Finance

Life Science

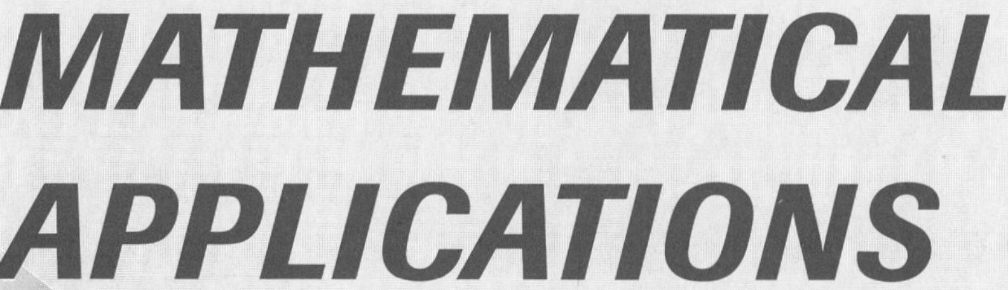

MATHEMATICAL APPLICATIONS

for the Management, Life, and Social Sciences

SIXTH EDITION

MATHEMATICAL APPLICATIONS

for the Management, Life, and Social Sciences

SIXTH EDITION

Ronald J. Harshbarger

University of South Carolina Beaufort

James J. Reynolds

Clarion University of Pennsylvania

HOUGHTON MIFFLIN COMPANY Boston New York

Cover designer: Walter Kopec

Cover image: Joseph Pobereskin/Tony Stone Images

Sponsoring Editor: Paul Murphy
Associate Editor: Mary Beckwith
Senior Project Editor: Maria A. Morelli
Editorial Assistant: Lauren M. Gagliardi
Senior Production/Design Coordinator: Jennifer Waddell
Senior Manufacturing Coordinator: Marie Barnes

Credits are listed at the back of the book.

Printed in the U.S.A.

Library of Congress Numer: 99-72016
ISBN: 0-395-96142-4
123456789-DW-03 02 01 00 99

Contents

Preface

To paraphrase English mathematician, philosopher, and educator Alfred North Whitehead, the purpose of education is not to fill a vessel but to kindle a fire. In particular, Whitehead encouraged students to be creative and imaginative in their learning and to continually form ideas into new and more exciting combinations. This desirable goal is not always an easy one to realize in mathematics with students whose primary interests are in areas other than mathematics. The purpose of this text, then, is to present mathematical skills and concepts and to apply them to ideas that are important to students in the management, life, and social sciences. We hope that this look at the relevance of mathematical ideas to a broad range of fields will help inspire the imaginative thinking and excitement for learning that Whitehead spoke of. The applications included allow students to view mathematics in a practical setting relevant to their intended careers. Almost every chapter of this book includes a section or two devoted to the applications of mathematical topics, and every section contains a number of application examples and problems. An index of these applications on the inside covers demonstrates the wide variety used in examples and exercises. Although intended for students who have completed two years of high school algebra or its equivalent, this text begins with a brief review of algebra which, if covered, will aid in preparing students for the work ahead.

Pedagogical Features

Important pedagogical features that have been retained in this new edition are the following.

Intuitive Viewpoint. The book is written from an intuitive viewpoint, with emphasis on concepts and problem solving rather than on mathematical theory. Each topic is carefully explained, and examples illustrate the techniques involved. Many exercises stress computation and drill, but there are enough challenging problems to stimulate students in thoughtful investigations.

Flexibility. At different colleges and universities, the coverage and sequencing of topics may vary according to the purpose of this course and the nature of the student audience. To accommodate alternative approaches, the text has a great deal of flexibility in the order in which topics may be presented and the degree to which they may be emphasized.

Chapter Warm-ups. A Warm-up appears at the beginning of each chapter and invites students to test themselves on the skills needed for that chapter. The Warm-ups present several prerequisite problem types that are keyed to the appropriate sections in the upcoming chapter where those skills are needed. Students who have difficulty with any particular skill are directed to specific sections of the text for review. Instructors may also find the Warm-ups useful in creating a course syllabus that includes an appropriate scope and sequence of topics.

Applications. We have found that integrating applied topics into the discussions and exercises helps to provide motivation within the sections and demonstrate the relevance of each topic. In addition, we have found that offering separate lessons on applied topics such as cost, revenue, and profit functions brings the preceding mathematical discussions into clear and concise focus and provides a thread of continuity as mathematical sophistication increases. Each application problem is identified, so the instructor or student can select applications that are of special interest. There are 10 such sections in the book, and more than 2000 exercises of the 5500 in the book are applied problems.

Extended Applications/Group Projects. Each of these case studies further illustrates how mathematics can be used in business and personal decision-making. In addition, each application is cumulative in nature in that its solution sometimes requires students to combine the mathematical concepts and techniques they learned in some chapters preceding the chapter presently being studied.

Graphing Utilities. After the introduction to graphing with technology in Section 1.4, discussion of the use of technology is placed in subsections and/or examples in many sections, so that graphing technology can be emphasized or de-emphasized at the option of the instructor.

The discussions of graphing utility technology highlight its most common features and uses, such as graphing, window setting, trace, zoom, tables, finding points of intersection, numerical derivatives, numerical integration, matrices, solving inequalities, and modeling (curve fitting). However, technology never replaces the mathematics, but rather supplements and extends it by providing opportunities for generalization and alternative ways of understanding, doing, and checking.

Some exercises that benefit by the use of technology, including graphics calculators, computer programs, and computer spreadsheets, are highlighted with a technology icon and modeling examples and exercises are highlighted with the icon shown at the left. Of course, many additional exercises can benefit from the use of technology, at the option of the instructor. Technology can be used to graph functions and to discuss the generalizations, applications and implications of problems being studied. In the Sixth Edition, we have increased the presence and depth of our treatment of spreadsheets. The basics of spreadsheet operations are introduced and opportunities to solve problems with spreadsheets are provided. In particular, spreadsheets are both illustrated and used as a problem solving tool when studying the Mathematics of Finance in Chapter 6.

Graphical, Numerical, and Symbolic Methods. A large number of real data and modeled applications are included in the examples and exercises throughout the text. Many sections include problems with functions modeled from real data, and some problems ask students to model functions from the data. These problems are solved by using one or more of graphical, numerical, and symbolic methods.

Checkpoints. These Checkpoints ask questions and pose problems to allow students to check their understanding of the skills and concepts under discussion before they proceed. Solutions to these checkpoints appear before the section exercises.

Application Previews. Each section begins with an Application Preview that establishes the context and direction for the concepts that will be presented.

Application Previews motivate the mathematics in the section and are then revisited in complete worked-out examples appearing later in the lesson.

Objective Lists. Every section begins with a brief list of objectives that outlines the goals of that section for the student.

Procedure/Example and Property/Example Tables. Appearing throughout the text, these tables aid student understanding by giving step-by-step descriptions of important procedures and properties with illustrative examples worked out beside them.

Boxed Information. All important information is boxed for easy reference, and key terms are highlighted in boldface.

Review Exercises. At the end of each chapter, a set of Review Exercises offers the student extra practice on the topics in that chapter. These exercises are annotated with section numbers so that any student who has difficulty can turn to the appropriate section for review.

Changes in the Sixth Edition

1. Numerous real data problems were added. A large number of real data and modeled applications have been added to the examples and exercises throughout the text.
2. The modeling of real data problems is introduced in Chapter 2 and used throughout the text. Many sections include problems with functions that are modeled from real data, as well as problems that require students to use graphing utilities to create equations from real data. The use of technology is also used for modeling functions from real data. This permits the inclusion of problems that are larger in scope and that contain more data points.
3. The sequence and scope of coverage has also changed significantly with the new edition of *Mathematical Applications.* The first four chapters of the text have been rearranged to allow for an earlier and more focused presentation of functions. Chapter 1, Linear Equations and Functions, is now followed by a new Chapter 2, Special Functions. In addition, the number and types of functions covered in Chapter 1 have been reduced in order to ensure a smooth transition into Chapter 2. Matrices are now covered in Chapter 3 and are followed by Inequalities and Linear Programming in Chapter 4. This new sequence represents a more logical presentation of algebra for both students and instructors.
4. Many of the real data problems introduced in the fifth edition have been updated. Students are asked to compare the previous models for the data with the current data to see if the same or different functions now model the data.
5. The number of Extended Applications/Group Projects has been doubled, with real life projects and/or real data projects added to each chapter. The instructor can chose projects that can be completed with graphing utilities or ones that can be done without graphing utilities. Most of the new projects use real data, and some require students to model the data as part of the solution.
6. A new guide to the use of the TI-82 and the TI-83 graphing calculators has been added. This supplement provides step-by-step procedures for the use of

the calculators alongside examples showing the screens and results of the steps.

7. Chapter Tests are included at the end of each chapter, so students can test their understanding of the major skills and concepts introduced in the chapter.
8. Exercises have been upgraded to increase the variety of problem types, to include some multi-step problems, and to use spreadsheets when appropriate.

Other significant improvements include the following.

CHAPTER 1 Linear Equations and Functions
The chapter has been reorganized to reduce the number of types of functions that are discussed. Topics included are limited to linear equations, functions, linear functions, systems of linear equations, and an introduction to graphing functions with graphing utilities. The power and limitations of graphing technology are discussed, paving the way for future discussions involving tabular, graphical, and analytic methods of solving problems.

CHAPTER 2 Special Functions
Quadratic functions, graphs, and applications are moved to this chapter from Chapter 4. Basic functions, polynomial functions, rational functions, and piecewise functions, including their graphs and applications, are moved to this chapter from Chapter 1. Section 2.5 discusses methods of fitting curves to data with graphing utilities. Real data applications are discussed in this section.

CHAPTER 3 Matrices
Many real data examples and problems that can be solved using matrices have been added to this chapter. Discussion of the use of technology to perform operations with large matrices and to find matrix inverses has been added.

CHAPTER 4 Inequalities and Linear Programming
A large number of new linear programming problems are introduced. Use of technology to find the corners of feasible regions has been added, and three Extended Applications/Group Projects are included., one of which incorporates LP Solvers accessible on the Internet. There is an increased emphasis in converting real data tables and figures to matrices.

CHAPTER 5 Exponential and Logarithmic Functions
More problems using equations created from real data are provided. Methods of modeling real data with exponential and logarithmic functions, and solving problems using these models, is included.

CHAPTER 6 Mathematics of Finance
This chapter still begins with simple and compound interest and a discussion of sequences. However, subsequent material now focuses separately on future value applications and present value applications. All discussions have been expanded to include both ordinary annuities and annuities due. Amortization and its related applications (such as unpaid balance of a loan) now form a single section.

CHAPTERS 9, 10, 11 Differential Calculus
Added to this edition are a large number of problems that require the modeling of real data before using calculus. Also, the use of technology has been updated and improved.

CHAPTER 14 Functions of Two or More Variables
The development of the linear regression formulas as an application of finding the minimum for a function of two variables is included.

Supplements

Complete Solutions Manual. This supplement contains complete solutions to all even- and odd-numbered exercises in the text.

Study & Solutions Guide, by Gordon Shilling. In addition to the solutions to all odd-numbered exercises in the text, this guide contains supplementary exercises that reinforce the concepts and techniques presented in the text. Answers to these problems are also provided.

Computerized Test Bank. This software supplement, available in Windows and Macintosh formats, allows for computerized testing. Select features of the program include support for multiple versions of the same test and the ability to output multiple-choice questions as free-response items.

Test Item File. This is a printed version of all test items appearing in the computerized testing program.

Easy Steps to Success Using the TI-82 and TI-83 Calculators. This technology guide gives step-by-step keystrokes for the use of these calculators, along with examples using these keystrokes to solve problems.

Acknowledgments

We would like to thank the many people who have helped us at various stages of revising this text. The encouragement, criticism, and suggestions that have been offered have been invaluable to us. Special thanks go to Dr. Jane Upshaw and Dr. Jon Beal for their assistance with the Extended Applications.

For their reviews of draft manuscript, responses to contents surveys, and many helpful comments, we would like to thank Bob Bradshaw, Ohlone College; Roxanne Byrne, University of Colorado-Denver; Mitzi Chaffer, Central Michigan University; Dr. John Collings, University of North Dakota; Ken Dodaro, Florida State University; Dr. Ewaugh Fields, Drexel University; Robert Hoburg, Western Connecticut State University; Lee H. LaRue, Paris Junior College; Charles Laws, Cleveland State Community College; Betty Liu, University of Rhode Island; Ho Kuen Ng, San Jose State University; Dr. J. Doug Richey, Northeast Texas Community College; Gordon Shilling, University of Texas at Arlington; and Patty Schovanec, Texas Tech University.

We would also like to express our appreciation to the editorial staff at Houghton Mifflin Company for their continued enthusiasm and support.

Ronald J. Harshbarger
James J. Reynolds

MATHEMATICAL APPLICATIONS

for the Management, Life, and Social Sciences

SIXTH EDITION

Chapter 0

Algebraic Concepts

This chapter provides a brief review of the algebraic concepts that will be used throughout the text.

The review begins with sets and the real numbers, the number system used throughout the text. Special subsets of the real numbers, including intervals, are also discussed.

Exponents and radicals, the building blocks for algebraic expressions, are then introduced. Operations with algebraic expressions and factoring are reviewed and then followed by operations with, and simplification of, algebraic fractions.

You should already be familiar with the topics covered in this chapter, but it may be helpful to spend some time reviewing them. In addition, each chapter after this one opens with a warm-up page that identifies prerequisite skills needed for that chapter. If algebraic skills are required, the warm-up cites their coverage in this chapter. Thus you will find that the following sections are a useful reference as you study later chapters.

0.1 Sets

We can begin to talk about sets by first looking at how blood is typed. The determination of blood types is based on the presence or absence of three antigens: the A antigen, the B antigen, and the antigen referred to as the Rh factor. Each different antigen combination determines a different blood type, and this scheme can be used to identify any individual as a member of a distinct group (or set) based on blood type (see Problem 59 in the exercises). For example, a person belongs to the set of people with blood type AB^+ if that person's blood contains antigen A, antigen B, and a positive Rh factor.

A **set** is a well-defined collection of objects. We may talk about a set of books, a set of dishes, a set of students, or, as described above, a set of individuals with a certain blood type. There are two ways to tell what a given set contains. One way is by listing the **elements** (or **members**) of the set (usually between braces). We may say that a set A contains 1, 2, 3, and 4 by writing $A = \{1, 2, 3, 4\}$. To say that 4 is a member of set A, we write $4 \in A$. Similarly, we write $5 \notin A$ to denote that 5 is not a member of set A.

If all the members of the set can be listed, the set is said to be a **finite set.** $A = \{1, 2, 3, 4\}$ and $B = \{x, y, z\}$ are examples of finite sets. When we do not wish to list all the elements of a finite set, we can use three dots to indicate the unlisted members of the set. For example, the set of even integers from 8 to 8952, inclusive, could be written as

$$\{8, 10, 12, 14, \ldots, 8952\}.$$

For an **infinite set,** we cannot list all the elements, so we use the three dots. For example, $N = \{1, 2, 3, 4, \ldots\}$ is an infinite set. This set N is called the set of **natural numbers.**

Another way to specify the elements of a given set is by description. For example, we may write $D = \{x\text{: } x \text{ is a Ford automobile}\}$ to describe the set of all Ford automobiles. $F = \{y\text{: } y \text{ is an odd natural number}\}$ is read "F is the set of all y such that y is an odd natural number."

EXAMPLE 1

Write the following sets in two ways.

(a) The set A of natural numbers less than 6.
(b) The set B of natural numbers greater than 10.
(c) The set C containing only 3.

Solution

(a) $A = \{1, 2, 3, 4, 5\}$ or $A = \{x\text{: } x \text{ is a natural number less than } 6\}$
(b) $B = \{11, 12, 13, 14, \ldots\}$ or $B = \{x\text{: } x \text{ is a natural number greater than } 10\}$
(c) $C = \{3\}$ or $C = \{x\text{: } x = 3\}$

Note that set C of Example 1 contains one member, 3; set A contains five members; and set B contains an infinite number of members. It is possible for a set to contain no members. Such a set is called the **empty set** or the **null set,** and it is denoted by $\varnothing$ or by { }. The set of living veterans of the War of 1812 is empty because there are no living veterans of that war. Thus

$$\{x: x \text{ is a living veteran of the War of 1812}\} = \varnothing.$$

Special relations that may exist between two sets are defined as follows.

Relations Between Sets

Definition	***Example***
1. Sets X and Y are **equal** if they contain the same elements.	1. If $X = \{1, 2, 3, 4\}$ and $Y = \{4, 3, 2, 1\}$, then $X = Y$.
2. $A \subseteq B$ if every element of A is an element of B. A is called a **subset** of B. The empty set is a subset of every set.	2. If $A = \{1, 2, c, f\}$ and $B = \{1, 2, 3, a, b, c, f\}$, then $A \subseteq B$.
3. If C and D have no elements in common, they are called **disjoint.**	3. If $C = \{1, 2, a, b\}$ and $D = \{3, e, 5, c\}$, then C and D are disjoint.

In the discussion of particular sets, the assumption is always made that the sets under discussion are all subsets of some larger set, called the **universal set** U. The choice of the universal set depends on the problem under consideration. For example, in discussing the set of all students and the set of all female students, we may use the set of all humans as the universal set.

We may use **Venn diagrams** to illustrate the relationships among sets. We use a rectangle to represent the universal set, and we use closed figures inside the rectangle to represent the sets under consideration. Figures 0.1–0.3 show such Venn diagrams.

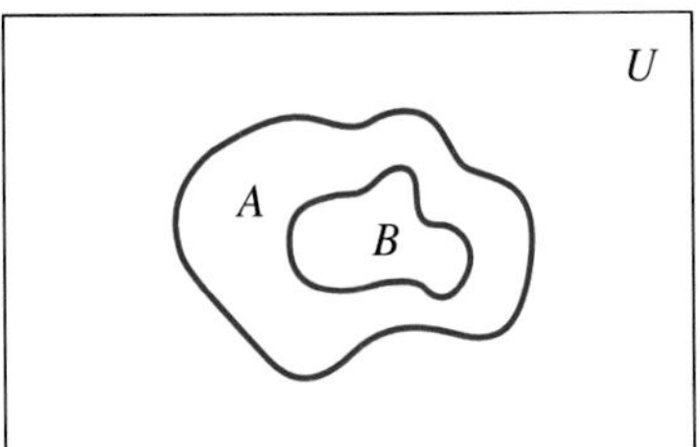

Figure 0.1
B is a subset of A; $B \subseteq A$.

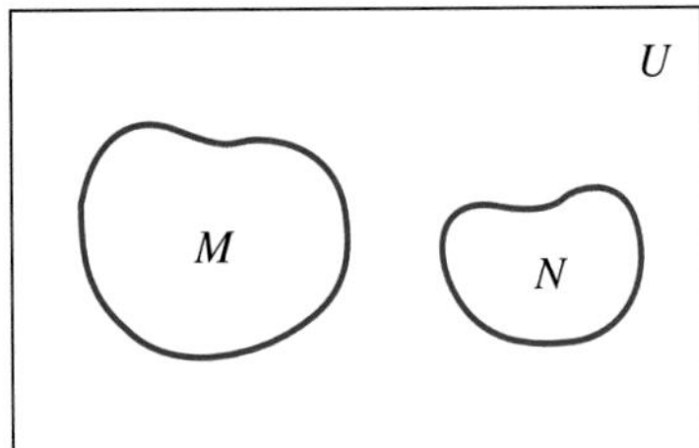

Figure 0.2
M and N are disjoint.

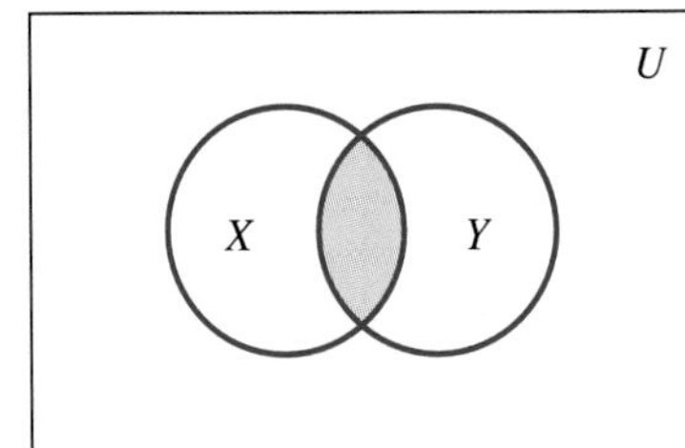

Figure 0.3
X and Y are not disjoint.

The shaded portion of Figure 0.3 indicates where the two sets overlap. The set containing the members that are common to two sets is said to be the **intersection** of the two sets.

Set Intersection The intersection of A and B, written $A \cap B$, is defined by

$$A \cap B = \{x: x \in A \text{ and } x \in B\}.$$

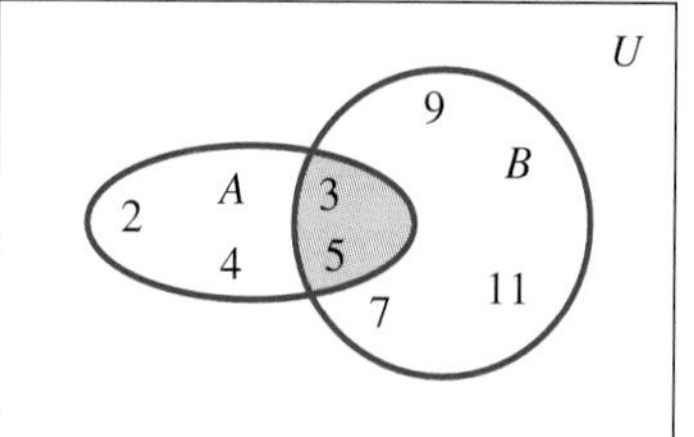

Figure 0.4

EXAMPLE 2

If $A = \{2, 3, 4, 5\}$ and $B = \{3, 5, 7, 9, 11\}$, find $A \cap B$.

Solution

$A \cap B = \{3, 5\}$ because 3 and 5 are the common elements of A and B. Figure 0.4 shows the sets and their intersection.

CHECKPOINT

Let $A = \{2, 3, 5, 7, 11\}$, $B = \{2, 4, 6, 8, 10\}$, and $C = \{6, 10, 14, 18, 22\}$. Use these sets to answer the following.

1. (a) Of which sets is 6 an element? (b) Of which sets is $\{6\}$ an element?
2. Which of the following are true?
   (a) $2 \in A$ (b) $2 \in B$ (c) $2 \in C$ (d) $5 \notin A$ (e) $5 \notin B$
3. Which pair of A, B, and C is disjoint?
4. Which of A, B, and C are subsets of
   (a) the set P of all prime numbers?
   (b) the set M of all multiples of 2?
5. Which of A, B, and C is equal to
   $D = \{x \mid x = 4n + 2 \text{ for natural numbers } 1 \le n \le 5\}$?

The **union** of two sets is the set that contains all members of the two sets.

Set Union The union of A and B, written $A \cup B$, is defined by

$$A \cup B = \{x: x \in A \text{ or } x \in B \text{ (or both)}\}.*$$

We can illustrate the intersection and union of two sets by the use of Venn diagrams. Figures 0.5 and 0.6 show Venn diagrams with universal set U represented by the rectangles and with sets A and B represented by the circles. The shaded region in Figure 0.5 represents $A \cap B$, the intersection of A and B, and the shaded region in Figure 0.6—which consists of all parts of both circles—represents $A \cup B$.

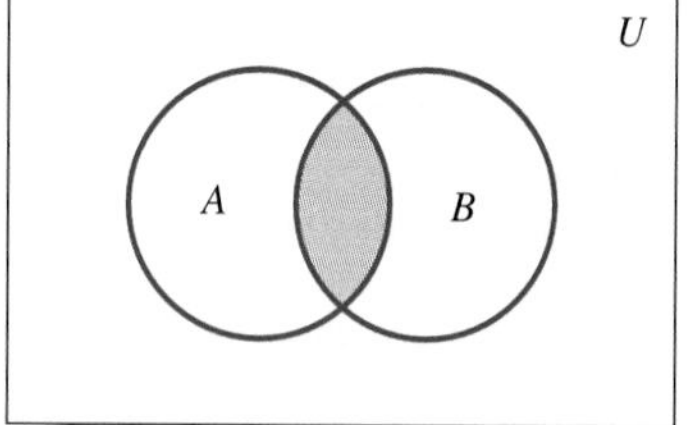

Figure 0.5

EXAMPLE 3

If $X = \{a, b, c, f\}$ and $Y = \{e, f, a, b\}$, find $X \cup Y$.

Solution

$X \cup Y = \{a, b, c, e, f\}$

* In mathematics, the word *or* means "one or the other or both."

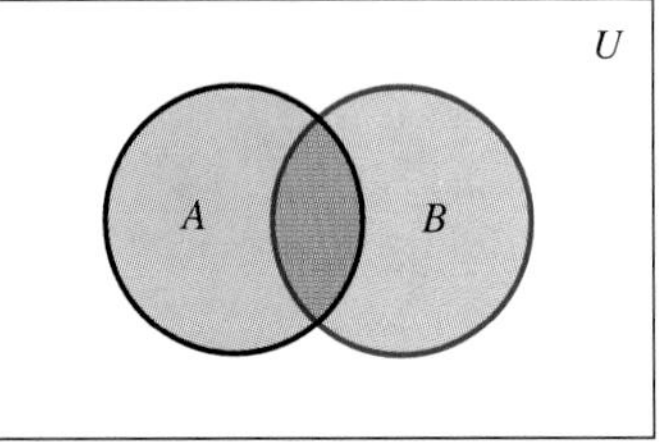

Figure 0.6

EXAMPLE 4

Let $A = \{x\text{: } x \text{ is a natural number less than } 6\}$ and $B = \{1, 3, 5, 7, 9, 11\}$.

(a) Find $A \cap B$.
(b) Find $A \cup B$.

Solution

(a) $A \cap B = \{1, 3, 5\}$
(b) $A \cup B = \{1, 2, 3, 4, 5, 7, 9, 11\}$

All elements of the universal set that are not contained in a set A form a set called the **complement** of A.

Set Complement The complement of A, written A', is defined by

$$A' = \{x\text{: } x \in U \text{ and } x \notin A\}$$

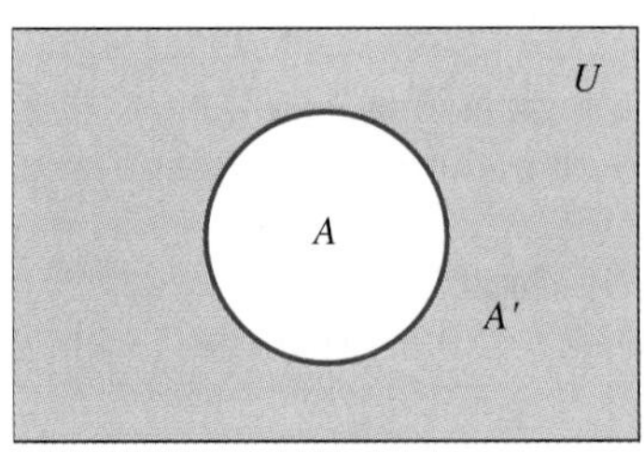

Figure 0.7

We can use a Venn diagram to illustrate the complement of a set. The shaded region of Figure 0.7 represents A', and the *un*shaded region of Figure 0.5 represents $(A \cap B)'$.

EXAMPLE 5

If $U = \{x \in N\text{: } x < 10\}$, $A = \{1, 3, 6\}$, and $B = \{1, 6, 8, 9\}$, find the following.

(a) A' (b) B' (c) $(A \cap B)'$ (d) $A' \cup B'$

Solution

(a) $U = \{1, 2, 3, 4, 5, 6, 7, 8, 9\}$ so $A' = \{2, 4, 5, 7, 8, 9\}$
(b) $B' = \{2, 3, 4, 5, 7\}$
(c) $A \cap B = \{1, 6\}$ so $(A \cap B)' = \{2, 3, 4, 5, 7, 8, 9\}$
(d) $A' \cup B' = \{2, 4, 5, 7, 8, 9\} \cup \{2, 3, 4, 5, 7\} = \{2, 3, 4, 5, 7, 8, 9\}$

CHECKPOINT

Given the sets $U = \{1, 2, 3, 4, 5, 6, 7, 8, 9, 10\}$, $A = \{1, 3, 5, 7, 9\}$, $B = \{2, 3, 5, 7\}$, and $C = \{4, 5, 6, 7, 8, 9, 10\}$, find the following.

6. $A \cup B$ 7. $B \cap C$ 8. A'

EXAMPLE 6

The local newspaper of Oil City, Pennsylvania, is called *The Derrick*. On the morning of June 30, 1998, *The Derrick* listed 23 stocks "of local interest." A prospective investor categorized these 23 stocks according to whether

- their closing price on June 29, 1998, was less than \$50/share (set C)
- their price-to-earnings ratio was less than 20 (set P)
- their dividend per share was at least \$1.50 (set D)

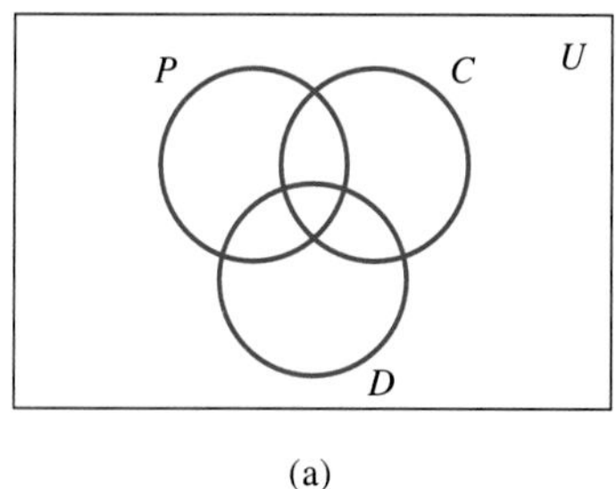

(a)

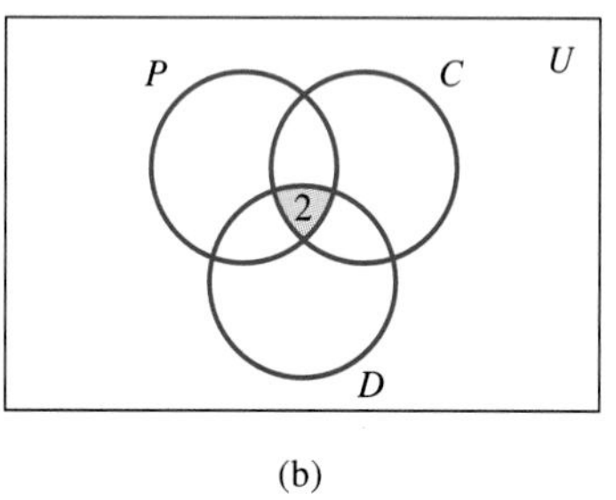

(b)

Figure 0.8

Of these 23 stocks,

16 belonged to set P
12 belonged to set C
8 belonged to set D
3 belonged to both C and D
10 belonged to both C and P
7 belonged to both D and P
2 belonged to all three sets

Draw a Venn diagram that represents this information. Use the diagram to answer the following.

(a) How many stocks had closing prices of less than \$50 per share or price-to-earnings ratios of less than 20?
(b) How many stocks had none of the characteristics of set C, P, or D?
(c) How many stocks had only dividends per share of at least \$1.50?

Solution

The Venn diagram for three sets has eight separate regions (see Figure 0.8(a)). To assign numbers from our data, we must begin with some information that refers to a single region, namely that two stocks belonged to all three sets (see Figure 0.8 (b)). Because the region common to all three sets is also common to any pair, we can next use the information about stocks that belonged to two of the sets (see Figure 0.8(c)). Finally, we can complete the Venn diagram (see Figure 0.8(d)).

(a) We need to add the numbers in the separate regions that lie within $C \cup P$. That is, 18 stocks closed under \$50 per share or had price-to-earnings ratios of less than 20.
(b) There are 5 stocks outside the three sets C, D, and P.
(c) Those stocks that had only dividends of at least \$1.50 per share are inside D but outside both C and P. There are no such stocks.

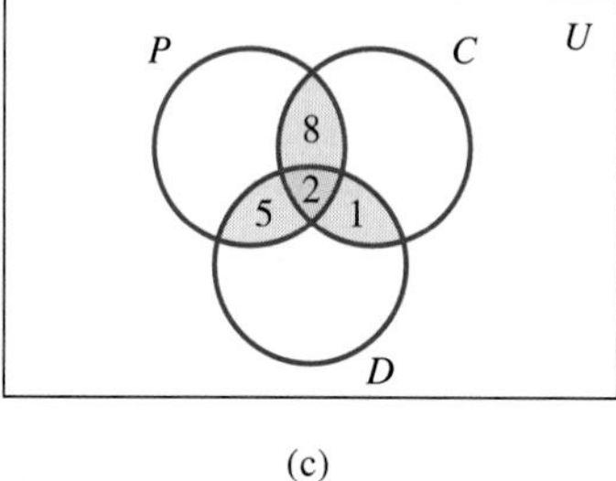

(c)

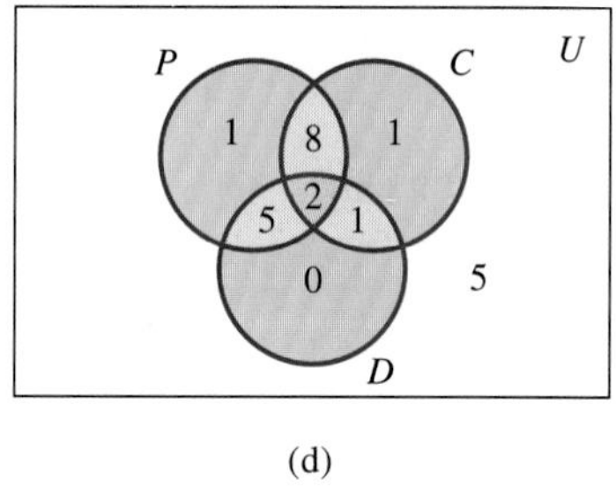

(d)

CHECKPOINT SOLUTIONS

1. (a) Sets B and C have 6 as an element.
   (b) None of A, B, or C has $\{6\}$ as an element; $\{6\}$ is itself a set, and the elements of A, B, and C are not sets.
2. (a) True (b) True (c) False; $2 \notin C$
   (d) False; $5 \in A$ (e) True
3. A and C are disjoint.
4. (a) $A \subseteq P$
   (b) $B \subseteq M \quad C \subseteq M$
5. $C = D$
6. $A \cup B = \{1, 2, 3, 5, 7, 9\}$
7. $B \cap C = \{5, 7\}$
8. $A' = \{2, 4, 6, 8, 10\}$

EXERCISE 0.1

Use $\in$ or $\notin$ to indicate whether the given object is an element of the given set in the following problems.

1. x $\{x, y, z, a\}$
2. 3 $\{1, 2, 4, 5, 6\}$
3. 12 $\{1, 2, 3, 4, \ldots\}$
4. 5 $\{x$: x is a natural number greater than $5\}$
5. 6 $\{x$: x is a natural number less than $6\}$
6. 3 $\varnothing$

In Problems 7–10, write the following sets a second way.

7. $\{x$: x is a natural number less than $8\}$
8. $\{x$: x is a natural number greater than 6 and less than $10\}$
9. $\{3, 4, 5, 6, 7\}$
10. $\{7, 8, 9, 10, \ldots\}$
11. If $A = \{1, 2, 3, 4\}$ and $B = \{1, 2, 3, 4, 5, 6\}$, is A a subset of B?
12. If $A = \{a, b, c, d\}$ and $B = \{c, d, a, b\}$, is A a subset of B?
13. Is $A \subseteq B$ if $A = \{a, b, c, d\}$ and $B = \{a, b, d\}$?
14. Is $A \subseteq B$ if $A = \{6, 8, 10, 12\}$ and $B = \{6, 8, 10, 14, 18\}$?

Use $\subseteq$ notation to indicate which set is a subset of the other in the following problems.

15. $C = \{a, b, 1, 2, 3\}$, $D = \{a, b, 1\}$
16. $E = \{x, y, a, b\}$, $F = \{x, 1, a, y, b, 2\}$
17. $A = \{1, \pi, e, 3\}$, $D = \{1, \pi, 3\}$
18. $A = \{2, 3, 4, e\}$, $\varnothing$
19. $A = \{6, 8, 7, 4\}$, $B = \{8, 7, 6, 4\}$
20. $D = \{a, e, 1, 3, c\}$, $F = \{e, a, c, 1, 3\}$

In Problems 21–24, indicate whether the following pairs of sets are equal.

21. $A = \{a, b, \pi, \sqrt{3}\}$, $B = \{a, \pi, \sqrt{3}, b\}$
22. $A = \{x, g, a, b\}$, $D = \{x, a, b, y\}$
23. $D = \{x$: x is a natural number less than $4\}$, $E = \{1, 2, 3, 4\}$
24. $F = \{x$: x is a natural number greater than $6\}$, $G = \{7, 8, 9, \ldots\}$
25. From the following list of sets, indicate which pairs of sets are disjoint.

$$A = \{1, 2, 3, 4\}$$

$$B = \{x: x \text{ is a natural number greater than } 4\}$$

$$C = \{4, 5, 6, \ldots\}$$

$$D = \{1, 2, 3\}$$

26. If A and B are disjoint sets, what does $A \cap B$ equal?

In Problems 27–30, find $A \cap B$, the intersection of sets A and B.

27. $A = \{2, 3, 4, 5, 6\}$ and $B = \{4, 6, 8, 10, 12\}$
28. $A = \{a, b, c, d, e\}$ and $B = \{a, d, e, f, g, h\}$
29. $A = \varnothing$ and $B = \{x, y, a, b\}$
30. $A = \{x$: x is a natural number less than $4\}$ and $B = \{3, 4, 5, 6\}$

In Problems 31–34, find $A \cup B$, the union of sets A and B.

31. $A = \{1, 2, 4, 5\}$ and $B = \{2, 3, 4, 5\}$
32. $A = \{a, e, i, o, u\}$ and $B = \{a, b, c, d\}$
33. $A = \varnothing$ and $B = \{1, 2, 3, 4\}$
34. $A = \{x$: x is a natural number greater than $5\}$ and $B = \{x$: x is a natural number less than $5\}$

In Problems 35–46, assume that

$$A = \{1, 3, 5, 8, 7, 2\} \qquad B = \{4, 3, 8, 10\}$$

$$C = \{2, 4, 6, 8, 10\}$$

and that U is the universal set of natural numbers less than 11. Find the following.

35. A'
36. B'
37. $A \cap B'$
38. $A' \cap B'$
39. $(A \cup B)'$
40. $(A \cap B)'$
41. $A' \cup B'$
42. $(A' \cup B)'$
43. $(A \cap B') \cup C'$
44. $A \cap (B' \cup C')$
45. $(A \cap B')' \cap C$
46. $A \cap (B \cup C)$

The difference of two sets, $A - B$, is defined as the set containing all elements of A except those in B. That is, $A - B = A \cap B'$. Find $A - B$ for each pair of sets in Problems 47–50 if $U = \{1, 2, 3, 4, 5, 6, 7, 8, 9\}$.

47. $A = \{1, 3, 7, 9\}$ and $B = \{3, 5, 8, 9\}$
48. $A = \{1, 2, 3, 6, 9\}$ and $B = \{1, 4, 5, 6, 7\}$
49. $A = \{2, 1, 5\}$ and $B = \{1, 2, 3, 4, 5, 6\}$
50. $A = \{1, 2, 3, 4, 5\}$ and $B = \{7, 8, 9\}$

Applications

51. ***Dow Jones Industrial Average*** The table on the next page shows information about yearly lows, highs, and percentage change for the years 1990–1997. Let L be the set of years where the low was greater than 3300. Let H be the set of years where the high was greater than 3300. Let C be the years when the percentage change (from low to high) exceeded 25%.
    (a) List the elements of L, H, and C.
    (b) Is any of L, H, or C a subset of one of the others (besides itself)?
    (c) Write a verbal description of C'.
    (d) Find $H' \cup C'$ and describe it in words.
    (e) Find $L' \cap C$ and describe it in words.

Dow Jones Industrial Average

Year	Low	High	% Change
97	6391.69	8259.31	29.2
96	5032.94	6569.91	30.5
95	3832.08	5216.47	36.1
94	3593.35	3978.36	10.7
93	3241.95	3794.33	17.0
92	3136.58	3413.21	8.8
91	2470.30	3168.83	28.3
90	2365.10	2999.75	26.8

Source: *The World Almanac, 1998*

52. ***Sticker shock*** The table below contains base prices for selected models of cars. Let A be the set of cars listed with a base price of at least \$20,000 in any year. Let B be the set of cars listed with a 1993 base price of at least \$20,000. Let C be the set of cars listed with an increase in the base price of at least 75% over 10 years.
    (a) List the elements of A, B, and C.
    (b) Is any of A, B, and C a subset of one of the others?
    (c) Find $A \cap C$ and describe it in words.
    (d) Write a verbal description for B'.
    (e) Find $A' \cap C'$.

Automobile Base Prices

Model	1983	1988	1993	10-year Increase
Cadillac Seville	\$21,440	\$27,627	\$36,990	73%
Chev. Camaro	\$8,036	\$10,995	\$13,500	68
Chev. Corvette	\$21,800	\$29,480	\$34,595	59
Ford Escort	\$5,639	\$6,586	\$8,355	48
Ford Mustang	\$6,727	\$8,835	\$10,719	59
Honda Accord	\$8,345	\$10,915	\$13,950	67
Honda Civic	\$4,899	\$6,195	\$8,400	71
Mazda RX-7	\$9,695	\$15,880	\$32,500	235
Nissan Maxima	\$10,869	\$17,449	\$20,960	93
Pont. Bonneville	\$8,899	\$14,099	\$19,444	118
Toyota Celica	\$7,299	\$11,348	\$14,198	95
Toyota Corolla	\$6,138	\$8,998	\$11,198	82

Source: *Motor Trend*, February 1993, p. 89.

National health care Suppose that the table below summarizes the opinions of various groups on the issue of national health care. Use this table for Problems 53 and 54.

	Whites		Non-whites		
Opinion	Rep.	Dem.	Rep.	Dem.	Total
Favor	100	250	30	200	580
Oppose	250	150	10	10	420
Total	350	400	40	210	1000

53. Identify the number of individuals in each of the following sets.
    (a) Republicans and those who favor national health care
    (b) Republicans or those who favor national health care
    (c) White Republicans or those who oppose national health care
54. Identify the number of individuals in each of the following sets.
    (a) Whites and those who oppose national health care
    (b) Whites or those who oppose national health care
    (c) Non-white Democrats and those who favor national health care

55. ***Languages*** A survey of 100 aides at the United Nations revealed that 65 could speak English, 60 could speak French, and 40 could speak both English and French.
    (a) Draw a Venn diagram representing the 100 aides. Use E to represent English-speaking and F to represent French-speaking aides.
    (b) How many aides are in $E \cap F$?
    (c) How many aides are in $E \cup F$?
    (d) How many aides are in $E \cap F'$?

56. ***Advertising*** Suppose that a survey of 100 advertisers in *U.S. News*, *These Times*, and *World* found the following.

    14 advertised in all three
    30 advertised in *These Times* and *U.S. News*
    26 advertised in *World* and *U.S. News*
    27 advertised in *World* and *These Times*
    60 advertised in *These Times*
    52 advertised in *U.S. News*
    50 advertised in *World*

    Draw a Venn diagram representing this information and use it to answer the following.
    (a) How many advertised in none of these publications?
    (b) How many advertised only in *These Times*?
    (c) How many advertised in *U.S. News* or *These Times*?

57. ***College enrollments*** Records at a small college show the following about the enrollments of 100 first-year students in mathematics, fine arts, and economics.

 38 take math
 42 take fine arts
 20 take economics
 4 take economics and fine arts
 15 take math and economics
 9 take math and fine arts
 12 take math and economics but not fine arts

 Draw a Venn diagram representing this information and label all the areas. Use this diagram to answer the following.
 (a) How many take none of these three courses?
 (b) How many take math or economics?
 (c) How many take exactly one of these three courses?

58. ***Survey analysis*** In a survey of the dining preferences of 110 dormitory students at the end of the spring semester, the following facts were discovered about Adam's Lunch (AL), Pizza Tower (PT), and the Student Union.

 30 liked AL but not PT
 21 liked AL only
 63 liked AL
 58 liked PT
 27 liked the Union
 25 liked PT and AL but not the Union
 18 liked PT and the Union

 Draw a Venn diagram representing this survey and label all the areas. Use this diagram to answer the following.
 (a) How many liked PT or the Union?
 (b) How many liked all three?
 (c) How many liked only the Union?

59. ***Blood types*** Blood types are determined by the presence or absence of three antigens: A antigen, B antigen, and an antigen called the Rh factor. The resulting blood types are classified as follows.

 type A if the A antigen is present
 type B if the B antigen is present
 type AB if both the A and B antigens are present
 type O if neither the A nor the B antigen is present

 These types are further classified as *Rh-positive* if the Rh-factor antigen is present and *Rh-negative* otherwise.
 (a) Draw a Venn diagram that illustrates this classification scheme.
 (b) Identify the blood type determined by each region of the Venn diagram (such as A^+ to indicate type A, Rh-positive).
 (c) Use a library or another source to find what percentage of the U.S. population has each blood type.

0.2 The Real Numbers

This text uses the set of **real numbers** as the universal set. We can represent the real numbers along a line called the **real number line.** This number line is a picture, or graph, of the real numbers. Each point on the number line corresponds to exactly one real number, and each real number can be located at exactly one point. Thus two real numbers are said to be equal whenever they are represented by the same point on the number line. The equation $a = b$ (a equals b) means that the symbols a and b represent the same real number. Thus $3 + 4 = 7$ means that $3 + 4$ and 7 represent the same number. Table 0.1 lists special subsets of the real numbers.

TABLE 0.1 Subsets of the Set of Real Numbers

	Description	Example (some elements shown)
Natural numbers	{1, 2, 3, . . .} The counting numbers.	−3 −2 −1 0 1 2 3 4 5
Integers	{. . . , −2, −1, 0, 1, 2, . . .} The natural numbers, 0, and the negatives of the natural numbers.	−4 −3 −2 −1 0 1 2 3 4
Rational numbers	All numbers that can be written as the ratio of two integers, a/b, with $b \neq 0$. These numbers have decimal representations that either terminate or repeat.	−2 −1 0 1 2; $-\frac{5}{4}$ $\frac{1}{2}$ $\frac{7}{4}$
Irrational numbers	Those real numbers that *cannot* be written as the ratio of two integers. Irrational numbers have decimal representations that neither terminate nor repeat.	−2 −1 0 1 2 3; $-\sqrt{3}$ $\sqrt{2}$ π
Real numbers	The set of all rational and irrational numbers (the entire number line).	−3 −2 −1 0 1 2 3

The properties of the real numbers are fundamental to the study of algebra. These properties follow.

Properties of the Real Numbers

Let a, b, and c denote real numbers.

1. Addition and multiplication are commutative.

$$a + b = b + a \qquad ab = ba$$

2. Addition and multiplication are associative.

$$(a + b) + c = a + (b + c) \qquad (ab)c = a(bc)$$

3. The additive identity is 0.

$$a + 0 = 0 + a = a$$

4. The multiplicative identity is 1.

$$a \cdot 1 = 1 \cdot a = a$$

5. Each element a has an additive inverse, denoted by $-a$.

$$a + (-a) = -a + a = 0$$

Note that there is a difference between a negative number and the negative of a number.

6. Each nonzero element a has a multiplicative inverse, denoted by a^{-1}.

$$a \cdot a^{-1} = a^{-1} \cdot a = 1$$

Note that $a^{-1} = 1/a$.

7. Multiplication is distributive over addition.

$$a(b + c) = ab + ac$$

Note that property 5 provides the means to subtract by defining $a - b = a + (-b)$ and property 6 provides a means to divide by defining $a \div b = a \cdot (1/b)$. The number 0 has no multiplicative inverse, so division by 0 is undefined.

We say that a is less than b (written $a < b$) if the point representing a is to the left of the point representing b on the real number line. For example, $4 < 7$ because 4 is to the left of 7 on the number line. We may also say that 7 is greater than 4 (written $7 > 4$). We may indicate that the number x is less than or equal to another number y by writing $x \leq y$. We may also indicate that p is greater than or equal to 4 by writing $p \geq 4$.

EXAMPLE 1

Use $<$ or $>$ notation to write

(a) 6 is greater than 5. (b) 10 is less than 15.
(c) 3 is to the left of 8 on the number line.
(d) x is at most 12.

Solution

(a) $6 > 5$ (b) $10 < 15$ (c) $3 < 8$
(d) "x is at most 12" means it must be less than or equal to 12. Thus, $x \leq 12$.

The subset of the real numbers consisting of all real numbers x that lie between a and b, excluding a and b, can be denoted by the *double inequality* $a < x < b$ or by the **open interval** (a, b). It is called an open interval because neither of the endpoints is included in the interval. The **closed interval** $[a, b]$ represents the set of all real numbers x satisfying $a \leq x \leq b$. Intervals containing one endpoint, such as $(a, b]$ and $[a, b)$, are called **half-open intervals.**

We can use $[a, +\infty)$ to represent the inequality $x \geq a$ and $(-\infty, a)$ to represent $x < a$. In each of these cases, the symbols $+\infty$ and $-\infty$ are not real numbers but represent the fact that x increases without bound ($+\infty$) or decreases without bound ($-\infty$). Table 0.2 summarizes three types of intervals.

TABLE 0.2 Intervals

Type of Interval	*Inequality Notation*	*Interval Notation*	*Graph*
Open interval	$x > a$	(a, ∞)	a
	$x < b$	$(-\infty, b)$	b
	$a < x < b$	(a, b)	a b
Half-open interval	$x \geq a$	$[a, \infty)$	a
	$x \leq b$	$(-\infty, b]$	b
	$a \leq x < b$	$[a, b)$	a b
	$a < x \leq b$	$(a, b]$	a b
Closed interval	$a \leq x \leq b$	$[a, b]$	a b

CHECKPOINT

1. Evaluate the following, if possible. For any that are meaningless, so state.
   (a) $\frac{4}{0}$ (b) $\frac{0}{4}$ (c) $\frac{4}{4}$ (d) $\frac{4-4}{4-4}$
2. For (a)–(d), write the inequality corresponding to the given interval and sketch its graph on a number line.
   (a) $(1, 3)$ (b) $(0, 3]$ (c) $[-1, \infty)$ (d) $(-\infty, 2)$

Sometimes we are interested in the *distance* a number is from the origin (0) of the number line, without regard to direction. The distance a number a is from 0 on the number line is the **absolute value** of a, denoted $|a|$. The absolute value of any nonzero number is positive, and the absolute value of 0 is 0.

EXAMPLE 2

Evaluate the following.

(a) $|-4|$

(b) $|+2|$

(c) $|0|$

(d) $|-5 - |-3||$

Solution

(a) $|-4| = +4 = 4$

(b) $|+2| = +2 = 2$

(c) $|0| = 0$

(d) $|-5 - |-3|| = |-5 - 3| = |-8| = 8$

Note that if a is a nonnegative number, then $|a| = a$, *but if a is negative, then* $|a|$ *is the positive number* $(-a)$. Thus

Absolute Value

$$|a| = \begin{cases} a & \text{if } a \geq 0 \\ -a & \text{if } a < 0 \end{cases}$$

In performing computations with real numbers, it is important to remember the rules for computations with signed numbers.

Operations with Real (Signed) Numbers

Procedure	Example
1. (a) To add two signed numbers with the same sign, add their absolute values and affix their common sign.	1. (a) $(+5) + (+6) = +11$ $\left(-\frac{1}{6}\right) + \left(-\frac{2}{6}\right) = -\frac{3}{6} = -\frac{1}{2}$
(b) To add two signed numbers with unlike signs, find the difference of their absolute values and affix the sign of the number with the larger absolute value.	(b) $(-4) + (+3) = -1$ $(+5) + (-3) = +2$ $\left(-\frac{11}{7}\right) + (+1) = -\frac{4}{7}$
2. To subtract one signed number from another, change the sign of the number being subtracted and proceed as in addition.	2. $(-9) - (-8) = (-9) + (+8) = -1$ $16 - (8) = 16 + (-8) = +8$
3. (a) The product of two numbers with like signs is positive.	3. (a) $(-3)(-4) = +12$ $\left(+\frac{3}{4}\right)(+4) = +3$
(b) The product of two numbers with unlike signs is negative.	(b) $5(-3) = -15$ $(-3)(+4) = -12$
4. (a) The quotient of two numbers with like signs is positive.	4. (a) $(-14) \div (-2) = +7$ $+36/4 = +9$
(b) The quotient of two numbers with unlike signs is negative.	(b) $(-28)/4 = -7$ $45 \div (-5) = -9$

When two or more operations with real numbers are indicated in an evaluation, it is important that everyone agree upon the order in which the operations are performed so that a unique result is guaranteed. The following **order of operations** is universally accepted.

1. Perform operations within parentheses.
2. Find indicated powers ($2^3 = 2 \cdot 2 \cdot 2 = 8$).
3. Perform multiplications and divisions from left to right.
4. Perform additions and subtractions from left to right.

EXAMPLE 3

Evaluate the following.

(a) $-4 + 3$ (b) $-4^2 + 3$
(c) $(-4 + 3)^2 + 3$ (d) $6 \div 2(2 + 1)$

Solution

(a) -1 (b) $-16 + 3 = -13$
(c) $(-1)^2 + 3 = 1 + 3 = 4$ (d) $6 \div 2(3) = 3 \cdot 3 = 9$

CHECKPOINT

True or false:

3. $-(-5)^2 = 25$
4. $|4 - 6| = |4| - |6|$
5. $9 - 2(2)(-10) = 7(2)(-10) = -140$

Operations with real numbers can be performed on a calculator. Calculator types vary widely, from those that perform only the four arithmetic operations to those that perform many operations automatically and those that have graphing capabilities and can be programmed.

The text assumes that you have a scientific or graphing calculator. Discussions of some of the capabilities of graphing calculators and graphing utilities may be found throughout the text.

Most scientific and graphing calculators use standard algebraic order when evaluating arithmetic expressions. Working outward from inner parentheses, calculations are performed from left to right. Powers and roots are evaluated first, followed by multiplications and divisions and then additions and subtractions.

CHECKPOINT SOLUTIONS

1. (a) Meaningless. A denominator of zero means division by zero, which is undefined.
   (b) $\frac{0}{4} = 0$. A numerator of zero (when the denominator is not zero) means the fraction has value 0.
   (c) $\frac{4}{4} = 1$
   (d) Meaningless. The denominator is zero.
2. (a) $1 < x < 3$

   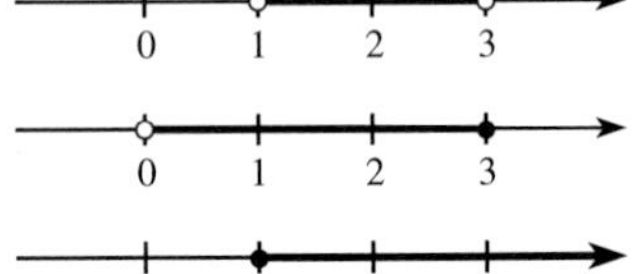

   (b) $0 < x \leq 3$

   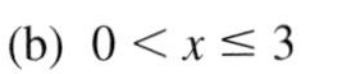

   (c) $-1 \leq x < \infty$ or $x \geq -1$

   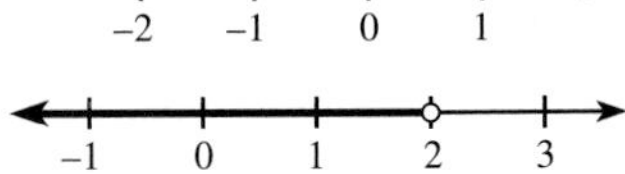

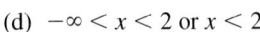

   (d) $-\infty < x < 2$ or $x < 2$
3. False. $-(-5)^2 = (-1)(-5)^2 = (-1)(25) = -25$. Exponentiation has priority and applies only to -5.
4. False. $|4 - 6| = |-2| = 2$ and $|4| - |6| = 4 - 6 = -2$.
5. False. Without parentheses, multiplication has priority over subtraction. $9 - 2(2)(-10) = 9 - 4(-10) = 9 + 40 = 49$.

EXERCISE 0.2

In Problems 1–2, indicate whether the given expression is one or more of the following types of numbers: rational, irrational, integer, natural. If the expression is meaningless, so state.

1. (a) $\frac{-\pi}{10}$ (b) -9 (c) $\frac{9}{3}$ (d) $\frac{4}{0}$
2. (a) $\frac{0}{6}$ (b) -1.2916 (c) 1.414 (d) $\frac{9}{6}$

Which property of real numbers is illustrated in each part of Problems 3 and 4?

3. (a) $8 + 6 = 6 + 8$
   (b) $5(3 + 7) = 5 \cdot 3 + 5 \cdot 7$ (c) $-e \cdot 1 = -e$
4. (a) $6(4 \cdot 5) = (6 \cdot 4)(5)$ (b) $-15 + 0 = -15$
   (c) $\left(\frac{3}{2}\right)\left(\frac{2}{3}\right) = 1$

Insert the proper sign $<$, $=$, or $>$ to replace $\square$ in Problems 5–10.

5. $-14 \square -3$
6. $\pi \square 3.14$
7. $0.333 \square \frac{1}{3}$
8. $\frac{1}{3} + \frac{1}{2} \square \frac{5}{6}$
9. $|-3| + |5| \square |-3 + 5|$
10. $|-9 - 3| \square |-9| + |3|$

In Problems 11–22, evaluate each expression.

11. $-3^2 + 10 \cdot 2$
12. $(-3)^2 + 10 \cdot 2$
13. $\frac{4 + 2^2}{2}$
14. $\frac{(4 + 2)^2}{2}$
15. $\frac{16 - (-4)}{8 - (-2)}$
16. $\frac{(-5)(-3) - (-2)(3)}{-9 + 2}$
17. $\frac{|5 - 2| - |-7|}{|5 - 2|}$
18. $\frac{|3 - |4 - 11||}{-|5^2 - 3^2|}$
19. $\frac{(-3)^2 - 2 \cdot 3 + 6}{4 - 2^2 + 3}$
20. $\frac{6^2 - 4(-3)(-2)}{6 - 6^2 \div 4}$
21. $\frac{-4^2 + 5 - 2 \cdot 3}{5 - 4^2}$
22. $\frac{3 - 2(5 - 2)}{(-2)^2 - 2^2 + 3}$
23. What part of the real number line corresponds to the interval $(-\infty, \infty)$?
24. Write the interval corresponding to $x \geq 0$.

In Problems 25–28, express each inequality or graph using interval notation.

25. $1 < x \leq 3$
26. $-4 \leq x \leq 3$
27. 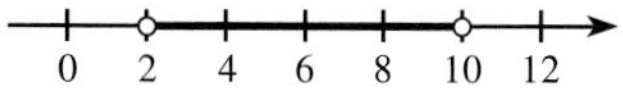

28. 

In Problems 29–32, write an inequality that describes each interval or graph.

29. $(-\infty, 5]$
30. $(-2, \infty)$
31. 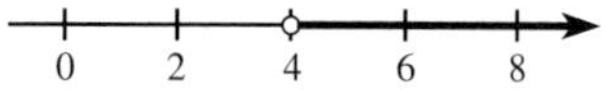

32. 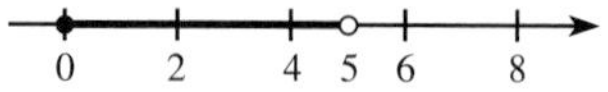

In Problems 33–40, graph the subset of the real numbers that is represented by each of the following and write your answer in interval notation.

33. $(-\infty, 4) \cap (-3, \infty)$
34. $[-4, 17) \cap [-20, 10]$
35. $x > 4$ and $x \geq 0$
36. $x < 10$ and $x < -1$
37. $[0, \infty) \cup [-1, 5]$
38. $(-\infty, 4) \cup (0, 2)$
39. $x > 7$ or $x < 0$
40. $x > 4$ and $x < 0$

In Problems 41–46, use your calculator to evaluate each of the following. List all the digits on your display in the answer.

41. $\frac{-1}{25916.8}$
42. $\frac{51.412}{127.01}$
43. $(3.679)^7$
44. $(1.28)^{10}$
45. $\frac{2500}{(1.1)^6 - 1}$
46. $100\left[\frac{(1.05)^{12} - 1}{0.05}\right]$

Applications

47. ***Take-home pay*** A sales representative's take-home pay is found by subtracting all taxes and retirement contributions from gross pay (which consists of salary plus commission). Given the following information, complete (a)–(c).

    Salary = \$300.00 Commission = \$788.91

    Retirement = 5% of gross pay

    Taxes: State = 5% of gross pay,

    Local = 1% of gross pay

    Federal withholding = 25% of (gross pay less retirement)

    Federal social security and Medicare = 7.65% of gross pay

    (a) Find the gross pay.
    (b) Find the amount of federal withholding.
    (c) Find the take-home pay.

48. ***Tax load*** By using data from 1955–1997, we can approximate the percentage P of average income used to pay federal state and local taxes by

$$P = 0.24627t + 25.96473$$

where t is the number of years past 1950.
(a) What t-value represents the year 1985?
(b) The actual tax load for 1985 was 34.8%. What does the formula give as an approximation?
(c) Approximate the tax load for 2005.
(d) What P-value signals that this approximation formula is no longer valid? Explain.

49. ***Cable rates*** By using data for 1982–1997 from Nielsen Media Research, we can approximate the average monthly cost C for basic cable TV service quite accurately by either
(1) $C = 1.2393t + 4.2594$ or
(2) $C = 0.0232t^2 + 0.7984t + 5.8271$
where t is the number of years past 1980.
(a) Both (1) and (2) closely fit all the data from 1982 to 1997, but which is more accurate for 1996 and for 1997? Use the fact that average cable rates for those years were as follows:

1996: \$24.41 1997: \$26.00

(b) Use the formula from (a) that was more accurate and predict the average cost for basic cable service for 2005.

0.3 Integral Exponents

If \$1000 is placed in a 5-year savings certificate that pays an interest rate of 10% per year, compounded annually, then the amount returned after 5 years is given by

$$1000(1.1)^5.$$

The 5 in this expression is an **exponent.** Exponents provide an easier way to denote certain multiplications. For example,

$$(1.1.)^5 = (1.1)(1.1)(1.1)(1.1)(1.1).$$

An understanding of the properties of exponents is fundamental to the algebra needed to study functions and solve equations. Furthermore, the definition of exponential and logarithmic functions and many of the techniques in calculus also require an understanding of the properties of exponents.

For any real number a,

$$a^2 = a \cdot a, \quad a^3 = a \cdot a \cdot a, \quad \text{and} \quad a^n = a \cdot a \cdot a \cdot \ldots \cdot a \quad (n \text{ factors})$$

for any positive integer n. The positive integer n is called the **exponent,** the number a is called the **base,** and a^n is read "a to the nth power."

Note that $4a^n$ means $4(a^n)$, which is different from $(4a)^n$. The 4 is the coefficient of a^n in $4a^n$. Note also that $-x^n$ is not equivalent to $(-x)^n$ when n is even. For example, $-3^4 = -81$, but $(-3)^4 = 81$.

Some of the rules of exponents follow.

Positive Integer Exponents For any real numbers a and b and positive integers m and n,

1. $a^m \cdot a^n = a^{m+n}$
2. For $a \neq 0$, $\dfrac{a^m}{a^n} = \begin{cases} a^{m-n} & \text{if } m > n \\ 1 & \text{if } m = n \\ 1/a^{n-m} & \text{if } m < n \end{cases}$
3. $(ab)^m = a^m b^m$
4. $\left(\dfrac{a}{b}\right)^m = \dfrac{a^m}{b^m} \quad (b \neq 0)$
5. $(a^m)^n = a^{mn}$

EXAMPLE 1

Use properties of positive integer exponents to rewrite each of the following. Assume all denominators are nonzero.

(a) $\dfrac{5^6}{5^4}$ (b) $\dfrac{x^2}{x^5}$ (c) $\left(\dfrac{x}{y}\right)^4$ (d) $(3x^2y^3)^4$ (e) $3^3 \cdot 3^2$

Solution

(a) $\dfrac{5^6}{5^4} = 5^{6-4} = 5^2$ (b) $\dfrac{x^2}{x^5} = \dfrac{1}{x^{5-2}} = \dfrac{1}{x^3}$

(c) $\left(\dfrac{x}{y}\right)^4 = \dfrac{x^4}{y^4}$ (d) $(3x^2y^3)^4 = 3^4(x^2)^4(y^3)^4 = 81x^8y^{12}$

(e) $3^3 \cdot 3^2 = 3^{3+2} = 3^5$

For certain calculus operations, use of negative exponents is necessary in order to write problems in the proper form. We can extend the rules for positive integer exponents to all integers by defining a^0 and a^{-n}. Clearly $a^m \cdot a^0$ should equal $a^{m+0} = a^m$, and it will if $a^0 = 1$.

Zero Exponent For any nonzero real number a, we define $a^0 = 1$. We leave 0^0 undefined.

In Section 0.2, we defined a^{-1} as $1/a$ for $a \neq 0$, so we define a^{-n} as $(a^{-1})^n$.

Negative Exponents

$$a^{-n} = (a^{-1})^n = \left(\frac{1}{a}\right)^n = \frac{1}{a^n} \qquad (a \neq 0)$$

$$\left(\frac{a}{b}\right)^{-n} = \left[\left(\frac{a}{b}\right)^{-1}\right]^n = \left(\frac{b}{a}\right)^n \qquad (a \neq 0, b \neq 0)$$

EXAMPLE 2

Write each of the following without exponents.

(a) $6 \cdot 3^0$ (b) 6^{-2}

(c) $\left(\frac{1}{3}\right)^{-1}$ (d) $-\left(\frac{2}{3}\right)^{-4}$

(e) $(-4)^{-2}$

Solution

(a) $6 \cdot 3^0 = 6 \cdot 1 = 6$ (b) $6^{-2} = \frac{1}{6^2} = \frac{1}{36}$

(c) $\left(\frac{1}{3}\right)^{-1} = \frac{3}{1} = 3$ (d) $-\left(\frac{2}{3}\right)^{-4} = -\left(\frac{3}{2}\right)^{4} = \frac{-81}{16}$

(e) $(-4)^{-2} = \frac{1}{(-4)^2} = \frac{1}{16}$

As we'll see in the chapter on the mathematics of finance, negative exponents arise in financial calculations when we have a future goal for an investment and want to know how much to invest now. For example, if money can be invested at 9%, compounded annually, then the amount we must invest now (which is called the present value) in order to have \$10,000 in the account after 7 years is given by $\$10,000(1.09)^{-7}$. Calculations such as this are often done directly with a scientific calculator.

Using the definitions of zero and negative exponents enables us to extend the rules of exponents to all integers and to express them more simply.

Rules of Exponents For real numbers a and b and *integers* m and n,

1. $a^m \cdot a^n = a^{m+n}$
2. $a^m/a^n = a^{m-n}$ $(a \neq 0)$
3. $(ab)^m = a^m b^m$
4. $(a^m)^n = a^{mn}$
5. $(a/b)^m = a^m/b^m$ $(b \neq 0)$
6. $a^0 = 1$ $(a \neq 0)$
7. $a^{-n} = 1/a^n$ $(a \neq 0)$
8. $(a/b)^{-n} = (b/a)^n$ $(a, b \neq 0)$

Throughout the remainder of the text, we will assume all expressions are defined.

EXAMPLE 3

Use the rules of exponents and the definitions of a^0 and a^{-n} to simplify each of the following with positive exponents.

(a) $2(x^2)^{-2}$ (b) $x^{-2} \cdot x^{-5}$ (c) $\frac{x^{-8}}{x^{-4}}$ (d) $\left(\frac{2x^3}{3x^{-5}}\right)^{-2}$

Solution

(a) $2(x^2)^{-2} = 2x^{-4} = 2\left(\frac{1}{x^4}\right) = \frac{2}{x^4}$

(b) $x^{-2} \cdot x^{-5} = x^{-2-5} = x^{-7} = \frac{1}{x^7}$

(c) $\frac{x^{-8}}{x^{-4}} = x^{-8-(-4)} = x^{-4} = \frac{1}{x^4}$

(d) $\left(\frac{2x^3}{3x^{-5}}\right)^{-2} = \left(\frac{2x^8}{3}\right)^{-2} = \left(\frac{3}{2x^8}\right)^2 = \frac{9}{4x^{16}}$

CHECKPOINT

1. Complete the following.

   (a) $x^3 \cdot x^8 = x^?$ (b) $x \cdot x^4 \cdot x^{-3} = x^?$ (c) $\frac{1}{x^4} = x^?$

   (d) $x^{24} \div x^{-3} = x^?$ (e) $(x^4)^2 = x^?$ (f) $(2x^4y)^3 = ?$

2. True or false:

   (a) $3x^{-2} = \frac{1}{9x^2}$ (b) $-x^{-4} = \frac{-1}{x^4}$ (c) $x^{-3} = -x^3$

3. Evaluate the following, if possible. For any that are meaningless, so state. Assume $x > 0$.

   (a) 0^4 (b) 0^0 (c) x^0

   (d) 0^x (e) 0^{-4} (f) -5^{-2}

EXAMPLE 4

Write $(x^2y)/(9wz^3)$ with all factors in the numerator.

Solution

$$\frac{x^2y}{9wz^3} = x^2y\left(\frac{1}{9wz^3}\right) = x^2y\left(\frac{1}{9}\right)\left(\frac{1}{w}\right)\left(\frac{1}{z^3}\right) = x^2y \cdot 9^{-1}w^{-1}z^{-3}$$

$$= 9^{-1}x^2yw^{-1}z^{-3}$$

EXAMPLE 5

Simplify the following so all exponents are positive.

(a) $(2^3x^{-4}y^5)^{-2}$ (b) $\frac{2x^4(x^2y)^0}{(4x^{-2}y)^2}$

Solution

(a) $(2^3x^{-4}y^5)^{-2} = 2^{-6}x^8y^{-10} = \frac{1}{2^6} \cdot x^8 \cdot \frac{1}{y^{10}} = \frac{x^8}{64y^{10}}$

(b) $\frac{2x^4(x^2y)^0}{(4x^{-2}y)^2} = \frac{2x^4 \cdot 1}{4^2x^{-4}y^2} = \frac{2}{4^2} \cdot \frac{x^4}{x^{-4}} \cdot \frac{1}{y^2} = \frac{2}{16} \cdot \frac{x^8}{1} \cdot \frac{1}{y^2} = \frac{x^8}{8y^2}$

CHECKPOINT SOLUTIONS

1. (a) $x^3 \cdot x^8 = x^{3+8} = x^{11}$ (b) $x \cdot x^4 \cdot x^{-3} = x^{1+4+(-3)} = x^2$
   (c) $\dfrac{1}{x^4} = x^{-4}$ (d) $x^{24} \div x^{-3} = x^{24-(-3)} = x^{27}$
   (e) $(x^4)^2 = x^{(4)(2)} = x^8$
   (f) $(2x^4y)^3 = 2^3(x^4)^3y^3 = 8x^{12}y^3$
2. (a) False. $3x^{-2} = 3\left(\dfrac{1}{x^2}\right) = \dfrac{3}{x^2}$
   (b) True. $-x^{-4} = (-1)\left(\dfrac{1}{x^4}\right) = \dfrac{-1}{x^4}$
   (c) False. $x^{-3} = \dfrac{1}{x^3}$
3. (a) $0^4 = 0$
   (b) Meaningless. 0^0 is undefined.
   (c) $x^0 = 1$ since $x \neq 0$
   (d) $0^x = 0$ because $x > 0$
   (e) Meaningless. 0^{-4} would be $\dfrac{1}{0^4}$, which is not defined.
   (f) $-5^{-2} = (-1)\left(\dfrac{1}{5^2}\right) = \dfrac{-1}{25}$

EXERCISE 0.3

Evaluate in Problems 1–8. Write all answers without using exponents.

1. $(-4)^3$
2. -5^3
3. -2^4
4. $(-2)^5$
5. 3^{-2}
6. 6^{-1}
7. $-\left(\dfrac{3}{2}\right)^2$
8. $\left(\dfrac{2}{3}\right)^3$

In Problems 9–16, use rules of exponents to simplify the expressions. Express answers with positive exponents.

9. $6^5 \cdot 6^3$
10. $\dfrac{7^8}{7^3}$
11. $\dfrac{10^8}{10^9}$
12. $\dfrac{5^4}{(5^{-2} \cdot 5^3)}$
13. $(3^3)^3$
14. $(2^{-3})^{-2}$
15. $\left(\dfrac{2}{3}\right)^{-2}$
16. $\left(\dfrac{-2}{5}\right)^{-4}$

In Problems 17–20, simplify by expressing answers with positive exponents.

17. $(x^2)^{-3}$
18. x^{-4}
19. $xy^{-2}z^0$
20. $(xy^{-2})^0$

In Problems 21–34, use the rules of exponents to simplify so that only positive exponents remain.

21. $x^3 \cdot x^4$
22. $a^5 \cdot a$
23. $x^{-5} \cdot x^3$
24. $y^{-5} \cdot y^{-2}$
25. $\dfrac{x^8}{x^4}$
26. $\dfrac{a^5}{a^{-1}}$
27. $\dfrac{y^5}{y^{-7}}$
28. $\dfrac{y^{-3}}{y^{-4}}$
29. $(x^4)^3$
30. $(y^3)^{-2}$
31. $(xy)^2$
32. $(2m)^3$
33. $\left(\dfrac{2}{x}\right)^4$
34. $\left(\dfrac{8}{a^3}\right)^3$

In Problems 35–46, compute and simplify so that only positive exponents remain.

35. $(2x^{-2}y)^{-4}$
36. $(-32x^5)^{-3}$
37. $(-8a^{-3}b^2)(2a^5b^{-4})$
38. $(-3m^2y^{-1})(2m^{-3}y^{-1})$
39. $(2x^{-2}) \div (x^{-1}y^2)$
40. $(-8a^{-3}b^2c) \div (2a^5b^4)$
41. $\left(\dfrac{x^3}{y^{-2}}\right)^{-3}$
42. $\left(\dfrac{x^{-2}}{y}\right)^{-3}$
43. $\left(\dfrac{a^{-2}b^{-1}c^{-4}}{a^4b^{-3}c^0}\right)^{-3}$
44. $\left(\dfrac{4x^{-1}y^{-40}}{2^{-2}x^4y^{-10}}\right)^{-2}$
45. (a) $\dfrac{2x^{-2}}{(2x)^2}$ (b) $\dfrac{(2x)^{-2}}{(2x)^2}$
    (c) $\dfrac{2x^{-2}}{2x^2}$ (d) $\dfrac{2x^{-2}}{(2x)^{-2}}$
46. (a) $\dfrac{2^{-1}x^{-2}}{(2x)^2}$ (b) $\dfrac{2^{-1}x^{-2}}{2x^2}$
    (c) $\dfrac{(2x^{-2})^{-1}}{(2x)^{-2}}$ (d) $\dfrac{(2x^{-2})^{-1}}{2x^2}$

In many applications it is often necessary to write expressions in the form cx^n, where c is a constant. In Problems 47–54, write the expressions in this form.

47. $\frac{1}{x}$
48. $\frac{1}{x^2}$
49. $(2x)^3$
50. $(3x)^2$
51. $\frac{1}{(4x^2)}$
52. $\frac{3}{(2x^4)}$
53. $\left(\frac{-x}{2}\right)^3$
54. $\left(\frac{-x}{3}\right)^2$

In Problems 55–58, use a scientific calculator to evaluate the indicated powers.

55. 1.2^4
56. $(-3.7)^3$
57. $(1.5)^{-5}$
58. $(-.8)^{-9}$

Applications

Compound interest If $P is invested for n years at rate i (as a decimal), compounded annually, the future value that accrues is given by $S = P(1 + i)^n$, and the interest earned is $I = S - P$. In Problems 59–62, find S and I for the given P, n, and i.

59. $1200 for 5 years at 12%
60. $1800 for 7 years at 10%
61. $5000 for 6 years at 11.5%
62. $800 for 20 years at 10.5%

Present value If an investment has a goal (future value) of $S after n years, invested at interest rate i (as a decimal), compounded annually, then the present value P that must be invested is given by $P = S(1 + i)^{-n}$. In Problems 63 and 64, find P for the given S, n, and i.

63. $15,000 after 6 years at 11.5%
64. $80,000 after 20 years at 10.5%

65. ***Investing*** Using data for 1978–1998 from First Union of Charlotte, the annual dividend D in dollars paid per share of First Union's common stock can be approximated with the formula

$$D = 0.091387(1.125013)^t$$

where t is the number of years past 1975.

(a) What t-values correspond to the years 1990, 1993, 1995, and 1997?

(b) The actual annual dividends per share for the years in (a) are as follows:

1990	1993	1995	1997
$0.54	$0.75	$0.98	$1.22

What does the formula predict (to the nearest cent) for these years?

(c) Use the formula to predict the dividend per share in 2005.

66. ***Dow Jones Industrial Average*** With data from 1979–1996, the high value H attained by the Dow Jones Industrial Average can be approximated by the formula

$$H = 0.1126601508(1.119409426)^t$$

where t is the number of years past 1900.

(a) What t-values correspond to the years 1980, 1985, 1991, and 1996?

(b) The actual Dow Jones highs for the years in (a) are as follows:

1980	1985	1991	1996
1000.17	1553.10	3168.83	6590.91

What does the formula predict for these years?

(c) Use the formula to predict the Dow Jones high for 2005.

67. ***U.S. debt*** With U.S. Department of Treasury data from 1900–1996 (published in *The World Almanac, 1998*), the U.S. debt, D, in billions of dollars, can be approximated by the formula

$$D = 1.531547528(1.089218396)^t$$

where t is the number of years past 1900.

(a) The actual U.S. debt, in billions of dollars, for selected years was as follows:

1940	1945	1991
43.0	258.7	3665.3

For each of these years, find the debt predicted by the formula.

(b) This debt formula is based on historical data and is sensitive to long-term trends in those data, but not necessarily to individual years with special circumstances. Why is the prediction for 1945 so inaccurate?

(c) Use available resources to find U.S. debt data for years since 1996. Then check the accuracy of the formula against these data to see whether recent federal efforts to control the debt have been successful. Is the debt for years after 1996 below what the formula predicts?

68. ***Investing*** Using data from the Investment Company Institute (published in *Investment Digest* of Valic Co., Summer 1992), the number of mutual funds N can be modeled with the formula

$$N = 4.8139(1.9387)^t$$

where t is the number of decades past 1900.
(a) What t-value corresponds to 1950?
(b) Approximate the number of mutual funds in 1950.
(c) Approximate the number of mutual funds in 1990.
(d) Estimate the number of mutual funds in the year 2000.

69. ***Health care expenditures*** Using data from the Congressional Budget Office (reported in *Newsweek*, October 4, 1993), the national health care expenditure H (in billions of dollars) can be modeled with the formula

$$H = 45(1.094)^t$$

where t is the number of years past 1960.
(a) What t-value corresponds to 1970?
(b) Approximate the national health care expenditure in 1970.
(c) Approximate the national health care expenditure in 1993.
(d) Estimate the national health care expenditure in 2005.

0.4 Radicals and Rational Exponents

A process closely linked to that of raising numbers to powers is that of extracting roots. From geometry we know that if an edge of a cube is x units, its volume is x^3 cubic units. Reversing this process, we determine that if the volume of a cube is V cubic units, the length of an edge is the cube root of V, which is denoted

$$\sqrt[3]{V} \text{ units.}$$

When we seek the **cube root** of a number such as 8 (written $\sqrt[3]{8}$), we are looking for a real number whose cube equals 8. Because $2^3 = 8$, we know that $\sqrt[3]{8} = 2$. Similarly, $\sqrt[3]{-27} = -3$ because $(-3)^3 = -27$. The expression $\sqrt[n]{a}$ is called a **radical,** where $\sqrt{}$ is the **radical sign,** n the **index,** and a the **radicand.** When no index is indicated, the index is assumed to be 2 and the expression is called a **square root;** thus $\sqrt{4}$ is the square root of 4 and represents the positive number whose square is 4.

Only one real number satisfies $\sqrt[n]{a}$ for a real number a and an odd number n; we call that number the **principal *n*th root,** or, more simply, the ***n*th root.**

For an even index n, there are two possible cases:

1. If a is negative, there is no real number equal to $\sqrt[n]{a}$. For example, there are no real numbers that equal $\sqrt{-4}$ or $\sqrt[4]{-16}$ because there is no real number b such that $b^2 = -4$ or $b^4 = -16$. In this case, we say $\sqrt[n]{a}$ is not a real number.
2. If a is positive, there are 2 real numbers whose nth power equals a. For example, $3^2 = 9$ and $(-3)^2 = 9$. In order to have a unique nth root, we define the (principal) nth root, $\sqrt[n]{a}$, as the *positive* number b that satisfies $b^n = a$.

We summarize this discussion as follows:

nth Root of a

The **(principal) nth root** of a real number is defined as

$$\sqrt[n]{a} = b \quad \text{only if} \quad a = b^n$$

subject to the following conditions:

	$a = 0$	$a > 0$	$a < 0$
n even	$\sqrt[n]{a} = 0$	$\sqrt[n]{a} > 0$	$\sqrt[n]{a}$ not real
n odd	$\sqrt[n]{a} = 0$	$\sqrt[n]{a} > 0$	$\sqrt[n]{a} < 0$

When we are asked for the root of a number, we give the principal root.

EXAMPLE 1

Find the roots, if they are real numbers.

(a) $\sqrt[6]{64}$ (b) $-\sqrt{16}$ (c) $\sqrt[3]{-8}$ (d) $\sqrt{-16}$

Solution

(a) $\sqrt[6]{64} = 2$ because $2^6 = 64$

(b) $-\sqrt{16} = -(\sqrt{16}) = -4$

(c) $\sqrt[3]{-8} = -2$

(d) $\sqrt{-16}$ is not a real number because an even root of a negative number is not real.

In order to perform evaluations on a scientific calculator or to perform calculus operations, it is sometimes necessary to rewrite radicals in exponential form with fractional exponents.

We have stated that for $a \geq 0$ and $b \geq 0$,

$$\sqrt{a} = b \quad \text{only if} \quad a = b^2$$

This means that $(\sqrt{a})^2 = b^2 = a$, or $(\sqrt{a})^2 = a$. In order to extend the properties of exponents to rational exponents, it is necessary to define

$$a^{1/2} = \sqrt{a} \quad \text{so that} \quad (a^{1/2})^2 = a$$

Exponent 1/n

For a positive integer n, we define

$$a^{1/n} = \sqrt[n]{a} \quad \text{if} \quad \sqrt[n]{a} \text{ exists}$$

Thus $(a^{1/n})^n = a^{(1/n) \cdot n} = a$.

Because we wish the properties established for integer exponents to extend to rational exponents, we make the following definitions.

Rational Exponents For positive integer n and any integer m (with $a \neq 0$ when $m \leq 0$):

1. $a^{m/n} = (a^{1/n})^m = (\sqrt[n]{a})^m$
2. $a^{m/n} = (a^m)^{1/n} = \sqrt[n]{a^m}$ if a is nonnegative when n is even.

Throughout the remaining discussion, we assume all expressions are real.

EXAMPLE 2

Write the following in radical form and simplify.

(a) $16^{3/4}$ (b) $y^{-3/2}$ (c) $(6m)^{2/3}$

Solution

(a) $16^{3/4} = \sqrt[4]{16^3} = (\sqrt[4]{16})^3 = (2)^3 = 8$

(b) $y^{-3/2} = \dfrac{1}{y^{3/2}} = \dfrac{1}{\sqrt{y^3}}$

(c) $(6m)^{2/3} = \sqrt[3]{(6m)^2} = \sqrt[3]{36m^2}$

EXAMPLE 3

Write the following without radical signs.

(a) $\sqrt{x^3}$ (b) $\dfrac{1}{\sqrt[3]{b^2}}$ (c) $\sqrt[3]{(ab)^3}$

Solution

(a) $\sqrt{x^3} = x^{3/2}$ (b) $\dfrac{1}{\sqrt[3]{b^2}} = \dfrac{1}{b^{2/3}} = b^{-2/3}$

(c) $\sqrt[3]{(ab)^3} = (ab)^{3/3} = ab$

Note that with many calculators, if we want to evaluate $\sqrt[10]{3.012}$, we must rewrite it as $(3.012)^{1/10}$. Our definition of $a^{m/n}$ guarantees that the rules for exponents will apply to fractional exponents. Thus we can perform operations with fractional exponents as we did with integer exponents.

EXAMPLE 4

Simplify the following expressions.

(a) $a^{1/2} \cdot a^{1/6}$ (b) $a^{3/4}/a^{1/3}$ (c) $(a^3b)^{2/3}$
(d) $(a^{3/2})^{1/2}$ (e) $a^{-1/2} \cdot a^{-3/2}$

Solution

(a) $a^{1/2} \cdot a^{1/6} = a^{1/2 + 1/6} = a^{3/6 + 1/6} = a^{4/6} = a^{2/3}$

(b) $a^{3/4}/a^{1/3} = a^{3/4 - 1/3} = a^{9/12 - 4/12} = a^{5/12}$

(c) $(a^3b)^{2/3} = (a^3)^{2/3}b^{2/3} = a^2b^{2/3}$

(d) $(a^{3/2})^{1/2} = a^{(3/2)(1/2)} = a^{3/4}$

(e) $a^{-1/2} \cdot a^{-3/2} = a^{-1/2 - 3/2} = a^{-2} = 1/a^2$

CHECKPOINT

1. Which of the following are *not* real numbers?
   (a) $\sqrt[3]{-64}$ (b) $\sqrt{-64}$
   (c) $\sqrt{0}$ (d) $\sqrt[4]{1}$
   (e) $\sqrt[5]{-1}$ (f) $\sqrt[8]{-1}$
2. (a) Write as radicals: $x^{1/3}$, $x^{2/5}$, $x^{-3/2}$
   (b) Write with fractional exponents: $\sqrt[4]{x^3} = x^?$, $\dfrac{1}{\sqrt{x}} = \dfrac{1}{x^?} = x^?$
3. Evaluate the following.
   (a) $8^{2/3}$ (b) $(-8)^{2/3}$ (c) $8^{-2/3}$
   (d) $-8^{-2/3}$ (e) $\sqrt[15]{71}$
4. Complete the following.
   (a) $x \cdot x^{1/3} \cdot x^3 = x^?$ (b) $x^2 \div x^{1/2} = x^?$
   (c) $(x^{-2/3})^{-3} = x^?$ (d) $x^{-3/2} \cdot x^{1/2} = x^?$
   (e) $x^{-3/2} \cdot x = x^?$ (f) $\left(\dfrac{x^4}{y^2}\right)^{3/2} = ?$
5. True or false:
   (a) $\dfrac{8x^{2/3}}{x^{-1/3}} = 4x$
   (b) $(16x^8y)^{3/4} = 12x^6y^{3/4}$
   (c) $\left(\dfrac{x^2}{y^3}\right)^{-1/3} = \left(\dfrac{y^3}{x^2}\right)^{1/3} = \dfrac{y}{x^{2/3}}$

We can perform operations with radicals by first rewriting in exponential form, performing the operations with exponents, and then converting the answer back to radical form. Another option is to apply directly the following rules for operations with radicals.

Rules for Radicals	**Examples**
Given that $\sqrt[n]{a}$ and $\sqrt[n]{b}$ are real,*	
1. $\sqrt[n]{a^n} = (\sqrt[n]{a})^n = a$	1. $\sqrt[5]{6^5} = (\sqrt[5]{6})^5 = 6$
2. $\sqrt[n]{a} \cdot \sqrt[n]{b} = \sqrt[n]{ab}$	2. $\sqrt[3]{2}\,\sqrt[3]{4} = \sqrt[3]{8} = \sqrt[3]{2^3} = 2$
3. $\dfrac{\sqrt[n]{a}}{\sqrt[n]{b}} = \sqrt[n]{\dfrac{a}{b}}$ $(b \neq 0)$	3. $\dfrac{\sqrt{18}}{\sqrt{2}} = \sqrt{\dfrac{18}{2}} = \sqrt{9} = 3$

*Note that this means $a \geq 0$ and $b \geq 0$ if n is even.

Let us consider Rule 1 for radicals more carefully. Note that if n is even and $a < 0$, then $\sqrt[n]{a}$ is not real, and Rule 1 does not apply. For example,

$$\sqrt{(-2)^2} \neq -2 \quad \text{because} \quad \sqrt{(-2)^2} = \sqrt{4} = 2 = -(-2).$$

We can generalize this observation as follows: If $a < 0$, then $\sqrt{a^2} = -a > 0$, so

$$\sqrt{a^2} = \begin{cases} a & \text{if } a \geq 0 \\ -a & \text{if } a < 0 \end{cases}$$

This means

$$\sqrt{a^2} = |a|$$

EXAMPLE 5

Simplify:

(a) $\sqrt[3]{8^3}$ (b) $\sqrt[5]{x^5}$ (c) $\sqrt{x^2}$

Solution

(a) $\sqrt[3]{8^3} = 8$ by Rule 1 for radicals

(b) $\sqrt[5]{x^5} = x$ (c) $\sqrt{x^2} = |x|$

Up to now, to *simplify* a radical has meant to find the indicated root. More generally, a radical expression $\sqrt[n]{x}$ is considered simplified if x has no nth powers as factors. Rule 2 for radicals ($\sqrt[n]{a} \cdot \sqrt[n]{b} = \sqrt[n]{ab}$) provides a procedure for simplifying radicals.

EXAMPLE 6

Simplify the following radicals; assume the expressions are real numbers.

(a) $\sqrt{48x^5y^6}$ $(y \geq 0)$

(b) $\sqrt[3]{72a^3b^4}$

Solution

(a) To simplify $\sqrt{48x^5y^6}$, we first factor $48x^5y^6$ into perfect-square factors and other factors. Then we apply Rule 2.

$$\sqrt{48x^5y^6} = \sqrt{16 \cdot 3 \cdot x^4xy^6} = \sqrt{16}\sqrt{x^4}\sqrt{y^6}\sqrt{3x} = 4x^2y^3\sqrt{3x}$$

(b) In this case, we factor $72a^3b^4$ into factors that are perfect cubes and other factors.

$$\sqrt[3]{72a^3b^4} = \sqrt[3]{8 \cdot 9a^3b^3b} = \sqrt[3]{8} \cdot \sqrt[3]{a^3} \cdot \sqrt[3]{b^3} \cdot \sqrt[3]{9b} = 2ab\sqrt[3]{9b}$$

Rule 2 for radicals also provides a procedure for multiplying two roots with the same index.

EXAMPLE 7

Multiply the following and simplify the answers, assuming nonnegative variables.

(a) $\sqrt[3]{2xy} \cdot \sqrt[3]{4x^2y}$ (b) $\sqrt{8xy^3z}\sqrt{4x^2y^3z^2}$

Solution

(a) $\sqrt[3]{2xy} \cdot \sqrt[3]{4x^2y} = \sqrt[3]{2xy \cdot 4x^2y} = \sqrt[3]{8x^3y^2} = \sqrt[3]{8} \cdot \sqrt[3]{x^3} \cdot \sqrt[3]{y^2} = 2x\sqrt[3]{y^2}$

(b) $\sqrt{8xy^3z}\,\sqrt{4x^2y^3z^2} = \sqrt{32x^3y^6z^3} = \sqrt{16x^2y^6z^2}\,\sqrt{2xz} = 4xy^3z\sqrt{2xz}$

Rule 3 for radicals ($\sqrt[n]{a}/\sqrt[n]{b} = \sqrt[n]{a/b}$) indicates how to find the quotient of two roots with the same index.

EXAMPLE 8

Find the quotients and simplify the answers, assuming nonnegative variables.

(a) $\dfrac{\sqrt[3]{32}}{\sqrt[3]{4}}$ (b) $\dfrac{\sqrt{16a^3x}}{\sqrt{2ax}}$

Solution

(a) $\dfrac{\sqrt[3]{32}}{\sqrt[3]{4}} = \sqrt[3]{\dfrac{32}{4}} = \sqrt[3]{8} = 2$

(b) $\dfrac{\sqrt{16a^3x}}{\sqrt{2ax}} = \sqrt{\dfrac{16a^3x}{2ax}} = \sqrt{8a^2} = 2a\sqrt{2}$

Occasionally, we wish to express a fraction containing radicals in an equivalent form that contains no radicals in the denominator. This is accomplished by multiplying the numerator *and* the denominator by the expression that will remove the radical. This process is called **rationalizing the denominator.**

EXAMPLE 9

Express each of the following with no radicals in the denominator. (Rationalize each denominator.)

(a) $\dfrac{15}{\sqrt{x}}$ (b) $\dfrac{2x}{\sqrt{18xy}}$ $(x, y > 0)$

(c) $\dfrac{3x}{\sqrt[3]{2x^2}}$

Solution

(a) We wish to create a perfect square under the radical in the denominator.

$$\frac{15}{\sqrt{x}} \cdot \frac{\sqrt{x}}{\sqrt{x}} = \frac{15\sqrt{x}}{x}$$

(b) $\dfrac{2x}{\sqrt{18xy}} \cdot \dfrac{\sqrt{2xy}}{\sqrt{2xy}} = \dfrac{2x\sqrt{2xy}}{\sqrt{36x^2y^2}} = \dfrac{2x\sqrt{2xy}}{6xy} = \dfrac{\sqrt{2xy}}{3y}$

(c) We wish to create a perfect cube under the radical in the denominator.

$$\frac{3x}{\sqrt[3]{2x^2}} \cdot \frac{\sqrt[3]{4x}}{\sqrt[3]{4x}} = \frac{3x\sqrt[3]{4x}}{\sqrt[3]{8x^3}} = \frac{3x\sqrt[3]{4x}}{2x} = \frac{3\sqrt[3]{4x}}{2}$$

CHECKPOINT

6. Simplify:
   (a) $\sqrt[7]{x^7}$ (b) $[\sqrt[5]{(x^2+1)^2}]^5$ (c) $\sqrt{12xy^2} \cdot \sqrt{3x^2y}$
7. Rationalize the denominator of $\dfrac{x}{\sqrt{5x}}$.

It is also sometimes useful, especially in calculus, to *rationalize the numerator* of a fraction. For example, we can rationalize the numerator of

$$\frac{\sqrt[3]{4x^2}}{3x}$$

by multiplying the numerator and denominator by $\sqrt[3]{2x}$, which creates a perfect cube under the radical:

$$\frac{\sqrt[3]{4x^2}}{3x} \cdot \frac{\sqrt[3]{2x}}{\sqrt[3]{2x}} = \frac{\sqrt[3]{8x^3}}{3x\sqrt[3]{2x}} = \frac{2x}{3x\sqrt[3]{2x}} = \frac{2}{3\sqrt[3]{2x}}.$$

CHECKPOINT SOLUTIONS

1. Only *even* roots of negatives are not real numbers. Thus $\sqrt{-64}$ and $\sqrt[8]{-1}$ are not real numbers.
2. (a) $x^{1/3} = \sqrt[3]{x}$, $x^{2/5} = \sqrt[5]{x^2}$, $x^{-3/2} = \dfrac{1}{x^{3/2}} = \dfrac{1}{\sqrt{x^3}}$
   (b) $\sqrt[4]{x^3} = x^{3/4}$, $\dfrac{1}{\sqrt{x}} = \dfrac{1}{x^{1/2}} = x^{-1/2}$
3. (a) $8^{2/3} = (\sqrt[3]{8})^2 = 2^2 = 4$
   (b) $(-8)^{2/3} = (\sqrt[3]{-8})^2 = (-2)^2 = 4$
   (c) $8^{-2/3} = \dfrac{1}{8^{2/3}} = \dfrac{1}{4}$
   (d) $-8^{-2/3} = -\left(\dfrac{1}{8^{2/3}}\right) = -\dfrac{1}{4}$
   (e) $\sqrt[15]{71} = (71)^{1/15} \approx 1.32867$
4. (a) $x \cdot x^{1/3} \cdot x^3 = x^{1+1/3+3} = x^{13/3}$ (b) $x^2 \div x^{1/2} = x^{2-1/2} = x^{3/2}$
   (c) $(x^{-2/3})^{-3} = x^{(-2/3)(-3)} = x^2$ (d) $x^{-3/2} \cdot x^{1/2} = x^{-3/2+1/2} = x^{-1}$
   (e) $x^{-3/2} \cdot x = x^{-3/2+1} = x^{-1/2}$ (f) $\left(\dfrac{x^4}{y^2}\right)^{3/2} = \dfrac{(x^4)^{3/2}}{(y^2)^{3/2}} = \dfrac{x^6}{y^3}$
5. (a) False. $\dfrac{8x^{2/3}}{x^{-1/3}} = 8x^{2/3} \cdot x^{1/3} = 8x^{2/3+1/3} = 8x$
   (b) False. $(16x^8y)^{3/4} = 16^{3/4}(x^8)^{3/4}y^{3/4} = (\sqrt[4]{16})^3x^6y^{3/4} = 8x^6y^{3/4}$
   (c) True.
6. (a) $\sqrt[7]{x^7} = x$ (b) $[\sqrt[5]{(x^2+1)^2}]^5 = (x^2+1)^2$
   (c) $\sqrt{12xy^2} \cdot \sqrt{3x^2y} = \sqrt{36x^3y^3} = \sqrt{36x^2y^2 \cdot xy} = \sqrt{36x^2y^2}\sqrt{xy} = 6xy\sqrt{xy}$
7. $\dfrac{x}{\sqrt{5x}} = \dfrac{x}{\sqrt{5x}} \cdot \dfrac{\sqrt{5x}}{\sqrt{5x}} = \dfrac{x\sqrt{5x}}{5x} = \dfrac{\sqrt{5x}}{5}$

EXERCISE 0.4

Unless stated otherwise, assume all variables are nonnegative and all denominators are nonzero.

In Problems 1–12, find the powers and roots, if they are real numbers.

1. $\sqrt{256/9}$
2. $\sqrt{1.44}$
3. $\sqrt[5]{-32^3}$
4. $\sqrt[4]{-16^5}$
5. $16^{3/4}$
6. $32^{3/5}$
7. $(-16)^{-3/2}$
8. $-27^{-1/3}$
9. $\left(\frac{8}{27}\right)^{-2/3}$
10. $\left(\frac{4}{9}\right)^{3/2}$
11. (a) $8^{2/3}$ (b) $(-8)^{-2/3}$
12. (a) $8^{-2/3}$ (b) $-8^{2/3}$

In Problems 13 and 14, rewrite each radical with a fractional exponent, and then approximate the value with a calculator.

13. $\sqrt[9]{(6.12)^4}$
14. $\sqrt[12]{4.96}$

In Problems 15–18, replace each radical with a fractional exponent. Do not simplify.

15. $\sqrt{m^3}$
16. $\sqrt[3]{x^5}$
17. $\sqrt[4]{m^2n^5}$
18. $\sqrt[5]{x^3}$

In Problems 19–22, write in radical form. Do not simplify.

19. $x^{7/4}$
20. $y^{11/5}$
21. $-(1/4)x^{-5/4}$
22. $-x^{-5/3}$

In Problems 23–36, use the properties of exponents to simplify each expression so that only positive exponents remain.

23. $y^{1/4} \cdot y^{1/2}$
24. $x^{2/3} \cdot x^{1/5}$
25. $z^{3/4} \cdot z^4$
26. $x^{-2/3} \cdot x^2$
27. $y^{-3/2} \cdot y^{-1}$
28. $z^{-2} \cdot z^{5/3}$
29. $\dfrac{x^{1/3}}{x^{-2/3}}$
30. $\dfrac{x^{-1/2}}{x^{-3/2}}$
31. $\dfrac{y^{-5/2}}{y^{-2/5}}$
32. $\dfrac{x^{4/9}}{x^{1/12}}$
33. $(x^{2/3})^{3/4}$
34. $(x^{4/5})^3$
35. $(x^{-1/2})^2$
36. $(x^{-2/3})^{-2/5}$

In Problems 37–42, simplify each expression by using the properties of radicals. Assume nonnegative variables.

37. $\sqrt{64x^4}$
38. $\sqrt[3]{-64x^6y^3}$
39. $\sqrt{128x^4y^5}$
40. $\sqrt[3]{54x^5x^8}$
41. $\sqrt[3]{40x^8y^5}$
42. $\sqrt{32x^5y}$

In Problems 43–50, perform the indicated operations and simplify.

43. $\sqrt{12x^3y} \cdot \sqrt{3x^2y}$
44. $\sqrt[3]{16x^2y} \cdot \sqrt[3]{3x^2y}$
45. $\sqrt{63x^5y^3} \cdot \sqrt{28x^2y}$
46. $\sqrt{10xz^{10}} \cdot \sqrt{30x^{17}z}$
47. $\dfrac{\sqrt{12x^3y^{12}}}{\sqrt{27xy^2}}$
48. $\dfrac{\sqrt{250xy^7z^4}}{\sqrt{18x^{17}y^2}}$
49. $\dfrac{\sqrt[4]{32a^9b^5}}{\sqrt[4]{162a^{17}}}$
50. $\dfrac{\sqrt[3]{-16x^3y^4}}{\sqrt[3]{128y^2}}$

In Problems 51–54, use properties of exponents and radicals to determine a value for x that makes each statement true.

51. $(A^9)^x = A$
52. $(B^{20})^x = B$
53. $(\sqrt[7]{R})^x = R$
54. $(\sqrt{T^3})^x = T$

In Problems 55–60, rationalize each denominator and then simplify.

55. $\sqrt{2/3}$
56. $\sqrt{5/8}$
57. $\dfrac{\sqrt{m^2x}}{\sqrt{mx^2}}$
58. $\dfrac{5x^3w}{\sqrt{4xw^2}}$
59. $\dfrac{\sqrt[3]{m^2x}}{\sqrt[3]{mx^5}}$
60. $\dfrac{\sqrt[4]{mx^3}}{\sqrt[4]{y^2z^5}}$

In calculus it is frequently important to write an expression in the form cx^n, where c is a constant. In Problems 61–64, write each expression in this form.

61. $\dfrac{-2}{3\sqrt[3]{x^2}}$
62. $\dfrac{-2}{3\sqrt[4]{x^3}}$
63. $3x\sqrt{x}$
64. $\sqrt{x} \cdot \sqrt[3]{x}$

In calculus problems, the answers are frequently expected to be in a simple form with a radical instead of an exponent. In Problems 65–68, write each expression with radicals.

65. $\dfrac{3}{2}x^{1/2}$
66. $\dfrac{4}{3}x^{1/3}$
67. $\dfrac{1}{2}x^{-1/2}$
68. $\dfrac{-1}{2}x^{-3/2}$

Applications

69. ***Richter scale*** The Richter scale reading for an earthquake measures its intensity (as a multiple of some minimum intensity used for comparison). The intensity I corresponding to a Richter scale reading R is given by

$$I = 10^R.$$

(a) A quake measuring 8.5 on the Richter scale would be severe. Express the intensity of such a quake in exponential form and in radical form.
(b) Find the intensity of a quake measuring 8.5.
(c) The San Francisco quake that occurred during the 1989 World Series measured 7.1, and the 1906 San Francisco quake (which devastated the city) measured 8.25. Calculate the ratio of these intensities (larger to smaller).

70. ***Sound intensity*** The intensity of sound I (as a multiple of the average minimum threshold of hearing intensity) is related to the decibel level D (or loudness of sound) according to

$$I = 10^{D/10}.$$

(a) Express $10^{D/10}$ using radical notation.
(b) The background noise level of a relatively quiet room has a decibel reading of 32. Find the intensity I_1 of this noise level.
(c) A decibel reading of 140 is at the threshold of pain. If I_2 is the intensity of this threshold and I_1 is the intensity found in (b), express the ratio I_2/I_1 as a power of 10. Then approximate this ratio.

Half-life In Problems 71 and 72, use the fact that the quantity of a radioactive substance after t years is given by $q = q_0(2^{-t/k})$, where q_0 is the original amount of radioactive material and k is its **half-life** (the number of years it takes for half the radioactive substance to decay).

71. The half-life of strontium-90 is 25 years. Find the amount of strontium-90 remaining after 10 years if $q_0 = 98$ kg.
72. The half-life of carbon-14 is 5600 years. Find the amount of carbon-14 remaining after 10,000 years if $q_0 = 40.0$ g.

73. ***Population growth*** Suppose the formula for the growth of the population of a city for the next 10 years is given by

$$P = P_0(2.5)^{ht}$$

where P_0 is the population of the city at the present time and P is the population t years from now. If $h = 0.03$ and $P_0 = 30{,}000$, find P when $t = 10$.

74. ***Advertising and sales*** Suppose it has been determined that the sales at Ewing Gallery decline after the end of an advertising campaign, with daily sales given by

$$S = 2000(2^{-0.1x})$$

where S is in dollars and x is the number of days after the campaign ends. What is the daily sales 10 days after the end of the campaign?

75. ***Company growth*** The growth of a company can be described by the equation

$$N = 500(0.02)^{0.7^t}$$

where t is the number of years the company has been in existence and N is the number of employees.
(a) What is the number of employees when $t = 0$? (This is the number of employees the company has when it starts.)
(b) What is the number of employees when $t = 5$?

0.5 Operations with Algebraic Expressions

In algebra we are usually dealing with combinations of real numbers (such as 3, 6/7, and $-\sqrt{2}$) and letters (such as x, a, and m). Unless otherwise specified, the letters are symbols used to represent real numbers and are sometimes called **variables.** An expression obtained by performing additions, subtractions, multiplications, divisions, or extraction of roots with one or more real numbers or variables is called an **algebraic expression.** Unless otherwise specified, the variables represent all real numbers for which the algebraic expression is a real number. Examples of algebraic expressions include

$$3x + 2y, \quad \frac{x^3y + y}{x - 1}, \quad \text{and} \quad \sqrt{x} - 3.$$

Note that the variable x cannot be negative in the expression $\sqrt{x} - 3$ and that $(x^3y + y)/(x - 1)$ is not a real number when $x = 1$, because division by 0 is undefined.

Any product of a real number (called the **coefficient**) and one or more variables to powers is called a **term.** The sum of a finite number of terms with nonnegative integer powers on the variables is called a **polynomial.** If a polynomial contains only one variable x, then it is called a polynomial in x.

Polynomial in x

The general form of a **polynomial in *x*** is

$$a_nx^n + a_{n-1}x^{n-1} + \cdots + a_1x + a_0$$

where each coefficient a_i is a real number and where $i = 0, 1, 2, \ldots, n$. If $a_n \neq 0$, the **degree** of the polynomial is n, and a_n is called the **leading coefficient.** The term a_0 is called the **constant term.**

Thus $4x^3 - 2x + 3$ is a third-degree polynomial in x with leading coefficient 4. If two or more variables are in a term, the degree of the term is the sum of the exponents of the variables. The degree of a nonzero constant term is zero. Thus the degree of $4x^2y$ is $2 + 1 = 3$, the degree of $6xy$ is $1 + 1 = 2$, and the degree of 3 is 0. The **degree of a polynomial** containing one or more variables is the degree of the term in the polynomial having the highest degree. Therefore $2xy - 4x + 6$ is a second-degree polynomial.

A polynomial containing two terms is called a **binomial,** and a polynomial containing three terms is called a **trinomial.** A single-term polynomial is a **monomial.**

Because monomials and polynomials represent real numbers, the properties of real numbers can be used to add, subtract, multiply, divide, and simplify polynomials. For example, we can use the Distributive Law to add $3x$ and $2x$.

$$3x + 2x = (3 + 2)x = 5x$$

Similarly, $9xy - 3xy = (9 - 3)xy = 6xy$.

Terms with exactly the same variable factors are called **like terms.** We can add or subtract like terms by adding or subtracting the coefficients of the variables.

EXAMPLE 1

Compute $(4xy + 3x) + (5xy - 2x)$.

Solution

$$(4xy + 3x) + (5xy - 2x) = 4xy + 3x + 5xy - 2x$$
$$= 9xy + x$$

Subtraction of polynomials uses the Distributive Law to remove the parentheses.

EXAMPLE 2

Compute $(3x^2 + 4xy + 5y^2 + 1) - (6x^2 - 2xy + 4)$.

Solution

Removing the parentheses gives

$$3x^2 + 4xy + 5y^2 + 1 - 6x^2 + 2xy - 4$$

which simplifies to

$$-3x^2 + 6xy + 5y^2 - 3.$$

Using the rules of exponents and the Commutative and Associative Laws for multiplication, we can multiply and divide monomials, as the following example shows.

EXAMPLE 3

Perform the indicated operations.

(a) $(8xy^3)(2x^3y)(-3xy^2)$ (b) $-15x^2y^3 \div (3xy^5)$

Solution

(a) $8 \cdot 2 \cdot (-3) \cdot x \cdot x^3 \cdot x \cdot y^3 \cdot y \cdot y^2 = -48x^5y^6$

(b) $\dfrac{-15x^2y^3}{3xy^5} = -\dfrac{15}{3} \cdot \dfrac{x^2}{x} \cdot \dfrac{y^3}{y^5} = -5 \cdot x \cdot \dfrac{1}{y^2} = -\dfrac{5x}{y^2}$

Symbols of grouping are used in algebra in the same way as they are used in the arithmetic of real numbers. We have removed parentheses in the process of adding and subtracting polynomials. Other symbols of grouping, such as brackets, [ ], are treated the same as parentheses.

When there are two or more symbols of grouping involved, we begin with the innermost and work outward.

EXAMPLE 4

Simplify $3x^2 - [2x - (3x^2 - 2x)]$.

Solution

$$\begin{aligned}3x^2 - [2x - (3x^2 - 2x)] &= 3x^2 - [2x - 3x^2 + 2x]\\ &= 3x^2 - [4x - 3x^2]\\ &= 3x^2 - 4x + 3x^2\\ &= 6x^2 - 4x\end{aligned}$$

By the use of the Distributive Law, we can multiply a binomial by a monomial. For example,

$$x(2x + 3) = x \cdot 2x + x \cdot 3 = 2x^2 + 3x.$$

We can extend the Distributive Law to multiply polynomials with more than two terms. For example,

$$5(x + y + 2) = 5x + 5y + 10.$$

EXAMPLE 5

Find the following products.

(a) $-4ab(3a^2b + 4ab^2 - 1)$
(b) $(4a + 5b + c)ac$

Solution

(a) $-4ab(3a^2b + 4ab^2 - 1) = -12a^3b^2 - 16a^2b^3 + 4ab$
(b) $(4a + 5b + c)ac = 4a \cdot ac + 5b \cdot ac + c \cdot ac = 4a^2c + 5abc + ac^2$

The Distributive Law can be used to show us how to multiply two polynomials. Consider the indicated multiplication $(a + b)(c + d)$. If we first treat the sum $(a + b)$ as a single quantity, then two successive applications of the Distributive Law give

$$(a + b)(c + d) = (a + b) \cdot c + (a + b) \cdot d = ac + bc + ad + bd.$$

Thus we see that the product can be found by multiplying $(a + b)$ by c, multiplying $(a + b)$ by d, and then adding the products. This is frequently set up as follows.

Product of Two Polynomials

Procedure	***Example***
To multiply two polynomials:	Multiply $(3x + 4xy + 3y)$ by $(x - 2y)$.
1. Write one of the polynomials above the other.	1. $\begin{array}{r}3x + 4xy + 3y\\ \underline{x - 2y}\end{array}$
2. Multiply each term of the top polynomial by each term of the bottom one, and write the similar terms of the product under one another.	2. $\begin{array}{l}3x^2 + 4x^2y + 3xy\\ \underline{ - 6xy - 8xy^2 - 6y^2}\end{array}$
3. Add like terms to simplify the product.	3. $3x^2 + 4x^2y - 3xy - 8xy^2 - 6y^2$

EXAMPLE 6

Multiply $(4x^2 + 3xy + 4x)$ by $(2x - 3y)$.

Solution

$$\begin{array}{rrrrr} 4x^2 + 3xy + 4x & & & & \\ 2x - 3y & & & & \\ \hline 8x^3 + 6x^2y + 8x^2 & & & & \\ -12x^2y & & -9xy^2 - 12xy & & \\ \hline 8x^3 - 6x^2y + 8x^2 - 9xy^2 - 12xy & & & & \end{array}$$

Because the multiplications we must perform often involve binomials, the following special products are worth remembering.

Special Products

A. $(x + a)(x + b) = x^2 + (a + b)x + ab$
B. $(ax + b)(cx + d) = acx^2 + (ad + bc)x + bd$

It is easier to remember these two special products if we note their structure. We can obtain these products by finding the products of the First terms, Outside terms, Inside terms, and Last terms, and then adding the results. This is called the FOIL method of multiplying two binomials.

EXAMPLE 7

Multiply the following.

(a) $(x - 4)(x + 3)$ (b) $(3x + 2)(2x + 5)$

Solution

First Outside Inside Last

(a) $(x - 4)(x + 3) = (x^2) + (3x) + (-4x) + (-12) = x^2 - x - 12$
(b) $(3x + 2)(2x + 5) = (6x^2) + (15x) + (4x) + (10)$
$= 6x^2 + 19x + 10$

Additional special products are as follows:

Additional Special Products

C. $(x + a)^2 = x^2 + 2ax + a^2$ binomial squared
D. $(x - a)^2 = x^2 - 2ax + a^2$ binomial squared
E. $(x + a)(x - a) = x^2 - a^2$ difference of two squares
F. $(x + a)^3 = x^3 + 3ax^2 + 3a^2x + a^3$ binomial cubed
G. $(x - a)^3 = x^3 - 3ax^2 + 3a^2x - a^3$ binomial cubed

EXAMPLE 8

Multiply the following.

(a) $(x + 5)^2$ (b) $(3x - 4y)^2$ (c) $(x - 2)(x + 2)$
(d) $(x^2 - y^3)^2$ (e) $(x + 4)^3$

Solution

(a) $(x + 5)^2 = x^2 + 2(5)x + 25 = x^2 + 10x + 25$
(b) $(3x - 4y)^2 = (3x)^2 - 2(3x)(4y) + (4y)^2 = 9x^2 - 24xy + 16y^2$
(c) $(x - 2)(x + 2) = x^2 - 4$
(d) $(x^2 - y^3)^2 = (x^2)^2 - 2(x^2)(y^3) + (y^3)^2 = x^4 - 2x^2y^3 + y^6$
(e) $(x + 4)^3 = x^3 + 3(4)(x^2) + 3(4^2)(x) + 4^3 = x^3 + 12x^2 + 48x + 64$

CHECKPOINT

1. Remove parentheses and combine like terms. $9x - 5x(x + 2) + 4x^2$
2. Find the following products.
   (a) $(2x + 1)(4x^2 - 2x + 1)$ (b) $(x + 3)^2$
   (c) $(3x + 2)(x - 5)$ (d) $(1 - 4x)(1 + 4x)$

All algebraic expressions can represent real numbers, so the techniques used to perform operations on polynomials and to simplify polynomials also apply to other algebraic expressions.

EXAMPLE 9

Perform the indicated operations.

(a) $3\sqrt{3} + 4x\sqrt{y} - 5\sqrt{3} - 11x\sqrt{y} - (\sqrt{3} - x\sqrt{y})$
(b) $x^{3/2}(x^{1/2} - x^{-1/2})$ (c) $(x^{1/2} - x^{1/3})^2$ (d) $(\sqrt{x} + 2)(\sqrt{x} - 2)$

Solution

(a) We remove parentheses and then combine the terms containing $\sqrt{3}$ and the terms containing $x\sqrt{y}$.

$$(3 - 5 - 1)\sqrt{3} + (4 - 11 + 1)x\sqrt{y} = -3\sqrt{3} - 6x\sqrt{y}$$

(b) $x^{3/2}(x^{1/2} - x^{-1/2}) = x^{3/2} \cdot x^{1/2} - x^{3/2} \cdot x^{-1/2} = x^2 - x$
(c) $(x^{1/2} - x^{1/3})^2 = (x^{1/2})^2 - 2x^{1/2}x^{1/3} + (x^{1/3})^2 = x - 2x^{5/6} + x^{2/3}$
(d) $(\sqrt{x} + 2)(\sqrt{x} - 2) = (\sqrt{x})^2 - (2)^2 = x - 4$

In later chapters we will need to write problems in a simplified form so that we can perform certain operations on them. We can often use division of one polynomial by another to obtain the simplification, as shown in the following.

Division of Polynomials

Procedure

To divide one polynomial by another:

1. Write with both polynomials in descending powers of a variable. Include missing terms with coefficient 0 in the dividend.
2. (a) Divide the highest power of the divisor into the highest power of the dividend, and write this partial quotient above the dividend. Multiply the partial quotient times the divisor, write the product under the dividend, and subtract, getting a new dividend.
   (b) Repeat until the degree of the new dividend is less than the degree of the divisor. Any remainder is written over the divisor and added to the quotient.

Example

Divide $4x^3 + 4x^2 + 5$ by $2x^2 + 1$.

1. $2x^2 + 1\overline{)4x^3 + 4x^2 + 0x + 5}$

2. (a)
$$\begin{array}{r} 2x \\ 2x^2+1\overline{)4x^3 + 4x^2 + 0x + 5} \\ \underline{4x^3 + 2x } \\ 4x^2 - 2x + 5 \end{array}$$

(b)
$$\begin{array}{r} 2x + 2 \\ 2x^2+1\overline{)4x^3 + 4x^2 + 0x + 5} \\ \underline{4x^3 + 2x } \\ 4x^2 - 2x + 5 \\ \underline{4x^2 + 2} \\ -2x + 3 \end{array}$$

Degree $(-2x + 3) <$ degree $(2x^2 + 1)$

Quotient: $2x + 2 + \dfrac{-2x + 3}{2x^2 + 1}$

EXAMPLE 10

Divide $(4x^3 - 13x - 22)$ by $(x - 3)$, $x \neq 3$.

Solution

$$\begin{array}{r} 4x^2 + 12x + 23 \\ x-3\overline{)4x^3 + 0x^2 - 13x - 22} \\ \underline{4x^3 - 12x^2 } \\ 12x^2 - 13x - 22 \\ \underline{12x^2 - 36x } \\ 23x - 22 \\ \underline{23x - 69} \\ 47 \end{array}$$

$0x^2$ is inserted so that each power of x is present.

The quotient is $4x^2 + 12x + 23$, with remainder 47, or

$$4x^2 + 12x + 23 + \frac{47}{x - 3}$$

CHECKPOINT

3. Use long division to find $(x^3 + 2x + 7) \div (x - 4)$.

Technology Note One important use of algebraic expressions is to describe relationships among quantities. For example, the expression "one more than a number" could be written as $n + 1$, where n represents an arbitrary number. This ability to represent quantities or their interrelationships algebraically is one of the keys to using spreadsheets.

TABLE 0.3

	A	B	C	D
1	cell A1	cell B1	cell C1	.
2	cell A2	cell B2	cell C1	
3	cell A3	cell B3	cell C1	
4	.			
5	.			

Each cell in a spreadsheet has an address based on its row and column (see Table 0.3). These cell addresses can act like variables in an algebraic expression, and the "fill down" or "fill across" capabilities update this cell referencing while maintaining algebraic relationships. For example, we noted previously that if \$1000 is invested in an account that earns 10%, compounded annually, then the future value of the account after n years is given by $\$1000(1.1)^n$. We can track the future value by starting with the spreadsheet shown in Table 0.4.

TABLE 0.4

	A	B
1	Year	Future value
2	1	"=1000* (1.1) ^ (A2)"
3	"=A2+1"	

Type what is in quotes (not the quotes).

gives →

	A	B
1	Year	Future value
2	1	1100
3	2	

The use of the = sign to begin a cell entry indicates an algebraic expression whose variables are other cells. In cell A3 of Table 0.4, the entry "=A2+1" creates a referencing scheme based on the algebraic expression $n + 1$, but with cell A2 acting as the variable. From this beginning, highlighting cell A3 and using the "fill down" command updates the referencing from cell to cell and creates a counter for the number of years (see Table 0.5 on the next page).

The future value is given by a formula, so we use that formula in the cell but replace the variable with the cell name where a value of that variable can be found. The "fill down" command updates the referencing and hence the values used. The first 15 years of the spreadsheet begun in Table 0.4 are shown in Table 0.5, along with a new column for interest earned. Because the interest earned is found by subtracting the original investment of \$1000 from the future value, cell C2 contains the entry "=B2−1000" and those below it are obtained by using "fill down." The entries for future value and interest earned also could be expressed in dollars and cents.

TABLE 0.5

	A	B	C
1	Year	Future value	Interest earned
2	1	1100	100
3	2	1210	210
4	3	1331	331
5	4	1464.1	464.1
6	5	1610.51	610.51
7	6	1771.561	771.561
8	7	1948.7171	948.7171
9	8	2143.58881	1143.58881
10	9	2357.947691	1357.947691
11	10	2593.7424601	1593.74246
12	11	2853.1167061	1853.116706
13	12	3138.4283767	2138.428377
14	13	3452.2712144	2452.271214
15	14	3797.4983358	2797.498336
16	15	4177.2481694	3177.248169

CHECKPOINT SOLUTIONS

1. $$\begin{aligned} 9x - 5x(x + 2) + 4x^2 &= 9x - 5x^2 - 10x + 4x^2 \\ &= -x^2 - x \end{aligned}$$

   Note that without parentheses around $9x - 5x$, multiplication has priority over subtraction.

2. (a) $$\begin{array}{r} 4x^2 - 2x + 1 \\ 2x + 1 \\ \hline 8x^3 - 4x^2 + 2x \\ 4x^2 - 2x + 1 \\ \hline 8x^3 \qquad\qquad + 1 \end{array}$$

   (b) $(x + 3)^2 = x^2 + 2(3x) + 3^2 = x^2 + 6x + 9$

   (c) $(3x + 2)(x - 5) = 3x^2 - 15x + 2x - 10 = 3x^2 - 13x - 10$

   (d) $(1 - 4x)(1 + 4x) = 1 - 16x^2$ Note that this is different from $16x^2 - 1$.

3. $$\begin{array}{r} x^2 + 4x + 18 \\ x - 4\overline{)x^3 + 0x^2 + 2x + 7} \\ x^3 - 4x^2 \qquad\qquad \\ \hline 4x^2 + 2x + 7 \\ 4x^2 - 16x \qquad \\ \hline 18x + 7 \\ 18x - 72 \\ \hline 79 \end{array}$$

   The answer is $x^2 + 4x + 18 + \dfrac{79}{x - 4}$.

EXERCISE 0.5

For each polynomial in Problems 1–4, (a) give the degree of the polynomial, (b) give the coefficient (numerical) of the highest-degree term, (c) give the constant term, and (d) decide whether it is a polynomial of one or several variables.

1. $10 - 3x - x^2$
2. $5x^4 - 2x^9 + 7$
3. $7x^2y - 14xy^3z$
4. $2x^5 + 7x^2y^3 - 5y^6$

The expressions in Problems 5 and 6 are polynomials (that is, they have the form $a_nx^n + a_{n-1}x^{n-1} + \cdots + a_1x + a_0$).

5. For $2x^5 - 3x^2 - 5$,
   (a) $2 = a_?$ (b) $a_3 = ?$
   (c) $-3 = a_?$ (d) $a_0 = ?$
6. For $5x^3 - 4x - 17$,
   (a) $a_3 = ?$ (b) $a_1 = ?$
   (c) $a_2 = ?$ (d) $-17 = a_?$

In Problems 7–10, evaluate each algebraic expression at the indicated values of the variables.

7. $4x - x^2$ at $x = -2$
8. $3x^2 - 4y^2 - 2xy$ at $x = 3$ and $y = -4$
9. $\dfrac{2x - y}{x^2 - 2y}$ at $x = -5$ and $y = -3$
10. $\dfrac{16y}{1 - y}$ at $y = -3$

In Problems 11–18, simplify by combining like terms.

11. $(16pq - 7p^2) + (5pq + 5p^2)$
12. $(3x^3 + 4x^2y^2) + (3x^2y^2 - 7x^3)$
13. $(4m^2 - 3n^2 + 5) - (3m^2 + 4n^2 + 8)$
14. $(4rs - 2r^2s - 11rs^2) - (11rs^2 - 2rs + 4r^2s)$
15. $-[8 - 4(q + 5) + q]$
16. $x^3 + [3x - (x^3 - 3x)]$
17. $x^2 - [x - (x^2 - 1) + 1 - (1 - x^2)] + x$
18. $y^3 - [y^2 - (y^3 + y^2)] - [y^3 + (1 - y^2)]$

In Problems 19–64, perform the indicated operations and simplify.

19. $(5x^3)(7x^2)$
20. $(-3x^2y)(2xy^3)(4x^2y^2)$
21. $(39r^3s^2) \div (13r^2s)$
22. $(-15m^3n) \div (5mn^4)$
23. $ax^2(2x^2 + ax + ab)$
24. $-3(3 - x^2)$
25. $(3y + 4)(2y - 3)$
26. $(4x - 1)(x - 3)$
27. $(1 - 2x^2)(2 - x^2)$
28. $(x^3 + 3)(2x^3 - 5)$
29. $(4x + 3)^2$
30. $(2y + 5)^2$
31. $\left(x^2 - \dfrac{1}{2}\right)^2$
32. $(x^3y^3 - 0.3)^2$
33. $(2x + 1)(2x - 1)$
34. $(5y + 2)(5y - 2)$
35. $(0.1 - 4x)(0.1 + 4x)$
36. $\left(\dfrac{2}{3} + x\right)\left(\dfrac{2}{3} - x\right)$
37. $(x - 2)(x^2 + 2x + 4)$
38. $(a + b)(a^2 - ab + b^2)$
39. $(x^3 + 5x)(x^5 - 2x^3 + 5)$
40. $(x^3 - 1)(x^7 - 2x^4 - 5x^2 + 5)$
41. $(18m^2n + 6m^3n + 12m^4n^2) \div (6m^2n)$
42. $(16x^2 + 4xy^2 + 8x) \div (4xy)$
43. $(24x^8y^4 + 15x^5y - 6x^7y) \div (9x^5y^2)$
44. $(27x^2y^2 - 18xy + 9xy^2) \div (6xy)$
45. $(x + 1)^3$
46. $(x - 3)^3$
47. $(2x - 3)^3$
48. $(3x + 4)^3$
49. $(0.1x - 2)(x + 0.05)$
50. $(6.2x + 4.1)(6.2x - 4.1)$
51. $(x^3 + x - 1) \div (x + 2)$
52. $(x^5 + 5x - 7) \div (x + 1)$
53. $(x^4 + 3x^3 - x + 1) \div (x^2 + 1)$
54. $(x^3 + 5x^2 - 6) \div (x^2 - 2)$
55. $x^{1/2}(x^{1/2} + 2x^{3/2})$
56. $x^{-2/3}(x^{5/3} - x^{-1/3})$
57. $(x^{1/2} + 1)(x^{1/2} - 2)$
58. $(x^{1/3} - x^{1/2})(4x^{2/3} - 3x^{3/2})$
59. $(\sqrt{x} + 3)(\sqrt{x} - 3)$
60. $(x^{1/5} + x^{1/2})(x^{1/5} - x^{1/2})$
61. $(2x + 1)^{1/2}[(2x + 1)^{3/2} - (2x + 1)^{-1/2}]$
62. $(4x - 3)^{-5/3}[(4x - 3)^{8/3} + 3(4x - 3)^{5/3}]$
63. (a) $(3x - 2)^2 - 3x - 2(3x - 2) + 5$
    (b) $(3x - 2)^2 - (3x - 2)(3x - 2) + 5$
64. (a) $(2x - 3)(3x + 2) - (5x - 2)(x - 3)$
    (b) $2x - 3(3x + 2) - 5x - 2(x - 3)$

Applications

65. ***Revenue*** A company sells its product for \$55 per unit. Write an expression for the amount of money received (revenue) from the sale of x units of the product.
66. ***Profit*** Suppose a company's revenue R (in dollars) from the sale of x units of its product is given by

$$R = 215x.$$

Suppose further that the total costs C (in dollars) of producing those x units is given by

$$C = 65x + 15{,}000.$$

(a) If profit is revenue minus cost, find an expression for the profit from the production and sale of x units.
(b) Find the profit received if 1000 units are sold.

67. ***Investments*** Suppose that you have \$4000 to invest, and you invest x dollars at 10% and the remainder at 8%. Write an expression that represents
    (a) the amount invested at 8%.
    (b) the interest earned on the x dollars at 10%.
    (c) the interest earned on the money invested at 8%.
    (d) the total interest earned.

68. ***Medications*** Suppose that a nurse needs 10 cc (cubic centimeters) of a 15.5% solution (that is, a solution that is 15.5% ingredient) of a certain medication, which must be obtained by mixing x cc of a 20% solution and y cc of a 5% solution. Write expressions involving x for
    (a) y, the amount of 5% solution.
    (b) the amount of ingredient in the x cc of 20% solution.
    (c) the amount of ingredient in the 5% solution.
    (d) the total amount of ingredient in the mixture.

69. ***Package design*** The volume of a rectangular box is given by $V =$ (length)(width)(height). If a rectangular piece of cardboard that is 10 in. by 15 in. has a square with sides of length x cut from each corner (see Figure 0.9), and if the sides are folded up along the dotted lines to form a box, what expression of x would represent the volume?

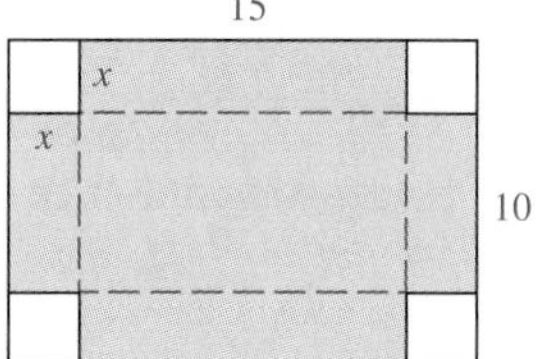

Figure 0.9

Problems 70–73 use spreadsheets.

70. ***Coin mix*** Suppose an individual has 125 coins, with some nickels and some dimes. We can use a spreadsheet to investigate the value for various numbers of nickels and dimes. Suppose we set up a spreadsheet with column A for the number of nickels, column B for the number of dimes, and column C for the value of the coins. Then the beginning of the spreadsheet might look like this:

	A	B	C
1	nickels	dimes	value
2	0	125	12.5
3	1	124	12.45
4	2	123	12.4

    (a) With 0 as the entry in cell A2, what cell A3 entry references A2 and could be used to fill column A with the possible numbers of nickels?
    (b) Knowing that the total number of coins is 125, how do we find the number of dimes if we know the number of nickels?
    (c) Use the result in (b) to write an entry for cell B2 that is referenced to A2 and that could be used to fill column B with the correct number of dimes for each possible number of nickels in column A.
    (d) How do you use the entries in columns A and B to find the corresponding value for column C? Write a rule.
    (e) Adapt the rule from (d) to write a C2 entry (referenced appropriately) that could be used to give the correct value for each different number of coins.
    (f) Use a spreadsheet to determine how many nickels and how many dimes (with 125 total coins) give the value \$7.95.

71. ***Area*** Suppose we were interested in different areas for rectangles for which the sum of their length and width is 50. If the length and width must be whole numbers, the beginning of a spreadsheet listing the possibilities might look like this:

	A	B	C
1	length	width	area = (l)(w)
2	49	1	49
3	48	2	96
4	47	3	141

    (a) With cell A2 as 49, what entry in cell A3 (referenced to A2) could be used to fill column A with the possible lengths?
    (b) Because the sum of the length and width must be 50, what rule would we use to find the width if we knew the length?
    (c) Use your result from (b) to write an entry for cell B2 that is referenced to A2 and that could be used to fill column B with the correct widths for each of the lengths in column A.
    (d) What entry in cell C2 (appropriately referenced) could be used to fill column C with the area for any pair of lengths and widths?
    (e) Use a spreadsheet to determine the (whole-number) length and width for which we get the smallest area that is still greater than 550.

72. ***Cable rates*** By using data for 1982–1997 from Nielsen Media Research, the average monthly cost C for basic cable TV service can be approximated quite accurately by

$$C = 0.0232t^2 + 0.7984t + 5.8271$$

where t is the number of years past 1980. Use a spreadsheet to track the predicted cost of basic cable TV service from 1995 to 2010.

73. ***Tax load*** By using data from 1955–1997, the percentage P of average income used to pay federal, state, and local taxes can be approximated by

$$P = 0.24627t + 25.96473$$

where t is the number of years past 1950.
(a) Use a spreadsheet to track the approximate tax loads from 1995 to 2010.
(b) In what year will the average tax load first surpass 40%?

0.6 Factoring

We can factor monomial factors out of a polynomial by using the Distributive Law in reverse; $ab + ac = a(b + c)$ is an example showing that a is a monomial factor of the polynomial $ab + ac$. But it is also a statement of the Distributive Law (with the sides of the equation interchanged). The monomial factor of a polynomial must be a factor of each term of the polynomial, so it is frequently called a **common monomial factor.**

EXAMPLE 1

Factor $-3x^2t - 3x + 9xt^2$.

Solution

1. We can factor out $3x$ and obtain

$$-3x^2t - 3x + 9xt^2 = 3x(-xt - 1 + 3t^2)$$

2. Or we can factor out $-3x$ (factoring out the negative will make the first term of the polynomial positive) and obtain

$$-3x^2t - 3x + 9xt^2 = -3x(xt + 1 - 3t^2).$$

If a factor is common to each term of a polynomial, we can use this procedure to factor it out, even if it is not a monomial. For example, we can factor $(a + b)$ out of the polynomial $2x(a + b) - 3y(a + b)$. If we factor $(a + b)$ from both terms, we get $(a + b)(2x - 3y)$. The following example demonstrates the **factoring by grouping** technique.

EXAMPLE 2

Factor $5x - 5y + bx - by$.

Solution

We can factor this polynomial by the use of grouping. The grouping is done so that common factors (frequently binomial factors) can be removed. We see that we can factor 5 from the first two terms and b from the last two, which gives

$$5(x - y) + b(x - y).$$

This gives two terms with the common factor $x - y$, so we get

$$(x - y)(5 + b).$$

We can use the formula for multiplying two binomials to factor certain trinomials. The formula

$$(x + a)(x + b) = x^2 + (a + b)x + ab$$

can be used to factor trinomials such as $x^2 - 7x + 6$.

EXAMPLE 3

Factor $x^2 - 7x + 6$.

Solution

If this trinomial can be factored into an expression of the form

$$(x + a)(x + b)$$

then we need to find a and b such that

$$x^2 - 7x + 6 = x^2 + (a + b)x + ab$$

That is, we need to find a and b such that $a + b = -7$ and $ab = 6$. The two numbers whose sum is -7 and whose product is 6 are -1 and -6. Thus

$$x^2 - 7x + 6 = (x - 1)(x - 6).$$

A similar method can be used to factor trinomials such as $9x^2 - 31x + 12$. Finding the proper factors for this type of trinomial may involve a fair amount of trial and error, because we must find factors a, b, c, and d such that

$$(ax + b)(cx + d) = acx^2 + (ad + bc)x + bd.$$

Another technique of factoring is used to factor trinomials such as those we have been discussing. It is useful in factoring more complicated trinomials, such as $9x^2 - 31x + 12$. This procedure for factoring second-degree trinomials follows.

Factoring a Trinomial

Procedure	***Example***
To factor a trinomial into the product of its binomial factors:	Factor $9x^2 - 31x + 12$.
1. Form the product of the second-degree term and the constant term.	1. $9x^2 \cdot 12 = 108x^2$
2. Determine whether there are any factors of the product of step 1 that will sum to the middle term of the trinomial. (If the answer is no, the trinomial will not factor into two binomials.)	2. The factors $-27x$ and $-4x$ give a sum of $-31x$.
3. Use the sum of these two factors to replace the middle term of the trinomial.	3. $9x^2 - 31x + 12 = 9x^2 - 27x - 4x + 12$
4. Factor this four-term expression by grouping.	4. $9x^2 - 31x + 12 = (9x^2 - 27x) + (-4x + 12)$ $= 9x(x - 3) - 4(x - 3)$ $= (x - 3)(9x - 4)$

In the example just completed, note that writing the middle term $(-31x)$ as $-4x - 27x$ rather than as $-27x - 4x$ (as we did) will also result in the correct factorization. (Try it.)

EXAMPLE 4

Factor $9x^2 - 9x - 10$.

Solution

The product of the second-degree term and the constant is $-90x^2$. Factors of $-90x^2$ that sum to $-9x$ are $-15x$ and $6x$. Thus

$$\begin{aligned} 9x^2 - 9x - 10 &= 9x^2 - 15x + 6x - 10 \\ &= (9x^2 - 15x) + (6x - 10) \\ &= 3x(3x - 5) + 2(3x - 5) \\ &= (3x - 5)(3x + 2) \end{aligned}$$

We can check this factorization by multiplying.

$$\begin{aligned} (3x - 5)(3x + 2) &= 9x^2 + 6x - 15x - 10 \\ &= 9x^2 - 9x - 10 \end{aligned}$$

Some special products that make factoring easier are the following.

Special Factorizations

The perfect-square trinomials:

$$x^2 + 2ax + a^2 = (x + a)^2$$
$$x^2 - 2ax + a^2 = (x - a)^2$$

The difference of two squares:

$$x^2 - a^2 = (x + a)(x - a)$$

EXAMPLE 5

Factor $25x^2 - 36y^2$.

Solution

The binomial $25x^2 - 36y^2$ is the difference of two squares, so we get

$$25x^2 - 36y^2 = (5x - 6y)(5x + 6y).$$

These two factors are called binomial **conjugates** because they differ in only one sign.

EXAMPLE 6

Factor $4x^2 + 12x + 9$.

Solution

Although we can use the technique we have learned to factor trinomials, the factors come quickly if we recognize that this trinomial is a perfect square. It has two square terms, and the remaining term ($12x$) is twice the product of the square roots of the squares ($12x = 2 \cdot 2x \cdot 3$). Thus

$$4x^2 + 12x + 9 = (2x + 3)^2.$$

Most of the polynomials we have factored have been second-degree polynomials, or **quadratic polynomials.** Some polynomials that are not quadratic are in a form that can be factored in the same manner as quadratics. For example, the polynomial $x^4 + 4x^2 + 4$ can be written as $a^2 + 4a + 4$, where $a = x^2$.

EXAMPLE 7

Factor $x^4 + 4x^2 + 4$ completely.

Solution

The trinomial is in the form of a perfect square, so letting $a = x^2$ will give us

$$x^4 + 4x^2 + 4 = a^2 + 4a + 4 = (a + 2)^2.$$

Thus

$$x^4 + 4x^2 + 4 = (x^2 + 2)^2.$$

EXAMPLE 8

Factor $x^4 - 16$ completely.

Solution

The binomial $x^4 - 16$ can be treated as the difference of two squares, $(x^2)^2 - 4^2$, so

$$x^4 - 16 = (x^2 - 4)(x^2 + 4).$$

But $x^2 - 4$ can be factored into $(x - 2)(x + 2)$, so

$$x^4 - 16 = (x - 2)(x + 2)(x^2 + 4).$$

CHECKPOINT

1. Factor the following.
   (a) $8x^3 - 12x$ (b) $3x(x^2 + 5) - 5(x^2 + 5)$ (c) $x^2 - 10x - 24$
   (d) $x^2 - 5x + 6$ (e) $4x^2 - 20x + 25$ (f) $100 - 49x^2$
2. Consider $10x^2 - 17x - 20$ and observe that $(10x^2)(-20) = -200x^2$.
   (a) Find two expressions whose product is $-200x^2$ and whose sum is $-17x$.
   (b) Replace $-17x$ in $10x^2 - 17x - 20$ with the two expressions in (a).
   (c) Factor (b) by grouping.
3. True or false:
   (a) $4x^2 + 9 = (2x + 3)^2$ (b) $x^2 - x + 12 = (x - 4)(x + 3)$
   (c) $5x^5 - 20x^3 = 5x^3(x^2 - 4) = 5x^3(x + 2)(x - 2)$

A polynomial is said to be factored completely if all possible factorizations have been completed. For example, $(2x - 4)(x + 3)$ is not factored completely because a 2 can still be factored out of $2x - 4$. Confining our attention to factors with integer coefficients, we can factor a number of polynomials completely by using the following guidelines.

Guidelines for Factoring Completely

Look for: Monomials first.
Then for: Difference of two squares.
Then for: Trinomial squares.
Then for: Other methods of factoring trinomials.

EXAMPLE 9

Factor completely $12x^2 - 36x + 27$.

Solution

$$\begin{aligned} 12x^2 - 36x + 27 &= 3(4x^2 - 12x + 9) && \text{Monomial} \\ &= 3(2x - 3)^2 && \text{Perfect square} \end{aligned}$$

EXAMPLE 10

Factor completely $16x^2 - 64y^2$.

Solution

$$\begin{aligned} 16x^2 - 64y^2 &= 16(x^2 - 4y^2) \\ &= 16(x + 2y)(x - 2y) \end{aligned}$$

Note that factoring the difference of two squares immediately would give $(4x + 8y)(4x - 8y)$, which is not factored completely (because we could still factor 4 from $4x + 8y$ and 4 from $4x - 8y$).

CHECKPOINT SOLUTIONS

1. (a) $8x^3 - 12x = 4x(2x^2 - 3)$
   (b) $3x(x^2 + 5) - 5(x^2 + 5) = (x^2 + 5)(3x - 5)$
   (c) $x^2 - 10x - 24 = (x - 12)(x + 2)$
   (d) $x^2 - 5x + 6 = (x - 3)(x - 2)$
   (e) $4x^2 - 20x + 25 = (2x - 5)^2$
   (f) $100 - 49x^2 = (10 + 7x)(10 - 7x)$
2. (a) $(-25x)(+8x) = -200x^2$ and $-25x + 8x = -17x$
   (b) $10x^2 - 17x - 20 = 10x^2 - 25x + 8x - 20$
   (c) $= (10x^2 - 25x) + (8x - 20)$
   $= 5x(2x - 5) + 4(2x - 5)$
   $= (2x - 5)(5x + 4)$
3. (a) False. $4x^2 + 9$ cannot be factored. In fact, sums of squares cannot be factored.
   (b) False. $x^2 - x + 12$ cannot be factored. We cannot find two numbers whose product is $+12$ and whose sum is -1.
   (c) True.

EXERCISE 0.6

In Problems 1–4, factor by finding the common monomial factor.

1. $9ab - 12a^2b + 18b^2$
2. $8a^2b - 160x + 4bx^2$
3. $4x^2 + 8xy^2 + 2xy^3$
4. $12y^3z + 4yz^2 - 8y^2z^3$

In Problems 5–8, factor by grouping.

5. $7x^3 - 14x^2 + 2x - 4$
6. $5y - 20 - x^2y + 4x^2$
7. $6x - 6m + xy - my$
8. $x^3 - x^2 - 5x + 5$

Factor each expression in Problems 9–20 as a product of binomials.

9. $x^2 + 8x + 12$
10. $x^2 + 6x + 8$
11. $x^2 - x - 6$
12. $x^2 - 2x - 8$
13. $7x^2 - 10x - 8$
14. $12x^2 + 11x + 2$
15. $x^2 - 10x + 25$
16. $4y^2 + 12y + 9$
17. $49a^2 - 144b^2$
18. $16x^2 - 25y^2$
19. (a) $9x^2 + 21x - 8$
    (b) $9x^2 + 22x + 8$
20. (a) $10x^2 - 99x - 63$
    (b) $10x^2 - 27x - 63$
    (c) $10x^2 + 61x - 63$
    (d) $10x^2 + 9x - 63$

In Problems 21–44, factor completely.

21. $4x^2 - x$
22. $2x^5 + 18x^3$
23. $x^3 + 4x^2 - 5x - 20$
24. $x^3 - 2x^2 - 3x + 6$
25. $2x^2 - 8x - 42$
26. $3x^2 - 21x + 36$
27. $2x^3 - 8x^2 + 8x$
28. $x^3 + 16x^2 + 64x$
29. $2x^2 + x - 6$
30. $2x^2 + 13x + 6$
31. $3x^2 + 3x - 36$
32. $4x^2 - 8x - 60$
33. $2x^3 - 8x$
34. $16z^2 - 81w^2$
35. $10x^2 + 19x + 6$
36. $6x^2 + 67x - 35$
37. $9 - 47x + 10x^2$
38. $10x^2 + 21x - 10$
39. $y^4 - 16x^4$
40. $x^8 - 81$
41. $x^4 - 8x^2 + 16$
42. $81 - 18x^2 + x^4$
43. $4x^4 - 5x^2 + 1$
44. $x^4 - 3x^2 - 4$

Use the following factorization formulas involving cubes to factor each expression in Problems 45–52.

Factorizations with Cubes

Perfect cube

$$a^3 + 3a^2b + 3ab^2 + b^3 = (a + b)^3$$

Perfect cube

$$a^3 - 3a^2b + 3ab^2 - b^3 = (a - b)^3$$

Difference of two cubes

$$a^3 - b^3 = (a - b)(a^2 + ab + b^2)$$

Sum of two cubes

$$a^2 + b^3 = (a + b)(a^2 - ab + b^2)$$

45. $x^3 + 3x^2 + 3x + 1$
46. $x^3 + 6x^2 + 12x + 8$
47. $x^3 - 12x^2 + 48x - 64$
48. $y^3 - 9y^2 + 27y - 27$
49. $x^3 - 64$
50. $8x^3 - 1$
51. $27 + 8x^3$
52. $a^3 + 216$

In Problems 53–58 determine the missing factor.

53. $x^{3/2} + x^{1/2} = x^{1/2}(?)$
54. $2x^{1/4} + 4x^{3/4} = 2x^{1/4}(?)$
55. $x^{-3} + x^{-2} = x^{-3}(?)$
56. $x^{-1} - x = x^{-1}(?)$
57. $(-x^3 + x)(3 - x^2)^{-1/2} + 2x(3 - x^2)^{1/2} = (3 - x^2)^{-1/2}(?)$
58. $4x(4x + 1)^{-1/3} - (4x + 1)^{2/3} = (4x + 1)^{-1/3}(?)$

Applications

59. ***Simple interest*** The future value of a simple-interest investment of P dollars at an annual interest rate r for t years is given by the expression $P + Prt$. Factor this expression.

60. ***Reaction to medication*** When medicine is administered, the reaction (measured in change of blood pressure or temperature) can be modeled by (that is, described by)

$$R = \frac{cm^2}{2} - \frac{m^3}{3}$$

where c is a positive constant and m is the amount of medicine absorbed into the blood.* Factor the expression for the reaction.

61. ***Sensitivity to medication*** From the formula for reaction to medication given in Problem 60, an expression for sensitivity S can be obtained, where

$$S = cm - m^2.$$

Factor this expression for sensitivity.

* R. M. Thrall *et al., Some Mathematical Models in Biology,* U.S. Department of Commerce, 1967.

62. ***Volume*** Suppose that squares of side x are cut from four corners of an 8-by-8-inch piece of cardboard and an open-top box is formed (see Figure 0.10). The volume of the box is given by $64x - 32x^2 + 4x^3$. Factor this expression.

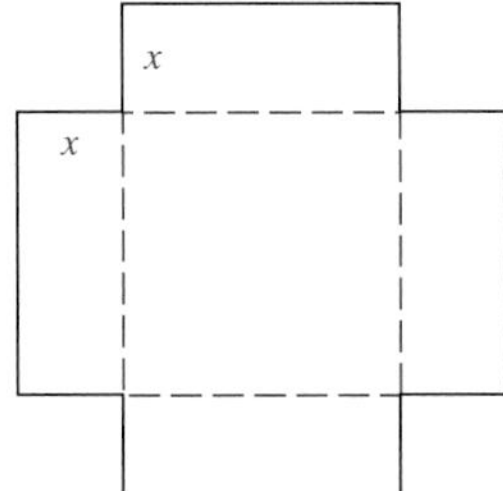

Figure 0.10

63. ***Consumer expenditure*** The consumer expenditure for a commodity is the product of its market price, p, and the number of units demanded, x—that is, px. Suppose that for a certain commodity, the consumer expenditure is given by

$$10{,}000p - 100p^2.$$

(a) Factor this in order to find an expression for the number of units demanded.
(b) Use (a) to find the number of units demanded when the market price is \$38.

64. ***Power in a circuit*** Factor the following expression for the maximum power in a certain electrical circuit.

$$(R + r)^2 - 2r(R + r)$$

0.7 Algebraic Fractions

Evaluating certain limits and graphing rational functions require an understanding of algebraic fractions. The fraction 6/8 can be reduced to 3/4 by dividing both the numerator and the denominator by 2. In the same manner, the algebraic fraction

$$\frac{(x+2)(x+1)}{(x+1)(x+3)}$$

can be reduced to

$$\frac{x+2}{x+3}$$

by dividing both the numerator and the denominator by $x + 1$, if $x \neq -1$.

Simplifying Fractions

We *simplify* algebraic fractions by factoring the numerator and denominator and then dividing both the numerator and the denominator by any common factors.*

EXAMPLE 1

Simplify $\dfrac{3x^2 - 14x + 8}{x^2 - 16}$, $\quad x^2 \neq 16$.

Solution

$$\frac{3x^2 - 14x + 8}{x^2 - 16} = \frac{(3x-2)(x-4)}{(x-4)(x+4)}$$

$$= \frac{(3x-2)\overset{1}{\cancel{(x-4)}}}{\underset{1}{\cancel{(x-4)}}(x+4)}$$

$$= \frac{3x-2}{x+4}$$

* We assume that all fractions are defined.

We can multiply fractions by writing the product as the product of the numerators divided by the product of the denominators. For example,

$$\frac{4}{5} \cdot \frac{10}{12} \cdot \frac{2}{5} = \frac{80}{300}$$

which reduces to $\frac{4}{15}$.

We can also find the product by reducing the fractions before we indicate the multiplication in the numerator and denominator. For example, in

$$\frac{4}{5} \cdot \frac{10}{12} \cdot \frac{2}{5}$$

we can divide the numerator and denominator by 5 and then by 4, which yields

$$\frac{\overset{1}{\cancel{4}}}{\underset{1}{\cancel{5}}} \cdot \frac{\overset{2}{\cancel{10}}}{\underset{3}{\cancel{12}}} \cdot \frac{2}{5} = \frac{1}{1} \cdot \frac{2}{3} \cdot \frac{2}{5} = \frac{4}{15}$$

Product of Fractions

We *multiply* algebraic fractions by writing the product of the numerators divided by the product of the denominators, and then reduce to lowest terms. We may also reduce prior to finding the product.

EXAMPLE 2

Multiply:

(a) $\frac{4x^2}{5y} \cdot \frac{10x}{y^2} \cdot \frac{y}{8x^2}$ (b) $\frac{-4x + 8}{3x + 6} \cdot \frac{2x + 4}{4x + 12}$

Solution

(a) $$\frac{4x^2}{5y} \cdot \frac{10x}{y^2} \cdot \frac{y}{8x^2} = \frac{\overset{1}{\cancel{4x^2}}}{\underset{1 \cdot 1}{\cancel{5y}}} \cdot \frac{\overset{2}{\cancel{10}}x}{y^2} \cdot \frac{\overset{1}{\cancel{y}}}{\underset{2}{\cancel{8x^2}}} = \frac{1}{1} \cdot \frac{\overset{1}{\cancel{2}}x}{y^2} \cdot \frac{1}{\underset{1}{\cancel{2}}} = \frac{x}{y^2}$$

(b) $$\frac{-4x + 8}{3x + 6} \cdot \frac{2x + 4}{4x + 12} = \frac{-4(x - 2)}{3(x + 2)} \cdot \frac{2(x + 2)}{4(x + 3)}$$

$$= \frac{-\overset{-1}{\cancel{4}}(x - 2)}{3\underset{1}{\cancel{(x + 2)}}} \cdot \frac{\overset{1}{2}\cancel{(x + 2)}}{\underset{1}{\cancel{4}}(x + 3)}$$

$$= \frac{-2(x - 2)}{3(x + 3)}$$

In arithmetic we learned to divide one fraction by another by inverting the divisor and multiplying. The same rule applies to division of algebraic fractions.

EXAMPLE 3

(a) Divide $\frac{a^2b}{c}$ by $\frac{ab}{c^2}$ (b) $\frac{6x^2-6}{x^2+3x+2} \div \frac{x-1}{x^2+4x+4}$

Solution

(a) $$\frac{a^2b}{c} \div \frac{ab}{c^2} = \frac{a^2b}{c} \cdot \frac{c^2}{ab} = \frac{\overset{a \cdot 1}{\cancel{a^2}\,\cancel{b}}}{\underset{1}{\cancel{c}}} \cdot \frac{\overset{c}{\cancel{c^2}}}{\underset{1 \cdot 1}{\cancel{ab}}} = \frac{ac}{1} = ac$$

(b) $$\begin{aligned} \frac{6x^2-6}{x^2+3x+2} \div \frac{x-1}{x^2+4x+4} &= \frac{6x^2-6}{x^2+3x+2} \cdot \frac{x^2+4x+4}{x-1} \\ &= \frac{6(x-1)(x+1)}{(x+2)(x+1)} \cdot \frac{(x+2)(x+2)}{x-1} \\ &= 6(x+2) \end{aligned}$$

CHECKPOINT

1. Reduce $\frac{2x^2-4x}{2x}$.
2. Multiply: $\frac{x^2}{x^2-9} \cdot \frac{x+3}{3x}$.
3. Divide: $\frac{5x^2(x-1)}{2(x+1)} \div \frac{10x^2}{(x+1)(x-1)}$.

If two fractions are to be added, it is convenient that both be expressed with the same denominator. If the denominators are not the same, we can write the equivalents of each of the fractions with a common denominator. We usually use the least common denominator (LCD) when we write the equivalent fractions. The **least common denominator** is the lowest-degree variable expression into which all denominators will divide. If the denominators are polynomials, then the LCD is the lowest-degree polynomial into which all denominators will divide. We can find the least common denominator as follows.

Finding the Least Common Denominator

Procedure	***Example***
To find the least common denominator of a set of fractions:	Find the LCD of $\frac{1}{x^2-x}, \frac{1}{x^2-1}, \frac{1}{x^2}$.
1. Completely factor each denominator.	1. The factored denominators are $x(x-1)$, $(x+1)(x-1)$, and $x \cdot x$.
2. Write the LCD as the product of each of these factors used the maximum number of times it occurs in any one denominator.	2. x occurs a maximum of twice in one denominator, $x-1$ occurs once, and $x+1$ occurs once. Thus the LCD is $x \cdot x(x-1)(x+1) = x^2(x-1)(x+1)$.

The procedure for combining (adding or subtracting) two or more fractions follows.

Adding or Subtracting Fractions

Procedure	Example
To combine fractions:	Combine $\frac{y-3}{y-5}+\frac{y-23}{y^2-y-20}$.
1. Find the LCD of the fractions.	1. $y^2-y-20=(y-5)(y+4)$, so the LCD is $(y-5)(y+4)$.
2. Write the equivalent of each fraction with the LCD as its denominator.	2. The sum is $\frac{(y-3)(y+4)}{(y-5)(y+4)}+\frac{y-23}{(y-5)(y+4)}$.
3. Add or subtract, as indicated, by combining like terms in the numerator over the LCD.	3. $=\frac{y^2+y-12+y-23}{(y-5)(y+4)}$ $=\frac{y^2+2y-35}{(y-5)(y+4)}$
4. Reduce the fraction, if possible.	4. $y-5$ is a factor of the numerator, so the sum is $\frac{y+7}{y+4}$, if $y \neq 5$.

EXAMPLE 4

Add $\frac{3x}{a^2}+\frac{4}{ax}$.

Solution

1. The LCD is a^2x.
2. $\frac{3x}{a^2}+\frac{4}{ax}=\frac{3x}{a^2}\cdot\frac{x}{x}+\frac{4}{ax}\cdot\frac{a}{a}=\frac{3x^2}{a^2x}+\frac{4a}{a^2x}$
3. $\frac{3x^2}{a^2x}+\frac{4a}{a^2x}=\frac{3x^2+4a}{a^2x}$
4. The sum is in lowest terms.

EXAMPLE 5

Combine $\frac{y-3}{(y-5)^2}+\frac{y-2}{y^2-y-20}$.

Solution

$y^2-y-20=(y-5)(y+4)$, so the LCD is $(y-5)^2(y+4)$. Writing the equivalent fractions and then combining them, we get

$$\frac{y-3}{(y-5)^2}+\frac{y-2}{(y-5)(y+4)}=$$

$$= \frac{(y-3)(y+4)}{(y-5)^2(y+4)} + \frac{(y-2)(y-5)}{(y-5)(y+4)(y-5)}$$

$$= \frac{(y^2+y-12)+(y^2-7y+10)}{(y-5)^2(y+4)}$$

$$= \frac{2y^2-6y-2}{(y-5)^2(y+4)}.$$

A fractional expression that contains one or more fractions in its numerator or denominator is called a **complex fraction.** An example of a complex fraction is

$$\frac{\frac{1}{3}+\frac{4}{x}}{3-\frac{1}{xy}}.$$

We can simplify fractions of this type using the property $\frac{a}{b} = \frac{ac}{bc}$, with c equal to the LCD of *all* the fractions contained in the numerator and denominator of the complex fraction.

For example, all fractions contained in

$$\frac{\frac{1}{3}+\frac{4}{x}}{3-\frac{1}{xy}}$$

have LCD $3xy$. We simplify this complex fraction by multiplying the numerator and denominator as follows:

$$\frac{3xy\left(\frac{1}{3}+\frac{4}{x}\right)}{3xy\left(3-\frac{1}{xy}\right)} = \frac{3xy\left(\frac{1}{3}\right)+3xy\left(\frac{4}{x}\right)}{3xy(3)-3xy\left(\frac{1}{xy}\right)} = \frac{xy+12y}{9xy-3}.$$

EXAMPLE 6

Simplify $\frac{x^{-3}+x^2y^{-3}}{(xy)^{-2}}$ so that positive exponents remain.

Solution

$$\frac{x^{-3}+x^2y^{-3}}{(xy)^{-2}} = \frac{\frac{1}{x^3}+\frac{x^2}{y^3}}{\frac{1}{(xy)^2}}, \quad \text{LCD} = x^3y^3$$

$$= \frac{x^3y^3\left(\frac{1}{x^3}+\frac{x^2}{y^3}\right)}{x^3y^3\left(\frac{1}{x^2y^2}\right)} = \frac{y^3+x^5}{xy}$$

CHECKPOINT

4. Add or subtract:
   (a) $\frac{5x-1}{2x-5} - \frac{x+9}{2x-5}$
   (b) $\frac{x+1}{x} + \frac{x}{x-1}$
5. Simplify $\frac{\frac{y}{x}-1}{\frac{y}{x}-\frac{x}{y}}$.

We can simplify algebraic fractions whose denominators contain sums and differences that involve square roots by rationalizing the denominators. Using the fact that $(x + y)(x - y) = x^2 - y^2$, we multiply the numerator and denominator of an algebraic fraction of this type by the conjugate of the denominator to simplify the fraction.

EXAMPLE 7

Rationalize the denominators.

(a) $\frac{1}{\sqrt{x}-2}$ (b) $\frac{3+\sqrt{x}}{\sqrt{x}+\sqrt{5}}$

Solution

Multiplying $\sqrt{x} - 2$ by $\sqrt{x} + 2$, its conjugate, gives the difference of two squares and removes the radical from the denominator in (a). We also use the conjugate in (b).

(a) $\frac{1}{\sqrt{x}-2} \cdot \frac{\sqrt{x}+2}{\sqrt{x}+2} = \frac{\sqrt{x}+2}{(\sqrt{x})^2-(2)^2} = \frac{\sqrt{x}+2}{x-4}$

(b) $\frac{3+\sqrt{x}}{\sqrt{x}+\sqrt{5}} \cdot \frac{\sqrt{x}-\sqrt{5}}{\sqrt{x}-\sqrt{5}} = \frac{3\sqrt{x}-3\sqrt{5}+x-\sqrt{5x}}{x-5}$

CHECKPOINT

6. Rationalize the denominator $\frac{\sqrt{x}}{2\sqrt{x}-3}$.

CHECKPOINT SOLUTIONS

1. $\frac{2x^2-4x}{2x} = \frac{2x(x-2)}{2x} = x-2$
2. $\frac{x^2}{x^2-9} \cdot \frac{x+3}{3x} = \frac{x^2 \cdot (x+3)}{(x+3)(x-3) \cdot 3x} = \frac{x}{3(x-3)} = \frac{x}{3x-9}$
3. $\frac{5x^2(x-1)}{2(x+1)} \div \frac{10x^2}{(x+1)(x-1)} = \frac{5x^2(x-1)}{2(x+1)} \cdot \frac{(x+1)(x-1)}{10x^2}$
   $= \frac{(x-1)^2}{4}$

4. (a) $\frac{5x-1}{2x-5} - \frac{x+9}{2x-5} = \frac{(5x-1)-(x+9)}{2x-5}$

$= \frac{5x-1-x-9}{2x-5}$

$= \frac{4x-10}{2x-5} = \frac{2(2x-5)}{2x-5} = 2$

(b) $\frac{x+1}{x} + \frac{x}{x-1}$ has LCD $= x(x-1)$

$\frac{x+1}{x} + \frac{x}{x-1} = \frac{x+1}{x}\cdot\frac{(x-1)}{(x-1)} + \frac{x}{x-1}\cdot\frac{x}{x}$

$= \frac{x^2-1}{x(x-1)} + \frac{x^2}{x(x-1)} = \frac{x^2-1+x^2}{x(x-1)}$

$= \frac{2x^2-1}{x(x-1)}$

5. $\frac{\frac{y}{x}-1}{\frac{y}{x}-\frac{x}{y}} = \frac{\left(\frac{y}{x}-1\right)}{\left(\frac{y}{x}-\frac{x}{y}\right)}\cdot\frac{xy}{xy} = \frac{\frac{y}{x}\cdot xy - 1\cdot xy}{\frac{y}{x}\cdot xy - \frac{x}{y}\cdot xy}$

$= \frac{y^2-xy}{y^2-x^2} = \frac{y(y-x)}{(y+x)(y-x)} = \frac{y}{y+x}$

6. $\frac{\sqrt{x}}{2\sqrt{x}-3} = \frac{\sqrt{x}}{2\sqrt{x}-3}\cdot\frac{2\sqrt{x}+3}{2\sqrt{x}+3}$

$= \frac{2x+3\sqrt{x}}{4x-9}$

EXERCISE 0.7

Simplify the following fractions.

1. $\frac{18x^3y^3}{9x^3z}$
2. $\frac{15a^4b^5}{30a^3b}$
3. $\frac{x-3y}{3x-9y}$
4. $\frac{x^2-6x+8}{x^2-16}$
5. $\frac{x^2-2x+1}{x^2-4x+3}$
6. $\frac{x^2-5x+6}{9-x^2}$
7. $\frac{6x^3y^3-15x^2y}{3x^2y^2+9x^2y}$
8. $\frac{x^2y^2-4x^3y}{x^2y-2x^2y^2}$

In Problems 9–38, perform the indicated operations and simplify.

9. $\frac{6x^3}{8y^3}\cdot\frac{16x}{9y^2}\cdot\frac{15y^4}{x^3}$
10. $\frac{25ac^2}{15a^2c}\cdot\frac{4ad^4}{15abc^3}$
11. $\frac{8x-16}{x-3}\cdot\frac{4x-12}{3x-6}$
12. $\frac{x^2+7x+12}{3x^2+13x+4}\cdot\frac{3x+1}{x+3}$
13. $(x^2-4)\cdot\frac{2x-3}{x+2}$
14. $\frac{4x+4}{x-4}\cdot\frac{x^2-6x+8}{8x^2+8x}$
15. $\frac{x^2-x-2}{2x^2-8}\cdot\frac{18-2x^2}{x^2-5x+4}\cdot\frac{x^2-2x-8}{x^2-6x+9}$
16. $\frac{x^2-5x-6}{x^2-5x+4}\cdot\frac{x^2-x-12}{x^3-6x^2}\cdot\frac{x-x^3}{x^2-2x+1}$
17. $\frac{15ac^2}{7bd}\div\frac{4a}{14b^2d}$
18. $\frac{16}{x-2}\div\frac{4}{3x-6}$
19. $\frac{y^2-2y+1}{7y^2-7y}\div\frac{y^2-4y+3}{35y^2}$
20. $\frac{6x^2}{4x^2y-12xy}\div\frac{3x^2+12x}{x^2+x-12}$
21. $(x^2-x-6)\div\frac{9-x^2}{x^2-3x}$
22. $\frac{2x^2+7x+3}{4x^2-1}\div(x+3)$

23. $\dfrac{2x}{x^2 - x - 2} - \dfrac{x + 2}{x^2 - x - 2}$

24. $\dfrac{4}{9 - x^2} - \dfrac{x + 1}{9 - x^2}$

25. $\dfrac{a}{a - 2} - \dfrac{a - 2}{a}$

26. $x - \dfrac{2}{x - 1}$

27. $\dfrac{x}{x + 1} - x + 1$

28. $\dfrac{x - 1}{x + 1} - \dfrac{2}{x^2 + x}$

29. $\dfrac{4a}{3x + 6} + \dfrac{5a^2}{4x + 8}$

30. $\dfrac{b - 1}{b^2 + 2b} + \dfrac{b}{3b + 6}$

31. $\dfrac{3x - 1}{2x - 4} + \dfrac{4x}{3x - 6} - \dfrac{x - 4}{5x - 10}$

32. $\dfrac{2x + 1}{4x - 2} + \dfrac{5}{2x} - \dfrac{x + 4}{2x^2 - x}$

33. $\dfrac{1}{x^2 - 4y^2} - \dfrac{1}{x^2 - 4xy + 4y^2}$

34. $\dfrac{3x^2}{x^2 - 4} + \dfrac{2}{x^2 - 4x + 4} - 3$

35. $\dfrac{x}{x^2 - 4} + \dfrac{4}{x^2 - x - 2} - \dfrac{x - 2}{x^2 + 3x + 2}$

36. $\dfrac{1}{y^2 + 2y - 35} - \dfrac{y - 1}{y^2 - 4y - 5} + \dfrac{2}{y^2 + 8y + 7}$

37. $\dfrac{-x^3 + x}{\sqrt{3 - x^2}} + 2x\sqrt{3 - x^2}$

38. $\dfrac{3x^2(x + 1)}{\sqrt{x^3 + 1}} + \sqrt{x^3 + 1}$

In Problems 39–48, simplify each complex fraction.

39. $\dfrac{3 - \dfrac{2}{3}}{14}$

40. $\dfrac{4}{\dfrac{1}{4} + \dfrac{1}{4}}$

41. $\dfrac{x + y}{\dfrac{1}{x} + \dfrac{1}{y}}$

42. $\dfrac{\dfrac{5}{2y} + \dfrac{3}{y}}{\dfrac{1}{4} + \dfrac{1}{3y}}$

43. $\dfrac{2 - \dfrac{1}{x}}{2x - \dfrac{3x}{x + 1}}$

44. $\dfrac{1 - \dfrac{2}{x - 2}}{x - 6 + \dfrac{10}{x + 1}}$

45. $\dfrac{\sqrt{a} - \dfrac{b}{\sqrt{a}}}{a - b}$

46. $\dfrac{\sqrt{x - 1} + \dfrac{1}{\sqrt{x - 1}}}{x}$

47. $\dfrac{\sqrt{x^2 + 9} - \dfrac{13}{\sqrt{x^2 + 9}}}{x^2 - x - 6}$

48. $\dfrac{\sqrt{x^2 + 3} - \dfrac{x + 5}{\sqrt{x^2 + 3}}}{x^2 + 5x + 4}$

In Problems 49–54, rewrite each of the following so that only positive exponents remain, and simplify.

49. (a) $(2^{-2} - 3^{-1})^{-1}$ (b) $(2^{-1} + 3^{-1})^2$

50. (a) $(3^2 + 4^2)^{-1/2}$ (b) $(2^2 + 3^2)^{-1}$

51. $\dfrac{2a^{-1} - b^{-1}}{(ab)^{-1}}$

52. $\dfrac{x^{-2} + y^{-2}}{x^{-2} - y^{-2}}$

53. $\dfrac{xy^{-2} + x^{-2}y}{x + y}$

54. $\dfrac{x^{-2} + xy^{-2}}{(x^2y)^{-2}}$

In Problems 55 and 56, rationalize the denominator of each fraction and simplify.

55. $\dfrac{1 - \sqrt{x}}{1 + \sqrt{x}}$

56. $\dfrac{x - 3}{x - \sqrt{3}}$

In Problems 57 and 58, rationalize the numerator of each fraction and simplify.

57. $\dfrac{\sqrt{x + h} - \sqrt{x}}{h}$

58. $\dfrac{\sqrt{9 + 2h} - 3}{h}$

Applications

59. ***Time study*** Workers A, B, and C can complete a job in a, b, and c hours, respectively. Working together, they can complete

$$\frac{1}{a} + \frac{1}{b} + \frac{1}{c}$$

of the job in 1 hour. Add these fractions to obtain an expression for what they can do in 1 hour, working together.

60. ***Focal length*** Two thin lenses with focal lengths p and q and separated by a distance d have their combined focal length given by the reciprocal of

$$\frac{1}{p} + \frac{1}{q} - \frac{d}{pq}.$$

(a) Combine these fractions.
(b) Use the reciprocal of your answer in (a) to find the combined focal length.

Average cost A company's average cost per unit when x units are produced is defined to be

$$\text{Average cost} = \frac{\text{Total cost}}{x}.$$

Use this equation in Problems 61 and 62.

61. Suppose a company's average costs are given by

$$\text{Average cost} = \frac{4000}{x} + 55 + 0.1x.$$

(a) Express the average-cost formula as a single fraction.
(b) Write the expression that gives the company's total costs.

62. Suppose a company's average costs are given by

$$\text{Average cost} = \frac{40{,}500}{x} + 190 + 0.2x.$$

(a) Express the average-cost formula as a single fraction.
(b) Write the expression that gives the company's total costs.

63. ***Advertising and sales*** Suppose that a company's daily sales volume attributed to an advertising campaign is given by

$$\text{Sales volume} = 1 + \frac{3}{t+3} - \frac{18}{(t+3)^2}$$

where t is the number of days since the campaign started. Express the sales volume as a single fraction.

KEY TERMS AND FORMULAS

Section	Key Terms	Formula
0.1	Sets and set membership	
	Natural numbers	$\{1, 2, 3, 4 \ldots\}$
	Empty set	$\varnothing$
	Set equality	
	Subset	$A \subseteq B$
	Universal set	U
	Venn diagrams	
	Set intersection	$A \cap B$
	Disjoint sets	$A \cap B = \varnothing$
	Set union	$A \cup B$
	Set complement	A'
0.2	Real numbers	
	Subsets and properties	
	Real number line	
	Inequalities	
	Intervals and interval notation	
	Absolute value	
	Order of operations	
0.3	Exponent and base	a^n has base a, exponent n
	Zero exponent	$a^0 = 1, a \neq 0$
	Negative exponent	$a^{-n} = \dfrac{1}{a^n}$
	Rules of exponents	

0.4	Radical	$\sqrt[n]{a}$
	Radicand, index	radicand $= a$; index $= n$
	Principal nth root	$\sqrt[n]{a} = b$ only if $b^n = a$ and $a \geq 0$ and $b \geq 0$ when n is even
	Fractional exponents	$a^{1/n} = \sqrt[n]{a}$
		$a^{m/n} = \sqrt[n]{a^m} = (\sqrt[n]{a})^m$
	Properties of radicals	
	Rationalizing the denominator	
0.5	Algebraic expression	
	Variable	
	Constant term	
	Coefficient; leading coefficient	
	Term	
	Polynomial	$a_nx^n + \cdots + a_1x + a_0$
	Degree	
	Monomial	
	Binomial	
	Trinomial	
	Like terms	
	Distributive Law	$a(b + c) = ab + ac$
	Binomial products	
	Division of polynomials	
0.6	Factor	
	Common factor	
	Factoring by grouping	
	Special factorizations	
	Conjugates	$a + b$; $a - b$
	Quadratic polynomials	$ax^2 + bx + c$
	Factoring completely	
0.7	Algebraic fractions	
	Numerator	
	Denominator	
	Reduce	
	Product of fractions	
	Quotient of fractions	
	Common denominator	
	Least common denominator (LCD)	
	Addition and subtraction of fractions	
	Complex fraction	
	Rationalize the denominator	

REVIEW EXERCISES

Section 0.1

1. Is $A \subseteq B$, if $A = \{1, 2, 5, 7\}$ and $B = \{x: x \text{ is a positive integer}, x \le 8\}$?
2. Is it true that $3 \in \{x: x > 3\}$?
3. Are $A = \{1, 2, 3, 4\}$ and $B = \{x: x \le 1\}$ disjoint?

In Problems 4–7, use sets $U = \{1, 2, 3, 4, 5, 6, 7, 8, 9, 10\}$, $A = \{1, 2, 3, 9\}$, and $B = \{1, 3, 5, 6, 7, 8, 10\}$ to find the elements of the sets described.

4. $A \cup B'$
5. $A' \cap B$
6. $(A' \cap B)'$
7. Does $(A' \cup B')' = A \cap B$?

Section 0.2

8. State the property of the real numbers that is illustrated in each case.
   (a) $6 + \frac{1}{3} = \frac{1}{3} + 6$ (b) $2(3 \cdot 4) = (2 \cdot 3)4$
   (c) $\frac{1}{3}(6 + 9) = 2 + 3$
9. Indicate whether each given expression is one or more of the following: rational, irrational, integer, natural, or meaningless.
   (a) π (b) $0/6$ (c) $6/0$
10. Insert the proper sign ($<$, $=$, or $>$) to replace each $\square$.
   (a) $\pi \,\square\, 3.14$ (b) $-100 \,\square\, 0.1$ (c) $-3 \,\square\, -12$

For Problems 11–18, evaluate each expression. Use a calculator when necessary.

11. $|5 - 11|$
12. $44 \div 2 \cdot 11 - 10^2$
13. $(-3)^2 - (-1)^3$
14. $\dfrac{(3)(2)(15) - (5)(8)}{(4)(10)}$
15. $2 - [3 - (2 - |-3|)] + 11$
16. $-4^2 - (-4)^2 + 3$
17. $\dfrac{4 + 3^2}{4}$
18. $\dfrac{(-2.91)^5}{\sqrt{3.29^5}}$
19. Write each inequality in interval notation, name the type of interval, and graph it on a real number line.
   (a) $0 \le x \le 5$ (b) $x \ge -3$ and $x < 7$
   (c) $(-4, \infty) \cap (-\infty, 0)$
20. Write an inequality that represents each of the following.
   (a) $(-1, 16)$ (b) $[-12, 8]$
   (c) 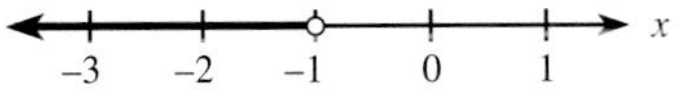

Section 0.3

21. Evaluate each of the following without a calculator.
   (a) $\left(\frac{3}{8}\right)^0$ (b) $2^3 \cdot 2^{-5}$
   (c) $\dfrac{4^9}{4^3}$ (d) $\left(\frac{1}{7}\right)^3\left(\frac{1}{7}\right)^{-4}$
22. Use the rules of exponents to simplify each of the following with positive exponents. Assume all variables are nonzero.
   (a) $x^5 \cdot x^{-7}$ (b) x^8/x^{-2} (c) $(x^3)^3$
   (d) $(y^4)^{-2}$ (e) $(-y^{-3})^{-2}$

For Problems 23–28 rewrite each expression so that only positive exponents remain. Assume all variables are nonzero.

23. $\dfrac{-(2xy^2)^{-2}}{(3x^{-2}y^{-3})^2}$
24. $\left(\frac{2}{3}x^2y^{-4}\right)^{-2}$
25. $\left(\dfrac{x^{-2}}{2y^{-1}}\right)^2$
26. $\dfrac{(-x^4y^{-2}z^2)^0}{-(x^4y^{-2}z^2)^{-2}}$
27. $\left(\dfrac{x^{-3}y^4z^{-2}}{3x^{-2}y^{-3}z^{-3}}\right)^{-1}$
28. $\left(\dfrac{x}{2y}\right)\left(\dfrac{y}{x^2}\right)^{-2}$

Section 0.4

29. Find the following roots.
   (a) $-\sqrt[3]{-64}$ (b) $\sqrt{4/49}$ (c) $\sqrt[7]{1.9487171}$
30. Write each of the following with an exponent and with the variable in the numerator.
   (a) $\sqrt{x}$ (b) $\sqrt[3]{x^2}$ (c) $1/\sqrt[4]{x}$
31. Write each of the following in radical form.
   (a) $x^{2/3}$ (b) $x^{-1/2}$ (c) $-x^{3/2}$
32. Rationalize each of the following denominators and simplify.
   (a) $\dfrac{5xy}{\sqrt{2x}}$ (b) $\dfrac{y}{x\sqrt[3]{xy^2}}$

In Problems 33–38, use the properties of exponents to simplify so that only positive exponents remain. Assume all variables are positive.

33. $x^{1/2} \cdot x^{1/3}$
34. $y^{-3/4}/y^{-7/4}$
35. $x^4 \cdot x^{1/4}$
36. $1/(x^{-4/3} \cdot x^{-7/3})$
37. $(x^{4/5})^{1/2}$
38. $(x^{1/2}y^2)^4$

In Problems 39–44, simplify each expression. Assume all variables are positive.

39. $\sqrt{12x^3y^5}$
40. $\sqrt{1250x^6y^9}$
41. $\sqrt[3]{24x^4y^4} \cdot \sqrt[3]{45x^4y^{10}}$
42. $\sqrt{16a^2b^3} \cdot \sqrt{8a^3b^5}$
43. $\dfrac{\sqrt{52x^3y^6}}{\sqrt{13xy^4}}$
44. $\dfrac{\sqrt{32x^4y^3}}{\sqrt{6xy^{10}}}$

Section 0.5

In Problems 45–62, perform the indicated operations and simplify.

45. $(3x + 5) - (4x + 7)$
46. $x(1 - x) + x[x - (2 + x)]$
47. $(3x^3 - 4xy - 3) + (5xy + x^3 + 4y - 1)$
48. $(4xy^3)(6x^4y^2)$
49. $(3x - 4)(x - 1)$
50. $(3x - 1)(x + 2)$
51. $(4x + 1)(x - 2)$
52. $(3x - 7)(2x + 1)$

53. $(2x - 3)^2$
54. $(4x + 3)(4x - 3)$
55. $(2x^2 + 1)(x^2 + x - 3)$
56. $(2x - 1)^3$
57. $(x - y)(x^2 + xy + y^2)$
58. $\dfrac{4x^2y - 3x^3y^3 - 6x^4y^2}{2x^2y^2}$
59. $(3x^4 + 2x^3 - x + 4) \div (x^2 + 1)$
60. $(x^4 - 4x^3 + 5x^2 + x) \div (x - 3)$
61. $x^{4/3}(x^{2/3} - x^{-1/3})$
62. $(\sqrt{x} + \sqrt{a - x})(\sqrt{x} - \sqrt{a - x})$

Section 0.6

In Problems 63–71, factor each expression completely.

63. $2x^4 - x^3$
64. $4(x^2 + 1)^2 - 2(x^2 + 1)^3$
65. $4x^2 - 4x + 1$
66. $16 - 9x^2$
67. $2x^4 - 8x^2$
68. $x^2 - 4x - 21$
69. $3x^2 - x - 2$
70. $12x^2 - 23x - 24$
71. $16x^4 - 72x^2 + 81$
72. Factor as indicated: $x^{-2/3} + x^{-4/3} = x^{-4/3}(?)$

Section 0.7

73. Reduce each of the following to lowest terms.
(a) $\dfrac{2x}{2x + 4}$ (b) $\dfrac{4x^2y^3 - 6x^3y^4}{2x^2y^2 - 3xy^3}$

In Problems 74–80, perform the indicated operations and simplify.

74. $\dfrac{x^2 - 4x}{x^2 + 4} \cdot \dfrac{x^4 - 16}{x^4 - 16x^2}$
75. $\dfrac{x^2 + 6x + 9}{x^2 - 7x + 12} \div \dfrac{x^2 + 4x + 3}{x^2 - 3x - 4}$
76. $\dfrac{x^4 - 2x^3}{3x^2 - x - 2} \div \dfrac{x^3 - 4x}{9x^2 - 4}$
77. $1 + \dfrac{3}{2x} - \dfrac{1}{6x^2}$
78. $\dfrac{1}{x - 2} - \dfrac{x - 2}{4}$
79. $\dfrac{x + 2}{x^2 - x} - \dfrac{x^2 + 4}{x^2 - 2x + 1} + 1$
80. $\dfrac{x - 1}{x^2 - x - 2} - \dfrac{x}{x^2 - 2x - 3} + \dfrac{1}{x - 2}$

In Problems 81 and 82, simplify each complex fraction.

81. $\dfrac{x - 1 - \dfrac{x - 1}{x}}{\dfrac{1}{x - 1} + 1}$
82. $\dfrac{x^{-2} - x^{-1}}{x^{-2} + x^{-1}}$

83. Rationalize the denominator of $\dfrac{3x - 3}{\sqrt{x} - 1}$ and simplify.

84. Rationalize the numerator of $\dfrac{\sqrt{x} - \sqrt{x - 4}}{2}$ and simplify.

Applications

Section 0.1

85. ***Job effectiveness factors*** In an attempt to determine some off-the-job factors that might be indicators of on-the-job effectiveness, a company made a study of 200 of its employees. It was interested in whether the employees had been recognized for superior work by their supervisors within the past year, whether they were involved in community activities, and whether they followed a regular exercise plan. The company found the following.

30 answered "yes" to all three
50 were recognized and they exercised
52 were recognized and were involved in the community
77 were recognized
37 were involved in the community but did not exercise
95 were recognized or were involved in the community
95 answered "no" to all three

(a) Draw a Venn diagram that represents this information.
(b) How many exercised only?
(c) How many exercised or were involved in the community?

Section 0.3

86. ***Future value*** If an individual makes monthly deposits of \$100 into an account that earns 9%, compounded monthly, then the future value S of the account after n months is given by the formula

$$S = \$100\left[\frac{(1.0075)^n - 1}{0.0075}\right].$$

(a) Find the future value after 36 months (3 years).
(b) Find the future value after 20 years.

Sections 0.3 and 0.7

87. ***Loan payment*** Suppose you borrow \$10,000 for n months to buy a car at an interest rate of 7.8%, compounded monthly. The size of each month's payment, R, is given by the formula

$$R = \$10{,}000\left[\frac{0.0065}{1 - (1.0065)^{-n}}\right].$$

(a) Rewrite the expression on the right-hand side of this formula as a fraction with only positive exponents.

(b) Find the monthly payment for a 48-month loan. Use both the original formula and your result from (a). (Both formulas should give the same payment.)

Section 0.4

88. ***Environment*** Suppose that in a study of water birds, the relationship between the acres of wetlands A and the number of species of birds S found in the wetlands area was given by

$$S = kA^{1/3}$$

where k is a constant.
(a) Express this formula using radical notation.
(b) If the area is expanded by a factor of 2.25 from 20,000 acres to 45,000 acres, find the expected increase in the number of species (as a multiple of the number of species on the 20,000 acres).

Section 0.7

89. ***Cost-benefit*** Suppose that the cost C (in dollars) of removing $p\%$ of the pollution from the waste water of a manufacturing process is given by

$$C = \frac{540{,}000}{100 - p} - 5400$$

(a) Express the right-hand side of this formula as a single fraction.
(b) Find the cost if $p = 0$. Write a sentence that explains the meaning of what you found.
(c) Find the cost of removing 98% of the pollution.
(d) What happens to this formula when $p = 100$? Explain why you think this happens.

CHAPTER TEST

1. Let $U = \{1, 2, 3, 4, 5, 6, 7, 8, 9\}$, $A = \{x\text{: } x \text{ is even and } x \geq 5\}$, and $B = \{1, 2, 5, 7, 8, 9\}$. Complete the following.
   (a) Find $A \cup B'$.
   (b) Find a two-element set that is disjoint from B.
   (c) Find a nonempty subset of A that is not equal to A.
2. Evaluate $(4 - 2^3)^2 - 3^4 \cdot 0^{15} + 12 \div 3 + 1$.
3. Use definitions and properties of exponents to complete the following.
   (a) $x^4 \cdot x^4 = x^?$ (b) $x^0 = ?$, if $x \neq 0$
   (c) $\sqrt{x} = x^?$ (d) $(x^{-5})^2 = x^?$
   (e) $a^{27} \div a^{-3} = a^?$ (f) $x^{1/2} \cdot x^{1/3} = x^?$
   (g) $\dfrac{1}{\sqrt[3]{x^2}} = \dfrac{1}{x^?}$ (h) $\dfrac{1}{x^3} = x^?$
4. Write each of the following as radicals.
   (a) $x^{1/4}$ (b) $x^{-3/4}$
5. Simplify each of the following so that only positive exponents remain.
   (a) x^{-5} (b) $\left(\dfrac{x^{-8}y^2}{x^{-1}}\right)^{-3}$
6. Simplify the following radical expressions, and rationalize any denominators.
   (a) $\dfrac{x}{\sqrt{5x}}$ (b) $\sqrt{24a^2b}\sqrt{a^3b^4}$ (c) $\dfrac{1 - \sqrt{x}}{1 + \sqrt{x}}$
7. Given the expression $2x^3 - 7x^5 - 5x - 8$, complete the following.
   (a) Find the degree.
   (b) Find the constant term.
   (c) Find the coefficient of x.
8. Express $(-2, \infty) \cap (-\infty, 3]$ using interval notation, and graph it.
9. Completely factor each of the following.
   (a) $8x^3 - 2x^2$ (b) $x^2 - 10x - 24$
   (c) $6x^2 - 13x + 6$ (d) $2x^3 - 32x^5$
10. Identify the quadratic polynomial from among (a)–(c), and evaluate it using $x = -3$.
    (a) $2x^2 - 3x^3 + 7$ (b) $x^2 + 3/x + 11$
    (c) $4 - x - x^2$
11. Use long division to find $(2x^3 + x^2 - 7) \div (x^2 - 1)$.
12. Perform the indicated operations and simplify.
    (a) $4y - 5(9 - 3y)$ (b) $-3t^2(2t^4 - 3t^7)$
    (c) $(4x - 1)(x^2 - 5x + 2)$
    (d) $(6x - 1)(2 - 3x)$
    (e) $(2m - 7)^2$ (f) $\dfrac{x^6}{x^2 - 9} \cdot \dfrac{x - 3}{3x^2}$
    (g) $\dfrac{x^4}{3^2} \div \dfrac{9x^3}{x^6}$ (h) $\dfrac{4}{x - 8} - \dfrac{x - 2}{x - 8}$
    (i) $\dfrac{x - 1}{x^2 - 2x - 3} - \dfrac{3}{x^2 - 3x}$
13. Simplify $\dfrac{\dfrac{1}{x} - \dfrac{1}{y}}{\dfrac{1}{x} + y}$.

14. In a nutrition survey of 320 students, the following information was obtained.

    145 ate breakfast
    270 ate lunch
    280 ate dinner
    125 ate breakfast and lunch
    110 ate breakfast and dinner
    230 ate lunch and dinner
    90 ate all three meals

    (a) How many students in the survey ate only breakfast?
    (b) How many students skipped breakfast?

15. If \$1000 is invested for x years at 8%, compounded quarterly, the future value of the investment is given by $S = 1000\left(1 + \frac{.08}{4}\right)^{4x}$. What will be the future value of this investment in 20 years?

Extended Applications / Group Projects

I. Campaign Management

A politician is trying to win election to the city council, and as his campaign manager, you need to decide how to promote the candidate. There are three ways you can do so: You can send glossy, full-color pamphlets to registered voters of the city, you can run a commercial during the television news on a local cable network, and/or you can buy a full-page ad in the newspaper.

Two hundred fifty thousand voters live in the city, and 36% of them read the newspaper. Fifty thousand voters watch the local cable network news, and 30% of them also read the newspaper.

You also know that the television commercial would cost $40,000, the newspaper ad $27,000, and the pamphlets mailed to voters 90 cents each, including printing and bulk-rate postage.

Suppose the success of the candidate depends on your campaign reaching at least 125,000 voters and that because your budget is limited, you must achieve this goal at a minimum cost. What would your plan and the cost of that plan be?

If you need help devising a method of solution for this problem, try answering the following questions first.

1. How many voters in the city read the newspaper but do not watch the local cable television news?
2. How many voters read the newspaper or watch the local cable television news or both?
3. Complete the following chart by indicating the number of voters reached by each promotional option, the total cost, and the cost per voter reached.

	Number of Voters Reached	*Total Cost*	*Cost per Voter Reached*
Pamphlet			
Television			
Newspaper			

4. Now explain your plan and the cost of that plan.

II. Pricing for Maximum Profit

Overnight Cleaners specializes in cleaning professional offices. The company is expanding into a new city and believes that its staff is capable of handling 300 office units at a weekly cost per unit of \$58 for labor and supplies. Preliminary pricing surveys indicate that if Overnight's weekly charge is \$100 per unit, then it would have clients for 300 units. However, for each \$5 price increase, it could expect to have 10 fewer units. By using a spreadsheet to investigate possible prices per unit and corresponding numbers of units, determine what weekly price Overnight Cleaners should charge in order to have the greatest profit.

The following steps will aid in creating a spreadsheet that can determine the price that gives maximum profit.

(a) Begin with columns for price per unit and number of units, and create a plan by hand to determine what your spreadsheet should look like. Use the information in the problem statement to determine the starting values for price per unit and number of units. Note that a price increase of \$5 resulting in a reduction of 10 units to clean can be equivalently expressed as a \$0.50 increase resulting in a 1-unit reduction. With this observation, determine the next three entries in the columns for price per unit and number of units. Write the rules used for each.

(b) Profit is found by subtracting expenses (or costs) from income. In the hand-done plan begun in (a), add a column or columns that would allow you to use information about price per unit and number of units to find profit. In your hand-done plan, complete entries for each new column (as you did for prices per unit and number of units). Write the rule(s) you used. (At this point, you should have a hand-done plan with at least four entries in each column and a rule for finding more entries in each column.)

(c) Use the starting values from (a) and the rules from (a) and (b) to create a computer spreadsheet (with appropriate cell references) that will duplicate and extend your hand-done plan.

(d) Fill the computer spreadsheet from (c) and find the price per unit that gives Overnight Cleaners its maximum profit. Find its maximum profit. (Print the spreadsheet.)

Warm-up

Prerequisite Problem Type	For Section	Answer	Section for Review
Evaluate: (a) $2(-1)^3 - 3(-1)^2 + 1$ (b) $3(-3) - 1$ (c) $14(10) - 0.02(10^2)$ (d) $\frac{3-1}{4-(-2)}$ (e) $\frac{-1-3}{2-(-2)}$ (f) $\frac{3(-8)}{4}$ (g) $2\left(\frac{23}{9}\right) - 5\left(\frac{2}{9}\right)$	1.1 1.2 1.3 1.5 1.6	(a) -4 (b) -10 (c) 138 (d) $\frac{1}{3}$ (e) -1 (f) -6 (g) 4	0.2 Signed numbers
(a) $\frac{1}{x}$ is *undefined* for which real numbers? (b) $\sqrt{x-4}$ is a real number for which values of x?	1.2	(a) Undefined for $x = 0$ (b) $x \geq 4$	0.2 Real numbers 0.4 Radicals
Identify the coefficient of x and the constant term for (a) $-9x + 2$ (b) $\frac{x}{2}$ (c) $x - 300$	1.1, 1.3 1.5, 1.6	*Coeff.* / *Const.* (a) -9 / 2 (b) $\frac{1}{2}$ / 0 (c) 1 / -300	0.5 Algebraic expressions
Simplify: (a) $4(-c)^2 - 3(-c) + 1$ (b) $[3(x + h) - 1] - [3x - 1]$ (c) $12\left(\frac{3x}{4} + 3\right)$ (d) $9x - (300 + 2x)$ (e) $-\frac{1}{5}[x - (-1)]$ (f) $2(2y + 3) + 3y$	1.1, 1.2 1.3, 1.5 1.6	(a) $4c^2 + 3c + 1$ (b) $3h$ (c) $9x + 36$ (d) $7x - 300$ (e) $-\frac{1}{5}x - \frac{1}{5}$ (f) $7y + 6$	0.5 Algebraic expressions
Find the LCD of $\frac{3x}{2x + 10}$ and $\frac{1}{x + 5}$	1.1	$2x + 10$	0.7 Algebraic fractions

Chapter 1

Linear Equations and Functions

A wide variety of problems from business, the social sciences, and the life sciences may be solved by using equations. Managers and economists use equations and their graphs to study costs, sales, national consumption, or supply and demand. Social scientists may plot demographic data or try to develop equations that predict population growth, voting behavior, or learning and retention rates. Life scientists use equations to model the flow of blood or the conduction of nerve impulses and to test theories or develop new ones by plotting experimental evidence.

In this chapter we begin by investigating the solution of linear equations. We then define the concepts of relation, function, domain, and range and introduce functional notation. Linear functions (including slope, equation writing, graphs, and systems) and their applications are the focus of most of the remainder of the chapter, along with a discussion of the uses of graphing utilities.

Numerous applications of mathematics are given throughout the text, but most chapters contain special sections emphasizing business and economics applications. In particular, this chapter introduces two important applications that will be expanded and used throughout the text as increased mathematical skills permit: supply and demand as functions of price (market analysis); and total cost, total revenue, and total profit as functions of the quantity produced or sold (theory of the firm).

Most businesses use linear equations to predict such things as future revenues and costs. Linear functions are used in economics, biology, and sociology to relate data, and they are also used in accounting courses. We also formulate linear supply and demand functions and then determine the price and quantity at which market equilibrium occurs. We use linear revenue and cost functions to obtain profit functions and to find break-even points, and we solve problems involving linear equations from the social and life sciences.

1.1 Solution of Linear Equations in One Variable

OBJECTIVES

- *To solve linear equations in one variable*
- *To solve applied problems by using linear equations*

APPLICATION PREVIEW

Woodwright Industries manufactures interior, raised-panel doors with a hardwood veneer from a pressed mixture of scrap hardwood chips. They break even when total revenue *equals* total costs. Suppose x represents the number of these doors that Woodwright sells and total revenue is given by $98x$ and total costs are given by $48x + 12{,}000$. Then Woodwright breaks even when total revenue = total cost, or when

$$98x = 48x + 12{,}000$$

In this section, we begin to study equality relationships such as this one.

An **equation** is a statement that two quantities or algebraic expressions are equal. The two quantities on either side of the equal sign are called **members** of the equation. For example, $2 + 2 = 4$ is an equation with members $2 + 2$ and 4; $3x - 2 = 7$ is an equation with $3x - 2$ as its left member and 7 as its right member. Note that the equation $7 = 3x - 2$ is the same statement as $3x - 2 = 7$. An equation such as $3x - 2 = 7$ is known as an equation in one variable. The x in this case is called a **variable** because its value determines whether the equation is true. For example, $3x - 2 = 7$ is true only for $x = 3$. Finding the value(s) of the variable(s) that make the equation true—that is, finding the **solutions**—is called **solving the equation.** The set of solutions of an equation is called a **solution set** of the equation. The variable in an equation is sometimes called the **unknown.**

Some equations involving variables are true only for certain values of the variables, whereas others are true for all values. Equations that are true for all values of the variables are called **identities.** The equation $2(x - 1) = 2x - 2$ is an example of an identity. Equations that are true only for certain values of the variables are called **conditional equations** or simply **equations.**

Two equations are said to be **equivalent** if they have exactly the same solution set. For example,

$$4x - 12 = 16$$
$$4x = 28$$
$$x = 7$$

are equivalent equations because they all have the same solution, namely 7. We can often solve a complicated linear equation by finding an equivalent equation whose solution is easily found. We use the following properties of equality to reduce an equation to a simple equivalent equation.

Property of Equality	***Examples***
Substitution Property The equation formed by substituting one expression for an equal expression is equivalent to the original equation.	$3(x-3)-\frac{1}{2}(4x-18)=4$ is equivalent to $3x-9-2x+9=4$ and to $x=4$. We say the solution set is $\{4\}$, or the solution is 4.
Addition Property The equation formed by adding the same quantity to both sides of an equation is equivalent to the original equation.	$x-4=6$ is equivalent to $x-4+4=6+4$, or to $x=10$. $x+5=12$ is equivalent to $x+5+(-5)=12+(-5)$, or to $x=7$.
Multiplication Property The equation formed by multiplying both sides of an equation by the same nonzero quantity is equivalent to the original equation.	$\frac{1}{3}x=6$ is equivalent to $3\left(\frac{1}{3}x\right)=3(6)$, or to $x=18$. $5x=20$ is equivalent to $(5x)/5=20/5$, or to $x=4$. (Dividing both sides by 5 is equivalent to multiplying both sides by $\frac{1}{5}$.)

If an equation contains one variable and if the variable occurs to the first degree, the equation is called a **linear equation in one variable.** The three properties above permit us to reduce any linear equation in one variable (also called an unknown) to an equivalent equation whose solution is obvious. We may solve linear equations in one variable by using the following procedure:

Solving a Linear Equation

Procedure	***Example***
To solve a linear equation in one variable:	Solve $\frac{3x}{4}+3=\frac{2(x-1)}{6}$.
1. If the equation contains fractions, multiply both sides by the least common denominator (LCD) of the fractions.	1. LCD is 12. $12\left(\frac{3x}{4}+3\right)=12\left(\frac{2(x-1)}{6}\right)$
2. Remove any parentheses in the equation.	2. $9x+36=4x-4$
3. Perform any additions or subtractions to get all terms containing the variable on one side and all other terms on the other side.	3. $9x+36-4x=4x-4-4x$ $5x+36=-4$ $5x+36-36=-4-36$ $5x=-40$
4. Divide both sides of the equation by the coefficient of the variable.	4. $\frac{5x}{5}=\frac{-40}{5}$ gives $x=-8$
5. Check the solution by substitution in the original equation.	5. $\frac{3(-8)}{4}+3\stackrel{?}{=}\frac{2(-8-1)}{6}$ gives $-3=-3$ ✓

EXAMPLE 1

(a) Solve for z: $\frac{2z}{3} = -6$ (b) Solve for x: $\frac{3x+1}{2} = \frac{x}{3} - 3$

Solution

(a) Multiply both sides by 3/2.

$\frac{3}{2}\left(\frac{2z}{3}\right) = \frac{3}{2}(-6)$ gives $z = -9$

Check: $\frac{2(-9)}{3} \stackrel{?}{=} -6$ gives $-6 = -6$ ✓

(b) $\frac{3x+1}{2} = \frac{x}{3} - 3$

$6\left(\frac{3x+1}{2}\right) = 6\left(\frac{x}{3} - 3\right)$ Multiplying both sides by the LCD, 6.

$3(3x+1) = 6\left(\frac{x}{3} - 3\right)$ Simplifying the fraction on the left side.

$9x + 3 = 2x - 18$ Distributing to remove parentheses.

$7x = -21$ Adding $(-2x) + (-3)$ to both sides.

$x = -3$ Dividing both sides by 7.

Check: $\frac{3(-3)+1}{2} \stackrel{?}{=} \frac{-3}{3} - 3$ gives $-4 = -4$ ✓

A **fractional equation** is an equation that contains a variable in a denominator. It is solved by first multiplying both sides of the equation by the least common denominator (LCD) of the fractions in the equation. Some fractional equations lead to linear equations. Note that the solution to any fractional equation *must* be checked in the original equation, because multiplying both sides of a fractional equation by a variable expression may result in an equation that is not equivalent to the original equation. Some fractional equations have no solutions.

EXAMPLE 2

Solve for x: (a) $\frac{3x}{2x+10} = 1 + \frac{1}{x+5}$ (b) $\frac{2x-1}{x-3} = 4 + \frac{5}{x-3}$

Solution

(a) First multiply both sides by the LCD $= 2x + 10$. Then simplify and solve.

$(2x+10)\left(\frac{3x}{2x+10}\right) = (2x+10)(1) + (2x+10)\left(\frac{1}{x+5}\right)$

$3x = (2x+10) + 2$ gives $x = 12$

Check: $\frac{3(12)}{2(12)+10} \stackrel{?}{=} 1 + \frac{1}{12+5}$ gives $\frac{36}{34} = \frac{18}{17}$ ✓

(b) First multiply both sides by the LCD $= x - 3$. Then simplify.

$(x-3)\left(\frac{2x-1}{x-3}\right) = (x-3)(4) + (x-3)\left(\frac{5}{x-3}\right)$

$2x - 1 = (4x - 12) + 5$ or $2x - 1 = 4x - 7$

Add $(-4x) + 1$ to both sides.

$-2x = -6$ gives $x = 3$

Check: The value $x = 3$ gives undefined expressions because the denominators equal 0. Hence the equation has no solution.

The steps used to solve linear equations in one variable can also be used to solve linear equations in more than one variable. Solving an equation like the one in the following example is important when using a graphing utility.

EXAMPLE 3

Solve $4x + 3y = 12$ for y.

Solution

No fractions or parentheses are present, so we subtract $4x$ from both sides to get only the term that contains y on one side.

$$3y = -4x + 12$$

Dividing both sides by 3 gives the solution.

$$y = -\frac{4}{3}x + 4$$

CHECKPOINT

1. Solve the following for x, and check.
   (a) $4x = 7$ (b) $4 - 5x = 4 + x$ (c) $3(x - 7) = 19 - x$
   (d) $\frac{-1}{3}x = 15$ (e) $\frac{5(x-3)}{6} - x = 1 - \frac{x}{9}$ (f) $\frac{x}{3x-6} = 2 - \frac{2x}{x-2}$
2. Solve for y: $7(y - 4) = -4(x + 2)$

EXAMPLE 4

Suppose that the relationship between a firm's profit P and the number x of items sold can be described by the equation

$$5x - 4P = 1200$$

(a) How many units must be produced and sold for the firm to make a profit of \$150?
(b) Solve this equation for P in terms of x.
(c) Find the profit when 240 units are sold.

Solution

(a)
$$\begin{aligned} 5x - 4(150) &= 1200 \\ 5x - 600 &= 1200 \\ 5x &= 1800 \\ x &= 360 \text{ units} \end{aligned}$$

Check: $5(360) - 4(150) = 1800 - 600 = 1200$ ✓

(b)
$$5x - 1200 = 4P$$
$$P = \frac{5}{4}x - 300$$

(c)
$$P = \frac{5}{4}x - 300$$
$$P = \frac{5}{4}(240) - 300 = 0$$

We say that the firm breaks even (profit $P = 0$) when $x = 240$ units.

We now return to the Application Preview problem.

EXAMPLE 5

Suppose Woodwright Industries' total revenue from the sale of x interior, raised-panel doors is given by

$$R = 98x$$

and its total costs are given by

$$C = 48x + 12{,}000$$

Woodwright breaks even when its total revenues equal its total costs. Find the number of doors that Woodwright must sell in order to break even.

Solution

The break-even point occurs when $R = C$, so we must solve for x in

$$98x = 48x + 12{,}000$$
$$50x = 12{,}000$$
$$x = \frac{12{,}000}{50} = 240$$

Thus Woodwright Industries breaks even if 240 doors are sold. This can be checked by noting that if $x = 240$ doors, Woodwright's revenue and cost both equal $23,250.

With an applied problem, it is frequently necessary to convert the problem from its stated form into one or more equations from which the problem's solution can be found. The following guidelines may be useful in solving stated problems.

Guidelines for Solving Stated Problems

1. Begin by reading the problem carefully to determine what you are to find. Use variables to represent the quantities to be found.
2. Reread the problem and use your variables to translate given information into algebraic expressions. Often, drawing a figure is helpful.
3. Use the algebraic expressions and the problem statement to formulate an equation (or equations).
4. Solve the equation.
5. Check the solution in the problem, not just in your equation or equations. The answer should satisfy the conditions.

EXAMPLE 6

Jill Bell has \$90,000 to invest. She has chosen one relatively safe investment fund that has an annual yield of 10% and another, riskier one that has a 15% annual yield. How much should she invest in each fund if she would like to earn \$10,000 in one year from her investments?

Solution

We want to find the amount of each investment, so we begin as follows:

Let x = the amount invested at 10%, then

$90{,}000 - x$ = the amount invested at 15% (because the two investments total \$90,000).

If P is the amount of an investment and r is the annual rate of yield (expressed as a decimal), then the annual earnings $I = Pr$. Using this relationship, we can summarize the information about these two investments in a table.

	P	r	I
10% investment	x	0.10	$0.10x$
15% investment	$90{,}000 - x$	0.15	$0.15(90{,}000 - x)$
Total investment	90,000		10,000

The column under I shows that the sum of the earnings is

$$0.10x + 0.15(90{,}000 - x) = 10{,}000$$

We solve this as follows.

$$0.10x + 13{,}500 - 0.15x = 10{,}000$$

$$-0.05x = -3500 \quad \text{or} \quad x = 70{,}000$$

Thus the amount invested at 10% is \$70,000, and the amount invested at 15% is $90{,}000 - 70{,}000 = 20{,}000$. Check: To check, we return to the problem and note that 10% of \$70,000 plus 15% of \$20,000 gives a yield of $\$7000 + \$3000 = \$10{,}000$.

The solutions of linear equations can also be checked by evaluating both sides of the equation with a calculator and verifying that the two results are equal.

CHECKPOINT SOLUTIONS

1. (a) $4x = 7$

$x = \frac{7}{4}$

Check: $4\left(\frac{7}{4}\right) = 7$ ✓

(b) $4 - 5x = 4 + x$

$-5x = x$

$0 = 6x$

$\frac{0}{6} = x$

$x = 0$

Check: $4 - 5(0) \stackrel{?}{=} 4 + 0$

$4 = 4$ ✓

(c) $3(x - 7) = 19 - x$

$3x - 21 = 19 - x$

$4x = 40$

$x = 10$

Check:

$3(10 - 7) \stackrel{?}{=} 19 - 10$

$9 = 9$ ✓

(d) $-\frac{1}{3}x = 15$

$x = (15)(-3)$

$x = -45$

Check: $-\frac{1}{3}(-45) = 15$ ✓

(e) $\frac{5(x - 3)}{6} - x = 1 - \frac{x}{9}$

LCD of 6 and 9 is 18.

$18\left(\frac{5x - 15}{6}\right) - 18(x) = 18(1) - 18\left(\frac{x}{9}\right)$

$3(5x - 15) - 18x = 18 - 2x$

$15x - 45 - 18x = 18 - 2x$

$-45 - 3x = 18 - 2x$

$-63 = x$

Check: $\frac{5(-63 - 3)}{6} - (-63) \stackrel{?}{=} 1 - \frac{(-63)}{9}$

$5(-11) + 63 \stackrel{?}{=} 1 + 7$

$8 = 8$ ✓

(f) $\frac{x}{3x - 6} = 2 - \frac{2x}{x - 2}$

LCD is $3x - 6$ or $3(x - 2)$.

$(3x - 6)\left(\frac{x}{3x - 6}\right) = (3x - 6)(2) - (3x - 6)\left(\frac{2x}{x - 2}\right)$

$x = 6x - 12 - 3(2x)$ or $x = 6x - 12 - 6x$

$x = -12$

Check: $\frac{-12}{3(-12) - 6} \stackrel{?}{=} 2 - \frac{2(-12)}{-12 - 2}$

$\frac{-12}{-42} \stackrel{?}{=} 2 - \frac{12}{7}$

$\frac{2}{7} = \frac{2}{7}$ ✓

2. $7(y - 4) = -4(x + 2)$

$7y - 28 = -4x - 8$

$7y = -4x + 20$

$y = \frac{-4}{7}x + \frac{20}{7}$

EXERCISE 1.1

In Problems 1–18, solve each equation.

1. $4x - 7 = 8x + 2$
2. $3x + 22 = 7x + 2$
3. $x + 8 = 8(x + 1)$
4. $x + x + x = x$
5. $8(x - 2) = 6(3x - 4)$
6. $3(x - 2) = 4(3 - x)$
7. $\frac{-3}{4}x = 24$
8. $\frac{-1}{6}x = 12$
9. $\frac{5x}{6} = \frac{8}{3}$
10. $\frac{17x}{2} = \frac{34}{3}$
11. $2(x - 7) = 5(x + 3) - x$
12. $3(x - 4) = 4 - 2(x + 2)$
13. $\frac{5x}{2} - 4 = \frac{2x - 7}{6}$
14. $\frac{2x}{3} - 1 = \frac{x - 2}{2}$
15. $\frac{5x - 1}{9} = \frac{5(x - 1)}{6}$
16. $\frac{6x + 5}{2} = \frac{5(2 - x)}{3}$
17. $x + \frac{1}{3} = 2\left(x - \frac{2}{3}\right) - 6x$
18. $\frac{3x}{4} - \frac{1}{3} = 1 - \frac{2}{3}\left(x - \frac{1}{6}\right)$

The equations in Problems 19–26 lead to linear equations. Because not all solutions to the linear equations are solutions to the original equations, be sure to check the solutions in the original equations.

19. $\frac{33 - x}{5x} = 2$
20. $\frac{3x + 3}{x - 3} = 7$
21. $\frac{3}{x - 5} = \frac{7}{2x + 13}$
22. $\frac{5}{3x + 8} + \frac{2}{x - 1} = 0$
23. $\frac{2x}{x - 1} + \frac{1}{3} = \frac{5}{6} + \frac{2}{x - 1}$
24. $\frac{2x}{x - 3} = 4 + \frac{6}{x - 3}$
25. $\frac{2x}{2x + 5} = \frac{2}{3} - \frac{5}{4x + 10}$
26. $\frac{3}{x} + \frac{1}{4} = \frac{2}{3} + \frac{1}{x}$

In Problems 27–30, use a calculator to solve each equation. Round your answer to three decimal places.

27. $3.259x - 8.638 = -3.8(8.625x + 4.917)$
28. $3.319(14.1x - 5) = 9.95 - 4.6x$
29. $0.000316x + 9.18 = 2.1(3.1 - 0.0029x) - 4.68$
30. $3.814x = 2.916(4.2 - 0.06x) + 5.3$

In Problems 31–34, solve for y in terms of x.

31. $3x - 4y = 15$
32. $3x - 5y = 25$
33. $9x + \frac{3}{2}y = 11$
34. $\frac{3x}{2} + 5y = \frac{1}{3}$
35. Solve $I = Prt$ for P.
36. Solve $S = P(1 + i)^n$ for P.

Applications

37. ***Course grades*** To earn an A in a course, a student must get at least a 90 average on four tests and a final exam, with the final exam weighted twice that of any one test. If the four test scores are 93, 69, 89, and 97, what is the lowest score the student can earn on the final exam and still get an A in the course?

38. ***Course grades*** Suppose a professor counts the final exam as being equal to each of the other tests in her course, and she will also change the lowest test score to match the final exam score if the final exam score is higher. If a student's four test scores are 83, 67, 52, and 90, what is the lowest score the student can earn on the final exam and still obtain at least an 80 average for the course?

39. ***Height-weight relationships*** It has been noted that for adults over 5 feet tall in the northeast United States, their weight is related to their height according to

$$3w + 110 = 11(h - 20)$$

where w is measured in pounds and h is measured in inches. Use the formula above to answer the following.

(a) Find the weight of an adult whose height is 5 feet, 6 inches.

(b) Find the height of adults weighing 160 pounds.

40. ***Fahrenheit-Celsius*** The equation $F = \frac{9}{5}C + 32$ describes the relation between temperature readings in Fahrenheit and Celsius. What Celsius temperature is equivalent to $-40°$F?

41. ***Cigarette use*** Data from the National Institute on Drug Abuse indicates that the percentage p of high school seniors who have tried cigarettes can be described (that is, predicted with some accuracy) by the equation

$$p = 75.4509 - 0.706948t$$

where t is the number of years past 1975.

(a) What percentage does this equation predict for 1998?

(b) This equation ceases to be effective when p reaches zero (and then becomes negative) and perhaps before that. Find the year when $p = 0$.

42. ***Heat and humidity index*** One way to measure people's discomfort during extremes of heat and humidity is with the apparent temperature A, which is given by

$$A = 2.70 + 0.885t - 78.7h + 1.20th^*$$

where t is the Fahrenheit temperature and h is the relative humidity (expressed as a decimal). On June 15, 1994, in Washington, D.C., the temperature reached 101°F; for this reading the apparent temperature formula becomes

$$A = 92.085 + 42.5h$$

If the local Washington, D.C., news reported that this 101°F reading was equivalent to an apparent temperature of 111°F, what was the relative humidity (as a decimal)? What was the percent relative humidity?

43. ***Sales projections*** Data in an article in the *Atlanta Journal* on July 14, 1993, indicated that the growth of sales, S (in millions of dollars), of the Chick-fil-A's restaurant chain can be described by

$$S = 241.33 + 29t$$

where t is the number of years past 1988. In what year does this equation project Chick-fil-A's sales will reach $1 billion?

44. ***Investing*** In Federal Signal Corporation's publication of March 7, 1994, *Notice of Annual Meeting,* data were given that compared the cumulative total return on a $100 investment made in 1988. If t is the number of years past 1988, then this cumulative total return can be described by R_{FS} for Federal Signal Corporation and by R_{SP} for the S & P 400 Index as follows:

$$R_{FS} = 85.714t + 88.381$$
$$R_{SP} = 17.1714t + 104.238$$

Assuming that these equations are valid, find the year in which the cumulative total return of
(a) Federal Signal is predicted to reach $850,
(b) S & P 400 Index is predicted to reach $300.

45. ***Organizational impact*** The Beaver-Castle Girl Scout Council of Beaver, Pennsylvania, determined the number of youths it had served by adding its active registrations and the number of dropouts. If the drop-out rate was one-third of the total served, find the total number of youths served when there were 6000 youths actively registered.

46. ***Population*** A city has a population P at the beginning of a year. The birth rate during the year was 10 per thousand, and the death rate was 12 per thousand (of the year's original population). During the year, 360 people moved into the city and 190 moved away. If the population at the end of the year is 30,110, what was the population at the beginning of the year?

47. ***Insect behavior*** The number of times n per minute that a cricket chirps is related to the Fahrenheit temperature T. This relationship can be approximated by

$$7n - 12T = 52$$

(a) Solve this equation for T.
(b) If you count 28 chirps in 15 seconds, find the approximate temperature.

48. ***Seawater pressure*** In seawater, the pressure p is related to the depth d according to

$$33p - 18d = 495$$

where d is in feet and p is in pounds per square inch.
(a) Solve this equation for p in terms of d.
(b) The *Titanic* was discovered at a depth of 12,460 ft. Find the pressure at this depth.

49. ***Investments*** The total amount of a simple interest investment is given by

$$A = P + Prt$$

What principal P must be invested for $t = 5$ years at the simple interest rate $r = 10\%$ so that the amount A grows to $6000?

50. ***Sales tax*** The total price of a new car (including 6% sales tax) is $21,041. How much of this is tax?

51. ***Investment mix*** A retired woman has $120,000 to invest. She has chosen one relatively safe investment fund that has an annual yield of 9% and another, riskier one that has a 13% annual yield. How much should she invest in each fund if she would like to earn $12,000 per year from her investments?

52. ***Investment yields*** One safe investment pays 10% per year, and a more risky investment pays 18% per year. A woman who has $145,600 to invest would like to have an income of $20,000 per year from her investments. How much should she invest at each rate?

*Bosch, W., and C. G. Cobb, "Temperature-Humidity Indices," UMAP Unit 691, *The UMAP Journal,* 10(3), Fall 1989, 237–256.

53. ***Salary increases*** A woman making \$2000 per month has her salary reduced by 10% because of sluggish sales. One year later, after a dramatic improvement in sales, she is given a 20% raise over her reduced salary. Find her salary after the raise. What percent change is this from the \$2000 per month?

54. ***Profit*** A car dealer purchases 20 new automobiles for \$8000 each. If he sells 16 of them at a profit of 20%, for how much must he sell the remaining 4 to obtain an average profit of 18%?

55. ***Profit and loss*** An antiques collector sold two pieces of furniture for \$480 each. For one of them, this represented a 20% loss; for the other, it was a 20% profit (based on the cost of each item). How much did he make or lose on the transaction?

56. ***Wildlife management*** In wildlife management, the capture-mark-recapture technique is used to estimate the populations of fish or birds in an area or to measure the infestation of insects such as Japanese beetles. Suppose 100 individuals of the species being studied are caught, marked, and released, and one week later 100 more are caught. To estimate the total number of individuals, the following relationship is used:

$$\frac{\text{Total marked found in 2nd capture}}{\text{Total in 2nd capture}} = \frac{\text{Total number marked}}{\text{Total population}}$$

(a) If in the second capture of 100, it is found that 3 are marked, what is the total population?

(b) Suppose that 1000 beetles are captured, marked, and released. Suppose further that in the second capture of 1000 it is found that 63 are marked. What is the population estimate?

57. ***Markups*** A retailer wants a 30% markup on the selling price of an item that costs him \$214.90. What selling price should he charge?

58. ***Markups*** A toaster costs the Ace Department Store \$22.74. If the store marks the toaster up by 40% of the selling price, what is the selling price?

59. ***Markups*** A room air conditioner costs the wholesaler \$154.98. If the wholesaler's markup is 10% of the wholesale selling price and if the retailer's markup is 30% of the retail selling price, for what price does the retailer sell the air conditioner?

60. ***Markups*** An electric mixer retails for \$48.54, which includes a markup of 40% for the retailer and a markup of 20% for the wholesaler. These markups are based on selling price.

(a) What did the mixer cost the retailer?

(b) What did the mixer cost the wholesaler?

1.2 Functions

OBJECTIVES

- *To determine whether a relation is a function*
- *To state the domains and ranges of certain functions*
- *To use function notation*
- *To perform operations with functions*
- *To find the composite of two functions*

APPLICATION PREVIEW

The dollar volume of transactions at automatic teller machines (ATMs) has increased as the number of machines has increased. This relationship can be described by the equation

$$y = 0.1369x - 5.091255^*$$

where y is billions of dollars of transactions and x is the number of terminals (in thousands). In this equation, y (the dollar volume of transactions) depends uniquely on x (the number of ATM machines), so we say y is a function of x. Understanding the mathematical meaning of the phrase *function of* and gaining the ability to interpret and apply such relationships are the goals of this section.

*Equation developed with data from the Electronic Funds Transfer Association published in *Statistical Abstract of the United States,* 1993, p. 517.

An equation or inequality containing two variables expresses a **relation** between those two variables. For example, the inequality $R \geq 35x$ expresses a relation between the two variables x and R, and the equation $y = 4x - 3$ expresses a relation between the two variables x and y.

In addition to defining a relation by an equation or rule of correspondence, we may also define it as any set of **ordered pairs** of real numbers (a, b). For example, the solutions to $y = 4x - 3$ are pairs of numbers (one for x and one for y). We write the pairs (x, y) so that the first number is the x-value and the second is the y-value, and these ordered pairs define the relation between x and y. Some relations may be defined by a table, a graph, or an equation.

Relation A **relation** is defined by a set of ordered pairs or by a rule that determines how the ordered pairs are found. It may also be defined by a table, a graph, or an equation.

For example, the set of ordered pairs

$$\{(1, 3), (1, 6), (2, 6), (3, 9), (3, 12), (4, 12)\}$$

expresses a relation between the set of first components, $\{1, 2, 3, 4\}$, and the set of second components, $\{3, 6, 9, 12\}$. The set of first components is called the **domain** of the relation, and the set of second components is called the **range** of the relation. Figure 1.1 uses arrows to indicate how the inputs from the domain (the first components) are associated with the outputs in the range (the second components). Because relations can also be defined by tables and graphs, Table 1.1 and Figure 1.2 are examples of relations.

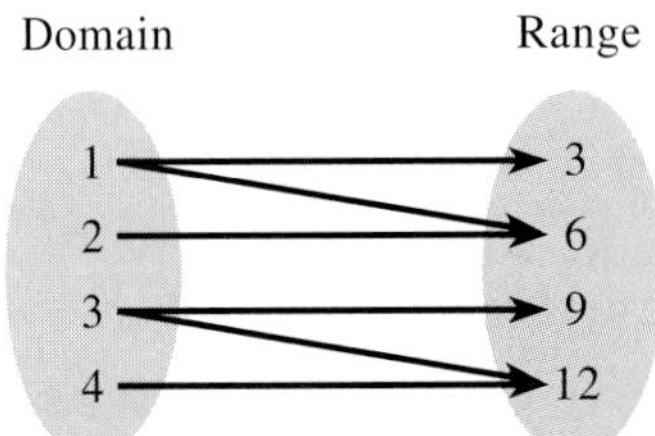

Figure 1.1

An equation frequently expresses how the second component (the output) is obtained from the first component (the input). For example, the equation

$$y = 4x - 3$$

expresses how the output y results from the input x. This equation expresses a special relation between x and y, because each value of x that is substituted into the equation results in only one value for y. If each value of x put into an equation results in one value of y, we say that the equation expresses y as a **function** of x.

Definition of a Function A **function** is a relation between two sets such that to each element of the domain (input) there corresponds exactly one element of the range (output).

TABLE 1.1 Single Filer's Tax Table

Taxable Income (Domain)		*Tax Due (Range)*
at least	less than	
30,000	**30,050**	5,203
30,050	**30,100**	5,217
30,100	**30,150**	5,231
30,150	**30,200**	5,245
30,200	**30,250**	5,259
30,250	**30,300**	5,273
30,300	**30,350**	5,287
30,350	**30,400**	5,301
30,400	**30,450**	5,315
30,450	**30,500**	5,329
30,500	**30,550**	5,343
30,550	**30,600**	5,357
30,600	**30,650**	5,371
30,650	**30,700**	5,385
30,700	**30,750**	5,399
30,750	**30,800**	5,413
30,800	**30,850**	5,427
30,850	**30,900**	5,441
30,900	**30,950**	5,455
30,950	**31,000**	5,469

Source: Internal Revenue Service, 1997 Form 1040, Instructions

When a function is defined, the variable that represents the numbers in the domain (input) is called the **independent variable** of the function, and the variable that represents the numbers in the range (output) is called the **dependent variable** (because its values depend on the values of the independent variable). The equation $y = 4x - 3$ defines y as a function of x, because only one value of y will result from each value of x that is substituted into the equation. Thus the equation defines a function in which x is the independent variable and y is the dependent variable.

We can also apply this idea to a relation defined by a table. For example, Table 1.1 defines tax due (the dependent variable) as a function of the taxable income of a single taxpayer (the independent variable).

EXAMPLE 1

Does $y^2 = 2x$ express y as a function of x?

Solution

No, because for some values of x there is more than one value for y. In fact, there are two y-values for each $x > 0$. For example, if $x = 8$, then $y = 4$ or $y = -4$, two different y-values for the same x-value. The equation $y^2 = 2x$ expresses a relation between x and y, but y is not a function of x.

It is possible to picture geometrically the relations and functions that we have been discussing by sketching their graphs on a rectangular coordinate system. We construct a rectangular coordinate system by drawing two real number lines (called **coordinate axes**) that are perpendicular to each other and intersect at their origins (called the **origin** of the system).

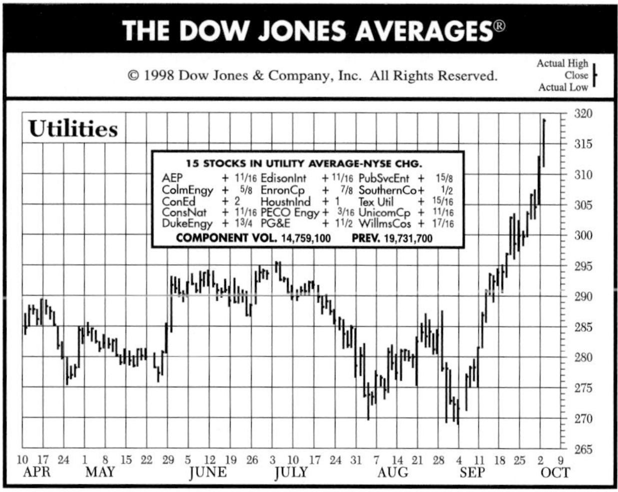

Figure 1.2 SOURCE: *Wall Street Journal*, October 6, 1998.

The ordered pair (a, b) represents the point P that is located a units along the x-axis and b units along the y-axis (see Figure 1.3). Similarly, any point has a unique ordered pair that describes it.

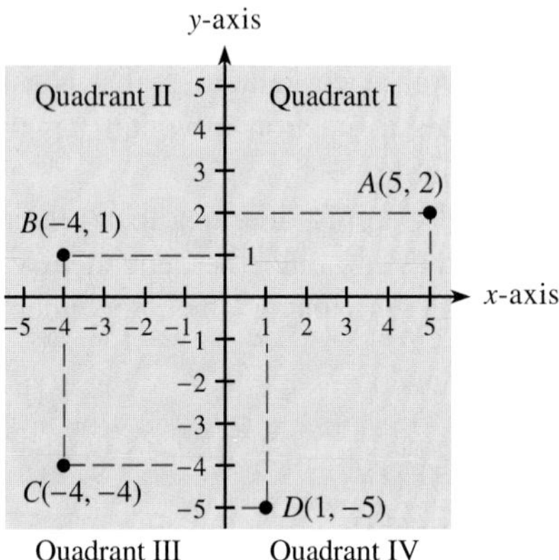

Figure 1.3

The values a and b in the ordered pair associated with the point P are called the **rectangular** (or **Cartesian**) **coordinates** of the point, where a is the **x-coordinate** (or **abscissa**), and b is the **y-coordinate** (or **ordinate**). The ordered pairs (a, b) and (c, d) are equal if and only if $a = c$ and $b = d$.

The **graph** of an equation that defines a function (or relation) is the picture that results when we plot the points whose coordinates (x, y) satisfy the equation. To sketch the graph, we plot enough points to suggest the shape of the graph and draw a smooth curve through the points. This is called the **point-plotting method** of sketching a graph.

EXAMPLE 2

Graph the function $y = 4x^2$.

Solution

We choose some sample values of x and find the corresponding values of y. Placing these in a table, we have sample points to plot. When we have enough to determine the shape of the graph, we connect the points to complete the graph. The table and graph are shown in Figure 1.4.

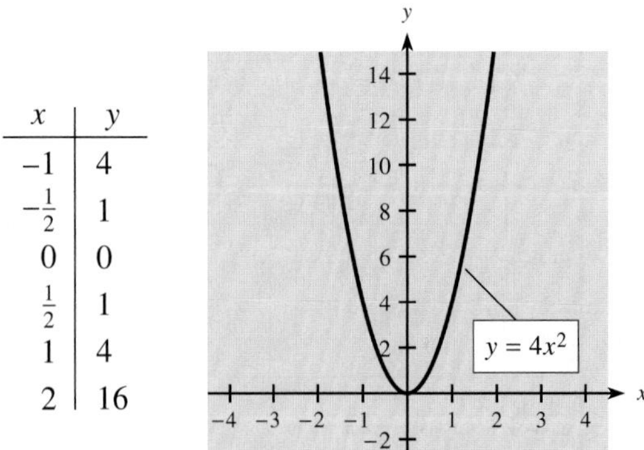

x	y
-1	4
$-\frac{1}{2}$	1
0	0
$\frac{1}{2}$	1
1	4
2	16

Figure 1.4

We can determine whether a relation is a function by inspecting its graph. If the relation is a function, no two points on the graph will have the same first coordinate (component). Thus no two points of the graph will lie on the same vertical line.

Vertical-Line Test If no vertical line exists that intersects the graph at more than one point, then the graph is that of a function.

Performing this test on the graph of $y = 4x^2$ (Figure 1.4), we easily see that this equation describes a function. The graph of $y^2 = 2x$ is shown in Figure 1.5, and we can see that the vertical-line test indicates that this is not a function (as we already saw in Example 1). For example, a vertical line at $x = 2$ intersects the curve at (2, 2) and (2, -2).

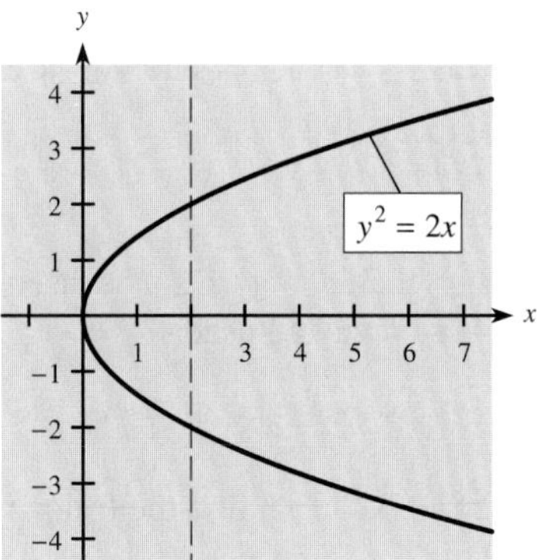

Figure 1.5

The graph in Figure 1.6 shows marijuana use (y) versus time (x). It represents a function, because for every x there is exactly one y. On the other hand, the graph in Figure 1.2 on page 77 does not represent a function, because for each input (day) there is an interval of output values (the Dow's range for that day).

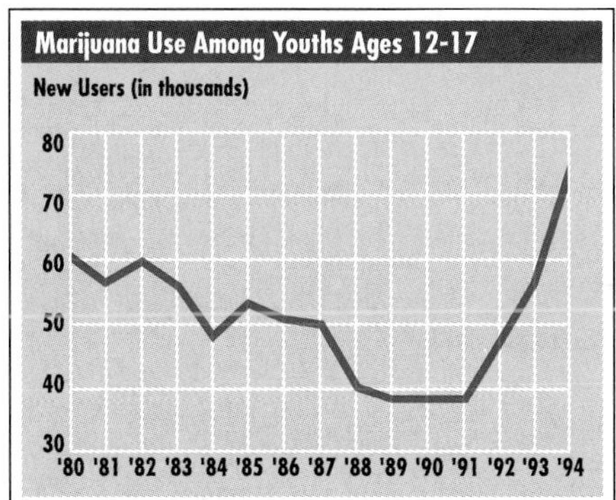

Source: Substance Abuse and Mental Health Services Administration Office of Applied Studies. U.S. Department of Health and Human Services

Figure 1.6

We can use functional notation to indicate that y is a function of x. The function is denoted by f, and we write $y = f(x)$. This is read "y is a function of x" or "y equals f of x." For specific values of x, $f(x)$ represents the values of the function (that is, outputs, or y-values) at those x-values. Thus if

$$f(x) = 3x^2 + 2x + 1$$

then $$f(2) = 3(2)^2 + 2(2) + 1 = 17$$

and $$f(-3) = 3(-3)^2 + 2(-3) + 1 = 22$$

Figure 1.7 represents this functional notation as (a) an operator on x and (b) a y-coordinate for a given x-value.

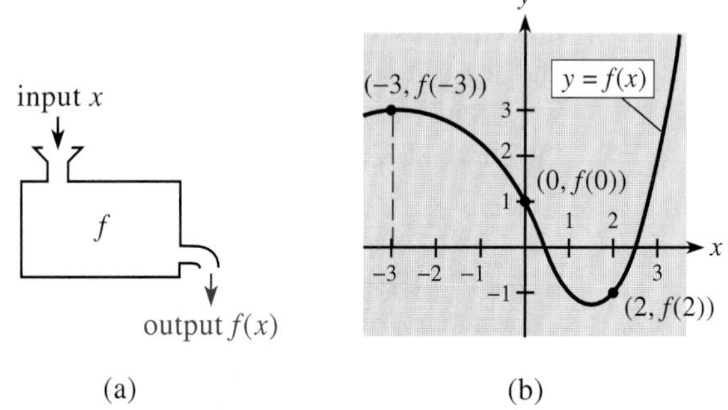

Figure 1.7 (a) (b)

Letters other than f may also be used to denote functions. For example, $y = g(x)$ or $y = h(x)$ may be used.

EXAMPLE 3

If $y = f(x) = 2x^3 - 3x^2 + 1$, find the following.

(a) $f(3)$ (b) $f(-1)$

Solution

(a) $f(3) = 2(3)^3 - 3(3)^2 + 1 = 2(27) - 3(9) + 1 = 28$
Thus $y = 28$ when $x = 3$.

(b) $f(-1) = 2(-1)^3 - 3(-1)^2 + 1 = 2(-1) - 3(1) + 1 = -4$
Thus $y = -4$ when $x = -1$.

EXAMPLE 4

If $g(x) = 4x^2 - 3x + 1$, find the following.

(a) $g(a)$ (b) $g(-a)$ (c) $g(b)$ (d) $g(a + b)$
(e) Does $g(a + b) = g(a) + g(b)$?

Solution

(a) $g(a) = 4(a)^2 - 3(a) + 1 = 4a^2 - 3a + 1$

(b) $g(-a) = 4(-a)^2 - 3(-a) + 1 = 4a^2 + 3a + 1$

(c) $g(b) = 4(b)^2 - 3(b) + 1 = 4b^2 - 3b + 1$

(d) $$g(a + b) = 4(a + b)^2 - 3(a + b) + 1$$
$$= 4(a^2 + 2ab + b^2) - 3a - 3b + 1$$
$$= 4a^2 + 8ab + 4b^2 - 3a - 3b + 1$$

(e) $$g(a) + g(b) = (4a^2 - 3a + 1) + (4b^2 - 3b + 1)$$
$$= 4a^2 + 4b^2 - 3a - 3b + 2$$

Thus $g(a + b) \neq g(a) + g(b)$.

EXAMPLE 5

Given $f(x) = x^2 - 3x + 8$, find $\dfrac{f(x+h) - f(x)}{h}$ and simplify (if $h \neq 0$).

Solution

$$\frac{f(x+h) - f(x)}{h} = \frac{[(x+h)^2 - 3(x+h) + 8] - [x^2 - 3x + 8]}{h}$$

$$= \frac{[(x^2 + 2xh + h^2) - 3x - 3h + 8] - x^2 + 3x - 8}{h}$$

$$= \frac{x^2 + 2xh + h^2 - 3x - 3h + 8 - x^2 + 3x - 8}{h}$$

$$= \frac{2xh + h^2 - 3h}{h} = \frac{h(2x + h - 3)}{h} = 2x + h - 3$$

EXAMPLE 6

As mentioned in the Application Preview, the dollar volume of transactions at ATMs can be described by the function

$$y = f(x) = 0.1369x - 5.091255$$

where y is billions of dollars of transactions and x is the number of terminals (in thousands).

(a) Find $f(100)$.
(b) Write a sentence that explains the meaning of the result in (a).

Solution

(a) $f(100) = 0.1369(100) - 5.091255$
$= 13.69 - 5.091255 = 8.598745.$

Thus the point (100, 8.598745) is on the graph of this function.

(b) The statement $f(100) = 8.598745$ means that when there were 100 thousand ATM terminals, there were (approximately) \$8.598745 billion in ATM transactions.

We will limit our discussion in this text to **real functions,** which are functions whose domains and ranges contain only real numbers. If the domain and range of a function are not specified, it is assumed that the domain consists of all real inputs (x-values) that result in real outputs (y-values), making the range a subset of the real numbers.

In general, if the domain of a function is unspecified, it will include all real numbers except

1. values that result in a denominator of 0, and
2. values that result in an even root of a negative number.

EXAMPLE 7

Find the domain of each of the following functions; find the range for the functions in (a) and (b).

(a) $y = 4x^2$ (b) $y = \sqrt{4 - x}$ (c) $y = 1 + \dfrac{1}{x - 2}$

Solution

(a) There are no restrictions on the numbers substituted for x, so the domain consists of all real numbers. Because the square of any real number is nonnegative, $4x^2$ must be nonnegative. Thus the range is $y \geq 0$. If we plot points or use a graphing utility, we will get the graph shown in Figure 1.8(a), which illustrates our conclusions about the domain and range.

(b) We note the restriction that $4 - x$ cannot be negative. Thus the domain consists of only numbers less than or equal to 4. That is, the domain is the set of real numbers satisfying $x \leq 4$. Because $\sqrt{4 - x}$ is always nonnegative, the range is all $y \geq 0$. Figure 1.8(b) shows the graph of $y = \sqrt{4 - x}$. Note that the graph is located only where $x \leq 4$ and on or above the x-axis (where $y \geq 0$).

(c) $y = 1 + \dfrac{1}{x - 2}$ is undefined at $x = 2$ because $\dfrac{1}{0}$ is undefined. Hence, the domain consists of all real numbers except 2. Figure 1.8(c) shows the graph of $y = 1 + \dfrac{1}{x - 2}$. The break where $x = 2$ indicates that $x = 2$ is not part of the domain.

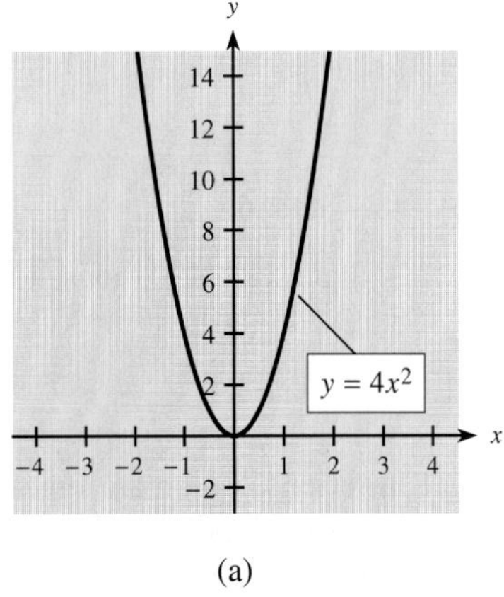

(a)

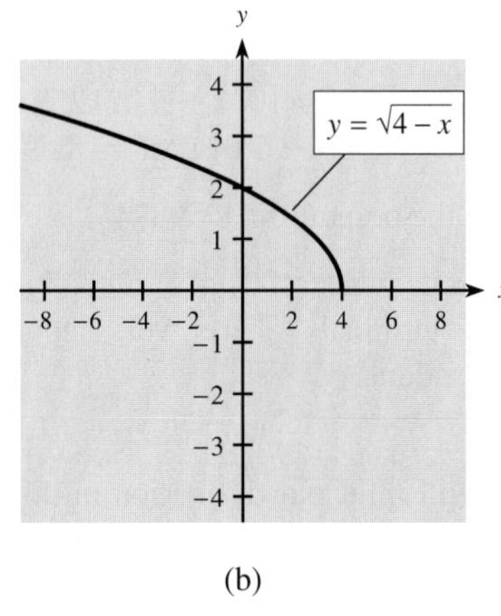

(b)

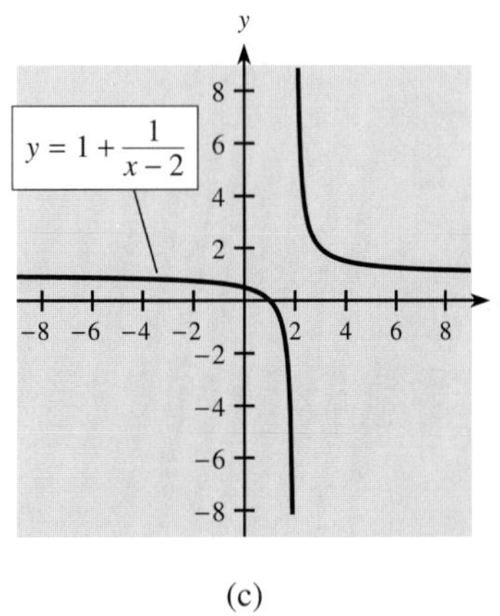

(c)

Figure 1.8

CHECKPOINT

1. If $y = f(x)$, the independent variable is ______ and the dependent variable is ______.
2. If (1, 3) is on the graph of $y = f(x)$, then $f(1) = ?$

3. If $f(x) = 1 - x^3$, find $f(-2)$.
4. If $f(x) = 2x^2$, find $f(x + h)$.
5. If $f(x) = \dfrac{1}{x+1}$, what is the domain of $f(x)$?

We can form new functions by performing algebraic operations with two or more functions. We define new functions that are the sum, difference, product, and quotient of two functions as follows:

Operations with Functions

Let f and g be functions of x, and define the following.

Sum	$(f + g)(x) = f(x) + g(x)$
Difference	$(f - g)(x) = f(x) - g(x)$
Product	$(f \cdot g)(x) = f(x) \cdot g(x)$
Quotient	$\left(\dfrac{f}{g}\right)(x) = \dfrac{f(x)}{g(x)}$ if $g(x) \neq 0$

EXAMPLE 8

If $f(x) = 3x + 2$ and $g(x) = x^2 - 3$, find the following functions.

(a) $(f + g)(x)$ (b) $(f - g)(x)$

(c) $(f \cdot g)(x)$ (d) $\left(\dfrac{f}{g}\right)(x)$

Solution

(a) $(f + g)(x) = f(x) + g(x) = (3x + 2) + (x^2 - 3) = x^2 + 3x - 1$

(b) $(f - g)(x) = f(x) - g(x) = (3x + 2) - (x^2 - 3) = -x^2 + 3x + 5$

(c) $(f \cdot g)(x) = f(x) \cdot g(x) = (3x + 2)(x^2 - 3) = 3x^3 + 2x^2 - 9x - 6$

(d) $\left(\dfrac{f}{g}\right)(x) = \dfrac{f(x)}{g(x)} = \dfrac{3x+2}{x^2-3}$, if $x^2 - 3 \neq 0$

We now consider a new way to combine two functions. Just as we can substitute a number for the independent variable in a function, we can substitute a second function for the variable. This creates a new function, called a **composite function.**

Composite Functions

Let f and g be functions. Then the **composite functions** g of f (denoted $g \circ f$) and f of g (denoted $f \circ g$) are defined as follows:

$$(g \circ f)(x) = g(f(x))$$
$$(f \circ g)(x) = f(g(x))$$

Note that the domain of $g \circ f$ is the subset of the domain of f for which $g \circ f$ is defined. Similarly, the domain of $f \circ g$ is the subset of the domain of g for which $f \circ g$ is defined.

EXAMPLE 9

If $f(x) = 2x^3 + 1$ and $g(x) = x^2$, find the following.

(a) $(g \circ f)(x)$ (b) $(f \circ g)(x)$

Solution

(a) $(g \circ f)(x) = g(f(x))$
$= g(2x^3 + 1)$
$= (2x^3 + 1)^2 = 4x^6 + 4x^3 + 1$

(b) $(f \circ g)(x) = f(g(x))$
$= f(x^2)$
$= 2(x^2)^3 + 1$
$= 2x^6 + 1$

Figure 1.9 illustrates both composite functions found in Example 9.

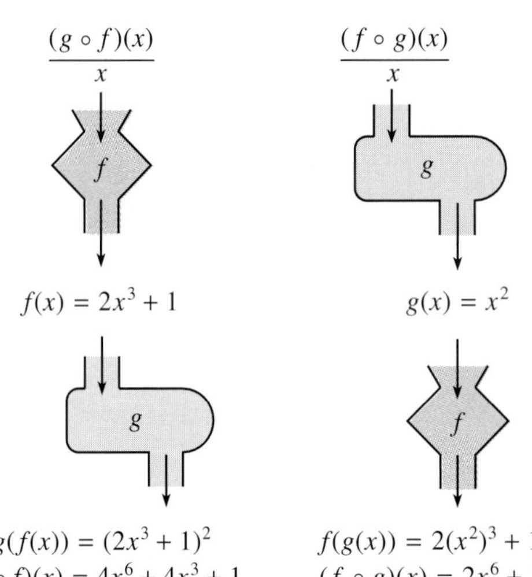

Figure 1.9

CHECKPOINT

6. If $f(x) = 1 - 2x$ and $g(x) = 3x^2$, find the following.
   (a) $(f \circ g)(x)$ (b) $(g \circ f)(x)$ (c) $(f \circ f)(x) = f(f(x))$

CHECKPOINT SOLUTIONS

1. Independent variable is x; dependent variable is y.
2. $f(1) = 3$
3. $f(-2) = 1 - (-2)^3 = 1 - (-8) = 9$
4. $f(x + h) = 2(x + h)^2$
   $= 2(x^2 + 2xh + h^2)$
5. The domain is all real numbers except $x = -1$, because $f(x)$ is undefined when $x = -1$.
6. (a) $(f \circ g)(x) = f(g(x)) = f(3x^2) = 1 - 2(3x^2) = 1 - 6x^2$
   (b) $(g \circ f)(x) = g(f(x)) = g(1 - 2x) = 3(1 - 2x)^2$
   (c) $(f \circ f)(x) = f(f(x)) = f(1 - 2x) = 1 - 2(1 - 2x) = 4x - 1$

EXERCISE 1.2

1. If $y = 3x^3$, is y a function of x?
2. If $y = 6x^2$, is y a function of x?
3. If $y^2 = 3x$, is y a function of x?
4. If $y^2 = 10x^2$, is y a function of x?

In Problems 5 and 6, are the relations defined by the tables functions? Explain why or why not and give the domain and range.

5.

x	1	2	3	8	9
y	-4	-4	5	16	5

6.

x	-1	0	1	3	1
y	0	2	4	6	9

7. If $R(x) = 8x - 10$, find the following.
   (a) $R(0)$ (b) $R(2)$
   (c) $R(-3)$ (d) $R(1.6)$
8. If $h(x) = 3x^2 - 2x$, find the following.
   (a) $h(3)$ (b) $h(-3)$
   (c) $h(2)$ (d) $h\left(\frac{1}{6}\right)$
9. If $C(x) = 4x^2 - 3$, find the following.
   (a) $C(0)$ (b) $C(-1)$
   (c) $C(-2)$ (d) $C\left(-\frac{3}{2}\right)$
10. If $R(x) = 100x - x^3$, find the following.
    (a) $R(1)$ (b) $R(10)$
    (c) $R(2)$ (d) $R(-10)$
11. If $f(x) = x^3 - 4/x$, find the following.
    (a) $f\left(-\frac{1}{2}\right)$ (b) $f(2)$ (c) $f(-2)$
12. If $C(x) = (x^2 - 1)/x$, find the following.
    (a) $C(1)$ (b) $C\left(\frac{1}{2}\right)$ (c) $C(-2)$
13. Let $f(x) = 1 + x + x^2$ and $h \neq 0$.
    (a) Is $f(2 + 1) = f(2) + f(1)$?
    (b) Find $f(x + h)$.
    (c) Does $f(x + h) = f(x) + f(h)$?
    (d) Does $f(x + h) = f(x) + h$?
    (e) Find $\dfrac{f(x + h) - f(x)}{h}$ and simplify.
14. Let $f(x) = 3x^2 - 6x$ and $h \neq 0$.
    (a) Is $f(3 + 2) = f(3) + 2$?
    (b) Find $f(x + h)$.
    (c) Does $f(x + h) = f(x) + h$?
    (d) Does $f(x + h) = f(x) + f(h)$?
    (e) Find $\dfrac{f(x + h) - f(x)}{h}$ and simplify.
15. If $f(x) = x - 2x^2$ and $h \neq 0$, find the following and simplify.
    (a) $f(x + h)$
    (b) $\dfrac{f(x + h) - f(x)}{h}$
16. If $f(x) = 2x^2 - x + 3$ and $h \neq 0$, find the following and simplify.
    (a) $f(x + h)$
    (b) $\dfrac{f(x + h) - f(x)}{h}$

17. Does either of the graphs in Figure 1.10 represent y as a function of x? Explain your choices.

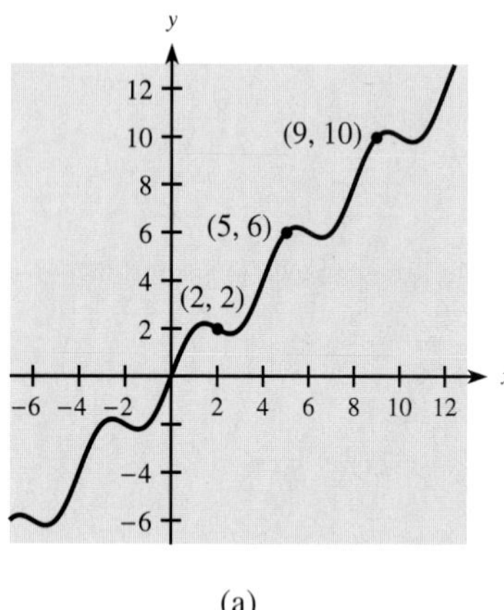

(a)

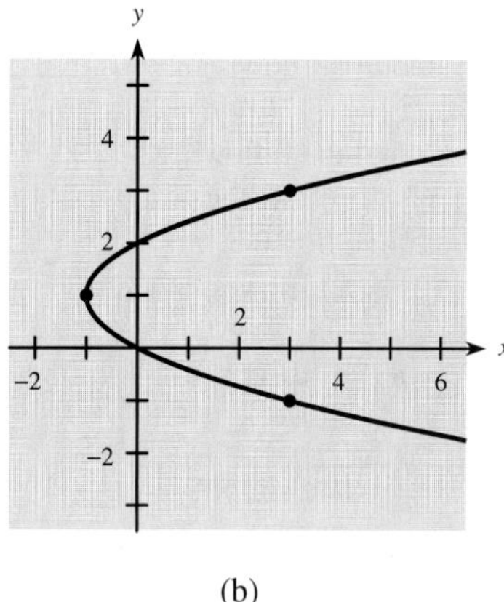

(b)

Figure 1.10

18. Does either of the graphs in Figure 1.11 represent y as a function of x? Explain your choices.

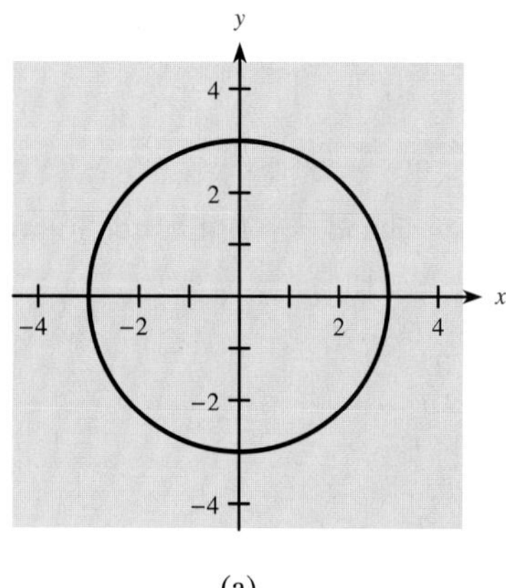

(a)

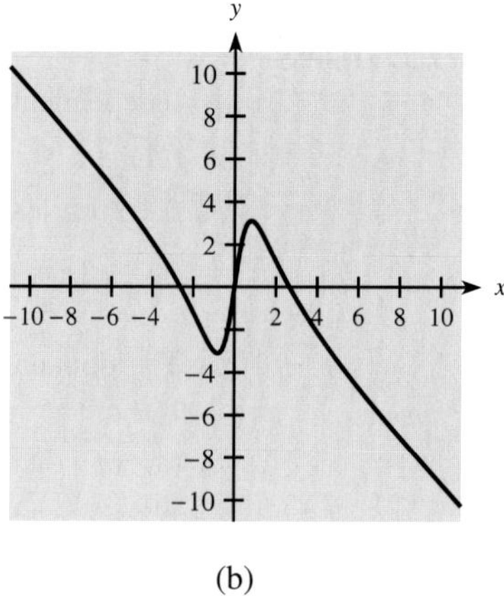

(b)

Figure 1.11

19. If $y = f(x)$ in Figure 1.10(a), find the following.
    (a) $f(9)$ (b) $f(5)$
20. Suppose $y = g(x)$ in Figure 1.11(b).
    (a) Find $g(0)$.
    (b) How many x-values in the domain of this function satisfy $g(x) = 0$?
21. The graph of $y = x^2 - 4x$ is shown in Figure 1.12.

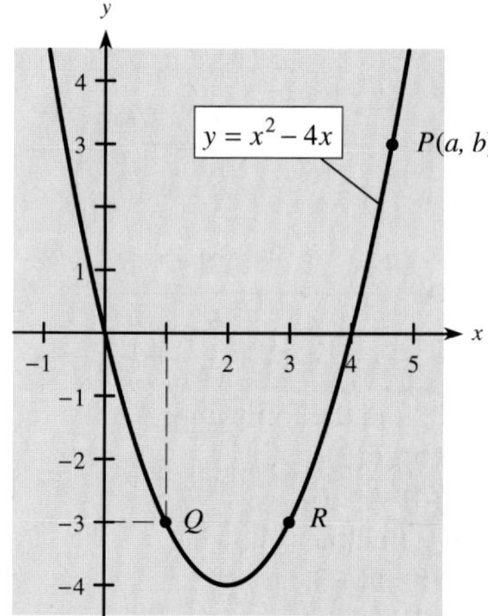

Figure 1.12

(a) If the coordinates of the point P on the graph are (a, b), how are a and b related?
(b) What are the coordinates of the point Q? Do they satisfy the equation?
(c) What are the coordinates of R? Do they satisfy the equation?
(d) What are the x-values of the points on the graph whose y-coordinates are 0? Are these x-values solutions to the equation $x^2 - 4x = 0$?

22. The graph of $y = 2x^2$ is shown in Figure 1.13.

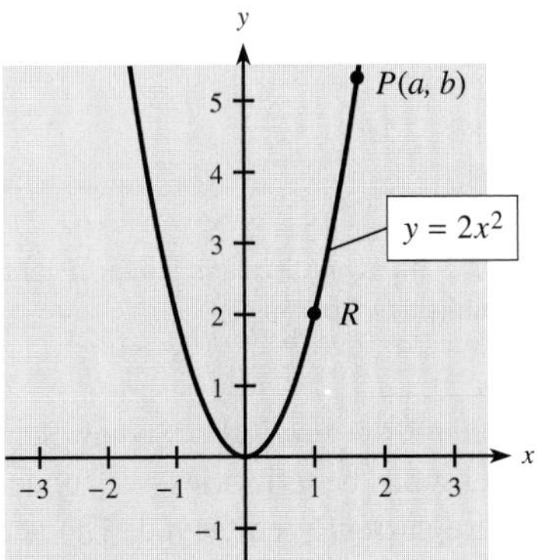

Figure 1.13

(a) If the point P, with coordinates (a, b), is on the graph, how are a and b related?
(b) Does the point $(1, 1)$ lie on the graph? Do the coordinates satisfy the equation?
(c) What are the coordinates of point R? Do they satisfy the equation?
(d) What is the x-value of the point whose y-coordinate is 0? Does this value of x satisfy the equation $0 = 2x^2$?

For $f(x)$ and $g(x)$ given in Problems 23–26, find
(a) $(f + g)(x)$
(b) $(f - g)(x)$
(c) $(f \cdot g)(x)$
(d) $(f/g)(x)$

23. $f(x) = 3x$ $\quad g(x) = x^3$
24. $f(x) = \sqrt{x}$ $\quad g(x) = 1/x$
25. $f(x) = \sqrt{2x}$ $\quad g(x) = x^2$
26. $f(x) = (x - 1)^2$ $\quad g(x) = 1 - 2x$

For $f(x)$ and $g(x)$ given in Problems 27–30, find
(a) $(f \circ g)(x)$
(b) $(g \circ f)(x)$
(c) $f(f(x))$
(d) $f^2(x) = (f \cdot f)(x)$

27. $f(x) = (x - 1)^3$ $\quad g(x) = 1 - 2x$
28. $f(x) = 3x$ $\quad g(x) = x^3 - 1$
29. $f(x) = 2\sqrt{x}$ $\quad g(x) = x^4 + 5$
30. $f(x) = \dfrac{1}{x^3}$ $\quad g(x) = 4x + 1$

State the domain and range of each of the functions in Problems 31–34.

31. $y = x^2 + 4$
32. $y = x^2 - 1$
33. $y = \sqrt{x + 4}$
34. $y = \sqrt{x^2 + 1}$

In Problems 35–38, a function and its graph are given. In each problem find the domain.

35. $f(x) = \dfrac{\sqrt{x - 1}}{x - 2}$

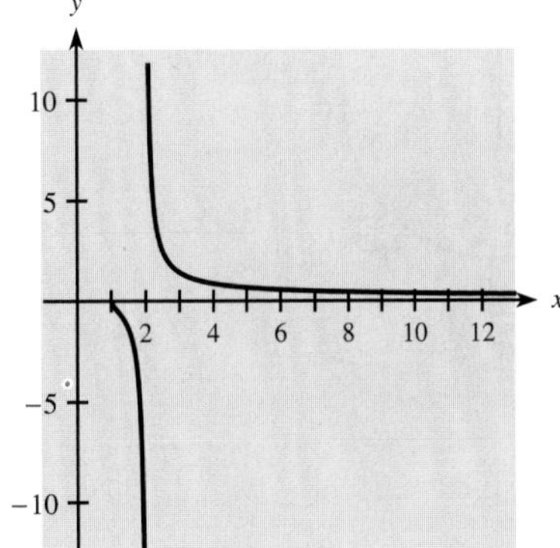

36. $f(x) = \dfrac{x + 1}{\sqrt{x + 3}}$

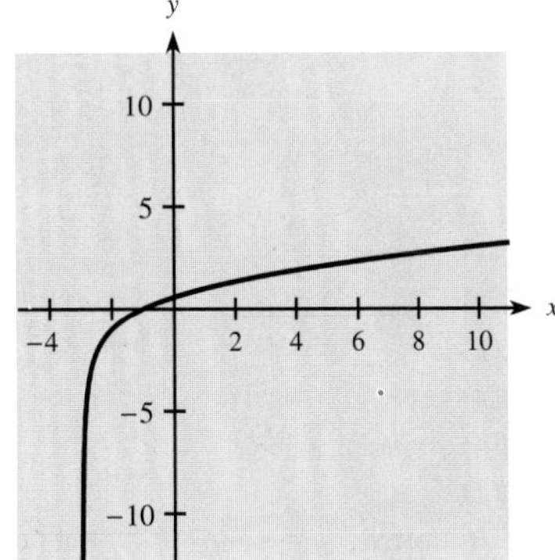

37. $f(x) = 4 + \sqrt{49 - x^2}$

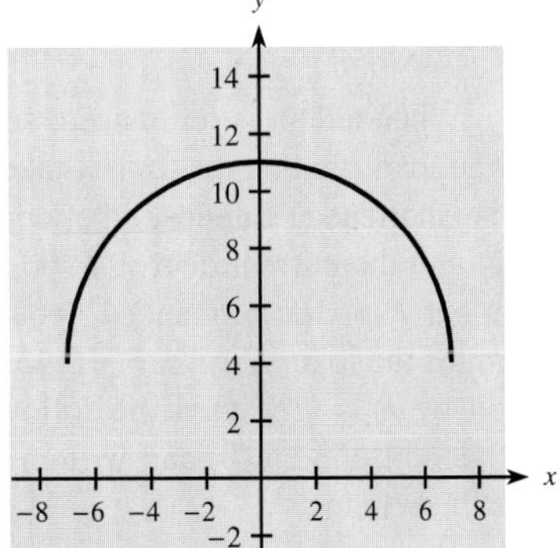

38. $f(x) = -2 - \sqrt{9 - x^2}$

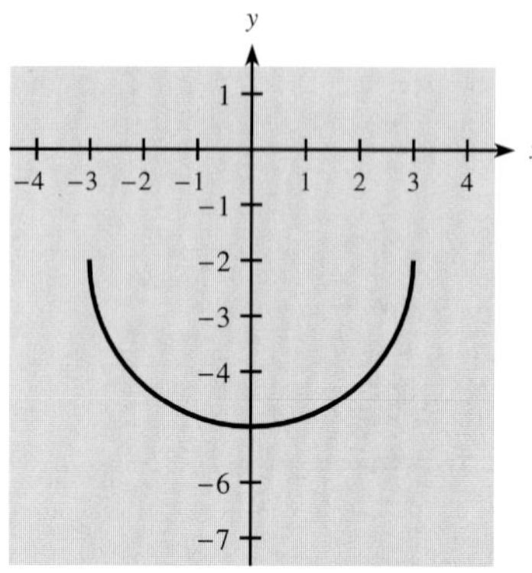

Applications

39. ***Mortgage*** A couple seeking to buy a home decides that a monthly payment of \$800 fits their budget. Their bank's interest rate is 7.5%. The amount they can borrow, A, is a function of the time t, in years, it will take to repay the debt. If we denote this function by $A = f(t)$, then the following table defines the function.

t	A
5	40,000
10	69,000
15	89,000
20	103,000
25	113,000
30	120,000

Source: *Comprehensive Mortgage Payment Tables,* Publication No. 492, Financial Publishing Co., Boston

(a) Find $f(20)$ and write a sentence that explains its meaning.
(b) Does $f(5 + 5) = f(5) + f(5)$?

40. ***Debt refinancing*** When a debt is refinanced, sometimes the term of the loan (that is, the time it takes to repay the debt) is shortened. Suppose the current interest rate is 7%, and the current debt is \$100,000. The monthly payment R of the refinanced debt is a function of the term of the loan, t, in years. If we represent this function by $R = f(t)$, then the following table defines the function. Find $f(10)$ and write a sentence that explains its meaning.

t	R
5	1980.12
10	1161.09
12	1028.39
15	898.83
20	775.30
25	706.78

Source: *Comprehensive Mortgage Payment Tables,* Publication No. 492, Financial Publishing Co., Boston

41. ***Social Security benefits funding*** Social Security benefits paid to eligible beneficiaries are funded by individuals who are currently employed. The accompanying graph, based on known data but showing projections into the future, defines a function that gives the number of workers, n, supporting each retiree as a function of time t (given by calendar year). Let us denote this function by $n = f(t)$.
(a) Find $f(1950)$ and explain its meaning.
(b) Find $f(1990)$.
(c) If after the year 2010, actual data through 2010 regarding workers per Social Security beneficiary were graphed, what parts of this graph *must* be the same and what parts *might* be the same? Explain.
(d) Find the domain and range of $n = f(t)$ if the function is defined by the graph.

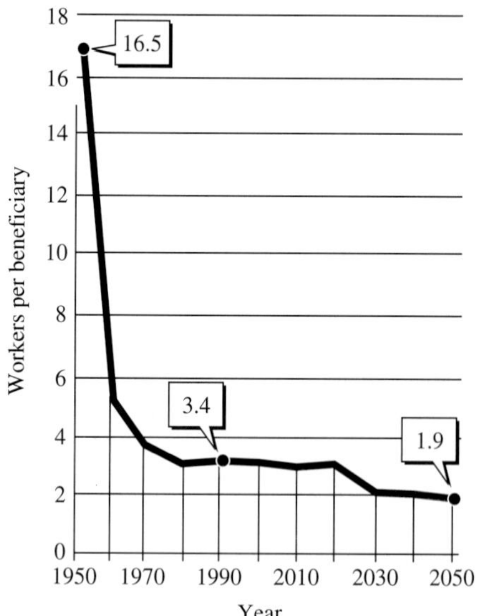

Source: Social Security Administration

42. ***Dow Jones average*** If t represents the number of hours after 9:30 A.M. on Thursday, March 31, 1994, then the graph defines the Dow Jones industrial average D as a function of time t. If we represent this function by $D = f(t)$, use the graph to complete the following.
(a) Find $f(0)$ and $f(6.5)$ and write a sentence that explains what each means.
(b) Find the domain and range for $D = f(t)$ as defined by the graph.
(c) How many t-values satisfy $f(t) = 3600$? Estimate one such t-value.

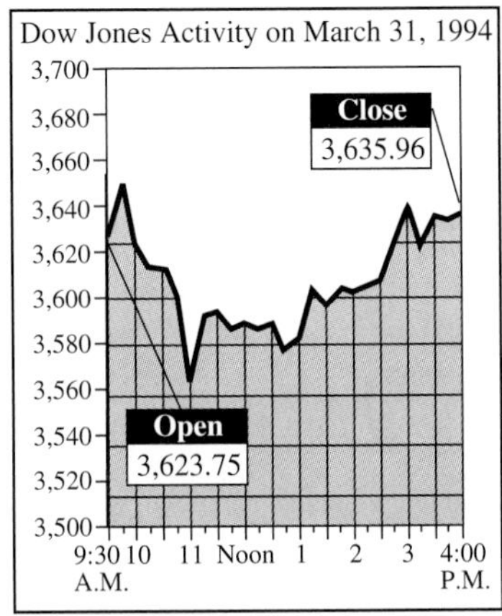

From *The Derrick*, Oil City, Pennsylvania, April 1, 1994
Source: Telerate Systems, Inc.

43. ***Wind chill factor*** Dr. Paul Siple conducted studies testing the effect of wind on the formation of ice at various temperatures and developed the concept of the wind chill factor, which we hear reported during winter weather reports. If the air temperature is $-5°$F, then the wind chill, WC, is a function of the wind speed, s (in mph), and is given by

$$WC = f(s) = 45.694 + 1.75s - 29.26\sqrt{s}$$

(a) Based on the formula for $f(s)$ and the physical context of the problem, what is the domain of $f(s)$?
(b) Find $f(10)$ and write a sentence that explains its meaning.
(c) The working domain of this wind chill function is actually $s \geq 4$. How can you tell that $s = 0$ is not in the working domain, even though it is in the mathematical domain?

44. ***Body-heat loss*** The description of body-heat loss due to convection involves a coefficient of convection K_c, which depends on wind speed v according to the equation

$$K_c = 4\sqrt{4v + 1}$$

(a) Is K_c a function of v?
(b) What is the domain of the function defined by this equation?
(c) What restrictions do nature and common sense put on v?

45. ***Efficiency of a muscle*** The efficiency E of a muscle performing a maximal contraction is related to the time t that the muscle is contracted according to

$$E = \frac{1 - 0.24t}{2 + t}$$

(a) Is E a function of t?
(b) What is the domain of the function defined by this equation?
(c) What restrictions do nature and common sense put on the domain?

46. ***Pressure of a gas*** The pressure P of a certain gas is related to volume V according to

$$P = \frac{100}{V}$$

(a) Is 0 in the domain of this function?
(b) What is $P(100)$? (c) What is $P(50)$?
(d) As volume decreases, what happens to pressure?

47. ***Fahrenheit-Celsius*** The equation

$$C = \frac{5}{9}F - \frac{160}{9}$$

gives the relation between temperature readings in Celsius and Fahrenheit.
(a) Is C a function of F?
(b) What is the domain?
(c) If we consider this equation as relating temperatures of water in its liquid state, what are the domain and range?
(d) What is C when $F = 40°$?

48. ***Cost*** The total cost of producing a product is given by

$$C = 300x + 0.1x^2 + 1200$$

where x represents the number of units produced. Give
(a) the total cost of producing 10 units
(b) the meaning of $C(100)$
(c) the value of $C(100)$

49. ***Cost-benefit*** Suppose that the cost C (in dollars) of removing p percent of the particulate pollution from the smokestacks of an industrial plant is given by

$$C(p) = \frac{7300p}{100 - p}$$

(a) Find the domain of this function. Recall that p represents the percentage of pollution that is removed.

In parts (b)–(e), find the functional values and explain what each means.

(b) $C(45)$ (c) $C(90)$
(d) $C(99)$ (e) $C(99.6)$

50. ***Test reliability*** If a test that has reliability r is lengthened by a factor n $(n \geq 1)$, the reliability R of the new test is given by

$$R(n) = \frac{nr}{1 + (n - 1)r} \qquad 0 < r \leq 1$$

If the reliability is $r = 0.6$, the equation becomes

$$R(n) = \frac{0.6n}{0.4 + 0.6n}$$

(a) Find $R(1)$.
(b) Find $R(2)$; that is, find R when the test length is doubled.
(c) What percentage improvement is there in the reliability when the test length is doubled?

51. ***Area*** If 100 feet of fence is to be used to fence in a rectangular yard, then the resulting area of the fenced yard is given by

$$A = x(50 - x)$$

where x is the width of the rectangle.
(a) Is A a function of x?
(b) If $A = A(x)$, find $A(2)$ and $A(30)$.
(c) What restrictions must be placed on x (the domain) so that the problem makes physical sense?

52. ***Postal restrictions*** If a box with square cross section is to be sent by the postal service, there are restrictions on its size such that its volume is given by $V = x^2(108 - 4x)$, where x is the length of each side of the cross section (in inches).
(a) Is V a function of x?
(b) If $V = V(x)$, find $V(10)$ and $V(20)$.
(c) What restrictions must be placed on x (the domain) so that the problem makes physical sense?

53. ***Profit*** Suppose that the profit from the production and sale of x units of a product is given by

$$P(x) = 180x - \frac{x^2}{100} - 200$$

In addition, suppose that for a certain month the number of units produced on day t of the month is

$$x = q(t) = 1000 + 10t$$

(a) Find $(P \circ q)(t)$ to express the profit as a function of the day of the month.
(b) Find the number of units produced, and the profit, on the fifteenth day of the month.

54. ***Fish species growth*** For many species of fish, the weight W is a function of the length L that can be expressed by

$$W = W(L) = kL^3 \qquad k = \text{constant}$$

Suppose that for a particular species $k = 0.02$, that for this species the length (in centimeters) is a function of the number of years t the fish has been alive, and that this function is given by

$$L = L(t) = 50 - \frac{(t - 20)^2}{10} \qquad 0 \leq t \leq 20$$

Find $(W \circ L)(t)$ in order to express W as a function of the age t of the fish.

55. ***Fencing a lot*** A farmer wishes to fence the perimeter of a rectangular lot with an area of 1600 square feet. If the lot is x feet long, express the amount L of fence needed as a function of x.

56. ***Cost*** A shipping crate has a square base with sides of length x feet, and it is half as tall as it is wide. If the material for the bottom and sides of the box costs \$2.00 per square foot and the material for the top costs \$1.50 per square foot, express the total cost of material for the box as function of x.

57. ***Revenue*** An agency charges \$10 per person for a trip to a concert if 30 people travel in a group. But for each person above the 30, the charge will be reduced by \$0.20. If x represents the number of people above the 30, write the agency's revenue R as a function of x.

58. ***Revenue*** A company handles an apartment building with 50 units. Experience has shown that if the rent for each of the units is \$360 per month, all of the units will be filled, but one unit will become vacant for each \$10 increase in the monthly rate. If x represents the number of \$10 increases, write the revenue R from the building as a function of x.

1.3 Linear Functions

OBJECTIVES

- *To find the intercepts of graphs*
- *To graph linear functions*
- *To find the slope of a line from its graph and from its equation*
- *To graph a line, given its slope and y-intercept or its slope and one point on the line*
- *To write the equation of a line, given information about its graph*
- *To find marginal cost from a linear total cost function*

APPLICATION PREVIEW

U.S. Census Bureau data from 15 different years showed that the average price p (in dollars) of a color television set depended on the number of sets sold N (in thousands). The data indicated that when N increased by 1000 units (a million more sets were sold), there was a decrease of \$10.40 in the average price per set. The data also showed that when 6485 (thousand) sets were sold one year, the average price per set was \$504.39. (Source: *Statistical Abstract of the United States,* Washington, D.C.) By using skills developed in this section, we could express p as a linear function of N and use that equation to understand better the relationship between p and N.

A special function, called the **linear function,** is defined as follows:

Linear Function A linear function is a function of the form

$$y = f(x) = ax + b$$

where a and b are constants.

Because the graph of a linear function is a line, only two points are necessary to determine its graph. It is frequently possible to use **intercepts** to graph a linear function. The point(s) where a graph intersects the x-axis are called the x-intercept points, and the x-coordinates of these points are the ***x*-intercepts.** Similarly, the point where the graph of a function intersects the y-axis is the y-intercept point, and the y-coordinate of the point is the ***y*-intercept.**

Intercepts

(a) To find the ***y*-intercept(s)** of the graph of an equation, set $x = 0$ in the equation and solve for y.

(b) To find the ***x*-intercept(s),** set $y = 0$ and solve for x.

EXAMPLE 1

Find the intercepts and graph the following.

(a) $3x + y = 9$ (b) $x = 4y$

Solution

(a) To find the y-intercept, we set $x = 0$ and solve for y. $3(0) + y = 9$ gives $y = 9$, so the y-intercept is 9. To find the x-intercept, we set $y = 0$ and solve for x. $3x + 0 = 9$ gives $x = 3$, so the x-intercept is 3. Using the intercepts gives the graph, shown in Figure 1.14.

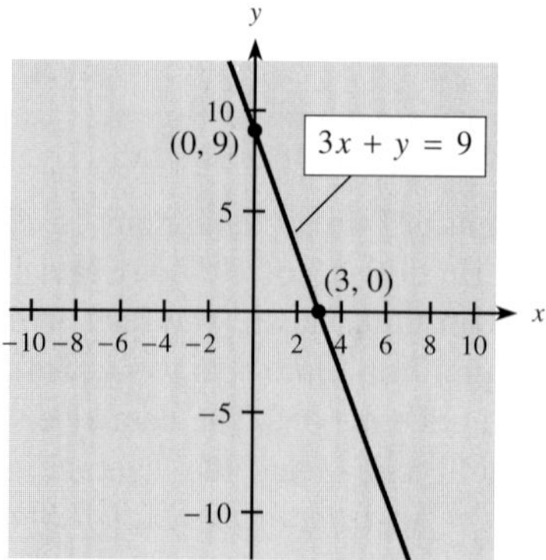

Figure 1.14

(b) Letting $x = 0$ gives $y = 0$, and letting $y = 0$ gives $x = 0$, so the only intercept of the graph of $x = 4y$ is at the point (0, 0). A second point is needed to graph the line. Hence, if we let $y = 1$ in $x = 4y$, we get $x = 4$ and have a second point (4, 1) on the graph. It is wise to plot a third point as a check. The graph is shown in Figure 1.15.

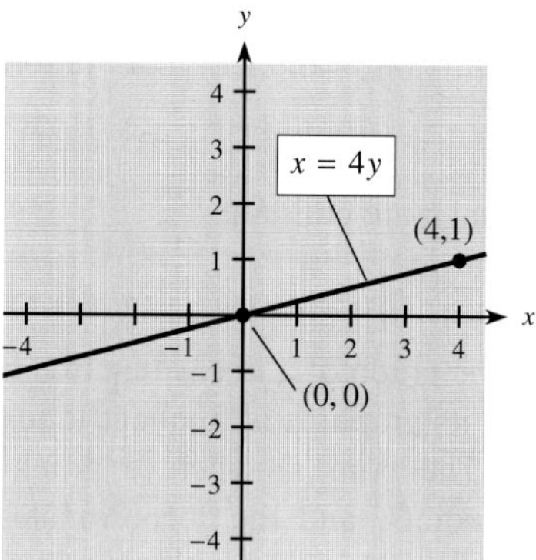

Figure 1.15

EXAMPLE 2

A business property is purchased for \$122,970 and depreciated over a period of 10 years. Its value y is related to the number of months of service x by the equation

$$4099x + 4y = 491{,}880$$

Find the x-intercept and the y-intercept and use them to sketch the graph of the equation.

Solution

$$x\text{-intercept: } y = 0 \quad \text{gives} \quad 4099x = 491{,}880$$
$$x = 120$$

Thus 120 is the x-intercept.

$$y\text{-intercept: } x = 0 \quad \text{gives} \quad 4y = 491{,}880$$
$$y = 122{,}970$$

Thus 122,970 is the y-intercept. The graph is shown in Figure 1.16. Note that the units on the x- and y-axes are different and that the y-intercept corresponds to the value of the property 0 months after purchase. That is, the y-intercept gives the purchase price. The x-intercept corresponds to the number of months that have passed before the value is 0; that is, the property is fully depreciated after 120 months, or 10 years. Note that only positive values for x and y make sense in this application, so only the Quadrant I portion of the graph is shown.

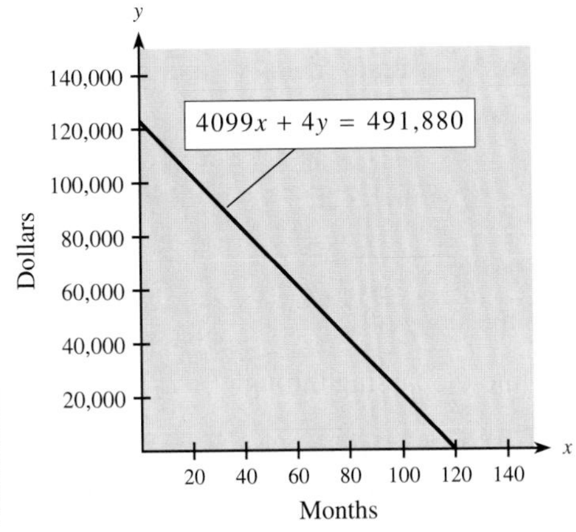

Figure 1.16

Despite the ease of using intercepts to graph linear equations, this method is not always the best. For example, vertical lines, horizontal lines, or lines that pass through the origin may have a single intercept, and if a line has both intercepts very close to the origin, using the intercepts may lead to an inaccurate graph.

Note that in Figure 1.16, as the graph moves from the y-intercept point (0, 122,970) to the x-intercept point (120, 0), the y-value on the line changes $-122,970$ units (from 122,970 to 0), whereas the x-value changes 120 units (from 0 to 120). The ratio of this change in y to the corresponding change in x is called the **slope** of the line. The slope measures the rate of change of y with respect to x (measured in dollars per month in Example 2). For any nonvertical line, the slope can be found by using any two points on the line, as follows.

Slope of a Line

If a nonvertical line passes through the points $P_1(x_1, y_1)$ and $P_2(x_2, y_2)$, its **slope,** denoted by m, is found by using either

$$m = \frac{y_2 - y_1}{x_2 - x_1}$$

or, equivalently,

$$m = \frac{y_1 - y_2}{x_1 - x_2}$$

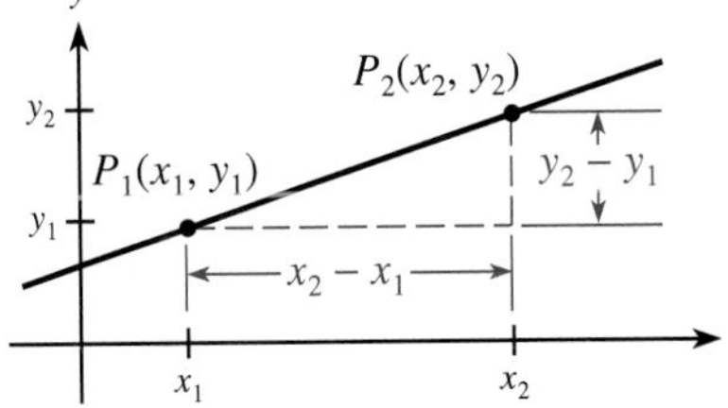

The slope of a vertical line is undefined.

Note that for a given line, the slope is the same regardless of which two points are used in the calculation; this is because corresponding sides of similar triangles are in proportion.

We may also write the slope by using the notation

$$m = \frac{\Delta y}{\Delta x} \qquad (\Delta y = y_2 - y_1 \quad \text{and} \quad \Delta x = x_2 - x_1)$$

where Δy is read "delta y" and means "change in y," and Δx means "change in x."

EXAMPLE 3

Find the slope of

(a) line ℓ_1, passing through $(-2, 1)$ and $(4, 3)$

(b) line ℓ_2, passing through $(3, 0)$ and $(4, -3)$

Solution

(a) $m = \frac{3 - 1}{4 - (-2)} = \frac{2}{6} = \frac{1}{3}$ or, equivalently, $m = \frac{1 - 3}{-2 - 4} = \frac{-2}{-6} = \frac{1}{3}$

This means that a point 3 units to the right and 1 unit up from any point on the line is also on the line. Line ℓ_1 is shown in Figure 1.17.

(b) $m = \frac{0 - (-3)}{3 - 4} = \frac{3}{-1} = -3$

This means that a point 1 unit to the right and 3 units down from any point on the line is also on the line. Line ℓ_2 is also shown in Figure 1.17.

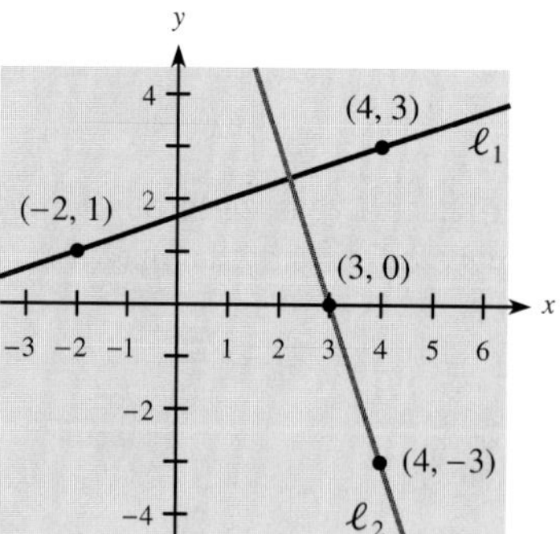

Figure 1.17

From the previous discussion, we see that the slope describes the direction of a line as follows.

Orientation of a Line and Its Slope

1. The slope is *positive* if the line slopes upward toward the right.

$$m = \frac{\Delta y}{\Delta x} > 0$$

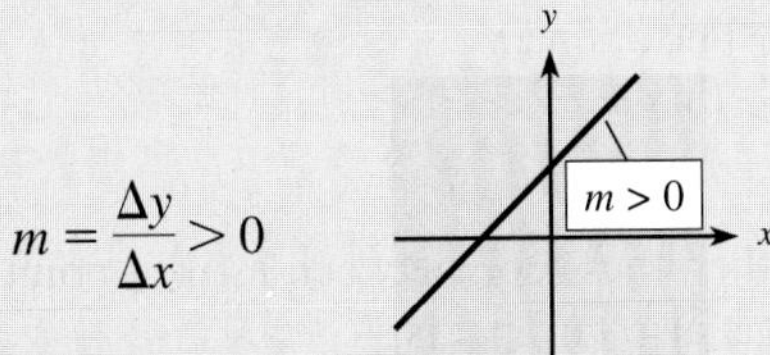

2. The slope is *negative* if the line slopes downward toward the right.

$$m = \frac{\Delta y}{\Delta x} < 0$$

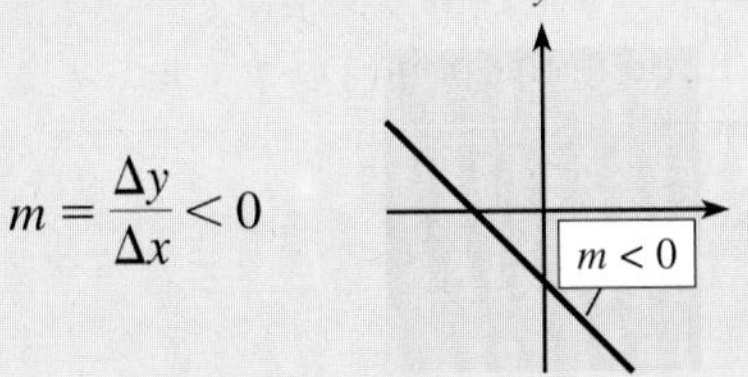

3. The slope of a *horizontal line* is 0, because $\Delta y = 0$.

$$m = \frac{\Delta y}{\Delta x} = 0$$

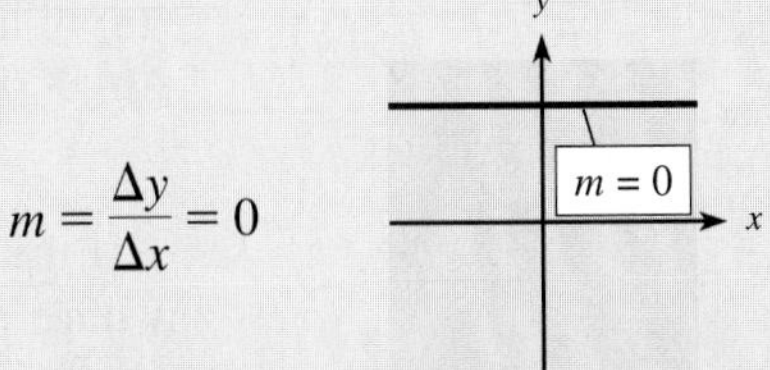

4. The slope of a *vertical line* is *undefined,* because $\Delta x = 0$.

$$m = \frac{\Delta y}{\Delta x} \text{ is undefined}$$

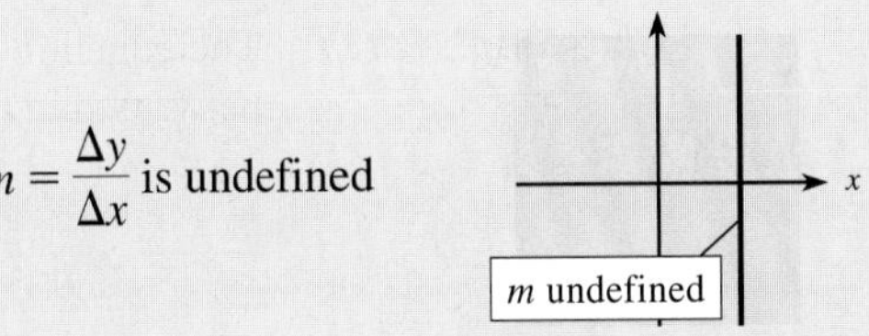

Two distinct nonvertical lines that have the same slope are parallel, and conversely, two parallel lines have the same slope.

Parallel Lines Two distinct nonvertical lines are *parallel* if and only if their slopes are *equal.*

In general, two lines are perpendicular if they form right angles where they intersect. Note that the lines ℓ_1 and ℓ_2 of Figure 1.17 appear to be perpendicular and that the slope of ℓ_1, $\frac{1}{3}$, is the negative reciprocal of the slope of ℓ_2, -3. In fact, any two nonvertical lines that are perpendicular have slopes that are negative reciprocals of each other.

Slopes of Perpendicular Lines A line ℓ_1 with slope m, where $m \neq 0$, is *perpendicular* to line ℓ_2 if and only if the slope of ℓ_2 is $-1/m$. (The slopes are *negative* reciprocals.)

Because the slope of a vertical line is undefined, we cannot use slope in discussing parallel and perpendicular relations that involve vertical lines. Two vertical lines are parallel, and any horizontal line is perpendicular to any vertical line.

CHECKPOINT

1. Find the slope of the line through (4, 6) and (28, −6).
2. If a line has slope $m = 0$, then the line is ________. If a line has slope m undefined, then the line is ________.
3. Suppose that line 1 has slope $m_1 = 3$ and line 2 has slope m_2.
   (a) If line 1 is perpendicular to line 2, find m_2.
   (b) If line 1 is parallel to line 2, find m_2.

If the slope of a line is m, then the slope between a fixed point (x_1, y_1) and any other point (x, y) on the line is also m. That is,

$$m = \frac{y - y_1}{x - x_1}$$

Solving for $y - y_1$ gives the point-slope form of the equation of a line.

Point-Slope Form The equation of the line passing through the point (x_1, y_1) and with slope m can be written in the **point-slope form**

$$y - y_1 = m(x - x_1)$$

EXAMPLE 4

Write equations for the lines that pass through (1, −2) and have

(a) slope $\frac{2}{3}$ (b) undefined slope (c) point (2, 3) also on the line

Solution

(a) Here $m = \frac{2}{3}$, $x_1 = 1$, and $y_1 = -2$. An equation of the line is

$$y - (-2) = \frac{2}{3}(x - 1)$$

This equation may be written as $y = \frac{2}{3}x - \frac{8}{3}$ or in the **general form** as $2x - 3y - 8 = 0$. Figure 1.18 shows the graph of this line; the point (1, −2) and the slope are highlighted.

(b) Because m is undefined, we cannot use the point-slope form. This line is vertical, so every point on it has x-coordinate 1. Thus the equation is $x = 1$. Note that $x = 1$ is not a function.

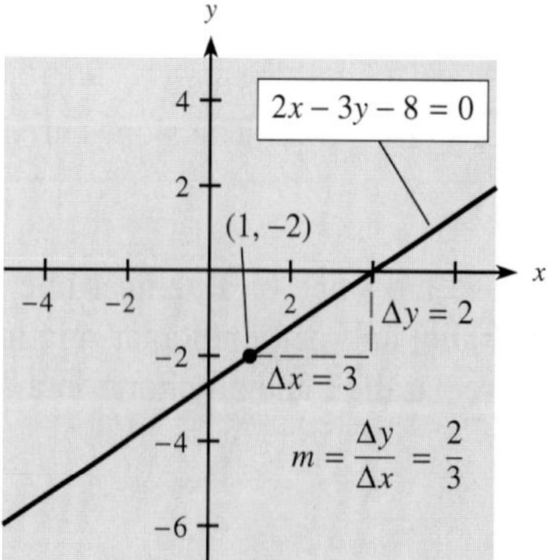

Figure 1.18

(c) First find

$$m = \frac{3 - (-2)}{2 - 1} = 5$$

Using $m = 5$ and the point $(1, -2)$ (the other point could also be used) gives

$$y - (-2) = 5(x - 1) \quad \text{or} \quad y = 5x - 7$$

The graph of $x = 1$ (from Example 4(b)) is a vertical line, as shown in Figure 1.19(a); the graph of $y = 1$ has slope 0, and its graph is a horizontal line, as shown in Figure 1.19(b).

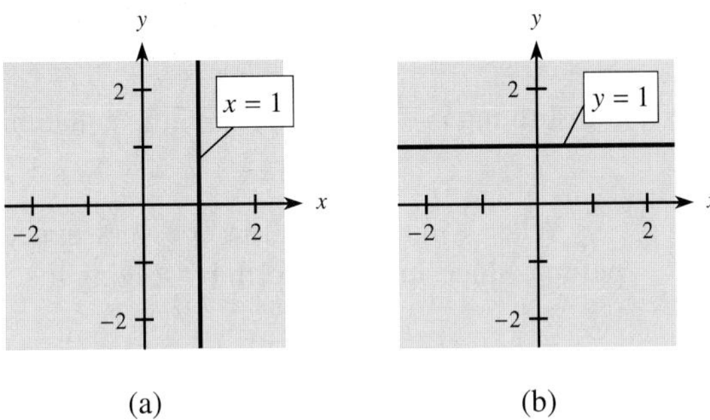

Figure 1.19

EXAMPLE 5

The U.S. Census Bureau data presented in the Application Preview indicates that the average price of color television sets can be expressed as a linear function of the number of sets sold N (in thousands). The relevant facts were that as N increased by 1000, p dropped by \$10.40 and that when 6485 (thousand) sets were sold, the average price per set was \$504.39. Write the equation of the line determined by this information.

Solution

We see that price p is a function of the number of sets N (in thousands), so the slope is given by

$$m = \frac{\text{change in } p}{\text{change in } N} = \frac{-10.40}{1000} = -0.0104$$

A point on the line is $(N_1, p_1) = (6485, 504.39)$. We use the point-slope form adapted to the variables N and p.

$$\begin{aligned} p - p_1 &= m(N - N_1) \\ p - 504.39 &= -0.0104(N - 6485) \\ p - 504.39 &= -0.0104N + 67.444 \\ p &= -0.0104N + 571.834 \end{aligned}$$

The point-slope form, with the intercept point $(0, b)$, can be used to derive a special form for the equation of a line.

$$\begin{aligned} y - b &= m(x - 0) \\ y &= mx + b \end{aligned}$$

Slope-Intercept Form The **slope-intercept form** of the equation of a line with slope m and y-intercept b is

$$y = mx + b$$

Note that if a linear equation has the form $y = mx + b$, then the coefficient of x is the slope and the constant is the y-intercept.

EXAMPLE 6

Write the equation of the line with slope $\frac{1}{2}$ and y-intercept 3.

Solution

Substituting $m = \frac{1}{2}$ and $b = 3$ in the equation $y = mx + b$ gives $y = \frac{1}{2}x + 3$.

When a linear equation does not appear in slope-intercept form, it can be put into slope-intercept form by solving the equation for y.

EXAMPLE 7

(a) Find the slope and y-intercept of the line whose equation is $x + 2y = 8$.
(b) Use this information to graph the equation.

Solution

(a) To put the equation in slope-intercept form, we must solve it for y.

$$2y = -x + 8 \quad \text{or} \quad y = -\frac{1}{2}x + 4$$

Thus the slope is $-\frac{1}{2}$ and the y-intercept is 4.

(b) First we plot the y-intercept point (0, 4). Because the slope is $-\frac{1}{2} = \frac{-1}{2}$, moving 2 units to the right and down 1 unit from (0, 4) gives the point (2, 3) on the line. A third point (for a check) is plotted at (4, 2). The graph is shown in Figure 1.20.

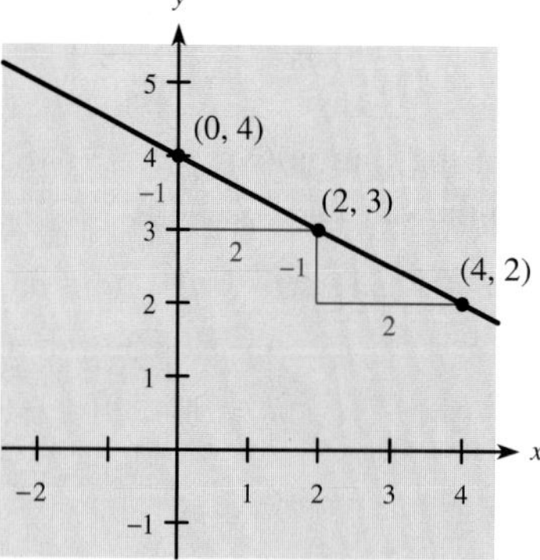

Figure 1.20

It is also possible to graph a straight line if we have its slope and any point that it passes through; we simply plot the point that is given and then use the slope to plot other points.

The following summarizes the forms of equations of lines.

Forms of Linear Equations

General form:	$ax + by + c = 0$
Point-slope form:	$y - y_1 = m(x - x_1)$
Slope-intercept form:	$y = mx + b$
Vertical line:	$x = a$
Horizontal line:	$y = b$

CHECKPOINT

4. Write the point-slope form of the equation of the line that has slope $-\frac{3}{4}$ and passes through $(4, -6)$.
5. What are the slope and y-intercept of the graph of $x = -4y + 1$?

CHECKPOINT SOLUTIONS

1. $m = \dfrac{y_2 - y_1}{x_2 - x_1} = \dfrac{-6 - (6)}{28 - (4)} = \dfrac{-12}{24} = -\dfrac{1}{2}$
2. If $m = 0$, then the line is horizontal. If m is undefined, then the line is vertical.
3. (a) $m_2 = \dfrac{-1}{m_1} = -\dfrac{1}{3}$ (b) $m_2 = m_1 = 3$
4. Use $y - y_1 = m(x - x_1)$. Hence $y - (-6) = \frac{-3}{4}(x - 4)$, or $3x + 4y = -12$.
5. If $x = -4y + 1$, then the slope-intercept form is $y = mx + b$ or $y = -\frac{1}{4}x + \frac{1}{4}$. Hence $m = -\frac{1}{4}$ and the y-intercept is $\frac{1}{4}$.

EXERCISE 1.3

Find the intercepts and graph the following functions.

1. $3x + 4y = 12$
2. $3x - 4y = 12$
3. $2x - 3y = 12$
4. $2x - y + 8 = 0$
5. $3x + 2y = 0$
6. $4x + 5y = 8$

In Problems 7–10, find the slope of the line passing through the given pair of points.

7. $(2, 1)$ and $(3, -4)$
8. $(-1, -2)$ and $(-2, -3)$
9. $(3, 2)$ and $(-1, 2)$
10. $(-4, 2)$ and $(-4, -2)$

Find the slopes and y-intercepts of the lines whose equations are given in Problems 11–18.

11. $y = \dfrac{7}{3}x - \dfrac{1}{4}$
12. $y = \dfrac{4}{3}x + \dfrac{1}{2}$
13. $y = 3$
14. $y = -2$
15. $x = -8$
16. $x = -\dfrac{1}{2}$
17. $2x + 3y = 6$
18. $3x - 2y = 18$

In Problems 19–22, write the equation and sketch the graph of each line with the given slope and y-intercept.

19. Slope $\frac{1}{2}$ and y-intercept 3
20. Slope 2 and y-intercept 3
21. Slope -2 and y-intercept $\frac{1}{2}$
22. Slope $\frac{2}{3}$ and y-intercept -1

In Problems 23–28, write the equation and graph the line that passes through the given point and has the slope indicated.

23. $(2, 0)$ with slope $\frac{1}{2}$
24. $(1, 1)$ with slope $-\frac{1}{3}$
25. $(-1, 3)$ with slope -2
26. $(3, -1)$ with slope 1
27. $(-1, 1)$ with undefined slope
28. $(1, 1)$ with 0 slope

In Problems 29–32, write the equations of the lines passing through the given pairs of points.

29. $(3, 2)$ and $(-1, -6)$
30. $(-4, 2)$ and $(2, 4)$
31. $(7, 3)$ and $(-6, 2)$
32. $(10, 2)$ and $(5, 7)$

33. Write the equation of the x-axis.
34. Write the equation of the y-axis.

In Problems 35–38, determine whether the following pairs of equations represent parallel lines, perpendicular lines, or neither of these.

35. $3x + 2y = 6$; $2x - 3y = 6$
36. $5x - 2y = 8$; $10x - 4y = 8$
37. $6x - 4y = 12$; $3x - 2y = 6$
38. $5x + 4y = 7$; $y = \frac{4}{5}x + 7$
39. Write the equation of the line through $(-2, -7)$ that is parallel to $3x + 5y = 11$.
40. Write the equation of the line through $(6, -4)$ that is parallel to $4x - 5y = 6$.
41. Write the equation of the line through $(3, 1)$ that is perpendicular to $5x - 6y = 4$.
42. Write the equation of the line through $(-2, -8)$ that is perpendicular to $x = 4y + 3$.

Applications

43. ***Depreciation*** A \$360,000 building is depreciated by its owner. The value y of the building after x months of use is $y = 360{,}000 - 1500x$.
    (a) Graph this function for $x \geq 0$.
    (b) How long is it until the building is completely depreciated (its value is zero)?
    (c) The point (60, 270,000) lies on the graph. Explain what this means.

44. ***Tax burden*** Using data from the Internal Revenue Service, the per capita tax burden T (in hundreds of dollars) can be described by

$$T(t) = 20.37 + 1.834t$$

where t is the number of years past 1980.
    (a) Find $T(20)$ and write a sentence that explains its meaning.
    (b) Graph the equation for per capita tax burden.

45. ***Investing*** The data below compare the cumulative total returns on two \$100 investments made in 1988: one in Federal Signal Corporation (FS) and one that represents the Standard and Poors 400 Index (S&P).

t	*1988*	*1989*	*1990*	*1991*	*1992*	*1993*
FS	\$100	\$170	\$248	\$357	\$402	\$539
S & P	\$100	\$129	\$128	\$168	\$171	\$187

Source: Federal Signal Corporation's *Notice of Annual Meeting*, March 7, 1994

If t is the number of years past 1988 (so $t = 0$ in 1988), then this cumulative return can be described by R_{FS} for Federal Signal Corporation and by R_{SP} for the S & P 400 Index as follows.

$$R_{FS} = 85.714t + 88.381$$
$$R_{SP} = 17.1714t + 104.238$$

    (a) Graph $R_{FS} = 85.714t + 88.381$ and the points (t, FS) from the data in the table.
    (b) Graph $R_{SP} = 17.1714t + 104.238$ and the points $(t, S\&P)$ from the data in the table.
    (c) What are the values of R_{FS} and R_{SP} when $t = 0$? Explain why they are different from the values in the tables.
    (d) What is the danger of predicting future investment returns from these equations?

46. ***Temperature-humidity models*** Two models for measuring the effects of high temperature and humidity are the Summer Simmer Index and the Apparent Temperature.* For an outside temperature of 100°F, these indices relate the relative humidity, H (expressed as a decimal), to the perceived temperature as follows.

Summer Simmer: $S = 141.1 - 45.78(1 - H)$

Apparent Temperature: $A = 91.2 + 41.3H$

    (a) For each index, find the point that corresponds to a relative humidity of 40%.
    (b) For each point in (a), write a sentence that explains its meaning.
    (c) Graph both equations for $0 \leq H \leq 1$.

47. ***ATM transactions*** The dollar volume of transactions at automatic teller machines (ATMs) has increased as the number of machines has increased. This relationship can be described by

$$y = 0.1369x - 5.091255$$

where y is billions of dollars of transactions and x is the number of terminals (in thousands). (Source: *Statistical Abstract of the United States,* 1993, p. 517.)
    (a) Find the slope and y-intercept of this equation.
    (b) What interpretation could be given to the y-intercept? Does this interpretation mean that there must be restrictions on x and y for the equation to make sense?
    (c) What interpretation could be given to the slope?

48. ***Cigarette use*** For selected years from 1975 to 1991, the percentage p of high school seniors who have tried cigarettes can be described by

*Bosch, W., and C. G. Cobb, "Temperature-Humidity Indices," UMAP Unit 691, *The UMAP Journal,* 10(3), Fall 1989, 237–256.

$$p = 75.4509 - 0.706948t$$

where t is the number of years past 1975 (Source: National Institute on Drug Abuse).

(a) Find the slope and p-intercept of this equation.

(b) Write a sentence that interprets the meaning of the slope.

(c) Write a sentence that interprets the meaning of the p-intercept.

49. ***Residential electric costs*** An electric utility company determines the monthly bill for a residential customer by adding an energy charge of 8.38 cents per kilowatt hour (kWh) to its base charge of $4.95 per month. Write an equation for the monthly charge y in terms of x, the number of kWh used.

50. ***Residential heating costs*** Residential customers who heat their homes with natural gas have their monthly bill calculated by adding a base service charge of $5.19 per month and an energy charge of 51.91 cents per hundred cubic feet (CCF). Write an equation for the monthly charge y in terms of x, the number of CCF used.

51. ***Gender differences in median salaries*** Data from selected years from 1973 to 1991 show that for each $1000 increase in the median annual salary of U.S. males, the median annual salary of females increased by about $838. Also, in 1985, the median annual salary for males was $22,300 and for females was $17,300 (Source: U.S. Equal Employment Opportunity Commission).

(a) Let m represent the median salary for males and f the median salary for females, and write the equation of the line that gives f as a linear function of m.

(b) When the median salary for males reaches $30,000, what does the equation in (a) predict for the median salary for females?

52. ***Retirement plans*** The retirement plan of the Public Service Company of Colorado as described on page 21 of their *Notice of Annual Meeting of Shareholders*, May 11, 1994, is based on the following formula: "1.5% of average final compensation multiplied by years of credited service." Let p represent annual retirement pension, y years of service, and c average final compensation.

(a) For someone with average final compensation of $80,000, write the linear equation that gives p in terms of y.

(b) For someone intending to retire after 30 years, write the linear equation that gives p in terms of c.

53. ***Pollution effects*** It has been estimated that a certain stream can support 85,000 fish if it is pollution-free. It has further been estimated that for each ton of pollutants in the stream, 1700 fewer fish can be supported. Assuming the relationship is linear, write the equation that gives the population of fish p in terms of the tons of pollutants x.

54. ***Age-sleep relationships*** Each day, a young person should sleep 8 hours plus $\frac{1}{4}$ hour for each year that the person is under 18 years of age. Assuming the relation is linear, write the equation relating hours of sleep y and age x.

55. ***Insulation R-values*** The R-value of insulation is a measure of its ability to resist heat transfer. For fiberglass insulation, $3\frac{1}{2}$ inches is rated at R-11 and 6 inches is rated at R-19. Assuming this relationship is linear, write the equation that gives the R-value of fiberglass insulation as a function of its thickness t (in inches).

56. ***Depreciation*** Suppose the cost of a business property is $960,000 and a company wants to use a straight-line depreciation schedule for a period of 240 months. If y is the value of this property after x months, then the company's depreciation schedule will be the equation of a line through (0, 960,000) and (240, 0). Write the equation of this depreciation schedule.

57. ***Cholesterol and coronary heart disease risk*** The Seven Countries Study, conducted by Ancel Keys, was a long-term study of the relationship of cholesterol to coronary heart disease (CHD) mortality in men. The relationship was approximated by the line shown in the accompanying figure. The line passes through the points (200, 25) and (250, 49), which means there were 25 CHD deaths per 1000 among men with 200 mg/dl (milligrams per deciliter) of cholesterol and 49 CHD deaths among men with 250 mg/dl of cholesterol. Using x to represent the cholesterol and y to represent CHD deaths, write the equation that represents this relationship.

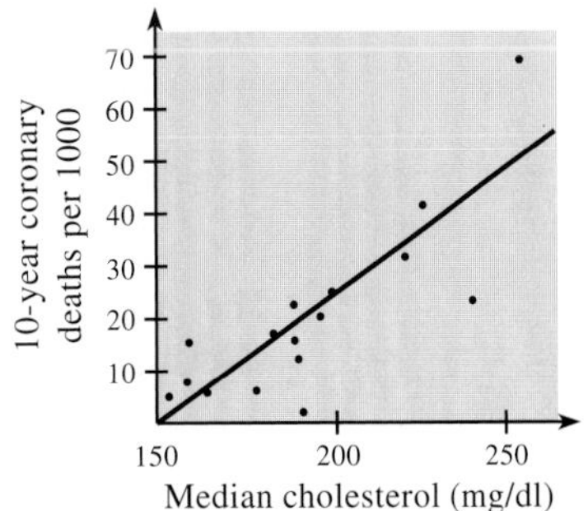

 1.4 Graphs and Graphing Utilities

OBJECTIVES

- *To use a graphing utility to graph equations in the standard viewing window*
- *To use a graphing utility and a specified range to graph equations*
- *To use a graphing utility to evaluate functions, to find intercepts, and to find zeros of a function*

APPLICATION PREVIEW

Suppose that for a certain city the cost C of obtaining drinking water with p percent impurities (by volume) is given by

$$C = \frac{120{,}000}{p} - 1200$$

where $0 < p \leq 100$. A graph of this equation would illustrate how the purity of water is related to the cost of obtaining it. Because this equation is not linear, graphing it would require many more than two points. Such a graph could be efficiently obtained with a graphing utility.

In the section "Linear Functions," we saw that a linear equation could easily be graphed by plotting points because only two points are required. When we want the graph of an equation that is not linear, we can still use point plotting, but we must have enough points to sketch an accurate graph.

Some computer software and all graphing calculators have **graphing utilities** (also called *graphics utilities*) that can be used to generate an accurate graph. All graphing utilities use the point-plotting method to plot scores of points quickly and thereby graph an equation. The computer monitor or calculator screen consists of a fine grid that looks like a piece of graph paper.

In this grid, each tiny area is called a pixel. Essentially, a pixel corresponds to a point on a piece of graph paper, and as a graphing utility plots points, the pixels corresponding to the points are lighted and then connected, revealing the graph. The graph is shown on a viewing window or viewing rectangle. The values that define the viewing window can be set individually or by using ZOOM keys. The important values are

x-min:	the smallest value on the x-axis (the leftmost x-value in the window)
x-max:	the largest value on the x-axis (the rightmost x-value in the window)
y-min:	the smallest value on the y-axis (the lowest y-value in the window)
y-max:	the largest value on the y-axis (the highest y-value in the window)

Most graphing utilities have a standard viewing window that gives a window with x-values and y-values between -10 and 10.

x-min: -10	y-min: -10
x-max: 10	y-max: 10

The graph of $y = \frac{1}{3}x^3 - x^2 - 3x + 2$ with the standard viewing window is shown in Figure 1.21.

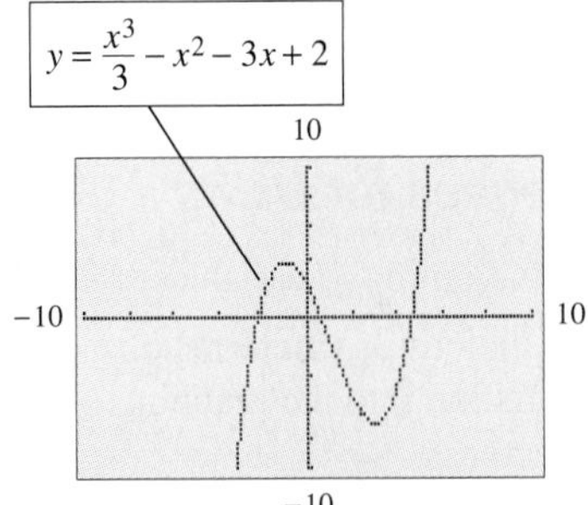

Figure 1.21

With a graphing utility, as with a graph plotted by hand, the appearance of the graph is determined by the part of the graph we are viewing, and the standard window does not always give a complete graph. (A complete graph shows the important parts of the graph and suggests the unseen parts.)

For some graphs, the standard viewing window may give a view of the graph that is misleading, incomplete, or perhaps even blank. For example, the graph of $y = 9x - 0.1x^2$ looks like a line (see Figure 1.22a) with the standard viewing window, but using a different viewing window shows that the standard window is inappropriate for graphing this function (see Figure 1.22b). A viewing window can be set manually. For example, the graph shown in Figure 1.22(b) was generated by setting the window (or range) as follows:

x-min: -25	y-min: -50
x-max: 100	y-max: 250

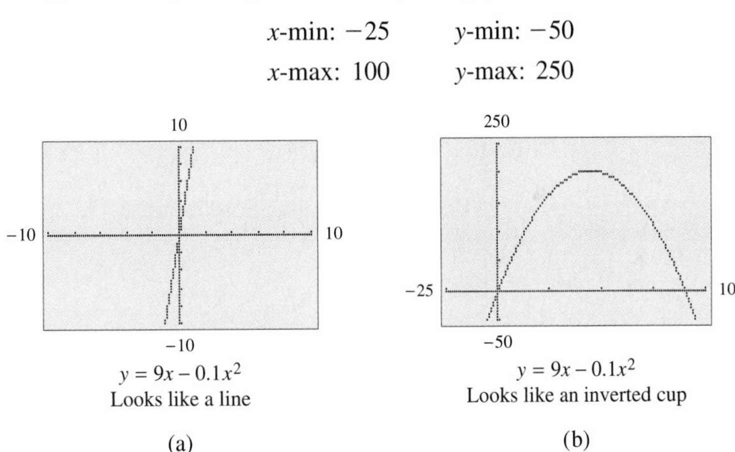

$y = 9x - 0.1x^2$
Looks like a line

(a)

$y = 9x - 0.1x^2$
Looks like an inverted cup

(b)

Figure 1.22

Figure 1.23 shows another example where the standard viewing window does not give the complete graph of a function.

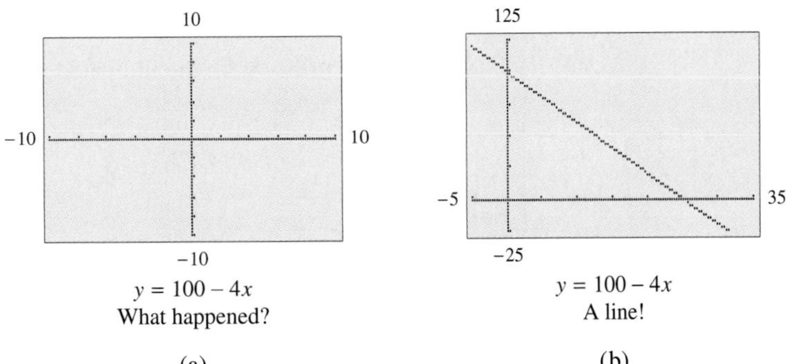

$y = 100 - 4x$
What happened?

(a)

$y = 100 - 4x$
A line!

(b)

Figure 1.23

The following guidelines may be used for most graphing utilities.

Guidelines for Using a Graphing Utility

To graph an equation in the variables x and y:

1. Solve the equation for y in terms of x.
2. Enter the equation into the graphing utility. Use parentheses as needed to ensure correctness.
3. Determine a viewing window.
4. Activate the graphing utility.

EXAMPLE 1

Graph $2x - 3y = 12$ with a graphing utility.

Solution

First note that this is a linear equation and that the graph is therefore a line. Begin by solving $2x - 3y = 12$ for y.

$$2x - 3y = 12$$
$$-3y = -2x + 12$$
$$y = \frac{-2x}{-3} + \frac{12}{-3}$$
$$y = \frac{2x}{3} - 4$$

Enter this last equation, and use the standard viewing window. The resulting graph is shown in Figure 1.24.

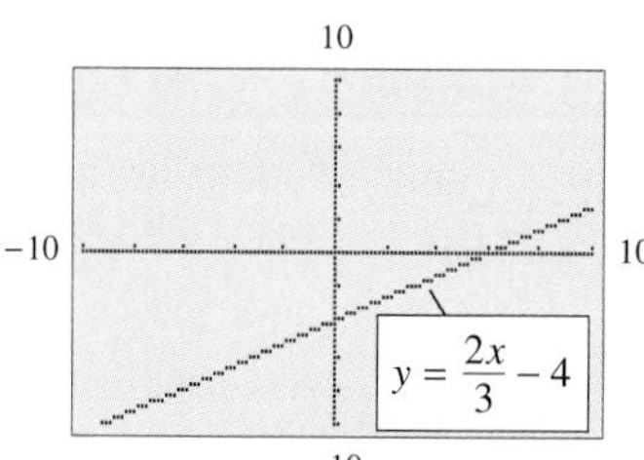

Figure 1.24

EXAMPLE 2

Use a graphing utility with the indicated range to graph $y = x^4 - 8x^2 - 10$.

x-min: -5	y-min: -35
x-max: 5	y-max: 10

Solution

The equation can be entered directly. Set the desired range and activate the graphing utility. The resulting graph is shown in Figure 1.25(a).

For comparison, graph the equation of Example 2 with the standard viewing window. How is it similar to the graph in Figure 1.25(a)? How is it different?

Most graphing utilities have a TRACE capability that allows the user to trace along a graph. Tracing highlights a point on the graph and simultaneously displays its coordinates. These coordinates can be plotted on paper to produce a copy of the graph of the equation. Even if the graph is not visible in the viewing window (because you're at a point beyond the y-range), the coordinates of the trace points are still displayed. These displayed coordinates can be used to help determine a more appropriate viewing window.

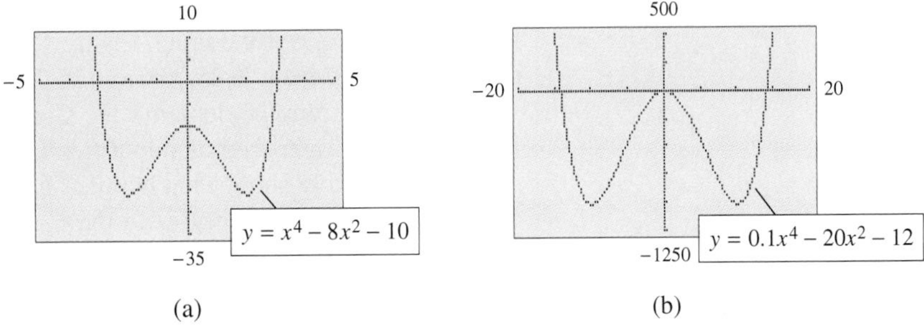

Figure 1.25

For example, if we graph $y = 0.1x^4 - 20x^2 - 12$ with the standard window, none of the graph appears in the window. When we trace along the graph with a standard window, the y-values range from -12 to -1012. Hence we could adjust the y-range to account for these y-values and also expand the x-range. If we use x-min $= -20$, x-max $= 20$, y-min $= -1250$, y-max $= 500$, we obtain the graph shown in Figure 1.25(b).

In addition to using a standard window, we can use windows that are called friendly. A "friendly" window causes TRACE to change each x-value by a "nice" amount, such as 0.1, 0.2, 1, etc., and it will occur if x-max $-$ x-min is nicely divisible by the number of pixels in the x-direction. For example, the TI-82 and TI-83 calculators have 94 pixels in the x-direction, so their windows will be "friendly" if x-max $-$ x-min gives a "nice" number when divided by 94. For example, if we set x-min $= -4.7$ and x-max $= 4.7$ on a T1-82 or T1-83 calculator, each step of TRACE will change x by

$$\frac{4.7 - (-4.7)}{94} = 0.1$$

On these calculators, ZOOM, 4 automatically gives x-min $= -4.7$ and x-max $=$ 4.7, and with ZOOM, 8, each step of TRACE changes by 1 unit for each press of an arrow. Using x-min $= -9.4$ and x-max $= 9.4$ with y-min $= -10$ and y-max $= 10$ gives a window that is "friendly" and close to the standard window.

We now return to the Application Preview problem.

EXAMPLE 3

For a certain city, the cost C of obtaining drinking water with p percent impurities (by volume) is given by

$$C = \frac{120{,}000}{p} - 1200$$

Because p is the percentage of impurities, we know that $0 \le p \le 100$. Use the restriction on p and a graphing utility to obtain an accurate graph of the equation.

Solution

To use a graphing utility, we identify C with y and p with x. Thus we enter

$$y = \frac{120{,}000}{x} - 1200$$

We will need an x-range on the graphing utility that corresponds to the p-range in the problem. Because $0 \leq p \leq 100$, we set the x-range to *include* these values, with the realization that only the portion of the graph above these values applies to our model and that $p = 0$ (i.e., $x = 0$) is excluded from the domain because it makes the denominator 0. With this x-range we can determine a y-, or C-, range from the equation or by tracing. Figure 1.26(a) shows the graph for x from -25 to 100 and for y from $-25{,}000$ to $125{,}000$. We see that when p is near 0 (the value excluded from the domain), the C-coordinates of the points are very large; indicating that water free of impurities is very costly. However, Figure 1.26(a) does not accurately show what happens for large p-values. Figure 1.26(b) shows another view of the equation with the C-range (that is, the y-range) from -500 to 3500. We see the p-intercept is $p = 100$, which indicates that water containing 100 percent impurities costs nothing ($C = 0$).

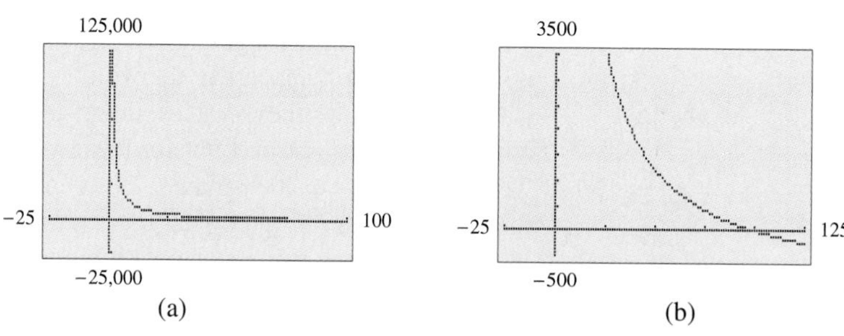

Figure 1.26 Two views of $C = \dfrac{120{,}000}{p} - 1200$

CHECKPOINT

Use a graphing utility and the standard viewing window to graph the following:

1. $x^2 + 4y = 0$
2. $y = \dfrac{4(x + 1)^2}{x^2 + 1}$

We can evaluate a function $y = f(x)$ with a graphing calculator by graphing the function, pressing TRACE, and moving the cursor to the value of the independent variable. We can also use TABLE to find the functional values corresponding to values of the independent variable in the table.

EXAMPLE 4

For the function $f(x) = x^4 - 8x^2 - 9$, use a graphing utility to evaluate

(a) $f(-2)$ and $f(3)$.

(b) Is 3 an x-intercept for the graph of $y = f(x)$?

(c) Find another x-intercept of the graph of $y = f(x)$.

Solution

(a) Graphing $y = x^4 - 8x^2 - 9$ and tracing to $x = -2$ gives a y-value of -25, so $f(-2) = -25$. See Figure 1.27(a). Tracing to $x = 3$ gives $y = 0$, so $f(3) = 0$.

(b) Because $f(3) = 0$, $x = 3$ is an x-intercept of the graph of $y = f(x)$. We also say that 3 is a zero of $f(x)$ and that 3 is a solution of $f(x) = 0$.

(c) It appears that the graph also crosses the x-axis at $x = -3$. Tracing to $x = -3$ on the graph gives $y = 0$, so -3 is another x-intercept of the graph and -3 is a zero of the function.

The table shown in Figure 1.27(b) shows selected values of x and the resulting values of $y = f(x)$. We can also see from the table that $f(-2) = -25$ and that 3 and -3 are zeros of the function, and thus x-intercepts of the graph of the function.

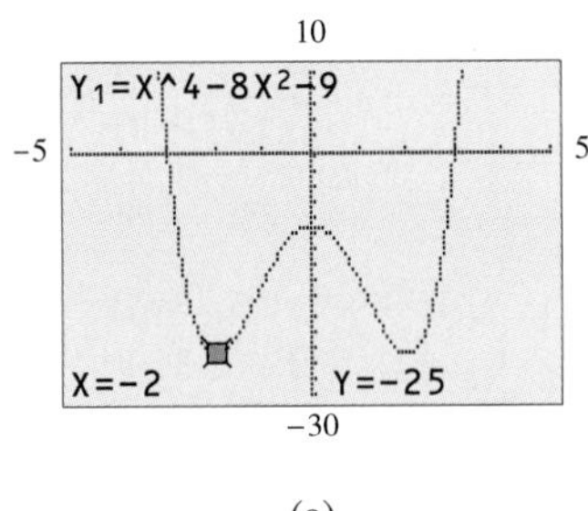

(a)

X	Y1
-3	0
-2	-25
-1	-16
0	-9
1	-16
2	-25
3	0

X=3

(b)

Figure 1.27

Technology Note

We can use a **spreadsheet,** such as Excel, to find the outputs of a function for given inputs. Table 1.2 shows a spreadsheet for the function $f(x) = 6x - 3$ for the input set $\{-2, -1, 0, 1, 3, 5, 10\}$, with these inputs listed as entries in the first column (column A). When using a formula with a spreadsheet, we use the cell location of the data to represent the variable. Thus if -2 is in cell A2, then $f(-2)$ can be found in cell B2 by typing $= 6*A2 - 3$ in cell B2. By using the fill-down capability, we can obtain all the functional values shown in column B.

TABLE 1.2

	A	B
1	x	f(x) = 6x–3
2	–2	–15
3	–1	–9
4	0	–3
5	1	3
6	3	15
7	5	27
8	10	57

CHECKPOINT SOLUTIONS

1. First solve $x^2 + 4y = 0$ for y.

$$4y = -x^2$$
$$y = \frac{-x^2}{4}$$

The graph is shown in Figure 1.28(a).

2. The equation $y = \frac{4(x+1)^2}{x^2+1}$ can be entered directly, but some care with parentheses is needed:

$$y = (4(x + 1)^2)/(x^2 + 1)$$

The graph is shown in Figure 1.28(b).

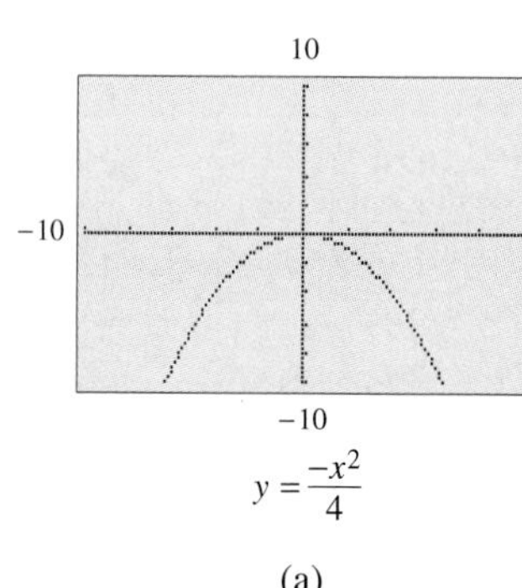

$y = \frac{-x^2}{4}$

(a)

10
−10
10
−10

$y = \frac{4(x+1)^2}{x^2+1}$

(b)

Figure 1.28

EXERCISE 1.4

In Problems 1–10, use a graphing utility with the standard viewing window to graph each function.

1. $y = 4 - x^2$
2. $y = x(x + 2)$
3. $y = x^2 - 6x + 5$
4. $y = 6 + x - x^2$
5. $y = x^3 - 3x$
6. $y = x^3 - 6x^2$
7. $y = \frac{1}{3}x^3 - x^2 - 3x + 2$
8. $y = \frac{1}{4}x^4 - \frac{1}{3}x^3 - 3x^2 + 8$
9. $y = \frac{12x}{x^2 + 1}$
10. $y = \frac{8}{x^2 + 1}$

In Problems 11 and 12, use a graphing utility with the specified range to graph each equation.

11. $y = x^3 - 12x - 1$
    x-min $= -5$, x-max $= 5$; y-min $= -20$, y-max $= 20$
12. $y = x^4 - 4x^3 + 6$
    x-min $= -4$, x-max $= 6$; y-min $= -40$, y-max $= 10$

In Problems 13–16, graph each equation with a graphing utility using
(a) the specified range
(b) the standard viewing window (for comparison)

13. $y = 0.01x^3 + 0.3x^2 - 72x + 150$
    x-min $= -100$, x-max $= 80$; y-min $= -2000$, y-max $= 4000$
14. $y = \frac{-0.01x^3 + 0.15x^2 + 60x + 700}{100}$
    x-min $= -100$, x-max $= 100$; y-min $= -50$, y-max $= 50$
15. $y = \frac{x + 15}{x^2 + 400}$
    x-min $= -200$, x-max $= 200$; y-min $= -0.02$, y-max $= 0.06$
16. $y = \frac{x - 80}{x^2 + 1700}$
    x-min $= -200$, x-max $= 400$; y-min $= -0.06$, y-max $= 0.01$

In Problems 17 and 18, do the following.

(a) Write a sentence that describes a plan to determine an appropriate viewing window for each graph.

(b) Determine a window that would show the graph.

(c) Use a graphing utility (with your window) to graph each equation.

17. $y = 0.001x - 0.03$

18. $y = 50{,}000 - 100x$

In Problems 19–22, a standard viewing window graph of a function is shown. Experiment with the viewing window to obtain a complete graph, adjusting the ranges where necessary.

19. $y = -0.15(x - 10.2)^2 + 10$

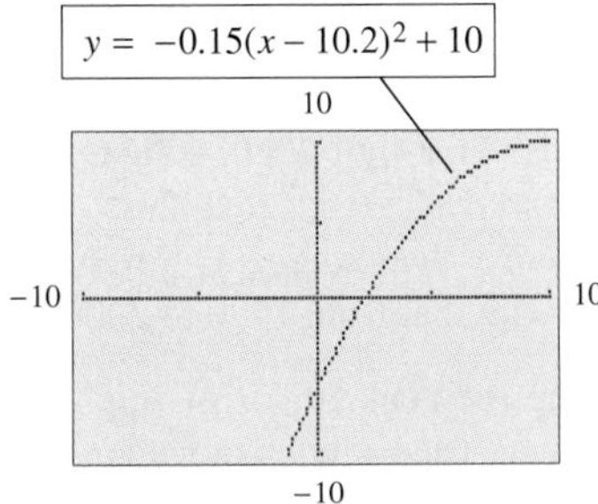

20. $y = x^2 - x - 42$

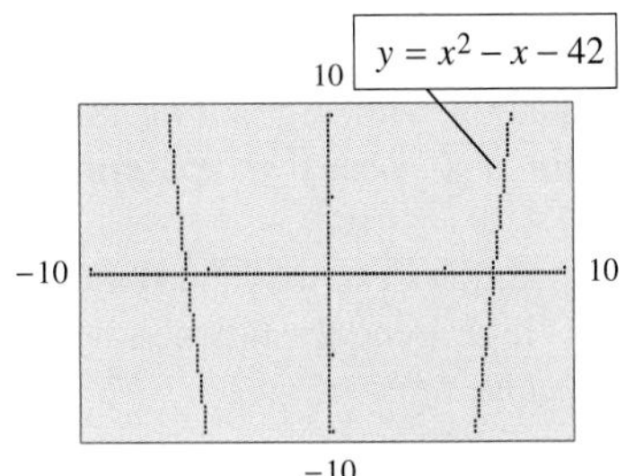

21. $y = \dfrac{x^3 + 19x^2 - 62x - 840}{20}$

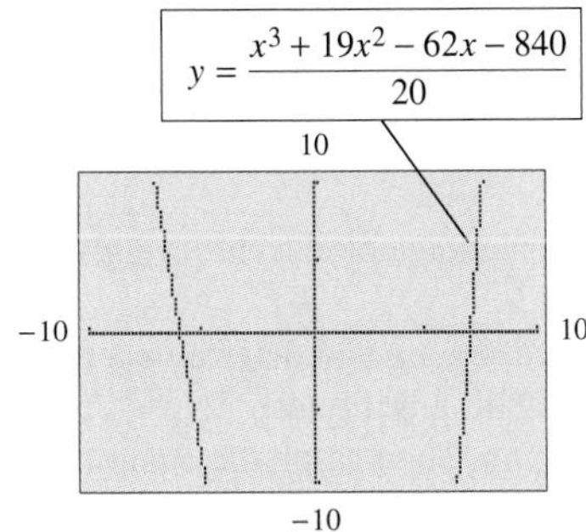

22. $y = \dfrac{-x^3 + 33x^2 + 120x}{20}$

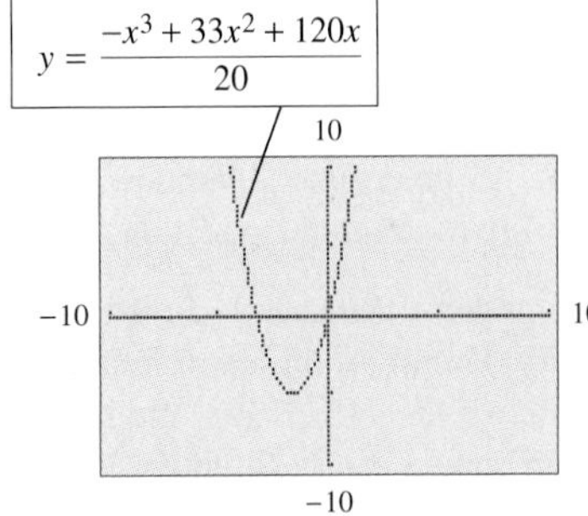

In Problems 23–28, graph the equations with a standard window on a graphing utility.

23. $5x - 3y = 5$

24. $4x - 2y = 5$

25. $4x^2 + 2y = 5$

26. $6x^2 + 3y = 8$

27. $x^2 + 2y = 6$

28. $3x^2 - 4y = 8$

29. If $f(x) = x^3 - 3x^2 + 2$, find $f(1)$ and $f(-3/2)$.

30. If $f(x) = \dfrac{x^2 - 2x}{x - 1}$, find $f(-2)$ and $f(3)$.

In Problems 31–34, graph each function using a window that gives a complete graph.

31. $y = \dfrac{12x^2 - 12}{x^2 + 1}$

32. $y = \dfrac{8}{x^2 + 1}$

33. $y = \dfrac{x^2 - x - 6}{x^2 + 5x + 6}$

34. $y = \dfrac{x^2 - 4}{x^2 - 9}$

In Problems 35–38, (a) find the x-intercepts to four decimal places of the graph of each function, and (b) find the zeros of each function.

35. $y = x^2 - 7x - 9$

36. $y = 2x^2 - 4x - 11$

37. $y = \dfrac{(x - 1)(x - 4)}{x(x + 2)}$

38. $y = \dfrac{(x - 3)^2(x + 7)}{(x - 9)(x + 1)}$

Applications

39. ***Advertising impact*** An advertising agency has found that when it promotes a new product in a city of 350,000 people, the rate of change R of the number of people x who are aware of the product is given by

$$R = 28{,}000 - 0.08x$$

Use the intercepts to determine a window, and then use a graphing utility to graph the equation for $x \geq 0$.

40. ***Learning rate*** In a study using 50 foreign-language vocabulary words, the learning rate L (in words per minute) was found to depend on the number of words already learned, x, according to the equation

$$L = 20 - 0.4x$$

Use the intercepts to determine a window, and then use a graphing utility to graph the equation for $x \geq 0$.

41. ***Earnings and gender*** With U.S. Equal Employment Opportunity Commission data found in the *Statistical Abstract of the U.S.*, 1993, the model that relates the median annual salary (in thousands of dollars) of females, F, and males, M, in the United States was found to be

$$F = 0.838M - 1.364$$

(a) Use a graphing utility to graph this equation for the range M-min $= 0$, M-max $= 35$; F-min $= 0$, F-max $= 30$.
(b) The point (50, 40.536) lies on the graph of this equation. Explain its meaning.

42. ***Earnings and minorities*** With U.S. Equal Employment Opportunity Commission data found in the *Statistical Abstract of the U.S.*, 1993, the model that relates the median annual salary (in thousands of dollars) of minorities, M, and whites, W, was found to be

$$M = 0.959W - 1.226$$

(a) Use a graphing utility to graph this equation for the range W-min $= 0$, W-max $= 30$; M-min $= 0$, M-max $= 30$.
(b) The point (40, 37.134) lies on the graph of this equation. Explain its meaning.

43. ***Consumer expenditure*** Suppose that the consumer expenditure E (in dollars) depends on the market price p per unit (in dollars) according to

$$E = 10{,}000p - 100p^2$$

(a) Graph this equation with a graphing utility and the range p-min $= -50$, p-max $= 150$; E-min $= -50{,}000$, E-max $= 300{,}000$.
(b) Because E represents consumer expenditure, we know $E \geq 0$. For what p-values is $E \geq 0$?

44. ***Rectilinear motion*** The height above ground (in feet) of a ball thrown vertically into the air is given by

$$S = 112t - 16t^2$$

where t is the time in seconds since the ball was thrown.

(a) Graph this equation with a graphing utility and the range t-min $= -2$, t-max $= 10$; S-min $= -20$, S-max $= 250$.
(b) Estimate the time at which the ball is at its highest point, and estimate the height of the ball at that time.

45. ***Sales revenue*** Sales revenue R (in billions of dollars) for Scott Paper Company can be modeled by

$$R = -0.031t^2 + 0.776t + 0.179$$

where t is the number of years past 1980 (Source: Scott Paper Company, *1993 Annual Report*).
(a) Graph this equation with a graphing utility and the range t-min $= 0$, t-max $= 15$; R-min $= 0$, R-max $= 8$, which corresponds to the years from 1980 to 1995.
(b) If this model is valid until 2020, adjust the range to obtain a graph that extends to 2020.
(c) Write a sentence that compares revenue from 1980 to 1995 with that from 1980 to 2020.

46. ***Drug usage*** The percentage p of high school seniors who have tried hallucinogens can be described by

$$p = 0.0241904(t + 70)^2 - 4.47459(t + 70) + 216.074$$

where t is the number of years past 1970 (Source: National Institute on Drug Abuse).
(a) Graph this equation with a graphing utility and the range t-min $= 0$, t-max $= 25$; p-min $= 0$, p-max $= 20$.
(b) Write a sentence that describes the trends in the percentage of high school students who have tried hallucinogens.

47. ***Cost-benefit*** The Millcreek watershed area was heavily strip-mined for coal during the late 1960s. Because of the resulting pollution, the streams can't support fish. Suppose the cost C of obtaining stream water that contains p percent of the current pollution levels is given by

$$C = \frac{285{,}000}{p} - 2850$$

Because p is the percentage of current levels, $0 \leq p \leq 100$.
(a) Use the restriction on p and determine a range for C so that a graphing utility can be used to obtain an accurate graph. Then graph the equation.
(b) Describe what happens to the cost as p takes on values near 0.

(c) The point (1, 282,150) lies on the graph of this equation. Explain its meaning.
(d) Explain the meaning of the p-intercept.

48. ***Cost-benefit*** Suppose the cost C of removing p percent of the particulate pollution from the exhaust gases at an industrial site is given by

$$C = \frac{8100p}{100 - p}$$

Because p is the percentage of particulate pollution, we know $0 \le p \le 100$.
(a) Use the restriction on p and experiment with a C-range to obtain an accurate graph of the equation with a graphing utility.
(b) Describe what happens to C as p gets close to 100.
(c) The point (98, 396,900) lies on the graph of this equation. Explain the meaning of the coordinates.
(d) Explain the meaning of the p-intercept.

49. ***Earning and gender*** The following table shows the 1986 median weekly earnings of men and women in various occupations.

	Earnings	
Occupation	*Men* (x)	*Women* (y)
Management	$666	$465
Technical sales	472	305
Service	299	208
Crafts, repair	446	302
Operators, fabricators	352	238
Fish, farm, forestry	234	201

Source: U.S. Bureau of Labor Statistics

(a) Plot the data in the table, using a graphing utility.
(b) Graph

$$y = 0.630526x + 27.0386 \quad \text{and}$$
$$y = 0.914811x + 20.4118$$

and decide which one fits the data better.
(c) Use the equation you chose in part (b) to see how well it predicts the 1986 median weekly earnings for a woman in a job where a man earned $472 (technical sales).

50. ***Training and productivity*** Suppose that a company purchased a new software package to handle its word processing, payroll, inventory, and accounting needs. After a few employees had received some training with the new system, their work with it was monitored and the following data were gathered.

Hours of Training (x)	*Number of Errors* (y)	*Hours of Training* (x)	*Number of Errors* (y)
1	8	12	0
2	7	7	3
10	2	6	3
8	2	4	4
6	8	4	6

(a) Plot the data in the table, using a graphing utility.
(b) Graph

$$y = -0.68868x + 8.4321 \quad \text{and}$$
$$y = -0.40116x + 6.3494$$

and decide which one fits the data better.
(c) Use the equation you chose in (b) to determine the number of hours of training needed for the predicted number of errors to equal 0.

1.5 Solutions of Systems of Linear Equations

OBJECTIVES

- *To solve systems of linear equations in two variables by graphing*
- *To solve systems of linear equations by substitution*
- *To solve systems of linear equations by elimination*
- *To solve systems of three linear equations in three variables*

APPLICATION PREVIEW

Suppose that a person has $200,000 invested, part at 9% and part at 8%, and that the yearly income from the two investments is $17,200. If x represents the amount invested at 9% and y represents the amount invested at 8%, then to find how much is invested at each rate we must find the values of x and y that satisfy both

$$x + y = 200{,}000 \quad \text{and} \quad 0.09x + 0.08y = 17{,}200$$

The terminology associated with problems like this and the skills needed to determine x and y are the topics of this section.

In the previous sections, we graphed linear equations in two variables and observed that the graphs are straight lines. Each point on the graph represents an ordered pair of values (x, y) that satisfies the equation, so a point of intersection of two (or more) lines represents a solution to both (or all) the equations of those lines.

The equations are referred to as a **system of equations,** and the ordered pairs (x, y) that satisfy all the equations are the **solutions** (or **simultaneous solutions)** to the system. We can use graphing to find the solution to a system of equations.

EXAMPLE 1

Use graphing to find the solution of the system

$$\begin{cases} 4x + 3y = 11 \\ 2x - 5y = -1 \end{cases}$$

Solution

The graphs of the two equations intersect (meet) at the point (2, 1). (See Figure 1.29.) The solution of the system is $x = 2$, $y = 1$. Note that these values satisfy both equations.

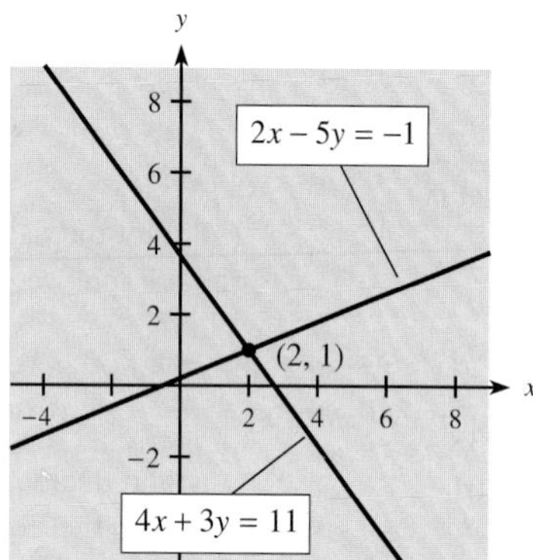

Figure 1.29 Two distinct nonparallel lines: one solution

If the graphs of two equations are parallel lines, they have no point in common and thus the system has no solution. Such a system of equations is called **inconsistent.** For example,

$$\begin{cases} 4x + 3y = 4 \\ 8x + 6y = 18 \end{cases}$$

is an **inconsistent system** (see Figure 1.30).

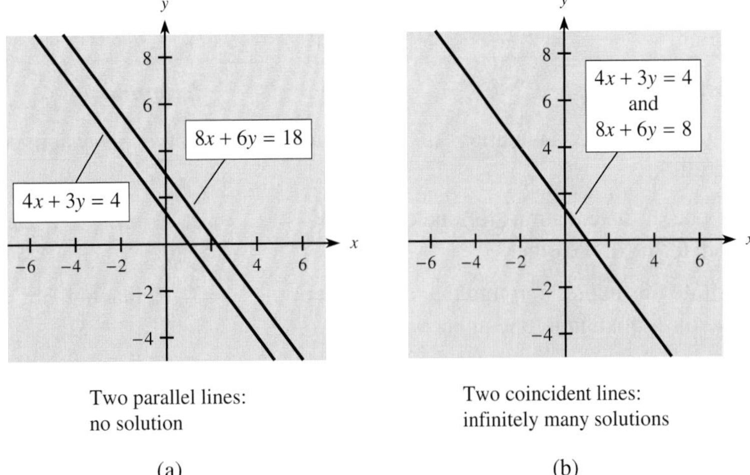

Two parallel lines: no solution

(a)

Two coincident lines: infinitely many solutions

(b)

Figure 1.30

It is also possible that two equations describe the same line. When this happens the equations are equivalent, and values that satisfy either equation are solutions to the system. For example,

$$\begin{cases} 4x + 3y = 4 \\ 8x + 6y = 8 \end{cases}$$

is called a **dependent system** because all points that satisfy one equation also satisfy the other (see Figure 1.30(b)).

Figures 1.29, 1.30(a) and 1.30(b) represent the three possibilities that can occur when we are solving a system of two linear equations in two variables.

Graphical solution methods may only yield approximate solutions to some systems. Exact solutions can be found using algebraic methods, which are based on the fact that equivalent systems result when any of the following operations are performed.

Equivalent Systems

Equivalent systems result when

1. One expression is replaced by an equivalent expression.
2. Two equations are interchanged.
3. A multiple of one equation is added to another equation.
4. An equation is multiplied by a nonzero constant.

The **substitution method** is based on operation (1).

Substitution Method for Solving Systems

Procedure	**Example**
To solve a system of two equations in two variables by substitution:	Solve the system containing $2x + 3y = 4$ and $x - 2y = 3$.
1. Solve one of the equations for one of the variables in terms of the other.	1. Solving $x - 2y = 3$ for x gives $x = 2y + 3$.
2. Substitute this expression into the other equation to give one equation in one unknown.	2. Replacing x by $2y + 3$ in $2x + 3y = 4$ gives $2(2y + 3) + 3y = 4$.
3. Solve this linear equation for the unknown.	3. $4y + 6 + 3y = 4$ $7y = -2 \Rightarrow y = -\frac{2}{7}$
4. Substitute this solution into the equation in step 1 or into one of the original equations to solve for the other variable.	4. $x = 2\left(-\frac{2}{7}\right) + 3 \Rightarrow x = \frac{17}{7}$
5. Check the solution by substituting for x and y in both original equations.	5. $2\left(\frac{17}{7}\right) + 3\left(-\frac{2}{7}\right) = 4$ ✔ $\frac{17}{7} - 2\left(-\frac{2}{7}\right) = 3$ ✔

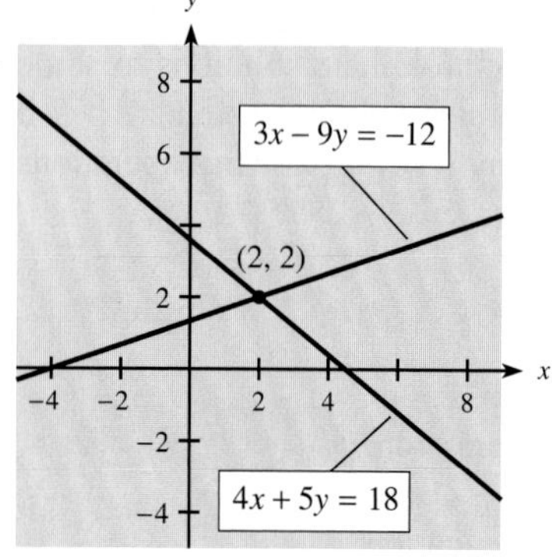

Figure 1.31

EXAMPLE 2

Solve the system

$$\begin{cases} 4x + 5y = 18 & (1) \\ 3x - 9y = -12 & (2) \end{cases}$$

Solution

1. $x = \frac{9y - 12}{3} = 3y - 4$ — Solving for x in equation (2)
2. $4(3y - 4) + 5y = 18$ — Substituting for x in equation (1)
3. $12y - 16 + 5y = 18$ — Solving for y
   $17y = 34$
   $y = 2$
4. $x = 3(2) - 4$ — Using $y = 2$ to find x
   $x = 2$
5. $4(2) + 5(2) = 18$ and $3(2) - 9(2) = -12$ — Checking

Thus the solution is $x = 2$, $y = 2$. This means that when the two equations are graphed simultaneously, their point of intersection is (2, 2). See Figure 1.31.

We can also eliminate one of the variables in a system by the **elimination method,** which uses addition or subtraction of equations.

Elimination Method for Solving Systems

Procedure	Example
To solve a system of two equations in two variables by the elimination method:	Solve the system $\begin{cases} 2x - 5y = 4 & (1) \\ x + 2y = 3 & (2) \end{cases}$
1. If necessary, multiply one or both equations by a nonzero number that will make the coefficients of one of the variables identical, except perhaps for signs.	1. Multiply Equation (2) by -2. $2x - 5y = 4$; $-2x - 4y = -6$
2. Add or subtract the equations to eliminate one of the variables.	2. Adding gives $0x - 9y = -2$
3. Solve for the variable in the resulting equation.	3. $y = \frac{2}{9}$
4. Substitute the solution into one of the original equations and solve for the other variable.	4. $2x - 5\left(\frac{2}{9}\right) = 4$; $2x = 4 + \frac{10}{9} = \frac{36}{9} + \frac{10}{9}$; $2x = \frac{46}{9}$ so $x = \frac{23}{9}$
5. Check the solutions in both original equations.	5. $2\left(\frac{23}{9}\right) - 5\left(\frac{2}{9}\right) = 4$ ✓; $\frac{23}{9} + 2\left(\frac{2}{9}\right) = 3$ ✓

We now return to the Application Preview problem.

EXAMPLE 3

A person has \$200,000 invested, part at 9% and part at 8%. If the total yearly income from the two investments is \$17,200, how much is invested at 9%, and how much at 8%?

Solution

If x represents the amount invested at 9% and y represents the amount invested at 8%, then $x + y$ is the total investment,

$$x + y = 200{,}000 \qquad (1)$$

and $0.09x + 0.08y$ is the total income earned.

$$0.09x + 0.08y = 17{,}200 \qquad (2)$$

We solve these equations as follows:

$$\begin{array}{rcrll} -8x - 8y &=& -1{,}600{,}000 & (3) & \text{Multiplying equation (1) by } -8 \\ 9x + 8y &=& 1{,}720{,}000 & (4) & \text{Multiplying equation (2) by } 100 \\ \hline x &=& 120{,}000 & & \text{Adding (3) and (4)} \end{array}$$

Thus \$120,000 is invested at 9%, and \$80,000 is invested at 8%.

EXAMPLE 4

Solve the systems:

(a) $\begin{cases} 4x + 3y = 4 \\ 8x + 6y = 18 \end{cases}$ (b) $\begin{cases} 4x + 3y = 4 \\ 8x + 6y = 8 \end{cases}$

Solution

(a) $\begin{cases} 4x + 3y = 4 \\ 8x + 6y = 18 \end{cases}$

$$\begin{array}{ll} \text{Multiply by } -2 \text{ to get:} & -8x - 6y = -8 \\ \text{Leave as is, which gives:} & 8x + 6y = 18 \\ \hline \text{Add the equations to get:} & 0x + 0y = 10 \\ & 0 = 10 \end{array}$$

The system is solved when $0 = 10$. This is impossible, so there are no solutions to the system. The equations are inconsistent. Their graphs are parallel lines; see Figure 1.30(a) on page 113.

(b) $\begin{cases} 4x + 3y = 4 \\ 8x + 6y = 8 \end{cases}$

$$\begin{array}{ll} \text{Multiply by } -2 \text{ to get:} & -8x - 6y = -8 \\ \text{Leave as is, which gives:} & 8x + 6y = 8 \\ \hline \text{Add the equations to get:} & 0x + 0y = 0 \\ & 0 = 0 \end{array}$$

This is an identity, so the two equations share infinitely many solutions. The equations are dependent. Their graphs coincide, and each point on this line represents a solution to the system; see Figure 1.30(b) on page 113.

EXAMPLE 5

A nurse has two solutions that contain different concentrations of a certain medication. One is a 12.5% concentration and the other is a 5% concentration. How many cubic centimeters of each should she mix to obtain 20 cubic centimeters of an 8% concentration?

Solution

Let x equal the number of cubic centimeters of the 12.5% solution, and let y equal the number of cubic centimeters of the 5% solution. The total amount of substance is

$$x + y = 20$$

and the total amount of medication is

$$0.125x + 0.05y = (0.08)(20) = 1.6$$

Solving this pair of equations simultaneously gives

$$\begin{array}{rcr} 50x + 50y &=& 1000 \\ -125x - 50y &=& -1600 \\ \hline -75x \qquad &=& -600 \end{array}$$

$$x = 8$$

$$8 + y = 20, \text{ so } y = 12$$

Thus 8 cubic centimeters of a 12.5% concentration and 12 cubic centimeters of a 5% concentration yield 20 cubic centimeters of an 8% concentration.

CHECKPOINT

1. Solve by substitution: $\begin{cases} 3x - 4y = -24 \\ x + \ \ y = -1 \end{cases}$
2. Solve by elimination: $\begin{cases} 2x + 3y = 5 \\ 3x + 5y = -25 \end{cases}$
3. In Problems 1 and 2, the solution method is given.
   (a) In each case, explain why you think that method is appropriate.
   (b) In each case, would the other method work as well? Explain.

Technology Note We can check the solution to a system of equations in two variables or solve the system by graphing the two equations on the same screen of a graphing utility and finding a point of intersection of the two graphs. Solutions of systems of equations, if they exist, can be found by using TRACE, or by using INTERSECT. TRACE may or may not give the solution exactly. INTERSECT will give the point of intersection exactly or to a large number of significant digits.

EXAMPLE 6

Solve the system

$$\begin{cases} 3x + 2y = 12 \\ 4x - 3y = -1 \end{cases}$$

Solution

Solving both of these equations for y gives $y = 6 - \dfrac{3x}{2}$ and $y = \dfrac{1}{3} + \dfrac{4x}{3}$.

Entering the equations in a graphing utility, graphing, and using TRACE or INTERSECT give the intersection of the graphs (see Figure 1.32). The point of intersection is (2, 3), so the solution to the system is $x = 2$, $y = 3$.

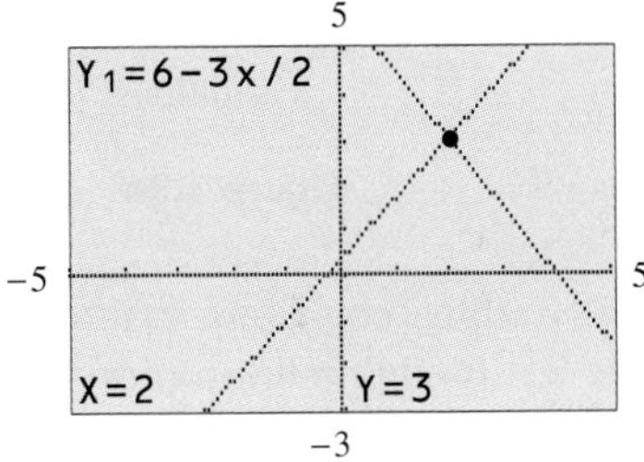

Figure 1.32

Solving a system of equations by graphing, whether by hand or with a graphing utility, is limited by two factors. (1) It may be difficult to determine a viewing window that contains the point of intersection and (2) the solution may be only approximate. With *some* systems of equations, the only practical method may be graphical approximation.

However, systems of *linear* equations can be consistently and accurately solved with algebraic methods. Computer algebra systems, some software packages (including spreadsheets), and some graphing calculators have the capability to solve systems of linear equations algebraically.

Three Equations in Three Variables

If a, b, c, and d represent constants, then

$$ax + by + cz = d$$

is a first-degree (linear) equation in three variables. When equations of this form are graphed in a three-dimensional coordinate system, their graphs are planes. Two different planes may intersect in a line (like two walls) or may not intersect at all (like a floor and ceiling). Three different planes may intersect in a single point (as when two walls meet the ceiling), may intersect in a line (as in a paddle wheel), or may not have a common intersection. (See Figures 1.33, 1.34, and 1.35.) Thus three linear equations in three variables may have a unique solution, infinitely many solutions, or no solution. For example, the solution to the system

$$\begin{cases} 3x + 2y + z = 6 \\ x - y - z = 0 \\ x + y - z = 4 \end{cases}$$

is $x = 1$, $y = 2$, $z = -1$, because these three values satisfy all three equations, and these are the only values that satisfy them. In this section, we will only discuss systems of three linear equations in three variables that have unique solutions. Additional systems will be discussed in Section 3.3, "Gauss–Jordan Elimination: Solving Systems of Equations."

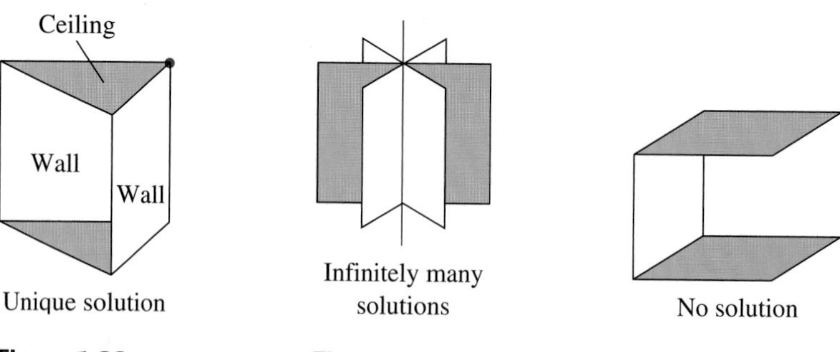

Unique solution

Figure 1.33

Infinitely many solutions

Figure 1.34

No solution

Figure 1.35

We can solve three equations in three variables using a systematic procedure, called the **left-to-right elimination method.**

Left-to-Right Elimination Method

Procedure	***Example***
To solve a system of three equations in three variables by the left-to-right elimination method:	Solve: $\begin{cases} 2x + 4y + 5z = 4 \\ x - 2y - 3z = 5 \\ x + 3y + 4z = 1 \end{cases}$
1. If necessary, interchange two equations or use multiplication to make the coefficient of the first variable in equation (1) a factor of the other first variable coefficients.	1. Interchange the first two equations: $\begin{aligned} x - 2y - 3z &= 5 \quad (1) \\ 2x + 4y + 5z &= 4 \quad (2) \\ x + 3y + 4z &= 1 \quad (3) \end{aligned}$
2. Add multiples of the first equation to each of the following equations so that the coefficients of the first variable in the second and third equations become zero.	2. Add $(-2) \times$ equation (1) to equation (2) and add $(-1) \times$ equation (1) to equation (3): $\begin{aligned} x - 2y - 3z &= 5 \quad (1) \\ 0x + 8y + 11z &= -6 \quad (2) \\ 0x + 5y + 7z &= -4 \quad (3) \end{aligned}$
3. Add a multiple of the second equation to the third equation so that the coefficient of the second variable in the third equation becomes zero.	3. Add $(-\frac{5}{8}) \times$ equation (2) to equation (3): $\begin{aligned} x - 2y - 3z &= 5 \quad (1) \\ 8y + 11z &= -6 \quad (2) \\ 0y + \frac{1}{8}z &= -\frac{2}{8} \quad (3) \end{aligned}$
4. Solve the third equation and *back substitute* from the bottom to find the remaining variables.	4. $z = -2$ from equation (3) $y = \frac{1}{8}(-6 - 11z) = 2$ from equation (2) $x = 5 + 2y + 3z = 3$ from equation (1) so $x = 3, y = 2, z = -2$.

EXAMPLE 7

$$\text{Solve: } \begin{cases} x + 2y + 3z = 6 & (1) \\ 2x + 3y + 2z = 6 & (2) \\ -x + y + z = 4 & (3) \end{cases}$$

Solution

Using equation (1) to eliminate x from the other equations gives the equivalent system:

$$\begin{cases} x + 2y + 3z = -6 & (1) \\ -y - 4z = -6 & (2) \quad (-2) \times \text{equation (1) added to equation (2)} \\ 3y + 4z = 10 & (3) \quad \text{Equation (1) added to equation (3)} \end{cases}$$

Using equation (2) to eliminate y from equation (3) gives

$$\begin{cases} x + 2y + 3z = 6 & (1) \\ -y - 4z = -6 & (2) \\ -8z = -8 & (3) \quad (3) \times \text{ equation (2) added to equation (3)} \end{cases}$$

Solving for each **lead variable** gives

$$x = 6 - 2y - 3z$$
$$y = 6 - 4z$$
$$z = 1$$

and using **back substitution** from the bottom gives

$$z = 1$$
$$y = 6 - 4 = 2$$
$$x = 6 - 4 - 3 = -1$$

Hence the solution is $x = -1$, $y = 2$, $z = 1$.

Although other methods for solving systems of equations in three variables may be useful, the left-to-right elimination method is important because it is systematic and can easily be extended to larger systems and to systems solved with matrices (see Section 3.3, "Gauss–Jordan Elimination: Solving Systems of Equations").

CHECKPOINT

4. Use left-to-right elimination to solve.

$$\begin{cases} x - y - z = 0 & (1) \\ y - 2z = -18 & (2) \\ x + y + z = 6 & (3) \end{cases}$$

CHECKPOINT SOLUTIONS

1. $x + y = -1$ means $x = -y - 1$, so $3x - 4y = -24$ becomes

$$3(-y - 1) - 4y = -24$$
$$-3y - 3 - 4y = -24$$
$$-7y = -21$$
$$y = 3$$

Hence $x = -3 - 1 = -4$, and the solution is $x = -4$, $y = 3$.

2. Multiply the first equation by (-3) and the second by (2). Add the resulting equations to eliminate the x-variable and solve for y.

$$\begin{array}{r} -6x - 9y = -15 \\ 6x + 10y = -50 \\ \hline y = -65 \end{array}$$

Hence $2x + 3(-65) = 5$ or $2x - 195 = 5$, so $2x = 200$. Thus $x = 100$ and $y = -65$.

3. (a) Substitution works well in Problem 1 because it is easy to solve for x (or for y). Substitution would not work well in Problem 2 because solving for x (or for y) would introduce fractions.
(b) Elimination would work well in Problem 1 if we multiplied the first equation by 4. As stated in (a), substitution wouldn't be as easy in Problem 2.

4.
$$\begin{cases} x - y - z = 0 & (1) \\ y - 2z = -18 & (2) \\ x + y + z = 6 & (3) \end{cases}$$

Add $(-1) \times$ equation (1) to equation (3).

$$\begin{cases} x - y - z = 0 & (1) \\ y - 2z = -18 & (2) \\ 2y + 2z = 6 & (3) \end{cases}$$

Add $(-2) \times$ equation (2) to equation (3).

$$\begin{cases} x - y - z = 0 \\ y - 2z = -18 \\ 6z = 42 \end{cases}$$

Solve the equations for x, y, and z.

$$x = y + z$$
$$y = 2z - 18$$
$$z = 7$$

Back substitution gives $y = 14 - 18 = -4$ and $x = -4 + 7 = 3$. Thus the solution is $x = 3$, $y = -4$, $z = 7$.

EXERCISE 1.5

In Problems 1–4, solve the systems of equations by using graphical methods.

1. $\begin{cases} 4x - 2y = 4 \\ x - 2y = -2 \end{cases}$
2. $\begin{cases} x - y = -2 \\ 2x + y = -1 \end{cases}$
3. $\begin{cases} 3x - y = 10 \\ 6x - 2y = 5 \end{cases}$
4. $\begin{cases} 2x - y = 3 \\ 4x - 2y = 6 \end{cases}$

In Problems 5–22, solve the systems of equations.

5. $\begin{cases} 4x - y = 3 \\ 2x + 3y = 19 \end{cases}$
6. $\begin{cases} 5x - 3y = 9 \\ x + 2y = 7 \end{cases}$
7. $\begin{cases} 2x - y = 2 \\ 3x + 4y = 6 \end{cases}$
8. $\begin{cases} x - y = 4 \\ 3x - 2y = 5 \end{cases}$
9. $\begin{cases} 3x - 2y = 6 \\ 4y = 8 \end{cases}$
10. $\begin{cases} 5x - 2y = 4 \\ 2x - 3y = 5 \end{cases}$
11. $\begin{cases} -4x + 3y = -5 \\ 3x - 2y = 4 \end{cases}$
12. $\begin{cases} x + 2y = 3 \\ 3x + 6y = 6 \end{cases}$
13. $\begin{cases} \frac{5}{2}x - \frac{7}{2}y = -1 \\ 8x + 3y = 11 \end{cases}$
14. $\begin{cases} 3m + 2n = 2 \\ m - 2n = 1 \end{cases}$
15. $\begin{cases} 0.3u - 0.2v = 0.5 \\ 0.9u - 0.6v = 0.1 \end{cases}$
16. $\begin{cases} x - \frac{1}{2}y = 1 \\ \frac{2}{3}x - \frac{1}{3}y = 1 \end{cases}$
17. $\begin{cases} 0.2x - 0.3y = 4 \\ 2.3x - y = 1.2 \end{cases}$
18. $\begin{cases} 0.5x + y = 3 \\ 0.3x + 0.2y = 6 \end{cases}$
19. $\begin{cases} 4x + 6y = 4 \\ 2x + 3y = 2 \end{cases}$
20. $\begin{cases} \dfrac{x}{4} + \dfrac{3y}{4} = 12 \\ \dfrac{y}{2} - \dfrac{x}{3} = -4 \end{cases}$
21. $\begin{cases} \dfrac{(x + y)}{4} = 2 \\ \dfrac{(y - 1)}{x} = 6 \end{cases}$
22. $\begin{cases} 3x + 4y = 8 \\ \dfrac{(16 - 6x)}{y} = 8 \end{cases}$

Use a graphing utility to find the solution to each system of equations in Problems 23–26.

23. $\begin{cases} y = 8 - \dfrac{3x}{2} \\ y = \dfrac{3x}{4} - 1 \end{cases}$

24. $\begin{cases} y = 9 - \dfrac{2x}{3} \\ y = 5 + \dfrac{2x}{3} \end{cases}$

25. $\begin{cases} 5x + 3y = -2 \\ 3x + 7y = 4 \end{cases}$

26. $\begin{cases} 4x - 5y = -3 \\ 2x - 7y = -6 \end{cases}$

Use the left-to-right elimination method to solve the systems in Problems 27–32.

27. $\begin{cases} x + 2y + z = 2 \\ -y + 3z = 8 \\ 2z = 10 \end{cases}$

28. $\begin{cases} x - 2y + 2z = -10 \\ y + 4z = -10 \\ -3z = 9 \end{cases}$

29. $\begin{cases} x - y - 8z = 0 \\ y + 4z = 8 \\ 3y + 14z = 22 \end{cases}$

30. $\begin{cases} x + 3y - 8z = 20 \\ y - 3z = 11 \\ 2y + 7z = -4 \end{cases}$

31. $\begin{cases} x + 4y - 2z = 9 \\ x + 5y + 2z = -2 \\ x + 4y - 28z = 22 \end{cases}$

32. $\begin{cases} x - 3y - z = 0 \\ x - 2y + z = 8 \\ 2x - 6y + z = 6 \end{cases}$

Applications

33. ***Investment yields*** One safe investment pays 10% per year, and a more risky investment pays 18% per year. A woman who has $145,600 to invest would like to have an income of $20,000 per year from her investments. How much should she invest at each rate?

34. ***Loans*** A bank lent $118,500 to a company for the development of two products. If the loan for product A was for $34,500 more than that for product B, how much was lent for each product?

35. ***Rental income*** A woman has $23,500 invested in two rental properties. One yields 10% on the investment, and the other yields 12%. Her total income from them is $2550. How much is her income from each property?

36. ***Loans*** Mr. Jackson borrowed money from his bank and on his life insurance to start a business. His interest rate on the bank loan was 10%, and his rate on the insurance loan was 12%. If the total amount borrowed was $10,000 and his total yearly interest payment was $1090, how much did he borrow from the bank?

37. ***Nutrition*** Each ounce of substance A supplies 5% of the nutrition a patient needs. Substance B supplies 12% of the required nutrition per ounce. If digestive restrictions require that the ratio of substance A to substance B be 3/5, how many ounces of each should be in the diet to provide 100% of the required nutrition?

38. ***Nutrition*** A glass of skim milk supplies 0.1 milligram of iron and 8.5 grams of protein. A quarter pound of lean red meat provides 3.4 mg of iron and 22 g of protein. If a person on a special diet is to have 7.15 mg of iron and 73.75 g of protein, how many glasses of skim milk and how many quarter-pound servings of meat would provide this?

39. ***Bacterial growth*** Bacteria of species A and species B are kept in a single test tube, where they are fed two nutrients. Each day the test tube is supplied with 10,600 units of the first nutrient and 19,650 units of the second nutrient. Each bacterium of species A requires 2 units of the first nutrient and 3 units of the second, and each bacterium of species B requires 1 unit of the first nutrient and 4 units of the second. What populations of each species can coexist in the test tube so that all the nutrients are consumed each day?

40. ***Botany*** A biologist has a 40% solution and a 10% solution of the same plant nutrient. How many cubic centimeters of each solution should be mixed to obtain 25 cc of a 28% solution?

41. ***Medications*** A nurse has two solutions that contain different concentrations of a certain medication. One is a 20% concentration and the other is a 5% concentration. How many cubic centimeters of each should he mix to obtain 10 cc of a 15.5% solution?

42. ***Medications*** Medication A is given every 4 hours and medication B is given twice each day. The total intake of the two medications is restricted to 50.6 mg per day, for a certain patient. If the ratio of the dosage of A to the dosage of B is 5 to 8, find the dosage for each administration of each medication.

43. ***Pricing*** A concert promoter needs to take in $380,000 on the sale of 16,000 tickets. If the promoter charges $20 for some tickets and $30 for others, how many of each type must be sold to yield the $380,000?

44. ***Pricing*** A nut wholesaler sells a mix of peanuts and cashews. He charges $2.80 per pound for peanuts and $5.30 per pound for cashews. If the mix is to sell for $3.30 per pound, how many pounds each of peanuts and cashews should be used to make 100 pounds of the mix?

45. ***Nutrient solutions*** How many cubic centimeters of a 20% solution of a nutrient must be added to 100 cc of a 2% solution of the same nutrient to make a 10% solution of the nutrient?

46. ***Mixtures*** How many gallons of washer fluid that is 13.5% antifreeze must a manufacturer add to 200 gallons of washer fluid that is 11% antifreeze to yield washer fluid that is 13% antifreeze?

Application Problems 47–50 require systems of equations in three variables.

47. ***Nutrition*** Each ounce of substance A supplies 5% of the nutrition a patient needs. Substance B supplies 15% of the required nutrition per ounce, and substance C supplies 12% of required nutrition per ounce. If digestive restrictions require that substances A and C be given in equal amounts, and the amount of substance B is one-fifth of either of these other amounts, find the number of ounces of each substance that should be in the meal to provide 100% of the required nutrition.

48. ***Dietary requirements*** A glass of skim milk supplies 0.1 mg of iron, 8.5 g of protein, and 1 g of carbohydrates. A quarter pound of lean red meat provides 3.4 mg of iron, 22 g of protein, and 20 g of carbohydrates. Two slices of whole grain bread supply 2.2 mg of iron, 10 g of protein, and 12 g of carbohydrates. If a person on a special diet must have 10.5 mg of iron, 94.5 g of protein, and 61 g of carbohydrates, how many glasses of skim milk, how many quarter-pound servings of meat, and how many two-slice servings of whole grain bread will supply this?

49. ***Social services*** A social agency is charged with providing services to three types of clients, A, B, and C. A total of 500 clients are to be served, with $150,000 available for counseling and $100,000 available for emergency food and shelter. Type A clients require an average of $200 for counseling and $300 for emergencies, Type B clients require an average of $500 for counseling and $200 for emergencies, and Type C clients require an average of $300 for counseling and $100 for emergencies. How many of each type client can be served?

50. ***Social services*** If funding for counseling is cut to $135,000 and funding for emergency food and shelter is cut to $90,000, only 450 clients can be served. How many of each type can be served in this case? (See Problem 49.)

1.6 Applications of Functions in Business and Economics

OBJECTIVES

- *To formulate and evaluate total cost, total revenue, and profit functions*
- *To find marginal cost, revenue, and profit, given linear total cost, total revenue, and profit functions*
- *To write the equations of linear total cost, total revenue, and profit functions by using information given about the functions*
- *To find break-even points*
- *To evaluate and graph supply and demand functions*
- *To find market equilibrium*

APPLICATION PREVIEW

We will discuss numerous applications of mathematics in this text, but in this section we will introduce two important applications, which will be expanded and used in different circumstances throughout the text as increased mathematical skills permit. These applications are total cost, total revenue, and profit as functions of quantity sold (theory of the firm) and supply and demand as functions of price (market analysis).

Total Cost, Total Revenue, and Profit

The **profit** a firm makes on its product is the difference between the amount it receives from sales (its revenue) and its cost. If x units are produced and sold, we can write

$$P(x) = R(x) - C(x)$$

where

$$P(x) = \text{profit from sale of } x \text{ units}$$
$$R(x) = \text{total revenue from sale of } x \text{ units}$$
$$C(x) = \text{total cost of production and sale of } x \text{ units*}$$

In general, **revenue** is found by using the equation

$$\text{Revenue} = (\text{price per unit})(\text{number of units})$$

The **cost** is composed of two parts, fixed costs and variable costs. **Fixed costs** (FC), such as depreciation, rent, utilities, and so on, remain constant regardless of the number of units produced. **Variable costs** (VC) are those directly related to the number of units produced. Thus the cost is found by using the equation

$$\text{Cost} = \text{variable costs} + \text{fixed costs}$$

EXAMPLE 1

Suppose that a firm manufactures radios and sells them for \$50 each. The costs incurred in the production and sale of the radios are \$200,000 plus \$10 for each radio produced and sold. Write the profit function for the production and sale of x radios.

Solution

The profit function is given by $P(x) = R(x) - C(x)$, where $R(x) = 50x$ and $C(x) = 10x + 200{,}000$. Thus,

$$P(x) = 50x - (10x + 200{,}000)$$
$$P(x) = 40x - 200{,}000$$

Figure 1.36 shows the graphs of $R(x)$, $C(x)$, and $P(x)$.

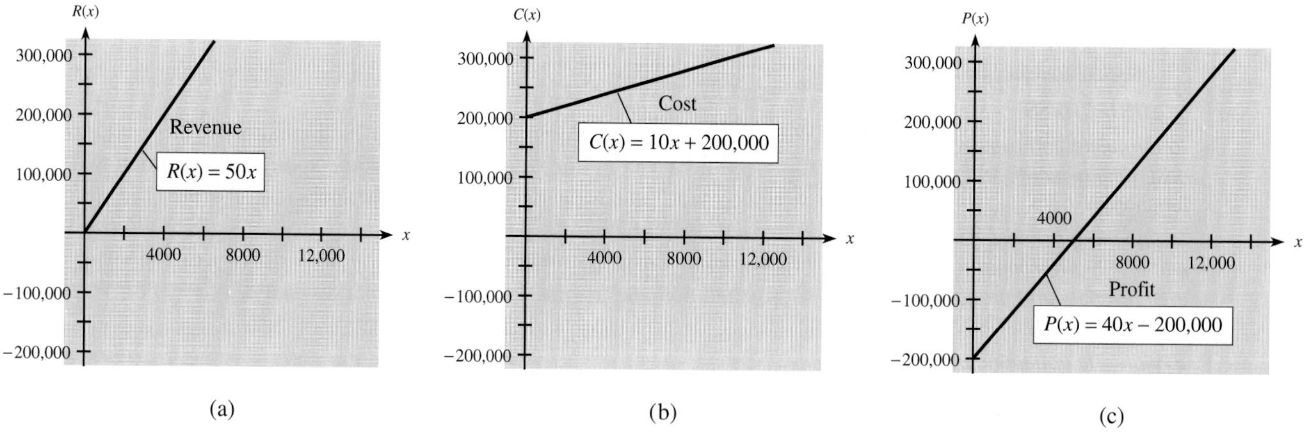

Figure 1.36

*The symbols generally used in economics for total cost, total revenue, and profit are TC, TR, and π, respectively. In order to avoid confusion, especially with the use of π as a variable, we do not use these symbols.

By observing the intercepts on the graphs in Figure 1.36, we note the following.

Revenue:	0 units produce 0 revenue.
Cost:	0 units' costs equal fixed costs = \$200,000.
Profit:	0 units yield a loss equal to fixed costs = \$200,000. 5000 units result in a profit of \$0 (no loss or gain).

In Example 1, both the revenue function and the cost function are linear, so their difference, the profit function, is also linear. The slope of the profit function represents the rate of change in profit with respect to the number of units produced and sold. This is called the **marginal profit** ($\overline{MP}$) for the product. Thus the marginal profit for the radios of Example 1 is \$40. Similarly, the **marginal cost** ($\overline{MC}$) for this product is \$10 (the slope of the cost function), and the **marginal revenue** ($\overline{MR}$) is \$50 (the slope of the revenue function).

EXAMPLE 2

Suppose that the cost (in dollars) for a product is $C = 21.75x + 4890$. What is the marginal cost for this product, and what does it mean?

Solution

The equation has the form $C = mx + b$, so the slope is 21.75. Thus the marginal cost is $\overline{MC} = 21.75$.

Because the marginal cost is the slope of the cost line, production of each additional unit will cost \$21.75 more, at any level of production.

Note that when total cost functions are linear, the marginal cost is the same as the variable cost. This is not the case if the functions are not linear, as we shall see later.

CHECKPOINT

1. Suppose that when a company produces its product, fixed costs are \$12,500 and variable costs are \$75 per item.
   (a) Write the total cost function if x represents the number of units.
   (b) Are fixed costs equal to $C(0)$?
2. Suppose the company in Problem 1 sells its products for \$175 per item.
   (a) Write the total revenue function.
   (b) Find $R(100)$ and give its meaning.
3. (a) Give the formula for profit in terms of revenue and cost.
   (b) Find the profit function for the company in Problems 1 and 2.

EXAMPLE 3

Suppose the profit function for a product is linear, and the marginal profit is \$5. If the profit is \$200 when 125 units are sold, write the equation of the profit function.

Solution

The marginal profit gives us the slope of the line representing the profit function. Using this slope and the point (125, 200), we have the equation

$$P - 200 = 5(x - 125)$$

or

$$P = 5x - 425$$

Break-Even Analysis

We can solve the equations for total revenue and total cost simultaneously to find the point where cost and revenue are equal. This point is called the **break-even point.** On the graph of these functions, we use x to represent the quantity produced and y to represent the dollar value of revenue *and* cost. The point where the total revenue line crosses the total cost line is the break-even point.

EXAMPLE 4

A manufacturer sells a product for \$10 per unit. The manufacturer's fixed costs are \$1200 per month, and the variable costs are \$2.50 per unit. How many units must the manufacturer produce each month to break even?

Solution

The total revenue for x units of the product is $10x$, so the equation for total revenue is $R = 10x$. The fixed costs are \$1200, so the total cost for x units is $2.50x + 1200$. Thus the equation for total cost is $C = 2.50x + 1200$. We find the break-even point by solving the two equations simultaneously ($R = C$ at the break-even point).

By substitution,

$$\begin{aligned} 10x &= 2.50x + 1200 \\ 7.5x &= 1200 \\ x &= 160 \end{aligned}$$

Thus the manufacturer will break even if 160 units are produced per month. The manufacturer will make a profit if more than 160 units are produced. Figure 1.37 shows that for $x < 160$, $R(x) < C(x)$ (resulting in a loss) and that for $x > 160$, $R(x) > C(x)$ (resulting in a profit).

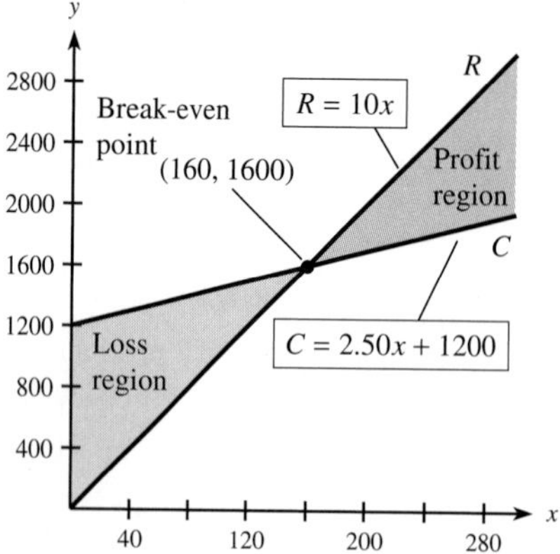

Figure 1.37

Using the fact that the profit function is found by subtracting the total cost function from the total revenue function, we can form the profit function for the previous example. The profit function is given by

$$P(x) = 10x - (2.50x + 1200) \quad \text{or} \quad P(x) = 7.50x - 1200$$

We can find the point where the profit is zero (the break-even point) by setting $P(x) = 0$ and solving for x.

$$0 = 7.50x - 1200$$
$$1200 = 7.50x$$
$$x = 160$$

Note that this is the same break-even point that we found by solving the total revenue and total cost equations simultaneously (see Figure 1.38).

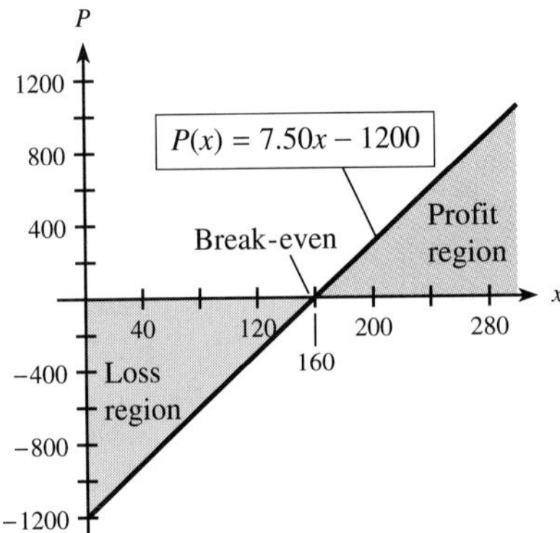

Figure 1.38

CHECKPOINT

4. Identify two ways in which break-even points can be found.

Supply, Demand, and Market Equilibrium

Economists and managers also use points of intersection to determine market equilibrium. **Market equilibrium** occurs when the quantity of commodity demanded is equal to the quantity supplied.

Demand by consumers for a commodity is related to the price of the commodity. The **law of demand** states that the quantity demanded will increase as price decreases, or that the quantity demanded will decrease as price increases. Figure 1.39 shows the graph of a typical linear demand function. Note that although quantity demanded is a function of price, economists have traditionally graphed the demand function with price on the vertical axis. Throughout this text, we will follow this tradition. Linear equations relating price p and quantity demanded q can be solved for either p or q, and we will have occasion to use the equations in both forms.

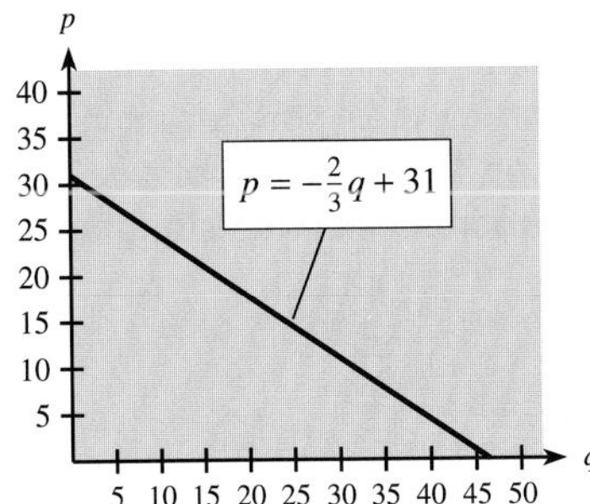

Figure 1.39

Just as a consumer's willingness to buy is related to price, a manufacturer's willingness to supply goods is also related to price. The **law of supply** states that the quantity supplied for sale will increase as the price of a product increases. Figure 1.40 on the next page shows the graph of a typical linear supply function.

As with demand, price is placed on the vertical axis. Note that negative prices and quantities have no meaning, so supply and demand curves are restricted to the first quadrant.

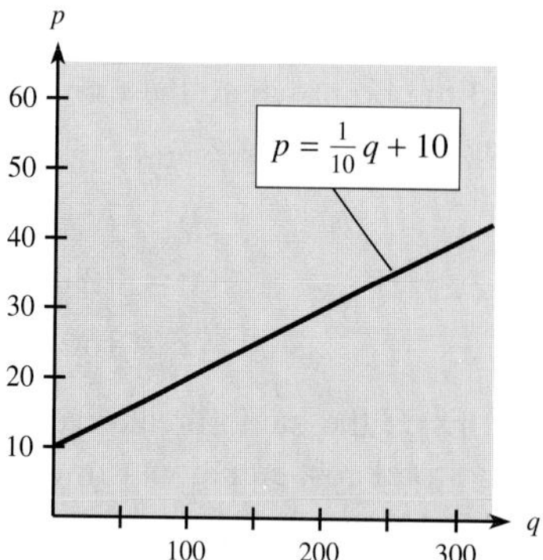

Figure 1.40

If the supply and demand curves for a commodity are graphed on the same coordinate system, with the same units, market equilibrium occurs at the point where the curves intersect. The price at that point is the **equilibrium price,** and the quantity at that point is the **equilibrium quantity.**

For the supply and demand functions shown in Figure 1.41, we see that the curves intersect at the point (30, 11). This means that when the price is $11, consumers are willing to purchase the same number of units (30) that producers are willing to supply.

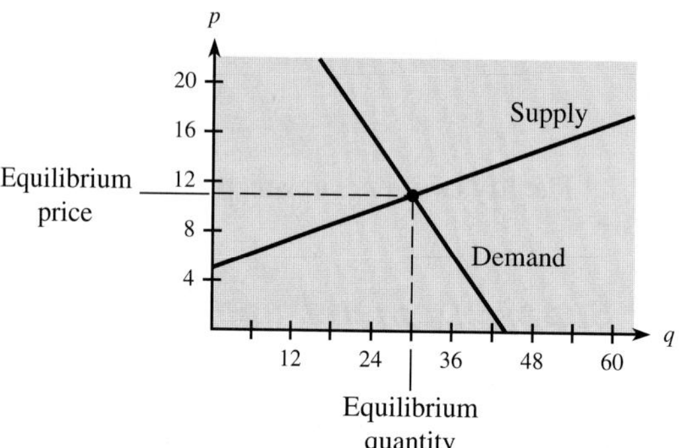

Figure 1.41

In general, the equilibrium price and the equilibrium quantity must both be positive for the market equilibrium to have meaning.

We can find the market equilibrium by graphing the supply and demand functions on the same coordinate system and observing their point of intersection. As we have seen, finding the point(s) common to the graphs of two (or more) functions is called **solving a system of equations** or **solving simultaneously.**

EXAMPLE 5

Find the market equilibrium point for the following supply and demand functions.

$$\text{Demand: } p = -3q + 36$$
$$\text{Supply: } p = 4q + 1$$

Solution

At market equilibrium, the demand price equals the supply price. Thus,

$$\begin{aligned} -3q + 36 &= 4q + 1 \\ 35 &= 7q \\ q &= 5 \\ p &= 21 \end{aligned}$$

The equilibrium point is (5, 21).

EXAMPLE 6

A group of wholesalers will buy 50 dryers per month if the price is \$200 and 30 per month if the price is \$300. The manufacturer is willing to supply 20 if the price is \$210 and 30 if the price is \$230. Assuming that the resulting supply and demand functions are linear, find the equilibrium point for the market.

Solution

Representing price by p and quantity by q, we have
Demand function:

$$\begin{aligned} m &= \frac{300 - 200}{30 - 50} = -5 \\ p - 200 &= -5(q - 50) \\ p &= -5q + 450 \end{aligned}$$

Supply function:

$$\begin{aligned} m &= \frac{230 - 210}{30 - 20} = 2 \\ p - 230 &= 2(q - 30) \\ p &= 2q + 170 \end{aligned}$$

Because the prices are equal at market equilibrium, we have

$$\begin{aligned} -5q + 450 &= 2q + 170 \\ 280 &= 7q \\ q &= 40 \\ p &= 250 \end{aligned}$$

The equilibrium point is (40, 250). See Figure 1.42 for the graphs of these functions.

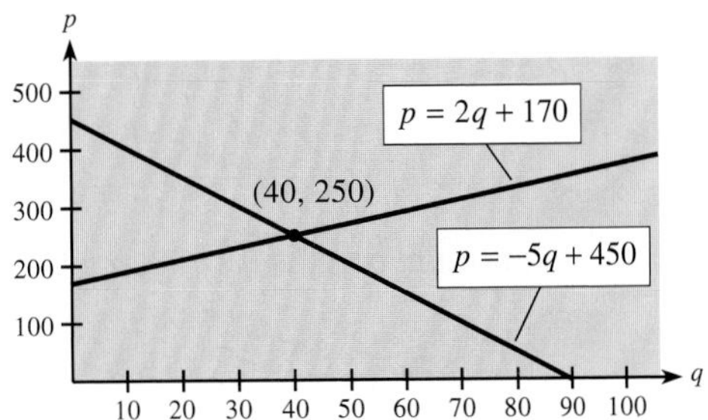

Figure 1.42

If a supplier is taxed \$$K$ per unit sold, then the tax is passed on to the consumer by adding \$$K$ to the selling price of the product. If the original supply function is $p = f(q)$, then passing the tax on gives a new supply function, $p = f(q) + K$. Because the value of the product is not changed by the tax, the demand function is unchanged. Figure 1.43 shows the effect that this has on market equilibrium.

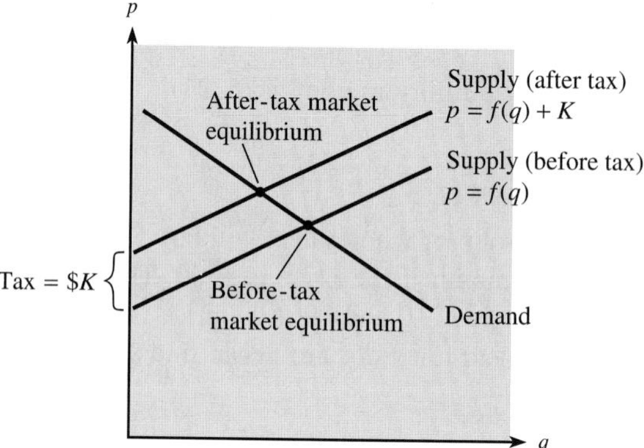

Figure 1.43

Note that the new market equilibrium point is the point of intersection of the original demand function and the new (after taxes) supply function.

EXAMPLE 7

In Example 6 the supply and demand functions for dryers were given as follows.

$$\begin{aligned} \text{Supply:} \quad & p = 2q + 170 \\ \text{Demand:} \quad & p = -5q + 450 \end{aligned}$$

The equilibrium point was $q = 40$, $p = \$250$. If the wholesaler is taxed \$14 per unit sold, what is the new equilibrium point?

Solution

The \$14 tax per unit is passed on by the wholesaler, so the new supply function is

$$p = 2q + 170 + 14$$

and the demand function is unchanged. Thus we solve the system

$$\begin{cases} p = 2q + 184 \\ p = -5q + 450 \end{cases}$$

$$\begin{aligned} 2q + 184 &= -5q + 450 \\ 7q &= 266 \\ q &= 38 \\ p &= 2(38) + 184 = 260 \end{aligned}$$

The new equilibrium point is $q = 38$, $p = \$260$.

CHECKPOINT

5. (a) Does a typical linear demand function have positive slope or negative slope? Why?
   (b) Does a typical linear supply function have positive slope or negative slope? Why?
6. (a) What do we call the point of intersection of a supply function and a demand function?
   (b) What algebraic technique is used to find the point named in (a)?

CHECKPOINT SOLUTIONS

1. (a) $C(x) = 75x + 12{,}500$
   (b) $C(0) = 12{,}500 =$ Fixed costs. In fact, fixed costs are defined to be $C(0)$.
2. (a) $R(x) = 175x$
   (b) $R(100) = 175(100) = \$17{,}500$, which means that revenue is \$17,500 when 100 units are sold.
3. (a) Profit = Revenue − Cost or $P(x) = R(x) - C(x)$
   (b) $P(x) = 175x - (75x + 12{,}500)$
   $= 175x - 75x - 12{,}500 = 100x - 12{,}500$
4. The break-even point occurs where revenue equals cost $[R(x) = C(x)]$ or where profit is zero $[P(x) = 0]$.
5. (a) Negative slope, because demand falls as price increases.
   (b) Positive slope, because supply increases as price increases.
6. (a) Market equilibrium (b) Solving simultaneously

EXERCISE 1.6

Total Cost, Total Revenue, and Profit Functions

1. For a certain period, a calculator manufacturer has fixed costs of \$3400 and variable costs of \$17 for each item produced.
   (a) Write the total cost function.
   (b) What will it cost to produce 200 units during the given period?
2. A computer manufacturer has a fixed cost of \$3300 and a variable cost of \$85 for each item produced.
   (a) Write the equation that represents total cost.
   (b) What is the cost if no units are produced? What is this called?
   (c) What is the cost if 300 units are produced?
3. A calculator is sold for \$34 per unit.
   (a) Write the total revenue function as an equation.
   (b) What will the revenue be if 300 units are sold?
4. A stereo receiver is sold for \$215 per item.
   (a) Write the equation that represents the revenue function.
   (b) What will the revenue be from the sale of 50 items?
5. Suppose a calculator manufacturer has the total cost function $C(x) = 17x + 3400$ and the total revenue function $R(x) = 34x$.
   (a) What is the equation of the profit function for the calculator?
   (b) What is the profit on 300 units?
6. Suppose a stereo receiver manufacturer has the total cost function $C(x) = 105x + 1650$ and the total revenue function $R(x) = 215x$.
   (a) What is the equation of the profit function for this commodity?
   (b) What is the profit on 50 items?

7. Suppose a radio manufacturer has the total cost function $C(x) = 43x + 1850$ and the total revenue function $R(x) = 80x$.
   (a) What is the equation of the profit function for this commodity?
   (b) What is the profit on 30 units? Interpret your result.
   (c) How many radios must be sold to avoid losing money?
8. Suppose a computer manufacturer has the total cost function $C(x) = 85x + 3300$ and the total revenue function $R(x) = 385x$.
   (a) What is the equation of the profit function for this commodity?
   (b) What is the profit on 351 items?
   (c) How many items must be sold to avoid losing money?
9. ***Cost*** A linear cost function is $C(x) = 5x + 250$.
   (a) What are the slope and the C-intercept?
   (b) What is the marginal cost, and what does it mean?
   (c) How are your answers to (a) and to (b) related?
   (d) What is the cost of producing *one more* item if 50 are currently being produced? What is it if 100 are currently being produced?
10. ***Cost*** A linear cost function is $C(x) = 21.75x + 4890$.
   (a) What are the slope and the C-intercept?
   (b) What is the marginal cost, and what does it mean?
   (c) How are your answers to (a) and to (b) related?
   (d) What is the cost of producing *one more* item if 50 are currently being produced? What is it if 100 are currently being produced?
11. ***Revenue*** A linear revenue function is $R = 27x$.
   (a) What is the slope?
   (b) What is the marginal revenue, and what does it mean?
   (c) What is the revenue received from selling *one more* item if 50 are currently being sold? If 100 are being sold?
12. ***Revenue*** A linear revenue function is $R = 38.95x$.
   (a) What is the slope?
   (b) What is the marginal revenue, and what does it mean?
   (c) What is the revenue received from selling *one more* item if 50 are currently being sold? If 100 are being sold?
13. ***Profit*** Let $C(x) = 5x + 250$ and $R(x) = 27x$.
   (a) Write the profit function $P(x)$.
   (b) What is the slope of the profit function?
   (c) What is the marginal profit?
   (d) Interpret the marginal profit.
14. ***Profit*** Given $C(x) = 21.95x + 1400$ and $R(x) = 20x$, find the profit function.
   (a) What is marginal profit, and what does it mean?
   (b) What should a firm with these cost, revenue, and profit functions do? (*Hint:* Graph the profit function and see where it goes.)
15. ***Profit*** A company charting its profits notices that the relationship between the number of units sold, x, and the profit, P, is linear. If 200 units sold results in \$3100 profit and 250 units sold results in \$6000 profit, write the profit function for this company. Find the marginal profit.
16. ***Cost*** Suppose that the total cost function for a radio is linear, that the marginal cost is \$27, and that the total cost for 50 radios is \$4350. Write the equation of this cost function and then graph it.

Break-Even Analysis

17. The figure shows graphs of the total cost function and the total revenue function for a commodity.

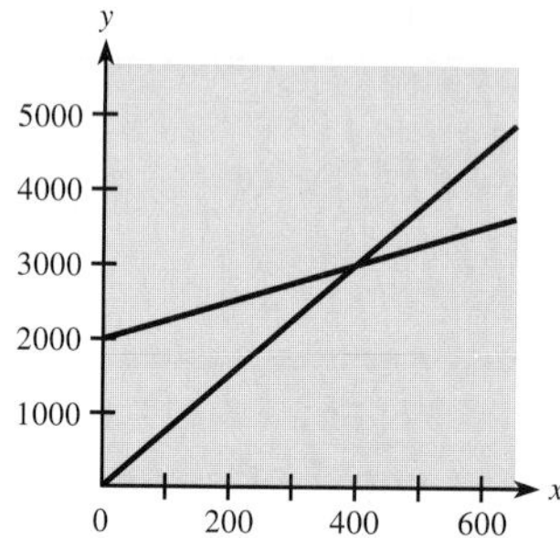

   (a) Label each function correctly.
   (b) Determine the fixed costs.
   (c) Locate the break-even point and determine the number of units sold to break even.
   (d) Estimate the marginal cost and marginal revenue.
18. A manufacturer of shower-surrounds has a revenue function of

$$R(x) = 81.50x$$

and a cost function of

$$C(x) = 63x + 1850$$

Find the number of units that must be sold to break even.

19. A jewelry maker incurs costs for a necklace according to

$$C(x) = 35x + 1650$$

If the revenue function for the necklaces is

$$R(x) = 85x$$

how many necklaces must be sold to break even?

20. A small business recaps and sells tires. If a set of four tires has the revenue function

$$R(x) = 89x$$

and the cost function

$$C(x) = 1400 + 75x$$

find the number of sets of recaps that must be sold to break even.

21. A manufacturer sells belts for \$12 per unit. The fixed costs are \$1600 per month, and the variable costs are \$8 per unit.
(a) Write the equations of the revenue and cost functions.
(b) Find the break-even point.

22. A manufacturer sells watches for \$50 per unit. The fixed costs related to this product are \$10,000 per month, and the variable costs are \$30 per unit.
(a) Write the equations of the revenue and cost functions.
(b) How many watches must be sold to break even?

23. (a) Write the profit function for Problem 21.
(b) Set profit equal to zero and solve for x. Compare this x-value with the break-even point from 21(b).

24. (a) Write the profit function for Problem 22.
(b) Set profit equal to zero and solve for x. Compare this x-value with the break-even point from 22(b).

25. A company manufactures and sells bookcases. The selling price is \$54.90 per bookcase. The total cost function is linear, and costs amount to \$50,000 for 2000 bookcases and \$32,120 for 800 bookcases.
(a) Write the equation for revenue.
(b) Write the equation for total costs.
(c) Find the break-even point.

26. A company distributes college logo sweatshirts and sells them for \$50 each. The total cost function is linear, and the total cost for 100 sweatshirts is \$4360, whereas the total cost for 250 sweatshirts is \$7060.
(a) Write the equation for the revenue function.
(b) Write the equation for the total cost function.
(c) Find the break-even point.

Supply and Demand and Market Equilibrium

27. As the price of a commodity increases, what happens to demand?

28. As the price of a commodity increases, what happens to supply?

Figure 1.44 is the graph of the demand function for a product, and Figure 1.45 is the graph of the supply function for the same product. Use these graphs to answer the questions in Problems 29 and 30.

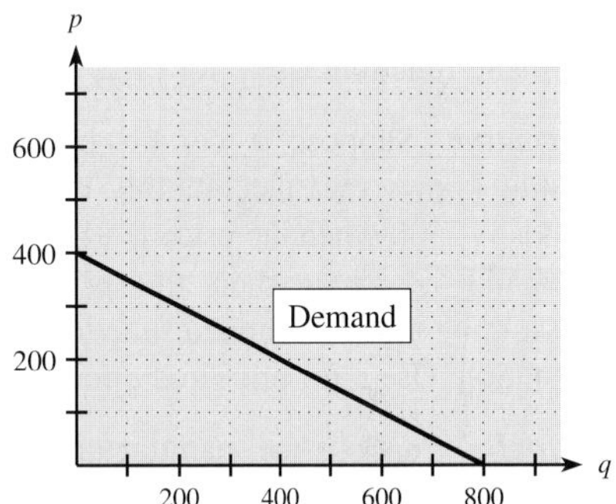

Figure 1.44

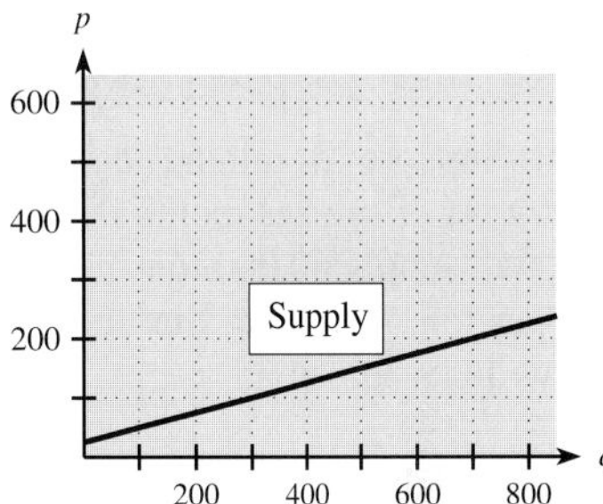

Figure 1.45

29. (a) How many units q are *demanded* when the price p is \$100?
(b) How many units q are *supplied* when the price p is \$100?
(c) Will there be a market surplus (more supplied) or shortage (more demanded) when $p = \$100$?

30. (a) How many units q are demanded when the price p is \$200?
(b) How many units q are supplied when the price p is \$200?
(c) Will there be a market surplus or shortage when the price p is \$200?

31. If the demand for a pair of shoes is given by $2p + 5q = 200$ and the supply function for it is $p - 2q = 10$, compare the quantity demanded and supplied when the price is \$60. Will there be a surplus or shortfall at this price?

32. If the demand function and supply function for Z-brand phones are $p + 2q = 100$ and $35p - 20q = 350$, respectively, compare the quantity demanded and the quantity supplied when $p = 14$. Are there surplus phones or not enough to meet demand?

33. ***Demand*** Suppose a certain outlet chain selling appliances has found that for one brand of stereo system, the monthly demand is 240 when the price is \$900. However, when the price is \$850, the monthly demand is 315. Assuming that the demand function for this system is linear, write the equation for the demand function. Use p for price and q for quantity.

34. ***Demand*** Suppose a certain home improvements outlet knows that the monthly demand for framing studs is 2500 when the price is \$1.00 each but that the demand is 3500 when the price is \$0.90 each. Assuming that the demand function is linear, write its equation. Use p for price and q for quantity.

35. ***Supply*** Suppose the manufacturer of a board game will supply 10,000 games if the wholesale price is \$1.50 each but will supply only 5000 if the price is \$1.00 each. Assuming that the supply function is linear, write its equation. Use p for price and q for quantity.

36. ***Supply*** Suppose a mining company will supply 100,000 tons of ore per month if the price is \$30 per ton but will supply 80,000 tons per month if the price is \$25 per ton. Assuming that the supply function is linear, write its equation.

Complete Problems 37–41 by using the accompanying figure, which shows a supply function and a demand function.

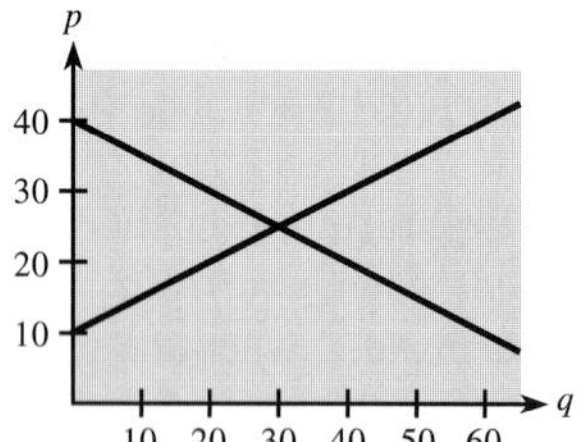

37. (a) Label each function as "demand" or "supply."
    (b) Label the equilibrium point and determine the price and quantity at which market equilibrium occurs.
38. (a) If the price is \$30, what quantity is demanded?
    (b) If the price is \$30, what quantity is supplied?
    (c) Is there a surplus or shortage when the price is \$30? How many units is this surplus or shortage?
39. (a) If the price is \$20, what quantity is supplied?
    (b) If the price is \$20, what quantity is demanded?
    (c) Is there a surplus or a shortage when the price is \$20? How many units is this surplus or shortage?
40. Will a price above the equilibrium price result in a market surplus or shortage?
41. Will a price below the equilibrium price result in a market surplus or shortage?
42. Find the market equilibrium point for these demand and supply functions.

    Demand: $p = -2q + 320$
    Supply: $p = 8q + 2$

43. Find the market equilibrium point for these demand and supply functions.

    Demand: $2p = -q + 56$
    Supply: $3p - q = 34$

44. Find the equilibrium point for the following supply and demand functions.

    Demand: $p = 480 - 3q$
    Supply: $p = 17q + 80$

45. Find the equilibrium point for the following supply and demand functions.

    Demand: $p = -4q + 220$
    Supply: $p = 15q + 30$

46. Retailers will buy 45 cordless phones from a wholesaler if the price is \$10 each and will buy 20 if the price is \$60. The wholesaler will supply 35 phones at \$30 each and 70 at \$50 each. Assuming the supply and demand functions are linear, find the market equilibrium point.
47. A group of retailers will buy 80 televisions from a wholesaler if the price is \$350 and 120 if the price is \$300. The wholesaler is willing to supply 60 if the price is \$280 and 140 if the price is \$370. Assuming the resulting supply and demand functions are linear, find the equilibrium point for the market.
48. A shoe store owner will buy 10 pairs of a certain shoe if the price is \$75 per pair, and 30 pairs if the price is \$25. The supplier of the shoes is willing to provide 35 pairs if the price is \$80 per pair, and 5 pairs if the price is \$20. Assuming the supply and demand functions for the shoes are linear, find the market equilibrium point.

Problems 49–56 involve market equilibrium after taxation. Use the figure to answer Problems 49 and 50.

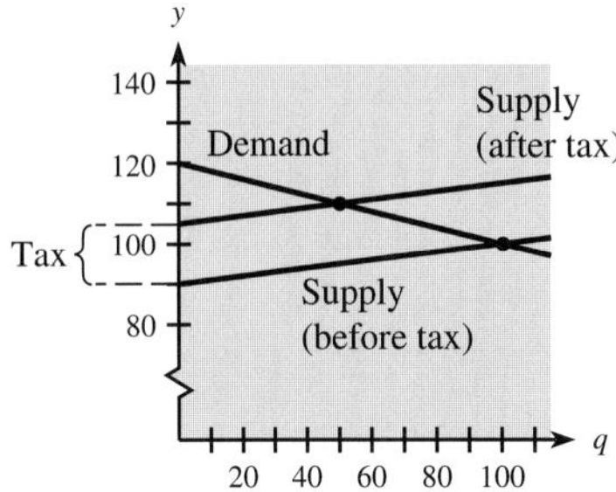

49. (a) What is the amount of the tax?
    (b) What is the original equilibrium price and quantity?
    (c) What is the new equilibrium price and quantity?
    (d) Does the supplier suffer from the tax even though it is passed on?
50. (a) If the tax is doubled, how many units will be sold?
    (b) Can a government lose money by increasing taxes?
51. If a \$38 tax is placed on each unit of the product of Problem 45, what are the new equilibrium price and quantity?
52. If a \$56 tax is placed on each unit of the product of Problem 44, what is the new equilibrium point?
53. Suppose that a certain product has the following demand and supply functions.

    Demand: $p = -0.05q + 65$
    Supply: $p = 0.05q + 10$

    If a \$5 tax per item is levied on the supplier and this tax is passed on to the consumer, find the market equilibrium point after the tax.
54. Suppose that a certain product has the following demand and supply functions.

    Demand: $p = -8q + 2800$
    Supply: $p = 3q + 35$

    If a \$15 tax per item is levied on the supplier, who passes it on to the consumer as a price increase, find the market equilibrium point after the tax.
55. Suppose that in a certain market the demand function for a product is given by $60p + q = 2100$ and the supply function is given by $120p - q = 540$. Then a tax of \$0.50 per item is levied on the supplier, who passes it on to the consumer as a price increase. Find the equilibrium price and quantity after the tax is levied.
56. Suppose that in a certain market the demand function for a product is given by $10p + q = 2300$ and the supply function is given by $45p - q = 360$. If the government levies a tax of \$2 per item on the supplier, who passes the tax on to the consumer as a price increase, find the equilibrium price and quantity after the tax is levied.

KEY TERMS AND FORMULAS

Section	Key Terms	Formula
1.1	Equation, members; variable; solution Identities; conditional equations Properties of equality Linear equation in one variable	
1.2	Relation Function Vertical-line test Domain, range Coordinate system Ordered pair, origin, x-axis, y-axis Graph Functional notation	

Section	Key Terms	Formula
1.3	Linear function	$y = ax + b$
	Intercepts	x-intercept: where $y = 0$
		y-intercept: where $x = 0$
	Slope of a line	$m = \dfrac{y_2 - y_1}{x_2 - x_1}$
	Parallel lines	$m_1 = m_2$
	Perpendicular lines	$m_2 = -1/m_1$
	Point-slope form	$y - y_1 = m(x - x_1)$
	Slope-intercept form	$y = mx + b$
	Vertical line	$x = a$
	Horizontal line	$y = b$
1.4	Graphing utilities	
	Standard viewing window	
	Range	
	Evaluating functions	
	x-intercept	
	Zeros of functions	
	Solutions of equations	
1.5	Systems of linear equations	
	Solutions	
	Equivalent systems	
	Substitution method	
	Elimination method	
	Left-to-right elimination method	
	Lead variable	
	Back substitution	
1.6	Cost and revenue functions	$C(x)$ and $R(x)$
	Profit functions	$P(x) = R(x) - C(x)$
	Marginal profit	Slope of linear profit function
	Marginal cost	Slope of linear cost function
	Marginal revenue	Slope of linear revenue function
	Break-even point	$C(x) = R(x)$ or $P(x) = 0$
	Supply and demand functions	
	Market equilibrium	

REVIEW EXERCISES

Section 1.1

Solve the equations in Problems 1–9.

1. $x + 7 = 14$
2. $3x - 8 = 23$
3. $2x - 8 = 3x + 5$
4. $\frac{6x + 3}{6} = \frac{5(x - 2)}{9}$
5. $2x + \frac{1}{2} = \frac{x}{2} + \frac{1}{3}$
6. $0.6x + 4 = x - 0.02$
7. $\frac{6}{3x - 5} = \frac{6}{2x + 3}$
8. Solve for y. $3(y - 2) = -2(x + 5)$
9. $\frac{2x + 5}{x + 7} = \frac{1}{3} + \frac{x - 11}{2x + 14}$

Section 1.2

10. If $p = 3q^3$, is p a function of q?
11. If $y^2 = 9x$, is y a function of x?
12. If $R = \sqrt[3]{x + 4}$, is R a function of x?
13. What are the domain and range of the function $y = \sqrt{9 - x}$?
14. If $f(x) = x^2 + 4x + 5$, find the following.
    (a) $f(-3)$ (b) $f(4)$ (c) $f\left(\frac{1}{2}\right)$
15. If $g(x) = x^2 + 1/x$, find the following.
    (a) $g(-1)$ (b) $g\left(\frac{1}{2}\right)$ (c) $g(0.1)$
16. If $f(x) = 9x - x^2$, find $\frac{f(x + h) - f(x)}{h}$.
17. Does the graph in Figure 1.46 represent y as a function of x?

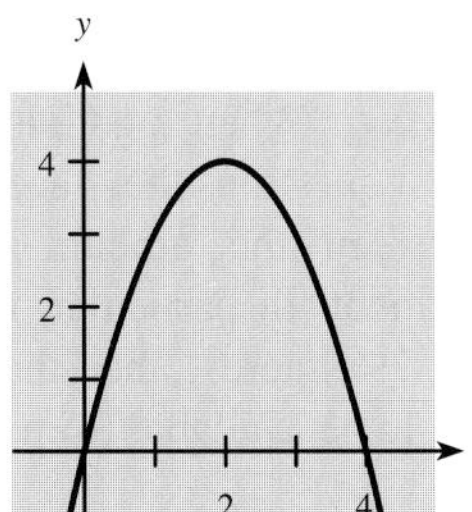

Figure 1.46

18. Does the graph in Figure 1.47 represent y as a function of x?

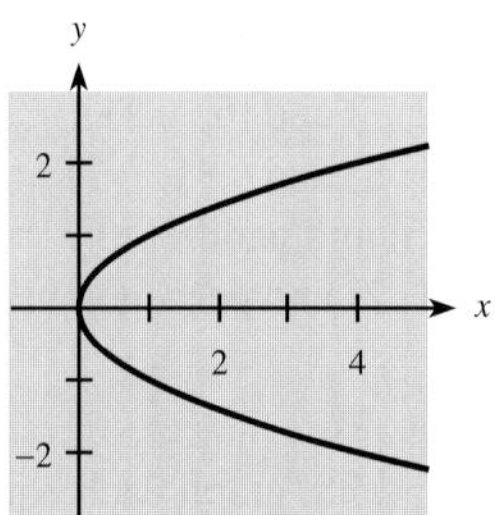

Figure 1.47

19. For the function f graphed in Figure 1.46, what is $f(2)$?
20. For the function f graphed in Figure 1.46, for what values of x does $f(x) = 0$?
21. The following table defines y as a function of x, denoted $y = f(x)$.

x	-2	-1	0	1	3	4
y	8	2	-3	4	2	7

Use the table to complete the following.
    (a) Find $f(4)$.
    (b) Find all x-values for which $f(x) = 2$.
    (c) Graph $y = f(x)$.
22. If $f(x) = 3x + 5$ and $g(x) = x^2$, find
    (a) $(f + g)(x)$ (b) $(f/g)(x)$
    (c) $f(g(x))$ (d) $(f \circ f)(x)$

Section 1.3

In Problems 23–25, find the intercepts and graph.

23. $5x + 2y = 10$
24. $6x + 5y = 9$
25. $x = -2$

In Problems 26 and 27, find the slope of the line that passes through each pair of points.

26. $(2, -1)$ and $(-1, -4)$
27. $(-3.8, -7.16)$ and $(-3.8, 1.16)$

In Problems 28 and 29 find the slope and y-intercept of each line.

28. $2x + 5y = 10$
29. $x = -\frac{3}{4}y + \frac{3}{2}$

In Problems 30–36, write the equation of each line described.

30. Slope 4 and y-intercept 2
31. Slope $-\frac{1}{2}$ and y-intercept 3
32. Through $(-2, 1)$ with slope $\frac{2}{5}$
33. Through $(-2, 7)$ and $(6, -4)$
34. Through $(-1, 8)$ and $(-1, -1)$
35. Through $(1, 6)$ and parallel to $y = 4x - 6$
36. Through $(-1, 2)$ and perpendicular to $3x + 4y = 12$

Section 1.4

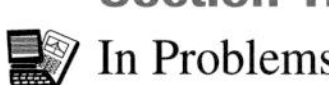

In Problems 37 and 38, graph each equation with a graphing utility and the standard viewing window.

37. $x^2 + y - 2x - 3 = 0$ 38. $y = \dfrac{x^3 - 27x + 54}{15}$

In Problems 39 and 40, use a graphing utility and
(a) graph each equation in the viewing window given.
(b) graph each equation in the standard viewing window.
(c) explain how the two views differ and why.

39. $y = (x + 6)(x - 3)(x - 15)$ with x-min $= -15$, x-max $= 25$; y-min $= -700$, y-max $= 500$
40. $y = x^2 - x - 42$ with x-min $= -15$, x-max $= 15$; y-min $= -50$, y-max $= 50$

41. What is the domain of $y = \dfrac{\sqrt{x+3}}{x}$? Check with a graphing utility.

Section 1.5

In Problems 42–48, solve each system of equations.

42. $\begin{cases} 4x - 2y = 6 \\ 3x + 3y = 9 \end{cases}$ 43. $\begin{cases} 2x + y = 19 \\ x - 2y = 12 \end{cases}$

44. $\begin{cases} 3x + 2y = 5 \\ 2x - 3y = 12 \end{cases}$ 45. $\begin{cases} 6x + 3y = 1 \\ y = -2x + 1 \end{cases}$

46. $\begin{cases} 4x - 3y = 253 \\ 13x + 2y = -12 \end{cases}$

47. $\begin{cases} x + 2y + 3z = 5 \\ y + 11z = 21 \\ 5y + 9z = 13 \end{cases}$ 48. $\begin{cases} x + y - z = 12 \\ 2y - 3z = -7 \\ 3x + 3y - 7z = 0 \end{cases}$

Applications

Section 1.1

49. ***Course grades*** In a certain course, grades are based on three tests worth 100 points each, three quizzes worth 50 points each, and a final exam worth 200 points. A student has test grades of 91, 82, and 88, and quiz grades of 50, 42, and 42. What is the lowest percent the student can get on the final and still earn an A (90% or more of the total points) in the course?

50. ***Cost analysis*** The owner of a small construction business needs a new truck. He can buy a diesel truck for \$18,000 and it will cost him \$0.16 per mile to operate. He can buy a gas engine truck for \$16,000 and it will cost him \$0.21 per mile to operate. Find the number of miles he must drive before the costs are equal. If he normally keeps a truck for 5 years, which is the better buy?

Section 1.2

51. ***Heart disease risk*** The Multiple Risk Factor Intervention Trial (MRFIT) used data from 356,222 men aged 35 to 57 to investigate the relationship between serum cholesterol and coronary heart disease (CHD) risk. Figure 1.48 shows the graph of the relationship of CHD risk and cholesterol, where a risk of 1 is assigned to 200 mg/dl of serum cholesterol and where the CHD risk is 4 times as high when serum cholesterol is 300 mg/dl.

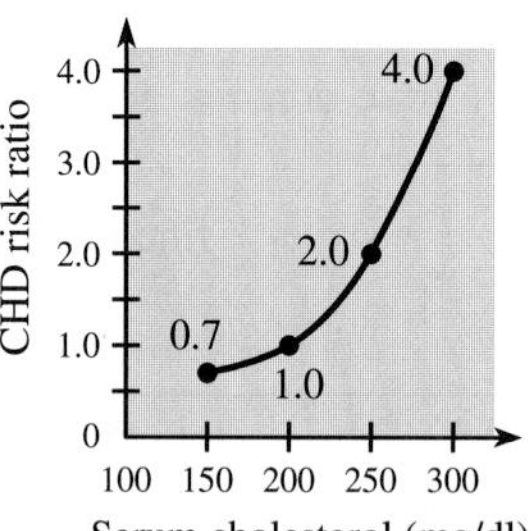

Figure 1.48

(a) Does this graph indicate that the CHD risk is a function of the serum cholesterol?
(b) Is the relationship a linear function?
(c) If CHD risk is a function f of serum cholesterol, what is $f(300)$?

52. ***Mortgage loans*** When a couple purchases a home, one of the first questions they face deals with the relationship between the amount borrowed and the monthly payment. In particular, if a bank offers 25-year loans at 7% interest, then the data in the following table would be available from the bank.

Amount Borrowed	*Monthly Payment*
\$40,000	\$282.72
50,000	353.39
60,000	424.07
70,000	494.75
80,000	565.44
90,000	636.11
100,000	706.78

(Source: *Comprehensive Mortgage Payment Tables,* Publication No. 492, Financial Publishing Co., Boston)

Assume that the monthly payment P is a function of the amount borrowed A (in thousands) and is denoted by $P = f(A)$, and answer the following.

(a) Find $f(80)$.

(b) Write a sentence that explains the meaning of $f(70) = 494.75$.

53. ***Profit*** Suppose that the profit from the production and sale of x units of a product is given by

$$P(x) = 180x - \frac{x^2}{100} - 200$$

In addition, suppose that for a certain month, the number of units produced on day t of the month is

$$x = q(t) = 1000 + 10t$$

(a) Find $(P \circ q)(t)$ to express the profit as a function of the day of the month.

(b) Find the number of units produced, and the profit, on the fifteenth day of the month.

54. ***Fish species growth*** For many species of fish, the weight W is a function of the length L that can be expressed by

$$W = W(L) = kL^3 \qquad k = \text{constant}$$

Suppose that for a particular species $k = 0.02$ and that for this species, the length (in cm) is a function of the number of years t the fish has been alive and that this function is given by

$$L = L(t) = 50 - \frac{(t-20)^2}{10} \qquad 0 \le t \le 20$$

Find $(W \circ L)(t)$ in order to express W as a function of the age t of the fish.

Section 1.3

55. ***Distance to a thunderstorm*** The distance d (in miles) to a thunderstorm is given by

$$d = \frac{t}{4.8}$$

where t is the number of seconds that elapse between seeing the lightning and hearing the thunder.

(a) Graph this function for $0 \le t \le 20$.

(b) The point (9.6, 2) satisfies the equation. Explain its meaning.

56. ***Body-heat loss*** Body-heat loss due to convection depends on a number of factors. If H_c is body-heat loss due to convection, A_c is the exposed surface area of the body, $T_s - T_a$ is skin temperature minus air temperature, and K_c is the convection coefficient (determined by air velocity and so on), then we have

$$H_c = K_c A_c (T_s - T_a)$$

When $K_c = 1$, $A_c = 1$, and $T_s = 90$, the equation is

$$H_c = 90 - T_a$$

Sketch the graph.

57. ***Profit*** A company charting its profits notices the relationship between the number of units sold, x, and the profit, P, is linear. If 200 units sold results in \$3100 profit and 250 units sold results in \$6000 profit, write the profit function for this company.

58. ***Fahrenheit-Celsius*** Write the equation of the linear relationship between temperature in Celsius (C) and Fahrenheit (F) if water freezes at 0°C and 32°F and boils at 100°C and 212°F.

Section 1.4

59. ***Photosynthesis*** The amount y of photosynthesis that takes place in a certain plant depends on the intensity x of the light present, according to

$$y = 120x^2 - 20x^3 \quad \text{for } x \ge 0$$

(a) Graph this function with a graphing utility. (Use y-min $= -100$ and y-max $= 700$.)

(b) The model is valid only when $f(x) \ge 0$ (that is, on or above the x-axis). For what x-values is this true?

60. ***Flow rates of water*** The speed at which water travels in a pipe can be measured by directing the flow through an elbow and measuring the height it spurts out the top. If the elbow height is 10 cm, the equation relating the height h (in centimeters) of the water above the elbow and its velocity v (in centimeters per second) is given by

$$v^2 = 1960(h + 10)$$

Solve this equation for h and graph the result, using the velocity as the independent variable.

Section 1.5

61. ***Investment mix*** A retired couple has \$150,000 to invest and wants to earn \$15,000 per year in interest. The safer investment yields 9.5%, but they can supplement their earnings by investing some of their money at 11%. How much should they invest at each rate to earn \$15,000 per year?

62. ***Botany*** A botanist has a 20% solution and a 70% solution of an insecticide. How much of each must be used to make 4.0 liters of a 35% solution?

Section 1.6

63. ***Supply and demand*** A certain product has supply and demand functions $p = 4q + 5$ and $p = -2q + 81$, respectively.
    (a) If the price is \$53, how many units are supplied and how many are demanded?
    (b) Does this give a shortfall or a surplus?
    (c) Is the price likely to increase from \$53 or decrease from it?

64. ***Market analysis*** Of the equations $p + 6q = 420$ and $p = 6q + 60$, one is the supply function for a product and one is the demand function for that product.
    (a) Graph these equations on the same set of axes.
    (b) Label the supply function and the demand function.
    (c) Find the market equilibrium point.

65. ***Cost, revenue, and profit*** The total costs and total revenues of a certain product are given by the following:

$$C(x) = 38.80x + 4500$$
$$R(x) = 61.30x$$

    (a) Find the marginal cost.
    (b) Find the marginal revenue.
    (c) Find the marginal profit.
    (d) Find the number of units required to break even.

66. ***Cost, revenue and profit*** A certain commodity has the following costs for a period.

Fixed costs:	\$1500
Variable costs:	\$22 per unit

The commodity is sold for \$52 per unit.
    (a) What is the total cost function?
    (b) What is the total revenue function?
    (c) What is the profit function?
    (d) What is the marginal cost?
    (e) What is the marginal revenue?
    (f) What is the marginal profit?
    (g) What is the break-even point?

67. ***Market analysis*** The supply function and the demand function for a product are linear and are determined by the tables that follow. Find the price that will give market equilibrium.

Supply Function

Price	*Quantity*
100	200
200	400
300	600

Demand Function

Price	*Quantity*
200	200
100	400
0	600

68. ***Market analysis*** Suppose that for a certain product the supply and demand functions prior to any taxation are

$$\text{Supply:} \quad p = \frac{q}{10} + 8$$
$$\text{Demand:} \quad 10p + q = 1500$$

If a tax of \$2 per item is levied on the supplier and is passed on to the consumer as a price increase, find the market equilibrium after the tax is levied.

CHAPTER TEST

In Problems 1–3, solve the following equations.

1. $4x - 3 = \frac{x}{2} + 6$
2. $\frac{3}{x} + 4 = \frac{4x}{x + 1}$
3. $\frac{3x - 1}{4x - 9} = \frac{5}{7}$
4. For $f(x) = 7 + 5x - 2x^2$, find and simplify $\frac{f(x + h) - f(x)}{h}$.

In Problems 7 and 8, find the intercepts and graph the functions.

5. $5x - 6y = 30$
6. $7x + 5y = 21$
7. Consider the function $f(x) = \sqrt{4x + 16}$.
   (a) Find the domain and range.
   (b) Find $f(3)$.
   (c) Find the y-coordinate on the graph of $y = f(x)$ when $x = 5$.
8. Write the equation of the line passing through $(-1, 2)$ and $(3, -4)$. Write your answer in slope-intercept form.
9. Find the slope and the y-intercept of the graph of $5x + 4y = 15$.
10. Write the equation of the line through $(-3, -1)$ that
    (a) has undefined slope.
    (b) is perpendicular to $x = 4y - 8$.
11. Which of the following relations [(a), (b), (c)] are functions? Explain.
    (a)

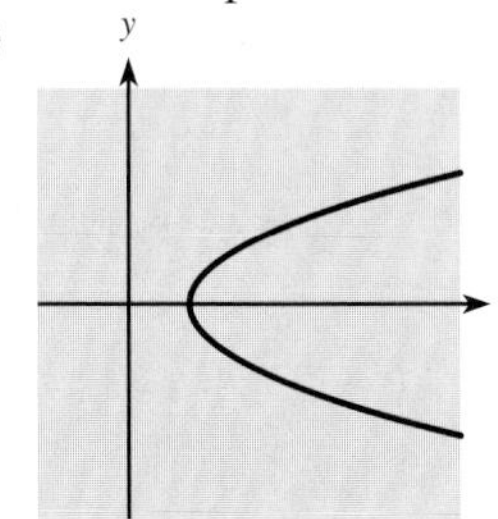

(b)

x	y
1	3
−1	3
4	2

(c) $y = \pm\sqrt{x^2 - 1}$

12. Solve the system

$$\begin{cases} 3x + 2y = -2 \\ 4x + 5y = 2 \end{cases}$$

13. Given $f(x) = 5x^2 - 3x$ and $g(x) = x + 1$, find
    (a) $(fg)(x)$
    (b) $g(g(x))$
    (c) $(f \circ g)(x)$
14. The total cost function for a product is $C(x) = 30x + 1200$, and the total revenue is $R(x) = 38x$, where x is the number of units produced and sold.
    (a) Find the marginal cost.
    (b) Find the profit function.
    (c) Find the number of units that gives break even.
    (d) Find the marginal profit and explain what it means.
15. The selling price for each item of a product is \$50, and the total cost is given by $C(x) = 10x + 18{,}000$, where x is the number of items.
    (a) Write the revenue function.
    (b) Find $C(100)$ and write a sentence that explains its meaning.
    (c) Find the number of units that gives break even.
16. The supply function for a product is $p = 5q + 1500$ and the demand function is $p = -3q + 3100$. Find the quantity and price that give market equilibrium.
17. A building is depreciated by its owner, with the value y of the building after x months given by $y = 360{,}000 - 1500x$.
    (a) Find the y-intercept of the graph of this function, and explain what it means.
    (b) Find the slope of the graph and tell what it means.

18. An airline has 360 seats on a plane for one of its flights. If 90% of the people making reservations actually buy a ticket, how many reservations should the airline accept to be confident that it will sell 360 tickets?

19. Amanda plans to invest $20,000, part of it at a 9% interest rate and part of it in a safer fund that pays 6%. How much should be invested in each fund to yield an annual return of $1560?

Extended Applications & Group Projects

I. Hospital Administration

Southwest Hospital has an operating room used only for eye surgery. The annual cost of rent, heat, and electricity for the operating room and its equipment is $180,000, and the annual salaries for the people who staff this room total $270,000.

Each surgery performed requires the use of $380 worth of medical supplies and drugs. To promote goodwill, every patient receives a bouquet of flowers the day after surgery. In addition, one-quarter of the patients require dark glasses, which the hospital provides free of charge. It costs the hospital $15 for each bouquet of flowers and $20 for each pair of glasses.

The hospital receives a payment of $1000 for each eye operation performed.

1. Identify the revenue per case and the annual fixed and variable costs for running the operating room.
2. How many eye operations must the hospital perform each year in order to break even?
3. Southwest Hospital currently averages 70 eye operations per month. One of the nurses has just learned about a machine that would reduce by $50 per patient the amount of medical supplies needed. It can be leased for $50,000 annually. Keeping in mind the financial cost and benefits, advise the hospital on whether it should lease this machine.
4. An advertising agency has proposed to the hospital's president that she spend $10,000 per month on television and radio advertising to persuade people that Southwest Hospital is the best place to have any eye surgery performed. Advertising account executives estimate that such publicity would increase business by 40 operations per month. If they are correct and if this increase is not big enough to affect fixed costs, what impact would this advertising have on the hospital's profits?
5. In case the advertising agency is being overly optimistic, how many extra operations per month are needed to cover the cost of the proposed ads?
6. If the ad campaign is approved and subsequently meets its projections, should the hospital review its decision about the machine discussed in Question 3?

II. Fund Raising

At most colleges and universities, student organizations conduct fund-raising activities, such as selling T-shirts or candy. If a club or organization decided to sell coupons for submarine sandwiches, it would want to find the best deal, based on the amount the club thought it could sell and how much profit it would make. In this project, you'll try to discover the best deal for this type of fund raiser conducted in your area.

1. Contact at least two different sub shops in your area from among local sub shops, a national chain, or a regional or national convenience store chain. From each contact, find the details of selling sub sandwich coupons as a fund raiser. In particular, determine the following for each sub shop, and present your findings in a chart.

   (a) The selling price for each coupon and its value to the coupon holder, including any expiration date

   (b) Your cost for each coupon sold and for each one returned

   (c) The total number of coupons provided

   (d) The duration of the sale
2. For each sub shop, determine a club's total revenue and total costs as linear functions of the number of coupons sold.
3. Form the profit function for each sub shop, and graph the profit functions together. For each function, determine the break-even point. Find the marginal profit for each function, and interpret its meaning.
4. The shop with the best deal will be the one whose coupons will provide the maximum profit for a club.

   (a) For each shop, determine a sales estimate that is based on location, local popularity, and customer value.

   (b) Use each estimate from (a) with that shop's profit function to determine which shop gives the maximum profit.

Fully explain and support your claims; make and justify a recommendation.

Warm-up

Prerequisite Problem Type	For Section	Answer	Section for Review
Find $b^2 - 4ac$ if (a) $a = 1, b = 2, c = -2$ (b) $a = -2, b = 3, c = -1$ (c) $a = -3, b = -3, c = -2$	2.1 2.2 2.3	(a) 12 (b) 1 (c) -15	0.2 Signed numbers
(a) Factor $6x^2 - x - 2$. (b) Factor $6x^2 - 9x$.	2.1 2.2 2.3	(a) $(3x - 2)(2x + 1)$ (b) $3x(2x - 3)$	0.6 Factoring
Is $\dfrac{1 + \sqrt{-3}}{2}$ a real number?	2.1	No	0.4 Radicals
(a) Find the y-intercept of $y = x^2 - 6x + 8$. (b) Find the x-intercepts of $y = 3x(2x - 3)$.	2.2 2.3	(a) 8 (b) $0, \frac{3}{2}$	1.2 x- and y-intercepts
If $f(x) = -x^2 + 4x$, find $f(2)$.	2.2, 2.3	4	1.2 Functions
Find the domain of $f(x) = \dfrac{12x + 8}{3x - 9}$.	2.4	All $x \neq 3$	
Assume revenue is $R(x) = 500x - 2x^2$ and cost is $C(x) = 3600 + 100x + 2x^2$. (a) Find the profit function $P(x)$. (b) Find $P(50)$.	2.3	(a) $P(x) = -3600 + 400x - 4x^2$ (b) 6400	1.6 Cost, revenue and profit

Chapter 2

Special Functions

In Chapter 1 we discussed functions in general and linear functions in particular. In this chapter we will discuss other types of functions, including identity, constant, reciprocal, power, absolute value, piecewise defined, quadratic, polynomial, and rational functions.

We will pay particular attention to quadratic equations and to quadratic functions, whose graphs are parabolas. Use of quadratic functions makes it possible to expand our study of the special business functions introduced in Chapter 1. In particular, we will see that the principles governing market equilibrium and break-even analysis still apply even when the functions are not linear. Furthermore, our analysis allows us to find maximum values for revenue and profit functions if they are quadratic functions.

We will also discuss, in an optional section on modeling, the creation of equations that approximately fit real data points.

2.1 Quadratic Equations

OBJECTIVES

- *To solve quadratic equations with factoring methods*
- *To solve quadratic equations with the quadratic formula*

APPLICATION PREVIEW

With fewer workers, future income from payroll taxes used to fund Social Security benefits won't keep pace with scheduled benefits. The function

$$B = -1.056785714t^2 + 8.259285714t + 74.07142857$$

describes the Social Security trust fund balance B, in billions of dollars, where t is the number of years past the year 2000 (Source: Social Security Administration). For planning purposes, it is important to know when the trust fund balance will be 0. That is, for what t-value does

$$0 = -1.056785714t^2 + 8.259285714t + 74.07142857?$$

In this section, we learn how to solve equations of this type, using factoring methods and using the quadratic formula.

Factoring Methods

It is important for us to know how to solve equations, such as the equation in the Application Preview. This was true for linear equations and is again true for quadratic equations.

A **quadratic equation** in one variable is an equation that can be written in the *general form*

$$ax^2 + bx + c = 0 \qquad (a \neq 0)$$

where a, b, and c represent constants. For example, the equations

$$3x^2 + 4x + 1 = 0 \qquad \text{and} \qquad 2x^2 + 1 = x^2 - x$$

are quadratic equations; the first of these is in general form, and the second can easily be rewritten in general form.

When we solve quadratic equations, we will be interested only in real number solutions and will consider two methods of solution: factoring and the quadratic formula. We will discuss solving quadratic equations by factoring first. (For a review of factoring, see Section 0.6, "Factoring.")

Solution by factoring is based on the following property of the real numbers.

Zero Product Property For real numbers a and b, $ab = 0$ if and only if $a = 0$ or $b = 0$ or both.

Hence, to solve by factoring, we must first write the equation with zero on one side.

EXAMPLE 1

Solve (a) $6x^2 + 3x = 4x + 2$. (b) $6x^2 = 9x$

Solution

(a)

$$6x^2 + 3x = 4x + 2$$
$$6x^2 - x - 2 = 0 \qquad \text{Proper form for factoring}$$
$$(3x - 2)(2x + 1) = 0 \qquad \text{Factored}$$

$$3x - 2 = 0 \quad \text{or} \quad 2x + 1 = 0 \qquad \text{Factors equal to zero}$$
$$3x = 2 \qquad 2x = -1$$
$$x = \frac{2}{3} \qquad x = -\frac{1}{2} \qquad \text{Solutions}$$

We now check that these values are, in fact, solutions to our original equation.

$$6\left(\frac{2}{3}\right)^2 + 3\left(\frac{2}{3}\right) \stackrel{?}{=} 4\left(\frac{2}{3}\right) + 2 \qquad 6\left(-\frac{1}{2}\right)^2 + 3\left(-\frac{1}{2}\right) \stackrel{?}{=} 4\left(-\frac{1}{2}\right) + 2$$
$$\frac{14}{3} = \frac{14}{3} \checkmark \qquad 0 = 0 \checkmark$$

(b)

$$6x^2 = 9x$$
$$6x^2 - 9x = 0$$
$$3x(2x - 3) = 0$$

$$3x = 0 \quad \text{or} \quad 2x - 3 = 0$$
$$x = 0 \qquad 2x = 3$$
$$x = \frac{3}{2}$$

CHECK: $6(0)^2 = 9(0) \checkmark \qquad 6\left(\frac{3}{2}\right)^2 = 9\left(\frac{3}{2}\right) \checkmark$

Thus the solutions are $x = 0$ and $x = \frac{3}{2}$.

Note that in the previous example it is incorrect to divide both sides of the equation by x, because this results in the loss of the solution $x = 0$. Never divide both sides of an equation by an expression containing the variable.

EXAMPLE 2

Solve (a) $(y - 3)(y + 2) = -4$ for y. (b) $\dfrac{x + 1}{3x + 6} = \dfrac{3}{x} + \dfrac{2x + 6}{x(3x + 6)}$

Solution

(a) Note that the left side of the equation is factored, but the right member is not 0. Therefore, we must multiply the factors before we can rewrite the equation in general form.

$$\begin{aligned}(y-3)(y+2) &= -4\\ y^2 - y - 6 &= -4\\ y^2 - y - 2 &= 0\\ (y-2)(y+1) &= 0\end{aligned}$$

$$y - 2 = 0 \quad \text{or} \quad y + 1 = 0$$
$$y = 2 \quad\quad\quad y = -1$$

CHECK: $(2-3)(2+2) = -4$ ✓ $\quad (-1-3)(-1+2) = -4$ ✓

(b) Multiplying both sides of the equation by the LCD of the fractions gives a quadratic equation that is equivalent to the original equation for $x \neq 0$ and $x \neq -2$. (The original equation is undefined for these values.)

$$\frac{x+1}{3x+6} = \frac{3}{x} + \frac{2x+6}{x(3x+6)}$$

$$x(3x+6)\frac{x+1}{3x+6} = x(3x+6)\left(\frac{3}{x} + \frac{2x+6}{x(3x+6)}\right) \qquad \text{Multiplying both sides by } x(3x+6)$$

$$x(x+1) = 3(3x+6) + (2x+6)$$

$$x^2 + x = 9x + 18 + 2x + 6$$

$$x^2 - 10x - 24 = 0$$

$$(x-12)(x+2) = 0$$

$$x - 12 = 0 \text{ or } x + 2 = 0$$

$$x = 12 \text{ or } x = -2$$

Checking $x = 12$ and $x = -2$ in the original equation, we see that -2 does not check, so the only solution is $x = 12$.

CHECKPOINT

1. The factoring method for solving a quadratic equation is based on the __________ product property. Hence, in order for us to solve a quadratic equation by factoring, one side of the equation must equal __________.
2. Solve the following equations by factoring.
   (a) $x^2 - 19x = 20$ (b) $2x^2 = 6x$

The Quadratic Formula

Factoring does not lend itself easily to solving quadratic equations like

$$x^2 - 5 = 0$$

However, we can solve this equation by writing

$$x^2 = 5$$
$$x = \pm\sqrt{5}$$

In general, we can solve quadratic equations of the form $x^2 = C$ (no x term) by taking the square root of both sides.

Square Root Property The solution of $x^2 = C$ is

$$x = \pm\sqrt{C}$$

This property also can be used to solve equations like those in the following example.

EXAMPLE 3

Solve the following equations.

(a) $4x^2 = 5$ (b) $(3x - 4)^2 = 9$

Solution

We can use the square root property for both parts.

(a) $4x^2 = 5$ is equivalent to $x^2 = \frac{5}{4}$. Thus,

$$x = \pm\sqrt{\frac{5}{4}} = \pm\frac{\sqrt{5}}{\sqrt{4}} = \pm\frac{\sqrt{5}}{2}$$

(b) $(3x - 4)^2 = 9$ is equivalent to $3x - 4 = \pm\sqrt{9}$. Thus

$3x - 4 = 3$	$3x - 4 = -3$
$3x = 7$	$3x = 1$
$x = \frac{7}{3}$	$x = \frac{1}{3}$

The solution of the general quadratic equation $ax^2 + bx + c = 0$, where $a \neq 0$, is called the **quadratic formula.** It can be derived by using the square root property, as follows:

$$ax^2 + bx + c = 0 \qquad \text{Standard form}$$
$$ax^2 + bx = -c \qquad \text{Subtracting } c \text{ from both sides}$$
$$x^2 + \frac{b}{a}x = -\frac{c}{a} \qquad \text{Dividing both sides by } a$$

We would like to make the left side of the last equation a perfect square of the form $(x + k)^2 = x^2 + 2kx + k^2$. If we let $2k = b/a$, then $k = b/(2a)$ and $k^2 = b^2/(4a^2)$. Hence we continue as follows:

$$x^2 + \frac{b}{a}x + \frac{b^2}{4a^2} = \frac{b^2}{4a^2} - \frac{c}{a} \qquad \text{Adding } \frac{b^2}{4a^2} \text{ to both sides}$$

$$\left(x + \frac{b}{2a}\right)^2 = \frac{b^2 - 4ac}{4a^2} \qquad \text{Simplifying}$$

$$x + \frac{b}{2a} = \pm\sqrt{\frac{b^2 - 4ac}{4a^2}} \qquad \text{Square root property}$$

$$x = \frac{-b}{2a} \pm \frac{\sqrt{b^2 - 4ac}}{2|a|} \qquad \text{Solve for } x$$

Because $\pm 2|a|$ represents the same number as $\pm 2a$, we obtain the following:

Quadratic Formula If $ax^2 + bx + c = 0$, where $a \neq 0$, then

$$x = \frac{-b \pm \sqrt{b^2 - 4ac}}{2a}$$

We may use the quadratic formula to solve all quadratic equations, but especially those in which factorization is difficult or impossible to see. The proper identification of values for a, b, and c to be substituted into the formula requires that the equation be in general form.

EXAMPLE 4

Use the quadratic formula to solve $2x^2 - 3x - 6 = 0$ for x.

Solution

The equation is already in general form, with $a = 2$, $b = -3$, and $c = -6$. Hence,

$$\begin{aligned} x &= \frac{-b \pm \sqrt{b^2 - 4ac}}{2a} \\ &= \frac{-(-3) \pm \sqrt{(-3)^2 - 4(2)(-6)}}{2(2)} \\ &= \frac{3 \pm \sqrt{9 + 48}}{4} \\ &= \frac{3 \pm \sqrt{57}}{4} \end{aligned}$$

Thus the solutions are

$$x = \frac{3 + \sqrt{57}}{4} \quad \text{and} \quad x = \frac{3 - \sqrt{57}}{4}$$

EXAMPLE 5

Using the quadratic formula, find the (real) solutions to $x^2 = x - 1$.

Solution

We must rewrite the equation in general form before we can determine the values of a, b, and c.

$$x^2 = x - 1$$

$$x^2 - x + 1 = 0 \qquad \text{Note: } a = 1,\ b = -1, \text{ and } c = 1.$$

$$x = \frac{-(-1) \pm \sqrt{(-1)^2 - 4(1)(1)}}{2(1)}$$

$$x = \frac{1 \pm \sqrt{1 - 4}}{2}$$

$$x = \frac{1 \pm \sqrt{-3}}{2}$$

Because $\sqrt{-3}$ is not a real number, the values of x are not real. Hence there are no real solutions to the given equation.

In Example 5 there were no real solutions to the quadratic equation because the radicand of the quadratic formula was negative. In general, when solving a quadratic equation, we can use the sign of the radicand in the quadratic formula—that is, the sign of $b^2 - 4ac$—to determine how many real solutions there are. Thus we refer to $b^2 - 4ac$ as the **quadratic discriminant.**

Quadratic Discriminant

Given $ax^2 + bx + c = 0$ and $a \neq 0$,

If $b^2 - 4ac > 0$, the equation has 2 distinct real solutions.
If $b^2 - 4ac = 0$, the equation has exactly 1 real solution.
If $b^2 - 4ac < 0$, the equation has no real solutions.

The quadratic formula is especially useful when the coefficients of a quadratic equation are decimal values that make factorization impractical. This occurs in many applied problems, such as the one in the Application Preview, to which we now return.

EXAMPLE 6

Solve $0 = -1.056785714t^2 + 8.259285714t + 74.07142857$ from the Application Preview to find when the Social Security trust fund balance is predicted to be zero. Recall that t is the number of years past the year 2000.

Solution

We use the quadratic formula with

$$a = -1.056785714, \quad b = 8.259285714, \quad c = 74.07142857$$

Thus

$$t = \frac{-b \pm \sqrt{b^2 - 4ac}}{2a}$$

$$= \frac{-8.259285714 \pm \sqrt{(8.259285714)^2 - 4(-1.056785714)(74.07142857)}}{2(-1.056785714)}$$

$$= \frac{-8.259285714 \pm \sqrt{381.3263106}}{-2.113571428}$$

$$\approx \frac{-8.259285714 \pm 19.52757821}{-2.113571428}$$

We are interested only in the positive solution, so

$$t = \frac{-8.259285714 - 19.52757821}{-2.113571428} \approx 13.14687393$$

Therefore, the trust fund balance is projected to reach zero slightly more than 13 years after 2000, during the year 2013.

Graphing Utilities

We can use graphing utilities to determine (or approximate) the solutions of quadratic equations. The solutions can be found by using commands (such as ZERO, ROOT, or SOLVER), programs, or TRACE and ZOOM. TRACE can be used to see where the graph of the function $y = f(x)$ intersects the x-axis. This **x-intercept** is also the value of x that makes the function zero, so it is called a **zero of the function.** Because this value of x makes $f(x) = 0$, it is also a **solution** (or **root**) of the equation $0 = f(x)$.

EXAMPLE 7

Find the solutions to the equation $0 = -7x^2 + 16x - 4$.

Solution

To solve the equation, we find the x-intercepts of the graph of $y = -7x^2 + 16x - 4$. We graph this function and use TRACE to find an x-value that gives $y = 0$. Tracing to $x = 2$ gives $y = 0$, so $x = 2$ is an x-intercept.

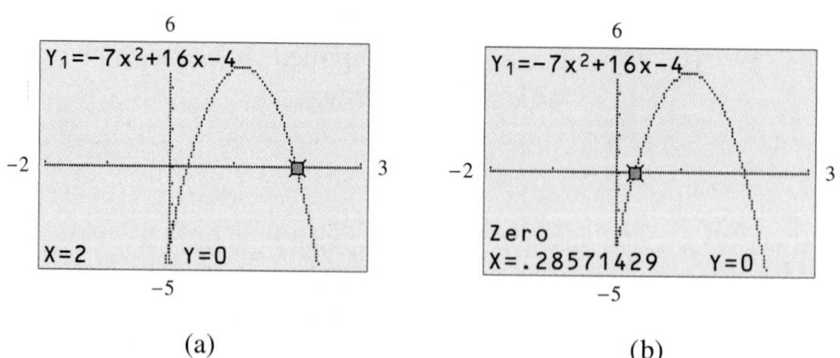

Figure 2.1 (a) (b)

Tracing near $x = 0.3$ gives a value of y near 0, and an (approximate) x-intercept near 0.3 can be found more accurately by using TRACE and ZOOM IN or ZERO (or ROOT). This gives $y = 0$ at $x = 0.28571429$ (approximately).

Solving $f(x) = 0$ algebraically gives solutions $x = 2$ and $x = \frac{2}{7}$.

$$0 = (-7x + 2)(x - 2)$$
$$-7x + 2 = 0 \quad \text{or} \quad x - 2 = 0$$
$$x = \frac{2}{7} \qquad\qquad x = 2$$

CHECKPOINT

3. The statement of the quadratic formula says that if ________, then $x =$ ________.
4. Solve $2x^2 - 5x = 9$ using the quadratic formula.

CHECKPOINT SOLUTIONS

1. Zero; zero
2. (a)

$$x^2 - 19x - 20 = 0$$
$$(x - 20)(x + 1) = 0$$

$x - 20 = 0$	$x + 1 = 0$
$x = 20$	$x = -1$

(b)

$$2x^2 - 6x = 0$$
$$2x(x - 3) = 0$$

$2x = 0$	$x - 3 = 0$
$x = 0$	$x = 3$

3. If $ax^2 + bx + c = 0$ with $a \neq 0$, then

$$x = \frac{-b \pm \sqrt{b^2 - 4ac}}{2a}$$

4. $2x^2 - 5x - 9 = 0$ so $a = 2, b = -5, c = -9$

$$x = \frac{-b \pm \sqrt{b^2 - 4ac}}{2a} = \frac{5 \pm \sqrt{25 - 4(2)(-9)}}{4}$$
$$= \frac{5 \pm \sqrt{25 + 72}}{4}$$
$$= \frac{5 \pm \sqrt{97}}{4}$$

$$x = \frac{5 + \sqrt{97}}{4} \approx 3.712 \quad \text{or} \quad x = \frac{5 - \sqrt{97}}{4} \approx -1.212$$

EXERCISE 2.1

In Problems 1–6, write the following equations in general form.

1. $x^2 + 5x = 3$
2. $y^2 + 4y = 2$
3. $2x^2 + 3 = x^2 - 2x + 4$
4. $x^2 - 2x + 5 = 2 - 2x^2$
5. $(y + 1)(y + 2) = 4$
6. $(z - 1)(z - 3) = 1$

In Problems 7–22, solve each equation by factoring.

7. $9 - 4x^2 = 0$
8. $25x^2 - 16 = 0$
9. $x = x^2$
10. $t^2 - 4t = 3t^2$
11. $x^2 + 5x = 21 + x$
12. $x^2 + 17x = 8x - 14$
13. $4t^2 - 4t + 1 = 0$
14. $49z^2 + 14z + 1 = 0$
15. $\frac{w^2}{8} - \frac{w}{2} - 4 = 0$
16. $\frac{y^2}{2} - \frac{11}{6}y + 1 = 0$
17. $\frac{x^2}{4} + \frac{3x}{2} + 2 = 0$
18. $\frac{2x^2}{27} + \frac{x}{3} - 3 = 0$
19. $(x - 1)(x + 5) = 7$
20. $(x - 3)(1 - x) = 1$
21. $(4x - 1)(x - 3) = 15$
22. $(2x + 1)(x - 1) = 14$

In Problems 23–26, multiply both sides of the equation by the LCD, and solve the resulting quadratic equation.

23. $x + \frac{8}{x} = 9$
24. $\frac{x}{x - 2} - 1 = \frac{3}{x + 1}$
25. $\frac{x}{x - 1} = 2x + \frac{1}{x - 1}$
26. $\frac{5}{z + 4} - \frac{3}{z - 2} = 4$

In Problems 27–34, solve each equation by using the quadratic formula. Give real answers (a) exactly and (b) rounded to two decimal places.

27. $x^2 - 4x - 4 = 0$
28. $x^2 - 6x + 7 = 0$
29. $2w^2 + w + 1 = 0$
30. $z^2 + 2z + 4 = 0$
31. $16z^2 + 16z - 21 = 0$
32. $10y^2 - y - 65 = 0$
33. $5x^2 = 2x + 6$
34. $3x^2 = -6x - 2$

In Problems 35–40, find the exact real solutions to each equation, if they exist.

35. $y^2 = 7$
36. $z^2 = 12$
37. $y^2 + 9 = 0$
38. $z^2 + 121 = 0$
39. $(x + 4)^2 = 25$
40. $(x + 1)^2 = 2$

In Problems 41 and 42, solve using quadratic methods.

41. $(x + 8)^2 + 3(x + 8) + 2 = 0$
42. $(s - 2)^2 - 5(s - 2) - 24 = 0$

In Problems 43–50, solve each equation by using a graphing utility.

43. $49x^2 + 28x + 4 = 0$
44. $25x^2 + 120x + 144 = 0$
45. $21x + 70 - 7x^2 = 0$
46. $3x^2 - 11x + 6 = 0$
47. $300 - 2x - 0.01x^2 = 0$
48. $-9.6 + 2x - 0.1x^2 = 0$
49. $25.6x^2 - 16.1x - 1.1 = 0$
50. $6.8z^2 - 4.9z - 2.6 = 0$

Applications

51. ***Profit*** If the profit from the sale of x units of a product is $P = 90x - 200 - x^2$, what level(s) of production yields a profit of \$1200?

52. ***Profit*** If the profit from the sale of x units of a product is $P = 16x - 0.1x^2 - 100$, what level(s) of production yields a profit of \$180?

53. ***Profit*** If the profit from the sale of x units of a product is $P = 6400x - 18x^2 - 400$, what level(s) of production yields a profit of \$61,800?

54. ***Profit*** If the profit from the sale of x units of a product is $P = 50x - 300 - 0.01x^2$, what level(s) of production yields a profit of \$250?

55. ***Wind chill*** Weather forecasters frequently report wind chill factors because a body exposed to wind loses heat due to convection. The amount of loss depends on many factors, but for a given situation there is a positive number called the coefficient of convection, K_c, which depends on the wind velocity v. The approximate relationship between K_c and v is given by

$$\frac{(K_c)^2}{64} - \frac{1}{4} = v$$

(a) Find the coefficient of convection when the wind velocity is 10 mph.
(b) Find K_c when $v = 40$ mph.
(c) What is the change in K_c for the change in v from 10 mph to 40 mph?

56. ***Velocity of blood*** The velocity of a blood corpuscle in a vessel depends on how far the corpuscle is from the center of the vessel. Let R be the constant radius of the vessel, v_m the constant maximum velocity of

the corpuscle, r the distance from the center to a particular blood corpuscle (variable), and v_r the velocity of that corpuscle. Then the velocity v_r is related to the distance r according to

$$v_r = v_m\left(1 - \frac{r^2}{R^2}\right)$$

(a) Find r when $v_r = \frac{1}{2}v_m$.
(b) Find r when $v_r = \frac{1}{4}v_m$.
(c) Find v_r when $r = R$.

57. ***Toyota Supra acceleration*** The time t, in seconds, that it takes a 1993 Toyota Supra to accelerate to x mph can be described by

$$t = 0.01(0.0839x^2 + 2.6536x + 92.3571)$$

(Source: *Motor Trend*, March 1993, p. 7). How fast is the Supra going after 5.5 seconds? Give your answer to the nearest tenth.

58. ***Social Security trust fund*** Social Security benefits are paid from a trust fund. As mentioned in the Application Preview, the trust fund balance, B, in billions of dollars, t years past the year 2000 is projected to be described by

$$B = -1.056785714t^2 + 8.259285714t + 74.07142857$$

(Source: Social Security Administration projections). Find in what year the trust fund balance is projected to be \$1000 billion in the red—that is, when $B = -1000$.

59. ***AT&T total revenues*** For the years 1985–1993, total revenues of AT&T can be described by

$$R = 0.253t^2 - 4.03t + 76.84$$

where R is in billions of dollars and t is the number of years past 1980 (Source: *AT&T Annual Report*, 1993).
(a) For what t-values will total revenue be \$70 billion?
(b) Use your answer to (a) to determine what year (after 1993) total revenue will reach \$70 billion.

60. ***Depth of a fissure*** A fissure in the earth appeared after an earthquake. To measure its vertical depth, a stone was dropped into it, and the sound of the stone's impact was heard 3.9 seconds later. The distance (in feet) the stone fell is given by $s = 16t_1^2$, and the distance (in feet) the sound traveled is given by $s = 1090t_2$. In these equations, the distances traveled by the sound and the stone are the same, but their times are not. Using the fact that the total time is 3.9 seconds, find the depth of the fissure.

61. ***Marijuana use*** For the years from 1975 to 1991, the percentage p of high school seniors who have tried marijuana can be considered as a function of time t according to

$$p = f(t) = -0.228596t^2 + 2.7783t + 49.813783$$

where t is the number of years past 1975 (Source: National Institute on Drug Abuse). In what year, after 1990, will the percentage predicted by the function reach 33% if this function remains valid?

62. ***Projectile motion*** Two projectiles are shot into the air over a lake. The paths of the projectiles are given by
(a) $y = -0.0013x^2 + x + 10$ and
(b) $y = -\frac{x^2}{81} + \frac{4}{3}x + 10$

where y is the height and x is the horizontal distance traveled. Determine which projectile travels farther by substituting $y = 0$ in each equation and finding x.

63. ***Percentage profit*** The Ace Jewelry Store sold a necklace for \$144. If the percentage of profit (based on cost) equals the cost of the necklace to the store, how much did the store pay for it? Use

$$P = \left(\frac{C}{100}\right) \cdot C$$

where P is profit and C is cost.

2.2 Quadratic Functions: Parabolas

OBJECTIVES

- *To find the vertex of the graph of a quadratic function*
- *To determine whether a vertex is a maximum point or a minimum point*
- *To find the zeros of a quadratic function*
- *To graph quadratic functions*

APPLICATION PREVIEW

Because additional equipment, raw materials, and labor may cause variable costs of some products to increase dramatically as more units are produced, total cost functions are not always linear functions. For example, suppose that the total cost of producing a product is given by the equation

$$C = C(x) = 300x + 0.1x^2 + 1200$$

where x represents the number of units produced. That is, the cost of this product is represented by a **second-degree function,** or a **quadratic function.** We can find the cost of producing 10 units by evaluating

$$C(10) = 300(10) + 0.1(10)^2 + 1200 = 4210$$

If the revenue function for this product is

$$R = R(x) = 600x$$

then the profit is also a quadratic function:

$$P = R - C = 600x - (300x + 0.1x^2 + 1200)$$
$$P = -0.1x^2 + 300x - 1200$$

It is natural to ask how many units give maximum profit and what that profit is. In this section we will describe ways to find the maximum point, or minimum point, for a quadratic function.

In Chapter 1, we studied functions of the form $y = ax + b$, called linear (or first-degree) functions. We now turn our attention to **quadratic** (or second-degree) **functions.** The general equation of a quadratic function has the form

$$y = f(x) = ax^2 + bx + c$$

where a, b, and c are real numbers and $a \neq 0$.

The graph of a quadratic function,

$$y = ax^2 + bx + c \qquad (a \neq 0)$$

has a distinctive shape called a **parabola.**

The basic function $y = x^2$ and a variation of it, $y = -\frac{1}{2}x^2$, are parabolas whose graphs are shown in Figure 2.2.

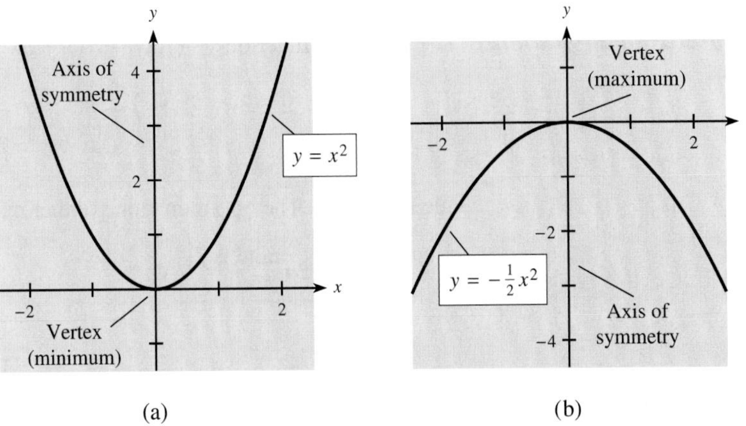

Figure 2.2

As these examples illustrate, the graph of $y = ax^2$ is a parabola that opens upward if $a > 0$ and downward if $a < 0$. The **vertex,** where the parabola turns, is a **minimum point** if $a > 0$ and a **maximum point** if $a < 0$. The vertical line through the vertex of a parabola is called the **axis of symmetry** because one half of the graph is a reflection of the other half through this line.

The graph of $y = (x - 2)^2 - 1$ is the graph of $y = x^2$ shifted to a new location that is 2 units to the right and 1 unit down; its vertex is shifted from $(0, 0)$ to $(2, -1)$ and its axis of symmetry is shifted 2 units to the right. (See Figure 2.3.) The graph of $y = -\frac{1}{2}(x + 1)^2 + 2$ is the graph of $y = -\frac{1}{2}x^2$ shifted 1 unit to the left and 2 units upward, with its vertex at $(-1, 2)$. (See Figure 2.4.)

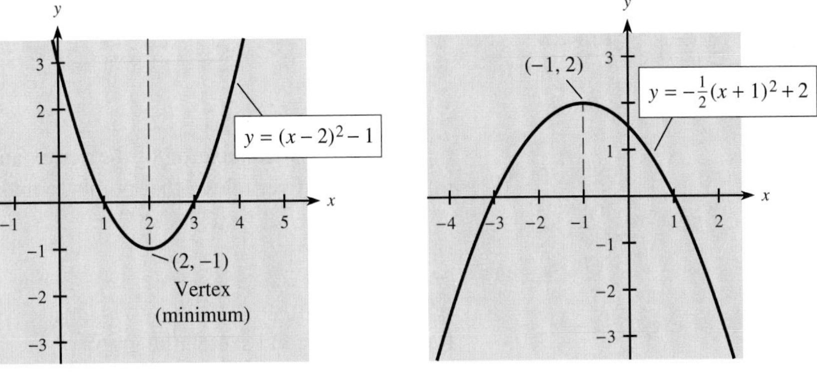

Figure 2.3 **Figure 2.4**

In general, if a quadratic function $y = f(x) = ax^2 + bx + c$ has the form

$$y = a(x - h)^2 + k$$

then its graph is that of $y = ax^2$ shifted h units in the x-direction and k units in the y-direction, with its vertex shifted to $(h, k) = (h, f(h))$.

In addition, $f(x) = ax^2 + bx + c$ can be written in the form

$$f(x) = a\left(x + \frac{b}{2a}\right)^2 + \left(c - \frac{b^2}{4a}\right)$$

so we have the following:

Vertex of a Parabola The quadratic function $y = f(x) = ax^2 + bx + c$ has its **vertex** at

$$\left(\frac{-b}{2a}, f\left(\frac{-b}{2a}\right)\right)$$

See Figure 2.5. The optimum value (either maximum or minimum) of the function occurs at $x = \frac{-b}{2a}$ and is $f\left(\frac{-b}{2a}\right)$.

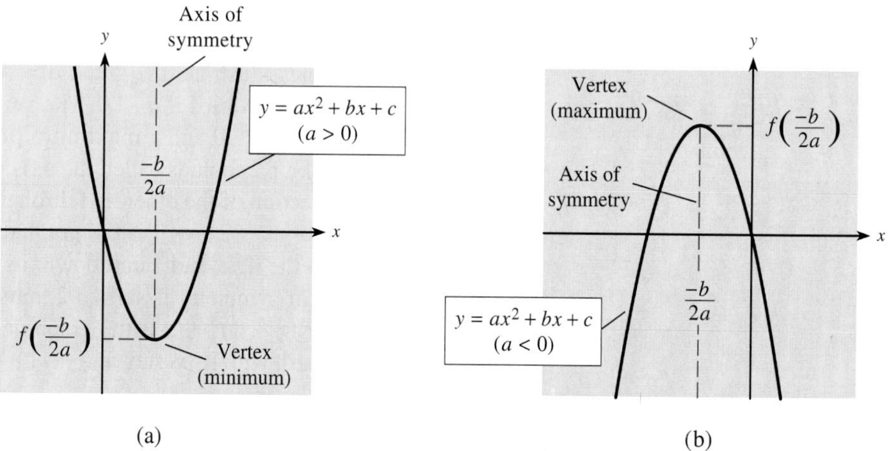

Figure 2.5 (a) (b)

If we know the location of the vertex and the direction in which the parabola opens, we need very few other points to make a good sketch.

EXAMPLE 1

Find the vertex and sketch the graph of

$$f(x) = 2x^2 - 4x + 4$$

Solution

Because $a = 2 > 0$ the graph of $f(x)$ opens upward and the vertex is the minimum point. We can calculate its coordinates as follows:

$$x = \frac{-b}{2a} = \frac{-(-4)}{2(2)} = 1$$
$$y = f(1) = 2$$

Thus the vertex is (1, 2). Using x-values on either side of the vertex to plot additional points allows us to sketch the graph accurately. (See Figure 2.6.)

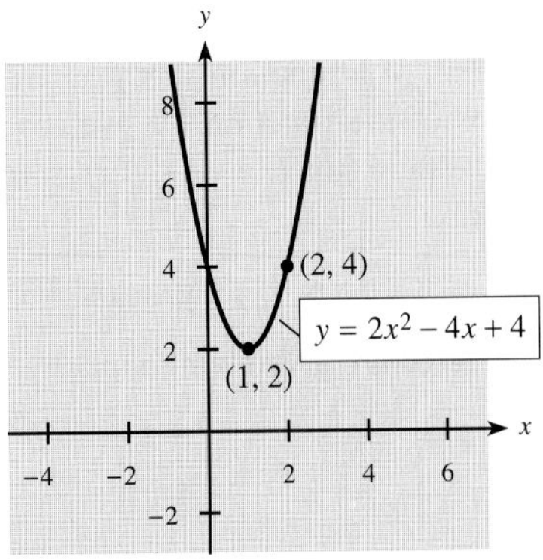

Figure 2.6

We can also use the coordinates of the vertex to find maximum or minimum values without a graph.

EXAMPLE 2

For the profit function in our Application Preview,

$$P(x) = -0.1x^2 + 300x - 1200$$

find the number of units that give maximum profit and find the maximum profit.

Solution

$P(x)$ is a quadratic function with $a < 0$. Thus the graph of $y = P(x)$ is a parabola that opens downward, so the vertex is a maximum point. The coordinates of the vertex are

$$x = \frac{-b}{2a} = \frac{-300}{2(-0.1)} = 1500$$

$$P = P(1500) = -0.1(1500)^2 + 300(1500) - 1200 = 223{,}800$$

Therefore, the maximum profit is \$223,800 when 1500 units are sold.

We can also use a graphing utility to graph a quadratic function. Even with a graphing tool, recognizing that the graph is a parabola and locating the vertex are important. They help us to set the range (to include the vertex) and to know when we have a complete graph. For example, suppose we wanted to graph the profit function from Example 2 (and the Application Preview). Because $x \geq 0$ and the vertex and axis of symmetry are at $x = 1500$, we can set the x-range so that it includes $x = 0$ and has $x = 1500$ near its center. For the P-range (or y-range), we know that the maximum is $P = 223{,}800$ and that $P(0) = -1200$, so the window should include these y-values. Figure 2.7 shows the graph. Note that the graph of this function has two x-intercepts.

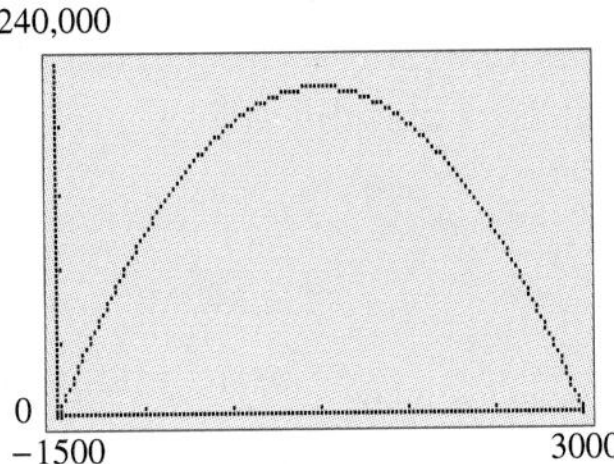

Figure 2.7

As we noted in Chapter 1, "Linear Equations and Functions," the x-intercepts of the graph of a function $y = f(x)$ are the values of x for which $f(x) = 0$, called the **zeros** of the function. As we saw earlier in the previous section, the zeros of the quadratic function $y = f(x) = ax^2 + bx + c$ are the solutions to the quadratic equation

$$ax^2 + bx + c = 0$$

Therefore, the zeros are given by the quadratic formula as

$$x = \frac{-b \pm \sqrt{b^2 - 4ac}}{2a}$$

The information that is useful in graphing quadratic functions is summarized as follows:

Graphs of Quadratic Functions

Form: $y = ax^2 + bx + c$

Graph: parabola (See Figure 2.5 on page 160.)

$a > 0$ parabola opens upward; vertex is a minimum point

$a < 0$ parabola opens downward; vertex is a maximum point

Coordinates of vertex: $x = \dfrac{-b}{2a}$, $y = f\left(\dfrac{-b}{2a}\right)$

Axis of symmetry equation: $x = \dfrac{-b}{2a}$

x-intercepts or zeros (if real*):

$$x = \frac{-b + \sqrt{b^2 - 4ac}}{2a}, \quad x = \frac{-b - \sqrt{b^2 - 4ac}}{2a}$$

y-intercept: Let $x = 0$; then $y = c$.

EXAMPLE 3

For the function $y = 4x - x^2$: determine whether its vertex is a maximum point or a minimum point and find the coordinates of this point; find the zeros, if any exist, and sketch the graph.

Solution

The proper form is $y = -x^2 + 4x + 0$, so $a = -1$. Thus the parabola opens downward, and the vertex is the highest (maximum) point.

The vertex occurs at $x = \dfrac{-b}{2a} = \dfrac{-4}{2(-1)} = 2$.

The y-coordinate of the vertex is $f(2) = -(2)^2 + 4(2) = 4$.

The zeros for the parabola are solutions to

$$-x^2 + 4x = 0$$

$$x(-x + 4) = 0$$

or $x = 0$ and $x = 4$.

*If the zeros are not real, the graph does not cross the x-axis.

Plotting these three points (or using a graphing utility) gives the graph (see Figure 2.8).

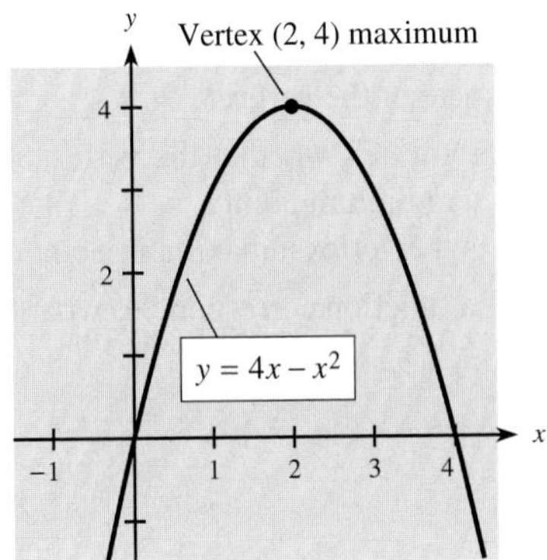

Figure 2.8

EXAMPLE 4

Figure 2.9 shows the graphs of two different quadratic functions. Use the figure to answer the following.

(a) Determine the vertex of each function.
(b) Determine the real solutions to $f_1(x) = 0$ and to $f_2(x) = 0$.
(c) One of the graphs in Figure 2.9 is the graph of $y = 7 + 6x - x^2$, and one is the graph of $y = x^2 - 6x + 10$. Determine which is which and why.

Solution

(a) For $y = f_1(x)$ the vertex is the maximum point, at (3, 16). For $y = f_2(x)$ the vertex is the minimum point, at (3, 1).
(b) For $y = f_1(x)$, the real solutions to $f_1(x) = 0$ are the zeros, or x-intercepts, at $x = -1$ and $x = 7$. For $y = f_2(x)$, the graph has no x-intercepts, so $f_2(x) = 0$ has no real solutions.
(c) Because the graph of $y = f_1(x)$ opens downward, the coefficient of x^2 must be negative. Hence Figure 2.9(a) shows the graph of $y = f_1(x) = 7 + 6x - x^2$. Similarly, the coefficient of x^2 in $y = f_2(x)$ must be positive, so Figure 2.9(b) shows the graph of $y = f_2(x) = x^2 - 6x + 10$.

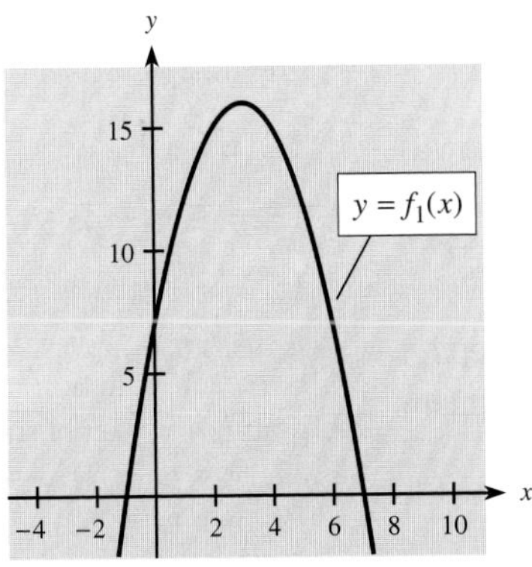

(a)

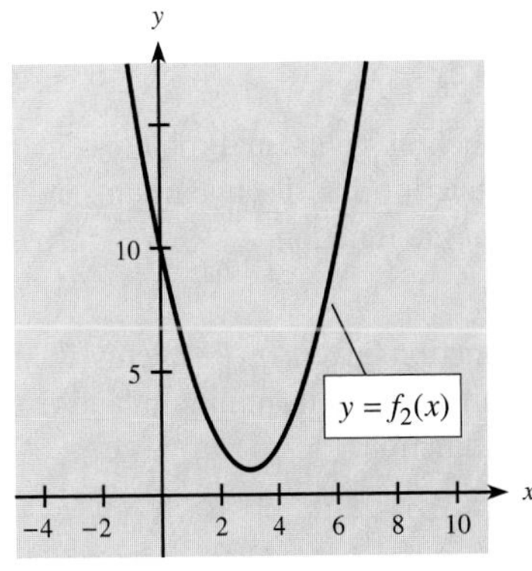

(b)

Figure 2.9

CHECKPOINT

1. Name the graph of a quadratic function.
2. (a) What is the x-coordinate of the vertex of $y = ax^2 + bx + c$?
   (b) For $y = 12x - \frac{1}{2}x^2$, what is the x-coordinate of the vertex? What is the y-coordinate of the vertex?
3. (a) How can you tell whether the vertex of $f(x) = ax^2 + bx + c$ is a maximum point or a minimum point?
   (b) In 2(b), is the vertex a maximum point or a minimum point?
4. The zeros of a function correspond to what feature of its graph?

CHECKPOINT SOLUTIONS

1. Parabola
2. (a) $x = \dfrac{-b}{2a}$
   (b) $x = \dfrac{-12}{2(-1/2)} = \dfrac{-12}{(-1)} = 12$
   $y = 12(12) - \frac{1}{2}(12)^2 = 144 - 72 = 72$
3. (a) Maximum point if $a < 0$; minimum point if $a > 0$.
   (b) (12, 72) is a maximum point because $a = -\frac{1}{2}$.
4. The x-intercepts

EXERCISE 2.2

In Problems 1–4, find the vertex of the graph and determine whether it is a maximum point or a minimum point.

1. $y = \frac{1}{2}x^2 + x$
2. $y = x^2 - 2x$
3. $y = 8 + 2x - x^2$
4. $y = 6 - 4x - 2x^2$

For each function given in Problems 5 and 6, find (a) where the function has its maximum and (b) the maximum value of the function.

5. $f(x) = 6x - x^2$
6. $f(x) = 4 + 3x - x^2$

For each function given in Problems 7 and 8, find (a) where the function has its minimum and (b) the minimum value of the function.

7. $f(x) = x^2 + 2x - 3$
8. $f(x) = \frac{1}{2}x^2 + 2x$

In Problems 9–16, determine whether each function's vertex is a maximum point or a minimum point and find the coordinates of this point. Find the zeros, if any exist, and sketch the graph of the function.

9. $y = x^2 - 4$
10. $y = x^2 - 2$
11. $y = x - \frac{1}{4}x^2$
12. $y = -2x^2 + 18x$
13. $y = x^2 + 4x + 4$
14. $y = x^2 - 6x + 9$
15. $\frac{1}{2}x^2 + x - y - 3 = 0$
16. $x^2 + x + 2y = 5$

For each function in Problems 17–20, (a) tell how the graph of $y = x^2$ is shifted, and (b) graph the function.

17. $y = (x - 3)^2 + 1$
18. $y = (x + 2)^2 - 2$
19. $y = (x - 10)^2 + 12$
20. $y = (x + 12)^2 - 8$

 In Problems 21–26, graph each function with a graphing utility. Use the graph to find the vertex and zeros. Check your results algebraically.

21. $y = \frac{1}{2}x^2 - x - \frac{15}{2}$
22. $y = 0.1(x^2 + 4x - 32)$
23. $y = -5x - x^2$
24. $y = 8 - \frac{1}{2}x^2$
25. $y = \frac{1}{4}x^2 + 3x + 12$
26. $y = x^2 - 2x + 5$

 In Problems 27–30, find the vertex and zeros and use them to determine a range for a graphing utility that includes these values; graph the function with that range.

27. $y = 63 + 0.2x - 0.01x^2$
28. $y = 0.2x^2 + 16x + 140$
29. $y = 0.0001x^2 - 0.01$
30. $y = 0.01x - 0.001x^2$

 In Problems 31 a–c, use a graphing utility to graph the basic function $f(x) = x^2$ with the requested variations.

31. (a) Graph $f(x) = x^2$ and $y = af(x) = ax^2$ for $a = \pm 3$, ± 2, $\pm\frac{1}{4}$, and $\pm\frac{1}{2}$.
    (b) Describe the differences between these graphs when $a > 0$ and when $a < 0$.
    (c) Describe how $af(x)$ differs from $f(x)$ when $0 < a < 1$ and when $a > 1$.
32. How does the discriminant $b^2 - 4ac$ affect the graph of $y = ax^2 + bx + c$ under each of these conditions.
    (a) $b^2 - 4ac = 0$
    (b) $b^2 - 4ac > 0$
    (c) $b^2 - 4ac < 0$

Applications

33. ***Profit*** The daily profit from the sale of a product is given by $P = 16x - 0.1x^2 - 100$.
    (a) What level of production maximizes profit?
    (b) What is the maximum possible profit?
34. ***Profit*** The daily profit from the sale of x units of a product is $P = 80x - 0.4x^2 - 200$.
    (a) What level of production maximizes profit?
    (b) What is the maximum possible profit?
35. ***Wind and pollution*** The amount of particulate pollution x depends on the wind velocity v, among other things. If the relationship between x and v can be approximated by

$$x = 20 - 0.01v^2$$

sketch the graph relating these quantities, with v on the horizontal axis.

36. ***Velocity of blood*** The velocity of a blood corpuscle in a vessel, v_r, depends on the distance of the corpuscle from the center of the vessel, r, according to

$$v_r = v_m\left(1 - \frac{r^2}{R^2}\right)$$

where v_m is the maximum velocity and R the radius of the vessel.

The blood pressure also affects the velocity of a corpuscle. If we make some simplifications in the formula (namely, let $v_m = 1$ and let $x^2 = r^2/R^2$), then we can observe the following:

Pressure	*Equation*
$p = 20$ mm	$v(x) = 8(1 - x^2)$
$p = 10$ mm	$v(x) = 3(1 - x^2)$
$p = 5$ mm	$v(x) = 1 - x^2$

Graph all of these equations on the same set of axes.

37. ***Crop yield*** The yield from a grove of orange trees is given by $Y = x(800 - x)$, where x is the number of orange trees per acre. How many trees will maximize the yield?
38. ***Stimulus-response*** One of the early results in psychology relating the magnitude of a stimulus x to the magnitude of a response y is expressed by the equation

$$y = kx^2$$

where k is an experimental constant. Sketch this graph for $k = 1$, $k = 2$, and $k = 4$.

39. ***Drug sensitivity*** The sensitivity S to a drug is related to the dosage x by

$$S = 1000x - x^2$$

Sketch the graph of this function and determine what dosage gives maximum sensitivity. Use the graph to determine the maximum sensitivity.

40. ***Maximizing an enclosed area*** If 100 feet of fence is used to fence in a rectangular yard, then the resulting area is given by

$$A = x(50 - x)$$

where x is the width of the rectangle and $50 - x$ is the length. Graph this equation and determine the length and width that give maximum area.

41. ***Photosynthesis*** The rate of photosynthesis R for a certain plant depends on the intensity of light x according to

$$R = 270x - 90x^2$$

Sketch the graph of this function, and determine the intensity that gives the maximum rate.

42. ***Projectiles*** A ball thrown vertically into the air has its height above ground given by

$$s = 112t - 16t^2$$

Find the maximum height of the ball.

43. ***Projectiles*** Two projectiles are shot into the air from the same location. The paths of the projectiles are parabolas and are given by
(a) $y = -0.0013x^2 + x + 10$ and
(b) $y = \frac{-x^2}{81} + \frac{4}{3}x + 10$
Determine which projectile goes higher by locating the vertex of each parabola.

44. ***Flow rates of water*** The speed at which water travels in a pipe can be measured by directing the flow through an elbow and measuring the height to which it spurts out the top. If the elbow height is 10 cm, the equation relating the height h (in centimeters) of the water above the elbow and its velocity v (in centimeters per second) is given by

$$v^2 = 1960(h + 10)$$

Solve this equation for h and graph the result, using the velocity as the independent variable.

45. ***U.S. trade deficit*** The figure to the right shows the U.S. trade deficit and the 12-month moving-average U.S. trade deficit.
    (a) What kind of function might be used to model the 12-month moving average?
    (b) If a function of the form $f(x) = ax^2 + bx + c$ were used to model the 12-month moving average, would $a > 0$ or $a < 0$? Why?

46. ***Projectile motion*** When a stone is thrown upward, it follows a parabolic path given by a form of the equation

$$y = ax^2 + bx + c$$

If $y = 0$ represents ground level, find the equation of a stone that is thrown from ground level at $x = 0$ and lands on the ground 40 units away if the stone reaches a maximum height of 40 units. (*Hint:* Find the coordinates of the vertex of the parabola and two other points.)

47. ***Health care*** Many politicians are discussing national health insurance because health care costs are increasing so rapidly. Health care costs in the United States are given in the following table.

Year	*Costs ($ millions)*
1960	27.1
1965	41.6
1970	74.4
1975	132.9
1980	249.1
1985	420.1
1989	539.9

Source: *World Almanac*, 1991

(a) Plot the points from the table. Use x to represent the number of years since 1960 and y to represent costs in millions of dollars.
(b) What type of function appears to be the best fit for these points?
(c) Graph different quadratic functions of the form $y = ax^2 + c$ until you find a curve that is a reasonable fit for the points.
(d) Use your "model" function to predict health care costs in 1990.
(e) Actual health care costs for 1990 were $666.2 million. Do health care costs seem to be increasing even faster than your model predicts? Does it appear that some measures are necessary to help slow the growth of health care costs?

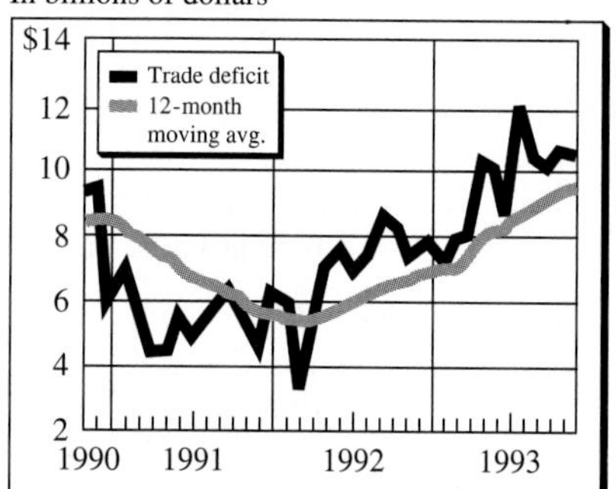

SOURCE: *The Wall Street Journal*, December 17, 1993

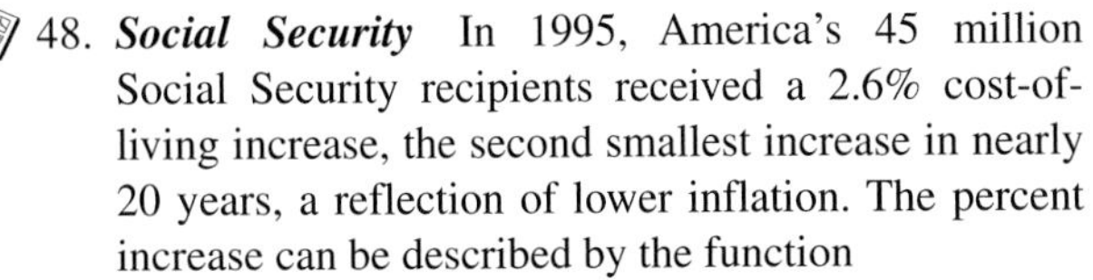

48. ***Social Security*** In 1995, America's 45 million Social Security recipients received a 2.6% cost-of-living increase, the second smallest increase in nearly 20 years, a reflection of lower inflation. The percent increase can be described by the function

$$p(t) = -0.3375t^2 + 7.3t - 34.3625$$

where t is the number of years past 1980. (Source: Social Security Administration)
(a) Graph the function $y = p(t)$.
(b) From the graph, identify t-values where $p(t) < 0$ and hence where the model is not valid.
(c) From the graph, identify the maximum point on the graph of $p(t)$.

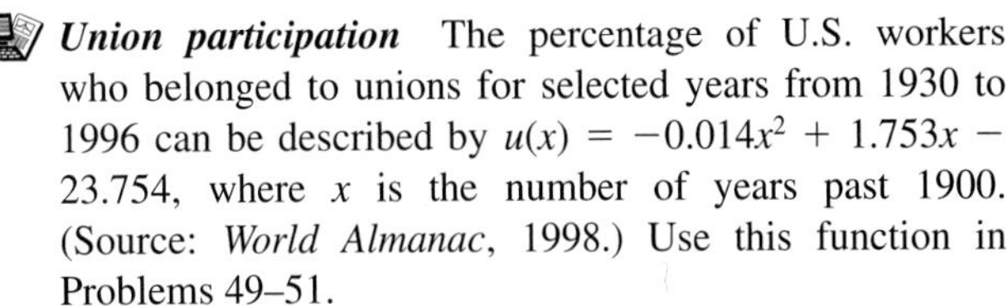

Union participation The percentage of U.S. workers who belonged to unions for selected years from 1930 to 1996 can be described by $u(x) = -0.014x^2 + 1.753x - 23.754$, where x is the number of years past 1900. (Source: *World Almanac*, 1998.) Use this function in Problems 49–51.

49. Graph the function $u(x)$.
50. For what year does the function $u(x)$ indicate a maximum percentage of workers belonged to unions?
51. (a) For what years does the function $u(x)$ predict that 0% of U.S. workers will belong to unions?
(b) When can you guarantee that $u(x)$ can no longer be used to describe the percentage of U.S. workers who belong to unions?

2.3 Business Applications of Quadratic Functions

OBJECTIVES

- *To graph quadratic supply and demand functions*
- *To find market equilibrium by using quadratic supply and demand functions*
- *To find break-even points by using quadratic cost and revenue functions*
- *To maximize quadratic revenue and profit functions*

APPLICATION PREVIEW

In this section, we revisit applications concerning market analysis (supply and demand) and theory of the firm (cost, revenue, and profit).

We extend our work to graphing supply and demand functions when they are quadratic. We also reconsider market equilibrium and find it by simultaneously solving supply and demand equations in situations when at least one of the equations is quadratic.

We consider graphs of cost, revenue, and profit functions when they are quadratic. We also revisit break-even points in a quadratic setting and notice that the notion of break-even points does not change even though the mathematics used to find them does. Also we use the vertex of quadratic functions to determine maximum and minimum points, such as for revenue and profit.

Supply, Demand, and Market Equilibrium

The first-quadrant parts of parabolas or other quadratic equations are frequently used to represent supply and demand functions. For example, the first-quadrant part of $p = q^2 + q + 2$ (Figure 2.10) may represent a supply curve, whereas the first-quadrant part of $q^2 + 2q + 6p - 23 = 0$ (Figure 2.11) may represent a demand curve.

When quadratic equations are used to represent supply or demand curves, we can solve their equations simultaneously to find the market equilibrium as we did with linear supply and demand functions. As in Section 1.5, we can solve two equations in two variables by eliminating one variable and obtaining an equation in the other variable. When the functions are quadratic, the substitution method of solution is perhaps the best, and the resulting equation in one unknown will usually be quadratic.

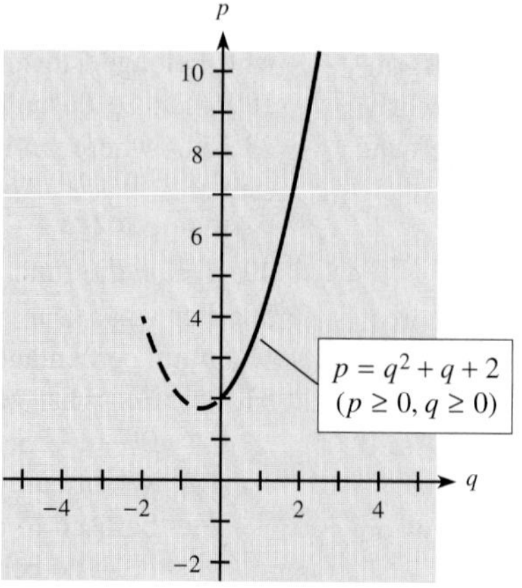

Figure 2.10

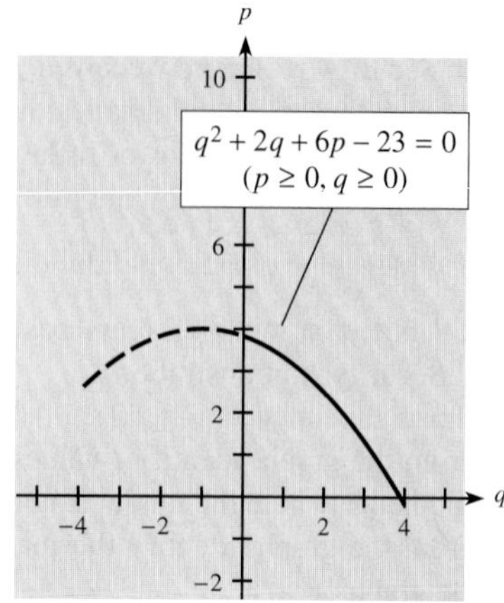

Figure 2.11

EXAMPLE 1

If the supply function for a commodity is given by $p = q^2 + 100$ and the demand function is given by $p = -20q + 2500$, find the point of market equilibrium.

Solution

At market equilibrium, both equations will have the same p-value. Thus substituting $q^2 + 100$ for p in $p = -20q + 2500$ yields

$$q^2 + 100 = -20q + 2500$$
$$q^2 + 20q - 2400 = 0$$
$$(q - 40)(q + 60) = 0$$
$$q = 40 \quad \text{or} \quad q = -60$$

Because a negative quantity has no meaning, the equilibrium point occurs when 40 units are sold, at (40, 1700). The graphs of the functions are shown (in the first quadrant only) in Figure 2.12.

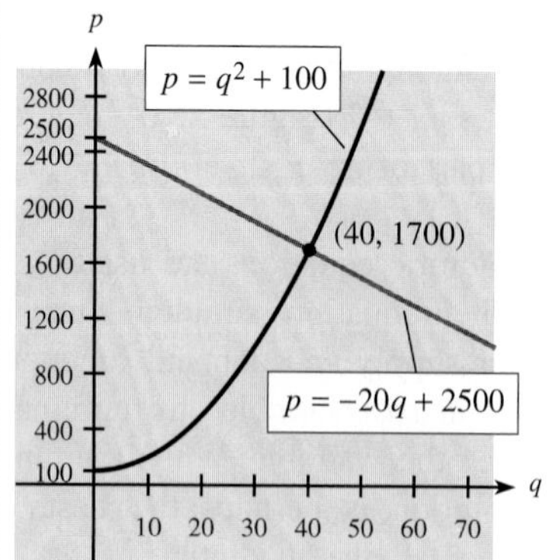

Figure 2.12

EXAMPLE 2

If the demand function for a commodity is given by $p(q + 4) = 400$ and the supply function is given by $2p - q - 38 = 0$, find the market equilibrium.

Solution

First note that the demand is defined by the function $p = 400/(q + 4)$. Solving $2p - q - 38 = 0$ for p gives $p = \frac{1}{2}q + 19$. Substituting for p in $p(q + 4) = 400$ gives

$$\left(\frac{1}{2}q + 19\right)(q + 4) = 400$$

$$\frac{1}{2}q^2 + 21q - 324 = 0$$

Multiplying both sides of the equation by 2 yields $q^2 + 42q - 648 = 0$. Factoring gives

$$(q - 12)(q + 54) = 0$$

$$q = 12 \quad \text{or} \quad q = -54$$

Thus the market equilibrium occurs when 12 items are sold, at a price of \$25 each. The graphs of the demand and supply functions are shown in Figure 2.13.

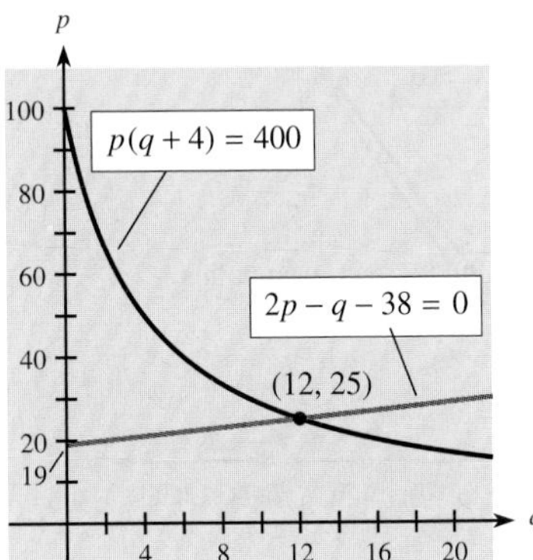

Figure 2.13

Technology Note Graphing utilities also can be used to sketch these graphs. TRACE and ZOOM, or a special command such as INTERSECT, could be used to determine points of intersection that give market equilibrium.

CHECKPOINT

1. The point of intersection of the supply and demand functions is called ________.
2. If the demand and supply functions for a product are

$$p + \frac{1}{10}q^2 = 1000 \quad \text{and} \quad p = \frac{1}{10}q + 10$$

respectively, finding the market equilibrium point requires solution of what equation?

Break-Even Points and Maximization

In Chapter 1, we discussed linear total cost and total revenue functions. Many total revenue functions may be linear, but costs tend to increase sharply after a certain level of production. Thus functions other than linear functions, including quadratic functions, are used to predict the total costs of products.

For example, the monthly total cost curve for a commodity may be the parabola with equation $C(x) = 360 + 40x + 0.1x^2$. If the total revenue function is $R(x) = 60x$, we can find the break-even point by finding the quantity x that makes $C(x) = R(x)$. (See Figure 2.14.)

Setting $C(x) = R(x)$, we have

$$360 + 40x + 0.1x^2 = 60x$$
$$0.1x^2 - 20x + 360 = 0$$
$$x^2 - 200x + 3600 = 0$$
$$(x - 20)(x - 180) = 0$$
$$x = 20 \qquad \text{or} \qquad x = 180$$

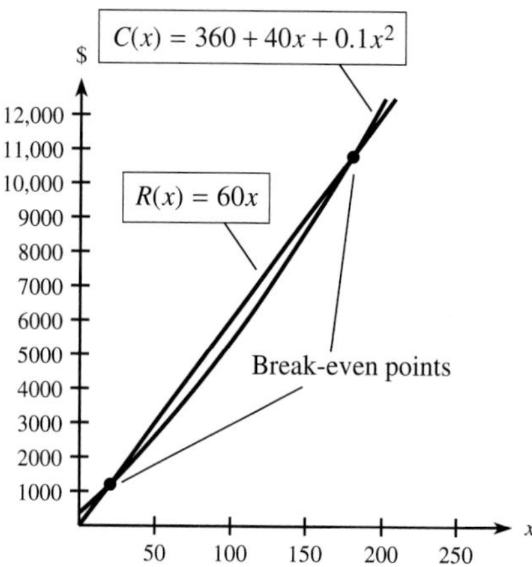

Figure 2.14

Thus $C(x) = R(x)$ at $x = 20$ and at $x = 180$. If 20 items are produced and sold, $C(x)$ and $R(x)$ are both \$1200; if 180 items are sold, $C(x)$ and $R(x)$ are both \$10,800. Thus there are two break-even points.

In a monopoly market, the revenue of a company is restricted by the demand for the product. In this case, the relationship between the price p of the product and the number of units sold x is described by the demand function $p = f(x)$, and the total revenue function for the product is given by

$$R = px = [f(x)]x$$

If, for example, the demand for a product is given by

$$p = 300 - x$$

where x is the number of units sold and p is the price, then the revenue function for this product is

$$R = px = (300 - x)x$$

or

$$R = 300x - x^2$$

a quadratic function.

EXAMPLE 3

Suppose that in a monopoly market the total cost per week of producing a high-tech product is given by $C = 3600 + 100x + 2x^2$. Suppose further that the weekly demand function for this product is $p = 500 - 2x$. Find the number of units that will give break-even for the product.

Solution

The total cost function is $C(x) = 3600 + 100x + 2x^2$, and the total revenue function is $R(x) = (500 - 2x)x = 500x - 2x^2$.

Setting $C(x) = R(x)$ and solving for x gives

$$3600 + 100x + 2x^2 = 500x - 2x^2$$
$$4x^2 - 400x + 3600 = 0$$
$$x^2 - 100x + 900 = 0$$
$$(x - 90)(x - 10) = 0$$
$$x = 90 \quad \text{or} \quad x = 10$$

Does this mean the firm will break even at 10 units and at 90 units? Yes. Figure 2.15 shows the graphs of $C(x)$ and $R(x)$. From the graph we can observe that the firm makes a profit after $x = 10$ *until* $x = 90$, because $R(x) > C(x)$ in that interval. At $x = 90$, the profit is 0, and the firm loses money if it produces more than 90 units per week.

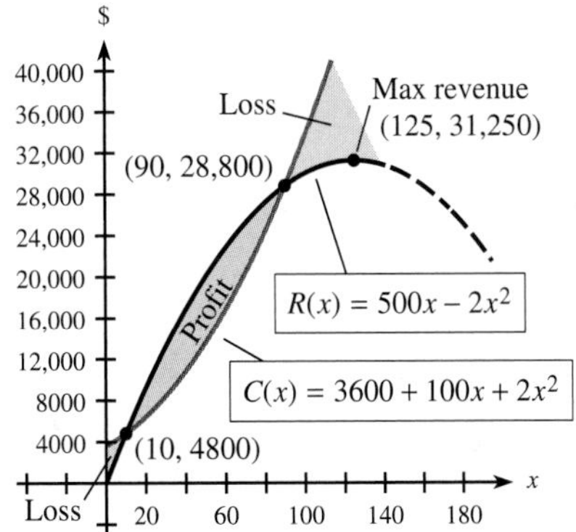

Figure 2.15

CHECKPOINT

3. The point of intersection of the revenue function and the cost function is called ________.

4. If $C(x) = 120x + 15{,}000$ and $R(x) = 370x - x^2$, finding the break-even points requires solution of what equation?

Note that for Example 3, the revenue function

$$R(x) = (500 - 2x)x = 500x - 2x^2$$

is a parabola that opens downward. Thus the vertex is the point at which revenue is maximum. We can locate this vertex by using the methods discussed in the previous section.

$$\text{Vertex: } x = \frac{-b}{2a} = \frac{-500}{2(-2)} = \frac{500}{4} = 125 \text{ (units)}$$

It is interesting to note that when $x = 125$, the firm achieves its maximum revenue of

$$R(125) = 500(125) - 2(125)^2 = 31{,}250 \text{ (dollars)}$$

but the costs when $x = 125$ are

$$C(125) = 3600 + 100(125) + 2(125)^2 = 47{,}350 \text{ (dollars)}$$

which results in a loss. This illustrates that maximizing revenue is not a good goal. We should seek to maximize profit.

Using

$$\text{Profit} = \text{revenue} - \text{cost}$$

we can determine the profit function:

$$P(x) = (500x - 2x^2) - (3600 + 100x + 2x^2)$$

or

$$P(x) = -3600 + 400x - 4x^2$$

This profit function is a parabola that opens downward, so the vertex will be the maximum point.

$$\text{Vertex: } x = \frac{-b}{2a} = \frac{-400}{2(-4)} = \frac{-400}{-8} = 50$$

Furthermore, when $x = 50$, we have

$$P(50) = -3600 + 400(50) - 4(50)^2 = 6400 \text{ (dollars)}$$

Thus, when 50 items are produced and sold, a maximum profit of \$6400 is made (see Figure 2.16).

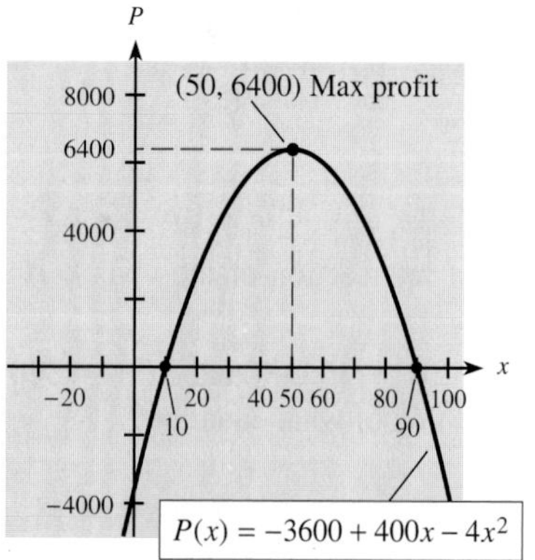

Figure 2.16

Comparing Figures 2.15 and 2.16, note first that the maximum profit of \$6400 at 50 units in Figure 2.16 corresponds to the vertical distance between revenue and cost at 50 units in Figure 2.15. In addition, the break-even points at $x = 10$ and $x = 90$ in Figure 2.15 correspond to x-intercepts (Profit = 0) in Figure 2.16.

It is important to note that the procedures for finding maximum revenue and profit in these examples depend on the fact that these functions are parabolas. Using methods discussed in Sections 2.1 and 2.2, we can use graphing utilities to locate maximum points, minimum points, and break-even points. For more general functions, procedures for finding maximum or minimum values are discussed in Chapter 10, "Applications of Derivatives."

CHECKPOINT SOLUTIONS

1. The market equilibrium point or market equilibrium
2. Demand: $p = -\frac{1}{10}q^2 + 1000$; supply: $p = \frac{1}{10}q + 10$

   Solution of $-\frac{1}{10}q^2 + 1000 = \frac{1}{10}q + 10$

   $$-q^2 + 10{,}000 = q + 100$$
   $$0 = q^2 + q - 9900$$
3. The break-even point
4. Solution of $C(x) = R(x)$. That is, solution of $120x + 15{,}000 = 370x - x^2$, or $x^2 - 250x + 15{,}000 = 0$.

EXERCISE 2.3

Supply, Demand, and Market Equilibrium

1. Sketch the first-quadrant portions of the following on the same set of axes.
   (a) The supply function whose equation is $p = \frac{1}{4}q^2 + 10$
   (b) The demand function whose equation is $p = 86 - 6q - 3q^2$
   (c) Label the market equilibrium point.
   (d) Algebraically determine the equilibrium point for the supply and demand functions.
2. Sketch the first-quadrant portions of the following on the same set of axes.
   (a) The supply function whose equation is $p = q^2 + 8q + 16$
   (b) The demand function whose equation is $p = 216 - 2q$
   (c) Label the market equilibrium point.
   (d) Algebraically determine the equilibrium point for the supply and demand functions.
3. Sketch the first-quadrant portions of the following on the same set of axes.
   (a) The supply function whose equation is $p = 0.2q^2 + 0.4q + 1.8$
   (b) The demand function whose equation is $p = 9 - 0.2q - 0.1q^2$
   (c) Label the market equilibrium point.
   (d) Algebraically determine the market equilibrium point.
4. Sketch the first-quadrant portions of the following on the same set of axes.
   (a) The supply function whose equation is $p = q^2 + 8q + 22$
   (b) The demand function whose equation is $p = 198 - 4q - \frac{1}{4}q^2$
   (c) Label the market equilibrium point.
   (d) Algebraically determine the market equilibrium point for the supply and demand functions.

5. If the supply function for a commodity is $p = q^2 + 8q + 16$ and the demand function is $p = -3q^2 + 6q + 436$, find the equilibrium quantity and equilibrium price.
6. If the supply function for a commodity is $p = q^2 + 8q + 20$ and the demand function is $p = 100 - 4q - q^2$, find the equilibrium quantity and equilibrium price.
7. If the demand function for a commodity is given by the equation $p^2 + 4q = 1600$ and the supply function is given by the equation $300 - p^2 + 2q = 0$, find the equilibrium quantity and equilibrium price.
8. If the supply and demand functions for a commodity are given by $4p - q = 42$ and $(p + 2)q = 2100$, respectively, find the price that will result in market equilibrium.
9. If the supply and demand functions for a commodity are given by $q = p - 10$ and $q = 1200/p$, what is the equilibrium price and what is the corresponding number of units supplied and demanded?
10. If the supply and demand functions for a certain product are given by the equations $2p - q + 6 = 0$ and $(p + q)(q + 10) = 3696$, respectively, find the price and quantity that give market equilibrium.
11. The supply function for a product is $2p - q - 10 = 0$, while the demand function for the same product is $(p + 10)(q + 30) = 7200$. Find the market equilibrium point.
12. The supply and demand for a product are given by $2p - q = 50$ and $pq = 100 + 20q$, respectively. Find the market equilibrium point.
13. For the product in Problem 11, if a \$22 tax is placed on production of the item, then the supplier passes this tax on by adding \$22 to his selling price. Find the new equilibrium point for this product when the tax is passed on. (The new supply function is given by $p = \frac{1}{2}q + 27$.)
14. For the product in Problem 12, if a \$12.50 tax is placed on production and passed through by the supplier, find the new equilibrium point.

Break-Even Points and Maximization

15. The total costs for a company are given by

$$C(x) = 2000 + 40x + x^2$$

and the total revenues are given by

$$R(x) = 130x$$

Find the break-even points.

16. If a firm has the following cost and revenue functions, find the break-even points.

$$C(x) = 3600 + 25x + \frac{1}{2}x^2$$

$$R(x) = \left(175 - \frac{1}{2}x\right)x$$

17. If a company has total costs $C(x) = 15{,}000 + 35x + 0.1x^2$ and total revenues given by $R(x) = 385x - 0.9x^2$, find the break-even points.
18. If total costs are $C(x) = 1600 + 1500x$ and total revenues are $R(x) = 1600x - x^2$, find the break-even points.
19. Given that $C(x) = 150 + x + 0.09x^2$ and $R(x) = 12.5x - 0.01x^2$, and given that production is restricted to fewer than 75 units, find the break-even points.
20. If the profit function for a firm is given by $P(x) = -1100 + 120x - x^2$ and limitations on space require that production is less than 100 units, find the break-even points.
21. Find the maximum revenue for the revenue function $R(x) = 385x - 0.9x^2$.
22. Find the maximum revenue for the revenue function $R(x) = 1600x - x^2$.
23. If, in a monopoly market, the demand for a product is $p = 175 - 0.50x$ and the revenue function is $R = px$, where x is the number of units sold, what price will maximize revenue?
24. If, in a monopoly market, the demand for a product is $p = 1600 - x$ and the revenue is $R = px$, where x is the number of units sold, what price will maximize revenue?
25. The profit function for a certain commodity is $P(x) = 110x - x^2 - 1000$. Find the level of production that yields maximum profit, and find the maximum profit.
26. The profit function for a firm making widgets is $P(x) = 88x - x^2 - 1200$. Find the number of units at which maximum profit is achieved and find the maximum profit.
27. (a) Form the profit function for the cost and revenue functions in Problem 17, and find maximum profit.
    (b) Compare the level of production to maximize profit with the level to maximize revenue (see Problem 21). Do they agree?
    (c) How do the break-even points compare with the zeros of $P(x)$?

28. (a) Form the profit function for the cost and revenue functions in Problem 18, and find maximum profit.
(b) Compare the level of production to maximize profit with the level to maximize revenue (see Problem 22). Do they agree?
(c) How do the break-even points compare with the zeros of $P(x)$?

29. Suppose a company has fixed costs of $28,000 and variable costs of $\frac{2}{5}x + 222$ dollars per unit, where x is the total number of units produced. Suppose further that the selling price of its product is $1250 - \frac{3}{5}x$ dollars per unit.
(a) Find the break-even points.
(b) Find the maximum revenue.
(c) Form the profit function from the cost and revenue functions and find maximum profit.
(d) What price will maximize the profit?

30. Suppose a company has fixed costs of $300 and variable costs of $\frac{3}{4}x + 1460$ dollars per unit, where x is the total number of units produced. Suppose further that the selling price of its product is $1500 - \frac{1}{4}x$ dollars per unit.
(a) Find the break-even points.
(b) Find the maximum revenue.
(c) Form the profit function from the cost and revenue functions and find maximum profit.
(d) What price will maximize the profit?

31. The following table gives the total revenues of AT&T for selected years.

Year	*Total revenues (billions)*
1985	$63.13
1986	$69.906
1987	$60.53
1989	$61.1
1990	$62.191
1991	$63.089
1992	$64.904
1993	$67.156

SOURCE: *AT&T Annual Report*, 1993

Suppose the data can be described by the equation

$$R(t) = 0.253t^2 - 4.03t + 76.84$$

where t is the number of years past 1980.
(a) Use the function to find the year in which revenue was minimum and find the minimum predicted revenue.
(b) Check the result from (a) against the data in the table.
(c) Graph $R(t)$ and the data points from the table.
(d) Write a sentence to describe how well the function fits the data.

The data in the table give sales revenues and costs and expenses for Scott Paper Co. for various years. Use this table in Problems 32 and 33.

Year	*Sales Revenue (billions)*	*Costs and Expenses (billions)*
1983	$2.6155	$2.4105
1984	2.7474	2.4412
1985	2.934	2.6378
1986	3.3131	2.9447
1987	3.9769	3.5344
1988	4.5494	3.8171
1989	4.8949	4.2587
1990	5.1686	4.8769
1991	4.9593	4.9088
1992	5.0913	4.6771
1993	4.7489	4.9025

SOURCE: Scott Paper Company, *1993 Annual Report*

32. Assume that sales revenues for Scott Paper can be described by

$$R(t) = -0.031t^2 + 0.776t + 0.179$$

where t is the number of years past 1980.
(a) Use the function to determine the year in which maximum revenue occurs and the maximum revenue it predicts.
(b) Check the result from (a) against the data in the table.
(c) Graph $R(t)$ and the data points from the table.
(d) Write a sentence to describe how well the function fits the data.

33. Assume that costs and expenses for Scott Paper can be described by

$$C(t) = -0.012t^2 + 0.492t + 0.725$$

where t is the number of years past 1980.
(a) Use $R(t)$ as given in Problem 32 and form the profit function (as a function of time).
(b) Use the function from (a) to find the year in which maximum profit occurs.
(c) Graph the profit function from (b) and the data points from the table.
(d) Through the mid to late 1990s, does the function project increasing or decreasing profits? Does the data support this trend (as far as it goes)?
(e) How might management respond to this kind of projection?

2.4 Special Functions and Their Graphs

OBJECTIVES

- *To graph and apply basic functions, including constant, power, rational, and absolute value functions*
- *To graph and apply piecewise defined functions*

APPLICATION PREVIEW

The average cost per item for a product is calculated by dividing the total cost by the number of items. Hence, if the total cost function for x units of a product is

$$C(x) = 900 + 3x + x^2$$

and if we denote the average cost function by $\overline{C}(x)$, then

$$\overline{C}(x) = \frac{C(x)}{x} = \frac{900 + 3x + x^2}{x}$$

This average cost function is a special kind of function, called a **rational function.** In this section, we discuss this and other types of functions and answer questions such as "What is the minimum average cost?"

In Section 1.3, "Linear Functions," we saw that linear functions have the form $y = ax + b$. The special linear function

$$y = f(x) = x$$

is called the **identity function** (Figure 2.17a), and a linear function defined by

$$y = f(x) = C \qquad C \text{ a constant}$$

is called a **constant function.** Figure 2.17(b) shows the graph of the constant function $y = f(x) = 2$. (Note that the slope of the graph of any constant function is 0.)

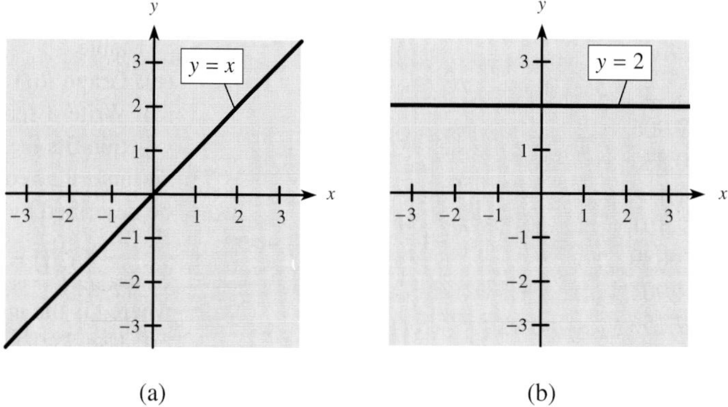

Figure 2.17 (a) (b)

The functions of the form $y = ax^b$, where $b > 0$, are called **power functions.** Examples of power functions include $y = x^2$, $y = x^3$, $y = \sqrt{x} = x^{1/2}$, and $y = \sqrt[3]{x} = x^{1/3}$. (See Figures 2.18a–d.) The functions $y = \sqrt{x}$ and $y = \sqrt[3]{x}$ are also called **root functions,** and $y = x^2$ and $y = x^3$ are basic **polynomial functions.** The general shape for the power function $y = ax^b$, where $b > 0$, depends on the value of b. Figure 2.19 shows the first-quadrant portion of typical graphs of $y = x^b$ for different values of b. Note how the direction in which the graph bends differs for $b > 1$ and for $0 < b < 1$. Getting accurate graphs of these functions requires plotting a number of points by hand or with a graphing utility. Our goal at this stage is to recognize the basic shapes of certain functions.

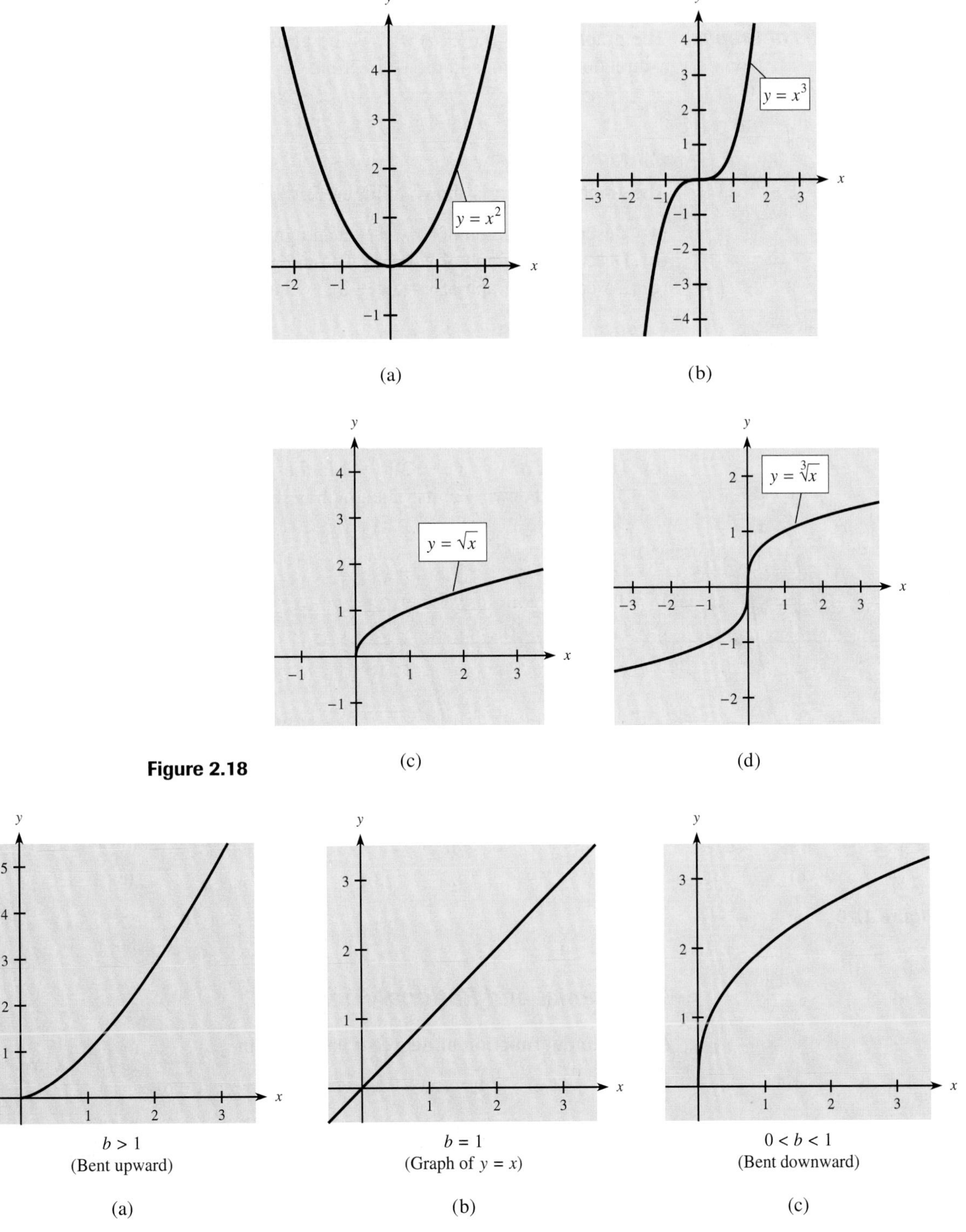

Figure 2.18

Figure 2.19

In Section 2.2, "Quadratic Functions: Parabolas," we noted that the graph of $y = (x - h)^2 + k$ is a parabola that is shifted h units in the x-direction and k units in the y-direction. In general, we have the following:

Shifts of Graphs The graph of $y = f(x - h) + k$ is the graph of $y = f(x)$ shifted h units in the x-direction and k units in the y-direction.

EXAMPLE 1

The graph of $y = x^3$ is shown in Figure 2.18(b).

(a) Describe the graph of $y = x^3 - 3$ and graph this function.
(b) Describe the graph of $y = (x - 2)^3$ and graph this function.
(c) Describe the graph of $y = (x - 2)^3 - 3$ and graph this function.

Solution

(a) The graph of $y = x^3 - 3$ is the graph of $y = x^3$ shifted down 3 units. The graph is shown in Figure 2.20(a).
(b) The graph of $y = (x - 2)^3$ is the graph of $y = x^3$ shifted to the right 2 units. The graph is shown in Figure 2.20(b).
(c) The graph of $y = (x - 2)^3 - 3$ is the graph of $y = x^3$ shifted to the right 2 units and down 3 units. The graph is shown in Figure 2.20(c).

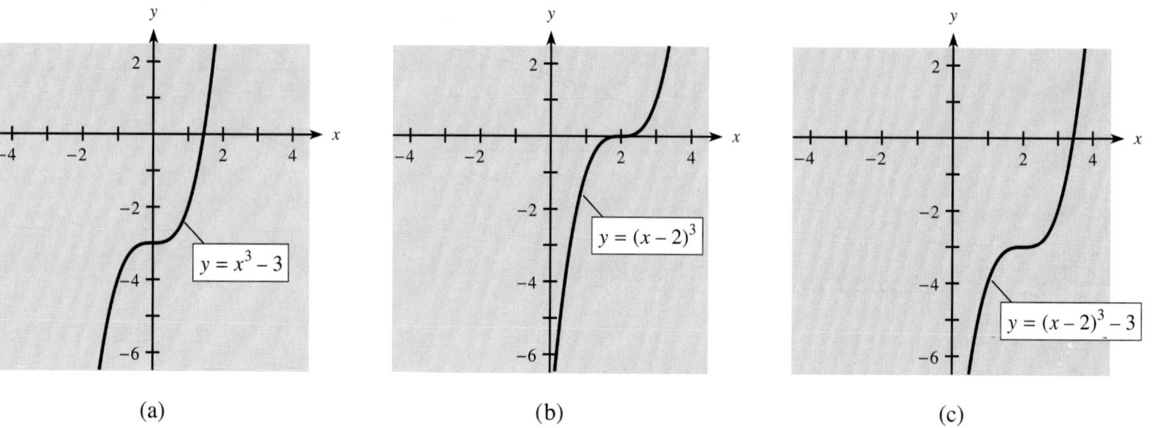

Figure 2.20

Polynomial and Rational Functions

A **polynomial function of degree n** has the form

$$y = a_n x^n + a_{n-1} x^{n-1} + \cdots + a_1 x + a_0$$

where $a_n \neq 0$, and n is an integer, $n \geq 0$.

A **linear function** is a polynomial function of degree 1, and a **quadratic function** is a polynomial function of degree 2.

Accurate graphing of a polynomial function of degree greater than 2 may require the methods of calculus; we will investigate these methods in Chapter 10, "Applications of Derivatives." For now, we will observe some characteristics of the graphs of polynomial functions of degrees 2, 3, and 4; these are summarized in Table 2.1. Using this information and point-plotting or a graphing utility yields the graphs of these functions.

TABLE 2.1 Graphs of Some Polynomials

	Degree 2	*Degree 3*	*Degree 4*
Turning points	1	0 or 2	1 or 3
x-intercepts	0, 1, or 2	1, 2, or 3	0, 1, 2, 3, or 4
Possible shapes	(a) or (b)	(a) (b) (c) (d)	(a) (b) (c) (d) (e) (f)

EXAMPLE 2

Graph $y = x^3 - 16x$.

Solution

This function is a third-degree polynomial function, so it has one of the four shapes shown in the "Degree 3" column in Table 2.1. By graphing the function with a large x-range and y-range, we obtain the graph shown in Figure 2.21(a). We see that the graph has a shape like (b) in Table 2.1 so we know that a smaller viewing window (using a smaller x-range and y-range) will provide a more detailed view of all the graph's turning points (see Figure 2.21b). Figure 2.21(b) indicates that the turning points are at $x = -2.3$ and $x = 2.3$ (to the nearest tenth).

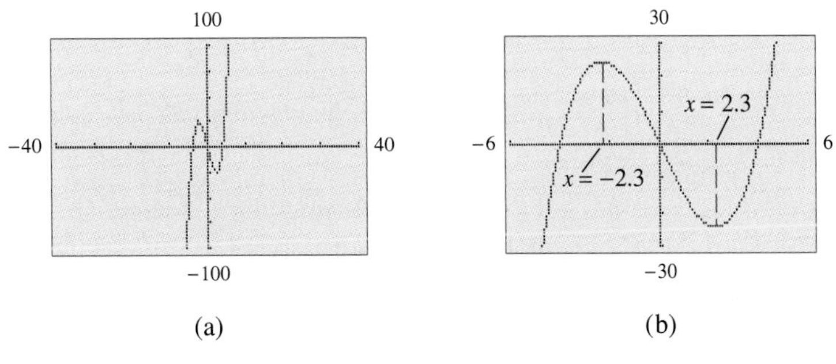

Figure 2.21 Two views of $y = x^3 - 16x$

EXAMPLE 3

Graph $y = x^4 - 2x^2$.

Solution

This is a polynomial function of degree 4. Thus it has one or three turning points and has one of the six shapes in the "Degree 4" column of Table 2.1. We begin by graphing this function with a large viewing window to obtain the graph in Figure 2.22(a). We see that the graph appears to fit either shape (a) or shape (f) in Table 2.1. Figure 2.22(b) shows the same graph with a smaller viewing window, near $x = 0$. We now see that the graph has shape (f) from Table 2.1, with turning points at $x = -1$, $x = 0$, and $x = 1$.

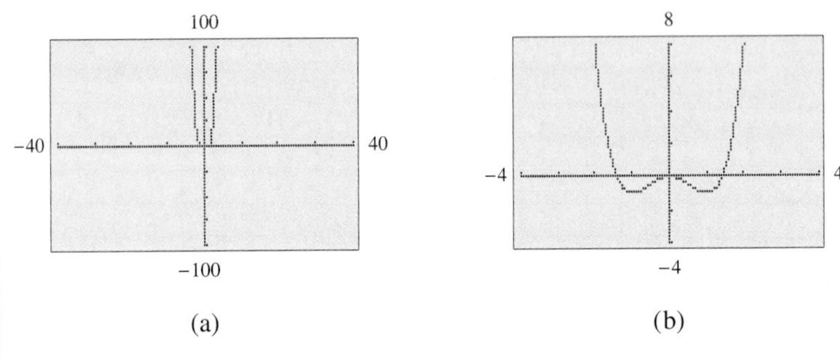

Figure 2.22 Two views of $y = x^4 - 2x^2$

CHECKPOINT

1. All constant functions (such as $f(x) = 8$) have graphs that are ________.
2. Which of the following are polynomial functions?
   (a) $f(x) = x^3 - x + 4$
   (b) $f(x) = \dfrac{x + 1}{4x}$
   (c) $f(x) = 1 + \sqrt{x}$
   (d) $g(x) = \dfrac{1 + \sqrt{x}}{1 + x + \sqrt{x}}$
   (e) $h(x) = 5$
3. A third-degree polynomial can have at most ________ turning points.

A **rational function** is a function that can be written as the ratio of two polynomials. The graph of the rational function

$$y = \frac{1}{x}$$

is shown in Figure 2.23. Because division by 0 is not possible, $x = 0$ is not in the domain of the function.

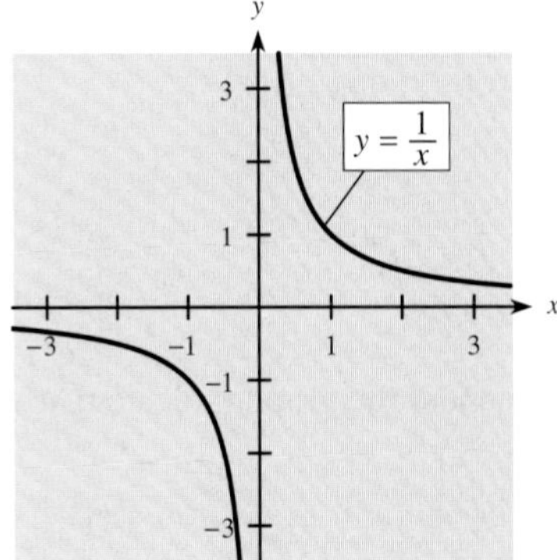

Figure 2.23

The graph of $y = 1/x$ approaches the y-axis but does not touch it. In this case, we call the y-axis a **vertical asymptote.**

On the graphs of polynomial functions, the turning points are usually the features of greatest interest. However, on the graphs of rational functions, vertical asymptotes frequently are the most interesting features. (Rational functions may or may not have turning points.)

In general, when a rational function has its denominator equal to 0 at a value $x = a$, we should include $x = a$ in our viewing rectangle to see what shape the graph is taking near $x = a$. When $|y|$ increases without bound as x gets close to a, we say that $x = a$ is a **vertical asymptote.**

Rational functions sometimes have **horizontal asymptotes** as well as vertical asymptotes. Whenever the values of y approach some finite number b as $|x|$ becomes very large, we say that there is a **horizontal asymptote** at $y = b$.

Note that the graph of $y = 1/x$ appears to get close to the x-axis as $|x|$ becomes large. Testing values of x for which $|x|$ is large, we see that y is close to 0. Thus, we say that $y = 0$ (or the x-axis) is a **horizontal asymptote** for the graph of $y = 1/x$. The graph in the following example has both a vertical and a horizontal asymptote.

EXAMPLE 4

Sketch the graph of

$$y = \frac{12x + 8}{3x - 9}$$

Determine any asymptotes.

Solution

Because $3x - 9 = 0$ when $x = 3$, it follows that $x = 3$ is not in the domain of the function. Thus, to search for a vertical asymptote, our window should include $x = 3$ and large values for y. Figure 2.24 shows a vertical asymptote at $x = 3$. Also, it looks like the graph has a horizontal asymptote. To verify this, we calculate y as $|x|$ becomes larger.

x	$y = f(x)$
$-10{,}000$	3.998
-1000	3.98
1000	4.01
10,000	4.001

This calculation and the graph indicate that $y = 4$ is a horizontal asymptote.

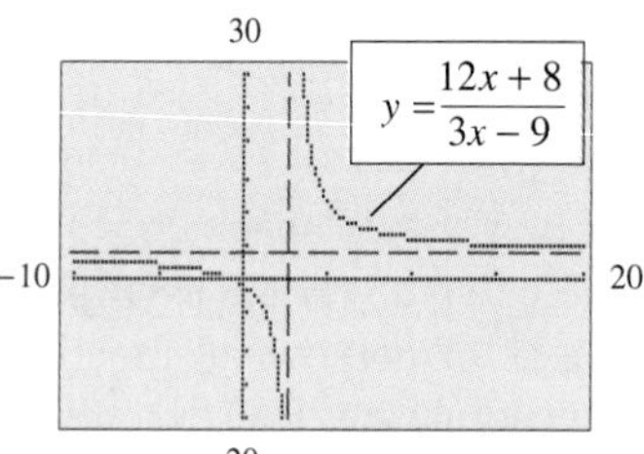

Figure 2.24

EXAMPLE 5

Suppose the total cost function for x units of a product is given by

$$C(x) = 900 + 3x + x^2$$

Graph the average cost function for this product, and determine the minimum average cost.

Solution

The average cost per unit is

$$\overline{C}(x) = \frac{C(x)}{x} = \frac{900 + 3x + x^2}{x}$$

or

$$\overline{C}(x) = \frac{900}{x} + 3 + x$$

Because x represents the number of units produced, $x \geq 0$. Because $x = 0$ cannot be in the domain of the function, we choose a window with $x \geq 0$. Figure 2.25 shows the graph of $\overline{C}(x)$. The graph appears to have a minimum near $x = 30$. By plotting points or using MINIMUM or TRACE on a graphing utility, we can verify that the minimum point occurs at $x = 30$ and $\overline{C} = 63$. Thus the minimum average cost is \$63 per unit when 30 units are produced.

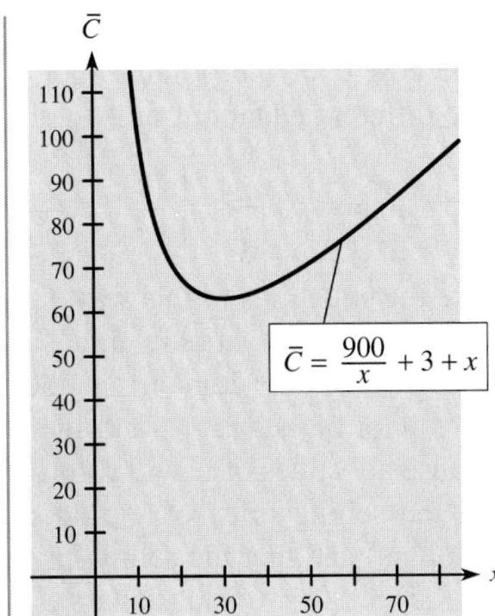

Figure 2.25

CHECKPOINT

4. Given $f(x) = \dfrac{3x}{x-4}$, decide whether the following are true or false.
   (a) $f(x)$ has a vertical asymptote at $x = 4$.
   (b) $f(x)$ has a horizontal asymptote at $x = 3$.

Piecewise Defined Functions

Another special function comes from the definition of $|x|$. The **absolute value function** can be written as

$$f(x) = |x| \quad \text{or} \quad f(x) = \begin{cases} x & \text{if } x \geq 0 \\ -x & \text{if } x < 0 \end{cases}$$

Note that restrictions on the domain of the absolute value function specify different formulas for different parts of the domain. To graph $f(x) = |x|$, we graph the portion of the line $y = x$ for $x \geq 0$ (see Figure 2.26a) and the portion of the line $y = -x$ for $x < 0$ (see Figure 2.26b). When we put these pieces on the same graph (Figure 2.26c), they give us the graph of $f(x) = |x|$. Because the absolute value function is defined by two equations, we say it is a **piecewise defined function.**

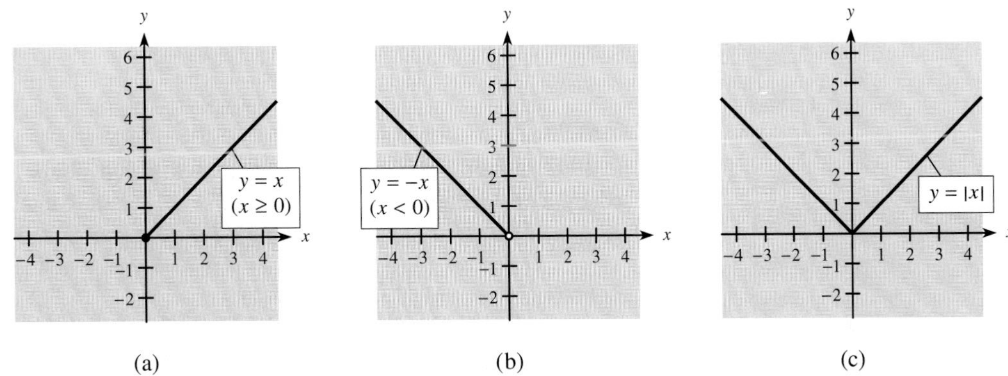

Figure 2.26 (a) (b) (c)

It is possible for the selling price S of a product to be defined as a piecewise function of the cost C of the product. For example, the selling price might be defined by two different equations on two different intervals, as follows:

$$S = f(C) = \begin{cases} 3C & \text{if } 0 \le C \le 20 \\ 1.5C + 30 & \text{if } C > 20 \end{cases}$$

When we write the equations in this way, the value of S depends on the value of C, so C is the independent variable and S is the dependent variable. The set of possible values of C (the domain) is the set of nonnegative real numbers (because cost cannot be negative), and the range is the set of nonnegative real numbers (because all values of C will result in nonnegative values for S).

Note that the selling price of a product that costs \$15 would be $f(15) = 3(15) = 45$ (dollars) and that the selling price of a product that costs \$25 would be $f(25) = 1.5(25) + 30 = 67.50$ (dollars).

EXAMPLE 6

Graph the selling price function

$$S = f(C) = \begin{cases} 3C & \text{if } 0 \le C \le 20 \\ 1.5C + 30 & \text{if } C > 20 \end{cases}$$

Solution

Each of the two pieces of the graph of this function is a line and is easily graphed. It remains only to graph each in the proper interval. The graph is shown in Figure 2.27(a).

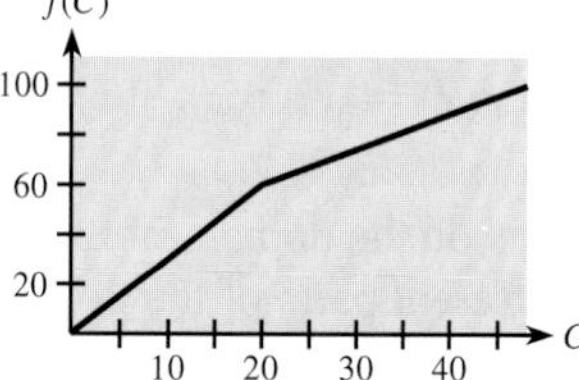

Figure 2.27(a)

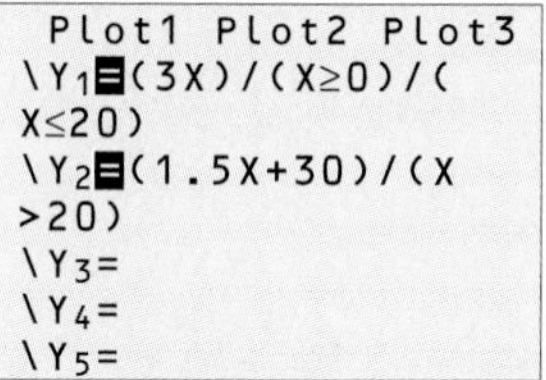

Figure 2.27(b)

Technology Note

Graphing utilities can be used to graph piecewise defined functions by entering each piece as a separate equation, along with the interval over which it is defined. Figure 2.27(b) shows how the function from Example 6 could be entered on a T1-83 calculator, using x to represent C.

EXAMPLE 7

The 1993 monthly charge in dollars for x kilowatt hours (kWh) of electricity used by a residential customer of Excelsior Electric Membership Corporation during the months of November through June is given by the function

$$C(x) = \begin{cases} 10 + 0.094x & \text{if } 0 \le x \le 100 \\ 19.40 + 0.075(x - 100) & \text{if } 100 < x \le 500 \\ 49.40 + 0.05(x - 500) & \text{if } x > 500 \end{cases}$$

(a) What is the monthly charge if 1100 kWh of electricity are consumed in a month?

(b) What is the monthly charge if 450 kWh are consumed in a month?
(c) Is there a charge if no electricity is used?

Solution

(a) We need to find $C(1100)$. Because $1100 > 500$, we use it in the third formula line.

$$C(1100) = 49.40 + 0.05(1100 - 500) = \$79.40$$

(b) We evaluate $C(450)$ by using the second formula line for $C(x)$.

$$C(450) = 19.40 + 0.075(450 - 100) = \$45.65$$

(c) If no electricity is used, the charge is $C(0)$; it is found by using the first formula line.

$$C(0) = 10 + 0.094(0) = \$10$$

There is a \$10 charge even when no electricity is used. This could be thought of as a service charge or line charge for the privilege and convenience of having electrical service available.

CHECKPOINT

5. If $f(x) = \begin{cases} 5 & \text{if } x \le 0 \\ 2x & \text{if } 0 < x < 5, \\ x + 6 & \text{if } x \ge 5 \end{cases}$ find the following.
   (a) $f(-5)$ (b) $f(4)$ (c) $f(20)$

CHECKPOINT SOLUTIONS

1. Horizontal lines
2. Polynomial functions are (a) and (e); also (e) is a constant function; function (b) is a quotient of polynomials (a rational function).
3. Two
4. (a) True.
   (b) False. The horizontal asymptote is $y = 3$.
5. (a) $f(-5) = 5$. In fact, for any negative value of x, $f(x) = 5$.
   (b) $f(5) = 5 + 6 = 11$
   (c) $f(20) = 20 + 6 = 26$

EXERCISE 2.4

In Problems 1–14, match each of the functions with one of the graphs labeled (*a*)–(*n*) shown following these functions. Recognizing special features of certain types of functions and plotting points for the functions will be helpful.

1. $y = x^3$
2. $y = \sqrt{x}$
3. $f(x) = -3$
4. $y = x$
5. $y = \sqrt[3]{x}$
6. $y = (\sqrt{x})^5$
7. $y = (x + 4)^3 + 1$
8. $y = \dfrac{1}{x}$
9. $y = |x|$
10. $y = |x - 2|$
11. $f(x) = \begin{cases} x^2 & \text{if } x \le 2 \\ 4 & \text{if } x > 2 \end{cases}$
12. $f(x) = \begin{cases} 2 & \text{if } x < 0 \\ x^3 & \text{if } x \ge 0 \end{cases}$
13. $y = \sqrt{x - 2}$
14. $y = \begin{cases} -x & \text{if } x < -1 \\ 1 & \text{if } -1 \le x \le 1 \\ x & \text{if } x > 1 \end{cases}$

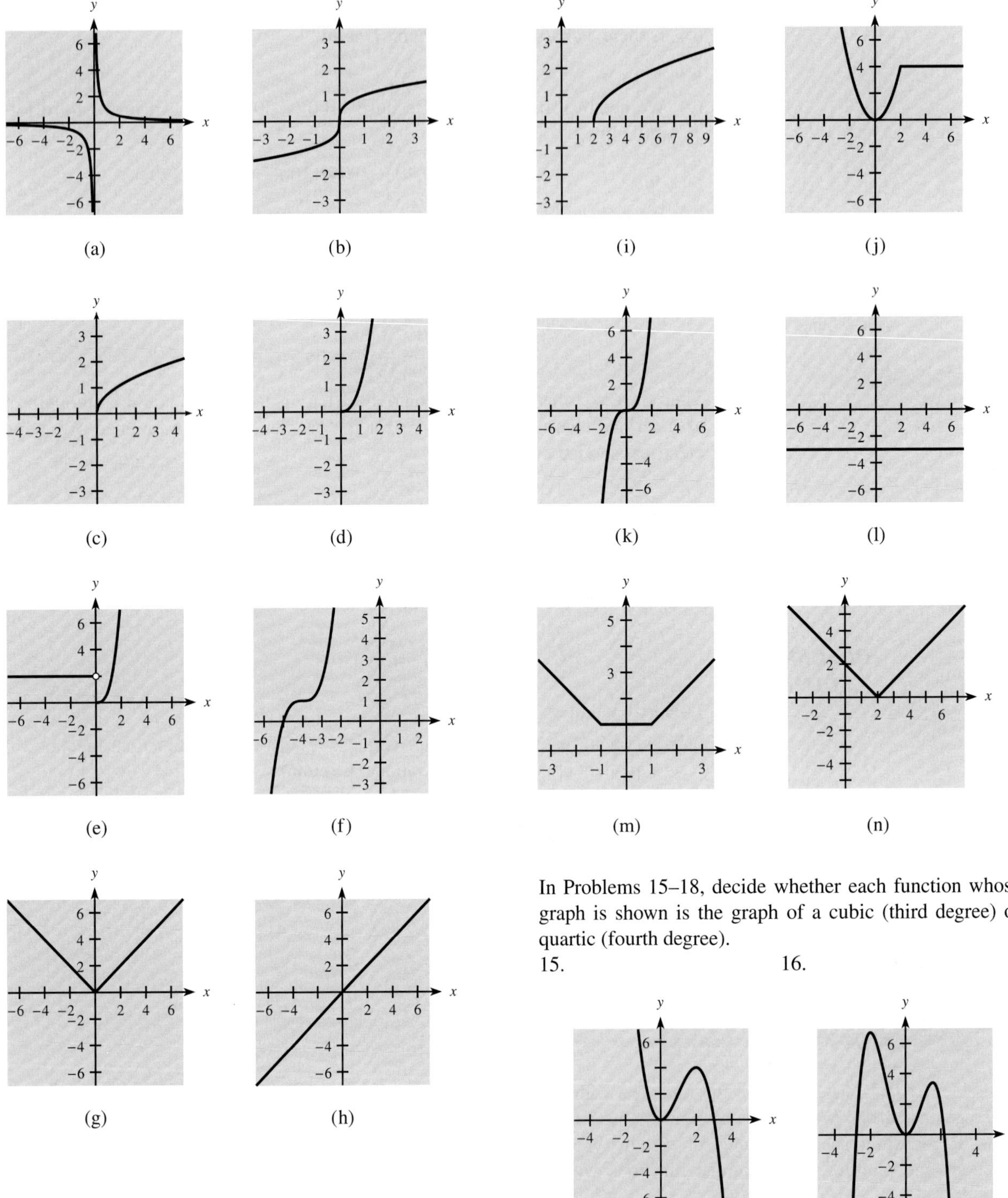

(a) (b) (i) (j)

(c) (d) (k) (l)

(e) (f) (m) (n)

(g) (h)

In Problems 15–18, decide whether each function whose graph is shown is the graph of a cubic (third degree) or quartic (fourth degree).

15.

16.

17.

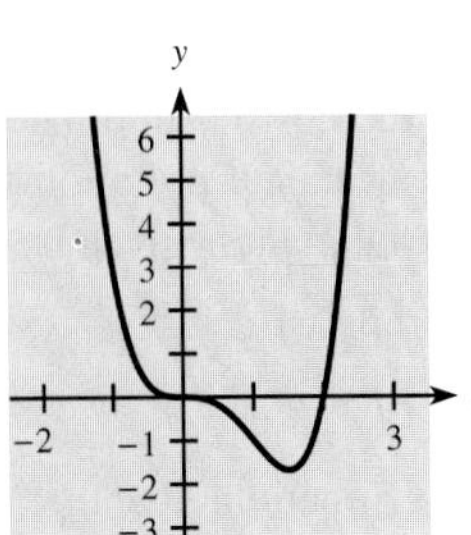

18.

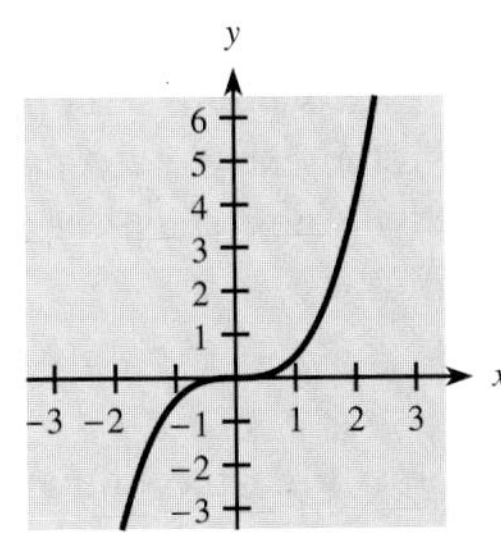

In Problems 19–28, match each equation with the correct graph among those labeled (*a*)–(*j*). Use a graphing utility to confirm your choice.

19. $y = x^3 - x$
20. $y = (x - 3)^2(x + 1)$
21. $y = 16x^2 - x^4$
22. $y = (x + 1)(x - 3)(x - 1)$
23. $y = x^2 + 7x$
24. $y = 7x - x^2$
25. $y = x^4 - 3x^2 - 4$
26. $y = \dfrac{1 - 2x}{x}$
27. $y = \dfrac{x - 3}{x + 1}$
28. $y = \dfrac{1 - 3x}{2x + 5}$

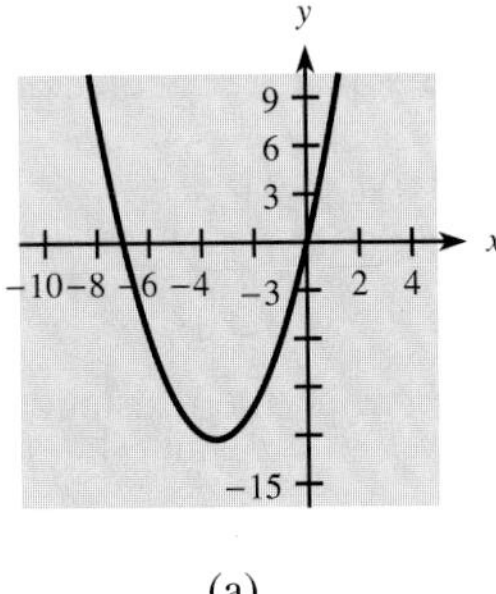

(a)

(b)

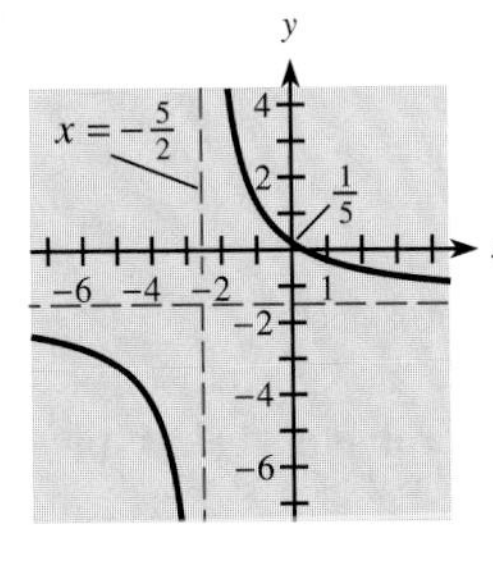

(c)

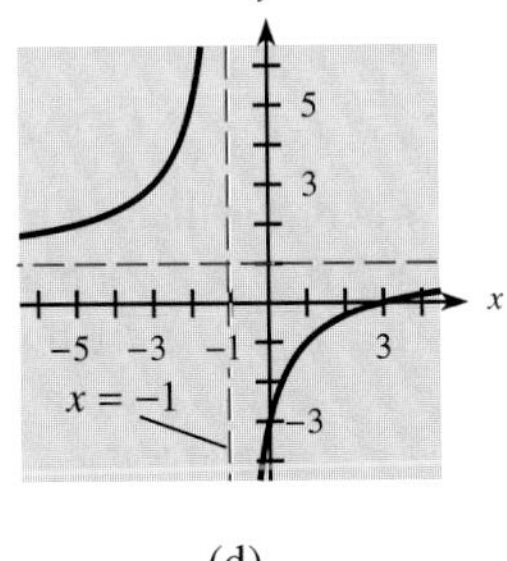

(d)

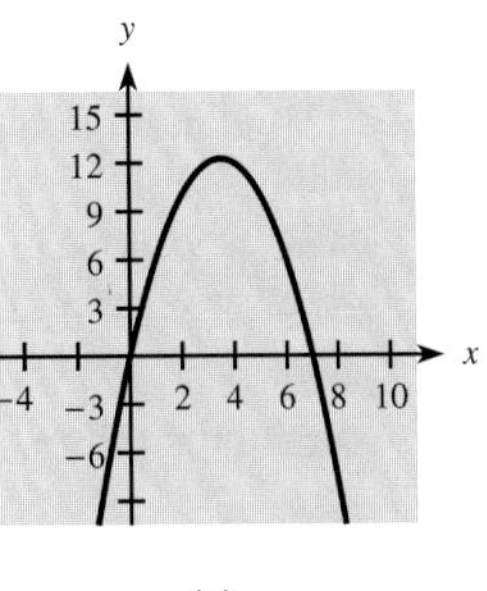

(e)

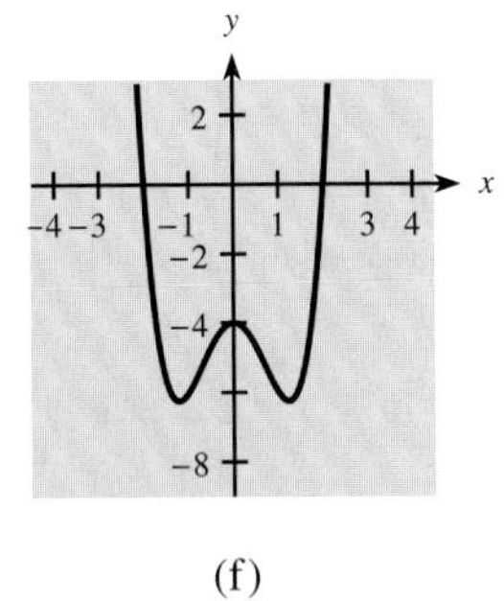

(f)

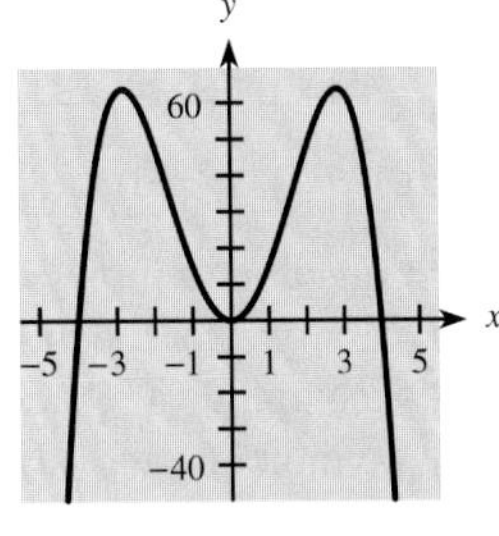

(g)

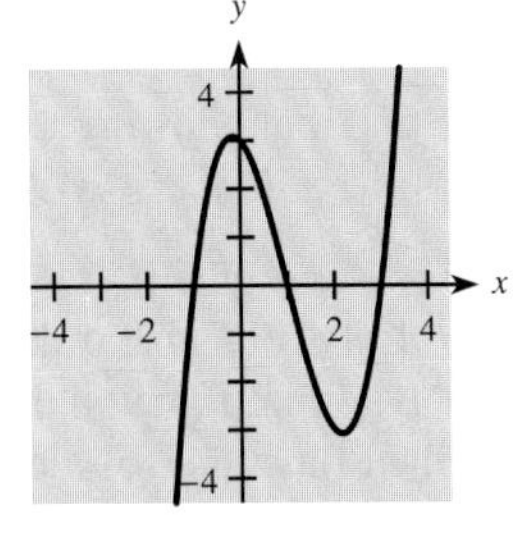

(h)

(i)

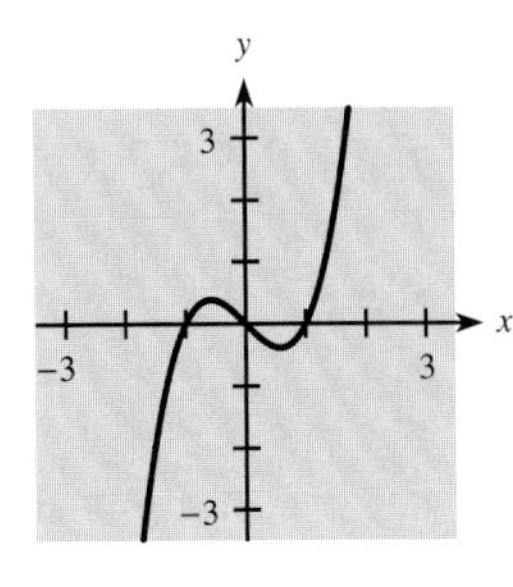

(j)

29. If $F(x) = \dfrac{x^2 - 1}{x}$, find the following.
    (a) $F(-\frac{1}{3})$ (b) $F(10)$
    (c) $F(0.001)$ (d) Is $F(0)$ defined?
30. If $H(x) = |x - 1|$, find the following.
    (a) $H(-1)$ (b) $H(1)$
    (c) $H(0)$ (d) Does $H(-x) = H(x)$?
31. If $f(x) = x^{3/2}$, find the following, if they exist.
    (a) $f(16)$ (b) $f(1)$ (c) $f(100)$ (d) $f(0.09)$

32. If $k(x) = \begin{cases} 2 & \text{if } x < 0 \\ x+4 & \text{if } 0 < x < 1 \\ |1-x| & \text{if } x \geq 1 \end{cases}$, find the following.
    (a) $k(-0.1)$ (b) $k(0.1)$ (c) $k(0.9)$ (d) $k(1.1)$

33. If $k(x) = \begin{cases} 2 & \text{if } x < 0 \\ x+4 & \text{if } 0 \leq x < 1 \\ 1-x & \text{if } x \geq 1 \end{cases}$, find the following.
    (a) $k(-5)$ (b) $k(0)$ (c) $k(1)$ (d) $k(-0.001)$

34. If $g(x) = \begin{cases} 0.5x+4 & \text{if } x < 0 \\ 4-x & \text{if } 0 \leq x < 4 \\ 0 & \text{if } x > 4 \end{cases}$,
    find the following.
    (a) $g(-4)$ (b) $g(1)$ (c) $g(7)$ (d) $g(3.9)$

In Problems 35–40, (a) graph each function (with a graphing utility if one is available); (b) classify each function as a polynomial function, a rational function, or a piecewise defined function; (c) identify any asymptotes; and (d) use the graphs to locate turning points.

35. $f(x) = 1.6x^2 - 0.1x^4$ 36. $f(x) = \dfrac{x^4 - 4x^3}{3}$

37. $f(x) = \dfrac{2x+4}{x+1}$

38. $f(x) = \dfrac{x-3}{x+2}$

39. $f(x) = \begin{cases} -x & \text{if } x < 0 \\ 5x & \text{if } x \geq 0 \end{cases}$

40. $f(x) = \begin{cases} 2x-1 & \text{if } x < 1 \\ -x & \text{if } x \geq 1 \end{cases}$

Applications

41. ***Postal restrictions*** If a box with square cross section is to be sent by the postal service, there are restrictions on its size such that its volume is given by $V = x^2(108 - 4x)$, where x is the length of each side of the cross section (in inches).
    (a) If $V = V(x)$, find $V(10)$ and $V(20)$.
    (b) What restrictions must be placed on x (the domain) so that the problem makes physical sense?

42. ***Fixed costs*** Fixed costs FC are business costs that remain constant regardless of the number of units produced. Some items that might contribute to fixed costs are rent and utilities.

$$FC = 2000$$

is an equation indicating that a business has fixed costs of \$2000. Graph $FC = 2000$ by putting x (the number of units produced) on the horizontal axis. (Note that FC does not mean the product of F and C.)

43. ***Cost-benefit*** Suppose that the cost C (in dollars) of removing p percent of the particulate pollution from the smokestacks of an industrial plant is given by

$$C(p) = \frac{7300p}{100 - p}$$

    (a) Find the domain of this function. Recall that p represents the percentage of pollution that is removed.
    In parts (b)–(e), find the functional values and explain what each means.
    (b) $C(45)$
    (c) $C(90)$
    (d) $C(99)$
    (e) $C(99.6)$

44. ***Test reliability*** If a test having reliability r is lengthened by a factor n ($n \geq 1$), the reliability R of the new test is given by

$$R(n) = \frac{nr}{1 + (n-1)r} \qquad 0 < r \leq 1$$

If the reliability is $r = 0.6$, the equation becomes

$$R(n) = \frac{0.6n}{0.4 + 0.6n}$$

    (a) Find $R(1)$.
    (b) Find $R(2)$; that is, find R when the test length is doubled.
    (c) What percentage improvement is there in the reliability when the test length is doubled?

45. ***Area*** If 100 feet of fence is to be used to fence in a rectangular yard, then the resulting area of the fenced yard is given by

$$A = x(50 - x)$$

where x is the width of the rectangle.
    (a) If $A = A(x)$, find $A(2)$ and $A(30)$.
    (b) What restrictions must be placed on x (the domain) so that the problem makes physical sense?

46. ***Water usage costs*** The monthly charge for water in a small town is given by

$$f(x) = \begin{cases} 18 & \text{if } 0 \leq x \leq 20 \\ 18 + 0.1(x - 20) & \text{if } x > 20 \end{cases}$$

where x is hundreds of gallons and $f(x)$ is in dollars. Find the monthly charge for each of the following usages: (a) 30 gallons, (b) 3000 gallons, and (c) 4000 gallons. (d) Graph the function for $0 \le x \le 100$.

47. ***Commercial electrical usage*** The monthly charge (in dollars) for x kilowatt hours (kWh) of electricity used by a commercial customer is given by the function

$$C(x) = \begin{cases} 7.52 + 0.1079x & \text{if } 0 \le x \le 5 \\ 19.22 + 0.1079x & \text{if } 5 < x \le 750 \\ 20.795 + 0.1058x & \text{if } 750 < x \le 1500 \\ 131.345 + 0.0321x & \text{if } x > 1500 \end{cases}$$

Find the monthly charges for the following usages.
(a) 5 kWh
(b) 6 kWh
(c) 3000 kWh

48. ***Residential power costs*** Excluding fuel adjustment costs and taxes, Georgia Power Company charges its residential customers during the months of June through September according to the function

$$C(x) = \begin{cases} 7.50 + 0.04783x & \text{if } 0 \le x \le 650 \\ 38.59 + 0.07948(x - 650) & \text{if } 650 < x \le 1000 \\ 66.41 + 0.08184(x - 1000) & \text{if } x > 1000 \end{cases}$$

where $C(x)$ is the charge in dollars for x kilowatt hours (kWh) of electricity.
(a) Find $C(780)$ and write a sentence that explains its meaning.
(b) Find the charge for using 1280 kWh in a month.

49. ***First-class postage*** The U.S. Postal Service fee for first-class postage P is a function of the weight w (in ounces) of the letter or package and is given by

$$P(w) = \begin{cases} 0.33 & 0 < w \le 1 \\ 0.55 & 1 < w \le 2 \\ 0.77 & 2 < w \le 3 \\ 0.99 & 3 < w \le 4 \\ \vdots & \vdots \end{cases}$$

(a) Find $P(2.3)$ and write a sentence that explains its meaning.
(b) Find the postage for a letter that weighs exactly 1 oz and for one that weighs 1.01 oz.
(c) Find the postage for a letter that weighs 3.3 oz.

50. ***Federal income tax*** The 1998 U.S. federal income tax owed by a married couple filing jointly can be found from the following schedule.

Schedule Y-1—Use if your filing status is Married filing jointly or Qualifying widow(er)

If line 5 is:		**The tax is:**	*of the*
Over—	*But not over—*		*amount over—*
\$0	\$42,350	.... 15%	\$0
42,350	102,300	\$6,352.50 + 28%	42,350
102,300	155,950	23,138.50 + 31%	102,300
155,950	278,450	39,770.00 + 36%	155,950
278,450	....	83,870.00 + 39.6%	278,450

SOURCE: Internal Revenue Service, 1998 Form 1040, ES/V

From this schedule, the tax due T (entered on Form 1040, line 38) is a function of income x (the amount on Form 1040, line 37) as follows:

$$T(x) = \begin{cases} 0.15x & \text{if } 0 < x \le 42{,}350 \\ 6352.50 + 0.28(x - 42{,}350) & \text{if } 42{,}350 < x \le 102{,}300 \\ 23{,}138.50 + 0.31(x - 102{,}300) & \text{if } 102{,}300 < x \le 155{,}950 \\ 39{,}770.00 + 0.36(x - 155{,}950) & \text{if } 155{,}950 < x \le 278{,}450 \\ 83{,}870.00 + 0.396(x - 278{,}450) & \text{if } 278{,}450 < x \end{cases}$$

(a) Find $T(58{,}676)$ and write a sentence that explains its meaning.
(b) When taxable income passes \$42,350, the tax rate changes from 15% to 28%. Does a couple whose taxable income is \$42,351 pay 13% more tax than a couple whose taxable income is \$42,349? Explain.

51. ***Demand*** The demand function for a product is given by

$$p = \frac{200}{2 + 0.1x}$$

where x is the number of units and p is the price. Graph this demand function for $0 \le x \le 250$, with x on the horizontal axis.

52. ***Minimum costs*** A printer has a contract to run 10,000 posters for a fire company benefit. He can use any number of printing plates from 1 to 10 to run the posters, with the cost of printing given by

$$C(x) = 3x + \frac{48}{x}$$

where x is the number of plates he uses.
(a) Graph this function for $1 \le x \le 10$.
(b) Identify and interpret any turning points for $1 \le x \le 10$.
(c) Does the domain extend beyond $1 \le x \le 10$? Explain.

53. ***Production costs*** A manufacturer estimates that the cost of a production run for a product is

$$C(x) = 30(x - 1) + \frac{3000}{x + 10}$$

where x is the number of machines used.

(a) Graph this total cost function for values of $x \geq 0$.
(b) Interpret any turning points.
(c) Interpret the y-intercept.

54. ***Mob behavior*** In studying lynchings between 1899 and 1946, psychologist Brian Mullin found that the size of a lynch mob relative to the number of victims predicted the level of brutality. He developed a formula for the other-total ratio (y) that predicts the level of self-attentiveness of people in a crowd of size x with 1 victim.

$$y = \frac{1}{x + 1}$$

The lower the value of y, the more likely an individual is to be influenced by "mob psychology." Graph this function; use positive integers as its domain.

M 2.5 Modeling; Fitting Curves to Data with Graphing Utilities (optional)

OBJECTIVES

- *To graph data points in a scatter plot*
- *To determine the function type that will best model data*
- *To use a graphing utility to create an equation that models the data*
- *To graph the data points and model on the same graph*

APPLICATION PREVIEW

As the data in Table 2.2 show, the amount of federal tax paid per capita has increased every year for the years from 1983 to 1998. We can use a technique called **linear regression** (or the **least-squares method**) to find the equation of the line that is the best fit for the data points. This equation **(model)** describes the data and gives a formula to plan for the future. We can use technology to fit linear functions and other functions to sets of data.

TABLE 2.2

Year	*Federal Tax per Capita*	*Year*	*Federal Tax per Capita*	*Year*	*Federal Tax per Capita*
1983	$2490	1989	$3884	1995	$5006
1984	2738	1990	4026	1996	5365
1985	2982	1991	4064	1997	5497
1986	3090	1992	4153	1998	5667 (est)
1987	3414	1993	4382		
1988	3598	1994	4728		

Source: Tax Foundation

We saw in Section 1.3, "Linear Functions," that it is possible to write the equation of a straight line if we have two points on the line. Business firms frequently like to treat demand functions as though they are linear, even when they are not exactly linear. They do this because linear functions are much easier to handle than other functions. If a firm has more than two points describing the demand for its product, it is likely that the points will not all lie on the same straight line. However, by using a technique called **linear regression,** the firm can determine the "best line" that fits these points.

Suppose we have the points shown in Figure 2.28. We seek the line that is the "best fit" for the points. We can "eyeball" a line that appears to lie along the points; we now find the equation of the line that fits best.

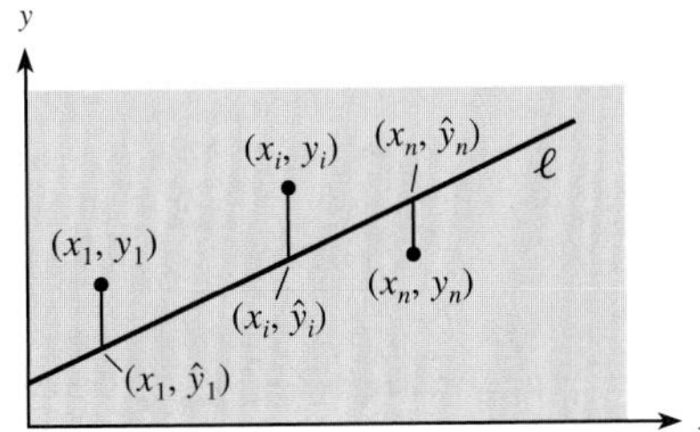

Figure 2.28

Corresponding to each point (x_1, y_1) that we know, we can find a point $(x_1, \hat{y}_1)$ on the line. If the line we have is a good fit for the points, the differences between the y-values of the given points and the corresponding $\hat{y}$-values on the line should be small. The line shown is the best line for the points only if the sum of the squares of these differences is a minimum.*

The formulas that give the best-fitting line for given data are developed in Chapter 14, "Functions of Two or More Variables," but graphing calculators, computer programs, and spreadsheets have built-in formulas or programs that give the equation of the best-fitting line for a set of data.

To find a function that models a set of data, the first step is to plot the data and determine the shape of the curve that would best fit the data points in the *scatter plot*. This shape determines the type of function that will be the best model for the data. The following example shows how a model can be developed for a set of data.

EXAMPLE 1

The following table represents the total return by the Standard and Poor's 400 Index (S & P) on a $100 investment made in 1988.

Year	1988	1989	1990	1991	1992	1993
S & P	100	129	128	168	171	187

(a) Use a graphing utility to create a scatter plot of the data.
(b) Determine what type of function will best model the data.
(c) Use the utility to create an equation that models the data.
(d) Graph the function and the data on the same graph to see how well the function models ("fits") the data.

Solution

(a) When we use the number of years past 1988 (so that $x = 0$ in 1988) and use y to represent the cumulative value of the $100 investment, the points representing this function are

$$(0,100), (1,129), (2,128), (3,168), (4,171), (5,187)$$

*Because linear regression involves minimizing the sum of squared numbers, it is also called the **least-squares method.**

Plotting these points on a coordinate plane gives the scatter plot shown in Figure 2.29(a). (These points can be plotted by hand or by using a graphing utility.)

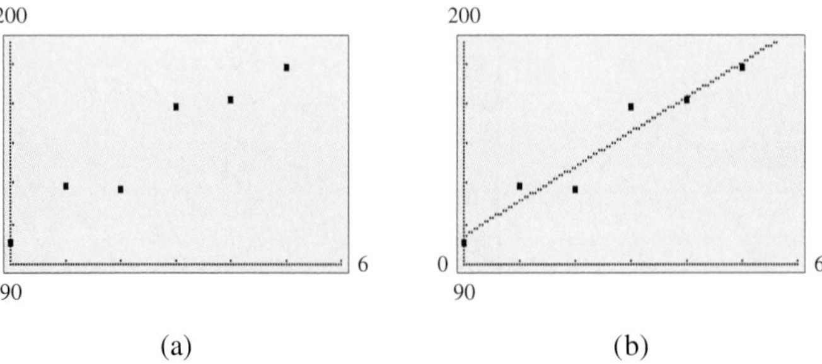

Figure 2.29 (a) (b)

(b) It appears that a line could be found that would fit close to the points in Figure 2.29(a). The points do not fit exactly on a line, because the S & P Index did not increase by exactly the same amount each year.

(c) The equation of the line that is the best fit for the data can be found by using a graphing utility programmed to perform the linear regression. The linear equation that is the best fit for the points is

$$y = 17.1714x + 104.238$$

(d) Figure 2.29(b) shows the graph of the function and the data on the same graph. Note that not all points fit on the graph of the equation, even though this is the line that is the best fit for the data.

Of course, many relations that occur are not linear. The use of graphing utilities permits us to use the same steps as those above to model nonlinear relations. Creating an equation, or **mathematical model,** for a set of data is a complex process that involves careful analysis of the data in accordance with a number of guidelines. Our approach to creating a model is much simplified and relies on a knowledge of the appearance of linear, quadratic, power, cubic, and quartic functions and the capabilities of computers and graphing calculators. We should note that the equations the computers and calculators provide are based on sophisticated formulas that are derived using calculus of several variables. These formulas provide the "best fit" for the data using the function type we choose, but we should keep the following limitations in mind.

1. The computer/calculator will give the best fit for whatever type of function we choose (even if the selected function is a bad fit for the points), so we must choose a function type carefully. To choose a function type, compare the scatter plot with the graphs of the functions discussed earlier in this chapter. The model will not fit the data if we choose a function type whose graph does not match the shape of the plotted data.
2. Some sets of data have no pattern, so they cannot be modeled by functions. Other data sets cannot be modeled by the functions we have studied.
3. Modeling provides a formula that relates the data, but it is not an absolute truth. In fact, even though a model may accurately fit a data set, it may not be a good predictor over a period of time.

EXAMPLE 2

The graphs of the data points shown in Figure 2.30(a–c) are called scatter plots (or stat plots). Determine what type of function is your choice as the best-fitting curve for each scatter plot. If it is a polynomial function, state the degree.

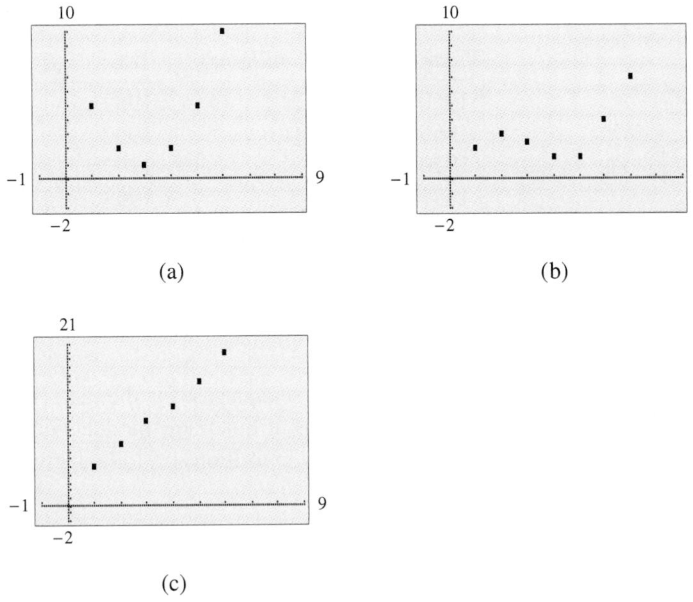

Figure 2.30

Solution

(a) It appears that a parabola will fit the data points in Figure 2.30(a), so the data points can be modeled by a second-degree (quadratic) function.

(b) The two "turns" in the scatter plot in Figure 2.30(b) suggest that it could be modeled by a cubic function.

(c) The data points in Figure 2.30(c) appear to lie (approximately) along a line, so the data can be modeled by a linear function.

After the type of function that gives the best fit for the data is chosen, a graphing utility or spreadsheet can be used to develop the best-fitting equation for the function chosen. The steps used to find an equation that models data follow.

M

Modeling Data

1. Enter the data and use a graphing utility or spreadsheet to plot the data points. (This gives a scatter plot.)
2. Visually determine what type of function (including degree, if it is a polynomial function) would have a graph that best fits the data.
3. Use a graphing utility or spreadsheet to determine the equation of the chosen type that gives the best fit for the data.
4. To see how well the equation models the data, graph the equation and the data points on the same set of axes. If the graph of the equation does not fit the data points well, another type of function may model the data better.

EXAMPLE 3

Politicians regularly discuss national health insurance because of the escalating cost of health care. The following table gives the health care costs in the United States. Find a model that fits the data points (with $x = 0$ in 1990), and use it to predict the cost of health care in the years 2000 and 2005.

Year	*Cost ($millions)*	*Year*	*Cost ($millions)*
1960	27.1	1989	539.9
1965	41.6	1990	697.5
1970	74.4	1992	834.2
1975	132.9	1993	892.1
1980	249.1	1994	937.1
1985	420.1	1995	988.5

Source: U.S. Department of Health and Human Services

Solution

A glance at the scatter plot for the data (Figure 2.31a) shows that the growth is not linear. Using a quadratic model, with x representing the number of years from 1900 and y in millions of dollars, gives an equation that fits the data.

$$y = 1.1310058x^2 - 148.682511x + 4907.772572$$

The graph in Figure 2.31(b) shows how well the model fits the data.

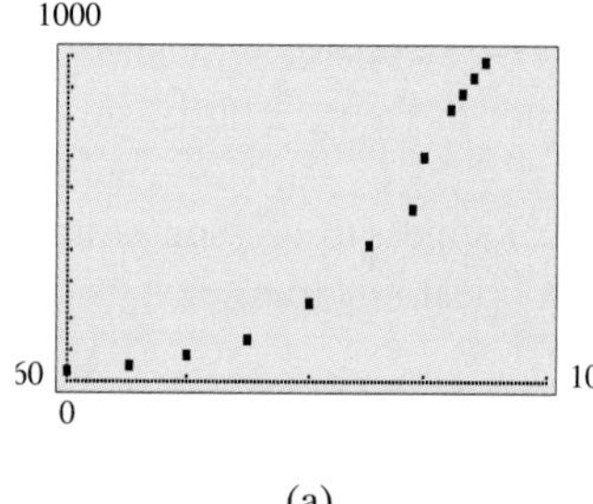

(a)

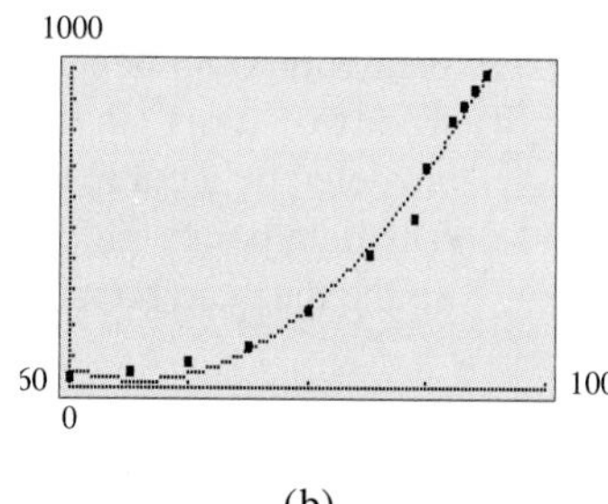

(b)

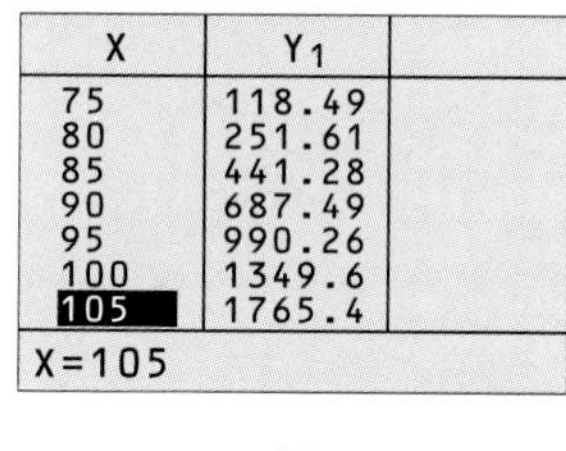

X	Y_1
75	118.49
80	251.61
85	441.28
90	687.49
95	990.26
100	1349.6
105	1765.4

X=105

(c)

Figure 2.31

The table of values for the equation (Figure 2.31c) can be used to predict future costs for health care. We can use $x = 100$ to predict the cost in the year 2000. The cost is $1349.6 million, or about $1.35 billion. Using $x = 105$ gives the predicted cost of $1,765,400,000 in 2005.

If it is not obvious what model will best fit a given set of data, several models can be developed and compared graphically with the data. For example, a power model also could have been used in Example 3.

EXAMPLE 4

The expected life span of people in the United States depends on their year of birth.

Year	*Life Span (years)*	Year	*Life Span (years)*
1920	54.1	1988	74.9
1930	59.7	1989	75.1
1940	62.9	1990	75.4
1950	68.2	1991	75.5
1960	69.7	1992	75.5
1970	70.8	1993	75.5
1975	72.6	1994	75.7
1980	73.7	1995	75.8
1987	75.0	1996	76.1

Source: National Center for Health Statistics

(a) Create linear and quadratic models that give life span as a function of birth year with $x = 0$ in 1900 and, by visual inspection, decide which model gives the best fit.
(b) Use both models to predict the life span of a person born in the year 2000.
(c) Which model's prediction for the lifespan in 2010 seems better?

Solution

(a) The scatter plot for the data is shown in Figure 2.32(a). It appears that a linear function could be used to model this data. The linear equation that is the best fit for the data is

$$y = 0.25807x + 52.24405$$

The graph in Figure 2.32(b) shows how well the line fits the data points. The quadratic function that is the best fit for the data is

$$y = -0.002654x^2 + 0.58567x + 44.03318$$

Its graph is shown in Figure 2.32(c).

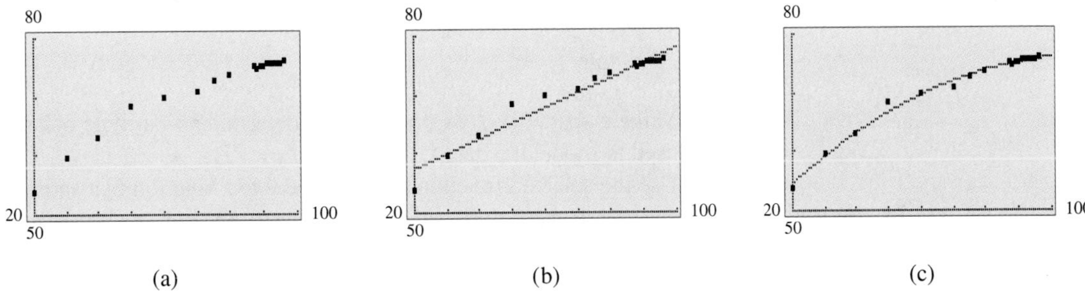

Figure 2.32

The quadratic model appears to fit the data points better than the linear model. The table can be used to compare the points from the model with the data points (see Table 2.3 with y_1 giving values from the quadratic model and y_2 giving values from the linear model).

X	Y1	Y2
65	70.889	69.019
70	72.025	70.309
75	73.03	71.599
80	73.901	72.89
85	74.64	74.18
90	75.246	75.47
95	75.719	76.761
X=65		

X	Y1	Y2
80	73.901	72.89
85	74.64	74.18
90	75.246	75.47
95	75.719	76.761
100	76.06	78.051
105	76.268	79.341
110	76.343	80.632
X=110		

Table 2.3 (a) (b)

(b) We can predict the life span of people born in the year 2000 with either model by evaluating the function at $x = 100$. For the linear model the expected life span is 78.051, and for the quadratic model the expected life span is 78.06.

(c) Looking at Table 2.3(b), we see that the linear model may be giving optimistic values in 2010, when $x = 110$, so the quadratic model may be better in the years up to 2010.

EXAMPLE 5

The following table shows the times that it takes a 1994 Porsche 911 to reach speeds from 0 mph to 100 mph, in increments of 10 mph after 30 mph.

Time (seconds)	*Speed (mph)*	*Time (seconds)*	*Speed (mph)*
1.9	30	7.1	70
2.7	40	9.2	80
4.1	50	11.9	90
5.5	60	14.0	100

Source: *Motor Trend*, January 1994

(a) Make a scatter plot of the data and determine what type of function can be used to model the data.

(b) Use the selected function type to model the function that satisfies the points, and graph the points and the function to see how well the function fits the points.

(c) What does the model indicate the speed is 5 seconds after the car starts to move?

(d) Use the model to determine the number of seconds until the Porsche reaches 110 mph.

Solution

(a) Using x as the time and y as the speed, the scatter plot is shown in Figure 2.33. The power function $y = ax^b$ can be used to model the function that satisfies these points.

(b) An equation found using this model (graphed in Figure 2.34) is

$$y = 21.637x^{0.586}$$

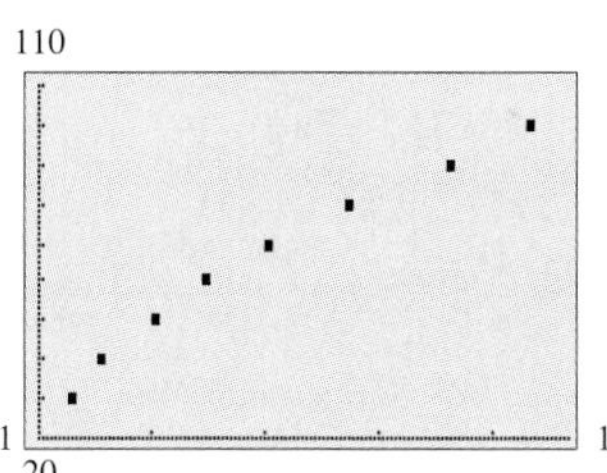

Figure 2.33

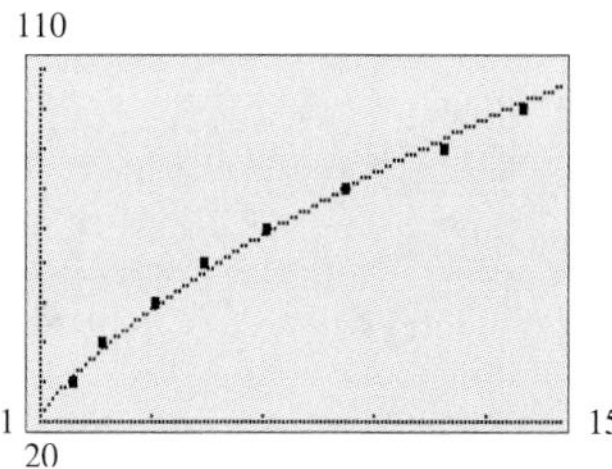

Figure 2.34

(c) Using $x = 5$ in this equation gives 55.56 mph as the speed of the Porsche.

(d) Tracing along the graph of $y_1 = 21.637x^{0.586}$ until the y-value is 110 gives 16.0, so the time required to reach 110 mph is 16.0 seconds.

EXERCISE 2.5

In Problems 1–8, determine whether the scatter plot should be modeled by a linear, power, quadratic, cubic, or quartic function.

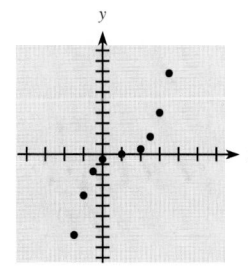

1.

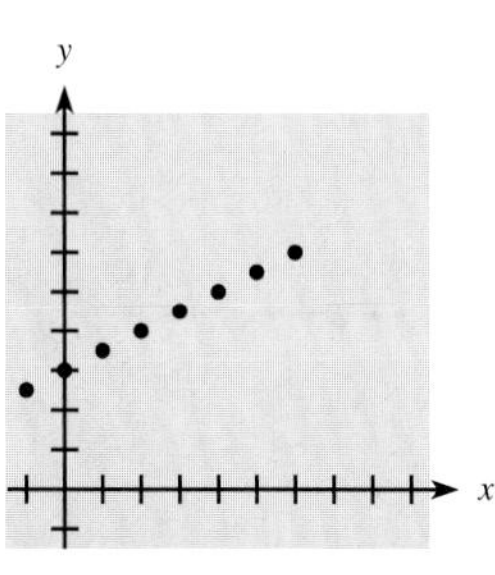

2.

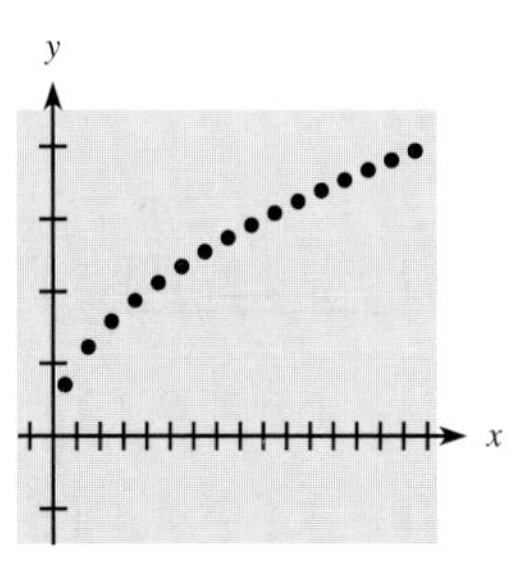

3.

4.

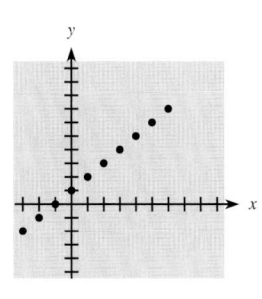

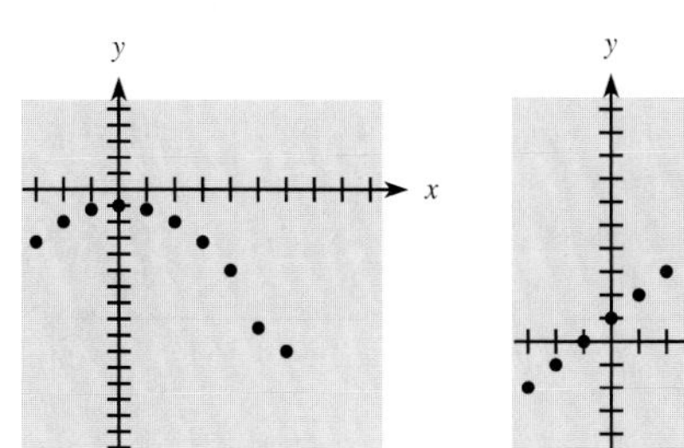

5.

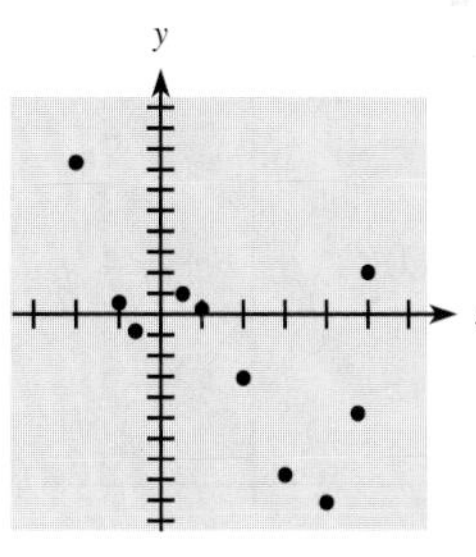

6.

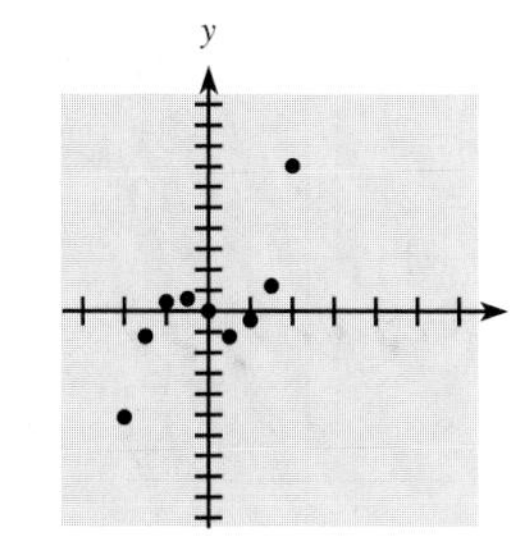

7.

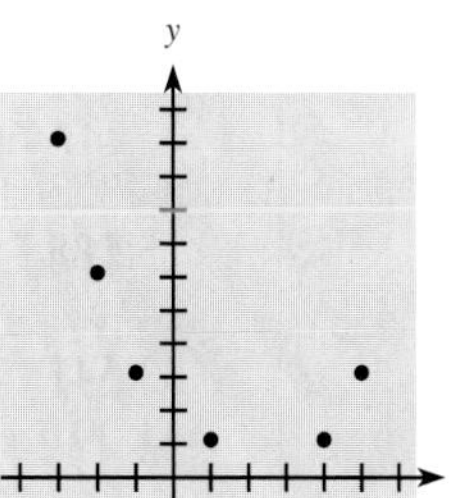

8.

In Problems 9–16, find the equation of the function of the specified type that is the best fit for the given data. Plot the data and the equation.

9. linear

x	y_1
−2	−7
−1	−5
0	−3
1	−1
2	1
3	3
4	5

10. linear

x	y_1
−1	−5.5
0	−4
1	−2.5
2	−1
3	.5
4	2
5	3.5

11. quadratic

x	y_1
−2	7
−1	−0.5
0	−4
1	−3.5
2	1
3	9.5
4	22

12. quadratic

x	y_1
−4	−8
−3	−4
−2	−2
−1	−2
0	−4
1	−8
2	−14

13. cubic

x	y_1
−4	−72
−3	−31
−2	−10
−1	−3
0	−4
1	−7
2	−6

14. cubic

x	y_1
−3	−22
−2	−6
−1	−2
0	−4
1	−6
2	−2
3	14

15. power

x	y_1
1	2
2	2.8284
3	3.4641
4	4
5	4.4721
6	4.899

16. power

x	y_1
1	3
2	8.4853
3	15.588
4	24
5	33.541
6	44.091

In Problems 17–24, (a) plot the given points, (b) determine what function best models the data, and (c) find the equation that is the best fit for the data.

17.

x	y_1
−1	−8
0	−3
1	2
2	7
3	12
4	17
5	22

18.

x	y_1
−1	−2
0	2
1	6
2	10
3	14
4	18
5	22

19.

x	y_1
−1	2
0	0
1	2
2	3.1748
3	4.1602
4	5.0397
5	5.848

20.

x	y_1
0.01	0.3
1	3
2	4.2426
3	5.1962
4	6
5	6.7082
6	7.3485

21.

x	y_1
−2	19
−1	8
0	1
1	−2
2	−1
3	4
4	13

22.

x	y_1
−4	37
−3	19
−2	7
−1	1
0	1
1	7
2	19

23.

x	y_1
−3	−11
−2	3
−1	5
0	1
1	−3
2	−1
3	13

24.

x	y_1
−3	−54
−2	−14
−1	0
0	0
1	−2
2	6
3	36

Applications

25. ***Earnings and gender*** The following data show the 1986 median weekly earnings of men (x) and women (y) in various occupations.
   (a) Find a linear model that gives women's earnings as a function of men's earnings.
   (b) Find the slope of the line that models the data.
   (c) What does the slope of the line tell us about the relationship between the earnings of men and the earnings of women?

Occupation	*Men*	*Women*
Management	\$666	\$465
Technical sales	472	305
Service	299	208
Crafts, repair	446	302
Operators, fabricators	352	238
Fish, farm, forestry	234	201

Source: U.S. Equal Opportunity Employment Commissions

26. ***Electric cooperative memberships*** The following table gives the membership of the Palmetto Electric Cooperative for 1987–1997. Plot the data points and decide what equation will best model the data.
   (a) Create the equation that gives the best fit using your model if x is the number of years from 1900.
   (b) Use the model to predict the membership in the year 2000.

Year	*Members of the Cooperative*
1987	30,847
1992	35,486
1993	36,451
1994	37,556
1995	39,938
1996	41,258
1997	42,739

Source: Palmetto Electric Cooperative Inc. *Annual Report*

27. ***AT&T revenue*** The following table gives the total revenues of AT&T for selected years.

Year	*Total Revenues (\$ billions)*
1985	63.130
1986	69.906
1987	60.530
1989	61.100
1990	62.191
1991	63.089
1992	64.904
1993	67.156

Source: AT&T Annual Report, 1993

   (a) What type function do the data points appear to follow?
   (b) Use x to represent the number of years since 1980 and y to represent the revenue, and find the equation which gives the best fit for the data.
   (c) Use this model to find the year in which revenue is at its minimum and find the minimum predicted revenue. Compare the outputs of your model with the real data in the table.
   (d) Write a sentence describing how the data compares with the predicted data from the model.

28. ***Stock prices*** The stock prices of the Anchor Bank for years 1993–1997 are given below.
   (a) Find the quadratic function that is the best fit for the data, where x is the number years from 1900.
   (b) Use the model to predict the stock price in the year 2000.

Year	*Closing Price of Stock*
1993	\$8.67
1994	9.17
1995	13.17
1996	21.67
1997	33.00

Source: The Anchor Bank Shareholder Information

29. ***Voluntary bumps from airline flights*** The table contains the numbers of people who gave up their seats in an overbooked airline flight (voluntary bumps) for 1993–1997. Plot the data points and decide what function type best models the data.
   (a) Create the equation that gives the best fit using your model if x is the number of years from 1990.
   (b) Look at a table of outputs from the model for the years 1993–1997, and determine which year's prediction is most accurate.
   (c) For what year is the prediction least accurate?

Year	*Voluntary Bumps*
1993	604,635
1994	770,017
1995	793,747
1996	898,876
1997	1,017,926

USA Today, Grant Jerding

30. ***Poverty*** The table shows the number of millions of people in the United States who lived below the poverty level for selected years from 1960 to 1996.
(a) Find a model that approximately fits the data, using x as the number of years from 1900.
(b) Use a graph of the model to determine when this model would no longer be valid. Explain why it would no longer be valid.

Year	*Persons Living Below the Poverty Level (millions)*
1960	39.9
1965	33.2
1970	25.4
1975	25.9
1980	29.3
1986	32.4
1989	31.5
1990	33.6
1991	35.7
1992	38
1993	39.3
1994	38.1
1995	36.4
1996	36.5

Bureau of the Census, U.S. Dept. of Commerce

31. ***Budget deficit*** The table gives the yearly budget deficit, in billions of dollars, for the years 1990–1997, with White House estimates for 1998 and 1999.
(a) Use x as the number of years from 1900 to create a model using the data from 1990 to 1999, and use this model to predict the values for 1998 and 1999.
(b) When does the model predict that the budget deficit will be 0?

Year	*Yearly Deficit ($billions)*
1990	221.2
1991	269.4
1992	290.4
1993	255.0
1994	203.1
1995	163.9
1996	107.3
1997	22.6
1998	22.0
1999	0

Source: *USA Today*, January 7, 1998

32. ***Tax load*** The percent of income used for federal, state, and local taxes for median-income households with one wage earner is given in the table.
(a) Using x as the number of years from 1950, write the equation of the function that is the best fit for the data.
(b) When is the model no longer valid?

Year	*Percent*
1955	26.7
1965	28.8
1975	34.2
1985	34.8
1997	36.6

Source: *USA Today*, Tax Foundation

33. ***Drug use*** The percent of high school seniors using marijuana is given in the table. If x is the years from 1900, find the equation of the function that is the best fit for the data.

Year	*Percent Using Marijuana/Hashish*	*Year*	*Percent Using Marijuana/Hashish*
1975	47.3	1986	38.8
1976	52.8	1987	36.3
1977	56.4	1988	33.1
1978	59.2	1989	29.6
1979	60.4	1990	27.0
1980	60.3	1991	23.9
1981	59.5	1992	21.9
1982	58.7	1993	26.0
1983	57	1994	30.7
1984	54.9	1995	34.7
1985	54.2	1996	35.8

Source: National Institute on Drug Abuse

34. ***Cable rates*** The average monthly rates for basic cable, without premium channels, for the years 1982–1997 are given in the table.
(a) Plot the data points, determine what type of function will model the data, and find the equation of the function that is the best fit, using that type of function. Use $x = 0$ in 1900.
(b) When will the rate be $50 per month if your model is accurate?

Year	Basic Cable Rate	Year	Basic Cable Rate
1982	$8.30	1990	16.78
1983	8.61	1991	18.10
1984	8.98	1992	19.08
1985	9.73	1993	19.39
1986	10.67	1994	21.62
1987	12.18	1995	23.07
1988	13.86	1996	24.41
1989	15.21	1997	26.00

Source: Nielsen Media Research and Paul Kagan Associates, *The Island Packet*, April 19, 1998

35. ***Catholics per priest*** The table gives the number of Catholics per Priest in the United States.
    (a) If x represents the number of years since 1900, find the equation of the function that is the best fit for the data.
    (b) Use a graph to determine the year when the model gives a minimum number.
    (c) When do the data give a minimum number?

Year	Number of Catholics per Priest in the United States
1905	899.39
1910	866.89
1915	858.66
1920	843.29
1925	787.19
1930	750.37
1935	678.45
1940	631.14
1945	623.23
1950	646.18
1955	693.54
1960	759.75
1965	778.43
1970	808.76
1975	826.73
1980	849.73
1985	912.23
1990	1073.58
1995	1214.72

Source: Dr. Robert G. Kennedy, University of St. Thomas

36. ***Microsoft revenue*** The net revenue for Microsoft for the years 1988–1997 is given in the table.
    (a) Plot the data points with x as the number of years since 1900, determine what function type fits the points, and create the model that is the best fit for the data.
    (b) What revenue does your model predict for 2005?
    (c) If your model is accurate, when will revenue reach $15 billion?

Year	Net Revenue ($million)	Year	Net Revenue ($million)
1988	$591	1993	3753
1989	804	1994	4649
1990	1183	1995	5940
1991	1843	1996	8671
1992	2759	1997	11,360

Source: Microsoft Press Pass: Microsoft Fast Facts

37. ***Church attendance*** The following table shows the percentage of adults and teens who attended church each year from 1980 to 1989.

Year	Percentage of Adults	Percentage of Teens
1980	41	50
1981	41	54
1982	41	50
1983	40	53
1984	40	52
1985	42	52
1986	40	54
1987	40	52
1988	42	51
1989	43	57

    (a) Find the equation that models the percentage of teenagers who attended church during these years, where x denotes the number of years since 1900. Also find the equation that models the percentage of adults who attended church during these years.
    (b) Assuming that your models are valid, in what year will 45% of adults attend church?
    (c) In what year will 60% of teenagers attend church?

KEY TERMS AND FORMULAS

Section	Key Terms	Formula
2.1	Quadratic equation	$ax^2 + bx + c = 0 \qquad (a \neq 0)$
	Quadratic formula	$x = \dfrac{-b \pm \sqrt{b^2 - 4ac}}{2a}$
	Quadratic discriminant	$b^2 - 4ac$
2.2	Quadratic function; parabola	$y = f(x) = ax^2 + bx + c$
	Vertex of a parabola	$(-b/2a, \quad f(-b/2a))$ Maximum point if $a < 0$ Minimum point if $a > 0$
	Axis of symmetry	$x = -b/2a$
	Zeros of a quadratic function	$x = \dfrac{-b \pm \sqrt{b^2 - 4ac}}{2a}$
2.3	Supply function	
	Demand function	
	Market equilibrium	
	Cost, revenue, and profit functions	$P(x) = R(x) - C(x)$
	Break-even points	$R(x) = C(x)$ or $P(x) = 0$
2.4	Basic functions	$f(x) = ax + b$ (linear) $f(x) = C, \quad C =$ constant $f(x) = ax^b$ (power) $f(x) = x$ (identity) $f(x) = x^2, \quad f(x) = x^3, f(x) = 1/x$
	Polynomial function of degree n	$f(x) = a_nx^n + a_{n-1}x^{n-1} + \cdots + a_1x + a_0$ $a_n \neq 0, n \geq 0$, n an integer
	Rational function	$f(x) = p(x)/q(x)$, where $p(x)$ and $q(x)$ are polynomials
	Vertical asymptote	
	Horizontal asymptote	
	Absolute value function	$f(x) = \lvert x \rvert = \begin{cases} x & \text{if } x \geq 0 \\ -x & \text{if } x < 0 \end{cases}$
	Piecewise defined functions	
2.5	Scatter plots	
	Function types	
	Mathematical model	
	Fitting curves to data points	
	Predicting from models	

REVIEW EXERCISES

Section 2.1

In Problems 1–10, find the real solutions to each quadratic equation.

1. $3x^2 + 10x = 5x$
2. $4x - 3x^2 = 0$
3. $x^2 + 5x + 6 = 0$
4. $11 - 10x - 2x^2 = 0$
5. $(x - 1)(x + 3) = -8$
6. $4x^2 = 3$
7. $20x^2 + 3x = 20 - 15x^2$
8. $8x^2 + 8x = 1 - 8x^2$
9. $7 = 2.07x - 0.02x^2$
10. $46.3x - 117 - 0.5x^2 = 0$

In Problems 11–14, solve each equation by using a graphing utility to find the zeros of a function. Solve the equation algebraically to check your results.

11. $4z^2 + 25 = 0$
12. $z(z + 6) = 27$
13. $3x^2 - 18x - 48 = 0$
14. $3x^2 - 6x - 9 = 0$
15. Solve $x^2 + ax + b = 0$ for x.
16. Solve $xr^2 - 4ar - x^2c = 0$ for r.

In Problems 17 and 18, approximate the real solutions to each quadratic equation to two decimal places.

17. $23.1 - 14.1x - 0.002x^2 = 0$
18. $1.03x^2 + 2.02x - 1.015 = 0$

Section 2.2

For each function in Problems 19–24, find the vertex and determine if it is a maximum or minimum point, find the zeros if they exist, and sketch the graph.

19. $y = \frac{1}{2}x^2 + 2x$
20. $y = 4 - \frac{1}{4}x^2$
21. $y = 6 + x - x^2$
22. $y = x^2 - 4x + 5$
23. $y = x^2 + 6x + 9$
24. $y = 12x - 9 - 4x^2$

In Problems 25–30, use a graphing utility to graph each function. Use the vertex and zeros to determine an appropriate range. Be sure to label the maximum or minimum point.

25. $y = \frac{1}{3}x^2 - 3$
26. $y = \frac{1}{2}x^2 + 2$
27. $y = x^2 + 2x + 5$
28. $y = -10 + 7x - x^2$
29. $y = 20x - 0.1x^2$
30. $y = 50 - 1.5x + 0.01x^2$

In Problems 31–34, a graph is given. Use the graph to
(a) locate the vertex,
(b) determine the zeros, and
(c) match the graph with one of the equations A, B, C, or D.

A. $y = 7x - \frac{1}{2}x^2$
B. $y = \frac{1}{2}x^2 - x - 4$
C. $y = 8 - 2x - x^2$
D. $y = 49 - x^2$

31.

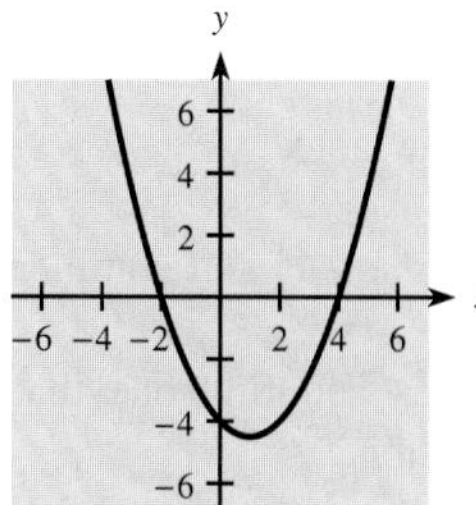

32.

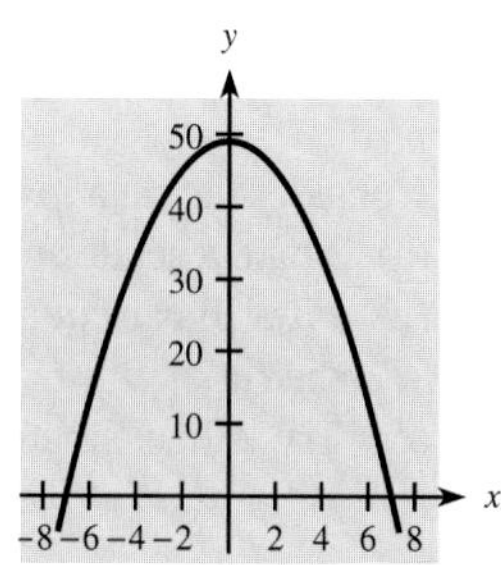

33.

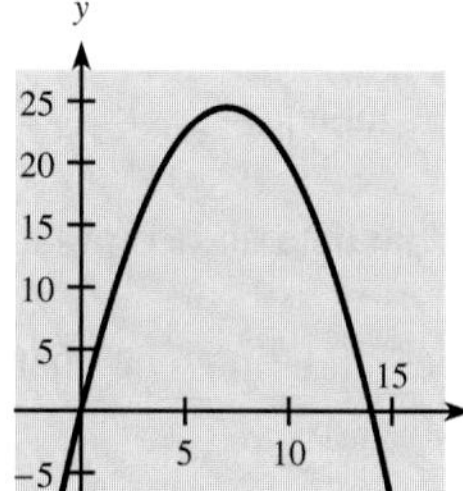

34.

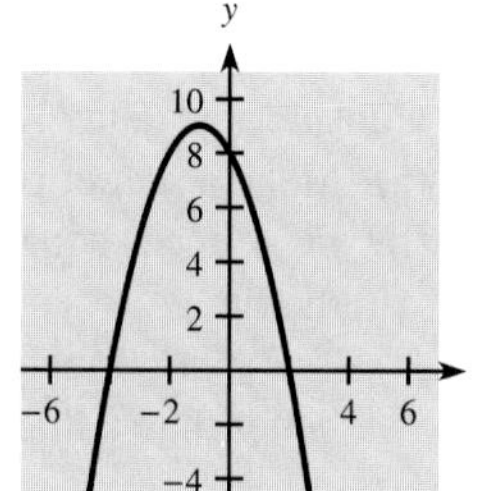

Section 2.4

35. Sketch a graph of each of the following basic functions.
(a) $f(x) = x^2$ (b) $f(x) = 1/x$ (c) $f(x) = x^{1/4}$

36. If $f(x) = \begin{cases} -x^2 & \text{if } x \le 0 \\ 1/x & \text{if } x > 0 \end{cases}$, find the following.
(a) $f(0)$ (b) $f(0.0001)$
(c) $f(-5)$ (d) $f(10)$

37. $f(x) = \begin{cases} x & \text{if } x \le 1 \\ 3x - 2 & \text{if } x > 1 \end{cases}$ find the following.
(a) $f(-2)$ (b) $f(0)$
(c) $f(1)$ (d) $f(2)$

In Problems 38 and 39, graph each function.

38. $f(x) = \begin{cases} x & \text{if } x \le 1 \\ 3x - 2 & \text{if } x > 1 \end{cases}$
39. (a) $f(x) = (x - 2)^2$ (b) $f(x) = (x + 1)^3$

In Problems 40 and 41, use a graphing utility to graph each function. Find any turning points.

40. $y = x^3 + 3x^2 - 9x$
41. $y = x^3 - 9x$

In Problems 42 and 43, use a graphing utility to graph each function. Find and identify any asymptotes.

42. $y = \dfrac{1}{x - 2}$

43. $y = \dfrac{2x - 1}{x + 3}$

Section 2.5

44. Consider the data given in the table.
   (a) Make a scatter plot.
   (b) Fit a linear equation to the data and comment on the fit.
   (c) Try other function types and find one that fits better than a linear function.

x	0	4	8	12	16	20	24
y	153	151	147	140	128	115	102

45. Consider the data given in the table.
   (a) Make a scatter plot.
   (b) Fit a linear equation to the data and comment on the fit.
   (c) Try other function types and find one that fits better than a linear function.

x	3	5	10	15	20	25	30
y	35	45	60	70	80	87	95

Applications

Section 2.1

46. ***Physics*** A ball is thrown into the air from a height of 96 ft above the ground, and its height is given by $S = 96 + 32t - 16t^2$, where t is the time in seconds.
   (a) Find the values of t that make $S = 0$.
   (b) Do both of the values of t have meaning for this application?
   (c) When will the ball strike the ground?

47. ***Profit*** The profit for a product is given by $P(x) = 82x - 0.10x^2 - 1600$, where x is the number of units produced and sold. Break-even points will occur at values of x where $P(x) = 0$. How many units will give a break-even point for the product?

Section 2.2

48. ***Drug use*** Data indicate that the percentage of high school seniors who have tried hallucinogens can be described by the function $f(t) = 0.0241904t^2 - 4.47459t + 216.074$, where t is the number of years since 1900.
   (a) For what years does the function predict the percentage will be 15%?
   (b) For what year does the function predict the percentage will be a minimum? Find that minimum predicted percentage.

49. ***Maximum area*** A rectangular lot is to be fenced in and then divided down the middle to create two identical fenced lots (see the figure). If 1200 ft of fence is to be used, the area of each lot is given by

$$A = x\left(\frac{1200 - 3x}{4}\right)$$

   (a) Find the x value that maximizes this area.
   (b) Find the maximum area.

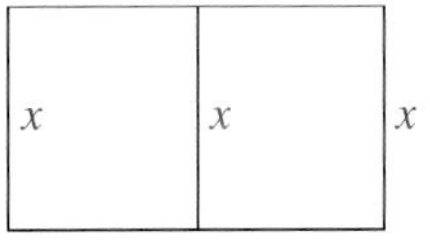

Section 2.3

50. ***Supply*** Graph the first-quadrant portion of the supply function

$$p = 2q^2 + 4q + 6$$

51. ***Demand*** Graph the first-quadrant portion of the demand function

$$p = 18 - 3q - q^2$$

52. ***Market equilibrium***
   (a) Suppose the supply function for a product is $p = 0.1q^2 + 1$ and the demand function is $p = 85 - 0.2q - 0.1q^2$. Sketch the first-quadrant portion of the graph of each function. Use the same set of axes for both and label the market equilibrium point.
   (b) Use algebraic methods to find the equilibrium price and quantity.

53. ***Market equilibrium*** The supply function for a product is given by $p = q^2 + 300$, and the demand is given by $p + q = 410$. Find the equilibrium quantity and price.

54. ***Market equilibrium*** If the demand function for a commodity is given by the equation $p^2 + 5q = 200$ and the supply function is given by $40 - p^2 + 3q = 0$, find the equilibrium quantity and price.

55. ***Break-even points*** If total costs for a product are given by $C(x) = 1760 + 8x + 0.6x^2$ and total rev-

enues are given by $R(x) = 100x - 0.4x^2$, find the break-even points.

56. ***Break-even points*** If total costs for a commodity are given by $C(x) = 900 + 25x$ and total revenues are given by $R(x) = 100x - x^2$, find the break-even points.

57. ***Maximizing revenue and profit*** Find the maximum revenue and maximum profit for the functions described in Problem 56.

58. ***Break-even points and profit maximization*** Given total profit $P(x) = 1.3x - 0.01x^2 - 30$, find maximum profit and the break-even points and sketch the graph.

59. ***Maximum profit*** Given $C(x) = 360 + 10x + 0.2x^2$ and $R(x) = 50x - 0.2x^2$, find the level of production that gives maximum profit and find the maximum profit.

60. ***Break-even points and profit maximization*** A certain company has fixed costs of \$15,000 for its product and variable costs given by $140 + 0.04x$ dollars per unit, where x is the total number of units. The selling price of the product is given by $300 - 0.06x$ dollars per unit.
    (a) Formulate the functions for total cost and total revenue.
    (b) Find the break-even points.
    (c) Find the level of sales that maximizes revenue.
    (d) Form the profit function and find the level of production and sales that maximizes profit.
    (e) Find the profit (or loss) at the production levels found in (c) and (d).

Section 2.4

61. ***Spread of AIDS*** The function $H(t) = 0.099t^{1.969}$, where t is the number of years since 1980 and $H(t)$ is the number of world HIV infections (in hundreds of thousands), has been used as one means of predicting the spread of AIDS.
    (a) What type of function is this?
    (b) What does this function predict as the number of HIV cases in the year 2005?
    (c) Find $H(15)$ and write a sentence that explains its meaning.

62. ***Photosynthesis*** The amount y of photosynthesis that takes place in a certain plant depends on the intensity x of the light present, according to

$$y = 120x^2 - 20x^3 \quad \text{for } x \geq 0$$

    (a) Graph this function with a graphing utility. (Use y-min $= -100$ and y-max $= 700$.)
    (b) The model is valid only when $f(x) \geq 0$ (that is, on or above the x-axis). For what x-values is this true?

63. ***Cost-benefit*** Suppose the cost C, in dollars, of eliminating p percent of the pollution from the emissions of a factory is given by

$$C(p) = \frac{4800p}{100 - p}$$

    (a) What type of function is this?
    (b) Given that p represents the percentage of pollution removed, what is the domain of $C(p)$?
    (c) Find $C(0)$ and interpret its meaning.
    (d) Find the cost of removing 99% of the pollution.

64. ***Municipal water costs*** The Borough Municipal Authority of Beaver, Pennsylvania, used the following function to determine charges for water.

$$C(x) = \begin{cases} 1.557x & 0 \leq x \leq 100 \\ 155.70 + 1.04(x - 100) & 100 < x \leq 1000 \\ 1091.70 + 0.689(x - 1000) & x > 1000 \end{cases}$$

where $C(x)$ is the cost in dollars for x thousand gallons of water.
    (a) Find the monthly charge for 12,000 gallons of water.
    (b) Find the monthly charge for 825,000 gallons of water.

Section 2.5

65. ***1998 Volkswagen Beetle*** The table shows the times that it takes a 1998 Volkswagen Beetle to accelerate from 0 mph to speeds of 30 mph, 40 mph, etc., up to 90 mph, in increments of 10 mph.

Time (seconds)	*Speed (mph)*	*Time (seconds)*	*Speed (mph)*
3.3	30	14.3	70
5.0	40	20.4	80
7.3	50	28.0	90
10.6	60		

Source: Motor Trend

    (a) Represent the times by x and the speeds by y, and model the function that is the best fit for the points.
    (b) Graph the points and the function to see how well the function fits the points.

(c) What does the model indicate the speed is 5 seconds after the car starts to move?
(d) According to the model, in how many seconds will the car reach 76.5 mph?

66. ***Beach nourishment*** Since the first beach nourishment on Rockaway Beach, New York, in 1923, the amount of sand being pumped onto barrier islands has grown, as shown in the table.
(a) Using $x = 3$ for the 1920s, $x = 4$ for the 1930s, etc, find a function that fits the data points.
(b) Predict how much sand will be pumped in the decade 2000–2010.

Decade	*Millions of Cubic Yards of Sand*
1920s	8
1930s	21
1940s	21
1950s	23
1960s	30
1970s	65
1980s	79
1990s	96

Source: *Island Packet*, July 13, 1997

CHAPTER TEST

1. Sketch a graph of each of the following functions.
(a) $f(x) = x^4$ (b) $g(x) = |x|$
(c) $h(x) = -1$ (d) $k(x) = \sqrt{x}$
2. The figures that follow show graphs of the power function $y = x^b$. Which is the graph for $b > 1$? Which is the graph for $0 < b < 1$?

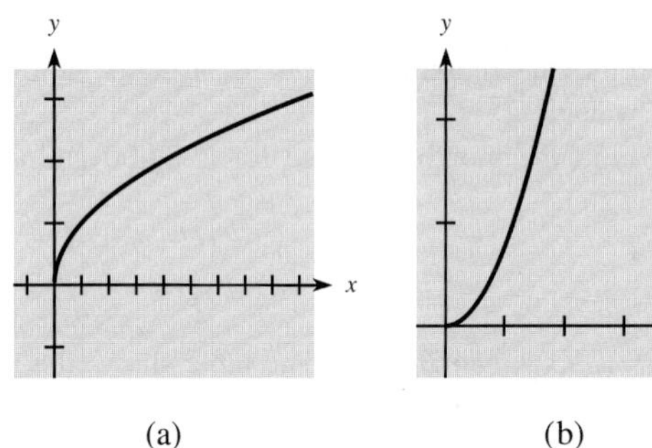

(a) (b)

3. If $f(x) = ax^2 + bx + c$ and $a < 0$, sketch the shape of the graph of $y = f(x)$.
4. Graph
(a) $f(x) = (x + 1)^2 - 1$ (b) $f(x) = (x - 2)^3 + 1$
5. Which of the following three graphs is the graph of $f(x) = x^3 - 4x^2$? Explain your choice.

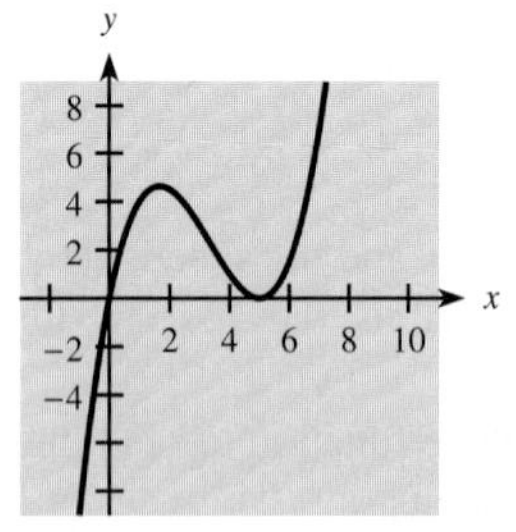

(a)

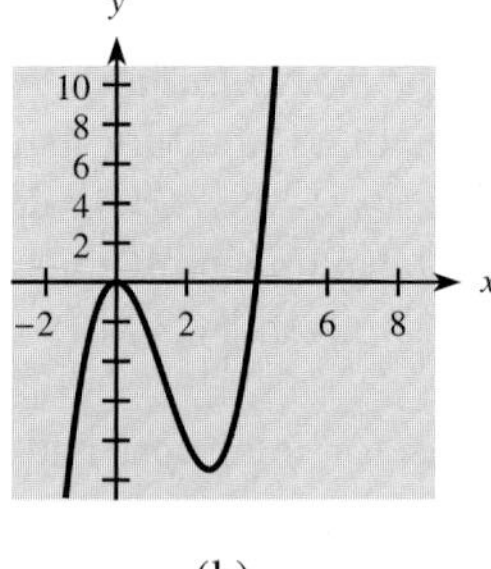

(b)

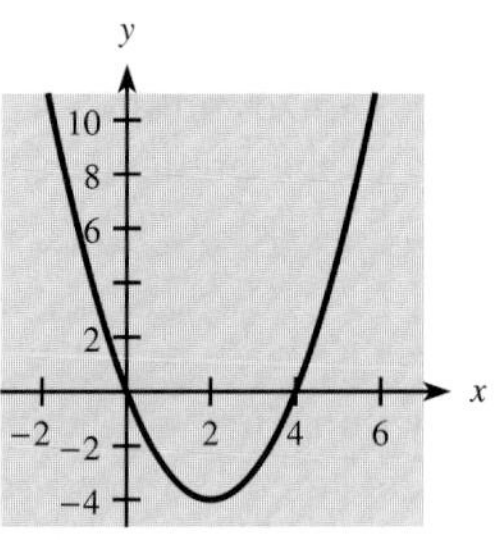

(c)

6. Let $f(x) = \begin{cases} 8x + 1/x & \text{if } x < 0 \\ 4 & \text{if } 0 \le x \le 2 \\ 6 - x & \text{if } x > 2 \end{cases}$

Find (a) the y-coordinate of the point on the graph of $y = f(x)$ where $x = 16$; (b) $f(-2)$; (c) $f(13)$.

7. Sketch the graph of $g(x) = \begin{cases} x^2 & \text{if } x \le 1 \\ 4 - x & \text{if } x > 1 \end{cases}$.
8. Find the vertex and zeros, if they exist, and sketch the graph of $f(x) = 21 - 4x - x^2$.

9. Solve $3x^2 + 2 = 7x$.
10. Solve $2x^2 + 6x = 9$.
11. Solve $\frac{1}{x} + 2x = \frac{1}{3} + \frac{x+1}{x}$.
12. Which of the following three graphs is that of $g(x) = \frac{3x - 12}{x + 2}$?

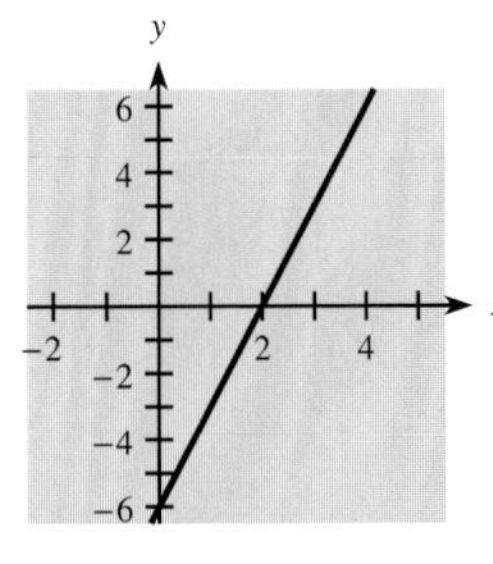

(a)

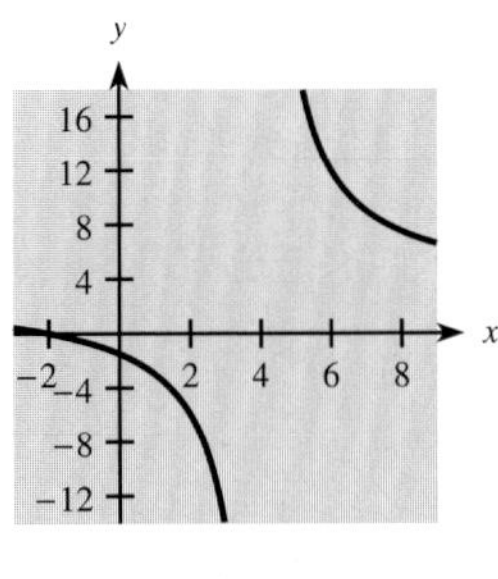

(b)

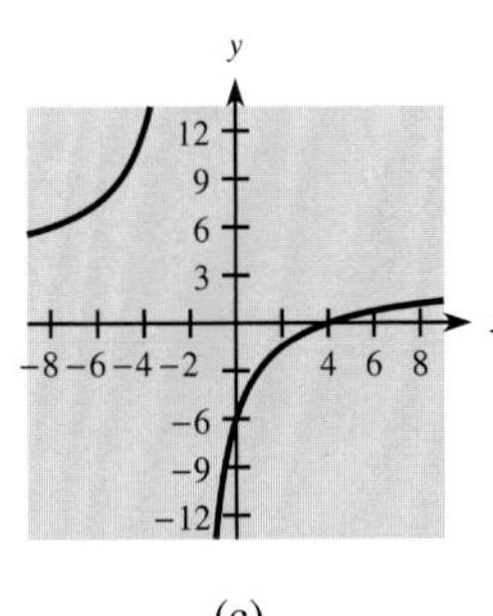

(c)

13. Choose the function type that models the following graphs.

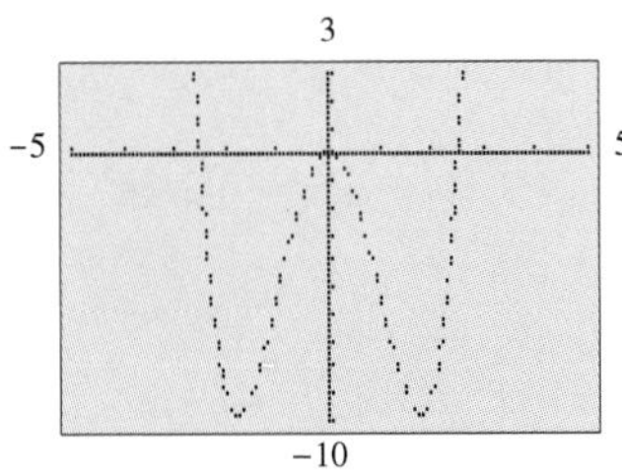

(a)

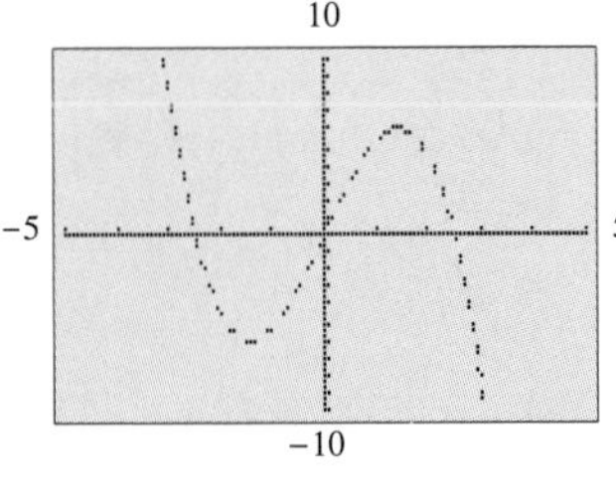

(b)

14. (a) Make a scatter plot, then develop a model, for the following data.
    (b) What does the model predict for $x = 40$?
    (c) When does the model predict $f(x) = 0$?

x	0	2	4	6	8	10	12	14	16	18	20
y	20.1	19.2	18.8	17.5	17.0	15.8	16	14.9	13.8	13.7	13.0

15. Suppose the supply and demand functions for a product are given by $6p - q = 180$ and $(p + 20)q = 30{,}000$, respectively. Find the equilibrium price and quantity.
16. Suppose a company's total cost for a product is given by $C(x) = 15{,}000 + 35x + 0.1x^2$ and the total revenue for the product is given by $R(x) = 285x - 0.9x^2$, where x is the number of units produced and sold.
    (a) Find the profit function.
    (b) Determine the number of units at which the profit for the product is maximized, and find the maximum possible profit.
    (c) Find the break-even point(s) for the product.
17. ***Wind chill*** The wind chill expresses the combined effects of low temperatures and wind speeds as a single temperature reading. When the outside temperature is 0°F, the wind chill, *WC*, is a function of the wind speed s (in mph) and is given by the following function.*

$$WC = f(s) = \begin{cases} 0 & \text{if } 0 \le s \le 4 \\ 91.4 - 7.46738(5.81 + 3.7\sqrt{s} - 0.25s) & \text{if } 4 < s \le 45 \\ -55 & \text{if } s > 45 \end{cases}$$

    (a) Find $f(15)$ and write a sentence that explains its meaning.
    (b) Find the wind chill when the wind speed is 48 mph.

National debt The table gives the national debt, in trillions of dollars, for the years 1990–1997, with White House estimates for 1998 and 1999. Use x as the number of years from 1900 in Problems 18 and 19.

Years	*National Debt*	*Years*	*National Debt*
1990	3.2	1995	4.9
1991	3.6	1996	5.2
1992	4.0	1997	5.4
1993	4.4	1998	5.5†
1994	4.6	1999	5.7†

Source: *USA Today*, January 7, 1998 †White House estimates

*Bosch, W., and L. G. Cobb, "Windchill," UMAP Unit 658, *The UMAP Journal,* 5(4), Winter 1984, 477–492.

18. (a) Plot the data for 1990 to 1997.
    (b) Find a linear model for the data.
    (c) Does this linear model predict that the debt will eventually decrease? Explain.
    (d) Why would the White House prefer to not use this model?
19. (a) The White House would be most interested in a model that would eventually predict a decrease in the debt. What function type could be used to do this and still fit the data?
    (b) Use the function type from (a) to create a model for the data.
    (c) Use this model to predict the debt for 1998 and 1999, and compare the predictions from the model with the White House predictions.
    (d) Find in what year this model predicts the debt will reach its maximum before starting to decrease.

Extended Applications / Group Projects

I. Maximizing Profit

The Digital Electronics Company has monopolistic control in the marketplace for its new product. Its average daily costs for the product are given by

$$\overline{C} = \frac{8000}{x} + 1000 + x$$

where x is the number of units produced and sold, and the daily demand for its product is

$$x = 150 - \frac{p}{40}$$

where x is the number of units demanded and p is the price.

For the short run, management wants to know how many units to produce per day to maximize its daily profit. In addition, Digital managers expect Congress to impose a new excise tax of \$100 per unit on its product, so they want to know how many units will maximize daily profit after the tax is imposed and how daily profit will be affected by the tax.

To assist the company, find the following.

1. The daily total cost function.
2. The daily revenue function. (Recall that $R = R(x) = px$.)
3. The daily profit function.
4. The number of units, x, that maximizes the profit function and the maximum daily profit.
5. The new total cost function after the tax is added and the new profit function.
6. The number of units that maximizes the new profit function and the new maximum daily profit.
7. The difference that the tax causes in daily profit.

II. International E-mail Usage

The instantaneous nature of telephone communication has always been important for business. However, differences in time zones have caused headaches for those who conduct business internationally. Recently, faxes and e-mail have helped alleviate some of this problem because messages are immediately transferred and made available, even though the recipient may not be there to receive them. Of these, e-mail is faster, more efficient with large documents, and more cost-effective. Thus the growing international economy is stimulating e-mail usage.

The following figure, published in *Newsweek* on January 26, 1998, shows some projections for international e-mail usage. (Data shown are estimates based on the figure.)

No Need to Phone
It's 5:00 in the morning at the Tokyo office and 8 p.m. in London. The international economy is creating a boom in e-mail.

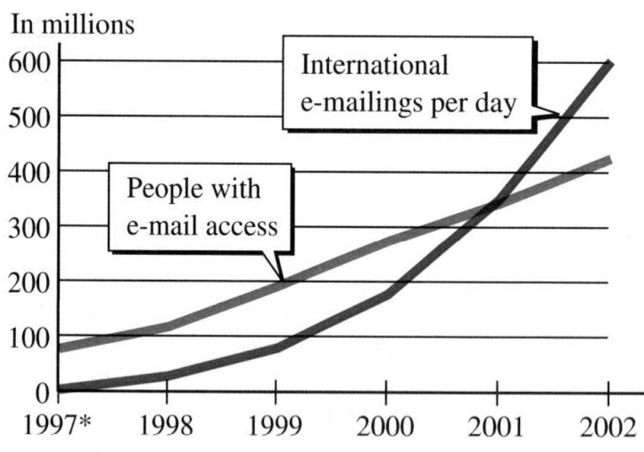

*1997 ESTIMATE. 1998–2002 PROJECTIONS. SOURCE: PIONEER CONSULTING

JANUARY 26, 1998 NEWSWEEK

Year	*People with E-mail Access (millions)*	*Daily International E-mailings*
1997	70	10
1998	130	30
1999	190	80
2000	270	190
2001	350	360
2002	430	600

A. 1. Decide what type of function would be most appropriate to model each of the following.
   (a) International e-mailings as a function of time in years, $I = I(t)$.
   (b) People with e-mail access as a function of time, $P = P(t)$.
2. (a) Identify the independent and dependent variables for $I(t)$ and $P(t)$,
   (b) Determine the domain and range for $I(t)$ and $P(t)$.

3. Develop a model for $I(t)$ and for $P(t)$ by using the given data points for each.
4. (a) Discuss the meaning of intercepts for each model.
   (b) Discuss the meaning of the slope for any linear model.
5. Consider the point of intersection in the figure.
   (a) Discuss its meaning.
   (b) Estimate its coordinates from the figure.
   (c) Calculate its coordinates using the equations from your models.
6. Use your models to find projections for 2005 and 2010, and discuss the implications of your findings.
7. Discuss the limitations of your model.

B. Next investigate international e-mailings as a function of the number of people with e-mail access.
   1. Make a scatter plot of the data points with the number of people as the independent variable. Then decide what type of function best models the data.
   2. Develop the function that models these data.
   3. What does this new model predict for the number of international e-mailings when the number of people with e-mail access is
      (a) 190,000,000 (as in 1999); (b) 650,000,000.
   4. Discuss the implications and limitations of this model.

C. Think of one other meaningful way to look at these data. Plot your data points and develop a model. Discuss the implications and limitations of this model.

Warm-up

Prerequisite Problem Type	For Section	Answer	Section for Review
Evaluate: (a) $4 - (-2)$ (b) $-4\left(\frac{2}{3}\right) + 3$ (c) $(-5)(4) + (8)(1) + (2)(5)$ (d) Multiply each of the numbers $0, -3, -2, 2$ by $-\frac{1}{3}$	3.1 3.2 3.3 3.4 3.5	(a) 6 (b) $\frac{1}{3}$ (c) -2 (d) $0, 1, -1, \frac{2}{3}, -\frac{2}{3}$	0.2 Signed numbers
Write the coefficients of x, y, and z and the constant term in each equation: (a) $2x + 5y + 4z = 4$ (b) $x - 3y - 2z = 5$ (c) $3x + y = 4$	3.3 3.4 3.5	*Coefficients* x y z / *Const. term* 2 5 4 / 4 1 −3 −2 / 5 3 1 0 / 4	0.5 Algebraic expressions
Solve (a) $x_1 - 2x_3 = 2$ for x_1 (b) $x_2 + x_3 = -3$ for x_2	3.3 3.4 3.5	(a) $x_1 = 2 + 2x_3$ (b) $x_2 = -3 - x_3$	1.1 Linear equations
Solve the system: $\begin{cases} x - 0.25y = 11{,}750 \\ -0.20x + y = 10{,}000 \end{cases}$	3.3 3.4 3.5	$x = 15{,}000$ $y = 13{,}000$	1.5 Systems of linear equations

Chapter 3

Matrices

A business may collect and store or analyze various types of data as a regular part of its record-keeping procedures. The data may be presented in tabular form. For example, a building contractor who builds four different styles of houses may catalog the number of units of certain materials needed to build each style in a table like Table 3.1.

TABLE 3.1

	House Style			
Required Units	*Ranch*	*Colonial*	*Split Level*	*Cape Cod*
Wood	20	25	22	18
Siding	27	40	31	25
Roofing	16	16	19	16

If we write the numbers from Table 3.1 in the rectangular array

$$A = \begin{bmatrix} 20 & 25 & 22 & 18 \\ 27 & 40 & 31 & 25 \\ 16 & 16 & 19 & 16 \end{bmatrix}$$

we say A is a ***matrix*** *(plural: matrices) representing Table 3.1. In addition to storing data in a matrix, we can analyze data and make business decisions by defining the operations of addition, subtraction, and multiplication for matrices. Matrices are also useful for analyzing many types of problems and for making comparisons of data. Matrices can be used to solve systems of linear equations, as we will see in this chapter; and they are useful in linear programming, which is discussed in Chapter 4 "Inequalities and Linear Programming." Matrix operations, discussed in this chapter, are the basis for computer spreadsheets, which are used extensively in education, research, and business.*

3.1 Matrices

OBJECTIVES

- *To organize and interpret data stored in matrices*
- *To add and subtract matrices*
- *To find the transpose of a matrix*

APPLICATION PREVIEW

Their organizational format makes matrices useful for solving an assortment of problems, including the following. Suppose the purchase prices and delivery charges (in dollars per unit) for wood, siding, and roofing used in construction are given by Table 3.2. If the supplier decides to raise the purchase prices by \$.60 per unit and the delivery charges by \$.05 per unit, the matrix that describes the new unit charges will be the sum of the charges matrix, formed from Table 3.2, and the matrix representing the increases.

TABLE 3.2

	Wood	*Siding*	*Roofing*
Purchase	6	4	2
Delivery	1	1	0.5

As we noted in the introduction to this chapter, matrices can be used to store data and to perform operations with the data. The matrix

$$C = \begin{bmatrix} 6 & 4 & 2 \\ 1 & 1 & 0.5 \end{bmatrix}$$

represents the original unit charges; it contains the heart of the information from Table 3.2, without the labels. The matrix representing the increases is given by

$$H = \begin{bmatrix} 0.60 & 0.60 & 0.60 \\ 0.05 & 0.05 & 0.05 \end{bmatrix}$$

We will discuss how to find the sum of these two and other pairs of matrices and how to find the difference of two matrices.

In general, we define a **matrix** as any rectangular array of numbers. In matrix A (representing Table 3.1 on the previous page), the *rows* correspond to the types of building materials, and the *columns* correspond to the types of houses. The rows of a matrix are numbered from the top to the bottom and the columns are numbered from left to right.

$$\begin{array}{r} \\ \text{Row 1} \\ \text{Row 2} \\ \text{Row 3} \end{array} \begin{array}{c} \begin{array}{cccc} \text{Column 1} & \text{Column 2} & \text{Column 3} & \text{Column 4} \end{array} \\ \begin{bmatrix} 20 & 25 & 22 & 18 \\ 27 & 40 & 31 & 25 \\ 16 & 16 & 19 & 16 \end{bmatrix} \end{array}$$

Matrices are classified in terms of the numbers of rows and columns they have. The matrix A has three rows and four columns, so we say this is a 3×4 (read "three by four") matrix.

The matrix

$$A = \begin{bmatrix} a_{11} & a_{12} & a_{13} & \cdots & a_{1n} \\ a_{21} & a_{22} & a_{23} & \cdots & a_{2n} \\ & \vdots & & & \vdots \\ a_{m1} & a_{m2} & a_{m3} & \cdots & a_{mn} \end{bmatrix}$$

has m rows and n columns, so it is an $m \times n$ matrix. When we designate A as an $m \times n$ matrix, we are indicating the size of the matrix. Two matrices are said to have the same **order** (be the same size) if they have the same number of rows and the same number of columns.

The numbers in a matrix are called its **entries** or **elements.** Note that the subscripts on an entry in matrix A correspond respectively to the row and column in which the entry is located. Thus a_{23} represents the entry in the second row and the third column, and we refer to it as the "two-three entry." In matrix B below, the entry denoted by b_{23} is 1.

Some matrices take special names because of their size. If the number of rows equals the number of columns, we say the matrix is a **square matrix.** Matrix B is a 3×3 square matrix.

$$B = \begin{bmatrix} 4 & 3 & 0 \\ 0 & 0 & 1 \\ -4 & 0 & 0 \end{bmatrix}$$

EXAMPLE 1

In a study of labor assignment rules for workers who are not perfectly interchangeable, Paul M. Bobrowski and Paul Sungchil Park used Figure 3.1 to describe the tardiness in completing a job (in hours per job) at six different locations for the job priorities as follows:

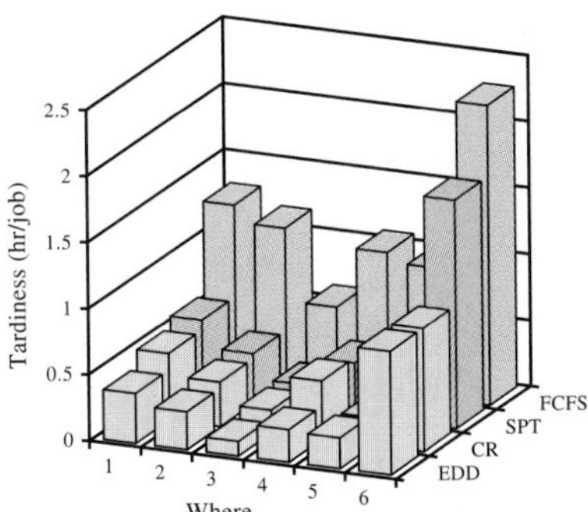

Figure 3.1

SOURCE: Adapted from Bobrowki, P. M. and P. S. Park, "An evaluation of labor assignment rules," *Journal of Operations Management,* Vol. 11, September 1993

(a) smallest ratio of time available until the due date to processing time remaining (the critical ratio, CR)
(b) first come, first served (FCFS)
(c) shortest processing time (SPT)
(d) earliest due date (EDD)

A partial table of values from the figure is on the next page. Use the figure to complete the table and to create a 4×6 matrix describing the information.

	1	*2*	*3*	*4*	*5*	*6*
EDD			0.1			
CR	0.5	0.4			0.2	1.0
SPT			0.2	0.4	0.4	1.5
FCFS		1.1	0.5	1.1	1.0	

Solution

Reading the data from the figure (to the nearest tenth) and completing the table, we get

	1	*2*	*3*	*4*	*5*	*6*
EDD	0.4	0.3	0.1	0.2	0.2	1.0
CR	0.5	0.4	0.2	0.3	0.2	1.0
SPT	0.5	0.4	0.2	0.4	0.4	1.5
FCFS	1.2	1.1	0.5	1.1	1.0	2.3

Writing these data in a matrix gives

$$T = \begin{bmatrix} 0.4 & 0.3 & 0.1 & 0.2 & 0.2 & 1.0 \\ 0.5 & 0.4 & 0.2 & 0.3 & 0.2 & 1.0 \\ 0.5 & 0.4 & 0.2 & 0.4 & 0.4 & 1.5 \\ 1.2 & 1.1 & 0.5 & 1.1 & 1.0 & 2.3 \end{bmatrix}$$

A matrix with one row, such as $[9 \quad 5]$ or $[3 \quad 2 \quad 1 \quad 6]$, is called a **row matrix,** and a matrix with one column, such as

$$\begin{bmatrix} 1 \\ 3 \\ 5 \end{bmatrix}$$

is called a **column matrix.** Row and column matrices are also called **vectors.**

Any matrix in which *every* entry is zero is called a **zero matrix;** examples include

$$\begin{bmatrix} 0 & 0 \\ 0 & 0 \end{bmatrix}, \quad \begin{bmatrix} 0 & 0 \\ 0 & 0 \\ 0 & 0 \end{bmatrix}, \quad \text{and} \quad \begin{bmatrix} 0 & 0 & 0 & 0 \\ 0 & 0 & 0 & 0 \\ 0 & 0 & 0 & 0 \end{bmatrix}$$

We define two matrices to be **equal** if they are the same order and if each entry in one equals the corresponding entry in the other.

EXAMPLE 2

Do the matrices $A = \begin{bmatrix} 1 & 3 & 2 \\ 4 & 1 & 2 \end{bmatrix}$ and $B = \begin{bmatrix} 1 & 4 \\ 2 & 1 \\ 3 & 0 \end{bmatrix}$ have the same order?

Solution

No. A is a 2×3 matrix and B is a 3×2 matrix.

CHECKPOINT

1. (a) Do matrices A and B have the same order?
   (b) Does matrix A equal matrix B?

$$A = \begin{bmatrix} 1 & 2 & 3 \\ 4 & 5 & 6 \\ 7 & 8 & 9 \end{bmatrix} \quad \text{and} \quad B = \begin{bmatrix} 1 & 4 & 7 \\ 2 & 5 & 8 \\ 3 & 6 & 9 \end{bmatrix}$$

Note that if the columns and rows of matrix A in the Checkpoint were interchanged the result is matrix B, and vice versa. When this relationship exists, we say that A and B are **transposes** of each other and write $A^T = B$ and $B^T = A$. We will see valuable uses for transposes of matrices in Chapter 4 "Inequalities and Linear Programming."

EXAMPLE 3

(a) Which element of

$$A = \begin{bmatrix} 1 & 0 & 3 \\ 3 & 4 & 2 \\ 7 & 8 & 3 \end{bmatrix}$$

is represented by a_{32}?

(b) Is A a square matrix?

(c) Find the transpose of matrix A.

Solution

(a) a_{32} represents the element in row 3 and column 2—that is, 8.

(b) Yes, it is a 3×3 (square) matrix.

(c) $\begin{bmatrix} 1 & 3 & 7 \\ 0 & 4 & 8 \\ 3 & 2 & 3 \end{bmatrix}$

If two matrices have the same number of rows and columns, we can add the matrices by adding their corresponding entries. More formally,

Sum of Two Matrices

If matrix A and matrix B are the same order and have elements a_{ij} and b_{ij}, respectively, then their **sum** $A + B$ is a matrix C whose elements are $c_{ij} = a_{ij} + b_{ij}$ for all i and j.

That is,

$$\begin{bmatrix} a_{11} & a_{12} & \cdots & a_{1n} \\ a_{21} & a_{22} & \cdots & a_{2n} \\ \vdots & & & \vdots \\ a_{m1} & a_{m2} & \cdots & a_{mn} \end{bmatrix} + \begin{bmatrix} b_{11} & b_{12} & \cdots & b_{1n} \\ b_{21} & b_{22} & \cdots & b_{2n} \\ \vdots & & & \vdots \\ b_{m1} & b_{m2} & \cdots & b_{mn} \end{bmatrix}$$

$$= \begin{bmatrix} a_{11} + b_{11} & a_{12} + b_{12} & \cdots & a_{1n} + b_{1n} \\ a_{21} + b_{21} & a_{22} + b_{22} & \cdots & a_{2n} + b_{2n} \\ \vdots & & & \vdots \\ a_{m1} + b_{m1} & a_{m2} + b_{m2} & \cdots & a_{mn} + b_{mn} \end{bmatrix}.$$

EXAMPLE 4

Find the sum of A and B if

$$A = \begin{bmatrix} 1 & 2 & 3 \\ 4 & -1 & -2 \end{bmatrix} \quad \text{and} \quad B = \begin{bmatrix} -1 & 2 & -3 \\ -2 & 0 & 1 \end{bmatrix}.$$

Solution

$$A + B = \begin{bmatrix} 1+(-1) & 2+2 & 3+(-3) \\ 4+(-2) & -1+0 & -2+1 \end{bmatrix} = \begin{bmatrix} 0 & 4 & 0 \\ 2 & -1 & -1 \end{bmatrix}.$$

The matrix $-B$ is called the **negative** of the matrix B, and each element of $-B$ is the negative of the corresponding element of B. For example, if

$$B = \begin{bmatrix} -1 & 2 & -3 \\ -2 & 0 & 1 \end{bmatrix}, \quad \text{then} \quad -B = \begin{bmatrix} 1 & -2 & 3 \\ 2 & 0 & -1 \end{bmatrix}.$$

Using the negative, we can define the difference $A - B$ (when A and B have the same order) by $A - B = A + (-B)$, or by subtracting corresponding elements.

EXAMPLE 5

For matrices A and B of Example 4, find $A - B$.

Solution

$A - B$ can be found by subtracting corresponding elements.

$$A - B = \begin{bmatrix} 1-(-1) & 2-2 & 3-(-3) \\ 4-(-2) & -1-0 & -2-1 \end{bmatrix} = \begin{bmatrix} 2 & 0 & 6 \\ 6 & -1 & -3 \end{bmatrix}$$

CHECKPOINT

2. If $A = \begin{bmatrix} 1 & 2 & 0 \\ 3 & 1 & 2 \end{bmatrix}$, $B = \begin{bmatrix} 3 & 6 & 1 \\ 0 & 2 & 1 \end{bmatrix}$, and $C = \begin{bmatrix} 1 & 0 & 2 \\ 0 & 1 & 1 \end{bmatrix}$, find $A + B - C$.

3. (a) What matrix D must be added to matrix A so that their sum is matrix Z?

$$A = \begin{bmatrix} 1 & -2 & 3 \\ -5 & 1 & 2 \end{bmatrix} \quad \text{and} \quad Z = \begin{bmatrix} 0 & 0 & 0 \\ 0 & 0 & 0 \end{bmatrix}$$

(b) Does $D = -A$?

Graphics calculators and computers have the capability to perform a number of operations on matrices. Computer spreadsheets are also very useful in performing operations with matrices.

EXAMPLE 6

Use a graphics calculator or computer to compute $A + B$ and $A - B$ for

$$A = \begin{bmatrix} -1 & 3 & 2 \\ 1 & 0 & 2 \\ 0 & 1 & -1 \\ -4 & 2 & 3 \end{bmatrix} \quad \text{and} \quad B = \begin{bmatrix} -3 & 2 & 1 \\ 0 & -2 & 2 \\ -1 & 6 & 8 \\ 3 & 2 & 4 \end{bmatrix}.$$

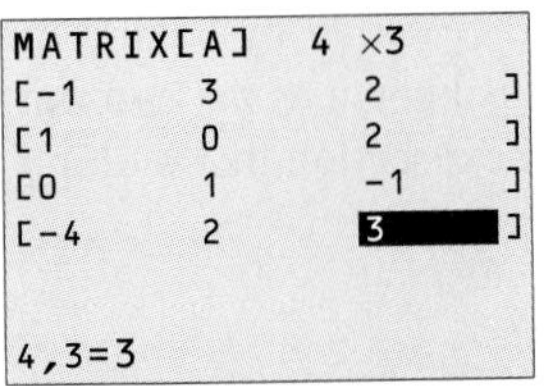

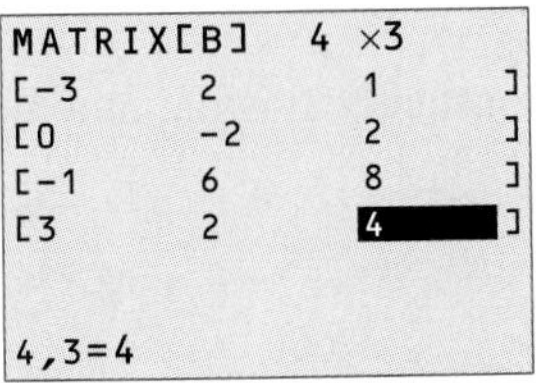

Figure 3.2

Solution

On most graphing calculators, you can use MATRIX and EDIT to define a matrix. To define matrix A, we select a name, A, select the dimensions, and enter the elements of matrix A. Note that the address (row and column) of each entry is shown when entering it. Figure 3.2 shows matrix A and matrix B displayed on screens of graphing calculators, with the dimensions of each matrix shown above it. We find the sum and difference of the two matrices by using the regular [+] and [−] keys (see Figure 3.3).

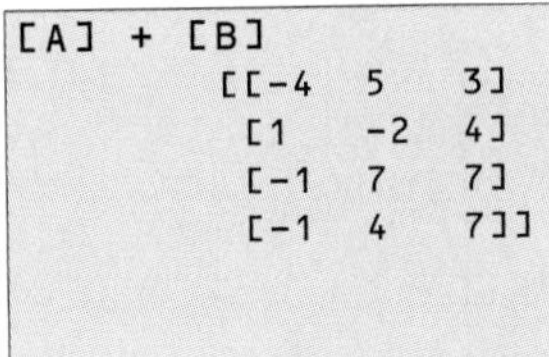


```
[A] - [B]
     [[2   1   1]
      [1   2   0]
      [1  -5  -9]
      [-7  0  -1]]
```

Figure 3.3

Note that if matrix A and matrix B have different orders, they cannot be added or subtracted, and a graphing calculator will give an error statement.

We now return to the Application Preview.

EXAMPLE 7

Suppose the purchase prices and delivery charges (in dollars per unit) for wood, siding, and roofing used in construction are given by Table 3.3. The supplier decides to raise the purchase prices by \$.60 per unit and the delivery charges by \$.05 per unit. Write the matrix that describes the new unit charges.

TABLE 3.3

	Wood	*Siding*	*Roofing*
Purchase	6	4	2
Delivery	1	1	0.5

Solution

The matrix representing the original unit charges is

$$C = \begin{bmatrix} 6 & 4 & 2 \\ 1 & 1 & 0.5 \end{bmatrix}$$

and the matrix representing the increases is given by

$$H = \begin{bmatrix} 0.60 & 0.60 & 0.60 \\ 0.05 & 0.05 & 0.05 \end{bmatrix}$$

The new unit charge for each item is its former charge plus the increase, so the new unit charge matrix is given by

$$C + H = \begin{bmatrix} 6 + 0.60 & 4 + 0.60 & 2 \ \ + 0.60 \\ 1 + 0.05 & 1 + 0.05 & 0.5 + 0.05 \end{bmatrix} = \begin{bmatrix} 6.60 & 4.60 & 2.60 \\ 1.05 & 1.05 & 0.55 \end{bmatrix}.$$

Note that the sum of C and H in Example 7 could be found by adding the matrices in either order. That is, $C + H = H + C$. This is known as the **commutative law of addition** for matrices. We will see in the next section that multiplication of matrices is *not* commutative.

EXAMPLE 8

Suppose that in a government agency, paperwork is constantly flowing among offices according to the diagram.

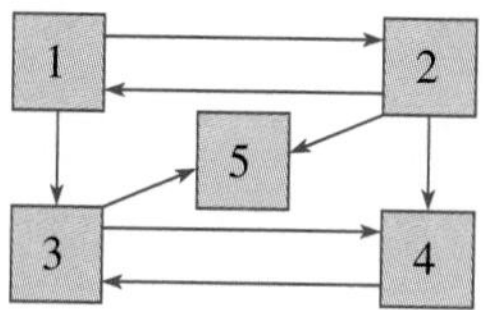

(a) Construct matrix A with elements

$$a_{ij} = \begin{cases} 1 & \text{if paperwork flows directly from } i \text{ to } j. \\ 0 & \text{if paperwork does not directly flow from } i \text{ to } j. \end{cases}$$

(b) Construct matrix B with elements

$$b_{ij} = \begin{cases} 1 & \text{if paperwork can flow from } i \text{ to } j \text{ through at most} \\ & 1 \text{ intermediary, with } i \neq j. \\ 0 & \text{if this is not true.} \end{cases}$$

(c) The person in office i has the most power to influence others if the sum of the elements of row i in the matrix $A + B$ is the largest. What is the office number of this person?

Solution

(a) $A = \begin{bmatrix} 0 & 1 & 1 & 0 & 0 \\ 1 & 0 & 0 & 1 & 1 \\ 0 & 0 & 0 & 1 & 1 \\ 0 & 0 & 1 & 0 & 0 \\ 0 & 0 & 0 & 0 & 0 \end{bmatrix}$ (b) $B = \begin{bmatrix} 0 & 1 & 1 & 1 & 1 \\ 1 & 0 & 1 & 1 & 1 \\ 0 & 0 & 0 & 1 & 1 \\ 0 & 0 & 1 & 0 & 1 \\ 0 & 0 & 0 & 0 & 0 \end{bmatrix}$

(c) $A + B = \begin{bmatrix} 0 & 2 & 2 & 1 & 1 \\ 2 & 0 & 1 & 2 & 2 \\ 0 & 0 & 0 & 2 & 2 \\ 0 & 0 & 2 & 0 & 1 \\ 0 & 0 & 0 & 0 & 0 \end{bmatrix}$

Because the sum of the elements in row 2 is the largest (7), the person in office 2 has the most power to influence others. Because row 5 has the smallest sum (0), the person in office 5 has the least power.

Graphing calculators, computer algebra systems, and computer spreadsheets have the capability to perform a number of complex operations on matrices. Many applications in business and industry call for repeated manipulations or operations with very large tables (matrices). Once the data are entered in a graphing calculator, computer program, or spreadsheet, they can be stored and then operated on or combined with other matrices.

The ability to store data in arrays (matrices) and the ability to add and subtract (and perform other operations) with matrices are at the heart of the development of computer spreadsheets such as *Lotus 1-2-3*, *Quattro Pro*, and *Excel*. Table 3.4 shows the financial results (in millions of dollars) from the *1993 Annual Report of the Horsham Corporation* in a spreadsheet.

TABLE 3.4

A	B	C	D	E
1		Financial	Results	($million)
2		1993	1992	1991
3	Earnings from Barrick	41.5	34.1	18.4
4	Dilution gains	2.4	0.9	16.7
5	Earnings from Clark	2.8	(3.0)	23.0
6	R&M Holdings parent	(11.6)	0	0
7	Horsham parent	5.0	4.2	3.3
8	Net earnings	40.1	36.2	61.4
9	Operating cash flow	62.3	31.8	91.6
10	Capital expenditures	100.9	81.1	90.7

One of the easiest and most frequently performed operations on a computer spreadsheet is adding columns, which is accomplished in the same manner as adding two (or more) $n \times 1$ column matrices. Similarly, adding all the rows of a spreadsheet is performed in the same manner as adding 2 (or more) $1 \times n$ row matrices.

EXAMPLE 9

Use the financial results in Table 3.4 to answer the following.

(a) Write the data from each of 1993 and 1992 as matrices, and find the gain (or loss) from 1992 to 1993 by using a matrix operation.

(b) Determine which row matrix is the sum of the row matrices above it on the spreadsheet.

Solution

(a) Writing the gains or losses (indicated by inclusion in parentheses in the annual report) for 1993 as column matrix A and for 1992 as B, we have

$$A = \begin{bmatrix} 41.5 \\ 2.4 \\ 2.8 \\ -11.6 \\ 5.0 \\ 40.1 \\ 62.3 \\ 100.9 \end{bmatrix} \quad \text{and} \quad B = \begin{bmatrix} 34.1 \\ 0.9 \\ -3.0 \\ 0 \\ 4.2 \\ 36.2 \\ 31.8 \\ 81.1 \end{bmatrix}$$

The gain (or loss) in 1993 is given by $A - B$.

$$\mathrm{A - B} = \begin{bmatrix} 7.4 \\ 1.5 \\ 5.8 \\ -11.6 \\ 0.8 \\ 3.9 \\ 30.5 \\ 19.8 \end{bmatrix}$$

This result indicates that gains were made in all categories but one.

(b) The row matrix that gives net earnings (row 8 of the spreadsheet) is the sum of the row matrices above it on the spreadsheet.

CHECKPOINT SOLUTIONS

1. (a) Yes, A and B have the same order (size).
   (b) No. They have the same elements, but they are not equal because their corresponding elements are not equal.

2. $\begin{bmatrix} 3 & 8 & -1 \\ 3 & 2 & 2 \end{bmatrix}$ 3. (a) $\begin{bmatrix} -1 & 2 & -3 \\ 5 & -1 & -2 \end{bmatrix}$ (b) Yes

EXERCISE 3.1

Use the following matrices for Problems 1–10.

$$A = \begin{bmatrix} 1 & 0 & 2 \\ 3 & 2 & 1 \\ 4 & 0 & 3 \end{bmatrix} \quad B = \begin{bmatrix} 1 & 1 & 3 & 0 \\ 4 & 2 & 1 & 1 \\ 3 & 2 & 0 & 1 \end{bmatrix}$$

$$C = \begin{bmatrix} 5 & 3 \\ 1 & 2 \end{bmatrix} \quad D = \begin{bmatrix} 4 & 2 \\ 3 & 5 \end{bmatrix} \quad E = \begin{bmatrix} 1 & 0 & 4 \\ 5 & 1 & 0 \end{bmatrix}$$

$$F = \begin{bmatrix} 1 & 2 & 3 \\ -1 & 0 & 1 \\ 2 & -3 & -4 \end{bmatrix} \quad Z = \begin{bmatrix} 0 & 0 & 0 \\ 0 & 0 & 0 \\ 0 & 0 & 0 \end{bmatrix}$$

1. How many rows does matrix B have?
2. What is the order of matrix E?
3. Write the negative of matrix F.
4. Write a zero matrix that is the same order as D.
5. Which of the matrices A, B, C, D, E, F, and Z are square?
6. Which pairs of matrices, if any, are equal?
7. Which pairs of matrices, if any, have the same order?
8. Write the matrix that is the negative of matrix B.
9. What is the element a_{23}?
10. What is element b_{24}?
11. Write the transpose of matrix A.
12. Write the transpose of matrix F.

Use the matrices M and N in Problems 13–16.

$$M = \begin{bmatrix} 1 & 0 & 2 \\ 3 & 2 & 1 \\ 4 & 0 & 3 \end{bmatrix} \quad \text{and} \quad N = \begin{bmatrix} 1 & 1 & 3 & 0 \\ 4 & 2 & 1 & 1 \\ 3 & 2 & 0 & 1 \end{bmatrix}$$

13. Can the sum of M and N be found?
14. What is the negative of N?
15. What is the sum of matrix M and its negative?
16. If matrix M has elements $m_{3j} = 0$, what is j?

For Problems 15–26, use the matrices below to compute the indicated sums and differences, if possible.

$$A = \begin{bmatrix} 1 & 0 & 2 \\ 3 & 2 & 1 \\ 4 & 0 & 3 \end{bmatrix} \quad B = \begin{bmatrix} 1 & 1 & 3 & 0 \\ 4 & 2 & 1 & 1 \\ 3 & 2 & 0 & 1 \end{bmatrix}$$

$$C = \begin{bmatrix} 5 & 3 \\ 1 & 2 \end{bmatrix} \quad D = \begin{bmatrix} 4 & 2 \\ 3 & 5 \end{bmatrix}$$

$$E = \begin{bmatrix} 2 & 1 & 1 \\ 1 & 0 & 4 \\ 5 & 1 & 0 \end{bmatrix} \quad F = \begin{bmatrix} 1 & 2 & 3 \\ -1 & 0 & 1 \\ 2 & -3 & -4 \end{bmatrix}$$

$$Z = \begin{bmatrix} 0 & 0 & 0 \\ 0 & 0 & 0 \\ 0 & 0 & 0 \end{bmatrix}$$

17. $C + D$
18. $A + F$
19. $A - F$
20. $C - D$
21. $A + A^T$
22. $A + F^T$
23. $D - E$
24. $B + F$
25. $A + E^T$
26. $Z + E^T$

In Problems 27–30 find x, y, z, and w.

27. $\begin{bmatrix} x & 1 & 0 \\ 0 & y & z \\ w & 2 & 1 \end{bmatrix} = \begin{bmatrix} 3 & 1 & 0 \\ 0 & 2 & 3 \\ 4 & 2 & 1 \end{bmatrix}$

28. $\begin{bmatrix} 0 & x & 1 \\ 3 & y & y \\ z & 0 & 2 \end{bmatrix} = \begin{bmatrix} 0 & 4 & 1 \\ 3 & 1 & y \\ 1 & 0 & w \end{bmatrix}$

29. $\begin{bmatrix} x & 3 & (2x-1) \\ y & 4 & 4y \end{bmatrix} = \begin{bmatrix} 2x-4 & z & 7 \\ 1 & (w+1) & (3y+1) \end{bmatrix}$

30. $\begin{bmatrix} x & y & (x+3) \\ z & 4 & 4y \end{bmatrix} = \begin{bmatrix} (2x-1) & -1 & w \\ x & (5+y) & -4 \end{bmatrix}$

31. Solve for x, y, and z if

$$\begin{bmatrix} x & y \\ y & z \end{bmatrix} + \begin{bmatrix} 2x & -y \\ 3y & -4z \end{bmatrix} = \begin{bmatrix} 6 & 0 \\ 8 & 9 \end{bmatrix}$$

32. Find x, y, z, and w if

$$\begin{bmatrix} x & 4 \\ 4y & w \end{bmatrix} - \begin{bmatrix} 4x & 2z \\ -3 & -2w \end{bmatrix} = \begin{bmatrix} 12 & 8 \\ y & 6 \end{bmatrix}$$

Applications

33. ***Endangered species*** The tables below give the number of some species of threatened and endangered wildlife in the United States and in foreign countries in 1996.

United States

	Mammals	*Birds*	*Reptiles*	*Amphibians*	*Fishes*
Endangered	55	74	14	7	65
Threatened	9	16	19	5	40

Foreign

	Mammals	*Birds*	*Reptiles*	*Amphibians*	*Fishes*
Endangered	252	178	65	8	11
Threatened	19	6	14	1	0

SOURCE: *Statistical Abstract of the United States,* U.S. Dept. of Commerce

(a) Write a matrix A that contains the number of each of these species in the United States in 1996 and a matrix B that contains the number of each of these species outside the United States in 1996.
(b) Find a matrix with the total number of these species, if U.S. and foreign species are different.
(c) Find the matrix $B - A$ and tell what the elements of this matrix mean. What do the negative entries in matrix $B - A$ mean?

34. ***Top-ten dog breeds*** The tables below give the rank and number of registered dogs for the top ten breeds for 1995 and 1996.
(a) Form a matrix for the 1995 data and for the 1996 data.
(b) Use a matrix operation to find the change from 1995 to 1996 in rank for each breed and in registration numbers for each breed.

1995

Breed	*Rank*	*Number Registered*
Labrador retriever	1	132,051
Rottweiler	2	93,656
German shepherd	3	76,088
Golden retriever	4	64,107
Beagle	5	57,063
Poodle	6	54,784
Cocker spaniel	7	48,065
Dachshund	8	44,680
Pomeranian	9	37,894
Yorkshire terrier	10	36,881

1996

	Rank	Number Registered
Labrador retriever	1	149,505
Rottweiler	2	89,867
German shepherd	3	79,076
Golden retriever	4	68,993
Beagle	5	56,946
Poodle	7	56,803
Cocker spaniel	6	45,305
Dachshund	8	48,426
Pomeranian	10	39,712
Yorkshire terrier	9	40,216

SOURCE: American Kennel Club

35. ***Pollution abatement costs*** The tables below give the capital expenditures and gross operating costs of manufacturing establishments for pollution abatement, in millions of dollars. Use a calculator or spreadsheet to find the sum of the matrices containing these data. This will yield the total output of manufacturing establishments for air, water, and solid contained waste pollution abatement for the years 1990–1993.

Pollution Abatement Capital Expenditures

	Air	Water	Solid Contained Waste
1990	6030.8	2562.0	817.5
1991	3706.3	2814.6	869.1
1992	4403.1	2509.8	953.9
1993	4122.0	2294.9	760.9

Pollution Abatement Gross Operating Costs

	Air	Water	Solid Contained Waste
1990	5010.9	6416.4	5643.5
1991	5033.5	6345.0	6008.2
1992	5395.0	6576.9	5494.5
1993	5574.6	6631.8	5348.6

SOURCE: *Statistical Abstract of the United States,* U.S. Dept. of Commerce

36. ***Death rates*** The tables below give the death rates, per 100,000 population, by age for selected years for males and females.
   (a) Use matrix operations to find the death rate per 100,000 for all people in the age categories given and for the given years.
   (b) If matrix M gives the male data and matrix F gives the female data, find matrix $M - F$ and describe what it means.

Males

	Under 1	1–4	5–14	15–24	25–34	35–44	45–54
1970	2410	93	51	189	215	403	959
1980	1429	73	37	172	196	299	767
1990	1083	52	29	147	204	310	610
1993	946	50	27	146	209	329	596
1994	899	52	26	151	207	337	585

Females

	Under 1	1–4	5–14	15–24	25–34	35–44	45–54
1970	1864	75	32	68	102	231	517
1980	1142	55	24	58	76	159	413
1990	856	41	19	49	74	138	343
1993	759	40	19	49	76	144	330
1994	719	37	19	47	76	144	326

SOURCE: *Statistical Abstract of the United States,* U.S. Dept. of Commerce

37. ***Sales*** Let matrix A represent the sales (in thousands of dollars) for the Walbash Company in 1994 in various cities, and let matrix B represent the sales (in thousands of dollars) for the same company in 1995 in the same cities.

$$A = \begin{array}{c} \text{Chicago} \quad \text{Atlanta} \quad \text{Memphis} \\ \begin{bmatrix} 450 & 280 & 850 \\ 400 & 350 & 150 \end{bmatrix} \end{array} \begin{array}{l} \text{Wholesale} \\ \text{Retail} \end{array}$$

$$B = \begin{array}{c} \text{Chicago} \quad \text{Atlanta} \quad \text{Memphis} \\ \begin{bmatrix} 375 & 300 & 710 \\ 410 & 300 & 200 \end{bmatrix} \end{array} \begin{array}{l} \text{Wholesale} \\ \text{Retail} \end{array}$$

   (a) Write the matrix that represents the total sales by type and city for both years.
   (b) Write the matrix that represents the change in sales by type and city from 1994 to 1995.

38. ***Opinion polls*** A poll of 3320 people revealed that of the respondents that are registered Democrats, 843 approved the president's job performance, 426 did not, and 751 had no opinion. Of the registered Republicans, 257 approved of the president's job performance, 451 did not, and 92 had no opinion. Of those registered as Independents, 135 approved, 127 did not approve, and 38 had no opinion. Of the remaining respondents, who were not registered, 92 approved, 64 did not approve, and 44 had no opinion. Represent these data in a 3×4 matrix.

39. ***Life expectancy*** The table below gives the years of life expected at birth for blacks and whites born in the United States in the years 1920, 1940, 1960, and 1980.

	Whites		*Blacks*	
	Males	*Females*	*Males*	*Females*
1920	54.4	55.6	45.5	45.2
1940	62.1	66.6	51.5	54.9
1960	67.4	74.1	61.1	67.4
1980	70.7	78.1	65.3	73.6

SOURCE: National Center for Health Statistics

Make a matrix A containing the information for whites and a matrix B for blacks. Use these to find how many more years whites in each category are expected to live than blacks.

40. ***Imports*** (a) From the data in the table below, make a matrix A that gives the value (in millions of dollars) of imports for the various country groupings in the years 1983–1985.
   (b) Make a matrix B that gives the value of exports (in millions of dollars) for the same groupings in the same years.
   (c) Find the trade balance for each country grouping in each year by finding matrix $B - A$.

		Imports	
Countries	*83*	*84*	*85*
Developed	122,822	135,884	134,018
Developing	72,342	74,421	72,673
Communist	5,085	7,214	7,091
Other	289	369	365
		Exports	
	83	*84*	*85*
Developed	152,117	200,714	223,314
Developing	102,266	119,790	116,161
Communist	3,604	5,221	5,801
Other	1	1	0

SOURCE: U.S. Bureau of Census

41. ***Nutrition*** In a computer simulation, three groups of baby laboratory animals were used. Group I had an enriched diet, group II (control) had the regular diet, and group III had a deficient diet. At the beginning of the simulation the animals were weighed and measured according to their group, with group I averaging 140 g with a length of 5.5 cm. Group II and group III weighed 151 g and 141 g, with lengths of 5.7 cm and 5.5 cm, respectively.
   (a) Make a 3×2 matrix that displays this information.
   (b) At the end of two weeks, the same measurements were made, with the following results: 12.5 cm, 250 g; 11.8 cm, 215 g; 9.8 cm, 190 g. Make a matrix that displays this information. Make sure the matrix is the same size as your answer in (a).

42. ***Nutrition*** Calculate the changes in weight and length from Problems 41 (a) and 41 (b) by using matrix subtraction.

43. ***Management*** Management is attempting to identify the most active person in labor's efforts to unionize. The following diagram shows how influence flows from one employee to another among the four most active employees.

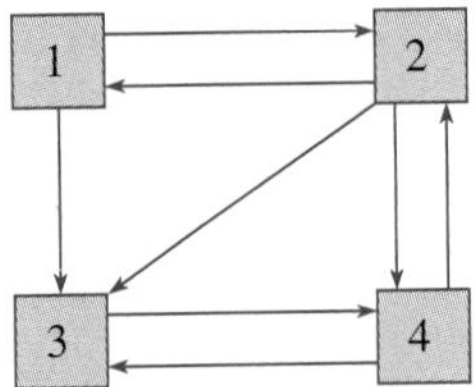

   (a) Construct matrix A with elements

$$a_{ij} = \begin{cases} 1 & \text{if } i \text{ influences } j \text{ directly.} \\ 0 & \text{otherwise.} \end{cases}$$

   (b) Construct matrix B with elements

$$b_{ij} = \begin{cases} 1 & \text{if } i \text{ influences } j \text{ through at most} \\ & 1 \text{ person with } i \neq j. \\ 0 & \text{otherwise.} \end{cases}$$

   (c) The person i is most active in influencing others if the sum of the elements in row i of the matrix $A + B$ is the largest. Who is the most active person?

44. ***Ranking*** In order to rank the five members of its chess team for play against another school, the coach draws the following diagram. An arrow from 1 to 2 means player 1 has defeated player 2.

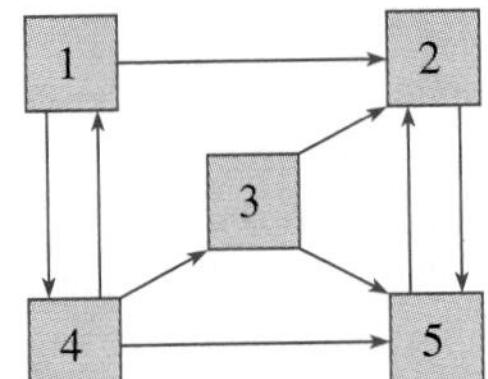

(a) Construct matrix A with elements

$$a_{ij} = \begin{cases} 1 & \text{if } i \text{ defeated } j. \\ 0 & \text{otherwise.} \end{cases}$$

(b) Construct matrix B with elements

$$b_{ij} = \begin{cases} 1 & \text{if } i \text{ defeated someone who defeated } j \text{ and } i \neq j. \\ 0 & \text{otherwise.} \end{cases}$$

(c) The player i is the top-ranked player if the sum of row i in the matrix $A + B$ is the largest. What is the number of this player?

45. ***Production and inventories*** Operating from two plants, the Book Equipment Company (BEC) produces bookcases and filing cabinets. Matrix A summarizes its production for a week, with row 1 representing the number of bookcases and row 2 representing the number of filing cabinets. Matrix B gives the production for the second week, and matrix C that of the third and fourth weeks combined.

$$A = \begin{bmatrix} 50 & 30 \\ 36 & 44 \end{bmatrix} \quad B = \begin{bmatrix} 30 & 45 \\ 22 & 62 \end{bmatrix} \quad C = \begin{bmatrix} 96 & 52 \\ 81 & 37 \end{bmatrix}$$

If column 1 in each matrix represents production from plant 1 and column 2 represents production from plant 2, answer the following.

(a) If production during the second week is given by matrix B, write a matrix that describes production for the first 2 weeks.

(b) If production over the next 2-week period is given by matrix C, describe production for the 4-week period.

(c) If matrix D

$$D = \begin{bmatrix} 40 & 26 \\ 29 & 42 \end{bmatrix}$$

describes the shipments made during the first week, write the matrix that describes the units added to the plants' inventories.

(d) If D also describes the shipments during the second week, describe the change in inventory. What happened at plant 1?

46. ***Management*** The figure that follows depicts the mean (average) flow time for a job when the critical-ratio rule is used to dispatch workers at several machines required to complete the job. In the figure, WHN 1, WHN 2, and WHN 3 represent three different rules for determining when a worker should be transferred to another machine, and Where represents six rules for determining the machine to which the worker is transferred. Construct a 3×6 matrix A to represent these data, with entries rounded to the nearest integer.

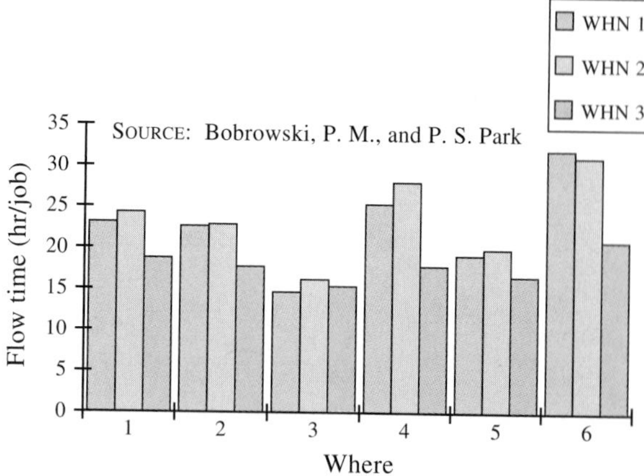

47. ***Management*** In an evaluation of labor assignment rules when workers are not perfectly interchangeable, Paul M. Bobrowski and Paul Sungchil Park created a dynamic job shop with 9 work centers and 9 workers, both numbered 1–9. The efficiency of each worker is specified in the labor efficiency matrix given below, which represents the degree of worker cross-training (Source: Bobrowski, P. M., and P. S. Park, "An evaluation of labor assignment rules," *Journal of Operations Management*, Vol. 11, September 1993).

(a) For what work center(s) is worker 7 least efficient?

(b) For what work center is worker 1 most efficient?

Work Centers

$$\begin{bmatrix} 1.00 & .95 & .95 & .95 & .95 & .85 & .85 & .85 & .85 \\ .85 & 1.00 & .95 & .95 & .95 & .95 & .85 & .85 & .85 \\ .85 & .85 & 1.00 & .95 & .95 & .95 & .95 & .85 & .85 \\ .85 & .85 & .85 & 1.00 & .95 & .95 & .95 & .95 & .85 \\ .85 & .85 & .85 & .85 & 1.00 & .95 & .95 & .95 & .95 \\ .95 & .85 & .85 & .85 & .85 & 1.00 & .95 & .95 & .95 \\ .95 & .95 & .85 & .85 & .85 & .85 & 1.00 & .95 & .95 \\ .95 & .95 & .95 & .85 & .85 & .85 & .85 & 1.00 & .95 \\ .95 & .95 & .95 & .95 & .85 & .85 & .85 & .85 & 1.00 \end{bmatrix} \text{Workers}$$

48. Suppose that for the workers and the work centers described in the matrix of Problem 47, changes in rules cause efficiency to decrease by .01 at work centers 2 and 5 and to decrease by .02 at work center 7. Write the matrix that describes the new efficiencies.

49. For the data of Problem 47, use a computer spreadsheet to find the average efficiency for each worker over the first four work centers. Which worker is the least efficient? Where does this worker perform best?

50. For the data of Problem 47, use a computer spreadsheet to find the average efficiency for each work center over the first five workers. Which work center is the most efficient? Which work center should be studied for improvement?

3.2 Multiplication of Matrices

OBJECTIVES

- *To multiply a matrix by a scalar (real number)*
- *To multiply two matrices*

APPLICATION PREVIEW

Pentico Industries is a manufacturing company that has two divisions, located at Clarion and Brooks, each of which needs different amounts (units) of production materials as described by the following table.

	Steel	*Wood*	*Plastic*
Clarion	20	30	8
Brooks	22	25	15

These raw materials are supplied by Western Supply and Coastal Supply, with prices given in dollars per unit in the following table.

	Western	*Coastal*
Steel	300	290
Wood	100	90
Plastic	145	180

If one supplier must be chosen to provide all materials for either or both divisions of Pentico, company officials can decide which supplier to choose by constructing matrices from these tables and using **matrix multiplication.**

In this section we will discuss **scalar multiplication** of a matrix and finding the **product of two matrices.**

Consider the matrix

$$A = \begin{bmatrix} 3 & 2 \\ 1 & 1 \\ 2 & 0 \end{bmatrix}$$

Because $2A$ is $A + A$, we see that

$$2A = \begin{bmatrix} 3 & 2 \\ 1 & 1 \\ 2 & 0 \end{bmatrix} + \begin{bmatrix} 3 & 2 \\ 1 & 1 \\ 2 & 0 \end{bmatrix} = \begin{bmatrix} 6 & 4 \\ 2 & 2 \\ 4 & 0 \end{bmatrix}$$

Note that $2A$ could have been found by multiplying each entry of A by 2. In the same manner,

$$\begin{aligned} 3A = A + A + A &= \begin{bmatrix} 3 & 2 \\ 1 & 1 \\ 2 & 0 \end{bmatrix} + \begin{bmatrix} 3 & 2 \\ 1 & 1 \\ 2 & 0 \end{bmatrix} + \begin{bmatrix} 3 & 2 \\ 1 & 1 \\ 2 & 0 \end{bmatrix} \\ &= \begin{bmatrix} 9 & 6 \\ 3 & 3 \\ 6 & 0 \end{bmatrix} = \begin{bmatrix} 3(3) & 3(2) \\ 3(1) & 3(1) \\ 3(2) & 3(0) \end{bmatrix} \end{aligned}$$

We can define **scalar multiplication** as follows.

Scalar Multiplication Multiplying a matrix by a real number (called a *scalar*) results in a matrix in which each entry of the original matrix is multiplied by the real number.

Thus, if

$$A = \begin{bmatrix} a_{11} & a_{12} \\ a_{21} & a_{22} \end{bmatrix}, \quad \text{then} \quad cA = \begin{bmatrix} ca_{11} & ca_{12} \\ ca_{21} & ca_{22} \end{bmatrix}$$

EXAMPLE 1

If

$$A = \begin{bmatrix} 4 & 1 & 4 & 0 \\ 2 & -7 & 3 & 6 \\ 0 & 0 & 2 & 5 \end{bmatrix}$$

find $5A$ and $-2A$.

Solution

$$5A = \begin{bmatrix} 5\cdot 4 & 5\cdot 1 & 5\cdot 4 & 5\cdot 0 \\ 5\cdot 2 & 5(-7) & 5\cdot 3 & 5\cdot 6 \\ 5\cdot 0 & 5\cdot 0 & 5\cdot 2 & 5\cdot 5 \end{bmatrix} = \begin{bmatrix} 20 & 5 & 20 & 0 \\ 10 & -35 & 15 & 30 \\ 0 & 0 & 10 & 25 \end{bmatrix}$$

$$-2A = \begin{bmatrix} -2\cdot 4 & -2\cdot 1 & -2\cdot 4 & -2\cdot 0 \\ -2\cdot 2 & -2(-7) & -2\cdot 3 & -2\cdot 6 \\ -2\cdot 0 & -2\cdot 0 & -2\cdot 2 & -2\cdot 5 \end{bmatrix}$$

$$= \begin{bmatrix} -8 & -2 & -8 & 0 \\ -4 & 14 & -6 & -12 \\ 0 & 0 & -4 & -10 \end{bmatrix}$$

Technology Note Graphing calculators, computer algebra systems, and computer spreadsheets have the capability to perform a number of complex operations on matrices. Figure 3.4 shows the product $-2A$ on a graphing calculator.

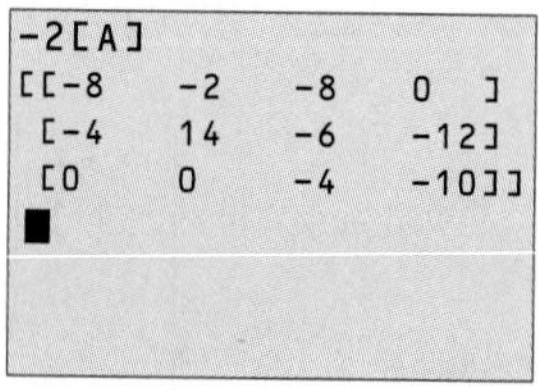

```
-2[A]
[[-8   -2   -8   0  ]
 [-4   14   -6   -12]
 [0    0    -4   -10]]
```

Figure 3.4

CHECKPOINT

1. (a) Find $3D - D$ if $D = \begin{bmatrix} 1 & 3 & 2 & 10 \\ 20 & 5 & 6 & 3 \end{bmatrix}$

   (b) Does $3D - D = 2D$?

EXAMPLE 2

Suppose the purchase prices and delivery costs (per unit) for wood, siding, and roofing used in construction are given by Table 3.5. Then the table of unit costs may be represented by the matrix C.

TABLE 3.5

	Wood	*Siding*	*Roofing*
Purchase	6	4	2
Delivery	1	1	0.5

$$C = \begin{bmatrix} 6 & 4 & 2 \\ 1 & 1 & 0.5 \end{bmatrix}$$

If the supplier announces a 10% increase on both purchase and delivery of these items, find the new unit cost matrix.

Solution

A 10% increase means that the new unit costs are the former costs plus 0.10 times the former cost. That is, the new costs are 1.10 times the former, so the new unit cost matrix is given by

$$\begin{aligned} 1.10C &= 1.10\begin{bmatrix} 6 & 4 & 2 \\ 1 & 1 & 0.5 \end{bmatrix} \\ &= \begin{bmatrix} 6.60 & 4.40 & 2.20 \\ 1.10 & 1.10 & 0.55 \end{bmatrix} \end{aligned}$$

Suppose one store of a firm has 30 washers, 20 dryers, and 10 dishwashers in its inventory. If the value of each washer is \$300, that of each dryer is \$250, and that of each dishwasher is \$350, then the value of this inventory is

$$30 \cdot 300 + 20 \cdot 250 + 10 \cdot 350 = \$17{,}500$$

If we write the value of each of the appliances in the *row matrix*

$$A = \begin{bmatrix} 300 & 250 & 350 \end{bmatrix}$$

and the number of each of the appliances in the *column matrix*

$$B = \begin{bmatrix} 30 \\ 20 \\ 10 \end{bmatrix}$$

then the value of the inventory may be represented by

$$\begin{aligned} AB &= \begin{bmatrix} 300 & 250 & 350 \end{bmatrix} \begin{bmatrix} 30 \\ 20 \\ 10 \end{bmatrix} \\ &= \begin{bmatrix} 300 \cdot 30 + 250 \cdot 20 + 350 \cdot 10 \end{bmatrix} \\ &= \begin{bmatrix} \$17{,}500 \end{bmatrix} \end{aligned}$$

This useful way of operating with a row matrix and a column matrix is called the product of A times B.

In general, we can multiply the row matrix

$$A = \begin{bmatrix} a_1 & a_2 & a_3 & \dots & a_n \end{bmatrix}$$

times the column matrix

$$B = \begin{bmatrix} b_1 \\ b_2 \\ b_3 \\ \vdots \\ b_n \end{bmatrix}$$

if A and B have the same number of elements. The product A times B is

$$AB = \begin{bmatrix} a_1 & a_2 & a_3 & \cdots & a_n \end{bmatrix} \begin{bmatrix} b_1 \\ b_2 \\ b_3 \\ \vdots \\ b_n \end{bmatrix} = \begin{bmatrix} a_1b_1 + a_2b_2 + a_3b_3 + \cdots + a_nb_n \end{bmatrix}$$

Suppose the firm has a second store with 40 washers, 25 dryers, and 5 dishwashers. We can use matrix C to represent the inventories of the two stores.

$$\begin{array}{cc} \text{Store I} & \text{Store II} \end{array}$$

$$C = \begin{bmatrix} 30 & 40 \\ 20 & 25 \\ 10 & 5 \end{bmatrix} \begin{array}{l} \text{Washers} \\ \text{Dryers} \\ \text{Dishwashers} \end{array}$$

Suppose the values of the appliances are \$300, \$250, and \$350, respectively. We have already seen that the value of the inventory of store I is \$17,500. The value of the inventory of store II is

$$300 \cdot 40 + 250 \cdot 25 + 350 \cdot 5 = \$20{,}000$$

Because the value of the inventory of store I can be found by multiplying the row matrix A times the first column of the matrix C, and the value of the inventory of store II can be found by multiplying matrix A times the second column of matrix C, we can write this result as

$$AC = \begin{bmatrix} 300 & 250 & 350 \end{bmatrix} \begin{bmatrix} 30 & 40 \\ 20 & 25 \\ 10 & 5 \end{bmatrix} = \begin{bmatrix} 17{,}500 & 20{,}000 \end{bmatrix}$$

The matrix we have represented as AC is called the **product** of the row matrix A and the 3×2 matrix C. We should note that matrix A is a 1×3 matrix, matrix C is a 3×2 matrix, and their product is a 1×2 matrix. In general, we can multiply an $m \times n$ matrix times an $n \times p$ matrix and the product will be an $m \times p$ matrix, as follows:

Product of Two Matrices Given an $m \times n$ matrix A and an $n \times p$ matrix B, the **matrix product** AB is an $m \times p$ matrix C, with the ij entry of C given by the formula

$$c_{ij} = a_{i1}b_{1j} + a_{i2}b_{2j} + \dots + a_{in}b_{nj}$$

which is illustrated in Figure 3.5.

$$\begin{bmatrix} a_{11} & a_{12} & \cdots & a_{1n} \\ a_{21} & a_{22} & \cdots & a_{2n} \\ & \vdots & & \\ a_{i1} & a_{i2} & \cdots & a_{in} \\ & \vdots & & \\ a_{m1} & a_{m2} & \cdots & a_{mn} \end{bmatrix} \begin{bmatrix} b_{11} & b_{12} & \cdots & b_{1j} & \cdots & b_{1p} \\ b_{21} & b_{22} & \cdots & b_{2j} & \cdots & b_{2p} \\ & \vdots & & & & \vdots \\ b_{n1} & b_{n2} & \cdots & b_{nj} & \cdots & b_{np} \end{bmatrix} = \begin{bmatrix} c_{11} & c_{12} & \cdots & c_{1j} & \cdots & c_{1p} \\ c_{21} & c_{22} & \cdots & c_{2j} & \cdots & c_{2p} \\ & \vdots & & & & \vdots \\ c_{i1} & c_{i2} & \cdots & c_{ij} & \cdots & c_{ip} \\ & \vdots & & & & \vdots \\ c_{m1} & c_{m2} & \cdots & c_{mj} & \cdots & c_{mp} \end{bmatrix}$$

Figure 3.5

EXAMPLE 3

Find AB if

$$A = \begin{bmatrix} 3 & 4 \\ 2 & 5 \\ 6 & 10 \end{bmatrix} \quad \text{and} \quad B = \begin{bmatrix} a & b & c & d \\ e & f & g & h \end{bmatrix}$$

Solution

A is a 3×2 matrix and B is a 2×4 matrix, so the number of columns of A equals the number of rows of B. Thus we can find the product AB, which is a 3×4 matrix, as follows:

$$\begin{aligned} AB &= \begin{bmatrix} 3 & 4 \\ 2 & 5 \\ 6 & 10 \end{bmatrix} \begin{bmatrix} a & b & c & d \\ e & f & g & h \end{bmatrix} \\ &= \begin{bmatrix} 3a + 4e & 3b + 4f & 3c + 4g & 3d + 4h \\ 2a + 5e & 2b + 5f & 2c + 5g & 2d + 5h \\ 6a + 10e & 6b + 10f & 6c + 10g & 6d + 10h \end{bmatrix} \end{aligned}$$

This example shows that if $AB = C$, element c_{32} is found by multiplying each entry of A's *third* row by the corresponding entry of B's *second* column and then adding these products.

EXAMPLE 4

Find AB and BA if

$$A = \begin{bmatrix} 3 & 2 \\ 1 & 0 \end{bmatrix} \quad \text{and} \quad B = \begin{bmatrix} 1 & 2 \\ 3 & 1 \end{bmatrix}$$

Solution

Both A and B are 2×2 matrices, so both AB and BA are defined.

$$AB = \begin{bmatrix} 3 & 2 \\ 1 & 0 \end{bmatrix}\begin{bmatrix} 1 & 2 \\ 3 & 1 \end{bmatrix} = \begin{bmatrix} 3 \cdot 1 + 2 \cdot 3 & 3 \cdot 2 + 2 \cdot 1 \\ 1 \cdot 1 + 0 \cdot 3 & 1 \cdot 2 + 0 \cdot 1 \end{bmatrix} = \begin{bmatrix} 9 & 8 \\ 1 & 2 \end{bmatrix}$$

$$BA = \begin{bmatrix} 1 & 2 \\ 3 & 1 \end{bmatrix}\begin{bmatrix} 3 & 2 \\ 1 & 0 \end{bmatrix} = \begin{bmatrix} 1 \cdot 3 + 2 \cdot 1 & 1 \cdot 2 + 2 \cdot 0 \\ 3 \cdot 3 + 1 \cdot 1 & 3 \cdot 2 + 1 \cdot 0 \end{bmatrix} = \begin{bmatrix} 5 & 2 \\ 10 & 6 \end{bmatrix}$$

Note that for these matrices, the two products AB and BA are matrices of the same size, but they are not equal. That is, $BA \neq AB$. Thus we say that *matrix multiplication is not commutative.*

CHECKPOINT

2. What is element c_{23} if $C = AB$ with

$$A = \begin{bmatrix} 1 & 2 & 0 \\ 2 & 1 & 3 \end{bmatrix} \quad \text{and} \quad B = \begin{bmatrix} 1 & 0 & 2 \\ 3 & 1 & 1 \\ -2 & 1 & 0 \end{bmatrix}?$$

3. Find the product AB.

EXAMPLE 5

Pentico Industries must choose a supplier for the raw materials that it uses in its two manufacturing divisions at Clarion and Brooks. Recall from the Application Preview on page 227 that each division uses different unit amounts of steel, wood, and plastic as shown in the table below.

	Steel	*Wood*	*Plastic*
Clarion	20	30	8
Brooks	22	25	15

The two supply companies being considered, Western and Coastal, can each supply all of these materials, but at different prices per unit, as described in the following table.

	Western	*Coastal*
Steel	300	290
Wood	100	90
Plastic	145	180

Use matrix multiplication to decide which supplier should be chosen to supply (a) the Clarion division, (b) the Brooks division, and (c) both divisions.

Solution

The different amounts of products needed can be placed in the matrix A.

$$A = \begin{array}{c} \\ \end{array}\begin{bmatrix} 20 & 30 & 8 \\ 22 & 25 & 15 \end{bmatrix}\begin{array}{l} \text{Clarion} \\ \text{Brooks} \end{array}$$

(columns: Steel, Wood, Plastic)

The prices charged by the suppliers are given in the matrix B.

$$B = \begin{array}{c} \\ \left[\begin{array}{rr} 300 & 290 \\ 100 & 90 \\ 145 & 180 \end{array}\right] \end{array} \begin{array}{l} \\ \text{Steel} \\ \text{Wood} \\ \text{Plastic} \end{array}$$

(columns: Western, Coastal)

The price from each supplier for each division is found from the product AB.

$$AB = \begin{bmatrix} 20 & 30 & 8 \\ 22 & 25 & 15 \end{bmatrix} \begin{bmatrix} 300 & 290 \\ 100 & 90 \\ 145 & 180 \end{bmatrix}$$

$$= \begin{bmatrix} 20 \cdot 300 + 30 \cdot 100 + \ 8 \cdot 145 & 20 \cdot 290 + 30 \cdot 90 + \ 8 \cdot 180 \\ 22 \cdot 300 + 25 \cdot 100 + 15 \cdot 145 & 22 \cdot 290 + 25 \cdot 90 + 15 \cdot 180 \end{bmatrix}$$

$$= \begin{bmatrix} 10{,}160 & 9940 \\ 11{,}275 & 11{,}330 \end{bmatrix}$$

(a) The price for the Clarion division for supplies from Western is in the first row, first column and is \$10,160; the price for Clarion for supplies from Coastal is in the first row, second column and is \$9940. Thus the best price for Clarion is from Coastal.

(b) The price for the Brooks division for supplies from Western is in the second row, first column and is \$11,275; the price for Brooks for supplies from Coastal is in the second row, second column and is \$11,330. Thus the best price for Brooks is from Western.

(c) The sum of the first column gives \$21,435, the price for supplies from Western for both divisions; the sum of the second column gives \$21,270, the price for supplies from Coastal for both divisions. Thus the best price is from Coastal if one supplier is used for both divisions.

An $n \times n$ (square) matrix (where n is any natural number) that has 1s down its diagonal and 0s everywhere else is called an **identity matrix.** The matrix

$$I = \begin{bmatrix} 1 & 0 & 0 \\ 0 & 1 & 0 \\ 0 & 0 & 1 \end{bmatrix}$$

is a 3×3 identity matrix. The matrix I is called an identity matrix because for any 3×3 matrix A, $AI = IA = A$. That is, if I is multiplied by a 3×3 matrix A, the product matrix is A. Note that when one of two square matrices being multiplied is an identity matrix, the product is *commutative.*

EXAMPLE 6

(a) Write the 2×2 identity matrix.

(b) Given $A = \begin{bmatrix} 4 & -7 \\ 13 & 2 \end{bmatrix}$, show that $AI = IA = A$.

Solution

(a) We denote the 2×2 identity matrix by

$$I = \begin{bmatrix} 1 & 0 \\ 0 & 1 \end{bmatrix}$$

(b) $AI = \begin{bmatrix} 4 & -7 \\ 13 & 2 \end{bmatrix}\begin{bmatrix} 1 & 0 \\ 0 & 1 \end{bmatrix} = \begin{bmatrix} 4+0 & 0-7 \\ 13+0 & 0+2 \end{bmatrix} = \begin{bmatrix} 4 & -7 \\ 13 & 2 \end{bmatrix}$

$IA = \begin{bmatrix} 1 & 0 \\ 0 & 1 \end{bmatrix}\begin{bmatrix} 4 & -7 \\ 13 & 2 \end{bmatrix} = \begin{bmatrix} 4 & -7 \\ 13 & 2 \end{bmatrix}$

Therefore, $AI = IA = A$.

EXAMPLE 7

Messages can be encoded through the use of a code and an encoding matrix. For example, given the code

a	b	c	d	e	f	g	h	i	j	k	l
1	2	3	4	5	6	7	8	9	10	11	12
m	n	o	p	q	r	s	t	u	v	w	x
13	14	15	16	17	18	19	20	21	22	23	24
y	z	blank									
25	26	27									

and the encoding matrix

$$A = \begin{bmatrix} 3 & 5 \\ 4 & 6 \end{bmatrix}$$

we can encode a message by separating it into number pairs (because A is a 2×2 matrix) and then multiplying each pair by A. Use this code and matrix to encode the message "good job."

Solution

The phrase "good job" is written

g	o	o	d	blank	j	o	b
7	15	15	4	27	10	15	2

and is encoded by writing each pair of numbers as a column matrix and multiplying each matrix by matrix A. Performing this multiplication gives the encoded message.

$$\begin{bmatrix} 3 & 5 \\ 4 & 6 \end{bmatrix}\begin{bmatrix} 7 \\ 15 \end{bmatrix} = \begin{bmatrix} 96 \\ 118 \end{bmatrix}$$

$$\begin{bmatrix} 3 & 5 \\ 4 & 6 \end{bmatrix}\begin{bmatrix} 15 \\ 4 \end{bmatrix} = \begin{bmatrix} 65 \\ 84 \end{bmatrix}$$

$$\begin{bmatrix} 3 & 5 \\ 4 & 6 \end{bmatrix}\begin{bmatrix} 27 \\ 10 \end{bmatrix} = \begin{bmatrix} 131 \\ 168 \end{bmatrix}$$

$$\begin{bmatrix} 3 & 5 \\ 4 & 6 \end{bmatrix}\begin{bmatrix} 15 \\ 2 \end{bmatrix} = \begin{bmatrix} 55 \\ 72 \end{bmatrix}$$

Thus the code sent is 96, 118, 65, 84, 131, 168, 55, 72.

Note in Example 7 that if we used a different encoding matrix, then the same message would be sent with a different sequence of code. We will discuss the methods of decoding messages in Section 3.4 "Inverse of a Square Matrix; Matrix Equations."

CHECKPOINT

4. (a) Compute AB if $A = \begin{bmatrix} 2 & 4 \\ 3 & 6 \end{bmatrix}$ and $B = \begin{bmatrix} 2 & -6 \\ -1 & 3 \end{bmatrix}$.
   (b) Can the product of two matrices be a zero matrix even if neither matrix is a zero matrix?

EXAMPLE 8

Use a graphing calculator and the matrices A, B, and C below to perform the operations (a) $B + 2C$, (b) AB, and (c) BA.

$$A = \begin{bmatrix} 2 & 3 & 4 & 5 \\ 3 & 1 & 3 & 5 \\ 3 & -4 & 3 & 1 \end{bmatrix} \quad B = \begin{bmatrix} 2 & 2 & 5 \\ 3 & 5 & 7 \\ 1 & 1 & 2 \\ 5 & 4 & 3 \end{bmatrix} \quad C = \begin{bmatrix} .5 & 3.5 & 2.5 \\ 2 & 7 & .8 \\ 5.5 & 3.5 & 7 \\ 2.5 & 4 & 3 \end{bmatrix}$$

Solution

We begin by entering matrices A, B, and C. The regular operational keys can then be used to perform the required operations.

(a) Figure 3.6 shows $B + 2C$. Because the original matrices are 4×3 matrices, the result of these operations is a 4×3 matrix.
(b) Figure 3.7 (a) shows the product AB. Because A is a 3×4 matrix and B is a 4×3 matrix, the product AB is a 3×3 matrix.
(c) Figure 3.7(b) shows the product BA. Because A is a 3×4 matrix and B is a 4×3 matrix, the product BA is a 4×4 matrix.

```
[B]+2[C]
    [[3    9    10 ]
     [7    19   8.6]
     [12   8    16 ]
     [10   12   9  ]]
```

Figure 3.6

```
[A][B]
     [[42   43   54]
      [37   34   43]
      [2    -7   -4]]
```

(a)

```
[B][A]
[[25   -12   29   25]
 [42   -14   48   47]
 [11   -4    13   12]
 [31   7     41   48]]
```

(b)

Figure 3.7

Like graphing calculators and software programs, modern computer spreadsheets can be used to multiply two matrices.

EXAMPLE 9

Suppose that a bank has three main sources of income—business loans, auto loans, and home mortgages—and that it draws funds from these sources for venture capital used to provide start-up funds for new businesses. Suppose the income from these sources for each of 3 years is given in Table 3.6, and the bank uses 45% of its income from business loans, 20% of its income from auto loans, and 30% of its income from home mortgages to get its venture capital funds. Write a matrix product that gives the venture capital for these years, and find the available venture capital in each of the 3 years.

TABLE 3.6

	A	B	C	D
1	Income	from	Loans	
2		Business	Auto	Home
3	1993	63,300	20,024	51,820
4	1994	48,305	15,817	63,722
5	1995	55,110	18,621	64,105

Solution

The matrix that describes the sources of income for the 3 years is

$$\begin{bmatrix} 63{,}300 & 20{,}024 & 51{,}820 \\ 48{,}305 & 15{,}817 & 63{,}722 \\ 55{,}110 & 18{,}621 & 64{,}105 \end{bmatrix}$$

and the matrix that describes the percent of each that is used for venture capital is

$$\begin{bmatrix} 0.45 \\ 0.20 \\ 0.30 \end{bmatrix}$$

The product of these matrices, which can be found with a calculator, spreadsheet, or by hand, gives the venture capital for the 3 years.

$$\begin{bmatrix} 63{,}300 & 20{,}024 & 51{,}820 \\ 48{,}305 & 15{,}817 & 63{,}722 \\ 55{,}110 & 18{,}621 & 64{,}105 \end{bmatrix} \begin{bmatrix} 0.45 \\ 0.20 \\ 0.30 \end{bmatrix} = \begin{bmatrix} 48{,}035.80 \\ 44{,}017.25 \\ 47{,}755.20 \end{bmatrix}$$

Thus the available venture capital for each of the 3 years is as follows.

1993	$48,035.80
1994	$44,017.25
1995	$47,755.20

CHECKPOINT SOLUTIONS

1. (a) $3D - D = \begin{bmatrix} 3 & 9 & 6 & 30 \\ 60 & 15 & 18 & 9 \end{bmatrix} - \begin{bmatrix} 1 & 3 & 2 & 10 \\ 20 & 5 & 6 & 3 \end{bmatrix}$

$= \begin{bmatrix} 2 & 6 & 4 & 20 \\ 40 & 10 & 12 & 6 \end{bmatrix}$

(b) Yes

2. $2 \cdot 2 + 1 \cdot 1 + 3 \cdot 0 = 5$

3. $\begin{bmatrix} 7 & 2 & 4 \\ -1 & 4 & 5 \end{bmatrix}$

4. (a) $AB = \begin{bmatrix} 0 & 0 \\ 0 & 0 \end{bmatrix}$ (b) Yes. See (a).

EXERCISE 3.2

In Problems 1–4, multiply the matrices.

1. $\begin{bmatrix} 1 & 2 & 3 \end{bmatrix} \begin{bmatrix} 4 \\ 5 \\ 6 \end{bmatrix}$

2. $\begin{bmatrix} 2 & 0 & 3 \end{bmatrix} \begin{bmatrix} 0 \\ 1 \\ -3 \end{bmatrix}$

3. $\begin{bmatrix} 1 & 2 \end{bmatrix} \begin{bmatrix} 3 & 5 \\ 4 & 6 \end{bmatrix}$

4. $\begin{bmatrix} 3 & 0 \end{bmatrix} \begin{bmatrix} 1 & 2 \\ 4 & 5 \end{bmatrix}$

In Problems 5–32, use matrices A through F to perform the indicated operations when possible.

$$A = \begin{bmatrix} 1 & 0 & 2 \\ 3 & 2 & 1 \\ 4 & 0 & 3 \end{bmatrix} \quad B = \begin{bmatrix} 1 & 1 & 3 & 0 \\ 4 & 2 & 1 & 1 \\ 3 & 2 & 0 & 1 \end{bmatrix}$$

$$C = \begin{bmatrix} 5 & 3 \\ 1 & 2 \end{bmatrix} \quad D = \begin{bmatrix} 4 & 2 \\ 3 & 5 \end{bmatrix}$$

$$E = \begin{bmatrix} 1 & 0 & 4 \\ 5 & 1 & 0 \end{bmatrix} \quad F = \begin{bmatrix} 1 & 0 & -1 & 3 \\ 2 & -1 & 3 & -4 \end{bmatrix}$$

5. $3A$
6. $4D$
7. $4C + 2D$
8. $5C - 3D$
9. $2A - 3B$
10. $3B + 4F$
11. CD
12. DE
13. DC
14. CF
15. DF
16. FB^T
17. AB
18. EC
19. BA
20. CE
21. EB
22. BE
23. EA^T
24. AE^T
25. A^2
26. A^3
27. C^3
28. F^2
29. Does $(AA^T)^T = A^TA$?
30. Are $(CD)E$ and $C(DE)$ equal?
31. Does $CD = DC$? (see Problems 11 and 13)
32. Are $\left(\frac{1}{4}A\right)B$ and $\frac{1}{4}(AB)$ equal?

In Problems 33–42, use the matrices below. Perform the indicated operations.

$$A = \begin{bmatrix} 2 & 5 & 4 \\ 1 & 4 & 3 \\ 1 & -3 & -2 \end{bmatrix} \quad B = \begin{bmatrix} -1 & 2 & 1 \\ -5 & 8 & 2 \\ -7 & 11 & -3 \end{bmatrix}$$

$$I = \begin{bmatrix} 1 & 0 & 0 \\ 0 & 1 & 0 \\ 0 & 0 & 1 \end{bmatrix} \quad C = \begin{bmatrix} 3 & 0 & 4 \\ 1 & 7 & -1 \\ 3 & 0 & 4 \end{bmatrix}$$

$$D = \begin{bmatrix} 4 & 4 & -8 \\ -1 & -1 & 2 \\ -3 & -3 & 6 \end{bmatrix} \quad F = \begin{bmatrix} 0 & 1 & 2 \\ 0 & 0 & -4 \\ 0 & 0 & 0 \end{bmatrix}$$

$$Z = \begin{bmatrix} 0 & 0 & 0 \\ 0 & 0 & 0 \\ 0 & 0 & 0 \end{bmatrix}$$

33. AB
34. BA
35. CD
36. DC
37. F^3
38. AZ
39. AI
40. IA
41. ZCI
42. IFZ

43. Is it true for matrices (as it is for real numbers) that AB equals a zero matrix if and only if either A or B equals a zero matrix? (Refer to Problems 33–42.)

44. Is it true for matrices (as it is for real numbers) that multiplication by a zero matrix gives a result of a zero matrix?

45. Find AB and BA if

$$A = \begin{bmatrix} a & b \\ c & d \end{bmatrix} \quad \text{and} \quad B = \frac{1}{ad - bc}\begin{bmatrix} d & -b \\ -c & a \end{bmatrix}$$

46. For B (in Problem 45) to exist, what restriction must $ad - bc$ satisfy?

In each of Problems 47–50, determine whether the given values of x, y, and z are the solution to the given matrix equation by substituting the given values into the matrix equation and by using matrix multiplication.

47. $\begin{bmatrix} 1 & 2 & 1 \\ 3 & 4 & -2 \\ 2 & 0 & -1 \end{bmatrix}\begin{bmatrix} x \\ y \\ z \end{bmatrix} = \begin{bmatrix} 2 \\ -2 \\ 2 \end{bmatrix}$ $\quad x = 2, y = -1, z = 2$

48. $\begin{bmatrix} 3 & 1 & 0 \\ 2 & -2 & 1 \\ 1 & 1 & 2 \end{bmatrix}\begin{bmatrix} x \\ y \\ z \end{bmatrix} = \begin{bmatrix} 4 \\ 9 \\ 2 \end{bmatrix}$ $\quad x = 2, y = -2, z = 1$

49. $\begin{bmatrix} 1 & 1 & 2 \\ 4 & 0 & 1 \\ 2 & 1 & 1 \end{bmatrix}\begin{bmatrix} x \\ y \\ z \end{bmatrix} = \begin{bmatrix} 5 \\ 5 \\ 5 \end{bmatrix}$ $\quad x = 1, y = 2, z = 1$

50. $\begin{bmatrix} 1 & 0 & 2 \\ 3 & 1 & 0 \\ 1 & 2 & 1 \end{bmatrix}\begin{bmatrix} x \\ y \\ z \end{bmatrix} = \begin{bmatrix} 0 \\ 7 \\ 3 \end{bmatrix}$ $\quad x = 2, y = 1, z = 1$

For matrices A and B in Problems 51 and 52, (a) find AB and BA and (b) tell whether $AB = BA$.

51. $A = \begin{bmatrix} 1 & 2 & 3 \\ 2 & -2 & -1 \\ 3 & 0 & 2 \end{bmatrix}$ $\quad B = \begin{bmatrix} 1 & 2 & 1 \\ 2 & 4 & 5 \\ 2 & -2 & 3 \end{bmatrix}$

52. $A = \begin{bmatrix} \frac{1}{2} & \frac{1}{2} & 1 \\ 1 & \frac{1}{2} & \frac{1}{2} \\ \frac{3}{2} & \frac{1}{2} & 1 \end{bmatrix}$ $\quad B = \begin{bmatrix} -1 & 0 & 1 \\ 1 & 4 & -3 \\ 1 & -2 & 1 \end{bmatrix}$

In Problems 55 and 56, use technology to find the product AB of the following matrices.

53. $A = \begin{bmatrix} 0.1 & 0.0 & 0.1 & 0.1 & 0.2 \\ 0.1 & 0.2 & -0.1 & 0.1 & -0.1 \\ 0.1 & -0.1 & 0.1 & -0.2 & 0.1 \\ 0.1 & 0.2 & 0.1 & 0.1 & 0.1 \\ 0.1 & 0.0 & 0.0 & 0.0 & 0.0 \end{bmatrix}$,

$$B = \begin{bmatrix} 0 & 0 & 0 & 0 & 10 \\ 0 & 5 & \frac{10}{3} & \frac{5}{3} & -10 \\ -10 & -15 & -\frac{20}{3} & \frac{35}{3} & 20 \\ 0 & -5 & -\frac{20}{3} & \frac{5}{3} & 10 \\ 10 & 10 & \frac{20}{3} & -\frac{20}{3} & -20 \end{bmatrix}$$

54. $A = \frac{1}{12}\begin{bmatrix} 3 & 15 & -9 & -15 & 6 \\ 3 & -9 & 3 & 9 & -6 \\ -1 & -17 & 3 & 23 & -4 \\ 0 & 12 & 0 & -12 & 0 \\ -1 & -5 & 3 & 5 & 2 \end{bmatrix}$

$$B = \begin{bmatrix} 2 & 3 & 0 & 1 & 3 \\ 1 & 0 & 2 & 3 & 1 \\ 0 & 1 & 0 & 2 & 3 \\ 1 & 0 & 2 & 2 & 1 \\ 1 & 0 & 0 & 0 & 3 \end{bmatrix}$$

Applications

55. ***Weekly earnings*** Below is a table showing median weekly earnings for men and women for different ages in 1995–1996.
    (a) Use the data to make a 6×2 matrix with each row representing an age category and each column representing a sex.
    (b) If union members' median weekly earnings are 20% more than the earnings given in the table, use a matrix operation to find a matrix giving the results by sex and age for union members.

Age	*16–24*	*25–34*	*35–44*	*45–54*	*55–64*	*65+*
Men	292	451	550	582	514	389
Women	275	403	453	464	403	353

SOURCE: Bureau of Labor Statistics, U.S. Dept. of Labor

56. ***Debt payment*** Ace, Baker, and Champ are being purchased by ALCO, Inc., and their outstanding debt must be paid by the purchaser. The matrix below gives the amount of debt and the terms for the companies being purchased.

	Due in 30 days	Due in 60 days
Ace	\$40,000	\$60,000
Baker	\$25,000	\$15,000
Champ	\$35,000	\$58,000

   (a) If ALCO pays 35% of the amount owed on each debt, write the matrix giving the remaining debt.
   (b) Suppose ALCO, Inc., decides to pay 80% of all debts due in 30 days and to increase the debts due in 60 days by 20%. Write a matrix that gives the debts after these transactions are made.

 57. ***Expense accounts*** A sales associate's expense account for the first week of a certain month has the daily expenses (in dollars) shown in matrix A.

$$A = \begin{array}{c} \\ \\ \\ \\ \\ \end{array}\begin{matrix} \text{Meals} & \text{Lodging} & \text{Travel} & \text{Other} \\ \end{matrix}$$

$$A = \begin{bmatrix} 22 & 40 & 100 & 5 \\ 20 & 40 & 20 & 0 \\ 28 & 70 & 45 & 0 \\ 15 & 70 & 20 & 10 \\ 20 & 0 & 100 & 5 \end{bmatrix} \begin{matrix} \text{Monday} \\ \text{Tuesday} \\ \text{Wednesday} \\ \text{Thursday} \\ \text{Friday} \end{matrix}$$

(a) The associate finds that expenses for the second week are 5% more (in each category) than for the first week. Find the expenses matrix for the second week (note that this is $1.05A$).

(b) Find the associate's expenses matrix for the third week if expenses for that week are 10% more (in each category) than they were in the first week.

 58. ***Debt payment*** When a firm buys another company, the company frequently has some outstanding debt that the purchaser must pay. Consider three companies, A, B, and C, that are purchased by Maxx Industries. The table below gives the amount of the debts of these companies, classified by the number of days remaining until the debt must be paid.

Company	*30 Days*	*60 Days*	*More than 60 Days*
A	$25,000	$26,000	$12,000
B	15,000	52,000	5,000
C	8,000	20,000	120,000

(a) Suppose Maxx Industries pays 20% of the amount owed on each loan. Write the matrix that would give the remaining debt in each category.

(b) What payment plan did Maxx Industries use if the outstanding debt is given by the following matrix?

$$\begin{bmatrix} 25{,}000 & 26{,}000 & 12{,}000 \\ 15{,}000 & 52{,}000 & 5{,}000 \\ 8{,}000 & 20{,}000 & 120{,}000 \end{bmatrix} - 0.50\begin{bmatrix} 0 & 26{,}000 & 0 \\ 0 & 52{,}000 & 0 \\ 0 & 20{,}000 & 0 \end{bmatrix} - \begin{bmatrix} 25{,}000 & 0 & 0 \\ 15{,}000 & 0 & 0 \\ 8{,}000 & 0 & 0 \end{bmatrix}$$

59. ***Car pricing*** A car dealer can buy midsize cars for 12% under the list price, and he can buy luxury cars for 15% under the list price. The table below gives the list prices for two midsize and two luxury cars. Write these data in a matrix and multiply it on the left by the matrix

$$\begin{bmatrix} 0.88 & 0 \\ 0 & 0.85 \end{bmatrix}$$

What does each entry in this product matrix represent?

Midsize	25,000	28,000
Luxury	36,000	42,000

60. ***Revenue*** A clothing manufacturer has factories in Atlanta, Chicago, and New York. Sales (in thousands) during the first quarter are summarized in the matrix below.

$$\begin{matrix} \text{Coats} \\ \text{Shirts} \\ \text{Pants} \\ \text{Ties} \end{matrix}\overset{\begin{matrix}\text{Atl.} & \text{Chi.} & \text{NY}\end{matrix}}{\begin{bmatrix} 40 & 63 & 18 \\ 85 & 56 & 42 \\ 6 & 18 & 8 \\ 7 & 10 & 8 \end{bmatrix}}$$

During this period the selling price of a coat is $100, of a shirt $10, of a pair of pants $25, and of a tie $5. Use matrix multiplication to find the total revenue received by each factory.

61. ***Egg production*** The tables below give the production of eggs (in millions) for 1995 and 1996 for the southern states and the average price in cents per dozen for these states for 1995 and 1996.

(a) Write the production data for 1995 and 1996 in the first table as matrix A, and write the transpose of matrix B, which contains the price data in the second table.

(b) What is the order of the product $\frac{1}{12}AB^T$?

(c) Where are the entries in the product found in (b) that give the total value of production of eggs for each of these states for the years 1995 and 1996?

Production of Eggs

State	*1995*	*1996*
Alabama	2703	3428
Arkansas	3608	2481
Florida	2383	2314
Georgia	4376	4584
Kentucky	679	664
Louisiana	472	477
Mississippi	1443	1523
North Carolina	3152	2988
South Carolina	1289	1224
Tennessee	254	257
Virginia	916	949
West Virginia	1455	1413

Price per Dozen

State	*1995*	*1996*
Alabama	96.1	109.0
Arkansas	97.9	105.0
Florida	48.1	63.3
Georgia	79.4	91.1
Kentucky	65.4	79.2
Louisiana	98.4	96.4
Mississippi	99.0	121.0
North Carolina	77.3	87.5
South Carolina	65.8	85.4
Tennessee	82.8	91.3
Virginia	89.5	95.3
West Virginia	76.9	76.4

SOURCE: National Agricultural Statistics Service, U.S. Dept. of Agriculture

62. ***Area and population*** Matrix A below gives the fraction of the Earth's area and the fraction of its population for five continents. Matrix B gives the Earth's area (in square miles) and its population. Find the area and population of each given continent by finding AB.

$$A = \begin{array}{c} \text{Fraction} \\ \text{of Area} \end{array} \begin{array}{c} \text{Fraction of} \\ \text{Population} \end{array}$$

$$A = \begin{bmatrix} 0.162 & 0.051 \\ 0.119 & 0.084 \\ 0.066 & 0.120 \\ 0.298 & 0.588 \\ 0.202 & 0.151 \end{bmatrix} \begin{array}{l} \text{North America} \\ \text{South America} \\ \text{Europe} \\ \text{Asia} \\ \text{Africa} \end{array}$$

$$B = \begin{array}{c} \begin{array}{cc} \text{Area} & \text{Population} \end{array} \\ \begin{bmatrix} 57{,}850{,}000 & 0 \\ 0 & 5{,}423{,}000{,}000 \end{bmatrix} \end{array} \begin{array}{l} \text{Area} \\ \text{Population} \end{array}$$

(SOURCE: Rand McNally and Co., *World Almanac* 1993)

63. ***Advertising*** A business plans to use three methods of advertising: cable TV, radio, and newspaper. The cost per ad (in thousands of dollars) is given by matrix C.

$$C = \begin{bmatrix} 20 \\ 9 \\ 6 \end{bmatrix} \begin{array}{l} \text{TV} \\ \text{Radio} \\ \text{Newspaper} \end{array}$$

Suppose the business has three target populations for its products: female teenagers, single women 20 to 35, and married women 35 to 60. Matrix T shows the number of ads per month directed at each of these groups.

$$T = \begin{array}{c} \begin{array}{ccc} \text{TV} & \text{Radio} & \text{Papers} \end{array} \\ \begin{bmatrix} 50 & 30 & 5 \\ 40 & 60 & 30 \\ 45 & 40 & 60 \end{bmatrix} \end{array} \begin{array}{l} \text{Female teens} \\ \text{Single women} \\ \text{Women 35–60} \end{array}$$

Find the matrix that gives the cost of each type of ad for each group of people.

64. ***Nutrition*** In problem 41 of Exercise 3.1, we found the weights and measures of the laboratory animals to be given by matrix A, where column 1 gives the measures.

$$A = \begin{bmatrix} 12.5 & 250 \\ 11.8 & 215 \\ 9.8 & 190 \end{bmatrix}$$

If the increase in both weight and length over the next two weeks is 20% for group I, 7% for group II, and 0% for group III, then the increases in the measures during the 2 weeks can be found by computing GA, where

$$G = \begin{bmatrix} 0.20 & 0 & 0 \\ 0 & 0.07 & 0 \\ 0 & 0 & 0 \end{bmatrix}.$$

(a) What are the increases in respective weights and measures at the end of these 2 weeks?

(b) Find the matrix that gives the new weights and measures at the end of this period by computing

$$(I + G)A$$

where I is the 3×3 identity matrix.

Encoding messages Multiplication by a matrix can be used to encode messages (and multiplication by its inverse can be used to decode messages). Given the code

a	b	c	d	e	f	g	h	i	j
1	2	3	4	5	6	7	8	9	10
k	l	m	n	o	p	q	r	s	t
11	12	13	14	15	16	17	18	19	20
u	v	w	x	y	z	blank			
21	22	23	24	25	26	27			

and the code matrix

$$A = \begin{bmatrix} 5 & 9 \\ 6 & 11 \end{bmatrix}$$

complete Problems 65 and 66.

65. Use matrix A to encode the message "The die is cast."

66. Use matrix A to encode the message "To be or not to be."

 67. ***Death rates*** The tables below give the death rates, per 100,000 population, by age for selected years for males and females. Use matrix operations to find the rate per 100,000 for all people in the age categories given and for the given years.

Males

	Under 1	*1–4*	*5–14*	*15–24*	*25–34*	*35–44*	*45–54*
1970	2410	93	51	189	215	403	959
1980	1429	73	37	172	196	299	767
1990	1083	52	29	147	204	310	610
1993	946	50	27	146	209	329	596
1994	899	52	26	151	207	337	585

Females

	Under 1	*1–4*	*5–14*	*15–24*	*25–34*	*35–44*	*45–54*
1970	1864	75	32	68	102	231	517
1980	1142	55	24	58	76	159	413
1990	856	41	19	49	74	138	343
1993	759	40	19	49	76	144	330
1994	719	37	19	47	76	144	326

SOURCE: *Statistical Abstract of the United States,* U.S. Dept. of Commerce

68. ***Manufacturing*** Suppose products A and B are made from plastic, steel, and glass, with the number of units of each raw material required for each product given by the table below.

	Plastic	*Steel*	*Glass*
Product A	3	1	0.50
Product B	4	0.50	2

Because of transportation costs to the firm's two plants, X and Y, the unit costs for some of the raw materials are different. The table below gives the unit costs for each of the raw materials at the two plants.

	Plant X	*Plant Y*
Plastic	10	9
Steel	22	26
Glass	14	14

Using the information just given, find the total cost of producing each of the products at each of the factories.

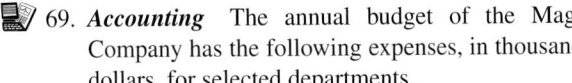

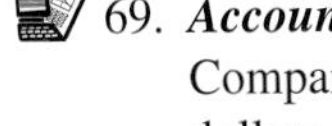 69. ***Accounting*** The annual budget of the Magnum Company has the following expenses, in thousands of dollars, for selected departments.

	Mfg.	*Office*	*Sales*	*Shp.*	*Act.*	*Mgt.*
Supplies	0.7	8.5	10.2	1.1	5.6	3.6
Phone	0.5	0.2	6.1	1.3	0.2	1.0
Transp.	2.2	0.4	8.8	1.2	1.2	4.8
Salaries	251.8	63.4	81.6	35.2	54.3	144.2
Utilities	30.0	1.0	1.0	1.0	1.0	1.0
Materials	788.9	0	0	0	0	0

Mfg. = Manufacturing; Shp. = Shipping;
Act. = Accounting; Mgt. = Management

Find the budget matrix for the following changes in the budget.
(a) a 5% decrease
(b) an 8% increase

70. ***Accounting*** Consider the original budget matrix of Problem 69. Assume there is a 20% increase in manufacturing, a 3% increase in office, a 5% increase in sales, a 20% increase in shipping, a 5% increase in accounting, and a 3% decrease in management. Find the new budget matrix by multiplying the following matrix times the original matrix from Problem 69.

$$\begin{bmatrix} 1.20 & 0 & 0 & 0 & 0 & 0 \\ 0 & 1.03 & 0 & 0 & 0 & 0 \\ 0 & 0 & 1.05 & 0 & 0 & 0 \\ 0 & 0 & 0 & 1.20 & 0 & 0 \\ 0 & 0 & 0 & 0 & 1.05 & 0 \\ 0 & 0 & 0 & 0 & 0 & 0.97 \end{bmatrix}$$

71. ***Management*** In an evaluation of labor assignment rules when workers are not perfectly interchangeable, Paul M. Bobrowski and Paul Sungchil Park created a dynamic job shop with 9 work centers and 9 workers. The efficiency of each worker is specified in the labor efficiency matrix E given here, which represents the degree of worker cross-training. If a worker slowdown causes a 10% decrease in efficiency at all work centers, write the matrix that describes the efficiencies for the 9 workers at the 9 work centers. (Use a graphing calculator, a computer program, or a spreadsheet.)

$$E = \begin{bmatrix} 1.00 & .95 & .95 & .95 & .95 & .85 & .85 & .85 & .85 \\ .85 & 1.00 & .95 & .95 & .95 & .95 & .85 & .85 & .85 \\ .85 & .85 & 1.00 & .95 & .95 & .95 & .95 & .85 & .85 \\ .85 & .85 & .85 & 1.00 & .95 & .95 & .95 & .95 & .85 \\ .85 & .85 & .85 & .85 & 1.00 & .95 & .95 & .95 & .95 \\ .95 & .85 & .85 & .85 & .85 & 1.00 & .95 & .95 & .95 \\ .95 & .95 & .85 & .85 & .85 & .85 & 1.00 & .95 & .95 \\ .95 & .95 & .95 & .85 & .85 & .85 & .85 & 1.00 & .95 \\ .95 & .95 & .95 & .95 & .85 & .85 & .85 & .85 & 1.00 \end{bmatrix}$$

3.3 Gauss-Jordan Elimination: Solving Systems of Equations

OBJECTIVES

- *To use matrices to solve systems of linear equations with unique solutions*
- *To use matrices to solve systems of linear equations with nonunique solutions*

APPLICATION PREVIEW

The Walters Manufacturing Company makes three types of metal storage sheds. The company has three departments: stamping, painting, and packaging. The following table gives the number of hours each division requires for each shed.

	Shed		
Departments	*Type I*	*Type II*	*Type III*
Stamping	2	3	4
Painting	1	2	1
Packaging	1	1	2

By using the information in the table and by solving a system of equations, we can determine how many of each type of shed can be produced if the stamping department has 3200 hours available, the painting department has 1700 hours, and the packaging department has 1300 hours. The set of equations used to solve this problem is as follows:

$$\begin{cases} 2x_1 + 3x_2 + 4x_3 = 3200 \\ x_1 + 2x_2 + x_3 = 1700 \\ x_1 + x_2 + 2x_3 = 1300 \end{cases}$$

In this section, we will solve such systems by using matrices and the Gauss-Jordan elimination method.

Systems with Unique Solutions

In solving systems with the left-to-right elimination method used in Example 7 of Section 1.5, "Solutions of Systems of Linear Equations," we operated on the coefficients of the variables x, y, and z and on the constants. If we keep the coefficients of the variables x, y, and z in distinctive columns, we do not need to write the equations. In solving a system of linear equations with matrices, we first write the coefficients and constants from the system (on the left) in the **augmented matrix** (on the right).

$$\begin{cases} a_1x + b_1y + c_1z = d_1 \\ a_2x + b_2y + c_2z = d_2 \\ a_3x + b_3y + c_3z = d_3 \end{cases} \qquad \left[\begin{array}{ccc|c} a_1 & b_1 & c_1 & d_1 \\ a_2 & b_2 & c_2 & d_2 \\ a_3 & b_3 & c_3 & d_3 \end{array}\right]$$

In the augmented matrix, the numbers on the left side of the solid line form the **coefficient matrix,** with each column containing the coefficients of a variable (0 represents any missing variable). The column on the right side of the line (called the *augment*) contains the constants. Each row of the matrix gives the corresponding coefficients of an equation.

The augmented matrix associated with the system

$$\begin{cases} x + 2y + 3z = 6 \\ x - z = 0 \\ x - y - z = -4 \end{cases} \text{ is } A = \left[\begin{array}{ccc|c} 1 & 2 & 3 & 6 \\ 1 & 0 & -1 & 0 \\ 1 & -1 & -1 & -4 \end{array}\right]$$

CHECKPOINT

1. (a) Write the augmented matrix for the following system of linear equations.

$$\begin{cases} 3x + 2y - z = 3 \\ x - y + 2z = 4 \\ 2x + 3y - z = 3 \end{cases}$$

(b) Write the coefficient matrix for the system.

We can use matrices to solve systems of linear equations by performing the same operations on the rows of a matrix to reduce it as we do on equations in a linear system. The three different operations we can use to reduce the matrix are called **elementary row operations** and are similar to the operations with equations that result in equivalent systems. These operations are

1. Interchange two rows.
2. Add a multiple of one row to another row.
3. Multiply a row by any nonzero constant.

When a new matrix results from one or more of these elementary row operations being performed on a matrix, the new matrix is said to be **equivalent** to the original because both these matrices represent equivalent systems of equations.

The process that we use to solve a system of equations using matrices (called the **elimination method)** is not always the fastest way, but it is a systematic procedure similar to the left-to-right elimination method discussed in Section 1.5 "Solutions of Systems of Linear Equations." This method uses these row operations to attempt to reduce the coefficient matrix to an identity matrix. Example 1 illustrates this process, which is called the **Gauss-Jordan elimination method.**

EXAMPLE 1

Use matrices to solve the system

$$\begin{cases} 2x + 5y + 4z = 4 \\ x + 4y + 3z = 1 \\ x - 3y - 2z = 5 \end{cases}$$

Solution

The augmented matrix for this system is

$$\left[\begin{array}{ccc|c} 2 & 5 & 4 & 4 \\ 1 & 4 & 3 & 1 \\ 1 & -3 & -2 & 5 \end{array}\right]$$

Note that this system has 3 equations in 3 variables, but the reduction procedure and row operations apply regardless of the size of the system.

Gauss-Jordan Elimination Method

Goal (for Each Step)	Row Operation	Equivalent Matrix
1. Get a 1 in row 1, column 1.	Interchange row 1 and row 2.	$\left[\begin{array}{rrr\|r} 1 & 4 & 3 & 1 \\ 2 & 5 & 4 & 4 \\ 1 & -3 & -2 & 5 \end{array}\right]$
2. Use row 1 *only* to get zeros in other entries of column 1.	Add -2 times row 1 to row 2; put the result in row 2. Add -1 times row 1 to row 3; put the result in row 3.	$\left[\begin{array}{rrr\|r} 1 & 4 & 3 & 1 \\ 0 & -3 & -2 & 2 \\ 0 & -7 & -5 & 4 \end{array}\right]$
3. Use rows below row 1 to get a 1 in row 2, column 2.	Multiply row 2 by $-\frac{1}{3}$; put the result in row 2.	$\left[\begin{array}{rrr\|r} 1 & 4 & 3 & 1 \\ 0 & 1 & \frac{2}{3} & -\frac{2}{3} \\ 0 & -7 & -5 & 4 \end{array}\right]$
4. Use row 2 *only* to get zeros as the other entries in column 2.	Add -4 times row 2 to row 1; put the result in row 1. Add 7 times row 2 to row 3; put the result in row 3.	$\left[\begin{array}{rrr\|r} 1 & 0 & \frac{1}{3} & \frac{11}{3} \\ 0 & 1 & \frac{2}{3} & -\frac{2}{3} \\ 0 & 0 & -\frac{1}{3} & -\frac{2}{3} \end{array}\right]$
5. Use row below row 2 to get a 1 in row 3, column 3.	Multiply row 3 by -3; put the result in row 3.	$\left[\begin{array}{rrr\|r} 1 & 0 & \frac{1}{3} & \frac{11}{3} \\ 0 & 1 & \frac{2}{3} & -\frac{2}{3} \\ 0 & 0 & 1 & 2 \end{array}\right]$
6. Use row 3 *only* to get zeros as the other entries in column 3.	Add $-\frac{1}{3}$ times row 3 to row 1; put the result in row 1. Add $-\frac{2}{3}$ times row 3 to row 2; put the result in row 2.	$\left[\begin{array}{rrr\|r} 1 & 0 & 0 & 3 \\ 0 & 1 & 0 & -2 \\ 0 & 0 & 1 & 2 \end{array}\right]$
7. Repeat the process until it cannot be continued.	All rows have been used. The matrix is in reduced form.	

The above display shows a series of step-by-step goals for reducing a matrix and shows a series of row operations that reduces this 3×3 matrix.

The system of equations corresponding to this final augmented matrix follows.

$$\begin{aligned} x + 0y + 0z &= 3 \\ 0x + y + 0z &= -2 \\ 0x + 0y + z &= 2 \end{aligned}$$

Thus the solution is $x = 3$, $y = -2$, and $z = 2$.

When a system of linear equations has a unique solution, the coefficient part of the reduced augmented matrix will be an identity matrix. Recall that the Gauss-Jordan elimination method (outlined in Example 1) may be used with systems of any size.

EXAMPLE 2

Solve the system

$$\begin{aligned} x_1 + x_2 + x_3 + 2x_4 &= 6 \\ x_1 + 2x_2 \qquad + x_4 &= -2 \\ x_1 + x_2 + 3x_3 - 2x_4 &= 12 \\ x_1 + x_2 - 4x_3 + 5x_4 &= -16. \end{aligned}$$

Solution

First note that there is a variable missing in the second equation. When we form the augmented matrix, we must insert a zero in this place so that each column represents the coefficients of one variable.

To solve this system, we must reduce the augmented matrix.

$$\left[\begin{array}{cccc|c} 1 & 1 & 1 & 2 & 6 \\ 1 & 2 & 0 & 1 & -2 \\ 1 & 1 & 3 & -2 & 12 \\ 1 & 1 & -4 & 5 & -16 \end{array}\right]$$

The reduction process follows. The entry in row 1, column 1 is 1. To get zeros in the first column, add -1 times row 1 to row 2, and put the result in row 2; add -1 times row 1 to row 3 and put the result in row 3; and add -1 times row 1 to row 4 and put the result in row 4.

$$\left[\begin{array}{cccc|c} 1 & 1 & 1 & 2 & 6 \\ 1 & 2 & 0 & 1 & -2 \\ 1 & 1 & 3 & -2 & 12 \\ 1 & 1 & -4 & 5 & -16 \end{array}\right] \xrightarrow[(-1)R_1 + R_4 \to R_4]{\substack{(-1)R_1 + R_2 \to R_2 \\ (-1)R_1 + R_3 \to R_3}} \left[\begin{array}{cccc|c} 1 & 1 & 1 & 2 & 6 \\ 0 & 1 & -1 & -1 & -8 \\ 0 & 0 & 2 & -4 & 6 \\ 0 & 0 & -5 & 3 & -22 \end{array}\right]$$

The entry in row 2, column 2 is 1. To get zeros in the second column, add -1 times row 2 to row 1, and put the result in row 1.

$$\xrightarrow{(-1)R_2 + R_1 \to R_1} \left[\begin{array}{cccc|c} 1 & 0 & 2 & 3 & 14 \\ 0 & 1 & -1 & -1 & -8 \\ 0 & 0 & 2 & -4 & 6 \\ 0 & 0 & -5 & 3 & -22 \end{array}\right]$$

The entry in row 3, column 3 is 2. To get a 1 in this position, multiply row 3 by $\frac{1}{2}$.

$$\xrightarrow{\frac{1}{2}R_3 \to R_3} \left[\begin{array}{cccc|c} 1 & 0 & 2 & 3 & 14 \\ 0 & 1 & -1 & -1 & -8 \\ 0 & 0 & 1 & -2 & 3 \\ 0 & 0 & -5 & 3 & -22 \end{array}\right]$$

Using row 3, add −2 times row 3 to row 1 and put the result in row 1; add row 3 to row 2 and put the result in row 2; and add 5 times row 3 to row 4 and put the result in row 4.

$$\xrightarrow[5R_3 + R_4 \to R_4]{\substack{(-2)R_3 + R_1 \to R_1 \\ R_3 + R_2 \to R_2}} \left[\begin{array}{cccc|c} 1 & 0 & 0 & 7 & 8 \\ 0 & 1 & 0 & -3 & -5 \\ 0 & 0 & 1 & -2 & 3 \\ 0 & 0 & 0 & -7 & -7 \end{array}\right]$$

Multiplying $-\frac{1}{7}$ times row 4 and putting the result in row 4 gives a 1 in row 4, column 4.

$$\xrightarrow{-\frac{1}{7}R_4 \to R_4} \left[\begin{array}{cccc|c} 1 & 0 & 0 & 7 & 8 \\ 0 & 1 & 0 & -3 & -5 \\ 0 & 0 & 1 & -2 & 3 \\ 0 & 0 & 0 & 1 & 1 \end{array}\right]$$

Adding appropriate multiples of row 4 to the other rows gives

$$\xrightarrow[2R_4 + R_3 \to R_3]{\substack{(-7)R_4 + R_1 \to R_1 \\ 3R_4 + R_2 \to R_2}} \left[\begin{array}{cccc|c} 1 & 0 & 0 & 0 & 1 \\ 0 & 1 & 0 & 0 & -2 \\ 0 & 0 & 1 & 0 & 5 \\ 0 & 0 & 0 & 1 & 1 \end{array}\right]$$

This corresponds to the system

$$\begin{aligned} x_1 \quad\quad\quad &= 1 \\ x_2 \quad\quad &= -2 \\ x_3 \quad &= 5 \\ x_4 &= 1 \end{aligned}$$

Thus the solution is $x_1 = 1$, $x_2 = -2$, $x_3 = 5$, and $x_4 = 1$.

Technology Note Spreadsheets, computer programs, and graphing calculators are especially useful in solving systems of equations, because they can be used to perform the row reductions necessary to obtain a reduced matrix. On some calculators, a direct command will give a reduced form for the augmented matrix. Figure 3.8(a) shows the augmented matrix for the system of equations in Example 2, and Figure 3.8(b) shows the reduced form of the augmented matrix (which is the same as the reduced matrix found in Example 2).

```
[A]
[[1  1  1   2   6  ]
 [1  2  0   1   -2 ]
 [1  1  3   -2  12 ]
 [1  1  -4  5   -16]]
```

(a)

```
rref([A])
  [[1  0  0  0  1 ]
   [0  1  0  0  -2]
   [0  0  1  0  5 ]
   [0  0  0  1  1 ]]
```

(b)

Figure 3.8

CHECKPOINT

2. Solve the system.

$$\begin{cases} 3x + 2y - z = 3 \\ x - y + 2z = 4 \\ 2x + 3y - z = 3 \end{cases}$$

EXAMPLE 3

As mentioned in the Application Preview, the Walters Manufacturing Company needs to know how best to use the time available within its three manufacturing departments in the construction and packaging of the three types of metal storage sheds. Each one must be stamped, painted, and packaged. Table 3.7 shows the number of hours required for the processing of each type of shed. Using the information in the table, determine how many of each type of shed can be produced if the stamping department has 3200 hours available, the painting department has 1700 hours, and the packaging department has 1300 hours.

TABLE 3.7

	Shed		
Department	*Type I*	*Type II*	*Type III*
Stamping	2	3	4
Painting	1	2	1
Packaging	1	1	2

Solution

If we let x_1 be the number of type I sheds, x_2 be the number of type II sheds, and x_3 be the number of type III sheds, the equation

$$2x_1 + 3x_2 + 4x_3 = 3200$$

represents the hours used by the stamping department. Similarly,

$$1x_1 + 2x_2 + 1x_3 = 1700$$

and

$$1x_1 + 1x_2 + 2x_3 = 1300$$

represent the hours used by the painting department and the packaging department, respectively.

The augmented matrix for this system of equations is

$$\left[\begin{array}{ccc|c} 2 & 3 & 4 & 3200 \\ 1 & 2 & 1 & 1700 \\ 1 & 1 & 2 & 1300 \end{array}\right]$$

Reducing this augmented matrix proceeds as follows:

$$\left[\begin{array}{ccc|c} 2 & 3 & 4 & 3200 \\ 1 & 2 & 1 & 1700 \\ 1 & 1 & 2 & 1300 \end{array}\right] \xrightarrow{\text{Switch } R_1 \text{ and } R_3} \left[\begin{array}{ccc|c} 1 & 1 & 2 & 1300 \\ 1 & 2 & 1 & 1700 \\ 2 & 3 & 4 & 3200 \end{array}\right]$$

$$\xrightarrow[(-2)R_1 + R_3 \to R_3]{(-1)R_1 + R_2 \to R_2} \left[\begin{array}{ccc|c} 1 & 1 & 2 & 1300 \\ 0 & 1 & -1 & 400 \\ 0 & 1 & 0 & 600 \end{array}\right]$$

$$\xrightarrow[(-1)R_2 + R_3 \to R_3]{(-1)R_2 + R_1 \to R_1} \left[\begin{array}{ccc|c} 1 & 0 & 3 & 900 \\ 0 & 1 & -1 & 400 \\ 0 & 0 & 1 & 200 \end{array}\right]$$

$$\xrightarrow[R_3 + R_2 \to R_2]{(-3)R_3 + R_1 \to R_1} \left[\begin{array}{ccc|c} 1 & 0 & 0 & 300 \\ 0 & 1 & 0 & 600 \\ 0 & 0 & 1 & 200 \end{array}\right]$$

Thus the solution to the system is

$$x_1 = 300$$
$$x_2 = 600$$
$$x_3 = 200$$

The company should make 300 type I, 600 type II, and 300 type III sheds.

Systems with Nonunique Solutions

All the systems considered so far had unique solutions, but it is also possible for a system of linear equations to have an infinite number of solutions or no solution at all. Although coefficient matrices for systems with an infinite number of solutions or no solution will not reduce to identity matrices, row operations can be used to obtain a reduced form from which the solutions, if they exist, can be determined.

A matrix is said to be in **reduced form** when it is in the following form:

1. The first nonzero element in each row is 1.
2. Every column containing a first nonzero element for some row has zeros everywhere else.
3. The first nonzero element of each row is to the right of the first nonzero element of every row above it.
4. All rows containing zeros are grouped together below the rows containing nonzero entries.

The following matrices are in reduced form because they satisfy these conditions.

$$\left[\begin{array}{cc|c} 1 & 0 & 4 \\ 0 & 1 & 2 \\ 0 & 0 & 0 \end{array}\right] \quad \left[\begin{array}{ccc|c} 1 & 0 & 0 & 0 \\ 0 & 1 & 0 & 3 \\ 0 & 0 & 1 & 1 \\ 0 & 0 & 0 & 0 \end{array}\right] \quad \left[\begin{array}{ccc|c} 1 & 4 & 0 & 1 \\ 0 & 0 & 1 & 2 \\ 0 & 0 & 0 & 0 \end{array}\right]$$

The following matrices are *not* in reduced form.

$$\left[\begin{array}{cc|c} 1 & \textcircled{1} & 0 \\ 0 & 1 & 3 \\ 0 & 0 & 0 \end{array}\right] \quad \left[\begin{array}{cc|c} 1 & 0 & 1 \\ 0 & \textcircled{2} & 1 \\ 0 & 0 & 0 \end{array}\right] \quad \left[\begin{array}{cc|c} 0 & 1 & 0 \\ 1 & 0 & 0 \\ 0 & 0 & 1 \end{array}\right]$$

In the first two matrices, the circled element must be changed to obtain a reduced form. Can you see what row operations would transform each of these matrices into reduced form? The third matrix does not satisfy point 3 above.

We can solve a system of linear equations by using row operations on the augmented matrix until the coefficient matrix is transformed to an equivalent matrix in reduced form.

EXAMPLE 4

Solve the system

$$\begin{cases} x_1 + x_2 - x_3 + x_4 = 3 \\ x_2 + x_3 + x_4 = 1 \\ x_1 - 2x_3 + x_4 = 6 \\ 2x_1 - x_2 - 5x_3 - 3x_4 = -5 \end{cases}$$

Solution

To solve this system we must reduce the augmented matrix

$$\left[\begin{array}{cccc|c} 1 & 1 & -1 & 1 & 3 \\ 0 & 1 & 1 & 1 & 1 \\ 1 & 0 & -2 & 1 & 6 \\ 2 & -1 & -5 & -3 & -5 \end{array}\right]$$

Attempting to reduce the coefficient matrix to an identity matrix requires the following.

$$\left[\begin{array}{cccc|c} 1 & 1 & -1 & 1 & 3 \\ 0 & 1 & 1 & 1 & 1 \\ 1 & 0 & -2 & 1 & 6 \\ 2 & -1 & -5 & -3 & -5 \end{array}\right] \xrightarrow[(-2)R_1 + R_4 \to R_4]{(-1)R_1 + R_3 \to R_3} \left[\begin{array}{cccc|c} 1 & 1 & -1 & 1 & 3 \\ 0 & 1 & 1 & 1 & 1 \\ 0 & -1 & -1 & 0 & 3 \\ 0 & -3 & -3 & -5 & -11 \end{array}\right]$$

$$\xrightarrow[3R_2 + R_4 \to R_4]{\substack{(-1)R_2 + R_1 \to R_1 \\ R_2 + R_3 \to R_3}} \left[\begin{array}{cccc|c} 1 & 0 & -2 & 0 & 2 \\ 0 & 1 & 1 & 1 & 1 \\ 0 & 0 & 0 & 1 & 4 \\ 0 & 0 & 0 & -2 & -8 \end{array}\right]$$

The entry in row 3, column 3 cannot be made 1 using rows *below* row 2. Moving to column 4, the entry in row 3, column 4 is a 1. Using row 3 gives

$$\xrightarrow[2R_3 + R_4 \to R_4]{(-1)R_3 + R_2 \to R_2} \left[\begin{array}{cccc|c} 1 & 0 & -2 & 0 & 2 \\ 0 & 1 & 1 & 0 & -3 \\ 0 & 0 & 0 & 1 & 4 \\ 0 & 0 & 0 & 0 & 0 \end{array}\right]$$

The augmented matrix is now reduced; this matrix corresponds to the system

$$\begin{cases} x_1 - 2x_3 = 2 \\ x_2 + x_3 = -3 \\ x_4 = 4 \\ 0 = 0 \end{cases}$$

If we solve each of the equations for the leading variable (the variable corresponding to the first 1 in each row of the reduced form of a matrix), we obtain the **general solution** to the system.

$$\begin{aligned} x_1 &= 2 + 2x_3 \\ x_2 &= -3 - x_3 \\ x_4 &= 4, \quad \text{where } x_3 \text{ is any real number} \end{aligned}$$

The general solution gives the values of x_1 and x_2 dependent on the value of x_3, so we can get many different solutions to the system by specifying different values of x_3. For example, if $x_3 = 1$, then $x_1 = 4$, $x_2 = -4$, $x_3 = 1$, and $x_4 = 4$ is a solution to the system; if we let $x_3 = -2$, then $x_1 = -2$, $x_2 = -1$, $x_3 = -2$, and $x_4 = 4$ is another solution.

We have seen two different possibilities for solutions to systems of linear equations. In Examples 1–3, there was only one solution, whereas in Example 4 we saw that there were many solutions. A third possibility exists: that the system has no solution. Recall that these possibilities also existed for two equations in two unknowns, as discussed in Chapter 1, "Linear Equations and Functions."

EXAMPLE 5

Solve the system

$$\begin{cases} x + 2y - z = 3 \\ 3x + y = 4 \\ 2x - y + z = 2 \end{cases}$$

Solution

The augmented matrix is

$$\left[\begin{array}{rrr|r} 1 & 2 & -1 & 3 \\ 3 & 1 & 0 & 4 \\ 2 & -1 & 1 & 2 \end{array}\right]$$

This can be reduced as follows:

$$\left[\begin{array}{rrr|r} 1 & 2 & -1 & 3 \\ 3 & 1 & 0 & 4 \\ 2 & -1 & 1 & 2 \end{array}\right] \xrightarrow[(-2)R_1 + R_3 \to R_3]{(-3)R_1 + R_2 \to R_2} \left[\begin{array}{rrr|r} 1 & 2 & -1 & 3 \\ 0 & -5 & 3 & -5 \\ 0 & -5 & 3 & -4 \end{array}\right]$$

$$\xrightarrow{-\frac{1}{5}R_2 \to R_2} \left[\begin{array}{rrr|r} 1 & 2 & -1 & 3 \\ 0 & 1 & -\frac{3}{5} & 1 \\ 0 & -5 & 3 & -4 \end{array}\right]$$

$$\xrightarrow[(5)R_2 + R_3 \to R_3]{(-2)R_2 + R_1 \to R_1} \left[\begin{array}{rrr|r} 1 & 0 & \frac{1}{5} & 1 \\ 0 & 1 & -\frac{3}{5} & 1 \\ 0 & 0 & 0 & 1 \end{array}\right]$$

The system of equations corresponding to the reduced matrix is

$$\begin{cases} x + \frac{1}{5}z = 1 \\ y - \frac{3}{5}z = 1 \\ \quad 0 = 1 \end{cases}$$

Thus we see that this system has $0 = 1$ as an equation. This is clearly an impossibility, so there is no solution.

Now that we have seen examples of each of the different solution possibilities, let us consider what we have learned. In all cases the solution procedure began by setting up and then reducing the augmented matrix. Once the matrix is reduced, we can write the corresponding reduced system of equations. The solutions, if they exist, are easily found from this reduced system. We summarize the possibilities in Table 3.8.

TABLE 3.8

Reduced Form of Augmented Matrix	*Solution to System*
1. Coefficient array is an identity matrix.	Unique solution (see Examples 1, 2, and 3).
2. Coefficient array is *not* an identity matrix with either:	
(a) A row of 0s in the coefficient array with a nonzero entry in the augment.	(a) No solution (see Example 5).
(b) Or otherwise.	(b) Infinitely many solutions. Solve for lead variables in terms of nonlead variables (see Example 4).

CHECKPOINT

For each system of equations, the reduced form of the augmented matrix is given. Give the solution to the system, if it exists.

3. $\begin{cases} 2x + 3y + 2z = 180 \\ -x + 2y + 4z = 180 \\ 2x + 6y + 5z = 270 \end{cases}$ $\left[\begin{array}{ccc|c} 1 & 0 & 0 & 100 \\ 0 & 1 & 0 & -80 \\ 0 & 0 & 1 & 110 \end{array}\right]$

4. $\begin{cases} x - 2y + 9z = 12 \\ 2x - y + 3z = 18 \\ x + y - 6z = 6 \end{cases}$ $\left[\begin{array}{ccc|c} 1 & 0 & -1 & 8 \\ 0 & 1 & -5 & -2 \\ 0 & 0 & 0 & 0 \end{array}\right]$

5. $\begin{cases} x - 4y + 3z = 4 \\ 2x - 2y + z = 6 \\ x + 2y - 2z = 4 \end{cases}$ $\left[\begin{array}{ccc|c} 1 & 0 & -\frac{1}{3} & 0 \\ 0 & 1 & -\frac{5}{6} & 0 \\ 0 & 0 & 0 & 1 \end{array}\right]$

Graphing Utilities

Whether or not a unique solution exists, we can perform the row operations necessary to reduce an augmented matrix to find any solution that exists. We can perform these row operations with a graphing calculator, and because computer spreadsheets are electronically stored matrices, we can solve systems of linear equations by instructing the spreadsheet to perform these row operations. As expected, the value of using a graphing calculator, a spreadsheet, or a computer algebra system to solve systems of linear equations increases as the size of the systems gets larger. Some calculators have commands that will directly give the reduced form for a matrix.

EXAMPLE 6

Solve the system

$$\begin{cases} 2x_1 + 3x_2 - x_3 + 3x_4 + x_5 = 20 \\ 3x_1 + x_2 - 4x_3 + 3x_4 - x_5 = 0 \\ x_1 + x_2 + 3x_3 - 4x_4 + 2x_5 = 0 \\ x_1 - 2x_2 - 3x_3 + 2x_4 - 2x_5 = -6 \\ 2x_1 + 4x_4 - 5x_5 = -7 \end{cases}$$

Solution

The augmented matrix for the system of equations is shown (in two parts, with two columns repeated).

```
[A]
[[2  3   -1   3   1...
 [3  1   -4   3   -...
 [1  1   3    -4  2...
 [1  -2  -3   2   -...
 [2  0   0    4   -...
```

```
[A]
... -1   3   1   20 ]
... -4   3   -1   0 ]
... 3    -4  2    0 ]
... -3   2   -2  -6]
... 0    4   -5  -7]]
```

The reduced matrix is shown (in two parts, with two columns repeated).

```
...     0                 ...
Ans▸Frac
[[1   0   0   0   -5/6   ...
 [0   1   0   0   -2     ...
 [0   0   1   0   -3/2   ...
 [0   0   0   1   -5/6   ...
 [0   0   0   0   0      ...
```

```
...      0                ...
Ans▸Frac
... 0   -5/6    -19/6]
... 0   -2      -8   ]
... 0   -3/2    -9/2 ]
... 1   -5/6    -1/6 ]
... 0   0       0    ]]
```

This resulting reduced augmented matrix is

$$\left[\begin{array}{ccccc|c} 1 & 0 & 0 & 0 & -\frac{5}{6} & -\frac{19}{6} \\ 0 & 1 & 0 & 0 & -2 & -8 \\ 0 & 0 & 1 & 0 & -\frac{3}{2} & -\frac{9}{2} \\ 0 & 0 & 0 & 1 & -\frac{5}{6} & -\frac{1}{6} \\ 0 & 0 & 0 & 0 & 0 & 0 \end{array}\right]$$

The resulting equations and the general solution are given below.

$$x_1 - \left(\frac{5}{6}\right)x_5 = -\frac{19}{6} \qquad x_1 = \left(\frac{5}{6}\right)x_5 - \frac{19}{6}$$

$$x_2 - 2x_5 = -8 \qquad x_2 = 2x_5 - 8$$

$$x_3 - \left(\frac{3}{2}\right)x_5 = -\frac{9}{2} \qquad x_3 = \left(\frac{3}{2}\right)x_5 - \frac{9}{2}$$

$$x_4 - \left(\frac{5}{6}\right)x_5 = -\frac{1}{6} \qquad x_4 = \left(\frac{5}{6}\right)x_5 - \frac{1}{6}$$

CHECKPOINT SOLUTIONS

1. (a) $\left[\begin{array}{ccc|c} 3 & 2 & -1 & 3 \\ 1 & -1 & 2 & 4 \\ 2 & 3 & -1 & 3 \end{array}\right]$ (b) $\left[\begin{array}{ccc} 3 & 2 & -1 \\ 1 & -1 & 2 \\ 2 & 3 & -1 \end{array}\right]$

2. $\left[\begin{array}{ccc|c} 3 & 2 & -1 & 3 \\ 1 & -1 & 2 & 4 \\ 2 & 3 & -1 & 3 \end{array}\right] \to \left[\begin{array}{ccc|c} 1 & -1 & 2 & 4 \\ 0 & 5 & -7 & -9 \\ 0 & 5 & -5 & -5 \end{array}\right] \to \left[\begin{array}{ccc|c} 1 & 0 & 1 & 3 \\ 0 & 1 & -1 & -1 \\ 0 & 0 & -2 & -4 \end{array}\right]$

   $\to \left[\begin{array}{ccc|c} 1 & 0 & 0 & 1 \\ 0 & 1 & 0 & 1 \\ 0 & 0 & 1 & 2 \end{array}\right]$ Thus $x = 1$, $y = 1$, $z = 2$.

3. $x = 100$, $y = -80$, $z = 110$
4. $x - z = 8$, $y - 5z = -2 \Rightarrow x = z + 8$, $y = 5z - 2$
5. No solution

EXERCISE 3.3

Unique Solutions

In Problems 1 and 2, use the indicated row operation to change matrix A, where

$$A = \left[\begin{array}{ccc|c} -1 & 2 & 1 & 7 \\ 3 & 1 & 2 & 0 \\ 4 & 2 & 2 & 1 \end{array}\right]$$

1. Add 3 times row 1 to row 2 of matrix A and place the result in row 2 to get 0 in row 2, column 1.
2. Add 4 times row 1 to row 3 of matrix A and place the result in row 3 to get 0 in row 3, column 1.

In Problems 3–6, write the augmented matrix associated with each system of linear equations.

3. $\begin{cases} 3x + 2y + 4z = 0 \\ 2x - y + 2z = 0 \\ x - 2y - 4z = 0 \end{cases}$

4. $\begin{cases} x - 3y + 4z = 0 \\ 2x + 2y + z = 1 \\ 3x - 4y + 2z = 9 \end{cases}$

5. $\begin{cases} x - 3y + 4z = 2 \\ 2x \qquad + 2z = 1 \\ x + 2y + z = 1 \end{cases}$

6. $\begin{cases} x + 2y + 2z = 3 \\ x - 2y \qquad = 4 \\ \qquad y - z = 1 \end{cases}$

In each of Problems 7–10, the given matrix is an augmented matrix used in the solution of a system of linear equations. What is the solution of the system?

7. $\left[\begin{array}{ccc|c} 1 & 0 & 0 & 2 \\ 0 & 1 & 0 & \frac{1}{2} \\ 0 & 0 & 1 & -5 \end{array}\right]$

8. $\left[\begin{array}{ccc|c} 1 & 0 & 0 & -8 \\ 0 & 1 & 0 & 1 \\ 0 & 0 & 1 & \frac{1}{3} \end{array}\right]$

9. $\left[\begin{array}{ccc|c} 1 & 0 & 0 & 18 \\ 0 & 1 & 0 & 10 \\ 0 & 0 & 1 & 0 \end{array}\right]$

10. $\left[\begin{array}{ccc|c} 1 & 0 & 0 & -1 \\ 0 & 1 & 0 & 7 \\ 0 & 0 & 1 & 0 \end{array}\right]$

In Problems 11–14, the given matrix is an augmented matrix representing a system of linear equations. Use the Gauss-Jordan elimination method to find the solution to the system.

11. $\left[\begin{array}{ccc|c} 1 & 1 & 2 & -1 \\ 0 & 3 & 1 & 7 \\ 0 & -2 & 4 & 0 \end{array}\right]$

12. $\left[\begin{array}{ccc|c} 1 & 5 & 2 & 6 \\ 0 & -2 & 3 & 9 \\ 0 & 1 & 3 & 0 \end{array}\right]$

13. $\left[\begin{array}{ccc|c} 1 & 2 & 5 & -4 \\ 2 & -2 & 4 & -2 \\ 0 & 1 & -3 & 7 \end{array}\right]$

14. $\left[\begin{array}{ccc|c} 1 & 1 & 1 & 3 \\ 3 & -2 & 4 & 5 \\ 1 & 2 & 1 & 4 \end{array}\right]$

In Problems 15–26, use row operations on augmented matrices to solve the given system of linear equations.

15. $\begin{cases} 7x - 2y = -1 \\ 3x + 6y = 11 \end{cases}$

16. $\begin{cases} 6x + 7y = 55 \\ 5x - 37y = 3 \end{cases}$

17. $\begin{cases} x + y - z = 0 \\ x + 2y + 3z = -5 \\ 2x - y - 13z = 17 \end{cases}$

18. $\begin{cases} x + 2y - z = 3 \\ 2x + 5y - 2z = 7 \\ -x + y + 5z = -12 \end{cases}$

19. $\begin{cases} 2x - 6y - 12z = 6 \\ 3x - 10y - 20z = 5 \\ 2x \qquad - 17z = -4 \end{cases}$

20. $\begin{cases} -3x + 6y - 9z = 3 \\ x - y - 2z = 0 \\ 5x + 5y - 7z = 63 \end{cases}$

21. $\begin{cases} x - 3y + 3z = 7 \\ x + 2y - z = -2 \\ 3x + 2y + 4z = 5 \end{cases}$

22. $\begin{cases} 3x + 2y - 4z = -12 \\ 3x - 3y + 2z = -5 \\ 4x + 6y + z = 0 \end{cases}$

23. $\begin{cases} x_1 + 3x_2 + 2x_3 + 2x_4 = 3 \\ x_1 + x_2 + 3x_3 \qquad = 4 \\ 2x_1 \qquad + 2x_3 - 3x_4 = 4 \\ x_1 - 3x_2 \qquad = 1 \end{cases}$

24. $\begin{cases} x_1 + x_2 + x_3 + 2x_4 = 1 \\ x_1 \qquad - 3x_3 \qquad = 2 \\ x_1 - 3x_2 \qquad + x_4 = -2 \\ \qquad x_2 - 4x_3 + x_4 = 0 \end{cases}$

25. $\begin{cases} x - 2y + 3z + w = -2 \\ x - 3y + z - w = -7 \\ x - y \qquad = -2 \\ x \qquad + z + w = 2 \end{cases}$

26. $\begin{cases} x + 4y - 6z - 3w = 3 \\ x - y \qquad + 2w = -5 \\ x \qquad + z + w = 1 \\ \qquad y + z + w = 0 \end{cases}$

Nonunique Solutions

In each of Problems 27–30, a system of linear equations and a reduced matrix for the system are given. Use the reduced matrix to find the general solution of the system, if one exists.

27. $\begin{cases} x + 2y + 3z = 1 \\ 2x - y = 3 \\ x + 2y + 3z = 2 \end{cases}$ $\left[\begin{array}{ccc|c} 1 & 0 & \frac{3}{5} & 0 \\ 0 & 1 & \frac{6}{5} & 0 \\ 0 & 0 & 0 & 1 \end{array}\right]$

28. $\begin{cases} 2x + 3y + 4z = 2 \\ x + 2y + 2z = 1 \\ x + y + 2z = 2 \end{cases}$ $\left[\begin{array}{ccc|c} 1 & 0 & 2 & 0 \\ 0 & 1 & 0 & 0 \\ 0 & 0 & 0 & -1 \end{array}\right]$

29. $\begin{cases} 2x + y - z = 7 \\ x - y - z = 4 \\ 3x + 3y - z = 10 \end{cases}$ $\left[\begin{array}{ccc|c} 1 & 0 & -\frac{2}{3} & \frac{11}{3} \\ 0 & 1 & \frac{1}{3} & -\frac{1}{3} \\ 0 & 0 & 0 & 0 \end{array}\right]$

30. $\begin{cases} x - y + z = 3 \\ 3x + 2z = 7 \\ x - 4y + 2z = 5 \end{cases}$ $\left[\begin{array}{ccc|c} 1 & 0 & \frac{2}{3} & \frac{7}{3} \\ 0 & 1 & -\frac{1}{3} & -\frac{2}{3} \\ 0 & 0 & 0 & 0 \end{array}\right]$

The systems of equations in Problems 31–42 may have unique solutions, an infinite number of solutions, or no solution. Use matrices to find the general solution to each system, if a solution exists.

31. $\begin{cases} x + y + z = 0 \\ 2x - y - z = 0 \\ -x + 2y + 2z = 0 \end{cases}$

32. $\begin{cases} 2x - y + 3z = 0 \\ x + 2y + 2z = 0 \\ x - 3y + z = 0 \end{cases}$

33. $\begin{cases} 3x + 2y + z = 0 \\ x + y + 2z = 1 \\ 2x + y - z = -1 \end{cases}$

34. $\begin{cases} x + 3y + 2z = 2 \\ 2x - y - 2z = 1 \\ 3x + 2y = 3 \end{cases}$

35. $\begin{cases} 2x + 2y + z = 2 \\ x - 2y + 2z = 1 \\ -x + 2y - 2z = -1 \end{cases}$

36. $\begin{cases} 2x + y - z = 2 \\ x - y + 2z = 3 \\ x + y - z = 1 \end{cases}$

37. $\begin{cases} 2x - 5y + z = -9 \\ x + 4y - 6z = 2 \\ 3x - 4y - 2z = -10 \end{cases}$

38. $\begin{cases} 2x - y + z = 2 \\ 3x + y - 6z = -7 \\ x - y + 2z = 3 \end{cases}$

39. $\begin{cases} x + y + z = 3 \\ x - y + z = 4 \\ x + y - z = 5 \end{cases}$

40. $\begin{cases} 3x + 2y + z = 3 \\ x - y - z = 2 \\ 2x + y - 2z = 1 \end{cases}$

41. $\begin{cases} -0.6x_1 + 0.1x_2 + 0.3x_3 = 0 \\ 0.4x_1 - 0.7x_2 + 0.2x_3 = 0 \\ 0.2x_1 + 0.6x_2 - 0.5x_3 = 0 \end{cases}$

42. $\begin{cases} 0.1x_1 - 0.1x_2 - 0.3x_3 = 0 \\ 0.2x_1 + 0.3x_2 + 0.1x_3 = -0.3 \\ 0.3x_1 + 0.7x_2 + 0.5x_3 = -0.6 \end{cases}$

In Problems 43–48, use technology to solve each system of equations, if a solution exists.

43. $\begin{cases} 3x + 2y + z - w = 3 \\ x - y - 2z + 2w = 2 \\ 2x + 3y - z + w = 1 \\ -x + y + 2z - 2w = -2 \end{cases}$

44. $\begin{cases} 2x_1 + 3x_2 - x_3 + 3x_4 + x_5 = 22 \\ 3x_1 + x_2 - 4x_3 + 3x_4 - x_5 = 0 \\ x_1 + x_2 + 3x_3 - 4x_4 + 2x_5 = 6 \\ x_1 + 2x_2 - 3x_3 + 2x_4 - 2x_5 = -6 \\ 2x_1 + 4x_4 - 5x_5 = -7 \end{cases}$

45. $\begin{cases} x_1 + 2x_2 - x_3 + x_4 = 3 \\ x_1 + 3x_2 + 4x_3 + x_4 = -2 \\ 2x_1 + 5x_2 + 2x_3 + 2x_4 = 1 \\ 2x_1 + 3x_2 - 6x_3 + 2x_4 = 3 \end{cases}$

46. $\begin{cases} x_1 + 2x_2 + 3x_3 + 4x_4 = 2 \\ x_1 + 2x_2 + 4x_3 + 3x_4 = 2 \\ 2x_1 + 4x_2 + 6x_3 + 8x_4 = 4 \\ x_1 + 4x_2 + 3x_3 = 4 \end{cases}$

47. $\begin{cases} x_1 + 3x_2 + 4x_3 - x_4 = 1 \\ x_1 - x_2 - 2x_3 + x_4 = 2 \\ x_1 + 2x_2 + 3x_3 + x_4 = 0 \\ 2x_1 + 2x_2 + 2x_3 = 3 \end{cases}$

48. $\begin{cases} 2x_1 - x_2 + x_3 = 6 \\ x_1 + x_2 + x_3 + 2x_4 = 0 \\ 2x_1 + x_2 - x_3 + x_4 = -3 \\ x_1 + 2x_2 + 2x_3 + x_4 = -9 \end{cases}$

49. Solve $\begin{cases} a_1x + b_1y = c_1 \\ a_2x + b_2y = c_2 \end{cases}$ for x. This gives a formula for solving two equations in two variables for x.

50. Solve $\begin{cases} a_1x + b_1y = c_1 \\ a_2x + b_2y = c_2 \end{cases}$ for y. This gives a formula for solving two equations in two variables for y.

Applications

Problems 51–60 involve systems with unique solutions.

51. ***Nutrition*** A preschool has Campbell's Chunky Beef soup, which contains 4 g of fat and 25 mg of cholesterol per serving (cup), and Campbell's Chunky Sirloin Burger soup, which contains 9 g of fat and 20 mg of cholesterol per serving. By combining the soups, it is possible to get 10 servings of soup that will have 80 g of fat and 210 mg of cholesterol. How many cups of each soup should be used?

52. ***Nutrition*** A psychologist studying the effects of nutrition on the behavior of laboratory rats is feeding one group a combination of three foods: I, II, and III. Each of these foods contains three additives: A, B, and C, that are being used in the study. Each additive is a certain percentage of each of the foods as follows:

	Foods		
	I	*II*	*III*
Additive A	10%	30%	60%
Additive B	0	4	5
Additive C	2	2	12

If the diet requires 53 g per day of A, 4.5 g per day of B, and 8.6 g per day of C, find the number of grams per day of each food that must be used.

53. ***Nutrition*** The following table gives the calories, fat, and carbohydrates per ounce for three brands of cereal. How many ounces of each brand should be combined to get 443 calories, 5.7 g of fat, and 113.4 g of carbohydrates?

Cereal	*Calories*	*Fat*	*Carbohydrates*
All Bran	50	0	22.0
Sugar Frosted Flakes	108	0.1	25.7
Natural Mixed Grain	127	5.5	18.0

54. ***Investment*** A brokerage house offers three stock portfolios. Portfolio I consists of 2 blocks of common stock and 1 municipal bond. Portfolio II consists of 4 blocks of common stock, 2 municipal bonds, and 3 blocks of preferred stock. Portfolio III consists of 7 blocks of common stock, 3 municipal bonds, and 3 blocks of preferred stock. A customer wants 21 blocks of common stock, 10 municipal bonds, and 9 blocks of preferred stock. How many of each portfolio should be offered?

55. ***Investment*** Suppose that portfolios I and II in Problem 54 are unchanged and portfolio III consists of 2 blocks of common stock, 2 municipal bonds, and 3 blocks of preferred stock. A customer wants 12 blocks of common stock, 6 municipal bonds, and 6 blocks of preferred stock. How many of each portfolio should be offered?

56. ***Nutrition*** Each ounce of substance A supplies 5% of the required nutrition a patient needs. Substance B supplies 15% of the required nutrition per ounce and substance C supplies 12% of required nutrition per ounce. If digestive restrictions require that substances A and C be given in equal amounts and that the amount of substance B be one-fifth of these other amounts, find the number of ounces of each substance that should be in the meal to provide 100% nutrition.

57. ***Nutrition*** A glass of skim milk supplies 0.1 mg of iron, 8.5 g of protein, and 1 g of carbohydrates. A quarter pound of lean red meat provides 3.4 mg of iron, 22 g of protein, and 20 g of carbohydrates. Two slices of whole-grain bread supply 2.2 mg of iron, 10 g of protein, and 12 g of carbohydrates. If a person on a special diet must have 12.1 mg of iron, 97 g of protein, and 70 g of carbohydrates, how many glasses of skim milk, how many quarter-pound servings of meat, and how many two-slice servings of whole-grain bread will supply this?

58. ***Loans*** A bank lent a company $118,500 for the development of two products. If the loan for product A was for $34,500 more than for product B, how much was lent for each product?

59. ***Investment*** A man has $23,500 invested in two rental properties. One earns 15% on the investment and the other yields 16%. His total income from the two properties is $3625. How much is his income from each property?

60. ***Transportation*** The King Trucking Company has an order for three products for delivery. The table below gives the particulars for the products.

	Type I	*Type II*	*Type III*
Unit volume (cubic feet)	10	8	20
Unit weight (pounds)	10	20	40
Unit value (dollars)	100	20	200

If the carrier can carry 6000 cu ft, can carry 11,000 lb, and is insured for $36,900, how many units of each type can be carried?

61. ***Nutrition*** A botanist can purchase plant food of four different types, I, II, III, and IV. Each food comes in the same size bag, and the following table summarizes the number of grams of each of three nutrients that each bag contains.

	Foods (grams)			
	I	*II*	*III*	*IV*
Nutrient A	5	5	10	5
Nutrient B	10	5	30	10
Nutrient C	5	15	10	25

The botanist wants to use a food that has these nutrients in a different proportion and determines that he will need a total of 10,000 g of A, 20,000 g of B, and

20,000 of C. Find the number of bags of each type of food that should be ordered.

Application Problems 62–67 involve systems with nonunique solutions.

62. ***Traffic flow*** In the analysis of traffic flow, a certain city estimates the following situation for an area of its downtown district. In the figure below the arrows indicate the flow of traffic. If x_1 represents the number of cars traveling between intersections A and B, x_2 represents the number of cars traveling between B and C, x_3 the number between C and D, and x_4 the number between D and A, we can formulate equations based on the principle that the number of vehicles entering an intersection equals the number leaving it. That is, for intersection A we obtain

$$200 + x_4 = 100 + x_1$$

(a) Formulate equations for the traffic at B, C, and D.
(b) Solve the system of these four equations.

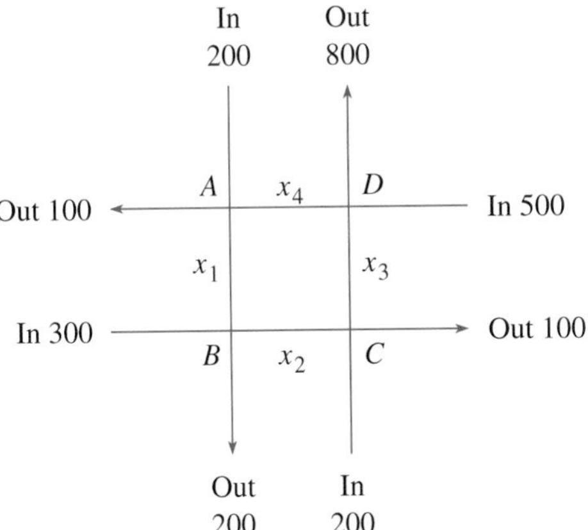

63. ***Investment*** An investment club has set a goal of earning 15% on the money they invest in stocks. The members are considering purchasing four possible stocks, with their cost per share (in dollars) and their projected growth per share (in dollars) summarized in the following table.

	Oil	*Bank*	*Computer*	*Retail*
Cost/share	100	40	30	20
Growth/share	10.00	3.60	6.00	2.40

If they have $102,300 to invest, how many shares of each stock should they buy to meet their goal?

64. ***Investment*** A trust account manager has $200,000 to be invested. The investment choices have current yields of 8%, 7%, and 10%. Suppose that the investment goal is to earn interest of $16,000, and risk factors make it prudent to invest some money in all three investments. What amount should be invested at each rate?

65. ***Nutrition*** Three different bacteria are cultured in one dish and feed on three nutrients. Each individual of species I consumes 1 unit of each of the first and second nutrients and 2 units of the third nutrient. Each individual of species II consumes 2 units of the first nutrient and 2 units of the third nutrient. Species III consumes 2 units of the first nutrient, 3 units of the second nutrient, and 5 units of the third nutrient. If the culture is given 5100 units of the first nutrient, 6900 units of the second nutrient, and 12,000 units of the third nutrient, how many of each species can be supported so all of the nutrients are consumed?

66. ***Irrigation*** An irrigation system allows water to flow in the pattern shown in the figure below. Water flows into the system at A and exits at B, C, D, and E with the amounts shown. Using the fact that at each point the water entering equals the water leaving, formulate an equation for water flow at each of the five points and solve the system.

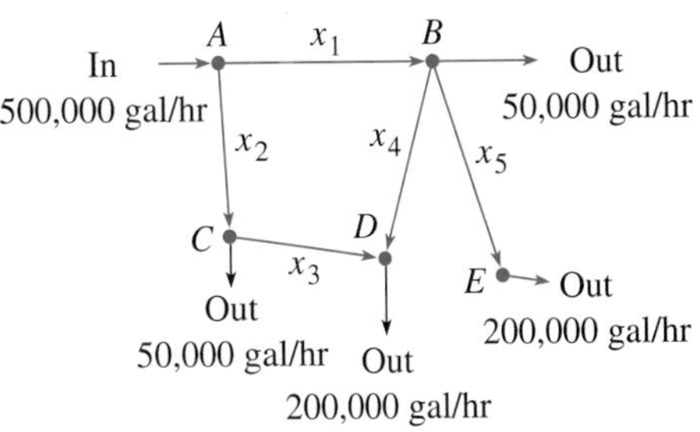

67. ***Investment*** A brokerage house offers three stock portfolios. Portfolio I consists of 2 blocks of common stock and 1 municipal bond. Portfolio II consists of 4 blocks of common stock, 2 municipal bonds, and 3 blocks of preferred stock. Portfolio III consists of 2 blocks of common stock, 1 municipal bond, and 3 blocks of preferred stock. A customer wants 16 blocks of common stock, 8 municipal bonds, and 6 blocks of preferred stock. If the numbers of each portfolio offered must be integers, find all possible offerings.

3.4 Inverse of a Square Matrix; Matrix Equations

OBJECTIVES

- *To find the inverse of a square matrix*
- *To use inverse matrices to solve systems of linear equations*
- *To find determinants of certain matrices*

APPLICATION PREVIEW

During World War II, the military commanders sent instructions that used simple substitution codes. These codes were easily broken, so they were further coded by using a coding matrix and matrix multiplication. The resulting coded messages, when received, could be unscrambled with a decoding matrix that is the **inverse matrix** of the coding matrix. In Section 3.2, "Multiplication of Matrices," we encoded messages by using the following code:

a	b	c	d	e	f	g	h	i	j	k	l
1	2	3	4	5	6	7	8	9	10	11	12

m	n	o	p	q	r	s	t	u	v	w	x
13	14	15	16	17	18	19	20	21	22	23	24

y	z	blank
25	26	27

and the encoding matrix

$$A = \begin{bmatrix} 3 & 5 \\ 4 & 6 \end{bmatrix}$$

which converts the numbers to the coded message (in pairs of numbers). To decode any message encoded by A, we must find the **inverse** of matrix A, denoted by A^{-1}, and multiply A^{-1} by the coded message (in pairs as 2×1 matrices). In this section, we will discuss how to find and apply the inverse of a given square matrix.

If the product of A and B is the identity matrix, I, we say that B is the inverse of A (and A is the inverse of B). The matrix B is called the **inverse matrix** of A, denoted A^{-1}.

Inverse Matrices

Two square matrices, A and B, are called **inverses** of each other if

$$AB = I \quad \text{and} \quad BA = I$$

In this case, $B = A^{-1}$ and $A = B^{-1}$.

EXAMPLE 1

Is B the inverse of A if $A = \begin{bmatrix} 1 & 2 \\ 1 & 1 \end{bmatrix}$ and $B = \begin{bmatrix} -1 & 2 \\ 1 & -1 \end{bmatrix}$?

Solution

$$AB = \begin{bmatrix} 1 & 2 \\ 1 & 1 \end{bmatrix} \cdot \begin{bmatrix} -1 & 2 \\ 1 & -1 \end{bmatrix} = \begin{bmatrix} 1 & 0 \\ 0 & 1 \end{bmatrix} = I$$

$$BA = \begin{bmatrix} -1 & 2 \\ 1 & -1 \end{bmatrix} \cdot \begin{bmatrix} 1 & 2 \\ 1 & 1 \end{bmatrix} = \begin{bmatrix} 1 & 0 \\ 0 & 1 \end{bmatrix} = I$$

Thus B is the inverse of A, or $B = A^{-1}$. We also say A is the inverse of B, or $A = B^{-1}$. Thus $(A^{-1})^{-1} = A$.

CHECKPOINT

1. Are the matrices A and B inverses if

$$A = \begin{bmatrix} 1 & 1 & -1 \\ 1 & -1 & 3 \\ 4 & 2 & 1 \end{bmatrix} \quad \text{and} \quad B = \begin{bmatrix} 3.5 & 1.5 & -1 \\ -5.5 & -2.5 & 2 \\ -3 & -1 & 1 \end{bmatrix}?$$

We have seen how to use elementary row operations on augmented matrices to solve systems of linear equations. We can also find the inverse of a matrix by using elementary row operations.

If the inverse exists for a square matrix A, we find A^{-1} as follows:

Finding the Inverse of a Square Matrix

Procedure	Example
To find the inverse of the square matrix A:	Find the inverse of matrix $A = \begin{bmatrix} 1 & 2 \\ 1 & 1 \end{bmatrix}$.
1. Form the augmented matrix $[A \mid I]$, where A is the $n \times n$ matrix and I is the $n \times n$ identity matrix.	1. $\left[\begin{array}{cc\|cc} 1 & 2 & 1 & 0 \\ 1 & 1 & 0 & 1 \end{array}\right]$
2. Perform elementary operations on $[A \mid I]$ until we have an augmented matrix of the form $[I \mid B]$—that is, until the matrix A on the left is transformed into the identity matrix.	2. $\left[\begin{array}{cc\|cc} 1 & 2 & 1 & 0 \\ 1 & 1 & 0 & 1 \end{array}\right] \xrightarrow{-R_1 + R_2 \to R_2} \left[\begin{array}{cc\|cc} 1 & 2 & 1 & 0 \\ 0 & -1 & -1 & 1 \end{array}\right] \xrightarrow[-R_2 \to R_2]{2R_2 + R_1 \to R_1} \left[\begin{array}{cc\|cc} 1 & 0 & -1 & 2 \\ 0 & 1 & 1 & -1 \end{array}\right]$
3. The matrix B (on the right) is the inverse of matrix A.	3. The inverse of A is $B = \begin{bmatrix} -1 & 2 \\ 1 & -1 \end{bmatrix}$. We verified this in Example 1.

Note that if A has no inverse, the reduction process on $[A \mid I]$ will yield a row of zeros in the left half.

EXAMPLE 2

Find the inverse of matrix A, below.

$$A = \begin{bmatrix} 2 & 5 & 4 \\ 1 & 4 & 3 \\ 1 & -3 & -2 \end{bmatrix}$$

Solution

To find the inverse of matrix A, we reduce A in $[A \mid I]$ (the matrix A augmented with the identity matrix I).

$$\left[\begin{array}{ccc|ccc} 2 & 5 & 4 & 1 & 0 & 0 \\ 1 & 4 & 3 & 0 & 1 & 0 \\ 1 & -3 & -2 & 0 & 0 & 1 \end{array}\right]$$

When the left side becomes an identity matrix, the inverse is in the augment.

$$\xrightarrow{\text{Switch } R_1 \text{ and } R_2} \left[\begin{array}{ccc|ccc} 1 & 4 & 3 & 0 & 1 & 0 \\ 2 & 5 & 4 & 1 & 0 & 0 \\ 1 & -3 & -2 & 0 & 0 & 1 \end{array}\right]$$

$$\xrightarrow[-R_1 + R_3 \to R_3]{-2R_1 + R_2 \to R_2} \left[\begin{array}{ccc|ccc} 1 & 4 & 3 & 0 & 1 & 0 \\ 0 & -3 & -2 & 1 & -2 & 0 \\ 0 & -7 & -5 & 0 & -1 & 1 \end{array}\right]$$

$$\xrightarrow{-\frac{1}{3}R_2 \to R_2} \left[\begin{array}{ccc|ccc} 1 & 4 & 3 & 0 & 1 & 0 \\ 0 & 1 & \frac{2}{3} & -\frac{1}{3} & \frac{2}{3} & 0 \\ 0 & -7 & -5 & 0 & -1 & 1 \end{array}\right]$$

$$\xrightarrow[7R_2 + R_3 \to R_3]{-4R_2 + R_1 \to R_1} \left[\begin{array}{ccc|ccc} 1 & 0 & \frac{1}{3} & \frac{4}{3} & -\frac{5}{3} & 0 \\ 0 & 1 & \frac{2}{3} & -\frac{1}{3} & \frac{2}{3} & 0 \\ 0 & 0 & -\frac{1}{3} & -\frac{7}{3} & \frac{11}{3} & 1 \end{array}\right]$$

$$\xrightarrow[-3R_3 \to R_3]{\substack{R_3 + R_1 \to R_1 \\ 2R_3 + R_2 \to R_2}} \left[\begin{array}{ccc|ccc} 1 & 0 & 0 & -1 & 2 & 1 \\ 0 & 1 & 0 & -5 & 8 & 2 \\ 0 & 0 & 1 & 7 & -11 & -3 \end{array}\right]$$

The inverse we seek is

$$\begin{bmatrix} -1 & 2 & 1 \\ -5 & 8 & 2 \\ 7 & -11 & -3 \end{bmatrix}$$

CHECKPOINT

2. Find the inverse of the matrix

$$A = \begin{bmatrix} 1 & 2 & 1 \\ 0 & 1 & 1 \\ 1 & 2 & 2 \end{bmatrix}$$

The technique of reducing $[A \mid I]$ can be used to find A^{-1} or to find that A^{-1}

doesn't exist for any square matrix A. However, a special formula can be used to find the inverse of a 2×2 matrix, if the inverse exists.

Inverse of a 2 × 2 Matrix

If $A = \begin{bmatrix} a & b \\ c & d \end{bmatrix}$, then $A^{-1} = \dfrac{1}{ad - bc}\begin{bmatrix} d & -b \\ -c & a \end{bmatrix}$ provided $ad - bc \neq 0$.

If $ad - bc = 0$, then A^{-1} does not exist.

This result can be verified by direct calculation or by reducing $[A \mid I]$.

EXAMPLE 3

Find the inverse, if it exists, for each of the following.

(a) $A = \begin{bmatrix} 3 & 7 \\ 2 & 5 \end{bmatrix}$ (b) $B = \begin{bmatrix} 2 & -1 \\ -4 & 2 \end{bmatrix}$

Solution

(a) For A, $ad - bc = (3)(5) - (2)(7) = 1 \neq 0$, so A^{-1} exists.

$$A^{-1} = \frac{1}{1}\begin{bmatrix} 5 & -7 \\ -2 & 3 \end{bmatrix} = \begin{bmatrix} 5 & -7 \\ -2 & 3 \end{bmatrix}$$

(b) For B, $ad - bc = (2)(2) - (-4)(-1) = 4 - 4 = 0$, so B^{-1} does not exist.

EXAMPLE 4

In the Application Preview we recalled how to encode messages with

a	b	c	d	e	f	g	h	i	j	k	l
1	2	3	4	5	6	7	8	9	10	11	12
m	n	o	p	q	r	s	t	u	v	w	x
13	14	15	16	17	18	19	20	21	22	23	24
y	z	blank									
25	26	27									

and with the encoding matrix

$$A = \begin{bmatrix} 3 & 5 \\ 4 & 6 \end{bmatrix}$$

which converts the numbers to the coded message (in pairs of numbers).

To decode any message encoded by A, we must find the inverse of A and multiply A^{-1} by the coded message (in pairs as 2×1 matrices). Use the inverse of A to decode the coded message 96, 118, 65, 84, 131, 168, 55, 72.

Solution

We first find the inverse of matrix A.

$$A^{-1} = \frac{1}{18-20}\begin{bmatrix} 6 & -5 \\ -4 & 3 \end{bmatrix} = -\frac{1}{2}\begin{bmatrix} 6 & -5 \\ -4 & 3 \end{bmatrix} = \begin{bmatrix} -3 & 2.5 \\ 2 & -1.5 \end{bmatrix}$$

Multiplying A^{-1} by the numbers, in pairs, gives

$$\begin{bmatrix} -3 & 2.5 \\ 2 & -1.5 \end{bmatrix}\begin{bmatrix} 96 \\ 118 \end{bmatrix} = \begin{bmatrix} 7 \\ 15 \end{bmatrix}$$

$$\begin{bmatrix} -3 & 2.5 \\ 2 & -1.5 \end{bmatrix}\begin{bmatrix} 65 \\ 84 \end{bmatrix} = \begin{bmatrix} 15 \\ 4 \end{bmatrix}$$

$$\begin{bmatrix} -3 & 2.5 \\ 2 & -1.5 \end{bmatrix}\begin{bmatrix} 131 \\ 168 \end{bmatrix} = \begin{bmatrix} 27 \\ 10 \end{bmatrix}$$

$$\begin{bmatrix} -3 & 2.5 \\ 2 & -1.5 \end{bmatrix}\begin{bmatrix} 55 \\ 72 \end{bmatrix} = \begin{bmatrix} 15 \\ 2 \end{bmatrix}$$

The result is 7, 15, 15, 4, 27, 10, 15, 2, which is the message "good job."

Graphing Utilities

Computer software programs and many types of calculators can be used to find the inverse of a matrix. Because of their ease of usage, it is logical to use graphing utilities to find inverses, especially for large matrices.

EXAMPLE 5

Use technology to find the inverse of the matrix

$$\begin{bmatrix} 1 & 2 & 4 \\ 3 & 1 & 3 \\ 5 & 2 & 4 \end{bmatrix}$$

Solution

Entering this matrix, using the inverse key or command, and converting the entries to fractions gives matrix A^{-1}.

```
[A]⁻¹▸Frac
[[-1/4    0     1/4  ]
 [3/8    -2     9/8  ]
 [1/8     1    -5/8]]
```

so $A^{-1} = \begin{bmatrix} -\frac{1}{4} & 0 & \frac{1}{4} \\ \frac{3}{8} & -2 & \frac{9}{8} \\ \frac{1}{8} & 1 & -\frac{5}{8} \end{bmatrix}$

We can verify that the product of matrix A and its inverse A^{-1} is the 3×3 identity matrix.

```
[A]⁻¹[A]
          [[1  0  0]
           [0  1  0]
           [0  0  1]]
```

It should be noted that roundoff error can occur when technology is used to find the inverse of a matrix and that the product of matrix A and its inverse may give a matrix that does not look exactly like the identity matrix, even though it is the identity matrix when the entries rounded to a reasonable number of decimal places.

Matrix Equations

In Example 1 of Section 3.3, "Gauss-Jordan Elimination: Solving Systems of Equations," we solved the system

$$\begin{cases} 2x + 5y + 4z = 4 \\ x + 4y + 3z = 1 \\ x - 3y - 2z = 5 \end{cases}$$

This system can be written as the following matrix equation.

$$\begin{bmatrix} 2x + 5y + 4z \\ x + 4y + 3z \\ x - 3y - 2z \end{bmatrix} = \begin{bmatrix} 4 \\ 1 \\ 5 \end{bmatrix}$$

The matrix equation is often written as follows:

$$\begin{bmatrix} 2 & 5 & 4 \\ 1 & 4 & 3 \\ 1 & -3 & -2 \end{bmatrix} \begin{bmatrix} x \\ y \\ z \end{bmatrix} = \begin{bmatrix} 4 \\ 1 \\ 5 \end{bmatrix}$$

We can see that this second form of a matrix equation is equivalent to the first if we carry out the multiplication. With the system written in this second form, we could also solve the system by multiplying both sides of the equation by the inverse of the coefficient matrix

$$\begin{bmatrix} 2 & 5 & 4 \\ 1 & 4 & 3 \\ 1 & -3 & -2 \end{bmatrix}$$

EXAMPLE 6

Solve the system

$$\begin{cases} 2x + 5y + 4z = 4 \\ x + 4y + 3z = 1 \\ x - 3y - 2z = 5 \end{cases}$$

by using inverse matrices.

Solution

To solve the system, we multiply both sides of the associated matrix equation, written above, by the inverse of the coefficient matrix, which we found in Example 2.

$$\begin{bmatrix} -1 & 2 & 1 \\ -5 & 8 & 2 \\ 7 & -11 & -3 \end{bmatrix} \begin{bmatrix} 2 & 5 & 4 \\ 1 & 4 & 3 \\ 1 & -3 & -2 \end{bmatrix} \begin{bmatrix} x \\ y \\ z \end{bmatrix} = \begin{bmatrix} -1 & 2 & 1 \\ -5 & 8 & 2 \\ 7 & -11 & -3 \end{bmatrix} \begin{bmatrix} 4 \\ 1 \\ 5 \end{bmatrix}$$

Note that we must be careful to multiply both sides *from the left* because matrix multiplication is not commutative. If we carry out the multiplications, we obtain

$$\begin{bmatrix} 1 & 0 & 0 \\ 0 & 1 & 0 \\ 0 & 0 & 1 \end{bmatrix}\begin{bmatrix} x \\ y \\ z \end{bmatrix} = \begin{bmatrix} 3 \\ -2 \\ 2 \end{bmatrix}, \quad \text{or} \quad \begin{bmatrix} x \\ y \\ z \end{bmatrix} = \begin{bmatrix} 3 \\ -2 \\ 2 \end{bmatrix}$$

which yields $x = 3$, $y = -2$, $z = 2$, the same solution as that found in Example 1 of the previous section.

Just as we wrote the system of three equations as a matrix equation of the form $AX = B$, this can be done in general. If A is an $n \times n$ matrix and if B and X are $n \times 1$ matrices, then

$$AX = B$$

is a **matrix equation.**

If the inverse of a matrix A exists, then we can use that inverse to solve the matrix equation for the matrix X. The general solution method follows.

$$AX = B$$

Multiplying by A^{-1} on the left of both sides of the equation gives

$$\begin{aligned} A^{-1}(AX) &= A^{-1}B \\ (A^{-1}A)X &= A^{-1}B \\ IX &= A^{-1}B \\ X &= A^{-1}B \end{aligned}$$

Thus inverse matrices can be used to solve systems of equations. Unfortunately, this method will work only if the solution to the system is unique. In fact, a system $AX = B$ has a unique solution if and only if A^{-1} exists. If the inverse of the coefficient matrix exists, the solution method above can be used to solve the system.

EXAMPLE 7

Use an inverse matrix to solve the system of equations.

$$\begin{cases} -x \phantom{{}+4y} + z = 1 \\ x + 4y - 3z = -3 \\ x - 2y + z = 3 \end{cases}$$

Solution

Writing the system as a matrix equation $AX = B$ gives

$$\begin{bmatrix} -1 & 0 & 1 \\ 1 & 4 & -3 \\ 1 & -2 & 1 \end{bmatrix}\begin{bmatrix} x \\ y \\ z \end{bmatrix} = \begin{bmatrix} 1 \\ -3 \\ 3 \end{bmatrix}$$

Using the methods of this section, the inverse of the coefficient matrix A is

$$A^{-1} = \begin{bmatrix} 0.5 & 0.5 & 1 \\ 1 & 0.5 & 0.5 \\ 1.5 & 0.5 & 1 \end{bmatrix}$$

Hence the solution is $X = A^{-1}B$.

$$X = \begin{bmatrix} 0.5 & 0.5 & 1 \\ 1 & 0.5 & 0.5 \\ 1.5 & 0.5 & 1 \end{bmatrix} \begin{bmatrix} 1 \\ -3 \\ 3 \end{bmatrix} = \begin{bmatrix} 2 \\ 1 \\ 3 \end{bmatrix}$$

Thus $x = 2$, $y = 1$, and $z = 3$.

CHECKPOINT

3. Use inverse matrices to solve the system.

$$\begin{cases} 7x - 5y = 12 \\ 2x - 3y = 6 \end{cases}$$

Graphing Utilities

Because technology can so easily find the inverse of a matrix, it is easy to use graphing utilities to solve a system of linear equations by using the inverse of the coefficient matrix.

EXAMPLE 8

Use the inverse of the coefficient matrix to solve the following system of linear equations.

$$\begin{cases} 2x_1 + 3x_2 - x_3 + 3x_4 + x_5 = 22 \\ 3x_1 + x_2 - 4x_3 + 3x_4 - x_5 = 0 \\ x_1 - 2x_2 + 3x_3 - 4x_4 + 2x_5 = 0 \\ x_1 + 2x_2 - 3x_3 + 2x_4 - 2x_5 = -6 \\ 2x_1 + \qquad\qquad 4x_4 + 5x_5 = -7 \end{cases}$$

Solution

The matrix equation for this system is $AX = B$, as follows:

$$AX = \begin{bmatrix} 2 & 3 & -1 & 3 & 1 \\ 3 & 1 & -4 & 3 & -1 \\ 1 & -2 & 3 & -4 & 2 \\ 1 & 2 & -3 & 2 & -2 \\ 2 & 0 & 0 & 4 & -5 \end{bmatrix} \begin{bmatrix} x_1 \\ x_2 \\ x_3 \\ x_4 \\ x_5 \end{bmatrix} = \begin{bmatrix} 22 \\ 0 \\ 0 \\ -6 \\ -7 \end{bmatrix}$$

We use technology to find the inverse of the coefficient matrix. The inverse of the coefficient matrix may be difficult to read on a small screen. However, we need not record the inverse. If we multiply both sides by A^{-1}, we get the solution, $X = A^{-1}B$ (see Figure 3.9).

```
[A]
[[2  3  -1  3   1 ]
 [3  1  -4  3  -1]
 [1  -2  3  -4  2 ]
 [1  2  -3  2  -2]
 [2  0   0  4  -5]]
```

```
[B]
[[22]
 [0 ]
 [0 ]
 [-6]
 [-7]]
```

```
[A]⁻¹[B]
[[1]
 [2]
 [3]
 [4]
 [5]]
```

Figure 3.9

Thus the solution is $x_1 = 1, x_2 = 2, x_3 = 3, x_4 = 4, x_5 = 5$.

Like graphing calculators and computer software programs, computer spreadsheets also can be used to find the inverse of a matrix, if it exists.

EXAMPLE 9

Suppose a bank draws its venture capital funds from three main sources of income—business loans, auto loans, and home mortgages. The income from these sources for each of 3 years is given in Table 3.9, and the venture capital for these years is given in Table 3.10. If the bank uses a fixed percent of its income from each of the business loans (x), auto loans (y), and home mortgages (z) to get its venture capital funds, find the percent of income used for each of these sources of income.

TABLE 3.9

	A	*B*	*C*	*D*
1	Income	from	Loans	
2		Business	Auto	Home
3	1993	68,210	23,324	57,234
4	1994	43,455	19,335	66,345
5	1995	58,672	15,654	69,334

TABLE 3.10

Venture	*Capital*
1993	53,448.74
1994	43,447.50
1995	50,582.88

Solution

The matrix equation describing this problem is $AX = V$, or

$$\begin{bmatrix} 68{,}210 & 23{,}324 & 57{,}234 \\ 43{,}455 & 19{,}335 & 66{,}345 \\ 58{,}672 & 15{,}654 & 69{,}334 \end{bmatrix} \begin{bmatrix} x \\ y \\ z \end{bmatrix} = \begin{bmatrix} 53{,}448.74 \\ 43{,}447.50 \\ 50{,}582.88 \end{bmatrix}$$

Finding the inverse of matrix of A and multiplying it times both sides of the matrix equation (on the left) gives the solution

$$\begin{bmatrix} x \\ y \\ z \end{bmatrix} = \begin{bmatrix} 0.47 \\ 0.23 \\ 0.28 \end{bmatrix}$$

Thus 47%, 23%, and 28% come from business, auto, and home loans, respectively.

Determinants

Recall that the inverse of $A = \begin{bmatrix} a & b \\ c & d \end{bmatrix}$ can be found with

$$A^{-1} = \frac{1}{ad - bc}\begin{bmatrix} d & -b \\ -c & a \end{bmatrix},$$

if $ad - bc \neq 0$.

The value $ad - bc$ used in finding the inverse of this 2×2 matrix is used so frequently that it is given a special name, the **determinant** of A (denoted *det* A or $|A|$).

Determinant of a 2 × 2 Matrix

$$\det\begin{bmatrix} a & b \\ c & d \end{bmatrix} = \begin{vmatrix} a & b \\ c & d \end{vmatrix} = ad - bc$$

EXAMPLE 10

Find det A if $A = \begin{bmatrix} 2 & 4 \\ 3 & -4 \end{bmatrix}$.

Solution

$$\det A = (2)(-4) - (3)(4) = -8 - 12 = -20$$

Formulas exist to find determinants of square matrices of orders larger than 2×2, but calculators and computers make it easy to find determinants. We have already seen that the inverse of a 2×2 matrix A does not exist if and only if det $A = 0$. In general, the inverse of an $n \times n$ matrix B does not exist if det $B = 0$. This also means that if the coefficient matrix of a system of linear equations has a determinant equal to 0, the system does not have a unique solution.

EXAMPLE 11

(a) Use a calculator to find the determinant of

$$A = \begin{bmatrix} 2 & 3 & 0 & 1 & 3 \\ 1 & 0 & 2 & 3 & 1 \\ 0 & 1 & 0 & 2 & 3 \\ 1 & 0 & 2 & 2 & 1 \\ 1 & 0 & 0 & 0 & 3 \end{bmatrix}$$

(b) Does the following system have a unique solution?

$$\begin{cases} 2x_1 + 3x_2 \qquad + x_4 + 3x_5 = 16 \\ x_1 \qquad + 2x_3 + 3x_4 + x_5 = 2 \\ \qquad x_2 \qquad + 2x_4 + 3x_5 = 9 \\ x_1 \qquad + 2x_3 + 2x_4 + x_5 = 3 \\ x_1 \qquad\qquad\qquad + 3x_5 = 10 \end{cases}$$

Solution

(a) Entering matrix A and using the determinant key gives $\det A = 24$.

(b) Because A is the coefficient matrix for the system and $\det A = 24$, the system has a unique solution. Using the inverse matrix gives the solution $x_1 = 1, x_2 = 2, x_3 = 0.5, x_4 = -1, x_5 = 3$.

CHECKPOINT SOLUTIONS

1. Yes, because both AB and BA give the 3×3 identity matrix.

2. $$\left[\begin{array}{ccc|ccc} 1 & 2 & 1 & 1 & 0 & 0 \\ 0 & 1 & 1 & 0 & 1 & 0 \\ 1 & 2 & 2 & 0 & 0 & 1 \end{array}\right] \to \left[\begin{array}{ccc|ccc} 1 & 2 & 1 & 1 & 0 & 0 \\ 0 & 1 & 1 & 0 & 1 & 0 \\ 0 & 0 & 1 & -1 & 0 & 1 \end{array}\right]$$

$$\to \left[\begin{array}{ccc|ccc} 1 & 0 & -1 & 1 & -2 & 0 \\ 0 & 1 & 1 & 0 & 1 & 0 \\ 0 & 0 & 1 & -1 & 0 & 1 \end{array}\right] \to \left[\begin{array}{ccc|ccc} 1 & 0 & 0 & 0 & -2 & 1 \\ 0 & 1 & 0 & 1 & 1 & -1 \\ 0 & 0 & 1 & -1 & 0 & 1 \end{array}\right]$$

The inverse is $A^{-1} = \begin{bmatrix} 0 & -2 & 1 \\ 1 & 1 & -1 \\ -1 & 0 & 1 \end{bmatrix}$.

3. This system can be written

$$\begin{bmatrix} 7 & -5 \\ 2 & -3 \end{bmatrix}\begin{bmatrix} x \\ y \end{bmatrix} = \begin{bmatrix} 12 \\ 6 \end{bmatrix}$$

and for the coefficient matrix A, $ad - bc = -21 - (-10) = -11$. Thus A^{-1} exists.

$$A^{-1} = -\frac{1}{11}\begin{bmatrix} -3 & 5 \\ -2 & 7 \end{bmatrix} = \begin{bmatrix} \frac{3}{11} & -\frac{5}{11} \\ \frac{2}{11} & -\frac{7}{11} \end{bmatrix}$$

so multiplying *from the left* gives the solution.

$$\begin{bmatrix} \frac{3}{11} & -\frac{5}{11} \\ \frac{2}{11} & -\frac{7}{11} \end{bmatrix}\begin{bmatrix} 7 & -5 \\ 2 & -3 \end{bmatrix}\begin{bmatrix} x \\ y \end{bmatrix} = \begin{bmatrix} \frac{3}{11} & -\frac{5}{11} \\ \frac{2}{11} & -\frac{7}{11} \end{bmatrix}\begin{bmatrix} 12 \\ 6 \end{bmatrix}$$

$$\begin{bmatrix} 1 & 0 \\ 0 & 1 \end{bmatrix}\begin{bmatrix} x \\ y \end{bmatrix} = \begin{bmatrix} \frac{6}{11} \\ -\frac{18}{11} \end{bmatrix}$$

Thus the solution is $x = \frac{6}{11}$, $y = -\frac{18}{11}$.

EXERCISE 3.4

1. If A is a 3×3 matrix and B is its inverse, what does the product AB equal?

2. If $C = \begin{bmatrix} 2 & -4 & 12 \\ 0 & 6 & -12 \\ 1 & -2 & 3 \end{bmatrix}$ and $D = \frac{1}{6}\begin{bmatrix} 1 & 2 & 4 \\ 2 & 1 & -4 \\ 1 & 0 & -2 \end{bmatrix}$, are C and D inverse matrices?

3. If $A = \begin{bmatrix} 1 & 2 & 1 \\ 0 & 0 & 3 \\ 1 & 0 & 1 \end{bmatrix}$ and $B = \begin{bmatrix} 0 & -\frac{1}{3} & 1 \\ \frac{1}{2} & 0 & -\frac{1}{2} \\ 0 & \frac{1}{3} & 0 \end{bmatrix}$, does $B = A^{-1}$?

4. If $D = \begin{bmatrix} 0 & 2 & -6 \\ -3 & 0 & 3 \\ 0 & -2 & 0 \end{bmatrix}$ and $C = -\frac{1}{3}\begin{bmatrix} 2 & 4 & 2 \\ 0 & 0 & 6 \\ 2 & 0 & 2 \end{bmatrix}$, does $C = D^{-1}$?

5. What is the inverse of $A = \begin{bmatrix} 3 & 0 & 0 \\ 0 & 3 & 0 \\ 0 & 0 & 3 \end{bmatrix}$?

6. What is the inverse of $B = \begin{bmatrix} 4 & 0 & 0 \\ 0 & 3 & 0 \\ 0 & 0 & 3 \end{bmatrix}$?

In Problems 7–14, find the inverse matrix for each matrix that has an inverse.

7. $\begin{bmatrix} 4 & 7 \\ 1 & 2 \end{bmatrix}$
8. $\begin{bmatrix} 4 & 5 \\ 7 & 9 \end{bmatrix}$
9. $\begin{bmatrix} 2 & -1 \\ -4 & 2 \end{bmatrix}$
10. $\begin{bmatrix} 3 & 12 \\ 1 & 4 \end{bmatrix}$
11. $\begin{bmatrix} 2 & 2 \\ 4 & 5 \end{bmatrix}$
12. $\begin{bmatrix} 6 & -4 \\ -1 & 1 \end{bmatrix}$
13. $\begin{bmatrix} 4 & 7 \\ 2 & 1 \end{bmatrix}$
14. $\begin{bmatrix} 1 & 2 \\ 3 & 0 \end{bmatrix}$

In Problems 15–22, find the inverse matrix for each matrix that has an inverse.

15. $\begin{bmatrix} 0 & 1 & 0 \\ 1 & 1 & 0 \\ 0 & 1 & 1 \end{bmatrix}$
16. $\begin{bmatrix} 0 & 2 & 1 \\ 3 & 0 & 1 \\ 1 & 1 & 1 \end{bmatrix}$
17. $\begin{bmatrix} 3 & 1 & 2 \\ 1 & 2 & 3 \\ 1 & 1 & 1 \end{bmatrix}$
18. $\begin{bmatrix} 1 & 2 & 3 \\ -1 & 5 & 6 \\ -1 & 3 & 3 \end{bmatrix}$
19. $\begin{bmatrix} 1 & 3 & 5 \\ -1 & -1 & 2 \\ 1 & 5 & 12 \end{bmatrix}$
20. $\begin{bmatrix} 1 & -1 & 4 \\ -1 & 0 & -2 \\ -1 & -3 & 4 \end{bmatrix}$
21. $\begin{bmatrix} 1 & 2 & 4 \\ 1 & -1 & -3 \\ 2 & 1 & 1 \end{bmatrix}$
22. $\begin{bmatrix} 3 & 4 & -1 \\ 4 & 2 & 2 \\ 2 & 6 & -4 \end{bmatrix}$

23. Find the inverse of

$$C = \begin{bmatrix} 0 & -1 & 0 & 1 & 0 \\ 1 & 1.5 & -1 & -1.5 & -0.5 \\ 0.5 & 1.75 & -1 & -1.25 & -0.25 \\ -0.5 & -1 & 1 & 1 & 0 \\ 0 & 0.5 & 0 & -0.5 & 0.5 \end{bmatrix}$$

24. Find the inverse of $B = \begin{bmatrix} 2 & 3 & 0 & 1 & 3 \\ 1 & 0 & 2 & 3 & 1 \\ 0 & 1 & 0 & 2 & 3 \\ 1 & 0 & 2 & 2 & 1 \\ 1 & 0 & 0 & 0 & 3 \end{bmatrix}$.

In each of Problems 25–28, the inverse of matrix A is given. Use the inverse to solve $AX = B$ for X.

25. $A^{-1} = \begin{bmatrix} 3 & 2 \\ 1 & 1 \end{bmatrix}$ Solve $AX = \begin{bmatrix} 3 \\ 2 \end{bmatrix}$.

26. $A^{-1} = \begin{bmatrix} 1 & -2 \\ 1 & 3 \end{bmatrix}$ Solve $AX = \begin{bmatrix} 5 \\ -1 \end{bmatrix}$.

27. $A^{-1} = \begin{bmatrix} 3 & 2 & 1 \\ 1 & 1 & 2 \\ 1 & 2 & 1 \end{bmatrix}$ Solve $AX = \begin{bmatrix} 3 \\ -1 \\ 2 \end{bmatrix}$.

28. $A^{-1} = \begin{bmatrix} 3 & 0 & -3 \\ 1 & 1 & 2 \\ 1 & 2 & 1 \end{bmatrix}$ Solve $AX = \begin{bmatrix} -1 \\ 2 \\ 3 \end{bmatrix}$.

29. Use the inverse found in Problem 15 to solve

$$\begin{bmatrix} 0 & 1 & 0 \\ 1 & 1 & 0 \\ 0 & 1 & 1 \end{bmatrix}\begin{bmatrix} x \\ y \\ z \end{bmatrix} = \begin{bmatrix} 1 \\ 2 \\ 3 \end{bmatrix}$$

30. Use the inverse found in Problem 16 to solve

$$\begin{bmatrix} 0 & 2 & 1 \\ 3 & 0 & 1 \\ 1 & 1 & 1 \end{bmatrix}\begin{bmatrix} x \\ y \\ z \end{bmatrix} = \begin{bmatrix} 1 \\ 0 \\ 4 \end{bmatrix}$$

In Problems 31–34, use inverse matrices to solve each system of linear equations.

31. $\begin{cases} x + 2y = 4 \\ 3x + 4y = 10 \end{cases}$ 32. $\begin{cases} 3x - 4y = 11 \\ 2x + 3y = -4 \end{cases}$

33. $\begin{cases} 2x + y = 4 \\ 3x + y = 5 \end{cases}$ 34. $\begin{cases} 5x - 2y = 6 \\ 3x + 3y = 12 \end{cases}$

In Problems 35–42, use inverse matrices to find the solution to the given matrix equation or system of equations.

35. $\begin{bmatrix} 6 & 3 & -8 \\ 1 & 0 & -3 \\ 20 & -12 & 2 \end{bmatrix}\begin{bmatrix} x \\ y \\ z \end{bmatrix} = \begin{bmatrix} 1 \\ -2 \\ 10 \end{bmatrix}$

36. $\begin{bmatrix} 3 & -1 & 0 \\ 2 & -1 & 1 \\ -1 & 1 & -2 \end{bmatrix}\begin{bmatrix} x \\ y \\ z \end{bmatrix} = \begin{bmatrix} 11 \\ 9 \\ -7 \end{bmatrix}$

37. $\begin{cases} x + y + z = 3 \\ 2x + y + z = 4 \\ 2x + 2y + z = 5 \end{cases}$ 38. $\begin{cases} 2x - y - 2z = 2 \\ 3x - y + z = -3 \\ x + y - z = 7 \end{cases}$

39. $\begin{cases} x + y + 2z = 8 \\ 2x + y + z = 7 \\ 2x + 2y + z = 10 \end{cases}$ 40. $\begin{cases} x - 2y + z = 0 \\ 2x + y - 2z = 2 \\ 3x + 2y - 3z = 2 \end{cases}$

41. $\begin{cases} -x_2 + x_4 = 0.7 \\ x_1 + x_2 - x_3 - 1.5x_4 - 0.5x_5 = -1.6 \\ 0.5x_1 + 1.75x_2 - x_3 - 1.25x_4 - 0.25x_5 = 1.275 \\ -0.5x_1 - x_2 + x_3 + x_4 = 1.15 \\ 0.5x_2 - 0.5x_4 + 0.5x_5 = -0.15 \end{cases}$

42. $\begin{cases} 2x_1 + 3x_2 + x_4 + 3x_5 = 0 \\ x_1 + 2x_3 + 3x_4 + x_5 = 27 \\ x_2 + 2x_4 + 3x_5 = -15 \\ x_1 + 2x_3 + 2x_4 + x_5 = 24 \\ x_1 + 3x_5 = -24 \end{cases}$

In Problems 43–46, evaluate each determinant.

43. $\begin{vmatrix} 1 & 2 \\ 3 & 4 \end{vmatrix}$ 44. $\begin{vmatrix} 4 & 5 \\ -1 & -2 \end{vmatrix}$

45. $\begin{vmatrix} 3 & -1 \\ 2 & 4 \end{vmatrix}$ 46. $\begin{vmatrix} -1 & 2 \\ 4 & -8 \end{vmatrix}$

In Problems 47–50, use technology to find each determinant.

47. $\begin{vmatrix} 3 & 2 & 1 \\ -1 & 0 & 2 \\ 0 & 1 & 1 \end{vmatrix}$ 48. $\begin{vmatrix} 1 & 2 & 2 \\ 3 & 0 & 1 \\ -1 & 0 & -2 \end{vmatrix}$

49. $\begin{vmatrix} 0 & 1 & 2 \\ 3 & 1 & 1 \\ 4 & -1 & 3 \end{vmatrix}$ 50. $\begin{vmatrix} 3 & 1 & 4 & 2 \\ -1 & 2 & 0 & 5 \\ 4 & 3 & 0 & -1 \\ 0 & 4 & 2 & 1 \end{vmatrix}$

In Problems 51–54, use determinants to decide whether each matrix has an inverse.

51. $\begin{bmatrix} 2 & 3 \\ -1 & 4 \end{bmatrix}$ 52. $\begin{bmatrix} 3 & 3 \\ 1 & -1 \end{bmatrix}$

53. $\begin{bmatrix} 1 & 3 & -2 \\ 2 & -1 & 5 \\ 3 & 2 & 3 \end{bmatrix}$ 54. $\begin{bmatrix} 4 & 1 & 0 \\ 0 & 5 & 0 \\ 0 & 0 & -1 \end{bmatrix}$

Applications

Decoding We have encoded messages by assigning the letters of the alphabet the numbers 1–26, with a blank assigned the number 27, and by using an encoding matrix A that converts the numbers to the coded message (in pairs of numbers). To decode any message encoded by A, we must find the inverse of A, denoted by A^{-1}, and multiply A^{-1} times the coded message. In Problems 55–58, use this code and the given encoding matrix A to decode the given message.

55. The code matrix is

$$A = \begin{bmatrix} 5 & 9 \\ 6 & 11 \end{bmatrix}$$

Use A^{-1} to decode the message

49, 59, 133, 161, 270, 327, 313, 381

56. If the code matrix is

$$A = \begin{bmatrix} 3 & 5 \\ 1 & 2 \end{bmatrix}$$

find A^{-1} and use it to decode

157, 59, 73, 29, 147, 58, 63, 24, 119, 44

57. If the code matrix is

$$A = \begin{bmatrix} 0 & 2 & 1 \\ 3 & 0 & 1 \\ 1 & 1 & 1 \end{bmatrix}$$

find A^{-1} and use it to decode

47, 22, 34, 28, 87, 46, 63, 66, 55, 56, 44, 43, 17, 14, 15

(Because A and A^{-1} are 3×3 matrices, use triples of numbers.)

58. The code matrix is

$$A = \begin{bmatrix} 1 & 2 & 1 \\ 5 & 8 & 2 \\ 7 & 11 & 3 \end{bmatrix}$$

Using A^{-1}, decode 49, 165, 231, 49, 154, 220, 39, 162, 226.

59. ***Competition*** A product is made by only two competing companies. If Company X retains two-thirds of its customers and loses one-third to Company Y each year, and Company Y retains three-quarters of its customers and loses one-quarter to Company X each year, then we can represent the fraction each company had last year by

$$\begin{bmatrix} x_0 \\ y_0 \end{bmatrix}$$

where x_0 is the fraction Company X had and y_0 is the fraction Company Y had. Then the fraction that each will have this year can be represented by

$$\begin{bmatrix} x \\ y \end{bmatrix} = \begin{bmatrix} \frac{2}{3} & \frac{1}{4} \\ \frac{1}{3} & \frac{3}{4} \end{bmatrix} \begin{bmatrix} x_0 \\ y_0 \end{bmatrix}$$

If Company X has 1900 customers and Company Y has 1700 customers this year, how many customers did each have last year?

60. ***Politics*** In a two-party system, it is observed that 70% of all Republicans vote for Republicans and the remainder for Democrats, whereas 60% of all Democrats vote for Democrats and the remainder for Republicans. (See the diagram below.)

The percent of each party that will win in this election is

$$\begin{bmatrix} r \\ d \end{bmatrix} = \begin{bmatrix} 0.70 & 0.40 \\ 0.30 & 0.60 \end{bmatrix} \begin{bmatrix} r_0 \\ d_0 \end{bmatrix}$$

where $\begin{bmatrix} r_0 \\ d_0 \end{bmatrix}$ represents the percents of Republicans and Democrats, respectively, that won in the last election. If 52% of those who win in this election are Republicans and 48% are Democrats, what were the percentages that won in the last election?

In Problems 61–65, set up each system of equations and then solve it by using inverse matrices.

61. ***Medication*** Medication A is given every 4 hours and medication B is given twice a day. For patient I, the total intake of the two medications is restricted to 50.6 mg per day, and for patient II, the total intake is restricted to 92 mg per day. The ratio of the dosage of A to the dosage of B is always 5 to 8. Find the dosage for each administration of each medication for both patient I and patient II.

62. ***Transportation*** The Ace Freight Company has an order for two products to be delivered to two stores of a company. The matrix below gives information regarding the two products.

Product	*I*	*II*
Unit volume (cu ft)	20	30
Unit weight (lb)	100	400

If truck A can carry 2350 cu ft and 23,000 lb, and truck B can carry 2500 cu ft and 24,500 lb, how many of each product can be carried by each truck?

63. ***Investment*** One safe investment pays 10% per year, and a more risky investment pays 18% per year. A woman has $145,600 to invest and would like to have an income of $20,000 per year from her investments. How much should she invest at each rate?

64. ***Nutrition*** A biologist has a 40% solution and a 10% solution of the same plant nutrient. How many cubic centimeters of each solution should be mixed to obtain 25 cc of a 28% solution?

65. ***Medication*** A nurse has two solutions that contain different concentrations of a certain medication. One is a 20% concentration and the other is a 5% concentration. How many cubic centimeters of each should he mix to obtain 10 cc of a 15.5% solution?

66. ***Bee ancestry*** Because a female bee comes from a fertilized egg and a male bee comes from an unfertilized egg, the number (n_{t+2}) of ancestors of a male bee $t + 2$ generations before the present generation is the sum of the number of ancestors t and $t + 1$ generations (n_t and n_{t+1}) before the present. If the number of ancestors of a male bee in a given generation t and in the previous generation is given by

$$N = \begin{bmatrix} n_t \\ n_{t+1} \end{bmatrix}$$

then there is a matrix M such that the number of ancestors in the two generations preceding generation t is given by

$$MN = \begin{bmatrix} 0 & 1 \\ 1 & 1 \end{bmatrix}\begin{bmatrix} n_t \\ n_{t+1} \end{bmatrix} = \begin{bmatrix} n_{t+1} \\ n_t + n_{t+1} \end{bmatrix}$$

For a given male bee, the number of ancestors 5 and 6 generations back is given by

$$\begin{bmatrix} 8 \\ 13 \end{bmatrix}$$

Find the number of ancestors 4 and 5 generations back by multiplying both sides of

$$MN = \begin{bmatrix} 8 \\ 13 \end{bmatrix}$$

by the inverse of M.

67. ***Attendance*** Suppose that during the first 2 weeks of a controversial art exhibit, the daily number of visitors to a gallery turns out to be the sum of the number of visitors on the previous 3 days. We represent the numbers of visitors on 3 successive days by

$$N = \begin{bmatrix} n_t \\ n_{t+1} \\ n_{t+2} \end{bmatrix}$$

(a) Find the matrix M so that

$$MN = \begin{bmatrix} n_{t+1} \\ n_{t+2} \\ n_t + n_{t+1} + n_{t+2} \end{bmatrix}$$

(b) If the number of visitors on 3 successive days is given by

$$\begin{bmatrix} 191 \\ 346 \\ 645 \end{bmatrix}$$

use M^{-1} to find the number of visitors 1 day earlier (before the 191-visitor day).

68. ***Bacteria growth*** The population of a colony of bacteria grows so that the population size at any hour t is the sum of the populations of the 3 previous hours. Suppose that the matrix

$$N = \begin{bmatrix} n_t \\ n_{t+1} \\ n_{t+2} \end{bmatrix}$$

describes the population size for 3 successive hours.

(a) Write matrix M so that MN gives the population size for 3 successive hours beginning 1 hour later—that is, so that

$$MN = \begin{bmatrix} n_{t+1} \\ n_{t+2} \\ n_t + n_{t+1} + n_{t+2} \end{bmatrix}$$

(b) If the population at the end of 3 successive 1-hour periods was 200 at the end of the first hour, 370 at the end of the second hour, and 600 at the end of the third hour, what was the population 1 hour before it was 200? Use M^{-1}.

3.5 Applications of Matrices: Leontief Input-Output Models

OBJECTIVES

- *To interpret Leontief technology matrices*
- *To use Leontief models to solve input-output problems*

APPLICATION PREVIEW

Suppose we consider a simple economy as being based on three commodities: agricultural products, manufactured goods, and fuels. Suppose further that production of 10 units of agricultural products requires 5 units of agricultural products, 2 units of manufactured goods, and 1 unit of fuels; that production of 10 units of manufactured goods requires 1 unit of agricultural products, 5 units of manufactured products, and 3 units of fuels; and that production of 10 units of fuels requires 1 unit of agricultural products, 3 units of manufactured goods, and 4 units of fuels.

Table 3.11 summarizes this information in terms of production of 1 unit. The first column represents the units of agricultural products, manufactured goods, and fuels, respectively, that are needed to produce 1 unit of agricultural products. Column 2 represents the units required to produce 1 unit of manufactured goods, and column 3 represents the units required to produce 1 unit of fuels.

TABLE 3.11

	Outputs		
Inputs	*Agricultural Products*	*Manufactured Goods*	*Fuels*
Agricultural products	0.5	0.1	0.1
Manufactured goods	0.2	0.5	0.3
Fuels	0.1	0.3	0.4

From Table 3.11 we can form matrix A, which is called a **technology matrix** or a **Leontief matrix.**

$$A = \begin{bmatrix} 0.5 & 0.1 & 0.1 \\ 0.2 & 0.5 & 0.3 \\ 0.1 & 0.3 & 0.4 \end{bmatrix}$$

The Leontief input-output model, named for Wassily Leontief, is an interesting application of matrices. Just as an airplane model approximates the real object, a *mathematical model* attempts to approximate a real-world situation. Models that are carefully constructed from past information may often provide remarkably accurate forecasts. The model Leontief developed was useful in predicting the effects on the economy of price changes or shifts in government spending.*

Leontief's work divided the economy into 500 sectors, which, in a subsequent article in *Scientific American* (October 1951), were reduced to a more manageable 42 departments of production. We can examine the workings of input-output analysis with a very simplified view of the economy.

In this section, we will interpret Leontief technology matrices for simple economies and will solve input-output problems for these economies by using Leontief models.

In Table 3.11, we note that the sum of the units of agricultural products, manufactured goods, and fuels required to produce 1 unit of fuels (column 3) does not add up to 1 unit. This is because not all commodities or industries are represented in the model. In particular, it is customary to omit labor from models of this type.

*Leontief's work dealt with a massive analysis of the American economy. He was awarded a Nobel Prize in economics in 1973 for his study.

EXAMPLE 1

Use Table 3.11 to answer the following questions.

(a) How many units of agricultural products and of fuels were required to produce 100 units of manufactured goods?
(b) Production of which commodity is least dependent on the other two?
(c) If fuel costs rise, which two industries will be most affected?

Solution

(a) Referring to column 2, manufactured goods, we see that 1 unit requires 0.1 unit of agricultural products and 0.3 unit of fuels. Thus 100 units of manufactured goods require 10 units of agricultural products and 30 units of fuels.
(b) Looking down the columns, we see that 1 unit of agricultural products requires 0.3 unit of the other two commodities; 1 unit of manufactured goods requires 0.4 unit of the other two; and 1 unit of fuels requires 0.4 unit of the other two. Thus production of agricultural products is least dependent on the others.
(c) A rise in the cost of fuels would most affect those industries that use the larger amount of fuels. One unit of agricultural products requires 0.1 unit of fuels, whereas a unit of manufactured goods requires 0.3 unit, and a unit of fuels requires 0.4 of its own units. Thus manufacturing and the fuel industry would be most affected by a cost increase in fuels.

CHECKPOINT

The following technology matrix for a simple economy describes the relationship of certain industries to each other in the production of one unit of product.

	Ag	Mfg	Fuels	Util
Agriculture	0.32	0.06	0.09	0.06
Manufacturing	0.11	0.41	0.19	0.31
Fuels	0.31	0.14	0.18	0.20
Utilities	0.18	0.26	0.21	0.25

1. What industry is most dependent on its own goods for its operation?
2. How many units of fuels are required to produce 100 units of utilities?

For a simplified model of the economy, such as that described with Table 3.11, not all information is contained within the technology matrix. In particular, each industry has a gross production. The **gross production matrix** for the economy can be represented by the column matrix

$$X = \begin{bmatrix} x_1 \\ x_2 \\ x_3 \end{bmatrix}$$

where x_1 is the gross production of agricultural products, x_2 is the gross production of manufactured goods, and x_3 is the gross production of fuels.

Those units of gross production not used by these industries are called **final demands** or **surpluses** and may be considered as being available for consumers, the government, or export. If we place these surpluses in a column matrix D, then the surplus can be represented by the equation

$$X - AX = D, \qquad \text{or} \qquad (I - A)X = D$$

where I is an identity matrix. This matrix equation is called the **technological equation** for an **open Leontief model.**

Open Leontief Model The technological equation for an open Leontief model is

$$(I - A)X = D$$

EXAMPLE 2

Use matrix A from Table 3.11 as the technology matrix. If we wish to have a surplus of 85 units of agricultural products, 65 units of manufactured goods, and 0 units of fuels, what should the gross outputs be?

Solution

Let X be the matrix as indicated above. Then the technological equation is $(I - A)X = D$.

$$I - A = \begin{bmatrix} 1 & 0 & 0 \\ 0 & 1 & 0 \\ 0 & 0 & 1 \end{bmatrix} - \begin{bmatrix} 0.5 & 0.1 & 0.1 \\ 0.2 & 0.5 & 0.3 \\ 0.1 & 0.3 & 0.4 \end{bmatrix} = \begin{bmatrix} 0.5 & -0.1 & -0.1 \\ -0.2 & 0.5 & -0.3 \\ -0.1 & -0.3 & 0.6 \end{bmatrix}$$

Hence we must solve the matrix equation

$$\begin{bmatrix} 0.5 & -0.1 & -0.1 \\ -0.2 & 0.5 & -0.3 \\ -0.1 & -0.3 & 0.6 \end{bmatrix} \begin{bmatrix} x_1 \\ x_2 \\ x_3 \end{bmatrix} = \begin{bmatrix} 85 \\ 65 \\ 0 \end{bmatrix}$$

The augmented matrix is

$$\left[\begin{array}{rrr|r} 0.5 & -0.1 & -0.1 & 85 \\ -0.2 & 0.5 & -0.3 & 65 \\ -0.1 & -0.3 & 0.6 & 0 \end{array}\right]$$

If we reduce by the methods of Section 3.3, "Gauss-Jordan Elimination: Solving Systems of Equations," we obtain

$$\left[\begin{array}{rrr|r} 1 & 0 & 0 & 300 \\ 0 & 1 & 0 & 400 \\ 0 & 0 & 1 & 250 \end{array}\right]$$

so the gross outputs for the industries are

Agriculture:	$x_1 = 300$
Manufacturing:	$x_2 = 400$
Fuels:	$x_3 = 250$

In Example 2 we solved the system of equations by using row operations on the augmented matrix. It is also possible to use the inverse of $I - A$ to solve the technology equation.

The technological equation for the open Leontief model can be solved by using the inverse of $I - A$, if the inverse exists. That is,

$$(I - A)X = D \quad \text{has the solution} \quad X = (I - A)^{-1}D$$

Technology Note If we have the technology matrix A and the surplus matrix D, we can solve the technological equation for an open Leontief model with a calculator or computer by defining matrices I, A, and D and evaluating $X = (I - A)^{-1}D$. Figure 3.10 shows the solution of Example 2 using a graphing calculator.

```
[A]    [[.5   .1   .1]
        [.2   .5   .3]
        [.1   .3   .4]]
[D]
              [[85]
               [65]
               [0 ]]
```

```
([I]-[A])-1*[D]
          [[300]
           [400]
           [250]]
```

Figure 3.10

EXAMPLE 3

A primitive economy with a lumber industry and a power industry has the following technological matrix.

$$\begin{array}{cc} & \begin{array}{cc}\text{Lumber} & \text{Power}\end{array} \\ A = & \begin{bmatrix} 0.1 & 0.2 \\ 0.2 & 0.4 \end{bmatrix} \begin{array}{l}\text{Lumber} \\ \text{Power}\end{array} \end{array}$$

If surpluses of 30 units of lumber and 70 units of power are desired, find the gross production of each industry.

Solution

Let x be the gross production of lumber, and let y be the gross production of power. Then

$$I - A = \begin{bmatrix} 0.9 & -0.2 \\ -0.2 & 0.6 \end{bmatrix}, \quad X = \begin{bmatrix} x \\ y \end{bmatrix}, \quad \text{and} \quad D = \begin{bmatrix} 30 \\ 70 \end{bmatrix}$$

We must solve the matrix equation $(I - A)X = D$, and we can do this by using

$$X = (I - A)^{-1}D$$

We can find $(I - A)^{-1}$ with the formula from Section 3.4: "Inverse of a Square Matrix; Matrix Equations" or with a graphing utility.

$$(I - A)^{-1} = \frac{1}{0.54 - 0.04}\begin{bmatrix} 0.6 & 0.2 \\ 0.2 & 0.9 \end{bmatrix} = 2\begin{bmatrix} 0.6 & 0.2 \\ 0.2 & 0.9 \end{bmatrix} = \begin{bmatrix} 1.2 & 0.4 \\ 0.4 & 1.8 \end{bmatrix}$$

Then we can find the gross production of each industry:

$$X = (I - A)^{-1}D = \begin{bmatrix} 1.2 & 0.4 \\ 0.4 & 1.8 \end{bmatrix}\begin{bmatrix} 30 \\ 70 \end{bmatrix} = \begin{bmatrix} 64 \\ 138 \end{bmatrix}$$

Hence the gross productions are

Lumber: $x = 64$

Power: $y = 138$

EXAMPLE 4

The economy of a developing nation is based on agricultural products, steel, and coal. An output of 1 ton of agricultural products requires an input of 0.1 ton of agricultural products, 0.02 ton of steel, and 0.05 ton of coal. An output of 1 ton of steel requires an input of 0.01 ton of agricultural products, 0.13 ton of steel, and 0.18 ton of coal. An output of 1 ton of coal requires an input of 0.01 ton of agricultural products, 0.20 ton of steel, and 0.05 ton of coal. Write the technological matrix for this economy. Find the necessary gross productions to provide surpluses of 2350 tons of agricultural products, 4552 tons of steel, and 911 tons of coal.

Solution

The technology matrix for the economy is

$$\begin{array}{c} \begin{array}{ccc} \text{Ag} & \text{Steel} & \text{Coal} \end{array} \\ \begin{bmatrix} 0.1 & 0.01 & 0.01 \\ 0.02 & 0.13 & 0.20 \\ 0.05 & 0.18 & 0.05 \end{bmatrix} \begin{array}{l} \text{Ag} \\ \text{Steel} \\ \text{Coal} \end{array} \end{array}$$

The technological equation for this economy is $(I - A)X = D$, or

$$\left(\begin{bmatrix} 1 & 0 & 0 \\ 0 & 1 & 0 \\ 0 & 0 & 1 \end{bmatrix} - \begin{bmatrix} 0.1 & 0.01 & 0.01 \\ 0.02 & 0.13 & 0.20 \\ 0.05 & 0.18 & 0.05 \end{bmatrix} \right) X = \begin{bmatrix} 2350 \\ 4552 \\ 911 \end{bmatrix}$$

Thus the solution is

$$X = \left(\begin{bmatrix} 1 & 0 & 0 \\ 0 & 1 & 0 \\ 0 & 0 & 1 \end{bmatrix} - \begin{bmatrix} 0.1 & 0.01 & 0.01 \\ 0.02 & 0.13 & 0.20 \\ 0.05 & 0.18 & 0.05 \end{bmatrix} \right)^{-1} \begin{bmatrix} 2350 \\ 4552 \\ 911 \end{bmatrix}$$

Thus the necessary gross productions of the industries are

$$X = \begin{bmatrix} 2700 \\ 5800 \\ 2200 \end{bmatrix} \begin{array}{l} \text{Ag} \\ \text{Steel} \\ \text{Coal} \end{array}$$

Note that we can use technology to find the gross productions in the previous example.

```
[A]  [[.1   .01   .01]
      [.02  .13   .2 ]
      [.05  .18   .05]]
[D]
           [[2350]
            [4552]
            [911 ]]
```

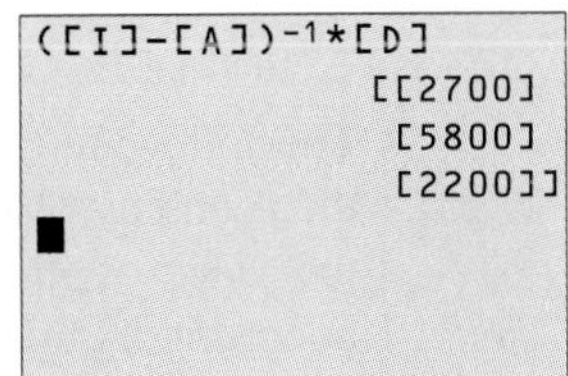

CHECKPOINT

3. Suppose a primitive economy has a wood products industry and a minerals industry, with technology matrix A (open Leontief model).

$$A = \begin{bmatrix} 0.12 & 0.11 \\ 0.09 & 0.15 \end{bmatrix} \begin{matrix} \text{Wood products} \\ \text{Minerals} \end{matrix}$$

(columns: W, M)

If surpluses of 1738 units of wood products and 1332 of minerals are desired, find the gross production of each industry.

A few comments are in order. First, the surplus matrix must contain positive or zero entries. If this is not the case, then the gross production of one or more industries is insufficient to supply the needs of the others. That is, the economy cannot use more of a certain commodity than it produces.

Second, the model we have considered is referred to as an **open Leontief model** because not all inputs and outputs were incorporated within the technology matrix. In particular, because labor was omitted, some (or all) of the surpluses must be used to support this labor. Problems that fit this open Leontief model have unique solutions, so the matrix $(I - A)^{-1}$ can be used to solve the technological equation.

If a model is developed in which all inputs and outputs are used within the system, then such a model is called a **closed Leontief model.** In such a model, labor (and perhaps other factors) must be included. Labor is included by considering a new industry, households, which produces labor. When such a closed model is developed, all outputs are used within the system, and the sum of the entries in each column equals 1. In this case, there is no surplus, so $D = 0$, and we have the following:

Closed Leontief Model

The technological equation for a closed Leontief model is

$$(I - A)X = 0$$

where 0 is a zero column matrix.

For closed models, the technological equation does *not* have a unique solution, so the matrix $(I - A)^{-1}$ does *not* exist and hence cannot be used to find the solution.

EXAMPLE 5

The following closed Leontief model with technology matrix A might describe the economy of the entire country, with x_1 equal to the government's budget, x_2 the value of industrial output (profit-making organizations), x_3 the budget of nonprofit organizations, and x_4 the household's budget.

$$A = \begin{bmatrix} 0.4 & 0.2 & 0.1 & 0.3 \\ 0.2 & 0.2 & 0.2 & 0.1 \\ 0.2 & 0 & 0.2 & 0.1 \\ 0.2 & 0.6 & 0.5 & 0.5 \end{bmatrix} \begin{matrix} \text{Government} \\ \text{Industry} \\ \text{Nonprofit} \\ \text{Household} \end{matrix}$$

(columns G, I, N, H)

Find the total budgets (or outputs) x_1, x_2, x_3, and x_4.

Solution

This is a closed Leontief model; therefore, we solve the technological equation $(I - A)X = 0$.

$$\left(\begin{bmatrix} 1 & 0 & 0 & 0 \\ 0 & 1 & 0 & 0 \\ 0 & 0 & 1 & 0 \\ 0 & 0 & 0 & 1 \end{bmatrix} - \begin{bmatrix} 0.4 & 0.2 & 0.1 & 0.3 \\ 0.2 & 0.2 & 0.2 & 0.1 \\ 0.2 & 0 & 0.2 & 0.1 \\ 0.2 & 0.6 & 0.5 & 0.5 \end{bmatrix}\right) X = \begin{bmatrix} 0 \\ 0 \\ 0 \\ 0 \end{bmatrix}$$

or

$$\begin{bmatrix} 0.6 & -0.2 & -0.1 & -0.3 \\ -0.2 & 0.8 & -0.2 & -0.1 \\ -0.2 & 0 & 0.8 & -0.1 \\ -0.2 & -0.6 & -0.5 & 0.5 \end{bmatrix} \begin{bmatrix} x_1 \\ x_2 \\ x_3 \\ x_4 \end{bmatrix} = \begin{bmatrix} 0 \\ 0 \\ 0 \\ 0 \end{bmatrix}.$$

The augmented matrix for this system is

$$\left[\begin{array}{cccc|c} 0.6 & -0.2 & -0.1 & -0.3 & 0 \\ -0.2 & 0.8 & -0.2 & -0.1 & 0 \\ -0.2 & 0 & 0.8 & -0.1 & 0 \\ -0.2 & -0.6 & -0.5 & 0.5 & 0 \end{array}\right]$$

If we multiply each entry by -5 and rearrange the rows by moving row 3 to row 1, row 1 to row 2, and row 2 to row 3, we have the equivalent matrix

$$\left[\begin{array}{cccc|c} 1 & 0 & -4 & 0.5 & 0 \\ -3 & 1 & 0.5 & 1.5 & 0 \\ 1 & -4 & 1 & 0.5 & 0 \\ 1 & 3 & 2.5 & -2.5 & 0 \end{array}\right]$$

Then, if we reduce this matrix by the methods of Section 3.3, "Gauss-Jordan Elimination: Solving Systems of Equations," or with a graphing utility, we obtain

$$\left[\begin{array}{cccc|c} 1 & 0 & 0 & -\frac{55}{82} & 0 \\ 0 & 1 & 0 & -\frac{15}{41} & 0 \\ 0 & 0 & 1 & -\frac{12}{41} & 0 \\ 0 & 0 & 0 & 0 & 0 \end{array}\right]$$

The system of equations that corresponds to this augmented matrix is

$$\begin{aligned} x_1 - \tfrac{55}{82}x_4 &= 0 \\ x_2 - \tfrac{15}{41}x_4 &= 0 \\ x_3 - \tfrac{12}{41}x_4 &= 0 \end{aligned} \qquad \text{or} \qquad \begin{aligned} x_1 &= \tfrac{55}{82}x_4 \\ x_2 &= \tfrac{15}{41}x_4 \\ x_3 &= \tfrac{12}{41}x_4 \end{aligned}$$

Note that the economy satisfies the given equation if the government's budget is $\frac{55}{82}$ times the household's budget, if the value of industrial output is $\frac{15}{41}$ times the household's budget, and if the budget of nonprofit organizations is $\frac{12}{41}$ times the household's budget. The dependency here is expected. The fact that the system is closed suggests the dependency, but even more obvious is the fact that industrial output is limited by labor supply.

Technology Note As noted previously, for a closed Leontief model, the inverse of $I - A$ does not exist. Thus, in Example 5, we must reduce $I - A$ by hand or with a calculator or computer (see Figure 3.11). Note that the existence or nonexistence of $(I - A)^{-1}$ dictates how the technology is used.

```
rref([I]-[A]▸Fr
ac
[[1   0   0   -55/82]
 [0   1   0   -15/41]
 [0   0   1   -12/41]
 [0   0   0   0      ]]
```

Figure 3.11

CHECKPOINT

4. For a closed Leontief model:
   (a) What is the sum of the column entries of the technology matrix?
   (b) State the technology equation.

We have examined two types of input-output models as they pertain to the economy as a whole. Since Leontief's original work with the economy, various other applications of input-output models have been developed. Consider the following parts-listing problem.

EXAMPLE 6

A storage shed consists of 4 walls and a roof. The walls and roof are made from stamped aluminum sheeting and are reinforced with 4 braces, each of which is held with 6 bolts. The final assembly joins the walls to each other and to the roof, using a total of 20 more bolts. The parts listing for these sheds can be described by the following matrix.

$$Q = \begin{array}{c} \begin{array}{cccccc} \text{S} & \text{W} & \text{R} & \text{Sht} & \text{Br} & \text{Bo} \end{array} \\ \begin{bmatrix} 0 & 0 & 0 & 0 & 0 & 0 \\ 4 & 0 & 0 & 0 & 0 & 0 \\ 1 & 0 & 0 & 0 & 0 & 0 \\ 0 & 1 & 1 & 0 & 0 & 0 \\ 0 & 4 & 4 & 0 & 0 & 0 \\ 20 & 24 & 24 & 0 & 0 & 0 \end{bmatrix} \end{array} \begin{array}{l} \text{Sheds} \\ \text{Walls} \\ \text{Roofs} \\ \text{Sheets} \\ \text{Braces} \\ \text{Bolts} \end{array}$$

This matrix is a form of input-output matrix, with the rows representing inputs that are used directly to produce the items that head the columns. We note that the entries here are quantities used rather than proportions used.

Thus to produce a shed requires 4 walls, 1 roof, and 20 bolts, and to produce a roof requires 1 aluminum sheet, 4 braces, and 24 bolts.

Suppose that an order is received for 4 completed sheds and spare parts including 2 walls, 1 roof, 8 braces, and 24 bolts. How many of each of the assembly items are required to fill the order?

Solution

If matrix D represents the order and matrix X represents the gross production, we have

$$D = \begin{bmatrix} 4 \\ 2 \\ 1 \\ 0 \\ 8 \\ 24 \end{bmatrix} \begin{matrix} \text{Sheds} \\ \text{Walls} \\ \text{Roofs} \\ \text{Sheets} \\ \text{Braces} \\ \text{Bolts} \end{matrix} \qquad X = \begin{bmatrix} x_1 \\ x_2 \\ x_3 \\ x_4 \\ x_5 \\ x_6 \end{bmatrix} = \begin{bmatrix} \text{total sheds required} \\ \text{total walls required} \\ \text{total roofs required} \\ \text{total sheets required} \\ \text{total braces required} \\ \text{total bolts required} \end{bmatrix}$$

Then X must satisfy $X - QX = D$, which is the technological equation for an open Leontief model.

We can find X by solving the system $(I - Q)X = D$, which is represented by the augmented matrix

$$\left[\begin{array}{cccccc|c} 1 & 0 & 0 & 0 & 0 & 0 & 4 \\ -4 & 1 & 0 & 0 & 0 & 0 & 2 \\ -1 & 0 & 1 & 0 & 0 & 0 & 1 \\ 0 & -1 & -1 & 1 & 0 & 0 & 0 \\ 0 & -4 & -4 & 0 & 1 & 0 & 8 \\ -20 & -24 & -24 & 0 & 0 & 1 & 24 \end{array}\right]$$

Although this matrix is quite large, it is easily reduced and yields

$$\left[\begin{array}{cccccc|c} 1 & 0 & 0 & 0 & 0 & 0 & 4 \\ 0 & 1 & 0 & 0 & 0 & 0 & 18 \\ 0 & 0 & 1 & 0 & 0 & 0 & 5 \\ 0 & 0 & 0 & 1 & 0 & 0 & 23 \\ 0 & 0 & 0 & 0 & 1 & 0 & 100 \\ 0 & 0 & 0 & 0 & 0 & 1 & 656 \end{array}\right]$$

This matrix tells us that the order will have 4 complete sheds. The 18 walls from the matrix include the 16 from the complete sheds, and the 5 roofs include the 4 from the complete sheds. The 23 sheets, 100 braces, and 656 bolts are the total number of *primary assembly items* (bolts, braces, and sheets) required to fill the order.

The inverse of $I - Q$ exists, so it could be used directly or with technology to find the solution to the parts-listing problem of Example 6.

CHECKPOINT SOLUTIONS

1. Manufacturing
2. Under the "Util" column and across from the "Fuels" row is 0.20, indicating that producing 1 unit of utilities requires 0.20 unit of fuels. Thus producing 100 units of utilities requires 20 units of fuels.
3. Let x = the gross production of wood products
   y = the gross production of minerals
   We must solve $(I - A)X = D$.

$$\left(\begin{bmatrix} 1 & 0 \\ 0 & 1 \end{bmatrix} - \begin{bmatrix} 0.12 & 0.11 \\ 0.09 & 0.15 \end{bmatrix}\right)\begin{bmatrix} x \\ y \end{bmatrix} = \begin{bmatrix} 1738 \\ 1332 \end{bmatrix}$$

$$\begin{bmatrix} 0.88 & -0.11 \\ -0.09 & 0.85 \end{bmatrix}\begin{bmatrix} x \\ y \end{bmatrix} = \begin{bmatrix} 1738 \\ 1332 \end{bmatrix}$$

$$\begin{bmatrix} 0.88 & -0.11 \\ -0.09 & 0.85 \end{bmatrix}^{-1} = \frac{1}{(0.88)(0.85) - (0.09)(0.11)}\begin{bmatrix} 0.85 & 0.11 \\ 0.09 & 0.88 \end{bmatrix}$$

   Hence

$$\begin{bmatrix} x \\ y \end{bmatrix} = \frac{1}{0.7381}\begin{bmatrix} 0.85 & 0.11 \\ 0.09 & 0.88 \end{bmatrix}\begin{bmatrix} 1738 \\ 1332 \end{bmatrix} = \begin{bmatrix} 2200 \\ 1800 \end{bmatrix}$$

4. (a) The sum of the column entries is 1.
   (b) $(I - A)X = 0$.

EXERCISE 3.5

1. The following technology matrix for a simple economy describes the relationship of certain industries to each other in the production of 1 unit of product.

A	M	F	U	
0.36	0.03	0.10	0.04	Agriculture
0.06	0.42	0.25	0.33	Manufacturing
0.18	0.15	0.10	0.41	Fuels
0.10	0.20	0.31	0.15	Utilities

   (a) For each 100 units of manufactured products produced, how many units of fuels are required?
   (b) How many units of utilities are required to produce 40 units of agriculture products?
2. For the economy of Problem 1, what industry is most dependent on utilities?

The following technology matrix describes the relationship of certain industries within the economy to each other. (A&F, Agriculture and food; RM, Raw materials; M, Manufacturing; F, Fuels industry; U, Utilities; SI, Service industries)

A&F	RM	M	F	U	SI	
0.410	0.008	0	0.002	0	0.006	A&F
0.025	0.493	0.190	0.024	0.030	0.150	RM
0.015	0.006	0.082	0.009	0.001	0.116	M
0.097	0.096	0.040	0.053	0.008	0.093	F
0.028	0.129	0.039	0.058	0.138	0.409	U
0.043	0.008	0.010	0.012	0.002	0.095	SI

Use this matrix in Problems 3–10.

3. For each 1000 units of raw materials produced, how many units of agricultural and food products were required?
4. For each 1000 units of raw materials produced, how many units of service were required?
5. How many units of fuels were required to produce 1000 units of manufactured goods?
6. How many units of fuels were required to produce 1000 units of power (utilities' goods)?
7. Which industry is most dependent on its own goods for its operations? Which industry is least dependent on its own goods?

8. Which industry is most dependent on the fuels industry?
9. Which three industries would be most affected by a rise in the cost of raw materials?
10. Which industry would be most affected by a rise in the cost of manufactured goods?
11. Given is the partially solved matrix for the gross production of each industry of an open Leontief model. Find the gross production for each industry.

$$\begin{array}{c} \begin{matrix} \text{U} & \text{A} & \text{M} & \end{matrix} \\ \left[\begin{array}{rrr|r} 1 & -2 & -1 & 80 \\ 0 & 8 & -4 & 60 \\ 0 & 0 & 10 & 50 \end{array}\right] \begin{matrix} \text{Utilities} \\ \text{Agriculture} \\ \text{Manufacturing} \end{matrix} \end{array}$$

12. Given is the partially solved matrix for the gross production of each industry of a closed Leontief model. Find the gross production for each industry.

$$\begin{array}{c} \begin{matrix} \text{U} & \text{A} & \text{H} & \end{matrix} \\ \left[\begin{array}{rrr|r} 3 & -10 & -7 & 0 \\ 0 & 20 & -10 & 0 \\ 0 & 0 & 0 & 0 \end{array}\right] \begin{matrix} \text{Utilities} \\ \text{Agriculture} \\ \text{Households} \end{matrix} \end{array}$$

13. Suppose a primitive economy consists of two industries, farm products and farm machinery. Suppose also that its technology matrix is

$$A = \begin{matrix} & \text{P} & \text{M} & \\ & \begin{bmatrix} 0.5 & 0.1 \\ 0.1 & 0.3 \end{bmatrix} & & \begin{matrix} \text{Products} \\ \text{Machinery} \end{matrix} \end{matrix}$$

If surpluses of 96 units of farm products and 8 units of farm machinery are desired, find the gross production of each industry.

14. Suppose an economy has two industries, agriculture and minerals. Suppose further that the technology matrix for this economy is A.

$$A = \begin{matrix} & \text{A} & \text{M} & \\ & \begin{bmatrix} 0.3 & 0.1 \\ 0.1 & 0.2 \end{bmatrix} & & \begin{matrix} \text{Agriculture} \\ \text{Minerals} \end{matrix} \end{matrix}$$

If a surplus of 60 agricultural units and 70 mineral units is desired, find the gross production for each industry.

15. Suppose the economy of an underdeveloped country has an agricultural industry and an oil industry, with technology matrix A.

$$A = \begin{matrix} & \text{A} & \text{O} & \\ & \begin{bmatrix} 0.3 & 0.1 \\ 0.2 & 0.1 \end{bmatrix} & & \begin{matrix} \text{Agriculture products} \\ \text{Oil products} \end{matrix} \end{matrix}$$

If a surplus of 0 units of agricultural products and 610 units of oil products are desired, find the gross production of each industry.

16. Suppose a simple economy with only an agricultural industry and a steel industry has the following technology matrix.

$$A = \begin{matrix} & \text{A} & \text{S} & \\ & \begin{bmatrix} 0.5 & 0.2 \\ 0.1 & 0.6 \end{bmatrix} & & \begin{matrix} \text{Agriculture} \\ \text{Steel} \end{matrix} \end{matrix}$$

If surpluses of 15 units of agricultural products and 33 units of steel are desired, find the gross production of each industry.

17. An economy is based on two industries: utilities and manufacturing. The technology matrix for these industries is A.

$$A = \begin{matrix} & \text{U} & \text{M} & \\ & \begin{bmatrix} 0.3 & 0.15 \\ 0.3 & 0.4 \end{bmatrix} & & \begin{matrix} \text{Utility} \\ \text{Manufacturing} \end{matrix} \end{matrix}$$

If surpluses of 80 units of utility output and 180 units of manufacturing are desired, find the gross production of each industry.

18. A primitive economy has a mining industry and a fishing industry, with technology matrix A.

$$A = \begin{matrix} & \text{M} & \text{F} & \\ & \begin{bmatrix} 0.25 & 0.05 \\ 0.05 & 0.40 \end{bmatrix} & & \begin{matrix} \text{Mining} \\ \text{Fishing} \end{matrix} \end{matrix}$$

If surpluses of 147 units of mining output and 26 units of fishing output are desired, find the gross production of each industry.

19. One sector of an economy consists of a mining industry and a manufacturing industry, with technology matrix A.

$$A = \begin{matrix} & \text{M} & \text{Mfg} & \\ & \begin{bmatrix} 0.2 & 0.4 \\ 0.3 & 0.3 \end{bmatrix} & & \begin{matrix} \text{Mining} \\ \text{Manufacturing} \end{matrix} \end{matrix}$$

Surpluses of 36 units from mining and 278 units from manufacturing are desired. Find the gross production of each industry.

20. An underdeveloped country has an agricultural industry and a manufacturing industry, with technology matrix A.

$$A = \begin{matrix} & \text{A} & \text{M} & \\ & \begin{bmatrix} 0.20 & 0.10 \\ 0.25 & 0.45 \end{bmatrix} & & \begin{matrix} \text{Agriculture} \\ \text{Manufacturing} \end{matrix} \end{matrix}$$

Surpluses of 8 units in agriculture and 620 units in manufacturing are desired. Find the gross production of each industry.

21. A simple economy has an electronics components industry and a computers industry, with technology matrix A.

$$A = \begin{bmatrix} 0.3 & 0.6 \\ 0.2 & 0.2 \end{bmatrix} \begin{matrix} \text{Electronics components} \\ \text{Computers} \end{matrix} \quad (\text{columns: EC, C})$$

Surpluses of 648 units of electronics components and 16 computers are desired. Find the gross production of each industry.

22. An economy has an agricultural industry and a textile industry, with technology matrix A.

$$A = \begin{bmatrix} 0.4 & 0.1 \\ 0.1 & 0.2 \end{bmatrix} \begin{matrix} \text{Agriculture} \\ \text{Textiles} \end{matrix} \quad (\text{columns: A, T})$$

Surpluses of 5 units in agriculture and 195 units of textiles are desired. Find the gross production of each industry.

23. A primitive economy consists of a fishing industry and an oil industry, with technology matrix A.

$$A = \begin{bmatrix} 0.30 & 0.04 \\ 0.35 & 0.10 \end{bmatrix} \begin{matrix} \text{Fishing} \\ \text{Oil} \end{matrix} \quad (\text{columns: F, O})$$

Surpluses of 20 units from fishing and 1090 units from oil are desired. Find the gross production of each industry.

24. An economy has a manufacturing industry and a banking industry, with technology matrix A.

$$A = \begin{bmatrix} 0.5 & 0.3 \\ 0.2 & 0.3 \end{bmatrix} \begin{matrix} \text{Manufacturing} \\ \text{Banking} \end{matrix} \quad (\text{columns: M, B})$$

If surpluses of 141 units from manufacturing and 106 units from banking are desired, find the gross production of each industry.

Interdepartment costs Within a company there is a (micro) economy that is monitored by the accounting procedures. In terms of the accounts, the various departments "produce" costs, some of which are internal and some of which are direct costs. Problems 25–28 show how an open Leontief model can be used to determine departmental costs.

25. Suppose the development department of a firm charges 10% of its total monthly costs to the promotional department, and the promotional department charges 5% of its total monthly costs to the development department. The direct costs of the development department are \$20,400, and the direct costs of the promotional department are \$9,900. The solution to the matrix equation

$$\begin{bmatrix} 20{,}400 \\ 9{,}900 \end{bmatrix} + \begin{bmatrix} 0 & 0.05 \\ 0.1 & 0 \end{bmatrix} \begin{bmatrix} x \\ y \end{bmatrix} = \begin{bmatrix} x \\ y \end{bmatrix}$$

for the column matrix $\begin{bmatrix} x \\ y \end{bmatrix}$ gives the total costs for each department. Note that this equation can be written in the form

$$D + AX = X, \qquad \text{or} \qquad X - AX = D$$

which is the form for an open economy. Find the total costs for each department in this (micro) economy.

26. Suppose the shipping department of a firm charges 20% of its total monthly costs to the printing department and that the printing department charges 10% of its total monthly costs to the shipping department. If the direct costs of the shipping department are \$16,500 and the direct costs of the printing department are \$11,400, find the total costs for each department.

27. Suppose the engineering department of a firm charges 20% of its total monthly costs to the computer department, and the computer department charges 25% of its total monthly costs to the engineering department. If during a given month the direct costs were \$11,750 for the engineering department and \$10,000 for the computer department, what are the total costs of each department?

28. The sales department of an auto dealership charges 10% of its total monthly costs to the service department, and the service department charges 20% of its total monthly costs to the sales department. During a given month, the direct costs were \$88,200 for sales and \$49,000 for service. Find the total costs of each department.

29. Suppose an economy has the same technological matrix as that used in Example 2. If surpluses of 110 units of agricultural goods and 50 units each of manufactured goods and fuels are desired, find the gross production of each industry required to satisfy this.

30. Suppose that an economy has the same technological matrix as that used in Example 2. If surpluses of 180 units of agricultural goods, 90 units of manufactured goods, and 40 units of fuels are desired, find the gross production of each industry required to satisfy this.

31. Suppose that the economy of a small nation has an electronics industry, a steel industry, and an auto industry, with the following technology matrix.

$$A = \begin{array}{c} \begin{matrix} E & S & A \end{matrix} \\ \begin{bmatrix} 0.6 & 0.2 & 0.2 \\ 0.1 & 0.4 & 0.5 \\ 0.1 & 0.2 & 0.2 \end{bmatrix} \end{array} \begin{matrix} \text{Electronics} \\ \text{Steel} \\ \text{Autos} \end{matrix}$$

If the nation wishes to have surpluses of 100 units of electronics production, 272 units of steel production, and 200 automobiles, find the gross production of each industry.

32. Suppose an economy has the same technology matrix as that in Problem 31. If surpluses of 540 units of electronics, 30 units of steel, and 140 autos are desired, find the gross production for each industry.

33. Suppose the technology matrix for a closed model of a simple economy is given by

$$A = \begin{array}{c} \begin{matrix} P & M & H \end{matrix} \\ \begin{bmatrix} 0.5 & 0.1 & 0.2 \\ 0.1 & 0.3 & 0 \\ 0.4 & 0.6 & 0.8 \end{bmatrix} \end{array} \begin{matrix} \text{Products} \\ \text{Machinery} \\ \text{Households} \end{matrix}$$

Find the gross productions for the industries.

34. Suppose the technology matrix for a closed model of a simple economy is given by matrix A. Find the gross productions for the industries.

$$A = \begin{array}{c} \begin{matrix} G & I & H \end{matrix} \\ \begin{bmatrix} 0.4 & 0.1 & 0.3 \\ 0.4 & 0.3 & 0.2 \\ 0.2 & 0.6 & 0.5 \end{bmatrix} \end{array} \begin{matrix} \text{Government} \\ \text{Industry} \\ \text{Households} \end{matrix}$$

35. Suppose the technology matrix for a closed model of a simple economy is given by matrix A. Find the gross productions for the industries.

$$A = \begin{array}{c} \begin{matrix} G & I & H \end{matrix} \\ \begin{bmatrix} 0.4 & 0.2 & 0.2 \\ 0.2 & 0.3 & 0.3 \\ 0.4 & 0.5 & 0.5 \end{bmatrix} \end{array} \begin{matrix} \text{Government} \\ \text{Industry} \\ \text{Households} \end{matrix}$$

36. Suppose the technology matrix for a closed model of an economy is given by matrix A. Find the gross productions for the industries.

$$A = \begin{array}{c} \begin{matrix} S & M & H \end{matrix} \\ \begin{bmatrix} 0.2 & 0.1 & 0.1 \\ 0.6 & 0.5 & 0.1 \\ 0.2 & 0.4 & 0.8 \end{bmatrix} \end{array} \begin{matrix} \text{Shipping} \\ \text{Manufacturing} \\ \text{Households} \end{matrix}$$

37. Suppose the technology matrix for a closed model of an economy is given by matrix A. Find the gross productions for the industries.

$$A = \begin{array}{c} \begin{matrix} M & U & H \end{matrix} \\ \begin{bmatrix} 0.5 & 0.4 & 0.3 \\ 0.4 & 0.5 & 0.3 \\ 0.1 & 0.1 & 0.4 \end{bmatrix} \end{array} \begin{matrix} \text{Manufacturing} \\ \text{Utilities} \\ \text{Households} \end{matrix}$$

38. Suppose the technology matrix for a closed model of an economy is given by matrix A. Find the gross productions for the industries.

$$A = \begin{array}{c} \begin{matrix} G & P & N & H \end{matrix} \\ \begin{bmatrix} 0.3 & 0.2 & 0.1 & 0.05 \\ 0.2 & 0.3 & 0.1 & 0.1 \\ 0.2 & 0.1 & 0.2 & 0.1 \\ 0.3 & 0.4 & 0.6 & 0.75 \end{bmatrix} \end{array} \begin{matrix} \text{Government} \\ \text{Profit sector} \\ \text{Nonprofit} \\ \text{Households} \end{matrix}$$

39. For the storage shed in Example 6, find the number of each primary assembly item required to fill an order for 24 sheds, 12 braces, and 96 bolts.

40. Card tables are made by joining 4 legs and a top using 4 bolts. The legs are each made from a steel rod. The top has a frame made from 4 steel rods. A cover and four special clamps that brace the top and hold the legs are joined to the frame using a total of 8 bolts. The parts-listing matrix for the card table assembly is given by

$$A = \begin{array}{c} \begin{matrix} CT & L & T & R & Co & Cl & B \end{matrix} \\ \begin{bmatrix} 0 & 0 & 0 & 0 & 0 & 0 & 0 \\ 4 & 0 & 0 & 0 & 0 & 0 & 0 \\ 1 & 0 & 0 & 0 & 0 & 0 & 0 \\ 0 & 1 & 4 & 0 & 0 & 0 & 0 \\ 0 & 0 & 1 & 0 & 0 & 0 & 0 \\ 0 & 0 & 4 & 0 & 0 & 0 & 0 \\ 4 & 0 & 8 & 0 & 0 & 0 & 0 \end{bmatrix} \end{array} \begin{matrix} \text{Card table} \\ \text{Legs} \\ \text{Top} \\ \text{Rods} \\ \text{Cover} \\ \text{Clamps} \\ \text{Bolts} \end{matrix}$$

If an order is received for 10 card tables, 4 legs, 1 top, 1 cover, 6 clamps, and 12 bolts, how many of each primary assembly item are required to fill the order?

41. A sawhorse is made from 2 pairs of legs and a top joined with 4 nails. The top is a 2 × 4-in. board 3 ft long. Each pair of legs is made from two 2 × 4-in. boards 3 ft long, a brace, and a special clamp, with the brace and the clamp joined using 8 nails. The parts-listing matrix for the sawhorse is given by

$$A = \begin{array}{c} \begin{matrix} \text{SH} & \text{T} & \text{LP} & 2\times4 & \text{B} & \text{C} & \text{N} \end{matrix} \\ \begin{bmatrix} 0 & 0 & 0 & 0 & 0 & 0 & 0 \\ 1 & 0 & 0 & 0 & 0 & 0 & 0 \\ 2 & 0 & 0 & 0 & 0 & 0 & 0 \\ 0 & 1 & 2 & 0 & 0 & 0 & 0 \\ 0 & 0 & 1 & 0 & 0 & 0 & 0 \\ 0 & 0 & 1 & 0 & 0 & 0 & 0 \\ 4 & 0 & 8 & 0 & 0 & 0 & 0 \end{bmatrix} \end{array} \begin{array}{l} \text{Sawhorse} \\ \text{Top} \\ \text{Leg pair} \\ 2 \times 4\text{s} \\ \text{Brace} \\ \text{Clamp} \\ \text{Nails} \end{array}$$

How many of each primary assembly item are required to fill an order for 10 sawhorses, 6 extra 2 × 4s, 6 extra clamps, and 100 nails?

42. A log carrier has a body made from a 4-ft length of reinforced material having a patch on each side and a dowel slid through each end to act as handles. The parts-listing matrix for the log carrier is given by

$$A = \begin{array}{c} \begin{matrix} \text{LC} & \text{B} & \text{H} & \text{C} & \text{P} \end{matrix} \\ \begin{bmatrix} 0 & 0 & 0 & 0 & 0 \\ 1 & 0 & 0 & 0 & 0 \\ 2 & 0 & 0 & 0 & 0 \\ 0 & 1 & 0 & 0 & 0 \\ 0 & 2 & 0 & 0 & 0 \end{bmatrix} \end{array} \begin{array}{l} \text{Log carrier} \\ \text{Body} \\ \text{Handles} \\ \text{Cloth} \\ \text{Patch} \end{array}$$

How many of each primary assembly item are required to fill an order for 500 log carriers and 20 handles?

KEY TERMS AND FORMULAS

Section	Key Terms	Formula

3.1	Matrices
	  Entries
	  Order
	  Square, row, column, and zero matrices	

	Sum of two matrices	
	Negative of a matrix	
3.2	Scalar multiplication	
	Matrix product	
	Identity matrix	2×2 is $\begin{bmatrix} 1 & 0 \\ 0 & 1 \end{bmatrix}$
3.3	Augmented matrix	
	Coefficient matrix	
	Elementary row operations	
	Solving systems of equations Gauss-Jordan elimination method	
	Nonunique solutions Reduced form	
3.4	Inverse matrices	
	Matrix equations	$AX = B$
	Determinants	$\begin{vmatrix} a & b \\ c & d \end{vmatrix} = ad - bc$
3.5	Technology matrix	
	Leontief input-output model	
	Gross production matrix Final demands (surpluses)	
	Leontief model	
	Technology equation	
	Open	$(I - A)X = D$
	Closed	$(I - A)X = 0$

REVIEW EXERCISES

Use the matrices

$$A = \begin{bmatrix} 4 & 4 & 2 & -5 \\ 6 & 3 & -1 & 0 \\ 0 & 0 & -3 & 5 \end{bmatrix}, B = \begin{bmatrix} 2 & -5 & -11 & 8 \\ 4 & 0 & 0 & 4 \\ -2 & -2 & 1 & 9 \end{bmatrix},$$

$$C = \begin{bmatrix} 4 & -2 \\ 5 & 0 \\ 6 & 0 \\ 1 & 3 \end{bmatrix}, D = \begin{bmatrix} 3 & 5 \\ 1 & 2 \end{bmatrix}, E = \begin{bmatrix} 1 & 1 \\ 1 & 1 \\ 4 & 6 \\ 0 & 5 \end{bmatrix},$$

$$F = \begin{bmatrix} -1 & 6 \\ 4 & 11 \end{bmatrix}, G = \begin{bmatrix} 2 & -5 \\ -1 & 3 \end{bmatrix}, \text{ and } I = \begin{bmatrix} 1 & 0 \\ 0 & 1 \end{bmatrix}$$

as needed to complete Problems 1–25.

Section 3.1

1. Find a_{12} in matrix A.
2. Find b_{23} in matrix B.
3. Which of the matrices, if any, are 3×4?
4. Which of the matrices, if any, are 2×4?
5. Which matrices are square?
6. Write the negative of matrix B.
7. If a matrix is added to its negative, what kind of matrix results?
8. Two matrices can be added if they have the same ________.

Again using the matrices above, perform the indicated operations in Problems 9–25.

9. $A + B$
10. $C - E$
11. $D^T - I$

Section 3.2

12. $3C$
13. $4I$
14. $-2F$
15. $4D - 3I$
16. $F + 2D$
17. $3A - 5B$
18. AC
19. CD
20. DF
21. FD
22. FI
23. IF
24. DG^T
25. $(DG)F$

Section 3.3

26. Given the following augmented matrix representing a system of linear equations, find the solution.

$$\left[\begin{array}{ccc|c} 1 & 1 & 2 & 5 \\ 4 & 0 & 1 & 5 \\ 2 & 1 & 1 & 5 \end{array}\right]$$

In Problems 27–32, solve each system using matrices.

27. $\begin{cases} x - 2y = 4 \\ -3x + 10y = 24 \end{cases}$

28. $\begin{cases} x + y + z = 4 \\ 3x + 4y - z = -1 \\ 2x - y + 3z = 3 \end{cases}$

29. $\begin{cases} -x + y + z = 3 \\ 3x - z = 1 \\ 2x - 3y - 4z = -2 \end{cases}$

30. $\begin{cases} x + y - 2z = 5 \\ 3x + 2y + 5z = 10 \\ -2x - 3y + 15z = 2 \end{cases}$

31. $\begin{cases} x - y = 3 \\ x + y + 4z = 1 \\ 2x - 3y - 2z = 7 \end{cases}$

32. $\begin{cases} x_1 + x_2 + x_3 + x_4 = 3 \\ x_1 - 2x_2 + x_3 - 4x_4 = -5 \\ x_1 - x_3 + x_4 = 0 \\ x_2 + x_3 + x_4 = 2 \end{cases}$

Section 3.4

33. Are D and G inverse matrices if

$$D = \begin{bmatrix} 3 & 5 \\ 1 & 2 \end{bmatrix} \quad \text{and} \quad G = \begin{bmatrix} 2 & -5 \\ -1 & 3 \end{bmatrix}?$$

In Problems 34–36, find the inverse of each matrix.

34. $\begin{bmatrix} 7 & -1 \\ -10 & 2 \end{bmatrix}$

35. $\begin{bmatrix} 1 & 0 & 2 \\ 3 & 4 & -1 \\ 1 & 1 & 0 \end{bmatrix}$

36. $\begin{bmatrix} 3 & 3 & 2 \\ -1 & 4 & 2 \\ 2 & 5 & 3 \end{bmatrix}$

In Problems 37–39, solve each system of equations by using inverse matrices.

37. $\begin{cases} x + 2z = 5 \\ 3x + 4y - z = 2 \\ x + y = -3 \end{cases}$ (See Problem 35.)

38. $\begin{cases} 3x + 3y + 2z = 1 \\ -x + 4y + 2z = -10 \\ 2x + 5y + 3z = -6 \end{cases}$ (See Problem 36.)

39. $\begin{cases} x + 3y + z = 0 \\ x + 4y + 3z = 2 \\ 2x - y - 11z = -12 \end{cases}$

40. Does the following matrix have an inverse?

$$\begin{bmatrix} 1 & 2 & 4 \\ 2 & 0 & -4 \\ 1 & 0 & -2 \end{bmatrix}$$

Applications

The Burr Cabinet Company manufactures bookcases and filing cabinets at two plants, A and B. Matrix M gives the production for the two plants during June, and matrix N gives the production for July. Use them in Problems 41–43.

$$M = \begin{matrix} & \text{A} & \text{B} & \\ & \left[\begin{matrix} 150 \\ 280 \end{matrix}\right. & \left.\begin{matrix} 80 \\ 300 \end{matrix}\right] & \begin{matrix} \text{Bookcases} \\ \text{Files} \end{matrix} \end{matrix} \qquad N = \begin{matrix} \text{A} \quad \text{B} \\ \begin{bmatrix} 100 & 60 \\ 200 & 400 \end{bmatrix} \end{matrix}$$

$$S = \begin{bmatrix} 120 & 80 \\ 180 & 300 \end{bmatrix} \qquad P = \begin{bmatrix} 1000 & 800 \\ 600 & 1200 \end{bmatrix}$$

Section 3.1

41. ***Production*** Write the matrix that represents total production at the two plants for the 2 months.

42. ***Production*** If matrix P represents the inventories at the plants at the beginning of June and matrix S represents shipments from the plants during June, write the matrix that represents the inventories at the end of June.

Section 3.2

43. ***Production*** If the company sells its bookcases to wholesalers for \$100 and its filing cabinets for \$120, for which month was the value of production higher at (a) plant A? (b) plant B?

A small church choir is made up of men and women who wear choir robes in the sizes shown in matrix A.

$$A = \begin{matrix} \text{Men} \quad \text{Women} & \\ \begin{bmatrix} 1 & 14 \\ 12 & 10 \\ 8 & 3 \end{bmatrix} & \begin{matrix} \text{Small} \\ \text{Medium} \\ \text{Large} \end{matrix} \end{matrix}$$

$$B = \begin{matrix} \text{S} \quad \text{M} \quad \text{L} & \\ \begin{bmatrix} 25 & 40 & 45 \\ 10 & 10 & 10 \end{bmatrix} & \begin{matrix} \text{Robes} \\ \text{Hoods} \end{matrix} \end{matrix}$$

Matrix B contains the prices (in dollars) of new robes and hoods according to size. Use these matrices in Problems 44 and 45.

44. ***Cost*** Find the product BA, and label the rows and columns to show what each entry represents.

45. ***Cost*** To find a matrix that gives the cost of new robes and the cost of new hoods, find

$$BA\begin{bmatrix} 1 \\ 1 \end{bmatrix}$$

46. ***Manufacturing*** Two departments of a firm, A and B, need differing amounts of the same products. The following table gives the amounts of the products needed by the departments.

	Steel	*Plastic*	*Wood*
Department A	30	20	10
Department B	20	10	20

These three products are supplied by two suppliers, Ace and Kink, with the unit prices given in the following table.

	Ace	*Kink*
Steel	300	280
Plastic	150	100
Wood	150	200

(a) Use matrix multiplication to find how much these two orders will cost at the two suppliers. The result should be a 2×2 matrix.

(b) From which supplier should each department make its purchase?

47. ***Investment*** Over the period of time 1984–1993, Chrysler's actual return (per dollar of stock) was 0.013469 per month, Ford's actual return was 0.013543 per month, and General Motors' actual return was 0.006504. Under certain assumptions, including that whatever underlying process generated the past set of observations will continue to hold in the present and near future, we can estimate the expected return for a stock portfolio containing these assets in some combination.

Suppose we have a portfolio that has 20% of our total wealth in Chrysler, 30% in Ford, and 50% in GM stock. Use the following steps to estimate the portfolio's historical return, which can be used to estimate the future return.

(a) Write a 1×3 matrix that defines the decimal part of our total wealth in Chrysler, in Ford, and in GM.

(b) Write a 3×1 matrix that gives the monthly returns from each of these companies.

(c) Find the historical return of the portfolio by computing the product of these two matrices.

(d) What is the expected monthly return of the stock portfolio?

Sections 3.3 and 3.4

48. ***Investment*** A woman has \$50,000 to invest. She has decided to invest all of it by purchasing some shares of stock in each of three companies: a fast-food chain that sells for \$50 per share and has an expected growth of 11.5% per year, a software company that sells for \$20 per share and has an expected growth of 15% per year, and a pharmaceutical company that sells for \$80 per share and has an expected growth of 10% per year. She plans to buy twice as many shares of stock in the fast-food chain as in the pharmaceutical company. If her goal is 12% growth per year, how many shares of each stock should she buy?

49. ***Nutrition*** A biologist is growing three different types of slugs (types A, B, and C) in the same laboratory environment. Each day, the slugs are given a nutrient mixture that contains three different ingredients (I, II, and III). Each type A slug requires 1 unit of I, 3 units of II, and 1 unit of III per day. Each type B slug requires 1 unit of I, 4 units of II, and 2 units of III per day. Each type C slug requires 2 units of I, 10 units of II, and 6 units of III per day. If the daily mixture contains 2000 units of I, 8000 units of II, and 4000 units of III, find the number of slugs of each type that can be supported.

50. ***Transportation*** An airline company has three types of aircraft that carry three types of cargo. The payload of each type is summarized in the table below.

	Plane Type		
Units Carried	*Passenger*	*Transport*	*Jumbo*
First-class mail	100	100	100
Passengers	150	20	350
Air freight	20	65	35

Suppose that on a given day the airline must move 1100 units of first-class mail, 460 units of air freight, and 1930 passengers. How many aircraft of each type should be scheduled?

Section 3.5

51. ***Economy models*** An economy has a shipping industry and an agricultural industry with technology matrix A.

$$A = \begin{array}{c} \\ \end{array}\begin{matrix} \text{S} & \text{A} \end{matrix} \qquad A = \begin{bmatrix} 0.2 & 0.1 \\ 0.1 & 0.2 \end{bmatrix} \begin{matrix} \text{Shipping} \\ \text{Agriculture} \end{matrix}$$

Surpluses of 4720 tons of shipping and 40 tons of agricultural output are desired. Find the gross production of each industry.

52. ***Economy models*** A simple economy has a shoe industry and a cattle industry with technology matrix A.

$$\begin{matrix} \text{S} & \text{C} \end{matrix} \qquad A = \begin{bmatrix} 0.1 & 0.1 \\ 0.2 & 0.05 \end{bmatrix} \begin{matrix} \text{Shoes} \\ \text{Cattle} \end{matrix}$$

Surpluses of 850 units of shoes and 275 units of cattle production are desired. Find the gross production of each industry.

53. ***Economy models*** A look at the industrial sector of an economy can be simplified to include three industries: the mining industry, the manufacturing industry, and the fuels industry. The technology matrix for this sector of the economy is given by

$$\begin{matrix} \text{M} & \text{Mf} & \text{F} \end{matrix} \qquad A = \begin{bmatrix} 0.4 & 0.4 & 0.2 \\ 0.2 & 0.4 & 0.2 \\ 0.1 & 0.2 & 0.4 \end{bmatrix} \begin{matrix} \text{Mining} \\ \text{Manufacturing} \\ \text{Fuels} \end{matrix}$$

Find the gross production of each industry if a surplus of 10 units of mined goods, 40 units of manufactured goods, and 140 units of fuels is desired.

54. ***Economy models*** Suppose a closed Leontief model for a nation's economy has the following technology matrix.

$$\begin{matrix} \text{G} & \text{A} & \text{M} & \text{H} \end{matrix} \qquad A = \begin{bmatrix} 0.4 & 0.2 & 0.2 & 0.2 \\ 0.2 & 0.4 & 0.1 & 0.2 \\ 0.2 & 0.1 & 0.3 & 0.1 \\ 0.2 & 0.3 & 0.4 & 0.5 \end{bmatrix} \begin{matrix} \text{Government} \\ \text{Agriculture} \\ \text{Manufacturing} \\ \text{Households} \end{matrix}$$

Find the gross production of each industry.

CHAPTER TEST

In Problems 1–6, perform the indicated matrix operations with the following matrices.

$$A = \begin{bmatrix} 1 & -2 \\ 3 & 2 \\ 4 & 1 \end{bmatrix} \qquad B = \begin{bmatrix} 2 & -2 & 1 \\ 3 & 1 & 5 \end{bmatrix}$$

$$C = \begin{bmatrix} 3 & -4 & -1 \\ 2 & 2 & -1 \end{bmatrix} \qquad D = \begin{bmatrix} 1 & 2 & 4 \\ 3 & 5 & 41 \\ 3 & 2 & 3 \end{bmatrix}$$

1. $A^T + B$
2. $B - C$
3. CD
4. DA
5. BA
6. ABD
7. Find the inverse of the matrix $\begin{bmatrix} 1 & 3 \\ 2 & 4 \end{bmatrix}$.
8. Find the inverse of the matrix

$$\begin{bmatrix} 1 & 2 & 4 \\ 1 & 2 & 2 \\ 1 & 1 & 4 \end{bmatrix}$$

9. If $AX = B$,

$$A^{-1} = \begin{bmatrix} 1 & 2 & 0 \\ 3 & 1 & 2 \\ 4 & 1 & 1 \end{bmatrix}, \quad \text{and} \quad B = \begin{bmatrix} 3 \\ 1 \\ 2 \end{bmatrix}$$

find the matrix X.

10. Solve

$$\begin{cases} x - y + 2z = 4 \\ x + 4y + z = 4 \\ 2x + 2y + 4z = 10 \end{cases}$$

11. Solve

$$\begin{cases} x - y + 2z = 4 \\ x + 4y + z = 4 \\ 2x + 3y + 3z = 8 \end{cases}$$

12. Solve

$$\begin{cases} x - y + 3z = 4 \\ x + 5y + 2z = 3 \\ 2x + 4y + 5z = 8 \end{cases}$$

13. Use an inverse matrix to solve

$$\begin{cases} x + y + 2z = 4 \\ x + 2y + z + w = 4 \\ 2x + 5y + 4z + 2w = 10 \\ 2y + z + 2w = 0 \end{cases}$$

14. Solve

$$\begin{cases} x - y + 2z - w = 4 \\ x + 4y + z + w = 4 \\ 2x + 2y + 4z + 2w = 10 \\ -y + z + 2w = 2 \end{cases}$$

15. ***Ecology*** In an ecological model, matrix A gives the fraction of several types of plants consumed by the herbivores in the ecosystem. Similarly, matrix B gives the fraction of each type of herbivore that is consumed by the carnivores in the system. Each row of matrix A refers to a type of plant, and each column refers to a kind of herbivore. Similarly, each row of matrix B is a kind of herbivore and each column is a kind of carnivore.

Herbivores

$$A = \begin{bmatrix} 0.2 & 0.1 & 0.4 \\ 0.3 & 0.3 & 0.1 \\ 0.2 & 0.1 & 0.1 \\ 0.1 & 0.2 & 0.5 \\ 0.4 & 0.4 & 0 \end{bmatrix} \text{Plants}$$

Carnivores

$$B = \begin{bmatrix} 0.1 & 0.1 & 0.2 \\ 0.2 & 0 & 0.4 \\ 0.1 & 0.5 & 0.1 \end{bmatrix} \text{Herbivores}$$

The matrix AB gives the fraction of each plant that is consumed by each carnivore.

(a) Find AB.

(b) What fraction of plant type 1 (row 1) is consumed by each carnivore?

(c) Identify the plant type that each carnivore consumes the most.

16. ***Manufacturing*** A furniture manufacturer produces four styles of chairs, and has orders for 1000 of style A, 4000 of style B, 2000 of style C, and 1000 of style D. The following table lists the numbers of units of raw material the manufacturer needs for each style. Suppose wood costs \$5 per unit, nylon costs \$3 per unit, velvet costs \$4 per unit, and springs cost \$4 per unit.

	Wood	*Nylon*	*Velvet*	*Springs*
Style A	10	5	0	0
Style B	5	0	20	10
Style C	5	20	0	10
Style D	5	10	10	10

(a) Form a 1×4 matrix containing the number of units of each style ordered.
(b) Use matrix multiplication to determine the total number of units of each raw material needed.
(c) Write a 4×1 matrix representing the costs for the raw materials.
(d) Use matrix multiplication to find the 1×1 matrix that represents the total investment to fill the orders.
(e) Use matrix multiplication to determine the cost for materials for each style of chair.

17. ***Investment*** A young couple with a \$120,000 inheritance wants to invest all this money. They plan to diversify their investments, choosing some of each of the following types of stock.

Type	*Cost/Share*	*Expected Total Growth/Share*
Growth	\$ 30	\$ 4.60
Blue-chip	100	11.00
Utility	50	5.00

Their strategy is to have the total investment in growth stocks equal to the sum of the other investments, and their goal is a 13% return on their investment. How many shares of each type of stock should they purchase?

18. Suppose an economy has two industries, agriculture and minerals, and the economy has the technology matrix

$$A = \begin{array}{c} \text{Ag} \quad \text{M} \\ \begin{bmatrix} 0.4 & 0.2 \\ 0.1 & 0.3 \end{bmatrix} \end{array} \begin{array}{l} \text{Ag} \\ \text{M} \end{array}$$

If surpluses of 140 agriculture units and 140 minerals units are desired, find the gross production of each industry.

19. Suppose the technology matrix for a closed model of a simple economy is given by matrix A. Find the gross productions of the industries.

$$A = \begin{array}{c} \text{P} \quad \text{NP} \quad \text{H} \\ \begin{bmatrix} 0.4 & 0.3 & 0.4 \\ 0.2 & 0.4 & 0.2 \\ 0.4 & 0.3 & 0.4 \end{bmatrix} \end{array} \begin{array}{l} \text{Profit} \\ \text{Nonprofit} \\ \text{Households} \end{array}$$

Use the following information in Problems 20 and 21. The national economy of Swiziland has four products: agricultural products, machinery, fuel, and steel. Producing 1 unit of agricultural products requires 0.2 unit of agricultural products, 0.3 unit of machinery, 0.2 unit of fuel, and 0.1 unit of steel. Producing 1 unit of machinery requires 0.1 unit of agricultural products, 0.2 unit of machinery, 0.2 unit of fuel, and 0.4 unit of steel. Producing 1 unit of fuels requires 0.1 unit of agricultural products, 0.2 unit of machinery, 0.3 unit of fuel, and 0.2 unit of steel. Producing 1 unit of steel requires 0.1 unit of agricultural products, 0.2 unit of machinery, 0.3 unit of fuel, and 0.2 unit of steel.

20. Create the technology matrix for this economy.

21. Determine how many units of each product will give surpluses of 1700 units of agriculture products, 1900 units of machinery, 900 units of fuel, and 300 units of steel.

22. A simple closed economy has the technology matrix A. Find the total output of the four sectors of the economy

$$\begin{array}{l} \\ \text{Agriculture} \\ \text{Steel} \\ \text{Fuel} \\ \text{Households} \end{array} \begin{array}{c} \text{Ag} \quad \text{S} \quad \text{F} \quad \text{H} \\ \begin{bmatrix} 0.25 & 0.25 & 0.3 & 0.1 \\ 0.3 & 0.25 & 0.2 & 0.4 \\ 0.15 & 0.2 & 0.1 & 0.3 \\ 0.3 & 0.3 & 0.4 & 0.2 \end{bmatrix} \end{array}$$

Extended Applications / Group Projects

I. Taxation

Federal income tax allows a deduction for any state income tax paid during the year. In addition, the state of Alabama allows a deduction from its state income tax for any federal income tax paid during the year. The federal corporate income tax rate is equivalent to a flat rate of 34% for taxable federal income between $335,000 and $10,000,000, and the Alabama rate is 5% of the taxable state income.*

Both the Alabama and federal taxable income for a corporation is $1,000,000 *before* either tax is paid. Because each tax is deductible on the other return, the taxable income will differ for the state and federal taxes. One procedure often used by tax accountants to find the tax due in this and similar situations is called *iteration* and is described by the first five steps below.†

1. Make an estimate of the federal taxes due by assuming no state tax is due, deduct this estimated federal tax due from the state taxable income, and calculate an estimate of the state taxes due on the basis of this assumption.
2. Deduct the estimated state tax due as computed in (1) from the federal taxable income, and calculate a new estimate of the federal tax due under this assumption.
3. Deduct the new estimated federal income tax due as computed in (2) from the state taxable income, and calculate a new estimate of the state taxes due on the basis of these calculations.
4. Repeat the process in (2) and (3) to get a better estimate of the taxes due to both governments.
5. Continue this process until the federal tax changes from one repetition to another by less than $1. What federal tax is due? What state tax is due?

To find the tax due each government directly, we can solve a matrix equation.

6. Create two linear equations that describe this taxation situation, convert the system to a matrix equation, and solve the system to see exactly what tax is due to each government.
7. What is the effective rate that this corporation pays to each government?

*Source: Federal Income Tax Tables, Alabama Income Tax Instructions.

†Performing these steps with an electronic spreadsheet is described by Kenneth H. Johnson in "A Simplified Calculation of Mutually Dependent Federal and State Tax Liabilities," in *The Journal of Taxation,* December 1994.

II. Company Profits after Bonuses and Taxes

A company earns $800,000 profit before paying bonuses and taxes. Suppose that a bonus of 2% is paid to each employee after all taxes are paid, that state taxes of 6% are paid on the profit after the bonuses are paid, and that federal taxes are 34% of the profit that remains after bonuses and state taxes are paid.

1. How much profit remains after all taxes and bonuses are paid? Give your answer to the nearest dollar.
2. If the employee bonus is increased to 3%, employees would receive 50% more bonus money. How much more profit would this cost the company than the 2% bonus?
3. Use a calculator or spreadsheet to determine the loss of profit associated with bonuses of 4%, 5%, etc., up to 10%, and make a management decision about what bonus to give that balances employee morale with loss of profit.
4. Use technology and the data points found above to create an equation that models the company's profit after expenses as a function of the employee bonus percent.

Warm-up

Prerequisite Problem Type	For Section	Answer	Section for Review
(a) Solve $3x - 2 = 7$. (b) Solve $2(x - 4) = \dfrac{x-3}{3}$.	4.1	(a) $x = 3$ (b) $x = \frac{21}{5}$	1.1 Linear equations
Graph the equation $y = \frac{3}{2}x - 2$.	4.1	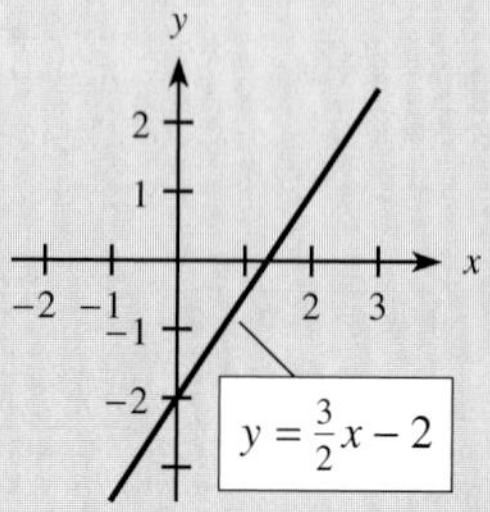	1.3 Graphing linear equations
Solve the systems: (a) $\begin{cases} x + 2y = 10 \\ 2x + y = 14 \end{cases}$	4.2	(a) $x = 6, y = 2$	1.5 Systems of linear equations
(b) $\begin{cases} x + 0.5y = 16 \\ x + y = 24 \end{cases}$	4.3	(b) $x = 8, y = 16$	
Write the system in an augmented matrix: $\begin{cases} x + 2y + s_1 = 10 \\ 2x + y + s_2 = 14 \\ -2x - 3y + f = 0 \end{cases}$	4.4 4.5 4.6	$\left[\begin{array}{ccccc\|c} 1 & 2 & 1 & 0 & 0 & 10 \\ 2 & 1 & 0 & 1 & 0 & 14 \\ -2 & -3 & 0 & 0 & 1 & 0 \end{array}\right]$	3.3 Gauss-Jordan elimination
Write a matrix equivalent to matrix A with the entry in row 1, column 2 equal to 1 and with all other entries in column 2 equal to 0. First multiply row 1 by $\frac{1}{2}$. $A = \left[\begin{array}{ccccc\|c} 1 & 2 & 1 & 0 & 0 & 10 \\ 2 & 1 & 0 & 1 & 0 & 14 \\ -2 & -3 & 0 & 0 & 1 & 0 \end{array}\right]$	4.4 4.5 4.6	$\left[\begin{array}{ccccc\|c} \frac{1}{2} & 1 & \frac{1}{2} & 0 & 0 & 5 \\ \frac{3}{2} & 0 & -\frac{1}{2} & 1 & 0 & 9 \\ -\frac{1}{2} & 0 & \frac{3}{2} & 0 & 1 & 15 \end{array}\right]$	3.3 Gauss-Jordan elimination

Chapter 4

Inequalities and Linear Programming

A firm frequently requires several components for the manufacture of the items it produces, and there are usually several stages for each item's assembly and final shipment. The company's costs and profits depend on the availability of the components (for example, labor and raw materials), the costs of these components, the unit profit for each product, and how many products are required. If the relationships among the various resources, the production requirements, the costs, and the profits are all linear, then these activities may be planned (or programmed) in the best possible (optimal) way by means of ***linear programming.***

Because linear programming is useful in solving the problem of allocating limited resources among various activities in the best possible way, its impact has been tremendous. Although it is a relatively recent development, it is a standard tool for companies of all sizes, and its application has been responsible for saving many thousands of dollars. Numerous textbooks have been written on the subject; our intention here is to provide an introduction to the method.

Because the restrictions (constraints) on most business operations can often be expressed as linear inequalities, we begin this chapter by introducing methods for solving and graphing linear inequalities. We will show how graphs of the constraint inequalities can be used to solve linear programming problems. The simplex method provides a technique for converting a system of inequalities into a system of equations that can be used to solve linear programming problems.

4.1 Linear Inequalities in One Variable

OBJECTIVE

- To solve and graph linear inequalities in one variable

APPLICATION PREVIEW

The height H in inches and age A in years for boys between 4 and 16 years of age are related according to

$$H = 2.31A + 31.26$$

To account for the normal variability among boys, normal height for a given age is $\pm 5\%$ of the height obtained in the equation.* We can express the range of normal height for a boy of a given age as an **inequality.** In this section we will solve inequalities involving one variable raised to the first power **(linear inequalities)**.

An **inequality** is a statement that one quantity is greater than (or less than) another quantity. We have already encountered some very simple inequalities. For example, the number of items a firm produces and sells, x, must be a nonnegative quantity. Thus x is greater than or equal to zero, which is written $x \geq 0$. The inequality $3x - 2 > 2x + 1$ is a first-degree (linear) inequality that states that the left member is greater than the right member. Certain values of the variable will satisfy the inequality. These values form the solution set of the inequality. For example, 4 is in the solution set of $3x - 2 > 2x + 1$ because $3 \cdot 4 - 2 > 2 \cdot 4 + 1$. On the other hand, 2 is not in the solution set because $3 \cdot 2 - 2 \not> 2 \cdot 2 + 1$. *Solving* an inequality means finding its solution set, and two inequalities are *equivalent* if they have the same solution set. As with equations, we find the solutions to inequalities by finding equivalent inequalities from which the solutions can be easily seen. We use the following properties to reduce an inequality to a simple equivalent inequality.

Inequalities

Properties	***Examples***
Substitution Property The inequality formed by substituting one expression for an equal expression is equivalent to the original inequality.	$5x - 4x < 6$ $x < 6$ The solution set is $\{x: x < 6\}$.
Addition Property The inequality formed by adding the same quantity to both sides of an inequality is equivalent to the original inequality.	$2x - 4 > x + 6$ $2x - 4 + 4 > x + 6 + 4$ $2x > x + 10$ $2x + (-x) > x + 10 + (-x)$ $x > 10$

Box continues on next page

*Adapted from data from the National Center for Health Statistics.

Inequalities

Properties	***Examples***	
Multiplication Property I The inequality formed by multiplying both sides of an inequality by the same *positive* quantity is equivalent to the original inequality.	$\frac{1}{2}x > 8$ $\frac{1}{2}x(2) > 8(2)$ $x > 16$	$3x < 6$ $3x\left(\frac{1}{3}\right) < 6\left(\frac{1}{3}\right)$ $x < 2$
Multiplication Property II The inequality formed by multiplying both sides of an inequality by the same *negative* number and reversing the sense of the inequality is equivalent to the original inequality.	$-x < 6$ $-x(-1) > 6(-1)$ $x > -6$	$-3x > -27$ $-3x\left(-\frac{1}{3}\right) < -27\left(-\frac{1}{3}\right)$ $x < 9$

We may graph the solution to inequalities in one unknown on the real number line. For example, the graph of $x < 2$ consists of all points to the left of 2 on the number line. The open circle on the graph in Figure 4.1 indicates that all points up to but *not* including 2 are in the solution set.

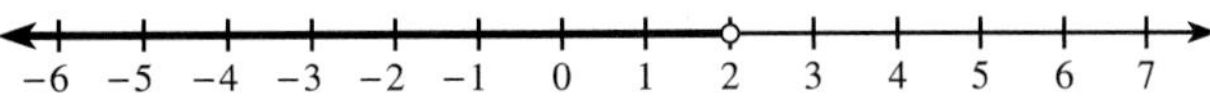

Figure 4.1

EXAMPLE 1

Solve the inequality

$$\frac{s}{-2} < -4$$

and graph the solution set.

Solution

$$\frac{s}{-2} < -4$$

Multiply both sides by -2 and reverse the inequality.

$$\frac{s}{-2}(-2) > -4(-2)$$

$$s > 8$$

The solution set is $\{s: s > 8\}$. The graph of the solution set is shown in Figure 4.2.

Figure 4.2

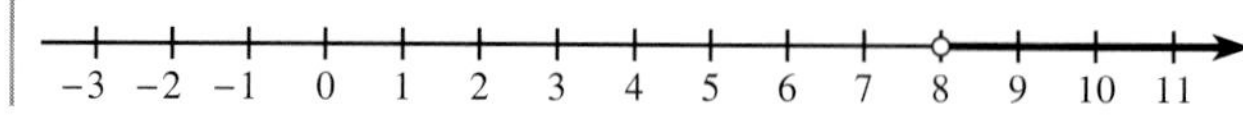

For some inequalities, it requires several operations to find their solution set. In this case, the order in which the operations are performed is the same as that used in solving linear equations.

EXAMPLE 2

Solve the inequality $2(x - 4) < \dfrac{x-3}{3}$.

Solution

$$\begin{aligned} 2(x-4) &< \frac{x-3}{3} \\ 6(x-4) &< x-3 && \text{Clear fractions} \\ 6x - 24 &< x - 3 && \text{Remove parentheses} \\ 5x &< 21 && \text{Perform additions and subtractions} \\ x &< \frac{21}{5} && \text{Multiply by } \frac{1}{5} \end{aligned}$$

Now, if we want to check that this solution is a reasonable one, we can substitute the integer values around $21/5$ into the original inequality. Note that $x = 4$ satisfies the inequality because

$$2[(4) - 4] < \frac{(4) - 3}{3}$$

but that $x = 5$ does not because

$$2[(5) - 4] \not< \frac{(5) - 3}{3}$$

Thus $x < 21/5$ is a reasonable solution.

We may also solve inequalities of the form $a \le b$. This means "a is less than b or $a = b$." The solution of $2x \le 4$ is $x \le 2$, because $x < 2$ is the solution of $2x < 4$ and $x = 2$ is the solution of $2x = 4$.

EXAMPLE 3

Solve the inequality $3x - 2 \le 7$.

Solution

This inequality states that $3x - 2 = 7$ or that $3x - 2 < 7$. By solving in the usual manner, we get $3x \le 9$, or $x \le 3$. Then $x = 3$ is the solution to $3x - 2 = 7$ and $x < 3$ is the solution to $3x - 2 < 7$, so the solution set for $3x - 2 \le 7$ is $\{x: x \le 3\}$.

The graph of the solution set includes the point $x = 3$ and all points $x < 3$ (see Figure 4.3).

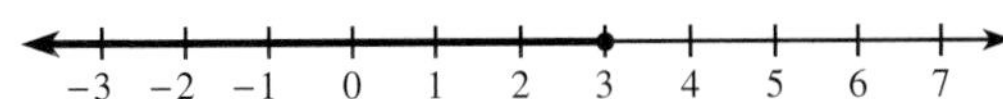

Figure 4.3

CHECKPOINT

Solve the following inequalities for y.

1. $3y - 7 \le 5 - y$ 2. $2y + 6 > 4y + 5$ 3. $4 - 3y \ge 4y + 5$

Recall from Section 0.2, "The Real Numbers," that the **compound inequality** $-2 < x < 4$ is called an **open interval** because neither endpoint is included. This interval (graphed in Figure 4.4) is denoted by $(-2, 4)$.

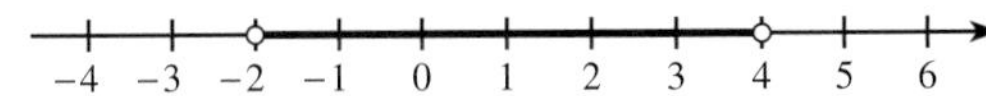

Figure 4.4

In Section 0.2 we also discussed **closed intervals** (both endpoints included) and **half-open intervals** (one endpoint excluded). Figure 4.5 shows the closed interval $-1 \le x \le 3$, denoted $[-1, 3]$. Figure 4.6 shows the half-open interval $-2 < x \le 4$, denoted $(-2, 4]$.

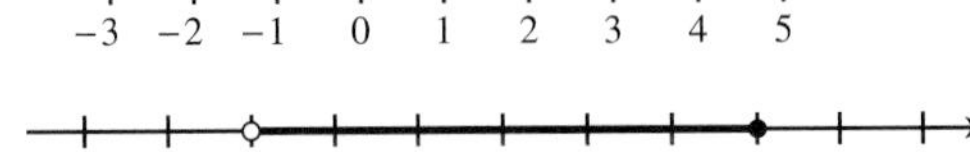

Figure 4.5

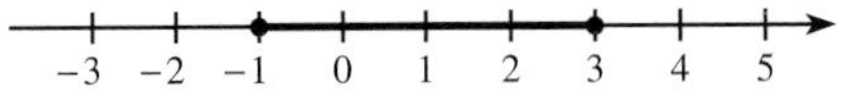

Figure 4.6

In general, we denote the closed interval $a \le x \le b$ by $[a, b]$, the open interval $a < x < b$ by (a, b), and the half-open interval $a \le x < b$ by $[a, b)$.

CHECKPOINT

Express each of the following inequalities in interval notation and name the type of interval.

4. $3 \le x \le 6$ 5. $-6 \le x < 4$

EXAMPLE 4

The Application Preview at the beginning of this section described how height and age are linearly related for boys between 4 and 16 years of age. That relation can be expressed as

$$H = 2.31A + 31.26$$

where H is height in inches and A is age in years. To account for natural variation among individuals, normal is considered to be any measure falling within $\pm 5\%$ of the height obtained from the equation. Write as an inequality the range of normal height for a boy who is 9 years old.

Solution

The boy's height from the formula is $H = 2.31(9) + 31.26 = 52.05$ inches. For a 9-year-old boy to be considered of normal height, H would have to be within $\pm 5\%$ of 52.05 inches. That is, the boy's height H is considered normal if $H \geq 52.05 - (0.05)(52.05)$ and $H \leq 52.05 + (0.05)(52.05)$. We can write this range of normal height by the compound inequality

$$52.05 - (0.05)(52.05) \leq H \leq 52.05 + (0.05)(52.05)$$

or

$$49.45 \leq H \leq 54.65.$$

CHECKPOINT SOLUTIONS

1. $4y - 7 \leq 5$; $4y \leq 12$; $y \leq 3$
2. $6 > 2y + 5$; $1 > 2y$; $y < \frac{1}{2}$
3. $4 \geq 7y + 5$; $-1 \geq 7y$; $y \leq -\frac{1}{7}$
4. [3,6]; closed interval
5. [−6, 4); half-open interval

EXERCISE 4.1

In Problems 1–10, solve each inequality.

1. $2x + 1 < x - 3$
2. $3x - 1 \geq 2x + 2$
3. $3(x - 1) < 2x - 1$
4. $2(x + 1) > x - 1$
5. $1 - 2x > 9$
6. $17 - x < -4$
7. $\frac{3(x - 1)}{2} \leq x - 2$
8. $\frac{x - 1}{2} + 1 > x + 1$
9. $2.9(2x - 4) \geq \frac{3.6x - 8.5}{5}$
10. $2.2(3.1x - 5) \leq \frac{3(4.6 - 6.1x)}{8}$

In Problems 11–22, solve each inequality and graph the solution.

11. $3(x - 1) < 2x$
12. $3(x + 2) \geq 4x + 1$
13. $2(x - 1) - 3 > 4x + 1$
14. $7x + 4 \leq 2(x - 1)$
15. $\frac{x}{3} > x - 1$
16. $\frac{x - 3}{4} \geq 2x$
17. $\frac{-3x}{2} > 9$
18. $\frac{-2x}{5} \leq -10$
19. $\frac{3x}{4} - \frac{1}{6} < x - \frac{2(x - 1)}{3}$
20. $\frac{4x}{3} - 3 > \frac{1}{2} + \frac{5x}{12}$
21. $\frac{3x}{4} - \frac{1}{3} < 1 - \frac{2}{3}\left(x - \frac{1}{2}\right)$
22. $\frac{x}{2} - \frac{4x}{5} > \frac{3(x - 1)}{10} - 2$

In Problems 23–30, write an inequality that describes each interval or graph.

23. $(-\frac{1}{2}, 3]$
24. $[-3, 5)$
25. $(1, 4)$
26. $[3, 7]$

27.

28.

29.

30.

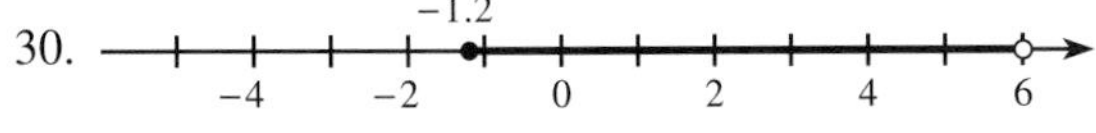

In Problems 31–40, express each inequality or graph using interval notation and name the type of interval.

31. $1 < x \le 3$
32. $-4 \le x \le 3$

33.

34.

35.

36.

37. $-4 < x < 3$
38. $-6 \le x \le -4$
39. $4 \le x \le 6$
40. $-2 \le x \le -1$

Applications

41. ***Profit*** For a certain product, the revenue function is $R(x) = 40x$ and the cost function is $C(x) = 20x + 1600$. To obtain a profit, the revenue must be greater than the cost. For what values of x will there be a profit? Graph the solution.

42. ***Car rental*** Thrifty rents a compact car for \$33 per day, and Budget rents a similar car for \$20 per day plus an initial fee of \$78. For how many days would it be cheaper to rent from Budget? Graph the solution.

43. ***Purchasing*** Sean can spend at most \$900 for a video camera and some video tapes. He plans to buy the camera for \$695 and tapes for \$5.75 each. Write an inequality that could be used to find the number of tapes (x) that he could buy. How many tapes could he buy?

44. ***Taxes*** In Sweetwater, Arizona, water bills are taxed on the basis of the amount of the monthly bill in order to encourage conservation. If the bill is more than \$0 but less than \$60, the tax is 2% of the bill; if the bill is \$60 but less than \$80, the tax is 4% of the bill; and if the bill is \$80 or more, the tax is 6% of the bill. Write the inequalities that represent the amount of tax owed in each of these three cases.

45. ***Income Taxes*** The 1998 tax brackets for a single person claiming one personal exemption are given in the following table.

Taxable Income I	*Tax T*
\$0–\$25,350	15% I
\$25,351–\$61,400	28% $(I - 25{,}350) + 3802.50$
\$61,401–\$128,100	31% $(I - 61{,}400) + 13{,}896.50$
\$128,101–\$278,450	36% $(I - 128{,}100) + 34{,}573.50$
Over \$278,450	39.6% $(I - 278{,}450) + 88{,}699.50$

SOURCE: 1040 Forms and Instructions, *IRS,* 1998

(a) Write the income ranges in the table as inequalities.
(b) For each income bracket, write the inequality that represents the amount of tax owed.

46. ***Health Statistics*** From data adapted from the National Center for Health Statistics, the height H in inches and age A in years for boys between 4 and 16 years of age are related according to

$$H = 2.31A + 31.26$$

To account for normal variability among boys, normal height for a given age is $\pm 5\%$ of the height obtained from the equation.
(a) Find the normal height range for a boy who is 10.5 years old, and write it as an inequality.
(b) Find the normal height range for a boy who is 5.75 years old, and write it as an inequality.

47. ***Heat index*** During the summer of 1998, Dallas, Texas, endured 29 consecutive days where the temperature was at least 110°F. On many of those days, the combination of heat and humidity made it feel even hotter than it was. When the temperature is 100°F, the apparent temperature A (or heat index) depends on the humidity h (expressed as a decimal) according to

$$A = 90.2 + 41.3h^*$$

(a) For what humidity levels is the apparent temperature at least 110°F? (Note that this answer will be a closed interval. Why?)
(b) For what humidity levels is the apparent temperature less than 100°F?

*Bosch, W., and C. G. Cobb, "Temperature-Humidity Indices," UMAP Unit 691, *The UMAP Journal,* 10(3), Fall 1989, 237–256.

48. ***Wind chill*** The combination of cold temperatures and wind speed determine what is called wind chill. The wind chill is a temperature that is the still-air equivalent of the combination of cold and wind. When the wind speed is 25 mph, the wind chill WC depends on the temperature t (in degrees Fahrenheit) according to

$$WC = 1.479t - 43.821$$

For what temperatures does it feel at least 30°F colder than the air temperature? That is, find t such that $WC \leq t - 30$.

49. ***Health statistics*** Normal body temperature is 98.6°F. What Celsius temperature C corresponds to $F \geq 98.6°$ if $F = \frac{9}{5}C + 32$?

50. ***Sales*** A saleswoman has monthly income I given by $I = 1000 + 0.062S$, where S is the monthly sales volume. How much must she sell to make at least \$3500 in a month?

51. ***Traffic flow*** The traffic flow problem illustrated in Figure 4.7 appeared as Problem 62 of Section 3.3, "Gauss-Jordan Elimination: Solving Systems of Equations," where x_1, x_2, x_3, and x_4 represent the number of cars on the indicated streets. The solution to that problem is described as follows:

$$x_1 = x_4 + 100$$
$$x_2 = x_4 + 200$$
$$x_3 = x_4 + 300$$

Because the total number of cars in the system is 1200 and each variable must be nonnegative, we can develop the following inequalities.

$$x_4 + 100 \leq 1200 \quad (1)$$
$$x_4 + 200 \leq 1200 \quad (2)$$
$$x_4 + 300 \leq 1200 \quad (3)$$
$$x_4 \geq 0 \quad (4)$$

(a) Solve each inequality for x_4.
(b) Determine an interval for x_4 that satisfies all these inequalities. (*Hint:* Find the intersection of the solution graphs.)
(c) Use the interval from (b) to develop an interval for each of the other variables.

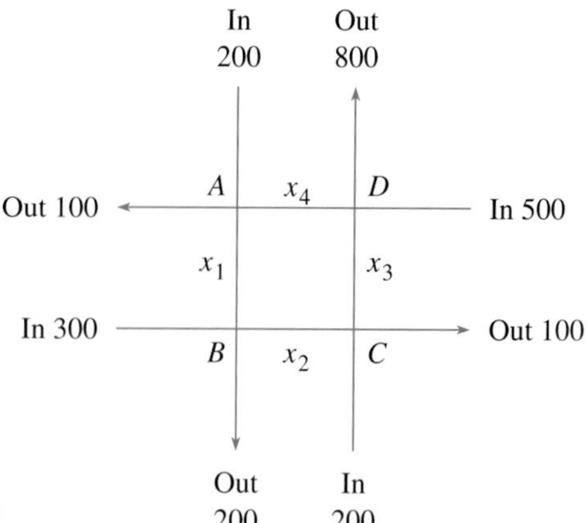

Figure 4.7

52. ***Nutrition*** Problem 61 of Section 3.3, "Gauss-Jordan Elimination: Solving Systems of Equations" described how a botanist could purchase four different plant foods with various nutrient values. If x_1 represents the number of bags of food I that were purchased, x_2 the number of bags of food II, and so on, then the number of bags of each food that the botanist can purchase and satisfy the requirements given in Problem 61 of that section can be described as follows:

$$x_1 = 160 - x_4$$
$$x_2 = 2x_4 - 220$$
$$x_3 = x_4 + 70$$

Because each variable must be nonnegative, we can develop the following inequalities.

$$160 - x_4 \geq 0$$
$$2x_4 - 220 \geq 0$$
$$x_4 + 70 \geq 0$$
$$x_4 \geq 0$$

(a) Solve each inequality for x_4.
(b) Determine an interval for x_4 that solves all these inequalities.
(c) Develop an interval for each of the other variables.

4.2 Linear Inequalities in Two Variables

OBJECTIVES

- *To graph linear inequalities in two variables*
- *To solve systems of linear inequalities in two variables*

APPLICATION PREVIEW

A rental agency has a maximum of \$240,000 to invest in the purchase of at most 13 cars of two different types, compact and full size. The cost per compact is \$15,000 and the cost per full-size car is \$24,000. The number of cars of each type is limited (constrained) by the budget available and the number of cars needed. These **constraints** can be expressed by a **system of inequalities** in two variables. Finding the values that satisfy all these constraints at the same time is called solving the system of inequalities.

Before we look at systems of inequalities, we will discuss solutions of one inequality in two variables, such as $y < x$. The solutions to this inequality are the ordered pairs (x, y) that satisfy the inequality. Thus $(1, 0)$, $(3, 2)$, $(0, -1)$, and $(-2, -5)$ are the solutions to $y < x$, but $(3, 7)$, $(-4, -3)$ and $(2, 2)$ are not.

The graph of $y < x$ consists of all points whose y-coordinate is less than the x-coordinate. The graph of the region $y < x$ can be found by graphing the line $y = x$ (as a dashed line, because the given inequality does not include $y = x$). This line separates the xy-plane into two **half-planes,** $y < x$ and $y > x$. We can determine which half-plane is the solution region by selecting as a **test point** any point **not on the line;** let's choose $(2, 0)$. Because the coordinates of this test point satisfy the inequality $y < x$, the half-plane containing this point is the solution region for $y < x$. (See Figure 4.8.) If the coordinates of the test point do not satisfy the inequality, then the other half-plane is the solution region. For example, say we had chosen $(0, 4)$ as our test point. Its coordinates do not satisfy $y < x$, so the half-plane that does *not* contain $(0, 4)$ is the solution region. (Note that we get the same region.)

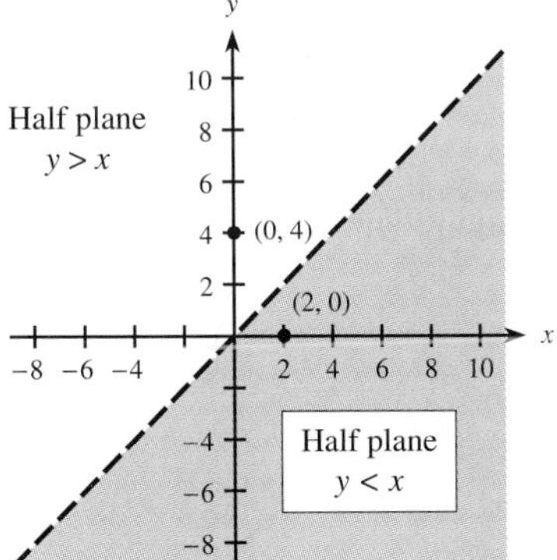

Figure 4.8

EXAMPLE 1

Graph the inequality $4x - 2y \le 6$.

Solution

First we graph the line $4x - 2y = 6$, or (equivalently) $y = 2x - 3$, as a solid line, because points lying on the line satisfy the given inequality. Next we pick a test point that is not on the line. If we use (0, 0), we see that its coordinates satisfy $4x - 2y \le 6$—that is, $y \ge 2x - 3$. Hence the solution region is the line $y = 2x - 3$ and the half-plane that contains the test point (0, 0). See Figure 4.9.

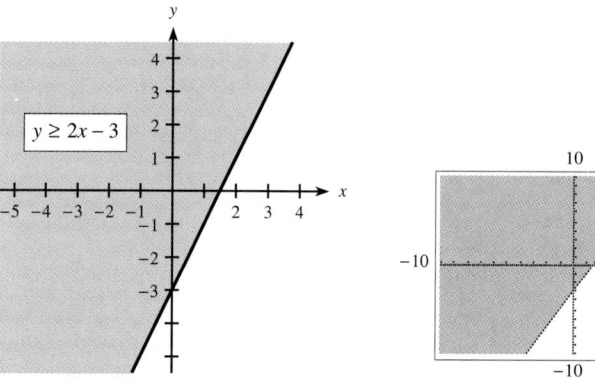

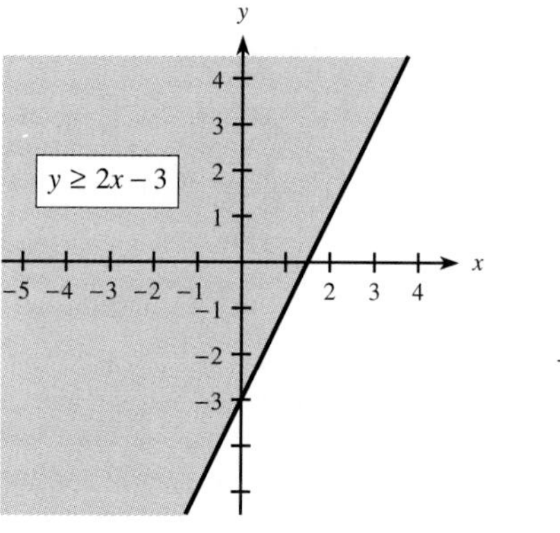

Figure 4.9

10
−10
10
−10

Figure 4.10

Technology Note

Graphing calculators can also be used to shade the solution region of an inequality. Figure 4.10 shows a graphing calculator window for the solution to $4x - 2y \le 6$, or $y \ge 2x - 3$.

If we have two inequalities in two variables, we can find the solutions that satisfy both inequalities. We call the inequalities a **system of inequalities,** and the solution to the system can be found by finding the intersection of the solution sets of the two inequalities.

The solution set of the system of inequalities can be found by graphing the inequalities on the same set of axes and noting their points of intersection.

EXAMPLE 2

Graph the solution to the system

$$\begin{cases} 3x - 2y \ge 4 \\ x + y - 3 > 0 \end{cases}$$

Solution

The inequalities can be written in the form

$$y \le \frac{3}{2}x - 2$$
$$y > -x + 3$$

We graph $y = \frac{3}{2}x - 2$ as a solid line and $y = -x + 3$ as a dashed line (see Figure 4.11a). We use any point not on either line as a test point; let's use (0, 0). Note that the coordinates of (0, 0) do not satisfy either $y > -x + 3$ or $y \leq \frac{3}{2}x - 2$. Thus the solution region for each individual inequality is the half-plane that does not contain the point (0, 0). Figure 4.11(b) indicates the half-plane solution for each inequality with arrows pointing from the line into the desired half-plane (away from the test point). The points that satisfy both of these inequalities lie in the intersection of the two individual solution regions, shown in Figure 4.11(c). This **solution region** is the graph of the solution to this system of inequalities.

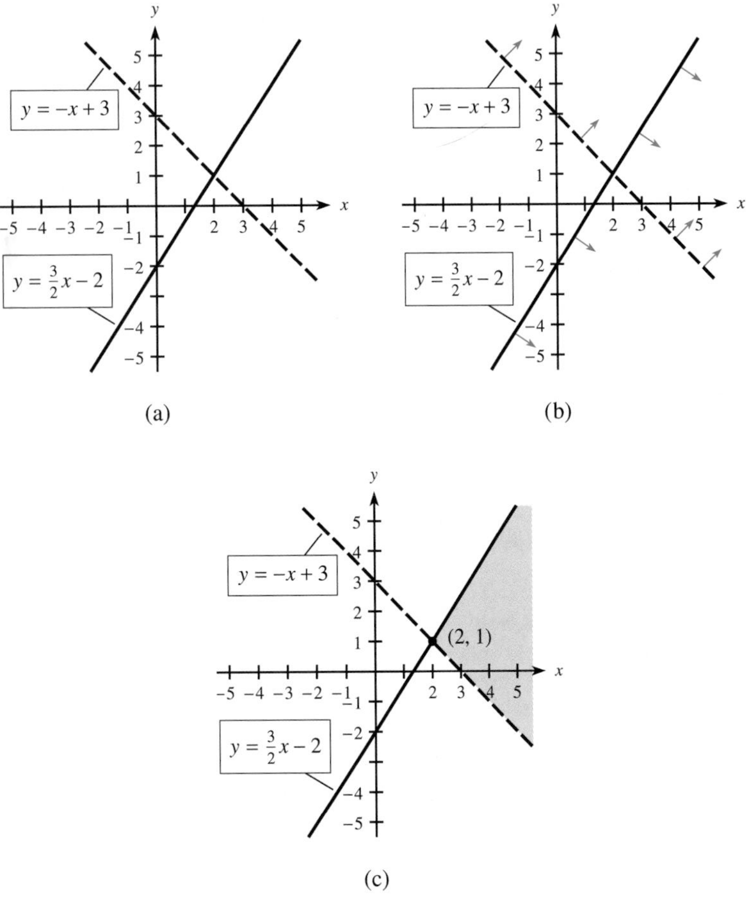

Figure 4.11

The point (2, 1) in Figure 4.11(c), where the two regions form a "corner," is found by solving the *equations* $y = \frac{3}{2}x - 2$ and $y = -x + 3$ simultaneously.

EXAMPLE 3

Graph the solution to the system

$$\begin{cases} x + 2y \leq 10 \\ 2x + y \leq 14 \\ x \geq 0 \\ y \geq 0 \end{cases}$$

Solution

The two inequalities $x \geq 0$ and $y \geq 0$ restrict the solution to quadrant I (and the axes bounding quadrant I).

We seek points in the first quadrant (on or above $y = 0$ and on or to the right of $x = 0$) that satisfy $x + 2y \leq 10$ *and* $2x + y \leq 14$. We can write these inequalities in their equivalent forms $y \leq 5 - \frac{1}{2}x$ and $y \leq 14 - 2x$. The points that satisfy these inequalities (in the first quadrant) are shown by the darkly shaded area in Figure 4.12. We can observe from the graph that the points (0, 0), (7, 0), and (0, 5) are corners of the solution region. The corner (6, 2) is found by solving the *equations* $y = 5 - \frac{1}{2}x$ and $y = 14 - 2x$ simultaneously as follows:

$$5 - \frac{1}{2}x = 14 - 2x$$
$$\frac{3}{2}x = 9$$
$$x = 6$$
$$y = 2$$

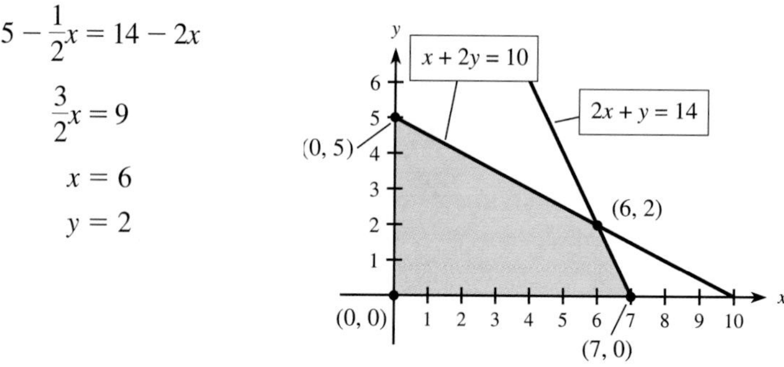

Figure 4.12

We will see that the corners of the solution region are important in solving linear programming problems.

Many applications restrict the variables to be nonnegative (such as $x \geq 0$ and $y \geq 0$ in Example 3). As we noted, the effect of this restriction is to limit the solution to quadrant I.

We now return to the Application Preview.

EXAMPLE 4

A rental agency has a maximum of \$240,000 to invest in the purchase of at most 13 cars of two different types, compact and full size. The cost per compact is \$15,000 and the cost per full-size car is \$24,000. Write the system of inequalities that describes the constraints, and graph the solution region for the system.

Solution

If x represents the number of compact cars and y represents the number of full-size cars, then the number of cars is limited by $x + y \leq 13$. The limitation on available funds is given by $15{,}000x + 24{,}000y \leq 240{,}000$. Because the number of cars must be positive, the system of inequalities is

$$\begin{aligned} x + y &\leq 13 \\ 15{,}000x + 24{,}000y &\leq 240{,}000 \\ x &\geq 0 \\ y &\geq 0 \end{aligned}$$

The solution region is shown in Figure 4.13. The solution region has corners at (0, 0), (13, 0), (8, 5), and (0, 10). The corner at (8, 5) is found by solving $x + y = 13$ and $15{,}000x + 24{,}000y = 240{,}000$ simultaneously.

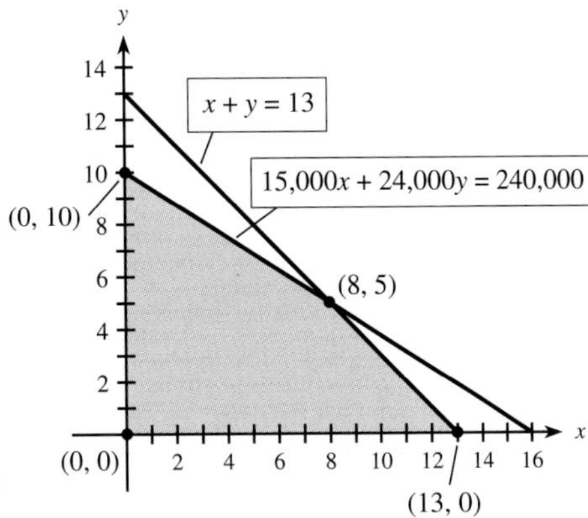

Figure 4.13

CHECKPOINT

1. Graph the region determined by the inequalities

$$\begin{aligned} 2x + 3y &\leq 12 \\ 4x + 2y &\leq 16 \\ x &\geq 0 \\ y &\geq 0 \end{aligned}$$

2. Determine the corners of the region.

Many computer programs save considerable time and energy in determining regions that satisfy a system of inequalities. Graphing calculators can also be used in the graphical solution of a system of inequalities.

EXAMPLE 5

Use a graphing utility to find the following.

(a) Find the region determined by the inequalities below.

$$\begin{aligned} 5x + 2y &\leq 54 \\ 2x + 4y &\leq 60 \\ x &\geq 0 \\ y &\geq 0 \end{aligned}$$

(b) Find the corners of this region.

Solution

(a) We write the inequalities above as equations, solved for y. Graphing these equations with a graphing calculator and using shading shows the region satisfying the inequalities (see Figure 4.14).

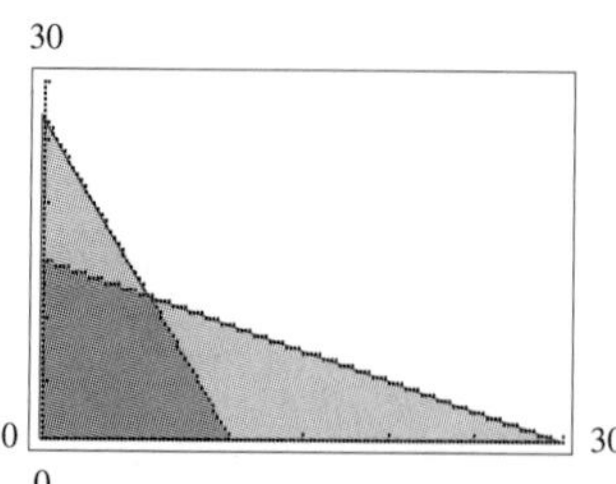

Figure 4.14

(b) Some utilities can determine the points of intersection of two equations with an INTERSECT command. This command can be used to find the intersection of the pair of equations from (a). With other graphing utilities, using the command SOLVER or INTERSECT, or using TRACE along the lines that form the borders of this region, will give the points where the boundaries intersect. These points, (0, 15), (6, 12), and (10.8, 0) can also be found algebraically. These points are the corners of the solution region. These corners will be important to us in the graphical solution to linear programming problems in the next section.

CHECKPOINT SOLUTIONS

1.

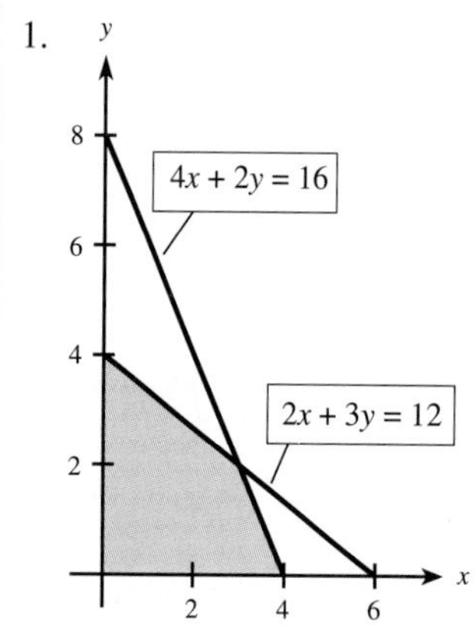

2. Corners occur at (0, 0) and where $x = 0$ or $y = 0$: $x = 0$ gives $y = 4$ and $y = 8$, so (0, 4) is a corner; $y = 0$ gives $x = 6$ and $x = 4$, so (4, 0) is a corner. A corner also occurs where $2x + 3y = 12$ and $4x + 2y = 16$ intersect. The point of intersection is $x = 3$, $y = 2$, so (3, 2) is a corner.

EXERCISE 4.2

In Problems 1–8, graph each inequality.

1. $y \le 2x - 1$
2. $y \ge 4x - 5$
3. $\frac{x}{2} + \frac{y}{4} < 1$
4. $x - \frac{y}{3} < \frac{-2}{3}$
5. $2(x - y) < y + 3$
6. $2(x - y) \ge x + 4$
7. $0.4x \ge 0.8$
8. $\frac{-y}{8} > \frac{1}{4}$

In Problems 9–14, the graph of the boundary equations for each system of inequalities is shown with that system.
(a) Locate the solution region.
(b) Find the corners of each solution region.

9. $\begin{cases} x + 4y \le 60 \\ 4x + 2y \le 100 \\ x \ge 0 \\ y \ge 0 \end{cases}$

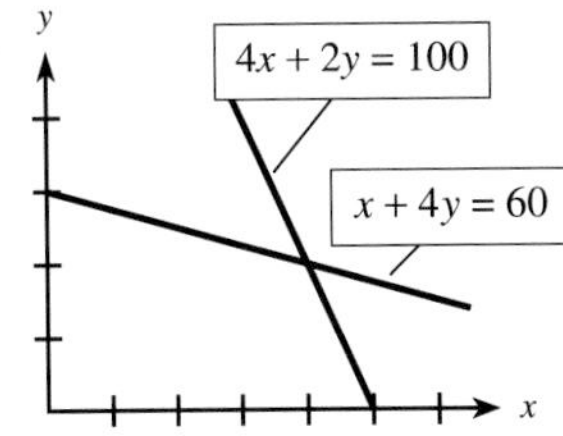

10. $\begin{cases} 4x + 3y \le 240 \\ 5x - y \le 110 \\ x \ge 0 \\ y \ge 0 \end{cases}$

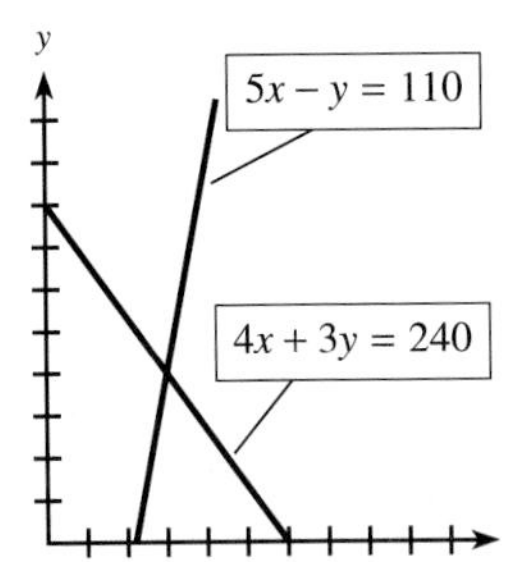

11. $\begin{cases} -x + y \le 4 \\ x + 3y \le 20 \\ 5x + 2y \le 35 \\ x \ge 0 \\ y \ge 0 \end{cases}$

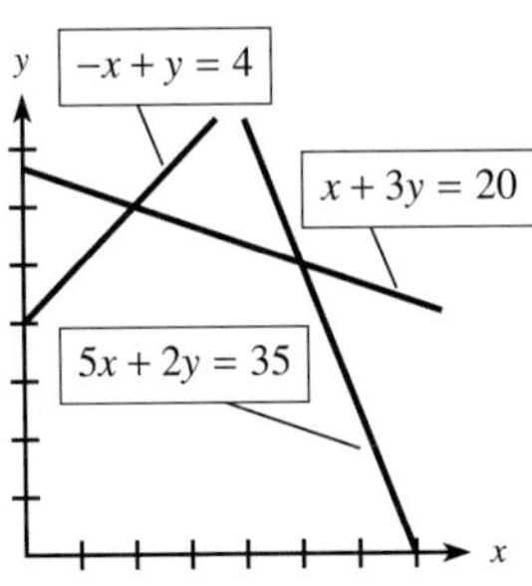

12. $\begin{cases} x + 2y \le 20 \\ 3x + 4y \le 48 \\ 3x + 2y \le 42 \\ x \ge 0 \\ y \ge 0 \end{cases}$

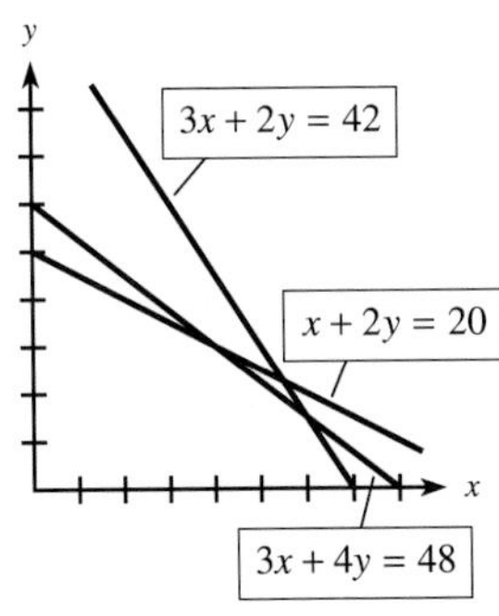

13. $\begin{cases} x + 3y \ge 6 \\ 2x + 4y \ge 10 \\ 3x + y \ge 5 \\ x \ge 0 \\ y \ge 0 \end{cases}$

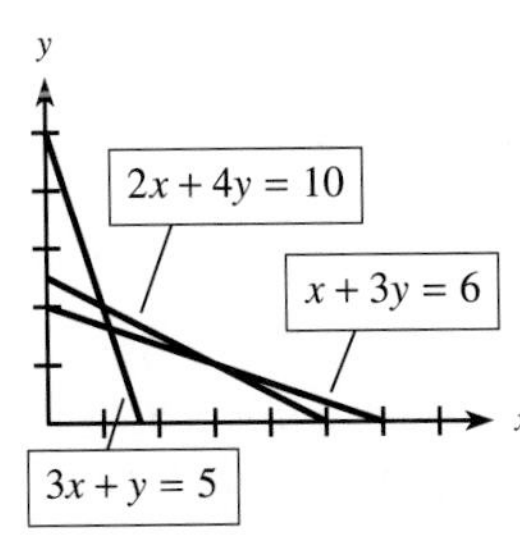

14. $\begin{cases} x + 4y \ge 10 \\ 4x + 2y \ge 10 \\ x + y \ge 4 \\ x \ge 0 \\ y \ge 0 \end{cases}$

y
x + y = 4
x + 4y = 10
x
4x + 2y = 10

In Problems 15–30, graph the solution to each system of inequalities.

15. $\begin{cases} y < 2x \\ y > x - 1 \end{cases}$

16. $\begin{cases} y > 3x - 4 \\ y < 2x + 3 \end{cases}$

17. $\begin{cases} 2x + y < 3 \\ x - 2y \ge -1 \end{cases}$

18. $\begin{cases} 3x + y > 4 \\ x - 2y < -1 \end{cases}$

19. $\begin{cases} y \ge 3 - 2x \\ y \ge \frac{1}{2}x + \frac{1}{2} \\ y \ge 3 \end{cases}$

20. $\begin{cases} y \le x + 1 \\ y \ge 2x - 1 \\ x \ge 0,\ y \ge 0 \end{cases}$

21. $\begin{cases} x + 5y \le 200 \\ 2x + 3y \le 134 \\ x \ge 0,\ y \ge 0 \end{cases}$

22. $\begin{cases} -x + y \le 2 \\ x + 2y \le 10 \\ 3x + y \le 15 \\ x \ge 0,\ y \ge 0 \end{cases}$

23. $\begin{cases} x + 2y \le 48 \\ x + y \le 30 \\ 2x + y \le 50 \\ x \ge 0,\ y \ge 0 \end{cases}$

24. $\begin{cases} 3x + y \le 9 \\ 3x + 2y \le 12 \\ x + 2y \le 8 \\ x \ge 0,\ y \ge 0 \end{cases}$

25. $\begin{cases} x + 2y \ge 19 \\ 3x + 2y \ge 29 \\ x \ge 0,\ y \ge 0 \end{cases}$

26. $\begin{cases} 4x + y \ge 12 \\ x + y \ge 9 \\ x + 3y \ge 15 \\ x \ge 0,\ y \ge 0 \end{cases}$

27. $\begin{cases} x + 3y \ge 3 \\ 2x + 3y \ge 5 \\ 2x + y \ge 3 \\ x \ge 0,\ y \ge 0 \end{cases}$

28. $\begin{cases} x + 2y \ge 10 \\ 2x + y \ge 11 \\ x + y \ge 9 \\ x \ge 0,\ y \ge 0 \end{cases}$

29. $\begin{cases} x + 2y \ge 20 \\ -3x + 2y \le 4 \\ x \ge 12 \\ x \ge 0,\ y \ge 0 \end{cases}$

30. $\begin{cases} 3x + 2y \ge 75 \\ -3x + 5y \ge 30 \\ y \le 40 \\ x \ge 0,\ y \ge 0 \end{cases}$

Applications

31. ***Management*** The Wellbuilt Company produces two types of wood chippers, Economy and Deluxe. The Deluxe model requires 3 hours to assemble and 1/2 hour to paint, and the Economy model requires 2 hours to assemble and 1 hour to paint. The maximum number of assembly hours available is 24 per day,

and the maximum number of painting hours available is 8 per day.

(a) Write the system of inequalities that describes the constraints on the production of these models.
(b) Graph the solution to the system of inequalities and find the corners of the solution region.

32. ***Learning environments*** An experiment that involves learning in animals requires placing white mice and rabbits into separate, controlled environments, Environment I and Environment II. The maximum amount of time available in Environment I is 500 minutes, and the maximum amount of time available in Environment II is 600 minutes. The white mice must spend 10 minutes in Environment I and 25 minutes in Environment II, and the rabbits must spend 15 minutes in Environment I and 15 minutes in Environment II.
(a) Write a system of inequalities that describes the constraints on the experiment.
(b) Graph the solution to the system of inequalities and find the corners of the solution region.

33. ***Production*** A company manufactures two types of electric hedge-trimmers, one of which is cordless. The cord-type trimmer requires 2 hours to make, and the cordless model requires 4 hours. The company has only 800 work hours to use in manufacturing each day, and the packing department can package only 300 trimmers per day.
(a) Write the inequalities that describe these constraints on production.
(b) Graph the region determined by these constraint inequalities.

34. ***Manufacturing*** A firm manufactures bumper bolts and fender bolts for cars. One machine can produce 130 fender bolts per hour, and another machine can produce 120 bumper bolts per hour. The combined number of fender bolts and bumper bolts the packaging department can handle is 230 per hour.
(a) Write the inequalities that describe these constraints on production.
(b) Graph the region determined by these constraint inequalities.

35. ***Advertising*** Apex Motors manufactures luxury cars and sport utility vehicles. The most likely customers are high-income men and women, and company managers want to initiate an advertising campaign targeting these groups. They plan to run 1-minute spots on business/investment programs, where they can reach 7 million women and 4 million men from their target groups. They also plan 1-minute spots during sporting events, where they can reach 2 million women and 12 million men from their target groups. Apex feels that the ads must reach at least 30 million women and at least 28 million men who are prospective customers.
(a) Write the inequalities that describe the constraints on reaching these target groups.
(b) Graph the region determined by these constraint inequalities.

36. ***Manufacturing*** The Video Star Company makes two different models of VCRs, which are assembled on two different assembly lines. Line 1 can assemble 30 units of the Star model and 40 units of the Prostar model per hour, and Line 2 can assemble 150 units of the Star model and 40 units of the Prostar model per hour. The company needs to produce at least 270 units of the Star model and 200 units of the Prostar model to fill an order.
(a) Write the inequalities that describe these production constraints.
(b) Graph the region determined by these constraint inequalities.

37. ***Politics*** A candidate wishes to use a combination of radio and television advertisements in her campaign. Research has shown that each 1-minute spot on television reaches 0.09 million people and each 1-minute spot on radio reaches 0.006 million. The candidate feels she must reach at least 2.16 million people, and she must buy a total of at least 80 minutes of advertisements.
(a) Write the inequalities that describe her needs.
(b) Graph the region determined by these constraint inequalities.

38. ***Nutrition*** In a hospital ward, the patients can be grouped into two general categories depending on their condition and the amount of solid foods they require in their diet. A combination of two diets is used for solid foods because they supply essential nutrients for recovery. The table below summarizes the patient groups and their minimum daily requirements.

	Diet A	*Diet B*	*Daily Requirements*
Group 1	4 oz per serving	1 oz per serving	26 oz
Group 2	2 oz per serving	1 oz per serving	18 oz

(a) Write the inequalities that describe these needs.
(b) Graph the region determined by these constraint inequalities.

39. ***Manufacturing*** A sausage company makes two different kinds of hot dogs, regular and all beef. Each pound of all-beef hot dogs requires 0.75 lb of beef and 0.2 lb of spices, and each pound of regular hot dogs requires 0.18 lb of beef, 0.3 lb of pork, and 0.2 lb of spices. Suppliers can deliver at most 1020 lb of beef, at most 600 lb of pork, and at least 500 lb of spices.
   (a) Write the inequalities that describe these needs.
   (b) Graph the region determined by these constraint inequalities.

40. ***Manufacturing*** A cereal manufacturer makes two different kinds of cereal, Senior Citizen's Feast and Kids Go. Each pound of Senior Citizen's Feast requires 0.6 lb of wheat and 0.2 lb of vitamin-enriched syrup, and each pound of Kids Go requires 0.4 lb of wheat, 0.2 lb of sugar, and 0.2 lb of vitamin-enriched syrup. Suppliers can deliver at most 2800 lb of wheat, at most 800 lb of sugar, and at least 1000 lb of the vitamin-enriched syrup.
   (a) Write the inequalities that describe these needs.
   (b) Graph the region determined by these constraint inequalities.

4.3 Linear Programming: Graphical Methods

OBJECTIVE

- *To use graphical methods to find the optimum value of a linear function subject to constraints*

APPLICATION PREVIEW

Many practical problems in business and economics involve complex relationships among capital, raw material, labor, and so forth. Consider the following example. Suppose that two chemical plants produce three types of fertilizer: low phosphorus (LP), medium phosphorus (MP), and high phosphorus (HP). The fertilizer is produced in a single production run, so the three types are produced in fixed proportions. The Macon plant produces 1 ton of LP, 2 tons of MP, and 3 tons of HP in a single operation, and it charges \$600 for what is produced in one operation, whereas one operation of the Jonesboro plant produces 1 ton of LP, 5 tons of MP, and 1 ton of HP, and it charges \$1000 for what it produces in one operation. If a customer needs a specific amount of each type of fertilizer, the minimum cost can be found by using a mathematical technique called **linear programming.** Linear programming can be used to solve problems such as this if the limits on the variables (called **constraints)** can be expressed as linear inequalities and if the function that is to be maximized or minimized (called the **objective function)** is a linear function.

Linear programming problems often involve many variables, but in this section we restrict our discussion to problems involving two variables. With two variables we can use graphical methods to help solve the problem. The constraints form a system of linear inequalities in two variables that we can solve by graphing. The solution to the system of constraint inequalities determines a region, any point of which may yield the *optimum* (maximum or minimum) value for the objective function.* Hence any point in the region determined by the constraints is called a **feasible solution,** and the region itself is called the **feasible region.**

In a linear programming problem, we seek the feasible solution that maximizes (or minimizes) the objective function. For example, suppose we wish to maximize the function $C = 2x + y$ subject to the constraints $x \geq 0$, $y \geq 0$, $x + 3y \leq 6$, and $x + y \leq 4$. The solution set (feasible region) for the system of constraints is shown in Figure 4.15(a). *Any* point inside the shaded region or on its boundary is a feasible solution to the problem. To determine which point will maximize the objective function, we graph $C = 2x + y$, for different values of C, on the same graph that contains the feasible region. The different values of C change the position of the line, *but the slope* of the line *does not change.* Letting $C = 0, 2, 3, 7$, and 8, we get the graphs shown in Figure 4.15(b). If any part of the line lies within the feasible region, we have feasible solutions.

Because we seek the values of x and y that will maximize $2x + y$, subject to the constraints, we keep trying larger values of C while keeping some part of the line intersecting the feasible region. It is clear from Figure 4.15(b) that any value

*The region determined by the constraints must be *convex* for the optimum to exist. A convex region is one such that for any two points in the region, the segment joining those points lies entirely within the region. We restrict our discussion to convex regions.

of C larger than 8 will cause the line $C = 2x + y$ to "miss" the region, and so the maximum value for $2x + y$, subject to the constraints, is 8. The point where the line $8 = 2x + y$ intersects the feasible region is (4, 0), so the function $2x + y$ is maximized when $x = 4$, $y = 0$. Note that the objective function was maximized at one of the "corners" (or vertices) of the feasible region.

Suppose we wanted to maximize $P = x + 2y$ over the constraint region shown in Figure 4.15(a). Again, we could choose several values for P and graph the corresponding lines (see Figure 4.15(c)). This figure shows that the maximum value is $P = 5$ and that it occurs at the corner (3, 1). Note in both examples that the objective function was maximized at one of the "corners" (or vertices) of the feasible region.

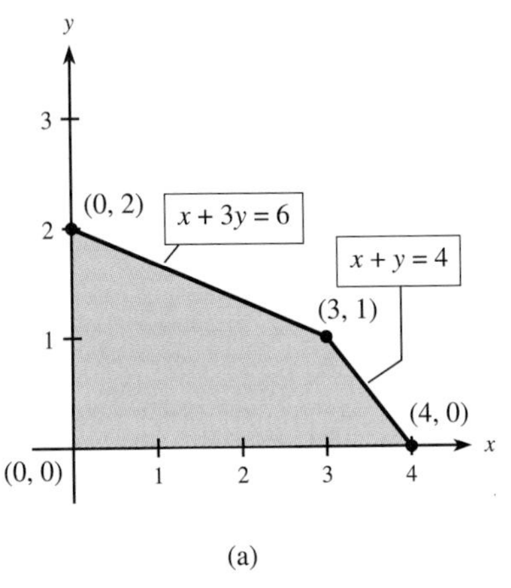

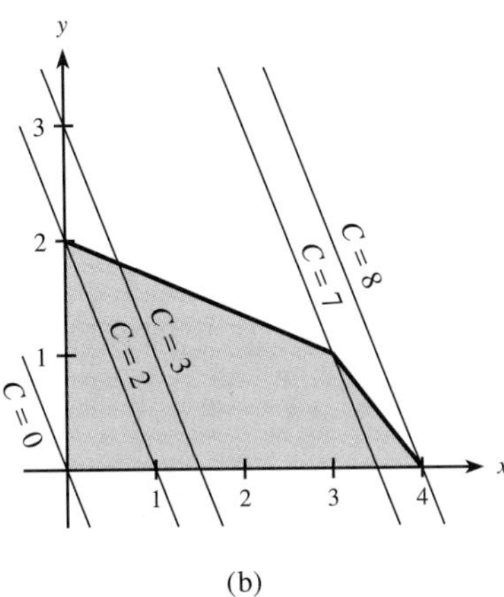

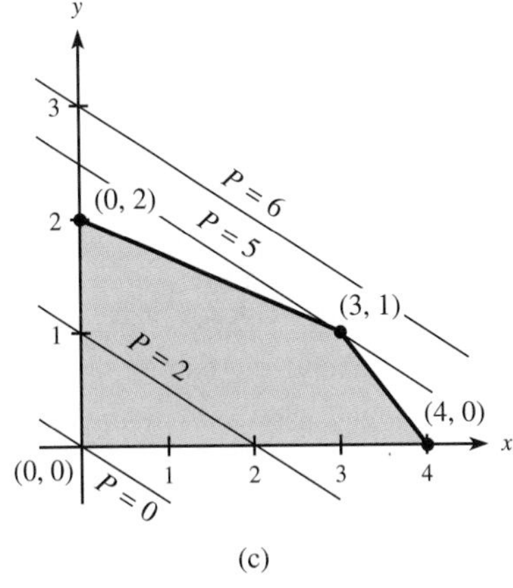

Figure 4.15

The feasible region in Figure 4.15 is an example of a **closed and bounded region** because it is entirely enclosed by, and includes, the lines associated with the constraints.

Solutions to Linear Programming Problems

1. When the feasible region for a linear programming problem is closed and bounded, the objective function has a maximum value and a minimum value.
2. When the feasible region is not closed and bounded, the objective function may have a maximum only, a minimum only, or no solution.
3. If a linear programming problem has a solution, then the optimum (maximum or minimum) value of an objective function occurs at a corner of the feasible region determined by the constraints.

Thus, for a closed and bounded region, we can find the maximum or minimum value of the objective function, by evaluating the function at each of the corners of the feasible region formed by the solution of the constraint inequalities. If the objective function has its optimum value at two corners, then it also has that optimum value at any point on the line (boundary) connecting those two corners. If the feasible region is not closed and bounded, we must check to make sure the objective function has an optimum value.

The steps involved in solving a linear programming problem are as follows.

Linear Programming (Graphical Method)

Procedure	Example
To find the optimal value of a function subject to constraints:	Maximize $C = 2x + 3y$ subject to the constraints $\begin{cases} x + 2y \le 10 \\ 2x + y \le 14 \\ x \ge 0 \\ y \ge 0 \end{cases}$
1. Write the objective function and constraint inequalities from the problem.	1. Objective function: $C = 2x + 3y$ Constraints: $x + 2y \le 10$ $2x + y \le 14$ $x \ge 0$ $y \ge 0$
2. Graph the solution to the constraint system. (a) If the feasible region is closed and bounded, proceed to step 3. (b) If the region is not closed and bounded, check whether an optimum value exists. If not, state this. If so, proceed to step 3.	2. See Figure 4.16. 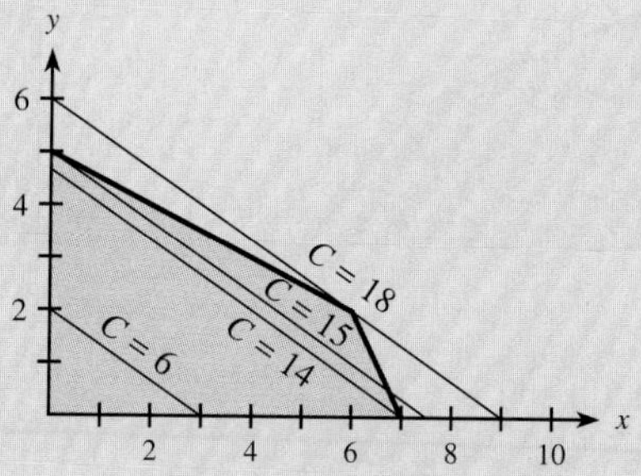 **Figure 4.16**
3. Find the corners of the resulting feasible region. This may require simultaneous solution of two boundary equations.	3. Corners are (0, 0), (0, 5), (6, 2), (7, 0).
4. Evaluate the objective function at each corner of the feasible region determined by the constraints.	4. At (0, 0), $2x + 3y = 0$ At (0, 5), $2x + 3y = 15$ At (6, 2), $2x + 3y = 18$ At (7, 0), $2x + 3y = 14$
5. If two corners give the optimum value for the objective function, then all points on the boundary line joining these two corners also optimize the function.	5. The function is maximized at $x = 6$, $y = 2$. The maximum value is 18.

EXAMPLE 1

Chairco manufactures two types of chairs, standard and plush. Standard chairs require 2 hours to construct and finish, and plush chairs require 3 hours to construct and finish. Upholstering takes 1 hour for standard chairs and 3 hours for plush chairs. There are 240 hours per day available for construction and finishing, and 150 hours per day are available for upholstering. If the revenue for standard chairs is \$89 and for plush chairs is \$133.50, how many of each type should be produced each day to maximize revenue?

Solution

Let x be the number of standard chairs produced each day, and let y be the number of plush chairs produced. Then the daily revenue function is given by $R = 89x + 133.5y$.

There are constraints for construction and finishing (no more than 240 hours/day) and for upholstering (no more than 150 hours/day). Thus we have the following.

$$\begin{aligned} \text{Construction/finishing constraint:} &\quad 2x + 3y \leq 240 \\ \text{Upholstering constraint:} &\quad x + 3y \leq 150 \end{aligned}$$

Because all quantities must be nonnegative, we also have the constraints $x \geq 0$ and $y \geq 0$.

Thus we seek to solve the following problem.

$$\text{Maximize } R = 89x + 133.5y \text{ subject to}$$

$$\begin{cases} 2x + 3y \leq 240 \\ x + 3y \leq 150 \\ x \geq 0,\ y \geq 0 \end{cases}$$

The feasible set is the closed and bounded region shaded in Figure 4.17. The corners of the feasible region are (0, 0), (120, 0), (0, 50), and (90, 20). All of these are obvious except (90, 20), which can be found by solving $2x + 3y = 240$ and $x + 3y = 150$ simultaneously. Testing the objective function at the corners gives the following.

$$\begin{aligned} \text{At } (0, 0), &\quad 89x + 133.5y = 0 \\ \text{At } (120, 0), &\quad 89x + 133.5y = 10{,}680 \\ \text{At } (0, 50), &\quad 89x + 133.5y = 6675 \\ \text{At } (90, 20), &\quad 89x + 133.5y = 10{,}680 \end{aligned}$$

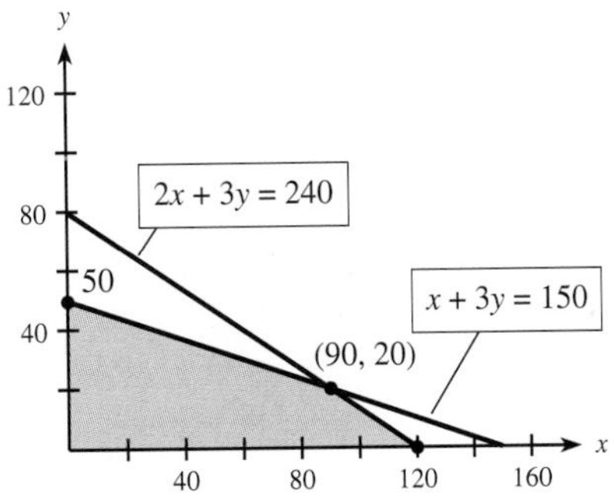

Figure 4.17

Thus the maximum revenue of \$10,680 occurs at either the point (120, 0) or the point (90, 20). This means that the revenue will be maximized not only at these two points but also at any point on the segment joining them. For example, the point (105, 10) is on this segment, and the revenue at this point is also \$10,680:

$$89x + 133.5y = 89(105) + 133.5(10) = 10,680$$

CHECKPOINT

1. The shaded region in the figure is determined by the following constraints.

$$\begin{cases} 2x + 3y \le 12 \\ 4x - 2y \le 8 \\ x \ge 0 \\ y \le 0 \end{cases}$$

Find the maximum value of the objective function $f = 4x + 3y$ on the region.

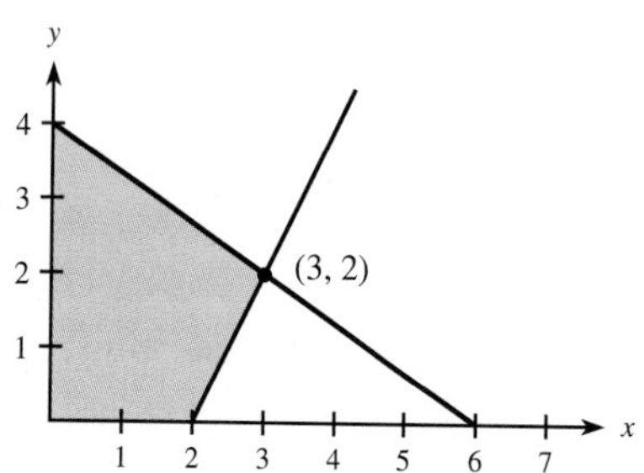

Although the examples so far have sought to maximize an objective function, the same procedures apply when a minimum is sought.

EXAMPLE 2

Minimize $C = x + y$ subject to the constraints

$$\begin{cases} 3x + 2y \ge 12 \\ x + 3y \ge 11 \\ x \ge 0, y \ge 0 \end{cases}$$

Solution

The graph of the constraint system is shown in Figure 4.18. Note that even though the feasible region is not closed and bounded, the objective function does have a minimum (but no maximum). Hence the corners still hold the key to the solution.

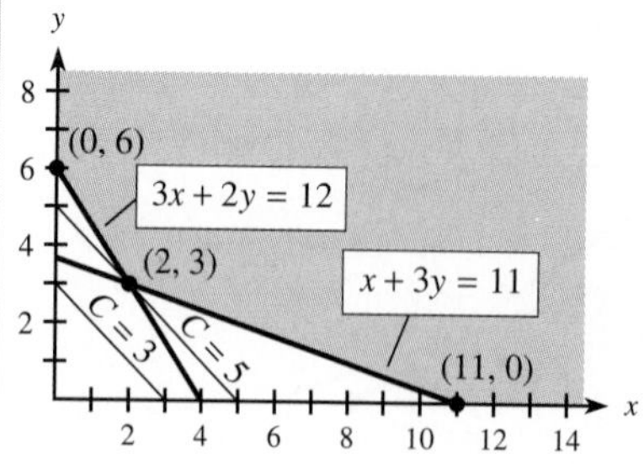

Figure 4.18

The corners (0, 6) and (11, 0) can be identified from the graph. The third corner, (2, 3), can be found by solving the equations of the two lines

$$3x + 2y = 12 \quad \text{and}$$
$$x + 3y = 11$$

simultaneously, as follows:

$$\begin{aligned} 3x + 2y &= 12 \\ -3x - 9y &= -33 \\ \hline -7y &= -21 \\ y &= 3 \\ x &= 2 \end{aligned}$$

Examining the value of C at each corner point, we have

$$\begin{aligned} &\text{At } (0, 6), \quad C = x + y = 6 \\ &\text{At } (11, 0), \quad C = x + y = 11 \\ &\text{At } (2, 3), \quad C = x + y = 5 \end{aligned}$$

Thus C is minimized at (2, 3) with minimum value $C = 5$. Note that for any smaller value of C, the graph of $C = x + y$ misses the feasible region.

CHECKPOINT

2. Minimize the objective function $g = 3x + 4y$ subject to the following constraints.

$$\begin{aligned} x + 2y \geq 12, \qquad & x \geq 0 \\ 3x + 4y \geq 30, \qquad & y \geq 2 \end{aligned}$$

EXAMPLE 3

Let's return to the Application Preview example of the two chemical plants that produce three types of fertilizer, low phosphorus (LP), medium phosphorus (MP), and high phosphorus (HP). Recall that the fertilizer is produced in a single production run, so the three types are produced in fixed proportions. The Macon plant produces 1 ton of LP, 2 tons of MP, and 3 tons of HP in a single operation, and it charges \$600 for what is produced in one operation, whereas one operation of the Jonesboro plant produces 1 ton of LP, 5 tons of MP, and 1 ton of HP, and it charges \$1000 for what it produces in one operation. If a customer needs 100 tons of LP, 260 tons of MP, and 180 tons of HP, how many production runs should be ordered from each plant to minimize costs?

Solution

If x represents the number of operations requested from the Macon plant and y represents the number of operations requested from the Jonesboro plant, then we seek to minimize cost

$$C = 600x + 1000y$$

The following table summarizes production capabilities and requirements.

	Macon Plant	*Jonesboro Plant*	*Requirements*
Units of LP	1	1	100
Units of MP	2	5	260
Units of HP	3	1	180

Using the number of operations requested and the fact that requirements must be met or exceeded, we can formulate the following constraints.

$$\begin{cases} x + y \geq 100 \\ 2x + 5y \geq 260 \\ 3x + y \geq 180 \\ x \geq 0, y \geq 0 \end{cases}$$

Graphing this system gives the feasible set shown in Figure 4.19. Again, the objective function has a minimum even though the feasible set is not closed and bounded. The corners are (0, 180), (40, 60), (80, 20), and (130, 0), where (40, 60) is obtained by solving $x + y = 100$ and $3x + y = 180$ simultaneously, and where (80, 20) is obtained by solving $x + y = 100$ and $2x + 5y = 260$ simultaneously.

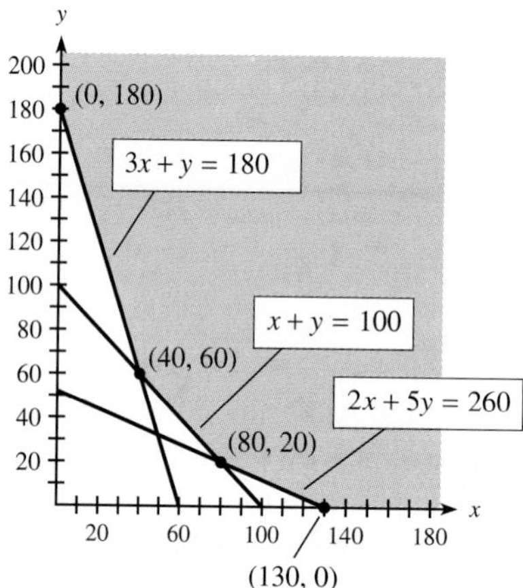

Figure 4.19

Evaluating $C = 600x + 1000y$ at each corner, we obtain

At (0, 180), $C = 180{,}000$
At (40, 60), $C = 84{,}000$
At (80, 20), $C = 68{,}000$
At (130, 0), $C = 78{,}000$

Thus, to minimize costs, the customer should place orders requiring 80 production runs from the Macon plant and 20 production runs from the Jonesboro plant.

EXAMPLE 4

Use a graphing utility to find the feasible region and to find the maximum value of $5x + 11y$ subject to the constraints

$$5x + 2y \le 54$$
$$2x + 4y \le 60$$
$$x \ge 0,\ y \ge 0$$

Solution

We write the inequalities above as equations, solved for y. Graphing these equations with a graphing calculator and using shading shows the closed and bounded region satisfying the inequalities (see Figure 4.20). By using an INTERSECT command or by using TRACE along the lines that form the borders of this region, we see that the points where the boundaries intersect are (0, 15), (6, 12), and (10.8, 0). These points can also be found algebraically. Testing the objective function at each of these corners gives the following values of f:

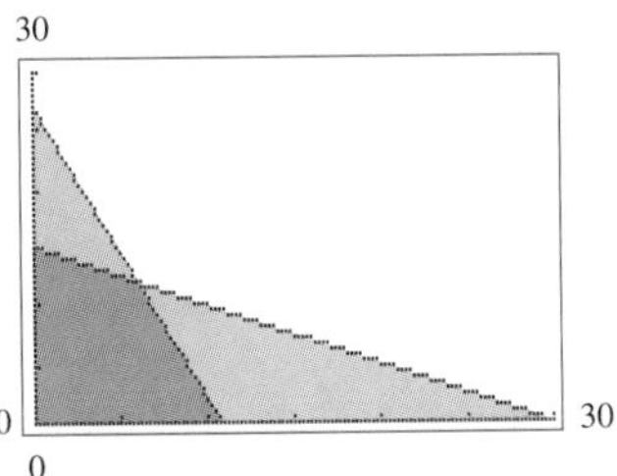

Figure 4.20

At (0, 15), $f = 165$
At (6, 12), $f = 162$
At (10.8, 0), $f = 54$

The maximum value of f is 165 at $x = 0$, $y = 15$.

CHECKPOINT SOLUTIONS

1. The values of f at the corners are found as follows.

   At (0, 0), $f = 0$
   At (2, 0), $f = 8$
   At (3, 2), $f = 12 + 6 = 18$
   At (0, 4), $f = 12$

   The maximum value of f is 18 at $x = 3$, $y = 2$.

2. The graph of the feasible region is shown in Figure 4.21. The values of g at the corners are found as follows:

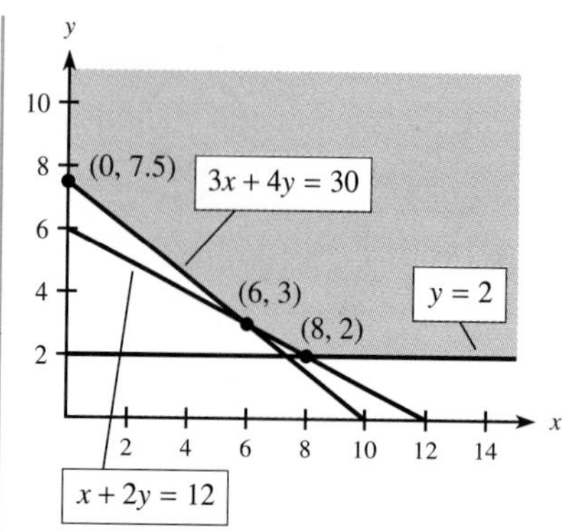

Figure 4.21

$$\begin{aligned} &\text{At } (0, 7.5), && g = 30 \\ &\text{At } (6, 3), && g = 18 + 12 = 30 \\ &\text{At } (8, 2), && g = 24 + 8 = 32 \end{aligned}$$

The minimum value of g is 30 at both (0, 7.5) and (6, 3). Thus any point on the border joining (0, 7.5) and (6, 3) will give the minimum value 30. For example, (2, 6) is on this border and gives the value $6 + 24 = 30$.

EXERCISE 4.3

1. The corners of a feasible region in a linear programming problem are (0, 0), (0, 8), (12, 0), and (6, 5). Which corner or corners give the maximum for the objective function $f = 6x + 3y$?
2. The corners of a feasible region in a linear programming problem are (0, 2), (3, 4), (5, 3), and (7, 0). Which corner or corners give the maximum for the objective function $f = 6x + 2y$?

In Problems 3–8, use the given feasible region determined by the constraint inequalities to find the maximum and minimum of the given objective function (if they exist).

3. $C = 2x + 3y$

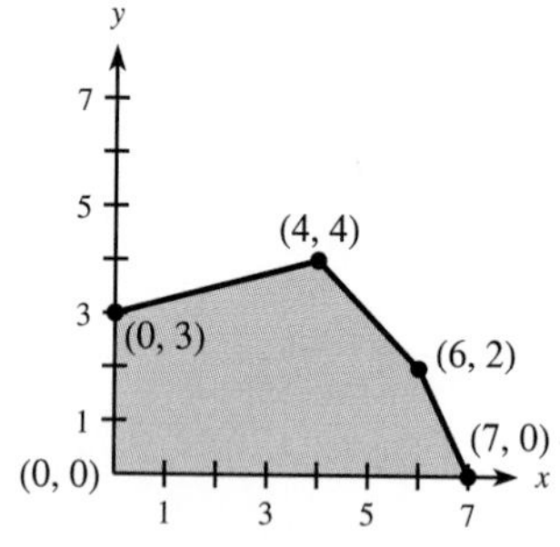

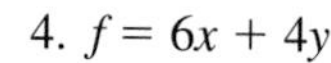

4. $f = 6x + 4y$

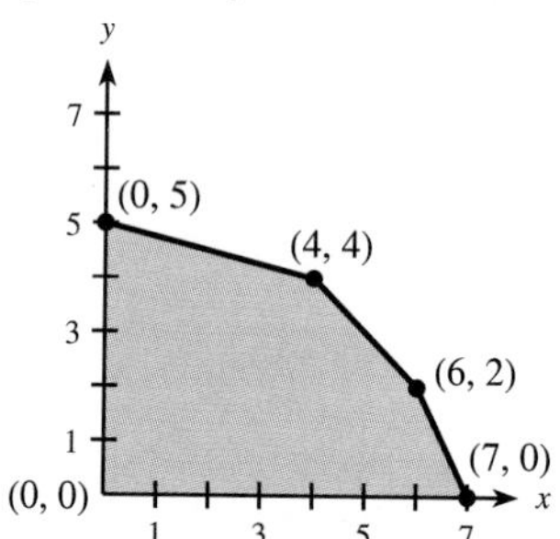

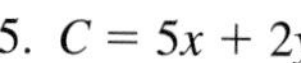

5. $C = 5x + 2y$

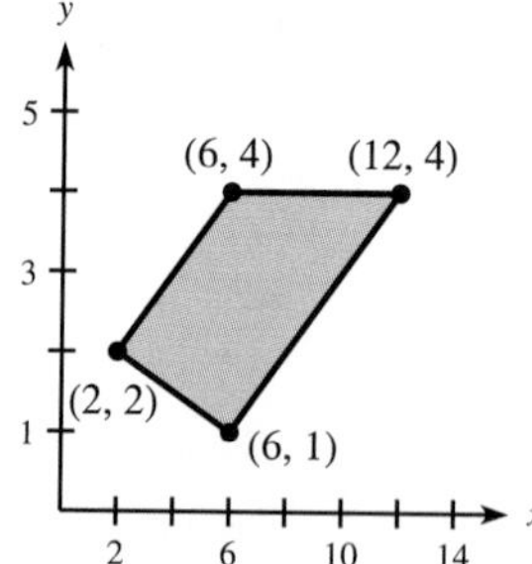

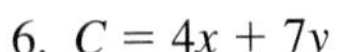

6. $C = 4x + 7y$

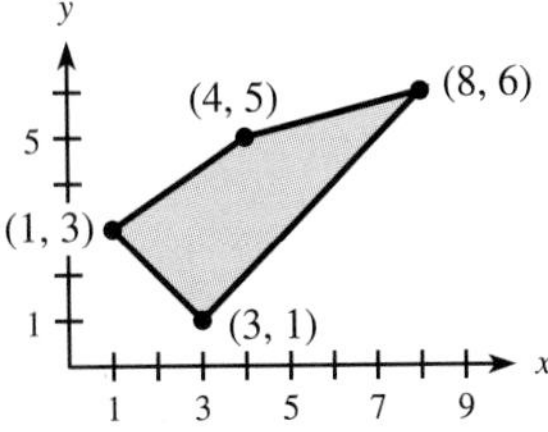

7. $f = 3x + 4y$

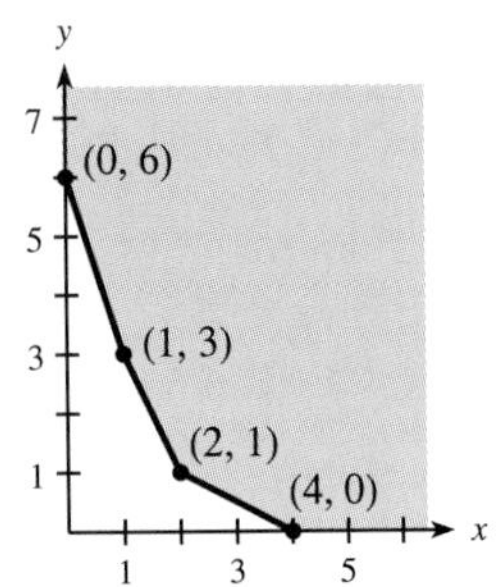

8. $f = 4x + 5y$

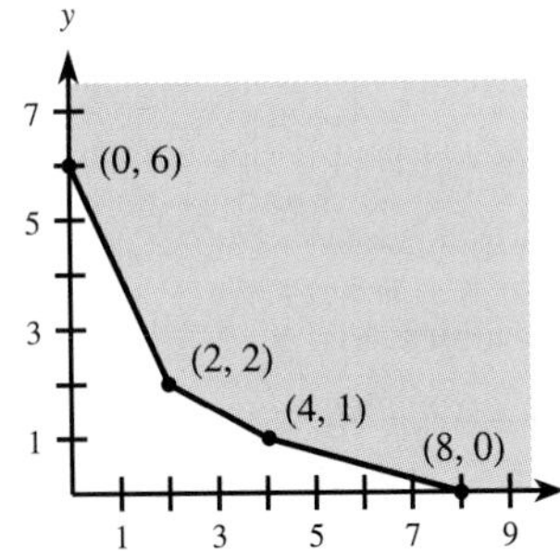

In each of Problems 9–12, the graph of the feasible region is shown. Find the corners of each feasible region, and maximize or minimize the function as directed.

9. Maximize $f = 3x + 2y$

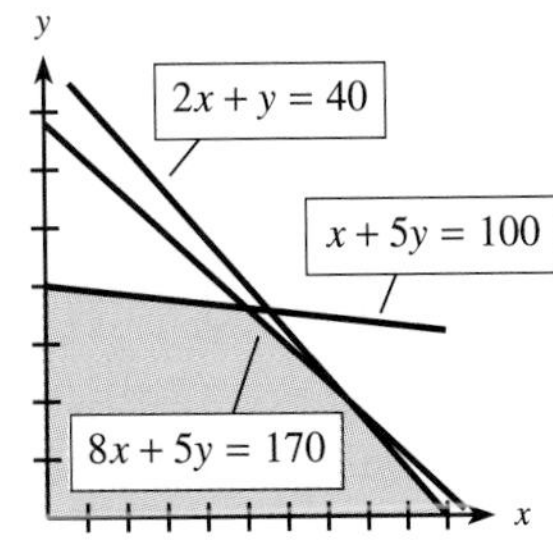

10. Maximize $f = 5x + 8y$

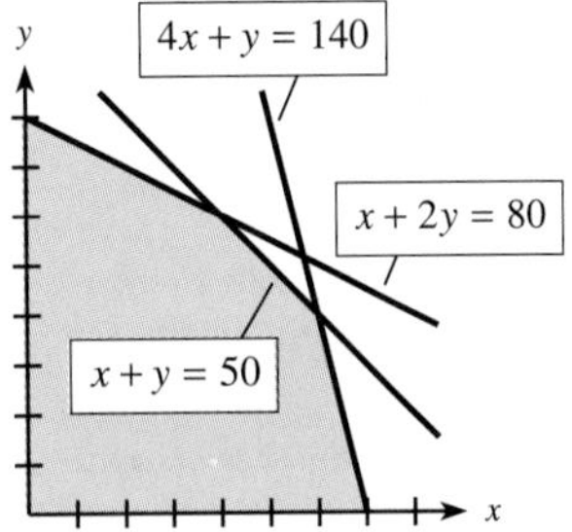

11. Minimize $g = 3x + 2y$

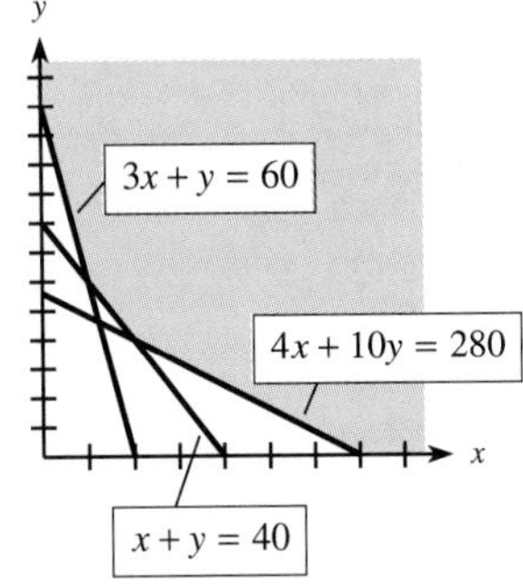

12. Minimize $g = x + 3y$

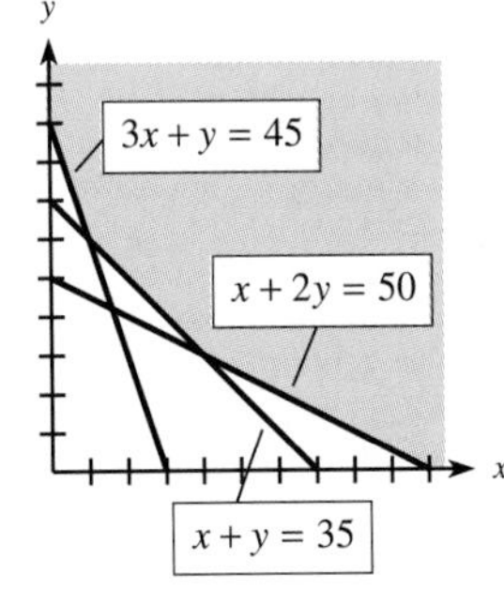

In Problems 13–20, find the indicated maximum or minimum value of the objective function in the linear programming problem. Note that the feasible regions for these problems are the solution regions sketched in Problems 21–28 of the previous set of exercises.

13. Maximize $f = 4x + 9y$ subject to

$$\begin{aligned} x + 5y &\le 200 \\ 2x + 3y &\le 134 \\ x \ge 0, y &\ge 0 \end{aligned}$$

14. Maximize $f = 2x + y$ subject to

$$\begin{aligned} -x + y &\le 2 \\ x + 2y &\le 10 \\ 3x + y &\le 15 \\ x \ge 0, y &\ge 0 \end{aligned}$$

15. Maximize $f = 3x + 2y$ subject to

$$\begin{aligned} x + 2y &\le 48 \\ x + y &\le 30 \\ 2x + y &\le 50 \\ x \ge 0, y &\ge 0 \end{aligned}$$

16. Maximize $f = 7x + 10y$ subject to

$$\begin{aligned} 3x + y &\le 9 \\ 3x + 2y &\le 12 \\ x + 2y &\le 8 \\ x \ge 0, y &\ge 0 \end{aligned}$$

17. Minimize $g = 9x + 10y$ subject to

$$\begin{aligned} x + 2y &\ge 19 \\ 3x + 2y &\ge 29 \\ x \ge 0, y &\ge 0 \end{aligned}$$

18. Minimize $g = 5x + 2y$ subject to

$$\begin{aligned} 4x + y &\ge 12 \\ x + y &\ge 9 \\ x + 3y &\ge 15 \\ x \ge 0, y &\ge 0 \end{aligned}$$

19. Minimize $g = 12x + 48y$ subject to

$$\begin{aligned} x + 3y &\ge 3 \\ 2x + 3y &\ge 5 \\ 2x + y &\ge 3 \\ x \ge 0, y &\ge 0 \end{aligned}$$

20. Minimize $g = 12x + 8y$ subject to

$$\begin{aligned} x + 2y &\ge 10 \\ 2x + y &\ge 11 \\ x + y &\ge 9 \\ x \ge 0, y &\ge 0 \end{aligned}$$

In Problems 21–36, solve the following linear programming problems.

21. Maximize $f = x + 3y$ subject to

$$\begin{aligned} x + 2y &\le 4 \\ 2x + y &\le 4 \\ x \ge 0, y &\ge 0 \end{aligned}$$

22. Maximize $f = 3x + 2y$ subject to

$$\begin{aligned} 2x + y &\le 8 \\ 2x + 3y &\le 12 \\ x \ge 0, y &\ge 0 \end{aligned}$$

23. Maximize $f = 3x + 4y$ subject to

$$\begin{aligned} x + y &\le 6 \\ 2x + y &\ge 10 \\ y &\le 4 \\ x \ge 0, y &\ge 0 \end{aligned}$$

24. Maximize $f = x + 3y$ subject to

$$\begin{aligned} x + 4y &\le 12 \\ y &\le 2 \\ x + y &\le 9 \\ x \ge 0, y &\ge 0 \end{aligned}$$

25. Maximize $f = 2x + 6y$ subject to

$$\begin{aligned} x + y &\le 7 \\ 2x + y &\le 12 \\ x + 3y &\le 15 \\ x \ge 0, y &\ge 0 \end{aligned}$$

26. Maximize $f = 4x + 2y$ subject to

$$\begin{aligned} x + 2y &\le 20 \\ x + y &\le 12 \\ 4x + y &\le 36 \\ x \ge 0, y &\ge 0 \end{aligned}$$

27. Minimize $g = 7x + 6y$ subject to

$$\begin{aligned} 5x + 2y &\ge 16 \\ 3x + 7y &\ge 27 \\ x \ge 0, y &\ge 0 \end{aligned}$$

28. Minimize $g = 22x + 17y$ subject to

$$\begin{aligned} 8x + 5y &\ge 100 \\ 12x + 25y &\ge 360 \\ x \ge 0, y &\ge 0 \end{aligned}$$

29. Minimize $g = 3x + y$ subject to

$$\begin{aligned} 4x + y &\ge 11 \\ 3x + 2y &\ge 12 \\ x \ge 0, y &\ge 0 \end{aligned}$$

30. Minimize $g = 50x + 70y$ subject to

$$\begin{aligned} 11x + 15y &\geq 225 \\ x + 3y &\geq 27 \\ x \geq 0, y &\geq 0 \end{aligned}$$

31. Minimize $f = x + 4y$ subject to

$$\begin{aligned} y &\leq 30 \\ 3x + 2y &\geq 75 \\ -3x + 5y &\geq 30 \\ x \geq 0, y &\geq 0 \end{aligned}$$

32. Minimize $f = 2x + y$ subject to

$$\begin{aligned} x &\leq 12 \\ x + 2y &\geq 20 \\ -3x + 2y &\leq 4 \\ x \geq 0, y &\geq 0 \end{aligned}$$

33. Maximize $f = x + 2y$ subject to

$$\begin{aligned} x + y &\geq 4 \\ 2x + y &\leq 8 \\ y &\leq 4 \end{aligned}$$

34. Maximize $f = 3x + 5y$ subject to

$$\begin{aligned} 2x + 4y &\geq 8 \\ 3x + y &\leq 7 \\ y &\leq 4 \end{aligned}$$

35. Minimize $g = 40x + 25y$ subject to

$$\begin{aligned} x + y &\geq 100 \\ -x + y &\leq 20 \\ -2x + 3y &\geq 30 \\ x \geq 0, y &\geq 0 \end{aligned}$$

36. Minimize $g = 3x + 8y$ subject to

$$\begin{aligned} 4x - 5y &\geq 50 \\ -x + 2y &\geq 4 \\ x + y &\leq 80 \\ x \geq 0, y &\geq 0 \end{aligned}$$

Applications

37. ***Manufacturing*** The Wellbuilt Company produces two types of wood chippers, Economy and Deluxe. The Deluxe model requires 3 hours to assemble and $\frac{1}{2}$ hour to paint, and the Economy model requires 2 hours to assemble and 1 hour to paint. The maximum number of assembly hours available is 24 per day, and the maximum number of painting hours available is 8 per day. If the profit on the Deluxe model is \$15 per unit and the profit on the Economy model is \$12 per unit, how many units of each model will maximize profit? (See Problem 31 of the previous exercise set.)

38. ***Learning environment*** An experiment involving learning in animals requires placing white mice and rabbits into separate, controlled environments, Environment I and Environment II. The maximum amount of time available in Environment I is 500 minutes, and the maximum amount of time available in Environment II is 600 minutes. The white mice must spend 10 minutes in Environment I and 25 minutes in Environment II, and the rabbits must spend 15 minutes in Environment I and 15 minutes in Environment II. Find the maximum possible number of animals that can be used in the experiment, and find the number of white mice and the number of rabbits that can be used. (See Problem 32 of the previous exercise set.)

39. ***Manufacturing*** A company manufactures two types of electric hedge-trimmers, one of which is cordless. The cord-type trimmer requires 2 hours to make, and the cordless model requires 4 hours. The company has only 800 work hours to use in manufacturing each day, and the packing department can package only 300 trimmers per day. If the company sells the cord-type model for \$30 and the cordless model for \$40, how many of each type should it produce per day to maximize its sales? (See Problem 33 of the previous exercise set.)

40. ***Manufacturing*** A firm manufactures bumper bolts and fender bolts for cars. One machine can produce 130 fender bolts per hour, and another machine can produce 120 bumper bolts per hour. The combined number of fender bolts and bumper bolts the packaging department can handle is 230 per hour. How many of each type of bolt should the firm produce hourly to maximize its sales if fender bolts sell for \$1 and bumper bolts sell for \$2? (See Problem 34 of the previous exercise set.)

41. ***Nutrition*** A privately owned lake contains two types of game fish, bass and trout. The owner provides two types of food, A and B, for these fish. Bass require 2 units of food A and 4 units of food B, and trout require 5 units of food A and 2 units of food B. If the owner has 800 units of each food, find the maximum number of fish that the lake can support.

42. ***Manufacturing*** A company manufactures two different sizes of boat lifts. Each size requires some time in the Welding and Assembly Department and some time in the Parts and Packaging Department. The smaller lift requires $\frac{3}{4}$ hour in Welding and Assembly and $1\frac{2}{3}$ hours in Parts and Packaging. The larger lift requires $1\frac{1}{2}$ hours in Welding and Assembly and 1 hour in Parts and Packaging. The factory has 156 hours/day available in Welding and Assembly and 174 hours/day available in Parts and Packaging. If demand for the lifts is at most 90 large and at most 100 small, and if profit is \$50 for each large lift and \$25 for each small lift, how many of each type should be produced each day to maximize profits?

43. ***Management*** A bank has two types of branches. A satellite branch employs 3 people, requires \$100,000 to construct and open, and generates an average daily revenue of \$10,000. A full-service branch employs 6 people, requires \$140,000 to construct and open, and generates an average daily revenue of \$18,000. The bank has up to \$2.98 million available to open new branches, and has decided to limit the new branches to a maximum of 25. If the bank further decides to hire at most 120 new employees, how many branches of each type should the bank open in order to maximize the average daily revenues?

44. ***Nutrition*** In a zoo, there is a natural habitat containing several feeding areas. One of these areas serves as a feeding area for two species, I and II, and it is supplied each day with 120 pounds of food A, 110 pounds of food B, and 57 pounds of food C. Each individual of species I requires 5 lb of A, 5 lb of B, and 2 lb of C, and each individual of species II requires 6 lb of A, 4 lb of B, and 3 lb of C. Find the maximum number of these species that can be supported.

45. ***Politics*** A candidate wishes to use a combination of radio and television advertisements in her campaign. Research has shown that each 1-minute spot on television reaches 0.09 million people and that each 1-minute spot on radio reaches 0.006 million. The candidate feels she must reach at least 2.16 million people, and she must buy a total of at least 80 minutes of advertisements. How many minutes of each medium should be used to minimize costs if television costs \$500/minute and radio costs \$100/minute? (See Problem 37 in the previous set of exercises.)

46. ***Nutrition*** In a hospital ward, the patients can be grouped into two general categories depending on their condition and the amount of solid foods they require in their diet. A combination of two diets is used for solid foods because they supply essential nutrients for recovery, but each diet has an amount of a substance deemed detrimental. The table summarizes the patient group, minimum diet requirements, and the amount of the detrimental substance. How many servings from each diet should be given each day in order to minimize the intake of this detrimental substance? (See Problem 38 in the previous set of exercises.)

	Diet A	*Diet B*	*Daily Reqs.*
Group 1	4 oz per serving	1 oz per serving	26 oz
Group 2	2 oz per serving	1 oz per serving	18 oz
Detrimental substance	0.18 oz per serving	0.07 oz per serving	

47. ***Manufacturing*** Two factories produce three different types of kitchen appliances. The table below summarizes the production capacity, the number of each type of appliance ordered, and the daily operating costs for the factories. How many days should each factory operate to fill the orders at minimum cost?

	Factory 1	*Factory 2*	*No. Ordered*
Appliance 1	80/day	20/day	1600
Appliance 2	10/day	10/day	500
Appliance 3	20/day	70/day	2000
Daily cost	\$10,000	\$20,000	

48. ***Nutrition*** In a laboratory experiment, two separate foods are given to experimental animals. Each food contains essential ingredients, A and B, for which the animals have a minimum requirement, and each food also has an ingredient C, which can be harmful to the animals. The table below summarizes this information.

	Food 1	*Food 2*	*Reqs.*
Ingredient A	10 units/g	3 units/g	49 units
Ingredient B	6 units/g	12 units/g	60 units
Ingredient C	3 units/g	1 unit/g	

How many grams of foods 1 and 2 should be given to the animals in order to satisfy the requirements for A and B while minimizing the amount of ingredient C ingested?

49. ***Manufacturing*** The Janie Gioffre Drapery Company makes three types of draperies at two different locations. At location I, it can make 10 pairs of deluxe drapes, 20 pairs of better drapes, and 13 pairs of standard drapes per day. At location II, it can make 20 pairs of deluxe, 50 pairs of better, and 6 pairs of standard per day. The company has orders for 2000 pairs of deluxe drapes, 4200 pairs of better drapes, and 1200 pairs of standard drapes. If the daily costs are $500 per day at location I and $800 per day at location II, how many days should Janie schedule at each location in order to fill the orders at minimum costs? Find the minimum costs.

50. ***Nutrition*** Two foods contain only proteins, carbohydrates, and fats. Food A costs $1 per pound and contains 30% protein and 50% carbohydrates. Food B costs $1.50 per pound and contains 20% protein and 75% carbohydrates. What combination of these two foods provides at least 1 pound of protein, $2\frac{1}{2}$ pounds of carbohydrates, and $\frac{1}{4}$ pound of fat at the lowest cost?

51. ***Manufacturing*** A sausage company makes two different kinds of hot dogs, regular and all beef. Each pound of all-beef hot dogs requires 0.75 lb of beef and 0.2 lb of spices, and each pound of regular hot dogs requires 0.18 lb of beef, 0.3 lb of pork, and 0.2 lb of spices. Suppliers can deliver at most 1020 lb of beef, at most 600 lb of pork, and at least 500 lb of spices. If the profit is $0.60 on each pound of all-beef hot dogs and $0.40 on each pound of regular hot dogs, how many pounds of each should be produced to obtain maximum profit? What is the maximum profit? (See Problem 39 of the previous exercise set.)

52. ***Manufacturing*** A cereal manufacturer makes two different kinds of cereal, Senior Citizen's Feast and Kids Go. Each pound of Senior Citizen's Feast requires 0.6 lb of wheat and 0.2 lb of vitamin-enriched syrup, and each pound of Kids Go requires 0.4 lb of wheat, 0.2 lb of sugar, and 0.2 lb of vitamin-enriched syrup. Suppliers can deliver at most 2800 lb of wheat, at most 800 lb of sugar, and at least 1000 lb of the vitamin-enriched syrup. If the profit is $0.90 on each pound of Senior Citizen's Feast and $1.00 on each pound of Kids Go, find the number of pounds of each cereal that should be produced to obtain maximum profit. Find the maximum profit. (See Problem 40 of the previous exercise set.)

53. ***Shipping costs*** TV Circuit has 30 large-screen televisions in a warehouse in Erie and 60 large-screen televisions in a warehouse in Pittsburgh. Thirty-five are needed in a store in Blairsville, and 40 are needed in a store in Youngstown. It costs $18 to ship from Pittsburgh to Blairsville and $22 to ship from Pittsburgh to Youngstown, whereas it costs $20 to ship from Erie to Blairsville and $25 to ship from Erie to Youngstown. How many televisions should be shipped from each warehouse to each store to minimize the shipping cost? *Hint:* If the number shipped from Pittsburgh to Blairsville is represented by x, then the number shipped from Erie to Blairsville is represented by $35 - x$.

54. ***Construction*** A contractor builds two types of homes. The Carolina requires one lot, $160,000 capital, and 160 worker-days of labor, whereas the Savannah requires one lot, $240,000 capital, and 160 worker-days of labor. The contractor owns 300 lots and has $48,000,000 available capital and 43,200 worker-days of labor. The profit on the Carolina is $40,000 and the profit on the Savannah is $50,000. List the corner points of the feasible region and find how many of each type of home should be built to maximize profit. Find the maximum possible profit.

4.4 The Simplex Method: Maximization

OBJECTIVE

- *To use the simplex method to maximize functions subject to constraints*

APPLICATION PREVIEW

Society has become increasingly aware of the hazards of toxic spill accidents and of pollution from dump sites, chemical plants, and nuclear power plants. Because it is difficult or impossible to eliminate a given facility, a desirable goal is to place it in a location that will maximize the distance from the facility to population centers and to environmentally sensitive locations. Solving such problems generally requires the use of more variables than can be handled with graphical linear programming methods. Consider the following business application.

The Solar Technology Company manufactures three different types of hand calculators and classifies them as small, medium, and large according to their calculating capabilities. The three types have production requirements given by the following table.

	Small	*Medium*	*Large*
Electronic circuit components	5	7	10
Assembly time (hours)	1	3	4
Cases	1	1	1

The firm has a monthly limit of 90,000 circuit components, 30,000 hours of labor, and 9000 cases. If the profit is \$6 for the small, \$13 for the medium, and \$20 for the large calculators, the number of each that should be produced to yield maximum profit and the amount of that maximum possible profit can be found by using the **simplex method** of linear programming.

The graphical method for solving linear programming problems is suitable only when there are three or fewer variables. If there are more than three variables, we could still find the corners of the convex region by simultaneously solving the equations of the boundaries. The objective function could then be evaluated at these corners. However, as the number of variables and the number of constraints increases, it becomes increasingly difficult to discover the corners and extremely time-consuming to evaluate the function. Furthermore, some problems have no solution, and this cannot be determined by using the method of solving the equations simultaneously.

The method discussed in this section is called the **simplex method** for solving linear programming problems. Basically, this method gives a systematic way of moving from one feasible corner of the convex region to another in such a way that the value of the objective function increases until an optimum value is reached or it is discovered that no solution exists.

The simplex method was developed by George Dantzig in 1947. One of the earliest applications of the method was to the scheduling problem that arose in connection with the Berlin airlift, begun in 1948. There the objective was to maximize the amount of goods delivered, subject to such constraints as the number of personnel, the number and size of available aircraft, and the number of runways. Since then, linear programming and the simplex method have been used to solve optimization problems in a wide variety of businesses. In fact, in a survey of Fortune 500 firms, 85% of the respondents indicated that they use linear programming.

In 1984, a researcher at Bell Labs named N. Karmarkar developed a new method (that uses interior points rather than corner points) for solving linear programming problems. When the number of variables is very large, Karmarkar's method is more efficient than the simplex method. For example, the Military Airlift Command and Delta Airlines have realized enormous savings by using Karmarkar's method on problems with thousands of variables. Despite the efficiency of Karmarkar's method with huge linear programming problems, the simplex method is just as effective for many applications and is still widely used.

In discussing the simplex method, we initially restrict ourselves to linear programming problems that satisfy the following conditions.

1. The objective function is to be maximized.
2. All variables are nonnegative.
3. The constraints are of the form

$$a_1x_1 + a_2x_2 + \dots + a_nx_n \le b$$

where $b > 0$.

These may seem to restrict the types of problems unduly, but in applied situations where the objective function is to be maximized, the constraints often satisfy conditions 2 and 3.

Before outlining a procedure for the simplex method, let us develop the procedure and investigate its rationale by seeing how the simplex method compares to the graphical method.

EXAMPLE 1

Maximize $f = 2x + 3y$ subject to

$$\begin{aligned} x + 2y &\le 10 \\ 2x + y &\le 14 \end{aligned}$$

Because we assume $x \ge 0$ and $y \ge 0$ in these problems, we shall no longer state these conditions.

Solution

The simplex method uses matrix methods on systems of equations, so we convert the constraint inequalities to equations by using **slack variables.** When we write $x + 2y \le 10$, this means that there is some nonnegative number—say s_1—such that s_1 added to $x + 2y$ equals 10. This variable s_1 is called a slack variable because it changes with x and y such that the sum is always 10. Thus

$$x + 2y + s_1 = 10$$

We can use another slack variable s_2 to write

$$2x + y \le 14 \quad \text{as} \quad 2x + y + s_2 = 14$$

If we write the objective function as an equation in the form

$$-2x - 3y + f = 0$$

then we have a system of equations that describes the problem.

$$\begin{cases} x + 2y + s_1 \qquad\qquad = 10 \\ 2x + y \qquad + s_2 \qquad = 14 \\ -2x - 3y \qquad\qquad + f = 0 \end{cases}$$

We seek to maximize f in the last equation subject to the constraints in the first two equations.

We can place this system of equations in a matrix called a **simplex matrix** or **simplex tableau.** Note that the objective function is in the last row of this matrix with all variables on the left side and that the first two rows correspond to the constraints.

$$A = \begin{array}{c} \\ \end{array}\left[\begin{array}{rrrrr|r} \overset{x}{1} & \overset{y}{2} & \overset{s_1}{1} & \overset{s_2}{0} & \overset{f}{0} & 10 \\ 2 & 1 & 0 & 1 & 0 & 14 \\ \hline -2 & -3 & 0 & 0 & 1 & 0 \end{array}\right] \begin{array}{l} \text{constraint 1} \\ \text{constraint 2} \\ \text{objective} \end{array}$$

From the graph in Figure 4.22, we know that (0, 0) is a feasible solution giving a value of 0 for f. Because the origin is always a feasible solution to the type of linear programming problems we are considering, the simplex method begins there and systematically moves to new corners while increasing the value of f.

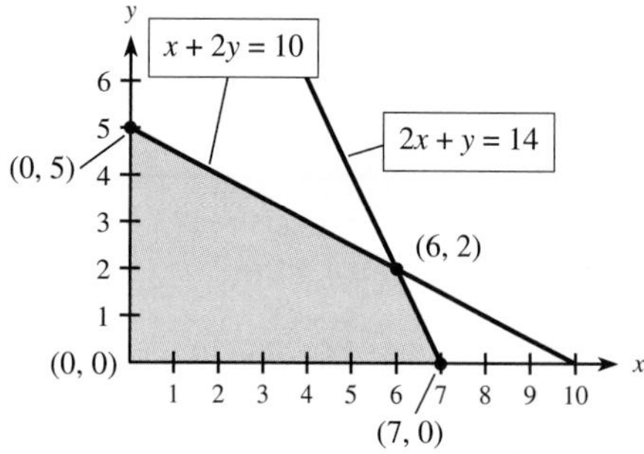

Figure 4.22

Note that if $x = 0$ and $y = 0$, then from rows 1, 2, and 3 we can read the values of s_1, s_2, and f, respectively.

$$s_1 = 10 \qquad s_2 = 14 \qquad f = 0$$

Recall that the last row still corresponds to the objective function $f = 2x + 3y$. If we seek to improve (that is, increase the value of) $f = 2x + 3y$ by changing the value of only one of the variables, then because the coefficient of y is larger, the greater improvement can be made by increasing y.

In general, we can find the variable that will improve f most by looking for the *most negative* value in the last row of the simplex matrix. The column that contains this value is called the **pivot** column.

The simplex matrix A is shown again below with the y-column indicated as the pivot column.

$$A = \left[\begin{array}{ccccc|c} x & y & s_1 & s_2 & f & \\ 1 & \boxed{②} & 1 & 0 & 0 & 10 \\ 2 & 1 & 0 & 1 & 0 & 14 \\ \hline -2 & -3 & 0 & 0 & 1 & 0 \end{array}\right]$$

↑ Most negative entry, so pivot column

The amount by which y can be increased is limited by the constraining equations (rows 1 and 2 of the simplex matrix). As the graph in Figure 4.22 shows, the constraints limit y to no larger than $y = 5$. Note that when the positive coefficients in the y-column of the simplex matrix are divided into the constants in the augment, the *smallest* quotient is $10/2 = 5$. The 2 (circled) in row 1, column 2 of the matrix A is called the **pivot entry.** The quotient $y = 5$ is found by setting $x = 0$ and $s_1 = 0$ in the row 1 equation. The point (0, 5) corresponds to a corner adjoining (0, 0) on the constraint region.

We will see that if we continue this method, then the next values for x and y will correspond to another corner of the constraint region.

To continue the solution, note that we used row 1 to determine the largest value for y. We can "transform" the simplex matrix into an equivalent simplex matrix by eliminating y from the remaining rows. If we also multiply the first row by $\frac{1}{2}$, we obtain the equivalent matrix B. We say that the variable y is now **basic.**

$$B = \left[\begin{array}{ccccc|c} x & y & s_1 & s_2 & f & \\ \frac{1}{2} & 1 & \frac{1}{2} & 0 & 0 & 5 \\ \textcircled{\tfrac{3}{2}} & 0 & -\frac{1}{2} & 1 & 0 & 9 \\ \hline -\frac{1}{2} & 0 & \frac{3}{2} & 0 & 1 & 15 \end{array}\right]$$

When we look at this transformed simplex matrix B, we notice that the bottom row represents $-\frac{1}{2}x + \frac{3}{2}s_1 + f = 15$, so $f = 15 + \frac{1}{2}x - \frac{3}{2}s_1$ depends on x and s_1 and $f = 15$ if $x = 0$ and $s_1 = 0$. Because x was held at 0 and s_1 must be nonnegative, the maximum value of f occurs at this point when $s_1 = 0$. Using $x = 0$ and $s_1 = 0$, rows, 1, 2, and 3 allow us to read the values of y, s_2, and f, respectively.

$$y = 5 \qquad s_2 = 9 \qquad f = 15$$

This solution corresponds to the corner (0, 5) in Figure 4.22. We also see that the value of f can be increased by increasing the value of x while holding the value s_1 at 0. This is clear from the equation

$$f = 15 + \frac{1}{2}x - \frac{3}{2}s_1$$

or by observing that the most negative number in the last row occurs in the x-column of the new matrix B.

The amount of the increase for x is limited by the constraints. We can discover the limitations on x by dividing the coefficient of x (if positive) into the constant term in the augment of each row and choosing the smaller quotient.

$$5 \div \frac{1}{2} = (5)(2) = 10 \qquad \text{from row 1}$$

$$9 \div \frac{3}{2} = (9)\left(\frac{2}{3}\right) = 6 \qquad \text{from row 2}$$

The smaller quotient is 6, so the pivot entry is in row 2, column 1 of B (circled). Note that if we move to a corner adjoining (0, 5) on the region in Figure 4.22, the largest x-value is 6 [at (6, 2)].

Let us again transform the simplex matrix B by eliminating x from all but the pivot row (the second row) in B and then multiplying the second row by $\frac{2}{3}$. The new simplex matrix is C.

$$C = \begin{array}{c} \begin{matrix} x & y & s_1 & s_2 & f & \end{matrix} \\ \left[\begin{array}{ccccc|c} 0 & 1 & \frac{2}{3} & -\frac{1}{3} & 0 & 2 \\ 1 & 0 & -\frac{1}{3} & \frac{2}{3} & 0 & 6 \\ \hline 0 & 0 & \frac{4}{3} & \frac{1}{3} & 1 & 18 \end{array}\right] \end{array}$$

The last row of this new matrix C corresponds to

$$\frac{4}{3}s_1 + \frac{1}{3}s_2 + f = 18 \quad \text{or} \quad f = 18 - \frac{4}{3}s_1 - \frac{1}{3}s_2$$

so any increase in s_1 or s_2 will cause a *decrease* in f. Observing this, or noting that all entries of the last row are nonnegative, tells us that we have the optimum solution when $s_1 = 0$ and $s_2 = 0$. Using this, rows 1, 2, and 3 give the values for y, x, and f, respectively.

$$y = 2 \qquad x = 6 \qquad f = 18$$

This corresponds to the corner (6, 2) on the graph and to the maximum value of 18 at this corner (which was found in the graphical solution in the Procedure/Example table in the previous section).

As we review the simplex solution in Example 1, observe that each of the feasible solutions corresponded to a corner (x, y) of the feasible region and that two of the four variables x, y, s_1, and s_2 were equal to 0 at each feasible solution.

At (0, 0): x and y were equal to 0.
At (0, 5): x and s_1 were equal to 0.
At (6, 2): s_1 and s_2 were equal to 0.

See Figure 4.23 for an illustration of how the simplex method moved us from corner to corner until the optimal solution was obtained.

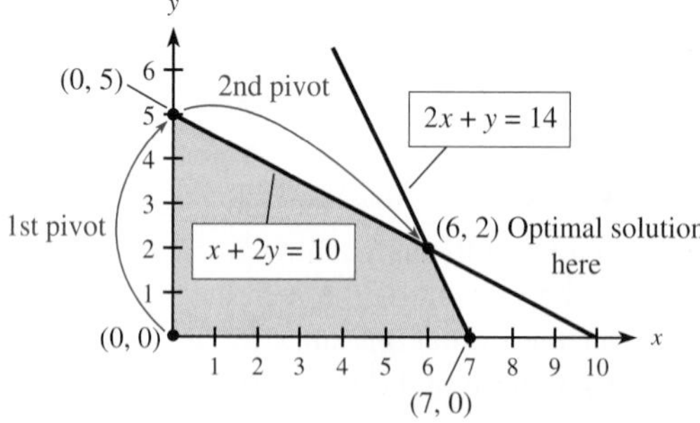

Figure 4.23

The values that were set equal to zero at each feasible solution are called the **nonbasic variables** for that feasible solution, and the other variables are called the **basic variables** for that feasible solution. At each step in the simplex method, we can identify the basic variables directly from the simplex matrix by noting that their columns contain a single entry of 1 with all other entries zeros. (Check this.) The remaining columns identify the nonbasic variables. Thus the values of the basic variables and of the objective function can be read from the rows of the simplex matrix, because the nonbasic variables are set equal to 0 at each step.

The simplex method involves a series of decisions and operations using matrices. It involves three major tasks.

Simplex Method Tasks

Task A: Setting up the matrix for the simplex method.
Task B: Determining necessary operations and implementing those to reach a solution.
Task C: Reading the solution from the simplex matrix.

From Example 1, it should be clear that the pivot selection and row operations of Task B are designed to move the system to the corner that achieves the greatest increase in the function and at the same time is sensitive to the nonnegative limitations on the variables. The procedure that combines these three tasks to maximize a function is given below.

Linear Programming (Simplex Method)

Procedure	Example
To use the simplex method to solve linear programming problems:	Maximize $f = 2x + 3y$ subject to $x + 2y \le 10$, $2x + y \le 14$.
1. Use slack variables to write the constraints as equations with positive constants on the right side.	1. $x + 2y \le 10$ becomes $x + 2y + s_1 = 10$. $2x + y \le 14$ becomes $2x + y + s_2 = 14$. Maximize f in $-2x - 3y + f = 0$.
2. Set up the simplex matrix. Put the objective function in the last row with all variables on the left and coefficient 1 for f.	2. $\begin{array}{c} \begin{matrix} x & y & s_1 & s_2 & f \end{matrix} \\ \left[\begin{array}{rrrrr\|r} 1 & 2 & 1 & 0 & 0 & 10 \\ 2 & 1 & 0 & 1 & 0 & 14 \\ \hline -2 & -3 & 0 & 0 & 1 & 0 \end{array}\right] \end{array}$
3. Find the pivot: (a) The pivot column has the most negative number in the last row. (If a tie occurs, use either column.)	3. (a) $\left[\begin{array}{rrrrr\|r} 1 & \boxed{2} & 1 & 0 & 0 & 10 \\ 2 & 1 & 0 & 1 & 0 & 14 \\ \hline -2 & -3 & 0 & 0 & 1 & 0 \end{array}\right]$ Most negative entry, so pivot column

Linear Programming (Simplex Method)

Procedure

(b) The positive coefficient in the pivot column that gives the smallest quotient when divided into the constant in the constraint rows of the augment is the pivot entry. If there is a tie, either coefficient may be chosen. If there are no positive coefficients in the pivot column, no solution exists. Stop.

Example

(b) Row 1 quotient: $\frac{10}{2} = 5$

Row 2 quotient: $\frac{14}{1} = 14$

Therefore, pivot row is row 1.

$$\left[\begin{array}{cc|ccc|c} 1 & \textcircled{2} & 1 & 0 & 0 & 10 \\ 2 & 1 & 0 & 1 & 0 & 14 \\ \hline -2 & -3 & 0 & 0 & 1 & 0 \end{array}\right]$$

Pivot entry is circled.

4. Use row operations with only the row containing the pivot entry to make the pivot entry a 1 and all other entries in the pivot column zeros. This makes the variable basic.

4. Row operations:
Multiply row 1 by $\frac{1}{2}$.
Add -1 times new row 1 to row 2.
Add 3 times new row 1 to row 3.

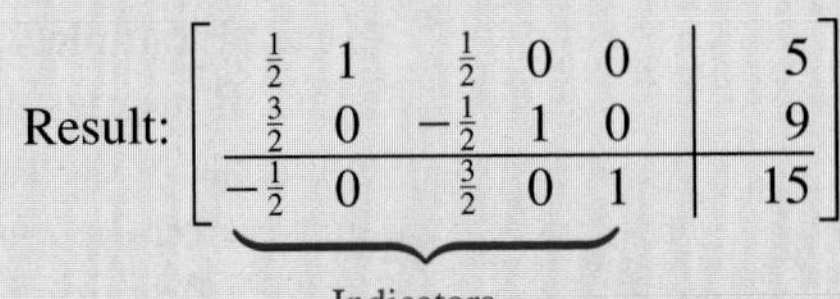

Result: $$\left[\begin{array}{ccccc|c} \frac{1}{2} & 1 & \frac{1}{2} & 0 & 0 & 5 \\ \frac{3}{2} & 0 & -\frac{1}{2} & 1 & 0 & 9 \\ \hline -\frac{1}{2} & 0 & \frac{3}{2} & 0 & 1 & 15 \end{array}\right]$$

Indicators

5. The numerical entries in the last row are the indicators of what to do next.
   (a) If there is a negative indicator, return to step 3.
   (b) If all indicators are positive or zero, an optimum value has been obtained for the objective function.

5. $-\frac{1}{2}$ is negative, so we identify a new pivot column and reduce again.

$$\left[\begin{array}{ccccc|c} \frac{1}{2} & 1 & \frac{1}{2} & 0 & 0 & 5 \\ \textcircled{\frac{3}{2}} & 0 & -\frac{1}{2} & 1 & 0 & 9 \\ \hline -\frac{1}{2} & 0 & \frac{2}{3} & 0 & 1 & 15 \end{array}\right] \quad \begin{array}{l} 5 \div \frac{1}{2} = 10 \\ 9 \div \frac{3}{2} = 6^* \end{array}$$

Pivot column

*6 is smaller quotient.

The new pivot entry is $\frac{3}{2}$ (circled). Now reduce, using the new pivot row, to obtain:

$$\left[\begin{array}{ccccc|c} 0 & 1 & \frac{2}{3} & -\frac{1}{3} & 0 & 2 \\ 1 & 0 & -\frac{1}{3} & \frac{2}{3} & 0 & 6 \\ \hline 0 & 0 & \frac{1}{2} & \frac{1}{3} & 1 & 18 \end{array}\right]$$

Indicators all 0 or positive, so the solution is complete.

6. Read the values of the basic variables and the objective function from the rows of the matrix (after setting the nonbasic variables equal to 0).

6. f is maximized at 18 when $y = 2$, $x = 6$, $s_1 = 0$, and $s_2 = 0$.

As we look back over our work, we can observe the following:

1. Steps 1 and 2 complete Task A.
2. Steps 3–5 complete Task B.
3. Step 6 completes Task C.

EXAMPLE 2

Complete Tasks A, B, and C to find the maximum value of $f = 4x + 3y$ subject to

$$
\begin{aligned}
x + 2y &\leq 8 \\
2x + y &\leq 10
\end{aligned}
$$

Solution

Task A: Introducing the slack variables gives the system of equations

$$
\begin{cases}
x + 2y + s_1 = 8 \\
2x + y + s_2 = 10 \\
-4x - 3y + f = 0
\end{cases}
$$

Writing this in a matrix with the objective function in the last row gives

$$
A = \begin{array}{c} \\ \end{array}
\left[\begin{array}{ccccc|c}
x & y & s_1 & s_2 & f & \\
1 & 2 & 1 & 0 & 0 & 8 \\
\textcircled{2} & 1 & 0 & 1 & 0 & 10 \\
\hline
-4 & -3 & 0 & 0 & 1 & 0
\end{array}\right]
\quad
\begin{array}{l}
8/1 = 8 \\
10/2 = 5^*
\end{array}
$$

↑ Most negative *Smallest quotient

Task B: The most negative entry in the last row is -4 (indicated by the arrow), so the pivot column is column 1. The smallest quotient formed by dividing the positive coefficients from column 1 into the constants in the augment is 5 (indicated by the asterisk), so row 2 is the pivot row. Thus the pivot entry is in row 2, column 1 (circled in matrix A above).

Next use row operations with row 2 only to make the pivot entry equal 1 and all other entries in column 1 equal zero. The row operations are

(a) Multiply row 2 by $\frac{1}{2}$.
(b) Add -1 times the new row 2 to row 1.
(c) Add 4 times the new row 2 to row 3.

This gives

$$
B = \left[\begin{array}{ccccc|c}
x & y & s_1 & s_2 & f & \\
0 & \textcircled{\tfrac{3}{2}} & 1 & -\frac{1}{2} & 0 & 3 \\
1 & \frac{1}{2} & 0 & \frac{1}{2} & 0 & 5 \\
\hline
0 & -1 & 0 & 2 & 1 & 20
\end{array}\right]
\quad
\begin{array}{l}
3 \div \frac{3}{2} = 2^* \\
5 \div \frac{1}{2} = 10
\end{array}
$$

↑ Most negative *Smallest quotient

The last row contains a negative entry, so we locate another pivot column and continue. The pivot column is indicated with an arrow, and the smallest quotient is indicated with an asterisk. The new pivot entry is circled.

The row operations using this pivot row are

(a) Multiply row 1 by $\frac{2}{3}$.
(b) Add $-\frac{1}{2}$ times the new row 1 to row 2.
(c) Add the new row 1 to row 3.

This gives

$$C = \begin{array}{c} \begin{matrix} x & y & s_1 & s_2 & f & \end{matrix} \\ \left[\begin{array}{ccccc|c} 0 & 1 & \frac{2}{3} & -\frac{1}{3} & 0 & 2 \\ 1 & 0 & -\frac{1}{3} & \frac{2}{3} & 0 & 4 \\ \hline 0 & 0 & \frac{2}{3} & \frac{5}{3} & 1 & 22 \end{array}\right] \end{array}$$

Because there are no negative indicators in the last row, the solution is complete.

Task C: s_1 and s_2 are nonbasic variables, so they equal zero. f is maximized at 22 when $y = 2$, $x = 4$, $s_1 = 0$, and $s_2 = 0$.

CHECKPOINT

1. Write the following constraints as equations by using slack variables.

$$\begin{aligned} 3x_1 + x_2 + x_3 &\le 9 \\ 2x_1 + x_2 + 3x_3 &\le 8 \\ 2x_1 + x_2 \quad\quad &\le 5 \end{aligned}$$

2. Set up the simplex matrix to maximize $f = 2x_1 + 5x_2 + x_3$ subject to the constraints discussed in Problem 1.

Using the simplex matrix from Problem 2, answer the following.

3. Find the first pivot entry.
4. Use row operations to make the variable of the first pivot entry basic.
5. Find the second pivot entry and make a second variable basic.
6. Find the solution.

EXAMPLE 3

The Solar Technology Company manufactures three different types of hand calculators and classifies them as small, medium, and large according to their calculating capabilities. The three types have production requirements given by the following table.

	Small	*Medium*	*Large*
Electronic circuit components	5	7	10
Assembly time (hours)	1	3	4
Cases	1	1	1

The firm has a monthly limit of 90,000 circuit components, 30,000 hours of labor, and 9000 cases. If the profit is \$6 for the small, \$13 for the medium, and \$20 for the large calculators, how many of each should be produced to yield maximum profit? What is the maximum profit?

Solution

Let x_1 be the number of small calculators produced, x_2 the number of medium calculators produced, and x_3 the number of large calculators produced. Then the problem is to maximize the profit $f = 6x_1 + 13x_2 + 20x_3$ subject to

Inequalities	*Equations with Slack Variables*
$5x_1 + 7x_2 + 10x_3 \le 90{,}000$	$5x_1 + 7x_2 + 10x_3 + s_1 = 90{,}000$
$x_1 + 3x_2 + 4x_3 \le 30{,}000$	$x_1 + 3x_2 + 4x_3 + s_2 = 30{,}000$
$x_1 + x_2 + x_3 \le 9{,}000$	$x_1 + x_2 + x_3 + s_3 = 9{,}000$

The simplex matrix, with the pivot entry circled, follows.

$$\begin{array}{c} \text{Slack variables} \\ \begin{array}{ccccccc|c} x_1 & x_2 & x_3 & s_1 & s_2 & s_3 & f & \\ \end{array} \\ \left[\begin{array}{ccccccc|c} 5 & 7 & 10 & 1 & 0 & 0 & 0 & 90{,}000 \\ 1 & 3 & \textcircled{4} & 0 & 1 & 0 & 0 & 30{,}000 \\ 1 & 1 & 1 & 0 & 0 & 1 & 0 & 9{,}000 \\ \hline -6 & -13 & -20 & 0 & 0 & 0 & 1 & 0 \end{array}\right] \\ \uparrow \text{ Most negative (column } x_3) \end{array} \quad \begin{array}{l} 90{,}000/10 = 9{,}000 \\ 30{,}000/4 = 7{,}500^* \\ 9{,}000/1 = 9{,}000 \\ \\ \text{*Smallest quotient} \end{array}$$

We now use row operations to change the pivot entry to 1 and create zeros elsewhere in the pivot column. The row operations are

1. Multiply row 2 by $\frac{1}{4}$ to convert the pivot entry to 1.
2. Using the new row 2,
   (a) Add -10 times new row 2 to row 1.
   (b) Add -1 times new row 2 to row 3.
   (c) Add 20 times new row 2 to row 4.

The result is the second simplex matrix.

$$\begin{array}{c} \text{Slack variables } (s_1, s_2, s_3) \\ \begin{array}{ccccccc|c} x_1 & x_2 & x_3 & s_1 & s_2 & s_3 & f & \\ \end{array} \\ \left[\begin{array}{ccccccc|c} \frac{5}{2} & -\frac{1}{2} & 0 & 1 & -\frac{5}{2} & 0 & 0 & 15{,}000 \\ \frac{1}{4} & \frac{3}{4} & 1 & 0 & \frac{1}{4} & 0 & 0 & 7500 \\ \textcircled{\frac{3}{4}} & \frac{1}{4} & 0 & 0 & -\frac{1}{4} & 1 & 0 & 1500 \\ \hline -1 & 2 & 0 & 0 & 5 & 0 & 1 & 150{,}000 \end{array}\right] \\ \uparrow \text{ Most negative (column } x_1) \end{array} \quad \begin{array}{l} 15{,}000 \div \frac{5}{2} = 6000 \\ 7500 \div \frac{1}{4} = 30{,}000 \\ 1500 \div \frac{3}{4} = 2000^* \\ \\ \text{*Smallest quotient} \end{array}$$

For this new simplex matrix, we check the last row indicators. Because there is an entry that is negative, we repeat the simplex process. Again the pivot entry is circled, and we apply row operations similar to the above. What are the row operations this time? The resulting matrix is

$$\begin{array}{c} \begin{array}{ccccccc|c} x_1 & x_2 & x_3 & s_1 & s_2 & s_3 & f & \\ \end{array} \\ \left[\begin{array}{ccccccc|c} 0 & -\frac{4}{3} & 0 & 1 & -\frac{5}{3} & -\frac{10}{3} & 0 & 10{,}000 \\ 0 & \frac{2}{3} & 1 & 0 & \frac{1}{3} & -\frac{1}{3} & 0 & 7000 \\ 1 & \frac{1}{3} & 0 & 0 & -\frac{1}{3} & \frac{4}{3} & 0 & 2000 \\ \hline 0 & \frac{7}{3} & 0 & 0 & \frac{14}{3} & \frac{4}{3} & 1 & 152{,}000 \end{array}\right] \end{array}$$

Checking the last row, we see all the entries are 0 or positive, so the solution is complete. Setting $x_2 = 0$, $s_2 = 0$, and $s_3 = 0$ (the nonbasic variables), we see from the matrix that $x_1 = 2000$, $x_3 = 7000$, $s_1 = 10{,}000$, and $f = 152{,}000$.

Thus the numbers of calculators that should be produced are

$$x_1 = 2000 \text{ small calculators}$$
$$x_2 = 0 \text{ medium calculators}$$
$$x_3 = 7000 \text{ large calculators}$$

in order to obtain a maximum profit of \$152,000 for the month. Note that s_1 being nonzero means that some circuit components are not used in this optimal situation.

Technology Note As our discussion indicates, the simplex method is quite complicated, even when the number of variables is relatively small (2 or 3). For this reason, computer software packages are often used to solve linear programming problems. It is useful to examine how the output from a typical simplex software package is similar to and different from the simplex matrices we've used.

EXAMPLE 4

In this example the calculator problem of maximizing $6x_1 + 13x_2 + 20x_3$ (Example 3) is solved again using a typical simplex software package. The constraints are the same:

$$\begin{aligned} 5x_1 + 7x_2 + 10x_3 &\le 90{,}000 && \text{Circuit components} \\ 1x_1 + 3x_2 + 4x_3 &\le 30{,}000 && \text{Assembly hours} \\ 1x_1 + 1x_2 + 1x_3 &\le 9000 && \text{Number of cases} \end{aligned}$$

A computer output gives the following (partial) tableau for this problem.

C		6	13	20	0	0	0	
		x_1	x_2	x_3	s_1	s_2	s_3	RHS
6	x_1	1	.33	0	0	−.33	1.33	2000
20	x_3	0	.67	1	0	.33	−.33	7000
0	s_1	0	−1.33	0	1	−1.67	−3.33	10,000
Z		6	15.33	20	0	4.67	1.33	152,000
C − Z		0	−2.33	0	0	−4.67	−1.33	

The C (first) row gives the coefficients from the objective function, and the second row identifies the variables in the problem.

(a) The solution is optimal when the values in the $C - Z$ (last) row are all nonpositive entries, and the optimal solution is the value of Z in the *RHS* (last) column of the output. What is the optimal solution to this problem?

(b) Row 3 is headed by x_1 and (the objective function coefficient of x_1) 6. The value of x_1 that gives the optimal solution is in the last (*RHS*) column of this row. What value of x_1 gives the optimal solution? What value of x_3 gives the optimal solution?

Solution

(a) The given solution in Row Z of the *RHS* column is optimal because all the values in the $C - Z$ (last) row are nonpositive entries. Row 6 (headed by Z) gives the maximum value of the objective function. This agrees with the solution in Example 3, where it was obtained with the simplex method.

(b) Row 3 (headed by x_1) gives $x_1 = 2000$ in the last column, so $x_1 = 2000$ gives the optimal solution. Row 4 (headed by x_3) gives $x_3 = 7000$, and row 5 (headed by s_1) gives $s_1 = 10{,}000$, so $x_3 = 7000$ gives the optimal solution. The absence of a row headed by x_2 tells us that $x_2 = 0$ gives the optimal solution. Note that these values agree with those found in Example 3, even though the computer output has a different form than the simplex matrix. Thus the profit is maximized at \$152,000 when 2000 small calculators, 0 medium calculators, and 7000 large calculators are produced.

All of the most popular spreadsheet programs (such as *Lotus 1-2-3*, *Microsoft Excel*, and *Quattro Pro*) offer a linear programming solver as a feature or as an add-in. Also, the mathematics software packages MATLAB and MAPLE have linear programming solvers. In addition, several Web sites are available that solve linear programming problems on-line, and others offer freeware linear programming solvers that can be downloaded. NEOS (Network-Enabled Optimization System) is a centralized governmental Web site (URL: http://www.mcs.anl.gov/home/otc/Server/) containing information and programs for solving optimization problems. (We should note that many of the materials available through NEOS require submission in a special form of both a problem and its dual, discussed in the next section. For more Web site information and application, see Extended Application/Group Project III at the end of this chapter.)

Nonunique Solutions

As we've seen, with graphical methods, a linear programming problem can have multiple solutions when the optimal value for the objective function occurs at two adjacent corners of the feasible region. Consider the following problem.

Maximize $f = 2x + y$ subject to

$$x + \frac{1}{2}y \leq 16$$

$$x + y \leq 24$$

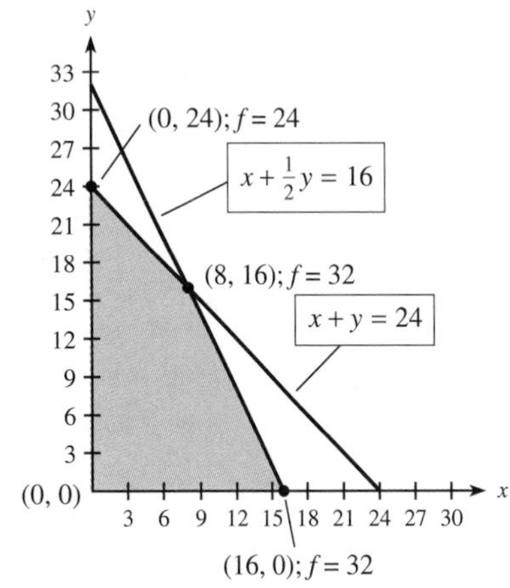

Figure 4.24

From Figure 4.24, we can see graphically that this problem has infinitely many solutions on the segment joining the points (8, 16) and (16, 0). Let's examine how we can discover these multiple solutions with the simplex method.

The simplex matrix for this problem (with the slack variables introduced) is

$$\begin{array}{c} \begin{matrix} x & y & s_1 & s_2 & f \end{matrix} \\ \left[\begin{array}{ccccc|c} \textcircled{1} & \frac{1}{2} & 1 & 0 & 0 & 16 \\ 1 & 1 & 0 & 1 & 0 & 24 \\ \hline -2 & -1 & 0 & 0 & 1 & 0 \end{array}\right] \end{array}$$

The most negative value in the last row is -2, so the x-column is the pivot column. The smallest quotient occurs in row 1; the pivot entry (circled) is 1. Making x a basic variable gives the following transformed simplex matrix.

$$\begin{array}{c} \begin{matrix} x & y & s_1 & s_2 & f & \end{matrix} \\ \left[\begin{array}{ccccc|c} 1 & \frac{1}{2} & 1 & 0 & 0 & 16 \\ 0 & \frac{1}{2} & -1 & 1 & 0 & 8 \\ \hline 0 & 0 & 2 & 0 & 1 & 32 \end{array}\right] \end{array}$$

This simplex matrix has no negative indicators, so the optimum value of f has been found ($f = 32$ when $x = 16$, $y = 0$). Note that one of the nonbasic variables (y) has a zero indicator. When this occurs, we can use the column with the 0 indicator as our pivot column and often obtain a new solution that has the same value for the objective function. In the present example, if we used the y-column as the pivot column, then we could find the second solution (at $x = 8$, $y = 16$). You will be asked to do this in Problem 41 of the exercises for this section.

Multiple Solutions When the simplex matrix for the optimum value of f has a nonbasic variable with a zero indicator in its column, there may be multiple solutions giving the *same* optimum value for f. We can discover whether another solution exists by using the column of that nonbasic variable as the pivot column.

It is also possible for a linear programming problem to have an unbounded solution (and thus no maximum value for f).

EXAMPLE 5

Maximize $f = 2x_1 + x_2 + 2x_3$ subject to

$$\begin{aligned} x_1 - 3x_2 + x_3 &\le 3 \\ x_1 - 6x_2 + 2x_3 &\le 6 \end{aligned}$$

if a maximum exists.

Solution

The simplex matrix for this problem (after introduction of slack variables) is

$$\begin{array}{c} \begin{matrix} x_1 & x_2 & x_3 & s_1 & s_2 & f & \end{matrix} \\ \left[\begin{array}{cccccc|c} 1 & -3 & \textcircled{1} & 1 & 0 & 0 & 3 \\ 1 & -6 & 2 & 0 & 1 & 0 & 6 \\ \hline -2 & -1 & -2 & 0 & 0 & 1 & 0 \end{array}\right] \end{array}$$

We see that the x_1 column and the x_3 column have the same most negative number, so we can use either column as the pivot column.

Using the x_3 column as the pivot column, we see that the smallest quotient is 3. Both coefficients give this quotient, so we can use either coefficient as the pivot entry. Using the element in row 1, we can make x_3 basic. The new simplex matrix is

$$\begin{array}{c} \begin{matrix} x_1 & x_2 & x_3 & s_1 & s_2 & f & \end{matrix} \\ \left[\begin{array}{cccccc|c} 1 & -3 & 1 & 1 & 0 & 0 & 3 \\ -1 & 0 & 0 & -2 & 1 & 0 & 0 \\ \hline 0 & -7 & 0 & 2 & 0 & 1 & 6 \end{array}\right] \end{array}$$

The most negative value in the last row of this matrix is -7, so the pivot column is the x_2-column. But there are no positive coefficients in the x_2-column; what does this mean? Even if we could pivot using the x_2-column, the variable x_1 and s_1 would have remained equal to 0. And the relationships among the remaining variables x_2, x_3, and s_2 would be the same.

$$-3x_2 + x_3 = 3 \quad \text{or} \quad x_3 = 3 + 3x_2$$
$$s_2 = 0$$

As x_2 is increased, x_3 also increases because all variables are nonnegative. This means that no matter how much we increase x_2, all other variables remain nonnegative. Furthermore, as we increase x_2 (and hence x_3), the value of f also increases. That is, there is no maximum value for f.

This example indicates that when a linear programming problem has an unbounded solution (and thus no maximum value for f), this is identified by the following conditions in the simplex method.

No Solution If, after finding the pivot column, there are no positive coefficients in that column, no maximum solution exists.

Some difficulties can arise in using the simplex method to solve linear programming problems. In the solution process, it is possible for the same simplex matrix to recur before an optimum value for the objective function is obtained. These "degeneracies" are discussed in more advanced courses. They rarely occur in applied linear programming problems, and we shall not encounter them in this text.

CHECKPOINT SOLUTIONS

1. $$\begin{aligned} 3x_1 + x_2 + x_3 + s_1 \qquad\qquad &= 9 \\ 2x_1 + x_2 + 3x_3 \qquad + s_2 \qquad &= 8 \\ 2x_1 + x_2 \qquad\qquad\qquad + s_3 &= 5 \end{aligned}$$

2. $$\left[\begin{array}{ccccccc|c} 3 & 1 & 1 & 1 & 0 & 0 & 0 & 9 \\ 2 & 1 & 3 & 0 & 1 & 0 & 0 & 8 \\ 2 & 1 & 0 & 0 & 0 & 1 & 0 & 5 \\ \hline -2 & -5 & -1 & 0 & 0 & 0 & 1 & 0 \end{array}\right]$$

3. The first pivot entry is the 1 in row 3, column 2.

4. $$\left[\begin{array}{ccccccc|c} 1 & 0 & 1 & 1 & 0 & -1 & 0 & 4 \\ 0 & 0 & 3 & 0 & 1 & -1 & 0 & 3 \\ 2 & 1 & 0 & 0 & 0 & 1 & 0 & 5 \\ \hline 8 & 0 & -1 & 0 & 0 & 5 & 1 & 25 \end{array}\right]$$

5. The second pivot entry is the 3 in row 2, column 3.
$$\left[\begin{array}{ccccccc|c} 1 & 0 & 0 & 1 & -\frac{1}{3} & -\frac{2}{3} & 0 & 3 \\ 0 & 0 & 1 & 0 & \frac{1}{3} & -\frac{1}{3} & 0 & 1 \\ 2 & 1 & 0 & 0 & 0 & 1 & 0 & 5 \\ \hline 8 & 0 & 0 & 0 & \frac{1}{3} & \frac{14}{3} & 1 & 26 \end{array}\right]$$

6. All indicators are 0 or positive, so f is maximized at 26 when $x_1 = 0$, $x_2 = 5$, and $x_3 = 1$.

EXERCISE 4.4

1. Convert the constraint inequalities $3x + 5y \le 15$ and $3x + 6y \le 20$ to equations containing slack variables.
2. Convert the constraint inequalities $2x + y \le 4$ and $3x + 2y \le 8$ to equations containing slack variables.

Task A: In Problems 3–8 set up the simplex matrix used to solve each linear programming problem. Assume all variables are nonnegative.

3. Maximize $f = 2x + 4y$ subject to
$$2x + 5y \le 30$$
$$x + 5y \le 25$$
4. Maximize $f = x + 5y$ subject to
$$3x + 7y \le 20$$
$$7x + 4y \le 44$$
5. Maximize $f = 4x + 9y$ subject to
$$x + 5y \le 200$$
$$2x + 3y \le 134$$
6. Maximize $f = x + 3y$ subject to
$$5x + y \le 50$$
$$x + 2y \le 50$$
7. Maximize $f = 2x + 5y + 2z$ subject to
$$2x + 7y + 9z \le 100$$
$$6x + 5y + z \le 145$$
$$x + 2y + 7z \le 90$$
8. Maximize $f = 3x + 5y + 11z$ subject to
$$2x + 3y + 4z \le 60$$
$$x + 4y + z \le 48$$
$$5x + y + z \le 48$$

Task B: In Problems 9–12, a simplex matrix is given. In each case, identify the pivot element.

9. $$\left[\begin{array}{ccccc|c} 2 & 4 & 1 & 0 & 0 & 24 \\ 1 & 1 & 0 & 1 & 0 & 5 \\ \hline -4 & -11 & 0 & 0 & 1 & 0 \end{array}\right]$$

10. $$\left[\begin{array}{cccccc|c} 3 & 1 & 1 & 1 & 0 & 0 & 100 \\ 4 & 3 & 6 & 0 & 1 & 0 & 250 \\ \hline -5 & -3 & -4 & 0 & 0 & 1 & 0 \end{array}\right]$$

11. $$\left[\begin{array}{cccccc|c} 10 & 27 & 1 & 0 & 0 & 0 & 200 \\ 4 & 51 & 0 & 1 & 0 & 0 & 400 \\ 15 & 27 & 0 & 0 & 1 & 0 & 350 \\ \hline -6 & -7 & 0 & 0 & 0 & 1 & 0 \end{array}\right]$$

12. $$\left[\begin{array}{ccccccc|c} 5 & 5 & 7 & 1 & 0 & 0 & 0 & 12 \\ 4 & 0 & 6 & 0 & 1 & 0 & 0 & 48 \\ 0 & 1 & 1 & 0 & 0 & 1 & 0 & 8 \\ \hline -1 & -3 & -1 & 0 & 0 & 0 & 1 & 0 \end{array}\right]$$

Task B: In Problems 13–20, a simplex matrix is given. Examine the indicators and determine whether the solution is complete. If it is not complete, find the next pivot element. If it is complete, indicate this.

13. $$\left[\begin{array}{ccccc|c} 2 & 0 & 1 & -\frac{3}{4} & 0 & 12 \\ 3 & 1 & 0 & \frac{1}{3} & 0 & 15 \\ \hline -4 & 0 & 0 & 3 & 1 & 15 \end{array}\right]$$

14. $$\left[\begin{array}{ccccc|c} 0 & 4 & 1 & -\frac{1}{5} & 0 & 2 \\ 1 & 2 & 0 & 1 & 0 & 4 \\ \hline 0 & 2 & 0 & 2 & 1 & 12 \end{array}\right]$$

15. $$\left[\begin{array}{ccccccc|c} 4 & 1 & 0 & 0 & \frac{3}{4} & 4 & 0 & 14 \\ -2 & 0 & 0 & 1 & -\frac{5}{8} & -2 & 0 & 6 \\ 3 & 0 & 1 & 0 & 2 & 6 & 0 & 11 \\ \hline 4 & 0 & 0 & 0 & 2 & \frac{1}{2} & 1 & 525 \end{array}\right]$$

16. $$\left[\begin{array}{ccccccc|c} 2 & 1 & 1 & 1 & 0 & 0 & 0 & 12 \\ -2 & 0 & -1 & 0 & 0 & 1 & 0 & 5 \\ 4 & 0 & 2 & 0 & 1 & 0 & 0 & 6 \\ \hline -2 & 0 & 5 & 0 & 0 & 0 & 1 & 30 \end{array}\right]$$

17. $$\left[\begin{array}{ccccccc|c} 4 & 4 & 1 & 0 & 0 & 2 & 0 & 12 \\ 2 & 4 & 0 & 1 & 0 & -1 & 0 & 4 \\ -3 & -11 & 0 & 0 & 1 & -1 & 0 & 6 \\ \hline -3 & -3 & 0 & 0 & 0 & 4 & 1 & 150 \end{array}\right]$$

18. $$\left[\begin{array}{ccccccc|c} 0 & 0 & -3 & 4 & -4 & 0 & 0 & 12 \\ 0 & 1 & -4 & 2 & 5 & 0 & 0 & 100 \\ 1 & 0 & -1 & -6 & -3 & 1 & 0 & 40 \\ \hline 0 & 0 & 6 & -1 & -3 & 0 & 1 & 380 \end{array}\right]$$

19. $$\left[\begin{array}{ccccccc|c} 4 & -1 & 0 & 1 & -5 & 0 & 0 & 5 \\ 1 & 0 & 1 & 0 & 1 & 0 & 0 & 12 \\ 3 & -3 & 0 & 0 & -2 & 1 & 0 & 6 \\ \hline -2 & -5 & 0 & 0 & 10 & 0 & 1 & 120 \end{array}\right]$$

20. $$\left[\begin{array}{ccccccc|c} 0 & 20 & -6 & 1 & 0 & -3 & 0 & 4 \\ 0 & 5 & -6 & 0 & 1 & -2 & 0 & 2 \\ 1 & -1 & 0 & 0 & 0 & 4 & 0 & 6 \\ \hline 0 & -9 & -12 & 0 & 0 & 10 & 1 & 20 \end{array}\right]$$

Task C: In Problems 21–26, a simplex matrix is given. In each case the solution is complete, so identify the maximum value of f and a set of values of the variables that gives this maximum value. If multiple solutions may exist, indicate this and locate the next pivot.

21. $$\left[\begin{array}{ccccc|c} 1 & 0 & \frac{3}{4} & -\frac{3}{4} & 0 & 11 \\ 0 & 1 & \frac{5}{12} & \frac{11}{12} & 0 & 9 \\ \hline 0 & 0 & 2 & 4 & 1 & 20 \end{array}\right]$$

22. $$\left[\begin{array}{ccccc|c} 0 & 4 & 1 & -\frac{1}{5} & 0 & 2 \\ 1 & 2 & 0 & 1 & 0 & 4 \\ \hline 0 & 2 & 0 & 2 & 1 & 12 \end{array}\right]$$

23. $$\left[\begin{array}{ccccccc|c} 4 & 1 & 0 & 0 & \frac{3}{4} & 4 & 0 & 14 \\ -2 & 0 & 0 & 1 & -\frac{8}{5} & -2 & 0 & 6 \\ 3 & 0 & 1 & 0 & 2 & 6 & 0 & 11 \\ \hline 4 & 0 & 0 & 0 & 2 & \frac{1}{2} & 1 & 525 \end{array}\right]$$

24. $$\left[\begin{array}{ccccccc|c} 0 & 1 & 0 & -3 & -1 & -4 & 0 & 12 \\ 1 & 0 & 0 & -2 & -1 & -5 & 0 & 16 \\ 0 & 0 & 1 & -1 & 2 & 1 & 0 & 22 \\ \hline 0 & 0 & 0 & 4 & 1 & 2 & 1 & 150 \end{array}\right]$$

25. $$\left[\begin{array}{cccccc|c} 1 & 0 & 3 & 0 & 6 & 0 & 50 \\ 0 & 0 & 4 & 1 & -4 & 0 & 6 \\ 0 & 1 & -2 & 0 & 2 & 0 & 10 \\ \hline 0 & 0 & 9 & 0 & 0 & 1 & 100 \end{array}\right]$$

26. $$\left[\begin{array}{cccccc|c} 0 & 1 & 2 & 1 & 0 & 0 & 20 \\ 1 & 0 & 6 & 1 & 0 & 0 & 50 \\ 0 & 0 & -1 & -2 & 1 & 0 & 10 \\ \hline 0 & 0 & 4 & 0 & 0 & 1 & 88 \end{array}\right]$$

In Problems 27–34, use the simplex method to maximize the given functions. Assume $x \geq 0$, $y \geq 0$ in all problems.

27. Maximize $f = 3x + 10y$ subject to
$$\begin{aligned} 14x + 7y &\leq 35 \\ 5x + 5y &\leq 50 \end{aligned}$$

28. Maximize $f = 7x + 10y$ subject to
$$\begin{aligned} 14x + 14y &\leq 98 \\ 8x + 10y &\leq 100 \end{aligned}$$

29. Maximize $f = 2x + 3y$ subject to
$$\begin{aligned} x + 2y &\leq 10 \\ x + y &\leq 7 \end{aligned}$$

30. Maximize $f = 5x + 30y$ subject to
$$\begin{aligned} 2x + 10y &\leq 96 \\ x + 10y &\leq 90 \end{aligned}$$

31. Maximize $f = x + 3y$ subject to
$$\begin{aligned} x + 2y &\leq 12 \\ y &\leq 5 \\ x + y &\leq 9 \end{aligned}$$

32. Maximize $f = 2x + 6y$ subject to
$$\begin{aligned} 2x + 5y &\leq 30 \\ x + 5y &\leq 25 \\ x + y &\leq 11 \end{aligned}$$

33. Maximize $f = 2x + y$ subject to
$$\begin{aligned} -x + y &\leq 2 \\ x + 2y &\leq 10 \\ 3x + y &\leq 15 \end{aligned}$$

34. Maximize $f = 3x + 2y$ subject to
$$\begin{aligned} x + 2y &\leq 48 \\ x + y &\leq 30 \\ 2x + y &\leq 50 \\ x + 10y &\leq 200 \end{aligned}$$

In Problems 35–40, assume $x \geq 0$, $y \geq 0$, and $z \geq 0$.

35. Maximize $f = 7x + 10y + 4z$ subject to
$$\begin{aligned} 3x + 5y + 4z &\leq 30 \\ 3x + 2y &\leq 4 \\ x + 2y &\leq 8 \end{aligned}$$

36. Maximize $f = 64x + 36y + 38z$ subject to

$$\begin{aligned} 4x + 2y + 2z &\leq 40 \\ 8x + 10y + 5z &\leq 90 \\ 4x + 4y + 3z &\leq 30 \end{aligned}$$

37. Maximize $f = 20x + 12y + 12z$ subject to

$$\begin{aligned} x + z &\leq 40 \\ x + y &\leq 30 \\ y + z &\leq 40 \end{aligned}$$

38. Maximize $f = 10x + 5y + 4z$ subject to

$$\begin{aligned} 2x + y + z &\leq 50 \\ x + 3y &\leq 20 \\ y + z &\leq 15 \end{aligned}$$

39. Maximize $f = 10x + 8y + 5z$ subject to

$$\begin{aligned} 2x + y + z &\leq 40 \\ x + 2y &\leq 10 \\ y + 3z &\leq 80 \end{aligned}$$

40. Maximize $f = x + 3y + z$ subject to

$$\begin{aligned} x + 4y &\leq 12 \\ 3x + 6y + 4z &\leq 48 \\ y + z &\leq 8 \end{aligned}$$

Problems 41–46 involve linear programming problems that have nonunique solutions.

The simplex matrices shown in Problems 41 and 42 indicate that an optimal solution has been found but that a second solution is possible. Find the second solution.

41. $\left[\begin{array}{ccccc|c} 1 & \frac{1}{2} & 1 & 0 & 0 & 16 \\ 0 & \frac{1}{2} & -1 & 1 & 0 & 8 \\ \hline 0 & 0 & 2 & 0 & 1 & 32 \end{array}\right]$

42. $\left[\begin{array}{ccccc|c} 1 & 2 & 0 & 1 & 0 & 40 \\ 0 & 1 & 1 & 2 & 0 & 15 \\ \hline 0 & 0 & 0 & 4 & 1 & 90 \end{array}\right]$

In Problems 43–46, use the simplex method to maximize each function (whenever possible) subject to the given constraints. If there is no solution, indicate this; if multiple solutions exist, find two of them.

43. Maximize $f = 3x + 2y$ subject to

$$\begin{aligned} x - 10y &\leq 10 \\ -x + y &\leq 40 \end{aligned}$$

44. Maximize $f = 10x + 15y$ subject to

$$\begin{aligned} -5x + y &\leq 24 \\ 2x - 3y &\leq 9 \end{aligned}$$

45. Maximize $f = 3x + 12y$ subject to

$$\begin{aligned} 2x + y &\leq 120 \\ x + 4y &\leq 200 \end{aligned}$$

46. Maximize $f = 8x + 6y$ subject to

$$\begin{aligned} 4x + 3y &\leq 24 \\ 10x + 9y &\leq 198 \end{aligned}$$

Applications

47. ***Manufacturing*** Newjet Inc. manufactures two types of printers, an inkjet printer and a laser printer. The company can make a total of 60 printers per day, and it has 120 labor-hours per day available. It takes 1 labor-hour to make an inkjet printer and 3 labor-hours to make a laser printer. The profit is \$40 per inkjet printer and \$60 per laser printer.
   (a) Write the simplex matrix to maximize the daily profit.
   (b) Find the maximum profit and the number of each type of printer that will give the maximum profit.

48. ***Construction*** A contractor builds two types of homes. The Carolina requires one lot, \$160,000 capital, and 160 worker-days of labor, and the Savannah requires one lot, \$240,000 capital, and 160 worker-days of labor. The contractor owns 300 lots and has \$48,000,000 available capital and 43,200 worker-days of labor. The profit on the Carolina is \$40,000 and the profit on the Savannah is \$50,000.
   (a) Write the simplex matrix to maximize the profit.
   (b) Use the simplex method to find how many of each type of home should be built to maximize profit, and find the maximum possible profit.

49. ***Manufacturing*** The Standard Steel Company produces two products, railroad car wheels and axles. Production requires processing in two departments: the smelting department (Department A) and the machining department (Department B). Department A has 50 hours available per week, and Department B has 43 hours available per week. Manufacturing one axle requires 4 hours in Department A and 3 hours in Department B, and manufacturing one wheel requires 3 hours in Department A and 5 hours in Department B. The profit is \$300 per axle and \$300 per wheel.
   (a) Set up the simplex matrix used to maximize the profit for the company.
   (b) Find the maximum possible profit and the number of axles and wheels that will maximize the profit.

50. ***Experimentation*** An experiment involves placing the males and females of a laboratory animal species in two separate controlled environments. There is a limited time available in these environments, and the experimenter wishes to maximize the number of animals subject to the constraints described.

	Males	*Females*	*Time Available*
Environment A	20 min	25 min	800 min
Environment B	20 min	15 min	600 min

How many males and how many females will maximize the total number of animals?

51. ***Shipping*** A produce wholesaler has determined that it takes $\frac{1}{2}$ hour of labor to sort and pack a crate of tomatoes and that it takes $1\frac{1}{4}$ hours to sort and pack a crate of peaches. The crate of tomatoes weighs 60 pounds, and the crate of peaches weighs 50 pounds. The wholesaler has 2500 hours of labor available each week and can ship 120,000 pounds per week. If profits are $1 per crate of tomatoes and $2 per crate of peaches, how many crates of each should be sorted, packed, and shipped to maximize overall profit? What is the maximum profit?

52. ***Manufacturing*** A manufacturer has two resources available, R_1 and R_2. These resources can be used to produce two different products, A and B, in the following way.

    2 units of R_1 and 5 units of R_2 give 1 unit of A.
    5 units of R_1 and 5 units of R_2 give 1 unit of B.
    30 units of R_1 and 25 units of R_2 are available.

    The profit for A is $2 per unit, and the profit for B is $4 per unit. How many units of products A and B should be produced to maximize overall profit?

53. ***Advertising*** Tire Corral has $6000 available per month for advertising. Newspaper ads cost $100 each and can occur a maximum of 21 times per month. Radio ads cost $300 each and can occur a maximum of 28 times per month at this price. Each newspaper ad reaches 6000 men over 20 years of age, and each radio ad reaches 8000 of these men. The company wants to maximize the number of ad exposures to this group. How many of each ad should it purchase?

54. ***Manufacturing*** A bicycle manufacturer makes a ten-speed and a regular bicycle. The ten-speed requires 2 units of steel and 6 units of aluminum in its frame and 12 special components for the hub, sprocket, and gear assembly. The regular bicycle requires 5 units each of steel and aluminum for its frame and 5 of the special components. Shipments are such that steel is limited to 100 units per day, aluminum is limited to 120 units per day, and the special components are limited to 180 units per day. If the profit is $30 on each ten-speed and $20 on each regular bike, how many of each should be produced to yield maximum profit? What is the maximum profit?

55. ***Advertising*** The total advertising budget for a firm is $200,000. The following table gives the costs per ad package for each medium and the number of exposures per ad package (with all numbers in thousands).

	Medium 1	*Medium 2*	*Medium 3*
Cost/package	10	4	5
Exposures/ package	3100	2000	2400

If the maximum numbers of medium 1, medium 2, and medium 3 packages that can be purchased are 18, 10, and 12, respectively, how many of each ad package should be purchased to maximize the number of ad exposures?

56. ***Advertising*** The Laposata Pasta Company has $12,000 available for advertising. The following table gives the cost per ad and the numbers of people exposed to its ads in three different media (with numbers in thousands).

Ad Packages	*Newspaper*	*Radio*	*TV*
Cost	2	2	4
Total audience	30	21	54
Working mothers	6	12	8

If the total available audience is 420,000, and if the company wishes to maximize the number of exposures to working mothers, how many ads of each type should it purchase?

57. ***Manufacturing*** A firm has decided to discontinue production of an unprofitable product. This will create excess capacity, and the firm is considering one or more of three possible new products, A, B, and C. The available weekly hours in the plant will be 477 hours in tool and die, 350 hours on the drill presses, and 150 hours on lathes. The hours of production required in each of these areas are as follows for each of the products.

	Tool and Die	Drill Press	Lathe
A	9	5	3
B	3	4	0
C	0.5	0	2

Furthermore, the sales department foresees no limitations on the sale of products A and C but anticipates sales of only 20 or fewer per week for B. If the unit profits expected are \$30 for A, \$9 for B, and \$15 for C, how many of each should be produced to maximize overall profit? What is the maximum profit?

58. ***Diagnostics*** A medical clinic performs three types of medical tests that use the same machines. Tests A, B, and C take 15 minutes, 30 minutes, and 1 hour, respectively, with respective profits of \$30, \$50, and \$100. The clinic has 4 machines available. One person is qualified to do test A, two to do test B, and one to do test C. If the clinic has a rush of customers for these tests, how many of each type should it schedule in a 12-hour day to maximize its profit?

59. ***Manufacturing*** A manufacturer of blenders produces three different sizes: Regular, Special, and Kitchen Magic. The production of each type requires the materials and amounts given in the table below. Suppose the manufacturer is opening a new plant and anticipates weekly supplies of 90,000 electrical components, 12,000 units of aluminum, and 9,000 containers. The unit profit is \$3 for the Regular, \$4 for the Special, and \$8 for the Kitchen Magic. A partial computer output for the problem is given below. Use the output to maximize $3x_1 + 4x_2 + 8x_3$ subject to the following constraints. How many of each type of blender should be produced to maximize overall profit? What is the maximum possible profit?

$$5x_1 + 12x_2 + 24x_3 \le 90{,}000 \quad \text{Electrical components}$$
$$1x_1 + 2x_2 + 4x_3 \le 12{,}000 \quad \text{Aluminum}$$
$$1x_1 + 1x_2 + 1x_3 \le 9{,}000 \quad \text{Containers}$$

	Regular	Special	Kitchen Magic
Electrical components	5	12	24
Units of aluminum	1	2	4
Blending containers	1	1	1

C		3	4	8	0	0	0	
		x_1	x_2	x_3	s_1	s_2	s_3	RHS
3	x_1	1	.67	0	0	−.33	1.33	8000
8	x_3	0	.33	1	0	.33	−.33	1000
0	s_1	0	.67	0	1	−6.33	1.33	26,000
Z		3	4.67	8	0	1.67	1.33	32,000
C − Z		0	−.67	0	0	−1.67	−1.33	

4.5 The Simplex Method: Duality and Minimization

OBJECTIVES

- *To formulate the dual for minimization problems*
- *To solve minimization problems using the simplex method on the dual*

APPLICATION PREVIEW

A beef producer is considering two different types of feed. Each feed contains some or all of the necessary ingredients for fattening beef. Brand 1 feed costs 20 cents per pound and brand 2 costs 30 cents per pound. The producer would like to determine how much of each brand to buy in order to satisfy the nutritional requirements for Ingredients A and B at minimum cost. Table 4.1 contains all the relevant data about nutrition and cost of each brand and the minimum requirements per unit of beef.

TABLE 4.1

	Brand 1	*Brand 2*	*Minimum Requirement*
Ingredient A	3 units/lb.	5 units/lb.	40 units
Ingredient B	4 units/lb.	3 units/lb.	46 units
Cost per pound	20¢	30¢	

In this section we will use the simplex method to solve minimization problems such as this one.

We have solved both maximization and minimization problems using graphical methods, but the simplex method, as discussed in the previous section, applies only to maximization problems. Now let us turn our attention to extending the use of the simplex method to solve minimization problems. Such problems might arise when a company seeks to minimize its production costs yet fill customer's orders or purchase items necessary for production.

As with maximization problems, we shall limit our discussion of minimization problems to those in which the constraints satisfy the following conditions.

1. All variables are nonnegative.
2. The constraints are of the form

$$a_1y_1 + a_2y_2 + \dots + a_ny_n \geq b$$

where b is positive.

In this section, we will show how minimization problems of this type can be solved using the simplex method.

EXAMPLE 1

Minimize $g = 11y_1 + 7y_2$ subject to

$$\begin{aligned} y_1 + 2y_2 &\geq 10 \\ 3y_1 + y_2 &\geq 15 \end{aligned}$$

Solution

Because the simplex method specifically seeks to increase the objective function, it does not apply to this minimization problem. Rather than solving this problem, let us solve a different but related problem called the **dual problem.**

In forming the dual problem, we write a matrix A that has a form similar to the simplex matrix, but without slack variables and with positive coefficients for the variables in the last row (and with the function to be minimized in the augment).

The matrix for the dual problem is formed by interchanging the rows and columns of matrix A. Matrix B is the result of this procedure, and we say that B is the **transpose** of A.

$$A = \left[\begin{array}{cc|c} 1 & 2 & 10 \\ 3 & 1 & 15 \\ \hline 11 & 7 & g \end{array}\right] \qquad B = \left[\begin{array}{cc|c} 1 & 3 & 11 \\ 2 & 1 & 7 \\ \hline 10 & 15 & g \end{array}\right]$$

Note that row 1 of A became column 1 of B, row 2 of A became column 2 of B, and row 3 of A became column 3 of B.

In the same way that we formed matrix A, we can now "convert back" from matrix B to the maximization problem that is the dual of the original minimization problem. We state this dual problem, using different letters to emphasize that this is a different problem from the original.

$$\text{Maximize } f = 10x_1 + 15x_2 \text{ subject to}$$
$$x_1 + 3x_2 \le 11$$
$$2x_1 + x_2 \le 7$$

The simplex method does apply to this maximization problem, with the following simplex matrix.

$$\begin{array}{ccccc} x_1 & x_2 & s_1 & s_2 & f \end{array}$$
$$\left[\begin{array}{ccccc|c} 1 & 3 & 1 & 0 & 0 & 11 \\ 2 & 1 & 0 & 1 & 0 & 7 \\ \hline -10 & -15 & 0 & 0 & 1 & 0 \end{array}\right]$$

Successive matrices that occur using the simplex method on the maximization problem are shown below with each pivot entry circled.

Beside the maximization problem is the solution to the original minimization problem, obtained using graphical methods.

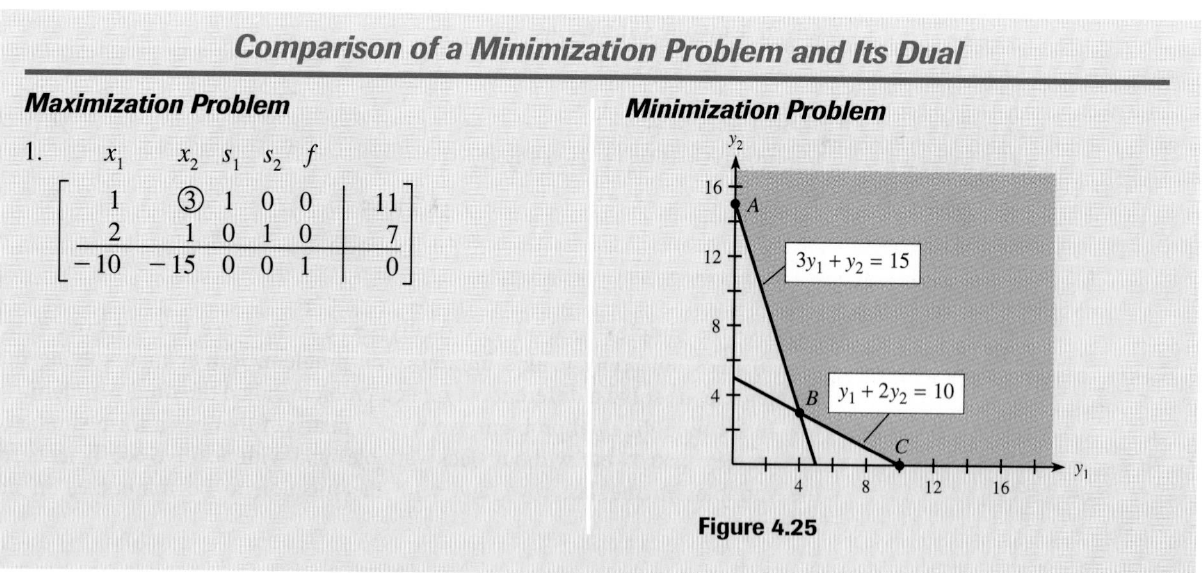

Comparison of a Minimization Problem and Its Dual

Maximization Problem

1. $\begin{array}{ccccc} x_1 & x_2 & s_1 & s_2 & f \end{array}$

$$\left[\begin{array}{ccccc|c} 1 & \textcircled{3} & 1 & 0 & 0 & 11 \\ 2 & 1 & 0 & 1 & 0 & 7 \\ \hline -10 & -15 & 0 & 0 & 1 & 0 \end{array}\right]$$

Minimization Problem

Figure 4.25

Maximization Problem	***Minimization Problem***
2. $\left[\begin{array}{ccccc|c} \frac{1}{3} & 1 & \frac{1}{3} & 0 & 0 & \frac{11}{3} \\ \textcircled{\frac{5}{3}} & 0 & -\frac{1}{3} & 1 & 0 & \frac{10}{3} \\ \hline -5 & 0 & 5 & 0 & 1 & 55 \end{array}\right]$	Corners: $A = (0, 15)$, $B = (4, 3)$, $C = (10, 0)$ $\left\{\text{obtained by solving simultaneously}\right.$
3. $\left[\begin{array}{ccccc|c} 0 & 1 & \frac{2}{5} & -\frac{1}{5} & 0 & 3 \\ 1 & 0 & -\frac{1}{5} & \frac{3}{5} & 0 & 2 \\ \hline 0 & 0 & 4 & 3 & 1 & 65 \end{array}\right]$	$g = 11y_1 + 7y_2$; At A, $g = 105$ At B, $g = 65$ At C, $g = 110$
Maximum $f = 65$ occurs at $x_1 = 2, x_2 = 3$.	Minimum $g = 65$ occurs at $y_1 = 4, y_2 = 3$.

In comparing the solutions to the previous two problems, the first thing to note is that the maximum value for f in the maximization problem and the minimum value for g in the minimization problem are the same. Furthermore, looking at the final simplex matrix, we can also find the values of y_1 and y_2 that give the minimum value for g by looking at the last row of the matrix.

$$\begin{array}{c} \begin{array}{cccccc} x_1 & x_2 & s_1 & s_2 & f & \end{array} \\ \left[\begin{array}{ccccc|c} 0 & 1 & \frac{2}{5} & -\frac{1}{5} & 0 & 3 \\ 1 & 0 & -\frac{1}{5} & \frac{3}{5} & 0 & 2 \\ \hline 0 & 0 & \underbrace{4 \quad 3} & & 1 & 65 \end{array}\right] \end{array}$$

These values give the values for y_1 and y_2 in the minimization problem.

From this example we see that the given minimization problem and its dual maximization problem are very closely related. Furthermore, it appears that when problems enjoy this relationship, the simplex method might be used to solve them both. The question is whether the simplex method will always solve them both. The answer is yes. This fact was proved by John von Neumann, and it is summarized by the Principle of Duality.

Principle of Duality

1. When a minimization problem and its dual have a solution, the maximum value of the function to be maximized is the same value as the minimum value of the function to be minimized.
2. When the simplex method is used to solve the maximization problem, the values for the variables that solve the corresponding minimization problem are *the last entries in the columns corresponding to the slack variables.*

EXAMPLE 2

Given the problem

$$\text{Minimize } 18y_1 + 12y_2 = g \text{ subject to}$$
$$2y_1 + y_2 \geq 8$$
$$6y_1 + 6y_2 \geq 36$$

(a) State the dual of this minimization problem.

(b) Use the simplex method to solve the given problem.

Solution

(a) The dual of this minimization problem will be a maximization problem. To form the dual, write the matrix A (without slack variables and with positive coefficients for the variables in the last row) for the given problem.

$$A = \left[\begin{array}{cc|c} 2 & 1 & 8 \\ 6 & 6 & 36 \\ \hline 18 & 12 & g \end{array}\right]$$

Transpose matrix A to yield matrix B.

$$B = \left[\begin{array}{cc|c} 2 & 6 & 18 \\ 1 & 6 & 12 \\ \hline 8 & 36 & g \end{array}\right]$$

Write the dual maximization problem, renaming the variables and the function.

Dual problem

$$\text{Maximize } f = 8x_1 + 36x_2 \text{ subject to}$$
$$2x_1 + 6x_2 \leq 18$$
$$x_1 + 6x_2 \leq 12$$

(b) To solve the given minimization problem, we use the simplex method on its dual maximization problem. The complete simplex matrix for each step is given, and each pivot entry is circled.

1. $$\left[\begin{array}{ccccc|c} 2 & 6 & 1 & 0 & 0 & 18 \\ 1 & \textcircled{6} & 0 & 1 & 0 & 12 \\ \hline -8 & -36 & 0 & 0 & 1 & 0 \end{array}\right]$$

2. $$\left[\begin{array}{ccccc|c} \textcircled{1} & 0 & 1 & -1 & 0 & 6 \\ \frac{1}{6} & 1 & 0 & \frac{1}{6} & 0 & 2 \\ \hline -2 & 0 & 0 & 6 & 1 & 72 \end{array}\right]$$

3. $$\begin{array}{c} \begin{array}{ccccc} x_1 & x_2 & s_1 & s_2 & f \end{array} \\ \left[\begin{array}{ccccc|c} 1 & 0 & 1 & -1 & 0 & 6 \\ 0 & 1 & -\frac{1}{6} & \frac{2}{6} & 0 & 1 \\ \hline 0 & 0 & 2 & 4 & 1 & 84 \end{array}\right] \end{array}$$

The last entries in the slack variable columns give the values of y_1 and y_2. Thus $y_1 = 2$, $y_2 = 4$, and $g = 84$.

CHECKPOINT

Perform the steps below to begin the process of finding the minimum value of $g = x + 4y$ subject to the following constraints.

$$2y_1 + 4y_2 \geq 18$$
$$y_1 + 5y_2 \geq 15$$

1. Form the matrix associated with the minimization problem.
2. Find the transpose of this matrix.
3. State the dual problem by using the matrix from Problem 2.
4. Write the simplex matrix of the dual of this problem.
5. The dual of this minimization problem has the following solution matrix. What is the solution to the minimization problem?

$$\left[\begin{array}{ccccc|c} 1 & 0 & \frac{5}{6} & -\frac{1}{6} & 0 & \frac{1}{6} \\ 0 & 1 & -\frac{2}{3} & \frac{1}{3} & 0 & \frac{2}{3} \\ \hline 0 & 0 & 5 & 2 & 1 & 13 \end{array}\right]$$

We now return to the Application Preview problem.

EXAMPLE 3

A beef producer is considering two different types of feed. Each feed contains some or all of the necessary ingredients for fattening beef. Brand 1 feed costs 20 cents per pound and brand 2 costs 30 cents per pound. How much of each brand should the producer buy in order to satisfy the nutritional requirements for Ingredients A and B at minimum cost? Table 4.2 contains all the relevant data about nutrition and cost of each brand and the minimum requirements per unit of beef.

TABLE 4.2

	Brand 1	*Brand 2*	*Minimum Requirement*
Ingredient A	3 units/lb.	5 units/lb.	40 units
Ingredient B	4 units/lb.	3 units/lb.	46 units
Cost per pound	20¢	30¢	

Solution

Let y_1 be the number of pounds of brand 1, and let y_2 be the number of pounds of brand 2. Then we can formulate the problem as follows:

$$\text{Minimize costs } C = 20y_1 + 30y_2 \text{ subject to}$$
$$3y_1 + 5y_2 \geq 40$$
$$4y_1 + 3y_2 \geq 46$$

The original linear programming problem is usually called the **primal problem.** If we write this in an augmented matrix, without the slack variables, then we can use the transpose to find the dual problem.

Primal | *Dual (with function renamed)*

$$\left[\begin{array}{cc|c} 3 & 5 & 40 \\ 4 & 3 & 46 \\ \hline 20 & 30 & C \end{array}\right] \qquad \left[\begin{array}{cc|c} 3 & 4 & 20 \\ 5 & 3 & 30 \\ \hline 40 & 46 & f \end{array}\right]$$

Thus the dual maximization problem is

$$\begin{aligned} \text{Maximize } f = 40x_1 + 46x_2 &\text{ subject to} \\ 3x_1 + 4x_2 &\le 20 \\ 5x_1 + 3x_2 &\le 30 \end{aligned}$$

The simplex matrix for this maximization problem, with slack variables included, is

1. $$\left[\begin{array}{ccccc|c} 3 & \textcircled{4} & 1 & 0 & 0 & 20 \\ 5 & 3 & 0 & 1 & 0 & 30 \\ \hline -40 & -46 & 0 & 0 & 1 & 0 \end{array}\right]$$

Solving the problem gives

2. $$\left[\begin{array}{ccccc|c} \frac{3}{4} & 1 & \frac{1}{4} & 0 & 0 & 5 \\ \textcircled{\frac{11}{4}} & 0 & -\frac{3}{4} & 1 & 0 & 15 \\ \hline -\frac{11}{2} & 0 & \frac{23}{2} & 0 & 1 & 230 \end{array}\right]$$

3. $$\left[\begin{array}{ccccc|c} 0 & 1 & \frac{5}{11} & -\frac{3}{11} & 0 & \frac{10}{11} \\ 1 & 0 & -\frac{3}{11} & \frac{4}{11} & 0 & \frac{60}{11} \\ \hline 0 & 0 & 10 & 2 & 1 & 260 \end{array}\right]$$

The solution to this problem is

$y_1 = 10$ lb of brand 1

$y_2 = 2$ lb of brand 2

Minimum cost $= 260$ cents or \$2.60 per unit of beef.

In a more extensive study of linear programming, the duality relationship that we have used in this section as a means to solve minimization problems can be shown to have more properties than we have mentioned. For example, in the Application Preview problem just completed, the dual maximization problem can be interpreted as follows:

$x_1 =$ cost per unit of Ingredient A

$x_2 =$ cost per unit of Ingredient B

$3x_1 + 4x_2 \le 20$	Cost per pound of the fattening ingredients in Brand 1
$5x_1 + 3x_2 \le 30$	Cost per pound of the fattening ingredients in Brand 2
$f = 40x_1 + 46x_2$	Maximize the money spent on ingredients for fattening

The values $x_1 = \frac{60}{11}$ cents/unit and $x_2 = \frac{10}{11}$ cents/unit from the solution to this dual maximization problem are called **shadow prices** (or marginal costs) for the desired ingredients. These values allow the beef producer to estimate additional costs associated with small changes in the minimum requirements for

Ingredients A and B. In particular, if the minimum requirement of each ingredient were increased by 1 unit, the approximate change in the feed cost per unit of beef would be

260 cents/unit	The current minimum cost
$+\ \frac{60}{11}$ cents/unit	Cost of one more unit of Ingredient A
$+\ \frac{10}{11}$ cents/unit	Cost of one more unit of Ingredient B

This gives the producer approximate new costs per unit of beef of $260\frac{70}{11}$ cents/unit or $\$2.66\frac{4}{11}$ per unit if each ingredient is increased by 1 unit.

Moreover, if the given problem is a profit maximization problem, then we can also form its dual, a minimization problem, by using matrix transposition, and we can think of the dual variables as representing "shadow prices" and interpret their values as marginal profits. A further discussion of shadow prices is given in Extended Application/Group Project II at the end of this chapter.

Technology Note

Many of the linear programming solvers available on the Internet require the submission of both a primal problem and its dual. Thus, if the primal problem is minimization, the dual will be maximization, and vice versa. Furthermore, we've seen that this duality relationship requires the constraints for both the maximization and minimization problems to have a special form. Of course, in applications the constraints may not be uniformly "≤" for a maximization problem or "≥" for a minimization problem; rather, they may be mixed. The next section discusses how to handle mixed constraints.

CHECKPOINT SOLUTIONS

1. $\left[\begin{array}{cc|c} 2 & 4 & 18 \\ 1 & 5 & 15 \\ \hline 1 & 4 & g \end{array}\right]$ 2. $\left[\begin{array}{cc|c} 2 & 1 & 1 \\ 4 & 5 & 4 \\ \hline 18 & 15 & g \end{array}\right]$

3. Find the maximum value of $f = 18x_1 + 15x_2$ subject to the constraints

$$2x_1 + x_2 \le 1$$
$$4x_1 + 5x_2 \le 4$$

4. The simplex matrix for this dual problem is

$$\left[\begin{array}{ccccc|c} 2 & 1 & 1 & 0 & 0 & 1 \\ 4 & 5 & 0 & 1 & 0 & 4 \\ \hline -18 & -15 & 0 & 0 & 1 & 0 \end{array}\right]$$

5. The minimum value is 13 when $y_1 = 5$ and $y_2 = 2$.

EXERCISE 4.5

In Problems 1–4, complete the following.

(a) Form the matrix associated with each given minimization problem and find its transpose.

(b) Write the dual maximization problem. Be sure to rename the variables.

1. Minimize $g = 3y_1 + y_2$ subject to

$$5y_1 + 2y_2 \ge 10$$
$$y_1 + 2y_2 \ge 6$$

2. Minimize $g = 3y_1 + 2y_2$ subject to

$$\begin{aligned} 3y_1 + y_2 &\geq 6 \\ 3y_1 + 4y_2 &\geq 12 \end{aligned}$$

3. Minimize $g = 5y_1 + 2y_2$ subject to

$$\begin{aligned} y_1 + y_2 &\geq 9 \\ y_1 + 3y_2 &\geq 15 \end{aligned}$$

4. Minimize $g = 9y_1 + 10y_2$ subject to

$$\begin{aligned} y_1 + 2y_2 &\geq 21 \\ 3y_1 + 2y_2 &\geq 27 \end{aligned}$$

In Problems 5 and 6, suppose a primal minimization problem and its dual maximization problem were solved by using the simplex method on the dual problem, and the final simplex method is given.

(a) Find the solution to the minimization problem. Use y_1, y_2, y_3 as the variables and g as the function.

(b) Find the solution to the maximization problem. Use x_1, x_2, x_3 as the variables and f as the function.

5. $$\left[\begin{array}{ccccccc|c} 1 & -\frac{2}{5} & 0 & 3 & -\frac{3}{5} & 0 & 0 & 5 \\ 0 & -\frac{3}{5} & 0 & -2 & -\frac{11}{5} & 1 & 0 & 3 \\ 0 & -\frac{4}{5} & 1 & -1 & -\frac{2}{5} & 0 & 0 & 9 \\ \hline 0 & 2 & 0 & 8 & 2 & 0 & 1 & 252 \end{array}\right]$$

6. $$\left[\begin{array}{ccccccc|c} 0 & 0 & 1 & \frac{2}{5} & \frac{5}{3} & \frac{4}{15} & 0 & 16 \\ 1 & 0 & 0 & -\frac{1}{5} & \frac{11}{3} & \frac{1}{5} & 0 & 19 \\ 0 & 1 & 0 & \frac{3}{5} & -\frac{2}{3} & \frac{2}{3} & 0 & 22 \\ \hline 0 & 0 & 0 & 12 & 15 & 20 & 1 & 554 \end{array}\right]$$

7–10. Use the simplex method to solve both the primal and dual problems from Problems 1–4.

In Problems 11–14, write the dual maximization problem, and then solve both the primal and the dual with the simplex method.

11. Minimize $g = 3y_1 + y_2$ subject to

$$\begin{aligned} 4y_1 + y_2 &\geq 11 \\ 3y_1 + 2y_2 &\geq 12 \\ 3y_1 + y_2 &\geq 6 \end{aligned}$$

12. Minimize $g = 2y_1 + 5y_2$ subject to

$$\begin{aligned} 4y_1 + y_2 &\geq 12 \\ y_1 + y_2 &\geq 9 \\ y_1 + 3y_2 &\geq 15 \end{aligned}$$

13. Minimize $g = 12y_1 + 48y_2 + 8y_3$ subject to

$$\begin{aligned} y_1 + 3y_2 \qquad &\geq 1 \\ 4y_1 + 6y_2 + y_3 &\geq 3 \\ 4y_2 + y_3 &\geq 1 \end{aligned}$$

14. Minimize $g = 12y_1 + 8y_2 + 10y_3$ subject to

$$\begin{aligned} y_1 + \qquad 2y_3 &\geq 10 \\ y_1 + y_2 \qquad &\geq 12 \\ 2y_1 + 2y_2 + y_3 &\geq 8 \end{aligned}$$

Use the simplex method in Problems 15–20.

15. Minimize $g = 2x + 10y$ subject to

$$\begin{aligned} 2x + y &\geq 11 \\ x + 3y &\geq 11 \\ x + 4y &\geq 16 \end{aligned}$$

16. Minimize $g = 8x + 4y$ subject to

$$\begin{aligned} 3x - 2y &\geq 6 \\ 2x + y &\geq 11 \end{aligned}$$

17. Minimize $g = 8x + 7y + 12z$ subject to

$$\begin{aligned} x + y + z &\geq 3 \\ y + 2z &\geq 2 \\ x &\geq 2 \end{aligned}$$

18. Minimize $g = 20x + 30y + 36z$ subject to

$$\begin{aligned} x + 2y + 3z &\geq 48 \\ 2x + 2y + 3z &\geq 70 \\ 2x + 3y + 4z &\geq 96 \end{aligned}$$

19. Minimize $g = 40y_1 + 90y_2 + 30y_3$ subject to

$$\begin{aligned} y_1 + 2y_2 + y_3 &\geq 16 \\ y_1 + 5y_2 + 2y_3 &\geq 18 \\ 2y_1 + 5y_2 + 3y_3 &\geq 38 \end{aligned}$$

20. Minimize $w = 48y_1 + 20y_2 + 8y_3$ subject to

$$\begin{aligned} 4y_1 + 2y_2 + y_3 &\geq 30 \\ 12y_1 + 4y_2 + 3y_3 &\geq 60 \\ 2y_1 + 3y_2 + y_3 &\geq 40 \end{aligned}$$

In Problems 21 and 22, a primal maximization problem is given.

(a) Form the dual minimization problem.

(b) Solve both the primal and the dual problem with the simplex method.

21. Maximize $f = 40x_1 + 20x_2$ subject to

$$\begin{aligned} 3x_1 + 2x_2 &\leq 120 \\ x_1 + x_2 &\leq 50 \end{aligned}$$

22. Maximize $f = 28x_1 + 12x_2$ subject to

$$\begin{aligned} 7x_1 + 12x_2 &\leq 50 \\ 2x_1 + 6x_2 &\leq 100 \end{aligned}$$

Applications

23. ***Manufacturing*** The Video Star Company makes two different models of VCRs, which are assembled on two different assembly lines. Line 1 can assemble 30 units of the Star model and 40 units of the Prostar model per hour, and Line 2 can assemble 150 units of the Star model and 40 units of the Prostar model per hour. The company needs to produce at least 270 units of the Star model and 200 units of the Prostar model to fill an order. If it costs $200 per hour to run Line 1 and $400 per hour to run Line 2, how many hours should each line be run to fill the order at the minimum cost? What is the minimum cost?

24. ***Nutrition*** A pork producer is considering two types of feed that contain the necessary ingredients for the nutritional requirements for fattening hogs. Red Star Brand contains 9 units of ingredient A and 12 units of ingredient B, and Blue Chip Brand contains 15 units of ingredient A and 9 units of ingredient B. The nutritional requirement for the hogs is at least 120 units of ingredient A and at least 138 units of ingredient B.
    (a) If Red Star Brand costs 50 cents per pound and Blue Chip Brand costs 75 cents, how many pounds of each brand should be bought to satisfy the nutritional requirements at the minimum cost?
    (b) What is the minimum cost?

25. ***Production*** A small company produces two products, I and II, at three facilities, A, B, and C. They have orders for 2000 of product I and 1200 of product II. The production capacity and cost per week to operate each facility are summarized in the following table.

	A	*B*	*C*
I	200	200	400
II	100	200	100
Cost/week	$1000	$3000	$4000

How many weeks should each facility operate to fill the company's orders at a minimum cost, and what is the minimum cost?

26. ***Nutrition*** In a hospital ward, the patients can be grouped into two general categories depending on their condition and the amount of solid foods they require in their diet. A combination of two diets is used for solid foods because they supply essential nutrients for recovery, but each diet has an amount of a substance deemed detrimental. The table below summarizes the patient group, diet requirements, and the amounts of the detrimental substance. How many servings from each diet should be given each day in order to minimize the intake of this detrimental substance?

	Diet A	*Diet B*	*Daily Reqs.*
Group 1	4 oz per serving	1 oz per serving	26 oz
Group 2	2 oz per serving	1 oz per serving	18 oz
Detrimental substance	0.18 oz per serving	0.07 oz per serving	

27. ***Production*** Two factories produce three different types of kitchen appliances. The following table summarizes the production capacity, the number of each type of appliance ordered, and the daily operating costs for the factories. How many days should each factory operate to fill the orders at minimum cost?

	Factory 1	*Factory 2*	*No. Ordered*
Appliance 1	80/day	20/day	1600
Appliance 2	10/day	10/day	500
Appliance 3	50/day	20/day	1900
Daily cost	$10,000	$20,000	

28. ***Nutrition*** In a laboratory experiment, two separate foods are given to experimental animals. Each food contains essential ingredients, A and B, for which the animals have minimum requirements, and each food also has an ingredient C, which can be harmful to the animals. The table below summarizes this information.

	Food 1	*Food 2*	*Reqs.*
Ingredient A	10 units/g	3 units/g	49 units
Ingredient B	6 units/g	12 units/g	60 units
Ingredient C	3 units/g	1 unit/g	

How many grams of foods 1 and 2 will satisfy the requirements for A and B and minimize the amount of ingredient C that is ingested?

29. ***Politics*** A political candidate wishes to use a combination of radio and TV advertisements in her campaign. Research has shown that each 1-minute spot

on TV reaches 0.9 million people and that each 1 minute on radio reaches 0.6 million. The candidate feels she must reach 63 million people, and she must buy at least 90 minutes of advertisements. How many minutes of each medium should be used if TV costs \$500 per minute, radio costs \$100 per minute, and the candidate wishes to minimize costs?

30. ***Production*** The James MacGregor Mining Company owns three mines, I, II, and III. Three ores, A, B, and C, are mined at these mines. For each ore, the number of tons per week available from each mine and the number of tons per week required to fill orders are given in the following table.

		I	*II*	*III*	*Required Tons/Week*
Ore Types	A	10	10	10	90
	B	0	10	10	50
	C	10	0	10	60
Cost/day		\$6000	\$8000	\$12,000	

Find the number of days the company should operate each mine so that orders are filled at minimum cost. Find the minimum cost.

31. ***Nutrition*** A hospital wishes to provide at least 24 units of nutrient A and 16 units of nutrient B in a meal, while minimizing the cost of the meal. If three types of food are available, with the nutritional value and costs (per ounce) given in the following table, how many ounces of each food should be served to minimize cost?

	Units of Nutrient A	*Units of Nutrient B*	*Cost*
Food I	2	1	1
Food II	2	5	5
Food III	2	1	2

32. ***Scheduling*** Each nurse works 8 consecutive hours at the Beaver Medical Center. The center has the following staffing requirements for each 4-hour work period.

Work Period	*Nurses Needed*
1 (7–11)	40
2 (11–3)	20
3 (3–7)	30
4 (7–11)	40
5 (11–3)	20
6 (3–7)	10

(a) If y_1 represents the number of nurses starting in period 1, y_2 the number starting in period 2, and so on, write the linear programming problem that will minimize the total number of nurses needed. (Note that the nurses who begin work in period 6, work periods 6 and 1 for their 8-hour shift.)
(b) Solve the problem in (a).

4.6 The Simplex Method with Mixed Constraints

OBJECTIVES

- *To solve maximization problems with mixed constraints*
- *To solve minimization problems with mixed constraints*

APPLICATION PREVIEW

The Laser Company manufactures two models of stereo systems, the Star and the Allstar, at two plants, located in Ashville and in Cleveland. The maximum daily output at the Ashville plant is 900, and the profit there is \$200 per unit of the Allstar and \$100 per unit of the Star. The maximum daily output at the Cleveland plant is 800, and the profit there is \$210 per unit of the Allstar and \$80 per unit of the Star. In addition, restrictions at the Ashville plant mean that the number of units of the Star model cannot exceed 100 more than the number of units of the Allstar model produced. If the company gets a rush order for 800 units of the Allstar model and 600 units of the Star model, finding the number of units of each model that should be produced at each location to fill the order and obtain the maximum profit is a **mixed constraint linear programming** problem. The term *mixed constraint* is used because some inequalities describing the constraints contain "≤" and some contain "≥" signs.

In the previous two sections, we used the simplex method to solve two different types of linear programming problems. These types are summarized as follows:

	Maximization Problems	*Minimization Problems*
Function	Maximized	Minimized
Variables	Nonnegative	Nonnegative
Constraints	$a_1x_1 + a_2x_2 + \cdots + a_nx_n \leq b\ (b \geq 0)$	$c_1y_1 + c_2y_2 + \cdots + c_ny_n \geq d\ (d \geq 0)$
Simplex method	Applied directly	Applied to dual

In this section we will show how to apply the simplex method to mixed constraint linear programming problems that do not exactly fit either of these types.

We begin by considering maximization problems with constraints that have a form different from those noted in the table above. This can happen in one of the following ways:

1. If the constraint has the form $a_1x_1 + a_2x_2 + \cdots + a_nx_n \leq b$, where $b < 0$, such as with $x - 2y \leq -8$
2. If the constraint has the form $a_1x_1 + a_2x_2 + \cdots + a_nx_n \geq b$, where $b \geq 0$, such as with $2x - 3y \geq 6$

Note in the latter case that multiplying both sides of the inequality by (-1) changes it as follows:

$$\begin{aligned} & a_1x_1 + a_2x_2 + \cdots + a_nx_n \geq b \\ \text{becomes} \quad & -a_1x_1 - a_2x_2 - \cdots - a_nx_n \leq -b \end{aligned}$$

That is,

$$2x - 3y \geq 6 \quad \text{becomes} \quad -2x + 3y \leq -6$$

Thus we see that the two possibilities for mixed constraints are actually one: constraints that have the form $a_1x_1 + a_2x_2 + \ldots + a_nx_n \leq b$, where b is any constant. We call these **less than or equal to constraints** and denote them as **$\leq$ constraints.**

Let's consider an example and see how we might proceed with a problem of this type.

EXAMPLE 1

Maximize $f = x + 2y$ subject to

$$\begin{aligned} x + y &\leq 13 \\ 2x - 3y &\geq 6 \\ x \geq 0, y &\geq 0 \end{aligned}$$

Solution

We begin by expressing all constraints as "$\leq$" constraints. In this case, we multiply $2x - 3y \geq 6$ by (-1) to obtain $-2x + 3y \leq -6$. We can now use slack variables to write the inequalities as equations and form the simplex matrix.

$$\begin{aligned} x + y + s_1 &= 13 \\ -2x + 3y + s_2 &= -6 \\ -x - 2y + f &= 0 \end{aligned} \quad \text{gives} \quad \begin{array}{c} \begin{matrix} x & y & s_1 & s_2 & f \end{matrix} \\ \left[\begin{array}{rrrrr|r} 1 & 1 & 1 & 0 & 0 & 13 \\ -2 & 3 & 0 & 1 & 0 & -6 \\ \hline -1 & -2 & 0 & 0 & 1 & 0 \end{array}\right] \end{array}$$

The -6 in the upper portion of the last column is a problem. If we compute values for the variables x, y, s_1, and s_2 that are associated with this matrix, we get $s_2 = -6$. This violates the condition of the simplex method that all variables be nonnegative. In order to use the simplex method, we must change the sign of any negative entry that appears in the upper portion of the last column (above the line separating the function from the constraints). There will always be another negative entry in the same row as this negative entry, but in a different column. We choose this column as the pivot column, because pivoting with it will give a positive entry in the last column.

$$\begin{array}{c} \begin{matrix} x & y & s_1 & s_2 & f \end{matrix} \\ \left[\begin{array}{rrrrr|r} 1 & 1 & 1 & 0 & 0 & 13 \\ -2 & 3 & 0 & 1 & 0 & -6 \\ \hline -1 & -2 & 0 & 0 & 1 & 0 \end{array}\right] \end{array} \leftarrow \text{Negative in last column}$$

↑ Pivot column (negative entry in the same row as the negative entry in the last column)

Once we have identified the pivot column, we choose the pivot entry by forming *all* quotients and choosing the entry that gives the *smallest positive quotient.* Often this will mean pivoting by a negative number, which is the case in this problem.

$$\begin{array}{c} \begin{matrix} x & y & s_1 & s_2 & f \end{matrix} \\ \left[\begin{array}{rrrrr|r} 1 & 1 & 1 & 0 & 0 & 13 \\ \textcircled{-2} & 3 & 0 & 1 & 0 & -6 \\ \hline -1 & -2 & 0 & 0 & 1 & 0 \end{array}\right] \end{array} \quad \begin{array}{l} 13/1 = 13 \\ (-6)/(-2) = 3^* \end{array}$$

↑ Pivot entry circled

*Smallest positive quotient

Pivoting with this entry gives the following matrix.

$$\begin{array}{c} \begin{matrix} x & y & s_1 & s_2 & f \end{matrix} \\ \left[\begin{array}{rrrrr|r} 0 & \textcircled{\frac{5}{2}} & 1 & \frac{1}{2} & 0 & 10 \\ 1 & -\frac{3}{2} & 0 & -\frac{1}{2} & 0 & 3 \\ \hline 0 & -\frac{7}{2} & 0 & -\frac{1}{2} & 1 & 3 \end{array}\right] \end{array} \Big\} \text{No negatives}$$

↑ Most negative

This new matrix has no negatives in the upper portion of the last column, so we proceed with the simplex method as we have used it previously. The new pivot column is found from the most negative entry in the last row (indicated), and the pivot is found from the smallest positive quotient (circled). Using this pivot completes the solution.

$$\begin{array}{c} \begin{matrix} x & y & s_1 & s_2 & f \end{matrix} \\ \left[\begin{array}{rrrrr|r} 0 & 1 & \frac{2}{5} & \frac{1}{5} & 0 & 4 \\ 1 & 0 & \frac{3}{5} & -\frac{1}{5} & 0 & 9 \\ \hline 0 & 0 & \frac{7}{5} & \frac{1}{5} & 1 & 17 \end{array}\right] \end{array}$$

We see that when $x = 9$ and $y = 4$, then the maximum value of $f = x + 2y$ is 17.

CHECKPOINT

1. Write the simplex matrix to maximize $f = 4x + y$ subject to the constraints

$$\begin{aligned} x + 2y &\le 12 \\ 3x - 4y &\ge 6 \\ x \ge 0, y &\ge 0 \end{aligned}$$

2. Using the matrix from Problem 1, find the maximum value of $f = 4x + y$ subject to the constraints.

If mixed constraints occur in a minimization problem, the simplex method does not apply to the dual problem. However, the same techniques used in Example 1 can be slightly modified and applied to minimization problems with mixed constraints. We could rewrite all constraints as "≥" constraints (by multiplying by −1 as needed) and then use the dual. Alternatively, if our objective is to minimize f, then we alter the problem so as to maximize $-f$ and then proceed as in Example 1.

EXAMPLE 2

Minimize the function $f = 3x + 4y$ subject to the constraints

$$\begin{aligned} x + y &\ge 20 \\ x + 2y &\ge 25 \\ -5x + y &\le 4 \\ x \ge 0, y &\ge 0 \end{aligned}$$

Solution

Because of the mixed constraints, we seek to maximize $-f = -3x - 4y$ subject to

$$\begin{aligned} -x - y &\le -20 \\ -x - 2y &\le -25 \\ -5x + y &\le 4 \end{aligned}$$

The simplex matrix for this problem is

$$\begin{array}{c} \begin{matrix} x & y & s_1 & s_2 & s_3 & -f & \end{matrix} \\ \left[\begin{array}{cccccc|c} -1 & -1 & 1 & 0 & 0 & 0 & -20 \\ -1 & -2 & 0 & 1 & 0 & 0 & -25 \\ -5 & 1 & 0 & 0 & 1 & 0 & 4 \\ \hline 3 & 4 & 0 & 0 & 0 & 1 & 0 \end{array}\right] \end{array} \begin{array}{l} \leftarrow \text{Negatives} \\ \swarrow \end{array}$$

In this case, there are two negative entries in the upper portion of the last column. When this happens, we can start with either one; we choose −20 (row 1). In row 1 (to the left of −20), we find negative entries in columns 1 and 2. We can choose our pivot column for either of these columns; we choose column 1. Once this choice is made, the pivot is determined from the quotients.

$$\begin{array}{cccccc|c} x & y & s_1 & s_2 & s_3 & -f & \\ \left[\begin{array}{cccccc|c} \textcircled{-1} & -1 & 1 & 0 & 0 & 0 & -20 \\ -1 & -2 & 0 & 1 & 0 & 0 & -25 \\ -5 & 1 & 0 & 0 & 1 & 0 & 4 \\ \hline 3 & 4 & 0 & 0 & 0 & 1 & 0 \end{array}\right] \end{array} \quad \begin{array}{l} -20/(-1) = 20^* \\ -25/(-1) = 25 \\ 4/(-5) = -4/5 \end{array}$$

↑ Pivot column (pivot entry circled) — *Smallest positive quotient

Pivoting with this entry gives the following matrix.

$$\begin{array}{c} x \quad y \quad s_1 \quad s_2 \quad s_3 \quad -f \\ \left[\begin{array}{cccccc|c} 1 & 1 & -1 & 0 & 0 & 0 & 20 \\ 0 & \textcircled{-1} & -1 & 1 & 0 & 0 & -5 \\ 0 & 6 & -5 & 0 & 1 & 0 & 104 \\ \hline 0 & 1 & 3 & 0 & 0 & 1 & -60 \end{array}\right] \end{array} \quad \begin{array}{l} \\ \leftarrow \text{Negative} \\ \\ \\ \end{array}$$

↑ Pivot column

Now there is only one negative in the upper portion of the last column. Looking for negatives in this same row, we see that we can choose either column 2 or column 3 for our pivot column. We choose column 2, and the pivot is circled. Pivoting gives the following matrix.

$$\begin{array}{c} x \quad y \quad s_1 \quad s_2 \quad s_3 \quad -f \\ \left[\begin{array}{cccccc|c} 1 & 0 & -2 & 1 & 0 & 0 & 15 \\ 0 & 1 & 1 & -1 & 0 & 0 & 5 \\ 0 & 0 & -11 & 6 & 1 & 0 & 74 \\ \hline 0 & 0 & 2 & 1 & 0 & 1 & -65 \end{array}\right] \end{array}$$

At this point we have all positives in the upper portion of the last column, so the simplex method can be applied. (The negative value in the lower portion is not a problem.) Looking at the indicators, we see that the solution is complete.

From the matrix we have $x = 15$, $y = 5$, and $-f = -65$. Thus the solution is $x = 15$ and $y = 5$, which gives the minimum value for the function $f = 3x + 4y = 65$.

A careful review of these examples allows us to summarize the key procedural steps for solving linear programming problems with mixed constraints.

Summary: Simplex Method for Mixed Constraints

1. If the problem is to minimize f, then maximize $-f$.
2. a. Make all constraints "$\leq$" constraints by multiplying both sides of any "$\geq$" constraints by (-1).
   b. Use slack variables and form the simplex matrix.
3. In the simplex matrix, scan the *upper portion* of the last column for any negative entries.
   a. If there are no negative entries, apply the simplex method.
   b. If there are negative entries, go to step 4.

4. When there is a negative value in the upper portion of the last column, proceed as follows:
   a. Select any negative entry in the same row and use its column as the pivot column.
   b. In the pivot column, compute all quotients for the entries above the line and determine the pivot from the *smallest positive quotient*.
   c. After completing the pivot operations, return to step 3.

Up to this point, we have discussed ways in which any problem can be expressed as a maximization problem with "≤" constraints (or as a minimization problem with "≥" constraints). If we have problems in either of these "standard" forms, then we can form the dual problem. As noted previously, the NEOS Web site (http://www.mcs.anl.gov/home/otc/Server/) has links to linear programming solvers, but many of these solvers require submission of both a primal and a dual problem. We are now equipped to transform any linear programming problem so that we can identify both a primal and a dual problem—and hence take advantage of on-line solvers.

EXAMPLE 3

The Laser Company manufactures two models of stereo systems, the Star and the Allstar, at two plants, located in Ashville and in Cleveland. The maximum daily output at the Ashville plant is 900, and the profit there is \$200 per unit of the Allstar and \$100 per unit of the Star. The maximum daily output at the Cleveland plant is 800, and the profit there is \$210 per unit of the Allstar and \$80 per unit of the Star. In addition, restrictions at the Ashville plant mean that the number of units of the Star model cannot exceed 100 more than the number of units of the Allstar model produced. If the company gets a rush order for 800 units of the Allstar model and 600 units of the Star model, how many units of each model should be produced at each location to fill the order and obtain the maximum profit?

Solution

The following table identifies our variables x and y and relates other important facts in the problem. In particular, we see that only variables x and y are required.

	Ashville Plant	*Total Needed*	*Cleveland Plant*
Allstar	x produced (profit = \$200 each)	800	$800 - x$ produced (profit = \$210 each)
Star	y produced (profit = \$100 each)	600	$600 - y$ produced (profit = \$80 each)
Capacity	900	—	800

We wish to maximize profit, and from the table we see that total profit is given by

$$P = 200x + 100y + 210(800 - x) + 80(600 - y)$$

or

$$P = 216{,}000 - 10x + 20y$$

We can read capacity constraints from our table.

$$x + y \leq 900 \quad \text{(Ashville)}$$

$$(800 - x) + (600 - y) \leq 800 \quad \text{or} \quad x + y \geq 600 \quad \text{(Cleveland)}$$

An additional Ashville plant constraint from the statement of the problem is

$$y \leq x + 100 \quad \text{or} \quad -x + y \leq 100$$

Thus our problem is to maximize

$$P = 216{,}000 - 10x + 20y$$

subject to

$$\begin{aligned} x + y &\leq 900 \\ x + y &\geq 600 \\ -x + y &\leq 100 \\ x \geq 0,\ y &\geq 0 \end{aligned}$$

We must express $x + y \geq 600$ as a "$\leq$" constraint; multiplying both sides by (-1) gives $-x - y \leq -600$. The simplex matrix is

$$\begin{array}{c} \\ \\ \\ \\ \\ \end{array}
\begin{array}{cccccc|c} x & y & s_1 & s_2 & s_3 & P & \\ \hline 1 & 1 & 1 & 0 & 0 & 0 & 900 \\ \textcircled{-1} & -1 & 0 & 1 & 0 & 0 & -600 \\ -1 & 1 & 0 & 0 & 1 & 0 & 100 \\ \hline 10 & -20 & 0 & 0 & 0 & 1 & 216{,}000 \\ \uparrow & & & & & & \\ \text{Pivot column} & & & & & & \end{array} \quad \begin{array}{l} \\ \\ \leftarrow \text{Negative} \\ \\ \\ \\ \end{array}$$

The negative in the upper portion of the last column is indicated. In row 2 we find negatives in both column 1 and column 2, so either of these may be our pivot column. Our choice is indicated, and the pivot entry is circled. The matrix that results from the pivot operations follows.

$$\begin{array}{cccccc|c} x & y & s_1 & s_2 & s_3 & P & \\ \hline 0 & 0 & 1 & 1 & 0 & 0 & 300 \\ 1 & 1 & 0 & -1 & 0 & 0 & 600 \\ 0 & \textcircled{2} & 0 & -1 & 1 & 0 & 700 \\ \hline 0 & -30 & 0 & 10 & 0 & 1 & 210{,}000 \\ & \uparrow & & & & & \\ & \text{Pivot column} & & & & & \end{array}$$

No entry in the upper portion of the last column is negative, so we can proceed with the simplex method. From the indicators we see that column 2 is the pivot column (indicated above), and the pivot entry is circled. The simplex matrix that results from pivoting follows.

$$\begin{array}{c} \\ \\ \\ \\ \\ \end{array}\begin{array}{cccccc|c} x & y & s_1 & s_2 & s_3 & P & \\ 0 & 0 & 1 & \textcircled{1} & 0 & 0 & 300 \\ 1 & 0 & 0 & -\frac{1}{2} & -\frac{1}{2} & 0 & 250 \\ 0 & 1 & 0 & -\frac{1}{2} & \frac{1}{2} & 0 & 350 \\ \hline 0 & 0 & 0 & -5 & 15 & 1 & 220{,}500 \\ & & & \uparrow & & & \\ & & & \text{Pivot column} & & & \end{array}$$

The new pivot column is indicated, and the pivot entry is circled. The pivot operation yields the following matrix.

$$\begin{array}{cccccc|c} x & y & s_1 & s_2 & s_3 & P & \\ 0 & 0 & 1 & 1 & 0 & 0 & 300 \\ 1 & 0 & \frac{1}{2} & 0 & -\frac{1}{2} & 0 & 400 \\ 0 & 1 & \frac{1}{2} & 0 & \frac{1}{2} & 0 & 500 \\ \hline 0 & 0 & 5 & 0 & 15 & 1 & 222{,}000 \end{array}$$

This matrix shows that the solution is complete. We see that $x = 400$ (so $800 - x = 400$), $y = 500$ (so $600 - y = 100$), and $P = \$222{,}000$.

Thus the company should operate the Ashville plant at capacity and produce 400 units of the Allstar model and 500 units of the Star model. The remainder of the order, 400 units of the Allstar model and 100 units of the Star model, should be produced at the Cleveland plant.

CHECKPOINT SOLUTIONS

1. $\left[\begin{array}{ccccc|c} 1 & 2 & 1 & 0 & 0 & 12 \\ -3 & 4 & 0 & 1 & 0 & -6 \\ \hline -4 & -1 & 0 & 0 & 1 & 0 \end{array}\right]$

2. $\left[\begin{array}{ccccc|c} 0 & \frac{10}{3} & 1 & \frac{1}{3} & 0 & 10 \\ 1 & -\frac{4}{3} & 0 & -\frac{1}{3} & 0 & 2 \\ \hline 0 & -\frac{19}{3} & 0 & -\frac{4}{3} & 1 & 8 \end{array}\right] \rightarrow \left[\begin{array}{ccccc|c} 0 & 1 & \frac{3}{10} & \frac{1}{10} & 0 & 3 \\ 1 & 0 & \frac{2}{5} & -\frac{1}{5} & 0 & 6 \\ \hline 0 & 0 & \frac{19}{10} & -\frac{7}{10} & 1 & 27 \end{array}\right]$

$\rightarrow \left[\begin{array}{ccccc|c} 0 & 10 & 3 & 1 & 0 & 30 \\ 1 & 2 & 1 & 0 & 0 & 12 \\ \hline 0 & 7 & 4 & 0 & 1 & 48 \end{array}\right]$

The maximum value of f is 48 at $x = 12$, $y = 0$, $s_1 = 0$, and $s_2 = 30$.

EXERCISE 4.6

In Problems 1–4, express each inequality as a "≤" constraint.

1. $3x - y \geq 5$
2. $4x - 3y \geq 6$
3. $y \geq 40 - 6x$
4. $x \geq 60 - 8y$
5. Write the simplex matrix used to maximize $f = 4x + 5y$ subject to the constraints $x \geq 0$, $y \geq 0$, $x + 2y \leq 6$, and $4x + 2y \geq 12$.
6. Write the simplex matrix used to maximize $f = 4x + y$ subject to the constraints $x \geq 0$, $y \geq 0$, $3x + 5y \leq 6$, and $x + y \geq 8$.
7. Write the simplex matrix used to minimize $g = 2x + 2y$ subject to the constraints $x \geq 0$, $y \geq 0$, $6x + 4y \leq 24$, and $5x + 2y \geq 16$.
8. Write the simplex matrix used to minimize $g = 10x + 2y$ subject to the constraints $x \geq 0$, $y \geq 0$, $2x + 3y \leq 16$, and $2x + y \geq 8$.

In Problems 9–12, complete both of the following.

(a) State the given problem in a form from which the simplex matrix can be formed (that is, as a maximization problem with "≤" constraints).

(b) Form the simplex matrix, and circle the first pivot.

9. Maximize $f = 2x + 3y$ subject to

$$\begin{aligned} 7x + 4y &\le 28 \\ -3x + y &\ge 2 \\ x \ge 0, y &\ge 0 \end{aligned}$$

10. Maximize $f = 5x + 11y$ subject to

$$\begin{aligned} x - 3y &\ge 3 \\ -x + y &\le 1 \\ x &\le 10 \\ x \ge 0, y &\ge 0 \end{aligned}$$

11. Minimize $g = 3x + 8y$ subject to

$$\begin{aligned} 4x - 5y &\le 50 \\ x + y &\le 80 \\ -x + 2y &\ge 4 \\ x \ge 0, y &\ge 0 \end{aligned}$$

12. Minimize $g = 40x + 25y$ subject to

$$\begin{aligned} x + y &\le 100 \\ -x + y &\le 20 \\ -2x + 3y &\ge 6 \\ x \ge 0, y &\ge 0 \end{aligned}$$

In Problems 13 and 14, a final simplex matrix for a minimization problem is given. In each case, find the solution.

13.
$$\begin{array}{c} \begin{matrix} x & y & z & s_1 & s_2 & s_3 & -f & \end{matrix} \\ \left[\begin{array}{ccccccc|c} 0 & 1 & 0 & \frac{7}{3} & \frac{1}{3} & -\frac{2}{3} & 0 & 8 \\ 0 & 0 & 1 & \frac{2}{3} & -\frac{1}{3} & \frac{8}{3} & 0 & 12 \\ 1 & 0 & 0 & -\frac{4}{3} & \frac{2}{3} & \frac{4}{3} & 0 & 6 \\ \hline 0 & 0 & 0 & 4 & \frac{4}{3} & 2 & 1 & -120 \end{array}\right] \end{array}$$

14.
$$\begin{array}{c} \begin{matrix} x & y & z & s_1 & s_2 & s_3 & -f & \end{matrix} \\ \left[\begin{array}{ccccccc|c} 1 & 0 & 0 & \frac{2}{5} & -\frac{2}{5} & 0 & 0 & 50 \\ 0 & 0 & 1 & \frac{1}{5} & \frac{1}{5} & \frac{11}{5} & 0 & 12 \\ 0 & 1 & 0 & 1 & -\frac{8}{5} & \frac{14}{5} & 0 & 30 \\ \hline 0 & 0 & 0 & 3 & 1 & 4 & 1 & -1200 \end{array}\right] \end{array}$$

In Problems 15–34, use the simplex method to find the optimal solution.

15. Maximize $f = 4x + y$ subject to

$$\begin{aligned} 5x + 2y &\le 84 \\ -3x + 2y &\ge 4 \\ x \ge 0, y &\ge 0 \end{aligned}$$

16. Maximize $f = x + 2y$ subject to

$$\begin{aligned} -x + 2y &\le 60 \\ -7x + 4y &\ge 20 \\ x \ge 0, y &\ge 0 \end{aligned}$$

17. Minimize $f = 2x + 3y$ subject to

$$\begin{aligned} x &\ge 5 \\ y &\le 13 \\ -x + y &\ge 2 \\ x \ge 0, y &\ge 0 \end{aligned}$$

18. Minimize $f = 3x + 2y$ subject to

$$\begin{aligned} x &\le 20 \\ y &\le 20 \\ x + y &\ge 21 \\ x \ge 0, y &\ge 0 \end{aligned}$$

19. Minimize $f = 2x + y$ subject to

$$\begin{aligned} x &\le 12 \\ x + 2y &\ge 20 \\ -3x + 2y &\le 4 \\ x \ge 0, y &\ge 0 \end{aligned}$$

20. Minimize $f = x + 4y$ subject to

$$\begin{aligned} y &\le 30 \\ 3x + 2y &\ge 75 \\ -3x + 5y &\ge 30 \\ x \ge 0, y &\ge 0 \end{aligned}$$

21. Maximize $f = 5x + 2y$ subject to

$$\begin{aligned} y &\le 20 \\ 2x + y &\le 32 \\ -x + 2y &\ge 4 \\ x \ge 0, y &\ge 0 \end{aligned}$$

22. Maximize $f = 3x + 4y$ subject to

$$\begin{aligned} 2x + 3y &\le 90 \\ x + y &\le 35 \\ -x + 2y &\ge 10 \\ x \ge 0, y &\ge 0 \end{aligned}$$

23. Maximize $f = 3x + 2y$ subject to

$$\begin{aligned} -x + y &\le 10 \\ x + y &\ge 20 \\ x + y &\le 35 \\ x \ge 0, y &\ge 0 \end{aligned}$$

24. Minimize $f = 4x + y$ subject to

$$\begin{aligned} -x + y &\le 4 \\ 3x + y &\ge 12 \\ x + y &\le 20 \\ x \ge 0, y &\ge 0 \end{aligned}$$

25. Maximize $f = 2x + 5y$ subject to

$$\begin{aligned} -x + y &\le 10 \\ x + y &\le 30 \\ -2x + y &\ge -24 \\ x \ge 0, y &\ge 0 \end{aligned}$$

26. Maximize $f = 3x + 2y$ subject to

$$\begin{aligned} -x + 2y &\le 20 \\ -3x + 2y &\le -36 \\ x + y &\le 22 \\ x \ge 0, y &\ge 0 \end{aligned}$$

27. Minimize $f = x + 2y + 3z$ subject to

$$\begin{aligned} x + z &\le 20 \\ x + y &\ge 30 \\ y + z &\le 20 \\ x \ge 0, y \ge 0, z &\ge 0 \end{aligned}$$

28. Minimize $f = x + 2y + z$ subject to

$$\begin{aligned} x + 3y + 2z &\ge 40 \\ x + y + z &\ge 30 \\ x + y + z &\le 100 \\ x \ge 0, y \ge 0, z &\ge 0 \end{aligned}$$

29. Maximize $f = 2x - y + 4z$ subject to

$$\begin{aligned} x + y + z &\le 8 \\ x - y + z &\ge 4 \\ x + y - z &\ge 2 \\ x \ge 0, y \ge 0, z &\ge 0 \end{aligned}$$

30. Maximize $f = 5x + 2y + z$ subject to

$$\begin{aligned} 2x + 3y + z &\le 30 \\ x - 2y + z &\ge 20 \\ 2x + 5y + 2z &\ge 25 \\ x \ge 0, y \ge 0, z &\ge 0 \end{aligned}$$

31. Maximize $f = 2x + 3y - z$ subject to

$$\begin{aligned} 3x + y + 2z &\le 240 \\ -2x - 2y + z &\ge 10 \\ x \ge 0, y \ge 0, z &\ge 0 \end{aligned}$$

32. Maximize $f = -x + 2y + 4z$ subject to

$$\begin{aligned} x + y + z &\le 40 \\ -x + y + z &\ge 20 \\ x + y - z &\ge 10 \\ x \ge 0, y \ge 0, z &\ge 0 \end{aligned}$$

33. Minimize $f = 10x + 30y + 35z$ subject to

$$\begin{aligned} x + y + z &\le 250 \\ x + y + 2z &\ge 150 \\ 2x + y + z &\le 180 \\ x \ge 0, y \ge 0, z &\ge 0 \end{aligned}$$

34. Minimize $f = 2x + 3y - z$ subject to

$$\begin{aligned} 3x + y + 2z &\le 120 \\ -2x - 2y + z &\ge 5 \\ x \ge 0, y \ge 0, z &\ge 0 \end{aligned}$$

Applications

35. ***Manufacturing*** A company manufactures commercial heating system components and domestic furnaces at its factories in Monaca, PA, and Hamburg, NY. The Monaca plant can produce no more than 1000 units per day, and the number of commercial components cannot exceed 100 more than half the number of domestic furnaces. The Hamburg plant can produce no more than 850 units per day. The profit on each commercial component is \$400 at the Monaca plant and \$390 at the Hamburg plant. The profit on each domestic furnace is \$200 at the Monaca plant and \$215 at the Hamburg plant. If there is a rush order for 500 commercial components and 750 domestic furnaces, how many of each should be produced at each plant in order to maximize profits? Find the maximum profit.

36. ***Manufacturing*** A manufacturer makes products A and B at locations I and II. Location I can produce at most 1800 products, and the production of B can be at most 200 units less than the production of A. Location II can produce at most 1200 products. The profit on product A is \$100 at location I and \$90 at location II, and the profit on product B is \$70 at location I and \$75 at location II. If the manufacturer gets

a rush order for 1500 units of product A and 1300 units of product B, how many units of each should be produced at each location so as to maximize profits? Find the maximum profit.

37. ***Manufacturing*** Refer to Problem 35. Suppose that the cost of each commercial component is \$380 at the Monaca plant and \$400 at the Hamburg plant. The cost of each domestic furnace is \$200 at the Monaca plant and \$185 at the Hamburg plant. If the same rush order for 500 commercial components and 750 domestic furnaces is received, how many of each should be produced at each location to minimize cost? Find the minimum cost.

38. ***Manufacturing*** Refer to Problem 36. Suppose the costs associated with product A are \$50 at location I and \$60 at location II, and the costs associated with product B are \$40 at location I and \$25 at location II. If the manufacturer gets the same rush order for 1500 units of A and 1300 units of B, how many units of each should be produced at each location in order to minimize costs? Find the minimum cost.

39. ***Production*** A sausage company makes two different kinds of hot dogs, regular and all beef. Each pound of all-beef hot dogs requires 0.75 lb of beef and 0.2 lb of spices, and each pound of regular hot dogs requires 0.18 lb of beef, 0.3 lb of pork, and 0.2 lb of spices. Suppliers can deliver at most 1020 lb of beef, at most 600 lb of pork, and at least 500 lb of spices. If the profit is \$0.60 on each pound of all-beef hot dogs and \$0.40 on each pound of regular hot dogs, how many pounds of each should be produced to obtain maximum profit? What is the maximum profit?

40. ***Manufacturing*** A cereal manufacturer makes two different kinds of cereal, Senior Citizen's Feast and Kids Go. Each pound of Senior Citizen's Feast requires 0.6 lb of wheat and 0.2 lb of vitamin-enriched syrup, and each pound of Kids Go requires 0.4 lb of wheat, 0.2 lb of sugar, and 0.2 lb of vitamin-enriched syrup. Suppliers can deliver at most 2800 lb of wheat, at most 800 lb of sugar, and at least 1000 lb of the vitamin-enriched syrup. If the profit is \$0.90 on each pound of Senior Citizen's Feast and \$1.00 on each pound of Kids Go, find the number of pounds of each cereal that should be produced to obtain maximum profit. Find the maximum profit.

41. ***Water purification*** Three water purification facilities can handle at most 10 million gallons in a certain time period. Plant 1 leaves 20% of certain impurities, and costs \$20,000 per million gallons. Plant 2 leaves 15% of these impurities and costs \$30,000 per million gallons. Plant 3 leaves 10% impurities and costs \$40,000 per million gallons. The desired level of impurities in the water from all three plants is at most 15%. If Plant 1 and Plant 3 combined must handle at least 6 million gallons, find the number of gallons each plant should handle so as to achieve the desired level of purity at minimum cost. Find the minimum cost.

42. ***Mixture*** A chemical storage tank has a capacity of 200 tons. Currently the tank contains 50 tons of mixture that has 10% of a certain active chemical and 1.8% of other inert ingredients. The owners of the tank want to replenish the supply in the tank and will purchase some combination of two available mixes. Mix 1 contains 70% of the active chemical and 3% of the inert ingredients; its cost is \$100 per ton. Mix 2 contains 30% of the active chemical and 1% of the inert ingredients; its cost is \$40 per ton. The desired final mixture should have at least 40% of the active chemical and at most 2% of the inert ingredients. How many tons of each mix should be purchased to obtain the desired final mixture at minimum cost? Find the minimum cost. Note that at least 40% of the active chemical means

$$70\% \text{ (Mix 1)} + 30\% \text{ (Mix 2)} + 10\% \text{ (mix on hand)} \geq 40\% \text{ (Mix 1 + Mix 2 + mix on hand)}$$

KEY TERMS

Section	Key Term
4.1	Linear inequalities Properties Solutions Graphs Intervals Open Closed Half-open
4.2	Linear inequalities in two variables Graphs Systems Solution region
4.3	Linear programming Objective function Constraints Optimum values Feasible region Graphical solution
4.4	Simplex method (maximization) Slack variables Simplex matrix Pivot entry Basic variable Nonunique solutions
4.5	Simplex method (minimization) Dual problem Transpose Primal problem
4.6	Simplex method with mixed constraints

REVIEW EXERCISES

Section 4.1

1. Solve $3x - 9 \le 4(3 - x)$ and graph the solution.
2. Solve $\frac{2}{5}x \le x + 4$ and graph the solution.
3. Solve $5x + 1 \ge \frac{2}{3}(x - 6)$ and graph the solution.
4. Solve $\frac{4(x-2)}{3} \ge 3x - \frac{1}{6}$ and graph the solution.
5. Determine whether the following represent open, closed, or half-open intervals. Write each in interval notation.
   (a) $0 \le x \le 5$ (b) $3 \le x < 7$
   (c) $-3 < x < 2$

6. Write the inequality represented by each of the following.
   (a) $(-1, 16)$ (b) $(-12, -8)$
   (c)

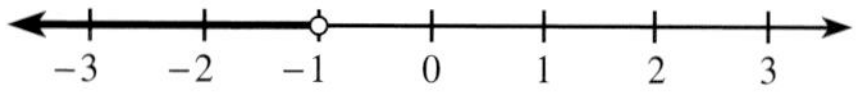

Section 4.2

In Problems 7–10, graph the solutions sets of each inequality or system of inequalities.

7. $2x + 3y \le 12$
8. $4x + 5y \ge 100$
9. $\begin{cases} x + 2y \le 20 \\ 3x + 10y \le 80 \\ x \ge 0, y \ge 0 \end{cases}$
10. $\begin{cases} 3x + y \ge 4 \\ x + y \ge 2 \\ -x + y \le 4 \\ x \le 5 \end{cases}$

Section 4.3

In Problems 11 and 12, a function and the graph of a feasible region are given. In each case, find both the maximum and the minimum value of the function and the point at which each occurs.

11. $f = -x + 3y$
12. $f = 6x + 4y$

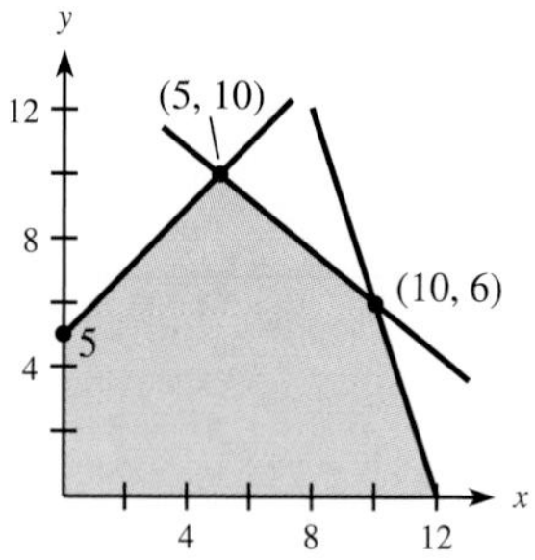

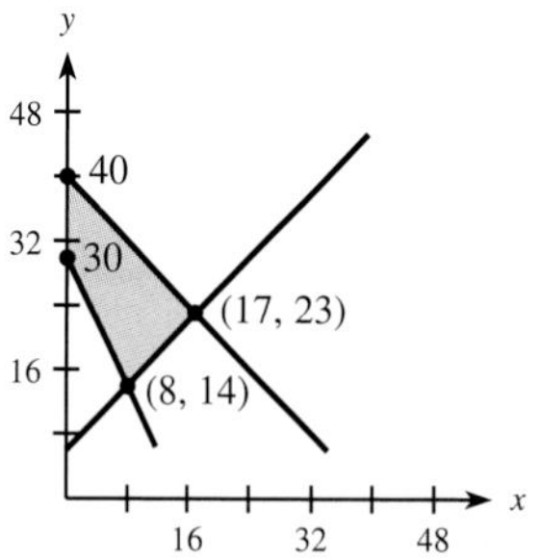

Solve the linear programming Problems 13–20 using graphical methods. Restrict $x \ge 0$ and $y \ge 0$.

13. Maximize $f = 5x + 6y$ subject to

$$\begin{aligned} x + 3y &\le 24 \\ 4x + 3y &\le 42 \\ 2x + y &\le 20 \end{aligned}$$

14. Maximize $f = 9x + 5y$ subject to the constraints in Problem 13.
15. Maximize $f = x + 4y$ subject to

$$\begin{aligned} 7x + 3y &\le 105 \\ 2x + 5y &\le 59 \\ x + 7y &\le 70 \end{aligned}$$

16. Maximize $f = 2x + y$ subject to the constraints in Problem 15.
17. Minimize $g = 5x + 3y$ subject to

$$\begin{aligned} 3x + y &\ge 12 \\ x + y &\ge 6 \\ x + 6y &\ge 11 \end{aligned}$$

18. Minimize $g = 3x + 5y$ subject to the constraints in Problem 17.
19. Minimize $g = x + 5y$ subject to

$$\begin{aligned} 8x + y &\ge 85 \\ x + y &\ge 50 \\ x + 4y &\ge 80 \\ x + 10y &\ge 94 \end{aligned}$$

20. Minimize $g = 7x + y$ subject to the constraints in Problem 19.

Section 4.4

Use the simplex method to solve the linear programming Problems 21–34. Assume all variables are nonnegative.

21. Maximize $f = 7x + 12y$ subject to the conditions in Problem 15.
22. Maximize $f = 3x + 4y$ subject to

$$\begin{aligned} x + 4y &\le 160 \\ x + 2y &\le 100 \\ 4x + 3y &\le 300 \end{aligned}$$

23. Maximize $f = 3x + 8y$ subject to the conditions in Problem 22.
24. Maximize $f = 39x + 5y + 30z$ subject to

$$\begin{aligned} x + z &\le 7 \\ 3x + 5y &\le 30 \\ 3x + y &\le 18 \end{aligned}$$

Problems 25 and 26 have nonunique solutions. If there is no solution, indicate this; if there are multiple solutions, find two different solutions. Use the simplex method.

25. Maximize $f = 4x + 4y$ subject to

$$\begin{aligned} x + 5y &\le 500 \\ x + 2y &\le 230 \\ x + y &\le 160 \end{aligned}$$

26. Maximize $f = 2x + 5y$ subject to

$$\begin{aligned} -4x + y &\le 40 \\ x - 7y &\le 70 \end{aligned}$$

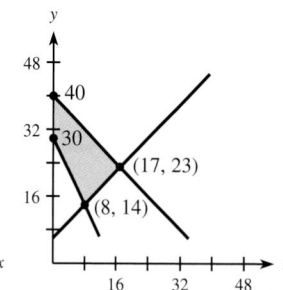

Section 4.5

Form the dual and use the simplex method to solve the minimization Problems 27–30.

27. Minimize $g = 7y_1 + 6y_2$ subject to

$$\begin{aligned} 5y_1 + 2y_2 &\geq 16 \\ 3y_1 + 7y_2 &\geq 27 \end{aligned}$$

28. Minimize $g = 3y_1 + 4y_2$ subject to

$$\begin{aligned} 3y_1 + y_2 &\geq 8 \\ y_1 + y_2 &\geq 6 \\ 2y_1 + 5y_2 &\geq 18 \end{aligned}$$

29. Minimize $g = 2y_1 + y_2$ subject to the conditions of Problem 28.

30. Minimize $g = 12y_1 + 48y_2 + 8y_3$ subject to

$$\begin{aligned} y_1 + 3y_2 &\geq 1 \\ 4y_1 + 6y_2 + y_3 &\geq 3 \\ 4y_2 + y_3 &\geq 1 \end{aligned}$$

Section 4.6

In Problems 31–34, use the simplex method.

31. Maximize $f = 3x + 5y$ subject to

$$\begin{aligned} x + y &\geq 19 \\ -x + y &\geq 1 \\ -x + 10y &\leq 190 \\ x \geq 0, y &\geq 0 \end{aligned}$$

32. Maximize $f = 4x + 6y$ subject to

$$\begin{aligned} 2x + 5y &\leq 37 \\ 5x - y &\leq 34 \\ -x + 2y &\geq 4 \\ x \geq 0, y &\geq 0 \end{aligned}$$

33. Minimize $f = 10x + 3y$ subject to

$$\begin{aligned} -x + 10y &\geq 5 \\ 4x + y &\geq 62 \\ x + y &\leq 50 \\ x \geq 0, y &\geq 0 \end{aligned}$$

34. Minimize $f = 4x + 3y$ subject to

$$\begin{aligned} -x + y &\geq 1 \\ x + y &\leq 45 \\ 10x + y &\geq 45 \\ x \geq 0, y &\geq 0 \end{aligned}$$

Applications

Section 4.3

35. ***Manufacturing*** A company manufactures backyard swing sets of two different sizes. The larger requires 5 hours of labor to complete, the smaller requires 2 hours, and there are 700 hours of labor available each week. The packing department can pack at most 185 swing sets per week. If the profit is \$100 on each larger set and \$50 on each smaller set, how many of each should be produced to yield maximum profit? What is the maximum profit? Use graphical methods.

36. ***Production*** A company produces two different grades of steel, A and B, at two different factories, 1 and 2. The table below summarizes the production capabilities of the factories, the cost per day, and the number of units of each grade of steel that is required to fill orders.

	Factory 1	*Factory 2*	*Required*
Grade A steel	1 unit	2 units	80 units
Grade B steel	3 units	2 units	140 units
Cost per day	\$5000	\$6000	

How many days should each factory operate in order to fill the orders at minimum cost? What is the minimum cost? Use graphical methods.

Use the simplex method to solve Problems 37–43.

Section 4.4

37. ***Production*** A small industry produces two items, I and II. It operates at capacity and makes a profit of \$6 on each item I and \$4 on each item II. The following table gives the hours required to produce each item and the hours available per day.

	I	*II*	*Hours Available*
Assembly	2 hours	1 hour	100
Packaging & inspection	1 hour	1 hour	60

Find the number of items that should be produced each day to maximize profits, and find the maximum daily profit.

38. ***Production*** Pinnochio Crafts makes two types of wooden crafts: Jacob's Ladders and locomotive engines. The manufacture of these crafts requires both carpentry and finishing. Each Jacob's Ladder

requires 1 hour of finishing and $\frac{1}{2}$ hour of carpentry. Each locomotive engine requires 1 hour of finishing and 1 hour of carpentry. Pinnochio Crafts can obtain all the necessary raw materials, but only 120 finishing hours and 75 carpentry hours per week are available. Also, demand for Jacob's Ladders is limited to at most 100 per week. If Pinnochio Crafts makes a profit of \$3 on each Jacob's Ladder and \$5 on each locomotive engine, how many of each should it produce each week to maximize profits? What is the maximum profit?

Section 4.5

39. ***Nutrition*** A nutritionist wants to find the least expensive combination of two foods that meet minimum daily vitamin requirements, which are 5 units of A and 30 units of B. Each ounce of food I provides 2 units of A and 1 unit of B, and each ounce of food II provides 10 units of A and 10 units of B. If food I costs 30 cents per ounce and food II costs 20 cents per ounce, find the number of ounces of each food that will provide the vitamins and minimize the cost.

40. ***Nutrition*** A laboratory wishes to purchase two different feeds, A and B, for its animals. The following table summarizes the nutritional content of the feeds, the required amounts of each ingredient, and the cost of each type of feed.

	Feed A	*Feed B*	*Reqs.*
Carbohydrates	1 unit/lb	4 units/lb	40 units
Protein	2 units/lb	1 unit/lb	80 units
Cost	14¢/lb	16¢/lb	

How many pounds of each type of feed should the laboratory buy in order to satisfy its needs at minimum cost?

41. ***Production*** A company makes three products, I, II, and III at three different factories. At factory A, it can make 10 units of each product per day. At factory B, it can make 20 units of II and 20 units of III per day. At factory C, it can make 20 units of I, 20 units of II, and 10 units of III per day. The company has orders for 200 units of I, 500 units of II, and 300 units of III. If the daily costs are \$200 at A, \$300 at B, and \$500 at C, find the number of days that each factory should operate in order to fill the company's orders at minimum cost. Find the minimum cost.

Section 4.6

42. ***Profit*** A company makes pancake mix and cake mix. Each pound of pancake mix uses 0.6 lb of flour and 0.1 lb of shortening. Each pound of cake mix uses 0.4 lb of flour, 0.1 lb of shortening, and 0.4 lb of sugar. Suppliers can deliver at most 6000 lb of flour, at least 500 lb of shortening, and at most 1200 lb of sugar. If the profit per pound is \$0.35 for pancake mix and \$0.25 for cake mix, how many pounds of each mix should be made to earn maximum profit? What is the maximum profit?

43. ***Manufacturing*** A company manufactures desks and computer tables at plants in Texas and Louisiana. At the Texas plant, production costs are \$12 for each desk and \$20 for each computer table, and the plant can produce at most 120 units per day. At the Louisiana plant, costs are \$14 for each desk and \$19 per computer table, and the plant can produce at most 150 units per day. The company gets a rush order for 130 desks and 130 computer tables at a time when the Texas plant is further limited by the fact that the number of computer tables it produces must be at least 10 more than the number of desks. How should production be scheduled at each location in order to fill the order at minimum cost? What is the minimum cost?

CHAPTER TEST

1. Solve $-\frac{7}{3}t \leq 21$ and graph the solution on a number line.
2. Write the interval $(-1, 4]$ as an inequality.
3. Graph the inequality $x > 2$ (a) on a number line and (b) in the xy-plane.
4. Graph the solution region for each of the following.

(a) $3x - 5y \leq 30$ (b) $\begin{cases} 5y \geq 2x \\ x + y > 7 \\ 2y \leq 5x \end{cases}$

5. Maximize $f = 3x + 5y$ subject to the constraints determined by the shaded region shown in the figure.

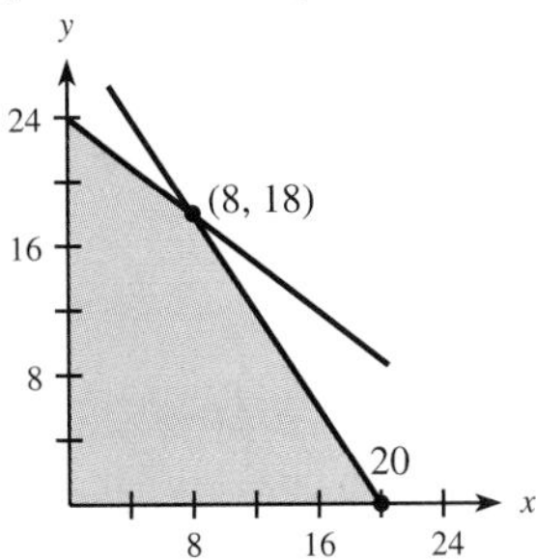

6. Use graphical methods to minimize $g = 3x + y$ subject to $5x - 4y \leq 40$, $x + y \geq 80$, and $2x - y \geq 4$.
7. Use the simplex method to maximize $f = 70x + 5y$ subject to $x + 1.5y \leq 150$, $x + 0.5y \leq 90$, $x \geq 0$, and $y \geq 0$.
8. Below are three simplex matrices in various stages of the simplex method solution of a maximization problem.
   (a) Identify the one for which multiple solutions exist. Circle the pivot and specify all row operations with that pivot that are needed to find a second solution.
   (b) Identify the one for which no solution exists. Explain how you can tell there is no solution.

$$(A)\ \left[\begin{array}{ccccccc|c} 2 & 0 & -4 & 1 & -2 & 0 & 0 & 6 \\ -1 & 1 & 0 & 0 & 1 & 0 & 0 & 3 \\ 4 & 0 & -6 & 0 & -1 & 1 & 0 & 10 \\ \hline -3 & 0 & -8 & 0 & 4 & 0 & 1 & 50 \end{array}\right]$$

$$(B)\ \left[\begin{array}{ccccccc|c} 0 & 1 & -2 & \frac{1}{4} & 0 & \frac{1}{5} & 0 & 12 \\ 0 & 0 & 4 & -\frac{3}{4} & 1 & -\frac{2}{5} & 0 & 20 \\ 1 & 0 & 3 & -\frac{5}{4} & 0 & \frac{4}{5} & 0 & 40 \\ \hline 0 & 0 & 4 & 6 & 0 & -\frac{1}{5} & 1 & 170 \end{array}\right]$$

$$(C)\ \left[\begin{array}{ccccccc|c} 1 & 2 & 0 & 1 & 0 & -\frac{3}{2} & 0 & 40 \\ 0 & 1 & 0 & -2 & 1 & \frac{1}{2} & 0 & 15 \\ 0 & 3 & 1 & -1 & 0 & \frac{1}{4} & 0 & 60 \\ \hline 0 & 0 & 0 & 4 & 0 & 6 & 1 & 220 \end{array}\right]$$

9. Formulate the dual maximization problem associated with the following:

Minimize $g = 2y_1 + 3y_2 + 5y_3$ subject to

$$3y_1 + 5y_2 + y_3 \geq 100,\ 4y_1 + 6y_2 + 3y_3 \geq 120,$$
$$y_1 \geq 0, y_2 \geq 0, y_3 \geq 0$$

10. Express the following linear programming problem as a maximization problem with "$\leq$" constraints. Assume $x \geq 0$, $y \geq 0$.

Minimize $g = 7x + 3y$ subject to

$$\begin{cases} -x + 4y \geq 4 \\ x - y \leq 5 \\ 2x + 3y \leq 30 \end{cases}$$

11. The final simplex matrix for a problem is given below. Give the solutions to *both* the maximization problem and the dual minimization problem. Carefully label each solution.

$$\left[\begin{array}{ccccccc|c} 0 & 1 & -0.5 & 3 & -0.7 & 0 & 0 & 15 \\ 1 & 0 & -0.8 & -2 & -0.5 & 0 & 0 & 17 \\ 0 & 0 & -0.6 & -1 & -2.4 & 1 & 0 & 11 \\ \hline 0 & 0 & 12 & 4 & 18 & 0 & 1 & 658 \end{array}\right]$$

12. River Brewery is a micro-brewery that produces an amber lager and an ale. Producing a barrel of lager requires 3 lb of corn and 2 lb of hops. Producing a barrel of ale requires 2 lb each of corn and hops. Profit from each barrel of lager is \$7 and from each barrel of ale is \$6. Suppliers can provide at most 1200 lb of corn and at most 1000 lb of hops per month. Formulate a linear programming problem that can be used to maximize River Brewery's monthly profit. Then solve it with the simplex method.
13. A marketing research group conducting a telephone survey must contact at least 150 wives and 120 husbands. It costs \$3 to make a daytime call and (because of higher labor costs) \$4 to make an evening call. On average, daytime calls reach wives 30% of the time, husbands 10% of the time, and neither of these 60% of the time, whereas evening calls reach wives 30% of the time, husbands 30% of the time, and neither of these 40% of the time. Staffing considerations mean that daytime calls must be less than or equal to half of the total calls made. Formulate a linear programming problem that can be used to minimize the cost of completing the survey. Then solve it graphically.

Extended Applications / Group Projects

I. Transportation

The Kimble Firefighting Equipment Company has two assembly plants, Plant A in Atlanta and Plant B in Buffalo. It has two distribution centers, Center I in Pittsburgh and Center II in Savannah. The company can produce at most 800 alarm valves per week at Plant A and at most 1000 per week at Plant B. Center I must have at least 900 alarm valves per week, and Center II must have at least 600 alarm valves per week. The costs per unit for transportation from the plants to the centers follow.

Plant A to Center I	\$2
Plant A to Center II	3
Plant B to Center I	4
Plant B to Center II	1

What weekly shipping plan will meet the market demands and minimize the total cost of shipping the valves from the assembly plants to the distribution centers? What is the minimum transportation cost? Should the company eliminate the route with the most expensive unit shipment cost?

1. To assist the company in making these decisions, use linear programming with the following variables.

y_1 = the number of units shipped from Plant A to Center I

y_2 = the number of units shipped from Plant A to Center II

y_3 = the number of units shipped from Plant B to Center I

y_4 = the number of units shipped from Plant B to Center II

2. Form the constraint inequalities that give the limits for the plants and distribution centers, and form the total cost function that is to be minimized.
3. Solve the mixed constraint problem for the minimum transportation cost.
4. Determine whether any shipment route is unnecessary.

II. Sensitivity Analysis

A manufacturer of television sets makes one or more of four models of color sets: a 19-inch portable, a 23-inch table model, a 35-inch table model, and a 35-inch floor model. The assembly and testing time requirements for each model are shown in the table below, together with the maximum amounts of time available for assembly and testing. In addition to these constraints, the supplier of picture tubes indicated that it would supply no more than 180 picture tubes in the next month and that of these, no more than 100 could be 35-inch picture tubes.

	19-Inch Portable	*23-Inch Table*	*35-Inch Table*	*35-Inch Floor*	*Total Available*
Assembly time (hours)	8	10	12	15	2000
Test time (hours)	2	2	4	5	500
Marginal profit (dollars)	40	60	80	100	

The computer output from a linear programming package analysis of this problem is given on page 000. Use it to answer the following questions.

1. What is the optimal production schedule for the television manufacturer for this month? (It is profitable to bring a variable into the solution only if its $C_j - Z_j$ entry in the output summary is greater than or equal to 0.)
2. By how much could marginal profit for 19-inch portable televisions increase before a new optimal solution should be sought? (The nonbasic variables will not affect the current optimal solution if the coefficients of these variables, C_j, are held in the range $-\infty < C_j \leq Z_j$, which is called the *range of insignificance*.)
3. If the marginal profit for the 35-inch table model increases beyond \$80, should a new optimal solution be sought?
4. Over what range of values can the marginal profit for the 23-inch table model vary without the optimal solution changing? This is the range of optimality for this model, with the lower and upper limits for the variables given in the table titled "Sensitivity Analysis of Objective Function Coefficients" (see the computer output).
5. Over what range of values can the marginal profit for the 35-inch floor model vary without the optimal solution changing?
6. The $C_j - Z_j$ values tell how much the objective function changes as one unit of a variable is introduced into the solution, and the negative of the $C_j - Z_j$ values for a slack variable associated with a constraint tells how much the objective function will increase if one additional unit of the resource is made available. This can be considered the price we would be willing to pay to obtain one additional unit of this resource, so we call it the **shadow price** of the resource. What is the marginal value of an additional hour of assembly time? Over what range of values is this marginal value valid?
7. What is the marginal value of an additional hour of test time? Over what range of values is this marginal value valid?
8. Management is considering the addition of a new 23-inch model that uses 10 hours of assembly and 3 hours of test time. If the marginal profit from this model would be \$70, can the model be produced at a profit? What would the profit or loss per unit be?

0 MAX 40X1+60X2+80X3+100X4
SUBJECT TO:
1 8X1+10X2+12X3+15X4<2000
2 2X1+2X2+4X3+5X4<500
3 X1+X2+X3+X4<180
4 X3+X4<100

OUTPUT SUMMARY:

C		40	60	80	100	0	0	0	0		
		X1	X2	X3	X4	S1	S2	S3	S4		RHS
60	X2	.5	1	0	0	.25	−.75	0	0	:	125
100	X4	.2	0	.8	1	−.1	.5	0	0	:	50
0	S3	.3	0	.2	0	−.15	.25	1	0	:	5
0	S4	−.2	0	.2	0	.1	−.5	0	1	:	50
Z		50	60	80	100	5	5	0	0	:	12500
C − Z		−10	0	0	0	−5	−5	0	0		

VARIABLE	QUANTITY	VARIABLE	SHADOW PRICE	
X2	125	X1	−10	
X4	50	X2	0	
S3	5	X3	0	
S4	50	X4	0	
		S1	−5	ASSEMBLY TIME
		S2	−5	TEST TIME
OPTIMAL Z = 12500		S3	0	PICTURE TUBES
		S4	0	35″ TUBES

SENSITIVITY ANALYSIS OF OBJECTIVE FUNCTION COEFFICIENTS

VARIABLE NAME	LOWER LIMIT	ORIGINAL VALUE	UPPER LIMIT
X2	40	60	66.67
X4	90	100	150
X1	NO LIMIT	40	50
X3	NO LIMIT	80	80

SENSITIVITY ANALYSIS OF RIGHT HAND SIDE RANGES

CONSTRAINT	LOWER LIMIT	ORIGINAL VALUE	UPPER LIMIT	
S1	1500	2000	2033.33	ASSEMBLY TIME
S2	480	500	600	TEST TIME
S3	175	180	NO LIMIT	PICTURE TUBES
S4	50	100	NO LIMIT	35″ TUBES

III. Minimizing Costs for Electrical Transmission

Dynamo Generation Systems has three electric power generation plants that supply electricity to four cities.* The table below summarizes the supply capacity of each plant, the peak power demand of each city, and the transmission costs for sending 1 million kilowatt hours (kWh) of electric power from each plant to each city.

	City 1	*City 2*	*City 3*	*City 4*	*Supply Capacity (million kWh)*
Plant 1	\$9	\$14	\$16	\$5	40
Plant 2	\$6	\$8	\$10	\$9	35
Plant 3	\$12	\$9	\$13	\$7	50
Peak demand (million kWh)	20	45	30	30	

Dynamo Generation Systems wants to know how to supply electricity from each plant to each of the various cities in order to minimize its transmission costs, subject to its own power generation capacities and subject to meeting the peak power demands of each city.

Dynamo has hired you to perform two tasks: (1) Carefully state this linear programming problem. (2) Solve it using a linear programming solver available on the Internet. To help you with these tasks, complete the following.

1. Let x_1 through x_4 represent the amount of power (in million kWh) sent from Plant 1 to Cities 1 through 4, respectively.
   Let x_5 through x_8 be the amount of power (in million kWh) sent from Plant 2 to Cities 1 through 4, respectively.
   Let x_9 through x_{12} be the amount of power (in million kWh) sent from Plant 3 to Cities 1 through 4, respectively.
   With these variables, formulate the objective cost function to be minimized.
2. (a) Set up the supply constraints for each plant. Note that the amount of power supplied by a plant cannot exceed its generation capacity.
   (b) Set up the peak demand constraint for each city. Note that the amount of power delivered to each city from all sources must be at least as much as its peak demand.
3. Use the results of (1) and (2) to state clearly Dynamo Generation Systems' linear programming problem.

*This project is based on Aarvik, O., and P. Randolph, "The Application of Linear Programming to the Determination of Transmission Line Fees in an Electrical Power Network," *Interfaces* 6 (1975):17–31.

4. NEOS (Network-Enabled Optimization System) is a centralized governmental Web site containing information and programs for solving optimization problems. Researchers and industries may submit linear programming problems through NEOS for free-of-charge solution. Within this NEOS site (NEOS URL: http://www.mcs.anl.gov/home/otc/Server/) are links to on-line solvers for relatively small problems and free ware to download for personal or educational use.
   (a) On-line Interactive Linear Programming Solver
   URL: ford.ieor.berkeley.edu:80/riot/Tolls/InteractLP/
   GeneralSolverInput.html
   (b) Freeware to download (others are available)
   —For PC (LINSOLVE)
   URL: http://archives.math.utk.edu/software/msdos/discrete.math/
   tslin/.html
   —For Mac (Lin Pro)
   URL: http://archives.math.utk.edu/software/mac/discreteMath/.
   directory.html

   Use one of these sites to minimize Dynamo Generation Systems' cost of power delivery.

Warm-up

Prerequisite Problem Type	For Section	Answer	Section for Review
Write the following with positive exponents: (a) x^{-3} (b) $\frac{1}{x^{-2}}$ (c) $\sqrt{x}$	5.1 5.2 5.3	(a) $\frac{1}{x^3}$ (b) x^2 (c) $x^{1/2}$	0.3, 0.4 Exponents and radicals
Simplify: (a) 2^0 (b) $x^0 \quad (x \neq 0)$ (c) $49^{1/2}$ (d) 10^{-2}	5.1 5.2 5.3	(a) 1 (b) 1 (c) 7 (d) $\frac{1}{100}$	0.3, 0.4 Exponents
Answer true or false. (a) $\left(\frac{1}{2}\right)^x = 2^{-x}$ (b) $\sqrt{50} = 50^{1/2}$ (c) If $8 = 2^y$, then $y = 4$. (d) If $x^3 = 8$, then $x = 2$.	5.1 5.2	(a) True (b) True (c) False; $y = 3$ (d) True	0.3, 0.4 Exponents and radicals
(a) If $f(x) = 2^{-2x}$, what is $f(-2)$? (b) If $f(x) = 2^{-2x}$, what is $f(1)$? (c) If $f(t) = (1 + 0.02)^t$, what is $f(0)$? (d) If $f(t) = 100(0.03)^{0.02^t}$, what is $f(0)$?	5.1 5.2 5.3	(a) 16 (b) $\frac{1}{4}$ (c) 1 (d) 3	1.2 Functional notation

Chapter 5

Exponential and Logarithmic Functions

In this chapter we study exponential and logarithmic functions, which provide models for many applications that at first seem remote and unrelated. For example, a business manager uses these functions to study the growth of money or corporations or the decay of new sales volume. A social scientist uses these functions to study the growth of organizations or populations or the dating of fossilized remains. And a biologist uses these functions to study the growth of microorganisms in a laboratory culture, the spread of disease, the measurement of pH, or the decay of radioactive material.

In our study of exponential and logarithmic functions, we will examine their descriptions, their properties, their graphs, and the special inverse relationship between these two functions. We will see how these functions are applied to some of the concerns of social scientists, business managers, and life scientists. In these applications, the inverse relationship of the exponential and logarithmic functions is used to solve some of the equations that arise. We will assume that you have a calculator that computes powers of e *and logarithms to the base* e *(denoted* e^x *and* ln x*, respectively).*

5.1 Exponential Functions

OBJECTIVES

- *To graph exponential functions*
- *To model exponential functions*

APPLICATION PREVIEW

Suppose a culture of bacteria has the characteristic that each minute, every microorganism splits into two new organisms. We can describe the number of bacteria in the culture as a function of time. That is, if we begin the culture with one microorganism, we know that after 1 minute we will have two organisms, after 2 minutes, four, and so on. Table 5.1 gives a few of the values that describe this growth. If x represents the number of minutes that have passed and y represents the number of organisms, the points (x, y) lie on the graph of the function with equation

$$y = 2^x$$

TABLE 5.1

Minutes Passed	*Number of Organisms*
0	1
1	2
2	4
3	8
4	16

The equation $y = 2^x$ in the Application Preview is an example of a special group of functions called **exponential functions.** In general, we define these functions as follows:

Exponential Functions If a is a positive real number and $a \neq 1$, then the function

$$f(x) = a^x$$

is an **exponential function.**

A table of some values satisfying $y = 2^x$ and the graph of this function are given in Figure 5.1. This function is said to model the growth of the number of organisms in the Application Preview discussion, even though some points on the graph do not correspond to a time and a number of organisms. For example, time x could not be negative, and the number of organisms y could not be fractional.

We earlier defined rational powers of x in terms of radicals in Chapter 0, so 2^x makes sense for any rational power x. It can also be shown that the laws of exponents apply for irrational numbers. We will assume that if we graphed $y = 2^x$ for irrational values of x, those points would lie on the curve in Figure 5.1. Thus, in general, we can graph an exponential function by plotting easily calculated points, such as those in the table in Figure 5.1, and drawing a smooth curve through the points.

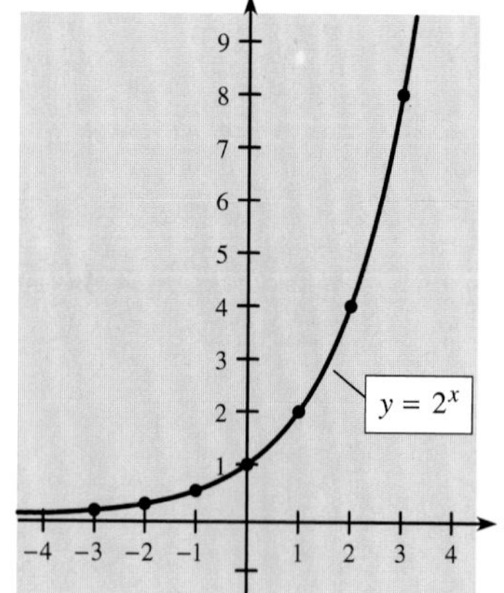

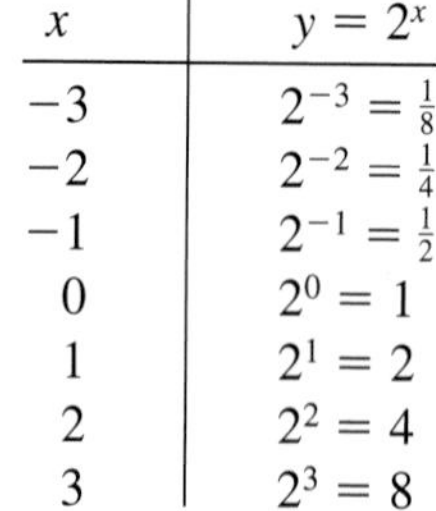

x	$y = 2^x$
-3	$2^{-3} = \frac{1}{8}$
-2	$2^{-2} = \frac{1}{4}$
-1	$2^{-1} = \frac{1}{2}$
0	$2^0 = 1$
1	$2^1 = 2$
2	$2^2 = 4$
3	$2^3 = 8$

Figure 5.1

EXAMPLE 1

Graph $y = 10^x$.

Solution

A table of values and the graph are given in Figure 5.2.

x	y
-3	$10^{-3} = 1/1000$
-2	$10^{-2} = 1/100$
-1	$10^{-1} = 1/10$
0	$10^0 = 1$
1	$10^1 = 10$
2	$10^2 = 100$
3	$10^3 = 1000$

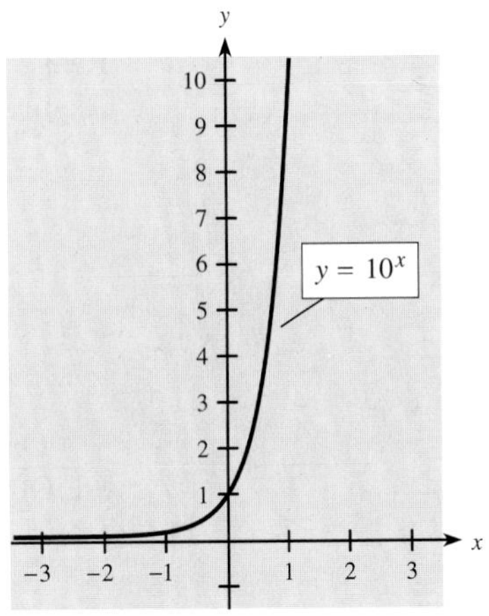

Figure 5.2

Technology Note We also could have used a graphing utility to graph $y = 2^x$ and $y = 10^x$. Figure 5.3 shows graphs of $f(x) = 1.5^x$, $f(x) = 4^x$, and $f(x) = 20^x$ that were obtained using a graphing utility. Note the range for each graph in the figure.

Figure 5.3

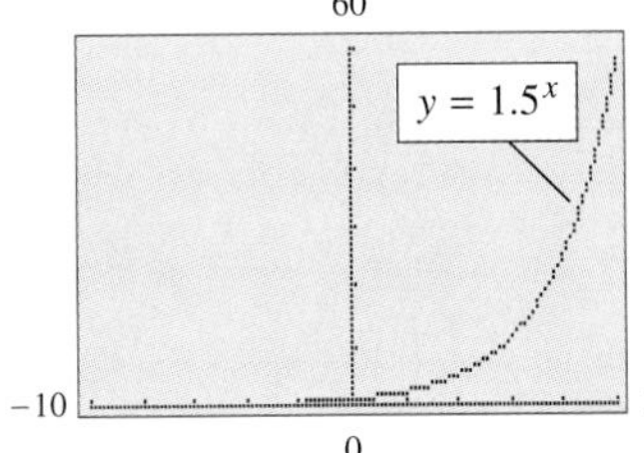

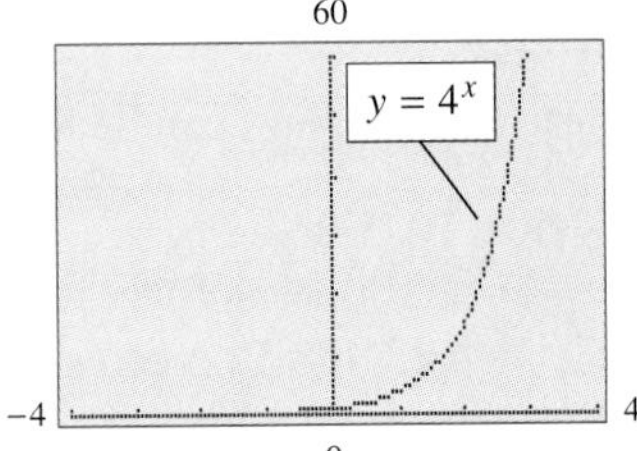

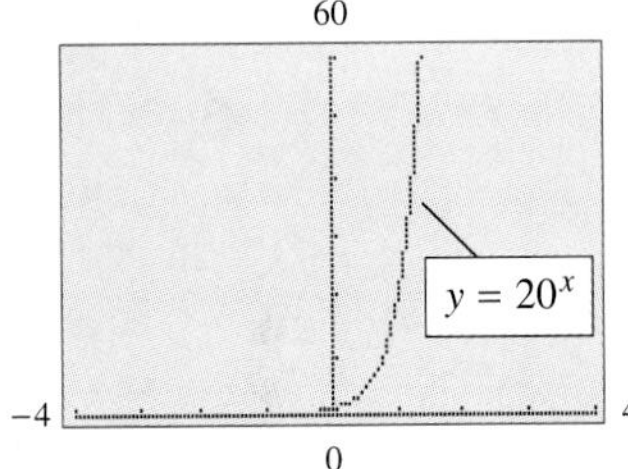

Note that the three graphs in the figure and the graphs of $y = 2^x$ and $y = 10^x$ are similar. The graphs of $y = 2^x$ and $y = 10^x$ clearly approach, but do not touch, the negative x-axis. However, the graphs in Figure 5.3 appear as if they might eventually merge with the negative x-axis. By adjusting the range, however, we would see that these graphs also approach, but do not touch, the negative x-axis. That is, the negative x-axis is an asymptote for these functions.

In fact, the shapes of the graphs of functions of the form $y = f(x) = a^x$, with $a > 1$, are similar to those in Figures 5.1, 5.2, and 5.3. Exponential functions of this type are called **exponential growth functions** because they are used to model growth in diverse applications. Their graphs have the basic shape shown in the following box.

Graphs of Exponential Growth Functions

Function: $y = f(x) = a^x \quad (a > 1)$
y-intercept: $(0, 1)$
Domain: All reals
Range: All positive reals
Asymptote: x-axis (negative half)

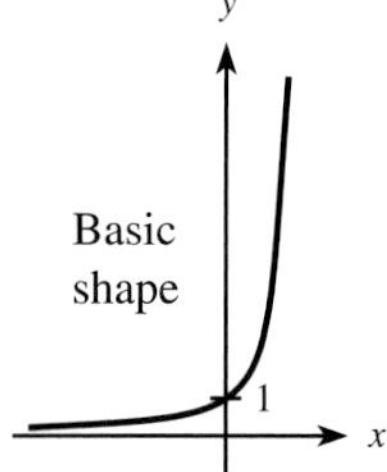

One exponential growth model concerns money invested at compound interest. Often we seek to evaluate rather than graph these functions.

EXAMPLE 2

If \$10,000 is invested at 6%, compounded monthly, then the future value of the investment S after x years is given by

$$S = 10{,}000(1.005)^{12x}$$

Find the future value of the investment after (a) 5 years and (b) 30 years.

Solution

These future values can be found with a calculator.

(a) $S = 10{,}000(1.005)^{12(5)}$
$= 10{,}000(1.005)^{60}$
$= \$13{,}488.50$ (nearest cent)

(b) $S = 10{,}000(1.005)^{12(30)}$
$= 10{,}000(1.005)^{360}$
$= \$60{,}225.75$ (nearest cent)

Note that the amount after 30 years is significantly more than the amount after 5 years, a result consistent with exponential growth models.

CHECKPOINT

1. Can any value of x give a negative value for y if $y = a^x$ and $a > 1$?
2. If $a > 1$, what asymptote does the graph of $y = a^x$ approach?

x	$y = e^x$
-2	0.14
-1	0.37
0	1.00
1	2.72
2	7.39

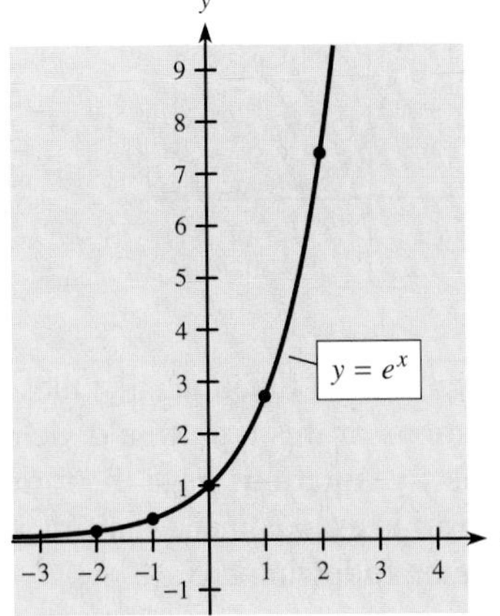

Figure 5.4

A special function that occurs frequently in economics and biology is $y = e^x$, where e is a fixed irrational number (approximately 2.71828 . . .). We will see how e arises when we discuss interest that is compounded continuously, and we will formally define e in Section 9.2, "Continuous Functions: Limits at Infinity."

Because $e > 1$, the graph of $y = e^x$ will have the same basic shape as other growth exponentials. We can calculate the y-coordinate for points on the graph of this function with a calculator. A table of some values (with y-values rounded to two places) and the graph are shown in Figure 5.4.

Exponentials whose bases are between 0 and 1, such as $y = \left(\frac{1}{2}\right)^x$, have graphs different from those of the exponentials just discussed. Using the properties of exponents, we have

$$y = \left(\frac{1}{2}\right)^x = (2^{-1})^x = 2^{-x}$$

This suggests that exponentials of the form $y = b^x$, where $0 < b < 1$, can be rewritten in the form $y = a^{-x}$, where $a > 1$.

EXAMPLE 3

Graph $y = 2^{-x}$.

Solution

A table of values and the graph are given in Figure 5.5.

x	$y = 2^{-x}$
-3	$2^3 = 8$
-2	$2^2 = 4$
-1	$2^1 = 2$
0	$2^0 = 1$
1	$2^{-1} = \frac{1}{2}$
2	$2^{-2} = \frac{1}{4}$
3	$2^{-3} = \frac{1}{8}$

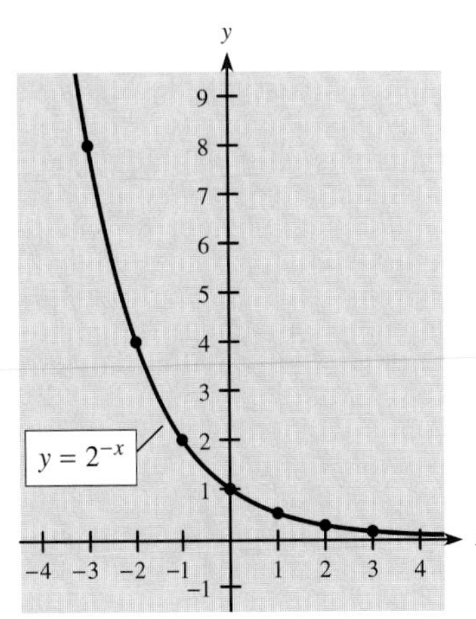

Figure 5.5

EXAMPLE 4

Graph $y = e^{-2x}$.

Solution

Using a calculator to find the values of powers of e (to two decimal places), we get the graph shown in Figure 5.6.

x	y
-3	$e^6 = 403.43$
-2	$e^4 = 54.60$
-1	$e^2 = 7.39$
0	$e^0 = 1.00$
1	$e^{-2} = 0.14$
2	$e^{-4} = 0.02$

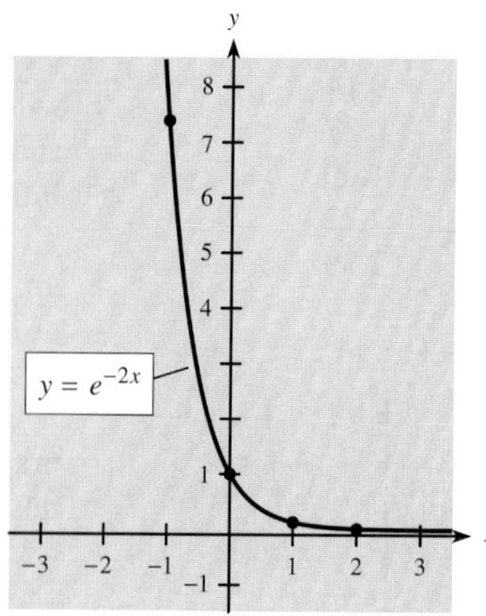

Figure 5.6

We can also use a graphing utility to graph exponential functions like those in Examples 3 and 4. Additional examples of functions of this type would yield graphs with the same basic shape. **Exponential decay functions** have the form $y = a^{-x}$, where $a > 1$. They model decay for various phenomena, and their graphs have the characteristics and shape shown in the following box.

Graphs of Exponential Decay Functions

Function: $y = f(x) = a^{-x} \quad (a > 1)$
y-intercept: $(0, 1)$
Domain: All reals
Range: All positive reals
Asymptote: x-axis (positive half)

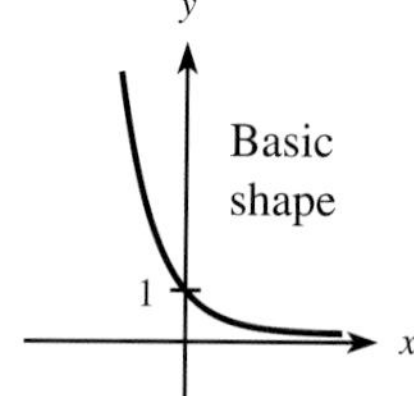

One example of exponential decay is the decrease in the number of atoms of a radioactive element. The number of atoms y present at an instant in time t is given by

$$y = w_0 e^{-h(t - t_0)}$$

where w_0 is the number of atoms at time t_0 and h is a constant that depends on the element. There are numerous other applications of decay exponentials.

EXAMPLE 5

The purchasing power P of a fixed income of \$30,000 per year (such as a pension) after t years of 4% inflation can be modeled by

$$P = 30{,}000e^{-0.04t}$$

Find the purchasing power after (a) 5 years and (b) 20 years.

Solution

We can use a calculator to answer both parts.

(a) $P = 30{,}000e^{-0.04(5)} = 30{,}000e^{-0.2} = \$24{,}562$ (nearest dollar)
(b) $P = 30{,}000e^{-0.04(20)} = 30{,}000e^{-0.8} = \$13{,}480$ (nearest dollar)

Note that the impact of inflation over time significantly erodes purchasing power and provides some insight into the plight of elderly people who live on a fixed income.

Exponential functions with base e often arise in natural ways. As we will see later in Section 6.2, the growth of money that is compounded continuously is given by $S = Pe^{rt}$, where P is the original principal, r the annual interest rate, and t the time in years. Certain populations (of insects, for example) grow exponentially, and the number of individuals can be closely approximated by the equation $y = P_0e^{ht}$, where P_0 is the original population size, h is a constant that depends on the type of population, and y is the population size at any instant t.

There are other important exponential functions that use base e but whose graphs are different from those we have discussed. For example, the standard normal probability curve (often referred to as a bell-shaped curve) is the graph of an exponential function with base e (see Figure 5.7). Later in this chapter we will study other exponential functions that model growth but whose graphs are also different from those discussed previously.

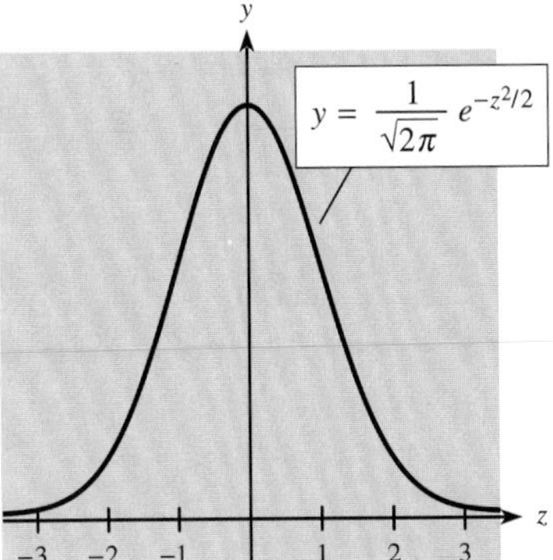

Figure 5.7 Standard normal probability curve

CHECKPOINT

3. True or false: The graph of $y = \left(\frac{1}{a}\right)^x$, with $a > 1$, is the same as the graph of $y = a^{-x}$.
4. True or false: The graph of $y = a^{-x}$, with $a > 1$, approaches the positive x-axis as an asymptote.
5. True or false: The graph of $y = 2^{-x}$ is the graph of $y = 2^x$ reflected about the y-axis.

Graphing Utilities Graphing utilities allow us to investigate variations of the growth and decay exponentials. For example, we can examine the similarities and differences between the graphs of $y = f(x) = a^x$ and $y = mf(x)$, $y = f(kx)$, $y = f(x + h)$, and $y = f(x) + C$ for constants m, k, h, and C. We also can explore these similarities and differences for $y = f(x) = a^{-x}$.

EXAMPLE 6

(a) Use a graphing utility to graph $y = f(x) = e^x$ and $y = mf(x) = me^x$ for $m = -6, -3, 2$, and 10.

(b) Explain the effect that m has on the shape of the graph, the asymptote, and the y-intercept.

Solution

(a) The graphs are shown in Figure 5.8.

(b) From the graphs, we see that when $m > 0$, the shape of the graph is still that of a growth exponential. When $m < 0$, the shape is that of a growth exponential turned upside down, or reflected through the x-axis (making the range $y < 0$). In all cases, the asymptote is unchanged and is the negative x-axis. In each case, the y-intercept is $(0, m)$.

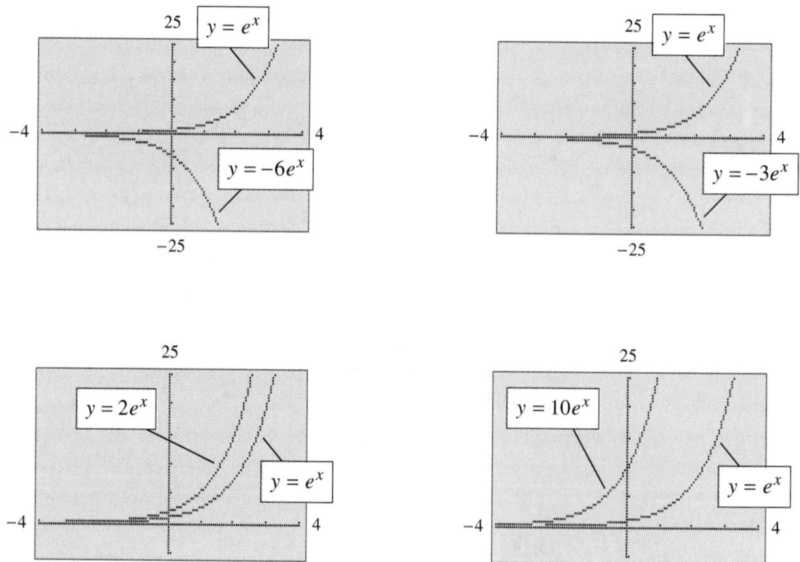

Figure 5.8

Modeling Exponential Functions

Many types of data can be modeled using an exponential growth function. Figure 5.9(a) shows a graph of amounts of carbon in the atmosphere due to emissions from the burning of fossil fuels. With curve-fitting tools available with some computer software or on some graphing calculators, we can develop an equation that models, or approximates, these data. The equation model and its graph are shown in Figure 5.9(b).

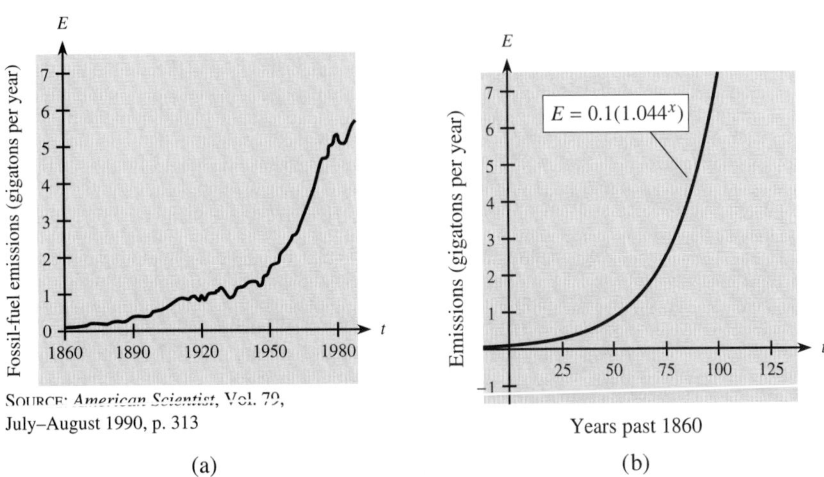

Figure 5.9 (a) (b)

The function in Figure 5.9 is a growth function. In general, an exponential growth function can be modeled with the general form $y = a(b^x)$ with $b > 1$, and an exponential decay function can be modeled with the general form $y = a(b^x)$ with $0 < b < 1$, or with the form $y = a(b^{-x})$ with $b > 1$. We now consider a real data problem that is modeled by an exponential decay function.

EXAMPLE 7

The purchasing power of \$1 is calculated by assuming that the Consumer Price Index (CPI) is \$1 in 1982. It is used to show the value of goods that could be purchased using a dollar in 1982. Table 5.2 gives the purchasing power of \$1 on the basis of consumer prices for 1963–1995.

(a) Use these data, with $x = 0$ in 1960, to find an exponential function that models the decay of the dollar.

(b) What does this model predict as the purchasing power of \$1 in 2005?

(c) When will the purchasing power of a 1982 dollar be \$0.40?

TABLE 5.2

Year	Purchasing Power of $1	Year	Purchasing Power of $1	Year	Purchasing Power of $1
1963	3.265	1974	2.029	1985	0.928
1964	3.22	1975	1.859	1986	0.913
1965	3.166	1976	1.757	1987	0.88
1966	3.08	1977	1.649	1988	0.846
1967	2.993	1978	1.532	1989	0.807
1968	2.873	1979	1.38	1990	0.766
1969	2.726	1980	1.215	1991	0.734
1970	2.574	1981	1.098	1992	0.713
1971	2.466	1982	1.035	1993	0.692
1972	2.391	1983	1.003	1994	0.675
1973	2.251	1984	0.961	1995	0.656

SOURCE: U.S. Bureau of Labor Statistics

Solution

(a) From a scatter plot of the points, with $x = 0$ in 1960, it appears that an exponential decay model is appropriate (see Figure 5.10a). Using the function of the form $a*b^x$, the data can be modeled by the equation $y = 4.2885(0.9441^x)$. Figure 5.10(b) shows how the graph of the equation fits the data. To write this equation with a base greater than 1, we can write 0.9441 as $(1/0.9441)^{-1} = 1.0592^{-1}$.

Thus the model could be written as $y = 4.2885(1.0592)^{-x}$.

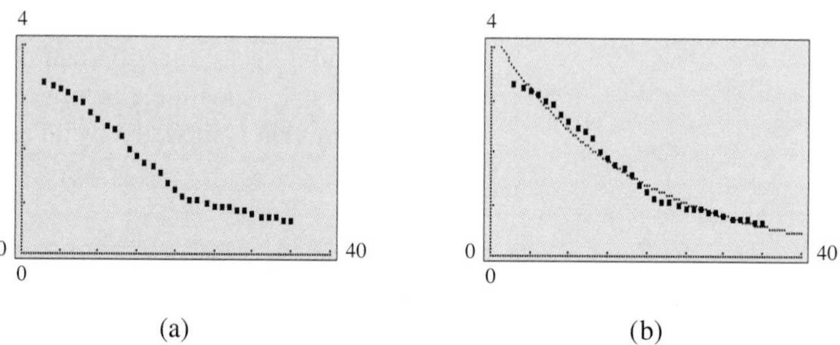

Figure 5.10

(b) For the year 2005, we use $x = 45$. Evaluating the function for $x = 45$ gives the purchasing power of $0.322.

(c) Tracing along the curve shows that the purchasing power is $0.41 when $x = 41$, or in 2001.

CHECKPOINT SOLUTIONS

1. No, all values of y are positive.
2. The left side of the x-axis ($y = 0$)
3. True 4. True 5. True

EXERCISE 5.1

In Problems 1–8, use a calculator to evaluate each expression.

1. $10^{0.5}$
2. $10^{3.6}$
3. $5^{-2.7}$
4. $8^{-2.6}$
5. $3^{1/3}$
6. $2^{11/6}$
7. e^2
8. e^{-3}

In Problems 9–18, graph each function.

9. $y = 4^x$
10. $y = 8^x$
11. $y = 2(x^3)$
12. $y = 3(2^x)$
13. $y = 5^x$
14. $y = 3^{x-1}$
15. $y = 2e^x$
16. $y = 3^{-x}$
17. $y = 3^{-2x}$
18. $y = e^{-x}$

In Problems 19–22, use a graphing utility to make the graphs.

19. Given $f(x) = e^{-x}$. Graph $y = f(x)$ and $y = f(kx) = e^{-kx}$ for each k, where $k = 0.1, 0.5, 2$, and 5. Explain the effect that different values of k have on the graph.
20. Given $f(x) = 2^{-x}$. Graph $y = f(x)$ and $y = mf(x) = m(2^{-x})$ for each m, where $m = -7, -2, 3$, and 8. Explain the effect that different values of m have on the graph.
21. Given $f(x) = 4^x$. Graph $y = f(x)$ and $y = f(x) + C = 4^x + C$ for each C, where $C = -5, -2, 3$, and 6. Explain the effect that C has on the graph.
22. Given $f(x) = 3^x$. Graph $y = f(x)$ and $y = f(x - h) = 3^{x-h}$ for each h, where $h = 4, 1, -2$, and -5. Explain the effect that h has on the graph.

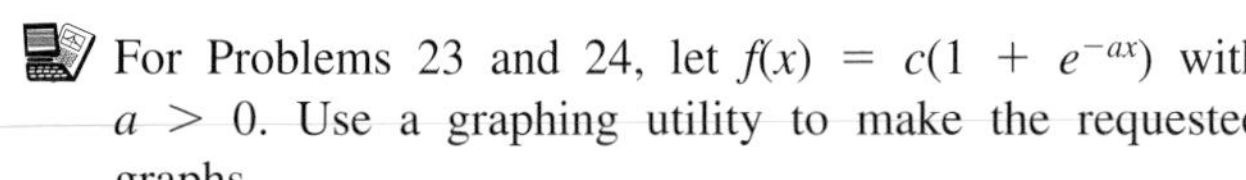

For Problems 23 and 24, let $f(x) = c(1 + e^{-ax})$ with $a > 0$. Use a graphing utility to make the requested graphs.

23. (a) Fix $a = 1$ and graph $y = f(x) = c(1 + e^{-x})$ for $c = 10, 50$, and 100.
    (b) What effect does c have on the graphs?
24. (a) Fix $c = 50$ and graph $y = f(x) = 50(1 + e^{-ax})$ for $a = 0.1, 1$, and 10.
    (b) What effect does a have on the graphs?

For Problems 25 and 26, let $f(x) = \dfrac{A}{1 + ce^{-x}}$. Use a graphing utility to make the requested graphs.

25. (a) Fix $A = 100$ and graph $y = f(x) = \dfrac{100}{1 + ce^{-x}}$ for $c = 0.25, 1, 9$, and 49.
    (b) What effect does c have on the graphs?
26. (a) Fix $c = 1$ and graph $y = f(x) = \dfrac{A}{1 + e^{-x}}$ for $A = 50, 100$, and 150.
    (b) What effect does A have on the graphs?

Applications

27. ***Compound interest*** If \$1000 is invested for x years at 8%, compounded quarterly, the future value that will result is

$$S = 1000(1.02)^{4x}$$

What amount will result in 8 years?

28. ***Compound interest*** If \$3200 is invested for x years at 8%, compounded quarterly, the interest earned is

$$I = 3200(1.02)^{4x} - 3200$$

What interest is earned in 5 years?

29. ***Compound interest*** We will show in the next chapter that if \$$P$ is invested for n years at 10% compounded continuously, the future value of the investment is given by

$$S = Pe^{0.1n}$$

Use $P = 1000$ and graph this function for $0 \le n \le 20$.

30. ***Compound interest*** If \$1000 is invested for x years at 10%, compounded continuously, the future value that results is

$$S = 1000e^{0.10x}$$

What amount will result in 5 years?

31. ***Drug in the bloodstream*** The concentration y of a certain drug in the bloodstream at any time t is given by the equation

$$y = 100(1 - e^{-0.462t})$$

Graph this equation for $0 \le t \le 10$. Write a sentence that interprets the graph.

32. ***Bacterial growth*** A single bacterium splits into two bacteria every half hour, so the number of bacteria in a culture quadruples every hour. Thus the equation by which a colony of 10 bacteria multiplies in t hours is given by

$$y = 10(4^t)$$

Graph this equation for $0 \le t \le 8$.

33. ***Product reliability*** A statistical study shows that the fraction of television sets of a certain brand that are still in service after x years is given by $f(x) = e^{-0.15x}$. Graph this equation for $0 \le x \le 10$. Write a sentence that interprets the graph.

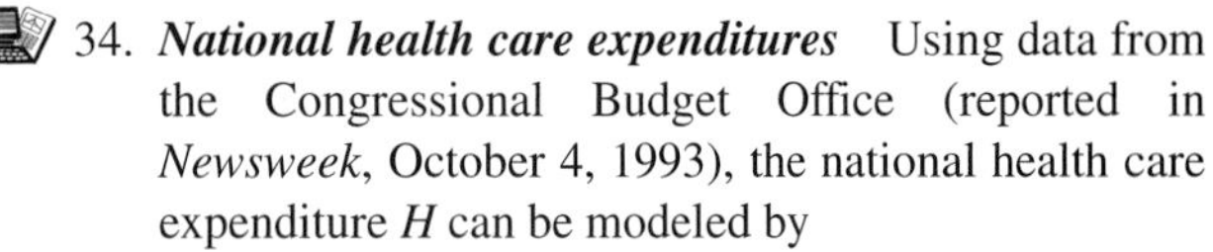

34. ***National health care expenditures*** Using data from the Congressional Budget Office (reported in *Newsweek*, October 4, 1993), the national health care expenditure H can be modeled by

$$H = 45e^{0.0898t}$$

where t is the number of years from 1960 and H is in billions of dollars. Graph this equation with a graphing utility to show the graph through the year 2010.

Population growth Use the following information to answer Problems 35–38. World population can be considered as growing according to the equation

$$N = N_0(1 + r)^t$$

where N_0 is the number of individuals at time $t = 0$, r is the yearly rate of growth, and t is the number of years.

35. Sketch the graph for $t = 0$ to $t = 10$ when the growth rate is 2% and N_0 is 4.1 billion.
36. Sketch the graph for $t = 0$ to $t = 10$ when the growth rate is 3% and N_0 is 4.1 billion.
37. Repeat Problem 35 when the growth rate is 5%.
38. Repeat Problem 36 when the growth rate is 7%.

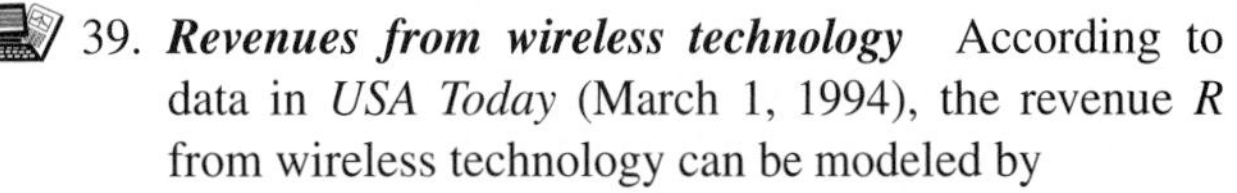

39. ***Revenues from wireless technology*** According to data in *USA Today* (March 1, 1994), the revenue R from wireless technology can be modeled by

$$R = 0.572e^{0.3860t}$$

where t is the number of years since 1985 and R is in billions of dollars. Graph this equation with a graphing utility to show the graph through the year 2010.

40. ***Advertising and sales*** Suppose that sales are related to advertising expenditures according to one of the following two models, where S_1 and S_n are sales and x is advertising, all in millions of dollars.

$$S_1 = 30 + 20x - 0.4x^2$$
$$S_n = 24.58 + 325.18(1 - e^{-x/14})$$

(a) Graph both of these functions on the same set of axes. Use a graphing utility.
(b) Do these two functions give approximately the same sales per million dollars of advertising for $0 \leq x \leq 20$?
(c) How do these functions differ for $x > 20$? Which more realistically represents the relationship between sales and advertising expenditures after \$20 million is spent on advertising? Why?

41. ***Industrial consolidation*** The figure below, from *Investor's Business Daily* (March 5, 1998) shows how quickly the U.S. metal processing industry is consolidating. The linear equation that is the best fit for the number of metal processors as a function of years from 1990 is $y_1 = -329.6970x + 5392.1212$, and the best exponential fit is $y_2 = 9933.7353e^{-0.1672x}$. The linear equation gives a much better fit for the data points than the exponential equation. Why, then, is the exponential equation a more useful model to predict the number of metal processors in 2007?

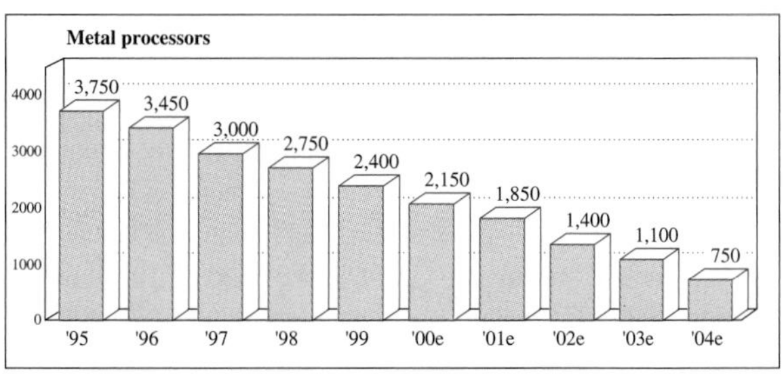

42. ***National debt*** The table below gives the U.S. national debt for selected years from 1900 to 1989.
(a) Using a function of the form a*b^x^, with $x = 0$ in 1900, model the data.
(b) Use the model to predict the debt in 1990.
(c) Predict when the debt will be \$5 trillion (\$5000 billion).

Year	*U.S. Debt (\$billions)*
1900	1.2
1910	1.1
1920	24.2
1930	16.1
1940	43.0
1945	258.7
1955	272.8
1965	313.8
1975	533.2
1985	1823.1
1987	2350.3
1989	2857.4

SOURCE: Bureau of Public Debt, U.S. Treasury

43. ***TV cable rates*** The average monthly rates for basic cable, without premium channels, for 1982–1997, are given in the table below.
   (a) These data can be modeled by an exponential function. Write the equation of the function, using x as the number of years since 1980.
   (b) If this model is accurate, what will basic cable cost in the year 2005?

Year	*Basic Cable Rate*	*Year*	*Basic Cable Rate*
1982	$8.30	1990	16.78
1983	8.61	1991	18.10
1984	8.98	1992	19.08
1985	9.73	1993	19.39
1986	10.67	1994	21.62
1987	12.18	1995	23.07
1988	13.86	1996	24.41
1989	15.21	1997	26.00

SOURCE: Nielsen Media Research and Paul Kagan Associates, *The Island Packet,* April 19, 1998

44. ***Dividend growth*** The dividend growth per share of First Union Bank from 1978 to 1997 (restated for stock splits) is shown in the table below. Use technology to create an exponential model for the data (with $x = 0$ in 1975), and use the model to predict the dividend in 2005.

Year	*Annual Dividend*	*Year*	*Annual Dividend*
1978	$0.145	1989	$.50
1979	0.155	1990	0.54
1980	0.165	1991	0.56
1981	0.18	1992	0.64
1982	0.20	1993	0.75
1983	0.225	1994	0.86
1984	0.245	1995	0.98
1985	0.29	1996	1.10
1986	0.325	1997	1.22
1987	0.385	1998	1.48
1988	0.43		

SOURCE: First Union Shareholder Information

45. ***Consumer Price Index*** The Consumer Price Index (CPI) is calculated by averaging the prices of various items after assigning a weight to each item. The following table gives the Consumer Price Indexes (CPI-U) for selected years 1940–1995, reflecting buying patterns of all urban consumers, with x representing years from 1900.
   (a) Find an equation that models these data.
   (b) Use the model to predict the Consumer Price Index in 2005.
   (c) During what year will the Consumer Price Index pass 200?

Year	*Consumer Price Index*
1940	14
1950	24.1
1960	29.6
1970	38.8
1980	82.4
1990	130.7
1995	152.4

SOURCE: Bureau of Labor Statistics

46. ***Compound interest*** The table below gives the value, at a bank, after intervals ranging from 0 to 7 years, of $20,000 invested at 10%, compounded annually.
   (a) Develop an exponential model for these data, accurate to four decimal places, with x in years and y in dollars.
   (b) Use the model to find the amount to which $20,000 will grow in 30 years if it is invested at 10%, compounded annually.

Year	*Amount Investment Grows to*
0	20,000
1	22,000
2	24,200
3	26,620
4	29,282
5	32,210.20
6	35,431.22
7	38,974.34

47. ***U.S. Postal rates*** The table below gives the U.S. postal rates for selected years from 1958 to 1995. Use an exponential equation to model the data, with x representing the number of years from 1900.

Year	*First-Class Postal Rate (1 oz)*
1958	4¢
1963	5
1968	6
1971	8
1974	10
1978	15
1981	18
1982	20
1985	22
1988	25
1991	29
1995	32

SOURCE: U.S. Postal Service

48. ***U.S. Postal rates UPDATE*** The table below gives the U.S. postal rates for selected years from 1958 to 1999. Does a cubic model fit the data better than an exponential model?

Year	*First-Class Postal Rate (1 oz)*
1958	4¢
1963	5
1968	6
1971	8
1974	10
1978	15
1981	18
1982	20
1985	22
1988	25
1991	29
1995	32
1999	33

5.2 Logarithmic Functions and Their Properties

OBJECTIVES

- *To convert equations for logarithmic functions from logarithmic to exponential form, and vice versa*
- *To evaluate some special logarithms*
- *To graph logarithmic functions*
- *To use properties of logarithmic functions to simplify expressions involving logarithms*
- *To use the change-of-base formula*
- *To model logarithmic functions*

APPLICATION PREVIEW

When interest is compounded continuously, if P dollars are invested at an annual rate r (as a decimal), then the future value of the investment after t years is given by

$$S = Pe^{rt}$$

A common question with investments such as this is "How long will it be before the investment doubles?" That is, when does $S = 2P$? To answer this question, and hence to develop a "doubling time" formula, requires that we solve for t, and this requires the use of **logarithmic functions.**

Logarithmic Functions and Graphs

Before the development and easy availability of calculators and computers, certain arithmetic computations, such as $(1.37)^{13}$ and $\sqrt[16]{3.09}$, were difficult to perform. The computations could be performed relatively easily using **logarithms,** which were developed in the 17th century by John Napier, or by using a slide rule, which is based on logarithms. The use of logarithms as a computing technique has all but disappeared today, but the study of **logarithmic functions** is still very important because of the many applications of these functions.

For example, let us again consider the culture of bacteria described at the beginning of the previous section. If we know that the culture is begun with one microorganism and that each minute every microorganism present splits into two new ones, then we can find the number of minutes it takes until there are 1024 organisms by solving

$$1024 = 2^y$$

The solution of this equation may be written in the form

$$y = \log_2 1024$$

which is read "y equals the logarithm of 1024 to the base 2."

In general, we may express the equation $x = a^y$ $(a > 0,\ a \neq 1)$ in the form $y = f(x)$ by defining a **logarithmic function.**

Logarithmic Function For $a > 0$ and $a \neq 1$, the **logarithmic function**

$$y = \log_a x \qquad \text{(logarithmic form)}$$

has domain $x > 0$ and is defined by

$$a^y = x \qquad \text{(exponential form).}$$

TABLE 5.3

Logarithmic Form	*Exponential Form*
$\log_{10} 100 = 2$	$10^2 = 100$
$\log_{10} 0.1 = -1$	$10^{-1} = 0.1$
$\log_2 x = y$	$2^y = x$
$\log_a 1 = 0$ $(a > 0)$	$a^0 = 1$
$\log_a a = 1$ $(a > 0)$	$a^1 = a$

The a is called the **base** in both $\log_a x = y$ and $a^y = x$, and y is the *logarithm* in $\log_a x = y$ and the *exponent* in $a^y = x$. Thus **a logarithm is an exponent.**

Table 5.3 shows some logarithmic equations and their equivalent exponential forms.

EXAMPLE 1

(a) Write $64 = 4^3$ in logarithmic form.
(b) Write $\log_4 \left(\frac{1}{64}\right) = -3$ in exponential form.
(c) If $4 = \log_2 x$, find x.

Solution

(a) $64 = 4^3$ is equivalent to $3 = \log_4 64$.
(b) $\log_4 \left(\frac{1}{64}\right) = -3$ is equivalent to $4^{-3} = \frac{1}{64}$.
(c) If $4 = \log_2 x$, then $2^4 = x$ and $x = 16$.

EXAMPLE 2

Evaluate:

(a) $\log_2 8$ (b) $\log_3 9$ (c) $\log_5\left(\frac{1}{25}\right)$

Solution

(a) If $y = \log_2 8$, then $8 = 2^y$. Because $2^3 = 8$, $\log_2 8 = 3$.
(b) If $y = \log_3 9$, then $9 = 3^y$. Because $3^2 = 9$, $\log_3 9 = 2$.
(c) If $y = \log_5\left(\frac{1}{25}\right)$, then $\frac{1}{25} = 5^y$. Because $5^{-2} = \frac{1}{25}$, $\log_5\left(\frac{1}{25}\right) = -2$.

EXAMPLE 3

Graph $y = \log_2 x$.

Solution

We may graph $y = \log_2 x$ by graphing $x = 2^y$. The table of values (found by substituting values for y and calculating x) and the graph are shown in Figure 5.11.

$x = 2^y$	y
$\frac{1}{8}$	-3
$\frac{1}{4}$	-2
$\frac{1}{2}$	-1
1	0
2	1
4	2
8	3

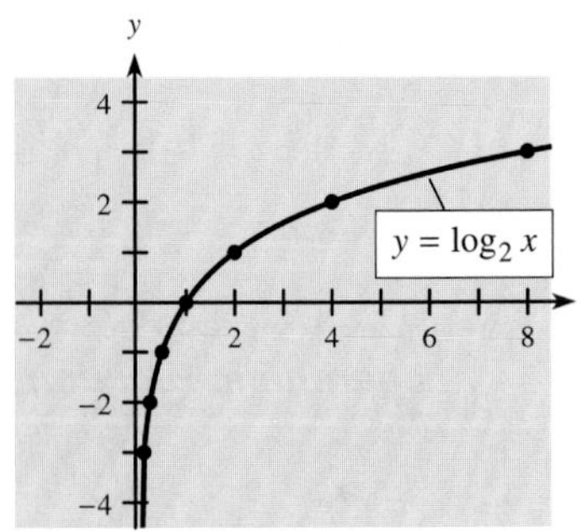

Figure 5.11

From the definition of logarithms, we see that every logarithm has a base. Most applications of logarithms involve logarithms to the base 10 (called **common logarithms**) or logarithms to the base e (called **natural logarithms**). In fact, logarithms to the base 10 and to the base e are the only ones that have function keys on scientific calculators. Thus it is important to be familiar with their names and designations.

Common and Natural Logarithms

Common logarithms: $\log x$ means $\log_{10} x$.
Natural logarithms: $\ln x$ means $\log_e x$.

Values of the natural logarithm function are usually found with a calculator. We now return to the Application Preview.

EXAMPLE 4

In the Application Preview we noted that the doubling time for a continuously compounded investment can be found by solving for t in $S = Pe^{rt}$ when $S = 2P$. That is, we must solve $2P = Pe^{rt}$, or (equivalently) $2 = e^{rt}$.

(a) Express $2 = e^{rt}$ in logarithmic form and then solve for t to find the doubling-time formula.
(b) If an investment earns 10% compounded continuously, in how many years will it double?

Solution

(a) In logarithmic form, $2 = e^{rt}$ is equivalent to $\log_e 2 = rt$. Solving for t gives the doubling-time formula

$$t = \frac{\log_e 2}{r} = \frac{\ln 2}{r}$$

(b) If the interest rate is $r = 10\%$, compounded continuously, the time required for the investment to double is

$$t = \frac{\ln 2}{0.10} = 6.93 \text{ years}$$

Note that we could write the doubling time for this problem as

$$t = \frac{\ln 2}{0.10} = \frac{0.693}{0.10} = \frac{69.3}{10}$$

In general we can approximate the doubling time for an investment at $r\%$, compounded continuously, with $\frac{70}{r}$.

EXAMPLE 5

Suppose that after a company introduces a new product, the number of months m before its market share is s percent can be modeled by

$$m = 20 \ln\left(\frac{40}{40 - s}\right)$$

When will its product have a 35% share of the market?

Solution

A 35% market share means $s = 35$. Hence

$$m = 20 \ln\left(\frac{40}{40 - s}\right)$$

$$= 20 \ln\left(\frac{40}{40 - 35}\right) = 20 \ln\left(\frac{40}{5}\right) = 20 \ln(8) \approx 41.6$$

Thus, the market share will be 35% after about 41.6 months.

EXAMPLE 6

Graph $y = \ln x$.

Solution

We can graph $y = \ln x$ by writing the equation in its exponential form, $x = e^y$, and then finding x for integer values of y. We can also graph by evaluating $y = \ln x$ for $x > 0$ (including some values $0 < x < 1$) with a calculator. The graph is shown in Figure 5.12.

y	$x = e^y$
−3	0.050
−2	0.135
−1	0.368
0	1.000
1	2.718
2	7.389
3	20.086

x	$y = \ln x$
0.05	−3.000
0.10	−2.303
0.50	−0.693
1	0.000
2	0.693
3	1.099
5	1.609
10	2.303

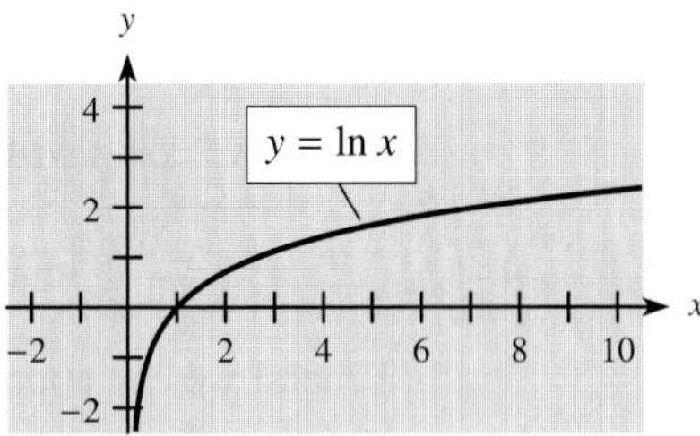

Figure 5.12

Note that because $\ln x$ is a standard function on a graphing utility, such a utility could be used to obtain its graph. Note also that the graphs of $y = \log_2 x$ (Figure 5.11) and $y = \ln x$ (Figure 5.12) are very similar. The shapes of graphs of equations of the form $y = \log_a x$ with $a > 1$ are similar to these two graphs.

Graphs of Logarithmic Functions

Equation: $y = \log_a x \quad (a > 1)$
x-intercept: $(1, 0)$
Domain: All positive reals
Range: All reals
Asymptote: y-axis (negative half)

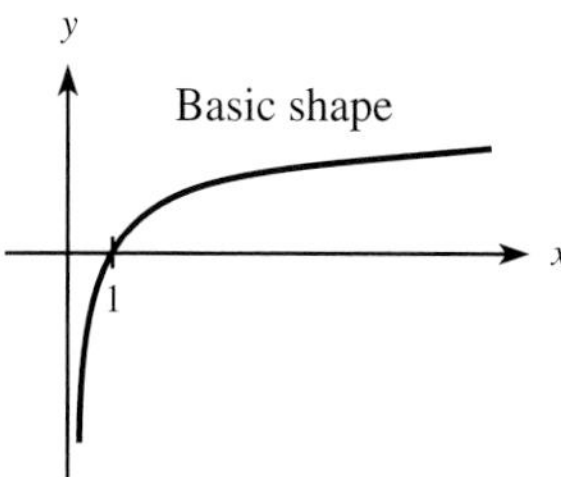

CHECKPOINT

1. What asymptote does the graph of $y = \log_a x$ approach when $a > 1$?
2. For $a > 1$, does the equation $y = \log_a x$ represent the same function as the equation $x = a^y$?
3. For what values of x is $y = \log_a x$, $a > 0$, $a \neq 1$, defined?

The basic shape of a logarithmic function is important for two reasons. First, when we graph a logarithmic function, we know that the graph should have this shape. Second, when data points have this basic shape, they suggest a logarithmic model.

M **EXAMPLE 7**

The expected life span of people in the United States depends on their year of birth (see the table and scatter plot, with $x = 0$ in 1900, in Figure 5.13). In Section 2.5, "Modeling; Fitting Curves to Data," we modeled these data with a linear function and with a quadratic function. However, the scatter plot suggests that the best model may be logarithmic.

(a) Use technology to find a logarithmic equation that models the data.
(b) The National Center for Health Statistics projects the expected life span for people born in 2000 to be 76.7 and that for those born in 2010 to be 77.9. Use your model to project the life span for those years.

Year	*Life Span (years)*	Year	*Life Span (years)*
1920	54.1	1988	74.9
1930	59.7	1989	75.2
1940	62.9	1990	75.4
1950	68.2	1991	75.5
1960	69.7	1992	75.8
1970	70.8	1993	75.5
1975	72.6	1994	75.7
1980	73.7	1996	76.1
1987	75.0		

SOURCE: National Center for Health Statistics

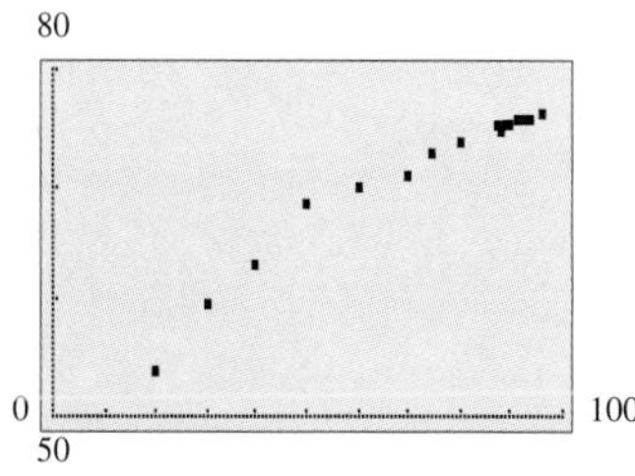

Figure 5.13

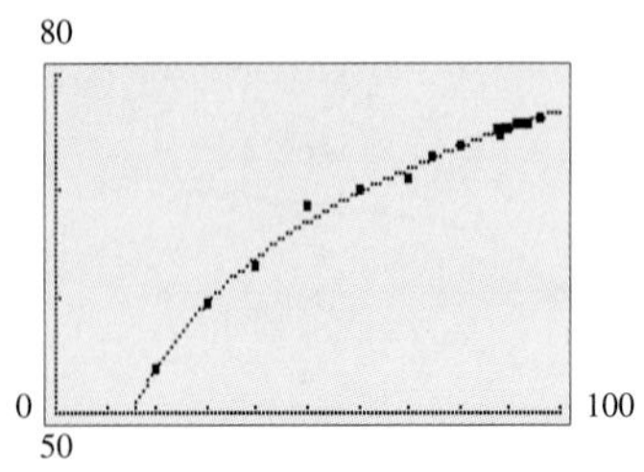

Figure 5.14

Solution

(a) A graphing calculator gives the logarithmic model

$$L(x) = 11.6439 + 14.1372 \ln x$$

where x is the number of years since 1900. Figure 5.14 shows the scatter plot and graph of the model.

(b) For the year 2000, the value of x is 100, so the projected life span is $L(100) = 76.75$. For 2010, the model gives 78.1 as the expected life span. These calculations closely approximate the projections from the National Center for Health Statistics.

Logarithm Properties

The definition of the logarithmic function and the previous examples suggest a special relationship between the logarithmic function $y = \log_a x$ and the exponential function $y = a^x$ $(a > 0, a \neq 1)$. Because we can write $y = \log_a x$ in exponential form as $x = a^y$, we see that the connection between

$$y = \log_a x \quad \text{and} \quad y = a^x$$

is that x and y have been interchanged from one function to the other. This is true for the functional description and hence for the ordered pairs that satisfy these functions. This is illustrated in Table 5.4 for the functions $y = \log_{10} x$ and $y = 10^x$.

TABLE 5.4

$y = \log_{10} x$		$y = 10^x$	
Coordinates	*Justification*	*Coordinates*	*Justification*
(1000, 3)	$3 = \log_{10} 1000$	(3, 1000)	$1000 = 10^3$
(100, 2)	$2 = \log_{10} 100$	(2, 100)	$100 = 10^2$
$\left(\frac{1}{10}, -1\right)$	$-1 = \log_{10} \frac{1}{10}$	$\left(-1, \frac{1}{10}\right)$	$\frac{1}{10} = 10^{-1}$

In general we say that $y = f(x)$ and $y = g(x)$ are **inverse functions** if, whenever the pair (a, b) satisfies $y = f(x)$, the pair (b, a) satisfies $y = g(x)$. Furthermore, because the values of the x- and y-coordinates are interchanged for inverse functions, their graphs are reflections of each other about the line $y = x$.

Thus for $a > 0$ and $a \neq 1$, the logarithmic function $y = \log_a x$ (also written $x = a^y$) and the exponential function $y = a^x$ are inverse functions.

The logarithmic function $y = \log x$ is the inverse of the exponential function $y = 10^x$. Thus the graphs of $y = \log_{10} x$ and $y = 10^x$ are reflections of each other about the line $y = x$. Some values of x and y for these functions are given in Table 5.4, and their graphs are shown in Figure 5.15.

This inverse relationship is a consequence of the definition of the logarithmic function. We can use this definition to discover several other properties of logarithms.

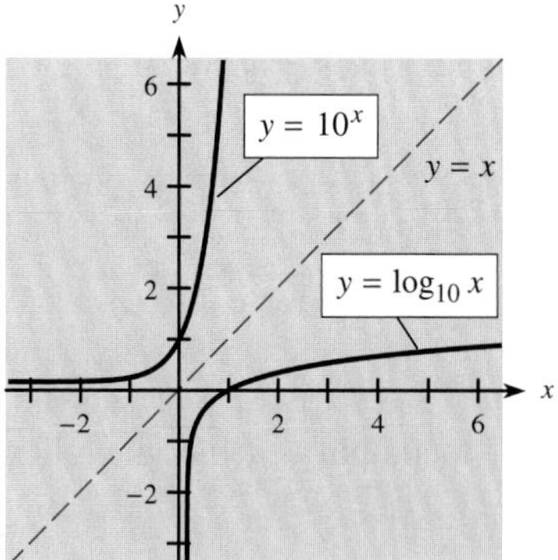

Figure 5.15

We know $y = \log_a x$ means $x = a^y$. This means that the logarithm y is an exponent. For example, $\log_3 81 = 4$ because $3^4 = 81$. In this case the logarithm, 4, was the exponent to which we had to raise the base to obtain 81. In general, if $y = \log_a x$, then y is the exponent to which the base a must be raised to obtain x.

Because logarithms are exponents, the properties of logarithms can be derived from the properties of exponents. (The properties of exponents are discussed in Chapter 0, "Algebraic Concepts." The following properties of logarithms are useful in simplifying expressions that contain logarithms.

Logarithm Property I If $a > 0$, $a \neq 1$, then $\log_a a^x = x$, for any real number x.

To prove this result, note that the exponential form of $y = \log_a a^x$ is $a^y = a^x$, so $y = x$. That is, $\log_a a^x = x$.

EXAMPLE 8

Use Property I to simplify each of the following.

(a) $\log_4 4^3$ (b) $\log_e e^x$

Solution

(a) $\log_4 4^3 = 3$
CHECK: The exponential form is $4^3 = 4^3$.

(b) $\log_e e^x = x$
CHECK: The exponential form is $e^x = e^x$.

We note that two special cases of Property I are used frequently; these are when $x = 1$ and $x = 0$.

Special Cases of Logarithm Property I

Because $a^1 = a$, we have $\log_a a = 1$.
Because $a^0 = 1$, we have $\log_a 1 = 0$.

The logarithmic form of $y = a^{\log_a x}$ is $\log_a y = \log_a x$, so $y = x$. This means that $a^{\log_a x} = x$ and proves Property II.

Logarithm Property II If $a > 0$, $a \neq 1$, then $a^{\log_a x} = x$, for any positive real number x.

EXAMPLE 9

Use Property II to simplify each of the following.

(a) $2^{\log_2 4}$ (b) $e^{\ln x}$

Solution

(a) $2^{\log_2 4} = 4$

CHECK: The logarithmic form is $\log_2 4 = \log_2 4$.

(b) $e^{\ln x} = x$

CHECK: The logarithmic form is $\log_e x = \ln x$.

If $u = \log_a M$ and $v = \log_a N$, then the exponential forms are $a^u = M$ and $a^v = N$. Thus

$$\log_a(MN) = \log_a(a^u \cdot a^v) = \log_a(a^{u+v}) = u + v = \log_a M + \log_a N$$

and Property III is established.

Logarithm Property III If $a > 0$, $a \neq 1$, and M and N are positive real numbers, then

$$\log_a(MN) = \log_a M + \log_a N$$

EXAMPLE 10

(a) Find $\log_2(4 \cdot 16)$, if $\log_2 4 = 2$ and $\log_2 16 = 4$.
(b) Find $\log_{10}(4 \cdot 5)$ if $\log_{10} 4 = 0.6021$ and $\log_{10} 5 = 0.6990$ (to four decimal places).

Solution

(a) $\log_2(4 \cdot 16) = \log_2 4 + \log_2 16 = 2 + 4 = 6$

CHECK: $\log_2(4 \cdot 16) = \log_2(64) = 6$, because $2^6 = 64$.

(b) $\log_{10}(4 \cdot 5) = \log_{10} 4 + \log_{10} 5 = 0.6021 + 0.6990 = 1.3011$

CHECK: Using a calculator, we can see that $10^{1.3011} = 20 = 4 \cdot 5$.

Logarithm Property IV If $a > 0$, $a \neq 1$, and M and N are positive real numbers, then

$$\log_a(M/N) = \log_a M - \log_a N$$

The proof of this property is left to the student in Problem 47.

EXAMPLE 11

(a) Evaluate $\log_3\left(\frac{9}{27}\right)$.

(b) Find $\log_{10}\left(\frac{16}{5}\right)$, if $\log_{10} 16 = 1.2041$ and $\log_{10} 5 = 0.6990$ (to four decimal places).

Solution

(a) $\log_3\left(\frac{9}{27}\right) = \log_3 9 - \log_3 27 = 2 - 3 = -1$

CHECK: $\log_3\left(\frac{1}{3}\right) = -1$ because $3^{-1} = \frac{1}{3}$.

(b) $\log_{10}\left(\frac{16}{5}\right) = \log_{10} 16 - \log_{10} 5 = 1.2041 - 0.6990 = 0.5051$

CHECK: Using a calculator, we can see that $10^{0.5051} = 3.2 = \frac{16}{5}$.

Logarithm Property V If $a > 0$, $a \neq 1$, M is a positive real number, and N is any real number, then

$$\log_a(M^N) = N \log_a M$$

The proof of this property is left to the student in Problem 48.

EXAMPLE 12

(a) Simplify $\log_3(9^2)$.

(b) Simplify $\ln 8^{-4}$, if $\ln 8 = 2.0794$ (to four decimal places).

Solution

(a) $\log_3(9^2) = 2 \log_3 9 = 2 \cdot 2 = 4$

CHECK: $\log_3 81 = 4$ because $3^4 = 81$.

(b) $\ln 8^{-4} = -4 \ln 8 = -4(2.0794) = -8.3176$

CHECK: Using a calculator, we can see that $8^{-4} \approx 0.000244$ and $\ln(0.000244) \approx -8.3176$.

CHECKPOINT

4. Simplify:
   (a) $6^{\log_6 x}$
   (b) $\log_7 7^3$
   (c) $\ln\left(\dfrac{1}{e^2}\right)$
   (d) $\log 1$

Change of Base

By using a calculator, we can directly evaluate only those logarithms with base 10 or base e. Also, logarithms with base 10 or base e are the only ones that are standard functions on a graphing utility. Thus, if we had a way to express a logarithmic function with any base in terms of a logarithm with base 10 or base e, we would be able to evaluate the original logarithmic function with a calculator and graph it with a graphing utility.

In general, if we use the properties of logarithms, we can write

$$y = \log_b x \quad \text{in the form} \quad b^y = x$$

If we take the base-a logarithms of both sides, we have

$$\begin{aligned} \log_a b^y &= \log_a x \\ y \log_a b &= \log_a x \\ y &= \frac{\log_a x}{\log_a b} \end{aligned}$$

This gives us the **change-of-base formula** from base b to base a.

Change-of-Base Formulas If $a \neq 1, b \neq 1, a > 0, b > 0$, then

$$\log_b x = \frac{\log_a x}{\log_a b}$$

For the purposes of calculation, we usually convert logarithms to base e.

$$\log_b x = \frac{\ln x}{\ln b}$$

EXAMPLE 13

Evaluate $\log_7 15$ by using a change-of-base formula.

Solution

$$\log_7 15 = \frac{\ln 15}{\ln 7} = \frac{2.70805}{1.94591} = 1.39166 \quad \text{(approximately)}$$

EXAMPLE 14

Use a change-of-base formula to rewrite each of the following logarithms in base e. Then graph each with a graphing utility.

(a) $y = \log_2 x$
(b) $y = \log_7 x$

Solution

(a) $y = \log_2 x = \dfrac{\ln x}{\ln 2}$. See Figure 5.16 for the graph. Note that it is exactly the same as the graph in Figure 5.11 in Example 3.

(b) $y = \log_7 x = \dfrac{\ln x}{\ln 7}$. See Figure 5.17 for the graph.

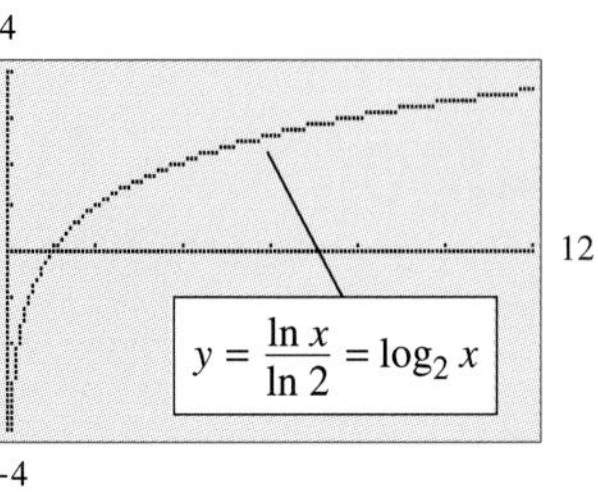

Figure 5.16

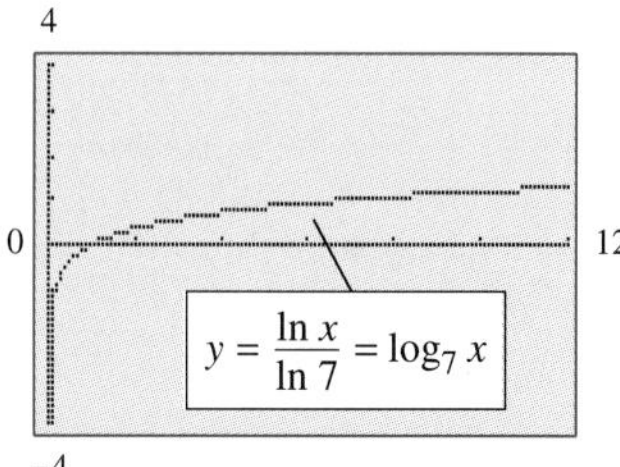

Figure 5.17

Natural logarithms, $y = \ln x$ (and the inverse exponential with base e), have many practical applications, some of which are considered in the next section. Common logarithms, $y = \log x$, were widely used for computation before computers and calculators became popular. They also have several applications to scaling variables, where the purpose is to reduce the scale of variation when a natural physical variable covers a wide range.

EXAMPLE 15

The Richter scale is used to measure the intensity of an earthquake. The magnitude on the Richter scale of an earthquake of intensity I is given by

$$R = \log(I/I_0)$$

where I_0 is a certain minimum intensity used for comparison.

(a) Find R if I is 3,160,000 times as great as I_0.

(b) The 1964 Alaskan earthquake measured 8.5 on the Richter scale. Find the intensity of the 1964 Alaskan earthquake.

Solution

(a) If $I = 3{,}160{,}000 I_0$, then $I/I_0 = 3{,}160{,}000$. Hence

$$\begin{aligned} R &= \log(3{,}160{,}000) \\ &= 6.5 \quad \text{(approximated to one decimal place)} \end{aligned}$$

(b) For $R = 8.5$, it follows that

$$8.5 = \log(I/I_0)$$

Rewriting this in exponential form gives

$$10^{8.5} = I/I_0$$

and from a calculator we obtain

$$I/I_0 = 316{,}000{,}000 \quad \text{(approximately)}$$

Thus the intensity is 316,000,000 times I_0.

Note that a Richter scale measurement that is 2 units larger means that the intensity is $10^2 = 100$ *times* greater.

CHECKPOINT SOLUTIONS

1. The (negative) y-axis
2. Yes
3. $x > 0$ is the domain of $\log_a x$.
4. (a) x, by Property II
   (b) 3, by Property I
   (c) $\ln\left(\frac{1}{e^2}\right) = \ln(e^{-2}) = -2$, by Property I
   (d) 0, because $\log_a 1 = 0$ for any base a.

EXERCISE 5.2

In Problems 1–4, write each equation in exponential form.

1. $4 = \log_2 16$
2. $4 = \log_3 81$
3. $\frac{1}{2} = \log_4 2$
4. $-2 = \log_3\left(\frac{1}{9}\right)$

In Problems 5–8, solve for x by writing the equations in exponential form.

5. $\log_2 x = 3$
6. $\log_4 x = -2$
7. $\log_8 x = -\frac{1}{3}$
8. $\log_{25} x = \frac{1}{2}$

In Problems 9–12, write each equation in logarithmic form.

9. $2^5 = 32$
10. $5^3 = 125$
11. $4^{-1} = \frac{1}{4}$
12. $9^{1/2} = 3$

In Problems 13–18, graph the functions.

13. $y = \log_3 x$
14. $y = \log_4 x$
15. $y = \ln x$
16. $y = \log_9 x$
17. $y = \log_2(-x)$
18. $y = \ln(-x)$

In Problems 19 and 20, use properties of logarithms or a definition to simplify each expression.

19. (a) $\log_3 27$ (b) $\log_5\left(\frac{1}{5}\right)$
20. (a) $\log_4 16$ (b) $\log_9 3$
21. If $f(x) = \ln(x)$, find $f(e^x)$.
22. If $f(x) = \ln(x)$, find $f(\sqrt{e})$.
23. If $f(x) = e^x$, find $f(\ln 3)$.
24. If $f(x) = 10^x$, find $f(\log 2)$.

In Problems 25 and 26, evaluate each logarithm by using properties of logarithms and the following facts.

$$\log_a x = 3.1 \qquad \log_a y = 1.8 \qquad \log_a z = 2.7$$

25. (a) $\log_a(xy)$ (b) $\log_a\left(\frac{x}{z}\right)$ (c) $\log_a(x^4)$ (d) $\log_a \sqrt{y}$
26. (a) $\log_a(yz)$ (b) $\log_a\left(\frac{z}{y}\right)$ (c) $\log_a(y^6)$ (d) $\log_a \sqrt[3]{z}$

Write each expression in Problems 27–30 as the sum or difference of two logarithmic functions containing no exponents.

27. $\log\left(\frac{x}{x+1}\right)$
28. $\ln[(x+1)(4x+5)]$
29. $\log_7(x\sqrt[3]{x+4})$
30. $\log_5\left(\frac{x^2}{\sqrt{x+4}}\right)$

Use the properties of logarithms to write each expression in Problems 31–34 as a single logarithm.

31. $\ln x - \ln y$
32. $\log_3(x+1) + \log_3(x-1)$
33. $\log_5(x+1) + \frac{1}{2}\log_5 x$
34. $\log(2x+1) - \frac{1}{3}\log(x+1)$

In Problems 35–38, use a calculator to determine whether expression (a) is equivalent to expression (b). If they are equivalent, state what properties are being illustrated. If they are not equivalent, rewrite expression (a) so that they are equivalent.

35. (a) $\ln \sqrt{4 \cdot 6}$ (b) $\frac{1}{2}(\ln 4 + \ln 6)$

36. (a) $\log \frac{\sqrt{56}}{23}$ (b) $\frac{1}{2}\log 56 - \log 23$

37. (a) $\log \sqrt[3]{\frac{8}{5}}$ (b) $\frac{1}{3}\log 8 - \log 5$

38. (a) $\frac{\log_2 34}{17}$ (b) $\log_2 34 - \log_2 17$

39. (a) Given $f(x) = \ln x$. Use a graphing utility to graph $f(x) = \ln x$ with $f(x - c) = \ln(x - c)$ for $c = -4, -2, 1$, and 5.
(b) For each c-value, identify the vertical asymptote and the domain of $y = \ln(x - c)$.
(c) For each c-value, find the x-intercept of $y = \ln(x - c)$.
(d) Write a sentence that explains precisely how the graphs of $y = f(x)$ and $y = f(x - c)$ are related.

40. Given $f(x) = \ln x$. Use a graphing utility to graph $f(ax) = \ln(ax)$ for $a = -2, -1, -0.5, 0.2$, and 3 with $f(x) = \ln x$. Explain the differences between the graphs
(a) when $a > 0$ and when $a < 0$.
(b) when $|a| > 1$ and when $0 < |a| < 1$.

In Problems 41 and 42, use a change-of-base formula to evaluate each logarithm.

41. (a) $\log_2 17$ (b) $\log_5(0.78)$

42. (a) $\log_3 12$ (b) $\log_8(0.15)$

In Problems 43–46, use a change-of-base formula to rewrite each logarithm. Then use a graphing utility to graph the function.

43. $y = \log_5 x$

44. $y = \log_6 x$

45. $y = \log_{13} x$

46. $y = \log_{16} x$

47. Prove logarithm Property IV.

48. Prove logarithm Property V.

49. The function $y = 4.2885(1.0592)^{-x}$ was used to model the purchasing power of \$1 in Example 7 of the previous section. Write this function as an exponential function with base e, so that it has the form $y = 4.2885e^{-kx}$. *Hint:* Solve $4.2885(1.0592)^{-x} = 4.2885e^{-kx}$ for k.

50. Write the function $y = 4(2.334)^x$ as an exponential function of the form $y = 4e^{kx}$.

Applications

Richter scale Use the formula $R = \log(I/I_0)$ in Problems 51–54.

51. In May 1983, an earthquake measuring 7.7 on the Richter scale occurred in Japan. This was the first major earthquake in Japan since 1948, when one registered 7.3 on the Richter scale. How many times more severe was the 1983 shock?

52. The strongest earthquake ever to strike Japan occurred in 1933 and measured 8.9 on the Richter scale. How many times more severe was this 1933 quake than the one in May 1983? (See Problem 51.)

53. The San Francisco earthquake of 1906 measured 8.25 on the Richter scale, and the San Francisco earthquake of 1989 measured 7.1. How much more intense was the 1906 quake?

54. The largest earthquakes ever recorded measured 8.9 on the Richter scale; the San Francisco earthquake of 1906 measured 8.25. Calculate the ratio of their intensities. (These readings correspond to devastating quakes.)

Decibel readings Problems 55–58. The loudness of sound (in decibels) perceived by the human ear depends on intensity levels according to

$$L = 10 \log(I/I_0)$$

where I_0 is the threshold of hearing for the average human ear.

55. Find the loudness when I is 10,000 times I_0. This is the intensity level of the average voice.

56. A sound that causes pain has intensity about 10^{14} times I_0. Find the decibel reading for this threshold.

57. Graph the equation for loudness. Use I/I_0 as the independent variable.

58. A relatively quiet room has a background noise level of about $L_1 = 32$ decibels, and a heated argument has a decibel level of about $L_2 = 66$. Find the ratio I_2/I_1 of the associated intensities.

pH levels Problems 59–62. Chemists use the pH (hydrogen potential) of a solution to measure its acidity or basicity. The pH is given by the formula

$$\text{pH} = -\log[\text{H}^+]$$

where $[\text{H}^+]$ is the concentration of hydrogen ions in moles per liter.

59. Most common solutions have a pH range between 1 and 14. What values of H^+ are associated with these extremes?

60. Find the approximate pH of each of the following.
(a) blood: $[H^+] = 3.98 \times 10^{-8}$
$= 0.0000000398$
(b) beer: $[H^+] = 6.31 \times 10^{-5} = 0.0000631$
(c) vinegar: $[H^+] = 6.3 \times 10^{-3} = 0.0063$

61. Sometimes pH is defined as the logarithm of the reciprocal of the concentration of hydrogen ions. Write an equation that represents this sentence, and explain how it and the equation given above can both represent pH.

62. Find the approximate hydrogen ion concentration $[H^+]$ for each of the following.
(a) apples: pH = 3.0
(b) eggs: pH = 7.79
(c) water (neutral): pH = 7.0

63. ***Life span UPDATE*** In Example 7 we used data from 1920 to 1996 and found that the life span in the United States depended on the year of birth according to the equation

$$L(x) = 11.6439 + 14.1372 \ln x$$

where x is the number of years from 1900. Using data from 1920 to 1989, the model

$$\ell(x) = 11.6164 + 14.1442 \ln x$$

where x is the number of years from 1900, predicts life span as a function of birth year. Use both models to predict the life span for people born in 1996 and in 2000. Did adding data from 1990 to 1996 give predictions that were quite different from or very similar to predictions based on the model found earlier?

64. ***Deforestation*** One of the major causes for rain forest deforestation is agricultural and residential development. The number of hectares destroyed in a particular year t can be modeled by

$$y = -3.91435 + 2.62196 \ln t$$

where $t = 0$ in 1950.
(a) Graph this function.
(b) Project the number of hectares that will be destroyed in the year 2000 because of agricultural and residential development.

65. ***Violent crime*** The following data represent the number of violent crimes (murders, rapes, and armed robberies) per 100,000 people for the years 1987 to 1992.
(a) Letting $t = 0$ in 1980, find a logarithmic equation (base e) that models the data.
(b) Use this model to predict the number of violent crimes in 2005.

Year	*Violent Crimes*
1987	610
1988	637
1989	663
1990	732
1991	758
1992	765

SOURCE: FBI Crime Report

66. ***Poverty threshold*** The table below gives the average poverty thresholds for 1987 to 1994. Use a logarithmic equation to model these data, with x the number of years from 1980.

Year	*Poverty Threshold*
1987	\$5778
1988	6022
1989	6310
1990	6652
1991	6932
1992	7143
1993	7363
1994	7547

SOURCE: U.S. Bureau of the Census

5.3 Solution of Exponential Equations; Applications of Exponential and Logarithmic Functions

OBJECTIVES

- *To solve exponential growth or decay equations when sufficient data are known*
- *To solve exponential equations representing demand, supply, total revenue, or total cost when sufficient data are known*

APPLICATION PREVIEW

In this section, we discuss applications of exponential and logarithmic functions to the solution of problems involving exponential growth and decay and supply and demand. With exponential decay, we consider various applications, including decay of radioactive materials and decay of sales following an advertising campaign. Exponential growth focuses on population growth and growth that can be modeled by Gompertz curves. Finally, we consider applications of demand functions that can be modeled by exponential functions.

Growth and Decay

Exponential decay models are those that can be described by a function of the following form.

Decay Models $f(x) = ma^{-x}$ with $a > 1$ and $m > 0$

We have already mentioned that radioactive decay has this form, as do the demand curves discussed later. Another example of this phenomenon occurs in the life sciences when the valves to the aorta are closed, and blood flows into the heart. During this period of time, the blood pressure in the aorta falls exponentially. This is illustrated in the following example.

EXAMPLE 1

Medical research has shown that over short periods of time when the valves to the aorta of a normal adult close, the pressure in the aorta is a function of time and can be modeled by the equation

$$P = 95e^{-0.491t}$$

where t is in seconds.

(a) What is the aortic pressure when the valves are first closed ($t = 0$)?
(b) What is the aortic pressure after 0.1 second?
(c) How long will it be before the pressure reaches 80?

Solution

(a) If $t = 0$, the pressure is given by

$$P = 95e^0 = 95$$

(b) If $t = 0.1$, the pressure is given by

$$\begin{aligned} P &= 95e^{-0.491(0.1)} \\ &= 95e^{-0.0491} \\ &= 95(0.952) \\ &= 90.44 \end{aligned}$$

(c) Setting $P = 80$ and solving for t will give us the length of time before the pressure reaches 80.

$$80 = 95e^{-0.491t}$$

To solve this equation for t, we must rewrite it so that t is not in an exponent. We first isolate the exponential containing t by dividing both sides by 95.

$$\frac{80}{95} = e^{-0.491t}$$

Rewriting this equation in logarithmic form gives

$$\ln\left(\frac{80}{95}\right) = -0.491t$$

Using a calculator, we have $-0.172 = -0.491t$. Thus,

$$\frac{-0.172}{-0.491} = t$$

$$t = 0.35 \text{ second} \qquad \text{(approximately)}$$

Note that in Example 1 we could have solved

$$\frac{80}{95} = e^{-0.491t}$$

for t by taking the logarithm, base e, of both sides and then using properties of logarithms.

$$\ln\left(\frac{80}{95}\right) = \ln e^{-0.491t}$$

By logarithm Property I, we have

$$\ln\left(\frac{80}{95}\right) = -0.491t$$

Note that taking the natural logarithm of both sides gives the same result as writing $80/95 = e^{-0.491t}$ in logarithmic form.

The advantage of taking the natural logarithm (or the common logarithm) of both sides of an exponential equation to solve the equation becomes apparent when the base of the exponential is not e (or 10).

To solve an exponential equation, we must be able to rewrite the equation so that the variable is not in an exponent. For equations involving a single exponential, we solve for that exponential. (That is, we isolate the exponential expression on one side of the equation.) For example,

$$Na^X = M \qquad \text{yields} \qquad a^x = \frac{M}{N}$$

Then taking the logarithm (base 10 or base e) of both sides of the equation, using logarithm Property V, and using algebraic methods, we can solve the equation. Part (b) of the following example illustrates this solution method.

EXAMPLE 2

A company finds that its daily sales begin to fall after the end of an advertising campaign, and the decline is such that the number of sales is $S = 2000(2^{-0.1x})$, where x is the number of days after the campaign's end.

(a) How many sales will be made 10 days after the end of the campaign?
(b) If the company does not want sales to drop below 500 per day, when should it start a new campaign?

Solution

(a) If $x = 10$, sales are given by $S = 2000(2^{-1}) = 1000$.
(b) Setting $S = 500$ and solving for x will give us the number of days after the end of the campaign when sales will reach 500.

$$500 = 2000(2^{-0.1x})$$

$$\frac{500}{2000} = 2^{-0.1x} \qquad \text{Isolate the exponential.}$$

$$0.25 = 2^{-0.1x}$$

Because the base of this exponential is 2 rather than e or 10, we choose to take the logarithm, base 10, of both sides of the equation instead of rewriting in logarithmic form.

$$\log(0.25) = \log(2^{-0.1x})$$

$$\log(0.25) = (-0.1x)(\log 2) \qquad \text{Property V}$$

$$\frac{\log(0.25)}{\log 2} = -0.1x$$

$$\frac{-0.6021}{0.3010} = -0.1x$$

$$-2 = -0.1x$$

$$x = 20$$

Thus sales will be 500 on the twentieth day after the end of the campaign. If a new campaign isn't begun on or before the 21st day, sales will drop below 500. (See Figure 5.18.)

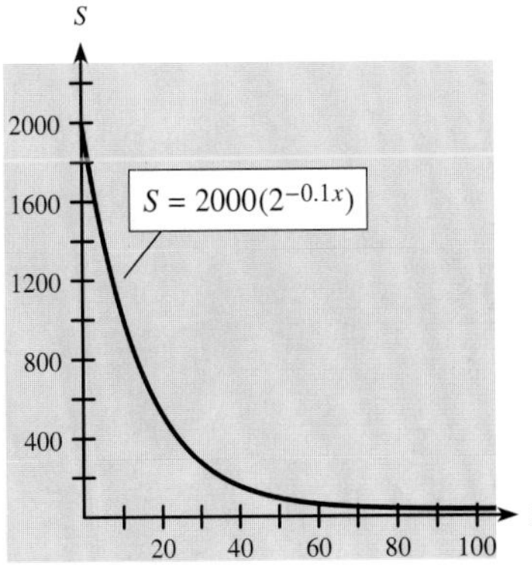

Figure 5.18

Example 2 is an example of *sales decay* and is typical of many exponential decay models. In particular, once some action is completed, such as an advertising campaign, its effect on sales volume diminishes, or decays, with time.

In business, economics, biology, and the social sciences, the growth of money, bacteria, or population is frequently of interest. **Exponential growth models** are those that can be described by a function of the following form.

Growth Models $f(x) = ma^x$ with $a > 1$ and $m > 0$

As we mentioned in Section 5.1 "Exponential and Logarithmic Functions," the curve that models the growth of some populations is given by $y = P_0e^{ht}$, where P_0 is the population size at a particular time t, h is a constant that depends on the population involved, and y is the total population at time t. This function may be used to model population growth for humans, insects, or bacteria.

EXAMPLE 3

The population of a certain city was 30,000 in 1970 and 40,500 in 1980. If the formula $y = P_0e^{ht}$ applies to the growth of the city's population, what should the population be in the year 2000?

Solution

We can first use the data from 1970 and 1980 to find the value of h in the formula. Letting $P_0 = 30{,}000$ and $t = 10$, we get

$$40{,}500 = 30{,}000e^{h(10)}$$

$$1.00035 = e^{10h} \qquad \text{Isolate the exponential.}$$

Taking the natural logarithm of both sides gives

$$\ln 1.35 = \ln e^{10h} = 10h(\ln e)$$

$$\ln 1.35 = 10h \qquad \text{Because } \ln e = 1$$

$$0.3001 = 10h$$

$$h = 0.0300 \qquad \text{(approximately)}$$

Thus the formula for this population is $y = P_0e^{0.03t}$. To predict the population for the year 2000, we use $P_0 = 40{,}500$ (for 1980) and $t = 20$. This gives

$$\begin{aligned} y &= 40{,}500e^{0.03(20)} \\ &= 40{,}500e^{0.6} \\ &= 40{,}500(1.8221) \\ &= 73{,}795 \qquad \text{(approximately)} \end{aligned}$$

One family of curves that has been used to describe human growth and development, the growth of organisms in a limited environment, and the growth

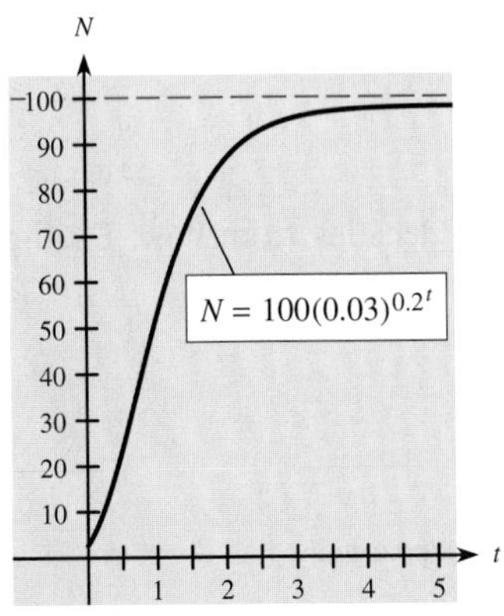

Figure 5.19

of many types of organizations is the family of **Gompertz curves.** These curves are graphs of equations of the form

$$N = Ca^{R^t}$$

where t represents the time, R $(0 < R < 1)$ is a constant that depends on the population, a represents the proportion of initial growth, C is the maximum possible number of individuals, and N is the number of individuals at a given time t.

For example, the equation $N = 100(0.03)^{0.2^t}$ could be used to predict the size of a deer herd introduced on a small island. Here the maximum number of deer C would be 100, the proportion of the initial growth a is 0.03, and R is 0.2. For this example, t represents time, measured in decades. The graph of this equation is given in Figure 5.19.

EXAMPLE 4

A hospital administrator predicts that the growth in the number of hospital employees will follow the Gompertz equation

$$N = 2000(0.6)^{0.5^t}$$

where t represents the number of years after the opening of a new facility.

(a) What is the number of employees when the facility opens $(t = 0)$?
(b) How many employees are predicted after 1 year of operation $(t = 1)$?
(c) Graph the curve.
(d) What is the maximum value for N that the curve will approach?

Solution

(a) If $t = 0$, $N = 2000(0.6)^1 = 1200$
(b) If $t = 1$, $N = 2000(0.6)^{0.5} = 2000\sqrt{0.6} = 1549$ (approximately)
(c) The graph is shown in Figure 5.20.
(d) From the graph we can see that as larger values of t are substituted in the function, the values of N approach, but never reach, 2000. We say that the line $N = 2000$ (dashed) is an **asymptote** for this curve and that 2000 is the maximum possible value.

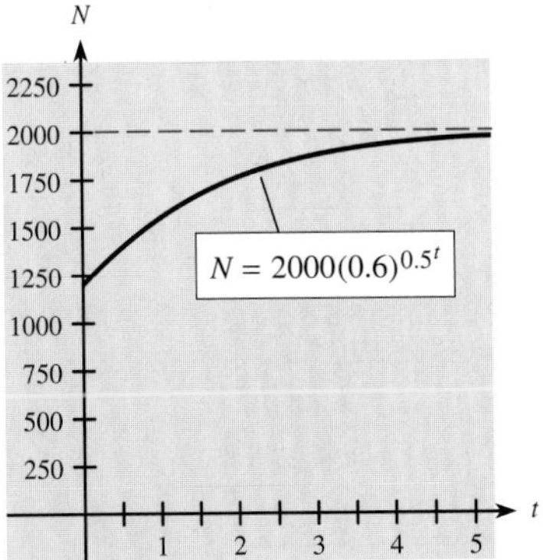

Figure 5.20

EXAMPLE 5

The Gompertz equation

$$N = 100(0.03)^{0.2^t}$$

predicts the size of a deer herd on a small island t decades from now. During what year will the deer population reach or exceed 70?

Solution

We solve the equation with $N = 70$.

$$70 = 100(0.03)^{0.2^t}$$

$$0.7 = 0.03^{0.2^t} \qquad \text{Isolate the exponential.}$$

$$\ln 0.7 = \ln 0.03^{0.2^t} = 0.2^t \ln 0.03$$

$$\frac{\ln 0.7}{\ln 0.03} = 0.2^t \qquad \text{Again, isolate the exponential.}$$

$$\ln\left(\frac{\ln 0.7}{\ln 0.03}\right) = t \ln 0.2$$

$$\ln(0.10172) = t(-1.6094)$$

$$\frac{-2.2855}{-1.6094} = t$$

$$t = 1.42 \text{ decades} \qquad \text{(approximately)}$$

The population will exceed 70 in just over 14 years, or during the 15th year.

Technology Note The value of a SOLVER feature or program is evident in Figure 5.21, where the solution to Example 5 is shown. Such a program may not yield a solution unless a reasonable guess is chosen to start the iteration process (which is used to solve the equations).

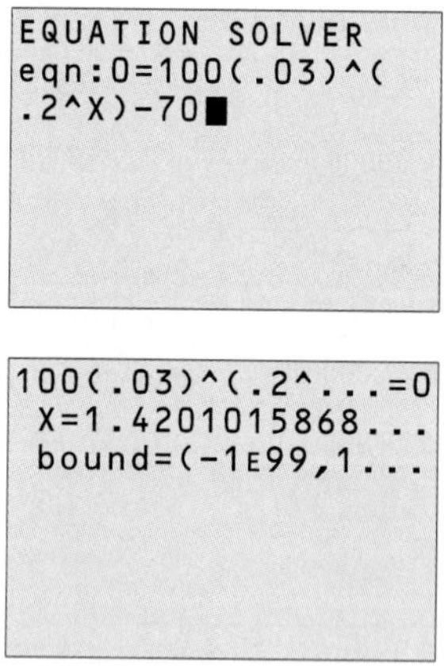

```
EQUATION SOLVER
eqn:0=100(.03)^(
.2^X)-70■
```

```
100(.03)^(.2^...=0
 X=1.4201015868...
 bound=(-1E99,1...
```

Figure 5.21

CHECKPOINT

1. Suppose the sales of a product, in dollars, are given by $S = 1000e^{-0.07x}$, where x is the number of days after the end of an advertising campaign.
   (a) What are sales 2 days after the end of the campaign?
   (b) How long will it be before sales are \$300?
2. Suppose the number of employees at a new regional hospital is predicted by the Gompertz curve

$$N = 3500(0.1)^{0.5^t}$$

   where t is the number of years after the hospital opens.
   (a) How many employees did the hospital have when it opened?
   (b) What is the expected upper limit on the number of employees?

Economic and Management Applications

In Section 2.3, "Business Applications of Quadratic Functions," we discussed quadratic cost, revenue, demand, and supply functions. But cost, revenue, demand, and supply may also be modeled by exponential or logarithmic equations. For example, suppose the demand for a product is given by $p = 30(3^{-q/2})$, where q is the number of units demanded at a price of p dollars per unit. Then the graph of the demand curve is as given in Figure 5.22.

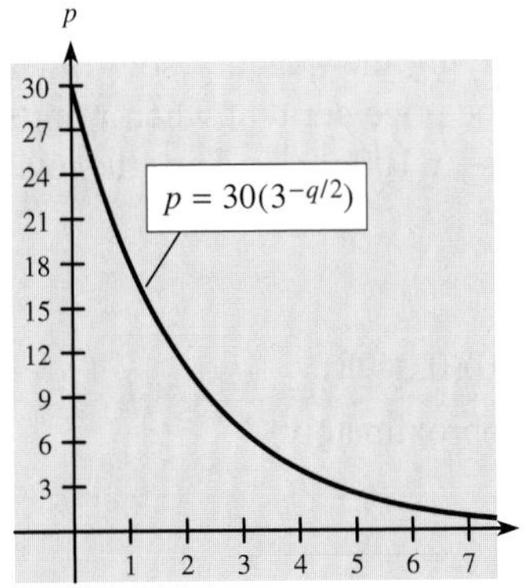

Figure 5.22

EXAMPLE 6

Suppose the demand function for a certain commodity is given by $p = 30(3^{-q/2})$.

(a) At what price per unit will the demand equal 4 units?
(b) How many units, to the nearest unit, will be demanded if the price is \$17.32?

Solution

(a) If $q = 4$, then
$$\begin{aligned} p &= 30(3^{-4/2}) \\ &= 30(0.1111) \\ &= 3.33 \text{ dollars} \quad \text{(approximately)} \end{aligned}$$

(b) If $p = 17.32$, then
$$\begin{aligned} 17.32 &= 30(3^{-q/2}) \\ 0.5773 &= 3^{-q/2} \quad \text{Isolate the exponential.} \\ \ln 0.5773 &= \ln 3^{-q/2} \\ \ln 0.5773 &= -\frac{q}{2}\ln 3 \\ \frac{-2\ln 0.5773}{\ln 3} &= q \\ 1 &= q \quad \text{(approximately)} \end{aligned}$$

The number of units demanded would be 1, to the nearest unit.

CHECKPOINT

3. Suppose the monthly demand for a product is given by $p = 400e^{-0.003x}$, where p is the price in dollars and x is the number of units. How many units will be demanded when the price is \$100?

EXAMPLE 7

Suppose the demand function for a commodity is given by $p = 100e^{-x/10}$, where p is the price per unit when x units are sold.

(a) What is the total revenue for the commodity?
(b) What would be the total revenue if 30 units were demanded and supplied?

Solution

(a) The total revenue can be computed by multiplying the quantity sold times the price per unit. The demand function gives the price per unit when x units are sold, so the total revenue for x units is $x \cdot p = x(100e^{-x/10})$. Thus the total revenue function is $R(x) = 100xe^{-x/10}$.
(b) If 30 units are sold, the total revenue is

$$\begin{aligned} R(30) &= 100(30)e^{-30/10} = 100(30)(0.0498) \\ &= 149.40 \text{ (dollars)} \qquad \text{(approximately)} \end{aligned}$$

CHECKPOINT SOLUTIONS

1. (a) $S = 100e^{-0.07(2)} = 1000e^{-0.14} \approx \$869.36$ Nearest cent
   (b)
$$\begin{aligned} 300 &= 1000e^{-0.07x} \\ 0.3 &= e^{-0.07x} \\ \ln(0.3) &= -0.07x \\ \frac{\ln(0.3)}{-0.07} &= x \approx 17 \quad \text{Nearest day} \end{aligned}$$

2. (a) $3500(0.1) = 350$
   (b) 3500

3.
$$\begin{aligned} 100 &= 400e^{-0.003x} \\ 0.25 &= e^{-0.003x} \\ \ln(0.25) &= -0.003x \\ \frac{\ln(0.25)}{-0.003} &= x \approx 462 \quad \text{Nearest unit} \end{aligned}$$

EXERCISE 5.3

Growth and Decay

1. ***Sales decay*** The sales decay for a product is given by $S = 50{,}000e^{-0.8x}$, where S is the monthly sales and x is the number of months that have passed since the end of a promotional campaign.
   (a) What will be the sales 4 months after the end of the campaign?
   (b) How many months after the end of the campaign will sales drop below 1000, if no new campaign is initiated?

2. ***Sales decay*** The sales of a product decline after the end of an advertising campaign, with the sales decay given by $S = 100{,}000e^{-0.5x}$, where S represents the weekly sales and x represents the number of weeks since the end of the campaign.
   (a) What will be the sales for the tenth week after the end of the campaign?
   (b) During what week after the end of the campaign will sales drop below 400?

3. ***Radioactive decay*** Radioactive carbon-14 can be used to determine the age of fossils. Carbon-14 decays according to the equation

$$y = y_0 e^{-0.00012378t}$$

   where y is the amount of carbon-14 at time t, in years, and y_0 is the original amount.
   (a) How much of the original amount would be left after 3000 years?
   (b) If a fossil is found to have $\frac{1}{100}$ the original amount of carbon-14, how old is the fossil?

4. ***Product reliability*** A statistical study shows that the fraction of television sets of a certain brand that are still in service after x years is given by $f(x) = e^{-0.15x}$.
   (a) What fraction of the sets are still in service after 5 years?
   (b) After how many years will the fraction still in service be 1/10?

5. ***Radioactive half-life*** An initial amount of 100 g of the radioactive isotope thorium-234 decays according to

$$Q(t) = 100e^{-0.02828t}$$

   where t is in years. How long before half of the initial amount has disintegrated? This time is called the half-life of this isotope.

6. ***Radioactive half-life*** A breeder reactor converts stable uranium-238 into the isotope plutonium-239. The decay of this isotope is given by

$$A(t) = A_0 e^{-0.00002876t}$$

   where $A(t)$ is the amount of the isotope at time t, in years, and A_0 is the original amount.
   (a) If $A_0 = 500$ lb, how much will be left after a human lifetime (use $t = 70$ years)?
   (b) Find the half-life of this isotope.

7. ***Population growth*** If the population of a certain county was 100,000 in 1980 and 110,517 in 1990, and if the formula $y = P_0 e^{ht}$ applies to the growth of the county's population, estimate the population of the county in 2005.

8. ***Population growth*** The population of a certain city grows according to the formula $y = P_0 e^{0.03t}$. If the population was 250,000 in 1980, estimate the population in the year 2100.

9. ***Sales growth*** The president of a company predicts that sales will increase after she assumes office and that the number of monthly sales will follow the curve given by $N = 3000(0.2)^{0.6^t}$, where t represents the months since she assumed office.
   (a) What will be the sales when she assumes office?
   (b) What will be the sales after 3 months?
   (c) What is the expected upper limit on sales?
   (d) Graph the curve.

10. ***Organizational growth*** Because of a new market opening, the number of employees of a firm is expected to increase according to the equation $N = 1400(0.5)^{0.3^t}$, where t represents the number of years after the new market opens.
   (a) What is the level of employment when the new market opens?
   (b) How many employees should be working at the end of 2 years?
   (c) What is the expected upper limit on the number of employees?
   (d) Graph the curve.

11. ***Organizational growth*** Suppose that the equation $N = 500(0.02)^{0.7^t}$ represents the number of employees working t years after a company begins operations.
   (a) How many employees are there when the company opens (at $t = 0$)?
   (b) After how many years will at least 100 employees be working?

12. ***Sales growth*** A firm predicts that sales will increase during a promotional campaign and that the number of daily sales will be given by $N = 200(0.01)^{0.8^t}$, where t represents the number of days after the campaign begins. How many days after the beginning of the campaign would the firm expect to sell at least 60 units per day?

Gompertz curves describe situations in which growth is limited. There are other equations that describe this phenomenon under different assumptions. Two examples are

(a) $y = c(1 - e^{-ax}),\quad a > 0$

(b) $y = \dfrac{A}{1 + ce^{-ax}},\quad a > 0$ (logistic curve)

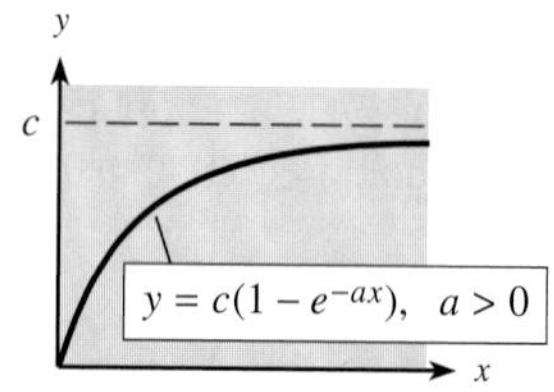

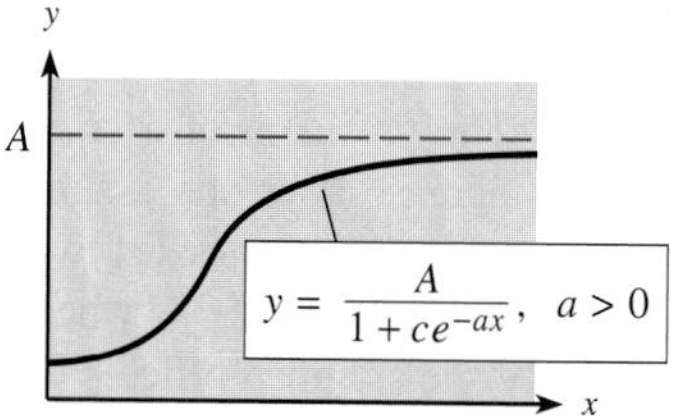

These equations have many applications. In general, both (a) and (b) can be used to describe learning, sales of new products, and population growth, and (b) can be used to describe the spread of epidemics. Problems 13–16 illustrate applications.

13. ***Drugs in the bloodstream*** The concentration y of a certain drug after t hours in the bloodstream (with $0 \le t \le 15$) is given by the equation

$$y = 100(1 - e^{-0.462t})$$

(a) What is y after 1 hour ($t = 1$)?
(b) How long does it take for y to reach 50?

14. ***Population growth*** Suppose that the size y of a deer herd t years after being introduced onto an island is given by

$$y = 2500 - 2490e^{-0.1t}$$

(a) Find the population when the herd was introduced (at $t = 0$).
(b) How long will it be before the herd numbers 1500?

15. ***Spread of disease*** On a college campus of 10,000 students, a single student returned to campus infected by a disease. The spread of the disease through the student body is given by

$$y = \frac{10{,}000}{1 + 9999e^{-0.99t}}$$

where y is the total number infected at time t (in days).
(a) How many are infected after 4 days?
(b) The school will shut down if 50% of the students are ill. During what day will it close?

16. ***Spread of a rumor*** The number of people $N(t)$ in a community who are reached by a particular rumor at time t is given by the equation

$$N(t) = \frac{50{,}500}{1 + 100e^{-0.7t}}$$

(a) Find $N(0)$.
(b) What is the upper limit on the number of people affected?
(c) How long before 75% of the upper limit is reached?

Economic and Management Applications

17. ***Demand*** The demand function for a certain commodity is given by $p = 100e^{-q/2}$.
(a) At what price per unit will the quantity demanded equal 6 units?
(b) If the price is \$1.83 per unit, how many units will be demanded, to the nearest unit?

18. ***Demand*** The demand function for a product is given by $p = 3000e^{-q/3}$.
(a) At what price per unit will the quantity demanded equal 6 units?
(b) If the price is \$149.40 per unit, how many units will be demanded, to the nearest unit?

19. ***Supply*** If the supply function for a product is given by $p = 100e^q/(q + 1)$, where q represents the number of hundreds of units, what will be the price when the producers are willing to supply 300 units?

20. ***Supply*** If the supply function for a product is given by $p = 200(2^q)$, where q represents the number of hundreds of units, what will be the price when the producers are willing to supply 500 units?

21. ***Total cost*** If the total cost function for a product is $C(x) = e^{0.1x} + 400$, where x is the number of items produced, what is the total cost of producing 30 units?

22. ***Total cost*** If the total cost for a product is given by $C(x) = 400 \ln(x + 10) + 100$, what is the total cost of producing 100 units?

23. ***Total revenue*** If the demand function for a product is given by $p = 200e^{-0.02x}$, where p is the price per unit when x units are demanded, what is the total revenue when 100 units are demanded and supplied?

24. ***Total revenue*** If the demand function for a product is given by $p = 4000/\ln(x + 10)$, where p is the price per unit when x units are demanded, what is the total revenue when 40 units are demanded and supplied?

25. ***Compound interest*** If \$8500 is invested at 11.5% compounded continuously, the future value A at any time t (in years) is given by

$$A = 8500e^{0.115t}$$

(a) What is the amount after 18 months?
(b) How long before the investment doubles?

26. ***Compound interest*** If \$1000 is invested at 10% compounded continuously, the future value A at any time t (in years) is given by $A = 1000e^{0.1t}$.
(a) What is the amount after 1 year?
(b) How long before the investment doubles?

27. ***Compound interest*** If \$5000 is invested at 9% per year compounded monthly, the future value A at any time t (in months) is given by $A = 5000(1.0075)^t$.
(a) What is the amount after 1 year?
(b) How long before the investment doubles?

28. ***Compound interest*** If \$10,000 is invested at 1% per month, the future value A at any time t (in months) is given by $A = 10{,}000(1.01)^t$.
(a) What is the amount after 1 year?
(b) How long before the investment doubles?

29. ***Market share*** Suppose that the market share y (as a percent) that a company expects t months after a new product is introduced is given by $y = 40 - 40e^{-0.05t}$.
(a) What is the market share after the first month (to the nearest percent)?
(b) How long (to the nearest month) before the market share is 25%?

30. ***Advertising*** An advertising agency has found that when it promotes a new product in a certain market of 350,000, the number of people x who are aware of the product t days after the ad campaign is initiated is given by

$$x = 350{,}000(1 - e^{-0.077t})$$

(a) How many people (to the nearest thousand) are aware after 1 week?
(b) How long (to the nearest day) before 300,000 are aware of the new product?

Additional Applications

31. ***Health care*** For selected years from 1960 to 1992, the national health care expenditures H, in billions of dollars, can be modeled by

$$H = 45e^{0.08988t}$$

where t is the number of years past 1960. (Source: Congressional Budget Office data as reported in *Newsweek*, October 4, 1993.)
(a) If this model remains accurate, how long will it be before national health care expenditures reach \$2.5 trillion (that is, \$2500 billion)?
(b) How do you think a national health care plan would alter this model?

32. ***U.S. debt*** For selected years from 1900 to 1989, the national debt d, in billions of dollars, can be modeled by

$$d = 1.53361\ e^{0.0853458t}$$

where t is the number of years past 1900 (Source: *World Almanac*, 1991). How long will it be before the debt is predicted to reach \$5 trillion (that is, \$5000 billion)?

33. ***Inflation*** The purchasing power P (in dollars) of an annual amount of A dollars after t years of 5% inflation is given by

$$P = Ae^{-0.05t}$$

(Source: *Viewpoints*, VALIC, Summer 1993.)
(a) How long will it be before a pension of \$20,000 per year has a purchasing power of \$8000?
(b) How much pension A would be needed so that the purchasing power P is \$20,000 after 15 years?

34. ***Mutual funds*** For selected years from 1940 to 1990, the number of mutual funds N, excluding money market funds, can be modeled by

$$N = 3.6992e^{0.662056t}$$

where t is the number of decades past 1900 (Source: Investment Company Institute data, published in *Investment Digest*, VALIC, Summer 1992).

(a) Use the model to estimate the number of mutual funds in 1970.
(b) Use the model to estimate the year when the number of mutual funds will reach 5000.

35. ***Doubling time*** If money is invested at 6% compounded monthly, then the doubling time t, in months, can be found from $2 = (1.005)^t$.
(a) Rewrite this equation in logarithmic form.
(b) Use your answer to (a) and a change-of-base formula to find the doubling time.

36. ***Demand*** Say the demand function for a product is given by $p = 100/\ln(q + 1)$.
(a) What will be the price if 19 units are demanded?
(b) How many units, to the nearest unit, will be demanded if the price is \$29.40?

37. ***Pollution*** Pollution levels in Lake Erie have been modeled by the equation

$$x = 0.05 + 0.18e^{-0.38t}$$

where x is the volume of pollutants (in cubic kilometers) and t is the time (in years). (Adapted from R. H. Rainey, *Science* 155 (1967), 1242–1243.)
(a) Find the initial pollution levels; that is, find x when $t = 0$.
(b) How long before x is 30% of that initial level?

38. ***Fish length*** Suppose that the length x (in centimeters) of an individual of a certain species of fish is given by

$$x = 50 - 40e^{-0.05t}$$

where t is its age in months.
(a) Find the length after 1 year.
(b) How long (to the nearest month) will it be until the length is 45 cm?

39. ***Chemical reaction*** When two chemicals, A and B, react to form another chemical C (such as in the digestive process), this is a special case of the **law of mass action,** which is fundamental to studying chemical reaction rates. Suppose that chemical C is formed from A and B according to

$$x = \frac{120[1 - (0.6)^{3t}]}{4 - (0.6)^{3t}}$$

where x is the number of pounds of C formed in t minutes.
(a) How much of C is present when the reaction begins?
(b) How much of C is formed in 4 minutes?
(c) How long does it take to form 10 lb of C?

40. ***Newton's law of cooling*** A certain object at 90°F that is allowed to cool in a room where the temperature is 20°F has its temperature T given as a function of time t (in minutes) according to

$$T = 20 + 70e^{-0.056t}$$

(a) Find the temperature of the object 30 seconds after it is in the room.
(b) How long does it take before the temperature of the object is 20.5°F?

41. ***Ventilation*** The ventilation system in a building operates when the concentration of carbon dioxide (CO_2) reaches a certain level. Suppose that when the ventilation system operates, the cubic feet of CO_2, x, in an 8000-cu-ft room depends on time t (in minutes) according to

$$x = 4.8 + 11.2e^{-t/4}$$

(a) Find the initial amount of CO_2 in the room, and find the concentration of CO_2 (as a percent) at this time.
(b) How long does it take to have a concentration of 0.07% CO_2?
(c) The steady-state or equilibrium concentration is what would result if the ventilation system were left on indefinitely (as t becomes larger and larger). Determine the steady-state concentration.

KEY TERMS AND FORMULAS

Section	Key Terms	Formula
5.1	Exponential functions	
	Growth functions	$f(x) = a^x \quad (a > 1)$
	Decay functions	$f(x) = a^{-x} \quad (a > 1)$
	e	$e \approx 2.71828$
5.2	Logarithmic function	$y = \log_a x$, defined by $x = a^y$
	Common logarithm	$\log x = \log_{10} x$
	Natural logarithm	$\ln x = \log_e x$
	Inverse functions	
	Logarithmic Properties I–V	$\log_a a^x = x$; $a^{\log_a x} = x$; $\log_a(MN) = \log_a M + \log_a N$; $\log_a(M/N) = \log_a M - \log_a N$; $\log_a(M^N) = N(\log_a M)$
	Change-of-base formulas	$\log_b x = \dfrac{\log_a x}{\log_a b}$ $\log_b x = \dfrac{\ln x}{\ln b}$
5.3	Exponential growth and decay	
	Decay models	$f(x) = ma^{-x} \quad (a > 1, m > 0)$
	Growth models	$f(x) = ma^x \quad (a > 1, m > 0)$
	Gompertz curves	$N = Ca^{R^t}$

REVIEW EXERCISES

Sections 5.1 and 5.2

1. Write each statement in logarithmic form.
   (a) $2^x = y$ (b) $3^y = 2x$
2. Write each statement in exponential form.
   (a) $\log_7\left(\dfrac{1}{49}\right) = -2$ (b) $\log_4 x = -1$

Graph the following functions.

3. $y = e^x$
4. $y = e^{-x}$
5. $y = \log_2 x$
6. $y = 2^x$
7. $y = 4^x$
8. $y = \ln x$
9. $y = \log_4 x$
10. $y = 3^{-2x}$
11. $y = \log x$
12. $y = \log(x + 2)$
13. $y = \ln(x - 3)$
14. $y = 10(2^{-x})$

In Problems 15–22 evaluate each logarithm without using a calculator.

15. $\log_5 1$
16. $\log_8 64$
17. $\log_{25} 5$
18. $\log_3\left(\dfrac{1}{3}\right)$
19. $\log_3 3^8$
20. $\ln e$
21. $e^{\ln 5}$
22. $10^{\log 3.15}$

In Problems 23–26, if $\log_a x = 1.2$ and $\log_a y = 3.9$, find each of the following by using the properties of logarithms.

23. $\log_a\left(\dfrac{x}{y}\right)$
24. $\log_a \sqrt{x}$
25. $\log_a (xy)$
26. $\log_a(y^4)$

In Problems 27 and 28, use the properties of logarithms to write each expression as the sum or difference of two logarithmic functions containing no exponents.

27. $\log(yz)$
28. $\ln\sqrt{\dfrac{x+1}{x}}$
29. Is it true that $\ln x + \ln y = \ln(x + y)$ for all values of x?
30. If $f(x) = \ln x$, find $f(e^{-2})$.
31. If $f(x) = 2^x + \log(7x - 4)$, find $f(2)$.
32. If $f(x) = e^x + \ln(x + 1)$, find $f(0)$.
33. If $f(x) = \ln(3e^x - 5)$, find $f(\ln 2)$.

In Problems 34 and 35, use a change-of-base formula to evaluate each logarithm.

34. $\log_9 2158$ 35. $\log_{12}(0.0195)$

In Problems 36 and 37, rewrite each logarithm by using a change-of-base formula, then graph the function with a graphing utility.

36. $y = \log_{\sqrt{3}} x$ 37. $f(x) = \log_{11}(2x - 5)$

Applications

Sections 5.1 and 5.2

38. ***Stock price history*** The graph in the accompanying figure shows the share price history of the stock of Federal Signal Corporation from 1978 to 1993. If the share price were modeled as a function of time, which of the following would give the best model: a growth exponential, a decay exponential, or a logarithm? Justify your choice.

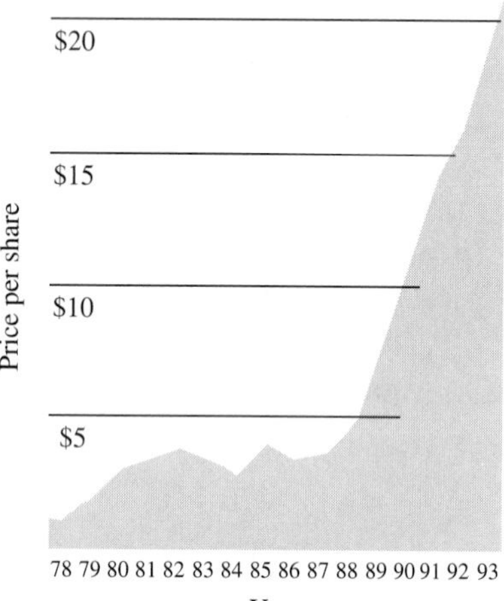

SOURCE: Federal Signal Corporation, 1993 Annual Report

39. ***HIV infections*** The figure shows the number of world HIV infections since AIDS was discovered in 1981.

(a) Suppose t is the number of years since 1980 and y is the number of world HIV infections (in hundreds of thousands). Graph each of the following functions to decide which one models the low estimate and which one models the high estimate. Use a graphing utility.

$$y = 0.099t^{1.969} \quad \text{and} \quad y = 0.08356(1.454)^t$$

(b) Consult the Internet or your library to see which model is more accurate.

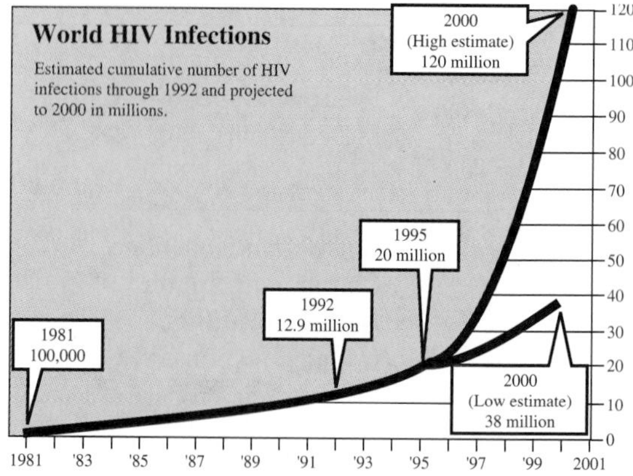

SOURCE: World Health Organization and the Global AIDS Policy Coalition, as reported in the *El Paso Times*, December 6, 1992.

40. ***Mutual funds*** The figure shows the history of the growth in the number of mutual funds. If the number of mutual funds were modeled by a function, do you think the best model would be linear, exponential, logarithmic, or quadratic? Explain.

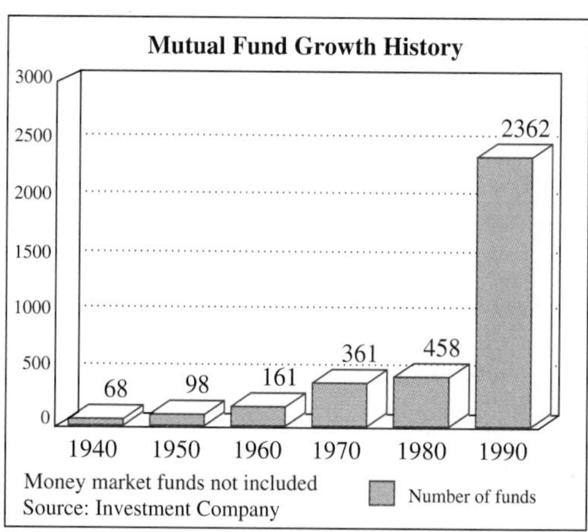

Published in *Investment Digest* of Valic Co., Vol. 5, No. 2, Summer 1992.

41. ***Prison population*** The prison population has grown exponentially from 1980 to 1995.

(a) Find an exponential function that models the data in the table, using $x = 0$ in 1900.

(b) What does your model predict as the prison population in 2005?

Year	*Prison Population*	*Year*	*Prison Population*
1980	319,598	1988	606,810
1981	360,029	1989	683,382
1982	402,914	1990	743,382
1983	423,898	1991	792,535
1984	448,264	1992	850,566
1985	487,593	1993	909,381
1986	526,436	1994	990,147
1987	562,814	1995	1,078,545

SOURCE: Bureau of Justice Statistics, U.S. Dept. of Justice

42. ***Taxes*** The percent of income claimed by federal, state, and local taxes for the median household with two incomes is given in the table below.
    (a) Using the number of years from 1950 as x and the percent as y, find a logarithmic equation that models the data.
    (b) What does this model predict as the percent claimed for taxes in 2000?
    (c) When is the tax load predicted to be 40%?

Year	*Percent of Income Paid in Taxes*
1955	27.9
1965	28.5
1975	36.2
1985	37.5
1997	38.2

SOURCE: Tax Foundation, *USA Today*, March 2, 1998

43. ***Stellar magnitude*** The stellar magnitude M of a star is related to its brightness B as seen from earth according to

$$M = -\frac{5}{2}\log(B/B_0)$$

where B_0 is a standard level of brightness (the brightness of the star Vega).
    (a) Find the magnitude of Venus if its brightness is 36.3 times B_0.
    (b) Find the brightness (as a multiple of B_0) of the North Star if its magnitude is 2.1.
    (c) If the faintest stars have magnitude 6, find their brightness (as a multiple of B_0).
    (d) Is a star with magnitude -1.0 brighter than a star with magnitude $+1.0$?

Section 5.3

44. ***Advertising and sales*** Because of a new advertising campaign, a company predicts that sales will increase and that the yearly sales will be given by the equation

$$N = 10{,}000(0.3)^{0.5^t}$$

where t represents the number of years after the start of the campaign.
    (a) What are the sales when the campaign begins?
    (b) What are the predicted sales for the third year?
    (c) What are the maximum predicted sales?

45. ***Sales decay*** The sales decay for a product is given by $S = 50{,}000e^{-0.1x}$, where S is the weekly sales (in dollars) and x is the number of weeks that have passed since the end of an advertising campaign.
    (a) What will sales be 6 weeks after the end of the campaign?
    (b) How many weeks will pass before sales drop below 15,000?

46. ***Sales decay*** The sales decay for a product is given by $S = 50{,}000e^{-0.6x}$, where S is the monthly sales (in dollars) and x is the number of months that have passed since the end of an advertising campaign. What will sales be 6 months after the end of the campaign?

47. ***Compound interest*** If \$1000 is invested at 12%, compounded monthly, the future value A at any time t (in years) is given by

$$A = 1000(1.01)^{12t}$$

How long will it take for the amount to double?

48. ***Compound interest*** If \$5000 is invested at 13.5%, compounded continuously, then the future value A at any time t (in years) is given by $A = 5000e^{0.135t}$.
    (a) What is the amount after 9 months?
    (b) How long will it be before the investment doubles?

CHAPTER TEST

In Problems 1–4, graph the functions.

1. $y = 5^x$
2. $y = 3^{-x}$
3. $y = \log_5 x$
4. $y = \ln x$

In Problems 5–8, use technology to graph the functions.

5. $y = 3^{0.5x}$
6. $y = \log(0.5x)$
7. $y = e^{2x}$
8. $y = \log_7 x$

In Problems 9–12, use a calculator to give a decimal approximation of the numbers.

9. e^4
10. $e^{-2.1}$
11. $\ln 4$
12. $\ln 21$
13. Write $\log_{17} x = 3.1$ in exponential form and find x to three decimal places.
14. Write $3^{2x} = 27$ in logarithmic form.

In Problems 15–18, simplify the expressions, using properties or definitions of logarithms.

15. $\log_2 8$
16. $e^{\ln x^4}$
17. $\log_7 7^3$
18. $\ln e^{x^2}$
19. Write $\ln(M \cdot N)$ as a sum involving M and N.
20. Write $\ln\left(\dfrac{x^3 - 1}{x + 2}\right)$ as a difference involving two binomials.
21. Write $\log_4(x^3 + 1)$ as a base-e logarithm using a change-of-base formula.
22. ***Social Security benefits*** The graph in the figure shows the number of active workers who will be (or have been) supporting each Social Security beneficiary. What type of function might be an appropriate model for this situation?

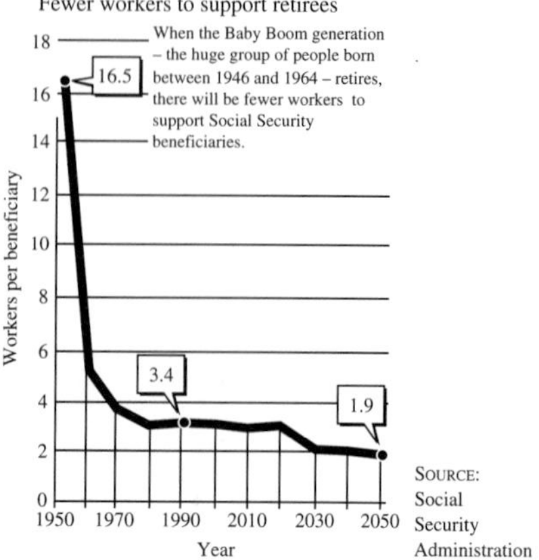

23. ***Postal rates*** The graph in the figure shows the history of first-class postage stamp rates from 1890 to 1995. If these data were modeled by a smooth curve, what type of function might be appropriate?

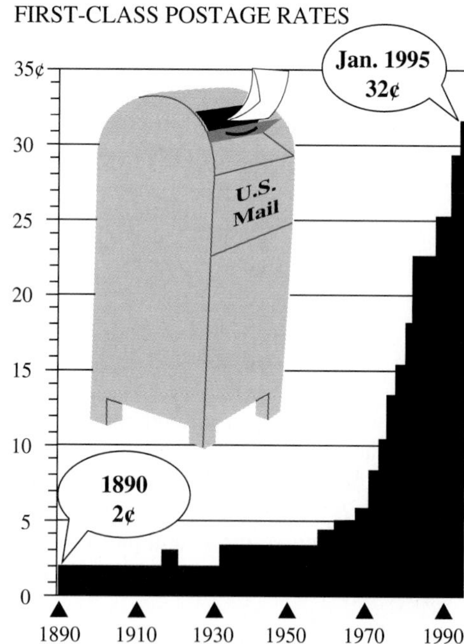

SOURCE: *Oil City Derrick*, Oil City, PA, December 1, 1994

24. ***Population*** The following table gives the world population, in millions, for selected years. If the equation that models these data is

$$y = 160e^{0.00648(x-1500)}$$

where x is the year, in what year will the population be double that of the world population in 1975?

Year	*World Population (millions)*
1	200
1650	500
1850	1000
1930	2000
1975	4000
1997	5800

SOURCE: *World Almanac,* 1997

M 25. ***Promise Keepers*** The following table gives the attendance at Promise Keepers rallies from 1991 to 1996.

(a) Using x as the number of years since 1990, model the data with an exponential equation.

(b) What would this model predict as the attendance for a rally in 2001? Does this seem reasonable?

Year	*Promise Keepers Rally Attendance*
1991	4,200
1992	32,000
1993	54,000
1994	300,000
1995	750,000
1996	1,100,000

SOURCE: *Time,* October 6, 1997

M 26. ***Dow Jones Industrial Average*** The following table shows the high values achieved by the Dow Jones Industrial Average for the years from 1979 to 1989. Create an exponential equation that models the data, where x is the number of years from 1900.

Year	*Dow Jones High*	*Year*	*Dow Jones High*
1979	897	1985	1553
1980	1000	1986	1955
1981	1024	1987	2722
1982	1070	1988	2183
1983	1287	1989	2791
1984	1286		

SOURCE: *World Almanac,* 1991

M 27. ***Dow Jones UPDATE*** The table below gives the high values achieved by the Dow Jones Industrial Average for the years from 1990 to 1997. Combine these data with those from Problem 26 to give the highs from 1979 to 1997. Create an exponential equation that models the data, where x is the number of years from 1900. By observing the equations and graphing them, decide whether the models are similar or radically different.

Year	*Dow Jones High*	*Year*	*Dow Jones High*
1990	3000	1994	3978
1991	3169	1995	5216
1992	3413	1996	6561
1993	3794	1997	8259

Extended Applications / Group Projects

I. Agricultural Business Management

A commercial vegetable and fruit grower carefully observes the relationship between the amount of fertilizer used on a certain variety of squash and the extra revenue made from the sales of the resulting squash crop. These observations are recorded in the table below.

Amount of fertilizer (pounds/acre)	0	250	500	750	1000	1250	1500	1750	2000
Extra revenue earned (dollars/acre)	0	96	145	172	185	192	196	198	199

The fertilizer costs 14¢ per pound. What would you advise the grower is the most profitable amount of fertilizer to use?

Check your advice by answering the following questions.

1. Is each pound of fertilizer equally effective?
2. Estimate the maximum amount of extra revenue that can be earned per acre by observing the data.
3. Graph the data points giving the extra revenue made from improved squash sales as a function of the amount of fertilizer used. What does the graph tell you about the growth of the extra revenue as a function of the amount of fertilizer?
4. Which of the following equations best models the graph you obtained in Question 3?

   (a) $N = Ca^{R^x}$ (b) $y = \dfrac{A}{1 + ce^{-ax}}$ (c) $y = c(1 - e^{-ax})$

   Determine a specific equation that fits the observed data. That is, find the value of the constants in the model you have chosen.
5. Graph the total cost of the fertilizer as a function of the amount of fertilizer used on the graph you created for Question 3. Determine an equation that models this relationship. For what amount of fertilizer is the extra revenue earned from squash sales offset by the cost of the fertilizer?
6. Use the graphs or the equations you have created to estimate the maximum extra profit that can be made and the amount of fertilizer per acre that would give that extra profit.

II. National Debt

For years politicians have talked about a balanced federal budget, and success in this effort was proclaimed in 1998. However, does this mean that there is no longer a national debt? Does it mean that the national debt is getting smaller? The table below gives the U.S. national debt for selected years from 1900 to 1989.

Year	*U.S. Debt ($billions)*	*Year*	*U.S. Debt ($billions)*
1900	1.2	1955	272.8
1910	1.1	1965	313.8
1920	24.2	1975	533.2
1930	16.1	1985	1823.1
1940	43.0	1987	2350.3
1945	258.7	1989	2857.4

SOURCE: Bureau of Public Debt, U.S. Treasury

To investigate this issue, complete the following.

1. Create a scatter plot with these data, and determine the best function type to model these data. (Let $x = 0$ in 1900.)
2. Model the data. Rewrite the equation with the base e.
3. Use the model to predict the debt in 1994 and in 1996.
4. Use this model to predict when the debt will be \$5 trillion (\$5000 billion).
5. Compare the results from Questions 3 and 4 with the actual results in the table below, which gives the national debt through 1996. Does the model you created predict the debt for these years fairly well?
6. Create a scatter plot that includes the data in the table below.
7. Graph the model developed in Question 2 on the same set of axes as the scatter plot for the new data. Does the model still represent the data?
8. Create an equation that models the data in the table below. Rewrite the equation with base e.
9. Look at the equations of the two models and graph both models on the same set of axes to determine whether the predictions generated by the new model differ greatly from those of the first model.

10. Use the Internet or a library to find the national debt for the last 3 years and determine whether it is falling or is still growing. Do the debts for these years fit the models you created?

Year	*U.S. Debt ($billions)*	*Year*	*U.S. Debt ($billions)*
1900	1.2	1987	2350.3
1910	1.1	1989	2857.4
1920	24.2	1990	3233.3
1930	16.1	1991	3665.3
1940	43.0	1992	4064.6
1945	258.7	1993	4411.5
1955	272.8	1994	4692.8
1965	313.8	1995	4974.0
1975	533.2	1996	5224.8
1985	1823.1		

SOURCE: Bureau of Public Debt, U.S. Treasury

Warm-up

Prerequisite Problem Type	For Section	Answer	Section for Review
What is $\frac{(-1)^n}{2n}$, if (a) $n = 1$? (b) $n = 2$? (c) $n = 3$?	**6.2**	(a) $-\frac{1}{2}$ (b) $\frac{1}{4}$ (c) $-\frac{1}{6}$	0.2 Signed numbers
(a) What is $(-2)^6$? (b) What is $\frac{\frac{1}{4}[1-(\frac{1}{2})^6]}{1-\frac{1}{2}}$?	**6.2**	(a) 64 (b) $\frac{63}{128}$	0.2 Signed numbers
If $f(x) = \frac{1}{2x}$, what is (a) $f(2)$? (b) $f(4)$?	**6.1**	(a) $\frac{1}{4}$ (b) $\frac{1}{8}$	1.2 Functional notation
Evaluate $e^{(.08)(20)}$	**6.2**	4.95303	5.1 Exponential functions
Evaluate (a) $\frac{1-(1.05)^{-16}}{0.05}$ (b) $5000 + 5000\left[\frac{1-(1.059)^{-9}}{0.059}\right]$	**6.3** **6.4**	(a) 10.83777 (b) 39,156.96575	0.3 Integral exponents
Solve: (a) $1100 = 1000 + 1000(0.058)t$ (b) $12{,}000 = P(1.03)^6$ (c) $850 = A_n\left[\frac{0.0065}{1-(1.0065)^{-360}}\right]$	**6.1–6.5**	(a) $t \approx 1.72$ (b) $P \approx 10{,}049.81$ (c) $A_n \approx 118{,}076.79$	1.1 Linear equations
Solve: (a) $2 = (1.08)^n$ (b) $20{,}000 = 10{,}000e^{0.08t}$	**6.2**	(a) $n \approx 9.01$ (b) $t \approx 8.66$	5.3 Solution of exponential equations

Chapter 6

Mathematics of Finance

Regardless of whether your career is in business, understanding how interest is computed on investments and loans is important to you as a consumer. The proliferation of personal finance and money management software for home computers attests to this importance. In this chapter we'll explain the mathematics that is the basis for such software packages.

We begin by studying simple interest, which is sometimes used on short-term investments or loans. Most investments (for example, savings accounts or bonds) pay compound interest, where the interest is calculated over short periods of time and added to the principal. Most loans are not simple interest loans because they are repaid by making partial payments of both principal and interest during the period of the loan; one such method of repaying a loan is called amortization.

The goal of this chapter is to provide some understanding of the methods used to determine the interest and the future value (principal plus interest) resulting from savings plans and the methods used in repayment of debts. We will also study sequences, and we will pay special attention to arithmetic and geometric sequences because of their applications in mathematics of finance. From these we will develop formulas for computing compound interest, the future and present values of annuities, amortization of debts, and the amount paid into a sinking fund to discharge a debt.

6.1 Simple Interest; Sequences

OBJECTIVES

- To find the future value and the amount of interest for simple interest loans and investments
- To find the simple interest rate earned on an investment
- To find the time required for a simple interest investment to reach a goal
- To write a specified number of terms of a sequence
- To find specified terms and sums of specified numbers of terms of arithmetic sequences

APPLICATION PREVIEW

In this section we begin our study of the mathematics of finance by considering **simple interest.** Simple interest forms the basis for all calculations involving interest that is paid on an investment or that is due on a loan. Also, we can evaluate short-term investments (such as in stocks or real estate) by calculating their simple interest rate. For example, on April 29, 1994, Mary Spaulding purchased 100 shares of Exxon stock for \$6125.00. After 6 months, the stock had risen in value by \$138.00 and had paid dividends totaling \$144.14. One way to evaluate this investment and compare it to a bank savings plan is to find the simple interest rate that this gain represents.

Simple Interest

If a sum of money (called the **principal**) P is invested for a time period t (frequently years) at an interest rate r per period, the **simple interest** is given by the following formula.

Simple Interest $I = Prt$,

where
I = interest (in dollars)
P = principal (in dollars)
r = annual interest rate (written as a decimal)
t = time (in years)*

Note that the time measurements for r and t must agree.

Simple interest is paid on investments involving time certificates issued by banks and on certain types of bonds, such as U.S. government series H bonds and municipal bonds. The interest for a given period is paid to the investor, and the principal remains the same.

If you borrow money from a friend or a relative, interest on your loan might be calculated with the simple interest formula. We'll consider some simple interest loans in this section, but interest on loans from banks and other lending institutions is calculated using methods discussed later.

*Periods of time other than years can be used, as can other monetary systems.

EXAMPLE 1

(a) If \$8000 is invested for 2 years at an annual interest rate of 9%, how much interest will be received at the end of the 2-year period?

(b) If \$4000 is borrowed for 39 weeks at an annual interest rate of 15%, how much interest is due at the end of the 39 weeks?

Solution

(a) The interest is $I = Prt = \$8000(0.09)(2) = \$1440$.

(b) Use $I = Prt$ with $t = 39/52 = 0.75$ year. Thus,

$$I = \$4000(0.15)(0.75) = \$450$$

The **future amount of an investment,** or its **future value,** at the end of an interest period is the sum of the principal and the interest. Thus in Example 1(a), the future value is

$$S = \$8000 + \$1440 = \$9440$$

Similarly, the **future amount of a loan,** or its **future value,** is the amount of money that must be repaid. In Example 1(b), the future value of the loan is the principal plus the interest, or

$$S = \$4000 + \$450 = \$4450$$

Future Value

If we use the letter S to denote the future value of an investment or a loan, then we have

$$\text{Future value of investment or loan:} \qquad S = P + I$$

where P is the principal (in dollars) and I is the interest (in dollars).

The principal P of a loan is also called the **face value** or the **present value** of the loan.

EXAMPLE 2

(a) If \$2000 is borrowed for 5 months at a simple interest rate of 1% per month, what is the future value of the loan at the end of the 5-month period?

(b) An investor wants to have \$20,000 in 9 months. If the best available interest rate is 6.05% per year, how much must be invested now to yield the desired amount?

Solution

(a) The interest for the 5-month period is $I = \$2000(0.01)(5) = \$100$. Thus the future value of the investment for the period is

$$S = P + I = \$2000 + \$100 = \$2100$$

(b) We know that $S = P + I = P + Prt$. In this case, we must solve for P, the present value.

$$\$20{,}000 = P + P(0.0605)(9/12) = P + 0.045375P$$

$$\$20{,}000 = 1.045375P$$

$$\frac{\$20{,}000}{1.045375} = P \quad \text{so} \quad P = \$19{,}132.89$$

We now return to the Application Preview problem.

EXAMPLE 3

Recall that Mary Spaulding bought Exxon stock for \$6125.00 and that after 6 months, the value of her shares had risen by \$138.00 and dividends totaling \$144.14 had been paid. Find the simple interest rate she earned on this investment if she sold the stock at this time.

Solution

We can view the principal as \$6125.00, the time as 1/2 year, and the interest earned as the total of all gains (that is, interest $= I =$ \$138.00 + \$144.14 = \$282.14). Hence

$$I = Prt$$

$$\$282.14 = (\$6125)r(0.5)$$

$$\$282.14 = \$3062.5r$$

$$r = \frac{\$282.14}{\$3062.5} \approx 0.092 = 9.2\%$$

Thus Mary's return was equivalent to an annual simple interest rate of about 9.2%.

EXAMPLE 4

If \$1000 is invested at 5.8% simple interest, how long will it take to grow to \$1100?

Solution

With $P = \$1000$, $S = \$1100$, and $r = 0.058$, we solve for t.

$$S = P + Prt$$

$$\$1100 = \$1000 + \$1000(0.058)t$$

$$\$100 = \$58t$$

$$\frac{\$100}{\$58} = t \quad \text{so} \quad t \approx 1.72 \text{ years}$$

CHECKPOINT

1. What is the simple interest formula?
2. If \$8000 is invested at 6% simple interest for 9 months, find the future value of the investment.
3. If a \$2500 investment grows to \$2875 in 15 months, what simple interest rate was earned?

Sequences

Let us look at the monthly future values of a \$2000 investment that earns 1% simple interest for each of 5 months.

Month	*Interest (I = Prt)*	*Future Value of the Investment*
1	(\$2000)(0.01)(1) = \$20	\$2000 + \$20 = \$2020
2	(\$2000)(0.01)(1) = \$20	\$2020 + \$20 = \$2040
3	(\$2000)(0.01)(1) = \$20	\$2040 + \$20 = \$2060
4	(\$2000)(0.01)(1) = \$20	\$2060 + \$20 = \$2080
5	(\$2000)(0.01)(1) = \$20	\$2080 + \$20 = \$2100

These future values are outputs that result when the inputs are positive integers that correspond to the number of months of the investment. Outputs (such as these future values) that arise uniquely from positive integer inputs define a function whose domain is the positive integers. Such a function is called a **sequence function.** Because the domain of a sequence function is the positive integers, the outputs form an ordered list called a **sequence.**

Sequence The set of functional values $a_1, a_2, a_3, \ldots$ of a sequence function is called a **sequence.** The values $a_1, a_2, a_3, \ldots$ are called **terms** of the sequence, with a_1 the first term, a_2 the second term, and so on.

Because calculations involving interest often result from using positive integer inputs, sequences are the basis for most of the financial formulas derived in this chapter.

EXAMPLE 5

Write the first four terms of the sequence whose nth term is $a_n = (-1)^n/(2n)$.

Solution

The first four terms of the sequence are as follows:

$$a_1 = \frac{(-1)^1}{2(1)} = -\frac{1}{2} \qquad a_2 = \frac{(-1)^2}{2(2)} = \frac{1}{4}$$

$$a_3 = \frac{(-1)^3}{2(3)} = -\frac{1}{6} \qquad a_4 = \frac{(-1)^4}{2(4)} = \frac{1}{8}$$

We usually write these terms in the form $-\frac{1}{2}, \frac{1}{4}, -\frac{1}{6}, \frac{1}{8}$.

Arithmetic Sequences

Earlier in this section, we described the sequence

$$2020, 2040, 2060, 2080, 2100, 2120, 2140, \ldots$$

This sequence can also be described in the following way:

$$a_1 = 2020, \qquad a_n = a_{n-1} + 20 \quad \text{for } n > 1$$

This sequence is an example of a special kind of sequence called an **arithmetic sequence.** In such a sequence, each term after the first can be found by adding a constant to the preceding term. Thus we have the following definition.

Arithmetic Sequence A sequence is called an **arithmetic sequence** (progression) if there exists a number d, called the **common difference,** such that

$$a_n = a_{n-1} + d \quad \text{for } n > 1$$

EXAMPLE 6

Write the next three terms of the following arithmetic sequences:

(a) $1, 3, 5, \ldots$ (b) $9, 6, 3, \ldots$ (c) $\frac{1}{2}, \frac{5}{6}, \frac{7}{6}, \ldots$

Solution

(a) The common difference is 2, so the next three terms are 7, 9, 11.
(b) The common difference is -3, so the next three terms are $0, -3, -6$.
(c) The common difference is $\frac{1}{3}$, so the next three terms are $\frac{3}{2}, \frac{11}{6}, \frac{13}{6}$.

Because each term after the first in an arithmetic sequence is obtained by adding d to the preceding term, the second term is $a_1 + d$, the third is $(a_1 + d) + d = a_1 + 2d, \ldots$, and the nth term is $a_1 + (n - 1)d$. Thus we have the following formula.

The ***n*th term of an arithmetic sequence** (progression) is given by

$$a_n = a_1 + (n - 1)d$$

where a_1 is the first term and d is the common difference between successive terms.

EXAMPLE 7

Find the 11th term of the arithmetic sequence with first term 3 and common difference -2.

Solution

The 11th term is $a_{11} = 3 + (11 - 1)(-2) = -17$.

EXAMPLE 8

If the first term of an arithmetic sequence is 4 and the ninth term is 20, find the fifth term.

Solution

Substituting the values $a_1 = 4$, $a_n = 20$, and $n = 9$ in $a_n = a_1 + (n - 1)d$ gives $20 = 4 + (9 - 1)d$. Solving this equation gives $d = 2$. Therefore, the fifth term is $a_5 = 4 + (5 - 1)2 = 12$.

Consider the arithmetic sequence with first term a_1, common difference d, and nth term a_n. The first n terms of an arithmetic sequence can be written from a_1 to a_n as

$$a_1, a_1 + d, a_1 + 2d, a_1 + 3d, \dots, a_1 + (n - 1)d$$

or backward from a_n to a_1 as

$$a_n, a_n - d, a_n - 2d, a_n - 3d, \dots, a_n - (n - 1)d$$

If we let s_n represent the sum of the first n terms of the sequence described above, then we have

$$s_n = a_1 + (a_1 + d) + (a_1 + 2d) + \cdots + [a_1 + (n - 1)d] \tag{1}$$

or, alternatively,

$$s_n = a_n + (a_n - d) + (a_n - 2d) + \cdots + [a_n - (n - 1)d] \tag{2}$$

If we add equations (1) and (2) together, term by term, we obtain

$$2s_n = (a_1 + a_n) + (a_1 + a_n) + (a_1 + a_n) + \cdots + (a_1 + a_n)$$

in which $(a_1 + a_n)$ appears as a term n times. Thus,

$$2s_n = n(a_1 + a_n)$$

$$s_n = \frac{n}{2}(a_1 + a_n)$$

We may state this result as follows:

The **sum of the first n terms of an arithmetic sequence** is given by the formula

$$s_n = \frac{n}{2}(a_1 + a_n)$$

where a_1 is the first term of the sequence and a_n is the nth term.

EXAMPLE 9

Find the sum of the first ten terms of the arithmetic sequence with first term 2 and common difference 4.

Solution

We are given the values $n = 10$, $a_1 = 2$, and $d = 4$. Thus the tenth term is $a_{10} = 2 + (10 - 1)4 = 38$, and the sum is

$$s_{10} = \frac{10}{2}(2 + 38) = 200$$

EXAMPLE 10

Find the sum of the first seven terms of the arithmetic sequence $\frac{1}{4}, \frac{7}{12}, \frac{11}{12}, \ldots$.

Solution

The first term is $\frac{1}{4}$ and the common difference is $\frac{1}{3}$. Therefore, the seventh term is $a_7 = \frac{1}{4} + (7 - 1)\left(\frac{1}{3}\right) = \frac{9}{4}$. The sum is $s_7 = \frac{7}{2}\left(\frac{1}{4} + \frac{9}{4}\right) = \frac{35}{4} = 8\frac{3}{4}$.

CHECKPOINT

4. Of the following sequences, identify the arithmetic sequences.
   (a) 1, 4, 9, 16, . . . (b) 1, 4, 7, 10, . . . (c) 1, 2, 4, 8, . . .
5. Given the arithmetic sequence $-10, -6, -2, \ldots$, find
   (a) the 51st term (b) the sum of the first 51 terms

Finally, we should note that many graphing calculators, some graphics software packages, and spreadsheets have the capability of defining sequence functions and then operating on them by finding additional terms, graphing the terms, or summing a fixed number of terms.

CHECKPOINT SOLUTIONS

1. $I = Prt$
2. $S = P + I = 8000 + 8000(0.06)\left(\frac{9}{12}\right) = \$8360$
3. $S = P + I$, so

$$2875 = 2500 + 2500r\left(\frac{15}{12}\right)$$

$$375 = 3125r$$

$$r = \frac{375}{3125} = 0.12, \quad \text{or} \quad 12\%$$

4. Only (b) is an arithmetic sequence; the common difference is 3.
5. (a) The common difference is $d = 4$. The 51st term ($n = 51$) is
   $a_n = a_1 + (n - 1)d = -10 + 50(4) = -10 + 200 = 190$
   (b) For $a_1 = -10$ and $a_n = 190$, the sum is $s_n = \frac{51}{2}(-10 + 190) = 4590$.

EXERCISE 6.1

Simple Interest

1. \$10,000 is invested for 6 years at an annual simple interest rate of 16%.
   (a) How much interest will be earned?
   (b) What is the future value of the investment at the end of the 6 years?
2. \$800 is invested for 5 years at an annual simple interest rate of 14%.
   (a) How much interest will be earned?
   (b) What is the future value of the investment at the end of the 5 years?
3. \$1000 is invested for 3 months at an annual simple interest rate of 12%.
   (a) How much interest will be earned?
   (b) What is the future value of the investment after 3 months?
4. \$1800 is invested for 9 months at an annual simple interest rate of 15%.
   (a) How much interest will be earned?
   (b) What is the future value of the investment after 9 months?
5. If you borrow \$800 for 6 months at 16% annual simple interest, how much must you repay at the end of the 6 months?
6. If you borrow \$1600 for 2 years at 14% annual simple interest, how much must you repay at the end of the 2 years?
7. If you lend \$3500 to a friend for 15 months at 8% annual simple interest, find the future value of the loan.
8. Mrs. Gonzalez lent \$2500 to her son Luis for 7 months at 9% annual simple interest. What is the future value of this loan?
9. A couple bought some stock for \$30 per share that pays an annual dividend of \$0.90 per share. After 1 year the price of the stock was \$33. Find the simple interest rate on the growth of their investment.
10. Jenny Reed bought SSX stock for \$16 per share. The annual dividend was \$1.50 per share, and after 1 year SSX was selling for \$35 per share. Find the simple interest rate of growth of her money.
11. (a) To buy a Treasury bill (T-bill) that matures to \$10,000 in 6 months, you must pay \$9685.23. What rate does this earn?
    (b) If the bank charges a fee of \$40 to buy a T-bill, what is the actual interest rate you earn?
12. Janie Christopher lent \$6000 to a friend for 90 days at 12%. After 30 days, she sold the note to a third party for \$6000. What interest rate did the third party receive? Use 360 days in a year.
13. A firm buys 12 file cabinets at \$140 each, with the bill due in 90 days. How much must the firm deposit now to have enough to pay the bill if money is worth 12% per year? Use 360 days in a year.
14. A student has a savings account earning 9% simple interest. She must pay \$1500 for first-semester tuition by September 1 and \$1500 for second-semester tuition on January 1. How much must she earn in the summer (by September 1) in order to pay the first-semester bill on time and still have the remainder of her summer earnings grow to \$1500 between September 1 and January 1?
15. If you want to earn 15% annual simple interest on an investment, how much should you pay for a note that will be worth \$13,500 in 10 months?
16. What is the present value of an investment at 6% annual simple interest if it is worth \$832 in 8 months?
17. If \$5000 is invested at 8% annual simple interest, how long does it take to be worth \$9000?
18. How long does it take for \$8500 invested at 11% annual simple interest to be worth \$13,000?
19. A retailer owes a wholesaler \$500,000 due in 45 days. If the payment is 15 days late, there is a 1% penalty charge. The retailer can get a 45-day certificate of deposit (CD) paying 6% or a 60-day certificate paying 7%. Is it better to take the 45-day certificate and pay on time or to take the 60-day certificate and pay late with the penalty?
20. An investor owns several apartment buildings. The taxes on these buildings total \$30,000 per year and are due before April 1. The late fee is 1/2% per month up to 6 months, at which time the buildings are seized by the authorities and sold for back taxes. If the investor has \$30,000 available on March 31, will he save money by paying the taxes at that time or by investing the money at 8% and paying the taxes and the penalty on September 30?
21. Bill Casler bought a \$2000, 9-month certificate of deposit (CD) that would earn 8% annual simple interest. Three months before the CD was due to mature, Bill needed his CD money, so a friend agreed to lend him money and receive the value of the CD when it matured.

(a) What is the value of the CD when it matures?

(b) If their agreement allowed the friend to earn a 10% annual simple interest return on his loan to Bill, how much did Bill receive from his friend?

22. Suppose you lent $5000 to Friend 1 for 18 months at an annual simple interest rate of 9%. After 1 year you need money for an emergency and decide to sell the note to Friend 2.

(a) How much does Friend 1 owe when the loan is due?

(b) If your agreement with Friend 2 means she earns simple interest at an annual rate of 12%, how much did Friend 2 pay you for the note?

Sequences

23. Write the first ten terms of the sequence defined by $a_n = 3n$.
24. Find the first six terms of the sequence defined by $a_n = 4n$.
25. Write the first eight terms of the sequence defined by $a_n = n/3$.
26. Write the first seven terms of the sequence defined by $a_n = 2/n$.
27. Write the first six terms of the sequence whose nth term is $(-1)^n/(4n)$.
28. Write the first five terms of the sequence whose nth term is $(-1)^n/(3n + 3)$.
29. Write the first six terms of the sequence whose nth term is $(-1)^n/(2n + 1)$.
30. Write the first five terms of the sequence whose nth term is

$$a_n = \frac{(-1)^n}{n^2}$$

31. Write the first four terms and the tenth term of the sequence whose nth term is

$$a_n = \frac{n - 4}{n(n + 2)}$$

32. Write the sixth term of the sequence whose nth term is

$$a_n = \frac{n(n - 1)}{n + 3}$$

Arithmetic Sequences

In Problems 33–36, (a) identify d and a_1 and (b) write the next 3 terms.

33. 2, 5, 8, . . .
34. 3, 9, 15, . . .
35. $3, \frac{9}{2}, 6, \ldots$
36. 2, 2.75, 3.5, . . .
37. Find the 28th term of the arithmetic sequence with first term -3 and common difference 4.
38. Find the 31st term of the arithmetic sequence with first term 7 and common difference 4.
39. Find the 83rd term of the arithmetic sequence with first term 6 and common difference $-\frac{1}{2}$.
40. Find the 66th term of the arithmetic sequence with first term $\frac{1}{2}$ and common difference $-\frac{1}{3}$.
41. Find the 100th term of the arithmetic sequence with first term 5 and eighth term 19.
42. Find the 73rd term of the arithmetic sequence with first term 20 and tenth term 47.
43. Find the sum of the first 38 terms of the arithmetic sequence with first term 2 and common difference 3.
44. Find the sum of the first 56 terms of the arithmetic sequence with first term 6 and common difference 4.
45. Find the sum of the first 70 terms of the arithmetic sequence with first term 10 and common difference $\frac{1}{2}$.
46. Find the sum of the first 80 terms of the arithmetic sequence with first term 12 and common difference -3.
47. Find the sum of the first 50 terms of the arithmetic sequence 2, 4, 6, ...
48. Find the sum of the first 60 terms of the arithmetic sequence 6, 9, 12, ...
49. Find the sum of the first 150 terms of the arithmetic sequence $6, \frac{9}{2}, 3, \ldots$.
50. Find the sum of the first 200 terms of the arithmetic sequence 12, 9, 6, ...

Applications

51. ***Bee reproduction*** A female bee hatches from a fertilized egg, whereas a male bee hatches from an unfertilized egg. Thus a female bee has a male parent and a female parent, whereas a male bee has only a female parent. Therefore, the number of ancestors of a male bee follows the *Fibonacci sequence*

$$1, 2, 3, 5, 8, 13, \ldots$$

Observe the pattern and write three more terms of the sequence.

52. ***Salaries*** Suppose you are offered a job with a relatively low starting salary but with a $1500 raise for each of the next 7 years. How much more than your starting salary would you be making in the eighth year?

53. ***Profit*** A new firm loses $2000 in its first month, but its profit increases by $400 in each succeeding month for the next year. What is its profit in the twelfth month?

54. ***Pay raises*** If you make $27,000 and get $1800 raises each year, in how many years will your salary double?

55. ***Salaries*** Suppose you are offered two identical jobs: one paying a starting salary of $20,000 with yearly raises of $1000 and one paying a starting salary of $18,000 with yearly raises of $1200. Which job will pay you more for your tenth year on the job?

56. ***Profit*** A new firm loses $2000 in its first month, but its profit increases by $400 in each succeding month for the next year. What is its profit for the year?

57. ***Pay raises*** If you are an employee, would you rather be given a raise of $1000 at the end of each year (Plan I) or a raise of $300 at the end of each 6-month period (Plan II)? Consider the following table for an employee whose base salary is $20,000 per year (or $10,000 per 6-month period), and answer parts (a)–(g).

	Salary Received per 6-Month Period	
Period (months)	*Plan I*	*Plan II*
0–6	$10,000	$10,000
6–12	10,000	10,300
12–18	10,500	10,600
18–24	10,500	10,900
24–30	11,000	11,200
30–36	11,000	11,500

(a) Find the sum of the raises for Plan I for the first 3 years.
(b) Find the sum of the raises for Plan II for the first 3 years.
(c) Which plan is better, and by how much?
(d) Find the sum of the raises in Plan I for 5 years.
(e) Find the sum of the raises in Plan II for 5 years.
(f) Which plan is better, and by how much?
(g) Do you want Plan I or Plan II?

58. ***Pay raises*** As an employee, would you prefer being given a $1200 raise each year for 5 years or a $200 raise each quarter for 5 years?

6.2 Compound Interest; Geometric Sequences

OBJECTIVES

- *To find the future value of a compound interest investment and the amount of interest earned when interest is compounded at regular intervals or continuously*
- *To find the annual percentage yield (APY) or the effective annual interest rate of money invested at compound interest*
- *To find the time it takes for an investment to reach a specified amount*
- *To find specified terms, and sums of specified numbers of terms, of geometric sequences*

APPLICATION PREVIEW

A compound interest investment is one in which interest is paid into an account at regular intervals throughout the duration of the investment. After an interest payment is made, that interest, as well as the principal, earns interest. Because of the seeming complexity of computing interest on the interest, interest on the interest on the interest, and so on, we need to develop formulas that allow us to find the **future value** of such an investment. It seems reasonable that this future value depends on the duration of the investment and on the interest rate, but it also depends on how the periodic interest payments are made (that is, on how the compounding is done).

Because the compounding method affects the future value, another application of this section develops formulas for the **annual percentage yield** (also called the effective annual rate) of an investment. By calculating the effective annual rates of investments with different compounding schemes, we can meaningfully compare those investments.

Finally, we should note that the methods we use with compound interest in finding periodic interest rates and numbers of interest periods are important for all subsequent financial applications of this chapter, including such things as savings plans (annuities) and debt repayment with regular payments (amortization).

Compound Interest

In the previous section we discussed simple interest. A second method of paying interest is the **compound interest** method, where the interest for each period is added to the principal before interest is calculated for the next period. With this method, the principal grows as the interest is added to it. This method is used in investments such as savings accounts and U.S. government series E bonds.

An understanding of compound interest is important not only for people planning careers with financial institutions but also for anyone planning to invest money. To see how compound interest is computed, we will first calculate the *future value* that will result if \$20,000 is invested for 3 years at 10%, *compounded annually* (each year).

The principal for the first year is \$20,000.

The interest at the end of the first year is

$$I = \$20{,}000(0.10)(1) = \$2000$$

The future value at the end of the first year is

$$S = P + I = \$20{,}000 + \$2000 = \$22{,}000$$

Thus the principal for the second year is \$22,000.

The interest at the end of the second year is

$$I = \$22{,}000(0.10)(1) = \$2200$$

The future value at the end of the second year is

$$S = \$22{,}000 + \$2200 = \$24{,}200$$

Thus the principal for the third year is \$24,200.

The interest at the end of the third year is

$$I = \$24{,}200(0.10)(1) = \$2420$$

The future value at the end of the third year is

$$S = \$24{,}200 + \$2420 = \$26{,}620$$

We see that calculating the future value of a compound interest investment is very tedious if we proceed in this manner. But note that the future value for each year can be found by multiplying the amount from the preceding year by $1 + 0.10$, or 1.10. That is, the future values at the end of the first, second, and third years are, respectively,

$$\$20{,}000(1.10) = \$22{,}000$$
$$[\$20{,}000(1.10)](1.10) = \$20{,}000(1.10)^2 = \$24{,}200 \quad \text{and}$$
$$[\$20{,}000(1.10)^2](1.10) = \$20{,}000(1.10)^3 = \$26{,}620$$

This suggests that if we maintained this investment for n years, the future value at the end of this time would be $\$20{,}000(1.10)^n$. Thus we have the following general formula.

If $\$P$ is invested at an interest rate of r per year, compounded annually, the future value S at the end of the nth year is

$$S = P(1 + r)^n$$

EXAMPLE 1

If \$3000 is invested for 4 years at 9%, compounded annually, how much interest is earned?

Solution

The future value is

$$\begin{aligned} S &= \$3000(1 + 0.09)^4 \\ &= \$3000(1.4115816) \\ &= \$4234.7448 \\ &= \$4234.74, \text{ to the nearest cent} \end{aligned}$$

Because \$3000 of this amount was the original investment, the interest earned is \$4234.74 − \$3000 = \$1234.74.

Some accounts have the interest compounded semiannually, quarterly, monthly, or daily. Unless specifically stated otherwise, a stated interest rate, called the **nominal annual rate,** is the rate per year and is denoted by r. The interest rate *per period,* denoted by i, is the nominal rate divided by the number of interest periods per year. The interest periods are also called *conversion periods,* and the number of periods is denoted by n. Thus, if \$100 is invested for 5 years at 6%, compounded semiannually (twice a year), it has been invested for $n = 10$ periods (5 years × 2 periods per year) at $i = 3\%$ per period (6% per year ÷ 2 periods per year). The future value of investments of this type is found using the following formula.

Future Value (Periodic Compounding)

If \$$P$ is invested for t years at a nominal interest rate r, compounded m times per year, then the total number of compounding periods is

$$n = mt$$

the interest rate per compounding period is

$$i = \frac{r}{m} \quad \text{(expressed as a decimal)}$$

and the future value is

$$S = P(1+i)^n = P\left(1 + \frac{r}{m}\right)^{mt}$$

EXAMPLE 2

For each of the following investments, identify the interest rate per period, i, and the number of compounding periods, n.

(a) 12% compounded monthly for 7 years
(b) 7.2% compounded quarterly for 11 quarters

Solution

(a) If the compounding is monthly and $r = 12\% = 0.12$, then $i = 0.12/12 = 0.01$. The number of compounding periods is $n =$ (7 yr)(12 periods/yr) = 84.
(b) $i = 0.072/4 = 0.018$, $n = 11$ (the number of quarters given)

Once we know i and n, we can calculate the future value from the formula using a calculator.

EXAMPLE 3

If \$8000 is invested for 6 years at 8%, compounded quarterly, find (a) the future value and (b) the compound interest.

Solution

(a) For this situation, $i = 0.08/4 = 0.02$ and $n = (4)(6) = 24$. Thus the future value of the \$8000 is given by

$$S = P(1 + i)^n = \$8000(1 + 0.02)^{24} = \$8000(1.608437) = \$12{,}867.50$$

(b) The compound interest is given by \$12,867.50 − \$8000 = \$4867.50.

We saw previously that compound interest calculations are based on those for simple interest, except that interest payments are added to the principal. In this way, interest is earned on both principal and previous interest payments. Let's examine the effect of this compounding by comparing compound interest and simple interest. If the investment in Example 3 had been at simple interest, the interest earned would have been $Prt = \$8000(0.08)(6) = \$3840$. This is over \$1000 less than the amount of compound interest earned. And this difference would have been magnified over a longer period of time. Try reworking Example 3's investment over 25 years and compare the compound interest earned with the simple interest earned. This comparison begins to shed some light on why Einstein characterized compound interest as "the most powerful force in the Universe."

EXAMPLE 4

What amount must be invested now in order to have \$12,000 after 3 years if money is worth 6% compounded semiannually?

Solution

We need to find the present value P, knowing that the future value is $S =$ \$12,000. Use $i = 0.06/2 = 0.03$ and $n = 3(2) = 6$.

$$\begin{aligned} S &= P(1 + i)^n \\ \$12{,}000 &= P(1 + 0.03)^6 = P(1.03)^6 \\ \$12{,}000 &= P(1.1940523) \\ P &= \frac{\$12{,}000}{1.1940523} = \$10{,}049.81\text{, to the nearest cent} \end{aligned}$$

EXAMPLE 5

Mutual fund advertisements sometimes have graphs similar to the one in Figure 6.1. For the investment in Figure 6.1, \$10,000 would have grown to \$30,118 in 10 years. What interest rate, compounded annually, would be earned?

Solution

We use $P = \$10{,}000$, $S = \$30{,}118$, and $n = 10$ in the formula $S = P(1 + i)^n$, and solve for i.

$$\$30{,}118 = \$10{,}000(1 + i)^{10}$$
$$\frac{30{,}118}{10{,}000} = (1 + i)^{10}$$
$$3.0118 = (1 + i)^{10}$$

At this point we take the tenth root of both sides (or, equivalently, raise both sides to the $1/10 = 0.1$ power).

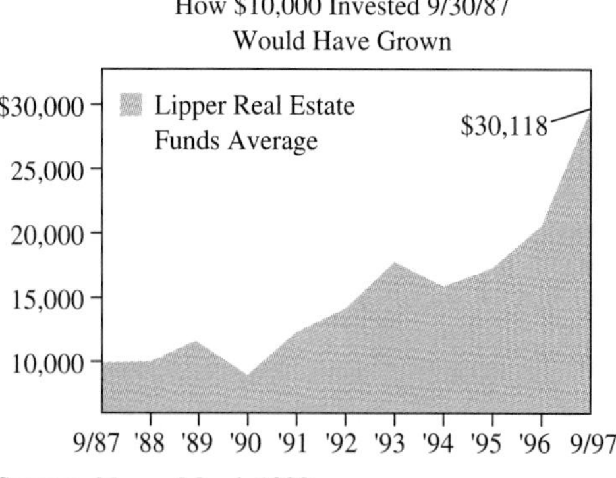

SOURCE: *Money*, March 1998

Figure 6.1

$$3.0118^{0.1} = [(1 + i)^{10}]^{0.01}$$
$$1.11656 = 1 + i$$
$$0.11656 = i$$

Thus this investment would earn about 11.66% compounded annually.

If we invest a sum of money, say \$100, then the higher the interest rate, the greater the future value. As follows, Figure 6.2 shows a graphical comparison of the future value when \$100 is invested at 5%, 8%, and 10%, all compounded annually over a period of 30 years. Note that higher interest rates yield consistently higher future values and have dramatically higher future values after 15–20 years. Note also that these graphs of the future values are growth exponentials.

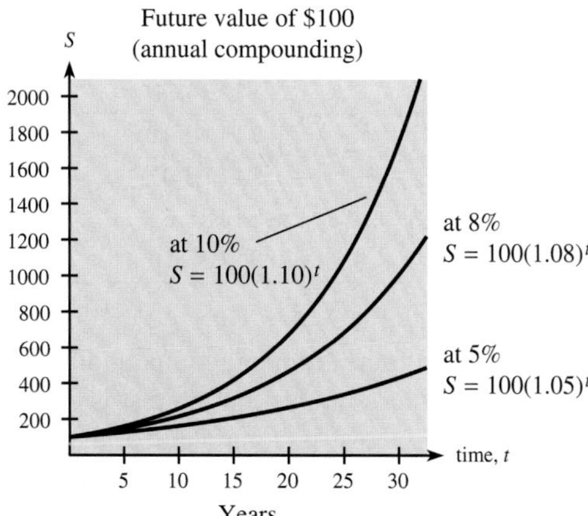

Figure 6.2

Technology Note Spreadsheets and financial calculators can also be used to investigate and analyze compound interest investments, such as tracking the future value of an investment or comparing investments with different interest rates.

Table 6.1 shows a portion of a spreadsheet that tracks the monthly growth of two investments, both compounded monthly and both with $P = \$1000$, but one at 6% (giving $i = 0.005$) and the other at 6.9% (giving $i = 0.00575$).

TABLE 6.1 First 24 Months of Two Investments at Different Interest Rates

	A	B	C
1	Month #	S=1000(1.005)^n	S=1000(1.00575)^n
2	1	$1005.00	$1005.75
3	2	$1010.03	$1011.53
4	3	$1015.08	$1017.35
5	4	$1020.15	$1023.20
6	5	$1025.25	$1029.08
7	6	$1030.38	$1035.00
8	7	$1035.53	$1040.95
9	8	$1040.71	$1046.94
10	9	$1045.91	$1052.96
11	10	$1051.14	$1059.01
12	11	$1056.40	$1065.10
13	12	$1061.68	$1071.22
14	13	$1066.99	$1077.38
15	14	$1072.32	$1083.58
16	15	$1077.68	$1089.81
17	16	$1083.07	$1096.08
18	17	$1088.49	$1102.38
19	18	$1093.93	$1108.72
20	19	$1099.40	$1115.09
21	20	$1104.90	$1121.50
22	21	$1110.42	$1127.95
23	22	$1115.97	$1134.44
24	23	$1121.55	$1140.96
25	24	$1127.16	$1147.52

Because more frequent compounding means that interest is paid more often (and hence more interest on interest is earned), it would seem that the more frequently the interest is compounded, the larger the future value will become. In order to determine the interest that results from *continuous* compounding (compounding every instant), consider an investment of \$1 for 1 year at a 100% interest rate. If the interest is compounded m times per year, the future value is given by

$$S = \left(1 + \frac{1}{m}\right)^m$$

Table 6.2 shows the future value that results as the number of compounding periods increases.

TABLE 6.2

Compounded	*Number of Periods per Year*	*Future Value*
Annually	1	$\left(1+\frac{1}{1}\right)^1 = 2$
Monthly	12	$\left(1+\frac{1}{12}\right)^{12} = 2.1630\ldots$
Daily	360 (business year)	$\left(1+\frac{1}{360}\right)^{360} = 2.7145\ldots$
Hourly	8640	$\left(1+\frac{1}{8640}\right)^{8640} = 2.71812\ldots$
Each minute	518,400	$\left(1+\frac{1}{518{,}400}\right)^{518{,}400} = 2.71827\ldots$

Table 6.2 shows that as the number of periods per year increases, the future value increases, although not very rapidly. In fact, no matter how often the interest is compounded, the future value will never exceed \$2.72. We say that as the number of periods increases, the future value approaches a **limit,** which is the number e:

$$e = 2.7182818\ldots.$$

We discussed the number e and the function $y = e^x$ in Chapter 5 "Exponential and Logarithmic Functions." The discussion here shows one way the number we call e may be derived. We will define e more formally later.

Future Value (Continuous Compounding) In general, if \$$P$ is invested for t years at a nominal rate r, compounded continuously, then the future value is given by the exponential function

$$S = Pe^{rt}$$

EXAMPLE 6

(a) Find the future value if \$1000 is invested for 20 years at 8%, compounded continuously.

(b) What amount must be invested at 6.5%, compounded continuously, so that it will be worth \$25,000 after 8 years?

Solution

(a) The future value is

$$\begin{aligned} S &= \$1000e^{(0.08)(20)} = \$1000e^{1.6} \\ &= \$1000(4.95303) \quad (\text{since } e^{1.6} = 4.95303) \\ &= \$4953.03 \end{aligned}$$

(b) Solve for the present value P in $\$25{,}000 = Pe^{(0.065)(8)}$.

$$\$25{,}000 = Pe^{(0.065)(8)} = Pe^{0.052}$$

$$\$25{,}000 = P(1.68202765)$$

$$\frac{\$25{,}000}{1.68202765} = P \quad \text{so} \quad P = \$14{,}863.01, \text{ to the nearest cent}$$

EXAMPLE 7

How much more will you earn if you invest $1000 for 5 years at 8% compounded continuously instead of at 8% compounded quarterly?

Solution

If the interest is compounded continuously, the future value at the end of the 5 years is

$$S = \$1000e^{(0.08)(5)} = \$1000e^{0.4} = \$1000(1.49182)$$
$$= \$1491.82$$

If the interest is compounded quarterly, the future value at the end of the 5 years is

$$S = \$1000(1.02)^{20} = \$1000(1.485947)$$
$$= \$1485.95, \text{ to the nearest cent}$$

Thus the extra interest earned by compounding continuously is

$$\$1491.82 - \$1485.95 = \$5.87$$

As Example 7 shows, when we invest money at a given compound interest rate, the method of compounding affects the amount of interest we earn. As a result, a rate of 8% can earn more than 8% interest if compounding is more frequent than annually.

For example, suppose $1 is invested for 1 year at 8%, compounded semiannually. Then $i = 0.08/2 = 0.04$, $n = 2$, the future value is $S = \$1(1.04)^2 = \$1.0816$, and the interest earned for the year is $\$1.0816 - \$1 = \$0.0816$. Note that this amount of interest represents an annual percentage yield of 8.16%, so we say that 8%, compounded semiannually, has an **annual percentage yield (APY),** or **effective annual rate,** of 8.16%. Similarly, if $1 is invested at 8% compounded continuously, then the interest earned is $\$1(e^{0.08}) - \$1 = \$0.0833$, for an APY of 8.33%.

Banks acknowledge this difference between stated nominal interest rates and annual percentage yields by posting both rates for their investments. Note that the annual percentage yield is equivalent to the stated rate when compounding is annual. In general, the annual percentage yield equals I/P, or just I if $P = \$1$. Hence we can calculate the APY with the following formula.

Annual Percentage Yield (APY)

Let r represent the annual (nominal) interest rate for an investment. Then the **annual percentage yield (APY)*** is found as follows.

Periodic Compounding. If m is the number of compounding periods per year, then $i = r/m$ is the interest rate per period, and

$$\text{APY} = \left(1 + \frac{r}{m}\right)^m - 1 = (1 + i)^m - 1 \quad \text{(as a decimal)}$$

Continuous Compounding

$$\text{APY} = e^r - 1 \quad \text{(as a decimal)}$$

*Note that the annual percentage yield is also called the **effective annual rate.**

Thus, although we cannot directly compare two nominal rates with different compounding periods, we can compare their corresponding APYs.

EXAMPLE 8

Find the annual percentage yield (or effective annual rate) for the nominal rates of 6% (a) compounded quarterly and (b) compounded continuously.

Solution

(a) Because the number of periods per year is $m = 4$ and the nominal rate is $r = 0.06$, the rate per period is $i = r/m = 0.06/4 = 0.015$. Thus,

$$\begin{aligned} \text{APY} = (1 + 0.015)^4 - 1 &= 1.061364 - 1 \\ &= 0.061364 \\ &= 6.14\% \quad \text{(to 2 decimal places)} \end{aligned}$$

(b) For continuous compounding, we have

$$\text{APY} = e^{0.06} - 1 = 0.06183655, \text{ or } 6.18\% \quad \text{(to 2 decimal places)}$$

EXAMPLE 9

How long does it take an investment of \$10,000 to double if it is invested at

(a) 8%, compounded annually?
(b) 8%, compounded continuously?

Solution

(a) We solve for n in $\$20{,}000 = \$10{,}000(1 + 0.08)^n$.

$$\begin{aligned} \$20{,}000 &= \$10{,}000(1 + 0.08)^n \\ 2 &= 1.08^n \end{aligned}$$

Taking the logarithm, base e, of both sides of the equation gives

$$\begin{aligned} \ln 2 &= \ln 1.08^n \\ \ln 2 &= n \ln 1.08 \qquad \text{[Logarithm Property IV]} \\ \frac{\ln 2}{\ln 1.08} &= n \\ n &\approx 9.0 \text{ (years)} \end{aligned}$$

(b) Solve for t in $\$20{,}000 = \$10{,}000e^{0.08t}$.

$$\begin{aligned} \$20{,}000 &= \$10{,}000e^{0.08t} \\ 2 &= e^{0.08t} \\ \ln 2 &= \ln e^{0.08t} \\ \ln 2 &= 0.08t \qquad \text{[Logarithm Property I]} \\ \frac{\ln 2}{0.08} &= t \\ t &\approx 8.7 \text{ (years)} \end{aligned}$$

Our examples and discussion have shown that investments with continuous compounding earn more interest if the time and rates are the same. Also, our development of the formula for future value of an investment with continuous compounding indicates that when the number of compounding periods is very large, there is little difference between periodic and continuous compounding. To understand better the effect of compounding periods on an investment, consider the following example.

EXAMPLE 10

If \$100 is invested at 10%, then the function

$$f(t) = 100\left(1 + \frac{0.10}{m}\right)^{mt}$$

gives the future value of the investment for t years, if the investment is compounded m times per year.

(a) Graph $f(t)$ for $m = 1$ (annual compounding), $m = 4$ (quarterly compounding), and $m = 365$ (daily compounding).
(b) Which graph lies above the others? What does this mean?

Solution

(a) The graphs are shown in Figure 6.3.
(b) The graph for $m = 365$ (daily compounding) lies above the others. This indicates that more compounding periods give a greater future value for a given interest rate. In addition, if $f(t)$ with $m = 365$ is graphed with $g(t) = 100e^{0.10t}$ (the function for continuous compounding), it is difficult to see any difference between the graphs. However, the graph for continuous compounding does lie above the one for daily compounding. This suggests, and it is indeed a fact, that at a given interest rate, more frequent compounding earns more interest (with continuous compounding earning the most interest).

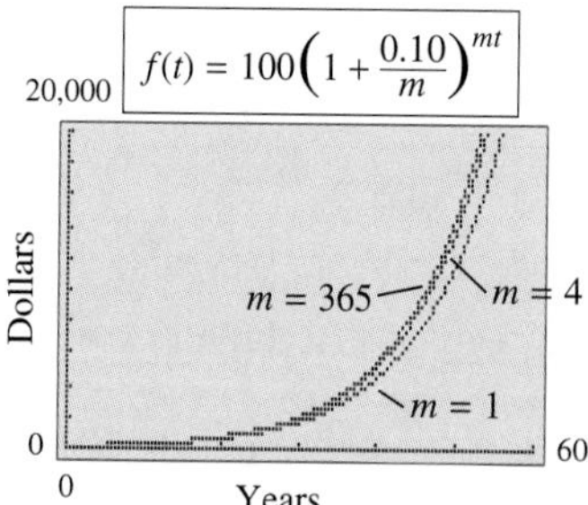

Figure 6.3

Technology Note We can also use a spreadsheet to compare different compounding schemes. Table 6.3 shows a spreadsheet that tracks the yearly growth of the Example 10 investment for quarterly compounding and continuous compounding.

TABLE 6.3 First 20 Years of Two Investments with Different Compounding Schemes

	A	B	C
1	Year	S=100(1.025)^(4n)	S=100*exp(.10n)
2	1	$110.38	$110.52
3	2	$121.84	$122.14
4	3	$134.49	$134.99
5	4	$148.45	$149.18
6	5	$163.86	$164.87
7	6	$180.87	$182.21
8	7	$199.65	$201.38
9	8	$220.38	$222.55
10	9	$243.25	$245.96
11	10	$268.51	$271.83
12	11	$296.38	$300.42
13	12	$327.15	$332.01
14	13	$361.11	$366.93
15	14	$398.60	$405.52
16	15	$439.98	$448.17
17	16	$485.65	$495.30
18	17	$536.07	$547.39
19	18	$591.72	$604.96
20	19	$653.15	$668.59
21	20	$720.96	$738.91

CHECKPOINT

1. Which future value formula below is used for interest that is compounded periodically, and which is used for interest that is compounded continuously?
   (a) $S = P(1 + i)^n$ (b) $S = Pe^{rt}$
2. If $5000 is invested at 6%, compounded quarterly, for 5 years, find
   (a) the number of compounding periods per year, m
   (b) the number of compounding periods for the investment, n
   (c) the interest rate for each compounding period, i
   (d) the future value of the investment
   (e) the effective annual rate
3. Find the present value of an investment that is worth $12,000 after 5 years at 9% compounded monthly.
4. If $5000 is invested at 6% compounded continuously for 5 years, find the future value of the investment.

Geometric Sequences

If \$P is invested at an interest rate of i per year, compounded at the end of each year, the future value at the end of each succeeding year is

$$P(1+i),\ P(1+i)^2,\ P(1+i)^3,\ \ldots,\ P(1+i)^n,\ \ldots$$

The future values for each of the succeeding years form a sequence in which each term (after the first) is found by multiplying the previous term by the same number. Such a sequence is called a **geometric sequence.**

A sequence is called a **geometric sequence** (progression) if there exists a number r, called the **common ratio,** such that

$$a_n = ra_{n-1} \quad \text{for } n > 1$$

Geometric sequences form the foundation for other applications involving compound interest.

EXAMPLE 11

Write the next three terms of the following geometric sequences:

(a) $1, 3, 9, \ldots$ (b) $4, 2, 1, \ldots$ (c) $3, -6, 12, \ldots$

Solution

(a) The common ratio is 3, so the next three terms are 27, 81, 243.
(b) The common ratio is $\frac{1}{2}$, so the next three terms are $\frac{1}{2}, \frac{1}{4}, \frac{1}{8}$.
(c) The common ratio is -2, so the next three terms are $-24, 48, -96$.

Because each term after the first in a geometric sequence is obtained by multiplying the previous term by r, the second term is a_1r, the third is $a_1r^2, \ldots$ and the nth term is a_1r^{n-1}. Thus we have the following formula.

The ***n*th term of a geometric sequence** (progression) is given by

$$a_n = a_1r^{n-1}$$

where a_1 is the first term of the sequence and r is the common ratio.

EXAMPLE 12

Find the seventh term of the geometric sequence with first term 5 and common ratio -2.

Solution

The seventh term is $a_7 = 5(-2)^{7-1} = 5(64) = 320$.

EXAMPLE 13

A ball is dropped from a height of 125 feet. If it rebounds $\frac{3}{5}$ of the height from which it falls every time it hits the ground, how high will it bounce after it strikes the ground for the fifth time?

Solution

The first rebound is $\frac{3}{5}(125) = 75$ feet; the second rebound is $\frac{3}{5}(75) = 45$ feet. The heights of the rebounds form a geometric sequence with first term 75 and common ratio $\frac{3}{5}$. Thus the fifth term is

$$a_5 = 75\left(\frac{3}{5}\right)^4 = 75\left(\frac{81}{625}\right) = \frac{243}{25} = 9\tfrac{18}{25} \text{ feet}$$

Next we develop a formula for the sum of a geometric sequence, a formula that is important in our study of annuities. The sum of the first n terms of a geometric sequence is

$$s_n = a_1 + a_1r + a_1r^2 + \cdots + a_1r^{n-1} \qquad (1)$$

If we multiply equation (1) by r, we have

$$rs_n = a_1r + a_1r^2 + a_1r^3 + \cdots + a_1r^n \qquad (2)$$

Subtracting equation (2) from equation (1), we obtain

$$s_n - rs_n = a_1 + (a_1r - a_1r) + (a_1r^2 - a_1r^2) + \ldots + (a_1r^{n-1} - a_1r^{n-1}) - a_1r^n$$

Thus

$$s_n(1 - r) = a_1 - a_1r^n$$

$$s_n = \frac{a_1 - a_1r^n}{1 - r} \qquad \text{if } r \neq 1$$

This gives the following.

The **sum of the first n terms of the geometric sequence** with first term a_1 and common ratio r is

$$s_n = \frac{a_1(1 - r^n)}{1 - r} \qquad \text{provided that } r \neq 1$$

EXAMPLE 14

Find the sum of the first five terms of the geometric progression with first term 4 and common ratio -3.

Solution

We are given that $n = 5$, $a_1 = 4$, and $r = -3$. Thus

$$s_5 = \frac{4[1 - (-3)^5]}{1 - (-3)} = \frac{4[1 - (-243)]}{4} = 244$$

EXAMPLE 15

Find the sum of the first six terms of the geometric sequence $\frac{1}{4}, \frac{1}{8}, \frac{1}{16}, \ldots$.

Solution

We know that $n = 6$, $a_1 = \frac{1}{4}$, and $r = \frac{1}{2}$. Thus

$$s_6 = \frac{\frac{1}{4}[1 - (\frac{1}{2})^6]}{1 - \frac{1}{2}} = \frac{\frac{1}{4}(1 - \frac{1}{64})}{\frac{1}{2}} = \frac{1 - \frac{1}{64}}{2} = \frac{64 - 1}{128} = \frac{63}{128}$$

CHECKPOINT

5. Identify any geometric sequences among the following.
   (a) $1, 4, 9, 16, \ldots$ (b) $1, 4, 7, 10, \ldots$ (c) $1, 4, 16, 64, \ldots$
6. (a) Find the 40th term of the geometric sequence $8, 6, \frac{9}{2}, \ldots$.
   (b) Find the sum of the first 20 terms of the geometric sequence $2, 6, 18, \ldots$.

CHECKPOINT SOLUTIONS

1. (a) Periodic compounding (b) Continuous compounding
2. (a) $m = 4$ (b) $n = m \cdot t = 4 \cdot 5 = 20$ (c) $i = \frac{r}{m} = \frac{0.06}{4} = 0.015$
   (d) $S = P(1 + i)^n = \$5000(1 + 0.015)^{20} \approx \$6734.28$, to the nearest cent
   (e) $r = (1 + 0.015)^4 - 1 \approx 0.0614$, or 6.14%
3. $i = 0.09/12 = 0.0075$ and $n = (5)(12) = 60$

$$\$12,000 = P(1 + 0.0075)^{60} = P(1.5656810270)$$
$$P = \$12,000/1.565681027 = \$7664.40, \text{ to the nearest cent}$$

4. $S = Pe^{rt} = \$5000e^{(0.06)(5)} = \$5000e^{0.3} \approx \$6749.29$, to the nearest cent
5. Only sequence (c) is geometric with common ratio $r = 4$.
6. (a) $a_{40} = a_1r^{40-1} = 8\left(\frac{3}{4}\right)^{39} \approx 0.0001$

   (b) $s_{20} = \frac{a_1(1 - r^{20})}{1 - r} = \frac{2(1 - 3^{20})}{1 - 3} = -(1 - 3^{20})$
   $= 3^{20} - 1 = 3,486,784,400$

EXERCISE 6.2

Compound Interest

For each investment situation in Problems 1–6, identify the (a) annual interest rate, (b) length of the investment in years, (c) periodic interest rate, and (d) number of periods of the investment.

1. 6% compounded semiannually for 6 years
2. 9% compounded semiannually for 10 years
3. 8% compounded quarterly for 7 years
4. 12% compounded monthly for 3 years
5. 9% compounded monthly for 5 years
6. 10% compounded quarterly for 8 years
7. Find the future value if \$8000 is invested for 10 years at 12%, compounded annually.
8. Find the interest that will be earned if \$5000 is invested for 3 years at 10%, compounded annually.
9. Find the interest that will be earned if \$10,000 is invested for 3 years at 9%, compounded monthly.
10. What is the future value if \$8600 is invested for 8 years at 10%, compounded semiannually?
11. What is the future value if \$3200 is invested for 5 years at 8%, compounded quarterly?
12. What interest will be earned if \$6300 is invested for 3 years at 12%, compounded monthly?

13. Find the interest that will be earned if \$8600 is invested for 6 years at 10%, compounded semiannually.
14. Find the future value if \$3500 is invested for 6 years at 8%, compounded quarterly.
15. What lump sum do parents need to deposit in an account earning 10%, compounded monthly, so that it will grow to \$40,000 for their son's college tuition in 18 years?
16. What lump sum should be deposited in an account that will earn 9%, compounded quarterly, to grow to \$100,000 for retirement in 25 years?
17. What present value amounts to \$10,000 if it was invested for 10 years at 6%, compounded annually?
18. What present value amounts to \$300,000 if it was invested at 7%, compounded semiannually, for 15 years?
19. Find the future value if \$5100 is invested for 4 years at 9%, compounded continuously.
20. Find the interest that will result if \$8000 is invested at 7%, compounded continuously, for 8 years.
21. What is the compound interest if \$410 is invested for 10 years at 8%, compounded continuously?
22. If \$8000 is invested at 8.5%, compounded continuously, find the future value after $4\frac{1}{2}$ years.
23. Grandparents want to make a gift of \$100,000 for their grandchild's 20th birthday. How much would have to be invested on the day of their grandchild's birth if their investment could earn:
    (a) 10.5% compounded continuously?
    (b) 11% compounded continuously?
    (c) Describe the effect that this slight change in the interest rate makes over the 20 years of this investment.
24. Suppose an individual wants to have \$200,000 available for her child's education. Find the amount that would have to be invested at 12%, compounded continuously, if the number of years until college is:
    (a) 7 years
    (b) 14 years
    (c) Does leaving the money invested twice as long mean that only half as much is needed initially? Explain why or why not.
25. Which investment will earn more money, a \$1000 investment for 5 years at 8%, compounded annually, or a \$1000 investment for 5 years, compounded continuously at 7%?
26. How much more interest will be earned if \$5000 is invested for 6 years at 7%, compounded continuously, instead of at 7%, compounded quarterly?
27. What is the annual percentage yield (or effective annual rate) for a nominal rate of 8.4%, compounded quarterly?
28. If money is invested at 9%, compounded monthly, what is the annual percentage yield (that is, the effective annual rate)?
29. Find the annual percentage yield for an investment at 7.3%, compounded monthly.
30. If money is invested at 6.6%, compounded semiannually, find the annual percentage yield.
31. If money is invested at 6%, compounded continuously, what is the annual percentage yield?
32. What is the annual percentage yield for an investment at 10%, compounded continuously?

In Problems 33 and 34, rank each interest rate and compounding scheme in order from highest yield to lowest yield.

33. 8% compounded quarterly, 8% compounded monthly, 8% compounded annually
34. 6% compounded continuously, 6% compounded semiannually, 6% compounded monthly
35. The figure below shows a graph of the future value of \$100 at 8% compounded annually, along with the graph of \$100 at 8% compounded continuously. Which is which? Explain.

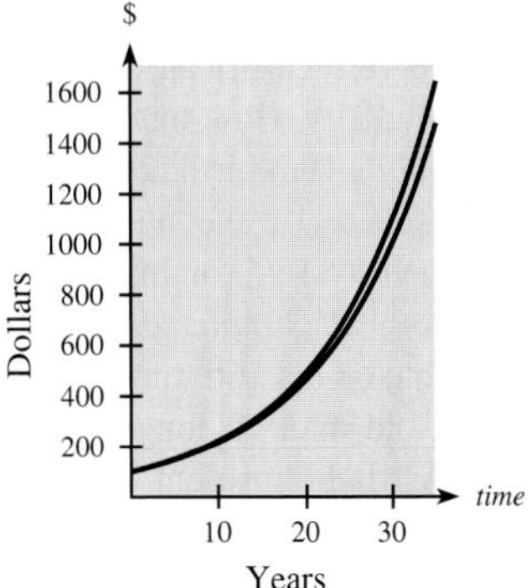

36. The figure below shows a graph of the future value of \$350 at 6% compounded monthly, along with the graph of \$350 at 6% compounded annually. Which is which? Explain.

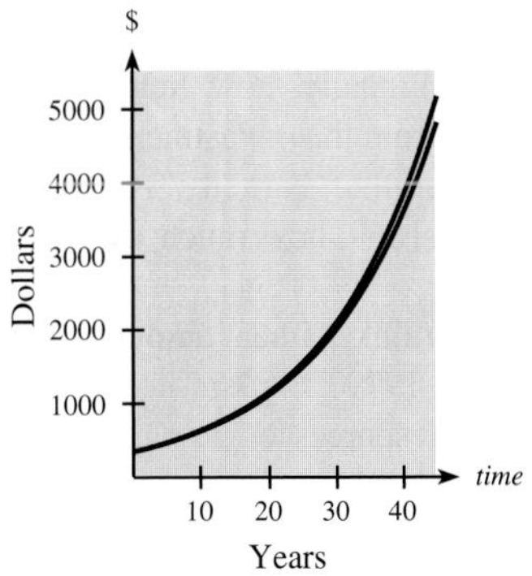

37. A shareholder who purchased 100 shares of Federal Signal Corporation's stock for \$1438 at the beginning of 1979 and reinvested the dividends would have had an investment worth \$50,620 at the end of 1993 (Source: Federal Signal Corporation, *1993 Annual Report*). What interest rate, compounded annually, did this investment earn?
38. An investment of \$10,000 in Templeton World Fund at the beginning of 1978 would have been worth \$93,575 in mid-1993. (Source: advertisement in *Newsweek,* November 8, 1993). What interest rate, compounded annually, did this investment earn?
39. How long (in years) would \$700 have to be invested at 11.9%, compounded continuously, to earn \$300 interest?
40. How long (in years) would \$600 have to be invested at 8%, compounded continuously, to amount to \$970?
41. At what nominal rate, compounded quarterly, would \$20,000 have to be invested to amount to \$26,425.82 in 7 years?
42. At what nominal rate, compounded annually, would \$10,000 have to be invested to amount to \$14,071 in 7 years?
43. For her first birthday, Polly's grandparents invested \$1000 in an 18-year certificate for her that pays 8% compounded annually. How much will the certificate be worth on Polly's 19th birthday?
44. To help their son buy a car on his 16th birthday, a boy's parents invest \$1500 on his 10th birthday. If the investment pays 9%, compounded continuously, how much is available on his 16th birthday?
45. (a) A 40-year-old man no longer qualifies for additional IRAs. If his present IRAs have a balance of \$12,860 and if he expects them to earn interest at 7.5%, compounded annually, how much does he expect to have when he retires at age 62?
    (b) How much more money would the man have if his investments earned 8.5%, compounded annually?
46. (a) The purchase of Alaska cost the United States \$7 million in 1869. If this money had been placed in a savings account paying 6%, compounded annually, how much money would be available from this investment in 1999?
    (b) If the \$7 million earned 7%, compounded annually since 1869, how much would be available in 1999?
    (c) Do you think either amount would purchase Alaska in 1999? Explain in light of the value of Alaska's resources or perhaps the price per acre of land.
47. A couple needs \$15,000 as a down payment for a home. If they invest the \$10,000 they have at 8%, compounded quarterly, how long will it take for the money to grow into \$15,000?
48. How long does it take for an account containing \$8000 to be worth \$15,000 if the money is invested at 9%, compounded monthly?
49. Mary Stahley invested \$2500 in a 36-month certificate of deposit (CD) that earned 8.5% annual simple interest. When the CD matured, she invested the full amount in a mutual fund that had an annual growth equivalent to 18%, compounded annually. How much was the mutual fund worth after 9 years?
50. Suppose Patrick Goldsmith deposited \$1000 in an account that earned simple interest at an annual rate of 7% and left it there for 4 years. At the end of the 4 years, Patrick deposited the entire amount from that account into a new account that earned 7% compounded quarterly. He left the money in this account for 6 years. How much did he have after the 10 years?

Problems 51 and 52 use a spreadsheet or financial program on a calculator or computer.

51. Track the future value of two investments of \$5000, one at 6.3% compounded quarterly and another at 6.3% compounded monthly for each interest payment period for 10 years.
    (a) How long does it take each investment to be worth more than \$7500?
    (b) What is the value of each investment after 3 years, 7 years, and 10 years?
52. Track the future value of two investments of \$1000, one at 6.0% compounded semiannually and one at 6.6% compounded semiannually for each interest payment period for 25 years.
    (a) How long before the difference between these investments is \$50?
    (b) How much sooner does the 6.6% investment reach \$1500?

Geometric Sequences

For each geometric sequence given in Problems 53 and 54, write the next three terms.

53. (a) $3, 6, 12, \ldots$ (b) $81, 54, 36, \ldots$
54. (a) $4, 12, 36, \ldots$ (b) $32, 40, 50, \ldots$
55. Find the 13th term of the geometric sequence with first term 10 and common ratio 2.
56. Find the 11th term of the geometric sequence with first term 6 and common ratio 3.
57. Find the 16th term of the geometric sequence with first term 4 and common ratio $\frac{3}{2}$.

58. Find the 20th term of the geometric sequence with first term 3 and common ratio -2.
59. Find the sum of the first 17 terms of the geometric sequence with first term 6 and common ratio 3.
60. Find the sum of the first 14 terms of the geometric sequence with first term 3 and common ratio 4.
61. Find the sum of the first 21 terms of the geometric sequence with first term 4 and common ratio $-\frac{1}{2}$.
62. Find the sum of the first 25 terms of the geometric sequence with first term 9 and common ratio $\frac{1}{3}$.
63. Find the sum of the first 35 terms of the geometric sequence 1, 3, 9, ...
64. Find the sum of the first 14 terms of the geometric sequence 16, 64, 256, ...
65. Find the sum of the first 18 terms of the geometric sequence 6, 4, $\frac{8}{3}$, ...
66. Find the sum of the first 31 terms of the geometric sequence 9, -6, 4, ...

Applications of Sequences

67. ***Inflation*** A house that 20 years ago was worth \$160,000 has increased in value by 4% each year because of inflation. What is its worth today?

68. ***Inflation*** If inflation causes the cost of automobiles to increase by 10% each year, what should a car that 6 years ago cost \$5000 cost today?

69. ***Population growth*** Suppose a country has a population of 20 million and projects a growth rate of 2% per year for the next 20 years. What will the population of this country be in 10 years?

70. ***Spread of AIDS*** Suppose a country is so devastated by the AIDS epidemic that its population decreases by 0.5% each year for a 4-year period. If the population was originally 10 million, what is the population at the end of the 4-year period?

71. ***Population growth*** If the rate of growth of a population continues at 2%, in how many years will the population double?

72. ***Population*** If a population of 8 million begins to increase at a rate of 0.1% each month, in how many months will it be 10 million?

73. ***Ball rebounding*** A ball is dropped from a height of 128 feet. If it rebounds $\frac{3}{4}$ of the height from which it falls every time it hits the ground, how high will it bounce after it strikes the ground for the fourth time?

74. ***Water pumping*** A pump removes $\frac{1}{3}$ of the water in a container with every stroke. What amount of water is still in a container after 5 strokes if it originally contained 81 cm^3?

75. ***Depreciation*** A machine is valued at \$10,000. If the depreciation at the end of each year is 20% of its value at the beginning of the year, find its value at the end of 4 years.

76. ***Profit*** Suppose a new business makes a \$1000 profit in its first month and has its profit increase by 10% each month for the next 2 years. How much profit will the business earn in its 12th month?

77. ***Bacterial growth*** The size of a certain bacteria culture doubles each hour. If the number of bacteria present initially is 5000, how many would be present at the end of 6 hours?

78. ***Bacterial growth*** If a bacteria culture increases by 20% every hour and 2000 are present initially, how many will be present at the end of 10 hours?

79. ***Profit*** If changing market conditions cause a company earning \$8,000,000 in 1992 to project a loss of 2% of its profit in each of the next 5 years, what profit does it project in 1997?

80. ***Profit*** Suppose a new business makes a \$1000 profit in its first month and has its profit increase by 10% each month for the next 2 years. How much profit will it earn in its first year?

81. ***Loans*** An interest-free loan of \$12,000 requires monthly payments of 10% of the outstanding balance. What is the outstanding balance after 18 payments?

82. ***Loans*** If Sherri must repay a \$9000 interest-free loan by making monthly payments of 15% of the unpaid balance, what is the unpaid balance after 1 year?

83. ***Chain letters*** Suppose you receive a chain letter with six names on it, and to keep the chain unbroken, you are to mail a dime to the person whose name is at the top, cross out the top name, add your name to the bottom, and mail it to five friends. If your friends mail out five letters each, and no one breaks the chain, you will eventually receive dimes. How many dimes would you receive? (This is a geometric sequence with first term 5.)

84. ***Chain letters*** Mailing chain letters that involve sending money has been declared illegal because most people would receive nothing while a comparative few would profit. Suppose the chain letter in Problem 83 were to go through 12 unbroken progressions.
    (a) How many people would receive money?
    (b) How much money would these people receive as a group?

85. ***Chain letters*** How many letters would be mailed if the chain letter in Problem 83 went through 12 unbroken progressions?

6.3 Future Value of Annuities

OBJECTIVES

- *To compute the future value of ordinary annuities and annuities due*
- *To compute the payment required in order for ordinary annuities and annuities due to have a specified future value*

APPLICATION PREVIEW

A pair of twins graduate from college together and start their careers. Twin 1 invests \$2000 at the end of each of 8 years in an account that earns 10%, compounded annually. After the initial 8 years, no additional contributions are made, but the investment continues to earn 10%, compounded annually. Twin 2 invests no money for 8 years but then contributes \$2000 at the end of each year for a period of 36 years (to age 65) to an account that pays 10%, compounded annually. How much money does each twin have at age 65?

Each twin's contributions form an **annuity.** An annuity is a financial plan characterized by regular payments. We can view an annuity as a savings plan in which the regular payments are contributions to the account, and then we can ask what the total value of the account will become (as in the Application Preview). Also, we can view an annuity as a payment plan (such as for retirement) in which regular payments are made from an account, often to an individual.

In this section we consider ordinary annuities and annuities due, and we develop a formula for each type of annuity that allows us to find its **future value**—that is, a formula for the value of the account after regular deposits have been made over a period of time.

Ordinary Annuities

In the previous section, we learned how to compute the future value and amount of interest if a fixed sum of money was deposited in an account that pays interest that is compounded periodically or continuously. But not many people are in a position to deposit a large sum of money at one time in an account. Most people save (or invest) money by depositing relatively small amounts at different times. If a depositor makes equal deposits at regular intervals, he or she is contributing to an **annuity.** The payments (deposits) may be made weekly, monthly, quarterly, yearly, or at any other interval of time. The sum of all payments plus all interest earned is called the **future amount of the annuity** or its **future value.**

Annuities may be classified into two categories—annuities certain and contingent annuities. An **annuity certain** is one in which the payments begin and end on fixed dates. In a **contingent annuity** the payments are related to events that cannot be paced regularly, so the payments are not regular. We will deal with annuities certain in this text, and we will deal first with annuities in which the payments are made at the end of each of the equal payment intervals. This type of annuity certain is called an **ordinary annuity** (and also an **annuity immediate**). The ordinary annuities we will consider have payment intervals that coincide with the compounding period of the interest.

Suppose you invested \$100 at the end of each year for 5 years in an account that paid interest at 10%, compounded annually. How much money would you have in the account at the end of the 5 years?

Because you are making payments at the end of each period (year), this annuity is called an ordinary annuity.

To find the future value of your annuity at the end of the 5 years, we compute the future value of each payment separately and add the amounts (see Figure 6.4). The \$100 invested *at the end* of the first year will draw interest for 4 years, so it will amount to $\$100(1.10)^4$. The \$100 invested at the end of the second year will draw interest for 3 years, so it will amount to $\$100(1.10)^3$. Similarly, the \$100 invested at the end of the third year will amount to $\$100(1.10)^2$, and the \$100 invested at the end of the fourth year will amount to \$100(1.10). The \$100 invested at the end of the fifth year will draw no interest, so it will amount to \$100.

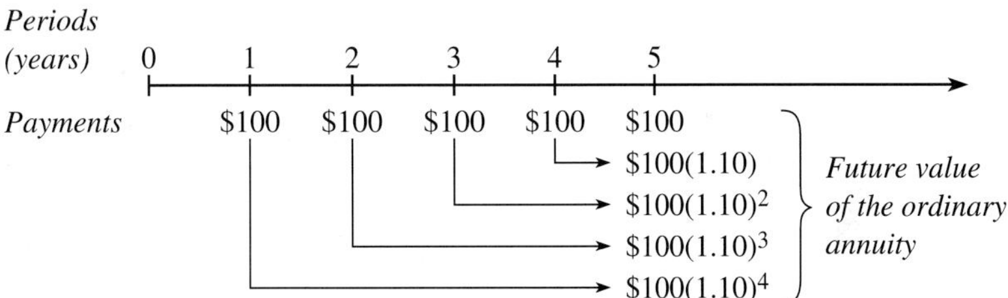

Figure 6.4

Thus the future value of the annuity is given by

$$S = 100 + 100(1.10) + 100(1.10)^2 + 100(1.10)^3 + 100(1.10)^4$$

The terms of this sum are the first five values of the geometric sequence having $a_1 = 100$ and $r = 1.10$. Thus

$$\begin{aligned} S &= \frac{100[1 - (1.10)^5]}{1 - 1.10} = \frac{100(-0.61051)}{-0.10} \\ &= 610.51 \end{aligned}$$

Thus the investment (\$100 at the end of each year for 5 years, at 10%, compounded annually) would return \$610.51.

Because every such annuity will take the same form, we can state that if a periodic payment R is made for n periods at an interest rate i *per period,* the **future amount of the annuity,** or its **future value,** will be given by

$$S = R \cdot \frac{1-(1+i)^n}{1-(1+i)}$$

This simplifies to the following.

Future Value of an Ordinary Annuity

If \$$R$ is deposited at the end of each period for n periods in an annuity that earns interest at a rate of i per period, the future value of the annuity will be

$$S = R \cdot s_{\overline{n}|i} = R \cdot \left[\frac{(1+i)^n - 1}{i}\right]$$

where $s_{\overline{n}|i}$ is read "*s, n* angle *i*" and represents the future value of an ordinary annuity of \$1 per period for n periods with an interest rate of i per period.

EXAMPLE 1

Richard Lloyd deposits \$100 at the end of each month in an account that pays 12%, compounded monthly. How much money will he have in his account in a year and a half?

Solution

The number of periods is $n = (12)(1.5) = 18$, and the rate *per period* is $i = 0.12/12 = 0.01$. At the end of 18 months the future value of the annuity will be

$$S = \$100 \cdot s_{\overline{18}|0.01} = \$100\left[\frac{(1+0.01)^{18} - 1}{0.01}\right]$$

$$= \$100(19.614748)$$

$$= \$1961.47$$

In the Application Preview, savings plans for twins were described. In the next example, we want to answer partially the question posed in the Application Preview. (See Example 5 for the rest of the "twin" problem.)

EXAMPLE 2

Twin 2 of the Application Preview invested \$2000 at the end of each year for 36 years (until age 65) in an account that paid 10% compounded annually. How much does twin 2 have at age 65?

Solution

This savings plan is an ordinary annuity with $i = 0.10$, $n = 36$, and $R = \$2000$. The future value (to the nearest dollar) is

$$S = R\left[\frac{(1+i)^n - 1}{i}\right] = \$2000\left[\frac{(1.10)^{36} - 1}{0.10}\right] = \$598{,}254$$

Figure 6.5 shows a comparison (each graphed as a smooth curve) of the future values of two annuities that are invested at 6% compounded monthly. One annuity has monthly payments of \$100, and one has payments of \$125. Note the impact of a slightly larger contribution on the future value of the annuity after 5 years, 10 years, 15 years, and 20 years. Of course, a slightly higher interest rate also can have a substantial impact over time; try a graphical comparison yourself.

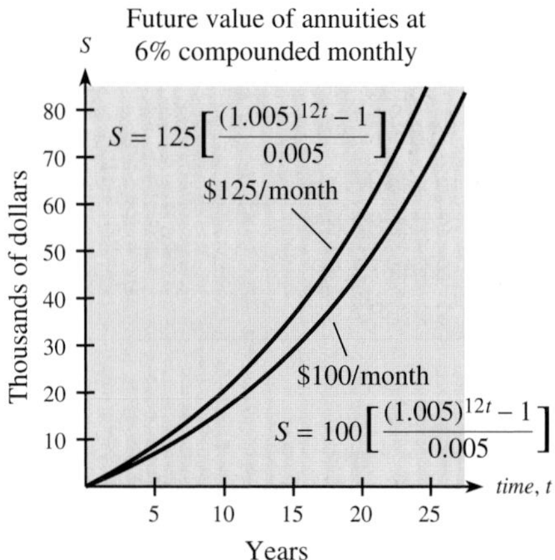

Figure 6.5

Technology Note A spreadsheet also can be used to compare investments. Table 6.4 shows the spreadsheet output for years 1–6 and 20–25 for the investments shown graphically in Figure 6.5. Table 6.5 on the next page shows the future value at the end of each of the first 10 years of an ordinary annuity of \$100 per month at two different interest rates, compounded monthly: 6% and 6.3%.

TABLE 6.4

	A	B	C	D
1		Future Value for R = \$100	Future Value for R = \$125	
2	End of Year #	at 6% compounded monthly	at 6% compounded monthly	
3	1	\$1233.56	\$1541.95	
4	2	\$2543.20	\$3178.99	
5	3	\$3933.61	\$4917.01	
6	4	\$5409.78	\$6762.23	
7	5	\$6977.00	\$8721.25	
8	6	\$8640.89	\$10801.11	
22	20	\$46204.09	\$57755.11	
23	21	\$50287.41	\$62859.27	
24	22	\$54622.59	\$68278.23	
25	23	\$59225.14	\$74031.43	
26	24	\$64111.58	\$80139.47	
27	25	\$69299.40	\$86624.25	

TABLE 6.5

	A	B	C	D
1		Future Value for R = $100	Future Value for R = $100	
2	End of Year #	at 6% compounded monthly	at 6.3% compounded monthly	
3	1	$1233.56	$1235.26	
4	2	$2543.20	$2550.64	
5	3	$3933.61	$3951.31	
6	4	$5409.78	$5442.82	
7	5	$6977.00	$7031.06	
8	6	$8640.89	$8722.30	
9	7	$10407.39	$10523.21	
10	8	$12282.85	$12440.92	
11	9	$14273.99	$14483.00	
12	10	$16387.93	$16657.50	

EXAMPLE 3

A young couple wants to save $15,000 over the next 5 years and then to use this amount as a down payment on a home. To reach this goal, how much money must they deposit at the end of each quarter into an account that earns interest at a rate of 5%, compounded quarterly?

Solution

This plan describes an ordinary annuity with a future value of $15,000 whose payment size, R, is to be determined. Quarterly compounding gives $n = 5(4) = 20$ and $i = 0.05/4 = 0.0125$.

$$S = R\left[\frac{(1+i)^n - 1}{i}\right]$$

$$\$15{,}000 = R\left[\frac{(1+0.0125)^{20} - 1}{0.0125}\right] = R(22.56297854)$$

$$R = \frac{\$15{,}000}{22.56297854} = \$664.81, \quad \text{to the nearest cent}$$

Sinking Funds

Just as the couple in Example 3 was saving for a future purchase, some borrowers, such as municipalities, may have a debt that must be paid in a single large sum on a specified date. If these borrowers make periodic deposits that will produce that sum on a specified date, we say that they have established a **sinking fund.** If the deposits are all the same size and are made regularly, they form an ordinary annuity whose future value (on a specified date) is the amount of the debt. To find the size of these periodic deposits, we solve the equation for the future value of an annuity,

$$S = R\left[\frac{(1+i)^n - 1}{i}\right]$$

for R. Thus we have the following.

Required Payment into an Ordinary Annuity (Sinking Fund Formula)

Suppose periodic payments are deposited at the end of each of n periods into an ordinary annuity (or sinking fund) earning interest at a rate of i per period, such that at the end of n periods, its value is \$$S$. The size of each required payment R is

$$R = \frac{S}{\left[\dfrac{(1+i)^n - 1}{i}\right]} = S\left[\frac{i}{(1+i)^n - 1}\right]$$

EXAMPLE 4

A company establishes a sinking fund to discharge a debt of \$100,000 due in 5 years by making equal semiannual deposits, the first due in 6 months. If the deposits are placed in an account that pays 6%, compounded semiannually, what is the size of the deposits?

Solution

For this sinking fund, we want to find the payment size, R, given that the future value is $S = \$100{,}000$, $n = 2(5) = 10$, and $i = 0.06/2 = 0.03$. Thus we have

$$R = S\left[\frac{i}{(1+i)^n - 1}\right] = \$100{,}000\left[\frac{0.03}{(1+0.03)^{10} - 1}\right]$$

$$= \$100{,}000\left[\frac{0.03}{(1.03)^{10} - 1}\right] = \$8723.10$$

The semiannual deposit is \$8723.10.

CHECKPOINT

1. Suppose that \$500 is deposited at the end of every quarter for 6 years into an account that pays 8%, compounded quarterly.
   (a) What is the total number of payments (periods)?
   (b) What is the interest rate per period?
   (c) What formula is used to find the future value of the annuity?
   (d) Find the future value of the annuity.
2. A sinking fund of \$100,000 is to be established with equal payments at the end of each half-year for 15 years. Find the amount of each payment into the fund if money is worth 10%, compounded semiannually.

Let's now complete the solution to the twin problem posed in the Application Preview and partially solved in Example 2.

EXAMPLE 5

Twin 1 invests \$2000 at the end of each of 8 years in an account that earns 10%, compounded annually. After the initial 8 years, no additional contributions are made, but the investment continues to earn 10%, compounded annually, for 36 more years (until twin 1 is age 65). How much does twin 1 have at age 65?

Solution

We seek the future values of two different investments. The first is an ordinary annuity with $R = \$2000$, $n = 8$ periods, and $i = 0.10$. The second is a compound interest investment with $n = 36$ periods and $i = 0.10$ and whose principal (that is, its present value) is the future value of the ordinary annuity above.

We first find the future value of the ordinary annuity.

$$S = R\left[\frac{(1+i)^n - 1}{i}\right] = \$2000\left[\frac{(1+0.10)^8 - 1}{0.10}\right]$$
$$= \$22{,}871.78, \text{ to the nearest cent}$$

This amount is the principal of the compound interest investment. If no deposits or withdrawals were made for the next 36 years, the future value of this investment would be

$$S = P(1 + i)^n = \$22{,}871.78(1 + 0.10)^{36} = \$707{,}028.03, \text{ to the nearest cent}$$

Thus at age 65, twin 1's investment is worth about \$707,028.

Looking back at Examples 2 and 5, we can extract the following summary and see which twin was the wiser.

	Contributions	*Account Value at Age 65*
Twin 1	\$2000/year for 8 years = \$16,000	\$707,028
Twin 2	\$2000/year for 36 years = \$72,000	\$598,254

Note that twin 1 contributed \$56,000 less than twin 2 but had \$108,774 more at age 65. This illustrates the powerful effect that time has on investments (and on loans, as we'll see in the section "Loans and Amortization").

Annuities Due

Deposits in savings accounts, rent payments, and insurance premiums are examples of **annuities due.** An annuity due differs from an ordinary annuity in that the periodic payments are made at the *beginning* of the period with an annuity due. The *term* of an annuity due is from the first payment to the end of one period after the last payment. Thus an annuity due draws interest for one period more than the ordinary annuity.

We can find the future value of an annuity due by treating each payment as though it were made at the *end* of the preceding period in an ordinary annuity. Thus we calculate $s_{\overline{n}|i}$ for one additional period. But increasing the number of periods also adds one more payment than should be paid. To compensate for this, we subtract the amount of one payment (see Figure 6.6).

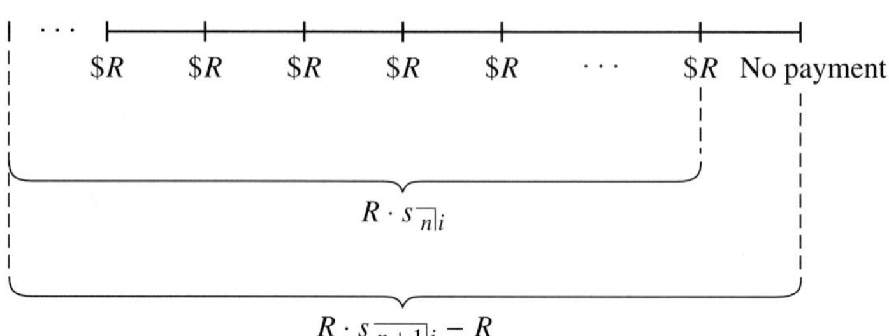

Figure 6.6

Thus the future value of an annuity due with n payments can be found by finding the future value of an ordinary annuity for $n + 1$ periods and subtracting one annuity payment.

That is,

$$\begin{aligned} S_{due} &= Rs_{\overline{n+1}|i} - R = R\left[\frac{(1+i)^{n+1}-1}{i}\right] - R \\ &= R\left[\frac{(1+i)^{n+1}}{i} - 1\right] = R\left[\frac{(1+i)^{n+1}-1}{i} - \frac{i}{i}\right] \\ &= R\left[\frac{(1+i)^{n+1}-(1+i)}{i}\right] = R(1+i)\left[\frac{(1+i)^{n}-1}{i}\right] \end{aligned}$$

Thus the formula for the future value of an annuity due is as follows:

Future Value of an Annuity Due

$$S_{due} = R\left[\frac{(1+i)^n - 1}{i}\right](1+i) = Rs_{\overline{n}|i}(1+i)$$

EXAMPLE 6

Find the future value of an investment if \$150 is deposited at the beginning of each month for 9 years and the interest rate is 7.2%, compounded monthly.

Solution

Because deposits are made at the *beginning* of each month, this is an annuity due with $R = \$150$, $n = 9(12) = 108$, and $i = 0.072/12 = 0.006$.

$$\begin{aligned} S_{due} &= R\left[\frac{(1+i)^n - 1}{i}\right](1+i) \\ &= \$150\left[\frac{(1+0.006)^{108} - 1}{0.006}\right](1+0.006) \\ &= \$150(151.3359308)(1.006) = \$22{,}836.59, \text{ to the nearest cent} \end{aligned}$$

We can also use the formula for the future value of an annuity due to determine the payment size required to reach an investment goal.

EXAMPLE 7

Suppose a company wants to have \$450,000 after $2\frac{1}{2}$ years to modernize its production equipment. How much of each previous quarter's profits should be deposited at the beginning of the current quarter to reach this goal, if the company's investment earns 6.8%, compounded quarterly?

Solution

We seek the payment size, R, for an annuity due with $S_{due} = \$450{,}000$, $n = (2.5)(4) = 10$, and $i = 0.068/4 = 0.017$.

$$S_{due} = R\left[\frac{(1+i)^n - 1}{i}\right](1+i)$$

$$\$450{,}000 = R\left[\frac{(1+0.017)^{10} - 1}{0.017}\right](1+0.017)$$

$$\$450{,}000 = R(10.80073308)(1.017) = (10.98434554)R$$

$$R = \frac{\$450{,}000}{10.98434554} = \$40{,}967.39, \text{ to the nearest cent}$$

Thus, the company needs to deposit about $40,967 at the beginning of each quarter for the next $2\frac{1}{2}$ years to reach its goal.

CHECKPOINT

3. Suppose $100 is deposited at the beginning of each month for 3 years into an account that pays 6%, compounded monthly.
   (a) What is the total number of payments (or periods)?
   (b) What is the interest rate per period?
   (c) What formula is used to find the future value of the annuity?
   (d) Find the future value.

CHECKPOINT SOLUTIONS

1. (a) $n = 4(6) = 24$ periods (b) $i = 0.08/4 = 0.02$ per period
   (c) $S = R\left[\frac{(1+i)^n - 1}{i}\right]$ for an ordinary annuity
   (d) $S = \$500\left[\frac{(1+0.02)^{24} - 1}{0.02}\right] = \$15{,}210.93$, to the nearest cent
2. Use $n = 2(15) = 30$, $i = 0.10/2 = 0.05$, and $S = \$100{,}000$ in the formula

$$R = S\left[\frac{i}{(1+i)^n - 1}\right] \quad \left(\text{or } S = R\left[\frac{(1+i)^n - 1}{i}\right]\right)$$

$$R = \$100{,}000\left[\frac{0.05}{(1.05)^{30} - 1}\right] = \$1505.14, \text{ to the nearest cent}$$

3. (a) $n = 3(12) = 36$ periods (b) $i = 0.06/12 = 0.005$ per period
   (c) $S_{due} = R\left[\frac{(1+i)^n - 1}{i}\right](1+i)$ for an annuity due
   (d) $S_{due} = \$100\left[\frac{(1+0.005)^{36} - 1}{0.005}\right](1+0.005) = \$3953.28$, to the nearest cent

EXERCISE 6.3

Ordinary Annuities and Sinking Funds

For each set of investment conditions in Problems 1–4, find (a) the periodic interest rate as a decimal and (b) the number of periods.

1. 7% compounded semiannually for 11 years
2. $5\frac{1}{4}$% compounded annually for 17 years
3. 3% compounded monthly for 5 years
4. 10% compounded quarterly for $7\frac{1}{2}$ years
5. The figure shows a graph that compares the future value, at 8% compounded annually, of an annuity of $1000 per year and one of $1120 per year.

(a) Decide which graph corresponds to which annuity.
(b) Verify your conclusion to (a) by finding the value of each annuity and the difference between them at $t = 25$ years.

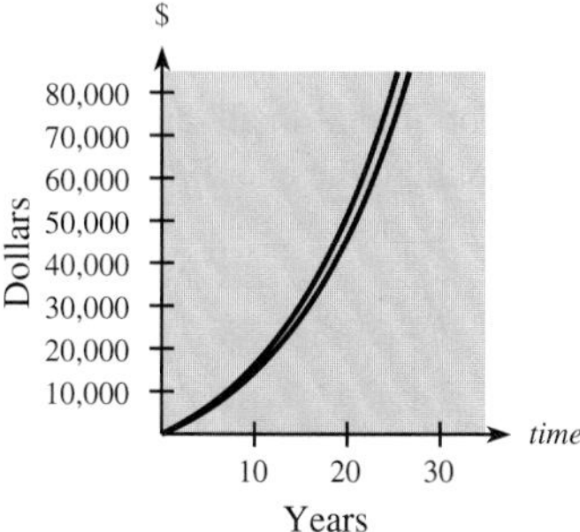

6. The figure below shows a graph that compares the future value, at 9% compounded monthly, of an annuity of \$50 per month and one of \$60 per month.
(a) Decide which graph corresponds to which annuity.
(b) Use the graph to estimate (to the nearest 10 months) how long it will be before the larger annuity is \$10,000 more than the smaller one.

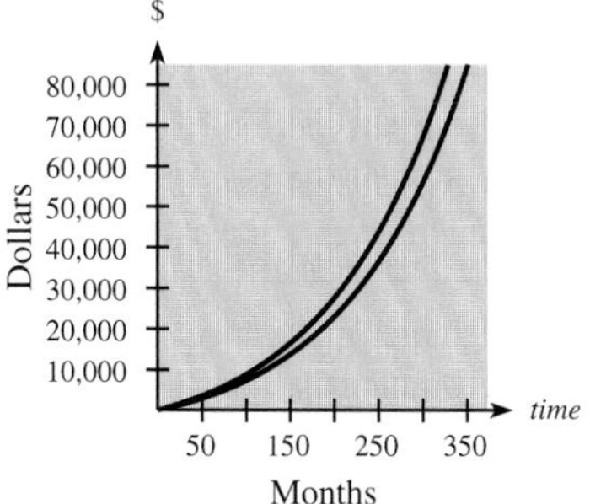

7. Find the future value of an annuity of \$1300 paid at the end of each year for 5 years, if interest is earned at a rate of 6%, compounded annually.
8. Find the future value of an annuity of \$5000 paid at the end of each year for 10 years, if it earns 9%, compounded annually.
9. Find the future value of an ordinary annuity of \$80 paid quarterly for 3 years, if the interest rate is 8%, compounded quarterly.
10. Find the future value of an ordinary annuity of \$300 paid quarterly for 5 years, if the interest rate is 12%, compounded quarterly.
11. Mr. Gordon plans to invest \$300 at the end of each month for 12 years. If the account pays 9%, compounded monthly, how much will he have at the end of the 12 years?
12. Sam deposits \$500 at the end of every 6 months into an account that pays 8%, compounded semiannually. How much will he have at the end of 8 years?
13. Parents agree to invest \$500 for their son (at 10%, compounded semiannually) on the December 31 or June 30 following each semester that he makes the Dean's list during his 4 years in college. If he makes the Dean's list in each of the 8 semesters, how much money will his parents have to give him when he graduates?
14. Jake Werkheiser qualifies to invest \$2000 in an IRA each April 15 for the next 10 years. If he makes these investments, and if the certificates pay 12%, compounded annually, how much will he have at the end of the 10 years?
15. How much will have to be invested at the end of each year at 10%, compounded annually, to pay off a debt of \$50,000 in 8 years?
16. How much will have to be invested at the end of each year at 12%, compounded annually, to pay off a debt of \$30,000 in 6 years?
17. A couple has computed that they need \$300,000 to establish an annuity when they retire in 25 years. How much money should they deposit at the end of each month into an investment plan that pays 10%, compounded monthly, so they will have the \$300,000 in 25 years?
18. How much money should a couple deposit at the end of each month into an investment plan that pays 7.5%, compounded monthly, so they will have \$800,000 in 30 years?
19. A family wants to have a \$200,000 college fund for their children at the end of 20 years. What contribution must be made at the end of each quarter if their investment pays 7.6%, compounded quarterly?
20. What size payments must be put into an account at the end of each quarter to establish an ordinary annuity that has a future value of \$50,000 in 14 years, if the investment pays 12%, compounded quarterly?
21. A sinking fund is established by a working couple so that they will have \$20,000 to pay for part of their daughter's education when she enters college. If they make deposits at the end of each 3-month period for 10 years, and if interest is paid at 12%, compounded quarterly, what size deposits must they make?
22. A company establishes a sinking fund to discharge a debt of \$75,000 due in 8 years by making equal semiannual deposits, the first due in 6 months. If its investment pays 12%, compounded semiannually, what is the size of the deposits?
23. The Weidmans want to save \$20,000 in 2 years for a down payment on a house. If they make monthly

deposits into an account paying 12%, compounded monthly, what size payments are required to meet their goal?

24. A sinking fund is established to discharge a debt of \$80,000 in 10 years. If deposits are made at the end of each 6-month period and interest is paid at the rate of 8%, compounded semiannually, what is the size of the deposits?
25. When you establish a sinking fund, which interest rate is better? Explain.
    (a) 10% (b) 6%
26. If you set up a sinking fund, which interest rate is better? Explain.
    (a) 12% compounded monthly
    (b) 12% compounded annually
27. Suppose a recent college graduate's first job allows her to deposit \$100 at the end of each month in a savings plan that earns 9%, compounded monthly. This savings plan continues for 8 years before new obligations make it impossible to continue. If the accrued amount remains in the plan for the next 15 years without deposits or withdrawals, how much money will be in the account 23 years after the plan began?
28. Suppose a young couple deposited \$1000 at the end of each quarter in an account that earned 7.6%, compounded quarterly, for a period of 8 years. After the 8 years, they start a family and find they can contribute only \$200 per quarter. If they leave the money from the first 8 years in the account and continue to contribute \$200 at the end of each quarter for the next $18\frac{1}{2}$ years, how much will they have in the account (to help with their child's college expenses)?
29. In this section's Application Preview, we considered the investment strategies of twins and found that starting early and stopping was a significantly better strategy than waiting, in terms of total contributions made as well as total value in the account at retirement. Suppose now that twin 1 invests \$2000 at the end of each year for 10 years only (until age 33) in an account that earns 8%, compounded annually. Suppose that twin 2 waits until turning 40 to think about investing. How much must twin 2 put aside at the end of each year for the next 25 years in an account that earns 8% compounded annually in order to have the same amount as twin 1 at the end of these 25 years (when they turn 65)?
30. (a) Patty Stacey deposited \$2000 at the end of each of the 5 years she qualified for an IRA. If she leaves the money that has accumulated in the IRA account for 25 additional years, how much is in her account at the end of the 30-year period? Assume an interest rate of 9%, compounded annually.
    (b) Suppose that Patty's husband delayed starting an IRA for the first 10 years he worked but then made \$2000 deposits at the end of each of the next 15 years. If the interest rate is 9%, compounded annually, and if he leaves the money in his account for 5 additional years, how much will be in his account at the end of the 30-year period?
    (c) Does Patty or her husband have more IRA money?

Annuities Due

31. Find the future value of an annuity due of \$100 each quarter for $2\frac{1}{2}$ years at 12%, compounded quarterly.
32. Find the future value of an annuity due of \$1500 each month for 3 years if the interest rate is 12%, compounded monthly.
33. Find the future value of an annuity due of \$200 paid at the beginning of each 6-month period for 8 years if the interest rate is 6%, compounded semiannually.
34. For 3 years \$400 is placed in a savings account at the beginning of each 6-month period. If the account pays interest at 10%, compounded semiannually, how much will be in the account at the end of the 3 years?
35. Jane Adele deposits \$500 in an account at the beginning of each 3-month period for 9 years. If the account pays interest at the rate of 8%, compounded quarterly, how much will she have in her account?
36. A house is rented for \$900 per quarter, with each quarter's rent payable in advance. If money is worth 8%, compounded quarterly, and the rent is deposited in an account, what is the future value of the rent for one year?
37. How much must be deposited at the beginning of each year into an account that pays 8%, compounded annually, so that the account will contain \$24,000 at the end of 5 years?
38. What size payments must be deposited at the beginning of each 6-month period into an account that pays 7.8%, compounded semiannually, so that the account will have a future value of \$120,000 at the end of 15 years?
39. Grandparents plan to open an account on their grandchild's birthday and contribute each month until she goes to college. How much must they contribute at the beginning of each month into an investment that pays 12%, compounded monthly, if they want the account to have \$180,000 at the end of 18 years?
40. A property owner has several rental units and wants to build more. How much of each month's rental

income should be deposited at the beginning of each month into an account that earns 6.6%, compounded monthly, if the goal is to have $100,000 at the end of 4 years?

Problems 41 and 42 use a spreadsheet or financial program on a calculator or computer.

41. Compare the future value of an ordinary annuity of $100 per month and that of an annuity due of $100 per month if each is invested at 7.2%, compounded monthly. Find the amount in each account at the end of each month for 10 years.
    (a) How much is contributed to each annuity?
    (b) Which account has more money after 10 years? Explain why.

42. Investigate the effect that small differences in payment size can have over time. Consider two ordinary annuities, both of which earn 8% compounded quarterly, one with payments of $100 at the end of each quarter and one with payments of $110 at the end of each quarter. Track the future value of each annuity at the end of each quarter for a period of 25 years.
    (a) How much is contributed to each annuity after 10 years, 20 years, and 25 years?
    (b) What is the difference in the future values after 10 years, 20 years, and 25 years? Why are these amounts more than the differences in the amounts contributed?

6.4 Present Value of Annuities

OBJECTIVES

- *To compute the present value of ordinary annuities, annuities due, and deferred annuities*
- *To compute the payments for a specified present value for an ordinary annuity, annuity due, and deferred annuity*

APPLICATION PREVIEW

If you wanted to receive, at retirement, $1000 at the end of each month for 16 years, what lump sum would you need to invest in an annuity that paid 9%, compounded monthly? We call this lump sum the **present value** of the annuity. Note that the annuity in this case is an account from which a person receives equal periodic payments (withdrawals).

We have discussed how contributing to an annuity program will result in a sum of money, and we have called that sum the future value of the annuity. Just as the term *annuity* is used to describe an account into which a person makes equal periodic payments (deposits), this term is also used to describe an account from which a person receives equal periodic payments (withdrawals). That is, if you invest a lump sum of money in an account today, so that at regular intervals you will receive a fixed sum of money, you have established an annuity.

Many people who are retiring purchase an annuity to supplement their income. This annuity pays them a fixed sum of money at regular intervals (usually each month). The single sum of money required to purchase an annuity that will provide these payments at regular intervals is the **present value** of the annuity.

Ordinary Annuities

Suppose we wish to invest a lump sum of money (denoted by A_n) in an annuity that earns interest at rate i per period in order to receive (withdraw) payments of size $\$R$ from this account at the end of each of n periods (after which time the account balance will be $0). Recall that our receiving payments at the end of each period means that this is an ordinary annuity.

To find a formula for A_n, we can find the present value of each future payment and then add these present values (see Figure 6.7).

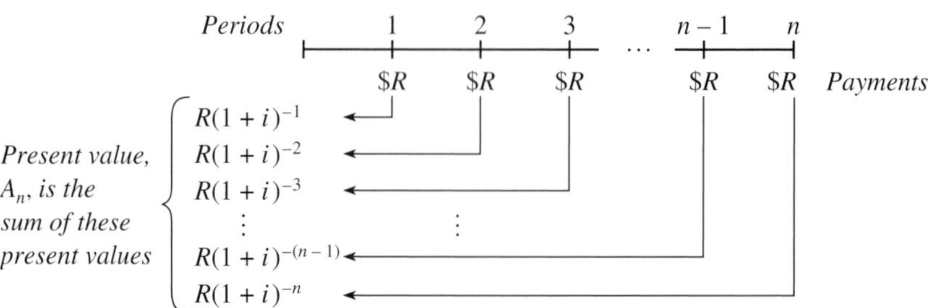

Figure 6.7

Figure 6.7 shows that we can express A_n as follows:

$$A_n = R(1+i)^{-1} + R(1+i)^{-2} + R(1+i)^{-3} + \cdots + R(1+i)^{-(n-1)} + R(1+i)^{-n} \quad (1)$$

Multiplying both sides of equation (1) by $(1+i)$ gives

$$(1+i)A_n = R + R(1+i)^{-1} + R(1+i)^{-2} + \cdots + R(1+i)^{-(n-2)} + R(1+i)^{-(n-1)} \quad (2)$$

If we subtract equation (1) from equation (2), we obtain $iA_n = R - R(1+i)^{-n}$. Then solving for A_n gives

$$A_n = R\left[\frac{1-(1+i)^{-n}}{i}\right]$$

Thus we have the following.

Present Value of an Ordinary Annuity If a payment of \$$R$ is to be made at the end of each period for n periods from an account that earns interest at a rate of i per period, then the account is an ordinary annuity, and the present value is

$$A_n = R \cdot a_{\overline{n}|i} = R \cdot \left[\frac{1-(1+i)^{-n}}{i}\right]$$

where $a_{\overline{n}|i}$ represents the present value of an ordinary annuity of \$1 per period for n periods, with an interest rate of i per period.

EXAMPLE 1

What is the present value of an annuity of \$1500 payable at the end of each 6-month period for 2 years if money is worth 8%, compounded semiannually?

Solution

We are given that $R = \$1500$ and $i = 0.08/2 = 0.04$. Because a payment is made twice a year for 2 years, the number of periods is $n = (2)(2) = 4$. Thus,

$$\begin{aligned} A_n = R \cdot a_{\overline{4}|0.04} &= \$1500\left[\frac{1-(1+0.04)^{-4}}{0.04}\right] \\ &= \$1500(3.69895) \\ &= \$5444.84, \text{ to the nearest cent} \end{aligned}$$

We now consider the annuity introduced in the Application Preview.

EXAMPLE 2

Find the lump sum that one must invest in an annuity in order to receive \$1000 at the end of each month for the next 16 years, if the annuity pays 9%, compounded monthly.

Solution

The sum we seek is the present value of an ordinary annuity, A_n, with $R =$ \$1000, $i = 0.09/12 = 0.0075$, and $n = (16)(12) = 192$.

$$A_n = R\left[\frac{1-(1+i)^{-n}}{i}\right]$$

$$= \$1000\left[\frac{1-(1.0075)^{-192}}{0.0075}\right] = \$1000(101.5727689) = \$101{,}572.77$$

Thus the required lump sum, to the nearest dollar, is \$101,573.

It is important to note that all annuities involve both periodic payments and a lump sum of money. It is whether this lump sum is in the present or in the future that distinguishes problems that use formulas for present value of annuities from those that use formulas for future value of annuities. In Example 2, the lump sum was needed now (in the present) to generate the \$1000 payments, so we used the present value formula.

Figure 6.8 shows a graph that compares the present value of an annuity of \$1000 per year at an interest rate of 6%, compounded annually, to the same annuity at an interest rate of 10%, compounded annually. Note the impact that the higher interest rate has on the present value needed to establish such an annuity for a longer period of time. Specifically, at 6% interest, a present value of more than \$12,500 is needed to generate 25 years of \$1000 payments, but at an interest rate of 10%, the necessary present value is less than \$10,000.

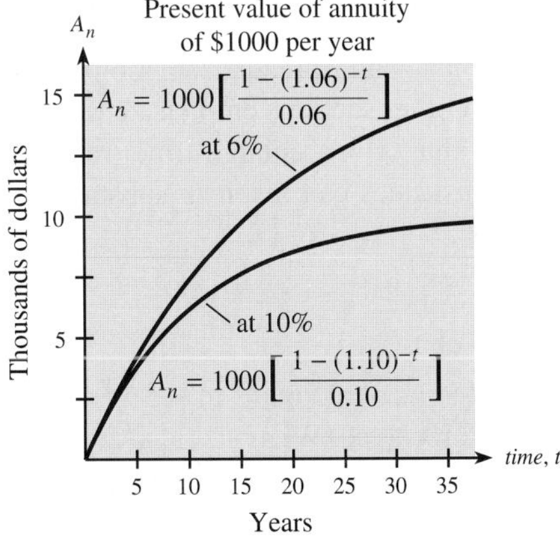

Figure 6.8

Technology Note The analysis shown graphically in Figure 6.8 can also be done with a spreadsheet, as Table 6.6 shows for the first 25 years.

TABLE 6.6

	A	B	C
1		Present Value giving	R = $1000 for n years
2	Year number	at 6% compounded annually	at 10% compounded annually
3	1	$943.40	$909.09
4	2	$1833.39	$1735.54
5	3	$2673.01	$2486.85
6	4	$3465.11	$3169.87
7	5	$4212.36	$3790.79
8	6	$4917.32	$4355.26
9	7	$5582.38	$4868.42
10	8	$6209.79	$5334.93
11	9	$6801.69	$5759.02
12	10	$7360.09	$6144.57
13	11	$7886.87	$6495.06
14	12	$8383.84	$6813.69
15	13	$8852.68	$7103.36
16	14	$9294.98	$7366.69
17	15	$9712.25	$7606.08
18	16	$10105.90	$7823.71
19	17	$10477.26	$8021.55
20	18	$10827.60	$8201.41
21	19	$11158.12	$8364.92
22	20	$11469.92	$8513.56
23	21	$11764.08	$8648.69
24	22	$12041.58	$8771.54
25	23	$12303.38	$8883.22
26	24	$12550.36	$8984.74
27	25	$12783.36	$9077.04

Our graphical and spreadsheet comparisons in Figure 6.8 and Table 6.6 and those from other sections emphasize the truth of the saying "Time is money." We have seen that, given enough time, relatively small differences in contributions or in interest rates can result in substantial differences in amounts (both present and future).

EXAMPLE 3

Suppose that a couple plans to set up an ordinary annuity with a $100,000 inheritance they received. What size quarterly payments will they receive for the next 6 years (while their children are in college) if the account pays 7%, compounded quarterly?

Solution

The \$100,000 is the amount the couple has now, so it is the present value of an ordinary annuity whose payment size, R, we seek. Using present value A_n = \$100,000, $n = 6(4) = 24$, and $i = 0.07/4 = 0.0175$, we solve for R.

$$A_n = R\left[\frac{1-(1+i)^{-n}}{i}\right]$$

$$\$100{,}000 = R\left[\frac{1-(1+0.0175)^{-24}}{0.0175}\right]$$

$$\$100{,}000 = R(19.46068565)$$

$$R = \frac{\$100{,}000}{19.46068565} = \$5138.57\text{, to the nearest cent}$$

CHECKPOINT

1. Suppose an annuity pays \$2000 at the end of each 3-month period for $3\frac{1}{2}$ years and money is worth 4%, compounded quarterly.
   (a) What is the total number of periods?
   (b) What is the interest rate per period?
   (c) What formula is used to find the present value of the annuity?
   (d) Find the present value.
2. An inheritance of \$400,000 will provide how much at the end of each year for the next 20 years, if money is worth 7%, compounded annually?

Annuities Due

Recall that an annuity due is one in which payments are made at the beginning of each period. This means that the present value of an annuity due of n payments (denoted $A_{(n,due)}$) of \$$R$ at interest rate i per period can be viewed as an initial payment of \$$R$ plus the payment program for an ordinary annuity of $n - 1$ payments of \$$R$ at interest rate i per period (see Figure 6.9).

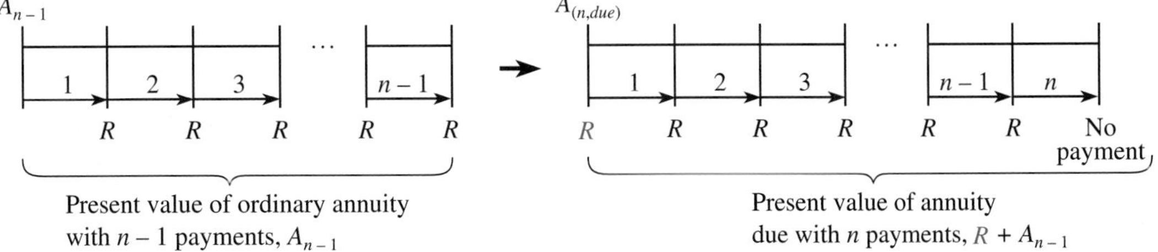

Figure 6.9

Thus from Figure 6.9, we have

$$
\begin{aligned}
A_{(n,due)} &= R + A_{n-1} = R + R\left[\frac{1-(1+i)^{-(n-1)}}{i}\right] \\
&= R\left[1 + \frac{1-(1+i)^{-(n-1)}}{i}\right] = R\left[\frac{i}{i} + \frac{1-(1+i)^{-(n-1)}}{i}\right] \\
&= R\left[\frac{(1+i)-(1+i)^{-(n-1)}}{i}\right] = R(1+i)\left[\frac{1-(1+i)^{-n}}{i}\right]
\end{aligned}
$$

Thus we have the following formula for the present value of an annuity due.

Present Value of an Annuity Due

If a payment of $\$R$ is to be made at the beginning of each period for n periods from an account that earns interest rate i per period, then the account is an annuity due, and its present value is given by

$$A_{(n,\,due)} = R\left[\frac{1-(1+i)^{-n}}{i}\right](1+i) = Ra_{\overline{n}|i}(1+i)$$

where $a_{\overline{n}|i}$ denotes the present value of an ordinary annuity of \$1 per period for n periods at interest rate i per period.

EXAMPLE 4

What lump sum would be needed on January 1 to generate annual payments of \$5000 at the beginning of each year for a period of 10 years if money is worth 5.9%, compounded annually?

Solution

Because the payments are made at the beginning of each year, this is an annuity due, and we seek its present value $A_{(n,due)}$. We use $R = \$5000$, $n = 10(1) = 10$, and $i = 0.059/1 = 0.059$.

$$
\begin{aligned}
A_{(n,due)} &= R\left[\frac{1-(1+i)^{-n}}{i}\right](1+i) = \$5000\left[\frac{1-(1+0.059)^{-10}}{0.059}\right](1+0.059) \\
&= \$5000(7.395083238)(1.059) = \$39{,}156.97, \text{ to the nearest cent}
\end{aligned}
$$

EXAMPLE 5

Suppose that a court settlement results in a \$250,000 award. If this is invested at 9%, compounded monthly, how much will it provide at the beginning of each month for a period of 4 years?

Solution

Because payments are made at the beginning of each month, this is an annuity due. We seek the payment size, R, and use the present value $A_{(n,due)} = \$250{,}000$, $n = 12(4) = 48$, and $i = 0.09/12 = 0.0075$.

$$A_{(n,due)} = R\left[\frac{1-(1+i)^{-n}}{i}\right](1+i)$$

$$\$250{,}000 = R\left[\frac{1-(1+0.0075)^{-48}}{0.0075}\right](1+0.0075) = R(40.48616775)$$

$$R = \frac{\$250{,}000}{40.48616775} = \$6174.95\text{, to the nearest cent}$$

CHECKPOINT

3. A lottery prize worth \$1,200,000 is awarded in payments of \$10,000 at the beginning of each month for 10 years. Money is worth 7.8%, compounded monthly.
   (a) What is the total number of periods?
   (b) What is the interest rate per period?
   (c) What formula is used to find the present value of this prize?
   (d) What is the *real* value of the prize? (That is, what is the present cash value of the prize?)

Deferred Annuities

A **deferred annuity** is one in which the first payment is made not at the beginning or end of the first period, but at some later date. An annuity that is deferred for k periods and then has payments of $\$R$ per period at the end of each of the next n periods is an ordinary deferred annuity and can be illustrated by Figure 6.10.

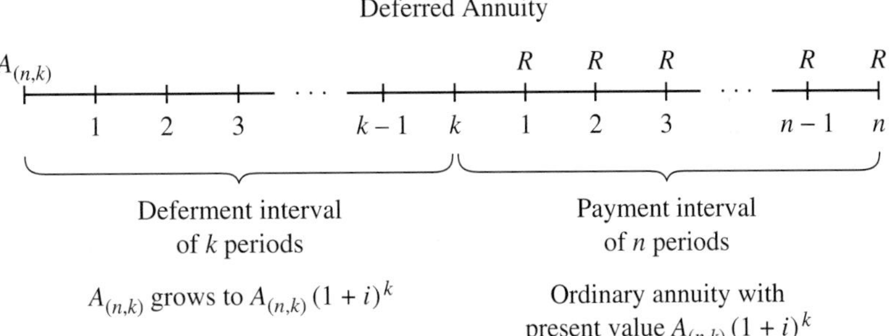

Figure 6.10

We now consider how to find the present value of such a deferred annuity when the interest rate is i per period. If payment is deferred for k periods, then the present value deposited now, denoted by $A_{(n,k)}$, is a compound interest investment for these k periods, and its future value is $A_{(n,k)}(1+i)^k$. This amount then becomes the present value for the ordinary annuity for the next n periods. From Figure 6.10, we see that we have two equivalent expressions for the amount at the beginning of the first payment period.

$$A_{(n,k)}(1+i)^k = R\left[\frac{1-(1+i)^{-n}}{i}\right]$$

Multiplying both sides by $(1+i)^{-k}$ gives the following.

Present Value of a Deferred Annuity The present value of a deferred annuity of $R per period for n periods, deferred for k periods with interest rate i per period, is given by

$$A_{(n,k)} = R\left[\frac{1-(1+i)^{-n}}{i}\right](1+i)^{-k} = Ra_{\overline{n}|i}(1+i)^{-k}$$

EXAMPLE 6

A deferred annuity is purchased that will pay $10,000 per quarter for 15 years after being deferred for 5 years. If money is worth 6%, compounded quarterly, what is the present value of this annuity?

Solution

We use $R = \$10{,}000$, $n = 4(15) = 60$, $k = 4(5) = 20$, and $i = 0.06/4 = 0.015$ in the formula for the present value of a deferred annuity.

$$\begin{aligned} A_{(60,20)} &= R\left[\frac{1-(1+i)^{-60}}{i}\right](1+i)^{-20} \\ &= \$10{,}000\left[\frac{1-(1.015)^{-60}}{0.015}\right](1.015)^{-20} \\ &= \$10{,}000(39.38026889)(0.7424704182) \\ &= \$292{,}386.85, \text{ to the nearest cent} \end{aligned}$$

EXAMPLE 7

Suppose a lottery prize of $50,000 is invested by a couple as their child's college fund. The family plans to use the money as 8 semiannual payments at the end of each 6-month period after payments are deferred for 10 years. How much would each payment be if the money can be invested at 8.6%, compounded semiannually?

Solution

We seek the payment R for a deferred annuity with $n = 8$ payment periods, deferred for $k = 2(10) = 20$ periods, $i = 0.086/2 = 0.043$, and $A_{(n,k)} = \$50{,}000$.

$$\begin{aligned} A_{(n,k)} &= R\left[\frac{1-(1+i)^{-n}}{i}\right](1+i)^{-k} \\ \$50{,}000 &= R\left[\frac{1-(1.043)^{-8}}{0.043}\right](1.043)^{-20} \\ \$50{,}000 &= R(6.650118184)(0.4308378316) \\ \$50{,}000 &= R(2.865122499) \\ R &= \frac{\$50{,}000}{2.865122499} = \$17{,}451.26, \text{ to the nearest cent} \end{aligned}$$

In Example 7, note the effect of the deferral time. The family receives 8($17,451.26) = $139,610.08 from the original $50,000 investment.

CHECKPOINT

4. Suppose an annuity at 6%, compounded semiannually, will pay $5000 at the end of each 6-month period for 5 years with the first payment deferred for 10 years.
   (a) What is the number of payment periods and the number of deferral periods?
   (b) What is the interest rate per period?
   (c) What formula is used to find the present value of this annuity?
   (d) Find the present value of this annuity.

CHECKPOINT SOLUTIONS

1. (a) $n = 3.5(4) = 14$ (b) $i = 0.04/4 = 0.01$
   (c) $A_n = R\left[\dfrac{1-(1+i)^{-n}}{i}\right]$ for an ordinary annuity
   (d) $A_n = \$2000\left[\dfrac{1-(1.01)^{-14}}{0.01}\right] = \$26{,}007.41$, to the nearest cent
2. We seek R for an ordinary annuity with $A_n = \$400{,}000$, $n = 20$, and $i = 0.07$.
$$A_n = R\left[\frac{1-(1+i)^{-n}}{i}\right]$$
$$\$400{,}000 = R\left[\frac{1-(1.07)^{-20}}{0.07}\right] = R(10.59401425)$$
$$R = \frac{\$400{,}000}{10.59401425} = \$37{,}757.17\text{, to the nearest cent}$$
3. (a) $n = 10(12) = 120$ (b) $i = 0.078/12 = 0.0065$
   (c) $A_{(n,due)} = R\left[\dfrac{1-(1+i)^{-n}}{i}\right](1+i)$ for an annuity due
   (d) $A_{(120,due)} = \$10{,}000\left[\dfrac{1-(1+0.0065)^{-120}}{0.0065}\right](1+0.0065)$
   $= \$836{,}843.55$, to the nearest cent
4. (a) $n = 5(2) = 10$ payment periods, and $k = 10(2) = 20$ deferral periods
   (b) $0.06/2 = 0.03$
   (c) $A_{(n,k)} = R\left[\dfrac{1-(1+i)^{-n}}{i}\right](1+i)^{-k}$ for a deferred annuity
   (d) $A_{(n,k)} = \$5000\left[\dfrac{1-(1.03)^{-10}}{0.03}\right](1.03)^{-20} = \$23{,}614.83$, to the nearest cent

EXERCISE 6.4

Ordinary Annuities

1. Find the present value of an annuity of $100 paid at the end of each year for 17 years if the interest rate is 7%, compounded annually.
2. Find the present value of an annuity of $800 paid at the end of each year for 15 years if the interest rate is 12%, compounded annually.
3. Find the present value of an annuity of $6000 paid at the end of each 6-month period for 8 years if the interest rate is 8%, compounded semiannually.
4. Find the present value of an annuity that pays $3000 at the end of each 6-month period for 6 years if the interest rate is 6%, compounded semiannually.

5. With a present value of \$135,000, what size withdrawals can be made at the end of each quarter for the next 10 years if money is worth 6.4%, compounded quarterly?
6. If \$88,000 is invested in an annuity that earns 5.8%, compounded quarterly, what size payments will it provide at the end of each quarter for the next $5\frac{1}{2}$ years?
7. An insurance settlement of \$750,000 must replace Trixie Eden's income for the next 40 years. What income will this settlement provide at the end of each month if it is invested in an annuity that earns 8.4%, compounded monthly?
8. A retiree inherits \$93,000 and invests it at 6.6%, compounded monthly, in an ordinary annuity that provides a monthly amount for the next 12 years. Find the monthly amount.
9. Juanita Domingo's parents want to establish a college trust for her. They want to make 16 quarterly withdrawals of \$2000, with the first withdrawal 3 months from now. If money is worth 7.2%, compounded quarterly, how much must be deposited now to provide for this trust?
10. Is it more economical to buy an automobile for \$3900 cash or to pay \$800 down and \$400 at the end of each quarter for 2 years, if money is worth 8%, compounded quarterly?
11. Suppose a state lottery prize of \$5 million is to be paid in 20 payments of \$250,000 each at the end of the next 20 years. If money is worth 10%, compounded annually, what is the present value of the prize?
12. Suppose Becky has her choice of \$10,000 at the end of each month for life or a single prize of \$1.5 million. She is 35 years old and her life expectancy is 40 more years.
    (a) Find the present value of the annuity if money is worth 7.2%, compounded monthly.
    (b) If she takes the \$1.5 million, spends \$700,000 of it, and invests the remainder at 7.2%, what monthly annuity will she receive?
13. Dr. Jane Kodiak plans to sell her practice to an HMO. The HMO will pay her \$750,000 now or will make a partial payment now and make additional payments at the end of each year for the next 10 years.
    (a) If money is worth 6.5%, compounded annually, what annual payment would the doctor receive if she took \$250,000 now?
    (b) Suppose the HMO can earn 6.82% compounded annually on its investments. How much can it save by paying \$250,000 now and the rest over the next 10 years rather than paying all \$750,000 now?
14. On March 20, 1998, *USA Today* reported that Buffalo Sabres Goalie Dominik Hasek, who led the Czech Republic to an Olympic gold medal, signed an \$18.5 million 2-year extension to his contract. This \$18.5 million included a \$4 million signing bonus. Suppose Hasek invested \$2.5 million of the signing bonus in an ordinary annuity that paid 7.2%, compounded monthly, for the next 50 years. How much would he receive at the end of each month during this time?
15. The figure below shows a graph that compares the present values of two ordinary annuities of \$1000 annually, one at 8% compounded annually and one at 10% compounded annually.
    (a) Determine which graph corresponds to the 8% rate and which to the 10% rate.
    (b) Use the graph to estimate the difference between the present values of these annuities for 25 years.
    (c) Write a sentence that explains this difference.

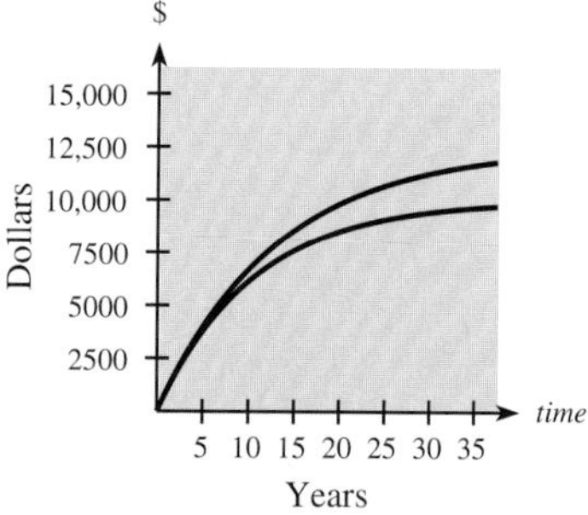

16. The figure below shows a graph that compares the present values of two ordinary annuities of \$800 quarterly, one at 6% compounded quarterly and one at 9% compounded quarterly.
    (a) Determine which graph corresponds to the 6% rate and which to the 9% rate.
    (b) Use the graph to estimate the difference between the present values of these annuities for 25 years (100 quarters).
    (c) Write a sentence that explains this difference.

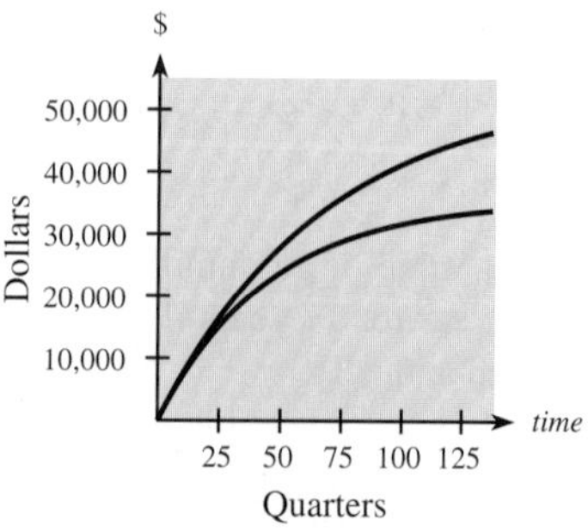

For each of Problems 17 and 18, answer the following questions.

(a) How much is in the account after the last deposit is made?

(b) How much was deposited?

(c) What is the amount of each withdrawal?

(d) What is the total amount withdrawn?

17. Suppose an individual makes an initial investment of \$2500 in an account that earns 7.8%, compounded monthly, and makes additional contributions of \$100 at the end of each month for a period of 12 years. After these 12 years, this individual wants to make withdrawals at the end of each month for the next 5 years (so that the account balance will be reduced to \$0).

18. Suppose that Nam Banh deposits his \$12,500 bonus in an account that earns 8%, compounded quarterly, and makes additional deposits of \$500 at the end of each quarter for the next $22\frac{1}{2}$ years, until he retires. To supplement his retirement, Nam wants to make withdrawals at the end of each quarter for the next 12 years (at which time the account balance will be \$0).

Annuities Due

19. Explain the difference between an ordinary annuity and an annuity due.

20. Is there any difference between the present values in (a) and (b)? Explain.
    (a) An annuity due that pays \$1000 at the beginning of each year for 10 years
    (b) Taking \$1000 now and establishing an ordinary annuity that pays \$1000 at the end of each year for 9 years

21. Find the present value of an annuity due that pays \$3000 at the beginning of each quarter for the next 7 years. Assume that money is worth 5.8%, compounded quarterly.

22. Find the present value of an annuity due that pays \$25,000 every 6 months for the next $2\frac{1}{2}$ years if money is worth 6.2%, compounded semiannually.

23. What amount must be set aside now to generate payments of \$50,000 at the beginning of each year for the next 12 years if money is worth 5.92%, compounded annually?

24. Suppose an annuity will pay \$15,000 at the beginning of each year for the next 7 years. How much money is needed to start this annuity if it earns 7.3%, compounded annually?

25. A company wants to have \$40,000 at the beginning of each 6-month period for the next $4\frac{1}{2}$ years. If an annuity is set up for this purpose, how much must be invested now if the annuity earns 6.68%, compounded semiannually?

26. A trust provides \$10,000 to a county library at the beginning of each 3-month period for the next $2\frac{1}{2}$ years. If money is worth 7.4%, compounded quarterly, find the amount in the trust when it began.

27. A year-end bonus of \$25,000 will generate how much money at the beginning of each month for the next year, if it can be invested at 6.48%, compounded monthly?

28. A couple inherits \$89,000. How much can this generate at the beginning of each month over the next 5 years, if money is worth 6.3%, compounded monthly?

29. Recent sales of some real estate and record profits make it possible for a manufacturer to set aside \$800,000 in a fund to be used for modernization and remodeling. How much can be withdrawn from this fund at the beginning of each half-year for the next 3 years if the fund earns 7.7%, compounded semiannually?

30. As a result of a court settlement, an accident victim is awarded \$1.2 million. The attorney takes one-third of this amount, another third is used for immediate expenses, and the remaining third is used to set up an annuity. What amount will this annuity pay at the beginning of each quarter for the next 5 years if the annuity earns 7.6%, compounded quarterly?

31. A used piece of rental equipment has $2\frac{1}{2}$ years of useful life remaining. When rented, the equipment brings in \$800 per month (paid at the beginning of the month). If the equipment is sold now and money is worth 4.8%, compounded monthly, what must the selling price be to recoup the income that the rental company loses by selling the equipment "early"?

32. A \$2.4 million state lottery pays \$10,000 at the beginning of each month for 20 years. How much money must the state actually have in hand to set up the payments for this prize if money is worth 6.3%, compounded monthly?

Deferred Annuities

33. Find the present value of an annuity of \$2000 per year at the end of each of 8 years after being deferred for 6 years, if money is worth 7%, compounded annually.

34. Find the present value of an annuity of \$2000, at the end of each quarter for 5 years after being deferred for 3 years, if money is worth 8%, compounded quarterly.

35. The terms of a single parent's will indicate that a child will receive an ordinary annuity of \$16,000 per year from age 18 to age 24 (so the child can attend college) and that the balance of the estate goes to a

niece. If the parent dies on the child's 14th birthday, how much money must be removed from the estate to purchase the annuity? (Assume an interest rate of 6%, compounded annually.)

36. On his 48th birthday, a man wants to set aside enough money to provide an income of $300 per month at the end of each month from his 60th birthday to his 65th birthday. If he earns 6%, compounded monthly, on his money, how much will this supplemental retirement plan cost him on his 48th birthday?
37. The semiannual tuition payment at a major university is expected to be $10,000 for the 4 years beginning in the year 2016. What lump sum payment should the university accept on January 1, 1998, in lieu of tuition payments beginning on July 1, 2016? Assume that money is worth 7%, compounded semiannually, and that tuition is paid on July 1 and January 1 of the 4 years.
38. A grateful alumnus wishes to provide a scholarship of $2000 per year for 5 years to his alma mater, with the first scholarship awarded on his 60th birthday. If money is worth 6%, compounded annually, how much money must he donate on his 50th birthday?
39. Danny Metzger's parents invested $1600 when he was born. This money is to be used for Danny's college education and is to be withdrawn in four equal annual payments beginning when Danny is age 19. Find the amount that will be available each year, if money is worth 6%, compounded annually.
40. Carol Goldsmith received a trust fund inheritance of $10,000 on her 30th birthday. She plans to use the money to supplement her income with 20 quarterly payments beginning on her 60th birthday. If money is worth 7.6%, compounded quarterly, how much will each quarterly payment be?
41. Hockey goalie Dominik Hasek, of the NHL's Buffalo Sabres and the Olympic gold medal-winning Czech Republic, received a $4 million signing bonus with his contract extension (as reported in *USA Today* on March 20, 1998). Suppose Hasek set aside $2.5 million in a fund that earns 9.6%, compounded monthly. If he deferred this amount for 15 years, how much per month would it provide for the 40 years after the deferral period?
42. A couple received a $134,000 inheritance the year they turned 48 and invested it in a fund that earns 7.7%, compounded semiannually. If this amount is deferred for 14 years (until they retire), how much will it provide at the end of each half-year for the next 20 years after they retire?

Problems 43 and 44 use a spreadsheet or financial program on a calculator or computer.

43. Suppose a couple had $100,000 at retirement that they can invest in an ordinary annuity that earns 7.8%, compounded monthly. Track the balance in this annuity account until it reaches $0, if the couple receives the following monthly payments.
    (a) $1000 (b) $2500
44. Suppose you invested $250,000 in an annuity that earned interest compounded monthly. This annuity paid $3000 at the end of each month. Experiment with the following different interest rates (compounded monthly) to see how long it will be, with each rate, until this annuity has an account balance of $0.
    (a) 6.5% (b) 9%

6.5 Loans and Amortization

OBJECTIVES

- *To find the regular payments required to amortize a debt*
- *To find the amount that can be borrowed for a specified payment*
- *To develop an amortization schedule*
- *To find the unpaid balance of a loan*

APPLICATIONS PREVIEW

When businesses, individuals, or families borrow money to make a major purchase, four questions commonly arise:

1. What will the payments be?
2. How much can be borrowed and still fit a budget?
3. What is the payoff amount of the loan before the final payment is due?
4. What is the total of all payments needed to pay off a loan?

In this section we discuss the most common way that consumers and businesses discharge debts: regular payments of fixed size made on a loan (called **amortization**). We study how amortization is related to the present value of an ordinary annuity and determine ways to answer the four questions posed above. We also study amortization schedules, which detail how loan payments are allocated to pay interest due and to discharge the debt.

Just as we invest money to earn interest, banks and lending institutions lend money and collect interest for its use. Although your aunt may lend you money with the understanding that you will repay the full amount of the money plus simple interest at the end of a year, financial institutions generally expect you to make partial payments on a regular basis (often monthly). Most consumer loans (for automobiles, appliances, televisions, and the like) are classed as **installment loans.**

Federal law now requires that the full cost of any loan and the **true annual percentage rate (APR)** be disclosed with the loan. The APR gives the interest rate as though the interest is always charged on the unpaid balance.

Prior to enactment of the law that requires "truth in lending," banks offered installment loans whose stated interest rates were different from the true APR. However, since the federal regulations have gone into effect, banks have phased out these types of loans and treat most consumer loans in the same manner as home mortgages and long-term commercial loans. That is, loans now are usually paid off by a series of partial payments with interest charged on the unpaid balance at the end of each payment period. In this case, the stated interest rate is the same as the APR.

A loan of this type can be made in one of two ways. One repayment plan applies an equal amount to the principal each payment period plus the interest for the period, which is the interest on the unpaid balance. For example, a loan of \$120,000 for 10 years at 12% could be repaid in 120 payments of \$1000 plus 1% of the remaining balance of the loan. When this payment method is used, the payments will decrease as the unpaid balance decreases.

This type of loan can also be repaid by making all payments (including principal and interest) of equal size. This process of repaying the loan is called **amortization.**

When a bank makes a loan of this type, it is purchasing from the borrower an ordinary annuity that pays a fixed return each payment period. The lump sum the bank gives to the borrower (the principal of the loan) is the present value of the ordinary annuity, and each payment the bank receives from the borrower is a payment from the annuity. Hence, to find the size of these periodic payments, we solve the formula for the present value of an ordinary annuity, $A_n = R \cdot a_{\overline{n}|i}$, for R. Thus we have the following.

Amortization Formula

If the debt of $\$A_n$, with interest at a rate of i per period, is amortized by n equal, periodic payments (each payment being made at the end of a period), the size of each payment is

$$R = A_n \cdot \frac{1}{a_{\overline{n}|i}} = A_n \cdot \left[\frac{i}{1-(1+i)^{-n}}\right]$$

where $1/a_{\overline{n}|i}$ represents the periodic payment necessary to amortize \$1 over n periods, at an interest rate of i per period.

EXAMPLE 1

A debt of \$1000 with interest at 16%, compounded quarterly, is to be amortized by 20 quarterly payments (all the same size) over the next 5 years. What will the size of these payments be?

Solution

The amortization of this loan is an ordinary annuity with present value \$1000. Therefore, $A_n = \$1000$, $n = 20$, and $i = 0.16/4 = 0.04$. Thus we have

$$R = A_n\left[\frac{i}{1-(1+i)^{-n}}\right] = \$1000\left[\frac{0.04}{1-(1.04)^{-20}}\right]$$
$$= \$1000(0.07358175) = \$73.58, \text{ to the nearest cent}$$

EXAMPLE 2

A man buys a house for \$200,000. He makes a \$50,000 down payment and agrees to amortize the rest of the debt with quarterly payments over the next 10 years. If the interest on the debt is 12%, compounded quarterly, find (a) the size of the quarterly payments, (b) the total amount of the payments, and (c) the total amount of interest paid.

Solution

(a) We know that $A_n = \$200{,}000 - \$50{,}000 = \$150{,}000$, $n = 4(10) = 40$, and $i = 0.12/4 = 0.03$. Thus, for the quarterly payment, we have

$$R = \$150{,}000\left[\frac{0.03}{1-(1.03)^{-40}}\right]$$
$$= (\$150{,}000)(0.043262378) = \$6489.36, \text{ to the nearest cent}$$

(b) The man made 40 payments of \$6489.36, so his payments totaled

$$(40)(\$6489.36) = \$259{,}574.40$$

plus the \$50,000 down payment, or \$309,574.40.

(c) Of the \$309,574.40 paid, \$200,000 was for payment of the house. The remaining \$109,574.40 was the total amount of interest paid.

EXAMPLE 3

A young couple is ready to buy their first home. They have \$30,000 for a down payment, and their budget can accommodate a monthly mortgage payment of \$850.00. What is the most expensive home they can buy if they can borrow money for 30 years at 7.8%, compounded monthly?

Solution

We seek the amount the couple can borrow, or A_n, knowing that $R = \$850$, $n = 30(12) = 360$, and $i = 0.078/12 = 0.0065$. We can use these values in the amortization formula and solve for A_n (or use the formula for the present value of an ordinary annuity).

$$R = A_n\left[\frac{i}{1-(1+i)^{-n}}\right]$$

$$\$850 = A_n\left[\frac{0.0065}{1-(1.0065)^{-360}}\right]$$

$$A_n = \$850\left[\frac{1-(1.0065)^{-360}}{0.0065}\right]$$

$$A_n = \$850(138.9138739) = \$118{,}076.79, \text{ to the nearest cent}$$

Thus, if they borrow \$118,077 (to the nearest dollar) and put down \$30,000, the most expensive home they can buy would cost \$118,077 + \$30,000 = \$148,077.

CHECKPOINT

1. A debt of \$25,000 is to be amortized with equal quarterly payments over 6 years, and money is worth 7%, compounded quarterly.
   (a) Find the total number of payments.
   (b) Find the interest rate per period.
   (c) Write the formula used to find the size of each payment.
   (d) Find the amount of each payment.
2. A new college graduate determines that she can afford a car payment of \$350 per month. If the auto manufacturer is offering a special 2.1% financing rate, compounded monthly for 4 years, how much can she borrow and still have a \$350 monthly payment?

We can construct an amortization schedule that summarizes all the information regarding the amortization of a loan.

For example, a loan of \$10,000 with interest at 10% could be repaid in 5 equal annual payments of size

$$R = \$10{,}000\left[\frac{0.10}{1-(1+0.10)^{-5}}\right] = \$10{,}000(0.263797) = \$2637.97$$

Each time this \$2637.97 payment is made, some is used to pay the interest on the unpaid balance, and some is used to reduce the principal. For the first payment, the unpaid balance is \$10,000, so the interest payment is 10% of \$10,000, or \$1000. The remaining \$1637.97 is applied to the principal. Hence, after this first payment, the unpaid balance is \$10,000 − \$1637.97 = \$8362.03.

For the second payment of \$2637.97, the amount used for interest is 10% of \$8362.03, or \$836.20; the remainder, \$1801.77, is used to reduce the principal.

This information for these two payments and for the remaining payments is summarized in the following amortization schedule, which gives the payment, the interest on the unpaid balance, the reduction of the balance, and the remaining unpaid balance for each payment period.

Period	*Payment*	*Interest*	*Balance Reduction*	*Unpaid Balance*
				10000.00
1	2637.97	1000.00	1637.97	8362.03
2	2637.97	836.20	1801.77	6560.26
3	2637.97	656.03	1981.94	4578.32
4	2637.97	457.83	2180.14	2398.18
5	2638.00	239.82	2398.18	0.00
Total	13189.88	3189.88	10000.00	

Note that the last payment was increased by 3¢ so that the balance was reduced to \$0 at the end of the 5 years. Such an adjustment is normal in amortizing a loan.

Technology Note The amortization schedule above involves only a few payments. For loans involving more payments, a spreadsheet program is an excellent tool to develop an amortization schedule. The spreadsheet, which is fast and accurate, requires the user to understand the interrelationships among the columns. The accompanying spreadsheet output in Table 6.7 shows the amortization schedule for the first 24 monthly payments on a \$100,000 loan amortized for 30 years at 6%, compounded monthly. It is interesting to observe how little the unpaid balance changes during these first 2 years—and, consequently, how much of each payment is devoted to paying interest due rather than to reducing the debt. In fact, the column labeled "Total Interest" shows a running total of how much has been paid toward interest. We see that after 12 payments of \$599.55, a total of \$7194.60 has been paid, and \$5966.59 of this has been interest payments.

Many people who borrow money, such as for a car or a home, do not pay on the loan for its entire term. Rather, they pay off the loan early by making a final lump sum payment. The unpaid balance found in the amortization schedule is the "payoff amount" of the loan and represents the lump sum payment that must be made to complete payment on the loan. When the number of payments is large, we may wish to find the unpaid balance of a loan without constructing an amortization schedule.

Recall that calculations for amortization of a debt are based on the present value formula for an ordinary annuity. Because of this, the **unpaid balance of a loan** (also called the **payoff amount** and the **outstanding principal of the loan**) is the present value needed to generate all the remaining payments.

Unpaid Balance or Payoff Amount of a Loan

For a loan of n payments of \$$R$ per period at interest rate i per period, the unpaid balance after k payments have been made is the present value of an ordinary annuity with $n - k$ payments. That is,

$$\text{Unpaid balance} = A_{n-k} = R\left[\frac{1-(1+i)^{-(n-k)}}{i}\right]$$

TABLE 6.7

	A	B	C	D	E	F
1	Payment Number	Payment Amount	Interest	Balance Reduction	Unpaid Balance	Total Interest
2	0	0	0	0	$100,000.00	$0.00
3	1	$599.55	$500.00	$99.55	$99,900.45	$500.00
4	2	$599.55	$499.50	$100.05	$99,800.40	$999.50
5	3	$599.55	$499.00	$100.55	$99,699.85	$1,498.50
6	4	$599.55	$498.50	$101.05	$99,598.80	$1,997.00
7	5	$599.55	$497.99	$101.56	$99,497.25	$2,495.00
8	6	$599.55	$497.49	$102.06	$99,395.18	$2,992.48
9	7	$599.55	$496.98	$102.57	$99,292.61	$3,489.46
10	8	$599.55	$496.46	$103.09	$99,189.52	$3,985.92
11	9	$599.55	$495.95	$103.60	$99,085.92	$4,481.87
12	10	$599.55	$495.43	$104.12	$98,981.80	$4,977.30
13	11	$599.55	$494.91	$104.64	$98,877.16	$5,472.21
14	12	$599.55	$494.39	$105.16	$98,771.99	$5,966.59
15	13	$599.55	$493.86	$105.69	$98,666.30	$6,460.45
16	14	$599.55	$493.33	$106.22	$98,560.09	$6,953.79
17	15	$599.55	$492.80	$106.75	$98,453.34	$7,446.59
18	16	$599.55	$492.27	$107.28	$98,346.05	$7,938.85
19	17	$599.55	$491.73	$107.82	$98,238.23	$8,430.58
20	18	$599.55	$491.19	$108.36	$98,129.87	$8,921.77
21	19	$599.55	$490.65	$108.90	$98,020.97	$9,412.42
22	20	$599.55	$490.10	$109.45	$97,911.53	$9,902.53
23	21	$599.55	$489.56	$109.99	$97,801.54	$10,392.09
24	22	$599.55	$489.01	$110.54	$97,690.99	$10,881.09
25	23	$599.55	$488.45	$111.10	$97,579.90	$11,369.55
26	24	$599.55	$487.90	$111.65	$97,468.25	$11,857.45

EXAMPLE 4

In Example 2, we found that the monthly payment for a loan of $150,000 at 12%, compounded quarterly, for 10 years is $6489.36 (to the nearest cent). Find the unpaid balance immediately after the 15th payment.

Solution

The unpaid balance after the 15th payment is the present value of the annuity with $40 - 15 = 25$ payments to be made. Thus we use $R = \$6489.36$, $i = 0.03$, and $n - k = 25$ in the formula for the unpaid balance of a loan.

$$A_{n-k} = R\left[\frac{1-(1+i)^{-(n-k)}}{i}\right] = \$6489.36\left[\frac{1-(1.03)^{-25}}{0.03}\right]$$

$$= (\$6489.36)(17.4131477) = \$113{,}000.18, \text{ to the nearest cent}$$

CHECKPOINT

3. A 42-month auto loan has monthly payments of \$411.35. If the interest rate is 8.1%, compounded monthly, find the unpaid balance immediately after the 24th payment.

The curve in Figure 6.11 shows the unpaid balance of a \$100,000 loan at 6%, compounded monthly, for 25 years (with monthly payments of \$644.30). In the figure, the straight line represents how the unpaid balance would decrease if each payment diminished the debt by the same amount. Note how the curve decreases slowly at first and more rapidly as the unpaid balance nears zero. The reason for this behavior is that when the unpaid balance is large, much of each payment is devoted to interest. Similarly, when the debt decreases, more of each payment goes toward the principal.

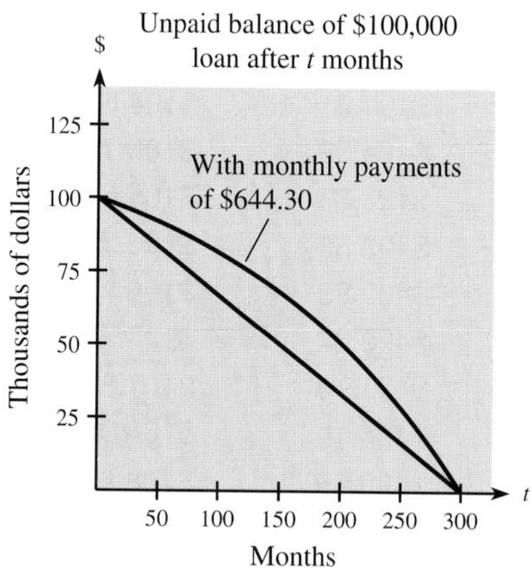

Figure 6.11

CHECKPOINT SOLUTIONS

1. (a) $n = 6(4) = 24$ payments (b) $i = 0.07/4 = 0.0175$ per period

(c) $R = A_n\left[\dfrac{i}{1-(1+i)^{-n}}\right]$

(d) $R = \$25{,}000\left[\dfrac{0.0175}{1-(1.0175)^{-24}}\right] = \$1284.64$, to the nearest cent

2. Use $R = A_n\left[\dfrac{i}{1-(1+i)^{-n}}\right]$ with $i = 0.021/12 = 0.00175$, $n = 48$, and $R = \$350$.

$$\$350 = A_n\left[\frac{0.00175}{1-(1.00175)^{-48}}\right] = A_n\,(0.021738795)$$

$$A_n = \frac{\$350}{0.021738795} = \$16{,}100.25\text{, to the nearest cent}$$

3. Use $R = \$411.35$, $i = 0.081/12 = 0.00675$, and $n - k = 42 - 24 = 18$.

$$A_{n-k} = R\left[\frac{1-(1+i)^{-(n-k)}}{i}\right] = \$411.35\left[\frac{1-(1.00675)^{-18}}{0.00675}\right]$$

$$= \$6950.13\text{, to the nearest cent}$$

EXERCISE 6.5

1. Answer parts (a) and (b) with respect to loans for 10 years and 25 years. Assume the same interest rate.
   (a) Which loan results in more of each payment being directed toward principal? Explain.
   (b) Which loan results in a lower periodic payment? Explain.
2. When a debt is amortized, which interest rate is better for the borrower, 10% or 6%? Explain.
3. A debt of \$10,000 is to be amortized by equal payments at the end of each year for 5 years. If the interest charged is 8%, compounded annually, find the periodic payment.
4. A debt of \$15,000 is to be amortized by equal payments at the end of each year for 10 years. If the interest charged is 9%, compounded annually, find the periodic payment.
5. A debt of \$8000 is to be amortized with 8 equal semiannual payments. If the interest rate is 12%, compounded semiannually, what is the size of each payment?
6. A loan of \$10,000 is to be amortized with 10 equal quarterly payments. If the interest rate is 6%, compounded quarterly, what is the periodic payment?
7. Develop an amortization schedule for a loan of \$100,000 to be repaid in 3 annual payments of equal size. The interest rate is 9%, compounded annually.
8. Develop an amortization schedule for a loan of \$30,000 to be repaid in 5 annual payments of equal size. The interest rate is 7%.
9. Develop an amortization schedule for a loan of \$20,000 with interest at 12%, compounded quarterly, if it is to be repaid in 4 quarterly payments of equal size.
10. Develop an amortization schedule for a loan of \$50,000 with interest at 10%, compounded semiannually, if it is to be repaid in $2\frac{1}{2}$ years by making equal semiannual payments.
11. A \$10,000 loan is to be amortized for 10 years with quarterly payments of \$334.27. If the interest rate is 6%, compounded quarterly, what is the unpaid balance immediately after the sixth payment?
12. A debt of \$8000 is to be amortized with 8 equal semiannual payments of \$1288.29. If the interest rate is 12%, compounded semiannually, find the unpaid balance immediately after the fifth payment.
13. When Maria Acosta bought a car $2\frac{1}{2}$ years ago, she borrowed \$14,000 for 48 months at 8.1% compounded monthly. Her monthly payments are \$342.44, but she'd like to pay off the loan. How much will she owe just after her payment at the $2\frac{1}{2}$-year mark?
14. Six and a half years ago, a small business borrowed \$50,000 for 10 years at 9%, compounded semiannually, in order to update some equipment. Now the company would like to pay off this loan. Find the payoff amount just after the company makes the 14th semiannual payment of \$3843.81.

Problems 15–18 describe a debt to be amortized. In each problem, find:

(a) the size of each payment,
(b) the total amount paid over the life of the loan, and
(c) the total interest paid over the life of the loan.

15. A man buys a house for \$75,000. He makes a \$25,000 down payment and amortizes the rest of the debt with semiannual payments over the next 10 years. The interest rate on the debt is 12%, compounded semiannually.
16. Sean Lee purchases \$20,000 worth of supplies for his restaurant by making a \$3000 down payment and amortizing the rest with quarterly payments over the next 5 years. The interest rate on the debt is 16%, compounded quarterly.
17. John Fare purchased \$10,000 worth of equipment by making a \$2000 down payment and promising to pay the remainder of the cost in semiannual payments over the next 4 years. The interest rate on the debt is 10%, compounded semiannually.
18. A woman buys an apartment house for \$250,000 by making a down payment of \$100,000 and amortizing the rest of the debt with semiannual payments over the next 10 years. The interest rate on the debt is 12%, compounded semiannually.
19. A man buys a car for \$12,000. If the interest rate on the loan is 12%, compounded monthly, and he wants to make monthly payments of \$300 for 36 months, how much must he put down?
20. A woman buys a car for \$15,000. If the interest rate on the loan is 12%, compounded monthly, and she wants to make monthly payments of \$500 for 3 years, how much must she have for a down payment?
21. A couple purchasing a home wants their payment to be \$500 per month. If they have \$15,000 available for a down payment and the mortgage rate on a 25-year loan is 12%, compounded monthly, how much can they spend on a house?

22. An investor interested in purchasing an apartment building determines that she can make payments of \$3000 per month. If a loan is available at 15%, compounded monthly for 20 years, how much can she afford to pay for the building?
23. A company is expanding its production capacity, and the budget for debt service is \$45,000 per quarter for 10 years. How much can the company borrow at each of the following interest rates?
    (a) 7.84%, compounded quarterly
    (b) 7.28%, compounded quarterly
24. A developer wants to buy a certain parcel of land. The developer feels she can afford payments of \$22,000 each half-year for the next 7 years. How much can she borrow and hold to this budget at each of the following interest rates?
    (a) 8.9%, compounded semiannually
    (b) 7.3%, compounded semiannually
25. A couple that borrows \$90,000 for 30 years at 7.2%, compounded monthly, must make monthly payments of \$610.91.
    (a) Find their unpaid balance after 1 year.
    (b) During that first year, how much interest did the couple pay?
26. A company that purchases a piece of equipment by borrowing \$250,000 for 10 years at 6%, compounded monthly, has monthly payments of \$2775.51.
    (a) Find the unpaid balance on this loan after 1 year.
    (b) During that first year, how much interest did the company pay?

A debt of \$100,000 is amortized at 6%, compounded monthly, over 25 years with 300 monthly payments of \$644.30 each. Figure 6.12 includes two graphs: one shows the total amount paid (in monthly payments) as a function of time (in months), and the other shows the amount paid toward the principal of the debt as a function of time. Use this figure to complete Problems 27–30.

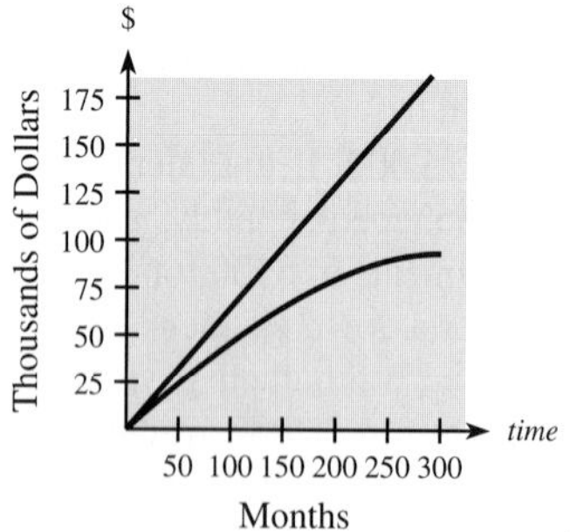

Figure 6.12

27. Correctly label each graph.
28. Draw a vertical segment whose length represents the outstanding principal of the debt (or the payoff amount of the loan) after 150 months.
29. Draw a vertical segment whose length represents the total amount of interest paid on the debt after 250 months.
30. If each year you made an extra payment devoted entirely toward payment of principal, would the total amount paid over the life of the loan be greater than, the same as, or less than the total amount that would be paid if you just made scheduled monthly payments? Explain.
31. What difference does 0.5% make on a loan? To answer this question, find (to the nearest dollar) the monthly payment and total interest paid over the life of the loan for each of the following.
    (a) An auto loan of \$15,000 at 8.0% versus 8.5%, compounded monthly, for 4 years.
    (b) A mortgage loan of \$80,000 at 6.75% versus 7.25%, compounded monthly, for 30 years.
    (c) In each of these 0.5% differences, what seems to have the greatest effect on the borrower: amount borrowed, interest rate, or duration of the loan? Explain.
32. Some banks now have biweekly mortgages (that is, with payments every other week). Compare a 20-year, \$100,000 loan at 8.1% by finding the payment size and the total interest paid over the life of the loan under each of the following conditions.
    (a) Payments are monthly, and the rate is 8.1%, compounded monthly.
    (b) Payments are biweekly, and the rate is 8.1%, compounded biweekly.
33. Many banks charge points on mortgage loans. Each point is the equivalent of a 1% charge on the amount borrowed and is paid before the loan is made as part of the closing costs of buying a home (closing costs include points, title fees, attorney's fees, assessor's fees, and so on).
    (a) If \$100,000 is borrowed for 25 years, for each of the following, find the payment size and the total paid over the life of the loan (including points).
        (i) $7\frac{1}{2}$%, compounded monthly, with 0 points
        (ii) $7\frac{1}{4}$%, compounded monthly, with 1 point
        (iii) 7%, compounded monthly, with 2 points
    (b) Which loan in (a) has the lowest total cost over the life of the loan?
34. Time shares provide an opportunity for vacationers to own a resort condo for 1 week (or more) each year forever. The owners may use their week at their own

condo or trade the week and vacation elsewhere. Time-share vacation sales usually require payment in full or financing through the time-share company, and interest rates are usually in the 13% to 18% range. Suppose the cost to buy a 1-week time share in a 3-bedroom condo is \$21,833. Also suppose a 10% down payment is required, with the balance financed for 15 years at 16.5%, compounded monthly.

(a) Find the monthly payment.
(b) Determine the total cost over the life of the loan.
(c) Suppose maintenance fees for this condo are \$400 per year. Find the annual cost of the condo over the life of the loan. Assume that the annual maintenance fees remain constant.
(d) Use (c) and the 10% down payment to determine the average annual cost for having this vacation condo for 1 week over the life of the loan.

Use a spreadsheet or financial program on a calculator or computer to complete Problems 35–38.

35. Develop an amortization schedule for a 4-year car loan if \$16,700 is borrowed at 8.2%, compounded monthly.
36. Develop an amortization schedule for a 10-year mortgage loan of \$80,000 at 7.2%, compounded monthly.
37. A recent college graduate bought a new car by borrowing \$18,000 at 8.4%, compounded monthly, for 5 years. She decides to pay an extra \$15 per payment. Use a spreadsheet to make an amortization schedule with this new payment. How many total payments will she have to make to pay off the loan?
38. A young couple buying their first home borrows \$85,000 for 30 years at 7.2%, compounded monthly. They make their required payment for the first 12 months. However, with the twelfth payment, they are able to pay an extra \$1000. After this payment how many of their original monthly payments will be required to pay off the loan (that is, how many payments do they save)? Use a spreadsheet amortization schedule from the thirteenth payment onward.

KEY TERMS AND FORMULAS

Section	Key Term	Formula
6.1	Simple interest	$I = Prt$
	Future value of a simple interest investment or loan	$S = P + I$
	Sequence function	
	Arithmetic sequence	$a_n = a_{n-1} + d \quad (n > 1)$
	Common difference	d
	nth term of	$a_n = a_1 + (n-1)d$
	Sum of first n terms	$s_n = \frac{n}{2}(a_1 + a_n)$
6.2	Future value of compound interest investment	
	Periodic compounding	$S = P(1+i)^n = P\left(1 + \frac{r}{m}\right)^{mt}$
	Continuous compounding	$S = Pe^{rt}$
	Annual percentage yield	
	Periodic compounding	$\text{APY} = \left(1 + \frac{r}{m}\right)^m - 1 = (1+i)^m - 1$
	Continuous compounding	$\text{APY} = e^r - 1$

	Geometric sequence	$a_n = ra_{n-1} \quad (n > 1)$
	Common ratio	r
	nth term of	$a_n = a_1 r^{n-1}$
	Sum of first n terms	$s_n = \dfrac{a_1(1-r^n)}{1-r} \quad (\text{if } r \neq 1)$
6.3	Future value of annuities	
	Ordinary annuity	$S = Rs_{\overline{n}\rvert i} = R\left[\dfrac{(1+i)^n - 1}{i}\right]$
	Required payment into an ordinary annuity (sinking fund)	$R = S\left[\dfrac{i}{(1+i)^n - 1}\right]$
	Annuity due	$S_{due} = R\left[\dfrac{(1+i)^n - 1}{i}\right](1+i) = Rs_{\overline{n}\rvert i}(1+i)$
6.4	Present value of annuities	
	Ordinary annuity	$A_n = R \cdot a_{\overline{n}\rvert i} = R\left[\dfrac{1-(1+i)^{-n}}{i}\right]$
	Annuity due	$A_{(n,due)} = R\left[\dfrac{1-(1+i)^{-n}}{i}\right](1+i) = Ra_{\overline{n}\rvert i}(1+i)$
	Deferred annuity	$A_{(n,k)} = R\left[\dfrac{1-(1+i)^{-n}}{i}\right](1+i)^{-k}$
6.5	True annual percentage rate (APR)	
	Amortization	$R = A_n\left(\dfrac{1}{a_{\overline{n}\rvert i}}\right) = A_n\left[\dfrac{i}{1-(1+i)^{-n}}\right]$
	Unpaid balance ($n - k$ payments left)	$A_{n-k} = R\left[\dfrac{1-(1+i)^{-(n-k)}}{i}\right]$
	Amortization schedule	

REVIEW EXERCISES

Section 6.1

1. Find the first 4 terms of the sequence with nth term

$$a_n = \frac{1}{n^2}$$

2. Identify any arithmetic sequences and find the common differences.
   (a) $12, 7, 2, -3, \ldots$ (b) $1, 3, 6, 10, \ldots$
   (c) $\frac{1}{6}, \frac{1}{3}, \frac{1}{2}, \frac{2}{3}, \ldots$

3. Find the 80th term of the arithmetic sequence with first term -2 and common difference 3.
4. Find the 36th term of the arithmetic sequence with third term 10 and eighth term 25.
5. Find the sum of the first 60 terms of the arithmetic sequence $\frac{1}{3}, \frac{1}{2}, \frac{2}{3}, \ldots$.

Section 6.2

6. Identify any geometric sequences and their common ratios.
   (a) $\frac{1}{4}, 2, 16, 128, \ldots$ (b) $16, -12, 9, -\frac{27}{4}, \ldots$
   (c) $4, 16, 36, 64, \ldots$
7. Find the fourth term of the geometric sequence with first term 64 and eighth term $\frac{1}{2}$.
8. Find the sum of the first 16 terms of the geometric sequence $\frac{1}{9}, \frac{1}{3}, 1, \ldots$.

Applications

Section 6.1

9. ***Simple interest*** If \$8000 is borrowed at 12% simple interest for 3 years, what is the future value of the loan at the end of the 3 years?

10. ***Simple interest*** Mary Toy borrowed \$2000 from her parents and repaid them \$2100 after 9 months. What simple interest rate did she pay?

11. ***Simple interest*** How much summer earnings must a college student deposit on August 31 in order to have \$3000 for tuition and fees on December 31 of the same year if the investment earns 6% simple interest?

12. ***Contest payments*** Suppose the winner of a contest receives \$10 on the first day of the month, \$20 on the second day, \$30 on the third day, and so on for a 30-day month. What is the total won?

13. ***Salaries*** Suppose you are offered two identical jobs: one paying a starting salary of \$20,000 with yearly raises of \$1000 and one paying a starting salary of \$18,000 with yearly raises of \$1200. Which job will pay more money over a 10-year period?

Section 6.2

14. ***Investments*** An investment is made at 8%, compounded quarterly for 10 years.
    (a) Find the number of periods.
    (b) Find the interest rate per period.

15. ***Future value*** Write the formula for the future value of a compound interest investment, if interest is compounded
    (a) periodically. (b) continuously.

16. ***Compound interest*** Which compounding method earns more money?
    (a) semiannually (b) monthly

17. ***Interest*** If \$1000 is invested for 4 years at 8%, compounded quarterly, how much interest will be earned?

18. ***Future value*** What is the future value if \$1000 is invested for 6 years at 8%, compounded semiannually?

19. ***Present value*** How much must one invest now in order to have \$18,000 in 4 years if the investment earns 5.4%, compounded monthly?

20. ***Future value*** What is the future value if \$1000 is invested for 6 years at 8%, compounded continuously?

21. ***Interest*** If \$8000 is invested at 12% compounded continuously for 3 years, what is the total interest earned at the end of the 3 years?

22. ***Present value*** A couple received an inheritance and plan to invest some of it for their grandchild's college education. How much must they invest if they would like the fund to have \$100,000 after 15 years, and their investment earns 10.31%, compounded continuously?

23. ***Continuous compounding*** If \$8000 is invested at 7%, compounded continuously, how long will it be before it grows to \$22,000?

24. ***Investments*** If \$15,000 is invested at 6%, compounded quarterly, how long will it be before it grows to \$25,000?

25. ***Compound rate*** If an initial investment of \$2500 grows to \$38,000 in 18 years, what annual interest rate, compounded annually, did this investment earn?

26. ***Continuous compounding*** (a) If an initial investment of \$35,000 grows to \$257,000 in 15 years, what annual interest rate, continuously compounded, was earned? (b) What is the annual percentage yield on this investment?

27. ***Annual percentage yield*** Find the annual percentage yield equivalent to a nominal rate of 7.2% (a) compounded quarterly and (b) compounded continuously.

28. ***Chess legend*** Legend has it that when the king of Persia wanted to reward the inventor of chess with whatever he desired, the inventor asked for one grain of wheat on the first square of the chessboard, with the number of grains doubled on each square thereafter for the remaining 63 squares. Find the number

of grains on the 64th square. (The sum of all the grains of wheat on all the squares would cover Alaska more than 3 inches deep with wheat!)

29. ***Chess legend*** If, in Problem 28, the king granted the inventor's wish for *half* a chessboard, how many grains would the inventor receive?

Section 6.3

30. ***Ordinary annuity*** Find the future value of an ordinary annuity of $800 paid at the end of every 6-month period for 10 years, if it earns interest at 12%, compounded semiannually.

31. ***Ordinary annuity*** If $100 is deposited at the end of each quarterly period for 4 years in an account that pays 8% interest, compounded quarterly, how much money will accrue in the account?

32. ***Payment into an ordinary annuity*** How much must be deposited at the end of each month into an account that earns 8.4%, compounded monthly, if the goal is to have $40,000 after 10 years?

33. ***Sinking fund*** How much would have to be invested at the end of each year at 6%, compounded annually, to pay off a debt of $80,000 in 10 years?

34. ***Annuity due*** Find the future value of an annuity due of $800 paid at the beginning of every 6-month period for 10 years, if it earns interest at 12%, compounded semiannually.

35. ***Payment into an annuity due*** A company wants to have $250,000 available in $4\frac{1}{2}$ years for new construction. How much must be deposited at the beginning of each quarter to reach this goal if the investment earns 10.2%, compounded quarterly?

Section 6.4

36. ***Present value of an ordinary annuity*** What lump sum would have to be invested at 9%, compounded semiannually, to provide an annuity of $10,000 at the end of each half-year for 10 years?

37. ***Deferred annuity*** A couple wishes to set up an annuity that will provide 6 monthly payments of $3000 while they take an extended cruise. How much of a $15,000 inheritance must be set aside if they plan to leave in 5 years and want the first payment before they leave? Assume that money is worth 7.8%, compounded monthly.

38. ***Divorce settlement*** A divorce settlement of $40,000 is paid in $1000 payments at the end of each of 40 months. What is the present value of this settlement if money is worth 12%, compounded monthly?

39. ***Powerball lottery*** Winners of lotteries receive the jackpot distributed over a period of years, usually 20 or 25 years. The winners of the Powerball lottery on July 29, 1998, elected to take a one-time cash payout rather than receive the $295.7 million jackpot in 25 annual payments beginning on the date the lottery was won.
    (a) How much money would the winners have received at the beginning of each of the 25 years?
    (b) If the value of money was 5.91%, compounded annually, what one-time payout did they receive in lieu of the annual payments?

40. ***Payment from an ordinary annuity*** A recent college graduate's gift from her grandparents is $20,000. How much will this provide at the end of each month for the next 12 months while the graduate travels? Assume that money is worth 6.6%, compounded monthly.

41. ***Payment from an annuity due*** An IRA of $250,000 is rolled into an annuity that pays a retired couple at the beginning of each quarter for the next 20 years. If the annuity earns 6.2%, compounded quarterly, how much does the couple receive each quarter?

42. ***Deferred annuity*** A young couple receives an inheritance of $72,000 that they want to set aside for a college fund for their two children. How much will this provide at the end of each half-year for a period of 9 years if it is deferred for 11 years and can be invested at 7.3%, compounded semiannually?

Section 6.5

43. ***Amortization*** A debt of $1000 with interest at 12%, compounded monthly, is amortized by 12 monthly payments (of equal size). What is the size of each payment?

44. ***Amortization*** A couple borrowed $92,000 to buy a condominium. Their loan was for 25 years and money is worth 6%, compounded monthly.
    (a) Find their monthly payment size.
    (b) Find the total amount they would pay over 25 years.
    (c) Find the total interest they would pay over 25 years.

45. ***Unpaid balance*** A debt of \$8000 is amortized with eight semiannual payments of \$1288.29 each. If money is worth 12%, compounded semiannually, find the unpaid balance after five payments have been made.

46. ***Cash value*** A woman paid \$10,000 down for a house and agreed to pay 18 quarterly payments of \$1500 each. If money is worth 4%, compounded quarterly, how much would the house have cost if she had paid cash?

47. ***Amortization schedule*** Complete the next two lines of the amortization schedule for a \$100,000 loan for 30 years at 7.5%, compounded monthly, with monthly payments of \$699.22.

Payment Number	*Payment Amount*	*Interest*	*Balance Reduction*	*Unpaid Balance*
56	\$699.22	\$594.67	\$104.55	\$95,042.20

CHAPTER TEST

1. If \$8000 is invested at 7%, compounded continuously, how long will it be before it grows to \$47,000?
2. How much must you put aside at the end of each half-year if, in 6 years, you want to have \$12,000, and the current interest rate is 6.2%, compounded semiannually?
3. To buy a bond that matures to \$5000 in 9 months, you must pay \$4756.69. What simple interest rate is earned?
4. A debt of \$280,000 is amortized with 40 equal semiannual payments of \$14,357.78. If interest is 8.2%, compounded semiannually, find the unpaid balance of the debt after 25 payments have been made.
5. If an investment of \$1000 grew to \$13,500 in 9 years, what interest rate, compounded annually, did this investment earn?
6. A couple borrowed \$97,000 at 7.2%, compounded monthly, for 25 years to purchase a condominium.
   (a) Find their monthly payment.
   (b) Over the 25 years, how much interest will they pay?
7. If you borrow \$2500 for 15 months at 4% simple interest, how much money must you repay after the 15 months?
8. Suppose you invest \$100 at the end of each month in an account that earns 6.9%, compounded monthly. What is the future value of the account after $5\frac{1}{2}$ years?
9. Find the annual percentage yield (the effective annual rate) for an investment that earns 8.4%, compounded monthly.
10. An accountant wants to withdraw \$3000 from an investment at the beginning of each quarter for the next 15 years. How much must be deposited originally if the investment earns 6%, compounded quarterly? Assume that after the 15 years, the balance is zero.
11. If \$10,000 is invested at 7%, compounded continuously, what is the value of the investment after 20 years?
12. A collector made a \$10,000 down payment for a classic car and agreed to make 18 quarterly payments of \$1500 at the end of each quarter. If money is worth 8%, compounded quarterly, how much would the car have cost if the collector had paid cash?
13. What amount must you invest in an account that earns 6.8%, compounded quarterly, if you want to have \$9500 after 5 years?
14. Suppose a salesman invests his \$12,500 bonus in a fund that earns 10.8%, compounded monthly. Suppose also that he makes contributions of \$150 at the end of each month to this fund.
    (a) Find the future value after $12\frac{1}{2}$ years.
    (b) If after the $12\frac{1}{2}$ years, the fund is used to set up an annuity, how much will it pay at the end of each month for the next 10 years?
15. A couple would like to have \$50,000 at the end of $4\frac{1}{2}$ years. They can invest \$10,000 now and make additional contributions at the end of each quarter over the $4\frac{1}{2}$ years. If the investment earns 7.7%, compounded quarterly, what size payments will allow them to reach their goal?
16. A woman makes \$3000 contributions at the end of each half-year to a retirement account for a period of 8 years. For the next 20 years, she makes no additional contributions and no withdrawals.
    (a) If the account earns 7.5%, compounded semiannually, find the value of the account after the 28 years.

(b) If this account is used to set up an annuity that pays her an amount at the beginning of each 6-month period for the next 20 years, how much will each payment be?

17. Maxine deposited \$80 at the beginning of each month for 15 years in an account that earned 6%, compounded monthly. Find the value of the account after the 15 years.

18. Grandparents plan to establish a college trust for their youngest grandchild. How much is needed now so the trust, which earns 6.4%, compounded quarterly, will provide \$4000 at the end of each quarter for 4 years after being deferred for 10 years?

19. The following sequence is arithmetic: 298.8, 293.3, 287.8, 282.3, ...
    (a) If you were not told that it was arithmetic, explain how you could tell.
    (b) Find the 51st term of this sequence.
    (c) Find the sum of the first 51 terms.

20. A 400-milligram dose of heart medicine is taken daily. During each 24-hour period, the body eliminates 40% of this drug (so that 60% remains in the body). Thus the amount of drug in the bloodstream just after the 31st dose is given by

$$400 + 400(0.6) + 400(0.6)^2 + 400(0.6)^3 + \cdots + 400(0.6)^{29} + 400(0.6)^{30}$$

Find the level of the drug in the bloodstream at this time.

Extended Applications Group Projects

I. Mail Solicitation Home Equity Loans: Are These Good Deals?

Discover® offers a Home Equity Line of Credit that lowers a hypothetical monthly payment by $366.12. The following table, published by Discover®, shows how.

The home equity loan payment is based on "10.49% APR annualized over a 10-year term." This means the loan is amortized at 10.49% with monthly payments.

(a) Using this interest rate and monthly payments for the 10 years, how much money will still be owed at the end of the 10 years?

(b) How much would the payment have to be to have the debt paid in the 10 years?

	Current Monthly Payment Example		*Discover Home Equity*	
Loan Type	*Loan Amount*	*Monthly Payment*	*Loan Amount*	*Monthly Payment*
Bank cards	$10,000	$186.00	PAID OFF	NONE
Auto loan	$12,500	$320.03	PAID OFF	NONE
Department store cards	$2,500	$78.63	PAID OFF	NONE
Discover Home Equity Line of Credit			$25,000	$218.54
Total	$25,000	$584.66	$25,000	$218.54

SOURCE: Discover Loan Center mailer

A second plan is offered by Direct Equity Lending, which gives the example shown in the table on the next page. The monthly savings with its plan are stated as $633. To see whether this plan actually gives savings:

(a) Find how long it will take to pay off the bills if the average interest rate for the bills is 18%.

(b) What is the total cost one incurs by paying the $31,500 in bills using the current payment plan?

(c) The company's example says the home equity loan will pay off the $31,500 in 25 years at an APR of 15.934%. Compute the payment that would do this, and compare your result with the payment the company states.

(d) What is the total cost of making the company's stated payments of $432.57?

CURRENT BILLS			DIRECT EQUITY LENDING LOAN		
Bills	*Balance*	*Payments*	*Payment*	*Monthly Saving*	*Yearly Saving*
Mastercard	5,850	135.00	0.00		
Visa	5,300	157.50	0.00		
Discover	10,090	262.50	0.00		
J.C. Penney	5,200	262.50	0.00		
R.H. Macy	5,060	249.00	0.00		
TOTAL	$31,500	1,066.50	432.57	633.93	7,607.16

SOURCE: Direct Equity Lending mailer

II. Profit Reinvestment

T. C. Hardware Store wants to construct a new building at a second location. If construction could begin immediately, the cost would be $250,000. At this time, however, the owners do not have the required 20% down payment, so they plan to invest $1000 per month of their profits until they have the necessary amount. They can invest their money in an annuity account that pays 6%, compounded monthly, but they are concerned about the 3% average inflation rate in the construction industry. They would like you to give them some projections about how 3% inflation will affect the time required to accrue the down payment and the building's eventual cost. They would also like to know how their projected profits, after the new building is complete, will affect their schedule for paying off the mortgage loan.

Specifically, the owners would like you to prepare a report that answers the following questions.

1. If the 3% annual inflation is accurate, how long will it take to get the down payment? *Hint:* Assume inflation is compounded monthly. Then the required down payment will be 20% of

$$250{,}000\left(1 + \frac{0.03}{12}\right)^n$$

   where n is in months. Thus you must find n such that this amount equals the future value of the owners' annuity. The solution to the resulting equation is difficult to obtain by algebraic methods but can be found with a graphing utility.
2. If the 3% annual inflation rate is accurate (compounded monthly), what will T. C. Hardware's projected construction costs be (to the nearest hundred dollars) when it has its down payment?
3. If T. C. Hardware borrows 80% of its construction costs and amortizes that amount at 7.8%, compounded monthly, for 15 years, what will its monthly payment be?
4. Finally, the owners believe that 2 years after this new building is begun (that is, after 2 years of payments on the loan), their profits will increase enough so that they'll be able to make double mortgage payments each month until the loan is paid. Under these assumptions, how long will it take T. C. Hardware to pay off the loan?

III. Purchasing a Home

Buying a home is the biggest single investment or purchase that most individuals make. This project is designed to give you some insight into the home-buying process and associated costs.

Find all the costs associated with buying a home by making a 20% down payment and borrowing 80% of the cost of the home.

1. Decide how much "your home" will cost (and how much you'll have to put down and how much to mortgage). Call a realtor to determine the cost of a home in your area (this could be a typical starter home, an average home for the area, or your dream house). Ask the realtor to identify and estimate all the closing costs associated with buying your home. Closing costs include your down payment, attorney's fees, title fees, and so on.
2. Call at least three different banks to determine mortgage loan rates for 15 years, 25 years, and 30 years and costs associated with obtaining your loan. These associated costs of a loan are paid at closing and include an application fee, appraisal fee, points, and so on.
3. For each bank and each different duration for the loan, develop a summary that contains
   (a) an itemization and explanation of all closing costs and the total amount due at closing
   (b) the monthly payment
   (c) the total paid over the life of the loan
   (d) the amortization schedule for each loan (use a spreadsheet or financial software package)
4. For each duration (15 years, 25 years, and 30 years), identify which bank gives the best loan rate and explain why you think it is best. Be sure to consider what is best for someone who is likely to remain in the home for 7 years or less compared with someone who is likely to stay 20 years or more.

Warm-up

Prerequisite Problem Type	For Section	Answer	Section for Review
(a) The sum of x, $\frac{1}{4}$, and $\frac{2}{3}$ is 1. What is x? (b) If the sum of x plus $\frac{1}{3}$ plus $\frac{1}{3}$ is 1, what is x?	7.1	(a) $x = \frac{1}{12}$ (b) $x = \frac{1}{3}$	1.1 Linear equations
If 12 numbers are weighted so that each of the 12 numbers has an equal weight and the sum of the weights is 1, what weight should be assigned to each number?	7.1	$\frac{1}{12}$	1.1 Linear equations
(a) List the set of integers that satisfies $2 \leq x < 6$. (b) List the set of integers that satisfies $4 \leq s \leq 12$.	7.1	(a) $\{2, 3, 4, 5\}$ (b) $\{4, 5, 6, 7, 8, 9, 10, 11, 12\}$	4.1 Inequalities
For the sets $E = \{1, 2, 3, 4, 5\}$ and $F = \{2, 4, 6, 8\}$, find (a) $E \cup F$ (b) $E \cap F$	7.1 7.2	(a) $\{1, 2, 3, 4, 5, 6, 8\}$ (b) $\{2, 4\}$	0.1 Sets
Find the product: $[0.9 \quad 0.1]\begin{bmatrix} 0.8 & 0.2 \\ 0.3 & 0.7 \end{bmatrix}$	7.7	$[0.75 \quad 0.25]$	3.2 Matrix multiplication
Solve the system using matrices. $\begin{cases} 0.5V_1 + 0.4V_2 + 0.3V_3 = V_1 \\ 0.4V_1 + 0.5V_2 + 0.3V_3 = V_2 \\ 0.1V_1 + 0.1V_2 + 0.4V_3 = V_3 \end{cases}$	7.7	$V_1 = 3V_3$ $V_2 = 3V_3$	3.3 Gauss-Jordan elimination

Chapter 7

Introduction to Probability

An economist cannot predict exactly how the gross national product will change, a sociologist cannot determine exactly how group behavior is affected by climatic conditions, and a psychologist cannot determine the exact effect of environment on behavior. Because the behavior of the economy and of people is subject to chance, its prediction involves probability. Answers to questions about the behavioral and life sciences, business, and economics are frequently found by conducting experiments with samples from the larger population and by using statistics (discussed in the next chapter). The experiments are designed to compare what actually happens with what probability theory predicted would happen. For example, if a rat runs a T-maze, can it learn which way to turn to get food? If the food is on the right and the rat turns right many more times than probability indicates that it should, then we have evidence that it has learned. The conclusions drawn from experiments are usually given in terms of probability because they are extended to larger populations. For example, substances that cause cancer in experimental animals may not have been proved to cause cancer in humans, but they are removed from the market because the probability is high that they will.

We will solve a number of problems involving games in this chapter because they lend themselves easily to probability. (In fact, probability theory is really an outgrowth of interest in solving gambling problems.) We will also see that the theory of probability can be useful in many applications in the management, social, and life sciences.

7.1 Probability; Odds

OBJECTIVES

- *To compute the probability of the occurrence of an event*
- *To construct a sample space for a probability experiment*
- *To compute the odds that an event will occur*
- *To compute the empirical probability that an event will occur*

APPLICATION PREVIEW

Before the General Standard Company can place a 3-year warranty on all of the faucets it produces, it needs to find the **probability** that each faucet it produces will function properly (not leak) for 3 years. In this section we will use **sample spaces** to find the probability that an event will occur by using assumptions or knowledge of the entire **population** of data under consideration. When knowledge of the entire population is not available, we can compute the **empirical probability** that an event will occur by using data collected about the probability experiment. This method can be used to compute the probability that the General Standard faucets will work properly if information about a large **sample** of the faucets is gathered.

When we toss a coin, it can land in one of two ways, heads or tails. If the coin is a "fair" coin, these two possible **outcomes** have an equal chance of occurring, and we say the outcomes of this **probability experiment** are **equally likely.** If we seek the probability that an experiment has a certain result, that result is called an **event.**

Suppose an experiment can have a total of n equally likely outcomes, and that k of these outcomes would be considered successes. Then the probability of achieving a success in the experiment is k/n. That is, the probability of a success in an experiment is the number of ways the experiment can result in a success divided by the total number of possible outcomes.

Probability of a Single Event

If an event E can happen in k ways out of a total of n equally likely possibilities, the probability of the occurrence of the event is denoted by

$$\Pr(E) = \frac{\text{Number of successes}}{\text{Number of possible outcomes}} = \frac{k}{n}$$

EXAMPLE 1

If we draw a ball from a bag containing 4 white balls and 6 black balls, what is the probability of

(a) getting a white ball?
(b) getting a black ball?
(c) not getting a white ball?

Solution

(a) A white ball can occur (be drawn) in 4 ways out of a total of 10 equally likely possibilities. Thus the probability of drawing a white ball is

$$\Pr(W) = \frac{4}{10} = \frac{2}{5}$$

(b) We have 6 chances to succeed at drawing a black ball out of a total of 10 possible outcomes. Thus the probability of getting a black ball is

$$\Pr(B) = \frac{6}{10} = \frac{3}{5}$$

(c) The probability of not getting a white ball is Pr(*B*), because not getting a white ball is the same as getting a black ball. Thus Pr(not *W*) = Pr(*B*) = 3/5.

Suppose a card is selected at random* from a box containing cards numbered 1 through 12. To find the probability that the card drawn has an even number on it, it is necessary to list (at least mentally) all the possible outcomes of the experiment (selecting a number at random). A set that contains all the possible outcomes of an experiment is called a **sample space.** For the experiment of drawing a numbered card, a sample space is

$$S = \{1, 2, 3, 4, 5, 6, 7, 8, 9, 10, 11, 12\}$$

Each element of the sample space is called a **sample point,** and an **event** is a subset of the sample space. Each sample point in the sample space is assigned a nonnegative **probability measure** or **probability weight** such that the sum of the weights in the sample space is 1. The probability of an event is the sum of the weights of the sample points in the event's **subspace** (subset of *S*).

For example, the experiment "drawing a number at random from the numbers 1 through 12" has the sample space *S* given above. Because each element of *S* is equally likely to occur, we assign a probability weight of 1/12 to each element. The event "drawing an even number" in this experiment has the subspace

$$E = \{2, 4, 6, 8, 10, 12\}$$

Because each of the 6 elements in the subspace has weight 1/12, the probability of event *E* is

$$\Pr(\text{even}) = \Pr(E) = \frac{1}{12} + \frac{1}{12} + \frac{1}{12} + \frac{1}{12} + \frac{1}{12} + \frac{1}{12} = \frac{6}{12} = \frac{1}{2}$$

Note that using the sample space *S* for this experiment gives the same result as using the formula $\Pr(E) = k/n$.

Some experiments can have an infinite number of outcomes, but we will concern ourselves only with experiments having finite sample spaces. We can usually construct more than one sample space for an experiment. The sample space in which each sample point is equally likely is called an **equiprobable sample space.** Suppose the number of elements in a sample space is $n(S) = n$, with each element representing an equally likely outcome of a probability experiment. Then the weight (probability) assigned to each element is

$$\frac{1}{n(S)} = \frac{1}{n}$$

If an event *E* that is a subset of an equiprobable sample space contains $n(E) = k$ elements of *S*, then we can restate the probability of an event *E* in terms of the number of elements of *E*.

*Selecting a card at random means every card has an equal chance of being selected.

Probability of an Event If an event E can occur in $n(E) = k$ ways out of $n(S) = n$ equally likely ways, then

$$\Pr(E) = \frac{n(E)}{n(S)} = \frac{k}{n}$$

EXAMPLE 2

If a number is to be selected at random from the integers 1 through 12, what is the probability that it is

(a) divisible by 3?
(b) even and divisible by 3?
(c) even or divisible by 3?

Solution

The set $S = \{1, 2, 3, 4, 5, 6, 7, 8, 9, 10, 11, 12\}$ is an equiprobable sample space for this experiment (as mentioned previously).

(a) The set $E = \{3, 6, 9, 12\}$ contains the numbers that are divisible by 3. Thus

$$\Pr(\text{divisible by } 3) = \frac{n(E)}{n(S)} = \frac{4}{12} = \frac{1}{3}$$

(b) The numbers 1 through 12 that are even *and* divisible by 3 are 6 and 12, so

$$\Pr(\text{even and divisible by } 3) = \frac{2}{12} = \frac{1}{6}$$

(c) The numbers that are even *or* divisible by 3 are 2, 3, 4, 6, 8, 9, 10, 12, so

$$\Pr(\text{even or divisible by } 3) = \frac{8}{12} = \frac{2}{3}$$

CHECKPOINT

1. If a coin is drawn from a box containing 4 gold, 10 silver, and 16 copper coins, all the same size, what is the probability of
   (a) getting a gold coin?
   (b) getting a copper coin?
   (c) not getting a copper coin?

If an event E is certain to occur, E contains all of the elements of the sample space, S. Hence the sum of the probability weights of E is the same as that of S, so

$$\Pr(E) = 1 \qquad \text{if } E \text{ is certain to occur}$$

If event E is impossible, $E = \varnothing$, so

$$\Pr(E) = 0 \qquad \text{if } E \text{ is impossible}$$

Because all probability weights of elements of a sample space are nonnegative,

$$0 \le \Pr(E) \le 1 \qquad \text{for any event } E$$

EXAMPLE 3

Suppose a coin is tossed 3 times.

(a) Construct an equiprobable sample space for the experiment.
(b) Find the probability of obtaining 0 heads.
(c) Find the probability of obtaining 2 heads.

Solution

(a) Perhaps the most obvious way to record the possibilities for this experiment is to list the number of heads that could result: {0, 1, 2, 3}. But the probability of obtaining 0 heads is different from the probability of obtaining 2 heads, so this sample space is not equiprobable. A sample space in which each outcome is equally likely is {HHH, HHT, HTH, THH, HTT, THT, TTH, TTT}, where HHT indicates that the first two tosses were heads and the third was a tail.

(b) Because there are 8 equally likely possible outcomes, $n(S) = 8$. Only one of the eight possible outcomes, $E = \{\text{TTT}\}$, gives 0 heads, so $n(E) = 1$. Thus

$$\Pr(0 \text{ heads}) = \frac{n(E)}{n(S)} = \frac{1}{8}$$

(c) The event "two heads" is $F = \{\text{HHT, HTH, THH}\}$, so

$$\Pr(2 \text{ heads}) = \frac{n(F)}{n(S)} = \frac{3}{8}$$

To find the probability of obtaining a given sum when a pair of dice is rolled, we need to determine how many outcomes are possible. If we distinguish between the two dice we are rolling, and we record all the possible outcomes for each die, we see that there are 36 possibilities, each of which is equally likely (see Table 7.1).

This list of possible outcomes for finding the sum of two dice is an equiprobable sample space for the experiment. Because each *element* of this sample space has the same chance of occurring, this is an equiprobable sample space. Because the 36 elements in the sample space are equally likely, we can find the probability that a given sum results by determining the number of ways

TABLE 7.1

	Second Die					
First Die	*1*	*2*	*3*	*4*	*5*	*6*
1	(1, 1)	(1, 2)	(1, 3)	(1, 4)	(1, 5)	(1, 6)
2	(2, 1)	(2, 2)	(2, 3)	(2, 4)	(2, 5)	(2, 6)
3	(3, 1)	(3, 2)	(3, 3)	(3, 4)	(3, 5)	(3, 6)
4	(4, 1)	(4, 2)	(4, 3)	(4, 4)	(4, 5)	(4, 6)
5	(5, 1)	(5, 2)	(5, 3)	(5, 4)	(5, 5)	(5, 6)
6	(6, 1)	(6, 2)	(6, 3)	(6, 4)	(6, 5)	(6, 6)

this sum can occur and dividing that number by 36. Thus the probability that sum 6 will occur is 5/36, and the probability that a 9 will occur is 4/36 = 1/9. (See the four ways the sum 9 can occur in Table 7.1.) Note that we could have made a sample space for this experiment (finding the sum of a pair of dice) by listing the sums: {2, 3, 4, 5, 6, 7, 8, 9, 10, 11, 12}. However, these outcomes are not equally likely; the probability of obtaining a 2 is different from the probability of obtaining a 6. We can see that the probabilities of events can be found more easily if an equiprobable sample space is used to determine the possibilities.

EXAMPLE 4

Use Table 7.1 to find the following probabilities if a pair of distinguishable dice is rolled.

(a) Pr(sum is 5) (b) Pr(sum is 2) (c) Pr(sum is 8)

Solution

(a) If E is the event "sum of 5," then $E = \{(4, 1), (3, 2), (2, 3), (1, 4)\}$.

Therefore, $\Pr(\text{sum is } 5) = \dfrac{n(E)}{n(S)} = \dfrac{4}{36} = \dfrac{1}{9}$.

(b) The sum 2 results only from $\{(1, 1)\}$. Thus Pr(sum is 2) = 1/36.

(c) The sample points that give a sum of 8 are in the subspace $F = \{(6, 2), (5, 3), (4, 4), (3, 5), (2, 6)\}$.

Thus $\Pr(\text{sum is } 8) = \dfrac{n(F)}{n(S)} = \dfrac{5}{36}$.

CHECKPOINT

2. A ball is to be drawn from a bag containing balls numbered 1, 2, 3, 4, and 5. To find the probability that a ball with an even number is drawn, we can use the sample space $S_1 = \{\text{even, odd}\}$ or $S_2 = \{1, 2, 3, 4, 5\}$.
   (a) Which of these sample spaces is an equiprobable sample space?
   (b) What is the probability of drawing a ball that is even-numbered?
3. A coin is tossed and a die is rolled.
   (a) Write an equiprobable sample space listing the possible outcomes for this experiment.
   (b) Find the probability of getting a head on the coin and an even number on the die.
   (c) Find the probability of getting a tail on the coin and a number divisible by 3 on the die.

If we know the probability that an event E will occur, we can compute the **odds** in favor of E occurring and the odds against E occurring.

Odds If the probability that event E occurs is $\Pr(E) \neq 1$, then the odds that E will occur are

$$\text{Odds in favor of } E = \frac{\Pr(E)}{1 - \Pr(E)}$$

and the odds that E will not occur, if $\Pr(E) \neq 0$, are

$$\text{Odds against } E = \frac{1 - \Pr(E)}{\Pr(E)}$$

The odds $\frac{a}{b}$ are stated as "a to b" or "$a{:}b$."

EXAMPLE 5

If the probability of drawing a queen from a deck of playing cards is 1/13, what are the odds

(a) in favor of drawing a queen? (b) against drawing a queen?

Solution

(a) $\dfrac{1/13}{1 - 1/13} = \dfrac{1/13}{12/13} = \dfrac{1}{12}$

The odds in favor of drawing a queen are 1 to 12, which we write as 1:12.

(b) $\dfrac{12/13}{1/13} = \dfrac{12}{1}$

The odds against drawing a queen are 12 to 1, denoted 12:1.

Odds and Probability

If the odds in favor of an event E occurring are $a{:}b$, the probability that E will occur is

$$\Pr(E) = \frac{a}{a + b}$$

If the odds against E are $b{:}a$, then

$$\Pr(E) = \frac{a}{a + b}$$

EXAMPLE 6

If the odds against dying in an industrial accident are 3992:3, what is the probability of dying in an industrial accident?

Solution

The probability is

$$\frac{3}{3 + 3992} = \frac{3}{3995}$$

Up to this point, the probabilities assigned to events were determined *theoretically,* either by assumption (a fair coin has probability 1/2 of obtaining a head in one toss) or by knowing the entire population under consideration (if 1000 people are in a room and 400 of them are males, the probability of picking a person at random and getting a male is 400/1000 = 2/5). We can also determine probabilities **empirically,** where the assignment of the probability of an event is frequently derived from our experiences or from data that have been gathered. Probabilities developed in this way are called **empirical probabilities.** An empirical probability is formed by conducting an experiment, or by observing a situation a number of times, and noting how many times a certain event occurs.

Empirical Probability of an Event

$$\Pr(\text{Event}) = \frac{\text{Number of observed occurrences of the event}}{\text{Total number of trials or observations}}$$

In some cases empirical probability is the only way to assign a probability to an event. For example, life insurance premiums are based on the probability that an individual will live to a certain age. These probabilities are developed empirically and organized into mortality tables.

In other cases, empirical probabilities may have a theoretical framework. For example, for tossing a coin,

$$\Pr(\text{Head}) = \Pr(\text{Tail}) = \frac{1}{2}$$

gives the theoretical probability. If a coin was actually tossed 1000 times, it might come up heads 517 times and tails 483 times, giving empirical probabilities for these 1000 tosses of

$$\Pr(\text{Head}) = \frac{517}{1000} = .517$$

$$\Pr(\text{Tail}) = \frac{483}{1000} = .483$$

On the other hand, if the coin was tossed only 4 times, it might come up heads 3 times and tails once, giving empirical probabilities vastly different from the theoretical ones. In general, empirical probabilities can differ from theoretical probabilities, but they are likely to reflect the theoretical probabilities when the number of observations is large. Moreover, when this is not the case, there may be reason to suspect that the theoretical model does not fit the event (for example, that a coin may be biased).

EXAMPLE 7

Experience has shown that 800 of the 500,000 microcomputer diskettes manufactured by a firm are defective. On the basis of this evidence, determine the probability that a diskette picked at random from those manufactured by this firm will be defective.

Solution

The 800 defective diskettes occurred in a very large number of diskettes manufactured, so the ratio of the number of defective diskettes to the total number of diskettes manufactured provides the *empirical* probability of a diskette picked at random being defective.

$$\Pr(\text{defective}) = \frac{800}{500{,}000} = \frac{1}{625}$$

EXAMPLE 8

As mentioned in the Application Preview, the General Standard Company can use information about a large sample of its faucets to compute the empirical probability that a faucet chosen at random will function properly (not leak) for 3 years. Of a sample of 12,316 faucets produced in 1992, 137 were found to fail within 3 years. Find the empirical probability that any faucet the company produces will work properly for 3 years.

Solution

If 137 of these faucets fail to work properly, then 12,316 − 137 = 12,179 will work properly. Thus the probability that any faucet selected at random will function properly is

$$\text{Pr(not leak)} = \frac{12{,}179}{12{,}316} \approx .989$$

CHECKPOINT SOLUTIONS

1. (a) The number of coins is 30 and the number of gold coins is 4, so Pr(gold) = 4/30 = 2/15.
   (b) The number of copper coins is 16, so Pr(copper) = 16/30 = 8/15.
   (c) The number of coins that are not copper is 14, so Pr(not copper) = 14/30 = 7/15.
2. (a) S_1 = {even, odd} is not an equiprobable sample space for this experiment because there are not equal numbers of even and odd numbers. S_2 = {1, 2, 3, 4, 5} is an equiprobable sample space.
   (b) Pr(even) = 2/5
3. (a) {H1, H2, H3, H4, H5, H6, T1, T2, T3, T4, T5, T6}
   (b) Pr(head and even number) = 3/12 = 1/4
   (c) Pr(tail and number divisible by 3) = 2/12 = 1/6

EXERCISE 7.1

1. If you draw one card at random from a deck of 12 cards numbered 1 through 12, inclusive, what is the probability that the number you draw is divisible by 4?
2. An ordinary die is tossed. What is the probability of getting a 3 or a 4?
3. A die is rolled. Find the probability of getting a number greater than 0.
4. A die is rolled. What is the probability that
   (a) a 4 will result?
   (b) a 7 will result?
   (c) an odd number will result?
5. One ball is drawn at random from a bag containing 4 red balls and 6 white balls. What is the probability that the ball is
   (a) red? (b) green? (c) red or white?
6. One ball is drawn at random from a bag containing 4 red balls and 6 white balls. What is the probability that the ball is
   (a) white? (b) white or red? (c) red and white?
7. An urn contains three red balls numbered 1, 2, 3, four white balls numbered 4, 5, 6, 7, and three black balls numbered 8, 9, 10. A ball is drawn from the urn. What is the probability that
   (a) it is red?
   (b) it is odd-numbered?
   (c) it is red and odd-numbered?
   (d) it is red or odd-numbered?
   (e) it is not black?
8. An urn contains three red balls numbered 1, 2, 3, four white balls numbered 4, 5, 6, 7, and three black balls

numbered 8, 9, 10. A ball is drawn from the urn. What is the probability that the ball is
(a) white? (b) white and odd?
(c) white or even? (d) black or white?
(e) black and white?

9. From a deck of 52 ordinary playing cards, one card is drawn. Find the probability that it is
(a) a queen. (b) a red card. (c) a spade.
10. From a deck of 52 ordinary playing cards, one card is drawn. Find the probability that it is
(a) a red king.
(b) a king or a black card.
11. Suppose a fair coin is tossed two times. Construct an equiprobable sample space for the experiment, and determine each of the following probabilities:
(a) Pr(0 heads) (b) Pr(1 head) (c) Pr(2 heads)
12. Suppose a fair coin is tossed four times. Construct an equiprobable sample space for the experiment, and determine each of the following probabilities:
(a) Pr(2 heads) (b) Pr(3 heads) (c) Pr(4 heads)
13. Use Table 7.1 to determine the following probabilities if a distinguishable pair of dice is rolled.
(a) Pr(sum is 4)
(b) Pr(sum is 10)
(c) Pr(sum is 12)
14. (a) When a pair of distinguishable dice is rolled, what sum is most likely to occur?
(b) When a pair of distinguishable dice is rolled, what is $\Pr(4 \leq S \leq 8)$, where S represents the sum rolled?
15. If a pair of dice, one green and one red, is rolled, what is
(a) $\Pr(4 \leq S \leq 7)$, where S is the sum rolled?
(b) $\Pr(8 \leq S \leq 12)$, where S is the sum rolled?
16. If a green die and a red die are rolled, find
(a) $\Pr(2 \leq S \leq 6)$, where S is the sum rolled on the two dice.
(b) $\Pr(4 \leq S)$, where S is the sum rolled on the two dice.
17. Suppose a die is tossed 1200 times and a 6 comes up 431 times.
(a) Find the empirical probability for a 6 to occur.
(b) On the basis of a comparison of the empirical probability and the theoretical probability, do you think the die is fair or biased?
18. Suppose that a coin is tossed 3000 times and 1800 heads result. What is the empirical probability that a head will occur with this coin? Is there evidence that the coin is a fair coin?
19. If the probability that an event will occur is 2/5, what are the odds
(a) in favor of the event occurring?
(b) against the event occurring?
20. If the probability that an event will occur is 7/15, what are the odds
(a) that the event will occur?
(b) that the event will not occur?
21. If the probability that an event E will not occur is 8/11, what are the odds
(a) that E will occur?
(b) that E will not occur?
22. If the probability that an event E will not occur is 4/19, what are the odds
(a) that E will not occur?
(b) that E will occur?
23. If the odds that a particular horse will win a race are 1:20, what is the probability
(a) that the horse will win the race?
(b) that the horse will lose the race?
24. If the odds that an event E will occur are 3:19, what is the probability
(a) that E will occur?
(b) that E will not occur?
25. If the odds that an event E will not occur are 23:57, what is the probability
(a) that event E will occur?
(b) that event E will not occur?
26. If the odds that an event F will not occur are 21:23, what is the probability
(a) that event F will occur?
(b) that event F will not occur?

Applications

27. ***Sales promotion*** In a sales promotion, a clothing store gives its customers a chance to draw a ticket from a box that contains a discount on their next purchase. The box contains 3000 tickets giving a 10% discount, 500 giving a 30% discount, 100 giving a 50% discount, and 1 giving a 100% discount. What is the probability that a given customer will randomly draw a ticket giving
(a) a 100% discount?
(b) a 50% discount?
(c) a discount of less than 50%?

28. ***Management*** A dry cleaning firm has 12 employees: 7 women and 5 men. Three of the women and

five of the men are 40 years old or older. The remainder are over 20 years of age and under 40. If a person is chosen at random from this firm, what is the probability that the person is
(a) a woman?
(b) under 40 years of age?
(c) 20 years old?

29. ***Blood types*** Human blood is classified by blood type, which indicates the presence or absence of the antigens A, B, and Rh, as follows.

A present	Type A
B present	Type B
Both A and B present	Type AB
Neither A nor B present	Type O

Each of these types is combined with a + or a − sign to indicate whether the Rh antigen is present or not. Write a sample space containing all possible blood types.

30. ***Genetics*** Construct an equiprobable sample space that gives the possible combinations of male and female children in a family with three children.

31. ***Drug use*** Forty-six percent of all drug use occurs in cities (Source: Partnership for a Drugfree America). If an instance of drug use is chosen at random, what is the probability that the use occurs in a city?

32. ***Breast cancer*** According to the American Cancer Society, 199 of 200 mammograms turn out to be normal. What is the probability that the mammogram of a woman chosen at random will be normal?

33. ***Blood type*** Four percent of the population of the United States has type AB blood (Source: *Encyclopedia Americana,* 1993 ed.). What is the probability that a U.S. resident chosen at random
(a) will have type AB blood?
(b) will not have type AB blood?

34. ***Blood type*** Fifteen percent of the population of the United States has a negative Rh (Rhesus) factor in their blood, and the remainder have a positive Rh factor (Source: American Medical Association, *Encyclopedia of Medicine,* 1989). What is the probability that a resident of the United States chosen at random
(a) will have a negative Rh factor?
(b) will have a positive Rh factor?

35. ***Minorities*** Among the residents of Los Angeles, 62.7% are classified as minorities (Source: *Upclose Census Digest,* City and Town Edition, 1992). If an L.A. resident is chosen at random, what is the probability that he or she is
(a) a member of a minority?
(b) not a member of a minority?

36. ***Minorities*** Among the population of Miami, 62.5% are of Hispanic descent (Source: *Upclose Census Digest,* City and Town Edition, 1992). What is the probability that a Miami resident chosen at random is
(a) of Hispanic descent?
(b) not of Hispanic descent?

37. ***Fraud*** A company selling substandard drugs to developing countries sold 2,000,000 capsules with 60,000 of them empty (Source: "60 Minutes," January 18, 1998). What is the probability that a person who takes a randomly chosen capsule from this company will get an empty capsule?

38. ***Inventory*** Forty percent of a company's total output consists of baseballs, 30% consists of softballs, and 10% consists of tennis balls. Its only remaining product is handballs. If we placed balls in a box in the same proportion as the company's output and selected a ball at random from the box, what is the probability that
(a) the ball is a baseball?
(b) the ball is a tennis ball?
(c) the ball is not a softball?
(d) the ball is a handball?

39. ***Lactose intolerance*** Lactose intolerance affects about 75% of African, Asian, and Native Americans (Source: Jean Carper, "Eat Smart," *USA Weekend,* Sept. 4, 1994). If a person is selected from this group of people, what is the probability that the person will have lactose intolerance?

40. ***Lactose intolerance*** Lactose intolerance affects about 20% of non-Hispanic white Americans (Source: Jean Carper, "Eat Smart," *USA Weekend,* Sept. 4, 1994). If a person is selected from this group of people, what is the probability that the person will be lactose-intolerant?

41. ***Management*** Because of a firm's growth, it is necessary to transfer one of its employees to one of its branch stores. Three of the nine employees are women, and all of the nine employees are equally qualified for the transfer. If the person to be transferred is chosen at random, what is the probability that the transferred person is a woman?

42. ***Quality control*** A supply of 500 television picture tubes has 6 defective tubes. What is the probability that a tube picked at random from the supply is not defective?

43. ***Education*** A professor assigns 6 homework projects and then, during the next class, rolls a die to determine which one of the 6 projects to grade. If a student has completed 2 of the assignments, what is the probability that the professor will grade a project that the student has completed?

44. ***Marketing*** Meow Mix offers a free coin in each bag of cat food. If each 10,000-bag shipment has one gold coin, 2 silver dollars, 5 quarters, 20 dimes, 50 nickels, and 9922 pennies, what is the probability that a bag of Meow Mix cat food selected at random will have a penny in it?

45. ***Minorities*** The NAACP claims that in Volusia County, Florida, police stop dark-skinned drivers on I-95 who are not violating the law much more frequently than they stop white drivers, and then they conduct searches of the cars for drugs or for large sums of money that could be related to drug sales (Source: *American Journal,* August 30, 1994). If police made routine stops of randomly selected motorists who are not violating a law, and if 22% of the drivers on I-95 passing through this county are dark-skinned, what is the probability that the police stop a driver who is not violating a law but who is dark-skinned? If 39% of the motorists stopped for routine checks are dark-skinned people, does the NAACP claim seem reasonable?

46. ***Quality control*** A computer store offered used 64K computers free to local middle schools. Of the 54 machines available, four have defective memories, six have defective keys, and the remainder have no defects. If a teacher picks one at random, what is the probability that she will select a defective computer?

47. ***Football*** Before the 1994 football season, a *USA Today* oddsmaker gave the San Francisco 49ers' odds against winning Super Bowl XXIX at 3:1, saying they were the favorite to win (Source: "Sheridan's Odds," *USA Today,* Sept. 2, 1994). If the odds were accurate, what probability would be assigned to the 49ers winning this Super Bowl? Did San Francisco win?

48. ***Football*** Before the 1994 football season, a *USA Today* oddsmaker gave the Cincinnati Bengals' odds against winning Super Bowl XXIX at 1 million:1, saying they were the least likely to win (Source: "Sheridan's Odds," *USA Today,* Sept. 2, 1994). If the odds were accurate, what probability would be assigned to the Bengals winning this Super Bowl?

49. ***Genetics*** A newly married couple plans to have three children. Assuming the probability of a girl being born equals that of a boy being born, what is the probability that exactly two of the three children born will be girls? *(Hint:* Construct the sample space.)

50. ***Management*** A frustrated store manager is asked to make four different yes-no decisions that have no relation to each other. Because he is impatient to leave work, he flips a coin for each decision. If the correct decision in each case was yes, what is the probability that
    (a) all of his decisions were correct?
    (b) none of his decisions was correct?
    (c) half of his decisions were correct?

51. ***Genetics*** A couple plans to have two children and wants to know the probability of having one boy and one girl.
    (a) Does {0, 1, 2} represent an equiprobable sample space for finding the probability of exactly one boy in two children?
    (b) Construct an equiprobable sample space for this problem.
    (c) What is the probability of getting one boy and one girl in two children?

52. ***Genetics*** What is the probability that a couple planning to have two children will have
    (a) at least one child of each sex?
    (b) two children of the same sex?

53. ***Testing*** A quiz consists of 6 multiple-choice questions with 5 possible answers for each question. If a student guesses on the first question, what is the probability that he will get the correct answer?

54. ***Testing*** In taking a quiz with 6 multiple-choice questions, a student knows that two of the possible five answers to the first question are not correct. What is the probability of her getting the correct answer to the first question?

55. ***Quality control*** On the average, 6 articles out of each 250 produced by a certain machine are defective. What is the probability that an article chosen at random is defective?

56. ***Baseball*** If a baseball player has 1150 at bats and 320 hits, what is the probability that he will get a hit his next time at bat?

57. ***Traffic safety*** In a certain town, citizens' groups have identified three intersections as potentially dangerous. They collected the following data.

	Intersections		
	A	*B*	*C*
Number of vehicles/day	25,500	3890	8580
Number of vehicles in accidents/year	182	51	118

For each intersection, find the probability that a vehicle entering the intersection will be involved in an accident. Which intersection is the most dangerous?

58. ***Traffic safety*** A traffic survey showed that 5680 cars entered an intersection of the main street of a city and that 1460 of them turned onto the intersecting street at this intersection.
    (a) What is the empirical probability a car will turn onto the intersecting street?
    (b) What is the probability a car will not turn onto the intersecting street?

59. ***Elections*** Prior to an election, a poll of 1200 citizens is taken. Suppose 557 plan to vote for A, 533 plan to vote for B, and the remainder indicate they will not vote.
    (a) If the poll is accurate, find the probability that a citizen chosen at random will vote for A.
    (b) If the poll is accurate, what is the probability that a person picked at random will not vote?

60. ***Elections*** Suppose a county has 800,000 registered voters, 480,000 of whom are Democrats and 290,000 of whom are Republicans. If a registered voter is chosen at random, what is the probability that
    (a) he or she is a Democrat?
    (b) he or she is neither a Democrat nor a Republican?

61. ***Births*** Suppose that at a certain hospital the births on a given day consisted of 8 girls and 2 boys. Suppose that at the same hospital, the births in a given year consisted of 1220 girls and 1194 boys.
    (a) Find the empirical probabilities for the birth of each sex on the given day.
    (b) Find the empirical probabilities for the given year.
    (c) What are the theoretical probabilities, and which empirical probabilities do you think more accurately reflect reality?

7.2 Unions and Intersections of Events; One-Trial Experiments

OBJECTIVES

- *To find the probability of the intersection of two events*
- *To find the probability of the union of two events*
- *To find the probability of the complement of an event*

APPLICATION PREVIEW

Suppose that the directors of the Chatham County School Board consist of 8 Democrats, 3 of whom are female, and 4 Republicans, 2 of whom are female. If a director is chosen at random to discuss an issue on television, we can find the probability that the person will be a Democrat or female by finding the probability of the union of the two events "a Democrat is chosen" and "a female is chosen." In this section we will discuss methods of finding the probability of the intersection and union of events by using sample spaces and formulas.

As we stated in the previous section, events determine subsets of the sample space for a probability experiment. The intersection, union, and complement of events are defined as follows:

If E and F are two events in a sample space S, then the **intersection of E and F** is

$$E \cap F = \{a\colon a \in E \quad \text{and} \quad a \in F\}$$

the **union of E and F** is

$$E \cup F = \{a\colon a \in E \quad \text{or} \quad a \in F\}$$

and the **complement of E** is

$$E' = \{a\colon a \in S \quad \text{and} \quad a \notin E\}$$

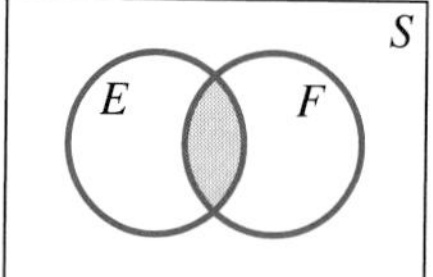

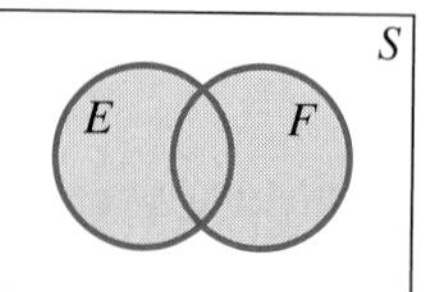

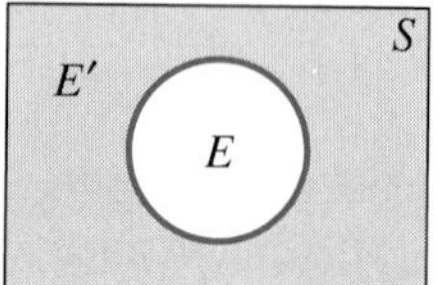

Using these definitions, we can write the probability that certain events occur as follows:

$$\Pr(E \textbf{ and } F \text{ both occur}) = \Pr(E \cap F)$$
$$\Pr(E \textbf{ or } F \text{ occurs}) = \Pr(E \cup F)$$
$$\Pr(E \text{ does } \textbf{not} \text{ occur}) = \Pr(E')$$

EXAMPLE 1

A card is drawn from a box containing 15 cards numbered 1 to 15. What is the probability that the card is

(a) not even?
(b) even and divisible by 3?
(c) even or divisible by 3?

Solution

The sample space S contains the 15 numbers.

$$S = \{1, 2, 3, 4, 5, 6, 7, 8, 9, 10, 11, 12, 13, 14, 15\}$$

If we let E represent "even-numbered" and D represent "number divisible by 3," we have

$$E = \{2, 4, 6, 8, 10, 12, 14\} \quad \text{and} \quad D = \{3, 6, 9, 12, 15\}$$

These sets are shown in the Venn diagram in Figure 7.1.

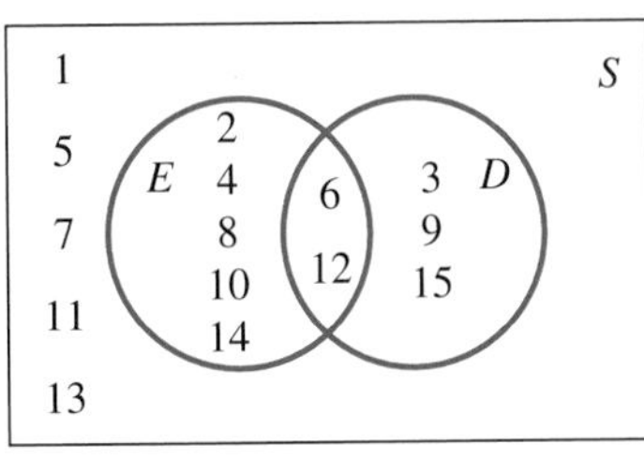

Figure 7.1

(a) The event "not even," or E', contains all elements of S not in E, so $E' = \{1, 3, 5, 7, 9, 11, 13, 15\}$, and the probability that the card is not even is

$$\Pr(\text{not } E) = \Pr(E') = \frac{n(E')}{n(S)} = \frac{8}{15}$$

(b) The event "even and divisible by 3" is $E \cap D = \{6, 12\}$, so

$$\Pr(\text{even and divisible by 3}) = \Pr(E \cap D) = \frac{n(E \cap D)}{n(S)} = \frac{2}{15}$$

(c) The event "even or divisible by 3" is $E \cup D = \{2, 3, 4, 6, 8, 9, 10, 12, 14, 15\}$, so

$$\Pr(\text{even or divisible by 3}) = \Pr(E \cup D) = \frac{n(E \cup D)}{n(S)} = \frac{10}{15} = \frac{2}{3}$$

CHECKPOINT

1. If a ball is drawn from a bag containing 4 red balls numbered 1, 2, 3, 4 and 3 white balls numbered 5, 6, 7, what is the probability that the ball is
   (a) red and even?
   (b) white and even?

Because the complement of an event contains all the elements of S *except* for those elements in E (see Figure 7.2), the number of elements in E' is $n(S) - n(E)$, so

$$\Pr(E') = \frac{n(S) - n(E)}{n(S)} = 1 - \frac{n(E)}{n(S)} = 1 - \Pr(E)$$

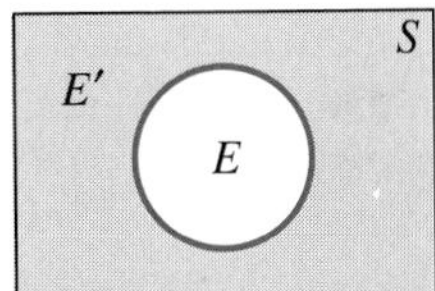

Figure 7.2

$$\Pr(\text{not } E) = \Pr(E') = 1 - \Pr(E)$$

EXAMPLE 2

A dry cleaning firm has 12 employees: 7 women and 5 men. Three of the women and 5 of the men are 40 years old or older. If a person is chosen at random from this firm, what is the probability that the person is under 40 years of age?

Solution

Eight people are 40 or older, so

$$\Pr(\text{under } 40) = 1 - \Pr(40+) = 1 - \frac{8}{12} = \frac{4}{12} = \frac{1}{3}$$

The number of elements in the set $E \cup F$ is the number of elements in E plus the number of elements in F, *minus* the number of elements in $E \cap F$ [because this number was counted twice in $n(E) + n(F)$]. Thus,

$$n(E \cup F) = n(E) + n(F) - n(E \cap F)$$

Thus,

$$\begin{aligned}\Pr(E \cup F) &= \frac{n(E \cup F)}{n(S)} = \frac{n(E)}{n(S)} + \frac{n(F)}{n(S)} - \frac{n(E \cap F)}{n(S)} \\ &= \Pr(E) + \Pr(F) - \Pr(E \cap F)\end{aligned}$$

If E and F are any two events, then the probability that one event or the other will occur, denoted $\Pr(E \text{ or } F)$, is given by

$$\Pr(E \text{ or } F) = \Pr(E \cup F) = \Pr(E) + \Pr(F) - \Pr(E \cap F)$$

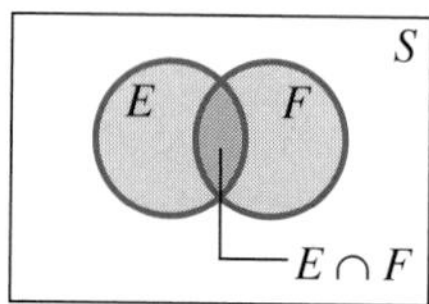

Suppose that of 30 people interviewed for a position, 18 had a business degree, 15 had previous experience, and 8 of those with experience also had a business degree. The Venn diagram in Figure 7.3 illustrates this situation.

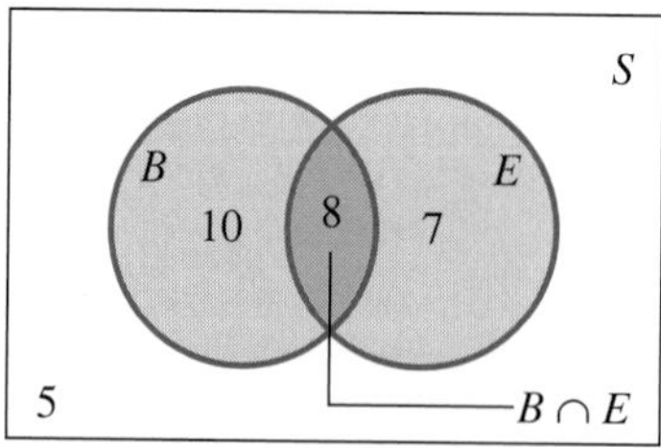

Figure 7.3

Because 8 of the 30 people have both a business degree and experience,

$$\Pr(B \cap E) = \frac{8}{30}$$

We also see that

$$\Pr(B) = \frac{18}{30} \quad \text{and} \quad \Pr(E) = \frac{15}{30}$$

Thus the probability of choosing a person at random who has a business degree or experience is

$$\begin{aligned}\Pr(B \cup E) &= \Pr(B) + \Pr(E) - \Pr(B \cap E)\\ &= \frac{18}{30} + \frac{15}{30} - \frac{8}{30} = \frac{25}{30} = \frac{5}{6}\end{aligned}$$

EXAMPLE 3

An integer between 1 and 6, inclusive, is chosen at random. What is the probability that it is even or divisible by 3?

Solution

Because Pr(even) = 3/6 = 1/2, Pr(divisible by 3) = 2/6 = 1/3, and Pr(even *and* divisible by 3) = Pr(6) = 1/6,

$$\Pr(\text{even or divisible by } 3) = \frac{1}{2} + \frac{1}{3} - \frac{1}{6} = \frac{5}{6} - \frac{1}{6} = \frac{4}{6} = \frac{2}{3}$$

We now consider the election problem mentioned in the Application Preview.

EXAMPLE 4

The directors of the Chatham County School Board consist of 8 Democrats, 3 of whom are female, and 4 Republicans, 2 of whom are female. If a director is chosen at random to discuss an issue on television, what is the probability that the person will be a Democrat or female?

Solution

There are 12 directors, of whom 8 are Democrats and 5 are female. But 3 of the directors are Democrats *and* females and cannot be counted twice. Thus

$$\begin{aligned}\Pr(D \text{ or } f) = \Pr(D \cup f) &= \Pr(D) + \Pr(f) - \Pr(D \text{ and } f)\\ &= \frac{8}{12} + \frac{5}{12} - \frac{3}{12} = \frac{10}{12}\end{aligned}$$

CHECKPOINT

2. If a ball is drawn from a bag containing 4 red balls numbered 1, 2, 3, 4 and 3 white balls numbered 5, 6, 7, what is the probability that the ball is
   (a) red or even? (b) white or even?

We say that events E and F are **mutually exclusive** if and only if $E \cap F = \varnothing$. Thus

$$\Pr(E \cup F) = \Pr(E) + \Pr(F) - 0 = \Pr(E) + \Pr(F)$$

if and only if E and F are mutually exclusive.

If E and F are **mutually exclusive,** then $\Pr(E \cap F) = 0$, and

$$\Pr(E \text{ or } F) = \Pr(E \cup F) = \Pr(E) + \Pr(F)$$

Mutually exclusive

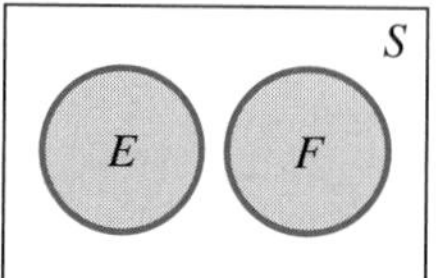

EXAMPLE 5

Find the probability of obtaining a 6 or a 4 in one roll of a die.

Solution

Rolling a 6 and rolling a 4 on one roll of a die are mutually exclusive events. Let E be the event "rolling a 6" and F be the event "rolling a 4."

$$\Pr(E) = \Pr(\text{rolling } 6) = \frac{1}{6} \quad \text{and} \quad \Pr(F) = \Pr(\text{rolling } 4) = \frac{1}{6}$$

Then

$$\Pr(E \cup F) = \Pr(E) + \Pr(F) = \frac{1}{6} + \frac{1}{6} = \frac{1}{3}$$

EXAMPLE 6

Sacco and Rosen are among 3 candidates running for a public office, and polls indicate that the probability that Sacco will win is 1/3 and the probability that Rosen will win is 1/2. Only one candidate can win.

(a) What is the probability that Sacco or Rosen will win?
(b) What is the probability that neither Sacco nor Rosen will win?

Solution

(a) The two events are mutually exclusive, so

$$\begin{aligned}\Pr(\text{Sacco or Rosen}) &= \Pr(\text{Sacco} \cup \text{Rosen}) \\ &= \Pr(\text{Sacco}) + \Pr(\text{Rosen}) = \frac{1}{3} + \frac{1}{2} = \frac{5}{6}\end{aligned}$$

(b) The probability that neither will win is $1 - 5/6 = 1/6$.

The formula for the probability of the union of mutually exclusive events can be extended beyond two events.

$$\Pr(E_1 \cup E_2 \cup \ldots \cup E_n) = \Pr(E_1) + \Pr(E_2) + \cdots + \Pr(E_n)$$

for mutually exclusive events $E_1, E_2, \ldots, E_n$.

EXAMPLE 7

Find the probability of tossing two distinct dice and obtaining a sum greater than 9.

Solution

Because the event "a sum greater than 9" consists of the mutually exclusive events "sum = 10," "sum = 11," and "sum = 12," and these probabilities can be found in Table 7.1 on page 501.

$$\begin{aligned}\Pr(10 \text{ or } 11 \text{ or } 12) &= \Pr(10 \cup 11 \cup 12)\\ &= \Pr(10) + \Pr(11) + \Pr(12)\\ &= \frac{3}{36} + \frac{2}{36} + \frac{1}{36} = \frac{6}{36} = \frac{1}{6}\end{aligned}$$

CHECKPOINT SOLUTIONS

1. (a) The sample space contains 7 balls: $S = \{R1, R2, R3, R4, W5, W6, W7\}$. The probability that a ball drawn at random is red and even is found by using the subspace that describes that event. Thus

$$\Pr(\text{red and even}) = \Pr(\text{red} \cap \text{even}) = 2/7.$$

   (b) $F = \{W6\}$ describes the event "white and even," so $\Pr(\text{white and even}) = \Pr(\text{white} \cap \text{even}) = 1/7$.

2. (a) Pr(red or even)

$$= \Pr(\text{red} \cup \text{even}) = \Pr(\text{red}) + \Pr(\text{even}) - \Pr(\text{red} \cap \text{even})$$
$$= \frac{4}{7} + \frac{3}{7} - \frac{2}{7} = \frac{5}{7}$$

   (b) Pr(white or even)

$$= \Pr(\text{white} \cup \text{even}) = \Pr(\text{white}) + \Pr(\text{even}) - \Pr(\text{white} \cap \text{even})$$
$$= \frac{3}{7} + \frac{3}{7} - \frac{1}{7} = \frac{5}{7}$$

EXERCISE 7.2

1. An ordinary die is tossed. What is the probability of getting a 3 or an odd number?
2. A card is drawn at random from a deck of 52 playing cards. Find the probability that it is either a club or a king.
3. If you draw one card at random from a deck of 12 cards numbered 1 through 12, inclusive, what is the probability that the number you draw is divisible by 3 and even?

4. If you draw one card at random from the deck of 12 cards numbered 1–12, what is the probability that the number you draw is even and divisible by 4?
5. If the probability that event E will occur is 3/5, what is the probability that E will not occur?
6. If the probability that an event will occur is 7/9, what is the probability that the event will not occur?
7. A bag contains 5 red balls numbered 1, 2, 3, 4, 5 and 9 white balls numbered 6, 7, 8, 9, 10, 11, 12, 13, 14. If a ball is drawn, what is the probability that
   (a) the ball is red and even-numbered?
   (b) the ball is red or even-numbered?
8. A bag contains 5 red balls numbered 1, 2, 3, 4, 5 and 9 white balls numbered 6, 7, 8, 9, 10, 11, 12, 13, 14. If a ball is drawn, what is the probability that it is
   (a) white and odd-numbered?
   (b) white or odd-numbered?
   (c) white or even-numbered?
9. If you draw one card from a deck of 12 cards, numbered 1 through 12, inclusive, what is the probability you will get an odd number or a number divisible by 4?
10. Suppose a die is biased (unfair) so that each odd-numbered face has probability 1/4 of resulting, and each even face has probability 1/12 of resulting. Find the probability of getting a number greater than 3.
11. If you draw one card from a deck of 12 cards numbered 1 through 12, inclusive, what is the probability that the card will be odd or divisible by 3?
12. A card is drawn from a deck of 52. What is the probability that it will be an ace, king, or jack?
13. A bag contains 4 white, 7 black, and 6 green balls. What is the probability that a ball drawn at random from the bag is white or green?
14. In a game where only one player can win, the probability that Jack will win is 1/5 and that Bill will win is 1/4. Find the probability that one of them will win.
15. A cube has 2 faces painted red, 2 painted white, and 2 painted blue. What is the probability of getting a red face or a white face in one roll?
16. A cube has 3 faces painted white, 2 faces painted red, and 1 face painted blue. What is the probability that a roll will result in a red or blue face?
17. A ball is drawn from a bag containing 13 red balls numbered 1–13 and 5 white balls numbered 14–18. What is the probability that
   (a) the ball is not even-numbered?
   (b) the ball is red and even-numbered?
   (c) the ball is red or even-numbered?
   (d) the ball is neither red nor even-numbered?
18. A ball is drawn from a bag containing 13 red balls numbered 1–13 and 5 white balls numbered 14–18. What is the probability that
   (a) the ball is not red?
   (b) the ball is white and odd-numbered?
   (c) the ball is white or odd-numbered?

Applications

19. ***Drug use*** Forty-six percent of marijuana use among youth occurs in the inner cities (Source: Partnership for a Drugfree America). If an instance of such marijuana use is chosen at random, what is the probability that the use does not occur in an inner city?

20. ***Breast cancer*** According to the American Cancer Society, 199 of 200 mammograms turn out to be normal. What is the probability that the mammogram of a woman chosen at random will show an abnormality?

21. ***Car maintenance*** A car rental firm has 425 cars. Sixty-three of these cars have defective turn signals and 32 have defective tires. What is the probability that one of these cars selected at random
   (a) does not have defective turn signals?
   (b) has no defects if no car has 2 defects?

22. ***Testing*** An unprepared student must take a 7-question multiple-choice test that has 5 possible answers per question. If the student guesses on the first question, what is the probability that she will answer that question incorrectly?

23. ***Linguistics*** Of 100 students, 24 can speak French, 18 can speak German, and 8 can speak both French and German. If a student is picked at random, what is the probability that he or she can speak French or German?

24. ***Management*** A company employs 65 people. Eight of the 30 men and 21 of the 35 women work in the business office. What is the probability that an employee picked at random is a woman or works in the business office?

25. ***Site selection*** A firm is considering three possible locations for a new factory. The probability that site A will be selected is 1/3 and the probability that site B will be selected is 1/5. Only one location will be chosen.
   (a) What is the probability that site A or site B will be chosen?
   (b) What is the probability that neither site A nor site B will be chosen?

26. ***Photography*** Rob Lee knows his camera will take a good picture unless the flash is defective or the batteries are dead. The probability of having a defective flash is 0.05, the probability of the batteries being dead is 0.3, and the probability that both these problems occur is 0.01. What is the probability that the picture will be good?

27. ***Cognitive complexity*** The cognitive complexity of a structure was studied by Scott* using a technique in which a person was asked to specify a number of objects and group them into as many groupings as he or she found meaningful. A person groups 12 objects into three groups in such a way that

    7 objects are in group A.
    7 objects are in group B.
    8 objects are in group C.
    3 objects are in both group A and group B.
    5 objects are in both group B and group C.
    4 objects are in both group A and group C.
    2 objects are in all three groups.

    (a) What is the probability that an object chosen at random has been placed into group A or group B?
    (b) What is the probability that an object chosen at random has been placed into group B or group C?

28. ***Politics*** A group of 100 people contains 60 Democrats and 35 Republicans. If there are 60 women and if 40 of the Democrats are women, what is the probability that a person selected at random is a Democrat or a woman?

29. ***Education*** A mathematics class has 16 engineering majors, 12 science majors, and 4 liberal arts majors.
    (a) What is the probability that a student selected at random will be a science or liberal arts major?
    (b) What is the probability that a student selected at random will be an engineering or science major?
    (c) Five of the engineering students, 6 of the science majors, and 2 of the liberal arts majors are female. What is the probability that a student selected at random is an engineering major or is a female?

30. ***Scheduling*** Maria has ordered a washer and dryer from two different companies. Both the washer and the dryer are to be delivered Thursday. The probability that the washer will be delivered in the morning is .6, and the probability that the dryer will be delivered in the morning is .8. If the probability that either the washer or dryer is delivered in the morning is .9, what is the probability that both will be delivered in the morning?

Military spending Suppose the table below summarizes the opinions of various groups on the issue of increased military spending. Use this table to calculate probabilities in Problems 31–34.

	Whites		*Non-Whites*		
Opinion	*Reps.*	*Dems.*	*Reps.*	*Dems.*	*Total*
Favor	300	100	25	10	435
Oppose	100	250	25	190	565
Total	400	350	50	200	1000

31. Find the probability that an individual chosen at random is a Republican or favors increased military spending.
32. Find the probability that an individual chosen at random is a Democrat or opposes increased military spending.
33. Find the probability that an individual chosen at random is white or opposes increased military spending.
34. Find the probability that an individual chosen at random is non-white or favors increased military spending.

35. ***Drinking age*** A survey questioned 1000 people regarding raising the legal drinking age from 18 to 21. Of the 560 who favored raising the age, 390 were female. Of the 440 opposition responses, 160 were female. A person is selected at random.
    (a) What is the probability the person is a female or favors raising the age?
    (b) What is the probability the person is a male or favors raising the age?
    (c) What is the probability the person is a male or opposes raising the age?

36. ***Job bids*** Three construction companies have bid for a job. Max knows that the two companies he is competing with have probabilities 1/3 and 1/6, respectively, of getting the job. What is the probability that Max will get the job?

*W. Scott, "Cognitive Complexity and Cognitive Flexibility," *Sociometry* 25 (1962), pp. 405–414.

37. ***Management*** A company employs 65 people. Eight of the 30 men and 21 of the 35 women work in the business office. What is the probability that an employee picked at random is a man or works in the business office?

38. ***Television*** The probability that a wife watches a certain television show is .55, that her husband watches it is .45, and that both watch the show is .30. What is the probability that the husband or the wife watches this show?

39. ***Cancer testing*** A long-term study has revealed that the prostate-specific antigen (PSA) test for prostate cancer in men is very effective. The study shows that 87% of the men for which the test is positive actually have prostate cancer (Source: *Journal of the American Medical Association,* January 1995). If a man selected at random tests positive for prostate cancer with this test, what is the probability that he does not have prostate cancer?

40. ***Cancer testing*** The study described in Problem 39 also revealed that 91% of the men for whom the PSA test was negative do not have prostate cancer (Source: *Journal of the American Medical Association,* January 1995). If a man selected at random tests negative for prostate cancer with this test, what is the probability that he does have prostate cancer?

7.3 Conditional Probability; the Product Rule

OBJECTIVES

- *To solve probability problems involving conditional probability*
- *To compute the probability that two or more independent events will occur*
- *To compute the probability that two or more dependent events will occur*

APPLICATION PREVIEW

Until recently, AT&T calling cards used each subscriber's home phone number plus a 4-digit personal identification number (PIN). Because criminals observe users of public phones to record and sell PINs for fraudulent use, AT&T has discontinued using home phone numbers and is now using 14-digit PINs. To see how this reduces the chance of someone obtaining a PIN, we can find the probability that someone can guess a PIN in these two cases if they have observed 3 of the digits and know the person's home phone number. To see which PIN would be safer, we can calculate the probability that someone can guess your PIN if you have a 4-digit PIN and if you have a 14-digit PIN by using the **Product Rule.** In this section we will use the Product Rule to find the probability that two events will occur, and we will solve problems involving **conditional probability.**

Each probability that we have computed has been relative to the sample space for the experiment. Sometimes information is given that reduces the sample space needed to solve a stated problem. For example, if a die is rolled, the probability that a 5 occurs, given that an odd number is rolled, is denoted by

$$\Pr(5 \text{ rolled} \mid \text{odd number rolled})$$

The knowledge that an odd number occurs reduces the sample space from the numbers on a die, $S = \{1, 2, 3, 4, 5, 6\}$, to the sample space for odd numbers on a die, $S_1 = \{1, 3, 5\}$, so the event E = "a 5 occurs" has probability

$$\Pr(5 \mid \text{odd number}) = \frac{n(5 \text{ and odd number rolled})}{n(\text{odd number rolled})} = \frac{n(\mathrm{E} \cap S_1)}{n(S_1)} = \frac{1}{3}$$

To find $\Pr(A \mid B)$, we seek the probability that A occurs, given that B occurs. Figure 7.4 shows the original sample space S with the reduced sample space B

shaded. Because all elements must be contained in B, we evaluate $\Pr(A \mid B)$ by dividing the number of elements in $A \cap B$ by the number of elements in B.

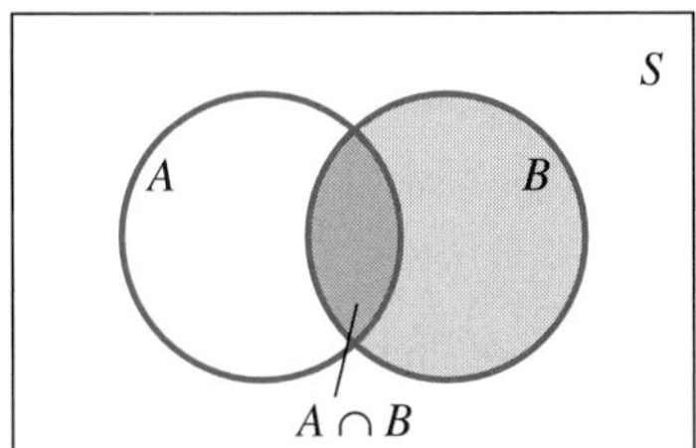

Figure 7.4

The probability that A occurs, given that B occurs, is denoted by $\Pr(A \mid B)$ and is given by

$$\Pr(A \mid B) = \frac{n(A \cap B)}{n(B)}$$

EXAMPLE 1

Suppose the table below summarizes the opinions of various groups on the issue of increased military spending. What is the probability that a person selected at random from this group of people

(a) is non-white?
(b) is non-white and favors increased military spending?
(c) favors increased military spending, given that the person is non-white?

	Whites		*Non-Whites*		
Opinion	*Republicans*	*Democrats*	*Republicans*	*Democrats*	*Totals*
Favor	300	100	25	10	435
Oppose	100	250	25	190	565
Total	400	350	50	200	1000

Solution

(a) Of the 1000 people in the group, $50 + 200 = 250$ are non-white, so

$$\Pr(\text{non-white}) = 250/1000 = 1/4$$

(b) There are 1000 people in the group and $25 + 10 = 35$ who both are non-white and favor increased spending, so

$$\Pr(\text{non-white and favor}) = 35/1000 = 7/200$$

(c) There are $50 + 200 = 250$ non-white and $25 + 10 = 35$ people who both are non-white and favor increased spending, so

$$\Pr(\text{favor} \mid \text{non-white}) = \frac{n(\text{favor and non-white})}{n(\text{non-white})} = \frac{35}{250} = \frac{7}{50}$$

The probability found in Example 1 is called **conditional probability.** The conditional probability $\Pr(A \mid B)$ can be found by using the number of elements in $A \cap B$ and in B or by using the probability of $A \cap B$ and B. To find the formula by using probabilities, we divide both the numerator and the denominator of the right-hand side of

$$\Pr(A \mid B) = \frac{n(A \cap B)}{n(B)}$$

by $n(S)$, the number of elements in the sample space, which gives

$$\Pr(A \mid B) = \frac{\dfrac{n(A \cap B)}{n(S)}}{\dfrac{n(B)}{n(S)}} = \frac{\Pr(A \cap B)}{\Pr(B)}$$

Conditional Probability Let A and B be events in the sample space S with $\Pr(B) > 0$. The conditional probability that event A occurs, given that event B occurs, is given by

$$\Pr(A \mid B) = \frac{\Pr(A \cap B)}{\Pr(B)}$$

EXAMPLE 2

A red die and a green die are rolled. What is the probability that the sum rolled on the dice is 6, given that the sum is less than 7?

Solution

Using the conditional probability formula and the sample space in Table 7.1 on page 501 we have

$$\Pr(\text{sum is } 6 \mid \text{sum} < 7) = \frac{\Pr(\text{sum is } 6 \cap \text{sum} < 7)}{\Pr(\text{sum} < 7)}$$

Because

$$\Pr(\text{sum is } 6 \cap \text{sum} < 7) = \Pr(\text{sum is } 6) = \frac{5}{36}$$

the probability is

$$\begin{aligned}\Pr(\text{sum is } 6 \mid \text{sum} < 7) &= \frac{5/36}{1/36 + 2/36 + 3/36 + 4/36 + 5/36} \\ &= \frac{5/36}{15/36} = \frac{1}{3}\end{aligned}$$

CHECKPOINT

1. Suppose that one ball is drawn from a bag containing 4 red balls numbered 1, 2, 3, 4 and 3 white balls numbered 5, 6, 7. What is the probability that the ball is white, given that it is an even-numbered ball?

We can solve the formula

$$\Pr(A \mid B) = \frac{\Pr(A \cap B)}{\Pr(B)}$$

for $\Pr(A \cap B)$ by multiplying both sides of the equation by $\Pr(B)$. This gives

$$\Pr(A \cap B) = \Pr(B) \cdot \Pr(A \mid B)$$

Similarly, we can solve

$$\Pr(B \mid A) = \frac{\Pr(B \cap A)}{\Pr(A)} = \frac{\Pr(A \cap B)}{\Pr(A)}$$

for $\Pr(A \cap B)$, getting

$$\Pr(A \cap B) = \Pr(A) \cdot \Pr(B \mid A)$$

Product Rule If A and B are probability events, then the probability of the event "A and B" is Pr(A and B), and it can be found by one or the other of these two formulas.

$$\Pr(A \text{ and } B) = \Pr(A \cap B) = \Pr(A) \cdot \Pr(B \mid A)$$

or

$$\Pr(A \text{ and } B) = \Pr(A \cap B) = \Pr(B) \cdot \Pr(A \mid B)$$

EXAMPLE 3

Suppose that from a deck of 52 cards, two cards are drawn in succession, without replacement. Find the probability that both cards are kings.

Solution

The probability we seek uses

$$\Pr(\text{2d king} \mid \text{1st king})$$

which is the probability that the second king will be drawn given that the first king has already been drawn. It is 3/51 because one king has been removed from the deck.

We can find the probability that two cards (drawn without replacement) are both kings as follows:

$$\Pr(\text{1st card is a king}) = \frac{4}{52}$$

$$\Pr(\text{2d card is a king} \mid \text{1st card is a king}) = \frac{3}{51}$$

$$\begin{aligned}\Pr(\text{2 kings}) &= \Pr(\text{1st king}) \cdot \Pr(\text{2d king} \mid \text{1st king}) \\ &= \frac{4}{52} \cdot \frac{3}{51} = \frac{1}{221}\end{aligned}$$

EXAMPLE 4

A box contains 3 black balls, 2 red balls, and 5 white balls. One ball is drawn, it is *not* replaced, and a second ball is drawn. Find the probability that the first ball is red and the second is black.

Solution

Let E_1 be "red ball first" and E_2 be "black ball second." Then

$$\Pr(E_1 \cap E_2) = \Pr(E_1)\cdot\Pr(E_2 \mid E_1) = \frac{2}{10}\cdot\frac{3}{9} = \frac{1}{15}$$

CHECKPOINT

2. Two balls are drawn, without replacement, from the box described in Example 4.
   (a) Find the probability that both balls are white.
   (b) Find the probability that the first ball is white and the second is red.

EXAMPLE 5

Polk's Department Store has observed that 80% of its charge accounts have men's names on them and that 16% of the accounts with men's names on them have been delinquent at least once, while 5% of the accounts with women's names on them have been delinquent at least once. What is the probability that an account selected at random

(a) is in a man's name and is delinquent?
(b) is in a woman's name and is delinquent?

Solution

(a) Pr(man's account) = .80
Pr(delinquent | man's) = .16
Pr(man's and delinquent) = Pr(man's $\cap$ delinquent)
= Pr(man's) · Pr(delinquent | man's)
= (.80)(.16) = .128

(b) Pr(woman's account) = .20
Pr(delinquent | woman's) = .05
Pr(woman's and delinquent) = (.20)(.05) = .01

EXAMPLE 6

All products on an assembly line must pass two inspections. It has been determined that the probability that the first inspector will miss a defective item is .09. If a defective item gets past the first inspector, the probability that the second inspector will not detect it is .01. What is the probability that a defective item will not be rejected by either inspector? (All good items pass both inspections.)

Solution

The second inspector inspects only items passed by the first inspector. The probability that a defective item will pass both inspections is as follows.

$$\begin{aligned}\Pr(\text{pass both}) &= \Pr(\text{pass first} \cap \text{pass second})\\ &= \Pr(\text{pass first}) \cdot \Pr(\text{pass second} \mid \text{passed first})\\ &= .09 \cdot .01 = .0009\end{aligned}$$

If a coin is tossed twice, the probability that both results will be heads if the first toss gives a head is

$$\Pr(2\text{H} \mid 1\text{st is H}) = \frac{\Pr(1\text{st is H} \cap 2\text{H})}{\Pr(1\text{st is H})} = \frac{\Pr(\text{Both H})}{\Pr(1\text{st is H})}$$

The sample space for this experiment is $S = \{\text{HH, HT, TH, TT}\}$, so

$$\Pr(2\text{H} \mid 1\text{st H}) = \frac{1/4}{2/4} = \frac{1}{2}$$

We can see that the probability that the second toss of a coin is a head, given that the first toss gave a head, was 1/2, which is the same as the probability that the second toss is a head, regardless of what happened on the first toss. This is because the results of the coin tosses are **independent events.** We can define *independent events* as follows:

Independent Events

The events A and B are independent if and only if

$$\Pr(A \mid B) = \Pr(A) \qquad \text{and} \qquad \Pr(B \mid A) = \Pr(B)$$

This means that the occurrence or nonoccurrence of one event does not affect the other.

If A and B are independent events, then this definition can be used to simplify the product rule:

$$\Pr(A \cap B) = \Pr(A) \cdot \Pr(B \mid A) = \Pr(A) \cdot \Pr(B)$$

Thus we have the following.

Product Rule for Independent Events

If A and B are independent events, then

$$\Pr(A \text{ and } B) = \Pr(A \cap B) = \Pr(A) \cdot \Pr(B)$$

EXAMPLE 7

A die is rolled and a coin is tossed. Find the probability of getting a 4 on the die and a head on the coin.

Solution

Let E_1 be "4 on the die" and E_2 be "head on the coin." The events are independent because what occurs on the die does not affect what happens to the coin.

$$\Pr(E_1) = \Pr(4 \text{ on die}) = \frac{1}{6}, \quad \text{and} \quad \Pr(E_2) = \Pr(\text{head on coin}) = \frac{1}{2}$$

Then

$$\Pr(E_1 \cap E_2) = \Pr(E_1) \cdot \Pr(E_2) = \frac{1}{6} \cdot \frac{1}{2} = \frac{1}{12}$$

The product rule can be expanded to include more than two events, as the next example shows.

EXAMPLE 8

A bag contains 3 red marbles, 4 white marbles, and 3 black marbles. Find the probability of getting a red marble on the first draw, a black marble on the second draw, and a white marble on the third draw (a) if the marbles are drawn with replacement, and (b) if the marbles are drawn without replacement.

Solution

(a) The marbles are replaced after each draw, so the events are independent. Let E_1 be "red on first," E_2 be "black on second," and E_3 be "white on third." Then

$$\Pr(E_1 \cap E_2 \cap E_3) = \Pr(E_1) \cdot \Pr(E_2) \cdot \Pr(E_3)$$
$$= \frac{3}{10} \cdot \frac{3}{10} \cdot \frac{4}{10} = \frac{36}{1000} = \frac{9}{250}$$

(b) The marbles are not replaced, so the events are dependent.

$$\Pr(E_1 \cap E_2 \cap E_3) = \Pr(E_1) \cdot \Pr(E_2 \mid E_1) \cdot \Pr(E_3 \mid E_1 \text{ and } E_2)$$
$$= \frac{3}{10} \cdot \frac{3}{9} \cdot \frac{4}{8} = \frac{1}{20}$$

We now turn to the problem posed in the Application Preview.

EXAMPLE 9

Suppose that a person knows your home phone number and has observed the first 3 digits of your PIN. Calculate the probability that the person can guess your PIN if

(a) you have a 4-digit PIN.
(b) you have a 14-digit PIN.

With which PIN would you feel safer?

Solution

(a) Because the person knows the first 3 digits, he or she need guess only the fourth digit, so the probability is

$$1 \cdot 1 \cdot 1 \cdot \frac{1}{10} = \frac{1}{10}$$

(b) The person knows the first 3 digits, so the person must guess the remaining 11 digits. The probability that he or she can do this is

$$1 \cdot 1 \cdot 1 \cdot \frac{1}{10} \cdot \frac{1}{10} \cdot \frac{1}{10} \cdot \frac{1}{10} \cdot \frac{1}{10} \cdot \frac{1}{10} \cdot \frac{1}{10} \cdot \frac{1}{10} \cdot \frac{1}{10} \cdot \frac{1}{10} \cdot \frac{1}{10}$$
$$= \frac{1}{100{,}000{,}000{,}000}$$

Clearly the 14-digit PIN is much safer than the 4-digit PIN. However, some professional criminals are able to steal even the 14-digit PIN if it is not guarded carefully.

EXAMPLE 10

A coin is tossed 4 times. Find the probability that

(a) no heads occur. (b) at least one head is obtained.

Solution

(a) Each toss is independent, and on each toss, the probability that a tail occurs is 1/2. Thus the probability of 4 tails in 4 tosses is

$$\Pr(\text{TTTT}) = \frac{1}{2} \cdot \frac{1}{2} \cdot \frac{1}{2} \cdot \frac{1}{2} = \frac{1}{16}$$

(b) The probability that at least 1 head will occur can be found by computing Pr(1 head or 2 heads or 3 heads or 4 heads).

The equiprobable sample space for this probability experiment is

$$S = \{\text{HHHH, HHHT, HHTH, HTHH, THHH, HHTT, HTHT, HTTH, THTH, TTHH, THHT, HTTT, THTT, TTHT, TTTH, TTTT}\}$$

It contains 16 elements, and 15 of these satisfy the conditions of the problem. The only element that does not satisfy the conditions of the problem is the element TTTT, which gives 0 heads. Thus the probability that at least one head occurs is

$$\Pr(\text{at least one head occurs}) = \Pr(\text{not 0 heads})$$
$$= 1 - \Pr(0 \text{ heads}) = 1 - \frac{1}{16} = \frac{15}{16}$$

In general, we can calculate the probability of at least one success in n trials as follows:

$$\text{Pr(at least 1 success in } n \text{ trials)} = 1 - \text{Pr(0 successes)}$$

CHECKPOINT SOLUTIONS

1. $\text{Pr(white} \mid \text{even-numbered)} = \dfrac{\text{Pr(white and even)}}{\text{Pr(even)}} = \dfrac{1/7}{3/7} = \dfrac{1}{3}$

2. (a) $\text{Pr(white on both draws)} = \Pr(W_1 \cap W_2) = \Pr(W_1) \cdot \Pr(W_2 \mid W_1)$
$= \dfrac{5}{10} \cdot \dfrac{4}{9} = \dfrac{2}{9}$

(b) $\text{Pr(white on first and red on second)} = \Pr(W_1 \cap R_2) = \Pr(W_1) \cdot \Pr(R_2 \mid W_1)$
$= \dfrac{5}{10} \cdot \dfrac{2}{9} = \dfrac{1}{9}$

EXERCISE 7.3

1. A card is drawn from a deck of 52 playing cards. Given that it is a red card, what is the probability that
   (a) it is a heart? (b) it is a king?
2. A card is drawn from a deck of 52 playing cards. Given that it is an ace, king, queen, or jack, what is the probability that it is a jack?
3. A die has been "loaded" so that the probability of rolling any even number is 2/9 and the probability of rolling any odd number is 1/9. What is the probability of
   (a) rolling a 6, given that an even number is rolled?
   (b) rolling a 3, given that a number divisible by 3 is rolled?
4. A die has been "loaded" so that the probability of rolling any even number is 2/9 and the probability of rolling any odd number is 1/9. What is the probability of
   (a) rolling a 5?
   (b) rolling a 5, given that an even number is rolled?
   (c) rolling a 5, given that an odd number is rolled?
5. A bag contains 9 red balls numbered 1, 2, 3, 4, 5, 6, 7, 8, 9 and 6 white balls numbered 10, 11, 12, 13, 14, 15. One ball is drawn from the bag. What is the probability that the ball is red, given that the ball is even-numbered?
6. If one ball is drawn from the bag of Problem 5, what is the probability that the ball is white, given that the ball is odd-numbered?
7. A bag contains 4 red balls and 6 white balls. Two balls are drawn without replacement.
   (a) What is the probability that the second ball is white, given that the first ball is red?
   (b) What is the probability that the second ball is red, given that the first ball is white?
   (c) Answer (a) if the first ball is replaced before the second is drawn.
8. A fair die is rolled. Find the probability that the result is a 4, given that the result is even.
9. A fair coin is tossed 3 times. Find the probability of
   (a) throwing 3 heads, given that the first toss is a head.
   (b) throwing 3 heads, given that the first two tosses result in heads.
10. A fair coin is tossed 14 times. What is the probability of tossing 14 heads, given that the first 13 tosses are heads?
11. A die is thrown twice. What is the probability that a 3 will result the first time and a 6 the second time?
12. A die is rolled and a coin is tossed. What is the probability that the die results in an even number and the coin results in a head?
13. (a) A coin is tossed three times. What is the probability of getting a head on all three tosses?
    (b) A coin is tossed three times. What is the probability of getting at least one tail? [*Hint:* Use the result from (a)].
14. The probability that Sam will win in a certain game whenever he plays is 2/5. If he plays two games, what is the probability that he will win just the first game?
15. (a) A box contains 3 red balls, 2 white balls, and 5 black balls. Two balls are drawn at random from the box (with replacement of the first before the second is drawn). What is the probability of getting a red ball on the first draw and a white ball on the second?
    (b) Answer the question if the first ball is not replaced before the second is drawn.

16. (a) One card is drawn at random from a deck of 52 cards. The first card is replaced, and a second card is drawn. Find the probability that both are hearts.
(b) Answer the question if the first card is not replaced before the second card is drawn.
17. Two balls are drawn from a bag containing 3 white balls and 2 red balls. If the first ball is replaced before the second is drawn, what is the probability that
(a) both balls are red?
(b) both balls are white?
(c) the first ball is red and the second is white?
(d) one of the balls is black?
18. Answer Problem 17 if the first ball is not replaced before the second is drawn.
19. A bag contains 9 nickels, 4 dimes, and 5 quarters. If you draw 3 coins at random from the bag, without replacement, what is the probability that you will get a nickel, a quarter, and a nickel, in that order?
20. A bag contains 6 red balls and 8 green balls. If two balls are drawn together, find the probability that
(a) both are red.
(b) both are green.
21. A red ball and 4 white balls are in a box. If two balls are drawn, without replacement, what is the probability
(a) of getting a red ball on the first draw and a white ball on the second?
(b) of getting 2 white balls?
(c) of getting 2 red balls?
22. From a deck of 52 playing cards two cards are drawn, one after the other without replacement. What is the probability that
(a) the first will be a king and the second will be a jack?
(b) the first will be a king and the second will be a jack of the same suit?
23. One card is drawn at random from a deck of 52 cards. The first card is not replaced, and a second card is drawn. Find the probability that
(a) both cards are spades.
(b) the first card is a heart and the second is a club.
24. Two cards are drawn from a deck of 52 cards. What is the probability that both are aces
(a) if the first card was replaced before the second was drawn?
(b) if the cards were drawn without replacement?
25. Two balls are drawn, without replacement, from a bag containing 13 red balls numbered 1–13 and 5 white balls numbered 14–18. What is the probability that
(a) the second ball is red, given that the first ball is white?
(b) both balls are even-numbered?
(c) the first ball is red and even-numbered and the second ball is even-numbered?
26. A red die and a green die are rolled. What is the probability that a 6 results on the green die and that a 4 results on the red die?

Applications

Universal health care The following table gives the result of a survey of 1000 people regarding the funding of universal health care by employers (employer mandate). Use the table to answer Problems 27–30.

	Favor	*Oppose*	*No Opinion*	*Total*
Democrat	310	150	60	520
Republican	125	345	10	480
Total	435	495	70	1000

27. What is the probability that a person selected at random from this group will favor universal health care with an employer mandate, given that the person is a Democrat?
28. What is the probability that a person selected at random from this group will oppose universal health care with an employer mandate, if the person is a Republican?
29. If the group surveyed represents the people of the United States, what is the probability that a citizen selected at random will favor universal health care with an employer mandate, given that the person is a Republican?
30. If the group surveyed represents the people of the United States, what is the probability that a citizen selected at random will oppose universal health care with an employer mandate, given that the person is a Democrat?

Military spending Suppose the table below summarizes the opinion of various groups on the issue of increased military spending. Use this table to calculate the empirical probabilities in Problems 31–38.

	Whites		*Non-Whites*		
Opinion	*Reps.*	*Dems.*	*Reps.*	*Dems.*	*Total*
Favor	300	100	25	10	435
Oppose	100	250	25	190	565
Total	400	350	50	200	1000

31. Given that a randomly selected individual is non-white, find the probability that he or she opposes increased military spending.
32. Given that a randomly selected individual is a Democrat, find the probability that he or she opposes increased military spending.
33. Given that a randomly selected individual is in favor of increased military spending, find the probability that he or she is a Republican.
34. Given that a randomly selected individual is opposed to increased military spending, find the probability that he or she is a Democrat.
35. Find the probability that a person who favors increased military spending is non-white.
36. Find the probability that an individual is white and opposes increased military spending.
37. Find the probability that an individual is a white Republican opposed to increased military spending.
38. Find the probability that an individual is a Democrat and opposes increased military spending.

39. ***Golf*** On June 16, 1997, two amateur golfers playing together hit back-to-back holes in one (Source: *The Island Packet,* June 19, 1997). According to the National Hole-in-One Association, the probability of an amateur golfer getting a hole-in-one is 1/12,000. If the golfer's shots are independent of each other, what is the probability that two amateur golfers will get back-to-back holes in one?

40. ***Lottery*** A name is drawn at random from the 96 entrants in a golf tournament. The person whose name is selected draws a ball from a bag containing four balls, one of which is embossed with the Peggos Company logo. If the person draws the ball with the logo, he or she wins a set of Ping irons. If you are one of the entrants in the tournament, what is the probability that you will win the Ping irons?

41. ***Blood types*** In the pretrial hearing of the O. J. Simpson case, the prosecution stated that Mr. Simpson's blood markers include type A blood, which 33.7% of the population has; blood SD subtype 1, which 79.6% of the population has; and PGM 2+2−, which 1.6% of the population has. If these blood markers are independent, what is the probability that a person selected at random will have the same blood markers as O. J. Simpson?

42. ***Education*** Fifty-one percent of the U.S. population is female and 18.6% of the female population has a college degree (Source: *Upclose Census Digest,* City and Town Edition, 1992). If a U.S. resident is chosen at random, what is the probability that the person is a female with a college degree?

43. ***Quality control*** Each computer component that the Peggos Company produces is tested twice before it is shipped. There is a .7 probability that a defective component will be so identified by the first test and a .8 probability that it will be identified as being defective by the second test. What is the probability that a defective component will not be identified as defective before it is shipped?

44. ***Lactose intolerance*** Lactose intolerance affects about 20% of non-Hispanic white Americans, and 75.6% of the residents of the United States are non-Hispanic whites (Source: Jean Carper, "Eat Smart," *USA Weekend,* Sept. 4, 1994). If a U.S. resident is selected at random, what is the probability that the person will be a non-Hispanic white and have lactose intolerance?

45. ***Lactose intolerance*** Lactose intolerance affects about 50% of Hispanic Americans, and 9% of the residents of the United States are Hispanic (Source: Jean Carper, "Eat Smart," *USA Weekend,* Sept. 4, 1994). If a U.S. resident is selected at random, what is the probability that the person will be Hispanic and have lactose intolerance?

46. ***Quality control*** If 3% of all light bulbs a company manufactures are defective, the probability of any one bulb being defective is .03. What is the probability that three bulbs drawn independently from the company's stock will be defective?

47. ***Quality control*** To test its shotgun shells, a company fires 5 of them. What is the probability that all 5 will fire properly if 5% of the company's shells are actually defective?

48. ***Quality control*** One machine produces 30% of a product for a company. If 10% of the products from this machine are defective and the other machines produce no defective items, what is the probability that an article produced by this company is defective?

49. ***Advertising*** A company estimates that 30% of the country has seen its commercial and that if a person sees its commercial, there is a 20% probability that the person will buy its product. What is the probability that a person chosen at random in the country will have seen the commercial and bought the product?

50. ***Employment*** Ronald Lee has been told by a company that the probability that he will be offered a job

in the quality control department is .6 and the probability that he will be asked to be foreman of the department, if he is offered the job, is .1. What is the probability that he will be offered the job and asked to be foreman?

51. ***Birth control*** Suppose a birth control pill is 99% effective in preventing pregnancy.
    (a) What is the probability that none of 100 women using the pill will become pregnant?
    (b) What is the probability that at least one woman per 100 users will become pregnant?

52. ***Combat*** If a fighter pilot has a 4% chance of being shot down on each mission during a war, is it certain he will be shot down in 25 missions?

53. ***Maintenance*** Twenty-three percent of the cars owned by a car rental firm have some defect. What is the probability that of 3 cars selected at random,
    (a) none has a defect?
    (b) at least one has defects?

54. ***Testing*** An unprepared student must take a 7-question multiple-choice test that has 5 possible answers per question. If the student guesses on every question, what is the probability that
    (a) she will answer every question correctly?
    (b) she will answer every question incorrectly?
    (c) she will answer at least one question correctly?

55. ***Testing*** An unprepared student must take a 7-question multiple-choice test that has 5 possible answers per question. If the student can eliminate two of the possible answers on the first three questions, and if she guesses on every question, what is the probability that
    (a) she will answer every question correctly?
    (b) she will answer every question incorrectly?
    (c) she will answer at least one question correctly?

56. ***Testing*** What is the probability that a student who guesses on every question of a 10-question true-or-false test will get at least one answer correct?

57. ***Racing*** Suppose that the odds in favor of the horse Portia winning a race are 3:8 and that the odds of the horse Trinka winning the same race are 1:10. What is the probability that Portia or Trinka will win the race? What are the odds that one or the other will win?

58. ***Racing*** Suppose that the odds that Blackjack will win a race are 1 to 3 and the odds that Snowball will win the same race are 1 to 5. If only one horse can win, what odds should be given that one of these two horses will win?

59. ***Birth dates*** Assuming there are 365 different birthdays, find the probability that two people chosen at random will have
    (a) different birthdays.
    (b) the same birthday.

60. ***Birth dates*** Assuming there are 365 different birthdays, find the probability that of three people chosen at random,
    (a) no two will have the same birthday.
    (b) at least two will have the same birthday.

61. ***Birth dates*** Assuming there are 365 different birthdays, find the probability that of 20 people chosen randomly,
    (a) no two will have the same birthday.
    (b) at least two will have the same birthday.

62. ***Crime*** In an actual case,* probability was used to convict a couple of mugging an elderly woman. Shortly after the mugging, a young, white woman with blonde hair worn in a ponytail was seen running from the scene of the crime and entering a yellow car that was driven away by a black man with a beard. A couple matching this description was arrested for the crime. A prosecuting attorney argued that the couple arrested had to be the couple at the scene of the crime because the probability of a second couple matching the description was very small. He estimated the probabilities of six events as follows:

    Probability of black-white couple: 1/1000
    Probability of black man: 1/3
    Probability of bearded man: 1/10
    Probability of blonde woman: 1/4
    Probability of hair in ponytail: 1/10
    Probability of yellow car: 1/10

    He multiplied these probabilities and concluded that the probability that another couple would have these characteristics is 1/12,000,000. On the basis of this circumstantial evidence, the couple was convicted and sent to prison. The conviction was overturned by the state supreme court because the prosecutor made an incorrect assumption. What error do you think he made?

**Time,* January 8, 1965, p. 42, and April 26, 1968, p. 41.

7.4 Probability Trees and Bayes' Formula

OBJECTIVES

- To use probability trees to solve problems
- To use Bayes' formula to solve probability problems

APPLICATION PREVIEW

A security system is manufactured with a "fail-safe" provision so that it functions properly if any two or more of its three main components, A, B, and C, are functioning properly. The probabilities that components A, B, and C are functioning properly are .95, .90, and .92, respectively. There are four different ways in which at least two components will function properly, so there are four mutually exclusive possible outcomes in which the system will function properly. We can find the probability that the system functions properly by using a **probability tree.** Probability trees are especially useful in dealing with probability problems that involve two or more stages.

Probability Trees

When a probability experiment involves more than one trial, it is frequently helpful to represent the possible outcomes in a tree diagram. We can think of the possible outcomes of the first trial as the beginning branches of the tree, as in Figure 7.5(a), where the first trial has three possible outcomes E_1, E_2, and E_3. The possible outcomes on the second trial can be treated as branches that emanate from each of the outcomes of the first trial, as in Figure 7.5(b), where there are two possible outcomes, F_1 and F_2, of the second trial. We know from the previous section that the probability that event E_1 occurs on the first trial and F_2 occurs on the second trial is

$$\Pr(E_1 \cap F_2) = \Pr(E_1) \cdot \Pr(F_2 \mid E_1)$$

If we put these probabilities on the branches of the tree, then the "event E_1 occurs on trial 1 and event F_2 occurs on trial 2" is represented by a "path" through the tree, and the probability that this occurs is the product of the probabilities on the branches.

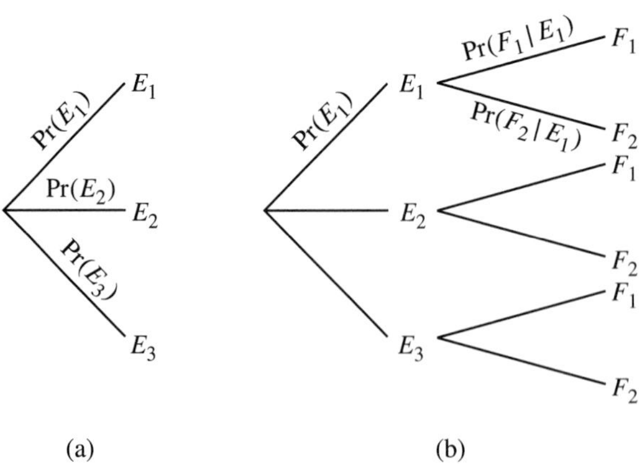

Figure 7.5

To illustrate how a probability tree is constructed, consider the following example.

EXAMPLE 1

A bag contains 5 red balls, 4 blue balls, and 3 white balls. Two balls are drawn, one after the other, without replacement. Draw a tree representing the experiment.

Solution

The first draw could result in a red ball (with probability 5/12), a blue ball (with probability 4/12), or a white ball (with probability 3/12). We can represent the results of the first draw by the tree shown in Figure 7.6.

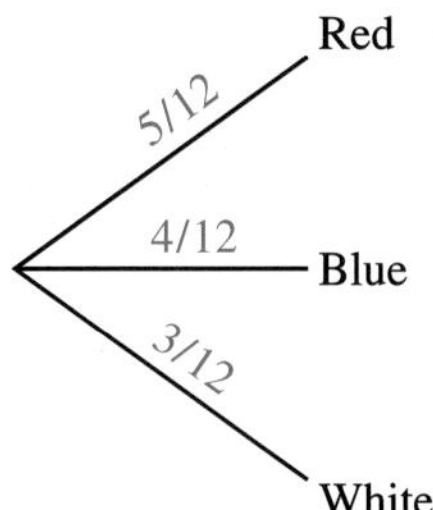

Figure 7.6

The probabilities for red, blue, and white balls occurring on the second draw depend on the result of the first draw. We can represent all the possibilities by adding a second stage to the tree we have drawn, with the conditional probabilities noted (see Figure 7.7).

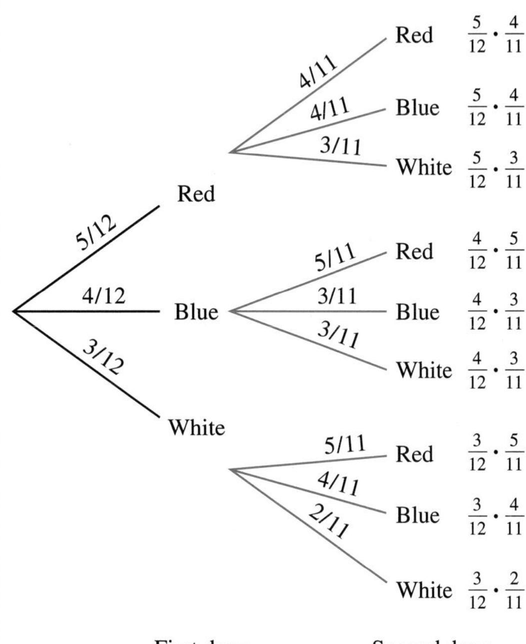

Figure 7.7

The first stage of the experiment in Example 1 has three possible outcomes, so the tree has three branches in the first stage. The second stage has three possible outcomes, so the tree has three branches in its second stage for each first outcome. The total number of possible outcomes in the compound experiment is the same as the total number of terminating branches, so each path through the tree represents a sample point for the experiment. For example, drawing a red ball and then drawing a blue ball is one possible outcome, and it is represented by one path through the tree—namely, the branch with red on the first draw and blue on the second draw.

Probability Tree The probability attached to *each branch* of the tree is the *conditional probability* that the specified event will occur, given that the events on the preceding branches have occurred. The probability of a **sample point** for the experiment described by a tree is the **product** of the probabilities attached to the branches **along the path** representing the sample point.

For example, the probability of drawing a red ball first and then a blue ball in the experiment described by Figure 7.7 is the product of 5/12 and 4/11, or 5/33.

EXAMPLE 2

In the experiment described by Example 1 (and Figure 7.7), find

(a) Pr(blue on first draw and white on second draw).
(b) Pr(white on both draws).
(c) Pr(drawing a blue ball and a white ball).
(d) Pr(second ball is red).

Solution

(a) By multiplying the probabilities along the path that represents blue on the first draw and white on the second draw, we obtain

$$\text{Pr(blue on the first draw and white on second draw)} = \frac{4}{12} \cdot \frac{3}{11} = \frac{1}{11}$$

(b) Multiplying the probabilities along the path that represents white on both draws gives

$$\text{Pr(white on both draws)} = \frac{3}{12} \cdot \frac{2}{11} = \frac{1}{22}$$

(c) This result can occur by drawing a blue ball first and then a white ball *or* by drawing a white ball and then a blue ball. Thus either of two paths leads to this result. Because the results represented by these paths are mutually exclusive, the probability is found by *adding* the probabilities from the two paths. That is, letting B represent a blue ball and W represent a white ball, we have

$$\begin{aligned}\Pr(B \text{ and } W) &= \Pr(B \cap W)\\ &= \Pr(B \text{ first, then } W) + \Pr(W \text{ first, then } B)\\ &= \frac{4}{12} \cdot \frac{3}{11} + \frac{3}{12} \cdot \frac{4}{11} = \frac{2}{11}\end{aligned}$$

(d) The second ball can be red after the first is red, black, or white.

$$\begin{aligned}\Pr(\text{2d ball is red}) &= \Pr(R \text{ then } R, \textit{ or } B \text{ then } R, \textit{ or } W \text{ then } R)\\ &= \Pr(R \text{ then } R) + \Pr(B \text{ then } R) + \Pr(W \text{ then } R)\\ &= \frac{5}{12}\cdot\frac{4}{11} + \frac{4}{12}\cdot\frac{5}{11} + \frac{3}{12}\cdot\frac{5}{11} = \frac{5}{12}\end{aligned}$$

CHECKPOINT

1. Urn I contains 3 gold coins, Urn II contains 1 gold coin and 2 silver coins, and Urn III contains 1 gold coin and 1 silver coin. If an urn is selected at random and a coin is drawn from the urn, what is the probability that a gold coin will be drawn?

EXAMPLE 3

As we mentioned in the Application Preview, a security system is manufactured with a "fail-safe" provision so that it functions properly if any two or more of its three main components, A, B, and C, are functioning properly. The probabilities that components A, B, and C are functioning properly are .95, .90, and .92, respectively. What is the probability that the system functions properly?

Solution

The probability tree in Figure 7.8 shows all of the possibilities for the components of the system. Each stage of the tree corresponds to a component, and G indicates that the component is good; N indicates it is not good. There are four paths through the probability tree that give at least two good (functioning) components, so there are four mutually exclusive possible outcomes. The probability that the system functions properly is

$$\begin{aligned}\Pr(\text{at least 2 } G) &= \Pr(GGG \cup GGN \cup GNG \cup NGG)\\ &= (.95)(.90)(.92) + (.95)(.90)(.08) + (.95)(.10)(.92) + (.05)(.90)(.92)\\ &= .9838\end{aligned}$$

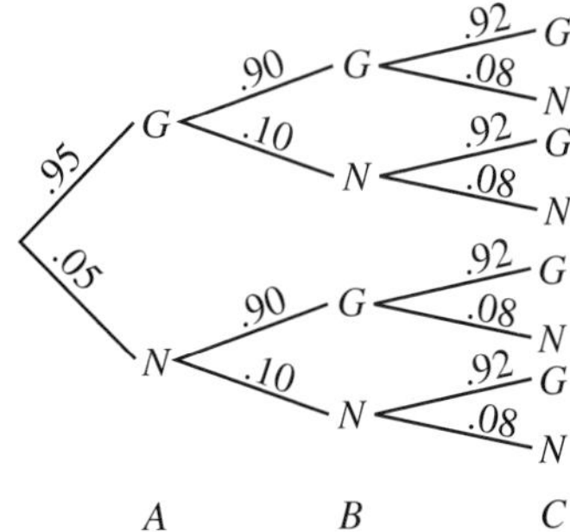

Figure 7.8

Bayes' Formula

Using a probability tree permits us to answer more difficult questions regarding conditional probability. For example, in the experiment described by Example 1, we can find the probability that the first ball is red, given that the second ball is blue. To find this probability, we recall the formula for conditional probability, which is

$$\Pr(E_2 \mid E_1)\frac{\Pr(E_1 \cap E_2)}{\Pr(E_1)}$$

so

$$\Pr(\text{1st is red} \mid \text{2d is blue}) = \frac{\Pr(\text{1st is red and 2d is blue})}{\Pr(\text{2d is blue})}$$

One path in Figure 7.7 (red-blue) describes "1st is red and 2d is blue" and three paths (red-blue, blue-blue, white-blue) describe "2d is blue," so

$$\Pr(\text{1st is red} \mid \text{2d is blue}) = \frac{\frac{5}{12}\cdot\frac{4}{11}}{\frac{5}{12}\cdot\frac{4}{11}+\frac{4}{12}\cdot\frac{3}{11}+\frac{3}{12}\cdot\frac{4}{11}}$$

$$= \frac{5/33}{5/33 + 1/11 + 1/11} = \frac{5}{11}$$

The preceding example illustrates a special type of problem, which we call a **Bayes problem.** In this type of problem, we know the result of the second stage of a two-stage experiment and wish to find the probability of a specified result in the first stage. In the preceding example we knew the second ball drawn was blue, and we sought the probability that the first ball was red.

Suppose there are n possible outcomes in the first stage of the experiment, denoted $E_1, E_2, \ldots, E_n$, and m possible outcomes in the second stage, denoted $F_1, F_2, \ldots, F_m$ (see Figure 7.9). Then the probability that event E_1 occurs in the first stage, given that F_1 occurred in the second stage, is

$$\Pr(E_1 \mid F_1) = \frac{\Pr(E_1 \cap F_1)}{\Pr(F_1)} \tag{1}$$

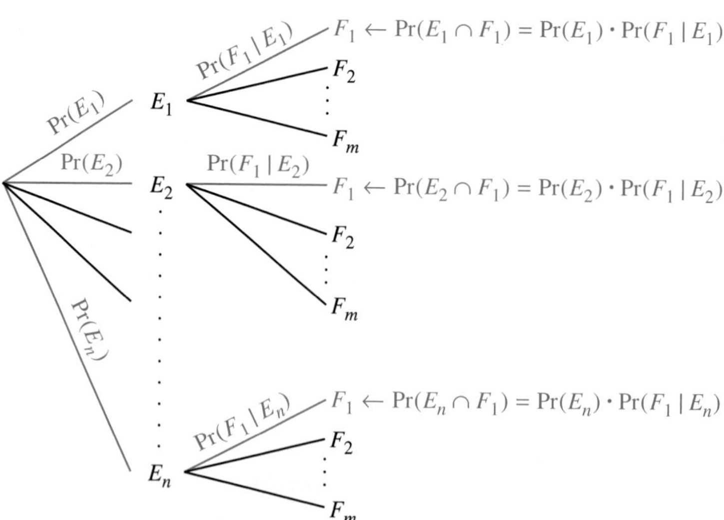

Figure 7.9

By looking at the probability tree in Figure 7.9, we see that

$$\Pr(E_1 \cap F_1) = \Pr(E_1)\cdot\Pr(F_1 \mid E_1)$$

and that

$$\Pr(F_1) = \Pr(E_1 \cap F_1) + \Pr(E_2 \cap F_1) + \cdots + \Pr(E_n \cap F_1)$$

Using these facts, equation (1) becomes

$$\Pr(E_1 \mid F_1) = \frac{\Pr(E_1) \cdot \Pr(F_1 \mid E_1)}{\Pr(E_1 \cap F_1) + \Pr(E_2 \cap F_1) + \cdots + \Pr(E_n \cap F_1)} \qquad (2)$$

Using the fact that for any E_i, $\Pr(E_i \text{ and } F_1) = \Pr(E_i) \cdot \Pr(F_1 \mid E_i)$, we can rewrite equation (2) in a form called **Bayes' formula.**

Bayes' Formula If a probability experiment has n possible outcomes in its first stage given by E_1, E_2, . . . , E_n, and if F_1 is an event in the second stage, then the probability that even E_1 occurs in the first stage, given that F_1 has occurred in the second stage, is

$$\Pr(E_1 \mid F_1) = \frac{\Pr(E_1) \cdot \Pr(F_1 \mid E_1)}{\Pr(E_1) \cdot \Pr(F_1 \mid E_1) + \Pr(E_2) \cdot \Pr(F_1 \mid E_2) + \cdots + \Pr(E_n) \cdot \Pr(F_1 \mid E_n)}$$

Note that Bayes problems can be solved either by this formula or by the use of a probability tree. Of the two methods, many students find using the tree easier because it is less abstract than the formula.

Bayes' Formula and Trees

$$\Pr(E_1 \mid F_1) = \frac{\text{Product of branch probabilities on path leading to } F_1 \text{ through } E_1}{\text{Sum of all branch products on paths leading to } F_1}$$

The following example illustrates both methods of solving a Bayes problem.

EXAMPLE 4

Suppose a test for diagnosing a certain serious disease is successful in detecting the disease in 95% of all persons infected but that it incorrectly diagnoses 4% of all healthy people as having the serious disease. Suppose also that it incorrectly diagnoses 12% of all people having another minor disease as having the serious disease. It is known that 2% of the population has the serious disease, 90% of the population is healthy, and 8% has the minor disease. Find the probability that a person selected at random has the serious disease if the test indicates that he or she does. Use H to represent healthy, M to represent having the minor disease, and D to represent having the serious disease.

Solution

The tree that represents the health condition of a person chosen at random and the results of the test on that person is shown in Figure 7.10. (A test that indicates that a person has the disease is called a *positive* test.)

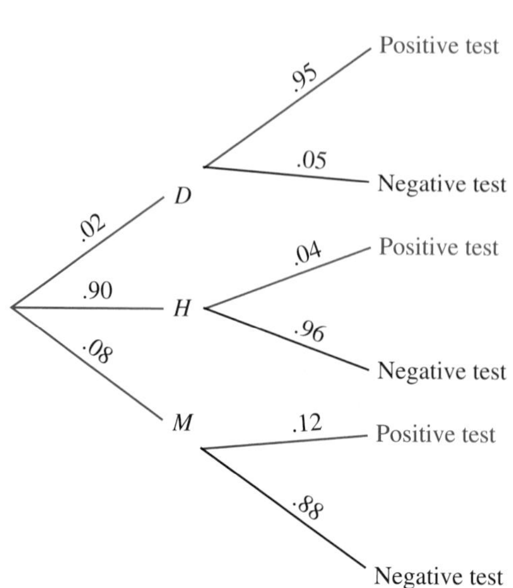

Figure 7.10

We seek $\Pr(D \mid \text{pos. test})$. Using the tree and the conditional probability formula gives

$$\begin{aligned}\Pr(D \mid \text{pos. test}) &= \frac{\Pr(D \cap \text{pos. test})}{\Pr(\text{pos. test})}\\ &= \frac{(.02)(.95)}{(.90)(.04) + (.02)(.95) + (.08)(.12)}\\ &= \frac{.0190}{.0360 + .0190 + .0096}\\ &= .2941\end{aligned}$$

Using Bayes' formula directly gives the same result.

$\Pr(D \mid \text{pos. test})$

$$= \frac{\Pr(D) \cdot \Pr(\text{pos. test} \mid D)}{\Pr(H) \cdot \Pr(\text{pos. test} \mid H) + \Pr(D) \cdot \Pr(\text{pos. test} \mid D) + \Pr(M) \cdot \Pr(\text{pos. test} \mid M)}$$

$$= \frac{(.02)(.95)}{(.90)(.04) + (.02)(.95) + (.08)(.12)} = .2941$$

CHECKPOINT

2. Use the information (and tree) from Example 4 to find the probability that a person chosen at random is healthy if the test is negative (a negative result indicates that the person does not have the disease).

Table 7.2 gives a summary of formulas and when they are used.

TABLE 7.2 Summary of Probability Formulas

		One Trial	*Two Trials*	*More Than Two Trials*
Pr(A and B)	Independent	Sample space	$\Pr(A) \cdot \Pr(B)$	Product of probabilities
	Dependent	Sample space	$\Pr(A) \cdot \Pr(B \mid A)$	Product of conditional probabilities
Pr(A or B)	Mutually exclusive	$\Pr(A) + \Pr(B)$	Probability tree	Probability tree
	Not mutually exclusive	$\Pr(A) + \Pr(B) - \Pr(A \text{ and } B)$	Probability tree	Probability tree

CHECKPOINT SOLUTIONS

1. The probability tree for this probability experiment is given below. From the tree we have

$$\begin{aligned}\Pr(\text{gold}) &= \Pr[(\text{I and } G) \text{ or } (\text{II and } G) \text{ or } (\text{III and } G)]\\ &= \frac{1}{3}\cdot\frac{3}{3}+\frac{1}{3}\cdot\frac{1}{3}+\frac{1}{3}\cdot\frac{1}{2}=\frac{11}{18}\end{aligned}$$

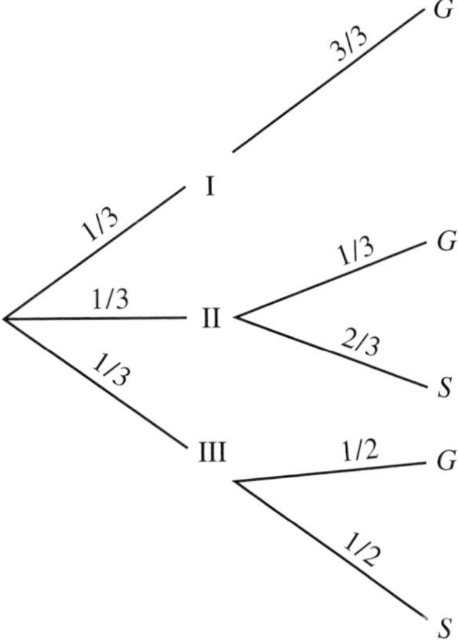

2. The probability that a person is healthy, given that the test is negative, is

$$\begin{aligned}\Pr(H \mid \text{neg. test}) &= \frac{\Pr(H \cap \text{neg. test})}{\Pr(\text{neg. test})}\\ &= \frac{(.90)(.96)}{(.02)(.05)+(.90)(.96)+(.08)(.88)}\\ &\approx .9237\end{aligned}$$

EXERCISE 7.4

In Problems 1–14, use probability trees to find the probabilities of the indicated outcomes.

1. Two balls are drawn, without replacement, from a bag containing 4 blue balls and 5 white balls. What is the probability that
   (a) both balls drawn are the same color?
   (b) one ball is blue and one ball is white?
2. Two cards are drawn, without replacement, from a regular deck of 52 cards. What is the probability that
   (a) both cards are the same suit?
   (b) both cards are the same suit, given that the first card is the king of hearts?
3. A bag contains 5 coins, 4 of which are fair and 1 that has a head on each side. If a coin is selected from the bag and tossed twice, what is the probability of obtaining 2 heads?
4. Three balls are drawn, without replacement, from a bag containing 4 red balls and 5 white balls. Find the probability that
   (a) three white balls are drawn.
   (b) two white balls and one red ball are drawn.
   (c) the third ball drawn is red.
5. Two cards are drawn, without replacement, from a regular deck of 52 cards. What is the probability that both cards are aces, given that they are the same color?
6. An urn contains 4 red, 5 white, and 6 black balls. One ball is drawn from the urn, it is replaced, and a second ball is drawn.
   (a) What is the probability that both balls are white?
   (b) What is the probability that one ball is white and one is red?
   (c) What is the probability that at least one ball is black?
7. An urn contains 4 red, 5 white, and 6 black balls. One ball is drawn from the urn, it is not replaced, and a second ball is drawn.
   (a) What is the probability that both balls are white?
   (b) What is the probability that one ball is white and one is red?
   (c) What is the probability that at least one ball is black?
8. An urn contains 4 red, 5 white, and 6 black balls. Three balls are drawn, without replacement, from the urn.
   (a) What is the probability that all three balls are red?
   (b) What is the probability that exactly two balls are red?
   (c) What is the probability that at least two balls are red?
9. A bag contains 4 white balls and 6 red balls. Three balls are drawn, without replacement, from the bag.
   (a) What is the probability that all three balls are white?
   (b) What is the probability that exactly one ball is white?
   (c) What is the probability that at least one ball is white?
10. A bag contains 5 coins, of which 4 are fair; the remaining coin has a head on both sides. If a coin is selected at random from the bag and tossed three times, what is the probability that heads will occur exactly twice?
11. A bag contains 5 coins, of which 4 are fair; the remaining coin has a head on both sides. If a coin is selected at random from the bag and tossed three times, what is the probability that heads will occur at least twice?
12. A bag contains 4 white balls and 7 red balls. If 2 balls are drawn at random from the bag, with the first ball being replaced before the second ball is drawn, what is the probability that
    (a) the first ball is red and the second is white?
    (b) one ball is red and the other is white?
13. A bag contains 4 white balls and 6 red balls. What is the probability that if 2 balls are drawn (with replacement),
    (a) the first ball is red and the second is white?
    (b) both balls are red?
    (c) one ball is red and one is white?
    (d) the first ball is red or the second is white?
14. Two balls are drawn, without replacement, from a bag containing 13 red balls numbered 1–13 and 5 white balls numbered 14–18. What is the probability that
    (a) the second ball is even-numbered, given that the first ball is even-numbered?
    (b) the first ball is red and the second ball is even-numbered?
    (c) the first ball is even-numbered and the second is white?

In Problems 15–18, use (a) a probability tree and (b) Bayes' formula to find the probabilities. Each of urns I and II has 5 red balls, 3 white balls, and 2 green balls. Urn III has 1 red ball, 1 white ball, and 8 green balls. Use this information in Problems 15 and 16.

15. An urn is selected at random and a ball is drawn. If the ball is green, find the probability that urn III was selected.

16. An urn is selected at random and a ball is drawn. If the ball is red, what is the probability that urn III was selected?
17. There are 3 urns containing coins. Urn I contains 3 gold coins, urn II contains 1 gold coin and 1 silver coin, and urn III contains 2 silver coins. An urn is selected and a coin is drawn from the urn. If the selected coin is gold, what is the probability that the urn selected was urn I?
18. In Problem 17, what is the probability that the urn selected was urn III if the coin selected was silver?

Applications

19. ***Lactose intolerance*** Lactose intolerance affects about 20% of non-Hispanic white Americans, 50% of Hispanic Americans, and 75% of African, Asian, and Native Americans (Source: Jean Carper, "Eat Smart," *USA Weekend,* Sept. 4, 1994). Seventy-six percent of U.S. residents are non-Hispanic whites, 9% of them are Hispanic, and 15% are African, Asian, or Native American. If a person is selected from this group of people, what is the probability that the person will have lactose intolerance?

20. ***Genetics*** What is the probability that a couple will have at least two sons if they plan to have 3 children and if the probability of having a son equals the probability of having a daughter?

21. ***Marksmanship*** Suppose that a marksman hits the bull's-eye 15,000 times in 50,000 shots. If the next 4 shots are independent, find the probability that
    (a) the next 4 shots hit the bull's-eye.
    (b) two of the next 4 shots hit the bull's-eye.

22. ***Majors*** In a random survey of students concerning student activities, 30 engineering majors, 25 business majors, 20 science majors, and 15 liberal arts majors were selected. If two students are selected at random, what is the probability of getting
    (a) two science majors?
    (b) a science major and an engineering major?

23. ***Quality control*** Suppose a box contains 3 defective transistors and 12 good transistors. If 2 transistors are drawn from the box without replacement, what is the probability that
    (a) the first transistor is good and the second transistor is defective?
    (b) the first transistor is defective and the second one is good?
    (c) one of the transistors drawn is good and one of them is defective?

24. ***Education*** The probability that an individual without a college education earns more than $20,000 is .2, whereas the probability that a person with a B.S. or higher degree will earn more than $20,000 is .6. The probability that a person chosen at random has a B.S. degree is .3. What is the probability a person has at least a B.S. degree if it is known that he or she earns more than $20,000?

25. ***Alcoholism*** A small town has 8000 adult males and 6000 adult females. A sociologist conducted a survey and found that 40% of the males and 30% of the females drink heavily. An adult is selected at random from the town.
    (a) What is the probability the person is a male?
    (b) What is the probability the person drinks heavily?
    (c) What is the probability the person is a male or drinks heavily?
    (d) What is the probability the person is a male, if it is known that the person drinks heavily?

26. ***TV violence*** One hundred boys and 100 girls were asked if they had ever been frightened by a television program. Thirty of the boys and 60 of the girls said they had been frightened. One of these children is selected at random.
    (a) What is the probability that he or she has been frightened?
    (b) What is the probability the child is a girl, given that he or she has been frightened?

27. ***Drinking age*** A survey questioned 1000 people regarding raising the legal drinking age from 18 to 21. Of the 560 who favored raising the age, 390 were female. Of the 440 opposition responses, 160 were female. If a person selected at random from this group is a man, what is the probability that the person favors raising the drinking age?

28. ***Politics*** A candidate for public office knows that he has a 60% chance of being elected to the office if he is nominated by his political party and that his party has a 50% chance of winning with another nominee. If the probability that he will be nominated by his party is .25, and his party wins the election, what is the probability that he is the winning candidate?

29. ***Pregnancy test*** A self-administered pregnancy test detects 85% of those who are pregnant but does not detect pregnancy in 15%. It is 90% accurate in indicating women who are not pregnant but indicates 10% of this group as being pregnant. Suppose it is known that 1% of the women in a neighborhood are pregnant. If a woman is chosen at random from those

living in this neighborhood, and if the test indicates she is pregnant, what is the probability that she really is?

30. ***Drinking age*** A survey questioned 1000 people regarding raising the legal drinking age from 18 to 21. Of the 560 who favored raising the age, 390 were female. Of the 440 who expressed opposition, 160 were female.
 (a) What is the probability that a person selected at random is a female?
 (b) What is the probability that a person selected at random favors raising the age if the person is a woman?

31. ***Survey*** In a SAPS (Student Activity Participation Study) survey, 30 engineering majors, 25 business majors, 20 science majors, 15 liberal arts majors, and 10 human development majors were selected. Ten of the engineering, 12 of the science, 6 of the human development, 13 of the business, and 8 of the liberal arts majors selected for the study were female.
 (a) What is the probability of selecting a female if one person from this group is selected randomly?
 (b) What is the probability that a student selected randomly from this group is a science major, given that she is female?

32. ***Survey*** If a student is selected at random from the group in Problem 31, what is the probability the student is
 (a) an engineering major, given that she is female?
 (b) an engineering major, given that he is male?

7.5 Counting; Permutations and Combinations

OBJECTIVES

- *To use the Fundamental Counting Principle and permutations to solve counting problems*
- *To use combinations to solve counting problems*

APPLICATION PREVIEW

During a national television advertising campaign, Little Caesar's Pizza stated that for $9.95, you could get 2 medium-sized pizzas, each with any of 0 to 5 toppings chosen from 11 that are available. The commercial asked the question, "How many different pairs of pizzas can you get?" Answering this question uses the computation of **combinations,** which are discussed in this section. Other counting techniques, including **permutations,** are also discussed in this section.

Suppose that you decide to dine at a restaurant that offers 3 appetizers, 8 entrees, and 6 desserts. How many different meals can you have? We can answer questions of this type using the **Fundamental Counting Principle.**

Fundamental Counting Principle

If there are $n(A)$ ways in which an event A can occur, and if there are $n(B)$ ways in which a second event B can occur after the first event has occurred, then the two events can occur in $n(A) \cdot n(B)$ ways.

The Fundamental Counting Principle can be extended to any number of events as long as they are independent. Thus the total number of possible meals at the restaurant mentioned above is

$$n(\text{appetizers}) \cdot n(\text{entrees}) \cdot n(\text{desserts}) = 3 \cdot 8 \cdot 6 = 144$$

EXAMPLE 1

The state of West Virginia has auto license plates that begin with a digit followed by two letters and three more digits. If no other plates are made for automobiles in this state, how many automobiles could be licensed in West Virginia?

Solution

The following numbers represent the number of possible digits or letters in each of the 6 spaces on the license plates.

$$\underline{10} \quad \underline{26} \quad \underline{26} \quad \underline{10} \quad \underline{10} \quad \underline{10}$$

The product of these numbers gives the number of license plates that can be made satisfying the given conditions.

$$\underline{10} \cdot \underline{26} \cdot \underline{26} \cdot \underline{10} \cdot \underline{10} \cdot \underline{10} = 6{,}760{,}000$$

Thus 6,760,000 automobiles can be licensed with these plates.

CHECKPOINT

1. If a state permits either a letter or a nonzero digit to be used in each of six places on its license plates, how many different plates can it issue?

EXAMPLE 2

Suppose a person planning a banquet cannot decide how to seat 6 honored guests at the head table. In how many arrangements can they be seated in the 6 chairs on one side of a table?

Solution

If we think of the 6 chairs as spaces, we can determine how many ways each space can be filled. The planner could place any one of the 6 people in the first space (say the first chair on the left). One of the 5 remaining persons could then be placed in the second space, one of the 4 remaining persons in the third space, and so on, as follows:

$$\underline{6} \quad \underline{5} \quad \underline{4} \quad \underline{3} \quad \underline{2} \quad \underline{1}$$

By the Fundamental Counting Principle, the total number of arrangements that can be made is the product of these numbers. That is, the total number of possible arrangements is

$$6 \cdot 5 \cdot 4 \cdot 3 \cdot 2 \cdot 1 = 720$$

Because special products such as $6 \cdot 5 \cdot 4 \cdot 3 \cdot 2 \cdot 1$ frequently occur in counting theory, we use special notation to denote them. We write 6! (read "6 factorial") to denote $6 \cdot 5 \cdot 4 \cdot 3 \cdot 2 \cdot 1$. Likewise, $4! = 4 \cdot 3 \cdot 2 \cdot 1$.

Factorial For any positive integer n, we define ***n*-factorial,** written $n!$, as

$$n! = n(n-1)(n-2) \cdots 3 \cdot 2 \cdot 1$$

We define $0! = 1$.

EXAMPLE 3

Suppose the planner in Example 2 knows that 8 people feel they should be at the head table, but only 6 spaces are available. How many arrangements can be made that place 6 of the 8 people at the head table?

Solution

There are again 6 spaces to fill, but any of 8 people can be placed in the first space, any of 7 people in the second space, and so on. Thus the total number of possible arrangements is

$$\underline{8} \cdot \underline{7} \cdot \underline{6} \cdot \underline{5} \cdot \underline{4} \cdot \underline{3} = 20{,}160$$

The number of possible arrangements in Example 3 is called the number of **permutations** of 8 things taken 6 at a time, and it is denoted ${}_8P_6$. Note that ${}_8P_6$ gives the first 6 factors of 8!, so we can use factorial notation to write the product.

$${}_8P_6 = \frac{8!}{2!} = \frac{8!}{(8-6)!}$$

Permutations The number of possible distinct arrangements of r objects chosen from a set of n objects is called the number of **permutations** of n objects taken r at a time, and it equals

$${}_nP_r = \frac{n!}{(n-r)!}$$

Most scientific and graphing calculators have keys that can be used to calculate permutations.

EXAMPLE 4

In how many ways can a president, vice president, secretary, and treasurer be selected from an organization with 20 members?

Solution

We seek the number of orders (arrangements) in which 4 people can be selected from a group of 20. This number is

$${}_{20}P_4 = \frac{20!}{(20-4)!} = 116{,}280$$

CHECKPOINT

2. In a 5-question matching test, how many different answer sheets are possible if no answer can be used twice and there are
   (a) 5 answers available for matching?
   (b) 7 answers available for matching?

When we talk about arrangements of people at a table or arrangements of digits on a license plate, order is important. We now consider counting problems in which order is not important. Suppose you are the president of a company and you want to pick 2 secretaries to work for you. If 5 people are qualified, how many different pairs of people can you select? If you select 2 people from the 5 in a specific order, the number of *arrangements* of the 5 people taken 2 at a time would be

$$_5P_2 = 20$$

But $_5P_2$ gives the number of ways you can select *and* order the pairs, and we seek only the number of ways you can select them. If the names of the people are A, B, C, D, and E, then two of the arrangements, AB and BA, would represent only one pair. Because each of the pairs of people can be arranged in $2! = 2$ ways, the number of pairs that can be selected without regard to order is

$$\frac{_5P_2}{2!} = \frac{20}{2} = 10$$

To find the number of ways you can select 3 people from the 5 without regard to order, we would first find $_5P_3$ and then divide by the number of ways the 3 people could be ordered (3!). Thus 3 people can be selected from 5 (without regard for order) in $_5P_3/3! = 10$ ways. We say that the number of combinations of 5 things taken 3 at a time is 10.

Combinations The number of ways in which r objects can be chosen from a set of n objects without regard to the order of selection is called the number of **combinations** of n objects taken r at a time, and it equals

$$_nC_r = \frac{_nP_r}{r!}$$

Note that the fundamental difference between permutations and combinations is that permutations are used when order is a factor in the selection, and combinations are used when it is not. Because $_nP_r = n!/(n-r)!$, we have

$$_nC_r = \frac{_nP_r}{r!} = \frac{\dfrac{n!}{(n-r)!}}{r!} = \frac{n!}{r!(n-r)!}$$

Combination Formula The number of combinations of n objects taken r at a time is also denoted by $\binom{n}{r}$; that is,

$$\binom{n}{r} = {_nC_r} = \frac{n!}{r!(n-r)!}$$

Most scientific and graphing calculators have keys that can be used to calculate combinations.

EXAMPLE 5

An auto dealer is offering any of 6 special options at one price on a specially equipped car being sold. If the dealer will sell as many of these options as you want, and you decide to buy the car with 4 options, how many different choices of specially equipped cars do you have?

Solution

The order in which you choose the options is not relevant, so we seek the number of combinations of 6 things taken 4 at a time. Thus the number of possible choices is

$$_6C_4 = \frac{6!}{4!(6-4)!} = \frac{6!}{4!2!} = 15$$

EXAMPLE 6

Suppose you are the junior class president, and you want to pick 4 people to serve on a committee. If 10 people are willing to serve, in how many different ways can you pick your committee?

Solution

Because the order in which the committee members are selected is not important, there are

$$_{10}C_4 = \frac{10!}{4!6!} = 210$$

ways in which the committee could be picked.

EXAMPLE 7

Six men and eight women have volunteered to serve on a committee. How many different committees can be formed containing three men and three women?

Solution

Three men can be selected from six in $\binom{6}{3}$ ways, and three women can be selected from eight in $\binom{8}{3}$ ways. Therefore, three men *and* three women can be selected in

$$\binom{6}{3}\binom{8}{3} = \frac{6!}{3!3!} \cdot \frac{8!}{5!3!} = 1120 \text{ ways}$$

CHECKPOINT

Determine whether each of the following counting problems can be solved by using permutations or combinations, and answer each question.

3. Find the number of ways in which 8 members of a space shuttle crew can be selected from 20 available astronauts.
4. The command structure on a space shuttle flight is determined by the order in which astronauts are selected for a flight. How many different command structures are possible if 8 astronauts are selected from the 20 that are available?
5. If 14 men and 6 women are available for a space shuttle flight, how many crews are possible that have 5 men and 3 women astronauts?

We now consider the television commercial mentioned in the Application Preview.

EXAMPLE 8

During a national television advertising campaign, Little Caesar's Pizza stated that for \$9.95, you could get 2 medium-sized pizzas, each with any of 0 to 5 toppings chosen from 11 that are available. The commercial asked the question, "How many different pairs of pizzas can you get?" Answer the question, if the first pizza has a thin crust and the second has a thick crust.

Solution

Because you can get 0, 1, 2, 3, 4, or 5 toppings from the 11 choices, the possible choices for the first pizza are

$$_{11}C_0 + {_{11}C_1} + {_{11}C_2} + {_{11}C_3} + {_{11}C_4} + {_{11}C_5}$$
$$= 1 + 11 + 55 + 165 + 330 + 462 = 1024$$

Because the same number of choices is available for the second pizza, the number of choices for two pizzas is

$$1024 \cdot 1024 = 1{,}048{,}576$$

CHECKPOINT SOLUTIONS

1. Any of 35 items (26 letters or 9 digits) can be used in any of the six places, so the number of different license plates is

$$\underline{35} \cdot \underline{35} \cdot \underline{35} \cdot \underline{35} \cdot \underline{35} \cdot \underline{35} = 35^6 = 1{,}838{,}265{,}625$$

2. (a) There are 5 answers available for the first question, 4 for the second, 3 for the third, 2 for the fourth, and 1 for the fifth, so there are

$$5 \cdot 4 \cdot 3 \cdot 2 \cdot 1 = 120$$

possible answer sheets. Note that this is the number of permutations of 5 things taken 5 at a time,

$$_5P_5 = 120$$

(b) There are 7 answers available for the first question, 6 for the second, 5 for the third, 4 for the fourth, and 3 for the fifth, so there are

$$7 \cdot 6 \cdot 5 \cdot 4 \cdot 3 = 2520$$

possible answer sheets. Note that this is the number of permutations of 7 things taken 5 at a time, $_7P_5 = 2520$.

3. The members of the crew are to be selected without regard to the order of selection, so we seek the number of combinations of 8 people selected from 20. There are

$$_{20}C_8 = 125{,}970$$

possible crews.

4. The order in which the members are selected is important, so we seek the number of permutations of 8 people selected from 20. There are

$$_{20}P_8 = 5{,}079{,}110{,}400$$

possible command structures.

5. As in Problem 3, the order of selection is not relevant, so we seek the number of combinations of 5 men out of 14 *and* 3 women out of 6. There are

$$_{14}C_5 \cdot {}_6C_3 = 2002 \cdot 20 = 40{,}040$$

possible crews that have 5 men and 3 women.

EXERCISE 7.5

Permutations

1. Compute $_6P_4$.
2. Compute $_8P_5$.
3. Compute $_{10}P_6$.
4. Compute $_7P_4$.
5. Compute $_5P_3$.
6. Compute $_4P_1$.
7. Compute $_5P_0$.
8. Compute $_nP_n$.
9. How many four-digit numbers can be formed from the digits 1, 3, 5, 7, 8, and 9
   (a) if each digit may be used once in each number?
   (b) if each digit may be used repeatedly in each number?
10. How many four-digit numbers can be formed from the digits 1, 3, 5, 7, 8, and 9
   (a) if the numbers must be even and digits are not used repeatedly?
   (b) if the numbers are less than 3000 and digits are not repeated?
   (*Hint:* In both parts, begin with the digit where there is a restriction on the choices.)
11. Compute $_nP_0$.
12. Compute $_nP_1$.
13. Find $\frac{(n+1)!}{n!}$.
14. Find $\frac{(2n+2)!}{(2n)!}$.
15. If $(n + 1)! = 17n!$, find n.
16. If $(n + 1)! = 30(n - 1)!$, find n.

Combinations

17. Compute $\binom{6}{2}$.
18. Compute $\binom{8}{6}$.
19. Compute $_{100}C_{98}$.
20. Compute $_{80}C_{76}$.
21. Compute $_4C_4$.
22. Compute $_3C_1$.
23. Compute $\binom{5}{0}$.
24. Compute $\binom{n}{0}$.
25. Compute $\binom{n}{n}$.
26. Compute $\binom{n}{1}$.
27. Verify $_{13}C_{10} = {}_{12}C_9 + {}_{12}C_{10}$.
28. Verify $_{20}C_{15} = {}_{19}C_{14} + {}_{19}C_{15}$.
29. If $_nC_6 = {}_nC_4$, find n.
30. If $_nC_8 = {}_nC_7$, find n.

Applications

31. ***License plates*** If a state wants each of its license plates to contain 7 different digits, how many different license plates can it make?

32. ***License plates*** If a state wants each of its license plates to contain any 7 digits, how many different license plates are possible?

33. ***Signaling*** A sailboat owner received 6 different signal flags with his new sailboat. If the order in which the flags are arranged on the mast determines the signal being sent, how many 3-flag signals can be sent?

34. ***Racing*** Eight horses are entered in a race. In how many ways can the horses finish?

35. ***Management*** Four candidates for manager of a department store are ranked according to the weighted average of several criteria. How many different rankings are possible if no two candidates receive the same rank?

36. ***Testing*** An examination consists of 12 questions. If 10 questions must be answered, find the number of different orders in which a student can answer the questions.

37. ***Molecules*** Biologists have identified 4 kinds of small molecules: adenine, cytosine, quanine, and thymine, which link together to form larger molecules in genes. How many 3-molecule chains can be formed if the order of linking is important and each small molecule can occur more than once in a chain?

38. ***Families*** A member of a family of four is asked to rank all members of the family in the order of their power in the family. How many possible rankings are there?

39. ***Management*** A department store manager wants to display 6 brands of a product along one shelf of an aisle. In how many ways can he arrange the brands?

40. ***Call letters*** The call letters for radio stations begin with K or W, followed by 3 additional letters. How many sets of call letters having 4 letters are possible?

41. ***Testing*** How many ways can a 10-question true-false test be answered?

42. ***Testing*** How many ways can a 10-question multiple-choice test be answered if each question has 4 possible answers?

43. ***Testing*** In a 10-part matching question, how many ways can the matches be made if no two parts are used twice and if there are 10 possible matches?

44. ***Testing*** Answer Problem 43 if there are 10 parts and 15 possible matches.

45. ***Committees*** In how many ways can a committee of 5 be selected from 10 people willing to serve?

46. ***Astronauts*** If 8 people are qualified for the next flight to a space station, how many different groups of 3 people can be chosen for the flight?

47. ***Sales*** A traveling salesperson has 30 products to sell but has room in her sample case for only 20 of the products. If all the products are the same size, in how many ways can she select 20 different products for her case?

48. ***Gambling*** A poker hand consists of 5 cards dealt from a deck of 52 cards. How many different poker hands are possible?

49. ***Psychology*** A psychological study of outstanding salespeople is to be made. If 5 salespeople are needed and 12 are qualified, in how many ways can the 5 be selected?

50. ***Banking*** A company wishes to use the services of 3 different banks in the city. If 15 banks are available, in how many ways can it choose the 3 banks?

51. ***Politics*** To determine voters' feelings regarding an issue, a candidate asks a sample of people to pick 4 words out of 10 that they feel best describes the issue. How many different groupings of 4 words are possible?

52. ***Sociology*** A sociologist wants to pick 3 fifth-grade students from each of four schools. If each of the schools has 50 fifth-grade students, how many different groups of 12 students can he select? (Set up in combination symbols; do not work out.)

53. ***Committees*** In how many ways can a committee consisting of 6 men and 6 women be selected from a group consisting of 20 men and 22 women?

54. ***Bridge*** In how many ways can a hand consisting of 6 spades, 4 hearts, 2 clubs, and 1 diamond be selected from a deck of 52 cards?

55. ***Sins*** The Iranian government issued an order requiring minibuses in Tehran to have separate sections to avoid millions of "sins," because every day 370,000 women ride the minibuses (Source: *USA Today,* August 29, 1995). If each of these women is brushed (intentionally or accidently) 10 times each day by one or more men, how many sins would be committed daily?

7.6 Permutations, Combinations, and Probability

OBJECTIVE

- *To use counting techniques to solve probability problems*

APPLICATION PREVIEW

Suppose that of the 20 prospective jurors for a trial, 12 favor the death penalty, whereas 8 do not, and that 12 are chosen at random from these 20. To find the probability that 7 of the 12 will favor the death penalty, we can use combinations to find the number of ways to choose 12 jurors from 20 prospective jurors and to find the number of ways to choose 7 from the 12 who favor the death penalty and 5 from the 8 who oppose it. We frequently use permutations, combinations, or other counting principles to determine the number of ways in which an event can occur, $n(E)$, and the total number of outcomes, $n(S)$.

As we saw in Section 7.1, "Probability; Odds," if an event E can happen in $n(E)$ ways out of a total of $n(S)$ equally likely possibilities, the probability of the occurrence of the event is

$$\Pr(E) = \frac{n(E)}{n(S)}$$

EXAMPLE 1

A psychologist claims she can teach four-year-old children to spell three-letter words very quickly. To test her, one of her students was given cards with the letters A, C, D, K, and T on them and told to spell CAT. What is the probability the child will spell CAT by chance?

Solution

The number of different three-letter "words" that can be formed using these 5 cards is $n(S) = {}_5P_3 = 5!/(5-3)! = 60$. Because only one of those 60 arrangements will give CAT, $n(E) = 1$, and the probability that the child will be successful by guessing is 1/60.

EXAMPLE 2

Suppose that license plates for a state each contain six digits. If a license plate from this state is selected at random, what is the probability that it will have six different digits?

Solution

The number of ways in which six different digits can be selected from 0, 1, 2, 3, 4, 5, 6, 7, 8, 9 is $n(E) = n(\text{all digits different}) = {}_{10}P_6$.

The total number of license plates that can be made with six digits will include those with digits repeated, so the number is

$$n(S) = 10^6$$

Thus the probability that a plate chosen at random will have six different digits is

$$\Pr(E) = \frac{n(E)}{n(S)} = \frac{{}_{10}P_6}{10^6} = \frac{10 \cdot 9 \cdot 8 \cdot 7 \cdot 6 \cdot 5}{1{,}000{,}000} = .1512$$

CHECKPOINT

1. If three wires (red, black, and white) are randomly attached to a 3-way switch (which has 3 poles to which wires can be attached), what is the probability that the wires will be attached at random in the one order that makes the switch work properly?

EXAMPLE 3

A box of 24 shotgun shells contains 4 shells that will not fire. If 8 shells are selected from the box, what is the probability

(a) that all 8 shells will be good?
(b) that 6 shells will be good and 2 will not fire?

Solution

(a) The 8 shells can be selected from 24 in $n(S) = {}_{24}C_8 = \binom{24}{8}$ ways, and 8 good shells can be selected from the 20 good ones in $n(E) = {}_{20}C_8 = \binom{20}{8}$ ways. Thus the probability that all 8 shells will be good is

$$\Pr(8 \text{ good}) = \frac{n(E)}{n(S)} = \frac{\frac{20!}{12!8!}}{\frac{24!}{16!8!}} = .17$$

(b) Six good shells *and* two defective shells can be selected in

$$n(E) = {}_{20}C_6 \cdot {}_4C_2 = \binom{20}{6} \cdot \binom{4}{2}$$

ways, so

$$\Pr(6 \text{ good and } 2 \text{ defective}) = \frac{n(E)}{n(S)} = \frac{\frac{20!}{14!6!} \cdot \frac{4!}{2!2!}}{\frac{24!}{16!8!}} = .316$$

EXAMPLE 4

A manufacturing process for computer chips is such that 5 out of 100 chips are defective. If 10 chips are chosen at random from a box containing 100 newly manufactured chips, what is the probability that none of the chips will be defective?

Solution

The number of ways in which any 10 chips can be chosen from 100 is

$$n(S) = {}_{100}C_{10} = \binom{100}{10} = \frac{100!}{10!90!}$$

The number of ways in which 10 chips can be chosen with none of them being defective is the same as the number of ways 10 good chips can be chosen from the 95 good chips in the box. Thus

$$n(E) = {}_{95}C_{10} = \binom{95}{10} = \frac{95!}{10!85!}$$

Thus the probability that none of the chips chosen will be defective is

$$\Pr(E) = \frac{n(E)}{n(S)} = \frac{\frac{95!}{10!85!}}{\frac{100!}{10!90!}} = .58375$$

We now consider the problem discussed in the Application Preview.

EXAMPLE 5

Suppose that of the 20 prospective jurors for a trial, 12 favor the death penalty and 8 do not. If 12 jurors are chosen at random from these 20, what is the probability that 7 of the jurors will favor the death penalty?

Solution

The number of ways in which 7 jurors of the 12 will favor the death penalty is the number of ways in which 7 favor and 5 do not favor, or

$$n(E) = {}_{12}C_7 \cdot {}_8C_5 = \frac{12!}{7!5!} \cdot \frac{8!}{5!3!}$$

The total number of ways in which 12 jurors can be selected from 20 people is

$$n(S) = {}_{20}C_{12} = \frac{20!}{12!8!}$$

Thus the probability we seek is

$$\Pr(E) = \frac{n(E)}{n(S)} = \frac{\frac{12!}{7!5!} \cdot \frac{8!}{5!3!}}{\frac{20!}{12!8!}} = .3521$$

CHECKPOINT

2. If 5 people are chosen out of a group that contains 4 men and 6 women, what is the probability that 3 men and 2 women will be chosen?

EXAMPLE 6

Pennsylvania auto license plates have three letters followed by three numbers. If Janie's initials are J. J. R. and she lives at 125 Spring Street, what is the probability that the license plate she draws at random will

(a) have her initials and house numbers on it?
(b) have her initials on it?

Solution

(a) The total number of possible license plates with three letters followed by three numbers is

$$26 \cdot 26 \cdot 26 \cdot 10 \cdot 10 \cdot 10 = 17{,}576{,}000$$

Of all these plates, only one will have Janie's initials and house number, so the probability that she will get this plate at random is 1/17,576,000.

(b) The total number of possible license plates is 17,576,000, and the number with Janie's initials and any numbers on it is $1 \cdot 1 \cdot 1 \cdot 10 \cdot 10 \cdot 10 = 1000$, so the probability that Janie will get a plate with her initials is $1000/17{,}576{,}000 = 1/17{,}576$.

EXAMPLE 7

Beginning in 1990, General Motors began to use a theft deterrent key on some of its cars. The key has six parts, with three patterns for each part, plus an electronic chip containing a code from 1 to 15. What is the probability that one of these keys selected at random will start a GM car requiring a key of this type?

Solution

Each of the six parts has three patterns, and the chip can have any of 15 codes, so there are

$$\underline{3} \cdot \underline{3} \cdot \underline{3} \cdot \underline{3} \cdot \underline{3} \cdot \underline{3} \cdot \underline{15} = 10{,}935$$

possible keys. Because there is only one key that will start the car, the probability that a key selected at random will start a given car is 1/10,935.

CHECKPOINT SOLUTIONS

1. To find the total number of orders in which the wires can be attached, we note that the first pole can have any of the 3 wires attached to it, the second pole can have any of 2, and the third pole can have only the remaining wire, so there are $3 \cdot 2 \cdot 1 = 6$ possible different orders. Only one of these orders permits the switch to work properly, so the probability is 1/6.
2. Five people can be chosen out of 10 people in ${}_{10}C_5 = 252$ ways. Three men and 2 women can be chosen in ${}_4C_3 \cdot {}_6C_2 = 4 \cdot 15 = 60$ ways. Thus the probability is $60/252 = 5/21$.

EXERCISE 7.6

Applications

1. ***Education*** If a child is given cards with A, C, D, G, O, and T on them, what is the probability he or she could spell DOG by guessing the correct arrangement of 3 cards from the 6?
2. ***Racing*** Eight horses in a race wear numbers 1, 2, 3, 4, 5, 6, 7, and 8. What is the probability that the first three horses to finish the race will be numbered 1, 2, and 3, respectively?
3. ***Politics*** A senator asks his constituents to rank five issues in order of importance.
   (a) How many rankings are possible?
   (b) What is the probability that one reply chosen at random will rank the issues in alphabetical order?
4. ***Photography*** Two men and a woman are lined up to have their picture taken. If they are arranged at random, what is the probability that
   (a) the woman will be on the left in the picture?
   (b) the woman will be in the middle in the picture?
5. ***Licenses*** Suppose that all license plates in a state have three letters and three digits. If a plate is chosen at random, what is the probability that all three letters and all three numbers on the plate will be different?
6. ***Spelling***
   (a) In how many different orders can the letters R, A, N, D, O, M be written?
   (b) If the letters R, A, N, D, O, M are placed in a random order, what is the probability that they will spell RANDOM?
7. ***ATMs***
   (a) An automatic teller machine requires that each customer enter a four-digit personal identification number (PIN) when he or she inserts a bank card. If a person finds a bank card and guesses at a PIN

to use the card fraudulantly, what is the probability that the person will succeed in one attempt?

(b) If the person knows that the PIN will not have any digit repeated, what is the probability the person will succeed in guessing in one attempt?

8. ***Keys*** Keys for older General Motors cars had six parts, with three patterns for each part.
   (a) How many different key designs are possible for these cars?
   (b) If you find an older GM key and own an older GM car, what is the probability that it will fit your trunk?

9. ***Telephones*** If the first digit of a seven-digit telephone number cannot be a 0 or a 1, what is the probability that a number chosen at random will have all seven digits the same?

10. ***Licenses*** If a license plate has seven different digits, what is the probability that it contains a zero?

11. ***Testing*** A quiz consists of 10 multiple-choice questions with 5 possible choices for each question. If a student guesses the answer to each question, what is the probability that
   (a) she gets all questions correct?
   (b) she gets no questions correct?
   (c) she gets at least one question correct?

12. ***Testing*** Suppose that the student in Problem 11 was certain of the first 5 questions, and on each of the last 5 she could narrow the answer to one of three choices. Find the probability that
   (a) she scores 50%.
   (b) she scores 100%.
   (c) she passes the test (at least 60%).

13. ***Testing*** In a 10-question matching test with 10 possible answers to match and no answers used more than once, what is the probability of guessing and getting every answer correct?

14. ***Testing*** In a 10-question matching test with 15 possible answers to match and no answers used more than once, what is the probability of guessing and getting every answer correct?

15. ***Quality control*** A box of 12 transistors has 3 defective ones. If 2 transistors are drawn from the box together, what is the probability
   (a) that both transistors are defective?
   (b) that neither transistor is defective?
   (c) that one transistor is defective?

16. ***Quality control*** Suppose that 6 transistors are drawn at random from a box containing 18 good transistors and 2 defective transistors. What is the probability that
   (a) all 6 of the transistors are good?
   (b) exactly 4 of the transistors are good?
   (c) exactly 2 of the transistors are good?

17. ***Quality control*** A retailer purchases 100 of a new brand of digital tape deck, of which 2 are defective. The purchase agreement says that if he tests 5 chosen at random and finds 1 or more defective, he receives all the tape decks free of charge. What is the probability that he will not have to pay for the tape decks?

18. ***Quality control*** A box containing 500 central processing units (CPUs) for microcomputers has 20 defective CPUs. If 10 CPUs are selected together from the box, what is the probability that
   (a) all 10 are defective?
   (b) none is defective?
   (c) half of them are defective?

19. ***Banking*** To see whether a bank has enough minority construction company loans, a social agency selects 30 loans to construction companies at random and finds that 2 of them are loans to minority companies. If the bank's claim that 10 of every 100 of its loans to construction companies are minority loans is true, what is the probability that 2 loans out of 30 are minority loans? Leave your answer with combination symbols.

20. ***Diversity*** A high school principal must select 12 girls at random from the freshman class to serve as hostesses at the junior-senior prom. There are 200 freshman girls, including 20 from minorities, and the principal would like at least one minority girl to have this honor. If he selects the girls at random, what is the probability that
   (a) he will select exactly one minority girl?
   (b) he will select no minority girls?
   (c) he will select at least one minority girl?

21. ***Diversity*** Suppose that an employer plans to hire four people from a group of nine equally qualified people, of whom three are minority candidates. If the employer does not know which candidates are minority candidates, and if she selects her employees at random, what is the probability that
   (a) no minority candidates are hired?
   (b) all three minority candidates are hired?
   (c) one minority candidate is hired?

22. ***Lottery*** Four men and three women are semifinalists in a lottery. From this group, three finalists are to be selected by a drawing. What is the probability that all three finalists will be men?

23. ***Management*** Suppose that a children's basketball coach knows the best 6 players to use on his team, but pressure from parents to give everyone a chance to start in a game causes him to pick the starting team by choosing 5 players at random from the 10 team members. What is the probability that this will give him a team with 5 of his 6 best players?

24. ***Sales*** A car dealer has 12 different cars that he would like to display, but he has room to display only 5.
    (a) In how many ways can he pick 5 cars to display?
    (b) Suppose 8 of the cars are the same color, with the remaining 4 having distinct colors. If the dealer tells a salesperson to display any 5 cars, what is the probability that all 5 cars will be the same color?

25. ***Diversity*** Suppose that two openings on an appellate court bench are to be filled from current municipal court judges. The municipal court judges consist of 23 men and 4 women. Find the probability
    (a) that both appointees are men.
    (b) that one man and one woman are appointed.
    (c) that at least one woman is appointed.

26. ***Politics*** Suppose that four of the eight students running for class officers (president, vice president, secretary, and treasurer) have grade-point averages (GPAs) above 3.0. If the officers are selected at random, what is the probability that all four officers will be students with GPAs above 3.0?

27. ***Management*** Suppose that an indecisive company owner has selected the three top officers of his company at random but claims that they earned their jobs because of ability.
    (a) What is the probability that the most able person is at the top?
    (b) What is the probability that the top three officers are ranked according to their ability?

28. ***Quality control*** Suppose that 10 computer chips are drawn from a box containing 12 good chips and 4 defective chips. What is the probability that
    (a) exactly 4 of the chips are defective?
    (b) all 10 of the chips are good?
    (c) exactly 8 of the chips are good?

29. ***Lotteries*** In New York state there is a state lottery game called Pick 10. The state chooses 20 numbers from 80 numbers, and each player chooses 10 numbers from the same 80 numbers. If all 10 of a player's numbers are among the 20 numbers that the state picked, then that player is a "big winner." Find the probability of being a big winner.

30. ***Diversity*** A task force studying sex discrimination wishes to establish a subcommittee. If the subcommittee is to consist of 11 members chosen from a group of 23 men and 19 women, find the probability that the subcommittee consists of
    (a) 6 men and 5 women.
    (b) all women.

31. ***Insurance*** An insurance company receives 25 claims in a certain month. Company policy mandates that a random selection of 5 of these claims be thoroughly investigated. If 2 of the 25 claims are fraudulent, find the probability that
    (a) both fraudulent claims are thoroughly investigated.
    (b) neither fraudulent claim is thoroughly investigated.

32. ***Evaluation*** Employees of a firm receive annual reviews. In a certain department, 4 employees received excellent ratings, 15 received good ratings, and 1 received a marginal rating. If 3 employees in this department are randomly selected to complete a form for an internal study of the firm, find the probability that
    (a) all 3 selected were rated excellent.
    (b) one from each category was selected.

33. ***Poker*** What is the probability of being dealt a poker hand of 5 cards containing
    (a) 5 spades? (b) 5 cards of the same suit?

34. ***Bridge*** A bridge deck contains 13 spades, 13 hearts, 13 diamonds, and 13 clubs. A bridge hand contains 13 cards selected at random from the 52-card deck. Find the probability that a hand contains
    (a) 13 hearts.
    (b) exactly 8 spades.

35. ***Bridge*** A bridge deck contains 13 spades, 13 hearts, 13 diamonds, and 13 clubs. A bridge hand contains 13 cards selected at random from the 52-card deck. Find the probability that a hand contains 4 spades, 4 hearts, 4 diamonds, and 1 club.

36. ***Bridge*** Find the probability of being dealt a hand containing 1 spade, 6 hearts, 4 diamonds, and 2 clubs. Assume that a normal deck of 52 cards is used.

7.7 Markov Chains

OBJECTIVES

- *To use transition matrices to find specific stages of Markov chains*
- *To write transition matrices for Markov chain problems*
- *To find steady-state vectors for certain Markov chain problems*

APPLICATION PREVIEW

Suppose a department store has determined that a woman who used a credit card last month will use it again this month with probability .8. That is,

$$\Pr(\text{woman will use a charge card this month} \mid \text{she used it last month}) = .8$$

and

$$\Pr(\text{woman will not use card this month} \mid \text{she used it last month}) = .2$$

Furthermore, suppose the store has determined that the probability that a woman who did not use a credit card last month will not use it this month is .7. That is,

$$\Pr(\text{woman will not use card this month} \mid \text{she did not use it last month}) = .7$$

and

$$\Pr(\text{woman will use card this month} \mid \text{she did not use it last month}) = .3$$

To find the probability that the woman will use the credit card in some later month, we use **Markov chains.** Markov chains are named for the famous Russian mathematician A. A. Markov, who developed this theory in 1906. They are useful in the analysis of price movements, consumer behavior, laboratory animal behavior, and many other processes in business, the social sciences, and the life sciences.

The Markov chain experiment described in the Application Preview has two states: using a credit card and not using a credit card. The probability of moving from one state to another is called the **transition** probability; these transition probabilities are the conditional probabilities stated in the preview.

Transition probabilities can be organized into a **transition matrix** P, illustrated below.

	Given month \ *Next month:*	Uses card	Card is not used
$P =$	Card used	.8	.2
	Card not used	.3	.7

In general, a transition matrix is characterized by the following.

1. The *ij*-entry in the matrix is Pr(moving from State i to State j). Note that this is really the probability of moving to State j given that you are in State i.
2. The entries are nonnegative.
3. The sum of the entries in each row equals 1.
4. The matrix is square.

Now suppose the department store data show that during the first month a credit card is received, the probabilities for each state are as follows:

$$\Pr(\text{a woman used a charge card}) = .9$$
$$\Pr(\text{a woman did not use a charge card}) = .1$$

These probabilities can be placed in the row matrix

$$A = \begin{bmatrix} .9 & .1 \end{bmatrix}$$

which is called the **initial-probability vector** for the problem. This vector gives the probabilities for each state at the outset of the trials. In general, a probability vector is defined as follows:

Probability Vector For a vector to be a probability vector, the sum of its entries must be 1 and each of its entries must be nonnegative.

Note that if a woman used her credit card in the initial month, the probability is .8 that she will use it next month, and if she didn't use the card, the probability is .3 that she will use it next month. Thus,

$$\begin{aligned} \Pr(\text{will use 2d month}) &= \Pr(\text{used 1st month}) \cdot \Pr(\text{will use 2d} \mid \text{used 1st}) \\ &\quad + \Pr(\text{not used 1st month}) \cdot \Pr(\text{will use 2d} \mid \text{not used 1st}) \\ &= (.9)(.8) + (.1)(.3) \\ &= .72 + .03 = .75 \end{aligned}$$

Note that this is one element of the product matrix AP. Similarly, the second element of the product matrix is the probability that a woman will not use the charge card in the second month.

$$AP = \underset{\text{Initial month}}{\begin{bmatrix} .9 & .1 \end{bmatrix}} \underset{\text{Transition matrix}}{\begin{bmatrix} .8 & .2 \\ .3 & .7 \end{bmatrix}} = \underset{\text{Second month}}{\begin{bmatrix} .75 & .25 \end{bmatrix}}$$

Thus

$$\Pr(\text{a woman will use a charge card in 2d month}) = .75$$

and

$$\Pr(\text{a woman will not use a charge card in 2d month}) = .25$$

To find the probabilities for each state for the third month, we multiply the second-month probability vector times the transition matrix.

$$\underset{\text{Second month}}{\begin{bmatrix} .75 & .25 \end{bmatrix}} \underset{\text{Transition matrix}}{\begin{bmatrix} .8 & .2 \\ .3 & .7 \end{bmatrix}} = \underset{\text{Third month}}{\begin{bmatrix} .675 & .325 \end{bmatrix}}$$

Continuing in this manner, we can find the probabilities for later months.

EXAMPLE 1

Suppose that in a certain city the Democratic, Republican, and Consumer parties always nominate candidates for mayor. The probability of winning in any election depends on the party in power and is given by the following transisiton matrix *P*.

Party in office now \ *Party in office next term*	D	R	C
Democrat	.5	.4	.1
$P =$ Republican	.4	.5	.1
Consumer	.3	.3	.4

Suppose the probability of winning the next election is given by $[.4 \quad .4 \quad .2]$; that is, Pr(D) = .4, Pr(R) = .4, Pr(C) = .2. Find the probability of each party winning the *fourth* election (the third election after this one).

Solution

The probability of each party winning the second election is found using the probabilities given for the first.

$$[.4 \quad .4 \quad .2]\begin{bmatrix} .5 & .4 & .1 \\ .4 & .5 & .1 \\ .3 & .3 & .4 \end{bmatrix} = [.42 \quad .42 \quad .16]$$

Using the probabilities for the second election, we can find the probabilities for the third.

$$[.42 \quad .42 \quad .16]\begin{bmatrix} .5 & .4 & .1 \\ .4 & .5 & .1 \\ .3 & .3 & .4 \end{bmatrix} = [.426 \quad .426 \quad .148]$$

Using the probabilities for the third election, we can find the probabilities for the fourth election.

$$[.426 \quad .426 \quad .148]\begin{bmatrix} .5 & .4 & .1 \\ .4 & .5 & .1 \\ .3 & .3 & .4 \end{bmatrix} = [.4278 \quad .4278 \quad .1444]$$

We can perform matrix multiplication with a graphing calculator, or we can use a computer spreadsheet to perform repeated multiplications. Continuing the multiplication process with the vectors in Example 1 eventually gives a vector with the probabilities

$$[.42857 \quad .42857 \quad .14286]$$

to five decimal places, and repeated multiplications will not change the probabilities, so the system appears to be stabilizing. This vector is the vector $[V_1 \quad V_2 \quad V_3]$ such that

$$[V_1 \quad V_2 \quad V_3]\cdot P = [V_1 \quad V_2 \quad V_3]$$

The exact probability vector for which this is true in Example 1 is $[\frac{3}{7} \quad \frac{3}{7} \quad \frac{1}{7}]$, because

$$\begin{bmatrix} \frac{3}{7} & \frac{3}{7} & \frac{1}{7} \end{bmatrix} \begin{bmatrix} .5 & .4 & .1 \\ .4 & .5 & .1 \\ .3 & .3 & .4 \end{bmatrix} = \begin{bmatrix} \frac{3}{7} & \frac{3}{7} & \frac{1}{7} \end{bmatrix}$$

This vector is called the **fixed-probability vector** for the given transition matrix. Because this vector is where the process eventually stabilizes, it is also called the **steady-state vector.**

Rather than trying to guess this steady-state vector by repeated multiplication, we can solve the equation

$$[V_1 \quad V_2 \quad V_3] \begin{bmatrix} .5 & .4 & .1 \\ .4 & .5 & .1 \\ .3 & .3 & .4 \end{bmatrix} = [V_1 \quad V_2 \quad V_3]$$

for V_1, V_2, and V_3 as follows:

$$[V_1 \quad V_2 \quad V_3] \begin{bmatrix} .5 & .4 & .1 \\ .4 & .5 & .1 \\ .3 & .3 & .4 \end{bmatrix} = [V_1 \quad V_2 \quad V_3]$$

$$\begin{aligned} .5V_1 + .4V_2 + .3V_3 &= V_1 \\ .4V_1 + .5V_2 + .3V_3 &= V_2 \\ .1V_1 + .1V_2 + .4V_3 &= V_3 \end{aligned}$$

Rewriting with zeros on the right and solving by Gauss-Jordan elimination give

$$\left[\begin{array}{ccc|c} -.5 & .4 & .3 & 0 \\ .4 & -.5 & .3 & 0 \\ .1 & .1 & -.6 & 0 \end{array}\right] \to \left[\begin{array}{ccc|c} 1 & 1 & -6 & 0 \\ 0 & -9 & 27 & 0 \\ 0 & 9 & -27 & 0 \end{array}\right]$$

$$\to \left[\begin{array}{ccc|c} 1 & 1 & -6 & 0 \\ 0 & 1 & -3 & 0 \\ 0 & 0 & 0 & 0 \end{array}\right]$$

$$\to \left[\begin{array}{ccc|c} 1 & 0 & -3 & 0 \\ 0 & 1 & -3 & 0 \\ 0 & 0 & 0 & 0 \end{array}\right]$$

so

$$V_1 = 3V_3 \qquad V_2 = 3V_3$$

Any vector that satisfies these conditions is a solution to the system. Because $[V_1 \quad V_2 \quad V_3]$ is a probability vector, we have $V_1 + V_2 + V_3 = 1$. Substituting gives $3V_3 + 3V_3 + V_3 = 1$, so $V_3 = \frac{1}{7}$, and the steady-state vector is

$$\begin{bmatrix} \frac{3}{7} & \frac{3}{7} & \frac{1}{7} \end{bmatrix}$$

Any transition matrix P such that some power of P, P^m, contains only positive entries is called a **regular transition matrix,** and it can be shown that each Markov chain using a regular transition matrix will have a steady-state vector.

Note that a regular transition matrix P may contain zeros; the only requirement is that some power of P contain all positive entries. In this text we will limit ourselves to regular transition matrices.

Furthermore, if a steady-state vector exists, then the effect of the initial probabilities diminishes as the number of steps in the process increases. This means that the probabilities for the elections would approach

$$\begin{bmatrix} \frac{3}{7} & \frac{3}{7} & \frac{1}{7} \end{bmatrix}$$

regardless of the probabilities for the initial election. For example, even if a Democrat were certain to be elected in the initial election (that is, the initial vector is $[1 \quad 0 \quad 0]$), the probabilities for the fifth election are $[.4292 \quad .4291 \quad .1417]$, which approximates the steady-state vector. (See Problem 17 of the exercise set for this section.)

CHECKPOINT

1. Use the transition matrix $\begin{bmatrix} .1 & .3 & .6 \\ .3 & .5 & .2 \\ .1 & .1 & .8 \end{bmatrix}$ and the initial probability vector $[.2 \quad .3 \quad .5]$ to
   (a) find the probability vector for the third stage (two steps after the initial stage).
   (b) find the steady-state vector associated with the matrix.

EXAMPLE 2

Suppose that in a certain city the probabilities that a woman with less than a high school education has a daughter with less than a high school education, with a high school degree, and with education beyond high school are .2, .6, and .2, respectively. The probabilities that a woman with a high school degree has a daughter with less than a high school education, with a high school degree, and with education beyond high school are .1, .5, and .4, respectively. The probabilities that a woman with education beyond high school has a daughter with less than a high school education, with a high school degree, and with education beyond high school are .1, .1, and .8, respectively. The population of women in the city is now 60% with less than a high school education, 30% with a high school degree, and 10% with education beyond high school.

(a) What is the transition matrix for this information?
(b) What is the probable distribution of women according to educational level two generations from now?
(c) What is the steady-state vector for this information?

Solution

(a) The transition matrix is

Mother's educational level \ *Daughter's educational level*	Less than H.S.	H.S. degree	More than H.S.
Less than H.S.	.2	.6	.2
H.S. degree	.1	.5	.4
More than H.S.	.1	.1	.8

(b) The probable distribution after one generation is

$$[.6 \quad .3 \quad .1]\begin{bmatrix} .2 & .6 & .2 \\ .1 & .5 & .4 \\ .1 & .1 & .8 \end{bmatrix} = [.16 \quad .52 \quad .32]$$

The probable distribution after two generations is

$$[.16 \quad .52 \quad .32]\begin{bmatrix} .2 & .6 & .2 \\ .1 & .5 & .4 \\ .1 & .1 & .8 \end{bmatrix} = [.116 \quad .388 \quad .496]$$

(c) The steady-state vector is found by solving

$$[V_1 \quad V_2 \quad V_3]\begin{bmatrix} .2 & .6 & .2 \\ .1 & .5 & .4 \\ .1 & .1 & .8 \end{bmatrix} = [V_1 \quad V_2 \quad V_3]$$

$$\begin{cases} .2V_1 + .1V_2 + .1V_3 = V_1 \\ .6V_1 + .5V_2 + .1V_3 = V_2 \\ .2V_1 + .4V_2 + .8V_3 = V_3 \end{cases} \quad \text{or} \quad \begin{cases} -.8V_1 + .1V_2 + .1V_3 = 0 \\ .6V_1 - .5V_2 + .1V_3 = 0 \\ .2V_1 + .4V_2 - .2V_3 = 0 \end{cases}$$

$$\left[\begin{array}{ccc|c} -.8 & .1 & .1 & 0 \\ .6 & -.5 & .1 & 0 \\ .2 & .4 & -.2 & 0 \end{array}\right] \to \left[\begin{array}{ccc|c} 1 & 2 & -1 & 0 \\ 0 & -17 & 7 & 0 \\ 0 & 17 & -7 & 0 \end{array}\right]$$

$$\to \left[\begin{array}{ccc|c} 1 & 0 & -\frac{3}{17} & 0 \\ 0 & 1 & -\frac{7}{17} & 0 \\ 0 & 0 & 0 & 0 \end{array}\right]$$

Thus $V_1 = \frac{3}{17}V_3$ and $V_2 = \frac{7}{17}V_3$. The probability vector that satisfies these conditions and $V_1 + V_2 + V_3 = 1$ must satisfy

$$\frac{3}{17}V_3 + \frac{7}{17}V_3 + V_3 = 1$$

Thus

$$\frac{27}{17}V_3 = 1, \quad \text{so} \quad V_3 = \frac{17}{27}$$

and the steady-state vector is

$$\left[\frac{3}{27} \quad \frac{7}{27} \quad \frac{17}{27}\right]$$

CHECKPOINT SOLUTIONS

1. (a) The second-stage probability vector is

$$[.2 \quad .3 \quad .5]\begin{bmatrix} .1 & .3 & .6 \\ .3 & .5 & .2 \\ .1 & .1 & .8 \end{bmatrix} = [.16 \quad .26 \quad .58]$$

and that for the third stage is

$$[.16 \quad .26 \quad .58]\begin{bmatrix} .1 & .3 & .6 \\ .3 & .5 & .2 \\ .1 & .1 & .8 \end{bmatrix} = [.152 \quad .236 \quad .612]$$

(b) To find the steady-state vector, we solve

$$[V_1 \quad V_2 \quad V_3]\begin{bmatrix} .1 & .3 & .6 \\ .3 & .5 & .2 \\ .1 & .1 & .8 \end{bmatrix} = [V_1 \quad V_2 \quad V_3]$$

Multiplying and combining like terms give the following system of equations.

$$\begin{cases} -.9V_1 + .3V_2 + .1V_3 = 0 \\ .3V_1 - .5V_2 + .1V_3 = 0 \\ .6V_1 + .2V_2 - .2V_3 = 0 \end{cases}$$

We solve this by using the augmented matrix

$$\left[\begin{array}{ccc|c} -.9 & .3 & .1 & 0 \\ .3 & -.5 & .1 & 0 \\ .6 & .2 & -.2 & 0 \end{array}\right]$$

The solution gives the steady-state vector $\begin{bmatrix} \frac{1}{7} & \frac{3}{14} & \frac{9}{14} \end{bmatrix}$.

EXERCISE 7.7

Which of the vectors in Problems 1–4 could *not* be probability vectors? Explain.

1. $\begin{bmatrix} \frac{1}{4} & \frac{3}{4} \end{bmatrix}$
2. $\begin{bmatrix} \frac{5}{6} & \frac{2}{3} \end{bmatrix}$
3. $\begin{bmatrix} \frac{1}{4} & \frac{3}{5} & \frac{1}{6} \end{bmatrix}$
4. $\begin{bmatrix} \frac{2}{3} & \frac{1}{12} & \frac{1}{4} \end{bmatrix}$

Which of the matrices in Problems 5–8 could *not* be transition matrices? Explain.

5. $\begin{bmatrix} .3 & .5 & .2 \\ .1 & .6 & .3 \end{bmatrix}$
6. $\begin{bmatrix} .6 & .2 & .2 \\ .1 & .5 & .4 \\ .3 & .5 & .1 \end{bmatrix}$
7. $\begin{bmatrix} .1 & .3 & .6 \\ .5 & .2 & .3 \\ .1 & .7 & .2 \end{bmatrix}$
8. $\begin{bmatrix} .6 & .5 & -.1 \\ .2 & .6 & .2 \\ .3 & .3 & .4 \end{bmatrix}$

In Problems 9–12, use the transition matrix and the initial-probability vector to find the probability vector for the third stage (two after the initial stage) of the Markov chain.

$$A = \begin{bmatrix} .1 & .9 \\ .3 & .7 \end{bmatrix} \quad B = \begin{bmatrix} .5 & .5 \\ .9 & .1 \end{bmatrix}$$

$$C = \begin{bmatrix} .5 & .3 & .2 \\ .3 & .5 & .2 \\ .1 & .1 & .8 \end{bmatrix} \quad D = \begin{bmatrix} .8 & .1 & .1 \\ .2 & .6 & .2 \\ .3 & .3 & .4 \end{bmatrix}$$

9. Transition matrix *A*, initial-probability vector $[.2 \quad .8]$.
10. Transition matrix *B*, initial-probability vector $[.6 \quad .4]$.
11. Transition matrix *C*, initial-probability vector $[.1 \quad .3 \quad .6]$.
12. Transition matrix *D*, initial-probability vector $[.9 \quad 0 \quad .1]$.

In Problems 13–16, find the steady-state vector associated with the given transition matrix from Problems 9–12.

13. *A*
14. *B*
15. *C*
16. *D*

Applications

Elections Use the following information for Problems 17 and 18. In a certain city the Democratic, Republican, and Consumer parties always nominate candidates for mayor. The probability of winning in any election depends on the party in power and is given by the transition matrix on the next page.

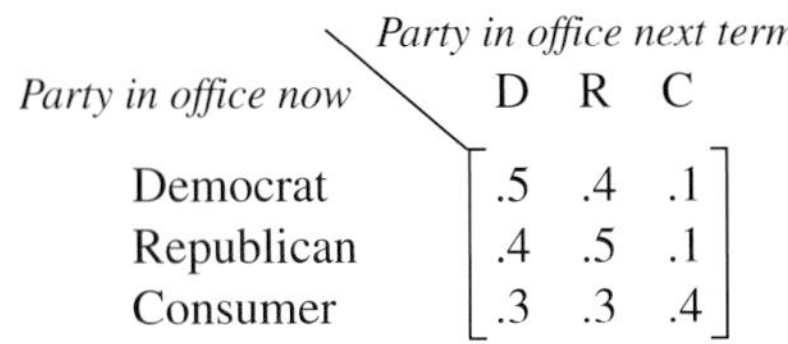

Party in office now \ *Party in office next term*	D	R	C
Democrat	.5	.4	.1
Republican	.4	.5	.1
Consumer	.3	.3	.4

17. Using the transition matrix given and assuming the initial-probability vector is [1 0 0], find the probability vectors for the next four steps of the Markov chain. (This initial-probability vector indicates that a Democrat is certain to win the initial election.)
18. Using the transition matrix given and assuming that a Republican is certain to win the initial election, find the probability vectors for the next four steps of the Markov chain.

Church attendance Use the following information for Problems 19–22. The probability that daughters of a mother who attends church regularly will also attend church regularly is .8, whereas the probability that daughters of a mother who does not attend regularly will attend regularly is .3.

19. What is the transition matrix for this information?
20. If a woman attends church regularly and has one daughter who in turn has one daughter, what is the probability that the granddaughter attends church regularly?
21. If a woman does not attend church, what is the probability that her granddaughter attends church regularly?
22. What is the steady-state vector for this information?

Car selection Use the following information for Problems 23–28. A man owns an Audi, a Ford, and a VW. He drives every day and never drives the same car two days in a row. This is the probability that he drives the other cars the next day:

Pr(Ford after Audi) = .7	Pr(VW after Audi) = .3
Pr(Audi after Ford) = .6	Pr(VW after Ford) = .4
Pr(Audi after VW) = .8	Pr(Ford after VW) = .2

23. Write the transition matrix for his selection of a car.
24. If he drove the Ford today, what are the probabilities for the cars that he will drive 2 days from today?
25. If he drove the Ford today, what are the probabilities for the cars that he will drive 4 days from today?
26. If he drove the Audi today, what are the probabilities for the cars that he will drive 4 days from now?
27. What is the steady-state vector for this problem?
28. Would the probabilities for which car he drives 100 days from now depend on whether he drove a Ford or an Audi today?
29. ***Population demographics*** Suppose a government study estimated that the probability of successive generations of a rural family remaining in a rural area was .7 and the probability of successive generations of an urban family remaining in an urban area was .9. Assuming a Markov chain applies to these facts, find the steady-state vector.
30. ***Psychology*** A psychologist found that a laboratory mouse placed in a T-maze will go down the same branch with probability .8 if there was food there on the previous trial and that it will go down the same branch with probability .6 if there was no food there on the previous trial. Find the steady-state vector for this experiment. If the mouse takes the branches with equal probability when no reward is found, does it appear that the mouse "learns" to choose the food path?
31. ***Charitable contributions*** The local community-service funding organization in a certain county has classified the population into those who did not give the previous year, those who gave less than their "fair share," and those who met or exceeded their "fair share." Suppose the organization developed the following transition matrix for these groups.

Last year \ *This year*	N	L	F
No contribution	.7	.2	.1
Less than fair share	.1	.6	.3
Fair share	0	.1	.9

Find the steady-state vector for this county's contribution habits.

Advertising Use the following information for Problems 32 and 33. A local business A has two competitors, B and C. No customer patronizes more than one of these businesses at the same time. Initially the probabilities that a customer patronizes A, B, or C are .2, .6, and .2, respectively. Suppose A initiates an advertising campaign to improve its business and finds the following transition matrix to describe the effect.

Last week's customers went to \ *This week's customers go to*	A	B	C
A	.7	.2	.1
B	.4	.4	.2
C	.4	.4	.2

32. If A runs the advertising campaign for 3 weeks, find the probability of a customer patronizing each business.

33. Find the steady-state vector for this market—that is, the long-range share of the market that each business can expect if the transition matrix holds.

Genetics Use the following information for Problems 34–36. For species that reproduce sexually, characteristics are determined by a gene from each parent. Suppose that for a certain trait there are two possible genes available from each parent: a dominant gene D and a recessive gene r. Then the different gene combinations (called *genotypes*) for the offspring are DD, Dr, rD, and rr, where Dr and rD produce the same trait. Suppose further that these genotypes are the states of a Markov chain with the following transition matrix.

Parent genotype \ *Offspring genotype*	DD	Dr	rr
DD	.7	.3	0
Dr	.35	.5	.15
rr	0	.7	.3

34. If the initial occurrence of those genotypes in the population is .5 for DD, .4 for Dr, and .1 for rr, find the distribution (probability) of each type after the first and second generations.
35. Find the steady-state vector for this genotype.
36. To which generation(s) does the steady-state vector correspond? This is called the **Hardy-Weinberg model.**

KEY TERMS AND FORMULAS

Section	Key Terms	Formula
7.1	Equally likely events	$\Pr(E) = k/n$
	Sample space	
	Certain event	$\Pr(E) = 1$
	Impossible event	$\Pr(E) = 0$
	Odds	
	Empirical probability	
7.2	Intersection of events	$\Pr(E \text{ and } F) = \Pr(E \cap F)$
	Union of events	$\Pr(E \text{ or } F) = \Pr(E \cup F) = \Pr(E) + \Pr(F) - \Pr(E \cap F)$
	Complement	$\Pr(\text{not } E) = \Pr(E') = 1 - \Pr(E)$
7.3	Conditional probability	$\Pr(E \mid F) = \dfrac{\Pr(E \cap F)}{\Pr(F)}$
	Product Rule	$\Pr(E \cap F) = \Pr(F) \cdot \Pr(E \mid F)$
	Independent events	$\Pr(A \mid B) = \Pr(A)$ and $\Pr(A \cap B) = \Pr(A) \cdot \Pr(B)$
7.4	Probability tree	
	Bayes' formula	
7.5	Fundamental Counting Principle	$n(A) \cdot n(B)$
	n-factorial	$n! = n(n-1) \cdots (3)(2)(1)$ and $0! = 1$
	Permutations	${}_nP_r = \dfrac{n!}{(n-r)!}$
	Combinations	${}_nC_r = \dfrac{n!}{r!(n-r)!}$

Section	Key Terms	Formula
7.6	Probability using permutations and combinations	
7.7	Markov chains Vectors Initial-probability vector Fixed-probability or steady-state vector	

REVIEW EXERCISES

Section 7.1

1. If one ball is drawn from a bag containing nine balls numbered 1 through 9, what is the probability that the ball's number is
   (a) odd?
   (b) divisible by 3?
   (c) odd and divisible by 3?
2. Suppose one ball is drawn from a bag containing nine red balls numbered 1 through 9 and three white balls numbered 10, 11, 12. What is the probability that
   (a) the ball is red?
   (b) the ball is odd-numbered?
   (c) the ball is white and even-numbered?
   (d) the ball is white or odd-numbered?
3. If the probability that an event E occurs is 3/7, what are the odds that
   (a) E will occur?
   (b) E will not occur?
4. Suppose that a fair coin is tossed two times. Construct an equiprobable sample space for the experiment and determine each of the following probabilities.
   (a) Pr(0 heads) (b) Pr(1 head) (c) Pr(2 heads)
5. Suppose that a fair coin is tossed three times. Construct an equiprobable sample space for the experiment and determine each of the following probabilities.
   (a) Pr(2 heads) (b) Pr(3 heads) (c) Pr(1 head)

Section 7.2

6. A card is drawn at random from an ordinary deck of 52 playing cards. What is the probability that it is a queen or a jack?
7. A deck of 52 cards is shuffled. A card is drawn, it is replaced, the pack is again shuffled, and a second card is drawn. What is the probability that each card drawn is an ace, king, queen, or jack?
8. If the probability of winning a game is 1/4 and there can be no ties, what is the probability of losing the game?
9. A card is drawn from a deck of 52 playing cards. What is the probability that it is an ace or a 10?
10. A card is drawn at random from a deck of playing cards. What is the probability that it is a king or a red card?
11. A bag contains 4 red balls numbered 1, 2, 3, 4 and 5 white balls numbered 5, 6, 7, 8, 9. A ball is drawn. What is the probability that the ball
    (a) is red and even-numbered?
    (b) is red or even-numbered?
    (c) is white or odd-numbered?

Section 7.3

12. A bag contains 4 red balls and 3 black balls. Two balls are drawn at random from the bag without replacement. Find the probability that both balls are red.
13. A box contains 2 red balls and 3 black balls. Two balls are drawn from the box without replacement. Find the probability that the second ball is red, given that the first ball is black.

Section 7.4

14. An urn contains 4 red and 6 white balls. One ball is drawn, it is not replaced, and a second ball is drawn. What is the probability that one ball is white and one is red?
15. Urn I contains 3 red and 4 white balls and urn II contains 5 red and 2 white balls. An urn is selected and a ball is drawn.
    (a) What is the probability that urn I is selected and a red ball is drawn?

(b) What is the probability that a red ball is selected?
(c) If a red ball is selected, what is the probability that urn I was selected?

Section 7.5

16. Compute ${}_6P_2$.
17. Compute ${}_7C_3$.
18. How many 3-letter sets of initials are possible?

Applications

Section 7.1

19. ***Senior citizens*** In a certain city, 30,000 citizens out of 80,000 are over 50 years of age. What is the probability that a citizen selected at random will be 50 years old or younger?

Section 7.2

20. ***United Nations*** Of 100 job applicants to the United Nations, 30 speak French, 40 speak German, and 12 speak both French and German. If an applicant is chosen at random, what is the probability that the applicant speaks French or German?

Productivity Suppose that in a study of leadership style versus industrial productivity, the following data were obtained. Use these empirical data to answer Problems 21–23.

	Leadership Style			
Productivity	*Demo-cratic*	*Authori-tarian*	*Laissez-faire*	*Total*
Low	40	15	40	95
Medium	25	75	10	110
High	25	30	20	75
Total	90	120	70	280

21. Find the probability that a person chosen at random has a democratic style and medium productivity.
22. Find the probability that an individual chosen at random has high productivity or an authoritarian style.

Section 7.3

23. Find the probability that an individual chosen at random has an authoritarian style, given that he or she has medium productivity.

Section 7.5

24. ***Management*** A personnel director ranks 4 applicants for a job. How many rankings are possible?

25. ***Management*** An organization wants to select a president, vice president, secretary, and treasurer. If 8 people are willing to serve and each of them is eligible for any of the offices, in how many different ways can the offices be filled?

26. ***Utilities*** A utility company sends teams of 4 people each to perform repairs. If it has 12 qualified people, how many different ways can people be assigned to a team?

27. ***Committees*** An organization wants to select a committee of 4 members from a group of 8 eligible members. How many different committees are possible?

28. ***Juries*** A jury can be deadlocked if one person disagrees with the rest. There are 12 ways in which a jury can be deadlocked if one person disagrees, because any one of the 12 jurors could disagree.
(a) In how many ways can a jury be deadlocked if 2 people disagree?
(b) If 3 people disagree?

29. ***License plates*** Because Atlanta was the site of the 1996 Olympics, the state of Georgia released vanity license plates containing the Olympic symbol, and a large number of these plates were sold. Because of the symbol, each plate could contain at most 5 digits and/or letters. How many possible plates could be produced using from 1 to 5 digits or letters?

30. ***Blood types*** In a book describing the mass murder of 18 people in Northern California, a policewoman was quoted as stating that there are blood groups O, A, B, and AB, positive and negative, blood types M, N, MN, and P, and 8 Rh blood types, so there are 288 unique groups (Source: Joseph Harrington and Robert Berger, *Eye of Evil,* St. Martin's Press, 1993). Comment on this conclusion.

Section 7.4

31. ***Color blindness*** Sixty men out of 1000 and 3 women out of 1000 are color blind. A person is picked at random from a group containing 10 men and 10 women.
(a) What is the probability that the person is color blind?
(b) What is the probability that the person is a man if the person is color blind?

32. ***Purchasing*** A regional survey found that 70% of all families who indicated an intention to buy a new car soon bought a new car within 3 months, that 10% of families who did not indicate an intention to buy a

new car soon bought one within 3 months, and that 22% indicated an intention to buy a new car. If a family chosen at random bought a car, find the probability that the family had not previously indicated an intention to buy a car.

Section 7.6

33. ***Scheduling*** A college registrar adds 4 new courses to the list of offerings for the spring semester. If he added the course names in random order at the end of the list, what is the probability that these 4 courses are listed in alphabetical order?

34. ***Lottery*** If a person "boxes" a 4-digit number with 4 different digits in Pennsylvania's Pick Four Lottery, what is the probability that the person will win? (See Problem 35 for the definition of "boxing.")

35. ***Lottery*** Pennsylvania's Daily Number pays 500 to 1 to people who play the winning 3-digit number exactly as it is drawn. However, players can "box" a number so they can win \$80 from a \$1 bet if the 3 digits they pick come out in any order. What is the probability that a "boxed" number with 3 different digits will be a winner?

36. ***Quality control*** A supplier has 200 compact discs, of which 10% are known to be defective. If a music store purchases 10 of these discs, what is the probability that
    (a) none of the discs is defective?
    (b) 2 of the discs are defective?

37. ***Stocks*** Mr. Way must sell stocks from 3 of the 5 companies whose stock he owns so that he can send his children to college. If he chooses the companies at random, what is the probability that the 3 companies will be the 3 with the best future earnings?

38. ***Quality control*** A sample of 6 fuses is drawn from a lot containing 10 good fuses and 2 defective fuses. Find the probability that the number of defective fuses is
    (a) exactly 1. (b) at least 1.

Section 7.7

Income level Use the following information for Problems 39 and 40. Suppose people in a certain community are classified as being low-income, middle-income, or high-income. Suppose further that the probabilities for children being in a given state depend on which state their parents were in according to the following Markov chain transition matrix.

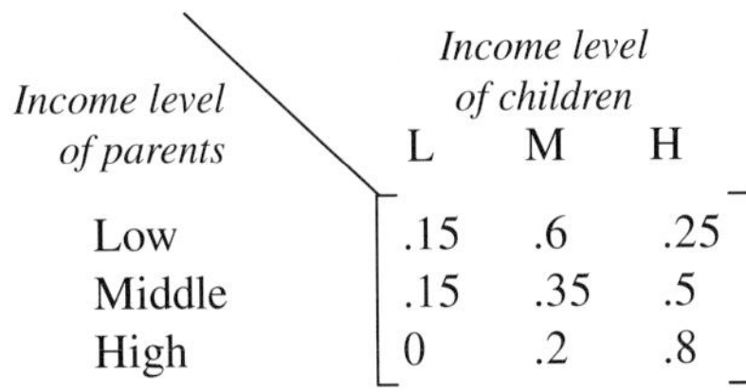

Income level of parents \ *Income level of children*	L	M	H
Low	.15	.6	.25
Middle	.15	.35	.5
High	0	.2	.8

39. If the initial distribution probabilities of families in a certain population are [.7 .2 .1], find the distribution for the next three generations.
40. Find the steady-state vector for the distribution of families by income level.

CHAPTER TEST

A bag contains three white balls numbered 1, 2, 3 and four black balls numbered 4, 5, 6, 7. If one ball is drawn at random, find the probability of each event described in Problems 1–3.

1. (a) The ball is odd-numbered.
   (b) The ball is white.
2. (a) The ball is black and even-numbered.
   (b) The ball is white or even-numbered.
3. (a) The ball is red.
   (b) The ball is numbered less than 8.

A bag contains three white balls numbered 1, 2, 3 and four black balls numbered 4, 5, 6, 7. Two balls are drawn without replacement. Find the probability of each event described in Problems 4–9.

4. The sum of the numbers is 7.
5. Both balls are white.
6. (a) The first ball is white, and the second is black.
   (b) A white ball and a black ball are drawn.
7. Both balls are odd-numbered.
8. The second ball is white. (Use a tree.)
9. The first ball is black, given that the second ball is white.
10. A cat hits the letters on a computer keyboard three times. What is the probability that the word RAT appears?

11. In a certain region of the country with a large number of cars, 45% of the cars are from Asian manufacturers, 35% are American, and 20% are European. If three cars are chosen at random, what is the probability that two American cars are chosen?
12. In a group of 20 people, 15 are right-handed, 4 are left-handed, and 1 is ambidextrous. What is the probability that a person selected at random is
    (a) left-handed? (b) ambidextrous?
13. In a group of 20 people, 15 are right-handed, 4 are left-handed, and 1 is ambidextrous. If 2 people are selected at random, without replacement, use a tree to find the following probabilities.
    (a) Both are left-handed.
    (b) One is right-handed, and one is left-handed.
    (c) Two are right-handed.
    (d) Two are ambidextrous.
14. The state of Arizona has a "6 of 42" game in which 6 balls are drawn without replacement from 42 balls numbered 1–42. The order in which the balls are drawn does not matter.
    (a) How many different drawing results are possible?
    (b) What is the probability that a person holding one ticket will win (match all the numbers drawn)?
15. The multi-state "5 of 50" game draws 5 balls without replacement from 50 balls numbered 1–50.
    (a) How many different drawing results are possible if order does not matter?
    (b) What is the probability that a person holding one ticket will win (match all the numbers drawn)?
16. Computer chips come from two suppliers, with 80% coming from supplier 1 and 20% coming from supplier 2. Six percent of the chips from supplier 1 are defective, and 8% of the chips from supplier 2 are defective. If a chip is chosen at random, what is the probability that it is defective?
17. A placement test is given by a university to predict student success in a calculus course. On average, 70% of students who take the test pass it, and 87% of those who pass the test also pass the course, whereas 8% of those who fail the test pass the course.
    (a) What is the probability that a student taking the placement test and the calculus course will pass the course?
    (b) If a student passed the course, what is the probability that he or she passed the test?
18. Lactose intolerance affects about 20% of non-Hispanic white Americans and 50% of white Hispanic Americans. Nine percent of Americans are white Hispanic and 75.6% are non-Hispanic whites (Source: Jean Carper, "Eat Smart," *USA Weekend*, Sept. 4, 1994). If a white American is chosen at random and is lactose-intolerant, what is the probability that he or she is Hispanic?
19. A car rental firm has 350 cars. Seventy of the cars have only defective windshield wipers and 25 have only defective tail lights. Two hundred of the cars have no defects; the remainder have other defects. What is the probability that a car chosen at random
    (a) has defective windshield wipers?
    (b) has defective tail lights?
    (c) does not have defective tail lights?
20. NACO Body Shops has found that 14% of the cars on the road need to be painted and that 3% need major body work and painting. What is the probability that a car selected at random needs major body work if it is known that it needs to be painted?
21. Garage door openers have ten on-off switches on the opener in the garage and on the remote control used to open the door. The door opens when the remote control is pressed and all the switches on both the opener and the remote agree.
    (a) How many different sequences of the on-off switches are possible on the door opener?
    (b) If the switches on the remote control are set at random, what is the probability that it will open a given garage door?
    (c) Remote controls and the openers in the garage for a given brand of door opener have the controls and openers matched in one of three sequences. If an owner does not change the sequence of switches on her or his remote control and opener after purchase, what is the probability that a different new remote control purchased from this company will open that owner's door?
    (d) What should the owner of a new garage door opener do to protect the contents of his or her garage (and home)?
22. A publishing company has determined that a new edition of an existing mathematics textbook will be re-adopted by 80% of its current users and will be adopted by 7% of the users of other texts if the text is not changed radically. To determine whether it should change the book radically to attract more sales, it uses Markov chains. Assume that the text in question currently has 25% of its possible market.
    (a) Create the transition matrix for this chain.
    (b) Find what percentage of the market the text will have three editions later.
    (c) Find the steady-state vector for this text to determine what percentage of its market this text will have if this policy is continued.

Extended Applications // Group Projects

I. Phone Numbers

Primarily because of the rapid growth in the use of cellular phones, the number of available phone numbers had to be increased. This was accomplished by increasing the number of available area codes in the United States. Investigate how this was done by completing the following, remembering that an area code cannot begin with a 0 or a 1.

1. Consult a phone book that was printed before 1995 and observe the other restrictions on the area codes at that time.
2. How many phone numbers were possible in the United States prior to 1995?
3. Consult a current phone book to see how the restrictions on area codes have been changed.
4. How many phone numbers are now possible because of the change? How many additional phone numbers are now possible because of the change?
5. What further changes could be made to the area code (other than adding a fourth digit) that would increase the available phone numbers? What would be the total number of available phone numbers with these changes? How many more numbers would be available?
6. After the area code, every number mentioned on television dramas has 555 as its first three digits. How many phone numbers are available in each area code of TV Land?
7. Los Angeles has two area codes. How many different phone numbers are available in TV Land's Los Angeles?
8. If the numbers containing 555 after the area code are not available for use by customers, how many numbers are currently available for customers?

II. Competition in the Telecommunications Industry

Of the 100,000 households in a particular city that have telephones, 85,000 use AT&T as their long-distance phone company, 10,000 use MCI, and 5000 use US Sprint. A research firm has surveyed residents to determine the level of customer satisfaction and to learn what company each expects to be using in a year's time. The results are depicted below.

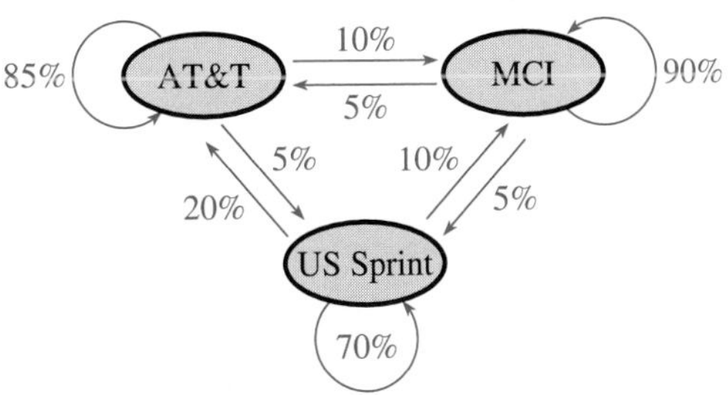

Although AT&T clearly dominates this city's long-distance service now, the situation could change in the future.

1. Calculate the percentage of the population that will be using each phone company in a year's time.
2. What will the long-term end result be if this pattern of change continues year after year?
3. You should have found, in your answer to Question 1, that the number of US Sprint customers increases in a year's time. Explain how this can happen when US Sprint loses 30% of its customers and gains only 5% of AT&T's and MCI's customers.
4. MCI managers have been presented with two proposals for increasing their market share, and each proposal has the same cost. It is estimated that Plan A would increase their customer retention rate from 90% to 95% through an enhanced customer services department. Plan B, an advertising campaign targeting AT&T customers, is intended to increase the percentage of people switching from AT&T to MCI from 10% to 11%. If the company has sufficient funds to implement one plan only, which one would you advise it to approve and why?
5. Suppose that because of budget constraints, MCI implements neither plan. A year later it faces the same choice. What should it do then?

Warm-up

Prerequisite Problem Type	For Section	Answer	Section for Review
If a fair coin is tossed 3 times, find Pr(2 heads and 1 tail).	**8.1**	$\frac{3}{8}$	7.4 Probability trees
Evaluate: (a) $\binom{4}{3}$ (b) $\binom{5}{2}$ (c) $\binom{24}{0}$	**8.3**	(a) 4 (b) 10 (c) 1	7.5 Counting; permutations and combinations
Expand: (a) $(p + q)^2$ (b) $(p + q)^3$	**8.3**	(a) $p^2 + 2pq + q^2$ (b) $p^3 + 3p^2q + 3pq^2 + q^3$	0.5 Special products
Graph all real numbers satisfying (a) $100 \leq x \leq 115$, (b) $-1 \leq z \leq 2$.	**8.5**	(a) 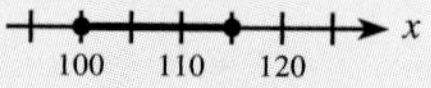 (b) 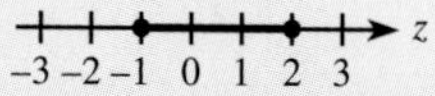	4.1 Linear inequalities
If 68% of the IQ scores of adults lie between 85 and 115, what is the probability that an adult chosen at random will have an IQ score between 85 and 115?	**8.5**	.68	7.1 Probability

Chapter 8

Further Topics in Probability; Statistics

In the previous chapter, we used sample spaces or probability trees to find the probability that the birth of three children resulted in 2 boys and 1 girl. But finding the probability that the birth of 6 children results in 4 boys and 2 girls is time-consuming using either of these methods. There is a more efficient method of finding this probability and other probabilities of events that occur in ***binomial experiments.***

In this chapter we continue our discussion of probability by considering binomial experiments, discrete probability distributions, binomial probability distributions, and normal probability distributions. We will also find the mean and standard deviation of discrete probability distributions and of normal probability distributions.

We will discuss how a set of data can be taken from a source of data for the purpose of learning about the source. This set of data is called a ***sample,*** *and the source of the data is called a* ***population.*** *The branch of statistics that deals with the collection, organization, summarization, and presentation of sample data is called* ***descriptive statistics.*** *Descriptive statistics is used by businesses to summarize data about advertising effectiveness, production costs, sales, wages, and profits. Behavioral scientists also collect statistical data about carefully selected samples in order to reach conclusions about general behavioral characteristics. The descriptive statistics methods in this chapter include graphical representation of data, measures of central tendency, and measures of dispersion.*

8.1 Binomial Probability Experiments

OBJECTIVE

- *To solve probability problems related to binomial experiments*

APPLICATION PREVIEW

Suppose it has been determined with probability .01 that any child whose mother lives in a specific area near a chemical dump will be born with a birth defect. To find the probability that 2 of 10 children whose mothers live in the area will be born with birth defects, we could use a probability tree that lists all the possibilities and then select the paths that give 2 children with birth defects. However, it is easier to find the probability by using a formula developed in this section. This formula can be used when the probability experiment is a **binomial probability experiment.**

Suppose you have a coin that is biased so that when the coin is tossed the probability of getting a head is 2/3 and the probability of getting a tail is 1/3. What is the probability of getting 2 heads in 3 tosses of this coin?

Suppose the question were "What is the probability that the first two tosses will be heads and the third will be tails?" In that case, the answer would be

$$\Pr(\text{HHT}) = \frac{2}{3} \cdot \frac{2}{3} \cdot \frac{1}{3} = \frac{4}{27}$$

However, the original question did not specify the order, so we must consider other orders that will give us 2 heads and 1 tail. We can use a tree to find all the possibilities (see Figure 8.1). We can see that there are 3 paths through the tree, which correspond to 2 heads and 1 tail in 3 tosses. Because the probability for each successful path is 4/27,

$$\Pr(\text{2H and 1T}) = 3 \cdot \frac{4}{27} = \frac{4}{9}$$

In this problem we can find the probability of 2 heads in 3 tosses by finding the probability for any one path of the tree (like HHT) and then multiplying that probability by the number of paths that result in 2 heads. We can also determine how many ways 2 heads and 1 tail can result by considering the 3 blanks below:

______ ______ ______

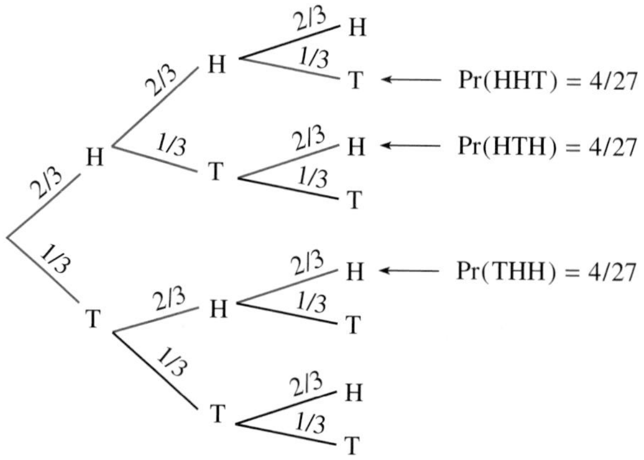

Figure 8.1

The number of ways we can pick 2 blanks (for the 2 heads) from the 3 blanks will be equivalent to the number of paths that result in 2 heads. But the number of ways we can select 2 blanks from the 3 blanks is ${}_3C_2 = \binom{3}{2}$. Thus the probability of 2 heads resulting from 3 tosses is

$$\Pr(2\text{H and }1\text{T}) = \binom{3}{2}\frac{2}{3}\cdot\frac{2}{3}\cdot\frac{1}{3} = \binom{3}{2}\left(\frac{2}{3}\right)^2\left(\frac{1}{3}\right)^1 = \frac{4}{9}$$

The experiment discussed above (tossing a coin 3 times) is an example of a general class of experiments called **binomial probability experiments.** These experiments are also called **Bernoulli experiments,** or **Bernoulli trials,** after the 18th-century mathematician Jacob Bernoulli.

A **binomial probability experiment** satisfies the following properties:

1. There are n repeated trials of the experiment.
2. Each trial results in one of two outcomes, with one denoted success (S) and the other failure (F).
3. The trials are independent, with the probability of success, p, the same for every trial. The probability of failure is $q = 1 - p$ for every trial.
4. The probability of x successes in n trials is

$$\Pr(x) = \binom{n}{x}p^x q^{n-x}$$

$\Pr(x)$ is called a **binomial probability.**

EXAMPLE 1

A die is rolled 4 times and the number of times a 6 results is recorded.

(a) Is the experiment a binomial experiment?
(b) What is the probability that three 6's will result?

Solution

(a) Yes. There are 4 trials, resulting in success (a 6) or failure (not a 6). The result of each roll is independent of previous rolls; the probability of success on each roll is 1/6.
(b) There are $n = 4$ trials with Pr(rolling a 6) $= p = 1/6$ and Pr(not rolling a 6) $= q = 5/6$, and we seek the probability of $x = 3$ successes (because we want three 6's). Thus

$$\Pr(\text{three 6's}) = \binom{4}{3}\left(\frac{1}{6}\right)^3\left(\frac{5}{6}\right)^1 = \frac{4!}{1!3!}\left(\frac{1}{216}\right)\left(\frac{5}{6}\right) = \frac{5}{324}$$

CHECKPOINT

Determine whether the following experiments satisfy the conditions for a binomial probability experiment.

1. A coin is tossed 8 times and the number of times a tail occurs is recorded.
2. A die is rolled twice and the sum that occurs is recorded.
3. A ball is drawn from a bag containing 4 white and 7 red balls and then replaced. This is done a total of 6 times, and the number of times a red ball is drawn is recorded.

We now consider the problem mentioned in the Application Preview.

EXAMPLE 2

Suppose it has been found that the probability is .01 that any child whose mother lives in an area near a chemical dump will be born with a birth defect. Suppose 10 children whose mothers live in the area are born in a given month.

(a) What is the probability that 2 of the 10 will be born with birth defects?
(b) What is the probability that at least 2 of them will have a birth defect?

Solution

(a) Note that we may consider a child having a birth defect as a success for this problem even though it is certainly not true in reality. Each of the 10 births is independent, with probability of success $p = .01$ and probability of failure $q = .99$. Therefore, the experiment is a binomial experiment, and

$$\Pr(2) = \binom{10}{2}(.01)^2(.99)^8 = .00415$$

(b) The probability of at least 2 of the children having a birth defect is

$$\begin{aligned}1 - [\Pr(0) + \Pr(1)] &= 1 - \left[\binom{10}{0}(.01)^0(.99)^{10} + \binom{10}{1}(.01)^1(.99)^9\right]\\ &= 1 - [.90438 + .09135] = .00427\end{aligned}$$

We interpret this to mean that there is a 0.427% likelihood that at least 2 of these children will have a birth defect.

EXAMPLE 3

A manufacturer of motorcycle parts guarantees that a box of 24 parts will contain at most 1 defective part. If the records show that the manufacturer's machines produce 1% defective parts, what is the probability that a box of parts will satisfy the guarantee?

Solution

The problem is a binomial experiment problem with $n = 24$ and (considering getting a defective part as a success) $p = .01$. Then the probability that the manufacturer will have no more than 1 defective part in a box is

$$\begin{aligned}\Pr(x \le 1) = \Pr(0) + \Pr(1) &= \binom{24}{0}(.01)^0(.99)^{24} + \binom{24}{1}(.01)^1(.99)^{23}\\ &= 1(1)(.7857) + 24(.01)(.7936) = .9762\end{aligned}$$

We interpret this to mean that there is a 97.62% likelihood that there will be no more than 1 defective part in a box of 24 parts.

CHECKPOINT

4. A bag contains 5 red balls and 3 white balls. If 3 balls are drawn, with replacement, from the bag, what is the probability that
   (a) two balls will be red?
   (b) at least two balls will be red?

CHECKPOINT SOLUTIONS

1. A tail is a success on each trial, and all of the conditions of a binomial probability distribution are satisfied.
2. Each trial in this experiment does not result in a success or failure, so this is not a binomial probability experiment.
3. A red ball is a success on each trial, and each condition of a binomial distribution is satisfied.
4. (a) $\Pr(2 \text{ red}) = \binom{3}{2}\left(\frac{5}{8}\right)^2\left(\frac{3}{8}\right)$
$= \frac{225}{512}$

   (b) $\Pr(\text{at least 2 red}) = \binom{3}{2}\left(\frac{5}{8}\right)^2\left(\frac{3}{8}\right) + \binom{3}{3}\left(\frac{5}{8}\right)^3\left(\frac{3}{8}\right)^0$
$= \frac{225}{512} + \frac{125}{512} = \frac{175}{256}$

EXERCISE 8.1

1. If the probability of success on each trial of an experiment is .3, what is the probability of 4 successes in 6 independent trials?
2. If the probability of success on each trial of an experiment is .4, what is the probability of 5 successes in 8 trials?
3. Suppose a fair coin is tossed 6 times. What is the probability that
   (a) 6 heads will occur? (b) 3 heads will occur?
   (c) 2 heads will occur?
4. If a fair die is rolled 3 times, what is the probability that
   (a) a 5 will result 2 times out of the 3 rolls?
   (b) an odd number will result 2 times?
   (c) a number divisible by 3 will occur 3 times?
5. Suppose a fair die is rolled 12 times. What is the probability that 5 "aces" (1's) will occur?
6. In Problem 5, what is the probability that no aces will occur?
7. A bag contains 6 red balls and 4 black balls. We draw 5 balls, with each one replaced before the next is drawn.
   (a) What is the probability that 2 balls drawn will be red?
   (b) What is the probability that 5 black balls will be drawn?
   (c) What is the probability that at least 3 black balls will be drawn?
8. A bag contains 6 red balls and 4 black balls. If we draw 5 balls, with each one replaced before the next is drawn, what is the probability that at least 2 balls drawn will be red?
9. Suppose a pair of dice is thrown 4 times. What is the probability that a sum of 9 occurs exactly 2 times?
10. If a pair of dice is thrown 4 times, find the probability that a sum of 7 occurs at least 3 times.
11. Suppose the probability that a marksman will hit a target each time he shoots is .85. If he fires 10 shots at a target, what is the probability he will hit it 8 times?
12. The probability that a sharpshooter will hit a target each time he shoots is .98. What is the probability that he will hit the target 10 times in 10 shots?

Applications

13. ***Genetics*** A family has 4 children. If the probability that each child is a girl is .5, what is the probability that
    (a) half of the children are girls?
    (b) all of the children are girls?
14. ***Testing*** A multiple-choice test has 10 questions and 5 choices for each question. If a student is totally unprepared and guesses on each question, what is the probability that
    (a) she will answer every question correctly?
    (b) she will answer half of the questions correctly?

15. ***Management*** The manager of a store buys portable radios in lots of 12. Suppose that, on the average, 2 out of each group of 12 are defective. The manager randomly selects 4 radios out of the group to test.
    (a) What is the probability that he will find 2 defective radios?
    (b) What is the probability that he will find no defective radios?

16. ***Quality control*** A hospital buys thermometers in lots of 1000. On the average, one out of 1000 is defective. If 10 are selected from one lot, what is the probability that none is defective?

17. ***Genetics*** The probability that a certain couple will have a blue-eyed child is 1/4, and they have 4 children. What is the probability that
    (a) 1 of their children has blue eyes?
    (b) 2 of their children have blue eyes?
    (c) none of their children has blue eyes?

18. ***Genetics*** If the probability that a certain couple will have a blue-eyed child is 1/2, and they have 4 children, what is the probability that
    (a) none of the children has blue eyes?
    (b) at least 1 child has blue eyes?

19. ***Health care*** Suppose that 10% of the patients who have a certain disease die from it. If 5 patients have the disease, what is the probability that
    (a) exactly 2 patients will die from it?
    (b) no patients will die from it?
    (c) no more than 2 patients will die from it?

20. ***Employee benefits*** In a certain school district, 3% of the faculty use none of their sick days in a school year. Find the probability that 5 faculty members selected at random used no sick days in a given year.

21. ***Genetics*** If the ratio of boys born to girls born is 105 to 100, and if 6 children are born in a certain hospital in a day, what is the probability that 4 of them are boys?

22. ***Biology*** It has been determined empirically that the probability that a given cell will survive for a given period of time is .4. Find the probability that 3 out of 6 of these cells will survive for this period of time.

23. ***Insurance*** If records indicate that 4 houses out of 1000 are expected to be damaged by fire in any year, what is the probability that a woman who owns 10 houses will have fire damage in 2 of them in a year?

24. ***Psychology*** Suppose 4 rats are placed in a T-maze in which they must turn right or left. If each rat makes a choice by chance, what is the probability that 2 of the rats will turn to the right?

25. ***Sports*** A baseball player has a lifetime batting average of .300. If he comes to bat 5 times in a given game, what is the probability that
    (a) he will get 3 hits?
    (b) he will get more than 3 hits?

26. ***Quality control*** If the probability of an automobile part being defective is .02, find the probability of getting exactly 3 defective parts in a sample of 12.

27. ***Quality control*** A company produces shotgun shells in batches of 300. A sample of 10 is tested from each batch, and if more than one defect is found, the entire batch is tested. If 1% of the shells are actually defective,
    (a) what is the probability of 0 defective shells in the sample?
    (b) what is the probability of 1 defective shell?
    (c) what is the probability of more than 1 defective shell?

28. ***Quality control*** A baseball pitching machine throws a bad pitch with probability .1. What is the probability that the machine will throw 5 or fewer bad pitches out of 20 pitches? (Set up only.)

29. ***Testing*** A quiz consists of 10 multiple-choice questions with 5 choices for each question. Suppose a student is sure of the first 5 answers and has each of the last 5 questions narrowed to 3 of the possible 5 choices. If the student guesses among the narrowed choices on the last 5 questions, find
    (a) the probability of passing the quiz (getting at least 60%).
    (b) the probability of getting at least a B (at least 80%).

30. ***Demographics*** In a certain community, 10% of the population is Jewish. A study shows that of 7 social service agencies, 4 have board presidents who are Jewish. Find the probability that this could happen by chance.

31. ***Suicide rates*** Suppose the probability of suicide among a certain age group is .003. If a randomly selected group of 100 Native Americans within this age group had no suicides, find the probability of this occurring by chance.

8.2 Discrete Probability Distributions; Decision Making

OBJECTIVES

- *To identify random variables*
- *To verify that a table or formula describes a discrete probability distribution*
- *To compute the expected value of a discrete probability distribution*
- *To make decisions by using expected value*
- *To find the variance and standard deviation of a discrete probability distribution*

APPLICATION PREVIEW

The T. J. Cooper Insurance Company insures 100,000 automobiles. Company records over a 5-year period indicate that during each 6-month period, it will pay out the following amounts for accidents.

$100,000 with probability .0001

$50,000 with probability .001

$25,000 with probability .002

$5000 with probability .008

$1000 with probability .02

To determine the premium the company should charge each driver, it must determine the **expected value** of its payments for accidents per car for each 6-month period. In this section we will find the expected value of **discrete probability distributions.**

If a player rolls a die and receives $1 for each dot on the face she rolls, the amount of money won on one roll can be represented by the variable x. If the die is rolled once, the following table gives the possible outcomes of the experiment and their probabilities.

x	1	2	3	4	5	6
$\Pr(x)$	1/6	1/6	1/6	1/6	1/6	1/6

Note that we have used the variable x to denote the possible numerical outcomes of this experiment. Because x results from a probability experiment, we call x a **random variable,** and because there are a finite number of possible values for x, we say that x is a discrete random variable. Whenever all possible values for a random variable can be listed (or counted), the random variable is a **discrete random variable.**

If x represents the possible number of heads that can occur in the toss of four coins, we can use the sample space $\{0, 1, 2, 3, 4\}$ to list the possible numerical values for x. Looking at Table 8.1, we see the values that the random variable of the coin-tossing experiment can assume and how the probability is distributed over these values.

TABLE 8.1

x	$\Pr(x)$
0	1/16
1	1/4
2	3/8
3	1/4
4	1/16

Discrete Probability Distribution A table, graph, or formula that assigns to each value of a discrete random variable x a probability $\Pr(x)$ describes a **discrete probability distribution** if the following two conditions hold:

(i) $0 \leq \Pr(x) \leq 1$, for any value of x.
(ii) The sum of all the probabilities is 1. We use Σ to denote "the sum of" and write $\Sigma\Pr(x) = 1$, where the sum is taken over all values of x.

EXAMPLE 1

Verify that the following table describes a discrete probability distribution.

x	$\Pr(x)$
80	1/16
120	3/8
160	5/16
200	1/4

Solution

Note that $0 \leq \Pr(x) \leq 1$ for each value of x and that

$$\sum \Pr(x) = \frac{1}{16} + \frac{3}{8} + \frac{5}{16} + \frac{1}{4} = 1$$

Thus the table describes a discrete probability distribution.

EXAMPLE 2

An experiment consists of selecting a ball from a bag containing 15 balls numbered 1 through 5. If the probability of selecting a ball with the number x on it is $\Pr(x) = x/15$, where x is an integer and $1 \leq x \leq 5$, verify that $f(x) = \Pr(x) = x/15$ describes a discrete probability distribution for the random variable x.

Solution

For each integer x $(1 \leq x \leq 5)$, $\Pr(x)$ satisfies $0 \leq \Pr(x) \leq 1$.

$$\begin{aligned}\sum \Pr(x) &= \Pr(1) + \Pr(2) + \Pr(3) + \Pr(4) + \Pr(5) \\ &= \frac{1}{15} + \frac{2}{15} + \frac{3}{15} + \frac{4}{15} + \frac{5}{15} = 1\end{aligned}$$

Hence $\Pr(x) = x/15$ describes a discrete probability distribution.

We can visualize the possible values of a discrete random variable and their associated probabilities by constructing a graph. This graph, called a **probability density histogram,** is designed so that centered over each value of the discrete random variable x along the horizontal axis is a bar having width equal to 1 unit and height (and thus, area) equal to $\Pr(x)$. Figure 8.2 gives the probability density histogram for the experiment described in Example 2.

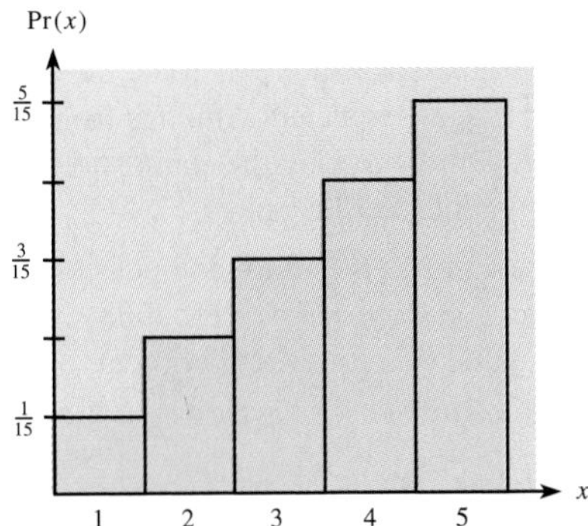

Figure 8.2

Many decisions in business and science are made on the basis of what the outcomes of specific decisions will be. One important property that is not a probability statement, but that can be developed using probability, is the **expected value,** or **mean,** of a distribution. Expected value is very useful in making decisions in business and the sciences.

Mean and Expected Value

If x is a discrete random variable with values $x_1, x_2, \ldots, x_n$, then the **mean, μ, of the distribution is the expected value of x,** denoted by **$E(x)$,** and is given by

$$\mu = E(x) = \Sigma x\text{Pr}(x) = x_1\text{Pr}(x_1) + x_2\text{Pr}(x_2) + \cdots + x_n\text{Pr}(x_n)$$

The mean μ is called a **measure of central tendency** because the probability density histogram can be balanced on a fulcrum located at $x = \mu$.

EXAMPLE 3

The probability distribution for the number of heads x that occur when a fair coin is tossed 4 times is given in the table below. What is the expected value (expected number of heads) for this experiment?

x	Pr(x)
0	1/16
1	1/4
2	3/8
3	1/4
4	1/16

Solution

The expected value is

$$E(x) = 0 \cdot \frac{1}{16} + 1 \cdot \frac{1}{4} + 2 \cdot \frac{3}{8} + 3 \cdot \frac{1}{4} + 4 \cdot \frac{1}{16} = 2$$

Note that an expected value of 2 heads in 4 tosses means that if 4 coins were tossed repeatedly, then the average (or mean) number of heads that would occur is expected to be 2.

CHECKPOINT

1. Consider a game in which you roll a die and receive \$1 for each spot that occurs.
   (a) What are the expected winnings from this game?
   (b) If you paid \$4 to play this game, how much would you lose, on average, each time you played the game?
2. An experiment has the following possible outcomes for the random variable x: 1, 2, 3, 4, 6, 7, 8, 9, 10. The probability that the value x occurs is $x/50$. Find the expected value of x for the experiment.

EXAMPLE 4

A fire company sells chances to win a new car, with each ticket costing \$5. If the car is worth \$8000 and the company sells 3000 tickets, what is the expected value for a person buying one ticket?

Solution

For a person buying one ticket, there are two possible outcomes from the drawing: winning or losing. The probability of winning is 1/3000, and the amount won would be \$8000 − \$5 (the cost of the ticket). The probability of losing is 2999/3000, and the amount lost is written as a winning of −\$5. Thus the expected value is

$$7995\left(\frac{1}{3000}\right) + (-5)\left(\frac{2999}{3000}\right) = \frac{7995}{3000} - \frac{14{,}995}{3000} = -2.33$$

Thus, on the average, a person can expect to lose \$2.33 on every ticket. It is not possible to lose exactly \$2.33 on a ticket—the person will either win \$8000 or lose \$5; the expected value means that a person who buys large numbers of tickets can expect to lose an average of \$2.33 on each ticket. That is, the fire company will make \$2.33 on each ticket it sells if it sells all 3000.

We now return to the Application Preview problem.

EXAMPLE 5

The T. J. Cooper Insurance Company insures 100,000 cars. Company records over a 5-year period indicate that during each 6-month period it will pay out the following amounts for accidents:

\$100,000 with probability .0001
\$50,000 with probability .001
\$25,000 with probability .002
\$5000 with probability .008
\$1000 with probability .02

How could the company use the expected payout per car for each 6-month period to help determine its rates?

Solution

The expected value of the company's payments is

$$\$100{,}000(.0001) + \$50{,}000(.001) + \$25{,}000(.002) + \$5000(.008) + \$1000(.02) = \$10 + \$50 + \$50 + \$40 + \$20 = \$170$$

Thus the average premium the company would charge each driver per car for a 6-month period would be

$$\$170 + \text{operating costs} + \text{profit} + \text{reserve for bad years}$$

Measures of Dispersion

Figure 8.3 shows two probability histograms that have the same mean (expected value). However, these histograms look very different. This is because the histograms differ in the **dispersion,** or **variation,** of the values of the random variables.

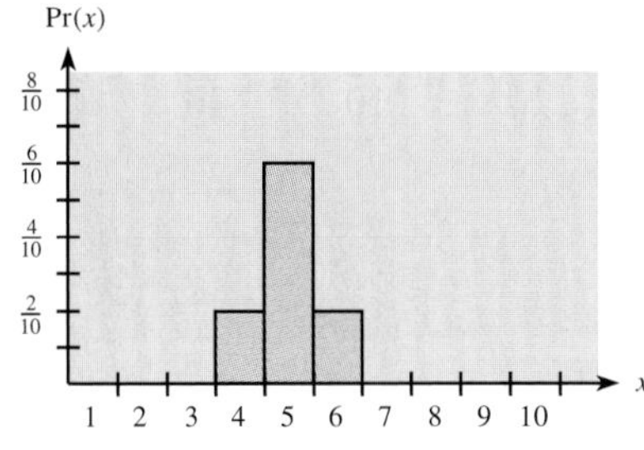

Pr(x)
6/10
4/10
2/10
x
1 2 3 4 5 6 7 8 9 10

Figure 8.3

(a)

$$\mu = E(x) = 4\left(\frac{2}{10}\right) + 5\left(\frac{6}{10}\right) + 6\left(\frac{2}{10}\right) = \frac{8 + 30 + 12}{10} = 5$$

(b)

$$\mu = E(x) = 1\left(\frac{2}{10}\right) + 2\left(\frac{1}{10}\right) + 4\left(\frac{1}{10}\right) + 5\left(\frac{2}{10}\right) + 6\left(\frac{1}{10}\right) + 8\left(\frac{1}{10}\right) + 9\left(\frac{2}{10}\right) = \frac{2 + 2 + 4 + 10 + 6 + 8 + 18}{10} = 5$$

Two useful measures of the variation of the values of the random variable about the mean are the **variance** of the distribution and the square root of the variance, which is called the **standard deviation.**

Variance and Standard Deviation

If x is a discrete random variable with values $x_1, x_2, \ldots, x_n$ and mean μ, then the variance of the distribution of x is

$$\sigma^2 = \sum (x - \mu)^2 \Pr(x)$$

The standard deviation is

$$\sigma = \sqrt{\sigma^2} = \sqrt{\sum (x - \mu)^2 \Pr(x)}$$

EXAMPLE 6

Find the standard deviation of the distributions described in (a) Figure 8.3(a) and (b) Figure 8.3(b).

Solution

(a)
$$\sigma^2 = \sum (x - \mu)^2 \Pr(x)$$
$$= (4-5)^2 \cdot \frac{2}{10} + (5-5)^2 \cdot \frac{6}{10} + (6-5)^2 \cdot \frac{2}{10}$$
$$= 1 \cdot \frac{2}{10} + 0 \cdot \frac{6}{10} + 1 \cdot \frac{2}{10}$$
$$= 0.4$$
$$\sigma = \sqrt{0.4} \approx 0.632$$

(b)
$$\sigma^2 = (1-5)^2 \cdot \frac{2}{10} + (2-5)^2 \cdot \frac{1}{10} + (4-5)^2 \cdot \frac{1}{10} + (5-5)^2 \cdot \frac{2}{10} +$$
$$(6-5)^2 \cdot \frac{1}{10} + (8-5)^2 \cdot \frac{1}{10} + (9-5)^2 \cdot \frac{2}{10}$$
$$= 16 \cdot \frac{2}{10} + 9 \cdot \frac{1}{10} + 1 \cdot \frac{1}{10} + 0 \cdot \frac{2}{10} + 1 \cdot \frac{1}{10} + 9 \cdot \frac{1}{10} + 16 \cdot \frac{2}{10}$$
$$= 8.4$$
$$\sigma = \sqrt{8.4} \approx 2.898$$

As we can see by referring to Figure 8.3 and to the standard deviations calculated in Example 6, smaller values of σ indicate that the values of x are clustered nearer μ. When σ is large, the values of x are more widely dispersed from μ.

CHECKPOINT

3. Find the standard deviation for the probability distribution in Example 3.

CHECKPOINT SOLUTIONS

1. (a) $E(x) = \$1\left(\frac{1}{6}\right) + \$2\left(\frac{1}{6}\right) + \$3\left(\frac{1}{6}\right) + \$4\left(\frac{1}{6}\right) + \$5\left(\frac{1}{6}\right) + \$6\left(\frac{1}{6}\right)$
$= \$3.50$

(b) $4 − $3.50 = $.50. You would lose $0.50, on average, from each play.

2. The expected value of x for this experiment is

$$E(x) = 1 \cdot \frac{1}{50} + 2 \cdot \frac{2}{50} + 3 \cdot \frac{3}{50} + 4 \cdot \frac{4}{50} + 6 \cdot \frac{6}{50} + 7 \cdot \frac{7}{50} + 8 \cdot \frac{8}{50} +$$
$$9 \cdot \frac{9}{50} + 10 \cdot \frac{10}{50} = 7.2$$

3. $\sigma^2 = (0-2)^2\left(\frac{1}{16}\right) + (1-2)^2\left(\frac{1}{4}\right) + (2-2)^2\left(\frac{3}{8}\right) + (3-2)^2\left(\frac{1}{4}\right) +$
$(4-2)^2\left(\frac{1}{16}\right) = 1$
$\sigma = \sqrt{1} = 1$

EXERCISE 8.2

Determine whether each table in Problems 1–4 describes a discrete probability distribution. Explain.

1.

x	$\Pr(x)$
1	$-1/5$
2	1/4
3	5/10
4	9/20

2.

x	$\Pr(x)$
0	$-4/5$
1	2/5
2	3/5
3	4/5

3.

x	$\Pr(x)$
-1	1/4
1	3/8
2	1/4
4	1/8

4.

x	$\Pr(x)$
0	1/3
4	1/6
8	1/2
12	0

Determine whether each formula in Problems 5–8 describes a discrete probability distribution.

5. $\Pr(x) = \frac{x}{21}; x = 0, 1, 2, 3, 4, 5, 6$
6. $\Pr(x) = \frac{x}{10}; x = 1, 2, 3, 4$
7. $\Pr(x) = \frac{10 - x}{10}; x = 1, 2, 3, 4, 5$
8. $\Pr(x) = \frac{x + 1}{15}; x = 1, 2, 3, 4$

Each of the tables in Problems 9–16 defines a discrete probability distribution. Find the expected value of each distribution.

9.

x	0	1	2	3
$\Pr(x)$	1/8	1/4	1/4	3/8

10.

x	1	2	3	4
$\Pr(x)$	1/10	1/5	3/10	2/5

11.

x	4	5	6	7
$\Pr(x)$	1/3	1/3	1/3	0

12.

x	1	2	3	4	5	6
$\Pr(x)$	1/12	1/4	1/12	1/4	1/12	1/4

13.

x	0	2	4
$\Pr(x)$	1/6	2/6	3/6

14.

x	3	4	5	6
$\Pr(x)$	3/10	1/5	3/10	1/5

15.

x	1	3	5
$\Pr(x)$	2/3	1/6	1/6

16.

x	1	2	3	4
$\Pr(x)$	1/15	4/15	1/5	7/15

In Problems 17–22, find the mean, variance, and standard deviation for each probability distribution.

17.

x	$\Pr(x)$
0	1/4
1	1/4
2	1/8
3	3/8

18.

x	$\Pr(x)$
3	1/4
4	1/4
5	1/4
6	1/4

19.

x	$\Pr(x)$
0	1/5
2	1/4
4	1/10
6	3/20
8	3/10

20.

x	$\Pr(x)$
1	1/8
3	1/8
5	1/8
7	1/4
9	3/8

21. $\Pr(x) = \frac{x}{21}; x = 0, 1, 2, 3, 4, 5, 6$
22. $\Pr(x) = \frac{x}{10}; x = 2, 3, 5$
23. Suppose an experiment has five possible outcomes for x: 0, 1, 2, 3, 4. The probability that each of these outcomes occurs is $x/10$. What is the expected value of x for the experiment?
24. Suppose an experiment has six possible outcomes for x: 0, 1, 2, 3, 4, 5. The probability that each of these outcomes occurs is $x/15$. What is the expected value of x for the experiment?
25. Five slips of paper containing the numbers 0, 1, 2, 3, 4 are placed in a hat. If the experiment consists of drawing one number, and if the experiment is repeated a large number of times, what is the expected value of the number drawn?
26. An experiment consists of rolling a die once. If the experiment is repeated a large number of times, what is the expected value of the number rolled?

Applications

27. ***Animal relocation*** In studying endangered species, scientists have found that when animals are relocated, it takes x years without offspring before the first young are born, where x and the probability of x are given below. What is the expected number of years before the first young are born?

x	0	1	2	3	4
$\Pr(x)$	.04	.35	.38	.18	.05

28. ***Raffle*** A charity sells 1000 raffle tickets for \$1 each. There is one grand prize of \$250, two prizes of \$75 each, and five prizes of \$10 each. If x is the net winnings for someone who buys a ticket, then the following table gives the probability distribution for x. Find the expected winnings of someone who buys a ticket.

x	Pr(x)
249	0.001
74	0.002
9	0.005
−1	0.992

29. ***Campaigning*** A candidate must decide whether he should spend his time and money on TV commercials or making personal appearances. His staff determines that by using TV he can reach 100,000 people with probability .01, 50,000 people with probability .47, and 25,000 with probability .52; by making personal appearances he can reach 80,000 people with probability .02, 50,000 people with probability .47, and 20,000 with probability .51. In the tables, x represents the number of people reached by each choice. In each case, find the expected value of x to decide which method will reach more people.

TV Commercials		*Personal Appearances*	
x	Pr(x)	x	Pr(x)
100,000	0.01	80,000	0.02
50,000	0.47	50,000	0.47
25,000	0.52	20,000	0.51

30. ***Gambling*** Suppose that a student is offered a chance to draw a card from an ordinary deck of 52 playing cards and win \$10 if he or she draws an ace, \$2 for a king, and \$1 for a queen. If \$2 must be paid to play the game, and if x is the net winnings (in dollars), use the table below to find the expected winnings each time the game is played.

x	Pr(x)
8	1/13
0	1/13
−1	1/13
−2	10/13

31. ***Genetics*** If the probability that a newborn child is a male is 1/2, what is the expected number of male children in a family having 4 children?

32. ***Testing*** A student is completely unprepared for a 4-choice multiple-choice test. If the test has 4 questions and all are guessed at, how many questions should he or she expect to get correct?

33. ***Testing*** Suppose the instructor of the student in Problem 32 tries to account for guessing on the test by counting the number of questions the student answers correctly and then subtracting from this number 1/4 of the number of questions a student answers. If the student guesses on all 4 problems on this test, for how many problems should he or she expect to get credit?

34. ***Gambling*** Suppose you are invited to play the following game: You toss 3 coins and receive \$1 if one head results, \$4 if two heads result, and \$9 if three heads result. How much should you pay to play if the game is fair?

35. ***Lottery*** A charity sells raffle tickets for \$1 each. First prize is \$500, second prize is \$100, third prize is \$10. If you bought one of the 1000 tickets sold, what are your expected winnings?

36. ***Gambling*** Suppose the coin used in Problem 34 has been altered so that the probability of a head on any toss is 1/3. How much should you pay to play if the game is fair?

37. ***Gambling*** Suppose a student is offered a chance to draw a card from an ordinary deck of 52 playing cards and win \$15 for an ace, \$10 for a king, and \$1 for a queen. If \$4 must be paid to play the game, what is the expected winnings every time the game is played by the student?

38. ***Revenue*** The Rent-to-Own Company estimates that 35% of its rentals result in the sale of the product, with an average revenue of 100%, 56% of the rentals are returned in good condition, with an average revenue of 35% on these rentals, and the remainder of the rentals are stolen or returned in poor condition, giving a loss of 15% on these rentals. If these estimates are accurate, what is the expected percent of revenue for this company?

39. ***Sales*** A young man plans to sell umbrellas at the city's Easter Parade. He knows that he can sell 180 umbrellas at \$5 each if it rains hard, he can sell 50 if it rains lightly, and he can sell 10 if it doesn't rain at all. Past records show it rains hard 25% of the time on Easter, rains lightly 20% of the time, and does not rain at all 55% of the time. If he can buy 0, 100, or 200 umbrellas at \$2 each and return the unsold ones for \$1 each, how many should he buy?

40. ***Sales*** If the young man in Problem 39 learns from a highly reliable weather forecaster that the probability of a hard rain on Easter is 15%, of a light rain is 20%, and of no rain is 65%, should he still plan to sell umbrellas at the Easter Parade?

41. ***Advertising*** A candidate must decide whether he should spend his advertising dollars on TV commercials or newspaper advertisements. His staff determines that by using TV he can reach 100,000 people with probability .01, reach 50,000 people with probability .47, and reach 25,000 with probability .52; by purchasing newspaper ads he can reach 70,000 people with probability .04, reach 50,000 people with probability .38, and reach 30,000 with probability .58. Which method will reach more people?

42. ***Insurance*** A car owner must decide whether she should take out a $100-deductible collision policy in addition to her liability insurance policy. Records show that each year, in her area, 8% of the drivers have an accident that is their fault or for which no fault is assigned, and that the average cost of repairs for these types of accidents is $1000. If the $100-deductible collision policy costs $100 per year, would she save money in the long run by buying the insurance or "taking the chance"? (*Hint:* Find the expected values if she has the policy and if she doesn't have the policy and compare them.)

43. ***Budgeting*** Suppose a youth has a part-time job in an ice cream shop. He receives $20 if he is called to work a full day and $10 if he is called to work a half-day. Over the past year he has been called to work a full day an average of 8 days per month and for a half-day an average of 14 days per month. How much can be expect to earn *per day* during a 30-day month?

44. ***Testing*** On a multiple-choice test with each question having five possible choices, a correct answer earns 4 points, an incorrect answer loses 1 point, and a blank receives 0 points.
    (a) Find the expected point-value earned when a student has no idea of the correct answer but makes a guess on every question.
    (b) Find the expected point-value earned when a student has narrowed the choice by eliminating two of the possible choices on each question but must guess among the remaining choices.

8.3 The Binomial Distribution

OBJECTIVES

- *To find the mean and standard deviation of a binomial distribution*
- *To graph a binomial distribution*
- *To expand a binomial to a power, using the binomial formula*

APPLICATION PREVIEW

Opponents to the distribution of condoms in high schools and colleges object on moral grounds and on the grounds that only abstinence is safe. If condoms are 95% effective in preventing pregnancy and the spread of disease (such as HIV), the probability of a failure is 5% each time a condom is used. Because the number of failures follows the **binomial probability distribution,** we can use the methods of this section to find the expected number of failures per thousand condom uses.

To begin our discussion of the binomial distribution, suppose a fair coin is tossed 6 times and x represents the number of heads. Then Table 8.2 gives all possible outcomes and their probabilities. This is an example of a special probability distribution called the **binomial probability distribution.**

TABLE 8.2

x	$\Pr(x)$	x	$\Pr(x)$
0	$\binom{6}{0}\left(\frac{1}{2}\right)^0\left(\frac{1}{2}\right)^6 = \frac{1}{64}$	4	$\binom{6}{4}\left(\frac{1}{2}\right)^4\left(\frac{1}{2}\right)^2 = \frac{15}{64}$
1	$\binom{6}{1}\left(\frac{1}{2}\right)^1\left(\frac{1}{2}\right)^5 = \frac{6}{64}$	5	$\binom{6}{5}\left(\frac{1}{2}\right)^5\left(\frac{1}{2}\right)^1 = \frac{6}{64}$
2	$\binom{6}{2}\left(\frac{1}{2}\right)^2\left(\frac{1}{2}\right)^4 = \frac{15}{64}$	6	$\binom{6}{6}\left(\frac{1}{2}\right)^6\left(\frac{1}{2}\right)^0 = \frac{1}{64}$
3	$\binom{6}{3}\left(\frac{1}{2}\right)^3\left(\frac{1}{2}\right)^3 = \frac{20}{64}$		

Binomial Probability Distribution

If x is a variable that assumes the values $0, 1, 2, \ldots, r, \ldots, n$ with probabilities $\binom{n}{0}p^0q^n, \binom{n}{1}p^1q^{n-1}, \binom{n}{2}p^2q^{n-2}, \ldots, \binom{n}{r}p^rq^{n-r}, \ldots, \binom{n}{n}p^nq^0$, respectively, then x is called a **binomial variable.**

The values of x and their corresponding probabilities described above form the **binomial probability distribution.**

Recall that $\binom{n}{r} = {}_nC_r$ can be evaluated easily on a calculator.

CHECKPOINT

1. Show that the values of x and the associated probabilities in Table 8.2 satisfy the conditions for a discrete probability distribution.

The expected value of the number of successes for the coin toss experiment above (expected value was discussed in Section 8.2) is given by

$$\begin{aligned} E(x) &= 0 \cdot \frac{1}{64} + 1 \cdot \frac{6}{64} + 2 \cdot \frac{15}{64} + 3 \cdot \frac{20}{64} + 4 \cdot \frac{15}{64} + 5 \cdot \frac{6}{64} + 6 \cdot \frac{1}{64} \\ &= \frac{192}{64} = 3 \end{aligned}$$

This expected number of successes seems reasonable. If the probability of success is 1/2, we would expect to succeed on half of the 6 trials. For any binomial distribution the expected number of successes is given by np, where n is the number of trials and p is the probability of success. Because the expected value of any probability distribution is defined as the mean of that distribution, we have the following.

Mean of a Binomial Distribution

The theoretical mean of any binomial distribution is

$$\mu = np$$

where n is the number of trials in the corresponding binomial experiment and p is the probability of success on each trial.

A simple formula can also be developed for the standard deviation of a binomial distribution.

Standard Deviation of a Binomial Distribution

The standard deviation of a binomial distribution is

$$\sigma = \sqrt{npq}$$

where n is the number of trials, p is the probability of success on each trial, and $q = 1 - p$.

The standard deviation of the binomial distribution corresponding to the number of heads resulting when a coin is tossed 16 times is

$$\sigma = \sqrt{16 \cdot \frac{1}{2} \cdot \frac{1}{2}} = \sqrt{4} = 2$$

EXAMPLE 1

A fair coin is tossed 4 times.

(a) Construct a table with each value of the binomial variable x, where x is the number of heads resulting, and the probability of each value of x.
(b) Graph the distribution.
(c) Find the mean of this binomial distribution.
(d) Find the standard deviation of this distribution.

Solution

(a)

x	$\Pr(x)$
0	$\binom{4}{0}\left(\frac{1}{2}\right)^0\left(\frac{1}{2}\right)^4 = \frac{1}{16}$
1	$\binom{4}{1}\left(\frac{1}{2}\right)^1\left(\frac{1}{2}\right)^3 = \frac{4}{16}$
2	$\binom{4}{2}\left(\frac{1}{2}\right)^2\left(\frac{1}{2}\right)^2 = \frac{6}{16}$
3	$\binom{4}{3}\left(\frac{1}{2}\right)^3\left(\frac{1}{2}\right)^1 = \frac{4}{16}$
4	$\binom{4}{4}\left(\frac{1}{2}\right)^4\left(\frac{1}{2}\right)^0 = \frac{1}{16}$

(b) The graph of the distribution is shown in Figure 8.4.

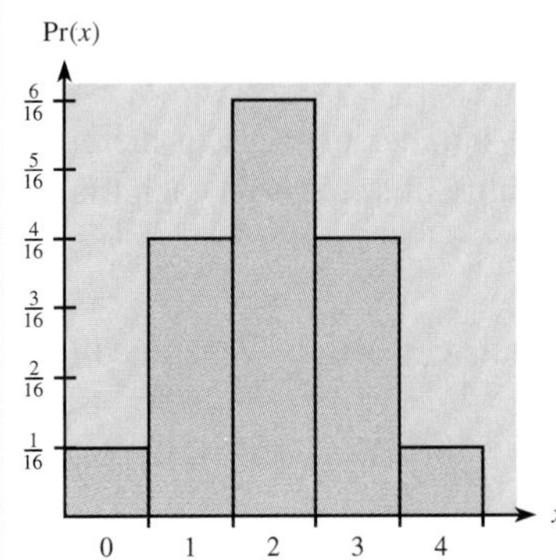

Figure 8.4

(c) The mean of this distribution is

$$\mu = np = 4 \cdot \frac{1}{2} = 2$$

(d) The standard deviation of this binomial distribution is

$$\sigma = \sqrt{npq} = \sqrt{4 \cdot \frac{1}{2} \cdot \frac{1}{2}} = 1$$

As we noted in the previous section, the mean of a distribution is also the expected value of the distribution. We use this information to solve the problem posed in the Application Preview.

EXAMPLE 2

Latex condoms are reported to be 95% effective in preventing the transmission of disease (such as HIV) and pregnancy. Find the expected number of failures per thousand condom uses.

Solution

In this case, we seek the expected number of failures where the probability of failure in each case is $1 - .95 = .05$. We find the expected number of failures by using $E(x) = \mu = np$, where $n = 1000$ and $p = .05$.

$$E(x) = \mu = 1000(.05) = 50$$

CHECKPOINT

2. A cube is colored so that 4 faces are green and 2 are red. The cube is tossed 9 times. If success is defined as a red face being up, find
   (a) the mean and
   (b) the standard deviation of the number of red faces that occur.

The binomial probability distribution is closely related to the powers of a binomial. For example, if a binomial experiment has 3 trials with probability of success p on each trial, then the binomial probability distribution is given by Table 8.3 with the values of x written from 3 to 0. Compare this to the expansion of $(p + q)^3$, which we introduced in the chapter on algebraic concepts.

$$(p + q)^3 = p^3 + 3p^2q + 3pq^2 + q^3$$

TABLE 8.3

x	$\Pr(x)$
3	$\binom{3}{3} p^3q^0 = p^3$
2	$\binom{3}{2} p^2q = 3p^2q$
1	$\binom{3}{1} pq^2 = 3pq^2$
0	$\binom{3}{0} p^0q^3 = q^3$

The formula that we can use to expand a binomial $(a + b)$ to any positive integer power n is

Binomial Formula

$$(a+b)^n = \binom{n}{n}a^n + \binom{n}{n-1}a^{n-1}b + \binom{n}{n-2}a^{n-2}b^2 + \cdots + \binom{n}{2}a^2b^{n-2} + \binom{n}{1}ab^{n-1} + \binom{n}{0}b^n$$

EXAMPLE 3

Expand $(x + y)^4$.

Solution

$$\begin{aligned}(x+y)^4 &= \binom{4}{4}x^4 + \binom{4}{3}x^3y + \binom{4}{2}x^2y^2 + \binom{4}{1}xy^3 + \binom{4}{0}y^4 \\ &= x^4 + 4x^3y + 6x^2y^2 + 4xy^3 + y^4\end{aligned}$$

Note that the coefficients in this expansion are related to the probabilities in Example 1.

EXAMPLE 4

(a) If a die is rolled 5 times, what is the probability that exactly 2 faces with 4 on them will result?

(b) What is the coefficient of p^2q^3 in the expansion of $(p + q)^5$?

Solution

(a) If getting a 4 is a success, then $\Pr(\text{success}) = \frac{1}{6}$ and we seek the probability of 2 successes in 5 trials. Thus,

$$\Pr(\text{two 4's}) = \binom{5}{2}\left(\frac{1}{6}\right)^2\left(\frac{5}{6}\right)^3 = 10 \cdot \frac{1}{36} \cdot \frac{125}{216} = \frac{625}{3888}$$

(b) By the Binomial Formula, the coefficient of p^2q^3 is $\binom{5}{2} = 10$.

CHECKPOINT SOLUTIONS

1. Each value of the random variable x has a probability, with $0 \le \Pr(x) \le 1$, and with $\sum \Pr(x) = 1$. Thus the conditions of a probability distribution are satisfied.
2. $\Pr(\text{red face}) = \frac{1}{3} = p$, so $q = \frac{2}{3}$ and $n = 9$.
   (a) $\mu = np = 9\left(\frac{1}{3}\right) = 3$
   (b) $\sigma = \sqrt{npq} = \sqrt{9\left(\frac{1}{3}\right)\left(\frac{2}{3}\right)} = \sqrt{2}$

EXERCISE 8.3

1. A die is rolled 3 times, and success is rolling a 5.
   (a) Construct the binomial distribution that describes this experiment.
   (b) Find the mean of this distribution.
   (c) Find the standard deviation of this distribution.
2. A die is rolled 3 times. Success is rolling a number divisible by 3.
   (a) Construct the binomial distribution that describes this experiment.
   (b) Find the mean of this distribution.
   (c) Find the standard deviation of this distribution.
3. A variable x has a binomial distribution with probability of success .7 for each trial and a total of 60 trials. What are
   (a) the mean and
   (b) the standard deviation of the distribution?
4. A variable x has a binomial distribution with probability of success .8 for each trial and a total of 40 trials. What are
   (a) the mean and
   (b) the standard deviation of the distribution?
5. Suppose a fair die is tossed 12 times. What are the mean and standard deviation of the number of 5's that occur?
6. Suppose a fair coin is tossed 8 times. What are the mean and standard deviation of the number of tails?
7. Suppose the probability that a certain coin will result in a head is 2/3 when it is tossed. If it is tossed 6 times, how many heads would we expect?
8. A coin is loaded so that the probability of tossing a head is 3/5. If it is tossed 50 times, what are
   (a) the mean and
   (b) the standard deviation of the number of heads that occur?
9. Suppose a pair of dice is thrown 12 times. How many times would we expect a sum of 7 to occur?
10. Suppose a pair of dice is thrown 20 times. How many times would we expect a sum of 6 to occur?
11. Suppose that a die is "loaded" so that the probability of getting each even number is 1/4 and the probability of getting each odd number is 1/12. If it is rolled 24 times, how many times would we expect to get a 3?
12. What is the coefficient of a^3b^5 in the expansion of $(a + b)^8$?
13. What is the coefficient of the term containing a^4b^2 in $(a + b)^6$?
14. (a) What is the coefficient of the term containing a^2b^4 in $(a + b)^6$?
    (b) Does $\binom{6}{4} = \binom{6}{2}$?
15. Expand $(a + b)^6$.
16. Expand $(x + y)^5$.
17. Expand $(x + h)^4$.
18. Write the general expression for $(x + h)^n$.

Applications

19. ***Health care*** Suppose that 10% of the patients who have a certain disease will die from it.
    (a) If 100 people have the disease, how many would we expect to die from it?
    (b) What is the standard deviation of the number of deaths that could occur?
20. ***Cancer research*** Suppose it has been determined that the probability that a rat injected with cancerous cells will live is .6. If 35 rats are injected, how many would be expected to die?
21. ***Voting*** A candidate claims 60% of the people in his district will vote for him.
    (a) If his district contains 100,000 voters, how many votes does he expect to get from his district?
    (b) What is the standard deviation of the number of his votes?

22. ***Genetics*** In a family with 2 children, the probability that both children will be boys is 1/4.
    (a) If 1200 families with two children are selected at random, how many of the families would we expect to have 2 boys?
    (b) What is the standard deviation for the number of families with 2 boys?

23. ***Voting*** If the number of votes the candidate in Problem 21 gets is 2 standard deviations below what he expected, how many votes did he get (approximately)?

24. ***Genetics*** If the 1200 families in Problem 22 had 315 families with two boys in it, how many standard deviations above the mean is the number of families?

25. ***Testing*** A multiple-choice test has 20 questions and 5 choices for each question. If a student is totally unprepared and guesses on each question, the mean is the number of questions she can expect to answer correctly.
    (a) How many questions can she expect to answer correctly?
    (b) What is the standard deviation of the distribution?

26. ***Birth control*** Suppose a birth control pill is 99% effective in preventing pregnancy. What number of women would be expected to get pregnant out of 100 women using this pill?

27. ***Quality control*** A certain calculator circuit board is manufactured in lots of 200. If 1% of the boards are defective, find the mean and standard deviation for the number of defects in each lot.

28. ***CPA exam*** Forty-eight percent of accountants taking exams to become a CPA fail the first time. If 1000 candidates take the exams for the first time, what is the expected number that will pass?

29. ***Seed germination*** A certain type of corn seed has an 85% germination rate. If 2000 seeds were planted, how many seeds would be expected not to germinate?

30. ***Quality control*** The probability that a manufacturer produces a defective medical thermometer is .001. What is the expected number of defective thermometers in a shipment of 3000?

8.4 Descriptive Statistics

OBJECTIVES

- *To set up frequency tables and construct frequency histograms for sets of data*
- *To find the mode of a set of scores (numbers)*
- *To find the median of a set of scores*
- *To find the mean of a set of scores*
- *To find the range of a set of data*
- *To find the variance and standard deviation of a set of data*

APPLICATION PREVIEW

In modern business, a vast amount of data is collected for use in making decisions about the production, distribution, and sale of merchandise. Businesses also collect and summarize data about advertising effectiveness, production costs, sales, wages, and profits. Behavioral scientists attempt to reach conclusions about general behavioral characteristics by studying the characteristics of a small sample of people. For example, election predictions are based on a careful sampling of the votes; correct predictions are frequently announced on television even though only 5% of the vote has been counted. Life scientists can use statistical methods with laboratory animals to detect substances that may be dangerous to humans.

For example, the table below gives the net income (or loss) in millions of dollars for the 11-year period 1983–1993 for the USF&G Corporation. These data can be summarized by using descriptive statistics. We will see in this section how to display these data with a histogram and how to summarize the data by finding the mean and standard deviation of the earnings over the 11-year period.

Year	*1983*	*1984*	*1985*	*1986*	*1987*	*1988*	*1989*	*1990*	*1991*	*1992*	*1993*
Net income	270	−64	−108	296	279	247	119	−569	−176	28	165

Source: USF&G *1993 Annual Report*

A set of data taken from some source of data for the purpose of learning about the source is called a **sample;** the source of the data is called a **population.** The branch of statistics that deals with the collection, organization, summarization, and presentation of sample data is called **descriptive statistics.** The descriptive statistics methods discussed in this section include graphical representation of data, measures of central tendency, and measures of dispersion.

The first step in statistical work is the collection of data. Data are regarded as statistics data if the numbers collected have some relationship. The heights of all female first-year students are statistical data because a definite relationship exists among the collected numbers.

A *graph* frequently is used to provide a picture of the statistical data that have been gathered and to show relationships among various quantities. The advantage of the graph for showing data is that a person can quickly get an idea of the relationships that exist. The disadvantage is that the data can be shown only approximately on a graph. One type of graph that is used frequently in statistical work is the **frequency histogram.** A histogram is really a bar graph. It is constructed by putting the scores (numbers collected) along the horizontal axis and the frequency with which they occur along the vertical axis. A frequency table is usually set up to prepare the data for the histogram.

EXAMPLE 1

Construct a frequency histogram for the following scores: 38, 37, 36, 40, 35, 40, 38, 37, 36, 37, 39, 38.

Solution

We first construct the frequency table and then use the table to construct the histogram shown in Figure 8.5.

Score	*Frequency*
35	1
36	2
37	3
38	3
39	1
40	2

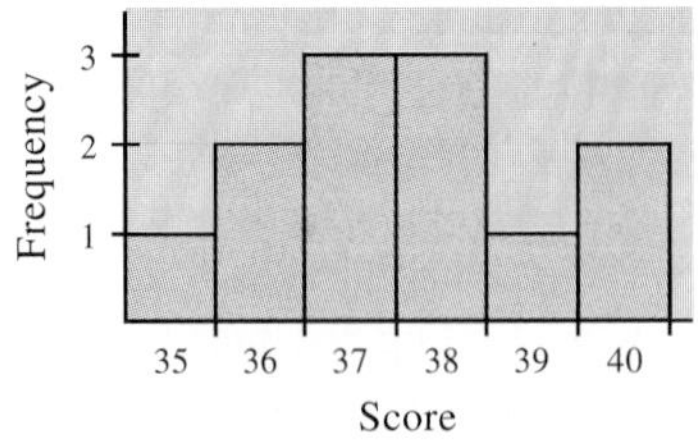

Figure 8.5

Graphing calculators, spreadsheets, and computer programs can be used to plot the histogram of a set of data. The width of each bar of the histogram can be determined by the calculator or adjusted by the user when the histogram is plotted.

EXAMPLE 2

Construct a histogram from the following frequency table.

Interval	*Frequency*
0–5	0
6–10	2
11–15	5
16–20	1
21–25	3

Solution

The histogram is shown in Figure 8.6.

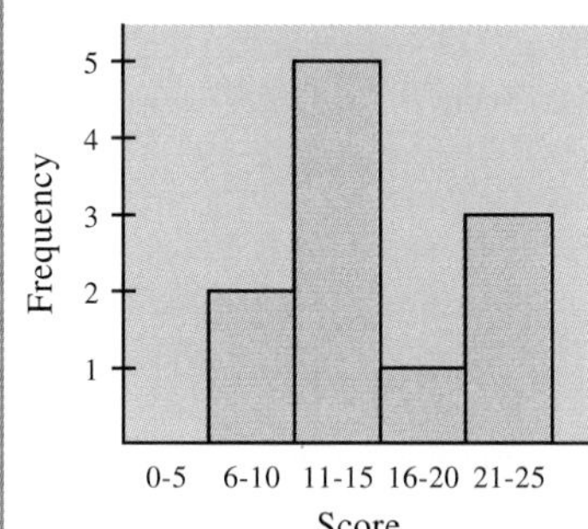

Figure 8.6

A set of data can be described by listing all the scores or by drawing a frequency histogram. But we often wish to describe a set of scores by giving the *average score* or the typical score in the set. There are three types of measures that are called *averages,* or measures of central tendency. They are the **mode,** the **median,** and the **mean.**

To determine what value represents the most "typical" score in a set of scores, we use the **mode** of the scores.

The **mode** of a set of scores is the score that occurs most frequently.

That is, the mode is the most popular score—the one that is most likely to occur. The mode can be readily determined from a frequency table or frequency histogram because it is determined according to the frequency of the scores.

EXAMPLE 3

Find the mode of the following scores: 10, 4, 3, 6, 4, 2, 3, 4, 5, 6, 8, 10, 2, 1, 4, 3.

Solution

The mode is 4, because it occurs more frequently than any other score.

EXAMPLE 4

Determine the mode of the scores shown in the histogram in Figure 8.7.

Solution

The most frequent score, as the histogram reveals, is 2.

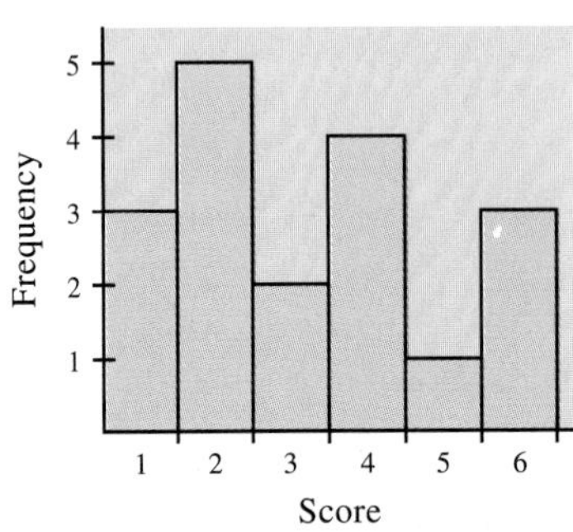

Figure 8.7

EXAMPLE 5

Find the mode of the following scores: 4, 3, 2, 4, 6, 5, 5, 7, 6, 5, 7, 3, 1, 7, 2.

Solution

Both 5 and 7 occur three times. Thus they are both modes. This set of data is said to be *bimodal* because it has two modes.

Although the mode of a set of scores tells us the most frequent score, it does not always represent the typical performance, or central tendency, of the set of scores. For example, the scores 2, 3, 3, 3, 5, 8, 9, 10, 13 have a mode of 3. But 3 is not a good measure of central tendency for this set of scores, for it is nowhere near the middle of the distribution.

One value that gives a better measure of central tendency is the **median.**

The **median** is the score or point above and below which an equal number of the scores lie.

It is the middle of the distribution of scores. If there is an odd number of scores, we find the median by ranking the scores from smallest to largest and picking the middle score. That score will be the median.

EXAMPLE 6

Find the median of the following scores: 2, 3, 16, 5, 15, 38, 18, 17, 12.

Solution

We rank the scores from smallest to largest: 2, 3, 5, 12, 15, 16, 17, 18, 38. The median is the middle score, which is 15. Note that there are 4 scores above and 4 below 15.

If the number of scores is even, there will be no middle score when the scores are ranked, so the *median* is the point that is the *mean of the two middle scores.*

EXAMPLE 7

Find the median of the following scores: 3, 2, 6, 8, 12, 4, 3, 2, 1, 6.

Solution

We rank the scores from smallest to largest: 1, 2, 2, 3, 3, 4, 6, 6, 8, 12. The two scores that lie in the middle of the distribution are 3 and 4. Thus the median is a point that is midway between 3 and 4; that is, it is the arithmetic average of 3 and 4. The median is

$$\frac{3+4}{2} = \frac{7}{2} = 3.5$$

In this case we say the median is a point because there is no *score* that is 3.5.

Some calculators have keys or functions for finding the median of a set of data. Such calculators, or computers, are valuable when the set of data is large.

The median is the most easily interpreted measure of central tendency, and it is the best indicator of central tendency when the set of scores contains a few extreme values. However, the most frequently used measure of central tendency is the **mean.**

The symbol $\bar{x}$ is used to represent the **mean** of a set of scores in a sample. Thus, if n is the number of scores, the formula for the mean is

$$\bar{x} = \frac{\Sigma x}{n} = \frac{\text{sum of scores}}{\text{number of scores}}$$

The mean is used most often as a measure of central tendency because it is more useful than the median in the general applications of statistics.

Scientific calculators and graphing calculators have keys or built-in functions that compute the mean of a set of data.

EXAMPLE 8

Find the mean of the following sample of numbers: 12, 8, 7, 10, 6, 14, 7, 6, 12, 9.

Solution

$$\bar{x} = \frac{\Sigma x}{10} = \frac{12+8+7+10+6+14+7+6+12+9}{10}$$

so $\bar{x} = 91/10 = 9.1$.

Note that the mean need not be one of the numbers (scores) given.

EXAMPLE 9

Find the mean of the following numbers: 2.1, 6.3, 7.1, 4.8, 3.2.

Solution

$$\bar{x} = \frac{\Sigma x}{5} = \frac{2.1 + 6.3 + 7.1 + 4.8 + 3.2}{5}$$

so $\bar{x} = 23.5/5 = 4.7$.

If we are given data in this frequency table,

x	1	2	3	4
f	3	1	4	2

we can find the mean $\bar{x}$ by using

$$\bar{x} = \frac{\Sigma x}{n}$$

where 1 is added 3 times, 2 is added once, 3 is added 4 times, and 4 is added twice. Or we can use $\Sigma(x \cdot f)$ and $n = \Sigma f$, so

$$\bar{x} = \frac{\Sigma(x \cdot f)}{\Sigma f}$$

where each value of x is used once.

The mean for the frequency table above is

$$\bar{x} = \frac{1 \cdot 3 + 2 \cdot 1 + 3 \cdot 4 + 4 \cdot 2}{3 + 1 + 4 + 2} = 2.5$$

EXAMPLE 10

Find the mean of the following sample of test scores for a math class.

Scores	*Class Marks*	*Frequencies*
40–49	44.5	2
50–59	54.5	0
60–69	64.5	6
70–79	74.5	12
80–89	84.5	8
90–99	94.5	2

Solution

For the purpose of finding the mean of interval data, we assume that all scores within an interval are represented by the class mark (midpoint of the interval). Thus

$$\bar{x} = \frac{(44.5)(2) + (54.5)(0) + (64.5)(6) + (74.5)(12) + (84.5)(8) + (94.5)(2)}{30}$$

$$= 74.5$$

CHECKPOINT

Figure 8.8 shows the payroll of the Ace Cap Company.

1. Find the mean, median, and mode of the salaries.
2. Which of the three measures of central tendency gives the lowest value?
3. Which of these measures gives the highest value?
4. Which of these measures best represents the "average" salary for the company?

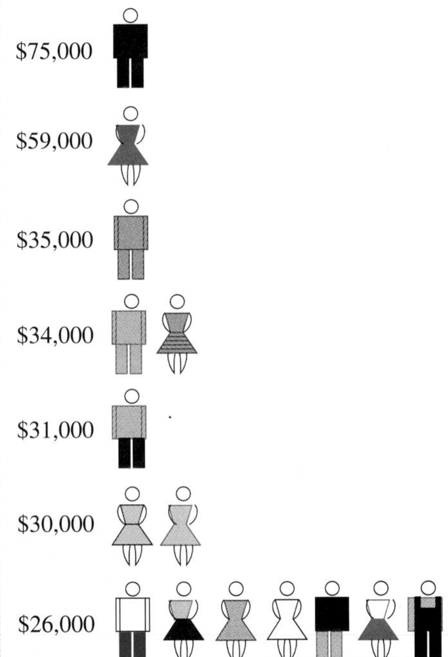

Figure 8.8

Although the mean of a set of data is useful in locating the center of the distribution of the data, it doesn't tell us as much about the distribution as we might think at first. For example, a basketball team with five 6-ft players is quite different from a team with one 6-ft, two 5-ft, and two 7-ft players. The distributions of the heights of these teams differ not in the mean height but in how the heights *vary* from the mean.

One measure of how a distribution varies is the **range** of the distribution.

Range The **range** of a set of numbers is the difference between the largest and smallest numbers in the set.

Consider the following set of heights of players on a basketball team, in inches: 69, 70, 75, 69, 73, 78, 74, 73, 78, 71. The range of this set of heights is $78 - 69 = 9$ inches. Note that this range is determined by only two numbers and does not give any information about how the other heights vary.

If we calculated the deviation of each score from the mean, $(x - \bar{x})$, and then attempted to average the deviations, we would see that the deviations add to 0. In an effort to find a meaningful measure of dispersion about the mean, statisticians developed the **variance,** which squares the deviations (to make them all positive) and then averages the squared deviations.

Frequently, we do not have all the data for a population and must use a sample of these data. To estimate the variance of a population from a sample, statisticians compensate for the fact that there is usually less variability in a sample than in the population itself by summing the squared deviations and dividing them by $n - 1$ rather than averaging them.

To get a measure comparable to the original deviations before they were squared, the **standard deviation,** which is the square root of the variance, was introduced. The formulas for the variance and standard deviation of sample data follow.

Variance and Standard Deviation

Sample Variance $$s^2 = \frac{\Sigma(x - \bar{x})^2}{n - 1}$$

Sample Standard Deviation $$s = \sqrt{\frac{\Sigma(x - \bar{x})^2}{n - 1}}$$

As with probability distributions, the standard deviation of a sample is a measure of the concentration of the scores about their mean. The smaller the standard deviation, the closer the scores lie to the mean. For example, the histograms in Figure 8.9 describe two sets of data with the same means and the same ranges. From looking at the histograms, we see that more of the data are concentrated about the mean $\bar{x} = 4$ in Figure 8.9(b) than in Figure 8.9(a). We will see that the standard deviation is smaller for the data in Figure 8.9(b) than in Figure 8.9(a).

EXAMPLE 11

Figure 8.9(a) is the histogram for the sample data 1, 1, 1, 3, 3, 4, 4, 5, 6, 6, 7, 7. Figure 8.9(b) is the histogram for the data 1, 2, 3, 3, 4, 4, 4, 4, 5, 5, 6, 7. Both sets of data have a range of 6 and a mean of 4. Find the standard deviations of these samples.

Figure 8.9

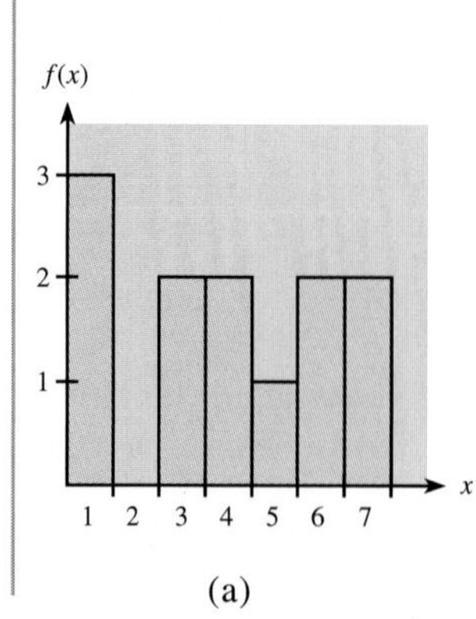

(a)

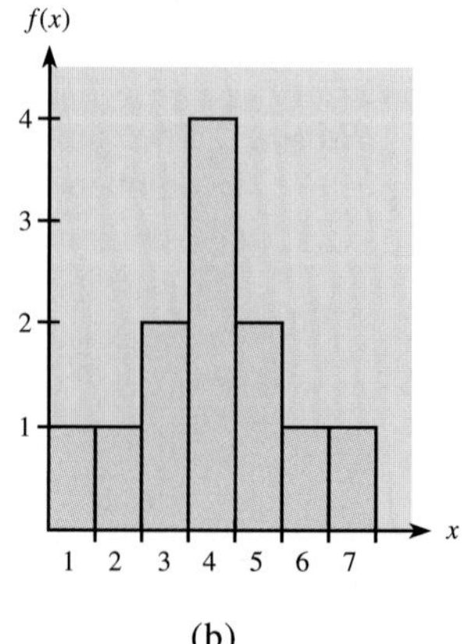

(b)

Solution

For the sample data of Figure 8.9(a), we have

$$\Sigma(x - \bar{x})^2 = (-3)^2 + (-3)^2 + (-3)^2 + (-1)^2 + (-1)^2 + 0^2 + 0^2 + 1^2 + 2^2 + 2^2 + 3^2 + 3^2$$

$$= 9 + 9 + 9 + 1 + 1 + 0 + 0 + 1 + 4 + 4 + 9 + 9 = 56$$

$$s^2 = \frac{\Sigma(x - \bar{x})^2}{n - 1} = \frac{56}{11} = 5.0909$$

$$s = \sqrt{s^2} = \sqrt{5.0909} = 2.26$$

For the sample data of Figure 8.9(b), we have

$$\Sigma(x - \bar{x})^2 = (-3)^2 + (-2)^2 + (-1)^2 + (-1)^2 + 0^2 + 0^2 + 0^2 + 0^2 + 1^2 + 1^2 + 2^2 + 3^2$$

$$= 9 + 4 + 1 + 1 + 0 + 0 + 0 + 0 + 1 + 1 + 4 + 9 = 30$$

$$s^2 = \frac{30}{11} = 2.7273$$

$$s = \sqrt{\frac{30}{11}} = 1.65$$

Technology Note

Scientific calculators and graphing calculators have keys or built-in functions that compute the mean and standard deviation of a set of data. The ease of using calculators to find the standard deviation makes it the preferred method.

EXAMPLE 12

The heights of the basketball team members mentioned earlier are 69, 70, 75, 69, 73, 78, 74, 73, 78, and 71 inches. Find the mean and standard deviation of this sample.

Solution

By using the formula or a calculator, we find that the mean is 73 and the sample standard deviation is

$$s = 3.333$$

We now return to the discussion of USF&G earnings begun in the Application Preview.

EXAMPLE 13

The accompanying table gives the net income (or loss), in millions of dollars, for the 11-year period 1983–1993 for the USF&G Corporation.

(a) Construct a bar graph to represent earnings over this period.
(b) What is the mean earnings for this 11-year period?
(c) What is the standard deviation of the earnings over this period?

Year	Net Income
1983	270
1984	−64
1985	−108
1986	296
1987	279
1988	247
1989	119
1990	−569
1991	−176
1992	28
1993	165

SOURCE: USF&G *1993 Annual Report*

Solution

(a) The bar graph displaying the earnings over this period is shown in Figure 8.10.

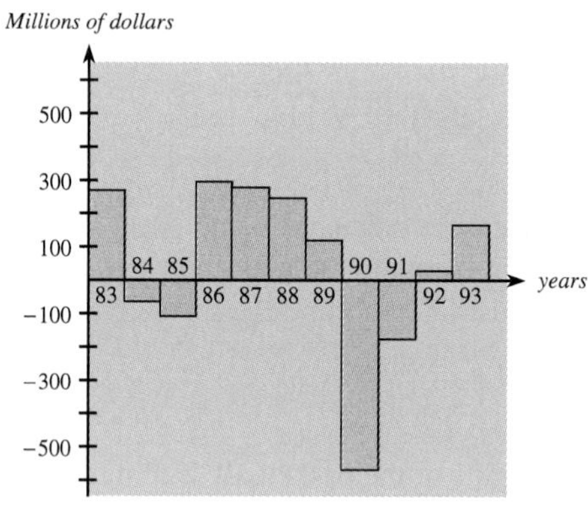

Figure 8.10

(b) The mean earnings over this period is

$$\bar{x} = \frac{\Sigma x}{n} = 44.27$$

(c) The standard deviation of the earnings over this period is

$$s = 262.77$$

This shows that the net income fluctuates widely.

CHECKPOINT SOLUTIONS

1. We can see that the mode (most frequent) of the salaries is \$26,000. Also, the figure has the 15 employees arranged in order, so we see that the median (middle) salary is \$30,000. The mean salary is

$$\begin{aligned}\bar{x} &= \frac{\Sigma(x \cdot f)}{\Sigma f}\\ &= \frac{75{,}000 + 59{,}000 + 35{,}000 + (34{,}000)2 + 31{,}000 + (30{,}000)2 + (26{,}000)7}{15}\\ &= \$34{,}000\end{aligned}$$

2. The mode gives the lowest salary for this company because the most people receive the lowest salary.
3. The mean gives the highest measure of central tendency because a few high salaries are added to many low salaries.
4. The median is most representative of the "average" salary because it is not affected by the few very high salaries or by the large number of very low salaries.

EXERCISE 8.4

Construct a frequency histogram for the data given in the frequency tables in Problems 1–6.

1.

Score	*Frequency*
12	2
13	3
14	4
15	3
16	1

2.

Score	*Frequency*
22	1
23	4
24	3
25	2
26	3

3.

x	3	4	5	6	7	8	9	10	11	12
$f(x)$	6	0	5	3	4	3	2	1	0	4

4.

x	5	6	8	9	11	12
$f(x)$	3	4	2	7	3	2

5.

Interval	*Frequency*
1–4	1
5–8	0
9–12	2
13–16	3
17–20	1

6.

Interval	*Frequency*
10–14	2
15–19	4
20–24	3
25–29	2
30–34	3

Construct a frequency histogram for the data in Problems 7 and 8.

7. 3, 2, 5, 6, 3, 2, 6, 5, 4, 2, 1, 6
8. 5, 4, 6, 5, 4, 6, 3, 6, 5, 4, 1, 7

Find the modes of the sets of scores in Problems 9–11.

9. 3, 4, 3, 2, 2, 3, 5, 7, 6, 2, 3
10. 5, 8, 10, 12, 5, 4, 6, 3, 5
11. 14, 17, 13, 16, 15, 12, 13, 12, 13

Find the medians of the sets of scores in Problems 12–15.

12. 1, 3, 6, 7, 5
13. 2, 1, 3, 4, 2, 1, 2
14. 4, 7, 9, 18, 36, 14, 12
15. 1, 0, 2, 1, 1, 0, 2, 14, 37

Find the mode, median, and mean of the scores in Problems 16–22.

16. 5, 7, 7, 7, 9, 12, 4
17. 3, 2, 1, 6, 8, 12, 14, 2
18. 5, 8, 7, 6, 1, 1, 31
19. 14, 17, 20, 31, 17, 42
20. 3.2, 3.2, 3.5, 3.7, 3.4
21. 2.8, 6.4, 5.3, 5.3, 6.8
22. 1.14, 2.28, 7.58, 6.32, 5.17
23. Use class marks to find the mean, mode, and median of the data in Problem 5.
24. Use class marks to find the mean, mode, and median of the data in Problem 6.

In Problems 25–28, find the range of the set of numbers given.

25. 3, 5, 7, 8, 2, 11, 6, 5
26. 5, 1, 3, 1.4, 6.3, 8
27. −1, 2, 4, 3, 6, 11, −3, 4, 2
28. 2, 3, 3, 3, 3, 3, 9

Find the mean, variance, and standard deviation of the following sets of sample data.

29. 5, 7, 1, 3, 0, 8, 6, 2
30. 7, 13, 5, 11, 8, 10, 9
31. 11, 12, 13, 14, 15, 16, 17
32. 4, 3, 6, 7, 8, 9, 12

In Problems 33–36, find the mean and standard deviation of the sample data in the table.

33.

x	1	2	3	4	5
$f(x)$	3	1	4	2	1

34.

x	3	4	5	6	8
$f(x)$	1	3	4	1	3

35.

x	3	4	5	6	7	8	9	10	11	12
$f(x)$	6	0	5	3	4	3	2	1	0	4

36.

x	5	6	7	8	9	10	11	12	13	14	15
$f(x)$	3	2	1	0	2	5	3	1	2	4	6

Applications

37. ***Death penalty*** The data in the table that follows give the total numbers of prisoners executed in the United States from 1950 to 1979 and the number of

black prisoners executed during the same period. Construct a histogram with x as the total number of executions (in groups of 10) and with $f(x)$ as the number of years in each group.

Year	Total	Blacks	Year	Total	Blacks
1950	82	42	1965	7	1
1951	105	47	1966	1	0
1952	83	47	1967	2	1
1953	62	31	1968	0	0
1954	81	42	1969	0	0
1955	76	32	1970	0	0
1956	65	43	1971	0	0
1957	65	31	1972	0	0
1958	49	28	1973	0	0
1959	49	33	1974	0	0
1960	56	35	1975	0	0
1961	42	22	1976	0	0
1962	47	19	1977	1	0
1963	21	8	1978	0	0
1964	15	7	1979	2	0

Source: U.S. Department of Justice, Bureau of Statistics

38. ***UPDATE*** The data in the table below give the total numbers of prisoners executed in the United States from 1983 to 1997 and the number of black prisoners executed during the same period. Construct a histogram with x as the years (in groups of 3) and with $f(x)$ as the number of black prisoners executed.

Year	Total	Blacks	Year	Total	Blacks
1983	5	4	1991	14	7
1984	21	13	1992	31	19
1985	18	11	1993	38	23
1986	18	11	1994	31	20
1987	25	13	1995	56	21
1988	11	6	1996	45	12
1989	16	8	1997	74	26
1990	23	16			

Source: U.S. Bureau of Justice Statistics

39. ***Salaries*** Suppose a company has 10 employees, 1 earning \$80,000, 1 earning \$60,000, 2 earning \$30,000, 1 earning \$20,000, and 5 earning \$16,000.
    (a) What is the mean salary for the company?
    (b) What is the median salary?
    (c) What is the mode of the salaries?

40. ***Salaries*** A survey revealed that of University of North Carolina alumni entering employment in 1984, those with a major in geography had the highest mean starting salary of all majors. Discuss this conclusion, using the fact that Michael Jordan was in this group and majored in geography.

Real estate Use the following information for Problems 41–43. Suppose you live in a neighborhood with a few expensive homes and many modest homes.

41. If you wanted to impress people with the neighborhood where you lived, which measure would you give as the "average" property value?
42. Which "average" would you cite to the property tax committee if you wanted to convince them that property values aren't very high in your neighborhood?
43. What measure of central tendency would give the most representative "average" property value for your neighborhood? Explain.

44. ***Salaries*** S & S Printing has 15 employees and 5 job classifications. Position and wages for these employees are given in the table.

Job	Number	Weekly Salary
Supervisor	1	\$1200
Printers	8	600
Camera-Darkroom	3	750
Secretaries	2	550
Delivery	1	820

    (a) What is the mean salary per job classification?
    (b) What is the mean salary per person?

45. ***Product testing*** A taxi company tests a new brand of tires by putting a set of four on each of three taxis. The following table indicates the number of miles the tires lasted. What is the mean number of miles the tires lasted?

Number of Miles	Number of Tires
30,000	4
32,000	2
34,000	3
36,000	2
38,000	1

46. ***Sales*** A new car with a \$19,000 list price can be bought for different prices from different dealers. In one city the car can be bought for \$18,200 from 2 dealers, for \$18,000 from 1 dealer, for \$17,800 from 3 dealers, for \$17,600 from 2 dealers, and for \$17,500 from 2 dealers. What are the mean and standard deviation of this sample of car prices?

47. ***Birth weights*** The birth weights (in kilograms) of a sample of 160 children are given in the following table. Find the mean and standard deviation of the weights.

Weight (kg)	*Frequency*	*Weight (kg)*	*Frequency*
2.0	4	3.5	26
2.3	12	3.8	20
2.6	20	4.1	16
2.9	26	4.4	10
3.2	20	4.7	6

48. ***Military presence*** The figure shows the number of military troops in Asia for each of ten countries. Add the number of U.S. troops in Japan to the number of Japanese troops, and add the number of U.S. troops in South Korea to the number of South Korean troops. Then compute the mean number of troops in Asia per country. How many of the countries have more troops than the mean number per country?

Military Troops in Asia

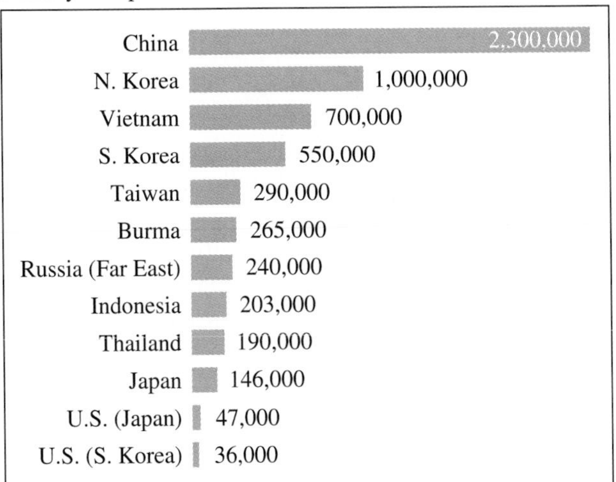

Sources: *Chicago Tribune*, Japan Self-Defense Agency, "Defense of Japan, 1994"; Knight-Ridder Tribune/Ken Marshall and Rick Tuma; *Savannah Morning News*, Jan. 5, 1995

49. ***Treasury bills*** The following table gives the annual percent return for Treasury bills from 1970 to 1989. Use a calculator or spreadsheet to compute the mean annual percent return over
(a) 1970–1979
(b) 1980–1989
(c) the two decades.
(d) Find the standard deviation of the returns over the two decades.

Year	*T-Bill Percent Return*	*Year*	*T-Bill Percent Return*
1970	6.5%	1980	11.2%
1971	4.4	1981	14.7
1972	3.8	1982	10.5
1973	6.9	1983	8.8
1974	8.0	1984	9.9
1975	5.8	1985	7.7
1976	5.1	1986	6.2
1977	5.1	1987	5.5
1978	7.2	1988	6.4
1979	10.4	1989	8.4

Source: *Mutual Fund Encyclopedia*

50. ***Earned capital*** The following table gives the percent earned total capital for Eli Lilly & Co. for the years 1987 to 1998. Use a calculator or spreadsheet to find the mean and standard deviation of the percent earned capital.

Year	*Percent Earned Total Capital*	*Year*	*Percent Earned Total Capital*
1987	18.8%	1993	25.4%
1988	21.5	1994	17.4
1989	23.6	1995	17.3
1990	30.4	1996	17.9
1991	25.0	1997	27.0
1992	25.8	1998	28.0

Source: Value Line Publishing Company, Oct. 31, 1997

51. ***Gold reserves*** American Barrick Resources Corporation has North America's largest gold reserves, 30.6 million ounces. The table gives the operating statistics for Barrick for the years 1991–1993. Use an electronic spreadsheet or other technology to find the mean of each category over the 3-year period.

Year Ended	*1991*	*1992*	*1993*
Gold production (thousand ounces)	789.8	1325.4	1632.0
Gold sales (thousand ounces)	787.7	1280.3	1633.7
Revenue per ounce ($)	438	422	409
Average spot price per ounce ($)	362	345	360
Operating costs per ounce ($)	205	164	170
Net income per ounce ($)	126	148	141

52. ***Gold reserves*** Use the gold production data for Barrick from the table in Problem 51 to find the standard deviation of gold production for 1991–1993.

53. ***Automotive testing*** The acceleration time from 0 to 60 mph and the braking distance from 60 mph to a stop for five 1995 subcompact cars are given in Table 8.4. Use the data in the accompanying table to find the mean and standard deviation of the braking distances of these cars.

TABLE 8.4 Acceleration and Braking—1995 Subcompacts

	0–60 mph	*Braking from 60 mph*
Honda	8.8 sec	165 ft
Mazda	9.1 sec	158 ft
Nissan	11.0 sec	148 ft
Plymouth	8.4 sec	142 ft
Toyota	10.2 sec	186 ft

54. ***Automotive testing*** Use Table 8.4 to find the mean and standard deviation of the acceleration time from 0 to 60 mph for the 5 cars.

8.5 Normal Probability Distribution

OBJECTIVES

- *To calculate the probability that a random variable following the standard normal distribution has values in a certain interval*
- *To convert normal distribution values to standard normal values (z-scores)*
- *To find the probability that normally distributed values lie in a certain interval*

APPLICATION PREVIEW

Discrete probability distributions are important, but they do not apply to many kinds of measurements. For example, the weights of people, the heights of trees, and the IQ scores of college students cannot be measured with whole numbers because each of them can assume any one of an infinite number of values on a measuring scale, and the values cannot be listed or counted, as is possible for a discrete random variable. The measurements mentioned above follow a special **continuous probability distribution** called the **normal distribution.**

The IQ scores of college students follow the normal distribution with a mean of 100 and a standard deviation of 15. We will see in this section that this information permits us to graph the distribution of the IQ scores and to find the probability that a person's IQ score lies within a given range of values.

The normal distribution is perhaps the most important probability distribution, because so many measurements that occur in nature follow this particular distribution.

The **normal distribution** has the following properties:

1. Its graph is a bell-shaped curve like that of Figure 8.11.* The graph is called the **normal curve.** It approaches but never touches the horizontal axis as it extends in both directions.

*Percentages shown are approximate.

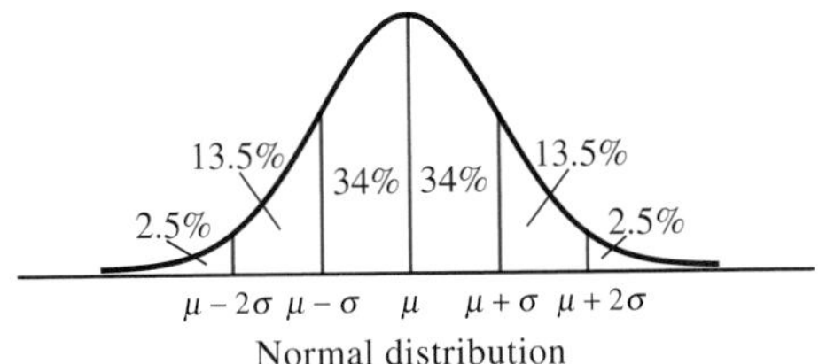

Figure 8.11

2. The curve is symmetric about a vertical center line that passes through the value that is the mean, the median, *and* the mode of the distribution. That is, the mean, median, and mode are the same for a normal distribution. (That is why they are all called average.)
3. A normal distribution is completely determined when its mean μ and its standard deviation σ are known.
4. Approximately 68% of all scores lie within 1 standard deviation of the mean. Approximately 95% of all scores lie within 2 standard deviations of the mean. More than 99.5% of all scores will lie within 3 standard deviations of the mean.

EXAMPLE 1

If IQ scores follow a normal distribution with mean 100 and standard deviation 15, what percentage of the scores will be

(a) between 100 and 115? (b) between 85 and 115?
(c) between 85 and 130? (d) between 70 and 130?
(e) greater than 130?

Solution

Because the mean is 100 and the standard deviation is 15, IQs of 85 and 115 are 1 standard deviation from 100, and IQs of 70 and 130 are 2 standard deviations from 100. The approximate percentages associated with these values are shown on the graph of a normal distribution in Figure 8.12. We can use this graph to answer the questions.

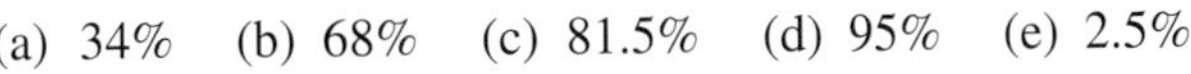
(a) 34% (b) 68% (c) 81.5% (d) 95% (e) 2.5%

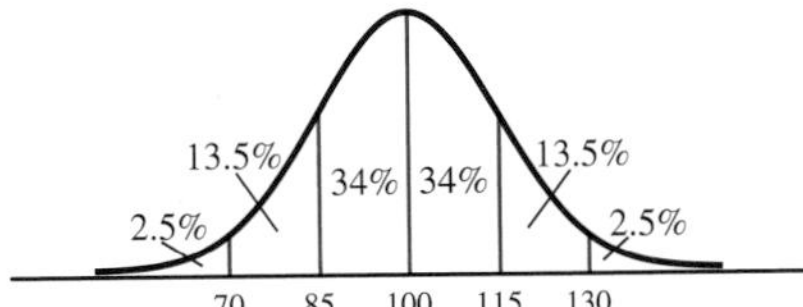

Figure 8.12

The total area under the normal curve is 1. The area under the curve from value x_1 to value x_2 represents the percentage of the scores that lie between x_1 and x_2. Thus the *area* under the curve from x_1 to x_2 *represents the probability* that a score chosen at random will lie between x_1 and x_2. Thus, in Example 1, the probability that a score chosen at random will lie between 85 and 115 is .68, and the probability that it will lie between 70 and 130 is .95.

CHECKPOINT

1. The mean of a normal distribution is 25 and the standard deviation is 3.
   (a) What percentage of the scores will be between 22 and 28?
   (b) What is the probability that a score chosen at random will be between 22 and 28?

We have seen that 34% of the scores lie between 100 and 115 in the normal distribution graph in Figure 8.12. As we have mentioned, this also means that the area under the curve from $x = 100$ to $x = 115$ is .34 and that the probability that a score chosen at random from this normal population lies between 100 and 115 is .34. We can write this as

$$\Pr(100 \leq x \leq 115) = .34$$

Because the probability of obtaining a score from a normal distribution is always related to how many standard deviations the score is away from the mean, it is desirable to convert all scores from a normal distribution to **standard scores,** or ***z*-scores.** The *z*-score for any score x is found by determining how many standard deviations x is from the mean μ. This is done using the formula

$$z = \frac{x - \mu}{\sigma}$$

If we convert all the scores from any normal distribution to *z*-scores, the distribution of *z*-scores will always be a normal distribution with mean 0 and standard deviation 1. This distribution is called the **standard normal distribution.** Figure 8.13 shows the graph of the standard normal distribution, with approximate percentages shown. The total area under the curve is 1, with .5 on either side of the mean 0.

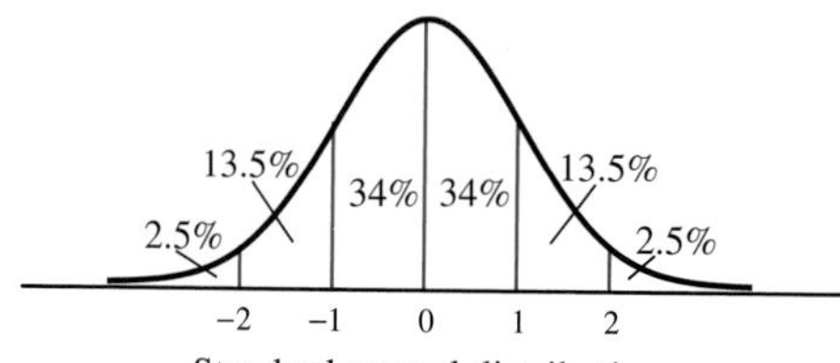

Figure 8.13 Standard normal distribution

By comparing Figure 8.13 with Figure 8.11, we see that each unit from 0 in the standard normal distribution corresponds to 1 standard deviation from the mean of the normal distribution.

We can use a table to determine more accurately the area under the standard normal curve between two *z*-scores. Table I in the appendix gives the area under the normal curve from $z = 0$ to $z = z_0$, for values of z_0 from 0 to 3. As with the normal curve, the area under the curve from $z = 0$ to $z = z_0$ is the probability that a *z*-score lies between 0 and z_0.

EXAMPLE 2

(a) Find the area under the standard normal curve from $z = 0$ to $z = 1.50$.
(b) Find $\Pr(0 \leq z \leq 1.50)$.

Solution

(a) Looking in the column headed by z in Table I, we see 1.50. Across from 1.50 in the column headed by A is .4332. Thus the area under the standard normal curve between $z = 0$ and $z = 1.50$ is $A_{1.50} = .4332$. (See Figure 8.14.)

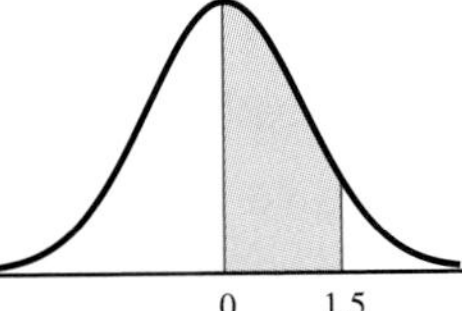

Figure 8.14

(b) The area under the curve from 0 to 1.50 equals the probability that z lies between 0 and 1.50. Thus

$$\Pr(0 \le z \le 1.50) = A_{1.50} = .4332$$

The following facts, which are direct results of the symmetry of the normal curve about m, are useful in calculating probabilities using Table I.

1. $\Pr(-z_0 \le z \le 0) = \Pr(0 \le z \le z_0) = A_{z_0}$
2. $\Pr(z \ge 0) = \Pr(z \le 0) = .5$

EXAMPLE 3

Find the following probabilities for the random variable z with standard normal distribution.

(a) $\Pr(-1 \le z \le 0)$ (b) $\Pr(-1 \le z \le 1.5)$ (c) $\Pr(1 \le z \le 1.5)$
(d) $\Pr(z > 2)$ (e) $\Pr(z < 1.35)$

Solution

(a) $\Pr(-1 \le z \le 0) = \Pr(0 \le z \le 1) = A_1 = .3413$
(b) We find $\Pr(-1 \le z \le 1.5)$ by using A_1 and $A_{1.5}$ as follows:

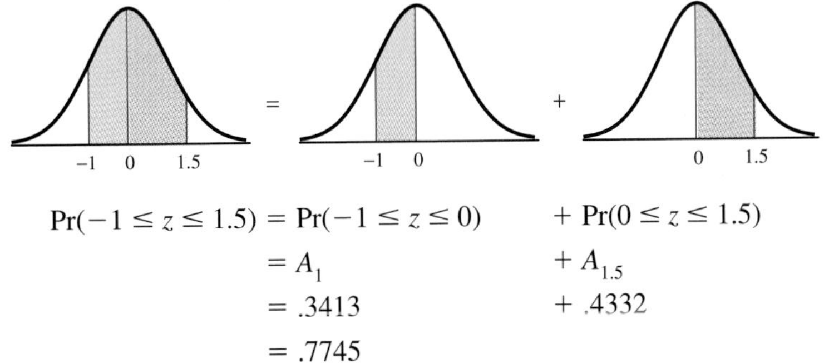

$$\begin{aligned}\Pr(-1 \le z \le 1.5) &= \Pr(-1 \le z \le 0) &&+ \Pr(0 \le z \le 1.5)\\ &= A_1 &&+ A_{1.5}\\ &= .3413 &&+ .4332\\ &= .7745\end{aligned}$$

(c) We find $\Pr(1 \le z \le 1.5)$ by using A_1 and $A_{1.5}$ as follows:

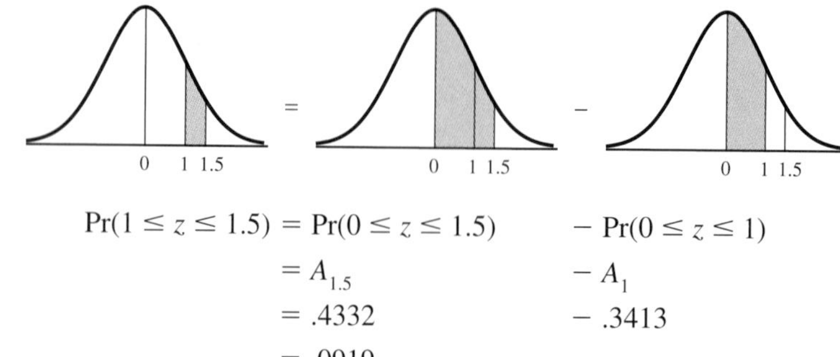

$$\begin{aligned}\Pr(1 \le z \le 1.5) &= \Pr(0 \le z \le 1.5) &&- \Pr(0 \le z \le 1)\\ &= A_{1.5} &&- A_1\\ &= .4332 &&- .3413\\ &= .0919\end{aligned}$$

(d) We find $\Pr(z > 2)$ by using .5 and A_2 as follows:

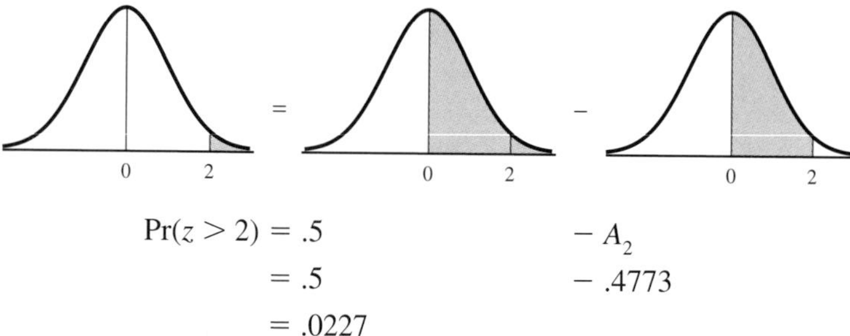

$$\begin{aligned}\Pr(z > 2) &= .5 &&- A_2\\ &= .5 &&- .4773\\ &= .0227\end{aligned}$$

(e) We find $\Pr(z < 1.35)$ by using .5 and $A_{1.35}$ as follows:

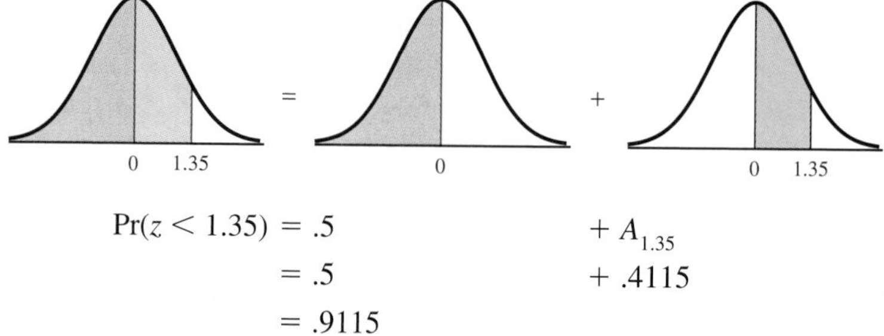

$$\begin{aligned}\Pr(z < 1.35) &= .5 &&+ A_{1.35}\\ &= .5 &&+ .4115\\ &= .9115\end{aligned}$$

CHECKPOINT

2. For the standard normal distribution, find the following probabilities.
   (a) $\Pr(0 \le z \le 2.5)$
   (b) $\Pr(z > 2.5)$
   (c) $\Pr(z \le 2.5)$

If a population follows a normal distribution, but with mean and/or standard deviation different from 0 and 1, respectively, it is desirable to convert all scores from a normal distribution to **standard scores,** or ***z*-scores.** The z-score for any score x is found by determining how many standard deviations x is from the mean μ. This is done using the formula

$$z = \frac{x - \mu}{\sigma}$$

Thus

$$\Pr(a \le x \le b) = \Pr(z_a \le z \le z_b)$$

where

$$z_a = \frac{a-\mu}{\sigma} \quad \text{and} \quad z_b = \frac{b-\mu}{\sigma}$$

and we can use Table I, the table of standard scores (z-scores), to find these probabilities. For example, the normal distribution of IQ scores has mean $\mu = 100$ and standard deviation $\sigma = 15$. Thus the z-score for $x = 115$ is

$$z = \frac{115-100}{15} = 1$$

A z-score of 1 indicates that 115 is 1 standard deviation above the mean (see Figure 8.15), and

$$\Pr(100 \le x \le 115) = \Pr(0 \le z \le 1) = .3413$$

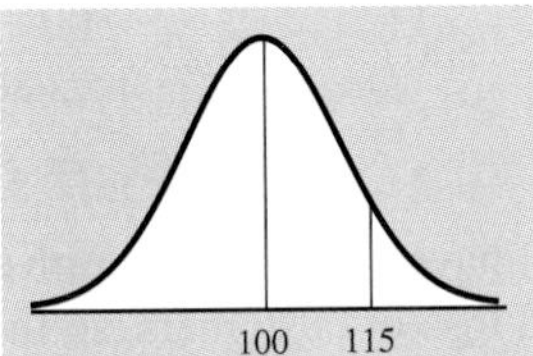

Figure 8.15

EXAMPLE 4

If the mean height of a population of students is $\mu = 68$ in. with standard deviation $\sigma = 3$ in., what is the probability that a person chosen at random from the population will be between

(a) 68 and 74 in. tall? (b) 65 and 74 in. tall?

Solution

(a) To find $\Pr(68 \le x \le 74)$, we convert 68 and 74 to z-scores.

$$\text{For 68: } z = \frac{68-68}{3} = 0 \qquad \text{For 74: } z = \frac{74-68}{3} = 2$$

Thus

$$\Pr(68 \le x \le 74) = \Pr(0 \le z \le 2) = A_2 = .4773$$

(b) The z-score for 65 is

$$z = \frac{65-68}{3} = -1$$

and the z-score for 74 is 2. Thus

$$\begin{aligned}\Pr(65 \le x \le 74) &= \Pr(-1 \le z \le 2)\\ &= A_1 + A_2\\ &= .3413 + .4773 = .8186\end{aligned}$$

CHECKPOINT

3. For the normal distribution with mean 70 and standard deviation 5, find the following probabilities.
   (a) $\Pr(70 \leq x \leq 77)$ (b) $\Pr(68 \leq x \leq 73)$ (c) $\Pr(x > 75)$

EXAMPLE 5

Recessionary pressures have forced an electrical supplies distributor to consider reducing its inventory. It has determined that, over the past year, the daily demand for 48-in. fluorescent light fixtures is normally distributed with a mean of 432 and a standard deviation of 86. If the distributor decides to reduce its daily inventory to 500 units, what percentage of the time will its inventory be insufficient to meet the demand? Will this cost the distributor business?

Solution

The distribution is normal, with $\mu = 432$ and $\sigma = 86$. The probability that more than 500 units will be demanded is

$$\Pr(x > 500) = \Pr\left(z > \frac{500 - 432}{86}\right) = \Pr(z > .79)$$
$$= .5 - A_{.79} = .5 - .2852 = .2148$$

The distributor will have an insufficient inventory 21.48% of the time. This will probably result in a loss of customers, so the distributor should not decrease the inventory to 500 units per day.

EXAMPLE 6

The function

$$f(x) = \frac{1}{\sigma\sqrt{2\pi}} e^{-(x-\mu)^2/(2\sigma^2)}$$

gives the equation for a normal curve with mean μ and standard deviation σ.

(a) Graph the following normal curves to see how different means yield different distributions.

$$\mu = 0, \sigma = 1, y = \frac{1}{\sqrt{2\pi}} e^{-x^2/2} \quad \text{and}$$
$$\mu = 3, \sigma = 1, y = \frac{1}{\sqrt{2\pi}} e^{-(x-3)^2/2}$$

If the standard deviations are the same, how does increasing the size of the mean affect the graph of the distribution?

(b) To see how normal distributions differ for different standard deviations, graph the normal distributions with

$$\mu = 0, \sigma = 3 \quad \text{and}$$
$$\mu = 0, \sigma = 2$$

What happens to the height of the curve as the standard deviation decreases? Do the data appear to be clustered nearer the mean when the standard deviation is smaller?

Solution

(a) The graphs are shown in Figure 8.16. The graphs are the same size but are centered over different means. A different mean causes the graph to shift to the right or left.

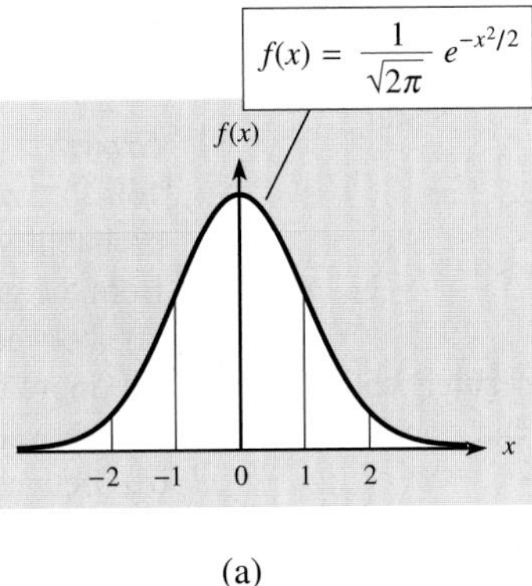

(a)

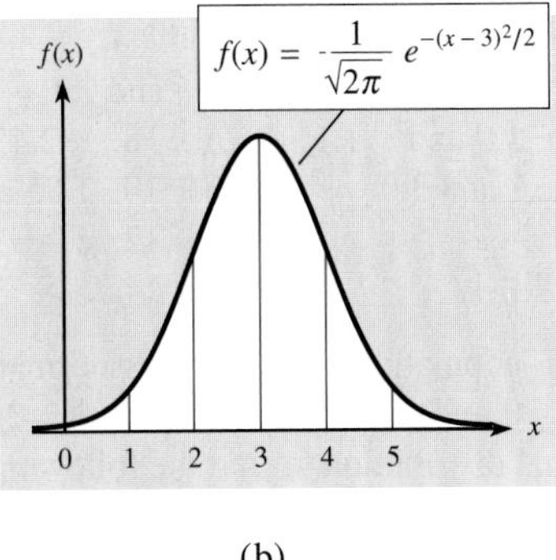

(b)

Figure 8.16

(b) The graphs are shown in Figure 8.17. The height of the graph will be larger if the standard deviation decreases. Yes, the data will be clustered closer to the mean when the standard deviation is smaller.

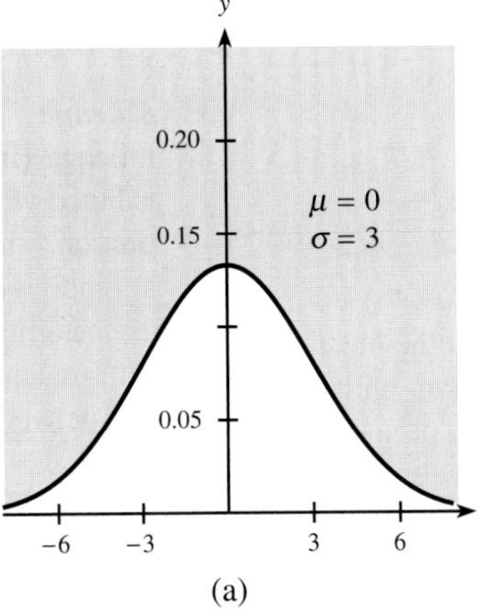

(a)

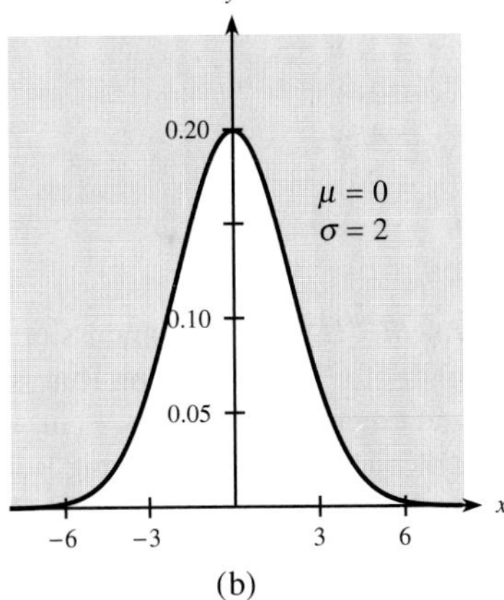

(b)

Figure 8.17

CHECKPOINT SOLUTIONS

1. (a) These scores lie within 1 standard deviation of the mean, 25, so 68% lie in this range.
   (b) The percentage of the scores in this range is 68%, so the probability that a score chosen at random from this normal distribution will be in this range is .68.
2. (a) $\Pr(0 \leq z \leq 2.5) = A_{2.5} = .4938$
   (b) $\Pr(z > 2.5) = .5 - A_{2.5} = .5 - .4938 = .0062$
   (c) $\Pr(z \leq 2.5) = .5 + A_{2.5} = .5 + .4938 = .9938$
3. (a) $\Pr(70 \leq x \leq 77) = \Pr(0 \leq z \leq 1.4) = A_{1.4} = .4192$
   (b) $\Pr(68 \leq x \leq 73) = \Pr(-.4 \leq z \leq .6) = A_{.4} + A_{.6}$
   $= .1554 + .2258 = .3812$
   (c) $\Pr(x > 75) = \Pr(z > 1) = .5 - A_1 = .5 - .3413 = .1587$

EXERCISE 8.5

Use Table I in the appendix to find the probability that a z-score from the standard normal distribution will lie within each of the intervals in Problems 1–14.

1. 0 and 1.8
2. 0 and 2.4
3. -1.8 and 0
4. -3 and 0
5. -1.5 and 2.1
6. -1.25 and 3
7. -1.9 and -1.1
8. -2.45 and -1.45
9. 2.1 and 3.0
10. 1.85 and 2.85
11. $z > 2$
12. $z > 1.5$
13. $z < 1.2$
14. $z > -1.6$

Say a population of scores x is normally distributed with $\mu = 20$ and $\sigma = 5$. In Problems 15–26, use the standard normal distribution to find the following.

15. $\Pr(20 \le x \le 22.5)$
16. $\Pr(20 \le x \le 21.25)$
17. $\Pr(13.75 \le x \le 20)$
18. $\Pr(18.5 \le x \le 20)$
19. $\Pr(12.25 \le x \le 21.25)$
20. $\Pr(18.5 \le x \le 27.75)$
21. $\Pr(13.75 \le x \le 18.75)$
22. $\Pr(21.25 \le x \le 25)$
23. $\Pr(x < 18)$
24. $\Pr(x < 14)$
25. $\Pr(x > 17)$
26. $\Pr(x < 24)$

Applications

27. ***Growth*** If the Fish Commission states that the mean length of all fish in Spring Run is $\mu = 15$ cm, with a standard deviation of $\sigma = 4$ cm, what is the probability that a fish caught in Spring Run will be between
   (a) 15 and 19 cm long? (b) 10 and 15 cm long?

28. ***Butterfat*** A quart of Parker's milk contains a mean of 39 grams of butterfat, with a standard deviation of 2 grams. Find the probability that a quart of this brand of milk chosen at random will contain between
   (a) 39 and 43 grams of butterfat.
   (b) 36 and 39 grams of butterfat.

29. ***Growth*** The heights of a certain species of plant are normally distributed with a mean $\mu = 20$ cm and $\sigma =$ 4 cm. What is the probability that a plant chosen at random will be between 10 and 30 cm tall?

30. ***Mating calls*** The mean duration of the mating call of a population of tree toads is 189 milliseconds (msec), with a standard deviation of 32 msec. What proportion of these calls would be expected to last between
   (a) 157 and 221 msec? (b) 205 and 253 msec?

31. ***Growth*** The mean weight of a group of boys is 160 lb, with a standard deviation of 15 lb. Find the probability that one of the boys picked at random from the group weighs
   (a) between 160 and 181 lb.
   (b) more than 190 lb.
   (c) between 181 and 190 lb.
   (d) between 130 and 181 lb.

32. ***Testing*** The Scholastic Aptitude Test (SAT) scores in mathematics at a certain high school are normally distributed with a mean of 500 and a standard deviation of 100. What is the probability that an individual chosen at random has a score
   (a) greater than 700?
   (b) less than 300?
   (c) between 550 and 600?

33. ***Mileage*** A certain model of automobile has its gas mileage (in miles per gallon, or mpg) normally distributed with a mean of 28 mpg and a standard deviation of 4 mpg. Find the probability that a car selected at random has gas mileage
   (a) less than 22 mpg.
   (b) greater than 30 mpg.
   (c) between 26 and 30 mpg.

34. ***Quality control*** A machine precision-cuts tubing such that $\mu = 40.00$ inches and $\sigma = 0.03$ inches. If a piece of tubing is selected at random, find the probability that it measures
   (a) greater than 40.05 inches.
   (b) less than 40.045 inches.
   (c) between 39.95 and 40.05 inches.

35. ***Blood pressure*** Systolic blood pressure for a group of women is normally distributed with a mean of 120 and a standard deviation of 12. Find the probability that a woman selected at random has blood pressure
   (a) greater than 140.
   (b) less than 110.
   (c) between 110 and 130.

36. ***Industrial waste*** In a certain state, the daily amounts of industrial waste are normally distributed with a mean of 8000 tons and a standard deviation of 2000 tons. Find the probability that on a randomly selected day the amount of industrial waste is
   (a) greater than 11,000 tons.
   (b) less than 6000 tons.
   (c) between 9000 and 11,000 tons.

37. ***Reaction time*** Reaction time is normally distributed with a mean of 0.7 second and a standard deviation of 0.1 second. Find the probability that an individual selected at random has a reaction time
   (a) greater than 0.9 second.
   (b) less than 0.6 second.
   (c) between 0.6 and 0.9 second.

38. ***Inventory*** Suppose that the electrical supplies distributor in Example 5 wants to reduce its daily inventory but wants to limit to 5% the number of days that it will have an insufficient inventory of 48-in. fluorescent light bulbs to meet daily demand. At what level should it maintain its daily inventory? (Recall that the daily demand over the past year had a mean of 432 and a standard deviation of 86; test values of 550, 575, and 600 units.)

KEY TERMS AND FORMULAS

Section	Key Term	Formula
8.1	Binomial probability experiments	$\Pr(x) = \binom{n}{x} p^x q^{n-x}$
8.2	Discrete probability distributions	
	Random variable	
	Probability density histogram	
	Expected value	$E(x) = \sum x\Pr(x)$
	Mean	$\mu = E(x) = \sum x\Pr(x)$
	Variance	$\sigma^2 = \sum (x - \mu)^2 \Pr(x)$
	Standard deviation	$\sigma = \sqrt{\sigma^2}$
8.3	Binomial probability distribution	
	Mean	$\mu = np$
	Standard deviation	$\sigma = \sqrt{npq}$
	Binomial formula	$(a + b)^n = {}_nC_n a^n + {}_nC_{n-1} a^{n-1} b + \cdots + {}_nC_1 a b^{n-1} + {}_nC_0 b^n$
8.4	Descriptive statistics	
	Frequency histogram	
	Mode	
	Median	
	Mean	$\bar{x} = \dfrac{\Sigma x}{n}$

Section	Key Term	Formula
	Range	
	Sample variance	$s^2 = \dfrac{\Sigma(x-\bar{x})^2}{n-1}$
	Sample standard deviation	$s = \sqrt{\dfrac{\Sigma(x-\bar{x})^2}{n-1}}$
8.5	Normal distribution Standard normal distribution	
	z-scores	$z = \dfrac{x-\mu}{\sigma}$

REVIEW EXERCISES

Section 8.1

1. Suppose 20% of the population are victims of crime. Out of 5 people selected at random, find the probability that 2 are crime victims.
2. If one person in 100,000 develops a certain disease, then calculate
   (a) Pr(exactly one person in 100,000 has the disease).
   (b) Pr(at least one person in 100,000 has the disease).
3. If a die is rolled 4 times, what is the probability that a number greater than 4 is rolled at least 2 times?

Section 8.2

4. For the following probability distribution, find the expected value $E(x)$.

x	Pr(x)
1	0.2
2	0.3
3	0.4
4	0.1

Determine whether each function or table in Problems 5–8 represents a discrete probability distribution.

5. $\Pr(x) = x/15$, $x = 1, 2, 3, 4, 5$
6. $\Pr(x) = x/55$, $x = 1, 2, 3, 4, 5, 6, 7, 8, 9$

7.

x	Pr(x)
3	1/6
4	1/3
5	1/3
6	1/6

8.

x	Pr(x)
2	1/2
3	1/4
4	−1/4
5	1/2

9. For the following probability distribution, find the expected value $E(x)$.

x	1	2	3	4
Pr(x)	.4	.3	.2	.1

10. For the discrete probability distribution described by $\Pr(x) = x/16$, $x = 1, 2, 3, 4, 6$, find the
    (a) mean. (b) variance. (c) standard deviation.
11. For the probability distribution described by the table, find the
    (a) mean.
    (b) variance.
    (c) standard deviation.

x	Pr(x)
1	1/12
2	1/6
3	1/3
4	5/12

Section 8.3

12. A coin has been altered so that the probability that a head will occur is 2/3. If the coin is tossed 6 times, give the mean and standard deviation of the distribution of the number of heads.
13. Refer to Problem 1.
    (a) What is the expected number of people in this sample that are victims?
    (b) Find the probability that exactly the expected number of people in this sample are victims of crime.
14. Expand $(x + y)^5$.

Section 8.4

Consider the data in the following frequency table, and use the data in Problems 15–18.

Score	*Frequency*
1	4
2	6
3	8
4	3
5	5

15. Construct a frequency histogram for the data.
16. What is the mode of the data?
17. What is the mean of the data?
18. What is the median of the data?
19. Construct a frequency histogram for the following set of test scores: 14, 16, 15, 14, 17, 16, 12, 12, 13, 14.
20. What is the median of the scores in Problem 19?
21. What is the mode of the scores in Problem 19?
22. What is the mean of the scores in Problem 19?
23. Find the mean, variance, and standard deviation of the following data: 4, 3, 4, 6, 8, 0, 2.
24. Find the mean, variance, and standard deviation of the following data: 3, 2, 1, 5, 1, 4, 0, 2, 1, 1.

Section 8.5

25. What is the area under the standard normal curve between $z = -1.6$ and $z = 1.9$?
26. If z is a standard normal score, find $\Pr(-1 \leq z \leq -.5)$.
27. Find $\Pr(1.23 \leq z \leq 2.55)$.
28. If a variable x is normally distributed with $\mu = 25$ and $\sigma = 5$, find $\Pr(25 \leq x \leq 30)$.
29. If a variable x is normally distributed with $\mu = 25$ and $\sigma = 5$, find $\Pr(20 \leq x \leq 30)$.
30. If a variable x is normally distributed with $\mu = 25$ and $\sigma = 5$, find $\Pr(30 \leq x \leq 35)$.

Applications

Section 8.1

31. ***Genetics*** Suppose the probability that a certain couple will have a blond child is 1/4. If they have 6 children, what is the probability that 2 of them will be blond?

32. ***Sampling*** Suppose 70% of a population opposes a proposal and a sample of size 5 is drawn from the population. What is the probability that the majority of the sample will favor the proposal?

Section 8.2

33. ***Lotteries*** A state lottery pays $500 to anyone who selects the correct 3-digit number from a random drawing. If it costs $1 to play, then the probability distribution for the amount won is given in the table. What are the expected winnings of a person who plays?

x	$\Pr(x)$
-1	0.999
499	0.001

Section 8.3

34. ***Cancer testing*** The probability that a man with prostate cancer will test positive with the prostate specific antigen (PSA) test is .91. If 500 men with prostate cancer are tested, how many would we expect to test positive?

35. ***Fraud*** A company selling substandard drugs to developing countries sold 2,000,000 capsules with 60,000 of them empty (Source: "60 Minutes," Jan. 18, 1998). If a person gets 100 randomly chosen capsules from this company, what is the expected number of empty capsules that this person will get?

36. ***Racing*** Suppose you paid $2 for a bet on a horse and as a result you will win $100 if your horse wins a given race. Find the expected value of the bet if the odds that the horse will win are 3 to 12.

Section 8.4

Farm families The distribution of farm families (as a percentage of the population) in a 50-county survey is given in the table. Use these data in Problems 37–39.

Percentage	*Number of Counties*
10–19	5
20–29	16
30–39	25
40–49	3
50–59	1

37. Make a frequency histogram for these data.
38. Find the mean percentage of farm families in these 50 counties.
39. Find the standard deviation of the percentage of farm families in these 50 counties.

Section 8.5

40. ***Testing*** Suppose the mean SAT score for students admitted to a university is 1000, with a standard deviation of 200. Suppose that a student is selected at random. Find the probability that the student's SAT score is
    (a) between 1000 and 1400.
    (b) between 1200 and 1400.
    (c) greater than 1400.

41. ***Net worth*** The mean net worth of the residents of Sun City, a retirement community, is \$611,000 (Source: *The Island Packet,* August 22, 1998). If their net worth is normally distributed with a standard deviation of \$96,000, what percentage of the residents have a net worth between \$700,000 and \$800,000?

CHAPTER TEST

1. If a coin is "loaded" so that the probability that a head will occur on each toss is 1/3, what is the probability that
   (a) 3 heads will result from 5 tosses of the coin?
   (b) at least 3 heads will occur in 5 tosses of the coin?
2. For the coin of Problem 1, what is the expected number of heads if the coin is tossed 12 times?
3. For the coin in Problem 2, find the mean, variance, and standard deviation of the distribution of heads.
4. Find the expected value of the variable x if the probability distribution is given in the table.

x	3	4	5	6	7	8
Pr(x)	.2	.3	.1	.1	.2	.1

5. For the probability distribution described by the table below, find the mean, variance, and standard deviation.

x	10	12	15	18	20	25
Pr(x)	.1	.3	.1	.2	.1	.2

6. For the data in the table below, find the mean, median, and mode.

Score	*Frequency*
20	5
21	7
22	3
23	4
24	2

7. If the variable x is normally distributed with a mean of 16 and a standard deviation of 6, find the following probabilities.
   (a) $\Pr(14 \le x \le 22)$
   (b) $\Pr(x \le 22)$
   (c) $\Pr(22 \le x \le 24)$
8. For the normal distribution with mean 70 and standard deviation 12, find:
   (a) $\Pr(73 \le x \le 97)$
   (b) $\Pr(65 \le x \le 84)$
   (c) $\Pr(x > 84)$

The following table gives the age of the householder in primary families. Use this table in Problems 9 and 10.

Age of Householder (years)	*Number (thousands)*
15–24	3,079
25–34	14,082
35–44	18,274
45–54	13,746
55–64	8,895
65+	11,236

SOURCE: U.S. Bureau of the Census

9. Construct a histogram for the data.
10. Use class marks and the given number (frequency) in each class to find the mean and standard deviation of the age of the householders. Use 72.5 as the class mark for the 65+ class.
11. The following table gives the 1995 resident population of the United States. Use class marks and the given number (frequency) in each class to find the mean of the age.

Age (years)	*Number (thousands)*	*Age (years)*	*Number (thousands)*
Under 5	19,591	50–54	13,630
5–9	19,220	55–59	11,085
10–14	18,915	60–64	10,046
15–19	18,065	65–69	9,928
20–24	17,882	70–74	8,831
25–29	19,005	75–79	1,652
30–34	21,868	80–84	4,464
35–39	22,249	85–89	2,321
40–44	20,219	90–94	991
45–49	17,449	95–99	263

12. The table below gives the number of cars (in millions) in use by age. Use the class marks and number to find the average age of cars in use in 1980 and in 1994. What has happened to the average age? What may be causing this?

	Class Mark	*1980 Number (millions)*	*1994 Number (millions)*
0–5	3	52.3	45.4
6–8	7	25.2	27.7
9–11	10	14.6	25.1
12+	15	12.5	31.4

Source: *Statistical Abstract of the United States,* 1996

13. Suppose that 2% of the computer chips shipped to an assembly plant are defective.
    (a) What is the probability that 10 of 100 chips chosen from the shipment will be defective?
    (b) What is the expected number of defective chips in the shipment if it contains 1500 chips?

The following table gives the percentage of women who become pregnant during 1 year of use with different types of contraceptives. The "Typical Use" column includes failures from incorrect or inconsistent use.

Failure Rates Over 1 Year

Contraceptive Type	*Correct and Consistent Use*	*Typical Use*
Spermicides	6%	21%
Diaphragm	6	18
Male latex condom	3	12
Unprotected	N/A	85

Source: Hatcher, R. A., Trussel, J., Stewart, F., *et al., Contraceptive Technology,* 1994, Irvington Pub.

14. If a group of 30 women use diaphragms correctly and consistently, how many of these 30 women could be expected to become pregnant in a year?
15. If a group of 30 women use diaphragms typically, how many of these 30 women could be expected to become pregnant in a year?
16. If spermicides and male latex condoms are used correctly by a group of 30 couples, how many of the 30 women could be expected to become pregnant in a year? Assume that these uses are independent events.
17. A new light bulb has an expected life of 18,620 hours, with a standard deviation of 4012 hours. Find the probability that one of these bulbs chosen at random will last
    (a) less than 10,000 hours.
    (b) more than 24,000 hours.
    (c) between 15,000 and 21,000 hours.

Extended Applications Group Projects

I. Lotteries

One of the reasons why the July 29, 1998, Powerball Lottery (sponsored by ten states and the District of Columbia) was so large ($295.7 million) is that the odds of winning are so small and the pot continues to grow until someone wins. To win the Powerball Jackpot, a player's 5 game balls and 1 red powerball must match those chosen. On November 2, 1997, the game was changed to lower the probability of winning, as shown in the following table.

Type of Game	*Old Powerball*	*New Powerball*
Game	5-of-45	5-of-49
Powerballs	1-of-45	1-of-42

Source: Lottery America

1. Use the information in the table to find the probability of winning the old 5-of-45 game and that of winning the new 5-of-49 game. Note that the order in which the first five numbers occur does not matter, and the Powerball is not used in this game.
2. Use the information in the table to find the probability of winning the Powerball Jackpot in the old game and in the new game. Note that the order in which the first five numbers occur does not matter. Approximate odds are given at http://bud.ica.net/powerball/powerodd.htm. Use these approximations to check your results for the new game.

The minimum payoff for the Jackpot has increased from $5 million to $10 million, and the table below gives the payoffs for each game.

Payoffs

Match	*Old Game Payout*	*New Game Payout*
5 + Powerball	Jackpot	Jackpot
5	$100,000	$100,000
4 + Powerball	5,000	5,000
4	100	100
3 + Powerball	100	100
3	5	7
2 + Powerball	5	7
1 + Powerball	2	4
Powerball	1	3

Source: Lottery America

3. Assume that a share of the Jackpot is \$10 million, and find the expected winnings from one ticket in the old game and in the new game. The sponsors of the Powerball game claim that making the probability lower improves the game. How do they justify that claim?

II. Statistics in Medical Research; Hypothesis Testing

Bering Research Laboratories is involved in a national study organized by the National Centers for Disease Control to determine whether a new birth control pill affects the blood pressure of women who are using it. A random sample of 100 women using the pill was selected, and the participants' blood pressures were measured. The mean blood pressure was $\bar{x} = 112.5$. If the blood pressure of all women is normally distributed with mean $\mu = 110$ and standard deviation $\sigma = 12$, is the mean of this sample sufficiently far from the mean of the population of all women to indicate that the birth control pill affects blood pressure?

To answer this question, researchers at Bering Labs first need to know how the means of samples drawn from a normal population are distributed. If all possible samples of size n are drawn from a normal population that has mean μ and standard deviation σ, then the distribution of the means of these samples will also be normally distributed with mean μ but with standard deviation given by

$$\sigma_{\bar{x}} = \frac{\sigma}{\sqrt{n}}$$

To conclude whether women using this birth control pill have blood pressure that is significantly different from that of all women in the population, researchers must decide whether the mean $\bar{x}$ of their sample is so far from the population mean μ that it is unlikely that the sample was chosen from the population.

To use a familiar analogy, the statistical proof that $\bar{x}$ does not come from the population of all women is like the trial of an accused criminal. In a criminal trial, we assume that the person who stands accused is innocent unless there is evidence "beyond all reasonable doubt" that the person is guilty. In this statistical test, the researchers will assume that the sample has been drawn from the population of women with mean $\mu = 110$ unless the blood pressures collected in the sample are so different from that of the population that it is not reasonable to make this assumption. In statistics, we can say that the assumption is unreasonable if the probability that the sample mean, $\bar{x} = 112.5$, could be drawn from this population, with mean 110, is less than .05. If this probability is less than .05, we say that the sample is drawn from a population with a mean different from 110, which strongly suggests that the women who have taken the birth control pill have, on average, different blood pressure from the women who are not using the pill. That is, there is statistically significant evidence that the pill affects blood pressure of women.

Because 95% of all normally distributed data points lie within 2 standard deviations of the mean, researchers at Bering Labs need to determine whether the sample mean $\bar{x} = 112.5$ is more than 2 standard deviations from $\mu = 110$. To determine how many standard deviations $\bar{x} = 112.5$ is from $\mu = 110$, they compute the z-score for $\bar{x}$ with the formula

$$z_{\bar{x}} = \frac{\bar{x} - \mu}{\sigma/\sqrt{n}}$$

Warm-up

Prerequisite Problem Type	For Section	Answer	Section for Review
If $f(x) = \dfrac{x^2 - x - 6}{x + 2}$, then find (a) $f(-3)$ (b) $f(-2.5)$ (c) $f(-2.1)$ (d) $f(-2)$	9.1–9.9	(a) -6 (b) -5.5 (c) -5.1 (d) undefined	1.2 Functional notation
Factor: (a) $x^2 - x - 6$ (b) $x^2 - 4$ (c) $x^2 + 3x + 2$	9.1 9.7	(a) $(x + 2)(x - 3)$ (b) $(x - 2)(x + 2)$ (c) $(x + 1)(x + 2)$	0.6 Factoring
Write as a power: (a) $\sqrt{t}$ (b) $\dfrac{1}{x}$ (c) $\dfrac{1}{\sqrt[3]{x^2 + 1}}$	9.4–9.8	(a) $t^{1/2}$ (b) x^{-1} (c) $(x^2 + 1)^{-1/3}$	0.3, 0.4 Exponents and radicals
Simplify: (a) $\dfrac{4(x + h)^2 - 4x^2}{h}$, if $h \neq 0$ (b) $(2x^3 + 3x + 1)(2x) + (x^2 + 4)(6x^2 + 3)$ (c) $\dfrac{x(3x^2) - x^3(1)}{x^2}$, if $x \neq 0$	9.3 9.5 9.7	(a) $8x + 4h$ (b) $10x^4 + 33x^2 + 2x + 12$ (c) $2x$	0.5 Simplifying algebraic expressions
Simplify: (a) $\dfrac{x^2 - x - 6}{x + 2}$ if $x \neq -2$ (b) $\dfrac{x^2 - 4}{x - 2}$ if $x \neq 2$	9.1 9.7	(a) $x - 3$ (b) $x + 2$	0.7 Simplifying fractions
If $f(x) = 3x^2 + 2x$, find $\dfrac{f(x + h) - f(x)}{h}$.	9.3	$6x + 3h + 2$	1.2 Functional notation
Find the slope of the line passing through (1, 2) and (2, 4).	9.3	2	1.3 Slopes
Write the equation of the line passing through (1, 5) with slope 8.	9.3 9.4 9.6	$y = 8x - 3$	1.3 Point-slope equation of a line

Chapter 9

Derivatives

If a firm receives \$30,000 in revenue during a 30-day month, its average revenue per day is \$30,000/30 = \$1000. This does not necessarily mean that the actual revenue was \$1000 on any one day, just that the average is \$1000 per day. Similarly, if a person drives a car 50 miles in an hour's time, the car's average velocity is 50 miles per hour, but the driver could have gotten a speeding ticket for traveling 70 miles per hour on this trip. When we say a car is moving at a velocity of 50 miles per hour, we are talking about the velocity of the car at an instant in time (the instantaneous velocity). We can use the average velocity to find the instantaneous velocity, as follows.

If a car travels in a straight line from position y_1 at time x_1 and arrives at position y_2 at time x_2, then it has traveled the distance $y_2 - y_1$ in the elapsed time $x_2 - x_1$. If we represent the distance traveled by Δy and the elapsed time by Δx, the average velocity is given by

$$V_{av} = \frac{\Delta y}{\Delta x}$$

The smaller the time interval, the nearer the average velocity will be to the instantaneous velocity. For example, knowing that a car traveled 50 miles in an hour does not tell us much about its instantaneous velocity at any time during that hour. But knowing that it traveled 1 mile in 1 minute, or 50 feet in one second, tells us much more about the velocity at a given time. Continuing to decrease the length of the time interval (Δx) will get us closer and closer to the instantaneous velocity.

Some police departments have equipment that measures how fast a car is traveling by measuring how much time elapses while the car travels between two sensors placed 60 inches apart on the road. This is not the instantaneous velocity of the car, but it is an excellent approximation.

We define the ***instantaneous velocity*** *to be the* limit *of $\Delta y/\Delta x$ as Δx approaches 0. We write this as*

$$V = \lim_{\Delta x \to 0} \frac{\Delta y}{\Delta x}$$

Thus we may think of velocity as the instantaneous rate of change of distance with respect to time.

This chapter is concerned with limits *and* rates of change. *We will see that the* derivative *of a function can be used to determine the rate of change of the dependent variable with respect to the independent variable. In this chapter the derivative will be used to find the marginal profit, marginal cost, and marginal revenue, given the respective profit, total cost, and total revenue functions, and we will find other rates of change, such as rates of change of populations and velocity. We will also use the derivative to determine the slope of a tangent to a curve at a point on the curve. In the next chapter, more applications of the derivative will be discussed. For example, we will use differentiation to minimize average cost, maximize total revenue, maximize profit, and find the maximum dosage for certain medications.*

9.1 Limits

OBJECTIVES

- *To use graphs and numerical tables to find limits of functions, when they exist*
- *To find limits of polynomial functions*
- *To find limits of rational functions*

APPLICATION PREVIEW

Although everyone recognizes the value of eliminating any and all particulate pollution from smokestack emissions of factories, company owners are concerned about the cost of removing this pollution. Suppose that USA Steel has shown that the cost C of removing p percent of the particulate pollution from the emissions at one of its plants is

$$C = C(p) = \frac{7300p}{100 - p}$$

To investigate the cost of removing as much of the pollution as possible, we can evaluate the **limit** as p (the percent) approaches 100 from values less than 100. Using a limit is important in this case, because we cannot evaluate this function at $p = 100$.

We have used the notation $f(c)$ to indicate the value of a function $f(x)$ at $x = c$. If we need to discuss a value that $f(x)$ approaches as x approaches c, we use the idea of a **limit.** For example, if

$$f(x) = \frac{x^2 - x - 6}{x + 2}$$

then we know that -2 is not in the domain of $f(x)$ so that $f(-2)$ does not exist. Figure 9.1 shows the graph of $y = f(x)$ with an open circle where $x = -2$. Even though $f(-2)$ is not defined, the figure shows that as x approaches -2 from either side of -2, the graph approaches the open circle at $(-2, -5)$ and the values of $f(x)$ approach -5. Thus -5 is the limit of $f(x)$ as x approaches -2, and we write

$$\lim_{x \to -2} f(x) = -5, \quad \text{or} \quad f(x) \to -5 \quad \text{as} \quad x \to -2$$

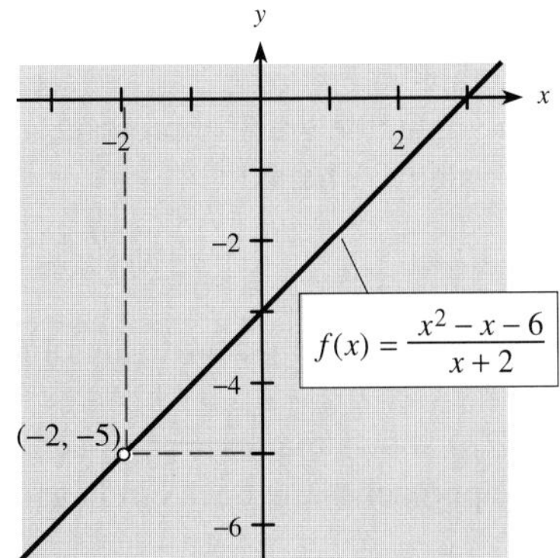

Figure 9.1

TABLE 9.1

x	$f(x) = \frac{x^2 - x - 6}{x + 2}$
Left of -2	
-3.000	-6.000
-2.500	-5.500
-2.100	-5.100
-2.010	-5.010
-2.001	-5.001
Right of -2	
-1.000	-4.000
-1.500	-4.500
-1.900	-4.900
-1.990	-4.990
-1.999	-4.999

This conclusion is fairly obvious from the graph, but it is not so obvious from the equation for $f(x)$.

We can use the values near $x = -2$ in Table 9.1 to help verify that $f(x) \to -5$ as $x \to -2$. Note that to the left of -2, the values of x increase from -3.000 to -2.001 in small increments, and in the corresponding column for $f(x)$, the values of the function $f(x)$ increase from -6.000 to -5.001. To the right of -2, the values of x decrease from -1.000 to -1.999 while the corresponding values of $f(x)$ decrease from -4.000 to -4.999. Hence, Table 9.1 and Figure 9.1 indicate that the value of $f(x)$ approaches -5 as x approaches -2 from both sides of $x = -2$.

From our discussion of the graph in Figure 9.1 and Table 9.1, we see that as x approaches -2 from either side of -2, the limit of the function is the value L that the function approaches. This limit L is not necessarily the value of the function at $x = -2$.

EXAMPLE 1

Figure 9.2 shows three functions for which the limit exists as x approaches 2. Use this figure to find the following.

(a) $\lim_{x \to 2} f(x)$ and $f(2)$ (if it exists)
(b) $\lim_{x \to 2} g(x)$ and $g(2)$ (if it exists)
(c) $\lim_{x \to 2} h(x)$ and $h(2)$ (if it exists)

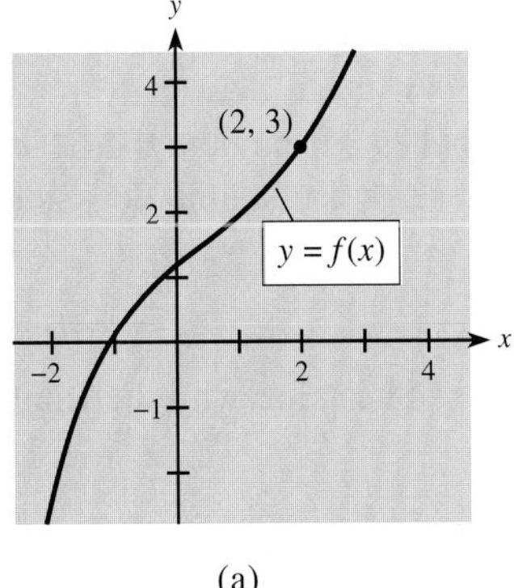

(a)

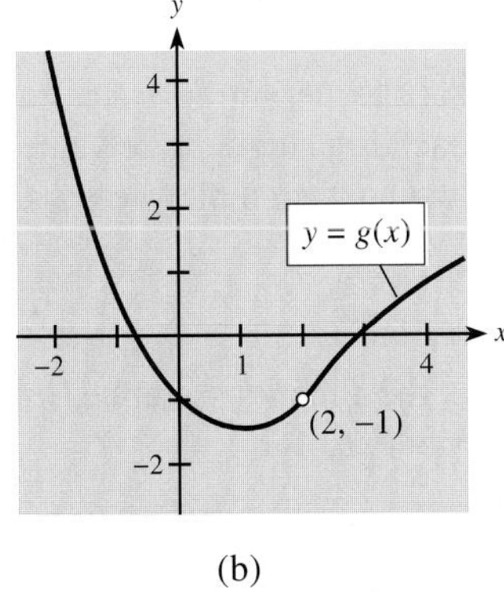

(b)

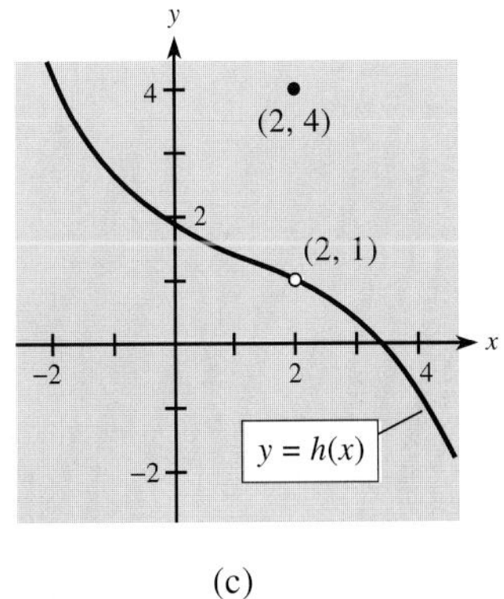

(c)

Figure 9.2

Solution

(a) From the graph in Figure 9.2(a), we see that as x approaches 2 from both the left and the right, the graph approaches the point (2, 3). Thus $f(x)$ approaches the single value 3. That is,

$$\lim_{x \to 2} f(x) = 3$$

The value of $f(2)$ is the y-coordinate of the point on the graph at $x = 2$. Thus $f(2) = 3$.

(b) Figure 9.2(b) shows that as x approaches 2 from both the left and the right, the graph approaches the open circle at $(2, -1)$. Thus

$$\lim_{x \to 2} g(x) = -1$$

The figure also shows that at $x = 2$ there is no point on the graph. Thus $g(2)$ is undefined.

(c) Figure 9.2(c) shows that

$$\lim_{x \to 2} h(x) = 1$$

The figure also shows that at $x = 2$ there is a point on the graph at (2, 4). Thus $h(2) = 4$, and we see that $\lim_{x \to 2} h(x) \neq h(2)$.

As Example 1 shows, the limit of the function as x approaches c may or may not be the same as the value of the function at $x = c$. This leads to our intuitive definition of *limit.*

Limit Let $f(x)$ be a function defined on an open interval containing c, except perhaps at c. Then

$$\lim_{x \to c} f(x) = L$$

is read "the limit of $f(x)$ as x approaches c equals L." The number L exists if we can make values of $f(x)$ as close to L as we desire by choosing values of x sufficiently close to c. When the values of $f(x)$ do not approach a single finite value L as x approaches c, we say the limit does not exist.

As the definition states, a limit as $x \to c$ can exist only if the function approaches a single finite value as x approaches c from both the left and right of c. In the next example, we consider some cases where a limit does not exist.

EXAMPLE 2

Using the functions graphed in Figure 9.3, determine why the limit as $x \to 2$ does not exist for

(a) $f(x)$ (b) $g(x)$ (c) $h(x)$

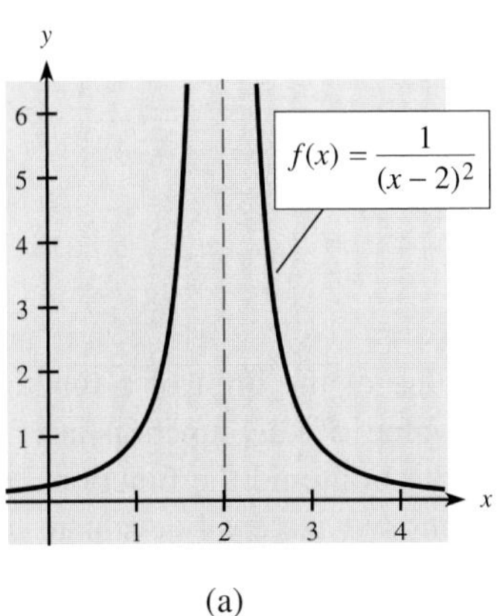

(a)

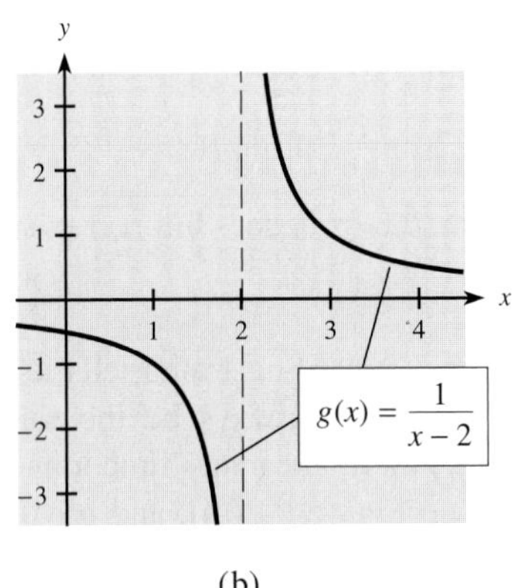

(b)

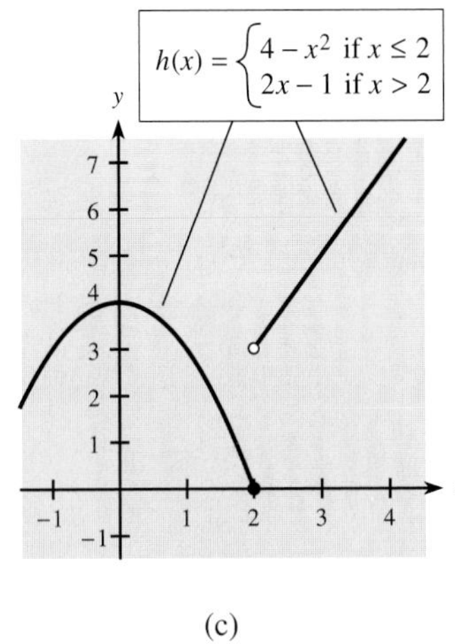

(c)

Figure 9.3

(a) As $x \to 2$ from the left side and the right side of $x = 2$, $f(x)$ increases without bound, which we denote by saying that $f(x)$ approaches $+\infty$ as $x \to 2$. In this case, $\lim_{x \to 2} f(x)$ does not exist (briefly noted by DNE) because $f(x)$ does not approach a finite value as $x \to 2$. The graph has a vertical asymptote at $x = 2$. We say that this is an **infinite limit.**

(b) As $x \to 2$ from the left, $g(x)$ approaches $-\infty$, and as $x \to 2$ from the right, $g(x)$ approaches $+\infty$, so $g(x)$ does not approach a finite value as $x \to 2$. Therefore, the limit does not exist. The graph of $y = g(x)$ has a vertical asymptote at $x = 2$.

(c) As $x \to 2$ from the left (while $x < 2$, denoted by $x \to 2^-$), the graph approaches the point at (2, 0), so $h(x)$ approaches the value 0, but as $x \to 2$ from the right (while $x > 2$, denoted by $x \to 2^+$), the graph approaches the open circle at (2, 3), so $h(x)$ approaches the value 3. Because $h(x)$ approaches two different numbers as x approaches 2, the limit does not exist.

Examples 1 and 2 illustrate the following two important facts regarding limits.

1. The limit of a function as x approaches c is independent of the value of the function at c. When $\lim_{x \to c} f(x)$ exists, the value of the function at c may be (i) the same as the limit, (ii) undefined, or (iii) defined but different from the limit (see Figure 9.2 and Example 1).

2. The limit is said to exist only if the following conditions are satisfied:

   (a) The limit L is a finite value (real number).

   (b) The limit as x approaches c from the left equals the limit as x approaches c from the right. That is, we must have

   $$\lim_{x \to c^-} f(x) = \lim_{x \to c^+} f(x)$$

   Figure 9.3 and Example 2 illustrate cases where $\lim_{x \to c} f(x)$ does not exist.

CHECKPOINT

1. Can $\lim_{x \to c} f(x)$ exist if $f(c)$ is undefined?
2. Does $\lim_{x \to c} f(x)$ exist if $f(c) = 0$?
3. Does $f(c) = 1$ if $\lim_{x \to c} f(x) = 1$?
4. If $\lim_{x \to c^-} f(x) = 0$, does $\lim_{x \to c} f(x)$ exist?

Fact 1 regarding limits tells us that the value of the limit of a function as $x \to c$ will not always be the same as the value of the function at $x = c$. However, there are many functions for which the limit and the functional value agree [see Figure 9.2(a)], and for these functions we can easily evaluate limits. The following properties of limits allow us to identify certain classes or types of functions for which $\lim_{x \to c} f(x)$ equals $f(c)$.

Properties of Limits

If k is a constant, $\lim_{x \to c} f(x) = L$ and $\lim_{x \to c} g(x) = M$, then the following are true.

I. $\lim_{x \to c} k = k$

II. $\lim_{x \to c} x = c$

III. $\lim_{x \to c} [f(x) \pm g(x)] = L \pm M$

IV. $\lim_{x \to c} [f(x) \cdot g(x)] = LM$

V. $\lim_{x \to c} \dfrac{f(x)}{g(x)} = \dfrac{L}{M}$ if $M \neq 0$

VI. $\lim_{x \to c} \sqrt[n]{f(x)} = \sqrt[n]{\lim_{x \to c} f(x)} = \sqrt[n]{L}$

If f is a polynomial function, then Properties I–IV imply that $\lim_{x \to c} f(x)$ can be found by evaluating $f(c)$. Moreover, if h is a rational function whose denominator is not zero at $x = c$, then Property V implies that $\lim_{x \to c} h(x)$ can be found by evaluating $h(c)$. The following summarizes these observations and recalls the definitions of polynomial and rational functions.

Function	***Definition***	***Limit***
Polynomial function	The function $f(x) = a_n x^n + a_{n-1} x^{n-1} + \cdots + a_1 x + a_0$, where n is a positive integer, is called a **polynomial function.**	$\lim_{x \to c} f(x) = f(c)$ for all values c (by Properties I–IV)
Rational function	The function $h(x) = \dfrac{f(x)}{g(x)}$ where both $f(x)$ and $g(x)$ are polynomial functions, is called a **rational function.**	$\lim_{x \to c} h(x) = \lim_{x \to c} \dfrac{f(x)}{g(x)} = \dfrac{f(c)}{g(c)}$ when $g(c) \neq 0$ (by Property V)

EXAMPLE 3

Find the following limits, if they exist.

(a) $\lim\limits_{x \to -1} (x^3 - 2x)$ (b) $\lim\limits_{x \to 4} \dfrac{x^2 - 4x}{x - 2}$

Solution

(a) Note that $f(x) = x^3 - 2x$ is a polynomial, so

$$\lim_{x \to -1} f(x) = f(-1) = (-1)^3 - 2(-1) = 1$$

Figure 9.4(a) shows the graph of $f(x) = x^3 - 2x$.

(b) Note that this limit has the form

$$\lim_{x \to c} \frac{f(x)}{g(x)}$$

where $f(x)$ and $g(x)$ are polynomials and $g(c) \neq 0$. Therefore, we have

$$\lim_{x \to 4} \frac{x^2 - 4x}{x - 2} = \frac{4^2 - 4(4)}{4 - 2} = \frac{0}{2} = 0$$

Figure 9.4(b) shows the graph of $g(x) = \dfrac{x^2 - 4x}{x - 2}$.

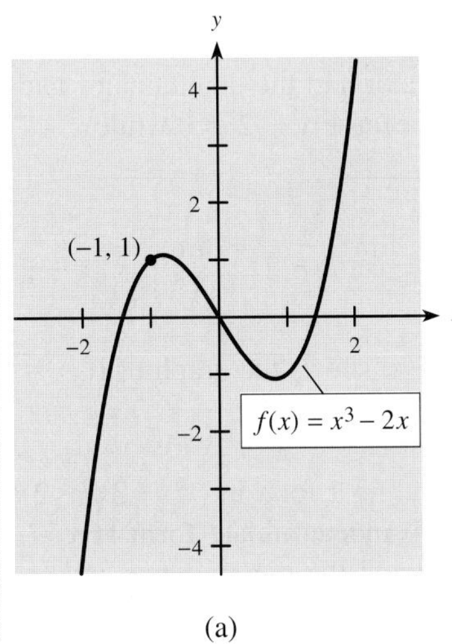

(a)

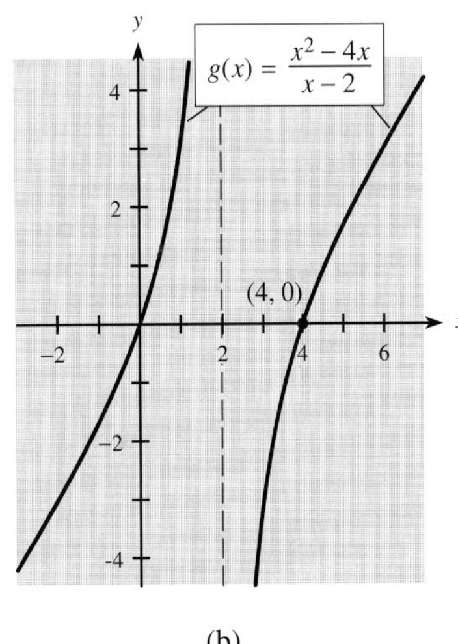

(b)

Figure 9.4

We have seen that we can use Property V to find the limit of a rational function $f(x)/g(x)$ as long as the denominator is *not* zero. If the limit of the denominator of $f(x)/g(x)$ *is* zero, then there are two possible cases.

I. Both $\lim\limits_{x \to c} g(x) = 0$ and $\lim\limits_{x \to c} f(x) = 0$, or

II. $\lim\limits_{x \to c} g(x) = 0$ and $\lim\limits_{x \to c} f(x) \neq 0$.

In case I we say that $f(x)/g(x)$ has the form 0/0 at $x = c$. We call this the **0/0 indeterminate form;** the limit cannot be evaluated until $x - c$ is factored from both $f(x)$ and $g(x)$ and the fraction is reduced. Example 4 will illustrate this case.

In case II, the limit has the form $a/0$, where a is a constant, $a \neq 0$. This expression is undefined, and the limit does not exist. Example 5 will illustrate this case.

EXAMPLE 4

Evaluate the following limits, if they exist.

(a) $\lim_{x \to 2} \frac{x^2 - 4}{x - 2}$

(b) $\lim_{x \to 1} \frac{x^2 - 3x + 2}{x^2 - 1}$

Solution

(a) We cannot find the limit by using Property V because the denominator is zero at $x = 2$. The numerator is also zero at $x = 2$, so the expression

$$\frac{x^2 - 4}{x - 2}$$

has the 0/0 indeterminate form at $x = 2$. Thus we can factor $x - 2$ from both the numerator and the denominator and reduce the fraction. (We can divide by $x - 2$ because $x - 2 \neq 0$ while $x \to 2$.)

$$\lim_{x \to 2} \frac{x^2 - 4}{x - 2} = \lim_{x \to 2} \frac{(x - 2)(x + 2)}{x - 2} = \lim_{x \to 2} (x + 2) = 4$$

Figure 9.5(a) shows the graph of $f(x) = (x^2 - 4)/(x - 2)$. Note the open circle at (2, 4).

(b) By substituting 1 for x in $(x^2 - 3x + 2)/(x^2 - 1)$, we see that the expression has the 0/0 indeterminate form at $x = 1$, so $x - 1$ is a factor of both the numerator and the denominator. (We can then reduce the fraction because $x - 1 \neq 0$ while $x \to 1$.)

$$\lim_{x \to 1} \frac{x^2 - 3x + 2}{x^2 - 1} = \lim_{x \to 1} \frac{(x - 1)(x - 2)}{(x - 1)(x + 1)}$$

$$= \lim_{x \to 1} \frac{x - 2}{x + 1}$$

$$= \frac{1 - 2}{1 + 1} = \frac{-1}{2} \quad \text{(by Property V)}$$

Figure 9.5(b) shows the graph of $g(x) = (x^2 - 3x + 2)/(x^2 - 1)$. Note the open circle at $\left(1, -\frac{1}{2}\right)$.

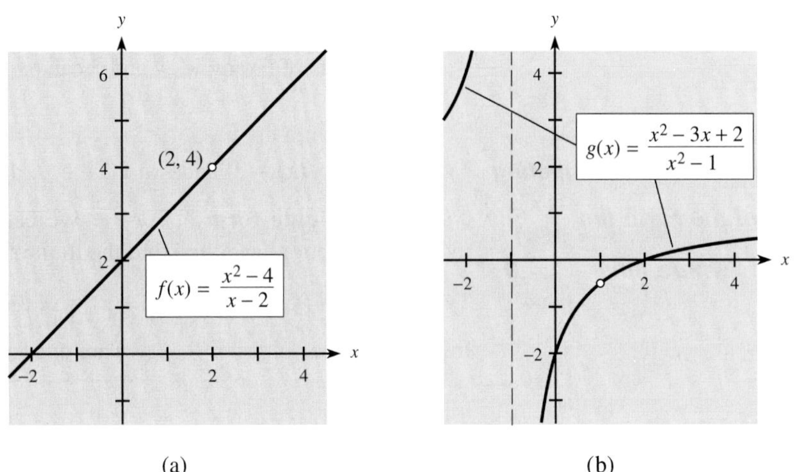

Figure 9.5

Note that although both problems in Example 4 had the 0/0 indeterminate form, they had different answers.

EXAMPLE 5

Find $\lim_{x \to 1} \frac{x^2 + 3x + 2}{x - 1}$, if it exists.

Solution

Substituting 1 for x in the function results in 6/0, so this limit has the form $a/0$, with $a \neq 0$, and is like case II discussed previously. Hence the limit does not exist. Because the numerator is not zero when $x = 1$, we know that $x - 1$ is *not* a factor of the numerator, and we cannot divide numerator and denominator as we did in Example 4. Table 9.2 confirms that this limit does not exist, because the values of the expression are unbounded near $x = 1$.

TABLE 9.2

Left of $x = 1$		*Right of* $x = 1$	
x	$\frac{x^2 + 3x + 2}{x - 1}$	x	$\frac{x^2 + 3x + 2}{x - 1}$
0	−2	2	12
0.5	−7.5	1.5	17.5
0.7	−15.3	1.2	35.2
0.9	−55.1	1.1	65.1
0.99	−595.01	1.01	605.01
0.999	−5,995.001	1.001	6,005.001
0.9999	−59,999.0001	1.001	60,005.0001
$\lim_{x \to 1^-} \frac{x^2 + 3x + 2}{x - 1} = -\infty$		$\lim_{x \to 1^+} \frac{x^2 + 3x + 2}{x - 1} = +\infty$	

The left-hand and right-hand limits are not finite, so they do not exist. Thus $\lim_{x\to 1} \frac{x^2 + 3x + 2}{x - 1}$ does not exist.

The results of Examples 4 and 5 can be summarized as follows:

Rational Functions: Evaluating Limits of the Form $\lim_{x\to c} \frac{f(x)}{g(x)}$ where $\lim_{x\to c} g(x) = 0$

Type I. If $\lim_{x\to c} f(x) = 0$ and $\lim_{x\to c} g(x) = 0$, then the fractional expression has the **0/0 indeterminate form** at $x = c$. We can factor $x - c$ from $f(x)$ and $g(x)$, reduce the fraction, and then find the limit of the resulting expression, if it exists.

Type II. If $\lim_{x\to c} f(x) \neq 0$ and $\lim_{x\to c} g(x) = 0$, then $\lim_{x\to c} \frac{f(x)}{g(x)}$ does not exist.
In this case, the values of $f(x)/g(x)$ are unbounded near $x = c$.

In Example 5, even though the left-hand and right-hand limits do not exist (see Table 9.2), knowledge that the functional values are unbounded (that is, that they become infinite) is helpful in graphing. The graph is shown in Figure 9.6. We see that $x = 1$ is a vertical asymptote.

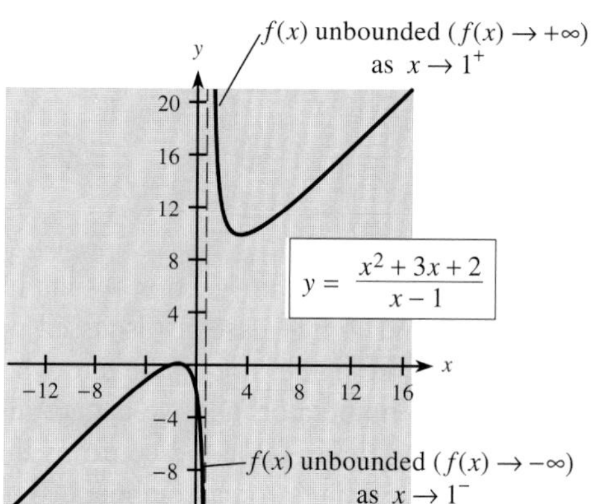

Figure 9.6

EXAMPLE 6

As mentioned in the Application Preview, USA Steel has shown that the cost C of removing p percent of the particulate pollution from the smokestack emissions at one of its plants is

$$C = C(p) = \frac{7300p}{100 - p}$$

To investigate the cost of removing as much of the pollution as possible, find:

(a) the cost of removing 50% of the pollution.
(b) the cost of removing 90% of the pollution.
(c) the cost of removing 99% of the pollution.
(d) the cost of removing 100% of the pollution.

Solution

(a) The cost of removing 50% of the pollution is \$7300 because

$$C(50) = \frac{7300(50)}{100 - 50} = \frac{365{,}000}{50} = 7300$$

(b) The cost of removing 90% of the pollution is \$65,700 because

$$C(90) = \frac{7300(90)}{100 - 90} = \frac{657{,}000}{10} = 65{,}700$$

(c) The cost of removing 99% of the pollution is \$722,700 because

$$C(99) = \frac{7300(99)}{100 - 99} = \frac{722{,}700}{1} = 722{,}700$$

(d) The cost of removing 100% of the pollution is undefined because the denominator of the function is 0 when $p = 100$. To see what the cost approaches as p approaches 100 from values smaller than 100, we evaluate $\lim_{x \to 100^-} \frac{7300p}{100 - p}$. This limit has the Type II form for rational functions. Thus $\lim_{x \to 100^-} \frac{7300}{100 - p} = +\infty$, which means that the cost of removing 100% of the pollution approaches infinity. (That is, it is impossible to remove 100% of the pollution.)

CHECKPOINT

5. Evaluate the following limits (if they exist).
   (a) $\lim_{x \to -3} \frac{2x^2 + 5x - 3}{x^2 - 9}$ (b) $\lim_{x \to 5} \frac{x^2 - 3x - 3}{x^2 - 8x + 1}$ (c) $\lim_{x \to -3/4} \frac{4x}{4x + 3}$

Assume that f, g, and h are polynomials.

6. Does $\lim_{x \to c} f(x) = f(c)$? 7. Does $\lim_{x \to c} \frac{g(x)}{h(x)} = \frac{g(c)}{h(c)}$?
8. If $g(c) = 0$ and $h(c) = 0$, can we be certain that
   (a) $\lim_{x \to c} \frac{g(x)}{h(x)} = 0$? (b) $\lim_{x \to c} \frac{g(x)}{h(x)}$ exists?
9. If $g(c) \neq 0$ and $h(c) = 0$, what can be said about $\lim_{x \to c} \frac{g(x)}{h(x)}$ and $\lim_{x \to c} \frac{h(x)}{g(x)}$?

As we noted in Section 2.4, "Special Functions and Their Graphs," many applications are modeled by piecewise defined functions. To see how we evaluate a limit involving a piecewise defined function, consider the following example.

EXAMPLE 7

Find $\lim_{x \to 1^-} f(x)$, $\lim_{x \to 1^+} f(x)$, and $\lim_{x \to 1} f(x)$, if they exist for

$$f(x) = \begin{cases} x^2 + 1 & \text{for } x \le 1 \\ x + 2 & \text{for } x > 1 \end{cases}$$

Solution

We select values of x that approach 1 from the left and right and evaluate f at these values (see Table 9.3). For values of x less than 1, we use $f(x) = x^2 + 1$ to find the value of the function (see the left-hand side of the table). For values of x that are greater than 1, we use $f(x) = x + 2$ to find the value of f (see the right-hand side of the table).

TABLE 9.3

Left of 1		*Right of* 1	
x	$f(x) = x^2 + 1$	x	$f(x) = x + 2$
0.1	1.01	1.2	3.2
0.9	1.81	1.01	3.01
0.99	1.98	1.001	3.001
0.999	1.998	1.0001	3.0001
0.9999	1.9998	1.00001	3.00001

In this case, we observe that $f(x)$ appears to be approaching 2 as x approaches 1 from the left, whereas $f(x)$ appears to be approaching 3 as x approaches 1 from the right. Figure 9.7 shows the graph of $f(x)$, with $\lim_{x \to 1^-} f(x) = 2$ and $\lim_{x \to 1^+} f(x) = 3$, as we determined from Table 9.3. We show this result algebraically as follows. Because $f(x)$ is defined by $x^2 + 1$ when $x < 1$,

$$\lim_{x \to 1^-} f(x) = \lim_{x \to 1^-} (x^2 + 1) = 2$$

Because $f(x)$ is defined by $x + 2$ when $x > 1$,

$$\lim_{x \to 1^+} f(x) = \lim_{x \to 1^+} (x + 2) = 3$$

And because

$$2 = \lim_{x \to 1^-} f(x) \neq \lim_{x \to 1^+} f(x) = 3$$

$\lim_{x \to 1} f(x)$ does not exist.

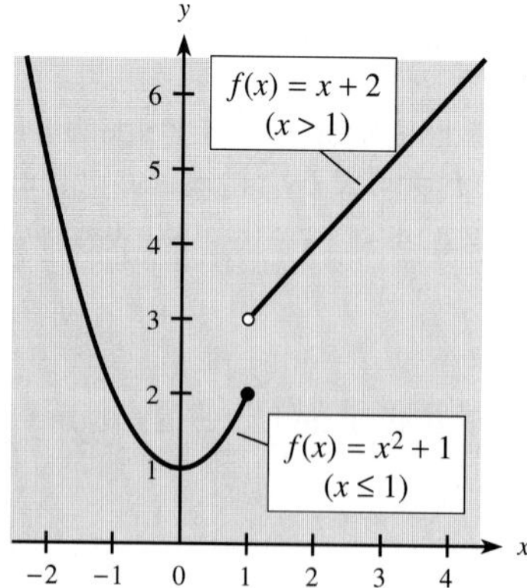

Figure 9.7

Graphing Utilities

We have used graphical, numerical, and algebraic methods to understand and evaluate limits. Graphing utilities can be especially effective when we are exploring limits graphically or numerically.

EXAMPLE 8

Consider the following limits.

(a) $\lim_{x \to 5} \frac{x^2 + 2x - 35}{x^2 - 6x + 5}$ (b) $\lim_{x \to -1} \frac{2x}{x + 1}$

Investigate each limit by using the following methods.

(i) Graphically: Graph the function with a graphing utility and trace near the limiting x-value.

(ii) Numerically: Use the table feature of a graphing utility to evaluate the function very close to the limiting x-value.

(iii) Algebraically: Use properties of limits and algebraic techniques.

Solution

(a) $\lim_{x \to 5} \frac{x^2 + 2x - 35}{x^2 - 6x + 5}$

(i) Figure 9.8(a) on the next page shows the graph of $y = (x^2 + 2x - 35)/(x^2 - 6x + 5)$. Tracing near $x = 5$ shows y-values getting close to 3.

(ii) Figure 9.8(b) on the next page shows a table from a graphing utility with $y_1 = (x^2 + 2x - 35)/(x^2 - 6x + 5)$ and x-values approaching 5 from both sides (note that the function is undefined at $x = 5$). Again, the y-values approach 3.

Both (i) and (ii) strongly suggest $\lim_{x \to 5} \frac{x^2 + 2x - 35}{x^2 - 6x + 5} = 3$.

(iii) Algebraic evaluation of this limit confirms what the graph and the table suggest.

$$\lim_{x \to 5} \frac{x^2 + 2x - 35}{x^2 - 6x + 5} = \lim_{x \to 5} \frac{(x + 7)(x - 5)}{(x - 1)(x - 5)} = \lim_{x \to 5} \frac{x + 7}{x - 1} = \frac{12}{4} = 3$$

(b) $\lim_{x \to -1} \frac{2x}{x + 1}$

(i) Figure 9.9(a) on the next page shows the graph of $y = 2x/(x + 1)$; it indicates that the graph is broken near $x = -1$. Tracing confirms that the break occurs at $x = -1$ and also suggests that the function becomes unbounded near $x = -1$. In addition, we can see that as x approaches -1 from either side, the function is headed in different directions. All this suggests that the limit does not exist.

(ii) Figure 9.9(b) shows a graphing utility table of values for $y_1 = 2x/(x + 1)$ and with x-values approaching $x = -1$. The table reinforces our preliminary conclusions from the graph that the limit does not exist, because the function is unbounded near $x = -1$.

(iii) Algebraically we see that this limit has the form $-2/0$. Thus $\lim_{x \to -1} \frac{2x}{x + 1}$ DNE.

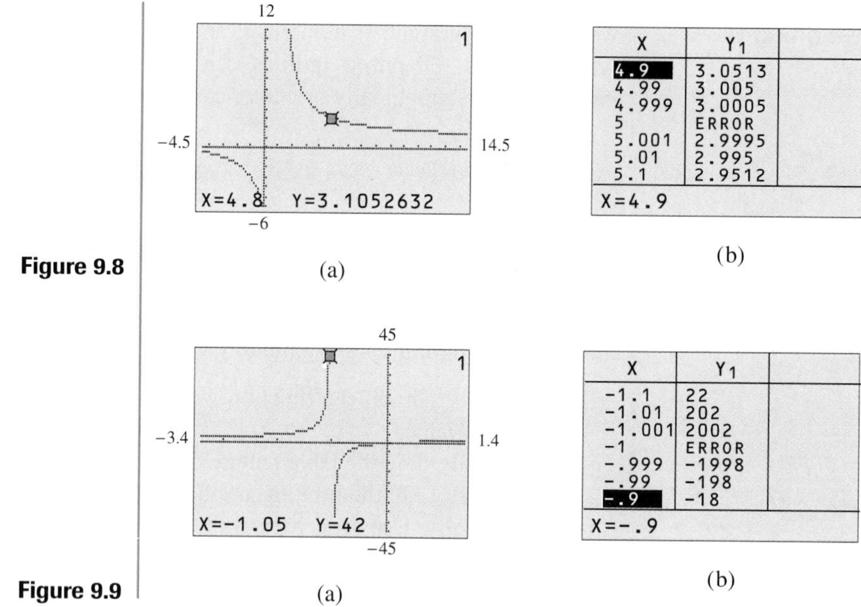

Figure 9.8

Figure 9.9

We could also use the graphing and table features of spreadsheets to explore limits.

CHECKPOINT SOLUTIONS

1. Yes. For example, Figure 9.1 on page 625 and Table 9.1 show that this is possible for $g(x) = \dfrac{x^2 - x - 6}{x + 2}$. Remember that $\lim\limits_{x \to c} f(x)$ does not depend on $f(c)$.
2. Not necessarily. Figure 9.3(c) on page 627 shows the graph of $y = h(x)$ with $h(2) = 0$, but $\lim\limits_{x \to 2} h(x)$ does not exist.
3. Not necessarily. Figure 9.2(c) on page 625 shows the graph of $y = h(x)$ with $\lim\limits_{x \to 2} f(x) = 1$ but $h(2) = 4$.
4. Not necessarily. For example, Figure 9.3(c) on page 627 shows the graph of $y = h(x)$ with $\lim\limits_{x \to 2^-} h(x) = 0$, but with $\lim\limits_{x \to 2^+} h(x) = 2$, so the limit doesn't exist. Recall that if $\lim\limits_{x \to c^-} f(x) = \lim\limits_{x \to c^+} f(x) = L$, then $\lim\limits_{x \to c} f(x) = L$.
5. (a) $\lim\limits_{x \to -3} \dfrac{2x^2 + 5x - 3}{x^2 - 9} = \lim\limits_{x \to -3} \dfrac{(2x - 1)(x + 3)}{(x + 3)(x - 3)} = \lim\limits_{x \to -3} \dfrac{2x - 1}{x - 3} = \dfrac{-7}{-6} = \dfrac{7}{6}$

   (b) $\lim\limits_{x \to 5} \dfrac{x^2 - 3x - 3}{x^2 - 8x + 1} = \dfrac{7}{-14} = -\dfrac{1}{2}$

   (c) Substituting $x = -3/4$ gives $-3/0$, so $\lim\limits_{x \to -3/4} \dfrac{4x}{4x + 3}$ does not exist.
6. Yes, Properties I–IV yield this result.
7. Not necessarily. If $h(c) \neq 0$, then this is true. Otherwise, it is not true.

8. For both (a) and (b), $g(x)/h(x)$ has the 0/0 indeterminate form at $x = c$. In this case we can make no general conclusion about the limit. It is possible for the limit to exist (and be zero or nonzero) or not to exist. Consider the following 0/0 indeterminate forms.

(i) $\lim_{x\to 0} \frac{x^2}{x} = \lim_{x\to 0} x = 0$ (ii) $\lim_{x\to 0} \frac{x(x+1)}{x} = \lim_{x\to 0} (x+1) = 1$

(iii) $\lim_{x\to 0} \frac{x}{x^2} = \lim_{x\to 0} \frac{1}{x}$, which does not exist

9. $\lim_{x\to c} \frac{g(x)}{h(x)}$ does not exist and $\lim_{x\to c} \frac{h(x)}{g(x)} = 0$

EXERCISE 9.1

In Problems 1–8, a graph of $y = f(x)$ is shown and a c-value is given. For each problem, use the graph to find the following, whenever they exist.

(a) $f(c)$ and (b) $\lim_{x\to c} f(x)$

1. $c = 2$

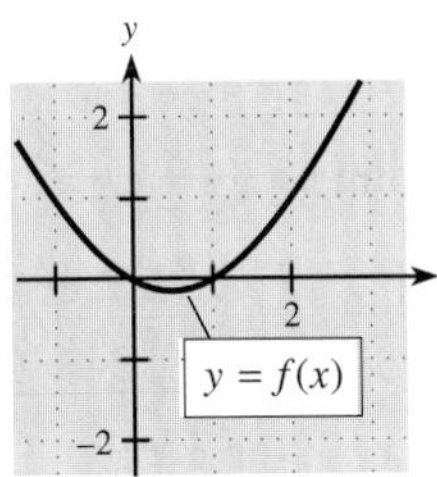

2. $c = 2$

3. $c = 4$

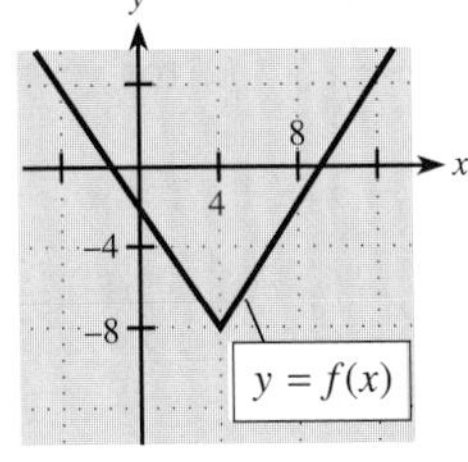

4. $c = 6$

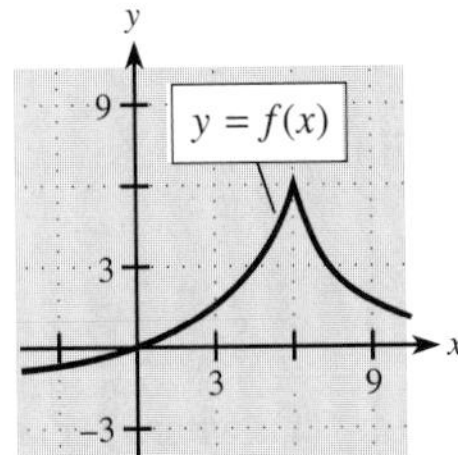

5. $c = 20$

6. $c = -10$

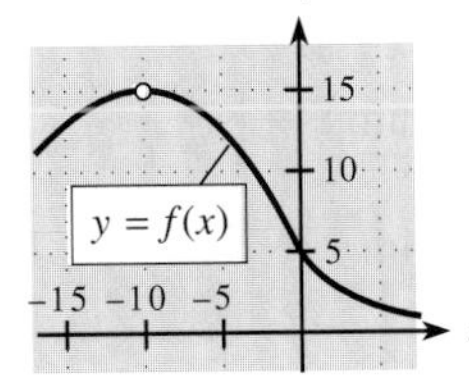

7. $c = -8$

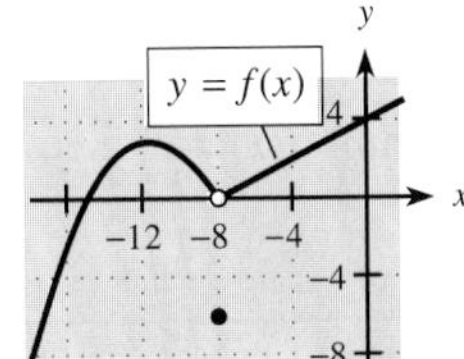

8. $c = -2$

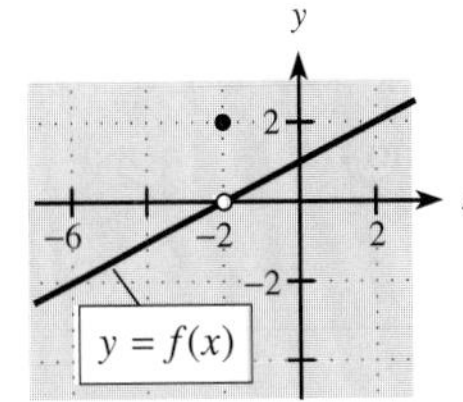

In Problems 9–12, use the graph of $y = f(x)$ and the given c-value to find the following, whenever they exist.

(a) $\lim_{x\to c^-} f(x)$ (b) $\lim_{x\to c^+} f(x)$

(c) $\lim_{x\to c} f(x)$ (d) $f(c)$

9. $c = -10$

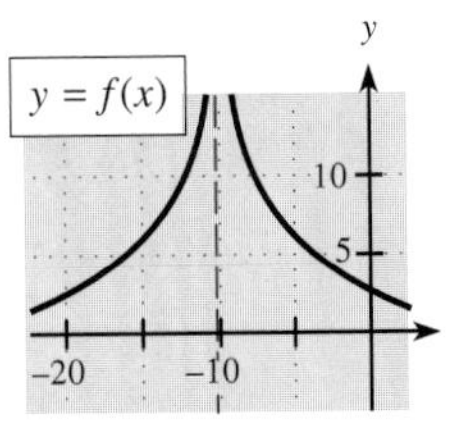

10. $c = 2$

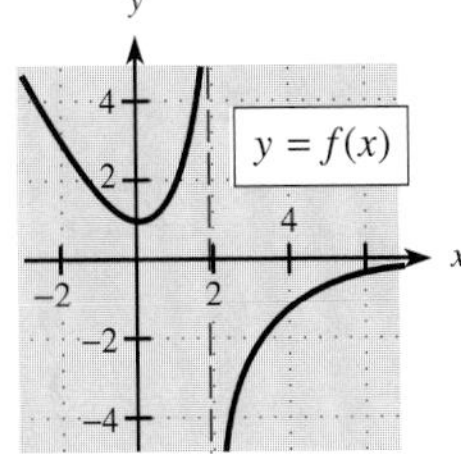

11. $c = -4\frac{1}{2}$

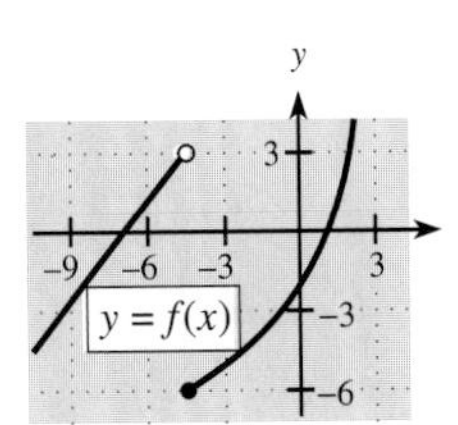

12. $c = 2$

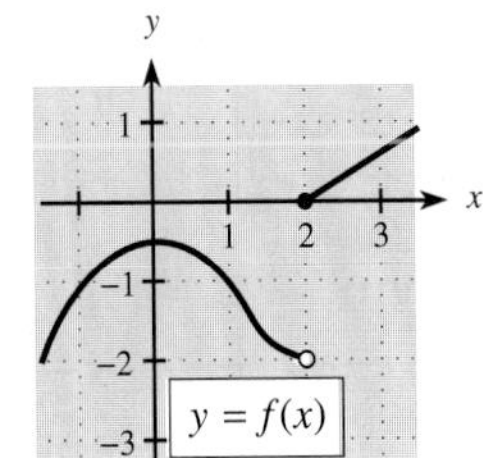

In Problems 13–18, complete each table and predict the limit, if it exists.

13. $f(x) = \dfrac{x^2 + 4x - 12}{x^2 - 2x}$

$\lim_{x \to 2} f(x) = ?$

x	$f(x)$
1.9	
1.99	
1.999	
↓	↓
2	?
↑	↑
2.001	
2.01	
2.1	

14. $f(x) = \dfrac{2x - 10}{x^2 - 25}$

$\lim_{x \to 5} f(x) = ?$

x	$f(x)$
4.9	
4.99	
4.999	
↓	↓
5	?
↑	↑
5.001	
5.01	
5.1	

15. $f(x) = \dfrac{2 - x - x^2}{x - 1}$

$\lim_{x \to 1} f(x) = ?$

x	$f(x)$
0.9	
0.99	
0.999	
↓	↓
1	?
↑	↑
1.001	
1.01	
1.1	

16. $f(x) = \dfrac{2x + 1}{\frac{1}{4} - x^2}$

$\lim_{x \to -0.5} f(x) = ?$

x	$f(x)$
−0.51	
−0.501	
−0.5001	
↓	↓
−0.5	?
↑	↑
−0.4999	
−0.499	
−0.49	

17. $f(x) = \begin{cases} 5x - 1 & \text{for } x < 1 \\ 8 - 2x - x^2 & \text{for } x \geq 1 \end{cases}$

$\lim_{x \to 1} f(x) = ?$

x	$f(x)$
0.9	
0.99	
0.999	
↓	↓
1	?
↑	↑
1.001	
1.01	
1.1	

18. $f(x) = \begin{cases} 4 - x^2 & \text{for } x \leq -2 \\ x^2 + 2x & \text{for } x > -2 \end{cases}$

$\lim_{x \to -2} f(x) = ?$

x	$f(x)$
−2.1	
−2.01	
−2.001	
↓	↓
−2	?
↑	↑
−1.999	
−1.99	

In Problems 19–40, use properties of limits and algebraic methods to find the limits, if they exist.

19. $\lim_{x \to -35} (34 + x)$

20. $\lim_{x \to 80} (82 - x)$

21. $\lim_{x \to -1} (4x^3 - 2x^2 + 2)$

22. $\lim_{x \to 3} (2x^3 - 12x^2 + 5x + 3)$

23. $\lim_{x \to -1/2} \dfrac{4x - 2}{4x^2 + 1}$

24. $\lim_{x \to -1/3} \dfrac{1 - 3x}{9x^2 + 1}$

25. $\lim_{x \to 3} \dfrac{x^2 - 9}{x - 3}$

26. $\lim_{x \to -4} \dfrac{x^2 - 16}{x + 4}$

27. $\lim_{x \to 7} \dfrac{x^2 - 8x + 7}{x^2 - 6x - 7}$

28. $\lim_{x \to -5} \dfrac{x^2 + 8x + 15}{x^2 + 5x}$

29. $\lim_{x \to -2} \dfrac{x^2 + 4x + 4}{x^2 + 3x + 2}$

30. $\lim_{x \to 10} \dfrac{x^2 - 8x - 20}{x^2 - 11x + 10}$

31. $\lim_{x \to 3} f(x)$, where $f(x) = \begin{cases} 10 - 2x & \text{for } x < 3 \\ x^2 - x & \text{for } x \geq 3 \end{cases}$

32. $\lim_{x \to 5} f(x)$, where $f(x) = \begin{cases} 7x - 10 & \text{for } x < 5 \\ 25 & \text{for } x \geq 5 \end{cases}$

33. $\lim_{x \to -1} f(x)$, where $f(x) = \begin{cases} x^2 + \dfrac{4}{x} & \text{for } x \leq -1 \\ 3x^3 - x - 1 & \text{for } x > -1 \end{cases}$

34. $\lim_{x \to 2} f(x)$, where $f(x) = \begin{cases} \dfrac{x^3 - 4}{x - 3} & \text{for } x \leq 2 \\ \dfrac{3 - x^2}{x} & \text{for } x > 2 \end{cases}$

35. $\lim_{x \to 2} \dfrac{x^2 + 6x + 9}{x - 2}$

36. $\lim_{x \to 5} \dfrac{x^2 - 6x + 8}{x - 5}$

37. $\lim_{x \to -1} \dfrac{x^2 + 5x + 6}{x + 1}$

38. $\lim_{x \to 3} \dfrac{x^2 + 2x - 3}{x - 3}$

39. $\lim_{h \to 0} \dfrac{(x + h)^3 - x^3}{h}$

40. $\lim_{h \to 0} \dfrac{2(x + h)^2 - 2x^2}{h}$

In Problems 41–46, graph each function with a graphing utility and use TRACE to predict the limit. Check your work either by using the table feature of the graphing utility or by finding the limit algebraically.

41. $\lim_{x\to 2} \frac{x^3 - 4x}{2x^2 - x^3}$

42. $\lim_{x\to 1/2} \frac{x^2 - \frac{1}{4}}{2x - 1}$

43. $\lim_{x\to 10} \frac{x^2 - 19x + 90}{3x^2 - 30x}$

44. $\lim_{x\to -3} \frac{x^4 + 3x^3}{2x^4 - 18x^2}$

45. $\lim_{x\to -1} \frac{x^3 - x}{x^2 + 2x + 1}$

46. $\lim_{x\to 5} \frac{x^2 - 7x + 10}{x^2 - 10x + 25}$

In Problems 47–52, use the table feature of a graphing utility to predict each limit. Check your work by using either a graphical or an algebraic approach.

47. $\lim_{x\to 6} \frac{x^2 - 2x - 24}{x^2 + 2x - 48}$

48. $\lim_{x\to 9} \frac{x^3 - 6x^2 - 27x}{x^2 - 12x - 27}$

49. $\lim_{x\to -2} \frac{x^4 - 4x^2}{x^2 + 8x + 12}$

50. $\lim_{x\to -4} \frac{x^3 + 4x^2}{2x^2 + 7x - 4}$

51. $\lim_{x\to 4} f(x)$, where $f(x) = \begin{cases} 12 - \frac{3}{4}x & \text{for } x \le 4 \\ x^2 - 7 & \text{for } x > 4 \end{cases}$

52. $\lim_{x\to 7} f(x)$,
where $f(x) = \begin{cases} 2 + x - x^2 & \text{for } x \le 7 \\ 23 - 9x & \text{for } x > 7 \end{cases}$

53. Use values 0.1, 0.01, 0.001, 0.0001, and 0.00001 with your calculator to approximate

$$\lim_{a\to 0}(1 + a)^{1/a}$$

to three decimal places. This limit equals the special number e that is discussed in Section 5.1, "Exponential Functions," and Section 6.2, "Compound Interest; Geometric Sequences."

54. If $\lim_{x\to 2} [f(x) + g(x)] = 5$ and $\lim_{x\to 2} g(x) = 11$, find
(a) $\lim_{x\to 2} f(x)$
(b) $\lim_{x\to 2} \{[f(x)]^2 - [g(x)]^2\}$
(c) $\lim_{x\to 2} \frac{3g(x)}{f(x) - g(x)}$

55. If $\lim_{x\to 3} f(x) = 4$ and $\lim_{x\to 3} g(x) = -2$, find
(a) $\lim_{x\to 3} [f(x) + g(x)]$ (b) $\lim_{x\to 3} [f(x) - g(x)]$
(c) $\lim_{x\to 3} [f(x) \cdot g(x)]$ (d) $\lim_{x\to 3} \frac{g(x)}{f(x)}$

56. (a) If $\lim_{x\to 2^+} f(x) = 5$, $\lim_{x\to 2^-} f(x) = 5$, and $f(2) = 0$, find $\lim_{x\to 2} f(x)$, if it exists. Explain your conclusions.
(b) If $\lim_{x\to 0^+} f(x) = 3$, $\lim_{x\to 0^-} f(x) = 0$, and $f(0) = 0$, find $\lim_{x\to 0} f(x)$, if it exists. Explain your conclusions.

Applications

57. ***Revenue*** The total revenue for a product is given by

$$R(x) = 1600x - x^2$$

where x is the number of units sold. What is $\lim_{x\to 100} R(x)$?

58. ***Profit*** If the profit function for a product is given by

$$P(x) = 92x - x^2 - 1760$$

find $\lim_{x\to 40} P(x)$.

59. ***Sales and training*** The average monthly sales volume (in thousands of dollars) for a firm depends on the number of hours x of training of its sales staff, according to

$$S(x) = \frac{4}{x} + 30 + \frac{x}{4}, \quad 4 \le x \le 100$$

(a) Find $\lim_{x\to 4^+} S(x)$. (b) Find $\lim_{x\to 100^-} S(x)$.

60. ***Sales and training*** During the first 4 months of employment, the monthly sales S (in thousands of dollars) for a new salesperson depends on the number of hours x of training, as follows:

$$S = S(x) = \frac{9}{x} + 10 + \frac{x}{4}, \quad x \ge 4$$

(a) Find $\lim_{x\to 4^+} S(x)$. (b) Find $\lim_{x\to 10} S(x)$.

61. ***Advertising and sales*** Suppose that the daily sales S (in dollars) t days after the end of an advertising campaign is

$$S = S(t) = 400 + \frac{2400}{t + 1}$$

(a) Find $S(0)$. (b) Find $\lim_{t\to 7} S(t)$.
(c) Find $\lim_{t\to 14} S(t)$.

62. ***Advertising and sales*** Sales y (in thousands of dollars) are related to advertising expenses x (in thousands of dollars) according to

$$y = y(x) = \frac{200x}{x + 10}, \quad x \ge 0$$

(a) Find $\lim_{x\to 10} y(x)$. (b) Find $\lim_{x\to 0^+} y(x)$.

63. ***Productivity*** During an 8-hour shift, the rate of change of productivity (in units per hour) of children's phonographs assembled after t hours on the job is

$$r(t) = \frac{128(t^2 + 6t)}{(t^2 + 6t + 18)^2}, \quad 0 \le t \le 8$$

(a) Find $\lim_{x \to 4} r(t)$. (b) Find $\lim_{x \to 8^-} r(t)$.

(c) Is the rate of productivity higher near the lunch break (at $t = 4$) or near quitting time (at $t = 8$)?

64. ***Revenue*** If the revenue for a product is $R(x) = 100x - 0.1x^2$, and the average revenue per unit is

$$\overline{R}(x) = \frac{R(x)}{x}, \quad x > 0$$

find (a) $\lim_{x \to 100} \frac{R(x)}{x}$ and (b) $\lim_{x \to 0^+} \frac{R(x)}{x}$.

65. ***Cost-benefit*** Suppose that the cost C of obtaining water that contains p percent impurities is given by

$$C(p) = \frac{120{,}000}{p} - 1200$$

(a) Find $\lim_{p \to 100^-} C(p)$, if it exists. Interpret this result.

(b) Find $\lim_{p \to 0^+} C(p)$, if it exists.

(c) Is complete purity possible? Explain.

66. ***Cost-benefit*** Suppose that the cost C of removing p percent of the particulate pollution from the smokestacks of an industrial plant is given by

$$C(p) = \frac{730{,}000}{100 - p} - 7300$$

(a) Find $\lim_{p \to 80} C(p)$.

(b) Find $\lim_{p \to 100^-} C(p)$, if it exists.

(c) Can 100% of the particulate pollution be removed? Explain.

67. ***Federal income tax*** Use the following tax rate schedule for single taxpayers, and create a table of values that could be used to find the following limits, if they exist. Let x represent the amount on Form 1040, line 37, and let $T(x)$ represent the tax due (entered on Form 1040, line 38).

(a) $\lim_{x \to 24{,}650^-} T(x)$ (b) $\lim_{x \to 24{,}650^+} T(x)$

(c) $\lim_{x \to 24{,}650} T(x)$

Schedule X—Use if your filing status is Single

If the amount on Form 1040, line 37, is: over—	*But not over—*	*Enter on Form 1040, line 38*	*of the amount over—*
$0	$24,650	—15%	$0
24,650	59,750	$3,697.50 + 28%	24,650
59,750	124,650	13,525.50 + 31%	59,750
124,650	271,050	33,644.50 + 36%	124,650
271,050	—	86,348.50 + 39.6%	271,050

Source: Internal Revenue Service, 1997 Form 1040 Instructions

68. ***Parking costs*** The Ace Parking Garage charges $2.00 for parking for 2 hours or less, and 50 cents for each extra hour or part of an hour after the 2-hour minimum. The parking charges for the first 5 hours could be written as a function of the time as follows:

$$f(t) = \begin{cases} \$2.00 & \text{if } 0 < t \le 2 \\ \$2.50 & \text{if } 2 < t \le 3 \\ \$3.00 & \text{if } 3 < t \le 4 \\ \$3.50 & \text{if } 4 < t \le 5 \end{cases}$$

(a) Find $\lim_{t \to 1} f(t)$, if it exists.

(b) Find $\lim_{t \to 2} f(t)$, if it exists.

69. ***Municipal water rates*** The Beaver, Pennsylvania, Borough Municipal Authority has the following rates per 1000 gallons of water used.

Usage (x)	*Cost per 1000 Gallons (C(x))*
First 100,000 gallons	1.557
Next 900,000 gallons	1.040
Over 1,000,000 gallons	0.689

Write a function $C = C(x)$ that models the charges, and find $\lim_{x \to 1000} C(x)$ (that is, as usage approaches 1,000,000 gallons).

70. ***Telephone charges*** A direct-dial call from Savannah, Georgia, to Atlanta, Georgia, costs $0.28 for the first minute and $0.24 for each additional minute or part of a minute. If $C = C(t)$ is the charge for a call lasting t minutes, create a table of charges for calls lasting close to 1 minute and use it to find the following limits, if they exist.

(a) $\lim_{t \to 1^-} C(t)$ (b) $\lim_{t \to 1^+} C(t)$ (c) $\lim_{t \to 1} C(t)$

Dow Jones average The graph in the figure shows the Dow Jones Industrial Average (DJIA) at 5-minute intervals for Monday, October 5, 1998. Use the graph for Problems 71 and 72, with t as the time of day and $D(t)$ as the DJIA at time t.

71. Estimate $\lim_{t \to 9{:}30\text{AM}^+} D(t)$, if it exists. Explain what this limit corresponds to.
72. Estimate $\lim_{t \to 4{:}00\text{PM}^-} D(t)$, if it exists. Explain what this limit corresponds to.

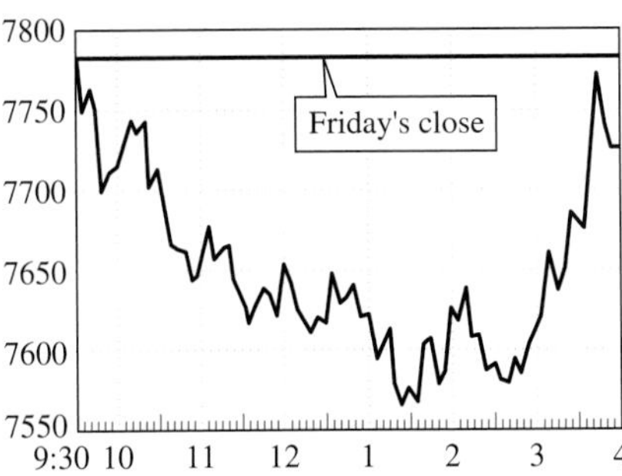

SOURCES: *Telerate, WSJ Statistics*
SOURCE: *Wall Street Journal*, October 6, 1998

Farm workers The percentage of U.S. workers in farm occupations during certain years is shown in the table.

Year	*Percent*	*Year*	*Percent*
1820	71.8	1950	11.6
1850	63.7	1960	6.1
1870	53	1970	3.6
1900	37.5	1980	2.7
1920	27	1985	2.8
1930	21.2	1990	2.4
1940	17.4		

Source: *The World Almanac and Book of Facts*, 1993

Assume that the percentage of U.S. workers in farm occupations can be modeled with the function

$$f(t) = 1000 \cdot \frac{-8.0912t + 1558.9}{1.09816t^2 - 122.183t + 21472.6}$$

where t is the number of years past 1800. (A graph of $f(t)$ along with the data in the table is shown in the figure.) Use the table and equation in Problems 73 and 74.

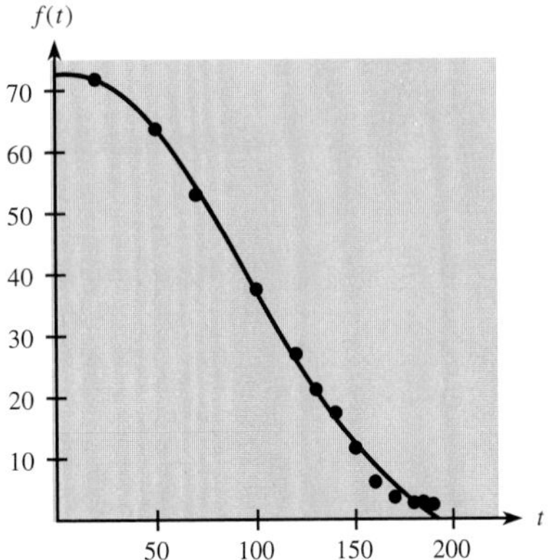

73. (a) Find $\lim_{t \to 200} f(t)$, if it exists.
    (b) What does this limit predict?
    (c) Is the equation accurate as $t \to 200$? Explain.
74. (a) Find $\lim_{t \to 100} f(t)$, if it exists.
    (b) What does this limit predict?
    (c) Is the equation accurate as $t \to 100$? Explain.

9.2 Continuous Functions; Limits at Infinity

OBJECTIVES

- *To determine whether a function is continuous or discontinuous*
- *To determine where a function is discontinuous*
- *To find limits at infinity*

APPLICATION PREVIEW

Suppose that a friend of yours and her husband have a taxable income of \$99,600, and she tells you that she doesn't want to make any more money because that would put them in a higher tax bracket. She makes this statement because the tax rate schedule for married taxpayers filing a joint return (shown in the table) appears to have a jump in taxes for taxable income at \$99,600.

Schedule Y-1—Use if your filing status is Married filing jointly or Qualifying widow(er)

If the amount on form 1040, line 37, is over—	*But not over—*	*Enter on Form 1040, line 38*	*of the amount over—*
\$0	\$41,200	15%	\$0
41,200	99,600	6,180 + 28%	41,200
99,600	151,750	22,532 + 31%	99,600
151,750	271,050	38,698.50 + 36%	151,750
271,050	—	81,646.50 + 39.6%	271,050

Source: Internal Revenue Service, 1997 Form 1040 Instructions

To see whether the couple's taxes would jump to some higher level, we will write the function that gives income tax for married taxpayers as a function of taxable income and show that the function is **continuous.** That is, we will see that the tax paid does not jump at \$99,600 even though the tax on income above \$99,600 is collected at a higher rate. In this section, we will show how to determine whether a function is continuous, and we will investigate some different types of discontinuous functions.

We have found that $f(c)$ is the same as the limit as $x \to c$ for any polynomial function $f(x)$ and any real number c. Any function for which this special property holds is called a **continuous function.** The graphs of such functions can be drawn without lifting the pencil from the paper, and graphs of others may have holes, vertical asymptotes, or jumps that make it impossible to draw them without lifting the pencil. Even though a function may not be continuous everywhere, it is likely to have some points where the limit of the function as $x \to c$ is the same as $f(c)$. In general, we define continuity of a function at the value $x = c$ as follows:

Continuity at a Point The function f is **continuous at $x = c$** if *all* of the following conditions are satisfied.

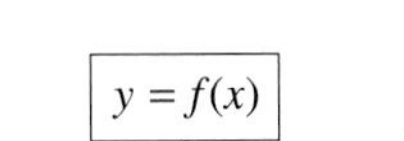

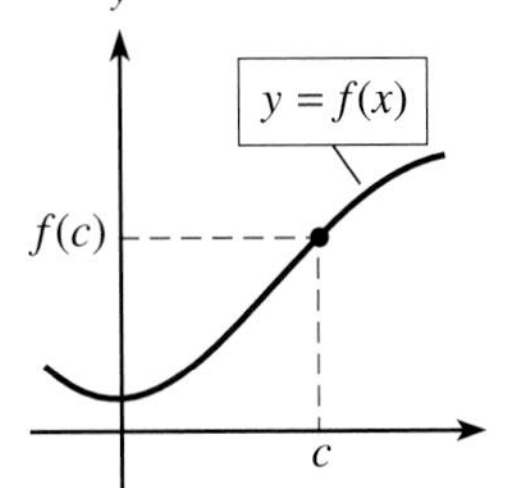

1. $f(c)$ exists 2. $\lim_{x \to c} f(x)$ exists 3. $\lim_{x \to c} f(x) = f(c)$

If one or more of the conditions above do not hold, we say the function is **discontinuous at $x = c$.** Figure 9.10 shows graphs of some functions that are discontinuous at $x = 2$.

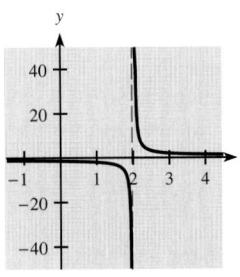

(a) $f(x) = \dfrac{1}{x-2}$

$\lim_{x \to 2} f(x)$ and $f(2)$ do not exist.

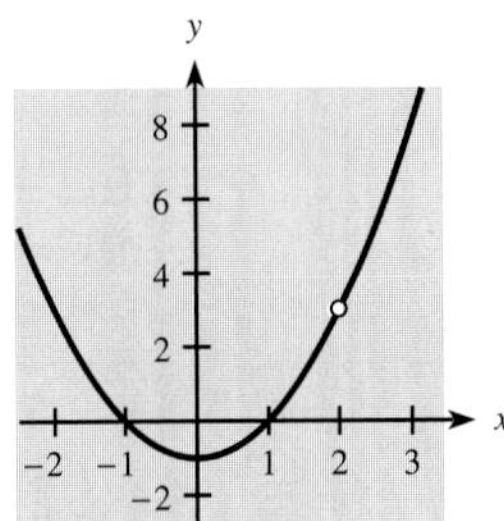

(b) $f(x) = \dfrac{x^3 - 2x^2 - x + 2}{x - 2}$

$f(2)$ does not exist.

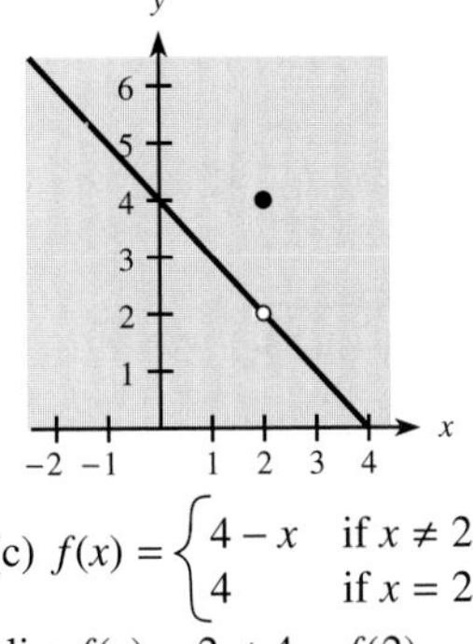

(c) $f(x) = \begin{cases} 4 - x & \text{if } x \neq 2 \\ 4 & \text{if } x = 2 \end{cases}$

$\lim_{x \to 2} f(x) = 2 \neq 4 = f(2)$

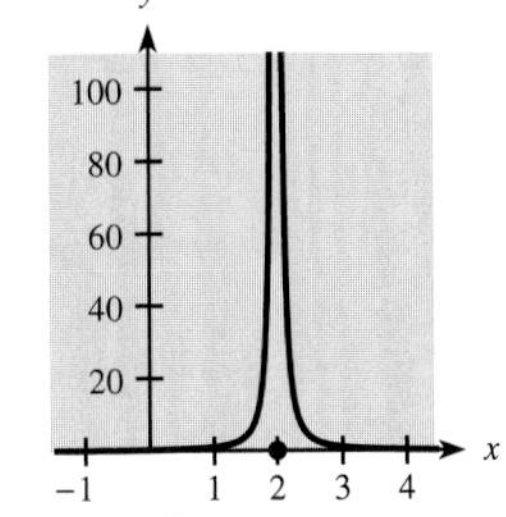

(d) $f(x) = \begin{cases} 1/(x-2)^2 & \text{if } x \neq 2 \\ 0 & \text{if } x = 2 \end{cases}$

$\lim_{x \to 2} f(x)$ does not exist.

Figure 9.10

In the previous section, we saw that if f is a polynomial function, then $\lim_{x \to c} f(x) = f(c)$ for every real number c, and also that $\lim_{x \to c} h(x) = h(c)$ if $h(x) = \dfrac{f(x)}{g(x)}$ is a rational function and $g(c) \neq 0$. Thus, by definition, we have the following.

Every polynomial function is continuous for all real numbers.

Every rational function is continuous at all values of x except those that make the denominator 0.

EXAMPLE 1

For what values of x, if any, is the function $h(x) = \dfrac{3x+2}{4x-6}$ continuous?

Solution

This is a rational function, so it is continuous for all values of x except for those that make the denominator, $4x - 6$, equal to 0. Because $4x - 6 = 0$ at $x = 3/2$, $h(x)$ is continuous for all real numbers except $x = 3/2$.

EXAMPLE 2

For what values of x, if any, is the function discontinuous if

$$f(x) = \frac{x^2 - x - 2}{x^2 - 4}$$

Solution

This is a rational function, so it is continuous everywhere except where the denominator is 0. To find the zeros of the denominator, we factor $x^2 - 4$.

$$f(x) = \frac{x^2 - x - 2}{x^2 - 4} = \frac{x^2 - x - 2}{(x-2)(x+2)}$$

Because the denominator is 0 for $x = 2$ and for $x = -2$, $f(2)$ and $f(-2)$ do not exist (recall that division by 0 is undefined). Thus the function is discontinuous at $x = 2$ and $x = -2$. The graph of this function (see Figure 9.11) shows a hole at $x = 2$ and a vertical asymptote at $x = -2$.

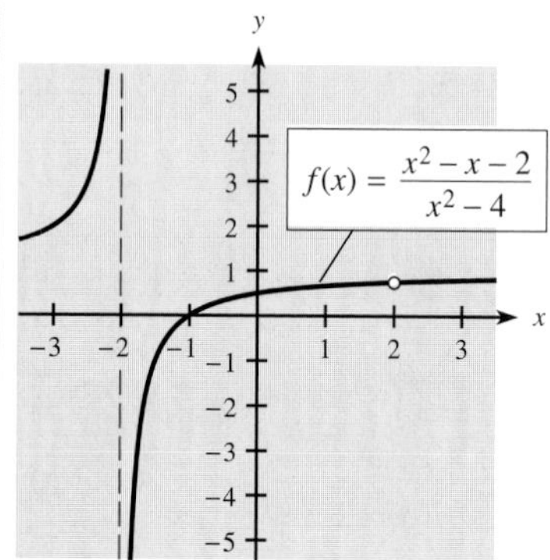

Figure 9.11

CHECKPOINT

1. Find any x-values where the following functions are discontinuous.

(a) $f(x) = x^3 - 3x + 1$ (b) $g(x) = \dfrac{x^3 - 1}{(x-1)(x+2)}$

If the pieces of a piecewise defined function are polynomials, the only values of x where the function might be discontinuous are those at which the definition of the function changes.

EXAMPLE 3

Determine the values of x, if any, for which the following functions are discontinuous.

(a) $g(x) = \begin{cases} (x+2)^3 + 1 & \text{if } x \le -1 \\ 3 & \text{if } x > -1 \end{cases}$ (b) $f(x) = \begin{cases} 4 - x^2 & \text{if } x < 2 \\ x - 2 & \text{if } x \ge 2 \end{cases}$

Solution

(a) $g(x)$ is a piecewise defined function in which each part is a polynomial. Thus, to see whether a discontinuity exists, we need only check the value of x for which the definition of the function changes—that is, at $x = -1$. Because $x = -1$ satisfies $x \le -1$, $g(-1) = (-1 + 2)^3 + 1 = 2$. Evaluating the left- and right-hand limits gives

$$\lim_{x\to -1^-} g(x) = \lim_{x\to -1^-} [(x+2)^3 + 1] = (-1+2)^3 + 1 = 2$$

and

$$\lim_{x\to -1^+} g(x) = \lim_{x\to -1^+} 3 = 3$$

Because the left- and right-hand limits differ, $\lim_{x\to -1} g(x)$ does not exist, so $g(x)$ is discontinuous at $x = -1$. This result is confirmed by examining the graph of g, shown in Figure 9.12.

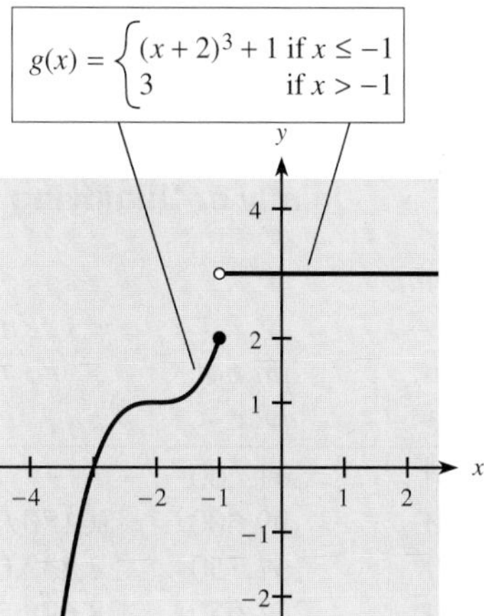

Figure 9.12

(b) As with $g(x)$, $f(x)$ is continuous everywhere except perhaps at $x = 2$, where the definition of $f(x)$ changes. Because $x = 2$ satisfies $x \ge 2$, $f(2) = 2 - 2 = 0$. The left- and right-hand limits are

$$\lim_{x\to 2^-} f(x) = \lim_{x\to 2^-} (4 - x^2) = 4 - 2^2 = 0$$

and

$$\lim_{x\to 2^+} f(x) = \lim_{x\to 2^+} (x - 2) = 2 - 2 = 0$$

Because the right- and left-hand limits are equal, we conclude that $\lim_{x\to 2} f(x) = 0$. The limit is equal to the functional value

$$\lim_{x\to 2} f(x) = f(2)$$

so we conclude that f is continuous at $x = 2$ and thus f is continuous for all values of x. This result is confirmed by the graph of f, shown in Figure 9.13.

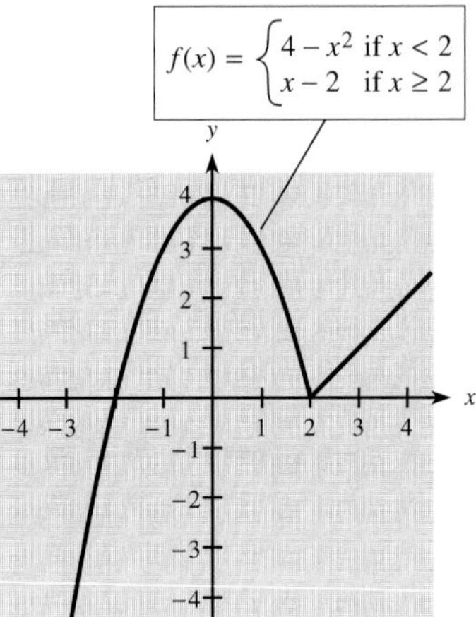

Figure 9.13

We now consider the problem posed in the Application Preview.

EXAMPLE 4

The tax rate schedule for married taxpayers filing a joint return (shown in the table) appears to have a jump in taxes for taxable income at $99,600.

Schedule Y-1—Use if your filing status is Married filing jointly or Qualifying widow(er)

If the amount on Form 1040, line 37, is: over—	*But not over—*	*Enter on Form 1040, line 38*	*of the amount over—*
$0	$41,200	15%	$0
41,200	99,600	$6180.00 + 28%	41,200
99,600	151,750	22,532.00 + 31%	99,600
151,750	271,050	38,698.50 + 36%	151,750
271,050	—	81,646.50 + 39.6%	271,050

Source: Internal Revenue Service, 1997 Form 1040 Instructions

(a) Use the table and write the function that gives income tax for married taxpayers as a function of taxable income, and graph the function.
(b) Is the function in (a) continuous at $x = 99{,}600$?
(c) A married friend of yours and her husband have a taxable income of $99,600, and she tells you that she doesn't want to make any more money because doing so would put her in a higher tax bracket. What would you tell her to do if she is offered a raise?

Solution

(a) The function that gives the tax due for married taxpayers is

$$T(x) = \begin{cases} 0.15x & \text{if } 0 \le x \le 41{,}200 \\ 6180 + 0.28(x - 41{,}200) & \text{if } 41{,}200 < x \le 99{,}600 \\ 22{,}532 + 0.31(x - 99{,}600) & \text{if } 99{,}600 < x \le 151{,}750 \\ 38{,}698.50 + 0.36(x - 151{,}750) & \text{if } 151{,}750 < x \le 271{,}050 \\ 81{,}646.50 + 0.396(x - 271{,}050) & \text{if } x > 271{,}050 \end{cases}$$

(b) This function is continuous at $x = 99{,}600$, because

(i) $T(99{,}600) = 22{,}532$, so $T(99{,}600)$ exists.

(ii) Because the function is piecewise defined near 99,600, we evaluate $\lim_{x\to 99{,}600} T(x)$ by evaluating $\lim_{x\to 99{,}600^-} T(x)$ and $\lim_{x\to 99{,}600^+} T(x)$.

$$\lim_{x\to 99{,}600^-} T(x) = \lim_{x\to 99{,}600^-} [6180 + 0.28(x - 41{,}200)] = 22{,}532$$

$$\lim_{x\to 99{,}600^+} T(x) = \lim_{x\to 99{,}600^+} [22{,}532 + 0.31(x - 99{,}600)] = 22{,}532$$

Because these limits are the same, $\lim_{x\to 99{,}600} T(x) = 22{,}532$, and so the limit exists.

(iii) Because $T(99{,}600) = \lim_{x\to 99{,}600} T(x) = 22{,}532$, the function is continuous at 99,600.

(c) If your friend earned more than $99,600, she and her husband would pay taxes at a higher rate on the money earned *above* the $99,600, but it would not increase the tax rate on any income *up to* $99,600. Thus she should take any raise that's offered.

CHECKPOINT

2. If $f(x)$ and $g(x)$ are polynomials, $h(x) = \begin{cases} f(x) & \text{if } x \le a \\ g(x) & \text{if } x > a \end{cases}$ is continuous everywhere except perhaps at ________.

Graphing Utilities

We noted earlier (in Chapter 2, "Special Functions") that the graph of $y = 1/x$ has a vertical asymptote at $x = 0$ [shown in Figure 9.14(a)]. By graphing $y = 1/x$ with a large x-range or by using TRACE to let x get very large, we can see that $y = 1/x$ never becomes negative for positive x-values regardless of how large the x-value is. Although no value of x makes $1/x$ equal to 0, it is easy to see that $1/x$ approaches 0 as x gets very large. This is denoted by

$$\lim_{x\to +\infty} \frac{1}{x} = 0$$

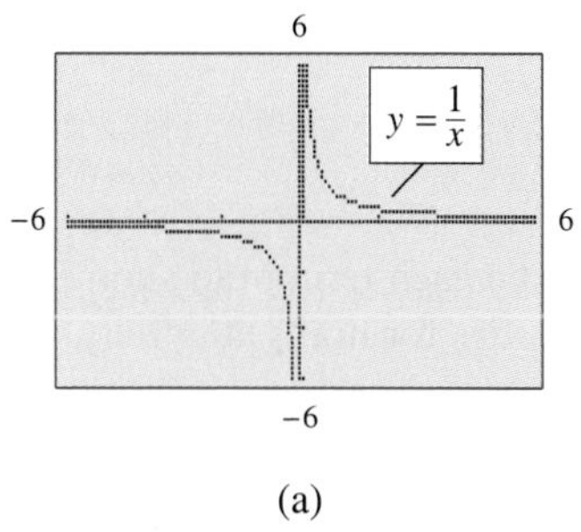

(a)

Figure 9.14

We say that $y = 0$ (the x-axis) is a horizontal asymptote for $y = 1/x$. We also see that $y = 1/x$ approaches 0 as x decreases without bound, and we denote this by

$$\lim_{x\to -\infty} \frac{1}{x} = 0$$

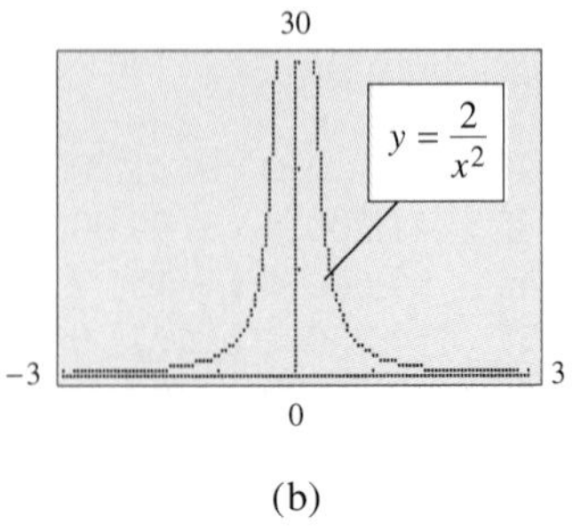

(b)

Figure 9.14 (continued)

These limits can also be established with numerical tables.

x	$f(x) = 1/x$	x	$f(x) = 1/x$
100	0.01	−100	−0.01
100,000	0.00001	−100,000	−0.00001
100,000,000	0.00000001	−100,000,000	−0.00000001
↓	↓	↓	↓
$+\infty$	0	$-\infty$	0

$$\lim_{x\to+\infty} \frac{1}{x} = 0 \qquad \lim_{x\to-\infty} \frac{1}{x} = 0$$

From Figure 9.14(b), we also see that

$$\lim_{x\to+\infty} \frac{2}{x^2} = 0 \quad \text{and} \quad \lim_{x\to-\infty} \frac{2}{x^2} = 0$$

By using graphs and/or tables of values, we can generalize the results for the functions shown in Figure 9.14 and conclude the following.

Limits at Infinity If c is any constant, then

1. $\lim_{x\to+\infty} c = c$ and $\lim_{x\to-\infty} c = c$.
2. $\lim_{x\to+\infty} \frac{c}{x^p} = 0$, where $p > 0$.
3. $\lim_{x\to-\infty} \frac{c}{x^n} = 0$, where $n > 0$ is any integer.

In order to use these properties for finding the limits of rational functions as x approaches $+\infty$ or $-\infty$, we first divide each term of the numerator and denominator by the highest power of x present and then determine the limit of the resulting expression.

EXAMPLE 5

Find each of the following limits, if they exist.

(a) $\lim_{x\to+\infty} \frac{2x-1}{x+2}$ (b) $\lim_{x\to-\infty} \frac{x^2+3}{1-x}$

Solution

(a) The highest power of x present is x^1, so we divide each term in the numerator and denominator by x and then use the properties for limits at infinity.

$$\lim_{x\to+\infty} \frac{2x-1}{x+2} = \lim_{x\to+\infty} \frac{\frac{2x}{x} - \frac{1}{x}}{\frac{x}{x} + \frac{2}{x}} = \lim_{x\to+\infty} \frac{2 - \frac{1}{x}}{1 + \frac{2}{x}}$$

$$= \frac{2-0}{1+0} = 2 \quad \text{(by Properties 1 and 2)}$$

Figure 9.15(a) shows the graph of this function with the y-coordinates of the graph approaching 2 as x approaches $+\infty$ and as x approaches $-\infty$. That is, $y = 2$ is a horizontal asymptote. Note also that there is a discontinuity (vertical asymptote) where $x = -2$.

(b) We divide each term in the numerator and denominator by x^2 and then use the properties.

$$\lim_{x\to-\infty} \frac{x^2+3}{1-x} = \lim_{x\to-\infty} \frac{\frac{x^2}{x^2}+\frac{3}{x^2}}{\frac{1}{x^2}-\frac{x}{x^2}} = \lim_{x\to-\infty} \frac{1+\frac{3}{x^2}}{\frac{1}{x^2}-\frac{1}{x}} = +\infty$$

This limit is $+\infty$ because the numerator approaches 1 and the denominator approaches 0 through positive values. Thus

$$\lim_{x\to-\infty} \frac{x^2+3}{1-x} \quad \text{does not exist}$$

The graph of this function, shown in Figure 9.15(b), has y-coordinates that increase without bound as x approaches $-\infty$ and that decrease without bound as x approaches $+\infty$. (There is no horizontal asymptote.) Note also that there is a vertical asymptote at $x = 1$.

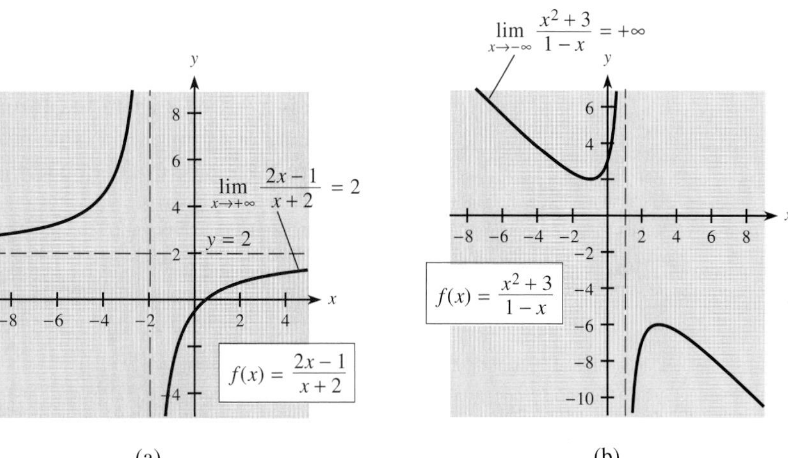

Figure 9.15

CHECKPOINT

3. Evaluate $\lim_{x\to+\infty} \dfrac{x^2-4}{2x^2-7}$.

Graphing Utilities

We can use the graphing and table features of a graphing utility to help locate and investigate discontinuities. The utility can be used to focus our attention on a possible discontinuity and to support or suggest appropriate algebraic calculations.

EXAMPLE 6

Use a graphing utility to investigate the continuity of the following functions.

(a) $f(x) = \dfrac{x^2 + 1}{x + 1}$ (b) $g(x) = \dfrac{x^2 - 2x - 3}{x^2 - 1}$

(c) $h(x) = \dfrac{|x + 1|}{x + 1}$ (d) $k(x) = \begin{cases} \dfrac{-x^2}{2} - 2x & \text{if } x \le -1 \\ \dfrac{x}{2} + 2 & \text{if } x > -1 \end{cases}$

Solution

(a) Figure 9.16(a) shows that $f(x)$ has a discontinuity (vertical asymptote) near $x = -1$. Because $f(-1)$ DNE, we know that $f(x)$ is not continuous at $x = -1$.

(b) Figure 9.16(b) shows that $g(x)$ is discontinuous (vertical asymptote) near $x = 1$, and this looks like the only discontinuity. However, the denominator of $g(x)$ is zero at $x = 1$ and $x = -1$, so $g(x)$ must have discontinuities at both of these x-values. Tracing, evaluating, or using the table feature confirms that $x = -1$ is a discontinuity (a hole, or missing point).

(c) Figure 9.16(c) shows a discontinuity (jump) at $x = -1$. We also see that $h(-1)$ DNE, which confirms the observations from the graph.

(d) The graph in Figure 9.16(d) appears to be continuous. The only "suspicious" x-value is $x = -1$, where the formula for $k(x)$ changes. Evaluating $k(-1)$ and tracing or examining a table near $x = -1$ indicates that $k(x)$ is continuous there. Algebraic evaluation of the two one-sided limits confirms this.

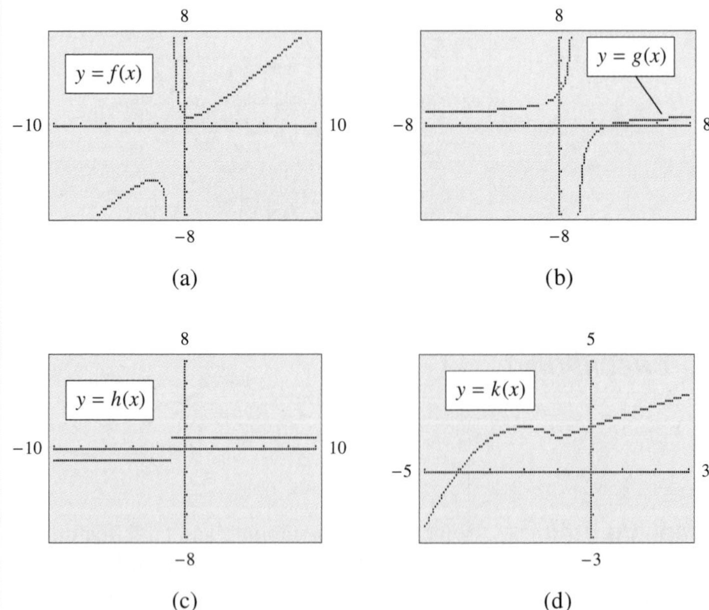

Figure 9.16

Summary

The following information is useful in discussing continuity of functions.

A. A polynomial function is continuous everywhere.

B. A rational function is a function of the form $\frac{f(x)}{g(x)}$, where $f(x)$ and $g(x)$ are polynomials.

1. If $g(x) \neq 0$ at any value of x, the function is continuous everywhere.
2. If $g(c) = 0$, the function is discontinuous at $x = c$.
   (a) If $g(c) = 0$ and $f(c) \neq 0$, then there is a vertical asymptote at $x = c$.
   (b) If $g(c) = 0$ and $\lim_{x \to c} \frac{f(x)}{g(x)} = L$, then the graph has an open circle at $x = c$.

C. A piecewise defined function *may* have a discontinuity at any x-value where the function changes its formula. One-sided limits must be used to see whether the limit exists.

The following steps are useful when we are evaluating limits at infinity for a rational function $f(x) = p(x)/q(x)$.

1. Divide both $p(x)$ and $q(x)$ by the highest power of x found in either polynomial.
2. Use the properties of limits at infinity to complete the evaluation.

CHECKPOINT SOLUTIONS

1. (a) This is a polynomial function, so it is continuous at all values of x (discontinuous at none).
   (b) This is a rational function. It is discontinuous at $x = 1$ and $x = -2$ because these values make its denominator 0.

2. $x = a$.

3. $$\lim_{x \to +\infty} \frac{x^2 - 4}{2x^2 - 7} = \lim_{x \to +\infty} \frac{1 - \frac{4}{x^2}}{2 - \frac{7}{x^2}} = \frac{1 - 0}{2 - 0} = \frac{1}{2}$$

EXERCISE 9.2

Problems 1 and 2 refer to the figure below. For each given x-value, use the figure to determine whether the function is continuous or discontinuous at that x-value. If the function is discontinuous, state which of the three conditions that define continuity is not satisfied.

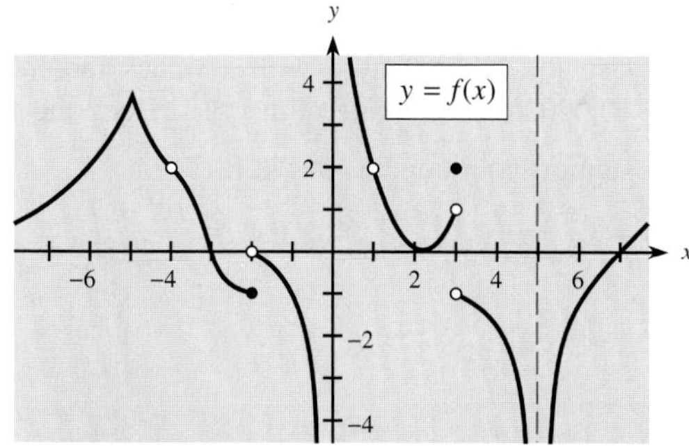

1. (a) $x = -5$ (b) $x = 1$ (c) $x = 3$ (d) $x = 0$
2. (a) $x = 2$ (b) $x = -4$ (c) $x = -2$ (d) $x = 5$

In Problems 3–14, determine whether each function is continuous or discontinuous at the given x-value. Examine the three conditions in the definition of continuity.

3. $f(x) = x^2 - 5x, \quad x = 0$
4. $f(x) = 3x - 5x^3, \quad x = 2$
5. $f(x) = \frac{x^2 - 4}{x - 2}, \quad x = -2$
6. $y = \frac{x^2 - 9}{x + 3}, \quad x = 3$
7. $y = \frac{x^2 - 9}{x + 3}, \quad x = -3$

8. $f(x) = \dfrac{x^2 - 4}{x - 2}, x = 2$

9. $y = \dfrac{x^2 + 5x - 6}{x + 1}, x = -1$

10. $y = \dfrac{x^2 - 2x - 3}{x - 1}, x = 1$

11. $f(x) = \begin{cases} 2 & \text{if } x \le 0 \\ x + 2 & \text{if } x > 0 \end{cases} \quad x = 0$

12. $f(x) = \begin{cases} x - 3 & \text{if } x \le 2 \\ 4x - 7 & \text{if } x > 2 \end{cases} \quad x = 2$

13. $f(x) = \begin{cases} x^2 + 1 & \text{if } x \le 1 \\ 2x^2 - 1 & \text{if } x > 1 \end{cases} \quad x = 1$

14. $f(x) = \begin{cases} x^2 - x & \text{if } x \le 2 \\ 8 - 3x & \text{if } x > 2 \end{cases} \quad x = 2$

In Problems 15–22, determine whether the given function is continuous. If it is not, identify where it is discontinuous and which condition fails to hold. You can verify your conclusions by graphing each function with a graphing utility, if one is available.

15. $f(x) = 4x^2 - 1$

16. $y = 5x^2 - 2x$

17. $g(x) = \dfrac{4x^2 + 3x + 2}{x + 2}$

18. $y = \dfrac{4x^2 + 4x + 1}{x + 1/2}$

19. $y = \dfrac{x}{x^2 + 1}$

20. $y = \dfrac{2x - 1}{x^2 + 3}$

21. $f(x) = \begin{cases} 3 & \text{if } x \le 1 \\ x^2 + 2 & \text{if } x > 1 \end{cases}$

22. $f(x) = \begin{cases} x^3 + 1 & \text{if } x \le 1 \\ 2 & \text{if } x > 1 \end{cases}$

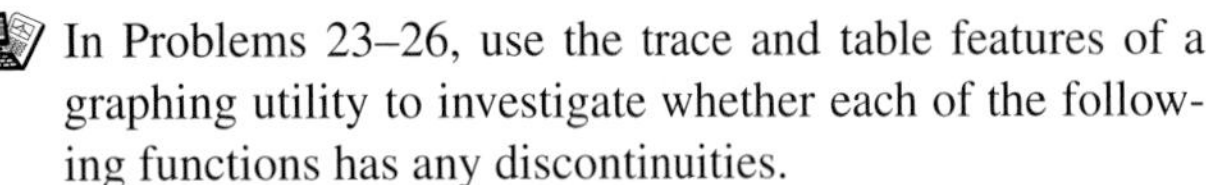

In Problems 23–26, use the trace and table features of a graphing utility to investigate whether each of the following functions has any discontinuities.

23. $y = \dfrac{x^2 - 5x - 6}{x + 1}$

24. $y = \dfrac{x^2 - 5x + 4}{x - 4}$

25. $f(x) = \begin{cases} x - 4 & \text{if } x \le 3 \\ x^2 - 8 & \text{if } x > 3 \end{cases}$

26. $f(x) = \begin{cases} x^2 + 4 & \text{if } x \ne 1 \\ 5 & \text{if } x = 1 \end{cases}$

Each of Problems 27–30 contains a function and its graph. For each problem, answer (a) and (b).

(a) Use the graph to determine, as well as you can, (i) vertical asymptotes, (ii) $\lim_{x\to+\infty} f(x)$, (iii) $\lim_{x\to-\infty} f(x)$

(b) Check your conclusions in (a) by using the functions to determine items (i)–(iii) analytically.

27. $f(x) = \dfrac{8}{x + 2}$

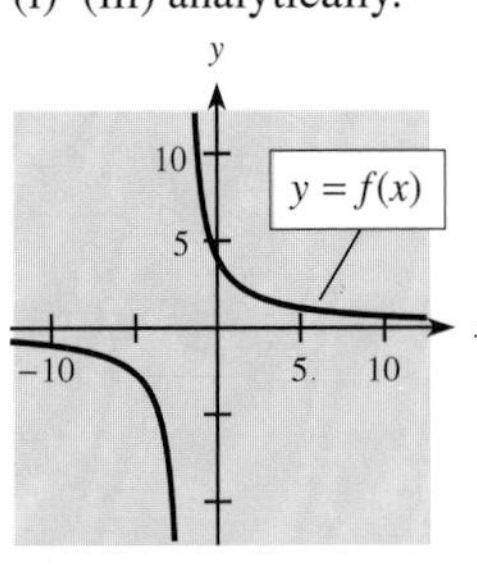

28. $f(x) = \dfrac{x - 3}{x - 2}$

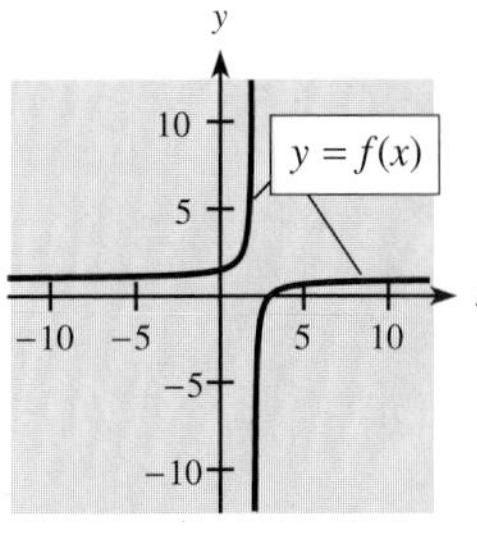

29. $f(x) = \dfrac{2(x + 1)^3(x + 5)}{(x - 3)^2(x + 2)^2}$

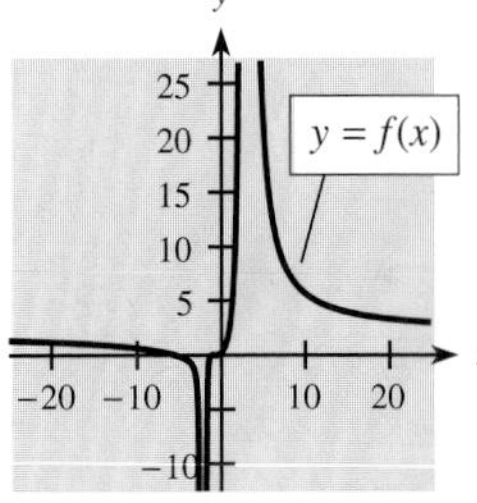

30. $f(x) = \dfrac{4x^2}{x^2 - 4x + 4}$

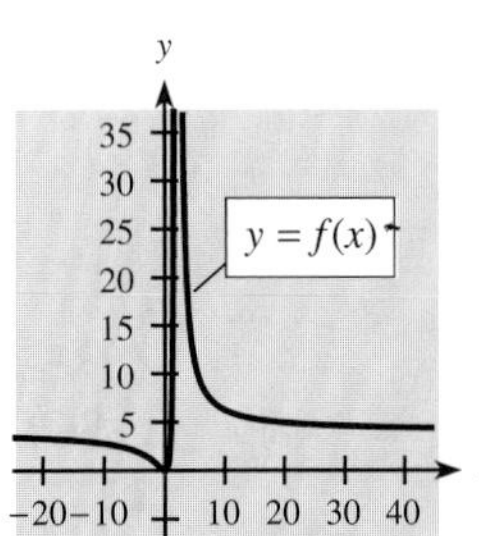

Use analytic methods to evaluate the limits in Problems 31–38. You can verify your conclusions by graphing the functions with a graphing utility, if one is available.

31. $\lim_{x\to+\infty} \dfrac{3}{x + 1}$

32. $\lim_{x\to-\infty} \dfrac{4}{x^2 - 2x}$

33. $\lim_{x\to+\infty} \dfrac{x^3 - 1}{x^3 + 4}$

34. $\lim_{x\to-\infty} \dfrac{3x^2 + 2}{x^2 - 4}$

35. $\lim_{x\to-\infty} \dfrac{5x^3 - 4x}{3x^3 - 2}$

36. $\lim_{x\to+\infty} \dfrac{4x^2 + 5x}{x^2 - 4x}$

37. $\lim_{x\to+\infty} \dfrac{3x^2 + 5x}{6x + 1}$

38. $\lim_{x\to-\infty} \dfrac{5x^3 - 8}{4x^2 + 5x}$

In Problems 39 and 40, use a graphing utility to complete (a) and (b).

(a) Graph each function in the window $0 \le x \le 300$ and $-2 \le y \le 2$. What does the graph indicate about $\lim_{x\to+\infty} f(x)$?

(b) Use the table feature with x-values larger than 10,000 to investigate $\lim_{x\to+\infty} f(x)$. Does the table support your conclusions in (a)?

39. $f(x) = \dfrac{x^2 - 4}{3 + 2x^2}$

40. $f(x) = \dfrac{5x^3 - 7x}{1 - 3x^3}$

In Problems 41 and 42, complete (a)–(c). Use analytic methods to locate (a) any points of discontinuity and (b) any horizontal asymptotes. (c) Then explain why, for these functions, a graphing utility is better as a support tool for the analytic methods than as the primary tool for investigation.

41. $f(x) = \dfrac{1000x - 1000}{x + 1000}$ 42. $f(x) = \dfrac{3000x}{4350 - 2x}$

Applications

43. ***Sales volume*** Suppose that the weekly sales volume (in thousands of units) for a product is given by

$$y = \frac{32}{(p + 8)^{2/5}}$$

where p is the price in dollars per unit. Is this function continuous

(a) for all values of p?
(b) at $p = 24$?
(c) for all $p \geq 0$?
(d) What is the domain for this application?

44. ***Worker productivity*** Suppose that the average number of minutes M that it takes a new employee to assemble one unit of a product is given by

$$M = \frac{40 + 30t}{2t + 1}$$

where t is the number of days on the job. Is this function continuous

(a) for all values of t?
(b) at $t = 14$?
(c) for all $t \geq 0$?
(d) What is the domain for this application?

45. ***Demand*** Suppose that the demand for a product is defined by the equation

$$p = \frac{200{,}000}{(q + 1)^2}$$

where p is the price and q is the quantity demanded.

(a) Is this function discontinuous at any value of q? What value?
(b) Because q represents quantity, we know that $q \geq 0$. Is this function continuous for $q \geq 0$?

46. ***Advertising and sales*** The sales volume y (in thousands of dollars) is related to advertising expenditures x (in thousands of dollars) according to

$$y = \frac{200x}{x + 10}$$

(a) Is this function discontinuous at any points?
(b) Advertising expenditures x must be nonnegative. Is this function continuous for these values of x?

47. ***Annuities*** If an annuity makes an infinite series of equal payments at the end of the interest periods, it is called a **perpetuity.** If a lump sum investment of A_n is needed to result in n periodic payments of R when the interest rate per period is i, then

$$A_n = R\left[\frac{1 - (1 + i)^{-n}}{i}\right]$$

(a) Evaluate $\lim_{n\to\infty} A_n$ to find a formula for the lump sum payment for a perpetuity.
(b) Find the lump sum investment needed to make payments of \$100 per month in perpetuity if interest is 12%, compounded monthly.

48. ***Response to adrenalin*** Experimental evidence suggests that the response y of the body to the concentration x of injected adrenalin is given by

$$y = \frac{x}{a + bx}$$

where a and b are experimental constants.

(a) Is this function continuous for all x?
(b) On the basis of your conclusion in (a) and the fact that in reality $x \geq 0$ and $y \geq 0$, must a and b be both positive, be both negative, or have opposite signs?

49. ***Cost-benefit*** Suppose that the cost C of removing p percent of the impurities from the waste water in a manufacturing process is given by

$$C(p) = \frac{9800p}{101 - p}$$

Is this function continuous for all those p-values for which the problem makes sense?

50. ***Cost-benefit*** Suppose that the cost C of removing p percent of the particulate pollution from the exhaust gases at an industrial site is given by

$$C(p) = \frac{8100p}{100 - p}$$

Describe any discontinuities for $C(p)$. Explain what each discontinuity means.

51. ***Cost-benefit*** The percentage p of particulate pollution that can be removed from the smokestacks of an industrial plant by spending C dollars is given by

$$p = \frac{100C}{7300 + C}$$

Find the percentage of the pollution that could be removed if spending C were allowed to increase without bound. Can 100% of the pollution be removed? Explain.

52. ***Cost-benefit*** The percentage p of impurities that can be removed from the waste water of a manufacturing process at a cost of C dollars is given by

$$p = \frac{100C}{8100 + C}$$

Find the percentage of the impurities that could be removed if cost were no object (that is, if cost were allowed to increase without bound). Can 100% of the impurities be removed? Explain.

53. ***Federal income tax*** The tax owed by a married couple filing jointly and their tax rates can be found in the following tax rate schedule.

Schedule Y-1—Use if your filing status is Married filing jointly or Qualifying widow(er)

If the amount on Form 1040, line 37, is: over—	*But not over—*	*Enter on Form 1040, line 38*	*of the amount over—*
$0	$41,200	15%	$0
41,200	99,600	$6180 + 28%	41,200
99,600	151,750	22,532 + 31%	99,600
151,750	271,050	38,698.50 + 36%	151,750
271,050	—	81,646.50 + 39.6%	271,050

Source: Internal Revenue Service, 1997 Form 1040 Instructions

From this schedule, the tax rate $R(x)$ is a function of income x (the amount on Form 1040, line 37) as follows.

$$R(x) = \begin{cases} 0.15 & \text{if} \quad 0 \le x \le 41{,}200 \\ 0.28 & \text{if} \quad 41{,}200 < x \le 99{,}600 \\ 0.31 & \text{if} \quad 99{,}600 < x \le 151{,}750 \\ 0.36 & \text{if} \quad 151{,}750 < x \le 271{,}050 \\ 0.396 & \text{if} \quad 271{,}050 < x \end{cases}$$

Identify any discontinuities in $R(x)$.

54. ***Calories and temperature*** Suppose that the number of calories of heat required to raise 1 gram of water (or ice) from $-40°$C to $x°$C is given by

$$f(x) = \begin{cases} \frac{1}{2}x + 20 & \text{if } -40 \le x < 0 \\ x + 100 & \text{if } 0 \le x \end{cases}$$

(a) What can be said about the continuity of the function $f(x)$?

(b) What accounts for the behavior of the function at 0°C?

55. ***Electrical usage costs*** The monthly charge in dollars for x kilowatt hours (kWh) of electricity used by a residential consumer of Excelsior Electric Membership Corporation from November through June is given by the function

$$C(x) = \begin{cases} 10 + .094x & \text{if} \quad 0 \le x \le 100 \\ 19.40 + .075(x - 100) & \text{if} \quad 100 < x \le 500 \\ 49.40 + .05(x - 500) & \text{if} \quad x > 500 \end{cases}$$

(a) What is the monthly charge if 1100 kWh of electricity is consumed in a month?

(b) Find $\lim_{x \to 100} C(x)$ and $\lim_{x \to 500} C(x)$, if the limits exist.

(c) Is C continuous at $x = 100$ and at $x = 500$?

56. ***Postage costs*** First-class postage is 33 cents for the first ounce or part of an ounce that a letter weighs and is an additional 22 cents for each additional ounce or part of an ounce above 1 ounce. Use the table or graph of the postage function, $f(x)$, to determine the following.

(a) $\lim_{x \to 2.5} f(x)$ (b) $f(2.5)$

(c) Is $f(x)$ continuous at 2.5?

(d) $\lim_{x \to 4} f(x)$ (e) $f(4)$

(f) Is $f(x)$ continuous at 4?

Weight x	*Postage f(x)*
$0 < x \le 1$	$0.33
$1 < x \le 2$	0.55
$2 < x \le 3$	0.77
$3 < x \le 4$	0.99
$4 < x \le 5$	1.21

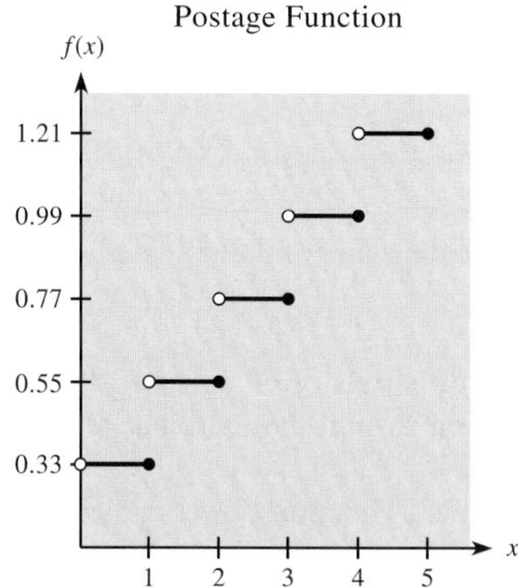

M 57. ***Public debt of the United States*** The interest paid on the public debt of the United States of America as a percentage of federal expenditures for selected years is shown in the following table.

Year	*Interest Paid as a Percentage of Federal Expenditures*	*Point Coordinates if t = 0 in 1900*
1930	0	(30, 0)
1940	10.5	(40, 10.5)
1950	13.4	(50, 13.4)
1955	9.4	(55, 9.4)
1960	10.0	(60, 10.0)
1965	9.6	(65, 9.6)
1970	9.9	(70, 9.9)
1975	9.8	(75, 9.8)
1980	12.7	(80, 12.7)
1985	18.9	(85, 18.9)
1990	21.1	(90, 21.1)
1995	22.0	(95, 22.0)

Source: Bureau of Public Debt, Department of the Treasury

If t is the number of years past 1900, use the table to complete the following.

(a) Use the data in the table to find a cubic and a fourth-degree function that model the percentage of federal expenditures devoted to payment of interest on the public debt. Let $d(t)$ be the one that better fits the data.

(b) Use $d(t)$ to predict the percentage of federal expenditures devoted to payment of interest in 2005.

(c) Calculate $\lim_{t \to +\infty} d(t)$.

(d) Can $d(t)$ be used to predict the percentage of federal expenditures devoted to payment of interest on the public debt for large values of t? Explain.

(e) For what years can you guarantee that $d(t)$ cannot be used to predict the percentage of federal expenditures devoted to payment of interest on the public debt? Explain.

58. ***Postal rates*** The following graphic shows the history of postal rates from 1890 to 1995. Use the figure to answer the following.

(a) If $P(t)$ represents first-class postage in year t, is $P(t)$ continuous?

(b) Identify the longest period of years after World War I when $P(t)$ was continuous.

(c) If $P(t)$ were modeled by a continuous curve, do you think the best model would be linear, exponential, or logarithmic? Explain.

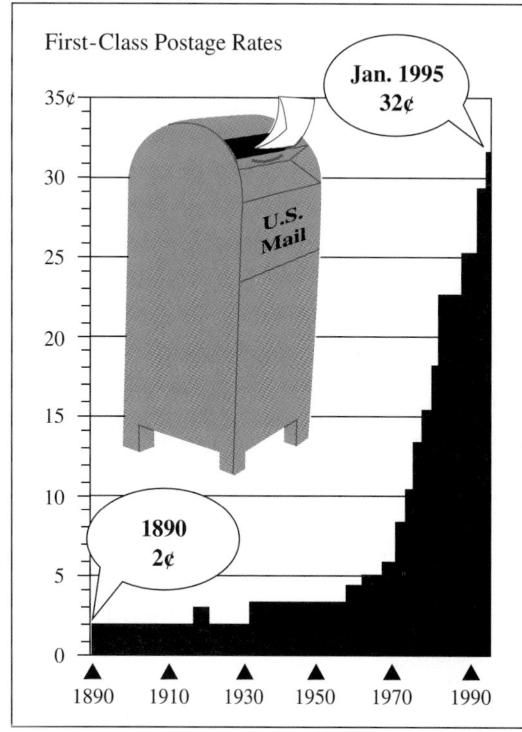

Source: *Oil City Derrick*, Oil City, PA, December 1, 1994.

9.3 The Derivative: Rates of Change; Tangent to a Curve

OBJECTIVES

- *To define the derivative as a rate of change*
- *To use the definition of derivative to find derivatives of functions*
- *To use derivatives to find slopes of tangents to curves*

APPLICATION PREVIEW

Suppose an oil company's revenue (in thousand of dollars) is given by

$$R = 100x - x^2, \quad x \geq 0$$

where x is the number of thousands of barrels of oil sold per day. Then we can use the **derivative** of this function to find the marginal revenue when 20,000 barrels are sold.

In Chapter 1, "Linear Equations and Functions," we studied linear revenue functions and defined the marginal revenue for a product as the rate of change of the revenue function. For linear revenue functions, this rate is also the slope of the line that is the graph of the revenue function. In this section, we will define **marginal revenue** as the rate of change of the revenue function, even when the revenue function is not linear. We will discuss the relationship between the marginal revenue at a given point and the slope of the line tangent to the revenue function at that point. We will see how the derivative of the revenue function can be used to find both the slope of this tangent line and the marginal revenue.

We will begin our study of rates of change (that is, derivatives) by investigating a common rate of change, velocity.

Suppose a ball is thrown straight upward at 64 feet per second from a spot 96 feet above ground level. The equation that describes the height y of the ball after x seconds is

$$y = f(x) = 96 + 64x - 16x^2$$

Figure 9.17 shows the graph of this function for $0 \leq x \leq 5$. The average velocity of the ball over a given time interval is the change in the height divided by the length of time that has passed. Table 9.4 shows some average velocities over time intervals beginning at $x = 1$.

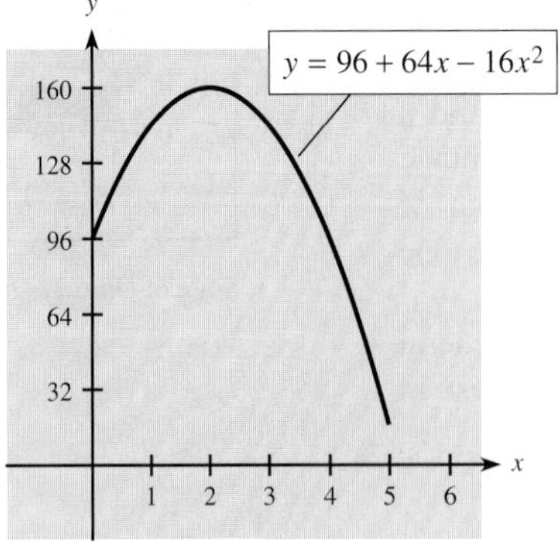

Figure 9.17

TABLE 9.4 Average Velocities

Time			*Height*			
Beginning	*Ending*	*Change* (Δx)	*Beginning*	*Ending*	*Change* (Δy)	*Average Velocity* ($\Delta y/\Delta x$)
1	2	1	144	160	16	16/1 = 16
1	1.5	0.5	144	156	12	12/0.5 = 24
1	1.1	0.1	144	147.04	3.04	3.04/01 = 30.4
1	1.01	0.01	144	144.3184	0.3184	0.3184/0.01 = 31.84

In Table 9.4, the smaller the time interval, the more closely the average velocity approximates the instantaneous velocity at $x = 1$. Thus the instantaneous velocity at $x = 1$ is closer to 31.84 ft/s than to 30.4 ft/s.

If we represent the change in time by h, then the average velocity from $x = 1$ to $x = 1 + h$ approaches the instantaneous velocity at $x = 1$ as h approaches 0. This is illustrated in the following example.

EXAMPLE 1

Suppose a ball is thrown straight upward so that its height $f(x)$ (in feet) is given by the equation

$$f(x) = 96 + 64x - 16x^2$$

where x is time (in seconds).

(a) Find the average velocity from $x = 1$ to $x = 1 + h$.
(b) Find the instantaneous velocity at $x = 1$.

Solution

(a) Let h represent the change in x (time) from 1 to $1 + h$. Then the corresponding change in $f(x)$ (height) is

$$\begin{aligned} f(1+h) - f(1) &= [96 + 64(1+h) - 16(1+h)^2] - [96 + 64 - 16] \\ &= 96 + 64 + 64h - 16(1 + 2h + h^2) - 144 \\ &= 16 + 64h - 16 - 32h - 16h^2 \\ &= 32h - 16h^2 \end{aligned}$$

The average velocity V_{av} is the change in height divided by the change in time.

$$\begin{aligned} V_{av} &= \frac{f(1+h) - f(1)}{h} \\ &= \frac{32h - 16h^2}{h} \\ &= 32 - 16h \end{aligned}$$

(b) The instantaneous velocity V is the limit of the average velocity as h approaches 0.

$$\begin{aligned} V = \lim_{h \to 0} V_{av} &= \lim_{h \to 0} (32 - 16h) \\ &= 32 \text{ ft/s} \end{aligned}$$

Note that average velocity is found over a time interval. Instantaneous velocity is usually called **velocity,** and it can be found at any time x, as follows:

Velocity Suppose that an object moving in a straight line has its position y at time x given by $y = f(x)$. Then the **velocity** of the object at time x is

$$V = \lim_{h\to 0} \frac{f(x+h) - f(x)}{h}$$

provided that this limit exists.

The instantaneous rate of change of any function (commonly called *rate of change*) can be found in the same way we find velocity. The function that gives this instantaneous rate of change of a function f is called the **derivative** of f.

Derivative If f is a function defined by $y = f(x)$, then the **derivative** of $f(x)$ at any value x, denoted $f'(x)$, is

$$f'(x) = \lim_{h\to 0} \frac{f(x+h) - f(x)}{h}$$

if this limit exists. If $f'(c)$ exists, we say that f is **differentiable** at c.

The following procedure illustrates how to find the derivative of a function $y = f(x)$ at any value x.

Derivative Using the Definition

Procedure	***Example***
To find the derivative of $y = f(x)$ at any value x:	Find the derivative of $f(x) = 4x^2$.
1. Let h represent the change in x from x to $x + h$.	1. The change in x from x to $x + h$ is h.
2. The corresponding change in $y = f(x)$ is $f(x + h) - f(x)$	2. The change in $f(x)$ is $f(x + h) - f(x) = 4(x + h)^2 - 4x^2$ $= 4(x^2 + 2xh + h^2) - 4x^2$ $= 4x^2 + 8xh + 4h^2 - 4x^2$ $= 8xh + 4h^2$
3. Form the difference quotient $\frac{f(x+h) - f(x)}{h}$ and simplify.	3. $\frac{f(x+h) - f(x)}{h} = \frac{8xh + 4h^2}{h}$ $= 8x + 4h$
4. Find $\lim_{h\to 0} \frac{f(x+h) - f(x)}{h}$ to determine $f'(x)$, the derivative of $f(x)$.	4. $f'(x) = \lim_{h\to 0} \frac{f(x+h) - f(x)}{h}$ $f'(x) = \lim_{h\to 0} (8x + 4h) = 8x$

Note that in the previous example, we could have found the derivative of the function $f(x) = 4x^2$ at a particular value of x, say $x = 3$, by evaluating the derivative formula at that value:

$$f'(x) = 8x \quad \text{so} \quad f'(3) = 8(3) = 24$$

In addition to $f'(x)$, the derivative at any point x may be denoted by

$$\frac{dy}{dx}, \quad y', \quad \frac{d}{dx}f(x), \quad D_x y, \quad \text{or} \quad D_x f(x)$$

We can, of course, use variables other than x and y to represent functions and their derivatives. For example, we can represent the derivative of the function defined by $p = 2q^2 - 1$ by dp/dq.

CHECKPOINT

1. For the function $y = f(x) = x^2 - x + 1$, find

   (a) $f(x + h) - f(x)$
   (b) $\dfrac{f(x + h) - f(x)}{h}$
   (c) $f'(x) = \lim_{h \to 0} \dfrac{f(x + h) - f(x)}{h}$
   (d) $f'(2)$

In Section 1.6, "Applications of Functions in Business and Economics," we defined the **marginal revenue** for a product as the rate of change of the total revenue function for the product. If the total revenue function for a product is not linear, we define the marginal revenue for the product as the instantaneous rate of change, or the derivative, of the revenue function.

Marginal Revenue

Suppose that the total revenue function for a product is given by $R = R(x)$, where x is the number of units sold. Then the **marginal revenue** at x units is

$$\overline{MR} = R'(x) = \lim_{h \to 0} \frac{R(x + h) - R(x)}{h}$$

provided that the limit exists.

Note that the marginal revenue (derivative of the revenue function) can be found by using the steps in the Procedure/Example above. These steps can also be combined, as they are in Example 2, which is the Application Preview problem.

EXAMPLE 2

Suppose that an oil company's revenue (in thousands of dollars) is given by the equation

$$R = R(x) = 100x - x^2, \quad x \geq 0$$

where x is the number of thousands of barrels of oil sold each day.

(a) Find the function that gives the marginal revenue at any value of x.
(b) Find the marginal revenue when 20,000 barrels are sold (that is, at $x = 20$).

Solution

(a) The marginal revenue function is found by evaluating

$$R'(x) = \lim_{h\to 0} \frac{R(x+h) - R(x)}{h}$$
$$= \lim_{h\to 0} \frac{[100(x+h) - (x+h)^2] - (100x - x^2)}{h}$$
$$= \lim_{h\to 0} \frac{100x + 100h - (x^2 + 2xh + h^2) - 100x + x^2}{h}$$
$$= \lim_{h\to 0} \frac{100h - 2xh - h^2}{h} = \lim_{h\to 0} (100 - 2x - h) = 100 - 2x$$

(b) The marginal revenue function found in (a) gives the marginal revenue at *any* value of x. To find the marginal revenue when 20 units are sold, we evaluate $R'(20)$.

$$R'(20) = 100 - 2(20) = 60$$

Hence the marginal revenue at $x = 20$ is \$60,000. Because the marginal revenue is used to approximate the revenue from the sale of one additional unit, we interpret $R'(20) = 60$ to mean that the expected revenue from the sale of the next thousand barrels (after 20,000) will be approximately \$60,000. [*Note:* The actual revenue from this sale is $R(21) - R(20) = 1659 - 1600 = 59$ (thousand dollars).]

As mentioned earlier, the rate of change of revenue (the marginal revenue) for a linear revenue function is given by the slope of the line. In fact, the slope of the revenue curve gives us the marginal revenue even if the revenue function is not linear. We will show that the slope of the graph of a function at any point is the same as the derivative at that point. In order to show this, we must define the slope of a curve at a point on the curve. We will define the slope of a curve at a point as the slope of the line tangent to the curve at the point.

In geometry, a **tangent** to a circle is defined as a line that has one point in common with the circle. [See Figure 9.18(a).] This definition does not apply to all curves, as Figure 9.18(b) shows. Many lines can be drawn through the point A that touch the curve only at A. One of the lines, line l, looks like it is tangent to the curve.

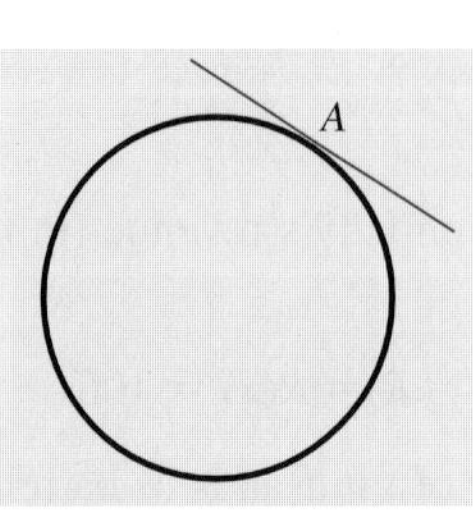

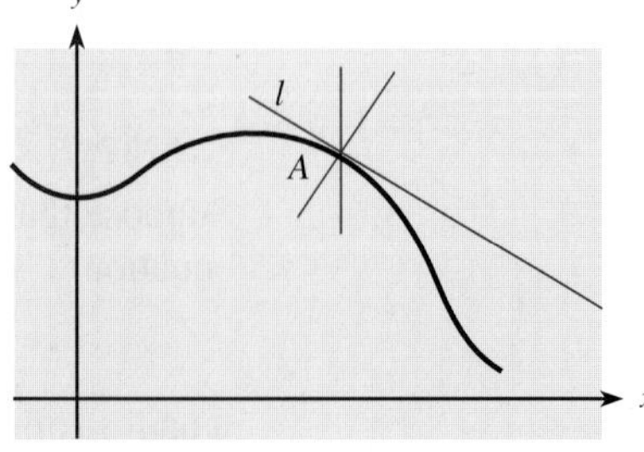

Figure 9.18 (a) (b)

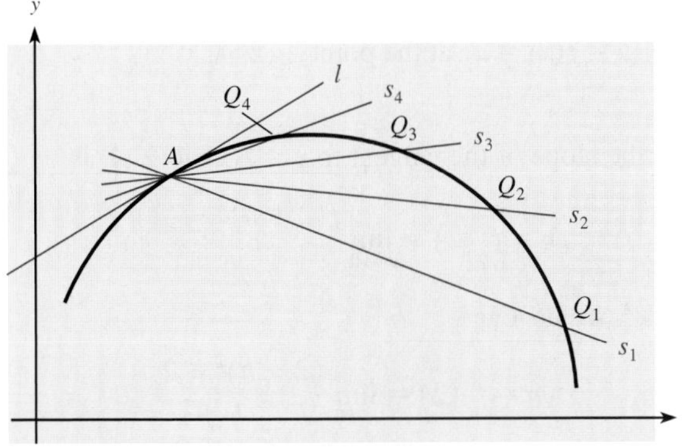

Figure 9.19

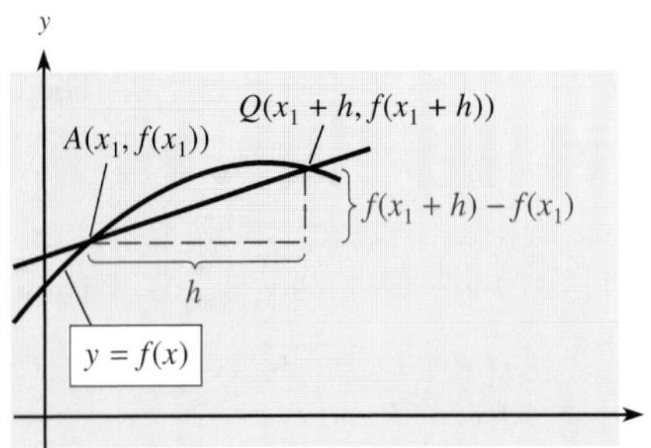

Figure 9.20

We can use **secant lines** (lines that intersect the curve at two points) to determine the tangent to a curve at a point. In Figure 9.19, we have a set of secant lines s_1, s_2, s_3, and s_4 that pass through a point A on the curve and points Q_1, Q_2, Q_3, and Q_4 on the curve near A. The line l represents the tangent line to the curve at point A. We can get a secant line as close as we wish to the tangent line l by choosing a "second point" Q sufficiently close to point A.

As we choose points on the curve closer and closer to A, the limiting position of the secant lines that pass through A is the **tangent line** to the curve at point A, and the slopes of those secant lines approach the slope of the tangent line at A. Thus we can find the slope of the tangent line by finding the slope of a secant line and taking the limit of this slope as the "second point" Q approaches A. To find the slope of the tangent to the graph of $y = f(x)$ at A $(x_1, f(x_1))$, we first draw a secant line from point A to a second point Q $(x_1 + h, f(x_1 + h))$ on the curve (see Figure 9.20).

The slope of this secant line is

$$m_{AQ} = \frac{f(x_1 + h) - f(x_1)}{h}$$

As Q approaches A, we see that the difference between the x-coordinates of these two points decreases, so h approaches 0. Thus the slope of the tangent is given by the following.

Slope of the Tangent

The **slope of the tangent** to the graph of $y = f(x)$ at point $A(x_1, f(x_1))$ is

$$m = \lim_{h \to 0} \frac{f(x_1 + h) - f(x_1)}{h}$$

if this limit exists. That is, $m = f'(x_1)$.

EXAMPLE 3

Find the slope of $y = f(x) = x^2$ at the point A (2, 4).

Solution

The formula for the slope of the tangent to $y = f(x)$ at (2, 4) is

$$m = f'(2) = \lim_{h \to 0} \frac{f(2+h) - f(2)}{h}$$

Thus for $f(x) = x^2$, we have

$$m = f'(2) = \lim_{h \to 0} \frac{(2+h)^2 - 2^2}{h}$$

Taking the limit immediately would result in both the numerator and the denominator approaching 0. To avoid this, we simplify the fraction before taking the limit.

$$m = \lim_{h \to 0} \frac{4 + 4h + h^2 - 4}{h} = \lim_{h \to 0} \frac{4h + h^2}{h} = \lim_{h \to 0} (4 + h) = 4$$

Thus the slope of the tangent to $y = x^2$ at (2, 4) is 4 (see Figure 9.21).

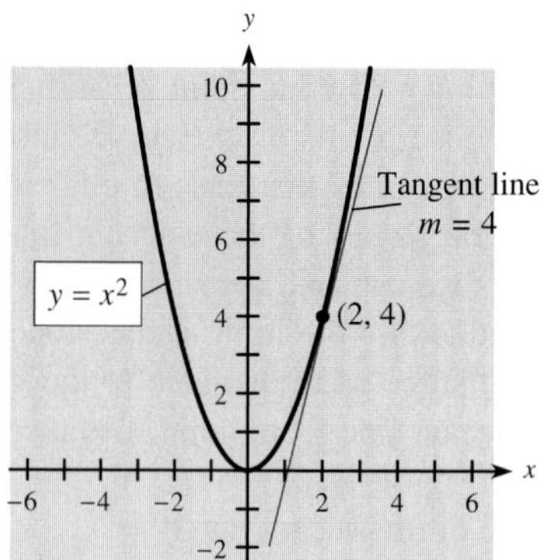

Figure 9.21

The statement "the slope of the tangent to the curve at (2, 4) is 4" is frequently simplified to the statement "the slope of the curve at (2, 4) is 4." Knowledge that the slope is a positive number on an interval tells us that the function is increasing on that interval, which means that a point moving along the graph of the function rises as it moves to the right on that interval. If the derivative (and thus the slope) is negative on an interval, the curve is decreasing on the interval; that is, a point moving along the graph falls as it moves to the right on that interval.

EXAMPLE 4

Given $y = f(x) = 3x^2 + 2x$, find

(a) the derivative of $f(x)$ at any point $(x, f(x))$.
(b) the slope of the curve at (1, 5).
(c) the equation of the line tangent to $y = 3x^2 + 2x$ at (1, 5).

Solution

(a) The derivative of $f(x)$ at any value x is denoted by $f'(x)$ and is

$$\begin{aligned} y' = f'(x) &= \lim_{h\to 0} \frac{f(x+h)-f(x)}{h} \\ &= \lim_{h\to 0} \frac{[3(x+h)^2 + 2(x+h)] - (3x^2+2x)}{h} \\ &= \lim_{h\to 0} \frac{3(x^2+2xh+h^2) + 2x + 2h - 3x^2 - 2x}{h} \\ &= \lim_{h\to 0} \frac{6xh + 3h^2 + 2h}{h} \\ &= \lim_{h\to 0} (6x + 3h + 2) \\ &= 6x + 2 \end{aligned}$$

(b) The derivative is $f'(x) = 6x + 2$, so the slope of the tangent to the curve at $(1, 5)$ is $f'(1) = 6(1) + 2 = 8$.

(c) The equation of the tangent line uses the given point (1, 5) and the slope $m = 8$. It is $y - 5 = 8(x - 1)$, or $y = 8x - 3$.

Technology Note Note in Figure 9.21 that near the point of tangency at (2, 4), the tangent line and the function look coincident. In fact, if we graphed both with a graphing utility and repeatedly zoomed in near the point (2, 4), the two graphs would eventually appear as one. Thus the derivative of $f(x)$ at the point where $x = a$ can be approximated by finding the slope between $(a, f(a))$ and a second point that is nearby.

In addition, we know that the slope of the tangent to $f(x)$ at $x = a$ is defined by

$$f'(a) = \lim_{h\to 0} \frac{f(a+h)-f(a)}{h}$$

Hence we could also estimate $f'(a)$—that is, the slope of the tangent at $x = a$—by evaluating

$$\frac{f(a+h)-f(a)}{h} \quad \text{when } h \approx 0$$

EXAMPLE 5

(a) Let $f(x) = 3x^2 + 2x$. Use $\frac{f(a+h)-f(a)}{h}$ and two values of h to make estimates of the slope of the tangent to $f(x)$ at $x = 3$ on opposite sides of $x = 3$.

(b) Use the following table of values of x and $g(x)$ to estimate $g'(3)$.

x	1	1.9	2.7	2.9	2.999	3	3.002	3.1	4	5
$g(x)$	1.6	4.3	11.4	10.8	10.513	10.5	10.474	10.18	6	−5

Solution

The table feature of a graphing utility can facilitate the following calculations.

(a) We can use $h = 0.0001$ and $h = -0.0001$ as follows:

$$\text{With } h = 0.0001: \quad f'(3) \approx \frac{f(3 + 0.0001) - f(3)}{0.0001} = \frac{f(3.0001) - f(3)}{0.0001} = 20.0003 \approx 20$$

$$\text{With } h = -0.0001: \quad f'(3) \approx \frac{f(3 + (-0.0001)) - f(3)}{-0.0001} = \frac{f(2.9999) - f(3)}{-0.0001} = 19.9997 \approx 20$$

(b) We use the given table and measure the slope between (3, 10.5) and another point that is nearby (the closer, the better). Using (2.999, 10.513), we obtain

$$g'(3) \approx \frac{y_2 - y_1}{x_2 - x_1} = \frac{10.5 - 10.513}{3 - 2.999} = \frac{-0.013}{0.001} = -13$$

Most graphing utilities have a feature called the **numerical derivative** (usually denoted by nDer or nDeriv) that can approximate the derivative of a function at a point. On most utilities this feature uses a calculation similar to our method in Example 5(a). The numerical derivative of $f(x) = 3x^2 + 2x$ with respect to x at $x = 3$ can be found as follows on many graphing utilities:

$$\text{nDeriv}(3x^2 + 2x, x, 3) = 20$$

The discussion in this section indicates that the derivative of a function can be used to accomplish the following.

1. Find the velocity of an object moving in a straight line.
2. Find the instantaneous rate of change of a function.
3. Find the marginal revenue function for a given revenue function.
4. Find the slope of the tangent to the graph of a function.

So far we have talked about how the derivative is defined, what it represents, and how to find it. However, there are functions for which derivatives do not exist at every value of x. Figure 9.22 shows some common cases where $f'(c)$ does not exist but where $f'(x)$ exists for all other values of x. These cases occur where there is a discontinuity, a corner, or a vertical tangent line.

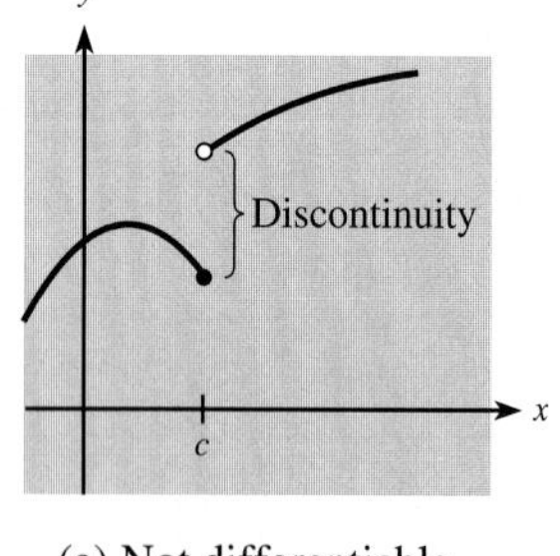

(a) Not differentiable at $x = c$

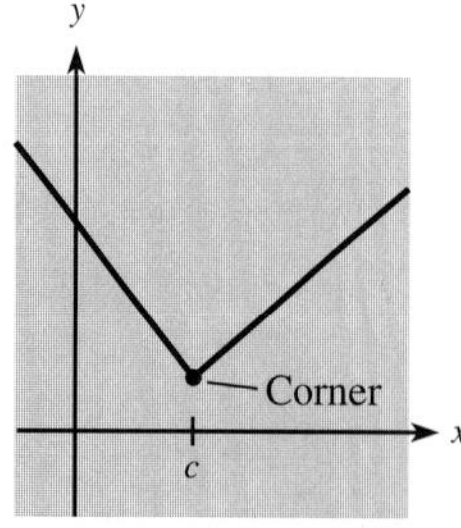

(b) Not differentiable at $x = c$

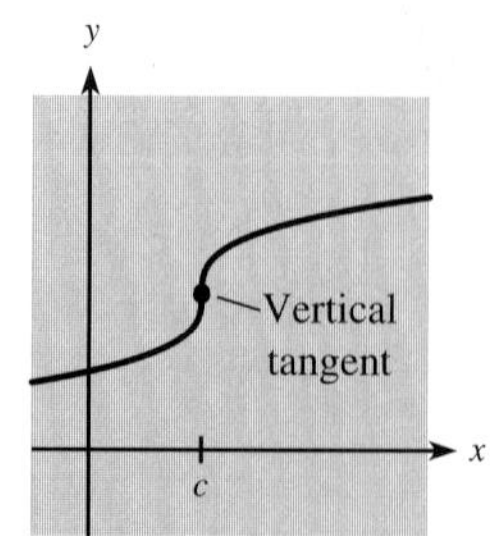

(c) Not differentiable at $x = c$

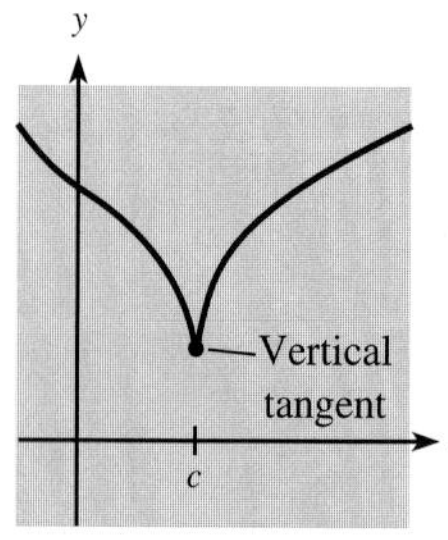

(d) Not differentiable at $x = c$

Figure 9.22

From Figure 9.22 we see that a function may be continuous at $x = c$ even though $f'(c)$ does not exist. Thus continuity does not imply differentiability at a point. However, differentiability does imply continuity.

Differentiability Implies Continuity

If a function f is differentiable at $x = c$, then f is continuous at $x = c$.

EXAMPLE 6

The monthly charge for water in a small town is given by

$$y = f(x) = \begin{cases} 18 & \text{if } 0 \le x \le 20 \\ 0.1x + 16 & \text{if } x > 20 \end{cases}$$

(a) Is this function continuous at $x = 20$?
(b) Is this function differentiable at $x = 20$?

Solution

(a) We must check the three properties for continuity.

1. $f(x) = 18$ for $x \le 20$, so $f(20) = 18$
2. $$\left.\begin{aligned} \lim_{x\to 20^-} f(x) &= \lim_{x\to 20^-} 18 = 18 \\ \lim_{x\to 20^+} f(x) &= \lim_{x\to 20^+} (0.1x + 16) = 18 \end{aligned}\right\} \Rightarrow \lim_{x\to 20} f(x) = 18$$
3. $\lim_{x\to 20} f(x) = f(20)$

Thus $f(x)$ is continuous at $x = 20$.

(b) Because the function is defined differently on either side of $x = 20$, we need to test to see whether $f'(20)$ exists by evaluating both

$$\text{(i)} \lim_{h\to 0^-} \frac{f(20+h) - f(20)}{h} \quad \text{and} \quad \text{(ii)} \lim_{h\to 0^+} \frac{f(20+h) - f(20)}{h}$$

and determining whether they are equal.

$$\text{(i)} \lim_{h\to 0^-} \frac{f(20+h) - f(20)}{h} = \lim_{h\to 0^-} \frac{18 - 18}{h} = \lim_{h\to 0^-} 0 = 0$$

$$\text{(ii)} \lim_{h\to 0^+} \frac{f(20+h) - f(20)}{h} = \lim_{h\to 0^+} \frac{[0.1(20+h) + 16] - 18}{h} = \lim_{h\to 0^+} \frac{0.1h}{h} = \lim_{h\to 0^+} 0.1 = 0.1$$

Because these limits are not equal, the derivative does not exist.

CHECKPOINT

2. Which of the following are given by $f'(c)$?
   (a) The slope of the tangent when $x = c$
   (b) The y-coordinate of the point where $x = c$
   (c) The instantaneous rate of change of $f(x)$ at $x = c$
   (d) The marginal revenue at $x = c$, if $f(x)$ is the revenue function
3. Must a graph that has no discontinuity, corner, or cusp at $x = c$ be differentiable at $x = c$?

EXAMPLE 7

If the point (a, b) lies on the graph of $y = x^2$, then the equation of the secant line to $y = x^2$ from $(1, 1)$ to (a, b) has the equation

$$y - 1 = \frac{b-1}{a-1}(x-1), \quad \text{or} \quad y = \frac{b-1}{a-1}(x-1) + 1$$

(a) Write the equation of the secant line from $(1, 1)$ to $(5, 25)$ and graph $y = x^2$ and this secant line.

(b) Write the equation of the secant line from $(1, 1)$ to $(3, 9)$ and graph $y = x^2$ and this secant line.

(c) Write the equation of the secant line from $(1, 1)$ to $(1.01, 1.0201)$ and graph $y = x^2$ and this secant line.

(d) Which secant line appears as if it might be closest to the tangent line at $(1, 1)$?

(e) Express the slope of the secant line from $(1, 1)$ to (a, b) in terms of a and find the limit of this slope as $a \to 1$. Is this limit the slope of the tangent line to $y = x^2$ at $(1, 1)$?

Solution

(a) The equation of the secant line from $(1, 1)$ to $a = 5$, $b = 25$ is

$$y = \frac{25-1}{5-1}(x-1) + 1, \quad \text{or} \quad y = 6x - 5$$

The graph of $y = x^2$ and the secant line are shown in Figure 9.23(a).

(b) The equation of the secant line from $(1, 1)$ to $a = 3$, $b = 9$ is

$$y = \frac{9-1}{3-1}(x-1) + 1, \quad \text{or} \quad y = 4x - 3$$

The graph of $y = x^2$ and the secant line are shown in Figure 9.23(b).

(c) The equation of the secant line from $(1, 1)$ to $a = 1.01$, $b = 1.0201$ is

$$y = \frac{1.0201 - 1}{1.01 - 1}(x-1) + 1, \quad \text{or} \quad y = 2.01x - 1.01$$

The graph of $y = x^2$ and the secant line are shown in Figure 9.23(c).

(d) The secant line from $(1, 1)$ to $(1.01, 1.0201)$ is closest to the tangent line at $(1, 1)$.

(e) The slope of the secant line from $(1, 1)$ to (a, b) is

$$\frac{b-1}{a-1} = \frac{a^2-1}{a-1}$$

The limit of this slope as a approaches 1, the x-value of the point $(1, 1)$, is

$$\lim_{a \to 1} \frac{b-1}{a-1} = \lim_{a \to 1} \frac{a^2-1}{a-1} = \lim_{a \to 1}(a+1) = 2$$

This limit, 2, is the slope of the tangent line at $(1, 1)$. That is, the derivative of $y = x^2$ at $(1, 1)$ is 2. [Note that a graphing utility's calculation of the numerical derivative of $f(x) = x^2$ with respect to x at $x = 1$ gives $f'(1) = 2$.]

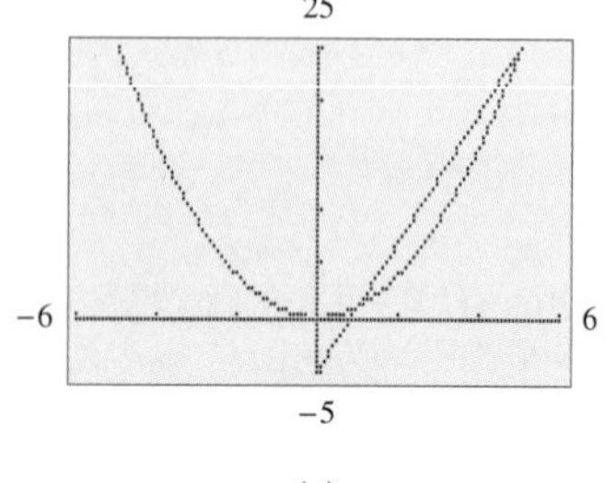

(a)

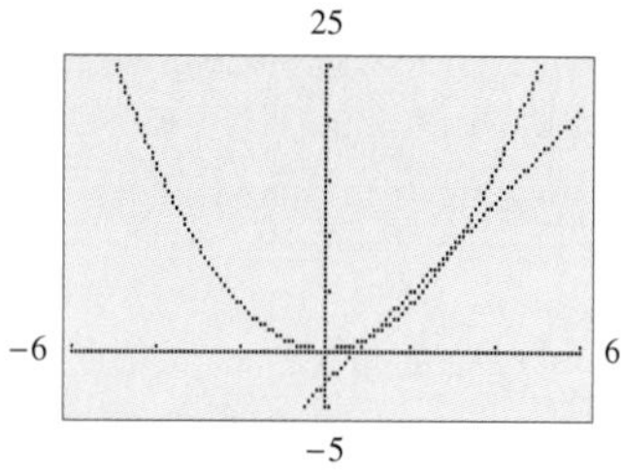

(b)

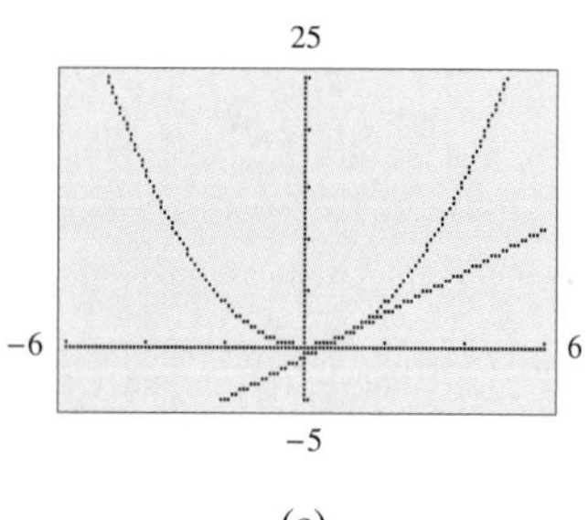

(c)

Figure 9.23

CHECKPOINT SOLUTIONS

1. (a) $f(x + h) - f(x) = [(x + h)^2 - (x + h) + 1] - (x^2 - x + 1)$
$= x^2 + 2xh + h^2 - x - h + 1 - x^2 + x - 1$
$= 2xh + h^2 - h$

(b) $\dfrac{f(x+h) - f(x)}{h} = \dfrac{2xh + h^2 - h}{h}$
$= 2x + h - 1$

(c) $f'(x) = \lim_{h \to 0} \dfrac{f(x+h) - f(x)}{h} = \lim_{h \to 0} (2x + h - 1)$
$= 2x - 1$

(d) $f'(x) = 2x - 1$, so $f'(2) = 3$.

2. Parts (a), (c), and (d) are given by $f'(c)$. The y-coordinate where $x = c$ is given by $f(c)$.
3. No. Figure 9.22(c) shows such an example.

EXERCISE 9.3

1. In the Procedure/Example table in this section we were given $f(x) = 4x^2$ and found $f'(x) = 8x$. Find
   (a) the instantaneous rate of change of $f(x)$ at $x = 4$.
   (b) the slope of the tangent to the graph of $y = f(x)$ at $x = 4$.
   (c) the point on the graph of $y = f(x)$ at $x = 4$.
2. In Example 4 of this section we were given $f(x) = 3x^2 + 2x$ and found $f'(x) = 6x + 2$. Find
   (a) the instantaneous rate of change of $f(x)$ at $x = 6$.
   (b) the slope of the tangent to the graph of $y = f(x)$ at $x = 6$.
   (c) the point on the graph of $y = f(x)$ at $x = 6$.
3. Let $f(x) = 2x^2 - x$.
   (a) Use the Procedure/Example in this section to verify that $f'(x) = 4x - 1$.
   (b) Find the instantaneous rate of change of $f(x)$ at $x = -1$.
   (c) Find the slope of the tangent to the graph of $y = f(x)$ at $x = -1$.
   (d) Find the point on the graph of $y = f(x)$ at $x = -1$.
4. Let $f(x) = 9 - \frac{1}{2}x^2$.
   (a) Use the Procedure/Example in this section to verify that $f'(x) = -x$.
   (b) Find the instantaneous rate of change of $f(x)$ at $x = 2$.
   (c) Find the slope of the tangent to the graph of $y = f(x)$ at $x = 2$.
   (d) Find the point on the graph of $y = f(x)$ at $x = 2$.

In Problems 5–8, the tangent line to the graph of $f(x)$ at $x = 1$ is shown. On the tangent line, P is the point of tangency and A is another point on the line.
(a) Find the coordinates of the points P and A.
(b) Use the coordinates of P and A to find the slope of the tangent line.
(c) Find $f'(1)$.
(d) Find the instantaneous rate of change of $f(x)$ at P.

5. 6.

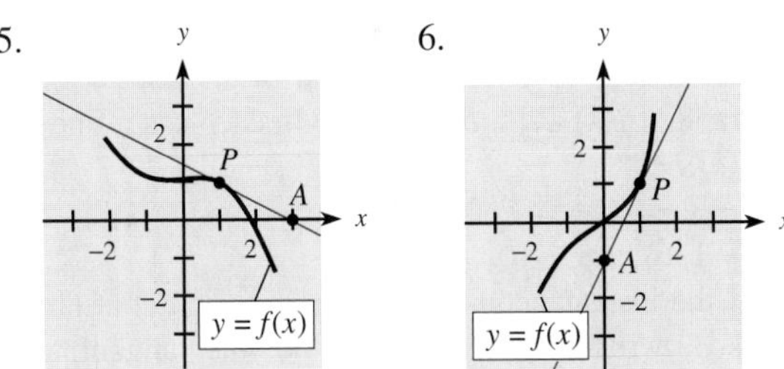

7. 8.

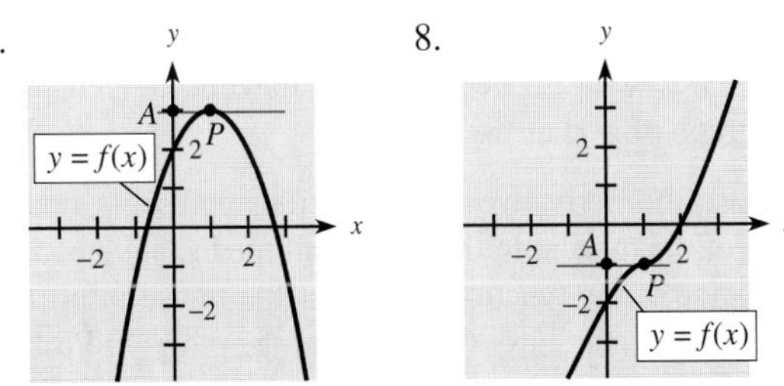

For each function in Problems 9–14, find
(a) the derivative, by using the definition.
(b) the instantaneous rate of change of the function at any value and at the given value.
(c) the slope of the tangent at the given value.

9. $f(x) = 1 - 6x;\quad x = 20$
10. $f(x) = 4 - 5x;\quad x = -8$
11. $f(x) = 4x^2 - 2x + 1;\quad x = -3$
12. $f(x) = 16x^2 - 4x + 2;\quad x = 1$
13. $p(q) = q^2 + 4q + 1;\quad q = 5$
14. $p(q) = 2q^2 - 4q + 5;\quad q = 2$

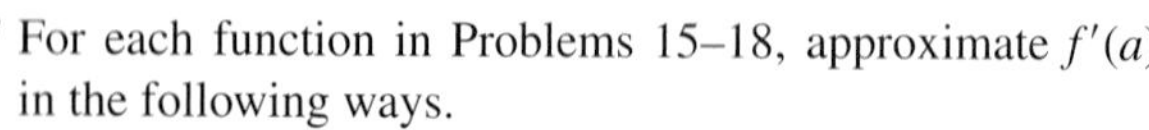

For each function in Problems 15–18, approximate $f'(a)$ in the following ways.

(a) Use the numerical derivative feature of a graphing utility.

(b) Use $\dfrac{f(a+h) - f(a)}{h}$ with $h = 0.0001$.

(c) Graph the function on a graphing utility. Then zoom in near the point until the graph appears straight, pick two points, and find the slope of the line you see.

15. $f'(2)$ for $f(x) = 3x^4 - 7x - 5$
16. $f'(-1)$ for $f(x) = 2x^3 - 11x + 9$
17. $f'(4)$ for $f(x) = (2x - 1)^3$
18. $f'(3)$ for $f(x) = \dfrac{3x + 1}{2x - 5}$

In Problems 19 and 20, use the given tables to approximate $f'(a)$ as accurately as you can.

19. $a = 13$

x	12.0	12.99	13	13.1
$f(x)$	1.41	17.42	17.11	22.84

20. $a = -7.5$

x	-7.4	-7.50	-7.51	-7
$f(x)$	22.12	22.351	22.38	24.12

In Problems 21 and 22, a point (a, b) on the graph of $y = f(x)$ is given, and the equation of the line tangent to the graph of $f(x)$ at (a, b) is given. In each case, find $f'(a)$ and $f(a)$.

21. $(-3, -9);\quad 5x - 2y = 3$
22. $(-1, 6);\quad x + 10y = 59$
23. If the instantaneous rate of change of $f(x)$ at $(1, -1)$ is 3, write the equation of the line tangent to the graph of $f(x)$ at $x = 1$.
24. If the instantaneous rate of change of $g(x)$ at $(-1, -2)$ is 1/2, write the equation of the line tangent to the graph of $g(x)$ at $x = -1$.

Because the derivative of a function represents both the slope of the tangent to the curve and the instantaneous rate of change of the function, it is possible to use information about one to gain information about the other. In Problems 25 and 26, use the graph of the function $y = f(x)$ given in Figure 9.24.

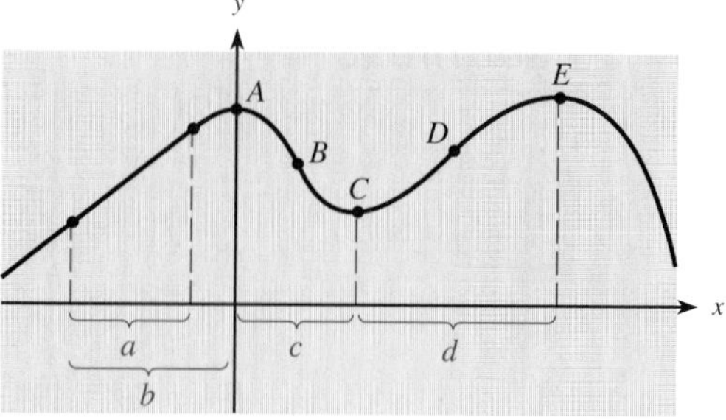

Figure 9.24

25. (a) Over what interval(s) (a) through (d) is the rate of change of $f(x)$ positive?
    (b) Over what interval(s) (a) through (d) is the rate of change of $f(x)$ negative?
    (c) At what point(s) A through E is the rate of change of $f(x)$ equal to zero?
26. (a) At what point(s) A through E does the rate of change of $f(x)$ change from positive to negative?
    (b) At what point(s) A through E does the rate of change of $f(x)$ change from negative to positive?
27. Given the graph of $y = f(x)$ in Figure 9.25, determine for which x-values A, B, C, D, or E the function is
    (a) continuous.
    (b) differentiable.
28. Given the graph of $y = f(x)$ in Figure 9.25, determine for which x-values F, G, H, I, or J the function is
    (a) continuous.
    (b) differentiable.

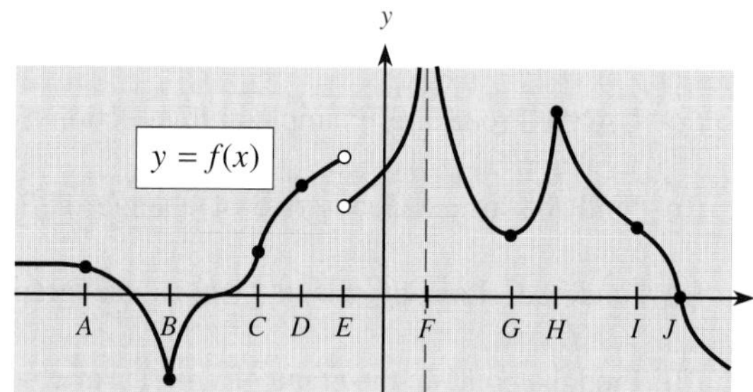

Figure 9.25

In Problems 29–32, (a) find the slope of the tangent to the graph of $f(x)$ at any point, (b) find the slope of the tangent at the given x-value, (c) write the equation of the line tangent to the graph of $f(x)$ at the given point, and (d) graph both $f(x)$ and its tangent line (use a graphing utility if one is available).

29. (a) $f(x) = x^2 + x$
(b) $x = 2$
(c) $(2, 6)$

30. (a) $f(x) = x^2 + 3x$
(b) $x = -1$
(c) $(-1, -2)$

31. (a) $f(x) = x^3 + 3$
(b) $x = 1$
(c) $(1, 4)$

32. (a) $f(x) = 5x^3 + 2$
(b) $x = -1$
(c) $(-1, -3)$

Applications

33. ***Average velocity*** When a ball is dropped from a height of 256 feet, its position (height above the ground) after x seconds is given by

$$S(x) = 256 - 16x^2$$

(a) What is the average velocity in the first 2 seconds of the fall?
(b) What does the negative average velocity in (a) mean?

34. ***Average velocity*** If an object is thrown upward at 64 ft/s from a height of 20 feet, its height S after x seconds is given by

$$S(x) = 20 + 64x - 16x^2$$

What is the average velocity in the
(a) first 2 seconds after it is thrown?
(b) next 2 seconds?

35. ***Demand*** If the demand for a product is given by

$$D(p) = \frac{1000}{\sqrt{p}} - 1$$

what is the average rate of change of demand when p increases from
(a) 1 to 25?
(b) 25 to 100?

36. ***Revenue*** If the total revenue function for a blender is

$$R(x) = 36x - 0.01x^2$$

where x is the number of units sold, what is the average rate of change in revenue $R(x)$ as x increases from 10 to 20 units?

37. ***Speed limit enforcement*** If 0.05 second elapses while a car on a road travels over two sensors that are 5 feet apart, what is the average velocity (in miles per hour) of the car as it travels between the sensors (88 feet per second is equivalent to 60 miles per hour)?

38. ***Speed limit enforcement*** One speed-check system used by local police measures the elasped time between two marks 0.1 mile apart. If 5.85 seconds elapse while a car on the highway travels between the two marks, find the average velocity of the car (in miles per hour) as it travels between the two marks.

39. ***Marginal revenue*** Say the revenue function for a stereo system is

$$R(x) = 300x - x^2$$

where x denotes the number of units sold.
(a) What is the function that gives marginal revenue?
(b) What is the marginal revenue if 50 units are sold and what does it mean?
(c) What is the marginal revenue if 200 units are sold and what does it mean?
(d) What is the marginal revenue if 150 units are sold and what does it mean?
(e) As the number of units sold passes through 150, what happens to revenue?

40. ***Marginal revenue*** Suppose the total revenue function for a blender is

$$R(x) = 36x - 0.01x^2$$

where x is the number of units sold.
(a) What function gives the marginal revenue?
(b) What is the marginal revenue when 600 units are sold and what does it mean?
(c) What is the marginal revenue when 2000 units are sold and what does it mean?
(d) What is the marginal revenue when 1800 units are sold and what does it mean?

41. ***Labor force and output*** The monthly output at the Olek Carpet Mill is

$$Q(x) = 15{,}000 + 2x^2 \text{ units,} \quad (40 \le x \le 60)$$

where x is the number of workers employed at the mill. If there are currently 50 workers, find the instantaneous rate of change of monthly output with respect to the number of workers. That is, find $Q'(50)$.

42. ***Consumer expenditure*** Suppose that the demand x for a product is

$$x = 10{,}000 - 100p$$

where p dollars is the price per unit. Then the consumer expenditure for the product is

$$E(p) = px = p(10{,}000 - 100p)$$
$$= 10{,}000p - 100p^2$$

What is the instantaneous rate of change of consumer expenditure with respect to price at
(a) any price p? (b) $p = 5$? (c) $p = 20$?

In Problems 43–46, find derivatives with the numerical derivative feature of a graphing utility.

43. ***Profit*** Suppose that the profit function for the monthly sales of a car by a dealership is

$$P(x) = 500x - x^2 - 100$$

where x is the number of cars sold. What is the instantaneous rate of change of profit when

(a) 200 cars are sold? Explain its meaning.
(b) 300 cars are sold? Explain its meaning.

44. ***Profit*** If the total revenue function for a toy is

$$R(x) = 2x$$

and the total cost function is

$$C(x) = 100 + 0.2x^2 + x$$

what is the instantaneous rate of change of profit if 10 units are produced and sold? Explain its meaning.

45. ***Heat index*** The highest recorded temperature in the state of Alaska was 100°F and occurred on June 27, 1915, at Fort Yukon. The *heat index* is the apparent temperature of the air at a given temperature and humidity level. If x denotes the relative humidity (in percent), then the heat index (in degrees Fahrenheit) for an air temperature of 100°F can be approximated with the function

$$f(x) = 0.009x^2 + 0.139x + 91.875$$

(a) At what rate is the heat index changing when the humidity is 50%?
(b) Write a sentence that explains the meaning of your answer in (a).

46. ***Receptivity*** In learning theory, receptivity is defined as the ability of students to understand a complex concept. Receptivity is highest when the topic is introduced and tends to decrease as time passes in a lecture. Suppose that the receptivity of a group of students in a mathematics class is given by

$$g(t) = -0.2t^2 + 3.1t + 32$$

where t is minutes after the lecture begins.

(a) At what rate is receptivity changing 10 minutes after the lecture begins?
(b) Write a sentence that explains the meaning of your answer in (a).

47. ***HIV infections*** The figure shows the number of world HIV infections since AIDS was discovered in 1981. The data are accurate through 1992, but are estimated beyond that time.

(a) Find the average rate of change in world HIV infections from 1981 to 1992.
(b) Find the average rates of change in world HIV infections from 1995 to 2000, using first the low estimate and then the high estimate.

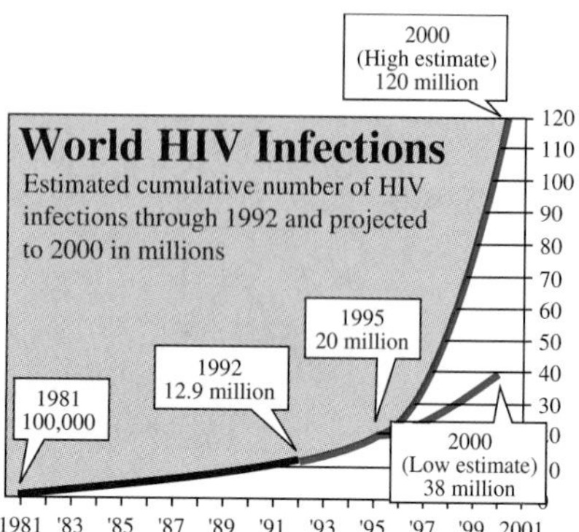

SOURCE: World Health Organization and the Global AIDS Policy Coalition, as reported in the *El Paso Times*, December 6, 1992.

48. ***Mutual funds*** The figure shows the history of the growth of mutual funds.

(a) Find the average annual rate of change in the number of funds from 1950 to 1960.
(b) Find the average rate of change in the number of funds from 1980 to 1990.

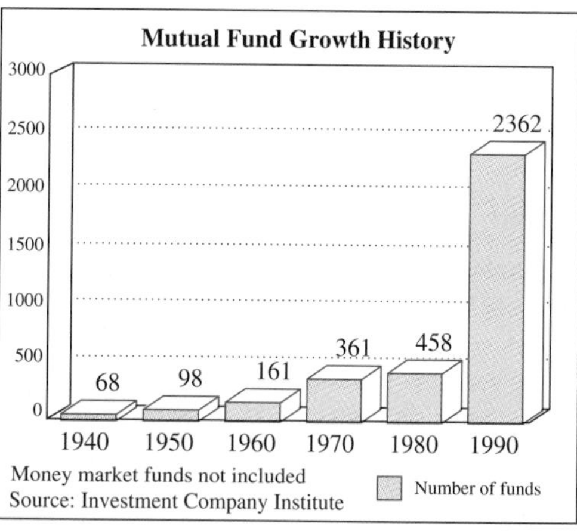

Published in *Investment Digest* of Valic Co., Vol. 5, No. 2, Summer, 1992.

9.4 Derivative Formulas

OBJECTIVES

- To find derivatives of powers of x
- To find derivatives of constant functions
- To find derivatives of functions involving constant coefficients
- To find derivatives of sums and differences of functions

APPLICATION PREVIEW

The killer bees bred in South America have entered the United States in spite of efforts to halt their spread. The first bees were recorded entering California in 1985 and were quickly destroyed. However, the first American was killed by the bees on July 15, 1993. Suppose that the bees enter a county in Texas and that the bee population in that county grows over a 6-week period, with the number of bees given by the equation

$$P(t) = 2t^2 + 10t + 1$$

where t is the number of weeks since the first bee is discovered. We can find the rate of growth of the bee population 2 weeks after the first bee is discovered by using the derivative $P'(t)$ of the growth function.

As we discussed in the previous section, the derivative of a function can be used to find the rate of change of the function. In this section we will develop formulas that will make it easier to find certain derivatives.

We can use the definition of derivative to show the following:

$$\text{If } f(x) = x^2, \text{ then } f'(x) = 2x.$$
$$\text{If } f(x) = x^3, \text{ then } f'(x) = 3x^2.$$
$$\text{If } f(x) = x^4, \text{ then } f'(x) = 4x^3.$$
$$\text{If } f(x) = x^5, \text{ then } f'(x) = 5x^4.$$

Do you recognize a pattern that could be used to find the derivative of $f(x) = x^6$? What is the derivative of $f(x) = x^n$? If you said that the derivative of $f(x) = x^6$ is $f'(x) = 6x^5$ and the derivative of $f(x) = x^n$ is $f'(x) = nx^{n-1}$, you're right. We can use the definition of derivative to show this. If n is a positive integer, then

$$\begin{aligned} f'(x) &= \lim_{h\to 0} \frac{f(x+h) - f(x)}{h} \\ &= \lim_{h\to 0} \frac{(x+h)^n - x^n}{h} \end{aligned}$$

Because we are assuming that n is a positive integer, we can use the binomial formula to expand $(x + h)^n$. You may recall from Section 8.3 that this formula is stated as follows:

$$(a+b)^n = a^n + na^{n-1}b + \frac{n(n-1)}{1 \cdot 2}a^{n-2}b^2 + \cdots + b^n$$

Thus replacing a with x and b with h gives

$$\begin{aligned} f'(x) &= \lim_{h\to 0} \frac{\left[x^n + nx^{n-1}h + \frac{n(n-1)}{1 \cdot 2}x^{n-2}h^2 + \cdots + h^n\right] - x^n}{h} \\ &= \lim_{h\to 0} \left[nx^{n-1} + \frac{n(n-1)}{1 \cdot 2}x^{n-2}h + \cdots + h^{n-1}\right] \end{aligned}$$

Now, each term after nx^{n-1} contains h as a factor, so all terms except nx^{n-1} will approach 0 as $h \to 0$. Thus

$$f'(x) = nx^{n-1}$$

Even though we proved this derivative rule only for the case when n is a positive integer, the rule applies for any real number n.

Powers of x Rule If $f(x) = x^n$, where n is a real number, then $f'(x) = nx^{n-1}$.

EXAMPLE 1

Find the derivatives of the following functions.

(a) $g(x) = x^6$ (b) $f(x) = x^{-2}$
(c) $y = x^4$ (d) $y = x^{1/3}$

Solution

(a) If $g(x) = x^6$, then $g'(x) = 6x^{6-1} = 6x^5$.
(b) The Powers of x Rule applies for all real values. Thus for $f(x) = x^{-2}$, we have

$$f'(x) = -2x^{-2-1} = -2x^{-3} = \frac{-2}{x^3}$$

(c) If $y = x^4$, then $dy/dx = 4x^{4-1} = 4x^3$.
(d) The Powers of x Rule applies to $y = x^{1/3}$.

$$\frac{dy}{dx} = \frac{1}{3}x^{1/3-1} = \frac{1}{3}x^{-2/3} = \frac{1}{3x^{2/3}}$$

In Example 1 we took the derivative with respect to x of *both sides* of each equation. We denote the operation "take the derivative with respect to x" by $\frac{d}{dx}$. Thus for $y = x^4$, in (c),

$$\frac{d}{dx}(y) = \frac{d}{dx}(x^4) \quad \text{gives} \quad \frac{dy}{dx} = 4x^3$$

Similarly, for $f(x) = x^{-2}$, in (b),

$$\frac{d}{dx}[f(x)] = \frac{d}{dx}(x^{-2}) \quad \text{gives} \quad f'(x) = -2x^{-3}$$

The differentiation rules are stated and proved for the independent variable x, but they also apply to other independent variables. The following examples illustrate differentiation with variables other than x.

EXAMPLE 2

Find the derivatives of the following functions.

(a) $u(s) = s^8$ (b) $p = q^{2/3}$ (c) $C(t) = \sqrt{t}$ (d) $s = \frac{1}{\sqrt{t}}$

Solution

(a) If $u(s) = s^8$, then $u'(s) = 8s^{8-1} = 8s^7$.

(b) If $p = q^{2/3}$, then

$$\frac{dp}{dq} = \frac{2}{3}q^{2/3-1} = \frac{2}{3}q^{-1/3} = \frac{2}{3q^{1/3}}$$

(c) Writing $\sqrt{t}$ in its equivalent form, $t^{1/2}$, permits us to use the derivative formula.

$$C'(t) = \frac{1}{2}t^{1/2-1} = \frac{1}{2}t^{-1/2}$$

Writing the derivative in radical form gives

$$C'(t) = \frac{1}{2} \cdot \frac{1}{t^{1/2}} = \frac{1}{2\sqrt{t}}$$

(d) Writing $1/\sqrt{t}$ as a power of t gives

$$s = \frac{1}{t^{1/2}} = t^{-1/2}, \quad \text{so} \quad \frac{ds}{dt} = -\frac{1}{2}t^{-1/2-1} = -\frac{1}{2}t^{-3/2}$$

Writing the derivative in a form similar to that of the original function gives

$$\frac{ds}{dt} = -\frac{1}{2} \cdot \frac{1}{t^{3/2}} = -\frac{1}{2\sqrt{t^3}}$$

EXAMPLE 3

Find the slope of the tangent to the curve $y = x^3$ at $x = 1$.

Solution

Finding the slope of the tangent line to $y = x^3$ at $x = 1$ involves two steps.

1. Find the derivative of $y = x^3$.

$$y' = 3x^2$$

2. Evaluate the derivative at $x = 1$.

$$m_{\tan} = y'\big|_{x=1} = y'(1) = 3(1)^2 = 3$$

The graph of $y = x^3$ and the tangent line to the graph at $x = 1$, $y = 1^3 = 1$ are shown in Figure 9.26.

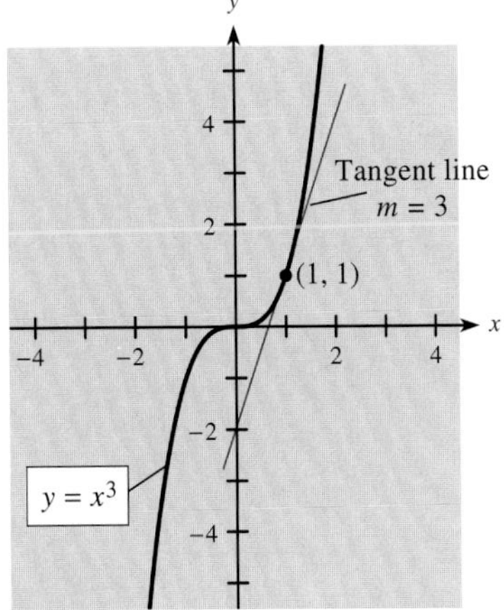

Figure 9.26

A function of the form $y = f(x) = c$, where c is a constant, is called a **constant function.** We can show that the derivative of a constant function is 0, as follows:

$$f'(x) = \lim_{h \to 0} \frac{f(x+h) - f(x)}{h} = \lim_{h \to 0} \frac{c - c}{h} = \lim_{h \to 0} 0 = 0$$

We can state this rule formally.

Constant Function Rule If $f(x) = c$, where c is a constant, then $f'(x) = 0$.

EXAMPLE 4

Find the derivative of the function defined by $y = 4$.

Solution

Because 4 is a constant, $\dfrac{dy}{dx} = 0$.

Recall that the function defined by $y = 4$ has a horizontal line as its graph. Thus the slope of the line (and the derivative of the function) is 0.

We now can take derivatives of constant functions and powers of x. But we do not yet have a rule for taking derivatives of functions of the form $f(x) = 4x^5$ or $g(t) = \frac{1}{2}t^2$. The following rule provides a method for handling functions of this type.

Coefficient Rule If $f(x) = c \cdot u(x)$, where c is a constant and $u(x)$ is a differentiable function of x, then $f'(x) = c \cdot u'(x)$.

The above formula says that the derivative of a constant times a function is the constant times the derivative of the function.

We can use the fact that

$$\lim_{h \to 0} c \cdot g(h) = c \cdot \lim_{h \to 0} g(h)$$

which was discussed in the section "Limits," to verify the coefficient rule. If $f(x) = c \cdot u(x)$, then

$$f'(x) = \lim_{h \to 0} \frac{f(x+h) - f(x)}{h} = \lim_{h \to 0} \frac{c \cdot u(x+h) - c \cdot u(x)}{h}$$

$$= \lim_{h \to 0} c \cdot \left[\frac{u(x+h) - u(x)}{h}\right] = c \cdot \lim_{h \to 0} \frac{u(x+h) - u(x)}{h}$$

$$\text{so } f'(x) = c \cdot u'(x)$$

EXAMPLE 5

Find the derivatives of the following functions.

(a) $f(x) = 4x^5$ (b) $f(t) = \frac{1}{2}t^2$ (c) $p = \frac{5}{\sqrt{q}}$

Solution

(a) $f'(x) = 4(5x^4) = 20x^4$ (b) $g'(t) = \frac{1}{2}(2t) = t$

(c) $p = \frac{5}{\sqrt{q}} = 5q^{-1/2}$, so

$$\frac{dp}{dq} = 5\left(-\frac{1}{2}q^{-3/2}\right) = -\frac{5}{2\sqrt{q^3}}$$

In Example 4 of Section 9.3, "The Derivative: Rates of Change; Tangent to a Curve," we found the derivative of $f(x) = 3x^2 + 2x$ to be $f'(x) = 6x + 2$. This result, along with the results of several of the derivatives calculated in the exercise for that section, suggest that we can find the derivative of a function by finding the derivatives of its terms and combining them. The following rules state this formally.

Sum Rule If $f(x) = u(x) + v(x)$, where u and v are differentiable functions of x, then $f'(x) = u'(x) + v'(x)$.

We can prove this rule as follows. If $f(x) = u(x) + v(x)$, then

$$\begin{aligned} f'(x) &= \lim_{h\to 0} \frac{f(x+h) - f(x)}{h} \\ &= \lim_{h\to 0} \frac{[u(x+h) + v(x+h)] - [u(x) + v(x)]}{h} \\ &= \lim_{h\to 0} \left[\frac{u(x+h) - u(x)}{h} + \frac{v(x+h) - v(x)}{h}\right] \\ &= \lim_{h\to 0} \frac{u(x+h) - u(x)}{h} + \lim_{h\to 0} \frac{v(x+h) - v(x)}{h} \\ &= u'(x) + v'(x) \end{aligned}$$

Difference Rule If $f(x) = u(x) - v(x)$, where u and v are differentiable functions of x, then $f'(x) = u'(x) - v'(x)$.

EXAMPLE 6

Find the derivatives of the following functions.

(a) $y = x^2 + 3$ (b) $p = q^2 - 4q$ (c) $y = 3x + 5$

Solution

(a) $y' = 2 \cdot x + 0 = 2x$
(b) $dp/dq = 2 \cdot q - 4 \cdot 1 = 2q - 4$
(c) $y' = 3 \cdot 1 + 0 = 3$

In Example 6(c) we saw that the derivative of $y = 3x + 5$ is 3. Because the slope of a line is the same at all points on the line, it is reasonable that the derivative of a linear equation is a constant. In particular, the slope of the graph of the equation $y = mx + b$ is m at all points on its graph because the derivative of $y = mx + b$ is $y' = f'(x) = m$.

The rules regarding the derivatives of sums and differences of two functions also apply if more than two functions are involved. For example, the derivative of $f(x) = 4x^3 - 2x^2 + 5x - 3$ is $f'(x) = 12x^2 - 4x + 5$. We may think of the functions that are added and subtracted as terms of the function f. Then it would be correct to say that we may take the derivative of a function term by term.

EXAMPLE 7

Find the derivatives of the following functions.

(a) $y = 3x^3 - 4x^2$
(b) $p = \frac{1}{3}q^3 + 2q^2 - 3$
(c) $u(x) = 5x^4 + x^{1/3}$
(d) $y = 4x^3 + \sqrt{x}$
(e) $s = 5t^6 - \dfrac{1}{t^2}$

Solution

(a) $y' = 3(3x^2) - 4(2x) = 9x^2 - 8x$
(b) $\dfrac{dp}{dq} = \frac{1}{3}(3q^2) + 2(2q) - 0 = q^2 + 4q$
(c) $u'(x) = 5(4x^3) + \frac{1}{3}x^{-2/3} = 20x^3 + \dfrac{1}{3x^{2/3}}$
(d) We may write the function as

$$y = 4x^3 + x^{1/2}$$

so

$$y' = 4(3x^2) + \frac{1}{2}x^{-1/2} = 12x^2 + \frac{1}{2x^{1/2}}$$
$$= 12x^2 + \frac{1}{2\sqrt{x}}$$

(e) We may write $s = 5t^6 - 1/t^2$ as

$$s = 5t^6 - t^{-2}$$

so

$$\frac{ds}{dt} = 5(6t^5) - (-2t^{-3}) = 30t^5 + 2t^{-3}$$

$$= 30t^5 + \frac{2}{t^3}$$

EXAMPLE 8

Find the slope of the tangent to $f(x) = \frac{1}{2}x^2 + 5x$ at each of the following.

(a) $x = 2$ (b) $x = -5$

Solution

The derivative of $f(x) = \frac{1}{2}x^2 + 5x$ is $f'(x) = x + 5$.

(a) At $x = 2$, the slope is $f'(2) = 2 + 5 = 7$.
(b) At $x = -5$, the slope is $f'(-5) = -5 + 5 = 0$. Thus when $x = -5$, the tangent to the curve is a horizontal line.

CHECKPOINT

1. True or false: The derivative of a constant times a function is equal to the constant times the derivative of the function.
2. True or false: The derivative of the sum of two functions is equal to the sum of the derivatives of the two functions.
3. True or false: The derivative of the difference of two functions is equal to the difference of the derivatives of the two functions.
4. Does the Coefficient Rule apply to $f(x) = x^n/c$, where c is a constant? Explain.
5. Find the derivative of each of the following functions.

   (a) $f(x) = x^{10} - 10x + 5$ (b) $s = \frac{1}{t^5} - 10^7 + 1$

6. Find the slope of the line tangent to $f(x) = x^3 - 4x^2 + 1$ at $x = -1$.

EXAMPLE 9

Find all points on the graph of $f(x) = x^3 + 3x^2 - 45x + 4$ where the tangent line is horizontal.

Solution

A horizontal line has slope equal to 0. Thus, to find the desired points, we solve $f'(x) = 0$.

$$f'(x) = 3x^2 + 6x - 45$$

We solve $3x^2 + 6x - 45 = 0$ as follows:

$$3x^2 + 6x - 45 = 0$$
$$3(x^2 + 2x - 15) = 0$$
$$3(x + 5)(x - 3) = 0$$

Solving $3(x + 5)(x - 3) = 0$ gives $x = -5$ and $x = 3$. The y-coordinates for these x-values come from $f(x)$. The desired points are $(-5, f(-5)) = (-5, 179)$ and $(3, f(3)) = (3, -77)$. Figure 9.27 shows the graph of $y = f(x)$ with these points and the tangent lines at them indicated.

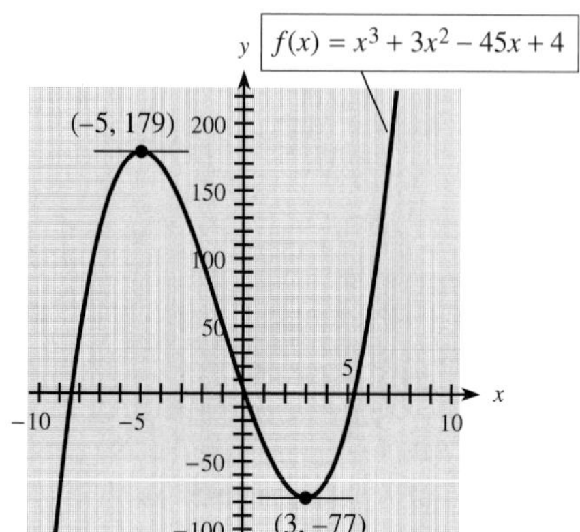

Figure 9.27

The marginal revenue $R'(x)$ is used to estimate the change in revenue caused by the sale of one additional unit.

EXAMPLE 10

Suppose that a manufacturer of a product knows that because of the demand for this product, his revenue is given by

$$R(x) = 1500x - 0.02x^2, \qquad 0 \leq x \leq 1000$$

where x is the number of units sold and $R(x)$ is in dollars.

(a) Find the marginal revenue at $x = 500$.
(b) Find the change in revenue caused by the increase in sales from 500 to 501 units.
(c) Find the difference between the marginal revenue found in (a) and the change in revenue found in (b).

Solution

(a) The marginal revenue for any value of x is

$$R'(x) = 1500 - 0.04x$$

The marginal revenue at $x = 500$ is

$$R'(500) = 1500 - 20 = 1480 \text{ (dollars)}$$

We can interpret this to mean that the approximate revenue from the sale of the 501st unit will be \$1480.

(b) The revenue at $x = 500$ is $R(500) = 745{,}000$, and the revenue at $x = 501$ is $R(501) = 746{,}479.98$, so the change in revenue is

$$R(501) - R(500) = 746{,}479.98 - 745{,}000 = 1479.98 \text{ (dollars)}$$

(c) The difference is $1480 - 1479.98 = 0.02$. Thus we see that the marginal revenue at $x = 500$ is a good estimate of the revenue from the 501st unit.

EXAMPLE 11

Suppose that the killer bees mentioned in the Application Preview enter a county in Texas and that the bee population in that county grows over a 6-week period, with the number of bees given by the equation

$$P(t) = 2t^2 + 10t + 1$$

where t is the number of weeks since the first bee is discovered. Find the rate of growth of the bee population 2 weeks after the first bee is discovered.

Solution

The rate of growth of the bees is given by $P'(t) = 4t + 10$, so the rate of growth 2 weeks after the first bee is discovered is $P'(2) = 18$ bees per week.

Graphing Utilities

We have mentioned that graphing utilities have a numerical derivative feature that can be used to estimate the derivative of a function at a specific value of x. This feature can also be used to check the derivative of a function that has been computed with a formula. We graph both the derivative calculated with a formula and the numerical derivative. If the two graphs lie on top of one another, the computed derivative agrees with the numerical derivative. Figure 9.28 illustrates this idea for the derivative of $f(x) = \frac{1}{3}x^3 - 2x^2 + 4$. Figure 9.28(a) shows $f'(x) = x^2 - 4x$ as y_1 and the calculator's numerical derivative of $f(x)$ as y_2. Figure 9.28(b) shows the graphs of both y_1 and y_2 (the graphs are coincident).

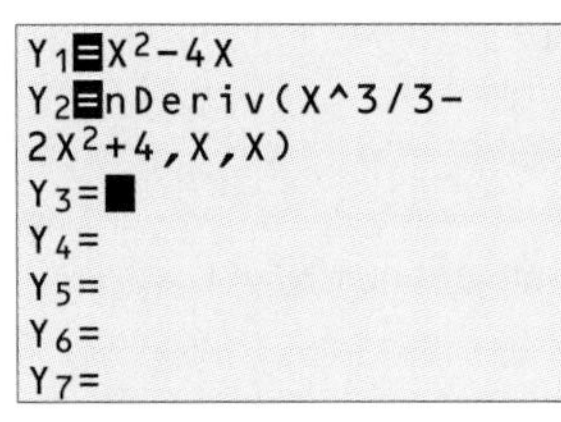

(a)

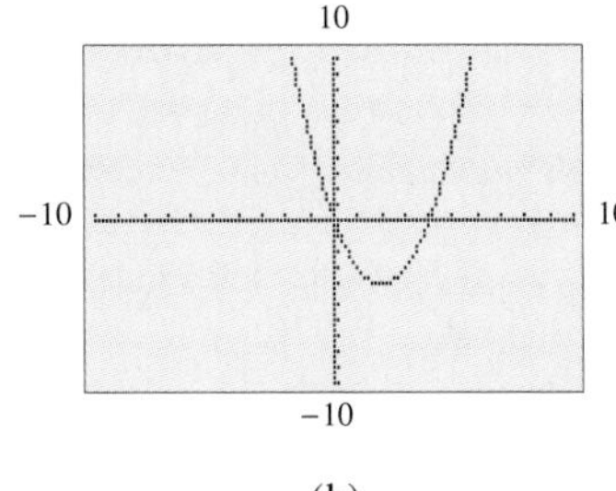

(b)

Figure 9.28

EXAMPLE 12

(a) Graph $f(x) = x^3 - 3x + 3$ and its derivative $f'(x)$ on the same set of axes so that all values of x that make $f'(x) = 0$ are in the x-range.

(b) Investigate the graph of $y = f(x)$ near values of x where $f'(x) = 0$. Does the graph of $y = f(x)$ appear to turn at values where $f'(x) = 0$?

(c) Compare the interval of x values where $f'(x) < 0$ with the interval where the graph of $y = f(x)$ is decreasing from left to right.

(d) What is the relationship between the intervals where $f'(x) > 0$ and where the graph of $y = f(x)$ is increasing from left to right?

Solution

(a) The graphs of $f(x) = x^3 - 3x + 3$ and $f'(x) = 3x^2 - 3$ are shown in Figure 9.29.

(b) The values where $f'(x) = 0$ are the x-intercepts, $x = -1$ and $x = 1$. The graph of $y = x^3 - 3x + 3$ appears to turn at both these values.

(c) $f'(x) < 0$ where the graph of $y = f'(x)$ is below the x-axis, for $-1 < x < 1$. The graph of $y = f(x)$ appears to be decreasing on this interval.

(d) They appear to be the same intervals.

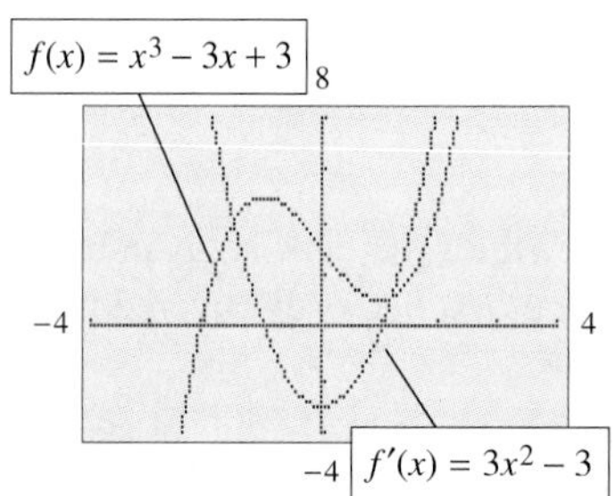

Figure 9.29

CHECKPOINT SOLUTIONS

1. True, by the Coefficient Rule.
2. True, by the Sum Rule.
3. True, by the Difference Rule.
4. Yes, $f(x) = x^n/c = (1/c)x^n$, so the coefficient is $(1/c)$.
5. (a) $f'(x) = 10x^9 - 10$
   (b) Note that $s = t^{-5} - 10^7 + 1$ and that 10^7 and 1 are constants.

$$\frac{ds}{dt} = -5t^{-6} = \frac{-5}{t^6}$$

6. The slope of the tangent at $x = -1$ is $f'(-1)$.

$$f'(x) = 3x^2 - 8x \qquad f'(-1) = 3(-1)^2 - 8(-1) = 11$$

EXERCISE 9.4

Find the derivatives of the functions in Problems 1–10.

1. $y = 4$
2. $f(s) = 6$
3. $y = x$
4. $s = t^2$
5. $f(x) = 2x^3 - x^5$
6. $f(x) = 3x^4 - x^9$
7. $y = 6x^4 - 5x^2 + x - 2$
8. $y = 3x^5 - 5x^3 - 8x + 8$
9. $g(x) = 10x^9 - 5x^5 + 7x^3 + 5x - 6$
10. $h(x) = 12x^{20} + 8x^{10} - 2x^7 + 17x - 9$

In Problems 11–14, at the indicated points, find
(a) the slope of the tangent to the curve, and
(b) the instantaneous rate of change of the function.

11. $y = 4x^2 + 3x, \quad x = 2$
12. $C(x) = 3x^2 - 5, \quad (3, 22)$
13. $P(x) = x^2 - 4x, \quad (2, -4)$
14. $R(x) = 16x + x^2, \quad x = 1$

In Problems 15–22, find the derivative of each function.

15. $y = x^{-5} + x^{-8} - 3$
16. $y = x^{-1} - x^{-2} + 13$
17. $y = 3x^{11/3} - 2x^{7/4} - x^{1/2} + 8$
18. $y = 5x^{8/5} - 3x^{5/6} + x^{1/3} + 5$
19. $f(x) = 5x^{-4/5} + 2x^{-4/3}$
20. $f(x) = 6x^{-8/3} - x^{-2/3}$
21. $g(x) = \dfrac{3}{x^4} + \dfrac{2}{x^5} + 5\sqrt[3]{x}$
22. $h(x) = \dfrac{7}{x^7} - \dfrac{3}{x^3} + 8\sqrt{x}$

In Problems 23–26, write the equations of the tangent lines to the curves at the indicated points.

23. $y = x^3 - 3x^2 + 5$ at $x = 1$
24. $y = x^4 - 4x^3 - 2$ at $x = 2$
25. $f(x) = 4x^2 - \dfrac{1}{x}$ at $x = -\dfrac{1}{2}$
26. $f(x) = \dfrac{x^3}{3} - \dfrac{3}{x^3}$ at $x = -1$

In Problems 27–30, find the coordinates of points where the graph of $f(x)$ has horizontal tangents.

27. $f(x) = -x^3 + 9x^2 - 15x + 6$
28. $f(x) = \dfrac{1}{3}x^3 - 3x^2 - 16x + 8$
29. $f(x) = x^4 - 4x^3 + 9$
30. $f(x) = 3x^5 - 5x^3 + 2$

In Problems 31 and 32, find each derivative at the given x-value (a) by finding the derivative with the appropriate rule and (b) with the numerical derivative feature of a graphing utility.

31. $y = 5 - 2\sqrt{x}$ at $x = 4$
32. $y = 1 + 3x^{2/3}$ at $x = -8$

In Problems 33–36, complete the following.

(a) Calculate the derivative of each function with the appropriate formula.
(b) Check your result from (a) by graphing your calculated derivative and the numerical derivative of the given function with respect to x evaluated at x.

33. $f(x) = 2x^3 + 5x - \pi^4 + 8$
34. $f(x) = 3x^2 - 8x + 2^5 - 20$
35. $h(x) = \dfrac{10}{x^3} - \dfrac{10}{\sqrt[5]{x^2}} + x^2 + 1$
36. $g(x) = \dfrac{5}{x^{10}} + \dfrac{4}{\sqrt[4]{x^3}} + x^5 - 4$

The tangent line to a curve at a point closely approximates the curve near the point. In fact, for x-values close enough to the point of tangency, the function and its tangent line are virtually indistinguishable. Problems 37 and 38 explore this relationship. Use each given function and the indicated point to complete the following.

(a) Write the equation of the tangent line to the curve at the indicated point.
(b) Use a graphing utility to graph both the function and its tangent line. Be sure your graph shows the point of tangency.
(c) Repeatedly zoom on the point of tangency until the function and the tangent line cannot be distinguished. Identify the x- and y-ranges in this window.

37. $f(x) = 3x^2 + 2x$ at $x = 1$
38. $f(x) = 4x - x^2$ at $x = 5$

For each function in Problems 39–42, do the following.

(a) Find $f'(x)$.
(b) Graph both $f(x)$ and $f'(x)$ with a graphing utility.
(c) Use the graph of $f'(x)$ to identify x-values where $f'(x) = 0$, $f'(x) > 0$, and $f'(x) < 0$.
(d) Use the graph of $f(x)$ to identify x-values where $f(x)$ has a maximum or minimum point, where the graph of $f(x)$ is rising, and where the graph of $f(x)$ is falling.

39. $f(x) = 8 - 2x - x^2$
40. $f(x) = x^2 + 4x - 12$
41. $f(x) = x^3 - 12x - 5$
42. $f(x) = 7 - 3x^2 - \dfrac{x^3}{3}$

Applications

43. ***Revenue*** Suppose that a wholesaler expects that his monthly revenue for small television sets will be

$$R(x) = 100x - 0.1x^2, \quad 0 \le x \le 800$$

where x is the number of units sold. Find his marginal revenue and interpret it when the quantity sold is

(a) $x = 300$ (b) $x = 600$

44. ***Revenue*** The total revenue for a commodity is described by the function

$$R = 300x - 0.02x^2$$

(a) What is the marginal revenue when 40 units are sold?
(b) Interpret your answer to (a).

45. ***Workers and output*** The weekly output of a certain product is

$$Q(x) = 200x + 6x^2$$

where x is the number of workers on the assembly line. There are presently 60 workers on the line.
(a) Find $Q'(x)$ and estimate the change in the weekly output caused by the addition of one worker.
(b) Calculate $Q(61) - Q(60)$ to see the actual change in the weekly output.

46. ***Capital investment and output*** The monthly output of a certain product is

$$Q(x) = 800x^{5/2}$$

where x is the capital investment in millions of dollars. Find dQ/dx, which can be used to estimate the effect on the output if an additional capital investment of $1 million is made.

47. ***Demand*** The demand q for a product depends on the price p (in dollars) according to

$$q = \frac{1000}{\sqrt{p}} - 1, \qquad \text{for } p > 0$$

Find and explain the meaning of the instantaneous rate of change of demand with respect to price when the price is
(a) $25 (b) $100

48. ***Demand*** Suppose that the demand for a product depends on the price p according to

$$D(p) = \frac{50{,}000}{p^2} - \frac{1}{2}, \qquad p > 0$$

where p is in dollars. Find and explain the meaning of the instantaneous rate of change of demand with respect to price when
(a) $p = 50$ (b) $p = 100$

49. ***Cost and average cost*** Suppose that the total cost function for the production of x units of a product is given by

$$C(x) = 4000 + 55x + 0.1x^2$$

Then the average cost of producing x items is

$$\overline{C(x)} = \frac{\text{total cost}}{x} = \frac{4000}{x} + 55 + 0.1x$$

(a) Find the instantaneous rate of change of average cost with respect to the number of units produced, at any level of production.
(b) Find the level of production at which this rate of change equals zero.

50. ***Cost and average cost*** Suppose that the total cost function for a certain commodity is given by

$$C(x) = 40{,}500 + 190x + 0.2x^2$$

where x is the number of units produced.
(a) Find the instantaneous rate of change of the average cost

$$\overline{C} = \frac{40{,}500}{x} + 190 + 0.2x$$

for any level of production.
(b) Find the level of production where this rate of change equals zero.

51. ***Cost-benefit*** Suppose that for a certain city the cost C of obtaining drinking water that contains p percent impurities (by volume) is given by

$$C = \frac{120{,}000}{p} - 1200$$

(a) Find the rate of change of cost with respect to p when impurities account for 1% (by volume).
(b) Write a sentence that explains the meaning of your answer in (a).

52. ***Cost-benefit*** Suppose that the cost C of processing the exhaust gases at an industrial site to ensure that only p percent of the particulate pollution escapes is given by

$$C(p) = \frac{8100(100 - p)}{p}$$

(a) Find the rate of change of cost C with respect to the percentage of particulate pollution that escapes when $p = 2$ (percent).
(b) Write a sentence interpreting your answer to (a).

53. ***Wind chill*** One form of the formula that meteorologists use to calculate wind chill temperature (WC) is

$$WC = 48.064 + 0.474t - 0.020ts - 1.85s + 0.304t\sqrt{s} - 27.74\sqrt{s}$$

where s is the wind speed in mph and t is the actual air temperature in degrees Fahrenheit. Suppose temperature is constant at 15°.
(a) Express wind chill WC as a function of wind speed s.
(b) Find the rate of change of wind chill with respect to wind speed when the wind speed is 25 mph.
(c) Interpret your answer to (b).

54. ***Allometric relationships—crabs*** For fiddler crabs, data gathered by Thompson* show that the allometric relationship between the weight C of the claw and the weight W of the body is given by

$$C = 0.11W^{1.54}$$

Find the function that gives the rate of change of claw weight with respect to body weight.

M 55. ***Union participation*** The following table shows the percentage of U.S. workers who belonged to unions for selected years from 1930 to 1996.

Year	*Percentage*
1930	11.6
1940	26.9
1950	31.5
1960	31.4
1970	27.3
1975	25.5
1980	21.9
1985	18
1990	16.1
1993	15.8
1994	15.5
1995	14.9
1996	14.5

Source: *World Almanac,* 1998

(a) Model these data with a cubic function, $u(x)$, where x is the number of years past 1900.
(b) For the period from 1980 to 1985, use the data points to find the average rate of change of the percentage of U.S. workers who belonged to unions.
(c) Find the instantaneous rate of change of the modeling function $u(x)$ for the year 1980.

M 56. ***Population below poverty level*** The table below shows the number of millions of people in the United States who lived below the poverty level for selected years between 1960 and 1996.

Year	*Persons Living Below the Poverty Level (millions)*
1960	39.9
1965	33.2
1970	25.4
1975	25.9
1980	29.3
1986	32.4
1990	33.6
1993	39.3
1994	38.1
1995	36.4
1996	36.5

Source: *The World Almanac and Book of Facts,* 1998

(a) Model these data with a cubic function $p = p(t)$, where p is the number of millions of people and t represents the number of years since 1900.
(b) Find the rate of growth in the number of persons living below the poverty level in 1980.
(c) Interpret your answer to (b).

M 57. ***Inflation rate*** The annual change in the consumer price index (CPI) for a 10-year period is shown in the table below. Assume that the annual change in the CPI can be modeled with a cubic function, $f(t)$, where t is the number of years past 1900.

Year	*Annual Percent Change in the CPI*
1987	3.6
1988	4.1
1989	4.8
1990	5.4
1991	4.2
1992	3.0
1993	3.0
1994	2.6
1995	2.8
1996	3.0

Source: *The World Almanac and Book of Facts,* 1998

(a) Find the function, $f(t)$, that models the data.
(b) Find the function that models the instantaneous rate of change of the CPI.
(c) Use the model found in (b) to find the instantaneous rate of change in 1989 and 1994.
(d) Interpret the two rates of change in (c).

*d'Arcy Thompson, *On Growth and Form* (Cambridge, England: Cambridge University Press, 1961).

58. ***Consumer debt*** The percentage of disposable income spent on consumer debt during certain years is shown in the table below.

Year	Consumer Debt as a Percentage of Disposable Income	Year	Consumer Debt as a Percentage of Disposable Income
1980	18.2	1990	20.0
1982	16.9	1991	18.9
1983	17.7	1994	19.7
1985	20.8	1995	21.3
1986	21.4	1996	23.7
1988	21.0		

Source: Federal Reserve System

Assume that consumer debt as a percentage of disposable income can be modeled with a fourth-degree function, $d(t)$, where t is the number of years past 1900.

(a) Find the function $d(t)$ that models the data.
(b) Find the function that models the instantaneous rate of change of consumer debt as a percentage of disposable income.
(c) Use the model found in (b) to find the instantaneous rate of change in 1985 and in 1995.
(d) Interpret the two rates of change found in (c).

9.5 Product and Quotient Rules

OBJECTIVES

- *To use the Product Rule to find the derivative of certain functions*
- *To use the Quotient Rule to find the derivative of certain functions*

APPLICATION PREVIEW

When medicine is administered, reaction (measured in change of blood pressure or temperature) can be modeled by

$$R = m^2\left(\frac{c}{2} - \frac{m}{3}\right)$$

where c is a positive constant and m is the amount of medicine absorbed into the blood.* The rate of change of R with respect to m is the sensitivity of the body to medicine. To find an expression for sensitivity as a function of m, we calculate dR/dm. We can find this derivative with the **Product Rule** for derivatives.

We have simple formulas for finding the derivatives of the sums and differences of functions. But we are not so lucky with products. The derivative of a product is *not* the product of the derivatives. To see this, we consider the function $f(x) = x \cdot x$. Because this function is $f(x) = x^2$, its derivative is $f'(x) = 2x$. But the product of the derivatives of x and x would give $1 \cdot 1 = 1 \neq 2x$. Thus we need a different formula to find the derivative of a product. This formula is given by the Product Rule.

Product Rule If $f(x) = u(x) \cdot v(x)$, where u and v are differentiable functions of x, then

$$f'(x) = u(x) \cdot v'(x) + v(x) \cdot u'(x)$$

*Source: Thrall, R. M., *et al.*, *Some Mathematical Models in Biology,* U.S. Department of Commerce, 1967.

Thus the derivative of a product of two functions is the first function times the derivative of the second plus the second function times the derivative of the first.

We can prove the Product Rule as follows. If $f(x) = u(x) \cdot v(x)$, then

$$\lim_{h \to 0} \frac{f(x+h) - f(x)}{h} = \lim_{h \to 0} \frac{u(x+h) \cdot v(x+h) - u(x) \cdot v(x)}{h}$$

Subtracting and adding $u(x + h) \cdot v(x)$ in the numerator gives

$$f'(x) = \lim_{h \to 0} \frac{u(x+h) \cdot v(x+h) - u(x+h) \cdot v(x) + u(x+h) \cdot v(x) - u(x) \cdot v(x)}{h}$$

$$= \lim_{h \to 0} \left(u(x+h) \left[\frac{v(x+h) - v(x)}{h} \right] + v(x) \left[\frac{u(x+h) - u(x)}{h} \right] \right)$$

Properties III and IV of limits give

$$f'(x) = \lim_{h \to 0} u(x+h) \cdot \lim_{h \to 0} \frac{v(x+h) - v(x)}{h} + \lim_{h \to 0} v(x) \cdot \lim_{h \to 0} \frac{u(x+h) - u(x)}{h}$$

Because u is differentiable and hence continuous, it follows that $\lim_{h \to 0} u(x + h) = u(x)$, so we have $f'(x) = u(x) \cdot v'(x) + v(x) \cdot u'(x)$.

EXAMPLE 1

Use the Product Rule to find the derivative of $f(x) = x^2 \cdot x$.

Solution

Using the formula with $u(x) = x^2$, $v(x) = x$, we have

$$\begin{aligned} f'(x) &= u(x) \cdot v'(x) + v(x) \cdot u'(x) \\ &= x^2 \cdot 1 + x(2x) \\ &= x^2 + 2x^2 \\ &= 3x^2 \end{aligned}$$

Note that we could have found the same result by multiplying the factors and finding the derivative of $f(x) = x^3$. But we will soon see how valuable the Product Rule is.

EXAMPLE 2

Find dy/dx if $y = (2x^3 + 3x + 1)(x^2 + 4)$.

Solution

Using the Product Rule with $u(x) = 2x^3 + 3x + 1$ and $v(x) = x^2 + 4$, we have

$$\begin{aligned} \frac{dy}{dx} &= (2x^3 + 3x + 1)(2x) + (x^2 + 4)(6x^2 + 3) \\ &= 4x^4 + 6x^2 + 2x + 6x^4 + 3x^2 + 24x^2 + 12 \\ &= 10x^4 + 33x^2 + 2x + 12 \end{aligned}$$

We could, of course, avoid using the Product Rule by multiplying the two factors before taking the derivative. But multiplying the factors first may involve more work than using the Product Rule.

EXAMPLE 3

Given $f(x) = (4x^3 + 5x^2 - 6x + 5)(x^3 - 4x^2 + 1)$, find the slope of the tangent to the graph of $y = f(x)$ at $x = 1$.

Solution

$$f'(x) = (4x^3 + 5x^2 - 6x + 5)(3x^2 - 8x) + (x^3 - 4x^2 + 1)(12x^2 + 10x - 6)$$

If we substitute $x = 1$ into $f'(x)$, we find that the slope of the curve at $x = 1$ is $f'(1) = 8(-5) + (-2)(16) = -72$.

The rule for finding the derivative of a function that is the quotient of two functions requires a new formula.

Quotient Rule If $f(x) = u(x)/v(x)$, where u and v are differentiable functions of x, with $v(x) \neq 0$, then

$$f'(x) = \frac{v(x) \cdot u''(x) - u(x) \cdot v''(x)}{[v(x)]^2}$$

The preceding formula says that the derivative of a quotient is the denominator times the derivative of the numerator minus the numerator times the derivative of the denominator, all divided by the square of the denominator.

To see that this rule is reasonable, consider the function $f(x) = x^3/x$, $x \neq 0$. Using the Quotient Rule, with $u(x) = x^3$ and $v(x) = x$, we get

$$f'(x) = \frac{x(3x^2) - x^3(1)}{x^2} = \frac{3x^3 - x^3}{x^2} = \frac{2x^3}{x^2} = 2x$$

Because $f(x) = x^3/x = x^2$ if $x \neq 0$, we see that $f'(x) = 2x$ is the correct derivative. The proof of the Quotient Rule is left for the student in Problem 41 of the exercises in this section.

EXAMPLE 4

If $f(x) = \dfrac{x^2 - 4x}{x + 5}$, find $f'(x)$.

Solution

Using the Quotient Rule with $u(x) = x^2 - 4x$ and $v(x) = x + 5$, we get

$$f'(x) = \frac{(x + 5)(2x - 4) - (x^2 - 4x)(1)}{(x + 5)^2}$$

$$= \frac{2x^2 + 6x - 20 - x^2 + 4x}{(x + 5)^2}$$

$$= \frac{x^2 + 10x - 20}{(x + 5)^2}$$

EXAMPLE 5

If $f(x) = \dfrac{x^3 - 3x^2 + 2}{x^2 - 4}$, find $f'(x)$.

Solution

Using the Quotient Rule with $u(x) = x^3 - 3x^2 + 2$ and $v(x) = x^2 - 4$, we get

$$\begin{aligned} f'(x) &= \frac{(x^2 - 4)(3x^2 - 6x) - (x^3 - 3x^2 + 2)(2x)}{(x^2 - 4)^2} \\ &= \frac{(3x^4 - 6x^3 - 12x^2 + 24x) - (2x^4 - 6x^3 + 4x)}{(x^2 - 4)^2} \\ &= \frac{x^4 - 12x^2 + 20x}{(x^2 - 4)^2} \end{aligned}$$

EXAMPLE 6

Use the Quotient Rule to find the derivative of $y = 1/x^3$.

Solution

Letting $u(x) = 1$ and $v(x) = x^3$, we get

$$\begin{aligned} y' &= \frac{x^3(0) - 1(3x^2)}{(x^3)^2} \\ &= -\frac{3x^2}{x^6} \\ &= -\frac{3}{x^4} \end{aligned}$$

Note that we could have found the derivative more easily by writing

$$y = 1/x^3 = x^{-3}$$

so

$$y' = -3x^{-4} = -\frac{3}{x^4}$$

Recall that we proved the Powers of x Rule for positive integer powers and assumed that it was true for all real number powers. In Problem 42 of the exercises in this section, you will be asked to use the Quotient Rule to show that the Powers of x Rule applies to negative integers.

It is not necessary to use the Quotient Rule when the denominator of the function in question contains only a constant. For example, the function $y = (x^3 - 3x)/3$ can be written $y = \frac{1}{3}(x^3 - 3x)$, so the derivative is $y' = \frac{1}{3}(3x^2 - 3) = x^2 - 1$.

CHECKPOINT

1. True or false: The derivative of the product of two functions is equal to the product of the derivatives of the two functions.
2. True or false: The derivative of the quotient of two functions is equal to the quotient of the derivatives of the two functions.

3. Find $f'(x)$ for each of the following.
   (a) $f(x) = (x^{12} + 8x^5 - 7)(10x^7 - 4x + 19)$ Do not simplify.
   (b) $f(x) = \dfrac{2x^4 + 3}{3x^4 + 2}$ Simplify.
4. If $y = \frac{4}{3}(x^2 + 3x - 4)$, does finding y' require the Product Rule? Explain.
5. If $y = f(x)/c$, where c is a constant, does finding y' require the Quotient Rule? Explain.

EXAMPLE 7

Suppose that the revenue function for a product is given by

$$R(x) = 10x + \frac{100x}{3x + 5}$$

where x is the number of units sold and R is in dollars.

(a) Find the marginal revenue function.
(b) Find the marginal revenue when $x = 15$.

Solution

(a) We must use the Quotient Rule to find the marginal revenue (the derivative).

$$\overline{MR} = R'(x) = 10 + \frac{(3x + 5)(100) - 100x(3)}{(3x + 5)^2}$$

$$= 10 + \frac{300x + 500 - 300x}{(3x + 5)^2} = 10 + \frac{500}{(3x + 5)^2}$$

(b) The marginal revenue when $x = 15$ is $R'(15)$.

$$R'(15) = 10 + \frac{500}{[(3)(15) + 5]^2} = 10 + \frac{500}{(50)^2}$$

$$= 10 + \frac{500}{2500} = 10.20$$

Recall that $R'(15)$ estimates the revenue from the sale of the 16th item.

We now consider the problem posed in the Application Preview.

EXAMPLE 8

When medicine is administered, reaction (measured in change of blood pressure or temperature) can be modeled by

$$R = m^2\left(\frac{c}{2} - \frac{m}{3}\right)$$

where c is a positive constant and m is the amount of medicine absorbed into the blood. The rate of change of R with respect to m is the sensitivity of the body to medicine. Find an expression for sensitivity s as a function of m.

Solution

The sensitivity is the rate of change of R with respect to m, or the derivative

$$\frac{dR}{dm} = m^2\left(0 - \frac{1}{3}\right) + \left(\frac{c}{2} - \frac{1}{3}m\right)(2m)$$

$$= -\frac{1}{3}m^2 + mc - \frac{2}{3}m^2 = mc - m^2$$

so sensitivity is given by

$$s = \frac{dR}{dm} = mc - m^2$$

EXAMPLE 9

(a) Graph $f(x) = \dfrac{8x}{x^2 + 4}$ and its derivative on the same set of axes over an interval that contains the x-values where $f'(x) = 0$.
(b) Determine the values of x where $f'(x) = 0$ and the intervals where $f'(x) > 0$ and where $f'(x) < 0$.
(c) What is the relationship between the derivative $f'(x)$ being positive or negative and the graph of $y = f(x)$ increasing (rising) or decreasing (falling)?

Solution

We find the derivative of $f(x)$ by using the Quotient Rule.

$$f'(x) = \frac{(x^2 + 4)(8) - (8x)(2x)}{(x^2 + 4)^2}$$

$$= \frac{32 - 8x^2}{(x^2 + 4)^2}$$

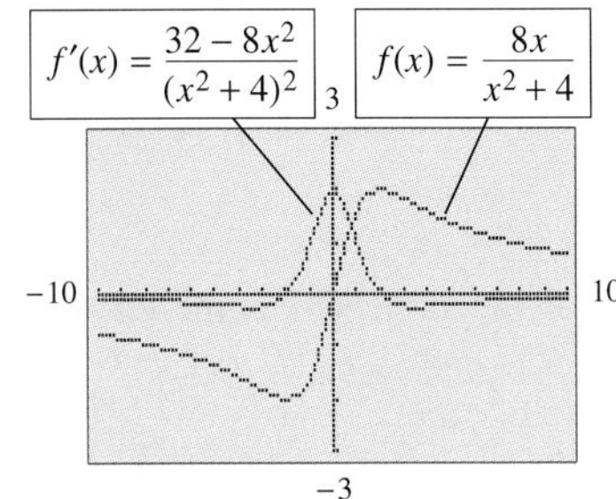

Figure 9.30

(a) The graphs of $f(x) = \dfrac{8x}{x^2 + 4}$ and $f'(x) = \dfrac{32 - 8x^2}{(x^2 + 4)^2}$ are shown in Figure 9.30.
(b) We can use TRACE or one of the solution features to find that $f'(x) = 0$ at $x = -2$ and $x = 2$. In the interval $-2 < x < 2$, the graph of $y = f'(x)$ is above the x-axis, so $f'(x)$ is positive for $-2 < x < 2$. In the intervals $x < -2$ and $x > 2$, the graph of $y = f'(x)$ is below the x-axis, so $f'(x)$ is negative there.
(c) The graph of $y = f(x)$ appears to turn at $x = -2$, and at $x = 2$, where $f'(x) = 0$. The graph of $y = f(x)$ is increasing where $f'(x) > 0$ and decreasing where $f'(x) < 0$.

CHECKPOINT SOLUTIONS

1. False. The derivative of a product is equal to the first function times the derivative of the second plus the second function times the derivative of the first. That is,

$$\frac{d}{dx}(fg) = f \cdot \frac{dg}{dx} + g \cdot \frac{df}{dx}$$

2. False. The derivative of a quotient is equal to the denominator times the derivative of the numerator minus the numerator times the derivative of the denominator, all divided by the square of the denominator. That is,

$$\frac{d}{dx}\left(\frac{f}{g}\right) = \frac{g \cdot f' - f \cdot g'}{g^2}$$

3. (a) $f'(x) = (x^{12} + 8x^5 - 7)(70x^6 - 4) + (10x^7 - 4x + 19)(12x^{11} + 40x^4)$

(b) $f'(x) = \dfrac{(3x^4 + 2)(8x^3) - (2x^4 + 3)(12x^3)}{(3x^4 + 2)^2}$

$= \dfrac{24x^7 + 16x^3 - 24x^7 - 36x^3}{(3x^4 + 2)^2} = \dfrac{-20x^3}{(3x^4 + 2)^2}$

4. No; y' can be found with the Coefficient Rule:

$$y' = \frac{4}{3}(2x + 3)$$

5. No; y' can be found with the Coefficient Rule:

$$y' = \left(\frac{1}{c}\right)f'(x)$$

EXERCISE 9.5

1. Find y' if $y = (x + 3)(x^2 - 2x)$.
2. Find $f'(x)$ if $f(x) = (3x - 1)(x^3 + 1)$.
3. Find $\dfrac{dp}{dq}$ if $p = (3q - 1)(q^2 + 2)$.
4. Find $\dfrac{ds}{dt}$ if $s = (t^4 + 1)(t^3 - 1)$.
5. If $f(x) = (x^{12} + 3x^4 + 4)(4x^3 - 1)$, find $f'(x)$.
6. If $y = (3x^7 + 4)(8x^6 - 6x^4 - 9)$, find $\dfrac{dy}{dx}$.

In Problems 7–10, find the derivative, but do not simplify your answer.

7. $y = (7x^6 - 5x^4 + 2x^2 - 1)(4x^9 + 3x^7 - 5x^2 + 3x)$
8. $y = (9x^9 - 7x^7 - 6x)(3x^5 - 4x^4 + 3x^3 - 8)$
9. $y = (x^2 + x + 1)(\sqrt[3]{x} - 2\sqrt{x} + 5)$
10. $y = (\sqrt[5]{x} - 2\sqrt[4]{x} + 1)(x^3 - 5x - 7)$

In Problems 11 and 12, at each indicated point find
(a) the slope of the tangent line, and
(b) the instantaneous rate of change of the function.

11. $y = (x^2 + 1)(x^3 - 4x)$ at $(-2, 0)$
12. $y = (x^3 - 3)(x^2 - 4x + 1)$ at $(2, -15)$

In Problems 13–24, find the indicated derivatives.

13. y' for $y = \dfrac{x}{x^2 - 1}$
14. $f'(x)$ for $f(x) = \dfrac{x^2}{x - 3}$
15. $\dfrac{dp}{dq}$ for $p = \dfrac{q^2 + 1}{q - 2}$
16. $C'(x)$ for $C(x) = \dfrac{x^2 + 1}{x^2 - 1}$
17. $\dfrac{dy}{dx}$ for $y = \dfrac{1 - 2x^2}{x^4 - 2x^2 + 5}$
18. $\dfrac{ds}{dt}$ for $s = \dfrac{t^3 - 4}{t^3 - 2t^2 - t - 5}$
19. $\dfrac{dz}{dx}$ for $z = x^2 + \dfrac{x^2}{1 - x - 2x^2}$
20. $\dfrac{dy}{dx}$ for $y = 200x - \dfrac{100x}{3x + 1}$
21. $\dfrac{dp}{dq}$ for $p = \dfrac{3\sqrt[3]{q}}{1 - q}$
22. $\dfrac{dy}{dx}$ for $y = \dfrac{2\sqrt{x} - 1}{1 - 4\sqrt{x^3}}$
23. y' for $y = \dfrac{x(x^2 + 4)}{x - 2}$
24. $f'(x)$ for $f(x) = \dfrac{(x + 1)(x - 2)}{x^2 + 1}$

In Problems 25 and 26, at the indicated point for each function, find
(a) the slope of the tangent line, and
(b) the instantaneous rate of change of the function.

25. $y = \dfrac{x^2 + 1}{x + 3}$ at $(2, 1)$
26. $y = \dfrac{x^2 - 4x}{x^2 + 2x}$ at $\left(2, -\dfrac{1}{2}\right)$

In Problems 27–30, write the equation of the tangent line to the graph of the function at the indicated point.

27. $y = (9x^2 - 6x + 1)(1 + 2x)$ at $x = 1$
28. $y = (4x^2 + 4x + 1)(7 - 2x)$ at $x = 0$
29. $y = \dfrac{3x^4 - 2x - 1}{4 - x^2}$ at $x = 1$
30. $y = \dfrac{x^2 - 4x}{2x - x^3}$ at $x = 2$

In Problems 31–34, use the numerical derivative feature of a graphing utility to find the derivative of each function at the given x-value.

31. $y = \left(4\sqrt{x} + \frac{3}{x}\right)\left(3\sqrt[3]{x} - \frac{5}{x^2} - 25\right)$ at $x = 1$

32. $y = (3\sqrt[4]{x^5} + \sqrt[5]{x^4} - 1)\left(\frac{2}{x^3} - \frac{1}{\sqrt{x}}\right)$ at $x = 1$

33. $f(x) = \frac{4x - 4}{3x^{2/3}}$ at $x = 1$

34. $f(x) = \frac{3\sqrt[3]{x} + 1}{x + 2}$ at $x = -1$

In Problems 35–38, complete the following.

(a) Find the derivative of each function, and check your work by graphing both your calculated derivative and the numerical derivative of the function.

(b) Use your graph of the derivative to find points where the original function has horizontal tangent lines.

(c) Use a graphing utility to graph the function and indicate the points found in (b) on the graph.

35. $f(x) = (x^2 + 4x + 4)(x - 7)$

36. $f(x) = (x^2 - 14x + 49)(2x + 1)$

37. $y = \frac{x^2}{x - 2}$

38. $y = \frac{x^2 - 7}{4 - x}$

In Problems 39 and 40,

(a) find $f'(x)$.

(b) graph both $f(x)$ and $f'(x)$ with a graphing utility.

(c) identify the x-values where $f'(x) = 0$, $f'(x) > 0$, and $f'(x) < 0$.

(d) identify x-values where $f(x)$ has a maximum point or a minimum point, where $f(x)$ is increasing, and where $f(x)$ is decreasing.

39. $f(x) = \frac{10x^2}{x^2 + 1}$

40. $f(x) = \frac{8 - x^2}{x^2 + 4}$

41. Prove the Quotient Rule for differentiation. (*Hint:* Add $[-u(x) \cdot v(x) + u(x) \cdot v(x)]$ to the expanded numerator and use steps similar to those used to prove the Product Rule.)

42. Use the Quotient Rule to show that the Powers of x Rule applies to negative integer powers. That is, show that $(d/dx)x^n = nx^{n-1}$ when $n = -k$, $k > 0$, by finding the derivative of $f(x) = 1/(x^k)$.

Applications

43. ***Cost-benefit*** If the cost C of removing p percent of the particulate pollution from the exhaust gases at an industrial site is given by

$$C(p) = \frac{8100p}{100 - p}$$

find the rate of change of C with respect to p.

44. ***Cost-benefit*** If the cost C of removing p percent of the impurities from the waste water in a manufacturing process is given by

$$C(p) = \frac{9800p}{101 - p}$$

find the rate of change of C with respect to p.

45. ***Revenue*** Suppose the revenue function for a product is given by

$$R(x) = \frac{60x^2 + 74x}{2x + 2}$$

Find the marginal revenue when 49 units are sold. Interpret your result.

46. ***Revenue*** The revenue from the sale of x units of a product is given by

$$R(x) = \frac{3000}{2x + 2} + 80x - 1500$$

Find the marginal revenue when 149 units are sold. Interpret your result.

47. ***Revenue*** A travel agency will plan a group tour for groups of size 25 or larger. If the group contains exactly 25 people, the cost is \$300 per person. If each person's cost is reduced by \$10 for each additional person above the 25, then the revenue is given by

$$R(x) = (25 + x)(300 - 10x)$$

where x is the number of additional people above 25. Find the marginal revenue if the group contains 30 people. Interpret your result.

48. ***Revenue*** McRobert's TV Shop sells 200 sets per month at a price of \$400 per unit. Market research indicates that the shop can sell one additional set for each \$1 it reduces the price, and in this case the total revenue is

$$R(x) = (200 + x)(400 - x)$$

where x is the number of additional sets beyond the 200. If the shop sells a total of 250 sets, find the marginal revenue. Interpret your result.

49. ***Response to a drug*** The reaction R to an injection of a drug is related to the dosage x according to

$$R(x) = x^2\left(500 - \frac{x}{3}\right)$$

where 1000 mg is the maximum dosage. If the rate of reaction with respect to the dosage defines the sensitivity to the drug, find the sensitivity.

50. ***Nerve response*** The number of action potentials produced by a nerve, t seconds after a stimulus, is given by

$$N(t) = 25t + \frac{4}{t^2 + 2} - 2$$

Find the rate at which the action potentials are produced by the nerve.

51. ***Test reliability*** If a test having reliability r is lengthened by a factor n, the reliability of the new test is given by

$$R = \frac{nr}{1 + (n - 1)r}, \quad 0 < r \leq 1$$

Find the rate at which R changes with respect to n.

52. ***Advertising and sales*** The sales of a product s (in thousands of dollars) are related to advertising expenses (in thousands of dollars) by

$$s = \frac{200x}{x + 10}$$

Find and interpret the meaning of the rate of change of sales with respect to advertising expenses when
(a) $x = 10$ (b) $x = 20$

53. ***Candidate recognition*** Suppose that the proportion P of voters who recognize a candidate's name t months after the start of the campaign is given by

$$P(t) = \frac{13t}{t^2 + 100} + 0.18$$

(a) Find the rate of change of P when $t = 6$ and explain its meaning.
(b) Find the rate of change of P when $t = 12$ and explain its meaning.
(c) One month prior to the election, is it better for $P'(t)$ to be positive or negative?

54. ***Endangered species population*** It is determined that a wildlife refuge can support a group of up to 120 of a certain endangered species. If 75 are introduced onto the refuge and their population after t years is given by

$$p(t) = 75\left(1 + \frac{4t}{t^2 + 16}\right)$$

find the rate of population growth after t years. Find the rate after each of the first 7 years.

55. ***Wind chill*** According to the National Climatic Data Center, during 1991, the lowest temperature recorded in Indianapolis, Indiana, was 0°F. If x is the wind speed in miles per hour and $x \geq 5$, then the wind chill (in degrees Fahrenheit) for an air temperature of 0°F can be approximated by the function

$$f(x) = \frac{289.173 - 58.5731x}{x + 1}$$

(a) At what rate is the wind chill changing when the wind speed is 20 mph?
(b) Explain the meaning of your answer to (a).

56. ***Response to injected adrenalin*** Experimental evidence has shown that the concentration of injected adrenaline x is related to the response y of a muscle according to the equation

$$y = \frac{x}{a + bx}$$

where a and b are constants. Find the rate of change of response with respect to the concentration.

57. ***Social Security*** America's 45 million Social Security recipients got a 2.6% cost-of-living increase in 1994. That was the second-smallest increase in nearly 20 years, a reflection of the low inflation rate. The following data show the percent increases for selected years. (We will assume that $t = 0$ in 1985.)

Year	*Percent Increase in Cost of Living*
1989	4.0
1990	4.9
1991	5.1
1992	4.1
1993	3.5
1994	2.6

Source: Social Security Administration

Assume these data can be modeled by the equation

$$C(t) = (t + 5)(-0.218t + 3.57) - 20.13$$

(a) Find the rate of change of the cost-of-living increase as predicted by the model for 1990.
(b) Interpret your answer to (a).
(c) Graph both the data and the model using a graphing utility.
(d) From the graph in (c), identify t-values for which the model must be invalid.

58. ***Drug use*** The percentage of high school seniors who have tried hallucinogens is shown in the table below for selected years.

Year of Graduation	*Hallucinogens*
1975	16.3
1978	14.3
1980	13.3
1983	11.9
1984	10.7
1985	10.3
1986	9.7
1987	10.3
1988	8.9
1989	9.4
1990	9.4
1991	9.6

Source: National Institute on Drug Abuse

The data from the table indicate that the percentage of high school seniors who have tried hallucinogens can be modeled with the function

$$f_h(t) = 216.074 - (t + 70)(2.782 - 0.0242t)$$

where t is the number of years past 1970.

(a) Find the function that models the instantaneous rate of change, with respect to time, of the percentage of seniors who have tried hallucinogens.
(b) Use the model in (a) to find the instantaneous rate of change in 1978 and in 1988.
(c) Interpret each rate of change found in (b).

59. ***Farm workers*** The percentage of U.S. workers in farm occupations during certain years is shown in the table.

Year	*Percent of All Workers in Farm Occupations*
1820	71.8
1850	63.7
1870	53
1900	37.5
1920	27
1930	21.2
1940	17.4
1950	11.6
1960	6.1
1970	3.6
1980	2.7
1985	2.8
1990	2.4

Source: *The World Almanac and Book of Facts,* 1993

Assume that the percentage of U.S. workers in farm occupations can be modeled with the function

$$f(t) = 1000 \cdot \frac{-8.0912t + 1558.9}{1.09816t^2 - 122.183t + 21472.6}$$

where t is the number of years past 1800.

(a) Find the function that models the instantaneous rate of change of the percentage of U.S. workers in farm occupations.
(b) Use the model in (a) to find the instantaneous rate of change in 1870 and in 1970.
(c) Interpret each of the rates of change in (b).

9.6 The Chain Rule and Power Rule

OBJECTIVES

- *To use the Chain Rule to differentiate functions*
- *To use the Power Rule to differentiate functions*

APPLICATION PREVIEW

The demand x for a product is given by

$$x = \frac{98}{\sqrt{2p + 1}} - 1$$

where p is the price per unit. To find how fast demand is changing when price is \$24, we take the derivative of x with respect to p. If we write this function with a power rather than a radical, it has the form

$$x = 98(2p + 1)^{-1/2} - 1$$

The formulas learned so far cannot be used to find this derivative. Rather than use the limit definition of derivative, we use a new formula, the **Power Rule,** to find this derivative. In this section we will discuss the **Chain Rule** and the Power Rule, which is one of the results of the Chain Rule, and we will use these formulas to solve applied problems.

Recall from Section 1.2, "Functions," that if f and g are functions, then the composite functions g of f (denoted $g \circ f$) and f of g (denoted $f \circ g$) are defined as follows:

$$(g \circ f)(x) = g(f(x)) \quad \text{and} \quad (f \circ g)(x) = f(g(x))$$

EXAMPLE 1

If $f(x) = 3x^2$ and $g(x) = 2x - 1$, find $F(x) = f(g(x))$.

Solution

Substituting $g(x) = 2x - 1$ for x in $f(x)$ gives

$$f(g(x)) = f(2x - 1) = 3(2x - 1)^2$$

Thus $F(x) = 3(2x - 1)^2$.

We could find the derivative of the function $F(x) = 3(2x - 1)^2$ by multiplying out the expression $3(2x - 1)^2$. Then

$$F(x) = 3(4x^2 - 4x + 1) = 12x^2 - 12x + 3$$

so $F'(x) = 24x - 12$. But we can also use a very powerful rule, called the **Chain Rule,** to find derivatives of functions of this type. If we write the composite function $y = f(g(x))$ in the form $y = f(u)$, where $u = g(x)$, we state the Chain Rule as follows:

Chain Rule If f and g are differentiable functions with $y = f(u)$ and $u = g(x)$, then y is a differentiable function of x, and

$$\frac{dy}{dx} = \frac{d}{du}f(u) \cdot \frac{d}{dx}g(x)$$

or, written another way,

$$\frac{dy}{dx} = \frac{dy}{du} \cdot \frac{du}{dx}$$

Note that dy/du represents the derivative of $y = f(u)$ *with respect to* u and du/dx represents the derivative of $u = g(x)$ *with respect to* x. For example, if $y = 3(2x - 1)^2$, we may write $y = f(u) = 3u^2$, where $u = 2x - 1$. Then the derivative is

$$\frac{dy}{dx} = \frac{dy}{du} \cdot \frac{du}{dx} = 6u \cdot 2 = 12u$$

To write this derivative in terms of x, we substitute $2x - 1$ for u. Thus

$$\frac{dy}{dx} = 12(2x - 1) = 24x - 12$$

Note that we get the same result by using the Chain Rule as we did by multiplying out $f(x) = 3(2x - 1)^2$. The Chain Rule is important because it is not always possible to rewrite the function as a polynomial. Consider the following example.

EXAMPLE 2

If $y = \sqrt{x^2 - 1}$, find $\frac{dy}{dx}$.

Solution

If we write this function as $y = f(u) = \sqrt{u}$, when $u = x^2 - 1$, we can find the derivative.

$$\frac{dy}{dx} = \frac{dy}{du} \cdot \frac{du}{dx} = \frac{1}{2} \cdot u^{-1/2} \cdot 2x = u^{-1/2} \cdot x = \frac{1}{\sqrt{u}} \cdot x = \frac{x}{\sqrt{u}}$$

To write this derivative in terms of x alone, we substitute $x^2 - 1$ for u. Then

$$\frac{dy}{dx} = \frac{x}{\sqrt{x^2 - 1}}$$

Note that we could not find the derivative of a function like that of Example 2 by the methods learned previously.

EXAMPLE 3

If $y = \frac{1}{(x^2 + 3x + 1)^2}$, find $\frac{dy}{dx}$.

Solution

If we let $u = x^2 + 3x + 1$, we can write $y = f(u) = \frac{1}{u^2}$, or $y = u^{-2}$. Then

$$\frac{dy}{dx} = \frac{dy}{du} \cdot \frac{du}{dx} = -2u^{-3}(2x + 3) = \frac{-4x - 6}{u^3}$$

Substituting for u gives

$$\frac{dy}{dx} = \frac{-4x - 6}{(x^2 + 3x + 1)^3}$$

EXAMPLE 4

The relationship between the length L (in meters) and weight W (in kilograms) of a species of fish in the Pacific Ocean is given by $W = 10.375L^3$. The rate of growth in length is given by $\frac{dL}{dt} = 0.36 - 0.18L$, where t is measured in years.

(a) Determine a formula for the rate of growth in weight $\frac{dW}{dt}$ in terms of L.

(b) If a fish weighs 30 kilograms, approximate its rate of growth in weight using the formula found in (a).

Solution

(a) The rate of change uses the Chain Rule, as follows:

$$\frac{dW}{dt} = \frac{dW}{dL} \cdot \frac{dL}{dt} = 31.125L^2(0.36 - 0.18L) = 11.205L^2 - 5.6025L^3$$

(b) From $W = 10.375L^3$, we have

$$L = \sqrt[3]{\frac{W}{10.375}}$$

If $W = 30$ kilograms, $L = \sqrt[3]{\frac{30}{10.375}} = 1.4247$ meters, so the rate of growth in weight is

$$\frac{dW}{dt} = 11.205(1.4247)^2 - 5.6025(1.4247)^3 = 6.542 \text{ kilograms/year}$$

The Chain Rule is very useful and will be extremely important with functions that we will study later, but a special case of the Chain Rule, called the **Power Rule,** is useful for the algebraic functions we have studied so far.

Power Rule If $y = u^n$, where u is a differentiable function of x, then

$$\frac{dy}{dx} = nu^{n-1} \cdot \frac{du}{dx}$$

EXAMPLE 5

If $y = (x^2 - 4x)^6$, find $\frac{dy}{dx}$.

Solution

The right side of the equation is in the form u^n, with $u = x^2 - 4x$. Thus, by the Power Rule,

$$\frac{dy}{dx} = nu^{n-1} \cdot \frac{du}{dx} = 6u^5(2x - 4)$$

Substituting for u gives

$$\frac{dy}{dx} = 6(x^2 - 4x)^5(2x - 4)$$
$$= (12x - 24)(x^2 - 4x)^5$$

EXAMPLE 6

If $y = 3\sqrt[3]{x^2 - 3x + 1}$, find y'.

Solution

Because $y = 3(x^2 - 3x + 1)^{1/3}$, we can make use of the Power Rule with $u = x^2 - 3x + 1$.

$$y' = 3\left(nu^{n-1}\frac{du}{dx}\right) = 3\left[\frac{1}{3}u^{-2/3}(2x - 3)\right]$$
$$= (x^2 - 3x + 1)^{-2/3}(2x - 3)$$
$$= \frac{2x - 3}{(x^2 - 3x + 1)^{2/3}}$$

EXAMPLE 7

If $p = \dfrac{4}{3q^2 + 1}$, find $\dfrac{dp}{dq}$.

Solution

We can use the Power Rule to find dp/dq if we write the equation in the form

$$p = 4(3q^2 + 1)^{-1}$$

Then

$$\frac{dp}{dq} = 4[-1(3q^2 + 1)^{-2}(6q)] = \frac{-24q}{(3q^2 + 1)^2}$$

The derivative of the function in Example 7 can also be found by using the Quotient Rule, but the Power Rule provides a more efficient method.

EXAMPLE 8

Find the derivative of $g(x) = \dfrac{1}{\sqrt{(x^2 + 1)^3}}$.

Solution

Writing $g(x)$ as a power gives

$$g(x) = (x^2 + 1)^{-3/2}$$

Then

$$g'(x) = -\frac{3}{2}(x^2 + 1)^{-5/2}(2x) = -3x \cdot \frac{1}{(x^2 + 1)^{5/2}}$$
$$= \frac{-3x}{\sqrt{(x^2 + 1)^5}}$$

CHECKPOINT

1. (a) If $f(x) = (3x^4 + 1)^{10}$, does $f'(x) = 10(3x^4 + 1)^9$?
   (b) If $f(x) = (2x + 1)^5$, does $f'(x) = 10(2x + 1)^4$?
   (c) If $f(x) = \dfrac{[u(x)]^n}{c}$, where c is a constant, does $f'(x) = \dfrac{n[u(x)]^{n-1} \cdot u'(x)}{c}$?
2. (a) If $f(x) = \dfrac{12}{2x^2 - 1}$, find $f'(x)$ by using the Power Rule (not the Quotient Rule).
   (b) If $f(x) = \dfrac{\sqrt{x^3 - 1}}{3}$, find $f'(x)$ by using the Power Rule (not the Quotient Rule).

EXAMPLE 9

The demand x for a product is given by

$$x = 98(2p + 1)^{-1/2} - 1$$

where p is the price per unit. Find the rate of change of the demand with respect to price when $p = 24$.

Solution

The rate of change of demand with respect to price is

$$\frac{dx}{dp} = 98\left[-\frac{1}{2}(2p+1)^{-3/2}(2)\right] = -98(2p+1)^{-3/2}$$

When $p = 24$, the rate of change is

$$\left.\frac{dx}{dp}\right|_{p=24} = -98(48+1)^{-3/2} = -98 \cdot \frac{1}{49^{3/2}}$$

$$= -98 \cdot \frac{1}{343}$$

$$= -\frac{2}{7}$$

CHECKPOINT SOLUTIONS

1. (a) No, $f'(x) = 10(3x^4 + 1)^9(12x^3)$. (b) Yes (c) Yes
2. (a) $f(x) = 12(2x^2 - 1)^{-1}$

   $f'(x) = -12(2x^2 - 1)^{-2}(4x) = \dfrac{-48x}{(2x^2 - 1)^2}$

   (b) $f(x) = \frac{1}{3}(x^3 - 1)^{1/2}$

   $f'(x) = \frac{1}{6}(x^3 - 1)^{-1/2}(3x^2) = \dfrac{x^2}{2\sqrt{x^3 - 1}}$

EXERCISE 9.6

In Problems 1–4, find the indicated derivatives.

1. $\frac{dy}{dx}$ for $y = u^3$ and $u = x^2 + 1$
2. $\frac{dp}{dq}$ for $p = u^4$ and $u = q^2 + 4q$
3. $\frac{dy}{dx}$ for $y = u^4$ and $u = 4x^2 - x + 8$
4. $\frac{dr}{ds}$ for $r = u^{10}$ and $u = s^2 + 5s$

Differentiate the functions in Problems 5–20.

5. $f(x) = \dfrac{1}{(x^2 + 2)^3}$
6. $g(x) = \dfrac{1}{4x^3 + 1}$
7. $g(x) = (x^2 + 4x)^{-2}$
8. $p = (q^3 + 1)^{-5}$
9. $c(x) = (x^2 + 3x + 4)^{-3}$
10. $y = (x^2 - 8x)^{2/3}$
11. $g(x) = \dfrac{1}{(2x^3 + 3x + 5)^{3/4}}$
12. $y = \dfrac{1}{(3x^3 + 4x + 1)^{3/2}}$
13. $y = \sqrt{x^2 + 4x + 5}$
14. $y = \sqrt{x^2 + 3x}$
15. $s = 4\sqrt{3x - x^2}$
16. $y = 3\sqrt[3]{(x - 1)^2}$

17. $y = \dfrac{8(x^2 - 3)^5}{5}$ 18. $y = \dfrac{5\sqrt{1 - x^3}}{6}$

19. $y = \dfrac{(3x + 1)^5 - 3x}{7}$

20. $y = \dfrac{\sqrt{2x - 1} - \sqrt{x}}{2}$

At the indicated point, for each function in Problems 21–24, find
(a) the slope of the tangent line, and
(b) the instantaneous rate of change of the function.

A graphing utility's numerical derivative feature can be used to check your work.

21. $y = (x^3 + 2x)^4$ at $x = 2$
22. $y = \sqrt{5x^2 + 2x}$ at $x = 1$
23. $y = \sqrt{x^3 + 1}$ at $(2, 3)$
24. $y = (4x^3 - 5x + 1)^3$ at $(1, 0)$

In Problems 25–28, write the equation of the line tangent to the graph of each function at the indicated point.

25. $y = (x^2 - 3x + 3)^3$ at $(1, 1)$
26. $y = (x^2 + 1)^3$ at $(2, 125)$
27. $y = \sqrt{3x^2 - 2}$ at $x = 3$
28. $y = \left(\dfrac{1}{x^3 - x}\right)^3$ at $x = 2$

In Problems 29 and 30, complete the following for each function.
(a) Find $f'(x)$.
(b) Check your result in (a) by graphing both it and the numerical derivative of the function.
(c) Find x-values for which the slope of the tangent is 0.
(d) Find points (x, y) where the slope of the tangent is 0.
(e) Use a graphing utility to graph the function and locate the points found in (d).

29. $f(x) = (x^2 - 4)^3 + 12$
30. $f(x) = 10 - (x^2 - 2x - 8)^2$

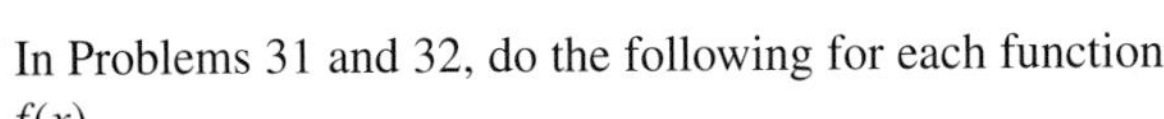

In Problems 31 and 32, do the following for each function $f(x)$.
(a) Find $f'(x)$,
(b) Graph both $f(x)$ and $f'(x)$ with a graphing utility,
(c) Determine x-values where $f'(x) = 0$, $f'(x) > 0$, $f'(x) < 0$.
(d) Determine x-values for which $f(x)$ has a maximum or minimum point, where the graph is increasing, and where it is decreasing.

31. $f(x) = 5 - 3(1 - x^2)^{4/3}$
32. $f(x) = 3 + \dfrac{1}{16}(x^2 - 4x)^4$

In Problems 33 and 34, find the derivative of each function.

33. (a) $y = \dfrac{2x^3}{3}$ (b) $y = \dfrac{2}{3x^3}$
(c) $y = \dfrac{(2x)^3}{3}$ (d) $y = \dfrac{2}{(3x)^3}$

34. (a) $y = \dfrac{3}{(5x)^5}$ (b) $y = \dfrac{3x^5}{5}$
(c) $y = \dfrac{3}{5x^5}$ (d) $y = \dfrac{(3x)^5}{5}$

Applications

35. ***Ballistics*** Ballistics experts are able to identify the weapon that fired a certain bullet by studying the markings on the bullet. Tests are conducted by firing into a bale of paper. If the distance s, in inches, that the bullet travels into the paper is given by

$$s = 27 - (3 - 10t)^3$$

for $0 \le t \le 0.3$ second, find the velocity of the bullet one-tenth of a second after it hits the paper.

36. ***Population of microorganisms*** Suppose that the population of a certain microorganism at time t (in minutes) is given by

$$P = 1000 - 1000(t + 10)^{-1}$$

Find the rate of change of population.

37. ***Revenue*** The revenue from the sale of x units of a product is

$$R = 1500x + 3000(2x + 3)^{-1} - 1000$$

where x is the number of units sold. Find the marginal revenue when 100 units are sold. Interpret your result.

38. ***Revenue*** The revenue from the sale of x units of a product is

$$R = 15(3x + 1)^{-1} + 50x - 15$$

Find the marginal revenue when 40 units are sold. Interpret your result.

39. ***Pricing and sales*** Suppose that the weekly sales volume y (in thousands of units sold) depends on the price per unit of the product according to

$$y = 32(3p + 1)^{-2/5}, \quad p > 0$$

where p is in dollars.
(a) What is the rate of change in sales volume when the price is \$21?
(b) Interpret your answer to (a).

40. ***Pricing and sales*** A chain of auto service stations has found that its monthly sales volume y (in thousands of dollars) is related to the price p (in dollars) of an oil change according to

$$y = \frac{90}{\sqrt{p+5}}, \quad p > 10$$

(a) What is the rate of change of sales volume when the price is \$20?
(b) Interpret your answer to (a).

41. ***Demand*** Suppose that the demand for a product is described by

$$p = \frac{200{,}000}{(q+1)^2}$$

(a) What is the rate of change of price with respect to the quantity demanded when $q = 49$?
(b) Interpret your answer to (a).

Stimulus-response The relation between the magnitude of a sensation y and the magnitude of the stimulus x is given by

$$y = k(x - x_0)^n$$

where k is a constant, x_0 is the threshold of effective stimulus, and n depends on the type of stimulus. Find the rate of change of sensation with respect to the amount of stimulus for each of Problems 42–44.

42. For the stimulus of visual brightness $y = k(x - x_0)^{1/3}$
43. For the stimulus of warmth $y = k(x - x_0)^{8/5}$
44. For the stimulus of electrical stimulation $y = k(x - x_0)^{7/2}$

45. ***Demand*** If the demand for a product is described by the equation

$$p = \frac{100}{\sqrt{2q+1}}$$

find the rate of change of p with respect to q.

46. ***Advertising and sales*** The daily sales S (in thousands of dollars) attributed to an advertising campaign are given by

$$S = 1 + \frac{3}{t+3} - \frac{18}{(t+3)^2}$$

where t is the number of weeks the campaign runs. What is the rate of change of sales at
(a) $t = 8$? (b) $t = 10$?
(c) Should the campaign be continued after the 10th week? Explain.

47. ***Body-heat loss*** The description of body-heat loss due to convection involves a coefficient of convection, K_c, which depends on wind velocity according to the following equation.

$$K_c = 4\sqrt{4v+1}$$

Find the rate of change of the coefficient with respect to the wind velocity.

48. ***Typing speed*** The typing speed (in words per minute) of a secretarial student is

$$S = 10\sqrt{0.8x+4}, \quad 0 \le x \le 100$$

where x is the number of hours of training he has had. What is the rate at which his speed is changing and what does this rate mean when he has had
(a) 15 hours of training?
(b) 40 hours of training?

49. ***Investments*** If an IRA is a variable-rate investment for 20 years at rate r percent per year, compounded monthly, then the future value S that accumulates from an initial investment of \$1000 is

$$S = 1000\left[1 + \frac{0.01r}{12}\right]^{240}$$

What is the rate of change of S with respect to r and what does it tell us if the interest rate is (a) 6%? (b) 12%?

50. ***Concentration of body substances*** The concentration C of a substance in the body depends on the quantity of the substance Q and the volume V through which it is distributed. For a static substance this is given by

$$C = \frac{Q}{V}$$

For a situation like that in the kidneys, where the fluids are moving, the concentration is the ratio of the rate of change of quantity with respect to time and the rate of change of volume with respect to time.
(a) Formulate the equation for concentration of a moving substance.
(b) Show that this is equal to the rate of change of quantity with respect to volume.

51. ***Public debt of the United States*** The interest paid on the public debt of the United States of America, as a percentage of federal expenditures for selected years, is shown in the table.

Assume that the percentage of federal expenditures devoted to payment of interest can be modeled with the function

$$d(t) = -0.1543(0.1t + 3)^4 + 4.2743(0.1t + 3)^3 - 42.1504(0.1t + 3)^2 + 175.805(0.1t + 3) - 251.334$$

where t is the number of years past 1930. Use this model to determine and interpret the instantaneous rate of change of the percentage of federal expenditures devoted to payment of interest on the public debt in 1960 and in 1990.

Year	*Interest Paid as a Percentage of Federal Expenditures*
1930	0
1940	10.5
1950	13.4
1955	9.4
1960	10.0
1965	9.6
1970	9.9
1975	9.8
1980	12.7
1985	18.9
1990	21.1
1995	22.0

Source: Bureau of Public Debt, Department of the Treasury

52. ***Union membership*** The table shows the percentage of U.S. workers who belonged to unions for selected years from 1930 to 1996.

Year	*Union Membership as a Percentage of the Labor Force*
1930	11.6
1935	13.2
1940	26.9
1945	35.5
1950	31.5
1955	33.2
1960	31.4
1965	28.4
1970	27.3
1975	25.5
1980	21.9
1985	18.0
1990	16.1
1993	15.8
1994	15.5
1995	14.9
1996	14.5

Source: Bureau of Labor Statistics, Department of Labor

Assume that the percentage of the labor force that belongs to unions can be modeled with the function

$$f(t) = 0.4711(0.1t + 2)^3 - 10.653(0.1t + 2)^2 + 73.874(0.1t + 2) - 129.237$$

where t is the number of years past 1920. Use the model to find and interpret the instantaneous rate of change of union membership in 1940 and in 1990.

53. ***Persons living below the poverty level*** The table below shows the number of millions of people in the United States who lived below the poverty level for selected years between 1960 and 1996.

Year	*Persons Living Below the Poverty Level (in Millions)*
1960	39.9
1965	33.2
1970	25.4
1975	25.9
1980	29.3
1986	32.4
1990	33.6
1993	39.3
1994	38.1
1995	36.4
1996	36.5

Source: *The World Almanac and Book of Facts,* 1998

Assume that the number of persons below the poverty level can be modeled by

$$p(t) = -0.001773(t + 60)^3 + 0.4506(t + 60)^2 - 37.44(t + 60) + 1047.8$$

where t is the number of years past 1960. Use the model to find and interpret the instantaneous rate of change of the number of persons below the poverty level in 1970 and in 1990.

 54. ***Inflation rate*** The annual change in the consumer price index (CPI) during certain years is shown in the table. Assume that the annual change in the CPI can be modeled with the function

$$f(t) = 0.0308469(t + 87)^3 - 8.4978(t + 87)^2 + 779.66(t + 87) - 23{,}820.3$$

where t is the number of years past 1987.

Year	*Consumer Price Index Annual Percent Change*
1987	3.6
1988	4.1
1989	4.8
1990	5.4
1991	4.2
1992	3.0
1993	3.0
1994	2.6
1995	2.8
1996	3.0

Source: *The World Almanac and Book of Facts,* 1998

Use the model to find and interpret the instantaneous rate of change of the CPI in 1990 and in 1995.

9.7 Using Derivative Formulas

OBJECTIVE

- *To use derivative formulas separately and in combination with each other*

APPLICATION PREVIEW

Suppose the weekly revenue function for a product is given by

$$R(x) = \frac{36{,}000{,}000x}{(2x + 500)^2}$$

where x is the number of units sold. We can find marginal revenue by finding the derivative of the revenue function. This revenue function contains both a quotient and a power, so its derivative is found by using both the Quotient Rule and the Power Rule. But before we do this, we must first decide the order in which to apply these formulas.

We have used the Power Rule to find the derivative of functions like

$$y = (x^3 - 3x^2 + x + 1)^5$$

but we have not found the derivative of functions like

$$y = [(x^2 + 1)(x^3 + x + 1)]^5$$

This function is different because the function u (which is raised to the fifth power) is the product of two functions, $(x^2 + 1)$ and $(x^3 + x + 1)$. The equation is of the form $y = u^5$, where $u = (x^2 + 1)(x^3 + x + 1)$. This means that the Product Rule should be used to find du/dx. Then

$$\begin{aligned}\frac{dy}{dx} &= 5u^4 \cdot \frac{du}{dx}\\ &= 5[(x^2+1)(x^3+x+1)]^4[(x^2+1)(3x^2+1)+(x^3+x+1)(2x)]\\ &= 5[(x^2+1)(x^3+x+1)]^4(5x^4+6x^2+2x+1)\\ &= (25x^4+30x^2+10x+5)[(x^2+1)(x^3+x+1)]^4\end{aligned}$$

A different type of problem involving the Power Rule and the Product Rule is finding the derivative of $y = (x^2 + 1)^5(x^3 + x + 1)$. We may think of y as the *product* of two functions, one of which is a power. Thus the fundamental formula we should use is the Product Rule. The two functions are $u(x) = (x^2 + 1)^5$ and $v(x) = x^3 + x + 1$. The Product Rule gives

$$\begin{aligned}\frac{dy}{dx} &= u(x)\cdot v'(x) + v(x)\cdot u'(x)\\ &= (x^2+1)^5(3x^2+1) + (x^3+x+1)[5(x^2+1)^4 2x]\end{aligned}$$

Note that the Power Rule was used to find $u'(x)$, since $u(x) = (x^2 + 1)^5$.

We can simplify dy/dx by factoring $(x^2 + 1)^4$ from both terms:

$$\begin{aligned}\frac{dy}{dx} &= (x^2+1)^4[(x^2+1)(3x^2+1)+(x^3+x+1)\cdot 5\cdot 2x]\\ &= (x^2+1)^4(13x^4+14x^2+10x+1)\end{aligned}$$

EXAMPLE 1

If $y = \left(\dfrac{x^2}{x-1}\right)^5$, find y'.

Solution

We again have an equation of the form $y = u^n$, but this time u is a quotient. Thus we will need the Quotient Rule to find du/dx.

$$y' = nu^{n-1}\cdot\frac{du}{dx} = 5u^4\,\frac{(x-1)\cdot 2x - x^2\cdot 1}{(x-1)^2}$$

Substituting for u and simplifying gives

$$\begin{aligned}y' &= 5\left(\frac{x^2}{x-1}\right)^4\cdot\frac{2x^2-2x-x^2}{(x-1)^2}\\ &= \frac{5x^8(x^2-2x)}{(x-1)^6} = \frac{5x^{10}-10x^9}{(x-1)^6}\end{aligned}$$

EXAMPLE 2

Find $f'(x)$ if $f(x) = \dfrac{(x-1)^2}{(x^4+3)^3}$.

Solution

This function is the quotient of two functions, $(x - 1)^2$ and $(x^4 + 3)^3$, so we must use the Quotient Rule to find the derivative of $f(x)$, but taking the derivatives of $(x - 1)^2$ and $(x^4 + 3)^3$ will require the Power Rule.

$$f'(x) = \frac{[v(x) \cdot u'(x) - u(x) \cdot v'(x)]}{[v(x)]^2}$$

$$= \frac{(x^4+3)^3[2(x-1)(1)] - (x-1)^2[3(x^4+3)^2 4x^3]}{[(x^4+3)^3]^2}$$

$$= \frac{2(x^4+3)^3(x-1) - 12x^3(x-1)^2(x^4+3)^2}{(x^4+3)^6}$$

We see that 2, $(x^4 + 3)^2$, and $(x - 1)$ are all factors in both terms of the numerator, so we can factor them from both terms and reduce the fraction.

$$f'(x) = \frac{2(x^4+3)^2(x-1)[(x^4+3) - 6x^3(x-1)]}{(x^4+3)^6}$$

$$= \frac{2(x-1)(-5x^4+6x^3+3)}{(x^4+3)^4}$$

EXAMPLE 3

Find $f'(x)$ if $f(x) = (x^2 - 1)\sqrt{3 - x^2}$.

Solution

The function is the product of two functions, $x^2 - 1$ and $\sqrt{3 - x^2}$. Therefore, we will use the Product Rule to find the derivative of $f(x)$, but the derivative of $\sqrt{3 - x^2} = (3 - x^2)^{1/2}$ will require the Power Rule.

$$f'(x) = u(x) \cdot v'(x) + v(x) \cdot u'(x)$$

$$= (x^2 - 1)\left[\frac{1}{2}(3 - x^2)^{-1/2}(-2x)\right] + (3 - x^2)^{1/2}(2x)$$

$$= (x^2 - 1)[-x(3 - x^2)^{-1/2}] + (3 - x^2)^{1/2}(2x)$$

$$= \frac{-x^3 + x}{(3 - x^2)^{1/2}} + 2x(3 - x^2)^{1/2}$$

We can combine these terms over the common denominator $(3 - x^2)^{1/2}$ as follows:

$$f'(x) = \frac{-x^3 + x}{(3 - x^2)^{1/2}} + \frac{2x(3 - x^2)^1}{(3 - x^2)^{1/2}} = \frac{-x^3 + x + 6x - 2x^3}{(3 - x^2)^{1/2}}$$

$$= \frac{-3x^3 + 7x}{(3 - x^2)^{1/2}}$$

We should note that in Example 3 we could have written $f'(x)$ in the form

$$f'(x) = (-x^3 + x)(3 - x^2)^{-1/2} + 2x(3 - x^2)^{1/2}$$

Now the factor $(3 - x^2)$, to different powers, is contained in both terms of the expression. Thus we can factor $(3 - x^2)^{-1/2}$ from both terms. (We choose the $-1/2$ power because it is the smaller of the two powers.) Dividing $(3 - x^2)^{-1/2}$ into the first term gives $(-x^3 + x)$, and dividing it into the second term gives $2x(3 - x^2)^1$. Why? Thus we have

$$f'(x) = (3 - x^2)^{-1/2}[(-x^3 + x) + 2x(3 - x^2)]$$
$$= \frac{-3x^2 + 7x}{(3 - x^2)^{1/2}}$$

which agrees with our previous answer.

CHECKPOINT

1. If a function has the form $y = [u(x)]^n \cdot v(x)$, where n is a constant, we begin to find the derivative by using the ________ Rule and then use the ________ Rule to find the derivative of $[u(x)]^n$.
2. If a function has the form $y = [u(x)/v(x)]^n$, where n is a constant, we begin to find the derivative by using the ________ Rule and then use the ________ Rule.
3. Find the derivative of each of the following and simplify.

   (a) $f(x) = 3x^4(2x^4 + 7)^5$ (b) $g(x) = \dfrac{(4x + 3)^7}{2x - 9}$

We now return to the Application Preview problem.

EXAMPLE 4

Suppose that the weekly revenue function for a product is given by

$$R(x) = \frac{36{,}000{,}000x}{(2x + 500)^2}$$

where x is the number of units sold.

(a) Find the marginal revenue function.
(b) Find the marginal revenue when 50 units are sold.

Solution

(a) $\overline{MR} = R'(x)$

$$= \frac{(2x + 500)^2(36{,}000{,}000) - 36{,}000{,}000x[2(2x + 500)^1(2)]}{(2x + 500)^4}$$
$$= \frac{36{,}000{,}000(2x + 500)(2x + 500 - 4x)}{(2x + 500)^4}$$
$$= \frac{36{,}000{,}000(500 - 2x)}{(2x + 500)^3}$$

(b) $\overline{MR}(50) = R'(50) = \dfrac{36{,}000{,}000(500 - 100)}{(100 + 500)^3}$

$$= \frac{36{,}000{,}000(400)}{(600)^3}$$
$$= \frac{200}{3} = 66.67$$

The marginal revenue is \$66.67 when 50 units are sold. That is, the predicted revenue from the sale of the 51st unit is approximately \$66.67.

It may be helpful to review the formulas needed to find the derivatives of various types of functions. Table 9.5 presents examples of different types of functions and the formulas needed to find their derivatives.

TABLE 9.5 Derivative Formulas Summary

Example	*Formula*
$f(x) = 14$	If $f(x) = c$, then $f'(x) = 0$.
$y = x^4$	If $f(x) = x^n$, then $f'(x) = nx^{n-1}$.
$g(x) = 5x^3$	If $g(x) = cf(x)$, then $g'(x) = cf'(x)$.
$y = 3x^2 + 4x$	If $f(x) = u(x) + v(x)$, then $f'(x) = u'(x) + v'(x)$.
$y = (x^2 - 2)(x + 4)$	If $f(x) = u(x) \cdot v(x)$, then $f'(x) = u(x) \cdot v'(x) + v(x) \cdot u'(x)$.
$f(x) = \dfrac{x^3}{x^2 + 1}$	If $f(x) = \dfrac{u(x)}{v(x)}$, then $f'(x) = \dfrac{v(x) \cdot u'(x) - u(x) \cdot v'(x)}{[v(x)]^2}$.
$y = (x^3 - 4x)^{10}$	If $y = u^n$ and $u = g(x)$, then $\dfrac{dy}{dx} = nu^{n-1} \cdot \dfrac{du}{dx}$.
$y = \left(\dfrac{x - 1}{x^2 + 3}\right)^3$	Power Rule, then Quotient Rule to find $\dfrac{du}{dx}$, where $u = \dfrac{x - 1}{x^2 + 3}$.
$y = (x + 1)\sqrt{x^3 + 1}$	Product Rule, then Power Rule to find $v'(x)$, where $v(x) = \sqrt{x^3 + 1}$.
$y = \dfrac{(x^2 - 3)^4}{x + 1}$	Quotient Rule, then Power Rule to find the derivative of the numerator.

CHECKPOINT SOLUTIONS

1. Product, Power
2. Power, Quotient
3. (a) $f'(x) = 3x^4[5(2x^4 + 7)^4(8x^3)] + (2x^4 + 7)^5(12x^3)$

$$= 120x^7(2x^4 + 7)^4 + 12x^3(2x^4 + 7)^5$$

$$= 12x^3(2x^4 + 7)^4[10x^4 + (2x^4 + 7)] = 12x^3(12x^4 + 7)(2x^4 + 7)^4$$

(b) $g'(x) = \dfrac{(2x - 9)[7(4x + 3)^6(4)] - (4x + 3)^7(2)}{(2x - 9)^2}$

$$= \frac{2(4x + 3)^6[14(2x - 9) - (4x + 3)]}{(2x - 9)^2} = \frac{2(24x - 129)(4x + 3)^6}{(2x - 9)^2}$$

EXERCISE 9.7

Find the derivatives of the functions in Problems 1–32. Simplify and express the answer using positive exponents only.

1. $f(x) = \pi^4$
2. $f(x) = \dfrac{1}{4}$
3. $g(x) = \dfrac{4}{x^4}$
4. $y = \dfrac{x^4}{4}$
5. $g(x) = 5x^3 + \dfrac{4}{x}$
6. $y = 3x^2 + 4\sqrt{x}$
7. $y = (x^2 - 2)(x + 4)$

8. $y = (x^3 - 5x^2 + 1)(x^3 - 3)$

9. $f(x) = \dfrac{x^3 + 1}{x^2}$ 10. $y = \dfrac{1 + x^2 - x^4}{1 + x^4}$

11. $y = \dfrac{(x^3 - 4x)^{10}}{10}$ 12. $y = \dfrac{5}{2}(3x^4 - 6x^2 + 2)^5$

13. $y = \dfrac{5}{3}x^3(4x^5 - 5)^3$ 14. $y = 3x^4(2x^5 + 1)^7$

15. $y = (x - 1)^2(x^2 + 1)$

16. $f(x) = (5x^3 + 1)(x^4 + 5x)^2$

17. $y = \dfrac{(x^2 - 4)^3}{x^2 + 1}$ 18. $y = \dfrac{(x^2 - 3)^4}{x}$

19. $p = [(q + 1)(q^3 - 3)]^3$

20. $y = [(4 - x^2)(x^2 + 5x)]^4$

21. $R(x) = [x^2(x^2 + 3x)]^4$ 22. $c(x) = [x^3(x^2 + 1)]^{-3}$

23. $y = \left(\dfrac{2x - 1}{x^2 + x}\right)^4$ 24. $y = \left(\dfrac{5 - x^2}{x^4}\right)^3$

25. $g(x) = (8x^4 + 3)^2(x^3 - 4x)^3$

26. $y = (3x^3 - 4x)^3(4x^2 - 8)^2$

27. $f(x) = \dfrac{\sqrt[3]{x^2 + 5}}{4 - x^2}$ 28. $g(x) = \dfrac{\sqrt[3]{2x - 1}}{2x + 1}$

29. $y = x^2\sqrt[4]{4x - 3}$ 30. $y = 3x\sqrt[3]{4x^4 + 3}$

31. $c(x) = 2x\sqrt{x^3 + 1}$ 32. $R(x) = x\sqrt[3]{3x^3 + 2}$

In Problems 33 and 34, find the derivative of each function.

33. (a) $F_1(x) = \dfrac{3(x^4 + 1)^5}{5}$ (b) $F_2(x) = \dfrac{3}{5(x^4 + 1)^5}$

(c) $F_3(x) = \dfrac{(3x^4 + 1)^5}{5}$ (d) $F_4(x) = \dfrac{3}{(5x^4 + 1)^5}$

34. (a) $G_1(x) = \dfrac{2(x^3 - 5)^3}{3}$ (b) $G_2(x) = \dfrac{(2x^3 - 5)^3}{3}$

(c) $G_3(x) = \dfrac{2}{3(x^3 - 5)^3}$ (d) $G_4(x) = \dfrac{2}{(3x^3 - 5)^3}$

Applications

35. ***Physical output*** The total physical output P of workers is a function of the number of workers, x. The function $P = f(x)$ is called the physical productivity function. Suppose that the physical productivity of x construction workers is given by

$$P = 10(3x + 1)^3 - 10$$

Find the marginal physical productivity, dP/dx.

36. ***Revenue*** Suppose that the revenue function for a certain product is given by

$$R(x) = 15(2x + 1)^{-1} + 30x - 15$$

where x is in thousands of units and R is in thousands of dollars.

(a) Find the marginal revenue when 2000 units are sold.

(b) How is revenue changing when 2000 units are sold?

37. ***Revenue*** Suppose that the revenue function for a computer is given by

$$R(x) = 60{,}000x + 40{,}000(10 + x)^{-1} - 4000$$

(a) Find the marginal revenue when 10 units are sold.

(b) How is revenue changing when 10 units are sold?

38. ***Production*** Suppose that the production of x items of a new line of products is given by

$$x = 200[(t + 10) - 400(t + 40)^{-1}]$$

where t is the number of weeks the line has been in production. Find the rate of production, dx/dt.

39. ***National consumption*** If the national consumption function is given by

$$C(y) = 2(y + 1)^{1/2} + 0.4y + 4$$

find the marginal propensity to consume, dC/dy.

40. ***Demand*** Suppose that the demand function for an appliance is given by

$$p = \frac{400(q + 1)}{(q + 2)^2}$$

Find the rate of change of price with respect to the number of appliances.

41. ***Volume*** When squares of side x are cut from the corners of a 12-inch-square piece of cardboard, an open-top box can be formed by folding up the sides. The volume of this box is given by

$$V = x(12 - 2x)^2$$

Find the rate of change of volume with respect to the size of the squares.

42. ***Advertising and sales*** Suppose that sales (in thousands of dollars) are directly related to an advertising campaign according to

$$S = 1 + \frac{3t - 9}{(t + 3)^2}$$

where t is the number of weeks of the campaign.

(a) Find the rate of change of sales after 3 weeks.

(b) Interpret the result in (a).

43. ***Advertising and sales*** An inferior product with an extensive advertising campaign does well when it is released, but sales decline as people discontinue use of the product. If the sales S after t weeks are given by

$$S(t) = \frac{200t}{(t+1)^2}, \quad t \geq 0$$

what is the rate of change of sales when $t = 9$? Interpret your result.

44. ***Advertising and sales*** An excellent film with a very small advertising budget must depend largely on word-of-mouth advertising. If attendance at the film after t weeks is given by

$$A = \frac{100t}{(t+10)^2}$$

what is the rate of change in attendance and what does it mean when (a) $t = 10$? (b) $t = 20$?

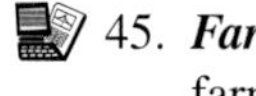
45. ***Farm workers*** The percentage of U.S. workers in farm occupations during certain years is shown in the table.

Year	*Percent*	*Year*	*Percent*
1820	71.8	1950	11.6
1850	63.7	1960	6.1
1870	53	1970	3.6
1900	37.5	1980	2.7
1920	27	1985	2.8
1930	21.2	1990	2.4
1940	17.4		

Source: *The World Almanac and Book of Facts*, 1993

Assume that the percentage of U.S. workers in farm occupations can be modeled with the function

$$f(t) = \frac{1000[-8.0912(t+20) + 1558.9]}{1.09816(t+20)^2 - 122.183(t+20) + 21472.6}$$

where t is the number of years past 1820.

(a) Find the function that models the instantaneous rate of change of the percentage of U.S. workers in farm occupations.
(b) Use the result of (a) to find the instantaneous rate of change in 1850 and in 1950.
(c) Interpret the two rates of change in (b).

9.8 Higher-Order Derivatives

OBJECTIVE

- *To find second derivatives and higher derivatives of certain functions*

APPLICATION PREVIEW

Suppose a particle travels according to the equation $s = 100t - 16t^2 + 200$, where s is the distance and t is the time. Then $\frac{ds}{dt}$ is the velocity, and the acceleration of the particle is the rate of change of the velocity of the particle, so acceleration is $\frac{dv}{dt}$. That is, acceleration is the derivative of the derivative of the distance function. This is called the second derivative. In this section we will discuss second-order and higher-order derivatives.

Because the derivative of a function is itself a function, we can take a derivative of the derivative. The derivative of a first derivative is called a **second derivative.** We can find the second derivative of a function f by differentiating it twice. If f' represents the first derivative of a function, then f'' represents the second derivative of that function.

EXAMPLE 1

If $f(x) = 3x^3 - 4x^2 + 5$, find $f''(x)$.

Solution

The first derivative is $f'(x) = 9x^2 - 8x$.
The second derivative is $f''(x) = 18x - 8$.

EXAMPLE 2

Find the second derivative of $y = x^4 - 3x^2 + x^{-2}$.

Solution

The first derivative is $y' = 4x^3 - 6x - 2x^{-3}$.
The second derivative, which we may denote by y'', is

$$y'' = 12x^2 - 6 + 6x^{-4}$$

It is also common to use $\frac{d^2y}{dx^2}$ and $\frac{d^2}{dx^2} f(x)$ to denote the second derivative of a function.

EXAMPLE 3

If $y = \sqrt{2x - 1}$, find d^2y/dx^2.

Solution

The first derivative is

$$\frac{dy}{dx} = \frac{1}{2}(2x - 1)^{-1/2}(2) = (2x - 1)^{-1/2}$$

The second derivative is

$$\frac{d^2y}{dx^2} = -\frac{1}{2}(2x - 1)^{-3/2}(2) = -(2x - 1)^{-3/2}$$

$$= \frac{-1}{(2x - 1)^{3/2}} = \frac{-1}{\sqrt{(2x - 1)^3}}$$

We can also find third, fourth, fifth, and higher derivatives, continuing indefinitely. The third, fourth, and fifth derivatives of a function f are denoted by f''', $f^{(4)}$, and $f^{(5)}$, respectively. Other notations for the third and fourth derivatives include

$$y''' = \frac{d^3y}{dx^3} = \frac{d^3f(x)}{dx^3} \qquad y^{(4)} = \frac{d^4y}{dx^4} = \frac{d^4f(x)}{dx^4}$$

EXAMPLE 4

Find the first four derivatives of $f(x) = 4x^3 + 5x^2 + 3$.

Solution

$f'(x) = 12x^2 + 10x$, $\quad f''(x) = 24x + 10$, $\quad f'''(x) = 24$, $\quad f^{(4)}(x) = 0$

Just as the first derivative, $f'(x)$, can be used to determine the rate of change of a function $f(x)$, the second derivative, $f''(x)$, can be used to determine the rate of change of $f'(x)$.

EXAMPLE 5

Let $f(x) = 3x^4 + 6x^3 - 3x^2 + 4$.

(a) How fast is $f(x)$ changing at (1, 10)?
(b) How fast is $f'(x)$ changing at (1, 10)?
(c) Is $f'(x)$ increasing or decreasing at (1, 10)?

Solution

(a) Because $f'(x) = 12x^3 + 18x^2 - 6x$, we have

$$f'(1) = 12 + 18 - 6 = 24$$

Thus the rate of change of $f(x)$ at (1, 10) is 24.

(b) Because $f''(x) = 36x^2 + 36x - 6$, we have

$$f''(1) = 66$$

Thus the rate of change of $f'(x)$ at (1, 10) is 66.

(c) Because $f''(1) = 66 > 0$, $f'(x)$ is increasing at (1, 10).

EXAMPLE 6

Suppose that a particle travels according to the equation

$$s = 100t - 16t^2 + 200$$

where s is the distance and t is the time. Then ds/dt is the velocity, and $d^2s/dt^2 = dv/dt$ is the acceleration of the particle. Find the acceleration.

Solution

The velocity is $v = ds/dt = 100 - 32t$, and the acceleration is

$$\frac{dv}{dt} = \frac{d^2s}{dt^2} = -32$$

CHECKPOINT

Suppose that the vertical distance a particle travels is given by

$$s = 4x^3 - 12x^2 + 6$$

where s is in feet and x is in seconds.

1. Find the function that describes the velocity of this particle.
2. Find the function that describes the acceleration of this particle.
3. Is the acceleration always positive?
4. When does the *velocity* of this particle increase?

We can use the numerical derivative feature of a graphing utility to find the second derivative of a function at a point.

EXAMPLE 7

Find $f''(2)$ if $f(x) = \sqrt{x^3 - 1}$.

Solution

We need the derivative of the derivative function, evaluated at $x = 2$. Figure 9.31 shows how the numerical derivative feature of a graphing utility can be used to obtain this result.

```
nDeriv(nDeriv(√(
X^3-1),X,X),X,2)

                .323969225
```

Figure 9.31

Thus Figure 9.31 shows that $f''(2) = 0.323969225 \approx 0.32397$. We can check this result by calculating $f''(x)$ with formulas.

$$f'(x) = \frac{1}{2}(x^3 - 1)^{-\frac{1}{2}}(3x^2)$$

$$f''(x) = \frac{1}{2}(x^3 - 1)^{-\frac{1}{2}}(6x) + (3x^2)\left[-\frac{1}{4}(x^3 - 1)^{-\frac{3}{2}}(3x^2)\right]$$

$$f''(2) = 0.3239695483 \approx 0.32397$$

Thus we see that the numerical derivative approximation is quite accurate.

EXAMPLE 8

(a) Given $f(x) = x^4 - 12x^2 + 2$, graph $f(x)$ and its second derivative on the same set of axes over an interval that contains all x-values where $f''(x) = 0$.
(b) When the graph of $y = f(x)$ is opening downward, is $f''(x) > 0, f''(x) < 0$, or $f''(x) = 0$?
(c) When the graph of $y = f(x)$ is opening upward, is $f''(x) > 0, f''(x) < 0$, or $f''(x) = 0$?

Solution

(a) $f'(x) = 4x^3 - 24x$ and $f''(x) = 12x^2 - 24 = 12(x^2 - 2)$. Because $f''(x) = 0$ at $x = -\sqrt{2}$ and at $x = \sqrt{2}$, we use an x-range that contains $x = -\sqrt{2}$ and $\sqrt{2}$. The graph of $f(x) = x^4 - 12x^2 + 2$ and its second derivative, $f''(x) = 12x^2 - 24$, are shown in Figure 9.32.
(b) The graph of $y = f(x)$ appears to be opening downward on the same interval for which $f''(x) < 0$.

(c) The graph appears to be opening upward on the same intervals for which $f''(x) > 0$.

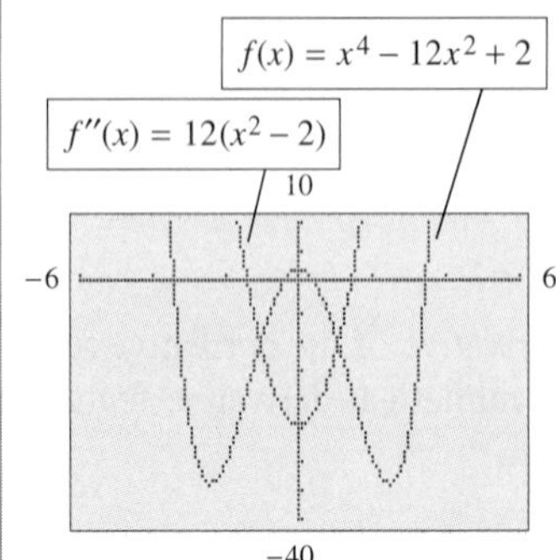

Figure 9.32

CHECKPOINT SOLUTIONS

1. The velocity is described by $s'(x) = 12x^2 - 24x$.
2. The acceleration is described by $s''(x) = 24x - 24$.
3. No; the acceleration is positive when $s''(x) > 0$—that is, when $24x - 24 > 0$. It is zero when $24x - 24 = 0$ and negative when $24x - 24 < 0$. Thus acceleration is negative when $x < 1$ second, zero when $x = 1$ second, and positive when $x > 1$ second.
4. The velocity increases when the acceleration is positive. Thus the velocity is increasing after 1 second.

EXERCISE 9.8

In Problems 1–8, find the second derivative.

1. $f(x) = 4x^3 - 15x^2 + 3x + 2$
2. $f(x) = 2x^{10} - 18x^5 - 12x^3 + 4$
3. $y = 10x^3 - x^2 + 14x + 3$
4. $y = 6x^5 - 3x^4 + 12x^2$
5. $g(x) = x^3 - \dfrac{1}{x}$
6. $h(x) = x^2 - \dfrac{1}{x^2}$
7. $y = x^3 - \sqrt{x}$
8. $y = 3x^2 - \sqrt[3]{x^2}$

In Problems 9–16, find the third derivative.

9. $y = x^5 - 16x^3 + 12$
10. $y = 6x^3 - 12x^2 + 6x$
11. $f(x) = 2x^9 - 6x^6$
12. $f(x) = 3x^5 - x^6$
13. $y = 1/x$
14. $y = 1/x^2$
15. $y = \sqrt{x}$
16. $y = \sqrt[3]{x}$

In Problems 17–28, find the indicated derivative.

17. If $y = x^5 - x^{1/2}$, find $\dfrac{d^2y}{dx^2}$.
18. If $y = x^4 + x^{1/3}$, find $\dfrac{d^2y}{dx^2}$.
19. If $f(x) = \sqrt{x+1}$, find $f'''(x)$.
20. If $f(x) = \sqrt{x-5}$, find $f'''(x)$.
21. Find $\dfrac{d^4y}{dx^4}$ if $y = 4x^3 - 16x$.
22. Find $y^{(4)}$ if $y = x^6 - 15x^3$.
23. Find $f^{(4)}(x)$ if $f(x) = \sqrt{x}$.
24. Find $f^{(4)}(x)$ if $f(x) = 1/x$.
25. Find $y^{(4)}$ if $y' = \sqrt{x-1}$.
26. Find $y^{(5)}$ if $\dfrac{d^2y}{dx^2} = \sqrt[3]{x+2}$.
27. Find $f^{(6)}(x)$ if $f^{(4)}(x) = x(x+1)^{-1}$.
28. Find $f^{(3)}(x)$ if $f'(x) = \dfrac{x^2}{x^2+1}$.
29. If $f(x) = 16x^2 - x^3$, what is the rate of change of $f'(x)$ at $(1, 15)$?
30. If $y = 36x^2 - 6x^3 + x$, what is the rate of change of y' at $(1, 31)$?

In Problems 31–34, use the numerical derivative feature of a graphing utility to approximate the given second derivatives.

31. $f''(3)$ for $f(x) = x^3 - \frac{27}{x}$

32. $f''(-1)$ for $f(x) = \frac{x^2}{4} - \frac{4}{x^2}$

33. $f''(21)$ for $f(x) = \sqrt{x^2 + 4}$

34. $f''(3)$ for $f(x) = \frac{1}{\sqrt{x^2 + 7}}$

In Problems 35–38, do the following for each function $f(x)$.

(a) Find $f'(x)$ and $f''(x)$.
(b) Graph $f(x)$, $f'(x)$, and $f''(x)$ with a graphing utility.
(c) Identify x-values where $f''(x) = 0$, $f''(x) > 0$, and $f''(x) < 0$.
(d) Identify x-values where $f'(x)$ has a maximum point or a minimum point, where $f'(x)$ is increasing, and where $f'(x)$ is decreasing.
(e) When $f(x)$ has a maximum point, is $f''(x) > 0$ or $f''(x) < 0$?
(f) When $f(x)$ has a minimum point, is $f''(x) > 0$ or $f''(x) < 0$?

35. $f(x) = x^3 - 3x^2 + 5$

36. $f(x) = 2 + 3x - x^3$

37. $f(x) = -\frac{1}{3}x^3 - x^2 + 3x + 7$

38. $f(x) = \frac{1}{3}x^3 - \frac{3x^2}{2} - 4x + 10$

Applications

39. ***Acceleration*** If a particle travels as a function of time according to the formula

$$s = 100 + 10t + 0.01t^3$$

find the acceleration of the particle when $t = 2$.

40. ***Acceleration*** If the formula describing the distance an object travels as a function of time is

$$s = 100 + 160t - 16t^2$$

what is the acceleration of the object when $t = 4$?

41. ***Revenue*** The revenue from sales of a certain product can be described by

$$R(x) = 100x - 0.01x^2$$

Find the instantaneous rate of change of the marginal revenue.

42. ***Revenue*** Suppose that the revenue from the sale of a product is given by

$$R = 70x + 0.5x^2 - 0.001x^3$$

where x is the number of units sold. How fast is the marginal revenue $\overline{MR}$ changing when $x = 100$?

43. ***Sensitivity*** When medicine is administered, reaction (measured in change of blood pressure or temperature) can be modeled by

$$R = m^2\left(\frac{c}{2} - \frac{m}{3}\right)$$

where c is a positive constant and m is the amount of medicine absorbed into the blood (Source: Thrall, R. M., *et al., Some Mathematical Models in Biology,* U.S. Department of Commerce, 1967). The sensitivity to the medication is defined to be the rate of change of reaction R with respect to the amount of medicine m absorbed in the blood.

(a) Find the sensitivity.
(b) Find the instantaneous rate of change of sensitivity with respect to the amount of medicine absorbed in the blood.
(c) Which order derivative of reaction gives the rate of change of sensitivity?

44. ***Photosynthesis*** The amount of photosynthesis that takes place in a certain plant depends on the intensity of light x according to the equation

$$f(x) = 145x^2 - 30x^3$$

(a) Find the rate of change of photosynthesis with respect to the intensity.
(b) What is the rate of change when $x = 1$? when $x = 3$?
(c) How fast is the rate found in (a) changing when $x = 1$? when $x = 3$?

45. ***Revenue*** The revenue (in thousands of dollars) from the sale of x units of a product is

$$R = 15x + 30(4x + 1)^{-1} - 30$$

where x is the number of units sold.

(a) At what rate is the marginal revenue $\overline{MR}$ changing when the number of units being sold is 25?
(b) Interpret your result in (a).

46. ***Advertising and sales*** The sales of a product S (in thousands of dollars) are given by

$$S = \frac{600x}{x + 40}$$

where x is the advertising expenditure (in thousands of dollars).
(a) Find the rate of change of sales with respect to advertising expenditure.
(b) Use the second derivative to find how this rate is changing at $x = 20$.
(c) Interpret your result in (b).

47. ***Advertising and sales*** The daily sales S (in thousands of dollars) that are attributed to an advertising campaign is given by

$$S = 1 + \frac{3}{t + 3} - \frac{18}{(t + 3)^2}$$

where t is the number of weeks the campaign runs.
(a) Find the rate of change of sales at any time t.
(b) Use the second derivative to find how this rate is changing at $t = 15$.
(c) Interpret your result in (b).

48. ***Advertising and sales*** A product with a large advertising budget has its sales S (in millions of dollars) given by

$$S = \frac{500}{t + 2} - \frac{1000}{(t + 2)^2}$$

where t is the number of months the product has been on the market.
(a) Find the rate of change of sales at any time t.
(b) What is the rate of change of sales at $t = 2$?
(c) Use the second derivative to find how this rate is changing at $t = 2$.
(d) Interpret your result from (b) and (c).

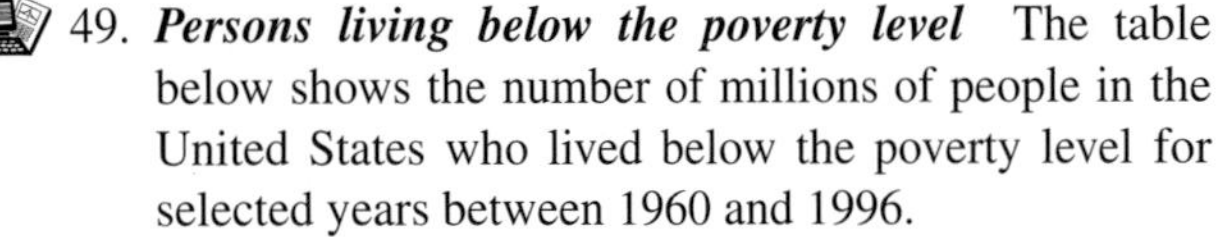

49. ***Persons living below the poverty level*** The table below shows the number of millions of people in the United States who lived below the poverty level for selected years between 1960 and 1996.

Year	*Persons Living Below the Poverty Level (millions)*
1960	39.9
1965	33.2
1970	25.4
1975	25.9
1980	29.3
1986	32.4
1990	33.6
1993	39.3
1994	38.1
1995	36.4
1996	36.5

Source: *The World Almanac and Book of Facts,* 1998

Assume that the number of millions of people in the United States who lived below the poverty level can be modeled by the function

$$p(t) = -0.001723t^3 + 0.1291t^2 - 2.5015t + 40.656$$

where t is the number of years past 1960.
(a) Find the function that models the instantaneous rate of change of $p(t)$.
(b) Use the second derivative to find how this rate is changing in 1970 and 1990.
(c) Interpret the meaning of $p'(10)$ and $p''(10)$. Write a sentence that explains each.

50. ***Consumer debt*** The percentage of disposable income spent on consumer debt during certain years is shown in the table below.

Year	*Consumer Debt as a Percentage of Disposable Income*
1980	18.2
1982	16.9
1983	17.7
1985	20.8
1986	21.4
1988	21.0
1990	20.0
1991	18.9
1994	19.7
1995	21.3
1996	23.7

Source: Federal Reserve System

Assume that consumer debt as a percentage of disposable income can be modeled with the function

$$d(t) = 0.0028t^4 - 0.081t^3 + 0.68t^2 - 1.3t + 17.9$$

where t is the number of years past 1980.

(a) Find the function that models the instantaneous rate of change of the percentage of disposable income spent on consumer debt.

(b) Use the second derivative to determine how this rate is changing in 1985 and 1990.

(c) Write a sentence that explains the meaning of $d'(15)$ and another that explains the meaning of $d''(15)$.

M 51. ***Consumer price index*** The annual change in the consumer price index (CPI) during certain years is shown in the table below. Assume that the annual change in the CPI can be modeled with a cubic function, $f(t)$, where t is the number of years past 1987.

Year	*Annual Percent Change in CPI*
1987	3.6
1988	4.1
1989	4.8
1990	5.4
1991	4.2
1992	3.0
1993	3.0
1994	2.6
1995	2.8
1996	3.0

Source: *The World Almanac and Book of Facts,* 1998

(a) Find the function $f(t)$ that models the CPI.

(b) Find the function that models the instantaneous rate of change of the annual change in the CPI.

(c) Use the second derivative to determine how this rate is changing in 1991 and 1995.

(d) Write sentences that explain the meanings of $f'(8)$ and $f''(8)$.

M 52. ***Union membership*** The table shows the percentage of U.S. workers who belonged to unions for selected years from 1930 to 1996.

Year	*Union Membership as a Percentage of the Labor Force*
1930	11.6
1935	13.2
1940	26.9
1945	35.5
1950	31.5
1955	33.2
1960	31.4
1965	28.4
1970	27.3
1975	25.5
1980	21.9
1985	18.0
1990	16.1
1993	15.8
1994	15.5
1995	14.9
1996	14.5

Source: Bureau of Labor Statistics, Department of Labor

Assume that the percentage of the labor force that belonged to unions can be modeled with a cubic function $f(t)$, where t is the number of years past 1930.

(a) Find the function $f(t)$ that models union membership as a percentage of the labor force.

(b) Find the function that models the instantaneous rate of change of the percentage of the U.S. labor force that belonged to unions.

(c) Use the second derivative to determine how this rate was changing in 1970 and in 1990.

(d) Write sentences that explain the meanings of $f'(40)$ and $f''(40)$.

9.9 Applications of Derivatives in Business and Economics

OBJECTIVES

- *To find the marginal cost and marginal revenue at different levels of production*
- *To find the marginal profit function, given information about cost and revenue*

APPLICATION PREVIEW

In Chapter 1, "Linear Equations and Functions," we defined marginal cost as the rate of change of the total cost function. For a linear total cost function, the marginal cost was defined as the slope of the function's graph. For any total cost function defined by an equation, we can find the instantaneous rate of change of cost (the marginal cost) at any level of production by finding the derivative of the function.

In the same way, we can find the marginal revenue from the total revenue function, and we can use total revenue and total cost functions to find marginal profit.

We begin this section by defining the marginal cost function.

Marginal Cost

If $C = C(x)$ is a total cost function for a commodity, then its derivative, $\overline{MC} = C'(x)$, is the **marginal cost function.**

The linear cost function with equation

$$C(x) = 300 + 6x \qquad \text{(in dollars)}$$

has marginal cost \$6 because its slope is 6. Taking the derivative of $C(x)$ gives

$$\overline{MC} = C'(x) = 6$$

which verifies that the marginal cost is \$6 at all levels of production.

The cost function

$$C(x) = 1000 + 6x + x^2$$

has derivative

$$C'(x) = 6 + 2x$$

Thus the *marginal cost* at $x = 10$ (when 10 units are produced) is

$$C'(10) = 6 + 2(10) = 26$$

and the marginal cost at 40 units is

$$C'(40) = 6 + 2(40) = 86$$

Note that when a cost function is linear, the marginal cost gives the amount by which cost would change if production were increased by 1 unit. When a cost function is *not* linear (as with $C(x) = 1000 + 6x + x^2$), the marginal cost is used to estimate the amount by which cost would change if production were increased by 1 unit. Thus at 10 units, the marginal cost is \$26, so costs would increase by approximately \$26 if 1 more unit were produced. (Note that $C(11) - C(10) = \$27$ gives the actual increase in costs.) Also, at 40 units, the marginal cost is \$86, so costs would increase by approximately \$86 if 1 more unit were produced.

As noted previously, when the derivative of a function is positive (and thus the slope of the tangent to the curve is positive), the function is increasing, and the value of the derivative gives us a measure of how fast it is increasing. As we saw, the marginal cost for

$$C(x) = 1000 + 6x + x^2$$

is 86 at $x = 40$ and 26 at $x = 10$. This tells us that the cost is increasing faster at $x = 40$ than it is at $x = 10$.

Because producing more units can never reduce the total cost of production, the following properties are valid:

1. The total cost can never be negative. If there are fixed costs, the cost of producing 0 units is positive; otherwise, the cost of producing 0 units is 0.
2. The total cost function is always increasing; the more units produced, the higher the total cost. Thus the marginal cost is always positive.
3. There may be limitations on the units produced, such as those imposed by plant space.

The graphs of many marginal cost functions tend to be U-shaped; they eventually will rise, even though there may be an initial interval where they decrease.

EXAMPLE 1

If the total cost function for a commodity is $C(x) = x^3 - 9x^2 + 33x + 30$, find the marginal cost.

Solution

The marginal cost is $\overline{MC} = C'(x) = 3x^2 - 18x + 33$.

The graph of the total cost function is shown in Figure 9.33(a), and the graph of the marginal cost function is shown in Figure 9.33(b).

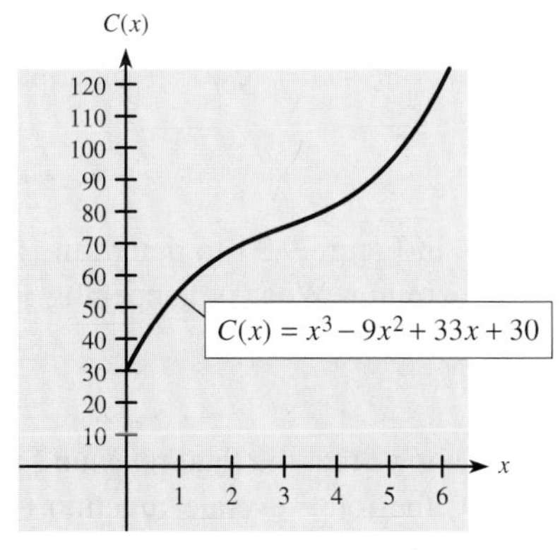

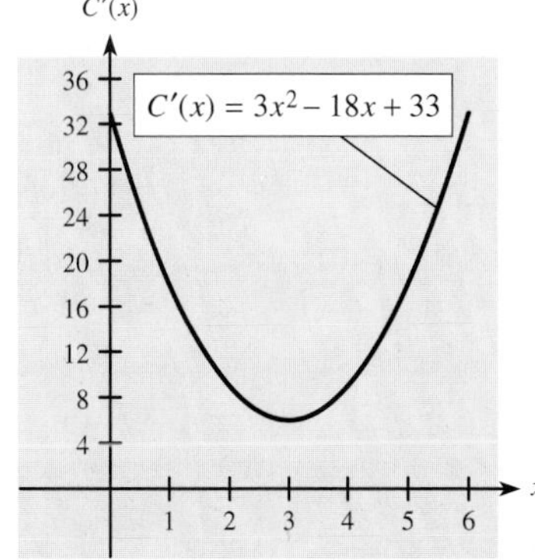

Figure 9.33 (a) Total cost function (b) Marginal cost function

As we saw in Section 9.3, "The Derivative," the instantaneous rate of change (the derivative) of the revenue function is the marginal revenue.

Marginal Revenue If $R = R(x)$ is the total revenue function for a commodity, then the **marginal revenue function** is $\overline{MR} = R'(x)$.

If the demand function for a product in a monopoly market is $p = f(x)$, then the total revenue from the sale of x units is

$$R(x) = px = f(x) \cdot x$$

EXAMPLE 2

If the demand for a product in a monopoly market is given by

$$p = 16 - 0.02x$$

where x is the number of units and p is the price per unit, (a) find the total revenue function, and (b) find the marginal revenue for this product at $x = 40$.

Solution

(a) The total revenue function is

$$\begin{aligned} R(x) &= px = (16 - 0.02x)x \\ &= 16x - 0.02x^2 \end{aligned}$$

(b) The marginal revenue function is

$$\overline{MR} = R'(x) = 16 - 0.04x$$

At $x = 40$, $R'(40) = 16 - 1.6 = 14.40$. Thus the 41st item sold will increase the total revenue by approximately \$14.40.

The marginal revenue is an approximation of the revenue gained from the sale of 1 additional unit. We have used marginal revenue in Example 2 to find that the revenue from the sale of the 41st item will be approximately \$14.40. The actual increase in revenue from the sale of the 41st item is

$$R(41) - R(40) = 622.38 - 608 = \$14.38$$

EXAMPLE 3

Use the graphs in Figure 9.34 to determine the x-value where the revenue function has its maximum. What is happening to the marginal revenue at and near this x-value?

Solution

Figure 9.34(a) shows that the total revenue function has a maximum value at $x = 400$. After that, the total revenue function decreases. This means that the total revenue will be reduced each time a unit is sold if more than 400 are produced and sold. The graph of the marginal revenue function in Figure 9.34(b) shows that the marginal revenue is positive to the left of 400. This indicates that the rate at which the total revenue is changing is positive until 400 units are sold; thus the total revenue is increasing. Then, at 400 units, the rate of change is 0. After

400 units are sold, the marginal revenue is negative, which indicates that the total revenue is now decreasing. It is clear from looking at either graph that 400 units should be produced and sold to maximize the total revenue function $R(x)$. That is, the *total revenue* function has its maximum at $x = 400$.

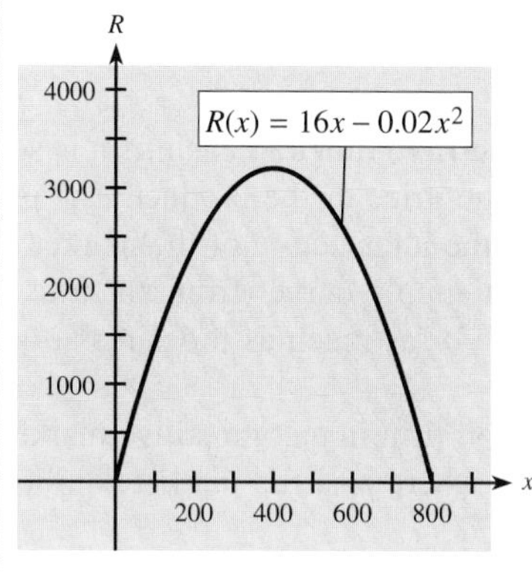

Figure 9.34 (a) Total revenue function (b) Marginal revenue function

CHECKPOINT

The total cost function for the commodity considered in Example 1 is $C(x) = x^3 - 9x^2 + 33x + 30$, and the marginal cost is $C'(x) = 3x^2 - 18x + 33$.

1. What is the marginal cost if $x = 10$ units are produced?
2. Use marginal cost to estimate the cost of producing the 11th unit.
3. Calculate $C(11) - C(10)$ to find the actual cost of producing the 11th unit.
4. True or false: For products that have linear cost functions, the actual cost of producing the $(x + 1)$st unit is equal to the marginal cost at x.

As with marginal cost and marginal revenue, the derivative of a profit function for a commodity will give us the marginal profit function for the commodity.

Marginal Profit If $P = P(x)$ is the profit function for a commodity, then the **marginal profit function** is $\overline{MP} = P'(x)$.

EXAMPLE 4

If the total profit, in thousands of dollars, for a product is given by $P(x) = 20\sqrt{x + 1} - 2x$, what is the marginal profit at a production level of 15 units?

Solution

The marginal profit function is

$$\overline{MP} = P'(x) = 20 \cdot \frac{1}{2}(x + 1)^{-1/2} - 2 = \frac{10}{\sqrt{x + 1}} - 2$$

If 15 units are produced, the marginal profit is

$$P'(15) = \frac{10}{\sqrt{15+1}} - 2 = \frac{1}{2}$$

This means that the profit from the sale of the 16th unit is approximately $\frac{1}{2}$ (thousand dollars), or \$500.

In a **competitive market,** each firm is so small that its actions in the market cannot affect the price of the product. The price of the product is determined in the market by the intersection of the market demand curve (from all consumers) and the market supply curve (from all firms that supply this product). The firm can sell as little or as much as it desires at the given market price, which it cannot change.

Therefore, a firm in a competitive market has a total revenue function given by $R(x) = px$, where p is the market equilibrium price for the product and x is the quantity sold.

EXAMPLE 5

A firm in a competitive market must sell its product for \$200 per unit. The cost per unit (per month) is $80 + x$, where x represents the number of units sold per month. Find the marginal profit function.

Solution

If the cost per unit is $80 + x$, then the total cost of x units is given by the equation $C(x) = (80 + x)x = 80x + x^2$. The revenue per unit is \$200, so the total revenue is given by $R(x) = 200x$. Thus the profit function is

$$P(x) = R(x) - C(x) = 200x - (80x + x^2), \quad \text{or} \quad P(x) = 120x - x^2$$

The marginal profit is $P'(x) = 120 - 2x$.

The marginal profit in Example 5 is not always positive, so producing and selling a certain number of items will maximize profit. Note that the marginal profit will be negative (that is, profit will decrease) if more than 60 items per month are produced. We will discuss methods of maximizing total revenue and profit, and for minimizing average cost, in the next chapter.

CHECKPOINT

If the total profit function for a product is $P(x) = 20\sqrt{x+1} - 2x$, then the marginal profit is

$$P'(x) = \frac{10}{\sqrt{x+1}} - 2 \quad \text{and} \quad P''(x) = \frac{-5}{\sqrt{(x+1)^3}}$$

5. Is $P''(x) < 0$ for all values of $x \geq 0$?
6. Is the marginal profit decreasing for all $x \geq 0$?

EXAMPLE 6

In Example 4, we found that the profit (in thousands of dollars) for a company's products is given by $P(x) = 20\sqrt{x+1} - 2x$ and its marginal profit is given by $P'(x) = \dfrac{10}{\sqrt{x+1}} - 2.$

(a) Use the graphs of $P(x)$ and $P'(x)$ to determine the relationship between the two functions.
(b) When is the marginal profit 0? What is happening to profit at this level of production?

Solution

(a) By comparing the graphs of the two functions (shown in Figure 9.35), we see that for $x > 0$, profit $P(x)$ is increasing over the interval where the marginal profit $P'(x)$ is positive, and profit is decreasing over the interval where the marginal profit $P'(x)$ is negative.
(b) By using SOLVER, INTERSECT, or TRACE, or by using algebra, we see that $P'(x) = 0$ when $x = 24$. This level of production ($x = 24$) is where profit is maximized, at 52 (thousand dollars).

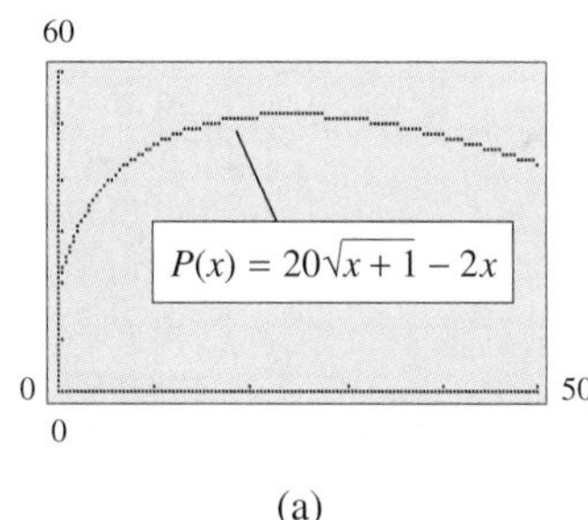

(a)

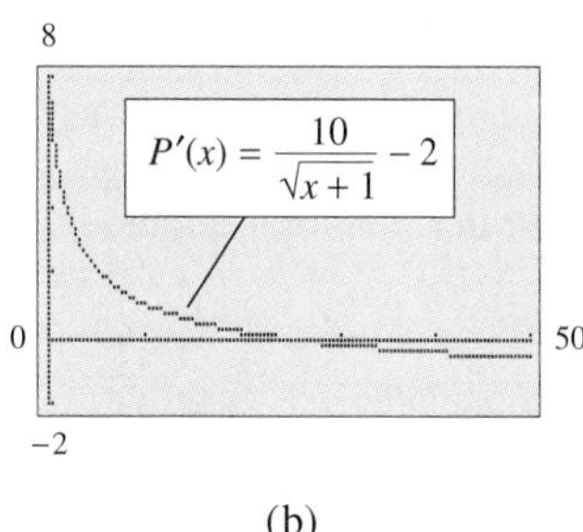

(b)

Figure 9.35

CHECKPOINT SOLUTIONS

1. $C'(10) = 153$
2. $C'(10) = 153$, so it will cost approximately \$153 to produce the 11th unit.
3. $C(11) - C(10) = 635 - 460 = 175$
4. True
5. Yes
6. Yes, because $P''(x) < 0$ for $x \geq 0$.

EXERCISE 9.9

Marginal Cost, Revenue, and Profit

Find the marginal cost functions related to the cost functions in Problems 1–8.

1. $C(x) = 40 + 8x$
2. $C(x) = 200 + 16x$
3. $C(x) = 500 + 13x + x^2$
4. $C(x) = 300 + 10x + \frac{1}{100}x^2$
5. $C = x^3 - 6x^2 + 24x + 10$
6. $C = x^3 - 12x^2 + 63x + 15$
7. $C = 400 + 27x + x^3$
8. $C(x) = 50 + 48x + x^3$
9. Suppose that the cost function for a commodity is

$$C(x) = 40 + x^2$$

(a) Find the marginal cost at $x = 5$ and tell what this predicts about the cost of producing 1 additional unit.
(b) Calculate $C(6) - C(5)$ to find the actual cost of producing 1 additional unit.

10. Suppose that the cost function for a commodity is

$$C(x) = 300 + 6x + \frac{1}{20}x^2$$

(a) Find the marginal cost at $x = 8$ and tell what this predicts about the cost of producing 1 additional unit.
(b) Calculate $C(9) - C(8)$ to find the actual cost of producing 1 additional unit.

11. If the cost function for a commodity is

$$C(x) = x^3 - 4x^2 + 30x + 20$$

find the marginal cost at $x = 4$ and tell what this predicts about the cost of producing 1 additional unit.

12. If the cost function for a commodity is

$$C(x) = \frac{1}{90}x^3 + 4x^2 + 4x + 10$$

find the marginal cost at $x = 3$ and tell what this predicts about the cost of producing 1 additional unit.

13. If the cost function for a commodity is

$$C(x) = 300 + 4x + x^2$$

graph the marginal cost function.

14. If the cost function for a commodity is

$$C(x) = x^3 - 12x^2 + 63x + 15$$

graph the marginal cost function.

15. (a) If the total revenue function for a product is $R(x) = 4x$, what is the marginal revenue function for that product?
(b) What does this marginal revenue function tell us?

16. If the total revenue function for a product is $R(x) = 32x$, what is the marginal revenue for the product? What does this mean?

17. Suppose that the total revenue function for a commodity is $R = 36x - 0.01x^2$.
(a) Find $R(100)$ and tell what it represents.
(b) Find the marginal revenue function.
(c) Find the marginal revenue at $x = 100$, and tell what it predicts about the sale of the next unit.
(d) Find $R(101) - R(100)$ and explain what this value represents.

18. Suppose that the total revenue function for a commodity is $R(x) = 25x - 0.05x^2$.
(a) Find $R(50)$ and tell what it represents.
(b) Find the marginal revenue function.
(c) Find the marginal revenue at $x = 50$, and tell what it predicts about the sale of the next unit.
(d) Find $R(51) - R(50)$ and explain what this value represents.

19. (a) Graph the marginal revenue function from Problem 17.
(b) At what value of x will total revenue be maximized for Problem 17?
(c) What is the maximum revenue?

20. (a) Graph the marginal revenue function from Problem 18.
(b) Determine the number of units that must be sold to maximize total revenue.
(c) What is the maximum revenue?

21. If the total profit function is $P(x) = 5x - 25$, find the marginal profit.

22. If the total profit function is $P(x) = 16x - 32$, find the marginal profit.

23. Suppose that the total revenue function for a product is $R(x) = 32x$ and that the total cost function is $C(x) = 200 + 2x + x^2$.
(a) Find the profit from the production and sale of 20 units.
(b) Find the marginal profit function.
(c) Find $\overline{MP}$ at $x = 20$ and explain what it predicts.
(d) Find $P(21) - P(20)$ and explain what this value represents.

24. Suppose that the total revenue function is given by

$$R(x) = 46x$$

and that the total cost function is given by

$$C(x) = 100 + 30x + \frac{1}{10}x^2$$

(a) Find $P(100)$.
(b) Find the marginal profit function.
(c) Find $\overline{MP}$ at $x = 100$ and explain what it predicts.
(d) Find $P(101) - P(100)$ and explain what this value represents.

25. (a) Graph the marginal profit function for the profit function $P(x) = 30x - x^2 - 200$.
(b) What level of production and sales will give a 0 marginal profit?
(c) At what level of production and sales will profit be at a maximum?
(d) What is the maximum profit?

26. (a) Graph the marginal profit function for the profit function $P(x) = 16x - 0.1x^2 - 100$.
(b) What level of production and sales will give a 0 marginal profit?
(c) At what level of production and sales will profit be at a maximum?
(d) What is the maximum profit?

27. The price of a product in a competitive market is \$300. If the cost per unit of producing the product is $160 + x$, where x is the number of units produced per month, how many units should the firm produce and sell to maximize its profit?
28. The cost per unit of producing a product is $60 + 2x$, where x represents the number of units produced per week. If the equilibrium price determined by a competitive market is \$220, how many units should the firm produce and sell each week to maximize its profit?
29. If the daily cost per unit of producing a product by the Ace Company is $10 + 2x$, and if the price on the competitive market is \$50, what is the maximum daily profit the Ace Company can expect on this product?
30. The Mary Ellen Candy Company produces chocolate Easter bunnies at a cost per unit of $0.10 + 0.01x$, where x is the number produced. If the price on the competitive market for a bunny this size is \$2.50, how many should the company produce to maximize its profit?
31. The following table gives the total revenues of AT&T for selected years.*

Year	Total Revenues (billions)
1985	\$63.13
1986	\$69.906
1987	\$60.53
1989	\$61.1
1990	\$62.191
1991	\$63.089
1992	\$64.904
1993	\$67.156

Source: *AT&T Annual Report*, 1993

Suppose the data can be modeled by the equation

$$R(t) = 0.253t^2 - 4.03t + 76.84$$

where t is the number of years past 1980.

(a) Find $R(7)$ from the data and compare it to $R(7)$ as found from the model. What does $R(7)$ represent?
(b) Find the function that gives the instantaneous rate of change of revenue.
(c) Find the instantaneous rate of change of revenue in 1992.
(d) Interpret your result to (c).

*Before AT&T split off NCR and Lucent

The data in the table give sales revenues and costs and expenses for Scott Paper Company for various years.[†] Use these data for Problems 32–34.

Year	Sales Revenue (billions)	Costs and Expenses (billions)
1983	\$2.6155	\$2.4105
1984	2.7474	2.4412
1985	2.934	2.6378
1986	3.3131	2.9447
1987	3.9769	3.5344
1988	4.5494	3.8171
1989	4.8949	4.2587
1990	5.1686	4.8769
1991	4.9593	4.9088
1992	5.0913	4.6771
1993	4.7489	4.9025

Source: Scott Paper Company, 1993 *Annual Report*

32. Assume that sales revenues for Scott Paper can be modeled by

$$R(t) = -0.031t^2 + 0.776t + 0.179$$

where t is the number of years past 1980.

(a) Use this model to find the instantaneous rate of change of revenue in 1990.
(b) Interpret your answer to (a).

33. Assume that costs and expenses for Scott Paper Company can be modeled by

$$C(t) = -0.012t^2 + 0.492t + 0.725$$

where t is the number of years past 1980.

(a) Use this model to find the instantaneous rate of change of costs and expenses in 1990.
(b) Interpret your answer to (a) and check this interpretation against the data in the table.

34. Let income from operations, $I(t)$, be revenues minus costs.
(a) Find $I(t)$.
(b) Find the instantaneous rate of change of income in 1990.
(c) Interpret your result in (b).
(d) Would the board of directors be interested in altering this model? Explain.

[†]Before Scott merged with Kimberly-Clark

KEY TERMS AND FORMULAS

Section	Key Terms	Formula
9.1	Limit, infinite limit 0/0 indeterminate form	
9.2	Continuous function Vertical asymptote Horizontal asymptote Limit at infinity	
9.3	Average velocity Velocity Instantaneous rate of change	
	Derivative	$f'(x) = \lim_{h \to 0} \frac{f(x+h) - f(x)}{h}$
	Marginal revenue	$\overline{MR} = R'(x)$
	Tangent line Secant line Slope of a curve Differentiability and continuity	
9.4	Powers of x Rule	$\frac{d(x^n)}{dx} = nx^{n-1}$
	Constant Function Rule	$\frac{d(c)}{dx} = 0$ for constant c
	Coefficient Rule	$\frac{d}{dx}[c \cdot f(x)] = c \cdot f'(x)$
	Sum Rule	$\frac{d}{dx}[u + v] = \frac{du}{dx} + \frac{dv}{dx}$
	Difference Rule	$\frac{d}{dx}[u - v] = \frac{du}{dx} - \frac{dv}{dx}$
9.5	Product Rule	$\frac{d}{dx}[uv] = uv' + vu'$
	Quotient Rule	$\frac{d}{dx}\left(\frac{u}{v}\right) = \frac{vu' - uv'}{v^2}$
9.6	Chain Rule	$\frac{dy}{dx} = \frac{dy}{du} \cdot \frac{du}{dx}$
	Power Rule	$\frac{d}{dx}(u^n) = nu^{n-1}\frac{du}{dx}$
9.8	Second derivative; third derivative; higher-order derivatives	
9.9	Marginal cost function Marginal revenue function Marginal profit function	$\overline{MC} = C'(x)$ $\overline{MR} = R'(x)$ $\overline{MP} = P'(x)$

REVIEW EXERCISES

Section 9.1

In Problems 1–6, use the graph of $y = f(x)$ in Figure 9.36 to find the functional values and limits, if they exist.

1. (a) $f(-2)$ (b) $\lim_{x\to -2} f(x)$
2. (a) $f(-1)$ (b) $\lim_{x\to -1} f(x)$
3. (a) $f(4)$ (b) $\lim_{x\to 4^-} f(x)$
4. (a) $\lim_{x\to 4^+} f(x)$ (b) $\lim_{x\to 4} f(x)$
5. (a) $f(1)$ (b) $\lim_{x\to 1} f(x)$
6. (a) $f(2)$ (b) $\lim_{x\to 2} f(x)$

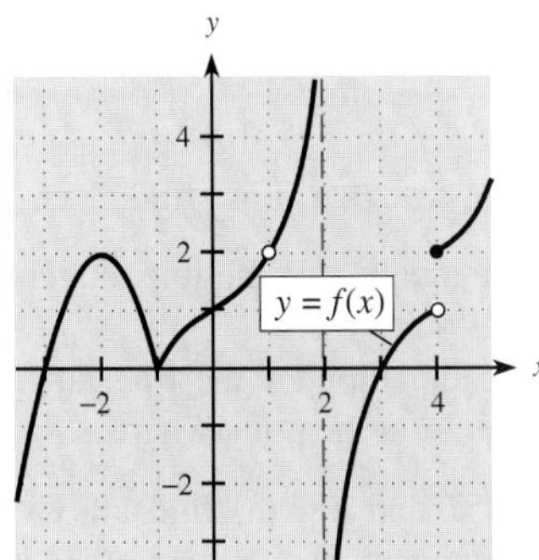

Figure 9.36

In Problems 7–20, find each limit, if it exists.

7. $\lim_{x\to 4} (3x^2 + x + 3)$
8. $\lim_{x\to 4} \dfrac{x^2 - 16}{x + 4}$
9. $\lim_{x\to -1} \dfrac{x^2 - 1}{x + 1}$
10. $\lim_{x\to 3} \dfrac{x^2 - 9}{x - 3}$
11. $\lim_{x\to 2} \dfrac{4x^3 - 8x^2}{4x^3 - 16x}$
12. $\lim_{x\to -\frac{1}{2}} \dfrac{x^2 - \frac{1}{4}}{6x^2 + x - 1}$
13. $\lim_{x\to 3} \dfrac{x^2 - 16}{x - 3}$
14. $\lim_{x\to -3} \dfrac{x^2 - 9}{x - 3}$
15. $\lim_{x\to 1} \dfrac{x^2 - 9}{x - 3}$
16. $\lim_{x\to 2} \dfrac{x^2 - 8}{x - 2}$
17. $\lim_{x\to 1} f(x)$ where $f(x) = \begin{cases} 4 - x^2 & \text{if } x < 1 \\ 4 & \text{if } x = 1 \\ 2x + 1 & \text{if } x > 1 \end{cases}$
18. $\lim_{x\to -2} f(x)$ where $f(x) = \begin{cases} x^3 - x & \text{if } x < -2 \\ 2 - x^2 & \text{if } x \ge -2 \end{cases}$
19. $\lim_{h\to 0} \dfrac{3(x + h)^2 - 3x^2}{h}$
20. $\lim_{h\to 0} \dfrac{[(x + h) - 2(x + h)^2] - (x - 2x^2)}{h}$

In Problems 21 and 22, use tables to investigate each limit. Check your result analytically or graphically.

21. $\lim_{x\to 2} \dfrac{x^2 + 10x - 24}{x^2 - 5x + 6}$
22. $\lim_{x\to -\frac{1}{2}} \dfrac{x^2 + \frac{1}{6}x - \frac{1}{6}}{x^2 + \frac{5}{6}x + \frac{1}{6}}$

Section 9.2

Use the graph of $y = f(x)$ in Figure 9.36 to answer the questions in Problems 23 and 24.

23. Is $f(x)$ continuous at
 (a) $x = -1$? (b) $x = 1$?
24. Is $f(x)$ continuous at
 (a) $x = -2$? (b) $x = 2$?

In Problems 25–30, suppose that

$$f(x) = \begin{cases} x^2 + 1 & \text{if } x \le 0 \\ x & \text{if } 0 < x < 1 \\ 2x^2 - 1 & \text{if } x \ge 1 \end{cases}$$

25. What is $\lim_{x\to -1} f(x)$?
26. What is $\lim_{x\to 0} f(x)$, if it exists?
27. What is $\lim_{x\to 1} f(x)$, if it exists?
28. Is $f(x)$ continuous at $x = 0$?
29. Is $f(x)$ continuous at $x = 1$?
30. Is $f(x)$ continuous at $x = -1$?

For the functions in Problems 31–34, determine which are continuous. Identify discontinuities for those that are not continuous.

31. $y = \dfrac{x^2 + 25}{x - 5}$
32. $y = \dfrac{x^2 - 3x + 2}{x - 2}$
33. $f(x) = \begin{cases} x + 2 & \text{if } x \le 2 \\ 5x - 6 & \text{if } x > 2 \end{cases}$
34. $y = \begin{cases} x^4 - 3 & \text{if } x \le 1 \\ 2x - 3 & \text{if } x > 1 \end{cases}$

In Problems 35 and 36, use the graphs to find (a) the points of discontinuity, (b) $\lim_{x \to +\infty} f(x)$, and (c) $\lim_{x \to -\infty} f(x)$.

35.

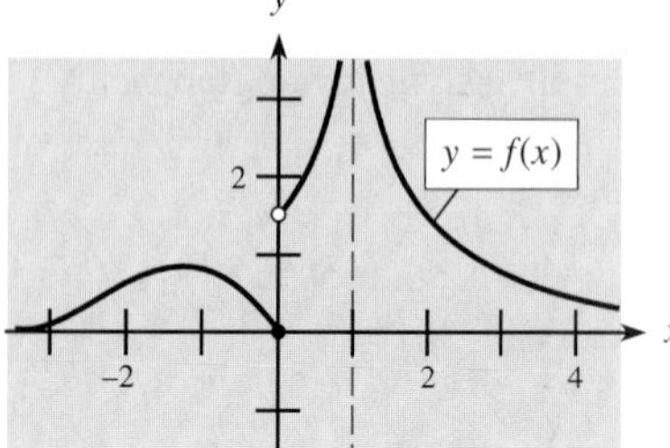

36.

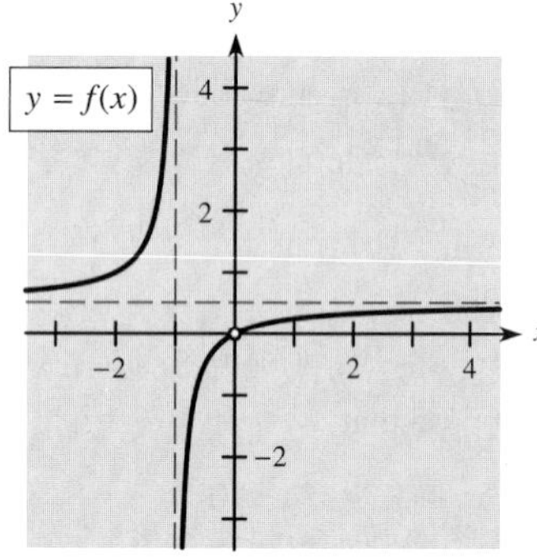

In Problems 37 and 38 evaluate the limits, if they exist.

37. $\lim_{x \to -\infty} \dfrac{2x^2}{1 - x^2}$

38. $\lim_{x \to +\infty} \dfrac{3x^{2/3}}{x + 1}$

Section 9.3

In Problems 39 and 40, decide whether the statements are true or false.

39. $\lim_{h \to 0} \dfrac{f(x+h) - f(x)}{h}$ gives the formula for the slope of the tangent and the instantaneous rate of change of $f(x)$ at any value of x.

40. $\lim_{h \to 0} \dfrac{f(c+h) - f(c)}{h}$ gives the equation of the tangent line to $f(x)$ at $x = c$.

41. Use the definition of derivative to find $f'(x)$ for $f(x) = 3x^2 + 2x - 1$.

42. Use the definition of derivative to find $f'(x)$ if $f(x) = x - x^2$.

Use the graph of $y = f(x)$ in Figure 9.36 on the previous page to answer the questions in Problems 43 and 44.

43. Is $f(x)$ differentiable at
    (a) $x = -1$? (b) $x = 1$?

44. Is $f(x)$ differentiable at
    (a) $x = -2$? (b) $x = 2$?

45. Let $f(x) = \dfrac{\sqrt[3]{4x}}{(3x^2 - 10)^2}$. Approximate $f'(2)$
    (a) by using the numerical derivative feature of a graphing utility, and
    (b) by evaluating $\dfrac{f(2+h) - f(2)}{h}$ with $h = 0.0001$.

46. Use the given table of values for $g(x)$ to approximate $g'(4)$ as accurately as possible.

x	2	2.3	3.1	4	4.3	5
$g(x)$	13.2	12.1	9.7	12.2	14.3	18.1

Section 9.4

47. If $c = 4x^5 - 6x^3$, find c'.
48. If $f(x) = 4x^2 - 1$, find $f'(x)$.
49. If $p = 3q + \sqrt{7}$, find dp/dq.
50. If $y = \sqrt{x}$, find y'.
51. If $f(z) = \sqrt[3]{z^4}$, find $f'(z)$.
52. If $v(x) = 4/\sqrt[3]{x}$, find $v'(x)$.
53. If $y = \dfrac{1}{x} - \dfrac{1}{\sqrt{x}}$, find y'.
54. If $f(x) = \dfrac{3}{2x^2} - \sqrt[3]{x} + 4^5$, find $f'(x)$.
55. Write the equation of the line tangent to the graph of $y = 3x^5 - 6$ at $x = 1$.
56. Write the equation of the line tangent to the curve $y = 3x^3 - 2x$ at the point where $x = 2$.

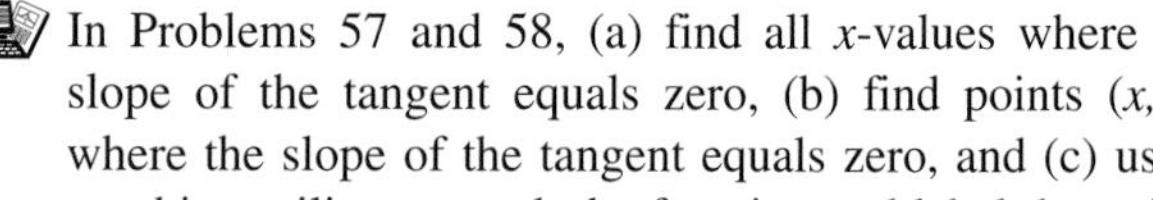

In Problems 57 and 58, (a) find all x-values where the slope of the tangent equals zero, (b) find points (x, y) where the slope of the tangent equals zero, and (c) use a graphing utility to graph the function and label the points found in (b).

57. $f(x) = x^3 - 3x^2 + 1$
58. $f(x) = x^6 - 6x^4 + 8$

Section 9.5

59. If $f(x) = (3x - 1)(x^2 - 4x)$, find $f'(x)$.
60. Find y' if $y = (x^2 + 1)(3x^3 + 1)$.
61. If $p = \dfrac{2q - 1}{q^2}$, find $\dfrac{dp}{dq}$.
62. Find $\dfrac{ds}{dt}$ if $s = \dfrac{\sqrt{t}}{(3t + 1)}$.
63. Find $\dfrac{dy}{dx}$ for $y = \sqrt{x}(3x + 2)$.
64. Find $\dfrac{dC}{dx}$ for $C = \dfrac{5x^4 - 2x^2 + 1}{x^3 + 1}$.

Section 9.6

65. If $y = (x^3 - 4x^2)^3$, find y'.
66. If $y = (5x^6 + 6x^4 + 5)^6$, find y'.
67. If $y = (2x^4 - 9)^9$, find $\dfrac{dy}{dx}$.
68. Find $g'(x)$ if $g(x) = \dfrac{1}{\sqrt{x^3 - 4x}}$.

Section 9.7

69. Find $f'(x)$ if $f(x) = x^2(2x^4 + 5)^8$.

70. Find S' if $S = \dfrac{(3x+1)^2}{x^2 - 4}$.

71. Find $\dfrac{dy}{dx}$ if $y = [(3x + 1)(2x^3 - 1)]^{12}$.

72. Find y' if $y = \left(\dfrac{x+1}{1-x^2}\right)^3$.

73. Find y' if $y = x\sqrt{x^2 - 4}$.

74. Find $\dfrac{dy}{dx}$ if $y = \dfrac{x}{\sqrt[3]{3x - 1}}$.

Section 9.8

In Problems 75 and 76, find the second derivatives.

75. $y = \sqrt{x} - x^2$

76. $y = x^4 - \dfrac{1}{x}$

In Problems 77 and 78, find the fifth derivatives.

77. $y = (2x + 1)^4$

78. $y = \dfrac{(1-x)^6}{24}$

79. If $\dfrac{dy}{dx} = \sqrt{x^2 - 4}$, find $\dfrac{d^3y}{dx^3}$.

80. If $\dfrac{d^2y}{dx^2} = \dfrac{x}{x^2 + 1}$, find $\dfrac{d^4y}{dx^4}$.

Applications

Section 9.4

81. ***Demand*** Suppose that the demand x for a product is given by $x = (100/p) - 1$, where p is the price per unit of the product. Find and interpret the rate of change of demand with respect to price if the price is
    (a) \$10.
    (b) \$20.

Section 9.6

82. ***Demand*** The demand q for a product at price p is given by

$$q = 10{,}000 - 50\sqrt{0.02p^2 + 500}$$

Find the rate of change of demand with respect to price.

83. ***Supply*** The number of units x of a product that is supplied at price p is given by

$$x = \sqrt{p - 1}, \quad p \geq 1$$

If the price p is \$10, what is the rate of change of the supply with respect to the price and what does it tell us?

Section 9.9

84. ***Cost*** If the cost function for a particular good is $C(x) = 3x^2 + 6x + 600$, what is the
    (a) marginal cost function?
    (b) marginal cost if 30 units are produced?
    (c) interpretation of your answer in (b)?

85. ***Cost*** If the total cost function for a commodity is $C(x) = 400 + 5x + x^3$, what is the marginal cost when 4 units are produced and what does it mean?

86. ***Revenue*** The total revenue function for a commodity is $R = 40x - 0.02x^2$, with x representing the number of units.
    (a) Find the marginal revenue function.
    (b) At what level of production will marginal revenue be 0?

87. ***Profit*** If the total revenue function for a product is given by $R(x) = 60x$ and the total cost function is given by $C = 200 + 10x + 0.1x^2$, what is the marginal profit at $x = 10$? What does the marginal profit at $x = 10$ predict?

88. ***Revenue*** The total revenue function for a commodity is given by $R = 80x - 0.04x^2$.
    (a) Find the marginal revenue function.
    (b) What is the marginal revenue at $x = 100$?
    (c) Interpret your answer in (b).

89. ***Revenue*** If the revenue function for a product is

$$R(x) = \frac{60x^2}{2x + 1}$$

find the marginal revenue.

90. ***Profit*** A firm has monthly costs given by

$$C = 45{,}000 + 100x + x^3$$

where x is the number of units produced per month. The firm can sell its product in a competitive market for \$4600 per unit. Find the marginal profit.

91. ***Profit*** A small business has weekly costs of

$$C = 100 + 30x + \frac{x^2}{10}$$

where x is the number of units produced each week. The competitive market price for this business's product is \$46 per unit. Find the marginal profit.

CHAPTER TEST

1. Evaluate the following limits, if they exist. Use algebraic methods.
   (a) $\lim_{x\to -2} \frac{4x - x^2}{4x - 8}$ (b) $\lim_{x\to\infty} \frac{8x^2 - 4x + 1}{2 + x - 5x^2}$
   (c) $\lim_{x\to 7} \frac{x^2 - 5x - 14}{x^2 - 6x - 7}$ (d) $\lim_{x\to -5} \frac{5x - 25}{x + 5}$
2. (a) Write the limit definition for $f'(x)$.
   (b) Use the definition from (a) to find $f'(x)$ for $f(x) = 3x^2 - x + 9$.
3. Let $f(x) = \frac{4x}{x^2 - 8x}$. Identify all x-values where $f(x)$ is *not* continuous.
4. Use derivative formulas to find the derivative of each of the following. Simplify, except for (b).
   (a) $y = \frac{3x^3}{2x^7 + 11}$
   (b) $f(x) = (3x^5 - 2x + 3)(4x^{10} + 10x^4 - 17)$
   (c) $g(x) = \frac{3}{4}(2x^5 + 7x^3 - 5)^{12}$
   (d) $y = (x^2 + 3)(2x + 5)^6$
   (e) $f(x) = 12\sqrt{x} - \frac{10}{x^2} + 17$
5. Find $\frac{d^3y}{dx^3}$ for $y = x^3 - x^{-3}$.
6. Let $f(x) = x^3 - 3x^2 - 24x - 10$.
   (a) Write the equation of the line tangent to the graph of $y = f(x)$ at $x = -1$.
   (b) Find all points (both x- and y-coordinates) where $f'(x) = 0$.
7. Use the given tables to evaluate the following limits, if they exist.
   (a) $\lim_{x\to 5} f(x)$ (b) $\lim_{x\to 5} g(x)$ (c) $\lim_{x\to 5^-} g(x)$

x	4.99	4.999	$\to 5 \leftarrow$	5.001	5.01
$f(x)$	2.01	2.001	$\to ? \leftarrow$	1.999	1.99

x	4.99	4.999	$\to 5 \leftarrow$	5.001	5.01
$g(x)$	-3.99	-3.999	$\to ? \leftarrow$	6.999	6.99

8. Use the definition of continuity to investigate whether $g(x)$ is continuous at $x = -2$. Show your work.

$$g(x) = \begin{cases} 6 - x & \text{if } x \le -2 \\ x^3 & \text{if } x > -2 \end{cases}$$

9. Suppose a company has its total cost for a product given by $C(x) = 200x + 10{,}000$ and its total revenue given by $R(x) = 250x - 0.01x^2$, where x is the number of units produced and sold.
   (a) Form the profit function for this product.
   (b) Find the marginal profit function.
   (c) Find the marginal profit when $x = 1000$, and then write a sentence that interprets this result.
10. Suppose that $f(x)$ is a differentiable function. Use the table of values to approximate $f'(3)$ as accurately as possible.

x	2	2.5	2.999	3	3.01	3.1
$f(x)$	0	18.4	44.896	45	46.05	56.18

11. Use the graph to perform the evaluations (a)–(e) and to answer (f)–(g). If no value exists, so indicate.
   (a) $f(1)$ (b) $\lim_{x\to 6} f(x)$
   (c) $\lim_{x\to 3^-} f(x)$ (d) $\lim_{x\to -4} f(x)$
   (e) $\lim_{x\to -\infty} f(x)$
   (f) Find all x-values where $f'(x)$ does not exist.
   (g) Find all x-values where $f(x)$ is not continuous.

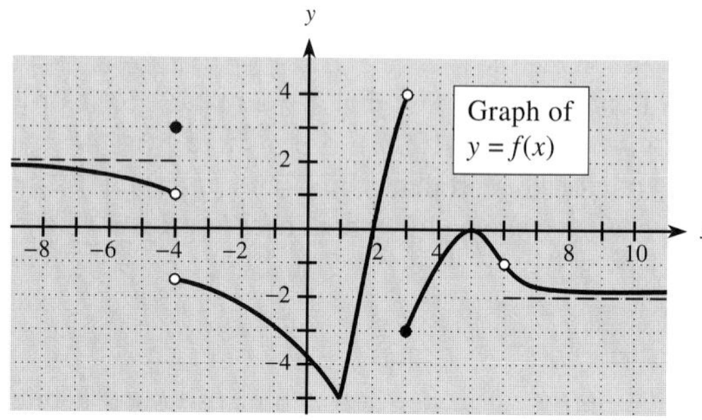

12. Given that the line $y = \frac{2}{3}x - 8$ is tangent to the graph of $y = f(x)$ at $x = 6$, find
   (a) $f'(6)$ (b) $f(6)$
   (c) the instantaneous rate of change of $f(x)$ with respect to x at $x = 6$

Extended Applications / Group Projects

I. Marginal Return to Sales

A tire manufacturer studying the effectiveness of television advertising and other promotions on sales of its GRIPPER-brand tires attempted to fit data it had gathered to the equation

$$S = a_0 + a_1x + a_2x^2 + b_1y$$

where S is sales revenue in millions of dollars, x is millions of dollars spent on television advertising, y is millions of dollars spent on other promotions, and a_0, a_1, a_2, and b_1 are constants. The data, gathered in two different regions of the country where expenditures for other promotions were kept constant (at B_1 and B_2), resulted in the following quadratic equations relating TV advertising and sales.

$$\text{Region 1:} \quad S_1 = 30 + 20x - 0.4x^2 + B_1$$
$$\text{Region 2:} \quad S_2 = 20 + 36x - 1.3x^2 + B_2$$

The company wants to know how to make the best use of its advertising dollars in the regions and whether the current allocation could be improved. Advise management about current advertising effectiveness, allocation of additional expenditures, and reallocation of current advertising expenditures by answering the following questions.

1. In the analysis of sales and advertising, **marginal return to sales** is usually used, and it is given by dS_1/dx for Region 1 and dS_2/dx for Region 2.
   (a) Find $\dfrac{dS_1}{dx}$ and $\dfrac{dS_2}{dx}$.
   (b) If \$10 million is being spent on TV advertising in each region, what is the marginal return to sales in each region?
2. Which region would benefit more from additional advertising expenditure, if \$10 million is currently being spent in each region?
3. If any additional money is made available for advertising, in which region should it be spent?
4. How could money already being spent be reallocated to produce more sales revenue?

II. Marginal Cost, Marginal Revenue, and Maximum Profit

In this chapter, we have seen how marginals for total cost, total revenue, and profit (that is, their derivatives) can be used to predict short-run future trends for each of these functions. In this project, we examine how a business might predict maximum profit by using the marginal cost and marginal revenue.

1. For each given pair of total cost and total revenue functions, complete the corresponding table. Assume that x represents the number of items produced and sold.

   (a) $C(x) = 3x + 6000$
   $R(x) = 12x - 0.001x^2$

   (b) $C(x) = 187x + 0.01x^2 + 15{,}750$
   $R(x) = 308x - 0.01x^2$

x	*Profit*	$\overline{MC} = C'(x)$	$\overline{MR} = R'(x)$
4000			
4100			
4200			
⋮			
4900			
5000			

x	*Profit*	$\overline{MC} = C'(x)$	$\overline{MR} = R'(x)$
3000			
3005			
3010			
⋮			
3095			
3100			

2. Examine the data you collected in each table.
   (a) Does each profit function seem to have a maximum value? If so, identify it in your table.
   (b) For each profit function, what are the values of marginal cost and marginal revenue at the x-value where profit has its maximum?
   (c) If a profit function has a maximum value at $x = a$, what seems to be the relationship between the values of marginal cost and marginal revenue at $x = a$? Answer in a summary sentence.
3. For each given total cost and total revenue pair from part 1, use a graphing utility to make the following graphs:
   (a) marginal cost and marginal revenue (graphed simultaneously)
   (b) profit

   These graphs should confirm your conclusions from part 2(c). Print or reproduce the graphs, and highlight the portion of the graphs that illustrates the relationship between the occurrence of maximum profit and the values of marginal cost and marginal revenue.
4. (a) In general, if $\overline{MR} > \overline{MC}$, is $P(x)$ increasing or decreasing? Justify your choice; use the interpretation of marginals as predictors for the next unit.
   (b) If $\overline{MR} < \overline{MC}$, is $P(x)$ increasing or decreasing? Justify your choice.
   (c) Explain how the observations in (a) and (b) provide general support for your statement relating the occurrence of maximum profit to the values of marginal cost and marginal revenue.

5. The conclusions relating marginal cost, marginal revenue, and maximum profit are valid as long as the total cost and total revenue functions are not both linear, such as with either of the following.

(a) $C(x) = 80x + 8000$
$R(x) = 120x$

(b) $C(x) = 52x + 7800$
$R(x) = 50x$

In each of these cases, find the marginal cost and marginal revenue. In light of your answers to parts 4(a) and 4(b) above, what do these calculations tell you about the corresponding profit function? Does the profit function for either (a) or (b) have a maximum value? Explain.

Warm-up

Prerequisite Problem Type	For Section	Answer	Section for Review
Write $\frac{1}{3}(x^2-1)^{-2/3}(2x)$ with positive exponents.	10.2	$\dfrac{2x}{3(x^2-1)^{2/3}}$	0.3, 0.4 Exponents and radicals
Factor: (a) $x^3 - x^2 - 6x$ (b) $8000 - 80x - 3x^2$	10.1 10.2 10.3	(a) $x(x-3)(x+2)$ (b) $(40-x)(200+3x)$	0.6 Factoring
(a) For what values of x is $\dfrac{2}{3\sqrt[3]{x+2}}$ undefined? (b) For what values of x is $\frac{1}{3}(x^2-1)^{-2/3}(2x)$ undefined?	10.1 10.2 10.5	(a) $x = -2$ (b) $x = -1, x = 1$	1.2 Domains of functions
If $f(x) = \frac{1}{3}x^3 - x^2 - 3x + 2$, and $f'(x) = x^2 - 2x - 3$, (a) find $f(-1)$. (b) find $f'(-2)$.	10.1	(a) $\dfrac{11}{3}$ (b) 5	1.2 Functional notation
(a) Solve $0 = x^2 - 2x - 3$. (b) If $f'(x) = 3x^2 - 3$, what values of x make $f'(x) = 0$?	10.1–10.5	(a) $x = -1, x = 3$ (b) $x = -1, x = 1$	2.1 Solving quadratic equations
Does $\displaystyle\lim_{x\to -2} \frac{2x-4}{3x+6}$ exist?	10.5	No; unbounded	9.1 Limits
(a) Find $f''(x)$ if $f(x) = x^3 - 4x^2 + 3$. (b) Find $P''(x)$ if $P(x) = 48x - 1.2x^2$.	10.2	(a) $f''(x) = 6x - 8$ (b) $P''(x) = -2.4$	9.8 Higher-order derivatives
Find the derivatives: (a) $y = \frac{1}{3}x^3 - x^2 - 3x + 2$ (b) $f = x + 2\left(\dfrac{80{,}000}{x}\right)$ (c) $p(t) = 1 + \dfrac{4t}{t^2+16}$ (d) $y = (x+2)^{2/3}$ (e) $y = \sqrt[3]{x^2-1}$	10.1 10.2 10.3 10.4	(a) $y' = x^2 - 2x - 3$ (b) $f' = 1 - \dfrac{160{,}000}{x^2}$ (c) $p'(t) = \dfrac{64-4t^2}{(t^2+16)^2}$ (d) $y' = \dfrac{2}{3(x+2)^{1/3}}$ (e) $y' = \dfrac{2x}{3(x^2-1)^{2/3}}$	9.4, 9.5, 9.6 Derivatives

Chapter 10

Applications of Derivatives

In this chapter we will discuss applications of the derivative. In particular, we will consider methods of determining when a function has a "turning point" on its graph, so that we can determine when the graph of the function reaches its highest or lowest point within a particular interval. These points are called ***relative maxima*** *and* ***relative minima,*** *respectively, and are useful in sketching the graph of a function whose equation is given and that has a maximum or minimum value within a particular interval. The endpoints of a given interval and the relative maxima and minima within the interval can be used to solve many types of applied problems. For example, we can use these points in determining the level of production that maximizes revenue or profit for a product and in minimizing the average cost of producing a product.*

In addition to using the first derivative to help graph a function, we can use the second derivative to determine where the graph will be concave up or concave down and where it will change from concave up to concave down or vice versa. Points where this change occurs are called ***points of inflection.*** *The second derivative uses concavity to determine where the graph of a function has a relative maximum or relative minimum. Knowledge of this information can be used to sketch the graph of a function or to determine the appropriate viewing window to use when graphing the function with a graphing utility. Because horizontal and vertical asymptotes are not always apparent when a graphing utility is used, we will discuss methods of finding and accounting for asymptotes when graphing functions.*

10.1 Relative Maxima and Minima; Curve Sketching

OBJECTIVES

- To find relative maxima and minima and horizontal points of inflection of functions
- To sketch graphs of functions by using information about maxima, minima, and horizontal points of inflection

APPLICATION PREVIEW

When a company initiates an advertising campaign, there is typically a surge in weekly sales. As the effect of the campaign lessens, sales attributable to it usually decrease. For example, suppose a company models its weekly sales during an advertising campaign by

$$S = \frac{100t}{t^2 + 100}$$

where t is the number of weeks since the beginning of the campaign. The company would like to determine accurately when the revenue function is increasing, when it is decreasing, and when sales revenue is maximized.

In this section we will use the derivative of a function to decide whether the function is increasing or decreasing on an interval and to find where the function has relative maximum points and relative minimum points. We will use the information about derivatives of functions to graph the functions and to solve applied problems.

Except for very simple graphs (lines and parabolas, for example), plotting points to sketch a graph may be tedious. Even when we use a graphing utility, special features of a graph of a function may be difficult to locate accurately. In addition to intercepts and asymptotes, we can use the first derivative as an aid in graphing. The first derivative identifies the "turning points" of the graph, which help us determine the general shape of the graph and choose a viewing window that includes the interesting points of the graph if a graphing utility is used.

In Figure 10.1(a) we see that the graph of $y = \frac{1}{3}x^3 - x^2 - 3x + 2$ has two "turning points," at $(-1, \frac{11}{3})$ and $(3, -7)$. The curve has a relative maximum at $(-1, \frac{11}{3})$ because this point is higher than any other point "near" it on the curve; the curve has a relative minimum at $(3, -7)$ because this point is lower than any other point "near" it on the curve. A formal definition follows.

Relative Maxima and Minima

The point $(x_1, f(x_1))$ is a **relative maximum point** for the function f if there is an interval around x_1 on which $f(x_1) \geq f(x)$ for all x in the interval. In this case, we say the relative maximum *occurs* at $x = x_1$ and the relative maximum is $f(x_1)$.

The point $(x_2, f(x_2))$ is a **relative minimum point** for the function f if there is an interval around x_2 on which $f(x_2) \leq f(x)$ for all x in the interval. In this case, we say the relative minimum *occurs* at $x = x_2$ and the relative minimum is $f(x_2)$.

In order to determine whether a turning point of a function is a maximum point or a minimum point, it is frequently helpful to know what the graph of the function does in intervals on either side of the turning point. We say a function is **increasing** on an interval if the functional values increase as the x-values increase (that is, if the graph rises as we move from left to right on the interval). Similarly, a function is **decreasing** on an interval if the functional values decrease as the x-values increase (that is, if the graph falls as we move from left to right on the interval).

We have seen that if the slope of a line is positive, then the linear function is increasing and its graph is rising. Similarly, if $f(x)$ is differentiable over an interval and if each tangent line to the curve over that interval has positive slope, then the curve is rising over the interval and the function is increasing. Because the derivative of the function gives the slope of the tangent to the curve, we see that if $f'(x) > 0$ on an interval, then $f(x)$ is increasing on that interval. A similar conclusion can be reached when the derivative is negative on the interval.

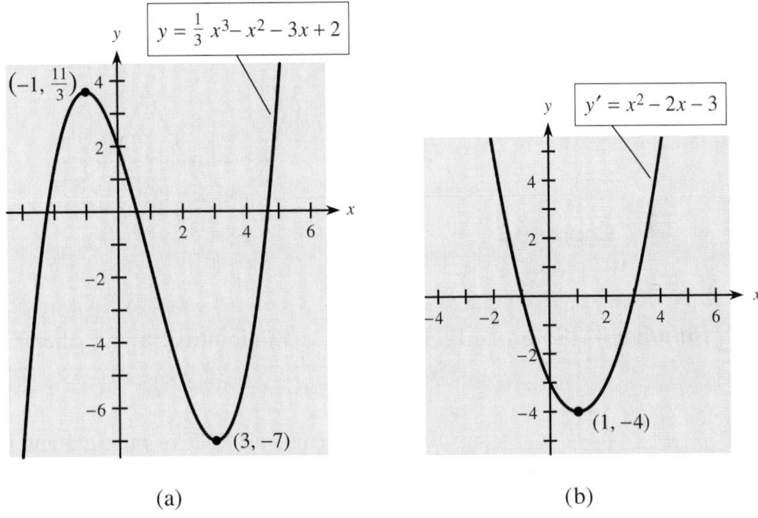

Figure 10.1

Increasing and Decreasing Functions

If f is a function that is differentiable on an interval (a, b), then

if $f'(x) > 0$ for all x in (a, b), f is increasing on (a, b).
if $f'(x) < 0$ for all x in (a, b), f is decreasing on (a, b).

Figure 10.1(a) shows the graph of a function, and Figure 10.1(b) shows the graph of its derivative. The figures show that the graph of $y = f(x)$ is increasing for the same x-values that the graph of $y' = f'(x)$ is above the x-axis (when $f'(x) > 0$). Similarly, the graph of $y = f(x)$ is decreasing for the same x-values $(-1 < x < 3)$ that the graph of $y' = f'(x)$ is below the x-axis (when $f'(x) < 0$).

The derivative $f'(x)$ can change signs only at values of x where $f'(x) = 0$ or $f'(x)$ is undefined. We call these values of x **critical values.** The point corresponding to a critical value for x is a **critical point.*** Because a curve changes from increasing to decreasing at a relative maximum (see Figure 10.1a), we have the following fact.

*There may be some critical values where $f'(x)$ and $f(x)$ are undefined. Critical points do not occur at these values, but studying the derivative on either side of such values may be of interest.

Relative Maximum If f has a relative maximum at $x = x_0$, then $f'(x_0) = 0$ or $f'(x_0)$ is undefined.

From Figure 10.2, we see that this function has two relative maxima, one at $x = x_1$ and the second at $x = x_3$. At $x = x_1$ the derivative is 0, and at $x = x_3$ the derivative does not exist. In Figure 10.2, we also see that a relative minimum occurs at $x = x_2$ and $f'(x_2) = 0$. As Figure 10.2 shows, the function changes from decreasing to increasing at a relative minimum. Thus we have the following fact.

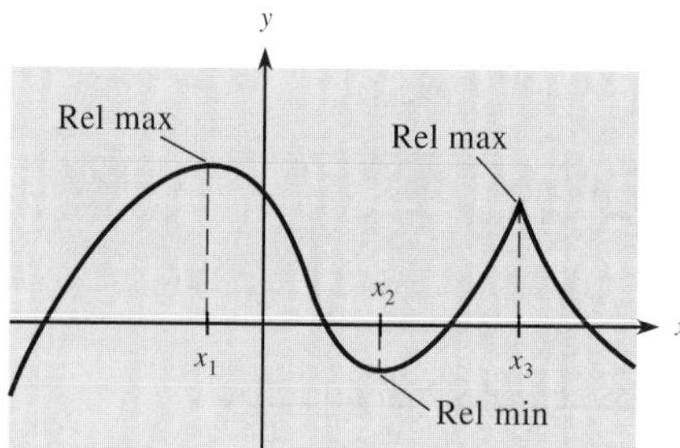

Figure 10.2

Relative Minimum If f has a relative minimum at $x = x_0$, then $f'(x_0) = 0$ or $f'(x_0)$ is undefined.

Thus we can find relative maxima and minima for a curve by finding values of x for which the function has critical points. The behavior of the derivative to the left and right of (and near) these points will tell us whether they are relative maxima, relative minima, or neither.

Because the critical values are the only values where the graph can have turning points, the derivative cannot change sign anywhere except at a critical value. Thus, in an interval between two critical values, the sign of the derivative at any value in the interval will be the sign of the derivative at all values in the interval.

Using the critical values of $f(x)$ and the sign of $f'(x)$ between those critical values, we can create a **sign diagram for $f'(x)$.** The sign diagram for the graph in Figure 10.2 is shown in Figure 10.3. This sign diagram was created from the graph of f, but it is also possible to predict the shape of a graph from a sign diagram.

Direction of graph of $f(x)$:

Signs and values of $f'(x)$:

x-axis with critical values:

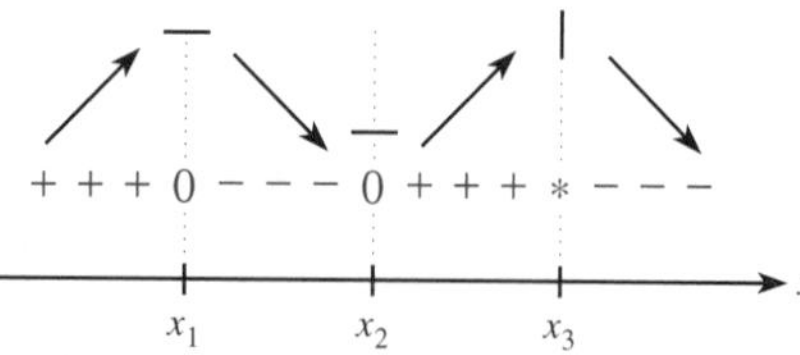

Figure 10.3

*means $f'(x_3)$ is undefined.

Suppose that the point (x_1, y_1) is a critical point. If $f'(x)$ is positive to the left of, and near, this critical point, and if $f'(x)$ is negative to the right of, and near, this critical point, then the curve is increasing to the left of the point and decreasing to the right. This means that a relative maximum occurs at the point. If we drew tangent lines to the curve on the left and right of this critical point, they would fit on the curve in one of the two ways shown in Figure 10.4.

Similarly, suppose that the point (x_2, y_2) is a critical point, $f'(x)$ is negative to the left of and near (x_2, y_2), and $f'(x)$ is positive to the right of and near this critical point. Then the curve is decreasing to the left of the point and increasing to the right, and a relative minimum occurs at the point. If we drew tangent lines to the curve to the left of and to the right of this critical point, they would fit on the curve in one of the two ways shown in Figure 10.5.

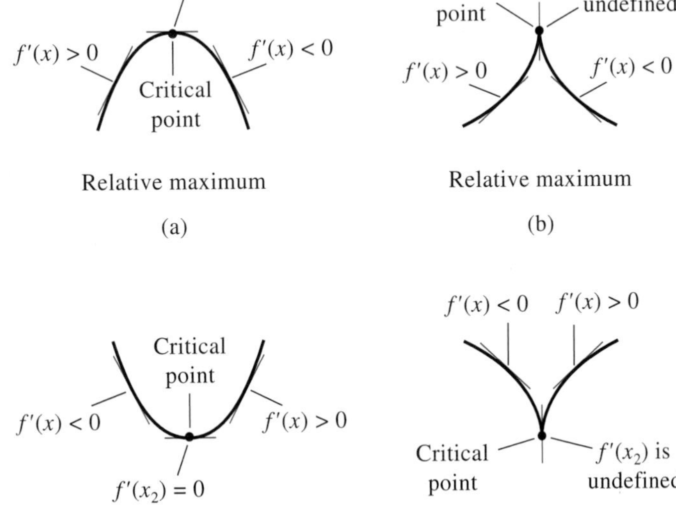

Figure 10.4

Figure 10.5

EXAMPLE 1

Show that the graph of $f(x) = 3x^2 - 2x^3$ has a relative maximum at $x = 1$.

Solution

We need to show that

$f'(1) = 0$ or $f'(1)$ is undefined.
$f'(x) > 0$ to the left of and near $x = 1$.
$f'(x) < 0$ to the right of and near $x = 1$.

Because $f'(x) = 6x - 6x^2 = 6x(x - 1)$ is 0 only at $x = 0$ and at $x = 1$, we can test (evaluate) the derivative at any value in the interval (0, 1) to see what the curve is doing to the left of $x = 1$, and we can test the derivative at any value to the right of $x = 1$ to see what the curve is doing for $x > 1$.

$$f'\left(\frac{1}{2}\right) = 3 - \frac{3}{2} = \frac{3}{2} > 0 \Rightarrow \text{increasing to left of } x = 1$$

$$f'(1) = 0 \Rightarrow \text{horizontal tangent at } x = 1$$

$$f'\left(\frac{3}{2}\right) = 9 - \frac{27}{2} = -\frac{9}{2} < 0 \Rightarrow \text{decreasing to right of } x = 1$$

A partial sign diagram for $f'(x)$ is given at the right. The sign diagram shows that the graph of the function has a relative maximum at $x = 1$.

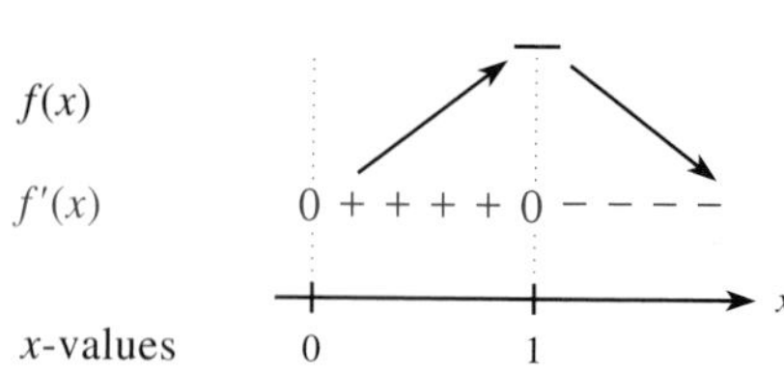

The preceding discussion suggests the following procedure for finding relative maxima and minima of a function.

First-Derivative Test

Procedure	**Example**
To find relative maxima and minima of a function:	Find the relative maxima and minima of $f(x) = \frac{1}{3}x^3 - x^2 - 3x + 2$.
1. Find the first derivative of the function.	1. $f'(x) = x^2 - 2x - 3$
2. Set the derivative equal to 0, and solve for values of x that satisfy $f'(x) = 0$. These are called **critical values.** Values that make $f'(x)$ undefined are also critical values.	2. $0 = x^2 - 2x - 3 = (x + 1)(x - 3)$ has solutions $x = -1, x = 3$. No values of x make $x^2 - 2x - 3$ undefined. Critical values are -1 and 3.
3. Substitute the critical values into the *original function* to find the **critical points.**	3. $f(-1) = \frac{11}{3}$ $f(3) = -7$ The critical points are $(-1, \frac{11}{3})$ and $(3, -7)$.
4. Evaluate $f'(x)$ at some value of x to the left and right of each critical point to develop a sign diagram. (a) If $f'(x) > 0$ to the left and $f'(x) < 0$ to the right of the critical value, the critical point is a relative maximum point. (b) If $f'(x) < 0$ to the left and $f'(x) > 0$ to the right of the critical value, the critical point is a relative minimum point.	4. $f'(-2) = 5 > 0$ and $f'(0) = -3 < 0$ Thus $(-1, 11/3)$ is a relative maximum point. $f'(2) = -3 < 0$ and $f'(4) = 5 > 0$ Thus $(3, -7)$ is a relative minimum point. The sign diagram for $f'(x)$ is 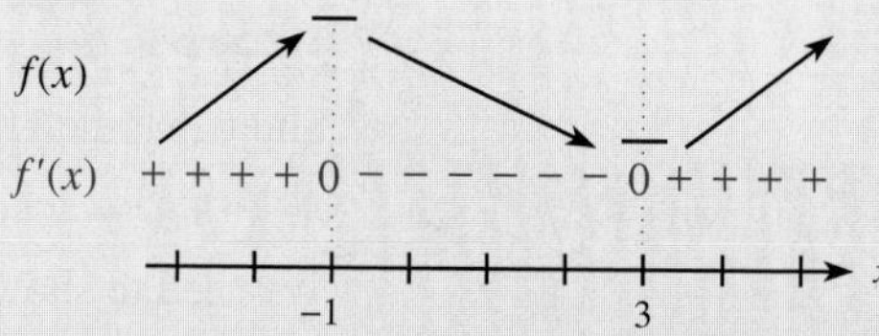
5. Use the information from the sign diagram and selected points to sketch the graph.	5. The information from this sign diagram is shown in Figure 10.6(a). Plotting additional points gives the graph of the function, which is shown in Figure 10.6(b).

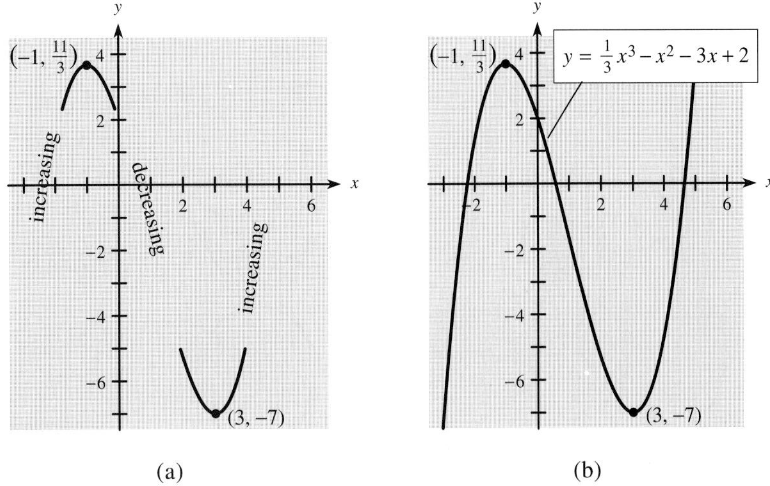

Figure 10.6

Because the critical values are the only x-values where the graph can have turning points, we can test to the left and right of each critical value by testing to the left of the smallest critical value, then testing a value *between* each two successive critical values, and then testing to the right of the largest critical value. The following example illustrates this procedure.

EXAMPLE 2

Find the relative maxima and minima of $f(x) = \frac{1}{4}x^4 - \frac{1}{3}x^3 - 3x^2 + 8$, and sketch its graph.

Solution

1. $f'(x) = x^3 - x^2 - 6x$
2. Setting $f'(x) = 0$ gives $0 = x^3 - x^2 - 6x$. Solving for x gives

$$0 = x(x - 3)(x + 2).$$

$x = 0$	$x - 3 = 0$	$x + 2 = 0$
	$x = 3$	$x = -2$

Thus the critical value are $x = 0$, $x = 3$, and $x = -2$.

3. Substituting the critical values into the original function gives the critical points:

$$f(-2) = \tfrac{8}{3}, \quad \text{so } (-2, \tfrac{8}{3}) \text{ is a critical point.}$$
$$f(0) = 8, \quad \text{so } (0, 8) \text{ is a critical point.}$$
$$f(3) = -\tfrac{31}{4}, \quad \text{so } (3, -\tfrac{31}{4}) \text{ is a critical point.}$$

4. Testing $f'(x)$ to the left of the smallest critical value, then between the critical values, and then to the right of the largest critical value will give the sign diagram. Evaluating $f'(x)$ at the test values $x = -3$, $x = -1$, $x = 1$, and $x = 4$ gives the signs to determine relative maxima and minima.

The sign diagram for $f'(x)$ is

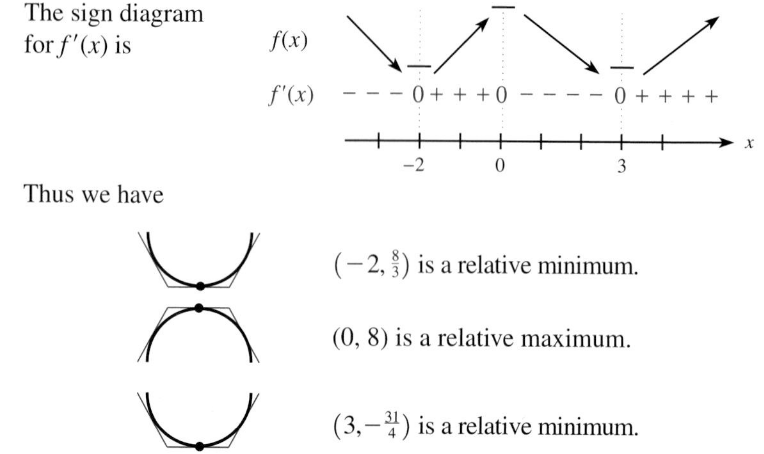

Thus we have

$(-2, \frac{8}{3})$ is a relative minimum.

$(0, 8)$ is a relative maximum.

$(3, -\frac{31}{4})$ is a relative minimum.

5. Figure 10.7(a) shows the graph of the function near the critical points, and Figure 10.7(b) shows the graph of the function.

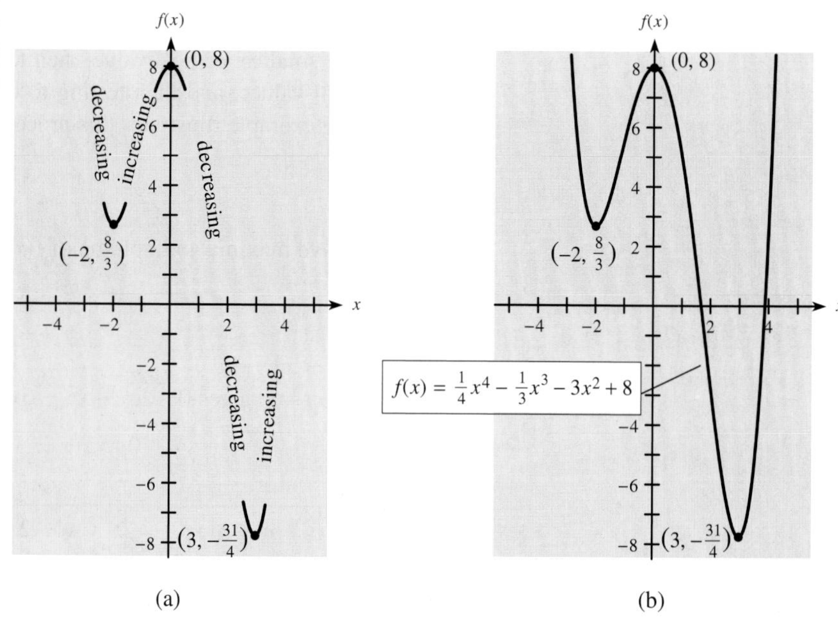

Figure 10.7

Note that we substitute the critical values into the *original function* $f(x)$ to find the y-values of the critical points, but we test for relative maxima and minima by substituting values near the critical values into the *derivative of the function,* $f'(x)$.

Only four values were needed to test three critical points in Example 2. This method will work *only if* the critical values are tested in order from smallest to largest.

If the first derivative of f is 0 at x_0 but does not change from positive to negative or from negative to positive as x passes through x_0, then the critical point at x_0 is neither a relative maximum nor a relative minimum. In this case we say that f has a **horizontal point of inflection** (abbreviated HPI) at x_0.

EXAMPLE 3

Find the relative maxima, relative minima, and horizontal points of inflection of $h(x) = \frac{1}{4}x^4 - \frac{2}{3}x^3 - 2x^2 + 8x + 4$, and sketch its graph.

Solution

1. $h'(x) = x^3 - 2x^2 - 4x + 8$
2. $0 = x^3 - 2x^2 - 4x + 8$ or $0 = x^2(x - 2) - 4(x - 2)$. Therefore, we have $0 = (x - 2)(x^2 - 4)$. Thus $x = -2$ and $x = 2$ are solutions.
3. The critical points are $(-2, -\frac{32}{3})$ and $(2, \frac{32}{3})$.
4. Using test values, such as $x = -3$, $x = 0$, and $x = 3$ gives the sign diagram for $h'(x)$.

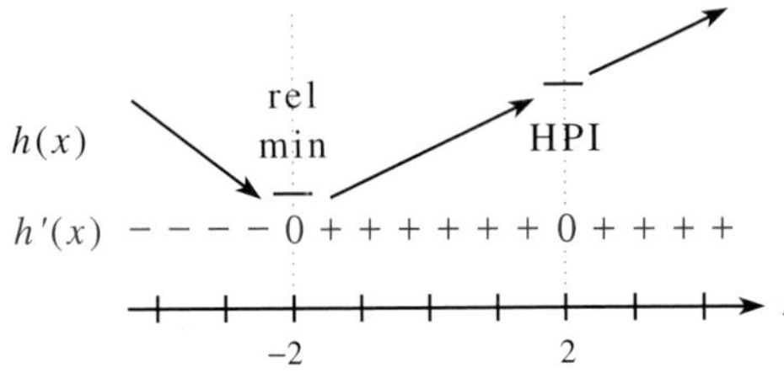

5. Figure 10.8(a) shows the graph of the function near the critical points, and Figure 10.8(b) shows the graph of the function.

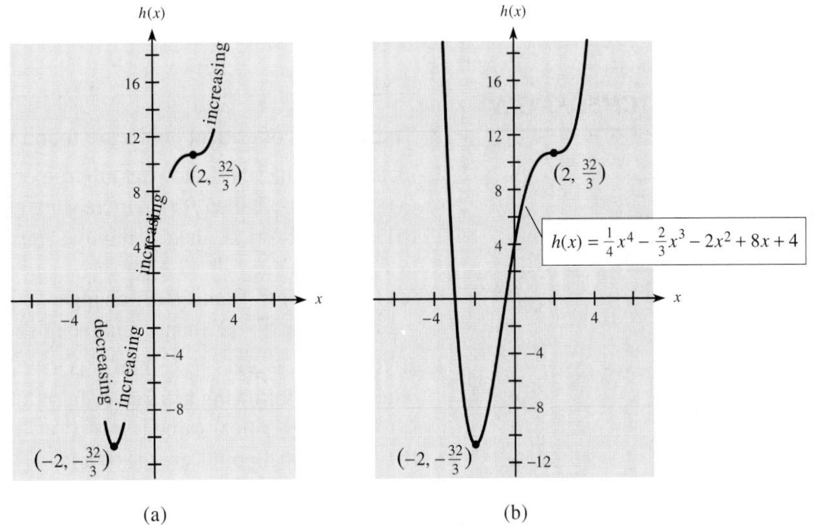

Figure 10.8

EXAMPLE 4

Find the relative maxima and minima (if any) of the graph of $y = (x + 2)^{2/3}$.

Solution

1. $y' = f'(x) = \frac{2}{3}(x + 2)^{-1/3} = \frac{2}{3\sqrt[3]{x + 2}}$

2. $0 = \dfrac{2}{3\sqrt[3]{x+2}}$ has no solutions; $f'(x)$ is undefined at $x = -2$.
3. $f(-2) = 0$, so the critical point is $(-2, 0)$.
4. The sign diagram for $f'(x)$ is

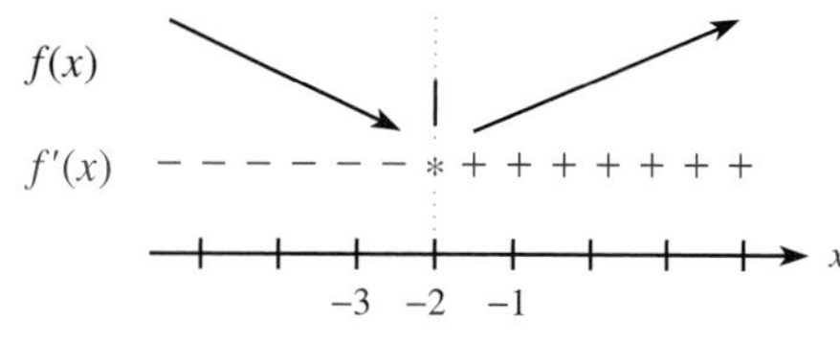

Thus a relative minimum occurs at $(-2, 0)$.

* means $f'(-2)$ is undefined.

5. Figure 10.9(a) shows the graph of the function near the critical point, and Figure 10.9(b) shows the graph.

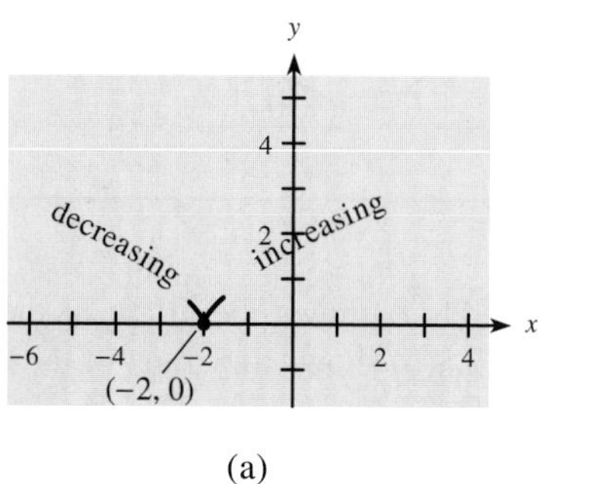

(a)

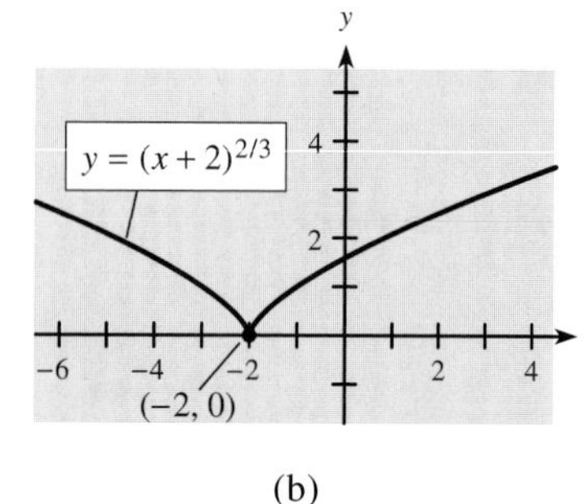

(b)

Figure 10.9

CHECKPOINT

1. The x-values of critical points are found where $f'(x)$ is ________ or ________.
2. Decide whether the following are true or false.
   (a) If $f'(1) = 7$, then $f(x)$ is increasing at $x = 1$.
   (b) If $f'(-2) = 0$, then a relative maximum or a relative minimum occurs at $x = -2$.
   (c) If $f'(-3) = 0$ and $f'(x)$ changes from positive on the left to negative on the right of $x = -3$, then a relative minimum occurs at $x = -3$.
3. If $f(x) = 7 + 3x - x^3$, then $f'(x) = 3 - 3x^2$. Use these functions to decide whether the following are true or false.
   (a) The only critical value is $x = 1$.
   (b) The critical points are $(1, 0)$ and $(-1, 0)$.
4. If $f'(x)$ has the following partial sign diagram, make a "stick-figure" sketch of $f(x)$ and label where any maxima and minima occur. Assume that $f(x)$ is defined for all real numbers.

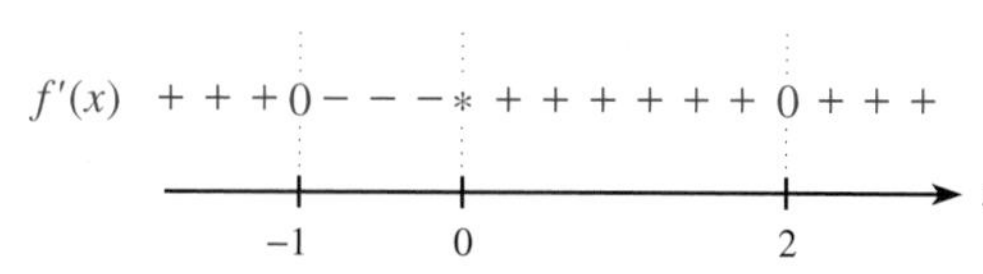

* means $f'(0)$ is undefined.

Let us now return to the discussion of advertising and sales revenue that we began in the Application Preview.

EXAMPLE 5

The weekly sales S of a product during an advertising campaign are given by

$$S = \frac{100t}{t^2 + 100}, \qquad 0 \le t \le 20$$

where t is the number of weeks since the beginning of the campaign and S is in thousands of dollars.

(a) Over what interval are sales increasing? decreasing?
(b) What is the maximum weekly sales?
(c) Sketch the graph for $0 \le t \le 20$.

Solution

(a) To find where S is increasing, we first find $S'(t)$.

$$S''(t) = \frac{(t^2 + 100)100 - (100t)2t}{(t^2 + 100)^2}$$
$$= \frac{10{,}000 - 100t^2}{(t^2 + 100)^2}$$

We see that $S'(t) = 0$ when $10{,}000 - 100t^2 = 0$, or

$$100(100 - t^2) = 0$$
$$(10 + t)(10 - t) = 0$$
$$t = -10 \quad \text{or} \quad t = 10$$

Because $S'(t)$ is never undefined ($t^2 + 100 \ne 0$ for any real t) and because $0 \le t \le 20$, our only critical value is $t = 10$. Testing $S'(t)$ to the left and right of $t = 10$ gives the sign diagram.

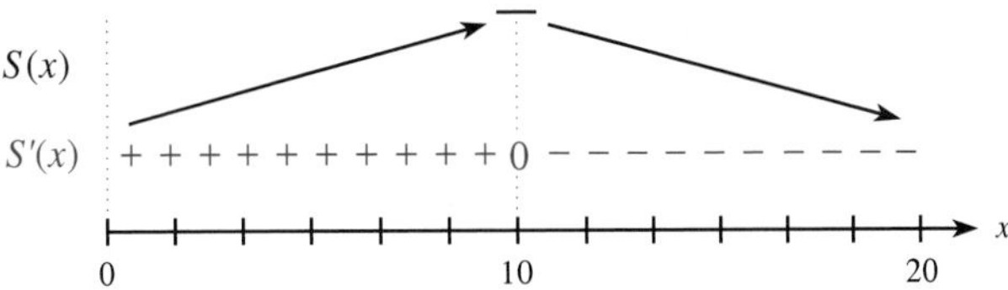

Hence, S is increasing on the interval $[0, 10)$ and decreasing on the interval $(10, 20]$.

(b) Because S is increasing to the left of $t = 10$ and S is decreasing to the right of $t = 10$, the maximum value of S occurs at $t = 10$ and is

$$S = S(10) = \frac{100(10)}{10^2 + 100} = \frac{1000}{200} = 5 \text{ (thousand dollars)}$$

(c) Plotting some additional points gives the graph; see Figure 10.10.

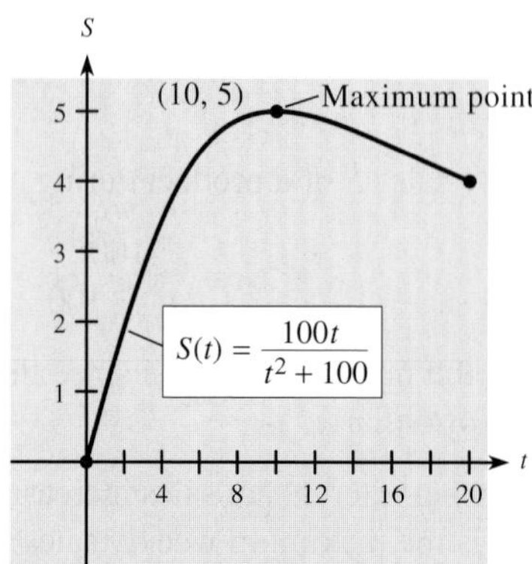

Figure 10.10

Graphing Utilities

With a graphing utility, choosing an appropriate window is the key to understanding the graph of a function. We saw that the derivative can be used to determine the critical values of a function and hence the interesting points of its graph. Therefore, the derivative can be used to determine the viewing window that provides an accurate representation of the graph. The derivative can also be used to discover graphical behavior that might be overlooked in a graph with a standard window.

EXAMPLE 6

Find the critical values for $f(x) = 0.0001x^3 + 0.003x^2 - 3.6x + 5$. Use them to determine an appropriate viewing window. Then sketch the graph.

Solution

Suppose that we first graph this function using a standard viewing window. The graph of this function, for $-10 \leq x \leq 10$, is shown in Figure 10.11(a). The graph looks like a line in this window, but the function is not linear. We could explore the function by tracing or zooming, but using the critical points is more helpful in graphing the function. We begin by finding $f'(x)$.

$$f'(x) = 0.0003x^2 + 0.006x - 3.6$$

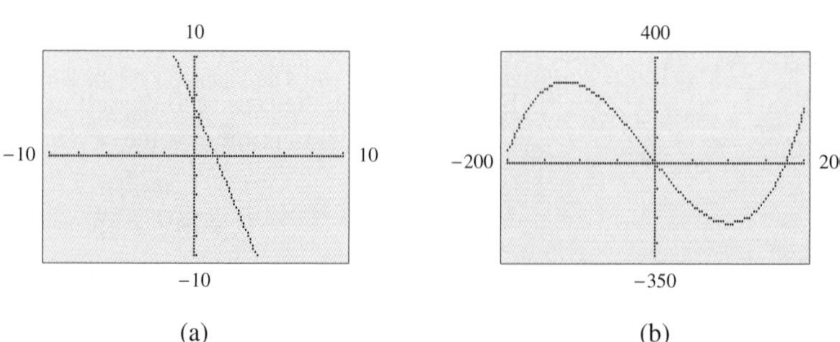

Figure 10.11

Solve $f'(x) = 0$ to find critical values.

$$0 = 0.0003x^2 + 0.006x - 3.6$$
$$0 = 0.0003(x^2 + 20x - 12000)$$
$$0 = 0.0003(x + 120)(x - 100)$$
$$x = -120 \quad \text{or} \quad x = 100$$

We choose a window that includes $x = -120$ and $x = 100$, graph, and use TRACE to find $f(-120) = 307.4$ and $f(100) = -225$. We then graph $y = f(x)$ on a window that includes $y = -225$ and $y = 307.4$ (see Figure 10.11b). On this graph we can verify that $(-120, 307.4)$ is a relative maximum and that $(100, -225)$ is a relative minimum.

We can also use a table of values of $f'(x)$ to create a sign diagram that will identify the relative maxima and minima. Figure 10.12(a) shows values of $y_1 = f(x)$ and $y_2 = f'(x)$ for different values of x at and near the critical values. Using the x- and y_2-values gives the sign diagram in Figure 10.12(b), which shows that $(-120, 307.4)$ is a relative maximum and that $(100, -225)$ is a relative minimum.

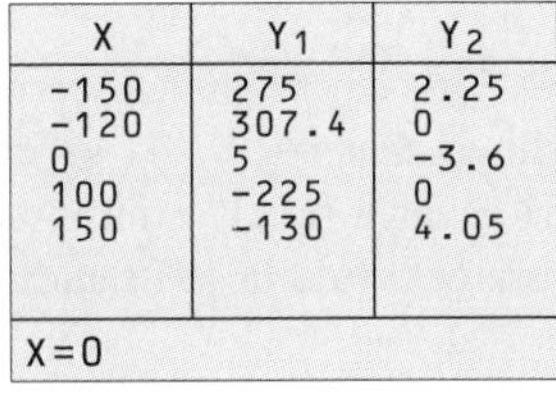

X	Y1	Y2
-150	275	2.25
-120	307.4	0
0	5	-3.6
100	-225	0
150	-130	4.05

X=0

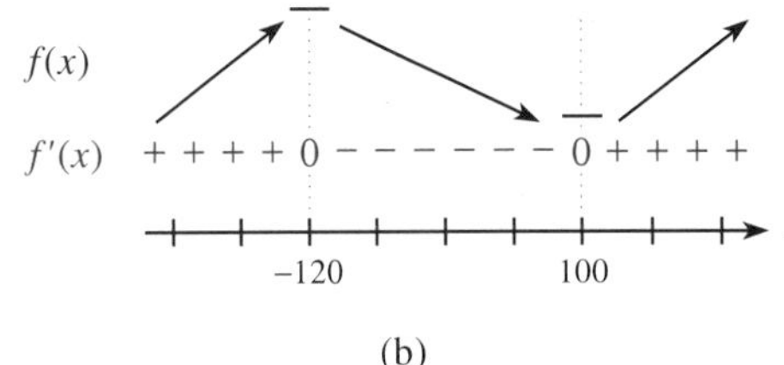

Figure 10.12 (a) (b)

M **EXAMPLE 7**

The table below gives the percentage of high school seniors using marijuana for the years 1975 to 1996.

(a) Using x as the number of years from 1970, develop an equation that models the percentage.

(b) During what years does the model indicate that the maximum and minimum use occurred?

Year	*Percent Using Marijuana*	*Year*	*Percent Using Marijuana*
1975	47.3	1986	38.8
1976	52.8	1987	36.3
1977	56.4	1988	33.1
1978	59.2	1989	29.6
1979	60.4	1990	27.0
1980	60.3	1991	23.9
1981	59.5	1992	21.9
1982	58.7	1993	26.0
1983	57	1994	30.7
1984	54.9	1995	34.7
1985	54.2	1996	35.8

Source: National Institute on Drug Abuse

Solution

An equation that models the percentage is $y = 0.03743x^3 - 1.7681x^2 + 23.3761x - 32.2371$. (See Figure 10.13.)

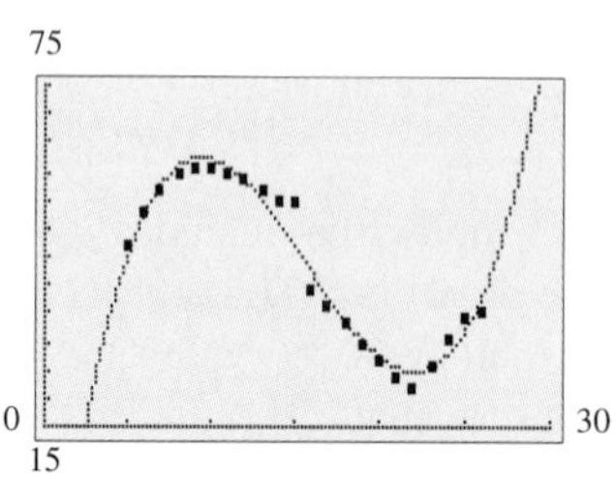

Figure 10.13

To find the critical values, we solve $y' = f'(x) = 0$, where $f'(x) = 0.11229x^2 - 3.5362x + 23.3761$. We can solve

$$0 = 0.11229x^2 - 3.5362x + 23.3761$$

with the quadratic formula or with a graphing utility. The two solutions are approximately

$$x = 9.441 \quad \text{or} \quad x = 22.051$$

The sign diagram shows that $x = 9.441$ gives a relative maximum (during 1979) and that $x = 22.051$ gives a relative minimum (during 1992). The data indicate that the maximum use occurred in 1979 and the minimum use in 1992.

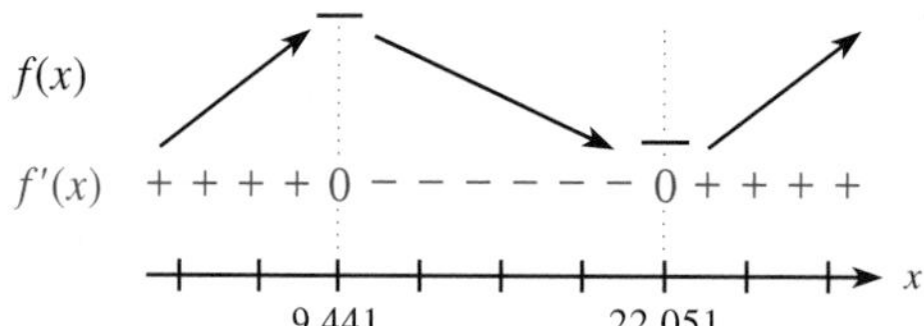

CHECKPOINT SOLUTIONS

1. $f'(x) = 0$ or $f'(x)$ is undefined.
2. (a) True, $f(x)$ is increasing when $f'(x) > 0$.
   (b) False. There may be a horizontal point of inflection at $x = -2$ (see Figure 10.8 on page 741).
   (c) False. A relative maximum occurs at $x = -3$.
3. (a) False. Critical values are solutions to $3 - 3x^2 = 0$, or $x = 1$ and $x = -1$.
   (b) False, y-coordinates of critical points come from $f(x) = 7 + 3x - x^3$. Thus critical points are $(1, 9)$ and $(-1, 5)$.
4.

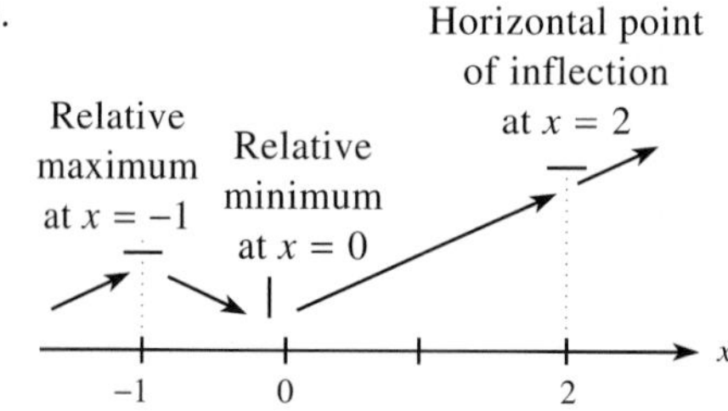

EXERCISE 10.1

In Problems 1 and 2, use the indicated points on the graph of $y = f(x)$ to identify points where $f(x)$ has (a) a relative maximum, (b) a relative minimum, and (c) a horizontal point of inflection.

1.

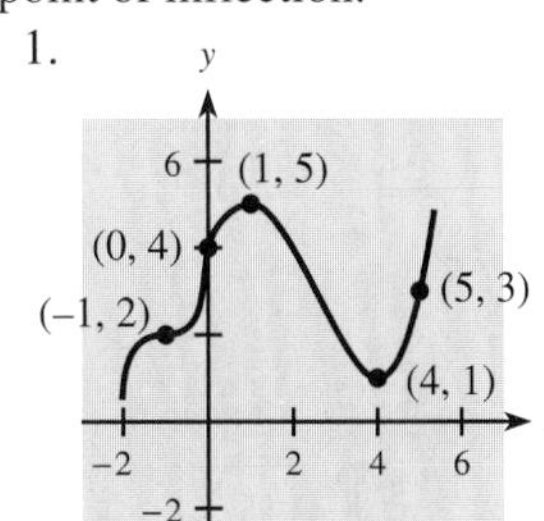

2.

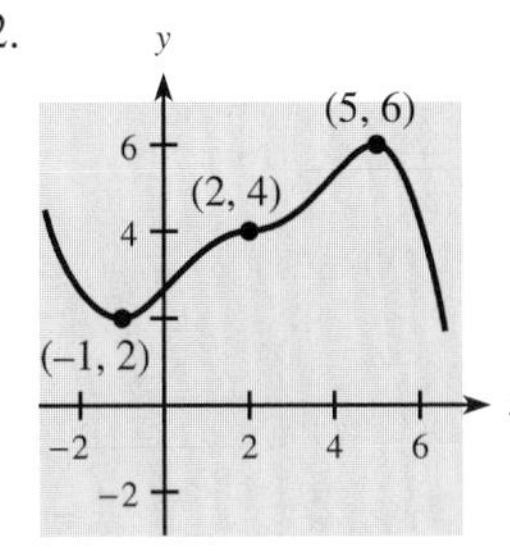

3. Use the graph of $y = f(x)$ in Problem 1 to identify at which of the indicated points the derivative $f'(x)$ (a) changes from positive to negative, (b) changes from negative to positive, and (c) does not change sign.
4. Use the graph of $y = f(x)$ in Problem 2 to identify at which of the indicated points the derivative $f'(x)$ (a) changes from positive to negative, (b) changes from negative to positive, and (c) does not change sign.

In Problems 5 and 6, use the sign diagram of $f'(x)$ to determine (a) the critical values of $f(x)$, (b) intervals where $f(x)$ increases, (c) intervals where $f(x)$ decreases, (d) x-values where relative maxima occur, and (e) x-values where relative minima occur.

5. $f'(x)$ $- - - 0 + + + + + 0 - - -$ (marks at 3 and 7 on the x-axis)

6. $f'(x)$ $+ + + + 0 + + + + + + 0 - - - -$ (marks at -5 and 8 on the x-axis)

In Problems 7 and 8, find the critical values of the function.

7. $y = 2x^3 - 12x^2 + 6$ 8. $y = x^3 - 3x^2 + 6x + 1$

In Problems 9 and 10, make a sign diagram for the function and determine the relative maxima and minima. (See Problems 7 and 8.)

9. $y = 2x^3 - 12x^2 + 6$ 10. $y = x^3 - 3x^2 + 6x + 1$

For each function and graph in Problems 11–14:

(a) Find $y' = f'(x) = \dfrac{dy}{dx}$.
(b) Use $y' = f'(x)$ to find the critical values.
(c) Find the critical points.
(d) Use the graph to classify each critical point as a relative maximum, relative minimum, or horizontal point of inflection.

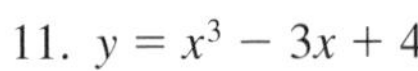
11. $y = x^3 - 3x + 4$

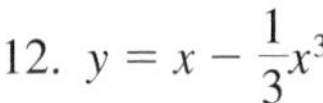
12. $y = x - \frac{1}{3}x^3$

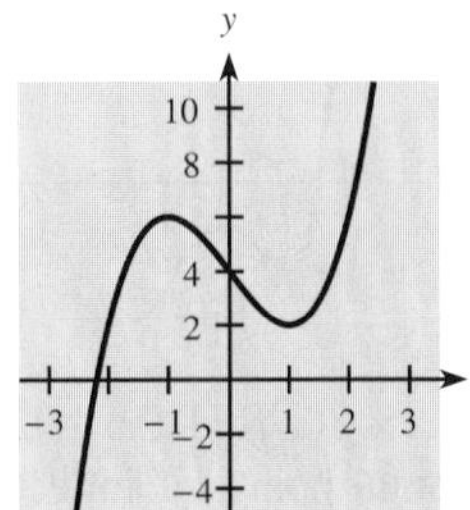

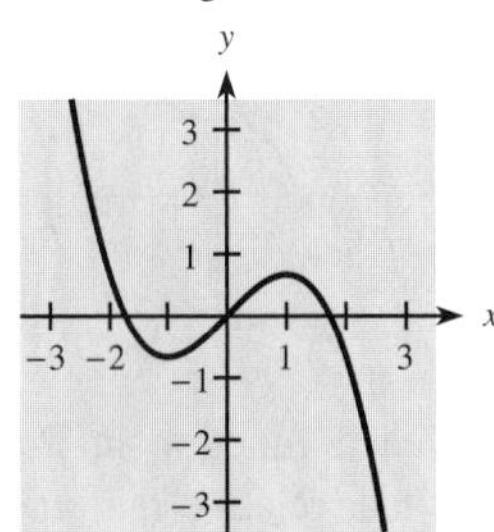

13. $y = x^3 + 3x^2 + 3x - 2$ 14. $y = x^3 - 6x^2 + 12x + 1$

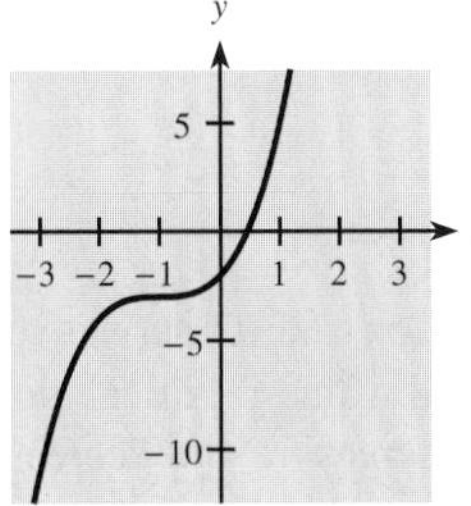

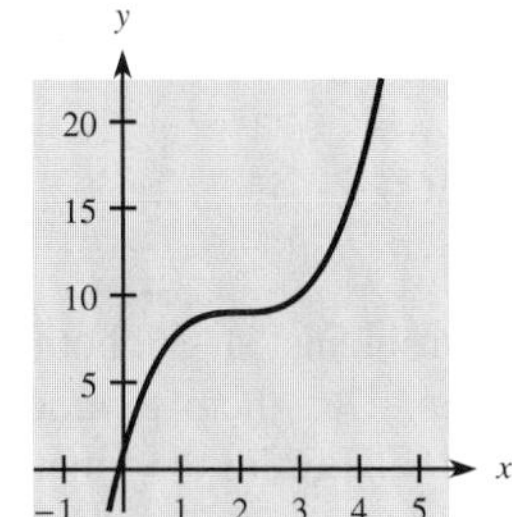

For each function in Problems 15–20:

(a) Find $y' = f'(x)$.
(b) Find the critical values.
(c) Find the critical points.
(d) Find intervals of x-values where the function is increasing and where it is decreasing.
(e) Classify the critical points as relative maxima, relative minima, or horizontal points of inflection. In each case, you may check your conclusions with a graphing utility.

15. $y = \frac{1}{2}x^2 - x$ 16. $y = x^2 + 4x$

17. $y = \dfrac{x^3}{3} + \dfrac{x^2}{2} - 2x + 1$ 18. $y = \dfrac{x^4}{4} - \dfrac{x^3}{3} - 2$

19. $y = x^{2/3}$ 20. $y = -(x - 3)^{2/3}$

For each function and graph in Problems 21–24:

(a) Use the graph to identify x-values for which $y' > 0$, $y' < 0$, $y' = 0$, and y' does not exist.
(b) Use the derivative to check your conclusions.

21. $y = 6 - x - x^2$ 22. $y = \frac{1}{2}x^2 - 4x + 1$

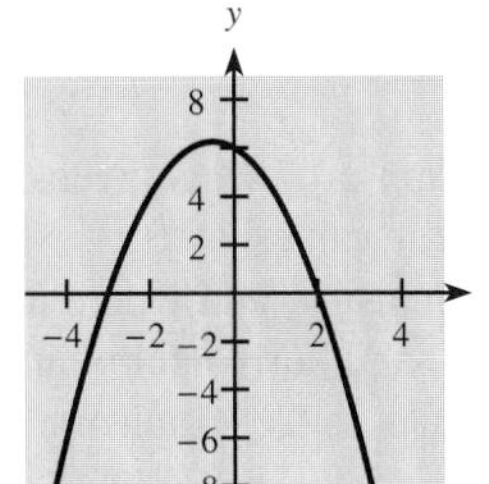

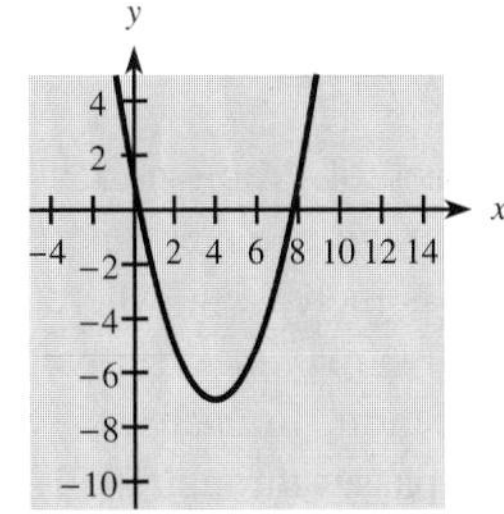

23. $y = 6 + x^3 - \frac{1}{15}x^5$

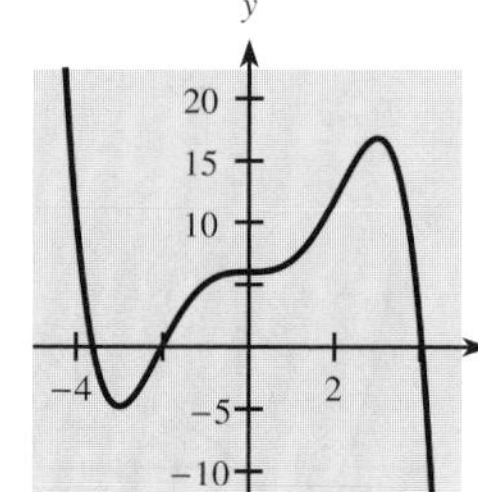

24. $y = x^4 - 2x^2 - 1$

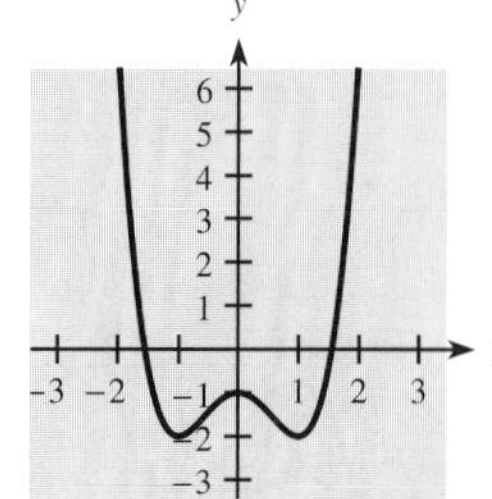

For each function in Problems 25–30, find the relative maxima, relative minima, horizontal points of inflection, and sketch the graph. You may check your graph with a graphing utility.

25. $y = \frac{1}{3}x^3 - x^2 + x + 1$
26. $y = \frac{1}{4}x^4 - \frac{2}{3}x^3 + \frac{1}{2}x^2 - 2$
27. $y = \frac{1}{3}x^3 + x^2 - 24x + 20$
28. $C(x) = x^3 - \frac{3}{2}x^2 - 18x + 5$
29. $y = 3x^5 - 5x^3 + 1$
30. $y = \frac{1}{6}x^6 - x^4 + 7$

In Problems 31–36, both a function and its derivative are given. Use them to find critical values, critical points, intervals where the function is increasing and decreasing, relative maxima, relative minima, and horizontal points of inflection; sketch the graph of each function.

31. $y = (x^2 - 2x)^2$ $\qquad \dfrac{dy}{dx} = 4x(x - 1)(x - 2)$

32. $f(x) = (x^2 - 4)^2$ $\qquad f'(x) = 4x(x + 2)(x - 2)$

33. $y = \dfrac{x^3(x - 5)^2}{27}$ $\qquad \dfrac{dy}{dx} = \dfrac{5x^2(x - 3)(x - 5)}{27}$

34. $y = \dfrac{x^2(x - 5)^3}{27}$ $\qquad \dfrac{dy}{dx} = \dfrac{5x(x - 2)(x - 5)^2}{27}$

35. $f(x) = x^{2/3}(x - 5)$ $\qquad f'(x) = \dfrac{5(x - 2)}{3x^{1/3}}$

36. $f(x) = x - 3x^{2/3}$ $\qquad f'(x) = \dfrac{x^{1/3} - 2}{x^{1/3}}$

In Problems 37–42, use the derivative to locate critical points and determine a viewing window that shows all features of the graph. Use a graphing utility to sketch a complete graph.

37. $f(x) = x^3 - 225x^2 + 15000x - 12000$
38. $f(x) = x^3 - 15x^2 - 16800x + 80000$
39. $f(x) = x^4 - 160x^3 + 7200x^2 - 40000$
40. $f(x) = x^4 - 240x^3 + 16200x^2 - 60000$
41. $y = 7.5x^4 - x^3 + 2$
42. $y = 2 - x^3 - 7.5x^4$

In each of Problems 43–46, a graph of $f'(x)$ is given. Use the graph to determine the critical values of $f(x)$, where $f(x)$ is increasing, where it is decreasing, and where it has relative maxima, relative minima, and horizontal points of inflection. In each case sketch a possible graph for $f(x)$ that passes through (0, 0).

43. $f'(x) = x^2 - x - 2$

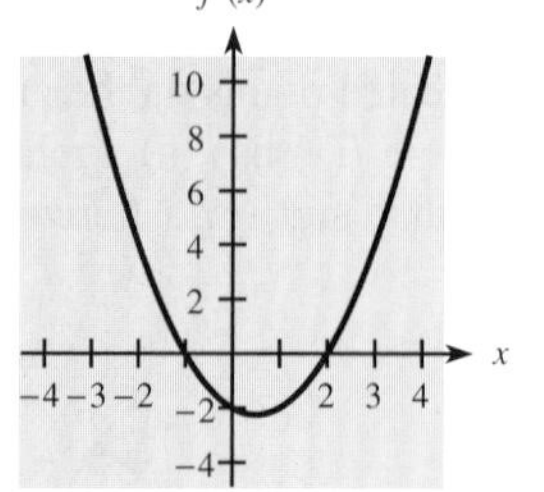

44. $f'(x) = 4x - x^2$

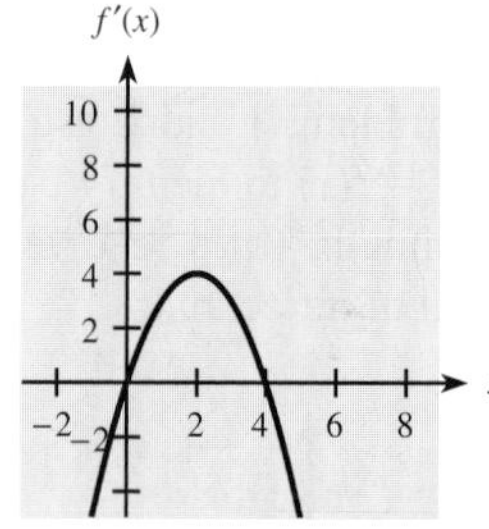

45. $f'(x) = x^3 - 3x^2$

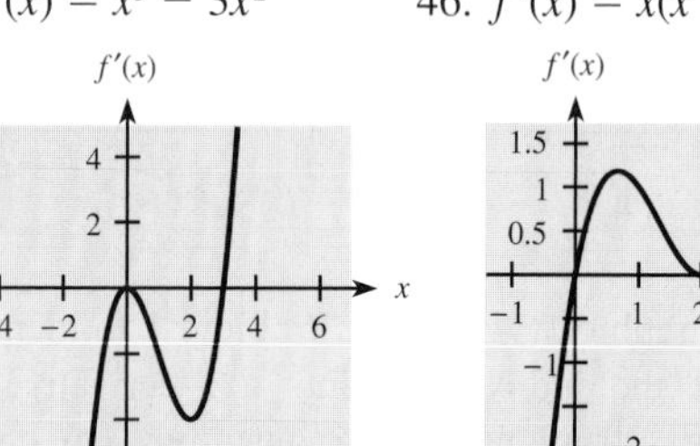

46. $f'(x) = x(x - 2)^2$

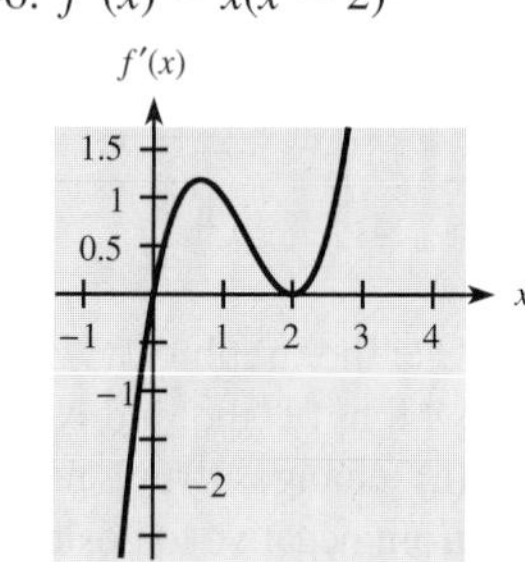

In Problems 47 and 48, two graphs are given. One is the graph of f and the other is the graph of f'. Decide which is which and explain your reasoning.

47.

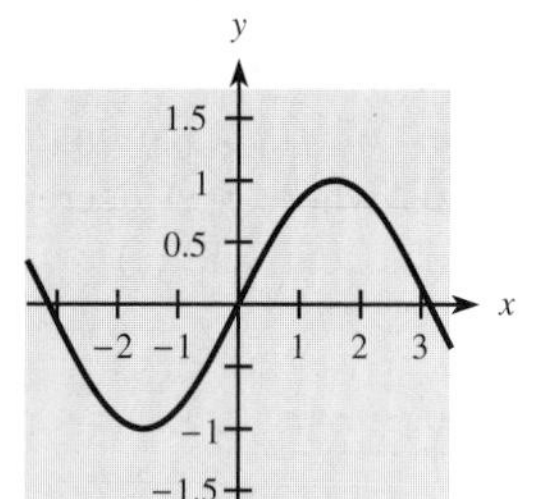

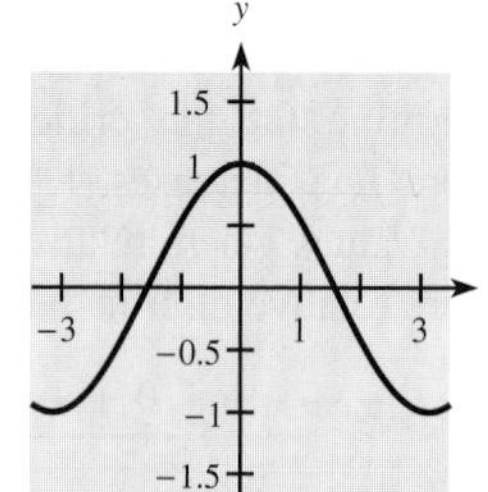

48.

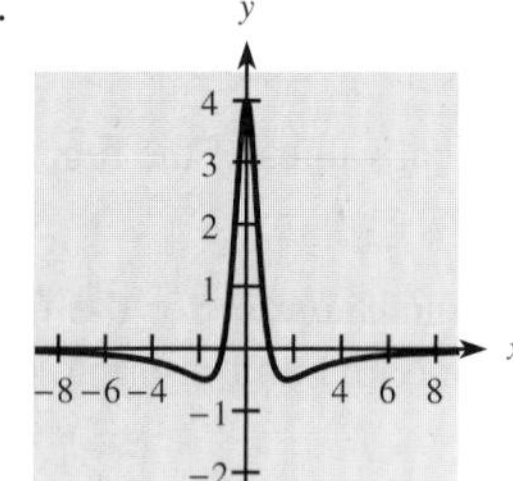

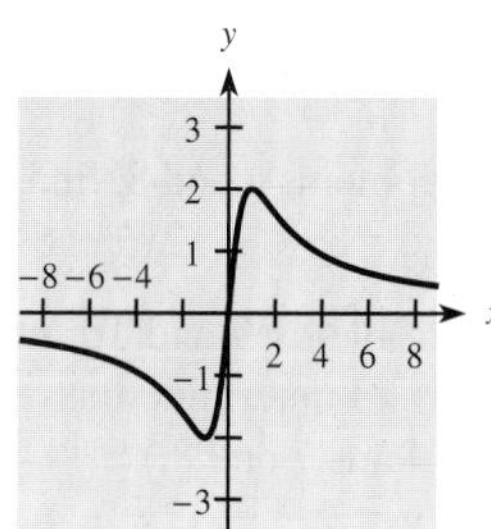

Applications

49. ***Advertising and sales*** Suppose that the daily sales (in dollars) t days after the end of an advertising campaign are given by

$$S = 1000 + \frac{400}{t + 1}, \qquad t \geq 0$$

Does S increase for all $t \geq 0$, decrease for all $t \geq 0$, or change direction at some point?

50. ***Pricing and sales*** Suppose that a chain of auto service stations, Quick-Oil, Inc., has found that its monthly sales volume y (in thousands of dollars) is related to the price p (in dollars) of an oil change by

$$y = \frac{90}{\sqrt{p+5}}, \qquad p > 10$$

Is y increasing or decreasing for all values of $p > 10$?

51. ***Productivity*** A time study showed that, on average, the productivity of a worker after t hours on the job can be modeled by

$$P(t) = 27t + 6t^2 - t^3, \quad 0 \leq t \leq 8$$

where P is the number of units produced per hour.
(a) Find the critical values for this function.
(b) Which critical value makes sense in this model?
(c) For what values of t is P increasing?
(d) Graph the function for $0 \leq t \leq 8$.

52. ***Production*** Analysis of daily output of a factory shows that, on average, the number of units per hour y produced after t hours of production is

$$y = 70t + \frac{1}{2}t^2 - t^3, \qquad 0 \leq t \leq 8$$

(a) Find the critical values for this function.
(b) Which critical values make sense in this particular problem?
(c) For which values of t, for $0 \leq t \leq 8$, is y increasing?
(d) Graph this function.

53. ***Production costs*** Suppose that the average cost of producing a shipment of a certain product is

$$\overline{C} = 5000x + \frac{125{,}000}{x}, \qquad x > 0$$

where x is the number of machines used in the production process.
(a) Find the critical values for this function.
(b) Over what interval does the average cost decrease?
(c) Over what interval does the average cost increase?

54. ***Average costs*** Suppose the average costs of a mining operation depend on the number of machines used, and average costs are given by

$$\overline{C}(x) = 2900x + \frac{1{,}278{,}900}{x}, \qquad x > 0$$

where x is the number of machines used.
(a) Find the critical values for $\overline{C}(x)$ that lie in the domain of the problem.
(b) Over what interval in the domain do average costs decrease?
(c) Over what interval in the domain do average costs increase?
(d) How many machines give minimum average costs?
(e) What are the minimum average costs?

55. ***Marginal revenue*** Suppose the weekly marginal revenue function for selling x units of a product is given by the graph in the figure.

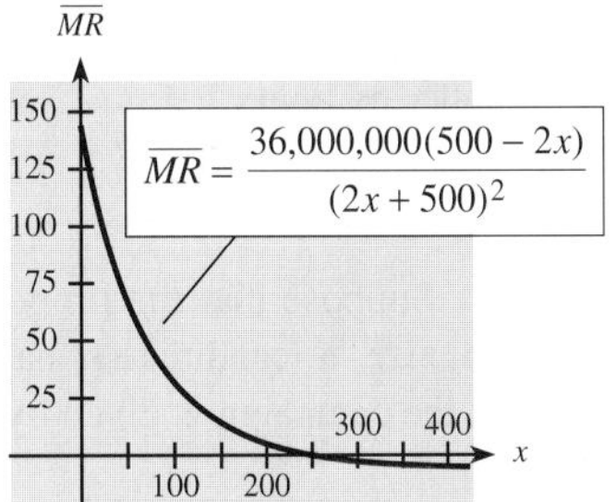

(a) At each of $x = 150$, $x = 250$, and $x = 350$, what is happening to revenue?
(b) Over what interval is revenue increasing?
(c) How many units must be sold to maximize revenue?

56. ***Earnings*** Suppose that the rate of change $f'(x)$ of the average annual earnings of new car salespersons is shown in the figure.

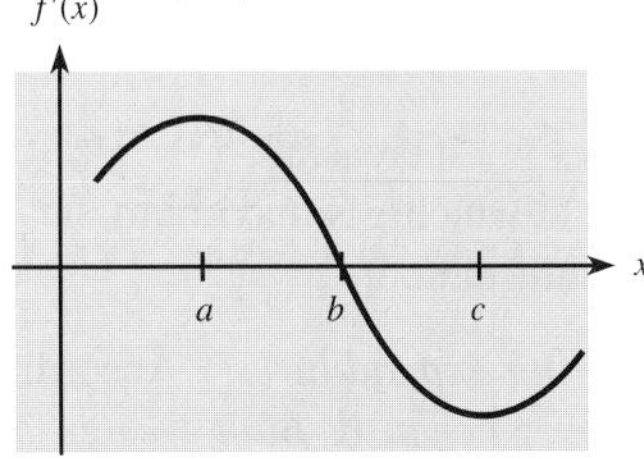

(a) If a, b, and c represent certain years, what is happening to $f(x)$, the average annual earnings of the salespersons, at a, b, and c?
(b) Over what interval (involving a, b, or c) is there an increase in $f(x)$, the average annual earnings of the salespersons?

57. ***Revenue*** The weekly revenue of a certain recently released film is given by

$$R(t) = \frac{50t}{t^2 + 36}, \qquad t \geq 0$$

where R is in millions of dollars and t is in weeks.
(a) Find the critical values.
(b) For how many weeks will weekly revenue increase?

58. ***Medication*** Suppose that the concentration C of a medication in the bloodstream t hours after an injection is given by

$$C(t) = \frac{0.2t}{t^2 + 1}$$

(a) Determine the number of hours before C attains its maximum.
(b) Find the maximum concentration.

59. ***Candidate recognition*** Suppose that the proportion P of voters who recognize a candidate's name t months after the start of the campaign is given by

$$P(t) = \frac{13t}{t^2 + 100} + 0.18$$

(a) How many months after the start of the campaign is recognition at its maximum?
(b) To have greatest recognition on November 1, when should a campaign be launched?

60. ***Medication*** The number of milligrams x of a medication in the bloodstream t hours after a dosage is taken can be modeled by

$$x(t) = \frac{2000t}{t^2 + 16}$$

(a) For what t-values is x increasing?
(b) Find the t-value at which x is maximum.
(c) Find the maximum value for x.

61. ***Residential remodeling expenditures*** According to the data provided by *Kitchen & Bath Design News,* the number of billions of dollars spent on residential remodeling (in 1992 dollars) can be modeled by

$$y = 0.003652x^4 - 0.1408x^3 + 1.6083x^2 - 4.4349x + 56.7387$$

where x is the number of years from 1980.
(a) To the nearest year from 1980 to 1999, find when this model predicts the highest expenditure for remodeling.
(b) Find the year in this interval when the model gives the minimum expenditure.

M 62. ***Poverty*** The table shows the number of millions of the people in the United States who lived below the poverty level for selected years from 1960 to 1996.
(a) Find a model that approximately fits the data, using x as years past 1900.
(b) Find the year when poverty is a minimum according to the model.

Year	*Persons Living Below the Poverty Level (millions)*
1960	39.9
1965	33.2
1970	25.4
1975	25.9
1980	29.3
1986	32.4
1989	31.5
1990	33.6
1991	35.7
1992	38
1993	39.3
1994	38.1
1995	36.4
1996	36.5

Source: Bureau of the Census, U.S. Dept. of Commerce

M 63. ***Budget deficit*** The table gives the yearly budget deficit, in billions of dollars, for the years 1990–1997, with White House estimates for 1998 and 1999.
(a) Use the data for 1990 to 1997 to find the equation that models the yearly deficit, with x as the number of years past 1990.
(b) Find the year when the model indicates that the deficit is a maximum, and compare it to the data.

Year	*Yearly Deficit (billions)*
1990	\$221.2
1991	269.4
1992	290.4
1993	255.0
1994	203.1
1995	163.9
1996	107.3
1997	22.6
1998	22.0
1999	0

Source: *USA Today,* Jan. 7, 1998

M 64. ***Total capital*** The table gives the percent earned total capital for Eli Lilly & Co. for the years 1987–1998.

(a) Find a cubic equation that models the data, using $x = 0$ in 1987.

(b) Use the model to find the year during which the maximum percent total return occurred.

Year	Percent Earned Total Capital	Year	Percent Earned Total Capital
1987	18.8	1993	25.4
1988	21.5	1994	17.4
1989	23.6	1995	17.3
1990	30.4	1996	17.9
1991	25.0	1997	27.0
1992	25.8	1998	28.0

Source: Value Line Publishing Company, Oct. 31, 1997

10.2 Concavity; Points of Inflection

OBJECTIVES

- *To find points of inflection of graphs of functions*
- *To use the second-derivative test to graph functions*

APPLICATION PREVIEW

Suppose that in 1996 a retailer wishes to sell his store and uses the graph in Figure 10.14 to show how profits have increased since he opened the store and the potential for profit in the future. Can we conclude that profits will continue to grow, or should we be concerned about future earnings?

Note that although profits are still increasing in 1995, they are increasing more slowly than in previous years. Indeed, they appear to have been growing at a decreasing rate since about 1990. This means that 1990 was the year that the rate of change of profits was maximum, and since then the rate of change of profits has been diminishing. For this reason, 1990 is called the *point of diminishing returns.* We say that this profit curve is **concave down** over the interval from 1990 to 1995. Judging by Figure 10.14, it would be unwise to expect a large increase in profit after 1995.

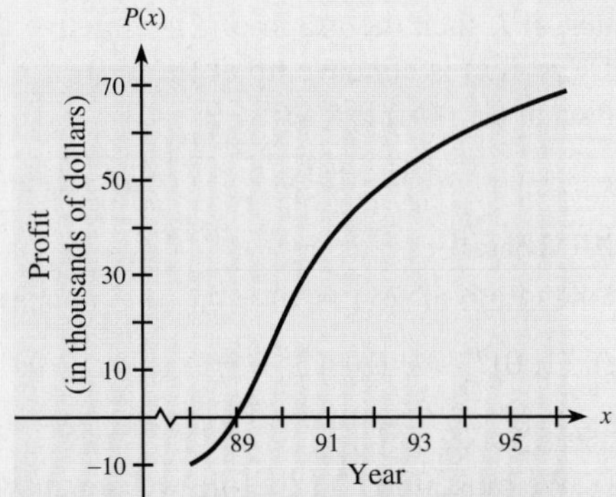

Figure 10.14

Just as we used the first derivative to determine whether a curve was increasing or decreasing on a given interval, we can use the second derivative to determine whether the curve is concave up or concave down on an interval.

A curve is said to be **concave up** on an interval $[a, b]$ if at each point on the interval the curve is above its tangent at the point (Figure 10.15a). If the curve is below all its tangents on a given interval, it is **concave down** on the interval (Figure 10.15b).

Looking at Figure 10.15(a), we see that the *slopes* of the tangent lines increase over the interval where the graph is concave up. Because $f'(x)$ gives the slopes of those tangents, it follows that $f'(x)$ is increasing over the interval where $f(x)$ is concave up. However, if we know that $f'(x)$ is increasing, then its derivative, $f''(x)$, must be positive. That is, the second derivative is positive if the curve is concave up. Conversely, it can be shown that the graph of a function is concave up if the second derivative is positive.

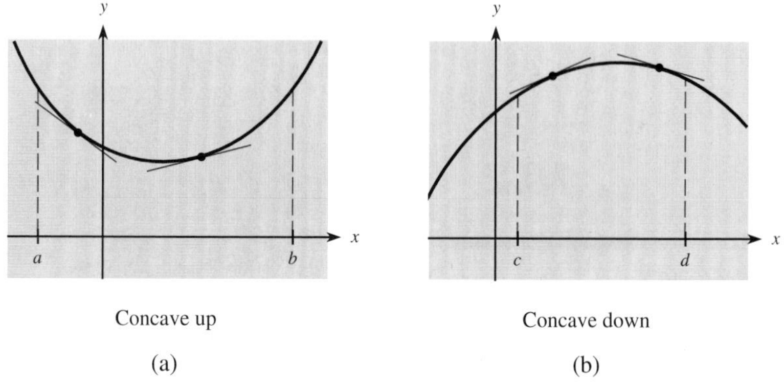

Concave up (a) Concave down (b)

Figure 10.15

Similarly, if the second derivative of a function is negative over an interval, the slopes of the tangents to the graph decrease over that interval. This happens when the tangent lines are above the graph, as in Figure 10.15(b), so the graph must be concave down on this interval.

Thus we see that the second derivative can be used to determine the concavity of a curve.

Concave Up and Concave Down

Assume that the first and second derivatives of function f exist. If $f''(x) > 0$ on an interval I, the graph of f is **concave up** on the interval. If $f''(x) < 0$ on an interval I, then the graph of f is **concave down** on I. We also say that the graph of $y = f(x)$ is concave up at $(a, f(a))$ if $f''(a) > 0$ and that the graph is concave down at $(b, f(b))$ if $f''(b) < 0$.

EXAMPLE 1

Is the graph of $f(x) = x^3 - 4x^2 + 3$ concave up or down at the point

(a) $(1, 0)$? (b) $(2, -5)$?

Solution

(a) We must find $f''(x)$ before we can answer this question.

$$f'(x) = 3x^2 - 8x \qquad f''(x) = 6x - 8$$

Then $f''(1) = 6(1) - 8 = -2$, so the graph is concave down at $(1, 0)$.

(b) Because $f''(2) = 6(2) - 8 = 4$, the graph is concave up at $(2, -5)$. The graph of $f(x) = x^3 - 4x^2 + 3$ is shown in Figure 10.16(a).

Looking at the graph of $y = x^3 - 4x^2 + 3$ (Figure 10.16a), we see that the curve is concave down on the left and concave up on the right. Thus it has changed from concave down to concave up. Figure 10.16(b) shows the graph of $y'' = f''(x) = 6x - 8$, and we can see that $y'' < 0$ for $x < \frac{4}{3}$ and $y'' > 0$ for $x > \frac{4}{3}$.

Thus the second derivative changes sign at $x = \frac{4}{3}$, so the concavity of the graph of $y = f(x)$ changes at $x = \frac{4}{3}$, $y = -\frac{47}{27}$. The point where concavity changes is called a **point of inflection.**

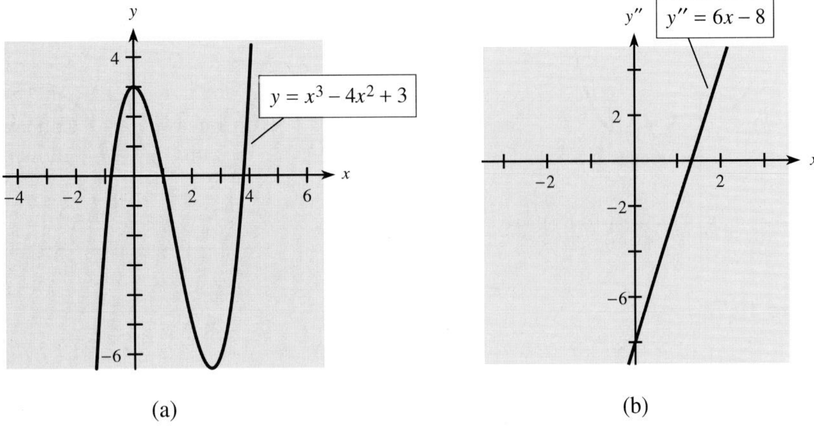

Figure 10.16 (a) (b)

Point of Inflection A point (x_0, y_0) on the graph of a function f is called a **point of inflection** if the curve is concave up on one side of the point and concave down on the other side. The second derivative at this point, $f''(x_0)$, will be 0 or undefined.

In general, we can find points of inflection and information about concavity as follows.

Finding Points of Inflection and Concavity

Procedure	***Example***
To find the point(s) of inflection of a curve and intervals where it is concave upward and where it is concave downward:	Find the points of inflection and concavity of the graph of $y = \frac{x^4}{2} - x^3 + 5$.
1. Find the second derivative of the function.	1. $y' = f'(x) = 2x^3 - 3x^2$ $y'' = f''(x) = 6x^2 - 6x$
2. Set the second derivative equal to 0, and solve for x. Potential points of inflection occur at these values of x or at values of x where $f(x)$ is defined and $f''(x)$ is undefined.	2. $0 = 6x^2 - 6x = 6x(x - 1)$ has solutions $x = 0, x = 1$. $f''(x)$ is defined everywhere.
3. Find the potential points of inflection.	3. $(0, 5)$ and $(1, \frac{9}{2})$ are potential points of inflection.
4. If the second derivative has opposite signs on the two sides of one of these values of x, a point of inflection occurs. The curve is concave upward where $f''(x) > 0$ and concave downward where $f''(x) < 0$.	4. $f''(-1) = 12 > 0$, $f''(0) = 0$, $f''(\frac{1}{2}) = -\frac{3}{2} < 0$ $\Rightarrow$ $(0, 5)$ is a point of inflection. $f''(\frac{1}{2}) = -\frac{3}{2} < 0$, $f''(1) = 0$, $f''(2) = 12 > 0$ $\Rightarrow$ $(1, \frac{9}{2})$ is a point of inflection. See the graph in Figure 10.17 and the sign diagram on the next page.

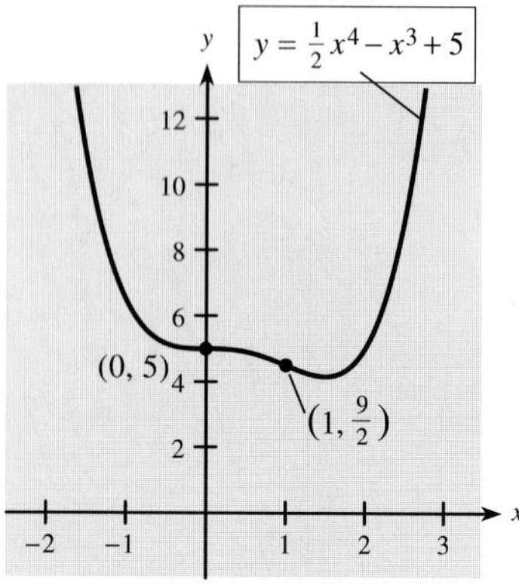

Figure 10.17

The graph of $y = \frac{1}{2}x^4 - x^3 + 5$ is shown in Figure 10.17. Note the points of inflection at (0, 5) and $(1, \frac{9}{2})$. The point of inflection at (0, 5) is a horizontal point of inflection because $f'(x)$ is also 0 at $x = 0$. A **sign diagram for $f''(x)$,** the second derivative of this function, is shown below. The changes in the sign of $f''(x)$ correspond to changes in concavity and occur at points of inflection.

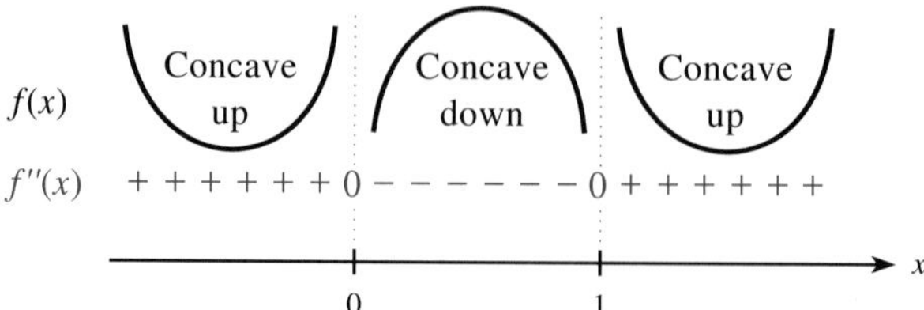

EXAMPLE 2

Suppose that a real estate developer wishes to remove pollution from a small lake so that she can sell lakefront homes on a "crystal clear" lake. The graph in Figure 10.18 shows the relation between dollars spent on cleaning the lake and the purity of the water. The point of inflection on the graph is called the **point of diminishing returns** on her investment because it is where the *rate* of return on her investment changes from increasing to decreasing. Show that the rate of change in the purity of the lake, $f'(x)$, is maximized at this point, $x = c$. Assume that $f(c)$, $f'(c)$, and $f''(c)$ are defined.

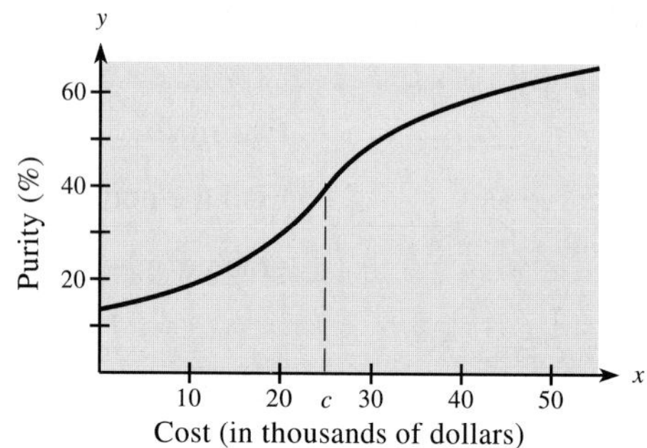

Figure 10.18

Solution

Because $x = c$ is a point of inflection for $f(x)$, we know that the concavity must change at $x = c$. From the figure we see the following.

$x < c$: $f(x)$ is concave up, so $f''(x) > 0$.
$f''(x) > 0$ means that $f'(x)$ is increasing.
$x > c$: $f(x)$ is concave down, so $f''(x) < 0$.
$f''(x) < 0$ means that $f'(x)$ is decreasing.

Thus $f'(x)$ has $f'(c)$ as its relative maximum.

EXAMPLE 3

Suppose that the daily sales (in thousands of dollars) of a product is given by

$$S = \frac{(-x^3 + 9x^2 + 6)}{6}$$

where x is thousands of dollars spent on advertising. Find the point of diminishing returns for money spent on advertising.

Solution

We seek the point where the graph of this function changes from concave up to concave down, if such a point exists.

$$\frac{dS}{dx} = S'(x) = \frac{1}{6}(-3x^2 + 18x)$$

$$S''(x) = \frac{1}{6}(-6x + 18) = -x + 3$$

$$S''(x) = 0 \quad \text{when} \quad 0 = -x + 3 \quad \text{or} \quad x = 3$$

Thus $x = 3$ is a possible point of inflection. We test $S''(x)$ to the left and right of $x = 3$.

$$S''(2) = 1 > 0 \Rightarrow \text{concave up to the left of } x = 3$$
$$S''(4) = -1 < 0 \Rightarrow \text{concave down to the right of } x = 3$$

Thus the point of diminishing returns occurs when $x = 3$ (thousand dollars) and $S = 10$ (thousand dollars). Figure 10.19 shows the graphs of S, S', and S''. At $x = 3$, we can see that the point of diminishing returns on the graph of S corresponds to the maximum point of the graph of S' and the zero (or x-intercept) of the graph of S''.

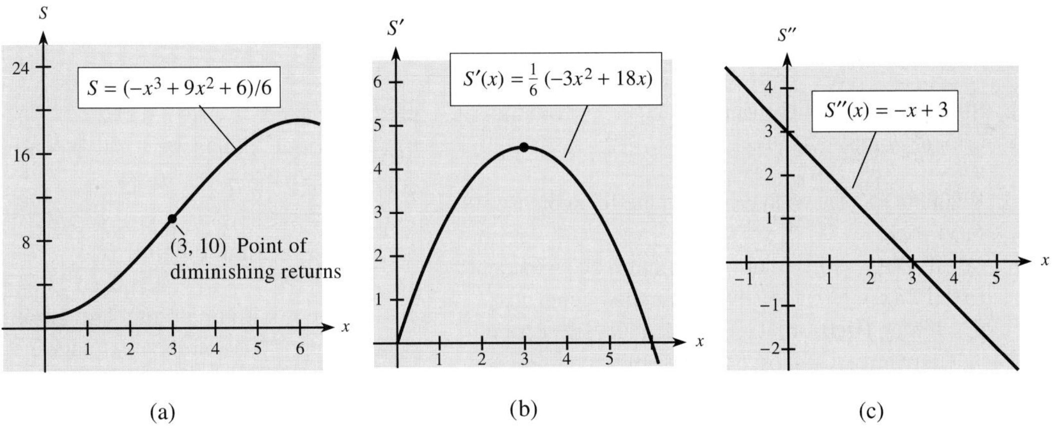

Figure 10.19

We can use information about points of inflection and concavity to help sketch graphs. For example, if we know that the curve is concave up at a critical point where $f'(x) = 0$, then the point must be a relative minimum because the tangent to the curve is horizontal at the critical point, and only a point at the bottom of a "concave up" curve could have a horizontal tangent (see Figure 10.20a).

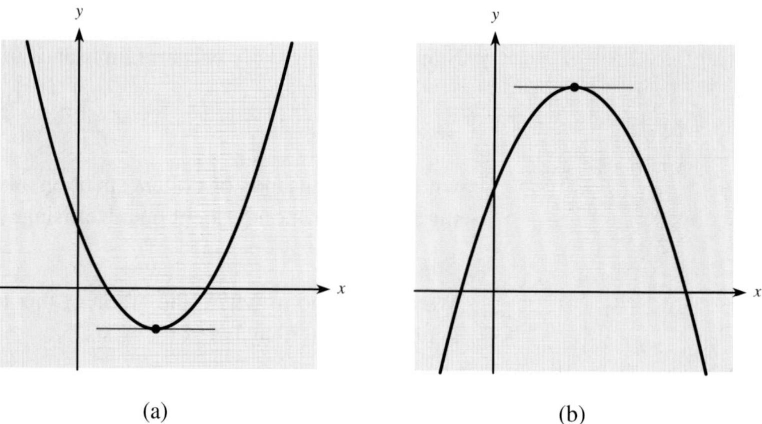

Figure 10.20 (a) (b)

On the other hand, if the curve is concave down at a critical point where $f'(x) = 0$, then the point is a relative maximum (see Figure 10.20b).

Thus we can use the **second-derivative test** to determine whether a critical point where $f'(x) = 0$ is a relative maximum or minimum.

Second-Derivative Test

Procedure	***Example***
To find relative maxima and minima of a function:	Find the relative maxima and minima of $y = f(x) = \frac{1}{3}x^3 - x^2 - 3x + 2$.
1. Find the critical values of the function.	1. $f'(x) = x^2 - 2x - 3$ $0 = (x - 3)(x + 1)$ has solutions $x = -1$ and $x = 3$. No values of x make $x^2 - 2x - 3$ undefined.
2. Substitute the critical values into $f(x)$ to find the critical points.	2. $f(-1) = \frac{11}{3}$ $\quad f(3) = -7$ The critical points are $(-1, \frac{11}{3})$ and $(3, -7)$.
3. Evaluate $f''(x)$ at each critical value for which $f'(x) = 0$. (a) If $f''(x_0) < 0$, a relative maximum occurs at x_0. (b) If $f''(x_0) > 0$, a relative minimum occurs at x_0. (c) If $f''(x_0) = 0$, or $f''(x_0)$ is undefined, the second-derivative test fails; use the first-derivative test.	3. $f''(x) = 2x - 2$ $f''(-1) = 2(-1) - 2 = -4 < 0$, so $(-1, \frac{11}{3})$ is a relative maximum point. $f''(3) = 2(3) - 2 = 4 > 0$, so $(3, -7)$ is a relative minimum point. (The graph is shown in Figure 10.21 on the next page.)

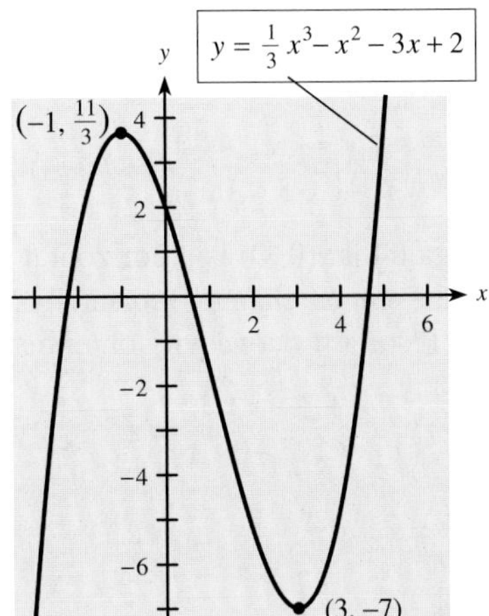

Figure 10.21

CHECKPOINT

1. If $f''(x) > 0$, then $f(x)$ is concave ________.
2. Where do possible points of inflection occur?
3. On the graph below, locate any points of inflection (approximately) and label where the curve satisfies $f''(x) > 0$ and $f''(x) < 0$.

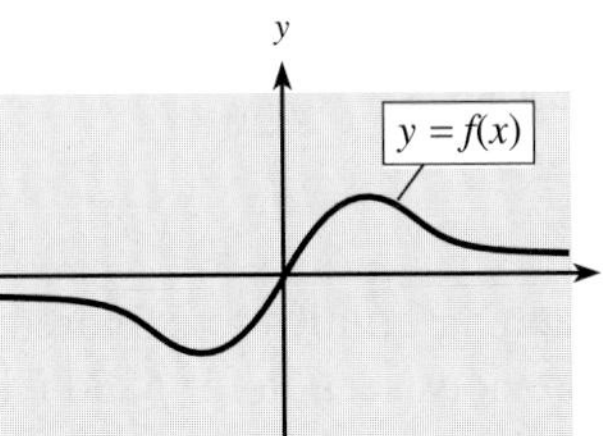

4. Determine whether the following is true or false. If $f''(0) = 0$, then $f(x)$ has a point of inflection at $x = 0$.

EXAMPLE 4

Find the relative maxima and minima and points of inflection of $y = 3x^4 - 4x^3$.

Solution

$$y' = f'(x) = 12x^3 - 12x^2$$

Solving $0 = 12x^3 - 12x^2 = 12x^2(x - 1)$ gives $x = 1$ and $x = 0$. Thus the critical points are $(1, -1)$ and $(0, 0)$.

$$y'' = f''(x) = 36x^2 - 24x$$

$$f''(1) = 12 > 0 \Rightarrow (1, -1) \text{ is a relative minimum point.}$$

$$f''(0) = 0 \Rightarrow \text{the second-derivative test fails.}$$

Because the second-derivative test fails, we must use the first-derivative test at the critical point (0, 0).

$$\left.\begin{aligned} f'(-1) &= -24 < 0 \\ f'(\tfrac{1}{2}) &= -\tfrac{3}{2} < 0 \end{aligned}\right\} \Rightarrow (0, 0) \text{ is a horizontal point of inflection.}$$

We look for points of inflection by setting $f''(x) = 0$ and solving for x. We find that $0 = 36x^2 - 24x$ has solutions $x = 0$ and $x = \frac{2}{3}$.

$$\left.\begin{array}{l} f''(-1) = 60 > 0 \\ f''(\frac{1}{2}) = -3 < 0 \end{array}\right\} \Rightarrow (0, 0) \text{ is a point of inflection.}$$

Thus we see again that (0, 0) is a horizontal point of inflection. This is a special point, where the curve changes concavity *and* has a horizontal tangent (see Figure 10.22). Testing for concavity on either side of $x = \frac{2}{3}$ gives

$$\left.\begin{array}{l} f''(\frac{1}{2}) = -3 < 0 \\ f''(1) = 12 > 0 \end{array}\right\} \Rightarrow (\tfrac{2}{3}, -\tfrac{16}{27}) \text{ is a point of inflection.}$$

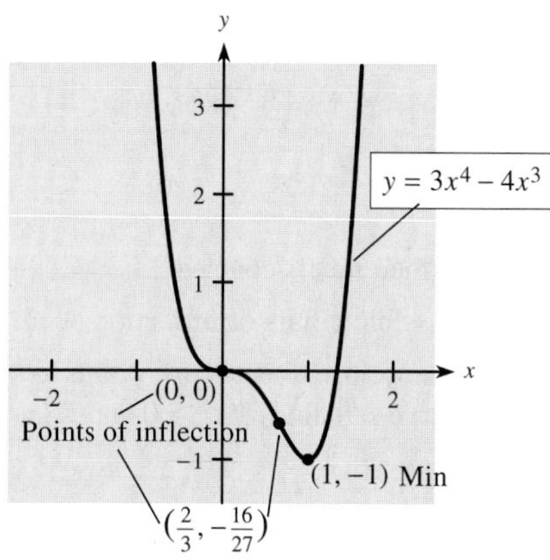

Figure 10.22

Graphing Utilities

We can use graphing utilities to explore the relationships among f, f', and f'', as we did in the previous section for f and f'.

EXAMPLE 5

Figure 10.23 shows the graph of $f(x) = \frac{1}{6}(2x^3 - 3x^2 - 12x + 12)$.

(a) From the graph, identify points where $f''(x) = 0$.
(b) From the graph, observe intervals where $f''(x) > 0$ and where $f''(x) < 0$.
(c) Check the conclusions from (a) and (b) by calculating $f''(x)$ and graphing it.

Solution

(a) From Figure 10.23, we can make an initial estimate of the x-value of the point of inflection. It appears to be near $x = \frac{1}{2}$, so we expect $f''(x) = 0$ at (or very near to) $x = \frac{1}{2}$.

(b) We see that the graph is concave downward (so $f''(x) < 0$) to the left of the point of inflection. That is, $f''(x) < 0$ when $x < \frac{1}{2}$. Similarly, $f''(x) > 0$ when $x > \frac{1}{2}$.

(c) $f(x) = \frac{1}{6}(2x^3 - 3x^2 - 12x + 12)$
$f'(x) = \frac{1}{6}(6x^2 - 6x - 12) = x^2 - x - 2$
$f''(x) = 2x - 1$
Thus $f''(x) = 0$ when $x = \frac{1}{2}$.

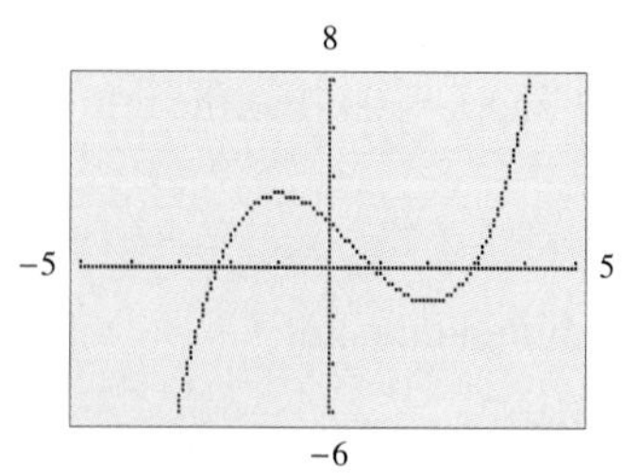

Figure 10.23

Figure 10.24 shows the graph of $f''(x) = 2x - 1$. We see that the graph crosses the x-axis ($f''(x) = 0$) when $x = \frac{1}{2}$, is below the x-axis ($f''(x) < 0$) when $x < \frac{1}{2}$, and is above the x-axis ($f''(x) > 0$) when $x > \frac{1}{2}$. This verifies our conclusions from (a) and (b).

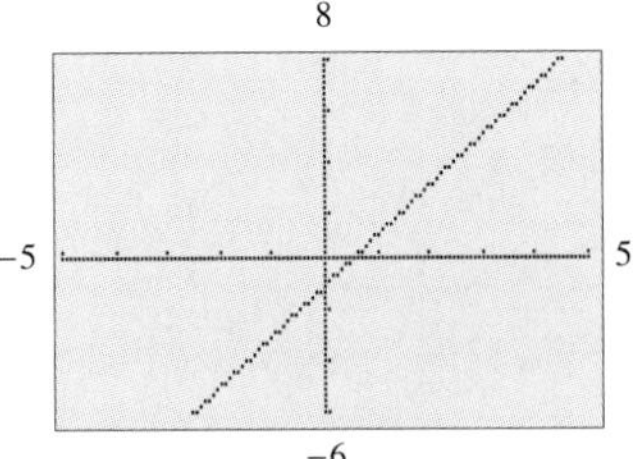

Figure 10.24

EXAMPLE 6

Figure 10.25 shows the graph of $f'(x) = -x^2 - 2x$. Use the graph of $f'(x)$ to do the following.

(a) Find intervals where $f(x)$ is concave downward and where it is concave upward.

(b) Find x-values where $f(x)$ has a point of inflection.

(c) Check the conclusions from (a) and (b) by finding $f''(x)$ and graphing it.

(d) For $f(x) = \frac{1}{3}(9 - x^3 - 3x^2)$, calculate $f'(x)$ to verify that this could be $f(x)$.

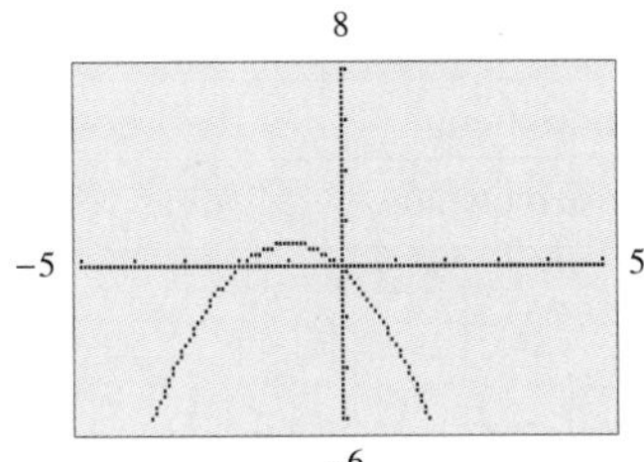

Figure 10.25

Solution

(a) Concavity for $f(x)$ can be found from the sign of $f''(x)$. Because $f''(x)$ is the first derivative of $f'(x)$, wherever the graph of $f'(x)$ is increasing, it follows that $f''(x) > 0$. Thus $f''(x) > 0$ and $f(x)$ is concave upward when $x < -1$. Similarly, $f''(x) < 0$, and $f(x)$ is concave downward when $f'(x)$ is decreasing—that is, when $x > -1$.

(b) From (a) we know that $f''(x)$ changes sign at $x = -1$, so $f(x)$ has a point of inflection at $x = -1$. Note that $f'(x)$ has its maximum at the x-value where $f(x)$ has a point of inflection. In fact, points of inflection for $f(x)$ will correspond to relative extrema for $f'(x)$.

(c) For $f'(x) = -x^2 - 2x$, we have $f''(x) = -2x - 2$. Figure 10.26 shows the graph of $y = f''(x)$ and verifies our conclusions from (a) and (b).

(d) If $f(x) = \frac{1}{3}(9 - x^3 - 3x^2)$, then $f'(x) = \frac{1}{3}(-3x^2 - 6x) = -x^2 - 2x$. Figure 10.27 shows the graph of $f(x) = \frac{1}{3}(9 - x^3 - 3x^2)$. Note that the point of inflection and the concavity correspond to what we discovered in (a) and (b).

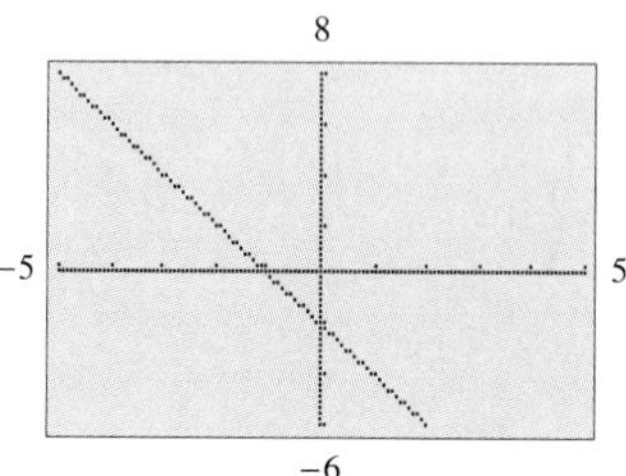

Figure 10.26

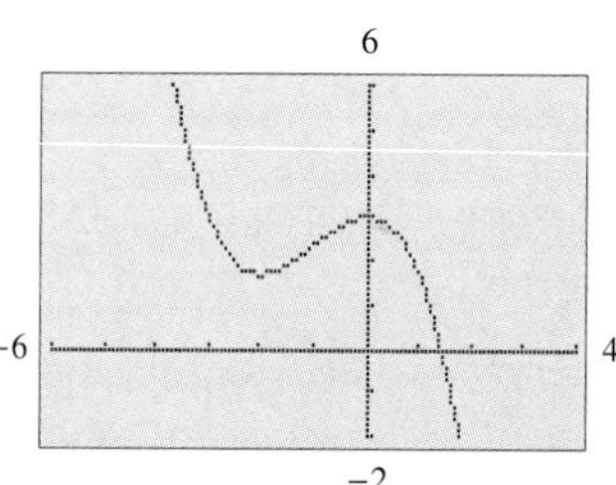

Figure 10.27

The relationship among $f(x)$, $f'(x)$, and $f''(x)$ that we explored in Example 6 can be summarized as follows.

$f(x)$	Concave Upward	Concave Downward	Point of Inflection	
$f'(x)$	increasing	decreasing	maximum	minimum
$f''(x)$	positive (+)	negative (−)	(+) to (−)	(−) to (+)

CHECKPOINT SOLUTIONS

1. up
2. Possible points of inflection occur where $f''(x) = 0$ or $f''(x)$ is undefined.
3. Points of inflection at A, B, and C
   $f''(x) < 0$ to the left of A and between B and C
   $f''(x) > 0$ between A and B and to the right of C

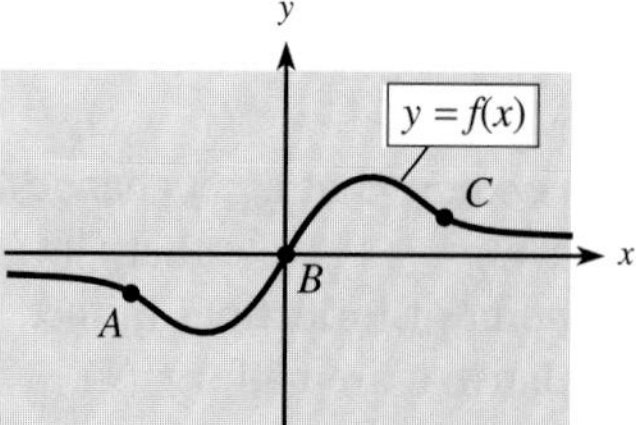

4. (a) False. For example, if $f(x) = x^4$, then $f'(x) = 4x^3$ and $f''(x) = 12x^2$. Note that $f''(x) = 12x^2$ does not change sign from $x < 0$ to $x > 0$.

EXERCISE 10.2

In Problems 1–4, determine whether each function is concave upward or concave downward at the indicated points.

1. $f(x) = x^3 - 3x^2 + 1$ at (a) $x = -2$ (b) $x = 3$
2. $f(x) = x^3 + 6x - 4$ at (a) $x = -5$ (b) $x = 7$
3. $y = 2x^3 + 4x - 8$ at (a) $x = -1$ (b) $x = 4$
4. $y = 4x^3 - 3x^2 + 2$ at (a) $x = 0$ (b) $x = 5$

In Problems 5–10, use the indicated x-values on the graph of $y = f(x)$ to find the following.

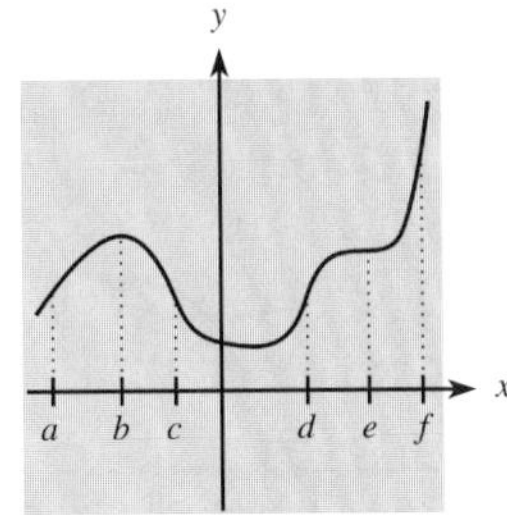

5. Find intervals over which the graph is concave down.
6. Find intervals over which the graph is concave up.
7. Find intervals where $f''(x) > 0$.
8. Find intervals where $f''(x) < 0$.
9. Find the x-coordinates of three points of inflection.
10. Find the x-coordinate of a horizontal point of inflection.

In Problems 11–14, a function and its graph are given. Use the second derivative to determine intervals where the function is concave upward, to determine intervals where it is concave downward, and to locate points of inflection. Check these results against the graph shown.

11. $f(x) = x^3 - 6x^2 + 5x + 6$

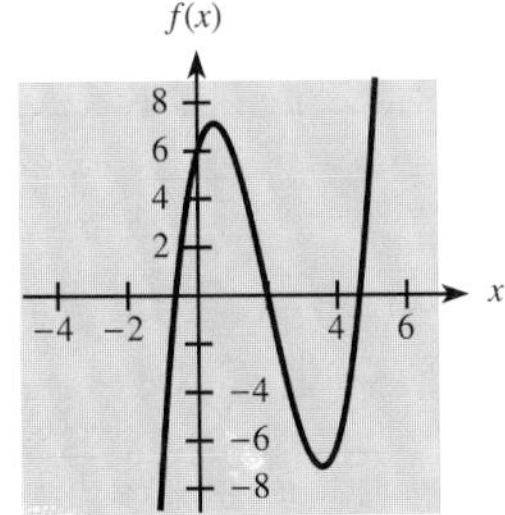

12. $y = x^3 - 9x^2$

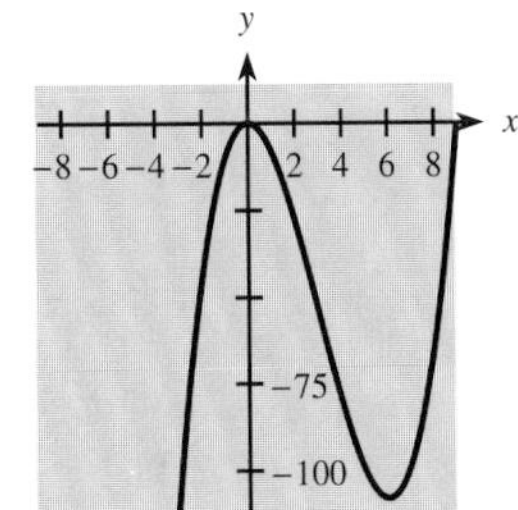

13. $f(x) = \frac{1}{4}x^4 + \frac{1}{2}x^3 - 3x^2 + 3$

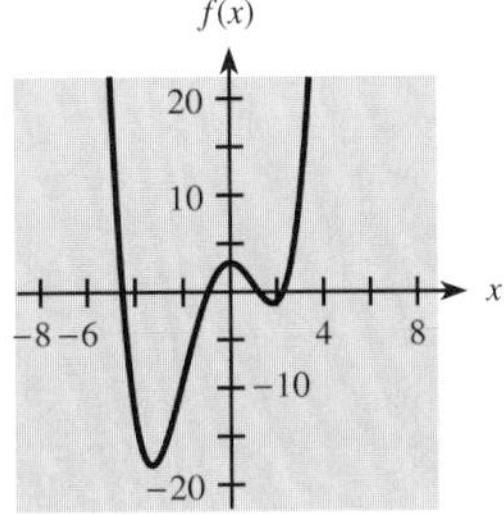

14. $y = 2x^4 - 6x^2 + 4$

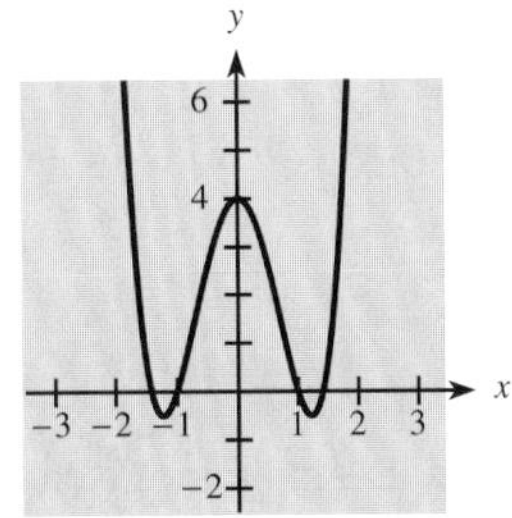

Find the relative maxima, relative minima, and points of inflection, and sketch the graph of the functions in Problems 15–20.

15. $y = x^2 - 4x + 2$
16. $y = x^3 - x^2$
17. $y = \frac{1}{3}x^3 - 2x^2 + 3x + 2$
18. $y = x^3 - 3x^2 + 6$
19. $y = x^4 - 16x^2$
20. $y = x^4 - 8x^3 + 16x^2$

In Problems 21–24, a function and its first and second derivatives are given. Use these to find critical values, relative maxima, relative minima, and points of inflection; sketch the graph of each function.

21. $f(x) = 3x^5 - 20x^3$
 $f'(x) = 15x^2(x - 2)(x + 2)$
 $f''(x) = 60x(x^2 - 2)$
22. $f(x) = x^5 - 5x^4$
 $f'(x) = 5x^3(x - 4)$
 $f''(x) = 20x^2(x - 3)$
23. $y = x^{1/3}(x - 4)$
 $y' = \dfrac{4(x-1)}{3x^{2/3}}$
 $y'' = \dfrac{4(x+2)}{9x^{5/3}}$
24. $y = x^{4/3}(x - 7)$
 $y' = \dfrac{7x^{1/3}(x-4)}{3}$
 $y'' = \dfrac{28(x-1)}{9x^{2/3}}$

In Problems 25 and 26, a function and its graph are given.

(a) From the graph, estimate where $f''(x) > 0$, where $f''(x) < 0$, and where $f''(x) = 0$.

(b) Use (a) to decide where $f'(x)$ has its relative maxima and relative minima.

(c) Verify your results in (a) and (b) by finding $f'(x)$ and $f''(x)$ and then graphing each with a graphing utility.

25. $f(x) = -\frac{1}{3}x^3 + x^2 + 8x - 12$

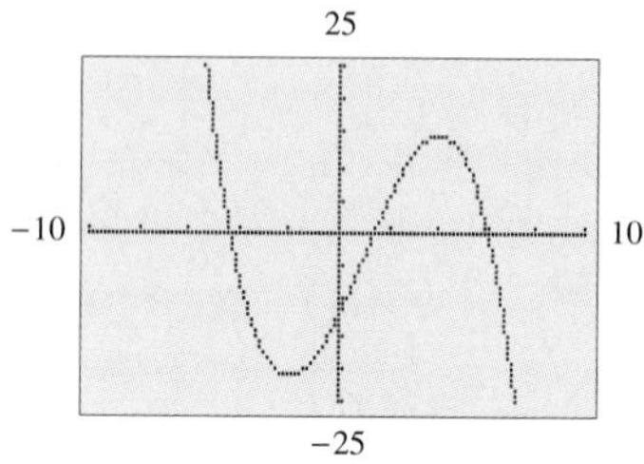

26. $f(x) = \frac{1}{3}x^3 + 2x^2 - 12x - 20$

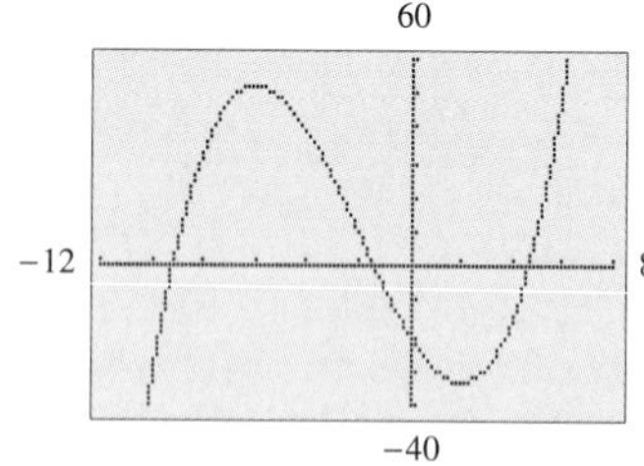

In Problems 27 and 28, $f'(x)$ and its graph are given. Use the graph of $f'(x)$ to determine the following.

(a) Where is the graph of $f(x)$ concave upward and where is it concave downward?
(b) Where does $f(x)$ have any points of inflection?
(c) Find $f''(x)$ and graph it. Then use that graph to check your conclusions from (a) and (b).
(d) Sketch a possible graph for $f(x)$.

27. $f'(x) = 4x - x^2$

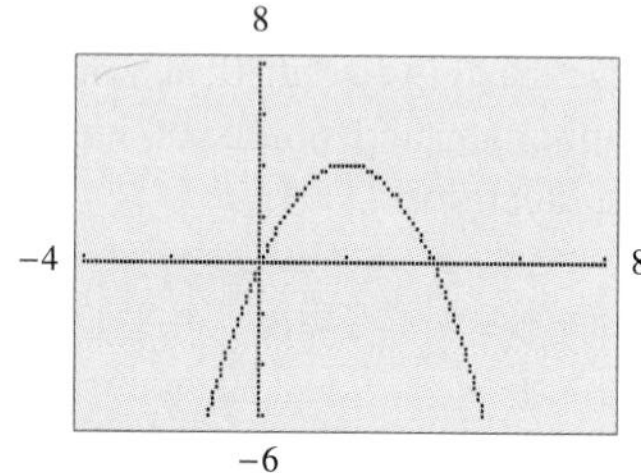

28. $f'(x) = x^2 - x - 2$

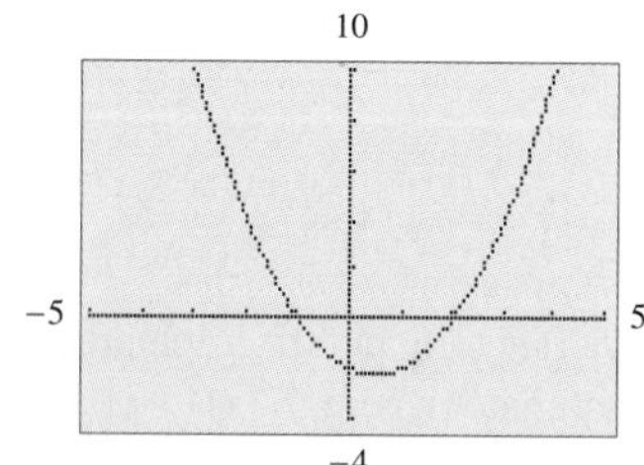

In Problems 29 and 30, use the graph shown in Figure 10.28 and identify points from A through I that satisfy the given conditions.

29. (a) $f'(x) > 0$ and $f''(x) > 0$
(b) $f'(x) < 0$ and $f''(x) < 0$
(c) $f'(x) = 0$ and $f''(x) > 0$
(d) $f'(x) > 0$ and $f''(x) = 0$
(e) $f'(x) = 0$ and $f''(x) = 0$

30. (a) $f'(x) > 0$ and $f''(x) < 0$
(b) $f'(x) < 0$ and $f''(x) > 0$
(c) $f'(x) = 0$ and $f''(x) < 0$
(d) $f'(x) < 0$ and $f''(x) = 0$

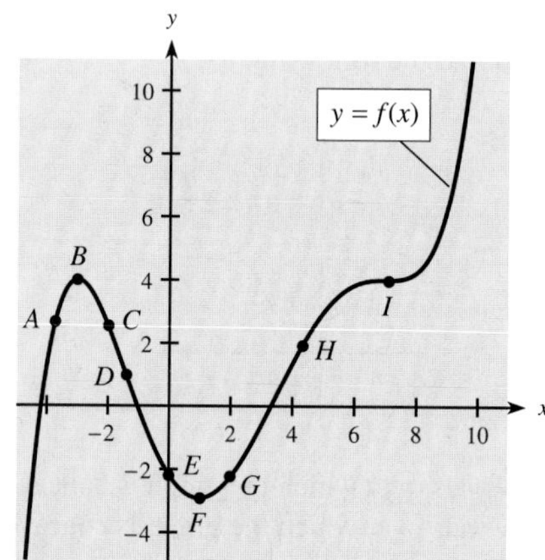

Figure 10.28

In Problems 31 and 32, a graph is given. Tell where $f(x)$ is concave upward, where it is concave downward, and where it has points of inflection on the interval $-2 < x < 2$, if the given graph is the graph of

(a) $f(x)$ (b) $f'(x)$ (c) $f''(x)$

31.

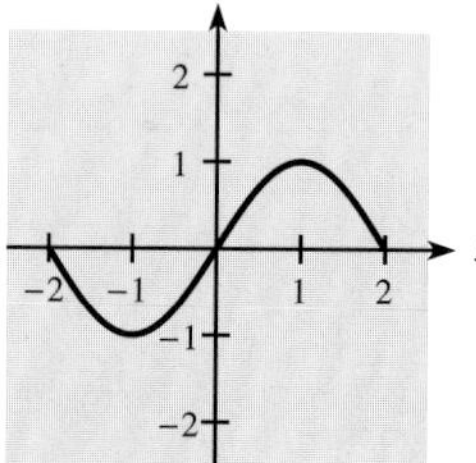

32.

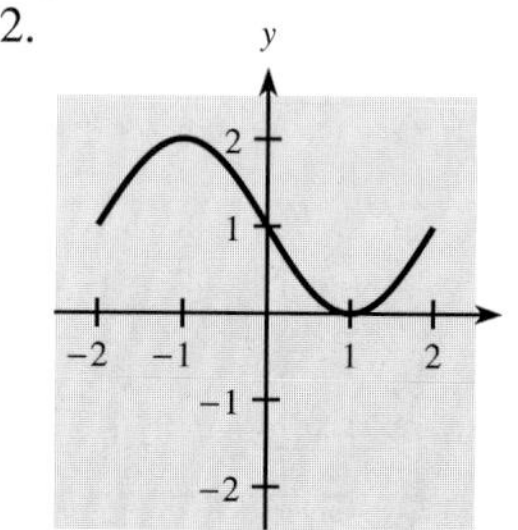

Applications

33. ***Productivity—diminishing returns*** The figure below is a typical graph of worker productivity as a function of time on the job.

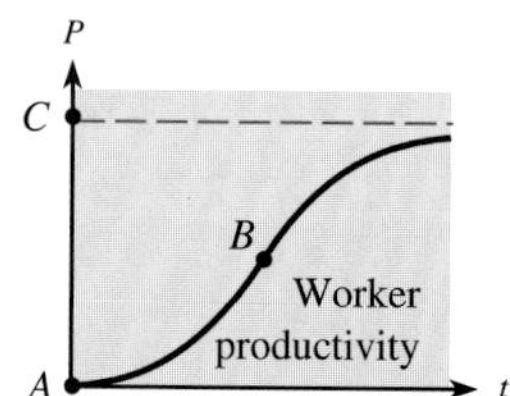

(a) If P represents the productivity and t represents the time, write a symbol that represents the rate of change of productivity with respect to time.
(b) Which of A, B, and C is the critical point for the rate of change found in (a)? This point actually corresponds to the point at which the rate of production is maximized, or the point for maximum worker efficiency. In economics, this is called the point of diminishing returns.
(c) Which of A, B, and C corresponds to the upper limit of production?

34. ***Population growth*** The figure below shows the growth of a population as a function of time.

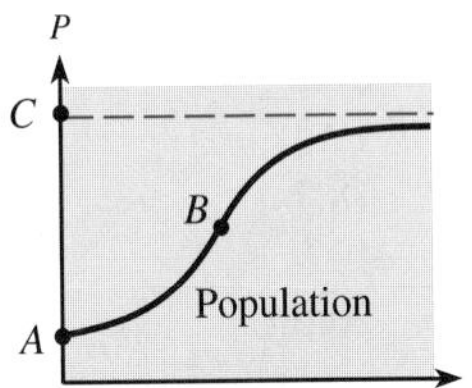

(a) If P represents the population and t represents the time, write a symbol that represents the rate of change (growth rate) of the population with respect to time.
(b) Which of A, B, and C corresponds to the point at which the growth *rate* attains its maximum?
(c) Which of A, B, and C corresponds to the upper limit of population?

35. ***Advertising and sales*** The figure below shows the daily sales volume S as a function of time t since an ad campaign began.

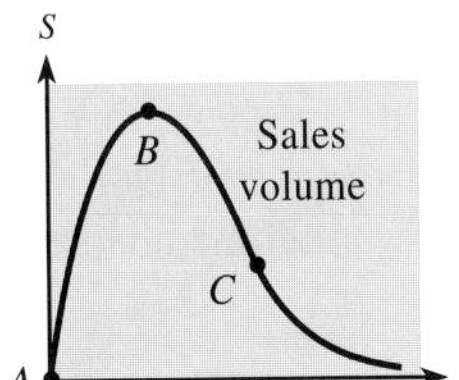

(a) Which of A, B, and C is the point of inflection for the graph?
(b) On which side of C is $d^2S/dt^2 > 0$?
(c) Does the *rate of change* of sales volume attain its minimum at C?

36. ***Oxygen purity*** The figure below shows the oxygen level P (for purity) in a lake t months after an oil spill.

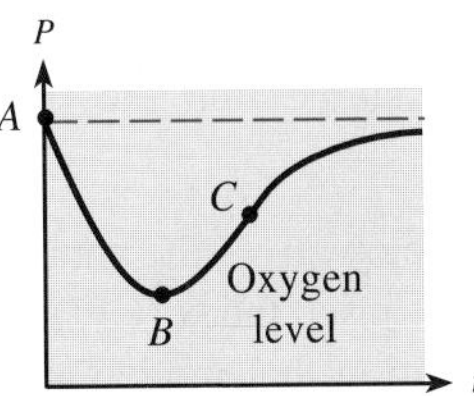

(a) Which of A, B, and C is the point of inflection for the graph?
(b) On which side of C is $d^2P/dt^2 < 0$?
(c) Does the *rate of change* of purity attain its maximum at C?

37. ***Production*** Suppose that the total number of units produced by a worker in t hours of an 8-hour shift can be modeled by the production function $P(t)$:

$$P(t) = 27t + 12t^2 - t^3$$

(a) Find the number of hours before production is maximized.
(b) Find the number of hours before the rate of production is maximized. That is, find the point of diminishing returns.

38. ***Poiseuille's law—velocity of blood*** According to Poiseuille's law, the speed S of blood through an artery of radius r at a distance x from the artery wall is given by

$$S = k[r^2 - (r - x)^2]$$

where k is a constant. Find the distance x that maximizes the speed.

39. ***Advertising and sales—diminishing returns*** Suppose that a company's daily sales volume attributed to an advertising campaign is given by

$$S(t) = \frac{3}{t+3} - \frac{18}{(t+3)^2} + 1$$

(a) Find how long it will be before sales volume is maximized.
(b) Find how long it will be before the rate of change of sales volume is minimized. That is, find the point of diminishing returns.

40. ***Oxygen purity—diminishing returns*** Suppose that the oxygen level P (for purity) in a body of water t months after an oil spill is given by

$$P(t) = 500\left[1 - \frac{4}{t+4} + \frac{16}{(t+4)^2}\right]$$

(a) Find how long it will be before the oxygen level reaches its minimum.

(b) Find how long it will be before the rate of change of P is maximized. That is, find the point of diminishing returns.

41. ***Inflation rate*** The table below gives the annual percent change in the consumer price index (CPI) for the years 1980–1996. The equation that models the data is

$$y = 0.002138x^4 - 0.08399x^3 + 1.12630x^2 - 5.9860x + 14.1484$$

where x is number of years past 1980. Use the model to find the year between 1985 and 1996 in which it indicates that the increase is a maximum.

Year	*Annual Percent Change in CPI*	Year	*Annual Percent Change in CPI*
1980	13.5	1989	4.8
1981	10.3	1990	5.4
1982	6.2	1991	4.2
1983	3.2	1992	3.0
1984	4.3	1993	3.0
1985	3.6	1994	2.6
1986	1.9	1995	2.8
1987	3.6	1996	2.9
1988	4.1		

Source: Bureau of Labor Statistics, U.S. Dept. of Labor

42. ***Personal savings*** The following table gives the personal savings (in billions of dollars) in the United States for selected years.

Year	*Personal Savings (billions)*	Year	*Personal Savings (billions)*
1980	\$153.8	1988	155.7
1985	189.3	1989	152.1
1986	187.5	1990	175.6
1987	142.0	1991	199.6

Source: *Survey of Current Business*, March 1993

Assume that personal savings is modeled by

$$PS(t) = 0.55t^3 - 25.6t^2 + 384t - 1678$$

where t is the number of years past 1970.

(a) Find a function that models the instantaneous rate of change of personal savings.

(b) Use the model from (a) to find when the rate of personal savings is decreasing, when it is increasing, and when it is a maximum.

(c) Find the point of inflection of the function $PS(t)$, and state to what in (b) it corresponds.

43. ***Murder rates*** The table below gives the murder rates per 1000 people from 1970 to 1994.

(a) Using x as the number of years past 1970, write the cubic function that models these data.

(b) Use this model to find the years during this period when the murders were at a minimum and when they were at a maximum.

(c) Does the model yield the same years for the minimum and maximum as the data?

Year	*Number of Murders per 1000 People*
1979	21.5
1980	23.0
1981	22.5
1982	21.0
1983	19.3
1984	18.7
1985	19.0
1986	20.6
1987	20.1
1988	20.7
1989	21.5
1990	23.4
1991	24.7
1992	23.8
1993	24.5
1994	23.3

Source: U.S. FBI

44. ***Union membership*** The data in the table can be modeled with the function

$$u(x) = 0.0003869x^3 - 0.08917x^2 + 6.2503x - 105.9$$

where $x = 0$ in 1900 and $u(x)$ is the percentage of U.S. workers who belonged to unions. During what year does the model indicate union membership was maximized?

Year	*Percent of Workers in Unions*	Year	*Percent of Workers in Unions*
1930	11.6	1989	16.4
1940	26.9	1990	16.1
1950	31.5	1991	16.1
1960	31.4	1992	15.8
1970	27.3	1993	15.8
1975	25.5	1994	15.5
1980	21.9	1995	14.9
1985	18.0	1996	14.5

Source: Bureau of Labor Statistics, U.S. Dept. of Labor

45. ***Gross national product*** The following table gives the U.S. gross national product (GNP) for the years 1913–1922.
 (a) Using $x = 0$ in 1912, find the function that models the data.
 (b) During what year in this time interval does the model predict a minimum GNP?
 (c) During what year in this time interval does the model predict a maximum GNP?

Year	*Gross National Product (GNP)*
1913	39.6
1914	38.6
1915	40.0
1916	48.3
1917	60.4
1918	76.4
1919	84.0
1920	91.5
1921	69.6
1922	74.1

Source: *National Debt in Perspective,* Oscar Falconi, Wholesale Nutrition (Internet)

10.3 Optimization in Business and Economics

OBJECTIVES

- *To find absolute maxima and minima*
- *To maximize revenue, given the total revenue function*
- *To minimize the average cost, given the total cost function*
- *To find the maximum profit from total cost and total revenue functions, or from a profit function*

APPLICATION PREVIEW

Most companies are interested in obtaining the greatest possible profit (instead of just making a profit that is relatively large). Similarly, manufacturers of products are concerned about producing their products for the lowest possible average cost per unit. Therefore, rather than just finding the relative maxima or relative minima of a function, we will consider where the **absolute maximum** or **absolute minimum** of a function occurs in a given interval. This requires evaluating the function at the endpoints of the given interval as well as finding the relative extrema.

As their name implies, **absolute extrema** are the functional values that are the largest or smallest values over the entire domain of the function (or over the interval of interest).

Absolute Extrema

The value $f(a)$ is the **absolute maximum** for f if $f(a) \geq f(x)$ for all x in the domain of f (or over the interval of interest).

The value $f(b)$ is the **absolute minimum** for f if $f(b) \leq f(x)$ for all x in the domain of f (or over the interval of interest).

In this section we will discuss how to find the absolute extrema of a function and then use these techniques to solve applications involving revenue, cost, and profit.

Let us begin by considering the graph of $y = (x - 1)^2$, shown in Figure 10.29(a). This graph has a relative minimum at (1, 0). Note that the relative minimum is the lowest point on the graph. In this case, the point (1, 0) is an **absolute minimum point,** and 0 is the absolute minimum for the function. Similarly, when there is a point that is the highest point on the graph over the domain of the function, we call the point an **absolute maximum point** for the function.

In Figure 10.29(a), we see that there is no relative maximum. However, if the domain of the function is restricted to the interval $\left[\frac{1}{2}, 2\right]$, then we get the graph shown in Figure 10.29(b). In this case, there is an absolute maximum of 1 at the point (2, 1), and the absolute minimum of 0 is still at (1, 0).

If the domain of $y = (x - 1)^2$ is restricted to the interval [2, 3], the resulting graph is that shown in Figure 10.29(c). In this case, the absolute minimum is 1 and occurs at the point (2, 1), and its absolute maximum is 4 and occurs at (3, 4).

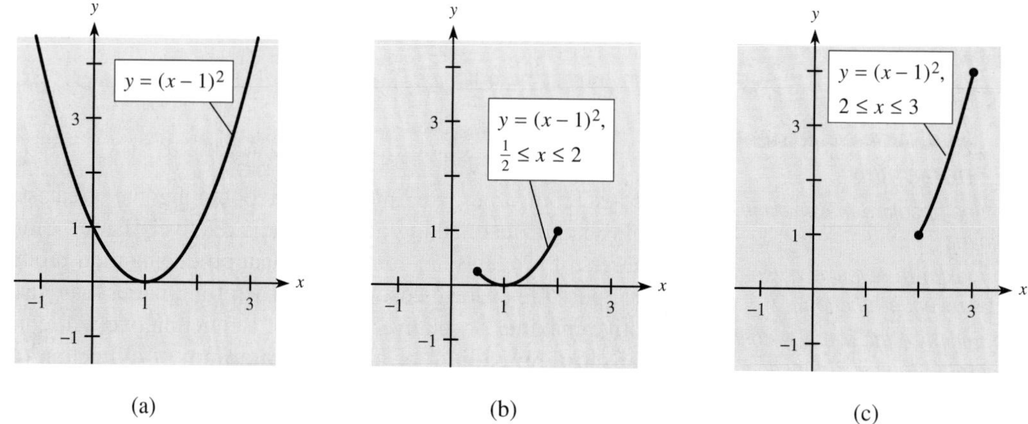

Figure 10.29 (a) (b) (c)

As the preceding discussion indicates, if the domain of a function is limited, an absolute maximum or minimum may occur at the endpoints of the domain. In testing functions with limited domains for absolute maxima and minima, we must compare the endpoints of the domain with the relative maxima and minima found by taking derivatives. In applications to the management, life, and social sciences, a limited domain occurs very often, because many quantities are required to be positive, or at least nonnegative.

Maximizing Revenue

Because the marginal revenue is the first derivative of the total revenue, it should be obvious that the total revenue function will have a critical point at the point where the marginal revenue equals 0. With the total revenue function $R(x) = 16x - 0.02x^2$, the point where $R'(x) = 0$ is clearly a maximum because $R(x)$ is a parabola that opens downward. But the revenue function may not always be a parabola, and the critical point may not always be a maximum, so it is wise to verify that the maximum value occurs at the critical point.

EXAMPLE 1

If total revenue for a firm is given by

$$R(x) = 8000x - 40x^2 - x^3$$

where x is the number of units sold, find the number of units that must be sold to maximize revenue. Find the maximum revenue.

Solution

$R'(x) = 8000 - 80x - 3x^2$, so we must solve $8000 - 80x - 3x^2 = 0$ for x.

$$(40 - x)(200 + 3x) = 0$$

$$40 - x = 0 \quad 200 + 3x = 0$$

so

$$x = 40 \quad \text{or} \quad x = -\frac{200}{3}$$

Now we reject the negative value for x, but we must verify that $x = 40$ will yield maximum revenue.

$$\left.\begin{aligned} R'(0) &= 8000 > 0 \\ R'(100) &= 8000 - 8000 - 30{,}000 < 0 \end{aligned}\right\} \Rightarrow \text{relative maximum}$$

This test shows that a relative maximum occurs at $x = 40$, giving revenue $R(40) = \$192{,}000$. Because $R'(x) < 0$ for all $x > 40$, the revenue function is decreasing to the right of $x = 40$. The revenue at $x = 0$ is $R(0) = 0$, so $R = \$192{,}000$ at $x = 40$ is the (absolute) maximum revenue.

EXAMPLE 2

A travel agency will plan tours for groups of 25 or larger. If the group contains exactly 25 people, the cost is \$300 per person. However, each person's cost is reduced by \$10 for each additional person above the 25. What size group will produce the largest revenue for the agency?

Solution

The total revenue is

$$R = (\text{number of people})(\text{cost per person})$$

If 25 people go, the total revenue will be

$$R = 25 \cdot \$300 = \$7500$$

But if x additional people go, the number of people will be $25 + x$, and the cost per person will be $(300 - 10x)$ dollars. Then the total revenue will be a function of x,

$$R = R(x) = (25 + x)(300 - 10x)$$

or

$$R(x) = 7500 + 50x - 10x^2$$

This function will have its maximum where $\overline{MR} = R'(x) = 0$; $R'(x) = 50 - 20x$, and the solution to $0 = 50 - 20x$ is $x = 2.5$. Thus adding 2.5 people to the group should maximize the total revenue. But we cannot add half a person, so we will test the total revenue function for 27 people and 28 people. This will determine the most profitable number because $R(x)$ is concave downward for all x.

For $x = 2$ (giving 27 people) we get $R(2) = 7500 + 50(2) - 10(2)^2 = 7560$. For $x = 3$ (giving 28 people) we get $R(3) = 7500 + 50(3) - 10(3)^2 = 7560$. Note that both 27 and 28 people give the same total revenue and that this revenue is greater than the revenue for 25 people. Thus the revenue is maximized at either 27 or 28 people in the group.

Minimizing Average Cost

Because the total cost function is always increasing, we cannot find the number of units that will make the total cost a minimum (except for producing 0 units, which is an absolute minimum). However, we usually can find the number of units that will make the average cost per unit a minimum.

Average Cost If the total cost is represented by $C = C(x)$, then the **average cost per unit** is

$$\overline{C} = \frac{C(x)}{x}$$

For example, if $C = 3x^2 + 4x + 2$ is the total cost function for a commodity, the **average cost function** is

$$\overline{C} = \frac{3x^2 + 4x + 2}{x} = 3x + 4 + \frac{2}{x}$$

Note that the average cost per unit is undefined if no units are produced.

We can use derivatives to find the minimum of the average cost function, as the following example shows.

EXAMPLE 3

If the total cost function for a commodity is given by $C = \frac{1}{4}x^2 + 4x + 100$, where x represents the number of units produced, producing how many units will result in a minimum *average cost* per unit? Find the minimum average cost.

Solution

The average cost function is given by

$$\overline{C} = \frac{\frac{1}{4}x^2 + 4x + 100}{x} = \frac{1}{4}x + 4 + \frac{100}{x}$$

Then

$$\overline{C}' = \overline{C}'(x) = \frac{1}{4} - \frac{100}{x^2}$$

Setting $\overline{C}' = 0$ gives

$$0 = \frac{1}{4} - \frac{100}{x^2}$$

$$0 = x^2 - 400, \quad \text{or} \quad x = \pm 20$$

Because the quantity produced must be positive, 20 units should minimize the average cost per unit. We show it is an absolute minimum by using the second derivative.

$$\overline{C}''(x) = \frac{200}{x^3} \quad \text{so} \quad \overline{C}''(x) > 0 \quad \text{when } x > 0$$

Thus the minimum average cost per unit occurs if 20 units are produced. The graph of the average cost per unit is shown in Figure 10.30. The minimum average cost per unit is $\overline{C}(20) = \$14$.

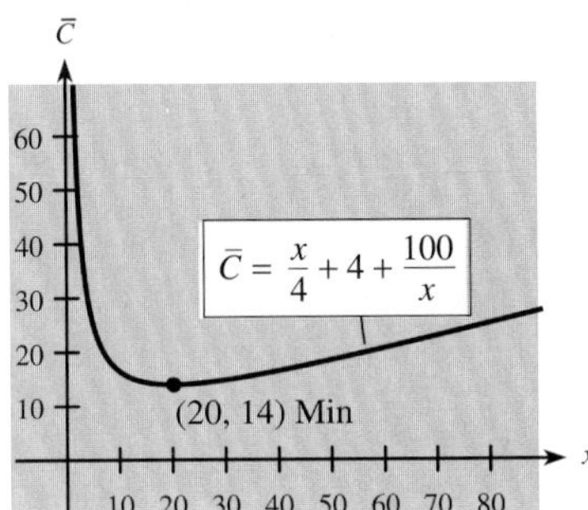

Figure 10.30

Maximizing Profit

In the previous chapter, we defined the marginal profit function as the derivative of the profit function. That is,

$$\overline{MP} = P'(x)$$

In this chapter we have seen how to use the derivative to find maxima and minima for various functions. Now we can apply those same techniques in the context of **profit maximization.** We can use marginal profit to maximize profit functions.

If there is a physical limitation on the number of units that can be produced in a given period of time, then the endpoints of the interval caused by these limitations should also be checked.

EXAMPLE 4

Suppose that the production capacity for a certain commodity cannot exceed 30. If the total profit function for this commodity is

$$P(x) = 4x^3 - 210x^2 + 3600x - 200$$

where x is the number of units sold, find the number of items that will maximize profit.

Solution

The restrictions on capacity mean that $P(x)$ is restricted by $0 \le x \le 30$. The marginal profit function is

$$P'(x) = 12x^2 - 420x + 3600$$

Setting $P'(x)$ equal to 0, we get

$$0 = 12(x - 15)(x - 20)$$

so $P'(x) = 0$ at $x = 15$ *and* $x = 20$. Testing to the right and left of these values (as in the first-derivative test), we get

$$\left.\begin{array}{l} P'(0) = 3600 > 0 \\ P'(18) = -72 < 0 \\ P'(25) = 600 > 0 \end{array}\right\} \begin{array}{l} \Rightarrow \text{relative maximum at } x = 15 \\ \Rightarrow \text{relative minimum at } x = 20 \end{array}$$

Thus, at (15, 20050) the total profit function has a *relative* maximum but we must check the endpoints (0 and 30) before deciding whether it is the absolute maximum.

$$P(0) = -200 \text{ and } \quad P(30) = 26{,}800$$

Thus the absolute maximum profit is \$26,800, and it occurs at the endpoint, $x = 30$. Figure 10.31 shows the graph of the profit function.

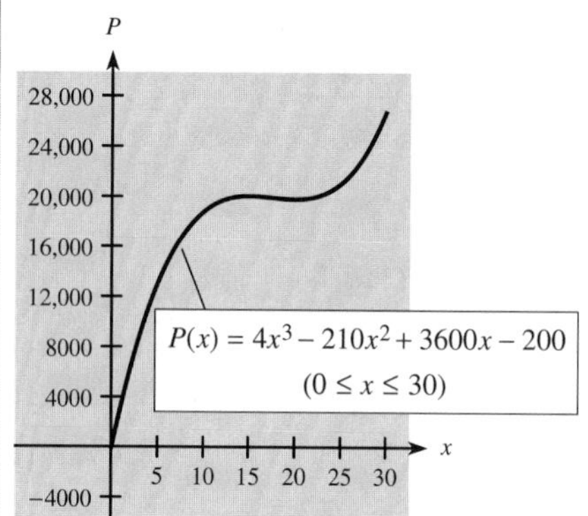

Figure 10.31

In a **monopolistic market,** the seller who has a monopoly can control the price by regulating the supply of the product. The seller controls the supply, so he or she can force the price higher by limiting supply.

If the demand function for the product is $p = f(x)$, total revenue for the sale of x units is $R(x) = px = f(x) \cdot x$. Note that the price p is fixed by the market in a competitive market but varies with output for the monopolist.

If $\overline{C} = \overline{C}(x)$ represents the average cost per unit sold, then the total cost for the x units sold is $C = \overline{C} \cdot x = \overline{C}x$. Because we have both total cost and total revenue as a function of the quantity, x, we can maximize the profit function, $P(x) = px - \overline{C}x$, where p represents the demand function $p = f(x)$ and $\overline{C}$ represents the average cost function $\overline{C} = \overline{C}(x)$.

EXAMPLE 5

The daily demand function for a product is

$$p = 168 - 0.2x$$

A monopolist finds that the average cost is

$$\overline{C} = 120 + x$$

(a) How many units must be sold to maximize profit?
(b) What is the selling price at this "optimal" level of production?
(c) What is the maximum possible profit?

Solution

(a) The total revenue function for the product is

$$R(x) = px = (168 - 0.2x)x = 168x - 0.2x^2$$

and the total cost function is

$$C(x) = \overline{C} \cdot x = (120 + x)x = 120x + x^2$$

Thus the profit function is

$$P(x) = R(x) - C(x) = 168x - 0.2x^2 - (120x + x^2)$$

or

$$P(x) = 48x - 1.2x^2$$

Then $P'(x) = 48 - 2.4x$, so $P'(x) = 0$ when $x = 20$. We see that $P''(20) = -2.4$, so by the second-derivative test, $P(x)$ has a maximum at $x = 20$. That is, selling 20 units will maximize profit.

(b) The selling price is determined by $p = 168 - 0.2x$, so the price that will result from supplying 20 units per day is $p = 168 - 0.2(20) = 164$. That is, the "optimal" selling price is \$164 per unit.

(c) The profit at $x = 20$ is $P(20) = 48(20) - 1.2(20)^2 = 960 - 480 = 480$. Thus the maximum possible profit is \$480 per day.

In a **competitive market,** each firm is so small that its actions in the market cannot affect the price of the product. The price of the product is determined in the market by the intersection of the market demand curve (from all consumers) and market supply curve (from all firms that supply this product). The firm can sell as little or as much as it desires at the market equilibrium price.

Therefore, a firm in a competitive market has a total revenue function given by $R(x) = px$, where p is the market equilibrium price for the product and x is the quantity sold.

EXAMPLE 6

A firm in a competitive market must sell its product for \$200 per unit. The average cost per unit (per month) is $\overline{C} = 80 + x$, where x is the number of units sold per month. How many units should be sold to maximize profit?

Solution

If the average cost per unit is $\overline{C} = 80 + x$, then the total cost of x units is given by $C(x) = (80 + x)x = 80x + x^2$. The revenue per unit is \$200, so the total revenue is given by $R(x) = 200x$. Thus the profit function is

$$P(x) = R(x) - C(x) = 200x - (80x + x^2), \quad \text{or} \quad P(x) = 120x - x^2$$

Then $P'(x) = 120 - 2x$. Setting $P'(x) = 0$ and solving for x gives $x = 60$. Because $P''(60) = -2$, the profit is maximized when the firm sells 60 units per month.

CHECKPOINT

1. True or false: If $R(x)$ is the revenue function, we find all possible points where $R(x)$ could be maximized by solving $\overline{MR} = 0$ for x.
2. If $C(x) = \dfrac{x^2}{20} + 10x + 2500$, form $\overline{C}(x)$, the average cost function.
3. (a) If $p = 5000 - x$ gives the demand function in a monopoly market, find $R(x)$, if it is possible with this information.
   (b) If $p = 5000 - x$ gives the demand function in a competitive market, find $R(x)$, if it is possible with this information.

Graphing Utilities

As we have seen, graphing utilities can be used to locate maximum values. In addition, if it is difficult to determine critical values algebraically, we may be able to find them graphically.

EXAMPLE 7

Suppose total revenue and total costs for a company are given by

$$R(x) = 3000 - \frac{3000}{x + 1}$$
$$C(x) = 500 + 12x + x^2$$

where x is thousands of units and revenue and costs are in thousands of dollars. Graph $P(x)$ and $P'(x)$ to determine the number of units that yields maximum profit and the amount of the maximum profit.

Solution

$$\begin{aligned} P(x) &= R(x) - C(x) \\ &= 3000 - \frac{3000}{x+1} - (500 + 12x + x^2) \\ &= 2500 - \frac{3000}{x+1} - 12x - x^2 \\ P'(x) &= \frac{3000}{(x+1)^2} - 12 - 2x \end{aligned}$$

Finding the critical values by solving $P'(x) = 0$ is very difficult in this case (try it). That is why we are using a graphing approach. Figure 10.32(a) shows the graph of $P(x)$ and Figure 10.32(b) shows the graph of $P'(x)$. These figures indicate that the maximum profit occurs near $x = 10$.

By adjusting the range for $P'(x)$, we obtain the graph in Figure 10.33. This shows that $P'(x) = 0$ when $x = 9$ (or when 9000 units are sold). The maximum profit is

$$\begin{aligned} P(9) &= 2500 - 300 - 108 - 81 \\ &= 2011 \text{ (thousands of dollars)} \end{aligned}$$

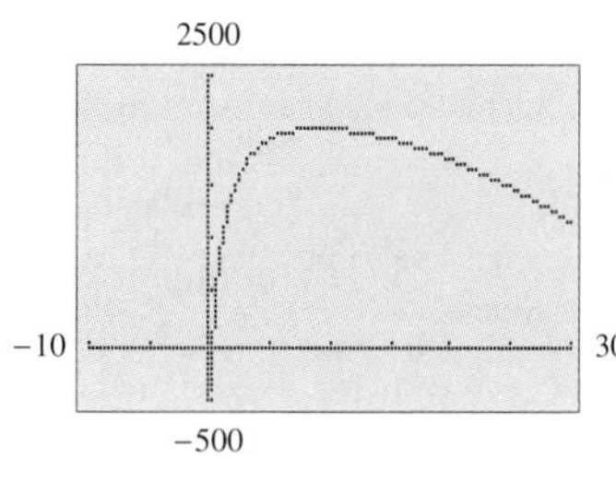

(a)

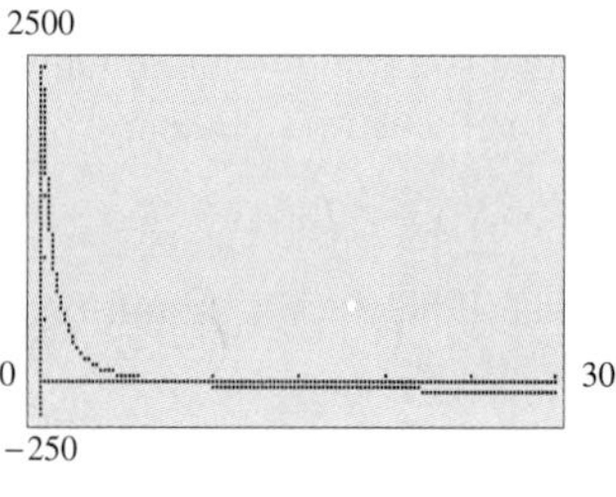

(b)

Figure 10.32

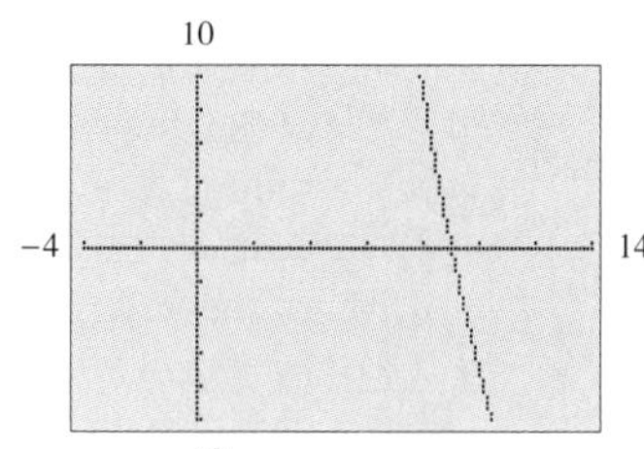

Figure 10.33

CHECKPOINT SOLUTIONS

1. False. $\overline{MR} = R'(x)$, but there may also be critical points where $R'(x)$ is undefined, or $R(x)$ may be maximized at endpoints of a restricted domain.
2. $\overline{C}(x) = \dfrac{C(x)}{x} = \dfrac{x^2/20 + 10x + 2500}{x} = \dfrac{x}{20} + 10 + \dfrac{2500}{x}$
3. (a) $R(x) = p \cdot x = (5000 - x)x = 5000x - x^2$
   (b) In a competitive market, $R(x) = p \cdot x$, where p is the constant equilibrium price. Thus we need to know the supply function and find the equilibrium price before we can form $R(x)$.

EXERCISE 10.3

In Problems 1–4, find the absolute maxima and minima for $f(x)$ on the interval $[a, b]$.

1. $f(x) = x^3 - 2x^2 - 4x + 2, [-1, 3]$
2. $f(x) = x^3 - 3x + 3, [-3, 1.5]$
3. $f(x) = x^3 + x^2 - x + 1, [-2, 0]$
4. $f(x) = x^3 - x^2 - x, [-0.5, 2]$

Maximizing Revenue

5. (a) If the total revenue function for a radio is $R = 36x - 0.01x^2$, then sale of how many units will maximize the total revenue? Find the maximum revenue.
   (b) Find the maximum revenue if production is limited to at most 1500 radios.

6. (a) If the total revenue function for a blender is $R(x) = 25x - 0.05x^2$, sale of how many units will provide the maximum total revenue? Find the maximum revenue.
   (b) Find the maximum revenue if production is limited to at most 200 blenders.
7. If the total revenue function for a computer is $R(x) = 2000x - 20x^2 - x^3$, find the level of sales that maximizes revenue and find the maximum revenue.
8. A firm has total revenues given by

$$R(x) = 2800x - 8x^2 - x^3$$

for a product. Find the maximum revenue from sales of that product.

9. An agency charges \$10 per person for a trip to a concert if 30 people travel in a group. But for each person above the 30, the charge will be reduced by \$.20. How many people will maximize the total revenue for the agency if the trip is limited to at most 50 people?
10. A company handles an apartment building with 50 units. Experience has shown that if the rent for each of the units is \$360 per month, all the units will be filled, but 1 unit will become vacant for each \$10 increase in the monthly rate. What rent should be charged to maximize the total revenue from the building if the upper limit on the rent is \$450 per month?
11. A cable TV company has 1000 customers paying \$20 each month. If each \$1 reduction in price attracts 100 new customers, find the price that yields maximum revenue. Find the maximum revenue.
12. If club members charge \$5 admission to a classic car show, 1000 people will attend, and for each \$1 increase in price, 100 fewer people will attend. What price will give the maximum revenue for the show? Find the maximum revenue.
13. The function $\overline{R}(x) = R(x)/x$ defines the average revenue for selling x units. For

$$R(x) = 2000x + 20x^2 - x^3$$

   (a) find the maximum average revenue.
   (b) show that $\overline{R}(x)$ attains its maximum at an x-value where $\overline{R}(x) = \overline{MR}$.
14. For the revenue function given by

$$R(x) = 2800x + 8x^2 - x^3$$

   (a) find the maximum average revenue.
   (b) show that $\overline{R}(x)$ attains its maximum at an x-value where $\overline{R}(x) = \overline{MR}$.

Minimizing Average Cost

15. If the total cost function for a lamp is $C(x) = 25 + 13x + x^2$, producing how many units will result in a minimum average cost per unit? Find the minimum average cost.
16. If the total cost function for a product is $C(x) = 300 + 10x + 0.03x^2$, producing how many units will result in a minimum average cost per unit? Find the minimum average cost.
17. If the total cost function for a product is $C(x) = 100 + x^2$, producing how many units will result in a minimum average cost per unit? Find the minimum average cost.
18. If the total cost function for a product is $C(x) = 250 + 6x + 0.1x^2$, producing how many units will minimize the average cost? Find the minimum average cost.
19. If the total cost function for a product is $C(x) = (x + 4)^3$, where x represents the number of hundreds of units produced, producing how many units will minimize average cost? Find the minimum average cost.
20. If the total cost function for a product is $C(x) = (x + 5)^3$, where x represents the number of hundreds of units produced, producing how many units will minimize average cost? Find the minimum average cost.
21. For the cost function $C(x) = 25 + 13x + x^2$, show that average costs are minimized at the x-value where

$$\overline{C}(x) = \overline{MC}$$

22. For the cost function $C(x) = 300 + 10x + 0.03x^2$, show that average costs are minimized at the x-value where

$$\overline{C}(x) = \overline{MC}$$

Maximizing Profit

23. If the profit function for a product is $P(x) = 5600x + 85x^2 - x^3 - 200{,}000$, selling how many items will produce a maximum profit? Find the maximum profit.
24. If the profit function for a commodity is $P = 6400x - 18x^2 - \frac{1}{3}x^3 - 40{,}000$, selling how many units will result in a maximum profit? Find the maximum profit.

25. A manufacturer estimates that x units of its product can be produced at a total cost of $C(x) = 45{,}000 + 100x + x^3$. If the manufacturer's total revenue from the sale of x units is $R(x) = 4600x$, determine the level of production x that will maximize the profit. Find the maximum profit.
26. A product can be produced at a total cost $C(x) = 800 + 100x^2 + x^3$, where x is the number produced. If the total revenue is given by $R(x) = 60{,}000x - 50x^2$, determine the level of production that will maximize the profit. Find the maximum profit.
27. A firm can produce only 1000 units per month. The monthly total cost is given by $C(x) = 300 + 200x$, where x is the number produced. If the total revenue is given by $R(x) = 250x - \frac{1}{100}x^2$, how many items should the firm produce for maximum profit? Find the maximum profit.
28. A firm can produce 100 units per week. If its total cost function is $C = 500 + 1500x$ and its total revenue function is $R = 1600x - x^2$, how many units should it produce to maximize its profit? Find the maximum profit.
29. A firm has monthly average costs given by

$$\overline{C} = \frac{45{,}000}{x} + 100 + x$$

where x is the number of units produced per month. The firm can sell its product in a competitive market for \$1600 per unit. If production is limited to 600 units per month, find the number of units that gives maximum profit, and find the maximum profit.
30. A small business has weekly average costs of

$$\overline{C} = \frac{100}{x} + 30 + \frac{x}{10}$$

where x is the number of units produced each week. The competitive market price for this business's product is \$46 per unit. If production is limited to 150 units per week, find the level of production that yields maximum profit, and find the maximum profit.
31. The weekly demand function for a product sold by only one firm is $p = 600 - \frac{1}{2}x$, and the average cost of production and sale is $\overline{C} = 300 + 2x$.
    (a) Find the quantity that will maximize profit.
    (b) Find the selling price at this optimal quantity.
    (c) What is the maximum profit?
32. The monthly demand function for a product sold by a monopoly is $p = 8000 - x$, and its average cost is $\overline{C} = 4000 + 5x$.
    (a) Determine the quantity that will maximize profit.
    (b) Determine the selling price at the optimal quantity.
    (c) Determine the maximum profit.
33. The monthly demand function for a product sold by a monopoly is $p = 1960 - \frac{1}{3}x^2$, and the average cost is $\overline{C} = 1000 + 2x + x^2$. Production is limited to 1000 units and x is in hundreds of units.
    (a) Find the quantity that will give maximum profit.
    (b) Find the maximum profit.
34. The monthly demand function for a product sold by a monopoly is $p = 5900 - \frac{1}{2}x^2$, and its average cost is $\overline{C} = 3020 + 2x$. If production is limited to 100 units, find the number of units that maximizes profit. Will the maximum profit result in a profit or loss?
35. An industry with a monopoly on a product has its average weekly costs given by

$$\overline{C} = \frac{10{,}000}{x} + 60 - 0.03x + 0.00001x^2$$

The weekly demand for the product is given by $p = 120 - 0.015x$. Find the price the industry should set and the number of units it should produce to obtain maximum profit. Find the maximum profit.
36. A large corporation with monopolistic control in the marketplace has its average daily costs given by

$$\overline{C} = \frac{800}{x} + 100x + x^2$$

The daily demand for its product is given by $p = 60{,}000 - 50x$. Find the quantity that gives maximum profit, and find the maximum profit. What selling price should the corporation set for its product?
37. A company handles an apartment building with 50 units. Experience has shown that if the rent for each of the units is \$360 per month, all of the units will be filled, but 1 unit will become vacant for each \$10 increase in this monthly rate. If the monthly cost of maintaining the apartment building is \$6 per rented unit, what rent should be charged per month to maximize the profit?
38. A travel agency will plan a tour for groups of size 25 or larger. If the group contains exactly 25 people, the cost is \$500 per person. However, each person's cost is reduced by \$10 for each additional person above the 25. If the travel agency incurs a cost of \$125 per person for the tour, what size group will give the agency the maximum profit?

Miscellaneous Applications

39. ***Sales revenue*** The data in the table give sales revenues for Scott Paper Company* for various years.

Year	Sales Revenue (billions)	Year	Sales Revenue (billions)
1983	$2.6155	1989	4.8949
1984	2.7474	1990	5.1686
1985	2.934	1991	4.9593
1986	3.3131	1992	5.0913
1987	3.9769	1993	4.7489
1988	4.5494		

Source: Scott Paper Company, 1993 *Annual Report*

Assume that sales revenues for Scott Paper can be modeled by

$$R(t) = -0.031t^2 + 0.776t + 0.179$$

where t is the number of years past 1980.

(a) Use the model to find when maximum revenue occurs and what maximum revenue it predicts.

(b) Check your result in (a) against the data in the table.

40. ***Revenue*** The following table gives the total revenues of AT&T[†] for selected years.

Year	Total Revenues (billions)	Year	Total Revenues (billions)
1985	$63.13	1990	$62.191
1986	$69.906	1991	$63.089
1987	$60.53	1992	$64.904
1989	$61.1	1993	$67.156

Source: *AT&T Annual Report,* 1993

Suppose the data can be modeled by the equation

$$R(t) = 0.253t^2 - 4.03t + 76.84$$

where t is the number of years past 1980.

(a) Use the model to find the year in which revenue was minimum and find the minimum predicted revenue.

(b) Check your result in (a) against the data in the table.

41. ***U.S. trade deficit*** The figure shows the U.S. trade deficit, in billions of dollars, from late 1990 to late 1993.

(a) Approximately when did the trade deficit reach its absolute minimum, and what was the minimum?

(b) Approximately when did the 12-month moving average reach its absolute minimum, and what was the minimum?

(c) Approximately when did the trade deficit reach its absolute maximum, and what was the maximum?

(d) Approximately when did the 12-month moving average reach its absolute maximum, and what was the maximum?

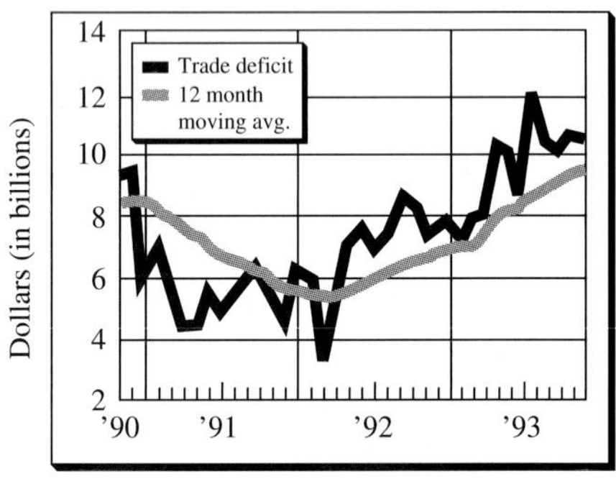

Source: *Wall Street Journal,* December 17, 1993

42. ***Dow Jones average*** The figures show the Dow Jones average for early 1994 (January through March) as well as the average through the day on March 31, 1994.

(a) Approximate the time of day on March 31, 1994, when the Dow Jones average reached its absolute maximum and when it reached its absolute minimum.

(b) Find the day of the year in early 1994 when the Dow Jones average reached its absolute maximum, and find its absolute maximum.

(c) Find the approximate day of the year in early 1994 when the Dow Jones average reached its absolute minimum, and find its absolute minimum.

Dow Jones Closing Averages—Early 1994

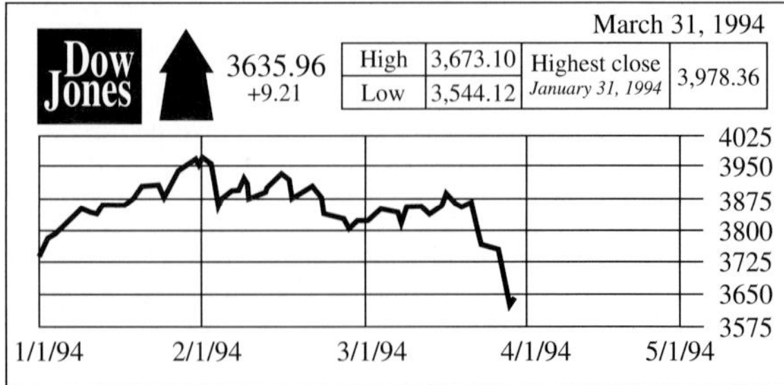

Source: *Oil City Derrick,* April 1, 1994

*Before Scott Paper merged with Kimberly Clark

[†]Before AT&T split off Lucent and NCR

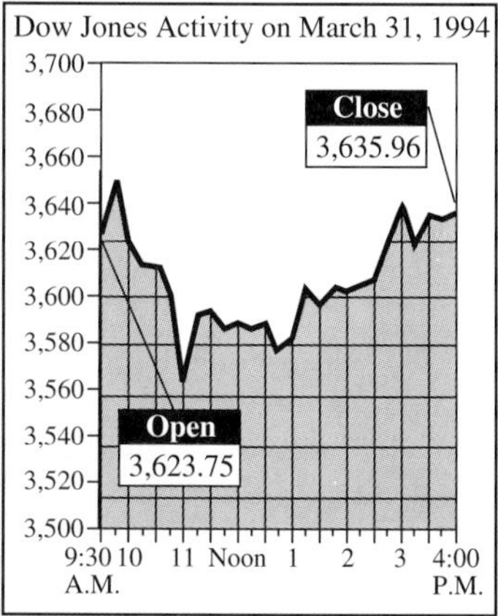

From *Oil City Derrick,* April 1, 1994
Source: Telerate Systems, Inc.

43. ***Crime concern*** The graph below shows that the percentage p of people in the United States who believe crime is the most serious domestic problem is a function of time t (in calendar years) and, of course, of the events that occur during those years. Denote this function by $p = f(t)$, and use the graph to answer the following.

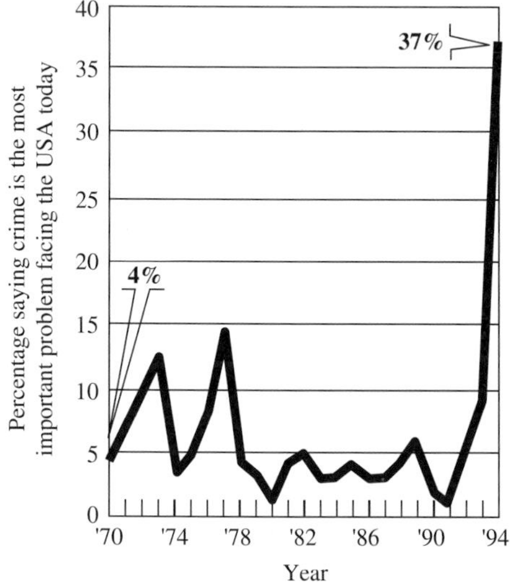

Source: *USA Today*, January 25, 1994

(a) When did $f(t)$ achieve its absolute minimum?
(b) What is the absolute maximum point for $f(t)$?
(c) Is it possible to have $f(t) < 0$? Explain.
(d) What effect do you think a foreign conflict (such as a war) would have on the graph?
(e) Do you think a candidate for public office should have based a campaign extensively on the "law and order" issue in 1980? in 1994? Explain.

44. ***Social Security support*** The graph below shows the number of workers, W, still in the workforce per Social Security beneficiary (historically and projected into the future) as a function of time t, in calendar years. Denote this function by $W = f(t)$, and use the graph to answer the following.

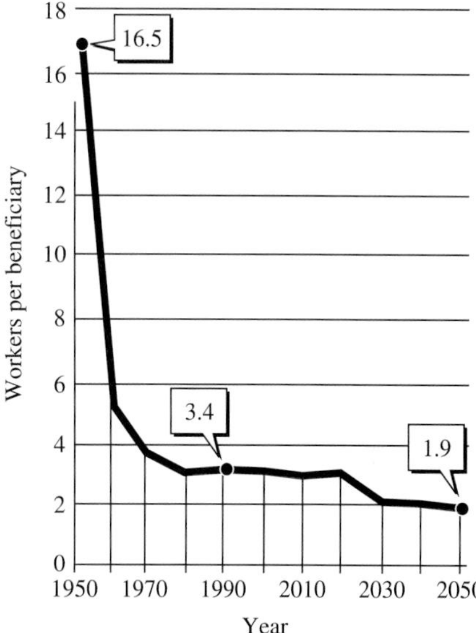

Source: Social Security Administration

(a) What is the absolute maximum for $f(t)$?
(b) What is the absolute minimum for $f(t)$?
(c) Does this graph suggest that Social Security taxes might rise or might fall in the early 21st century? Explain.

10.4 Applications of Maxima and Minima

OBJECTIVE

- *To apply the procedures for finding maxima and minima to solve problems from the management, life, and social sciences*

APPLICATION PREVIEW

Manufacturers make production runs to restock their inventories. Because costs are associated with both the production of items and their storage (placement into inventory), a typical question in these **inventory cost models** is "How many items should be produced in each production run to minimize the total costs of production and storage?" This question is a typical example of the kinds of questions and important business applications that require the use of the derivative for finding maxima and minima. As managers, workers, or consumers, we may be interested in such things as maximum revenue, maximum profit, minimum cost, maximum medical dosage, maximum utilization of resources, and so on.

If we have functions that model cost, revenue, or population growth, we can apply the methods of this chapter to find the maxima and minima of the functions.

EXAMPLE 1

Suppose that a new company begins production in 1995 with eight employees and the growth of the company over the next 10 years is predicted by

$$N = N(t) = 8\left(1 + \frac{160t}{t^2 + 16}\right), \qquad 0 \le t \le 10$$

where N is the number of employees t years after 1995.

(a) In what year will the number of employees be maximized?
(b) What will be the maximum number of employees?

Solution

This function will have a relative maximum when $N'(t) = 0$.

$$\begin{aligned} N'(t) &= 8\left[\frac{(t^2 + 16)(160) - (160t)(2t)}{(t^2 + 16)^2}\right] \\ &= 8\left[\frac{160t^2 + 2560 - 320t^2}{(t^2 + 16)^2}\right] \\ &= 8\left[\frac{2560 - 160t^2}{(t^2 + 16)^2}\right] \end{aligned}$$

Because $N'(t) = 0$ when its numerator is 0 (note that the denominator is never 0), we must solve

$$\begin{aligned} 2560 - 160t^2 &= 0 \\ 160(4 + t)(4 - t) &= 0 \end{aligned}$$

so

$$t = -4 \quad \text{or} \quad t = 4$$

We are interested only in positive t-values, so we test $t = 4$.

$$\left.\begin{aligned} N'(0) &= 8\left[\frac{2560}{256}\right] > 0 \\ N'(10) &= 8\left[\frac{-13{,}440}{(116)^2}\right] < 0 \end{aligned}\right\} \Rightarrow \textit{relative maximum}$$

The relative maximum is

$$N(4) = 8\left(1 + \frac{640}{32}\right) = 168$$

At $t = 0$, the number of employees is $N(0) = 8$, and it increases to $N(4) = 168$. After $t = 4$, $N(t)$ decreases to $N(10) = 118$ (approximately), so $N(4) = 168$ is the maximum number of employees. Figure 10.34 verifies these conclusions.

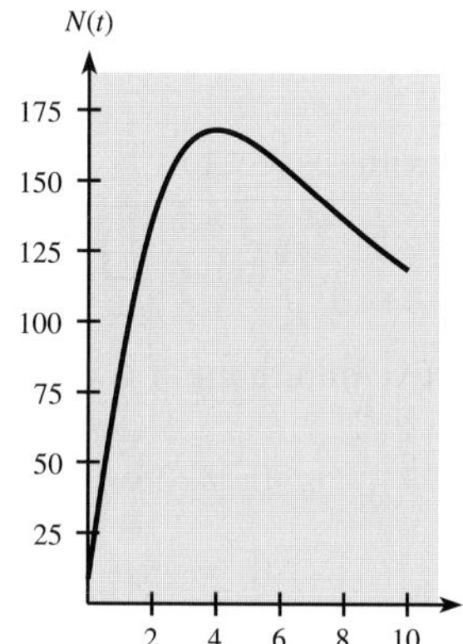

Figure 10.34

Sometimes we must develop the function we need from the statement of the problem. In this case, it is important to understand what is to be maximized or minimized and to express that quantity as a function of *one* variable.

EXAMPLE 2

A farmer needs to enclose a rectangular pasture containing 1,600,000 square feet. Suppose that along the road adjoining his property he wants to use a more expensive fence and that he needs no fence on one side perpendicular to the road because a river bounds his property on that side. If the fence costs \$15 per foot along the road and \$10 per foot along the two remaining sides that must be fenced, what dimensions of his rectangular field will minimize his cost? (See Figure 10.35.)

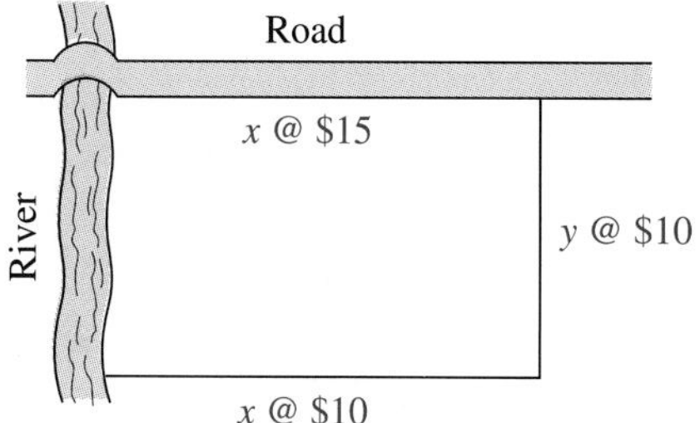

Figure 10.35

Solution

In Figure 10.35, x represents the length of the pasture along the road (and parallel to the road) and y represents the width. The cost function for the fence used is

$$C = 15x + 10y + 10x = 25x + 10y$$

We cannot use a derivative to find where C is minimized unless we write C as a function of x or y only. Because the area of the rectangular field must be 1,600,000 square feet, we have

$$A = xy = 1{,}600{,}000$$

Solving for y in terms of x and substituting give

$$y = \frac{1{,}600{,}000}{x}$$

$$C = 25x + 10\left(\frac{1{,}600{,}000}{x}\right) = 25x + \frac{16{,}000{,}000}{x}$$

The derivative of C with respect to x is

$$C'(x) = 25 - \frac{16{,}000{,}000}{x^2}$$

and we find the relative minimum of C as follows:

$$\begin{aligned} 0 &= 25 - \frac{16{,}000{,}000}{x^2} \\ 0 &= 25x^2 - 16{,}000{,}000 \\ 25x^2 &= 16{,}000{,}000 \\ x^2 &= 640{,}000 \\ x &= 800 \text{ feet} \end{aligned}$$

Testing to see if $x = 800$ gives the minimum cost, we find

$$C''(x) = \frac{32{,}000{,}000}{x^3}$$

$C''(x) > 0$ for $x > 0$, so $C(x)$ is concave up for all positive x. Thus $x = 800$ gives the absolute minimum, and $C(800)$ is the minimum cost. The other dimension of the rectangular field is $x = 1{,}600{,}000/800 = 2000$ feet. Figure 10.36 verifies that $C(x)$ reaches its minimum (of 40,000) at $x = 800$.

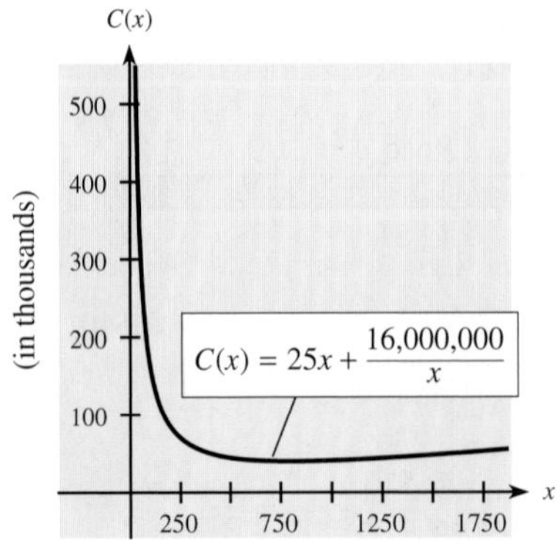

Figure 10.36

EXAMPLE 3

Postal restrictions limit the size of packages sent through the mail. If the restrictions are that the length plus the girth may not exceed 108 in., find the volume of the largest box with square cross section that can be mailed.

Solution

Let l equal the length of the box, and let s equal a side of the square end. See Figure 10.37(a). The volume we seek to maximize is given by

$$V = s^2 l$$

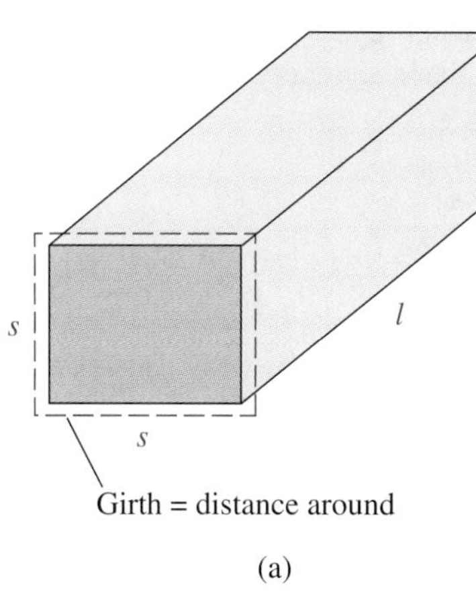

(a)

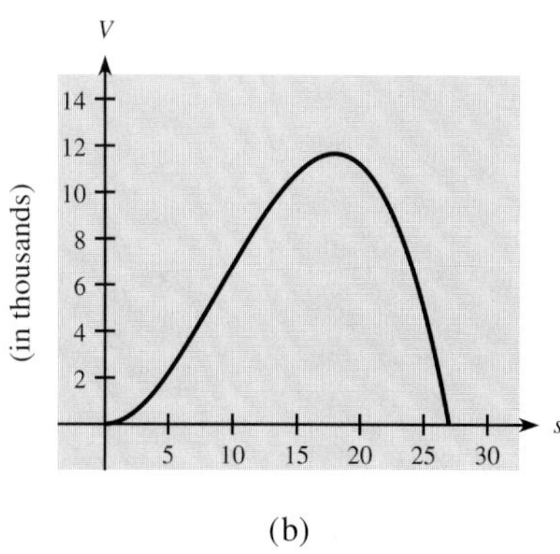

(b)

Figure 10.37

We can use the restriction that girth plus length equals 108,

$$4s + l = 108$$

to express V as a function of s or l. Because $l = 108 - 4s$, the equation for V becomes

$$V = s^2(108 - 4s)$$

or

$$V = 108s^2 - 4s^3$$

Thus we can use dV/ds to find the critical values.

$$\frac{dV}{ds} = 216s - 12s^2$$

$$0 = s(216 - 12s)$$

The critical values are $s = 0$, $s = \frac{216}{12} = 18$. The critical value $s = 0$ will not maximize the volume, for in this case, $V = 0$. Testing to the left and right of $s = 18$ gives

$$V'(17) > 0 \quad \text{and} \quad V'(19) < 0$$

Thus $s = 18$ in. and $l = 108 - 4(18) = 36$ in. yield a maximum volume of 11,664 cubic inches. Once again we can verify our results graphically. Figure 10.37(b) shows that $V = s^2(108 - 4s)$ achieves its maximum when $s = 18$.

CHECKPOINT

Suppose we want to find the minimum value of $C = 5x + 2y$ and we know that x and y must be positive and that $xy = 1000$.

1. What equation do we differentiate to solve this problem?
2. Find the critical values.
3. Find the minimum value of C.

Consider the **inventory cost model** in which x items are produced in each run and items are removed from inventory at a fixed constant rate. Then the number of units in storage changes with time and is illustrated in Figure 10.38. To see how these inventory cost models work, consider the following example.

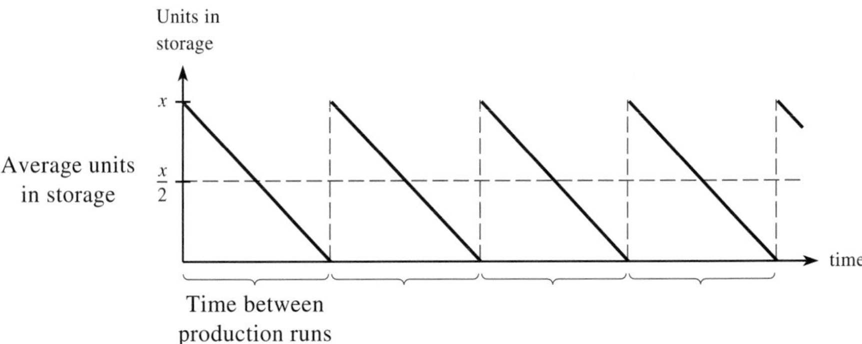

Figure 10.38

EXAMPLE 4

Suppose that a company needs 1,000,000 items during a year and that preparation costs are \$800 for each production run. Suppose further that it costs the company \$6 to produce each item and \$1 to store an item for up to a year. If each production run consists of x items, find x so that the total costs of production and storage are minimized.

Solution

The total production costs are given by

$$\begin{pmatrix}\text{No. of}\\ \text{runs}\end{pmatrix}\begin{pmatrix}\text{cost}\\ \text{per run}\end{pmatrix} + \begin{pmatrix}\text{no. of}\\ \text{items}\end{pmatrix}\begin{pmatrix}\text{cost}\\ \text{per item}\end{pmatrix}$$

$$= \left(\frac{1{,}000{,}000}{x}\right)(\$800) + (1{,}000{,}000)(\$6)$$

The total storage costs are

$$\begin{pmatrix}\text{Average}\\ \text{no. stored}\end{pmatrix}\begin{pmatrix}\text{storage cost}\\ \text{per item}\end{pmatrix} = \left(\frac{x}{2}\right)(\$1)$$

Thus the total costs of production and storage are

$$C = \left(\frac{1{,}000{,}000}{x}\right)(800) + 6{,}000{,}000 + \frac{x}{2}$$

We wish to find x so that C is minimized.

$$C' = \frac{-800{,}000{,}000}{x^2} + \frac{1}{2}$$

If $x > 0$, critical values occur when $C' = 0$.

$$0 = \frac{-800{,}000{,}000}{x^2} + \frac{1}{2}$$

$$\frac{800{,}000{,}000}{x^2} = \frac{1}{2}$$

$$1{,}600{,}000{,}000 = x^2$$

$$x = \pm\ 40{,}000$$

Because x must be positive, we test $x = 40{,}000$ with the second derivative.

$$C''(x) = \frac{1{,}600{,}000{,}000}{x^3}, \quad \text{so} \quad C''(40{,}000) > 0$$

Note that $x = 40{,}000$ yields an absolute minimum value for C, because $C'' > 0$ for all $x > 0$. That is, production runs of 40,000 items yield minimum total costs for production and storage.

Graphing Utilities

Problems of the types we've studied in this section could also be solved (at least approximately) with a graphing utility. With this approach, our first goal is still to express the quantity to be maximized or minimized as a function of one variable. Then that function can be graphed, and the (at least approximate) optimal value can be obtained from the graph.

EXAMPLE 5

A homeowner has to pay for utility line installation to her house from a transformer on the street at the corner of her property. Because of local restrictions, the lines must be underground on her property. Suppose that the costs are \$5/foot along the street and \$10/foot underground. How far from the transformer should she enter the property to minimize installation costs? See Figure 10.39(a).

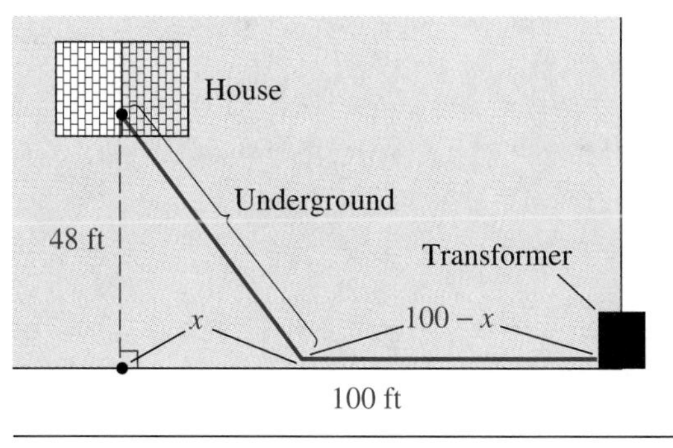

(a)

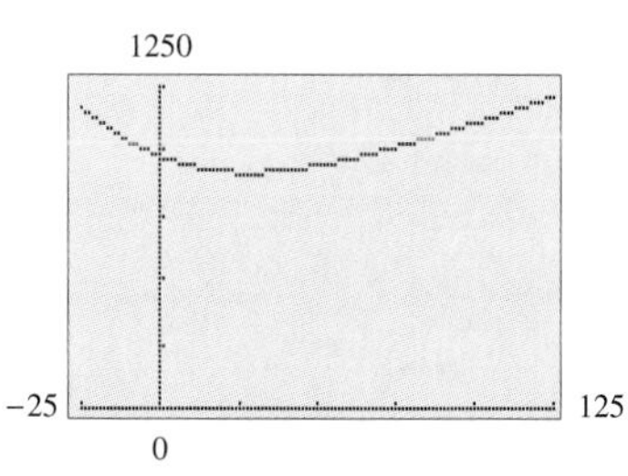

(b)

Figure 10.39

Solution

If the homeowner had the cable placed underground from the house perpendicular to the street and then to the transformer, the cost would be

$$\$10(48) + \$5(100) = \$980$$

It may be possible to save some money by placing the cable on a diagonal to the street. By using the Pythagorean Theorem, we find that the length of the underground cable that meets the street x feet closer to the transformer is

$$\sqrt{48^2 + x^2} = \sqrt{2304 + x^2} \quad \text{feet}$$

Thus the cost C of installation is given by

$$C = 10\sqrt{2304 + x^2} + 5(100 - x) \quad \text{dollars}$$

Figure 10.39(b) shows the graph of this function over the possible interval for x $(0 \le x \le 100)$. Because any extrema must occur in this interval, we can find the minimum by using TRACE or the MIN command on a calculator or computer. The minimum cost is \$915.70, when $x = 28$ feet (that is, when the cable meets the street 72 feet from the transformer). We can use the derivative of C to verify that $x = 28$ gives the minimum cost. By finding the derivative and graphing it, we can see that the only relative extremum occurs at $x = 28$. (Try this.) Because the cost is lower at $x = 28$ than at either endpoint of the interval, the lowest possible cost occurs there.

CHECKPOINT SOLUTIONS

1. We must differentiate C, but first C must be expressed as a function of one variable: $xy = 1000$ means that $y = 1000/x$. If we substitute $1000/x$ for y in $C = 5x + 2y$, we get

$$C(x) = 5x + 2\left(\frac{1000}{x}\right) = 5x + \frac{2000}{x}$$

   Find $C'(x)$ to solve the problem.

2. $C'(x) = 5 - 2000/x^2$, so $C'(x) = 0$ when

$$5 - \frac{2000}{x^2} = 0, \quad \text{or} \quad x = \pm 20$$

   Because x must be positive, the only critical value is $x = 20$.

3. $C''(x) = 4000/x^3$, so $C''(x) > 0$ for all $x > 0$. Thus $x = 20$ yields the minimum value. Also, when $x = 20$, we have $y = 50$, so the minimum value of C is $C = 5(20) + 2(50) = 200$.

EXERCISE 10.4

Applications

1. ***Return to sales*** The manufacturer of GRIPPER tires modeled its return on sales from television advertising expenditures in two regions, as follows:

$$\text{Region 1: } S_1 = 30 + 20x_1 - 0.4x_1^2$$
$$\text{Region 2: } S_2 = 20 + 36x_2 - 1.3x^2$$

where S_1 and S_2 are the sales revenue in millions of dollars and x_1 and x_2 are millions of dollars of expenditures for television advertising.
   (a) What advertising expenditures would maximize sales revenue in each district?
   (b) How much money will be needed to maximize sales revenue in both districts?

2. ***Projectiles*** A ball thrown into the air from a building 100 ft high travels along a path described by

$$y = \frac{-x^2}{110} + x + 100$$

where y is its height and x is the horizontal distance from the building. What is the maximum height the ball will reach?

3. ***Profit*** The profit from a grove of orange trees is given by $x(800 - x)$, where x is the number of orange trees per acre. How many trees per acre will maximize the profit?

4. ***Reaction rates*** The velocity v of an autocatalytic reaction can be represented by the equation

$$v = x(a - x)$$

where a is the amount of material originally present and x is the amount that has been decomposed at any given time. Find the maximum velocity of the reaction.

5. ***Productivity*** Analysis of daily output of a factory during an 8-hour shift shows that the hourly number of units y produced after t hours of production is

$$y = 70t + \tfrac{1}{2}t^2 - t^3, \qquad 0 \le t \le 8$$

   (a) After how many hours will the hourly number of units be maximized?
   (b) What is the maximum hourly output?

6. ***Productivity*** A time study showed that, on average, the productivity of a worker after t hours on the job can be modeled by

$$P = 27t + 6t^2 - t^3, \quad 0 \le t \le 8$$

where P is the number of units produced per hour. After how many hours will productivity be maximized? What is the maximum productivity?

7. ***Consumer expenditure*** Suppose that the demand x for a product is $x = 10{,}000 - 100p$, where p dollars is the market price per unit. Then the consumer expenditure for the product is

$$E = px = 10{,}000p - 100p^2$$

For what market price will expenditure be greatest?

8. ***Production costs*** Suppose that the monthly cost of mining a certain ore is related to the number of pieces of equipment purchased, according to

$$C = 25{,}000x + \frac{870{,}000}{x}, \qquad x > 0$$

where x is the number of pieces of equipment used. Using how many pieces of equipment will minimize the cost?

Medication For Problems 9 and 10, consider that when medicine is administered, reaction (measured in change of blood pressure or temperature) can be modeled by

$$R = m^2\left(\frac{c}{2} - \frac{m}{3}\right)$$

where c is a positive constant and m is the amount of medicine absorbed into the blood (Source: Thrall, R. M., *et al., Some Mathematical Models in Biology,* U.S. Dept. of Commerce, 1967).

9. Find the amount of medicine that is being absorbed into the blood when the reaction is maximum.
10. The rate of change of reaction R with respect to the amount of medicine m is defined to be the sensitivity.
    (a) Find the sensitivity, S.
    (b) Find the amount of medicine that is being absorbed into the blood when the sensitivity is maximum.

11. ***Advertising and sales*** An inferior product with a large advertising budget sells well when it is introduced, but sales fall as people discontinue use of the product. Suppose that the weekly sales S are given by

$$S = \frac{200t}{(t+1)^2}, \qquad t \geq 0$$

where S is in millions of dollars and t is in weeks. After how many weeks will sales be maximized?

12. ***Revenue*** A newly released film has its weekly revenue given by

$$R(t) = \frac{50t}{t^2 + 36}, \qquad t \geq 0$$

where R is in millions of dollars and t is in weeks.
(a) After how many weeks will the weekly revenue be maximized?
(b) What is the maximum weekly revenue?

13. ***News impact*** Suppose that the percentage p (as a decimal) of people who could correctly identify two of eight defendants in a drug case t days after their trial began is given by

$$p(t) = \frac{6.4t}{t^2 + 64} + 0.05$$

Find the number of days before the percentage is maximized, and find the maximum percentage.

14. ***Candidate recognition*** Suppose that in an election year the proportion p of voters who recognize a certain candidate's name t months after the campaign started is given by

$$p(t) = \frac{7.2t}{t^2 + 36} + 0.2$$

After how many months is the proportion maximized?

15. ***Minimum fence*** Two equal rectangular lots are enclosed by fencing the perimeter of a rectangular lot and then putting a fence across its middle. If each lot is to contain 1200 square feet, what is the minimum amount of fence needed to enclose the lots (include the fence across the middle)?

16. ***Minimum fence*** The running yard for a dog kennel must contain at least 900 square feet. If a 20-foot side of the kennel is used as part of one side of a rectangular yard with 900 square feet, what dimensions will require the least amount of fencing?

17. ***Minimum cost*** A rectangular field with one side along a river is to be fenced. Suppose that no fence is needed along the river, the fence on the side opposite the river costs \$20 per foot, and the fence on the other sides costs \$5 per foot. If the field must contain 45,000 square feet, what dimensions will minimize costs?

18. ***Minimum cost*** From a tract of land a developer plans to fence a rectangular region and then divide it into two identical rectangular lots by putting a fence down the middle. Suppose that the fence for the outside boundary costs \$5 per foot and the fence for the middle costs \$2 per foot. If each lot contains 13,500 square feet, find the dimensions of each lot that yield the minimum cost for the fence.

19. ***Optimization at a fixed cost*** A rectangular area is to be enclosed and divided into thirds. The family has \$800 to spend for the fencing material. The outside fence costs \$10 per running foot installed, and the dividers cost \$20 per running foot installed. What are the dimensions that will maximize the area enclosed? (The answer contains a fraction.)

20. ***Minimum cost*** A kennel of 640 square feet is to be constructed as shown. The cost is \$4 per running foot for the sides and \$1 per running foot for the ends and dividers. What are the dimensions of the kennel that will minimize the cost?

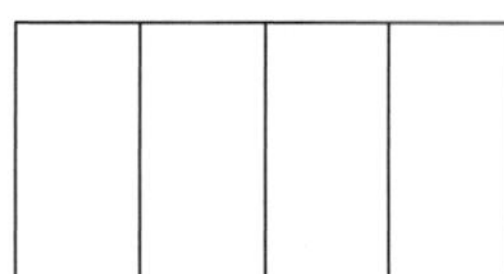

21. ***Minimum cost*** The base of a rectangular box is to be twice as long as it is wide. The volume of the box is 256 cubic inches. The material for the top costs \$0.10 per square inch and the material for the sides and bottom costs \$0.05 per square inch. Find the dimensions that will make the cost a minimum.

22. ***Velocity of air during a cough*** According to B. F. Visser, the velocity v of air in the trachea during a cough is related to the radius r of the trachea according to

$$v = ar^2(r_0 - r)$$

where a is a constant and r_0 is the radius of the trachea in a relaxed state. Find the radius r that produces the maximum velocity of air in the trachea during a cough.

23. ***Inventory cost model*** Suppose that a company needs 1,500,000 items during a year and that preparation for each production run costs \$600. Suppose also that it costs \$15 to produce each item and \$2 per year to store an item. Use the inventory cost model to find the number of items in each production run so that the total costs of production and storage are minimized.

24. ***Inventory cost model*** Suppose that a company needs 60,000 items during a year and that preparation for each production run costs \$400. Suppose further that it costs \$4 to produce each item and \$0.75 to store an item for one year. Use the inventory cost model to find the number of items in each production run that will minimize the total costs of production and storage.

25. ***Inventory cost model*** A company needs 150,000 items per year. It costs the company \$360 to prepare a production run of these items and \$7 to produce each item. If it also costs the company \$0.75 per year for each item stored, find the number of items that should be produced in each run so that total costs of production and storage are minimized.

26. ***Inventory cost model*** A company needs 450,000 items per year. Production costs are \$500 to prepare for a production run and \$10 for each item produced. Inventory costs are \$2 per item per year. Find the number of items that should be produced in each run so that the total costs of production and storage are minimized.

27. ***Volume*** A rectangular box with a square base is to be formed from a square piece of metal with 12-inch sides. If a square piece with side x is cut from the corners of the metal and the sides are folded up to form an open box, the volume of the box is $V = (12 - 2x)^2x$. What value of x will maximize the volume of the box?

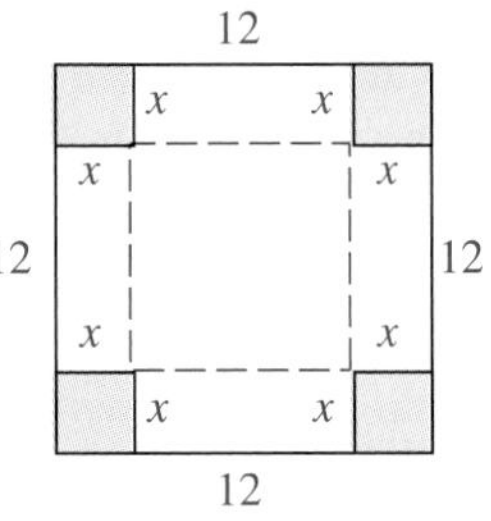

28. ***Volume*** A square piece of cardboard 36 centimeters on a side is to be formed into a rectangular box by cutting squares with length x from each corner and folding up the sides. What is the maximum volume possible for the box?

29. ***Revenue*** The owner of an orange grove must decide when to pick one variety of oranges. She can sell them for \$8 a bushel if she sells them now, with each tree yielding an average of 5 bushels. The yield increases by half a bushel per week for the next 5 weeks, but the price per bushel decreases by \$0.50 per bushel each week. When should the oranges be picked for maximum return?

30. ***Minimum material*** A box with an open top and a square base is to be constructed to contain 4000 cubic inches. Find the dimensions that will require the minimum amount of material to construct the box.

31. ***Minimum cost*** A printer has a contract to print 100,000 posters for a political candidate. He can run the posters by using any number of plates from 1 to 30 on his press. If he uses x metal plates, they will produce x copies of the poster with each impression of the press. The metal plates cost \$2.00 to prepare, and it costs \$12.50 per hour to run the press. If the press can make 1000 impressions per hour, how many metal plates should the printer make to minimize costs?

32. ***Shortest time*** A vacationer on an island 8 miles offshore from a point that is 48 miles from town must travel to town occasionally. (See the figure.) The vacationer has a boat capable of traveling 30 mph and can go by auto along the coast at 55 mph. At what point should the car be left to minimize the time it takes to get to town?

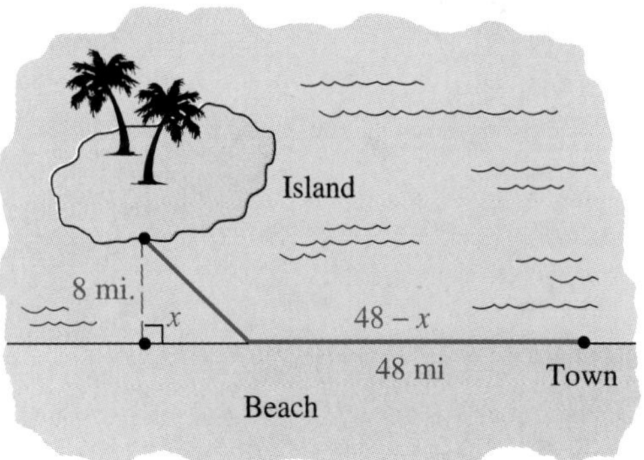

10.5 Asymptotes; More Curve Sketching

OBJECTIVES

- *To locate horizontal asymptotes*
- *To locate vertical asymptotes*
- *To sketch graphs of functions that have vertical and/or horizontal asymptotes*

APPLICATION PREVIEW

If the total daily cost of producing plastic rafts for swimming pools is given by

$$C = 500 + 8x + 0.05x^2$$

where x is the number of units produced per day, then the average cost per unit produced is given by

$$\overline{C} = \frac{500 + 8x + 0.05x^2}{x} = \frac{500}{x} + 8 + 0.05x, \qquad \text{for } x > 0$$

The graph of this function is very useful in understanding how many rafts should be produced to keep the average daily cost under control. The graph of this function contains a vertical asymptote at $x = 0$. We will discuss graphs and applications involving asymptotes in this section.

The procedures for using the first-derivative test and the second-derivative test are given in previous sections, but none of the graphs discussed in those sections contains vertical asymptotes or horizontal asymptotes. In this section, we consider how to use information about asymptotes along with the first and second derivatives, and we present a unified approach to curve sketching.

Asymptotes

In Section 2.4, "Special Functions and Their Graphs," we first discussed asymptotes and saw that they are important features of the graphs that have them. Then, in our discussion of limits, we discovered the relationship between certain limits and asymptotes. Limits are used to define and locate asymptotes precisely.

Because a horizontal asymptote tells us the behavior of the functional values (y-coordinates) when x increases or decreases without bound, we use limits at infinity to determine the existence of horizontal asymptotes.

Horizontal Asymptote

The graph of a rational function $y = f(x)$ will have a **horizontal asymptote** at $y = b$, for a constant b, if

$$\lim_{x \to +\infty} f(x) = b \quad \text{or} \quad \lim_{x \to -\infty} f(x) = b$$

Otherwise, the graph has no horizontal asymptote.

For a rational function f, $\lim_{x\to+\infty} f(x) = b$ if and only if $\lim_{x\to-\infty} f(x) = b$, so we only need to find one of these limits to locate a horizontal asymptote. Just as with horizontal asymptotes, the formal definition of vertical asymptotes uses limits.

Vertical Asymptote

The line $x = x_0$ is a **vertical asymptote** of the graph of $y = f(x)$ if the values of $f(x)$ approach $+\infty$ or $-\infty$ as x approaches x_0 (from the left or the right).

From our work with limits, recall that a vertical asymptote will occur on the graph of a function at an x-value where the function has its denominator (but not its numerator) equal to zero. These observations allow us to determine where vertical asymptotes occur.

Vertical Asymptote of a Rational Function

The graph of the rational function

$$h(x) = \frac{f(x)}{g(x)}$$

has a vertical asymptote at $x = c$ if $g(c) = 0$ and $f(c) \neq 0$.

EXAMPLE 1

Find any vertical and horizontal asymptotes for

(a) $f(x) = \dfrac{2x-1}{x+2}$ (b) $f(x) = \dfrac{x^2+3}{1-x}$

Solution

(a) The denominator of this function is 0 at $x = -2$, and because this value does not make the numerator 0, there is a vertical asymptote at $x = -2$.

Because the function is rational, we can find horizontal asymptotes by evaluating

$$\lim_{x\to+\infty} \frac{2x-1}{x+2} \quad \text{or} \quad \lim_{x\to-\infty} \frac{2x-1}{x+2}$$

We will evaluate both.

$$\lim_{x\to+\infty} \frac{2x-1}{x+2} = \lim_{x\to+\infty} \frac{2-1/x}{1+2/x} = \frac{2-0}{1+0} = 2$$

$$\lim_{x\to-\infty} \frac{2x-1}{x+2} = \lim_{x\to-\infty} \frac{2-1/x}{1+2/x} = \frac{2-0}{1+0} = 2$$

Thus there is a horizontal asymptote at $y = 2$. The graph is shown in Figure 10.40(a).

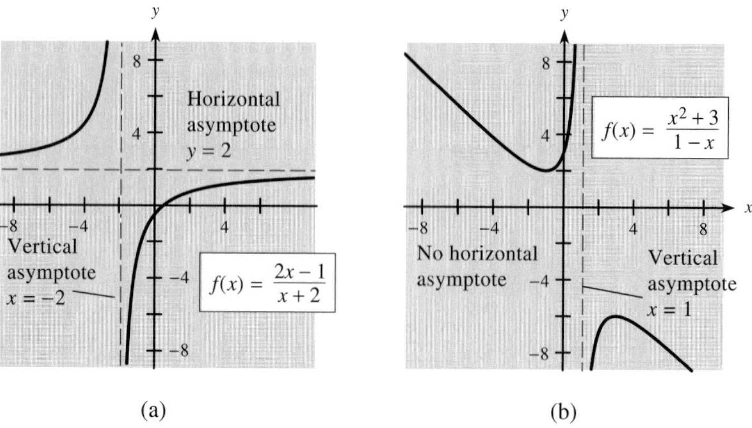

Figure 10.40

(b) The denominator of this function is 0 at $x = 1$, and because this value does not make the numerator 0, there is a vertical asymptote at $x = 1$. To find horizontal asymptotes, we evaluate the following.

$$\lim_{x\to+\infty} \frac{x^2+3}{1-x} = \lim_{x\to+\infty} \frac{1+3/x^2}{1/x^2 - 1/x} = \frac{1+0}{0-0} = -\infty$$

This limit is $-\infty$ because the numerator approaches 1 and the denominator approaches 0 through negative values. Thus

$$\lim_{x\to+\infty} \frac{x^2+3}{1-x} \text{ does not exist}$$

and the graph has no horizontal asymptotes. Note also that

$$\lim_{x\to-\infty} \frac{x^2+3}{1-x} \text{ does not exist}$$

The graph is shown in Figure 10.40(b).

More Curve Sketching

We now extend our first- and second-derivative techniques of curve sketching to include functions that have asymptotes.

In general, the following steps are helpful when we sketch the graph of a function.

1. Determine the domain of the function. The domain may be restricted by the nature of the problem or by the equation.
2. Look for vertical asymptotes, especially if the function is a rational function.
3. Look for horizontal asymptotes, especially if the function is a rational function.
4. Find the relative maxima and minima by using the first-derivative test or the second-derivative test.
5. Use the second derivative to find the points of inflection if this derivative is easily found.

6. Use other information (intercepts, for example) and plot additional points to complete the sketch of the graph.

EXAMPLE 2

Sketch the graph of the function $f(x) = \dfrac{x^2}{(x+1)^2}$.

Solution

1. The domain is the set of all real numbers except $x = -1$.
2. Because $x = -1$ makes the denominator 0 and does not make the numerator 0, there is a vertical asymptote at $x = -1$.
3. Because $\displaystyle\lim_{x\to+\infty} \frac{x^2}{(x+1)^2} = \lim_{x\to+\infty} \frac{x^2}{x^2+2x+1}$

$$= \lim_{x\to+\infty} \frac{1}{1+\dfrac{2}{x}+\dfrac{1}{x^2}}$$

$$= \frac{1}{1+0+0} = 1$$

there is a horizontal asymptote at $y = 1$.
4. To find any maxima and minima, we first find $f'(x)$.

$$f'(x) = \frac{(x+1)^2(2x) - x^2[2(x+1)]}{(x+1)^4}$$

$$= \frac{2x(x+1)[(x+1) - x]}{(x+1)^4}$$

$$= \frac{2x}{(x+1)^3}$$

Thus $f'(x) = 0$ when $x = 0$ (and $y = 0$), and $f'(x)$ is undefined at $x = -1$ (where the vertical asymptote occurs). Testing $f'(x)$ on either side of $x = 0$ and $x = -1$ gives the following sign diagram. The sign diagram for f' shows that the critical point (0, 0) is a relative minimum and shows how the graph approaches the vertical asymptote at $x = -1$.

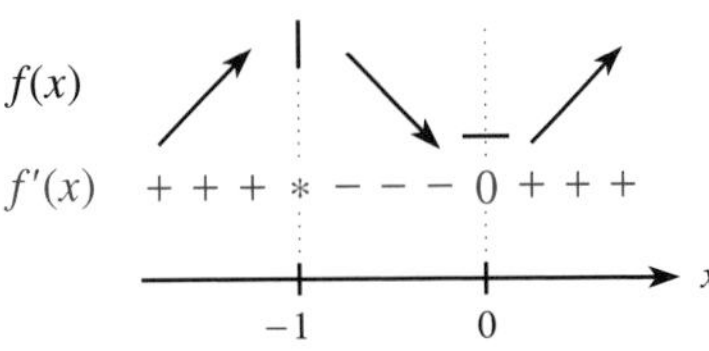

* $x = -1$ is a vertical asymptote.

5. The second derivative is

$$f''(x) = \frac{(x+1)^3(2) - 2x[3(x+1)^2]}{(x+1)^6}$$

Factoring $(x+1)^2$ from the numerator and simplifying give

$$f''(x) = \frac{2-4x}{(x+1)^4}$$

We can see that $f''(0) = 2 > 0$, so the second-derivative test also shows that $(0, 0)$ is a relative minimum. We see that $f''(x) = 0$ when $x = \frac{1}{2}$. Checking $f''(x)$ between $x = -1$ (where it is undefined) and $x = \frac{1}{2}$ shows that the graph is concave up on this interval. Note too that $f''(x) < 0$ for $x > \frac{1}{2}$, so the point $(\frac{1}{2}, \frac{1}{9})$ is a point of inflection. Also see the sign diagram for $f''(x)$.

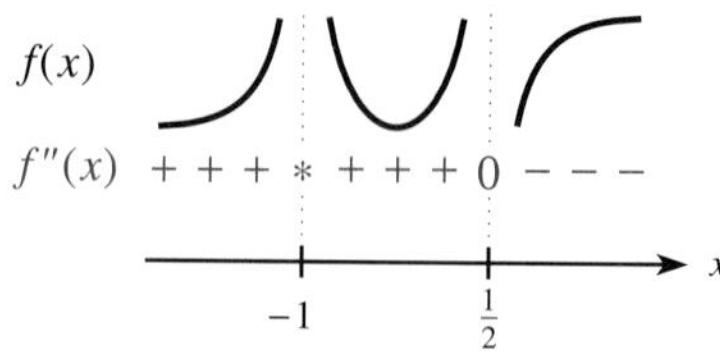

* $x = -1$ is a vertical asymptote.

6. To see how the graph approaches the horizontal asymptote, we check $f(x)$ for large values of $|x|$.

$$f(-100) = \frac{(-100)^2}{(-99)^2} = \frac{10{,}000}{9{,}801} > 1, \quad f(100) = \frac{100^2}{101^2} = \frac{10{,}000}{10{,}201} < 1$$

Thus the graph has the characteristics shown in Figure 10.41(a). The graph is shown in Figure 10.41(b).

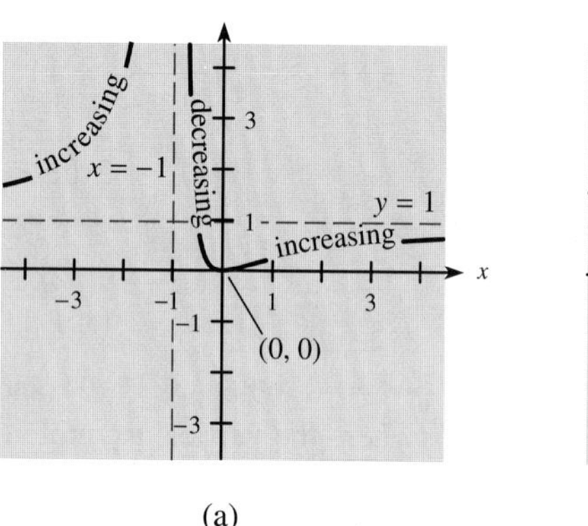

(a)

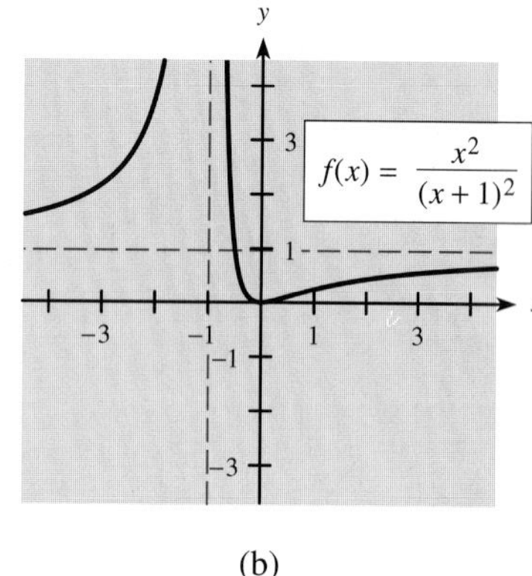

(b)

Figure 10.41

When we wish to learn about a function $f(x)$ or sketch its graph, it is important to understand what information we obtain from $f(x)$, from $f'(x)$, and from $f''(x)$. The following summary may be helpful.

Summary

Source	*Information Provided*
$f(x)$	y-coordinates; horizontal asymptotes, vertical asymptotes; domain restrictions
$f'(x)$	Increasing $[f'(x) > 0]$; decreasing $[f'(x) < 0]$; critical points $[f'(x) = 0$ or $f'(x)$ undefined]; sign-diagram tests for maxima and minima
$f''(x)$	Concave up $[f''(x) > 0]$; concave down $[f''(x) < 0]$; possible points of inflection $[f''(x) = 0$ or $f''(x)$ undefined]; sign-diagram tests for points of inflection; second-derivative test for maxima and minima

CHECKPOINT

1. Let $f(x) = \dfrac{2x + 10}{x - 1}$ and decide whether the following are true or false.
   (a) $f(x)$ has a vertical asymptote at $x = 1$.
   (b) $f(x)$ has $y = 2$ as its horizontal asymptote.
2. Let $f(x) = \dfrac{x^3 - 16}{x} + 1$; then $f'(x) = \dfrac{2x^3 + 16}{x^2}$ and $f''(x) = \dfrac{2x^3 - 32}{x^3}$.
   Use these to determine whether the following are true or false.
   (a) There are no asymptotes.
   (b) $f'(x) = 0$ when $x = -2$
   (c) A partial sign diagram for $f'(x)$ is

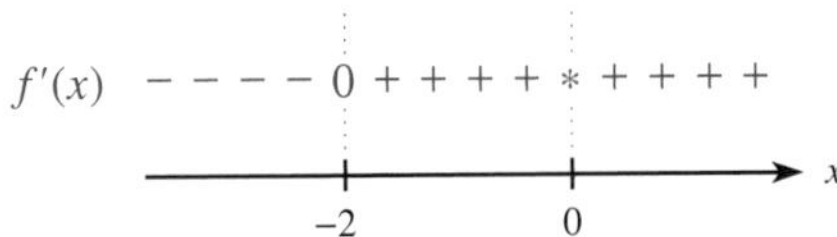

   * means $f'(0)$ is undefined.

   (d) There is a relative minimum at $x = -2$.
   (e) A partial sign diagram for $f''(x)$ is

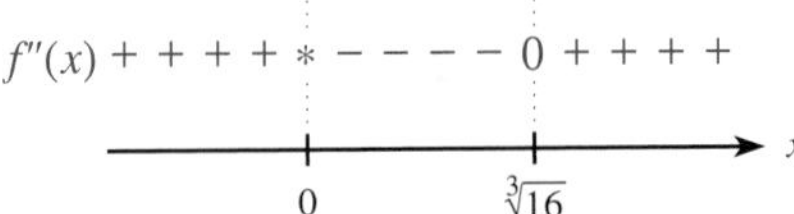

   * means $f''(0)$ is undefined.

   (f) There are points of inflection at $x = 0$ and $x = \sqrt[3]{16}$.

We now consider the problem introduced in the Application Preview.

EXAMPLE 3

If the total daily cost of producing plastic rafts for swimming pools is given by

$$C = 500 + 8x + 0.05x^2$$

where x is the number of units produced per day, then the average cost per unit produced is given by

$$\overline{C} = \frac{500 + 8x + 0.05x^2}{x} = \frac{500}{x} + 8 + 0.05x \qquad \text{for } x > 0$$

Graph this function.

(a) Discuss what happens to average cost as the number of units decreases, approaching 0.
(b) Find the level of production that minimizes average cost.

Solution

(a) The domain of $\overline{C}(x)$ does not include 0, and $\lim_{x \to 0^+} \overline{C} = +\infty$, so there is a vertical asymptote at $x = 0$. Thus the average cost per unit increases without bound as the number of units produced approaches zero. (b) Finding the derivative of $\overline{C}(x)$ gives

$$\overline{C}' = -\frac{500}{x^2} + 0.05$$

Setting $\overline{C}' = 0$ and solving for x gives the critical values of x.

$$0 = -\frac{500}{x^2} + 0.05$$
$$0.05x^2 = 500$$
$$x = -100 \quad \text{or} \quad x = 100$$

The second derivative, $\overline{C}'' = 1000x^{-3} = \dfrac{1000}{x^3}$, is positive at $x = 100$, so $\overline{C}(100) = 18$ is the minimum possible average daily cost. The graph of this function is shown in Figure 10.42. The graph confirms that average cost is minimized when 100 units are produced and shows the asymptotic behavior of the function at 0.

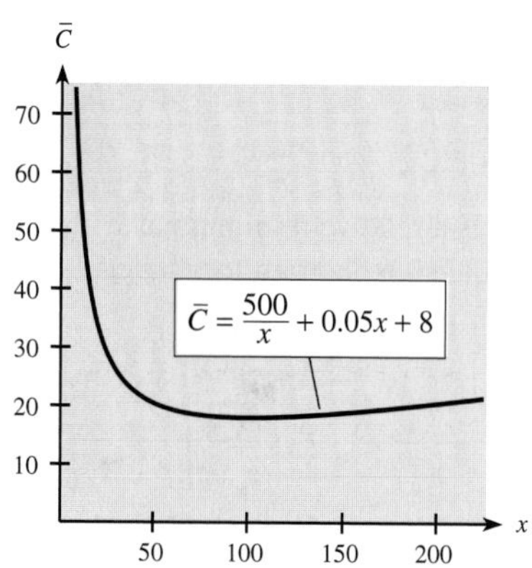

Figure 10.42

Graphing Utilities

If a graphing utility is not available, the procedures previously outlined in this section are necessary to generate a complete and accurate graph. With a graphing utility, the graph of a function is easily generated as long as the viewing window dimensions are appropriate. However, although a graphing utility may reveal the existence of asymptotes, it cannot always precisely locate them. Also, we sometimes need information provided by derivatives to obtain a window that shows all features of a graph.

EXAMPLE 4

Figure 10.43 shows the graph of $f(x) = \dfrac{71x^2}{28(3 - 2x^2)}$.

(a) Determine whether the function has horizontal or vertical asymptotes, and estimate where they occur.
(b) Check your conclusions to (a) analytically.
(c) Discuss which method is more accurate.

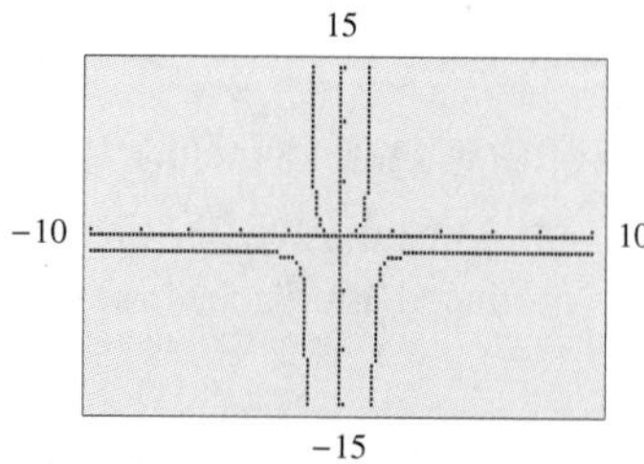

Figure 10.43

Solution

(a) The graph appears to have a horizontal asymptote somewhere between $y = -1$ and $y = -2$, perhaps near $y = -1.5$. Also, there are two vertical asymptotes located approximately at $x = 1.25$ and $x = -1.25$.

(b) We locate the horizontal asymptote by finding the limit of the function as x approaches infinity.

$$\lim_{x\to\infty} \frac{71x^2}{28(3-2x^2)} = \lim_{x\to\infty} \frac{\frac{71x^2}{x^2}}{\frac{84}{x^2} - \frac{56x^2}{x^2}} = \frac{71}{-56} \approx -1.268$$

Vertical asymptotes occur at x-values where $28(3 - 2x^2) = 0$, or

$$3 - 2x^2 = 0$$
$$3 = 2x^2$$
$$\frac{3}{2} = x^2$$
$$\pm\sqrt{\frac{3}{2}} = x \quad \text{or} \quad x \approx \pm 1.225$$

(c) The analytic method is more accurate, of course, because asymptotes reveal extreme behavior of the function either vertically or horizontally. An accurate graph shows all features, but not necessarily the details of any feature. Despite this, our estimates from the graph were not too bad.

Even with a graphing utility, sometimes analytic methods are needed to determine an appropriate viewing window.

EXAMPLE 5

The standard viewing window of the graph of $f(x) = \dfrac{x+10}{x^2+300}$ appears blank (check and see). Find any asymptotes, maxima, and minima, and determine an appropriate viewing window. Sketch the graph.

Solution

Because $x^2 + 300 = 0$ has no real solution, there are no vertical asymptotes. We locate the horizontal asymptotes by finding the limit of the function as x approaches infinity.

$$\lim_{x\to\infty} \frac{x+10}{x^2+300} = \lim_{x\to\infty} \frac{\frac{x}{x^2} + \frac{10}{x^2}}{\frac{x^2}{x^2} + \frac{300}{x^2}} = \frac{0+0}{1+0} = 0$$

Hence $y = 0$ (the x-axis) is a horizontal asymptote. We then find an appropriate viewing window by locating the critical points.

$$f'(x) = \frac{(x^2+300)(1) - (x+10)(2x)}{(x^2+300)^2}$$
$$= \frac{x^2 + 300 - 2x^2 - 20x}{(x^2+300)^2} = \frac{300 - 20x - x^2}{(x^2+300)^2}$$

$f'(x) = 0$ when the numerator is zero. Thus

$$\begin{aligned} 300 - 20x - x^2 &= 0 \\ 0 &= x^2 + 20x - 300 \\ 0 &= (x + 30)(x - 10) \\ x + 30 = 0 \qquad & \qquad x - 10 = 0 \\ x = -30 \qquad & \qquad x = 10 \end{aligned}$$

The critical points are $x = -30$, $y = -\frac{1}{60} \approx -0.01666667$ and $x = 10$, $y = \frac{1}{20} = 0.05$. A sign diagram for $f'(x)$ is shown at the right.

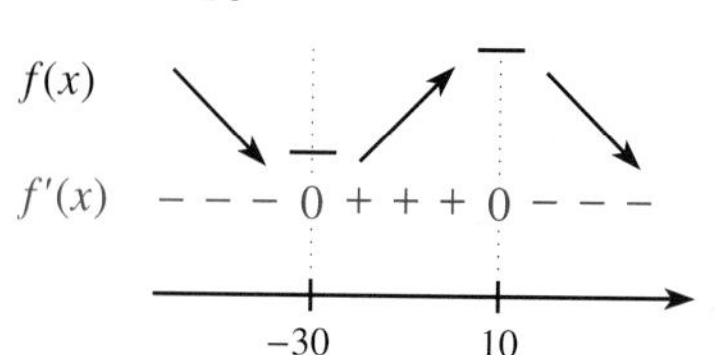

Without using the information above, a graphing utility may not give a useful graph. An x-range that includes -30 and 10 is needed. Because $y = 0$ is a horizontal asymptote, these relative extrema are absolute, and the y-range must be quite small for the shape of the graph to be seen clearly. Figure 10.44 shows the graph.

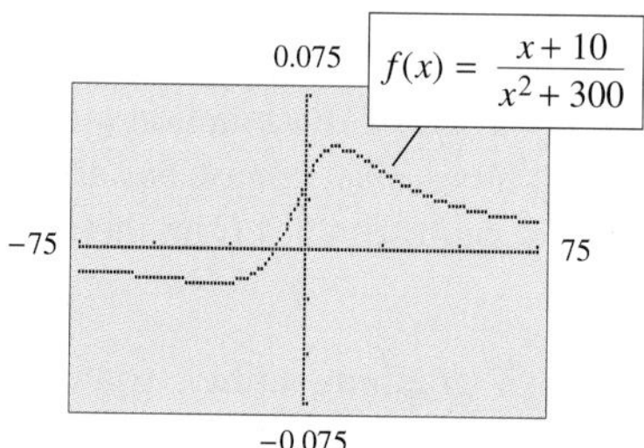

Figure 10.44

EXAMPLE 6

A profit function for a product is given by

$$P(x) = \frac{-x^2 + 16x - 4}{4x^2 + 16}, \qquad \text{for } x \geq 0$$

where x is in thousands of units and $P(x)$ is in billions of dollars. Because of fixed costs, profit is negative when fewer than 254 units are produced and sold. Will a loss occur at any other level of production and sales?

Solution

Looking at the graph of this function over the range $0 \leq x \leq 10$ (see Figure 10.45a), it is not clear whether the graph will eventually cross the x-axis. To see whether a loss ever occurs and to see what profit is approached as the number of units produced and sold becomes large, we evaluate the limit of $P(x)$ as $x \to \infty$.

$$\lim_{x\to\infty} \frac{-x^2 + 16x - 4}{4x^2 + 16} = \lim_{x\to\infty} \frac{-1 + \frac{16}{x} - \frac{4}{x^2}}{4 + \frac{16}{x^2}} = -\frac{1}{4}$$

Thus a loss of $\$\frac{1}{4}$ billion is approached as the number of units increases without bound. Figure 10.45(b) shows that the graph does cross the x-axis, where $-x^2 + 16x - 4 = 0$. Using TRACE, SOLVER, the Quadratic Formula, or a program gives $P(x) = 0$ at $x \approx 0.254$ and $x \approx 15.7459$. Thus if 15,746 units or more are produced and sold, the profit is negative.

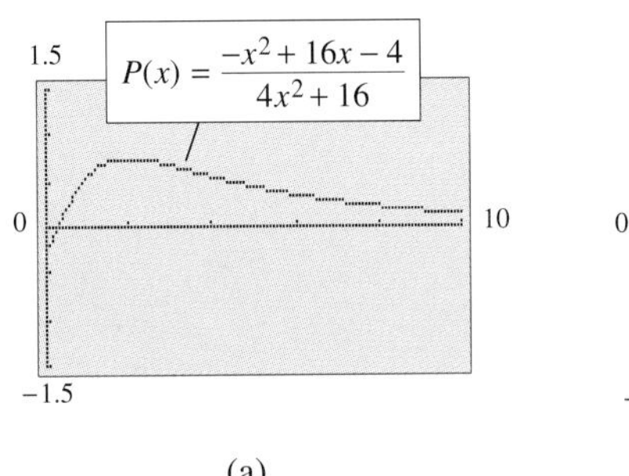

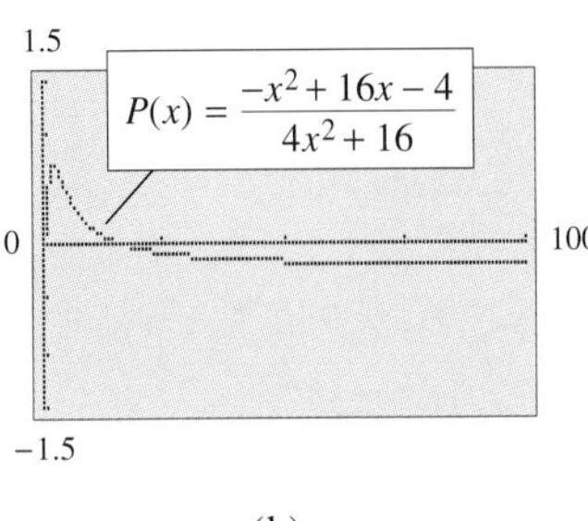

Figure 10.45 (a) (b)

CHECKPOINT SOLUTIONS

1. (a) True, $x = 1$ makes the denominator of $f(x)$ equal to zero, whereas the numerator is nonzero.
   (b) True, $\lim_{x\to\infty} \frac{2x+10}{x-1} = 2$ and $\lim_{x\to\infty} \frac{2x+10}{x-1} = 2$.
2. (a) False. There are no horizontal asymptotes, but $x = 0$ is a vertical asymptote.
   (b) True
   (c) True
   (d) True. The relative minimum point is $(-2, f(-2)) = (-2, 13)$.
   (e) True
   (f) False. There is a point of inflection only at $(\sqrt[3]{16}, 1)$. At $x = 0$ the vertical asymptote occurs, so there is no point on the graph and hence no point of inflection.

EXERCISE 10.5

In Problems 1–4, a function and its graph are given. Use the graph to find each of the following, if they exist. Then confirm your results analytically.

(a) vertical asymptotes
(b) $\lim_{x\to\infty} f(x)$
(c) horizontal asymptotes
(d) $\lim_{x\to-\infty} f(x)$

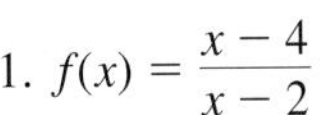

1. $f(x) = \frac{x-4}{x-2}$

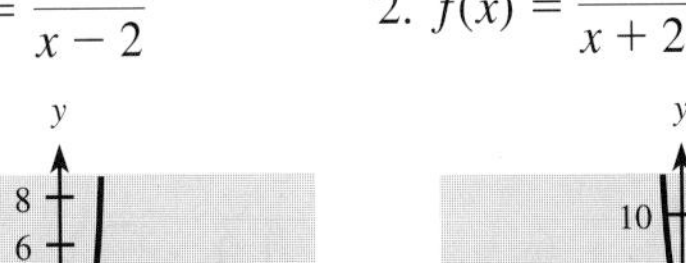

2. $f(x) = \frac{8}{x+2}$

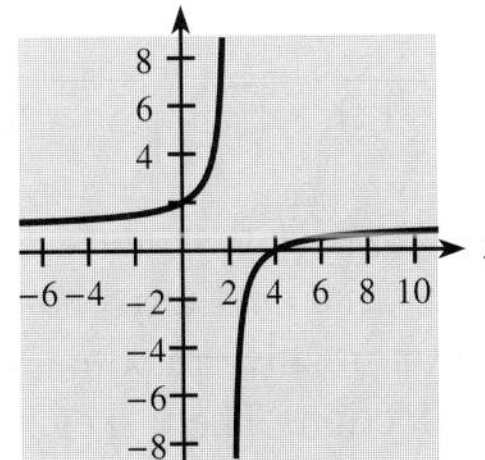

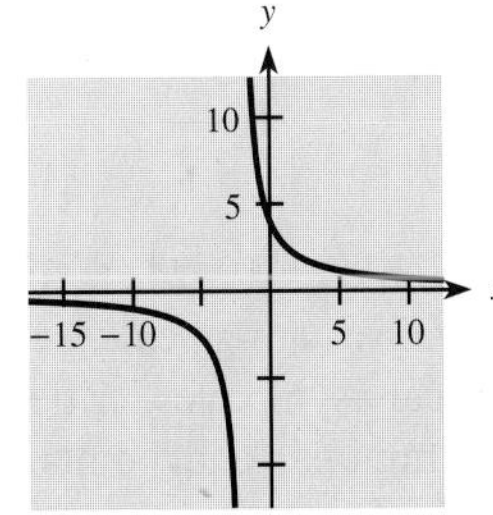

3. $f(x) = \dfrac{x^2}{(x-2)^2}$

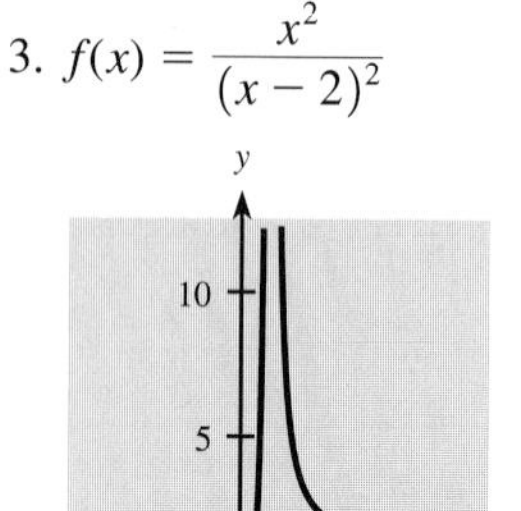

4. $f(x) = \dfrac{3(x^4 + 2x^3 + 6x^2 + 2x + 5)}{(x^2 - 4)^2}$

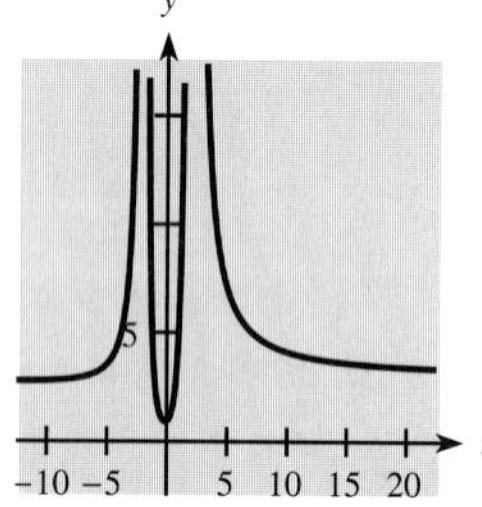

In Problems 5–10, find any horizontal and vertical asymptotes for each function.

5. $y = \dfrac{2x}{x-3}$

6. $y = \dfrac{3x-1}{x+5}$

7. $y = \dfrac{x+1}{x^2-4}$

8. $y = \dfrac{4x}{9-x^2}$

9. $y = \dfrac{3x^3-6}{x^2+4}$

10. $y = \dfrac{6x^3}{4x^2+9}$

For each function in Problems 11–18, find any horizontal and vertical asymptotes, and use information from the first derivative to sketch the graph.

11. $f(x) = \dfrac{2x+2}{x-3}$

12. $f(x) = \dfrac{5x-15}{x+2}$

13. $y = \dfrac{x^2+4}{x}$

14. $y = \dfrac{x^2+4}{x^2}$

15. $y = \dfrac{27x^2}{(x+1)^3}$

16. $y = \left(\dfrac{x+2}{x-3}\right)^2$

17. $f(x) = \dfrac{16x}{x^2+1}$

18. $f(x) = \dfrac{4x^2}{x^4+1}$

In Problems 19–24, a function and its first and second derivatives are given. Use these to find any horizontal and vertical asymptotes, critical points, relative maxima, relative minima, and points of inflection. Then sketch the graph of each function.

19. $y = \dfrac{x}{(x-1)^2}$

$y' = -\dfrac{x+1}{(x-1)^3}$

$y'' = \dfrac{2x+4}{(x-1)^4}$

20. $y = \dfrac{(x-1)^2}{x^2}$

$y' = \dfrac{2(x-1)}{x^3}$

$y'' = \dfrac{6-4x}{x^4}$

21. $y = x + \dfrac{3}{\sqrt[3]{x-3}}$

$y' = \dfrac{(x-3)^{2/3}-1}{(x-3)^{2/3}}$

$y'' = \dfrac{2}{3(x-3)^{5/3}}$

22. $y = 3\sqrt[3]{x} + \dfrac{1}{x}$

$y' = \dfrac{x^{4/3}-1}{x^2}$

$y'' = \dfrac{6-2x^{4/3}}{3x^3}$

23. $f(x) = \dfrac{9(x-2)^{2/3}}{x^2}$

$f'(x) = \dfrac{12(3-x)}{x^3(x-2)^{1/3}}$

$f''(x) = \dfrac{4(7x^2-42x+54)}{x^4(x-2)^{4/3}}$

24. $f(x) = \dfrac{3x^{2/3}}{x+1}$

$f'(x) = \dfrac{2-x}{x^{1/3}(x+1)^2}$

$f''(x) = \dfrac{2(2x^2-8x-1)}{3x^{4/3}(x+1)^3}$

In Problems 25–28, a function and its graph are given.

(a) Use the graph to estimate the location of any horizontal or vertical asymptotes.

(b) Use the function to precisely determine the location of any asymptotes.

25. $f(x) = \dfrac{9x}{17-4x}$

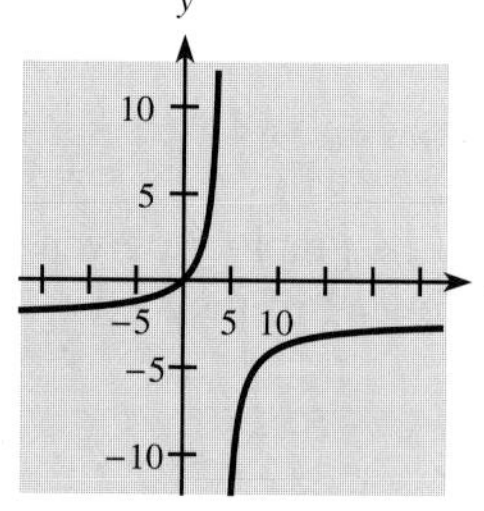

26. $f(x) = \dfrac{5-13x}{3x+20}$

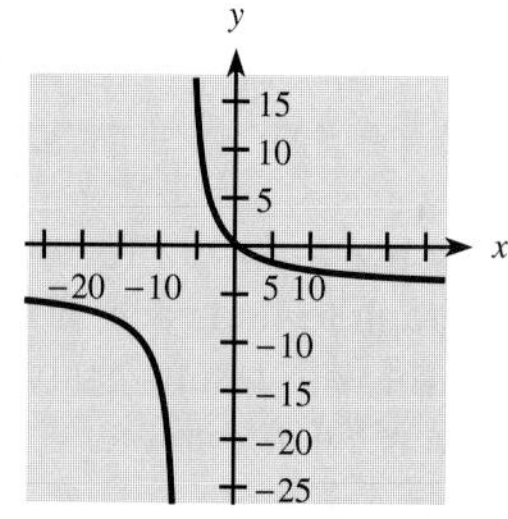

27. $f(x) = \dfrac{20x^2+98}{9x^2-49}$

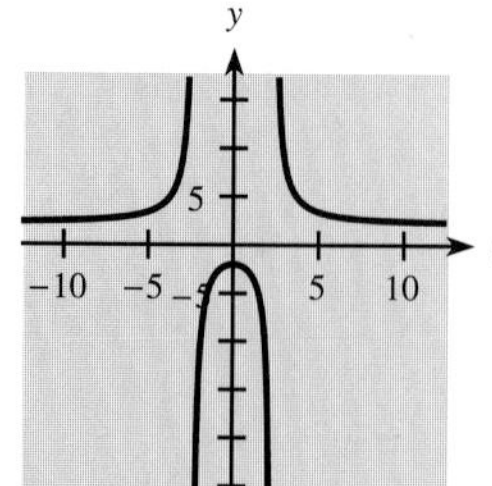

28. $f(x) = \dfrac{15x^2-x}{7x^2-35}$

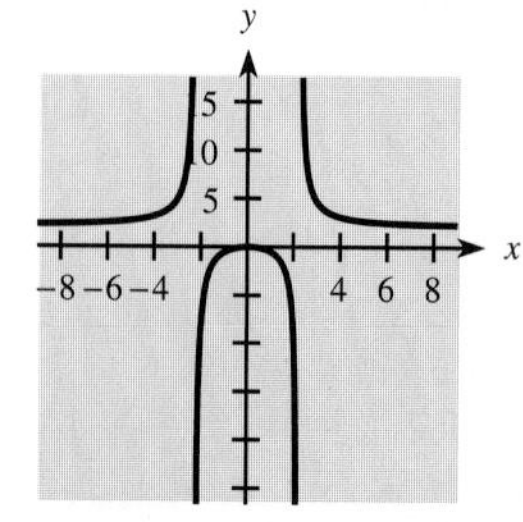

For each function in Problems 29–34, complete the following steps.

(a) Use a graphing utility to graph the function in the standard viewing window.
(b) Analytically determine the location of any asymptotes and extrema.
(c) Graph the function in a viewing window that shows all features of the graph. State the ranges for x-values and y-values for your viewing window.

29. $f(x) = \dfrac{x + 25}{x^2 + 1400}$ 30. $f(x) = \dfrac{x - 50}{x^2 + 1100}$

31. $f(x) = \dfrac{100(9 - x^2)}{x^2 + 100}$ 32. $f(x) = \dfrac{200x^2}{x^2 + 100}$

33. $f((x) = \dfrac{1000x - 4000}{x^2 - 10x - 2000}$

34. $f(x) = \dfrac{900x + 5400}{x^2 - 30x - 1800}$

Applications

35. ***Cost-benefit*** The percentage p of particulate pollution that can be removed from the smokestacks of an industrial plant by spending C dollars is given by

$$p = \frac{100C}{7300 + C}$$

(a) Find any C-values where the rate of change of p with respect to C does not exist. Make sure that these make sense in the problem.
(b) Find C-values for which p is increasing.
(c) If there is a horizontal asymptote, find it.
(d) Can 100% of the pollution be removed?

36. ***Cost-benefit*** The percentage p of impurities that can be removed from the waste water of a manufacturing process at a cost of C dollars is given by

$$p = \frac{100C}{8100 + C}$$

(a) Find any C-values where the rate of change of p with respect to C does not exist. Make sure that these make sense in the problem.
(b) Find C-values for which p is increasing.
(c) Find any horizontal asymptotes.
(d) Can 100% of the pollution be removed?

37. ***Revenue*** A recently released film has its weekly revenue given by

$$R(t) = \frac{50t}{t^2 + 36}, \qquad t \geq 0$$

where $R(t)$ is in millions of dollars and t is in weeks.

(a) Graph $R(t)$.
(b) When will revenue be maximized?
(c) Suppose that if revenue decreases for 4 consecutive weeks, the film will be removed from theaters and will be released as a video 12 weeks later. When will the video come out?

38. ***Production costs*** Suppose that the total cost of producing a shipment of a certain product is

$$C = 5000x + \frac{125{,}000}{x}, \qquad x > 0$$

where x is the number of machines used in the production process.

(a) Graph this total cost function.
(b) Using how many machines will minimize the total cost?

39. ***Wind chill*** If x is the wind speed in miles per hour and is greater than or equal to 5, then the wind chill (in degrees Fahrenheit) for an air temperature of 0°F can be approximated by the function

$$f(x) = \frac{289.173 - 58.5731x}{x + 1}, \qquad x \geq 5$$

(a) Does $f(x)$ have a vertical asymptote? If so, what is it?
(b) Does $f(x)$ have a vertical asymptote within its domain?
(c) Does $f(x)$ have a horizontal asymptote? If so, what is it?
(d) In the context of wind chill, does $\lim_{x\to\infty} f(x)$ have a physical interpretation? If so, what is it, and is it meaningful?

40. ***Profit*** An entrepreneur starts new companies and sells them when their growth is maximized. Suppose that the annual profit for a new company is given by

$$P(x) = 22 - \frac{1}{2}x - \frac{18}{x + 1}$$

where P is in thousands of dollars and x is the number of years after the company is formed. If she wants to sell the company before profits begin to decline, after how many years should she sell it?

41. ***Productivity*** The figure is a typical graph of worker productivity per hour P as a function of time t on the job.

(a) What is the horizontal asymptote?
(b) What is $\lim_{t\to\infty} P(t)$?
(c) What is the horizontal asymptote for $P'(t)$?
(d) What is $\lim_{t\to\infty} P'(t)$?

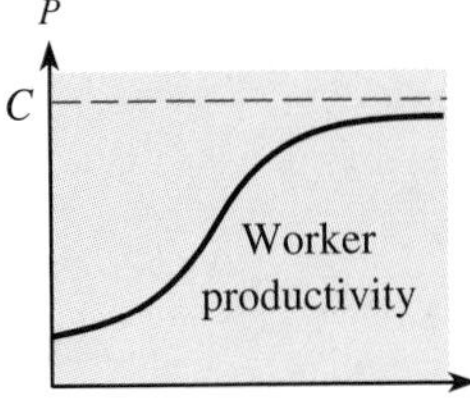

42. ***Sales volume*** The figure shows a typical curve that gives the volume of sales S as a function of time t after an ad campaign.
   (a) What is the horizontal asymptote?
   (b) What is $\lim_{x\to\infty} S(t)$?
   (c) What is the horizontal asymptote for $S'(t)$?
   (d) What is $\lim_{x\to\infty} S'(t)$?

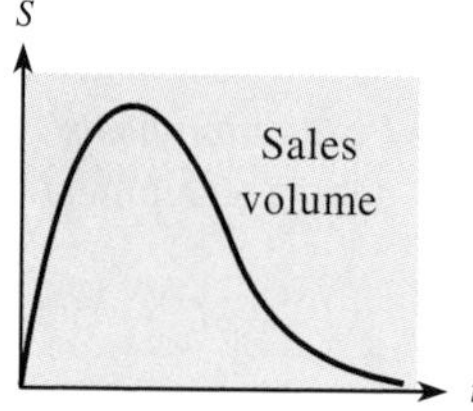

Farm workers The percentage of U.S. workers in farm occupations during certain years is shown in the table.

Year	*Percent of All Workers in Farm Occupations*
1820	71.8
1850	63.7
1870	53
1900	37.5
1920	27
1930	21.2
1940	17.4
1950	11.6
1960	6.1
1970	3.6
1980	2.7
1985	2.8
1990	2.4

Source: *The World Almanac and Book of Facts,* 1993

Assume that the percentage of U.S. workers in farm occupations can be modeled with the function

$$f(t) = 1000 \cdot \frac{-8.0912t + 1558.9}{1.09816t^2 - 122.183t + 21472.6}$$

where t is the number of years past 1800. Use this model in Problems 43 and 44.

43. (a) Find $\lim_{t\to\infty} f(t)$.
   (b) Interpret your answer to (a).
   (c) Does $f(t)$ have any vertical asymptotes within its domain $t \geq 0$?

44. (a) Use a graphing utility to graph $f(t)$ for $t = 0$ to $t = 220$.
   (b) From the graph, identify t-values where the model is inappropriate, and explain why it is inappropriate.

45. ***Barometric pressure*** The figure shows a barograph readout of the barometric pressure as recorded by Georgia Southern University's meteorological equipment. The figure shows a tremendous drop in barometric pressure on Saturday morning, March 13, 1993.
   (a) If $B(t)$ is barometric pressure expressed as a function of time, as shown in the figure, does $B(t)$ have a vertical asymptote sometime after 8 A.M. on Saturday, March 13, 1993? Explain why or why not.
   (b) Consult your library or some other resource to find out what happened in Georgia (and in the eastern United States) on March 13, 1993, to cause such a dramatic drop in barometric pressure.

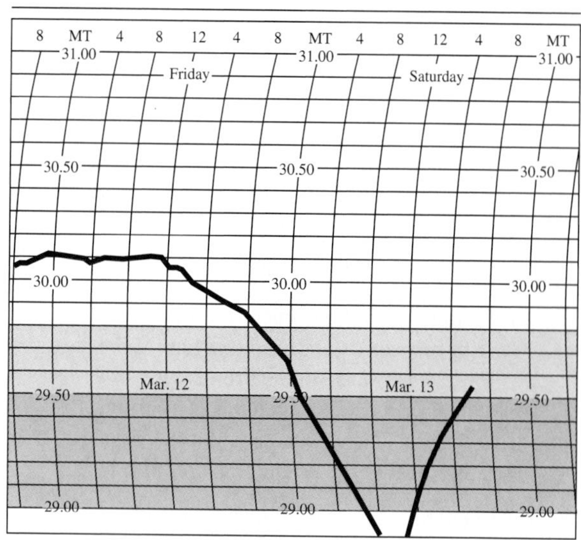

Source: *Statesboro Herald,* March 14, 1993

KEY TERMS AND FORMULAS

Section	Key Terms	Formula
10.1	Relative maxima and minima	
	Increasing	$f'(x) > 0$
	Decreasing	$f'(x) < 0$
	Critical points	$f'(x) = 0$ or $f'(x)$ undefined
	Sign diagram for $f'(x)$	
	First-derivative test	
	Horizontal point of inflection	
10.2	Concave up	$f''(x) > 0$
	Concave down	$f''(x) < 0$
	Point of inflection	May occur where $f''(x) = 0$ or $f''(x)$ undefined
	Sign diagram for $f''(x)$	
	Second-derivative test	
10.3	Absolute extrema	
	Average cost	$\overline{C}(x) = C(x)/x$
	Profit maximization	
	Competitive market	$R(x) = p \cdot x$ where $p =$ equilibrium price
	Monopolistic market	$R(x) = p \cdot x$ where $p = f(x)$ is the demand function
10.4	Inventory cost models	
10.5	Asymptotes	
	Horizontal: $y = b$	$\lim_{x \to +\infty} f(x) = b$ or $\lim_{x \to -\infty} f(x) = b$
	Vertical: $x = c$ for rational function $y = f(x)/g(x)$	y unbounded near $x = c$ if $g(c) = 0$ and $f(c) \neq 0$

REVIEW EXERCISES

Section 10.1

In Problems 1–4, find all critical points and determine whether they are relative maxima, relative minima, or horizontal points of inflection.

1. $y = -x^2$
2. $p = q^2 - 4q - 5$
3. $f(x) = 1 - 3x + 3x^2 - x^3$
4. $f(x) = \dfrac{3x}{x^2 + 1}$

In Problems 5–10.

(a) Find all critical values, including those where $f'(x)$ is undefined.
(b) Find the relative maxima and minima, if any exist.
(c) Find the horizontal points of inflection, if any exist.
(d) Sketch the graph.

5. $y = x^3 + x^2 - x - 1$
6. $f(x) = 4x^3 - x^4$

7. $f(x) = x^3 - \frac{15}{2}x^2 - 18x + \frac{3}{2}$
8. $y = 5x^7 - 7x^5 - 1$
9. $y = x^{2/3} - 1$
10. $y = x^{2/3}(x - 4)^2$

Section 10.2

11. Is the graph of $y = x^4 - 3x^3 + 2x - 1$ concave up or concave down at $x = 2$?
12. Find intervals where the graph of $y = x^4 - 2x^3 - 12x^2 + 6$ is concave upward and intervals where it is concave downward, and find points of inflection.
13. Find the relative maxima, relative minima, and points of inflection of the graph of $y = x^3 - 3x^2 - 9x + 10$.

In Problems 14 and 15, find any relative maxima, relative minima, and points of inflection, and sketch each graph.

14. $y = x^3 - 12x$
15. $y = 2 + 5x^3 - 3x^5$

Section 10.3

16. Given $R = 280x - x^2$, find the absolute maximum and minimum for R when (a) $0 \le x \le 200$ and (b) $0 \le x \le 100$.
17. Given $y = 6400x - 18x^2 - \frac{x^3}{3}$, find the absolute maximum and minimum for y when (a) $0 \le x \le 50$ and (b) $0 \le x \le 100$.

Section 10.5

In Problems 18 and 19, use the graphs to find the following items.

(a) vertical asymptotes
(b) horizontal asymptotes
(c) $\lim_{x \to +\infty} f(x)$
(d) $\lim_{x \to -\infty} f(x)$

18.

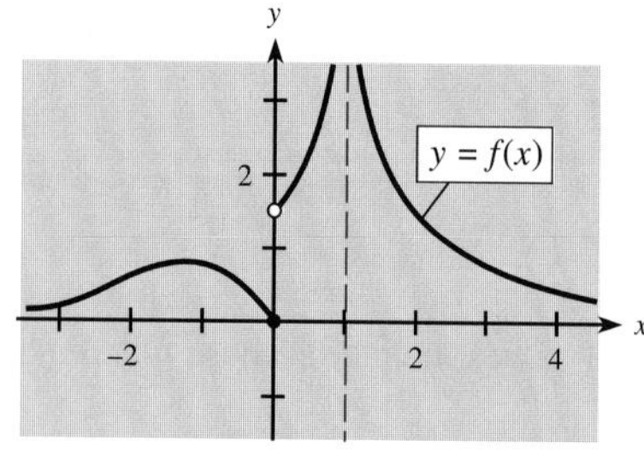

19.

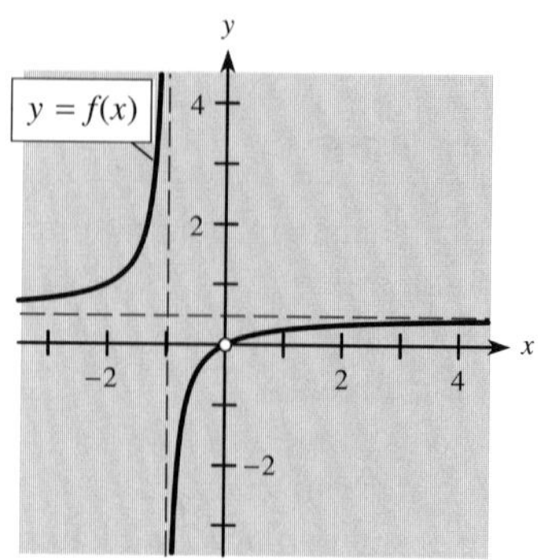

In Problems 20 and 21, find any horizontal asymptotes and any vertical asymptotes.

20. $y = \frac{3x + 2}{2x - 4}$
21. $y = \frac{x^2}{1 - x^2}$

In Problems 22–24:

(a) Find any horizontal and vertical asymptotes.
(b) Find any relative maxima and minima.
(c) Sketch each graph.

22. $y = \frac{3x}{x + 2}$
23. $y = \frac{8(x - 2)}{x^2}$
24. $y = \frac{x^2}{x - 1}$

Sections 10.1 and 10.2

In Problems 25 and 26, a function and its graph are given.

(a) Use the graph to determine (estimate) x-values where $f'(x) > 0$, where $f'(x) < 0$, and where $f'(x) = 0$.
(b) Use the graph to determine x-values where $f''(x) > 0$, where $f''(x) < 0$, and where $f''(x) = 0$.
(c) Check your conclusions to (a) by finding $f'(x)$ and graphing it with a graphing utility.
(d) Check your conclusions to (b) by finding $f''(x)$ and graphing it with a graphing utility.

25. $f(x) = x^3 - 4x^2 + 4x$

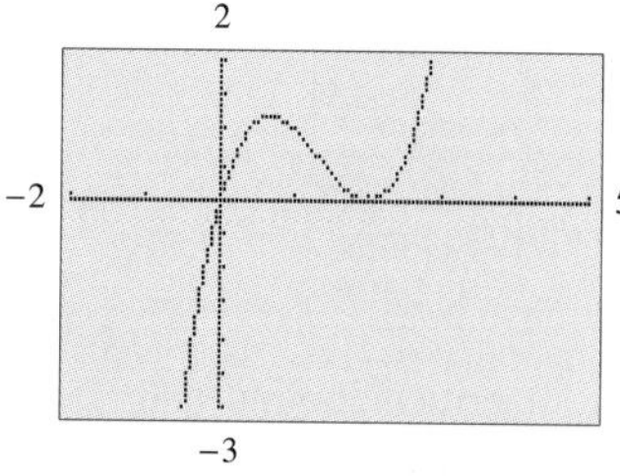

26. $f(x) = 0.0025x^4 + 0.02x^3 - 0.48x^2 + 0.08x + 4$

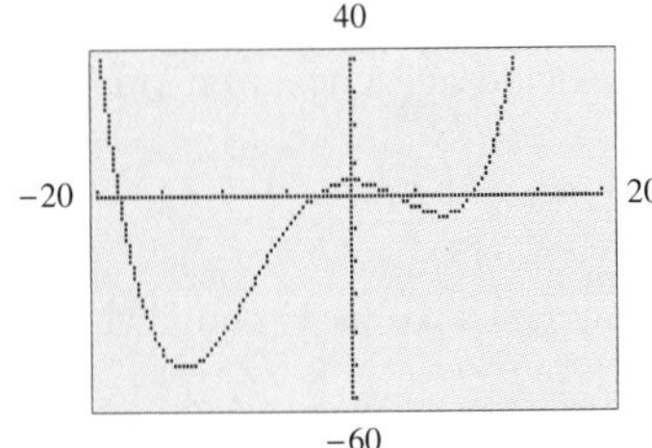

In Problems 27 and 28, $f'(x)$ and its graph are given.

(a) Use the graph of $f'(x)$ to determine (estimate) where the graph of $f(x)$ is increasing, where it is decreasing, and where it has relative extrema.

(b) Use the graph of $f'(x)$ to determine where $f''(x) > 0$, where $f''(x) < 0$, and where $f''(x) = 0$.

(c) Verify that the given $f(x)$ has $f'(x)$ as its derivative, and graph $f(x)$ to check your conclusions in (a).

(d) Calculate $f''(x)$ and graph it to check your conclusions in (b).

27. $f'(x) = x^2 + 4x - 5 \quad \left(\text{for } f(x) = \frac{x^3}{3} + 2x^2 - 5x\right)$

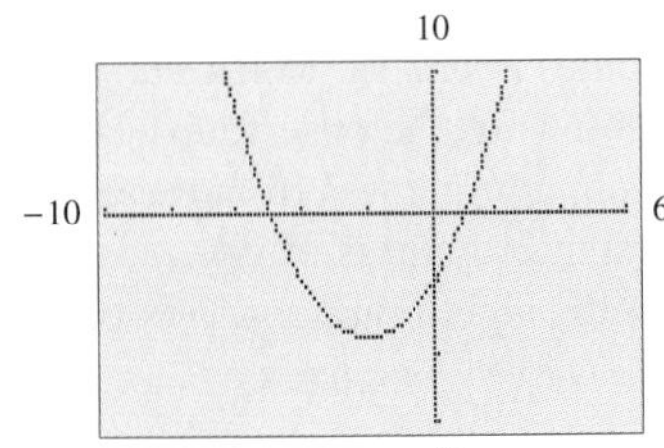

28. $f'(x) = 6x^2 - x^3 \quad \left(\text{for } f(x) = 2x^3 - \frac{x^4}{4}\right)$

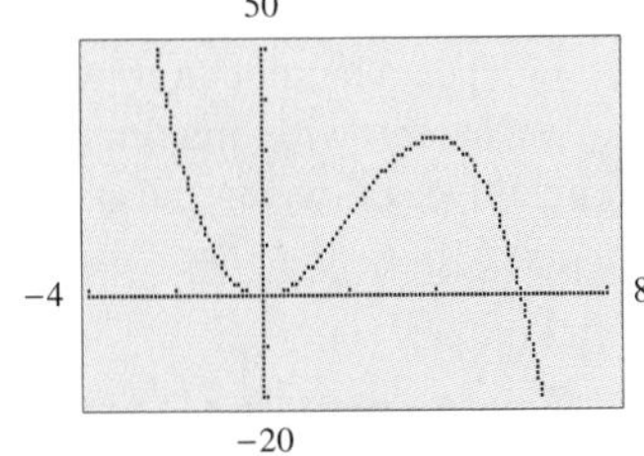

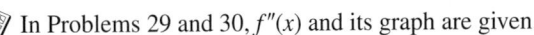

In Problems 29 and 30, $f''(x)$ and its graph are given.

(a) Use the graph to determine (estimate) where the graph of $f(x)$ is concave upward, where it is concave downward, and where it has points of inflection.

(b) Verify that the given $f(x)$ has $f''(x)$ as its second derivative, and graph $f(x)$ to check your conclusions in (a).

29. $f''(x) = 4 - x \quad \left(\text{for } f(x) = 2x^2 - \frac{x^3}{6}\right)$

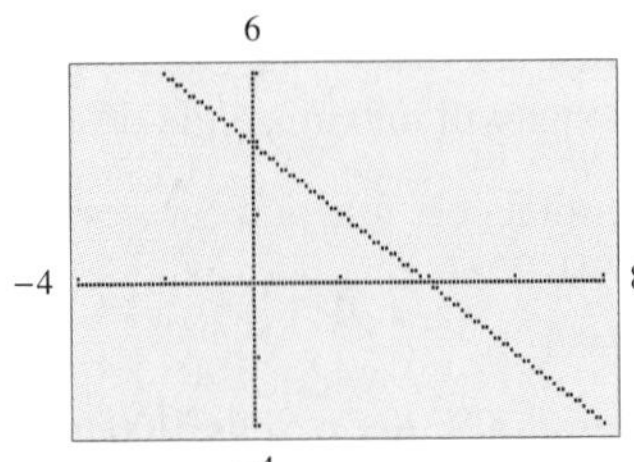

30. $f''(x) = 6 - x - x^2 \quad \left(\text{for } f(x) = 3x^2 - \frac{x^3}{6} - \frac{x^4}{12}\right)$

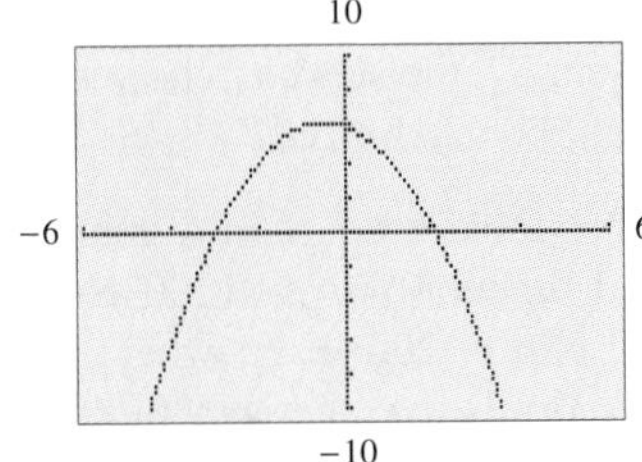

Applications

Section 10.3

31. ***Cost*** Suppose the total cost function for a product is

$$C(x) = 3x^2 + 15x + 75$$

How many units minimize the average cost? Find the minimum average cost.

32. ***Revenue*** Suppose the total revenue function for a product is given by

$$R(x) = 32x - 0.01x^2$$

(a) How many units will maximize the total revenue? Find the maximum revenue.

(b) If production is limited to 2500 units, how many units will maximize the total revenue? Find the maximum revenue.

33. ***Profit*** Suppose the profit function for a product is

$$P(x) = 1080x + 9.6x^2 - 0.1x^3 - 50{,}000$$

Find the maximum profit.

34. ***Profit*** How many units (x) will maximize profit if $R(x) = 46x - x^2$ and $C(x) = 5x^2 + 10x + 3$?

35. ***Profit*** A product can be produced at a total cost $C(x) = 800 + 4x$, where x is the number produced and is limited to at most 150 units. If the total revenue is given by $R(x) = 80x - \frac{1}{4}x^2$, determine the level of production that will maximize the profit.

36. ***Average cost*** The total cost function for a product is $C = 2x^2 + 54x + 98$. Producing how many units will minimize average cost?

37. ***Revenue*** McRobert's TV Shop sells 200 sets per month at a price of \$400 per unit. Market research indicates that the shop can sell one additional set for each \$1 that it reduces the price. At what selling price will the shop maximize revenue?

38. ***Profit*** If, in Problem 37, the sets cost the shop \$250 each, when will profit be maximized?

39. ***Profit*** Suppose that for a product in a competitive market, the demand function is $p = 1200 - 2x$ and the supply function is $p = 200 + 2x$, where x is the number of units. A firm's average cost function for this product is

$$\overline{C}(x) = \frac{12{,}000}{x} + 50 + x$$

Find the maximum profit. *Hint:* First find the equilibrium price.

40. ***Profit*** The monthly demand function for a product sold by a monopoly is $p = 800 - x$, and its average cost is $\overline{C} = 200 + x$.
    (a) Determine the quantity that will maximize profit.
    (b) Find the selling price at the optimal quantity.

41. ***Profit*** Suppose that in a monopolistic market, the demand function for a commodity is

$$p = 7000 - 10x - \frac{x^2}{3}$$

If a company's average cost function for this commodity is

$$\overline{C}(x) = \frac{40{,}000}{x} + 600 + 8x$$

find the maximum profit.

Section 10.4

42. ***Reaction to a drug*** The reaction R to an injection of a drug is related to the dosage x according to

$$R(x) = x^2\left(500 - \frac{x}{3}\right)$$

Find the dosage that yields the maximum reaction.

43. ***Productivity*** The number of parts produced per hour by a worker is given by

$$N = 4 + 3t^2 - t^3$$

where t is the number of hours on the job without a break. If the worker starts at 8 A.M., when will she be at maximum productivity during the morning?

44. ***Population*** Population estimates show that the equation $P = 300 + 10t - t^2$ represents the size of the graduating class of a high school, where t represents the number of years after 1990, $0 \le t \le 10$. What will be the largest graduating class in the decade?

45. ***Night brightness*** Suppose that an observatory is to be built between cities A and B, which are 30 miles apart. For the best viewing, the observatory should be located where the night brightness from these cities is minimum. If the night brightness of city A is 8 times that of city B, then the night brightness b between the two cities and x miles from A is given by

$$b = \frac{8k}{x^2} + \frac{k}{(30 - x)^2}$$

where k is a constant. Find the best location for the observatory; that is, find x that minimizes b.

46. ***Product design*** A playpen manufacturer wants to make a rectangular enclosure with maximum play area. To remain competitive, he wants the perimeter of the base to be only 16 feet. What dimensions should the playpen have?

47. ***Printing design*** A printed page is to contain 56 square inches and have a $\frac{3}{4}$-inch margin at the bottom and 1-inch margins at the top and on both sides. Find the dimensions that minimize the size of the page (and hence the costs for paper).

48. ***Drug sensitivity*** The reaction R to an injection of a drug is related to the dosage x according to

$$R(x) = x^2\left(500 - \frac{x}{3}\right)$$

The sensitivity to the drug is defined by dR/dx. Find the dosage that maximizes sensitivity.

49. ***Photosynthesis*** The amount of photosynthesis that takes place in a certain plant depends on the intensity of light x according to the equation

$$f(x) = 145x^2 - 30x^3$$

The rate of change of the amount of photosynthesis with respect to the intensity is $f'(x)$. Find the intensity that maximizes the rate of change.

50. ***Inventory cost model*** A company needs 288,000 items per year. Production costs are \$1500 to prepare for a production run and \$30 for each item produced. Inventory costs are \$1.50 per year for each item stored. Find the number of items that should be produced in each run so that the total costs of production and storage are minimum.

CHAPTER TEST

Find the local maxima, local minima, horizontal points of inflection, and asymptotes, if they exist, for each of the functions in Problems 1–3. Graph each function.

1. $f(x) = x^3 + 6x^2 + 9x + 3$
2. $y = x^3 - 3x^2 + 3x + 4$
3. $y = \dfrac{x^2 - 3x + 6}{x - 2}$

In Problems 4–6, use the function $y = 3x^5 - 5x^3 + 2$.

4. Over what intervals is the graph of this function concave upward?
5. Find the points of inflection of this function.
6. Find the relative maxima and minima of this function.
7. Find the absolute maximum and minimum for $f(x) = 2x^3 - 15x^2 + 3$ on the interval $[-2, 8]$.
8. Find all horizontal and vertical asymptotes for the function $f(x) = \dfrac{200x - 500}{x + 300}$.
9. Use the graph of $y = f(x)$ and the indicated points to complete the following chart. Enter +, −, or 0, according to whether f, f', and f'' are positive, negative, or zero at each point.

Point	f	f'	f''
A			
B			
C			

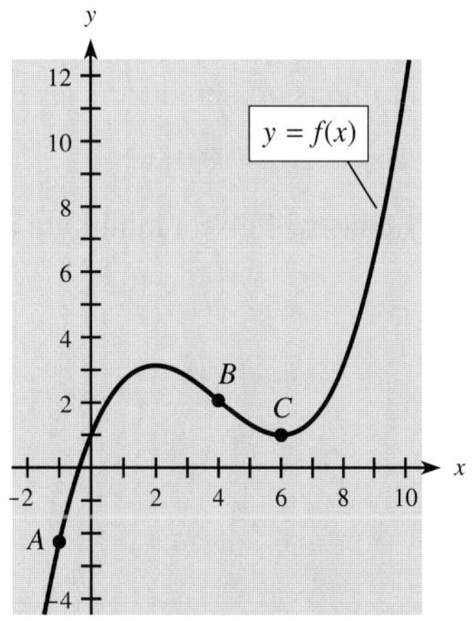

10. Use the figure to complete the following.
    (a) $\lim_{x \to -\infty} f(x) = ?$
    (b) What is the vertical asymptote?

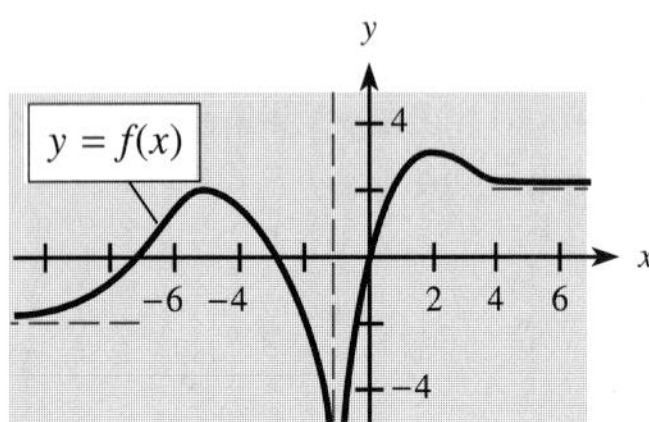

11. If $f(6) = 10$, $f'(6) = 0$, and $f''(6) = -3$, what can we conclude about the point on the graph of $y = f(x)$ where $x = 6$? Explain.
12. If x represents the number of years past 1900, the equation that models the number of Catholics per priest in the United States is

$$y = 0.19x^2 - 16.59x + 1038.29$$

(Source: Dr. Robert G. Kennedy, University of St. Thomas)
    (a) During which year does this model indicate that the number of Catholics per priest is a minimum?
    (b) What does this mean about the number of priests relative to the number of Catholics in the United States?
13. The revenue function for a product is $R(x) = 164x$ and the cost function for the product is

$$C(x) = 0.01x^2 + 20x + 300$$

where x is the number of units produced and sold.
    (a) Selling how many units of the product gives maximum profit?
    (b) What is the maximum possible profit?
14. The cost of producing x units of a product is given by

$$C(x) = 100 + 20x + 0.01x^2$$

Producing how many units will give a minimum average cost?
15. A firm sells 100 TV sets per month at $300 each, but market research indicates that it can sell 1 more set per month for each $2 reduction of the price. At what price will the revenue be maximized?
16. An open-top box is made by cutting squares from the corners of a piece of tin and folding up the sides. If the piece of tin was originally 20 centimeters on a side, how long should the sides of the removed squares be to maximize the resulting volume?
17. A company estimates that it will need 784,000 items during the coming year. It costs $420 to manufacture each item, $2500 to prepare for each production run, and $5 per year for each item stored. How many units should be in each production run so that the total costs of production and storage are minimized?
18. The table below gives the national debt, in trillions of dollars, for the years 1990–1997, with White House estimates for 1998 and 1999.
    (a) Use x as the number of years past 1900 to create a model using the data from 1990 to 1997.
    (b) Find what year the model indicates that the national debt reaches its maximum.

Year	*National Debt ($ Trillions)*
1990	3.2
1991	3.6
1992	4.0
1993	4.4
1994	4.6
1995	4.9
1996	5.2
1997	5.4
1998	5.5
1999	5.7

Source: *USA Today,* Jan. 7, 1998

Extended Applications / Group Projects

I. Production Management

Metal Containers, Inc., is reviewing the way it submits bids on U.S. Army contracts. The army often requests open-top boxes, with square bases and of specified volumes. The army also specifies the materials for the boxes, and the base is usually made of a different material than the sides. The box is put together by riveting a bracket at each of the eight corners. For Metal Containers, the total cost of producing a box is the sum of the cost of the materials for the box and the labor costs associated with affixing each bracket.

Instead of estimating each job separately, the company wants to develop an overall approach that will allow it to cost out proposals more easily. To accomplish this, company managers need you to devise a formula for the total cost of producing each box and determine the dimensions that allow a box of specified volume to be produced at minimum cost. Use the following notation to help you solve this problem.

Cost of the material for the base $= A$ per square unit

Cost of the material for the sides $= B$ per square unit

Cost of each bracket $= C$

Cost to affix each bracket $= D$

Length of the sides of the base $= x$

Height of the box $= h$

Volume specified by the army $= V$

1. Write an expression for the company's total cost in terms of these quantities.
2. At the time an order is received for boxes of a specified volume, the costs of the materials and labor will be fixed and only the dimensions will vary. Find a formula for each dimension of the box so that the total cost is a minimum.
3. The army requests bids on boxes of 48 cubic feet with base material costing the container company \$12 per square foot and side material costing \$8 per square foot. Each bracket costs \$5, and the associated labor cost is \$1 per bracket. Use your formulas to find the dimensions of the box that meet the army's requirements at a minimum cost. What is this cost?

Metal Containers asks you to determine how best to order the brackets it uses on its boxes. You are able to obtain the following information: The company uses approximately 100,000 brackets a year, and the purchase price of each is \$5. It buys the same number of brackets (say, n) each time it places an order with the supplier, and it costs \$60 to process each order. Metal Containers also has additional costs associated with storing, insuring, and financing its inventory of brackets. These carrying costs amount to 15% of the average value of inventory annually. The brackets are used steadily and deliveries are made just as inventory reaches zero, so that inventory fluctuates between zero and n brackets.

4. If the total annual cost associated with the bracket supply is the sum of the annual purchasing cost and the annual carrying costs, what order size n would minimize the total cost?
5. In the general case of the bracket-ordering problem, the order size n that minimizes the total cost of the bracket supply is called the economic order quantity, or EOQ. Use the following notations to determine a general formula for the EOQ.

Fixed cost per order $= F$	Unit cost $= C$
Quantity purchased per year $= P$	Carrying cost (as a decimal rate) $= r$

M II. Room Pricing in the Off Season

The data in the table below, from a survey of resort hotels with comparable rates on Hilton Head Island, show that room occupancy during the off season (November through February) is related to the price charged for a basic room.

Price per Day	*Occupancy Rate, %*
\$ 69	53
89	47
95	46
99	45
109	40
129	32

The goal is to use these data to help answer the following questions.

A. What price per day will maximize the daily off season revenue for a typical hotel in this group if it has 200 rooms available?
B. Suppose that for this typical hotel the daily cost is \$4592 plus \$30 per occupied room. What price will maximize the profit for this hotel in the off season?

The price per day that will maximize the off season profit for this typical hotel applies to this group of hotels. To find the room price per day that will maximize the daily revenue and the room price per day that will maximize the profit for this hotel (and thus the group of hotels) in the off season, complete the following.

1. Multiply each occupancy rate by 200 to get the hypothetical room occupancy. Create the revenue data points that compare the price with the revenue, R, which is equal to price times the room occupancy.
2. Use technology to create an equation that models the revenue, R, as a function of the price per day, x.
3. Use maximization techniques to find the price that these hotels should charge to maximize the daily revenue.
4. Use technology to get the occupancy as a function of the price, and use the occupancy function to create a daily cost function.
5. Form the profit function.
6. Use maximization techniques to find the price that will maximize the profit.

Warm-up

Prerequisite Problem Type	For Section	Answer	Section for Review
(a) Simplify: $\frac{1}{(3x)^{1/2}} \cdot \frac{1}{2}(3x)^{-1/2} \cdot 3$ (b) Write with a positive exponent: $\sqrt{x^2 - 1}$	11.1	(a) $\frac{1}{2x}$ (b) $(x^2 - 1)^{1/2}$	0.4 Rational exponents
(a) Vertical lines have _____ slopes. (b) Horizontal lines have _____ slopes.	11.3	(a) Undefined (b) 0	1.3 Slopes
Write the equation of the line passing through $(-2, -2)$ with slope 5.	11.3	$y = 5x + 8$	1.3 Equations of lines
Solve: $x^2 + y^2 - 9 = 0$, for y.	11.3	$y = \pm\sqrt{9 - x^2}$	2.1 Quadratic equations
(a) Write $\log_a(x + h) - \log_a x$ as an expression involving one logarithm. (b) Does $\ln x^4 = 4 \ln x$? (c) Does $\frac{x}{h} \log_a\left(\frac{x+h}{x}\right) = \log_a\left(1 + \frac{h}{x}\right)^{x/h}$? (d) Expand $\ln(xy)$ to separate x and y. (e) If $y = a^x$, then $x =$ _____. (f) Simplify $\ln e^x$.	11.1 11.2	(a) $\log_a\left(\frac{x+h}{x}\right)$ (b) Yes (c) Yes (d) $\ln x + \ln y$ (e) $\log_a y$ (f) x	5.2 Logarithms
Find the derivative of (a) $y = x^2 - 2x - 2$ (b) $T(q) = 400q - \frac{4}{3}q^2$ (c) $y = \sqrt{9 - x^2}$	11.1–11.5	(a) $y' = 2x - 2$ (b) $T'(q) = 400 - \frac{8}{3}q$ (c) $y' = -x(9 - x^2)^{-1/2}$	9.4–9.6 Derivatives
If $\frac{dy}{dx} = \frac{-x}{\sqrt{9 - x^2}}$, find the slope of the tangent to $y = f(x)$ at $x = \sqrt{5}$.	11.3	Slope $= -\frac{\sqrt{5}}{2}$	9.3–9.7 Derivatives

Chapter 11

Derivatives Continued

In this chapter we will develop derivative formulas for logarithmic and exponential functions and apply them to problems in the management, life, and social sciences. We will also develop methods for finding derivatives of one variable with respect to another even though the relationship between them may not be functional. This method is called ***implicit differentiation.***

We will use implicit derivatives with respect to time to solve problems involving the rates of change of two or more variables. These problems are called ***related-rates problems.***

The special business and economics applications include elasticity of demand and maximization of taxation revenue.

11.1 Derivatives of Logarithmic Functions

OBJECTIVE

- *To find derivatives of logarithmic functions*

APPLICATION PREVIEW

The table below shows the expected life span at birth for people born in certain years in the United States. Assume that for these years, the expected life span at birth can be modeled with the function $l(x) = 11.64 + 14.14 \ln x$, where x is the number of years past 1900. The graph of this function is shown in Figure 11.1. If we wanted to use this model to find the rate of change of life span with respect to the number of years past 1900, we would need the derivative of this function and hence the derivative of the logarithmic function $\ln x$.

Year	Life Span (years)	Year	Life Span (years)
1920	54.1	1987	75.0
1930	59.7	1988	74.9
1940	62.9	1989	75.2
1950	68.2	1990	75.4
1960	69.7	1991	75.5
1970	70.8	1993	75.5
1975	72.6	1994	75.7
1980	73.7	1996	76.1

Source: National Center for Health Statistics, 1997

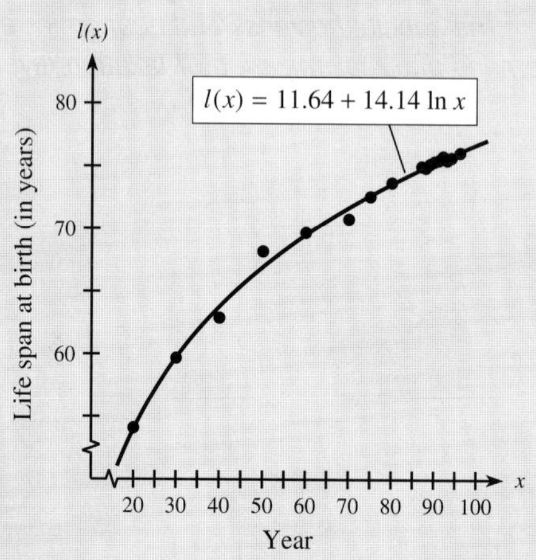

Figure 11.1

In Chapter 5, "Exponential and Logarithmic Functions," we defined the logarithmic function $y = \log_a x$ and developed properties of logarithms that will be important in this section. Although logarithmic functions can involve logarithms of any base, including base 10 (called **common logarithms**), most of the problems in calculus and many of the applications to the management, life, and social sciences involve logarithms with base e, called **natural logarithms.** We state the properties of logarithms for natural logarithms next.

Properties of Natural Logarithms

I. $\ln e^x = x$, for any real number x
II. $e^{\ln x} = x$, for $x > 0$
III. $\ln(MN) = \ln M + \ln N$, for M and N positive real numbers
IV. $\ln(M/N) = \ln M - \ln N$, for M and N positive real numbers
V. $\ln(M^N) = N \ln M$, for M a positive real number and N any real number

The formula for the derivative of $y = \ln x$ follows.

Derivative of $y = \ln x$

If $y = \ln x$, then $\dfrac{dy}{dx} = \dfrac{1}{x}$.

The proof follows.

If $y = f(x) = \ln x$, then

$$\begin{aligned}
\frac{dy}{dx} &= \lim_{h\to 0} \frac{f(x+h) - f(x)}{h} \\
&= \lim_{h\to 0} \frac{\ln(x+h) - \ln x}{h} \\
&= \lim_{h\to 0} \frac{\ln\left(\dfrac{x+h}{x}\right)}{h} && \text{Property IV} \\
&= \lim_{h\to 0} \frac{x}{x} \cdot \frac{1}{h} \ln\left(\frac{x+h}{x}\right) && \text{Introduce } \frac{x}{x}. \\
&= \lim_{h\to 0} \frac{1}{x} \cdot \frac{x}{h} \ln\left(1 + \frac{h}{x}\right) \\
&= \lim_{h\to 0} \frac{1}{x} \ln\left(1 + \frac{h}{x}\right)^{x/h} && \text{Property V}
\end{aligned}$$

The natural logarithmic function is continuous when it is defined, so

$$\frac{dy}{dx} = \frac{1}{x} \ln\left[\lim_{h\to 0}\left(1 + \frac{h}{x}\right)^{x/h}\right]$$

We can evaluate

$$\lim_{h\to 0}\left(1 + \frac{h}{x}\right)^{x/h} \tag{1}$$

by recalling from Problem 51 of Exercise 9.1 that

$$\lim_{a\to 0}(1 + a)^{1/a} = e$$

and noting that equation (1) has this form. Thus

$$\lim_{h\to 0}\left(1 + \frac{h}{x}\right)^{x/h} = e$$

and

$$\frac{dy}{dx} = \frac{1}{x} \ln e = \frac{1}{x}$$

EXAMPLE 1

If $y = x^3 + 3 \ln x$, find dy/dx.

Solution

$$\frac{dy}{dx} = 3x^2 + 3\left(\frac{1}{x}\right) = 3x^2 + \frac{3}{x}$$

EXAMPLE 2

If $y = x^2 \ln x$, find y'.

Solution

By the Product Rule,

$$y' = x^2 \cdot \frac{1}{x} + (\ln x)(2x) = x + 2x \ln x$$

We can use the Chain Rule to find the formula for the derivative of $y = \ln u$, where $u = f(x)$.

Derivatives of the Natural Logarithmic Functions

If $y = \ln u$, where u is a differentiable function of x, then

$$\frac{dy}{dx} = \frac{1}{u} \cdot \frac{du}{dx}$$

EXAMPLE 3

Find $f'(x)$ for each of the following.

(a) $f(x) = \ln(x^2)$ (b) $f(x) = \ln(x^4 - 3x + 7)$

Solution

(a) $f'(x) = \frac{1}{x^2}(2x) = \frac{2x}{x^2} = \frac{2}{x}$

(b) $f'(x) = \frac{1}{x^4 - 3x + 7}(4x^3 - 3) = \frac{4x^3 - 3}{x^4 - 3x + 7}$

EXAMPLE 4

(a) Find $f'(x)$ if $f(x) = \frac{1}{3}\ln(2x^6 - 3x + 2)$.

(b) Find $g'(x)$ if $g(x) = \frac{\ln(2x + 1)}{2x + 1}$.

Solution

(a) $f'(x)$ is $\frac{1}{3}$ of the derivative of $\ln(2x^6 - 3x + 2)$.

$$\begin{aligned} f'(x) &= \frac{1}{3} \cdot \frac{1}{2x^6 - 3x + 2}(12x^5 - 3) \\ &= \frac{4x^5 - 1}{2x^6 - 3x + 2} \end{aligned}$$

(b) We begin with the Quotient Rule.

$$g'(x) = \frac{(2x+1)\frac{1}{2x+1}(2) - [\ln(2x+1)]2}{(2x+1)^2}$$
$$= \frac{2 - 2\ln(2x+1)}{(2x+1)^2}$$

EXAMPLE 5

Use logarithm properties to find dy/dx when

$$y = \ln[x(x^5-2)^{10}]$$

Solution

We use logarithm Properties III and V to rewrite the function.

$$y = \ln x + \ln(x^5-2)^{10} \qquad \text{Property III}$$
$$y = \ln x + 10\ln(x^5-2) \qquad \text{Property V}$$

We now take the derivative.

$$\frac{dy}{dx} = \frac{1}{x} + 10 \cdot \frac{1}{x^5-2} \cdot 5x^4$$
$$= \frac{1}{x} + \frac{50x^4}{x^5-2}$$

CHECKPOINT

1. If $y = \ln(3x^2+2)$, find y'.
2. If $y = \ln x^6$, find y'.
3. If $y = \ln \sqrt[3]{x^2+1}$, find y'.

We now return to the life span model in the Application Preview.

EXAMPLE 6

Assume that the average life span (in years) for people born in 1920–1989 can be modeled by

$$l(x) = 11.64 + 14.14\ln x$$

where x is the number of years past 1900.

(a) Find the function that models the rate of change of life span.
(b) Does $l(x)$ have a maximum value for $x > 0$?
(c) Evaluate $\lim_{x\to\infty} l'(x)$.
(d) What do (b) and (c) tell us about the average life span?

Solution

(a) The rate of change of life span is given by the derivative.

$$l'(x) = 0 + 14.14\left(\frac{1}{x}\right) = \frac{14.14}{x}$$

(b) For $x > 0$, we see that $l'(x) > 0$. Hence $l(x)$ is increasing for all values of $x > 0$, so $l(x)$ never achieves a maximum value. That is, there is no maximum life span.

(c) $\lim_{x \to \infty} l'(x) = \lim_{x \to \infty} \frac{14.14}{x} = 0$

(d) If this model is accurate, life span will continue to increase, but at an ever slower rate.

EXAMPLE 7

Suppose the cost function for x units of a product is given by

$$C(x) = 18{,}250 + 615 \ln(4x + 10)$$

where $C(x)$ is in dollars. Find the marginal cost when 100 units are produced, and explain what it means.

Solution

Marginal cost is given by $C'(x)$.

$$\overline{MC} = C'(x) = 615\left(\frac{1}{4x + 10}\right)(4) = \frac{2460}{4x + 10}$$

$$\overline{MC}(100) = \frac{2460}{4(100) + 10} = \frac{2460}{410} = 6$$

When 100 units are produced, the marginal cost is 6. This means that the approximate cost of producing the 101st item is \$6.

The **change-of-base formula,** introduced in Section 5.2, "Logarithmic Functions and Their Properties," can be used to express logarithms with base a as natural logarithms (that is, logarithms with base e):

$$\log_a x = \frac{\ln x}{\ln a}$$

We can apply this change-of-base formula to find the derivative of a logarithm with any base, as the following example illustrates.

EXAMPLE 8

If $y = \log_4(x^3 + 1)$, find dy/dx.

Solution

By using the change-of-base formula, we have

$$y = \log_4(x^3 + 1) = \frac{\ln(x^3 + 1)}{\ln 4} = \frac{1}{\ln 4} \cdot \ln(x^3 + 1)$$

Thus

$$\frac{dy}{dx} = \frac{1}{\ln 4} \cdot \frac{1}{x^3 + 1} \cdot 3x^2 = \frac{3x^2}{(x^3 + 1)\ln 4}$$

EXAMPLE 9

Let $f(x) = x \ln x - x$. Use the graph of the derivative of $f(x)$ for $x > 0$ to answer the following questions.

(a) At what value a does the graph of $f'(x)$ cross the x-axis (that is, where is $f'(x) = 0$)?
(b) What value a is a critical value for $y = f(x)$?
(c) Does $f(x)$ have a relative maximum or a relative minimum at $x = a$?

Solution

$f'(x) = x \cdot \frac{1}{x} + \ln x - 1 = \ln x$. The graph of $f'(x) = \ln x$ is shown in Figure 11.2.

(a) The graph crosses the x-axis at $x = 1$, so $f'(a) = 0$ if $a = 1$.
(b) $a = 1$ is a critical value of $f(x)$.
(c) Because $f'(x)$ is negative for $x < 1$, $f(x)$ is decreasing for $x < 1$.
Because $f'(x)$ is positive for $x > 1$, $f(x)$ is increasing for $x > 1$.
Therefore, $f(x)$ has a relative minimum at $x = 1$.
The graph of $y = x \ln x - x$ is shown in Figure 11.3. It has a relative minimum at $x = 1$.

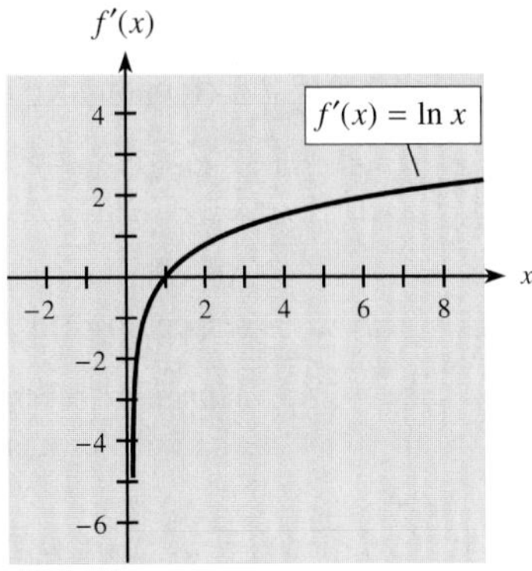

Figure 11.2

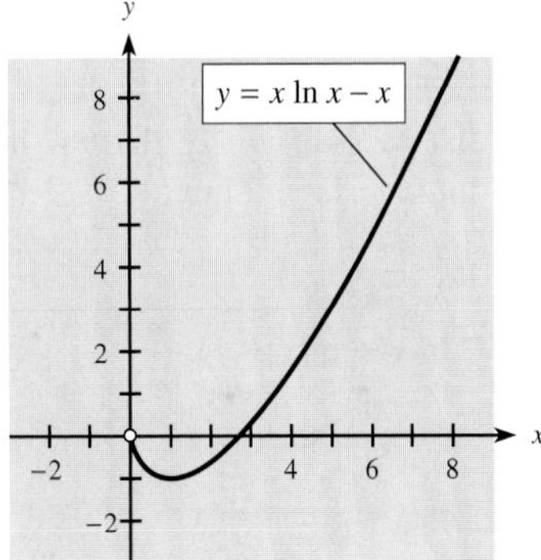

Figure 11.3

CHECKPOINT SOLUTIONS

1. $y' = \dfrac{6x}{3x^2 + 2}$

2. $y' = \dfrac{1}{x^6} \cdot 6x^5 = \dfrac{6}{x}$; or, by first using logarithm Property V, we get $y = 6 \ln x$, so $y' = 6\left(\dfrac{1}{x}\right) = \dfrac{6}{x}$.

3. By first using logarithm Property V, we get $y = \ln(x^2 + 1)^{1/3} = \frac{1}{3}\ln(x^2 + 1)$. Then

$$y' = \frac{1}{3} \cdot \frac{2x}{x^2 + 1} = \frac{2x}{3(x^2 + 1)}$$

EXERCISE 11.1

Find the derivatives of the functions in Problems 1–10.

1. $f(x) = 4 \ln x$
2. $y = 3 \ln x$
3. $y = \ln 8x$
4. $y = \ln 5x$
5. $y = \ln x^4$
6. $f(x) = \ln x^3$
7. $f(x) = \ln(4x + 9)$
8. $y = \ln(6x + 1)$
9. $y = \ln(2x^2 - x) + 3x$
10. $y = \ln(8x^3 - 2x) - 2x$
11. Find dp/dq if $p = \ln(q^2 + 1)$.
12. Find $\dfrac{ds}{dq}$ if $s = \ln\left(\dfrac{q^2}{4} + 1\right)$.

In each of Problems 13–20, find the derivative of the function in (a). Then find the derivative of the function in (b) or show that the function in (b) is the same function as that in (a).

13. (a) $y = \ln x - \ln(x - 1)$
    (b) $y = \ln \dfrac{x}{x - 1}$
14. (a) $y = \ln(x - 1) + \ln(2x + 1)$
    (b) $y = \ln[(x - 1)(2x + 1)]$
15. (a) $y = \frac{1}{3} \ln(x^2 - 1)$
    (b) $y = \ln \sqrt[3]{x^2 - 1}$
16. (a) $y = 3 \ln(x^4 - 1)$
    (b) $y = \ln(x^4 - 1)^3$
17. (a) $y = \ln(4x - 1) - 3 \ln x$
    (b) $y = \ln\left(\dfrac{4x - 1}{x^3}\right)$
18. (a) $y = 3 \ln x - \ln(x + 1)$
    (b) $y = \ln\left(\dfrac{x^3}{x + 1}\right)$
19. (a) $y = 3 \ln x + \frac{1}{2} \ln(x + 1)$
    (b) $y = \ln(x^3\sqrt{x + 1})$
20. (a) $y = 2 \ln x + \ln(x^4 - x + 1)$
    (b) $y = \ln[x^2(x^4 - x + 1)]$
21. Find $\dfrac{dp}{dq}$ if $p = \ln\left(\dfrac{q^2 - 1}{q}\right)$.
22. Find $\dfrac{ds}{dt}$ if $s = \ln[t^3(t^2 - 1)]$.
23. Find $\dfrac{dy}{dt}$ if $y = \ln\left(\dfrac{t^2 + 3}{\sqrt{1 - t}}\right)$.
24. Find $\dfrac{dy}{dx}$ if $y = \ln\left(\dfrac{3x + 2}{x^2 - 5}\right)^{1/4}$.

In Problems 25–38, find dy/dx.

25. $y = x - \ln x$
26. $y = x^2 \ln(2x + 3)$
27. $y = \dfrac{\ln x}{x}$
28. $y = \dfrac{1 + \ln x}{x^2}$
29. $y = \ln(x^4 + 3)^2$
30. $y = \ln(3x + 1)^{1/2}$
31. $y = (\ln x)^4$
32. $y = (\ln x)^{-1}$
33. $y = [\ln(x^4 + 3)]^2$
34. $y = \sqrt{\ln(3x + 1)}$
35. $y = \log_4 x$
36. $y = \log_5 x$
37. $y = \log_6(x^4 - 4x^3 + 1)$
38. $y = \log_2(1 - x - x^2)$

In Problems 39–42, find the relative maxima and relative minima, and sketch the graph with a graphing utility to check your results.

39. $y = x \ln x$
40. $y = x^2 \ln x$
41. $y = x^2 - 8 \ln x$
42. $y = \ln x - x$

Applications

43. ***Marginal cost*** Suppose that the total cost (in dollars) for a product is given by

$$C(x) = 1500 + 200 \ln(2x + 1)$$

where x is the number of units produced.
(a) Find the marginal cost function.
(b) Find the marginal cost when 200 units are produced, and interpret your result.

44. ***Investing*** The number of years t that it takes for an investment to double is a function of the interest rate r, compounded continuously, according to

$$t = \frac{\ln 2}{r}$$

At what rate is the required time changing with respect to r if $r = 10\%$, compounded continuously?

45. ***Marginal revenue*** The total revenue from the sale of x units of a product is given by

$$R(x) = \frac{2500x}{\ln(10x + 10)}$$

(a) Find the marginal revenue function.
(b) Find the marginal revenue when 100 units are sold, and interpret your result.

46. ***Supply*** Suppose that the supply of q units of a product at price x dollars is given by

$$q = 10 + 50 \ln(3x + 1)$$

Find the rate of change of supply with respect to price.

47. ***Demand*** The demand function for a product is given by $p = 4000/\ln(x + 10)$, where p is the price per unit when x units are demanded.
(a) Find the rate of change of price with respect to the number of units sold when 40 units are sold.
(b) Find the rate of change of price with respect to the number of units sold when 90 units are sold.
(c) Find the second derivative to see whether the rate at which the price is changing at 40 units is increasing or decreasing.

48. ***pH level*** The pH of a solution is given by

$$pH = -\log[H^+]$$

where $[H^+]$ is the concentration of hydrogen ions (in gram atoms per liter). What is the rate of change of pH with respect to $[H^+]$?

49. ***Reynolds number*** If the Reynolds number relating to the flow of blood exceeds R, where

$$R = A\ln(r) - Br$$

and r is the radius of the aorta and A and B are positive constants, the blood flow becomes turbulent. What is the radius r that makes R a maximum?

50. ***Decibels*** The loudness of sound (L, measured in decibels) perceived by the human ear depends on intensity levels (I) according to

$$L = 10\log(I/I_0)$$

where I_0 is the standard threshold of audibility. Using the change-of-base formula, we get

$$L = \frac{10\ln(I/I_0)}{\ln 10}$$

At what rate is the loudness changing with respect to the intensity when the intensity is 100 times the standard threshold of audibility?

51. ***Richter scale*** The Richter scale reading, R, used for measuring the magnitude of an earthquake with intensity I is determined by

$$R = \frac{\ln(I/I_0)}{\ln 10}$$

where I_0 is a standard minimum threshold of intensity. If $I_0 = 1$, what is the rate of change of the Richter scale reading with respect to intensity?

52. ***Violent crime*** The following data represent the number of violent crimes (murders, rapes, and armed robberies) per 100,000 people for the years 1987–1992.

Year	*Violent Crimes (per 100,000)*
1987	610
1988	637
1989	663
1990	732
1991	758
1992	765

Source: FBI Crime Report

Letting $t = 0$ in 1980, these data can be modeled by $y = -22.9 + 321\ln t$. If this model is accurate, at what rate did violent crimes change in 1990?

53. ***Poverty threshold*** The table below gives the average poverty thresholds for individuals for 1987–1994.
(a) Use a logarithmic equation to model these data, with x equal to the number of years past 1980.
(b) Use this model to predict the rate at which the poverty threshold will be growing in 2003.

Year	*Poverty Threshold Income*
1987	\$5778
1988	6022
1989	6310
1990	6652
1991	6932
1992	7143
1993	7363
1994	7547

Source: U.S. Bureau of the Census

54. ***Grade point averages*** The core grade point averages (based on 15 college-preparatory high school courses) of University of South Carolina freshman classes from 1992 to 1997 are shown in the table below.
(a) Find a logarithmic equation that models GPA as a function of the year of entering USC. (That is, use $x = 0$ as 1900.)
(b) Using this model, find the rate at which the grade point average was changing in 1997.

Year	*Core H.S. Grade Point Average*
1992	2.81
1993	2.91
1994	2.93
1995	3.02
1996	3.10
1997	3.11

Source: University of South Carolina

11.2 Derivatives of Exponential Functions

OBJECTIVE

- *To find derivatives of exponential functions*

APPLICATION PREVIEW

We saw in Chapter 6, "Mathematics of Finance," that the amount that accrues when \$100 is invested at 8%, compounded continuously, is

$$S(t) = 100e^{0.08t}$$

where t is the number of years. If we want to find the rate at which the money in this account is growing at the end of 1 year, then we need to find the derivative of this function, which is an exponential function.

In the previous section we found derivatives of logarithmic functions. In this section we turn our attention to exponential functions. The formula for the derivative of $y = e^x$ is developed as follows.

From Property I of logarithms, we know that

$$\ln e^x = x$$

Taking the derivative, with respect to x, of both sides of this equation, we have

$$\frac{d}{dx} \ln e^x = \frac{d}{dx} x$$

Using the Chain Rule for logarithms gives

$$\frac{1}{e^x} \cdot \frac{d}{dx} e^x = 1$$

and solving for $\frac{d}{dx} e^x$ yields

$$\frac{d}{dx} e^x = e^x$$

Thus we can conclude the following.

Derivative of y = eˣ

If $y = e^x$, then $\frac{dy}{dx} = e^x$.

EXAMPLE 1

If $p = e^q$, find dp/dq.

Solution

$$\frac{dp}{dq} = e^q$$

As with logarithmic functions, the Chain Rule permits us to expand our derivative formulas.

Derivatives of Exponential Functions

If $y = e^u$, where u is a differentiable function of x, then

$$\frac{dy}{dx} = e^u \cdot \frac{du}{dx}$$

EXAMPLE 2

If $f(x) = e^{4x^3}$, find $f'(x)$.

Solution

$$f'(x) = e^{4x^3} \cdot 12x^2 = 12x^2e^{4x^3}$$

EXAMPLE 3

If $p = qe^{5q}$, find dp/dq.

Solution

Using the Product Rule, we get

$$\frac{dp}{dq} = (q)(e^{5q} \cdot 5) + (e^{5q})(1) = 5qe^{5q} + e^{5q}$$

EXAMPLE 4

If $s = 3te^{3t^2 + 5t}$, find ds/dt.

Solution

$$\frac{ds}{dt} = 3t \cdot e^{3t^2+5t}(6t + 5) + e^{3t^2+5t} \cdot 3$$
$$= (18t^2 + 15t)e^{3t^2 + 5t} + 3e^{3t^2 + 5t}$$

CHECKPOINT

1. If $y = 2e^{4x}$, find y'.
2. If $y = e^{x^2 + 6x}$, find y'.
3. If $s = te^{t^2}$, find ds/dt.

EXAMPLE 5

If $y = e^{\ln x^2}$, find y'.

Solution

$$y' = e^{\ln x^2} \cdot \frac{1}{x^2} \cdot 2x = \frac{2}{x} e^{\ln x^2}$$

By Property II of logarithms (see the previous section and Section 5.2, "Logarithmic Functions and Their Properties"), $e^{\ln u} = u$, and we can simplify the derivative to

$$y' = \frac{2}{x} \cdot x^2 = 2x$$

Note that if we had used this property *before* taking the derivative, we would have had

$$y = e^{\ln x^2} = x^2$$

Then the derivative is $y' = 2x$.

Let us now return to the problem described in the Application Preview.

EXAMPLE 6

The future value when \$100 is invested at 8%, compounded continuously, is $S(t) = 100e^{0.08t}$, where t is the number of years. At what rate is the money in this account growing

(a) at the end of 1 year?
(b) at the end of 10 years?

Solution

The rate of growth of the money is given by

$$S'(t) = 100e^{0.08t}(0.08) = 8e^{0.08t}$$

(a) The rate of growth of the money at the end of 1 year is

$$S'(1) = 8e^{0.08} = 8.666$$

Thus the future value will change by about \$8.67 during the next year.

(b) The rate of growth of the money at the end of 10 years is

$$S'(10) = 8e^{0.08(10)} = 17.804$$

Thus the future value will change by about \$17.80 during the next year.

EXAMPLE 7

If $u = w/e^{3w}$, find u'.

Solution

The function is a quotient, with the denominator equal to e^{3w}. Using the Quotient Rule gives

$$u' = \frac{e^{3w} \cdot 1 - w \cdot e^{3w} \cdot 3}{(e^{3w})^2} = \frac{e^{3w} - 3we^{3w}}{e^{6w}} = \frac{1 - 3w}{e^{3w}}$$

EXAMPLE 8

North Forty, Inc. is a wilderness camping equipment manufacturer. The revenue function for its best-selling backpack, the Sierra, can be modeled by the function

$$R(x) = 25xe^{(1 - 0.01x)}$$

where $R(x)$ is the revenue in thousands of dollars from the sale of x thousand Sierra backpacks. Find the marginal revenue when 75,000 packs are sold, and explain what it means.

Solution

The marginal revenue function is given by $R'(x)$, and to find this derivative we use the Product Rule.

$$R'(x) = \overline{MR} = 25x[e^{(1-0.01x)} \cdot (-0.01)] + e^{(1-0.01x)}(25)$$
$$\overline{MR} = 25e^{(1-0.01x)}(1 - 0.01x)$$

To find the marginal revenue when 75,000 packs are sold, we use $x = 75$.

$$\overline{MR}\,(75) = 25e^{(1-0.75)}(1 - 0.75) \approx 8.025$$

This means that the sale of one (thousand) more Sierra backpacks will yield approximately \$8.025 (thousand) in additional revenue.

CHECKPOINT

4. If the sales of a product are given by $S = 1000e^{-0.2x}$, where x is the number of days after the end of an advertising campaign, what is the rate of decline in sales 20 days after the end of the campaign?

In a manner similar to that used to find the derivative of $y = e^x$, we can develop a formula for the derivative of $y = a^x$ for any base a.

Derivative of $y = a^u$

If $y = a^x$, then

$$\frac{dy}{dx} = a^x \ln a$$

If $y = a^u$, where u is a differentiable function of x, then

$$\frac{dy}{dx} = a^u \frac{du}{dx} \ln a$$

EXAMPLE 9

If $y = 4^x$, find dy/dx.

Solution

$$\frac{dy}{dx} = 4^x \ln 4$$

We can make use of a graphing utility to study the behavior of an exponential function and its derivative.

EXAMPLE 10

For the function $y = e^x - 3x^2$, complete the following.

(a) Approximate the critical values of the function to four decimal places.

(b) Determine whether relative maxima or relative minima occur at the critical values.

Solution

(a) The derivative is $y' = e^x - 6x$. Using the built-in features of a graphing utility, we find that $y' = 0$ at $x = 0.2045$ (see Figure 11.4a) and at $x = 2.8331$.

(b) From the graph of $y' = e^x - 6x$ in Figure 11.4(a), we can observe where $y' > 0$ and where $y' < 0$. From this we can make a sign diagram to determine relative maxima and relative minima.

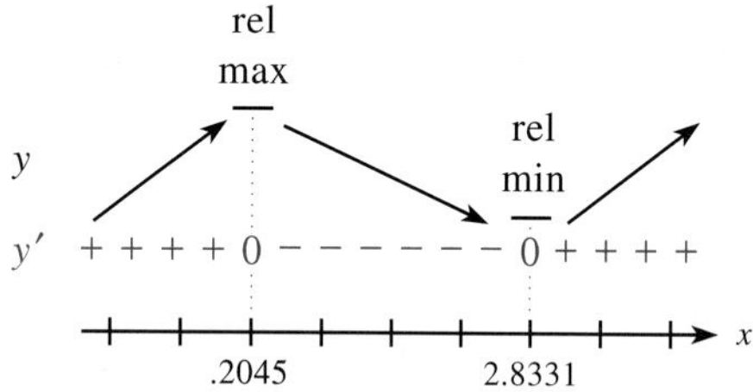

The graph of $y = e^x - 3x^2$ in Figure 11.4(b) shows that the relative maximum point is (0.2045, 1.1015) and that the relative minimum point is (2.8331, −7.0813).

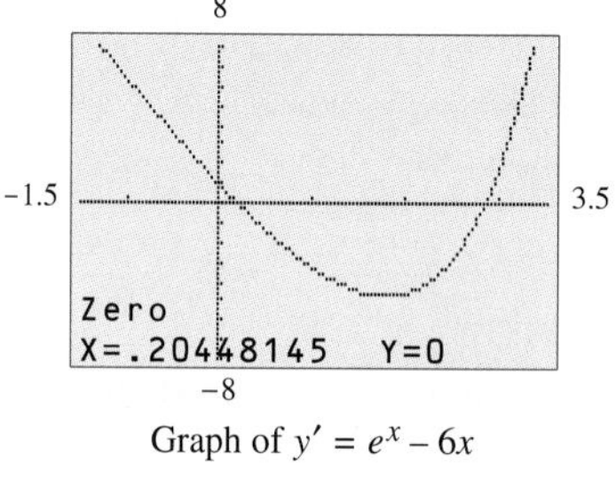

Graph of $y' = e^x - 6x$

(a)

6
−2
6
−8

Graph of $y = e^x - 3x^2$

(b)

Figure 11.4

CHECKPOINT SOLUTIONS

1. $y' = 2e^{4x}(4) = 8e^{4x}$
2. $y' = (2x + 6)e^{x^2 + 6x}$
3. By the Product Rule, $\frac{ds}{dt} = e^{t^2}(1) + t[e^{t^2}(2t)] = e^{t^2} + 2t^2e^{t^2}$.
4. The rate of decline is given by dS/dx.

$$\frac{dS}{dx} = 1000e^{-0.2x}(-0.2) = -200e^{-0.2x}$$

$$\left.\frac{dS}{dx}\right|_{x=20} = -200e^{(-0.2)(20)}$$

$$= -200e^{-4} \approx -3.663 \text{ sales/day}$$

EXERCISE 11.2

Find the derivatives of the functions in Problems 1–32.

1. $y = 5e^x - x$
2. $y = x^2 - 3e^x$
3. $f(x) = e^x - x^e$
4. $f(x) = 4e^x - \ln x$
5. $y = e^{x^3}$
6. $y = e^{x^2 - 1}$
7. $y = 6e^{3x^2}$
8. $y = 1 - 2e^{-x^3}$
9. $y = 2e^{(x^2 + 1)^3}$
10. $y = e^{\sqrt{x^2 - 9}}$
11. $y = e^{\ln x^3}$
12. $y = e^3 + e^{\ln x}$
13. $y = e^{-1/x}$
14. $y = 2e^{\sqrt{x}}$
15. $y = e^{-1/x^2} + e^{-x^2}$
16. $y = \dfrac{2}{e^{2x}} + \dfrac{e^{2x}}{2}$
17. $s = t^2e^t$
18. $p = 4qe^{q^3}$
19. $y = e^{x^4} - (e^x)^4$
20. $y = 4(e^x)^3 - 4e^{x^3}$
21. $y = \ln(e^{4x} + 2)$
22. $y = \ln(e^{2x} + 1)$
23. $y = e^{-3x}\ln(2x)$
24. $y = e^{2x^2}\ln(4x)$
25. $y = \dfrac{1 + e^{5x}}{e^{3x}}$
26. $y = \dfrac{x}{1 + e^{2x}}$
27. $y = (e^{3x} + 4)^{10}$
28. $y = \dfrac{e^x - e^{-x}}{e^x + e^{-x}}$
29. $y = 6^x$
30. $y = 3^x$
31. $y = 4^{x^2}$
32. $y = 5^{x - 1}$
33. (a) What is the slope of the line tangent to $y = xe^{-x}$ at $x = 1$?
    (b) Write the equation of the line tangent to the graph of $y = xe^{-x}$ at $x = 1$.
34. (a) What is the slope of the line tangent to $y = e^{-x}/(1 + e^{-x})$ at $x = 0$?
    (b) Write the equation of the line tangent to the graph of $y = e^{-x}/(1 + e^{-x})$ at $x = 0$.

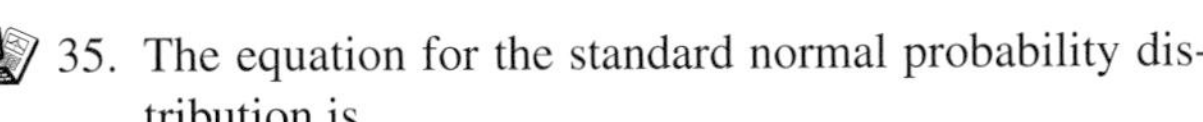

35. The equation for the standard normal probability distribution is

$$y = \frac{1}{\sqrt{2\pi}}e^{-z^2/2}$$

    (a) At what value of z will the curve be at its highest point?
    (b) Graph this function with a graphing utility to verify your answer.

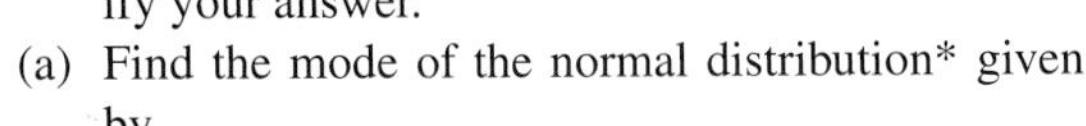

36. (a) Find the mode of the normal distribution* given by

$$y = \frac{1}{\sqrt{2\pi}}e^{-(x-10)^2/2}$$

    (b) What is the mean of this normal distribution?
    (c) Use a graphing utility to verify your answers.

*The mode occurs at the highest point on normal curves and equals the mean.

In Problems 37–40, find any relative maxima and minima. Use a graphing utility to check your results.

37. $y = \dfrac{e^x}{x}$
38. $y = \dfrac{x}{e^x}$
39. $y = x - e^x$
40. $y = \dfrac{x^2}{e^x}$

Applications

41. ***Future value*** If $P is invested for n years at 10%, compounded continuously, the future value is given by the function

$$S = Pe^{0.1n}$$

    (a) At what rate is the future value growing at any time (for any n)?
    (b) At what rate is the future value growing after 1 year ($n = 1$)?
    (c) Is the rate of growth of the future value after 1 year greater than 10%? Why?

42. ***Future value*** The future value that accrues when $700 is invested at 9%, compounded continuously, is

$$S(t) = 700e^{0.09t}$$

where t is the number of years.
    (a) At what rate is the money in this account growing when $t = 4$?
    (b) At what rate is it growing when $t = 10$?

43. ***Sales decay*** After the end of an advertising campaign, the sales of a product are given by

$$S = 100{,}000e^{-0.5t}$$

where S is weekly sales and t is the number of weeks since the end of the campaign. Find the rate of change of S (that is, the rate of *sales decay*).

44. ***Sales decay*** The sales decay for a product is given by

$$S = 50{,}000e^{-0.8t}$$

where S is the daily sales and t is the number of days since the end of a promotional campaign. Find the rate of sales decay.

45. ***Marginal cost*** Suppose that the total cost in dollars of producing x units of a product is given by

$$C(x) = 10{,}000 + 600xe^{x/600}$$

Find the marginal cost when 600 units are produced.

46. ***Marginal revenue*** Suppose that the revenue in dollars from the sale of x units of a product is given by

$$R(x) = 1000xe^{-x/50}$$

Find the marginal revenue function.

47. ***Drugs in a bloodstream*** The concentration y of a certain drug in the bloodstream at any time t (in hours) is given by

$$y = 100(1 - e^{-0.462t})$$

Find the rate of change of the concentration after 1 hour. Give your answer to three decimal places.

48. ***Radioactive decay*** The amount of the radioactive isotope thorium-234 present at time t is given by

$$Q(t) = 100e^{-0.02828t}$$

Find the rate of radioactive decay of the isotope.

49. ***Spread of disease*** Suppose that the spread of a disease through the student body at an isolated college campus can be modeled by

$$y = \frac{10{,}000}{1 + 9999e^{-0.99t}}$$

where y is the total number affected at time t (in days). Find the rate of change of y.

50. ***Spread of a rumor*** The number of people $N(t)$ in a community who are reached by a particular rumor at time t (in days) is given by

$$N(t) = \frac{50{,}500}{1 + 100e^{-0.7t}}$$

Find the rate of change of $N(t)$.

51. ***Chemical reaction*** The number of molecules of a certain substance that have enough energy to activate a reaction is given by

$$y = 100{,}000e^{-1/x}$$

where y is the number of molecules and x is the (absolute) temperature of the substance. What is the rate of change of y with respect to temperature?

52. ***Newton's law of cooling*** When a body is moved from one medium to another, its temperature T will change according to the equation

$$T = T_0 + Ce^{kt}$$

where T_0 is the temperature of the new medium, C is the temperature difference between the mediums (old − new), t is the time in the new medium, and k is a constant. If T_0, C, and k are held constant, what is the rate of change of T with respect to time?

53. ***World population*** Suppose that world population can be considered to be growing according to the equation

$$N = N_0(1 + r)^t$$

where N_0 and r are constants. Find the rate of change of N with respect to t.

54. ***Blood pressure*** Medical research has shown that between heartbeats, the pressure in the aorta of a normal adult is a function of time and can be modeled by the equation

$$P = 95e^{-0.491t}$$

(a) Use the derivative to find the rate at which the pressure changes at any time t.
(b) Use the derivative to find the rate at which the pressure changes after 0.1 second.
(c) Is the pressure increasing or decreasing?

55. ***Richter scale*** The intensity of an earthquake is related to the Richter scale reading R by

$$\frac{I}{I_0} = 10^R$$

where I_0 is a standard minimum intensity. If $I_0 = 1$, what is the rate of change of the intensity I with respect to the Richter scale reading?

56. ***Decibel readings*** The intensity level of sound, I, is given by

$$\frac{I}{I_0} = 10^{L/10}$$

where L is the decibel reading and I_0 is the standard threshold of audibility. At what rate is I/I_0 changing with respect to L when $L = 20$?

57. ***National health care*** Using data from the Congressional Budget Office (reported in *Newsweek*, October 4, 1993), the national health expenditure H can be modeled by

$$H = 45e^{0.0898t}$$

where t is the number of years past 1960 and H is in billions of dollars. If this model is accurate, at what rate did health care expenditures change in 1997?

58. ***Revenue*** According to data published in *USA Today* (March 1, 1994), the revenue R from wireless technology can be modeled by

$$R = 0.572e^{0.3860t}$$

where t is the number of years past 1985 and R is in billions of dollars. If this model is accurate, at what rate did revenue change in 1999?

59. ***U.S. debt*** The following table shows the U.S. national debt and the percent of federal expenditures devoted to payment of the interest on this debt for selected years from 1900 to 1989.

Year	*U.S. Debt (billions)*	*Percent to Interest*
1900	\$1.2	0.0
1910	1.1	0.0
1920	24.2	0.0
1930	16.1	0.0
1940	43.0	10.5
1945	258.7	4.1
1955	272.8	9.4
1965	313.8	9.6
1975	533.2	9.8
1985	1823.1	18.9
1987	2350.3	19.5
1989	2857.4	25.0

Source: *World Almanac,* 1991

Assume that the amount of the national debt (in billions of dollars) can be modeled by the function

$$d(x) = 1956.16 + 29x - 2018.91e^{x/100} + 0.44e^{x/10}$$

where x is the number of years past 1900. This function and the data points are graphed in the following figure.

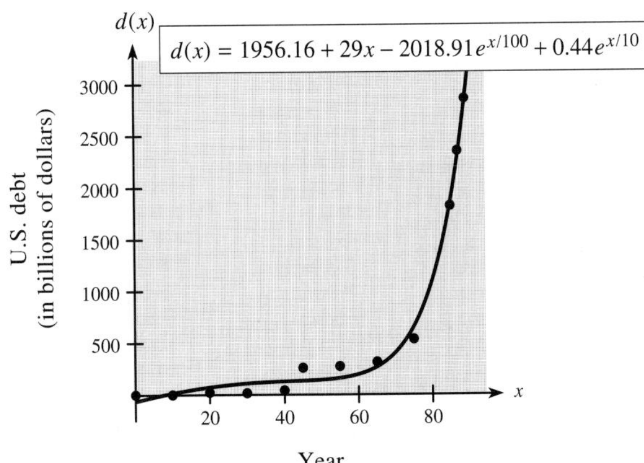

(a) What function describes how fast the national debt is changing?

(b) Find the instantaneous rate of change of the national debt model $d(x)$ in 1940 and 1985.

(c) The Gramm-Rudman Balanced Budget Act (passed in 1985) was designed to limit the growth of the national debt and hence change the function that modeled it. Why was this legislation necessary?

60. ***Social Security*** The number of workers available to support Social Security beneficiaries is declining, as the following table indicates.

Year	*No. of Workers per Retiree*
1950	16.5
1960	5.0
1980	3.7
1990	3.4

Source: Social Security Administration

These data can be modeled by the equation

$$y = 16.03e^{-0.0346t}$$

where $t = 0$ in 1940. Using this model, find the rate at which the number of workers was changing in 1990 (when $t = 50$).

61. ***Purchasing power*** The following table gives the purchasing power of \$1 based on consumer prices for 1963–1995. Using these data with $x = 0$ in 1960, the dollar's purchasing power, P, can be modeled by

$$P = 4.2885e^{-0.05751x}$$

Use the model to find the rate of decay of the purchasing power of \$1 in 1990.

Year	*Purchasing Power of \$1*	*Year*	*Purchasing Power of \$1*	*Year*	*Purchasing Power of \$1*
1963	3.265	1974	2.029	1985	0.928
1964	3.22	1975	1.859	1986	0.913
1965	3.166	1976	1.757	1987	0.880
1966	3.08	1977	1.649	1988	0.846
1967	2.993	1978	1.532	1989	0.807
1968	2.873	1979	1.38	1990	0.766
1969	2.726	1980	1.215	1991	0.734
1970	2.574	1981	1.098	1992	0.713
1971	2.466	1982	1.035	1993	0.692
1972	2.391	1983	1.003	1994	0.675
1973	2.251	1984	0.961	1995	0.656

Source: U.S. Bureau of Labor Statistics

62. ***Postal rates*** Using U.S. Postal Service data, the cost of first-class postage, y, for a letter weighing 1 ounce or less can be modeled by

$$y = 2.5206e^{0.05789x}$$

where x is the number of years past 1950. Use the model to find the rate of change of postal costs in 1974 and in 1995, with correct units.

63. ***TV cable rates*** The average monthly rates for basic cable, without premium channels, for the years 1982–1997 are given in the table.
   (a) These data can be modeled by an exponential function. Write the equation of the function, using x as the number of years past 1980.
   (b) At what rate does this model indicate that the basic cable rate will be growing in 2003? Give the correct units.

Year	*Basic Cable Rate*	*Year*	*Basic Cable Rate*
1982	\$8.30	1990	\$16.78
1983	8.61	1991	18.10
1984	8.98	1992	19.08
1985	9.73	1993	19.39
1986	10.67	1994	21.62
1987	12.18	1995	23.07
1988	13.86	1996	24.41
1989	15.21	1997	26.00

Source: Nielsen Media Research and Paul Kagan Associates, *The Island Packet,* April 19, 1998

64. ***Consumer price index*** The consumer price index (CPI) is calculated by finding the total price of various items that have been averaged according to a prescribed formula. The following table gives the consumer price indexes of all urban consumers (CPI-U) for selected years from 1940 to 1995.
   (a) With x representing years past 1900, find an exponential equation that models these data.
   (b) Use your model to predict the rate of growth in this price index in 2005.

Year	*Consumer Price Index*
1940	14
1950	24.1
1960	29.6
1970	38.8
1980	82.4
1990	130.7
1995	152.4

Source: Bureau of Labor Statistics

65. ***Prison population*** The prison population has grown exponentially from 1980 to 1995.
   (a) Use $x = 0$ in 1900 to find an exponential function that models the data given in the table below.
   (b) What does your model predict as the rate of growth of the prison population in the year 2000?

Year	*Prison Population*	*Year*	*Prison Population*
1980	319,598	1988	606,810
1981	360,029	1989	683,382
1982	402,914	1990	743,382
1983	423,898	1991	792,535
1984	448,264	1992	850,566
1985	487,593	1993	909,381
1986	526,436	1994	990,147
1987	562,814	1995	1,078,545

Source: Bureau of Justice Statistics, U.S. Dept. of Justice

11.3 Implicit Differentiation

OBJECTIVES

- *To find derivatives by using implicit differentiation*
- *To find slopes of tangents by using implicit differentiation*

APPLICATION PREVIEW

In the retail electronics industry, suppose the monthly demand for Precision, Inc., stereo headphones is given by

$$p = \frac{10,000}{(x+1)^2}$$

where p is the price in dollars per set of headphones and x is demand in hundreds of sets of headphones. If we want to find the rate of change of the quantity demanded with respect to the price, then we need to find dx/dp. Although we can (with some difficulty) solve this equation for x so that dx/dp can be found, the resulting equation does not define x as a function of p. In this case, and in other cases where we cannot solve equations for the variable we need, we can find derivatives with a technique called **implicit differentiation.**

When an equation has the form $F(x, y) = 0$, we can say that y is defined implicitly as a function of x whether or not we can solve for y. For example, the equation $xy - 4x + 1 = 0$ is in the form $F(x, y) = 0$, but we can solve for y to write the equation in the form $y = (4x - 1)/x$. We can say that $xy - 4x + 1 = 0$ defines y **implicitly** as a function of x, whereas $y = (4x - 1)/x$ defines the function explicitly.

The equation $x^2 + y^2 - 9 = 0$ has a circle as its graph. If we solve the equation for y, we get $y = \pm\sqrt{9 - x^2}$, which indicates that y is not a function of x. We can, however, consider the equation as defining *two* functions, $y = \sqrt{9 - x^2}$ and $y = -\sqrt{9 - x^2}$ (see Figure 11.5). We say that the equation $x^2 + y^2 - 9 = 0$ defines the two functions implicitly.

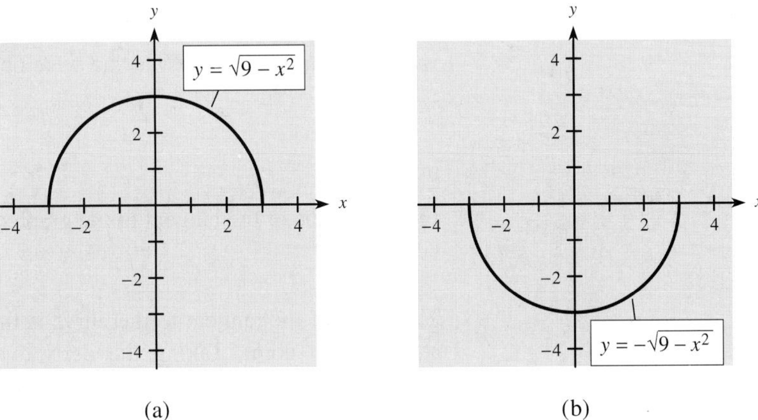

Figure 11.5

Even though an equation such as

$$\ln xy + xe^y + x - 3 = 0$$

may be difficult or even impossible to solve for y, and even though the equation may not represent y as a single function of x, we can use the technique of **implicit differentiation** to find the derivative of y with respect to x. The word *implicit* means we are implying that y is a function of x without verifying it. We simply take the derivative of both sides of $f(x, y) = 0$ and then solve algebraically for dy/dx.

For example, we can find the derivative dy/dx from $x^2 + y^2 - 9 = 0$ by taking the derivative of both sides of the equation.

$$\frac{d}{dx}(x^2 + y^2 - 9) = \frac{d}{dx}(0)$$

$$\frac{d}{dx}(x^2) + \frac{d}{dx}(y^2) + \frac{d}{dx}(-9) = \frac{d}{dx}(0)$$

We have assumed that y is a function of x; the derivative of y^2 is treated like the derivative of u^n, where u is a function of x. Thus the derivative is

$$2x + 2y^1 \cdot \frac{dy}{dx} + 0 = 0$$

Solving for dy/dx gives

$$\frac{dy}{dx} = -\frac{2x}{2y} = -\frac{x}{y}$$

Let us now compare this derivative with the derivative of the two functions $y = \sqrt{9 - x^2}$ and $y = -\sqrt{9 - x^2}$. The derivative of $y = \sqrt{9 - x^2}$ is

$$\frac{dy}{dx} = \frac{1}{2}(9 - x^2)^{-1/2}(-2x) = \frac{-x}{\sqrt{9 - x^2}}$$

and the derivative of $y = -\sqrt{9 - x^2}$ is

$$\frac{dy}{dx} = -\frac{1}{2}(9 - x^2)^{-1/2}(-2x) = \frac{x}{\sqrt{9 - x^2}}$$

Note that if we substituted $\pm\sqrt{9 - x^2}$ for y in our "implicit" derivative, we would get the two derivatives that were obtained from the "explicit" functions.

EXAMPLE 1

Find the slope of the tangent to the graph of $x^2 + y^2 - 9 = 0$ at $(\sqrt{5}, 2)$.

Solution

The slope of the tangent to the curve is the derivative of the equation, evaluated at the given point. Taking the derivative implicitly gives us $dy/dx = -x/y$. Evaluating the derivative at $(\sqrt{5}, 2)$ gives the slope of the tangent as $-\sqrt{5}/2$.

We also found the derivative of $x^2 + y^2 - 9 = 0$ by solving for y explicitly. The function whose graph contains $(\sqrt{5}, 2)$ is $y = \sqrt{9 - x^2}$, and its derivative is

$$\frac{dy}{dx} = \frac{-x}{\sqrt{9 - x^2}}$$

Evaluating the derivative at $(\sqrt{5}, 2)$, we get the slope of the tangent: $-\sqrt{5}/2$. Thus we see that both methods give us the same slope for the tangent but that the implicit method is easier to use.

EXAMPLE 2

Find dy/dx if $x^2 + 4x - 3y^2 + 4y = 0$.

Solution

Taking the derivative implicitly gives the following.

$$\frac{d}{dx}(x^2) + \frac{d}{dx}(4x) + \frac{d}{dx}(-3y^2) + \frac{d}{dx}(4y) = \frac{d}{dx}(0)$$

$$2x + 4 - 6y\frac{dy}{dx} + 4\frac{dy}{dx} = 0$$

Next we solve for $\frac{dy}{dx}$.

$$(-6y+4)\frac{dy}{dx} = -2x-4$$
$$\frac{dy}{dx} = \frac{-2x-4}{-6y+4}$$
$$\frac{dy}{dx} = \frac{x+2}{3y-2}$$

EXAMPLE 3

Write the equation of the tangent to the graph of $x^3 + xy + 4 = 0$ at the point $(2, -6)$.

Solution

Taking the derivative implicitly gives

$$\frac{d}{dx}(x^3) + \frac{d}{dx}(xy) + \frac{d}{dx}(4) = \frac{d}{dx}(0)$$

The $\frac{d}{dx}(xy)$ indicates that we should take the derivative of the *product* of x and y. Because we are assuming that y is a function of x, and because x is a function of x, we must use the Product Rule to find $\frac{d}{dx}(xy)$.

$$\frac{d}{dx}(xy) = x \cdot 1\frac{dy}{dx} + y \cdot 1 = x\frac{dy}{dx} + y$$

Thus we have $$3x^2 + \left(x\frac{dy}{dx} + y\right) + 0 = 0$$

Solving for dy/dx gives $$\frac{dy}{dx} = \frac{-3x^2 - y}{x}$$

The slope of the tangent to the curve at $x = 2$, $y = -6$ is

$$m = \frac{-3(2)^2 - (-6)}{2} = -3$$

The equation of the tangent line is

$$y - (-6) = -3[x - (2)], \quad \text{or} \quad y = -3x$$

A graphing utility can be used to graph the function of Example 3 and the line that is tangent to the curve at $(2, -6)$. To graph the equation, we solve the equation for y, getting

$$y = \frac{-x^3 - 4}{x}$$

The graph of the equation and the line that is tangent to the curve at $(2, -6)$ are shown in Figure 11.6.

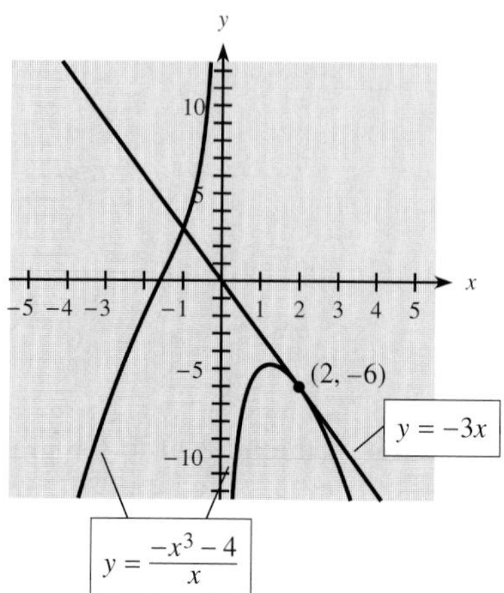

Figure 11.6

CHECKPOINT

1. Find the following:

   (a) $\frac{d}{dx}(x^3)$ (b) $\frac{d}{dx}(y^4)$ (c) $\frac{d}{dx}(x^2y^5)$

2. Find $\frac{dy}{dx}$ for $x^3 + y^4 = x^2y^5$.

EXAMPLE 4

At what point(s) does $x^2 + 4y^2 - 2x + 4y - 2 = 0$ have a horizontal tangent? At what point(s) does it have a vertical tangent?

Solution

First we find the derivative implicitly.

$$2x + 8y \cdot y' - 2 + 4y' - 0 = 0$$

$$y' = \frac{2 - 2x}{8y + 4} = \frac{1 - x}{4y + 2}$$

Horizontal tangents will occur where $y' = 0$—that is, where $x = 1$. We can now find the corresponding y-value(s) by substituting 1 for x in the original equation and solving.

$$1 + 4y^2 - 2 + 4y - 2 = 0$$

$$4y^2 + 4y - 3 = 0$$

$$(2y - 1)(2y + 3) = 0$$

$$y = \frac{1}{2} \quad \text{or} \quad y = -\frac{3}{2}$$

Thus horizontal tangents occur at $(1, \frac{1}{2})$, and $(1, -\frac{3}{2})$; see Figure 11.7.

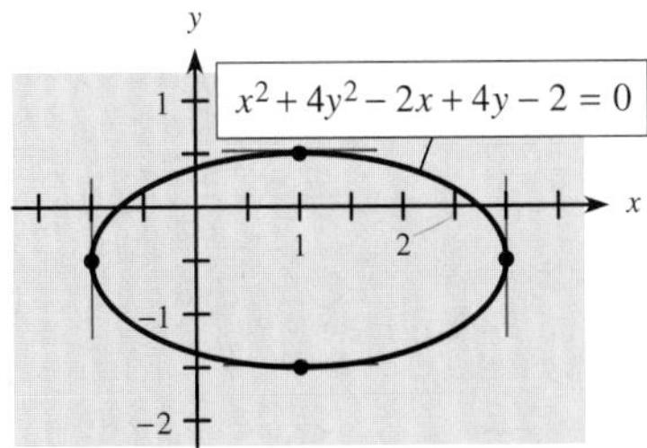

Figure 11.7

Vertical tangents will occur where the derivative is undefined—that is, where $y = -\frac{1}{2}$. To find the corresponding x-value(s), we substitute $-\frac{1}{2}$ in the equation for y and solve for x.

$$x^2 + 4\left(-\frac{1}{2}\right)^2 - 2x + 4\left(-\frac{1}{2}\right) - 2 = 0$$

$$x^2 - 2x - 3 = 0$$

$$(x - 3)(x + 1) = 0$$

$$x = 3 \quad \text{or} \quad x = -1$$

Thus vertical tangents occur at $(3, -\frac{1}{2})$ and $(-1, -\frac{1}{2})$; see Figure 11.7.

EXAMPLE 5

If $\ln xy = 6$, find dy/dx.

Solution

Using the properties of logarithms, we have

$$\ln x + \ln y = 6$$

which leads to the implicit derivative:

$$\frac{1}{x} + \frac{1}{y}\frac{dy}{dx} = 0$$

Solving gives

$$\frac{dy}{dx} = -\frac{y}{x}$$

EXAMPLE 6

Find dy/dx if $4x^2 + e^{xy} = 6y$.

Solution

We take the derivative of both sides.

$$\begin{aligned}
\frac{d}{dx}(4x^2) + \frac{d}{dx}(e^{xy}) &= \frac{d}{dx}(6y) \\
8x + e^{xy} \cdot \frac{d}{dx}(xy) &= 6\frac{dy}{dx} \\
8x + e^{xy}\left(x\frac{dy}{dx} + y\right) &= 6\frac{dy}{dx} \\
8x + xe^{xy}\frac{dy}{dx} + ye^{xy} &= 6\frac{dy}{dx} \\
8x + ye^{xy} &= 6\frac{dy}{dx} - xe^{xy}\frac{dy}{dx} \\
8x + ye^{xy} &= (6 - xe^{xy})\frac{dy}{dx} \\
\frac{8x + ye^{xy}}{6 - xe^{xy}} &= \frac{dy}{dx}
\end{aligned}$$

EXAMPLE 7

In the Application Preview, the demand for Precision, Inc., stereo headphones was given by

$$p = \frac{10{,}000}{(x+1)^2}$$

where p was the price per set in dollars and x was hundreds of headphone sets demanded. Find the rate of change of demand with respect to price when 19 (hundred) sets are demanded.

Solution

The rate of change of demand with respect to price is dx/dp. Using implicit differentiation, we get the following.

$$\begin{aligned}
\frac{d}{dp}(p) = \frac{d}{dp}\left[\frac{10{,}000}{(x+1)^2}\right] &= \frac{d}{dp}[10{,}000(x+1)^{-2}] \\
1 &= 10{,}000\left[-2(x+1)^{-3}\frac{dx}{dp}\right] \\
1 &= \frac{-20{,}000}{(x+1)^3}\frac{dx}{dp} \\
\frac{(x+1)^3}{-20{,}000} &= \frac{dx}{dp}
\end{aligned}$$

When 19 (hundred) headphone sets are demanded we use $x = 19$, and the rate of change of demand with respect to price is

$$\left.\frac{dx}{dp}\right|_{x=19} = \frac{(19+1)^3}{-20{,}000} = \frac{8000}{-20{,}000} = -0.4$$

This result means that when 19 (hundred) headphone sets are demanded, if the price per set is increased by \$1, then the expected change in demand is a decrease of 0.4 hundred, or 40, headphone sets.

Graphing Utilities To graph a function with a graphing utility, we need to write y as an *explicit* function of x (such as $y = \sqrt{4 - x^2}$). If an equation defines y as an *implicit* function of x, we have to solve for y in terms of x before we can use the graphing utility. Sometimes we cannot solve for y, and other times, such as in

$$x^{2/3} + y^{2/3} = 8^{2/3}$$

y cannot be written as a single function of x. If this equation is solved for y and a graphing utility is used to graph that function, the resulting graph usually shows only the portion of the graph that lies in quadrants I and II, and sometimes only the part in quadrant I. The complete graph is shown in Figure 11.8. Thus, for the graph of an implicitly defined function, a graphing utility must be used carefully (and sometimes cannot be used at all).

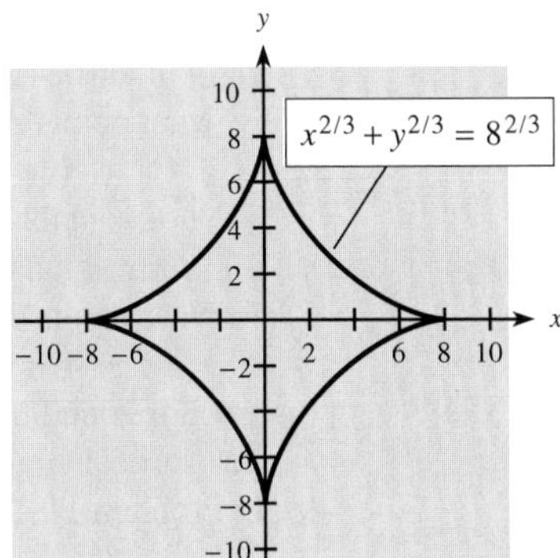

Figure 11.8

CHECKPOINT SOLUTIONS

1. (a) $\dfrac{d}{dx}(x^3) = 3x^2$ (b) $\dfrac{d}{dx}(y^4) = 4y^3\dfrac{dy}{dx}$

   (c) $\dfrac{d}{dx}(x^2y^5) = x^2\left(5y^4\dfrac{dy}{dx}\right) + y^5(2x)$ (by the Product Rule)

2. For $x^3 + y^4 = x^2y^5$, we can use the answers to Question (1) to obtain

$$3x^2 + 4y^3\frac{dy}{dx} = x^2\left(5y^4\frac{dy}{dx}\right) + y^5(2x)$$

$$3x^2 + 4y^3\frac{dy}{dx} = 5x^2y^4\frac{dy}{dx} + 2xy^5$$

$$3x^2 - 2xy^5 = (5x^2y^4 - 4y^3)\frac{dy}{dx}$$

$$\frac{3x^2 - 2xy^5}{5x^2y^4 - 4y^3} = \frac{dy}{dx}$$

EXERCISE 11.3

In Problems 1–6, find dy/dx at the given point without first solving for y.

1. $x^2 - 4y - 17 = 0$ at $(1, -4)$
2. $3x^2 - 10y + 400 = 0$ at $(10, 70)$
3. $xy^2 = 8$ at $(2, 2)$
4. $e^y = x$ at $(1, 0)$
5. $x^2 + 3xy - 4 = 0$ at $(1, 1)$
6. $x^2 + 5xy + 4 = 0$ at $(1, -1)$

Find dy/dx for the functions in Problems 7–10.

7. $x^2 + 2y^2 - 4 = 0$
8. $x + y^2 - 4y + 6 = 0$
9. $x^2 + 4x + y^2 - 3y + 1 = 0$
10. $x^2 - 5x + y^3 - 3y - 3 = 0$
11. If $x^2 + y^2 = 4$, find y'.
12. If $p^2 + 4p - q = 4$, find dp/dq.
13. If $xy^2 - y^2 = 1$, find y'.
14. If $p^2 - q = 4$, find dp/dq.
15. If $p^2q = 4p - 2$, find dp/dq.
16. If $x^2 - 3y^4 = 2x^5 + 7y^3 - 5$, find dy/dx.
17. If $3x^5 - 5y^3 = 5x^2 + 3y^5$, find dy/dx.
18. If $x^2 + 3x^2y^4 = y + 8$, find dy/dx.
19. If $x^4 + 2x^3y^2 = x - y^3$, find dy/dx.
20. If $(x + y)^2 = 5x^4y^3$, find dy/dx.
21. Find dy/dx for $x^4 + 3x^3y^2 - 2y^5 = (2x + 3y)^2$.
22. Find y' for $2x + 2y = \sqrt{x^2 + y^2}$.

For Problems 23–26, find the slope of the tangent to the curve.

23. $x^2 + 4x + y^2 + 2y - 4 = 0$ at $(1, -1)$
24. $x^2 - 4x + 2y^2 - 4 = 0$ at $(2, 2)$
25. $x^2 + 2xy + 3 = 0$ at $(-1, 2)$
26. $y + x^2 = 4$ at $(0, 4)$
27. Write the equation of the line tangent to the curve $x^2 - 2y^2 + 4 = 0$ at $(2, 2)$.
28. Write the equation of the line tangent to the curve $x^2 + y^2 + 2x - 3 = 0$ at $(-1, 2)$.
29. Write the equation of the line tangent to the curve $4x^2 + 3y^2 - 4y - 3 = 0$ at $(-1, 1)$.
30. Write the equation of the line tangent to the curve $xy + y^2 = 0$ at $(3, 0)$.
31. If $\ln x = y^2$, find dy/dx.
32. If $\ln(x + y) = y^2$, find dy/dx.
33. If $y^2 \ln x = 4$, find dy/dx.
34. If $\ln xy = 2$, find dy/dx.
35. Find the slope of the tangent to the curve $x^2 + \ln y = 4$ at the point $(2, 1)$.
36. Write the equation of the line tangent to the curve $x \ln y + 2xy = 2$ at the point $(1, 1)$.
37. If $xe^y = 6$, find dy/dx.
38. If $x + e^{xy} = 10$, find dy/dx.
39. If $e^{xy} = 4$, find dy/dx.
40. If $x - xe^y = 3$, find dy/dx.
41. If $ye^x - y = 3$, find dy/dx.
42. If $x^2y = e^{x+y}$, find dy/dx.
43. Find the slope of the line tangent to the graph of $y = xe^{-x}$ at $x = 1$.
44. Find the slope of the line tangent to the curve $ye^x = 4$ at $(0, 4)$.
45. Write the equation of the line tangent to the curve $xe^y = 3$ at $(3, 0)$.
46. Write the equation of the line tangent to the curve $ye^x = 4$ at $(0, 4)$.
47. At what points does the curve defined by $x^2 + 4y^2 - 4x - 4 = 0$ have
    (a) horizontal tangents?
    (b) vertical tangents?
48. At what points does the curve defined by $x^2 + 4y^2 - 4 = 0$ have
    (a) horizontal tangents?
    (b) vertical tangents?
49. In Problem 11, the derivative y' was found to be
$$y' = \frac{-x}{y}$$
when $x^2 + y^2 = 4$. Take the implicit derivative of the equation for y' to show that
$$y'' = \frac{-y + xy'}{y^2}$$
50. Find y' implicitly for $x^3 - y^3 = 8$. Then, by taking derivatives implicitly, use it to show that
$$y'' = \frac{2x(y - xy')}{y^3}$$
51. Use the result of Problem 49 in (a) and (b) below.
    (a) Substitute $-x/y$ for y' in the expression for y'' and simplify to show that
$$y'' = -\frac{(x^2 + y^2)}{y^3}$$
    (b) Does $y'' = -4/y^3$? Why or why not?

52. Use the result of Problem 50.
   (a) Substitute x^2/y^2 for y' in the expression for y'' and simplify to show that

$$y'' = \frac{2x(y^3 - x^3)}{y^5}$$

   (b) Does $y'' = -16x/y^5$? Why or why not?

53. Find y'' for $\sqrt{x} + \sqrt{y} = 1$ and simplify.

54. Find y'' for $\frac{1}{x} - \frac{1}{y} = 1$.

In Problems 55 and 56, find the maximum and minimum values of y. Use a graphing utility to verify your conclusion.

55. $x^2 + y^2 - 9 = 0$

56. $4x^2 + y^2 - 8x = 0$

Applications

57. ***Advertising and sales*** Suppose that a company's sales volume y (in thousands of dollars) is related to its advertising expenditures x (in thousands of dollars) according to

$$xy - 20x + 10y = 0$$

Find the rate of change of sales volume with respect to advertising expenditures when $x = 10$ (thousand dollars).

58. ***Insect control*** Suppose that the number of mosquitoes N (in thousands) in a certain swampy area near a community is related to the number of pounds of insecticide x sprayed on the nesting areas according to

$$Nx - 10x + N = 300$$

Find the rate of change of N with respect to x when 49 pounds of insecticide is used.

59. ***Production*** Suppose that a company can produce 12,000 units when the number of hours of skilled labor y and unskilled labor x satisfy

$$384 = (x + 1)^{3/4}(y + 2)^{1/3}$$

Find the rate of change of skilled-labor hours with respect to unskilled-labor hours when $x = 255$ and $y = 214$. This can be used to approximate the change in skilled-labor hours required to maintain the same production level when unskilled-labor hours are increased by 1 hour.

60. ***Production*** Suppose that production of 10,000 units of a certain agricultural crop is related to the number of hours of labor x and the number of acres of the crop y according to

$$300x + 30{,}000y = 11xy - 0.0002x^2 - 5y$$

Find the rate of change of the number of hours with respect to the number of acres.

61. ***Demand*** If the demand function for a product is given by

$$p(q + 1)^2 = 200{,}000$$

find the rate of change of quantity with respect to price when $p = \$80$. Interpret this result.

62. ***Demand*** If the demand function for a commodity is given by

$$p^2(2q + 1) = 100{,}000$$

find the rate of change of quantity with respect to price when $p = \$50$. Interpret this result.

63. ***Radioactive decay*** The number of grams of radium, y, that will remain after t years if 100 grams existed originally can be found by using the equation

$$-0.000436t = \ln\left(\frac{y}{100}\right)$$

Use implicit differentiation to find the rate of change of y with respect to t—that is, the rate at which the radium will decay.

64. ***Disease control*** Suppose the proportion of people affected by a certain disease is described by

$$\ln\left(\frac{P}{1 - P}\right) = 0.5t$$

where t is the time in months. Find dP/dt, the rate at which P grows.

65. ***Temperature-humidity index*** The temperature-humidity index (THI) is given by

$$\text{THI} = t - 0.55(1 - h)(t - 58)$$

where t is the air temperature in degrees Fahrenheit and h is the relative humidity. If the THI remains constant, find the rate of change of humidity with respect to temperature if the temperature is 70°F (Source: "Temperature-Humidity Indices," *UMAP Journal,* Fall 1989).

11.4 Related Rates

OBJECTIVE

- To use implicit differentiation to solve problems that involve related rates

APPLICATION PREVIEW

According to Poiseuille's law, the flow of blood F is related to the radius r of the vessel according to

$$F = kr^4$$

where k is a constant. When the radius of a blood vessel is reduced, such as by cholesterol deposits, the flow of blood is also restricted. Drugs can be administered that increase the radius of the blood vessel and, hence, the flow of blood. The rate of change of the blood flow and the rate of change of the radius of the blood vessel are time rates of change that are related to each other, so they are called **related rates.** We can use these related rates to find the **percentage rate of change** in the blood flow that corresponds to the percentage rate of change in the radius of the blood vessel caused by the drug.

We have seen that the derivative represents the instantaneous rate of change of one variable with respect to another. When the derivative is taken with respect to time, it represents the rate at which that variable is changing with respect to time (or the velocity). For example, if distance x is measured in miles and time t in hours, then dx/dt is measured in miles per hour and indicates how fast x is changing. Similarly, if V represents the volume (in cubic feet) of water in a swimming pool and t is time (in minutes), then dV/dt is measured in cubic feet per minute (ft^3/min) and might measure the rate at which the pool is being filled with water or being emptied.

Sometimes, two (or more) quantities that depend on time are also related to each other. For example, the height of a tree h (in feet) is related to the radius r (in inches) of its trunk, and this relationship can be modeled by

$$h = kr^{2/3}$$

where k is a constant.* Of course, both h and r are also related to time, so the rates of change dh/dt and dr/dt are related to each other. Thus they are called **related rates.**

The specific relationship between dh/dt and dr/dt can be found by differentiating $h = kr^{2/3}$ implicitly with respect to time t.

EXAMPLE 1

Suppose that for a certain type of tree, the height of the tree (in feet) is related to the radius of its trunk (in inches) by

$$h = 15r^{2/3}$$

Suppose that the rate of change of r is $\frac{3}{4}$ inch per year. Find how fast the height is changing when the radius is 8 inches.

*T. McMahon, "Size and Shape in Biology," *Science* 179 (1979): 1201.

Solution

To find how the rates dh/dt and dr/dt are related, we differentiate $h = 15r^{2/3}$ implicitly with respect to time t.

$$\frac{dh}{dt} = 10r^{-1/3}\frac{dr}{dt}$$

Using $r = 8$ inches and $dr/dt = \frac{3}{4}$ inch per year gives

$$\frac{dh}{dt} = 10(8)^{-1/3}(3/4) = \frac{15}{4} = 3\tfrac{3}{4} \text{ feet per year}$$

The work in Example 1 shows how to obtain related rates, but the different units (feet per year and inches per year) may be somewhat difficult to interpret. For this reason, many applications in the life sciences deal with **percentage rates of change.** The percentage rate of change of a quantity is the rate of change of the quantity divided by the quantity.

EXAMPLE 2

As mentioned in the Application Preview, Poiseuille's law expresses the flow of blood F as a function of the radius r of the vessel according to

$$F = kr^4$$

where k is a constant. When the radius of a blood vessel is restricted, such as by cholesterol deposits, drugs can be administered that will increase the radius of the blood vessel (and hence the blood flow). Find the percentage rate of change of the flow of blood that corresponds to the percentage rate of change of the radius of a blood vessel caused by the drug.

Solution

We seek the percentage rate of change of flow, $(dF/dt)/F$, that results from a given percentage rate of change of the radius $(dr/dt)/r$. We first find the related rates of change by differentiating

$$F = kr^4$$

implicitly with respect to time.

$$\frac{dF}{dt} = k\left(4r^3\frac{dr}{dt}\right)$$

Then the percentage rate of change of flow can be found by dividing both sides of the equation by F.

$$\frac{\frac{dF}{dt}}{F} = \frac{4kr^3\frac{dr}{dt}}{F}$$

If we replace F on the right side of the equation with kr^4 and reduce, we get

$$\frac{\frac{dF}{dt}}{F} = \frac{4kr^3\frac{dr}{dt}}{kr^4} = 4\left(\frac{\frac{dr}{dt}}{r}\right)$$

Thus we see that the percentage rate of change of the flow of blood is 4 times the corresponding percentage rate of change of the radius of the blood vessel. This means that a drug that would cause a 12% increase in the radius of a blood vessel at a certain time would produce a corresponding 48% increase in blood flow through that vessel at that time.

In these examples, the equation relating the time-dependent variables has been given. For some problems, the original equation relating the variables must first be developed from the statement of the problem. These problems can be solved with the aid of the following procedure.

Solving Related-Rates Problems

Procedure	**Example**
To solve related-rates problems:	Sand falls at a rate of 5 ft^3/min on a conical pile, with the diameter always equal to the height of the pile. At what rate is the height increasing when it is 10 ft high?
1. Use geometric and/or physical conditions to write an equation that relates the time-dependent variables.	1. The conical pile has its volume given by $$V = \frac{1}{3}\pi r^2 h$$
2. Substitute into the equation values or relationships that are true *at all times*.	2. The radius $r = \frac{1}{2}h$ at all times, so $$V = \frac{1}{3}\pi\left[\frac{1}{4}h^2\right]h = \frac{\pi}{12}h^3$$
3. Differentiate both sides of the equation implicitly with respect to time. This equation is valid for all times.	3. $\frac{dV}{dt} = \frac{\pi}{12}\left[3h^2\frac{dh}{dt}\right] = \frac{\pi}{4}h^2\frac{dh}{dt}$
4. Substitute the values that are known at the instant specified, and solve the equation.	4. $\frac{dV}{dt} = 5$ at all times, so when $h = 10$, $$5 = \frac{\pi}{4}(10^2)\frac{dh}{dt}$$
5. Solve for the specified quantity at the given time.	5. $\frac{dh}{dt} = \frac{20}{100\pi} = \frac{1}{5\pi}$ (ft/min)

Note that you should *not* substitute numerical values for any quantity that varies with time until after the derivative is taken. If values are substituted before the derivative is taken, that quantity will have the constant value resulting from the substitution and hence will have a derivative equal to zero.

EXAMPLE 3

A hot air balloon has a velocity of 50 ft/min and is flying at a constant height of 500 ft. An observer on the ground is watching the balloon approach. How fast is the distance between the balloon and the observer changing when the balloon is 1000 ft from the observer?

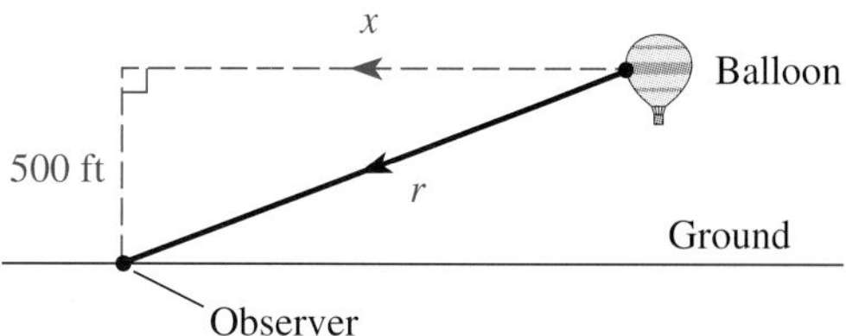

Figure 11.9

Solution

If we let r be the distance between the balloon and the observer and x be the horizontal distance from the balloon to a point directly above the observer, then we see that these quantities are related by the equation

$$x^2 + 500^2 = r^2 \qquad \text{(See Figure 11.9)}$$

Because the distance x is decreasing, we know that dx/dt must be negative. Thus we are given that $dx/dt = -50$ at all times, and we need to find dr/dt when $r = 1000$. Taking the derivative with respect to t of both sides of the equation $x^2 + 500^2 = r^2$ gives

$$2x\frac{dx}{dt} + 0 = 2r\frac{dr}{dt}$$

Using $dx/dt = -50$ and $r = 1000$, we get

$$2x(-50) = 2000\frac{dr}{dt}$$

$$\frac{dr}{dt} = \frac{-100x}{2000} = \frac{-x}{20}$$

Using $r = 1000$ in $x^2 + 500^2 = r^2$ gives $x^2 = 750{,}000$. Thus $x = 500\sqrt{3}$, and

$$\frac{dr}{dt} = \frac{-500\sqrt{3}}{20} = -25\sqrt{3}\text{ ft/min}$$

CHECKPOINT

1. If V represents volume, write a mathematical symbol that represents "the rate of change of volume with respect to time."
2. (a) Differentiate $x^2 + 64 = y^2$ implicitly with respect to time.
   (b) Suppose that we know that y is increasing at 2 units per minute. Use (a) to find the rate of change of x at the instant when $x = 6$ and $y = 10$.
3. True or false: In solving a related-rates problem, we substitute all numerical values into the equation before we take derivatives.

EXAMPLE 4

Suppose that oil is spreading in a circular pattern from a leak at an offshore rig. If the rate at which the radius of the oil slick is growing is 1 ft/min, at what rate is the area of the oil slick growing when the radius is 600 ft?

Solution

The area of the circular oil slick is given by

$$A = \pi r^2$$

where r is the radius. The rate at which the area is changing is

$$\frac{dA}{dt} = 2\pi r \frac{dr}{dt}$$

Using $r = 600$ ft and $dr/dt = 1$ ft/min gives

$$\frac{dA}{dt} = 2\pi(600 \text{ ft})(1 \text{ ft/min}) = 1200\pi \text{ ft}^2/\text{min}$$

Thus when the radius of the oil slick is 600 ft, the area is growing at the rate of 1200π ft²/min, or approximately 3770 ft²/min.

CHECKPOINT SOLUTIONS

1. dV/dt
2. (a) $2x\dfrac{dx}{dt} + 0 = 2y\dfrac{dy}{dt}$

   $$x\frac{dx}{dt} = y\frac{dy}{dt}$$

   (b) Use $\dfrac{dy}{dt} = 2$, $x = 6$, and $y = 10$ in (a) to obtain

   $$6\frac{dx}{dt} = 10(2) \quad \text{or} \quad \frac{dx}{dt} = \frac{20}{6} = \frac{10}{3}$$
3. False. The numerical values for any quantity that is varying with time should not be substituted until after the derivative is taken.

EXERCISE 11.4

In Problems 1–4, find dy/dt using the given values.

1. $y = x^3 - 3x$, for $x = 2$, $dx/dt = 4$
2. $y = 3x^3 + 5x^2 - x$ for $x = 4$, $dx/dt = 3$
3. $xy = 4$, for $x = 8$, $dx/dt = -2$
4. $xy = x + 3$, for $x = 3$, $dx/dt = -1$

In Problems 5–8, assume that x and y are differentiable functions of t. In each case, find dx/dt given that $x = 5$, $y = 12$, and $dy/dt = 2$.

5. $x^2 + y^2 = 169$
6. $y^2 - x^2 = 119$
7. $y^2 = 2xy + 24$
8. $x^2(y - 6) = 12y + 6$
9. If $x^2 + y^2 = z^2$, find dy/dt when $x = 3$, $y = 4$, $dx/dt = 10$, and $dz/dt = 2$.
10. If $s = 2\pi r(r + h)$, find dr/dt when $r = 2$, $h = 8$, $dh/dt = 3$, and $ds/dt = 10\pi$.
11. A point is moving along the graph of the equation $y = -4x^2$. At what rate is y changing when $x = 5$ and is changing at a rate of 2 units/sec?
12. A point is moving along the graph of the equation $y = 5x^3 - 2x$. At what rate is y changing when $x = 4$ and is changing at a rate of 3 units/sec?

13. The radius of a circle is increasing at a rate of 2 ft/min. At what rate is its area changing when the radius is 3 ft? (Recall that for a circle, $A = \pi r^2$.)
14. The area of a circle is changing at a rate of 1 in^2/sec. At what rate is its radius changing when the radius is 2 in.?
15. The volume of a cube is increasing at a rate of 64 in^3/sec. At what rate is the length of each edge of the cube changing when the edges are 6 in. long? (Recall that for a cube, $V = x^3$.)
16. The lengths of the edges of a cube are increasing at a rate of 8 ft/min. At what rate is the surface area changing when the edges are 24 ft long? (Recall that for a cube, $S = 6x^2$.)

Applications

17. ***Profit*** Suppose that the daily profit (in dollars) from the production and sale of x units of a product is given by

$$P = 180x - \frac{x^2}{1000} - 2000$$

At what rate is the profit changing when the number of units produced and sold is 100 and is increasing at a rate of 10 units per day?

18. ***Profit*** Suppose that the monthly revenue and cost (in dollars) for x units of a product are

$$R = 400x - \frac{x^2}{20} \quad \text{and} \quad C = 5000 + 70x$$

At what rate is the profit changing if the number of units produced and sold is 100 and is increasing at a rate of 10 units per month?

19. ***Demand*** Suppose that the price p (in dollars) of a product is given by the demand function

$$p = \frac{1000 - 10x}{400 - x}$$

where x represents the quantity demanded. If the daily demand is *decreasing* at a rate of 20 units per day, at what rate is the price changing when the demand is 20 units?

20. ***Supply*** The supply function for a product is given by $p = 40 + 100\sqrt{2x + 9}$, where x is the number of units supplied and p is the price in dollars. If the price is increasing at a rate of \$1 per month, at what rate is the supply changing when $x = 20$?

21. ***Capital investment and production*** Suppose that for a particular product, the number of units x produced per month depends on the number of thousands of dollars y invested, with $x = 30y + 20y^2$. At what rate will production increase if \$10,000 is invested and if the investment capital is increasing at a rate of \$1000 per month?

22. ***Boyle's law*** Boyle's law for enclosed gases states that at a constant temperature, the pressure is related to the volume by the equation

$$P = \frac{k}{V}$$

where k is a constant. If the volume is increasing at a rate of 5 cubic inches per hour, at what rate is the pressure changing when the volume is 30 cubic inches and $k = 2$ inch-pounds?

Tumor growth For Problems 23 and 24, suppose that a tumor in a person's body has a spherical shape and that treatment is causing the radius of the tumor to decrease at a rate of 1 millimeter per month.

23. At what rate is the volume decreasing when the radius is 3 millimeters? Recall that $V = \frac{4}{3}\pi r^3$.
24. At what rate is the surface area of the tumor decreasing when the radius is 3 mm? (Recall that for a sphere, $S = 4\pi r^2$.)

25. ***Allometric relationships—fish*** For many species of fish, the allometric relationship between the weight W and the length L is approximately $W = kL^3$, where k is a constant. Find the percentage rate of change of the weight as a corresponding percentage rate of change of the length.

26. ***Blood flow*** The resistance R of a blood vessel to the flow of blood is a function of the radius r of the blood vessel and is given by

$$R = \frac{k}{r^4}$$

where k is a constant. Find the percentage rate of change of the resistance of a blood vessel in terms of the percentage rate of change in the radius of the blood vessel.

27. ***Allometric relationships—crabs*** For fiddler crabs, data gathered by Thompson* show that the allometric relationship between the weight C of the claw and the weight W of the body is given by

$$C = 0.11W^{1.54}$$

Find the percentage rate of change of the claw weight in terms of the percentage rate of change of the body weight for fiddler crabs.

28. ***Body weight and surface area*** For human beings, the surface area S of the body is related to the body's weight W according to

$$S = kW^{2/3}$$

where k is a constant. Find the percentage rate of change of the body's surface area in terms of the percentage rate of change of the body's weight.

29. ***Cell growth*** A bacterial cell has a spherical shape. If the volume of the cell is increasing at a rate of 4 cubic micrometers per day, at what rate is the radius of the cell increasing when it is 2 micrometers? (Recall that for a sphere, $V = \frac{4}{3}\pi r^3$.)

30. ***Water purification*** Assume that water is being purified by flowing through a conical filter that has a height of 15 inches and a radius of 5 inches. If the depth of the water is decreasing at a rate of 1 inch per minute when the depth is 6 inches, at what rate is the (volume of) water flowing out of the filter at this instant?

31. ***Volume and radius*** Suppose that air is being pumped into a spherical balloon at a rate of 5 in^3/min. At what rate is the radius of the balloon increasing when the radius is 5 in.?

32. ***Boat docking*** Suppose that a boat is being pulled toward a dock by a winch that is 5 ft above the level of the boat deck. If the winch is pulling the cable at a rate of 3 ft/min, at what rate is the boat approaching the dock when it is 12 ft from the dock? Use the figure below.

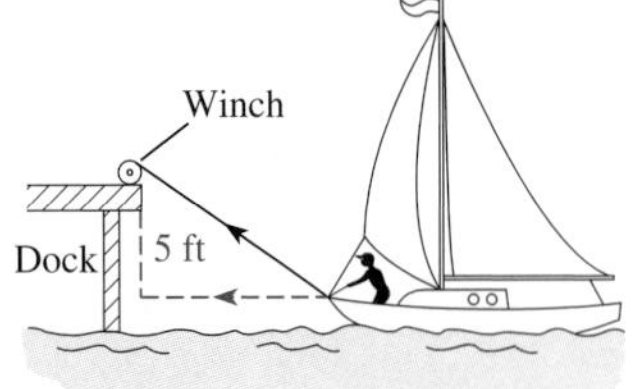

*d'Arcy Thompson, *On Growth and Form* (Cambridge, England: Cambridge University Press, 1961).

33. ***Ladder safety*** A 30-ft ladder is leaning against a wall. If the bottom is pulled away from the wall at a rate of 1 ft/sec, at what rate is the top of the ladder sliding down the wall when the bottom is 18 ft from the wall?

34. ***Flight*** A kite is 30 ft high and is moving horizontally at a rate of 10 ft/min. If the kite string is taut, at what rate is the string being played out when 50 ft of string is out?

35. ***Flight*** A plane is flying at a constant altitude of 1 mile and a speed of 300 mph. If it is flying toward an observer on the ground, how fast is the plane approaching the observer when it is 5 miles from the observer?

36. ***Distance*** Two boats leave the same port at the same time, with boat A traveling north at 15 knots and boat B traveling east at 20 knots. How fast is the distance between them changing when boat A is 30 nautical miles from port?

37. ***Distance*** Two cars are approaching an intersection on roads that are perpendicular to each other. Car A is north of the intersection and traveling south at 40 mph. Car B is east of the intersection and traveling west at 55 mph. How fast is the distance between the cars changing when car A is 15 miles from the intersection and car B is 8 miles from the intersection?

38. ***Water depth*** Water is flowing into a barrel in the shape of a right circular cylinder at the rate of 200 in^3/min. If the radius of the barrel is 18 in., at what rate is the depth of the water changing when the water is 30 in. deep?

39. ***Water depth*** Suppose that water is being pumped into a rectangular swimming pool of uniform depth at 10 ft^3/hr. If the pool is 10 ft wide and 25 ft long, at what rate is the water rising when it is 4 ft deep?

11.5 Applications in Business and Economics

OBJECTIVES

- *To find the elasticity of demand*
- *To find the tax per unit that will maximize tax revenue*

APPLICATION PREVIEW

In this section, we consider two applications: **elasticity of demand** and **taxation in a competitive market.** Elasticity of demand measures how sensitive the demand for a product is to price changes, and it can be used to measure the effect that price changes have on total revenue. Taxation in a competitive market examines how a tax levied on goods produces shifts in market equilibrium, and it also addresses the associated problem of finding the tax per unit that, despite changes in market equilibrium, maximizes tax revenues.

Elasticity of Demand

We know from the law of demand that consumers will respond to changes in prices; if prices increase, the quantity demanded will decrease. But the degree of responsiveness of the consumers to price changes will vary widely for different products. For example, a price increase in insulin will not greatly decrease the demand for it by diabetics, but a price increase in clothes may cause consumers to buy less and wear their old clothes longer. When the response to price changes is considerable, we say the demand is *elastic.* When price changes cause relatively small changes in demand for a product, the demand is said to be *inelastic* for that product.

Economists measure the **elasticity of demand** on an interval by dividing the rate of change of demand by the rate of change of price. We may write this as

$$E_d = -\frac{\text{change in quantity demanded}}{\text{original quantity demanded}} \div \frac{\text{change in price}}{\text{original price}}$$

or

$$E_d = -\frac{\Delta q}{q} \div \frac{\Delta p}{p}$$

The demand curve usually has a negative slope, so we have introduced a negative sign into the formula to give us a positive elasticity.

We can write the equation for elasticity as

$$E_d = -\frac{p}{q} \cdot \frac{\Delta q}{\Delta p}$$

If q is a function of p, then we can write

$$\frac{\Delta q}{\Delta p} = \frac{f(p + \Delta p) - f(p)}{\Delta p}$$

and the limit as Δp approaches 0 gives the **point elasticity of demand:**

$$\eta = \lim_{\Delta p \to 0}\left(-\frac{p}{q} \cdot \frac{\Delta q}{\Delta p}\right) = -\frac{p}{q} \cdot \frac{dq}{dp}$$

Elasticity The **elasticity of demand** at the point (q_A, p_A), is

$$\eta = -\frac{p}{q} \cdot \frac{dq}{dp}\bigg|_{(q_A,\, p_A)}$$

EXAMPLE 1

Find the elasticity of the demand function $p + 5q = 100$ when

(a) the price is \$40, (b) the price is \$60, and (c) the price is \$50.

Solution

Solving the demand function for q gives $q = 20 - \frac{1}{5}p$. Then $dq/dp = -\frac{1}{5}$ and

$$\eta = -\frac{p}{q}\left(-\frac{1}{5}\right)$$

(a) When $p = 40$, $q = 12$ and $\eta = -\frac{p}{q}\left(-\frac{1}{5}\right)\bigg|_{(12,\,40)} = -\frac{40}{12}\left(-\frac{1}{5}\right) = \frac{2}{3}$.

(b) When $p = 60$, $q = 8$ and $\eta = -\frac{p}{q}\left(-\frac{1}{5}\right)\bigg|_{(8,\,60)} = -\frac{60}{8}\left(-\frac{1}{5}\right) = \frac{3}{2}$.

(c) When $p = 50$, $q = 10$ and $\eta = -\frac{p}{q}\left(-\frac{1}{5}\right)\bigg|_{(10,\,50)} = -\frac{50}{10}\left(-\frac{1}{5}\right) = 1$.

Note that in Example 1 the demand equation was $p + 5q = 100$, so the demand "curve" is a straight line, with slope $m = -5$. But the elasticity was $\eta = \frac{2}{3}$ at (12, 40), $\eta = \frac{3}{2}$ at (8, 60), and $\eta = 1$ at (10, 50). This illustrates that the elasticity of demand may be different at different points on the demand curve, even though the slope of the demand "curve" is constant. (See Figure 11.10.) Economists use η to measure how responsive demand is to price at different points on the demand curve for a product.

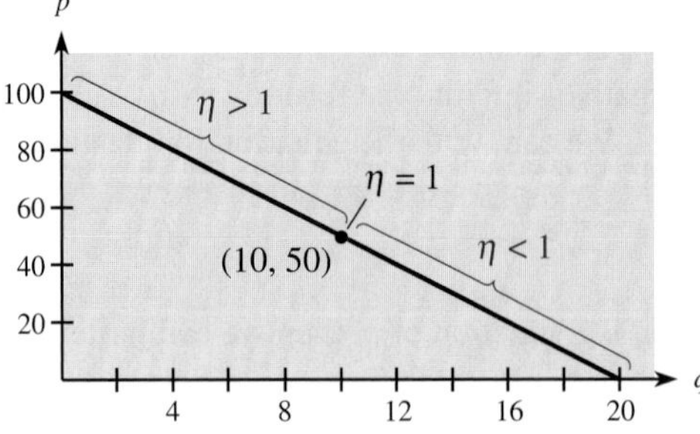

Figure 11.10

This example shows that the elasticity of demand is more than just the slope of the demand curve, which is the rate at which the demand is changing. Recall that the elasticity measures the consumers' degree of responsiveness to a price change.

Economists classify demand curves according to how responsive demand is to price changes by using elasticity.

Elasticity of Demand

- If $\eta > 1$, the demand is **elastic,** and the percent decrease in demand is greater than the corresponding percent increase in price.
- If $\eta < 1$, the demand is **inelastic,** and the percent decrease in demand will be less than the corresponding percent increase in price.
- If $\eta = 1$, the demand is **unitary elastic,** and the percent decrease in demand is approximately equal to the corresponding percent increase in price.

We can also use implicit differentiation to find dq/dp in evaluating the point elasticity of demand.

EXAMPLE 2

The demand for a certain product is given by

$$p = \frac{1000}{(q+1)^2}$$

where p is the price per unit in dollars and q is demand in units of the product. Find the elasticity of demand with respect to price when $q = 19$.

Solution

To find the elasticity, we need to find dq/dp. Using implicit differentiation, we get the following:

$$\frac{d}{dp}(p) = \frac{d}{dp}[1000(q+1)^{-2}]$$

$$1 = 1000\left[-2(q+1)^{-3}\frac{dq}{dp}\right]$$

$$1 = \frac{-2000}{(q+1)^3}\frac{dq}{dp}$$

$$\frac{(q+1)^3}{-2000} = \frac{dq}{dp}$$

When $q = 19$, we have $p = 1000/(19+1)^2 = 1000/400 = 5/2$ and

$$\left.\frac{dq}{dp}\right|_{(q=19)} = \frac{(19+1)^3}{-2000} = \frac{8000}{-2000} = -4$$

The elasticity of demand when $q = 19$ is

$$\eta = \frac{-p}{q}\cdot\frac{dq}{dp} = -\frac{(5/2)}{19}\cdot(-4) = \frac{10}{19} < 1$$

Thus the demand for this product is inelastic.

Elasticity is related to revenue in a special way. We can see how by computing the derivative of the revenue function

$$R = pq$$

with respect to p.

$$\begin{aligned}\frac{dR}{dp} &= p \cdot \frac{dq}{dp} + q \cdot 1 \\ &= \frac{q}{q} \cdot p \cdot \frac{dq}{dp} + q = q \cdot \frac{p}{q} \cdot \frac{dq}{dp} + q \\ &= q(-\eta) + q \\ &= q(1 - \eta)\end{aligned}$$

From this we can summarize the relationship of elasticity and revenue.

Elasticity and Revenue The rate of change of revenue R with respect to price p is related to elasticity in the following way.

- Elastic ($\eta > 1$) means $\frac{dR}{dp} < 0$. $\begin{cases}\text{Hence if price increases, revenue decreases,} \\ \text{and if price decreases, revenue increases.}\end{cases}$
- Inelastic ($\eta < 1$) means $\frac{dR}{dp} > 0$. $\begin{cases}\text{Hence if price increases, revenue increases,} \\ \text{and if price decreases, revenue decreases.}\end{cases}$
- Unitary elastic ($\eta = 1$) means $\frac{dR}{dp} = 0$. Hence an increase or decrease in price will not change revenue. Revenue is optimized at this point.

EXAMPLE 3

The demand for a product is given by

$$p = 10\sqrt{100 - q}, \qquad 0 \le q \le 100$$

(a) Find the point at which demand is of unitary elasticity, and find intervals in which the demand is inelastic and intervals in which it is elastic.

(b) Find where revenue is increasing, where it is decreasing, and where it is maximized.

(c) Use a graphing utility to show the graph of the revenue function $R = pq$, with $0 \le q \le 100$, and confirm the results from (b).

Solution

The elasticity is

$$\eta = -\frac{10\sqrt{100 - q}}{q} \cdot \frac{dq}{dp}$$

Finding dq/dp implicitly, we have

$$1 = 10\left[\frac{1}{2}(100-q)^{-1/2}\left(-\frac{dq}{dp}\right)\right]$$

so

$$\frac{dq}{dp} = -\frac{1}{5}\sqrt{100-q}$$

Thus

$$\eta = -\frac{10\sqrt{100-q}}{q}\left[-\frac{1}{5}\sqrt{100-q}\right] = \frac{200-2q}{q}$$

(a) Unitary elasticity occurs where $\eta = 1$.

$$1 = \frac{200-2q}{q}$$
$$q = 200 - 2q$$
$$3q = 200$$
$$q = 66\tfrac{2}{3}$$

so unitary elasticity occurs when $66\frac{2}{3}$ units are sold, at a price of \$57.74. For values of q between 0 and $66\frac{2}{3}$, $\eta > 1$ and demand is elastic. For values of q between $66\frac{2}{3}$ and 100, $\eta < 1$ and demand is inelastic.

(b) When q increases over $0 < q < 66\frac{2}{3}$, p decreases, so $\eta > 1$ means R increases. Similarly, when q increases over $66\frac{2}{3} < q < 100$, p decreases, so $\eta < 1$ means R decreases. Revenue is maximized where $\eta = 1$, at $q = 66\frac{2}{3}$, $p = 57.74$.

(c) The graph of this revenue function,

$$R = 10q\sqrt{1-q}$$

is shown in Figure 11.11 and confirms our conclusions from (b).

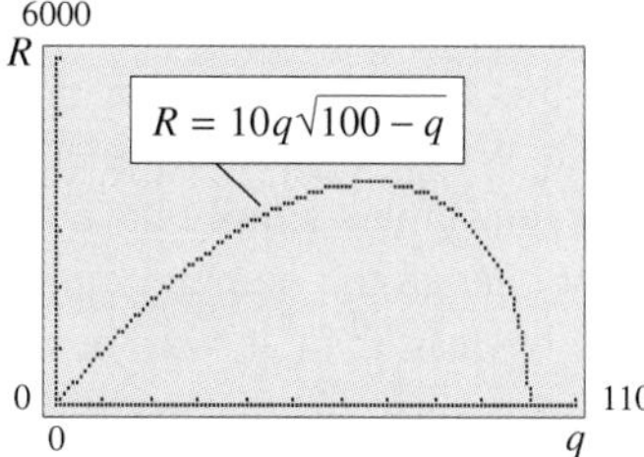

Figure 11.11

CHECKPOINT

1. Write the formula for point elasticity, η.
2. (a) If $\eta > 1$, the demand is called ________.
   (b) If $\eta < 1$, the demand is called ________.
   (c) If $\eta = 1$, the demand is called ________.
3. Find the elasticity of demand for $q = \dfrac{100}{p} - 1$ when $p = 10$ and $q = 9$.

Taxation in a Competitive Market

Many taxes imposed by governments are "hidden." That is, the tax is levied on goods produced, and the producers must pay the tax. Of course, the tax becomes a cost to the producers, and they pass that cost on to the consumer in the form of higher prices for goods.

Suppose the government imposes a tax of t dollars on each unit produced and sold by producers. If we are in pure competition in which the consumers' demand depends only on price, the *demand function* will not change. The tax will change the supply function, of course, because at each level of output q, the firm will want to charge a price higher by the amount of the tax.

The graphs of the market demand function, the original market supply function, and the market supply function after taxes are shown in Figure 11.12. Because the tax added to each item is constant, the graph of the supply function is t units above the original supply function. If $p = f(q)$ defines the original supply function, then $p = f(q) + t$ defines the supply function after taxation.

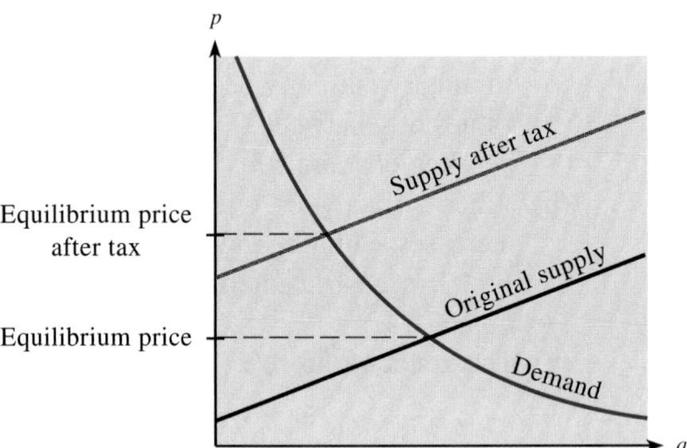

Figure 11.12

Note in this case that after the taxes are imposed, *no* items are supplied at the price that was the equilibrium price before taxation. After the taxes are imposed, the consumers simply have to pay more for the product. Because taxation does not change the demand curve, the quantity purchased at market equilibrium will be less than it was before taxation. Thus governments planning taxes should recognize that they will not collect taxes on the original equilibrium quantity. They will collect on the *new* equilibrium quantity, a quantity reduced by their taxation. Thus a large tax on each item may reduce the quantity demanded at the new market equilibrium so much that very little revenue results from the tax!

If the tax revenue is represented by $T = tq$, where t is the tax per unit and q is the equilibrium quantity of the supply and demand functions after taxation, we can use the following procedure for maximizing the total tax revenue in a competitive market.

Maximizing Total Tax Revenue

Procedure	Example
To find the tax per item (under pure competition) that will maximize total tax revenue:	If the demand and supply functions are given by $p = 600 - q$ and $p = 200 + \frac{1}{3}q$, respectively, find the tax rate t that will maximize the total tax revenue T.
1. Write the supply function after taxation.	1. $p = 200 + \frac{1}{3}q + t$
2. Set the demand function and the new supply function equal, and solve for t.	2. $600 - q = 200 + \frac{1}{3}q + t$ $400 - \frac{4}{3}q = t$
3. Form the total tax revenue function, $T = tq$, by multiplying the equation for t by q, and then take its derivative with respect to q.	3. $T = tq = 400q - \frac{4}{3}q^2$ $T'(q) = \frac{dT}{dq} = 400 - \frac{8}{3}q$
4. Set $T' = 0$, and solve for q. This is the q that should maximize T. Use the second-derivative test to verify it.	4. $0 = 400 - \frac{8}{3}q$ $q = 150$ $T''(q) = -\frac{8}{3}$. Thus T is maximized at $q = 150$.
5. Substitute the value of q into the equation for t (in step 2). This is the value of t that will maximize T.	5. $t = 400 - \frac{4}{3}(150) = 200$ A tax of \$200 per item will maximize the total tax revenue. The total tax revenue for the period would be \$200 · (150) = \$30,000.

Note that in the example just given, if a tax of \$300 were imposed, the total tax revenue the government would receive would be

$$(\$300)(75) = \$22{,}500$$

This means that consumers would spend \$100 more for each item, suppliers would sell 75 fewer items, and the government would lose \$7500 in tax revenue. Thus everyone would suffer if the tax rate were raised to \$300.

CHECKPOINT

4. For problems involving taxation in a competitive market, if supply is $p = f(q)$ and demand is $p = g(q)$, is the tax t added to $f(q)$ or to $g(q)$?

EXAMPLE 4

The demand and supply functions for a product are $p = 900 - 20q - \frac{1}{3}q^2$ and $p = 200 + 10q$, respectively. Find the tax per unit that will maximize the tax revenue T.

Solution

After taxation, the supply function is $p = 200 + 10q + t$, where t is the tax per unit. The demand function will meet the new supply function where

$$900 - 20q - \frac{1}{3}q^2 = 200 + 10q + t$$

so

$$t = 700 - 30q - \frac{1}{3}q^2$$

Then the total tax T is $T = tq = 700q - 30q^2 - \frac{1}{3}q^3$, and we maximize T as follows:

$$\begin{aligned} T'(q) &= 700 - 60q - q^2 \\ 0 &= -(q + 70)(q - 10) \\ q &= 10 \quad \text{or} \quad q = -70 \end{aligned}$$

Because $q = -70$ is meaningless, we test $q = 10$.

$$\left.\begin{aligned} &T(0) = 0 \\ &T'(0) = 700 > 0 \\ &T'(q) < 0 \text{ for all } q > 10 \end{aligned}\right\} \Rightarrow \text{absolute maximum at } q = 10$$

The maximum possible tax revenue is

$$T(10) = \$3666.67$$

The tax per unit that maximizes T is

$$t = 700 - 30(10) - \frac{1}{3}(10)^2 = \$366.67$$

An infamous example of a tax increase that resulted in decreased tax revenue and economic disaster is the "luxury tax" that went into effect in 1991. This was a 10% excise tax on the sale of more expensive jewelry, furs, airplanes, certain expensive boats, and automobiles costing over \$30,000. The Congressional Joint Tax Committee had estimated that the luxury tax would raise \$6 million from airplanes alone, but it raised only \$53,000 while it destroyed the small-airplane market (one company lost \$130 million and 480 jobs in a single year). It also capsized the boat market. The luxury tax was repealed at the end of 1993 (except for automobiles).*

***Fortune,* Sept. 6, 1993; *Motor Trend,* December 1993.

CHECKPOINT SOLUTIONS

1. $\eta = \dfrac{-p}{q} \cdot \dfrac{dq}{dp}$
2. (a) elastic (b) inelastic (c) unitary elastic
3. $\dfrac{dq}{dp} = \dfrac{-100}{p^2}$ and $\dfrac{dq}{dp} = -1$ when $p = 10, q = 9$

   $\eta = \dfrac{-10}{9}(-1) = \dfrac{10}{9}$ (elastic)
4. Tax t is added to supply: $p = f(q) + t$.

EXERCISE 11.5

Elasticity of Demand

1. (a) Find the elasticity of the demand function $p + 4q = 80$ at $(10, 40)$.
   (b) How will a price increase affect total revenue?
2. (a) Find the elasticity of the demand function $2p + 3q = 150$ at the price $p = 15$.
   (b) How will a price increase affect total revenue?
3. (a) Find the elasticity of the demand function $p^2 + 2p + q = 49$ at $p = 6$.
   (b) How will a price increase affect total revenue?
4. (a) Find the elasticity of the demand function $pq = 81$ at $p = 3$.
   (b) How will a price increase affect total revenue?
5. Suppose that the demand for a product is given by $pq + p = 5000$.
   (a) Find the elasticity when $p = \$50$ and $q = 99$.
   (b) Tell what type of elasticity this is: unitary, elastic, or inelastic.
   (c) How would revenue be affected by a price increase?
6. Suppose that the demand for a product is given by $2p^2q = 10{,}000 + 9000p^2$.
   (a) Find the elasticity when $p = \$50$ and $q = 4502$.
   (b) Tell what type of elasticity this is: unitary, elastic, or inelastic.
   (c) How would revenue be affected by a price increase?
7. Suppose that the demand for a product is given by $pq + p + 100q = 50{,}000$.
   (a) Find the elasticity when $p = \$401$.
   (b) Tell what type of elasticity this is.
   (c) How would a price increase affect revenue?
8. Suppose that the demand for a product is given by

   $$(p + 1)\sqrt{q + 1} = 1000$$

   (a) Find the elasticity when $p = \$39$.
   (b) Tell what type of elasticity this is.
   (c) How would a price increase affect revenue?
9. Suppose the demand function for a product is given by

   $$p = \frac{1}{2}\ln\left(\frac{5000 - q}{q + 1}\right)$$

   where p is in hundreds of dollars and q is the number of tons.
   (a) What is the elasticity of demand when the quantity demanded is 2 tons and the price is \$371?
   (b) Is the demand elastic or inelastic?
10. Suppose the weekly demand function for a product is

    $$q = \frac{5000}{1 + e^{2p}} - 1$$

    where p is the price in thousands of dollars and q is the number of units demanded. What is the elasticity of demand when price is \$1000 and the quantity demanded is 595?

 In Problems 11 and 12, the demand functions for specialty steel products are given. For both problems:

(a) Find the elasticity of demand as a function of quantity demanded, q.

(b) Find the point at which demand is of unitary elasticity and find intervals in which the demand is inelastic and intervals in which it is elastic.

(c) Use information about elasticity in (b) to decide where the revenue is increasing, where it is decreasing, and where it is maximized.

(d) Graph the revenue function $R = pq$, and use it to find where revenue is maximized. Is it at the same quantity as that determined in (c)?

11. $p = 120\sqrt[3]{125 - q}$
12. $p = 30\sqrt{49 - q}$

Taxation in a Competitive Market

13. If the weekly demand function is $p = 30 - q$ and the supply function before taxation is $p = 6 + 2q$, what tax per item will maximize the total tax revenue?
14. If the demand function for a fixed period of time is given by $p = 38 - 2q$ and the supply function before taxation is $p = 8 + 3q$, what tax per item will maximize the total tax revenue?
15. If the demand and supply functions for a product are $p = 800 - 2q$ and $p = 100 + 0.5q$, respectively, find the tax per unit t that will maximize the tax revenue T.
16. If the demand and supply functions for a product are $p = 2100 - 3q$ and $p = 300 + 1.5q$, respectively, find the tax per unit t that will maximize the tax revenue T.
17. If the weekly demand function is $p = 200 - 2q^2$ and the supply function before taxation is $p = 20 + 3q$, what tax per item will maximize the total tax revenue?
18. If the monthly demand function is $p = 7230 - 5q^2$ and the supply function before taxation is $p = 30 + 30q^2$, what tax per item will maximize the total revenue?
19. Suppose the weekly demand for a product is given by $p + 2q = 840$ and the weekly supply before taxation is given by $p = 0.02q^2 + 0.55q + 7.4$. Find the tax per item that produces maximum tax revenue. Find the tax revenue.
20. If the daily demand for a product is given by the function $p + q = 1000$ and the daily supply before taxation is $p = q^2/30 + 2.5q + 920$, find the tax per item that maximizes tax revenue. Find the tax revenue.
21. If the demand and supply functions for a product are $p = 2100 - 10q - 0.5q^2$ and $p = 300 + 5q + 0.5q^2$, respectively, find the tax per unit t that will maximize the tax revenue T.
22. If the demand and supply functions for a product are $p = 5000 - 20q - 0.7q^2$ and $p = 500 + 10q + 0.3q^2$, respectively, find the tax per unit t that will maximize the tax revenue T.

KEY TERMS AND FORMULAS

Section	Key Terms	Formula
11.1	Logarithmic function	$y = \log_a x$, defined by $x = a^y$
	Natural logarithm	$\ln x = \log_e x$
	Logarithmic Properties I–V for natural logarithms	$\ln e^x = x$; $e^{\ln x} = x$; $\ln(MN) = \ln M + \ln N$; $\ln(M/N) = \ln M - \ln N$; $\ln(M^N) = N(\ln M)$
	Change-of-base formula	$\log_a x = \dfrac{\ln x}{\ln a}$
	Derivatives of logarithmic functions	$\dfrac{d}{dx}(\ln x) = \dfrac{1}{x}$
		$\dfrac{d}{dx}(\ln u) = \dfrac{1}{u} \cdot \dfrac{du}{dx}$
11.2	Exponential function	$f(x) = a^x$
	e	$e = \lim\limits_{a \to 0} (1 + a)^{1/a}$
	Derivatives of exponential functions	$\dfrac{d}{dx}(e^x) = e^x$
		$\dfrac{d}{dx} e^u = e^u \dfrac{du}{dx}$
		$\dfrac{d}{dx} a^u = a^u \dfrac{du}{dx} \ln a$
11.3	Implicit differentiation	
11.4	Related rates	
	Percentage rates of change	
11.5	Elasticity of demand	$\eta = \dfrac{-p}{q} \cdot \dfrac{dq}{dp}$
	Elastic	$\eta > 1$
	Inelastic	$\eta < 1$
	Unitary elastic	$\eta = 1$
	Taxation in competitive market	
	Supply function after taxation	$p = f(q) + t$

REVIEW EXERCISES

Sections 11.1 and 11.2

In Problems 1–8, find the indicated derivative.

1. If $y = e^{3x^2 - x}$, find dy/dx.
2. If $y = \ln e^{x^2}$, find y'.
3. If $p = \ln\left(\frac{q}{q^2 - 1}\right)$, find $\frac{dp}{dq}$.
4. If $y = xe^{x^2}$, find dy/dx.
5. If $y = 3^{3x-4}$, find dy/dx.
6. If $y = 1 + \ln x^{10}$, find dy/dx.
7. If $y = \frac{\ln x}{x}$, find $\frac{dy}{dx}$.
8. If $y = \frac{1 + e^{-x}}{1 - e^{-x}}$, find $\frac{dy}{dx}$.
9. Write the equation of the line tangent to $y = 4e^{x^3}$ at $x = 1$.
10. Write the equation of the line tangent to $y = x \ln x$ at $x = 1$.

Section 11.3

In Problems 11–16, find the indicated derivative.

11. If $y \ln x = 5$, find dy/dx.
12. Find dy/dx for $e^{xy} = y$.
13. Find dy/dx for $y^2 = 4x - 1$.
14. Find dy/dx if $x^2 + 3y^2 + 2x - 3y + 2 = 0$.
15. Find dy/dx for $3x^2 + 2x^3y^2 - y^5 = 7$.
16. Find the second derivative of $x^2 + y^2 = 1$.
17. Find the slope of the tangent to the curve $x^2 + 4x - 3y^2 + 6 = 0$ at $(3, 3)$.
18. Find the points where tangents to the graph of the equation in Problem 17 are horizontal.

Section 11.4

19. Suppose $3x^2 - 2y^3 = 10y$, where x and y are differentiable functions of t. If $dx/dt = 2$, find dy/dt when $x = 10$ and $y = 5$.
20. A right triangle with legs of lengths x and y has its area given by

$$A = \frac{1}{2}xy$$

If the rate of change of x is 2 units per minute and the rate of change of y is 5 units per minute, find the rate of change of the area when $x = 4$ and $y = 1$.

Applications

Section 11.2

21. ***Compound interest*** If the future value of \$1000 invested for n years at 12%, compounded continuously, is given by

$$S = 1000e^{0.12n}$$

find the rate at which the future value is growing after 1 year.

22. ***Compound interest***
    (a) In Problem 21, find the rate of growth of the future value after 2 years.
    (b) How much faster is the future value growing at the end of 2 years than after 1 year?
23. ***Radioactive decay*** A breeder reactor converts stable uranium-238 into the isotope plutonium-239. The decay of this isotope is given by

$$A(t) = A_0e^{-0.00002876t}$$

where $A(t)$ is the amount of isotope at time t, in years, and A_0 is the original amount. This isotope has a half-life of 24,101 years (that is, half of it will decay away in 24,101 years).
    (a) At what rate is $A(t)$ decaying at this point in time?
    (b) At what rate is $A(t)$ decaying after 1 year?
    (c) Is the rate of decay at its half-life greater or less than after 1 year?
24. ***Marginal cost*** The average cost of producing x units of a product is $\overline{C} = 600e^{x/600}$. What is the marginal cost when 600 units are produced?
25. ***Inflation*** The impact of inflation on a \$20,000 pension can be measured by the purchasing power P of \$20,000 after t years. For an inflation rate of 5% per year, compounded annually, P is given by

$$P = 20{,}000e^{-0.0495t}$$

At what rate is purchasing power changing when $t = 10$? (Source: *Viewpoints*, VALIC, Summer 1993)

Section 11.4

26. ***Evaporation*** A spherical droplet of water evaporates at a rate of 1 mm^3/min. Find the rate of change of the radius when the droplet has a radius of 2.5 mm.

27. ***Worker safety*** A sign is being lowered over the side of a building at the rate of 2 ft/min. A worker handling a guide line is 7 ft away from a spot directly below the sign. How fast is the worker taking in the guide line at the instant the sign is 25 ft from the worker's hands? See the figure below.

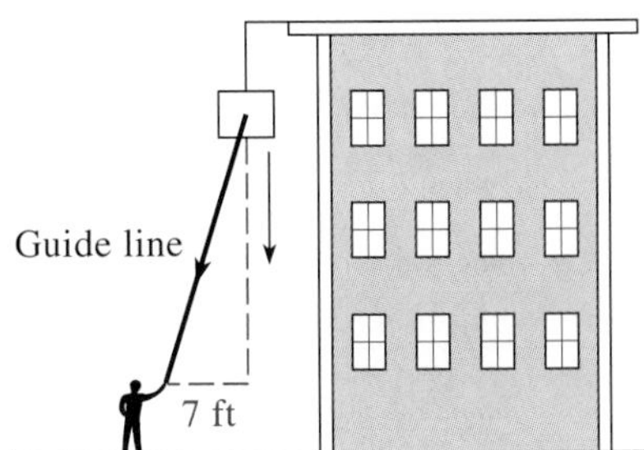

28. ***Environment*** Suppose that in a study of water birds, the relationship between the square miles of wetlands A and the number of different species S of birds found in the area was determined to be

$$S = kA^{1/3}$$

where k is constant. Find the percentage rate of change of the number of species in terms of the percentage rate of change of the area.

Section 11.5

29. ***Taxes*** Can increasing the tax per unit sold actually lead to a decrease in tax revenues?

30. ***Taxes*** If the demand and supply functions for a product are

$$p = 2800 - 8q - \frac{q^2}{3} \quad \text{and} \quad p = 400 + 2q$$

respectively, find the tax per unit t that will maximize the tax revenue T.

31. ***Taxes*** If the supply and demand functions for a product are

$$p = 40 + 20q \quad \text{and} \quad p = \frac{5000}{q + 1}$$

respectively, find the tax t that maximizes the tax revenue T.

32. ***Elasticity*** A demand function is given by

$$pq = 27$$

(a) Find the elasticity of demand at (9, 3).
(b) How will a price increase affect total revenue?

33. ***Elasticity*** Suppose the demand for a product is given by

$$p^2(2q + 1) = 10{,}000$$

Find the elasticity of demand when $p = \$20$.

34. ***Elasticity*** Suppose the weekly demand function for a product is given by

$$p = 100e^{-0.1q}$$

where p is the price in dollars and q is the number of tons demanded. What is the elasticity of demand when price is \$36.79 and the quantity demanded is 10?

35. ***Revenue*** A product has the demand function

$$p = 100 - 0.5q$$

(a) Find the elasticity $\eta(q)$ as a function of q, and graph the function

$$f(q) = \eta(q) - 1$$

(b) Find the q-intercept of $f(q)$, which gives the quantity for which the product has unitary elasticity.
(c) The revenue function for this product is

$$R(q) = pq = (100 - 0.5q)q$$

Graph $R(q)$ and find the q-value for which the maximum revenue occurs.
(d) What is the relationship between elasticity and maximum revenue?

CHAPTER TEST

Find the derivatives of the following functions.

1. $y = 4\ln(x^3 + 1)$
2. $y = \ln(x^4 + 1)^3$
3. $y = \dfrac{\ln x}{x}$
4. $y = 5e^{x^3} + x^2$
5. $S = te^{t^4}$
6. $y = \dfrac{e^{x^3+1}}{x}$
7. $f(x) = 10(3^{2x})$
8. $g(x) = 2\log_5(4x + 7)$
9. Find y' if $3x^4 + 2y^2 + 10 = 0$.
10. Find y' if $xe^y = 10y$.
11. Let $x^2 + y^2 = 100$. If $\dfrac{dx}{dt} = 2$, find $\dfrac{dy}{dt}$ when $x = 6$ and $y = 8$.
12. The sales of a product are given by $S = 80{,}000e^{-0.4t}$, where S is the daily sales and t is the number of days after the end of an advertising campaign. Find the rate of sales decay 10 days after the end of the ad campaign.
13. Suppose the demand function for a product is given by $(p + 1)q^2 = 10{,}000$, where p is the price and q is the quantity. Find the rate of change of quantity with respect to price when $p = \$99$.
14. Suppose the weekly revenue and weekly cost for a product are given by $R(x) = 300x - 0.001x^2$ and $C(x) = 4000 + 30x$, respectively, where x is the number of units produced and sold. Find the rate at which profit is changing with respect to time when the number of units produced and sold is 50 and is increasing at a rate of 5 units per week.
15. If the demand for a product is $p^2 + 3p + q = 1500$, find the elasticity of demand at $p = 30$. If the price is raised to \$31, does revenue increase or decrease?
16. The following table gives the world population, in millions, for selected years. If the equation that models these data is

$$y = 160e^{0.00648(x - 1500)}$$

where $x = 1$ in year 1, what does this model give as the rate of growth of world population (a) in 1930? (b) in 1997?

Year	*World Population (millions)*
1	200
1650	500
1850	1000
1930	2000
1975	4000
1997	5800

Source: *World Almanac*, 1997

17. The percent of income claimed by federal, state, and local taxes for median households with two incomes is given in the table below. Using the number of years from 1950 as x and the percent as y, the logarithmic equation that models the data is

$$y = 19.08889 + 4.3271\ln x$$

What does this model predict as the rate of increase of the percent claimed for taxes in 2000?

Year	*Percent of Income Paid in Taxes*
1955	26.7
1965	28.8
1975	34.2
1985	34.8
1997	35.6

Source: Tax Foundation

18. The U.S. national debt can be modeled by the equation

$$y = 1.5336(1.08909)^x$$

where x is the number of years from 1900 and y is in billions of dollars. Use the model to predict the rate of growth of the national debt in 2005.

19. If the demand and supply functions for a product are $p = 1100 - 5q$ and $p = 20 + 0.4q$, respectively, find the tax per unit t that will maximize the tax revenue $T = tq$.

Extended Applications / Group Projects

I. Inflation

Hollingsworth Pharmaceuticals specializes in manufacturing generic medicines. Recently it developed an antibiotic with outstanding profit potential. The new antibiotic's total costs, sales, and sales growth, as well as projected inflation, are described as follows.

Total monthly costs to produce x *units (1 unit is 100 capsules):*

$$C(x) = \begin{cases} 15{,}000 + 10x & 0 \le x \le 11{,}000 \\ 15{,}000 + 10x + 0.001(x - 11{,}000)^2 & x \ge 11{,}000 \end{cases}$$

Sales: 10,000 units per month and growing at 1.25% per month
Selling price: \$17 per unit
Inflation: Approximately 0.25% per month, affecting both total costs and selling price

Company owners are pleased with the sales growth but are concerned about the projected increase in variable costs when production levels exceed 11,000 units per month. The consensus is that improvements eventually can be made that will reduce costs at higher production levels, thus altering the current cost function model. To plan properly for these changes. Hollingsworth Pharmaceuticals would like you to determine when the company's profits will begin to decrease. To help you determine this, answer the following.

1. If inflation is assumed to be compounded continuously, the selling price and total costs must be multiplied by the factor $e^{0.0025t}$. In addition, if sales growth is assumed to be compounded continuously, then sales must be multiplied by a factor of the form e^{rt}, where r is the monthly sales growth rate (expressed as a decimal) and t is time in months. Use these factors to write each of the following as a function of time t:
   (a) selling price p per unit (including inflation)
   (b) number of units x sold per month (including sales growth)
   (c) total revenue (Recall that $R = px$.)
2. Determine how many months it will be before monthly sales exceed 11,000 units.
3. If you restrict your attention to total costs when $x \ge 11{,}000$, then, after expanding and collecting like terms, $C(x)$ can be written as follows:

$$C(x) = 136{,}000 - 12x + 0.001x^2 \quad \text{for } x \ge 11{,}000$$

   Use this form for $C(x)$ with your result from Question 1(b) and with the inflationary factor $e^{0.0025t}$ to express these total costs as a function of time.

4. Form the profit function that would be used when monthly sales exceed 11,000 units by using the total revenue function from Question 1(c) and the total cost function from Question 3. This profit function should be a function of time t.
5. Find how long it will be before the profit is maximized. You may have to solve $P'(t) = 0$ by using a graphing calculator or computer to find the t-intercept of the graph of $P'(t)$. In addition, because $P'(t)$ has large numerical coefficients, you may want to divide both sides of $P'(t) = 0$ by 1000 before solving or graphing.

M II. Knowledge Workers

In January 1997, *Working Woman* made the following points about today's economy and the place of women in the economy.

- The telecommunications industry employs more people than the auto and auto parts industries combined.
- More Americans make semiconductors than make construction equipment.
- Almost twice as many Americans make surgical and medical instruments as make plumbing and heating products.
- The ratio of male to female knowledge workers (engineers, scientists, technicians, professionals, and senior managers) was 3 to 2 in 1983. The following table, which gives the number (in millions) of male and female knowledge workers from 1983 to 1997, shows how that ratio is changing.

Year	*Female Knowledge Workers (millions)*	*Male Knowledge Workers (millions)*
1983	11.0	15.4
1984	11.6	15.9
1985	12.3	16.3
1986	12.9	16.7
1987	13.6	16.8
1988	14.3	17.6
1989	15.3	18.1
1990	15.9	18.6
1991	16.1	18.4
1992	16.7	18.66
1993	17.3	18.7
1994	18.0	19.0
1995	18.5	19.8
1996	19.0	19.6
1997	19.5	19.8

Source: *Working Woman,* January 1997

To compare how the growth in the number of female knowledge workers compares with that of male knowledge workers, do the following.

1. Find a logarithmic equation (with $x = 0$ in 1980) that models the number of females, and find a logarithmic equation (with $x = 0$ in 1980) that models the number of males.
2. Find the rate of growth with respect to time of the number of female knowledge workers by taking the derivative of the equation that models the number.
3. Find the rate of growth with respect to time of the number of male knowledge workers by taking the derivative of the equation that models the number.
4. Compare the two rates of growth in the year 2000 and determine which rate is larger.
5. If these models indicate that it is possible for the number of females to equal the number of males, during what year do they indicate that this will occur?

Warm-up

Prerequisite Problem Type	For Section	Answer	Section for Review
Write as a power: (a) $\sqrt{x}$ (b) $\sqrt{x^2 - 9}$	12.1–12.4	(a) $x^{1/2}$ (b) $(x^2 - 9)^{1/2}$	0.4 Radicals
Expand $(x^2 + 4)^2$.	12.2	$x^4 + 8x^2 + 16$	0.5 Special powers
Divide $x^4 - 2x^3 + 4x^2 - 7x - 1$ by $x^2 - 2x$.	12.3	$x^2 + 4 + \dfrac{x - 1}{x^2 - 2x}$	0.5 Division
Find the derivative of (a) $f(x) = 2x^{1/2}$ (b) $u = x^3 - 3x$	12.1 12.2 12.3	(a) $f'(x) = x^{-1/2}$ (b) $u' = 3x^2 - 3$	9.4 Derivatives
If $y = \dfrac{(x^2 + 4)^6}{6}$, what is y'?	12.2	$(x^2 + 4)^5 2x$	9.6 Derivatives
(a) If $y = \ln u$, what is y'? (b) If $y = e^u$, what is y'?	12.3	(a) $y' = \dfrac{1}{u} \cdot u'$ (b) $y' = e^u \cdot u'$	11.1, 11.2 Derivatives
Solve for y: $\ln y = kt + C$	12.5	$y = e^{kt + C}$	5.2 Logarithmic functions
Solve for k: $0.5 = e^{5600k}$	12.5	$k \approx -0.00012378$	5.3 Exponential equations

Chapter 12

Indefinite Integrals

If the marginal cost for a product is \$36 at all levels of production, we know that the total cost function is a linear function. In particular, $C(x) = 36x + FC$*, where* FC *is the fixed cost. But if the marginal cost changes at different levels of production, the total cost function cannot be linear. In this chapter we will use integration to find total cost functions, given information about marginal costs and fixed costs.*

Accountants can use linear regression to translate information about marginal cost into a linear equation defining (approximately) the marginal cost function. By integrating this marginal cost function, it is possible to find an (approximate) function that defines the total cost.

We can also use integration to find total revenue functions from marginal revenue functions, to optimize profit from information about marginal cost and marginal revenue, and to find national consumption functions from information about marginal propensity to consume.

Integration can be used in the social and life sciences to predict growth or decay from expressions giving rates of change. For example, we can determine equations for population size from the rate of growth; we can write equations for the number of radioactive atoms remaining in a substance if we know the rate of disintegration of the substance; and we can determine the volume of blood flow from information about the rate of flow.

12.1 The Indefinite Integral

OBJECTIVE

- To find certain indefinite integrals

APPLICATION PREVIEW

In our study of the theory of the firm, we have worked with total cost, total revenue, and profit functions and have found their marginal functions. In practice, it is often easier for a company to measure marginal cost, revenue, and profit and use these data to form marginal functions from which it can find total cost, revenue, and profit functions. For example, Jarus Technologies manufactures computer memory boards, and the company's sales records show that the marginal revenue for its 64 MB memory board is given by

$$\overline{MR} = 300 - 0.2x$$

where x is the number of units sold. If we want to use this function to find the total revenue function for Jarus Technologies' 64 MB memory board, we need to find $R(x)$ from $\overline{MR} = R'(x)$. In this situation, we need to be able to reverse the process of differentiation.

We have studied procedures for and applications of finding derivatives of a given function. We now turn our attention to reversing this process of differentiation. When we know the derivative of a function, the process of finding the function itself is called **antidifferentiation.** For example, if the derivative of a function is $2x$, we know that the function could be $f(x) = x^2$ because $\frac{d}{dx}(x^2) = 2x$. But the function could also be $f(x) = x^2 + 4$ because $\frac{d}{dx}(x^2 + 4) = 2x$. It is clear that any function of the form $f(x) = x^2 + C$, where C is a constant, will have $f'(x) = 2x$ as its derivative. Thus we say that the **general antiderivative** of $f'(x) = 2x$ is $f(x) = x^2 + C$, where C is an arbitrary constant.

Antiderivative A function $F(x)$ is called an **antiderivative** of a function $f(x)$ if, for every x in the domain of f, $F'(x) = f(x)$. If C is an arbitrary constant, then $F(x) + C$ is called the **general antiderivative** of $f(x)$ because all antiderivatives of $f(x)$ have this form.

EXAMPLE 1

If $f'(x) = 3x^2$, what is $f(x)$?

Solution

The derivative of the function $f(x) = x^3$ is $f'(x) = 3x^2$. But other functions also have this derivative. They will all be of the form $f(x) = x^3 + C$, where C is a constant. Thus we say that $f(x) = x^3 + C$ is the general antiderivative of $f'(x) = 3x^2$.

EXAMPLE 2

If $f'(x) = x^3$, what is $f(x)$?

Solution

We know that the derivative of $f(x) = x^4$ is $4x^3$, so the derivative of $f(x) = \frac{1}{4}x^4$ is $f'(x) = x^3$. Thus any function of the form $f(x) = \frac{1}{4}x^4 + C$ will have the derivative $f'(x) = x^3$.

It is easily seen that

$$\text{if } f'(x) = x^4, \quad \text{then} \quad f(x) = \frac{x^5}{5} + C \text{ is the general antiderivative;}$$

$$\text{if } f'(x) = x^5, \quad \text{then} \quad f(x) = \frac{x^6}{6} + C \text{ is the general antiderivative.}$$

In general, we have the following.

Antiderivative of $f(x) = x^n$

If $f'(x) = x^n$, then $f(x) = \dfrac{x^{n+1}}{n+1} + C, \quad \text{for } n \neq -1.$

We can see that this general formula applies for any $n \neq -1$ by noting that the derivative of

$$f(x) = \frac{x^{n+1}}{n+1} + C \quad \text{is} \quad f'(x) = \frac{(n+1)x^n}{n+1} + 0 = x^n$$

Later, we will discuss the case where $n = -1$.

EXAMPLE 3

What is the general antiderivative of $f'(x) = x^{-1/2}$?

Solution

Using the formula, we get

$$f(x) = \frac{x^{1/2}}{1/2} + C = 2x^{1/2} + C$$

We can check by noting that the derivative of $2x^{1/2} + C$ is $x^{-1/2}$.

Note that the general antiderivative in Example 3 is a function (actually a number of functions, one for each value of C). Several members of this family of functions are shown in Figure 12.1 on the next page. Note that at any given x-value, the tangent line to each curve would have the same slope, indicating that all family members have the same derivative.

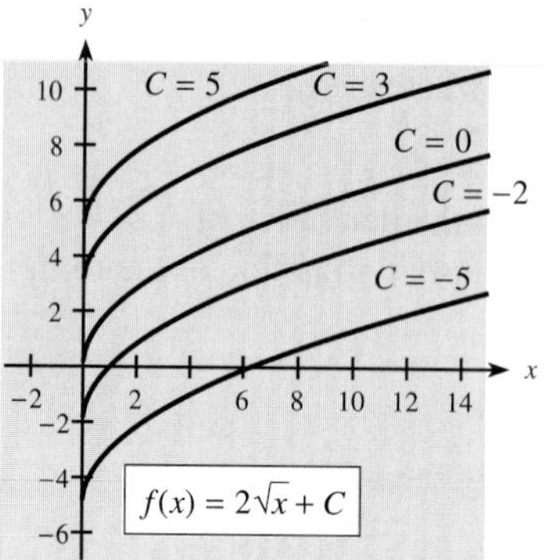

Figure 12.1

The process of finding an antiderivative is called **integration.** The function that results when integration takes place is called an **indefinite integral** or, more simply, an **integral.** We can denote the indefinite integral (that is, the general antiderivative) of a function $f(x)$ by $\int f(x)\,dx$. Thus we can write $\int x^2\,dx$ to indicate the general antiderivative of the function $f(x) = x^2$. The expression is read as "the integral of x^2 with respect to x." In this case, x^2 is called the **integrand.** The **integral sign,** $\int$, indicates the process of integration, and the dx indicates that the integral is to be taken with respect to x. Because the antiderivative of x^2 is $(x^3/3) + C$, we can write

$$\int x^2\,dx = \frac{x^3}{3} + C$$

We can now use the integral sign and rewrite the formula for integrating powers of x.

Powers of x Formula

$$\int x^n\,dx = \frac{x^{n+1}}{n+1} + C \quad (\text{for } n \neq -1)$$

EXAMPLE 4

Find (a) $\int \sqrt[3]{x}\,dx$ and (b) $\int \frac{1}{x^2}\,dx$.

Solution

(a) $$\int \sqrt[3]{x}\,dx = \int x^{1/3}\,dx = \frac{x^{4/3}}{4/3} + C$$
$$= \frac{3}{4}x^{4/3} + C = \frac{3}{4}\sqrt[3]{x^4} + C$$

(b) We write the power of x in the numerator so that the integral has the form in the formula above.

$$\int \frac{1}{x^2}\,dx = \int x^{-2}\,dx = \frac{x^{-2+1}}{-2+1} + C = \frac{x^{-1}}{-1} + C = \frac{-1}{x} + C$$

Other formulas will be useful in evaluating integrals. The following table shows how some new integration formulas result from differentiation formulas.

Integration Formulas

Derivative	**Resulting Integral**
$\frac{d}{dx}(x) = 1$	$\int 1\,dx = \int dx = x + C$
$\frac{d}{dx}[c \cdot u(x)] = c \cdot \frac{d}{dx}u(x)$	$\int c\,u(x)\,dx = c\int u(x)\,dx$
$\frac{d}{dx}[u(x) \pm v(x)] = \frac{d}{dx}u(x) \pm \frac{d}{dx}v(x)$	$\int [u(x) \pm v(x)]\,dx = \int u(x)\,dx \pm \int v(x)\,dx$

The formulas above indicate that we can integrate functions term by term just as we were able to take derivatives term by term.

EXAMPLE 5

Evaluate $\int 4\,dx$.

Solution

$$\int 4\,dx = 4\int dx = 4(x + C_1) = 4x + C$$

(Because C_1 is an unknown constant, we can write $4C_1$ as the unknown constant C.)

EXAMPLE 6

Evaluate $\int 8x^5\,dx$.

Solution

$$\int 8x^5\,dx = 8\int x^5\,dx = 8\left(\frac{x^6}{6} + C_1\right) = \frac{4x^6}{3} + C$$

EXAMPLE 7

Evaluate $\int (x^3 + 4x)\,dx$.

Solution

$$\begin{aligned}\int (x^3 + 4x)\,dx &= \int x^3dx + \int 4x\,dx\\ &= \left(\frac{x^4}{4} + C_1\right) + \left(4 \cdot \frac{x^2}{2} + C_2\right)\\ &= \frac{x^4}{4} + 2x^2 + C_1 + C_2\\ &= \frac{x^4}{4} + 2x^2 + C\end{aligned}$$

Note that we need only one constant because the sum of C_1 and C_2 is just a new constant.

EXAMPLE 8

Evaluate $\int (x^2 - 4)^2\, dx$.

Solution

We expand $(x^2 - 4)^2$ so that the integrand is in a form that fits the basic integration formulas.

$$\int (x^2 - 4)^2\, dx = \int (x^4 - 8x^2 + 16)\, dx = \frac{x^5}{5} - \frac{8x^3}{3} + 16x + C$$

CHECKPOINT

1. True or false:
   (a) $\int (4x^3 - 2x)\, dx = \int 4x^3 dx - \int 2x\, dx$
   $= (x^4 + C) - (x^2 + C) = x^4 - x^2$
   (b) $\displaystyle\int \frac{1}{3x^2}\, dx = \frac{1}{3(x^3/3)} + C = \frac{1}{x^3} + C$
2. Evaluate $\int (2x^3 + x^{-1/2} - 4x^{-5})\, dx$.

We now return to the Application Preview problem and consider how to find total revenue from marginal revenue.

EXAMPLE 9

Sales records at Jarus Technologies show that the rate of change of the revenue (that is, the marginal revenue) for an 64 MB memory board is $\overline{MR} = 300 - 0.2x$, where x represents the quantity sold. Find the total revenue function for the product.

Solution

We know that the marginal revenue can be found by differentiating the total revenue function. That is,

$$R'(x) = 300 - 0.2x$$

Thus integrating the marginal revenue function gives the total revenue function.

$$R(x) = \int (300 - 0.2x)\, dx = 300x - 0.1x^2 + K^*$$

We can use the fact that there is no revenue when no units are sold to evaluate K. Setting $x = 0$ and $R = 0$ gives $0 = 300(0) - 0.1(0)^2 + K$, so $K = 0$. Thus the total revenue function is

$$R(x) = 300x - 0.1x^2$$

*Here we are using K rather than C to represent the constant of integration to avoid confusion between the constant C and the cost function $C = C(x)$.

Graphing Utilities We can check that the $R(x)$ we found in Example 9 is correct by verifying that $R'(x) = 300 - 0.2x$ and $R(0) = 0$. Also, graphs can help us check the reasonableness of our result. Figure 12.2 shows the graphs of $\overline{MR} = 300 - 0.2x$ and of the $R(x)$ we found. Note that $R(x)$ passes through the origin, indicating $R(0) = 0$. Also, reading both graphs from left to right, we see that $R(x)$ increases when $\overline{MR} > 0$, attains its maximum when $\overline{MR} = 0$, and decreases when $\overline{MR} < 0$.

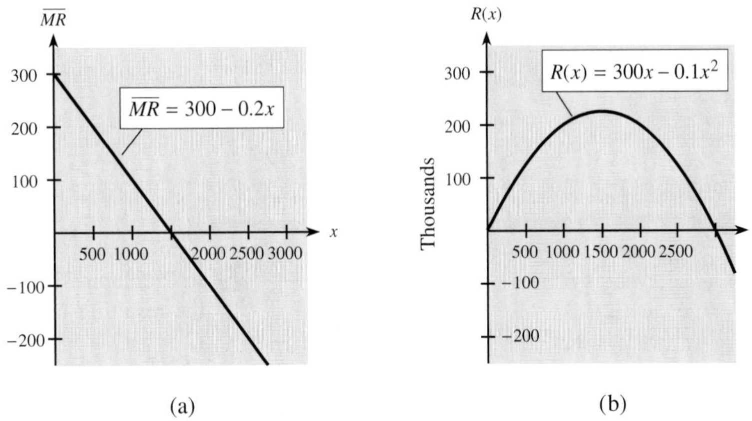

Figure 12.2 (a) (b)

We mentioned in Chapter 9, "Derivatives," that graphing utilities have a numerical derivative feature that can be used to check graphically the derivative of a function that has been calculated with a formula. We can also use the numerical integration feature on graphing utilities to check our integration (if we assume temporarily that the constant of integration is 0). We do this by graphing the integral calculated with a formula and the numerical integral from the graphing utility on the same set of axes. If the graphs lie on top of one another, the integrals agree. Figure 12.3 illustrates this for the function $f(x) = 3x^2 - 2x + 1$. Its integral, with the constant of integration set equal to 0, is shown as $y_1 = x^3 - x^2 + x$ in Figure 12.3(a). Of course, it is often easier to use the derivative to check integration.

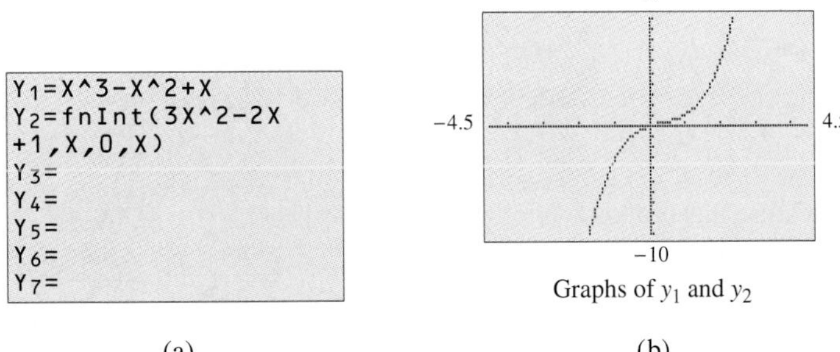

Graphs of y_1 and y_2

Figure 12.3 (a) (b)

CHECKPOINT SOLUTIONS

1. (a) False, $\int (4x^3 - 2x)\,dx = \int 4x^3\,dx - \int 2x\,dx = (x^4 + C_1) - (x^2 + C_2)$
$= x^4 - x^2 + C.$

(b) False, $\displaystyle\int \frac{1}{3x^2}\,dx = \int \frac{1}{3}\cdot\frac{1}{x^2}\,dx = \frac{1}{3}\int x^{-2}\,dx$
$\displaystyle = \frac{1}{3}\cdot\frac{x^{-1}}{-1} + C = \frac{-1}{3x} + C.$

2. $\displaystyle\int (2x^3 + x^{-1/2} - 4x^{-5})\,dx = \frac{2x^4}{4} + \frac{x^{1/2}}{1/2} - \frac{4x^{-4}}{-4} + C$
$\displaystyle = \frac{x^4}{2} + 2x^{1/2} + x^{-4} + C$

EXERCISE 12.1

1. If $f'(x) = 4x^3$, what is $f(x)$?
2. If $f'(x) = 5x^4$, what is $f(x)$?
3. If $f'(x) = x^6$, what is $f(x)$?
4. If $g'(x) = x^4$, what is $g(x)$?

Evaluate the integrals in Problems 5–26. Check your answers by differentiating.

5. $\int x^7\,dx$
6. $\int x^5\,dx$
7. $\int 8x^5\,dx$
8. $\int 16x^9\,dx$
9. $\int (3^3 + x^{13})\,dx$
10. $\int (5^2 + x^{10})\,dx$
11. $\int (3 - x^{3/2})\,dx$
12. $\int (8 + x^{2/3})\,dx$
13. $\int (x^4 - 9x^2 + 3)\,dx$
14. $\int (3x^2 - 4x - 4)\,dx$
15. $\int (2 + 2\sqrt{x})\,dx$
16. $\int (17 + \sqrt{x^3})\,dx$
17. $\int 6\sqrt[4]{x}\,dx$
18. $\int 3\sqrt[3]{x^2}\,dx$
19. $\displaystyle\int \frac{5}{x^4}\,dx$
20. $\displaystyle\int \frac{6}{x^5}\,dx$
21. $\displaystyle\int \frac{dx}{2\sqrt[3]{x^2}}$
22. $\displaystyle\int \frac{2\,dx}{5\sqrt{x^3}}$
23. $\displaystyle\int \left(x^3 - 4 + \frac{5}{x^6}\right) dx$
24. $\displaystyle\int \left(x^3 - 7 - \frac{3}{x^4}\right) dx$
25. $\displaystyle\int \left(x^9 - \frac{1}{x^3} + \frac{2}{\sqrt[3]{x}}\right) dx$
26. $\displaystyle\int \left(3x^8 + \frac{4}{x^8} - \frac{5}{\sqrt[5]{x}}\right) dx$

In Problems 27–32, use algebra to rewrite the integrands; then integrate and simplify.

27. $\int (x + 5)^2 x\,dx$
28. $\int (2x + 1)^2 x\,dx$
29. $\int (4x^2 - 1)^2 x^3\,dx$
30. $\int (x^3 + 1)^2 x\,dx$
31. $\displaystyle\int \frac{x + 1}{x^3}\,dx$
32. $\displaystyle\int \frac{x - 3}{\sqrt{x}}\,dx$

In Problems 33 and 34, find the antiderivatives and graph the resulting family members that correspond to $C = 0$, $C = 4$, $C = -4$, $C = 8$, and $C = -8$.

33. $\int (2x + 3)\,dx$
34. $\int (4 - x)\,dx$

In each of Problems 35–38, a family of functions is given and graphs of some members of the family are shown. Write the indefinite integral that gives the family.

35. $F(x) = 5x - \dfrac{x^2}{4} + C$

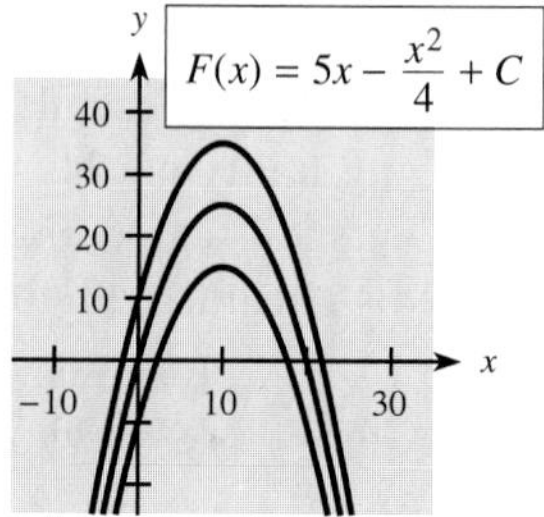

36. $F(x) = \dfrac{x^2}{2} + 3x + C$

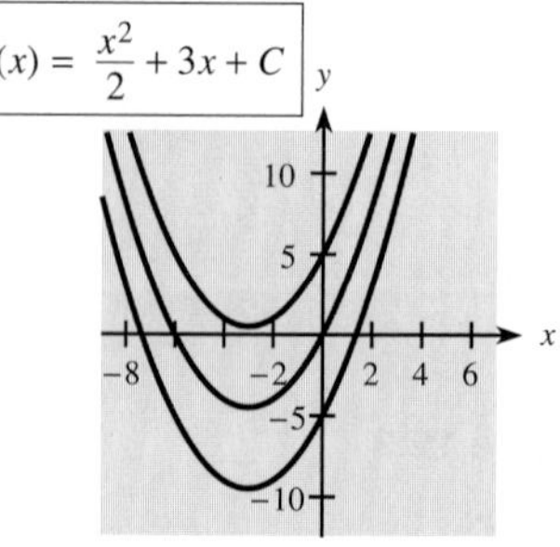

37. $F(x) = x^3 - 3x^2 + C$

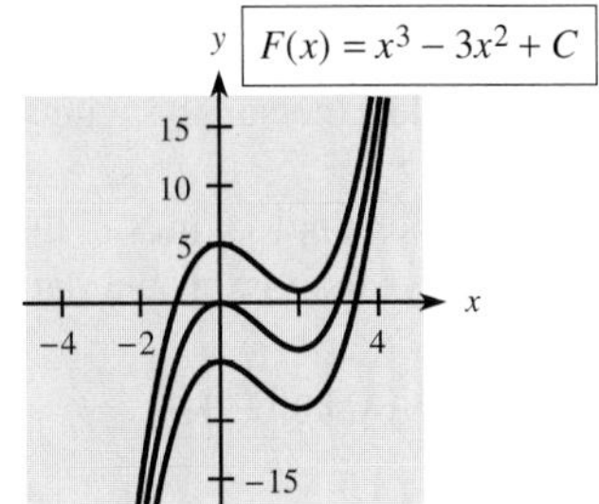

38. $F(x) = 12x - x^3 + C$

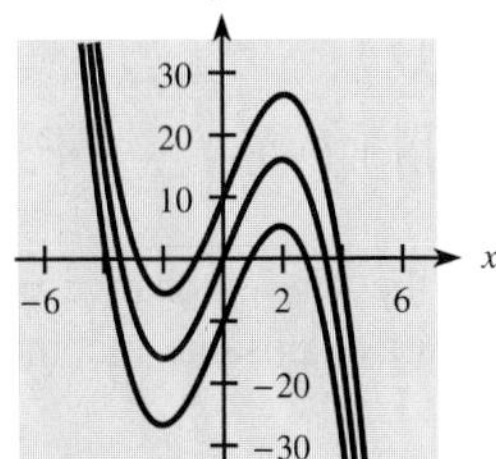

Applications

39. ***Revenue*** If the marginal revenue for a month for a commodity is $\overline{MR} = 3$, what is the total revenue function?

40. ***Revenue*** If the marginal revenue for a month for a commodity is $\overline{MR} = 5$, what is the total revenue function?

41. ***Revenue*** If the marginal revenue for a month for a commodity is $\overline{MR} = 0.4x + 3$, find the total revenue function.

42. ***Revenue*** If the marginal revenue for a month for a commodity is $\overline{MR} = 0.5x + 2$, find the total revenue function.

43. ***Revenue*** If the marginal revenue for a month is given by $\overline{MR} = 3x + 1$, what is the total revenue from the production and sale of 50 units?

44. ***Revenue*** If the marginal revenue for a month is given by $\overline{MR} = 5x + 3$, find the total revenue from the sale of 75 units.

45. ***Stimulus-response*** Suppose that when a sense organ receives a stimulus at time t, the total number of action potentials is $P(t)$. If the rate at which action potentials are produced is $t^3 + 4t^2 + 6$, and if there are 0 action potentials when $t = 0$, find the formula for $P(t)$.

46. ***Projectiles*** Suppose that a particle has been shot into the air in such a way that the rate at which its height is changing is $v = 320 - 32t$, in feet per second, and suppose that it is 1600 feet high when $t = 10$. Write the equation that describes the height of the particle at any time t.

47. ***Pollution*** A factory is dumping pollutants into a river at a rate given by $dx/dt = t^{3/4}/600$ tons per week, where t is the time in weeks since the dumping began and x is the number of tons of pollutants.
    (a) Find the equation for total tons of pollutants dumped.
    (b) How many tons were dumped during the first year?

48. ***Population growth*** The rate of growth of the population of a city is predicted to be

$$\frac{dp}{dt} = 1000t^{1.08}$$

where p is the population at time t, and t is measured in years from the present. Suppose that the current population is 100,000. What is the predicted
    (a) rate of growth 5 years from the present?
    (b) population 5 years from the present?

49. ***Average cost*** The DeWitt Company has found that the rate of change of its average cost for a product is

$$\overline{C}'(x) = \frac{1}{4} - \frac{100}{x^2}$$

in dollars, where x is the number of units. The average cost of producing 20 units is $40.00.
    (a) Find the average cost function for the product.
    (b) Find the average cost of 100 units of the product.

50. ***Oil leakage*** An oil tanker hits a reef and begins to leak. The efforts of the workers repairing the leak cause the rate at which the oil is leaking to decrease. The oil was leaking at a rate of 31 barrels per hour at the end of the first hour after the accident, and the rate is decreasing at a rate of one barrel per hour.
    (a) What function describes the rate of loss?
    (b) How many barrels of oil will leak in the first 6 hours?
    (c) When will the oil leak be stopped? How much will have leaked altogether?

51. ***Revenue*** Assume that the rate of change of sales revenue for Scott Paper Company can be modeled by

$$\frac{dR}{dt} = -0.062t + 0.776$$

where t is the number of years past 1980, and sales revenues are in billions of dollars.

(a) If sales revenue in 1987 was \$3.9769 billion, find the function that gives sales revenue. Graph this function.

(b) The data in the table show Scott Paper Company's sales revenue in billions of dollars for selected years. Graph the data in the table.

(c) Compare the graphs in parts (a) and (b)

Year	*Sales Revenue (billions)*	*Year*	*Sales Revenue (billions)*
1983	\$2.6155	1989	4.8949
1984	2.7474	1990	5.1686
1985	2.934	1991	4.9593
1986	3.3131	1992	5.0913
1987	3.9769	1993	4.7489
1988	4.5494		

Source: Scott Paper Company, *1993 Annual Report*

52. ***Revenue*** Assume that the rate of change of AT&T's total revenues can be modeled by

$$\frac{dR}{dt} = 0.506t - 4.03$$

where t is the number of years past 1980, and total revenues are in billions of dollars.

(a) If AT&T's total revenue in 1985 was \$63.13 billion, find the function for total revenue. Graph this function.

(b) The data in the table give AT&T's total revenue for selected years. Graph the data in the table.

(c) Compare the graphs in parts (a) and (b).

Year	*Total Revenue (billions)*	*Year*	*Total Revenue (billions)*
1985	\$63.13	1990	62.191
1986	69.906	1991	63.089
1987	60.53	1992	64.904
1989	61.1	1993	67.156

Source: AT&T *Annual Report, 1993*

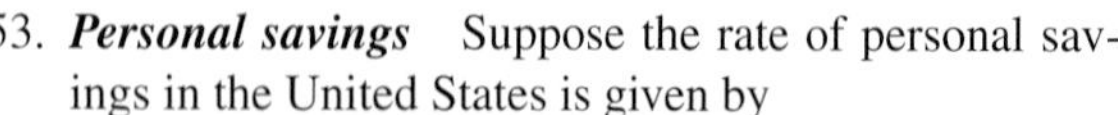

53. ***Personal savings*** Suppose the rate of personal savings in the United States is given by

$$\frac{dPS}{dt} = 2.1t^2 - 65.4t + 491.6$$

where t is the number of years past 1970, and personal savings PS is in billions of dollars.

(a) If personal savings in 1980 was \$153.8 billion, find the function that models personal savings. Graph this function.

(b) The data in the table show personal savings in the United States for selected years. Graph the data in the table.

(c) Compare the graphs in parts (a) and (b).

Year	*Personal Savings (billions)*	*Year*	*Personal Savings (billions)*
1980	\$153.8	1988	155.7
1985	189.3	1989	152.1
1986	187.5	1990	175.6
1987	142.0	1991	199.6

Source: *Survey of Current Business,* March 1993

54. ***Consumer debt*** Assume that the rate of change of consumer debt as a percentage of disposable income can be modeled by

$$\frac{dD}{dt} = 0.0112t^3 - 0.243t^2 + 1.36t - 1.3$$

where t is the number of years past 1980.

(a) If consumer debt in 1980 was 18.2% of disposable income, find the function that models consumer debt as a percentage of disposable income. Graph this function.

(b) The data in the table show consumer debt as a percentage of disposable income for selected years. Graph the data in the table.

(c) Compare the graphs in parts (a) and (b).

Year	*Consumer Debt as a Percentage of Disposable Income*	*Year*	*Consumer Debt as a Percentage of Disposable Income*
1980	18.2	1990	20.0
1982	16.9	1991	18.9
1983	17.7	1994	19.7
1985	20.8	1995	21.3
1986	21.4	1996	23.7
1988	21.0		

Source: Federal Reserve System

12.2 The Power Rule

OBJECTIVE

- *To evaluate integrals of the form $\int u^n \cdot u' \, dx = \int u^n \, du$ if $n \neq -1$*

APPLICATION PREVIEW

In the previous section, we saw that total revenue could be found by integrating marginal revenue. That is,

$$R(x) = \int \overline{MR} \, dx$$

For example, if the marginal revenue for a product is given by

$$\overline{MR} = \frac{600}{\sqrt{3x+1}} + 2$$

then

$$R(x) = \int \left[\frac{600}{\sqrt{3x+1}} + 2 \right] dx$$

To evaluate this integral, however, we need a more general formula than the Powers of x Formula.

In this section, we will extend the Powers of x Formula to a rule for powers of a function of x.

Our goal in this section is to extend the Powers of x Formula,

$$\int x^n \, dx = \frac{x^{n+1}}{n+1} + C \quad (n \neq -1)$$

to powers of a function of x. In order to do this, we must understand the symbol dx.

Recall from Section 9.3, "The Derivative," that the derivative of $y = f(x)$ with respect to x can be denoted by dy/dx. As we will see, there are advantages to using dy and dx as separate quantities whose ratio dy/dx equals $f'(x)$.

Differentials

If $y = f(x)$ is a differentiable function with derivative $dy/dx = f'(x)$, then the **differential of x** is dx, and the **differential of y** is dy, where

$$dy = f'(x) \, dx$$

Although differentials are useful in certain approximation problems, we are interested in the differential notation at this time.

EXAMPLE 1

Find the differential dy if $y = x^3 - 4x^2 + 5$.

Solution

$$dy = f'(x)\,dx = (3x^2 - 8x)\,dx$$

If the dependent variable in a function is u, then $du = u'(x)\,dx$.

EXAMPLE 2

If $u = x^2 + 4$, find du.

Solution

Because u is a function of x (that is, $u = u(x)$),

$$du = u'(x)\,dx = 2x\,dx$$

In terms of our goal of extending the Powers of x Formula, we would suspect that if x is replaced by a function of x, then dx should be replaced by the differential of that function. Let's see whether this is true.

Recall that if $y = [u(x)]^n$, the derivative of y is

$$\frac{dy}{dx} = n[u(x)]^{n-1} \cdot u'(x)$$

Using this formula for derivatives, we can see that

$$\int n[u(x)]^{n-1} \cdot u'(x)\,dx = [u(x)]^n + C$$

It is easy to see that this formula is equivalent to the following formula, which is called the **Power Rule for Integration.**

Power Rule for Integration

$$\int [u(x)]^n \cdot u'(x)\,dx = \frac{[u(x)]^{n+1}}{n+1} + C, \qquad n \neq -1$$

Using the fact that

$$du = u'(x)\,dx \quad \text{or} \quad du = u'\,dx$$

we can write the Power Rule in the following alternative form.

Power Rule (Alternative Form)

$$\int u^n\,du = \frac{u^{n+1}}{n+1} + C, \qquad n \neq -1$$

Note that this formula has the same form as the formula

$$\int x^n\,dx = \frac{x^{n+1}}{n+1} + C, \qquad n \neq -1$$

with the function u *substituted for* x *and* du *substituted for* dx.

EXAMPLE 3

Evaluate $\int (x^2 + 4)^5 \cdot 2x\,dx$.

Solution

To use the Power Rule, we must be sure that we have the function $u(x)$, its derivative $u'(x)$, and n.

$$u = x^2 + 4, \qquad n = 5$$
$$u' = 2x$$

All required parts are present, so the integral is of the form

$$\int (x^2+4)^5 2x\,dx = \int u^5 \cdot u'\,dx = \int u^5 du$$
$$= \frac{u^6}{6} + C = \frac{(x^2+4)^6}{6} + C$$

We can check the integration by noting that the derivative of

$$\frac{(x^2+4)^6}{6} + C \quad \text{is} \quad (x^2+4)^5 \cdot 2x$$

EXAMPLE 4

Evaluate $\int \sqrt{2x+3} \cdot 2\,dx$.

Solution

If we let $u = 2x + 3$, then $u' = 2$, and so we have

$$\int \sqrt{2x+3} \cdot 2\,dx = \int \sqrt{u}\,u'\,dx = \int \sqrt{u}\,du$$
$$= \int u^{1/2} du = \frac{u^{3/2}}{3/2} + C$$

Because $u = 2x + 3$, we have

$$\int \sqrt{2x+3} \cdot 2\,dx = \frac{2}{3}(2x+3)^{3/2} + C$$

CHECK: The derivative of $\frac{2}{3}(2x + 3)^{3/2} + C$ is $(2x + 3)^{1/2} \cdot 2$.

Some members of the family of functions given by

$$\int \sqrt{2x+3} \cdot 2\,dx = \frac{2}{3}(2x+3)^{3/2} + C$$

are shown in Figure 12.4 on the next page. Note from the graphs that the domain of each function is $x \geq -3/2$. This is because $2x + 3$ must be nonnegative so that $(2x + 3)^{3/2} = (\sqrt{2x+3})^3$ is a real number.

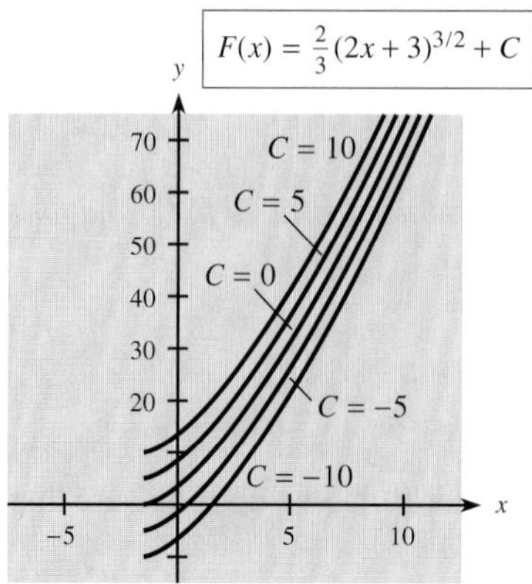

Figure 12.4

EXAMPLE 5

Evaluate $\int (x^2 + 4)^4 \cdot x\,dx$.

Solution

If we let $u = x^2 + 4$, then $u' = 2x$. Thus we do not have an integral of the form $\int u^n \cdot u'\,dx$, as we had in Example 3 and Example 4; the factor 2 is not in the integrand. To get the integrand in the correct form, we can multiply by 2 and divide it out as follows:

$$\int (x^2 + 4)^4 \cdot x\,dx = \int (x^2 + 4)^4 \cdot \frac{1}{2}(2x)\,dx$$

Because $\frac{1}{2}$ is a constant factor, we can factor it outside the integral sign, getting

$$\frac{1}{2}\int (x^2 + 4)^4 \cdot 2x\,dx$$

Now the integral is in the form $\frac{1}{2}\int u^4 \cdot u'\,dx$. Thus

$$\begin{aligned}\int (x^2 + 4)^4 \cdot x\,dx &= \frac{1}{2}\int (x^2 + 4)^4 \cdot 2x\,dx \\ &= \frac{1}{2}\,\frac{(x^2 + 4)^5}{5} + C \\ &= \frac{1}{10}(x^2 + 4)^5 + C\end{aligned}$$

EXAMPLE 6

Evaluate $\int \sqrt{x^3 - 4} \cdot 5x^2\,dx$.

Solution

If we let $u = x^3 - 4$, then $u' = 3x^2$. Thus we need the factor 3, rather than 5, in the integrand. If we multiply by the constant factor 3 (and divide it out), we have

$$\begin{aligned}\int \sqrt{x^3 - 4} \cdot 5x^2\,dx &= \int \sqrt{x^3 - 4} \cdot \frac{5}{3}(3x^2)\,dx \\ &= \frac{5}{3}\int (x^3 - 4)^{1/2} \cdot 3x^2\,dx\end{aligned}$$

This integral is of the form $\frac{5}{3} \int u^{1/2} \cdot u' \, dx$, resulting in

$$\frac{5}{3} \cdot \frac{u^{3/2}}{3/2} + C = \frac{5}{3} \cdot \frac{(x^3 - 4)^{3/2}}{3/2} + C = \frac{10}{9}(x^3 - 4)^{3/2} + C$$

Note that we can factor a constant outside the integral sign to obtain the integrand in the form we seek, but if the integral requires the introduction of a variable to obtain the form $u^n \cdot u' \, dx$, we *cannot* use this form and must try something else.

EXAMPLE 7

Evaluate $\int (x^2 + 4)^2 \, dx$.

Solution

If we let $u = x^2 + 4$, then $u' = 2x$. Because we would have to introduce a variable to get u' in the integral, we cannot solve this problem by using the Power Rule. We must find another method. We can evaluate this integral by squaring and then integrating term by term.

$$\begin{aligned} \int (x^2 + 4)^2 \, dx &= \int (x^4 + 8x^2 + 16) \, dx \\ &= \frac{x^5}{5} + \frac{8x^3}{3} + 16x + C \end{aligned}$$

Note that if we tried to introduce the factor $2x$ into the integral of Example 7, we would get

$$\begin{aligned} \int (x^2 + 4)^2 \, dx &= \int (x^2 + 4)^2 \cdot \frac{1}{2x}(2x) \, dx \\ &= \frac{1}{2} \int (x^2 + 4)^2 \cdot \frac{1}{x}(2x) \, dx \end{aligned}$$

But we cannot factor the $1/x$ outside the integral, so we do not have the proper form. Again, we can only introduce *a constant factor* to get an integral in the proper form.

EXAMPLE 8

Evaluate $\int (2x^2 - 4x)^2 (x - 1) \, dx$.

Solution

If we want to treat this as an integral of the form $\int u^n \, u' \, dx$, we will have to let $u = 2x^2 - 4x$. Then u' will be $4x - 4$. Multiplying and dividing by 4 will give us this form, as follows.

$$\begin{aligned}\int (2x^2 - 4x)^2 (x - 1)\, dx &= \int (2x^2 - 4x)^2 \cdot \frac{1}{4} \cdot 4(x - 1)\, dx \\ &= \frac{1}{4}\int (2x^2 - 4x)^2 (4x - 4)\, dx \\ &= \frac{1}{4}\int u^2 u'\, dx = \frac{1}{4} \cdot \frac{u^3}{3} + C \\ &= \frac{1}{4}\frac{(2x^2 - 4x)^3}{3} + C \\ &= \frac{1}{12}(2x^2 - 4x)^3 + C\end{aligned}$$

EXAMPLE 9

Evaluate $\int \frac{x^2 - 1}{(x^3 - 3x)^3}\, dx.$

Solution

This integral can be treated as $\int u^{-3}u'\, dx$ if we let $u = x^3 - 3x$. Then we can multiply and divide by 3 to get $u' = 3(x^2 - 1)$.

$$\begin{aligned}\int \frac{x^2 - 1}{(x^3 - 3x)^3}\, dx &= \int (x^3 - 3x)^{-3} \cdot \frac{1}{3} \cdot 3(x^2 - 1)\, dx \\ &= \frac{1}{3}\int (x^3 - 3x)^{-3} (3x^2 - 3)\, dx \\ &= \frac{1}{3}\left[\frac{(x^3 - 3x)^{-2}}{-2}\right] + C \\ &= \frac{-1}{6(x^3 - 3x)^2} + C\end{aligned}$$

CHECKPOINT

1. Which of the following can be evaluated with the Power Rule?
   (a) $\int (4x^2 + 1)^{10} (8x\, dx)$ (b) $\int (4x^2 + 1)^{10} (x\, dx)$
   (c) $\int (4x^2 + 1)^{10} (8\, dx)$ (d) $\int (4x^2 + 1)^{10}\, dx$
2. Which of the following is equal to $\int (2x^3 + 5)^{-2} (6x^2\, dx)$?
   (a) $\frac{[(2x^4)/4 + 5x]^{-1}}{-1} \cdot \frac{6x^3}{3} + C$ (b) $\frac{(2x^3 + 5)^{-1}}{-1} \cdot \frac{6x^3}{3} + C$
   (c) $\frac{(2x^3 + 5)^{-1}}{-1} + C$
3. True or false: Constants can be factored outside the integral sign.
4. Evaluate:
   (a) $\int (x^3 + 9)^5 (3x^2\, dx)$ (b) $\int (x^3 + 9)^{15} (x^2\, dx)$ (c) $\int (x^3 + 9)^2 (x\, dx)$

We now return to the problem introduced in the Application Preview.

EXAMPLE 10

Suppose that the marginal revenue for a product is given by

$$\overline{MR} = \frac{600}{\sqrt{3x+1}} + 2$$

Find the total revenue function.

Solution

$$\begin{aligned} R(x) &= \int \overline{MR}\,dx = \int \left[\frac{600}{(3x+1)^{1/2}} + 2\right] dx \\ &= \int 600(3x+1)^{-1/2}\,dx + \int 2\,dx \\ &= 600\left(\frac{1}{3}\right)\int (3x+1)^{-1/2}(3\,dx) + 2\int dx \\ &= 200\,\frac{(3x+1)^{1/2}}{1/2} + 2x + K \\ &= 400\sqrt{3x+1} + 2x + K \end{aligned}$$

We know that $R(0) = 0$, so we have

$$0 = 400\sqrt{1} + 0 + K \quad \text{or} \quad K = -400$$

Thus the total revenue function is

$$R(x) = 400\sqrt{3x+1} + 2x - 400$$

Note in Example 10 that even though $R(0) = 0$, the constant of integration K was *not* 0. This is because $x = 0$ does not necessarily mean that $u(x)$ will also be 0.

CHECKPOINT SOLUTIONS

1. Expressions (a) and (b) can be evaluated with the Power Rule. For (a), we let $u = 4x^2 + 1$ so that the integral becomes

$$\int u^{10}\,u'\,dx = \int u^{10}\,du$$

For (b), we let $u = 4x^2 + 1$ again, and the integral becomes

$$\frac{1}{8}\int u^{10}\,u'\,dx = \frac{1}{8}\int u^{10}\,du$$

Expressions (c) and (d) do not fit the format of the Power Rule, because neither integral has an x with the dx, outside the power (so they need to be multiplied out before integrating).

2. $u = 2x^3 + 5$, so $u' = 6x^2$

$$\begin{aligned} \int (2x^3+5)^{-2}(6x^2)\,dx &= \int u^{-2}u'\,dx = \int u^{-2}\,du \\ &= -u^{-1} + C = -(2x^3+5)^{-1} + C \end{aligned}$$

Thus (c) is the correct choice.

3. True

4. (a) $\int (x^3+9)^5(3x^2)\,dx = \int u^5u'\,dx = \dfrac{u^6}{6} + C = \dfrac{(x^3+9)^6}{6} + C$

(b) $\int (x^3+9)^{15}(x^2\,dx) = \dfrac{1}{3}\int (x^3+9)^{15}(3x^2\,dx)$

$= \dfrac{1}{3}\cdot\dfrac{(x^3+9)^{16}}{16} + C = \dfrac{(x^3+9)^{16}}{48} + C$

(c) The Power Rule does not fit, so we expand the integrand.

$$\int (x^3+9)^2(x\,dx) = \int (x^6+18x^3+81)x\,dx = \int (x^7+18x^4+81x)\,dx = \frac{x^8}{8} + \frac{18x^5}{5} + \frac{81x^2}{2} + C$$

EXERCISE 12.2

Evaluate the integrals in Problems 1–32. Check your results by differentiation.

1. $\int (x^2+3)^3 2x\,dx$
2. $\int (3x^3+1)^4 9x^2\,dx$
3. $\int (15x^2+10)^4(30x)\,dx$
4. $\int (8x^4+5)^3(32x^3)\,dx$
5. $\int (3x-x^3)^2(3-3x^2)\,dx$
6. $\int (4x^2-3x)^4(8x-3)\,dx$
7. $\int (x^2+5)^3 x\,dx$
8. $\int (3x^2-4)^6 x\,dx$
9. $\int 7(4x-1)^6\,dx$
10. $\int 3(5-x)^{-3}\,dx$
11. $\int (x^2+1)^{-3}\,x\,dx$
12. $\int (3x^4+7)^{-4}(5x^3\,dx)$
13. $\int (x-1)(x^2-2x+5)^4\,dx$
14. $\int (2x^3-x)(x^4-x^2)^6\,dx$
15. $\int 2(x^3-1)(x^4-4x+3)^{-5}\,dx$
16. $\int 3(x^5-2x)(x^6-6x^2+7)^{-2}\,dx$
17. $\int 7x^3\sqrt{x^4+6}\,dx$
18. $\int 3x\sqrt{5-x^2}\,dx$
19. $\int (x^3+1)^2(3x\,dx)$
20. $\int (x^2-5)^2(2x^2\,dx)$
21. $\int (3x^2-1)^2(8x^2\,dx)$
22. $\int (2x^4+3)^2(8x\,dx)$
23. $\int \sqrt{x^3-3x}\,(x^2-1)\,dx$
24. $\int \sqrt[3]{x^2+2x}\,(x+1)\,dx$
25. $\displaystyle\int \frac{x^2\,dx}{(x^3-1)^2}$
26. $\displaystyle\int \frac{x\,dx}{(x^2-1)^3}$
27. $\displaystyle\int \frac{3x^4\,dx}{(2x^5-5)^4}$
28. $\displaystyle\int \frac{5x^3\,dx}{(x^4-8)^3}$
29. $\displaystyle\int \frac{x^3-1}{(x^4-4x)^3}\,dx$
30. $\displaystyle\int \frac{3x^5-2x^3}{(x^6-x^4)^5}\,dx$
31. $\displaystyle\int \frac{x^2-4x}{\sqrt{x^3-6x^2+2}}\,dx$
32. $\displaystyle\int \frac{x^2+1}{\sqrt{x^3+3x+10}}\,dx$

In Problems 33 and 34, (a) evaluate each integral and (b) graph the members of the solution family for $C=-5$, $C=0$, and $C=5$.

33. $\int x(x^2-1)^3\,dx$
34. $\int (3x-11)^{1/3}\,dx$

Each of Problems 35 and 36 has the form $\int f(x)\,dx$.

(a) Evaluate each integral to obtain a family of functions.
(b) Find and graph the family member that passes through the point (0, 2). Call that function $F(x)$.
(c) Find any x-values where $f(x)$ is not defined but $F(x)$ is.
(d) At the x-values found in (c), what kind of tangent line does $F(x)$ have?

35. $\displaystyle\int \frac{3dx}{(2x-1)^{3/5}}$
36. $\displaystyle\int \frac{x^2\,dx}{(x^3-1)^{1/3}}$

In each of Problems 37 and 38, a family of functions is given, together with the graphs of some functions in the family. Write the indefinite integral that gives the family.

37. $F(x) = (x^2-1)^{4/3} + C$

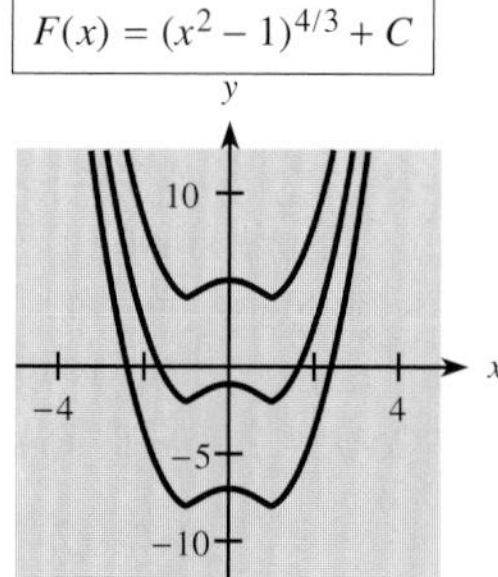

38. $F(x) = 54(4x^2 + 9)^{-1} + C$

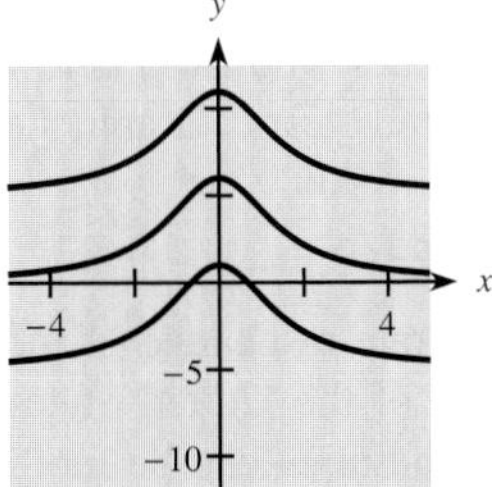

Applications

39. ***Revenue*** Suppose that the marginal revenue for a product is given by

$$\overline{MR} = \frac{-30}{(2x + 1)^2} + 30$$

Find the total revenue.

40. ***Revenue*** The marginal revenue for a new calculator is given by

$$\overline{MR} = 60{,}000 - \frac{40{,}000}{(10 + x)^2}$$

where x represents hundreds of calculators. Find the total revenue function for these calculators.

41. ***Physical productivity*** The total physical output of a number of machines or workers is called *physical productivity* and is a function of the number of machines or workers. If $P = f(x)$ is the productivity, dP/dx is the marginal physical productivity. If the marginal physical productivity for bricklayers is $dP/dx = 90(x + 1)^2$, where P is the number of bricks laid per day, find the physical productivity of 4 bricklayers. *Note:* $P = 0$ when $x = 0$.

42. ***Production*** The rate of production of a new line of products is given by

$$\frac{dx}{dt} = 200\left[1 + \frac{400}{(t + 40)^2}\right]$$

where x is the number of items and t is the number of weeks the product has been in production.

(a) Assuming that $x = 0$ when $t = 0$, find the total number of items produced as a function of time t.

(b) How many items were produced in the fifth week?

43. ***Typing speed*** The rate of change in typing speed of the average student is $ds/dx = 5(x + 1)^{-1/2}$, where x is the number of typing lessons the student has had.

(a) Find the typing speed as a function of the number of lessons if the average student can type 10 words per minute with no lessons ($x = 0$).

(b) How many words per minute can the average student type after 24 lessons?

44. ***Productivity*** Because a new employee must learn an assigned task, production will increase with time. Suppose that for the average new employee, the rate of performance is given by

$$\frac{dN}{dt} = \frac{1}{2\sqrt{t + 1}}$$

where N is the number of units completed t hours after beginning a new task. If 2 units are completed after 3 hours, how many units are completed after 8 hours?

45. ***Film attendance*** An excellent film with a very small advertising budget must depend largely on word-of-mouth advertising. In this case, the rate at which weekly attendance might grow can be given by

$$\frac{dA}{dt} = \frac{-100}{(t + 10)^2} + \frac{2000}{(t + 10)^3}$$

where t is the time in weeks since release and A is attendance in millions.

(a) Find the function that describes weekly attendance at this film.

(b) Find the attendance at this film in the tenth week.

46. ***Product quality and advertising*** An inferior product with a large advertising budget does well when it is introduced, but sales decline as people discontinue use of the product. Suppose that the rate of weekly sales is given by

$$S'(t) = \frac{400}{(t + 1)^3} - \frac{200}{(t + 1)^2}$$

where S is sales in millions and t is time in weeks.

(a) Find the function that describes the weekly sales.

(b) Find the sales for the first week and the ninth week.

47. ***Demographics*** Because of the decline of the steel industry, a western Pennsylvania town predicts that its public school population will decrease at the rate

$$\frac{dN}{dx} = \frac{-300}{\sqrt{x+9}}$$

where x is the number of years and N is the total school population. If the present population ($x = 0$) is 8000, what population size is planned for in 7 years?

48. ***Franchise growth*** A new fast-food firm predicts that the number of franchises for its products will grow at the rate

$$\frac{dn}{dt} = 9\sqrt{t+1}$$

where t is the number of years, $0 \le t \le 10$. If there is one franchise ($n = 1$) at present ($t = 0$), how many franchises are predicted for 8 years from now?

49. ***Poverty line*** Suppose the rate of change of the number of people (in millions) in the United States who lived below the poverty level can be modeled by

$$\frac{dp}{dt} = -0.004635(t-60)^2 + 0.2412(t-60) - 2.4165$$

where t is the number of years past 1900.

Year	*Persons Below the Poverty Level (millions)*	Year	*Persons Below the Poverty Level (millions)*
1960	39.9	1990	33.6
1965	33.2	1991	35.7
1970	25.4	1992	38
1975	25.9	1993	39.3
1980	29.3	1994	38.1
1986	32.4	1995	36.4
1989	31.5	1996	36.5

Source: Bureau of the Census, U.S. Dept. of Commerce

(a) If 33.6 million people lived below the poverty level in 1990, find the function that models the number of people, in millions, in the United States who lived below the poverty level. Graph this function.

(b) The data in the table show the number of people, in millions, in the United States who lived below the poverty level for selected years. Graph the data in the table with $t = 0$ in 1900.

(c) Compare the graphs in (a) and (b).

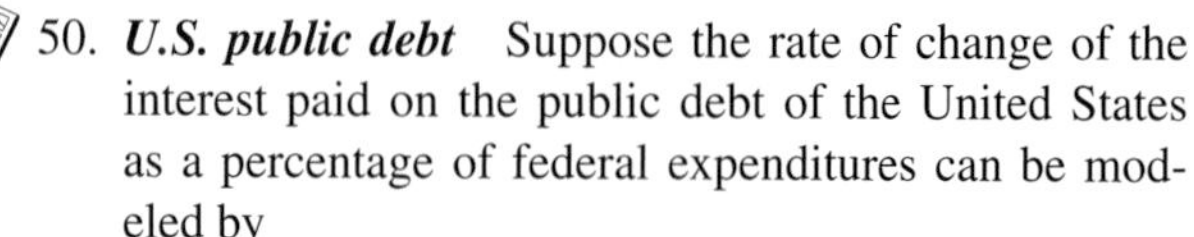

50. ***U.S. public debt*** Suppose the rate of change of the interest paid on the public debt of the United States as a percentage of federal expenditures can be modeled by

$$\frac{dD}{dt} = -0.06172(0.1t+3)^3 + 1.2823(0.1t+3)^2 - 8.43(0.1t+3) + 17.58$$

where t is the number of years past 1930.

Year	*Interest Paid as a Percentage of Federal Expenditures*	*Year*	*Interest Paid as a Percentage of Federal Expenditures*
1930	0.0	1970	9.9
1940	10.5	1975	9.8
1950	13.4	1980	12.7
1955	9.4	1985	18.9
1960	10.0	1990	21.1
1965	9.6	1995	22.0

Source: Bureau of Public Debt, U.S. Dept. of the Treasury

(a) If 9.6% of federal expenditures was devoted to interest payments on the public debt in 1965, find the function $D(t)$ that models the interest paid on the public debt of the United States as a percentage of federal expenditures. Graph this function.

(b) The data in the table give the interest paid on the public debt of the United States as a percentage of federal expenditures for selected years. Graph the data in the table with $t = 0$ in 1930.

(c) Compare the graphs in (a) and (b).

51. ***Union membership*** Suppose the rate of change of the percentage of U.S. workers who belonged to unions can be modeled by

$$\frac{dU}{dt} = 0.1415(0.1t + 2)^2 - 2.1317(0.1t + 2) + 7.3900$$

where t is the number of years past 1920.

(a) If 18.0% of U.S. workers belonged to unions in 1985, find the function $U(t)$ that models the percentage of U.S. workers who belonged to unions. Graph this function.

(b) The data in the table show the percentage of U.S. workers who belonged to unions for selected years. Graph the data ($t = 0$ in 1920).

(c) Compare the graphs in (a) and (b).

Year	*Union Membership as a Percentage of U.S. Labor Force*
1930	11.6
1935	13.2
1940	26.9
1945	35.5
1950	31.5
1955	33.2
1960	31.4
1965	28.4
1970	27.3
1975	25.5
1980	21.9
1985	18.0
1990	16.1
1993	15.8
1994	15.5
1995	14.9
1996	14.5

Source: Bureau of Labor Statistics, U.S. Dept. of Labor

12.3 Integrals Involving Logarithmic and Exponential Functions

OBJECTIVES

- *To evaluate integrals of the form* $\int \frac{u'}{u}\,dx$ *or, equivalently,* $\int \frac{1}{u}\,du$
- *To evaluate integrals of the form* $\int e^u u'\,dx$ *or, equivalently,* $\int e^u\,du$

APPLICATION PREVIEW

The rate of growth of the market value of a home has typically exceeded the inflation rate. Suppose, for example, that the real estate market has an average annual inflation rate of 8%. Then the rate of change of the value of a house that cost \$100,000 can be modeled by

$$\frac{dV}{dt} = 7.7e^{0.077t}$$

where V is the value of the home in hundreds of thousands of dollars and t is the time in years since the home was purchased. To find the market value of such a home 10 years after it was purchased, we would first have to integrate dV/dt. That is, we must be able to integrate an exponential.

In this section, we consider integration formulas that result in natural logarithms and formulas for integrating exponentials.

Recall that the Power Rule for integrals applies only if $n \neq -1$. That is,

$$\int u^n u'\,dx = \frac{u^{n+1}}{n+1} + C \qquad \text{if } n \neq -1$$

The following formula applies when $n = -1$.

Logarithmic Formula If u is a function of x, then

$$\int u^{-1}u'\,dx = \int \frac{u'}{u}\,dx = \int \frac{1}{u}\,du = \ln|u| + C$$

This formula is a direct result of the fact that

$$\frac{d}{dx}(\ln|u|) = \frac{1}{u}\cdot u'$$

We use the absolute value of u because the logarithm is defined only when the quantity is positive. We can see this result by considering the following.

For $u > 0$: $$\frac{d}{dx}(\ln|u|) = \frac{d}{dx}(\ln u) = \frac{1}{u}\cdot u'$$

For $u < 0$: $$\frac{d}{dx}(\ln|u|) = \frac{d}{dx}[\ln(-u)] = \frac{1}{(-u)}\cdot(-u') = \frac{1}{u}\cdot u'$$

In addition to this verification, we can graphically illustrate the need for the absolute value sign. Figure 12.5(a) shows that $f(x) = 1/x$ is defined for $x \neq 0$, and from Figures 12.5(b) and 12.5(c), we see that $F(x) = \int 1/x\,dx = \ln|x|$ is also defined for $x \neq 0$. But $y = \ln x$ is defined only for $x > 0$.

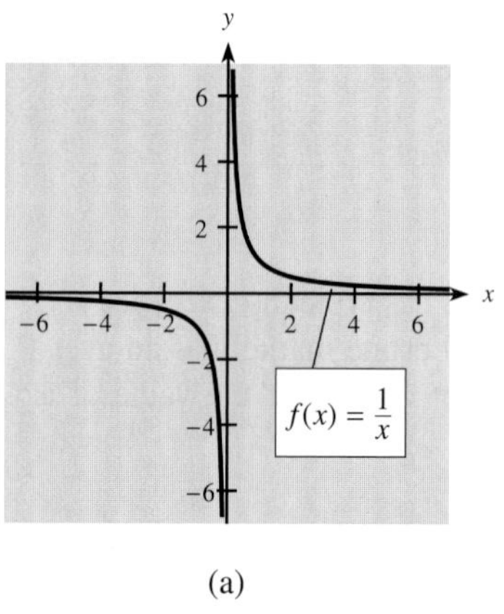

(a)

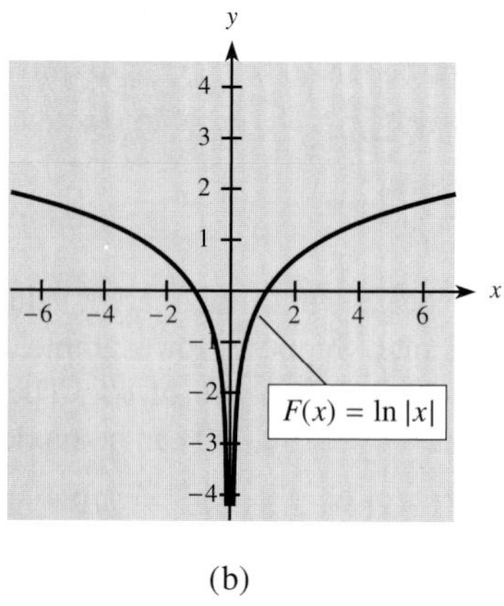

(b)

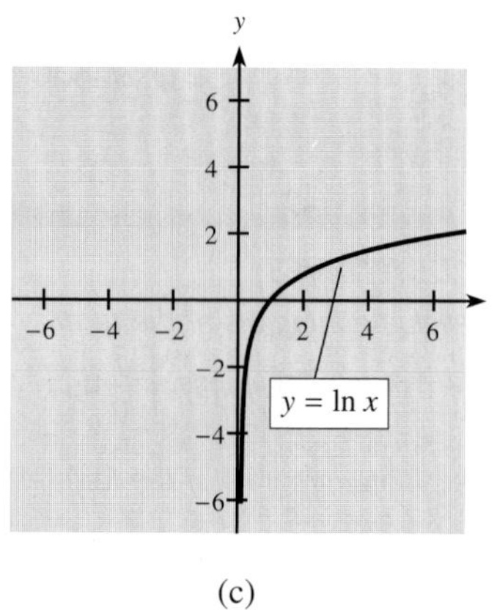

(c)

Figure 12.5

EXAMPLE 1

Evaluate $\int \frac{4}{4x+8}\,dx$.

Solution

This integral is of the form

$$\int \frac{u'}{u}\,dx = \ln|u| + C$$

with $u = 4x + 8$ and $u' = 4$. Thus

$$\int \frac{4}{4x + 8}\, dx = \ln |4x + 8| + C$$

Figure 12.6 shows several members of the family

$$F(x) = \int \frac{4\, dx}{4x + 8} = \ln |4x + 8| + C$$

We can choose different values for C and use a graphing utility to graph families of curves such as those in Figure 12.6.

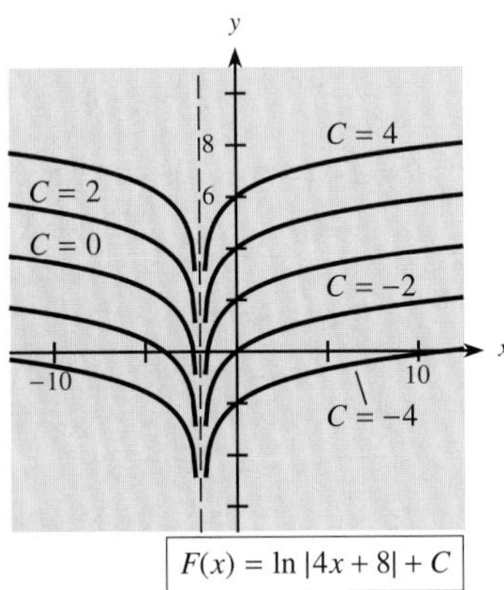

Figure 12.6

EXAMPLE 2

Evaluate $\displaystyle\int \frac{x - 3}{x^2 - 6x + 1}\, dx.$

Solution

This integral is of the form $\int (u'/u)\, dx$, *almost.* If we let $u = x^2 - 6x + 1$, then $u' = 2x - 6$. If we multiply (and divide) the numerator by 2, we get

$$\begin{aligned}
\int \frac{x - 3}{x^2 - 6x + 1}\, dx &= \frac{1}{2}\int \frac{2(x - 3)}{x^2 - 6x + 1}\, dx \\
&= \frac{1}{2}\int \frac{2x - 6}{x^2 - 6x + 1}\, dx \\
&= \frac{1}{2}\int \frac{u'}{u}\, dx = \frac{1}{2}\ln |u| + C \\
&= \frac{1}{2}\ln |x^2 - 6x + 1| + C
\end{aligned}$$

If an integral contains a fraction in which the degree of the numerator is equal to or greater than that of the denominator, we should divide the denominator into the numerator as a first step.

EXAMPLE 3

Evaluate $\displaystyle\int \frac{x^4 - 2x^3 + 4x^2 - 7x - 1}{x^2 - 2x}\,dx.$

Solution

Because the numerator is of higher degree than the denominator, we begin by dividing $x^2 - 2x$ into the numerator.

$$\begin{array}{r} x^2 \qquad\quad + 4 \\ x^2 - 2x\,\overline{)\,x^4 - 2x^3 + 4x^2 - 7x - 1} \\ \underline{x^4 - 2x^3 \qquad\qquad\qquad} \\ 4x^2 - 7x - 1 \\ \underline{4x^2 - 8x \quad\;} \\ x - 1 \end{array}$$

Thus

$$\begin{aligned}\int \frac{x^4 - 2x^3 + 4x^2 - 7x - 1}{x^2 - 2x}\,dx &= \int \left(x^2 + 4 + \frac{x - 1}{x^2 - 2x}\right) dx \\ &= \int (x^2 + 4)\,dx + \frac{1}{2}\int \frac{2(x - 1)\,dx}{x^2 - 2x} \\ &= \frac{x^3}{3} + 4x + \frac{1}{2}\ln|x^2 - 2x| + C\end{aligned}$$

CHECKPOINT

1. True or false:

(a) $\displaystyle\int \frac{3x^2\,dx}{x^3 + 4} = \ln|x^3 + 4| + C$ (b) $\displaystyle\int \frac{2x\,dx}{\sqrt{x^2 + 1}} = \ln\left|\sqrt{x^2 + 1}\right| + C$

(c) $\displaystyle\int \frac{2}{x}\,dx = 2\ln|x| + C$

(d) $\displaystyle\int \frac{x}{x + 1}\,dx = x\int \frac{1}{x + 1}\,dx = x\ln|x + 1| + C$

(e) To evaluate $\displaystyle\int \frac{4x}{4x + 1}\,dx$, our first step is to divide $4x + 1$ into $4x$.

2. (a) Divide $4x + 1$ into $4x$. (b) Evaluate $\displaystyle\int \frac{4x}{4x + 1}\,dx.$

We know that

$$\frac{d}{dx}(e^u) = e^u \cdot u'$$

The corresponding integral is given by the following.

Exponential Formula If u is a function of x,

$$\int e^u \cdot u'\,dx = \int e^u\,du = e^u + C$$

EXAMPLE 4

Evaluate $\int 5e^x\,dx$.

Solution

$\int 5e^x\,dx = 5\int e^x\,dx = 5e^x + C$

EXAMPLE 5

Evaluate $\int 2xe^{x^2}\,dx$.

Solution

Letting $u = x^2$ implies that $u' = 2x$, and the integral is of the form $\int e^u \cdot u'\,dx$. Thus

$$\int 2xe^{x^2}\,dx = \int e^{x^2}(2x)\,dx = \int e^u \cdot u'\,dx = e^u + C = e^{x^2} + C$$

EXAMPLE 6

Evaluate $\displaystyle\int \frac{x^2 dx}{e^{x^3}}$.

Solution

In order to use $\int e^u \cdot u'\,dx$, we write the exponential in the numerator. Thus

$$\int \frac{x^2\,dx}{e^{x^3}} = \int e^{-x^3}\,(x^2\,dx)$$

This is *almost* of the form $\int e^u \cdot u'\,dx$. Letting $u = -x^3$ gives $u' = -3x^2$. Thus

$$\int e^{-x^3}\,(x^2\,dx) = -\frac{1}{3}\int e^{-x^3}\,(-3x^2\,dx) = -\frac{1}{3}e^{-x^3} + C = \frac{-1}{3e^{x^3}} + C$$

CHECKPOINT

3. True or false:

(a) $\int e^{x^2}(2x\,dx) = e^{x^2} \cdot x^2 + C$ (b) $\int e^{-3x}\,dx = -\frac{1}{3}e^{-3x} + C$

(c) $\displaystyle\int \frac{dx}{e^{3x}} = \frac{1}{3}\left(\frac{1}{e^{3x}}\right) + C$ (d) $\displaystyle\int e^{3x+1}\,(3\,dx) = \frac{e^{3x+2}}{3x+2} + C$

We now return to the real estate inflation rate problem in the Application Preview.

EXAMPLE 7

Suppose the rate of change of the value of a house that cost \$100,000 can be modeled by

$$\frac{dV}{dt} = 7.7e^{0.077t}$$

where V is the market value of the home in hundreds of thousands of dollars and t is the time in years since the home was purchased.

(a) Find the function that expresses the value V in terms of t.
(b) Find the predicted value after 10 years.

Solution

(a) $V = \int \frac{dV}{dt}\,dt = \int 7.7e^{0.077t}\,dt$

$$V = 7.7 \int e^{0.077t}\left(\frac{1}{0.077}\right)(0.077\,dt)$$

$$V = 7.7\left(\frac{1}{0.077}\right) \int e^{0.077t}(0.077\,dt)$$

$$V = 100e^{0.077t} + C$$

Using $V = 100$ (thousand) when $t = 0$, we have

$$100 = 100 + C$$
$$0 = C$$

Thus we have the value as a function of time given by

$$V = 100e^{0.077t}$$

(b) The value after 10 years is found by using $t = 10$.

$$V = 100e^{0.077(10)} = 100e^{0.77} \approx 215.98$$

Thus, after 10 years, the predicted value of the home is \$215,980.

EXAMPLE 8

Figure 12.7 shows the graphs of $g(x) = 5e^{-x^2}$ and $h(x) = -10xe^{-x^2}$. One of these functions is $f(x)$ and the other is $\int f(x)\,dx$ with $C = 0$.

(a) Decide which of $g(x)$ and $h(x)$ is $f(x)$ and which is $\int f(x)\,dx$.
(b) How can the graph of $f(x)$ be used to locate and classify the extrema of $\int f(x)\,dx$?
(c) What feature of the graph of $f(x)$ occurs at the same x-values as the inflection points of the graph of $\int f(x)\,dx$?

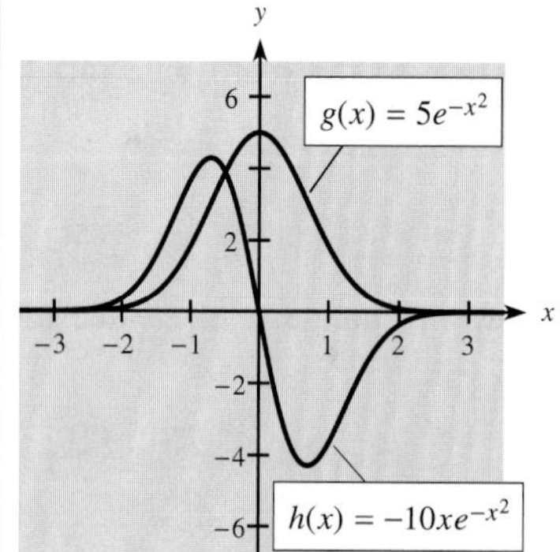

Figure 12.7

Solution

(a) The graph of $h(x)$ looks like the graph of $g'(x)$ because $h(x) > 0$ where $g(x)$ is increasing, $h(x) < 0$ where $g(x)$ is decreasing, and $h(x) = 0$ where $g(x)$ has its maximum. However, if $h(x) = g'(x)$, then, equivalently,

$$\int h(x)\,dx = \int g'(x)\,dx = g(x) + C$$

so $h(x) = f(x)$ and $g(x) = \int f(x)\,dx$. We can verify this by noting that

$$\int -10xe^{-x^2}\,dx = 5\int e^{-x^2}(-2x\,dx) = 5e^{-x^2} + C$$

(b) We know that $f(x)$ is the derivative of $\int f(x)\,dx$, so, as we saw in (a), the x-intercepts of $f(x)$ locate the critical values and extrema of $\int f(x)\,dx$.

(c) The first derivative of $\int f(x)\,dx$ is $f(x)$, and its second derivative is $f'(x)$. Hence the inflection points of $\int f(x)\,dx$ occur where $f'(x) = 0$. But $f(x)$ has its extrema where $f'(x) = 0$. Thus the extrema of $f(x)$ occur at the same x-values as the inflection points of $\int f(x)\,dx$.

CHECKPOINT SOLUTIONS

1. (a) True

   (b) False; $\displaystyle\int \frac{2x\,dx}{\sqrt{x^2+1}} = \int (x^2+1)^{-1/2}(2x\,dx)$

   $$= \frac{(x^2+1)^{1/2}}{1/2} + C = 2(x^2+1)^{1/2} + C$$

   (c) True

   (d) False. We cannot factor the variable x outside the integral sign.

   (e) True

2. (a)
$$4x+1\,\overline{)\,4x\,}\quad\text{quotient } 1$$
$$\begin{array}{r} 4x \\ \underline{4x+1} \\ -1 \end{array}$$

   so $\dfrac{4x}{4x+1} = 1 - \dfrac{1}{4x+1}$.

   (b) $\displaystyle\int \frac{4x\,dx}{4x+1} = \int\left(1 - \frac{1}{4x+1}\right)dx = x - \frac{1}{4}\ln|4x+1| + C$

3. (a) False. The correct solution is $e^{x^2} + C$ (see Example 5).

   (b) True

   (c) False; $\displaystyle\int \frac{dx}{e^{3x}} = \int e^{-3x}\,dx = -\frac{1}{3}e^{-3x} + C = \frac{-1}{3e^{3x}} + C$

   (d) False; $\displaystyle\int e^{3x+1}(3\,dx) = e^{3x+1} + C$

EXERCISE 12.3

Evaluate the integrals in Problems 1–32.

1. $\int \frac{3x^2}{x^3+4}\,dx$
2. $\int \frac{8x^7}{x^8-1}\,dx$
3. $\int \frac{dz}{4z+1}$
4. $\int \frac{y}{y^2+1}\,dy$
5. $\int \frac{x^3}{x^4+1}\,dx$
6. $\int \frac{x^2}{x^3-9}\,dx$
7. $\int \frac{4x}{x^2-4}\,dx$
8. $\int \frac{5x^2}{x^3-1}\,dx$
9. $\int \frac{3x^2-2}{x^3-2x}\,dx$
10. $\int \frac{4x^3+2x}{x^4+x^2}\,dx$
11. $\int \frac{z^2+1}{z^3+3z+17}\,dz$
12. $\int \frac{(x+2)dx}{x^2+4x-9}$
13. $\int \frac{x^3-x^2+1}{x-1}\,dx$
14. $\int \frac{2x^3+x^2+2x+3}{2x+1}\,dx$
15. $\int \frac{x^2+x+3}{x^2+3}\,dx$
16. $\int \frac{x^4-2x^2+x}{x^2-2}\,dx$
17. $\int 3e^{3x}\,dx$
18. $\int 4e^{4x}\,dx$
19. $\int e^{-x}\,dx$
20. $\int e^{2x}\,dx$
21. $\int 1000e^{0.1x}\,dx$
22. $\int 1600e^{0.4x}\,dx$
23. $\int 840e^{-0.7x}\,dx$
24. $\int 250e^{-0.5x}\,dx$
25. $\int x^3e^{3x^4}\,dx$
26. $\int xe^{2x^2}\,dx$
27. $\int \frac{3}{e^{2x}}\,dx$
28. $\int \frac{4}{e^{1-2x}}\,dx$
29. $\int \frac{x^5}{e^{2-3x^6}}\,dx$
30. $\int \frac{x^3}{e^{4x^4}}\,dx$
31. $\int \left(e^{4x}-\frac{3}{e^{x/2}}\right)dx$
32. $\int \left(xe^{3x^2}-\frac{5}{e^{x/3}}\right)dx$

In Problems 33 and 34, graphs of two functions labeled $g(x)$ and $h(x)$ are given. Decide which is the graph of $f(x)$ and which is one member of the family $\int f(x)\,dx$. Check your conclusions by evaluating the integral.

33.

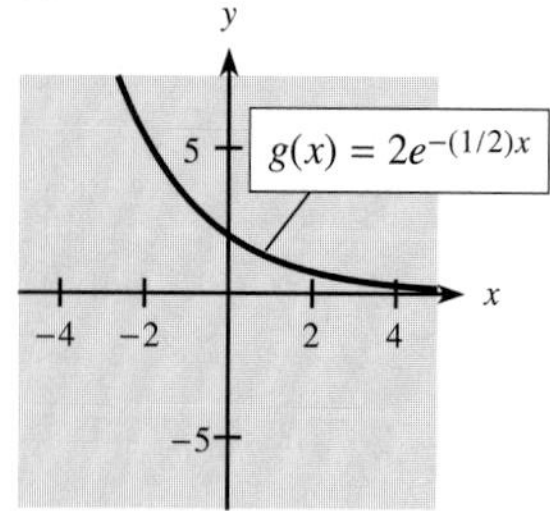

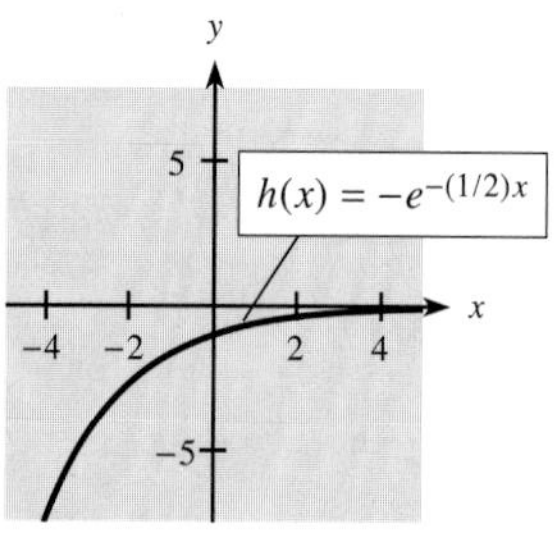

34.

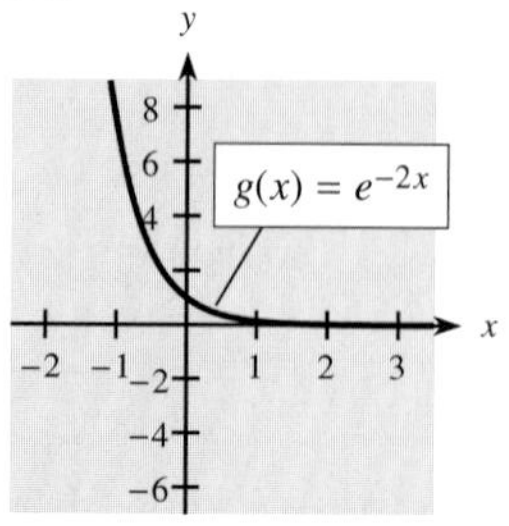

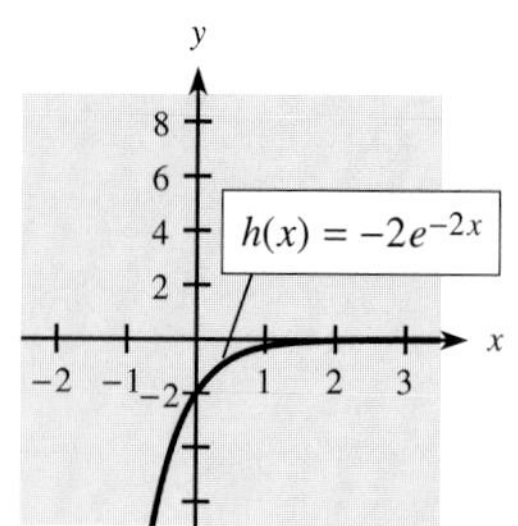

In Problems 35 and 36, a function $f(x)$ and its graph are given. Find the family $F(x) = \int f(x)\,dx$ and graph the member that satisfies $F(0) = 0$.

35.

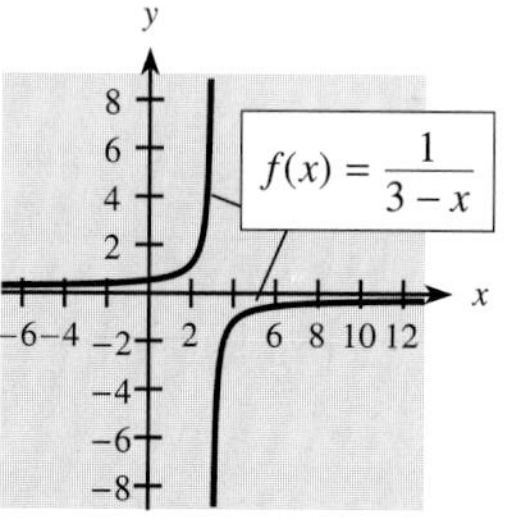

36.

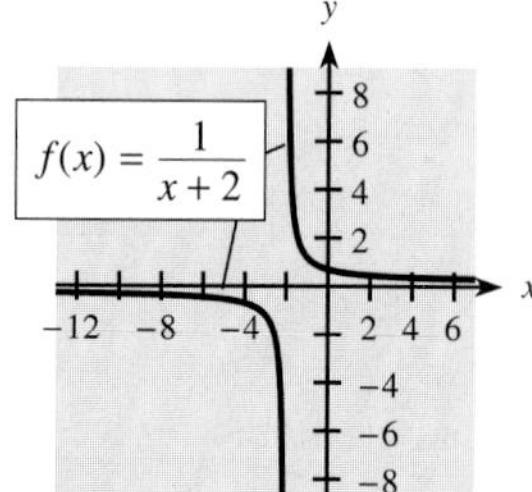

In Problems 37–40, a family of functions is given and graphs of some members are shown. Find the function $f(x)$ such that the family is given by $\int f(x)\,dx$.

37. $F(x) = x + \ln|x| + C$

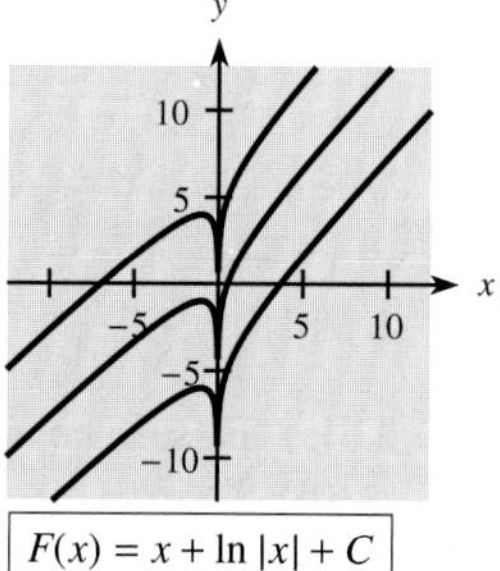

$F(x) = x + \ln|x| + C$

38. $F(x) = -\ln(x^2+4) + C$

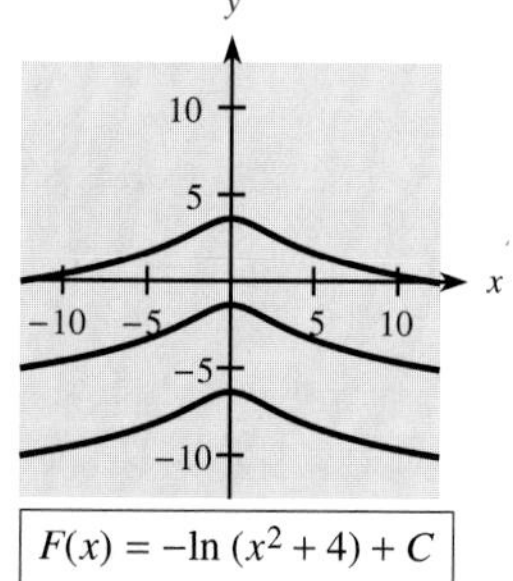

$F(x) = -\ln(x^2+4) + C$

39. $F(x) = 5xe^{-x} + C$

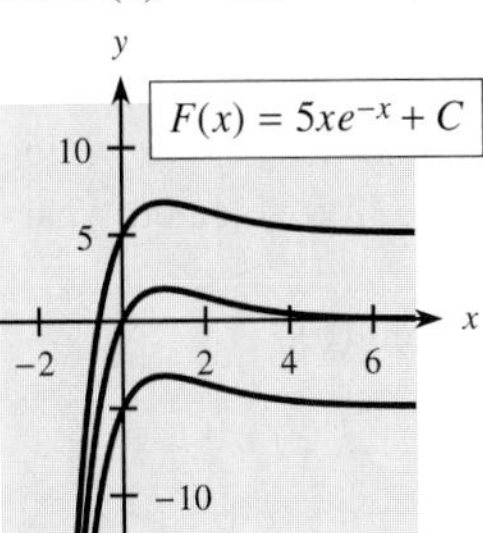

40. $F(x) = e^{0.4x} + e^{-0.4x} + C$

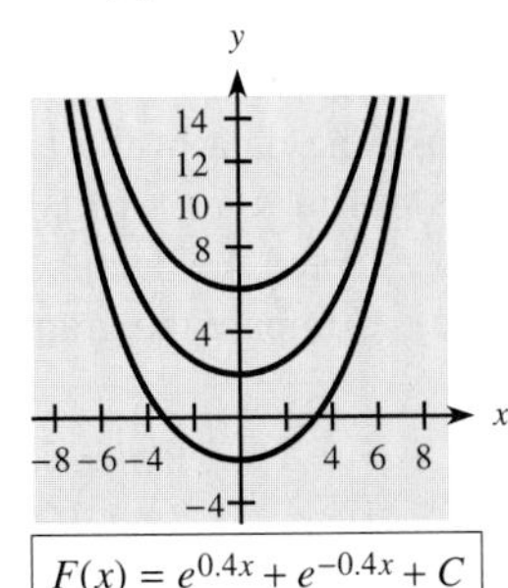

Applications

41. ***Revenue*** Suppose that the marginal revenue from the sale of a product is $\overline{MR} = R'(x) = 6e^{0.01x}$. What is the revenue on the sale of 100 units of the product?

42. ***Concentration of a drug*** Suppose that the rate at which the concentration of a drug in the blood changes with respect to time t is given by

$$C'(t) = \frac{c}{b-a}(be^{-bt} - ae^{-at}), \qquad t \geq 0$$

where a, b, and c are constants depending on the drug administered, with $b > a$. Assuming that $C(t) = 0$ when $t = 0$, find the formula for the concentration of the drug in the blood at any time t.

43. ***Radioactive decay*** The rate of disintegration of a radioactive substance can be described by

$$\frac{dn}{dt} = n_0(-K)e^{-Kt}$$

where n_0 is the number of radioactive atoms present when time t is 0, and K is a positive constant that depends on the substance involved. Using the fact that the constant of integration is 0, integrate dn/dt to find the number of atoms n that are still radioactive after time t.

44. ***World population*** Because the world contains only about 10 billion acres of arable land, world population is limited. Suppose that the world population is limited to 40 billion people and that the rate of population growth is proportional to how close the world is to this upper limit. Then the rate of growth would be given by $dP/dt = K(40 - P)$, where K is a positive constant. This means that

$$t = \frac{1}{K}\int \frac{1}{40 - P}\,dP$$

(a) Evaluate this integral to find an expression relating P and t.
(b) Use the properties relating logarithms and exponential functions to write P as a function of t.

45. ***Memorization*** The rate of vocabulary memorization of the average student in a foreign language is given by

$$\frac{dv}{dt} = \frac{40}{t+1}$$

where t is the number of continuous hours of study, $0 < t \leq 4$. How many words would the average student memorize in 3 hours?

46. ***Population growth*** The rate of growth of world population can be modeled by

$$\frac{dn}{dt} = N_0(1+r)^t \ln(1+r), \qquad r < 1$$

where t is the time in years from the present and N_0 and r are constants. What function describes world population if the present population is N_0? Use the formula $\int (a^u \ln a)u'\,dx = a^u + C$.

47. ***Compound interest*** If \$$P$ is invested for n years at 10%, compounded continuously, the rate at which the future value is growing is

$$\frac{dS}{dn} = 0.1Pe^{0.1n}$$

(a) What function describes the future value at the end of n years?
(b) In how many years will the future value double?

48. ***Temperature changes*** When an object is moved from one environment to another, its temperature T changes at a rate given by

$$\frac{dT}{dt} = kCe^{kt}$$

where t is the time in the new environment (in hours), C is the temperature difference (old − new) between the two environments, and k is a constant. If the temperature of the body (and the old environment) is 70°F, and $C = -10$°F, what function describes the temperature T of the object t hours after it is moved?

49. ***Blood pressure in the aorta*** The rate at which blood pressure decreases in the aorta of a normal adult after a heartbeat is

$$\frac{dp}{dt} = -46.645e^{-0.491t}$$

where t is time in seconds.

(a) What function describes the blood pressure in the aorta if $p = 95$ when $t = 0$?

(b) What is the blood pressure 0.1 second after a heartbeat?

50. ***Sales and advertising*** A store finds that its sales decline after the end of an advertising campaign, with its daily sales for the period declining at the rate $S'(t) = -147.78e^{-0.2t}$, $0 \le t \le 100$, where t is the number of days since the end of the campaign. Suppose that $S = 7389$ when $t = 0$.

(a) Find the function that describes the number of daily sales t days after the end of the campaign.

(b) Find the total number of sales 10 days after the end of the advertising campaign.

51. ***Life expectancy*** Suppose the rate of change of the expected life span l at birth of people born in the United States can be modeled by

$$\frac{dl}{dt} = \frac{14.1372}{t + 20}$$

where t is the number of years past 1920.

(a) If the expected life span was 72.6 years for people born in 1975, find the function that models the life span. Graph this function.

(b) The data in the table give the expected life span for people born in various years. Graph the data.

(c) Compare the graphs in (a) and (b).

Year	*Life Span (years)*	*Year*	*Life Span (years)*
1920	54.1	1988	74.9
1930	59.7	1989	75.2
1940	62.9	1990	75.4
1950	68.2	1991	75.5
1960	69.7	1992	75.8
1970	70.8	1993	75.5
1975	72.6	1994	75.7
1980	73.7	1996	76.1
1987	75.0		

Source: *World Almanac*, 1991

52. ***Violent crime*** Suppose the rate of change of the number of violent crimes per 100,000 people in the United States can be modeled by

$$\frac{dc}{dt} = \frac{321}{t + 5}$$

when t is the number of years past 1985.

(a) If there were 758 violent crimes per 100,000 people in 1991, find the function $c(t)$ that models the number of violent crimes. Graph this function.

(b) The data in the table give the number of violent crimes per 100,000 people for selected years. Graph the data in the table.

(c) Compare the graphs in (a) and (b).

Year	*Violent Crimes (per 100,000)*
1987	610
1988	637
1989	663
1990	732
1991	758
1992	765

Source: FBI Crime Report, 1993

53. ***Revenue*** Suppose the rate of change of revenue from wireless technology can be modeled by

$$\frac{dR}{dt} = 0.1994e^{0.3486t}$$

where t is the number of years past 1984 and R is revenue in billions of dollars.

(a) Find a model for revenue, $R(t)$, if revenue from wireless technology was \$11 billion in 1993.

(b) Graph the function you found in (a) and compare it to the graph in the figure.

(c) What does your model predict for revenue from wireless technology in 1995?

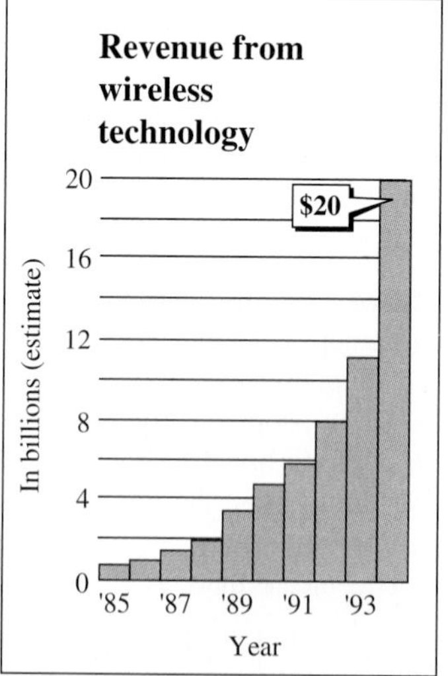

Source: *USA Today*, March 1, 1994

54. ***Health care costs*** Suppose the rate of change of national health care costs in the United States can be modeled by

$$\frac{dHC}{dt} = 4.0428e^{0.08984t}$$

where t is the number of years past 1960 and HC is the national health care expenditure in billions of dollars.

(a) Use the fact that 1991 health care costs totaled \$752 billion to find a function $HC(t)$ that models health care costs.

(b) Graph the function you found in (a) and compare it to the graph in the figure.

(c) What does your model project for health care costs for the year 2000?

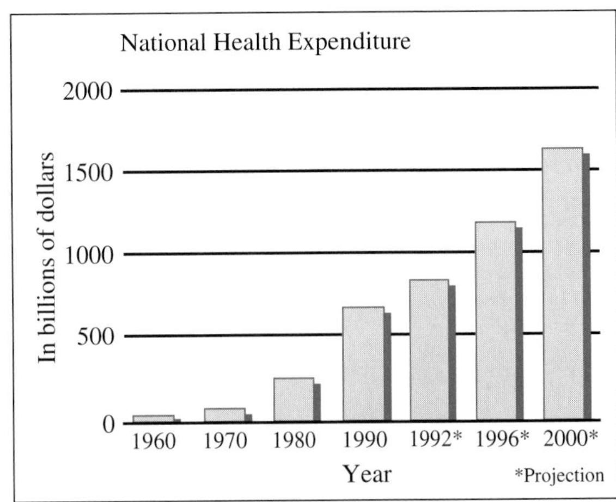

Source: Congressional Budget Office, printed in *Newsweek*, October 4, 1993

12.4 Applications of the Indefinite Integral in Business and Economics

OBJECTIVES

- *To use integration to find total cost functions from information involving marginal cost*
- *To optimize profit, given information regarding marginal cost and marginal revenue*
- *To use integration to find national consumption functions from information about marginal propensity to consume and marginal propensity to save*

APPLICATION PREVIEW

We turn our attention to applications involving cost, profit, and consumption. A thorough understanding of these concepts is imperative for success in any study of economics and business.

In this section, we will use integration to derive total cost and profit functions from the marginal cost and marginal revenue functions. One of the reasons for the marginal approach in economics is that firms can observe marginal changes in real life. If they know the marginal cost and the total cost when a given quantity is sold, they can develop their total cost function.

In this section, we also use integration to obtain the national consumption function from functions that describe the marginal propensity to consume or the marginal propensity to save.

Total Cost and Profit

We know that the marginal cost for a commodity is $\overline{MC} = C'(x)$, where $C(x)$ is the total cost function. Thus if we have the marginal cost function, we can integrate to find the total cost. That is, $C(x) = \int \overline{MC}\, dx$.

If, for example, the marginal cost is $\overline{MC} = 4x + 3$, the total cost is given by

$$\begin{aligned} C(x) &= \int \overline{MC}\, dx \\ &= \int (4x + 3)\, dx \\ &= 2x^2 + 3x + K \end{aligned}$$

where K represents the constant of integration. Now, we know that the total revenue is 0 if no items are produced, but the total cost may not be 0 if nothing is produced. The fixed costs accrue whether goods are produced or not. Thus the value for the constant of integration depends on the fixed costs FC of production.

Thus we cannot determine the total cost function from the marginal cost unless additional information is available to help us determine the fixed costs.

EXAMPLE 1

If the marginal cost function for a month for a certain product is $\overline{MC} = 3x + 50$, and if the fixed costs related to the product amount to \$100 per month, find the total cost function for the month.

Solution

The total cost function is

$$C(x) = \int (3x + 50)\,dx$$
$$= \frac{3x^2}{2} + 50x + K$$

But the constant of integration K is found by using the fact that $C(0) = FC = 100$. Thus

$$3(0)^2 + 50(0) + K = 100, \quad \text{so } K = 100$$

and the total cost for the month is given by

$$C(x) = \frac{3x^2}{2} + 50x + 100$$

EXAMPLE 2

Suppose monthly records show that the rate of change of the cost (that is, the marginal cost) for a product is $\overline{MC} = 3(2x + 25)^{1/2}$ and that the fixed costs for the month are \$11,125. What would be the total cost of producing 300 items per month?

Solution

We can integrate the marginal cost to find the total cost function.

$$C(x) = \int \overline{MC}\,dx = \int 3(2x + 25)^{1/2}\,dx$$
$$= 3 \cdot \left(\frac{1}{2}\right) \int (2x + 25)^{1/2}(2\,dx)$$
$$= \left(\frac{3}{2}\right) \frac{(2x + 25)^{3/2}}{3/2} + K$$
$$= (2x + 25)^{3/2} + K$$

We can find K by using the fact that fixed costs are \$11,125.

$$C(0) = 11{,}125 = (25)^{3/2} + K$$
$$11{,}125 = 125 + K, \quad \text{or} \quad K = 11{,}000$$

Thus the total cost function is

$$C(x) = (2x + 25)^{3/2} + 11{,}000$$

and the cost of producing 300 items per month is

$$C(300) = (625)^{3/2} + 11{,}000$$
$$= 26{,}625 \quad \text{(dollars)}$$

It can be shown that the profit is usually maximized when $\overline{MR} = \overline{MC}$. To see that this does not always give us a maximum *positive* profit, consider the following facts concerning the manufacture of widgets over the period of a month:

1. The marginal revenue is $\overline{MR} = 400 - 30x$.
2. The marginal cost is $\overline{MC} = 20x + 50$.
3. When 5 widgets are produced and sold, the total cost is \$1750. The profit *should* be maximized when $\overline{MR} = \overline{MC}$, or when $400 - 30x = 20x + 50$. Solving for x gives $x = 7$. To see whether our profit is maximized when 7 units are produced and sold, let us examine the profit function.

The profit function is given by $P(x) = R(x) - C(x)$, where

$$R(x) = \int \overline{MR}\, dx \qquad \text{and} \qquad C(x) = \int \overline{MC}\, dx$$

Integrating, we get

$$R(x) = \int (400 - 30x)\, dx = 400x - 15x^2 + K$$

but $K = 0$ for this total revenue function, so

$$R(x) = 400x - 15x^2$$

The total cost function is

$$C(x) = \int (20x + 50)\, dx = 10x^2 + 50x + K$$

The value of fixed cost can be determined by using the fact that 5 widgets cost \$1750. This tells us that $C(5) = 1750 = 250 + 250 + K$, so $K = 1250$.

Thus the total cost is $C(x) = 10x^2 + 50x + 1250$. Now, the profit is

$$P(x) = R(x) - C(x)$$

or

$$P(x) = (400x - 15x^2) - (10x^2 + 50x + 1250)$$

Simplifying gives

$$P(x) = 350x - 25x^2 - 1250$$

We have found that $\overline{MR} = \overline{MC}$ if $x = 7$, and the graph of $P(x)$ is a parabola that opens downward, so profit is maximized at $x = 7$. But if $x = 7$, profit is

$$P(7) = 2450 - 1225 - 1250 = -25$$

That is, the production and sale of 7 items result in a loss of \$25.

The preceding discussion indicates that, although setting $\overline{MR} = \overline{MC}$ may optimize profit, it does not indicate the level of profit or loss, as forming the profit function does.

If the widget firm is in a competitive market, and its optimal level of production results in a loss, it has two options. It can continue to produce at the optimal level in the short run until it can lower or eliminate its fixed costs, even though it is losing money; or it can take a larger loss (its fixed cost) by stopping production. Producing 7 units causes a loss of \$25 per month, and ceasing production results in a loss of \$1250 (the fixed cost) per month. If this firm and many others like it cease production, the supply will be reduced, causing an eventual increase in price. The firm can resume production when the price increase indicates that it can make a profit.

EXAMPLE 3

Given that $\overline{MR} = 200 - 4x$, $\overline{MC} = 50 + 2x$, and the total cost of producing 10 Wagbats is \$700, at what level should the Wagbat firm hold production in order to maximize the profits?

Solution

Setting $\overline{MR} = \overline{MC}$, we can solve for the production level that maximizes profit.

$$\begin{aligned} 200 - 4x &= 50 + 2x \\ 150 &= 6x \\ 25 &= x \end{aligned}$$

The level of production that should optimize profit is 25 units. To see whether 25 units maximizes profits or minimizes the losses (in the short run), we must find the total revenue and total cost functions.

$$\begin{aligned} R(x) &= \int (200 - 4x)\,dx = 200x - 2x^2 + K \\ &= 200x - 2x^2, \quad \text{because } K = 0 \\ C(x) &= \int (50 + 2x)\,dx = 50x + x^2 + K \end{aligned}$$

We find K by noting that $C(x) = 700$ when $x = 10$.

$$700 = 50(10) + (10)^2 + K$$

so $K = 100$.

Thus the cost is given by $C = C(x) = 50x + x^2 + 100$. At $x = 25$, $R = R(25) = 200(25) - 2(25)^2 = \$3750$ and $C = C(25) = 50(25) + (25)^2 + 100 = \$1975$.

We see that the total revenue is greater than the total cost, so production should be held at 25 units, which results in a maximum profit.

Graphing Utilities

If it is difficult to solve $\overline{MC} = \overline{MR}$ analytically, we could use a graphing utility to solve this equation by finding the point of intersection of the graphs of $\overline{MC}$ and $\overline{MR}$. We may also be able to integrate $\overline{MC}$ and $\overline{MR}$ to find the functions $C(x)$ and $R(x)$ and then use a graphing utility to graph them. From the graphs of $C(x)$ and $R(x)$ we can learn about these functions—and hence about profit.

EXAMPLE 4

Suppose that $\overline{MC} = 1.01(x + 190)^{0.01}$ and $\overline{MR} = (1/\sqrt{2x+1}) + 2$, where x is the number of thousands of units and both revenue and cost are in thousands of dollars. Suppose further that fixed costs are \$100,236 and that production is limited to at most 180 thousand units.

(a) Determine $C(x)$ and $R(x)$ and graph them to determine whether a profit can be made.

(b) Estimate the level of production that yields maximum profit, and find the maximum profit.

Solution

(a) $C(x) = \int \overline{MC}\, dx = \int 1.01\,(x + 190)^{0.01}\, dx$

$$= 1.01\,\frac{(x+190)^{1.01}}{1.01} + K$$

When we say that fixed costs equal \$100,236, we mean $C(0) = 100.236$.

$$\begin{aligned} 100.236 &= C(0) = (190)^{1.01} + K \\ 100.236 &= 200.236 + K \\ -100 &= K \end{aligned}$$

Thus $C(x) = (x + 190)^{1.01} - 100$.

$$\begin{aligned} R(x) &= \int \overline{MR}\, dx = \int \left[(2x+1)^{-1/2} + 2\right] dx \\ &= \frac{1}{2}\int (2x+1)^{-1/2}\,(2\,dx) + \int 2\, dx \\ &= \frac{\frac{1}{2}(2x+1)^{1/2}}{1/2} + 2x + K \end{aligned}$$

$R(0) = 0$ means

$$0 = R(0) = (1)^{1/2} + 0 + K, \quad \text{or} \quad K = -1$$

Thus $R(x) = (2x + 1)^{1/2} + 2x - 1$.

The graphs of $C(x)$ and $R(x)$ are shown in Figure 12.8 on the next page. (The x-range is chosen to include the production range from 0 to 180 (thousand) units. The y-range is chosen to extend beyond fixed costs of about 100 thousand dollars.)

From the figure we see that a profit can be made as long as the number of units sold exceeds about 95 (thousand). We could locate this value more precisely by using INTERSECT or TRACE and ZOOM.

(b) From the graph we also see that $R(x) - C(x) = P(x)$ is at its maximum at the right edge of the graph. Because production is limited to at most 180 thousand units, profit will be maximized when $x = 180$ and the maximum profit is

$$\begin{aligned} P(180) &= R(180) - C(180) \\ &= [(361)^{1/2} + 360 - 1] - [(370)^{1.01} - 100] \\ &\approx 85.46 \quad \text{(thousand dollars)} \end{aligned}$$

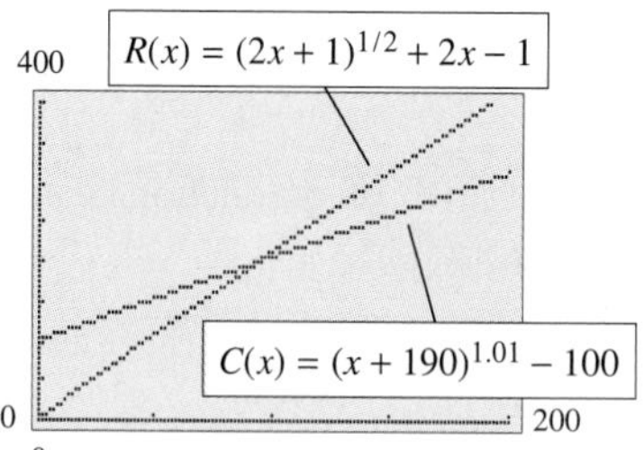

Figure 12.8

CHECKPOINT

1. True or false:
   (a) If $C(x) = \int \overline{MC}\, dx$, then the constant of integration equals the fixed costs.
   (b) If $R(x) = \int \overline{MR}\, dx$, then the constant of integration equals 0.
2. Find $C(x)$ if $\overline{MC} = \dfrac{100}{\sqrt{x + 1}}$ and fixed costs are \$8000.

National Consumption and Savings

The consumption function is one of the basic ingredients in a larger discussion of how an economy can have persistent high unemployment or persistent high inflation. This study is often called **Keynesian analysis,** after its founder John Maynard Keynes.

If C represents national consumption (in billions of dollars), then a **national consumption function** has the form $C = f(y)$, where y is disposable national income (also in billions of dollars). The **marginal propensity to consume** is the derivative of the national consumption function with respect to y, or $dC/dy = f'(y)$. For example, suppose that

$$C = f(y) = 0.8y + 6$$

is a national consumption function; then the marginal propensity to consume is $f'(y) = 0.8$.

If we know the marginal propensity to consume, we can integrate with respect to y to find national consumption:

$$C = \int f'(y)\, dy = f(y) + K$$

We can find the unique national consumption function if we have additional information to help us determine the value of K, the constant of integration.

EXAMPLE 5

If consumption is \$6 billion when disposable income is 0, and if the marginal propensity to consume is $dC/dy = 0.3 + 0.4/\sqrt{y}$ (in billions of dollars), find the national consumption function.

Solution

If

$$\frac{dC}{dy} = 0.3 + \frac{0.4}{\sqrt{y}}$$

then

$$C = \int \left(0.3 + \frac{0.4}{\sqrt{y}}\right) dy = 0.3y + 0.8y^{1/2} + K$$

Now, if $C = 6$ when $y = 0$, then $6 = 0.3(0) + 0.8\sqrt{0} + K$. Thus the constant of integration is $K = 6$, and the consumption function is

$$C = 0.3y + 0.8\sqrt{y} + 6 \qquad \text{(billions of dollars)}$$

If S represents national savings, we can assume that the disposable national income is given by $y = C + S$, or $S = y - C$. Then the **marginal propensity to save** is $dS/dy = 1 - dC/dy$.

EXAMPLE 6

If the consumption is \$9 billion when income is 0, and if the marginal propensity to save is 0.25, find the consumption function.

Solution

If $dS/dy = 0.25$, then $0.25 = 1 - dC/dy$, or $dC/dy = 0.75$. Thus

$$C = \int 0.75\, dy = 0.75y + K$$

If $C = 9$ when $y = 0$, then $9 = 0.75(0) + K$, or $K = 9$. Then the consumption function is $C = 0.75y + 9$ (billions of dollars).

CHECKPOINT

3. If the marginal propensity to save is

$$\frac{dS}{dy} = 0.7 - \frac{0.4}{\sqrt{y}}$$

find the marginal propensity to consume.

4. Find the national consumption function if the marginal propensity to consume is

$$\frac{dC}{dy} = \frac{1}{\sqrt{y+4}} + 0.2$$

and national consumption is \$6.8 billion when disposable income is 0.

CHECKPOINT SOLUTIONS

1. (a) False. $C(0)$ equals the fixed costs. It may or may not be the constant of integration. [In Problem 2, $C(0) = 8000$, but $K = 7800$.]
   (b) False. We use $R(0) = 0$ to determine the constant of integration, but it may be nonzero. See Example 4.
2. $C(x) = 100\int (x + 1)^{-1/2}\, dx = 100\left[\dfrac{(x + 1)^{1/2}}{1/2}\right] + K$

   $C(x) = 200\sqrt{x + 1} + K$

   When $x = 0$, $C(x) = 8000$, so

   $$8000 = 200\sqrt{1} + K$$
   $$7800 = K$$
   $$C(x) = 200\sqrt{x + 1} + 7800$$
3. $\dfrac{dC}{dy} = 1 - \dfrac{dS}{dy} = 1 - \left(0.7 - \dfrac{0.4}{\sqrt{y}}\right) = 0.3 + \dfrac{0.4}{\sqrt{y}}$
4. $C(y) = \int \left(\dfrac{1}{\sqrt{y + 4}} + 0.2\right) dy$

   $= \int [(y + 4)^{-1/2} + 0.2]\, dy$

   $= \dfrac{(y + 4)^{1/2}}{1/2} + 0.2y + K$

   $C(y) = 2\sqrt{y + 4} + 0.2y + K$

   Using $C(0) = 6.8$ gives $6.8 = 2\sqrt{4} + 0 + K$, or $K = 2.8$.

   Thus $C(y) = 2\sqrt{y + 4} + 0.2y + 2.8$.

EXERCISE 12.4

Total Cost and Profit

1. If the monthly marginal cost for a product is $\overline{MC} = 2x + 100$, with fixed costs amounting to \$200, find the total cost function for the month.
2. If the monthly marginal cost for a product is $\overline{MC} = x + 30$, and the related fixed costs are \$50, find the total cost function for the month.
3. If the marginal cost for a product is $\overline{MC} = 4x + 2$, and the production of 10 units results in a total cost of \$300, find the total cost function.
4. If the marginal cost for a product is $\overline{MC} = 3x + 50$, and the total cost of producing 20 units is \$2000, what will be the total cost function?
5. If the marginal cost for a product is $\overline{MC} = 4x + 40$, and the total cost of producing 25 units is \$3000, what will be the cost of producing 30 units?
6. If the marginal cost for producing a product is $\overline{MC} = 5x + 10$, with a fixed cost of \$800, what will be the cost of producing 20 units?
7. A firm knows that its marginal cost for a product is $\overline{MC} = 3x + 20$, that its marginal revenue is $\overline{MR} = 44 - 5x$, and that the cost of production and sale of 80 units is \$11,400.
   (a) Find the optimal level of production.
   (b) Find the profit function.
   (c) Find the profit or loss at the optimal level.
8. A certain firm's marginal cost for a product is $\overline{MC} = 6x + 60$, its marginal revenue is $\overline{MR} = 180 - 2x$, and its total cost of production of 10 items is \$1000.
   (a) Find the optimal level of production.
   (b) Find the profit function.
   (c) Find the profit or loss at the optimal level of production.
   (d) Should production be continued for the short run?
   (e) Should production be continued for the long run?
9. Suppose that the marginal revenue for a product is $\overline{MR} = 900$ and the marginal cost is $\overline{MC} = 30\sqrt{x + 4}$, with a fixed cost of \$1000.

(a) Find the profit or loss from the production and sale of 5 units.
(b) How many units will result in a maximum profit?

10. Suppose that the marginal cost for a product is $\overline{MC} = 60\sqrt{x+1}$ and its fixed cost is \$340.00. If the marginal revenue for it is $\overline{MR} = 80x$, find the profit or loss from production and sale of:
(a) 3 units.
(b) 8 units.

11. The average cost of a product changes at the rate

$$\overline{C}'(x) = -6x^{-2} + 1/6$$

and the average cost of 6 units is \$10.00.
(a) Find the average cost function.
(b) Find the average cost of 12 units.

12. The average cost of a product changes at the rate

$$\overline{C}'(x) = \frac{-10}{x^2} + \frac{1}{10}$$

and the average cost of 10 units is \$20.00.
(a) Find the average cost function.
(b) Find the average cost of 20 units.

13. Suppose for a certain product that marginal cost is given by $\overline{MC} = 1.05(x + 180)^{0.05}$ and marginal revenue is given by $\overline{MR} = (1/\sqrt{0.5x+4}) + 2.8$, where x is in thousands of units and both revenue and cost are in thousands of dollars. Fixed costs are \$200,000 and production is limited to at most 200 thousand units.
(a) Find $C(x)$ and $R(x)$.
(b) Graph $C(x)$ and $R(x)$ to determine whether a profit can be made.
(c) Determine the level of production that yields maximum profit, and find the maximum profit (or minimum loss).

14. Suppose for a certain product that the marginal cost is given by $\overline{MC} = 1.02(x + 200)^{0.02}$ and marginal revenue is given by $\overline{MR} = (2/\sqrt{4x+1}) + 1.75$, where x is in thousands of units and revenue and cost are in thousands of dollars. Suppose further that fixed costs are \$150,000 and production is limited to at most 200 thousand units.
(a) Find $C(x)$ and $R(x)$.
(b) Graph $C(x)$ and $R(x)$ to determine whether a profit can be made.
(c) Determine what level of production yields maximum profit, and find the maximum profit (or minimum loss).

National Consumption and Savings

15. If consumption is \$5 billion when disposable income is 0, and if the marginal propensity to consume is

$$\frac{dC}{dy} = 0.4 + \frac{0.3}{\sqrt{y}} \quad \text{(in billions of dollars)}$$

find the national consumption function.

16. If consumption is \$7 billion when disposable income is 0, and if the marginal propensity to consume is 0.80, find the national consumption function (in billions of dollars).

17. If consumption is \$8 billion when income is 0, and if the marginal propensity to consume is

$$\frac{dC}{dy} = 0.3 + \frac{0.2}{\sqrt{y}} \quad \text{(in billions of dollars)}$$

find the national consumption function.

18. If national consumption is \$9 billion when income is 0, and if the marginal propensity to consume is 0.30, what is consumption when disposable income is \$20 billion?

19. If consumption is \$6 billion when disposable income is 0, and if the marginal propensity to consume is

$$\frac{dC}{dy} = \frac{1}{\sqrt{y+1}} + 0.4 \quad \text{(in billions of dollars)}$$

find the national consumption function.

20. If consumption is \$5.8 billion when disposable income is 0, and if the marginal propensity to consume is

$$\frac{dC}{dy} = \frac{1}{\sqrt{2y+9}} + 0.8 \quad \text{(in billions of dollars)}$$

find the national consumption function.

21. Suppose that the marginal propensity to consume is

$$\frac{dC}{dy} = 0.7 - e^{-2y} \quad \text{(in billions of dollars)}$$

and that consumption is \$5.65 billion when disposable income is 0. Find the national consumption function.

22. Suppose that the marginal propensity to consume is

$$\frac{dC}{dy} = 0.04 + \frac{\ln(y+1)}{y+1} \quad \text{(in billions of dollars)}$$

and that consumption is \$6.04 billion when disposable income is 0. Find the national consumption function.

23. Suppose that the marginal propensity to save is

$$\frac{dS}{dy} = 0.15 \qquad \text{(in billions of dollars)}$$

and that consumption is \$5.15 billion when disposable income is 0. Find the national consumption function.

24. Suppose that the marginal propensity to save is

$$\frac{dS}{dy} = 0.22 \qquad \text{(in billions of dollars)}$$

and that consumption is \$8.6 billion when disposable income is 0. Find the national consumption function.

25. Suppose that the marginal propensity to save is

$$\frac{dS}{dy} = 0.2 - \frac{1}{\sqrt{3y+7}} \qquad \text{(in billions of dollars)}$$

and that consumption is \$6 billion when disposable income is 0. Find the national consumption function.

26. If consumption is \$3 billion when disposable income is 0, and if the marginal propensity to save is

$$\frac{dS}{dy} = 0.2 + e^{-1.5y} \qquad \text{(in billions of dollars)}$$

find the national consumption function.

12.5 Differential Equations

OBJECTIVES

- To show that a function is the solution to a differential equation
- To use integration to find the general solution of a differential equation
- To find particular solutions of differential equations using given conditions
- To solve separable differential equations
- To solve applied problems involving separable differential equations

APPLICATION PREVIEW

Carbon-14 dating, used to determine the age of fossils, is based on three facts. First, the half-life of carbon-14 is 5600 years. Second, the amount of carbon-14 in any living organism is essentially constant. Third, when an organism dies, the rate of change of carbon-14 in the organism is proportional to the amount present. If y represents the amount of carbon-14 present in the organism, then we can express the rate of change of carbon-14 by the **differential equation**

$$\frac{dy}{dt} = ky$$

where k is a constant and t is time in years. In this section, we study methods that allow us to find a function y that satisfies this differential equation, and then we use that function to date a fossil.

Recall that we introduced the derivative as an instantaneous rate of change and denoted the instantaneous rate of change of y with respect to time as dy/dt. For many growth or decay processes, such as carbon-14 decay, the rate of change of the amount of a substance with respect to time is proportional to the amount present. As we noted above, this can be represented by the equation

$$\frac{dy}{dt} = ky \qquad (k = \text{constant})$$

An equation of this type, where y is an unknown function of x or t, is called a **differential equation** because it contains derivatives (or differentials). In this section, we restrict ourselves to differential equations where the highest derivative present in the equation is the first derivative. These differential equations are called **first-order differential equations.** Examples are

$$f'(x) = \frac{1}{x+1}, \qquad \frac{dy}{dt} = 2t, \quad \text{and} \quad x\,dy = (y+1)\,dx$$

Solution of Differential Equations

The solution to a differential equation is a function [say $y = f(x)$] that, when used in the differential equation, results in an identity.

EXAMPLE 1

Show that $y = 4e^{-5t}$ is a solution of $dy/dt + 5y = 0$.

Solution

We must show that substituting $y = 4e^{-5t}$ into the equation $dy/dt + 5y = 0$ results in an identity:

$$\begin{aligned} \frac{d}{dt}(4e^{-5t}) + 5(4e^{-5t}) &= 0 \\ -20e^{-5t} + 20e^{-5t} &= 0 \\ 0 &= 0 \end{aligned}$$

Thus $y = 4e^{-5t}$ is a solution.

Now that we know what it means for a function to be a solution to a differential equation, let us consider how to find solutions.

The most elementary differential equations are of the form

$$\frac{dy}{dx} = f(x)$$

where $f(x)$ is a continuous function. These equations are elementary to solve because the solutions are found by integration:

$$y = \int f(x)\,dx$$

EXAMPLE 2

Find the solution of

$$f'(x) = \frac{1}{x+1}$$

Solution

The solution is

$$f(x) = \int f'(x)\,dx = \int \frac{1}{x+1}\,dx = \ln|x+1| + C$$

The solution in Example 2, $f(x) = \ln|x + 1| + C$, is called the **general solution** because every solution to the equation has this form, and different values of C give different **particular solutions.** Figure 12.9 on the next page shows the graphs of several members of the family of solutions to this differential equation. (We cannot, of course, show all of them.)

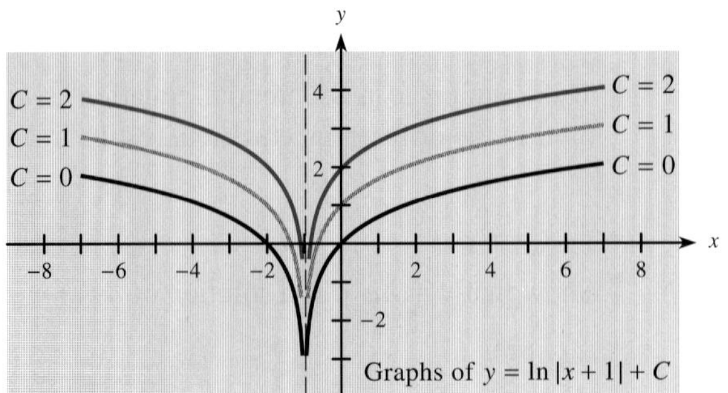

Figure 12.9

We can find a particular solution to a differential equation when we know that the solution must satisfy additional conditions, such as **initial conditions** or **boundary conditions.** For instance, to find the particular solution to

$$f'(x) = \frac{1}{x+1} \qquad \text{with the condition that} \qquad f(-2) = 2$$

we use $f(-2) = 2$ in the general solution, $f(x) = \ln|x + 1| + C$.

$$2 = f(-2) = \ln|-2 + 1| + C$$
$$2 = \ln|-1| + C \quad \text{so} \quad C = 2$$

Thus the particular solution is

$$f(x) = \ln|x + 1| + 2$$

and is shown in Figure 12.9 with $C = 2$.

We frequently denote the value of the solution function $y = f(t)$ at the initial time $t = 0$ as $y(0)$ instead of $f(0)$.

CHECKPOINT

1. Given $f'(x) = 2x - [1/(x + 1)], f(0) = 4$,
   (a) find the general solution to the differential equation.
   (b) find the particular solution that satisfies $f(0) = 4$.

Just as we can find the differential of both sides of an equation, we can find the solution to a differential equation of the form

$$G(y)\,dy = f(x)\,dx$$

by integrating both sides.

EXAMPLE 3

Solve $3y^2\,dy = 2x\,dx$, if $y(1) = 2$.

Solution

We find the general solution by integrating both sides.

$$\begin{aligned}\int 3y^2\,dy &= \int 2x\,dx\\ y^3 + C_1 &= x^2 + C_2\\ y^3 &= x^2 + C, \qquad \text{where } C = C_2 - C_1\end{aligned}$$

By using $y(1) = 2$, we can find C.

$$\begin{aligned}2^3 &= 1^2 + C\\ 7 &= C\end{aligned}$$

Thus the particular solution is given implicitly by

$$y^3 = x^2 + 7$$

Separable Differential Equations

It is frequently necessary to change the form of a differential equation before it can be solved by integrating both sides.

For example, the equation

$$\frac{dy}{dx} = y^2$$

cannot be solved by simply integrating both sides of the equation with respect to x because we cannot evaluate $\int y^2\,dx$.

However, we can multiply both sides of $dy/dx = y^2$ by dx/y^2 to obtain an equation that has all terms containing y on one side of the equation and all terms containing x on the other side. That is, we obtain

$$\frac{dy}{y^2} = dx$$

Separable Differential Equations

When a differential equation can be equivalently expressed in the form

$$g(y)\,dy = f(x)\,dx$$

we say that the equation is **separable.**

The solution of a separable differential equation is obtained by integrating both sides of the equation after the variables have been separated.

EXAMPLE 4

Solve the differential equation

$$(x^2y + x^2)\,dy = x^3\,dx$$

Solution

To write the equation in separable form, we first factor x^2 from the left side and divide both sides by it.

$$\begin{aligned} x^2(y+1)\,dy &= x^3\,dx \\ (y+1)\,dy &= \frac{x^3}{x^2}\,dx \end{aligned}$$

The equation is now separated, so we integrate both sides.

$$\begin{aligned} \int (y+1)\,dy &= \int x\,dx \\ \frac{y^2}{2} + y + C_1 &= \frac{x^2}{2} + C_2 \end{aligned}$$

This equation, as well as the equation

$$y^2 + 2y - x^2 = C, \quad \text{where } C = 2(C_2 - C_1)$$

gives the solution implicitly.

Note that we need not write both C_1 and C_2 when we integrate, because it is always possible to combine the two constants into one.

EXAMPLE 5

Solve the differential equation

$$\frac{dy}{dt} = ky \qquad (k = \text{constant})$$

Solution

To solve the equation, we write it in separated form as

$$\frac{dy}{y} = k\,dt$$

and integrate both sides as follows:

$$\begin{aligned} \int \frac{dy}{y} &= \int k\,dt \\ \ln|y| &= kt + C_1 \end{aligned}$$

Assuming that $y > 0$ and writing this equation in exponential form gives

$$\begin{aligned} y &= e^{kt + C_1} \\ y &= e^{kt} \cdot e^{C_1} = Ce^{kt}, \quad \text{where } C = e^{C_1} \end{aligned}$$

This solution,

$$y = Ce^{kt}$$

is the general solution of the differential equation $dy/dt = ky$ because all solutions have this form, with different values of C giving different particular solutions. The case of $y < 0$ is covered by values of $C < 0$.

CHECKPOINT

2. True or false:
   (a) The general solution to $dy = (x/y)\,dx$ can be found from
   $$\int y\,dy = \int x\,dx$$
   (b) The first step in solving $dy/dx = -2xy^2$ is to separate it.
   (c) The equation $dy/dx = -2xy^2$ separates as $y^2\,dy = -2x\,dx$.
3. Suppose that $(xy + x)(dy/dx) = x^2y + y$.
   (a) Separate this equation. (b) Find the general solution.

In many applied problems that can be modeled with differential equations, we know of conditions that allow us to obtain a particular solution.

Applications of Differential Equations

We now consider two applications that can be modeled by differential equations. These are radioactive decay (as introduced in the Application Preview) and one-container mixture problems (as a model for drugs in an organ).

EXAMPLE 6

In the Application Preview, we introduced carbon-14 dating and the facts that the process is based on. We said that when an organism dies, the rate of change of the amount of carbon-14 present is proportional to the amount present and is represented by the differential equation

$$\frac{dy}{dt} = ky$$

where y is the amount present, k is a constant, and t is time in years. If we denote the initial amount of carbon-14 in an organism as y_0, then $y = y_0$ represents the amount present at time $t = 0$ (when the organism died). Suppose that anthropologists discover a fossil that contains 1% of the initial amount of carbon-14. Find the age of the fossil. (Recall that the half-life of carbon-14 is 5600 years.)

Solution

We must find a particular solution to

$$\frac{dy}{dt} = ky$$

subject to the fact that when $t = 0$, $y = y_0$, and we must determine the value of k on the basis of the half-life of carbon-14 ($t = 5600$ years, $y = \frac{1}{2}y_0$ units.) From Example 5, we know that the general solution to the differential equation $dy/dt = ky$ is $y = Ce^{kt}$. Using $y = y_0$ when $t = 0$, we obtain $y_0 = C$, so the equation becomes $y = y_0e^{kt}$. Using $t = 5600$ and $y = \frac{1}{2}y_0$ in this equation gives

$$\frac{1}{2}y_0 = y_0e^{5600k} \qquad \text{or} \qquad 0.5 = e^{5600k}$$

Rewriting this equation in logarithmic form and then solving for k, we get

$$\begin{aligned} \ln(0.5) &= 5600k \\ -0.69315 &= 5600k \\ -0.00012378 &= k \end{aligned}$$

Thus the equation we seek is

$$y = y_0 e^{-0.00012378t}$$

Using the fact that $y = 0.01y_0$ when the fossil was discovered, we can find its age t by solving

$$0.01y_0 = y_0 e^{-0.00012378t} \quad \text{or} \quad 0.01 = e^{-0.00012378t}$$

Rewriting this in logarithmic form and then solving gives

$$\begin{aligned} \ln(0.01) &= -0.00012378t \\ -4.6051702 &= -0.00012378t \\ 37{,}204 &= t \end{aligned}$$

Thus the fossil is approximately 37,200 years old.

Another application of differential equations comes from a group of applications called *one-container mixture problems.* In problems of this type, there is a substance whose amount in a container is changing with time, and the goal is to determine the amount of the substance at any time t. The differential equations that model these problems are of the following form:

$$\begin{bmatrix} \text{Rate of change} \\ \text{of the amount} \\ \text{of the substance} \end{bmatrix} = \begin{bmatrix} \text{Rate at which} \\ \text{the substance} \\ \text{enters the container} \end{bmatrix} - \begin{bmatrix} \text{Rate at which} \\ \text{the substance} \\ \text{leaves the container} \end{bmatrix}$$

We consider this application as it applies to the amount of a drug in an organ.

EXAMPLE 7

A liquid carries a drug into an organ of volume 300 cc at a rate of 5 cc/s, and the liquid leaves the organ at the same rate. If the concentration of the drug in the entering liquid is 0.1 g/cc, and if x represents the amount of drug in the organ at any time t, then using the fact that the rate of change of the amount of the drug in the organ, dx/dt, equals the rate at which the drug enters minus the rate at which it leaves, we have

$$\frac{dx}{dt} = \left(\frac{5 \text{ cc}}{\text{s}}\right)\left(\frac{0.1 \text{ g}}{\text{cc}}\right) - \left(\frac{5 \text{ cc}}{\text{s}}\right)\left(\frac{x \text{ g}}{300 \text{ cc}}\right)$$

or

$$\frac{dx}{dt} = 0.5 - \frac{x}{60} = \frac{30}{60} - \frac{x}{60} = \frac{30 - x}{60}, \quad \text{in g/s}$$

Find the amount of the drug in the organ as a function of time t.

Solution

Multiplying both sides of the equation $\frac{dx}{dt} = \frac{30 - x}{60}$ by $\frac{dt}{(30 - x)}$ gives

$$\frac{dx}{30 - x} = \frac{1}{60}\,dt$$

The equation is now separated, so we can integrate both sides.

$$\int \frac{dx}{30 - x} = \int \frac{1}{60}\,dt$$

$$-\ln(30 - x) = \frac{1}{60}t + C_1 \qquad (30 - x > 0)$$

$$\ln(30 - x) = -\frac{1}{60}t - C_1$$

Rewriting this in exponential form gives

$$30 - x = e^{-t/60 - C_1} = e^{-t/60} \cdot e^{-C_1}$$

Letting $C = e^{-C_1}$ yields

$$30 - x = Ce^{-t/60}$$

so

$$x = 30 - Ce^{-t/60}$$

and we have the desired function.

CHECKPOINT SOLUTIONS

1. (a) $f(x) = \int \left[2x - \frac{1}{x + 1}\right] dx = x^2 - \ln|x + 1| + C$

   (b) If $f(0) = 4$, then $4 = 0^2 - \ln|1| + C$, so $C = 4$.
   Thus $f(x) = x^2 - \ln|x + 1| + 4$ is the particular solution.

2. (a) True, and the solution is

$$\int y\,dy = \int x\,dx \qquad \frac{y^2}{2} = \frac{x^2}{2} + C$$

   (b) True

   (c) False. It separates as $dy/y^2 = -2x\,dx$, and the solution is

$$\int \frac{-dy}{y^2} = \int 2x\,dx \qquad y^{-1} = x^2 + C$$

$$\frac{1}{y} = x^2 + C$$

$$y = \frac{1}{x^2 + C}$$

3. (a) $x(y + 1)\frac{dy}{dx} = (x^2 + 1)y$

$$\frac{y + 1}{y}\,dy = \frac{x^2 + 1}{x}\,dx$$

$$\left(1 + \frac{1}{y}\right) dy = \left(x + \frac{1}{x}\right) dx$$

   (b) $\int \left(1 + \frac{1}{y}\right) dy = \int \left(x + \frac{1}{x}\right) dx$

$$y + \ln|y| = \frac{x^2}{2} + \ln|x| + C$$

   This is the general solution.

EXERCISE 12.5

In Problems 1–4, show that the given function is a solution of the differential equation.

1. $y = x^2$; $4y - 2xy' = 0$
2. $y = x^3$; $3y - xy' = 0$
3. $y = 3x^2 + 1$; $2y\,dx - x\,dy = 2\,dx$
4. $y = 4x^3 + 2$; $3y\,dx - x\,dy = 6\,dx$

In Problems 5–10, use integration to find the general solution to each differential equation.

5. $dy = xe^{x^2} + 1\,dx$
6. $dy = x^2e^{x^3 - 1}\,dx$
7. $2y\,dy = 4x\,dx$
8. $4y\,dy = 4x^3\,dx$
9. $3y^2\,dy = (2x - 1)\,dx$
10. $4y^3\,dy = (3x^2 + 2x)\,dx$

In Problems 11–14, find the particular solutions.

11. $y' = e^{x-3}$, $y(0) = 2$
12. $y' = e^{2x+1}$, $y(0) = e$
13. $dy = \left(\dfrac{1}{x} - x\right) dx$, $y(1) = 0$
14. $dy = \left(x^2 - \dfrac{1}{x+1}\right) dx$, $y(0) = \dfrac{1}{3}$

In Problems 15–28, find the general solution to the given differential equation.

15. $\dfrac{dy}{dx} = \dfrac{x^2}{y}$
16. $y^3\,dx = \dfrac{dy}{x^3}$
17. $dx = x^3y\,dy$
18. $dy = x^2y^3\,dx$
19. $dx = (x^2y^2 + x^2)\,dy$
20. $dy = (x^2y^3 + xy^3)\,dx$
21. $y^2\,dx = x\,dy$
22. $y\,dx = x\,dy$
23. $\dfrac{dy}{dx} = \dfrac{x}{y}$
24. $\dfrac{dy}{dx} = \dfrac{x^2 + x}{y + 1}$
25. $(x + 1)\dfrac{dy}{dx} = y$
26. $x^2y\dfrac{dy}{dx} = y^2 + 1$
27. $e^{2x}y\,dy = (y + 1)\,dx$
28. $e^{4x}(y + 1)\,dx + e^{2x}y\,dy = 0$

In Problems 29–36, find the particular solution to each differential equation.

29. $\dfrac{dy}{dx} = \dfrac{x^2}{y^3}$, when $x = 1, y = 1$
30. $\dfrac{dy}{dx} = \dfrac{x + 1}{xy}$, when $x = 1, y = 3$
31. $2y^2\,dx = 3x^2\,dy$, when $x = 2, y = -1$
32. $(x + 1)\,dy = y^2\,dx$, when $x = 0, y = 2$
33. $x^2e^{2y}\,dy = (x^3 + 1)\,dx$, when $x = 1, y = 0$
34. $y' = \dfrac{1}{xy}$, when $x = 1, y = 3$
35. $2xy\dfrac{dy}{dx} = y^2 + 1$, when $x = 1, y = 2$
36. $xe^y\,dx = (x + 1)dy$, when $x = 0, y = 0$

Applications

37. ***Allometric growth*** If x and y are measurements of certain parts of an organism, then the rate of change of y with respect to x is proportional to the ratio of y to x. That is, these measurements satisfy

$$\frac{dy}{dx} = k\frac{y}{x}$$

which is referred to as an allometric law of growth. Solve this differential equation.

38. ***Bimolecular chemical reactions*** A bimolecular chemical reaction is one in which two chemicals react to give another substance. Suppose that one molecule of each of the two chemicals react to form two molecules of a new substance. If x represents the number of molecules of the new substance at time t, then the rate of change of x is proportional to the product of the numbers of molecules of the original chemicals available to be converted. That is, if each of the chemicals initially contained A molecules, then

$$\frac{dx}{dt} = k(A - x)^2$$

If 40% of the initial amount A is converted after 1 hour, how long will it be before 90% is converted?

Compound interest In Problems 39 and 40, use the following information.

When interest is compounded continuously, the rate of change of the amount x of the investment is proportional to the amount present. In this case, the proportionality constant is the annual interest rate r (as a decimal); that is,

$$\frac{dx}{dt} = rx$$

39. (a) If \$10,000 is invested at 6%, compounded continuously, find an equation for the future value of the investment as a function of time t in years.
    (b) What is the future value of the investment after 1 year? After 5 years?
    (c) How long will it take for investment to double?
40. (a) If \$2000 is invested at 8%, compounded continuously, find an equation for the future value of the investment as a function of time t, in years.
    (b) How long will it take for investment to double?
    (c) What will be the future value of this investment after 35 years?

41. ***Bacterial growth*** Suppose that the growth of a certain population of bacteria satisfies

$$\frac{dy}{dt} = ky$$

where y is the number of organisms and t is the number of hours. If initially there are 10,000 organisms and the number triples after 2 hours, how long will it be before there is 100 times the original population?

42. ***Bacterial growth*** Suppose that, for a certain population of bacteria, growth occurs according to

$$\frac{dy}{dt} = ky \quad (t \text{ in hours})$$

If the doubling rate depends on temperature, find how long it takes for the number of bacteria to reach 50 times the original number at each given temperature in (a) and (b).
(a) At 90°F, the number doubles after 30 minutes $\left(\frac{1}{2}\text{ hour}\right)$.
(b) At 40°F, the number doubles after 3 hours.

43. ***Half-life*** A breeder reactor converts uranium-238 into an isotope of plutonium-239 at a rate proportional to the amount present at any time. After 10 years, 0.03% of the radioactivity has dissipated (that is, 0.9997 of the initial amount remains). Suppose that initially there is 100 pounds of this substance. Find the half-life.

44. ***Radioactive decay*** A certain radioactive substance has a half-life of 50 hours. Find how long it will take for 90% of the radioactivity to be dissipated if the amount of material x satisfies

$$\frac{dx}{dt} = kx \quad (t \text{ in hours})$$

45. ***Drug in an organ*** Suppose that a liquid carries a drug into a 100-cc organ at a rate of 5 cc/s and leaves the organ at the same rate. Suppose that the concentration of the drug entering is 0.06 g/cc. If initially there is no drug in the organ, find the amount of drug in the organ as a function of time t.

46. ***Drug in an organ*** Suppose that a liquid carries a drug into a 250-cc organ at a rate of 10 cc/s and leaves the organ at the same rate. Suppose that the concentration of the drug entering is 0.15 g/cc. Find the amount of drug in the organ as a function of time t if initially there is none in the organ.

47. ***Drug in an organ*** Suppose that a liquid carries a drug with concentration 0.1 g/cc into a 200-cc organ at a rate of 5 cc/s and leaves the organ at the same rate. If initially there is 10 g of the drug in the organ, find the amount of drug in the organ as a function of time t.

48. ***Drug in an organ*** Suppose that a liquid carries a drug with concentration 0.05 g/cc into a 150-cc organ at a rate of 6 cc/s and leaves at the same rate. If initially there is 1.5 g of drug in the organ, find the amount of drug in the organ as a function of time t.

49. ***Sales and pricing*** Suppose that in a certain company, the relationship between the price per unit p of its product and the weekly sales volume y, in thousands of dollars, is given by

$$\frac{dy}{dp} = -\frac{2}{5}\left(\frac{y}{p+8}\right)$$

Solve this differential equation if $y = 8$ when $p =$ \$24.

50. ***Sales and pricing*** Suppose that a chain of auto service stations, Quick-Oil, Inc., has found that the relationship between its price p for a oil change and its monthly sales volume y, in thousands of dollars, is

$$\frac{dy}{dp} = -\frac{1}{2}\left(\frac{y}{p+5}\right)$$

Solve this differential equation if $y = 18$ when $p =$ \$20.

51. ***Tumor volume*** Let V denote the volume of a tumor, and suppose that the growth rate of the tumor satisfies

$$\frac{dV}{dt} = 0.2Ve^{-0.1t}$$

If the initial volume of the tumor is 1.86 units, find an equation for V as a function of t.

52. ***Gompertz curves*** The differential equation

$$\frac{dx}{dt} = x(a - b\ln x)$$

where x represents the number of objects at time t, and a and b are constants, is the model for Gompertz curves. Recall from Section 5.3, "Solution of Exponential Equations," that Gompertz curves can be used to study growth or decline of populations, organizations, and revenue from sales of a product, as well as forecast equipment maintenance costs. Solve the differential equation to obtain the Gompertz curve formula

$$x = e^{a/b}e^{-ce^{-bt}}$$

53. ***Cell growth*** If V is the volume of a spherical cell, then in certain cell growth and for some fetal growth models, the rate of change of V is given by

$$\frac{dV}{dt} = kV^{2/3}$$

where k is a constant depending on the organism. If $V = 0$ when $t = 0$, find V as a function of t.

54. ***Atmospheric pressure*** The rate of change of atmospheric pressure P with respect to the altitude above sea level h is proportional to the pressure. That is,

$$\frac{dP}{dh} = kP$$

Suppose that the pressure at sea level is denoted by P_0, and at 18,000 ft the pressure is half what it is at sea level. Find the pressure, as a percentage of P_0, at 25,000 ft.

55. ***Newton's law of cooling*** Newton's law of cooling states that the rate of change of temperature $u = u(t)$ of an object is proportional to the temperature difference between the object and its surroundings, where T is the constant temperature of the surroundings. That is,

$$\frac{du}{dt} = k(u - T)$$

Suppose an object at 0°C is placed in a room where the temperature is 20°C. If the temperature of the object is 8°C after 1 hour, how long will it take for the object to reach 18°C?

56. ***Newton's law of cooling*** Newton's law of cooling can be used to estimate time of death. (Actually the estimate may be quite rough, because cooling does not begin until metabolic processes have ceased.) Suppose a corpse is discovered at noon in a 70°F room, and at that time the body temperature is 96.1°F. If at 1:00 P.M. the body temperature is 94.6°F, use Newton's law of cooling to estimate the time of death.

57. ***Fossil-fuel emissions*** The amount of carbon in the atmosphere has been estimated to have increased at a rate of about 4.3% per year from 1860 until 1973, except during the Great Depression and the world wars.* This increase is due to carbon emissions from fossil-fuel burning and can be modeled by the differential equation

$$\frac{dE}{dt} = 0.043E$$

where t is the number of years since 1860 and E is fossil-fuel emissions in gigatons per year.

(a) Solve this differential equation, and find a particular solution that satisfies $E(0) = 0.1$.

(b) Graph your solution and compare it with the graph below, which shows actual emissions since 1860.

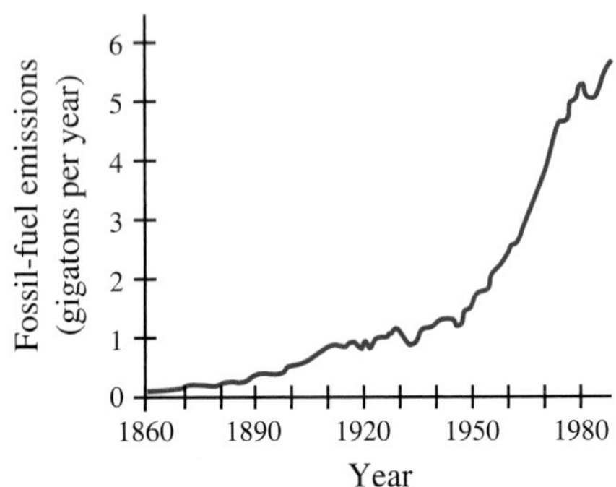

Source: *American Scientist*, Vol. 79, (July–August 1990), p. 313

58. ***Mutual fund growth*** The number of mutual funds established since 1940 can be modeled by

$$\frac{dy}{dt} = 0.662y, \quad y(4) = 68$$

where t is the number of decades since 1900 and y is the number of mutual funds.

(a) Find the particular solution to this differential equation.

(b) Check your model against the data in the figure for 1960 and 1990.

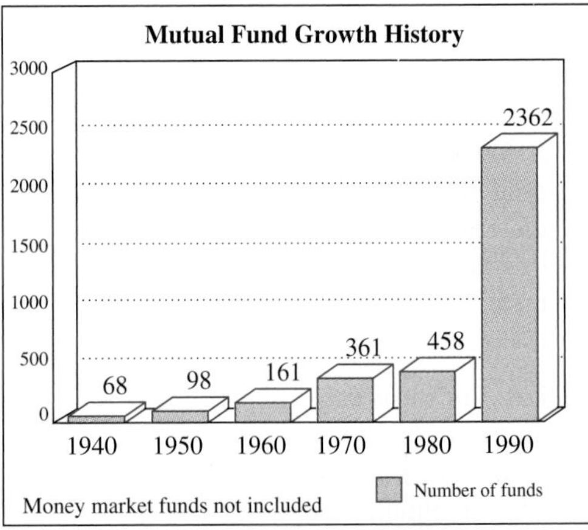

Source: Investment Company Institute, published in *Investment Digest of Valic Co.*, Vol. 5, No. 2 (Summer), 1992.

*American Scientist, Vol. 79 (July–August 1990), p. 313.

59. ***Impact of inflation*** The impact of a 5% inflation rate on a \$20,000-per-year pension is shown in the accompanying figure. If P represents the dollars of purchasing power of a \$20,000 pension, then the effect of a 5% inflation rate can be modeled by the differential equation

$$\frac{dP}{dt} = -0.05P, \quad P(0) = 20{,}000$$

where t is in years.

(a) Find the particular solution to this differential equation.

(b) Check your model against the graph in the figure. In particular, find the purchasing power after 30 years. Does your calculation match the graph?

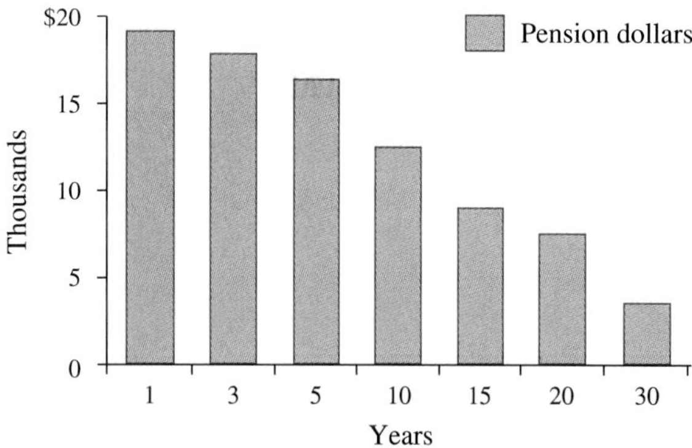

Source: *Viewpoints*, Financial news from Valic Co., Summer 1993

KEY TERMS AND FORMULAS

Section	Key Terms	Formula
12.1	General antiderivative of $f'(x)$	$f(x) + C$
	Integral	$\int f(x)\,dx$
	Powers of x Formula	$\int x^n\,dx = \frac{x^{n+1}}{n+1} + C \quad (n \neq -1)$
	Integration Formulas	$\int dx = x + C$
		$\int cu(x)\,dx = c\int u(x)\,dx$; c = a constant
		$\int [u(x) \pm v(x)]\,dx = \int u(x)\,dx \pm \int v(x)\,dx$
12.2	Power Rule	$\int [u(x)]^n u'(x)\,dx = \frac{[u(x)]^{n+1}}{n+1} + C \quad (n \neq -1)$
12.3	Logarithmic Formula	$\int \frac{u'}{u}\,dx = \int \frac{1}{u}\,du = \ln\lvert u\rvert + C$
	Exponential Formula	$\int e^u u'\,dx = \int e^u\,du = e^u + C$
12.4	Total cost	$C(x) = \int \overline{MC}\,dx$
	Total revenue	$R(x) = \int \overline{MR}\,dx$
	Profit	$P(x) = R(x) - C(x)$
	Marginal propensity to consume	$\frac{dC}{dy}$
	Marginal propensity to save	$\frac{dS}{dy} = 1 - \frac{dC}{dy}$

Section	Key Terms	Formula
	National consumption	$C = \int f'(y)\,dy = \int \frac{dC}{dy}\,dy$
12.5	Differential equations Solutions General Particular	
	First order	$\frac{dy}{dx} = f(x) \Rightarrow y = \int f(x)\,dx$
	Separable	$g(y)\,dy = f(x)\,dx$ $\Rightarrow \int g(y)\,dy = \int f(x)\,dx$
	Radioactive decay	$\frac{dy}{dt} = ky$
	Drugs in an organ	Rate = (rate in) − (rate out)

REVIEW EXERCISES

Sections 12.1–12.3

Evaluate the integrals in Problems 1–26.

1. $\int x^6 dx$
2. $\int x^{1/2}\,dx$
3. $\int (x^3 - 3x^2 + 4x + 5)\,dx$
4. $\int (x^2 - 1)^2\,dx$
5. $\int (x^2 - 1)^2 x\,dx$
6. $\int (x^3 - 3x^2)(x^2 - 2x)\,dx$
7. $\int (x^3 + 4)^2 x\,dx$
8. $\int (x^3 + 4)^6 x^2\,dx$
9. $\int \frac{x^2}{x^3 + 1}\,dx$
10. $\int \frac{x^2}{(x^3 + 1)^2}\,dx$
11. $\int \frac{x^2\,dx}{\sqrt[3]{x^3 - 4}}$
12. $\int \frac{x^2\,dx}{x^3 - 4}$
13. $\int \frac{x^3 + 1}{x^2}\,dx$
14. $\int \frac{x^3 - 3x + 1}{x - 1}\,dx$
15. $\int y^2 e^{y^3}\,dy$
16. $\int (x - 1)^2\,dx$
17. $\int \frac{3x^2}{2x^3 - 7}\,dx$
18. $\int \frac{5\,dx}{e^{4x}}$
19. $\int (x^3 - e^{3x})\,dx$
20. $\int xe^{1 + x^2}\,dx$
21. $\int \frac{6x^7}{(5x^8 + 7)^3}\,dx$
22. $\int \frac{7x^3}{\sqrt{1 - x^4}}\,dx$
23. $\int \left(\frac{e^{2x}}{2} + \frac{2}{e^{2x}}\right)dx$
24. $\int \left[x - \frac{1}{(x + 1)^2}\right]dx$
25. (a) $\int (x^2 - 1)^4 x\,dx$ (b) $\int (x^2 - 1)^{10} x\,dx$
    (c) $\int (x^2 - 1)^7 3x\,dx$ (d) $\int (x^2 - 1)^{-2/3} x\,dx$
26. (a) $\int \frac{2x\,dx}{x^2 - 1}$ (b) $\int \frac{2x\,dx}{(x^2 - 1)^2}$
    (c) $\int \frac{3x\,dx}{\sqrt{x^2 - 1}}$ (d) $\int \frac{3x\,dx}{x^2 - 1}$

Section 12.5

In Problems 27–32, find the general solution to each differential equation.

27. $\frac{dy}{dt} = 4.6e^{-0.05t}$
28. $dy = (64 + 76x - 36x^2)\,dx$
29. $\frac{dy}{dx} = \frac{4x}{y - 3}$
30. $t\,dy = \frac{dt}{y + 1}$
31. $\frac{dy}{dx} = \frac{x}{e^y}$
32. $\frac{dy}{dt} = \frac{4y}{t}$

In Problems 33 and 34, find the particular solution to each differential equation.

33. $y' = \frac{x^2}{y + 1}$ $y(0) = 4$
34. $y' = \frac{2x}{1 + 2y}$ $y(2) = 0$

Applications

Section 12.1

35. ***Revenue*** If the marginal revenue for a month for a product is $\overline{MR} = 6x + 12$, find the total revenue from the sale of 4 units of the product.

36. ***Productivity*** Suppose that the rate of change of production of the average worker at a factory is given by

$$\frac{dp}{dt} = 27 + 24t - 3t^2, \qquad 0 \le t \le 8$$

where p is the number of units the worker produces in t hours. How many units will the average worker produce in an 8-hour shift? (Assume that $p = 0$ when $t = 0$.)

Section 12.2

37. ***Oxygen levels in water*** The rate of change of the oxygen level in a body of water after an oil spill is given by

$$P'(t) = 400\left[\frac{5}{(t+5)^2} - \frac{50}{(t+5)^3}\right]$$

where t is the number of months after the spill. What function gives the oxygen level P at any time t if $P = 400$ when $t = 0$?

38. ***Bacterial growth*** A population of bacteria grows at the rate

$$r = \frac{100{,}000}{(t+100)^2}$$

where t is time. If the population is 1000 when $t = 1$, write the equation that gives the size of the population at any time t.

Section 12.3

39. ***Market share*** The rate of change of the market share (as a percentage) a firm expects for a new product is

$$\frac{dy}{dt} = 2.4e^{-0.04t}$$

where t is the number of months after the product is introduced.
   (a) Write the equation that gives the expected market share y at any time t. (Note that $y = 0$ when $t = 0$.)
   (b) What market share does the firm expect after one year?

40. ***Revenue*** If the marginal revenue for a product is $\overline{MR} = \frac{800}{x+1}$, find the total revenue function.

Section 12.4

41. ***Cost*** The marginal cost for a product is $\overline{MC} = 6x + 4$ and the cost of producing 100 items is \$31,400.
   (a) Find the fixed costs.
   (b) Find the total cost function.

42. ***Profit*** Suppose a product has a daily marginal revenue $\overline{MR} = 46$ and a daily marginal cost $\overline{MC} = 30 + \frac{1}{5}x$. If the daily fixed cost is \$200.00, how many units will give maximum profit and what is the maximum profit?

43. ***National consumption*** If consumption is \$8.5 billion when disposable income is 0, and if the marginal propensity to consume is

$$\frac{dC}{dy} = \frac{1}{\sqrt{2y+16}} + 0.6 \quad \text{(in billions of dollars)}$$

find the national consumption function.

44. ***National consumption*** Suppose that the marginal propensity to save is

$$\frac{dS}{dy} = 0.2 - 0.1e^{-2y} \quad \text{(in billions of dollars)}$$

and consumption is \$7.8 billion when disposable income is 0. Find the national consumption function.

Section 12.5

45. ***Allometric growth*** For many species of fish, the length L and weight W of a fish are related by

$$\frac{dW}{dL} = \frac{3W}{L}$$

The general solution to this differential equation expresses the allometric relationship between the length and weight of a fish. Find the general solution.

46. ***Fossil dating*** Radioactive beryllium is sometimes used to date fossils found in deep-sea sediment. The amount of radioactive material x satisfies

$$\frac{dx}{dt} = kx$$

Suppose that 10 units of beryllium are present in a living organism and that the half-life of beryllium is 4.6 million years. Find the age of a fossil if 20% of the original radioactivity is present when the fossil is discovered.

47. ***Drug in an organ*** Suppose that a liquid carries a drug into a 120-cc organ at a rate of 4 cc/s and leaves the organ at the same rate. If initially there is no drug in the organ and if the concentration of drug in the liquid is 3 g/cc, find the amount of drug in the organ as a function of time.

48. ***Chemical mixture*** A 300-gal tank initially contains a solution with 100 lb of a chemical. A mixture containing 2 lb/gal of the chemical enters the tank at 3 gal/min, and the well-stirred mixture leaves at the same rate. Find an equation that gives the amount of the chemical in the tank as a function of time. How long will it be before there are 500 lb of chemical in the tank?

CHAPTER TEST

Evaluate the integrals in Problems 1–8.

1. $\int (6x^2 + 8x - 7)\, dx$
2. $\int \left(4 + \sqrt{x} - \frac{1}{x^2}\right) dx$
3. $\int 5x^2 (4x^3 - 7)^9\, dx$
4. $\int (3x^2 - 6x + 1)^9 (2x - 2)\, dx$
5. $\int \frac{s^3}{2s^4 - 5}\, ds$
6. $\int 100e^{-0.01x}\, dx$
7. $\int 5y^3 e^{2y^4 - 1}\, dy$
8. $\int \left(e^x + \frac{5}{x} - 1\right) dx$
9. Evaluate $\int \frac{x^2}{x+1}\, dx$. Use long division.
10. If $\int f(x)dx = 2x^3 - x + 5e^x + C$, find $f(x)$.

In Problems 11 and 12, find the particular solution to each differential equation.

11. $y' = 4x^3 + 3x^2$, if $y(0) = 4$
12. $\frac{dy}{dx} = e^{4x}$, if $y(0) = 2$
13. Find the general solution of the separable differential equation $\frac{dy}{dx} = x^3y^2$.
14. Suppose the rate of growth of the population of a city is predicted to be
$$\frac{dp}{dt} = 2000t^{1.04}$$
where p is the population and t is the number of years past 2000. If the population in the year 2000 is 50,000, what is the predicted population in the year 2010?
15. Suppose that the marginal cost for a product is $\overline{MC} = 4x + 50$, the marginal revenue is $\overline{MR} = 500$, and the cost of the production and sale of 10 units is \$1000. What is the profit function for this product?
16. Suppose the marginal propensity to save is given by
$$\frac{dS}{dy} = 0.22 - \frac{0.25}{\sqrt{0.5y + 1}} \quad \text{(in billions of dollars)}$$
and national consumption is \$6.6 billion when disposable income is \$0. Find the national consumption function.
17. A certain radioactive material has a half-life of 100 days. If the amount of material present, x, satisfies $\frac{dx}{dt} = kx$, where t is in days, how long will it take for 90% of the radioactivity to dissipate?

Extended Applications Group Projects

I. Employee Production Rate

The manager of a plant has been instructed to hire and train additional employees to manufacture a new product. She must hire a sufficient number of new employees so that within 30 days they will be producing 2500 units of the product each day.

Because a new employee must learn an assigned task, production will increase with training. Suppose that research on similar projects indicates that production increases with training according to the learning curve, so that for the average employee, the rate of production per day is given by

$$\frac{dN}{dt} = be^{-at}$$

where N is the number of units produced per day after t days of training. Because of experience with a very similar project, the manager expects the rate for this project to be

$$\frac{dN}{dt} = 2.5e^{-0.05t}$$

The manager tested her training program with 5 employees and learned that the average employee could produce 11 units per day after 5 days of training. On the basis of this information, she must decide how many employees to hire and begin to train so that a month from now they will be producing 2500 units of the product per day. She estimates that it will take her 10 days to hire the employees, and thus she will have 15 days remaining to train them. She also expects a 10% attrition rate during this period.

How many employees would you advise the plant manager to hire? Check your advice by answering the following questions.

1. Use the expected rate of production and the results of the manager's test to find the function relating N and t—that is, $N = N(t)$.
2. Find the number of units the average employee can produce after 15 days of training. How many such employees would be needed to maintain a production rate of 2500 units per day?
3. Explain how you would revise this last result to account for the expected 10% attrition rate. How many new employees should the manager hire?

II. Supply and Demand

If p is the price of a given commodity at time t, then we can think of price as a function of time. Similarly, the number of units demanded by consumers q_d at any time, and the number of units supplied by producers q_s at any time, may also be considered as functions of time as well as functions of price.

Both the quantity demanded and the quantity supplied depend not only on the price at the time, but also on the direction and rate of change that consumers and producers ascribe to prices. For example, even when prices are high, if consumers feel that prices are rising, the demand may rise. Similarly, if prices are low but producers feel they may go lower, the supply may rise.

If we assume that prices are determined in the marketplace by supply and demand, then the equilibrium price is the one we seek.

Suppose the supply and demand functions for a certain commodity in a competitive market are given, in hundreds of units, by

$$q_s = 30 + p + 5\frac{dp}{dt}$$

$$q_d = 51 - 2p + 4\frac{dp}{dt}$$

where dp/dt denotes the rate of change of the price with respect to time. If, at $t = 0$, the market equilibrium price is 12, we can express the market equilibrium price as a function of time.

Our goals are

A. To express the market equilibrium price as a function of time.
B. To determine whether there is price stability in the marketplace for this item (that is, to determine whether the equilibrium price approaches a constant over time).

To achieve these goals, do the following.

1. Set the expressions for q_s and q_d equal to each other.
2. Solve this equation for $\frac{dp}{dt}$.
3. Write this equation in the form $f(p)\,dp = g(t)\,dt$.
4. Integrate both sides of this separated differential equation.
5. Solve the resulting equation for p in terms of t.
6. Use the fact that $p = 12$ when $t = 0$ to find C, the constant of integration, and write the market equilibrium price p as a function of time t.
7. Find the $\lim_{t\to\infty} p$, which gives the price we can expect this product to approach. If this limit is finite, then for this item there is price stability in the marketplace. If $\lim_{t\to\infty} p = \infty$, then price will continue to increase until economic conditions change.

Warm-up

Prerequisite Problem Type	For Section	Answer	Section for Review
Simplify: $\frac{1}{n^3}\left[\frac{n(n+1)(2n+1)}{6} - \frac{2n(n+1)}{2} + n\right]$	13.1	$\frac{2n^2 - 3n + 1}{6n^2}$	0.7 Fractions
(a) If $F(x) = \frac{x^4}{4} + 4x + C$, what is $F(4) - F(2)$? (b) If $F(x) = -\frac{1}{9}\ln\frac{9+\sqrt{81-9x^2}}{3x}$, what is $F(3) - F(2)$?	13.2 13.5	(a) 68 (b) $\frac{1}{9}\ln\left(\frac{3+\sqrt{5}}{2}\right)$	1.2 Functional notation
Find the limit: (a) $\lim_{n\to+\infty}\frac{n^2+n}{2n^2}$ (b) $\lim_{n\to+\infty}\frac{2n^2-3n+1}{6n^2}$ (c) $\lim_{b\to\infty}\left(1-\frac{1}{b}\right)$ (d) $\lim_{b\to\infty}\left(\frac{-100{,}000}{e^{0.10b}} + 100{,}000\right)$	13.1 13.7	(a) $\frac{1}{2}$ (b) $\frac{1}{3}$ (c) 1 (d) 100,000	9.2 Limits at infinity
Find the derivative of $y = \ln x$.	13.6	$\frac{1}{x}$	11.1 Derivatives of logarithmic functions
Integrate: (a) $\int (x^3 + 4)\,dx$ (b) $\int x\sqrt{x^2-9}\,dx$ (c) $\int e^{2x}\,dx$	13.2– 13.7	(a) $\frac{x^4}{4} + 4x + C$ (b) $\frac{1}{3}(x^2-9)^{3/2} + C$ (c) $\frac{1}{2}e^{2x} + C$	12.1, 12.2, 12.3 Integration

Chapter 13

Definite Integrals; Techniques of Integration

We saw some applications of the indefinite integral in Chapter 12. In this chapter we define the definite integral and discuss a theorem that is useful for evaluating it. We will also see how it can be used to find the areas under certain curves. The definite integral is used to solve many interesting types of problems from economics, finance, and probability.

Definite integrals can be used to approximate the ***total value,*** *the* ***present value,*** *and the* ***future value*** *of a* ***continuous income stream.*** *Improper integrals can be used to find the* ***capital values*** *of continuous income streams.*

Some consumers in a competitive market are willing to pay more than the market equilibrium price, and some producers are willing to sell for less than this price. The savings are called ***consumer's surplus*** *and* ***producer's surplus,*** *respectively, and areas under the demand and supply curves are used to calculate them.*

Evaluation of improper integrals is one of three new techniques of integration introduced in this chapter. The other two are use of integral tables and use of integration by parts.

13.1 Area Under a Curve

OBJECTIVES

- To use the sum of areas of rectangles to approximate the area under a curve
- To use Σ notation to denote sums
- To find the exact area under a curve

APPLICATION PREVIEW

One way to find the accumulated production (such as the production of ore from a mine) over a period of time is to graph the rate of production as a function of time and find the area under the resulting curve over a specified time interval. For example, if a coal mine produces at a rate of 30 tons per day, the production over 10 days ($30 \cdot 10 = 300$) could be represented by the area under the line $y = 30$ between $x = 0$ and $x = 10$ (see Figure 13.1).

Using area to determine the accumulated production is very useful when the rate-of-production function varies at different points in time. For example, if the rate of production is represented by

$$y = 100e^{-0.1x}$$

where x represents the number of days, then the area under the curve (and above the x-axis) from $x = 0$ to $x = 10$ represents the total production over the 10-day period (see Figure 13.2a). In order to determine the accumulated production and to solve other types of problems, we need a method for finding areas under curves. That is the goal of this section.

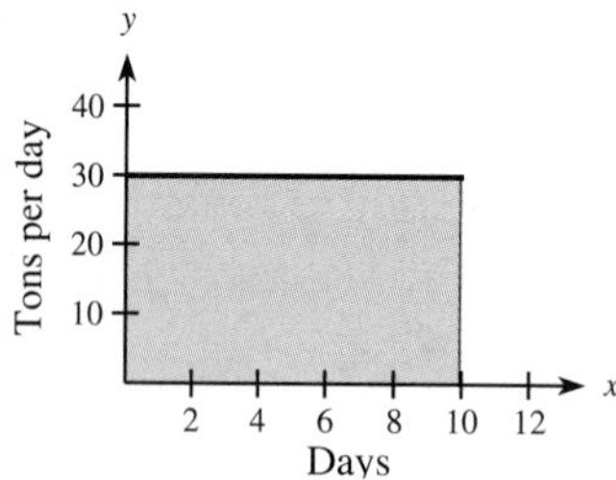

Figure 13.1

To estimate the accumulated production for the example in the Application Preview, we approximate the area under the graph of the production rate function. We can find a rough approximation of the area under this curve by fitting two rectangles to the curve as shown in Figure 13.2(b). The area of the first rectangle is $5 \cdot 100 = 500$ square units, and the area of the second rectangle is $(10 - 5)[100e^{-0.1(5)}] \approx 5(60.65) = 303.25$ square units, so this rough approximation is 803.25 square units. This approximation is clearly larger than the exact area under the curve. Why?

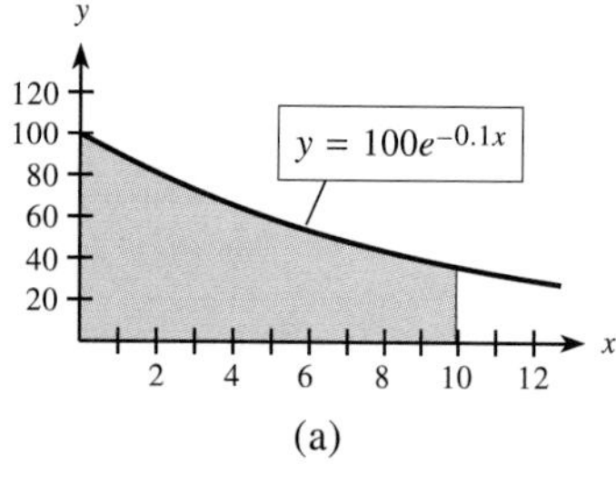

(a)

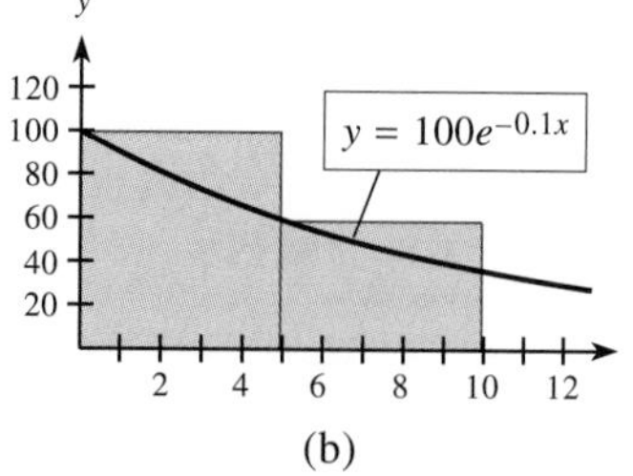

(b)

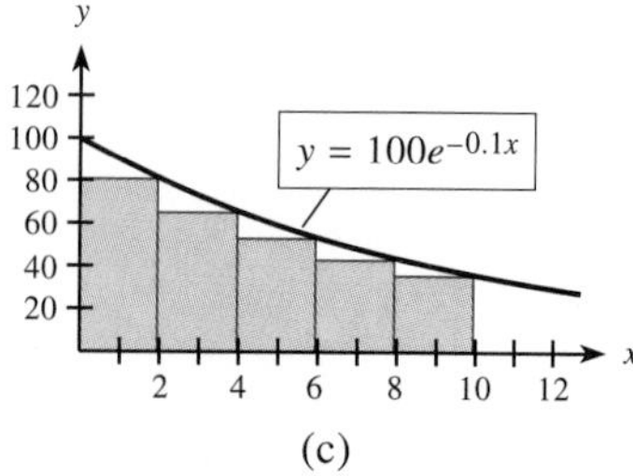

(c)

Figure 13.2

EXAMPLE 1

Fit five rectangles with equal bases inside the area under the curve $y = 100e^{-0.1x}$, and use them to approximate the area under the curve from $x = 0$ to $x = 10$ (see Figure 13.2c).

Solution

Each of the five rectangles has base 2, and the height of each rectangle is the value of the function at the right-hand endpoint of the interval forming its base. Thus the areas of the rectangles are as follows:

Rectangle	*Base*	*Height*	*Area = Base × Height*
1	2	$100e^{-0.1(2)} \approx 81.87$	$2(81.87) = 163.74$
2	2	$100e^{-0.1(4)} \approx 67.03$	$2(67.03) = 134.06$
3	2	$100e^{-0.1(6)} \approx 54.88$	$2(54.88) = 109.76$
4	2	$100e^{-0.1(8)} \approx 44.93$	$2(44.93) = 89.86$
5	2	$100e^{-0.1(10)} \approx 36.79$	$2(36.79) = 73.58$

The area under the curve is approximately equal to

$$163.74 + 134.06 + 109.76 + 89.86 + 73.58 = 571$$

The area is actually 632.12, to two decimal places, so this approximation is much better than the one we obtained with just two rectangles. In general, if we use bases of equal width, the approximation of the area under a curve improves when more rectangles are used.

Suppose that we wish to find the area between the curve $y = x$ and the x-axis from $x = 0$ to $x = 1$ (see Figure 13.3). One way to approximate this area is to use the areas of rectangles whose bases are on the x-axis and whose heights are the vertical distances from points on their bases to the curve. We can divide the interval [0, 1] into n equal subintervals and use them as the bases of n rectangles whose heights are determined by the curve (see Figure 13.4). The width of each of these rectangles is $1/n$. Using the functional value at the right-hand endpoint of each subinterval as the height of the rectangle, we get n rectangles as shown in Figure 13.4. Because part of each rectangle lies above the curve, the sum of the areas of the rectangles will overestimate the area.

Then, with $y = f(x) = x$ and subinterval width $1/n$, the areas of the rectangles are as follows:

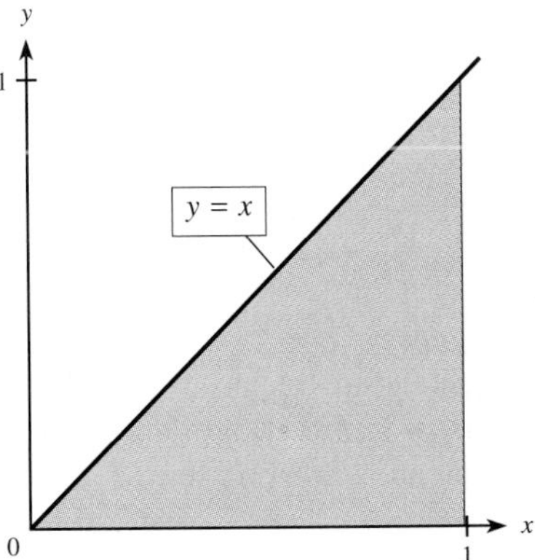

Figure 13.3

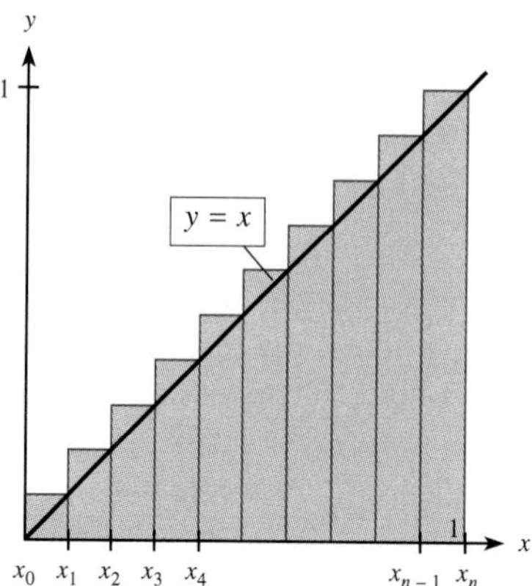

Figure 13.4

Rectangle	*Base*	*Endpoint*	*Height*	*Area* = *Base* × *Height*
1	$\frac{1}{n}$	$x_1 = \frac{1}{n}$	$f(x_1) = \frac{1}{n}$	$\frac{1}{n} \cdot \frac{1}{n} = \frac{1}{n^2}$
2	$\frac{1}{n}$	$x_2 = \frac{2}{n}$	$f(x_2) = \frac{2}{n}$	$\frac{1}{n} \cdot \frac{2}{n} = \frac{2}{n^2}$
3	$\frac{1}{n}$	$x_3 = \frac{3}{n}$	$f(x_3) = \frac{3}{n}$	$\frac{1}{n} \cdot \frac{3}{n} = \frac{3}{n^2}$
⋮				
i	$\frac{1}{n}$	$x_i = \frac{i}{n}$	$f(x_i) = \frac{i}{n}$	$\frac{1}{n} \cdot \frac{i}{n} = \frac{i}{n^2}$
⋮				
n	$\frac{1}{n}$	$x_n = \frac{n}{n}$	$f(x_n) = \frac{n}{n}$	$\frac{1}{n} \cdot \frac{n}{n} = \frac{n}{n^2}$

Note that i/n^2 gives the area of the ith rectangle for *any* value of i. Thus for any value of n, this area can be approximated by the sum

$$A \approx \frac{1}{n^2} + \frac{2}{n^2} + \frac{3}{n^2} + \cdots + \frac{i}{n^2} + \cdots + \frac{n}{n^2}$$

In particular, we have the following approximations of this area for specific values of n (the number of rectangles).

$$n = 5: \quad A \approx \frac{1}{25} + \frac{2}{25} + \frac{3}{25} + \frac{4}{25} + \frac{5}{25} = \frac{15}{25} = 0.60$$

$$n = 10: \quad A \approx \frac{1}{100} + \frac{2}{100} + \frac{3}{100} + \cdots + \frac{10}{100} = \frac{55}{100} = 0.55$$

$$n = 100: \quad A \approx \frac{1}{10{,}000} + \frac{2}{10{,}000} + \frac{3}{10{,}000} + \cdots + \frac{100}{10{,}000}$$

$$= \frac{5050}{10{,}000} = 0.505$$

We can find this sum for any n more easily if we observe that the common denominator is n^2 and that the numerator is the sum of the first n terms of an arithmetic sequence with first term 1 and last term n. As you may recall from

Section 6.1, "Simple Interest; Sequences," the first n terms of this arithmetic sequence add to $(n + 1)/2$. Thus the area is approximated by

$$A \approx \frac{1 + 2 + 3 + \cdots + n}{n^2} = \frac{n(n+1)/2}{n^2} = \frac{n+1}{2n}$$

Using this formula, we see the following.

$$n = 5: \quad A \approx \frac{5+1}{2(5)} = \frac{6}{10} = 0.60$$

$$n = 10: \quad A \approx \frac{10+1}{2(10)} = \frac{11}{20} = 0.55$$

$$n = 100: \quad A \approx \frac{100+1}{2(100)} = \frac{101}{200} = 0.505$$

Note that as n gets larger, the number of rectangles increases, the area of each rectangle decreases, and the approximation becomes more accurate. If we let n increase without bound, the approximation approaches the exact area.

$$A = \lim_{n\to+\infty} \frac{n+1}{2n} = \lim_{n\to+\infty} \frac{1 + 1/n}{2} = \frac{1}{2}$$

We can see that this area is correct, for we are computing the area of a triangle with base 1 and height 1. The formula for the area of a triangle gives

$$A = \frac{1}{2}bh = \frac{1}{2} \cdot 1 \cdot 1 = \frac{1}{2}$$

A special notation exists that uses the Greek letter Σ (sigma) to express the sum of numbers or expressions. (We used sigma notation informally in Chapter 8, "Further Topics in Probability; Statistics.") We may indicate the sum of the n numbers $a_1, a_2, a_3, a_4, \ldots, a_n$ by

$$\sum_{i=1}^{n} a_i = a_1 + a_2 + a_3 + \cdots + a_n$$

This may be read as "The sum of a_i as i goes from 1 to n." The subscript i in a_i is replaced first by 1, then by 2, then by 3, . . . , until it reaches the value above the sigma. The i is called the **index of summation,** and it starts with the lower limit, 1, and ends with the upper limit, n. For example, if $x_1 = 2$, $x_2 = 3$, $x_3 = -1$, and $x_4 = -2$, then

$$\sum_{i=1}^{4} x_i = x_1 + x_2 + x_3 + x_4 = 2 + 3 + (-1) + (-2) = 2$$

The area of the triangle under $y = x$ that we discussed above was approximated by

$$A \approx \frac{1}{n^2} + \frac{2}{n^2} + \frac{3}{n^2} + \cdots + \frac{i}{n^2} + \cdots + \frac{n}{n^2}$$

Using **sigma notation,** we can write this sum as

$$A \approx \sum_{i=1}^{n} \left(\frac{i}{n^2}\right)$$

Sigma notation allows us to represent the sums of the areas of the rectangles in an abbreviated fashion. Some formulas that simplify computations involving sums follow.

Sum Formulas

I. $\sum_{i=1}^{n} 1 = n$

II. $\sum_{i=1}^{n} cx_i = c \sum_{i=1}^{n} x_i \quad (c = \text{constant})$

III. $\sum_{i=1}^{n} (x_i + y_i) = \sum_{i=1}^{n} x_i + \sum_{i=1}^{n} y_i$

IV. $\sum_{i=1}^{n} i = \frac{n(n+1)}{2}$

V. $\sum_{i=1}^{n} i^2 = \frac{n(n+1)(2n+1)}{6}$

We have found that the area of the triangle discussed above was approximated by

$$A \approx \sum_{i=1}^{n} \frac{i}{n^2}$$

We can use these formulas to simplify this sum as follows.

$$\begin{aligned}\sum_{i=1}^{n} \frac{i}{n^2} &= \frac{1}{n^2} \sum_{i=1}^{n} i && \text{(Formula II)}\\ &= \frac{1}{n^2}\left[\frac{n(n+1)}{2}\right] && \text{(Formula IV)}\\ &= \frac{n+1}{2n}\end{aligned}$$

Note that this is the same formula we obtained previously using other methods.

The following example shows that we can find the area by evaluating the function at the left-hand endpoints of the subintervals.

EXAMPLE 2

Use rectangles to find the area under $y = x^2$ (and above the x-axis) from $x = 0$ to $x = 1$.

Solution

We again divide the interval [0, 1] into n equal subintervals of length $1/n$. If we evaluate the function at the left-hand endpoints of these subintervals to determine the heights of the rectangles, the sum of the areas of the rectangles will underestimate the area (see Figure 13.5). Thus we have the following:

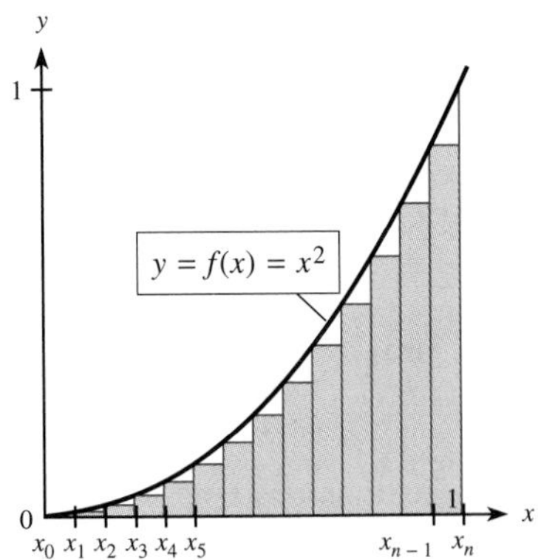

Figure 13.5

Rectangle	*Base*	*Endpoint*	*Height*	*Area = Base × Height*
1	$\frac{1}{n}$	$x_0 = 0$	$f(x_0) = 0$	$\frac{1}{n} \cdot 0 = 0$
2	$\frac{1}{n}$	$x_1 = \frac{1}{n}$	$f(x_1) = \frac{1}{n^2}$	$\frac{1}{n} \cdot \frac{1}{n^2} = \frac{1}{n^3}$
3	$\frac{1}{n}$	$x_2 = \frac{2}{n}$	$f(x_2) = \frac{4}{n^2}$	$\frac{1}{n} \cdot \frac{4}{n^2} = \frac{4}{n^3}$
4	$\frac{1}{n}$	$x_3 = \frac{3}{n}$	$f(x_3) = \frac{9}{n^2}$	$\frac{1}{n} \cdot \frac{9}{n^2} = \frac{9}{n^3}$
⋮				
i	$\frac{1}{n}$	$x_{i-1} = \frac{i-1}{n}$	$\frac{(i-1)^2}{n}$	$\frac{(i-1)^2}{n^3}$
⋮				
n	$\frac{1}{n}$	$x_{n-1} = \frac{n-1}{n}$	$\frac{(n-1)^2}{n^2}$	$\frac{(n-1)^2}{n^3}$

Note that $(i - 1)^2/n^3 = (i^2 - 2i + 1)/n^3$ gives the area of the ith rectangle for *any* value of i. The sum of these areas may be written as

$$S = \sum_{i=1}^{n} \frac{i^2 - 2i + 1}{n^3} = \frac{1}{n^3}\left(\sum_{i=1}^{n} i^2 - 2 \sum_{i=1}^{n} i + \sum_{i=1}^{n} 1 \right) \qquad \text{(Formulas II and III)}$$

$$= \frac{1}{n^3}\left[\frac{n(n+1)(2n+1)}{6} - \frac{2n(n+1)}{2} + n \right] \qquad \text{(Formulas V, IV, and I)}$$

$$= \frac{2n^3 + 3n^2 + n}{6n^3} - \frac{n^2 + n}{n^3} + \frac{n}{n^3} = \frac{2n^2 - 3n + 1}{6n^2}$$

If n is large, there will be a large number of smaller rectangles, and the approximation of the area under the curve will be good; the larger n, the better the approximation. For example, if $n = 10$, the area approximation is

$$S(10) = \frac{200 - 30 + 1}{600} = 0.285$$

whereas if $n = 100$,

$$S(100) = \frac{20{,}000 - 300 + 1}{60{,}000} = 0.328$$

If we let n increase without bound, we find the exact area.

$$A = \lim_{n\to\infty}\left(\frac{2n^2 - 3n + 1}{6n^2}\right) = \lim_{n\to\infty}\left(\frac{2 - \frac{3}{n} + \frac{1}{n^2}}{6}\right) = \frac{1}{3}$$

Note that the approximations with $n = 10$ and $n = 100$ were less than $\frac{1}{3}$. This is because all the rectangles were *under* the curve (see Figure 13.5 on the previous page).

Thus we see that we can determine the area under a curve $y = f(x)$ from $x = a$ to $x = b$ by dividing the interval $[a, b]$ into n equal subintervals of width $(b - a)/n$ and evaluating

$$A = \lim_{n\to\infty} S_R = \lim_{n\to\infty} \sum_{i=1}^{n} f(x_i)\left(\frac{b-a}{n}\right) \qquad \text{(using right-hand endpoints)}$$

or

$$A = \lim_{n\to\infty} S_L = \lim_{n\to\infty} \sum_{i=1}^{n} f(x_{i-1})\left(\frac{b-a}{n}\right) \qquad \text{(using left-hand endpoints)}$$

CHECKPOINT

1. For the interval [0, 2], determine whether the following are true or false.
   (a) For 4 subintervals, each subinterval has width $\frac{1}{2}$.
   (b) For 200 subintervals, each subinterval has width $\frac{1}{100}$.
   (c) For n subintervals, each subinterval has width $\frac{2}{n}$.
   (d) For n subintervals, $x_0 = 0, x_1 = \frac{2}{n}, x_2 = 2\left(\frac{2}{n}\right), \ldots, x_i = i\left(\frac{2}{n}\right), \ldots, x_n = 2$.
2. If $\frac{b-a}{n} = \frac{2}{n}$, $x_i = \frac{2i}{n}$, and $f(x) = 3x - x^2$, find:
   (a) $f(x_i)$
   (b) $f(x_i)\,\frac{b-a}{n}$
   (c) $\sum_{i=1}^{n} f(x_i)\,\frac{b-a}{n}$, and simplify
   (d) $\lim_{n\to\infty} \sum_{i=1}^{n} f(x_i)\,\frac{b-a}{n}$

EXAMPLE 3

To approximate the area under the graph of $y = \sqrt{x}$ on the interval [0, 4], do the following.

(a) Use a calculator or computer to find S_L and S_R for $n = 8$ on the interval [0, 4].
(b) Predict the area under the curve in the graph by evaluating S_L and S_R for larger values of n.

Solution

Figure 13.6(a) shows the graph of $y = \sqrt{x}$ with 8 rectangles whose heights are determined by evaluating the function at the left-hand endpoint of each interval (the first of these rectangles has height 0). Figure 13.6(b) shows the same graph with 8 rectangles whose heights are determined at the right-hand endpoints.

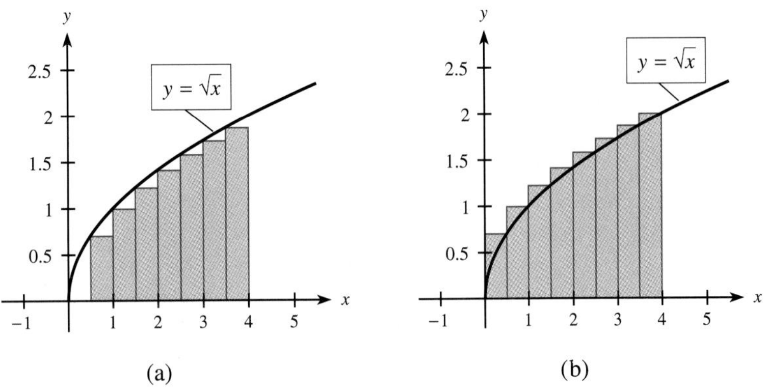

Figure 13.6

(a) The formula for the sum of the areas of n rectangles over the interval [0, 4], using left-hand endpoints, is

$$S_L = \sum_{i=1}^{n} \sqrt{\frac{4(i-1)}{n}}\,\frac{4}{n} = \frac{8}{n^{3/2}} \sum_{i=1}^{n} \sqrt{i-1}$$

and the formula for the sum of the areas of n rectangles over the interval [0, 4], using right-hand endpoints, is

$$S_R = \sum_{i=1}^{n} \sqrt{\frac{4(i)}{n}}\,\frac{4}{n} = \frac{8}{n^{3/2}} \sum_{i=1}^{n} \sqrt{i}$$

We have no formula to write either of these summations in a simpler form, but we can use a graphing calculator, computer program, or spreadsheet to compute the sum for any given n. For $n = 8$, the sums are

$$S_L = \frac{1}{\sqrt{8}} \sum_{i=1}^{8} \sqrt{i-1} \quad \text{and} \quad S_R = \frac{1}{\sqrt{8}} \sum_{i=1}^{8} \sqrt{i}$$

The following spreadsheet output shows the sum for $n = 8$ rectangles.

	A	B	C	D
1	Area for	n=8		
2		i	SL	SR
3		1	0	.35355
4		2	.35355	.5
5		3	.5	.61237
6		4	.61237	.70711
7		5	.70711	.79057
8		6	.79057	.86603
9		7	.86603	.93541
10		8	.93541	1
11	Total		4.76504	5.76504

As line 11 shows, $S_L = 4.76504$ and $S_R = 5.76504$ for $n = 8$.

(b) Larger values of n give better approximations of the area under the curve (try some). For example, $n = 1000$ gives $S_L = 5.3293$ and $S_R = 5.3373$, and because the area under the curve is between these values, the area is approximately 5.33, to two decimal places. By using values of n larger than 1000, we can get better approximations of the area.

CHECKPOINT SOLUTIONS

1. All parts are true.

2. (a) $f(x_i) = f\left(\frac{2i}{n}\right) = 3\left(\frac{2i}{n}\right) - \left(\frac{2i}{n}\right)^2 = \frac{6i}{n} - \frac{4i^2}{n^2}$

(b) $f(x_i)\frac{b-a}{n} = \left(\frac{6i}{n} - \frac{4i^2}{n^2}\right)\left(\frac{2}{n}\right) = \frac{12i}{n^2} - \frac{8i^2}{n^3}$

(c)
$$\begin{aligned}\sum_{i=1}^{n} f(x_i)\frac{b-a}{n} &= \sum_{i=1}^{n}\left(\frac{12i}{n^2} - \frac{8i^2}{n^3}\right)\\ &= \sum_{i=1}^{n}\frac{12i}{n^2} - \sum_{i=1}^{n}\frac{8i^2}{n^3}\\ &= \frac{12}{n^2}\sum_{i=1}^{n} i - \frac{8}{n^3}\sum_{i=1}^{n} i^2\\ &= \frac{12}{n^2}\left[\frac{n(n+1)}{2}\right] - \frac{8}{n^3}\left[\frac{n(n+1)(2n+1)}{6}\right]\\ &= \frac{6(n+1)}{n} - \frac{4(n+1)(2n+1)}{3n^2}\end{aligned}$$

(d) $\lim_{n\to\infty}\sum_{i=1}^{n} f(x_i)\frac{b-a}{n} = \lim_{n\to\infty}\left(\frac{6n+6}{n} - \frac{8n^2+12n+4}{3n^2}\right) = 6 - \frac{8}{3} = \frac{10}{3}$

EXERCISE 13.1

In Problems 1–4, approximate the area under each curve over the interval specified by using the indicated number of subintervals (or rectangles) and evaluating the function at the *right-hand* endpoints of the subintervals.

1. $f(x) = 4x - x^2$ from $x = 0$ to $x = 2$; 2 subintervals
2. $f(x) = x^3$ from $x = 0$ to $x = 3$; 3 subintervals
3. $f(x) = 9 - x^2$ from $x = -1$ to $x = 3$; 4 subintervals
4. $f(x) = x^2 + x + 1$ from $x = -1$ to $x = 1$; 4 subintervals

In Problems 5–8, approximate the area under each curve by evaluating the function at the *left-hand* endpoints of the subintervals.

5. $f(x) = 4x - x^2$ from $x = 0$ to $x = 2$; 2 subintervals
6. $f(x) = x^3$ from $x = 0$ to $x = 3$; 3 subintervals
7. $f(x) = 9 - x^2$ from $x = -1$ to $x = 3$; 4 subintervals
8. $f(x) = x^2 + x + 1$ from $x = -1$ to $x = 1$; 4 subintervals

When the area under $f(x) = x^2 + x$ from $x = 0$ to $x = 2$ is approximated, the formulas for the sum of n rectangles using *left-hand* endpoints and *right-hand* endpoints are

Left-hand endpoints: $S_L = \frac{14}{3} - \frac{6}{n} + \frac{4}{3n^2}$

Right-hand endpoints: $S_R = \frac{14n^2 + 18n + 4}{3n^2}$

Use these formulas to answer Problems 9–13.

9. Find $S_L(10)$ and $S_R(10)$.
10. Find $S_L(100)$ and $S_R(100)$.
11. Find $\lim_{n\to\infty} S_L$ and $\lim_{n\to\infty} S_R$.
12. Compare the right-hand and left-hand values by finding $S_R - S_L$ for $n = 10$, for $n = 100$, and as $n \to \infty$. (Use Problems 9–11.)
13. Because $f(x) = x^2 + x$ is increasing over the interval from $x = 0$ to $x = 2$, functional values at the right-

hand endpoints are maximum values for each subinterval, and functional values at the left-hand endpoints are minimum values for each subinterval. How would the approximate area using $n = 10$ and *any* other point within each subinterval compare with $S_L(10)$ and $S_R(10)$? What would happen to the area result as $n \to \infty$ if any other point in each subinterval were used?

In Problems 14–23, find the value of each sum.

14. $\sum_{k=1}^{3} x_k$, if $x_1 = 1, x_2 = 3, x_3 = -1, x_4 = 5$
15. $\sum_{i=1}^{4} x_i$, if $x_1 = 3, x_2 = -1, x_3 = 3, x_4 = -2$
16. $\sum_{i=3}^{5} (i^2 + 1)$
17. $\sum_{j=2}^{5} (j^2 - 3)$
18. $\sum_{i=4}^{7} \left(\frac{i-3}{i^2}\right)$
19. $\sum_{j=0}^{4} (j^2 - 4j + 1)$
20. $\sum_{k=1}^{50} 1$
21. $\sum_{j=1}^{60} 3$
22. $\sum_{k=1}^{50} (6k^2 + 5)$
23. $\sum_{k=1}^{30} (k^2 + 4k)$

In Problems 24 and 25, use the Sum Formulas I–V to express each of the following without the summation symbol.

24. $\sum_{i=1}^{n} \left(1 - \frac{i^2}{n^2}\right)\left(\frac{2}{n}\right)$
25. $\sum_{i=1}^{n} \left(1 - \frac{2i}{n} + \frac{i^2}{n^2}\right)\left(\frac{3}{n}\right)$

Use the function $y = x$ from $x = 0$ to $x = 1$ and n equal subintervals with the function evaluated at the *left-hand* endpoint of each subinterval for Problems 26–27.

26. What is the area of the
    (a) first rectangle? (b) second rectangle?
    (c) ith rectangle?
27. (a) Find a formula for the sum of the areas of the n rectangles (call this S). Then find
    (b) $S(10)$. (c) $S(100)$. (d) $S(1000)$. (e) $\lim_{n\to\infty} S$.
28. How do your answers to Problems 27(a)–(e) compare with the corresponding calculations in the discussion (after Example 1) of the area under $y = x$ using *right-hand* endpoints?

For Problems 29(a)–(e), use the function $y = x^2$ from $x = 0$ to $x = 1$ and n equal subintervals with the function evaluated at the *right-hand* endpoints.

29. (a) Find a formula for the sum of the areas of the n rectangles (call this S). Then find
    (b) $S(10)$. (c) $S(100)$. (d) $S(1000)$. (e) $\lim_{n\to\infty} S$.
30. How do your answers to Problems 29(a)–(e) compare with the corresponding calculations in Example 2?
31. Use rectangles to find the area between $y = x^2 - 6x + 8$ and the x-axis from $x = 0$ to $x = 2$. Divide the interval [0, 2] into n equal subintervals so that each subinterval has length $2/n$.
32. Use rectangles to find the area between $y = 4x - x^2$ and the x-axis from $x = 0$ to $x = 4$. Divide the interval [0, 4] into n equal subintervals so that each subinterval has length $4/n$.

Applications

33. ***Speed trials*** The graph in the figure gives the times that it takes a 1994 Porsche 911 to reach speeds from 0 mph to 100 mph, in increments of 10 mph, with a curve connecting them. The area under this curve from $t = 0$ seconds to $t = 14$ seconds represents the total amount of distance traveled over the 14-second period. Count the squares under the curve to estimate this distance. This car will travel 1/4 mile in 14 seconds, to a speed of 100.2 mph. Is your estimate close to this result? (Be careful with time units.)

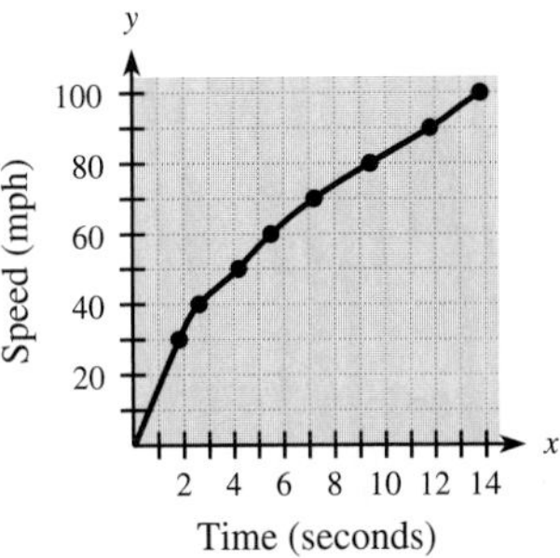

Source: *Motor Trend,* January 1994

34. ***Speed trials*** The graph in the figure gives the times that it takes a 1995 Mitsubishi Eclipse GSX to reach speeds from 0 mph to 100 mph, in increments of 10 mph, with a curve connecting them. The area under this curve from $t = 0$ seconds to $t = 21.1$ seconds represents the total amount of distance traveled over the 21.1-second period. Count the squares under the curve to estimate this distance. This car will travel 1/4 mile in 15.4 seconds, to a speed of 89.0 mph, so your estimate should be more than 1/4 mile. Is it? (Be careful with time units.)

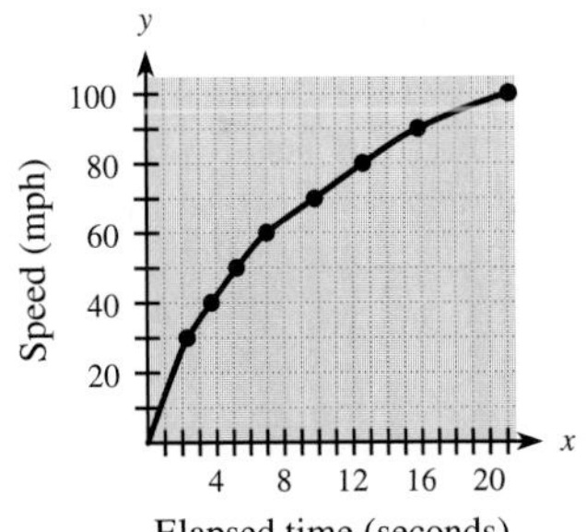

Source: *Road & Track,* June 1994

35. ***Pollution monitoring*** Suppose the presence of phosphates in certain waste products dumped into a lake promotes the growth of algae. Rampant growth of algae affects the oxygen supply in the water, so an environmental group wishes to estimate the area of algae growth. The group measures the length across the algae growth (see the figure) and obtains the following data (in feet).

x	*Length*	x	*Length*
0	0	50	27
10	15	60	24
20	18	70	23
30	18	80	0
40	30		

Use 8 rectangles with bases of 10 feet and lengths measured at the left-hand endpoints to approximate the area of the algae growth.

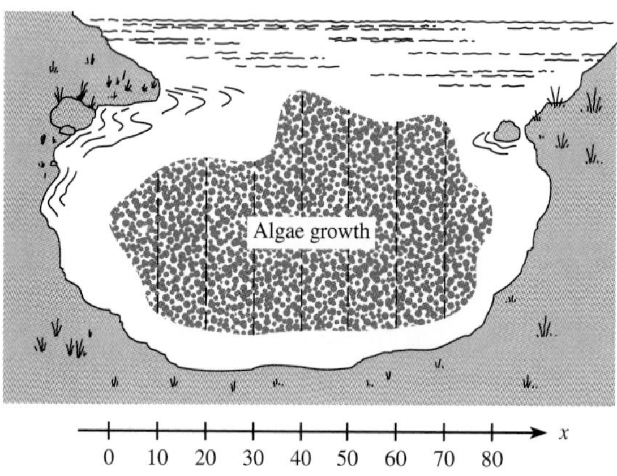

36. ***Drug levels in the blood*** The manufacturer of a medicine wants to test how a new 300-milligram capsule is released into the bloodstream. After a volunteer is given a capsule, blood samples are drawn every half-hour, and the number of milligrams of the drug in the bloodstream is calculated. The results obtained follow.

Time t (*hr*)	$N(t)$ (*mg*)	*Time t* (*hr*)	$N(t)$ (*mg*)
0	0	2.0	178.3
0.5	247.3	2.5	113.9
1.0	270	3.0	56.2
1.5	236.4	3.5	19.3

Use 7 rectangles, each with height $N(t)$ and with width 0.5 hr, to estimate the area under the graph representing these data. Divide this area by 3.5 hr to estimate the average drug level over this time period.

13.2 The Definite Integral; The Fundamental Theorem of Calculus

OBJECTIVES

- *To evaluate definite integrals using the Fundamental Theorem of Calculus*
- *To use definite integrals to find the area under a curve*

APPLICATION PREVIEW

Suppose that money flows continuously into a slot machine at a casino and grows at a rate given by

$$A'(t) = 100e^{0.1t}$$

where t is the time in hours and $0 \le t \le 10$. Then the total amount of money that accumulates over the 10-hour period, if no money is paid out, is given by the **definite integral**

$$\int_0^{10} 100e^{0.1t}\,dt$$

In the previous section, we used the sum of areas of rectangles to approximate the area under curves. In this section, we will see how this sum is related to the definite integral and how to evaluate definite integrals. In addition, we will see how definite integrals will be used to solve several types of applied problems.

In the previous section, we saw that we could determine the area under a curve using equal subintervals and the functional values at either the left-hand endpoints or the right-hand endpoints of the subintervals. In fact, we can use subintervals that are not of equal length, and we can use any point within each subinterval to determine the height of each rectangle. Suppose that we wish to find the area above the x-axis and under the curve $y = f(x)$ over a closed interval $[a, b]$. We can divide the interval into n subintervals (not necessarily equal), with the endpoints of these intervals at $x_0 = a, x_1, x_2 \ldots, x_n = b$. We now choose a point (*any* point) in each subinterval, and denote the points $x_1^*, x_2^*, \ldots, x_i^*, \ldots, x_n^*$. Then the ith rectangle (for any i) has height $f(x_i^*)$ and width $x_i - x_{i-1}$, so its area is $f(x_i^*)(x_i - x_{i-1})$. Then the sum of the areas of the n rectangles is

$$\begin{aligned} S &= \sum_{i=1}^{n} f(x_i^*)(x_i - x_{i-1}) \\ &= \sum_{i=1}^{n} f(x_i^*)\Delta x_i, \qquad \text{where } \Delta x_i = x_i - x_{i-1} \end{aligned}$$

Because the points in the subinterval may be chosen anywhere in the subinterval, we cannot be sure whether the rectangles will underestimate or overestimate the area under the curve. But increasing the number of subintervals (increasing n) and making sure that every interval becomes smaller (just increasing n will not guarantee this if the subintervals are unequal) will in the long run improve the estimation. Thus for any subdivision of $[a, b]$ and any x_i^*, the area is given by

$$A = \lim_{\substack{n\to\infty \\ \max \Delta x_i \to 0}} \sum_{i=1}^{n} f(x_i^*)\Delta x_i, \qquad \text{provided that this limit exists}$$

The preceding discussion takes place in the context of finding the area under a curve. However, if f is any function (not necessarily nonnegative) defined on $[a, b]$, then for each subdivision of $[a, b]$ and each choice of x_i^*, we define the sum S above as the **Riemann sum** of f for the subdivision of $[a, b]$. In addition, the limit of the Riemann sum (as $\max \Delta x_i \to 0$) has other important applications and is called the **definite integral** of $f(x)$ over interval $[a, b]$.

Definite Integral If f is a function on the interval $[a, b]$, then the *definite integral* of f from a to b is

$$\int_a^b f(x)dx = \lim_{\substack{n\to\infty \\ \max \Delta x_i \to 0}} \sum_{i=1}^{n} f(x_i^*)\Delta x_i$$

If the limit exists, then the definite integral exists and we say that f is integrable on $[a, b]$.

The obvious question is how this definite integral is related to the indefinite integral (antiderivative) we have been studying. The answer to this question is given by the **Fundamental Theorem of Calculus.**

Fundamental Theorem of Calculus

Let f be a continuous function on the closed interval $[a, b]$; then the definite integral of f exists on this interval, and

$$\int_a^b f(x)\,dx = F(b) - F(a)$$

where F is any function such that $F'(x) = f(x)$ for all x in $[a, b]$.

Stated differently, the theorem says that if the function F is an indefinite integral of a function f that is continuous on the interval $[a, b]$, then

$$\int_a^b f(x)\,dx = F(b) - F(a)$$

We denote $F(b) - F(a)$ by $F(x)\Big|_a^b$.

EXAMPLE 1

Evaluate $\int_2^4 (x^3 + 4)\,dx$.

Solution

$$\begin{aligned}\int_2^4 (x^3 + 4)\,dx &= \frac{x^4}{4} + 4x + C\Big|_2^4 \\ &= \left[\frac{(4)^4}{4} + 4(4) + C\right] - \left[\frac{(2)^4}{4} + 4(2) + C\right] \\ &= (64 + 16 + C) - (4 + 8 + C) \\ &= 68 \qquad \text{(Note that the } C\text{'s subtract out.)}\end{aligned}$$

Note that the Fundamental Theorem states that F can be *any* indefinite integral of f, so we need not add the constant of integration to the integral.

EXAMPLE 2

Evaluate $\int_1^3 (3x^2 + 6x)\,dx$.

Solution

$$\begin{aligned}\int_1^3 (3x^2 + 6x)\,dx &= x^3 + 3x^2\Big|_1^3 \\ &= (3^3 + 3 \cdot 3^2) - (1^3 + 3 \cdot 1^2) \\ &= 54 - 4 = 50\end{aligned}$$

The properties of definite integrals given below follow from properties of summations.

1. $\int_a^b [f(x) \pm g(x)]\,dx = \int_a^b f(x)\,dx \pm \int_a^b g(x)\,dx$
2. $\int_a^b kf(x)\,dx = k\int_a^b f(x)\,dx,$ where k is a constant

The following example uses both of these properties.

EXAMPLE 3

Evaluate $\int_3^5 (\sqrt{x^2 - 9} + 2)x\,dx.$

Solution

$$\begin{aligned}\int_3^5 (\sqrt{x^2-9}+2)x\,dx &= \int_3^5 \sqrt{x^2-9}(x\,dx) + \int_3^5 2x\,dx \\ &= \frac{1}{2}\int_3^5 (x^2-9)^{1/2}(2x\,dx) + \int_3^5 2x\,dx \\ &= \frac{1}{2}\left[\frac{2}{3}(x^2-9)^{3/2}\right]\Bigg|_3^5 + x^2\Big|_3^5 \\ &= \frac{1}{3}[(16)^{3/2} - (0)^{3/2}] + (25 - 9) \\ &= \frac{64}{3} + 16 = \frac{64}{3} + \frac{48}{3} = \frac{112}{3}\end{aligned}$$

In the integral $\int_a^b f(x)\,dx$, we call a the *lower limit* and b the *upper limit* of integration. Although we developed the definite integral with the assumption that the lower limit was less than the upper limit, the following properties permit us to evaluate the definite integral even when that is not the case.

3. $\int_a^a f(x)\,dx = 0$
4. If f is integrable on $[a, b]$, then

$$\int_b^a f(x)\,dx = -\int_a^b f(x)\,dx$$

The following examples illustrate these properties.

EXAMPLE 4

Evaluate $\int_4^4 x^2\,dx.$

Solution

$$\int_4^4 x^2\,dx = \frac{x^3}{3}\Bigg|_4^4 = \frac{4^3}{3} - \frac{4^3}{3} = 0$$

Note that because the limits of integration are the same, it is not necessary to integrate to see that the value of the integral is 0.

EXAMPLE 5

Compare $\int_2^4 3x^2\,dx$ and $\int_4^2 3x^2\,dx$.

Solution

$$\int_2^4 3x^2\,dx = x^3\Big|_2^4 = 4^3 - 2^3 = 56$$

$$\int_4^2 3x^2\,dx = x^3\Big|_4^2 = 2^3 - 4^3 = -56$$

Thus

$$\int_4^2 3x^2\,dx = -\int_2^4 3x^2\,dx$$

Another property of definite integrals is called the additive property.

5. If f is continuous on some interval containing a, b, and c,* then

$$\int_a^b f(x)\,dx = \int_a^c f(x)\,dx + \int_c^b f(x)\,dx$$

EXAMPLE 6

Show that $\int_2^3 4x\,dx + \int_3^5 4x\,dx = \int_2^5 4x\,dx$.

Solution

$$\int_2^3 4x\,dx = 2x^2\Big|_2^3 = 18 - 8 = 10$$

$$\int_3^5 4x\,dx = 2x^2\Big|_3^5 = 50 - 18 = 32$$

$$\int_2^5 4x\,dx = 2x^2\Big|_2^5 = 50 - 8 = 42$$

Thus

$$\int_2^3 4x\,dx + \int_3^5 4x\,dx = \int_2^5 4x\,dx$$

Let us now return to area problems, to see the relationship between the definite integral and the area under a curve. By the formula for the area of a triangle and by summing areas of rectangles, we found the area under the curve (line) $y = x$ from $x = 0$ to $x = 1$ to be $\frac{1}{2}$ (see Figure 13.7a). Using the definite integral to find the area gives

*Note that c need not be between a and b.

$$A = \int_0^1 x\,dx = \frac{x^2}{2}\bigg|_0^1 = \frac{1}{2} - 0 = \frac{1}{2}$$

In Example 2 of the previous section, we used rectangles to find that the area under $y = x^2$ from $x = 0$ to $x = 1$ was $\frac{1}{3}$ (see Figure 13.7b). Using the definite integral, we get

$$A = \int_0^1 x^2\,dx = \frac{x^3}{3}\bigg|_0^1 = \frac{1}{3} - 0 = \frac{1}{3}$$

which agrees with the answer obtained in Example 2.

However, not every definite integral represents the area between the curve and the x-axis over an integral. For example,

$$\int_0^2 (x-2)\,dx = \frac{x^2}{2} - 2x\bigg|_0^2 = (2-4) - (0) = -2$$

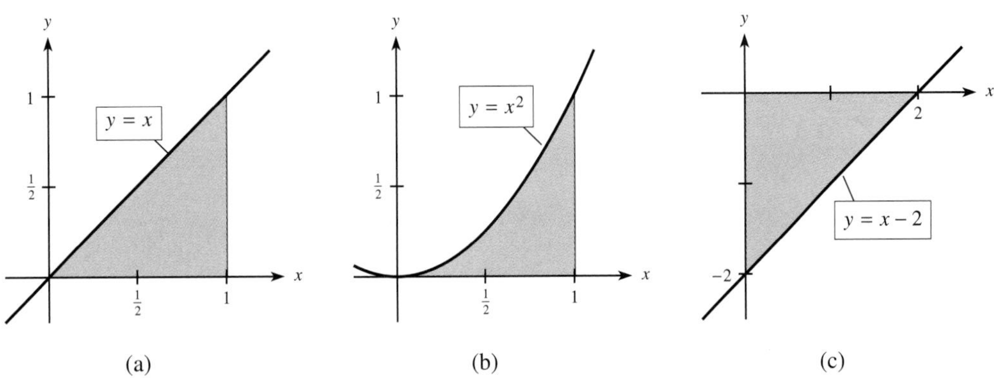

Figure 13.7 (a) (b) (c)

This would indicate that the area between the curve and the x-axis is negative, but area must be positive. A look at the graph of $y = x - 2$ (see Figure 13.7c) shows us what is happening. The region bounded by $y = x - 2$ and x-axis between $x = 0$ and $x = 2$ is a triangle whose base is 2 and height is 2, so its area is $\frac{1}{2}bh = \frac{1}{2}(2)(2) = 2$. The integral has value -2 because $y = x - 2$ lies below the x-axis from $x = 0$ to $x = 2$, and the functional values over the interval $[0, 2]$ are negative. Thus the value of the definite integral over *this* interval does not represent the area between the curve and the x-axis.

In general, the definite integral will give the area under the curve and above the x-axis only when $f(x) \geq 0$ for all x in $[a, b]$.

Area Under a Curve If f is a continuous function on $[a, b]$ and $f(x) \geq 0$ on $[a, b]$, then the exact area between $y = f(x)$ and the x-axis from $x = a$ to $x = b$ is given by

$$\text{Area (shaded)} = \int_a^b f(x)\,dx$$

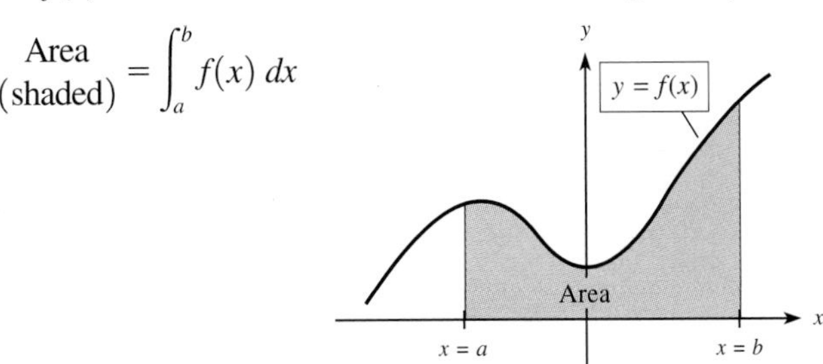

CHECKPOINT

1. True or false:
   (a) For any integral, we can omit the constant of integration (the $+C$).
   (b) $-\int_{-1}^{3} f(x)\,dx = \int_{3}^{-1} f(x)\,dx$, if f is integrable on $[-1, 3]$.
   (c) The area between $f(x)$ and the x-axis on the interval $[a, b]$ is given by $\int_{a}^{b} f(x)\,dx$.
2. Evaluate:
   (a) $\int_{0}^{3} (x^2 + 1)\,dx$ (b) $\int_{0}^{3} (x^2 + 1)^4 x\,dx$

If the rate of growth of some function with respect to time t is $f'(t)$, then the total growth of the function during the period from $t = 0$ to $t = k$ can be found by evaluating the definite integral

$$\int_{0}^{k} f'(t)\,dt = f(t)\Big|_{0}^{k} = f(k) - f(0)$$

For nonnegative rates of growth, this definite integral (and thus growth) is the same as the area under the graph of $f'(t)$ from $t = 0$ to $t = k$.

We now return to the problem introduced in the Application Preview.

EXAMPLE 7

Suppose that money flows continuously into a slot machine at a casino and grows at a rate given by

$$A'(t) = 100e^{0.1t}$$

where t is time in hours and $0 \le t \le 10$. Find the total amount that accumulates in the machine during the 10-hour period, if no money is paid out.

Solution

The total amount is given by

$$\begin{aligned} A = \int_{0}^{10} 100e^{0.1t}\,dt &= \frac{100}{0.1}\int_{0}^{10} e^{0.1t}\,(0.1)\,dt \\ &= 1000e^{0.1t}\Big|_{0}^{10} \\ &= 1000e - 1000 \\ &\approx 1718.28 \quad \text{(dollars)} \end{aligned}$$

In Section 8.5, "Normal Probability Distribution," we stated that the total area under the normal curve is 1 and that the area under the curve from value x_1 to value x_2 represents the probability that a score chosen at random will lie between x_1 and x_2.

The normal distribution is an example of a **continuous distribution** because the values of the random variable are considered over intervals rather than at discrete values. The above statements relating probability and area under the graph apply to other continuous probability distributions determined by **probability density functions.** Thus we can use the definite integral to find the probability that a random variable lies within a given interval. Consider the following example.

EXAMPLE 8

Suppose the probability density function for the life of a computer component is $f(x) = 0.10e^{-0.10x}$, where $x \geq 0$ is the number of years the component is in use. Find the probability that the component will last between 3 and 5 years.

Solution

The probability that the component will last between 3 and 5 years is the area under the graph of the function between $x = 3$ and $x = 5$. The probability is given by the integral

$$\begin{aligned}\int_3^5 0.10e^{-0.10x}\,dx &= -e^{-0.10x}\Big|_3^5 \\ &= -e^{-0.5} + e^{-0.3} \\ &\approx -0.6065 + 0.7408 \\ &= 0.1343\end{aligned}$$

Most graphing calculators and graphing utilities have a numerical integration feature that can be used to get very accurate approximations of definite integrals. This feature can be used to evaluate definite integrals directly or to check those done with the Fundamental Theorem. Figure 13.8 shows the numerical integration feature applied to the integral in Example 8. Note that when this answer is rounded to four decimal places, the results agree.

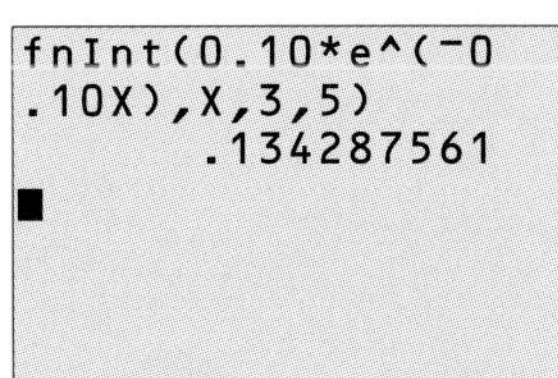

Figure 13.8

CHECKPOINT SOLUTIONS

1. (a) False. We can omit the constant of integration ($+C$) only for definite integrals.
   (b) True
   (c) False. Only if $f(x) \geq 0$ on $[a, b]$ is this true.
2. (a) $\int_0^3 (x^2 + 1)\,dx = \frac{x^3}{3} + x \Big|_0^3 = \left(\frac{27}{3} + 3\right) - (0 + 0) = 12$

   (b) $$\begin{aligned}\int_0^3 (x^2 + 1)^4 x\,dx &= \frac{1}{2}\int_0^3 (x^2 + 1)^4 (2x\,dx)\\ &= \frac{1}{2} \cdot \frac{(x^2 + 1)^5}{5}\Big|_0^3\\ &= \frac{1}{10}[(3^2 + 1)^5 - (1)^5]\\ &= \frac{1}{10}(10^5 - 1) = 9999.9\end{aligned}$$

EXERCISE 13.2

Evaluate the definite integrals in Problems 1–26.

1. $\int_0^3 4x\,dx$
2. $\int_0^1 8x\,dx$
3. $\int_2^4 dx$
4. $\int_1^5 2\,dy$
5. $\int_2^4 x^3\,dx$
6. $\int_0^5 x^2\,dx$
7. $\int_0^5 4\sqrt[3]{x^2}\,dx$
8. $\int_2^4 3\sqrt{x}\,dx$
9. $\int_2^4 (4x^3 - 6x^2 - 5x)\,dx$
10. $\int_0^2 (x^4 - 5x^3 + 2x)\,dx$
11. $\int_2^3 (x - 4)^2\,dx$
12. $\int_{-1}^3 (x + 2)^3\,dx$
13. $\int_2^4 (x^2 + 2)^3\,x\,dx$
14. $\int_0^3 (2x - x^2)^4(1 - x)\,dx$
15. $\int_{-1}^2 (x^3 - 3x^2)^3(x^2 - 2x)\,dx$
16. $\int_0^4 (3x^2 - 2)^4\,x\,dx$
17. $\int_2^3 x\sqrt{x^2 + 3}\,dx$
18. $\int_{-1}^2 x\sqrt[3]{x^2 - 5}\,dx$
19. $\int_1^3 \frac{3}{y^2}\,dy$
20. $\int_1^2 \frac{5}{z^3}\,dz$
21. $\int_0^1 e^{3x}\,dx$
22. $\int_0^2 e^{4x}\,dx$
23. $\int_1^e \frac{4}{z}\,dz$
24. $\int_1^e 3y^{-1}\,dy$
25. $\int_4^4 \sqrt{x^2 - 2}\,dx$
26. $\int_2^2 (x^3 - 4x)\,dx$

In Problems 27–30, evaluate each integral (a) with the Fundamental Theorem and (b) with a graphing utility (as a check).

27. $\int_3^6 \frac{x}{3x^2 + 4}\,dx$
28. $\int_0^2 \frac{x}{x^2 + 4}\,dx$
29. $\int_1^2 \frac{x^2 + 3}{x}\,dx$
30. $\int_1^4 \frac{4\sqrt{x} + 5}{\sqrt{x}}\,dx$
31. In the figure, which of the shaded regions (A, B, C, or D) has the area given by
   (a) $\int_a^b f(x)\,dx$?
   (b) $-\int_a^b f(x)\,dx$?

A
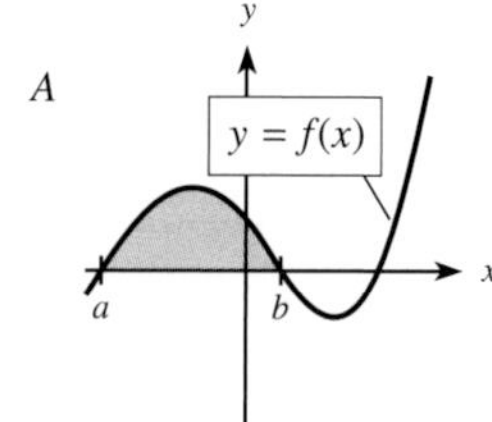

B
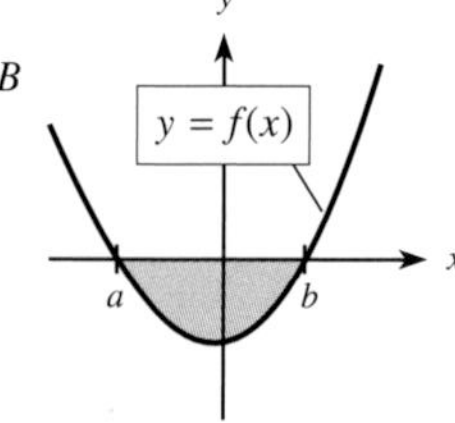

C
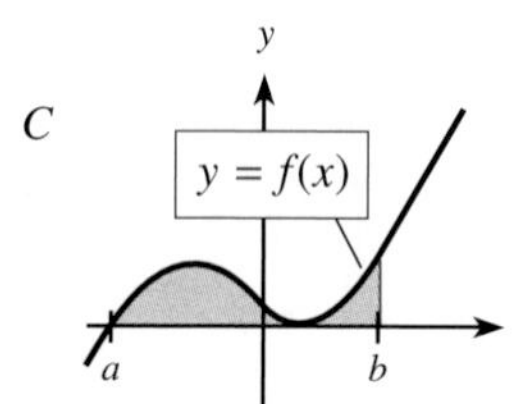

D
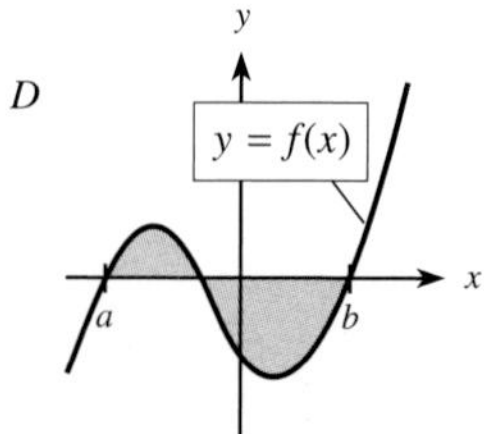

32. For which of the following functions $f(x)$ does

$$\int_0^2 f(x)\,dx$$

give the area between the graph of $f(x)$ and the x-axis from $x = 0$ to $x = 2$?
(a) $f(x) = x^2 + 1$
(b) $f(x) = -x^2$
(c) $f(x) = x - 1$

In Problems 33–36,
(a) write the integral that describes the area of the shaded region.
(b) find the area.

33.
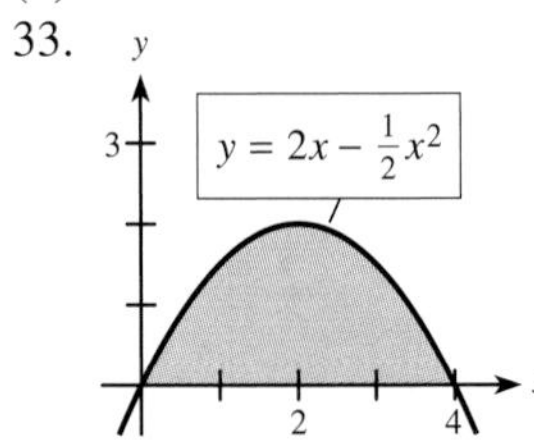

34.
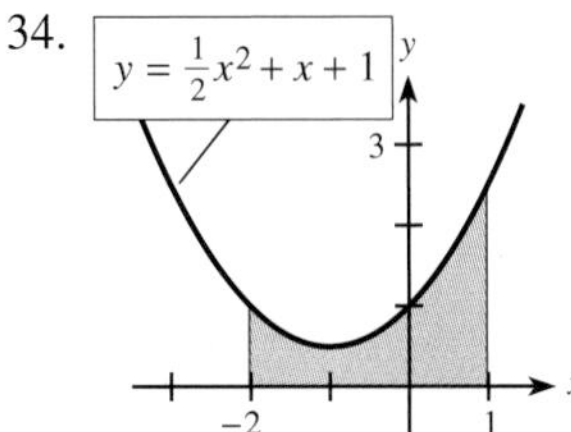

35.
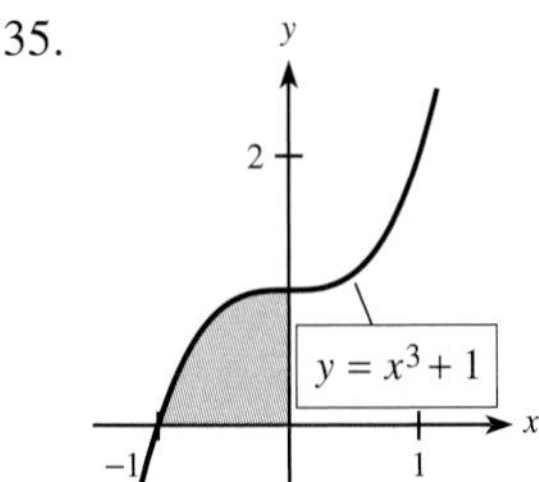

36.
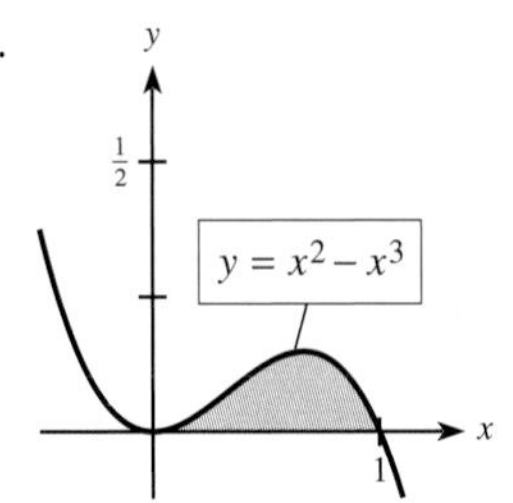

37. Find the area between the curve $y = -x^2 + 3x - 2$ and the x-axis from $x = 1$ to $x = 2$.
38. Find the area between the curve $y = x^2 + 3x + 2$ and the x-axis from $x = -1$ to $x = 3$.
39. Find the area between the curve $y = xe^{x^2}$ and the x-axis from $x = 1$ to $x = 3$.
40. Find the area between the curve $y = e^{-x}$ and the x-axis from $x = -1$ to $x = 1$.
41. How does $\int_{-1}^{-3} x\sqrt{x^2+1}\,dx$ compare with $\int_{-3}^{-1} x\sqrt{x^2+1}\,dx$?
42. If $\int_{-1}^{0} x^3\,dx = -\frac{1}{4}$ and $\int_0^1 x^3\,dx = \frac{1}{4}$, what does $\int_{-1}^{1} x^3\,dx$ equal?
43. If $\int_1^2 (2x - x^2)\,dx = \frac{2}{3}$ and $\int_2^4 (2x - x^2)\,dx = -\frac{20}{3}$, what does $\int_1^4 (x^2 - 2x)\,dx$ equal?
44. If $\int_1^2 (2x - x^2)\,dx = \frac{2}{3}$, what does $\int_1^2 6(2x - x^2)\,dx$ equal?

Applications

45. ***Total income*** The income from an oil change service chain can be considered as flowing continuously at an annual rate given by

$$f(t) = 10{,}000e^{0.02t} \qquad \text{(dollars/year)}$$

Find the total income for this chain over the first 2 years (from $t = 0$ to $t = 2$).

46. ***Total income*** Suppose that a vending machine service company models its income by assuming that money flows continuously into the machines, with the annual rate of flow given by

$$f(t) = 120e^{0.01t}$$

in thousands of dollars. Find the total income from the machines over the first 3 years.

Velocity of blood In Problems 47 and 48, the velocity of blood through a vessel is given by $v = K(R^2 - r^2)$, where K is the (constant) maximum velocity of the blood, R the (constant) radius of the vessel, and r the distance of the particular corpuscle from the center of the vessel. The rate of flow can be found by measuring the volume of blood that flows past a point in a given time period. This volume, V, is given by

$$V = \int_0^R v(2\pi r\,dr)$$

47. If $R = 0.30$ cm and $v = (0.30 - 3.33r^2)$ cm/s, find the volume.
48. Develop a general formula for V by evaluating

$$V = \int_0^R v(2\pi r\,dr)$$

using $v = K(R^2 - r^2)$.

Production In Problems 49 and 50, the rate of production of a new line of products is given by

$$\frac{dx}{dt} = 200\left[1 + \frac{400}{(t+40)^2}\right]$$

where x is the number of items produced and t is the number of weeks the products have been in production.
49. How many units were produced in the first 5 weeks?
50. How many units were produced in the sixth week?

51. ***Depreciation*** If the rate of depreciation of a building is given by $D'(t) = 3000(20 - t)$, $0 \le t \le 20$, what is the total depreciation of the building over the first 10 years ($t = 0$ to $t = 10$)?

52. ***Depreciation*** What is the total depreciation of the building in Problem 51 during the next 10 years ($t = 10$ to $t = 20$)?

53. ***Sales and advertising*** A store finds that its sales change at a rate given by

$$S'(t) = -3t^2 + 300t$$

where t is the number of days after an advertising campaign ends and $0 \le t \le 30$.
   (a) Find the total sales for the first week after the campaign ends ($t = 0$ to $t = 7$).
   (b) Find the total sales for the second week after the campaign ends ($t = 7$ to $t = 14$).

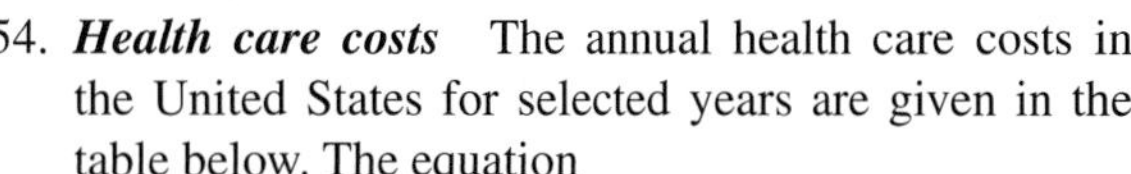

54. ***Health care costs*** The annual health care costs in the United States for selected years are given in the table below. The equation

$$y = 1.094x^2 - 11.103x + 49.562$$

models the annual health care costs, y (in billions of dollars), as a function of the years past 1960, x. Use a definite integral and this model to find the total cost of health care over the period 1960–1995.

Year	1960	1965	1970	1975	1980	1985
Cost	27.1	41.1	73.2	130.7	247.2	428.2

Year	1990	1992	1993	1994	1995
Cost	697.5	834.2	892.1	937.1	988.5

($ million)

Source: *World Almanac,* 1998

M 55. ***Dodge Viper acceleration*** Table 13.1(a) shows the time in seconds that a 1996 Dodge Viper GTS requires to reach various speeds up to 100 mph. Table 13.1(b) shows the same data, but with speeds in miles per second.
   (a) Fit a power model to the data in Table 13.1(b).
   (b) Use a definite integral from 0 to 9.2 of the function you found in (a) to find the distance traveled by the Viper as it went from 0 mph to 100 mph in 9.2 seconds.

TABLE 13.1

(a)		(b)	
Time (seconds)	*Speed (mph)*	*Time (seconds)*	*Speed (mi/s)*
1.7	30	1.7	0.00833
2.4	40	2.4	0.01111
3.2	50	3.2	0.01389
4.1	60	4.1	0.01667
5.8	70	5.8	0.01944
6.2	80	6.2	0.02222
7.8	90	7.8	0.02500
9.2	100	9.2	0.02778

Source: *Motor Trend,* April 1998

13.3 Area Between Two Curves

OBJECTIVES

- *To find the area between two curves*
- *To find the average value of a function*

APPLICATION PREVIEW

In economics, the **Lorenz curve** is used to represent the inequality of income distribution among different groups in the population of a country. The curve is constructed by plotting the cumulative percentage of families at or below a given income level and the cumulative percentage of total personal income received by these families. For example, the table shows the coordinates of some points on the Lorenz curve $y = L(x)$ that divide the income (for the United States in 1996) into 5 equal income levels (quintiles). The point (0.40, 0.142) is on the Lorenz curve because the families with income in the bottom 40% of the country received 14.2% of the total income in 1996. The graph of the Lorenz curve $y = L(x)$ is shown in Figure 13.9.

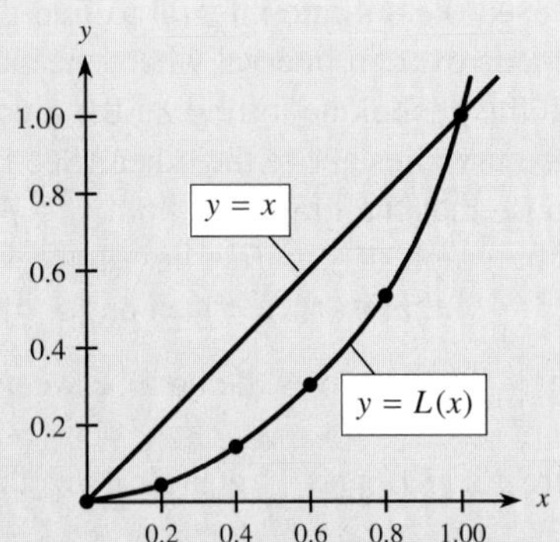

Figure 13.9

U.S. Income Distribution for 1996 (Points on the Lorenz Curve)

x, Cumulative Proportion of Families Below Income Level	*y = L(x), Cumulative Proportion of Total Income*
0	0
0.20	0.042
0.40	0.142
0.60	0.302
0.80	0.532
1	1

Source: *Statistical Abstract of the United States,* 1997

Equality of income would result if each family received an equal proportion of the total income, so that the bottom 20% would receive 20% of the total income, the bottom 40% would receive 40%, and so on. The Lorenz curve representing this would have the equation $y = x$.

The inequality of income distribution is measured by the **Gini coefficient** of income, which measures how far the Lorenz curve falls below $y = x$. It is defined as

$$\frac{\text{Area between } y = x \text{ and } y = L(x)}{\text{Area below } y = x}$$

Because the area of the triangle below $y = x$ and above the x-axis from $x = 0$ to $x = 1$ is $1/2$, the Gini coefficient of income is

$$\frac{\text{Area between } y = x \text{ and } y = L(x)}{1/2} = 2\cdot[\text{area between } y = x \text{ and } y = L(x)]$$

In this section we will use the definite integral to find the area between two curves. We will use the area between two curves to find the Gini coefficient of income and to find average cost, average revenue, average profit, and average inventory.

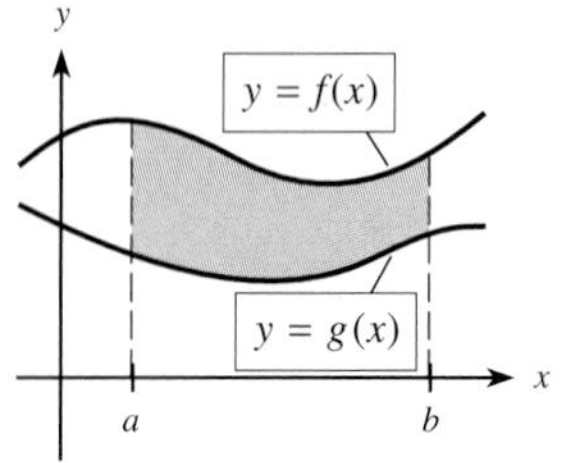

Figure 13.10

We have used the definite integral to find the area of the region between a curve and the x-axis over an interval where the curve lies above the x-axis. We can easily extend this technique to finding the area between two curves over an interval where one curve lies above the other. (See Figure 13.10.)

Suppose that the graphs of both $y = f(x)$ and $y = g(x)$ lie above the x-axis and that the graph of $y = f(x)$ lies above $y = g(x)$ throughout the interval from $x = a$ to $x = b$; that is, $f(x) \geq g(x)$ on $[a, b]$.

Then $\int_a^b f(x)\, dx$ gives the area between the graph of $y = f(x)$ and the x-axis (see Figure 13.11a), and $\int_a^b g(x)\, dx$ gives the area between the graph of $y = g(x)$ and the x-axis (see Figure 13.11b). As Figure 13.11(c) shows, the area of the region between the graphs of $y = f(x)$ and $y = g(x)$ is the difference of these two areas. That is,

$$\text{Area between the curves} = \int_a^b f(x)\, dx - \int_a^b g(x)\, dx$$

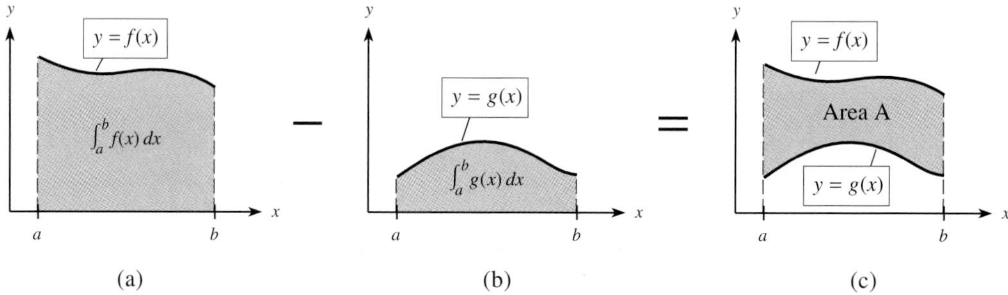

Figure 13.11 (a) (b) (c)

Although Figure 13.11(c) shows the graphs of both $y = f(x)$ and $y = g(x)$ lying above the x-axis, this difference of their integrals will always give the area between their graphs if both functions are continuous and if $f(x) \geq g(x)$ on the interval $[a, b]$. Using the fact that

$$\int_a^b f(x)\, dx - \int_a^b g(x)\, dx = \int_a^b [f(x) - g(x)]\, dx$$

we have the following result.

Area Between Two Curves If f and g are continuous functions on $[a, b]$ and if $f(x) \geq g(x)$ on $[a, b]$, then the area of the region bounded by $y = f(x)$, $y = g(x)$, $x = a$, and $x = b$ is

$$A = \int_a^b [f(x) - g(x)]\, dx$$

EXAMPLE 1

Find the area of the region bounded by $y = x^2 + 4$, $y = x$, $x = 0$, and $x = 3$.

Solution

We first sketch the graphs of the functions. The graphs of the region is shown in Figure 13.12. Because $y = x^2 + 4$ lies above $y = x$ in the interval from $x = 0$ to $x = 3$, the area is

$$A = \int_0^3 [(x^2 + 4) - x]\, dx = \frac{x^3}{3} + 4x - \frac{x^2}{2}\bigg|_0^3$$

$$= \left(9 + 12 - \frac{9}{2}\right) - (0 + 0 - 0)$$

$$= 16\frac{1}{2} \text{ square units}$$

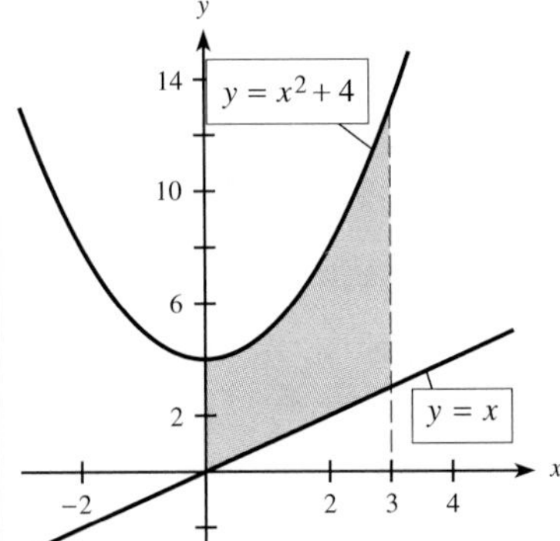

Figure 13.12

We are sometimes asked to find the area enclosed by two curves. In this case, we find the points of intersection of the curves to determine a and b.

EXAMPLE 2

Find the area enclosed by $y = x^2$ and $y = 2x + 3$.

Solution

We first find a and b by finding the x-coordinates of the points of intersection of the graphs. Setting the y-values equal gives

$$x^2 = 2x + 3$$
$$x^2 - 2x - 3 = 0$$
$$(x - 3)(x + 1) = 0$$
$$x = 3, \quad x = -1$$

Thus $a = -1$ and $b = 3$.

We next sketch the graphs of these functions on the same set of axes. Because the graphs do not intersect on the interval $(-1, 3)$, we can determine which function is larger on this interval by evaluating $2x + 3$ and x^2 at any value c where $-1 < c < 3$. Figure 13.13 on the next page shows the region between the graphs, with $2x + 3 \geq x^2$ from $x = -1$ to $x = 3$. The area of the enclosed region is

$$A = \int_{-1}^{3} [(2x + 3) - x^2]\,dx = x^2 + 3x - \frac{x^3}{3}\bigg|_{-1}^{3}$$

$$= (9 + 9 - 9) - \left(1 - 3 + \frac{1}{3}\right)$$

$$= 10\tfrac{2}{3} \text{ square units}$$

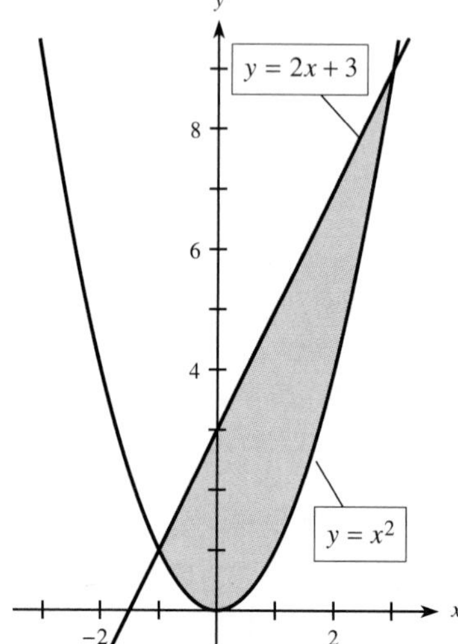

Figure 13.13

Some graphs enclose two or more regions because they have more than two points of intersection.

EXAMPLE 3

Find the area of the region enclosed by the graphs of

$$y = f(x) = x^3 - x^2 \quad \text{and} \quad y = g(x) = 2x$$

Solution

To find the points of intersection of the graphs, we set the y-values equal and solve for x.

$$x^3 - x^2 = 2x$$

$$x^3 - x^2 - 2x = 0$$

$$x(x - 2)(x + 1) = 0$$

$$x = 0, \quad x = 2, \quad x = -1$$

Graphing these functions between $x = -1$ and $x = 2$, we see that for any x-value in the interval $(-1, 0)$, $f(x) \geq g(x)$, so $f(x) \geq g(x)$ for the region enclosed by the curves from $x = -1$ to $x = 0$. But evaluating the functions for any x-value in the interval $(0, 2)$ shows that $f(x) \leq g(x)$ for the region enclosed by the curves from $x = 0$ to $x = 2$. See Figure 13.14.

Thus we need one integral to find the area of the region from $x = -1$ to $x = 0$ and a second integral to find the area from $x = 0$ to $x = 2$. The area is found by summing these two integrals.

$$\begin{aligned} A &= \int_{-1}^{0} [(x^3 - x^2) - (2x)]\, dx + \int_{0}^{2} [(2x) - (x^3 - x^2)]\, dx \\ &= \int_{-1}^{0} (x^3 - x^2 - 2x)\, dx + \int_{0}^{2} (2x - x^3 + x^2)\, dx \\ &= \left(\frac{x^4}{4} - \frac{x^3}{3} - x^2\right)\Bigg|_{-1}^{0} + \left(x^2 - \frac{x^4}{4} + \frac{x^3}{3}\right)\Bigg|_{0}^{2} \\ &= \left[(0) - \left(\frac{1}{4} - \frac{-1}{3} - 1\right)\right] + \left[\left(4 - \frac{16}{4} + \frac{8}{3}\right) - (0)\right] = \frac{37}{12} \end{aligned}$$

Thus the area between the curves is $\frac{37}{12}$ square units.

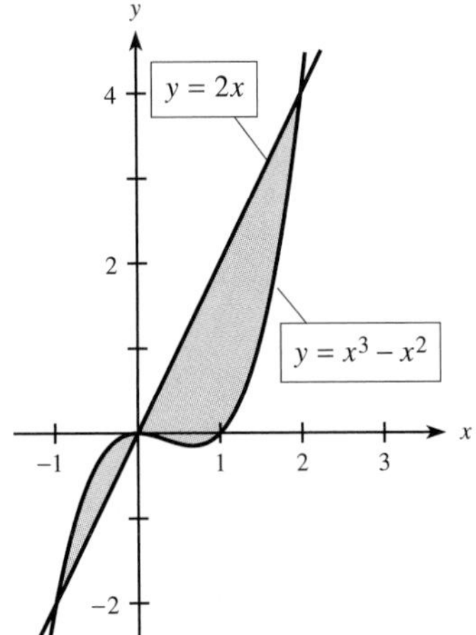

Figure 13.14

Graphing Utilities

Most graphing utilities have a numerical integration feature, and some can perform both symbolic and numerical integration.

A graphing utility could be used to find the area enclosed by the graphs of $y = f(x) = x^3 - x^2$ and $y = g(x) = 2x$, found in Example 3. By using SOLVER, INTERSECT, or ZOOM and TRACE, we can find that the curves intersect at $x = -1$, $x = 0$, and $x = 2$. From the graph, shown in Figure 13.13, we see that the curves enclose two regions, with $f(x) > g(x)$ for $-1 < x < 0$ and $g(x) > f(x)$ for $0 < x < 2$. Thus the area is given by

$$\int_{-1}^{0} [(x^3 - x^2) - (2x)]\, dx + \int_{0}^{2} [(2x) - (x^3 - x^2)]\, dx$$

By using the numerical integration feature of a graphing utility, we can find the integral (and the area) to be $0.4166666667 + 2.666666667 = 3.083333333$. This is the same (to nine decimal places) as the decimal representation of $37/12$, found in Example 3.

CHECKPOINT

1. True or false:
   (a) Over the interval $[a, b]$, the area between the continuous functions $f(x)$ and $g(x)$ is

$$\int_a^b [f(x) - g(x)]\,dx$$

   (b) If $f(x) \geq g(x)$ and the area between $f(x)$ and $g(x)$ is given by

$$\int_a^b \Big[f(x) - g(x)\Big]\,dx$$

   then $x = a$ and $x = b$ represent the left and right boundaries, respectively, of the region.
   (c) To find points of intersection of $f(x)$ and $g(x)$, solve $f(x) = g(x)$.
2. Consider the functions $f(x) = x^2 + 3x - 9$ and $g(x) = \frac{1}{4}x^2$.
   (a) Find the points of intersection of $f(x)$ and $g(x)$.
   (b) Determine which function is greater than the other between the points found in (a).
   (c) Set up the integral used to find the area between the curves in the interval between the points found in (a).
   (d) Find the area.

EXAMPLE 4

As we noted in the Application Preview, the inequality of income distribution is measured by the Gini coefficient of income, which is defined as

$$\frac{\text{Area between } y = x \text{ and } y = L(x)}{\text{Area below } y = x} = \frac{\int_0^1 [x - L(x)]\,dx}{1/2} = 2\int_0^1 [x - L(x)]\,dx$$

We can use the data in the Application Preview and a graphing utility to model the Lorenz curve for United States income in 1996. By using a power model, we obtain $L(x) = 0.8743x^{1.9262}$.

(a) Use this $L(x)$ to find the Gini coefficient of income for 1996.
(b) If the Gini coefficient of income for 1991 is 0.374, during which year is the distribution of income more nearly equal?

Solution

(a) The Gini coefficient of income for 1996 is

$$2\int_0^1 (x - L(x))\,dx = 2\int_0^1 (x - 0.8743x^{1.9262})\,dx$$

$$= 2\left[\frac{x^2}{2} - \frac{0.8743x^{2.9262}}{2.9262}\right]\Bigg|_0^1 = 2\left[\frac{1}{2} - 0.2988\right] = 0.4024$$

(b) Absolute equality of income would occur if the Gini coefficient of income were 0; and smaller coefficients indicate more nearly equal incomes. Thus the distribution of income was more nearly equal in 1991.

If the graph of $y = f(x)$ lies on or above the x-axis from $x = a$ to $x = b$, then the area between the graph and the x-axis is

$$A = \int_a^b f(x)\,dx \qquad \text{(See Figure 13.15a.)}$$

The area A is also the area of a rectangle with base equal to $b - a$ and height equal to the *average value* (or average height) of the function $y = f(x)$ (see Figure 13.15b).

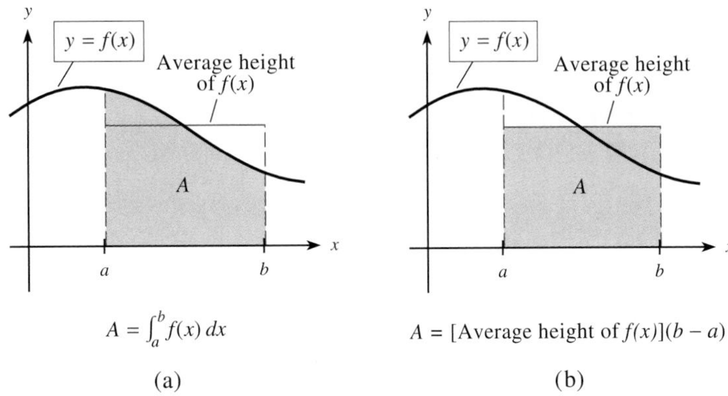

Figure 13.15

Thus the average value of the function is

$$\frac{A}{b-a} = \frac{1}{b-a}\int_a^b f(x)\,dx$$

Even if $f(x) \le 0$ on all or part of the interval $[a, b]$, we can find the average value by using the integral. Thus we have the following.

Average Value The average value of a continuous function $y = f(x)$ over the interval $[a, b]$ is

$$\text{Average value} = \frac{1}{b-a}\int_a^b f(x)\,dx$$

EXAMPLE 5

Suppose that the cost function for a product is $C(x) = 400 + x + 0.3x^2$.

(a) What is the average value of $C(x)$ for $x = 10$ to $x = 20$ units?
(b) Find the average cost per unit if 40 units are produced.

Solution

(a) The average value of $C(x)$ is

$$\frac{1}{20-10}\int_{10}^{20} (400 + x + 0.3x^2)\,dx = \frac{1}{10}\left(400x + \frac{x^2}{2} + 0.1x^3\right)\Big|_{10}^{20}$$

$$= \frac{1}{10}[(8000 + 200 + 800) - (4000 + 50 + 100)]$$

$$= 485 \quad \text{(dollars)}$$

(b) The average cost per unit if 40 units are produced is the average cost function evaluated at $x = 40$. The average cost function is

$$\overline{C}(x) = \frac{C(x)}{x} = \frac{400}{x} + 1 + 0.3x$$

Thus the average cost per unit if 40 units are produced is

$$\overline{C}(40) = \frac{400}{40} + 1 + 0.3(40) = 23 \quad \text{(dollars)}$$

CHECKPOINT

3. Find the average value of $f(x) = x^2 - 4$ over $[-1, 3]$.

EXAMPLE 6

Consider the functions $f(x) = x^2 - 4$ and $g(x) = x^3 - 4x$. For each function, do the following.

(a) Graph the function on the interval $[-2, 2]$.

(b) On the graph, "eyeball" the average value (height) of each function on $[-2, 2]$.

(c) Compute the average value of the function over the interval $[-2, 2]$.

Solution

For $f(x) = x^2 - 4$.

(a) The graph of $f(x) = x^2 - 4$ is shown in Figure 13.16(a).

(b) The average height of $f(x)$ may be near -2.

(c) The average value of $f(x)$ over the integral is given by

$$\begin{aligned}\frac{1}{2-(-2)}\int_{-2}^{2}(x^2-4)\,dx &= \frac{1}{4}\left[\frac{x^3}{3} - 4x\right]_{-2}^{2} \\ &= \left(\frac{8}{12} - 2\right) - \left(-\frac{8}{12} + 2\right) \\ &= \frac{4}{3} - 4 = -\frac{8}{3} = -2\tfrac{2}{3}\end{aligned}$$

For $g(x) = x^3 - 4x$.

(a) The graph of $g(x) = x^3 - 4x$ is shown in Figure 13.16(b).

(b) The average height of $g(x)$ graph may be approximately 0.

(c) The average value of $g(x)$ is given by

$$\begin{aligned}\frac{1}{2-(-2)}\int_{-2}^{2}(x^3-4x)\,dx &= \frac{1}{4}\left[\frac{x^4}{4} - \frac{4x^2}{2}\right]_{-2}^{2} \\ &= \left(\frac{16}{16} - \frac{16}{8}\right) - \left(\frac{16}{16} - \frac{16}{8}\right) \\ &= -1 + 1 = 0\end{aligned}$$

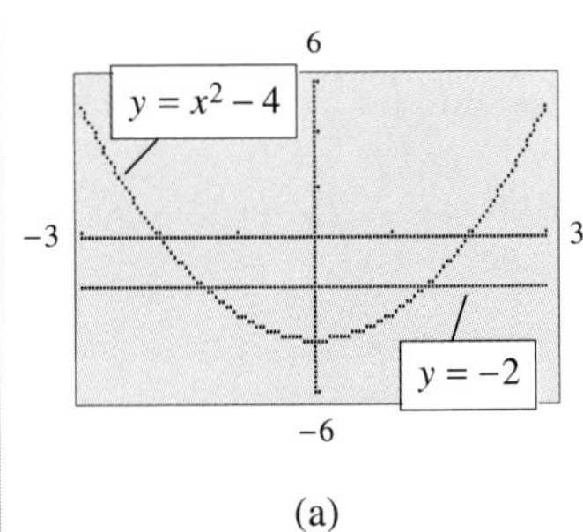

(a)

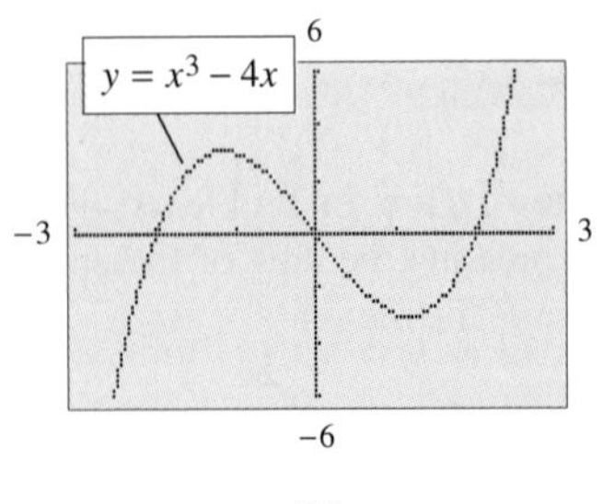

(b)

Figure 13.16

CHECKPOINT SOLUTIONS

1. (a) False. This is true only if $f(x) \geq g(x)$ over $[a, b]$.

   (b) True (c) True

2. (a) Solve $f(x) = g(x)$, or $x^2 + 3x - 9 = \frac{1}{4}x^2$.

$$\frac{3}{4}x^2 + 3x - 9 = 0$$
$$3x^2 + 12x - 36 = 0$$
$$x^2 + 4x - 12 = 0$$
$$(x + 6)(x - 2) = 0$$
$$x = -6 \mid x = 2$$

   The points of intersection are $(-6, 9)$ and $(2, 1)$.

   (b) Evaluating $g(x)$ and $f(x)$ at any point in the interval $(-6, 2)$ shows that $g(x) > f(x)$, so $g(x) \geq f(x)$ on $[-6, 2]$.

   (c) $A = \displaystyle\int_{-6}^{2} \left[\frac{1}{4}x^2 - (x^2 + 3x - 9)\right] dx$

   (d) $A = \displaystyle\frac{x^3}{12} - \frac{x^3}{3} - \frac{3x^2}{2} + 9x\Big|_{-6}^{2} = 64$ square units

3. $$\frac{1}{3 - (-1)} \int_{-1}^{3} (x^2 - 4)\, dx = \frac{1}{4}\left(\frac{x^3}{3} - 4x\right)\Bigg|_{-1}^{3}$$
$$= \frac{1}{4}\left[(9 - 12) - \left(-\frac{1}{3} + 4\right)\right]$$
$$= \frac{1}{4}\left(\frac{-20}{3}\right) = -\frac{5}{3}$$

EXERCISE 13.3

For each shaded region in Problems 1–6, (a) form the integral that represents the area of the shaded region and (b) find the area of the region.

1.

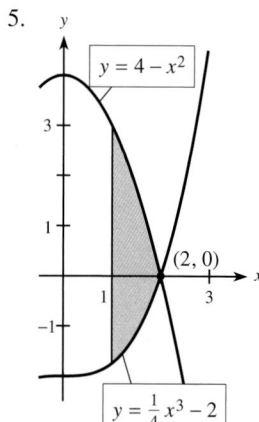

2.

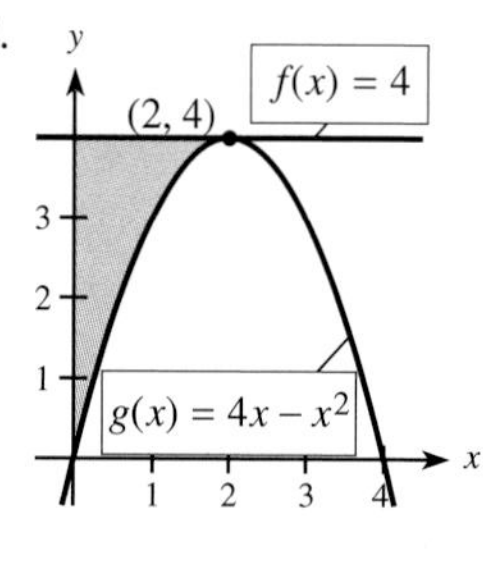

3.

4.

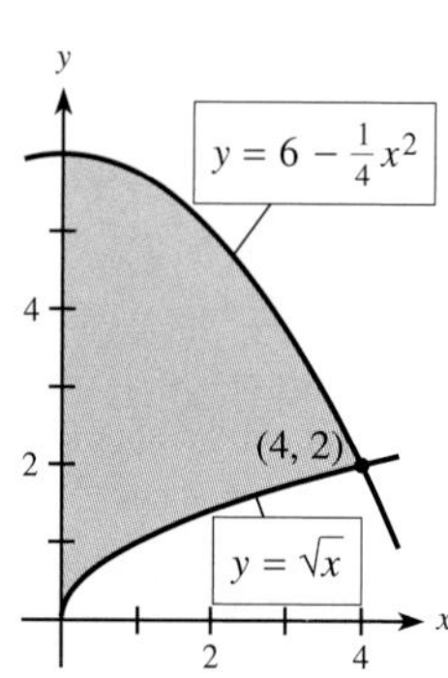

5.

6.

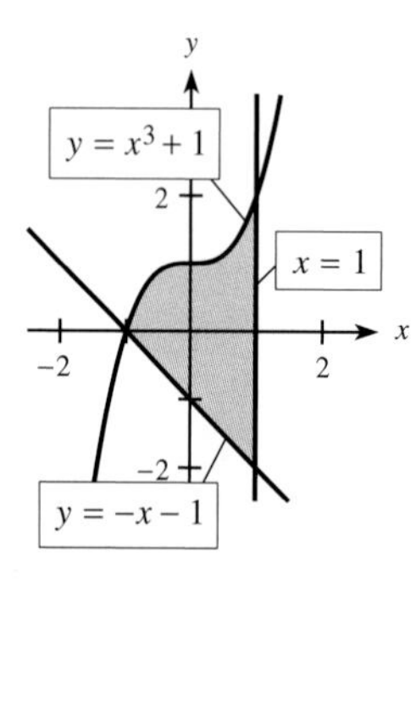

For each shaded region in Problems 7–12, (a) find the points of intersection of the curves, (b) form the integral that represents the area of the shaded region, and (c) find the area of the shaded region.

7.

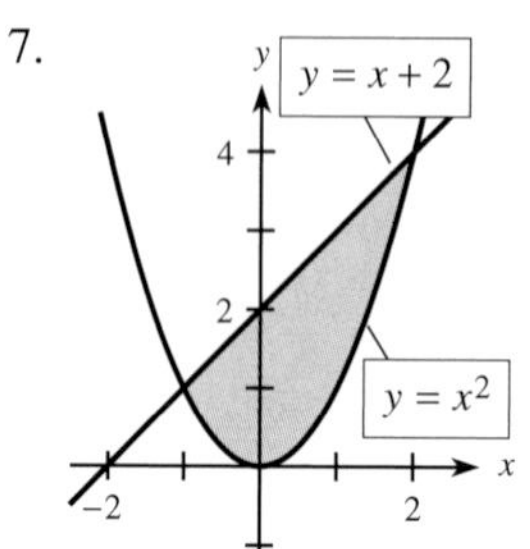

8.

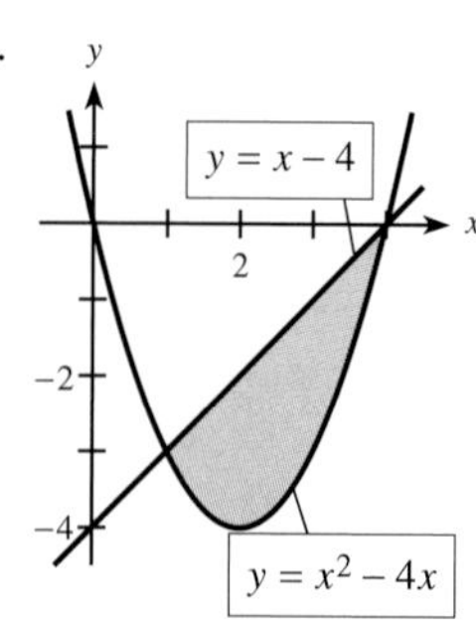

9.

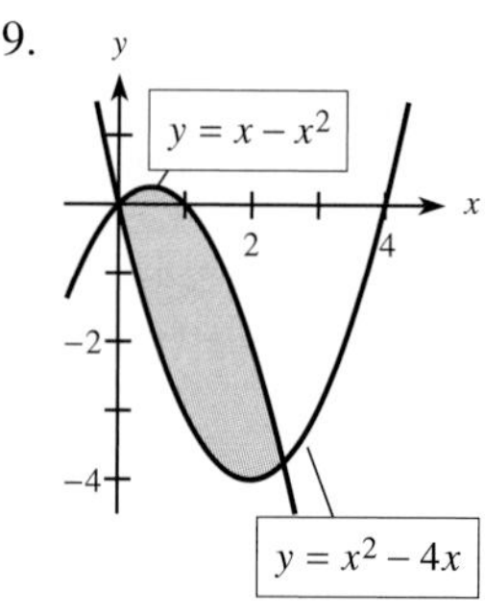

10.

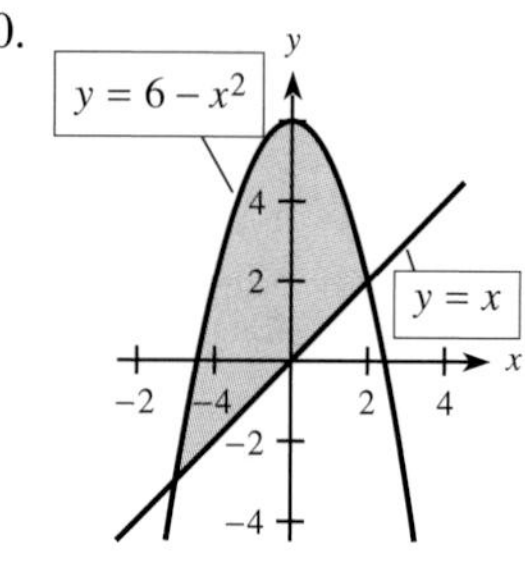

11.

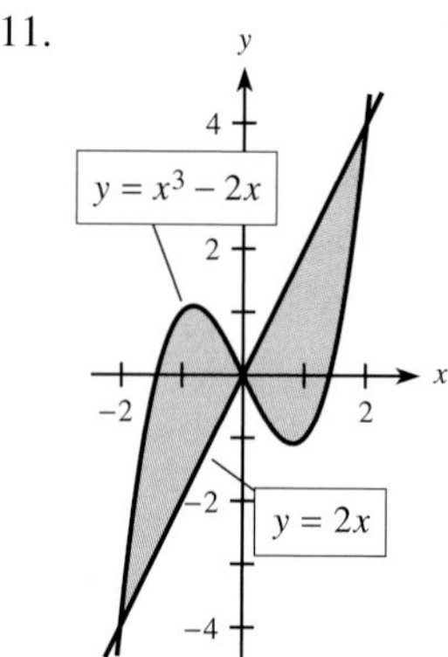

12.

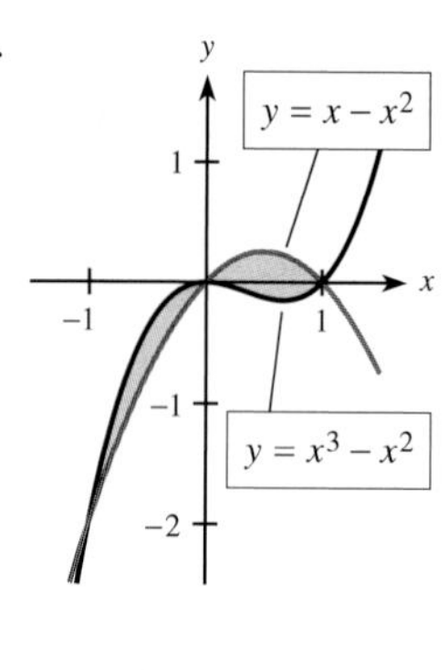

In Problems 13–26, equations are given whose graphs enclose a region. In each problem, find the area of the region.

13. $f(x) = x^2 + 2;\quad g(x) = -x^2;\quad x = 0; x = 2$
14. $f(x) = x^2;\quad g(x) = -\frac{1}{10}(10 + x);\quad x = 0; x = 3$
15. $y = x^3 - 1;\quad y = x - 1;$ to the right of the y-axis
16. $y = x^2 - 2x + 1;\quad y = x^2 - 5x + 4;\quad x = 2$
17. $y = \frac{1}{2}x^2;\quad y = x^2 - 2x$
18. $y = x^2;\quad y = 4x - x^2$
19. $h(x) = x^2;\quad k(x) = \sqrt{x}$
20. $g(x) = 1 - x^2;\quad h(x) = x^2 + x$
21. $f(x) = x^3;\quad g(x) = x^2 + 2x$
22. $f(x) = x^3;\quad g(x) = 2x - x^2$
23. $f(x) = \frac{3}{x};\quad g(x) = 4 - x$

24. $f(x) = \frac{6}{x}$; $g(x) = -x - 5$

25. $y = \sqrt{x + 3}$; $x = -3; y = 2$

26. $y = \sqrt{4 - x}$; $x = 4$; $y = 3$

In Problems 27–32, find the average value of each function over the given interval.

27. $f(x) = 9 - x^2$ over $[0, 3]$

28. $f(x) = 2x - x^2$ over $[0, 2]$

29. $f(x) = x^3 - x$ over $[-1, 1]$

30. $f(x) = \frac{1}{2}x^3 + 1$ over $[-2, 0]$

31. $f(x) = \sqrt{x} - 2$ over $[1, 4]$

32. $f(x) = \sqrt[3]{x}$ over $[-8, -1]$

33. Use a graphing calculator or computer to find the area between the curves $y = f(x) = x^3 - 4x$ and $y = g(x) = x^2 - 4$.

34. Use a graphing calculator or computer software to find the area between the curves $f(x) = \sqrt[3]{x}$ and $g(x) = x^3 - x$.

Applications

35. ***Income distribution*** Changes in tax laws in the United States during the 1980s were widely thought to help the wealthy at the expense of the poor. The Lorenz curves for the income distribution in 1980 and in 1990 are given below. Find the Gini coefficient of income for both years and determine whether the distribution of income is more or less nearly equal in 1990 than it was in 1980. What was the effect of the tax laws?

1980	$y = 0.916x^{1.821}$
1990	$y = 0.896x^{1.878}$

36. ***Income distribution*** The Lorenz curves for the income distribution in the United States in 1950 and in 1970 are given below. Find the Gini coefficient of income for both years and compare the distributions of income for these years.

1950	$y = 0.925x^{1.891}$
1970	$y = 0.920x^{1.783}$

37. ***Income distribution*** Data from the U.S. Bureau of the Census yields the Lorenz curves given below for income distribution among blacks and among whites in the United States for 1996. Find the Gini coefficient of income for both groups and determine in which the group income is more nearly equally distributed.

Whites	$y = 0.8693x^{1.8556}$
Blacks	$y = 0.8693x^{2.0982}$

38. ***Income distribution*** In an effort to make the distribution of income more nearly equal, the government of a country passes a tax law that changes the Lorenz curve for one year from $y = 0.99x^{2.1}$ to $y = 0.32x^2 + 0.68x$ for the next year. Find the Gini coefficient of income for both years and compare the distributions of income before and after the tax law is passed. Interpret the result.

39. ***Average profit*** For the product whose total cost and total revenue are shown in the figure, represent total revenue by $R(x)$ and total cost by $C(x)$ and write an integral that gives the average profit for the product over the interval from x_0 to x_1.

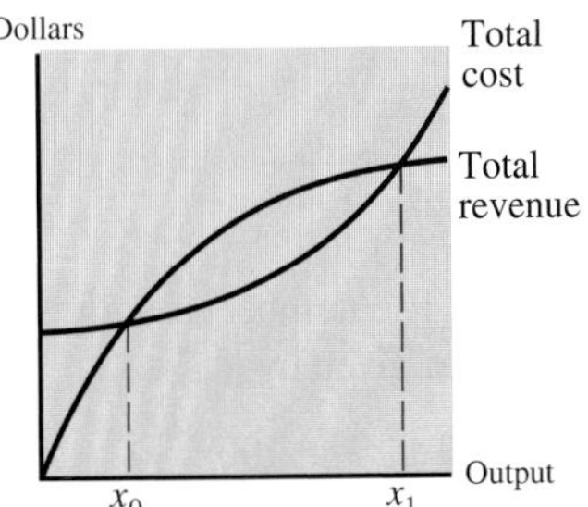

40. ***Sales and advertising*** The figure shows the sales growth rates under different levels of distribution and advertising from a to b. Set up an integral to determine the extra sales growth if \$4 million is used in advertising rather than \$2 million.

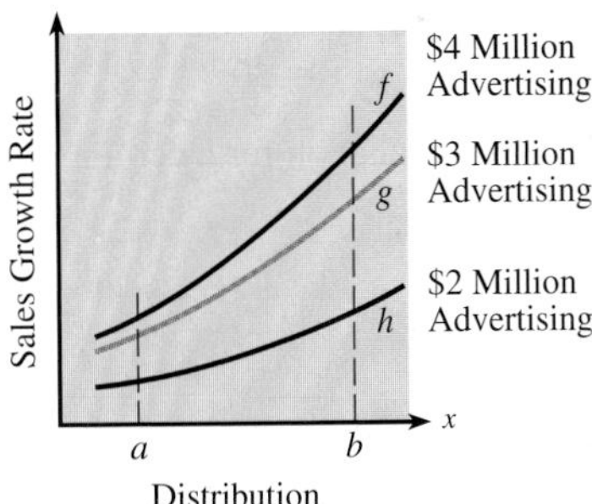

41. ***Cost*** The cost of producing x units of a certain item is $C(x) = x^2 + 400x + 2000$.

(a) Use $C(x)$ to find the average cost of producing 1000 units.

(b) Find the average value of the cost function $C(x)$ over the interval from 0 to 1000.

42. ***Inventory management*** The figure shows how an inventory of a product is depleted each quarter of a given year. What is the average inventory per month for the first 3 months for this product? (Assume that the graph is a line joining (0, 1300) and (3,100).)

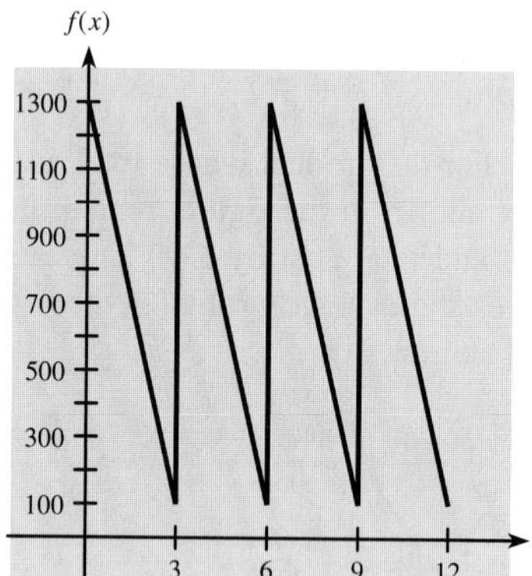

43. ***Sales and advertising*** The number of daily sales of a product was found to be given by

$$S = 100xe^{-x^2} + 100$$

x days after the start of an advertising campaign for this product.
   (a) Find the average daily sales during the first 20 days of the campaign—that is, from $x = 0$ to $x = 20$.
   (b) If no new advertising campaign is begun, what is the average number of sales per day for the next 10 days (from $x = 20$ to $x = 30$)?

44. ***Demand*** The demand function for a certain product is given by

$$p = 500 + \frac{1000}{q + 1}$$

where p is the price and q is the number of units demanded. Find the average price as demand ranges from 49 to 99 units.

45. ***Interest rates*** The equation $y = 0.00144100x^4 - 0.0593608x^3 + 0.740741x^2 - 2.573133x + 6.941375$ describes interest rates for the years 1970 to 1989, where $x = 0$ in 1970. Use a definite integral from 0 to 19 to compute the average interest rate over the period.

46. ***Total income*** Suppose that the income from a slot machine in a casino flows continuously at a rate

$$f(t) = 100e^{0.1t}$$

where t is the time in hours since the casino opened. Then the total income during the first 10 hours is given by

$$\int_0^{10} 100e^{0.1t}\,dt$$

Find the average income over the first 10 hours.

47. ***Drug levels in the blood*** A drug manufacturer has developed a time-release capsule with the number of milligrams of the drug in the bloodstream given by

$$S = 30x^{18/7} - 240x^{11/7} + 480x^{4/7}$$

where x is in hours and $0 \le x \le 4$. Find the average number of milligrams of the drug in the bloodstream for the first 4 hours after a capsule is taken.

13.4 Applications of Definite Integrals in Business and Economics

OBJECTIVES

- *To use definite integrals to find total income, present value, and future value of continuous income streams*
- *To use definite integrals to find the consumer's surplus*
- *To use definite integrals to find the producer's surplus*

APPLICATION PREVIEW

The definite integral can be used in a number of applications in business and economics. For example, the definite integral can be used to find the total income over a fixed number of years from a **continuous income stream.** The definite integral can also be used to find the **present value** and the **future value** of a continuous income stream. And it can be used to find the **consumer's surplus** and **producer's surplus** when the demand function and the supply function for a product are known.

Continuous Income Streams

An oil company's profits depend on the amount of oil that can be pumped from a well. Thus we can consider a pump at an oil field as producing a **continuous stream of income** for the owner. Because both the pump and the oil field "wear out" with time, the continuous stream of income is a function of time. Suppose $f(t)$ is the (annual) *rate* of flow of income from this pump; then we can find the total income from the rate of income by using integration. In particular, the total income for k years is given by

$$\text{Total income} = \int_0^k f(t)\,dt$$

EXAMPLE 1

A small oil company considers the continuous pumping of oil from a well as a continuous income stream with its annual rate of flow at time t given by

$$f(t) = 600e^{-0.2t}$$

in thousands of dollars. Find an estimate of the total income from this well over the next 10 years.

Solution

$$\begin{aligned}\text{Total income} &= \int_0^{10} f(t)\,dt = \int_0^{10} 600e^{-0.2t}\,dt \\ &= \frac{600}{-0.2}e^{-0.2t}\Big|_0^{10} \\ &\approx 2594 \quad \text{(to the nearest integer)}\end{aligned}$$

Thus the total income is approximately \$2,594,000.

In addition to the total income from a continuous income stream, the **present value** of the stream is also important. The present value is the value today of a continuous income stream that will be providing income in the future. The present value is useful in deciding when to replace machinery (such as the oil pump in the example) or what new equipment to select.

To find the present value of a continuous stream of income with rate of flow $f(t)$, we first graph the function $f(t)$ and divide the time interval from 0 to k into n subintervals of width Δt_i, $i = 1$ to n.

The total amount of income is the area under this curve between $t = 0$ and $t = k$. We can approximate the amount of income in each subinterval by finding the area of the rectangle in that subinterval. (See Figure 13.17.)

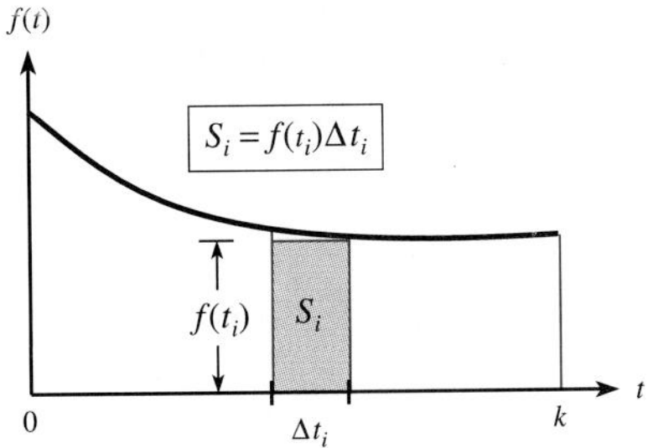

Figure 13.17

We have shown that the future value S that accrues if $\$P$ is invested for t years at an annual rate r, compounded continuously, is

$$S = Pe^{rt}$$

Thus the present value of the investment that yields the single payment of $\$S$ after t years is

$$P = \frac{S}{e^{rt}} = Se^{-rt}$$

The contribution to S in the ith subinterval is $S_i = f(t_i)\ \Delta t_i$, and the present value of this amount is

$$P_i = f(t_i)\ \Delta t_i e^{-rt_i}$$

Thus the total present value of S can be approximated by

$$\sum_{i=1}^{n} f(t_i)\ \Delta t_i e^{-rt_i}$$

This approximation improves as $\Delta t_i \to 0$ with the present value given by

$$\lim_{\Delta t_i \to 0} \sum_{i=1}^{n} f(t_i)\ \Delta t_i e^{-rt_i}$$

This limit gives the present value as a definite integral.

Present Value of a Continuous Income Stream

If $f(t)$ is the rate of continuous income flow earning interest at rate r, compounded continuously, then the present value of the continuous income stream is

$$\text{Present value} = \int_0^k f(t)e^{-rt}\,dt$$

where $t = 0$ to $t = k$ is the time interval.

EXAMPLE 2

Suppose that the oil company in Example 1 is planning to sell the well because of its remote location. Suppose further that the company wants to use the present value of the well over the next 10 years to help establish its selling price. If the company determines that the annual rate of flow is

$$f(t) = 600e^{-0.2(t+5)}$$

in thousands of dollars, and if money is worth 10%, compounded continuously, find this present value.

Solution

$$\begin{aligned}\text{Present value} &= \int_0^{10} f(t)e^{-rt}\,dt \\ &= \int_0^{10} 600e^{-0.2(t+5)}e^{-0.1t}\,dt \\ &= \int_0^{10} 600e^{-0.3t-1}\,dt \\ &= \frac{600}{-0.3}e^{-0.3t-1}\Big|_0^{10} \\ &= -2000(e^{-4} - e^{-1}) \\ &\approx 699 \quad \text{(to the nearest integer)}\end{aligned}$$

Thus the present value is \$699,000.

Future Value of a Continuous Income Stream If $f(t)$ is the rate of continuous income flow for k years earning interest at rate r, compounded continuously, then the future value of the continuous income stream is

$$FV = e^{rk}\int_0^k f(t)e^{-rt}\,dt$$

EXAMPLE 3

If the rate of flow of income from an asset is $1000e^{0.02t}$ and if the income is invested at 6%, compounded continuously, find the future value of the asset 4 years from now.

Solution

The future value is given by

$$\begin{aligned}FV &= e^{rk}\int_0^k f(t)e^{-rt}\,dt \\ &= e^{(0.06)4}\int_0^4 1000e^{0.02t}e^{-0.06t}\,dt = e^{0.24}\int_0^4 1000e^{-0.04t}\,dt \\ &= e^{0.24}\left[-25000e^{-0.04t}\right]_0^4 = -25000e^{0.24}\left[e^{-0.16} - 1\right] \\ &\approx 4699.05\end{aligned}$$

CHECKPOINT

1. Suppose that a continuous income stream has an annual rate of flow given by $f(t) = 5000e^{-0.01t}$, and suppose that money is worth 7%, compounded continuously. Create the integral used to find
   (a) the total income for the next 5 years.
   (b) the present value for the next 5 years.
   (c) the future value 5 years from now.

Consumer's Surplus

Suppose that the demand for a product is given by $p = f(x)$ and that the supply of the product is described by $p = g(x)$. The price p_1 where the graphs of these functions intersect is the **equilibrium price** (see Figure 13.18). As the demand curve shows, some consumers (but not all) would be willing to pay more than $\$p_1$ for the product.

For example, some consumers would be willing to buy x_3 units if the price were $\$p_3$. Those consumers willing to pay more than $\$p_1$ are benefiting from the lower price. The total gain for all those consumers willing to pay more than $\$p_1$ is called the **consumer's surplus,** and under proper assumptions the area of the shaded region in Figure 13.18 represents this consumer's surplus.

Looking at Figure 13.19, we see that if the demand curve has equation $p = f(x)$, the consumer's surplus is given by the area between $f(x)$ and the x-axis from 0 to x_1, *minus* the area of the rectangle denoted *TR:*

$$CS = \int_0^{x_1} f(x)\,dx - p_1x_1$$

Note that p_1x_1 is the area of the rectangle that represents the total revenue (see Figure 13.19).

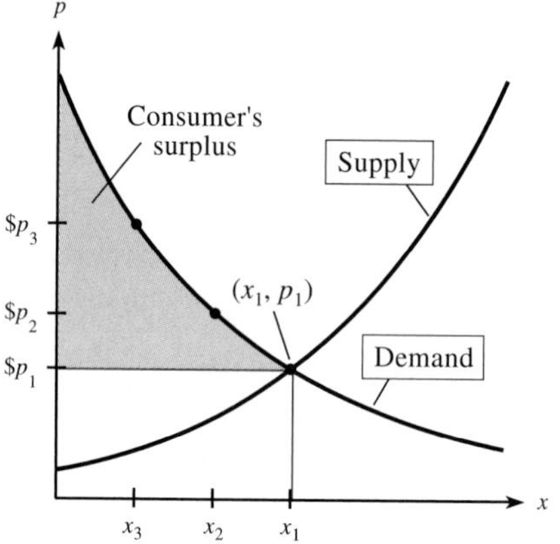

Figure 13.18

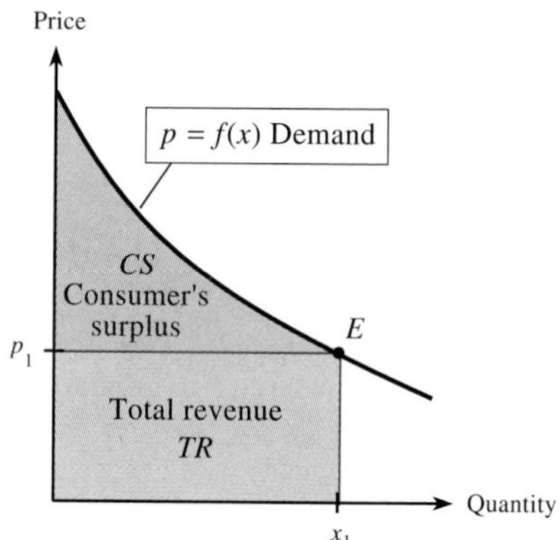

Figure 13.19

EXAMPLE 4

The demand function for a product is $p = 100/(x + 1)$. If the equilibrium price is \$20, what is the consumer's surplus?

Solution

We must first find the quantity that will be purchased at this price. Letting $p = 20$ and solving for x, we get

$$\begin{aligned} 20 &= \frac{100}{x + 1} \\ 20(x + 1) &= 100 \\ x + 1 &= 5 \\ x &= 4 \end{aligned}$$

Thus the equilibrium point is (4, 20). The consumer's surplus is given by the formula

$$CS = \int_0^{x_1} f(x)\,dx - p_1x_1 = \int_0^4 \frac{100}{x+1}\,dx - 20 \cdot 4$$

$$= 100 \ln|x+1|\Big|_0^4 - 80$$

$$= 100(\ln 5 - \ln 1) - 80$$

$$\approx 100(1.6094 - 0) - 80$$

$$= 160.94 - 80$$

$$= 80.94$$

The consumer's surplus is \$80.94.

EXAMPLE 5

A product's demand function is $p = \sqrt{49 - 6x}$ and its supply function is $p = x + 1$. Find the equilibrium point and the consumer's surplus there.

Solution

We can determine the equilibrium point by solving the two equations simultaneously.

$$\sqrt{49 - 6x} = x + 1$$

$$49 - 6x = (x + 1)^2$$

$$0 = x^2 + 8x - 48$$

$$0 = (x + 12)(x - 4)$$

$$x = 4 \quad \text{or} \quad x = -12$$

Thus the equilibrium quantity is 4 and the equilibrium price is \$5 (because $x = -12$ is not a solution). The graphs of the supply and demand functions are shown in Figure 13.20.

The consumer's surplus is given by

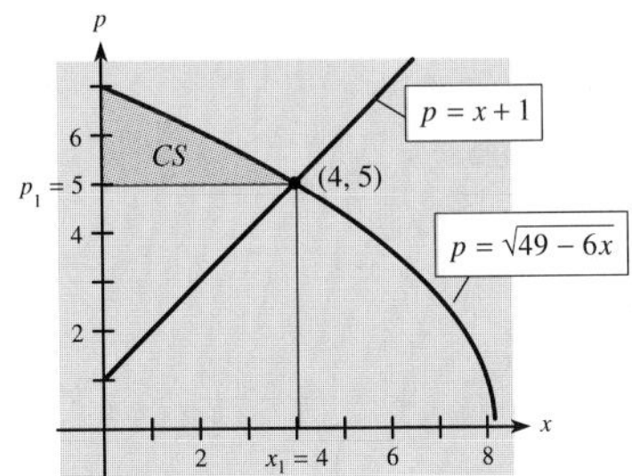

Figure 13.20

$$CS = \int_0^4 f(x)\,dx - p_1x_1$$

$$= \int_0^4 \sqrt{49 - 6x}\,dx - 5 \cdot 4$$

$$= -\frac{1}{6}\int_0^4 \sqrt{49 - 6x}\,(-6\,dx) - 20$$

$$= -\frac{1}{9}(49 - 6x)^{3/2}\Big|_0^4 - 20$$

$$= -\frac{1}{9}[(25)^{3/2} - (49)^{3/2}] - 20$$

$$= -\frac{1}{9}(125 - 343) - 20$$

$$\approx 24.22 - 20 = 4.22$$

The consumer's surplus is \$4.22.

EXAMPLE 6

If a monopoly has a total cost function $C(x) = 60 + 2x^2$ for a product whose demand is given by $p = 30 - x$, find the consumer's surplus at the point where the monopoly has maximum profit.

Solution

We must first find the point where the profit function is maximized. Because the demand for x units is $p = 30 - x$, the total revenue is

$$R(x) = (30 - x)x = 30x - x^2$$

Thus the profit function is

$$P(x) = R(x) - C(x)$$
$$P(x) = 30x - x^2 - (60 + 2x^2)$$
$$P(x) = 30x - 60 - 3x^2$$

Then $P'(x) = 30 - 6x$. So $0 = 30 - 6x$ has the solution $x = 5$.

Because $P''(5) = -6 < 0$, the profit for the monopolist is maximized when $x = 5$ units are sold at price $p = 30 - x = 25$.

The consumer's surplus at $x = 5$, $p = 25$ is given by

$$CS = \int_0^5 f(x)\,dx - 5 \cdot 25$$

where $f(x)$ is the demand function.

$$\begin{aligned} CS &= \int_0^5 (30 - x)\,dx - 125 \\ &= 30x - \frac{x^2}{2}\bigg|_0^5 - 125 \\ &= \left(150 - \frac{25}{2}\right) - 125 \\ &= \frac{25}{2} = 12.50 \end{aligned}$$

The consumer's surplus is \$12.50.

Producer's Surplus

When a product is sold at the equilibrium price, some producers will also benefit, for they would have sold the product at a lower price. The area between the line $p = p_1$ and the supply curve (from $x = 0$ to $x = x_1$) gives the producer's surplus (see Figure 13.21).

If the supply function is $p = g(x)$, the **producer's surplus** is given by the area between the graph of $p = g(x)$ and the x-axis from 0 to x_1 *subtracted from* the area of the rectangle $0x_1Ep_1$.

$$PS = p_1x_1 - \int_0^{x_1} g(x)\,dx$$

Note that p_1x_1 represents the total revenue at the equilibrium point.

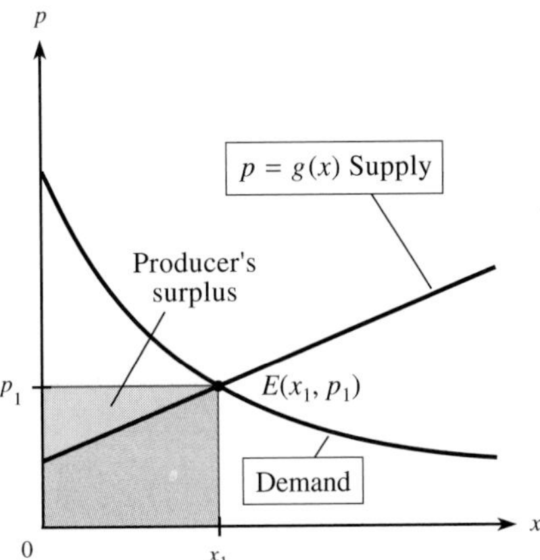

Figure 13.21

EXAMPLE 7

Suppose that the supply function for a product is $p = x^2 + x$. If the equilibrium price is \$20, what is the producer's surplus?

Solution

Because $p = 20$, we can find x as follows:

$$\begin{aligned} 20 &= x^2 + x \\ 0 &= x^2 + x - 20 \\ 0 &= (x + 5)(x - 4) \\ x &= -5, \quad x = 4 \end{aligned}$$

The equilibrium point is $x = 4$, $p = 20$. The producer's surplus is given by

$$\begin{aligned} PS &= 20 \cdot 4 - \int_0^4 (x^2 + x)\,dx \\ &= 80 - \left(\frac{x^3}{3} + \frac{x^2}{2}\right)\Big|_0^4 \\ &= 80 - \left(\frac{64}{3} + 8\right) \\ &\approx 50.67 \end{aligned}$$

The producer's surplus is \$50.67. See Figure 13.22.

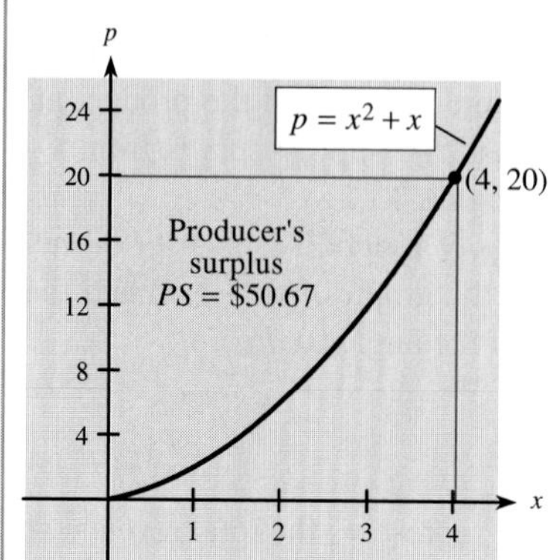

Figure 13.22

EXAMPLE 8

The demand function for a product is $p = \sqrt{49 - 6x}$ and the supply function is $p = x + 1$. Find the producer's surplus.

Solution

We found the equilibrium point for these functions to be (4, 5) in Example 5 (see Figure 13.20). The producer's surplus is

$$PS = 5 \cdot 4 - \int_0^4 (x + 1)\,dx$$

$$= 20 - \left(\frac{x^2}{2} + x\right)\Big|_0^4$$

$$= 20 - (8 + 4) = 8$$

The producer's surplus is \$8.

CHECKPOINT

2. Suppose that for a certain product, the supply function is $p = f(x)$, the demand function is $p = g(x)$, and the equilibrium point is (x_1, p_1). Decide whether the following are true or false.
   (a) $CS = \int_0^{x_1} f(x)\,dx - p_1x_1$ (b) $PS = \int_0^{x_1} f(x)dx - p_1x_1$
3. If demand is $p = \dfrac{100}{x + 1}$, supply is $p = x + 1$, and the market equilibrium is (9, 10), create the integral used to find the
   (a) consumer's surplus. (b) producer's surplus.

EXAMPLE 9

Suppose that for a certain product, the demand function is $p = 200e^{-0.01x}$ and the supply function is $p = \sqrt{200x + 49}$.

(a) Use a calculator or computer to find the market equilibrium point.
(b) Find the consumer's surplus.
(c) Find the producer's surplus.

Solution

(a) Solving $200e^{-0.01x} = \sqrt{200x + 49}$ requires solving

$$40000e^{-0.02x} = 200x + 49$$

which is very difficult using algebraic techniques. Using SOLVER or TRACE gives $x = 60$, to the nearest unit, with a price of \$109.76. (See Figure 13.23a.)

(b) The consumer's surplus is

$$\begin{aligned}\int_0^{60} 200e^{-0.01x}\,dx - 109.76(60) &= [-20{,}000e^{-0.01x}]_0^{60} - 6585.60\\ &= -20{,}000e^{-0.6} + 20{,}000 - 6585.60\\ &= 2438.17\end{aligned}$$

(c) The producer's surplus is

$$\begin{aligned}&60(109.76) - \int_0^{60} \sqrt{200x + 49}\,dx\\ &\quad = 6585.60 - \frac{1}{200}\left[\frac{(200x + 49)^{3/2}}{3/2}\right]_0^{60}\\ &\quad = 6585.60 - \frac{1}{300}[(12049^{3/2} - 49^{3/2})]\\ &\quad \approx 2178.10\end{aligned}$$

Note that we also could have evaluated these definite integrals with the numerical integration feature of a graphing utility, and we would have obtained the same results.

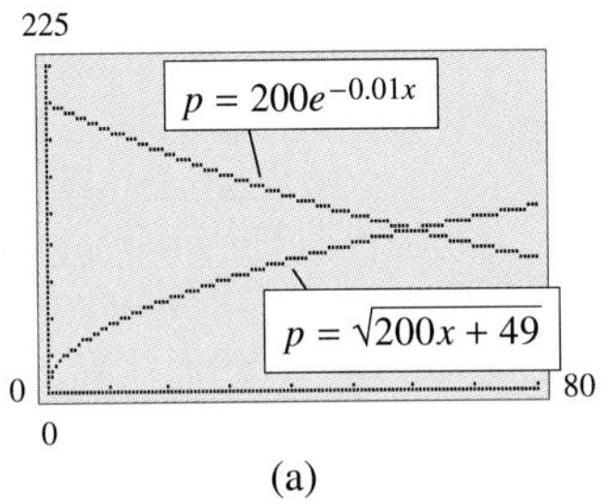

(a)

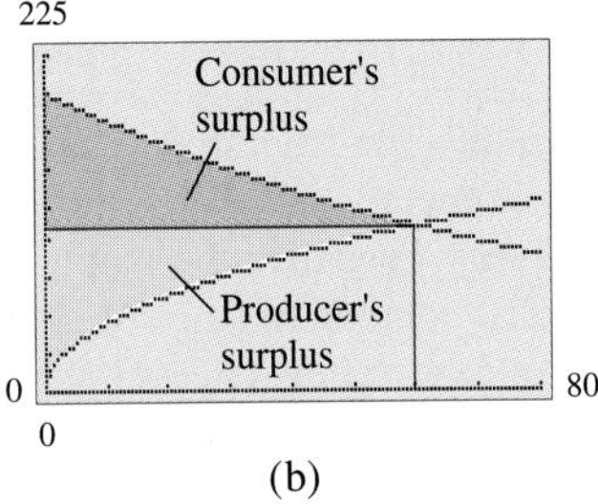

(b)

Figure 13.23

CHECKPOINT SOLUTIONS

1. (a) $\int_0^5 5000e^{-0.01t}\,dt$

   (b) $\int_0^5 (5000e^{-0.01t})(e^{-0.07t})\,dt = \int_0^5 5000e^{-0.08t}\,dt$

   (c) $e^{(0.07)(5)} \int_0^5 (5000e^{-0.01t})(e^{-0.07t})\,dt = e^{0.35} \int_0^5 5000e^{-0.08t}\,dt$

2. (a) False. Consumer's surplus uses the demand function, so

$$CS = \int_0^{x_1} g(x)\,dx - p_1x_1$$

(b) False. Producer's surplus uses the supply function, but the formula is

$$PS = p_1x_1 - \int_0^{x_1} f(x)\,dx$$

3. (a) $CS = \int_0^9 \frac{100}{x+1}\,dx - 90$ (b) $PS = 90 - \int_0^9 (x+1)\,dx$

EXERCISE 13.4

Continuous Income Streams

1. Find the total income over the next 10 years from a continuous income stream that has an annual rate of flow at time t given by $f(t) = 12{,}000$ (dollars).
2. Find the total income over the next 8 years from a continuous income stream with an annual rate of flow at time t given by $f(t) = 8500$ (dollars).
3. Suppose that a steel company views the production of its continuous caster as a continuous income stream with a monthly rate of flow at time t given by

$$f(t) = 24{,}000e^{0.03t} \quad \text{(dollars)}$$

Find the total income from this caster in the first year.

4. Suppose that the Quick-Fix Car Service franchise finds that the income generated by its stores can be modeled by assuming that the income is a continuous stream with a monthly rate of flow at time t given by

$$f(t) = 10{,}000e^{0.02t} \quad \text{(dollars)}$$

Find the total income from a Quick-Fix store for the first 2 years of operation.

5. A small brewery considers the output of its bottling machine as a continuous income stream with an annual rate of flow at time t given by

$$f(t) = 80e^{-0.1t}$$

in thousands of dollars. Find the income from this stream for the next 10 years.

6. A company that services a number of vending machines considers its income as a continuous stream with an annual rate of flow at time t given by

$$f(t) = 120e^{-0.4t}$$

in thousands of dollars. Find the income from this stream over the next 5 years.

7. A franchise models the profit from its store as a continuous income stream with a monthly rate of flow at time t given by

$$f(t) = 3000e^{0.004t} \quad \text{(dollars)}$$

When a new store opens, its manager is judged against the model, with special emphasis on the second half of the first year. Find the total profit for the second 6-month period ($t = 6$ to $t = 12$).

8. The Quick-Fix Car Service franchise has a continuous income stream with a monthly rate of flow modeled by $f(t) = 10{,}000e^{0.02t}$ (dollars). Find the total income for years 2 through 5.
9. A continuous income stream has an annual rate of flow at time t given by

$$f(t) = 12{,}000e^{0.04t} \quad \text{(dollars)}$$

If money is worth 8%, compounded continuously, find the present value of this stream for the next 8 years.

10. A continuous income stream has an annual rate of flow at time t given by

$$f(t) = 9000e^{0.12t} \quad \text{(dollars)}$$

Find the present value of this income stream for the next 10 years, if money is worth 6%, compounded continuously.

11. The income from an established chain of laundromats is a continuous stream with its annual rate of flow at time t given by $f(t) = 63{,}000$ (dollars). If money is worth 7%, compounded continuously, find the present value and future value of this chain over the next 5 years.
12. The profit from an insurance agency can be considered as a continuous income stream with an annual rate of flow at time t given by $f(t) = 84{,}000$ (dollars). Find the present value and future value of this agency over the next 12 years, if money is worth 8%, compounded continuously.
13. Suppose that a printing firm considers the production of its presses as a continuous income stream. If the annual rate of flow at time t is given by

$$f(t) = 97.5e^{-0.2(t+3)}$$

in thousands of dollars, and if money is worth 6%, compounded continuously, find the present value and future value of the presses over the next 10 years.

14. Suppose that a vending machine company is considering selling some of its machines. Suppose further that the income from these particular machines is a continuous stream with an annual rate of flow at time t given by

$$f(t) = 12e^{-0.4(t+3)}$$

in thousands of dollars. Find the present value and future value of the machines over the next 5 years if money is worth 10%, compounded continuously.

15. A 58-year-old couple is considering opening a business of their own. They will either purchase an established Gift and Card Shoppe or open a new Video Rental Palace. The Gift Shoppe has a continuous income stream with an annual rate of flow at time t given by

$$G(t) = 30{,}000 \quad \text{(dollars)}$$

and the Video Palace has a continuous income stream with a projected annual rate of flow at time t given by

$$V(t) = 21{,}600e^{0.08t} \quad \text{(dollars)}$$

The initial investment is the same for both businesses, and money is worth 10%, compounded continuously. Find the present value of each business over the next 7 years (until the couple reaches age 65) to see which is the better buy.

16. If the couple in Problem 15 plans to keep the business until age 70 (for the next 12 years), find each present value to see which business is the better buy in this case.

Consumer's Surplus

17. The demand function for a product is $p = 34 - x^2$. If the equilibrium price is \$9, what is the consumer's surplus?
18. The demand function for a product is $p = 100 - 4x$. If the equilibrium price is \$40, what is the consumer's surplus?
19. The demand function for a product is $p = 200/(x + 2)$. If the equilibrium quantity is 8 units, what is the consumer's surplus?
20. The demand function for a certain product is $p = 100/(1 + 2x)$. If the equilibrium quantity is 12 units, what is the consumer's surplus?
21. The demand function for a certain product is $p = 81 - x^2$ and the supply function is $p = x^2 + 4x + 11$. Find the equilibrium point and the consumer's surplus there.
22. The demand function for a product is $p = 49 - x^2$ and the supply function is $p = 4x + 4$. Find the equilibrium point and the consumer's surplus there.
23. If the demand function for a product is $p = 12/(x + 1)$ and the supply function is $p = 1 + 0.2x$, find the consumer's surplus under pure competition.
24. If the demand function for a good is $p = 110 - x^2$ and the supply function for it is $p = 2 - \frac{6}{5}x + \frac{1}{5}x^2$, find the consumer's surplus under pure competition.
25. A monopoly has a total cost function $C = 1000 + 120x + 6x^2$ for its product, which has demand function $p = 360 - 3x - 2x^2$. Find the consumer's surplus at the point where the monopoly has maximum profit.
26. A monopoly has a total cost function $C = 500 + 2x^2 + 10x$ for its product, which has demand function $p = -\frac{1}{3}x^2 - 2x + 30$. Find the consumer's surplus at the point where the monopoly has maximum profit.

Producer's Surplus

27. Suppose that the supply function for a good is $p = 4x^2 + 2x + 2$. If the equilibrium price is \$422, what is the producer's surplus there?
28. Suppose that the supply function for a good is $p = 0.1x^2 + 3x + 20$. If the equilibrium price is \$36, what is the producer's surplus there?
29. If the supply function for a commodity is $p = 10e^{x/3}$, what is the producer's surplus when 15 units are sold?
30. If the supply function for a commodity is $p = 40 + 100(x + 1)^2$, what is the producer's surplus at $x = 20$?
31. Find the producer's surplus for a product if its demand function is $p = 81 - x^2$ and its supply function is $p = x^2 + 4x + 11$.
32. Find the producer's surplus for a product if its demand function is $p = 49 - x^2$ and its supply function is $p = 4x + 4$.
33. Find the producer's surplus for a product with demand function $p = 12/(x + 1)$ and supply function $p = 1 + 0.2x$.
34. Find the producer's surplus for a product with demand function $p = 110 - x^2$ and supply function $p = 2 - \frac{6}{5}x + \frac{1}{5}x^2$.
35. The demand function for a certain product is $p = 144 - 2x^2$ and the supply function is $p = x^2 + 33x + 48$. Find the producer's surplus at the equilibrium point.
36. The demand function for a product is $p = 280 - 4x - x^2$ and the supply function for it is $p = 160 + 4x + x^2$. Find the producer's surplus at the equilibrium point.

13.5 Using Tables of Integrals

OBJECTIVE

- *To use tables of integrals to evaluate certain integrals*

APPLICATION PREVIEW

In 1982, the rate of increase of new AIDS cases was predicted to be

$$\frac{dN}{dt} = 2^t(700)$$

where t is the number of years after 1982. The number of AIDS cases was 1012 in 1982 ($t = 0$). Finding the predicted number of cases for 1986 ($t = 4$) requires evaluating the integral

$$\int 2^t(700)\,dt$$

Evaluating this integral is made easier by the existence of a formula such as those given in Table 13.2. The formulas in this table, and others listed in other resources, such as the Chemical Rubber Company's *Standard Mathematical Tables,* extend the number of integrals that can be evaluated. Using the formulas is not quite as easy as it may sound because finding the correct formula and using it properly may present problems. The examples in this section illustrate how some of these formulas are used.

TABLE 13.2 Integration Formulas

1. $\int u^n\,du = \frac{u^{n+1}}{n+1} + C, \quad \text{for } n \neq -1$
2. $\int \frac{du}{u} = \int u^{-1}\,du = \ln|u| + C$
3. $\int a^u\,du = a^u \log_a e + C = \frac{a^u}{\ln a} + C$
4. $\int e^u\,du = e^u + C$
5. $\int \frac{du}{a^2 - u^2} = \frac{1}{2a}\ln\left|\frac{a+u}{a-u}\right| + C$
6. $\int \sqrt{u^2 + a^2}\,du = \frac{1}{2}(u\sqrt{u^2+a^2} + a^2\ln|u + \sqrt{u^2+a^2}|) + C$
7. $\int \sqrt{u^2 - a^2}\,du = \frac{1}{2}(u\sqrt{u^2-a^2} - a^2\ln|u + \sqrt{u^2-a^2}|) + C$
8. $\int \frac{du}{\sqrt{u^2+a^2}} = \ln|u + \sqrt{u^2+a^2}| + C$
9. $\int \frac{du}{u\sqrt{a^2-u^2}} = -\frac{1}{a}\ln\left|\frac{a+\sqrt{a^2-u^2}}{u}\right| + C$
10. $\int \frac{du}{\sqrt{u^2-a^2}} = \ln|u + \sqrt{u^2-a^2}| + C$
11. $\int \frac{du}{u\sqrt{a^2+u^2}} = -\frac{1}{a}\ln\left|\frac{a+\sqrt{a^2+u^2}}{u}\right| + C$

12. $\displaystyle\int \frac{u\,du}{au+b} = \frac{u}{a} - \frac{b}{a^2}\ln|au+b| + C$

13. $\displaystyle\int \frac{du}{u(au+b)} = \frac{1}{b}\ln\left|\frac{u}{au+b}\right| + C$

14. $\displaystyle\int \ln u\,du = u(\ln u - 1) + C$

15. $\displaystyle\int \frac{u\,du}{(au+b)^2} = \frac{1}{a^2}\left(\ln|au+b| + \frac{b}{au+b}\right) + C$

16. $\displaystyle\int u\sqrt{au+b}\,du = \frac{2(3au-2b)(au+b)^{3/2}}{15a^2} + C$

17. $\displaystyle\int u\,dv = uv - \int v\,du$

EXAMPLE 1

Evaluate $\displaystyle\int \frac{dx}{\sqrt{x^2+4}}$.

Solution

We must find a formula in Table 13.2 that is of the same form as this integral. We see that formula 8 has the desired form, *if* we let $u = x$ and $a = 2$. Thus

$$\int \frac{dx}{\sqrt{x^2+4}} = \ln|x + \sqrt{x^2+4}| + C$$

EXAMPLE 2

Evaluate $\displaystyle\int_1^2 \frac{dx}{x^2+2x}$.

Solution

There does not appear to be any formula having exactly the same form as our integral. But if we rewrite our integral as

$$\int_1^2 \frac{dx}{x(x+2)}$$

we see that formula 13 will work. Letting $u = x$, $a = 1$, and $b = 2$, we get

$$\begin{aligned}\int_1^2 \frac{dx}{x(x+2)} = \frac{1}{2}\ln\left|\frac{x}{x+2}\right|\Bigg|_1^2 &= \frac{1}{2}\ln\left|\frac{2}{4}\right| - \frac{1}{2}\ln\left|\frac{1}{3}\right| \\ &= \frac{1}{2}\left(\ln\frac{1}{2} - \ln\frac{1}{3}\right) \\ &= \frac{1}{2}\ln\frac{3}{2} \\ &= \frac{1}{2}\ln 1.5\end{aligned}$$

Although the formulas in Table 13.2 are given in terms of the variable u, they may be used with any variable.

EXAMPLE 3

Evaluate $\int \frac{dq}{9 - q^2}$.

Solution

The formula that applies in this case is formula 5, with $a = 3$ and $u = q$. Then

$$\int \frac{dq}{9 - q^2} = \frac{1}{2 \cdot 3} \ln \left|\frac{3 + q}{3 - q}\right| + C = \frac{1}{6} \ln \left|\frac{3 + q}{3 - q}\right| + C$$

EXAMPLE 4

Evaluate $\int \ln (2x + 1)\, dx$.

Solution

This integral has the form of formula 14, with $u = 2x + 1$. But if $u = 2x + 1$, du must be represented by the differential of $2x + 1$ (that is, $2\, dx$). Thus

$$\int \ln (2x + 1)\, dx = \frac{1}{2} \int \ln (2x + 1)(2\, dx)$$
$$= \frac{1}{2}(2x + 1)[\ln (2x + 1) - 1] + C$$

CHECKPOINT

1. Can both $\int \frac{dx}{\sqrt{x^2 - 4}}$ and $-\int \frac{dx}{\sqrt{4 - x^2}}$ be evaluated with formula 10 in Table 13.2?
2. Determine the formula used to evaluate $\int \frac{3x}{4x - 5}\, dx$, and show how the formula would be applied.
3. True or false: In order for us to use a formula, the given integral must correspond exactly to the formula, including du.
4. True or false: $\int \frac{dx}{x^2(3x^2 - 7)}$ can be evaluated with formula 13.
5. True or false: $\int \frac{dx}{(6x + 1)^2}$ can be evaluated with either formula 1 or formula 15.
6. True or false: $\int \sqrt{x^2 + 4}\, dx$ can be evaluated with formula 1, formula 6, or formula 16.

EXAMPLE 5

Evaluate $\int_1^2 \frac{dx}{x\sqrt{81 - 9x^2}}$.

Solution

This integral is similar to that of formula 9 in Table 13.2. Letting $a = 9$, letting $u = 3x$, and multiplying the numerator and denominator by 3 give the proper form.

$$\begin{aligned}\int_1^2 \frac{dx}{x\sqrt{81-9x^2}} &= \int_1^2 \frac{3\,dx}{3x\sqrt{81-9x^2}}\\ &= -\frac{1}{9}\ln\left|\frac{9+\sqrt{81-9x^2}}{3x}\right|\Bigg|_1^2\\ &= \left[-\frac{1}{9}\ln\left(\frac{9+\sqrt{45}}{6}\right)\right]-\left[-\frac{1}{9}\ln\left(\frac{9+\sqrt{72}}{3}\right)\right]\\ &\approx 0.0889\end{aligned}$$

Remember that the formulas given in Table 13.2 represent only a very small sample of all possible integration formulas. Additional formulas may be found in books of mathematical tables.

EXAMPLE 6

In 1982, the rate of increase of new AIDS cases was predicted to be

$$\frac{dN}{dt} = 2^t(700)$$

where t is the number of years since 1982. If the number of AIDS cases was 1012 in 1982 ($t = 0$), what would be the predicted number of cases for 1986 ($t = 4$), before added attention was given to prevention?

Solution

Finding the predicted number of cases N in 1986 requires evaluating the integral

$$N = \int 2^t(700)\,dt = 700\int 2^t\,dt$$

and using the fact that $N = 1012$ when $t = 0$ to find the constant C. The integral has the form of formula 3 with $a = 2$ and $u = t$.

$$N = 700\int 2^t\,dt = 700\cdot\frac{2^t}{\ln 2} + C$$

Using $N = 1012$ when $t = 0$ gives

$$\begin{aligned}1012 &= 700\cdot\frac{2^0}{\ln 2} + C\\ 1012 &= 1010 + C\\ C &= 2,\end{aligned}$$

so

$$N = 700\cdot\frac{2^t}{\ln 2} + 2$$

The number of cases in 1986 (when $t = 4$) is predicted by this model to be

$$N = 700\cdot\frac{2^4}{\ln 2} + 2 = 16{,}160$$

The actual number of cases in 1986 was 13,898. Paying more attention to prevention has reduced the rate of spread of the disease somewhat, changing the function that models the number of cases. Models using more recent data are discussed in other sections of the text.

Graphing Utilities

Numerical integration with a graphing calculator or computer software is especially useful in evaluating definite integrals when the formulas for the integral are difficult to use or are not available. For example, evaluating the definite integral in Example 5 above with the numerical integration feature of a graphing utility gives 0.08892484. The decimal approximation of the answer in Example 5 is 0.088924836, so the numerical approximation of the answer agrees for the first eight decimal places.

CHECKPOINT SOLUTIONS

1. No. Although $\int \frac{dx}{\sqrt{x^2-4}}$ can be evaluated with formula 10 from Table 13.2, $-\int \frac{dx}{\sqrt{4-x^2}}$ cannot, because $\sqrt{4-x^2}$ cannot be rewritten in the form $\sqrt{u^2-a^2}$ as is needed for this formula to be used.
2. Use formula 12 with $u = x$, $a = 4$, $b = -5$, and $du = dx$.
$$\int \frac{3x}{4x-5}\,dx = 3\int \frac{x\,dx}{4x-5} = 3\int \frac{u\,du}{au+b} = 3\left(\frac{u}{a} - \frac{b}{a^2}\ln|au+b| + C\right)$$
$$= \frac{3x}{4} + \frac{15}{16}\ln|4x-5| + C$$
3. True. An exact correspondence with the formula and du is necessary.
4. False. With $u = x^2$, we must have $du = 2x\,dx$. In this problem there is no x with dx, so the problem cannot correspond to formula 13.
5. False. The integral can be evaluated only with formula 1, not with formula 15. The correspondence is $u = 6x + 1$, $du = 6\,dx$, and $n = -2$.
6. False. The integral can be evaluated only with formula 6, with $u = x$, $du = dx$, and $a = 2$.

EXERCISE 13.5

Evaluate the integrals in Problems 1–32.

1. $\int \frac{dx}{16-x^2}$
2. $\int \frac{dx}{x(3x+5)}$
3. $\int_1^4 \frac{dx}{x\sqrt{9+x^2}}$
4. $\int \frac{dx}{x\sqrt{9-x^2}}$
5. $\int \ln w\,dw$
6. $\int \frac{dv}{v(3v+8)}$
7. $\int_0^2 \frac{q\,dq}{6q+9}$
8. $\int_1^5 \frac{dq}{q\sqrt{25+q^2}}$
9. $\int 3^x\,dx$
10. $\int_0^3 \sqrt{x^2+16}\,dx$
11. $\int_5^7 \sqrt{x^2-25}\,dx$
12. $\int \frac{x\,dx}{(3x+2)^2}$
13. $\int w\sqrt{4w+5}\,dw$
14. $\int \frac{dy}{\sqrt{9+y^2}}$
15. $\int x\,5^{x^2}\,dx$
16. $\int \sqrt{9x^2+4}\,dx$
17. $\int_0^3 x\sqrt{x^2+4}\,dx$
18. $\int x\sqrt{x^4-36}\,dx$
19. $\int \frac{5\,dx}{x\sqrt{4-9x^2}}$
20. $\int x\,e^{x^2}\,dx$
21. $\int \frac{dx}{\sqrt{9x^2-4}}$
22. $\int \frac{dx}{25-4x^2}$
23. $\int_5^6 \frac{dx}{x^2-16}$
24. $\int_0^1 \frac{x\,dx}{6-5x}$

25. $\int \frac{dx}{\sqrt{(3x+1)^2+1}}$ 26. $\int \frac{dx}{9-(2x+3)^2}$

27. $\int_0^3 x\sqrt{(x^2+1)^2+9}\, dx$ 28. $\int_1^e x \ln x^2\, dx$

29. $\int \frac{x\, dx}{7-3x^2}$ 30. $\int_0^1 \frac{e^x}{1+e^x}\, dx$

31. $\int \frac{dx}{\sqrt{4x^2+7}}$ 32. $\int e^{2x}\sqrt{3e^x+1}\, dx$

Use formulas or numerical integration with a graphing calculator or computer to evaluate the definite integral in Problems 33–36.

33. $\int_2^3 \frac{e^{\sqrt{x-1}}}{\sqrt{x-1}}\, dx$ 34. $\int_2^4 \frac{3x}{\sqrt{x^4-9}}\, dx$

35. $\int_0^1 \frac{x^3\, dx}{(4x^2+5)^2}$ 36. $\int_0^1 (e^x+1)^3 e^x\, dx$

Applications

37. ***Producer's surplus*** If the supply function for a commodity is $p = 40 + 100 \ln (x+1)^2$, what is the producer's surplus at $x = 20$?

38. ***Consumer's surplus*** If the demand function for a good is $p = 5000e^{-x} + 4$, where x is the number of hundreds of bushels of wheat, what is the consumer's surplus at $x = 7, p = 9.10$?

39. ***Cost*** (a) If the marginal cost for a good is $\overline{MC} = \sqrt{x^2+9}$ and if the fixed cost is \$300, what is the total cost function of the good?
    (b) What is the total cost of producing 4 units of this good?

40. ***Consumer's surplus*** Suppose that the demand function for an appliance is

$$p = \frac{400q + 400}{(q+2)^2}$$

What is the consumer's surplus if the equilibrium price is \$19 and the equilibrium quantity is 18?

41. ***Income streams*** Suppose that when a new oil well is opened, its production is viewed as a continuous income stream with monthly rate of flow

$$f(t) = 10 \ln (t+1) - 0.1t$$

where t is time in months and $f(t)$ is in thousands of dollars. Find the total income over the next 10 years (120 months).

42. ***Spread of disease*** An isolated community of 1000 people susceptible to a certain disease is exposed when one member returns carrying the disease. If x represents the number infected with the disease at time t (in days), then the rate of change of x is proportional to the product of the number infected, x, and the number still susceptible, $1000 - x$. That is,

$$\frac{dx}{dt} = kx(1000-x) \quad \text{or} \quad \frac{dx}{x(1000-x)} = k\, dt$$

(a) If $k = 0.001$, integrate both sides to solve this differential equation.
(b) Find how long it will be before half the population of the community is affected.
(c) Find the rate of new cases, dx/dt, every other day for the first 13 days.

13.6 Integration by Parts

OBJECTIVE

- *To evaluate integrals using the method of integration by parts*

APPLICATION PREVIEW

If the value of oil produced by a piece of oil extraction equipment is considered a continuous income stream with an annual rate of flow at time t given by

$$f(t) = 300{,}000 - 2500t, \qquad 0 \le t \le 10$$

and if money can be invested at 8%, compounded continuously, then the present value of the piece of equipment is

$$\int_0^{10} (300{,}000 - 2500t)e^{-0.08t}\, dt$$

$$= 300{,}000 \int_0^{10} e^{-0.08t}\, dt - 2500 \int_0^{10} te^{-0.08t}\, dt$$

The first integral can be evaluated with the formula for the integral of $e^u\,du$. Evaluating the second integral can be done by using **integration by parts,** which is a special technique that uses formula 17 from Table 13.2. This technique involves rewriting an integral in a form that can be evaluated.

Formula 17 in Table 13.2 follows from the Product Rule for derivatives (actually differentials) as follows.

$$\frac{d}{dx}(uv) = u\frac{dv}{dx} + v\frac{du}{dx} \quad \text{so} \quad d(uv) = u\,dv + v\,du$$

Rearranging the differential form and integrating both sides give the following.

$$u\,dv = d(uv) - v\,du$$
$$\int u\,dv = \int d(uv) - \int v\,du$$
$$\int u\,dv = uv - \int v\,du$$

Integration by Parts

$$\int u\,dv = uv - \int v\,du$$

Integration by parts is very useful if the integral we seek to evaluate can be treated as the product of one function, u, and the differential dv of a second function, so that the two integrals $\int dv$ and $\int v \cdot du$ can be found. Let us consider an example using this method.

EXAMPLE 1

Evaluate $\int xe^x\,dx$.

Solution

We cannot evaluate this integral using methods we have learned. But we can "split" the integrand into two parts, setting one part equal to u and the other part equal to dv. This "split" must be done in such a way that $\int dv$ and $\int v\,du$ can be evaluated. Letting $u = x$ and letting $dv = e^x\,dx$ are possible choices. If we make these choices, we have

$$u = x \qquad dv = e^x\,dx$$
$$du = 1\,dx \qquad v = \int e^x\,dx = e^x$$

Then

$$\begin{aligned}\int xe^x\,dx &= u \cdot v - \int v\,du \\ &= x \cdot e^x - \int e^x\,dx \\ &= xe^x - e^x + C\end{aligned}$$

We see that choosing $u = x$ and $dv = e^x\,dx$ worked in evaluating $\int xe^x\,dx$ in Example 1. If we had chosen $u = e^x$ and $dv = x\,dx$, the results would not have been so successful.

How can we select u and dv to make integration by parts work? There are no general rules for separating the integrand into u and dv, but the goal is to select a dv that is integrable and will result in an $\int v\,du$ that is also integrable. There are usually just two reasonable choices, and it may be necessary to try both. Practice will enhance your insight and lead to increasingly successful choices. Consider the following examples.

EXAMPLE 2

Evaluate $\int x \ln x\,dx$.

Solution

Let $u = \ln x$ and $dv = x\,dx$. Then

$$du = \frac{1}{x}\,dx \quad \text{and} \quad v = \frac{x^2}{2}$$

so

$$\begin{aligned}\int x \ln x\,dx &= u \cdot v - \int v\,du \\ &= (\ln x)\frac{x^2}{2} - \int \frac{x^2}{2} \cdot \frac{1}{x}\,dx \\ &= \frac{x^2}{2}\ln x - \int \frac{x}{2}\,dx \\ &= \frac{x^2}{2}\ln x - \frac{x^2}{4} + C\end{aligned}$$

Note that letting $dv = \ln x\,dx$ would lead to great difficulty in evaluating $\int dv$ and $\int v\,du$, so it would not be wise choice.

EXAMPLE 3

Evaluate $\int \ln x^2\,dx$.

Solution

It is frequently good practice to let expressions involving logarithms be part of u in integrating by parts, because the derivatives of logarithmic expressions are usually simple.

In this problem, we can let $u = \ln x^2$. Thus

$$u = \ln x^2 \qquad dv = dx$$

$$du = \frac{2x}{x^2}\,dx = \frac{2}{x}\,dx \qquad v = x$$

Then

$$\begin{aligned}\int \ln x^2\,dx &= x \ln x^2 - \int x \cdot \frac{2}{x}\,dx \\ &= x \ln x^2 - 2x + C\end{aligned}$$

Note that if we write $\ln x^2$ as $2 \ln x$, we can also evaluate this integral using formula 14 in Table 13.2, so integration by parts would not be needed.

CHECKPOINT

1. True or false: In evaluating $\int u\,dv$ by parts,
   (a) the parts u and dv are selected and the parts du and v are calculated.
   (b) the differential (often dx) is always chosen as part of dv.
   (c) the parts du and v are found from u and dv as follows:
   $$du = u'\,dx \qquad \text{and} \qquad v = \int dv$$
   (d) For $\displaystyle\int \frac{3x}{e^{2x}}\,dx$, we could choose $u = 3x$ and $dv = e^{2x}\,dx$.
2. For $\displaystyle\int \frac{\ln x}{x^4}\,dx$,
   (a) identify u and dv.
   (b) find du and v.
   (c) complete the evaluation of the integral.

Sometimes it is necessary to repeat the integration by parts to complete the evaluation. As before, the goal is to produce a new integral that is simpler.

EXAMPLE 4

Evaluate $\int x^2e^{2x}\,dx$.

Solution

Let $u = x^2$ and $dv = e^{2x}\,dx$, so $du = 2x\,dx$ and $v = \frac{1}{2}e^{2x}$. Then

$$\int x^2e^{2x}\,dx = \frac{1}{2}x^2e^{2x} - \int xe^{2x}\,dx$$

We cannot evaluate $\int xe^{2x}\,dx$ directly, but this new integral is simpler than the original, and a second integration by parts will be successful. Letting $u = x$ and $dv = e^{2x}\,dx$ gives $du = dx$ and $v = \frac{1}{2}e^{2x}$. Thus

$$\begin{aligned}\int x^2e^{2x}\,dx &= \frac{1}{2}x^2e^{2x} - \left(\frac{1}{2}xe^{2x} - \int \frac{1}{2}e^{2x}\,dx\right)\\ &= \frac{1}{2}x^2e^{2x} - \frac{1}{2}xe^{2x} + \frac{1}{4}e^{2x} + C\\ &= \frac{1}{4}e^{2x}(2x^2 - 2x + 1) + C\end{aligned}$$

The most obvious choices for u and dv are not always the correct ones, as the following example shows. Integration by parts still requires some trial and error.

EXAMPLE 5

Evaluate $\int x^3\sqrt{x^2 + 1}\,dx$.

Solution

Because x^3 can be integrated easily, it may appear that the new integral would be simplified if we let $u = \sqrt{x^2+1}$ and $dv = x^3$. But then du would be $\frac{1}{2}(x^2 + 1)^{-1/2}2x\,dx$, making $\int v\,du$ more complicated than the original integral. However, we can use $\sqrt{x^2+1}$ as part of dv, and we can evaluate $\int dv$ if we let $dv = x\sqrt{x^2+1}\,dx$. Then

$$u = x^2 \qquad dv = (x^2 + 1)^{1/2}x\,dx$$

$$du = 2x\,dx \qquad v = \int (x^2+1)^{1/2}(x\,dx) = \frac{1}{2}\int (x^2+1)^{1/2}(2x\,dx)$$

$$v = \frac{1}{2}\,\frac{(x^2+1)^{3/2}}{3/2} = \frac{1}{3}(x^2+1)^{3/2}$$

Then $\displaystyle\int x^3\sqrt{x^2+1}\,dx = \frac{x^2}{3}(x^2+1)^{3/2} - \int \frac{1}{3}(x^2+1)^{3/2}(2x\,dx)$

$$= \frac{x^2}{3}(x^2+1)^{3/2} - \frac{1}{3}\,\frac{(x^2+1)^{5/2}}{5/2} + C$$

$$= \frac{x^2}{3}(x^2+1)^{3/2} - \frac{2}{15}(x^2+1)^{5/2} + C$$

We now consider the problem in the Application Preview.

EXAMPLE 6

Suppose that the value of oil produced by a piece of oil extraction equipment is considered a continuous income stream with an annual rate of flow at time t given by

$$f(t) = 300{,}000 - 2500t, \qquad 0 \le t \le 10$$

and that money is worth 8%, compounded continuously. Find the present value of the piece of equipment.

Solution

The present value of the piece of equipment is given by

$$\int_0^{10} (300{,}000 - 2500t)e^{-0.08t}\,dt$$

$$= 300{,}000\int_0^{10} e^{-0.08t}\,dt - 2500\int_0^{10} te^{-0.08t}\,dt$$

$$= \frac{300{,}000}{-0.08}e^{-0.08t}\Big|_0^{10} - 2500\int_0^{10} te^{-0.08t}\,dt$$

The value of the first integral is

$$= \frac{300{,}000}{-0.08}e^{-0.08t}\Big|_0^{10} = \frac{300{,}000}{-0.08}e^{-0.8} - \frac{300{,}000}{-0.08}$$

$$\approx -1{,}684{,}983.615 + 3{,}750{,}000$$

$$= 2{,}065{,}016.385$$

The second of these integrals can be evaluated by using integration by parts, with $u = t$ and $dv = e^{-0.08t}\,dt$. Then $du = 1\,dt$ and $v = \dfrac{e^{-0.08t}}{-0.08}$, and this integral is

$$\begin{aligned}
-2500\int_0^{10} te^{-0.08t}\,dt &= -2500\left.\frac{te^{-0.08t}}{-0.08}\right|_0^{10} + 2500\int_0^{10}\frac{e^{-0.08t}}{-0.08}dt \\
&= \frac{2500}{0.08}\left.te^{-0.08t}\right|_0^{10} + \frac{2500}{0.0064}\left.e^{-0.08t}\right|_0^{10} \\
&= \frac{2500}{0.08}10e^{-0.8} + \frac{2500}{0.0064}e^{-0.8} - \frac{2500}{0.0064} \\
&\approx -74{,}690.572
\end{aligned}$$

Thus the sum of the integrals is

$$2{,}065{,}016.385 + (-74{,}690.572) = 1{,}990{,}325.823$$

and the present value of this piece of equipment is $1,990,325.82.

One further note about integration by parts. It can be very useful on certain types of problems, but not on all types. Don't attempt to use integration by parts when easier methods are available.

Graphing Utilities Using numerical integration with a graphing calculator or computer software to evaluate the integral in Example 6 gives the present value of $1,990,325.80, so this answer is the same as that found in Example 6, to the nearest dollar.

CHECKPOINT SOLUTIONS

1. (a) True (b) True (c) True
   (d) False. The product of u and dv must equal the original integrand. Rewrite as

$$\int \frac{3x}{e^{2x}}\,dx = \int 3xe^{-2x}\,dx$$

   and then choose $u = 3x$ and $dv = e^{-2x}\,dx$.
2. (a) $u = \ln x$ and $dv = x^{-4}\,dx$
   (b) $du = \dfrac{1}{x}\,dx$ and $v = \displaystyle\int x^{-4}\,dx = \frac{x^{-3}}{-3}$
   (c) $$\begin{aligned}
\int \frac{\ln x}{x^4}\,dx &= uv - \int v\,du \\
&= (\ln x)\left(\frac{x^{-3}}{-3}\right) - \int \frac{x^{-3}}{-3}\cdot\frac{1}{x}\,dx \\
&= \frac{-\ln x}{3x^3} + \frac{1}{3}\int x^{-4}\,dx \\
&= -\frac{\ln x}{3x^3} + \frac{1}{3}\left(\frac{x^{-3}}{-3}\right) + C \\
&= -\frac{\ln x}{3x^3} - \frac{1}{9x^3} + C
\end{aligned}$$

EXERCISE 13.6

In Problems 1–16, use integration by parts to evaluate the integral.

1. $\int xe^{2x}\,dx$
2. $\int xe^{-x}\,dx$
3. $\int x^2 \ln x\,dx$
4. $\int x^3 \ln x\,dx$
5. $\int_4^6 q\sqrt{q-4}\,dq$
6. $\int_0^1 y(1-y)^{3/2}\,dy$
7. $\int \frac{\ln x}{x^2}\,dx$
8. $\int \frac{\ln(x-1)}{\sqrt{x-1}}\,dx$
9. $\int_1^e \ln x\,dx$
10. $\int \frac{x}{\sqrt{x-3}}\,dx$
11. $\int x\ln(2x-3)\,dx$
12. $\int x\ln(4x)\,dx$
13. $\int q^3\sqrt{q^2-3}\,dq$
14. $\int \frac{x^3}{\sqrt{9-x^2}}\,dx$
15. $\int_0^4 x^3\sqrt{x^2+9}\,dx$
16. $\int \sqrt{x}\ln x\,dx$

In Problems 17–24, use integration by parts to evaluate the integral. Note that evaluation will require integration by parts more than once.

17. $\int x^2e^{-x}\,dx$
18. $\int_0^1 x^2e^x\,dx$
19. $\int_0^2 x^3e^{x^2}\,dx$
20. $\int x^3e^x\,dx$
21. $\int x^3\ln^2 x\,dx$
22. $\int \frac{x^2}{\sqrt{x-3}}\,dx$
23. $\int e^{2x}\sqrt{e^x+1}\,dx$
24. $\int_1^2 (\ln x)^2\,dx$

In Problems 25–30, match each of the integrals with the formula or method (I–IV) that should be used to evaluate it. Then evaluate the integral.

I. Integration by parts II. $\int e^u\,du$ III. $\int \frac{du}{u}$ IV. $\int u^n\,du$

25. $\int xe^{x^2}\,dx$
26. $\int \frac{x}{\sqrt{9-x^2}}\,dx$
27. $\int e^x\sqrt{e^x+1}\,dx$
28. $\int 4x^2e^{x^3}\,dx$
29. $\int_0^4 \frac{t}{e^t}\,dt$
30. $\int x^2\sqrt{x-1}\,dx$

Applications

31. ***Producer's surplus*** If the supply function for a commodity is $p = 30 + 50\ln(2x+1)^2$, what is the producer's surplus at $x = 30$?

32. ***Cost*** If the marginal cost function for a product is $\overline{MC} = 1 + 3\ln(x+1)$, and if the fixed cost is \$100, find the total cost function.

33. ***Present value*** Suppose that a machine's production can be considered as a continuous income stream with annual rate of flow at time t given by

$$f(t) = 10{,}000 - 500t \quad \text{(dollars)}$$

If money is worth 10%, compounded continuously, find the present value of the machine over the next 5 years.

34. ***Present value*** Suppose that the production of a machine used to mine coal is considered as a continuous income stream with annual rate of flow at time t given by

$$f(t) = 280{,}000 - 14{,}000t \quad \text{(dollars)}$$

If money is worth 7%, compounded continuously, find the present value of this machine over the next 8 years.

35. ***Income distribution*** Suppose the Lorenz curve for the distribution of income of a certain country is given by

$$y = xe^{x-1}$$

Find the Gini coefficient of income.

36. ***Income streams*** Suppose the income from an Internet access business is a continuous income stream with annual rate of flow given by

$$f(t) = 100te^{-0.1t}$$

in thousands of dollars. Find the total income over the next 10 years.

13.7 Improper Integrals and Their Applications

OBJECTIVES

- To evaluate improper integrals
- To apply improper integrals to continuous income streams and to probability density functions

APPLICATION PREVIEW

We saw in Section 13.4, "Applications of Definite Integrals in Business and Economics," that the present value of a continuous income stream over a fixed number of years can be found by using a definite integral. When this notion is extended to an infinite time interval, the result is called the **capital value** of the income stream and is given by

$$\textbf{Capital value} = \int_0^\infty f(t)e^{-rt}\,dt$$

where $f(t)$ is the annual rate of flow at time t, and r is the annual interest rate, compounded continuously. This is called an **improper integral.**

Other applications of calculus to business and to statistics use improper integrals to find the area of a region that extends infinitely to the left or right along the x-axis (see Figure 13.24).

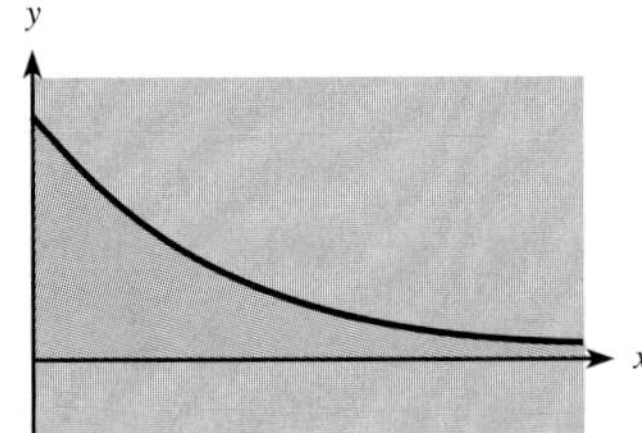

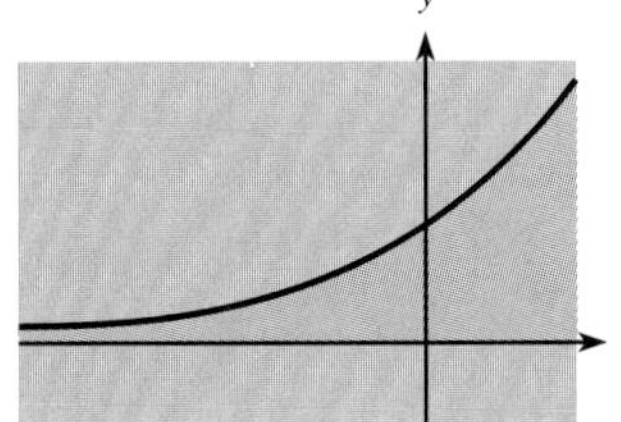

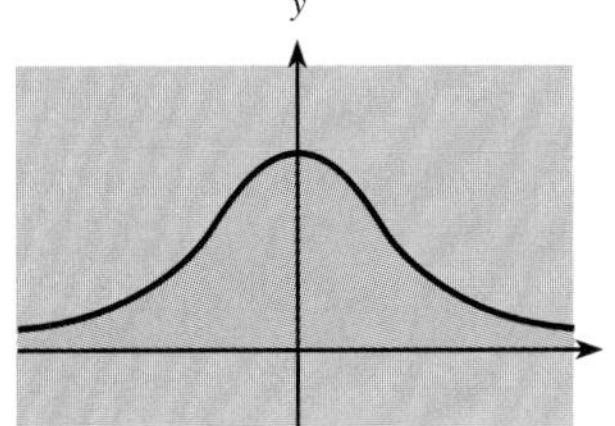

Figure 13.24

Let us consider how to find the area between the curve $y = 1/x^2$ and x-axis to the right of $x = 1$.

To find the area under this curve from $x = 1$ to $x = b$, where b is any number greater than 1 (see Figure 13.25), we evaluate

$$A = \int_1^b \frac{1}{x^2}\,dx = \left.\frac{-1}{x}\right|_1^b = \frac{-1}{b} - \left(\frac{-1}{1}\right) = 1 - \frac{1}{b}$$

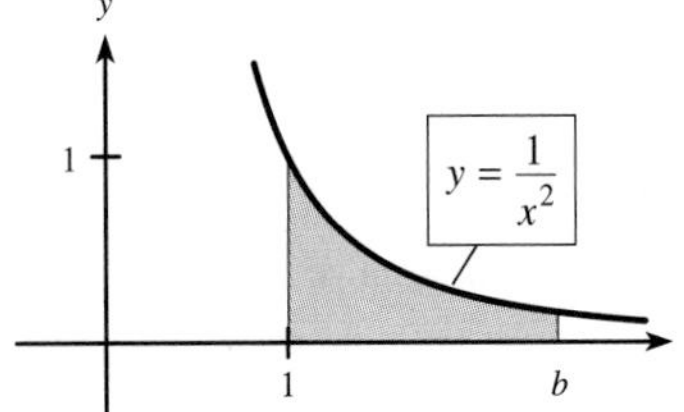

Figure 13.25

Note that the larger b is, the closer the area is to 1. If $b = 100$, $A = 0.99$; if $b = 1000$, $A = 0.999$; and if $b = 1{,}000{,}000$, $A = 0.999999$.

We can represent the area of the region under $1/x^2$ to the right of 1 using the notation

$$\lim_{b\to\infty} \int_1^b \frac{1}{x^2}\,dx = \lim_{b\to\infty} \left(1 - \frac{1}{b}\right)$$

where $\lim\limits_{b\to\infty}$ represents the limit as b gets larger without bound. Clearly,

$$\lim_{b\to\infty} \frac{1}{b} = 0$$

so

$$\lim_{b\to\infty} \left(1 - \frac{1}{b}\right) = 1$$

Thus the area under the curve $y = 1/x^2$ to the right of $x = 1$ is 1.

In general, we define the area under a curve $y = f(x)$ to the right of $x = a$, with $f(x) \geq 0$, to be

$$\text{Area} = \lim_{b\to\infty} (\text{area from } a \text{ to } b) = \lim_{b\to\infty} \int_a^b f(x)\, dx$$

This motivates the definition that follows.

Improper Integral

$$\int_a^\infty f(x)\, dx = \lim_{b\to\infty} \int_a^b f(x)\, dx$$

If the limit defining the improper integral is a unique finite number, we say that the integral *converges;* otherwise, we say that the integral *diverges.*

EXAMPLE 1

Evaluate the following improper integrals, if they converge.

(a) $\int_1^\infty \frac{1}{x^3}\, dx$ (b) $\int_1^\infty \frac{1}{x}\, dx$

Solution

(a)
$$\int_1^\infty \frac{1}{x^3}\, dx = \lim_{b\to\infty} \int_1^b \frac{1}{x^3}\, dx = \lim_{b\to\infty} \left[\frac{x^{-2}}{-2}\right]_1^b$$
$$= \lim_{b\to\infty} \left[\frac{-1}{2b^2} - \left(\frac{-1}{2(1)^2}\right)\right]$$
$$= \lim_{b\to\infty} \left(\frac{-1}{2b^2} + \frac{1}{2}\right)$$

Now as $b \to \infty$, $\frac{-1}{2b^2} \to 0$, so the limit, and the integral, converge to $\frac{1}{2}$. That is,

$$\int_1^\infty \frac{1}{x^3}\, dx = \frac{1}{2}$$

(b)
$$\int_1^\infty \frac{1}{x}\, dx = \lim_{b\to\infty} \int_1^b \frac{1}{x}\, dx = \lim_{b\to\infty} \Big[\ln|x|\Big]_1^b$$
$$= \lim_{b\to\infty} (\ln b - \ln 1)$$

Now $\ln b$ increases without bound as $b \to \infty$, so the limit, and the integral, diverge. We write this as

$$\int_1^\infty \frac{1}{x}\, dx = \infty$$

From Example 1 we can conclude that the area under the curve $y = 1/x^3$ to the right of $x = 1$ is $\frac{1}{2}$, whereas the corresponding area under the curve $y = 1/x$ is infinite. (We have already seen that the corresponding area under $y = 1/x^2$ is 1.)

As Figure 13.26 shows, the graphs of the curves look similar, but the graph of $1/x^2$ gets "close" to the x-axis much more rapidly than the graph of $1//x$. The area under $y = 1/x$ does not converge to a finite number because as $x \to \infty$ the graph of $1/x$ does not approach the x-axis rapidly enough.

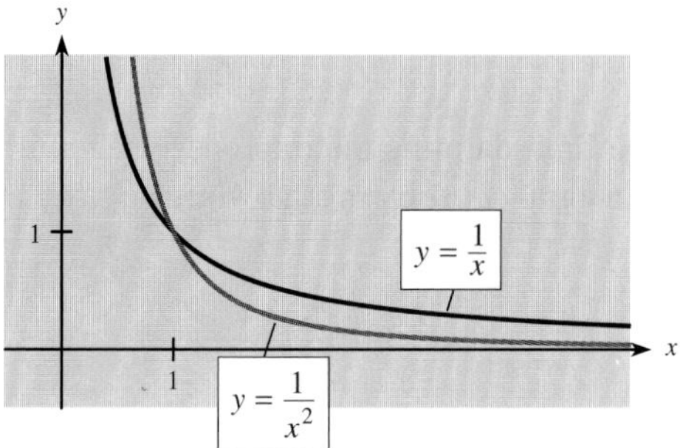

Figure 13.26

EXAMPLE 2

Suppose that an organization wants to establish a trust fund that will provide a continuous income stream with an annual rate of flow at time t given by $f(t) = 10{,}000$. If the interest rate remains at 10%, compounded continuously, find the capital value of the fund.

Solution

The capital value of the fund is given by

$$\int_0^\infty f(t)\, e^{-rt}\, dt$$

where $f(t)$ is the annual rate of flow at time t, and r is the annual interest rate, compounded continuously.

$$\begin{aligned}
\int_0^\infty 10{,}000e^{-0.10t}\, dt &= \lim_{b\to\infty} \int_0^b 10{,}000e^{-0.10t}\, dt \\
&= \lim_{b\to\infty} \Big[-100{,}000e^{-0.10t}\Big]_0^b \\
&= \lim_{b\to\infty} \left(\frac{-100{,}000}{e^{0.10b}} + 100{,}000\right) \\
&= 100{,}000
\end{aligned}$$

Thus the capital value of the fund is \$100,000.

Another term for a fund such as the one in Example 2 is a **perpetuity.** Usually the rate of flow of a perpetuity is a constant. If the rate of flow is a constant A, it can be shown that the capital value is given by A/r (see Problem 37 in the exercise set).

CHECKPOINT

1. True or false:

   (a) $\lim_{b\to+\infty} \frac{1}{b^p} = 0$ if $p > 0$ (b) $\lim_{b\to+\infty} b^p = +\infty$ if $p > 0$

   (c) $\lim_{b\to+\infty} e^{-pb} = 0$ if $p > 0$

2. Evaluate the following (if they exist).

   (a) $\int_1^{\infty} \frac{1}{x^{4/3}}\,dx = \lim_{b\to\infty} \int_1^b x^{-4/3}\,dx$ (b) $\int_0^{\infty} \frac{dx}{\sqrt{x+1}}$

A second improper integral has the form

$$\int_{-\infty}^{b} f(x)\,dx$$

and is defined by

$$\int_{-\infty}^{b} f(x)\,dx = \lim_{a\to\infty} \int_{-a}^{b} f(x)\,dx$$

The integral converges if the limit is finite. In addition, the improper integral

$$\int_{-\infty}^{\infty} f(x)\,dx$$

is defined by

$$\int_{-\infty}^{\infty} f(x)\,dx = \lim_{a\to\infty} \int_{-a}^{c} f(x)\,dx + \lim_{b\to\infty} \int_{c}^{b} f(x)\,dx$$

for any finite constant c. (0 is often used for c.) If both limits are finite, the improper integral converges; otherwise, it diverges.

EXAMPLE 3

Evaluate the following integrals.

(a) $\int_{-\infty}^{4} e^{3x}\,dx$ (b) $\int_{-\infty}^{\infty} \frac{x^3}{(x^4+3)^2}\,dx$

Solution

(a)
$$\int_{-\infty}^{4} e^{3x}\,dx = \lim_{a\to\infty} \int_{-a}^{4} e^{3x}\,dx$$
$$= \lim_{a\to\infty} \left[\left(\frac{1}{3}\right)e^{3x}\right]_{-a}^{4}$$
$$= \lim_{a\to\infty} \left[\left(\frac{1}{3}\right)e^{12} - \left(\frac{1}{3}\right)e^{-3a}\right]$$

$$= \lim_{a\to\infty}\left[\left(\frac{1}{3}\right)e^{12} - \left(\frac{1}{3}\right)\left(\frac{1}{e^{3a}}\right)\right]$$

$$= \frac{1}{3}e^{12} \qquad \text{(because } 1/e^{3a} \to 0 \text{ as } a \to \infty\text{)}$$

(b) $$\int_{-\infty}^{\infty} \frac{x^3}{(x^4+3)^2}\,dx = \lim_{a\to\infty}\int_{-a}^{0} \frac{x^3}{(x^4+3)^2}\,dx + \lim_{b\to\infty}\int_{0}^{b} \frac{x^3}{(x^4+3)^2}\,dx$$

$$= \lim_{a\to\infty}\left[\frac{1}{4}\frac{(x^4+3)^{-1}}{-1}\right]_{-a}^{0} + \lim_{b\to\infty}\left[\frac{1}{4}\frac{(x^4+3)^{-1}}{-1}\right]_{0}^{b}$$

$$= \lim_{a\to\infty}\left[-\frac{1}{4}\left(\frac{1}{3} - \frac{1}{a^4+3}\right)\right] + \lim_{b\to\infty}\left[-\frac{1}{4}\left(\frac{1}{b^4+3} - \frac{1}{3}\right)\right]$$

$$= -\frac{1}{12} + 0 + 0 + \frac{1}{12} = 0$$

$$\left(\text{since } \lim_{a\to\infty}\frac{1}{a^4+3} = 0 \quad \text{and} \quad \lim_{b\to\infty}\frac{1}{b^4+3} = 0\right)$$

In Section 13.2, "The Definite Integral," we calculated the probability that a computer component will last between 3 and 5 years when the probability density function for the life span is $f(x) = 0.10e^{-0.10x}$, where x is the number of years, $x \geq 0$. The probability that the component will last more than 3 years is given by the improper integral

$$\int_{3}^{\infty} 0.10e^{-0.10x}\,dx$$

which gives

$$\lim_{b\to\infty}\int_{3}^{b} 0.10e^{-0.10x}\,dx = \lim_{b\to\infty}\left[-e^{-0.10x}\right]_{3}^{b}$$

$$= \lim_{b\to\infty}\left(-e^{-0.10b} + e^{-0.3}\right)$$

$$= e^{-0.3} = 0.7408$$

We noted in Chapter 8, "Further Topics in Probability" that the sum of the probabilities for a probability distribution (a probability density function) equals 1. In particular, we stated that the area under the normal probability curve is 1.

Probability Density Function

In general, if $f(x) \geq 0$ for all x, then f is a probability density function for a continuous random variable if and only if

$$\int_{-\infty}^{\infty} f(x)\,dx = 1$$

The complete probability density function for the life span of the computer component mentioned above is

$$f(x) = \begin{cases} 0.10e^{-0.10x} & \text{if } x \geq 0 \\ 0 & \text{if } x < 0 \end{cases}$$

We can verify that

$$\int_{-\infty}^{\infty} f(x)\,dx = 1$$

for this function.

$$\int_{-\infty}^{\infty} f(x)\,dx = \int_{-\infty}^{0} f(x)\,dx + \int_{0}^{\infty} f(x)\,dx = \int_{-\infty}^{0} 0\,dx + \int_{0}^{\infty} 0.10e^{-0.10x}\,dx$$

$$= 0 + \lim_{b\to\infty} \int_{0}^{b} 0.1e^{-0.10x}\,dx$$

$$= \lim_{b\to\infty} \left[-e^{-0.10x}\right]_0^b$$

$$= \lim_{b\to\infty} \left[-e^{-0.10b} + 1\right]$$

$$= 1$$

In Section 8.2, "Discrete Probability Distributions," we found the expected value (mean) of a discrete probability distribution using the formula

$$E(x) = \sum x\text{Pr}(x)$$

For continuous probability distributions, such as the normal probability distribution, the expected value, or mean, can be found by evaluating the improper integral

$$\int_{-\infty}^{\infty} xf(x)\,dx$$

Mean (Expected Value) If x is a continuous random variable with probability density function f, then the mean of the probability distribution is

$$\mu = \int_{-\infty}^{\infty} xf(x)\,dx$$

The normal distribution density function, in standard form, is

$$f(x) = \frac{1}{\sqrt{2\pi}}e^{-x^2/2}$$

so the mean of the normal probability distribution is given by

$$\mu = \int_{-\infty}^{\infty} x\left(\frac{1}{\sqrt{2\pi}}e^{-x^2/2}\right)dx$$

$$= \lim_{a\to\infty} \int_{-a}^{0} \frac{1}{\sqrt{2\pi}}xe^{-x^2/2}\,dx + \lim_{b\to\infty} \int_{0}^{b} \frac{1}{\sqrt{2\pi}}xe^{-x^2/2}\,dx$$

$$= \lim_{a\to\infty} \frac{1}{\sqrt{2\pi}}\left[-e^{-x^2/2}\right]_{-a}^{0} + \lim_{b\to\infty} \frac{1}{\sqrt{2\pi}}\left[-e^{-x^2/2}\right]_{0}^{b}$$

$$= \frac{1}{\sqrt{2\pi}}(-1+0) + \frac{1}{\sqrt{2\pi}}(0+1) = 0$$

This verifies the statement in Chapter 8 that the mean of the standard normal distribution is 0.

EXAMPLE 4

Find the area of the region below the graph of $f(x) = \frac{\ln x}{x^2}$ and above the x-axis. This can be done by evaluating the integral over the interval where $f(x)$ is above the x-axis. It is not possible to find two points where $f(x) = 0$, so we will use graphs to find the interval and to evaluate the integral.

(a) Graph $y = f(x)$ to find the interval over which to integrate.
(b) Evaluate the integral of $f(x)$ over this interval.

Solution

(a) The graph of $y = f(x)$ is shown in Figure 13.27(a). Using TRACE and ZOOM shows that $f(1) = 0$ and $f(x)$ approaches 0 as x approaches $+\infty$. Thus the area is found by evaluating

$$\int_1^{+\infty} f(x)\, dx = \int_1^{+\infty} \frac{\ln x}{x^2}\, dx = \lim_{b \to +\infty} \int_1^b \frac{\ln x}{x^2}\, dx$$

(b) Evaluating $\int_1^b \frac{\ln x}{x^2}\, dx$ requires the use of integration by parts, with

$$u = \ln x \quad \text{and} \quad dv = x^{-2}$$

Thus

$$du = \frac{1}{x}\, dx \quad \text{and} \quad v = \frac{x^{-1}}{-1}$$

$$\int_1^b \frac{\ln x}{x^2}\, dx = (\ln x)\left(\frac{x^{-1}}{-1}\right)\Big|_1^b - \int_1^b (-x^{-1})\frac{1}{x}\, dx = -\frac{\ln x}{x}\Big|_1^b + \frac{x^{-1}}{-1}\Big|_1^b$$

$$= \left(-\frac{\ln b}{b} + \frac{\ln 1}{1}\right) - \left(\frac{1}{b} - 1\right) = 1 - \frac{1}{b} - \frac{\ln b}{b}$$

$$\int_1^{\infty} f(x)\, dt = \lim_{b \to \infty} \left(1 - \frac{1}{b} - \frac{\ln b}{b}\right) = \lim_{b \to \infty} 1 - \lim_{b \to \infty} \frac{1}{b} - \lim_{b \to \infty} \frac{\ln b}{b}$$

We have not developed a method to evaluate $\lim_{b \to \infty} \left(\frac{\ln b}{b}\right)$ or, equivalently, $\lim_{x \to \infty} \left(\frac{\ln x}{x}\right)$. But we can use ZOOM and TRACE to see that the graph of $y = 1 - \frac{1}{x} - \frac{\ln x}{x}$ approaches 1 as x approaches $+\infty$ (see Figure 13.27b).

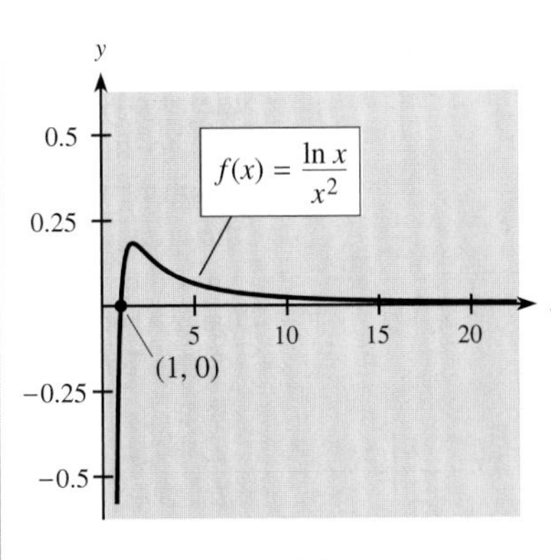

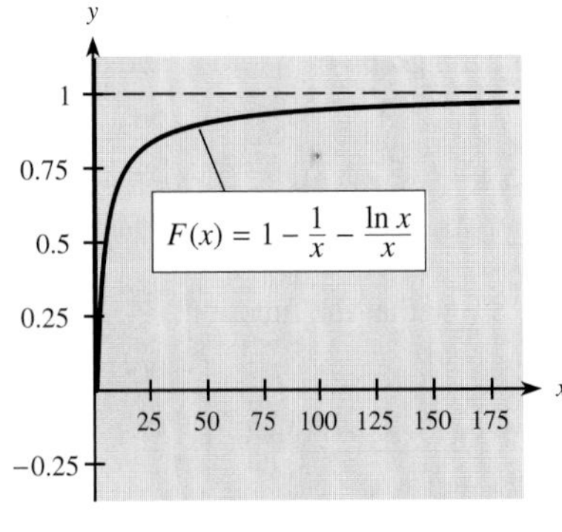

Figure 13.27 (a) (b)

Thus $\lim_{b\to\infty}\left(1 - \frac{1}{b} - \frac{\ln b}{b}\right) = 1$, and the area under $y = f(x)$ is $\int_1^\infty f(x)\,dx = \lim_{b\to\infty}\left(1 - \frac{1}{b} - \frac{\ln b}{b}\right) = 1$.

CHECKPOINT SOLUTIONS

1. (a) True (b) True (c) True
2. (a) $\lim_{b\to\infty}\int_1^b x^{-4/3}\,dx = \lim_{b\to\infty}\left.\frac{x^{-1/3}}{-1/3}\right|_1^b = \lim_{b\to\infty}\left(\frac{-3}{b^{1/3}} - \frac{-3}{1}\right) = 0 + 3 = 3$

   (b) $\lim_{b\to\infty}\int_0^b (x+1)^{-1/2}\,dx = \lim_{b\to\infty}\left.\frac{(x+1)^{1/2}}{1/2}\right|_0^b = \lim_{b\to\infty} 2\sqrt{x+1}\Big|_0^b$

   $= \lim_{b\to\infty}(2\sqrt{b+1} - 2) = \infty$

   (Integral diverges)

EXERCISE 13.7

In Problems 1–20, evaluate the improper integrals that converge.

1. $\int_1^\infty \frac{dx}{x^6}$
2. $\int_1^\infty \frac{1}{x^4}\,dx$
3. $\int_1^\infty \frac{dt}{t^{3/2}}$
4. $\int_5^\infty \frac{dx}{(x-1)^3}$
5. $\int_1^\infty e^{-x}\,dx$
6. $\int_0^\infty x^2 e^{-x^3}\,dx$
7. $\int_1^\infty \frac{dt}{t^{1/3}}$
8. $\int_1^\infty \frac{1}{\sqrt{x}}\,dx$
9. $\int_0^\infty e^{3x}\,dx$
10. $\int_1^\infty x e^{x^2}\,dx$
11. $\int_{-\infty}^{-1} \frac{10}{x^2}\,dx$
12. $\int_{-\infty}^{-2} \frac{x}{\sqrt{x^2-1}}\,dx$
13. $\int_{-\infty}^{0} x^2 e^{-x^3}\,dx$
14. $\int_{-\infty}^{0} \frac{x}{(x^2+1)^2}\,dx$
15. $\int_{-\infty}^{-1} \frac{6}{x}\,dx$
16. $\int_{-\infty}^{-2} \frac{3x}{x^2+1}\,dx$
17. $\int_{-\infty}^{\infty} \frac{2x}{x^2+1}\,dx$
18. $\int_{-\infty}^{\infty} \frac{x}{(x^2+1)^2}\,dx$
19. $\int_{-\infty}^{\infty} x^3 e^{-x^4}\,dx$
20. $\int_{-\infty}^{\infty} x^4 e^{-x^5}\,dx$
21. For what value of c does
$$\int_0^\infty \frac{c}{e^{0.5t}}\,dt = 1?$$
22. For what value of c does
$$\int_{10}^\infty \frac{c}{x^3}\,dx = 1?$$

In Problems 23–26, find the area, if it exists, of the region under the graph of $y = f(x)$ and to the right of $x = 1$.

23. $f(x) = \dfrac{x}{e^{x^2}}$

24. $f(x) = \dfrac{1}{\sqrt[5]{x^3}}$

25. $f(x) = \dfrac{1}{\sqrt[3]{x^5}}$

26. $f(x) = \dfrac{1}{x\sqrt{x}}$

27. Show that the function

$$f(x) = \begin{cases} \dfrac{200}{x^3} & \text{if } x \geq 10 \\ 0 & \text{otherwise} \end{cases}$$

is a probability density function.

28. Show that

$$f(x) = \begin{cases} 3e^{-3t} & \text{if } t \geq 0 \\ 0 & \text{if } t < 0 \end{cases}$$

is a probability density function.

29. For what value of c is the function

$$f(x) = \begin{cases} c/x^2 & \text{if } x \geq 1 \\ 0 & \text{otherwise} \end{cases}$$

a probability density function?

30. For what value of c is the function

$$f(x) = \begin{cases} c/x^3 & \text{if } x \geq 100 \\ 0 & \text{otherwise} \end{cases}$$

a probability density function?

31. If

$$f(x) = \begin{cases} ce^{-x/4} & x \geq 0 \\ 0 & x < 0 \end{cases}$$

is a probability density function, what must be the value of c?

32. If

$$f(x) = \begin{cases} ce^{-kx} & \text{if } x \geq 0 \\ 0 & \text{if } x < 0 \end{cases}$$

is a probability density function, what must be the value of c?

33. Find the mean of the probability distribution if the probability density function is

$$f(x) = \begin{cases} \dfrac{200}{x^3} & \text{if } x \geq 10 \\ 0 & \text{otherwise} \end{cases}$$

34. Find the mean of the probability distribution if the probability density function is

$$f(x) = \begin{cases} c/x^3 & \text{if } x \geq 100 \\ 0 & \text{otherwise} \end{cases}$$

35. Find the area below the graph of $y = f(x)$ and above the x-axis for $f(x) = 24xe^{-3x}$. Use the graph of $y = f(x)$ to find the interval for which $f(x) \geq 0$ and the graph of the integral of $f(x)$ over this interval to find the area.

36. Find the area below the graph of $y = f(x)$ and above the x-axis for $f(x) = x^2e^{-x}$ and $x \geq 0$. Use the graph of the integral of $f(x)$ over this interval to find the area.

Applications

37. ***Capital value*** Suppose that a continuous income stream has an annual rate of flow at time t given by $f(t) = A$, where A is a constant. If the interest rate is r (as a decimal, $r > 0$), compounded continuously, show that the capital value of the stream is A/r.

38. ***Capital value*** Suppose that a donor wishes to provide a cash gift to a hospital that will generate a continuous income stream with an annual rate of flow at time t given by $f(t) = \$20{,}000$. If the annual interest rate is 12%, compounded continuously, find the capital value of this perpetuity.

39. ***Capital value*** Suppose that a business provides a continuous income stream with an annual rate of flow at time t given by $f(t) = 120e^{0.04t}$ in thousands of dollars. If the interest rate is 9%, compounded continuously, find the capital value of the business.

40. ***Capital value*** Suppose that the output of the machinery in a factory can be considered as a continuous income stream with annual rate of flow at time t given by $f(t) = 450e^{-0.09t}$ in thousands of dollars. If the annual interest rate is 6%, compounded continuously, find the capital value of the machinery.

41. ***Capital value*** A business has a continuous income stream with an annual rate of flow at time t given by $f(t) = 56{,}000e^{0.02t}$ (dollars). If the interest rate is 10%, compounded continuously, find the capital value of the business.

42. ***Capital value*** Suppose that a business provides a continuous income stream with an annual rate of flow at time t given by $f(t) = 10{,}800e^{0.06t}$ (dollars). If money is worth 12%, compounded continuously, find the capital value of the business.

43. ***Radioactive waste*** Suppose that the rate at which a nuclear power plant produces radioactive waste is proportional to the number of years it has been operating, according to $f(t) = 500t$ in pounds per year. Suppose also that the waste decays exponentially at a

rate of 3% per year. Then the amount of radioactive waste that will accumulate in b years is given by

$$\int_0^b 500te^{-0.03(b-t)}\,dt$$

(a) Evaluate this integral.
(b) How much waste will accumulate in the long run? Take the limit as $b \to \infty$ in (a).

44. ***Field intensity*** The field intensity around an infinitely long, straight electrical wire is

$$F = \frac{\mu IM}{10} \int_{-\infty}^{\infty} \frac{dx}{(x^2 + r^2)^{3/2}}$$

where μ, r, I, and M are constants. Evaluate this integral.

45. ***Quality control*** The probability density function for the life span of an electronics part is $f(t) = 0.08e^{-0.08t}$, where t is the number of months in service. Find the probability that any given part of this type lasts longer than 24 months.

46. ***Warranties*** A transmission repair firm that wants to offer a lifetime warranty on its repairs has determined that the probability density function for transmission failure after repair is $f(t) = 0.3e^{-0.3t}$, where t is the number of months after repair. What is the probability that a transmission chosen at random will last
(a) 3 months or less?
(b) more than 3 months?

KEY TERMS AND FORMULAS

Section	Key Terms	Formula
13.1	Sigma notation	$\sum_{i=1}^{n} 1 = n;\quad \sum_{i=1}^{n} i = \frac{n(n+1)}{2}$
		$\sum_{i=1}^{n} i^2 = \frac{n(n+1)(2n+1)}{6}$
	Area	
	Right-hand endpoints	$\lim_{n\to\infty} \sum_{i=1}^{n} f(x_i)\frac{b-a}{n}$
	Left-hand endpoints	$\lim_{n\to\infty} \sum_{i=1}^{n} f(x_{i-1})\frac{b-a}{n}$
13.2	Riemann sum	$\sum_{i=1}^{n} f(x_i^*)\Delta x_i$
	Definite integral	$\int_a^b f(x)\,dx = \lim_{\substack{\max \Delta x_i \to 0 \\ (n\to\infty)}} \sum_{i=1}^{n} f(x_i^*)\,\Delta x_i$
	Fundamental Theorem of Calculus	$\int_a^b f(x)\,dx = F(b) - F(a)$, where $F'(x) = f(x)$
	Definite Integral Properties	$\int_a^a f(x)\,dx = 0$
		$\int_a^b f(x)\,dx = -\int_b^a f(x)\,dx$
		$\int_a^b [f(x) \pm g(x)]\,dx = \int_a^b f(x)\,dx \pm \int_a^b g(x)\,dx$
		$\int_a^b c \cdot f(x)\,dx = c\int_a^b f(x)\,dx$

Section	Key Terms	Formula
		$\int_a^c f(x)\,dx + \int_c^b f(x)\,dx = \int_a^b f(x)\,dx$
	Area under $f(x)$, where $f(x) \geq 0$	$A = \int_a^b f(x)\,dx$
13.3	Area between $f(x)$ and $g(x)$, where $f(x) \geq g(x)$	$A = \int_a^b [f(x) - g(x)]\,dx$
	Average value over $[a, b]$	$\frac{1}{b-a}\int_a^b f(x)\,dx$
	Lorenz curve	
	Gini coefficient	$2\int_0^1 [x - L(x)]\,dx$
13.4	Continuous income streams	
	Total income	$\int_0^k f(t)\,dt$ (for k years)
	Present value	$\int_0^k f(t)e^{-rt}\,dt$, where r is the interest rate
	Future value	$e^{rk}\int_0^k f(t)e^{-rt}\,dt$
	Consumer's surplus [demand is $f(x)$]	$CS = \int_0^{x_1} f(x)\,dx - p_1x_1$
	Producer's surplus [supply is $g(x)$]	$PS = p_1x_1 - \int_0^{x_1} g(x)\,dx$
13.5	Integration from tables	See Table 13.2.
13.6	Integration by parts	$\int u\,dv = uv - \int v\,du$
13.7	Improper integrals	$\int_a^{\infty} f(x)\,dx = \lim_{b\to\infty}\int_a^b f(x)\,dx$
		$\int_{-\infty}^{b} f(x)\,dx = \lim_{a\to\infty}\int_{-a}^b f(x)\,dx$
		$\int_{-\infty}^{\infty} f(x)\,dx = \int_{-\infty}^{c} f(x)\,dx + \int_c^{\infty} f(x)\,dx$
	Capital value of a continuous income stream	$\int_0^{\infty} f(t)\,e^{-rt}\,dt$
	Probability distribution	$\int_{-\infty}^{\infty} f(x)\,dx = 1$
	Mean	$\mu = \int_{-\infty}^{\infty} xf(x)\,dx$

REVIEW EXERCISES

Section 13.1

1. Calculate $\sum_{k=1}^{8} (k^2 + 1)$.
2. Use formulas to simplify
$$\sum_{i=1}^{n} \frac{3i}{n^3}$$
3. Use 6 subintervals of the same size to approximate the area under the graph of $y = 3x^2$ from $x = 0$ to $x = 1$. Use the right-hand endpoints of the subintervals to find the heights of the rectangles.
4. Use rectangles to find the area under the graph of $y = 3x^2$ from $x = 0$ to $x = 1$. Use n equal subintervals.

Section 13.2

5. Use a definite integral to find the area under the graph of $y = 3x^2$ from $x = 0$ to $x = 1$.
6. Find the area between the graph of $y = x^3 - 4x + 5$ and the x-axis from $x = 1$ to $x = 3$.

Evaluate the integrals in Problems 7–18.

7. $\int_1^4 4\sqrt{x^3}\, dx$
8. $\int_{-3}^{2} (x^3 - 3x^2 + 4x + 2)\, dx$
9. $\int_0^5 (x^3 + 4x)\, dx$
10. $\int_{-2}^{3} (x + 2)^2\, dx$
11. $\int_{-3}^{-1} (x + 1)\, dx$
12. $\int_2^3 \frac{x^2}{2x^3 - 7}\, dx$
13. $\int_{-1}^{2} (x^2 + x)\, dx$
14. $\int_1^4 \left(\frac{1}{x} + \sqrt{x}\right) dx$
15. $\int_0^4 (2x + 1)^{1/2}\, dx$
16. $\int_0^1 \frac{x}{x^2 + 1}\, dx$
17. $\int_0^1 e^{-2x}\, dx$
18. $\int_0^1 xe^{x^2}\, dx$

Section 13.3

Find the area between the curves in Problems 19–22.

19. $y = x^2 - 3x + 2$ and $y = x^2 + 4$ from $x = 0$ to $x = 5$
20. $y = x^2$ and $y = 4x + 5$
21. $y = x^3$ and $y = x$ from $x = -1$ to $x = 0$
22. $y = x^3 - 1$ and $y = x - 1$

Section 13.5

Evaluate the integrals in Problems 23–26, using Table 13.2.

23. $\int \sqrt{x^2 - 4}\, dx$
24. $\int_0^1 3^x\, dx$
25. $\int x \ln x^2\, dx$
26. $\int \frac{dx}{x(3x + 2)}$

Section 13.6

In Problems 27–30, use integration by parts to evaluate.

27. $\int x^5 \ln x\, dx$
28. $\int xe^{-2x}\, dx$
29. $\int \frac{x\, dx}{\sqrt{x + 5}}$
30. $\int_1^e \ln x\, dx$

Section 13.7

Evaluate the improper integrals in Problems 31–34.

31. $\int_1^{\infty} \frac{1}{x}\, dx$
32. $\int_{-\infty}^{-1} \frac{200}{x^3}\, dx$
33. $\int_0^{\infty} 5e^{-3x}\, dx$
34. $\int_{-\infty}^{0} \frac{x}{(x^2 + 1)^2}\, dx$

Applications

Section 13.2

35. ***Maintenance*** Maintenance costs for buildings increase as the buildings age. If the rate of increase in maintenance costs for a building is
$$M'(t) = \frac{14{,}000}{\sqrt{t + 16}}$$
where M is in dollars and t is time in years, $0 \le t \le 15$, find the total maintenance cost for the first 9 years ($t = 0$ to $t = 9$).

36. ***Quality control*** Suppose the probability density function for the life expectancy of a "disposable" telephone is
$$f(x) = \begin{cases} 1.4e^{-1.4x} & x \ge 0 \\ 0 & x < 0 \end{cases}$$
Find the probability that the telephone lasts 2 years.

Section 13.3

37. ***Savings*** The future value of \$1000 invested in a savings account at 10%, compounded continuously, is $S = 1000e^{0.1t}$, where t is in years. Find the average amount in the savings account during the first 5 years.

38. ***Income streams*** Suppose the total income from a video machine is given by
$$I = 50e^{0.2t}, \quad 0 \le t \le 4,\ t \text{ in hours}$$
Find the average income over this 4-hour period.

Section 13.4

39. ***Consumer's surplus*** The demand function for a product under pure competition is $p = \sqrt{64 - 4x}$, and the supply function is $p = x - 1$.
    (a) Find the market equilibrium.
    (b) Find the consumer's surplus at market equilibrium.

40. ***Producer's surplus*** Find the producer's surplus at market equilibrium for Problem 39.

41. ***Income streams*** Find the total income over the next 10 years from a continuous income stream that has an annual flow rate at time t given by
$$f(t) = 125e^{0.05t}$$
in thousands of dollars.

42. ***Income streams*** Suppose that a machine's production is considered a continuous income stream with an annual rate of flow at time t given by
$$f(t) = 150e^{-0.2t}$$
in thousands of dollars. If money is worth 8%, compounded continuously, find (a) the present value of the machine's production over the next 5 years, and (b) the future value of the production 5 years from now.

Section 13.5

43. ***Average cost*** Suppose the cost function for a product is given by $C(x) = \sqrt{40{,}000 + x^2}$. Find the average cost over the first 150 units.

Section 13.6

44. ***Income streams*** Suppose the present value of a continuous income stream over the next 5 years is given by
$$P = 9000\int_0^5 te^{-0.08t}\,dt, \quad P \text{ in dollars, } t \text{ in years}$$
Find the present value.

45. ***Cost*** If the marginal cost for a product is $\overline{MC} = 3 + x\ln(x + 1)$ and if the fixed cost is \$2000, find the total cost function.

Section 13.7

46. ***Quality control*** Find the probability that a telephone lasts more than 1 year if the probability density function for its life expectancy is given by
$$f(x) = \begin{cases} 1.4e^{-1.4x} & x \geq 0 \\ 0 & x < 0 \end{cases}$$

47. ***Capital value*** Find the capital value of a business if its income is considered a continuous income stream with annual rate of flow given by
$$f(t) = 120e^{0.03t}$$
in thousands of dollars, and the current interest rate is 6%, compounded continuously.

CHAPTER TEST

1. Use left-hand endpoints and $n = 4$ subdivisions to approximate the area under $f(x) = \sqrt{4 - x^2}$ on the interval $[0, 2]$.
2. Consider $f(x) = 5 - 2x$ from $x = 0$ to $x = 1$ with n equal subdivisions.
   (a) If $f(x)$ is evaluated at right-hand endpoints, find a formula for the sum, S, of the areas of the n rectangles.
   (b) Find $\lim_{n\to\infty} S$.
3. Express the area in quadrant I under $y = 12 + 4x - x^2$ (shaded in the figure) as an integral. Then evaluate the integral to find the area.

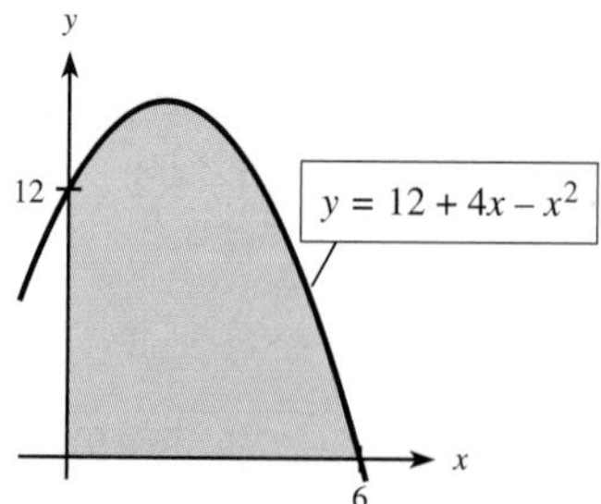

4. Evaluate the following integrals with the Fundamental Theorem.
   (a) $\int_0^4 (9 - 4x)\,dx$ (b) $\int_0^3 x(8x^2 + 9)^{-1/2}dx$

(c) $\int_1^4 \frac{5}{4x-1}\,dx$ (d) $\int_1^\infty \frac{7}{x^2}\,dx$

5. Use integration by parts to evaluate the following.

(a) $\int 3xe^x\,dx$ (b) $\int x\ln(2x)\,dx$

6. If $\int_1^4 f(x)\,dx = 3$ and $\int_3^4 f(x)\,dx = 7$, find $\int_1^3 2f(x)\,dx$.

7. Use Table 13.2 on page 966 to evaluate each of the following.

(a) $\int \ln(2x)\,dx$ (b) $\int x\sqrt{3x-7}\,dx$

8. Use the numerical integration feature of a graphing utility to approximate $\int_1^4 \sqrt{x^3+10}\,dx$.

9. Suppose the supply function for a product is $p = 40 + 0.001x^2$ and the demand function is $p = 120 - 0.2x$, where x is the number of units and p is the price in dollars. If the market equilibrium price is \$80, find (a) the consumer's surplus and (b) the producer's surplus.

10. Suppose a continuous income stream has an annual rate of flow $f(t) = 85e^{-0.01t}$, in thousands of dollars, and the current interest rate is 7%, compounded continuously.

(a) Find the total income over the next 12 years.

(b) Find the present value over the next 12 years.

(c) Find the capital value of the stream.

11. Find the area between $y = 2x + 4$ and $y = x^2 - x$.

12. The figure shows typical supply and demand curves. On the figure, sketch and shade the region whose area represents the consumer's surplus.

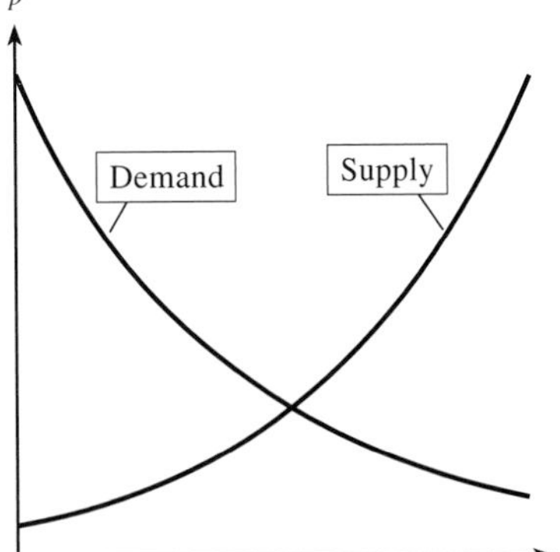

13. In an effort to make the distribution of income more nearly equal, the government of a country passes a tax law that changes the Lorenz curve from $y = 0.998x^{2.6}$ for one year to $y = 0.57x^2 + 0.43x$ for the next year. Find the Gini coefficient of income for both years and determine whether the distribution of income is more or less equitable after the tax law is passed. Interpret the result.

14. With data from the U.S. Department of Energy, the number of millions of barrels of oil imported from the Persian Gulf each year from 1985 to 1996 can be modeled by the function

$$f(t) = 1.107t^3 - 46.288t^2 + 614.913t - 1952.342$$

where t is the number of years past 1980 (*Monthly Energy Review*, June 1997).

(a) Find the average number of barrels imported per year from 1985 to 1996 (that is, from $t = 5$ to $t = 16$).

(b) Find the average number of barrels imported per year during the 1990s (from $t = 10$ to $t = 16$).

Extended Applications / Group Projects

I. Retirement Planning

A 52-year-old client asks an accountant how to plan for his future retirement at age 62. He expects income from Social Security in the amount of \$16,000 per year and a retirement pension of \$30,000 per year from his employer. He wants to make monthly contributions to an investment plan that pays 8%, compounded monthly, for 10 years so that he will have a total income of \$62,000 per year for 30 years. What will the size of the monthly contributions have to be to accomplish this goal, if it is assumed that money will be worth 8%, compounded continuously throughout the period after he is 62?

To help you answer this question, complete the following.

1. How much money must the client withdraw annually from his investment plan during his retirement so that his total income goal is met?
2. How much money S must the client's account contain when he is 62 so that it will generate this annual amount for 30 years? *Hint:* S can be considered the present value over 30 years of a continuous income stream with the amount you found in question 1 as its annual rate of flow.
3. The monthly contribution R that would, after 10 years, amount to the present value S found in question 2 can be obtained from the formula

$$R = S\left[\frac{i}{(1+i)^n - 1}\right]$$

where i represents the monthly interest rate and n the number of months. Find the client's monthly contribution, R.

II. Purchasing Electrical Power

In order to plan its purchases of electrical power from suppliers over the next 5 years, the PAC Electric Company needs to model its load data (demand for power by its customers) and use this model to predict future loads. The company pays for the electrical power each month on the basis of the peak load (demand) at any point during the month. The table gives, for the years 1980–1998, the load in megawatts (million watts) for the month when the maximum load occurred and the load in megawatts for the month when the minimum load occurred. The maximum loads occurred in summer, and the minimum loads occurred in spring or fall.

Year	*Maximum Monthly Load*	*Minimum Monthly Load*
1980	40.9367	19.4689
1981	45.7127	22.1504
1982	48.0460	25.3670
1983	56.1712	28.7254
1984	55.5793	31.0460
1985	62.4285	31.3838
1986	76.6536	34.8426
1987	73.8214	38.4544
1988	74.8844	40.6080
1989	83.0590	47.3621
1990	88.3914	45.8393
1991	88.7704	48.7956
1992	94.2620	48.3313
1993	105.1596	52.7710
1994	95.8301	54.4757
1995	97.8854	55.2210
1996	102.8912	55.1360
1997	109.5541	57.2162
1998	111.2516	58.3216

The company wishes to predict the average monthly load over the next 5 years so that it can plan its future monthly purchases. To assist the company, proceed as follows.

1. (a) Using the years and the maximum monthly load given for each year, graph the data, with x representing the number of years from 1980 and y representing the load in megawatts.
   (b) Find the equation that best fits the data, using both a quadratic model and a cubic model.
   (c) Graph the data and both of these models from 1980 to 2000 (that is, from $x = 0$ to $x = 20$).
2. Do the two models appear to fit the data equally well in the interval 1980–2000? Which model appears to be a better predictor for the next decade?
3. Use the quadratic model to predict the maximum monthly load in the year 2003. How can this value be used by the company? Should this number be used to plan monthly power purchases for each month in 2003?
4. To create a "typical" monthly load function:
   (a) Create a table with the year as the independent variable and the average of the maximum and minimum monthly loads as the dependent variable.
   (b) Find the quadratic model that best fits these data points, using $x = 0$ in 1980.
5. Use a definite integral with the typical monthly load function to predict the average monthly load over the years 2000–2005.
6. What factors in addition to the average monthly load should be considered when the company plans future purchases of power?

Warm-up

Prerequisite Problem Type	For Section	Answer	Section for Review
If $y = f(x)$, x is the independent variable and y is the ______ variable.	14.1	Dependent	1.2 Functions
What is the domain of $f(x) = \dfrac{3x}{x-1}$?	14.1	All reals except $x = 1$	1.2 Domains
If $C(x) = 5 + 5x$, what is $C(0.20)$?	14.1	6	1.2 Functional notation
(a) Solve for x and y: $\begin{cases} 0 = 50 - 2x - 2y \\ 0 = 60 - 2x - 4y \end{cases}$ (b) Solve for x and y: $\begin{cases} x = 2y \\ x + y - 9 = 0 \end{cases}$	14.4 14.5	(a) $x = 20,\ y = 5$ (b) $x = 6,\ y = 3$	1.5 Systems of equations
If $z = 4x^2 + 5x^3 - 7$, what is $\dfrac{dz}{dx}$?	14.2 14.3 14.4 14.5	$\dfrac{dz}{dx} = 8x + 15x^2$	9.4 Derivatives
If $f(x) = (x^2 - 1)^2$, what is $f'(x)$?	14.2	$f'(x) = 4x(x^2 - 1)$	9.6 Derivatives
If $z = 10y - \ln y$, what is $\dfrac{dz}{dy}$?	14.2	$\dfrac{dz}{dy} = 10 - \dfrac{1}{y}$	11.1 Derivatives of logarithmic functions
If $z = 5x^2 + e^x$, what is $\dfrac{dz}{dx}$?	14.2	$\dfrac{dz}{dx} = 10x + e^x$	11.2 Derivatives of exponential functions
Find the slope of the tangent to $y = 4x^3 - 4e^x$ at $(0, -4)$.	14.2	-4	9.3, 9.4, 11.2 Derivatives

Chapter 14

Functions of Two or More Variables

Although we have been dealing primarily with functions of one variable, many real-life situations involve quantities that are functions of two or more variables. For example, the grade you receive in a course is a function of several test grades. The cost of manufacturing a product may involve the cost of labor, the cost of materials, and overhead expenses. The concentration of a substance at any point in a vein after an injection is a function of time since the injection t, *the velocity of the blood* v, *and the distance the point is from the point of injection.*

In this chapter we will extend our study to functions of two or more variables. We will extend the derivative concept to functions of several variables by taking partial derivatives, and we will learn how to maximize functions of two variables. We will use these concepts to solve problems in the management, social, and life sciences. In particular, we will discuss joint cost functions, marginal cost, marginal productivity, and marginal demand functions. We will use Lagrange multipliers to optimize functions of two variables subject to a condition that constrains the variables.

14.1 Functions of Two or More Variables

OBJECTIVES

- To find the domain and range of a function of two or more variables
- To evaluate a function of two or more variables given values for the independent variables

APPLICATION PREVIEW

The relations we have studied up to this point have been limited to two variables, with one of the variables assumed to be a function of the other. But there are many instances where one variable may depend on two or more other variables. For example, the output or production Q (for quantity) of a company can be modeled according to the equation

$$Q = AK^{\alpha}L^{1-\alpha}$$

where A is a constant, K is the company's capital investment, L is the size of the labor force (in work-hours), and α is a constant with $0 < \alpha < 1$. Functions of this type are called **Cobb-Douglas production functions,** and they are frequently used in economics. For example, suppose the Cobb-Douglas production function for a company is given by

$$Q = 4\,K^{0.4}L^{0.6}$$

where Q is dollars of production value. We could use this function to determine the production value for a given amount of capital investment and available work-hours of labor. We could also find how production is affected by changes in capital investment or available work-hours.

In addition, the demand function for a commodity frequently depends on the price of the commodity, available income, and prices of competing goods. Other examples from economics will be presented later in this chapter.

We write $z = f(x, y)$ to state that z is a function of both x and y. The variables x and y are called the **independent variables** and z is called the **dependent variable.** Thus the function f associates with each pair of possible values for the independent variables (x and y) exactly one value of the dependent variable (z).

The equation $z = x^2 - xy$ defines z as a function of x and y. We can denote this by writing $z = f(x, y) = x^2 - xy$. The domain of the function is the set of all ordered pairs (of real numbers), and the range is the set of all real numbers.

EXAMPLE 1

Give the domain of the function

$$g(x, y) = \frac{x^2 - 3y}{x - y}$$

Solution

The domain of the function is the set of ordered pairs that do not give a 0 denominator. That is, the domain is the set of all ordered pairs where the first and second elements are not equal (that is, where $x \neq y$).

CHECKPOINT

1. Find the domain of the function

$$f(x, y) = \frac{2}{\sqrt{x^2 - y^2}}$$

We graph the function $z = f(x, y)$ by using three dimensions. We can construct a three-dimensional coordinate space by drawing three mutually perpendicular axes, as in Figure 14.1. By setting up a scale of measurement along the three axes from the origin 0, we can determine the three coordinates (x, y, z) for any point P. The point shown in Figure 14.1 is $+2$ units in the x-direction, $+3$ units in the y-direction, and $+4$ units in the z-direction, so the coordinates of the point are (2, 3, 4).

The pairs of axes determine the three **coordinate planes;** the xy-plane, the yz-plane, and the xz-plane. The planes divide the space into eight **octants.** The point $P(2, 3, 4)$ is in the first octant.

If we are given a function $z = f(x, y)$, we can find the z-value corresponding to $x = a$ and $y = b$ by evaluating $f(a, b)$.

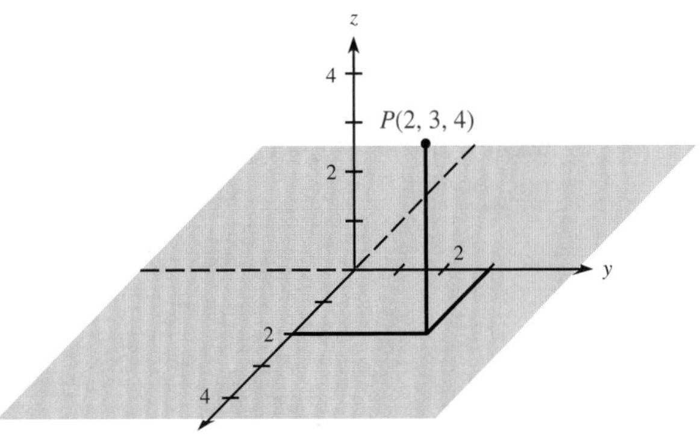

Figure 14.1

EXAMPLE 2

If $z = f(x, y) = x^2 - 4xy + xy^3$, find the following.

(a) $f(1, 2)$ (b) $f(2, 5)$ (c) $f(-1, 3)$

Solution

(a) $f(1, 2) = 1^2 - 4(1)(2) + (1)(2)^3 = 1$
(b) $f(2, 5) = 2^2 - 4(2)(5) + (2)(5)^3 = 214$
(c) $f(-1, 3) = (-1)^2 - 4(-1)(3) + (-1)(3)^3 = -14$

CHECKPOINT

2. If $f(x, y, z) = x^2 + 2y - z$, find $f(2, 3, 4)$.

EXAMPLE 3

A small furniture company's cost (in dollars) to manufacture 1 unit of several different all-wood items is given by

$$C(x, y) = 5 + 5x + 22y$$

where x represents the number of board feet of material used and y represents the number of work-hours of labor required for assembly and finishing. A certain bookcase uses 20 board feet of material and requires 2.5 work-hours for assembly and finishing. Find the cost of manufacturing this bookcase.

Solution

The desired cost is

$$C(20, 2.5) = 5 + 5(20) + 22(2.5) = 160 \quad \text{(dollars)}$$

For a given function $z = f(x, y)$, we can construct a table of values by assigning values to x and y and finding the corresponding values of z. To each pair of values for x and y there corresponds a unique value of z, and thus a unique point in a three-dimensional coordinate system. From a table of values such as this, a finite number of points can be plotted. All points that satisfy the equation form a "surface" in space. Because z is a function of x and y, lines parallel to the z-axis will intersect such a surface in at most one point. The graph of the equation $z = 4 - x^2 - y^2$ is a surface like that shown in Figure 14.2. The portion of the surface above the xy-plane resembles a bullet and is called a **paraboloid.**

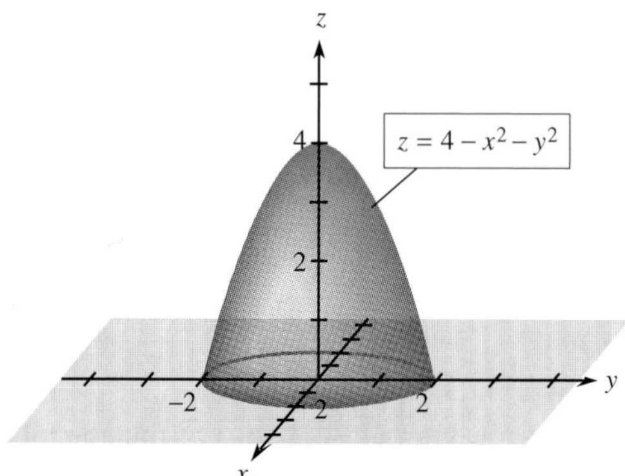

Figure 14.2

In practical applications of functions of two variables, we will have little need to construct the graphs of the surfaces. For this reason, we will not discuss methods of sketching the graphs. Although you will not be asked to sketch graphs of these surfaces, the fact that the graphs do *exist* will be used in studying relative maxima and minima of functions of two variables.

The properties of functions of one variable can be extended to functions of two variables. The precise definition of continuity for functions of two variables is technical and may be found in more advanced books. We will limit our study to functions that are continuous and have continuous derivatives in the domain of interest to us. We may think of continuous functions as functions whose graphs consist of surfaces without "holes" or "breaks" in them.

Let the function $U = f(x, y)$ represent the **utility** (that is, satisfaction) derived by a consumer from the consumption of two goods, X and Y, where x and y represent the amounts of X and Y, respectively. Because we will assume that the utility function is continuous, a given level of utility can be derived from an infinite number of combinations of x and y. The graph of all points (x, y) that give the same utility is called an **indifference curve.** A set of indifference curves corresponding to different levels of utility is called an **indifference map** (see Figure 14.3).

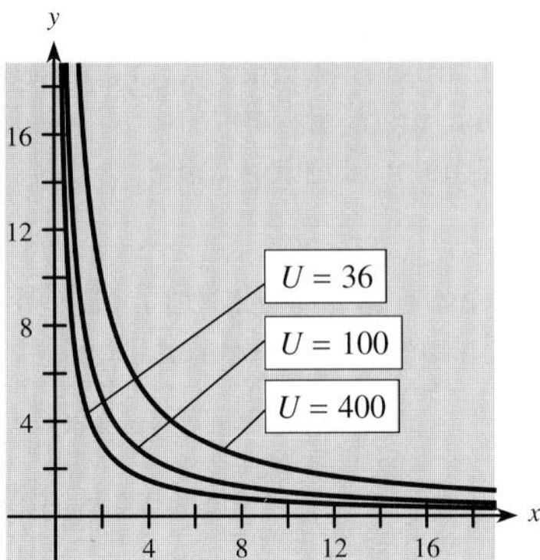

Figure 14.3

EXAMPLE 4

Suppose that the utility function for two goods, X and Y, is $U = x^2y^2$ and a consumer purchases 10 units of X and 2 units of Y.

(a) If the consumer purchases 5 units of X, how many units of Y must be purchased to retain the same level of utility?
(b) Graph the indifference curve for this level of utility.
(c) Graph the indifference curves for this utility function if $U = 100$ and if $U = 36$.

Solution

(a) If $x = 10$ and $y = 2$ satisfy the utility function, then $U = 10^2 \cdot 2^2 = 400$. Thus if x is 5, y must satisfy $400 = 5^2y^2$, so $y = 4$.
(b) The indifference curve for $U = 400$ is $400 = x^2y^2$. The graph for positive x and y is shown in Figure 14.3.
(c) The indifference map in Figure 14.3 contains these indifference curves.

Graphing Utilities

A graphing utility can be used to graph each indifference curve in the indifference map shown in Figure 14.3. To graph the indifference curve for a given value of U, we must recognize that y will be positive and solve for y to express it as a function of x. Try it for $U = 100$. Does your graph agree with the one shown in Figure 14.3?

Sometimes functions of two variables are studied by fixing a value for one variable and graphing the resulting function of a single variable. We'll do this in Section 14.3 with production functions.

EXAMPLE 5

Suppose a company has the Cobb-Douglas production function introduced in the Application Preview,

$$Q = 4K^{0.4}L^{0.6}$$

where Q is thousands of dollars of production value, K is hundreds of dollars of capital investment per week, and L is work-hours of labor per week.

(a) If current capital investment is \$72,900 per week and work-hours are 3072 per week, find the current weekly production value.
(b) If weekly capital investment is increased to \$97,200 and new employees are hired so that there are 4096 total weekly work-hours, find the percentage increase in the production value.

Solution

(a) Capital investment of \$72,900 means that $K = 729$. We use this value and $L = 3072$ in the production function.

$$Q = 4(729)^{0.4}(3072)^{0.6} = 6912$$

Thus the weekly production value is \$6,912,000.

(b) In this case we use $K = 972$ and $L = 4096$.

$$Q = 4(972)^{0.4}(4096)^{0.6} = 9216$$

This is an increase in production value of $9216 - 6912 = 2304$, which is equivalent to a weekly increase of

$$\frac{2304}{6912} = 0.33\tfrac{1}{3} = 33\tfrac{1}{3}\%$$

CHECKPOINT SOLUTIONS

1. The domain is the set of ordered pairs of real numbers where $x^2 - y^2 > 0$ or $x^2 > y^2$—that is, where $|x| > |y|$.
2. $f(2, 3, 4) = 4 + 6 - 4 = 6$

EXERCISE 14.1

Give the domain of each function in Problems 1–8.

1. $z = x^2 + y^2$
2. $z = 4x - 3y$
3. $z = \dfrac{4x - 3}{y}$
4. $z = \dfrac{x + y^2}{\sqrt{x}}$
5. $z = \dfrac{4x^3y - x}{2x - y}$
6. $z = \sqrt{x - y}$
7. $q = \sqrt{p_1} + 3p_2$
8. $q = \dfrac{p_1 + p_2}{\sqrt{p_1}}$

In Problems 9–22, evaluate the following functions at the given values of the independent variables.

9. $z = x^3 + 4xy + y^2$; $x = 1, y = -1$
10. $z = 4x^2 - 3xy^3$; $x = 2, y = 2$
11. $z = \dfrac{x - y}{x + y}$; $x = 4, y = -1$
12. $z = \dfrac{x^2 + xy}{x - y}$; $x = 3, y = 2$
13. $C(x_1, x_2) = 600 + 4x_1 + 6x_2$; $x_1 = 400, x_2 = 50$
14. $C(x_1, x_2) = 500 + 5x_1 + 7x_2$; find $C(200, 300)$.
15. $q_1(p_1, p_2) = \dfrac{p_1 + 4p_2}{p_1 - p_2}$; find $q_1(40, 35)$.
16. $q_1(p_1, p_2) = \dfrac{5p_1 - p_2}{p_1 + 3p_2}$; find $q_1(50, 10)$.
17. $z(x, y) = xe^{x+y}$; find $z(3, -3)$.

18. $f(x, y) = ye^{2x} + y^2$; find $f(0, 7)$.

19. $f(x, y) = \dfrac{\ln(xy)}{x^2 + y^2}$; find $f(-3, -4)$.

20. $z(x, y) = x \ln y - y \ln x$; find $z(1, 1)$.

21. $w = \dfrac{x^2 + 4yz}{xyz}$ at (1, 3, 1)

22. $u = f(w, x, y, z) = \dfrac{wx - yz^2}{xy - wz}$ at (2, 3, 1, −1)

Applications

23. ***Investment*** The future value S of an investment earning 6%, compounded continuously, is a function of the principal P and the length of time t that the principal has been invested. It is given by

$$S = f(P, t) = Pe^{0.06t}$$

Find $f(2000, 20)$, and interpret your answer.

24. ***Amortization*** If \$100,000 is borrowed to purchase a home, then the monthly payment R is a function of the interest rate i (expressed as a percent) and the number of years n before the mortgage is paid. It is given by

$$R = f(i, n) = 100{,}000\left[\frac{0.01(i/12)}{1 - (1 + 0.01(i/12))^{-12n}}\right]$$

Find $f(7.25, 30)$ and interpret your answer.

25. ***Wilson's lot size formula*** In economics, the most economical quantity Q of goods (TVs, dresses, gallons of paint, etc.) for a store to order is given by Wilson's lot size formula

$$Q = f(K, M, h) = \sqrt{2KM/h}$$

where K is the cost of placing the order, M is the number of items sold per week, and h is the weekly holding cost for each item (the cost of storage space, utilities, taxes, security, etc.). Find $f(200, 625, 1)$ and interpret your answer.

26. ***Gas law*** Suppose that a gas satisfies the universal gas law, $V = nRT/P$, with n equal to 10 moles of the gas and R, the universal gas constant, equal to 0.082054. What is V if $T = 10$ K (kelvins, the units in which temperature is measured on the Kelvin scale) and $P = 1$ atmosphere?

Temperature-humidity models There are different models for measuring the effects of high temperature and humidity. Two of these are the Summer Simmer Index (S) and the Apparent Temperature (A),* and they are given by

$$S = 1.98T - 1.09(1 - H)(T - 58) - 56.9$$
$$A = 2.70 + 0.885T - 78.7H + 1.20TH$$

where T is the air temperature (in degrees Fahrenheit) and H is the relative humidity (expressed as a decimal). Use these models in Problems 27 and 28.

27. At the Dallas–Fort Worth Airport, the average daily temperatures and humidities for July are

Maximum: 97.8°F with 44% humidity
Minimum: 74.7°F with 80% humidity**

Calculate the Summer Simmer Index S and the Apparent Temperature A for both the average daily maximum and the average daily minimum temperature.

28. In Orlando, Florida, the following represent the average daily temperatures and humidities for August.

Maximum: 91.6°F with 60% humidity
Minimum: 73.4°F with 92% humidity**

Calculate the Summer Simmer Index S and the Apparent Temperature A for both the average daily maximum and the average daily minimum temperature.

29. The tables below and on the next page show that a monthly mortgage payment, R, is a function of the amount financed, A, in thousands of dollars; the duration of the loan, n, in years; and the annual interest rate, r, as a percent. If $R = f(A, n, r)$, use the tables to find the following, and then write a sentence of explanation for each.
(a) $f(90, 20, 8)$ (b) $f(160, 15, 9)$

8% Annual Percentage Rate
Monthly Payments (Principal and Interest)

Amount Financed	*10 Years*	*15 Years*	*20 Years*	*25 Years*	*30 Years*
\$50,000	\$606.64	\$477.83	\$418.22	\$385.91	\$366.88
60,000	727.97	573.39	501.86	463.09	440.26
70,000	849.29	668.96	585.51	540.27	513.64
80,000	970.62	764.52	669.15	617.45	587.01
90,000	1091.95	860.09	752.80	694.63	660.39
100,000	1213.28	955.65	836.44	771.82	733.76
120,000	1455.94	1146.78	1003.72	926.18	880.52
140,000	1698.58	1337.92	1171.02	1080.54	1027.28
160,000	1941.24	1529.04	1338.30	1234.90	1174.02
180,000	2183.90	1720.18	1505.60	1389.26	1320.78
200,000	2426.56	1911.30	1672.88	1543.64	1467.52

*Bosch, W., and L. G. Cobb, "Temperature-Humidity Indices," UMAP Unit 691, *The UMAP Journal,* 10(3), Fall 1989, 237–256.

**Ruffner, James, and Frank Bair (eds.), *Weather of U.S. Cities,* Gale Research Co., Detroit, MI, 1987.

9% Annual Percentage Rate
Monthly Payments (Principal and Interest)

Amount Financed	*10 Years*	*15 Years*	*20 Years*	*25 Years*	*30 Years*
\$50,000	\$633.38	\$507.13	\$449.86	\$419.60	\$402.31
60,000	760.05	608.56	539.84	503.52	482.77
70,000	886.73	709.99	629.81	587.44	563.24
80,000	1013.41	811.41	719.78	671.36	643.70
90,000	1140.08	912.84	809.75	755.28	724.16
100,000	1266.76	1014.27	899.73	839.20	804.62
120,000	1520.10	1217.12	1079.68	1007.04	965.54
140,000	1773.46	1419.96	1259.62	1174.88	1126.48
160,000	2026.82	1622.82	1439.56	1342.72	1287.40
180,000	2280.16	1825.68	1619.50	1510.56	1448.32
200,000	2533.52	2028.54	1799.46	1678.40	1609.24

Source: *The Mortgage Money Guide,* Federal Trade Commission

30. Wind and cold temperatures combine to make the air temperature feel colder than it actually is. This combination is reported as wind chill. The table below shows that wind chill temperatures, WC, are a function of wind speed, s, and air temperature, t. If $WC = f(s, t)$, use the table to find the following, and then write a sentence of explanation for each.
(a) $f(25, 5)$ (b) $f(15, -15)$

Wind Speed (mph)	*Air Temperature (°F)* 35	25	15	5	−5	−15	−25
5	33	21	12	0	−10	−21	−31
15	16	2	−11	−25	−38	−51	−65
25	8	−7	−22	−36	−51	−66	−81
35	4	−12	−27	−43	−58	−74	−89
45	2	−14	−30	−46	−62	−78	−93

Source: *World Almanac,* 1998

31. ***Utility*** Suppose that the utility function for two goods X and Y is given by $U = xy^2$, and a consumer purchases 9 units of X and 6 units of Y.
(a) If the consumer purchases 9 units of Y, how many units of X must be purchased to retain the same level of utility?
(b) If the consumer purchases 81 units of X, how many units of Y must be purchased to retain the same level of utility?
(c) Graph the indifference curve for the utility level found in (a) and (b). Use the graph to confirm your answers to (a) and (b).

32. ***Utility*** Suppose that an indifference curve for two goods, X and Y, has the equation $xy = 400$.
(a) If 25 units of X are purchased, how many units of Y must be purchased to remain on this indifference curve?
(b) Graph this indifference curve and confirm your results in (a).

33. ***Production*** Suppose that a company's production is given by the Cobb-Douglas production function

$$Q = 30K^{1/4}L^{3/4}$$

where K is dollars of capital investment and L is labor hours.
(a) Find Q if $K = \$10{,}000$ and $L = 625$ hours.
(b) Show that if *both* K and L are doubled, then the output is doubled.
(c) If capital investment is held at \$10,000, graph Q as a function of L.

34. ***Production*** Suppose that a company's production is given by the Cobb-Douglas production function

$$Q = 70K^{2/3}L^{1/3}$$

where K is dollars of capital investment and L is labor hours.
(a) Find Q if $K = \$64{,}000$ and $L = 512$ hours.
(b) Show that if both K and L are halved, then Q is also halved.
(c) If capital investment is held at \$64,000, graph Q as a function of L.

35. ***Production*** Suppose that the number of units of a good produced, z, is given by $z = 20xy$, where x is the number of machines working properly and y is the average number of work-hours per machine. Find the production for a week in which
(a) 12 machines are working properly and the average number of work-hours per machine is 30.
(b) 10 machines are working properly and the average number of work-hours per machine is 25.

36. ***Profit*** The Kirk Kelly Kandy Company makes two kinds of candy, Kisses and Kreams. The profit function for the company is

$$P(x, y) = 100x + 64y - 0.01x^2 - 0.25y^2$$

where x is the number of pounds of Kisses sold per week and y is the number of pounds of Kreams. What is the company's profit if it sells
(a) 20 pounds of Kisses and 10 pounds of Kreams?
(b) 100 pounds of Kisses and 16 pounds of Kreams?
(c) 10,000 pounds of Kisses and 256 pounds of Kreams?

37. ***Epidemic*** The cost per day to society of an epidemic is

$$C(x, y) = 20x + 200y$$

where C is in dollars, x is the number of people infected on a given day, and y is the number of people who die on a given day. If 14,000 people are infected and 20 people die on a given day, what is the cost to society?

38. ***Pesticide*** An area of land is to be sprayed with two brands of pesticide: x liters of brand 1 and y liters of brand 2. If the number of insects killed is given by

$$f(x, y) = 10{,}000 - 6500e^{-0.01x} - 3500e^{-0.02y}$$

how many insects would be killed if 80 liters of brand 1 and 120 liters of brand 2 were used?

14.2 Partial Differentiation

OBJECTIVES

- *To find partial derivatives of functions of two or more variables*
- *To evaluate partial derivatives of functions of two or more variables at given points*
- *To use partial derivatives to find slopes of tangents to surfaces*
- *To find and evaluate second- and higher-order partial derivatives of functions of two variables*

APPLICATION PREVIEW

We used derivatives to find the rate of change of cost with respect to the quantity produced. If cost is given as a function of two variables (such as material costs x and labor costs y), then we can find the rate of change of cost with respect to *one* of these independent variables. This is done by finding the **partial derivative** of the function with respect to one variable, while holding the other one constant.

For example, suppose that the cost of manufacturing a good is given by

$$C(x, y) = 5 + 5x + 2y$$

where x represents the cost of 1 ounce of material used and y represents the labor cost in dollars per hour. To find the rate at which the cost changes with respect to material used x, we treat the y-variable as though it were a constant and take the derivative of C with respect to x. This derivative is

$$\frac{\partial C}{\partial x} = 0 + 5 + 0, \quad \text{or} \quad \frac{\partial C}{\partial x} = 5$$

Note that the partial derivative of $2y$ with respect to x is 0 because y is treated as a constant. The partial derivative is a new type of derivative, so we use a new symbol to denote it, $\partial C/\partial x$.

The partial derivative $\partial C/\partial x = 5$ tells us that a change of \$1 in the cost of materials will cause an increase of \$5 in total costs, *if* labor costs remain constant. To see the rate at which total cost changes with respect to labor costs, we find $\partial C/\partial y = 0 + 0 + 2$, or $\partial C/\partial y = 2$. Thus if the cost of materials is held constant, an increase of \$1 in labor costs will cause an increase of \$2 in the total cost of the good.

First-Order Partial Derivatives

In general, if $z = f(x, y)$ we denote the partial derivative of z with respect to x as $\partial z/\partial x$ and the partial derivative of z with respect to y as $\partial z/\partial y$. Note that dz/dx represents the derivative of a function of one variable, x, and that $\partial z/\partial x$ represents the partial derivative of a function of two or more variables.

As mentioned in the Application Preview, $\partial z/\partial x$ is found by treating y as a constant and taking the derivative of $z = f(x, y)$ with respect to x. Other notations used to represent the partial derivative of $z = f(x, y)$ with respect to x are

$$\frac{\partial z}{\partial x}, \quad \frac{\partial f}{\partial x}, \quad \frac{\partial}{\partial x}f(x, y), \quad f_x(x, y), \quad f_x, \quad \text{and} \quad z_x$$

If x is held constant in the function $z = f(x, y)$ and the derivative is taken with respect to y, we have the partial derivative of z with respect to y, denoted by

$$\frac{\partial z}{\partial y}, \quad \frac{\partial f}{\partial y}, \quad \frac{\partial}{\partial y}f(x, y), \quad f_y(x, y), \quad f_y, \quad \text{or} \quad z_y$$

EXAMPLE 1

If $z = 4x^2 + 5x^2y^2 + 6y^3 - 7$, find $\partial z/\partial x$ and $\partial z/\partial y$.

Solution

$$\frac{\partial z}{\partial x} = 8x + 10y^2x$$

$$\frac{\partial z}{\partial y} = 10x^2y + 18y^2$$

EXAMPLE 2

If $z = x^2y + e^x - \ln y$, find z_x and z_y.

Solution

$$z_x = \frac{\partial z}{\partial x} = 2yx + e^x$$

$$z_y = \frac{\partial z}{\partial y} = x^2 - \frac{1}{y}$$

EXAMPLE 3

If $f(x, y) = (x^2 - y^2)^2$, find the following.

(a) f_x (b) f_y

Solution

(a) $f_x = 2(x^2 - y^2)2x = 4x^3 - 4xy^2$
(b) $f_y = 2(x^2 - y^2)(-2y) = -4x^2y + 4y^3$

CHECKPOINT

1. If $z = 100x + 10xy - y^2$, find the following.
   (a) z_x
   (b) $\dfrac{\partial z}{\partial y}$

EXAMPLE 4

If $q = \dfrac{p_1p_2 + 2p_1}{p_1p_2 - 2p_2}$, find $\partial q/\partial p_1$.

Solution

$$\begin{aligned}\frac{\partial q}{\partial p_1} &= \frac{(p_1p_2 - 2p_2)(p_2 + 2) - (p_1p_2 + 2p_1)p_2}{(p_1p_2 - 2p_2)^2} \\ &= \frac{p_1p_2^2 + 2p_1p_2 - 2p_2^2 - 4p_2 - p_1p_2^2 - 2p_1p_2}{(p_1p_2 - 2p_2)^2} \\ &= \frac{-2p_2^2 - 4p_2}{p_2^2(p_1 - 2)^2} \\ &= \frac{-2p_2(p_2 + 2)}{p_2^2(p_1 - 2)^2} \\ &= \frac{-2(p_2 + 2)}{p_2(p_1 - 2)^2}\end{aligned}$$

We may evaluate partial derivatives by substituting values for x *and* y in the same way we did with derivatives of functions of one variable. For example, if $\partial z/\partial x = 2x - xy$, the value of the partial derivative with respect to x at $x = 2$, $y = 3$ is

$$\left.\frac{\partial z}{\partial x}\right|_{(2,\,3)} = 2(2) - 2 \cdot 3 = -2$$

Other notations used to denote evaluation of partial derivatives with respect to x at (a, b) are

$$\frac{\partial}{\partial x} f\,(a, b) \quad \text{and} \quad f_x(a, b)$$

We denote the evaluation of partial derivatives with respect to y at (a, b) by

$$\left.\frac{\partial z}{\partial y}\right|_{(a,\,b)}, \quad \frac{\partial}{\partial y} f(a, b), \quad \text{or} \quad f_y(a, b)$$

EXAMPLE 5

Find the partial derivative of $f(x, y) = x^2 + 3xy + 4$ with respect to x at the point $(1, 2, 11)$.

Solution

$$f_x(x, y) = 2x + 3y$$
$$f_x(1, 2) = 2(1) + 3(2) = 8$$

CHECKPOINT

2. If $g(x, y) = 4x^2 - 3xy + 10y^2$, find the following.
   (a) $\dfrac{\partial g}{\partial x}(1, 3)$
   (b) $g_y(4, 2)$

EXAMPLE 6

Suppose that a company's sales are related to its television advertising by

$$s = 20{,}000 + 10nt + 20n^2$$

where n is the number of commercials per day and t is the length of the commercials in seconds. Find the partial derivative of s with respect to n, and use the result to find the instantaneous rate of change of sales with respect to the number of commercials per day, if the company is currently running ten 30-second commercials.

Solution

The partial derivative of s with respect to n is $\partial s/\partial n = 10t + 40n$. At $n = 10$ and $t = 30$, the rate of change in sales is approximately

$$\left.\frac{\partial s}{\partial n}\right|_{\substack{n=10\\ t=30}} = 10(30) + 40(10) = 700$$

Thus increasing the number of commercials by 1 would result in approximately 700 additional sales.

We have seen that the partial derivative $\partial z/\partial x$ is found by holding y constant and taking the derivative of z with respect to x and that the partial derivative $\partial z/\partial y$ is found by holding x constant and taking the derivative of z with respect to y. We now give formal definitions of these partial derivatives.

Partial Derivatives

The partial derivative of $z = f(x, y)$ with respect to x at the point (x, y) is

$$\frac{\partial z}{\partial x} = \frac{\partial}{\partial x} f(x, y) = \lim_{h \to 0} \frac{f(x + h, y) - f(x, y)}{h}$$

provided this limit exists.

The partial derivative of $z = f(x, y)$ with respect to y at the point (x, y) is

$$\frac{\partial z}{\partial y} = \frac{\partial}{\partial y} f(x, y) = \lim_{h \to 0} \frac{f(x, y + h) - f(x, y)}{h}$$

provided this limit exists.

We have already stated that the graph of $z = f(x, y)$ is a surface in three dimensions. The partial derivative with respect to x of such a function may be thought of as the slope of the tangent to the surface at a point (x, y, z) on the surface in the *positive direction of the* x*-axis.* That is, if a plane parallel to the xz-plane cuts the surface, passing through the point (x_0, y_0, z_0), the line in the plane that is tangent to the surface will have a slope equal to $\partial z/\partial x$ evaluated at the point. Thus

$$\left.\frac{\partial z}{\partial x}\right|_{(x_0, y_0)}$$

represents the slope of the tangent to the surface in the positive direction of the x-axis (see Figure 14.4).

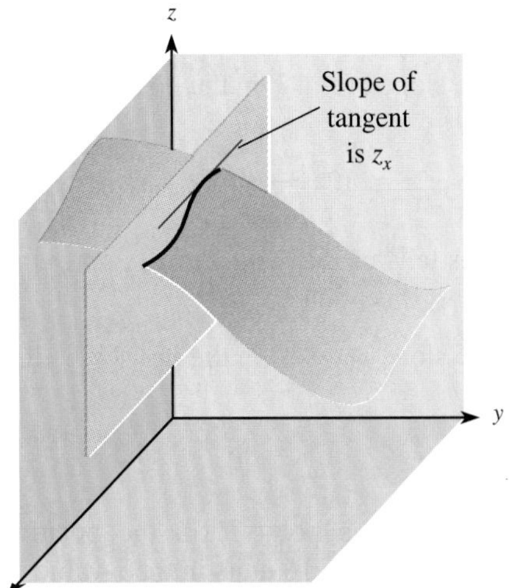

Figure 14.4

Similarly,

$$\left.\frac{\partial z}{\partial y}\right|_{(x_0,\, y_0)} = \frac{\partial}{\partial y} f(x_0, y_0)$$

represents the slope of the tangent to the surface at (x_0, y_0, z_0) in the positive direction of the y-axis (see Figure 14.5).

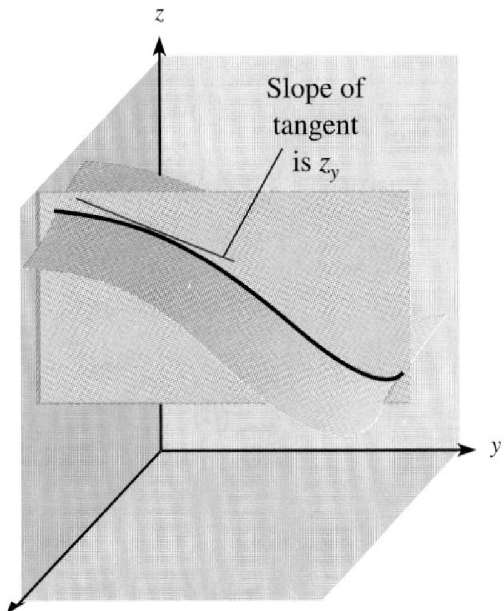

Figure 14.5

EXAMPLE 7

Let $z = 4x^3 - 4e^x + 4y^2$ and let P be the point (0, 2, 12). Find the slope of the tangent to z at the point P in the positive direction of (a) the x-axis and (b) the y-axis.

Solution

(a) The slope of z at P in the positive x-direction is given by $\frac{\partial z}{\partial x}$, evaluated at P.

$$\frac{\partial z}{\partial x} = 12x^2 - 4e^x \quad \text{and} \quad \left.\frac{\partial z}{\partial x}\right|_{(0,\,2)} = 12(0)^2 - 4e^0 = -4$$

This tells us that z *decreases* approximately 4 units for an increase of 1 unit in x at this point.

(b) The slope of z at P in the positive y-direction is given by $\frac{\partial z}{\partial y}$, evaluated at P.

$$\frac{\partial z}{\partial y} = 8y \quad \text{and} \quad \left.\frac{\partial z}{\partial y}\right|_{(0,\,2)} = 8(2) = 16$$

Thus, at the point P, in the positive y-direction the function *increases* approximately 16 units in the z-value for a unit increase in y.

Up to this point, we have considered derivatives of functions of two variables. We can easily extend the concept to functions of three or more variables. We can find the partial derivative with respect to any one independent variable by taking the derivative of the function with respect to that variable while holding all other independent variables constant.

EXAMPLE 8

If $u = f(w, x, y, z) = 3x^2y + w^3 - 4xyz$, find the following.

(a) $\frac{\partial u}{\partial w}$ (b) $\frac{\partial u}{\partial x}$ (c) $\frac{\partial u}{\partial y}$ (d) $\frac{\partial u}{\partial z}$

Solution

(a) $\frac{\partial u}{\partial w} = 3w^2$ (b) $\frac{\partial u}{\partial x} = 6xy - 4yz$

(c) $\frac{\partial u}{\partial y} = 3x^2 - 4xz$ (d) $\frac{\partial u}{\partial z} = -4xy$

EXAMPLE 9

If $C = 4x_1 + 2x_1^2 + 3x_2 - x_1x_2 + x_3^2$, find the following.

(a) $\frac{\partial C}{\partial x_1}$ (b) $\frac{\partial C}{\partial x_2}$ (c) $\frac{\partial C}{\partial x_3}$

Solution

(a) $\frac{\partial C}{\partial x_1} = 4 + 4x_1 - x_2$ (b) $\frac{\partial C}{\partial x_2} = 3 - x_1$ (c) $\frac{\partial C}{\partial x_3} = 2x_3$

CHECKPOINT

3. If $f(w, x, y, z) = 8xy^2 + 4yz - xw^2$, find

(a) $\frac{\partial f}{\partial x}$. (b) $\frac{\partial f}{\partial w}$. (c) $\frac{\partial f}{\partial y}(1, 2, 1, 3)$. (d) $\frac{\partial f}{\partial z}(0, 2, 1, 3)$.

Higher-Order Partial Derivatives

Just as we have taken derivatives of derivatives to obtain higher-order derivatives of functions of one variable, we may also take partial derivatives of partial derivatives to obtain higher-order partial derivatives of a function of more than one variable. If $z = f(x, y)$, then the partial derivative functions z_x and z_y are called *first partials.* Partial derivatives of z_x and z_y are called *second partials,* so $z = f(x, y)$ has *four* **second partial derivatives.** The notations for these second partial derivatives follow.

Second Partial Derivatives

$z_{xx} = \frac{\partial^2 z}{\partial x^2} = \frac{\partial}{\partial x}\left(\frac{\partial z}{\partial x}\right)$: both derivatives taken with respect to x.

$z_{yy} = \frac{\partial^2 z}{\partial y^2} = \frac{\partial}{\partial y}\left(\frac{\partial z}{\partial y}\right)$: both derivatives taken with respect to y.

$z_{xy} = \frac{\partial^2 z}{\partial y\, \partial x} = \frac{\partial}{\partial y}\left(\frac{\partial z}{\partial x}\right)$: first derivative taken with respect to x, second with respect to y.

$z_{yx} = \frac{\partial^2 z}{\partial x\, \partial y} = \frac{\partial}{\partial x}\left(\frac{\partial z}{\partial y}\right)$: first derivative taken with respect to y, second with respect to x.

EXAMPLE 10

If $z = x^3y - 3xy^2 + 4$, find each of the second partial derivatives of the function.

Solution

Because

$$z_x = 3x^2y - 3y^2 \quad \text{and} \quad z_y = x^3 - 6xy,$$

$$z_{xx} = \frac{\partial}{\partial x}(3x^2y - 3y^2) = 6xy$$

$$z_{xy} = \frac{\partial}{\partial y}(3x^2y - 3y^2) = 3x^2 - 6y$$

$$z_{yy} = \frac{\partial}{\partial y}(x^3 - 6xy) = -6x$$

$$z_{yx} = \frac{\partial}{\partial x}(x^3 - 6xy) = 3x^2 - 6y$$

Note that z_{xy} and z_{yx} are equal for the function in Example 10. This will always occur if the derivatives of this function are continuous.

$z_{xy} = z_{yx}$ If the second partial derivatives z_{xy} and z_{yx} of a function $z = f(x, y)$ are continuous at a point, they are equal there.

EXAMPLE 11

Find each of the second partial derivatives of $z = x^2y + e^{xy}$.

Solution

Because $z_x = 2xy + e^{xy} \cdot y = 2xy + ye^{xy}$,

$$\begin{aligned} z_{xx} &= 2y + e^{xy} \cdot y^2 = 2y + y^2e^{xy} \\ z_{xy} &= 2x + (e^{xy} \cdot 1 + ye^{xy} \cdot x) \\ &= 2x + e^{xy} + xye^{xy} \end{aligned}$$

Because $z_y = x^2 + e^{xy} \cdot x = x^2 + xe^{xy}$,

$$\begin{aligned} z_{yx} &= 2x + (e^{xy} \cdot 1 + xe^{xy} \cdot y) \\ &= 2x + e^{xy} + xye^{xy} \\ z_{yy} &= 0 + xe^{xy} \cdot x = x^2e^{xy} \end{aligned}$$

CHECKPOINT

4. If $z = 4x^3y^4 + 4xy$, find the following.
   (a) z_{xx} (b) z_{yy} (c) z_{xy} (d) z_{yx}
5. If $z = x^2 + 4e^{xy}$, find z_{xy}.

We can find partial derivatives of order higher than the second. For example, we can find the third-order partial derivatives z_{xyx} and z_{xyy} for the function in Example 10 from the second derivative $z_{xy} = 3x^2 - 6y$.

$$\begin{aligned} z_{xyx} &= 6x \\ z_{xyy} &= -6 \end{aligned}$$

EXAMPLE 12

If $z = x^3y^2 + 4 \ln x$, find z_{xyy}.

Solution

$$\begin{aligned} z_x &= 3x^2y^2 + 4 \cdot \frac{1}{x} \\ z_{xy} &= 3x^2(2y) + 0 = 6x^2y \\ z_{xyy} &= 6x^2 \end{aligned}$$

CHECKPOINT SOLUTIONS

1. (a) $z_x = 100 + 10y$
   (b) $\frac{\partial z}{\partial y} = 10x - 2y$
2. (a) $\frac{\partial g}{\partial x} = 8x - 3y$ and $\frac{\partial g}{\partial x}(1, 3) = 8(1) - 3(3) = -1$
   (b) $g_y = -3x + 20y$ and $g_y(4, 2) = -3(4) + 20(2) = 28$
3. (a) $\frac{\partial f}{\partial x} = 8y^2 - w^2$ (b) $\frac{\partial f}{\partial w} = -2xw$
   (c) $\frac{\partial f}{\partial y} = 16xy + 4z$ and $\frac{\partial f}{\partial y}(1, 2, 1, 3) = 16(2)(1) + 4(3) = 44$
   (d) $\frac{\partial f}{\partial z} = 4y$ and $\frac{\partial f}{\partial z}(0, 2, 1, 3) = 4(1) = 4$
4. $z_x = 12x^2y^4 + 4y$ and $z_y = 16x^3y^3 + 4x$
   (a) $z_{xx} = 24xy^4$ (b) $z_{yy} = 48x^3y^2$
   (c) $z_{xy} = 48x^2y^3 + 4$ (d) $z_{yx} = 48x^2y^3 + 4$
5. $z_x = 2x + 4e^{xy}(y) = 2x + 4ye^{xy}$
   Calculation of z_{xy} requires the Product Rule.
   $$z_{xy} = 0 + (4y)(e^{xy}x) + (e^{xy})(4) = 4xye^{xy} + 4e^{xy}$$

EXERCISE 14.2

1. If $z = x^4 - 5x^2 + 6x + 3y^3 - 5y + 7$, find $\frac{\partial z}{\partial x}$ and $\frac{\partial z}{\partial y}$.
2. If $z = x^5 - 6x + 4y^4 - y^2$, find $\frac{\partial z}{\partial x}$ and $\frac{\partial z}{\partial y}$.
3. If $z = x^3 + 4x^2y + 6y^2$, find z_x and z_y.
4. If $z = 3xy + y^2$, find z_x and z_y.
5. If $f(x, y) = (x^3 + 2y^2)^3$, find $\frac{\partial f}{\partial x}$ and $\frac{\partial f}{\partial y}$.
6. If $f(x, y) = (xy^3 + y)^2$, find $\frac{\partial f}{\partial x}$ and $\frac{\partial f}{\partial y}$.
7. If $f(x, y) = \sqrt{2x^2 - 5y^2}$, find f_x and f_y.
8. If $g(x, y) = x\sqrt{y - x}$, find g_x and g_y.
9. If $C(x, y) = 600 - 4xy + 10x^2y$, find $\frac{\partial C}{\partial x}$ and $\frac{\partial C}{\partial y}$.
10. If $C(x, y) = 1000 - 4x + xy^2$, find $\frac{\partial C}{\partial x}$ and $\frac{\partial C}{\partial y}$.
11. If $Q(s, t) = \frac{2s - 3t}{s^2 + t^2}$, find $\frac{\partial Q}{\partial s}$ and $\frac{\partial Q}{\partial t}$.
12. If $q = \frac{5p_1 + 4p_2}{p_1 + p_2}$, find $\frac{\partial q}{\partial p_1}$ and $\frac{\partial q}{\partial p_2}$.
13. If $z = e^{2x} + y \ln x$, find z_x and z_y.
14. If $z = \ln(1 + x^2y) - ye^{-x}$, find z_x and z_y.
15. If $f(x, y) = 100e^{xy}$, find $\frac{\partial f}{\partial x}$ and $\frac{\partial f}{\partial y}$.
16. If $f(x, y) = \ln(xy + 1)$, find $\frac{\partial f}{\partial x}$ and $\frac{\partial f}{\partial y}$.
17. Find the partial derivative of
    $$f(x, y) = 4x^3 - 5xy + y^2$$
    with respect to x at the point $(1, 2, -2)$.
18. Find the partial derivative of
    $$f(x, y) = 3x^2 + 4x + 6xy$$
    with respect to y at $x = 2$, $y = -1$.
19. Find the slope of the tangent in the positive x-direction to the surface $z = 5x^3 - 4xy$ at the point $(1, 2, -3)$.
20. Find the slope of the tangent in the positive y-direction to the surface $z = x^3 - 5xy$ at $(2, 1, -2)$.
21. Find the slope of the tangent in the positive y-direction to the surface $z = e^{xy}$ at $(0, 1, 1)$.
22. Find the slope of the tangent in the positive x-direction to the surface $z = \ln(xy)$ at $(1, 1, 0)$.
23. If $u = f(w, x, y, z) = y^2 - x^2z + 4x$, find the following.
    (a) $\frac{\partial u}{\partial w}$ (b) $\frac{\partial u}{\partial x}$ (c) $\frac{\partial u}{\partial y}$ (d) $\frac{\partial u}{\partial z}$
24. If $u = x^2 + 3xy + xz$, find the following.
    (a) u_x (b) u_y (c) u_z

25. If $C(x_1, x_2, x_3) = 4x_1^2 + 5x_1x_2 + 6x_2^2 + x_3$, find the following.
   (a) $\frac{\partial C}{\partial x_1}$ (b) $\frac{\partial C}{\partial x_2}$ (c) $\frac{\partial C}{\partial x_3}$
26. If $f(x, y, z) = 2x\sqrt{yz - 1} + x^2z^3$, find the following.
   (a) $\frac{\partial f}{\partial x}$ (b) $\frac{\partial f}{\partial y}$ (c) $\frac{\partial f}{\partial z}$
27. If $z = x^2 + 4x - 5y^3$, find the following.
   (a) z_{xx} (b) z_{xy} (c) z_{yx} (d) z_{yy}
28. If $z = x^3 - 5y^2 + 4y + 1$, find the following.
   (a) z_{xx} (b) z_{xy} (c) z_{yx} (d) z_{yy}
29. If $z = x^2y - 4xy^2$, find the following.
   (a) z_{xx} (b) z_{xy} (c) z_{yx} (d) z_{yy}
30. If $z = xy^2 + 4xy - 5$, find the following.
   (a) z_{xx} (b) z_{xy} (c) z_{yx} (d) z_{yy}
31. If $z = x^2 - xy + 4y^3$, find z_{xyx}.
32. If $z = x^3 - 4x^2y + 5y^3$, find z_{yyx}.
33. If $f(x, y) = x^3y + 4xy^4$, find $\left.\frac{\partial^2}{\partial x^2} f(x, y)\right|_{(1, -1)}$.
34. If $f(x, y) = x^4y^2 + 4xy$, find $\left.\frac{\partial^2}{\partial y^2} f(x, y)\right|_{(1, 2)}$.
35. If $f(x, y) = \frac{2x}{x^2 + y^2}$, find the following.
   (a) $\left.\frac{\partial^2 f}{\partial x^2}\right|_{(-1, 4)}$ (b) $\left.\frac{\partial^2 f}{\partial y^2}\right|_{(-1, 4)}$
36. If $f(x, y) = \frac{2y^2}{3xy + 4}$, find the following.
   (a) $\left.\frac{\partial^2 f}{\partial x^2}\right|_{(1, -2)}$ (b) $\left.\frac{\partial^2 f}{\partial y^2}\right|_{(1, -2)}$
37. If $z = x^2y + ye^{x^2}$, find $z_{yx}\big|_{(1, 2)}$.
38. If $z = xy^3 + x \ln y^2$, find $z_{xy}\big|_{(1, 2)}$.
39. If $f(x, y) = x^2 + e^{xy}$, find the following.
   (a) $\frac{\partial^2 f}{\partial x^2}$ (b) $\frac{\partial^2 f}{\partial y\, \partial x}$ (c) $\frac{\partial^2 f}{\partial x\, \partial y}$ (d) $\frac{\partial^2 f}{\partial y^2}$
40. If $z = xe^{xy}$, find the following.
   (a) z_{xx} (b) z_{yy} (c) z_{xy} (d) z_{yx}
41. If $f(x, y) = y^2 - \ln xy$, find the following.
   (a) $\frac{\partial^2 f}{\partial x^2}$ (b) $\frac{\partial^2 f}{\partial y\, \partial x}$
   (c) $\frac{\partial^2 f}{\partial x\, \partial y}$ (d) $\frac{\partial^2 f}{\partial y^2}$
42. If $f(x, y) = x^3 + \ln(xy - 1)$, find the following.
   (a) $\frac{\partial^2 f}{\partial x^2}$ (b) $\frac{\partial^2 f}{\partial y^2}$
   (c) $\frac{\partial^2 f}{\partial x\, \partial y}$ (d) $\frac{\partial^2 f}{\partial y\, \partial x}$
43. If $w = 4x^3y + y^2z + z^3$, find the following.
   (a) w_{xxy} (b) w_{xyx} (c) w_{xyz}
44. If $w = 4xyz + x^3y^2z + x^3$, find the following.
   (a) w_{xyz} (b) w_{xzz} (c) w_{yyz}

Applications

45. ***Mortgage*** When a homeowner has a 25-year variable-rate mortgage loan, the monthly payment R is a function of the amount of the loan A and the current interest rate i (as a percent); that is, $R = f(A, i)$. Interpret each of the following.
   (a) $f(100{,}000, 8) = 1289$
   (b) $\frac{\partial f}{\partial i}(100{,}000, 8) = 62.51$
46. ***Mass transportation ridership*** Suppose that in a certain city, the number of people N using the mass transportation system is a function of the fare f and the daily cost of downtown parking p, so that $N = N(f, p)$. Interpret each of the following.
   (a) $N(5, 10) = 6500$
   (b) $\frac{\partial N}{\partial f}(5, 10) = -400$
   (c) $\frac{\partial N}{\partial p}(5, 10) = 250$
47. ***Wilson's lot size formula*** In economics, the most economical quantity Q of goods (TVs, dresses, gallons of paint, etc.) for a store to order is given by Wilson's lot size formula

$$Q = \sqrt{2KM/h}$$

   where K is the cost of placing the order, M is the number of items sold per week, and h is the weekly holding costs for each item (the cost of storage space, utilities, taxes, security, etc.).
   (a) Explain why $\frac{\partial Q}{\partial M}$ will be positive.
   (b) Explain why $\frac{\partial Q}{\partial h}$ will be negative.
48. ***Cost*** Suppose that the total cost of producing a product is $C(x, y) = 25 + 2x^2 + 3y^2$, where x is the cost per pound for material and y is cost per hour for labor.
   (a) If material costs are held constant, at what rate will the total cost increase for each \$1-per-hour increase in labor?
   (b) If the labor costs are held constant, at what rate will the total cost increase for each increase of \$1 in material cost?

49. ***Pesticide*** Suppose that the number of insects killed by two brands of pesticide is given by

$$f(x, y) = 10{,}000 - 6500e^{-0.01x} - 3500e^{-0.02y}$$

where x is the number of liters of brand 1 and y is the number of liters of brand 2.
(a) What is the rate of change of insect deaths with respect to the number of liters of brand 1?
(b) What is the rate of change of insect deaths with respect to the number of liters of brand 2?

50. ***Profit*** Suppose that the profit from the sale of Kisses and Kreams is given by

$$P(x, y) = 100x + 64y - 0.01x^2 - 0.25y^2$$

where x is the number of pounds of Kisses and y is the number of pounds of Kreams.
(a) Find $\partial P/\partial x$, and give the approximate rate of change of profit with respect to the number of pounds of Kisses if present sales are 20 pounds of Kisses and 10 pounds of Kreams.
(b) Find $\partial P/\partial y$, and give the approximate rate of change of profit with respect to the number of pounds of Kreams that are sold if 100 pounds of Kisses and 16 pounds of Kreams are currently being sold.

51. ***Utility*** If $U = f(x, y)$ is the utility function for goods X and Y, the *marginal utility* of X is $\partial U/\partial x$ and the *marginal utility* of Y is $\partial U/\partial y$. If $U = x^2y^2$, find the marginal utility of
(a) X. (b) Y.

52. ***Utility*** If the utility function for goods X and Y is $U = xy + y^2$, find the marginal utility of
(a) X. (b) Y.

53. ***Production*** Suppose that the output Q (in thousands of units) of a certain company is $Q = 75K^{1/3}L^{2/3}$, where K is the capital expenditures in thousands of dollars and L is the number of labor hours. Find $\partial Q/\partial K$ and $\partial Q/\partial L$ when capital expenditures are \$729,000 and the labor hours total 1728. Interpret each answer.

54. ***Production*** Suppose that the production Q (in hundreds of gallons of paint) of a paint manufacturer can be modeled by $Q = 140K^{1/2}L^{1/2}$, where K is the company's capital expenditures in thousands of dollars and L is the size of the labor force (in hours worked). Find $\partial Q/\partial K$ and $\partial Q/\partial L$ when capital expenditures are \$250,000 and the labor hours are 1225. Interpret each answer.

Wind chill factor Dr. Paul Siple conducted studies testing the effect of wind on the formation of ice at various temperatures and developed the concept of the wind chill factor, which we hear reported during winter weather reports. The wind chill temperatures for selected air temperatures and wind speeds are shown in the table below. For example, the table shows that an air temperature of 15°F together with a wind speed of 35 mph feels the same as an air temperature of −27°F when there is no wind.

Wind Speed (mph)	*Air Temperature (°F)*						
	35	*25*	*15*	*5*	*−5*	*−15*	*−25*
5	33	21	12	0	−10	−21	−31
15	16	2	−11	−25	−38	−51	−65
25	8	−7	−22	−36	−51	−66	−81
35	4	−12	−27	−43	−58	−74	−89
45	2	−14	−30	−46	−62	−78	−93

Source: *World Almanac*, 1991

One form of the formula that meteorologists use to calculate wind chill temperatures is

$$WC = 48.064 + 0.474t - 0.020ts - 1.85s + 0.304t\sqrt{s} - 27.74\sqrt{s}$$

where s is wind speed and t is the actual air temperature. Use this equation to answer Problems 55 and 56.

55. (a) To see how the wind chill temperature changes with wind speed, find $\partial WC/\partial s$.
(b) Find $\partial WC/\partial s$ when the temperature is 10°F and the wind speed is 25 mph. What does this mean?

56. (a) To see how wind chill temperature changes with temperature, find $\partial WC/\partial t$.
(b) Find $\partial WC/\partial t$ when the temperature is 10°F and the wind speed is 25 mph. What does this mean?

14.3 Applications of Functions of Two Variables in Business and Economics

OBJECTIVES

- To evaluate cost functions at given levels of production
- To find marginal costs from total cost and joint cost functions
- To find marginal productivity for given production functions
- To find marginal demand functions from demand functions for a pair of related products

APPLICATION PREVIEW

In this section, we consider three classes of applications of functions of two variables and their partial derivatives. We begin with joint cost functions and their marginals. Next, we consider production functions and revisit Cobb-Douglas production functions. Marginal productivity is introduced and its meaning is explained. Finally, we consider demand functions for two products in a competitive market. Partial derivatives are used to define marginal demands, and these marginals are used to classify the products as competitive or complementary.

Joint Cost and Marginal Cost

Suppose that a firm produces two commodities using the same inputs in different proportions. In such a case the **joint cost function** is of the form $C = Q(x, y)$, where x and y represent the quantities of each commodity and C represents the total cost for the two commodities. Then $\partial C/\partial x$ is the **marginal cost** with respect to product x, and $\partial C//\partial y$ is the **marginal cost** with respect to product y.

EXAMPLE 1

If the joint cost function for two products is

$$C = Q(x, y) = 50 + x^2 + 8xy + y^3$$

find the marginal cost with respect to the following.

(a) x (b) y (c) x at (5, 3) (d) y at (5, 3)

Solution

(a) The marginal cost with respect to x is $\partial C/\partial x = 2x + 8y$.

(b) The marginal cost with respect to y is $\dfrac{\partial C}{\partial y} = 8x + 3y^2$.

(c) $\left.\dfrac{\partial C}{\partial x}\right|_{(5,\,3)} = 2(5) + 8(3) = 34$

Thus if 5 units of product x and 3 units of product y are produced, the total cost will increase approximately \$34 for each unit increase in product x if y is held constant.

(d) $\left.\dfrac{\partial C}{\partial y}\right|_{(5,\,3)} = 8(5) + 3(3)^2 = 67$

Thus if 5 units of product x and 3 units of product y are produced, the total cost will increase approximately \$67 for each unit increase in product y if x is held constant.

Production Functions

An important problem in economics concerns how the factors necessary for production determine the output of a product. For example, the output of a product depends on available labor, land, capital, material, and machines. If the amount of output z of a product depends on the amounts of two inputs x and y, then the quantity z is given by the **production function** $z = f(x, y)$.

EXAMPLE 2

Suppose that it is known that z bushels of a crop can be harvested according to the function

$$z = (21)\frac{6xy - 4x^2 - 3y}{2x + 0.01y}$$

when $100x$ work-hours of labor are employed on y acres of land. What would be the output (in bushels) if 200 work-hours were used on 300 acres?

Solution

Because $z = f(x, y)$,

$$\begin{aligned} f(2, 300) &= (21)\frac{6(2)(300) + 4(2)^2 - 3(300)}{2(2) + 3} \\ &= (21)\frac{3600 + 16 - 900}{7} = 8148 \quad \text{(bushels)} \end{aligned}$$

If $z = f(x, y)$ is a production function, $\partial z/\partial x$ represents the change in the output z with respect to input x while input y remains constant. This partial derivative is called the **marginal productivity of *x*.** The partial derivative $\partial z/\partial y$ is the **marginal productivity of *y*** and measures the rate of change of z with respect to input y.

Marginal productivity (for either input) will be positive over a wide range of inputs, but it increases at a decreasing rate, and it may eventually reach a point where it no longer increases and begins to decrease.

EXAMPLE 3

If a production function is given by $z = 5x^{1/2}y^{1/4}$, find the marginal productivity of

(a) x. (b) y.

Solution

(a) $\dfrac{\partial z}{\partial x} = \dfrac{5}{2}x^{-1/2}y^{1/4}$ (b) $\dfrac{\partial z}{\partial y} = \dfrac{5}{4}x^{1/2}y^{-3/4}$

Note that the marginal productivity of x is positive for all values of x but that it decreases as x gets larger (because of the negative exponent). The same is true for the marginal productivity of y.

Graphing Utilities

If we have a production function and fix a value for one variable, then we can use a graphing utility to analyze the marginal productivity with respect to the other variable.

EXAMPLE 4

Suppose the Cobb-Douglas production function for a company is given by

$$z = 100x^{1/4}y^{3/4}$$

where x is the company's capital investment and y is the size of the labor force (in work-hours).

(a) Find the marginal productivity of x.
(b) If the current labor force is 625 work-hours, substitute $y = 625$ in your answer to (a) and graph the result.
(c) From the graph in (b), what can be said about the effect on production of additional capital investment when the work-hours remain at 625?
(d) Find the marginal productivity of y.
(e) If current capital investment is \$10,000, substitute $x = 10{,}000$ in your answer to (d) and graph the result.
(f) From the graph in (e), what can be said about the effect on production of additional work-hours when capital investment remains at \$10,000?

Solution

(a) $z_x = 25x^{-3/4}y^{3/4}$
(b) If $y = 625$, then z_x becomes

$$z_x = 25x^{-3/4}(625)^{3/4} = 25\left(\frac{1}{x^{3/4}}\right)(125) = \frac{3125}{x^{3/4}}$$

The graph of z_x can be limited to quadrant I because the capital investment is $x > 0$, and hence $z_x > 0$. Knowledge of asymptotes can help us determine range values for x and z_x that give an accurate graph. See Figure 14.6.

(c) Figure 14.6 shows that $z_x > 0$ for $x > 0$. This means that any increases in capital investment will result in increases in productivity. However, Figure 14.6 also shows that z_x is decreasing for $x > 0$, which means that increases in capital investment have a diminishing impact on productivity.

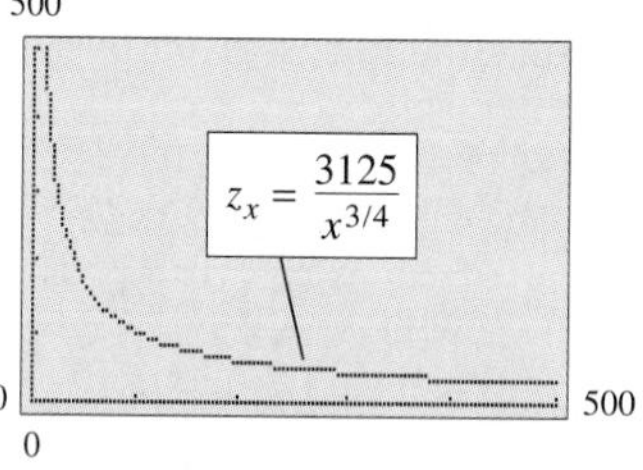

Figure 14.6

(d) $z_y = 75x^{1/4}y^{-1/4}$
(e) If $x = 10{,}000$, then z_y becomes

$$z_y = 75(10{,}000)^{1/4}\left(\frac{1}{y^{1/4}}\right) = \frac{750}{y^{1/4}}$$

The graph is shown in Figure 14.7.

(f) Figure 14.7 also shows that $z_y > 0$ when $y > 0$, so increasing work-hours increases productivity. Note that z_y is decreasing for $y > 0$ but that it does so more slowly than z_x. This indicates that increases in work-hours have a diminishing impact on productivity, but still a more significant one than increases in capital expenditures (as seen in Figure 14.6).

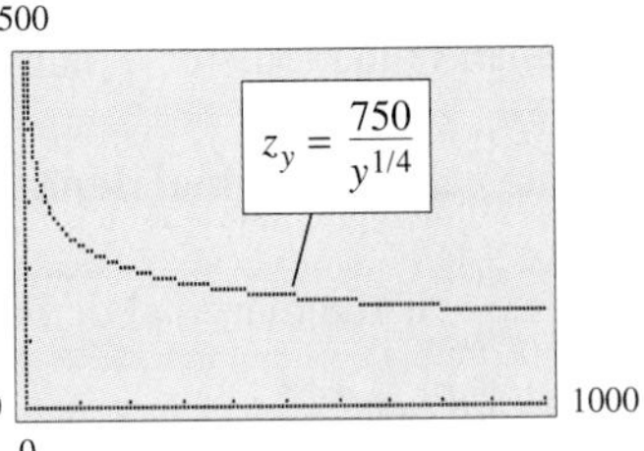

Figure 14.7

Demand Functions

Suppose that two products are sold at prices p_1 and p_2, respectively, in a competitive market consisting of a fixed number of consumers with given tastes and incomes. Then the amount of each *one* of the products demanded by the consumers is dependent on the prices of *both* products on the market. If q_1 represents the demand for the first product, then $q_1 = f(p_1, p_2)$ is the **demand function** for that product. The graph of such a function is called a **demand surface.** An example of a demand function in two variables is $q_1 = 400 - 2p_1 - 4p_2$. Here q_1 is a function of two variables p_1 and p_2. If $p_1 = \$10$ and $p_2 = \$20$, the demand would equal $400 - 2(10) - 4(20) = 300$.

EXAMPLE 5

The demand functions for two products are

$$q_1 = 50 - 5p_1 - 2p_2$$
$$q_2 = 100 - 3p_1 - 8p_2$$

(a) What is the demand for each of the products if the price of the first is $p_1 = \$5$ and the price of the second is $p_2 = \$8$?

(b) Find a pair of prices p_1 and p_2 such that the demands for product 1 and product 2 are equal.

Solution

(a)
$$q_1 = 50 - 5(5) - 2(8) = 9$$
$$q_2 = 100 - 3(5) - 8(8) = 21$$

Thus if these are the prices, the demand for product 2 is higher than the demand for product 1.

(b) We want q_1 to equal q_2. Setting $q_1 = q_2$, we see that

$$50 - 5p_1 - 2p_2 = 100 - 3p_1 - 8p_2$$
$$6p_2 - 50 = 2p_1$$
$$p_1 = 3p_2 - 25$$

Now, any pair of positive values that satisfies this equation will make the demands equal. Letting $p_2 = 10$, we see that $p_1 = 5$ will satisfy the equation. Thus the prices $p_1 = 5$ and $p_2 = 10$ will make the demands equal. The prices $p_1 = 2$ and $p_2 = 9$ will also make the demands equal. Many pairs of values (that is, all those satisfying $p_1 = 3p_2 - 25$) will equalize the demands.

If the demand functions for a pair of related products, product 1 and product 2, are $q_1 = f(p_1, p_2)$ and $q_2 = g(p_1, p_2)$, respectively, then the partial derivatives of q_1 and q_2 are called **marginal demand functions.**

$\dfrac{\partial q_1}{\partial p_1}$ is the marginal demand of q_1 with respect to p_1.

$\dfrac{\partial q_1}{\partial p_2}$ is the marginal demand of q_1 with respect to p_2.

$\dfrac{\partial q_2}{\partial p_1}$ is the marginal demand of q_2 with respect to p_1.

$\dfrac{\partial q_2}{\partial p_2}$ is the marginal demand of q_2 with respect to p_2.

For typical demand functions, if the price of product 2 is fixed, the demand for product 1 will decrease as its price p_1 increases. In this case the marginal demand of q_1 with respect to p_1 will be negative; that is, $\partial q_1/\partial p_1 < 0$. Similarly, $\partial q_2/\partial p_2 < 0$.

But what about $\partial q_2/\partial p_1$ and $\partial q_1/\partial p_2$? If $\partial q_2/\partial p_1$ and $\partial q_1/\partial p_2$ are both positive, the two products are **competitive** because an increase in price p_1 will result in an increase in demand for product 2 (q_2) if the price p_2 is held constant, and an increase in price p_2 will increase the demand for product 1 (q_1) if p_1 is held constant. Stated more simply, an increase in the price of one of the two products will result in an increased demand for the other, so the products are in competition. For example, an increase in the price of a Japanese automobile will result in an increase in demand for an American automobile if the price of the American automobile is held constant.

If $\partial q_2/\partial p_1$ and $\partial q_1/\partial p_2$ are both negative, the products are **complementary** because an increase in the price of one product will cause a decrease in demand for the other product if the price of the second product doesn't change. Under these conditions, a *decrease* in the price of product 1 will result in an *increase* in the demand for product 2, and a decrease in the price of product 2 will result in an increase in the demand for product 1. For example, a decrease in the price of gasoline will result in an increase in the demand for large automobiles.

If the signs of $\partial q_2/\partial p_1$ and $\partial q_1/\partial p_2$ are different, the products are neither competitive nor complementary. This situation rarely occurs but is possible.

EXAMPLE 6

The demand functions for two related products, product 1 and product 2, are given by

$$q_1 = 400 - 5p_1 + 6p_2 \qquad q_2 = 250 + 4p_1 - 5p_2$$

(a) Determine the four marginal demands.
(b) Are product 1 and product 2 complementary or competitive?

Solution

(a) $\frac{\partial q_1}{\partial p_1} = -5 \qquad \frac{\partial q_2}{\partial p_2} = -5 \qquad \frac{\partial q_1}{\partial p_2} = 6 \qquad \frac{\partial q_2}{\partial p_1} = 4$

(b) Because $\partial q_1/\partial p_2$ and $\partial q_2/\partial p_1$ are positive, products 1 and 2 are competitive.

CHECKPOINT

1. If the joint cost function for two products is

$$C = 100 + 3x + 10xy + y^2$$

find the marginal cost with respect to
(a) x. (b) y at (7, 3).

2. If the production function for a product is

$$P = 10x^{1/4}y^{1/2}$$

find the marginal productivity of x.

3. If the demand functions for two products are

$$q_1 = 200 - 3p_1 - 4p_2 \quad \text{and} \quad q_2 = 50 - 6p_1 - 5p_2$$

find the marginal demand of
(a) q_1 with respect to p_1. (b) q_2 with respect to p_2.

CHECKPOINT SOLUTIONS

1. (a) $\frac{\partial C}{\partial x} = 3 + 10y$ (b) $\frac{\partial C}{\partial y} = 10x + 2y$

$\frac{\partial C}{\partial y}(7, 3) = 10(7) + 2(3) = 76$

2. $\frac{\partial P}{\partial x} = \frac{2.5y^{1/2}}{x^{3/4}}$

3. (a) $\frac{\partial q_1}{\partial p_1} = -3$ (b) $\frac{\partial q_2}{\partial p_2} = -5$

EXERCISE 14.3

Joint Cost and Marginal Cost

1. The cost of manufacturing one item is given by

$$C(x, y) = 30 + 3x + 5y$$

where x is the cost of 1 hour of labor and y is the cost of 1 pound of material. If the hourly cost of labor is \$4, and the material costs \$3 per pound, what is the cost of manufacturing one of these items?

2. The manufacture of 1 unit of a product has a cost given by

$$C(x, y, z) = 10 + 8x + 3y + z$$

where x is the cost of 1 pound of one raw material, y is the cost of 1 pound of a second material, and z is the cost of 1 work-hour of labor. If the cost of the first raw material is \$16 per pound, the cost of the second raw material is \$8 per pound, and labor costs \$8 per work-hour, what will it cost to produce 1 unit of the product?

3. The total cost of producing 1 unit of a product is

$$C(x, y) = 30 + 2x + 4y + \frac{xy}{50}$$

where x is the cost per pound of raw materials and y is the cost per hour of labor.
   (a) If labor costs are held constant, at what rate will the total cost increase for each increase of \$1 per pound in material cost?
   (b) If material costs are held constant, at what rate will the total cost increase for each \$1 per hour increase in labor costs?
4. The total cost of producing an item is

$$C(x, y) = 40 + 4x + 6y + \frac{x^2y}{100}$$

where x is the cost per pound of raw materials and y is the cost per hour for labor. How will an increase of
   (a) \$1 per pound of raw materials affect the total cost?
   (b) \$1 per hour in labor costs affect the total cost?
5. The total cost of producing 1 unit of a product is given by

$$C(x, y) = 20x + 70y + \frac{x^2}{1000} + \frac{xy^2}{100}$$

where x represents the cost per pound of raw materials and y represents the hourly rate for labor. The present cost for raw materials is \$10 per pound and the present hourly rate for labor is \$4. How will an increase of
   (a) \$1 per pound for raw materials affect the total cost?
   (b) \$1 per hour in labor costs affect the total cost?
6. The total cost of producing 1 unit of a product is given by

$$C(x, y) = 30 + 10x^2 + 20y - xy$$

where x is the hourly labor rate and y is the cost per pound of raw materials. The current hourly rate is \$5, and the raw materials cost \$6 per pound. How will an increase of
   (a) \$1 per pound for the raw materials affect the total cost?
   (b) \$1 in the hourly labor rate affect the total cost?
7. The joint cost function for two products is

$$C(x, y) = 30 + x^2 + 3y + 2xy$$

where x represents the quantity of product X produced and y represents the quantity of product Y produced.
   (a) Find the marginal cost with respect to x if 8 units of product X and 10 units of product Y are produced.
   (b) Find the marginal cost with respect to y if 8 units of product X and 10 units of product Y are produced.
8. The joint cost function for products X and Y is

$$C(x, y) = 40 + 3x^2 + y^2 + xy$$

where x represents the quantity of X and y represents the quantity of Y.
   (a) Find the marginal cost with respect to x if 20 units of product X and 15 units of product Y are produced.
   (b) Find the marginal cost with respect to y if 20 units of X and 15 units of Y are produced.
9. If the joint cost function for two products is

$$C(x, y) = x\sqrt{y^2 + 1}$$

   (a) Find the marginal cost (function) with respect to x.
   (b) Find the marginal cost with respect to y.
10. Suppose the joint cost function for x units of product X and y units of product Y is given by

$$C(x, y) = 2500\sqrt{xy + 1}$$

Find the marginal cost with respect to
   (a) x. (b) y.
11. Suppose that the joint cost function for two products is

$$C(x, y) = 1200 \ln(xy + 1) + 10{,}000$$

Find the marginal cost with respect to
   (a) x. (b) y.
12. Suppose that the joint cost function for two products is

$$C(x, y) = y \ln(x + 1)$$

Find the marginal cost with respect to
   (a) x. (b) y.

Production Functions

13. Suppose that the production function for a product is $z = \sqrt{4xy}$, where x represents the number of work-hours per month and y is the number of available machines. Determine the marginal productivity of
   (a) x. (b) y.

14. Suppose the production function for a product is

$$z = 60x^{2/5}y^{3/5}$$

where x is the capital expenditures and y is the number of work-hours. Find the marginal productivity of
(a) x. (b) y.

15. Suppose that the production function for a product is $z = \sqrt{x}\ln(y + 1)$, where x represents the number of work-hours and y represents the available capital (per week). Find the marginal productivity of
(a) x. (b) y.

16. Suppose that a company's production function for a certain product is

$$z = (x + 1)^{1/2}\ln(y^2 + 1)$$

where x is the number of work-hours of unskilled labor and y is the number of work-hours of skilled labor. Find the marginal productivity of
(a) x. (b) y.

For Problems 17–19, suppose that the production function for an agricultural product is given by

$$z = \frac{11xy - 0.0002x^2 - 5y}{0.03x + 3y}$$

where x is the number of hours of labor and y is the number of acres of the crop.

17. Find the output when $x = 300$ and $y = 500$.
18. Find the marginal productivity of the number of acres of the crop (y) when $x = 300$ and $y = 500$.
19. Find the marginal productivity of the number of hours of labor (x) when $x = 300$ and $y = 500$.
20. If a production function is given by $z = 12x^{3/4}y^{1/3}$, find the marginal productivity of
(a) x. (b) y.
21. Suppose the Cobb-Douglas production function for a company is given by

$$z = 400x^{3/5}y^{2/5}$$

where x is the company's capital investment and y is the size of the labor force (in work-hours).
(a) Find the marginal productivity of x.
(b) If the current labor force is 1024 work-hours, substitute $y = 1024$ in your answer to (a) and graph the result.
(c) Find the marginal productivity of y.
(d) If the current capital investment is \$59,049, substitute $x = 59{,}049$ in your answer to (c) and graph the result.
(e) Interpret the graphs in (b) and (d) with regard to what they say about the effects on productivity of an increased capital investment (part b) and of an increased labor force (part d).

22. Suppose the Cobb-Douglas production function for a company is given by

$$z = 300x^{2/3}y^{1/3}$$

where x is the company's capital investment and y is the size of the labor force (in work-hours).
(a) Find the marginal productivity of x.
(b) If the current labor force is 729 work-hours, substitute $y = 729$ in your answer to (a) and graph the result.
(c) Find the marginal productivity of y.
(d) If the current capital investment is \$27,000, substitute $x = 27{,}000$ in your answer to (c) and graph the result.
(e) Interpret the graphs in (b) and (d) with regard to what they say about the effects on productivity of an increased capital investment (part b) and of an increased labor force (part d).

Demand Functions

23. The demand functions for two products are given by

$$q_1 = 300 - 8p_1 - 4p_2$$
$$q_2 = 400 - 5p_1 - 10p_2$$

Find the demand for each of the products if the price of the first is $p_1 = 10$ and the price of the second is $p_2 = 8$.

24. The demand functions for two products are given by

$$q_1 = 900 - 9p_1 + 2p_2$$
$$q_2 = 1200 + 6p_1 - 10p_2$$

Find the demands q_1 and q_2 if $p_1 = \$10$ and $p_2 = \$12$.

25. Find a pair of prices p_1 and p_2 such that the demands for the two products in Problem 23 will be equal.
26. Find a pair of prices p_1 and p_2 such that the demands for the two products in Problem 24 will be equal.

In Problems 27–30, the demand functions for two related products, A and B, are given. Complete (a)–(e) for each problem.
(a) Find the marginal demand of q_A with respect to p_A.
(b) Find the marginal demand of q_A with respect to p_B.
(c) Find the marginal demand of q_B with respect to p_B.
(d) Find the marginal demand of q_B with respect to p_A.
(e) Are the two goods competitive or complementary?

27. $\begin{cases} q_A = 400 - 3p_A - 2p_B \\ q_B = 250 - 5p_A - 6p_B \end{cases}$

28. $\begin{cases} q_A = 600 - 4p_A + 6p_B \\ q_B = 1200 + 8p_A - 4p_B \end{cases}$

29. $\begin{cases} q_A = 5000 - 50p_A - \dfrac{600}{p_B + 1} \\ q_B = 10{,}000 - \dfrac{400}{p_A + 4} + \dfrac{400}{p_B + 4} \end{cases}$

30. $\begin{cases} q_A = 2500 + \dfrac{600}{p_A + 2} - 40p_B \\ q_B = 3000 - 100p_A + \dfrac{400}{p_B + 5} \end{cases}$

14.4 Maxima and Minima

OBJECTIVES

- *To find relative maxima, minima, and saddle points of functions of two variables*
- *To apply linear regression formulas*

APPLICATION PREVIEW

Adele Lighting manufactures 20-inch lamps and 31-inch lamps. Suppose that x is the number of thousands of 20-inch lamps and that the demand for these is given by $p_1 = 50 - x$, where p_1 is in dollars. Similarly, suppose that y is the number of thousands of 31-inch lamps and that the demand for these is given by $p_2 = 60 - 2y$, where p_2 is also in dollars. Adele Lighting's joint cost function for these lamps is $C = 2xy$ (in thousands of dollars). Therefore, Adele Lighting's profit (in thousands of dollars) is a function of the two variables x and y. In order to determine Adele's maximum profit, we need to develop methods for finding maximum values for a function of two variables.

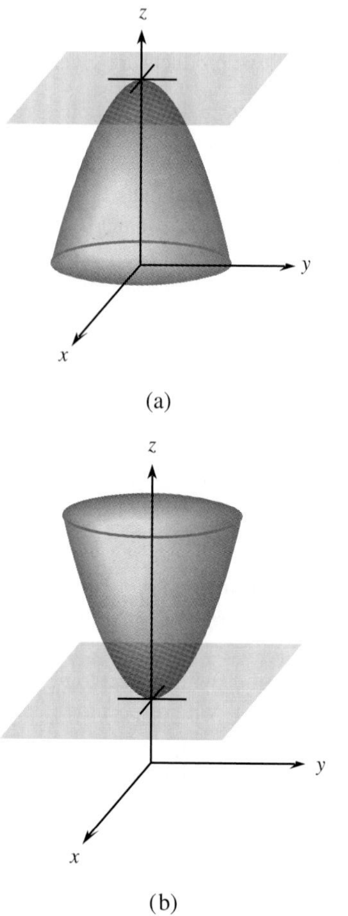

Figure 14.8

In our study of differentiable functions of one variable, we saw that for a relative maximum or minimum to occur at a point, the tangent line to the curve had to be horizontal at that point. The function $z = f(x, y)$ describes a surface in three dimensions. If all partial derivatives of $f(x, y)$ exist, then the surface described by $z = f(x, y)$ must have a horizontal plane tangent to the surface at a point in order to have a relative maximum or minimum at that point (see Figure 14.8). But if the plane tangent to the surface at the point is horizontal, then all the tangent lines to the surface at that point must also be horizontal, for they lie in the tangent plane. In particular, the tangent line in the direction of the x-axis will be horizontal, so $\partial z/\partial x = 0$ at the point; and the tangent line in the direction of the y-axis will be horizontal, so $\partial z/\partial y = 0$ at the point. Thus we can determine the *critical points* for a surface by finding those points where *both* $\partial z/\partial x = 0$ and $\partial z/\partial y = 0$.

How can we determine whether a critical point is a relative maximum, a relative minimum, or neither of these? Finding that $\partial^2 z/\partial x^2 < 0$ and $\partial^2 z/\partial y^2 < 0$ is not enough to tell us that we have a relative maximum. The "second derivative" test we must use involves the values of the second partial derivatives and the value of D at the critical point (a, b), where D is defined as follows:

$$D = \frac{\partial^2 z}{\partial x^2} \cdot \frac{\partial^2 z}{\partial y^2} - \left(\frac{\partial^2 z}{\partial x \partial y}\right)^2$$

We shall state, without proof, the result that determines whether there is a relative maximum, a relative minimum, or neither at the critical point (a, b).

Test for Maxima and Minima

Let $z = f(x, y)$ be a function for which

$$\frac{\partial z}{\partial x} = \frac{\partial z}{\partial y} = 0 \quad \text{at a point } (a, b)$$

and suppose that all second partial derivatives are continuous there. Evaluate

$$D = \frac{\partial^2 z}{\partial x^2} \cdot \frac{\partial^2 z}{\partial y^2} - \left(\frac{\partial^2 z}{\partial x \partial y}\right)^2$$

at the critical point (a, b), and conclude the following.

a. If $D > 0$ and $\partial^2 z/\partial x^2 > 0$ at (a, b), then a relative minimum occurs at (a, b). In this case, $\partial^2 z/\partial y^2 > 0$ at (a, b) also.
b. If $D > 0$ and $\partial^2 z/\partial x^2 < 0$ at (a, b), then a relative maximum occurs at (a, b). In this case, $\partial^2 z/\partial y^2 < 0$ at (a, b) also.
c. If $D < 0$ at (a, b), there is neither a relative maximum nor a relative minimum at (a, b).
d. If $D = 0$ at (a, b), the test fails; investigate the function near the point.

We can test for relative maxima and minima by using the following procedure.

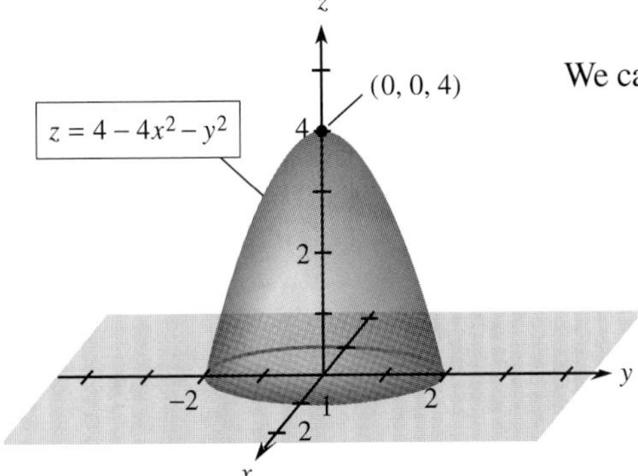

Figure 14.9

Maxima and Minima of z = f(x, y)

Procedure	**Example**
To find relative maxima and minima of $z = f(x, y)$:	Test $z = 4 - 4x^2 - y^2$ for relative maxima and minima.
1. Find $\partial z/\partial x$ and $\partial z/\partial y$.	1. $\dfrac{\partial z}{\partial x} = -8x$; $\dfrac{\partial z}{\partial y} = -2y$
2. Find the point(s) that satisfy *both* $\partial z/\partial x = 0$ and $\partial z/\partial y = 0$. These are the critical points.	2. $\dfrac{\partial z}{\partial x} = 0$ if $x = 0$. $\dfrac{\partial z}{\partial y} = 0$ if $y = 0$. The critical point is (0, 0, 4).
3. Find all second partial derivatives.	3. $\dfrac{\partial^2 z}{\partial x^2} = -8$; $\dfrac{\partial^2 z}{\partial y^2} = -2$; $\dfrac{\partial^2 z}{\partial x\, \partial y} = \dfrac{\partial^2 z}{\partial y\, \partial x} = 0$
4. Evaluate D at each critical point.	4. At (0, 0), $D = (-8)(-2) - 0^2 = 16$.
5. Use the test for maxima and minima to determine whether relative maxima or minima occur.	5. $D > 0$, $\partial^2 z/\partial x^2 < 0$, and $\partial^2 z/\partial y^2 < 0$. A relative maximum occurs at (0, 0). See Figure 14.9.

EXAMPLE 1

Test $z = x^2 + y^2 - 2x + 1$ for relative maxima and minima.

Solution

1. $\dfrac{\partial z}{\partial x} = 2x - 2; \quad \dfrac{\partial z}{\partial y} = 2y$
2. $\dfrac{\partial z}{\partial x} = 0$ if $x = 1$. $\quad \dfrac{\partial z}{\partial y} = 0$ if $y = 0$.
   Both are 0 if $x = 1$ *and* $y = 0$, so the critical point is $(1, 0, 0)$.
3. $\dfrac{\partial^2 z}{\partial x^2} = 2; \quad \dfrac{\partial^2 z}{\partial y^2} = 2; \quad \dfrac{\partial^2 z}{\partial x\, \partial y} = \dfrac{\partial^2 z}{\partial y\, \partial x} = 0$
4. At $(1, 0)$, $D = 2 \cdot 2 - 0^2 = 4$.
5. $D > 0$, $\partial^2 z/\partial x^2 > 0$, and $\partial^2 z/\partial y^2 > 0$. A relative minimum occurs at $(1, 0)$. (See Figure 14.10.)

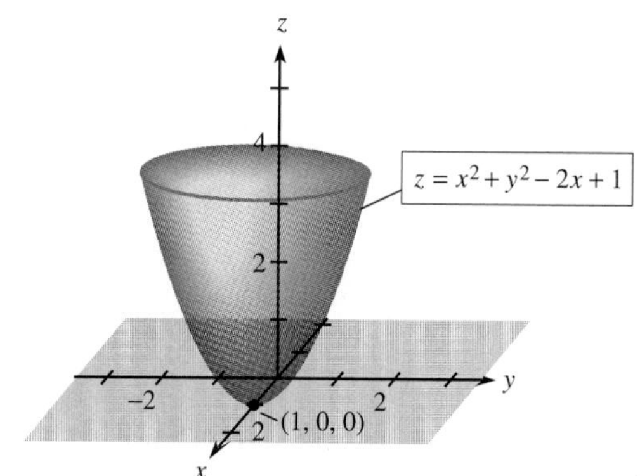

Figure 14.10

EXAMPLE 2

Test $z = y^2 - x^2$ for relative maxima and minima.

Solution

1. $\dfrac{\partial z}{\partial x} = -2x; \quad \dfrac{\partial z}{\partial y} = 2y$
2. $\dfrac{\partial z}{\partial x} = 0$ if $x = 0$; $\quad \dfrac{\partial z}{\partial y} = 0$ if $y = 0$.
   Thus both equal 0 if $x = 0$, $y = 0$. The critical point is $(0, 0, 0)$.
3. $\dfrac{\partial^2 z}{\partial x^2} = -2; \quad \dfrac{\partial^2 z}{\partial y^2} = 2; \quad \dfrac{\partial^2 z}{\partial x\, \partial y} = \dfrac{\partial^2 z}{\partial y\, \partial x} = 0$
4. $D = (-2)(2) - 0 = -4$
5. $D < 0$, so the critical point is neither a relative maximum nor a relative minimum. As Figure 14.11 shows, the surface formed has the shape of a saddle. For this reason, critical points that are neither relative maxima nor relative minima are called **saddle points.**

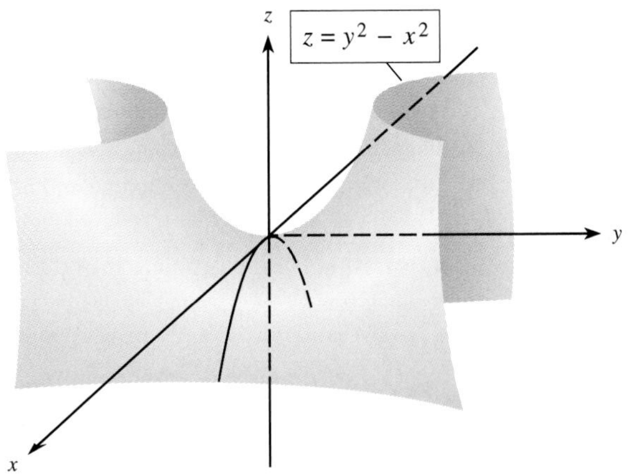

Figure 14.11

The following example involves a surface with two critical points.

EXAMPLE 3

Test $z = x^3 + y^3 + 6xy$ for relative maxima and minima.

Solution

1. $\dfrac{\partial z}{\partial x} = 3x^2 + 6y;\quad \dfrac{\partial z}{\partial y} = 3y^2 + 6x$
2. $\dfrac{\partial z}{\partial x} = 0$ if $0 = 3x^2 + 6y$—that is, if $y = -\frac{1}{2}x^2$.

   $\dfrac{\partial z}{\partial y} = 0$ if $0 = 3y^2 + 6x$—that is, if $x = -\frac{1}{2}y^2$.

   Because *both* conditions must be satisfied, we can substitute $-\frac{1}{2}y^2$ for x in $y = -\frac{1}{2}x^2$ and obtain

$$y = -\frac{1}{2}\left(-\frac{1}{2}y^2\right)^2$$
$$y = -\frac{1}{8}y^4$$
$$y + \frac{1}{8}y^4 = 0$$
$$y(8 + y^3) = 0$$

   Hence $y = 0$ or $y^3 = -8$; thus, $y = 0$ or $y = -2$. If $y = 0$, $x = -\frac{1}{2}(0)^2 = 0$, so one critical point is $(0, 0, 0)$. If $y = -2$, $x = -\frac{1}{2}(-2)^2 = -2$, so the second critical point is $(-2, -2, 8)$.
3. $\dfrac{\partial^2 z}{\partial x^2} = 6x;\quad \dfrac{\partial^2 z}{\partial y^2} = 6y;\quad \dfrac{\partial^2 z}{\partial x\,\partial y} = \dfrac{\partial^2 z}{\partial y\,\partial x} = 6$

   Thus $D = (6x)(6y) - (6)^2$.
4. At $(0, 0)$, $D = 0 \cdot 0 - (6)^2 = -36 < 0$.
   At $(-2, -2)$, $D = (-12)(-12) - 36 = 108 > 0$.
5. At $(0, 0)$, $D < 0$, so a saddle point occurs at $(0, 0, 0)$.
   At $(-2, -2)$, $D > 0$, $\partial^2 z/\partial x^2 = 6(-2) = -12$, and $\partial^2 z/\partial y^2 = 6(-2) = -12$, so a relative maximum occurs at $(-2, -2)$. It is 8.

CHECKPOINT

Suppose that $z = 4 - x^2 - y^2 + 2x - 4y$.

1. Find z_x and z_y.
2. Solve $z_x = 0$ and $z_y = 0$ simultaneously to find the critical point(s) for the graph of this function.
3. Test the point(s) for relative maxima and minima.

Let us now solve the problem introduced in the Application Preview.

EXAMPLE 4

Maximize Adele Lighting's profit if the demand functions are $p_1 = 50 - x$ for 20-inch lamps and $p_2 = 60 - 2y$ for 31-inch lamps, and if the joint cost function is $C = 2xy$. Recall that x and y are in thousands of lamps, p_1 and p_2 are in dollars, and C is in thousands of dollars.

Solution

The profit function is $P(x, y) = p_1x + p_2y - C(x, y)$. Thus,

$$\begin{aligned} P(x, y) &= (50 - x)x + (60 - 2y)y - 2xy \\ &= 50x - x^2 + 60y - 2y^2 - 2xy \end{aligned}$$

gives the profit in thousands of dollars. To maximize the profit, we proceed as follows.

$$P_x = 50 - 2x - 2y \quad \text{and} \quad P_y = 60 - 4y - 2x$$

Solving simultaneously $P_x = 0$ and $P_y = 0$, we have

$$\begin{cases} 0 = 50 - 2x - 2y \\ 0 = 60 - 2x - 4y \end{cases}$$

Subtraction gives $-10 + 2y = 0$, so $y = 5$. Thus $0 = 40 - 2x$, so $x = 20$. Now

$$P_{xx} = -2, \quad P_{yy} = -4, \quad \text{and} \quad P_{xy} = -2, \quad \text{and}$$
$$D = (P_{xx})(P_{yy}) - (P_{xy})^2 = (-2)(-4) - (-2)^2 = +4$$

Because $P_{xx} < 0$, $P_{yy} < 0$, and $D > 0$, the values $x = 20$ and $y = 5$ yield maximum profit. Therefore, when $x = 20$ and $y = 5$, $p_1 = 30$, $p_2 = 50$, and the maximum profit is

$$P(20, 5) = 600 + 250 - 200 = 650$$

That is, Adele Lighting's maximum profit is \$650,000 when the company sells 20,000 of the 20-inch lamps at \$30 each and 5000 of the 31-inch lamps at \$50 each.

Linear Regression

We have used different types of functions to model cost, revenue, profit, demand, supply, and other real-world relationships. Sometimes we have used calculus to study the behavior of these functions, finding, for example, marginal cost, marginal revenue, producer's surplus, and so on. We now have the mathematical tools to understand and develop the formulas that graphing utilities and other technology use to find the equations for linear models.

The formulas used to find the equation of the straight line that is the best fit for a set of data are developed using max-min techniques for functions of two variables. This line is called the **regression line.** In Figure 14.12, we define line ℓ to be the best fit for the data points (that is, the regression line) if the sum of the squares of the differences between the actual y-values of the data points and the y-values of the points on the line is a minimum.

In general, to find the equation of the regression line, we assume that the relationship between x and y is approximately linear and that we can find a straight line with equation

$$\hat{y} = a + bx$$

where the values of $\hat{y}$ will approximate the y-values of the points we know. That is, for each given value of x, the point $(x, \hat{y})$ will be on the line. For any given x-value, x_i, we are interested in the deviation between the y-value of the data point (x_i, y_i) and the $\hat{y}$-value from the equation, $\hat{y}_i$, that results when x_i is substituted for x. These deviations are of the form

$$d_i = \hat{y}_i - y_i, \qquad \text{for } i = 1, 2, \ldots, n$$

(See Figure 14.12 for a general case with the deviations exaggerated.)

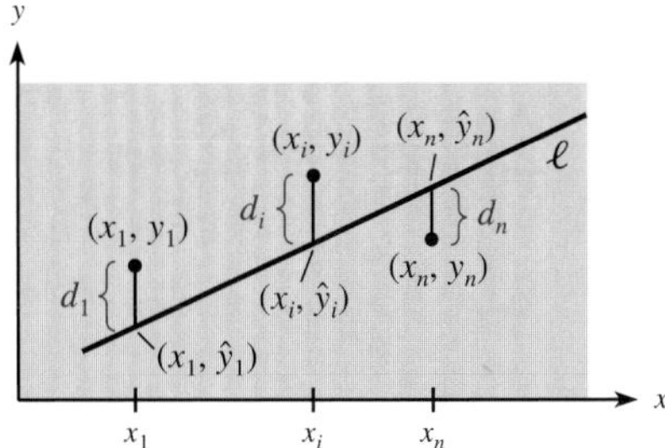

Figure 14.12

To measure the deviations in a way that accounts for the fact that some of the y-values will be above the line and some below, we will say that the line that is the best fit for the data is the one for which the sum of the squares of the deviations is a minimum. That is, we seek the a and b in the equation

$$\hat{y} = a + bx$$

such that the sum of the squares of the deviations,

$$\begin{aligned} S &= \sum_{i=1}^{n} (\hat{y}_i - y_i)^2 = \sum_{i=1}^{n} [(bx_i + a) - y_i]^2 \\ &= (bx_1 + a - y_1)^2 + (bx_2 + a - y_2)^2 + \cdots + (bx_n + a - y_n)^2 \end{aligned}$$

is a minimum. The procedure for determining a and b is called the **method of least squares.**

We seek the values of b and a that make S a minimum, so we find the values that make

$$\frac{\partial S}{\partial b} = 0 \quad \text{and} \quad \frac{\partial S}{\partial a} = 0$$

$$\frac{\partial S}{\partial b} = 2(bx_1 + a - y_1)x_1 + 2(bx_2 + a - y_2)x_2 + \cdots + 2(bx_n + a - y_n)x_n$$

$$\frac{\partial S}{\partial a} = 2(bx_1 + a - y_1) + 2(bx_2 + a - y_2) + \cdots + 2(bx_n + a - y_n)$$

Setting each equation equal to 0, dividing by 2, and using sigma notation give

$$0 = b\sum_{i=1}^{n} x_i^2 + a\sum_{i=1}^{n} x_i - \sum_{i=1}^{n} x_iy_i \tag{1}$$

$$0 = b\sum_{i=1}^{n} x_i + a\sum_{i=1}^{n} 1 - \sum_{i=1}^{n} y_i \tag{2}$$

We can write equations (1) and (2) as follows:

$$\sum_{i=1}^{n} x_iy_i = a\sum_{i=1}^{n} x_i + b\sum_{i=1}^{n} x_i^2 \tag{3}$$

$$\sum_{i=1}^{n} y_i = an + b\sum_{i=1}^{n} x_i \tag{4}$$

Multiplying equation (3) by n and equation (4) by $\sum_{i=1}^{n} x_i$ permits us to begin to solve for b.

$$n\sum_{i=1}^{n} x_iy_i = na\sum_{i=1}^{n} x_i + nb\sum_{i=1}^{n} x_i^2 \tag{5}$$

$$\sum_{i=1}^{n} x_i \sum_{i=1}^{n} y_i = na\sum_{i=1}^{n} x_i + b\left(\sum_{i=1}^{n} x_i\right)^2 \tag{6}$$

Subtracting equation (5) from equation (6) gives

$$\begin{aligned}\sum_{i=1}^{n} x_i \sum_{i=1}^{n} y_i - n\sum_{i=1}^{n} x_iy_i &= b\left(\sum_{i=1}^{n} x_i\right)^2 - nb\sum_{i=1}^{n} x_i^2 \\ &= b\left[\left(\sum_{i=1}^{n} x_i\right)^2 - n\sum_{i=1}^{n} x_i^2\right]\end{aligned}$$

Thus

$$b = \frac{\sum_{i=1}^{n} x_i \sum_{i=1}^{n} y_i - n\sum_{i=1}^{n} x_iy_i}{\left(\sum_{i=1}^{n} x_i\right)^2 - n\sum_{i=1}^{n} x_i^2}$$

and, from equation (4),

$$a = \frac{\sum_{i=1}^{n} y_i - b\sum_{i=1}^{n} x_i}{n}$$

It can be shown that these values for b and a give a minimum value for S, so we have the following.

Linear Regression Equation

Given a set of data points $(x_1, y_1), (x_2, y_2), \ldots, (x_n, y_n)$, the equation of the line that is the best fit for these data is

$$\hat{y} = a + bx$$

where

$$b = \frac{\Sigma x \cdot \Sigma y - n\Sigma xy}{(\Sigma x)^2 - n\Sigma x^2}, \qquad a = \frac{\Sigma y - b\Sigma x}{n}$$

and each summation is taken over the entire data set (that is, from 1 to n).

EXAMPLE 5

The following data show the relation between the diameter of a partial roll of blue denim material at MacGregor Mills and the actual number of yards remaining on the roll. Use linear regression to find the linear equation that gives the number of yards as a function of the diameter.

Diameter (inches)	*Yards/Roll*	*Diameter (inches)*	*Yards/Roll*
14.0	120	22.5	325
15.0	145	24.0	360
16.5	170	24.5	380
17.75	200	25.25	405
18.5	220	26.0	435
19.8	255	26.75	460
20.5	270	27.0	470
22.0	305	28.0	500

Solution

Let x be the diameter of the partial rolls and y be the yards on a roll. Before finding the values for a and b, we evaluate some parts of the formulas:

$$\begin{aligned}
n &= 16 \\
\Sigma x &= 348.05 \\
\Sigma x^2 &= 7871.48 \\
\Sigma y &= 5020 \\
\Sigma xy &= 117{,}367.75 \\
b &= \frac{\Sigma x \Sigma y - n\Sigma xy}{(\Sigma x)^2 - n\Sigma x^2} \\
&= \frac{(348.05)(5020) - 16(117{,}367.75)}{(348.05)^2 - 16(7871.48)} = 27.1959 \\
a &= \frac{\Sigma y - b\Sigma x}{n} \\
&= \frac{(5020) - (27.1959)(348.05)}{16} = -277.8458
\end{aligned}$$

Thus the linear equation that can be used to estimate the number of yards of denim remaining on a roll is

$$\hat{y} = -277.85 + 27.20x$$

Note that if we use the linear regression capability of a graphing utility, we obtain exactly the same equation.

CHECKPOINT

4. Use linear regression to write the equation of the line that is the best fit for the following points.

x	50	25	10	5
y	2	4	10	20

Finally, we note that formulas for models other than linear ones, such as power models ($y = ax^b$), exponential models ($y = ab^x$), and logarithmic models [$y = a + b\ln(x)$], can also be developed with the least-squares method. That is, we apply max-min techniques for functions of two variables to minimize the sum of the squares of the deviations.

CHECKPOINT SOLUTIONS

1. $z_x = -2x + 2$, $z_y = -2y - 4$
2. $-2x + 2 = 0$ gives $x = 1$.
   $-2y - 4 = 0$ gives $y = -2$.
   Thus the critical point is $(1, -2, 9)$.
3. $z_{xx} = -2$, $z_{yy} = -2$, and $z_{xy} = 0$, so

$$D(x, y) = (-2)(-2) - (0)^2 = 4$$

Hence, at $(1, -2, 9)$ we have $D > 0$ and $z_{xx} < 0$, so $(1, -2, 9)$ is a relative minimum.

4.

$$\begin{aligned}\Sigma x &= 50 + 25 + 10 + 5 = 90\\ \Sigma x^2 &= 2500 + 625 + 100 + 25 = 3250\\ \Sigma y &= 2 + 4 + 10 + 20 = 36\\ \Sigma xy &= 100 + 100 + 100 + 100 = 400\end{aligned}$$

Then

$$b = \frac{\Sigma x \cdot \Sigma y - n\Sigma xy}{(\Sigma x)^2 - n\Sigma x^2} = \frac{90 \cdot 36 - 4 \cdot 400}{90^2 - 4 \cdot 3250} = \frac{1640}{-4900} \approx -0.33,$$

and

$$a = \frac{\Sigma y - b\Sigma x}{n} = \frac{36 - (-0.33)(90)}{4} = \frac{65.7}{4} = 16.43$$

Thus the line that gives the best fit to these points is

$$\hat{y} = -0.33x + 16.43$$

EXERCISE 14.4

In Problems 1–16, test for relative maxima and minima.

1. $z = 9 - x^2 - y^2$
2. $z = 16 - 4x^2 - 9y^2$
3. $z = x^2 + y^2 + 4$
4. $z = x^2 + y^2 - 4$
5. $z = x^2 + y^2 - 2x + 4y + 5$
6. $z = 4x^2 + y^2 + 4x + 1$
7. $z = x^2 + 6xy + y^2 + 16x$
8. $z = x^2 - 4xy + y^2 - 6y$
9. $z = \dfrac{x^2 - y^2}{9}$
10. $z = \dfrac{y^2}{4} - \dfrac{x^2}{9}$
11. $z = 24 - x^2 + xy - y^2 + 36y$
12. $z = 46 - x^2 + 2xy - 4y^2$
13. $z = x^2 + xy + y^2 - 4y + 10x$
14. $z = x^2 + 5xy + 10y^2 + 8x - 40y$
15. $z = x^3 + y^3 - 6xy$
16. $z = x^3 + y^3 + 3xy$

In Problems 17 and 18, use the points given in the tables to write the equation of the line that is the best fit for the points.

17.

x	3	4	5	6
y	15	22	28	32

18.

x	10	20	30	40
y	2	6	5	6

Applications

19. ***Profit*** Suppose that the profit from the sale of Kisses and Kreams is given by

$$P(x, y) = 100x + 64y - 0.01x^2 - 0.25y^2$$

where x is the number of pounds of Kisses and y is the number of pounds of Kreams. Selling how many pounds of Kisses and Kreams will maximize profit?

20. ***Profit*** The profit from the sales of two products is given by

$$P(x, y) = 20x + 70y - x^2 - y^2$$

where x is the number of units of product 1 sold and y is the number of units of product 2. Selling how much of each product will maximize profit?

21. ***Nutrition*** A new food is designed to add weight to mature beef cattle. The increase in weight is given by $W = xy(20 - x - 2y)$, where x is the number of units of the first ingredient and y is the number of units of the second ingredient. How many units of each ingredient will maximize the weight gain?

22. ***Profit*** The profit for a grain crop is related to fertilizer and labor. The profit per acre is

$$P = 100x + 40y - 5x^2 - 2y^2$$

where x is the number of units of fertilizer and y is the number of work-hours. What values of x and y will maximize the profit?

23. ***Production*** Suppose that

$$P = 3.78x^2 + 1.5y^2 - 0.09x^3 - 0.01y^3$$

is the production function for a product with x units of one input and y units of a second input. Find the values of x and y that will maximize production.

24. ***Production*** Suppose that x units of one input and y units of a second input result in

$$P = 40x + 50y - x^2 - y^2 - xy$$

units of a product. Determine the inputs x and y that will maximize P.

25. ***Production*** Suppose that a manufacturer produces two brands of a product, brand 1 and brand 2. If the demand for brand 1 is $p_1 = 10 - x$, the demand for brand 2 is $p_2 = 40 - 2y$, and the joint cost function is $C = xy$, how many of each brand should be produced to maximize profit?

26. ***Production*** Suppose that a firm produces two products, A and B, that sell for \$$a$ and \$$b$, respectively, with the total cost of producing x units of A and y units of B equal to $C(x, y)$. Show that profit from these products is maximized when

$$\frac{\partial C}{\partial x}(x, y) = a \quad \text{and} \quad \frac{\partial C}{\partial y}(x, y) = b$$

27. ***Manufacturing*** Find the values for each of the dimensions of an open-top box of length x, width y, and height $500{,}000/(xy)$ such that the box requires the least amount of material to make.

28. ***Manufacturing*** Find the values for each of the dimensions of a closed-top box of length x, width y, and height z if the volume equals 27,000 cubic inches and the box requires the least amount of material to make. *Hint:* First write z in terms of x and y, as in Problem 27.

29. ***Profit*** A company manufactures two products, A and B. If x is the number of units of A and y is the number of units of B, then the cost and revenue functions are

$$C(x, y) = 2x^2 - 2xy + y^2 - 7x + 10y + 11$$
$$R(x, y) = 5x + 4y$$

Find the number of each type of product that should be manufactured to maximize profit.

30. ***Production*** Let x be the number of work-hours required and y be the amount of capital required to produce z units of a product. Show that the average production per work-hour, z/x, is maximized when

$$\frac{\partial z}{\partial x} = \frac{z}{x}$$

Use $z = f(x, y)$ and assume a maximum exists.

31. ***Retirement benefits*** The following table gives the approximate benefits for PepsiCo executives who earned an average of $250,000 per year during the last 5 years of service, based on the number of years of service, from 15 years to 45 years.
    (a) Use linear regression to find the linear equation that is the best fit for the data.
    (b) Use the equation to find the expected annual retirement benefits after 40 years of service.

Years of Service	*Annual Retirement*
25	$109,280
30	121,130
35	132,990
40	145,490
45	160,790

Source: TRICON Salaried Employees Retirement Plan

32. ***Health insurance*** The number of people enrolled in health insurance plans (in millions) is given in the table below for the years 1987 to 1995.
    (a) Use linear regression to find the equation of the line of best fit. Use $x = 0$ in 1987.
    (b) Use the equation to estimate the number enrolled in 1996.

Year	*Number of People Enrolled in Health Insurance Plans (millions)*
1987	241.2
1988	243.7
1989	246.2
1990	248.9
1991	251.4
1992	256.8
1993	259.8
1994	262.1
1995	264.3

Source: Bureau of the Census, 1996

33. ***Hourly earnings*** The following table shows the average hourly earnings for full-time workers in various industries for selected years.
    (a) Find the linear regression equation for hourly earnings as a function of time (with $x = 0$ in 1970).
    (b) What does this model predict for the average hourly earnings in 2005?
    (c) Write a sentence that interprets the slope of the linear regression equation.

Year	*Hourly Earnings*
1970	$3.23
1975	4.53
1980	6.66
1985	8.57
1986	8.76
1987	8.98
1988	9.28
1989	9.66
1990	10.01
1991	10.32
1992	10.57
1993	10.83
1994	11.12
1995	11.43
1996	11.81

Source: Bureau of Labor Statistics, U.S. Dept. of Labor

34. ***Tuition*** The following table gives the annual tuition for universities in Georgia from 1985 to 1994.

Year	*Tuition*	*Year*	*Tuition*
1985	$377	1990	528
1986	424	1991	552
1987	460	1992	574
1988	487	1993	597
1989	506	1994	615

Source: University System of Georgia

(a) Let $t = 0$ in 1985 and use linear regression to write a linear equation representing the annual tuition at universities in Georgia as a function of the number of years past 1985.
(b) Use the equation to predict tuition in 2003.

14.5 Maxima and Minima of Functions Subject to Constraints; Lagrange Multipliers

OBJECTIVE

- *To find the maximum or minimum value of a function of two or more variables subject to a condition that constrains the variables*

APPLICATION PREVIEW

Many practical problems require that a function of two or more variables be maximized or minimized subject to certain conditions, or constraints, that limit the variables involved. For example, a firm will want to maximize its profits within the limits (constraints) imposed by its production capacity. Similarly, a city planner may want to locate a new building to maximize access to public transportation yet may be constrained by the availability and cost of building sites.

Specifically, suppose that the utility function for commodities X and Y is given by $U = x^2y^2$, where x and y are the amounts of X and Y, respectively. If p_1 and p_2 represent the prices of X and Y, respectively, and I represents the consumer's income available to purchase these two commodities, the equation $p_1x + p_2y = I$ is called the *budget constraint.* If the price of X is \$2, the price of Y is \$4, and the income available is \$40, then the budget constraint is $2x + 4y = 40$. Thus we seek to maximize the consumer's utility $U = x^2y^2$ subject to the budget constraint $2x + 4y = 40$. In this section we develop methods to solve this type of constrained maximum or minimum.

We can obtain maxima and minima for a function $z = f(x, y)$ subject to the constraint $g(x, y) = 0$ by using the method of **Lagrange multipliers,** named for the famous eighteenth-century mathematician Joseph Louis Lagrange. Lagrange multipliers can be used with functions of two or more variables when the constraints are given by an equation.

In order to find the critical values of a function $f(x, y)$ subject to the constraint $g(x, y) = 0$, we will use the new variable λ to form the **objective function**

$$F(x, y, \lambda) = f(x, y) + \lambda g(x, y)$$

It can be shown that the critical values of $F(x, y, \lambda)$ will satisfy the constraint $g(x, y)$ and will also be critical points of $f(x, y)$. Thus we need only find the critical points of $F(x, y, \lambda)$ to find the required critical points.

To find the critical points of $F(x, y, \lambda)$, we must find the points that make all the partial derivatives equal to 0. That is, the points must satisfy

$$\partial F/\partial x = 0, \quad \partial F/\partial y = 0, \quad \text{and} \quad \partial F/\partial \lambda = 0$$

Because $F(x, y, \lambda) = f(x, y) + \lambda g(x, y)$, these equations may be written as

$$\frac{\partial f}{\partial x} + \lambda\frac{\partial g}{\partial x} = 0$$

$$\frac{\partial f}{\partial y} + \lambda\frac{\partial g}{\partial y} = 0$$

$$g(x, y) = 0$$

Finding the values of x and y that satisfy these three equations simultaneously gives the critical values.

This method will not tell us whether the critical points correspond to maxima or minima, but this can be determined either from the physical setting for the problem or by testing according to a procedure similar to that used for unconstrained maxima and minima. The following examples illustrate the use of Lagrange multipliers.

EXAMPLE 1

Find the maximum value of $z = x^2y$ subject to $x + y = 9$, $x \geq 0$, $y \geq 0$.

Solution

The function to be maximized is $f(x, y) = x^2y$.
The constraint is $g(x, y) = 0$, where $g(x, y) = x + y - 9$.
The objective function is

$$F(x, y, \lambda) = f(x, y) + \lambda g(x, y)$$

or

$$F(x, y, \lambda) = x^2y + \lambda(x + y - 9)$$

Thus

$$\frac{\partial F}{\partial x} = 2xy + \lambda(1) = 0, \quad \text{or} \quad 2xy + \lambda = 0$$

$$\frac{\partial F}{\partial y} = x^2 + \lambda(1) = 0, \quad \text{or} \quad x^2 + \lambda = 0$$

$$\frac{\partial F}{\partial \lambda} = 0 + 1(x + y - 9) = 0, \quad \text{or} \quad x + y - 9 = 0$$

Solving the first two equations for λ and substituting gives

$$\lambda = -2xy$$
$$\lambda = -x^2$$
$$2xy = x^2$$
$$2xy - x^2 = 0$$
$$x(2y - x) = 0$$

so

$$x = 0 \quad \text{or} \quad x = 2y$$

Because $x = 0$ could not make $z = x^2y$ a maximum, we substitute $x = 2y$ into $x + y - 9 = 0$.

$$2y + y = 9$$
$$y = 3$$
$$x = 6$$

Thus the function $z = x^2y$ is maximized at 108 when $x = 6$, $y = 3$, if the constraint is $x + y = 9$. Testing values near $x = 6$, $y = 3$, and satisfying the constraint shows that the function is maximized there. (Try $x = 5.5$, $y = 3.5$; $x = 7$, $y = 2$; and so on.)

EXAMPLE 2

Find the minimum value of the function $z = x^3 + y^3 + xy$ subject to the constraint $x + y - 4 = 0$.

Solution

The function to be minimized is $f(x, y) = x^3 + y^3 + xy$.
The constraint function is $g(x, y) = x + y - 4$.
The objective function is

$$F(x, y, \lambda) = f(x, y) + \lambda g(x, y)$$

or

$$F(x, y, \lambda) = x^3 + y^3 + xy + \lambda(x + y - 4)$$

Then

$$\frac{\partial F}{\partial x} = 3x^2 + y + \lambda = 0$$

$$\frac{\partial F}{\partial y} = 3y^2 + x + \lambda = 0$$

$$\frac{\partial F}{\partial \lambda} = x + y - 4 = 0$$

Solving the first two equations for λ and substituting, we get

$$\lambda = -(3x^2 + y)$$
$$\lambda = -(3y^2 + x)$$
$$3x^2 + y = 3y^2 + x$$

Solving $x + y - 4 = 0$ for y gives $y = 4 - x$. Substituting for y in the equation above, we get

$$3x^2 + (4 - x) = 3(4 - x)^2 + x$$
$$3x^2 + 4 - x = 48 - 24x + 3x^2 + x$$
$$22x = 44 \quad \text{or} \quad x = 2$$

Thus when $x + y - 4 = 0$, $x = 2$ and $y = 2$ give the minimum value $z = 20$.

CHECKPOINT

Find the minimum value of $f(x, y) = x^2 + y^2 - 4xy$, subject to the constraint $x + y = 10$, by:

1. forming the objective function $F(x, y, \lambda)$;
2. finding $\frac{\partial F}{\partial x}$, $\frac{\partial F}{\partial y}$, and $\frac{\partial F}{\partial \lambda}$;
3. setting the three partial derivatives (from question 2) equal to 0, and solving the equations simultaneously for x and y;
4. finding the value of $f(x, y)$ at the critical values of x and y.

We can also use Lagrange multipliers to find the maxima and minima of functions of three (or more) variables, subject to two (or more) constraints. The method involves using two multipliers, one for each constraint, to form an objective function $F = f + \lambda g_1 + \mu g_2$. We leave further discussion for more advanced courses.

We can easily extend the method to functions of three or more variables, as the following example shows.

EXAMPLE 3

Find the minimum value of the function $w = x + y^2 + z^2$, subject to the constraint $x + y + z = 1$.

Solution

The function to be minimized is $f(x, y, z) = x + y^2 + z^2$. The constraint is $g(x, y, z) = 0$, where $g(x, y, z) = x + y + z - 1$.
The objective function is

$$F(x, y, z, \lambda) = f(x, y, z) + \lambda g(x, y, z)$$

or

$$F(x, y, z, \lambda) = x + y^2 + z^2 + \lambda(x + y + z - 1)$$

Then

$$\frac{\partial F}{\partial x} = 1 + \lambda = 0$$

$$\frac{\partial F}{\partial y} = 2y + \lambda = 0$$

$$\frac{\partial F}{\partial z} = 2z + \lambda = 0$$

$$\frac{\partial F}{\partial \lambda} = x + y + z - 1 = 0$$

Solving the first three equations simultaneously gives

$$\lambda = -1 \qquad y = \frac{1}{2} \qquad z = \frac{1}{2}$$

Substituting these values in the fourth equation (which is the constraint), we get $x + \frac{1}{2} + \frac{1}{2} - 1 = 0$, so $x = 0$, $y = \frac{1}{2}$, $z = \frac{1}{2}$. Thus $w = \frac{1}{2}$ is the minimum value because other values of x, y, and z that satisfy $x + y + z = 1$ give larger values for w.

Let us now return to the utility problem posed in the Application Preview.

EXAMPLE 4

Find x and y that maximize the utility function $U = x^2y^2$ subject to the budget constraint $2x + 4y = 40$.

Solution

First we rewrite the constraint as $2x + 4y - 40 = 0$. Then the objective function is

$$F(x, y, \lambda) = x^2y^2 + \lambda(2x + 4y - 40)$$

$$\frac{\partial F}{\partial x} = 2xy^2 + 2\lambda, \qquad \frac{\partial F}{\partial y} = 2x^2y + 4\lambda, \qquad \frac{\partial F}{\partial \lambda} = 2x + 4y - 40$$

Setting these partial derivatives equal to 0 and solving gives

$$-\lambda = xy^2 = x^2y/2, \quad \text{or} \quad xy^2 - x^2y/2 = 0$$

so

$$xy(y - x/2) = 0$$

yields $x = 0$, $y = 0$, or $x = 2y$. Neither $x = 0$ nor $y = 0$ maximizes utility. If $x = 2y$, we have

$$0 = 4y + 4y - 40$$

Thus $y = 5$ and $x = 10$.

Testing values near $x = 10$, $y = 5$ shows that these values maximize utility at $U = 2500$.

Figure 14.13 shows the budget constraint $2x + 4y = 40$ from Example 4 graphed with the indifference curves for $U = x^2y^2$ that correspond to $U = 500$, $U = 2500$, and $U = 5000$.

Whenever an indifference curve intersects the budget constraint, that utility level is attainable within the budget. Note that the highest attainable utility (such as $U = 2500$, found in Example 4) corresponds to the indifference curve that touches the budget constraint at exactly one point—that is, the curve that has the budget constraint as a tangent line. Note also that utility levels greater than $U = 2500$ are not attainable within the budget because the indifference curve "misses" the budget constraint line (as for $U = 5000$).

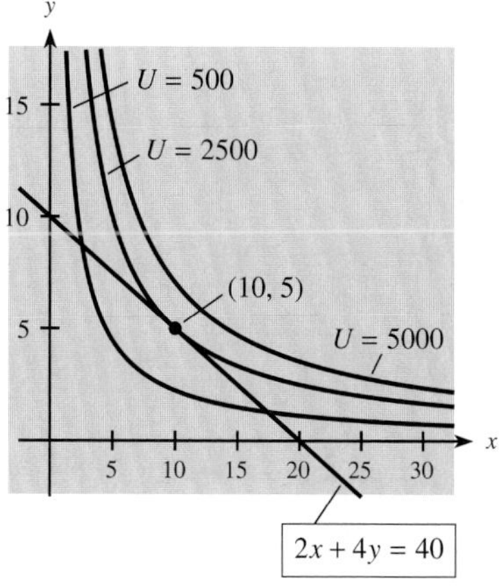

Figure 14.13

EXAMPLE 5

Suppose that the Cobb-Douglas production function for a certain manufacturer gives the number of units of production z according to

$$z = f(x, y) = 100x^{4/5}y^{1/5}$$

where x is the number of units of labor and y is the number of units of capital. Suppose further that labor costs \$160 per unit, capital costs \$200 per unit, and the total cost for capital and labor is limited to \$100,000, so that production is constrained by

$$160x + 200y = 100{,}000$$

Find the number of units of labor and the number of units of capital that maximize production.

Solution

The objective function is

$$F(x, y, \lambda) = 100x^{4/5}y^{1/5} + \lambda(160x + 200y - 100{,}000)$$

$$\frac{\partial F}{\partial x} = 80x^{-1/5}y^{1/5} + 160\lambda, \qquad \frac{\partial F}{\partial y} = 20x^{4/5}y^{-4/5} + 200\lambda$$

$$\frac{\partial F}{\partial \lambda} = 160x + 200y - 100{,}000$$

Setting these partial derivatives equal to 0 and solving gives

$$\lambda = \frac{-80x^{-1/5}y^{1/5}}{160} = \frac{-20x^{4/5}y^{-4/5}}{200} \qquad \text{or} \qquad \frac{y^{1/5}}{2x^{1/5}} = \frac{x^{4/5}}{10y^{4/5}}$$

This means $5y = x$. Using this in $\frac{\partial F}{\partial \lambda} = 0$ gives

$$\begin{aligned} 160(5y) + 200y - 100{,}000 &= 0 \\ 1000y &= 100{,}000 \\ y &= 100 \\ x = 5y &= 500 \end{aligned}$$

Thus production is maximized at $z = 100(500)^{4/5}(100)^{1/5} \approx 36{,}239$ when $x = 500$ (units of labor) and $y = 100$ (units of capital). See Figure 14.14.

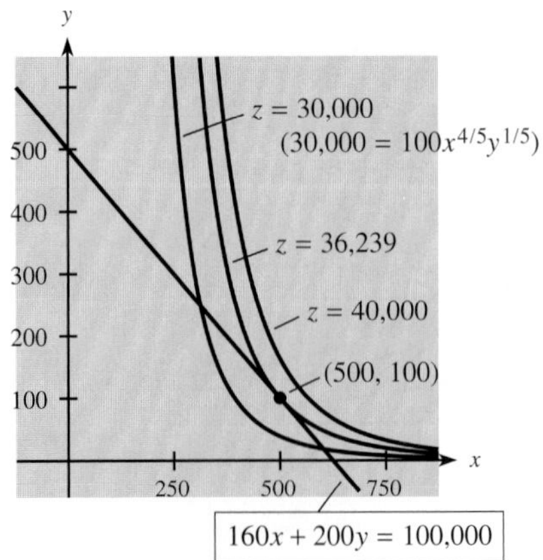

Figure 14.14

In problems of this type, economists call the value of $-\lambda$ the **marginal productivity of money.** In this case

$$-\lambda = \frac{y^{1/5}}{2x^{1/5}} = \frac{(100)^{0.2}}{2(500)^{0.2}} \approx 0.362$$

This means that each additional dollar spent on production results in approximately 0.362 additional unit produced.

Finally, Figure 14.14 shows the graph of the constraint together with some production function curves that correspond to different production levels.

CHECKPOINT SOLUTIONS

1. $F(x, y, \lambda) = x^2 + y^2 - 4xy + \lambda(10 - x - y)$
2. $\dfrac{\partial F}{\partial x} = 2x - 4y - \lambda, \quad \dfrac{\partial F}{\partial y} = 2y - 4x - \lambda, \quad \dfrac{\partial F}{\partial \lambda} = 10 - x - y$
3. $0 = 2x - 4y - \lambda \quad (1)$
   $0 = 2y - 4x - \lambda \quad (2)$
   $0 = 10 - x - y \quad (3)$
   From equations (1) and (2) we have the following:
   $$\lambda = 2x - 4y \quad \text{and} \quad \lambda = 2y - 4x, \quad \text{so}$$
   $$2x - 4y = 2y - 4x$$
   $$6x = 6y, \quad \text{or} \quad x = y$$
   Using $x = y$ in equation (3) gives $0 = 10 - x - x$, or $2x = 10$. Thus $x = 5$ and $y = 5$.
4. $f(5, 5) = 25 + 25 - 100 = -50$ is the minimum because other values that satisfy the constraint give larger z-values.

EXERCISE 14.5

1. Find the minimum value of $z = x^2 + y^2$ subject to the condition $x + y = 6$.
2. Find the minimum value of $z = 4x^2 + y^2$ subject to the constraint $x + y = 5$.
3. Find the minimum value of $z = 3x^2 + 5y^2 - 2xy$ subject to the constraint $x + y = 5$.
4. Find the maximum value of $z = 2xy - 3x^2 - 5y^2$ subject to the constraint $x + y = 5$.
5. Find the maximum value of $z = x^2y$ subject to $x + y = 6$, $x \geq 0$, $y \geq 0$.
6. Find the maximum value of the function $z = x^3y^2$ subject to $x + y = 10$, $x \geq 0$, $y \geq 0$.
7. Find the maximum value of the function $z = 2xy - 2x^2 - 4y^2$ subject to the condition $x + 2y = 8$.
8. Find the minimum value of $z = 2x^2 + y^2 - xy$ subject to the constraint $2x + y = 8$.
9. Find the minimum value of $z = x^2 + y^2$ subject to the condition $2x + y + 1 = 0$.
10. Find the minimum value of $z = x^2 + y^2$ subject to the condition $xy = 1$.
11. Find the minimum value of $w = x^2 + y^2 + z^2$ subject to the constraint $x + y + z = 3$.
12. Find the minimum value of $w = x^2 + y^2 + z^2$ subject to the condition $2x - 4y + z = 21$.
13. Find the maximum value of $w = xz + y$ subject to the constraint $x^2 + y^2 + z^2 = 1$.
14. Find the maximum value of $w = x^2yz$ subject to the constraint $4x + y + z = 4$, $x \geq 0$, $y \geq 0$, and $z \geq 0$.

Applications

15. ***Utility*** Suppose that the utility function for two commodities is given by $U = x^2y$ and that the budget constraint is $3x + 6y = 18$. What values of x and y will maximize utility?

16. ***Utility*** Suppose that the budget constraint in Problem 15 is $5x + 20y = 80$. What values of x and y will maximize $U = x^2y$?

17. ***Utility*** Suppose that the utility function for two products is given by $U = x^2y$, and the budget constraint is $2x + 3y = 120$. Find the values of x and y that maximize utility. Check by graphing the budget constraint with the indifference curve for maximum utility and with two other indifference curves.

18. ***Utility*** Suppose that the utility function for two commodities is given by $U = x^2y^3$, and the budget constraint is $10x + 15y = 250$. Find the values of x and y that maximize utility. Check by graphing the budget constraint with the indifference curve for maximum utility and with two other indifference curves.

19. ***Production*** A company has the Cobb-Douglas production function

$$z = 400x^{0.6}y^{0.4}$$

where x is the number of units of labor and y is the number of units of capital. Suppose labor costs \$150 per unit, capital costs \$100 per unit, and the total cost of labor and capital is limited to \$100,000.
   (a) Find the number of units of labor and the number of units of capital that maximize production.
   (b) Find the marginal productivity of money and interpret it.
   (c) Graph the constraint with the optimal value for production and with two other z-values (one smaller than the optimal value and one larger).

20. ***Production*** Suppose a company has the Cobb-Douglas production function

$$z = 100^{0.75}y^{0.25}$$

where x is the number of units of labor and y is the number of units of capital. Suppose further that labor costs \$90 per unit, capital costs \$150 per unit, and the total costs of labor and capital are limited to \$90,000.
   (a) Find the number of units of labor and the number of units of capital that maximize production.
   (b) Find the marginal productivity of money and interpret it.
   (c) Graph the constraint with the optimal value for production and with two other z-values (one smaller than the optimal value and one larger).

21. ***Cost*** A firm has two plants, X and Y. Suppose that the cost of producing x units at plant X is $x^2 + 1200$ and the cost of producing y units of the same product at plant Y is given by $3y^2 + 800$. If the firm has an order for 1200 units, how many should it produce at each plant to fill this order and minimize the cost of production?

22. ***Cost*** Suppose that the cost of producing x units at plant X is $(3x + 4)x$ and that the cost of producing y units of the same product at plant Y is $(2y + 8)y$. If the firm that owns the plants has an order for 149 units, how many should it produce at each plant to fill this order and minimize its cost of production?

23. ***Revenue*** On the basis of past experience a company has determined that its sales revenue is related to its advertising according to the formula $s = 20x + y^2 + 4xy$, where x is the amount spent on radio advertising and y is the amount spent on television advertising. If the company plans to spend \$30,000 on these two means of advertising, how much should it spend on each method to maximize its sales revenue?

24. ***Manufacturing*** Find the dimensions x, y, and z of the rectangular box with the largest volume that satisfies

$$3x + 4y + 12z = 12$$

25. ***Manufacturing*** Find the dimensions of the box with square base, open top, and volume 500,000 cubic centimeters that requires the least materials.

26. ***Manufacturing*** Show that a box with a square base, an open top, and a fixed volume requires the least material to build if it has a height equal to one-half the length of one side of the base.

KEY TERMS AND FORMULAS

Section	Key Terms	Formula
14.1	Function of two variables	
	Variables: independent, dependent	
	Domain	
	Coordinate planes	
	Utility	
	Indifference curve	
	Indifference map	
14.2	First-order partial derivative	
	With respect to x	$z_x = \frac{\partial z}{\partial x}$
	With respect to y	$z_y = \frac{\partial z}{\partial y}$
	Higher-order partial derivatives	
	Second partial derivatives	z_{xx}, z_{yy}, z_{xy}, and z_{yx}
14.3	Joint cost function	$C = Q(x, y)$
	Marginal cost	
	Marginal productivity	
	Demand function	
	Marginal demand function	
	Competitive products	
	Complementary products	
14.4	Critical values for maxima and minima	Solve simultaneously $\begin{cases} z_x = 0 \\ z_y = 0 \end{cases}$.
	Test for critical values	Use $D(x, y) = (z_{xx})(z_{yy}) - (z_{xy})^2$.
	Linear regression	$\hat{y} = a + bx$
		$b = \frac{\sum x \cdot \sum y - n\sum xy}{(\sum x)^2 - n\sum x^2}$
		$a = \frac{\sum y - b\sum x}{n}$
14.5	Maxima and minima subject to constraints	
	Lagrange multipliers	
	Objective function	$F(x, y, \lambda) = f(x, y) + \lambda g(x, y)$

REVIEW EXERCISES

Section 14.1

1. What is the domain of $z = \dfrac{3}{2x - y}$?
2. What is the domain of $z = \dfrac{3x + 2\sqrt{y}}{x^2 + y^2}$?
3. If $w(x, y, z) = x^2 - 3yz$, find $w(2, 3, 1)$.
4. If $Q(K, L) = 70K^{2/3}L^{1/3}$, find $Q(64{,}000, 512)$.

Section 14.2

5. Find $\frac{\partial z}{\partial x}$ if $z = 5x^3 + 6xy + y^2$.
6. Find $\frac{\partial z}{\partial y}$ if $z = 12x^5 - 14x^3y^3 + 6y^4 - 1$.

In Problems 7–12, find z_x and z_y.

7. $z = 4x^2y^3 + \frac{x}{y}$
8. $z = \sqrt{x^2 + 2y^2}$
9. $z = (xy + 1)^{-2}$
10. $z = e^{x^2y^3}$
11. $z = e^{xy} + y \ln x$
12. $z = e^{\ln xy}$
13. Find the partial derivative of $f(x, y) = 4x^3 - 5xy^2 + y^3$ with respect to x at the point $(1, 2, -8)$.
14. Find the slope of the tangent in the x-direction to the surface $z = 5x^4 - 3xy^2 + y^2$ at $(1, 2, -3)$.

In Problems 15–18, find the second partials.

(a) z_{xx} (b) z_{yy} (c) z_{xy} (d) z_{yx}

15. $z = x^2y - 3xy$
16. $z = 3x^3y^4 - \frac{x^2}{y^2}$
17. $z = x^2e^{y^2}$
18. $z = \ln(xy + 1)$

Section 14.4

19. Test $z = 16 - x^2 - xy - y^2 + 24y$ for maxima and minima.
20. Test $z = x^3 + y^3 - 3xy$ for maxima and minima.

Section 14.5

21. Find the minimum value of $z = 4x^2 + y^2$ subject to the constraint $x + y = 10$.
22. Find the maximum value of $z = x^4y^2$ subject to the constraint $x + y = 9$, $x \geq 0$, $y \geq 0$.

Applications

Section 14.1

23. ***Utility*** Suppose that the utility function for two goods X and Y is given by $U = x^2y$, and a consumer purchases 6 units of X and 15 units of Y. If the consumer purchases 60 units of Y, how many units of X must be purchased to retain the same level of utility?

24. ***Utility*** Suppose that an indifference curve for two products, X and Y, has the equation $xy = 1600$. If 80 units of X are purchased, how many units of Y must be purchased?

Section 14.2

25. ***Concorde sonic booms*** The width of the region on the ground on either side of the path of France's Concorde jet in which people hear the sonic boom is given by

$$w = f(T, h, d) = 2\sqrt{Th/d}$$

where T is the air temperature at ground level in kelvins (K), h is the Concorde's altitude in kilometers, and d is the vertical temperature gradient (the temperature drop in kelvins per kilometer).*

(a) Suppose the Concorde approaches Washington, D.C., from Europe on a course that takes it south of Nantucket Island at an altitude of 16.8 km. If the surface temperature is 293 K and the vertical temperature gradient is 5 K/km, how far south of Nantucket must the plane pass to keep the sonic boom off the island?

(b) Interpret $f(287, 17.1, 4.9) \approx 63.3$.

(c) Find $\frac{\partial f}{\partial h}(293, 16.8, 5)$ and interpret the result.

(d) Find $\frac{\partial f}{\partial d}(293, 16.8, 5)$ and interpret the result.

Section 14.3

26. ***Cost*** The joint cost function for two products is $C(x, y) = x^2\sqrt{y^2 + 13}$. Find the marginal cost with respect to

(a) x if 20 units of x and 6 units of y are produced.

(b) y if 20 units of x and 6 units of y are produced.

27. ***Production*** Suppose that the production function for a company is given by

$$Q = 80K^{1/4}L^{3/4}$$

where Q is the output (in hundreds of units), K is the capital expenditures (in thousands of dollars), and L is the work-hours. Find $\partial Q/\partial K$ and $\partial Q/\partial L$ when expenditures are \$625,000 and total work-hours are 4096. Interpret the results.

28. ***Marginal demand*** The demand functions for two related products, product A and product B, are given by

$$q_A = 400 - 2p_A - 3p_B$$
$$q_B = 300 - 5p_A - 6p_B$$

(a) Find the marginal demand of q_A with respect to p_A.

(b) Find the marginal demand of q_B with respect to p_B.

(c) Are the products complementary or competitive?

*Balachandra, N. K., W. L. Donn, and D. H. Rind, "Concorde Sonic Booms as an Atmospheric Probe," *Science,* 1 July 1977, Vol. 197, p. 47.

29. ***Marginal demand*** Suppose that the demand functions for two related products, A and B, are given by

$$q_A = 800 - 40p_A - \frac{2}{p_B + 1}$$

$$q_B = 1000 - \frac{10}{p_A + 4} - 30p_B$$

Determine whether the products are competitive or complementary.

Section 14.4

30. ***Profit*** The profit from the sale of two products is given by $P(x, y) = 40x + 80y - x^2 - y^2$, where x is the number of units of product 1 and y is the number of units of product 2. Selling how much of each product will maximize profit?

Section 14.5

31. ***Utility*** If the utility function for two commodities is $U = x^2y$, and the budget constraint is $4x + 5y = 60$, find the values of x and y that maximize utility.

32. ***Production*** Suppose a company has the Cobb-Douglas production function

$$z = 300x^{2/3}y^{1/3}$$

where x is the number of units of labor and y is the number of units of capital. Suppose labor costs are \$50 per unit, capital costs are \$50 per unit, and total costs are limited to \$75,000.
   (a) Find the number of units of labor and the number of units of capital that maximize production.
   (b) Find the marginal productivity of money and interpret your result.
   (c) Graph the constraint with the production function when $z = 180{,}000$, $z = 300{,}000$, and when the z-value is optimal.

33. ***Taxes*** The following data show U.S. national personal income and personal taxes for selected years.
   (a) Write the linear regression equation that best fits these data.
   (b) Use the equation found in (a) to predict the taxes when national personal income reaches \$7000 billion.

Income (x) (*billions*)	*Taxes* (y) (*billions*)
\$2285.7	\$312.4
2560.4	360.2
2718.7	371.4
2891.7	369.3
3205.5	395.5
3439.6	437.7
3647.5	459.9
3877.3	514.2
4172.8	532.0
4489.3	594.9
4791.6	624.8
4968.5	624.8
5264.2	650.5
5480.1	689.9
5753.1	731.4
6150.8	795.1
6495.2	886.9

Source: Bureau of Economic Analysis, U.S. Commerce Dept.

34. ***Supply and demand*** The table gives the number of color television sets (in thousands) sold in 15 different years, along with the corresponding average price per set for the year. Use linear regression to find the best linear equation defining the demand function $q = f(p)$.

Price (p) (*dollars*)	*Quantity Sold* (q) (*thousands*)
471.56	9793
487.79	10236
487.32	9107
510.78	7700
504.39	6485
466.29	8411
449.81	10071
509.86	7908
524.96	6349
514.10	4822
515.43	5962
520.82	5981
525.01	5777
462.32	5892
560.09	2646

Source: U.S. Bureau of the Census, *Statistical Abstract of the United States,* Washington, D.C.

CHAPTER TEST

1. Consider the function $f(x, y) = \dfrac{2x + 3y}{\sqrt{x^2 - y}}$.
   (a) Find the domain of $f(x, y)$.
   (b) Evaluate $f(-4, 12)$.
2. Find all first and second partial derivatives of
$$z = f(x, y) = 5x - 9y^2 + 2(xy + 1)^5$$
3. Let $z = 6x^2 + x^2y + y^2 - 4y + 9$. Find the pairs (x, y) that are critical points for z, and then classify each as a relative maximum, a relative minimum, or a saddle point.
4. Suppose a company's monthly production value Q, in thousands of dollars, is given by the Cobb-Douglas production function
$$Q = 10K^{0.45}L^{0.55}$$
where K is thousands of dollars of capital investment per month and L is the total hours of labor per month. Capital investment is currently \$10,000 per month and monthly work-hours of labor total 1590.
   (a) Find the monthly production value (to the nearest thousand dollars).
   (b) Find the marginal productivity with respect to capital investment, and interpret your result.
   (c) Find the marginal productivity with respect to total hours of labor, and interpret your result.
5. The monthly payment R on a loan is a function of the amount borrowed, A, in thousands of dollars; the length of the loan, n, in years; and the annual interest rate, r, as a percent. Thus $R = f(A, n, r)$. In (a) and (b), write a sentence that explains the practical meaning of each mathematical statement.
   (a) $f(94.5, 25, 7) = \$667.91$
   (b) $\dfrac{\partial f}{\partial r}(94.5, 25, 7) = \$49.76$
   (c) Would $\dfrac{\partial f}{\partial n}(94.5, 25, 7)$ be positive, negative, or zero? Explain.
6. Let $f(x, y) = 2e^{x^2y^2}$. Find $\dfrac{\partial^2 f}{\partial x\, \partial y}$.
7. Suppose the demand functions for two products are
$$q_1 = 300 - 2p_1 - 5p_2 \quad \text{and} \quad q_2 = 150 - 4p_1 - 7p_2$$
where q_1 and q_2 represent quantities demanded and p_1 and p_2 represent prices. What calculations enable us to decide whether the products are competitive or complementary? Are these products competitive or complementary?
8. Suppose a store sells two brands of disposable cameras and the profit for these is a function of their two selling prices. The type 1 camera sells for \$$x$, the type 2 sells for \$$y$, and profit is given by
$$P = 915x - 30x^2 - 45xy + 975y - 30y^2 - 3500$$
Find the selling prices that maximize profit.
9. Find x and y that maximize the utility function $U = x^3y$ subject to the budget constraint $30x + 20y = 8000$.
10. For a middle-income family, the estimated annual expenditures associated each year of raising a child through age 11 are given in the table below.
   (a) Find the linear regression line for these data.
   (b) Use the regression equation to project the annual expenditure associated with raising a child during his or her 14th year.
   (c) If this linear model had the form $E = f(x)$, where E is the expenditures and x is the age, would $f(35)$ make sense? Explain.

Age	Expenditures	Age	Expenditures
0	\$7,880	6	\$11,020
1	8,270	7	11,590
2	8,700	8	12,200
3	9,380	9	12,780
4	9,870	10	13,450
5	10,390	11	14,150

Source: Family Economics Research Group, U.S. Dept. of Agriculture, 1996

Extended Applications / Group Projects

I. Advertising

To model sales of its tires, the manufacturer of GRIPPER tires used the quadratic equation $S = a_0 + a_1x + a_2x^2 + b_1y$, where S is regional sales in millions of dollars, x is TV advertising expenditures in millions of dollars, and y is other promotional expenditures in millions of dollars. (See the Extended Application/Group Project "Marginal Return to Sales," on page 729.)

Although this model represents the relationship between advertising and sales dollars for small changes in advertising expenditures, it is clear to the vice president of advertising that it does not apply to large expenditures for TV advertising on a national level. He knows from experience that increased expenditures for TV advertising do result in more sales, but at a decreasing rate of return for the product.

The vice president is aware that some advertising agencies model the relationship between advertising and sales by the function

$$S = b_0 + b_1(1 - e^{-ax}) + c_1y$$

where $a > 0$, S is sales in millions of dollars, x is TV advertising expenditures in millions of dollars, and y is other promotional expenditures in millions of dollars.* The equation

$$S_n = 24.58 + 325.18(1 - e^{-x/14}) + b_1y$$

has the form mentioned previously as being used by some advertising agencies. For TV advertising expenditures up to \$20 million, this equation closely approximates

$$S_1 = 30 + 20x - 0.4x^2 + b_1y$$

which, in the Extended Application/Group Project on page 729, was used with fixed promotional expenses to describe advertising and sales in Region 1.

To help the vice president decide whether this is a better model for large expenditures, answer the following questions.

1. What is $\partial S_1/\partial x$? Does this indicate that sales might actually decline after some amount is spent on TV advertising? If so, what is this amount?
2. Does the quadratic model $S_1(x, y)$ indicate that sales will become negative after some amount is spent on TV advertising? Does this model cease to be useful in predicting sales after a certain point?
3. What is $\partial S_n/\partial x$? Does this indicate that sales will continue to rise if additional money is devoted to TV advertising? Is S_n growing at a rate that is increasing or decreasing when promotional sales are held constant? Is S_n a better model for large expenditures?

*Mansfield, Edwin, *Managerial Economics* (New York: Norton, 1990).

4. If this model does describe the relationship between advertising and sales, and if promotional expenditures are held constant at y_0, is there an upper limit to the sales, even if an unlimited amount of money is spent on TV advertising? If so, what is it?

II. Competitive Pricing

Often retailers sell different brands of competing products. Depending on the joint demand for the products, the retailer may be able to set prices that regulate demand and, therefore, influence profits.

Suppose HOME-ALL, Inc., a national chain of home improvement retailers, sells two competing brands of interior flat paint, En-Dure 100 and Croyle & James, which the chain purchases for $8 per gallon and $10 per gallon, respectively. HOME-ALL's research department has determined the following two monthly demand equations for these paints:

$$D = 120 - 40d + 30c \qquad \text{and} \qquad C = 680 + 30d - 40c$$

where D is hundreds of gallons of En-Dure 100 demanded at $\$d$ per gallon and C is hundreds of gallons of Croyle & James demanded at $\$c$ per gallon. For what prices should HOME-ALL sell these paints in order to maximize its monthly profit on these items?

To answer this question, complete the following.

1. Recall that revenue is a product's selling price per item times the number of items sold. With this in mind, formulate HOME-ALL's total revenue function for the two paints as a function of their prices.
2. Form HOME-ALL's profit function for the two paints (in terms of their selling prices).
3. Determine the price of each type of paint that will maximize HOME-ALL's profit.
4. Write a brief report to management that details your pricing recommendations and justifies them.

Appendix

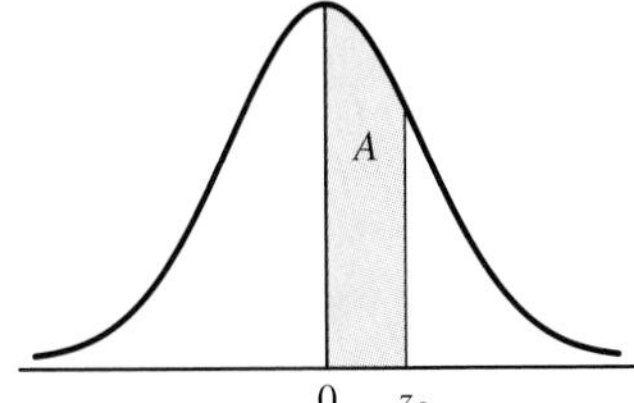

Areas Under the Standard Normal Curve

The value of A is the area under the standard normal curve between $z = 0$ and $z = z_0$, for $z_0 \geq 0$. Areas for negative values of z_0 are obtained by symmetry.

z_0	A	z_0	A	z_0	A	z_0	A
.00	.0000	.36	.1406	.72	.2642	1.08	.3599
.01	.0040	.37	.1443	.73	.2673	1.09	.3621
.02	.0080	.38	.1480	.74	.2704	1.10	.3643
.03	.0120	.39	.1517	.75	.2734	1.11	.3665
.04	.0160	.40	.1554	.76	.2764	1.12	.3686
.05	.0199	.41	.1591	.77	.2794	1.13	.3708
.06	.0239	.42	.1628	.78	.2823	1.14	.3729
.07	.0279	.43	.1664	.79	.2852	1.15	.3749
.08	.0319	.44	.1700	.80	.2881	1.16	.3770
.09	.0359	.45	.1736	.81	.2910	1.17	.3790
.10	.0398	.46	.1772	.82	.2939	1.18	.3810
.11	.0438	.47	.1808	.83	.2967	1.19	.3830
.12	.0478	.48	.1844	.84	.2996	1.20	.3849
.13	.0517	.49	.1879	.85	.3023	1.21	.3869
.14	.0557	.50	.1915	.86	.3051	1.22	.3888
.15	.0596	.51	.1950	.87	.3079	1.23	.3907
.16	.0636	.52	.1985	.88	.3106	1.24	.3925
.17	.0675	.53	.2019	.89	.3133	1.25	.3944
.18	.0714	.54	.2054	.90	.3159	1.26	.3962
.19	.0754	.55	.2088	.91	.3186	1.27	.3980
.20	.0793	.56	.2123	.92	.3212	1.28	.3997
.21	.0832	.57	.2157	.93	.3238	1.29	.4015
.22	.0871	.58	.2190	.94	.3264	1.30	.4032
.23	.0910	.59	.2224	.95	.3289	1.31	.4049
.24	.0948	.60	.2258	.96	.3315	1.32	.4066
.25	.0987	.61	.2291	.97	.3340	1.33	.4082
.26	.1026	.62	.2324	.98	.3365	1.34	.4099
.27	.1064	.63	.2357	.99	.3389	1.35	.4115
.28	.1103	.64	.2389	1.00	.3413	1.36	.4131
.29	.1141	.65	.2422	1.01	.3438	1.37	.4147
.30	.1179	.66	.2454	1.02	.3461	1.38	.4162
.31	.1217	.67	.2486	1.03	.3485	1.39	.4177
.32	.1255	.68	.2518	1.04	.3508	1.40	.4192
.33	.1293	.69	.2549	1.05	.3531	1.41	.4207
.34	.1331	.70	.2580	1.06	.3554	1.42	.4222
.35	.1368	.71	.2612	1.07	.3577	1.43	.4236

Areas Under the Standard Normal Curve (Continued)

z_0	A	z_0	A	z_0	A	z_0	A
1.44	.4251	1.93	.4732	2.42	.4922	2.91	.4982
1.45	.4265	1.94	.4738	2.43	.4925	2.92	.4983
1.46	.4279	1.95	.4744	2.44	.4927	2.93	.4983
1.47	.4292	1.96	.4750	2.45	.4929	2.94	.4984
1.48	.4306	1.97	.4756	2.46	.4931	2.95	.4984
1.49	.4319	1.98	.4762	2.47	.4932	2.96	.4985
1.50	.4332	1.99	.4767	2.48	.4934	2.97	.4985
1.51	.4345	2.00	.4773	2.49	.4936	2.98	.4986
1.52	.4357	2.01	.4778	2.50	.4938	2.99	.4986
1.53	.4370	2.02	.4783	2.51	.4940	3.00	.4987
1.54	.4382	2.03	.4788	2.52	.4941	3.01	.4987
1.55	.4394	2.04	.4793	2.53	.4943	3.02	.4987
1.56	.4406	2.05	.4798	2.54	.4945	3.03	.4988
1.57	.4418	2.06	.4803	2.55	.4946	3.04	.4988
1.58	.4430	2.07	.4808	2.56	.4948	3.05	.4989
1.59	.4441	2.08	.4812	2.57	.4949	3.06	.4989
1.60	.4452	2.09	.4817	2.58	.4951	3.07	.4989
1.61	.4463	2.10	.4821	2.59	.4952	3.08	.4990
1.62	.4474	2.11	.4826	2.60	.4953	3.09	.4990
1.63	.4485	2.12	.4830	2.61	.4955	3.10	.4990
1.64	.4495	2.13	.4834	2.62	.4956	3.11	.4991
1.65	.4505	2.14	.4838	2.63	.4957	3.12	.4991
1.66	.4515	2.15	.4842	2.64	.4959	3.13	.4991
1.67	.4525	2.16	.4846	2.65	.4960	3.14	.4992
1.68	.4535	2.17	.4850	2.66	.4961	3.15	.4992
1.69	.4545	2.18	.4854	2.67	.4962	3.16	.4992
1.70	.4554	2.19	.4857	2.68	.4963	3.17	.4992
1.71	.4564	2.20	.4861	2.69	.4964	3.18	.4993
1.72	.4573	2.21	.4865	2.70	.4965	3.19	.4993
1.73	.4582	2.22	.4868	2.71	.4966	3.20	.4993
1.74	.4591	2.23	.4871	2.72	.4967	3.21	.4993
1.75	.4599	2.24	.4875	2.73	.4968	3.22	.4994
1.76	.4608	2.25	.4878	2.74	.4969	3.23	.4994
1.77	.4616	2.26	.4881	2.75	.4970	3.24	.4994
1.78	.4625	2.27	.4884	2.76	.4971	3.25	.4994
1.79	.4633	2.28	.4887	2.77	.4972	3.26	.4994
1.80	.4641	2.29	.4890	2.78	.4973	3.27	.4995
1.81	.4649	2.30	.4893	2.79	.4974	3.28	.4995
1.82	.4656	2.31	.4896	2.80	.4974	3.29	.4995
1.83	.4664	2.32	.4898	2.81	.4975	3.30	.4995
1.84	.4671	2.33	.4901	2.82	.4976	3.31	.4995
1.85	.4678	2.34	.4904	2.83	.4977	3.32	.4996
1.86	.4686	2.35	.4906	2.84	.4977	3.33	.4996
1.87	.4693	2.36	.4909	2.85	.4978	3.34	.4996
1.88	.4700	2.37	.4911	2.86	.4979	3.35	.4996
1.89	.4706	2.38	.4913	2.87	.4980	3.36	.4996
1.90	.4713	2.39	.4916	2.88	.4980	3.37	.4996
1.91	.4719	2.40	.4918	2.89	.4981	3.38	.4996
1.92	.4726	2.41	.4920	2.90	.4981	3.39	.4997

Areas Under the Standard Normal Curve (Continued)

z_0	A	z_0	A	z_0	A	z_0	A
3.40	.4997	3.52	.4998	3.64	.4999	3.76	.4999
3.41	.4997	3.53	.4998	3.65	.4999	3.77	.4999
3.42	.4997	3.54	.4998	3.66	.4999	3.78	.4999
3.43	.4997	3.55	.4998	3.67	.4999	3.79	.4999
3.44	.4997	3.56	.4998	3.68	.4999	3.80	.4999
3.45	.4997	3.57	.4998	3.69	.4999	3.81	.4999
3.46	.4997	3.58	.4998	3.70	.4999	3.82	.4999
3.47	.4997	3.59	.4998	3.71	.4999	3.83	.4999
3.48	.4998	3.60	.4998	3.72	.4999	3.84	.4999
3.49	.4998	3.61	.4999	3.73	.4999	3.85	.4999
3.50	.4998	3.62	.4999	3.74	.4999	3.86	.4999
3.51	.4998	3.63	.4999	3.75	.4999		

Answers

Below are the answers to odd-numbered Section Exercises and all the Chapter Review and Chapter Test problems.

Exercise 0.1 (page 7)

1. $\in$ **3.** $\in$ **5.** $\notin$ **7.** $\{1, 2, 3, 4, 5, 6, 7\}$
9. $\{x: x$ is a natural number greater than 2 and less than $8\}$
11. yes **13.** no **15.** $D \subseteq C$ **17.** $D \subseteq A$
19. $A \subseteq B$ or $B \subseteq A$ **21.** yes **23.** no
25. A and B, B and D, C and D
27. $A \cap B = \{4, 6\}$ **29.** $A \cap B = \varnothing$
31. $A \cup B = \{1, 2, 3, 4, 5\}$
33. $A \cup B = \{1, 2, 3, 4\} = B$
35. $A' = \{4, 6, 9, 10\}$
37. $A \cap B' = \{1, 2, 5, 7\}$ **39.** $(A \cup B)' = \{6, 9\}$
41. $A' \cup B' = \{1, 2, 4, 5, 6, 7, 9, 10\}$
43. $\{1, 2, 3, 5, 7, 9\}$ **45.** $\{4, 6, 8, 10\}$
47. $A - B = \{1, 7\}$ **49.** $A - B = \varnothing$ or $\{\ \}$
51. **(a)** $L = \{94, 95, 96, 97\}$;
$H = \{92, 93, 94, 95, 96, 97\}$;
$C = \{90, 91, 95, 96, 97\}$
(b) $L \subseteq H$
(c) C' is the set of years when the percentage change from low to high was 25% or less.
(d) $\{90, 91, 92, 93, 94\}$ = the set of years when the high was 3300 or less or the percentage change was 25% or less.
(e) $\{90, 91\}$ = the set of years when the low was 3300 or less and the percentage change exceeded 25%.
53. **(a)** 130 **(b)** 840 **(c)** 520
55. **(a)**

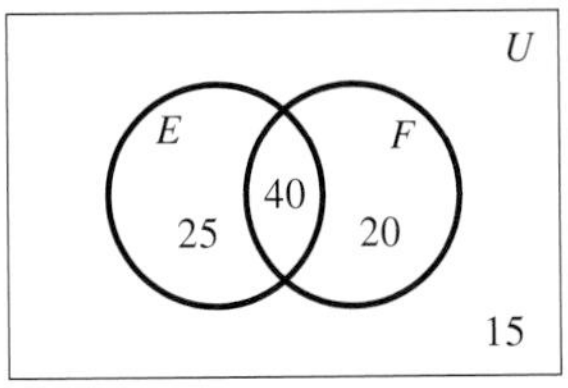

(b) 40
(c) 85
(d) 25

57.

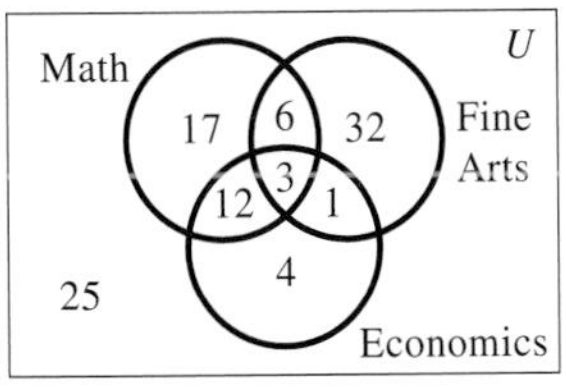

(a) 25
(b) 43
(c) 53

59. **(a)** and **(b)**

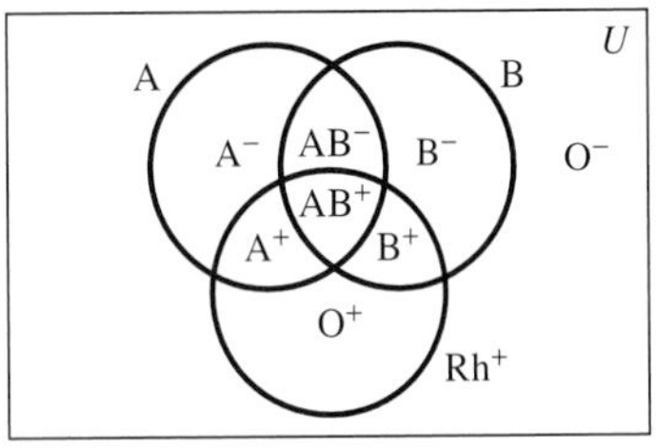

Exercise 0.2 (page 15)

1. **(a)** irrational **(b)** rational, integer
(c) rational, integer, natural **(d)** meaningless
3. **(a)** Commutative **(b)** Distributive
(c) Multiplicative identity
5. $<$ **7.** $<$ **9.** $>$ **11.** 11 **13.** 4
15. 2 **17.** $\frac{-4}{3}$ **19.** 3 **21.** $\frac{17}{11}$ **23.** entire line
25. $(1, 3]$ **27.** $(2, 10)$ **29.** $x \le 5$ **31.** $x > 4$
33. $(-3, 4)$

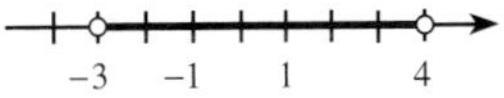

35. $(4, +\infty)$

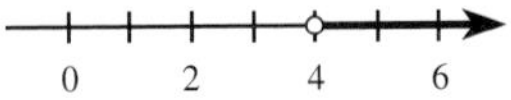

37. $[-1, +\infty)$

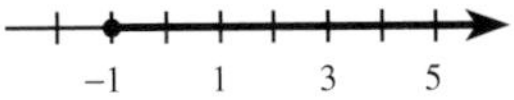

39. $(-\infty, 0) \cup (7, +\infty)$

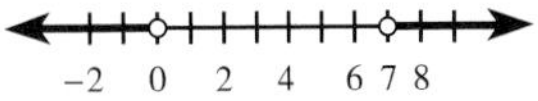

41. -0.000038585
43. 9122.387471 **45.** 3240.184509
47. **(a)** \$1088.91 **(b)** \$258.62 **(c)** \$627.20
49. **(a)** Formula (2) is more accurate; 1996: \$24.54; 1997: \$26.10
(b) \$40.29

Exercise 0.3 (page 20)

1. -64 **3.** -16 **5.** $\frac{1}{9}$ **7.** $-\frac{9}{4}$ **9.** 6^8
11. $\frac{1}{10}$ **13.** 3^9 **15.** $(\frac{3}{2})^2 = \frac{9}{4}$ **17.** $1/x^6$
19. x/y^2 **21.** x^7 **23.** $x^{-2} = 1/x^2$ **25.** x^4
27. y^{12} **29.** x^{12} **31.** x^2y^2 **33.** $16/x^4$
35. $x^8/(16y^4)$ **37.** $-16a^2/b^2$ **39.** $2/(xy^2)$
41. $1/(x^9y^6)$ **43.** $(a^{18}c^{12})/b^6$
45. **(a)** $1/(2x^4)$ **(b)** $1/(16x^4)$ **(c)** $1/x^4$ **(d)** 8
47. x^{-1} **49.** $8x^3$ **51.** $\frac{1}{4}x^{-2}$ **53.** $-\frac{1}{8}x^3$
55. 2.0736 **57.** 0.1316872428
59. $S = \$2114.81$; $I = \$914.81$
61. $S = \$9607.70$; $I = \$4607.70$
63. \$7806.24

65. **(a)** 15, 18, 20, 22 **(b)** \$0.53, \$0.76, \$0.96, \$1.22 **(c)** \$3.13
67. **(a)** 46.7, 71.7, 3652.5 all in billions of \$ **(b)** World War II had just ended.
69. **(a)** 10 **(b)** \$110.5 billion **(c)** \$872.5 billion **(d)** \$2564.4 billion

Exercise 0.4 (page 29)

1. $\frac{16}{3}$ **3.** -8 **5.** 8 **7.** not real **9.** $\frac{9}{4}$
11. **(a)** 4 **(b)** $\frac{1}{4}$ **13.** $(6.12)^{4/9} \approx 2.237$
15. $m^{3/2}$ **17.** $(m^2n^5)^{1/4}$ **19.** $\sqrt[4]{x^7}$ **21.** $-1/(4\sqrt[4]{x^5})$
23. $y^{3/4}$ **25.** $z^{19/4}$ **27.** $1/y^{5/2}$ **29.** x
31. $1/y^{21/10}$ **33.** $x^{1/2}$ **35.** $1/x$ **37.** $8x^2$
39. $8x^2y^2\sqrt{2y}$ **41.** $2x^2y\sqrt[3]{5x^2y^2}$ **43.** $6x^2y\sqrt{x}$
45. $42x^3y^2\sqrt{x}$ **47.** $2xy^5/3$ **49.** $2b\sqrt[4]{b}/(3a^2)$
51. 1/9 **53.** 7 **55.** $\sqrt{6}/3$ **57.** $\sqrt{mx}/x$
59. $\sqrt[3]{mx^2}/x^2$ **61.** $-\frac{2}{3}x^{-2/3}$ **63.** $3x^{3/2}$
65. $(3\sqrt{x})/2$ **67.** $1/(2\sqrt{x})$
69. **(a)** $10^{8.5} = 10^{17/2} = \sqrt{10^{17}}$ **(b)** 316,227,766 **(c)** 14.125
71. 74 kg **73.** 39,491 **75.** **(a)** 10 **(b)** 259

Exercise 0.5 (page 39)

1. **(a)** 2 **(b)** -1 **(c)** 10 **(d)** one
3. **(a)** 5 **(b)** -14 **(c)** 0 **(d)** several
5. **(a)** 5 **(b)** 0 **(c)** 2 **(d)** -5
7. -12 **9.** $\frac{-7}{31}$ **11.** $21pq - 2p^2$
13. $m^2 - 7n^2 - 3$ **15.** $3q + 12$
17. $x^2 - 1$ **19.** $35x^5$ **21.** $3rs$
23. $2ax^4 + a^2x^3 + a^2bx^2$ **25.** $6y^2 - y - 12$
27. $2 - 5x^2 + 2x^4$ **29.** $16x^2 + 24x + 9$
31. $x^4 - x^2 + \frac{1}{4}$ **33.** $4x^2 - 1$ **35.** $0.01 - 16x^2$
37. $x^3 - 8$ **39.** $x^8 + 3x^6 - 10x^4 + 5x^3 + 25x$
41. $3 + m + 2m^2n$ **43.** $8x^3y^2/3 + 5/(3y) - 2x^2/(3y)$
45. $x^3 + 3x^2 + 3x + 1$ **47.** $8x^3 - 36x^2 + 54x - 27$
49. $0.1x^2 - 1.995x - 0.1$
51. $x^2 - 2x + 5 - 11/(x + 2)$
53. $x^2 + 3x - 1 + (-4x + 2)/(x^2 + 1)$
55. $x + 2x^2$ **57.** $x - x^{1/2} - 2$ **59.** $x - 9$
61. $4x^2 + 4x$
63. **(a)** $9x^2 - 21x + 13$ **(b)** 5 **65.** $55x$
67. **(a)** $4000 - x$ **(b)** $0.10\,x$ **(c)** $0.08(4000 - x)$ **(d)** $0.10x + 0.08(4000 - x)$
69. $(15 - 2x)(10 - 2x)x$
71. **(a)** = A2 − 1 **(b)** width = 50 − length **(c)** = 50 − A2 **(d)** = A2*B2 **(e)** length = 33, width = 17

73. **(a)**

	A	B
1	Year	Tax load
2	1995	37.04688
3	1996	37.29315
4	1997	37.53942
5	1998	37.78569
6	1999	38.03196
7	2000	38.27823
8	2001	38.5245
9	2002	38.77077
10	2003	39.01704
11	2004	39.26331
12	2005	39.50958
13	2006	39.75585
14	2007	40.00212
15	2008	40.24839
16	2009	40.49466
17	2010	40.74093

(b) 2007

Exercise 0.6 (page 46)

1. $3b(3a - 4a^2 + 6b)$ **3.** $2x(2x + 4y^2 + y^3)$
5. $(7x^2 + 2)(x - 2)$ **7.** $(6 + y)(x - m)$
9. $(x + 2)(x + 6)$ **11.** $(x - 3)(x + 2)$
13. $(7x + 4)(x - 2)$ **15.** $(x - 5)^2$
17. $(7a + 12b)(7a - 12b)$
19. **(a)** $(3x - 1)(3x + 8)$ **(b)** $(9x + 4)(x + 2)$
21. $x(4x - 1)$ **23.** $(x^2 - 5)(x + 4)$
25. $2(x - 7)(x + 3)$ **27.** $2x(x - 2)^2$
29. $(2x - 3)(x + 2)$ **31.** $3(x + 4)(x - 3)$
33. $2x(x + 2)(x - 2)$ **35.** $(5x + 2)(2x + 3)$
37. $(5x - 1)(2x - 9)$
39. $(y^2 + 4x^2)(y + 2x)(y - 2x)$
41. $(x + 2)^2(x - 2)^2$
43. $(2x + 1)(2x - 1)(x + 1)(x - 1)$
45. $(x + 1)^3$ **47.** $(x - 4)^3$
49. $(x - 4)(x^2 + 4x + 16)$
51. $(3 + 2x)(9 - 6x + 4x^2)$ **53.** $x + 1$
55. $1 + x$ **57.** $7x - 3x^3$ **59.** $P(1 + rt)$
61. $m(c - m)$
63. **(a)** $p(10{,}000 - 100p)$; $x = 10{,}000 - 100p$ **(b)** 6200

Exercise 0.7 (page 53)

1. $2y^3/z$ **3.** $\frac{1}{3}$ **5.** $(x - 1)/(x - 3)$
7. $(2xy^2 - 5)/(y + 3)$ **9.** $20x/y$ **11.** $\frac{32}{3}$
13. $2x^2 - 7x + 6$
15. $-(x + 1)(x + 3)/[(x - 1)(x - 3)]$
17. $15bc^2/2$ **19.** $5y/(y - 3)$
21. $\dfrac{-x(x - 3)(x + 2)}{x + 3}$ **23.** $\dfrac{1}{x + 1}$ **25.** $\dfrac{4a - 4}{a(a - 2)}$

27. $\dfrac{-x^2+x+1}{x+1}$ **29.** $\dfrac{16a+15a^2}{12(x+2)}$ **31.** $\dfrac{79x+9}{30(x-2)}$

33. $\dfrac{-4y}{(x-2y)^2(x+2y)}$ **35.** $\dfrac{9x+4}{(x-2)(x+2)(x+1)}$

37. $(7x-3x^3)/\sqrt{3-x^2}$ **39.** $\frac{1}{6}$ **41.** xy

43. $\dfrac{x+1}{x^2}$ **45.** $\dfrac{1}{\sqrt{a}}=\dfrac{\sqrt{a}}{a}$ **47.** $\dfrac{x-2}{(x-3)\sqrt{x^2+9}}$

49. **(a)** -12 **(b)** $\frac{25}{36}$ **51.** $2b-a$

53. $\dfrac{x^3+y^3}{x^2y^2(x+y)}=\dfrac{x^2-xy+y^2}{x^2y^2}$

55. $(1-2\sqrt{x}+x)/(1-x)$ **57.** $1/(\sqrt{x+h}+\sqrt{x})$

59. $(bc+ac+ab)/abc$

61. **(a)** $\dfrac{0.1x^2+55x+4000}{x}$ **(b)** $0.1x^2+55x+4000$

63. $\dfrac{t^2+9t}{(t+3)^2}$

Chapter 0 Review Exercises (page 57)

1. yes **2.** no **3.** no **4.** $\{1, 2, 3, 4, 9\}$
5. $\{5, 6, 7, 8, 10\}$ **6.** $\{1, 2, 3, 4, 9\}$
7. yes, $(A' \cup B')' = \{1, 3\} = A \cap B$
8. **(a)** commutative property of addition
(b) associative property of multiplication
(c) distributive law
9. **(a)** irrational **(b)** rational, integer
(c) meaningless
10. **(a)** $>$ **(b)** $<$ **(c)** $>$ **11.** 6
12. 142 **13.** 10 **14.** 5/4 **15.** 9 **16.** -29
17. 13/4 **18.** -10.62857888
19. **(a)** $[0, 5]$, closed

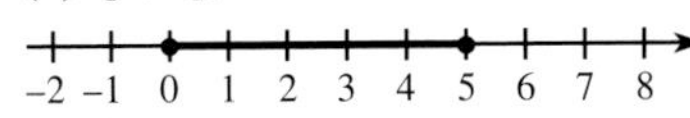

(b) $[-3, 7)$, half-open

−6 −4 −2 0 2 4 6 8 10

(c) $(-4, 0)$, open

−6 −5 −4 −3 −2 −1 0 1 2

20. **(a)** $-1<x<16$ **(b)** $-12 \le x \le 8$
(c) $x<-1$
21. **(a)** 1 **(b)** $2^{-2}=1/4$ **(c)** 4^6 **(d)** 7
22. **(a)** $1/x^2$ **(b)** x^{10} **(c)** x^9 **(d)** $1/y^8$ **(e)** y^6
23. $-x^2y^2/36$ **24.** $9y^8/(4x^4)$ **25.** $y^2/(4x^4)$
26. $-x^8z^4/y^4$ **27.** $3x/(y^7z)$ **28.** $x^5/(2y^3)$
29. **(a)** 4 **(b)** 2/7 **(c)** 1.1
30. **(a)** $x^{1/2}$ **(b)** $x^{2/3}$ **(c)** $x^{-1/4}$
31. **(a)** $\sqrt[3]{x^2}$ **(b)** $1/\sqrt{x}=\sqrt{x}/x$ **(c)** $-x\sqrt{x}$
32. **(a)** $5y\sqrt{2x}/2$ **(b)** $\sqrt[3]{x^2y}/x^2$ **33.** $x^{5/6}$ **34.** y
35. $x^{17/4}$ **36.** $x^{11/3}$ **37.** $x^{2/5}$ **38.** x^2y^8

39. $2xy^2\sqrt{3xy}$ **40.** $25x^3y^4\sqrt{2y}$ **41.** $6x^2y^4\sqrt[3]{5x^2y^2}$
42. $8a^2b^4\sqrt{2a}$ **43.** $2xy$ **44.** $4x\sqrt{3xy}/(3y^4)$
45. $-x-2$ **46.** $-x^2-x$
47. $4x^3+xy+4y-4$ **48.** $24x^5y^5$
49. $3x^2-7x+4$ **50.** $3x^2+5x-2$
51. $4x^2-7x-2$ **52.** $6x^2-11x-7$
53. $4x^2-12x+9$ **54.** $16x^2-9$
55. $2x^4+2x^3-5x^2+x-3$
56. $8x^3-12x^2+6x-1$
57. x^3-y^3 **58.** $(2/y)-(3xy/2)-3x^2$
59. $3x^2+2x-3+(-3x+7)/(x^2+1)$
60. $x^3-x^2+2x+7+21/(x-3)$
61. x^2-x **62.** $2x-a$ **63.** $x^3(2x-1)$
64. $2(x^2+1)^2(1+x)(1-x)$ **65.** $(2x-1)^2$
66. $(4+3x)(4-3x)$ **67.** $2x^2(x+2)(x-2)$
68. $(x-7)(x+3)$ **69.** $(3x+2)(x-1)$
70. $(4x+3)(3x-8)$ **71.** $(2x+3)^2(2x-3)^2$

72. $x^{2/3}+1$ **73.** **(a)** $\dfrac{x}{(x+2)}$ **(b)** $\dfrac{2xy(2-3xy)}{2x-3y}$

74. $\dfrac{x^2-4}{x(x+4)}$ **75.** $\dfrac{(x+3)}{(x-3)}$ **76.** $\dfrac{x^2(3x-2)}{(x-1)(x+2)}$

77. $(6x^2+9x-1)/(6x^2)$ **78.** $\dfrac{4x-x^2}{4(x-2)}$

79. $-\dfrac{x^2+2x+2}{x(x-1)^2}$ **80.** $\dfrac{x(x-4)}{(x-2)(x+1)(x-3)}$

81. $\dfrac{(x-1)^3}{x^2}$ **82.** $\dfrac{1-x}{1+x}$ **83.** $3(\sqrt{x}+1)$

84. $2/(\sqrt{x}+\sqrt{x-4})$
85. **(a)**

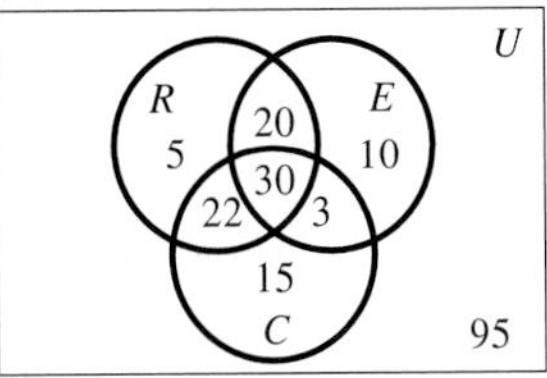

(b) 10 **(c)** 100
86. **(a)** 4115.27 **(b)** $66,788.69
87. **(a)** $10{,}000\left[\dfrac{(0.0065)(1.0065)^n}{(1.0065)^n-1}\right]$
(b) $243.19 (for both)
88. **(a)** $5=k\sqrt[3]{A}$ **(b)** $\sqrt[3]{2.25}\times 1.31$
89. **(a)** $\dfrac{5400p}{100-p}$
(b) $0. It costs nothing if no effort is made to remove pollution.
(c) $264,600
(d) Undefined. Removing 100% would be impossible, and the cost of getting close would be enormous.

Chapter 0 Test (page 59)

1. **(a)** $\{3, 4, 6, 8\}$ **(b)** $\{3, 4\}$; $\{3, 6\}$; or $\{4, 6\}$
(c) $\{6\}$ or $\{8\}$
2. 21
3. **(a)** 8 **(b)** 1 **(c)** $\frac{1}{2}$ **(d)** -10 **(e)** 30
(f) $\frac{5}{6}$ **(g)** $\frac{2}{3}$ **(h)** -3
4. **(a)** $\sqrt[4]{x}$ **(b)** $\frac{1}{\sqrt[4]{x^3}}$ **5.** **(a)** $\frac{1}{x^5}$ **(b)** $\frac{x^{21}}{y^6}$
6. **(a)** $\frac{\sqrt{5x}}{5}$ **(b)** $2a^2b^2\sqrt{6ab}$ **(c)** $\frac{1 - 2\sqrt{x} + x}{1 - x}$
7. **(a)** 5 **(b)** -8 **(c)** -5
8. $(-2, 3]$ 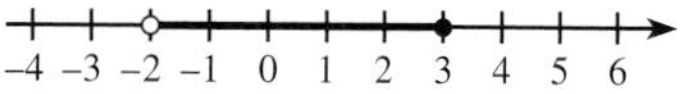

9. **(a)** $2x^2(4x - 1)$ **(b)** $(x - 12)(x + 2)$
(c) $(3x - 2)(2x - 3)$ **(d)** $2x^3(1 + 4x)(1 - 4x)$
10. (c); -2
11. $2x + 1 + \frac{2x - 6}{x^2 - 1}$
12. **(a)** $19y - 45$ **(b)** $-6t^6 + 9t^9$
(c) $4x^3 - 21x^2 + 13x - 2$ **(d)** $-18x^2 + 15x - 2$
(e) $4m^2 - 28m + 49$ **(f)** $\frac{x^4}{3x + 9}$ **(g)** $\frac{x^7}{81}$
(h) $\frac{6 - x}{x - 8}$ **(i)** $\frac{x^2 - 4x - 3}{x(x - 3)(x + 1)}$
13. $\frac{y - x}{y + xy^2}$
14. **(a)** 0 **(b)** 175
15. \$4875.44 (nearest cent)

Exercise 1.1 (page 73)

1. $x = -9/4$ **3.** $x = 0$ **5.** $x = 4/5$
7. $x = -32$ **9.** $x = 16/5$ **11.** $x = -29/2$
13. $x = 17/13$ **15.** $x = 13/5$ **17.** $x = -1/3$
19. $x = 3$ **21.** $x = 74$ **23.** No solution
25. $x = 5/4$ **27.** $x \approx -0.279$
29. $x \approx -1147.362$ **31.** $y = \frac{3}{4}x - \frac{15}{4}$
33. $y = -6x + \frac{22}{3}$ **35.** $P = I/(rt)$
37. 96 **39.** **(a)** 132 pounds **(b)** 73.6 inches
41. **(a)** 59.2% (approx.) **(b)** $t \approx 107$, in 2082
43. $t \approx 26$ in 2014 **45.** 9000
47. **(a)** $T = (7n - 52)/12$ **(b)** 61°
49. \$4000 **51.** \$90,000 at 9%; \$30,000 at 13%
53. \$2160/month; 8% increase
55. lost \$40 **57.** \$307 **59.** \$246

Exercise 1.2 (page 85)

1. yes **3.** no
5. yes; each x-value has exactly one y-value;
$D = \{1, 2, 3, 8, 9\}$, $R = \{-4, 5, 16\}$
7. **(a)** -10 **(b)** 6 **(c)** -34 **(d)** 2.8
9. **(a)** -3 **(b)** 1 **(c)** 13 **(d)** 6
11. **(a)** 63/8 **(b)** 6 **(c)** -6
13. **(a)** no, $f(2 + 1) = f(3) = 13$ but $f(2) + f(1) = 10$
(b) $1 + x + h + x^2 + 2xh + h^2$
(c) no, $f(x) + f(h) = 2 + x + h + x^2 + h^2$
(d) no, $f(x) + h = 1 + x + x^2 + h$
(e) $1 + 2x + h$
15. **(a)** $-2x^2 - 4xh - 2h^2 + x + h$
(b) $-4x - 2h + 1$
17. The vertical-line test shows that graph (a) represents a function of x, but graph (b) does not.
19. **(a)** 10 **(b)** 6
21. **(a)** $b = a^2 - 4a$ **(b)** $(1, -3)$, yes
(c) $(3, -3)$, yes **(d)** $x = 0$, $x = 4$, yes
23. **(a)** $3x + x^3$ **(b)** $3x - x^3$ **(c)** $3x^4$ **(d)** $\frac{3}{x^2}$
25. **(a)** $\sqrt{2x} + x^2$ **(b)** $\sqrt{2x} - x^2$
(c) $x^2\sqrt{2x}$ **(d)** $\frac{\sqrt{2x}}{x^2}$
27. **(a)** $-8x^3$ **(b)** $1 - 2(x - 1)^3$
(c) $[(x - 1)^3 - 1]^3$ **(d)** $(x - 1)^6$
29. **(a)** $2\sqrt{x^4 + 5}$ **(b)** $16x^2 + 5$
(c) $2\sqrt{2\sqrt{x}}$ **(d)** $4x$
31. D: all reals; R: reals $y \geq 4$
33. D: reals $x \geq -4$: R: reals $y \geq 0$
35. $x \geq 1, x \neq 2$ **37.** $-7 \leq x \leq 7$
39. **(a)** $f(20) = 103{,}000$ means that if \$103,000 is borrowed, it can be repaid in 20 years (of \$800-per-month payments).
(b) no; $f(5 + 5) = f(10) = 69{,}000$, but $f(5) + f(5) = 80{,}000$
41. **(a)** $f(1950) = 16.5$ means that in 1950 there were 16.5 workers supporting each Social Security beneficiary.
(b) 3.4
(c) The parts of the graph that correspond to data prior to 1995 would be the same, but data on this graph beyond 1995 are predictions for the future and only might be accurate.
(d) Domain: $1950 \leq t \leq 2050$
Range: $1.9 \leq n \leq 16.5$
43. **(a)** $s \geq 0$
(b) $f(10) \approx -29.33$ means that if the air temperature is -5°F and there is a 10 mph wind, then the temperature feels like -29.33°F.
(c) $f(0) = 45.694$ from the formula, but $f(0)$ should equal the air temperature, -5°F.
45. **(a)** yes **(b)** $t \neq -2$ **(c)** $t \geq 0$
47. **(a)** yes **(b)** all reals
(c) D: $32° \leq F \leq 212°$; R: $0° \leq C \leq 100°$
(d) 4.44°C

49. **(a)** $0 \le p < 100$
(b) \$5972.73; to remove 45% of the particulate pollution would cost \$5972.73.
(c) \$65,700; to remove 90% of the particulate pollution would cost \$65,700.
(d) \$722,700; to remove 99% of the particulate pollution would cost \$722,700.
(e) \$1,817,700; to remove 99.6% of the particulate pollution would cost \$1,817,700.

51. **(a)** yes **(b)** $A(2) = 96$; $A(30) = 600$
(c) $0 < x < 50$

53. **(a)** $(p \circ x)(t) = 180(1000 + 10t) - \dfrac{(1000 + 10t)^2}{100} - 200$
(b) $x = 1150$, $p = \$193,575$

55. $L = 2x + 3200/x$ **57.** $R = (30 + x)(10 - 0.20x)$

Exercise 1.3 (page 99)

1. x-intercept 4
y-intercept 3

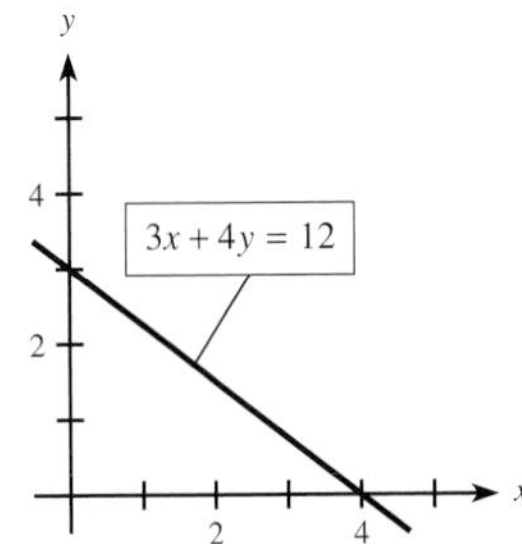

3. x-intercept 6
y-intercept −4

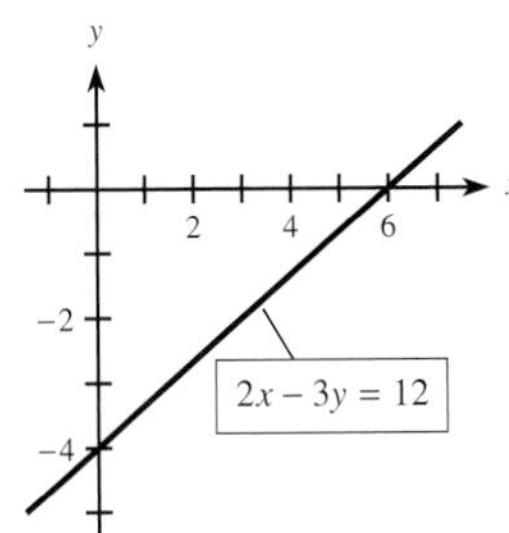

5. x-intercept 0
y-intercept 0

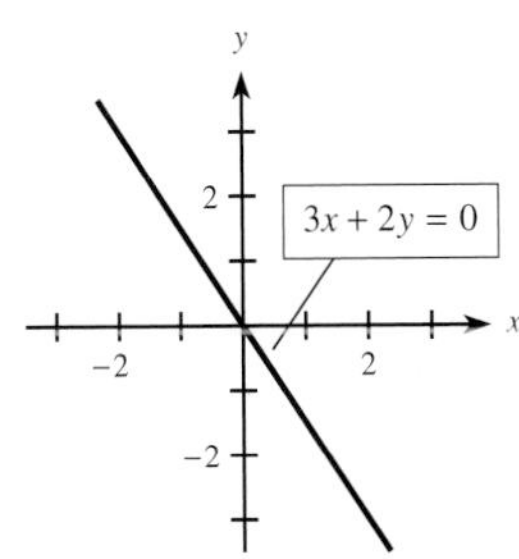

7. $m = -5$ **9.** $m = 0$

11. $m = 7/3$, $b = -1/4$ **13.** $m = 0$, $b = 3$

15. no slope, no y-intercept

17. $m = -2/3$, $b = 2$

19.

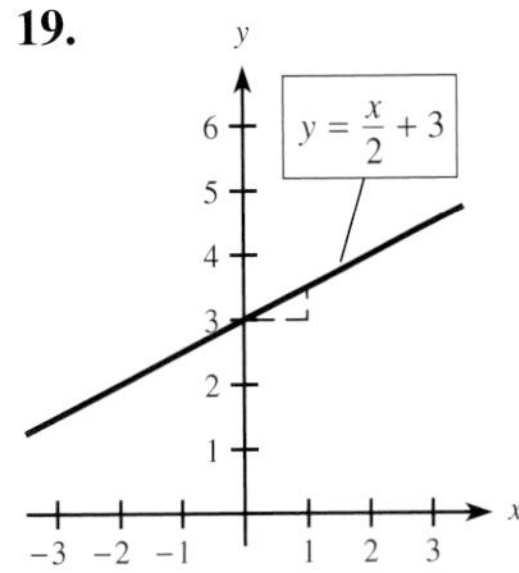

21.

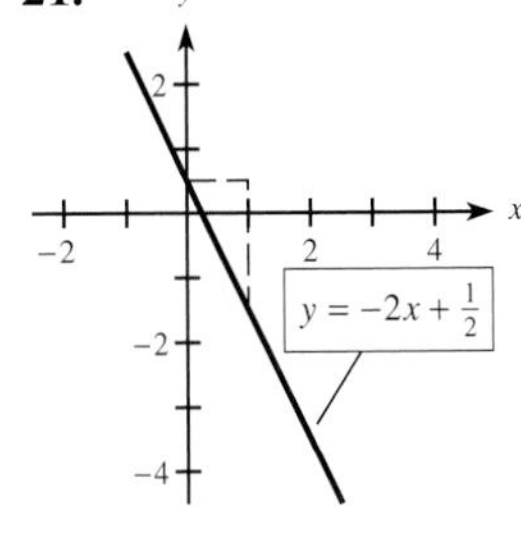

23.

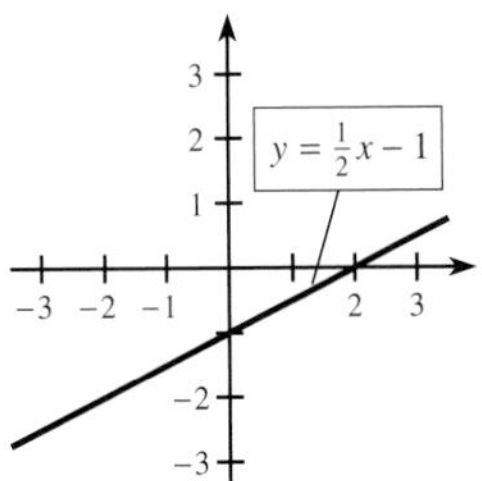

25.

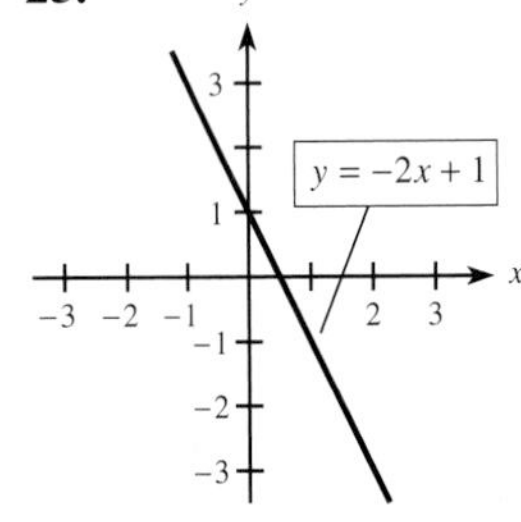

27.

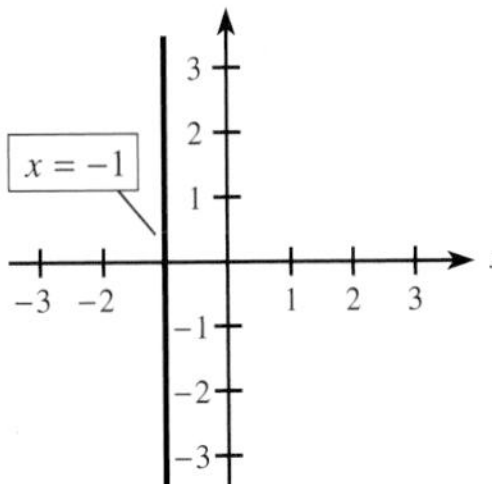

29. $y = 2x - 4$ **31.** $-x + 13y = 32$ **33.** $y = 0$

35. perpendicular

37. neither; same line

39. $y = -\frac{3}{5}x - \frac{41}{5}$ **41.** $y = -\frac{6}{5}x + \frac{23}{5}$

43. **(a)**

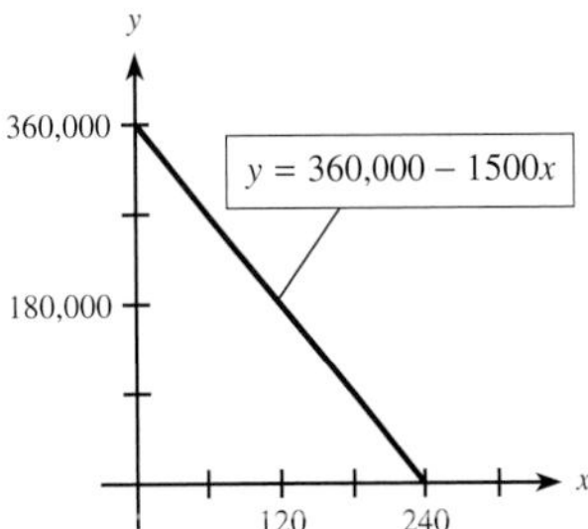

(b) 240 months
(c) After 60 months, the value of the building is \$270,000.

45. (a)

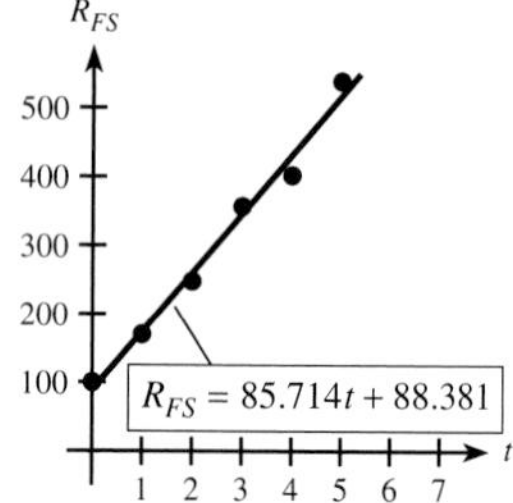

(b)

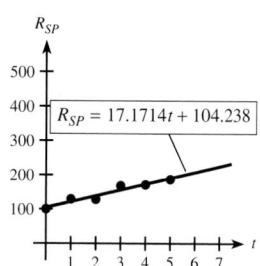

$R_{SP} = 17.1714t + 104.238$

(c) At $t = 0$, $R_{FS} = 88.381$ and $R_{SP} = 104.238$. These are different from 100 because these equations are a good fit to the data but are not perfect; the data are not exactly linear.

(d) These equations show past performance; future sales, market conditions, profits, and confidence cannot be measured and are not part of these equations.

47. (a) $m = 0.1369$ $b = -5.091255$

(b) The y-intercept indicates that when there were 0 terminals, the amount transacted was negative. This is impossible. The model must be restricted to when both $x > 0$ and $y \geq 0$.

(c) The slope means that the transaction amount increases by \$0.1369 (billion) when the number of ATMs increases by 1 (thousand).

49. $y = 4.95 + 0.0838x$

51. (a) $f = 0.838m - 1387.4$ **(b)** \$23,752.60

53. $p = 85{,}000 - 1700x$

55. $R = 3.2t - 0.2$

57. $y = 0.48x - 71$

Exercise 1.4 (page 108)

1.

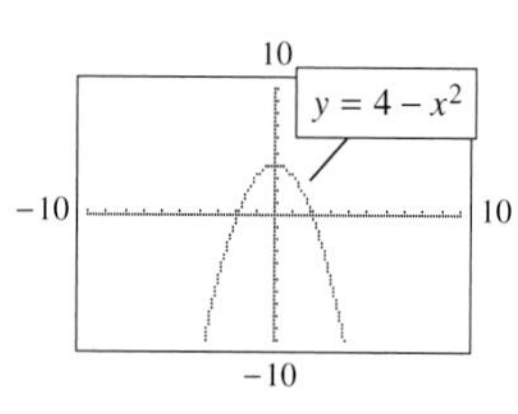

3.

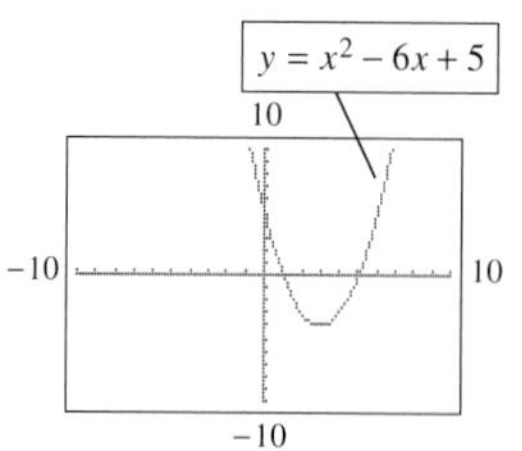

5.

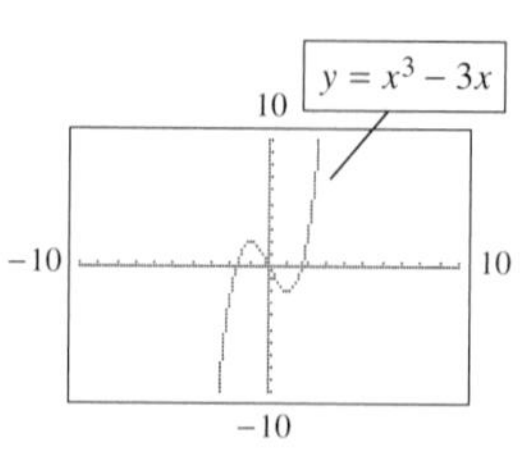

7.

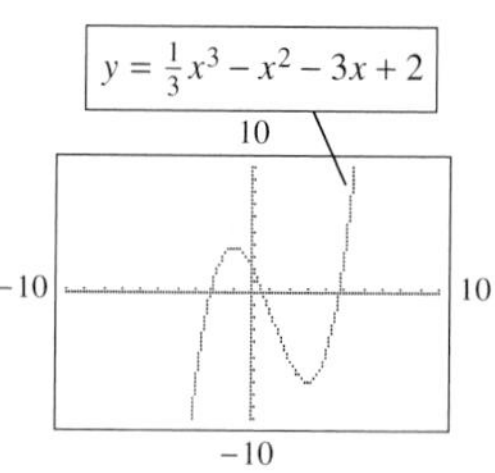

9.

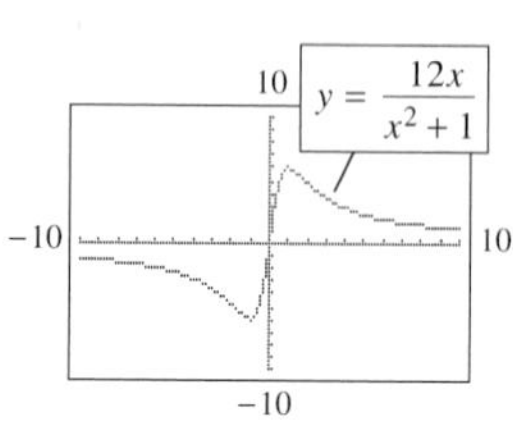

11.

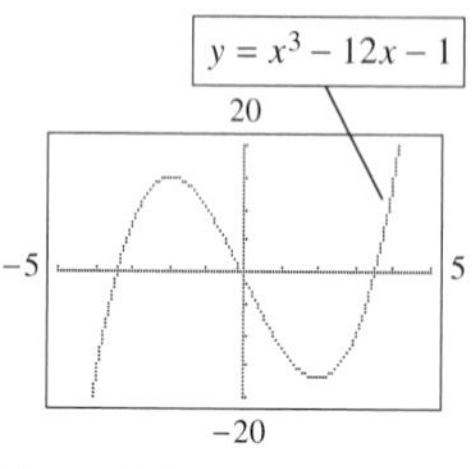

13. (a) $y = 0.01\,x^3 + 0.3x^2 - 72x + 150$

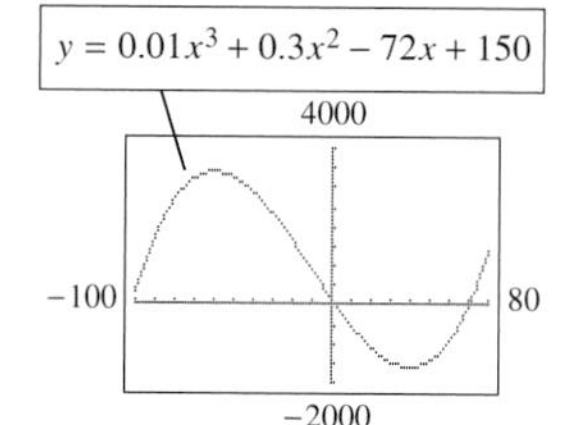

(b) Standard window

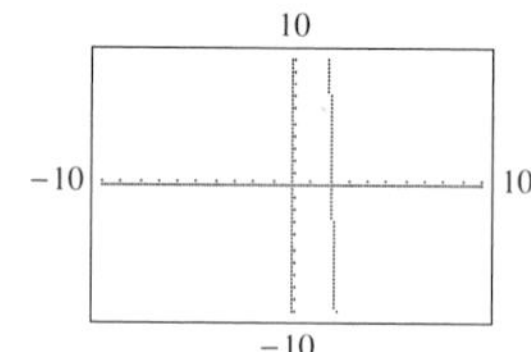

15. (a) $y = \dfrac{x + 15}{x^2 + 400}$

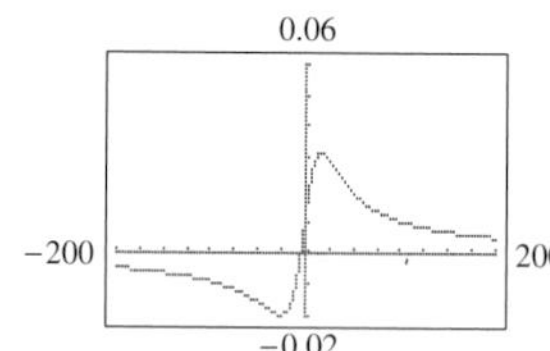

(b) Standard window

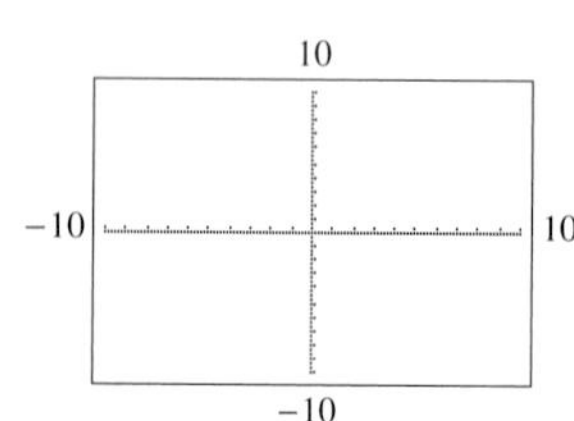

17. **(a)** The equation is linear, so the graph will be a line. Use the intercepts to determine a window.
(b) Window: x-min $= -5$, x-max $= 35$, y-min $= -0.06$, y-max $= 0.02$
(c) $y = 0.001x - 0.03$

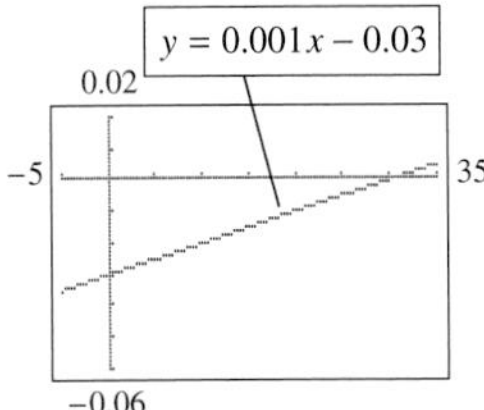

19.

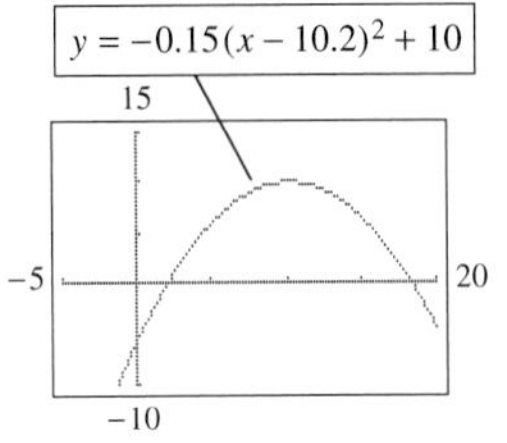

21. $y = \dfrac{(x^3 + 19x^2 - 62x - 840)}{20}$

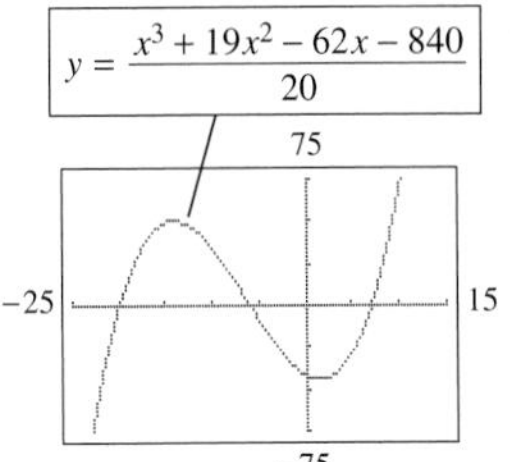

23.

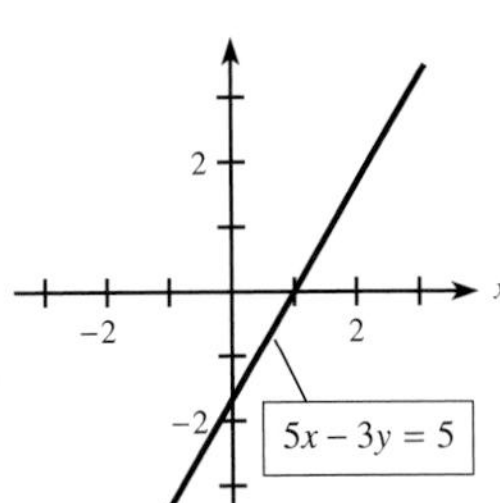

25.

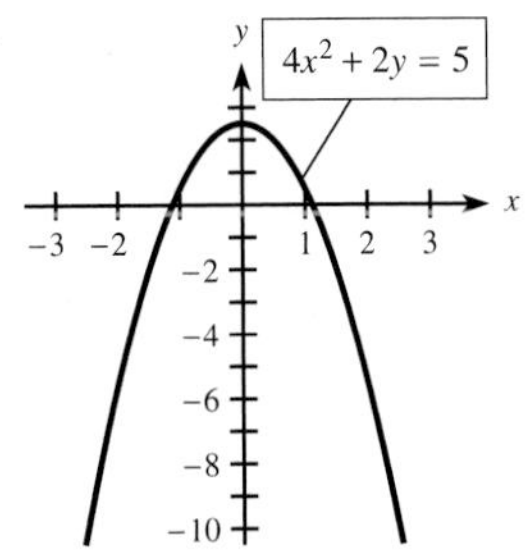

27.

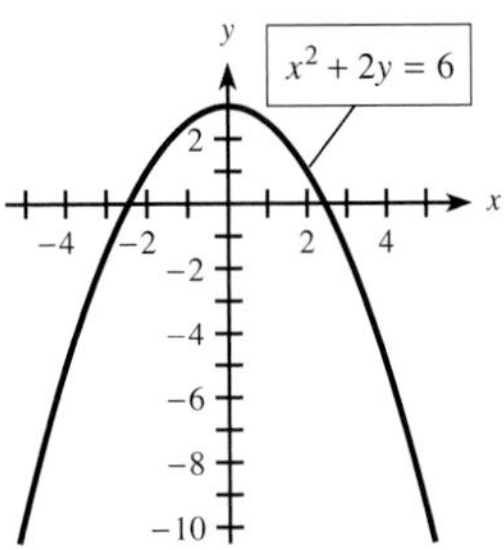

29. $f(1) = 0, f(-3/2) = -65/8$

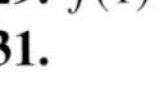

31.

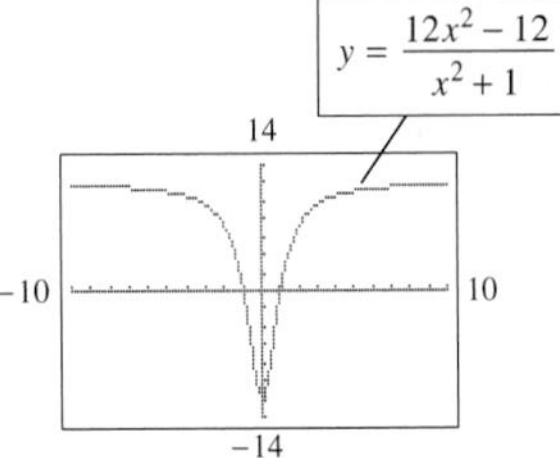

33.

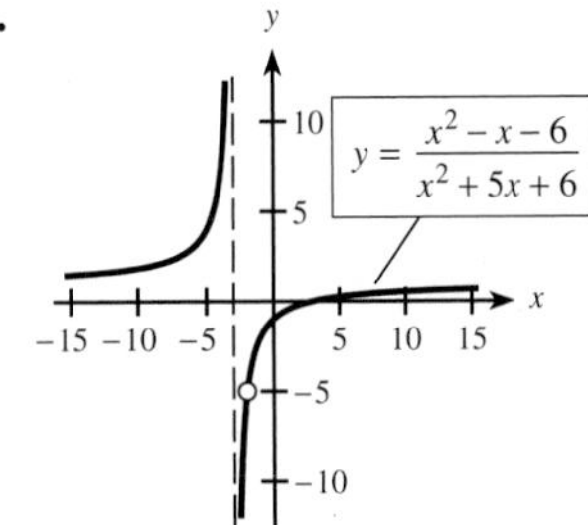

35. **(a)** $(-1.11, 8.11)$ **(b)** $(-1.11, 8.11)$

37. **(a)** 1.000, 4.000 **(b)** 1.000, 4.000

39.

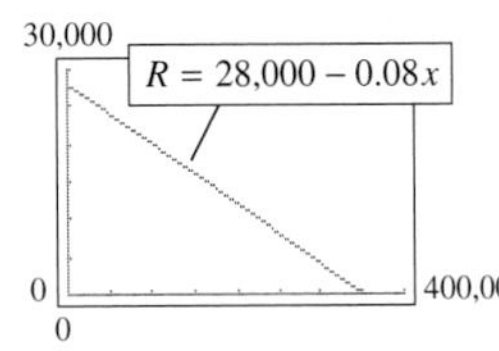

41. **(a)**

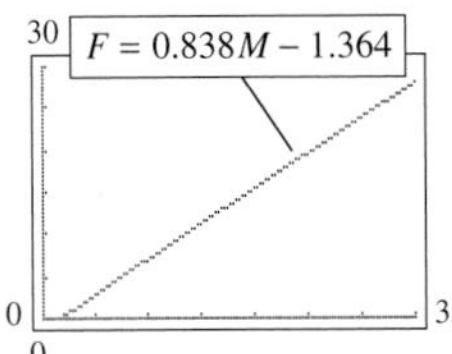

(b) The coordinates mean that when a male's salary is \$50 thousand, a female's is \$40.536 thousand.

43. **(a)**

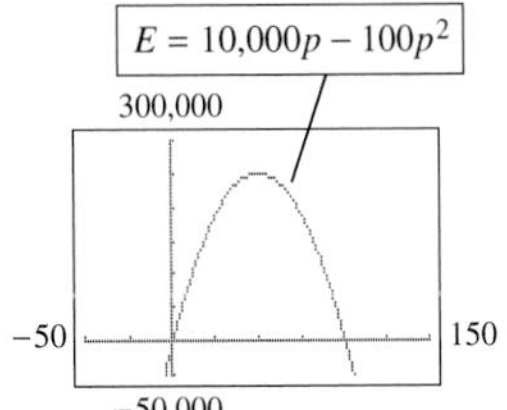

(b) $E \geq 0$ when $0 \leq p \leq 100$

45. (a)

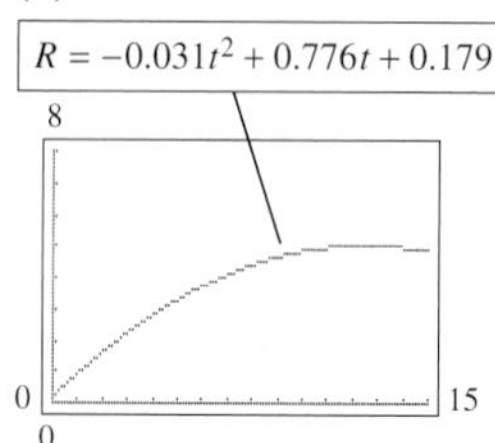

(b)

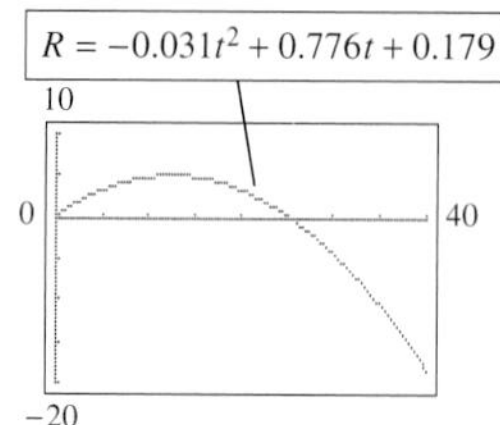

(c) From 1980 to 1995 revenues were rising to what appears to be a maximum. From 1980 to 2020 revenues rose then declined below 0. This model cannot be valid for this period, because revenue cannot be negative.

47. (a)

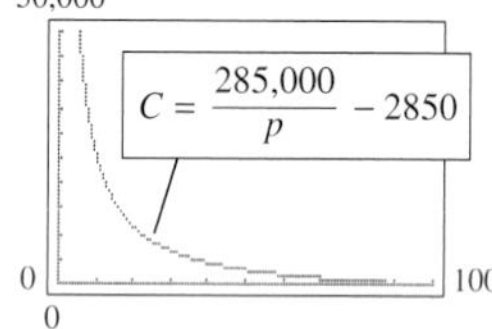

(b) Near $p = 0$, cost grows without bound.
(c) The coordinates of the point mean that obtaining stream water with 1% of the current pollution levels would cost $282,150.
(d) The p-intercept means that stream water with 100% of the current pollution levels would cost $0.

49. (a)

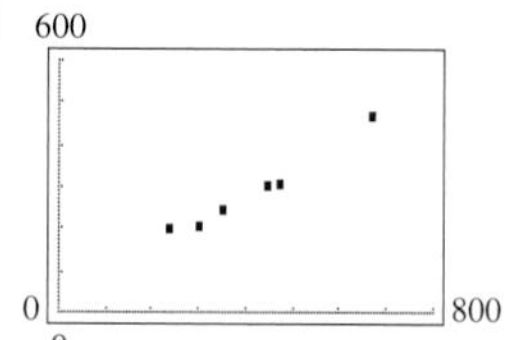

(b)

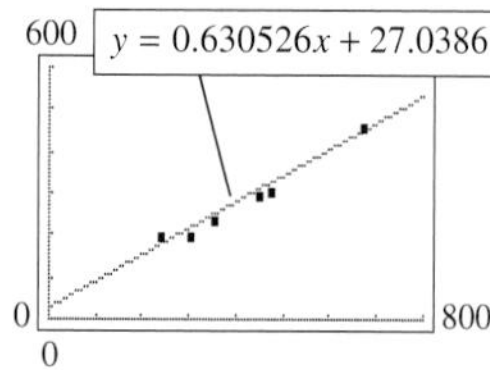

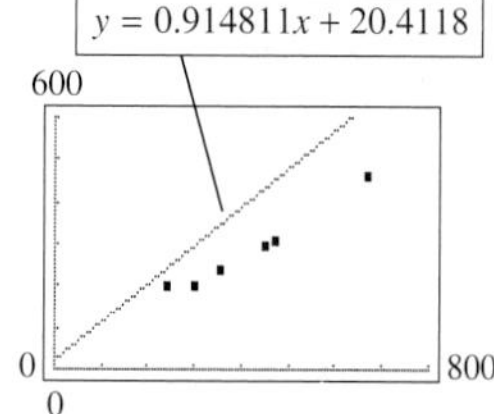

The first equation fits better.

(c) predicts $325 (nearest dollar), actual is $305.

Exercise 1.5 *(page 121)*

1. $x = 2, y = 2$ 3. no solution 5. $x = 2, y = 5$
7. $x = 14/11, y = 6/11$ 9. $x = 10/3, y = 2$
11. $x = 2, y = 1$ 13. $x = 1, y = 1$ 15. no solution
17. $x = -52/7, y = -128/7$ 19. dependent
21. $x = 1, y = 7$
23. $x = 4, y = 2$ 25. $x = -1, y = 1$
27. $x = -17, y = 7, z = 5$
29. $x = 4, y = 12, z = -1$
31. $x = 44, y = -9, z = -1/2$
33. $68,000 at 18%; $77,600 at 10%
35. $13,500 at 10%; $10,000 at 12%
37. $A = 4$ oz, $B = 6\frac{2}{3}$ oz 39. $A = 4550, B = 1500$
41. 7 cc of 20%; 3 cc of 5%
43. 10,000 at $20; 6000 at $30 45. 80 cc
47. 5 oz of *A*, 1 oz of *B*, 5 oz of *C*
49. $A = 200, B = 100, C = 200$

Exercise 1.6 *(page 131)*

1. (a) $C(x) = 17x + 3400$ (b) $6800
3. (a) $R(x) = 34x$ (b) $10,200
5. (a) $P(x) = 17x - 3400$ (b) $1700
7. (a) $P(x) = 37x - 1850$ (b) −$740, loss of $740
   (c) 50
9. (a) $m = 5, b = 250$
   (b) $\overline{MC} = 5$ means each additional unit produced costs $5.
   (c) 250
   (d) Slope = marginal cost; c-intercept = fixed costs
   (e) 5, 5
11. (a) 27
   (b) $\overline{MR} = 27$ means each additional unit sold brings in $27.
   (c) 27, 27
13. (a) $P(x) = 22x - 250$
   (b) 22
   (c) $\overline{MP} = 22$
   (d) Each unit sold adds $22 to profits at all levels of production, so produce and sell as much as possible.
15. $P = 58x - 8500$, $\overline{MP} = 58$

17. **(a)** Revenue passes through the origin. **(b)** \$2000
(c) 400 units **(d)** $\overline{MC} = 2.5$; $\overline{MR} = 7.5$
19. 33
21. **(a)** $R(x) = 12x$; $C(x) = 8x + 1600$ **(b)** 400 units
23. **(a)** $P(x) = 4x - 1600$
(b) $x = 400$ units to break even
25. **(a)** $R(x) = 54.90x$ **(b)** $C(x) = 14.90x + 20{,}200$
(c) 505
27. demand decreases
29. **(a)** 650 (approx) **(b)** 300 **(c)** shortage
31. 16 demanded, 25 supplied; surplus
33. $p = -2q/3 + 1060$ **35.** $p = 0.0001q + 0.5$
37. **(a)** demand falls; supply rises **(b)** (30, \$25)
39. **(a)** $q = 20$ **(b)** $q = 40$
(c) shortage, 20 units short
41. shortage **43.** $q = 20, p = \$18$
45. $q = 10, p = \$180$ **47.** $q = 100, p = \$325$
49. **(a)** \$15 **(b)** $q = 100, p = \$100$
(c) $q = 50, p = \$110$ **(d)** yes
51. $q = 8; p = \$188$ **53.** $q = 500, p = 40$
55. $q = 1200, p = \$15$

Chapter 1 Review Exercises (page 137)

1. $x = 7$ **2.** $x = \frac{31}{3}$ **3.** $x = -13$
4. $-\frac{29}{8}$ **5.** $x = -\frac{1}{9}$ **6.** $x = 10.05$
7. $x = 8$ **8.** $y = -\frac{2}{3}x - \frac{4}{3}$ **9.** no solution
10. yes **11.** no **12.** yes
13. D: reals $x \le 9$; R: reals $y \ge 0$
14. **(a)** 2 **(b)** 37 **(c)** 29/4
15. **(a)** 0 **(b)** 9/4 **(c)** 10.01
16. $9 - 2x - h$ **17.** yes **18.** no **19.** 4
20. $x = 0, x = 4$
21. **(a)** 7 **(b)** $x = -1, x = 3$
(c)

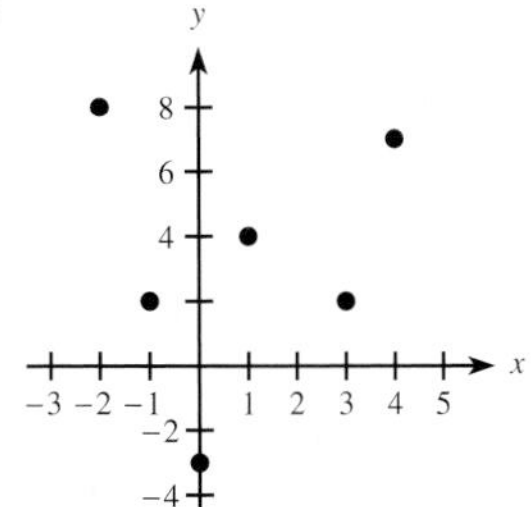

22. **(a)** $x^2 + 3x + 5$ **(b)** $(3x + 5)/x^2$
(c) $3x^2 + 5$ **(d)** $9x + 20$

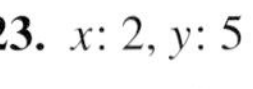

23. x: 2, y: 5

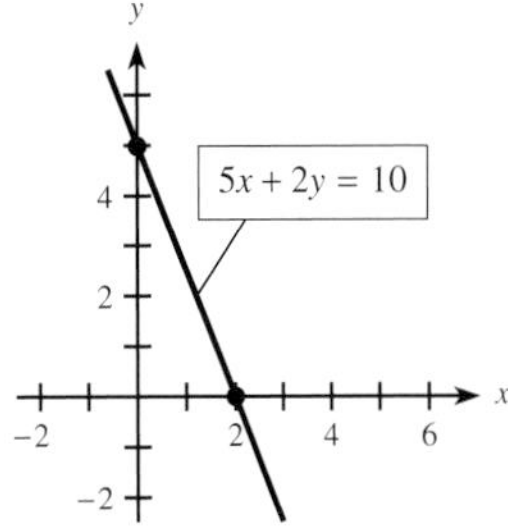

24. x: 3/2, y: 9/5

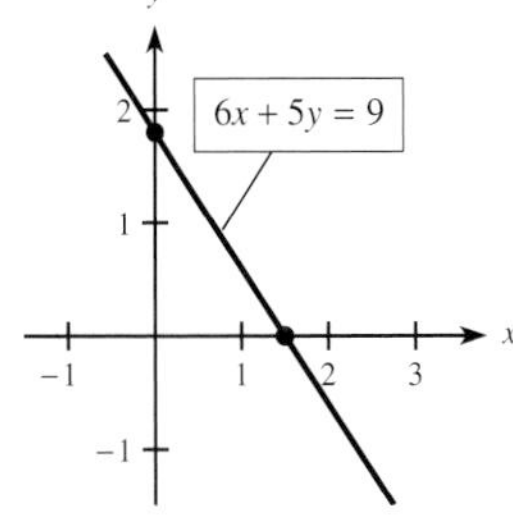

25. x: −2, y: none

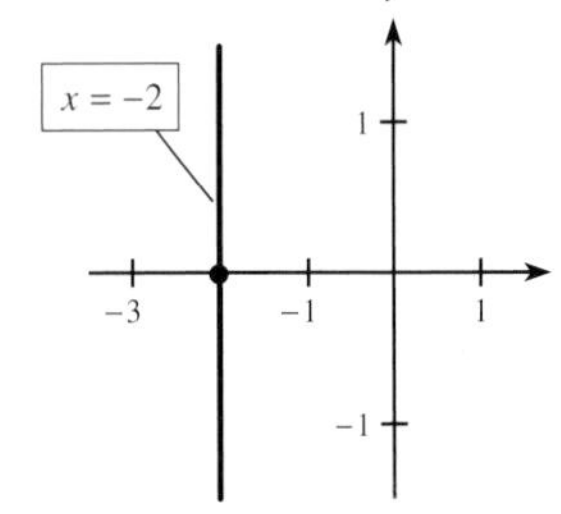

26. $m = 1$ **27.** undefined
28. $m = -\frac{2}{5}, b = 2$ **29.** $m = -\frac{4}{3}, b = 2$
30. $y = 4x + 2$ **31.** $y = -\frac{1}{2}x + 3$
32. $y = \frac{2}{5}x + \frac{9}{5}$ **33.** $y = -\frac{11}{8}x + \frac{17}{4}$ **34.** $x = -1$
35. $y = 4x + 2$ **36.** $y = \frac{4}{3}x + \frac{10}{3}$
37.

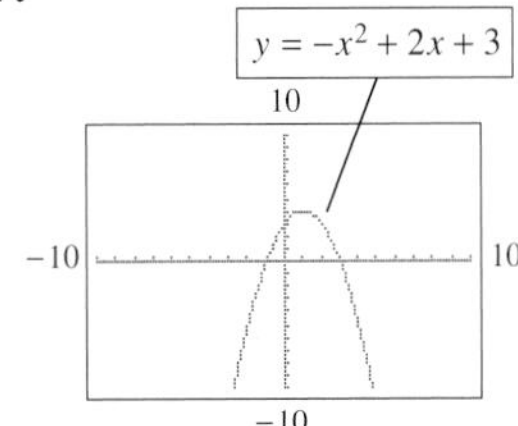

38.

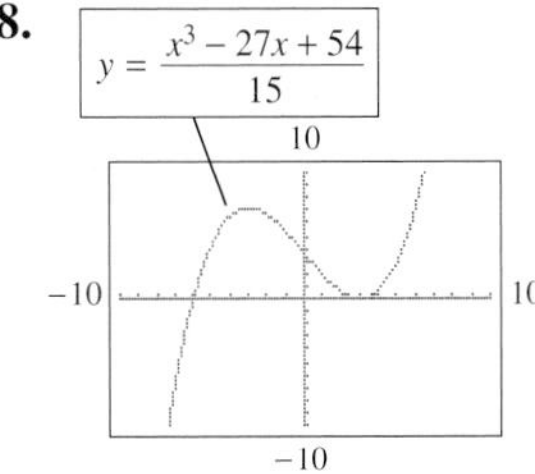

39. **(a)**

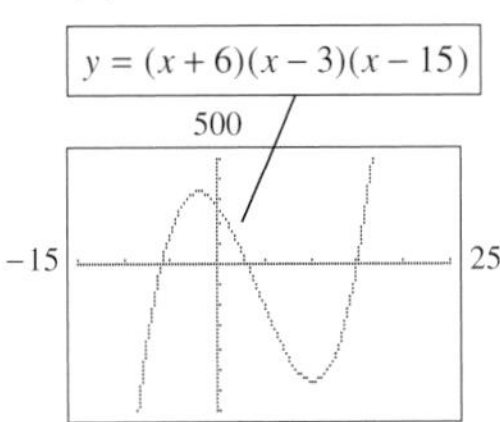

(b)

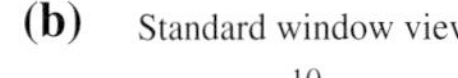

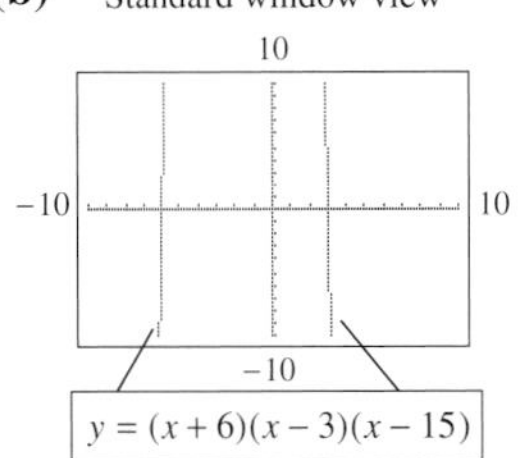

(c) The graph in (a) shows the complete graph. The graph in (b) shows a piece that rises toward the high point and a piece between the high and low points.

40. **(a)**

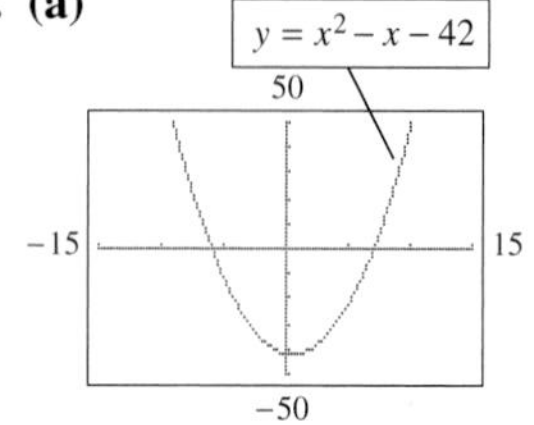

(b) Standard window view

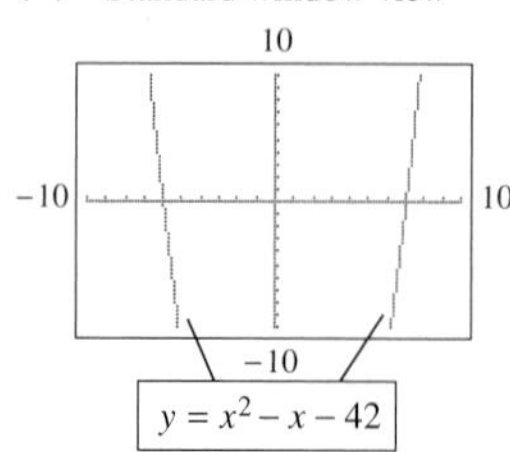

(c) The graph in (a) is a complete graph. The one in (b) shows pieces that fall toward the minimum point and rise from it.

41. reals $x \geq -3$ with $x \neq 0$ **42.** $x = 2, y = 1$

43. $x = 10, y = -1$ **44.** $x = 3, y = -2$

45. no solution **46.** $x = 10, y = -71$

47. $x = 1, y = -1, z = 2$

48. $x = 11, y = 10, z = 9$ **49.** 95%

50. 40,000 mi. He would normally drive more than 40,000 miles in 5 years, so he should buy diesel.

51. **(a)** yes **(b)** no **(c)** 4

52. **(a)** \$565.44

(b) The monthly payment on a \$70,000 loan is \$494.75.

53. **(a)** $(p \circ q)(t) = 180(1000 + 10t) - \dfrac{(1000 + 10t)^2}{100} - 200$

(b) $x = 1150, p = \$193{,}575$

54. $(W \circ L)(t) = k\left(50 - \dfrac{(t - 20)^2}{10}\right)^3$

55. **(a)**

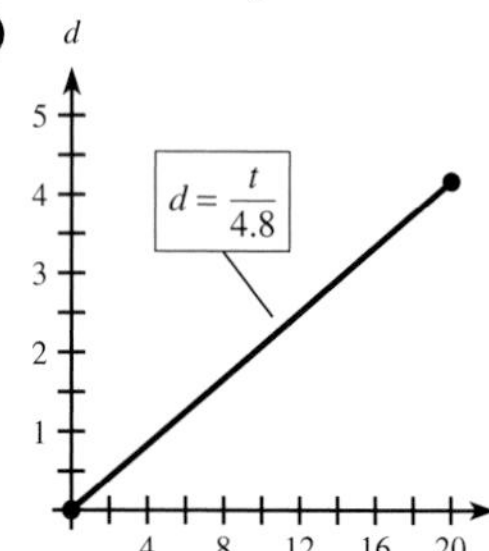

(b) When the time between lightning flashes is 9.6 seconds the storm is 2 miles away.

56.

57. $P = 58x - 8500$

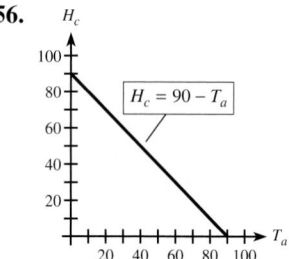

58. $F = \frac{9}{5}C + 32$ or $C = \frac{5}{9}(F - 32)$

59. **(a)**

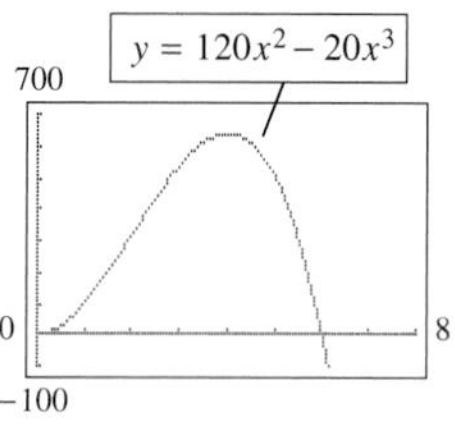

(b) $0 \leq x \leq 6$

60.

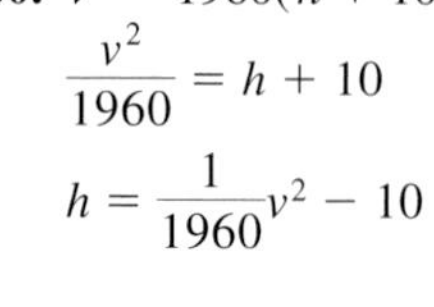

$$v^2 = 1960(h + 10)$$
$$\frac{v^2}{1960} = h + 10$$
$$h = \frac{1}{1960}v^2 - 10$$

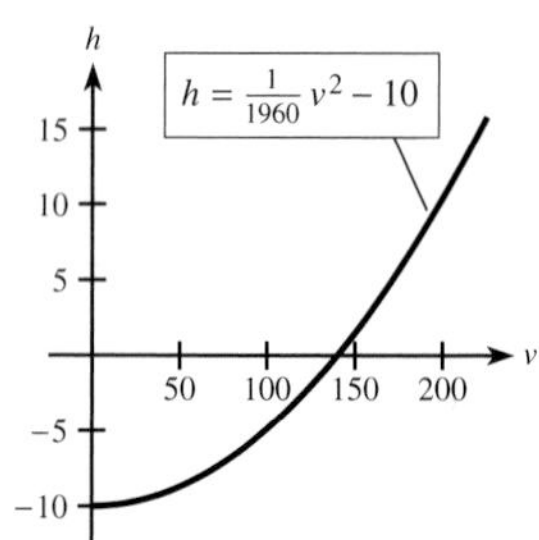

61. \$100,000 at 9.5%; \$50,000 at 11%

62. 2.8 liters of 20%; 1.2 liters of 70%

63. **(a)** 12 supplied; 14 demanded **(b)** shortage **(c)** increase

64.

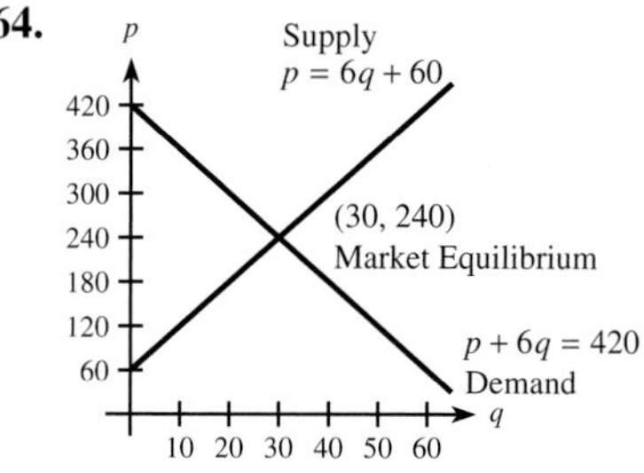

65. **(a)** 38.80 **(b)** 61.30 **(c)** 22.50 **(d)** 200

66. **(a)** $C(x) = 22x + 1500$ **(b)** $R(x) = 52x$

(c) $P(x) = 30x - 1500$ **(d)** $\overline{MC} = 22$

(e) $\overline{MR} = 52$ **(f)** $\overline{MP} = 30$ **(g)** $x = 50$

67. $q = 300, p = \$150$ **68.** $q = 700, p = 80$

Chapter 1 Test (page 141)

1. $x = 18/7$ **2.** $x = -3/7$

3. $x = -38$ **4.** $5 - 4x - 2h$

5. x: 6 y: −5

6. x: 3 y: 21/5

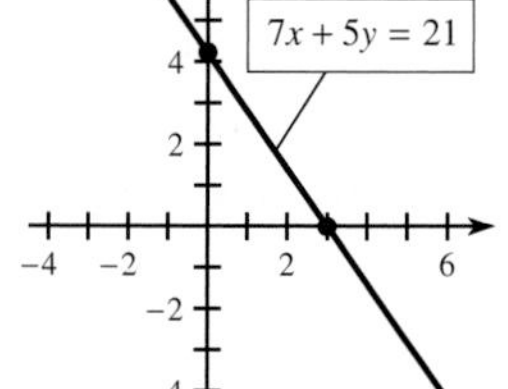

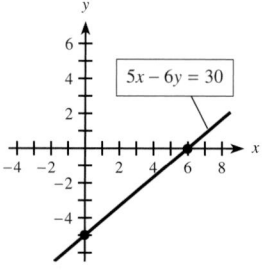

7. **(a)** Domain: $x \geq -4$ Range: $f(x) \geq 0$ **(b)** $2\sqrt{7}$ **(c)** 6

8. $y = -\frac{3}{2}x + \frac{1}{2}$ **9.** $m = -\frac{5}{4}; b = \frac{15}{4}$
10. (a) $x = -3$ **(b)** $y = -4x - 13$
11. (a) No; a vertical line intersects the curve twice.
(b) Yes; there is one y-value for each x-value.
(c) No; one value of x gives two y-values.
12. $x = -2, y = 2$
13. (a) $5x^3 + 2x^2 - 3x$ **(b)** $x + 2$ **(c)** $5x^2 + 7x + 2$
14. (a) 30 **(b)** $P = 8x - 1200$ **(c)** 150
(d) 8; the sale of each additional unit gives \$8 more profit.
15. (a) $R = 50x$
(b) 19,000; it costs \$19,000 to produce 100 units.
(c) 450
16. $q = 200, p = \$2500$
17. (a) 360,000; original value of the building
(b) -1500; building depreciates \$1500 per month.
18. 400 **19.** 12,000 at 9%, 8000 at 6%

Exercise 2.1 *(page 156)*

1. $x^2 + 5x - 3 = 0$ **3.** $x^2 + 2x - 1 = 0$
5. $y^2 + 3y - 2 = 0$ **7.** $\frac{3}{2}, -\frac{3}{2}$ **9.** 0, 1
11. $-7, 3$ **13.** $\frac{1}{2}$ **15.** $8, -4$ **17.** $-4, -2$
19. $-6, 2$ **21.** $4, -3/4$ **23.** 8, 1 **25.** 1/2
27. (a) $2 + 2\sqrt{2}, 2 - 2\sqrt{2}$ **(b)** $4.83, -0.83$
29. no real solutions
31. (a) $-\frac{7}{4}, \frac{3}{4}$ **(b)** $-1.75, 0.75$
33. (a) $(1 \pm \sqrt{31})/5$ **(b)** $1.31, -0.91$
35. $\sqrt{7}, -\sqrt{7}$ **37.** no real solutions **39.** $1, -9$
41. $-9, -10$ **43.** $-\frac{2}{7} \approx -0.29$
45. $-2, 5$ **47.** $-300, 100$ **49.** $0.69, -0.06$
51. $x = 20$ or $x = 70$ **53.** $x = 10$ or $x = 345\frac{5}{9}$
55. (a) $4\sqrt{41} \approx 25.6$ **(b)** $4\sqrt{161} \approx 50.8$
(c) 25.2; K_c is approximately doubled
57. 59.7 mph **59. (a)** 1.93, 13.99 **(b)** 1994
61. $x = 16.59$; 1991 **63.** \$80

Exercise 2.2 *(page 164)*

1. $(-1, -\frac{1}{2})$; min **3.** (1, 9); max
5. (a) $x = 3$ **(b)** $f(3) = 9$
7. (a) $x = -1$ **(b)** $f(-1) = -4$
9. min $(0, -4)$; zeros $(-2, 0)$, $(2, 0)$

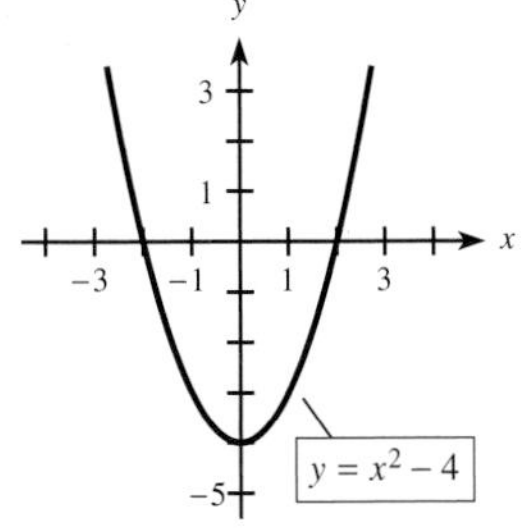

11. max (2, 1); zeros (0, 0), (4, 0)

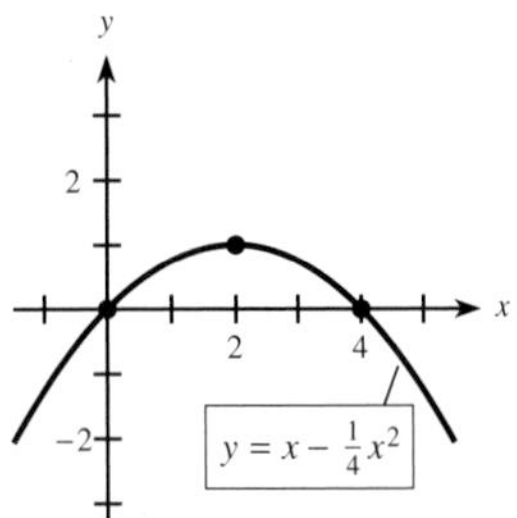

13. min $(-2, 0)$; zero $(-2, 0)$

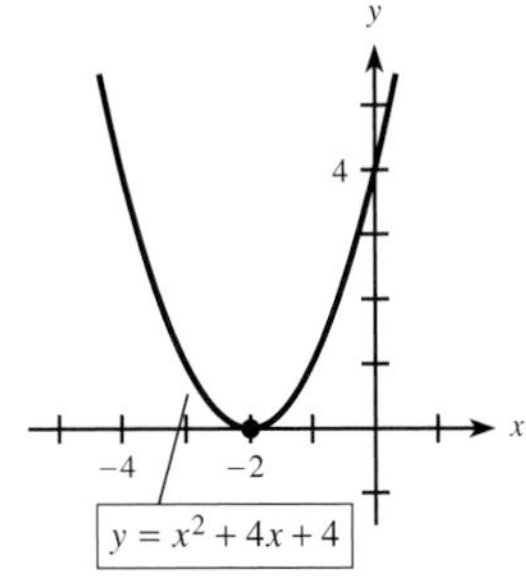

15. min $(-1, -3\frac{1}{2})$; zeros $(-1 + \sqrt{7}, 0)$, $(-1 - \sqrt{7}, 0)$

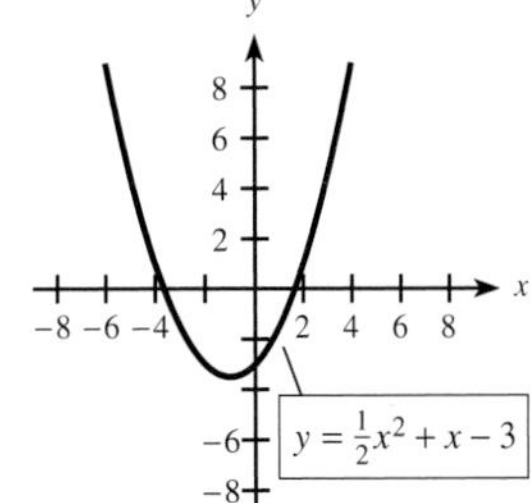

17. (a) 3 units to the right and 1 unit up
(b)

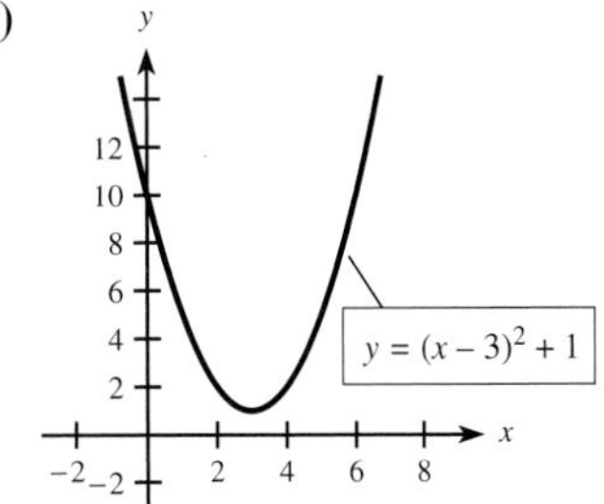

19. (a) 10 units to the right and 12 units up
(b)

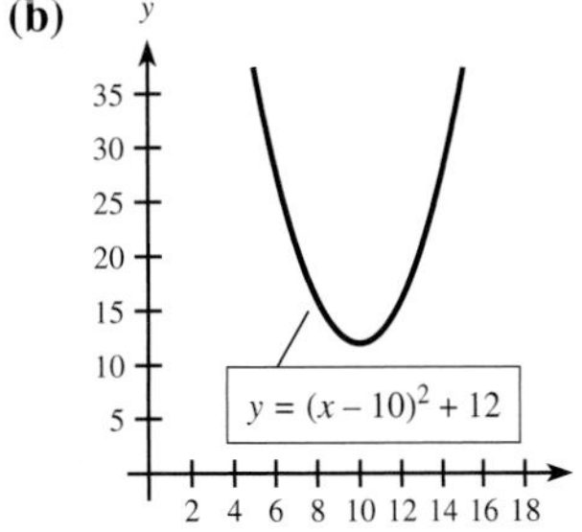

21. vertex $(1, -8)$; zeros $(-3, 0)$, $(5, 0)$

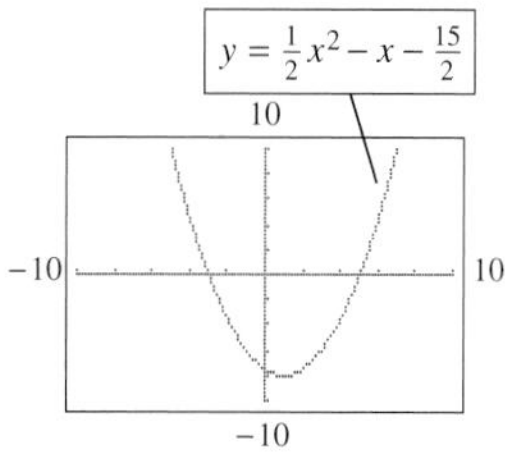

23. vertex $\left(-\frac{5}{2}, \frac{25}{4}\right)$; zeros $(-5, 0)$, $(0, 0)$

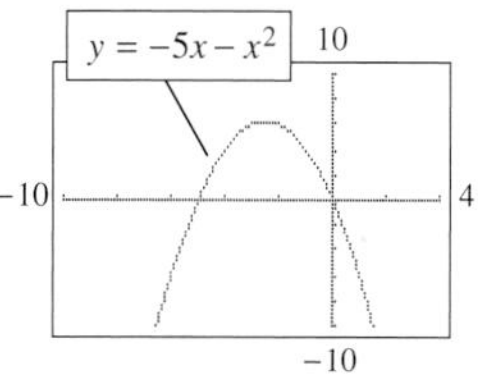

25. vertex $(-6, 3)$; no real zeros

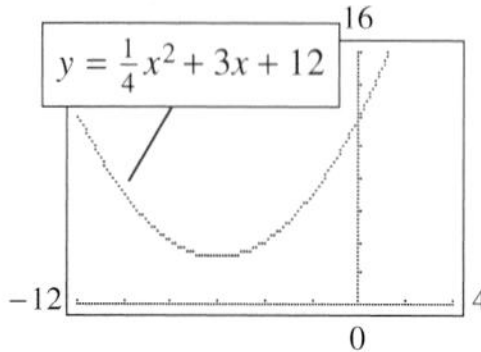

27. vertex $(10, 64)$; zeros $(90, 0)$, $(-70, 0)$

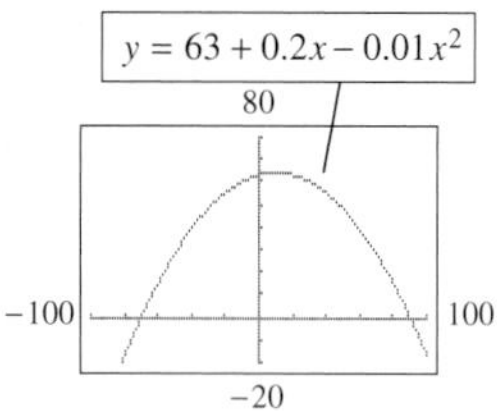

29. vertex $(0, -0.01)$; zeros $(10, 0)$, $(-10, 0)$

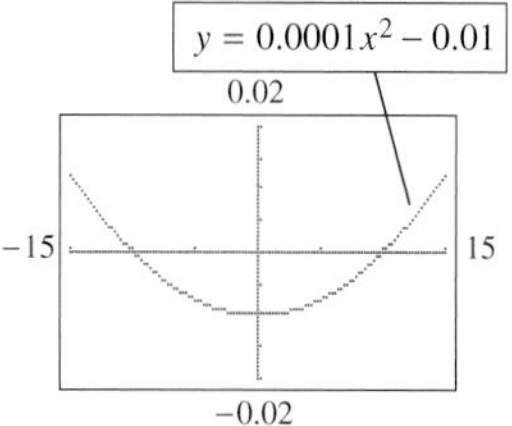

31. (a) $y = 2x^2$, $y = 3x^2$, $y = x^2$, $y = \frac{1}{2}x^2$, $y = \frac{1}{4}x^2$ (window −6 to 6, −4 to 12); $y = -\frac{1}{2}x^2$, $y = x^2$, $y = -2x^2$, $y = -\frac{1}{4}x^2$, $y = -3x^2$ (window −6 to 6, −10 to 6)

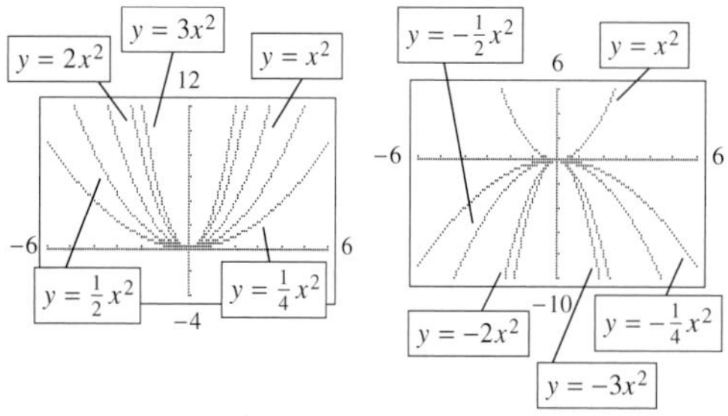

(b) When $a > 0$, the graph of $y = ax^2$ has the same basic shape as $y = x^2$. When $a < 0$, the graph is turned upside down.

(c) When $0 < a < 1$, the graph of $y = ax^2$ opens more gradually than $y = x^2$. When $a > 1$, the opening is narrower than for $y = x^2$.

33. (a) 80 units **(b)** \$540

35.

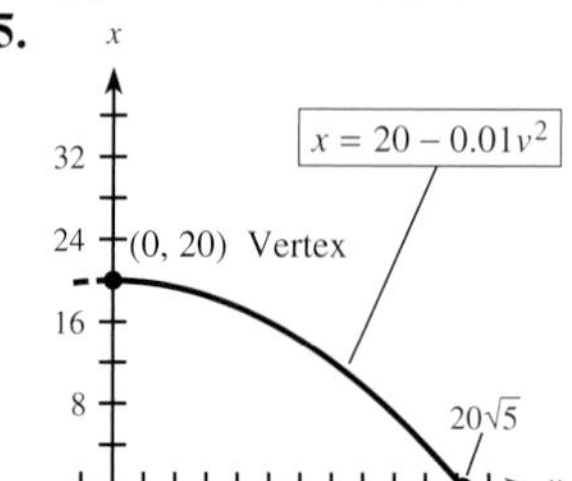

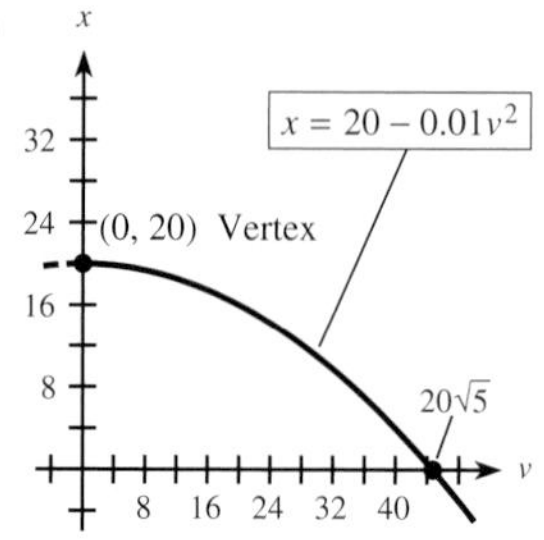

37. 400

39. Dosage = 500

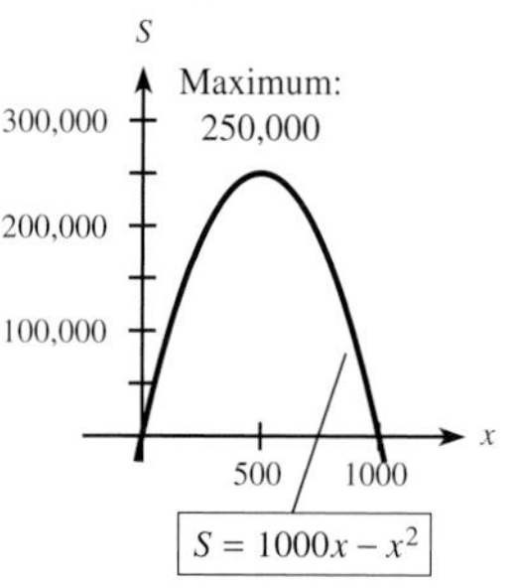

41. Intensity = 1.5

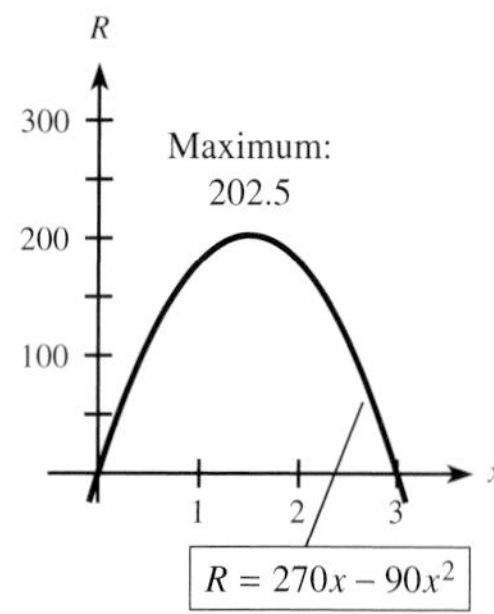

43. Equation **(a)** (384.62, 202.31) **(b)** (54, 46)
Projectile **(a)** goes higher.

45. (a) quadratic
(b) $a > 0$ because the graph opens upward.

47. (a)

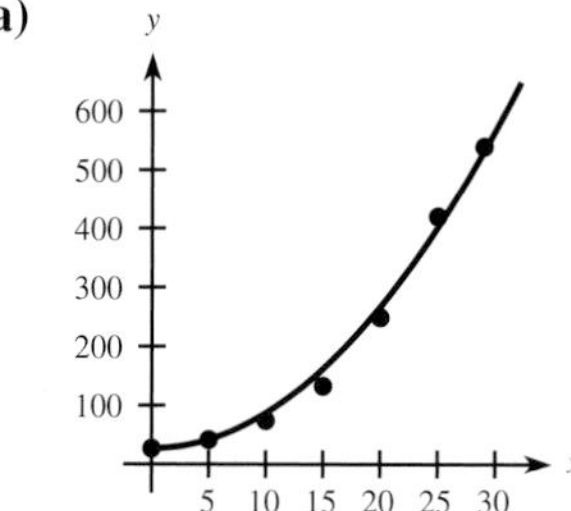

x	y
0	27.1
5	41.6
10	74.4
15	132.9
20	249.1
25	420.1
29	539.9

(b) The shape appears to be quadratic.
(c) $y = 0.6x^2 + 27.1$ fits fairly well.
(d) The model predicts 567.1.
(e) Yes, \$666.2 million is significantly larger than the predicted \$567.1 million. Yes.

49.

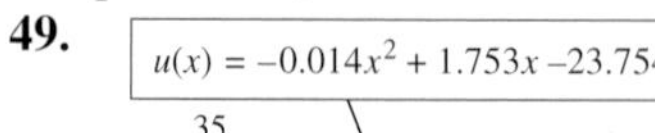

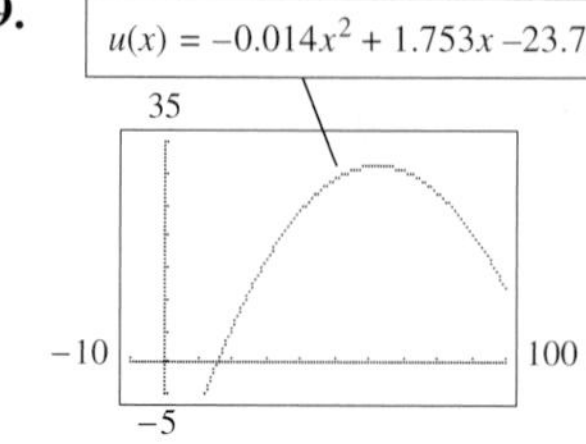

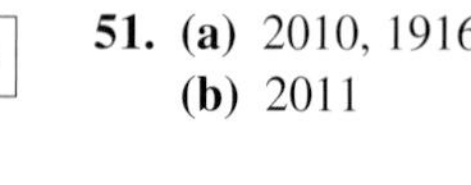

51. (a) 2010, 1916
(b) 2011

Exercise 2.3 (page 173)

1. **(a)** and **(b)**

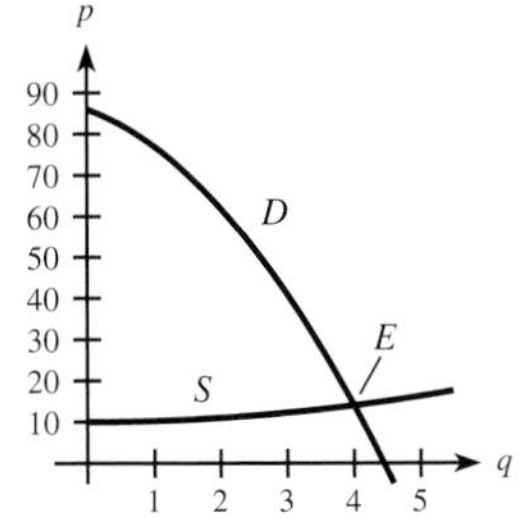

(c) See E on graph. **(d)** $q = 4$, $p = \$14$

3. **(a)** and **(b)**

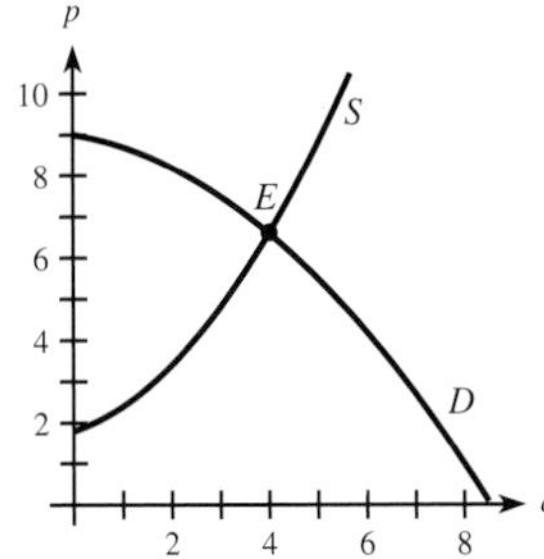

(c) See E on graph. **(d)** $q = 4$, $p = \$6.60$

5. $q = 10$, $p = \$196$ **7.** $p = \$27.08$, $q = 216\frac{2}{3}$

9. $p = \$40$, $q = 30$ **11.** $q = 90$, $p = \$50$

13. $q = 70$, $p = \$62$ **15.** $x = 40$ units, $x = 50$ units

17. $x = 50$, $x = 300$

19. $x = 15$ units; reject $x = 100$ **21.** \$41,173.61

23. \$87.50 **25.** $x = 55$, $P(55) = 2025$

27. **(a)** $P(x) = -x^2 + 350x - 15{,}000$; max is \$15,625
(b) no **(c)** x-values agree

29. **(a)** $x = 28$ units, $x = 1000$ units **(b)** \$651,041.67
(c) $P(x) = -x^2 + 1028x - 28{,}000$; max is \$236,196
(d) \$941.60

31. **(a)** $t \approx 8$; 1988 $R = \$60.792$ billion
(b) The data show a smaller revenue, $R = \$60.53$ billion, in 1987
(c)

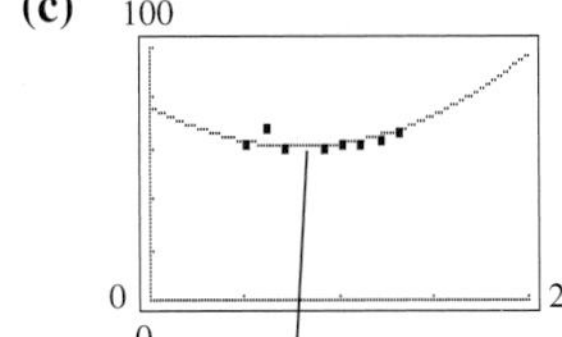

(d) Except for 1986 ($t = 6$), the model fits the data quite well.

33. **(a)** $P(t) = -0.019t^2 + 0.284t - 0.546$ **(b)** 1987
(c)

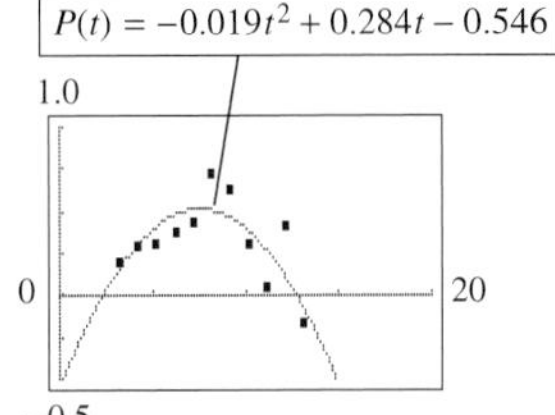

(d) The model projects decreasing profits, and except for 1992, the data support this.
(e) Management would be interested in increasing revenues or reducing costs (or both) to improve profits.

Exercise 2.4 (page 185)

1. k **3.** l **5.** b **7.** f **9.** g
11. j **13.** i **15.** 3rd **17.** 4th **19.** j
21. g **23.** a **25.** f **27.** d
29. **(a)** 8/3 **(b)** 9.9 **(c)** -999.999 **(d)** no
31. **(a)** 64 **(b)** 1 **(c)** 1000 **(d)** 0.027
33. **(a)** 2 **(b)** 4 **(c)** 0 **(d)** 2
35. **(a)**

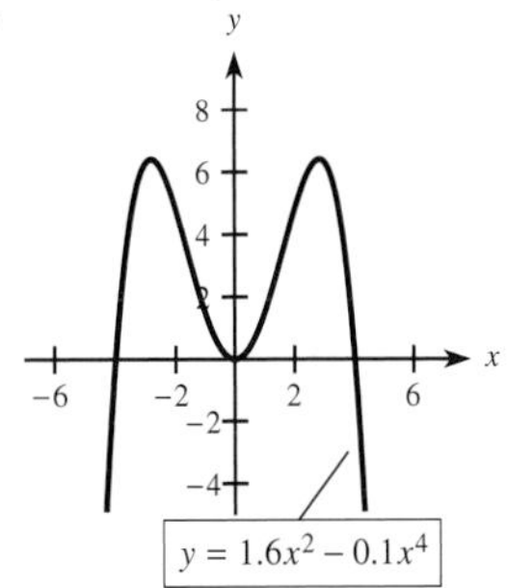

(b) polynomial **(c)** no asymptotes
(d) turning points at $x = 0$ and approximately $x = -2.8$ and $x = 2.8$

37. **(a)**

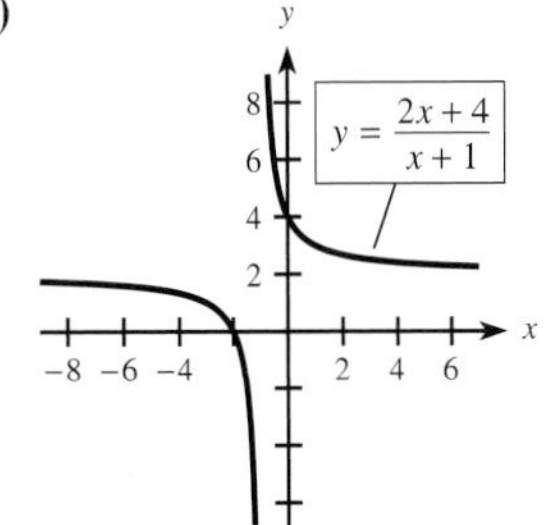

(b) rational
(c) vertical: $x = -1$, horizontal: $y = 2$
(d) no turning points

39. **(a)**

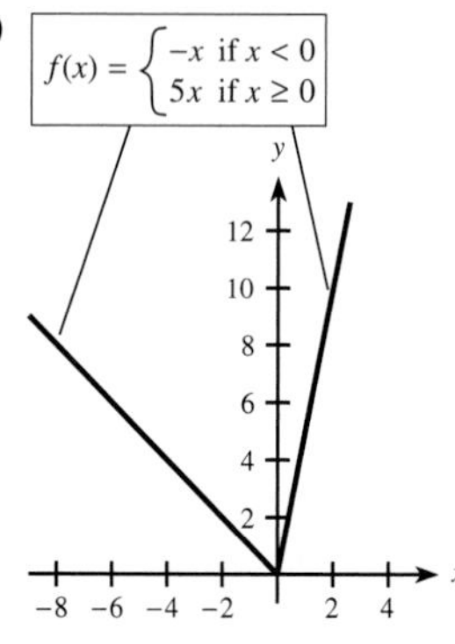

(b) piecewise
(c) no asymptotes
(d) turning point at $x = 0$

41. **(a)** 6800; 11,200 **(b)** $0 < x < 27$

43. **(a)** $0 \leq p < 100$
(b) \$5972.73; to remove 45% of the particulate pollution would cost \$5972.73.
(c) \$65,700; to remove 90% of the particulate pollution would cost \$65,700.
(d) \$722,700; to remove 99% of the particulate pollution would cost \$722,700.
(e) \$1,817,700; to remove 99.6% of the particulate pollution would cost \$1,817,700.

45. **(a)** $A(2) = 96$; $A(30) = 600$
(b) $0 < x < 50$

47. **(a)** $C(5) = \$8.06$ **(b)** $C(6) = \$19.87$
(c) $C(3000) = \$227.65$

49. **(a)** $P(2.3) = 0.77$
The first-class postage for mailing 2.3 oz is 77¢.
(b) $p(1) = 0.33$ $p(1.01) = 0.55$

51. $p = \dfrac{200}{2 + 0.1x}$

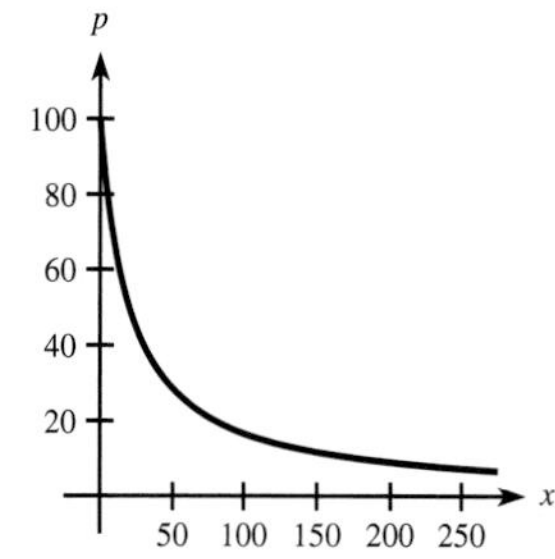

53. $C(x) = 30(x - 1) + \dfrac{3000}{x + 10}$

(a)

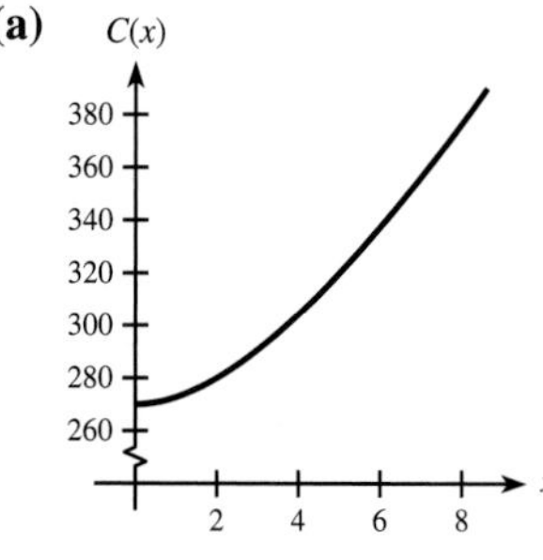

(b) Any turning point would indicate the minimum or the maximum cost. In this case, $x = 0$ gives a minimum.
(c) The y-intercept is the fixed cost of production.

Exercise 2.5 *(page 197)*

1. linear **3.** quadratic **5.** quartic **7.** quadratic

9. $y = 2x - 3$

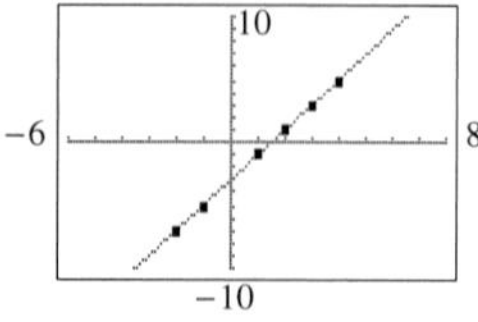

11. $y = 2x^2 - 1.5x - 4$

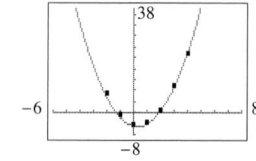

13. $y = x^3 - x^2 - 3x - 4$

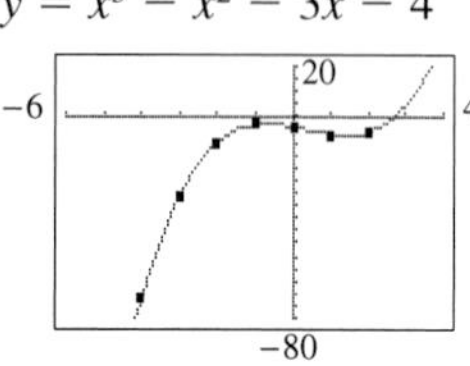

15. $y = 2x^{0.5}$

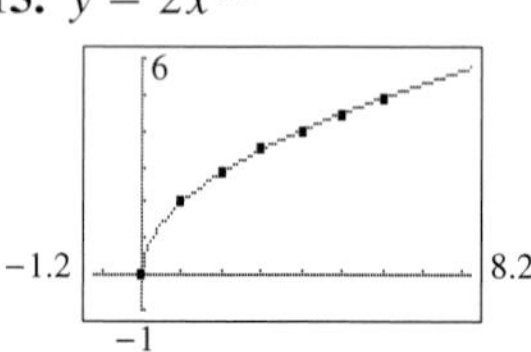

17. **(a)**
(b) linear
(c) $y = 5x - 3$

19. **(a)**
(b) quadratic
(c) $y = 0.09595x^2 + 0.4656x + 1.4758$

21. **(a)**

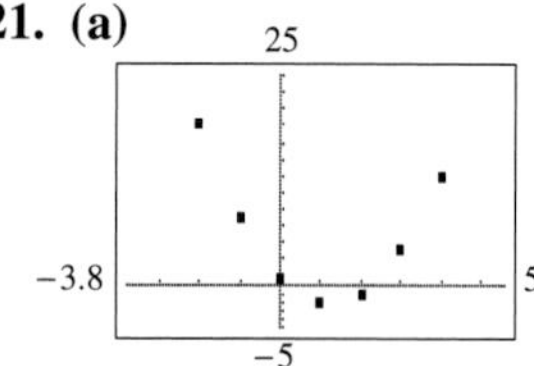

(b) quadratic
(c) $y = 2x^2 - 5x + 1$

23. **(a)**

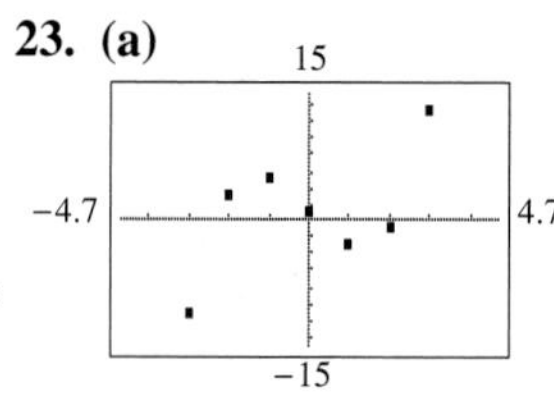

(b) cubic
(c) $y = x^3 - 5x + 1$

25. **(a)** $y = 0.6305x + 27.0386$ **(b)** 0.6305
(c) For each additional dollar earned by men, \$0.63 will be earned by women.

27. **(a)** quadratic **(b)** $0.2912x^2 - 5.1350x + 84.4689$
(c) $x = 8.8$, during 1988; minimum is \$61.83 billion
(d) Predicted data are higher, but close to actual data.

29. **(a)** $-804.6429x^2 + 103{,}590.5286x + 320{,}812.9143$
(b) 1997 **(c)** 1994

31. **(a)** $y = -3.628x^2 + 652.1699x - 29{,}042.0332$
(b) $x = 98.39$, in 1998

33. $y = 0.03743x^3 - 9.6275x^2 + 821.0664x - 23{,}169.1694$

35. **(a)** $y = 0.19x^2 - 16.59x + 1038.29$
(b) $x = 43.7$, 1944 **(c)** 1945

37. **(a)** $T(x) = 0.33x + 24.33$, $A(x) = 0.13x + 29.73$
(b) $x = 117.5$, during 2017
(c) $x = 108.1$, during 2008

Chapter 2 Review Exercises (page 203)

1. $x = 0, x = -\frac{5}{3}$ **2.** $x = 0, x = \frac{4}{3}$
3. $x = -2, x = -3$
4. $x = (-5 + \sqrt{47})/2, x = (-5 - \sqrt{47})/2$
5. no real solutions **6.** $x = \frac{\sqrt{3}}{2}, x = -\frac{\sqrt{3}}{2}$
7. $\frac{5}{7}, -\frac{4}{5}$ **8.** $(-1 + \sqrt{2})/4, (-1 - \sqrt{2})/4$
9. 7/2, 100 **10.** 13/5, 90
11. no real solutions **12.** $z = -9, z = 3$
13. $x = 8, x = -2$ **14.** $x = 3, x = -1$
15. $x = (-a \pm \sqrt{a^2 - 4b})/2$
16. $r = (2a \pm \sqrt{4a^2 + x^3c})/x$
17. 1.64, −7051.64 **18.** 0.41, −2.38
19. vertex (−2, −2); zeros (0, 0), (−4, 0)

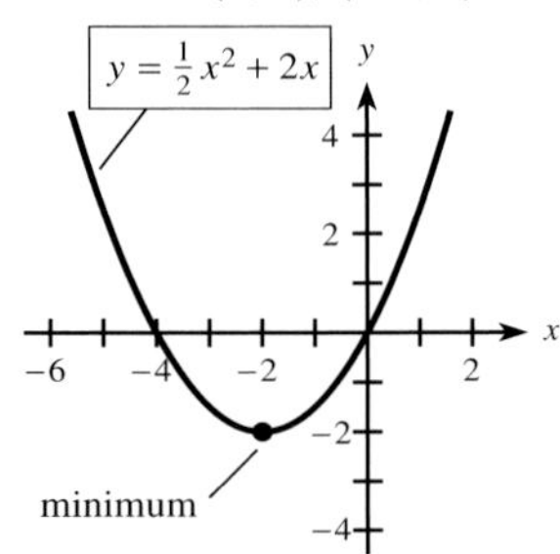

20. vertex (0, 4); zeros (4, 0), (−4, 0)

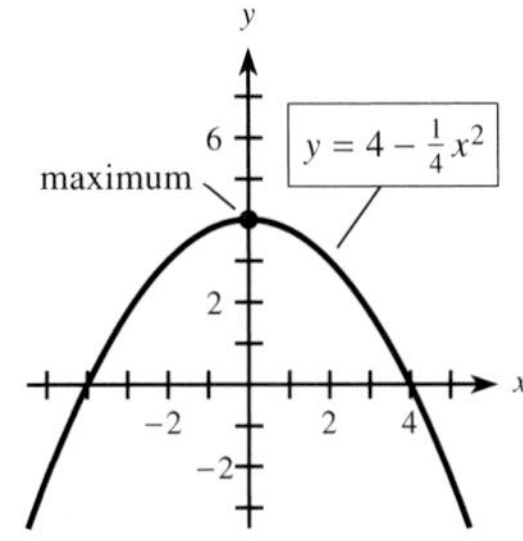

21. vertex $(\frac{1}{2}, \frac{25}{4})$; zeros (−2, 0), (3, 0)

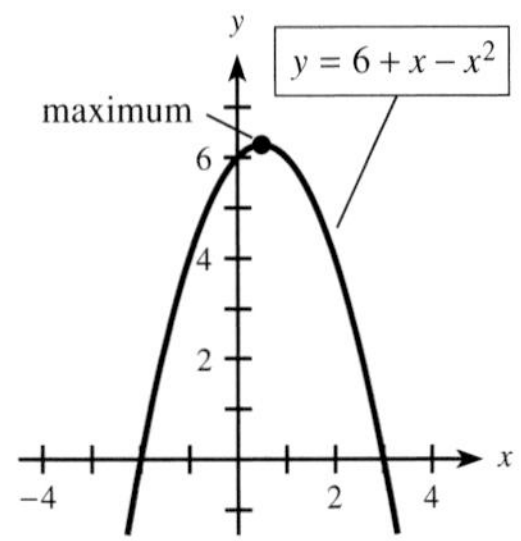

22. vertex (2, 1); no real zeros

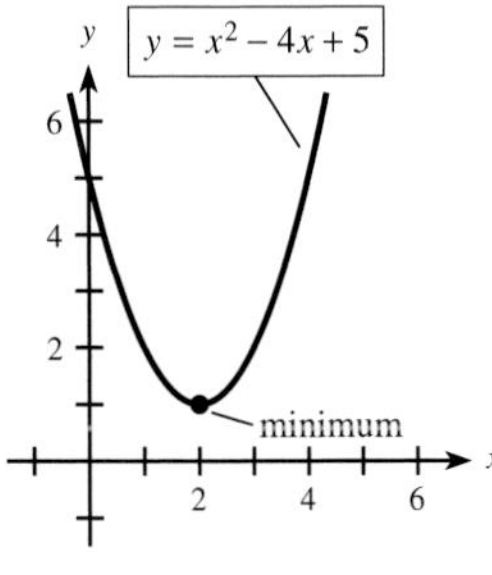

23. vertex (−3, 0); zero (−3, 0)

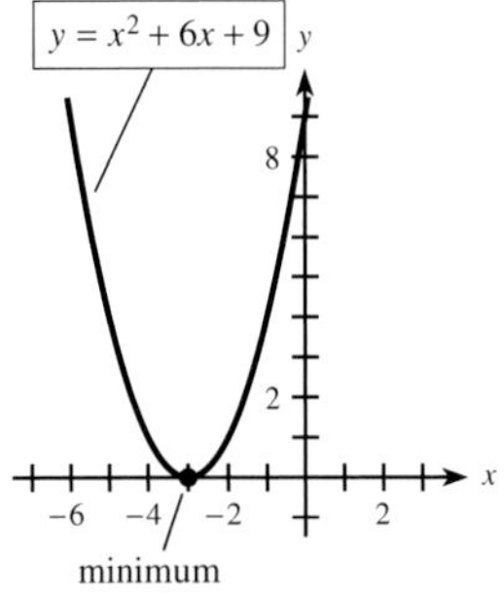

24. vertex $(\frac{3}{2}, 0)$; zero $(\frac{3}{2}, 0)$

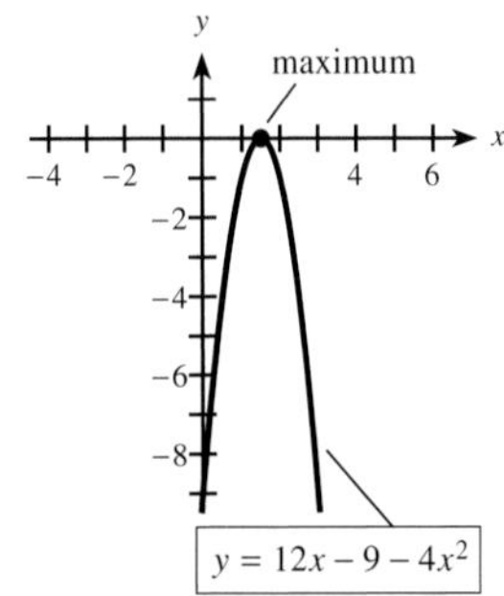

25. vertex (0, −3); zeros (−3, 0), (3, 0)

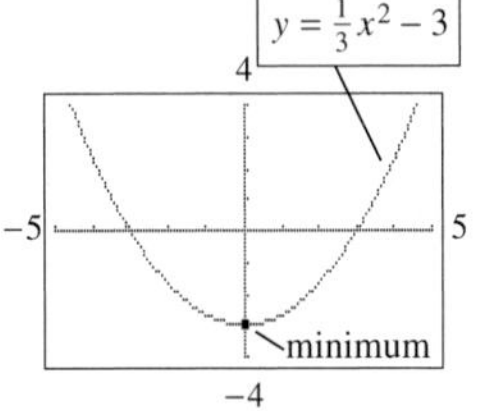

26. vertex (0, 2); no real zeros

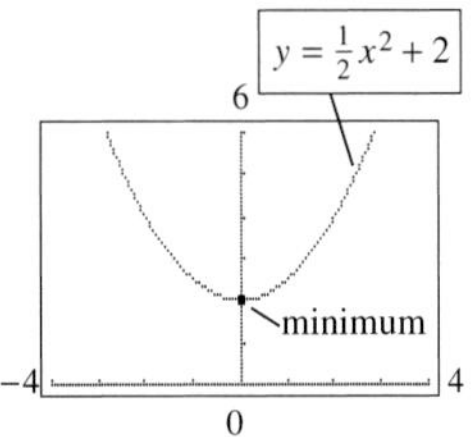

27. vertex (−1, 4); no real zeros

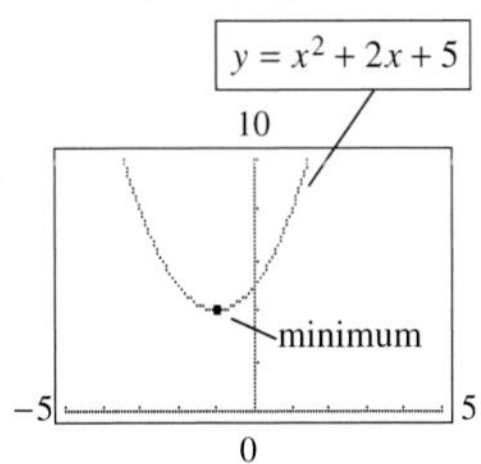

28. vertex $\left(\frac{7}{2}, \frac{9}{4}\right)$; zeros $(2, 0)$, $(5, 0)$

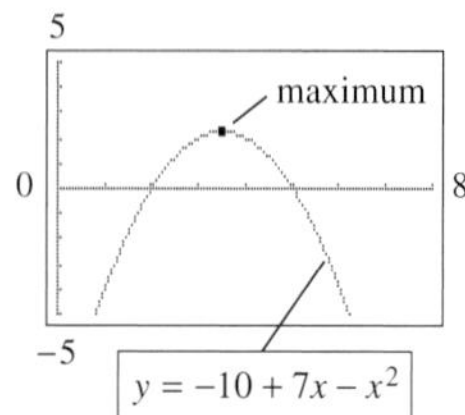

29. vertex $(100, 1000)$; zeros $(0, 0)$, $(200, 0)$

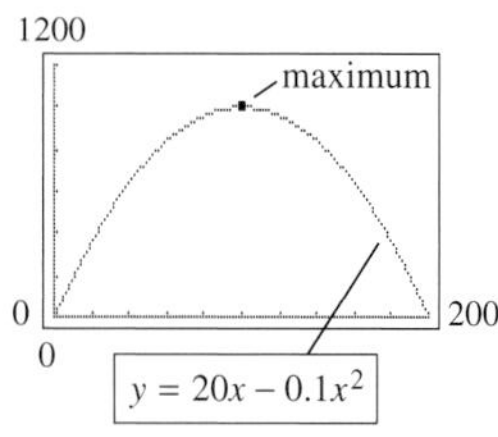

30. vertex $(75, -6.25)$, zeros $(50, 0)$, $(100, 0)$

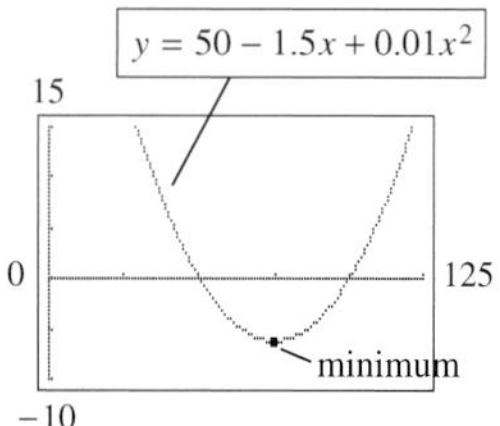

31. **(a)** $(1, -4\frac{1}{2})$ **(b)** $x = -2, x = 4$ **(c)** B

32. **(a)** $(0, 49)$ **(b)** $x = -7, x = 7$ **(c)** D

33. **(a)** $(7, 25)$ approximately, actual is $(7, 24\frac{1}{2})$ **(b)** $x = 0, x = 14$ **(c)** A

34. **(a)** $(-1, 9)$ **(b)** $x = -4, x = 2$ **(c)** C

35. **(a)**

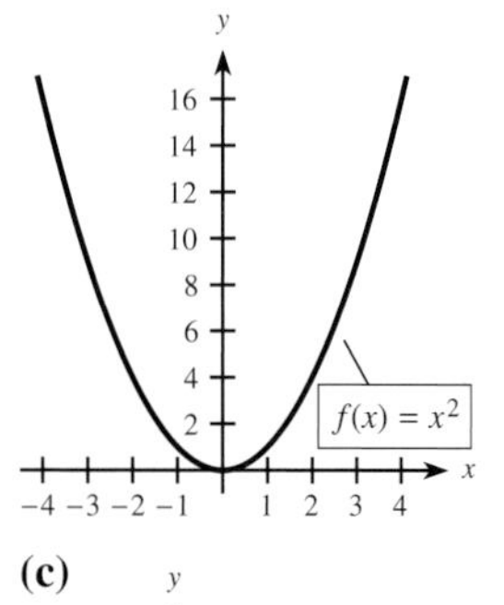

(b)

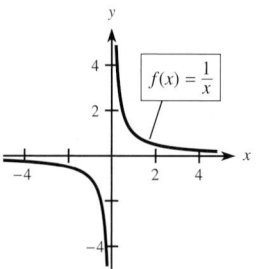

(c)

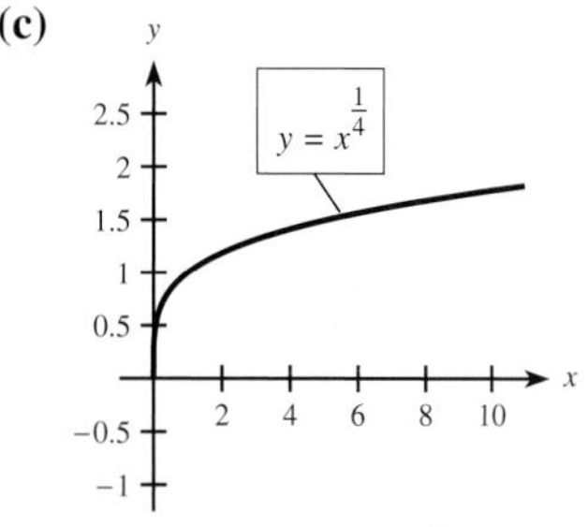

36. **(a)** 0 **(b)** 10,000 **(c)** −25 **(d)** 0.1

37. **(a)** −2 **(b)** 0 **(c)** 1 **(d)** 4

38.

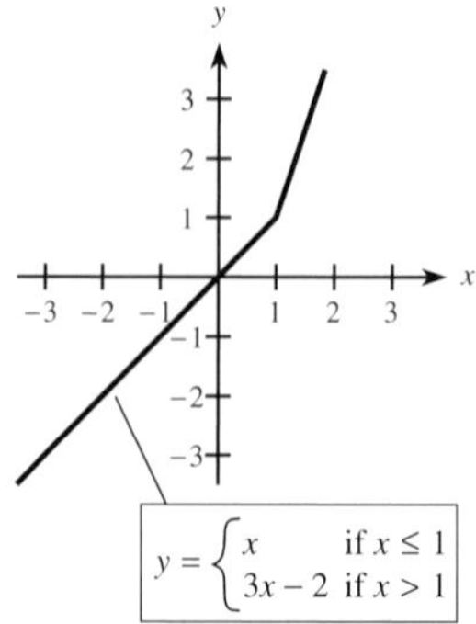

39. **(a)** **(b)**

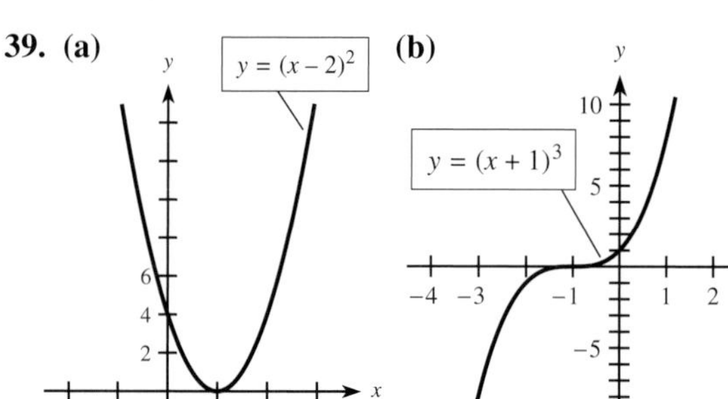

40. Turns: $(1, -5)$, $(-3, 27)$

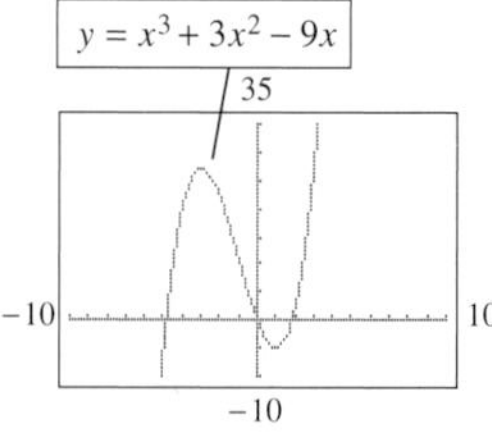

41. Turns: $(1.7, -10.4)$, $(-1.7, 10.4)$

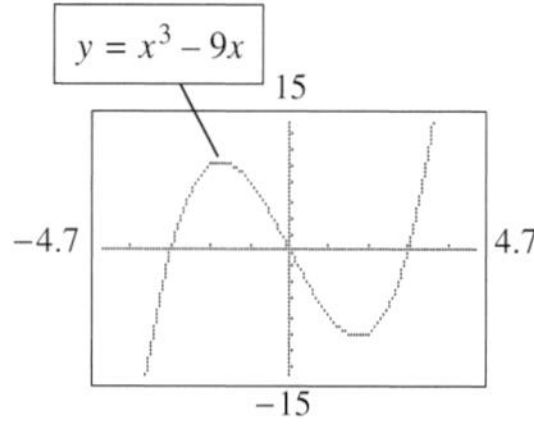

42. VA: $x = 2$; HA: $y = 0$

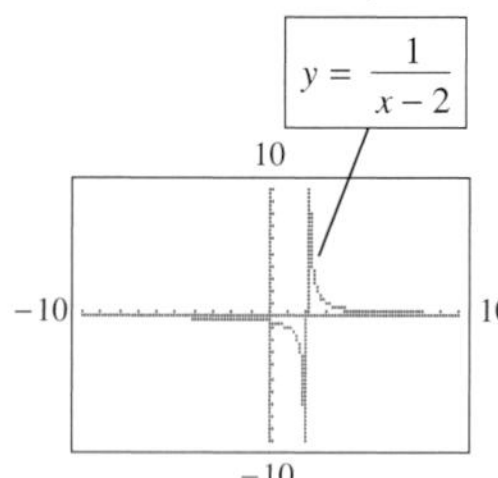

43. VA: $x = -3$; HA: $y = 2$

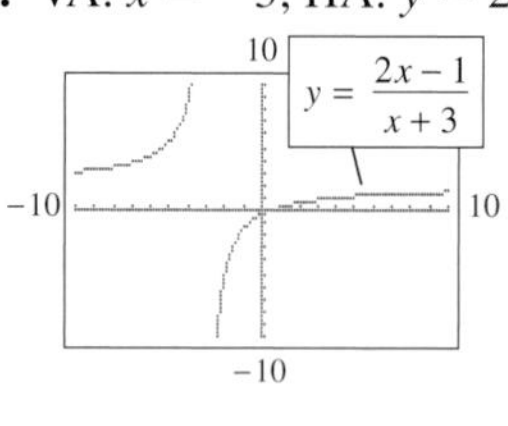

44. **(a)**

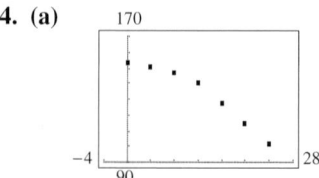

(b) $y = -2.1786x + 159.8571$

(c) $y = -0.0818x^2 - 0.2143x + 153.3095$

45. **(a)**

(b) $y = 2.1413x + 34.3913$

(c) $y = 22.2766x^{0.4259}$

46. **(a)** $t = -1.65$ $t = 3.65$ **(b)** Just $t = 3.65$

(c) At 3.65 seconds

47. $x = 20, x = 800$

48. **(a)** $x = 76.9$, in 1977; $x = 108$, in 2008

(b) $x = 92.5$, in 1992; 9.15%

49. **(a)** $x = 200$ **(b)** $A = 30{,}000$ square feet

50.

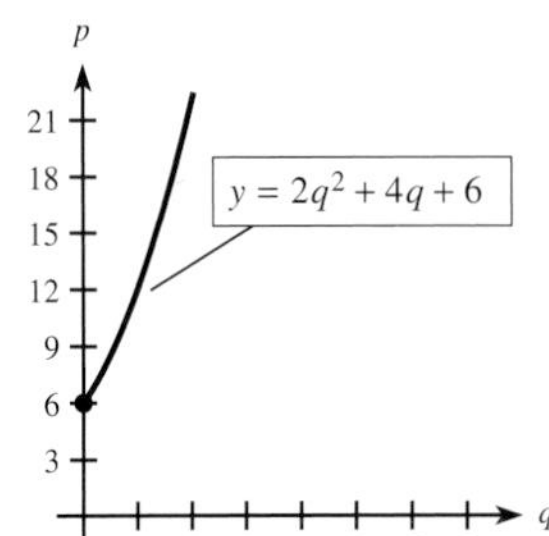

51.

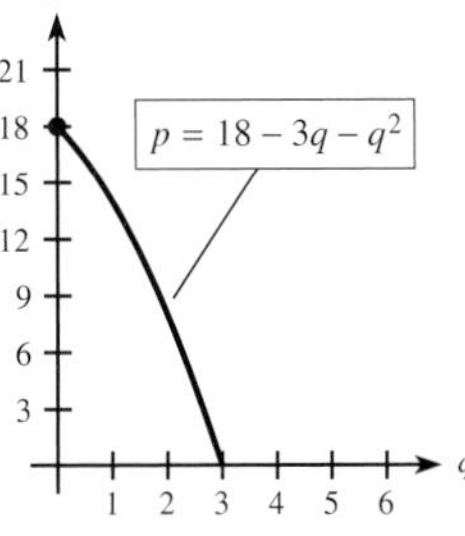

52. **(a)**

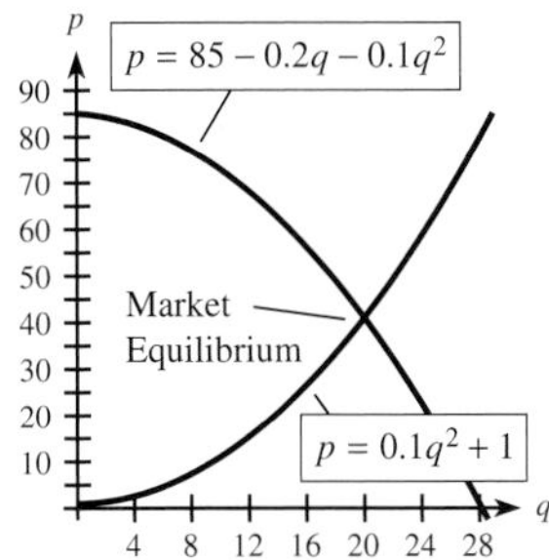

(b) $p = 41, q = 20$

53. $p = 400, q = 10$ **54.** $p = 10, q = 20$

55. $x = 46 + 2\sqrt{89}, x = 46 - 2\sqrt{89}$

56. (15, 1275), (60, 2400)

57. max revenue = \$2500; max profit = \$506.25

58. max profit = 12.25; break-even $x = 100, x = 30$

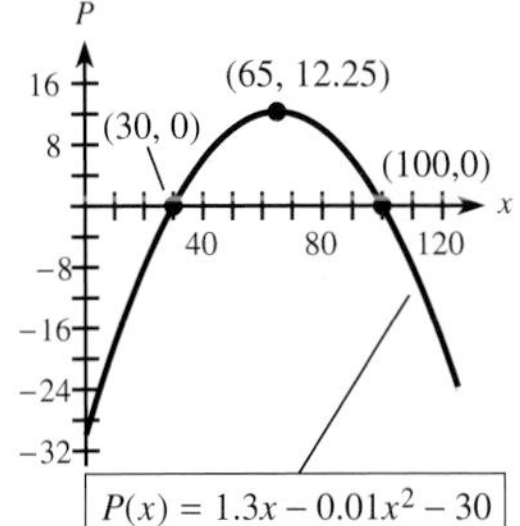

59. $x = 50, P(50) = 640$

60. **(a)** $C = 15{,}000 + 140x + 0.04x^2$; $R = 300x - 0.06x^2$

(b) 100, 1500 **(c)** 2500

(d) $P = 160x - 15{,}000 - 0.1x^2$; max at 800

(e) at 2500: $P = -240{,}000$; at 800: $P = 49{,}000$

61. **(a)** power **(b)** 5,599,885

(c) 20.481364; this model predicts 2,048,136 cases in 1995.

62. **(a)**

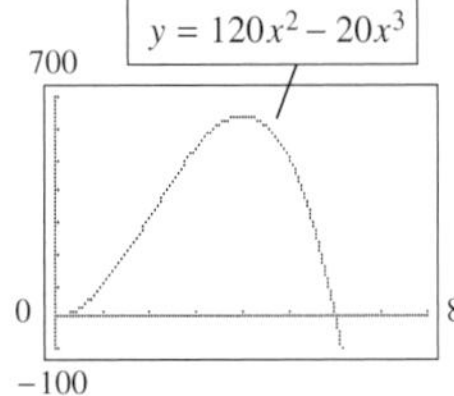

(b) $0 \le x \le 6$

63. **(a)** rational **(b)** $0 \le p < 100$

(c) 0; it costs \$0 to remove no pollution.

(d) \$475,200

64. **(a)** $x = 12$; $C(12) = \$18.68$

(b) $x = 825$; $C(825) = \$909.70$

65. **(a)** $y = 17.3969x^{0.5094}$

(b)

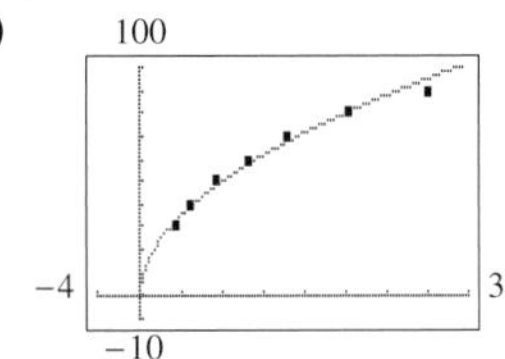

(c) 39.5 mph

(d) 18.3 seconds

66. **(a)** $y = 1.8155x^2 - 11.1607x + 29.1845$

(b) Use $x = 11$; 126 million cubic yards

Chapter 2 Test *(page 206)*

1. **(a)**

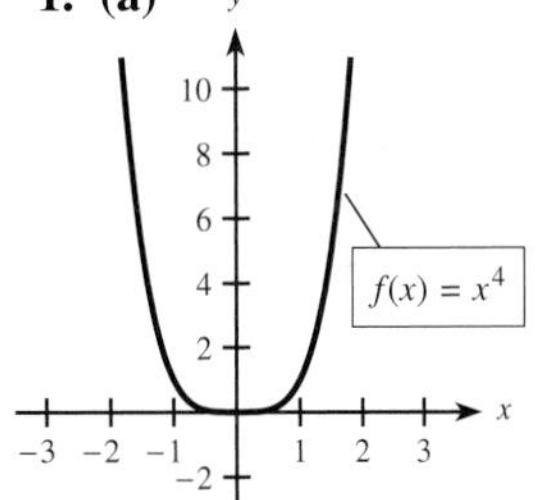

(b)

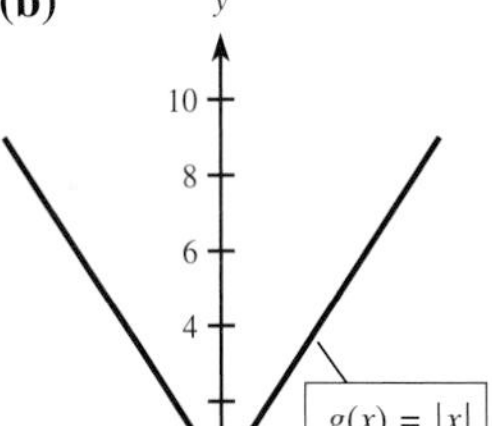

(c)

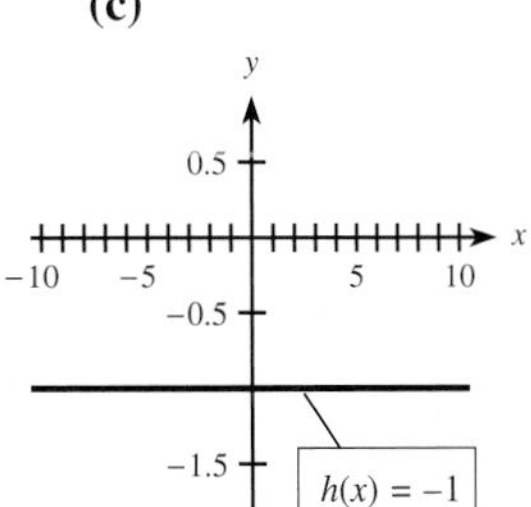

(d)

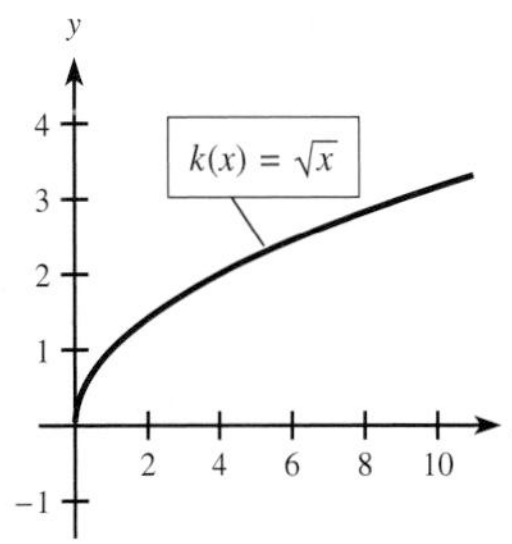

2. *b; a*

3.

4. (a) **(b)**

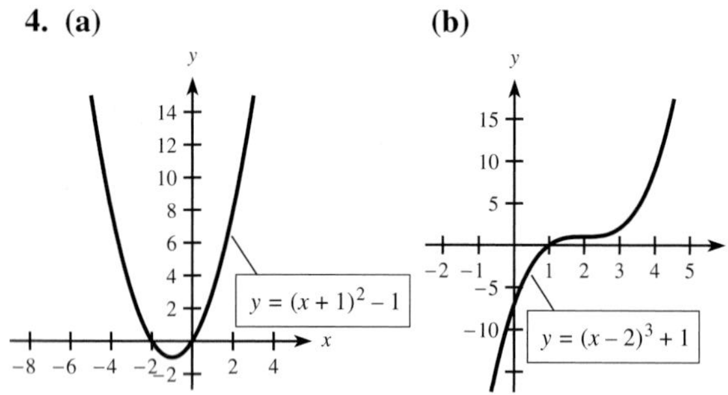

5. *b;* is cubic; $f(1) < 0$

6. (a) -10 **(b)** $-15\frac{1}{2}$ **(c)** -7

7.

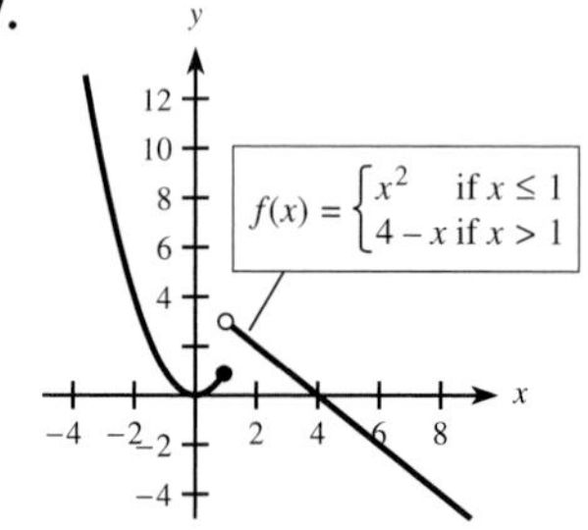

8. vertex $(-2, 25)$; zeros $-7, 3$

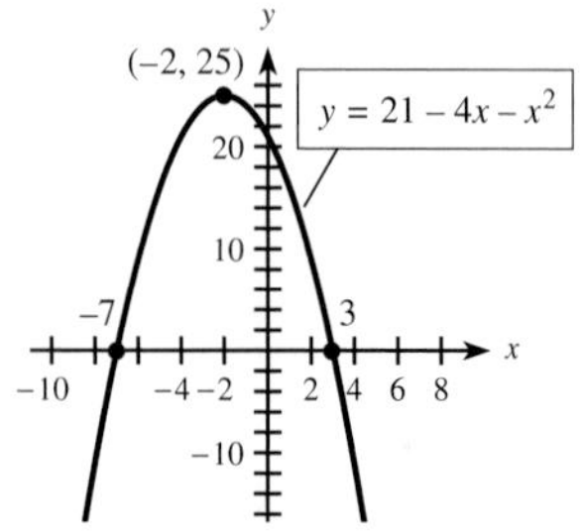

9. $x = 2, x = 1/3$

10. $x = \dfrac{-3 + 3\sqrt{3}}{2}, x = \dfrac{-3 - 3\sqrt{3}}{2}$

11. $x = 2/3$ **12.** c

13. (a) quartic (fourth-degree) **(b)** cubic

14. (a)

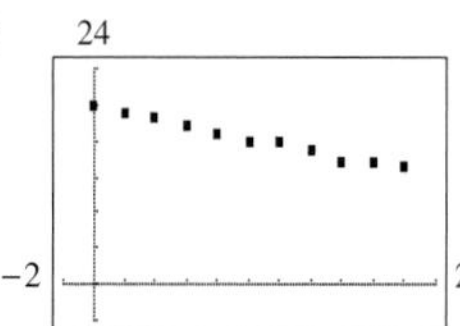

Model: $y = -0.3577x + 19.9227$

(b) 5.6

(c) At $x = 55.7$

15. $q = 300, p = \$80$

16. (a) $P(x) = -x^2 + 250x - 15{,}000$

(b) 125 units, \$625

(c) 100 units, 150 units

17. (a) $f(15) = -31$ means that when the air temperature is 0°F and the wind speed is 15 mph, the air temperature feels like -31°F.

(b) -55°F

18. (a)

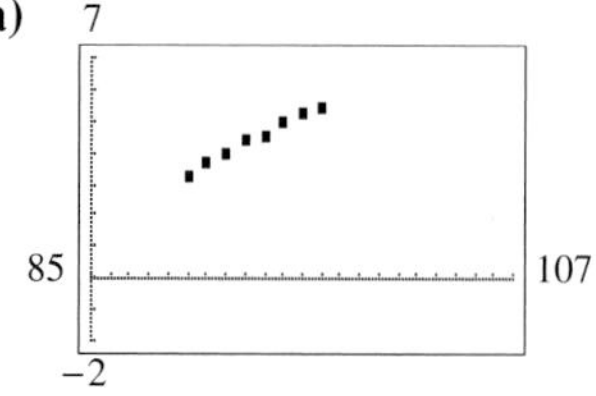

(b) $y = 0.3131x - 24.8619$

(c) No; the slope is positive.

(d) The predicted debt continues to increase.

19. (a) quadratic

(b) $y = -0.01607x^2 + 3.31845x - 165.27798$

(c) 5.59 in 1998; 5.75 in 1999

(d) $x = 103.2$ or after 2003

Exercise 3.1 *(page 222)*

1. 3 **3.** $\begin{bmatrix} -1 & -2 & -3 \\ 1 & 0 & -1 \\ -2 & 3 & 4 \end{bmatrix}$ **5.** *A, C, D, F, Z*

7. *A, F,* and *Z* are 3×3; *C* and *D* are 2×2

9. 1 **11.** $\begin{bmatrix} 1 & 3 & 4 \\ 0 & 2 & 0 \\ 2 & 1 & 3 \end{bmatrix}$ **13.** no

15. $\begin{bmatrix} 0 & 0 & 0 \\ 0 & 0 & 0 \\ 0 & 0 & 0 \end{bmatrix}$ **17.** $\begin{bmatrix} 9 & 5 \\ 4 & 7 \end{bmatrix}$ **19.** $\begin{bmatrix} 0 & -2 & -1 \\ 4 & 2 & 0 \\ 2 & 3 & 7 \end{bmatrix}$

21. $\begin{bmatrix} 2 & 3 & 6 \\ 3 & 4 & 1 \\ 6 & 1 & 6 \end{bmatrix}$ **23.** impossible **25.** $\begin{bmatrix} 3 & 1 & 7 \\ 4 & 2 & 2 \\ 5 & 4 & 3 \end{bmatrix}$

27. $x = 3, y = 2, z = 3, w = 4$

29. $x = 4, y = 1, w = 3, z = 3$

31. $x = 2, y = 2, z = -3$

33. (a) $A = \begin{bmatrix} 55 & 74 & 14 & 7 & 65 \\ 9 & 16 & 19 & 5 & 40 \end{bmatrix}$

$B = \begin{bmatrix} 252 & 178 & 65 & 8 & 11 \\ 19 & 6 & 14 & 1 & 0 \end{bmatrix}$

(b) $A + B = \begin{bmatrix} 307 & 252 & 79 & 15 & 76 \\ 28 & 22 & 33 & 6 & 40 \end{bmatrix}$

(c) $\begin{bmatrix} 197 & 104 & 51 & 1 & -54 \\ 10 & -10 & -5 & -4 & -40 \end{bmatrix}$

more species in U.S.

35. $\begin{bmatrix} 11{,}041.7 & 8978.4 & 6461 \\ 8739.8 & 9159.6 & 6877.3 \\ 9798.1 & 9086.7 & 6448.4 \\ 9696.6 & 8926.7 & 6109.5 \end{bmatrix}$

37. **(a)** $\begin{bmatrix} 825 & 580 & 1560 \\ 810 & 650 & 350 \end{bmatrix}$

(b) $\begin{bmatrix} -75 & 20 & -140 \\ 10 & -50 & 50 \end{bmatrix}$

39. $A = \begin{array}{c} \begin{array}{cc} \text{M} & \text{F} \end{array} \\ \begin{bmatrix} 54.4 & 55.6 \\ 62.1 & 66.6 \\ 67.4 & 74.1 \\ 70.7 & 78.1 \end{bmatrix} \end{array}$ $B = \begin{array}{c} \begin{array}{cc} \text{M} & \text{F} \end{array} \\ \begin{bmatrix} 45.5 & 45.2 \\ 51.5 & 54.9 \\ 61.1 & 67.4 \\ 65.3 & 73.6 \end{bmatrix} \end{array}$

$A + B = C = \begin{bmatrix} 8.9 & 10.4 \\ 10.6 & 11.7 \\ 6.3 & 6.7 \\ 5.4 & 4.5 \end{bmatrix}$

41. **(a)** $\begin{array}{c} \\ \text{I} \\ \text{II} \\ \text{III} \end{array} \begin{array}{c} \begin{array}{cc} \text{wt} & \text{l} \end{array} \\ \begin{bmatrix} 140 & 5.5 \\ 151 & 5.7 \\ 141 & 5.5 \end{bmatrix} \end{array}$ **(b)** $\begin{array}{c} \\ \text{I} \\ \text{II} \\ \text{III} \end{array} \begin{array}{c} \begin{array}{cc} \text{wt} & \text{l} \end{array} \\ \begin{bmatrix} 250 & 12.5 \\ 215 & 11.8 \\ 190 & 9.8 \end{bmatrix} \end{array}$

43. **(a)** $A = \begin{bmatrix} 0 & 1 & 1 & 0 \\ 1 & 0 & 1 & 1 \\ 0 & 0 & 0 & 1 \\ 0 & 1 & 1 & 0 \end{bmatrix}$ **(b)** $B = \begin{bmatrix} 0 & 1 & 1 & 1 \\ 1 & 0 & 1 & 1 \\ 0 & 1 & 0 & 1 \\ 1 & 1 & 1 & 0 \end{bmatrix}$

(c) person 2

45. **(a)** $\begin{bmatrix} 80 & 75 \\ 58 & 106 \end{bmatrix}$ **(b)** $\begin{bmatrix} 176 & 127 \\ 139 & 143 \end{bmatrix}$

(c) $\begin{bmatrix} 10 & 4 \\ 7 & 2 \end{bmatrix}$ **(b)** $\begin{bmatrix} -10 & 19 \\ -7 & 20 \end{bmatrix}$ Shortage, taken from inventory.

47. **(a)** 3, 4, 5, 6 **(b)** 1

49. Worker 1: 0.9625 Worker 2: 0.9375
Worker 3: 0.9125 Worker 4: 0.8875
Worker 5: 0.85 Worker 6: 0.875
Worker 7: 0.90 Worker 8: 0.925
Worker 9: 0.95

Worker 5 is least efficient; at center 5

Exercise 3.2 *(page 237)*

1. [32] **3.** [11 17] **5.** $\begin{bmatrix} 3 & 0 & 6 \\ 9 & 6 & 3 \\ 12 & 0 & 9 \end{bmatrix}$

7. $\begin{bmatrix} 28 & 16 \\ 10 & 18 \end{bmatrix}$ **9.** impossible **11.** $\begin{bmatrix} 29 & 25 \\ 10 & 12 \end{bmatrix}$

13. $\begin{bmatrix} 22 & 16 \\ 20 & 19 \end{bmatrix}$ **15.** $\begin{bmatrix} 8 & -2 & 2 & 4 \\ 13 & -5 & 12 & -11 \end{bmatrix}$

17. $\begin{bmatrix} 7 & 5 & 3 & 2 \\ 14 & 9 & 11 & 3 \\ 13 & 10 & 12 & 3 \end{bmatrix}$ **19.** impossible

21. $\begin{bmatrix} 13 & 9 & 3 & 4 \\ 9 & 7 & 16 & 1 \end{bmatrix}$ **23.** $\begin{bmatrix} 9 & 7 & 16 \\ 5 & 17 & 20 \end{bmatrix}$

25. $\begin{bmatrix} 9 & 0 & 8 \\ 13 & 4 & 11 \\ 16 & 0 & 17 \end{bmatrix}$ **27.** $\begin{bmatrix} 161 & 126 \\ 42 & 35 \end{bmatrix}$ **29.** no

31. no **33.** $\begin{bmatrix} -55 & 88 & 0 \\ -42 & 67 & 0 \\ 28 & -44 & 1 \end{bmatrix}$ **35.** $\begin{bmatrix} 0 & 0 & 0 \\ 0 & 0 & 0 \\ 0 & 0 & 0 \end{bmatrix}$

37. $\begin{bmatrix} 0 & 0 & 0 \\ 0 & 0 & 0 \\ 0 & 0 & 0 \end{bmatrix}$ **39.** A **41.** Z

43. no (see Problem 35)

45. $AB = \begin{bmatrix} 1 & 0 \\ 0 & 1 \end{bmatrix}$, $BA = \begin{bmatrix} 1 & 0 \\ 0 & 1 \end{bmatrix}$

47. $\begin{bmatrix} 2-2+2 \\ 6-4-4 \\ 4+0-2 \end{bmatrix} = \begin{bmatrix} 2 \\ -2 \\ 2 \end{bmatrix}$; is solution

49. $\begin{bmatrix} 1+2+2 \\ 4+0+1 \\ 2+2+1 \end{bmatrix} = \begin{bmatrix} 5 \\ 5 \\ 5 \end{bmatrix}$; is solution

51. **(a)** $AB = \begin{bmatrix} 11 & 4 & 20 \\ -4 & -2 & -11 \\ 7 & 2 & 9 \end{bmatrix}$ $BA = \begin{bmatrix} 8 & -2 & 3 \\ 25 & -4 & 12 \\ 7 & 8 & 14 \end{bmatrix}$

(b) no

53. $\begin{bmatrix} 1 & 0 & 0 & 0 & 0 \\ 0 & 1 & 0 & 0 & 0 \\ 0 & 0 & 1 & 0 & 0 \\ 0 & 0 & 0 & 1 & 0 \\ 0 & 0 & 0 & 0 & 1 \end{bmatrix}$ (Some entries may appear as decimal approximations of 0.)

55. **(a)** $\begin{bmatrix} 292 & 275 \\ 451 & 403 \\ 550 & 453 \\ 582 & 464 \\ 514 & 403 \\ 389 & 353 \end{bmatrix}$ **(b)** $\begin{bmatrix} 350.40 & 330 \\ 541.20 & 483.60 \\ 660 & 543.60 \\ 698.40 & 556.80 \\ 616.80 & 483.60 \\ 466.80 & 423.60 \end{bmatrix}$

57. **(a)** $\begin{bmatrix} 23.10 & 42.00 & 105.00 & 5.25 \\ 21.00 & 42.00 & 21.00 & 0.00 \\ 29.40 & 73.50 & 47.25 & 0.00 \\ 15.75 & 73.50 & 21.00 & 10.50 \\ 21.00 & 0.00 & 105.00 & 5.25 \end{bmatrix}$

(b) $\begin{bmatrix} 24.20 & 44.00 & 110.00 & 5.50 \\ 22.00 & 44.00 & 22.00 & 0.00 \\ 30.80 & 77.00 & 49.50 & 0.00 \\ 16.50 & 77.00 & 22.00 & 11.00 \\ 22.00 & 0.00 & 110.00 & 5.50 \end{bmatrix}$

59. $\begin{bmatrix} 22{,}000 & 24{,}640 \\ 30{,}600 & 35{,}700 \end{bmatrix}$

61. **(a)** $\begin{bmatrix} 2703 & 3428 \\ 3608 & 2481 \\ 2383 & 2314 \\ 4376 & 4584 \\ 679 & 664 \\ 472 & 477 \\ 1443 & 1523 \\ 3152 & 2988 \\ 1289 & 1224 \\ 254 & 257 \\ 916 & 949 \\ 1455 & 1413 \end{bmatrix}$ $\begin{bmatrix} 96.1 & 97.9 & 48.1 & 79.4 & 65.4 & 98.4 & 99.0 & 77.3 & 65.8 & 82.8 & 89.5 & 76.9 \\ 109.0 & 105.0 & 63.3 & 91.1 & 79.2 & 96.4 & 121.0 & 87.5 & 85.4 & 91.3 & 95.3 & 76.4 \end{bmatrix}$

(b) 12×12 **(c)** the diagonal entries

63. 1300 teens, 1520 single, 1620 married

65. 172, 208, 268, 327, 101, 123, 268, 327, 216, 263, 162, 195, 176, 215, 343, 417

67. $\begin{bmatrix} 2137 & 84 & 41.5 & 128.5 & 158.5 & 317 & 738 \\ 1285.5 & 64 & 30.5 & 115 & 136 & 229 & 590 \\ 969.5 & 46.5 & 24 & 98 & 139 & 224 & 476.5 \\ 852.5 & 45 & 23 & 97.5 & 142.5 & 236.5 & 463 \\ 809 & 44.5 & 22.5 & 99 & 141.5 & 240.5 & 455.5 \end{bmatrix}$

69. **(a)** $\begin{bmatrix} 0.665 & 8.075 & 9.690 & 1.045 & 5.320 & 3.420 \\ 0.475 & 0.190 & 5.795 & 1.235 & 0.190 & 0.950 \\ 2.090 & 0.380 & 8.360 & 1.140 & 1.140 & 4.560 \\ 239.210 & 60.230 & 77.520 & 33.440 & 51.585 & 136.990 \\ 28.500 & 0.950 & 0.950 & 0.950 & 0.950 & 0.950 \\ 749.455 & 0 & 0 & 0 & 0 & 0 \end{bmatrix}$

(b) $\begin{bmatrix} 0.7560 & 9.1800 & 11.0160 & 1.1880 & 6.0480 & 3.8880 \\ 0.5400 & 0.2160 & 6.5880 & 1.4040 & 0.2160 & 1.0800 \\ 2.3760 & 0.4320 & 9.5040 & 1.2960 & 1.2960 & 5.1840 \\ 271.9440 & 68.4720 & 88.1280 & 38.0160 & 58.6440 & 155.7360 \\ 32.4000 & 1.0800 & 1.0800 & 1.0800 & 1.0800 & 1.0800 \\ 852.0121 & 0 & 0 & 0 & 0 & 0 \end{bmatrix}$

71. $\begin{bmatrix} 0.900 & 0.855 & 0.855 & 0.855 & 0.855 & 0.765 & 0.765 & 0.765 & 0.765 \\ 0.765 & 0.900 & 0.855 & 0.855 & 0.855 & 0.855 & 0.765 & 0.765 & 0.765 \\ 0.765 & 0.765 & 0.900 & 0.855 & 0.855 & 0.855 & 0.855 & 0.765 & 0.765 \\ 0.765 & 0.765 & 0.765 & 0.900 & 0.855 & 0.855 & 0.855 & 0.855 & 0.765 \\ 0.765 & 0.765 & 0.765 & 0.765 & 0.900 & 0.855 & 0.855 & 0.855 & 0.855 \\ 0.855 & 0.765 & 0.765 & 0.765 & 0.765 & 0.900 & 0.855 & 0.855 & 0.855 \\ 0.855 & 0.855 & 0.765 & 0.765 & 0.765 & 0.765 & 0.900 & 0.855 & 0.855 \\ 0.855 & 0.855 & 0.855 & 0.765 & 0.765 & 0.765 & 0.765 & 0.900 & 0.855 \\ 0.855 & 0.855 & 0.855 & 0.855 & 0.765 & 0.765 & 0.765 & 0.765 & 0.900 \end{bmatrix}$

Exercise 3.3 *(page 254)*

1. $\left[\begin{array}{rrr|r} -1 & 2 & 1 & 7 \\ 0 & 7 & 5 & 21 \\ 4 & 2 & 2 & 1 \end{array}\right]$ **3.** $\left[\begin{array}{rrr|r} 3 & 2 & 4 & 0 \\ 2 & -1 & 2 & 0 \\ 1 & -2 & -4 & 0 \end{array}\right]$

5. $\left[\begin{array}{rrr|r} 1 & -3 & 4 & 2 \\ 2 & 0 & 2 & 1 \\ 1 & 2 & 1 & 1 \end{array}\right]$ **7.** $x = 2, y = 1/2, z = -5$

9. $x = 18, y = 10, z = 0$ **11.** $x = -5, y = 2, z = 1$

13. $x = 4, y = 1, z = -2$ **15.** $x = \frac{1}{3}, y = \frac{5}{3}$

17. $x = 15, y = -13, z = 2$

19. $x = 15, y = 0, z = 2$ **21.** $x = 1, y = -1, z = 1$

23. $x_1 = 1, x_2 = 0, x_3 = 1, x_4 = 0$

25. $x = 1, y = 3, z = 1, w = 0$

27. no solution **29.** $x = (11 + 2z)/3, y = (-1 - z)/3$

31. $x = 0, y = -z$ **33.** $x = 3z - 2, y = 3 - 5z$

35. $x = 1 - z, y = \frac{1}{2}z$ **37.** $x = 2z - 2, y = 1 + z$
39. $x = \frac{9}{2}, y = -\frac{1}{2}, z = -1$ **41.** $x_1 = \frac{23}{38}x_3, x_2 = \frac{12}{19}x_3$
43. $x = 7/5, y = -3/5, z = w$ **45.** no solution
47. $x_1 = 0.5 - 2x_4, x_2 = 3.5 + 5x_4, x_3 = -2.5 - 3x_4$
49. $x = (b_2c_1 - b_1c_2)/(a_1b_2 - a_2b_1)$
51. Beef: 2 cups; sirloin: 8 cups
53. AB: 2 oz, SFF: 2 oz, NMG: 1 oz
55. 2 of Portfolio I, 2 of Portfolio II
57. $\frac{3}{8}$ pound of red meat, 6 slices of bread, 4 glasses of milk
59. \$ 13,500 @ 15% and \$ 10,000 @ 16%
61. Type I = 3 Type IV, Type II = 1000 − 2(Type IV), Type III = 500 − Type IV, Type IV = any amount (less than 500 bags)
63. Oil = 6138 − 3.3C − 0.6R, Bank = 7.5C + R − 12,787.50; R = any amount (less than 10,230); C = any amount (less than 1705)
65. Bacteria III = any amount (between 1800 and 2300)
Bacteria I = 6900 − 3 (Bacteria III)
Bacteria II = $\frac{1}{2}$ (Bacteria III) − 900
67. There are three possibilities:
(1) 4 of I and 2 of II
(2) 5 of I, 1 of II, and 1 of III
(3) 6 of I and 2 of III

Exercise 3.4 (page 269)

1. $\begin{bmatrix} 1 & 0 & 0 \\ 0 & 1 & 0 \\ 0 & 0 & 1 \end{bmatrix}$ **3.** yes **5.** $\begin{bmatrix} \frac{1}{3} & 0 & 0 \\ 0 & \frac{1}{3} & 0 \\ 0 & 0 & \frac{1}{3} \end{bmatrix}$

7. $\begin{bmatrix} 2 & -7 \\ -1 & 4 \end{bmatrix}$ **9.** no inverse **11.** $\begin{bmatrix} \frac{5}{2} & -1 \\ -2 & 1 \end{bmatrix}$

13. $\begin{bmatrix} -\frac{1}{10} & \frac{7}{10} \\ \frac{1}{5} & -\frac{2}{5} \end{bmatrix}$ **15.** $\begin{bmatrix} -1 & 1 & 0 \\ 1 & 0 & 0 \\ -1 & 0 & 1 \end{bmatrix}$

17. $\begin{bmatrix} \frac{1}{3} & -\frac{1}{3} & \frac{1}{3} \\ -\frac{2}{3} & -\frac{1}{3} & \frac{7}{3} \\ \frac{1}{3} & \frac{2}{3} & -\frac{5}{3} \end{bmatrix}$ **19.** no inverse **21.** no inverse

23. $\begin{bmatrix} 2 & 2 & 0 & 2 & 2 \\ 1 & 0 & 2 & 2 & 1 \\ 0 & 1 & 0 & 2 & 1 \\ 2 & 0 & 2 & 2 & 1 \\ 1 & 0 & 0 & 0 & 2 \end{bmatrix}$ **25.** $\begin{bmatrix} 13 \\ 5 \end{bmatrix}$ **27.** $\begin{bmatrix} 9 \\ 6 \\ 3 \end{bmatrix}$

29. $\begin{bmatrix} x \\ y \\ z \end{bmatrix} = \begin{bmatrix} 1 \\ 1 \\ 2 \end{bmatrix}$ **31.** $x = 2, y = 1$

33. $x = 1, y = 2$ **35.** $x = 1, y = 1, z = 1$
37. $x = 1, y = 1, z = 1$ **39.** $x = 1, y = 3, z = 2$
41. $x_1 = 5.6, x_2 = 5.4, x_3 = 3.25, x_4 = 6.1, x_5 = 0.4$
43. −2 **45.** 14 **47.** −5 **49.** −19 **51.** yes
53. no **55.** Hang on **57.** Answers in back

59. $x_0 = 2400, y_0 = 1200$
61. A = 5.5 mg and B = 8.8 mg for Patient I; A = 10 mg and B = 16 mg for Patient II
63. \$68,000 at 18%, \$77,600 at 10%
65. 7 cc of 20%, 3 cc of 5%
67. (a) $\begin{bmatrix} 0 & 1 & 0 \\ 0 & 0 & 1 \\ 1 & 1 & 1 \end{bmatrix}$ **(b)** 108

Exercise 3.5 (page 282)

1. (a) 15 **(b)** 4 **3.** 8 **5.** 40
7. most: raw materials; least: fuels
9. raw materials, manufacturing, service
11. 5 units of manufacturing; 10 units of ag; 105 units of utilities
13. farm products = 200; machinery = 40
15. agricultural products = 100; oil = 700
17. utilities = 200; manufacturing = 400
19. mining = 310; manufacturing = 530
21. electronic components = 1200; computers = 320
23. fishing = 100; oil = 1250
25. development = \$21,000; promotional = \$12,000
27. engineering = \$15,000; computer = \$13,000
29. agricultural goods = 400; manufactured goods = 500; fuels = 400
31. electronics = 1240; steel = 1260; autos = 720
33. products = $\frac{7}{17}$ households; machinery = $\frac{1}{17}$ household
35. government = $\frac{10}{19}$ households; industry = $\frac{11}{19}$ households
37. manufacturing = 3 households; utilities = 3 households
39. $\begin{bmatrix} 24 \\ 96 \\ 24 \\ 120 \\ 492 \\ 3456 \end{bmatrix}$ 3456 bolts, 492 braces, 120 sheets
41. $\begin{bmatrix} 10 \\ 10 \\ 20 \\ 56 \\ 20 \\ 26 \\ 300 \end{bmatrix}$ 56 2 × 4's, 20 braces, 26 clamps, 300 nails

Chapter 3 Review Exercises (page 287)

1. 4 **2.** 0 **3.** A, B **4.** none
5. D, F, G, I **6.** $\begin{bmatrix} -2 & 5 & 11 & -8 \\ -4 & 0 & 0 & -4 \\ 2 & 2 & -1 & -9 \end{bmatrix}$

7. Zero matrix **8.** Order

9. $\begin{bmatrix} 6 & -1 & -9 & 3 \\ 10 & 3 & -1 & 4 \\ -2 & -2 & -2 & 14 \end{bmatrix}$ **10.** $\begin{bmatrix} 3 & -3 \\ 4 & -1 \\ 2 & -6 \\ 1 & -2 \end{bmatrix}$

11. $\begin{bmatrix} 2 & 1 \\ 5 & 1 \end{bmatrix}$ **12.** $\begin{bmatrix} 12 & -6 \\ 15 & 0 \\ 18 & 0 \\ 3 & 9 \end{bmatrix}$ **13.** $\begin{bmatrix} 4 & 0 \\ 0 & 4 \end{bmatrix}$

14. $\begin{bmatrix} 2 & -12 \\ -8 & -22 \end{bmatrix}$ **15.** $\begin{bmatrix} 9 & 20 \\ 4 & 5 \end{bmatrix}$ **16.** $\begin{bmatrix} 5 & 16 \\ 6 & 15 \end{bmatrix}$

17. $\begin{bmatrix} 2 & 37 & 61 & -55 \\ -2 & 9 & -3 & -20 \\ 10 & 10 & -14 & -30 \end{bmatrix}$ **18.** $\begin{bmatrix} 43 & -23 \\ 33 & -12 \\ -13 & 15 \end{bmatrix}$

19. $\begin{bmatrix} 10 & 16 \\ 15 & 25 \\ 18 & 30 \\ 6 & 11 \end{bmatrix}$ **20.** $\begin{bmatrix} 17 & 73 \\ 7 & 28 \end{bmatrix}$ **21.** $\begin{bmatrix} 3 & 7 \\ 23 & 42 \end{bmatrix}$

22. F **23.** F **24.** $\begin{bmatrix} -19 & 12 \\ -8 & 5 \end{bmatrix}$

25. F **26.** (1, 2, 1)

27. $x = 22, y = 9$ **28.** $x = -3, y = 3, z = 4$

29. $x = -\frac{3}{2}, y = 7, z = -\frac{11}{2}$ **30.** no solution

31. $x = 2 - 2z;\ y = -1 - 2z$

32. $x_1 = 1, x_2 = 11, x_3 = -4, x_4 = -5$

33. yes **34.** $\begin{bmatrix} \frac{1}{2} & \frac{1}{4} \\ \frac{5}{2} & \frac{7}{4} \end{bmatrix}$ **35.** $\begin{bmatrix} -1 & -2 & 8 \\ 1 & 2 & -7 \\ 1 & 1 & -4 \end{bmatrix}$

36. $\begin{bmatrix} 2 & 1 & -2 \\ 7 & 5 & -8 \\ -13 & -9 & 15 \end{bmatrix}$ **37.** $\begin{bmatrix} -33 \\ 30 \\ 19 \end{bmatrix}$ **38.** $\begin{bmatrix} 4 \\ 5 \\ -13 \end{bmatrix}$

39. $A^{-1} = \begin{bmatrix} -41 & 32 & 5 \\ 17 & -13 & -2 \\ -9 & 7 & 1 \end{bmatrix}$; $x = 4, y = -2, z = 2$

40. No **41.** $\begin{bmatrix} 250 & 140 \\ 480 & 700 \end{bmatrix}$ **42.** $\begin{bmatrix} 1030 & 800 \\ 700 & 1200 \end{bmatrix}$

43. **(a)** higher in June **(b)** higher in July

44. $\begin{array}{c} \text{men} \quad \text{women} \\ \begin{bmatrix} 865 & 885 \\ 210 & 270 \end{bmatrix} \end{array} \begin{array}{l} \text{Robes} \\ \text{Hoods} \end{array}$ **45.** $\begin{bmatrix} 1750 \\ 480 \end{bmatrix} \begin{array}{l} \text{Robes} \\ \text{Hoods} \end{array}$

46. **(a)** $\begin{bmatrix} 13{,}500 & 12{,}400 \\ 10{,}500 & 10{,}600 \end{bmatrix}$

(b) Dept. A buys from Kink; Dept. B buys from Ace

47. **(a)** [0.20 0.30 0.50] **(b)** $\begin{bmatrix} 0.013469 \\ 0.013543 \\ 0.006504 \end{bmatrix}$

(c) $[0.20\ 0.30\ 0.50] \begin{bmatrix} 0.013469 \\ 0.013543 \\ 0.006504 \end{bmatrix} = 0.20(0.013469) + 0.30(0.013543) + 0.50(0.006504) = 0.0100087$

(d) The historical return of the portfolio, 0.0100087, is the estimated expected monthly return of the portfolio. This is roughly 1% per month.

48. 400 fast food, 700 computer, 200 pharmaceutical

49. $A = 2C,\ b = 2000 - 4C$

50. 3 passenger, 4 transport, 4 jumbo

51. $S = 6000;\ A = 800$ **52.** $S = 1000;\ C = 500$

53. $X = \begin{bmatrix} 366 \\ 322 \\ 402 \end{bmatrix}$, approximately

54. $\text{G} = \frac{64}{93}\text{H};\ \text{A} = \frac{59}{93}\text{H};\ \text{M} = \frac{40}{93}\text{H}$

Chapter 3 Test *(page 290)*

1. $\begin{bmatrix} 3 & 1 & 5 \\ 1 & 3 & 6 \end{bmatrix}$ **2.** $\begin{bmatrix} -1 & 2 & 2 \\ 1 & -1 & 6 \end{bmatrix}$

3. $\begin{bmatrix} -12 & -16 & -155 \\ 5 & 12 & 87 \end{bmatrix}$ **4.** $\begin{bmatrix} 23 & 6 \\ 182 & 45 \\ 21 & 1 \end{bmatrix}$

5. $\begin{bmatrix} 0 & -7 \\ 26 & 1 \end{bmatrix}$ **6.** $\begin{bmatrix} -43 & -46 & -207 \\ 39 & 30 & -77 \\ 17 & 5 & -216 \end{bmatrix}$

7. $\begin{bmatrix} -2 & 3/2 \\ 1 & -1/2 \end{bmatrix}$ **8.** $\begin{bmatrix} -3 & 2 & 2 \\ 1 & 0 & -1 \\ 1/2 & -1/2 & 0 \end{bmatrix}$

9. $\begin{bmatrix} 5 \\ 14 \\ 15 \end{bmatrix}$ **10.** $x = -0.5, y = 0.5, z = 2.5$

11. $x = 4 - 1.8z,\ y = 0.2z$ **12.** no solution

13. $x = 2, y = 2, z = 0, w = -2$

14. $x = 6w - 0.5,\ y = 0.5 - w,\ z = 2.5 - 3w$

15. **(a)** $\begin{bmatrix} .08 & .22 & .12 \\ .10 & .08 & .19 \\ .05 & .07 & .09 \\ .10 & .26 & .15 \\ .12 & .04 & .24 \end{bmatrix}$

(b) 0.08, 0.22, 0.12 consumed by carnivores 1, 2, 3

(c) Plant 5 by 1, Plant 4 by 2, Plant 5 by 3

16. **(a)** [1000 4000 2000 1000]

(b) [45,000 55,000 90,000 70,000]

(c) $\begin{bmatrix} 5 \\ 3 \\ 4 \\ 4 \end{bmatrix}$ **(d)** [$1,030,000] **(e)** $\begin{array}{c} \\ A \\ B \\ C \\ D \end{array} \begin{array}{c} \$ \\ \begin{bmatrix} 65 \\ 145 \\ 125 \\ 135 \end{bmatrix} \end{array}$

17. Growth, 2000; blue-chip, 400, utility, 400
18. Ag: 315, M: 245
19. Profit = Households
Non-profit = $\frac{2}{3}$ Households
20.

	Ag	M	F	S
Ag	0.2	0.1	0.1	0.1
M	0.3	0.2	0.2	0.2
F	0.2	0.2	0.3	0.3
S	0.1	0.4	0.2	0.2

21. Ag: 35,725, M: 101,000, F: 121,189, S: 115,752
22. Ag: 5000, M: 8000, F: 8000, S: 7000

Exercise 4.1 (page 302)

1. $x < -4$ **3.** $x < 2$ **5.** $x < -4$
7. $x \le -1$ **9.** $x \ge 1.949$
11. $x < 3$

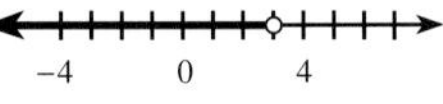

13. $x < -3$

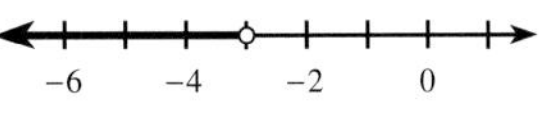

15. $x < \frac{3}{2}$

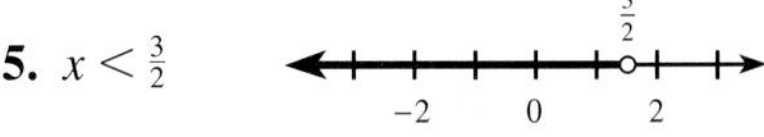

17. $x < -6$

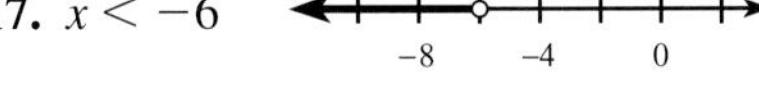

19. $x < 2$

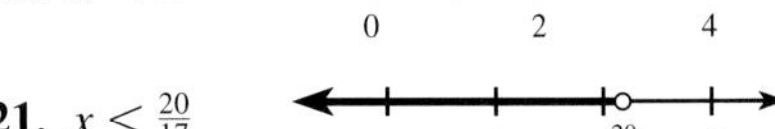

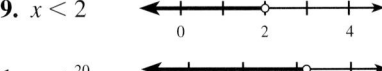

21. $x < \frac{20}{17}$

23. $-\frac{1}{2} < x \le 3$ **25.** $1 < x < 4$
27. $x > 4$ **29.** $-50 \le x < -22$
31. (1, 3], half-open
33. (2, 10), open **35.** [−3, 2], closed
37. (−4, 3), open **39.** [4, 6], closed
41. $x > 80$
43. $695 + 5.75x \le 900$; 35 or fewer
45. (a) $0 \le I \le 25{,}350$
$25{,}351 \le I \le 61{,}400$
$61{,}401 \le I \le 128{,}100$
$128{,}101 \le I \le 278{,}450$
$I \ge 278{,}451$
(b) $0 \le T \le 3802.50$
$3802.50 < T \le 13{,}896.50$
$13{,}896.50 < T \le 34{,}573.50$
$34{,}573.50 < T \le 88{,}699.50$
$T > 88{,}699.50$
47. (a) $0.237 \le h \le 1$; $h = 1$ means 100% humidity
(b) $0 \le h \le 0.237$
49. $C \ge 37$
51. (a) (1) $x_4 \le 1100$, (2) $x_4 \le 1000$, (3) $x_4 \le 900$, (4) $x_4 \ge 0$
(b) $0 \le x_4 \le 900$
(c) $x_1 \in [100, 1000]$, $x_2 \in [200, 1100]$, $x_3 \in [300, 1200]$

Exercise 4.2 (page 310)

1.

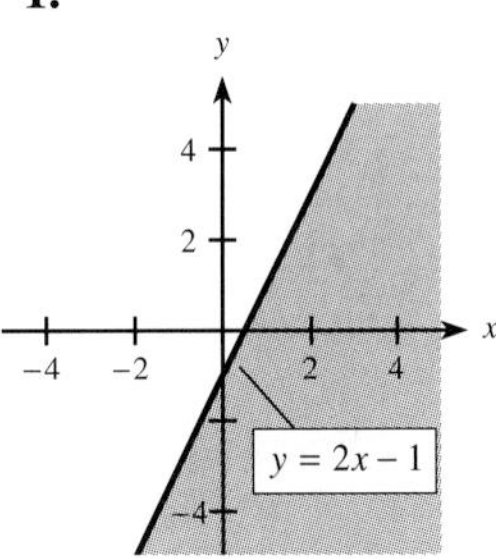

3.

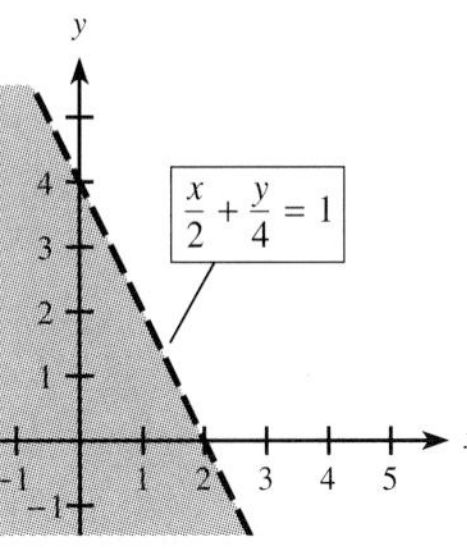

5.

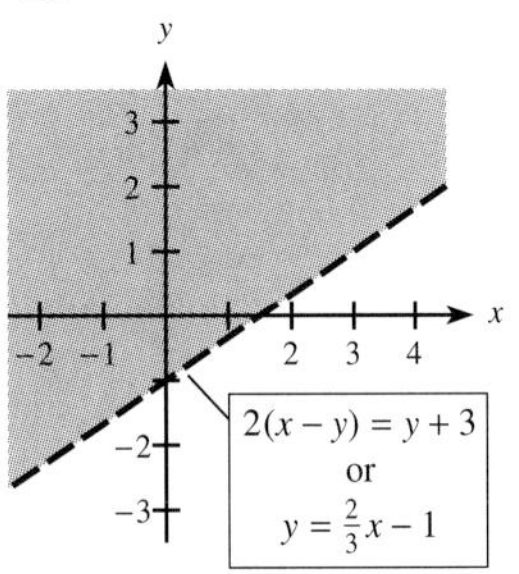

7.

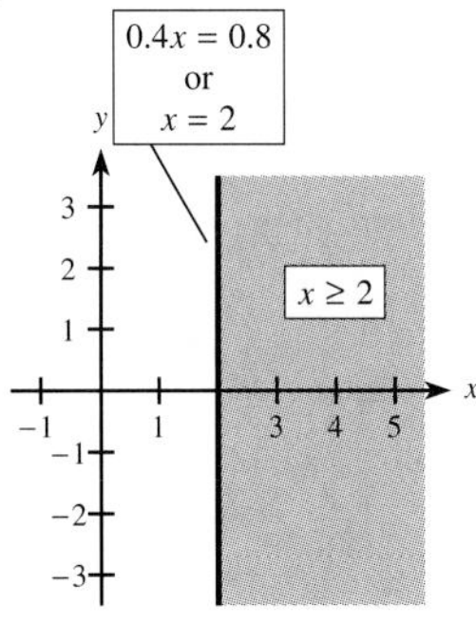

9. (0, 0), (20, 10), (0, 15), (25, 0)
11. (0, 0), (7, 0), (5, 5), (2, 6), (0, 4)
13. (0, 5), (1, 2), (3, 1), (6, 0)
15.

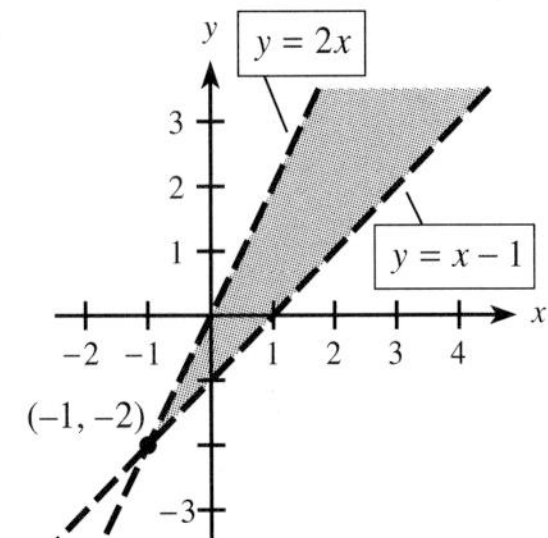

17.

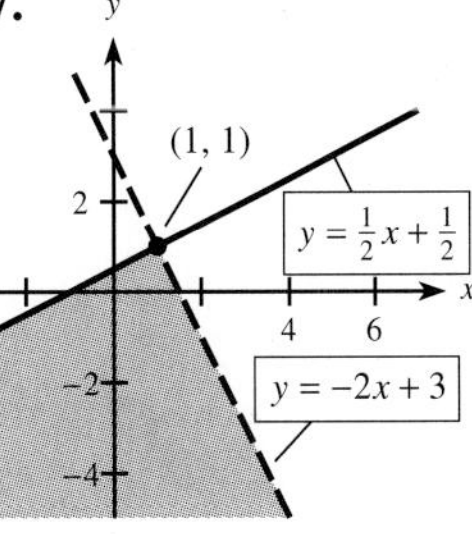

19.

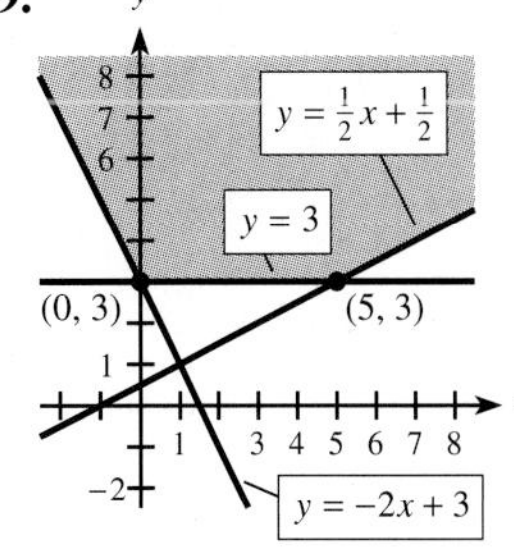

21.

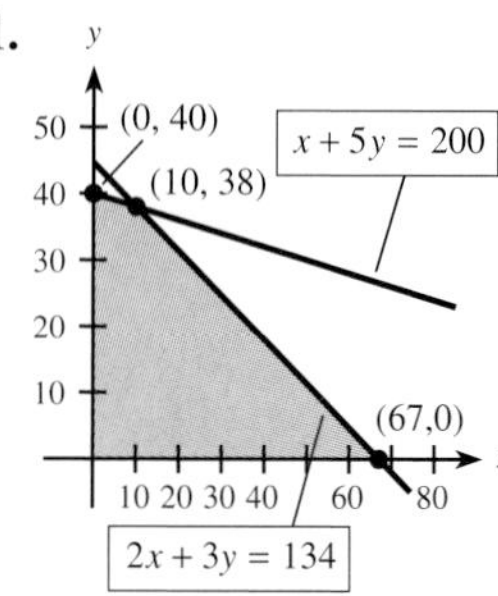

23.

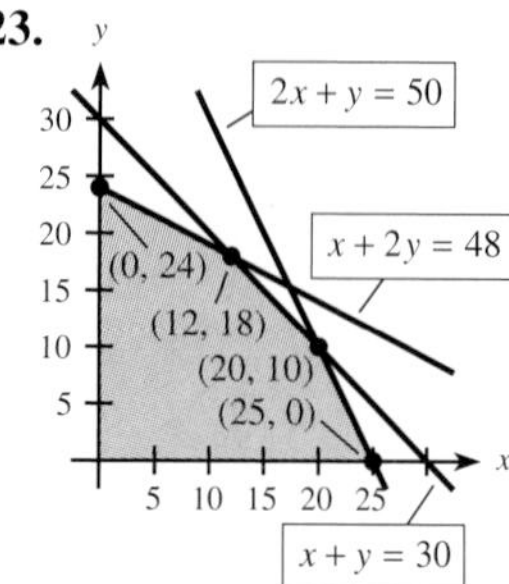

25.

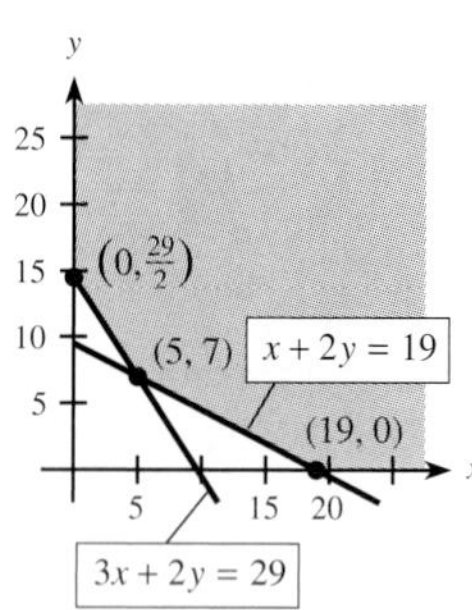

27.

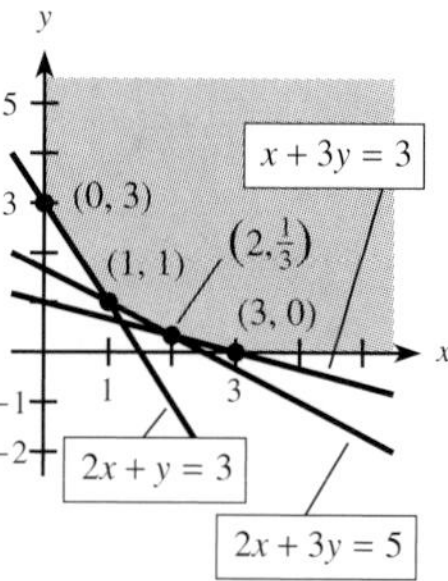

29.

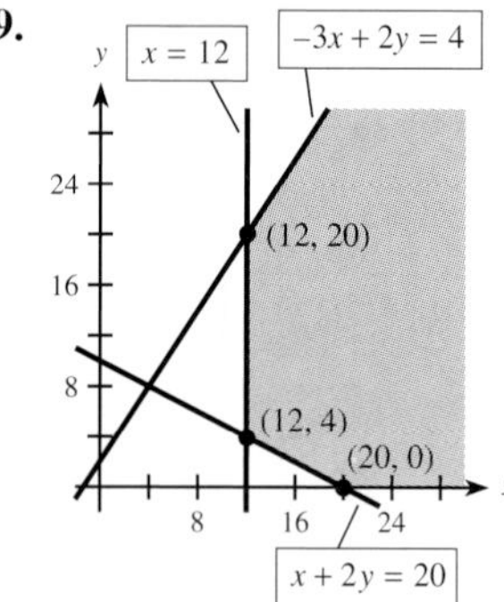

31. **(a)** Let x = deluxe model and y = economy model

$3x + 2y \leq 24$

$\frac{1}{2}x + y \leq 8$

$x \geq 0, y \geq 0$

(b)

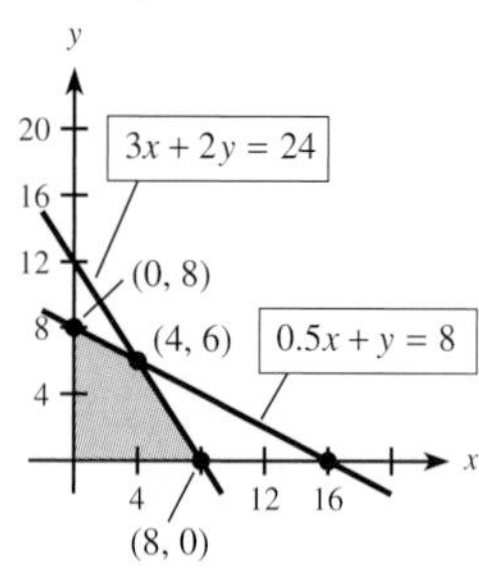

Corners: (0, 0), (8, 0), (4, 6), (0, 8)

33. Let x = cord-type trimmer and y = cordless trimmer.

(a) Constraints are

$x + y \leq 300$

$2x + 4y \leq 800$

$x \geq 0, y \geq 0$

(b)

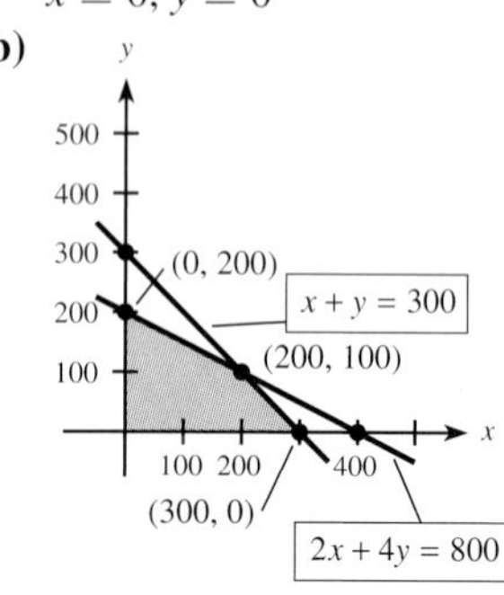

35. **(a)** $7x + 2y \geq 30$

$4x + 12y \geq 28$

$x \geq 0, y \geq 0$

(b)

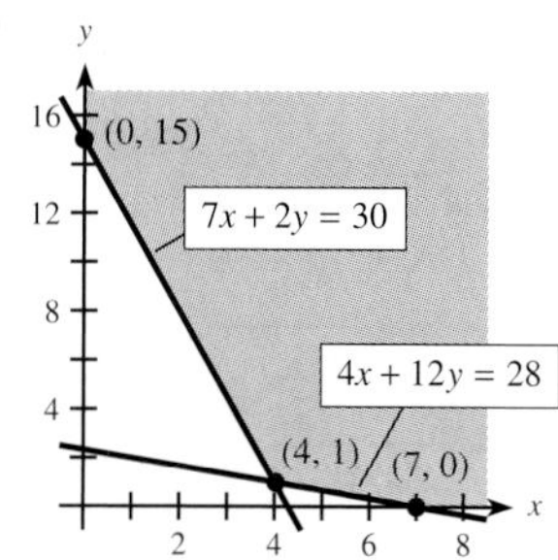

37. Let x = minutes of radio and y = minutes of television

(a) Constraints are

$x + y \geq 80$

$0.006x + 0.09y \geq 2.16$

$x \geq 0, y \geq 0$

(b)

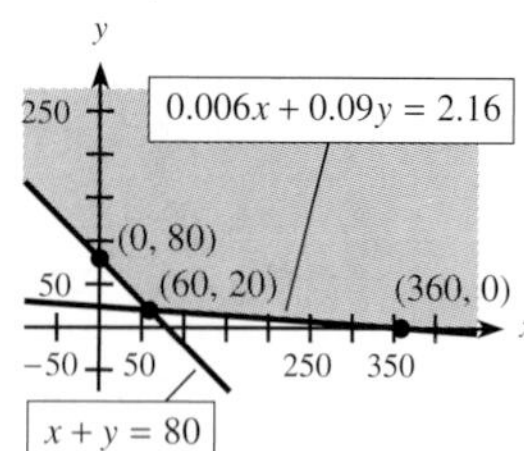

39. **(a)** x = pounds of regular

y = pounds of all beef

$0.18x + 0.75y \leq 1020$

$0.2x + 0.2y \geq 500$

$0.3x \qquad \leq 600$

(b)

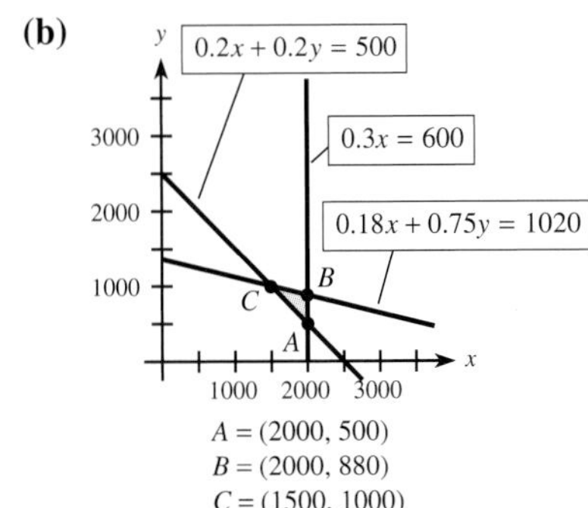

Exercise 4.3 *(page 322)*

1. (12, 0) **3.** max = 20 at (4, 4); min = 0 at (0, 0)
5. max = 68 at (12, 4); min = 14 at (2, 2)
7. no max; min = 10 at (2, 1)
9. (0, 0), (0, 20), (10, 18), (15, 10), (20, 0); max = 66 at (10, 18)
11. (0, 60), (10, 30), (20, 20), (70, 0); min = 90 at (10, 30)
13. max = 382 at $x = 10$, $y = 38$
15. max = 80 at $x = 20$, $y = 10$
17. min = 115 at $x = 5$, $y = 7$
19. min = 36 at $x = 3$, $y = 0$
21. max = 6 at (0, 2)
23. max = 22 at (2, 4)
25. max = 30 on line between (0, 5) and (3, 4)
27. min = 32 at (2, 3) **29.** min = 9 at (2, 3)
31. min = 75 at (15, 15)
33. max = 10 at (2, 4)
35. min = 3100 at (40, 60)
37. max = \$132 at (4, 6)
39. max = \$10,000 at 200 cord-type, 100 cordless
41. 250 fish: 150 bass and 100 trout
43. R = \$366,000 with 6 satellite and 17 full-service branches
45. radio = 60, TV = 20, C = \$16,000
47. 30 days for Factory 1 and 20 days for Factory 2; cost = \$700,000
49. 60 days for Location I and 70 days for Location II; cost = \$86,000
51. reg = 2000 lb; beef = 880 lb; profit = \$1328
53. From Pittsburgh: 20 to Blairsville, 40 to Youngstown
From Erie: 15 to Blairsville, 0 to Youngstown
Minimum cost = \$1550

Exercise 4.4 *(page 342)*

1. $3x + 5y + s_1 = 15$, $3x + 6y + s_2 = 20$

3. $\left[\begin{array}{rrrrr|r} 2 & 5 & 1 & 0 & 0 & 30 \\ 1 & 5 & 0 & 1 & 0 & 25 \\ \hline -2 & -4 & 0 & 0 & 1 & 0 \end{array}\right]$

5. $\left[\begin{array}{rrrrr|r} 1 & 5 & 1 & 0 & 0 & 200 \\ 2 & 3 & 0 & 1 & 0 & 134 \\ \hline -4 & -9 & 0 & 0 & 1 & 0 \end{array}\right]$

7. $\left[\begin{array}{rrrrrrr|r} 2 & 7 & 9 & 1 & 0 & 0 & 0 & 100 \\ 6 & 5 & 1 & 0 & 1 & 0 & 0 & 145 \\ 1 & 2 & 7 & 0 & 0 & 1 & 0 & 90 \\ \hline -2 & -5 & -2 & 0 & 0 & 0 & 1 & 0 \end{array}\right]$

9. $\left[\begin{array}{rrrrr|r} 2 & 4 & 1 & 0 & 0 & 24 \\ 1 & \textcircled{1} & 0 & 1 & 0 & 5 \\ \hline -4 & -11 & 0 & 0 & 1 & 0 \end{array}\right]$

11. $\left[\begin{array}{rrrrrr|r} 10 & \textcircled{27} & 1 & 0 & 0 & 0 & 200 \\ 4 & 51 & 0 & 1 & 0 & 0 & 400 \\ 15 & 27 & 0 & 0 & 1 & 0 & 350 \\ \hline -6 & -7 & 0 & 0 & 0 & 1 & 0 \end{array}\right]$

13. $\left[\begin{array}{rrrrr|r} 2 & 0 & 1 & -\frac{3}{4} & 0 & 12 \\ \textcircled{3} & 1 & 0 & \frac{1}{3} & 0 & 15 \\ \hline -4 & 0 & 0 & 3 & 1 & 15 \end{array}\right]$

15. Solution is complete

17. $\left[\begin{array}{rrrrrrr|r} 4 & 4 & 1 & 0 & 0 & 2 & 0 & 12 \\ \textcircled{2} & \textcircled{4} & 0 & 1 & 0 & -1 & 0 & 4 \\ -3 & -11 & 0 & 0 & 1 & -1 & 0 & 6 \\ \hline -3 & -3 & 0 & 0 & 0 & 4 & 1 & 150 \end{array}\right]$

Either circled number may act as the next pivot, but only one of them.

19. No solution is possible. **21.** $x = 11, y = 9; f = 20$
23. $x = 0, y = 14, z = 11; f = 525$
25. $x = 50, y = 10, z = 0, f = 100$. Multiple solutions are possible.
Next pivot is circled.

$\left[\begin{array}{rrrrrr|r} 1 & 0 & 3 & 0 & 6 & 0 & 50 \\ 0 & 0 & 4 & 1 & -4 & 0 & 6 \\ 0 & 1 & -2 & 0 & \textcircled{2} & 0 & 10 \\ \hline 0 & 0 & 9 & 0 & 0 & 1 & 100 \end{array}\right]$

27. $x = 0, y = 5, f = 50$ **29.** $x = 4, y = 3, f = 17$
31. $x = 2, y = 5, f = 17$ **33.** $x = 4, y = 3, f = 11$
35. $x = 0, y = 2, z = 5, f = 40$
37. $x = 15, y = 15, z = 25, f = 780$
39. $x = 6, y = 2, z = 26, f = 206$
41. $x = 8, y = 16, f = 32$ **43.** no solution
45. $x = 0, y = 50$ or $x = 40, y = 40, f = 600$

47. (a) $\left[\begin{array}{rrrrr|r} 1 & 1 & 1 & 0 & 0 & 60 \\ 1 & 3 & 0 & 1 & 0 & 120 \\ \hline -40 & -60 & 0 & 0 & 1 & 0 \end{array}\right]$

(b) Max profit is \$3000 with 30 ink jet and 30 laser printers

49. (a) $\left[\begin{array}{rrrrr|r} 4 & 3 & 1 & 0 & 0 & 50 \\ 3 & 5 & 0 & 1 & 0 & 43 \\ \hline -300 & -300 & 0 & 0 & 1 & 0 \end{array}\right]$

(b) Max profit is \$3900 with 11 axles and 2 wheels

51. 500 tomatoes, 1800 peaches; $P = 4100$

53. 21 newspaper, 13 radio

55. Medium 1 = 10, Medium 2 = 10, Medium 3 = 12

57. \$1650 profit with 46A, 20B, 6C

59. 8000 Regular, 0 Special, and 1000 Kitchen Magic; \$32,000

Exercise 4.5 (page 353)

1. (a) $\left[\begin{array}{rr|r} 5 & 2 & 10 \\ 1 & 2 & 6 \\ \hline 3 & 1 & g \end{array}\right]$ transpose $= \left[\begin{array}{rr|r} 5 & 1 & 3 \\ 2 & 2 & 1 \\ \hline 10 & 6 & g \end{array}\right]$

(b) Maximize $f = 10x + 6y$ subject to $5x + y \le 3$, $2x + 2y \le 1$, $x \ge 0$, $y \ge 0$.

3. (a) $\left[\begin{array}{rr|r} 1 & 1 & 9 \\ 1 & 3 & 15 \\ \hline 5 & 2 & g \end{array}\right]$ transpose $= \left[\begin{array}{rr|r} 1 & 1 & 5 \\ 1 & 3 & 2 \\ \hline 9 & 15 & g \end{array}\right]$

(b) Maximize $f = 9x_1 + 15x_2$ subject to
$x_1 + x_2 \le 5$
$x_1 + 3x_2 \le 2$

5. (a) $y_1 = 8, y_2 = 2, y_3 = 0$; min: $g = 252$
(b) $x_1 = 5, x_2 = 0, x_3 = 9$; max $f = 252$

7. $g = 5$ at $y_1 = 0, y_2 = 5$; $f = 5$ at $x_1 = 1/2, x_2 = 0$

9. $y_1 = 0, y_2 = 9$; $g = 18$ (min); $x_1 = 2, x_2 = 0$; $f = 18$ (max)

11. Maximize $f = 11x_1 + 12x_2 + 6x_3$
subject to $4x_1 + 3x_2 + 3x_3 \le 3$
$x_1 + 2x_2 + x_3 \le 1$
Primal: $y_1 = 2, y_2 = 3$; $g = 9$ (min)
Dual: $x_1 = 3/5, x_2 = 1/5, x_3 = 0$; $f = 9$ (max)

13. Maximize $f = x_1 + 3x_2 + x_3$
subject to $x_1 + 4x_2 \le 12$
$3x_1 + 6x_2 + 4x_3 \le 48$
$x_2 + x_3 \le 8$
$y_1 = \frac{2}{5}, y_2 = \frac{1}{5}, y_3 = \frac{1}{5}$; $g = 16$ (min)
Dual: $x_1 = 4, x_2 = 2, x_3 = 6$; $f = 16$ (max)

15. min = 32 at $x = 16, y = 0$

17. min = 28 at $x = 2, y = 0, z = 1$

19. min = 480 at $y_1 = 0, y_2 = 0, y_3 = 16$

21. (a) Minimize $g = 120y_1 + 50y_2$
subject to $3y_1 + y_2 \ge 40$
$2y_1 + y_2 \ge 20$

(b) Primal: $x_1 = 40, x_2 = 0$; $f = 1600$ (max)
Dual: $y_1 = 40/3, y_2 = 0$; $g = 1600$ (min)

23. Line 1 for 4 hours, Line 2 for 1 hour; \$1200

25. $A = 12$ weeks, $B = 0$ weeks, $C = 0$ weeks; cost = \$12,000

27. Factory 1: 50 days, Factory 2: 0 days. Minimum cost \$500,000.

29. 105 minutes on radio, nothing on TV. Minimum cost \$10,500.

31. 16 oz of Food I, 0 oz of Food II, 0 oz of Food III, Minimum cost = \$16.

Exercise 4.6 (page 363)

1. $-3x + y \le -5$ **3.** $-6x - y \le -40$

5. $\left[\begin{array}{rrrrr|r} 1 & 2 & 1 & 0 & 0 & 6 \\ -4 & -2 & 0 & 1 & 0 & -12 \\ \hline -4 & -5 & 0 & 0 & 1 & 0 \end{array}\right]$

7. $\left[\begin{array}{rrrrr|r} 6 & 4 & 1 & 0 & 0 & 24 \\ -5 & -2 & 0 & 1 & 0 & -16 \\ \hline -2 & -2 & 0 & 0 & 1 & 0 \end{array}\right]$

9. (a) Maximize $f = 2x + 3y$ subject to
$7x + 4y \le 28$
$3x - y \le -2$
$x \ge 0, y \ge 0$

(b) $\left[\begin{array}{rrrrr|r} 7 & 4 & 1 & 0 & 0 & 28 \\ 3 & \textcircled{-1} & 0 & 1 & 0 & -2 \\ \hline -2 & -3 & 0 & 0 & 1 & 0 \end{array}\right]$

11. (a) Maximize $-g = -3x - 8y$ subject to
$4x - 5y \le 50$
$x + y \le 80$
$x - 2y \le -4$
$x \ge 0, y \ge 0$

(b) $\left[\begin{array}{rrrrrr|r} 4 & -5 & 1 & 0 & 0 & 0 & 50 \\ 1 & 1 & 0 & 1 & 0 & 0 & 80 \\ 1 & \textcircled{-2} & 0 & 0 & 1 & 0 & -4 \\ \hline 3 & 8 & 0 & 0 & 0 & 1 & 0 \end{array}\right]$

13. $x = 6, y = 8, z = 12$; $f = 120$

15. $x = 10, y = 17$; $f = 57$ **17.** $x = 5, y = 7$; $f = 31$

19. $x = 4, y = 8$; $f = 16$ **21.** $x = 12, y = 8$; $f = 76$

23. $x = 5, y = 15$; $f = 45$

25. $x = 10, y = 20$; $f = 120$

27. $x = 20, y = 10, z = 0$; $f = 40$

29. $x = 5, y = 0, z = 3$; $f = 22$

31. $x = 0, y = 44, z = 98$; $f = 34$

33. $x = 70, y = 0, z = 40$; $f = 2100$

35. Produce 200 of each at Monaca; produce 300 commercial components and 550 domestic furnaces at Hamburg; profit = \$355,250

37. Produce 200 of each at Monaca; produce 300 commercial components and 550 domestic furnaces at Hamburg; cost = \$337,750
39. regular = 2000 lb; beef = 880 lb; profit = \$1328
41. I = 3 million, II = 0, III = 3 million; cost = \$180,000

Chapter 4 Review Exercises *(page 367)*

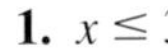

1. $x \leq 3$

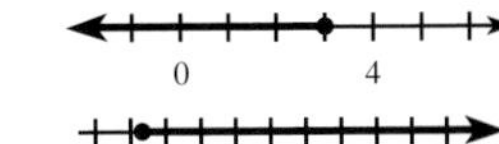

2. $x \geq -20/3$

3. $x \geq -15/13$

4. $x \leq -3/2$

5. **(a)** closed; $[0, 5]$ **(b)** half-open; $[3, 7)$
(c) open; $(-3, 2)$
6. **(a)** $-1 < x < 16$ **(b)** $-12 < x \leq 8$
(c) $x < -1$

7.

8.

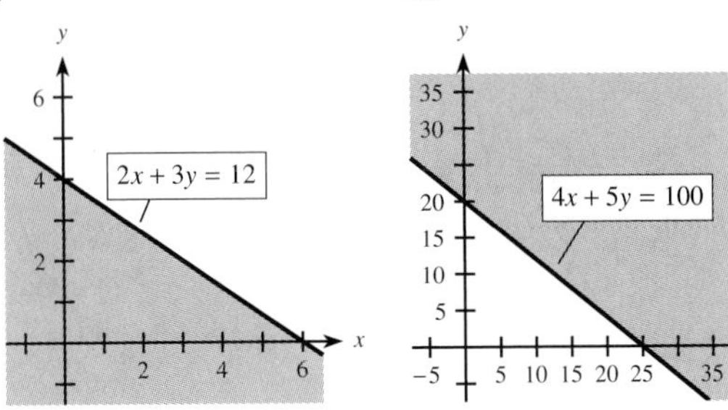

9.

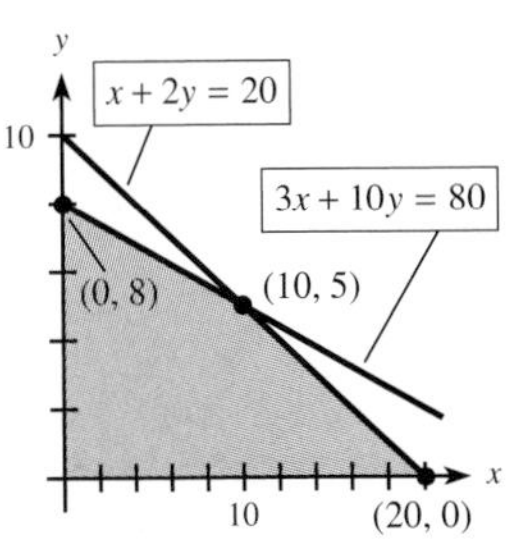

10.

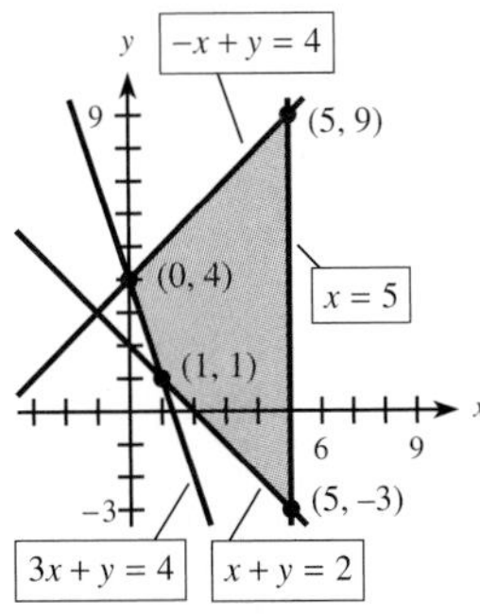

11. max = 25 at (5, 10); min = −12 at (12, 0)
12. max = 194 at (17, 23); min = 104 at (8, 14)
13. $f = 66$ at (6, 6) **14.** $f = 91$ at (9, 2)
15. $f = 43$ at (7, 9) **16.** $f = 31$ at (12, 7)
17. $g = 24$ at (3, 3) **18.** $g = 20$ at (5, 1)
19. $g = \frac{247}{3}$ at $(\frac{212}{3}, \frac{7}{3})$ **20.** $g = 80$ at (5, 45)
21. $f = 168$ at (12, 7) **22.** $f = 260$ at (60, 20)
23. $f = 360$ at (40, 30) **24.** $f = 270$ at (5, 3, 2)
25. $f = 640$ on the line between (160, 0) and (90, 70)
26. no solution **27.** $g = 32$ at $y_1 = 2, y_2 = 3$
28. $g = 20$ at $y_1 = 4, y_2 = 2$
29. $g = 7$ at $y_1 = 1, y_2 = 5$
30. $g = 16$ at $y_1 = \frac{2}{5}, y_2 = \frac{1}{5}, y_3 = \frac{1}{5}$
31. $f = 165$ at $x = 20, y = 21$
32. $f = 54$ at $x = 6, y = 5$
33. $f = 156$ at $x = 15, y = 2$
34. $f = 31$ at $x = 4, y = 5$
35. $p = \$14,750$ when 110 large and 75 small swingsets are made
36. $C = \$300,000$ when #1 operates 30 days, #2 operates 25 days
37. $P = \$320$; I = 40, II = 20
38. $P = \$420$; Jacob's ladders = 90, locomotives = 30
39. food I = 0 oz, food II = 3 oz; $C = \$0.60$ (min)
40. Cost = \$5.60; $A = 40$ lb, $B = 0$ lb
41. Cost = \$8500; $A = 20$ days, $B = 15$ days, $C = 0$ days
42. pancake mix = 8000 lb; cake mix = 3000 lb; profit = \$3550
43. Texas: 55 desks, 65 computer tables; Louisiana: 75 desks, 65 computer tables; cost = \$4245

Chapter 4 Test *(page 370)*

1. $t \geq -9$

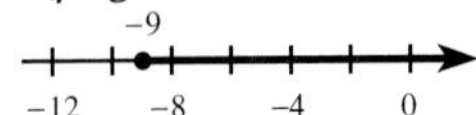

2. $-1 < x \leq 4$
3. **(a)**

(b)

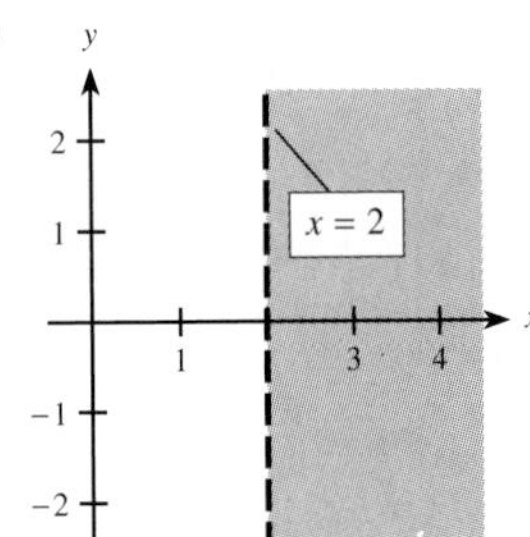

4. **(a)**

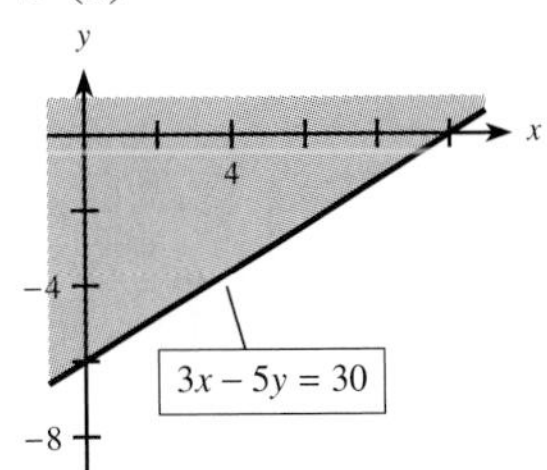

(b)

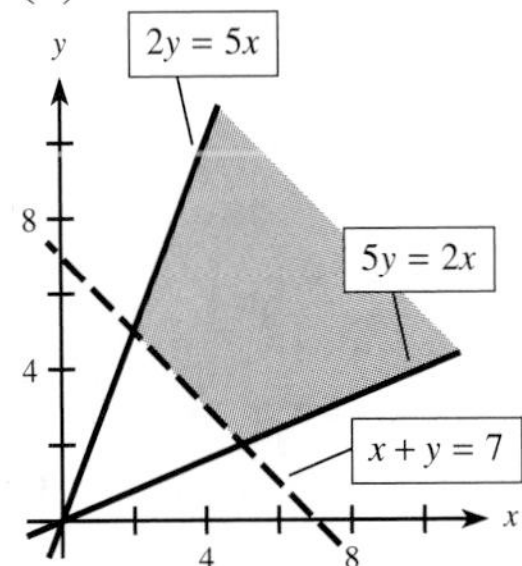

5. Max = 120 at (0, 24) **6.** Min = 136 at (28, 52)
7. Max = 6300 at $x = 90$, $y = 0$

8. (a) C; $\left[\begin{array}{ccccccc|c} 1 & 2 & 0 & 1 & 0 & -3/2 & 0 & 40 \\ 0 & \textcircled{1} & 0 & -2 & 1 & 1/2 & 0 & 15 \\ 0 & 3 & 1 & -1 & 0 & 1/4 & 0 & 60 \\ \hline 0 & 0 & 0 & 4 & 0 & 6 & 1 & 220 \end{array}\right]$

$-2R_2 + \text{to } R_1 \quad -3R_2 + \text{to } R_3$

(b) A; Pivot column is column 3, but new pivot is undefined.

9. Maximize $f = 100x_1 + 120x_2$
subject to $3x_1 + 4x_2 \le 2$
$5x_1 + 6x_2 \le 3$
$x_1 + 3x_2 \le 5$
$x_1 \ge 0, x_2 \ge 0$

10. Maximize $-g = -7x - 3y$
subject to $x - 4y \le -4$
$x - y \le 5$
$2x + 3y \le 30$

11. Max: $x_1 = 17, x_2 = 15, x_3 = 0; f = 658$ (max)
Min: $y_1 = 4, y_2 = 18, y_3 = 0; g = 658$ (min)

12. Maximize $P = 7x + 6y$ $\quad x$ = barrels of lager;
subject to $3x + 2y \le 1200$ $\quad y$ = barrels of ale
$2x + 2y \le 1000$
$P = \$3200$ (max) at $x = 200$, $y = 300$

13. x = number of day calls, y = number of evening calls
Minimize $C = 3x + 4y$
subject to $0.3x + 0.3y \ge 150$
$0.1x + 0.3y \ge 120$
$C = \$1850$ (min) at $x = 150$, $y = 350$

Exercise 5.1 *(page 389)*

1. 3.162278 **3.** 0.01296525 **5.** 1.44225
7. 7.3891 **9.**

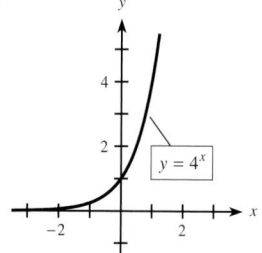

11. **13.**

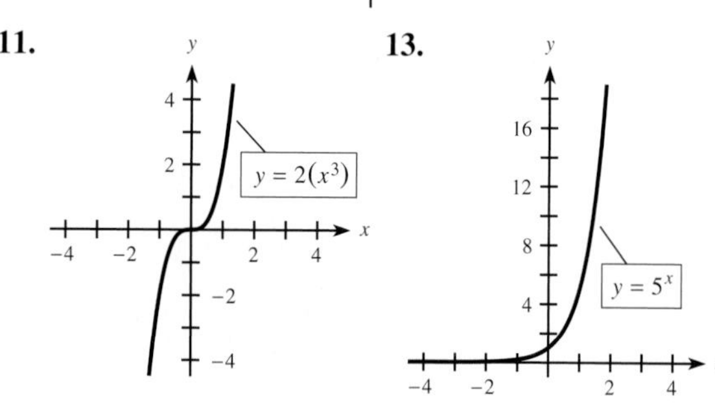

15. **17.**

19. All graphs have the same basic shape. For larger positive k, the graphs fall more sharply. For positive k nearer 0, the graphs fall more slowly.

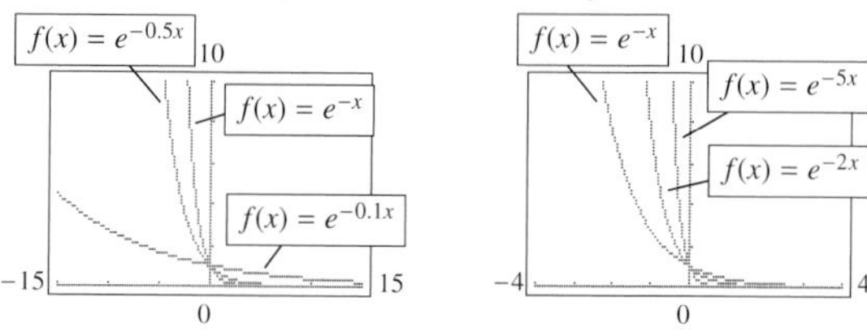

21. $y = f(x) + C$ is the same graph as $y = f(x)$ but shifted C units on the y-axis.

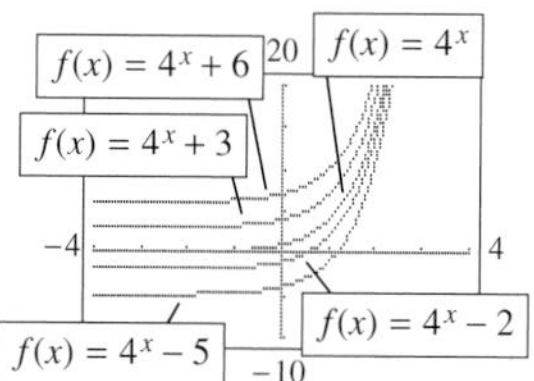

23. (a)

(b) As c changes, the y-intercept and the asymptote change.

25. (a)

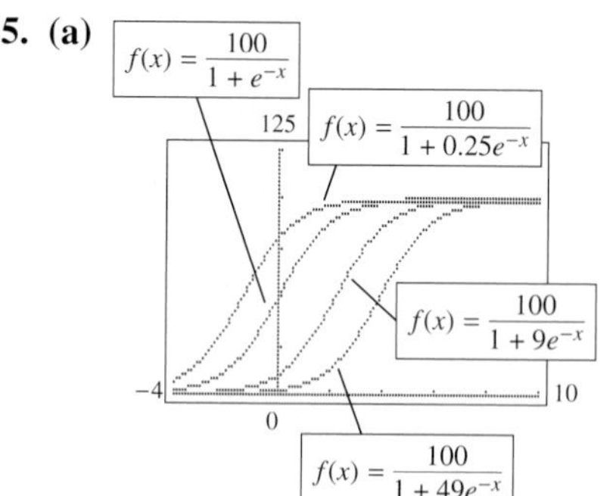

(b) Different c values change the y-intercept and how the graph approaches the asymptote.

27. \$1884.54

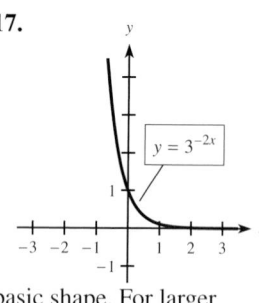

29.

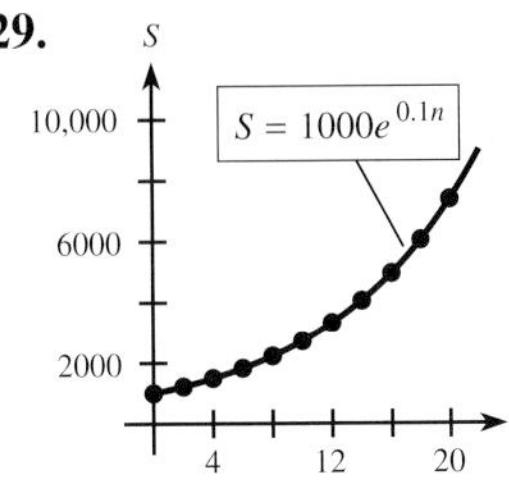

31.

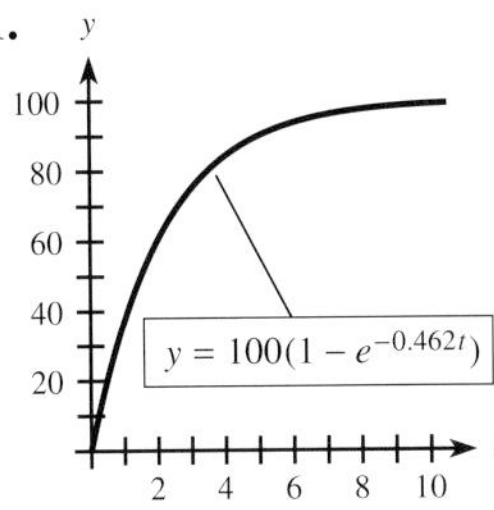

33.

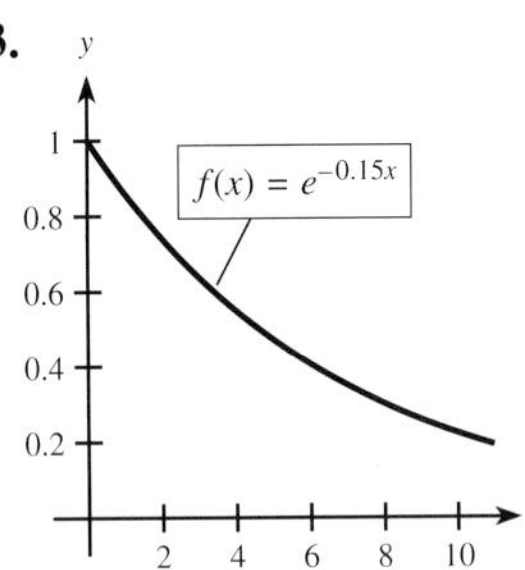

35.

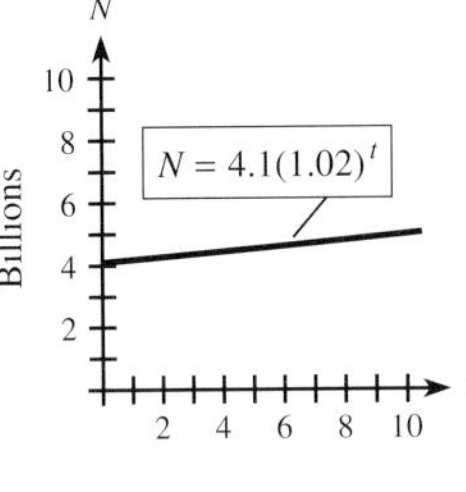

37.

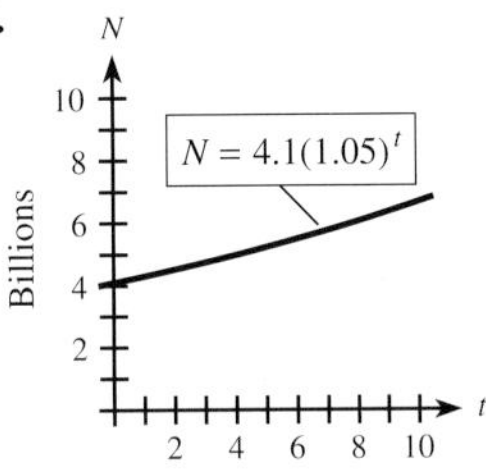

39.

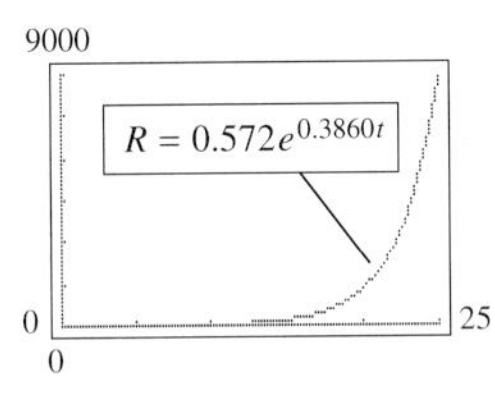

41. The linear model fails in 2007, giving a negative number of processors.

43. **(a)** $y = 6.869(1.0851)^x$ **(b)** \$52.92

45. **(a)** $y = 2.366(1.04457)^x$ **(b)** 230.39 **(c)** $x = 101.76$, 2001

47. $y = 0.1018(1.06443)^x$

Exercise 5.2 *(page 402)*

1. $2^4 = 16$ **3.** $4^{1/2} = 2$ **5.** $x = 8$ **7.** $x = \frac{1}{2}$ **9.** $\log_2 32 = 5$ **11.** $\log_4(\frac{1}{4}) = -1$

13.

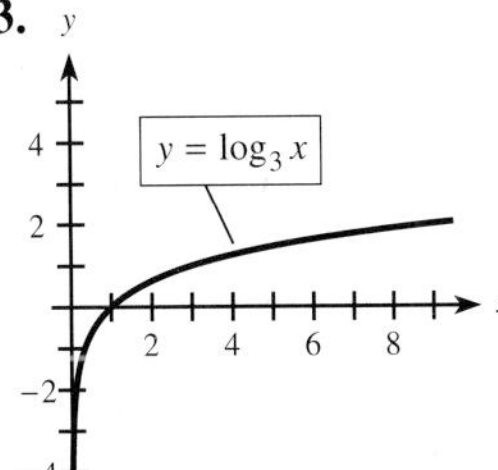

15.

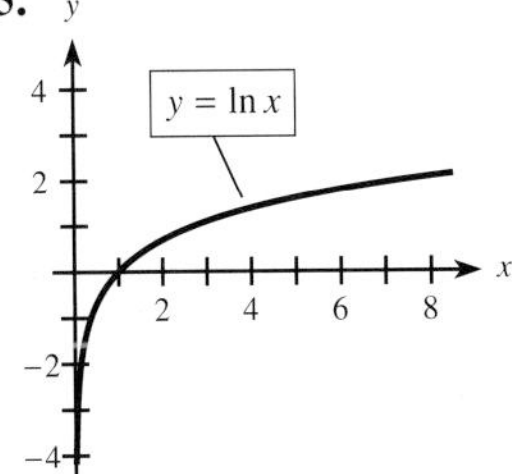

17.

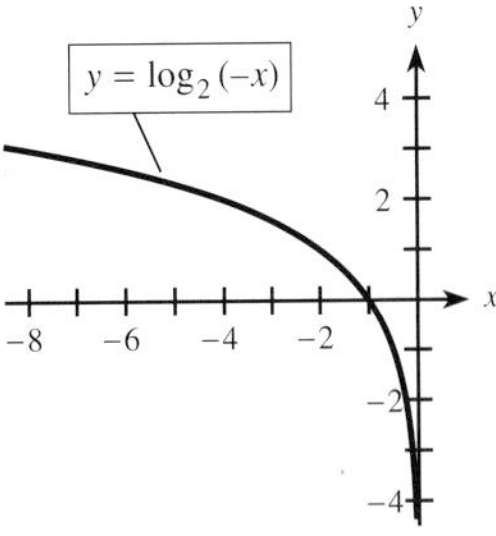

19. **(a)** 3 **(b)** -1

21. x **23.** 3

25. **(a)** 4.9 **(b)** 0.4 **(c)** 12.4 **(d)** 0.9

27. $\log x - \log(x + 1)$ **29.** $\log_7 x + \frac{1}{3}\log_7(x + 4)$

31. $\ln(x/y)$ **33.** $\log_5[x^{1/2}(x + 1)]$

35. equivalent; Properties V and III

37. not equivalent; $\log(\sqrt[3]{8}/5)$

39. **(a)**

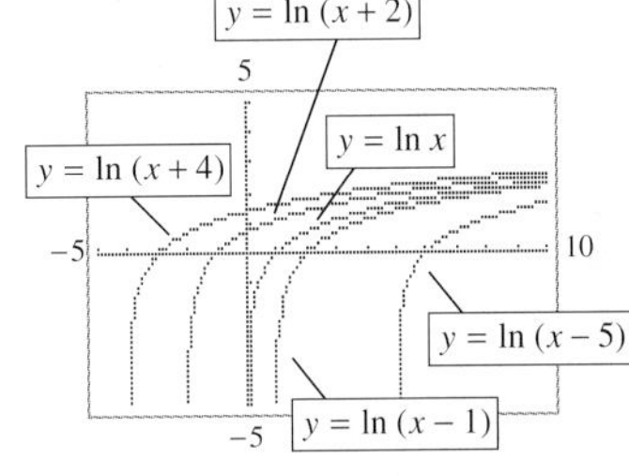

(b) For each c, the domain is $x > c$ and the vertical asymptote is at $x = c$.

(c) Each x-intercept is at $x = c + 1$.

(d) The graph of $y = f(x - c)$ is the graph of $y = f(x)$ shifted c units on the x-axis.

41. **(a)** 4.0875 **(b)** -0.1544

43.

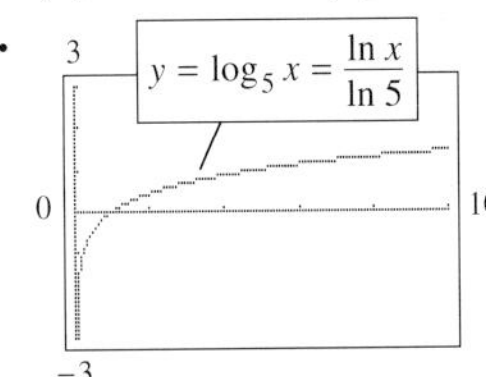

45.

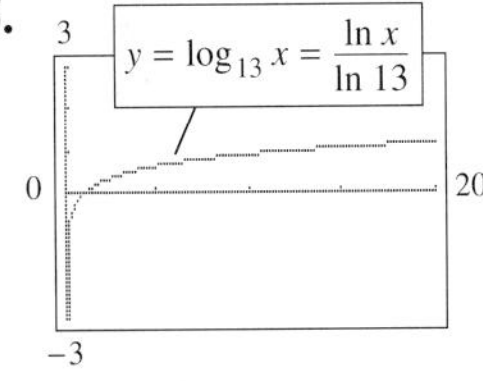

47. If $\log_a M = u$ and $\log_a N = v$, then $a^u = M$ and $a^v = N$. Therefore, $\log_a(M/N) = \log_a(a^u/a^v) = \log_a(a^{u-v}) = u - v = \log_a M - \log_a N$.

49. $y = 4.2885e^{-0.05751x}$ **51.** 2.5 times as severe

53. 14 times as severe **55.** 40

57. $L = 10\log(I/I_0)$

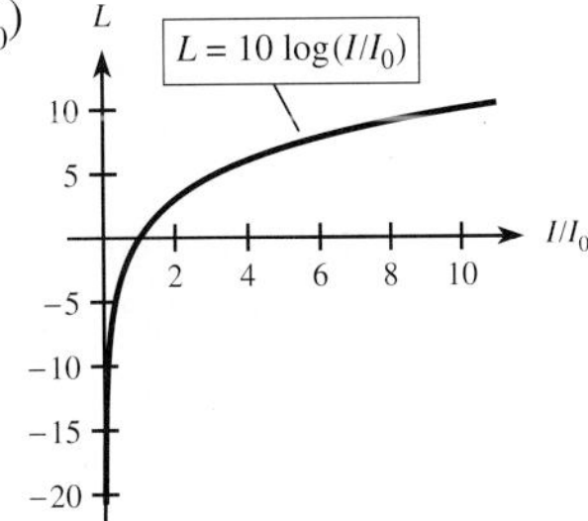

59. 0.1 and 1×10^{-14}

61. $\text{pH} = \log \dfrac{1}{[\text{H}^+]} = \log 1 - \log[\text{H}^+] = -\log[\text{H}^+]$

63.

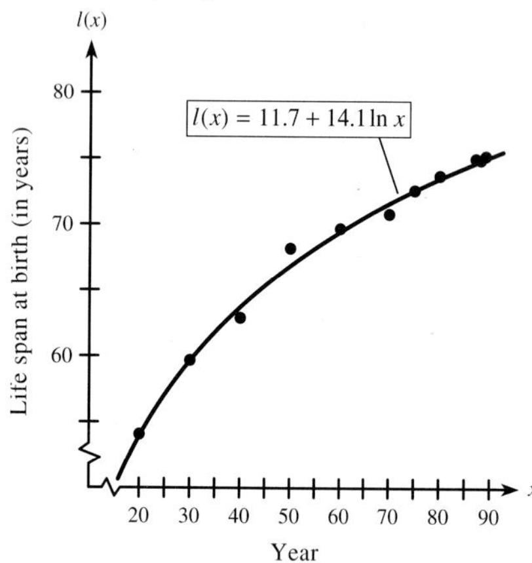

Very similar.

65. **(a)** $y = -22.88 + 320.88 \ln x$ **(b)** 1010

Exercise 5.3 *(page 415)*

1. **(a)** 2038 **(b)** 4.9 months

3. **(a)** 69% **(b)** 37,204 years (approximately)

5. 24.5 years **7.** 128,402

9. **(a)** 600 **(b)** 2119 **(c)** 3000

(d)

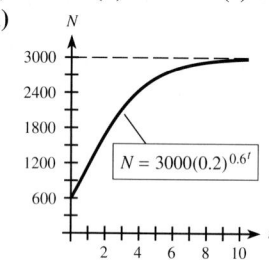

11. **(a)** 10 **(b)** 2.5 years

13. **(a)** 37 **(b)** 1.5 hours **15.** **(a)** 52 **(b)** tenth

17. **(a)** \$4.98 **(b)** 8 **19.** \$502 **21.** \$420.09

23. \$2706.71 **25.** **(a)** \$10,100.31 **(b)** 6.03 years

27. **(a)** \$5469.03

(b) 7 years, 9 months (approximately)

29. **(a)** 2% **(b)** 20 months ($x = 19.6$)

31. **(a)** 44.7 years (approximately)

(b) The intent of such a plan would be to reduce future increases in health care expenditures. A new model might not be exponential, or, if it were, it would be one that rose more gently.

33. **(a)** 18.3 yrs (approx.) **(b)** \$42,340.00

35. **(a)** $\log_{1.005}(2) = t$

(b) 139 months (approximately)

37. **(a)** 0.23 km^3 **(b)** 5.9 years

39. **(a)** 0 lb **(b)** 29.95 lb **(c)** 0.21 min

41. **(a)** $x = 16$; 0.2% **(b)** 10.56 min **(c)** 0.06%

Chapter 5 Review Exercises *(page 419)*

1. **(a)** $\log_2 y = x$ **(b)** $\log_3 2x = y$

2. **(a)** $7^{-2} = \frac{1}{49}$ **(b)** $4^{-1} = x$

3.

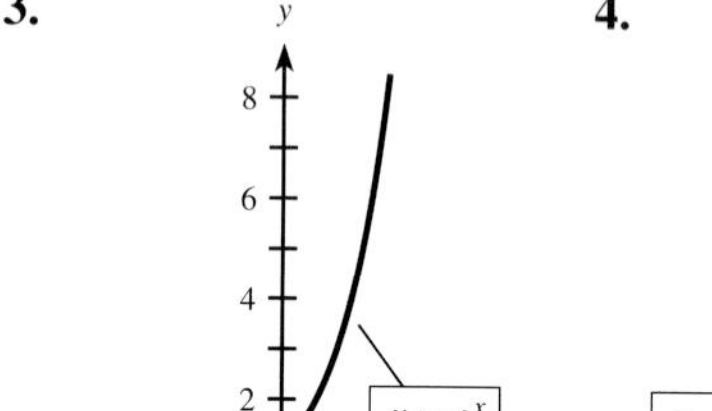

4.

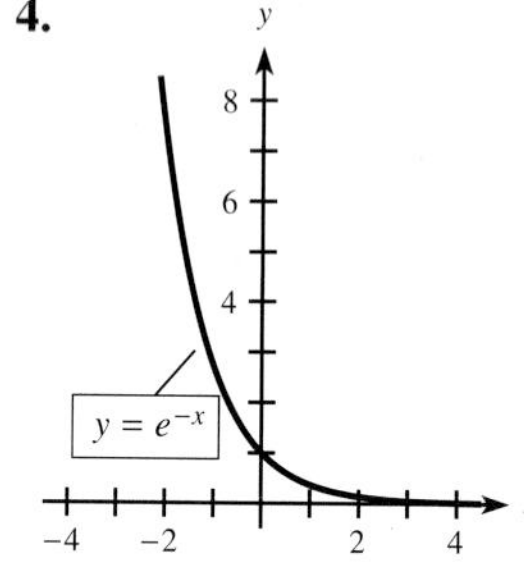

5.

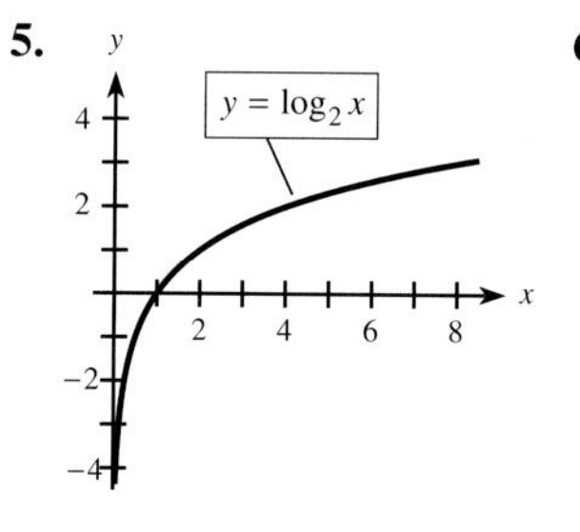

6.

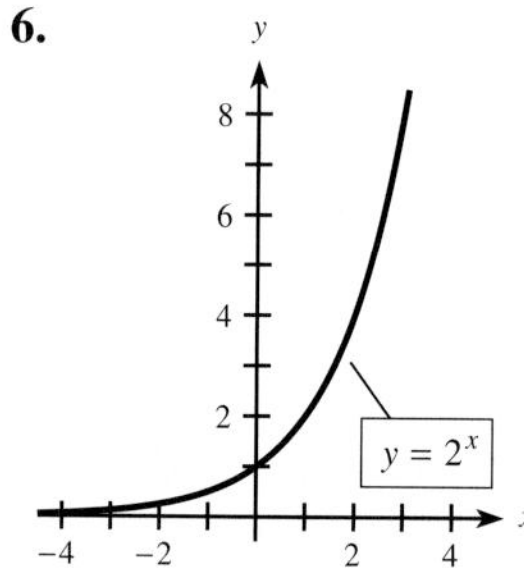

7.

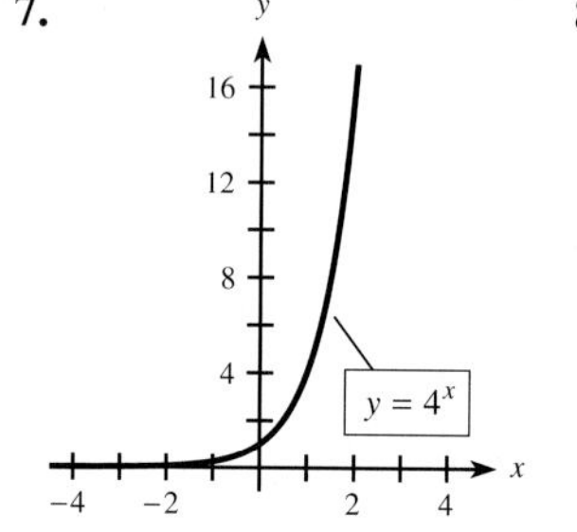

8.

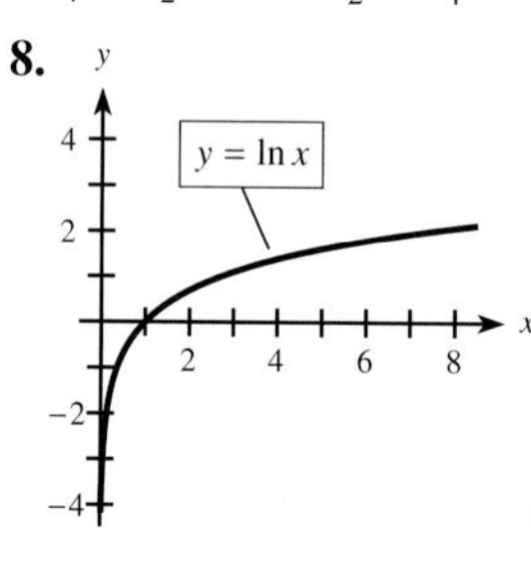

9. **10.**

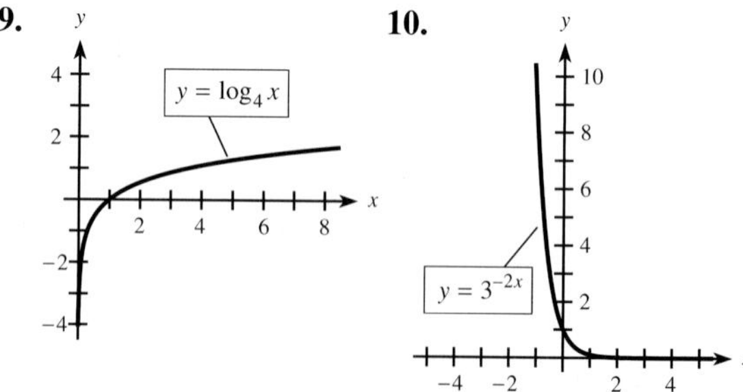

11.

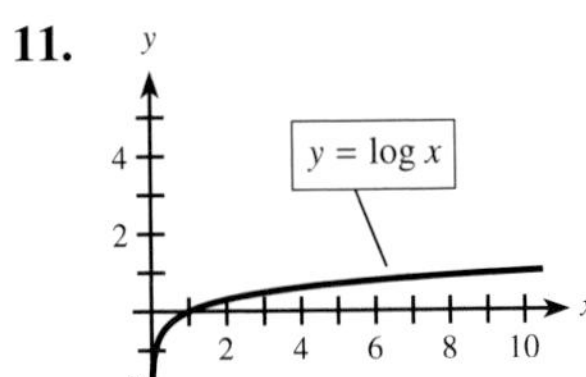

12.

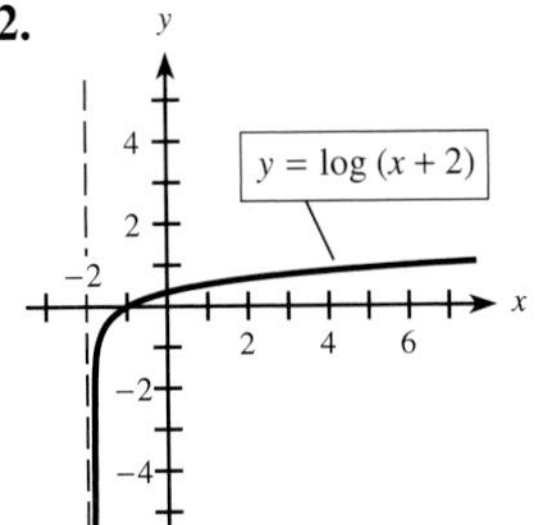

13.

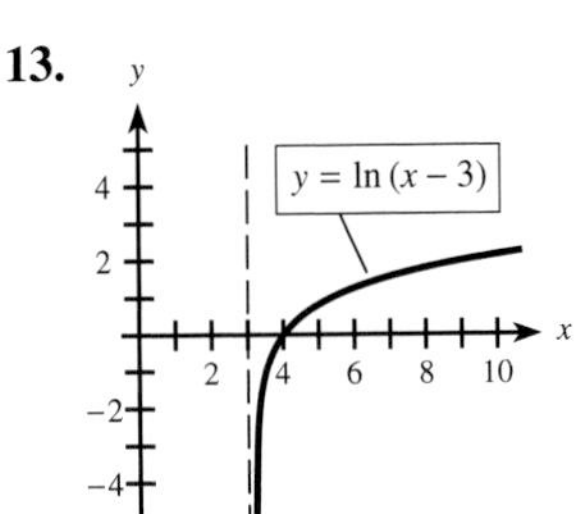

14.

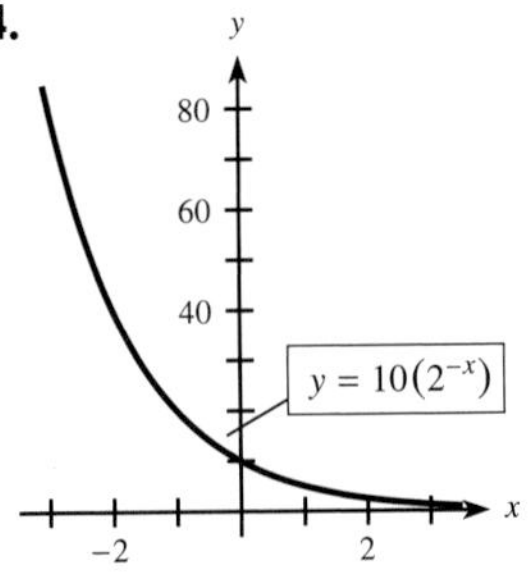

15. 0 **16.** 2 **17.** $\frac{1}{2}$ **18.** -1 **19.** 8
20. 1 **21.** 5 **22.** 3.15 **23.** -2.7
24. 0.6 **25.** 5.1 **26.** 15.6 **27.** $\log y + \log z$
28. $\frac{1}{2}\ln(x+1) - \frac{1}{2}\ln x$ **29.** no
30. -2 **31.** 5 **32.** 1 **33.** 0
34. 3.4939 **35.** -1.5845

36.

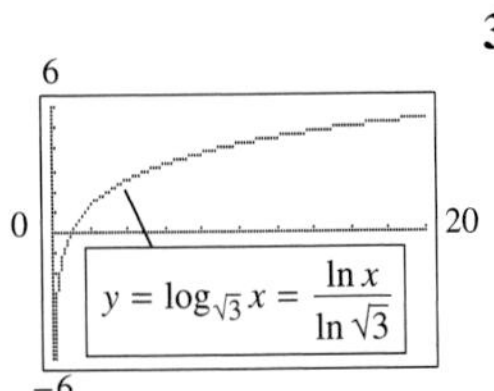

37.

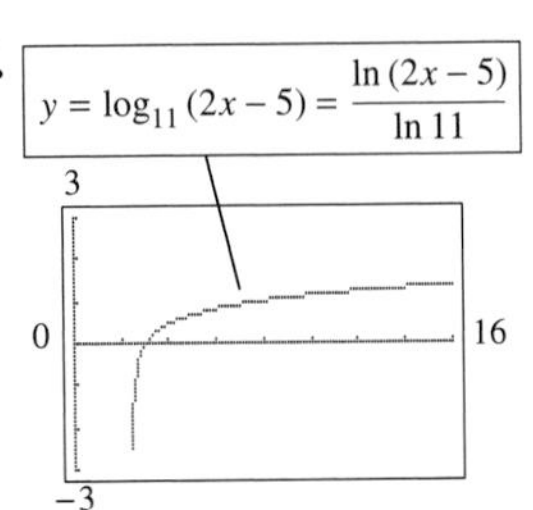

38. Growth exponential because the general outline has the same shape as a growth exponential.

39. (a)

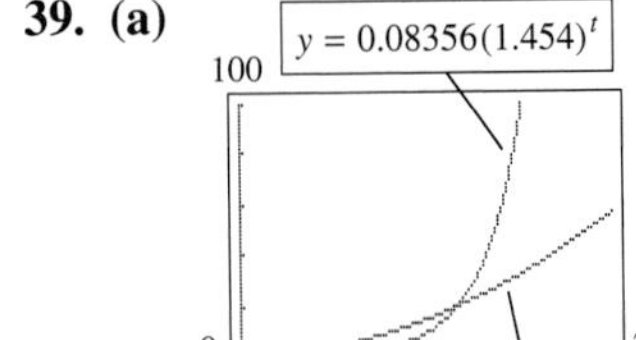

(b) The function $y = 0.099t^{1.969}$

40. Exponential because the general shape is similar to the graph of a growth exponential.
41. (a) $y = 604.9211\,(1.082)^x$, $x = 0$ in 1900
(b) 2,374,440
42. (a) $y = 18.1607 + 5.1865\ln(x)$
(b) 38.4% **(c)** 2017 ($x = 67$)
43. (a) -3.9 **(b)** $0.14B_0$ **(c)** $0.004B_0$ **(d)** yes
44. (a) 3000 **(b)** 8603 **(c)** 10,000
45. (a) 27,441 **(b)** 12 weeks
46. 1366 **47.** 5.8 years
48. (a) \$5532.77 **(b)** 5.13 years

Chapter 5 Test *(page 422)*

1. **2.**

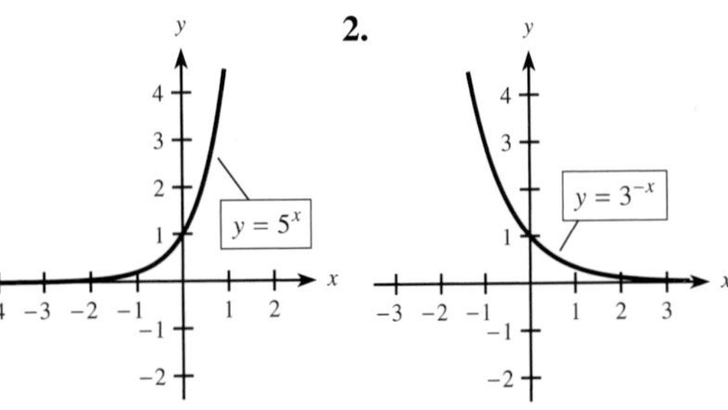

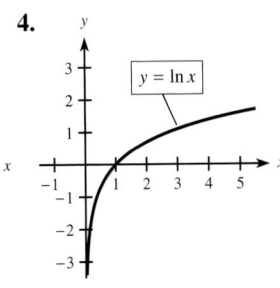

3. **4.**

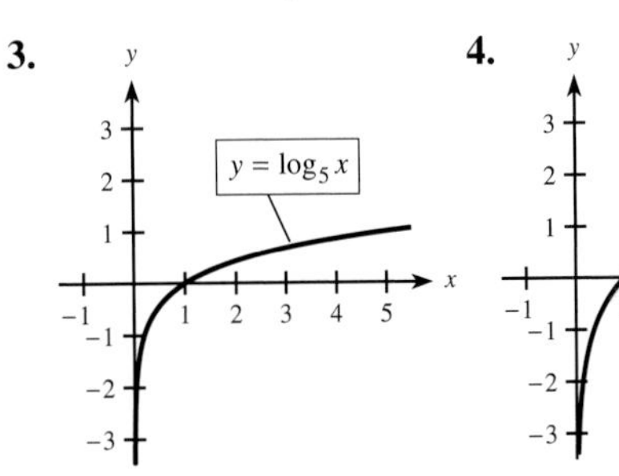

5. **6.**

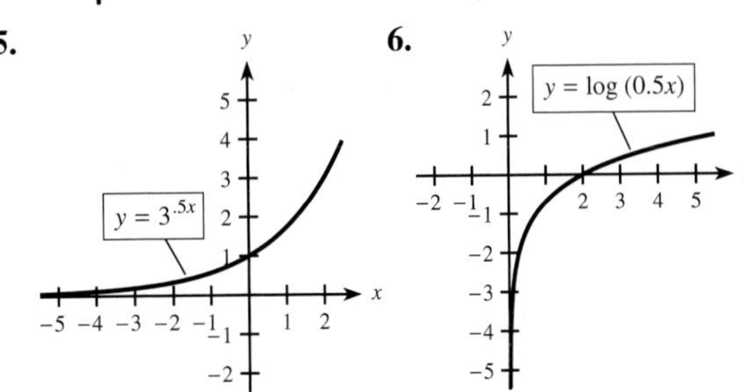

7. **8.**

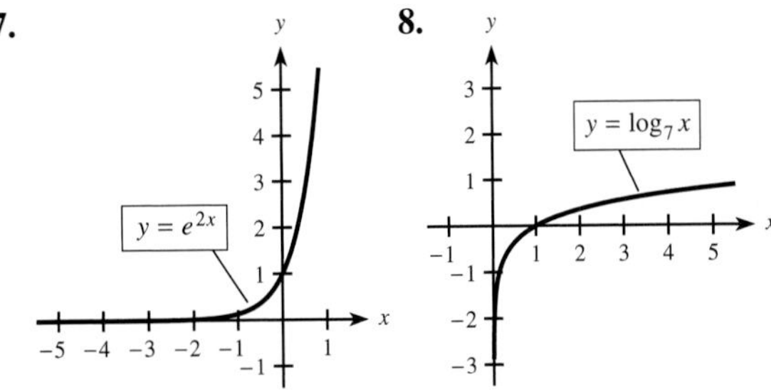

9. 54.59815 **10.** 0.122456 **11.** 1.38629
12. 3.04452 **13.** $x = 17^{3.1} \approx 6522.163$
14. $\log_3(27) = 2x$, so $x = \frac{3}{2}$ **15.** 3
16. x^4 **17.** 3 **18.** x^2
19. $\ln(M) + \ln(N)$ **20.** $\ln(x^3 - 1) - \ln(x + 2)$
21. $\dfrac{\ln(x^3+1)}{\ln 4} \approx 0.721\ln(x^3+1)$
22. A decay exponential

23. A growth exponential **24.** During 2103
25. **(a)** $y = 2227.8035\ (3.0489)^x$
(b) Model predicts attendance for 2001 will be about 471,468,000 (to the nearest thousand). It seems unlikely that this many people would attend one rally.
26. $y = 0.06626\ (1.126724)^x$
27. $y = 0.080073\ (1.12401)^x$; The models are quite similar.

Exercise 6.1 *(page 437)*

1. **(a)** \$9600 **(b)** \$19,600
3. **(a)** \$30 **(b)** \$1030
5. \$864 **7.** \$3850 **9.** 13%
11. **(a)** 6.5% **(b)** 5.65% **13.** \$1631.07
15. \$12,000 **17.** 10 years **19.** pay on time
21. **(a)** \$2120 **(b)** \$2068.29 (nearest cent)
23. 3, 6, 9, 12, 15, 18, 21, 24, 27, 30
25. $\frac{1}{3}, \frac{2}{3}, 1, \frac{4}{3}, \frac{5}{3}, 2, \frac{7}{3}, \frac{8}{3}$ **27.** $-\frac{1}{4}, \frac{1}{8}, -\frac{1}{12}, \frac{1}{16}, \frac{-1}{20}, \frac{1}{24}$
29. $-\frac{1}{3}, \frac{1}{5}, \frac{-1}{7}, \frac{1}{9}, \frac{-1}{11}, \frac{1}{13}$ **31.** $-1, -\frac{1}{4}, -\frac{1}{15}, 0; a_{10} = \frac{1}{20}$
33. **(a)** $d = 3, a_1 = 2$ **(b)** 11, 14, 17
35. **(a)** $d = \frac{3}{2}, a_1 = 3$ **(b)** $\frac{15}{2}, 9, \frac{21}{2}$ **37.** 105
39. −35 **41.** 203 **43.** 2185 **45.** 1907.5
47. 2550 **49.** −15,862.5 **51.** 21, 34, 55
53. \$2400 **55.** the job starting at \$20,000
57. **(a)** \$3000 **(b)** \$4500 **(c)** Plan II, by \$1500
(d) \$10,000 **(e)** \$13,500 **(f)** Plan II, by \$3500
(g) Plan II

Exercise 6.2 *(page 452)*

(Minor differences may occur because of rounding.)
1. **(a)** 6% **(b)** 6 **(c)** 3% = 0.03 **(d)** 12
3. **(a)** 8% **(b)** 7 **(c)** 2% = 0.02 **(d)** 28
5. **(a)** 9% **(b)** 5 **(c)** $(\frac{9}{12})\% = .0075$ **(d)** 60
7. \$24,846.78 **9.** \$3086.45 **11.** \$4755.03
13. \$6844.36 **15.** \$6661.46 **17.** \$5583.95
19. \$7309.98 **21.** \$502.47
23. **(a)** \$12,245.64 **(b)** \$11,080.32
(c) A $\frac{1}{2}$% increase in the interest rate reduces the amount required by \$1165.32.
25. \$50.26 more at 8% **27.** 8.67% **29.** 7.55%
31. 6.18%
33. 8% compounded monthly, 8% compounded quarterly, 8% compounded annually
35. The higher graph is for continuous compounding because its yield (its effective annual rate) is higher.
37. 26.8% **39.** 3 years **41.** 4% **43.** \$3996.02
45. **(a)** \$63,128.75 **(b)** \$14,263.10
47. 5.12 years (approximately) **49.** \$13,916.24
51.

	A	B	C
1		Future Value	(Yearly)
2	End of Year	Quarterly	Monthly
3	0	\$5000.00	\$5000.00
4	1	\$5322.52	\$5324.26
5	2	\$5665.84	\$5669.54
6	3	\$6031.31	\$6037.22
7	4	\$6420.36	\$6428.74
8	5	\$6834.50	\$6845.65
9	6	\$7275.35	\$7289.60
10	7	\$7744.64	\$7762.34
11	8	\$8244.20	\$8265.74
12	9	\$8775.99	\$8801.79
13	10	\$9342.07	\$9372.59

(a) From quarterly and monthly spreadsheets: after $6\frac{1}{2}$ years (26 quarters or 78 months)
(b) See the spreadsheet.
53. **(a)** 24, 48, 96 **(b)** 24, 16, $\frac{32}{3}$
55. 40,960 **57.** $4 \cdot (\frac{3}{2})^{15}$ **59.** $\frac{6(1 - 3^{17})}{-2}$
61. $\frac{8}{3}[1 + (\frac{1}{2})^{21}]$ **63.** $\frac{3^{35} - 1}{2}$ **65.** $18[1 - (\frac{2}{3})^{18}]$
67. \$350,580 (approx.) **69.** 24.4 million (approx.)
71. 35 years **73.** 40.5 ft **75.** \$4096
77. 320,000 **79.** \$7,231,366 **81.** \$1801.14
83. 15,625 **85.** 305,175,780

Exercise 6.3 *(page 464)*

1. $3\frac{1}{2}\% = 0.035$ for 22 periods
3. $\frac{1}{4}\% = 0.0025$ for 60 periods
5. **(a)** The higher graph is \$1120 per year.
(b) \$8772.71
7. \$7328.22 **9.** \$1072.97 **11.** \$77,313.47
13. \$4774.55 **15.** \$4372.20 **17.** \$226.10
19. \$1083.40 **21.** \$265.25 **23.** \$741.47
25. A sinking fund is a savings plan, so the 10% rate (a) is better.
27. \$53,677.40 **29.** \$4651.61 **31.** \$1180.78
33. \$4152.32 **35.** \$26,517.13 **37.** \$3787.92
39. \$235.16

41. The spreadsheet shows the amount at the end of each of the first 12 months and the last 12 months. Amount after 10 years is shown.

	A	B	C
1		Future Value	
2	End of Month	Ordinary Ann.	Annuity Due
3	0	0	100
4	1	$100.00	$100.60
5	2	$200.60	$201.80
6	3	$301.80	$303.61
7	4	$403.61	$406.04
8	5	$506.04	$509.07
9	6	$609.07	$612.73
10	7	$712.73	$717.00
11	8	$817.00	$821.91
12	9	$921.91	$927.44
13	10	$1027.44	$1033.60
14	11	$1133.60	$1140.40
15	12	$1240.40	$1247.85
	⋮	⋮	⋮
112	109	$15324.39	$15416.34
113	110	$15516.34	$15609.44
114	111	$15709.44	$15803.70
115	112	$15903.70	$15999.12
116	113	$16099.12	$16195.71
117	114	$16295.71	$16393.49
118	115	$16493.49	$16592.45
119	116	$16692.45	$16792.60
120	117	$16892.60	$16993.96
121	118	$17093.96	$17196.52
122	119	$17296.52	$17400.30
123	120	$17500.30	$17605.30

(b) $12,000

(c) Annuity due. Each payment for an annuity due earns 1 month's interest more than that for an ordinary annuity.

Exercise 6.4 (page 475)

1. $976.32 **3.** $69,913.77 **5.** $4595.46
7. $5541.23 **9.** $27,590.62 **11.** $2,128,391

13. (a) $69,552.35
(b) $1045.23 per annual payment; $10,452.30 over 10 years

15. (a) The higher graph corresponds to 8%.
(b) $1500 (approximately)
(c) With an interest rate of 10%, a present value of about $9000 is needed to purchase an annuity of $1000 for 25 years. If the interest rate is 8%, about $10,500 is needed.

17. (a) $30,078.99 **(b)** $16,900
(c) $607.02 **(d)** $36,421.20

19. Ordinary annuity—payments at the end of each period
Annuity due—payments at the beginning of each period

21. $69,632.02 **23.** $445,962.23 **25.** $316,803.61
27. $2145.59 **29.** $146,235.06 **31.** $22,663.74
33. $7957.86 **35.** $74,993.20 **37.** $19,922.97
39. $1317.98 **41.** $85,804.29

43. (a) The spreadsheet shows the payments for first 12 months and the last 12 months. Full payments for $13\frac{1}{2}$ years.

	A	B	C	D
1	End of Month	Acct. Value	Payment	New Balance
2	0	$100000.00	$0.00	$100000.00
3	1	$100650.00	$1000.00	$99650.00
4	2	$100297.73	$1000.00	$99297.73
5	3	$99943.16	$1000.00	$98943.16
6	4	$99586.29	$1000.00	$98586.29
7	5	$99227.10	$1000.00	$98227.10
8	6	$98865.58	$1000.00	$97865.58
9	7	$98501.70	$1000.00	$97501.70
10	8	$98135.47	$1000.00	$97135.47
11	9	$97766.85	$1000.00	$96766.85
12	10	$97395.83	$1000.00	$96395.83
13	11	$97022.40	$1000.00	$96022.40
14	12	$96646.55	$1000.00	$95646.55
	⋮	⋮	⋮	⋮
154	152	$10684.71	$1000.00	$9684.71
155	153	$9747.66	$1000.00	$8747.66
156	154	$8804.52	$1000.00	$7804.52
157	155	$7855.25	$1000.00	$6855.25
158	156	$6899.81	$1000.00	$5899.81
159	157	$5938.16	$1000.00	$4938.16
160	158	$4970.25	$1000.00	$3970.25
161	159	$3996.06	$1000.00	$2996.06
162	160	$3015.53	$1000.00	$2015.53
163	161	$2028.64	$1000.00	$1028.64
164	162	$1035.32	$1000.00	$35.32
165	163	$35.55	$35.55	$0.00

43. **(b)** The spreadsheet shows the payments for the first 12 months and the last 12 months. Full payments for almost 4 years.

	A	B	C	D
1	End of Month	Acct. Value	Payment	New Balance
2	0	$100000.00	$0.00	$100000.00
3	1	$100650.00	$2500.00	$98150.00
4	2	$98787.98	$2500.00	$96287.98
5	3	$96913.85	$2500.00	$94413.85
6	4	$95027.54	$2500.00	$92527.54
7	5	$93128.97	$2500.00	$90628.97
8	6	$91218.05	$2500.00	$88718.05
9	7	$89294.72	$2500.00	$86794.72
10	8	$87358.89	$2500.00	$84858.89
11	9	$85410.47	$2500.00	$82910.47
12	10	$83449.39	$2500.00	$80949.39
13	11	$81475.56	$2500.00	$78975.56
14	12	$79488.90	$2500.00	$76988.90
	⋮	⋮	⋮	⋮
38	36	$27734.95	$2500.00	$25234.95
39	37	$25398.98	$2500.00	$22898.98
40	38	$23047.83	$2500.00	$20547.83
41	39	$20681.39	$2500.00	$18181.39
42	40	$18299.57	$2500.00	$15799.57
43	41	$15902.26	$2500.00	$13402.26
44	42	$13489.38	$2500.00	$10989.38
45	43	$11060.81	$2500.00	$8560.81
46	44	$8616.45	$2500.00	$6116.45
47	45	$6156.21	$2500.00	$3656.21
48	46	$3679.98	$2500.00	$1179.98
49	47	$1187.65	$1187.65	$0.00

Exercise 6.5 *(page 485)*

1. **(a)** The 10-year loan because the loan must be paid more quickly.
(b) The 25-year loan because the loan is paid more slowly.

3. $2504.56 **5.** $1288.29

7.

Period	*Payment*	*Interest*
1	$39,505.50	$9000.00
2	39,505.50	6254.51
3	39,505.43	3261.92
	118,516.43	18,516.43

Period	*Balance Reduction*	*Unpaid Balance*
		$100,000.00
1	$30,505.50	69,494.50
2	33,250.99	36,243.51
3	36,243.51	0.00
	100,000.00	

9.

Period	*Payment*	*Interest*
1	$5380.54	$600.00
2	5380.54	456.58
3	5380.54	308.87
4	5380.54	156.71
	21,522.16	1522.16

Period	*Balance Reduction*	*Unpaid Balance*
		$20,000.00
1	$4780.54	15,219.46
2	4923.96	10,295.50
3	5071.67	5,223.83
4	5223.83	0.00
	20,000.00	

11. $8852.05 **13.** $5785.83
15. **(a)** $4359.23 **(b)** $87,184.60 **(c)** $37,184.60
17. **(a)** $1237.78 **(b)** $9902.24 **(c)** $1902.24
19. $2967.75 **21.** $62,473.28
23. **(a)** $1,239,676.52 **(b)** $1,270,768.38
25. **(a)** $89,120.53 **(b)** $6451.45
27. The line is the total amount paid ($644.30 per month × the number of months). The other curve is the total amount paid toward the principal.

29.

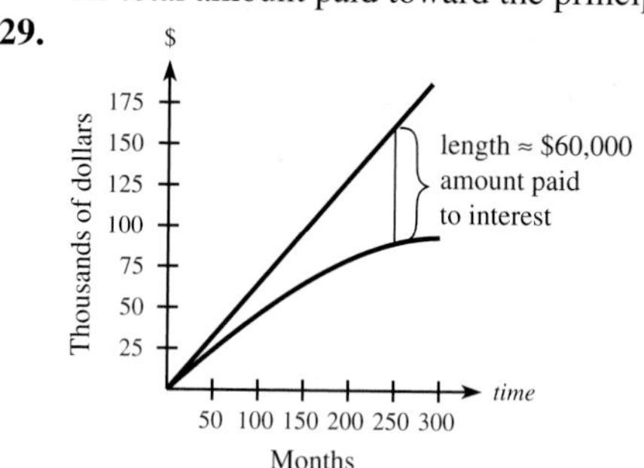

31.

	Payment	*Total Interest*
(a) 8%	$366.19	$2577.12
8.5%	$369.72	$2746.56
(b) 6.75%	$518.88	$106,796.80
7.25%	$545.74	$116,466.40

(c) The duration of the loan seems to have the greatest effect. It greatly influences payment size (for a $15,000 loan vs. one for $80,000), and it also affects total interest paid.

33.

	Payment	*Points*	*Total paid*
(a) (i)	$738.99	—	$221,697
(ii)	$722.81	$1000	$217,843
(iii)	$706.78	$2000	$214,034

(b) The 7% loan with 2 points.

35. The spreadsheet shows the amortization schedule for the first 12 and the last 12 payments.

	A	B	C	D	E
1	Period	Payment	Interest	Bal. Reduction	Unpaid Bal.
2	0				$16700.00
3	1	$409.27	$114.12	$295.15	$16404.85
4	2	$409.27	$112.10	$297.17	$16107.68
5	3	$409.27	$110.07	$299.20	$15808.48
6	4	$409.27	$108.02	$301.25	$15507.23
7	5	$409.27	$105.97	$303.30	$15203.93
8	6	$409.27	$103.89	$305.38	$14898.55
9	7	$409.27	$101.81	$307.46	$14591.09
10	8	$409.27	$99.71	$309.56	$14281.52
11	9	$409.27	$97.59	$311.68	$13969.84
12	10	$409.27	$95.46	$313.81	$13656.03
13	11	$409.27	$93.32	$315.95	$13340.08
14	12	$409.27	$91.16	$318.11	$13021.97
	⋮	⋮	⋮	⋮	⋮
39	37	$409.27	$32.11	$377.16	$4322.49
40	38	$409.27	$29.54	$379.73	$3942.75
41	39	$409.27	$26.94	$382.33	$3560.43
42	40	$409.27	$24.33	$384.94	$3175.49
43	41	$409.27	$21.70	$387.57	$2787.91
44	42	$409.27	$19.05	$390.22	$2397.70
45	43	$409.27	$16.38	$392.89	$2004.81
46	44	$409.27	$13.70	$395.57	$1609.24
47	45	$409.27	$11.00	$398.27	$1210.97
48	46	$409.27	$8.27	$401.00	$809.97
49	47	$409.27	$5.53	$403.74	$406.24
50	48	$409.02	$2.78	$406.24	$0.00

37. The spreadsheet shows the amortization schedule for the first 12 payments and the last 12. Paying an extra $15 takes 57 full payments (rather than 60) and a final payment of $44.07.

	A	B	C	D	E
1	Period	Payment	Interest	Bal. Reduction	Unpaid Bal.
2	0				$18000.00
3	1	$383.43	$126.00	$257.43	$17742.57
4	2	$383.43	$124.20	$259.23	$17483.34
5	3	$383.43	$122.38	$261.05	$17222.29
6	4	$383.43	$120.56	$262.87	$16959.42
7	5	$383.43	$118.72	$264.71	$16694.70
8	6	$383.43	$116.86	$266.57	$16428.14
9	7	$383.43	$115.00	$268.43	$16159.70
10	8	$383.43	$113.12	$270.31	$15889.39
11	9	$383.43	$111.23	$272.20	$15617.19
12	10	$383.43	$109.32	$274.11	$15343.08
13	11	$383.43	$107.40	$276.03	$15067.05
14	12	$383.43	$105.47	$277.96	$14789.09
	⋮	⋮	⋮	⋮	⋮
48	46	$383.43	$31.07	$352.36	$4086.36
49	47	$383.43	$28.60	$354.83	$3731.53
50	48	$383.43	$26.12	$357.31	$3374.22
51	49	$383.43	$23.62	$359.81	$3014.41
52	50	$383.43	$21.10	$362.33	$2652.09
53	51	$383.43	$18.56	$364.87	$2287.22
54	52	$383.43	$16.01	$367.42	$1919.80
55	53	$383.43	$13.44	$369.99	$1549.81
56	54	$383.43	$10.85	$372.58	$1177.23
57	55	$383.43	$8.24	$375.19	$802.04
58	56	$383.43	$5.61	$377.82	$424.22
59	57	$383.43	$2.97	$380.46	$43.76
60	58	$44.07	$0.31	$43.76	$0.00

Chapter 6 Review Exercises (page 488)

1. $1, \frac{1}{4}, \frac{1}{9}, \frac{1}{16}$
2. Arithmetic: (a) and (c) **(a)** $d = -5$ **(c)** $d = \frac{1}{6}$
3. 235 **4.** 109 **5.** 315
6. Geometric: (a) and (b) **(a)** $r = 8$ **(b)** $r = \frac{-3}{4}$
7. 8 **8.** $2{,}391{,}484\frac{4}{9}$ **9.** $10,880 **10.** $6\frac{2}{3}\%$
11. $2941.18 **12.** $4650
13. the $20,000 job ($245,000 vs $234,000)
14. (a) 40 **(b)** 2% = 0.02
15. (a) $S = P(1 + i)^n$ **(b)** $S = Pe^{rt}$
16. (b) monthly **17.** $372.79 **18.** $1601.03
19. $14,510.26 **20.** $1616.07 **21.** $3,466.64
22. $21,299.21 **23.** 14.5 years **24.** 34.3 months
25. (a) 16.32% **(b)** 14.22% **26.** 13.29%
27. (a) 7.40% **(b)** 7.47%
28. 2^{63} **29.** $2^{32} - 1$ **30.** $29,428.47
31. $1863.93 **32.** $213.81 **33.** $6069.44
34. $31,194.18 **35.** $10,841.24 **36.** $130,079.36
37. $12,007.09 **38.** $32,834.69
39. (a) $11.828 million **(b)** $161.5 million
40. $1726.85 **41.** $5390.77 **42.** $12,162.06
43. $88.85
44. (a) $592.76 **(b)** $177,828 **(c)** $85,828
45. $3443.61 **46.** $34,597.40

47.

Payment Number	*Payment Amount*	*Interest*	*Balance Reduction*	*Unpaid Balance*
57	$699.22	$594.01	$105.21	$94,936.99
58	$699.22	$593.36	$105.86	$94,831.13

Chapter 6 Test (page 491)

1. 25.3 years (approximately) **2.** \$840.75
3. 6.82% **4.** \$158,524.90 **5.** 33.53%
6. **(a)** \$698.00 **(b)** \$112,400 **7.** \$2625
8. \$7999.41 **9.** 8.73% **10.** \$119,912.92
11. \$40,552.00 **12.** \$32,488 (to the nearest dollar)
13. \$6781.17 **14.** **(a)** \$95,164.21 **(b)** \$1300.14
15. \$1688.02 **16.** **(a)** \$279,841.35 **(b)** \$13,124.75
17. \$23,381.82 **18.** \$29,716.47
19. **(a)** The difference between successive terms is always -5.5.
(b) 23.8
(c) 8226.3
20. 1000 mg (correct to three decimal places)

Exercise 7.1 (page 505)

1. $\frac{1}{4}$ **3.** 1 **5.** **(a)** $\frac{2}{5}$ **(b)** 0 **(c)** 1
7. **(a)** $\frac{3}{10}$ **(b)** $\frac{1}{2}$ **(c)** $\frac{1}{5}$ **(d)** $\frac{3}{5}$ **(e)** $\frac{7}{10}$
9. **(a)** $\frac{1}{13}$ **(b)** $\frac{1}{2}$ **(c)** $\frac{1}{4}$
11. {HH, HT, TH, TT}; **(a)** $\frac{1}{4}$ **(b)** $\frac{1}{2}$ **(c)** $\frac{1}{4}$
13. **(a)** $\frac{1}{12}$ **(b)** $\frac{1}{12}$ **(c)** $\frac{1}{36}$ **15.** **(a)** $\frac{1}{2}$ **(b)** $\frac{5}{12}$
17. **(a)** 431/1200
(b) If fair, $\Pr(6) = \frac{1}{6}$; 431/1200 not close to $\frac{1}{6}$, so not a fair die
19. **(a)** 2:3 **(b)** 3:2 **21.** **(a)** 3:8 **(b)** 8:3
23. **(a)** $\frac{1}{21}$ **(b)** $\frac{20}{21}$ **25.** **(a)** $\frac{57}{80}$ **(b)** $\frac{23}{80}$
27. **(a)** 1/3601 **(b)** 100/3601 **(c)** 3500/3601
29. S = {A+, A−, B+, B−, AB+, AB−, O+, O−}
31. .46 **33.** **(a)** .04 **(b)** .96
35. **(a)** .627 **(b)** .373 **37.** .03 **39.** .75
41. $\frac{1}{3}$ **43.** $\frac{1}{3}$
45. .22; yes, .39 is much higher than .22
47. $\frac{1}{4}$ **49.** $\frac{3}{8}$
51. **(a)** no **(b)** {BB, BG, GB, GG} **(c)** $\frac{1}{2}$
53. $\frac{1}{5}$ **55.** $\frac{3}{125}$
57. Pr(A) = 0.000019554 or about 1.9 accidents per 100,000
Pr(B) = 0.000035919 or about 3.6 accidents per 100,000
Pr(C) = 0.000037679 or about 3.8 accidents per 100,000
Intersection C is the most dangerous.
59. **(a)** 557/1200 **(b)** 11/120
61. **(a)** boy: 1/5; girl: 4/5
(b) boy: .4946; girl: .5054 **(c)** part b

Exercise 7.2 (page 515)

1. $\frac{1}{2}$ **3.** $\frac{1}{6}$ **5.** $\frac{2}{5}$ **7.** **(a)** $\frac{1}{7}$ **(b)** $\frac{5}{7}$
9. $\frac{3}{4}$ **11.** $\frac{2}{3}$ **13.** $\frac{10}{17}$ **15.** $\frac{2}{3}$
17. **(a)** $\frac{1}{2}$ **(b)** $\frac{1}{3}$ **(c)** $\frac{8}{9}$ **(d)** $\frac{1}{9}$
19. .54 **21.** **(a)** 362/425 **(b)** $\frac{66}{85}$
23. $\frac{17}{50}$ **25.** **(a)** $\frac{8}{15}$ **(b)** $\frac{7}{15}$
27. **(a)** $\frac{11}{12}$ **(b)** $\frac{5}{6}$ **29.** **(a)** $\frac{1}{2}$ **(b)** $\frac{7}{8}$ **(c)** $\frac{3}{4}$
31. .56 **33.** .965 **35.** **(a)** .72 **(b)** .84 **(c)** $\frac{61}{100}$
37. $\frac{51}{65}$ **39.** .13

Exercise 7.3 (page 526)

1. **(a)** $\frac{1}{2}$ **(b)** $\frac{1}{13}$ **3.** **(a)** $\frac{1}{3}$ **(b)** $\frac{1}{3}$ **5.** $\frac{4}{7}$
7. **(a)** $\frac{2}{3}$ **(b)** $\frac{4}{9}$ **(c)** $\frac{3}{5}$ **9.** **(a)** $\frac{1}{4}$ **(b)** $\frac{1}{2}$
11. $\frac{1}{36}$ **13.** **(a)** $\frac{1}{8}$ **(b)** $\frac{7}{8}$ **15.** **(a)** $\frac{3}{50}$ **(b)** $\frac{1}{15}$
17. **(a)** $\frac{4}{25}$ **(b)** $\frac{9}{25}$ **(c)** $\frac{6}{25}$ **(d)** 0 **19.** $\frac{5}{68}$
21. **(a)** $\frac{1}{5}$ **(b)** $\frac{3}{5}$ **(c)** 0 **23.** **(a)** $\frac{1}{17}$ **(b)** 13/204
25. **(a)** $\frac{13}{17}$ **(b)** $\frac{4}{17}$ **(c)** $\frac{8}{51}$ **27.** $\frac{31}{52}$ **29.** $\frac{25}{96}$
31. $\frac{43}{50}$ **33.** $\frac{65}{87}$ **35.** $35/435 = \frac{7}{87}$ **37.** $\frac{1}{10}$
39. 1/144,000,000 **41.** .004292 **43.** .06
45. .045 **47.** $(.95)^5 = .774$ **49.** .06
51. **(a)** .366 **(b)** .634 **53.** **(a)** .4565 **(b)** .5435
55. **(a)** $(\frac{1}{3})^3(\frac{1}{5})^4 = 1/16{,}875$
(b) $(\frac{2}{3})^3(\frac{4}{5})^4 = 2048/16{,}875$ **(c)** 14,827/16,875
57. 4/11; 4:7
59. **(a)** 364/365 **(b)** $\frac{1}{365}$ **61.** **(a)** .59 **(b)** .41

Exercise 7.4 (page 538)

1. **(a)** $\frac{4}{9}$ **(b)** $\frac{5}{9}$ **3.** $\frac{2}{5}$ **5.** $\frac{1}{325}$
7. **(a)** $\frac{2}{21}$ **(b)** $\frac{4}{21}$ **(c)** $\frac{23}{35}$
9. **(a)** $\frac{1}{30}$ **(b)** $\frac{1}{2}$ **(c)** $\frac{5}{6}$ **11.** $\frac{3}{5}$
13. **(a)** $\frac{6}{25}$ **(b)** $\frac{9}{25}$ **(c)** $\frac{12}{25}$ **(d)** $\frac{19}{25}$
15. $\frac{2}{3}$ **17.** $\frac{2}{3}$ **19.** 0.3095
21. **(a)** 81/10,000 **(b)** 1323/5000
23. **(a)** $\frac{6}{35}$ **(b)** $\frac{6}{35}$ **(c)** $\frac{12}{35}$
25. **(a)** $\frac{4}{7}$ **(b)** $\frac{5}{14}$ **(c)** $\frac{7}{10}$ **(d)** $\frac{16}{25}$
27. $\frac{17}{45}$ **29.** 0.079 **31.** **(a)** 49/100 **(b)** $\frac{12}{49}$

Exercise 7.5 (page 546)

1. 360 **3.** 151,200 **5.** 60 **7.** 1
9. **(a)** $6 \cdot 5 \cdot 4 \cdot 3 = 360$ **(b)** $6^4 = 1296$
11. 1 **13.** $n + 1$ **15.** 16 **17.** 15
19. 4950 **21.** 1 **23.** 1 **25.** 1
27. ${}_{13}C_{10} = 286$; ${}_{12}C_9 + {}_{12}C_{10} = 220 + 66 = 286$
29. 10 **31.** 604,800 **33.** 120 **35.** 24 **37.** 64
39. 720 **41.** $2^{10} = 1024$ **43.** $10! = 3{,}628{,}800$
45. 252 **47.** 30,045,015 **49.** 792 **51.** 210
53. 2,891,999,880 **55.** 3,700,000

Exercise 7.6 (page 551)

1. $\frac{1}{120}$ **3.** **(a)** 120 **(b)** $\frac{1}{120}$ **5.** .639
7. **(a)** 1/10,000 **(b)** 1/5040 **9.** $1/10^6$
11. **(a)** $(\frac{1}{5})^{10} = 1/9{,}765{,}625$ **(b)** $(\frac{4}{5})^{10} = .107$
(c) .893
13. $1/10! = 1/3{,}628{,}800$

15. (a) $\frac{1}{22}$ (b) $\frac{6}{11}$ (c) $\frac{9}{22}$ 17. .098

19. $\frac{{}_{90}C_{28} \cdot {}_{10}C_2}{{}_{100}C_{30}}$ 21. (a) .119 (b) .0476 (c) .476

23. .0238 25. (a) .721 (b) .262 (c) .279

27. (a) $\frac{1}{3}$ (b) $\frac{1}{6}$ 29. $\frac{{}_{20}C_{10}}{{}_{80}C_{10}} = .00000011$

31. (a) .033 (b) .633 33. (a) .0005 (b) .002

35. .00748

Exercise 7.7 (page 560)

1. can 3. cannot, sum $\neq 1$ 5. cannot, not square

7. can 9. [.248 .752] 11. [.228 .236 .536]

13. $[\frac{1}{4}\ \frac{3}{4}]$ 15. $[\frac{1}{4}\ \frac{1}{4}\ \frac{1}{2}]$

17. [.5 .4 .1]; [.44 .43 .13]; [.431 .43 .139]; [.4292 .4291 .1417]

19.
$$\begin{array}{c} \\ R \\ N \end{array}\begin{array}{c} \begin{array}{cc} R & N \end{array} \\ \begin{bmatrix} 0.8 & 0.2 \\ 0.3 & 0.7 \end{bmatrix} \end{array}$$

21. 0.45

23.
$$\begin{array}{c} \\ A \\ F \\ V \end{array}\begin{array}{c} \begin{array}{ccc} A & F & V \end{array} \\ \begin{bmatrix} 0 & .7 & .3 \\ .6 & 0 & .4 \\ .8 & .2 & 0 \end{bmatrix} \end{array}$$

25. [.3928 .37 .2372]

27. [46/113 38/113 29/113]

29.
$$\begin{array}{c} \\ r \\ u \end{array}\begin{array}{c} \begin{array}{cc} r & u \end{array} \\ \begin{bmatrix} .7 & .3 \\ .1 & .9 \end{bmatrix} \end{array}; [1/4 \quad 3/4]$$

31. $[\frac{1}{14}\ \frac{3}{14}\ \frac{5}{7}]$ 33. $[\frac{4}{7}\ \frac{2}{7}\ \frac{1}{7}]$

35. [49/100 42/100 9/100]

Chapter 7 Review Exercises (page 563)

1. (a) $\frac{5}{9}$ (b) $\frac{1}{3}$ (c) $\frac{2}{9}$

2. (a) $\frac{3}{4}$ (b) $\frac{1}{2}$ (c) $\frac{1}{6}$ (d) $\frac{2}{3}$

3. (a) 3:4 (b) 4:3 4. (a) $\frac{1}{4}$ (b) $\frac{1}{2}$ (c) $\frac{1}{4}$

5. (a) $\frac{3}{8}$ (b) $\frac{1}{8}$ (c) $\frac{3}{8}$ 6. $\frac{2}{13}$ 7. 16/169

8. $\frac{3}{4}$ 9. $\frac{2}{13}$ 10. $\frac{7}{13}$ 11. (a) $\frac{2}{9}$ (b) $\frac{2}{3}$ (c) $\frac{7}{9}$

12. $\frac{2}{7}$ 13. $\frac{1}{2}$ 14. $\frac{8}{15}$ 15. (a) $\frac{3}{14}$ (b) $\frac{4}{7}$ (c) $\frac{3}{8}$

16. 30 17. 35 18. 26^3 19. $\frac{5}{8}$ 20. $\frac{29}{50}$

21. $\frac{5}{56}$ 22. $\frac{33}{56}$ 23. $\frac{15}{22}$ 24. $4! = 24$

25. ${}_8P_4 = 1680$ 26. ${}_{12}C_4 = 495$ 27. ${}_8C_4 = 70$

28. (a) ${}_{12}C_2 = 66$ (b) ${}_{12}C_3 = 220$ 29. 62,193,780

30. If her assumption about blood groups is accurate, there would be $4 \cdot 2 \cdot 4 \cdot 8 = 256$, not 288, unique groups.

31. (a) 63/2000 (b) $\frac{60}{63}$ 32. 39/116

33. $\frac{1}{24}$ 34. $\frac{3}{1250}$ 35. $\frac{3}{500}$

36. (a) .3398 (b) .1975 37. $\frac{1}{10}$

38. (a) $({}_{10}C_5)({}_2C_1)/{}_{12}C_6$
(b) $\frac{({}_{10}C_5)({}_2C_1) + ({}_{10}C_4)({}_2C_2)}{{}_{12}C_6}$

39. [.135 .51 .355], [.09675 .3305 .57275], [.0640875 .288275 .6476375]

40. [12/265 68/265 37/53]

Chapter 7 Test (page 565)

1. (a) $\frac{4}{7}$ (b) $\frac{3}{7}$ 2. (a) $\frac{2}{7}$ (b) $\frac{5}{7}$

3. (a) 0 (b) 1 4. $\frac{1}{7}$ 5. $\frac{1}{7}$

6. (a) $\frac{2}{7}$ (b) $\frac{4}{7}$ 7. $\frac{2}{7}$ 8. $\frac{3}{7}$ 9. $\frac{2}{3}$

10. 1/17,576 11. .2389 12. (a) $\frac{1}{5}$ (b) $\frac{1}{20}$

13. (a) $\frac{3}{95}$ (b) $\frac{6}{19}$ (c) $\frac{21}{38}$ (d) 0

14. 1/5,245,786

15. (a) 2,118,760 (b) 1/2,118,760

16. .064 17. (a) .633 (b) .962 18. .229

19. (a) $\frac{1}{5}$ (b) $\frac{1}{14}$ (c) $\frac{13}{14}$ 20. $\frac{3}{14}$

21. (a) 2^{10} (b) $\frac{1}{2^{10}}$ (c) $\frac{1}{3}$ (d) Change the code.

22. (a) $A = \begin{bmatrix} .80 & .20 \\ .07 & .93 \end{bmatrix}$ (b) [.25566 .74434]
(c) $\frac{7}{27}$; 25.9% of market

Exercise 8.1 (page 575)

1. .0595 3. (a) $\frac{1}{64}$ (b) $\frac{5}{16}$ (c) $\frac{15}{64}$

5. .0284 7. (a) .2304 (b) .0102 (c) .3174

9. .0585 11. .2759 13. (a) .375 (b) .0625

15. (a) .1157 (b) .4823

17. (a) $\frac{27}{64}$ (b) $\frac{27}{128}$ (c) $\frac{81}{256}$

19. (a) .0729 (b) .5905 (c) .9914

21. .2457 23. .0007

25. (a) .1323 (b) .0308

27. (a) .9044 (b) .0914 (c) .0043

29. (a) .8683 (b) .2099 31. .740

Exercise 8.2 (page 583)

1. no; $\Pr(x) \ngeq 0$ 3. yes 5. yes

7. no; $\Sigma \Pr(x) > 1$ 9. $\frac{15}{8}$ 11. 5 13. $\frac{8}{3}$ 15. 2

17. $\mu = \frac{13}{8}, \sigma^2 = 1.48, \sigma = 1.22$

19. $\mu = 4.2, \sigma^2 = 9.56, \sigma = 3.09$

21. $\mu = \frac{13}{3}, \sigma^2 = 2.22, \sigma = 1.49$

23. 3 25. 2 27. 1.85

29. TV, 37,500; P.A., 35,300 31. 2

33. 0 35. −\$0.39 37. Expect to lose \$2 each time.

39. If he buys 0, his profit is 0. If the buys 100, his expected profit is \$3(100)(.25) + \$3(50)(.20) + \$3(10)(.55) + \$(−1) 50(.20) + \$(−1)90(.55) = \$62. If he buys 200, his expected profit is \$3(180)(.25) + \$3(50)(.20) + \$3(10)(.55) + (−\$1)(20)(.25) + (−\$1)(150)(.20) + (−\$1)(190)(.55) = \$42. He should buy 100.

41. TV: 37,500; newspapers: 39,200; newspapers

43. \$10

Exercise 8.3 (page 590)

1. (a)

x	$\Pr(x)$
0	125/216
1	25/72
2	5/72
3	1/216

(b) $3(\frac{1}{6}) = \frac{1}{2}$ **(c)** $\sqrt{3(\frac{1}{6})(\frac{5}{6})} = (\frac{1}{6})\sqrt{15}$
3. (a) 42 **(b)** 3.55 **5.** 2, 1.29
7. 4 **9.** 2 **11.** 2 **13.** 15
15. $a^6 + 6a^5b + 15a^4b^2 + 20a^3b^3 + 15a^2b^4 + 6ab^5 + b^6$
17. $x^4 + 4x^3h + 6x^2h^2 + 4xh^3 + h^4$
19. (a) $100(.10) = 10$ **(b)** $\sqrt{100(.10)(90)} = 3$
21. (a) 60,000 **(b)** $\sqrt{24{,}000} = 154.919$
23. 59,690 **25. (a)** 4 **(b)** 1.79
27. 2, 1.41 **29.** 300

Exercise 8.4 (page 601)

1. **3.**

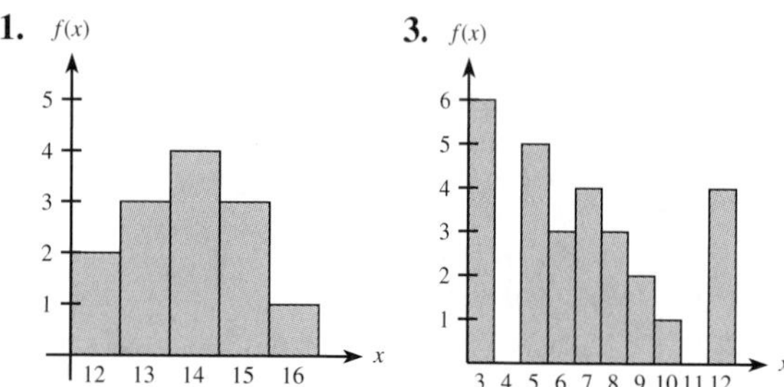

5. **7.**

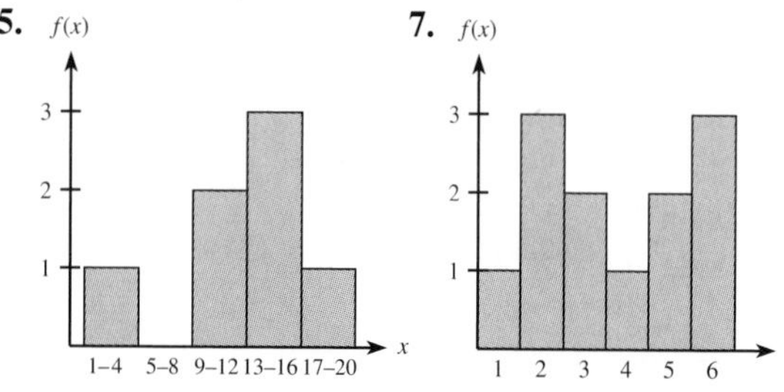

9. 3 **11.** 13 **13.** 2 **15.** 1
17. mode = 2, median = 4.5, mean = 6
19. mode = 17, median = 18.5, mean = 23.5
21. mode = 5.3, median = 5.3, mean = 5.32
23. 12.21, 14.5, 14.5 **25.** 9 **27.** 14
29. 4, 8.5714, 2.9277 **31.** 14, 4.6667, 2.1602
33. 2.73, 1.35 **35.** 6.75, 2.96

37.

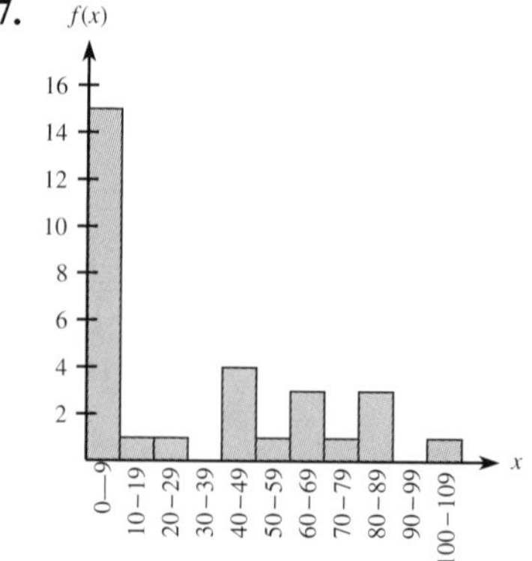

Total Executions in U.S. 1950–79

39. (a) \$30,000 **(b)** \$18,000 **(c)** \$16,000
41. The mean will give the highest measure.
43. The median will give the most representative average.
45. 33,000 **47.** 3.32, 0.677
49. (a) 6.32% **(b)** 8.93% **(c)** 7.625% **(d)** 2.69%
51.

Gold Prod.	1249.1	thousand oz
Gold Sales	1233.9	thousand oz
Revenue	423	\$/oz
Aver. price	355.7	\$/oz
Oper. cost	179.7	\$/oz
Net income	138.3	\$/oz

53. $\bar{x} = 159.8$, $s = 17.123$

Exercise 8.5 (page 612)

1. 0.4641 **3.** 0.4641 **5.** 0.9153 **7.** 0.1070
9. 0.0166 **11.** 0.0227 **13.** 0.8849 **15.** 0.1915
17. 0.3944 **19.** 0.5381 **21.** 0.2957 **23.** 0.3446
25. 0.7258 **27. (a)** 0.3413 **(b)** 0.3944
29. 0.9876
31. (a) 0.4192 **(b)** 0.0227 **(c)** 0.0581
(d) 0.8965
33. (a) 0.0668 **(b)** 0.3085 **(c)** 0.3830
35. (a) 0.0475 **(b)** 0.2033 **(c)** 0.5934
37. (a) 0.0227 **(b)** 0.1587 **(c)** 0.8186

Chapter 8 Review Exercises (page 614)

1. $\frac{11}{27}$ **2. (a)** $(99{,}999/100{,}000)^{99{,}999} \approx 0.37$
(b) $1 - (99{,}999/100{,}000)^{100{,}000} \approx 0.63$
3. .2048 **4.** 2.4 **5.** yes **6.** no; $\Sigma\Pr(x) \neq 1$
7. yes **8.** no; $\Pr(x) \not\geq 0$ **9.** 2
10. (a) 4.125 **(b)** 2.7344 **(c)** 1.654
11. (a) $\frac{37}{12}$ **(b)** .9097 **(c)** .9538
12. $\mu = 4$, $\sigma = (2\sqrt{3})/3$ **13. (a)** 1 **(b)** $(\frac{4}{5})^4$
14. $x^5 + 5x^4y + 10x^3y^2 + 10x^2y^3 + 5xy^4 + y^5$

15.

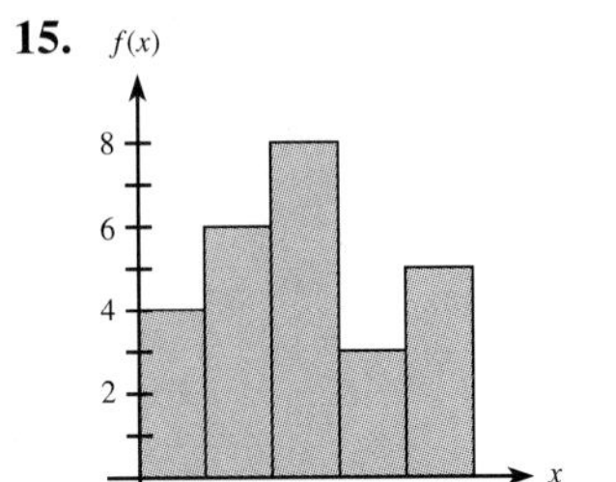

16. 3
17. $\frac{77}{26} = 2.96$
18. 3

19. f(x)

3
2
1
12 13 14 15 16 17
x

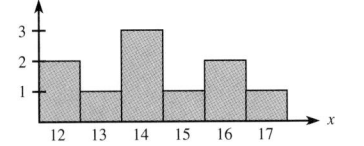

20. 14
21. 14
22. 14.3

23. $\bar{x} = 3.86$; $s^2 = 6.81$; $s = 2.61$
24. $\bar{x} = 2$; $s^2 = 2.44$; $s = 1.56$
25. .9165 **26.** .1498 **27.** .1039 **28.** .3413
29. .6826 **30.** .1360 **31.** .297 **32.** .16308
33. \$−.50 **34.** 455 **35.** 3 **36.** \$18.00
37.

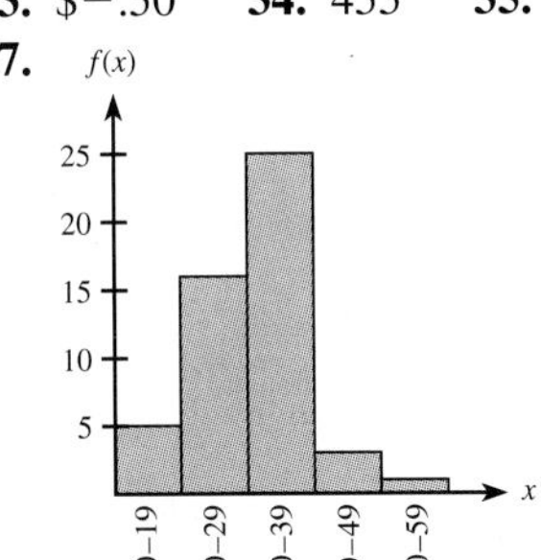

38. 30.3%
39. 8.35%
40. **(a)** .4773 **(b)** .1360 **(c)** .0227
41. 15%

Chapter 8 Test (page 616)

1. **(a)** $\frac{40}{243}$ **(b)** $\frac{51}{243}$ **2.** 4
3. $\mu = 4, \sigma^2 = \frac{8}{3}, \sigma = \frac{2}{3}\sqrt{6}$ **4.** 5.1
5. $\mu = 16.7, \sigma^2 = 26.61, \sigma = 5.16$
6. $\mu = 21.57$, median = 21, mode = 21
7. **(a)** .4706 **(b)** .8413 **(c)** .0669
8. **(a)** .3891 **(b)** .5418 **(c)** .1210
9.

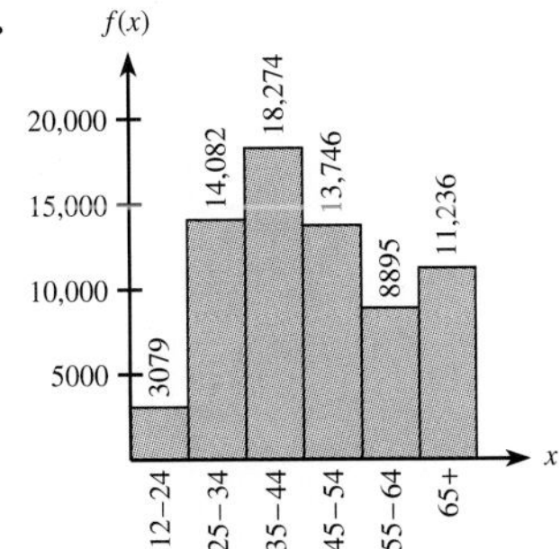

10. $\bar{x} = 46.48, s = 15.40$

11. 34.47
12. 1980, 6.4 yr; 1994, 8.1 yr; Higher prices, better quality
13. **(a)** .00003 **(b)** 30 **14.** 2 (1.8) **15.** 5 (5.4)
16. 0 (.054) with correct use; 1 (.76) with typical use
17. **(a)** .0158 **(b)** .0901 **(c)** .5213

Exercise 9.1 (page 637)

1. **(a)** 1 **(b)** 1 **3.** **(a)** −8 **(b)** −8
5. **(a)** 10 **(b)** does not exist **7.** **(a)** 0 **(b)** −6
9. **(a)** does not exist $(+\infty)$ **(b)** does not exist $(+\infty)$
(c) does not exist $(+\infty)$ **(d)** does not exist
11. **(a)** 3 **(b)** −6 **(c)** does not exist **(d)** −6

13.

x	$f(x)$
1.9	4.1579
1.99	4.0151
1.999	4.0015
↓	↓
2	4
↑	↑
2.001	3.9985
2.01	3.9851
2.1	3.8571

$\lim_{x\to 2} f(x) = 4$

15.

x	$f(x)$
0.9	−2.9
0.99	−2.99
0.999	−2.999
↓	↓
1	−3
↑	↑
1.001	−3.001
1.01	−3.01
1.1	−3.1

$\lim_{x\to 1} f(x) = -3$

17.

x	$f(x)$
0.9	3.5
0.99	3.95
0.999	3.995
↓	↘ 4
1	} Different
↑	↗ 5
1.001	4.995999
1.01	4.9599
1.1	4.59

$\lim_{x\to 1} f(x)$ does not exist

19. −1 **21.** −4 **23.** −2 **25.** 6
27. 3/4 **29.** 0 **31.** does not exist
33. −3 **35.** does not exist **37.** does not exist
39. $3x^2$ **41.** −2 **43.** $\frac{1}{30}$ **45.** does not exist
47. 5/7 **49.** −4 **51.** 9

53.

a	$(1 + a)^{1/a}$
0.1	2.5937
0.01	2.7048
0.001	2.7169
0.0001	2.7181
0.00001	2.71827
↓	↓
0	≈2.718

55. **(a)** 2 **(b)** 6 **(c)** −8 **(d)** $-\frac{1}{2}$

57. \$150,000
59. **(a)** \$32 (thousands) **(b)** \$55.04 (thousands)
61. **(a)** \$2800 **(b)** \$700 **(c)** \$560
63. **(a)** 1.52 units/hr **(b)** 0.85 units/hr **(c)** lunch
65. **(a)** 0; $p \to 100^-$ means the water approaches not being treated (containing 100% or all of its impurities); the associated costs of nontreatment approach zero.
(b) ∞ **(c)** No, because $C(0)$ is undefiined.
67. **(a)** \$3697.50 **(b)** \$3697.50 **(c)** \$3697.50
69. \$1091.70
71. 7780 (approximately) This corresponds to the Dow Jones opening average on October 5, 1998.
73. **(a)** -1.449 (approximately)
(b) This predicts the percentage of U.S. workers in farm occupations as the year approaches 2000.
(c) The model is inaccurate because the percentage must be 0 or positive; it cannot be negative.

Exercise 9.2 (page 651)

1. **(a)** continuous
(b) discontinuous; $f(1)$ does not exist
(c) discontinuous; $\lim_{x\to 3} f(x)$ does not exist
(d) discontinuous; $f(0)$ does not exist and $\lim_{x\to 0} f(x)$ does not exist
3. continuous **5.** continuous
7. discontinuous; $f(-3)$ does not exist
9. discontinuous; $f(-1)$ and $\lim_{x\to -1} f(x)$ do not exist
11. continuous
13. discontinuous; $\lim_{x\to 1} f(x)$ does not exist
15. continuous
17. discontinuity at $x = -2$; $g(-2)$ and $\lim_{x\to -2} g(x)$ do not exist
19. continuous **21.** continuous
23. discontinuity at $x = -1$; $f(-1)$ does not exist
25. discontinuity at $x = 3$; $\lim_{x\to 3} f(x)$ does not exist
27. vertical asymptote: $x = -2$; $\lim_{x\to +\infty} f(x) = 0$; $\lim_{x\to -\infty} f(x) = 0$
29. vertical asymptotes: $x = -2, x = 3$; $\lim_{x\to +\infty} f(x) = 2$; $\lim_{x\to -\infty} f(x) = 2$
31. 0 **33.** 1 **35.** $\frac{5}{3}$ **37.** does not exist $(+\infty)$
39. **(a)**

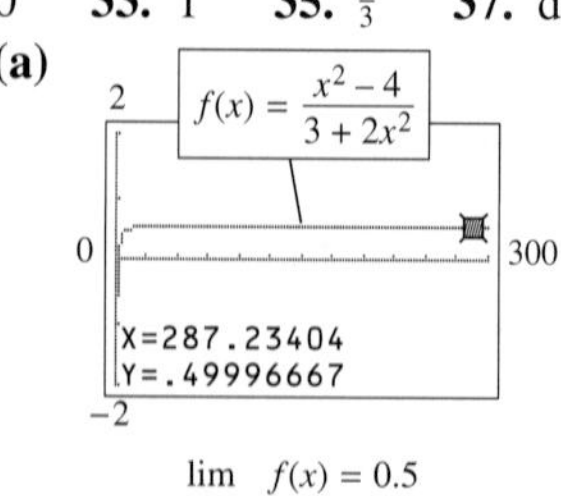

$\lim_{x\to +\infty} f(x) = 0.5$

(b) The table indicates $\lim_{x\to +\infty} f(x) = 0.5$.

41. **(a)** $x = -1000$ **(b)** $y = 1000$
(c) These values are so large that experimenting with windows to discover asymptotes may never locate them.
43. **(a)** no, not at $p = -8$ **(b)** yes **(c)** yes
(d) $p > 0$
45. **(a)** yes, $q = -1$ **(b)** yes
47. **(a)** R/i **(b)** \$10,000 **49.** yes, $0 \le p \le 100$
51. 100%; No, for p to approach 100% (as a limit) requires spending to increase without bound, which is impossible.
53. $R(x)$ is discontinuous at $x = 41{,}200$; $x = 99{,}600$; $x = 151{,}750$; and $x = 271{,}050$.
55. **(a)** \$79.40 **(b)** $\lim_{x\to 100} C(x) = 19.40$; $\lim_{x\to 500} C(x) = 49.40$ **(c)** yes
57. **(a)** Fourth degree is better.
$d(t) = -0.00001543t^4 + 0.004274t^3 - 0.4215t^2 + 17.58t - 251.274$
(b) 19.75% **(c)** $-\infty$
(d) No, $d(t)$ is negative for t-values greater than 114.
(e)

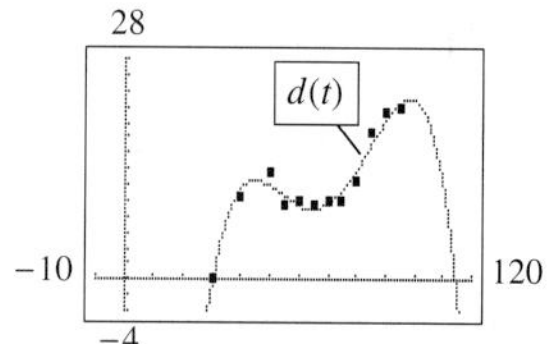

Exercise 9.3 (page 667)

1. **(a)** 32 **(b)** 32 **(c)** (4, 64)
3. **(a)** verification **(b)** -5 **(c)** -5 **(d)** $(-1, 3)$
5. **(a)** $P(1, 1), A(3, 0)$ **(b)** $-\frac{1}{2}$ **(c)** $-\frac{1}{2}$ **(d)** $-\frac{1}{2}$
7. **(a)** $P(1, 3), A(0, 3)$ **(b)** 0 **(c)** 0 **(d)** 0
9. **(a)** $f'(x) = -6$ **(b)** -6; -6 **(c)** -6
11. **(a)** $f'(x) = 8x - 2$ **(b)** $8x - 2$; -26 **(c)** -26
13. **(a)** $p'(q) = 2q + 4$ **(b)** $2q + 4$; 14 **(c)** 14
15. **(a)** 89.000024 **(b)** 89.0072 **(c)** ≈ 89
17. **(a)** 294.000008 **(b)** 294.0084 **(c)** ≈ 294
19. -31 **21.** $f'(-3) = \frac{5}{2}$; $f(-3) = -9$
23. $y = 3x - 4$
25. **(a)** a, b, d **(b)** c **(c)** A, C, E
27. **(a)** A, B, C, D **(b)** A, D
29. **(a)** $f'(x) = 2x + 1$ **(b)** $f'(2) = 5$ **(c)** $y = 5x - 4$
(d)

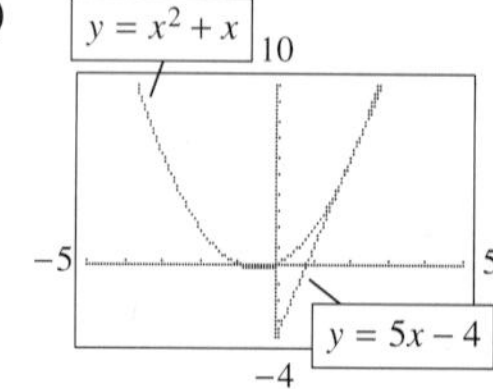

31. **(a)** $f'(x) = 3x^2$ **(b)** $f'(1) = 3$ **(c)** $y = 3x + 1$

(d)

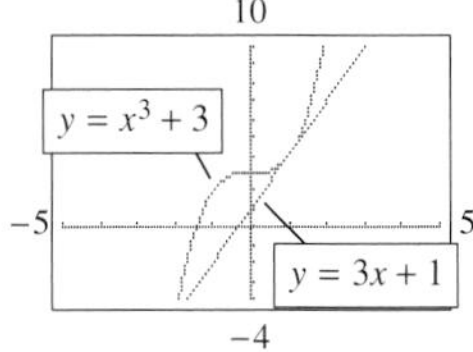

33. **(a)** -32 ft/s **(b)** the ball is falling, so S is decreasing

35. **(a)** $\frac{-100}{3}$ **(b)** $\frac{-4}{3}$ **37.** 68 mph

39. **(a)** $R'(x) = \overline{MR} = 300 - 2x$ **(b)** 200

(c) -100 **(d)** 0

(e) It changes from increasing to decreasing

41. 200

43. **(a)** 100; the expected profit from the sale of the 201st car is \$100.

(b) -100; the expected profit from the sale of the 301st car is a loss of \$100.

45. **(a)** 1.039

(b) If humidity changes by 1%, the heat index will change by about 1.039°F.

47. **(a)** 1.164 million infections per year (approximately)

(b) Low: 3.6 million infections per year

High: 20 million infections per year

Exercise 9.4 *(page 680)*

1. $y' = 0$ **3.** $y' = 1$ **5.** $f'(x) = 6x^2 - 5x^4$

7. $y' = 24x^3 - 10x + 1$

9. $g'(x) = 90x^8 - 25x^4 + 21x^2 + 5$

11. **(a)** 19 **(b)** 19 **13.** **(a)** 0 **(b)** 0

15. $y' = -5x^{-6} - 8x^{-9}$

17. $y' = 11x^{8/3} - \frac{7}{2}x^{3/4} - \frac{1}{2}x^{-1/2}$

19. $f'(x) = -4x^{-9/5} - \frac{8}{3}x^{-7/3}$

21. $g'(x) = -\dfrac{12}{x^5} - \dfrac{10}{x^6} + \dfrac{5}{3\sqrt[3]{x^2}}$

23. $y = -3x + 6$ **25.** $y = 3$

27. $(1, -1), (5, 31)$ **29.** $(0, 9), (3, -18)$

31. **(a)** $-1/2$ **(b)** -0.5000 (to four decimal places)

33. **(a)** $f'(x) = 6x^2 + 5$

(b)

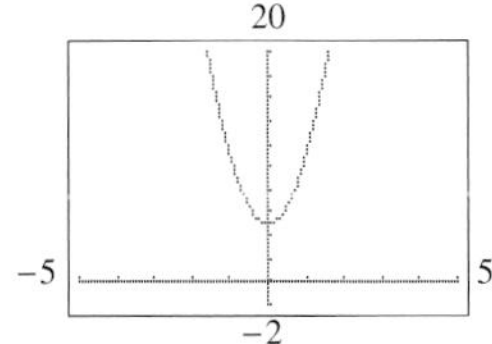

Graph of $f'(x)$ and numerical derivative of $f(x)$

35. **(a)** $h'(x) = \dfrac{-30}{x^4} + \dfrac{4}{\sqrt[5]{x^7}} + 2x$

(b)

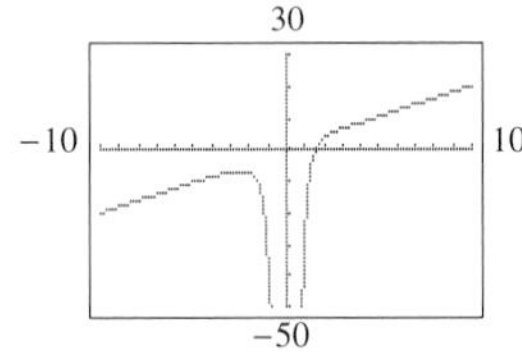

Graph of $h'(x)$ and numerical derivative of $h(x)$

37. **(a)** $y = 8x - 3$

(b)

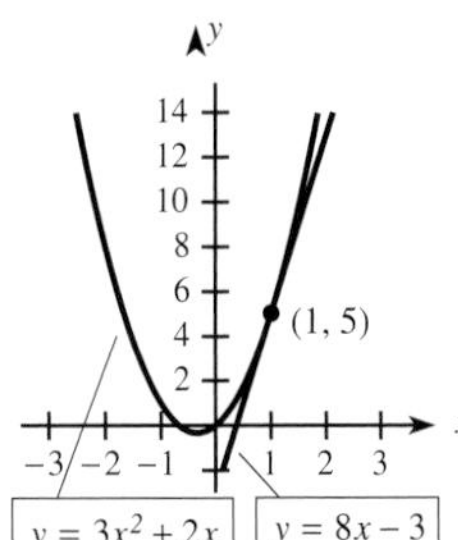

(c) x:0.7 → 1.6; y:3.0 → 7.9

39. **(a)** $f'(x) = -2 - 2x$

(b)

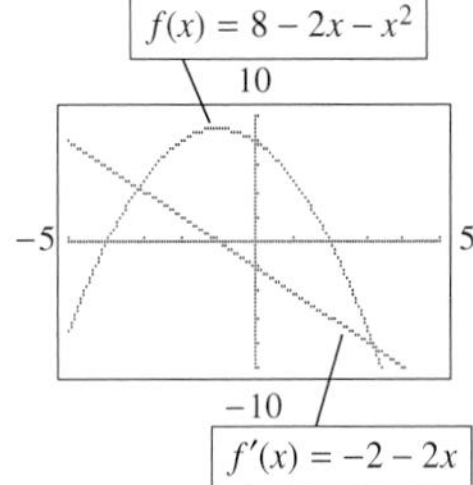

(c) $f'(x) = 0$ at $x = -1$; $f'(x) > 0$ for $x < -1$

$f'(x) < 0$ for $x > -1$

(d) $f(x)$ has max when $x = -1$

$f(x)$ rises for $x < -1$

$f(x)$ falls for $x > -1$

41. **(a)** $f'(x) = 3x^2 - 12$

(b)

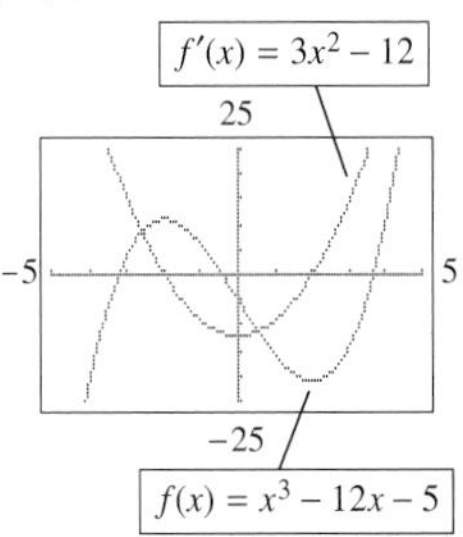

(c) $f'(x) = 0$ at $x = -2$ and $x = 2$

$f'(x) > 0$ for $x < -2$ and $x > 2$

$f'(x) < 0$ for $-2 < x < 2$

(d) $f(x)$ has max when $x = -2$, min when $x = 2$

$f(x)$ rises when $x < -2$ and when $x > 2$

$f(x)$ falls when $-2 < x < 2$

43. **(a)** 40, revenue increasing

(b) -20, revenue decreasing

45. **(a)** 920 **(b)** 926

47. **(a)** -4; if the price changes to \$26, the quantity demanded will change by approximately -4 units
(b) $-\frac{1}{2}$; if the price changes to \$101, the quantity demanded will change by approximately $-\frac{1}{2}$ unit

49. **(a)** $\overline{C(x)}' = (-4000/x^2) + 0.1$ **(b)** 200

51. **(a)** $-120{,}000$
(b) If the impurities change from 1% to 2%, then the expected change in cost is $-120{,}000$ (dollars).

53. **(a)** $WC = 55.174 - 2.15s - 23.18\sqrt{s}$
(b) -4.468
(c) If the wind speed changes by + 1 mph (to 26 mph), then the wind chill will change by approximately -4.468°F.

55. **(a)** $u(x) = 0.0003836x^3 - 0.08853x^2 + 6.2132x - 105.247$
(b) -0.78% per year
(c) $u'(80) \approx -0.59\%$ per year

57. **(a)** $f(t) = 0.030847t^3 - 8.49779t^2 + 779.66274t - 23{,}820.28$
(b) $f'(t) = 0.092541t^2 - 16.99558t + 779.66274$
(c) $f'(89) \approx 0.07$; $f'(94) \approx -0.23$
(d) $f'(89) \approx 0.07$ means that from 1989 to 1990, the consumer price index rose about 0.07%. $f'(94) \approx -0.23$ means that from 1994 to 1995, the consumer price index fell about 0.23%.

Exercise 9.5 *(page 690)*

1. $y' = 3x^2 + 2x - 6$ **3.** $dp/dq = 9q^2 - 2q + 6$

5. $f'(x) = (x^{12} + 3x^4 + 4)(12x^2) + (4x^3 - 1)(12x^{11} + 12x^3)$

7. $y' = (7x^6 - 5x^4 + 2x^2 - 1)(36x^8 + 21x^6 - 10x + 3) + (4x^9 + 3x^7 - 5x^2 + 3x)(42x^5 - 20x^3 + 4x)$

9. $y' = (x^2 + x + 1)(\frac{1}{3}x^{-2/3} - x^{-1/2}) + (x^{1/3} - 2x^{1/2} + 5)(2x + 1)$

11. **(a)** 40 **(b)** 40 **13.** $y' = (-x^2 - 1)/(x^2 - 1)^2$

15. $\dfrac{dp}{dq} = (q^2 - 4q - 1)/(q - 2)^2$

17. $\dfrac{dy}{dx} = \dfrac{4x^5 - 4x^3 - 16x}{(x^4 - 2x^2 + 5)^2}$

19. $\dfrac{dz}{dx} = 2x + \dfrac{2x - x^2}{(1 - x - 2x^2)^2}$

21. $\dfrac{dp}{dq} = \dfrac{2q + 1}{\sqrt[3]{q^2}\,(1 - q)^2}$

23. $y' = \dfrac{2x^3 - 6x^2 - 8}{(x - 2)^2}$

25. **(a)** $\frac{3}{5}$ **(b)** $\frac{3}{5}$ **27.** $y = 44x - 32$

29. $y = \frac{10}{3}x - \frac{10}{3}$ **31.** 104

33. 1.3333 (to four decimal places)

35. **(a)** $f'(x) = 3x^2 - 6x - 24$

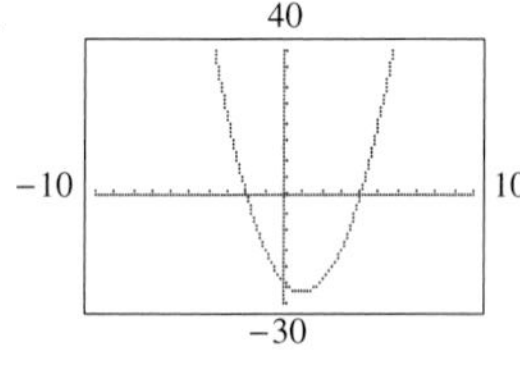

Graph of both $f'(x)$ and numerical derivative of $f(x)$

(b) Horizontal tangents where $f'(x) = 0$; at $x = -2$ and $x = 4$
(c)

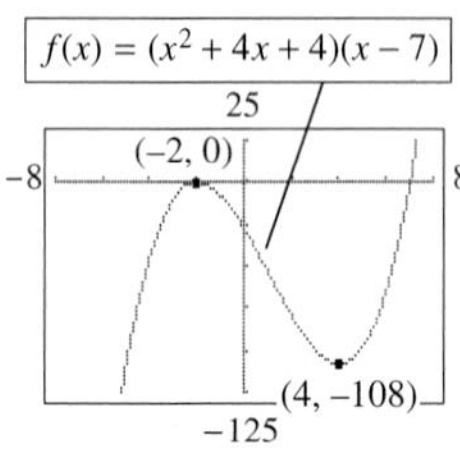

37. **(a)** $y' = \dfrac{x^2 - 4x}{(x - 2)^2}$

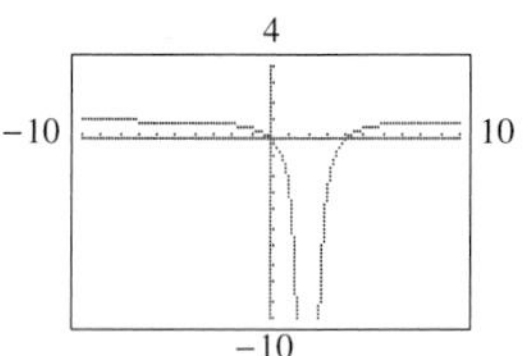

Graph of both y' and the numerical derivative of y

(b) Horizontal tangents where $y' = 0$; at $x = 0$ and $x = 4$
(c) (0, 0) (4, 8)

39. **(a)** $f'(x) = \dfrac{20x}{(x^2 + 1)^2}$ **(b)**

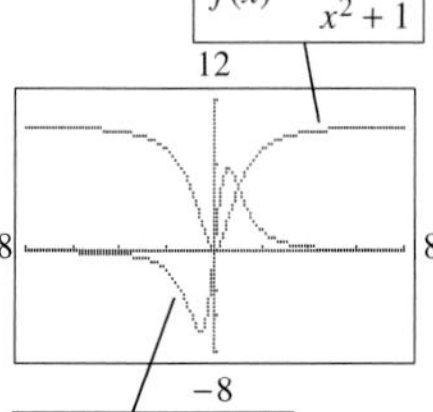

(c) $f' = 0$ at $x = 0$
$f' > 0$ for $x > 0$
$f' < 0$ for $x < 0$
(d) f has a min at $x = 0$
f increasing for $x > 0$
f decreasing for $x < 0$

41.

$$f'(x) = \lim_{h\to0} \frac{\dfrac{u(x+h)}{v(x+h)} - \dfrac{u(x)}{v(x)}}{h}$$

$$= \lim_{h\to0} \frac{u(x+h)\,v(x) - u(x)\,v(x+h)}{h\cdot v(x)\,v(x+h)}$$

$$= \lim_{h\to0} \frac{u(x+h)\,v(x) - u(x)\,v(x) + u(x)v(x) - u(x)v(x+h)}{h\cdot v(x)\,v(x+h)}$$

$$= \lim_{h\to0} \frac{v(x)\left[\dfrac{u(x+h) - u(x)}{h}\right] - u(x)\left[\dfrac{v(x+h) - v(x)}{h}\right]}{v(x)\,v(x+h)}$$

$$= \frac{v(x)\,u'(x) - u(x)\,v'(x)}{[v(x)]^2}$$

43. $C'(p) = 810{,}000/(100 - p)^2$

45. $R'(49) \approx 30.00$ The expected revenue from the sale of the next unit (the 50th) is about \$30.00.

47. $R'(5) = -50$ As the group changes by 1 person (to 31), the revenue will drop by about \$50.

49. $S = 1000x - x^2$ **51.** $\dfrac{dR}{dn} = \dfrac{r(1-r)}{[1+(n-1)r]^2}$

53. **(a)** $P'(6) \approx 0.045$ During the next (7th) month of the campaign, the proportion of voters who recognize the candidate will change by about 0.045, or 4.5%.

(b) $P'(12) \approx -0.010$ During the next (13th) month of the campaign, the proportion of voters who recognize the candidate will drop by about 0.010, or 1%.

(c) It is better for $P'(t)$ to be positive—that is, to have increasing recognition.

55. **(a)** $f'(20) \approx -0.79$

(b) At 0°F, if the wind speed changes by 1 mph (to 21 mph), the wind chill will change by about -0.79°F.

57. **(a)** $C'(5) = 0.3$

(b) From 1990 to 1991, the model predicts a change of 0.3% in the cost of living.

(c)

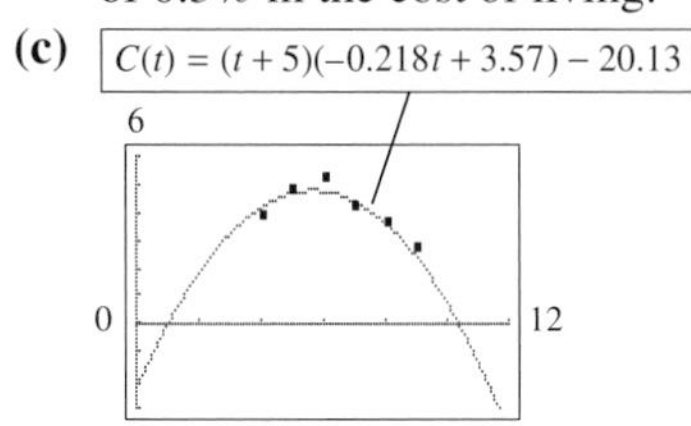

(d) The model fails to be valid when $C(t) < 0$, below the t-axis.

59. **(a)** $f'(t) =$

$$1000\left[\frac{8.88543219t^2 - 3423.843248t + 16{,}731.97758}{(1.09816t^2 - 122.183t + 21{,}472.6)^2}\right]$$

(b) $f'(70) \approx -0.5356$ $f'(170) \approx -0.2932$

(c) $f'(70)$ means that from 1870 to 1871, the model predicts a change of about -0.5356% in U.S. workers in farm occupations. $f'(170)$ means that from 1970 to 1971, the model predicts a change of about -0.2932% in U.S. workers in farm occupations.

Exercise 9.6 *(page 698)*

1. $6x(x^2+1)^2$ **3.** $4(8x-1)(4x^2-x+8)^3$

5. $-6x/(x^2+2)^4$ **7.** $-4(x+2)(x^2+4x)^{-3}$

9. $-3(2x+3)(x^2+3x+4)^{-4}$

11. $\dfrac{-3(6x^2+3)}{4(2x^3+3x+5)^{7/4}}$ **13.** $(x+2)/\sqrt{x^2+4x+5}$

15. $(6-4x)/\sqrt{3x-x^2}$ **17.** $16x(x^2-3)^4$

19. $\dfrac{15(3x+1)^4 - 3}{7}$ **21.** **(a)** and **(b)** 96,768

23. **(a)** and **(b)** 2 **25.** $y = -3x + 4$

27. $9x - 5y = 2$

29. **(a)** $f'(x) = 6x(x^2-4)^2$

(b)

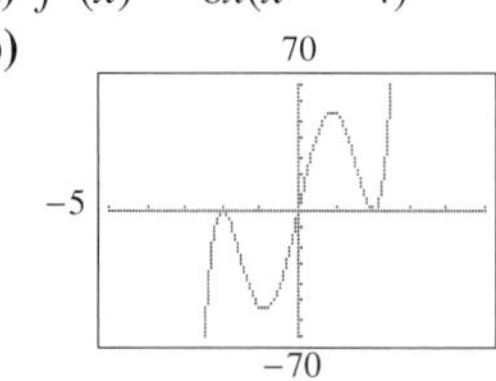

Graph of both $f'(x)$ and numerical derivative of $f(x)$

(c) $x = 0, x = 2, x = -2$

(d) $(0, -52), (2, 12), (-2, 12)$

(e) $f(x) = (x^2-4)^3 + 12$

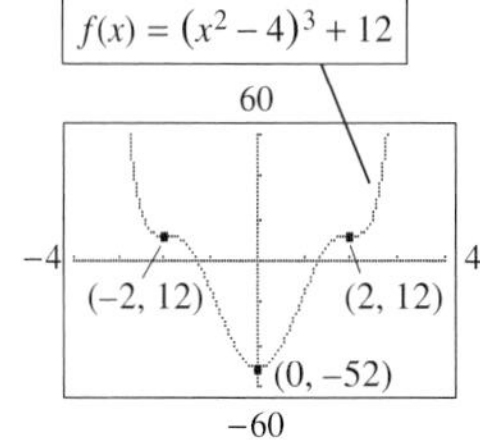

31. **(a)** $f'(x) = 8x(1-x^2)^{1/3}$

(b)

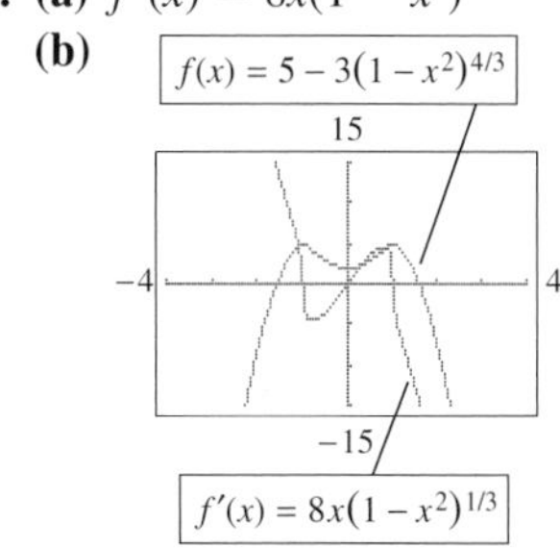

(c) $f'(x) = 0$ at $x = -1, x = 0, x = 1$
$f'(x) > 0$ for $x < -1$ and $0 < x < 1$
$f'(x) < 0$ for $-1 < x < 0$ and $x > 1$

(d) $f(x)$ has a maximum at $x = -1$ and $x = 1$, a minimum at $x = 0$
$f(x)$ increasing for $x < -1$ and $0 < x < 1$
$f(x)$ decreasing for $-1 < x < 0$ and $x > 1$

33. **(a)** $y' = 2x^2$ **(b)** $y' = -2/x^4$

(c) $y' = 2(2x)^2$ **(d)** $y' = \dfrac{-18}{(3x)^4}$

35. 10 ft/sec **37.** \$1499.85 (approximately)

39. **(a)** -0.114 (approximately)

(b) If the price changes by \$1, to \$22, the weekly sales volume will change by approximately -0.114 thousand unit.

41. **(a)** $-\$3.20$ per unit

(b) If the quantity demanded changes from 49 to 50 units, the change in price will be about $-\$3.20$.

43. $\dfrac{dy}{dx} = \left(\dfrac{8k}{5}\right)(x - x_0)^{3/5}$

45. $\dfrac{dp}{dq} = \dfrac{-100}{(2q+1)^{3/2}}$ **47.** $\dfrac{dK_c}{dv} = \dfrac{8}{\sqrt{4v+1}}$

49. **(a)** \$658.75. If the interest changed from 6% to 7%, the amount of the investment would change by about \$658.75
(b) \$2156.94. If the interest rate changed from 12% to 13%, the amount of the investment would change by about \$2156.94.

51. $d'(30) \approx -0.170$ means from 1960 to 1961, the model predicts that the interest paid as a percent of federal expenditures would change by -0.17%. $d'(60) \approx 0.58$ means from 1990 to 1991, the model predicts that the interest paid as a percent of federal expenditures would change by 0.58%.

53. $p'(10) \approx -0.42$ means from 1970 to 1971, the model predicts that the number of persons below the poverty level would change by -0.42 million (a decrease). $p'(30) \approx 0.58$ means from 1990 to 1991, the model predicts that the number of persons below the poverty level would change by 0.58 million (an increase).

Exercise 9.7 *(page 706)*

1. 0 **3.** $4(-4x^{-5})$; $-16/x^5$
5. $15x^2 + 4(-x^{-2})$; $15x^2 - 4/x^2$
7. $(x^2-2)1 + (x+4)(2x)$; $3x^2 + 8x - 2$
9. $\dfrac{x^2(3x^2) - (x^3+1)(2x)}{(x^2)^2}$; $(x^3-2)/x^3$
11. $(3x^2-4)(x^3-4x)^9$
13. $\frac{5}{3}x^3[3(4x^5-5)^2(20x^4)] + (4x^5-5)^3(5x^2)$; $5x^2(4x^5-5)^2(24x^5-5)$
15. $(x-1)^2(2x) + (x^2+1)2(x-1)$; $2(x-1)(2x^2-x+1)$
17. $\dfrac{(x^2+1)\,3(x^2-4)^2(2x) - (x^2-4)^3(2x)}{(x^2+1)^2}$; $\dfrac{2x(x^2-4)^2(2x^2+7)}{(x^2+1)^2}$
19. $3[(q+1)(q^3-3)]^2[(q+1)3q^2 + (q^3-3)1]$; $3(4q^3+3q^2-3)[(q+1)(q^3-3)]^2$
21. $4[x^2(x^2+3x)]^3[x^2(2x+3) + (x^2+3x)(2x)]$; $4x^2(4x+9)[x^2(x^2+3x)]^3$
23. $4\left(\dfrac{2x-1}{x^2+x}\right)^3\left[\dfrac{(x^2+x)2 - (2x-1)(2x+1)}{(x^2+x)^2}\right]$; $\dfrac{4(-2x^2+2x+1)(2x-1)^3}{(x^2+x)^5}$
25. $(8x^4+3)^2 3(x^3-4x)^2(3x^2-4) + (x^3-4x)^3 2(8x^4+3)(32x^3)$; $(8x^4+3)(x^3-4x)^2(136x^6-352x^4+27x^2-36)$
27. $\dfrac{(4-x^2)\frac{1}{3}(x^2+5)^{-2/3}(2x) - (x^2+5)^{1/3}(-2x)}{(4-x^2)^2}$; $\dfrac{2x(2x^2+19)}{3\sqrt[3]{(x^2+5)^2}\,(4-x^2)^2}$
29. $(x^2)\frac{1}{4}(4x-3)^{-3/4}(4) + (4x-3)^{1/4}(2x)$; $(9x^2-6x)/\sqrt[4]{(4x-3)^3}$
31. $(2x)\frac{1}{2}(x^3+1)^{-1/2}(3x^2) + (x^3+1)^{1/2}(2)$; $(5x^3+2)/\sqrt{x^3+1}$
33. **(a)** $F_1'(x) = 12x^3(x^4+1)^4$
(b) $F_2'(x) = \dfrac{-12x^3}{(x^4+1)^6}$
(c) $F_3'(x) = 12x^3(3x^4+1)^4$
(d) $F_4'(x) = \dfrac{-300x^3}{(5x^4+1)^6}$
35. $dP/dx = 90(3x+1)^2$
37. **(a)** \$59,900
(b) It is increasing.
39. $C'(y) = 1/\sqrt{y+1} + 0.4$
41. $dV/dx = 144 - 96x + 12x^2$
43. **(a)** -1.6 This means that from the 9th to the 10th week, sales are expected to change by -1.6 (decrease).
45. **(a)** $f'(t) =$
$$1000\left[\frac{8.885432192\,(t+20)^2 - 3423.843248\,(t+20) - 16{,}731.9776}{[1.09816\,(t+20)^2 - 122.183\,(t+20) + 21472.6]^2}\right]$$
(b) 1850: $f'(30) \approx -0.4033$
1950: $f'(130) \approx -0.3827$
(c) $f'(30)$ means that from 1850 to 1851, the model predicts a change of $-.4033\%$ in U.S. workers in farm occupations.
$f'(130)$ means that from 1950 to 1951, the model predicts a change of -0.3827% in U.S. workers in farm occupations.

Exercise 9.8 *(page 712)*

1. $24x - 30$ **3.** $60x - 2$ **5.** $6x - 2x^{-3}$
7. $6x + \frac{1}{4}x^{-3/2}$ **9.** $60x^2 - 96$
11. $1008x^6 - 720x^3$ **13.** $-6/x^4$ **15.** $\frac{3}{8}x^{-5/2}$
17. $20x^3 + \frac{1}{4}x^{-3/2}$ **19.** $\frac{3}{8}(x+1)^{-5/2}$ **21.** 0
23. $-15/(16x^{7/2})$ **25.** $\frac{3}{8}(x-1)^{-5/2}$
27. $-2(x+1)^{-3}$ **29.** 26
31. 16.0000 (to four decimal places)
33. 0.0004261
35. **(a)** $f'(x) = 3x^2 - 6x$ $f''(x) = 6x - 6$
(b)

$f'(x) = 3x^2 - 6x$
$f''(x) = 6x - 6$
15
-3
5
-10
$f(x) = x^3 - 3x^2 + 5$

(c) $f''(x) = 0$ at $x = 1$
$f''(x) > 0$ for $x > 1$
$f''(x) < 0$ for $x < 1$
(d) $f'(x)$ has a min at $x = 1$
$f'(x)$ is increasing for $x > 1$
$f'(x)$ is decreasing for $x < 1$
(e) $f''(x) < 0$ **(f)** $f''(x) > 0$

37. (a) $f'(x) = -x^2 - 2x + 3$ $f''(x) = -2x - 2$
(b)
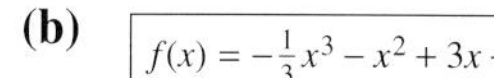

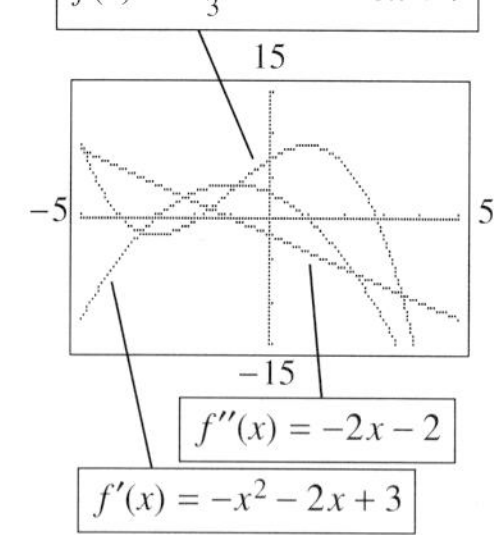

(c) $f''(x) = 0$ at $x = -1$
$f''(x) > 0$ for $x < -1$
$f''(x) < 0$ for $x > -1$
(d) $f'(x)$ has a max at $x = -1$
$f'(x)$ is increasing for $x < -1$
$f'(x)$ is decreasing for $x > -1$
(e) $f''(x) < 0$ **(f)** $f''(x) > 0$

39. $a = 0.12$ **41.** -0.02

43. (a) $\dfrac{dR}{dm} = mc - m^2$ **(b)** $\dfrac{d^2R}{dm^2} = c - 2m$
(c) Second

45. (a) 0.0009 (approximately)
(b) When 1 more unit is sold (beyond 25), the marginal revenue will change by about 0.0009 thousand dollars per unit, or \$0.90 per unit.

47. (a) $S' = \dfrac{-3}{(t+3)^2} + \dfrac{36}{(t+3)^3}$ **(b)** $S''(15) = 0$
(c) After 15 weeks, the rate of change of the rate of sales is zero because the rate of sales reaches a minimum value.

49. (a) $p'(t) = -0.005169t^2 + 0.2582t - 2.5015$
(b) 1970: $p''(10) = 0.155$
1990: $p''(30) = -0.052$
(c) $p'(10) = -0.44$ means that from 1970 to 1971, the number of people who lived below the poverty level was expected to change by about -0.44 million.
$p''(10) = 0.155$ means that from 1970 to 1971, the rate of change of the number of people who lived below the poverty level was expected to change by about 0.155 million per year.
Thus, from 1970 to 1971, the number of people who lived below the poverty level was increasing at an increasing rate.

51. (a) $f(t) = 0.030847t^3 - 0.44674t^2 + 1.48932t + 3.43357$
(b) $f'(t) = 0.092541t^2 - 0.89348t + 1.48932$
(c) 1991: $f''(4) \approx -0.153$
1995: $f''(8) \approx 0.587$
(d) $f'(8) \approx 0.264$ means that from 1995 to 1996, the expected change in the CPI was about 0.264%.
$f''(8) \approx 0.587$ means that from 1995 to 1996, the expected change in the rate of change of the CPI was about 0.587% per year.

Exercise 9.9 *(page 721)*

1. $\overline{MC} = 8$ **3.** $\overline{MC} = 13 + 2x$
5. $\overline{MC} = 3x^2 - 12x + 24$ **7.** $\overline{MC} = 27 + 3x^2$
9. (a) \$10; the cost will increase by \$10. **(b)** \$11
11. \$46; the cost will increase by \$46.
13.
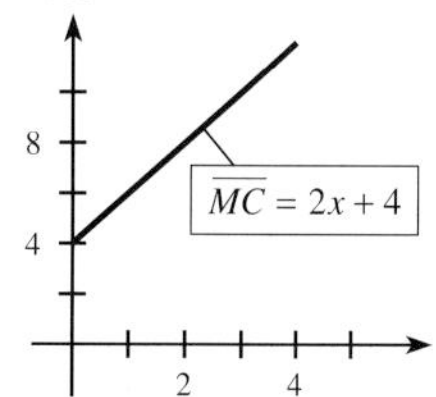

15. (a) $\overline{MR} = 4$
(b) The sale of each additional item brings in \$4 revenue at all levels of production.

17. (a) \$3500; this is revenue from the sale of 100 units.
(b) $\overline{MR} = 36 - 0.02x$
(c) \$34; Revenue will increase by \$34.
(d) Actual revenue from the sale of the 101st item is \$33.99.

19. (a) $\overline{MR} = 36 - 0.02x$ **(b)** $x = 1800$
(c) \$32,400
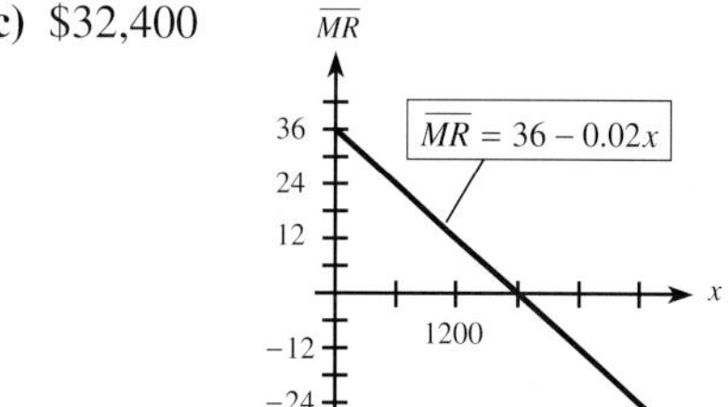

21. $\overline{MP} = 5$

23. (a) \$0 **(b)** $\overline{MP} = 30 - 2x$
(c) $-\$10$; profit will decrease by \$10 if 1 additional unit is sold.
(d) The sale of the 21st item results in a loss of \$11.

25. (a)

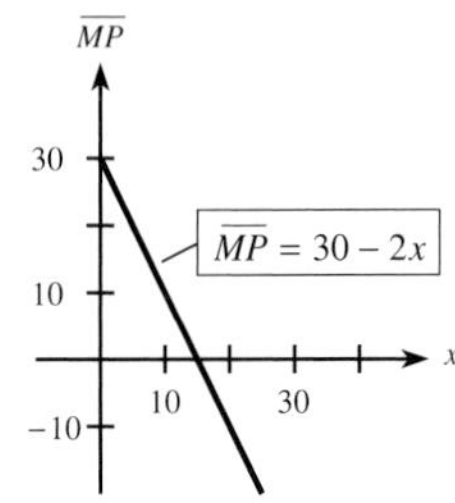

(b) 15
(c) 15
(d) \$25

27. 70 **29.** \$200

31. (a) Data: $R(7) = 60.53$ Model: $R(7) = 61.027$
$R(7)$ represents AT&T's total revenues (in billions of dollars) for 1987.
(b) $R'(t) = 0.506\,t - 4.03$
(c) $R'(12) = 2.042$
(d) $R'(12) = 2.042$ means that from 1992 to 1993, the model predicts that AT&T's revenue rose by about \$2.042 billion.

33. (a) $C'(10) = 0.252$
(b) This means that from 1990 to 1991, the model predicts that Scott Paper Company's costs and expenses will change by about \$0.252 billion. (The actual change for that period was \$0.0319 billion.)

Chapter 9 Review Exercises *(page 725)*

1. (a) 2 **(b)** 2 **2. (a)** 0 **(b)** 0
3. (a) 2 **(b)** 1 **4. (a)** 2 **(b)** does not exist
5. (a) does not exist **(b)** 2
6. (a) does not exist **(b)** does not exist
7. 55 **8.** 0 **9.** -2 **10.** 6 **11.** $\frac{1}{2}$ **12.** $\frac{1}{5}$
13. no limit **14.** 0 **15.** 4 **16.** no limit
17. 3 **18.** no limit **19.** $6x$ **20.** $1 - 4x$
21. -14 **22.** 5 **23. (a)** yes **(b)** no
24. (a) yes **(b)** no **25.** 2 **26.** no limit
27. 1 **28.** no **29.** yes **30.** yes
31. discontinuity at $x = 5$ **32.** discontinuity at $x = 2$
33. continuous **34.** discontinuity at $x = 1$
35. (a) $x = 0, x = 1$ **(b)** 0 **(c)** 0
36. (a) $x = -1, x = 0$ **(b)** $\frac{1}{2}$ **(c)** $\frac{1}{2}$
37. -2 **38.** 0 **39.** true **40.** false
41. $f'(x) = 6x + 2$ **42.** $f'(x) = 1 - 2x$
43. (a) no **(b)** no **44. (a)** yes **(b)** no
45. (a) -5.9171 (to four decimal places) **(b)** -5.9
46. 7 **47.** $20x^4 - 18x^2$ **48.** $8x$ **49.** 3
50. $1/(2\sqrt{x})$ **51.** 0 **52.** $-4/(3\sqrt[3]{x^4})$
53. $\dfrac{-1}{x^2} + \dfrac{1}{2\sqrt{x^3}}$ **54.** $\dfrac{-3}{x^3} - \dfrac{1}{3\sqrt[3]{x^2}}$
55. $y = 15x - 18$ **56.** $y = 34x - 48$

57. (a) $x = 0, x = 2$ **(b)** $(0, 1)\ (2, -3)$

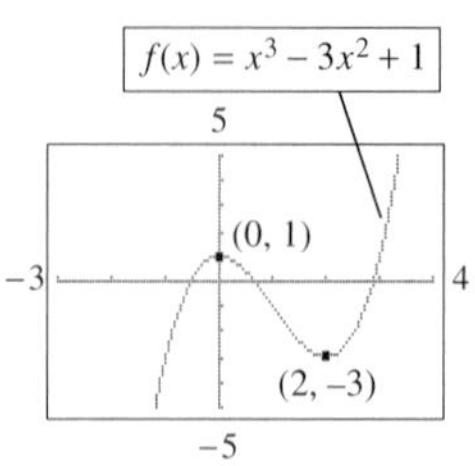

58. (a) $x = 0, x = 2, x = -2$
(b) $(0, 8)\ (2, -24)\ (-2, -24)$

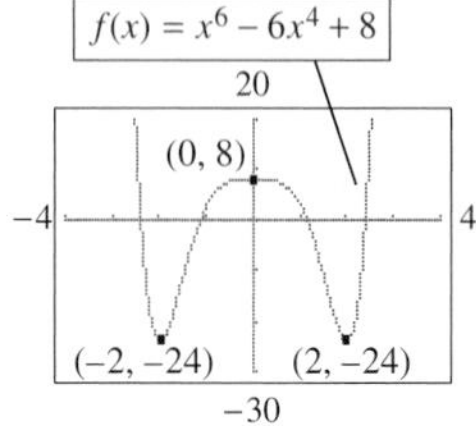

59. $9x^2 - 2x + 12$ **60.** $15x^4 + 9x^2 + 2x$
61. $\dfrac{2(1-q)}{q^3}$ **62.** $\dfrac{1-3t}{[2\sqrt{t}(3t+1)^2]}$ **63.** $\dfrac{9x+2}{2\sqrt{x}}$
64. $\dfrac{5x^6 + 2x^4 + 20x^3 - 3x^2 - 4x}{(x^3+1)^2}$
65. $(9x^2 - 24x)(x^3 - 4x^2)^2$
66. $6(30x^5 + 24x^3)(5x^6 + 6x^4 + 5)^5$
67. $72x^3(2x^4 - 9)^8$ **68.** $\dfrac{-(3x^2-4)}{2\sqrt{(x^3-4x)^3}}$
69. $2x(2x^4 + 5)^7\,(34x^4 + 5)$ **70.** $\dfrac{-2(3x+1)(x+12)}{(x^2-4)^2}$
71. $36[(3x + 1)(2x^3 - 1)]^{11}(8x^3 + 2x^2 - 1)$
72. $\dfrac{3}{(1-x)^4}$ **73.** $\dfrac{(2x^2-4)}{\sqrt{x^2-4}}$ **74.** $\dfrac{2x-1}{(3x-1)^{4/3}}$
75. $y'' = \frac{-1}{4}x^{-3/2} - 2$ **76.** $y'' = 12x^2 - 2/x^3$
77. $\dfrac{d^5y}{dx^5} = 0$ **78.** $\dfrac{d^5y}{dx^5} = -30(1 - x)$
79. $\dfrac{d^3y}{dx^3} = -4/[(x^2 - 4)^{3/2}]$ **80.** $\dfrac{d^4y}{dx^4} = \dfrac{2x(x^2-3)}{(x^2+1)^3}$
81. (a) $x'(10) = -1$ means if price changes from \$10 to \$11, the number of units demanded will change by about -1.
(b) $x'(20) = -\frac{1}{4}$ means if price changes from \$20 to \$21, the number of units demanded will change by about $-\frac{1}{4}$.
82. $\dfrac{dq}{dp} = \dfrac{-p}{\sqrt{0.02p^2 + 500}}$
83. $x'(10) = \frac{1}{6}$ means if price changes from \$10 to \$11, the number of units supplied will change by about $\frac{1}{6}$.
84. (a) $\overline{MC} = 6x + 6$ **(b)** 186
(c) If a 31st unit is produced, costs will change by about \$186.

85. $C'(4) = 53$ means that a 5th unit produced would change total costs by about \$53.

86. **(a)** $\overline{MR} = 40 - 0.04x$ **(b)** $x = 1000$ units

87. $\overline{MP}(10) = 48$ means if an 11th unit is sold, profit will change by about \$48.

88. **(a)** $\overline{MR} = 80 - 0.08x$ **(b)** 72
(c) If a 101st unit is sold, revenue will change by about \$72.

89. $\dfrac{120x(x+1)}{(2x+1)^2}$ 90. $\overline{MP} = 4500 - 3x^2$

91. $\overline{MP} = 16 - 0.2x$

Chapter 9 Test (page 728)

1. **(a)** $\frac{3}{4}$ **(b)** $-8/5$ **(c)** $9/8$ **(d)** does not exist

2. **(a)** $f'(x) = \lim_{h\to 0} \dfrac{f(x+h) - f(x)}{h}$
(b) $f'(x)$
$= \lim_{h\to 0} \dfrac{[3(x+h)^2 - (x+h) + 9] - [3x^2 - x + 9]}{h}$
$= \lim_{h\to 0} \dfrac{[3x^2 + 6xh + 3h^2 - x - h + 9] - [3x^2 - x + 9]}{h}$
$= \lim_{h\to 0} \dfrac{6xh + 3h^2 - h}{h} = \lim_{h\to 0} [6x + 3h - 1] = 6x - 1$

3. $x = 0, x = 8$

4. **(a)** $\dfrac{99x^2 - 24x^9}{(2x^7 + 11)^2}$
(b) $(3x^5 - 2x + 3)(40x^9 + 40x^3) + (4x^{10} + 10x^4 - 17)(15x^4 - 2)$
(c) $9(10x^4 + 21x^2)(2x^5 + 7x^3 - 5)^{11}$
(d) $2(8x^2 + 5x + 18)(2x + 5)^5$
(e) $\dfrac{6}{\sqrt{x}} + \dfrac{20}{x^3}$

5. $\dfrac{d^3y}{dx^3} = 6 + 60x^{-6}$

6. **(a)** $y = -15x - 5$ **(b)** $(4, -90), (-2, 18)$

7. **(a)** 2 **(b)** does not exist **(c)** -4

8. $g(-2) = 8;\ \lim_{x\to -2^-} g(x) = 8,\ \lim_{x\to -2^+} g(x) = -8$
$\therefore \lim_{x\to -2} g(x)$ does not exist and $g(x)$ is not continuous at $x = -2$.

9. **(a)** $P(x) = 50x - 0.01x^2 - 10{,}000$
(b) $\overline{MP} = 50 - 0.02x$
(c) $\overline{MP}(1000) = 30$ means the predicted profit from the sale of the 1001st unit is approximately \$30.

10. 104

11. **(a)** -5 **(b)** -1 **(c)** 4 **(d)** does not exist
(e) 2 **(f)** $-4, 1, 3, 6$ **(g)** $-4, 3, 6$

12. **(a)** $\frac{2}{3}$ **(b)** -4 **(c)** $\frac{2}{3}$

Exercise 10.1 (page 747)

1. **(a)** $(1, 5)$ **(b)** $(4, 1)$ **(c)** $(-1, 2)$

3. **(a)** $(1, 5)$ **(b)** $(4, 1)$ **(c)** $(-1, 2)$

5. **(a)** 3, 7 **(b)** $3 < x < 7$
(c) $x < 3, x > 7$ **(d)** 7 **(e)** 3

7. $x = 0, x = 4$

9.

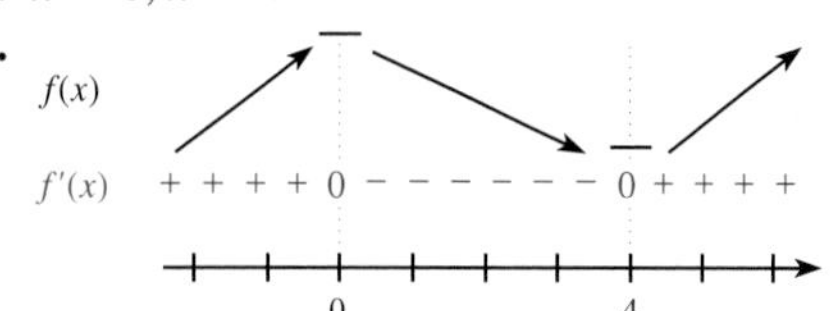

Min: $(4, -58)$; Max: $(0, 6)$

11. **(a)** $dy/dx = 3x^2 - 3$ **(b)** $x = 1, x = -1$
(c) $(1, 2), (-1, 6)$
(d) relative max at $(-1, 6)$, relative min at $(1, 2)$

13. **(a)** $dy/dx = 3x^2 + 6x + 3$ **(b)** $x = -1$
(c) $(-1, -3)$
(d) horizontal point of inflection at $(-1, -3)$

15. **(a)** $\dfrac{dy}{dx} = x - 1$ **(b)** $x = 1$ **(c)** $(1, -\frac{1}{2})$
(d) decreasing: $x < 1$
increasing: $x > 1$
(e)

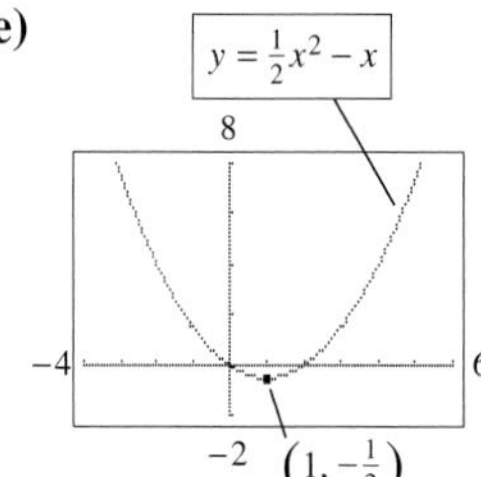

17. **(a)** $dy/dx = x^2 + x - 2$
(b) $x = -2, x = 1$ **(c)** $(-2, \frac{13}{3}), (1, -\frac{1}{6})$
(d) increasing: $x < -2$ and $x > 1$
decreasing: $-2 < x < 1$
(e)

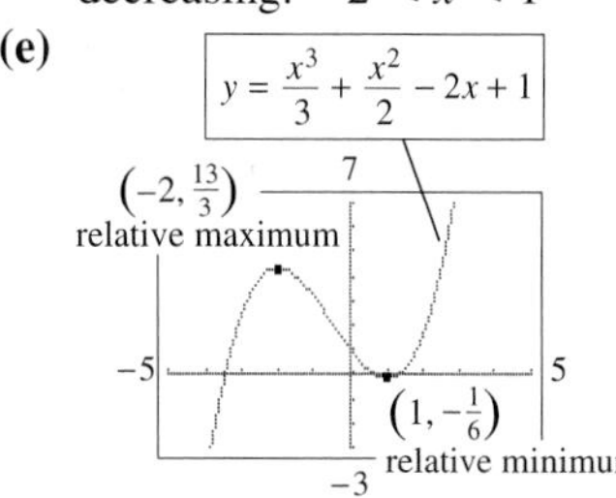

19. **(a)** $\dfrac{dy}{dx} = \dfrac{2}{3x^{1/3}}$ **(b)** $x = 0$ **(c)** $(0, 0)$
(d) decreasing: $x < 0$
increasing: $x > 0$
(e)

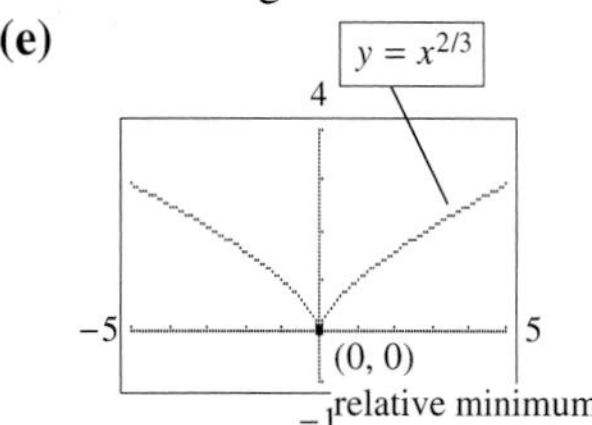

21. **(a)** $f'(x) = 0$ at $x = -\frac{1}{2}$
$f'(x) > 0$ for $x < -\frac{1}{2}$
$f'(x) < 0$ for $x > -\frac{1}{2}$
(b) $f'(x) = -1 - 2x$ verifies these conclusions

23. **(a)** $f'(x) = 0$ at $x = 0, x = -3, x = 3$
$f'(x) > 0$ for $-3 < x < 3, x \neq 0$
$f'(x) < 0$ for $x < -3$ and $x > 3$
(b) $f'(x) = \frac{1}{3}x^2(9 - x^2)$ verifies these conclusions.

25. HPI $(1, \frac{4}{3})$
no max or min

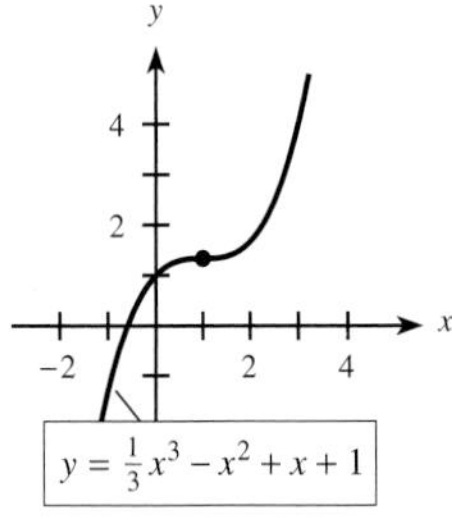

27. $(-6, 128)$ rel max;
$(4, -38\frac{2}{3})$ rel min

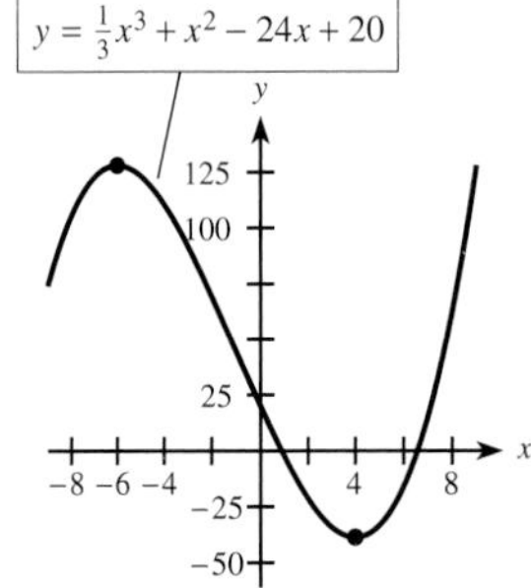

29. $(-1, 3)$ rel max;
$(1, -1)$ rel min;
HPI $(0, 1)$

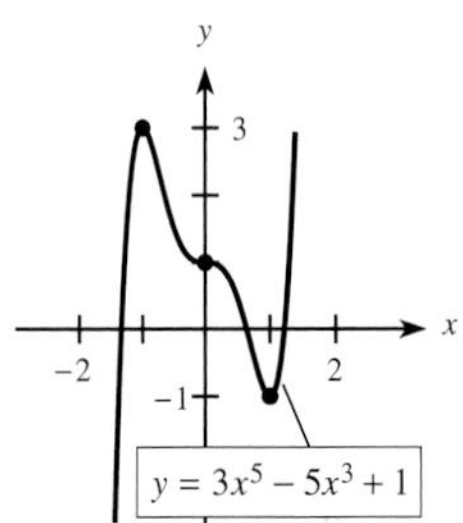

31. $(1, 1)$ rel max;
$(0, 0)$, $(2, 0)$ rel min

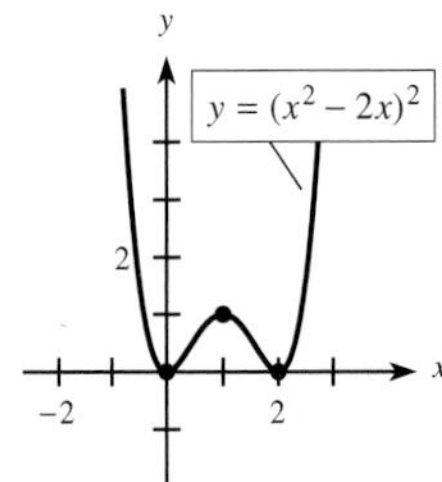

33. $(3, 4)$ rel max;
$(5, 0)$ rel min;
HPI $(0, 0)$

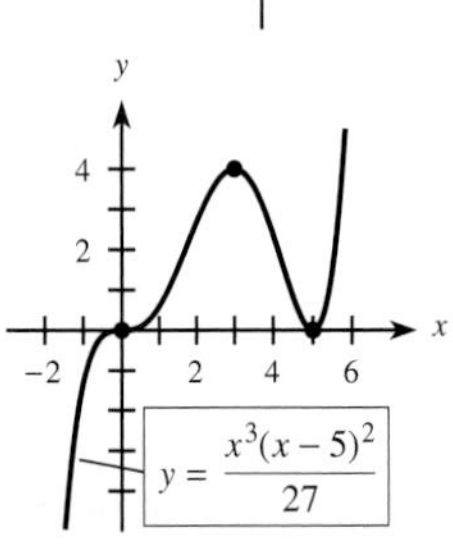

35. $(0, 0)$ rel max;
$(2, -4.8)$ rel min

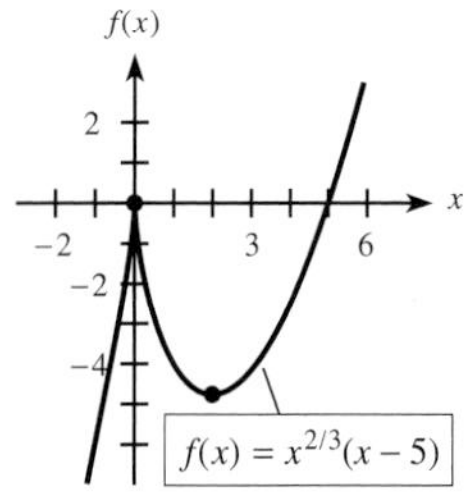

37. $(50, 300{,}500)$, $(100, 238{,}000)$
$0 \leq x \leq 150$, $0 \leq y \leq 301{,}000$

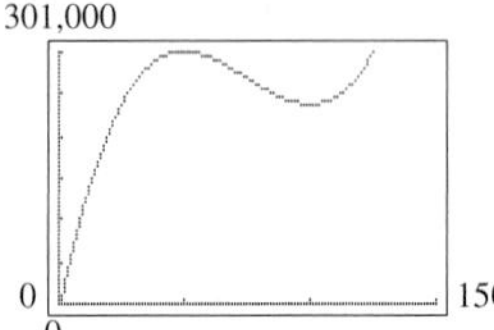

39. $(0, -40{,}000)$ $(60, 4{,}280{,}000)$
$-20 \leq x \leq 90$ $\quad -50{,}000 \leq y \leq 5{,}000{,}000$

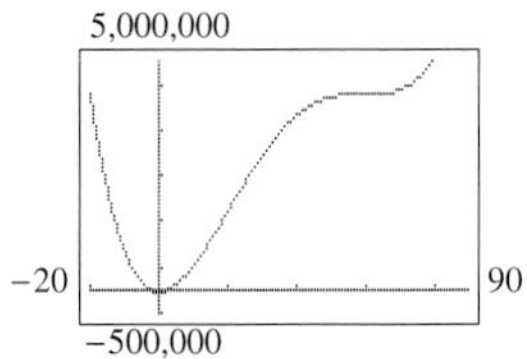

41. $f(x) = 7.5x^4 - x^3 + 2$
$(0, 2)$ HPI
$(0.1, 1.99975)$ rel min
$-0.1 \leq x \leq 0.2$
$1.9997 \leq y \leq 2.007$

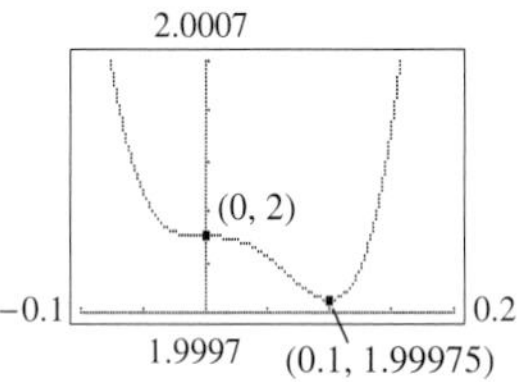

43. critical values: $x = -1, x = 2$
$f(x)$ increasing for $x < -1$ and $x > 2$
$f(x)$ decreasing for $-1 < x < 2$
rel max at $x = -1$; rel min at $x = 2$

possible graph for $f(x)$

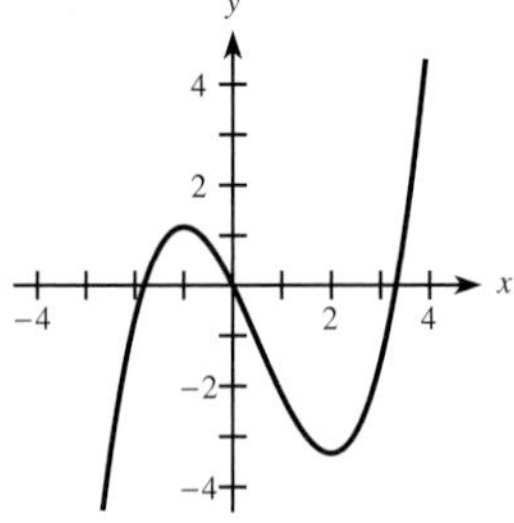

45. critical values: $x = 0, x = 3$
$f(x)$ increasing for $x > 3$
$f(x)$ decreasing for $x < 3, x \neq 0$
rel min at $x = 3$; HPI at $x = 0$

possible graph of $f(x)$

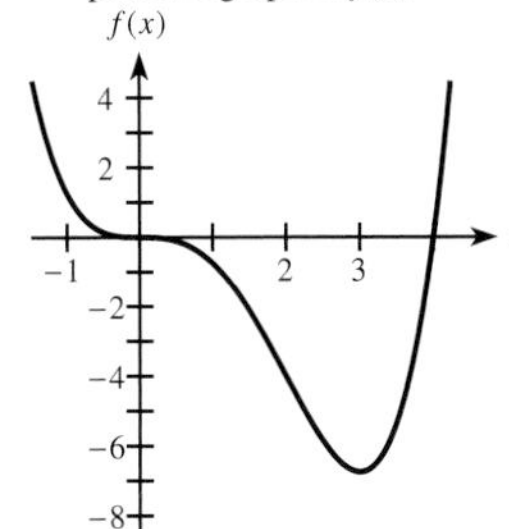

47. Graph on left is $f(x)$; on right is $f'(x)$ because $f(x)$ is increasing when $f'(x) > 0$ (i.e., above the x-axis) and $f(x)$ is decreasing when $f'(x) < 0$ (i.e., below the x-axis).

49. decreasing for $t \geq 0$

51. **(a)** $2 \pm \sqrt{13}$
(b) $2 + \sqrt{13} \approx 5.6$
(c) $0 \leq t < 2 + \sqrt{13}$

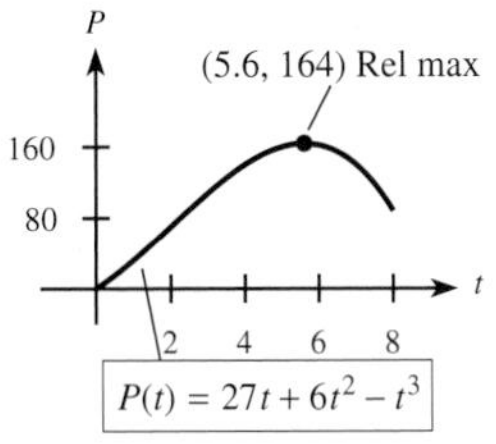

53. **(a)** $x = 5$ **(b)** decreasing for $0 < x < 5$
(c) $x > 5$

55. **(a)** at $x = 150$, increasing; at $x = 250$, changing from increasing to decreasing; at $x = 350$, decreasing
(b) increasing for $x < 250$ **(c)** 250 units

57. **(a)** $t = 6$ **(b)** 6 weeks

59. **(a)** 10 **(b)** January 1

61. **(a)** $x \approx 10.12$, during 1990
(b) $x \approx 1.76$, during 1981

63. **(a)** $y = -9.3458x^2 + 34.0863x + 235.8625$
(b) at $x = 1.82$, late in 1991; data indicate the maximum in 1992.

Exercise 10.2 *(page 761)*

1. **(a)** concave down **(b)** concave up

3. **(a)** concave down **(b)** concave up

5. (a, c) and (d, e) **7.** (c, d) (e, f) **9.** c, d, e

11. concave up when $x > 2$; concave down when $x < 2$; POI at $x = 2$

13. concave up when $x < -2$ and $x > 1$
concave down when $-2 < x < 1$
points of inflection at $x = -2$ and $x = 1$

15. no points of inflection; $(2, -2)$ min

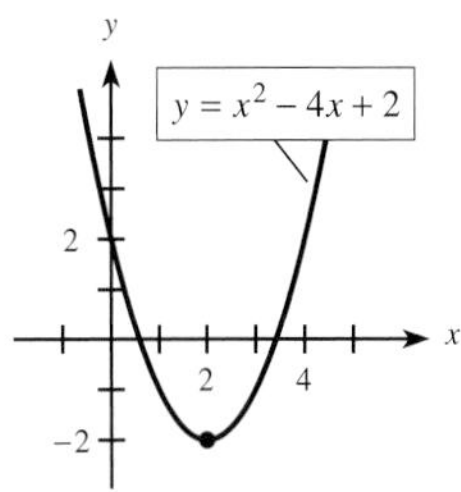

17. $(1, \frac{10}{3})$ max; $(3, 2)$ min; $(2, \frac{8}{3})$ point of inflection

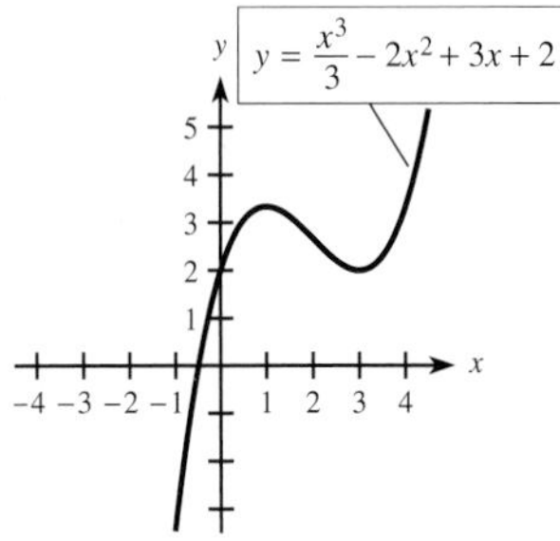

19. $(0, 0)$ rel max; $(2\sqrt{2}, -64)$, $(-2\sqrt{2}, -64)$ min; points of inflection: $(2\sqrt{6}/3, -320/9)$ and $(-2\sqrt{6}/3, -320/9)$

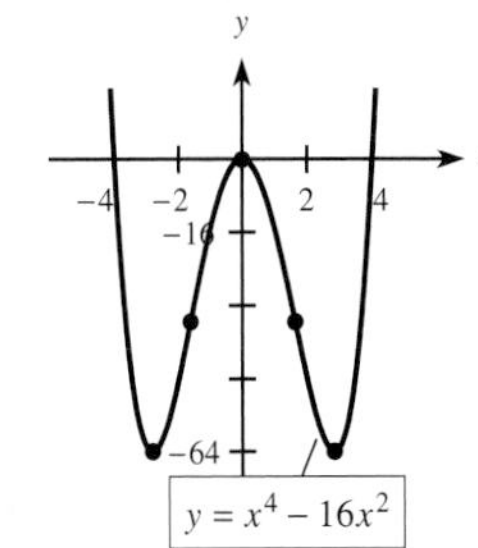

21. $(-2, 64)$ rel max; $(2, -64)$ rel min; and points of inflection: $(-\sqrt{2}, 39.6)$, $(0, 0)$ and $(\sqrt{2}, -39.6)$

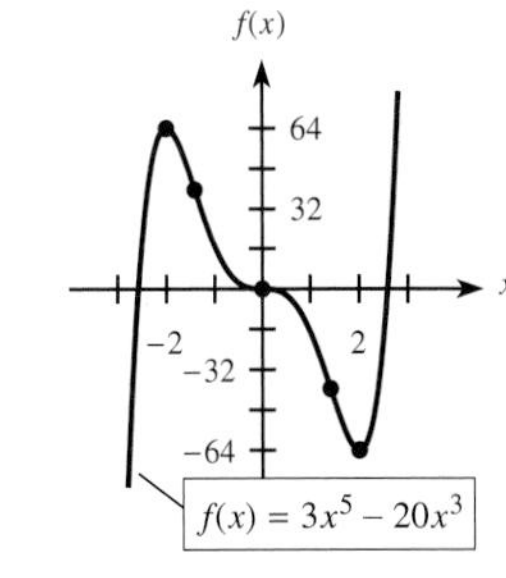

23. $(1, -3)$ min; points of inflection: $(-2, 7.6)$ and $(0, 0)$

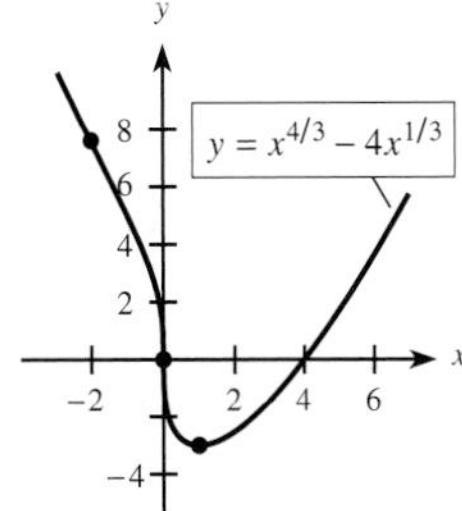

25. **(a)** $f''(x) = 0$ when $x = 1$
$f''(x) > 0$ when $x < 1$
$f''(x) < 0$ when $x > 1$
(b) rel max for $f'(x)$ at $x = 1$
no rel min
(c)

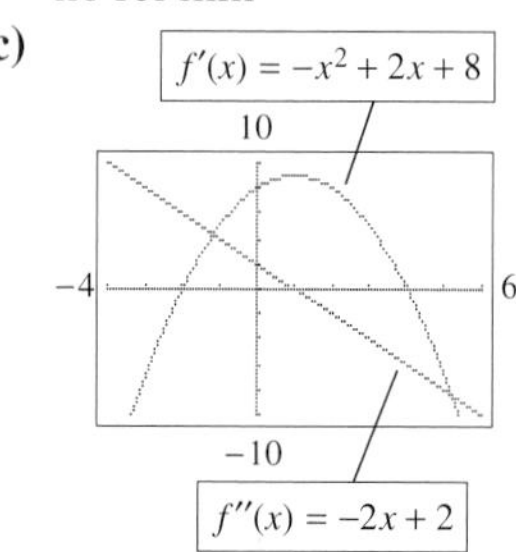

27. **(a)** concave up when $x < 2$ concave down when $x > 2$
(b) point of inflection at $x = 2$
(c) **(d)**

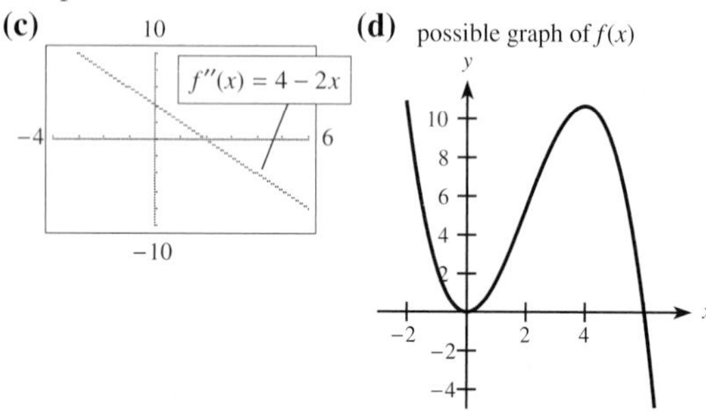

29. **(a)** G **(b)** C **(c)** F **(d)** H **(e)** I
31. **(a)** concave up when $x < 0$
concave down when $x > 0$
point of inflection at $x = 0$
(b) concave up when $-1 < x < 1$
concave down when $x < -1$ and $x > 1$
POI at $x = -1$ and $x = 1$
(c) concave up when $x > 0$
concave down when $x < 0$
point of inflection at $x = 0$
33. **(a)** $P'(t)$ **(b)** B **(c)** C
35. **(a)** C **(b)** right **(c)** yes
37. **(a)** in an 8-hour shift, max when $t = 8$ **(b)** 4 hr
39. **(a)** 9 days **(b)** 15 days
41. $x = 9.53$, during 1989
43. **(a)** $y = -0.00816x^3 + 0.4635x^2 - 8.1122x + 64.7785$
(b) min at $x = 13.73$, during 1983
max at $x = 24.14$, during 1994
(c) no; data give min in 1984, max in 1991.
45. **(a)** $y = -0.4052x^3 + 6.0161x^2 - 17.9605x + 52$
(b) $x = 1.83$, during 1913
(c) $x = 8.07$, during 1920

Exercise 10.3 *(page 773)*

1. min -6 at $x = 2$, max $3.\overline{481}$ at $x = -2/3$
3. min -1 at $x = -2$, max 2 at $x = -1$
5. **(a)** $x = 1800$ units, $R = \$32{,}400$
(b) $x = 1500$ units, $R = \$31{,}500$
7. $x = 20$ units, $R = \$24{,}000$ **9.** 40 people
11. $p = \$15$, $R = \$22{,}500$
13. **(a)** max $= \$2100$ **(b)** $x = 10$
15. $x = 5$ units, $\overline{C} = \$23$
17. $x = 10$ units, $\overline{C} = \$20$
19. 200 units ($x = 2$), $\overline{C} = \$108$ **21.** $x = 5$
23. $x = 80$ units, $P = \$280{,}000$
25. $x = 10\sqrt{15} \approx 39$ units, $P \approx \$71{,}181$ (using $x = 39$)
27. $x = 1000$ units, $P = \$39{,}700$
29. $x = 600$ units, $P = \$495{,}000$
31. **(a)** 60 **(b)** \$570 **(c)** \$9000
33. **(a)** 1000 units **(b)** \$8066.67 (approximately)
35. 2000 units priced at \$90/unit; max profit is \$90,000/wk
37. rent $= \$430$
39. **(a)** $t \approx 12.5$ (in 1992), $R = \$5.0353$ billion
(b) the data show a max in 1990 of \$5.1686 billion
41. **(a)** August 1991; approximately \$3 billion
(b) August or September 1991; approximately \$5.5 billion
(c) December 1992 or January 1993; approximately \$12 billion
(d) May 1993; approximately \$9.5 billion
43. **(a)** About January 1980 or January 1991
(b) (January 25, 1994, 37%)
(c) No, either some citizens think crime is the most important problem or no citizens think that.
(d) Crime would be a less important issue, and the percent would drop.
(e) Not 1980, because crime was a low priority. Yes, in 1994, as seen from the high concern about crime then.

Exercise 10.4 *(page 785)*

1. **(a)** $x_1 = \$25$ million, $x_2 = \$13.846$ million
(b) \$38.846 million
3. 400 trees **5.** **(a)** 5 **(b)** 237.5 **7.** \$50
9. $m = c$ **11.** 1 week **13.** $t = 8$, $p = 45\%$
15. 240 ft **17.** 300 ft × 150 ft
19. 20 ft long, $6\frac{2}{3}$ ft across (dividers run across)
21. 4 in. × 8 in. × 8 in. high **23.** 30,000
25. 12,000 **27.** $x = 2$ **29.** 3 weeks from now
31. 25 plates

Exercise 10.5 (page 797)

1. **(a)** $x = 2$ **(b)** 1 **(c)** $y = 1$ **(d)** 1
3. **(a)** $x = 2$ **(b)** 1 **(c)** $y = 1$ **(d)** 1
5. HA: $y = 2$; VA: $x = 3$
7. HA: $y = 0$; VA: $x = -2, x = 2$
9. HA: none; VA: none
11. HA: $y = 2$; VA: $x = 3$
    no max, min, or points of inflection

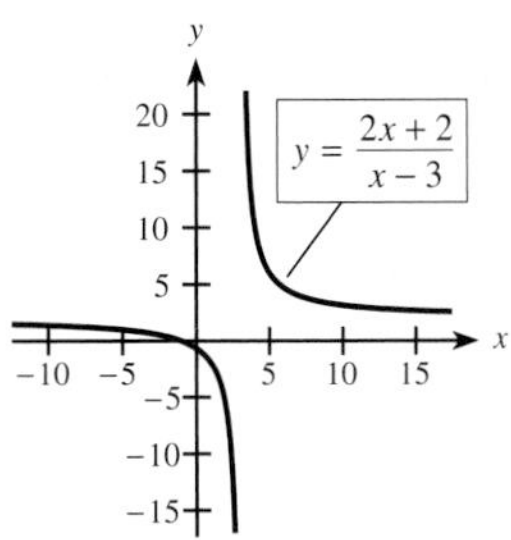

13. VA: $x = 0$;
    $(-2, -4)$ rel max;
    $(2, 4)$ rel min

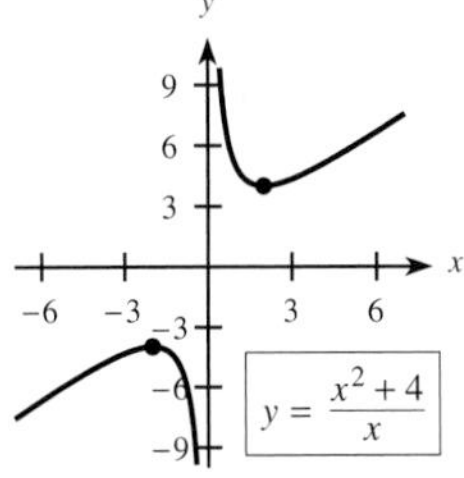

15. VA: $x = -1$; HA: $y = 0$;
    $(0, 0)$ rel min; $(2, 4)$ rel max;

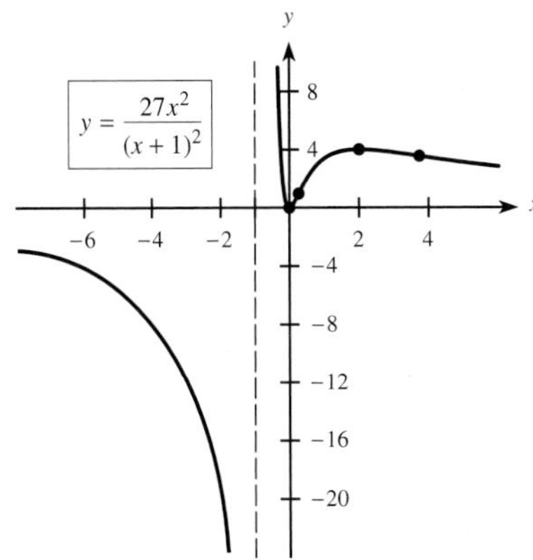

17. HA: $y = 0$; $(1, 8)$ rel max;
    $(-1, -8)$ rel min;
    points of inflection:
    $(0, 0)$, $(-\sqrt{3}, -4\sqrt{3})$,
    and $(\sqrt{3}, 4\sqrt{3})$

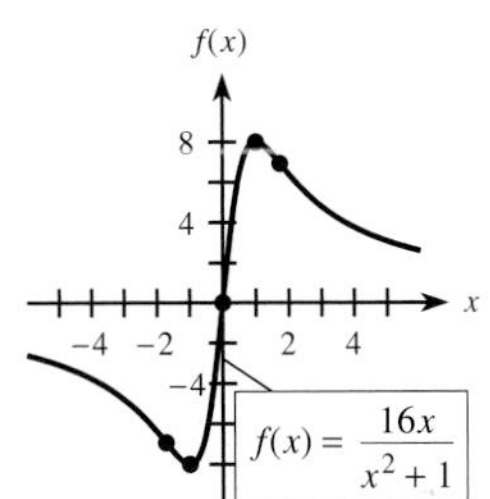

19. HA: $y = 0$; VA: $x = 1$; $(-1, -\frac{1}{4})$ rel min; point of inflection: $(-2, -\frac{2}{9})$

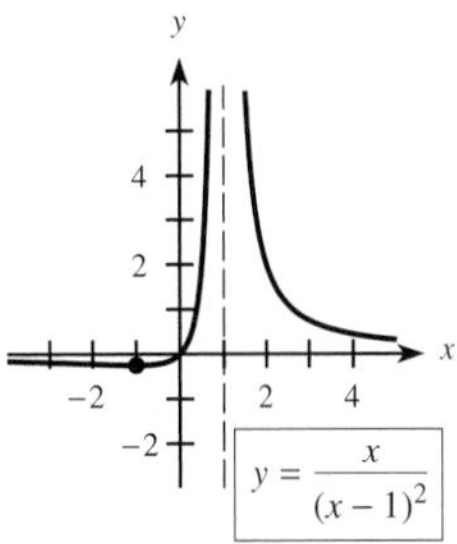

21. VA: $x = 3$;
    $(2, -1)$ rel max;
    $(4, 7)$ rel min

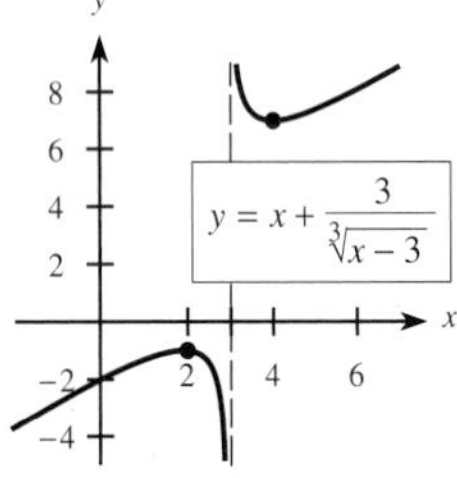

23. HA: $y = 0$; VA: $x = 0$
    $(2, 0)$ rel min; $(3, 1)$ rel max
    points of inflection: $(1.87, 0.66)$, $(4.13, 0.87)$

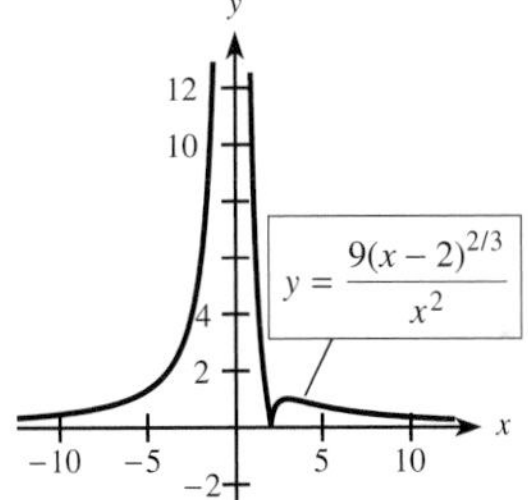

25. **(a)** HA: approx. $y = -2$; VA: approx. $x = 4$
    (b) HA: $y = -\frac{9}{4}$; VA: $x = \frac{17}{4}$
27. **(a)** HA: approx. $y = 2$;
    VA: approx. $x = 2.5$, $x = -2.5$
    (b) HA: $y = \frac{20}{9}$; VA: $x = \frac{7}{3}$, $x = -\frac{7}{3}$
29. $f(x) = \dfrac{x + 25}{x^2 + 1400}$
    (a)

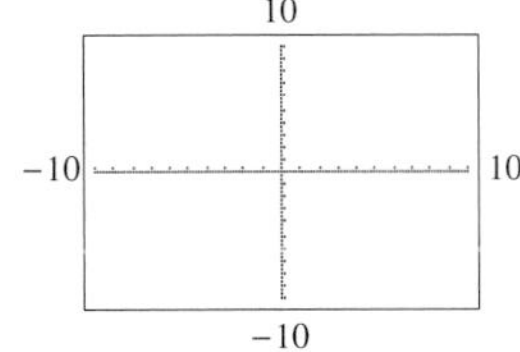

    (b) HA: $y = 0$; rel min $(-70, -0.0071)$
    rel max $(20, 0.025)$

(c) x: -500 to 400
y: -0.01 to 0.03

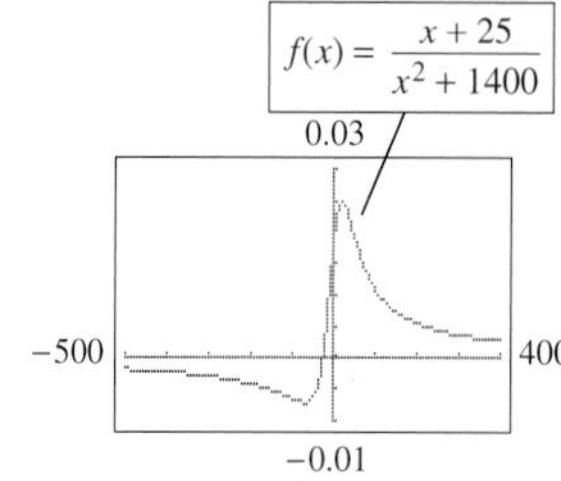

31. $f(x) = \dfrac{100(9 - x^2)}{x^2 + 100}$

(a)

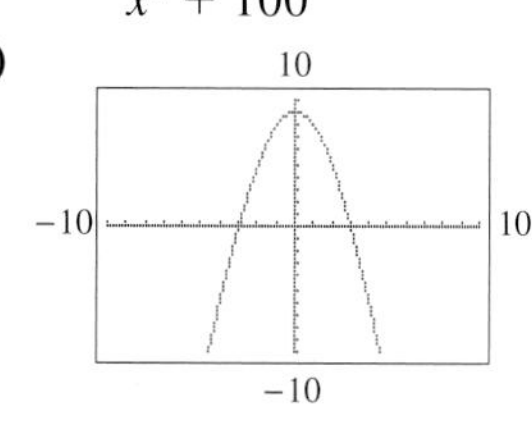

(b) HA: $y = -100$; rel max $(0, 9)$

(c) x: -75 to 75
y: -120 to 20

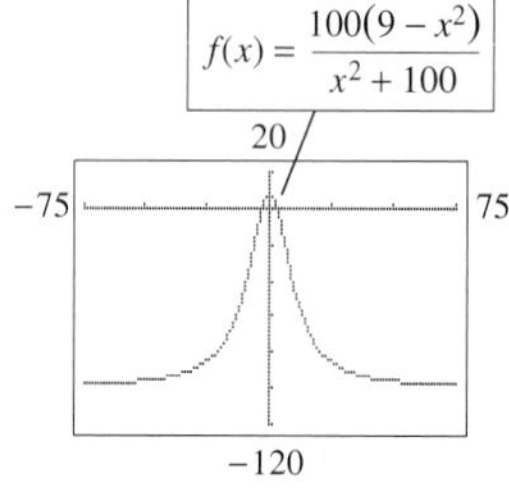

33. $f(x) = \dfrac{1000x - 4000}{x^2 - 10x - 2000}$

(a)

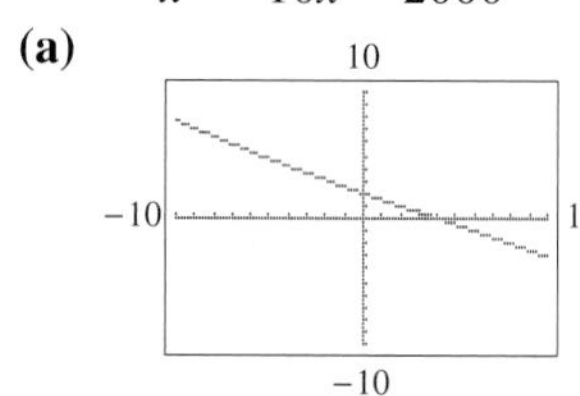

(b) HA: $y = 0$
VA: $x = -40$, $x = 50$
no max or min

(c) x: -200 to 200
y: -200 to 200

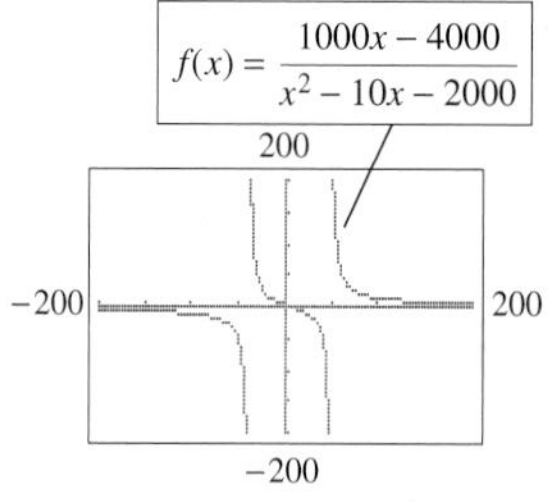

35. **(a)** none **(b)** $C \geq 0$ **(c)** $p = 100$ **(d)** no

37. **(a)**

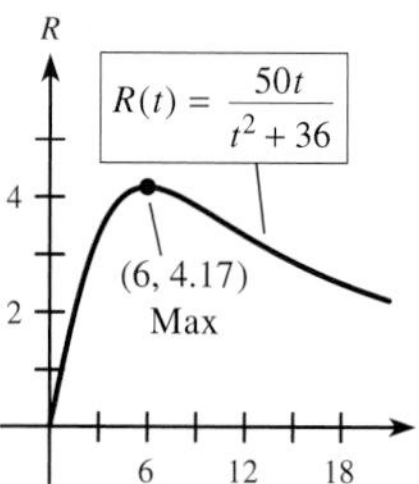

(b) 6 weeks
(c) 22 weeks after its release

39. **(a)** yes, $x = -1$ **(b)** no; domain is $x \geq 5$
(c) yes, $y = -58.5731$
(d) At 0°F, as the wind speed increases, there is a limiting wind chill of about -58.6°F. This is meaningful because at high wind speeds, additional wind probably has little noticeable effect.

41. **(a)** $P = C$ **(b)** C **(c)** $P' = 0$ **(d)** 0

43. **(a)** 0
(b) As the years past 1800 increase, the percentage of workers in farm occupations approaches 0.
(c) no

45. **(a)** No. Barometric pressure can drop off the scale (as shown), but it cannot decrease without bound. In fact, it must always be positive.
(b) See your library with regard to the "Storm of the Century" in March 1993.

Chapter 10 Review Exercises *(page 801)*

1. $(0, 0)$ max **2.** $(2, -9)$ min **3.** HPI $(1, 0)$

4. $(1, \frac{3}{2})$ max, $(-1, -\frac{3}{2})$ min

5. **(a)** $\frac{1}{3}, -1$ **(b)** $(-1, 0)$ rel max, $(\frac{1}{3}, -\frac{32}{27})$ rel min
(c) none **(d)**

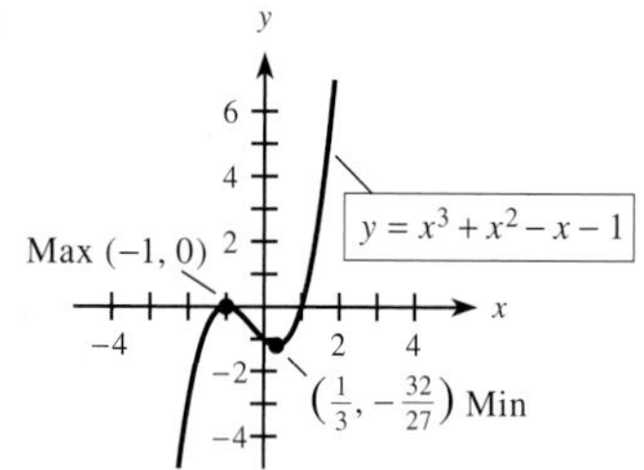

6. **(a)** 3, 0 **(b)** $(3, 27)$ max **(c)** $(0, 0)$
(d)

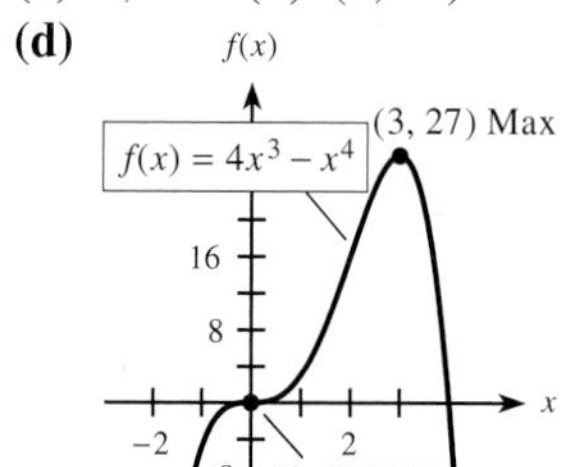

7. **(a)** $-1, 6$
(b) $(-1, 11)$ rel max, $(6, -160.5)$ rel min
(c) none **(d)**

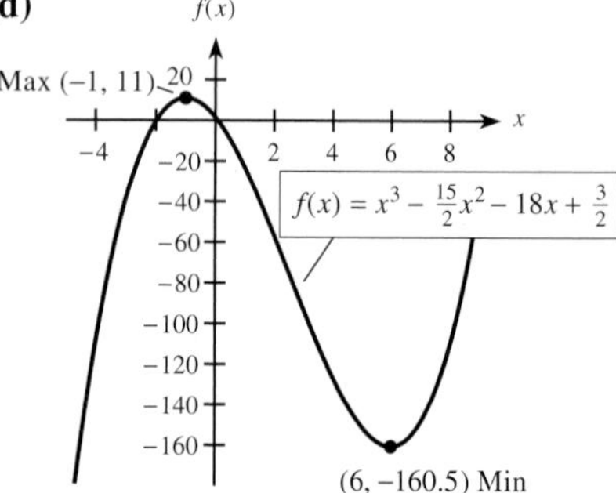

8. **(a)** 0, ± 1 **(b)** $(-1, 1)$ rel max, $(1, -3)$ rel min
(c) $(0, -1)$ **(d)**

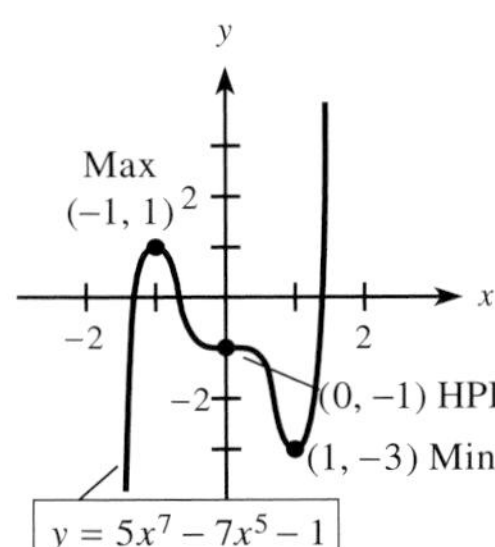

9. **(a)** 0 **(b)** $(0, -1)$ min **(c)** none
(d)

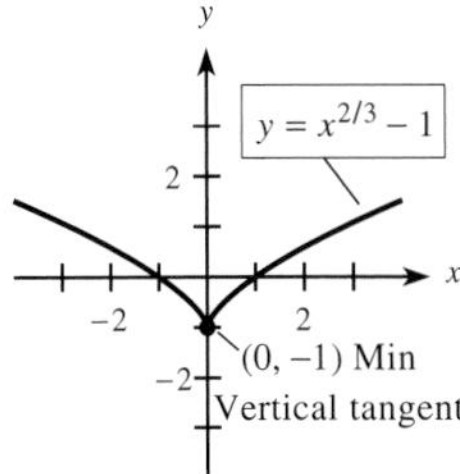

10. **(a)** 0, 1, 4
(b) $(0, 0)$ rel min, $(1, 9)$ rel max, $(4, 0)$ rel min
(c) none **(d)**

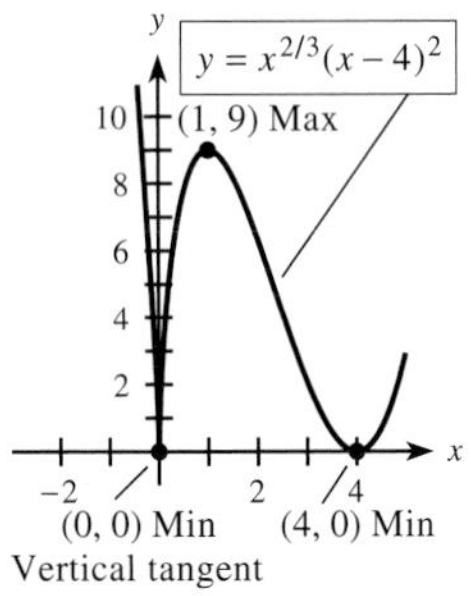

11. concave up

12. concave up when $x < -1$ and $x > 2$
concave down when $-1 < x < 2$
points of inflection at $(-1, -3)$ and $(2, -42)$

13. $(-1, 15)$ rel max; $(3, -17)$ rel min;
point of inflection $(1, -1)$

14. $(-2, 16)$ rel max; $(2, -16)$ rel min;
point of inflection $(0, 0)$

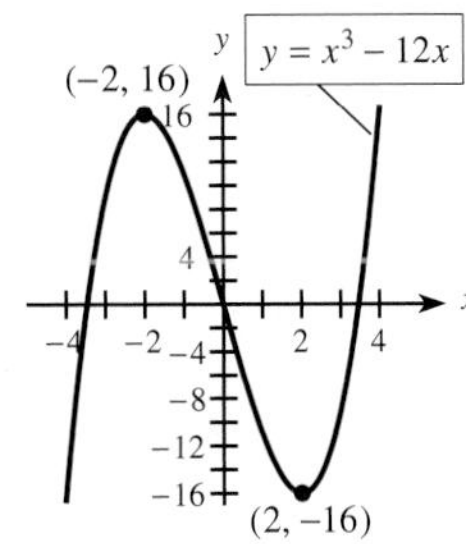

15. $(1, 4)$ rel max; $(-1, 0)$ rel min; points of inflection:
$\left(\frac{1}{\sqrt{2}}, 2 + \frac{7}{4\sqrt{2}}\right)$, $(0, 2)$, and $\left(-\frac{1}{\sqrt{2}}, 2 - \frac{7}{4\sqrt{2}}\right)$

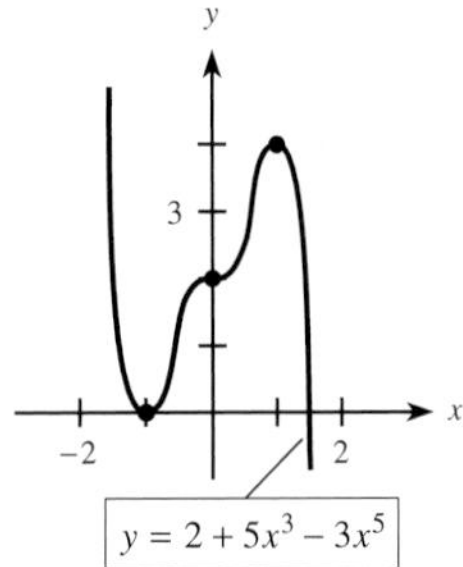

16. **(a)** $(0, 0)$ absolute min; $(140, 19{,}600)$ absolute max
(b) $(0, 0)$ absolute min; $(100, 18{,}000)$ absolute max

17. **(a)** $(50, 233{,}333)$ absolute max; $(0, 0)$ absolute min
(b) $(64, 248{,}491)$ absolute max; $(0, 0)$ absolute min

18. **(a)** $x = 1$ **(b)** $y = 0$ **(c)** 0 **(d)** 0

19. **(a)** $x = -1$ **(b)** $y = \frac{1}{2}$ **(c)** $\frac{1}{2}$ **(d)** $\frac{1}{2}$

20. HA: $y = \frac{3}{2}$; VA: $x = 2$

21. HA: $y = -1$; VA: $x = 1, x = -1$

22. **(a)** HA: $y = 3$; VA: $x = -2$
(b) no max or min **(c)**

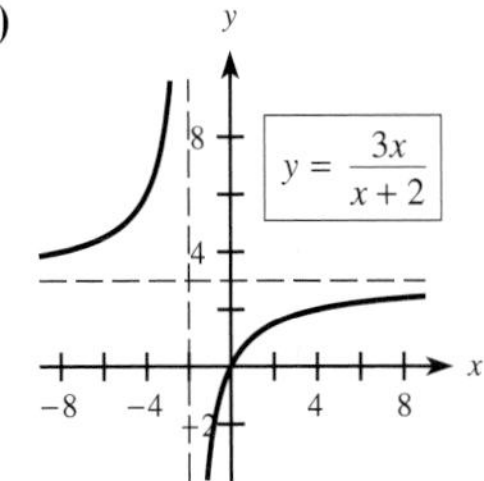

23. **(a)** HA: $y = 0$; VA: $x = 0$
(b) $(4, 1)$ max **(c)**

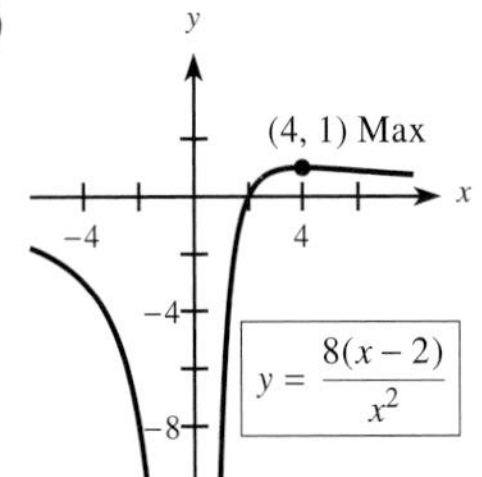

24. **(a)** HA: none; VA: $x = 1$
(b) $(0, 0)$ rel max; $(2, 4)$ rel min
(c)

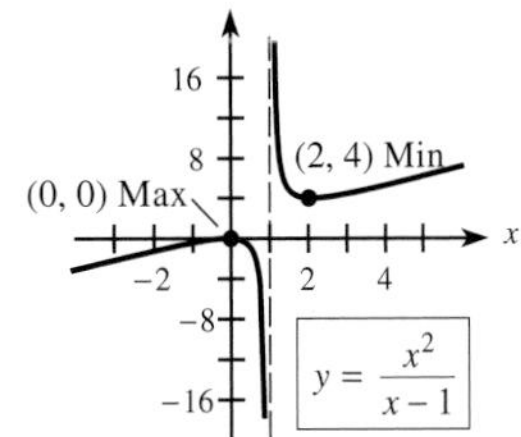

25. (a) $f'(x) > 0$ for $x < \frac{2}{3}$ (approximately) and $x > 2$
$f'(x) < 0$ for about $\frac{2}{3} < x < 2$
$f'(x) = 0$ about $x = \frac{2}{3}$ and $x = 2$

(b) $f''(x) > 0$ for $x > \frac{4}{3}$
$f''(x) < 0$ for $x < \frac{4}{3}$
$f''(x) = 0$ at $x = \frac{4}{3}$

(c)
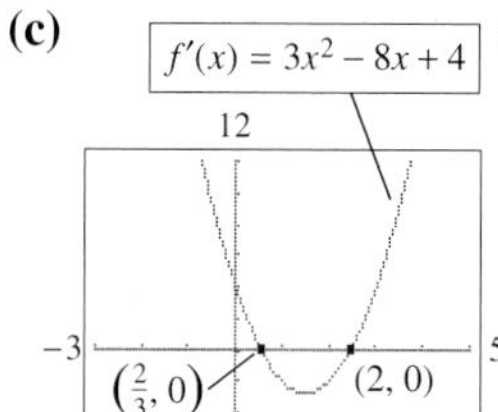

(d)
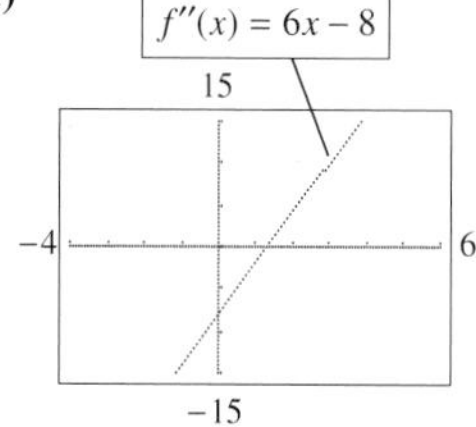

26. (a) $f'(x) > 0$ for about $-13 < x < 0$ and $x > 7$
$f'(x) < 0$ for about $x < -13$ and $0 < x < 7$
$f'(x) = 0$ for about $x = 0, x = -13, x = 7$

(b) $f''(x) > 0$ for about $x < -8$ and $x > 4$
$f''(x) < 0$ for about $-8 < x < 4$
$f''(x) = 0$ for about $x = -8$ and $x = 4$

(c)
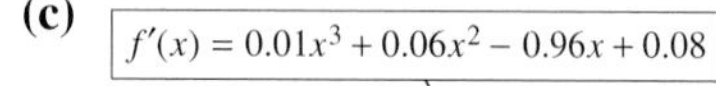

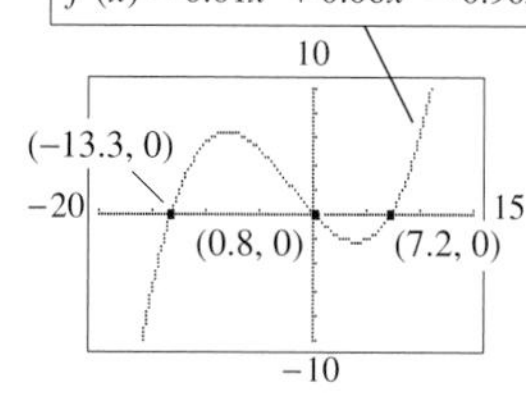

(d)
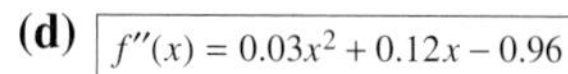

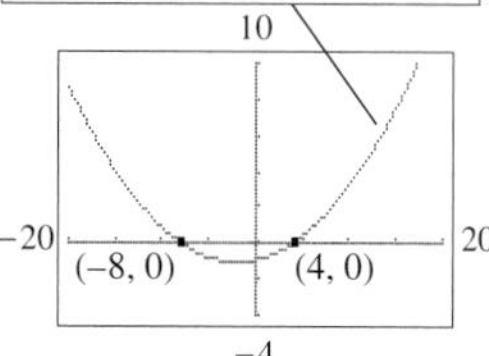

27. (a) $f(x)$ increasing for $x < -5$ and $x > 1$
$f(x)$ decreasing for $-5 < x < 1$
$f(x)$ has rel max at $x = -5$, rel min at $x = 1$

(b) $f''(x) > 0$ for $x > -2$ (where $f'(x)$ increases)
$f''(x) < 0$ for $x < -2$ (where $f'(x)$ decreases)
$f''(x) = 0$ for $x = -2$

(c)
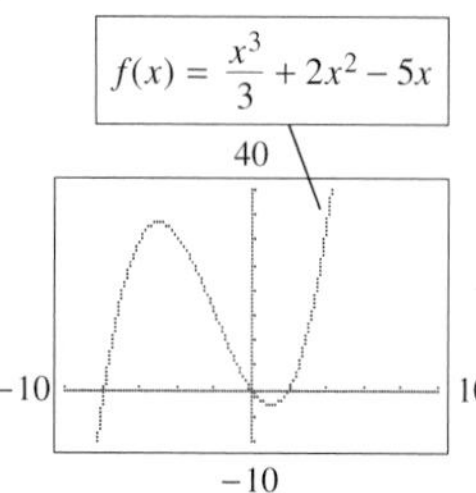

(d)
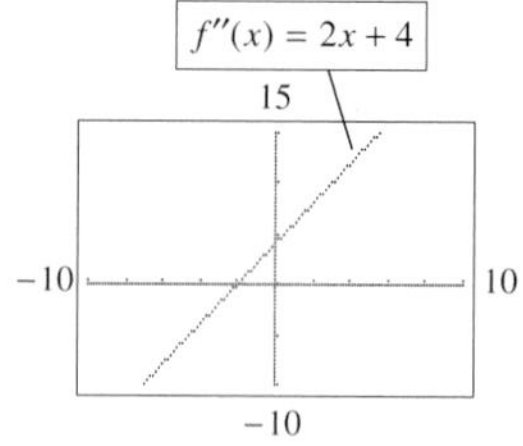

28. (a) $f(x)$ increasing for $x < 6, x \neq 0$
$f(x)$ decreasing for $x > 6$
$f(x)$ has max at $x = 6$, point of inflection at $x = 0$

(b) $f''(x) > 0$ for $0 < x < 4$
$f''(x) < 0$ for $x < 0$ and $x > 4$
$f''(x) = 0$ at $x = 0$ and $x = 4$

(c)
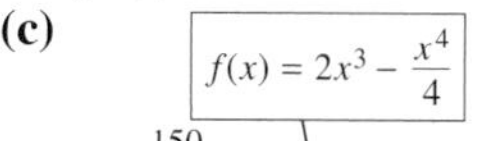

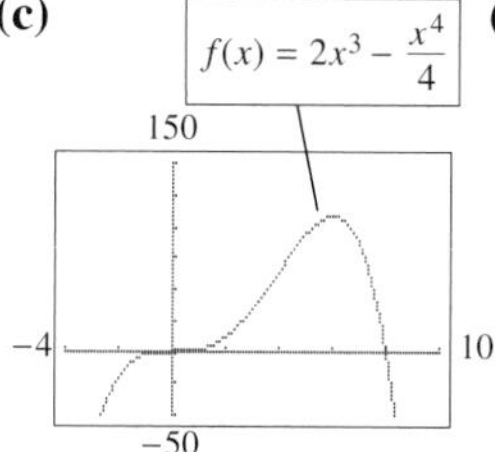

(d)
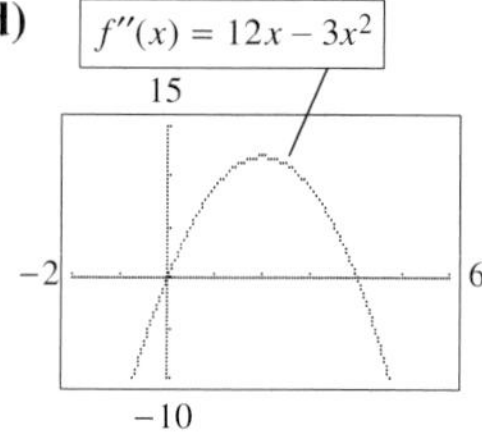

29. (a) $f(x)$ is concave up for $x < 4$
$f(x)$ is concave down for $x > 4$
$f(x)$ has point of inflection at $x = 4$

(b) $f''(x) = 4 - x$

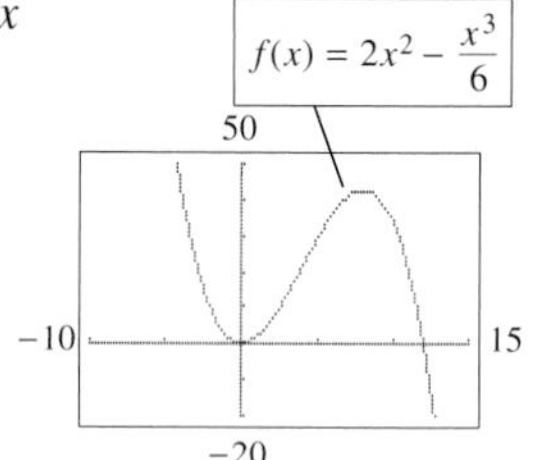

30. (a) $f(x)$ is concave up for $-3 < x < 2$
$f(x)$ is concave down for $x < -3$ and $x > 2$
$f(x)$ has points of inflection at $x = -3$ and $x = 2$

(b) $f''(x) = 6 - x - x^2$

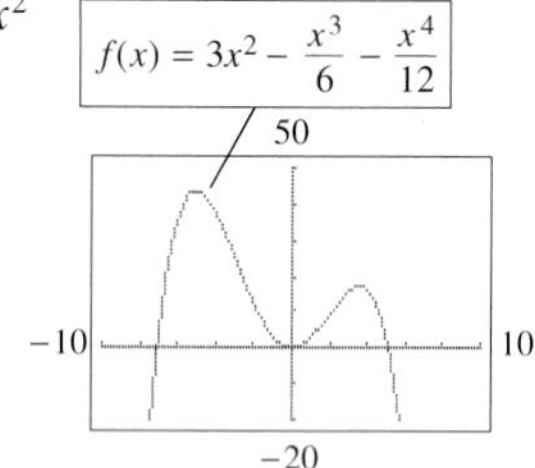

31. $x = 5$ units, $\overline{C} = \$45$

32. (a) $x = 1600$ units, $R = \$25,600$
(b) $x = 1600$ units, $R = \$25,600$

33. $P = \$54,000$ at $x = 100$ units **34.** $x = 3$ units

35. $x = 152$ units **36.** $x = 7$ units **37.** \$300

38. selling 175 sets at \$425 each

39. \$93,625 at 325 units
40. **(a)** 150 **(b)** \$650
41. \$208,490.67 at 64 units **42.** $x = 1000$ units
43. 10:00 A.M. **44.** 325 in 1995
45. 20 mi from A, 10 mi from B **46.** 4 ft × 4 ft
47. $8\frac{3}{4}$ in. × 10 in. **48.** 500 **49.** $\frac{29}{18}$ **50.** 24,000

Chapter 10 Test (page 805)

1. max $(-3, 3)$; min $(-1, -1)$

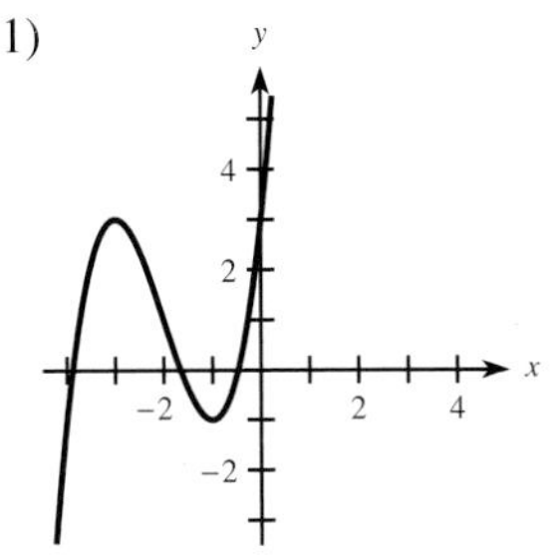

2. HPI $(1, 5)$

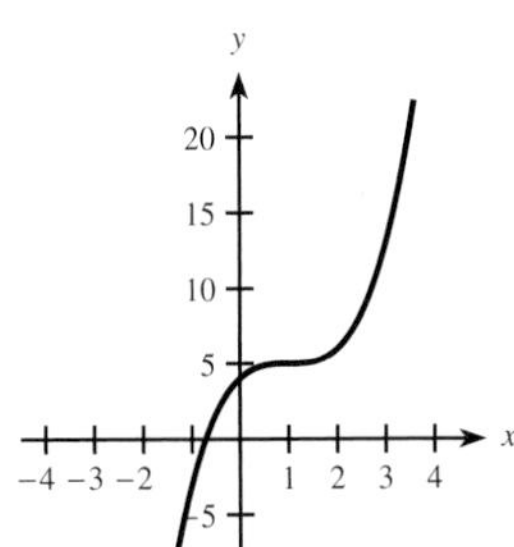

3. max $(0, -3)$; min $(4, 5)$; vert asym $x = 2$

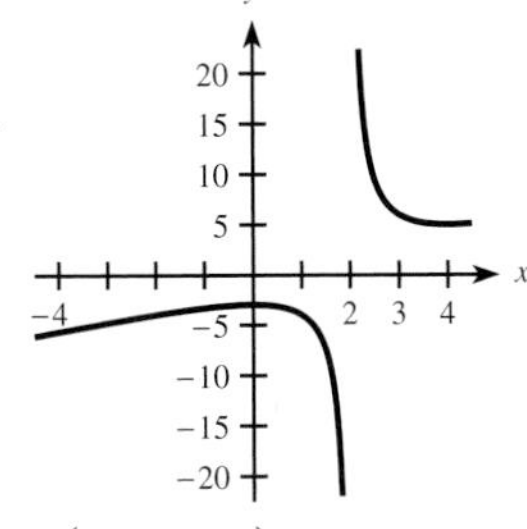

4. $\left(-\frac{1}{\sqrt{2}}, 0\right)$ and $\left(\frac{1}{\sqrt{2}}, \infty\right)$

5. $(0, 2)$, HPI; $\left(-\frac{1}{\sqrt{2}}, 3.237\right), \left(\frac{1}{\sqrt{2}}, 0.763\right)$

6. max $(-1, 4)$; min $(1, 0)$

7. max 67 at $x = 8$; min -122 at $x = 5$

8. horiz asym $y = 200$; vert asym $x = -300$

9.

Point	f	f'	f''
A	$-$	$+$	$-$
B	$+$	$-$	0
C	$+$	0	$+$

10. **(a)** -2 **(b)** $x = -1$

11. local max at $(6, 10)$

12. **(a)** $x = 43.7$, during 1943
(b) The ratio of priests to Catholics was highest in 1943.

13. **(a)** $x = 7200$ **(b)** \$518,100 **14.** 100 units

15. \$250 **16.** $\frac{10}{3}$ centimeter **17.** 28,000 units

18. **(a)** $y = -0.0161x^2 + 3.3185x - 165.278$
(b) $x = 103.059$; during 2003

Exercise 11.1 (page 818)

1. $f'(x) = 4/x$ **3.** $y' = 1/x$

5. $y' = 4/x$ **7.** $f'(x) = \dfrac{4}{4x + 9}$

9. $y' = \dfrac{4x - 1}{2x^2 - x} + 3$ **11.** $dp/dq = 2q/(q^2 + 1)$

13. **(a)** $y' = \dfrac{1}{x} - \dfrac{1}{x - 1}$ **(b)** $y' = \dfrac{-1}{x(x - 1)}$

15. **(a)** $y' = \dfrac{2x}{3(x^2 - 1)}$ **(b)** $y' = \dfrac{2x}{3(x^2 - 1)}$

17. **(a)** $y' = \dfrac{4}{4x - 1} - \dfrac{3}{x}$ **(b)** $y' = \dfrac{-8x + 3}{x(4x - 1)}$

19. **(a)** $y' = \dfrac{3}{x} + \dfrac{1}{2(x + 1)}$ **(b)** $y' = \dfrac{7x + 6}{2x(x + 1)}$

21. $\dfrac{dp}{dq} = \dfrac{(q^2 + 1)}{q(q^2 - 1)}$ **23.** $\dfrac{dy}{dt} = -\dfrac{(3t^2 - 4t - 3)}{2(1 - t)(t^2 + 3)}$

25. $\dfrac{dy}{dx} = 1 - \dfrac{1}{x}$ **27.** $y' = (1 - \ln x)/x^2$

29. $y' = 8x^3/(x^4 + 3)$ **31.** $y' = \dfrac{4(\ln x)^3}{x}$

33. $y' = \dfrac{8x^3 \ln(x^4 + 3)}{x^4 + 3}$ **35.** $y' = \dfrac{1}{x \ln 4}$

37. $y' = \dfrac{4x^3 - 12x^2}{(x^4 - 4x^3 + 1) \ln 6}$

39. rel min $(e^{-1}, -e^{-1})$

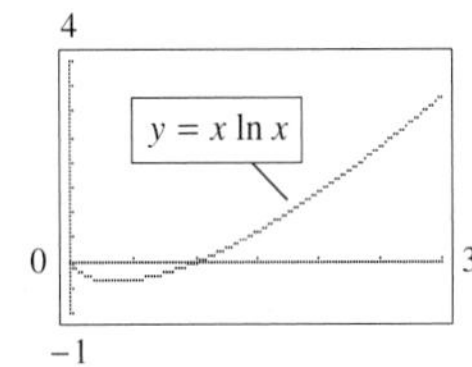

41. rel min $(2, 4 - 8 \ln 2)$

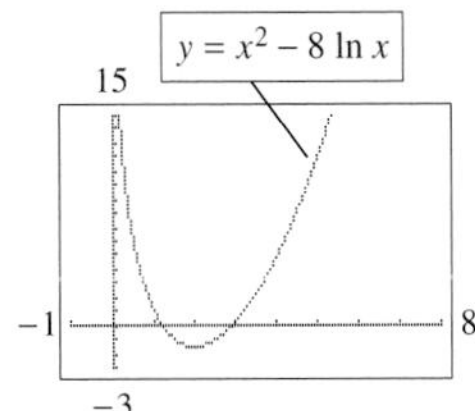

43. **(a)** $\overline{MC} = \dfrac{400}{2x + 1}$

(b) $\overline{MC} = \dfrac{400}{401} \approx 1.0$; the approximate cost of the 201st unit is \$1.00

45. **(a)** $\overline{MR} = \dfrac{2500[(x + 1)\ln(10x + 10) - x]}{(x + 1)\ln^2(10x + 10)}$
(b) 309.67; at 100 units, selling 1 additional unit yields \$309.67.
47. **(a)** -5.23 **(b)** -1.89 **(c)** increasing
49. A/B **51.** $dR/dI = 1/(I \ln 10)$
53. **(a)** $y = 561.2713 + 2646.5135 \ln(x)$
(b) \$115/year

Exercise 11.2 (page 825)

1. $y' = 5e^x - 1$ **3.** $f'(x) = e^x - ex^{e-1}$
5. $y' = 3x^2e^{x^3}$ **7.** $y' = 36xe^{3x^2}$
9. $y' = 12x(x^2 + 1)^2e^{(x^2+1)^3}$ **11.** $y' = 3x^2$
13. $y' = e^{-1/x}/x^2$ **15.** $y' = \dfrac{2}{x^3}e^{-1/x^2} - 2xe^{-x^2}$
17. $ds/dt = te^t(t + 2)$ **19.** $4x^3 e^{x^4} - 4e^{4x}$
21. $\dfrac{4e^{4x}}{e^{4x} + 2}$ **23.** $y' = e^{-3x}/x - 3e^{-3x}\ln(2x)$
25. $y' = (2e^{5x} - 3)/e^{3x} = 2e^{2x} - 3e^{-3x}$
27. $y' = 30e^{3x}(e^{3x} + 4)^9$ **29.** $y' = 6^x \ln 6$
31. $y' = 4^{x^2}(2x \ln 4)$
33. **(a)** $y'(1) = 0$ **(b)** $y = e^{-1}$
35. **(a)** $z = 0$ **(b)**

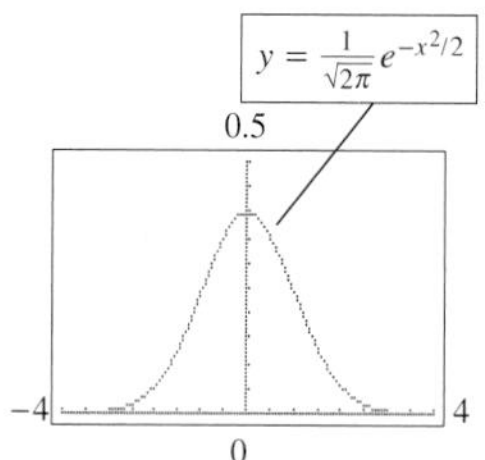

37. rel min at $x = 1, y = e$
39. rel max at $x = 0, y = -1$
41. **(a)** $(0.1)Pe^{0.1n}$ **(b)** $(0.1)Pe^{0.1}$
(c) Yes, because $e^{0.1n} > 1$ for any $n \geq 1$.
43. $\dfrac{dS}{dt} = -50{,}000e^{-0.5t}$ **45.** $1200e$
47. 29.107 **49.** $y' = \dfrac{98{,}990{,}100e^{-0.99t}}{(1 + 9999e^{-0.99t})^2}$
51. $y' = 100{,}000e^{-1/x}/x^2$
53. $\dfrac{dN}{dt} = N_0(1 + r)^t \cdot \ln(1 + r)$
55. $\dfrac{dI}{dR} = 10^R \ln 10$ **57.** 112.066 (\$billion/year)
59. **(a)** $d'(x) = 29 - 20.1891e^{x/100} + 0.044e^{x/10}$
(b) \$1.28 (billion/year), \$198.01 (billion/year)
(c) The United States could not tolerate exponential growth in its debt. The percent of federal expenditures devoted to payment of interest on the debt was too high.
61. $\dfrac{dP}{dx} = -0.044$ dollars per year
63. **(a)** $y = 6.8690\,(1.08514)^x$
(b) $y'(23) = 3.68$; \$3.68 per year
65. **(a)** $y = 604.9211\,(1.0820)^x$
(b) 126,186 per year

Exercise 11.3 (page 836)

1. $\frac{1}{2}$ **3.** $-\frac{1}{2}$ **5.** $-\frac{5}{3}$ **7.** $-x/(2y)$
9. $-(2x + 4)/(2y - 3)$ **11.** $-x/y$
13. $y' = \dfrac{-y}{2(x - 1)}$ **15.** $\dfrac{dp}{dq} = \dfrac{p^2}{4 - 2pq}$
17. $\dfrac{dy}{dx} = \dfrac{x(3x^3 - 2)}{3y^2(1 + y^2)}$ **19.** $\dfrac{dy}{dx} = \dfrac{4x^3 + 6x^2y - 1}{-4x^3y - 3y^2}$
21. $\dfrac{(4x^3 + 9x^2y^2 - 8x - 12y)}{(18y + 12x - 6x^3y + 10y^4)}$ **23.** undefined
25. 1 **27.** $y = \frac{1}{2}x + 1$ **29.** $y = 4x + 5$
31. $y' = \dfrac{1}{2xy}$ **33.** $y' = \dfrac{-y}{2x \ln x}$ **35.** -4
37. $-1/x$ **39.** $-y/x$
41. $ye^x/(1 - e^x)$ **43.** 0 **45.** $y = -\frac{1}{3}x + 1$
47. horizontal: $(2, \sqrt{2})$, $(2, -\sqrt{2})$;
vertical: $(2 + 2\sqrt{2}, 0)$, $(2 - 2\sqrt{2}, 0)$
51. **(b)** yes, because $x^2 + y^2 = 4$ **53.** $1/(2x\sqrt{x})$
55. max at $(0, 3)$; min at $(0, -3)$ **57.** $\frac{1}{2}$ **59.** $-\frac{243}{128}$
61. At $p = \$80$, $q = 49$ and $dq/dp = -\frac{5}{16}$, which means that if the price is increased to \$81, quantity demanded will decrease by approximately $\frac{5}{16}$ unit.
63. $-0.000436\,y$ **65.** $\dfrac{dh}{dt} = -\dfrac{3}{44} - \dfrac{h}{12}$

Exercise 11.4 (page 842)

1. 36 **3.** $\frac{1}{8}$ **5.** $-\frac{24}{5}$ **7.** $\frac{7}{6}$
9. -5 if $z = 5$, -10 if $z = -5$
11. -80 units/sec **13.** 12π ft²/min
15. $\frac{16}{27}$ in./sec **17.** \$1798/day **19.** \$0.42/day
21. 430 units/month **23.** 36π mm³/month
25. $\dfrac{\frac{dW}{dt}}{W} = 3\left(\dfrac{\frac{dL}{dt}}{L}\right)$ **27.** $\dfrac{\frac{dC}{dt}}{C} = 1.54\left(\dfrac{\frac{dW}{dt}}{W}\right)$
29. $\dfrac{1}{4\pi}$ micrometers/day **31.** $1/(20\pi)$ in./min
33. -0.75 ft/sec **35.** $-120\sqrt{6}$
37. 61.18 mph **39.** $\frac{1}{25}$ ft/hr

Exercise 11.5 (page 853)

1. **(a)** 1 **(b)** no change **3.** **(a)** 84
(b) Revenue will decrease.
5. **(a)** $\frac{100}{99}$ **(b)** elastic **(c)** decrease
7. **(a)** 0.81 **(b)** inelastic **(c)** increase
9. **(a)** $\eta = 11.1$ (approximately) **(b)** elastic

11. **(a)** $\eta = \dfrac{375 - 3q}{q}$

(b) unitary: $q = 93.75$; inelastic: $q > 93.75$; elastic: $q < 93.75$

(c) As q increases over $0 < q < 93.75$, p decreases, so elastic demand means R increases. Similarly, R decreases for $q > 93.75$.

(d) Maximum for R when $q = 93.75$; yes.

13. \$12/item **15.** $t = \$350$ **17.** \$115/item
19. \$483 per item; \$40,100 **21.** \$1100/item

Chapter 11 Review Exercises (page 856)

1. $(6x - 1)e^{3x^2 - x}$ **2.** $2x$ **3.** $1/q - \left[\dfrac{2q}{(q^2 - 1)}\right]$
4. $dy/dx = e^{x^2}(2x^2 + 1)$ **5.** $dy/dx = 3^{3x - 3}\ln 3$
6. $dy/dx = \dfrac{10}{x}$ **7.** $y' = \dfrac{1 - \ln x}{x^2}$
8. $dy/dx = -2e^{-x}/(1 - e^{-x})^2$
9. $y = 12ex - 8e$, or $y = 32.62x - 21.75$
10. $y = x - 1$ **11.** $y' = \dfrac{-y}{x \ln x}$
12. $dy/dx = ye^{xy}/(1 - xe^{xy})$ **13.** $dy/dx = 2/y$
14. $\dfrac{dy}{dx} = \dfrac{2(x + 1)}{3(1 - 2y)}$ **15.** $y' = \dfrac{6x(1 + xy^2)}{y(5y^3 - 4x^3)}$
16. $d^2y/dx^2 = -(x^2 + y^2)/y^3 = -1/y^3$ **17.** 5/9
18. $(-2, \pm\sqrt{\tfrac{2}{3}})$ **19.** 3/4 **20.** 11 square units/min
21. 135.3 **22.** **(a)** 152.5 **(b)** 1.13 times faster
23. **(a)** $-0.00002876A_0$ **(b)** $-0.00002876A_0$ **(c)** less
24. $\$1200e \approx \$3261.94$ **25.** $-\$603.48$
26. $1/(25\pi)$ mm/min **27.** $\frac{48}{25}$ ft/min
28. $\dfrac{dS/dt}{S} = \dfrac{1}{3}\left(\dfrac{dA/dt}{A}\right)$ **29.** yes
30. $t = 1446.67$ **31.** \$880
32. **(a)** 1 **(b)** no change **33.** $\frac{25}{12}$, elastic **34.** 1
35. **(a)**

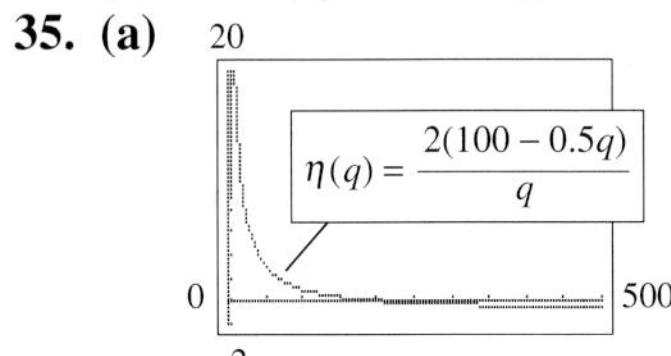

(b) $q = 100$

(c) max revenue at $q = 100$

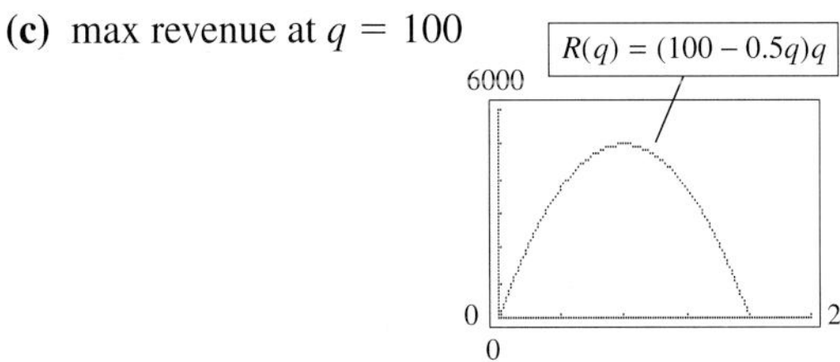

(d) Revenue is maximized where elasticity is unitary.

Chapter 11 Test (page 858)

1. $y' = \dfrac{12x^2}{x^3 + 1}$ **2.** $y' = \dfrac{12x^3}{x^4 + 1}$ **3.** $y' = \dfrac{1 - \ln x}{x^2}$
4. $y' = 15x^2 e^{x^3} + 2x$ **5.** $\dfrac{dS}{dt} = e^{t^4}(4t^4 + 1)$
6. $y' = \dfrac{e^{x^3 + 1}(3x^3 - 1)}{x^2}$ **7.** $f'(x) = 20(3^{2x})\ln 3$
8. $g'(x) = \dfrac{8}{(4x + 7)\ln 5}$ **9.** $y' = \dfrac{-3x^3}{y}$
10. $y' = \dfrac{-e^y}{xe^y - 10}$ **11.** $-\frac{3}{2}$ **12.** -586 per day
13. -0.05 unit per dollar **14.** \$1349.50 per week
15. $n = 3.71$; decreases
16. **(a)** 16.82 **(b)** 25.96 **17.** 0.087%
18. 1020 (\$ billions/year) **19.** \$540

Exercise 12.1 (page 870)

1. $x^4 + C$ **3.** $\frac{1}{7}x^7 + C$ **5.** $\frac{1}{8}x^8 + C$
7. $\frac{4}{3}x^6 + C$ **9.** $27x + \frac{1}{14}x^{14} + C$
11. $3x - \frac{2}{5}x^{5/2} + C$ **13.** $\frac{1}{5}x^5 - 3x^3 + 3x + C$
15. $2x + \frac{4}{3}x\sqrt{x} + C$ **17.** $\frac{24}{5}x\sqrt[4]{x} + C$
19. $-5/(3x^3) + C$ **21.** $\frac{3}{2}\sqrt[3]{x} + C$
23. $\dfrac{1}{4}x^4 - 4x - \dfrac{1}{x^5} + C$ **25.** $\dfrac{1}{10}x^{10} + \dfrac{1}{2x^2} + 3x^{2/3} + C$
27. $\frac{1}{4}x^4 + \frac{10}{3}x^3 + \frac{25}{2}x^2 + C$ **29.** $2x^8 - \frac{4}{3}x^6 + \frac{1}{4}x^4 + C$
31. $-1/x - 1/(2x^2) + C$ **33.**

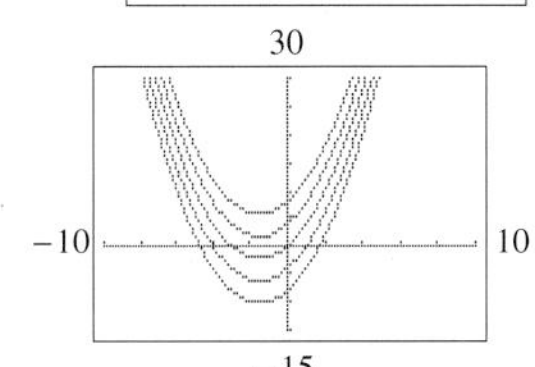

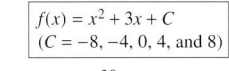

35. $\int(5 - \frac{1}{2}x)\,dx$ **37.** $\int(3x^2 - 6x)\,dx$ **39.** $R(x) = 3x$
41. $R(x) = 0.2x^2 + 3x$ **43.** \$3800
45. $P(t) = \frac{1}{4}t^4 + \frac{4}{3}t^3 + 6t$
47. **(a)** $x = t^{7/4}/1050$ **(b)** 0.96 tons
49. **(a)** $x/4 + 100/x + 30$ **(b)** \$56
51. **(a)** $R = -0.031t^2 + 0.776t + 0.0639$

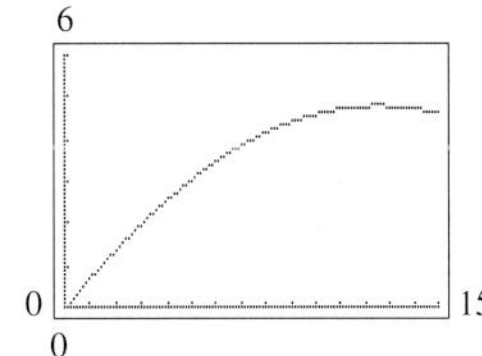

(b) Graph of $R(t)$ with data points

Graph of $R(t)$ with data points

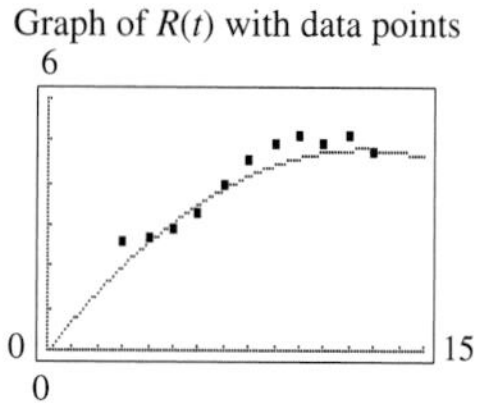

53. (a) $PS = 0.7t^3 - 32.7t^2 + 491.6t - 2192.2$

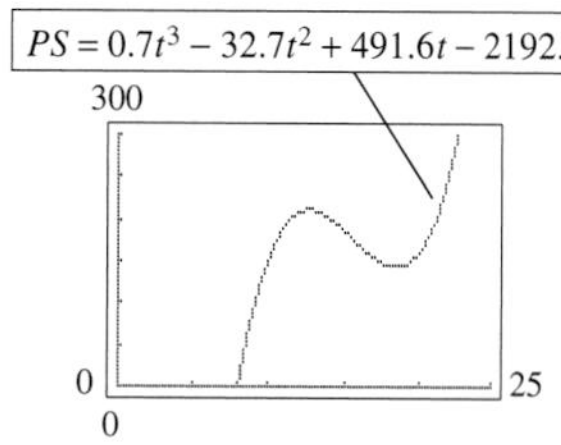

(b) Graph of $PS(t)$ with data points

Graph of $PS(t)$ with data points

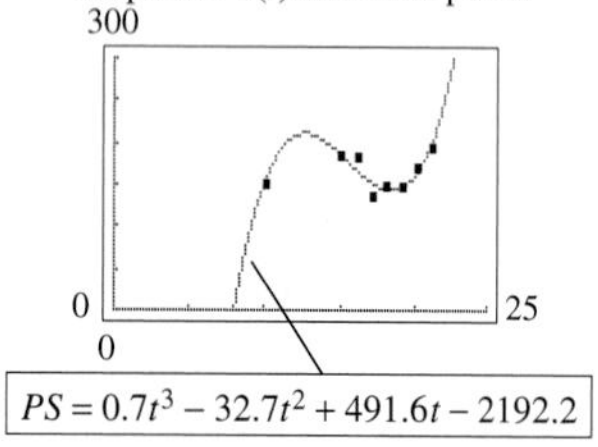

Exercise 12.2 (page 850)

1. $\frac{1}{4}(x^2 + 3)^4 + C$ **3.** $(15x^2 + 10)^5/5 + C$
5. $\frac{1}{3}(3x - x^3)^3 + C$ **7.** $\frac{1}{8}(x^2 + 5)^4 + C$
9. $\frac{1}{4}(4x - 1)^7 + C$ **11.** $-\frac{1}{4}(x^2 + 1)^{-2} + C$
13. $\frac{1}{10}(x^2 - 2x + 5)^5 + C$ **15.** $-\frac{1}{8}(x^4 - 4x + 3)^{-4} + C$
17. $\frac{7}{6}(x^4 + 6)^{3/2} + C$ **19.** $\frac{3}{8}x^8 + \frac{6}{5}x^5 + \frac{3}{2}x^2 + C$
21. $\frac{72}{7}x^7 - \frac{48}{5}x^5 + \frac{8}{3}x^3 + C$ **23.** $\frac{2}{9}(x^3 - 3x)^{3/2} + C$
25. $\dfrac{-1}{[3(x^3 - 1)]} + C$ **27.** $\dfrac{-1}{[10(2x^5 - 5)^3]} + C$
29. $\dfrac{-1}{[8(x^4 - 4x)^2]} + C$ **31.** $\frac{2}{3}\sqrt{x^3 - 6x^2 + 2} + C$
33. (a) $f(x) = \frac{1}{8}(x^2 - 1)^4 + C$
(b)

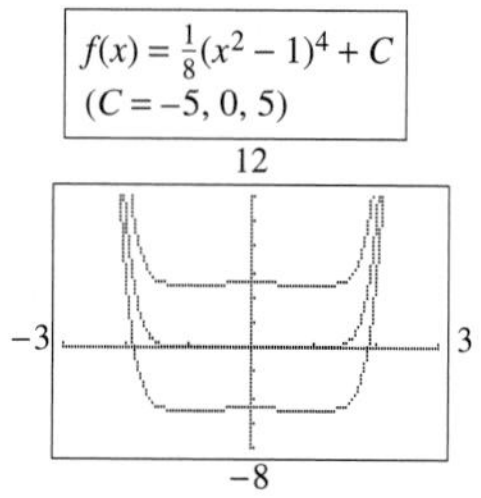

35. (a) $F(x) = \frac{15}{4}(2x - 1)^{2/5} + C$
(b)

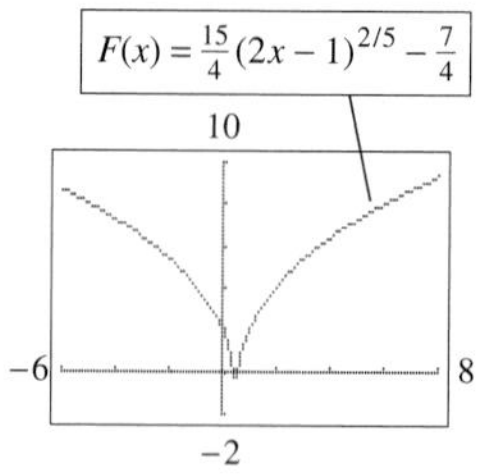

(c) $x = \frac{1}{2}$
(d) vertical

37. $\displaystyle\int \frac{8x(x^2 - 1)^{1/3}}{3}\,dx$

39. $R(x) = \dfrac{15}{2x + 1} + 30x - 15$

41. 3720 **43. (a)** $s = 10\sqrt{x + 1}$ **(b)** 50

45. (a) $100/(t + 10) - 1000/(t + 10)^2$
(b) 2.5 million

47. 7400

49. (a) $p = -0.001545(t - 60)^3 + 0.1206(t - 60)^2 - 2.4165t + 184.26$

(a)

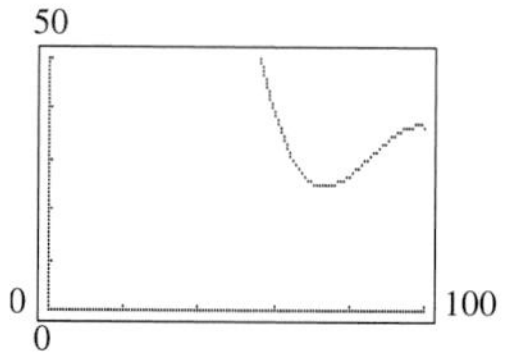

(b) Graph of $p(t)$ with data points

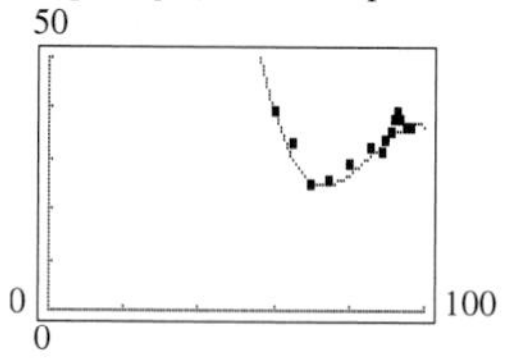

(c) a good fit

51. (a) $U = 0.4717(0.1t + 2)^3 - 10.6585(0.1t + 2)^2 + 7.3900t + 18.044$

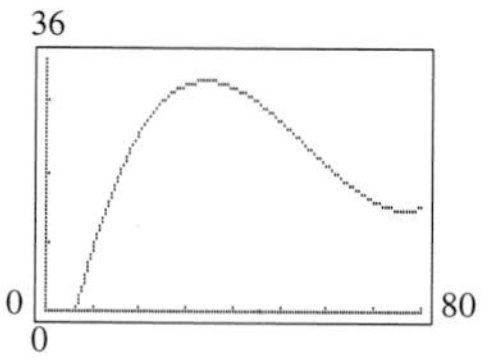

(b) Graph of $U(t)$ with data points

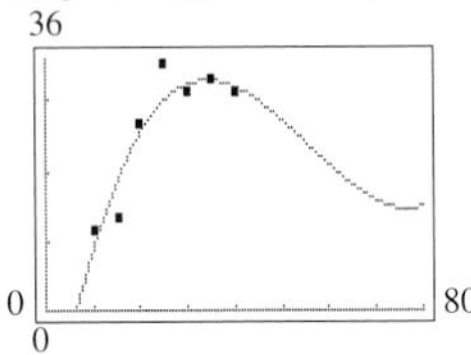

(c) The data fit fairly well, but the maximum points do not agree.

53. (a) $R = 0.572\, e^{0.3486t} + C$
(b)

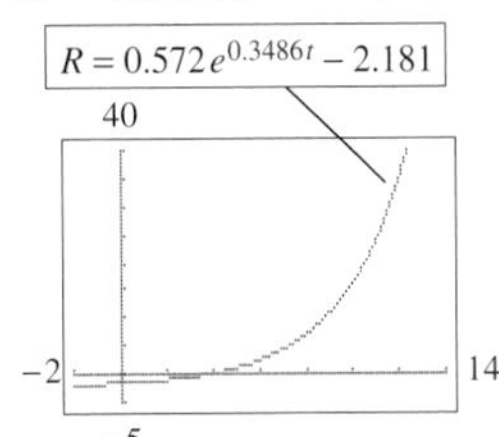

(c) \$24.29 billion

Exercise 12.3 *(page 890)*

1. $\ln|x^3 + 4| + C$ **3.** $\frac{1}{4}\ln|4z + 1| + C$
5. $\frac{1}{4}\ln|x^4 + 1| + C$ **7.** $2\ln|x^2 - 4| + C$
9. $\ln|x^3 - 2x| + C$ **11.** $\frac{1}{3}\ln|z^3 + 3z + 17| + C$
13. $\frac{1}{3}x^3 + \ln|x - 1| + C$ **15.** $x + \frac{1}{2}\ln|x^2 + 3| + C$
17. $e^{3x} + C$ **19.** $-e^{-x} + C$ **21.** $10{,}000\, e^{0.1x} + C$
23. $-1200\, e^{-0.7x} + C$ **25.** $\frac{1}{12}e^{3x^4} + C$
27. $-\frac{3}{2}e^{-2x} + C$ **29.** $\frac{1}{18}e^{3x^6 - 2} + C$
31. $\frac{1}{4}e^{4x} + 6/e^{x/2} + C$ **33.** $f(x) = h(x), \int f(x)\, dx = g(x)$
35. $F(x) = -\ln|3 - x| + C$

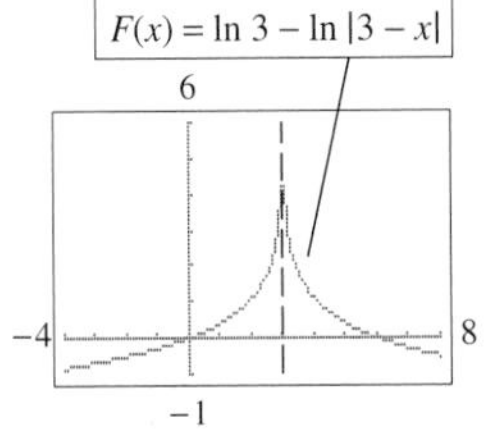

37. $\int \left(1 + \frac{1}{x}\right) dx$ **39.** $5e^{-x} - 5xe^{-x}$
41. \$1030.97 **43.** $n = n_0 e^{-Kt}$ **45.** 55
47. (a) $Pe^{0.1n}$ **(b)** approx. 7 yrs
49. (a) $p = 95e^{-0.491t}$ **(b)** ≈ 90.45
51. $l = 14.1372 \ln|t + 20| + 11.6$
(a)

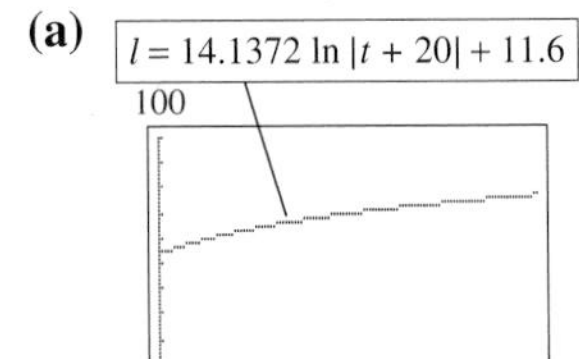

(b) Graph of $l(t)$ with data points

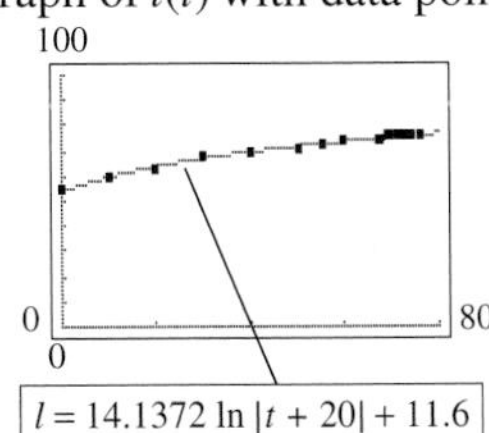

(c) They are a good match.

Exercise 12.4 *(page 900)*

1. $C(x) = x^2 + 100x + 200$
3. $C(x) = 2x^2 + 2x + 80$ **5.** \$3750
7. (a) $x = 3$ units is optimal level
(b) $P(x) = -4x^2 + 24x - 200$ **(c)** loss of \$164
9. (a) \$2960 **(b)** 896
11. (a) $\overline{C}(x) = \frac{6}{x} + \frac{x}{6} + 8$ **(b)** \$10.50
13. (a) and **(b)**

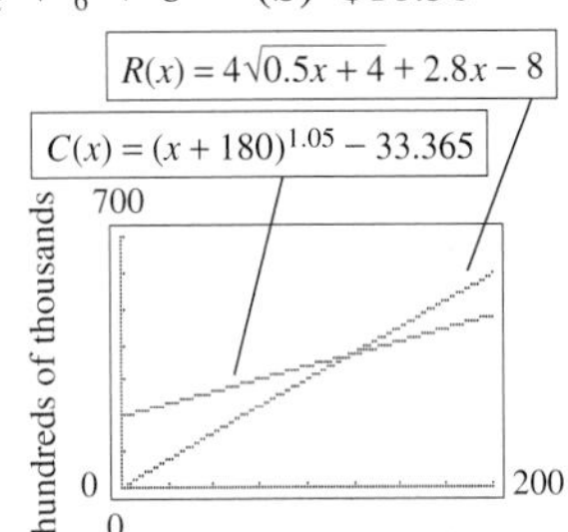

(c) Maximum profit is \$114.743 thousand at $x = 200$ thousand units.

15. $C(y) = 0.4y + 0.6\sqrt{y} + 5$
17. $C(y) = 0.3y + 0.4\sqrt{y} + 8$
19. $C = 2\sqrt{y + 1} + 0.4y + 4$
21. $C = 0.7y + 0.5e^{-2y} + 5.15$
23. $C = 0.85y + 5.15$
25. $C = 0.8y + \dfrac{2\sqrt{3y + 7}}{3} + 4.24$

Exercise 12.5 *(page 910)*

1. $4y - 2xy' = 4x^2 - 2x(2x) = 0$ ✓
3. $2y\, dx - x\, dy = 2(3x^2 + 1)\, dx - x(6x\, dx) = 2\, dx$ ✓
5. $y = \frac{1}{2}e^{x^2 + 1} + C$ **7.** $y^2 = 2x^2 + C$
9. $y^3 = x^2 - x + C$ **11.** $y = e^{x - 3} - e^{-3} + 2$
13. $y = \ln|x| - \dfrac{x^2}{2} + \dfrac{1}{2}$ **15.** $\dfrac{y^2}{2} = \dfrac{x^3}{3} + C$
17. $\dfrac{1}{2x^2} + \dfrac{y^2}{2} = C$ **19.** $\dfrac{1}{x} + y + \dfrac{y^3}{3} = C$
21. $\dfrac{1}{y} + \ln|x| = C$ **23.** $x^2 - y^2 = C$
25. $y = C(x + 1)$ **27.** $y - \ln|y + 1| = -\frac{1}{2}e^{-2x} + C$
29. $3y^4 = 4x^3 - 1$ **31.** $2y = 3x + 4xy$ or $y = \dfrac{3x}{2 - 4x}$

33. $e^{2y} = x^2 - \frac{2}{x} + 2$ **35.** $y^2 + 1 = 5x$

37. $y = Cx^k$

39. **(a)** $x = 10{,}000e^{0.06t}$ **(b)** \$10,618.37; \$13,498.59
(c) 11.55 years

41. $\approx$ 8.4 hours **43.** $\approx$ 23,100 years

45. $x = 6(1 - e^{-0.05t})$ **47.** $x = 20 - 10e^{-0.025t}$

49. $y = \frac{32}{(p+8)^{2/5}}$ **51.** $V = 1.86e^{2-2e^{-0.1t}}$

53. $V = \frac{k^3t^3}{27}$ **55.** $t \approx 4.5$ hours

57. **(a)** $E(t) = 0.1e^{0.043t}$
(b) The graph is a similar, but smooth, representation of the given data.

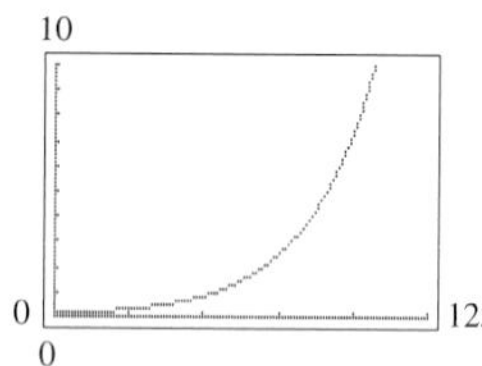

59. **(a)** $P(t) = 20{,}000e^{-0.05t}$
(b) $P(30) \approx \$4463$. This result corresponds quite well to the graph.

Chapter 12 Review Exercises *(page 914)*

1. $\frac{1}{7}x^7 + C$ **2.** $\frac{2}{3}x^{3/2} + C$

3. $\frac{1}{4}x^4 - x^3 + 2x^2 + 5x + C$

4. $\frac{1}{5}x^5 - \frac{2}{3}x^3 + x + C$

5. $\frac{1}{6}(x^2-1)^3 + C$ **6.** $\frac{1}{6}(x^3-3x^2)^2 + C$

7. $\frac{1}{8}x^8 + \frac{8}{5}x^5 + 8x^2 + C$ **8.** $\frac{1}{21}(x^3+4)^7 + C$

9. $\frac{1}{3}\ln|x^3+1| + C$ **10.** $\frac{-1}{3(x^3+1)} + C$

11. $\frac{1}{2}(x^3-4)^{2/3} + C$ **12.** $\frac{1}{3}\ln|x^3-4| + C$

13. $\frac{1}{2}x^2 - \frac{1}{x} + C$ **14.** $\frac{1}{3}x^3 + \frac{1}{2}x^2 - 2x - \ln|x-1| + C$

15. $\frac{1}{3}e^{y^3} + C$ **16.** $x^3/3 - x^2 + x + C$

17. $\frac{1}{2}\ln|2x^3-7| + C$ **18.** $\frac{-5}{4e^{4x}} + C$

19. $x^4/4 - e^{3x}/3 + C$ **20.** $\frac{1}{2}e^{x^2+1} + C$

21. $\frac{-3}{40(5x^8+7)^2} + C$ **22.** $-\frac{7}{2}\sqrt{1-x^4} + C$

23. $\frac{1}{4}e^{2x} - e^{-2x} + C$ **24.** $x^2/2 + 1/(x+1) + C$

25. **(a)** $\frac{1}{10}(x^2-1)^5 + C$ **(b)** $\frac{1}{22}(x^2-1)^{11} + C$
(c) $\frac{3}{16}(x^2-1)^8 + C$ **(d)** $\frac{3}{2}(x^2-1)^{1/3} + C$

26. **(a)** $\ln|x^2-1| + C$ **(b)** $\frac{-1}{x^2-1} + C$
(c) $3\sqrt{x^2-1} + C$ **(d)** $\frac{3}{2}\ln|x^2-1| + C$

27. $y = C - 92e^{-0.05t}$

28. $y = 64x + 38x^2 - 12x^3 + C$

29. $(y-3)^2 = 4x^2 + C$ **30.** $(y+1)^2 = 2\ln|t| + C$

31. $e^y = \frac{x^2}{2} + C$ **32.** $y = Ct^4$

33. $3(y+1)^2 = 2x^3 + 75$

34. $x^2 = y + y^2 + 4$ **35.** 96 **36.** 472

37. $400[1 - 5/(t+5) + 25/(t+5)^2]$

38. $p = 1990.099 - 100{,}000/(t+100)$

39. **(a)** $y = -60e^{-0.04t} + 60$ **(b)** 23%

40. $R = 800\ln(x+1)$

41. **(a)** \$1000 **(b)** $C(x) = 3x^2 + 4x + 1000$

42. 80 units, \$440 **43.** $C = \sqrt{2y+16} + 0.6y + 4.5$

44. $C = 0.8y - 0.05e^{-2y} + 7.85$ **45.** $W = CL^3$

46. $\approx$ 10.7 million years **47.** $x = 360(1 - e^{-t/30})$

48. $x = 600 - 500e^{-0.01t}$; $\approx$ 161 min

Chapter 12 Test *(page 916)*

1. $2x^3 + 4x^2 - 7x + C$ **2.** $4x + \frac{2}{3}x\sqrt{x} + \frac{1}{x} + C$

3. $\frac{(4x^3-7)^{10}}{24} + C$ **4.** $\frac{(3x^2-6x+1)^{10}}{30} + C$

5. $\frac{\ln|2x^4-5|}{8} + C$ **6.** $-10{,}000e^{-0.01x} + C$

7. $\frac{5}{8}e^{2y^4-1} + C$ **8.** $e^x + 5\ln|x| - x + C$

9. $\frac{x^2}{2} - x + \ln|x+1| + C$ **10.** $6x^2 - 1 + 5e^x$

11. $y = x^4 + x^3 + 4$ **12.** $y = \frac{1}{4}e^{4x} + \frac{7}{4}$

13. $y = \frac{4}{C - x^4}$ **14.** 104,824

15. $P(x) = 450x - 2x^2 - 300$

16. $C(y) = 0.78y + \sqrt{0.5y+1} + 5.6$ **17.** 332.3 days

Exercise 13.1 *(page 930)*

1. 7 square units **3.** 22 square units

5. 3 square units **7.** 30 square units

9. $S_L(10) = 4.08$; $S_R(10) = 5.28$

11. Both equal 14/3.

13. It would lie between $S_L(10)$ and $S_R(10)$. It would equal 14/3.

15. 3 **17.** 42 **19.** -5 **21.** 180 **23.** 11,315

25. $3 - \frac{3(n+1)}{n} + \frac{(n+1)(2n+1)}{2n^2} = \frac{2n^2-3n+1}{2n^2}$

27. **(a)** $S = (n-1)/2n$ **(b)** 9/20 **(c)** 99/200
(d) 999/2000 **(e)** $\frac{1}{2}$

29. **(a)** $S = \frac{(n+1)(2n+1)}{6n^2}$
(b) $77/200 = 0.385$ **(c)** $6767/20{,}000 \approx 0.3384$
(d) $667{,}667/2{,}000{,}000 \approx 0.3338$ **(e)** $\frac{1}{3}$ **31.** $\frac{20}{3}$

33. There are approximately 90 squares under the curve, each representing 1 second by $10 \text{ mph} = 10 \times \frac{1 \text{ hr}}{3600} \times \frac{1 \text{ mile}}{\text{hour}} = \frac{1}{360}$ mile. The area under the curve is approximately $90\frac{1}{360} = \frac{1}{4}$ mile.
35. 1550 square feet

Exercise 13.2 (page 940)

1. 18 **3.** 2 **5.** 60 **7.** $12\sqrt[3]{25}$ **9.** 98
11. $\frac{7}{3}$ **13.** 12,960 **15.** 0 **17.** $8\sqrt{3} - \frac{7}{3}\sqrt{7}$
19. 2 **21.** $e^3/3 - 1/3$ **23.** 4 **25.** 0
27. **(a)** $\frac{1}{6}\ln(112/31) \approx 0.2140853$ **(b)** 0.2140853
29. **(a)** $\frac{3}{2} + 3\ln 2 \approx 3.5794415$ **(b)** 3.5794415
31. **(a)** A, C **(b)** B
33. $\int_0^4 (2x - \frac{1}{2}x^2)\,dx$ **(b)** 16/3
35. **(a)** $\int_{-1}^0 (x^3 + 1)\,dx$ **(b)** 3/4
37. $\frac{1}{6}$ **39.** $\frac{1}{2}(e^9 - e)$
41. same absolute values, opposite signs
43. 6 **45.** 20,405.39 **47.** 0.04 cm^3
49. 1222 (approximately) **51.** $450,000
53. **(a)** $7007 **(b)** $19,649
55. **(a)** $y = 0.005963x^{0.701215}$
(b) 0.15 mile (approximately)

Exercise 13.3 (page 952)

1. **(a)** $\int_0^2 (4 - x^2)\,dx$ **(b)** $\frac{16}{3}$
3. **(a)** $\int_1^8 [\sqrt[3]{x} - (2 - x)]\,dx$ **(b)** 28.75
5. **(a)** $\int_1^2 [(4 - x^2) - (\frac{1}{4}x^3 - 2)]\,dx$ **(b)** 131/48
7. **(a)** $(-1, 1), (2, 4)$ **(b)** $\int_{-1}^2 [(x + 2) - x^2]\,dx$
(c) 9/2
9. **(a)** $(0, 0), (\frac{5}{2}, -\frac{15}{4})$
(b) $\int_0^{5/2} [(x - x^2) - (x^2 - 4x)]\,dx$ **(c)** $\frac{125}{24}$
11. **(a)** $(-2, -4), (0, 0), (2, 4)$
(b) $\int_{-2}^0 [(x^3 - 2x) - 2x]\,dx + \int_0^2 [2x - (x^3 - 2x)\,dx]$
(c) 8
13. $\frac{28}{3}$ **15.** $\frac{1}{4}$ **17.** $\frac{16}{3}$ **19.** $\frac{1}{3}$ **21.** $\frac{37}{12}$
23. $4 - 3\ln 3$ **25.** $\frac{8}{3}$ **27.** 6 **29.** 0 **31.** $-\frac{4}{9}$
33. $11.8\overline{3}$
35. 1980, 0.351; 1990, 0.377. The difference in incomes widened.
37. Whites, 0.391; Blacks, 0.439. In 1996, income was more equally distributed among whites.
39. average profit $= \frac{1}{x_1 - x_0}\int_{x_0}^{x_1} (R(x) - C(x))\,dx$
41. **(a)** $1402 **(b)** $535,333.33
43. **(a)** 102.5 units **(b)** 100 units
45. 7.40% **47.** 147 milligrams

Exercise 13.4 (page 964)

1. $120,000 **3.** $346,664 (nearest dollar)
5. $506,000 (nearest thousand)
7. $18,660 (nearest dollar)
9. $82,155 (nearest dollar)
11. PV = $265,781 (nearest dollar), FV = $377,161 (nearest dollar)
13. PV = $190,519 (nearest dollar), FV = $347,148 (nearest dollar)
15. Gift shop, $151,024; Video rental, $141,093. Gift shop is a better buy.
17. $83.33 **19.** $161.89 **21.** (5, 56); $83.33
23. $11.50 **25.** $204.17 **27.** $2766.67
29. $17,839.58 **31.** $133.33 **33.** $2.50
35. $103.35

Exercise 13.5 (page 970)

1. $\frac{1}{8}\ln|(4 + x)/(4 - x)| + C$
3. $\frac{1}{3}\ln[(3 + \sqrt{10})/2]$ **5.** $w(\ln w - 1) + C$
7. $\frac{1}{3} + \frac{1}{4}\ln(\frac{3}{7})$ **9.** $3^x \log_3 e + C$ or $3^x/\ln 3 + C$
11. $\frac{1}{2}[7\sqrt{24} - 25\ln(7 + \sqrt{24}) + 25\ln 5]$
13. $\frac{(6w - 5)(4w + 5)^{3/2}}{60} + C$ **15.** $\frac{1}{2}(5^{x^2})\log_5 e + C$
17. $\frac{1}{3}(13^{3/2} - 8)$ **19.** $-\frac{5}{2}\ln\left|\frac{2 + \sqrt{4 - 9x^2}}{3x}\right| + C$
21. $\frac{1}{3}\ln|3x + \sqrt{9x^2 - 4}| + C$ **23.** $\frac{1}{8}\ln(\frac{9}{5})$
25. $\frac{1}{3}\ln|3x + 1 + \sqrt{(3x + 1)^2 + 1}| + C$
27. $\frac{1}{4}[10\sqrt{109} - \sqrt{10} + 9\ln(10 + \sqrt{109}) - 9\ln(1 + \sqrt{10})]$
29. $-\frac{1}{6}\ln|7 - 3x^2| + C$
31. $\frac{1}{2}\ln|2x + \sqrt{4x^2 + 7}| + C$
33. $2(e^{\sqrt{2}} - e) \approx 2.7899$
35. $\frac{1}{32}[\ln(9/5) - 4/9] \approx .004479$ **37.** $3391.10
39. **(a)** $C = \frac{1}{2}x\sqrt{x^2 + 9} + \frac{9}{2}\ln\left|\frac{x + \sqrt{x^2 + 9}}{3}\right| + 300$
(b) $314.94
41. $3882.9 thousand

Exercise 13.6 (page 977)

1. $\frac{1}{2}xe^{2x} - \frac{1}{4}e^{2x} + C$ **3.** $\frac{1}{3}x^3\ln x - \frac{1}{9}x^3 + C$
5. $\frac{104\sqrt{2}}{15}$ **7.** $-(1 + \ln x)/x + C$ **9.** 1
11. $\frac{x^2}{2}\ln(2x - 3) - \frac{1}{4}x^2 - \frac{3}{4}x - \frac{9}{8}\ln(2x - 3) + C$
13. $\frac{1}{5}(q^2 - 3)^{3/2}(q^2 + 2) + C$ **15.** 282.4
17. $-e^{-x}(x^2 + 2x + 2) + C$ **19.** $(3e^4 + 1)/2$
21. $\frac{1}{4}x^4\ln^2 x - \frac{1}{8}x^4\ln x + \frac{1}{32}x^4 + C$

23. $\frac{2}{15}(e^x + 1)^{3/2}(3e^x - 2) + C$ **25.** $\frac{1}{2}e^{x^2} + C$
27. $\frac{2}{3}(e^x + 1)^{3/2} + C$ **29.** $-5e^{-4} + 1$ **31.** \$2794.46
33. \$34,836.73 **35.** 0.264

Exercise 13.7 (page 985)

1. 1/5 **3.** 2 **5.** $1/e$ **7.** diverges **9.** diverges
11. 10 **13.** diverges **15.** diverges **17.** diverges
19. 0 **21.** 0.5 **23.** $1/(2e)$ **25.** $\frac{3}{2}$
27. $\int_{-\infty}^{\infty} f(x)\,dx = 1$ **29.** $c = 1$ **31.** $c = \frac{1}{4}$
33. 20 **35.** area $= \frac{8}{3}$ **37.** $\int_0^{\infty} Ae^{-rt}\,dt = A/r$
39. \$2,400,000 **41.** \$700,000
43. **(a)** $500\left[\dfrac{e^{-0.03b} + 0.03b - 1}{0.0009}\right]$ **(b)** limit $= \infty$
45. 0.147

Chapter 13 Review Exercises (page 989)

1. 212 **2.** $\dfrac{3(n+1)}{2n^2}$ **3.** $\frac{91}{72}$ **4.** 1 **5.** 1
6. 14 **7.** $\frac{248}{5}$ **8.** $-\frac{205}{4}$ **9.** $\frac{825}{4}$ **10.** $\frac{125}{3}$
11. -2 **12.** $\frac{1}{6}\ln 47 - \frac{1}{6}\ln 9$ **13.** $\frac{9}{2}$
14. $\ln 4 + \frac{14}{3}$ **15.** $\frac{26}{3}$ **16.** $\frac{1}{2}\ln 2$ **17.** $(1 - e^{-2})/2$
18. $(e - 1)/2$ **19.** 95/2 **20.** 36 **21.** $\frac{1}{4}$ **22.** $\frac{1}{2}$
23. $\frac{1}{2}x\sqrt{x^2 - 4} - 2\ln\left|x + \sqrt{x^2 - 4}\right| + C$
24. $2\log_3 e$ **25.** $\frac{1}{2}x^2(\ln x^2 - 1) + C$
26. $\frac{1}{2}\ln|x| - \frac{1}{2}\ln|3x + 2| + C$
27. $\frac{1}{6}x^6\ln x - \frac{1}{36}x^6 + C$
28. $(-xe^{-2x}/2) - (e^{-2x}/4) + C$
29. $2x\sqrt{x + 5} - \frac{4}{3}(x + 5)^{3/2} + C$ **30.** 1 **31.** ∞
32. -100 **33.** $\frac{5}{3}$ **34.** $-\frac{1}{2}$ **35.** \$28,000
36. $e^{-2.8} \approx 0.061$ **37.** \$1297.44 **38.** \$76.60
39. **(a)** (7, 6) **(b)** \$7.33 **40.** \$24.50
41. \$1,621,803 **42.** **(a)** \$403,609 **(b)** \$602,114
43. \$217.42 **44.** \$86,557.41
45. $-\dfrac{x^2}{4} + \dfrac{7}{2}x + \dfrac{x^2 - 1}{2}\ln(x + 1) + 2000$
46. $e^{-1.4} \approx 0.247$
47. \$4000 thousand, or \$4 million

Chapter 13 Test (page 990)

1. 3.496 (approximately) **2.** **(a)** $5 - \dfrac{n+1}{n}$ **(b)** 4
3. $\int_0^6 (12 + 4x - x^2)\,dx$; 72
4. **(a)** 4 **(b)** 3/4 **(c)** $\frac{5}{4}\ln 5$ **(d)** 7
5. **(a)** $3xe^x - 3e^x + C$ **(b)** $\dfrac{x^2}{2}\ln(2x) - \dfrac{x^2}{4} + C$
6. -8
7. **(a)** $x[\ln(2x) - 1] + C$
(b) $\dfrac{2(9x + 14)(3x - 7)^{3/2}}{135} + C$
8. 16.089
9. **(a)** \$4000 **(b)** \$16,000/3
10. **(a)** \$961.18 thousand **(b)** \$655.68 thousand
(c) \$1062.5 thousand
11. 125/6 **12.**

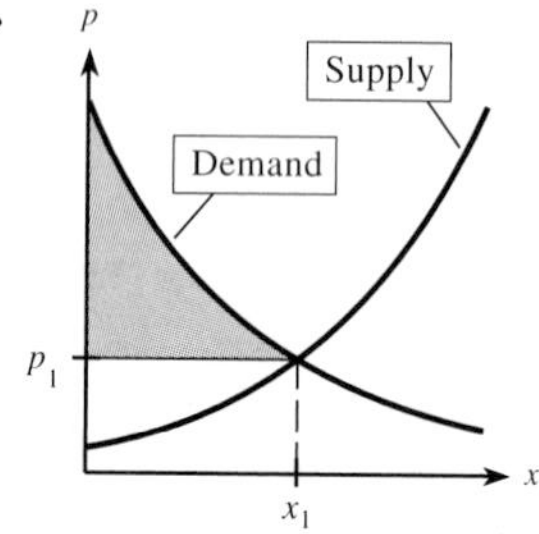

13. Before, 0.446; After, 0.19. The change decreases the difference in income.
14. **(a)** 567.357 million barrels per year
(b) 641.589 million barrels per year

Exercise 14.1 (page 1002)

1. $\{(x, y): x$ and y are real numbers$\}$
3. $\{(x, y): x$ and y are real numbers and $y \neq 0\}$
5. $\{(x, y): x$ and y are real numbers and $2x - y \neq 0\}$
7. $\{(p_1, p_2): p_1$ and p_2 are real numbers and $p_1 \geq 0\}$
9. -2 **11.** $\frac{5}{3}$ **13.** 2500 **15.** 36 **17.** 3
19. $\frac{1}{25}\ln(12)$ **21.** $\frac{13}{3}$
23. \$6640.23; the amount that results when \$2000 is invested for 20 years
25. 500; if the cost of placing an order is \$200, the number of items ordered is 625, and the weekly holding cost per item is \$1, then the most economical order size is 500.
27. Max: $S \approx 112.5$°F; $A \approx 106.3$°F
Min: $S \approx 87.4$°F; $A \approx 77.6$°F
29. **(a)** \$752.80; when \$90,000 is borrowed for 20 years at 8%, the monthly payment is \$752.80.
(b) \$1622.82; when \$160,000 is borrowed for 15 years at 9%, the monthly payment is \$1622.82.
31. **(a)** $x = 4$ **(b)** $y = 2$
(c)

33. (a) 37,500
(b) $30(2K)^{1/4}(2L)^{3/4} = 30(2^{1/4})(2^{3/4})K^{1/4}L^{3/4} = 2[30\,K^{1/4}L^{3/4}]$
(c)

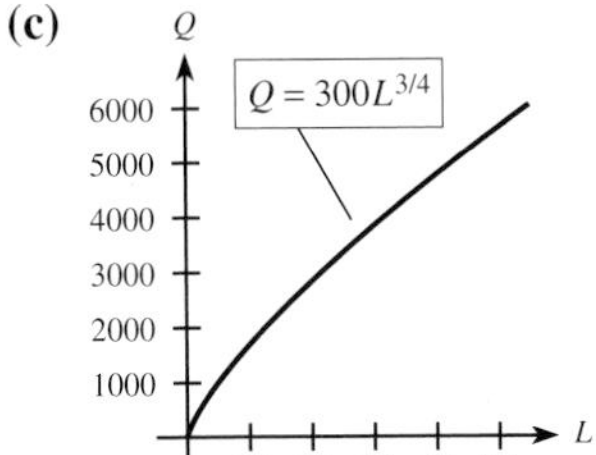

35. (a) 7200 (b) 5000 37. $284,000

Exercise 14.2 *(page 1013)*

1. $\frac{\partial z}{\partial x} = 4x^3 - 10x + 6$ $\frac{\partial z}{\partial y} = 9y^2 - 5$
3. $z_x = 3x^2 + 8xy$ $z_y = 4x^2 + 12y$
5. $\frac{\partial f}{\partial x} = 9x^2(x^3 + 2y^2)^2$ $\frac{\partial f}{\partial y} = 12y(x^3 + 2y^2)^2$
7. $f_x = 2x(2x^2 - 5y^2)^{-1/2}$ $f_y = -5y(2x^2 - 5y^2)^{-1/2}$
9. $\frac{\partial C}{\partial x} = -4y + 20xy$ $\frac{\partial C}{\partial y} = -4x + 10x^2$
11. $\frac{\partial Q}{\partial s} = \frac{2(t^2 + 3st - s^2)}{(s^2 + t^2)^2}$ $\frac{\partial Q}{\partial t} = \frac{3t^2 - 4st - 3s^2}{(s^2 + t^2)^2}$
13. $z_x = 2e^{2x} + \frac{y}{x}$ $z_y = \ln x$
15. $\frac{\partial f}{\partial x} = 100ye^{xy}$ $\frac{\partial f}{\partial y} = 100xe^{xy}$ 17. 2
19. 7 21. 0
23. (a) 0 (b) $-2xz + 4$ (c) $2y$ (d) $-x^2$
25. (a) $8x_1 + 5x_2$ (b) $5x_1 + 12x_2$ (c) 1
27. (a) 2 (b) 0 (c) 0 (d) $-30y$
29. (a) $2y$ (b) $2x - 8y$ (c) $2x - 8y$ (d) $-8x$
31. 0 33. -6
35. (a) $\frac{188}{4913}$ (b) $\frac{-188}{4913}$ 37. $2 + 2e$
39. (a) $2 + y^2e^{xy}$ (b) $xye^{xy} + e^{xy}$
(c) $xye^{xy} + e^{xy}$ (d) x^2e^{xy}
41. (a) $1/x^2$ (b) 0 (c) 0 (d) $2 + 1/y^2$
43. (a) $24x$ (b) $24x$ (c) 0
45. (a) For a mortgage of $100,000 and an 8% interest rate, the monthly payment is $1289.
(b) The rate of change of the payment with respect to the interest rate is $62.51. That is, if the rate goes from 8% to 9% on a $100,000 mortgage, the approximate increase in the monthly payment is $62.51.

47. (a) If the number of items sold per week changes by 1, the most economical order quantity should also increase. $\frac{\partial Q}{\partial M} = \sqrt{\frac{K}{2Mh}} > 0$
(b) If the weekly storage costs change by 1, the most economical order quantity should decrease.
$$\frac{\partial Q}{\partial h} = -\sqrt{\frac{KM}{2h^3}} < 0$$
49. (a) $65e^{-0.01x}$ (b) $70e^{-0.02y}$
51. (a) $2xy^2$ (b) $2x^2y$
53. $\frac{\partial Q}{\partial K} = 44\frac{4}{9}$; If labor hours are held constant at 1728 and K changes by $1 (thousand) to $730,000, Q will change by about $44\frac{4}{9}$ thousand units. $\frac{\partial Q}{\partial L} = 37\frac{1}{2}$; If capital expenditures are held constant at $729,000 and L changes by 1 hour (to 1729), Q will change by about $37\frac{1}{2}$ thousand units.
55. (a) $\frac{\partial WC}{\partial s} = -0.020t - 1.85 + \frac{0.152t}{\sqrt{s}} - \frac{13.87}{\sqrt{s}}$
(b) at $t = 10$, $s = 25$, $\frac{\partial WC}{\partial s} = -0.20 - 1.85 + 0.304 - 2.774 = -4.52$
This means that the rate of change of WC with respect to s is -4.52 if $t = 10°$F, $s = 25$ mph. An increase in speed causes a decrease in wind chill temperature.

Exercise 14.3 *(page 1021)*

1. 57 3. (a) $2 + y/50$ (b) $4 + x/50$
5. (a) 20.18 (b) 70.80 7. (a) 36 (b) 19
9. (a) $\sqrt{y^2 + 1}$ (b) $xy/\sqrt{y^2 + 1}$
11. (a) $1200y/(xy + 1)$ (b) $1200x/(xy + 1)$
13. (a) $\sqrt{y/x}$ (b) $\sqrt{x/y}$
15. (a) $\ln(y + 1)/(2\sqrt{x})$ (b) $\sqrt{x}/(y + 1)$
17. $z = 1092$ (approximately) 19. $z_x = 3.6$
21. (a) $z_x = \frac{240y^{2/5}}{x^{2/5}}$ (b)

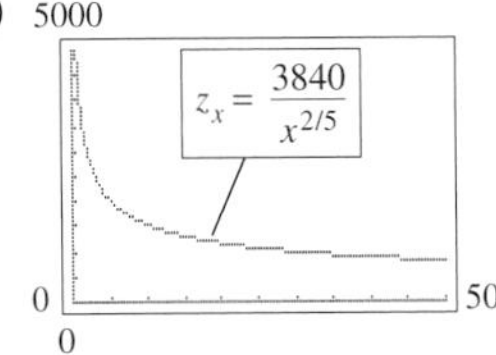

(c) $z_y = \frac{160x^{3/5}}{y^{3/5}}$ (d)

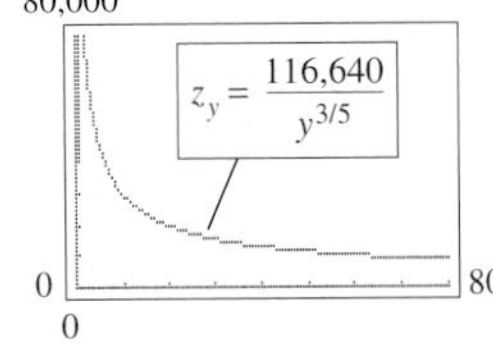

(e) Both z_x and z_y are positive, so increases in both capital investment and work-hours result in increases in productivity. However, both are decreasing, so such increases have a diminishing effect on productivity. Also, z_y decreases more slowly than z_x, so that increases in work-hours have a more significant impact on productivity than do increases in capital investment.

23. $q_1 = 188$; $q_2 = 270$

25. any values for p_1 and p_2 that satisfy $6p_2 - 3p_1 = 100$ and that make q_1 and q_2 nonnegative, such as $p_1 = 10$, $p_2 = 21\frac{2}{3}$

27. **(a)** -3 **(b)** -2 **(c)** -6 **(d)** -5
(e) complementary

29. **(a)** -50 **(b)** $600/(p_B + 1)^2$
(c) $-400/(p_B + 4)^2$
(d) $400/(p_A + 4)^2$ **(e)** competitive

Exercise 14.4 (page 1032)

1. max (0, 0, 9) **3.** min(0, 0, 4) **5.** min(1, −2, 0)
7. saddle (1, −3, 8) **9.** saddle (0, 0, 0)
11. max(12, 24, 456) **13.** min(−8, 6, −52)
15. saddle (0, 0, 0); min(2, 2, −8)
17. $\hat{y} = 5.7x - 1.4$
19. $x = 5000$, $y = 128$ **21.** $x = \frac{20}{3}$, $y = \frac{10}{3}$
23. $x = 28$, $y = 100$ **25.** $x = 0$, $y = 10$
27. length = 100, width = 100, height = 50
29. $x = 3$, $y = 0$
31. **(a)** $\hat{y} = 2547.6x + 44{,}770$ **(b)** \$146,674
33. **(a)** $\hat{y} = 0.334x + 3.235$ **(b)** \$14.925
(c) Slope = 0.334 means that hourly earnings change at the rate of \$0.334 per year.

Exercise 14.5 (page 1041)

1. 18 at (3, 3) **3.** 35 at (3, 2) **5.** 32 at (4, 2)
7. -28 at $(3, \frac{5}{2})$ **9.** $\frac{1}{5}$ at $(-\frac{2}{5}, -\frac{1}{5})$ **11.** 3 at (1, 1, 1)

13. 1 at (0, 1, 0) **15.** $x = 4$, $y = 1$
17. $x = 40$, $y = \frac{40}{3}$

50
$U = \frac{80{,}000}{3}$
$U = \frac{64{,}000}{3}$
$U = \frac{50{,}000}{3}$
$(40, \frac{40}{3})$
0
80
0
$2x + 3y = 120$

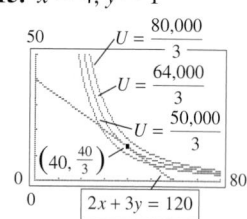

19. **(a)** $x = 400$, $y = 400$
(b) $-\lambda = 1.6$ means that each additional dollar spent on production results in approximately 1.6 additional units produced.

(c)

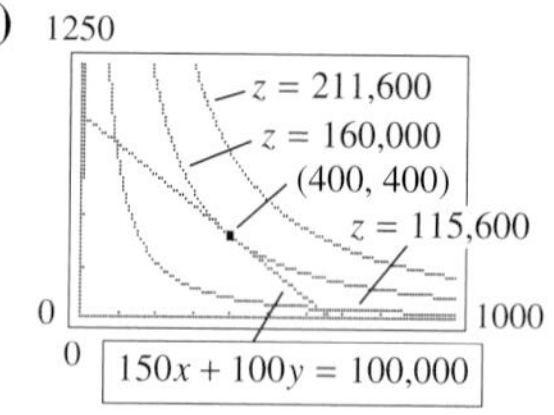

21. $x = 900$, $y = 300$
23. $x = \$10{,}003.33$, $y = \$19{,}996.67$
25. length = 100 cm, width = 100 cm, height = 50 cm

Chapter 14 Review Exercises (page 1043)

1. $\{(x, y)$: x and y are real numbers and $y \neq 2x\}$
2. $\{(x, y)$: x and y are real numbers with $y \geq 0$ and $(x, y) \neq (0, 0)\}$
3. -5 **4.** 896,000
5. $15x^2 + 6y$ **6.** $24y^3 - 42x^3y^2$
7. $z_x = 8xy^3 + 1/y$; $z_y = 12x^2y^2 - x/y^2$
8. $z_x = x/\sqrt{x^2 + 2y^2}$; $z_y = 2y/\sqrt{x^2 + 2y^2}$
9. $z_x = -2y/(xy + 1)^3$; $z_y = -2x/(xy + 1)^3$
10. $z_x = 2xy^3e^{x^2y^3}$; $z_y = 3x^2y^2e^{x^2y^3}$
11. $z_x = ye^{xy} + y/x$; $z_y = xe^{xy} + \ln x$
12. $z_x = y$; $z_y = x$ **13.** -8 **14.** 8
15. **(a)** $2y$ **(b)** 0 **(c)** $2x - 3$ **(d)** $2x - 3$
16. **(a)** $18xy^4 - 2/y^2$ **(b)** $36x^3y^2 - 6x^2/y^4$
(c) $36x^2y^3 + 4x/y^3$ **(d)** $36x^2y^3 + 4x/y^3$
17. **(a)** $2e^{y^2}$ **(b)** $4x^2y^2e^{y^2} + 2x^2e^{y^2}$
(c) $4xye^{y^2}$ **(d)** $4xye^{y^2}$
18. **(a)** $-y^2/(xy + 1)^2$ **(b)** $-x^2/(xy + 1)^2$
(c) $1/(xy + 1)^2$ **(d)** $1/(xy + 1)^2$
19. max(−8, 16, 208)
20. saddle(0, 0, 0); min(1, 1, −1)
21. 80 at (2, 8) **22.** 11,664 at (6, 3)
23. 3 units **24.** 20
25. **(a)** 62.8 km (approximately)
(b) If the surface temperature is 287 K, the Concorde's altitude is 17.1 km, and the vertical temperature gradient is 4.9 K/km, then the width of the sonic boom is about 63.3 km.
(c) $\frac{\partial f}{\partial h} \approx 1.87$ If the surface temperature is 293 K and the temperature gradient is 5 K/km, then a change of 1 km in the Concorde's altitude would result in a change in the width of the sonic boom of about 1.87 km.
(d) $\frac{\partial f}{\partial d} \approx -6.28$ If the surface temperature is 293 K and the Concorde's altitude is 16.8 km, then a change of 1 K/km in the temperature gradient would result in a change in the width of the sonic boom of about −6.28 km.

26. (a) 280 (b) 2400/7
27. $\partial Q/\partial K = 81.92$; $\partial Q/\partial L = 37.5$
28. (a) -2 (b) -6 (c) complementary
29. competitive 30. $x = 20, y = 40$
31. $x = 10, y = 4$
32. (a) $x = 1000, y = 500$
(b) $-\lambda \approx 3.17$ means that each additional dollar spent on production results in approximately 3 additional units.
(c)

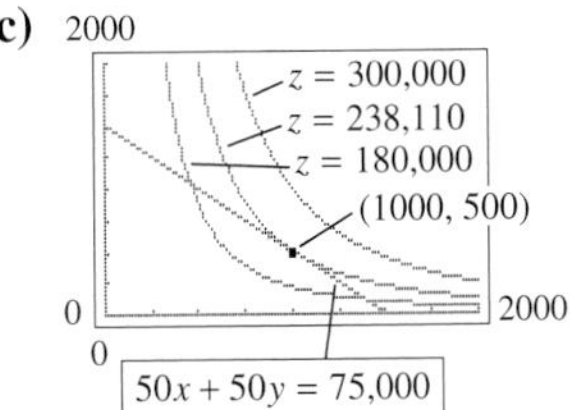

33. (a) $\hat{y} = 0.127x + 8.926$ (b) $897.926 billion
34. $\hat{q} = 34{,}726 - 55.09p$

Chapter 14 Test (page 1046)

1. (a) all pairs (x, y) with $y < x^2$ (b) 14
2. $z_x = 5 + 10y(xy + 1)^4$ $z_y = -18y + 10x(xy + 1)^4$
$z_{xx} = 40y^2(xy + 1)^3$ $z_{yy} = -18 + 40x^2(xy + 1)^3$
$z_{xy} = z_{yx} = 10(5xy + 1)(xy + 1)^3$
3. (0, 2), a relative minimum; (4, −6) and (−4, −6), saddle points
4. (a) $1625 thousand
(b) 73.11 means that if capital investment increases from $10,000 to $11,000, the expected change in monthly production value will be $73.11 thousand, if labor hours remain at 1590.
(c) 0.56 means that if labor hours increase by 1 to 1591, the expected change in monthly production value will be $0.56 thousand, if capital investment remains at $10,000.
5. (a) When $94,500 is borrowed for 25 years at 7%, the monthly payment is $667.91
(b) If the percent goes from 7% to 8%, the expected change in the monthly payment is $49.76, if the loan amount remains at $94,500 for 25 years.
(c) Negative. If the loan amount remains at $94,500 and the percent remains at 7%, increasing the time to pay off the loan will decrease the monthly payment, and vice versa.
6. $8xy\, e^{x^2y^2}(x^2y^2 + 1)$
7. Find $\dfrac{\partial q_1}{\partial p_2}$ and $\dfrac{\partial q_2}{\partial p_1}$ and compare their signs.
Both positive means competitive. Both negative means complementary. These products are complementary.
8. $x = \$7, y = \$11$
9. $x = 200, y = 100$
10. (a) $\hat{y} = 573.57x + 7652.05$
(b) $15,682 (nearest dollar)
(c) No; a child cannot be 35 years old.

Index

Credits

Chapter 1 "The Dow Jones Averages" chart: Republished by permission of Dow Jones, Inc. *via* Copyright Clearance Center, Inc. (c) Dow Jones and Company, Inc. All Rights Reserved Worldwide. "Dow Jones Activity: AP/Wide World Photos. **Chapter 2** "U.S. Trade Deficit" chart: Republished by permission of Dow Jones, Inc. *via* Copyright Clearance Center, Inc. (c) Dow Jones and Company, Inc. All Rights Reserved Worldwide. "E-mail Usage Chart": *Newsweek* (c) January 26, 1998 Newsweek, Inc. All rights reserved. Reprinted by permission. **Chapter 3** "Tardiness" and "Time" charts: Adapted from Bobrowki, P.M. and P.S. Part, "An evaluation of labor assignment rules," *Journal of Operations Management,* Vol. II Sept., 1993. **Chapter 5** "Price per share 1978–1993" chart: Republished by permission of Federal Signal Corporation. Metal Processors chart: Reprinted by permission of *Investor's Business Daily* "For People Who Chose To Succeed." "Basic Cable Rate" chart: Source: Nielsen Media Research. "World HIV Infections" chart: World Health Organization and the Global AIDS Policy Coalition, as reported in the *El Paso Times,* December 6, 1992. "Tax Foundation" chart: Copyright 1998, USA TODAY. Reprinted by permission. "First-Class Postage Rates": AP/Wide World Photos. "Rally Attendance" Chart: (c) 1997 Time Inc. Reprinted by permission. **Chapter 8** "Military Troops in Asia" chart: Reprinted with permission. Knight-Ridder/Tribune Information Services. **Chapter 9** "DIJA" at 5-Min Intervals Yesterday" chart: Republished by permission of Dow Jones, Inc. *via* Copyright Clearance Center, Inc. (c) Dow Jones and Company, Inc. All Rights Reserved Worldwide. "First-Class Postage Rates": AP/Wide World Photos. "Mutual Fund Growth History" chart: Reprinted with permission of VALIC (The Variable Annuity Life Insurance Company) from *Investment Digest* Vol 5 No 2. **Chapter 10** "Trade Deficit" chart: Republished by permission of Dow Jones, Inc. *via* Copyright Clearance Center, Inc. (c) Dow Jones and Company, Inc. All Rights Reserved Worldwide. "Dow Jones Closing Averages" and "Dow Jones Activity on March 31, 1994": AP/Wide World Photos. "Crime Statistics" chart: Copyright 1998, USA TODAY. Reprinted by permission. "Barometric Pressure" chart: By permission of the *Statesboro Herald.* **Chapter 12** "Revenue from Wireless Technology" chart: Copyright 1998, USA TODAY. Reprinted by permission. "National Health Expenditures" chart: *Newsweek* 4 October 1993 (c) Newsweek, Inc. All rights reserved. Reprinted by permission. "Fossil-Fuel Emissions" chart: *American Scientist,* Vol. 79, July-August 1990, p. 313. "Mutual Fund Growth History" chart: Reprinted with permission of VALIC (The Variable Annuity Life Insurance Company) from *Investment Digest* Vol 5 No 2. "Pension Dollars" graph: Reprinted with permission of VALIC (The Variable Annuity Life Insurance Company) from *Viewpoints* Summer 1993.

Social Science